THE HUMAN–COMPUTER

INTERACTION HANDBOOK

Book

*Fundamentals, Evolving Technologies
and Emerging Applications*

Human Factors and Ergonomics
Gavriel Salvendy, Series Editor

Hendrick, H., and Kleiner, B. (Eds.): *Macroergonomics: Theory, Methods, and Applications.*

Jacko, J. A., and Sears, A. (Eds.): *The Human–Computer Interaction Handbook: Fundamentals, Evolving Technologies and Emerging Applications.*

Meister, D., and Enderwick, T. (Eds.): *Human Factors in System Design, Development, and Testing.*

Stanney, Kay M. (Ed.): *Handbook of Virtual Environments Technology: Design, Implementation, and Applications.*

Stephanidis, C. (Ed.): *User Interfaces for All: Concepts, Methods, and Tools.*

Also in this Series

HCI 1999 Proceedings   2-Volume Set

- **Bullinger, H.-J., and Ziegler, J.** (Eds.): *Human-Computer Interaction: Ergonomics and User Interfaces.*
- **Bullinger, H.-J., and Ziegler, J.** (Eds.): *Human-Computer Interaction: Communication, Cooperation, and Application Design.*

HCI 2001 Proceedings   3-Volume Set

- **Smith, M. J., Salvendy, G., Harris, D., and Koubek, R. J.** (Eds.): *Usability Evaluation and Interface Design: Cognitive Engineering, Intelligent Agents and Virtual Reality*
- **Smith, M. J., and Salvendy, G.** (Eds.): *Systems, Social, and Internationalization Design Aspects of Human-Computer Interactions*
- **Stephanidis, C.** (Ed.): *Universal Access in HCI: Towards an Information Society for All*

For more information on LEA titles, please contact Lawrence Erlbaum Associates, Publishers, at www.erlbaum.com.

THE HUMAN–COMPUTER INTERACTION HANDBOOK

*Fundamentals, Evolving Technologies
and Emerging Applications*

JULIE A. JACKO, Editor
Georgia Institute of Technology

ANDREW SEARS, Editor
UMBC

LEA
2003
LAWRENCE ERLBAUM ASSOCIATES, PUBLISHERS
Mahwah, New Jersey   London

Senior Acquisitions Editor:	Anne Duffy
Editorial Assistant:	Kristin Duch
Cover Design:	Kathryn Houghtaling Lacey
Textbook Production Manager:	Paul Smolenski
Full-Service Compositor:	TechBooks
Text and Cover Printer:	Edwards Brothers, Inc.

This book was typeset in 9/11 pt. ITC Garamond Roman, Bold, and Italic.
The heads were typeset in Novarese, Novarese Medium, and Novarese Bold Italic.

Lawrence Erlbaum Associates, Inc., Publishers
10 Industrial Avenue
Mahwah, New Jersey 07430
www.erlbaum.com

Library of Congress Cataloging-in-Publication Data

The human-computer interaction handbook : fundamentals, evolving technologies, and emerging
   applications / Julie A. Jacko, Andrew Sears, editors.
      p.   cm.—(Human factors and ergonomics)
   Includes bibliographical references and index.
   ISBN 0-8058-3838-4 (case)—ISBN 0-8058-4468-6 (pbk.)
   1. Human-computer interaction—Handbooks, manuals, etc.   I. Jacko, Julie A.   II. Sears, Andrew.
III. Series.

QA76.9.H85 H8568   2002
004′.01′9—dc21   2002011381

Books published by Lawrence Erlbaum Associates are printed
on acid-free paper, and their bindings are chosen for strength
and durability.

Printed in the United States of America
10 9 8 7 6 5 4 3 2

For François, Beth, Nicole, and Kristen.

CONTENTS

Part
I

HUMANS IN HUMAN–COMPUTER INTERACTION
Mary Czerwinski 19

Part II

COMPUTERS IN HUMAN–COMPUTER INTERACTION
Robert J. K. Jacob......147

Part III

HUMAN–COMPUTER INTERACTION......241

A. Interaction Fundamentals
Julie A. Jacko......243

B. Interaction for Diverse Users
Gregg Vanderheiden......397

C. Interaction Issues for Special Applications
Jenny Preece......523

Part IV
APPLICATION DOMAINS
Arnold M. Lund......753

Part V
THE DEVELOPMENT PROCESS......907

A. Requirements Specification
Michael J. Muller......909

B. Design and Development
Tom Stewart......987

C. Testing and Evaluation
Andrew Sears......1091

Part VI

MANAGING HUMAN–COMPUTER INTERACTION AND EMERGING ISSUES
Hans-Jörg Bullinger and Jürgen Ziegler......1165

SERIES FOREWORD

With the rapid introduction of highly sophisticated computers, (tele)communication, service, and manufacturing systems, a major shift has occurred in the way people use and work with technology. The objectives of this series on human factors and ergonomics is to provide researchers and practitioners a platform for discusson of important issues related to these changes, where methods and recommendations can be presented for ensuring that emerging technologies provide increased productivity, quality, satisfaction, safety, and health in the new workplace and the Information Society.

The Human–Computer Interaction Handbook: Fundamentals, Evolving Technologies, and Emerging Applications, is published at a very opportune time, when the Information Society technologies are emerging as a dominant force both in the workplace and in everyday life. For these new technologies to be truly effective, they must provide communication modes and interaction modalities across different languages and culture. They should accommodate the requirements of a diverse user population, including the aging and the disabled.

The *Handbook* provides comprehensive coverage of human–computer interaction as it has been evolving in the international arena of research, development, and practice. The book's 64 chapters were authored by leading international human–computer interaction gurus including academicians,

researchers, and industry practitioners. For further in-depth coverage of the discipline, over 3,000 references are cited; and more than 125 tables and 450 figures facilitate the communication of knowledge and enliven the reader's experience.

This work should be of special value to software developers for interactive systems and to human–computer interaction professionals. For graduate students, this book provides a wealth of theoretically based and practically oriented information on understanding the human–computer interactive design process. Further, it addresses itself to design-diversified user populations, design-diversified development processes, and design-diversified applications.

The reader will discover a systematic, structured approach to the design, experimentation, operation, and evaluation of user interfaces so that emerging information technologies can be easily and joyfully accessible to all individuals.

The editors, in cooperation with their distinguished Advisory Board, have produced a most comprehensive human–computer handbook that establishes the boundaries of the discipline and which will become an invaluable resource in the professional's library.

Gavriel Salvendy
Series Editor
April 2002

FOREWORD

This Handbook celebrates the success of human–computer interaction (HCI). Readers from within the discipline can take satisfaction in their substantial accomplishments: Basic research on menus, pointing devices, and screen design as well as design innovations in hot HCI topics online communities, information visualization, mobile devices, and interactive games. Practitioners can take pride in their contributions to better products and services that have gained growing acceptance by decision makers.

Newcomers to HCI will find informative section introductions that highlight the key ideas and guide them to the details. This Handbook provides them with grounding in the underlying cognitive principles (Part II), an appreciation of the range of design possibilities (Parts III and IV-A), and recognition of the growing sophistication of modern development processes (Part VI).

Outsiders can dip into specific topics of timely value to them, and expand their reading as their needs and interests grow. They will likely begin with chapters on applications for which they must build or improve user interfaces (Section IVC and V). Visionary readers will search for forward-looking ideas, especially the reflections on human values, ethics, and trust (Section VII). Every reader will benefit from the numerous references that connect to the foundational research.

It is satisfying to recognize the breadth and depth of HCI, and to reflect on its future. I have wondered if this volume marks HCI's emergence from childhood, its youthful adolescence, or young adult status.

The case for "childhood" is that HCI is just beginning to be accepted in academic departments and corporate boardrooms. Only a few years have passed since the topic became a funding area within governmental research support agencies (e.g., U. S. National Science Foundation and other federal counterparts) and the first doctoral programs were announced. On the professional side, usability engineers fight for acceptance, authors propose competing design guidelines, and analysts squabble over appropriate metrics of success.

The argument for "adolescence" is that the uncertainty of childhood has passed. HCI has already made its mark, as these chapters confirm, but this energetic adolescent still has to choose from many possible futures. It could become a basic science like physics and psychology; it could remain an eclectic interdiscipline like biophysics and sociolinguistics; or it could mature into a professional discipline with scientific foundations like architecture and medicine.

The rationale for the "young adult" status is that HCI has already produced interesting theories and stimulated important innovations such as graphical user interfaces, the World Wide Web, email, online communities, instant messaging, and multimedia. It has successfully done its part in bringing information and communications technologies to a diverse set of users and in fashioning devices that are well on their way to universality, ubiquity, and invisibility.

Different readers may come to different conclusions and all be correct. Like any sufficiently large intellectual domain, there is room for many concurrent viewpoints. Those who think HCI is still in childhood can guide its development by engaging in new applications and innovative technologies. They can try to influence HCI's maturation by refining, extending, and applying predictive and explanatory theories, while proposing new theories and integrative frameworks.

Those who see HCI in the adolescent stage can engage in broadening the field by introducing new educational programs, expanding participation in professional societies, and enlarging existing journals. Energetic adolescents should take some risks and explore affiliations with other disciplines such as sociology, semiotics, or nanotechnology; they can expand their experiences by working in cultures other than their own.

Those who see HCI as a young adult can promote its integration into traditional social structures. They can help usability professionals rise through corporate and government ranks and be more influential in controlling product development plans, budgets, and schedules. They can publicize HCI by engaging journalists to report HCI stories and make HCI issues more visible in related disciplines such as software engineering, networking, or security. HCI advocates still have work to do to gain acceptance by top corporate executives, key government officials, and leading science policy makers.

As for the future, I have high expectations and great hopes for HCI as a mature discipline and profession. While preserving our "youthful" infatuation with new and cool technologies, however, we must also address "adult" responsibilities for *usable*, *universal*, and *useful* technologies (CHI'99 Research Agenda Workshop; see Chapter 40). These three concepts are essential as we look to the future of HCI.

1. **Usable.** Existing user interfaces and their underlying systems are just too frustrating for most users. A major effort for software developers should be to increase the reliability and usability of every system. The frequency of system reboots, application crashes, and incompatible file formats that stop users in the middle of their tasks must be reduced. Then the incomprehensible instructions, mysterious menus, and troubling dialog boxes need to be revised to enable users to complete their work promptly and confidently. In our study of the frustrations encountered by more than 100 users, we received frequent comments about long download times for web pages, dropped sessions for networked applications, and the disruption caused by unsolicited email ("spam") and destructive viruses. As the number of users has grown, all these problems have become more serious. Novice users want the benefits of email or web services, but they are poorly equipped and motivated to overcome the problems that high-tech early adopters proudly conquered. Improved training can help, but improved designs are an important component of making the next generation of interfaces more appealing and usability more successful.

2. **Universal.** After systems are made more reliable, we will be ready to address the challenge of making them universally usable. We must enable people with diverse backgrounds to become successful and satisfied users—old and young, men and women, novice and expert; users with varied disabilities; and users with low motivation, low self-confidence, and poor reading skills. We will have to make it possible for users with differing language skills, cultures, and nontraditional social norms to participate in online communities, e-commerce, and e-learning. We must also enable people with limited or outdated technologies—slow modems, small screens, or older processors—to benefit from information and communications services. Advanced technical solutions will be needed to convert web pages to fit small mobile devices and provide web services by cellphone. Finally, we must learn to bridge the gap between what people know and what they need to know.

Improved online help, tutorials, email assistance, and customer service are essential. These contributions will often come first from software producers in developed nations, but they need to be pressured to include adequate concerns for the needs and opportunities of users in developing nations.

3. **Useful.** Usable and universal technologies will make it possible to deliver vital services: Life-critical medical systems, military and police protection, safe transportation, and effective job training. Then content providers can create truly useful e-business, e-learning, e-healthcare, e-government, and many other applications. HCI professionals and researchers can promote greater respect for the discipline by addressing urgent political and global issues such as economic growth, community development, counterterrorism, and conflict resolution. By taking responsibility for difficult questions such as the contemporary issues of free speech, intellectual property, voting technologies, pornography control, and privacy, we will win respect within and outside our field. Better theories regarding human needs are required to understand what is missing and where the greatest opportunities lie. The next major direction I foresee is the enhancement of creativity support tools to make more people more creative more of the time. This ambitious goal will produce a sustainable chain reaction of further creativity and even better tools to effectively harness each individual's capacity to contribute and participate in shaping the future.

A hallmark of adulthood will be the willingness of HCI researchers and professionals to take on issues that transcend technology. We will gain respect as we contribute to systems that empower individuals, enrich human relationships, encourage cooperation, build trust, and support empathy. Our acceptance of responsibility for the future will be more than a model for other disciplines; it can bring hope to many people for a better life.

Ben Shneiderman
University of Maryland

PREFACE

This Handbook marks a critical juncture in the field of human-computer interaction (HCI). Over the last two decades, HCI emerged as a major discipline in computing, and was quickly energized by substantial contributions from other fields, such as engineering, psychology, education, and graphic design. The effective utilization of the resulting interdisciplinary breakthroughs has contributed to a heightened global awareness of the importance of HCI not just as an academic discipline. HCI in practice helps us achieve increased productivity, enhances our quality of life, envigorates competitiveness, and improves the quality of services through the effective design and use of interactive computing systems. HCI now has a home in every application, environment, and device, and is routinely used as a tool for inclusion. Therefore, this Handbook should be of tremendous value to practitioners, researchers, students, and academicians regardless of enterprise, because HCI, like computing itself, is now ubiquitous.

Composed of eight sections which include 64 chapters, the 1242 pages of this work were authored by over 120 of the most eminent researchers and professionals in the discipline from around the world. Each contribution was carefully reviewed by experts in the field, as well as by ourselves. The Advisory Board was instrumental throughout the process as they helped shape and refine our vision for this volume. We thank the contributing authors and the Advisory Board. Their commitment to the success of this volume was unwavering, and we are honored to have worked with each one of them.

The Handbook opens with an evolutionary perspective on HCI, which provides a sense of perspective on the field, recognizes significant visionary developments, and directs the reader to seminal literature. The book then explores the topics of humans in HCI, computers in HCI, interaction fundamentals, interaction issues for diverse users, and interaction issues for special applications. The 37 chapters within these sections provide a comprehensive knowledge base that distinguishes this volume from others in its field. The following section, Application Domains, is another unique feature of this work. In this section, HCI principles are applied to 9 specialized domains. The final 13 chapters explore the state of the art in the development process and the management of HCI. The volume concludes with an exploration of the future of HCI.

The editing of the Handbook was made possible with the very valuable assistance of Brynley Zorich, who so effectively coordinated the management of incoming contributions, reviews, and outgoing correspondence. It was truly a pleasure working with Gavriel Salvendy, Anne Duffy, and, more recently, Bill Webber, our brilliant editorial team whose insight and graceful facilitation was priceless to this effort.

Julie A. Jacko and Andrew Sears
May 2002

HANDBOOK WEB SITE

Additional material related to this handbook can be found on the world wide web including:

- Electronic versions of all figures, including high-quality, color images where appropriate.
- An expanded index with additional keywords.

To view this, and other supplementary material, please visit:

www.isrc.umbc.edu/HCIHandbook

ADVISORY BOARD

ABOUT THE EDITORS

Julie A. Jacko is an Associate Professor of Industrial & Systems Engineering (ISyE) at the Georgia Institute of Technology and is the author or coauthor of over 70 research publications including journal articles, books, book chapters, and conference proceedings. Dr. Jacko's research activities focus on human–computer interaction, human aspects of computing, and universal access to electronic information technologies. Her research has been supported by the Intel Corporation, Microsoft Corporation, the National Science Foundation, and NASA. Dr. Jacko received a National Science Foundation CAREER Award for her research titled, "Universal Access to the Graphical User Interface: Design For The Partially Sighted." She is also a recipient of the National Science Foundation's Presidential Early Career Award for Scientists and Engineers (PECASE), the highest honor bestowed on young scientists and engineers by the United States government. She is on the editorial board of the *International Journal of Human-Computer Interaction* and *Universal Access in the Information Society*. She served as Conference and Technical Program Cochair of the Association for Computing Machinery's Conference on Human Factors in Computing Systems (CHI 2001). She also served as Program Chair for the Fifth ACM SIGCAPH Conference on Assistive Technologies (ASSETS 2002). Dr. Jacko routinely consults for organizations and corporations on systems usability and human aspects of interactive systems design. She holds the B.S., M.S., and Ph.D. degrees in Industrial Engineering from Purdue University.

Andrew Sears is an Associate Professor and Chair of the Information Systems Department at UMBC. Dr. Sears' research explores issues related to human–computer interaction. His recent projects investigate issues associated with mobile computing, speech recognition, and the difficulties information technology users experience as a result of the environment in which they are working or the tasks in which they are engaged. His research projects have been supported by Aether Systems, Intel Corporation, Microsoft Corporation, Motorola, Platinum Technologies, NASA, and the National Science Foundation. Dr. Sears is the author or coauthor of over 70 research publications including journal articles, books, book chapters, and conference proceedings. He is on the editorial board of the *International Journal of Human-Computer Interaction* and *Universal Access in the Information Society*. He served as Conference and Technical Program Cochair of the Association for Computing Machinery's Conference on Human Factors in Computing Systems (CHI 2001), and has been on the program board for a variety of events including the Assets, ERCIM User Interfaces for All, HCII ad UAIS conferences and workshops. Dr. Sears routinely consults for organizations and corporations on issues related to interactive systems design and evaluation. He holds the B.S. degree in Computer Science from Rensselaer Polytechnic Institute and the Ph.D. degree in Computer Science with an emphasis on human–computer interaction from the University of Maryland-College Park.

LIST OF CONTRIBUTORS

Norman Alm
Department of Applied Computing
University of Dundee
Scotland

Alisa Bandlow
College of Computing
Georgia Institute of Technology
USA

Michel Beaudouin-Lafon
Laboratoire de Recherche en Informatique
Université Paris–Sud/CNRS
France

Randolph G. Bias
Austin Usability
USA

Jeanette Blomberg
Department of Human Work Science
   and Media Technology
Blekinge Institute of Technology
Sweden

Bridget C. Booske
University of Wisconsin—Madison
USA

Scott Brave
Department of Communication
Stanford University
USA

Stephen Brewster
Department of Computing Science
University of Glasgow
UK

Amy Bruckman
College of Computing
Georgia Institute of Technology
USA

Hans-Jörg Bullinger
Fraunhofer IAO
Germany

Mark Burrell
Sapient Corporation
USA

Michael D. Byrne
Department of Psychology
Rice University
USA

Pascale Carayon
Department of Industrial Engineering
University of Wisconsin—Madison
USA

Stuart K. Card
User Interface Research Group
PARC
USA

Alex Carmichael
Department of Psychology
Manchester University
UK

John M. Carroll
Center for Human–Computer Interaction
   and Department of Computer Science
Virginia Tech
USA

Justine Cassell
MIT Media Lab
Massachusetts Institute of Technology
USA

Romeo Chua
School of Human Kinetics
University of British Columbia
Canada

Gilbert Cockton
School of Computing and Technology
University of Sunderland
UK

Thomas A. Cofino
IBM T. J. Watson Research Center
USA

William J. Cohen
Consultant
USA

Sara J. Czaja
Department of Psychiatry and Behavioral Sciences
University of Miami School of Medicine
USA

Mary Czerwinski
Microsoft Research
Microsoft Corporation
USA

Steve Diller
Cheskin
USA

Alan Dix
Computing Department
Lancaster University
UK

Joseph S. Dumas
Oracle Corporation
USA

Ashley G. Durham
Centers for Medicare and Medicaid Services
USA

Henry H. Emurian
Information Systems Department
UMBC
USA

B. J. Fogg
Stanford University
USA

Batya Friedman
The Information School
University of Washington
USA

Thomas Fuller
Microsoft Game Studios
Microsoft Corporation
USA

Norman D. Geddes
Applied Systems Intelligence, Inc.
USA

David Goodman
School of Kinesiology
Simon Fraser University
Canada

Paul Green
Transportation Research Institute, Human Factors Division
Industrial and Operations Engineering
    and Mechanical Engineering
University of Michigan
USA

Peter Gregor
Department of Applied Computing
University of Dundee
Scotland

Greg Guest
Sapient Corporation
USA

Ken Hinckley
Microsoft Research
Microsoft Corporation
USA

Karen Holtzblatt
InContext Enterprises
USA

Hiroo Iwata
Institute of Engineering Mechanics and Systems
University of Tsukuba
Japan

Edmond Israelski
Abbott Laboratories
USA

Julie A. Jacko
School of Industrial and Systems Engineering
Georgia Institute of Technology
USA

Robert J. K. Jacob
Department of Electrical Engineering
    and Computer Science
Tufts University
USA

Anthony Jameson
DFKI, German Research Center for Artificial Intelligence, and
International University in Germany
Germany

Peter H. Kahn, Jr.
Department of Psychology
University of Washington
USA

Clare-Marie Karat
IBM T. J. Watson Research Center
USA

John Karat
IBM T. J. Watson Research Center
USA

Waldemar Karwowski
Center for Industrial Ergonomics
University of Louisville
USA

Kevin Keeker
Microsoft Game Studios
Microsoft Corporation
USA

David Kieras
Electrical Engineering and Computer Science Department
University of Michigan
USA

Joseph Krajcik
Center for Highly Interactive Computing in Education
University of Michigan
USA

Joseph Kramer
IBM T. J. Watson Research Center
USA

Diane Maloney-Krichmar
UMBC
USA

Jennifer Lai
IBM T. J. Watson Research Center
USA

Darryn Lavery
Microsoft Corporation
USA

Jonathan Lazar
Department of Computer and Information Sciences
Towson University
USA

Chin Chin Lee
Department of Psychiatry
   and Behavioral Sciences
University of Miami School of Medicine
USA

Jon Lenchner
IBM T. J. Watson Research Center
USA

Lynn Lin
Cheskin
USA

Holger Luczak
Institute of Industrial Engineering and Ergonomics
Aachen University of Technology
Germany

Arnold M. Lund
Sapient
USA

Wendy E. Mackay
Institut National de Recherche en Informatique
   et en Automatique (INRIA)
France

Aaron Marcus
Aaron Marcus and Associates, Inc.
USA

Deborah J. Mayhew
Deborah J. Mayhew & Associates
USA

Brad Mehlenbacher
Department of English, and
Department of Psychology
North Carolina State University
USA

Maria del Carmen Puerta Melguizo
Department of Information Systems
   and Software Engineering
Vrije Universiteit
The Netherlands

Michael J. Muller
IBM Research
USA

David Nahamoo
IBM T. J. Watson Research Center
USA

Clifford Nass
Department of Communications
Stanford University
USA

Alan F. Newell
Department of Applied Computing
University of Dundee
Scotland

Cathleen Norris
Department of Cognition and Technology
University of North Texas
USA

Sunil Noronha
IBM T. J. Watson Research
USA

Olaf Oehme
Institute of Industrial Engineering and Ergonomics
Aachen University of Technology
Germany

Gary M. Olson
School of Information
The University of Michigan
USA

Judith S. Olson
School of Information
The University of Michigan
USA

Sharon Oviatt
Department of Computer Science
Oregon Health and Science University
USA

Randy J. Pagulayan
Microsoft Game Studios
Microsoft Corporation
USA

Richard W. Pew
BBN Technologies
USA

Jenny Preece
Information Systems Department
UMBC
USA

Amy R. Pritchett
Schools of Industrial and Systems
    Engineering and Aerospace Engineering
Georgia Institute of Technology
USA

Robert W. Proctor
Department of Psychological Sciences
Purdue University
USA

Chris Quintana
Center for Highly Interactive Computing in Education
University of Michigan
USA

Janice Redish
Redish & Associates, Inc.
USA

Matthias Roetting
Liberty Mutual Research Center
USA

Ramon L. Romero
Microsoft Game Studios
Microsoft Corporation
USA

Mary Beth Rosson
Center for Human–Computer Interaction
    and Department of Computer Science
Virginia Tech
USA

François Sainfort
School of Industrial and Systems Engineering
Georgia Institute of Technology
USA

Gavriel Salvendy
School of Industrial Engineering,
    Purdue University, USA, and
Department of Industrial Engineering,
    Tsinghua University, China

Anthony Savidis
Institute of Computer Science,
    Foundation for Research
    and Technology—Hellas
Greece

Jesse Schell
VR Studio
Walt Disney Corporation
USA

Kevin M. Schofield
Microsoft Research
Microsoft Corporation
USA

Jean C. Scholtz
Visualization and Usability Group
National Institute of Standards and Technology
USA

Ingrid U. Scott
Bascom Palmer Eye Institute
University of Miami School of Medicine
USA

Andrew Sears
Interactive Systems Research Center
Information Systems Department
UMBC
USA

Bill Sharpe
The Appliance Studio
UK

Daniel P. Siewiorek
Human–Computer Interaction Institute
Carnegie Mellon University
USA

Asim Smailagic
Institute for Complex Engineered Systems,
    College of Engineering
Carnegie Mellon University
USA

Michael J. Smith
Department of Industrial Engineering
University of Wisconsin—Madison
USA

Philip J. Smith
Institute for Ergonomics
Ohio State University
USA

Elliot Soloway
Center for Highly Interactive
    Computing in Education
University of Michigan
USA

Neville A. Stanton
Department of Design
Brunel University
UK

Kay M. Stanney
Industrial Engineering & Management Systems
University of Central Florida
USA

Phil Stenton
Hewlett-Packard Laboratories
UK

Constantine Stephanidis
Department of Computer Science,
    University of Crete
Institute of Computer Science, Foundation
    for Research and Technology—Hellas
Greece

Tom Stewart
Systems Concepts
UK

Alistair Sutcliffe
Department of Computation
UMIST
UK

Vania Tashjian
Cheskin
USA

David Travis
Systems Concepts
UK

Dilip Upmanyu
IBM
USA

Gregg Vanderheiden
Trace Research & Development Center
Departments of Industrial Engineering
    and BioMedical Engineering
University of Wisconsin—Madison
USA

Gerrit C. van der Veer
Department of Information Systems
    and Software Engineering
Vrije Universiteit
The Netherlands

John Vergo
IBM T. J. Watson Research Center
USA

Holly S. Vitense
Department of Industrial Engineering
University of Wisconsin—Madison
USA

Kim-Phuong L. Vu
Department of Psychological Sciences
Purdue University
USA

Suzanne Watzman
Watzman Information Design
USA

Daniel J. Weeks
Department of Psychology
Simon Fraser University
Canada

Dennis Wixon
Microsoft Game Studios
Microsoft Corporation
USA

Alan Woolrych
School of Computing
   and Technology
University of Sunderland
UK

Nicole Yankelovich
Sun Microsystems
USA

Hidekazu Yoshikawa
Department of Socio-Environmental Energy Science,
   Graduate School of Energy Science
Kyoto University
Japan

Mark A. Young
The Maryland Rehabilitation Center &
   Workforce Technology Center
State of Maryland and Department of Education
USA

Jürgen Ziegler
Fraunhofer IAO
Germany

THE HUMAN–COMPUTER

INTERACTION HANDBOOK

·Introduction·

EVOLUTION OF HUMAN–COMPUTER INTERACTION: FROM MEMEX TO BLUETOOTH AND BEYOND

Richard W. Pew
BBN Technologies

> If your cursor finds a menu item followed by a dash,
> and the double-clicking icon puts your window in the trash,
> and your data is corrupted cause the index doesn't hash,
> then your situation's hopeless and your system's gonna crash!
> —Author Unknown

PREAMBLE

The editors decided to begin this handbook with a "history" chapter, not because we all like to read history, but because researchers and practitioners can benefit from a chance to step back from the bits and bites of the moment to reflect on the many different ways in which contributions have been and will continue to be made. Successful innovation is not so much the result of lucky guesses as it is understanding the need and the context of what is possible . . . so thoroughly that a solution seems obvious.

Some readers will have seen the anonymous e-mail that begins, "If an unnamed [automobile company] had developed technology like unnamed [software vendors], we would all be driving cars with the following characteristics":

1. Occasionally, your car would die on the freeway unexpectedly, and you would just accept this, restart and drive on.
2. The oil, water temperature, and alternator warning lights would be replaced by a single "general car default" warning light.
3. Occasionally, your car would lock you out and refuse to let you in until you simultaneously lifted the door handle, turned the key, and grabbed hold of the radio antenna.
4. Every time [automobile company] introduced a new model, car buyers would have to learn to drive all over again because none of the controls would operate in the same manner as the old car.
5. Etc.

We find these and many other similar net-icisms funny because on the one hand, they are absurd but on the other, they contain more than a grain of truth. It is the thesis of this chapter that, although we have made amazing progress—this handbook as a whole provides ample evidence of the progress we have made—in fact, our visions continue to outstrip our capacity to deliver, and our technology continues to evolve so rapidly that the symphony of well-orchestrated hardware and software in seamless support of the user's requirements seems to remain just out of reach. As Landauer (1995) so skillfully demonstrated, although computers are supposed to help us *solve* problems, in too many instances their interfaces still *are* the problem.

It is difficult to draw boundaries between the history of computers and the history of human–computer interaction (HCI). The history of computers and software has set the context in which the field of HCI has emerged. With this perspective in mind, this chapter will seek to provide a sense of context for the field as it exists today, in 2001; introduce a collection of remarkable and significant visionary developments; provide references to the seminal work in the field, with perhaps a little nostalgia thrown in; and, because the author does not consider himself a visionary, provide some modest hints at directions for the future. There is not space to do justice to all these topics. For those who would like more complete coverage of the history

of HCI than is possible here, fortunately there are a number of other reviews of the historical developments to which the reader may refer, such as Association for Computing Machinery (1986), Baecker and Buxton (1987), Baecker, Grudin, Buxton, and Greenberg (1995), Carroll (1997), Gaines and Shaw (1986), Myers (1998), Myers, Hollan, and Cruz (1996), Nickerson and Landauer (1997), Rheingold (2000), and Shackel (1997, 2000).

No better way to organize this material than chronologically has emerged. The sections that follow will break the discourse into reasonably coherent epochs and for each epoch to consider:

- The visions that emerged
- The dominant hardware and software technology
- The important research and/or design methodologies
- The significant application innovations.

1966 AND BEFORE

State-of-the-Art

Throughout the history of HCI there has been a need to distinguish the interface requirements for programmers from those for users. In the period before 1963 or so, there were a great many constraints. Programmers entered programs by keyboard or prepared punched cards on special machines; to review and understand their programs, they read printouts produced on line printers. The human factors issues for programmers were how to format the printouts of their programs so they were easy to read and debug, because the only feedback from running a batch job was in the form of a rejected batch of cards and/or a printout of their program with one or more abstruse error messages. Initial submissions usually had formatting errors, because the formats were inexorably fixed and the program would not execute with even one such error. It often took several submissions to get to the point were the first substantive error was flagged, at which point the execution hung, and further correction and job resubmission were required to find the next bug.[1] Woe-be-unto the programmer who accidentally dropped a deck of cards and had to reorder them before the next submittal.

The human factors issues for the users were how to prepare program printouts in a form that they could be read and interpreted easily, especially given the constraint of line printers that could only present alphanumeric characters in a single, often difficult-to-read font with 80-character lines and limited formatting capabilities. The specification of how they were to appear was a problem for the programmers, of course. In the case of scientific computing, programmers and users were most often the same individuals; but, for accounting departments, communication between users and programmers was even rarer than it is today. In 1971, Gerald Weinberg published the first book documenting the problems of programmers and users, and describing the research under way to solve them. It took a strongly user-oriented perspective (Weinberg, 1971).

---

[1] Grace Murray Hopper (1906-1992) is credited with coining the term, "bug," when she discovered an error in the Harvard Mark II output was attributable to a moth that had infiltrated a set of relay contacts in the computer (Dickason, 2001).

Initially, programs were written exclusively in machine language until compilers and higher level programming languages, such as COBOL (Common Business Oriented Language) and FORTRAN, were developed. The COBOL language was designed to use English statements and be self-documenting. It was specifically intended for business and accounting, and was first introduced in 1960. Grace Hopper created FLOWMATIC, its immediate predecessor, and is considered the "mother" of COBOL (LegacyJ Corp., 2001). FORTRAN, on the other hand was developed specifically for the programmer/scientist at IBM and was first announced in 1957 for the IBM 704. FORTRAN made it possible to code very sophisticated mathematical models and calculations without resorting to assembly language, but was somewhat slow to be widely adopted (Wilkes, 1993).

From the point of view of HCI, the most significant developments of this period were the first visions of what the world of computing might be like, if only we were not limited to punched cards and batch processing on monolithic machines. The five visionaries that stand out during this period were Vannevar Bush, J. C. R. Licklider, Douglas Englebart, Ivan Sutherland, and Ted Nelson.

Vannevar Bush

Vannevar Bush was the Director of the Office of Scientific Research and Development and served as an advisor to Presidents Truman and Roosevelt. At the end of World War II, he wrote an article for the *Atlantic Monthly* in which he summarized his thinking and sought to redirect the attention of scientists from the instrumentalities of war to the need for improved storage, management, and access to information, both scientific and otherwise (Bush, 1945).

The summation of human experience is being expanded at a prodigious rate, and the means we use for threading through the consequent maze to the momentarily important item is the same as was used in the days of square-rigged ships. (p. 102)

He carefully documented his vision for the kinds of technology that he thought could contribute to the solution of this problem, including photocell-based electronic cameras mounted on one's forehead with which to document what was of particular interest, dry photography, speech recording that could be time-stamped to synchronize it with the pictures being recorded by his advanced camera, electronic calculating machines that could arithmetically *and logically* manipulate data, and speech-controlled typewriters. However, his most interesting ideas were conceptual, relating to the processes of information retrieval:

Selection by association, rather than indexing, may yet be mechanized. One cannot hope thus to equal the speed and flexibility with which the mind follows an associative trail, but it should be possible to beat the mind decisively in regard to the permanence and clarity of the items resurrected from storage. (p. 104)

Thus, Bush was advocating that we should organize and manipulate information in the same ways we think about it. He used the term Memex to describe a mechanized, associatively indexed, specialized workstation from which to access private files and libraries, complete with screens that projected automatically retrieved microfilm images. Ted Nelson would later call this nonsequential, associative method of cataloging "hypertext" (Nelson, 1965).

J. C. R. Licklider

In 1959, J. C. R. Licklider purchased Bolt Beranek and Newman.[2] (BBN)—its first computer. It was a Royal McBee LGP 30. Having been influenced by the MIT Whirlwind computer and the Lincoln Lab TX-0 and TX-2, it was not an IBM 650, or even a 704, the dominant machines in use at the time, particularly in accounting departments or scientific laboratories. Lick, as he was called by anyone who knew him, was not interested in batch processing and punched cards. He was interested in how individuals could use computers to support their work. The LGP-30 received its input by reading a punched paper tape that was prepared offline by a Flexowriter, a special typewriter that looked a little like a teletype machine. Although the LGP-30 cost $25,000, it was a small machine by today's standards. It had a magnetic drum memory of 8,000 characters and was coded in a primitive machine language. One full rotation of the drum took approximately 1 ms, so that if processing speed or efficiency was of interest, the programmer had to plan the sequence of literal memory locations assigned in the program so that sequential read instructions would not consume extra drum rotations. The machine was definitely one a single user could get his or her arms around, but it was marginally useful and prone to errors, and Lick soon replaced it with Digital Equipment Corporation PDP-1, serial number 1. The PDP-1 was about the capacity of a Radioshack TRS 80, of 1970s vintage, but was attractive because it featured a Cathode-Ray Tube display, and a keyboard, light pen, and paper tape reader as input devices. Even though it cost more than $120,000, it was truly a computer with which an individual could interact.

The year 1960 was also the year Lick published his seminal paper, "Man–Computer Symbiosis" (Licklider, 1960). In it, he laid out his vision for what HCI could be.

Man-computer symbiosis . . . will involve very close coupling between the human and the electronic member of the partnership. The main aims are 1) to let computers facilitate formulative thinking as they now facilitate the solution of formulated problems, and 2) to enable men and

---

[2]BBN, originally Bolt Beranek and Newman, is a commercial company that was founded by and named for Richard Bolt, Leo Beranek, and Robert Newman, three MIT academicians whose specialty was acoustics. In 1994, the name was officially shortened to BBN; in 1997, it was acquired by GTE and renamed BBN Technologies. As a result of GTE's merger in 2000 with Bell Atlantic, it is now a business unit of Verizon Communications. Licklider was hired by Leo Beranek because of his interest in psychoacoustics, but, as will become evident, his interests were much broader and he launched BBN into the computer age.

computers to cooperate in making decisions and controlling complex situations without inflexible dependence on predetermined programs. In the anticipated symbiotic partnership, men will set the goals, formulate the hypotheses, determine the criteria and perform the evaluations. Computing machines will do the routinizable work that must be done to prepare the way for insights and decisions in technical and scientific thinking. (Licklider, 1960, Summary, p. 4)

Lick wanted to think of computers as aids to human thinking, not as monolithic, number-crunching machines. He laid out the technological advances required to achieve these goals—developments in:

1. Computer time-sharing, because use of one machine for one knowledge worker was not, at the time, cost-effective.
2. Hardware memory requirements, because he foresaw the need for the user to have access to large quantities of data and reference material, a virtual library at one's fingertips.
3. Memory organization, because serial search through sequentially organized databases was too time-consuming and inefficient.
4. Programming languages, because of the extreme mismatch between languages the computer could understand and those the human could understand.
5. Input–output equipment, because he envisioned the time when input and output should match the "flexibility and convenience of the pencil and doodle pad or the chalk and blackboard" (Licklider, 1960).

In this vein, he also discussed the then-current state of speech recognition technology (claimed to be 98% correct recognition of decimal digits at Bell Laboratories and MIT Lincoln Labs.) and the need for computer-posted wall displays. He was well aware of the fledgling state of artificial intelligence (AI) research. He cited an Air Force study group conclusion that it would be at least 1980, that is 20 years hence, before developments in AI would make it possible for machines alone to do much thinking or problem-solving of military significance. To that, Lick added,

That would leave, say, five years to develop man-computer symbiosis and 15 years to use it. The 15 may be 10 or 500, but those years should be intellectually the most creative and exciting in the history of mankind.

In 1962, Lick became the Director of the Information Processing Techniques Office at the Advanced Research Projects Agency (ARPA) of the Department of Defense and actually served during two periods of time; 1962–1964 and 1974–1975. In that role, he developed a far-reaching program of contract research in all of the elements required to achieve his vision. By 1963, that vision had broadened to include the concept of computer networking and the Internet. In a memo in 1963 to his ARPA research community, addressed as "the members and affiliates of the Intergelactic Computer Network," he outlined an important piece of his strategy—to connect all the community's computers and time-sharing systems into a single computer network spanning the continent. Then, in 1965, in a book entitled *Libraries of the Future* (Licklider, 1965), he expanded on this idea:

The concept of a "desk" may have changed from passive to active: a desk may be primarily a display-and-control station in a telecommunication-telecomputation system and its most vital part may be the cable ("umbilical cord") that connects it, via a wall socket, into the precognitive utility net. (Licklider, 1965, p. 33)

Lick passed away in 1990 at age 75. Today, he is widely credited with having been the first to have a vision of computer networking and the Internet. For a comprehensive record of Lick's Contributions see Waldrop (2001).

Douglas Englebart

One of the key contributors to the ARPA program that Lick originated was Douglas Englebart. In 1950, Englebart began to shape his vision for what became his life's work—augmenting human intellect. The first version in the open literature was published in 1963 (Englebart, 1963).

By "augmenting human intellect" we mean increasing the capability of a man to approach a complex problem situation, to gain comprehension to suit his particular needs, and to derive solutions to problems . . .. We do not speak of isolated clever tricks that help in particular situations. We refer to a way of life in an integrated domain where hunches, cut-and-try, intangibles, and the human "feel for a situation" usefully co-exist with powerful concepts, streamlined terminology and notation, sophisticated methods, and high-powered electronic aids. (Englebart, 1963, p. 1)

This is a clear statement of his vision; however, there is much more to say about Englebart's accomplishments. In 1965, Bob Taylor took over for Ivan Sutherland as Director of the Information Processing Techniques at ARPA and substantially increased Englebart's funding over that supplied by Air Force Office of Scientific Research. With the improved support, Englebart created the Augmentation Research Center at the Stanford Research Institute.

In 4 years, he made giant steps toward implementing his vision for a computer-supported environment for accomplishing intellectual work. Just as Lick had conceived of computers supporting an individual knowledge worker, Englebart surrounded the user with cathode-ray tube displays and a variety of experimental input–output devices beyond a keyboard. He wrote software for his ARPA-supplied, time-sharing computers that transformed them into a visionary prototype system for the support of intellectual collaboration called NLS, a somewhat forced acronym for oNLine System. He is credited with inventing the mouse, but in addition to the mouse and a standard keyboard, the environment included a five-key control box used to control information presentation. At the heart of NLS was a hierarchical information management capability that could deal with text, diagrams, pictures, even video conference imagery, could manipulate these objects symbolically, and display them in "windows" laid out on the multiple screens. He thought of text and pictures in hypertext terms. NLS could outline, expand, collapse, reorganize, and link paragraphs, illustrations, and sections hierarchically.

In the most dramatic computer demonstration of the 1960s, some would say the most dramatic ever, at the Fall Joint Computer Conference in 1968 in San Francisco, Englebart assembled all his gear on the stage. The tangle of equipment required to carry all this off might be characterized as kluggy, but the reader must appreciate the state of the early time-shared computers and displays he had to work with. He was using character generators and analog vector graphics, combined with television imagery in standard raster-scan formats. He piped the TV imagery from his Stanford laboratory and achieved switching among various hierarchical representations of these artifacts reasonably easily using his mouse and control box. He presented the results of 4 years work in 90 minutes to a large crowd who appreciatively gave him a standing ovation (Rheingold, 2000).

Englebart's most lasting contributions were the mouse, his working hypertext system as a part of NLS, and the vision for a personal workstation that can legitimately be thought of as one of the sources of ideas for the personal computer. He has continued his mission in life to provide computer support to the human intellect, but never with the funding and momentum of the NLS project. In 1989, he and his daughter, Christine, formed the Bootstrap Institute that continues to elaborate hypertext concepts well beyond what is found in current internet applications, and to consult with industry about how they can increase the efficiency and productivity of their knowledge workers.

Ted Nelson

The history of hypertext would not be complete without mention of the guru who invented the term, and typifies the 1960s and 1970s generation of hackers, Ted Nelson.

Let me introduce the word *hypertext* to mean a body of written or pictorial material interconnected in such a way that it could not conveniently be presented or reproduced on paper. It may contain summaries, or maps, of its contents and their interrelations; it may contain annotations, additions and footnotes from scholars who have examined it. Let me suggest that such an object and system, properly designed and administered, could have great potential for education, increasing the student's range of choice, his sense of freedom, his motivation and his intellectual grasp. (Nelson, 1965)

Nelson writes copiously, with great vision, at times irreverently, but with great humanity, continuing to elaborate and refine his vision for international connectedness through hypertext (Nelson, 1965, 1973, 1974a, 1974b, 1981). He also tried, for more than 30 years, to create an uncompromised implementation of his vision, Xanadu. Although it never coalesced beyond beta versions, it has spawned a number of creative contributors to the field and served to focus Nelson's innovative thinking about hypertext.

Ivan Sutherland

While Licklider was talking and writing, Ivan Sutherland was producing code to describe his own visions. His MIT PhD thesis, Sketchpad, captured the attention of the fledgling computer science world immediately (Sutherland, 1963). The Sketchpad system not only opened the door to computer-aided design systems, but also became the basis for many of the graphics construction and editing capabilities that are commonplace today. Baecker and Buxton (1987) provide a nice summary of the far-reaching visions his code demonstrated:

1. The concept of a hierarchic structure to a computer-represented picture.
2. The concept of a master drawing and its instances, a forerunner of object-oriented programming.
3. The use of the method of constraints to specify the details of drawing geometry.
4. The ability to display and manipulate iconic representations of constraints.
5. The ability to copy, as well as "instance" both pictures and constraints.
6. Some elegant techniques for drawing construction using a light pen.
7. Separation of the coordinate system in which a drawing is defined from that on which it is displayed, permitting interactive three-dimensional manipulation of the drawing as an object, and recursive operations, such as move and delete, applied to hierarchically defined drawings.

The author personally remembers a remarkable early video from the project showing interactive rotation of three-dimensional objects.

It should be mentioned in passing on to the next epoch that the single-user features of the PDP-1 inevitably led to the development of the first computer game, Space War. It consisted of little more than missile-firing space vehicles with unidirectional thrusters in the inertial environment of space impacted only by the gravity of the sun. By orienting the vehicle and specifying the amount of thrust to be applied, each player moved about the screen and attempted to fire deadly missiles at each other. It started out as an MIT hack in the early 1960s, but was widely distributed to any organization that had the appropriate hardware. Little did the developers expect that such games would spawn a thriving industry.

1967–1977

State-of-the-Art

The period beginning in the late 1960s marked the end of the era of small independent machines and introduced the era of time-shared computing resources. With the impetus from Licklider and others, software and hardware infrastructures were introduced to make it possible to swap several individuals' programs and memory allocations between relatively fast magnetic core memory and relatively slow but much larger memory recorded on a magnetic drum so quickly that, when the machine was lightly loaded, users were barely aware that they did not have the full resources of the machine at their command. However, under heavy use, the machine began spending more time swapping users' work in and out than it did getting the actual users' processing accomplished. The users began experiencing significant lags between when they typed and when the result appeared on their screen. This led to a flurry of HCI activity characterizing the taxonomy of system-induced time delays and research on how time delays affected user behavior. The seminal taxonomic paper, which also suggested acceptable delays for various conditions, was Miller (1968). Recently, it has seen a revival of interest directed toward web-based quality of service issues (Sears & Jocko, 2000). An example of the early user-delay interaction research is Morfield, Wiesen, Grossberg, and Yntema (1969).

Soon, geographically dispersed organizations with a frequent requirement for transaction processing, such as the insurance industry, the U.S. Department of Agriculture, the Social Security Administration, and other government organizations, became interested in taking advantage of time-sharing to implement "interactive" online transaction processing in the field and to connect their diverse locations to their central records repositories. The path was via leased telephone lines, through a series of regional data concentrators, and then to the central database machines. Only a few such systems were actually implemented. "Interactive" is in quotes because, by today's standards, there were usually substantial delays before a reply to an interactive query was received, and usually the central databases themselves were only updated overnight. These constraints launched the concept of the "screen" of information. The remotely connected systems, using a store-and-forward paradigm, would accumulate a "screen" or page of information and then transmit it to the central system via the concentrator. The concentrator would then download a new form or "screen" and set it up for response from the remote terminal. Initially, only the most advanced systems offered full duplex interaction, that is, interaction in which each ASCII character was transmitted and echoed by the host machine or concentrator as it is typed, rather than transmitted a line at a time or a "screen" at a time.

Human–Computer Interaction and Screen Design

In those days, HCI design was conceptually much easier than today, because the interface was defined anew for each application and was limited to a countable set of "screens" having a well-defined sequentially-branching structure. This was the state of HCI through much of the 1970s. Nevertheless, this stage of user requirements development, where the emphasis was on deciding what a user would want to see and what a user needed to type, led to much confusion and user frustration that was only occasionally addressed directly by usability studies. Martin (1973) and Galitz (1981) document the state-of-the-art of interface design during this time period.

Interactive Programming

Even more significant from the point of view of programmers' HCI was the development of interactive programming that replaced punched cards and seriously reduced the requirement for and the pain of batch compiling and code execution. Although programmers were already writing code at time-sharing consoles, the introduction of incremental compilers, infrastructure for writing programs in which each program statement was assembled incrementally, able to be verified in small sections and built up incrementally into a full application, made the process much more interactive and signaled the final end to the tedium and inefficiency of punched cards and batch program submissions. The first such programming language was JOSS, created at the RAND Corp. in the early 1960s (Marks & Armerding, 1967), but it was soon followed by Kemeny and Kurtz's BASIC (Kemeny & Kurtz, 1966) and BBN's TELCOMP (Myer, 1967).

Although it is hard to imagine today, the groups funding programming, in both commercial and government organizations, were skeptical at first that it was actually more cost-effective to program interactively than to use the batch systems. Since, it was argued, programmers sitting at interactive consoles would waste time trying different things, and the time delays of the then-current timesharing systems would inhibit timely results. Since, eventually, the entire program had to be tested as a whole anyway, why would it be more cost-effective to program interactively? Several HCI researchers were funded to evaluate the relative cost-effectiveness of the two approaches, for example, Gold (1969), Grant and Sackman (1967), and Sackman (1970).

As it turned out, the adoption of online programming techniques, most of which used incremental compilers, made punched cards and traditional batch submissions obsolete, regardless of the outcome of the experiments. In the 1970s, the first programming environments, such as INTERLISP and Smalltalk, were introduced. They included a whole range of programming support from integrated editors and debugging routines to property inheritance that supported the reuse of selective code segments. Following these developments, there was no longer even an argument that could be made supporting the batch mode of operation.

Human Factors Engineering

A brief diversion is needed to introduce the role of human factors engineering into this discussion. During the first half of the 20th century, the fields of industrial engineering and psychology both became involved in issues of engineering design to take account of human performance capacities and limitations. Industrial engineering came to it by way of Frederick W. Taylor's views of "scientific management" and time and motion study in industrial settings (Taylor, 1911). Psychology came to it by way of their involvement in personnel selection, training, and eventually design of military systems. The psychologists who entered this field, sometimes referred to as Engineering Psychologists, were interested in design to accommodate cognitive or intellectual capacities and limitations, as well as perceptual-motor skills. The Human Factors Society of America (now the Human Factors and Ergonomics Society) was founded in 1957 and had about 600 members by 1960 (5,000 members today). Because this group had always been interested in the design of systems to take account of their human users, it is only natural that they would become involved with the design of computers, software, and HCI more generally as the field emerged. In fact, through the early 1970s, this group played the dominant role in introducing the concepts of cognitive psychology and psychological methodology into the design and evaluation of human–computer systems.

The notion that to begin man–machine system (sic) design, the engineer had to understand the tasks the system was required to perform, the context in which those tasks were required, and to make explicit the assignment of functions to various system components, including humans, was fundamental to the development of systems involving humans and machines. Furthermore, it became clear quite early that the programming of computers and the use of computers involved human cognitive and intellectual capacities and limitations that psychologists might be able to understand. Then, in the 1980s, when it came time to think seriously about the evaluation of user interfaces, the early efforts were based largely on experiment design methodology derived directly from experimental psychology. These facts account for the early and continued involvement of cognitive psychologists in the HCI field.

The first computer company to have a human factors group was IBM in their Federal Systems Division supporting military systems development during World War II. By 1956, there was a group in computer research at San Jose and soon thereafter at Yorktown Heights. However, at that time, IBM was largely a computer hardware manufacturer, and the research the group undertook was primarily concerned with factors that would influence how users would interact with their hardware, through control panels, keyboards, and hardware displays. It was not until 1979 that IBM, recognizing its larger role in software as well as hardware, undertook a serious initiative to build on human factors efforts already underway in a number of its labs, and introduced usability concepts and usability laboratories into its software design and development organizations (R. Hirsh, 2001, personal communication).

Alan Kay and the Palo Alto Research Center Nexus

During the 1970s, Alan Kay was carrying forward the visions initiated by Licklider, Englebart, and Sutherland. In fact, Kay got his start in computer science at the University of Utah in 1966 under Donald Evans and Ivan Sutherland. At Utah, he began dreaming of the idea of the personal computer. His first realization was embodied in his PhD thesis on the "Reactive Engine" (Kay, 1969). Also known as the FLEX machine, his software was truly innovative. It introduced some of the concepts of parallel processing, windows, and the message passing that is the foundation of object-oriented software. Kay was not interested in programming languages to support computation. He envisioned the computer as a tool for logical simulation and modeling of conceptual worlds. Accordingly, his object-oriented approach created structures whose complexity could be built up incrementally using a language that emphasized symbolic

logic rather than computation. His vision, encouraged by his belief in Moore's Law,[3] was that we should eventually have a powerful, inexpensive, easy-to-use computer on everyone's desk.

In 1972, Kay moved from the Stanford AI Lab to join other innovators at the Xerox Palo Alto Research Center (PARC)— such as Butler Lampson, Bob Metcalf, Bill Newman, Charles Simonyi, Bill English, Chuck Thacker, Ed McCreight, Skip Ellis, and Bob Sproull—seeking to realize the dreams of Licklider and Englebart. The PARC was created in 1970, initially under the direction of George Pake (Waldrop, 2001). Here, Kay continued the evolution of the concepts in FLEX, which by now, with the collaboration of Adele Goldberg, had morphed into the Smalltalk programming system. Kay, influenced by the language, Simula, together with Goldberg, cast all computation in Smalltalk in the form of self-contained objects, each with its own interface, to the world that allowed partitioning into local and "public" information (S. Card, 2001, personal communication). As it evolved at PARC, Smalltalk became a programming environment in the sense that the programming language and the supporting software to facilitate development and debugging were seamlessly interconnected. These innovations, which are widely emulated today, were a giant step beyond incremental compilers toward object-oriented programming languages and interactive support for programmers.

At PARC, a team was developing the ALTO, the first truly powerful bit-mapped graphics personal machine (Lampson & Taft, 1979). The ALTO was a box the size of a two-drawer file cabinet and fit under the desk rather than on it, but it was clearly a step along the way toward Alan Kay's vision, for which by this time he had coined the term "Dynabook." It had a 256-kb core memory and a 2-megabyte removable hard disk required to support the object-oriented features of Smalltalk and INTERLISP. PARCers populated their offices with them and interconnected them through the PARC-developed Ethernet.

Information Processing Theory, GOMS, and Verbal Protocols

Now for a second digression. One significant stimulus for advancement in HCI and for the work at PARC came much earlier and from a somewhat unexpected source. In the early 1970s, Allan Newell and Herbert Simon began thinking quantitatively about the possibility of building an information processing theory of human problem solving and, more importantly, the possibility of embodying that theory in computer simulations of thought (Newell & Simon, 1972). This work may be considered to have spawned the whole field of AI, but for our purposes, it had two other salient impacts. First, to generate their theory, these authors began having the subjects of their experiments think aloud as they performed a problem-solving task and the resulting verbal protocols were considered to be data for their

---

[3] Moore's Law: The observation made in 1965 by Gordon Moore, cofounder of Intel, that the number of transistors per square inch on integrated circuits had doubled every year since the integrated circuit was invented. Moore predicted that this trend would continue for the foreseeable future. In subsequent years, the pace decreased a bit, but data density has doubled approximately every 18 months, and this is the current definition of Moore's Law. http://webopedia.internet.com/TERM/M/Moores_Law.html

theories (Ericsson & Simon, 1980). Simon and Newell's use of think-aloud protocols led HCI researchers to suggest that the *thinking-aloud method* could be useful for cognitive–interface design (Lewis, 1982). In fact, the method has, over the succeeding years, been widely adopted as an extremely profitable method for identifying the difficulties a user is having and pinpointing the causes of those difficulties and their remedies. (See Chapter 56, this Handbook.)

Second, in 1971, Allan Newell was consulting at PARC and proposed a project on applied information–processing psychology that would translate into practice some of his groundbreaking work with Simon. The project was begun in 1974 with the hiring of Stuart Card and Thomas Moran. Newell continued as a consultant. It is no wonder that with the generous financial support of the Xerox Corporation and the culture of the organization committed to the partnership between humans and computers that this HCI group became one of the premiere sites for ground-breaking HCI research.

They set as a goal to create a psychological theory that would have an influence on the ongoing technological developments of their colleagues. They thought it might be possible to use what was known in cognitive and perceptual psychology to create predictive models that could influence conceptual design and be used during the user interface design cycle to evaluate alternative designs before they were implemented (Card, Moran, & Newell, 1980a, 1980b). Their work culminated in the publication of *The Psychology of Human–Computer Interaction* (Card, Moran, & Newell, 1983). Not only did they develop useful predictive models, the book was also a tour-de-force that summarized all of the relevant human information processing literature in the behavioral sciences in the "Model Human Processor" and elaborated the keystroke model into a useful engineering application called GOMS (Goals, Objects, Methods, Selection Rules). It was a significant development that helped ground this emerging field in the early 1980s.

Human Sciences and Advanced Technology Centre and the British Community

Meanwhile, also in 1970, in Great Britain, Brian Shackel formed the Human Sciences and Advanced Technology Centre (HUSAT) at Loughborough University (Shackel, 1997). Although not exclusively interested in computer-related issues, under his leadership the Centre (now Institute) has played a leading role, along with T. R. G Green at the Computer-Based Learning Unit at the University of Leeds and the Applied Psychology Research Unit at Cambridge, during the tenure of D. E. Broadbent and later Alan Baddley, in applying the concerns, methods, and knowledge traditional to the fields of psychology and ergonomics (the term used in Europe for human factors engineering) to the study of computer design and use.

Design Guidelines and Style Guides

The period of the early 1970s, when time-shared remote terminal systems were becoming popular, was also the time when the first notions that there might be some user-oriented design principles that could be applied to the design of the "screens." Hanson is usually credited with the first statement of "User Engineering Principles" (Hanson, 1971). Hanson was an engineer at the Argonne National Laboratory who became interested in making Emily, a text editing system for programmers, as well adapted to his users as possible. In the course of constructing Emily, he derived four top-level principles:

- Know the user
- Minimize memorization
- Optimize operations
- Engineer for errors.

His paper then describes several subtopics designed to accomplish these higher level goals. Then, in 1975, Engle and Granda (1975), at IBM, using a variety of sources, assembled a much more comprehensive set of guidelines covering the topics:

- Display formats
- Frame content
- Command language
- Recovery procedures
- User entry techniques
- Response time requirements.

The development of guidelines generally evolved into the most comprehensive set developed by Sid Smith and his colleagues. His guidelines went through several revisions during the 1970s and early 1980s, culminating in the final version, Smith and Mosier (1986). This set of guidelines was so comprehensive that it was difficult to decide which specific recommendations apply to particular design requirements and to choose between two guidelines that in a specific context might suggest conflicting solutions for the same design detail.

As everyone now accepts, guidelines are limited to recommendations that can be applied without regard to the context of the specific application. This is a strength in the sense that they can apply broadly, but a weakness in that they are necessarily watered down to be general and unable to address issues that are conceptual rather than physically explicit, or that arise in a specific task context. Today, it is more common to generate style guides. A style guide usually is developed in reference to one product or product line for which the guidelines can be considerably more specific and contextually constrained. It often provides explicit examples of how the guidelines apply to the product(s) in question.

A very early style guide, perhaps the first, was developed by Pew and Rollins for the Department of Agriculture (Pew & Rollins, 1975; Pew, Rollins, & Williams, 1976). Although the system was never implemented, the Department was preparing to place interactive computers in their county field offices that would be connected to the central office databases in Washington, DC. They had a cadre of programmers who were experienced in producing software for batch operations, and the Department wished to retrain them to develop software with which users, county agricultural agents, having no knowledge of how the computers worked, could interact and enter data as they conducted interviews with their farmer-clients. The management was concerned that the modules being developed by different teams needed to look the same to their users, who would be distributed all over the US. Accordingly, they commissioned the development of a Dialog Specification Procedure. In addition to providing specifications for what the screens would look like for different purposes, what terminology would be used, and how the user would interact with the systems, the Dialog Specification Procedures provided specific instructions on how to decide what information was required to be collected by the user in the field, that is, how to conduct task analyses, and specific procedures and templates for specifying the design of the screen layouts for the coders who would be implementing the programs.

Final 1970s Developments

As the 1970s were winding down, packet-switched networks were beginning to proliferate, and e-mail over such networks was just beginning to take hold with users in the research community. Frankston and Bricklin introduced the spreadsheet paradigm on an Apple II in the VisiCalc application in 1979 (Stranahan, 1996). It was a significant breakthrough because it blurred the line between the programmer and the user, making it possible for moderately sophisticated users to program simple repetitive mathematical and logical operations and to share the programs with other users. Finally, the first commercially successful video arcade game, PONG, produced by Atari, hit the market in several versions around 1973 (Winter, 1999).

1978–1988

State-of-the-Art

The beginning of this period ushered in a paradigm shift in the nature of computer interfaces. The approach to interface design went from a keyboard-controlled *glass teletype* running software having a frame-driven dialog imbedded in the application with rigid format and coding constraints to the fundamental features of the graphic user interface (GUI), well known today. Stimulated by the demonstrations of NLS, conceptions of the Dynabook, the Smalltalk object-oriented software programming environment, together with significant hardware improvements enabling much higher resolution bit-mapped displays and faster multiprocess hardware, the era of GUIs began. Bit-map graphics supported selective redisplay of screen images and sophisticated font and icon design. The improved hardware also included memory cashe systems that, in turn, supported multiple window displays.

The Xerox-developed display software, D-Lisp and the BRAVO text editor, implemented on the ALTO were the first instantiations (Teitelman, 1977) soon to be incorporated in the Xerox Star secretarial workstation in 1982 and then, of course, next in the Apple Lisa in 1983. and in the original Macintosh in

1984 (Bewley, Roberts, Schroit, & Verplank, 1983). No longer were abstruse command-line formatting instructions required in word processors, The text could be portrayed in literal page-layout format and printed without transformation on laser printers, another significant PARC development, so that "What you see is what you get!" (WYSIWYG). On the basis of quantitative human factors studies, Englebart's mouse was selected as the interactive device of choice to supplement the keyboard (Card, English, & Burr, 1978). The Xerox Star introduced the desktop metaphor that has become virtually a standard in today's GUIs. The concept of the Icon, also a Xerox invention, to represent desktop objects, was more than a simple pictogram. It provided a visual manifestation of the products of computing that had mnemonic value and could be buttoned, dragged, and dropped into other icons representing destinations as a means for carrying out intended actions on them (Bewley et al., 1983).

Perhaps even more importantly, facilitated by the object-oriented conception of software, the advent of the GUI brought with it the clean separation of the user interface code from the application code it was designed to support. Objects essentially had their own interface and made it possible to hide information internally that was not relevent to other objects. The user interface became a technology to be developed on its own. Ben Shneiderman coined the term, "Direct Manipulation Interfaces," to capture this paradigm shift (Shneiderman, 1983), but the term GUI has persisted in the literature.

On the one hand, the GUI made the computer much more accessible to nonprogrammer users who just wanted to do their work. On the other hand, it increased tremendously the sophistication of the things that a nonprogrammer user might want to do. On the one hand, it made the user interface much more intuitive. On the other hand, it increased the complexity and thereby increased the potential for user confusion and the challenge of user interface design because it relaxed many of the design constraints that previously existed. No longer could one layout a series of display frames in advance and be confident of all the actions a user could do next. No longer was one restricted to blinking and inverse video as text coding dimensions. The envelope of possibilities, manifest in multiple windows and applications, each with its own icons, menus, and widgets, seldom could be fully understood. Once again, the technological capabilities were outstripping the ability of the HCI community to accommodate user requirements.

Iterative System Development

It was during this time also that the industrial world was beginning to recognize the limitations of the waterfall system development or product development process and began experimenting with a more spiral process in which you build-a-little-test-a-little. This iterative approach had great appeal to software interface developers because it was much easier in software to develop prototypes that represented successive approximations to a final design. Furthermore, it created opportunities for the HCI and user communities to evaluate and critique designs in the formative stages before they were cast in stone. As early as 1986, Marilyn Mantei described how to incorporate typical

HCI methods into the software life cycle processes of the time (see Mantei & Teorey, 1989). John Carroll (1997) attributes the fundamental idea of iterative cycles of prototyping and design to the industrial designer Henry Dreyfuss (1955). It was Fred Brooks (1975), referring to software programs, who said, "Plan to throw one away, you will anyhow" (p. 116).

Interface Builders

Once the interface had been disentangled from applications and the technology of GUIs began to mature, there was interest in simplifying the implementation of GUIs. In 1979–1985, Albert Stevens, Bruce Roberts, and their colleagues at BBN together with James Hollan at the Naval Personnel Research and Development Center (NPRDC) built a prototype graphics-based simulation system for training Navy crewmen to operate a steam propulsion plant (Stevens, Roberts, & Stead, 1983). The software demonstrated many of the ideas later incorporated into "Interface Builders," as they came to be called. It had a library of widgets corresponding to components in a steam plant (valves, turbines, pumps, gauges, switches, piping) that could be introduced into a diagram by simply dragging them to the proper location and dropping them in place, then adjusting their visual properties and connecting them to the underlying simulation through a menu interface. As the simulation ran, the widgets were automatically updated to reflect the changes in the simulation (gauges moved, pipes showed animated flow) and the user controlled the simulation by interacting with the widgets using a mouse (opening and closing valves). Brad Myers suggests that Stevens and Roberts' "Steamer" was possibly the first object-oriented graphics system (Myers, 1998). In 1981, Austin Henderson built Trillium (Henderson, 1986), which he referred to as a user interface design environment, building on and greatly extending the Steamer work. The real power of this approach became evident when the Macintosh was introduced. The Macintosh incorporated widgets directly on a computer chip, which allowed Apple to control their use by constraining software developers to use a well-defined and limited set. In addition, the Macintosh developer tools included a Resource Editor that further simplified the interface development task. Widgets-on-a-chip and the Resource Editor, together with carefully constructed guidelines for developers, were very successful in ensuring that each new application, particularly applications developed by a third party, would retain the look and feel intended by Apple's design staff. Users benefited because the Macintosh developer tools enforced a consistent look and feel across applications.

Donald A. Norman

Don Norman has identified with the field of HCI at least since 1981 when he authored an irreverent paper criticizing the UNIX interface (Norman, 1981a). Having first been a leader in the field of human information processing, then in cognitive science, at the University of California, San Diego, he brought to the applied world in general and the HCI world in particular a profound knowledge of the relevant cognitive psychology and behavioral science. The book he co-edited with Stephen Draper and contributed to in 1986 on "User-Centered System Design" focused specifically on HCI and set the tone for how behavioral science could impact HCI. In it, he used the term Cognitive Engineering, a term that, in the author's opinion, he coined (Norman, 1981b). He spoke of the Gulf of Execution and the Gulf of Evaluation, referring to the gulf between human goals and the physical interfaces through which humans must communicate with technology:

The user of the systems starts off with goals expressed in psychological terms. The system, however, presents its current state in physical terms. Goals and system state differ significantly in form and content, creating the Gulfs that need to be bridged if the system can be used. The Gulfs can be bridged by starting in either direction. The designer can bridge the Gulfs by starting at the system side and moving closer to the person by constructing the input and output characteristics of the interface so as to make better matches to the psychological needs of the user. The user can bridge the Gulfs by creating plans, action sequences and interpretations that move the normal description of the goals and intentions closer to the description required by the physical system. (Norman & Draper, 1986, pp. 38-39)

Over the next several years, he became more committed to working on applied problems that could be informed by cognitive science. He worked on understanding typewriting skill. He undertook a deep analysis of human error and his landmark book, *The Psychology of Everyday Things*, later renamed *The Design of Everyday Things* (Norman, 1988), sensitized the public to the way people think and its implications for design. In 1993, he placed a foot solidly in the applied camp when he joined Apple Computer, Inc., and since then has been a consistent advocate for better HCI educational preparation for work in industry and better understanding of what cognitive science has to contribute to design.

Human–Computer Interaction as a Professional Field

It was during this period that specialists began to speak of a professional field of usability and HCI. The first textbook, (Shneiderman, 1980) was published, to be followed by many others, for example, Foley and Van Dam (1982) and Smith and Green (1980). In 1978, Ramsay and Attwood published a landmark bibliography abstracting the existing publications in the then nascent field (Ramsey & Atwood, 1979). Gary Perlman deserves credit for initiating a still more extensive bibliography project in 1988 that is now the primary reference source for HCI materials (Perlman, 2001).

Brian Shackel, quoting Brian Gaines, attributes the first HCI conference to the International Conference on Man–Machine Systems in 1969 at Cambridge, England, Shackel (1997), but that event was really a human systems conference at which HCI was highlighted. In 1981, the Association for Computing Machinery (ACM) Special Interest Group on Social and Behavioral Computing (SIGSOC) held a conference in Ann Arbor, MI, entitled, "Conference on Easier and More Productive Use of Computing." The first general conference devoted to HCI was held in Gaithersburg, MD, in 1982 and at that meeting it was announced that SICSOC would change its name to the Special

Interest Group on Computer–Human Interaction (ACM SIGCHI; http://www.acm.org/sigchi/). It was officially founded later that year (Borman, 1996). The series of CHI conferences officially began in 1983. The first international conference, IFIP Interact '84, was held in London. The annual SIGCHI Conference and the biannual INTERACT Conference remain the premiere HCI forms in the field.

The *International Journal of Man-Machine Systems*, founded in 1968, began publishing articles concerned with HCI in the 1970s and 1980s, and changed its name to *International Journal of Human-Computer Studies* in 1994. It was followed soon after by *Behaviour and Information Technology* in 1982, *Human-Computer Interaction* in 1985, *International Journal of Human-Computer Interaction*, and *Interacting with Computers* in 1989.

Usability Testing

Usability testing, as a part of the research on interactive technologies, was already going on in laboratories around the world, but it was during this period that the first application usability testing was initiated. IBM established formal usability labs. Considerable formative testing and evaluation went into the development of the Xerox Star Workstation (Bewley et al., 1983). In 1978, the Social Security Administration was undertaking a serious design exercise focused on the "Future Process," a major software development that was to replace Social Security hardware and software with a distributed interactive system that provided for source data entry and retrieval in Social Security Offices around the country. BBN set up a usability laboratory at Social Security headquarters in Baltimore in support of this project. In the course of the work, prototype interface designs were developed on Xerox Altos[4] and more than 60 Social Security Claims Representatives from around the country spent a week at headquarters participating as users in alternative design evaluations. A year later, management of Social Security changed and plans for the Future Process were abandoned. It took at least another 5 years before Social Security began the serious implementation of large-scale interactive systems in the field.

Naturalistic Observation and Ethnography

Converging forces derived from perceived limitations of cognitive task analysis based on interviews, and laboratory experimentation with alternative interface designs, led to new emphasis on field, naturalistic observation and contextual research as a means for better understanding the user, and the environment in which work is accomplished. John Whiteside led this movement, deriving his arguments from the philosophy of Heidegger (1962) (Whiteside, Bennett, & Holtzblatt, 1988). The argument was that truly useful applications depended on understanding in depth the social and environmental context in which the software would be used, and for that purpose there was no

substitute for extensive in situ observation in the workplace. At about the same time, anthropologists and ethnographers whose specific expertise is in just this kind of field observation, albeit normally in more culturally remote field sites, also began supporting HCI requirements analysis. Suchman (1987) specifically refers to *situated action*, suggesting that activity is goal driven and is tied to the current knowledge state of the individual as well as the task-specific state of the workplace. It is now quite common to find anthropologists or ethnographers working side by side with HCI designers.

1989–1999

State-of-the-Art

The scene now shifts to the period that we can expect to be familiar to most readers. Without a doubt, this should be characterized as the era of the Internet. Although much of the preparation occurred earlier, it was in the 1990s that its major impact began to be felt. The complete timeline is documented by The Hobbes' Internet Timeline (Zakon, 2001). Only a few relevant highlights are presented here. Packet-switching technology was first introduced in 1969 with the building of the ARPANET by the ARPA. The TCP/IP protocol was adopted as the standard protocol for network transmissions in 1982. The first long-haul network e-mail was sent over the ARPANET in 1972 by Ray Tomlinson, a colleague of the author at BBN. The introduction of a Domain Name System, a common address space that eliminated the need to know the IP address of the recipient, took place in 1984. The NSFnet was created in 1984 to link the five major supercomputer centers that were created by the National Science Foundation, and this network became a national backbone to which other networks soon linked. At this point it could really be called an Internet.

In 1999, Tim Berners-Lee, a researcher at CERN, the European Organization for Nuclear Physics Research in France, first described the idea for a hypertext/hypermedia information management system that would be available at CERN for the management of myriad documents to communicate, share information, and share software across groups (Berners-Lee, 1989). It had to work across platforms. It had to move documents across networks. It needed a scheme for identifying local and remote documents (evolved into Uniform Resource Locators, (URLs)) and required a formatting language (Hypertext Markup Language, (HTML)) (Wilson, 1999). Berners-Lee imagined being able to summarize and analyze data across these very large hypertext distributed databases. He initially referred to the concept as *Mesh*, but changed the name to World Wide Web when he began to implement it. Quickly, there was a flurry of browser developments, but MOSAIC, the internet information browser and www client created by the University of Illinios National Center for Supercomputing Applications (NCSA) in 1993 quickly became the dominant technology (NCSA, 2000). Marc Andreessen and some colleagues left the University of Illinois and founded Netscape, thus privatizing the technology

---

[4]Special arrangements were made with the Xerox Corporation to lease Altos, because none were ever made available for sale.

developed at NCSA. Subsequently, the technology was licensed by many others wishing to get into the Internet game. The NCSA technology provided the foundation for Microsoft's Internet Explorer as well (NCSA, 2000).

There is no need to document here the impact the Internet has had on the business, commercial, and educational worlds. However, the jury may still be out on whether e-commerce will be the major economic force that was once forecast. The major strengths of the technology—(1) near-universal access, (2) relative ease of locating unique data, and (3) data that may need to be backed up, but never replicated—must be balanced against continuing concerns about privacy, security, and impersonalization.

Web Usability

Of course the advent of the Internet and the World Wide Web substantially changed the HCI game again, just when its specialists were beginning to think they had a handle on things. Those inexperienced users who first explored the web via an HTML hypertext browser were wildly enthusiastic. Military officers asked why all software couldn't be this easy to use? It truly reduced the task to point, click, and fill in the blanks. But, contrary to early popular belief, it soon became clear that it was rather easy to get lost in myriad hypertext layers and not realize how you reached that point. The sources of information were so vast that it became difficult to know how to reach a site of particular interest, giving birth to search engines, history lists, and bookmarks. In the drive to make economical use of page space, too many things were (and still are) presented on a single page, making it difficult to find and review material of interest. Efficiency of information retrieval (now often referred to as the quality of service) depended on the capacity of the user's computer, the capacity of the modem in use, the capacity of network communication links accessed, the load on the network and the complexity of the retrieval request and graphics of the to-be-downloaded page. User-system response time became an issue all over again.

Thus, *design* of web pages and the architecture of web sites has become a major occupation for specialists drawn from the technical writing community, the industrial design community, the advertising community, not to mention the usability and HCI communities.

The 1990s also signaled the mushrooming growth of interest and activity in HCI in general. The man on the street began to complain about how difficult it was to program a videorecorder (*Business Week*, 1991). Usability and ease of learning claims were found in advertisements. In some organizations, usability became an informal, but important product requirement. Other major software companies such as Oracle and Fidelity formally incorporated usability requirements into their product development process and created such positions as Vice President, User Interface Design, and Senior Director, Advanced User Interfaces. Promising dot-coms with extensive venture capital created an insatiable market for web pages. A small proportion of these companies actually became concerned about the usability of their pages.

Speech Technology

Speech has been a proverbial interface medium since first envisioned by Licklider in 1961. Unfortunately, the technology to support it only began to reach maturity in the late 1990s. The first serious research support of speech understanding was sponsored by ARPA in 1971. It brought together speech recognition and language specialists as a result of the insight that speech recognition was not going to advance without incorporating information derived from syntax and semantics. Limited progress was made, and after 3 years, ARPA discontinued financing because their management believed there was little real hope that it could ever be accomplished in real time. Serious DARPA research funding resumed in 1984, largely after the promise of Hidden Markoff Models (HMMs) for this application was identified (Baker, 1979). By 1987, virtually all the research was focused on the application of HMMs. It was only in the 1990s that the confluence of improved processor speeds and more efficient HMM pattern matching algorithms led to feasible, software-only (no requirement for special purpose hardware), real time, continuous, speaker independent, large vocabulary recognition on personal computers.

The first successful application of a speech recognition interface was by the United Parcel Service in the 1980s for hands-free package sorting, and at that time required speaking isolated words (digits). It was the 1990s breakthroughs that have opened the gates to more widespread application, and even today cost-effectiveness is limited to special domains, such as telephone call directors and information services.

Cost-Justifying Usability

Marketing the need for usability requirements has become an important part of the HCI job, and in the late 1980s, there was increased attention to the cost-effectiveness justification for including usability as an integral part of design cycle. The idea was first proposed in Mantei and Teorey (1988) and is well documented in Bias and Mayhew (1994). It can now be argued, sometimes quite convincingly, that attention to usability can result in significant cost savings. A frequently cited example is Project Ernestine, in which a GOMS model predicted that a new telephone toll assistance operator console the company was considering buying from an independent vendor would actually cost the company $2 million per year more to operate than the one it was replacing. This argument convinced the client not to purchase it. Subsequent user testing verified these results (Gray, John, & Attwood, 1993). One lesson learned from the early efforts to calculate cost benefits is the importance of documenting the usability and effectiveness of an application in its original form and as development progresses so that savings can be attributed to the changes.

Continued Professional Development

There was a flurry of new textbooks—Dix, Finlay, Abowd, and Beale (1993), Hix & Hartson (1993), Mayhew (1992), Preece et al. (1994), and Shneiderman (1992)—perhaps the most influential being Jacob Nielsen's (Nielsen, 1993), Neilsen was a pioneer in advocating discount usability methods, by which he meant shortcuts from the more formal (and costly) usability testing designs derived from experimental psychology. There was some research underway at the time to support his recommendations. Methods he advocated include:

1. Prototypes on paper or in simple prototyping environments that only follow a single scenario thread at a time and allow for rapid revision and re-evaluation.
2. Simplified thinking aloud based on notes taken at the time of a user's walkthrough rather than detailed protocol analysis of videotaped sessions.
3. Heuristic evaluation that involves experts exercising the software and applying a limited set of broad user guidelines or heuristics to identify usability problems (Neilsen, 1993).

He made an important contribution by quantifying the results of heuristic evaluations in terms of the number of problems found as a function of the number of evaluators and, for actual usability tests, the recommended number of users required to achieve a given level of confidence in the results (Nielsen, 1993). While Nielsen did not directly advocate it, another useful technique championed during this period was the cognitive walkthrough, which is similar to Nielsen's heuristic evaluation (Lewis, Polson, & Rieman, 1991). In a cognitive walkthrough, the investigator steps through one or more typical tasks for which the interface would be used. At each step, the analysts ask whether, from the perspective of how the user thinks about it, the task is being supported. The current status of these approaches and others is presented and elaborated in other chapters of this handbook.

Computer-Supported Cooperative Work

When the ARPANET was first conceived, the motivation that sold the program was the idea that university scientists would be able, via the computer network, to transfer files back and forth and to collaborate at a distance. This was a part of Bush and Licklider's original vision that was carried forward by Englebart, Nelson, and a host of others. It came as a complete surprise that e-mail was the first widely accepted application of the ARPANET; however, it certainly supported the use of a computer network for collaboration independent of geography and identity (Hafner & Lyon, 1996).

Computer-to-computer communication on a packet-switched network began with the first ARPANET connections in 1969. The TELENET protocol was created soon thereafter and was widely used by specialists in the field to interconnect machines. The first e-mail on a packet-switched network was sent in 1971, and the first person-to-person real-time chat over a long-haul network took place in 1972 (Zacon, 2001). Files were exchanged with difficulty initially, but then more easily through the FTP protocol. However, it has only been with the introduction of e-mail attachments that file transfers have been widely accessible to the public.

Video teleconferencing, desktop video, even use of the telephone, should legitimately be considered under the aegis of computer-supported cooperative work (CSCW), but their discussion is beyond the scope of this chapter.

There were two software threads that could legitimately be called groupware: Lotus Notes Domino and GroupSystems. Lotus Notes originally supported mainly asynchronous collaboration. The original concept for notes was developed at the University of Illinois in the Computer-based Education Research Laboratory, in 1976 then called Plato Notes and later Plato Group Notes, and was used strictly for intragroup communication. The first commercial release of Lotus Notes was in 1989 and was the product of Ray Ozzie and his colleagues, several of whom had a history that included the Plato group. It included online discussion, e-mail, phone books, and document databases (Mathers, 2001). At first it was sold as an internal collaboration product, but as the Internet emerged, Notes quickly adapted to the broader application and has been widely successful. Lotus Notes has not won any awards for usability. Its success is a function of the extent to which it satisfies a corporate need, regardless of its ease of use. (See http://www.iarchitect.com/mtarget.htm for an extensive list of specific interface problems.)

At about the same time, Skip Ellis at Microelectronics and Computer Technology Corporation and Jay Nunamaker and his colleagues, working at the University of Arizona, were building and experimenting with synchronous face-to-face collaborative meeting room computer support for large group meetings. This work built on earlier efforts of Joyner and Tunstall (1970) and others (Kraemer & King, 1986; Jessup & Valacich, 1993). Nunamaker's developments achieved a critical mass in 1989 with the incorporation of the Ventana Corp., a company that offers the product, GroupSystems, initially to support collaborative decision making in a single meeting room setting, but now broadened to include remote collaboration (Nunamaker, Dennis, Valacich, Vogel, & George, 1991; Ventana, 2001).

Remote collaboration, which may still be regarded as a developing technology, creates a whole new set of HCI challenges. Productive work requires embedding the collaborators in the same context and minimizing the time spent in facilitating the communication itself (Are we on the same page? Who just wrote that? Who is controlling the pointer now?), as contrasted with accomplishing useful communication. Making turn-taking transparent and easily understood, and striving to capture the richness of face-to-face communication are challenges that continue to be addressed (Olson et al., 1993).

1999 AND BEYOND

State-of-the-Art

We appear to be heading into the much heralded epoch of ubiquitous computing. Mark Weiser, of PARC, who died in 1999 at the age of 46, coined the term, championed the vision, and led serious development efforts toward this goal. He said,

For thirty years most interface design, and most computer design, has been headed down the path of the "dramatic" machine. Its highest ideal is to make a computer so exciting, so wonderful, so interesting, that we never want to be without it. A less-traveled path I call the "invisible"; its highest ideal is to make a computer so imbedded, so fitting, so natural, that we use it without even thinking about it. (I have also called this notion "Ubiquitous Computing.") (Weiser, 1996)

The PARC developments were focused on ubiquitous computing in an office, including seamless connections among wearable, "deskable" and wall-mounted computers (Weiser, 1993). There is a natural extension to the consumer world in which more and more of our products actually contain computer chips. Donald Norman has gone so far as to predict transparently supplemented hearing, sight, memory, and reasoning via computer(s) implanted under the skin (Norman, 2001a).

The enabling technologies to accomplish much of this are already available. Many of them are listed in Table I.1. The three things that are missing are the multiindustry cooperation, the systems engineering required to make them all work together seamlessly, and, most importantly, from the point of view of this chapter, the human factors knowledge and experience to understand just what it would mean for them to be transparent to the human user. Research is still needed to further interpret this concept.

TABLE I.1. Technologies That Are Impacting Current and Future HCI Developments

Technologies
Global positioning systems (GPSs)
Personal digital assistants (PDAs)
Cellular telephones
Wireless application protocol (WAP)
Bluetooth short-range radio links
Miniature digital television cameras
Liveboards
Wearable computers

Note. HCI = human–computer interface.

Human–Computer Interaction Challenges and Ubiquitous Computing

On the one hand, we want to bury the complexity, but on the other, the users want to have control of the devices with which they are interacting. In the case of the automobile, the computer engine controls of today are transparent because they involve little or no modification of the existing controls. But, as our products become more sophisticated, for example, auto navigation computers and cruise and headway control sensors and computers, there will surely be additional controls. Perhaps an even better example, highlighted in a cogent critique by Donald Norman, is how to make home theater transparent to the user (Norman, 2001b). These two dimensions of future systems, level of sophistication and control, cannot be independently manipulated.

An important feature of such transparency is context sensitivity. Specialists already recognize that one of the keys to successful, natural, human interaction is detecting and accommodating the context in which the user is operating. A very significant part of that context is to know or be able to infer the user's intent. The GPS systems give us the location context. Clever system design can support some level of intent inferencing. Speech recognition has the potential to support more. Nevertheless, one of the challenges of the next several years will be to improve our methods and techniques for intent inferencing.

Portability goes hand in hand with small size. Even if the technical issues associated with seamless and timely downloading of megabits of information into pocket-sized devices are solved, an additional significant challenge is to understand how, indeed even whether, it is possible to provide, in the size of handheld devices, all the information management, display, and manipulation capabilities that a user would like to have. There is general agreement that the Apple Newton and the Palm Pilot have done an excellent job with the basic capabilities they provide, but the potential sophistication of such devices has not begun to be realized, and it is not clear that Palm's solutions will scale to these new requirements. We can expect to be overwhelmed with information and that it will be available to us anywhere via mobile connectivity. There is more work to be done to decide where to draw the line, in the range from wearable devices to liveboards; to decide what to make available in different scale devices (who wants to read a lengthy e-mail or order a new computer on their personal data unit?), and just how much of it should be available on mobile devices.

Interacting through speech has been the holy grail for many years. We can expect continued improvement in the performance of speech recognition systems and the naturalness of text-to-speech systems. However, making these systems more widely useful requires more than improved technology. Most specialists now realize that human-to-human communication is imperfect and is facilitated by a very wide range of sources of contextual cues that will need to be mimicked and interpreted in some way to make human-to-computer communication as facile.

In 2001, the final version of Section 508 of the Rehabilitation Act was approved, requiring equal access for disabled persons to electronic and information technology provided to federal employees and for public access to federal electronic and information technology. This requirement, although acknowledged for a number of years, is now the law. It represents significant challenges for HCI developers, but such efforts should be broadly rewarding. (See chapters 25 and 26 in this handbook.) The fact is that making information accessible to many types of disabled persons usually also enhances its accessibility for the rest of us. Alphonse Chapanis once noted that, when he revised some insurance regulations so that they met the standards for eighth grade readability, college graduates found them more accessible as well.

New Methodologies

As the scope and sophistication of digital systems become ubiquitous, the pressure for improved HCI methodologies will continue to increase. On the design side, specialists and funding organizations have long sought the ultimate design methodology that would provide seamless integration of the elements of product usability into a single design support tool. Such a tool would include means for documenting the results of cognitive task analyses and naturalistic observation at the time they were collected. The analysis would directly support the generation of representative scenarios of use that, in turn, provide the data for user system requirement specification and system architecture design. The scenarios and requirements then could be converted easily into quantitative models that will be useful for evaluating alternative design concepts. Prototyping tools for formative evaluation by experts and by potential users would build from the scenarios and models, and make multiple levels of prototyping detail realizable as the design matures. In this idealized methodology, the models developed along the way would continue to be available for support of further exploration and evaluation of detailed design alternatives. The final results of the prototyping studies would be directly convertible into efficient application code.

On the usability evaluation side, the scenarios and prototypes would be readily converted into a simulation that would support human-in-the-loop studies of the effectiveness of the resultant designs. The tools would include online performance measures, fully indexed digital video recording, and spoken interaction automatically synchronized into a single visualization and analysis package for summary and reporting.

The Last Word

Over the last four decades, the profession of HCI has emerged. Although it began with relatively mundane human factors questions about the format and layout of programs and data, what has emerged is a collection of activities that can be regarded as central to the society of the 21st century. We are the gatekeepers to ensure that the technological capabilities that are thrust on us will be manageable and responsive to human requirements. This responsibility does not rest with social planners, sociologists, or politicians. It rests with us because we have the opportunity to design the interfaces that translate that technology into tools and products that are useful, usable, meaningful, accessible, rewarding, and fun.

ACKNOWLEDGMENTS

I thank Stuart Card, Richard Granda, Richard Hirsh, Bonnie John, Clayton Lewis, John Makhoul, Bruce Roberts, Marilyn Tremaine, and Ray Tomlinson for many helpful suggestions.

References

Baecker, R. M., & Buxton, W. A. S. (1987). *Readings in human-computer interaction: A multidisciplinary approach*. Los Altos, CA: Morgan Kaufmann.

Baecker, R. M., Grudin, J., Buxton, W. A. S., & Greenberg, S. (1995). *Readings in human-computer interaction: Toward the year 2000* (2nd ed.). San Francisco: Morgan Kaufmann.

Baker, J. (1979). Trainable grammars for speech recognition. In D. Klatt and J. Wolf (Eds.), *Speech Communication Papers for the 97th Meeting of the Acoustical Society of America* (pp. 547-550). Cambridge, MA: MIT Press.

Bewley, W. L., Roberts, T. L., Schroit, D., & Verplank, W. L. (1983). Human factors testing in the design of Xerox's 8010 "Star" Office Workstation. In *Proceedings of the CHI '83 Conference* (pp. 72-77). New York: Association of Computing Machinery.

Berners-Lee, T. (1989). Information management: A proposal. CERN. Available: http://www.w3.org/History/1989/proposal.html

Bias, R. G., & Mayhew, D. J. (1994). *Cost justifying usability*. Cambridge, MA: Academic Press.

Borman, L. (1996). SIGCHI: The early years. *SIGCHI Bulletin* [On-line], *28*(1), 4-6. Available: http://www.acm.org/sigchi/bulletin/1996.1

Brooks, F. P. (1975). *The mythical man-month: Essays on software engineering*. Reading, MA: Addison-Wesley.

Bush, V. (1945). As we may think. *Atlantic Monthly, 176*, 101-108.

Card, S. K., English, W. K., & Burr, B. J. (1978). Evaluation of mouse, rate-controlled isometric joystick, step keys and text keys for text selection on a CRT. *Ergonomics, 21*(8), 601-613.

Card, S. K., Moran, T. P., & Newell, A. (1980a). The keystroke-level model for user performance time with interactive systems. *Communications of the ACM, 23*(7), 396-410.

Card, S. K., Moran, T. P., & Newell, A. (1980b). Computer text editing: An information-processing analysis of a routine cognitive skill. *Cognitive Psychology, 12*, 32-74.

Card, S. K., Moran, T. P., & Newell, A. (1983). *The psychology of human-computer interaction*, Hillsdale, NJ: Erlbaum.

Carroll, J. M. (1997). Human-computer interaction: Psychology as a science of design. In *Annual review of psychology* (Vol. 48, pp. 61-83). Palo Alto, CA: Annual Reviews.

Dickason, E. (2001). Remembering Grace Murray Hopper: A legend in her own time. Available: http://www.norfolk.navy.mil/chips/grace_hopper/file2.htm

Dix, A., Finlay, J., Abowd, G., & Beale, R. (1993). *Human-computer interaction*. Hillsdale, NJ: Prentice Hall.

Dreyfuss, H. (1955). *Designing for people*. New York: Simon & Schuster.

Engle, S. E., & Granda, R. E. (1975). Guidelines for man/display interfaces. *IBM Technical Report TR 00.2720*. Poughkeepsie, NY: IBM.

Englebart, D. (1963). A conceptual framework for the augmentation of man's intellect. In P. W. Howerton, & D. C. Weeks (Eds.), *Vistas in information handling* (Vol. 1, pp. 1-29). Washington, DC: Spartan.

Ericsson, K. A., & Simon, H. A. (1980). Verbal reports as data. *Psychological Review, 3*, 215-251.

Foley, J. D., & Van Dam, A. (1982). *Fundamentals of interactive computer graphics*. Reading, MA: Addison-Wesley.

Gaines, B. R., & Shaw, M. L. G. (1986). From timesharing to the sixth generation: The development of human-computer interaction. Part I. *International Journal of Man-Machine Studies, 24*, 1-27.

Galitz, W. O. (1981). *Handbook of screen format design*. Wellesley, MA: Q. E. D. Information Sciences.

Gold, M. (1969). Time sharing and batch processing: An experimental comparison of their values in a problem-solving situation. *Communications of the ACM, 12*(5), 249-259.

Goldberg, A. (1998). *A history of personal workstations.* Reading, MA Addison-Wesley.

Grant, E. E., & Sackman, H. (1967). An exploratory investigation of programmer performance under on-line and off-line conditions. *IEEE Transactions on Human Factors in Electronics, HFE-8*(1), 33-48.

Gray, W. D., John, B. E., & Atwood, M. E. (1993). Project Ernestine: Validating a GOMS analysis for predicting and explaining real-work task performance. *Human-Computer Interaction, 8,* 237-309.

Hafner, K., & Lyon, M. (1996). *Where wizards stay up late: The origins of the Internet.* New York: Simon & Schuster.

Hanson, W. (1971). User engineering principles for interactive systems. In *AFIPS Conference Proceedings 39, Fall Joint Computer Conference* (pp. 523-532). Montvale, New Jersey: AFIPS Press.

Heidegger, M. (1962). *Being and time* (J. Macquarrie and E. Robinson, Trans.). New York: Harper Row.

Henderson, D. A., Jr. (1986). The trillium user interface design environment. In *Proceedings SIGCHI'86: Human Factors in Computing Systems* (pp. 221-227). New York: Association for Computing Machinery.

Hix, D., & Hartson, H. R. (1993). *Developing user interfaces: Ensuring usability through product and process.* New York: John Wiley & Sons, Inc.

Jessup, L., & Valacich, J. (Eds.). (1993). *Group support systems: New perspectives.* New York: Macmillan.

Joyner, R., & Tunstall, K. (1970). Computer augmented organizational problem solving. *Management Science, 17*(4), 212-225.

Kay, A. (1969). The reactive engine. PhD Thesis, University of Utah, Salt Lake City.

Kemeny, J. G., & Kurtz, T. E. (1966). *BASIC* (3rd ed.). Hanover, NH: Dartmouth College.

Kraemer, K. L., & King, J. L. (1986). Computer-based systems for group decision support: Status of use and problems in development. *Proceedings of the Conference on Computer-Supported Cooperative Work* (pp. 353-375). Austin, TX.

Lampson, B., & Taft, E. (1979). *Alto user's handbook.* Palo Alto, CA: Xerox Palo Alto Research Center.

Landauer, T. (1995). *The trouble with computers.* Cambridge, MA: The MIT Press.

LegacyJ Corp. (2001). COBOL milestones. Available: http://www.legacyj.com/cobol/cobol_history.html

Lewis, C. (1982). Using the "thinking-aloud" method in cognitive interface design. *IBM Research Report RC 9265.* Yorktown Heights, NY: IBM Thomas J. Watson Research Center.

Lewis, C., Polson, P. G., & Rieman, J. (1991). Cognitive walkthrough forms and instructions. *Institute of Cognitive Science Technical Report #ICS 91-14.* Boulder, CO: University of Colorado.

Licklider, J. C. R. (1960). Man-computer symbiosis. *IRE Transaction on Human Factors in Electronics, HFE-1,* 4-11.

Licklider, J. C. R. (1965). *Libraries of the future.* Cambridge, MA: MIT Press.

Mantei, M., & Teorey, T. J. (1988). Cost/benefit analysis for incorporating human factors in the software lifecycle. *Communications of the ACM, 31*(4), 428-439.

Mantei, M., & Teorey, T. J. (1989). Incorporating behavioral techniques into the system development life cycle. *Management Information Systems Quarterly, 13*(3), 257-276.

Marks, S. L., & Armerding, G. W. (1967). The JOSS primer *Memorandum RM-5220-PR.* Santa Monica, CA: The Rand Corp.

Martin, J. (1973). *Design of man-computer dialogues.* Edgewood Cliffs, NJ: Prentice-Hall.

Mathers, B. (2001). Westford, MA: Iris Associates. Available: http://www.notes.net/history.nsf/text?OpenPage

Mayhew, D. (1992). *Principles and guidelines in software user interface design.* Edgewood Cliffs, NJ: Prentice Hall.

Miller, R. B. (1968). Response time in man-computer conversational transactions. *AFIPS Conference Proceedings, 33,* 267-277.

Morfield, M. A., Wiesen, R. A., Grossberg, M., & Yntema, D. B. (1969). Initial experiments on the effects of system delay on on-line problem-solving. *MIT Lincoln Laboratory Technical Note 1969-5.* Lexington, MA: MIT.

Myer, T. H. (1967). *TELCOMP manual.* Cambridge, MA: Bolt Beranek and Newman.

Myers, B. A. (1998). A brief history of human computer interaction technology. *ACM Interactions, 5*(2), 44-54.

Myers, B., Hollan, J., & Cruz I. (1996). Strategic directions in human-computer interaction. *ACM Computing Surveys, 28*(4), 794-809.

National Center for Supercomputing Applications [NCSA]. (2000). Board of Trustees, University of Illinois Associates. Available: http://www.ncsa.uiuc.edu/Divisions/Communications/MosaicHistory/impact.html

Nelson, T. H. (1965). A file structure for the complex, the changing and the indeterminate. *Proceedings of the ACM National Conference* (pp. 84-100). New York: Association for Computing Machinery.

Nelson, T. H. (1973). A conceptual framework for man-machine everything. *Proceedings of the ACM National Conference* (pp. m21-m26). New York: Association for Computing Machinery.

Nelson, T. H. (1974a). *Computer lib.* South Bend, IN: The Distributors.

Nelson, T. H. (1974b). *Dream machines.* South Bend, IN: The Distributors.

Nelson, T. H. (1981). *Literary machines.* San Antonio, TX: Project Xanadu.

Newell, A., & Simon, H. A. (1972). *Human problem solving.* Englewood Cliffs, NJ: Prentice-Hall.

Nickerson, R. S., & Landauer, T. K. (1997). Human-computer interaction: Background and issues. In M. Helender, T. K. Landauer, & P. Prabhu (Eds.), *Handbook of human-computer interaction: Second, completely revised edition* (pp. 3-31). Amsterdam: Elsevier Science B.V.

Nielsen, J. (1993). *Usability engineering.* Cambridge, MA: Academic Press.

Norman, D. (2001a). Cyborgs. *Communications of the ACM, 44*(3), 36-37.

Norman, D. (2001b). The perils of home theater. Available: http://www.jnd.org/dn.mss/ProblemsOfHomeTheater.html

Norman, D. A. (1981a). The trouble with UNIX: The user interface is horrid. *Datamation, 27*(12), 139-150.

Norman, D. A. (1981b). Comments on cognitive engineering, the need for clear "system images," and the safety of nuclear power plants. *Conference Record for 1981 IEEE Standards Workshop on Human Factors and Nuclear Safety* (pp. 91-92, 211-212). New York: IEEE.

Norman, D. A., & Draper, S. (Eds.). (1986). User-centered system design: New perspectives on human-computer interaction. Hillsdale, NJ: Erlbaum.

Norman, D. A. (1988). The psychology of everyday things. New York: Basic Books.

Nunamaker, J. F., Dennis, A. R., Valacich, J. S., Vogel, D. R., & George, A. F. (1991). Electronic meeting systems to support group work. *Communications of the ACM, 34*(7), 40-61.

Nussbaum, B., & Neff, R. (1991). I can't work this thing. *Business Week,* 58-66.

Olson, J. S., Card, S. K., Landauer, T. K., Olson, G. M., Malone, T., & Leggett, J. (1993). Computer-supported co-operative work: Research issues for the 90s. *Behaviour & Information Technology, 12*(2), 115–129.

Perlman, G. (2001). HCI Bibliography: free access to human-computer interaction resources. Available: www.hcibib.org

Pew, R. W., & Rollins, A. M. (1975). Dialog specification procedures (Rev. ed.). *BBN Report 3129.* Cambridge, MA: Bolt Beranek and Newman.

Pew, R. W., Rollins, A. M., & Williams, G. A. (1976). Generic man-computer dialogue specification: An alternative to dialogue specialists. *Proceeding of the 6th Congress of the International Ergonomics Association* (pp. 251–254). Santa Monica, CA: Human Factors Society.

Preece, J., Rogers, Y., Sharp, H., Benyon, D., Holland, S., & Carey, T. (1994). *Human-computer interaction.* Wokingham, UK: Addison-Wesley.

Ramsey, H. R., & Atwood, M. E. (1979). Human factors in computer systems: A review of the literature. *Technical Report SAI-79-111-DEN.* Englewood, CO: Science Applications, Inc.

Rheingold, H. (2000). *Tools for thought: MIT Press edition.* Cambridge, MA: MIT Press.

Sackman, H. (1970). *Man-computer problem solving: Experimental evaluation of time sharing and batch processing.* Princeton, NJ: Auerback.

Sears, A., & Jacko, J. A. (2000). Understanding the relationship between network quality of service and the usability of distributed multimedia documents. *Human Computer Interaction, 15,* 43–68.

Shackel, B. (1997). Human-computer interaction—Whence and whither? *Journal of the American Society for Information Science, 48*(11), 970–986.

Shackel, B. (2000). People and computers—Some recent highlights. *Applied Ergonomics, 31,* 595–608.

Shneiderman, B. (1980). *Software psychology: Human factors in computer and information systems.* Cambridge, MA: Winthrop Publishers.

Shneiderman, B. (1983). Direct manipulation: A step beyond programming languages. *IEEE Computers, 16*(8), 57–62.

Shneiderman, B. (1992). *Designing the user interface: Strategies for effective human-computer interaction.* Reading, MA: Addison-Wesley.

Smith, H. T., & Green, T. R. G. (Eds.). (1980). *Human interaction with computers.* London: Academic Press.

Smith, S. L., & Mosier J. N. (1986). Guidelines for designing user interface software. *Technical Report ESD TR-86-278.* Hanscom AFB, MA: USAF Electronic Systems Division.

Stevens, A., Roberts, B., & Stead, L. (1983). The use of a sophisticated graphics interface in computer-assisted instruction. *IEEE Computer Graphics and Applications, 3*(2), 25–31.

Stranahan, P. (1996). Dan Bricklin. *Jones Telecommunications and Multimedia Encyclopedia.* Available: http://ei.cs.vt.edu/~history/BRICKLIN.Fleming.HTML

Suchman, L. (1987). *Plans and situated actions: The problem of human-machine communication.* Cambridge, UK: Cambridge University Press.

Sutherland, I. E. (1963). Sketchpad: A man-machine graphical communication system. *AFIOS Conference Proceedings, 23,* 329–346.

Taylor, F. W. (1911). *Principles of scientific management.* New York: Harper.

Teitelman, W. (1979). A display oriented programmer's assistant. *International Journal of Man-Machine Studies, 11,* 157–187.

Ventana (2001). Available: http://www.ventana.com.html

Waldrop, M. M. (2001). *The dream machine: J. C. R. Licklider and the revolution that made computing personal.* New York, NY: Viking.

Weinberg, G. (1971). *The psychology of computer programming.* New York: Van Nostrand Reinhold.

Weiser, M. (1993). Some computer science issues in ubiquitous computing. *Communications of the ACM, 36*(7), 75–84.

Weiser, M. (1996). Mark Weiser. Available: http://nano.xerox.com/hypertext/weiser/UbiHome.html

Whiteside, J., Bennett, J., & Holtzblatt, K. (1988). Usability engineering: Our experience and evolution. In Helander M. (Ed.), *Handbook of human-computer interaction* (pp. 791–817). Amsterdam: North Holland.

Wilkes, M. (1993). From Fortran and Algol to object-oriented languages. *Communications of the ACM, 36*(7), 21–23.

Wilson, B. (1999). HTML overview. Available: http://www.zdnet.com/devhead/resources/tag_library/history/html.html

Winter, D. (1999). Pong-story. Available: http://pong-story.com/atpong1.htm

Zakon, R. H. (2001). Hobbes' Internet Timeline. Available: http://www.zakon.org/robert/internet/timeline/

Part

· I ·

HUMANS IN
HUMAN–COMPUTER
INTERACTION

Mary Czerwinski
Microsoft Research

We are fortunate to be alive at a time when research and invention in the computing domain flourishes, and many industrial, government, and research institutions are aggressively funding creative research in the field of human–computer interaction (HCI). There exists today a deep level of understanding that HCI is a multidisciplinary field, incorporating research and theories from Computer Science, Psychology, Anthropology, Education, Design, Engineering, Math, and even Physics. Partly driven by a flourishing economy and certainly aided by a relatively stable global political situation, much innovation has occurred in HCI since the last edited volume on this topic. In particular, our understanding of how our knowledge about the user, the user's context, and culture all fit together to determine optimal performance and satisfaction during HCI tasks has grown immensely. Rapid advancements in our understanding of the laws governing input devices (e.g., Chua, Weeks, & Goodman, chapter 1, this volume; Accot & Zhai, 2001), the mapping of perceptual codes to display design (e.g., Proctor & Vu, chapter 2, this volume; Ware, 2000), a deeper understanding of how humans process information to the point of automating various aspects in computer simulations (e.g., Byrne, chapter 5, this volume; Pirolli, Card, & Van Der Wege, 2001), and a much better understanding of the emotional and social aspects of HCI (e.g., Brave & Nass, chapter 4, this volume; Picard, 1997) have occurred. In addition, the artificial intelligence community's collaboration with HCI researchers has produced some fascinating examples of how to capture information about a user's context to make better decisions about how to interact with that user (e.g., Horvitz, 1999).

Given these rich opportunities for research and exploration, it would seem fit for any book entitled *Human-Computer Interaction Handbook* to cover in depth the recent progress surrounding the human aspects of interacting with computers. It would be impractical to assemble such a book without addressing those human aspects relevant to computing in the foremost section of the book, much as the editors have done for this volume. The chapters chosen to comprise this section of the book cover most of the human characteristics necessary for optimal HCI design, including perceptual, cognitive, emotional, motoric, and social aspects of computing. Issues related to sociocultural influences on cognition (e.g., Hutchins, 1995) are addressed in various other chapters in this volume. In addition, the authors of the following chapters cover myriad research techniques and methods to provide guidance by way of example to the reader, whether a practitioner, scientist, or student. Methodologies such as chronometric methods and subtraction logic from psychology's early history are covered, in addition to more advanced modeling techniques, physiological recording approaches, and computer simulation efforts.

All of the authors of this section emphasize the need to study the human's perceptual and cognitive processing capabilities to design systems that are more natural, less error-prone, and generally more satisfying to use. Psychologists who work in the software industry typically find themselves designing and evaluating complex software systems to aid humans in a wide range of problem domains, such as word processing, interpersonal communications, information access, finance, remote meeting support, air traffic control, or even gaming situations. In these domains, the technologies and the users' tasks are in a constant state of flux, evolution, and coevolution. Cognitive psychologists working in HCI design may try to start from first principles developing these systems, but they often encounter novel usage scenarios for which no guidance is available. It was encouraging to see the authors of this section describe applications of their theories, models, and specific findings from basic psychological research to user interface design. Of special interest to the reader should be the various analysis techniques and guidelines generated presented throughout the chapters. The chapter authors outline some efforts in HCI research from their own applied research experience, demonstrating at what points in the design cycle HCI practitioners typically draw from the psychological literature. This is especially valuable to students and current practitioners in HCI.

To begin the section, Chua, Weeks, and Goodman introduce two basic theoretical and analytical frameworks as part of their approach to studying perceptual–motor interaction. Integrating aspects of information processing theory and the use of analytical tools to investigate both static and dynamic interactions, the group presents a series of elegant arguments about the limitations of Fitts' Law in predicting movement time for input devices in HCI. Specifically, the authors argue that simply basing estimations of movement time on distance and target size ignores factors such as the number/type of distractor items (as often accompany a target in any display or graphical user interface), in addition to top-down or cognitive decision processes that might influence the selection of a target, and hence, movement to that target. The authors detail the movement process approach (Kelso, 1982), which supplements chronometric techniques in the study of dynamic, perceptual–motor interaction. In addition, they discuss the phenomena of compatibility and some recent data from their laboratory to push their framework further. Given the emphasis on Fitts' Law in most HCI publications and studies, this chapter provides an excellent alternative view and focus in the area of perception and motoric behavior and HCI.

Proctor and Vu provide an overview of the human information processing approach that has so long been the mainstay of perceptual, cognitive, and HCI research. The premise of this chapter is that only through understanding the user's information processing capabilities can the user interface to a computing system be optimally designed. Taking a historical approach, the authors walk the reader through many of the prominent approaches and methods over the last 50 years of psychology, and more recent applications of psychology to computing systems. The authors survey methods used to study human information processing and then provide a summarization of the key findings and theories that guide our explanations of those findings. From chronometric methods, including the subtractive and additive factors techniques, through electrophysiological measurements and recordings of magnetic fields, Proctor and Vu attempt to inform the reader about the need to distinguish between serial and parallel processing stages, and between local and distributed processing models. In addition, excellent coverage of recent models of attention and memory is provided, including a discussion of the effects of practice, analogy, and decision-making

heuristics. This chapter paves the way for a subsequent discussion of Cognitive Architectures by Byrne.

The chapter by van der Veer and del Carmen Puerta Melguizo is an absolute must read for anyone new to the field of HCI, or who has come to HCI from an intense background of psychology or computer science. The authors not only define the meaning of the term "mental model," but they also cover the modern techniques for capturing users' system models and provide a practical approach to applying that modeling with an eye toward system design. Any system designer who has had to sit and watch a real end user attempt to understand a novel user interface will appreciate the guidance presented in this chapter!

Brave and Nass cover the definition of emotions, including how they are measured, the course of emotions, and their causal basis. They review their concept of the media equation, asserting that if the user considers computers as social actors, then the social aspects of computing need to be part of optimal HCI design. This would include capturing the user's emotional state and responding appropriately, if possible. In their discussion of the effects of emotions during HCI, they discuss how emotions tend to alter attention and memory, bias judgment, motivate behavior, and can short circuit cognitive processing and lead to irrationality. In addition, they discuss how emotions can interact with a user's cognitive style, and that emotional interactions with a computing situation could alter future interactions with a system. They argue that capturing a user's emotional state can be useful for diffusing user frustration, providing help when needed, or simply making a user feel satisfied while performing a necessary task.

Byrne's chapter defines cognitive architecture to be "a broad theory of human cognition based on a wide selection of human experimental data, implemented as a running computer simulation program." Cognitive architectures are also described as simulating aspects of cognition that are fairly constant over time and task domain. They are an attempt to bring theoretical unification to cognitive psychology via computer simulation. Speaking from experience, the unification effort itself is of fundamental importance to designers, who may have to rely on their psychology partners to inform them of typically piecemeal empirical findings that are relevant to a computer system's task domain or target end user. Other benefits of using cognitive simulation models are discussed, including quantitative predictions (e.g., error rates, task times, learning curves) and the identification of alternative strategies and comparative findings as a provision for some level of automation in design. Byrne covers a brief history of such systems, leading up to current-day architectures. The shortcomings of these systems is acknowledged, but the chapter ends with a call for more research and development, because the broad importance of these architectures will most likely grow in the future.

The section on humans in HCI closes with a chapter by Yoshikawa also on modeling HCI. The motivation to Yoshikawa's model is to generate good system performance, while avoiding human error in the complex task domains of process control, education, and training. Yoshikawa also discusses the application of findings from human memory, including the various modes of information processing, and what is referred to as the balance sheet of human capabilities. Focusing on the problem of human error, and frameworks for minimizing it through good design, the author provides an overview of Rasmussen's skill, rule, and knowledge-based modes of human action. An important focus in this chapter is the importance of context in the manner in which a user will interpret the task or procedures to carry out given a set of system input signals. In addition, an overview of the common types of errors made during complex system interaction and their classification is described. Similar to Byrne's chapter, methods for error identification and prevention via modeling are discussed. In addition, Yoshikawa describes the importance of observing and recognizing system and user state information through artificial intelligence techniques. Especially in the

aviation and military domains, these are becoming increasingly important techniques for proficiency in our field. The chapter closes with a discussion of application domains.

In all, this section on the human in HCI provides an important overview of many of the methods available to HCI researchers and practitioners as they attempt to understand their users with an eye toward optimal design. A new understanding of motoric behavior, information processing, and user modeling during HCI has been provided and detailed. The compiled works provide the proper tools and techniques for the next generation of advanced human–computer systems. Now it is up to each of us to put these thoughtful, scholarly principles to good use.

References

Accot, J., & Zhai, S. (2001). Scale effects in steering law tasks. In J. Jacko, A. Sears, M. Beaudouin-Lafon, & R. J. Jacobs (Eds.), *Proceedings of CHI 2001 ACM SIGCHI Conference on Human Factors in Computing Systems, 3*, 1–8.

Horvitz, E. (1999). Principles of mixed-initiative user interfaces. *Proceedings of CHI '99, ACM SIGCHI Conference on Human Factors in Computing Systems*, Pittsburgh, PA, May 1999, 159–166.

Hutchins, E. (1995). *Cognition in the wild*. Cambridge, MA: MIT Press.

Kelso, J. A. S. (1982). The process approach to understanding human motor behavior: An introduction. In J. A. S. Kelso (Ed.), *Human motor behavior: An introduction* (pp. 3–19). Hillsdale, NJ: Lawrence Erlbaum Associates.

Picard, R. W. (1997). *Affective computing*. Cambridge, MA: MIT Press.

Pirolli, P., Card, S. K., & Van Der Wege, M. M. (2001). Visual information foraging in a focus + context visualization. In J. Jacko, A. Sears, M. Beaudouin-Lafon, & R. J. Jacobs (Eds.), *Proceedings of CHI 2001 ACM SIGCHI Conference on Human Factors in Computing Systems, 3*, 506–513.

Ware, C. (2000). *Information visualization: Perception for design*. San Francisco, CA: Morgan Kaufman Publishers.

•1•

PERCEPTUAL–MOTOR INTERACTION: SOME IMPLICATIONS FOR HUMAN–COMPUTER INTERACTION

Romeo Chua
University of British Columbia

Daniel J. Weeks and David Goodman
Simon Fraser University

PERCEPTUAL–MOTOR INTERACTION:
A BEHAVIORAL EMPHASIS

Two of us (D.W., D.G.) can still remember purchasing our first computers to be used for research purposes. The primary attributes of these new tools were their utility in solving relatively complex mathematical problems and performing computer-based experiments. However, it was not long after that word processing brought about the demise of the typewriter, and our department secretaries no longer prepared our research manuscripts and reports (but that story is for another time). It is interesting to us that computers are not so substantively different from other tools such that we should disregard much of what the study of human factors and engineering psychology has contributed to our understanding of human behavior in simple and complex systems. Rather, it is the computer's capacity for displaying, storing, and processing information that has led us to the point at which the manner with which we interact with such systems has become a research area in itself.

In our studies of human–computer interaction (HCI) and perceptual–motor interactions in general, we have adopted two basic theoretical and analytical frameworks as part of an integrated approach. In the first framework, we view perceptual–motor interactions in the context of an information–processing model. In the second, we use analytical tools that allow detailed investigations of both static and dynamic interactions. The purpose of this chapter is to outline this approach and some of the current empirical work on perceptual-motor behavior we believe have considerable implications for those working in HCI.

Human Information Processing
and Perceptual-Motor Behavior

For many scientists interested in perceptual-motor behavior, the information-processing framework has traditionally provided a major theoretical and empirical platform for the study of these interactions. The study of perceptual-motor behavior within this framework has inquired into such issues as the information capacity of the motor system (e.g., Fitts, 1954), the attentional demands of movements (e.g., Posner & Keele, 1969), motor memory (e.g., Adams & Dijkstra, 1966), and processes of motor learning (e.g., Adams, 1971). The language of information processing (e.g., Broadbent, 1958) has provided the vehicle for discussions of mental and computational operations of the cognitive and perceptual-motor system (Posner, 1982). Of interest in the study of perceptual-motor behavior is the nature of the cognitive processes that underlie perception and action.

The information-processing approach describes the human as an active processor of information, in terms that are now commonly used to describe complex computing mechanisms. An information-processing analysis describes observed behavior in terms of the encoding of perceptual information, the manner in which internal psychological subsystems utilize the encoded information, and the functional organization of these subsystems. At the heart of the human processing system are processes of information transmission, translation, reduction, collation, and storage (e.g., Fitts, 1964; Marteniuk, 1976; Stelmach, 1982; Welford, 1968).

Consistent with a general model of human information processing (e.g., Fitts & Posner, 1967), three basic processes have been distinguished historically. For our purposes, we refer to these processes as stimulus identification, response selection, and response programming. Briefly, stimulus identification is associated with processes responsible for the perception of information. Response selection pertains to the translation between stimuli and responses and the selection of a response. Response programming is associated with the organization of the final output (see Proctor & Vu, this volume, for a more detailed discussion of these processes).

A key feature of models of information processing is the emphasis on the cognitive activities that precede action (Marteniuk, 1976; Stelmach, 1982). From this perspective, action is viewed only as the end result of a complex chain of information-processing activities (Marteniuk, 1976). Thus, chronometric measures, such as reaction time and movement time, as well as other global outcome measures, are often the predominant dependent measures. However, even a cursory examination of the literature indicates that time to engage a target has been a primary measure of interest. For example, a classic assessment of perceptual-motor behavior in the context of HCI and input devices was conducted by Card, English, and Burr (1978; see also English, Engelhart, & Berman, 1967). Using measures of error and speed, Card et al. (1978) had subjects complete a cursor positioning task using four different control devices (mouse, joystick, step keys, and text keys). Data revealed the now well-known advantage for the mouse. Of interest is that the speed measure was decomposed into homing time and positioning time. The former denoted the time that it took to engage the control device and initiate cursor movement, and the latter the time to complete the cursor movement. Although the mouse was actually the poorest device in terms of the homing time measure, the advantage in positioning time produced the faster overall time. That these researchers sought to glean more information from the time measure acknowledges the importance of the movement itself in perceptual–motor interactions such as these.

The fact that various pointing devices depend on hand movement to control cursory movement has led to the emphasis that researchers in HCI have placed on Fitts' law (Fitts, 1954) as a predictive model of time to engage a target. The law predicts pointing (movement) time as a function of the distance to and width of the target. The impact of Fitts' law is most evident by its inclusion in the battery of tests to evaluate computer pointing devices in ISO 9241-9. We argue that there are number of important limitations to an exclusive reliance on Fitts' law in this context. First, although the law predicts movement time, it does so on the basis of distance and target size. Consequently, it does not allow for determining what other factors may influence movement time. Specifically, Fitts' law is often based on a movement to a single target at any given time (although it was originally developed using reciprocal movements between two targets). However, in most HCI and graphic user interface contexts, there is an array of potential targets that can be engaged by an operator. As we will discuss later in this chapter, the influence of these distractor targets on movements to the imperative target can be significant. Second, we suggest that the emphasis on Fitts' law has diverted attention from the fact that cognitive processes involving the selection of a potential target from an array are an important, and time-consuming, information processing activity that must precede movement to that target. Indeed, the Hick-Hyman law (Hick, 1952; Hyman, 1953) predicts the decision time required to select a target response from a set of potential responses. In fact, if an operator executes the decision and movement components sequentially, then the time to complete the task will be the sum of the times predicted by the Hick-Hyman and Fitts' laws. However, an operator may opt to make a general movement first and select the final target destination concurrently. Under such conditions, Hoffman and Lim (1997) reported interference between the decision and movement component that was dependent on their respective difficulties. Finally, although Fitts' law predicts movement time given a set of movement parameters, it does not actually reveal much about the underlying movement itself. Indeed, considerable research effort has been directed toward revealing the movement processes that give rise to Fitts' law. For example, theoretical models of limb control have been forwarded that propose that Fitts' law emerges as a result of multiple submovements (e.g., Crossman & Goodeve, 1983), or as a function of both initial movement impulse variability and subsequent corrective processes late in the movement (Meyer, Abrams, Kornblum, Wright, & Smith, 1988). These models highlight the importance of conducting detailed examination of movements themselves as a necessary complement to chronometric explorations only.

Translation, Coding, and Mapping. As outlined previously, the general model of human information processing (e.g., Fitts & Posner, 1967) distinguishes three basic processes: stimulus identification, response selection, and response programming. Although stimulus identification and response programming are functions of stimulus and response properties, respectively, response selection is associated with the translation between stimuli and responses (Welford, 1968).

Translation is the seat of the human interface between perception and action. Moreover, the effectiveness of translation processes at this interface is influenced to a large extent by the relation between perceptual inputs (e.g., stimuli) and motor outputs (e.g., responses). Indeed, since the seminal work of Fitts and colleagues (Fitts & Deninger, 1954; Fitts & Seeger, 1953). It has been repeatedly demonstrated that errors and choice reaction times to stimuli in a spatial array are shorter when the stimuli are mapped onto responses in a spatially compatible manner. Fitts and Seeger (1953) referred to this finding as stimulus-response (S-R) compatibility and ascribed it to cognitive codes associated with the spatial locations of elements in the stimulus and response arrays. Presumably, it is the degree of coding and recoding required to map the locations of stimulus and response elements that determine the speed and accuracy of translation and thus response selection (e.g., Wallace, 1971).

The relevance of studies of S-R compatibility to the domain of human factors engineering is paramount. It is now well understood that the design of an optimal interface in which effective S-R translation facilitates fast and accurate responses is largely determined by the manner in which stimulus and response arrays are arranged and mapped onto each other (e.g., Bayerl, Millen, & Lewis, 1988; Chapanis & Lindenbaum, 1959; Proctor & Van Zandt, 1994).

Movement Dynamics and Perceptual-Motor Behavior

As discussed previously, many HCI situations involve dynamic perceptual–motor interactions that may not be best indexed merely by chronometric methods (cf. Card, English, & Burr, 1978). Indeed, as HCI moves beyond the simple key press interfaces characteristic of early systems to include virtual and augmented reality, teleoperation, gestural and haptic interfaces, among others, the dynamic nature of perceptual–motor interactions are even more evident. Consequently, assessment of the actual movement required to engage such interfaces may be more revealing.

To supplement chronometric explorations of basic perceptual–motor interactions, motor behavior researchers have also advocated a movement process approach (Kelso, 1982). The argument is that to understand the nature of movement organization and control, analyses should also encompass the movement itself, and not just the activities preceding it (e.g., Kelso, 1982; Marteniuk, MacKenzie, & Leavitt, 1988). Thus, investigators have examined the kinematics of movements in attempts to further understand the underlying organization involved (e.g., Brooks, 1974; Chua & Elliott, 1993; Elliott, Carson, Goodman, & Chua, 1991; Kelso, Southard, & Goodman, 1979; MacKenzie, Marteniuk, Dugas, Liske, & Eickmeier, 1987; Marteniuk, MacKenzie, Jeannerod, Athenes, & Dugas, 1987).

This emphasis on movement dynamics is evident in the dynamical systems approach to the study of perception and action. This theoretical perspective seeks to explain perceptual–motor behavior in terms of fundamental, physical laws and principles (Jeka & Kelso, 1989; Kelso, 1995; Turvey, 1990). The dynamical systems framework is characterized by the application of the tools and principles from physical biology, synergetics (self-organization), and nonlinear dynamics. In the language of dynamics, movement systems are thought of as self-organizing systems, whereby patterns emerge from the interaction of the many variables inherent in the system. Thus, the theoretical and analytical tools and principles of synergetics and nonlinear dynamics become relevant to the study of perceptual-motor systems and perceptual–motor interactions (e.g., Schöner & Kelso, 1988). Furthermore, and of greatest utility in our work, is that with the application of these tools, the set of measurable, dependent, variables are extended to capture the richness of movement dynamics more adequately.

A key element of the dynamical systems approach is the identification of patterns (e.g., perceptual-motor patterns or interactions) relevant to the system under study. The primary strategy for identifying these patterns is to find transitions, situations in which one observes qualitative changes in the system's behavior. The transition demarcates one pattern from another, and the qualitative change allows one not only to distinguish between the patterns, but also to identify the relevant dimension of the pattern (Jeka & Kelso, 1989; Kelso, 1995). It is also the change about the transition that helps to identify the relevant variable that characterizes the pattern itself. A second important element is the study of stability and loss of stability of the patterns. The study of the system's stability or instability allows a determination of the system's dynamics. It is the stability of a given pattern that distinguishes it from others, characterizing the state in which the system resides. Moreover, the loss of stability is hypothesized to be a mechanism that effects a change in pattern (Jeka & Kelso, 1989; Kelso, 1995). It is the loss of stability of a pattern that may lead to a transition to a new pattern, one distinguished by its greater stability. In the section that follows, we describe our efforts to integrate compatibility phenomena and dynamic perceptual–motor interactions under one theoretical framework. This is followed by some recent data from our lab that has attempted to apply the framework.

Relative Organizational Mapping

Our studies of perceptual–motor interactions, and compatibility phenomena more generally, has led us to propose a relative organizational mapping model (Chua & Weeks, 1997). Consistent with other conceptual divisions forwarded by Fitts and colleagues, among others (e.g., Fitts & Deninger, 1954; Fitts & Posner, 1967; Kantowitz, Triggs, & Barnes, 1990; Kornblum, Hasbroucq, & Osman, 1990), the model proposes three levels of organization: global relation, configuration, and mapping (see Fig. 1.1). The mapping relation refers to how stimulus events are mapped onto response events. The configuration relation refers to the orientation of the stimulus and response arrays with respect to the other. The global relation refers to the overall relation in space between the stimulus display and response array. For example, in a prototypical two-choice reaction time task, the assignment of a given stimulus event to a response event would be categorized as a mapping relation. Whether or not the stimulus display and response array are arranged along the same spatial dimension (e.g., parallel or orthogonal) would be a configuration relation. Lastly, where the response array is

physically located with respect to the stimulus display would be subsumed under global relation.

The three levels of organization can be considered as a nested hierarchy. Mapping bears directly on the response (i.e., stimuli are assigned to responses on the basis of the mapping). Mapping is nested within a configuration, which may affect performance indirectly through its influence on the relation between the mapping and the response. Both mapping and configuration are, in turn, subsumed within a global spatial relation. The global relation may affect performance indirectly through its constraint on the relation between the configuration, mapping, and the response action. The hierarchy is not necessarily meant to suggest that one level has precedence over the other. Rather, it provides a means to conceptualize potential constraints on perceptual–motor interactions and to provide a framework for our experimental manipulations. These levels of organization reveal the complex factors that a human operator must contend with to translate between perceptual and motor workspaces. Moreover, not only does the model provide a useful taxonomy for standard compatibility phenomena involving discrete stimulus and response events (as do a number of models and theories—Kantowitz et al., 1990; Kornblum et al., 1990), together with the analytical tools used to study movement dynamics, the model also provides a platform for examining the manner in which response selection and response execution become connected in dynamic perceptual–motor interactions.

In a simple and elegant study that acted as a catalyst for our work on relative organizational mapping, Worringham and Beringer (1989) examined the influence of operator orientation on visual-motor performance, thereby providing us with the requisite experimental conditions for the application of our distinction between levels of spatial organization. In their study, participants used a joystick control to move a cursor on a monitor from a central location to targets positioned radially about a central location. The authors noted three types of compatibility relations that could exist between movement of the joystick and cursor, and the different orientations of the operator with respect to the display and control. Control-display compatibility referred to a relation in which the direction of control motion corresponded spatially to motion of the cursor on the display, independent of operator orientation. Visual-motor compatibility referred to the relation in which the directions of control motion and cursor motion were referenced to the visual field axis of the operator, if the operator was looking at either control or display. Visual-trunk compatibility referred to the relation in which the cursor motion was referenced to the visual field axis, and control motion was referenced to the body midline. Depending on the orientation of the operator, and the spatial position of the display with respect to the control, different levels of either one, two, or all three of these compatibility relations could exist (see Worringham & Beringer, 1989). Worringham and Beringer found that participants' visual-motor task performance was influenced by the spatial relations between the display, control, and operator. Moreover, they also found that situations that yielded compatible relations according to a visual-motor reference (visual-motor compatibility) led to superior performance compared with other compatibility relations (display-control and visual-trunk compatibility).

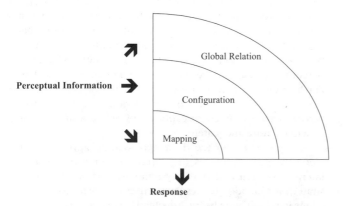

FIGURE 1.1. Relative organizational mapping: levels of organization in perceptual–motor interaction.

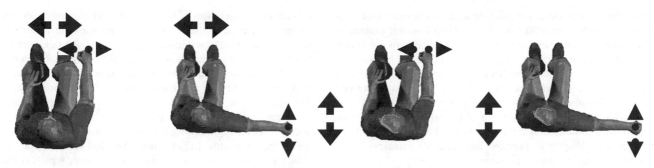

FIGURE 1.2. Orientation of operator with respect to display and control from Chua et al. (2001). See text for description.

Using Worringham and Beringer's (1989) study as an empirical starting point, we (Chua, Weeks, Ricker, & Poon, 2001) imported a subset of their task conditions and examined these within the context of examining interactions between the organizational levels of mapping, configuration, and global relation. Specifically, in our set-up, participants sat with the stimulus display (a rectangular panel with a light-emitting diode toward each end) located either directly in front of them, aligned with their midline, or directly to their left. The display panel could be oriented either vertically or horizontally. A control lever (allowing leftward and rightward rotational movements of the forearm) was located either in front, in right ipsilateral space, or directly to the participant's right side. In the first case, participants could reach forward to grasp the control, whereas in the second, the participants reached to their right. We assigned the relation between display-stimulus events and control-response events to the mapping level of organization, the orientation of the display with respect to the control to the configuration level, and the spatial relation between the display, control, and operator to the global relation level. Thus, with the set-up shown in Fig. 1.2, we manipulated the orientation of the participant (operator) with respect to the locations of the display and control to examine the influence of global spatial relations. We varied display orientation to examine the influence of spatial configuration. Finally, we varied mapping rules to examine the effects of spatial mapping.

In an initial experiment (Chua et al., 2001, Experiment 1), we used a choice reaction time paradigm in which participants had to make rapid, discrete, rotational movements in response to the brief onset of a visual stimulus. In a second experiment (Chua et al., 2001, Experiment 2), we used a coordination paradigm in which participants had to synchronize continuous rhythmic movements with an oscillating visual display. Thus, we examined independently the impact of the three levels of spatial organization on both discrete actions that emphasized selection of a response, and continuous actions that emphasized dynamic perceptual-motor coordination. Our results showed that, for discrete responses, the spatial mapping that yielded faster responding was dependent on the display-control array configuration and the global relation. When the display and control dimensions were parallel to one another (e.g., movement in horizontal plane), participants appeared to code the spatial aspects of the display and control in a manner that was unaffected by the global spatial relation. Thus, one mapping rule yielded an

advantage that was not affected by the global relation. When the display and control dimensions were orthogonal to one another (e.g., vertical display, horizontal control movement), there was a tendency for the direction of mapping effects to be influenced by the global relation. The results for the coordination task showed that spatial configuration had an impact on whether or not performance differences between spatial mapping rules emerged. Differences in coordination performance between mapping conditions was evident only under a parallel configuration. At the global relation level, there was no influence on accuracy and stability of perceptual-motor coordination under different configuration or mapping conditions. In both the discrete task (Experiment 1) and the coordination task (Experiment 2), our findings pointed to the consistent adoption of a visual-motor reference (e.g., see Worringham & Beringer, 1989), even if the end result was that the direction of motion of the stimulus display and the motion of the response were in actuality spatially incompatible (see Chua et al., 2001). The results thus support the proposal that a visual-motor frame of reference may dominate over others (Worringham & Beringer, 1989). More importantly, these findings suggested that, in determining the degree of compatibility between a stimulus display and a control array, the determination must include consideration of the fact that a human operator is physically oriented, and must translate between stimulus and response events.

In our initial proposal of the relative organizational mapping framework (Chua & Weeks, 1997; Chua, Weeks, & Goodman, 1996), we presented a series of experiments that examined spatial compatibility effects in coordinative actions. Consistent with our most recent work described previously (Chua et al., 2001), we showed that spatial compatibility effects that were observed in discrete tasks could also be demonstrated during performance of continuous coordination tasks. These spatial compatibility effects were sufficiently robust so as to influence performance even when the spatial stimulus dimension was irrelevant for the task. In one study (Chua, 1995), participants coordinated rhythmic leftward and rightward movements of a control lever with a visual stimulus that oscillated between two spatial locations (cf. Chua et al., 2001). The stimulus changed colors such that the two oscillation points were distinguished by a particular color. The position of the two oscillation points was also varied, such that the oscillation would begin along a horizontal plane, then rotate such that the two oscillation points would exchange positions (e.g., a counter-clockwise rotation of 180 degrees). The changing orientation of the stimulus resulted in a change in the stimulus display–control configuration. The endpoints of the leftward and rightward movements were mapped onto the stimulus colors—the actual position of the stimulus was irrelevant. Nevertheless, the results showed that coordination performance (in both accuracy and variability) was influenced by the orientation of the stimulus display relative to the spatial dimension of the control action. When both motion of the display and control happened to correspond, coordination performance was improved.

This impact of spatial stimulus information, even when it is not directly relevant to the task, has led us to speculate about the influence that symbolic or implied spatial information might have on the execution of continuous tasks. Symbolic spatial information has been shown to exert effects on reaction time in prototypical compatibility tasks (e.g., Weeks & Proctor, 1990). Presently, we have turned our focus toward symbolic cues to investigate whether or not similar effects can be found in tasks that require continuous perceptual–motor interactions. In addition, the effects described previously reminded us of the Simon effect in choice reaction time tasks (see later in this chapter). Briefly, the Simon effect (Simon, 1968) occurs when an irrelevant spatial stimulus attribute interferes with response selection. As described later in this chapter, the Simon effect has been linked to the operation of attention-based mechanisms. Indeed, attention-based accounts of the Simon effect has become an important point of contact in considerations of the role of attention in perceptual–motor interactions.

PERCEPTUAL–MOTOR INTERACTION: ATTENTION AND PERFORMANCE

The vast literature on selective attention and its role in the filtering and selection of information (e.g., Cherry, 1953; Deutsch & Deutch, 1963; Treisman, 1964a, 1964b, 1986; Treisman & Gelade, 1980) has no doubt been informative in the resolution of issues in HCI pertaining to stimulus displays and inputs (e.g., the use of color and sound). However, attention can be thought of as not a unitary function, but rather as a set of information-processing activities that are important for perceptual, cognitive, and motor skills. Indeed, the evolution of HCI into the realm of augmented reality, teleoperation, gestural interfaces, and other areas that highlight the importance of dynamic perceptual–motor interactions necessitates a greater consideration of the role of attention in the selection and execution of action. Recent developments in the study of how selective attention mediates perception and action, and more importantly, how action in turn influences attentional processes, are poised to make just such a contribution to HCI. We will turn to a brief review of these developments and some thoughts on their potential relevance to HCI.

Action-Centered Attention

It has been suggested (Allport, 1987) that one role for human selective attention is to provide the motor system with the relevant stimulus characteristics necessary for selecting and executing an appropriate action (e.g., reaching and grasping a particular selected object). Tipper and colleagues (e.g., Tipper, Lortie, & Baylis, 1992) have suggested that, for actions such as selective reaching to an item in a cluttered multi-item environment, attention is mediated by an action-centered cognitive representation. The architecture of this representation is such that distractor items are included in the response selection computations associated with programming the motor output (e.g., the reach) for a target stimulus. In their now oft-cited study, Tipper and colleagues (1992) used a reaching paradigm in which participants reached and pointed to a target located within a 3 × 3 matrix of push buttons that denoted potential target locations. On any given trial, light emitting diodes alongside the buttons were used to cue participants to the position of the target only, or the target and to-be-ignored distractor. Tipper and colleagues found that responses were slowed in the presence of distractors, with the effect even more pronounced when distractors were located along the path of movement (Tipper et al., 1992).

Several investigators have both corroborated and qualified the work of Tipper and colleagues (1992). Pratt and Abrams (1994) confirmed that responses are slower in the presence of distractors, particularly when distractors are positioned along the path of the movement. Moreover, the same authors demonstrated that the interference effect of distractors were found during both response preparation (indexed by reaction time) and response execution (indexed by movement time). In addition, interference effects during the execution of the response had its locus in the terminal, corrective phase, of the movement.

The cluttered environment of response buttons used by Tipper and colleagues (1992) struck us as being analogous to the array of icons present in a typical graphical user interface. In a recent study, Lyons, Elliott, Ricker, Weeks, and Chua (1999) sought to determine whether the paradigm developed by Tipper and colleagues (1992) could be imported into a virtual environment and ultimately serve as a test bed for investigations of perceptual-motor interactions in an HCI context. The task space utilized a 3 × 3 matrix similar to that used by Tipper et al. (1992). The matrix, made up of nine blue circles, was displayed on a monitor placed vertically in front of the participant. Consistent with Tipper and colleagues' paradigm, the target was presented either in isolation or in the presence of a distractor. Consequently, at trial onset, a target circle would turn red in color, and for distractor trials a yellow circle would appear simultaneously in one of the other eight locations. The participants were required to move the mouse on the graphics tablet, which would in turn move a cursor on the monitor in the desired direction toward the target circle. The participants were unable to view their hand; the only visual feedback of their progress was from the monitor. The graphics tablet allowed the researchers to record displacement and time data of the mouse throughout the trial. In contrast to previous experiments (e.g., Meegan & Tipper, 1998; Tipper et al., 1992), the presence of a distractor had relatively little influence on performance. Lyons et al. postulated that, in a task environment in which perceptual–motor interaction is less direct (e.g., using a mouse to move a cursor on a remote display), perceptual and motor workspaces are misaligned, and the increased translation processing owing to the misalignment serves to limit the impact of distractor items.

To test this idea, Lyons et al. (1999) modified the task environment so as to align the perceptual and motor workspaces. The monitor was turned and held screen down inside a support frame. The same 3 × 3 matrix was displayed on the monitor and reflected into a half-silvered mirror positioned above the graphics tablet allowing for sufficient space for the participant to manipulate the mouse and move the cursor to the target without vision of the hand. With this configuration, the stimulus display was presented and superimposed on the same plane as the motor workspace (i.e., the graphics tablet). Under this setup, distractor effects became evident and were consistent with an action-centered framework of attention. Taken together, these findings underscore the influence of translation requirements demanded by relative alignment of perceptual and motor workspaces. More importantly, these findings suggest that even relatively innocuous changes to the layout of the task environment may have significant impact on processes associated with selective attention in the mediation of action in an HCI context.

The behavioral consequence of selecting and executing target-directed actions in the presence of potential distractors is not limited simply to the time taken to prepare and execute the movement. Indeed, further recent investigations have provided evidence showing that increases in movement time due to the presence of distractors may be a result of deviations in the movement trajectory itself. Tipper and colleagues (Howard & Tipper, 1997; Tipper, Howard, & Jackson, 1997) have reported that the trajectory of the movement toward the target deviates away so as to avoid (or inhibit a competing response to) a distractor object that is located in proximity to the hand. Recently, Welsh and Elliott (2001) examined cursor movement trajectories using a virtual environment setup similar to Lyons et al. (1999). Interestingly, the results revealed that participants were actually attracted to, as opposed to repelled from, the competing stimulus. Welsh and Elliott (2001) suggested that subtle methodological differences may account for the discrepancy in results. Specifically, in the task used by Howard and Tipper (1997), the location of the nontarget stimulus was known before the location of the target. This may have facilitated early inhibition of the response to the nontarget location. In contrast, in the virtual environment used by Welsh and Elliott (2001), presentation of the target and distractor was simultaneous. This may have resulted in the preparation of an initial averaged response that was subsequently corrected online during execution. Regardless, these results again reveal the impact that even subtle changes to the stimulus array can have on task performance and the need for a detailed examination of the execution of the response itself.

Action Requirements and Attention. The first round of basic research into the action-centered model of attention has been focused primarily on the examination of the spatial locations of distractors with respect to the target. In that context, an action-centered framework could offer a useful perspective for the spatial organization of perceptual information presented in an HCI context. However, often the reason for engaging a target in an HCI task is because the target symbolically represents an outcome or operation to be achieved. Indeed, this is what defines a target as an icon—target features symbolically carry a meaning that defines it as the appropriate target. An emerging

line of research investigations also under the umbrella of the action-centered framework have turned to issues pertaining to the impact of intrinsic features of targets and distractors. For example, Jervis, Bennett, Thomas, Lim, and Castiello (1999) have examined the relation between semantic categories to which targets and distractors belong (fruits vs. three-dimensional shapes). They have shown distractor interference effects on grasping the target object when the target and distractor belong to different semantic categories. When distractors and targets come from similar semantic categories (e.g., both fruits), distractors may not be of behavioral importance to the task, and therefore can be inhibited (Castiello, 1996; Jervis et al., 1999).

Although using objects from different semantic categories may effectively distinguish the visual similarity between targets and distractors, the physical features of these objects may still evoke similar response requirements (i.e., a similar reach and grasp). An interest in the application of the action-centered model to human factors and HCI led Weir et al. (2002) to consider whether or not distractor effects could be elicited based on the specific actions required to engage a target and distractor object. The question was whether the engagement properties of target and distractor objects (i.e., turn or pull) in a control array would mediate the influence of the distractor on the control of movement. In that study, participants executed their movements on a control panel that was located directly in front of them. On some trials, the control panel consisted of a single pull-knob or right-turn dial located at the midline either near or far from a starting position located proximal to the participant. On other trials, a second control device (pull-knob or dial) was placed into the other position on the display. If this second device was present, it served as a distractor object and was to be ignored. The findings suggested that, when moving in an environment with distracting stimuli or objects, competing responses may be programmed in parallel. When the distractor is different from the target object, greater interference is present when the competing responses are being programmed, resulting in an increased response time. The implication is that the terminal action required to engage a target object can also be important to movement planning and execution.

One potential area of relevance to HCI is whether terminal action requirements will have a similar impact on perceptual–motor interactions within a virtual environment. Current work in our labs (Ibbotson, Chua, & Weeks, in preparation) is focusing on the influence the action and engagement properties of target and distractor objects on response competition and selection for action in the same task environment we have used previously (cf. Lyons et al., 1999). Thus, in our studies of the action-centered model of attention and its relevance to HCI, we have progressed to considering the response execution requirements of the task environment, in addition to the spatial layout of the environment.

Attention and Stimulus-Response Compatibility

To this point, we have separated the discussion of attention from the issues related to translation and perceptual–motor interaction. However, the action-centered model of selective attention clearly advocates the view that attention and action are intimately linked. The fundamental premise is that attention mediates perceptual–motor interactions, and these, in turn, influence attention. In line with this perspective, the role of attention in the translation between perceptual inputs and motor outputs has also received considerable interest over the past decade. As discussed previously, a key element in the selection of an action is the translation between stimuli and responses, the effectiveness of which is influenced to a large extent by the spatial relation between the stimuli and responses. The degree of coding and recoding required to map the locations of stimulus and response elements has been proposed to be a primary determinant of the speed and accuracy of translation (e.g., Wallace, 1971). Attentional processes have been implicated in the issue of how relative spatial stimulus information is coded. Specifically, the orienting of attention to the location of a stimulus has been proposed to result in the generation of the spatial stimulus code.

Initial interest in the link between attention orienting and spatial translation have emerged as a result of attempts to explain the Simon effect. The Simon effect (Simon, 1968; Simon & Rudell, 1969), often considered a variant of spatial S-R compatibility, occurs in a situation in which a nonspatial stimulus attribute indicates the correct response and the spatial attribute is irrelevant to the task. Thus, the spatial dimension of the stimulus is an irrelevant attribute, and a symbolic stimulus feature constitutes the relevant attribute. Although the spatial stimulus attribute is irrelevant to the task, faster responding is found when the position of the stimulus and the position of the response happen to correspond. A number of researchers (e.g., Umiltà & Nicoletti, 1992) have suggested that attentional processes may be a unifying link between the Simon effect and the spatial compatibility effect proper. Specifically, the link between attention and action in these cases is that a shift in attention is postulated to be the mechanism that underlies the generation of the spatial stimulus code (e.g., Nicoletti & Umiltà, 1989, 1994; Proctor & Lu, 1994; Rubichi, Nicoletti, Iani, & Umiltà, 1997; Stoffer, 1991; Stoffer & Umiltà, 1997; Umiltà & Nicoletti, 1992). According to an attention-shift account, when a stimulus is presented to the left or right of the current focus of attention, a reorienting of attention occurs toward the location of the stimulus. This attention shift is associated with the generation of a spatial code that specifies the position of the stimulus with respect to the last attended location. If this spatial stimulus code is congruent with the spatial code of the response, then S-R translation, and, hence response selection, is facilitated. If the two codes are incongruent, response selection is hindered.

Recent work in our lab has also implicated a role for attention shifts in compatibility effects and object recognition. In these studies, Lyons, Weeks, and Chua (2000a, 2000b) sought to examine the influence of spatial orientation on the speed of object identification. Participants were presented with video images of common objects that possessed a graspable surface (e.g., a tea cup, frying pan) and were instructed to make a left or right key press under two distinct mapping rules, depending on whether the object was in an upright or inverted vertical orientation. The first mapping rule required participants to respond with a left key press when the object was inverted and a right key press when the object was upright. The opposite was true for the second mapping rule. The orientation of the object's graspable surface was irrelevant to the task. The results showed that identification of object orientation was facilitated when the graspable surface of the object was also oriented to the same side of space as the response (see also Tucker & Ellis, 1998). In contrast, when participants were presented with objects that possessed symmetrical graspable surfaces on both sides (e.g., a sugar bowl with two handles), identification of object orientation was not facilitated. Lyons et al. (2000a) also showed that response facilitation was evident when the stimuli consisted simply of objects that, though may not inherently be graspable, possessed a left-right asymmetry. Taken together, these results were interpreted in terms of an attentional mechanism. Specifically, Lyons et al. (2000a, 2000b) proposed that a left-right object asymmetry (e.g., a protruding handle) may serve to capture spatial attention (cf. Tucker & Ellis, 1998). If attention is thus oriented toward the same side of space as the ensuing action, the spatial code associated with the attention shift (e.g., see previous discussion) would lead to facilitation of the response. In situations in which no such object asymmetry exists, attentional capture and orienting may be hindered, and, as a result, there is no facilitation of the response.

Taken into the realm of HCI, it is our position that the interplay between shifts of attention, spatial compatibility, and object recognition will be a central human performance factor as technological developments continue to enhance the directness of direct-manipulation systems (cf. Shneiderman, 1983, 1992). Specifically, as interactive environments become better abstractions of reality with greater transparency (Rutkowski, 1982), the potential influence of these features of human information-processing will likely increase. Thus, it is somewhat ironic that the view toward virtual reality, as the solution to the problem of creating the optimal display representation, may bring with it an unintended consequence (Tenner, 1996). Indeed, the operator in such an HCI environment will be subject to the same constraints that are present in everyday life.

A goal of human factors research is to guide technological design to optimize perceptual–motor interactions between human operators and the systems they use. Thus, the design of machines, tools, interfaces, and other sorts of devices utilizes knowledge about the characteristics, capabilities, as well as limitations, of the human perceptual-motor system. In computing, the development of input devices such as the mouse and graphical user interfaces was intended to improve HCI. As technology has continued to advance, the relatively simple mouse and graphical displays have begun to give way to exploration of complex gestural interfaces and virtual environments. This development may perhaps, in part, be a desire to move beyond the artificial nature of such devices as the mouse to ones that provide a better mimic of reality. Why move an arrow on a monitor using a hand-held device to point to a displayed object, when instead, you can reach and interact with the object. Perhaps such an interface would provide a closer reflection of real-world interactions, and the seeming ease with which we interact with our environments, but also subject to the constraints of the human system.

PERCEPTUAL–MOTOR INTERACTION IN APPLIED TASKS: A FEW EXAMPLES

As we described at the outset of this chapter, the evolution of computers and computer-related technology has brought us to the point at which the manner with which we interact with such systems has become a research area in itself. Current research in motor behavior and experimental psychology pertaining to attention, perception, action, and spatial cognition is poised to make significant contributions to the area of HCI. In addition to the continued development of a knowledge base of fundamental information pertaining to the perceptual-motor capabilities of the human user, these contributions will include new theoretical and analytical frameworks that can guide the study of HCI in various settings. In this final section, we highlight just a few specific examples of HCI situations that offer a potential arena for the application of the basic research that we have outlined in this chapter.

Remote and Endoscopic Surgery

Recent work by Hanna and colleagues (Hanna, Shimi, & Cuschieri, 1998) examined task performance of surgeons as a function of the location of the image display used during endoscopic surgical procedures. In their study, the display was located either in front, to the left, or to the right of the surgeon. In addition, the display was placed either at eye level or at the level of the surgeon's hands. The surgeons' task performance was observed with the image display positioned at each of these locations. Hanna et al. (1998) showed that the surgeons' performance was affected by the location of the display. Performance was facilitated when the surgeons were allowed to view their actions with the monitor positioned in front and at the level of the immediate workspace (the hands). In terms of our relative organizational mapping framework (see earlier in this chapter), the position of the display relative to the surgeon and the workspace is a global relation. We would suggest that the optimal display location (or global relation) in this task environment placed less translation demands on the surgeon during task performance. Given that surgeons work with a team of support personnel each with a different vantage point relative to the operating field, an interesting empirical question will be the manner in which perceptual-motor workspace can be effectively optimized for each team member.

Similar findings have also been demonstrated by Mandryk and MacKenzie (1999). These investigators also examined the impact of display location on endoscopic surgical performance. In addition to the frontal image display location used by Hanna et al. (1998), Mandryk and MacKenzie also investigated the benefits of projecting and superimposing the image from the endoscopic camera directly over the workspace. Their results showed that performance was superior when participants were initially exposed to the superimposed viewing conditions. This finding was attributed to the superimposed view allowing the participants to better calibrate the display space with the workspace. These findings are consistent with our own investigations of action-centered attention in virtual environments (Lyons et al., 1999). We would suggest that the alignment of perceptual and motor workspaces in the superimposed viewing condition facilitated performance due to the decreased translation requirements demanded by such a situation. However, the findings of Lyons et al. (1999) would also lead us to suspect that this alignment may have additional implications with respect to processes associated with selective attention in the mediation of action. Although the demands on perceptual-motor translation may be reduced, the potential intrusion of processes related to selective attention and action selection may now surface.

Remote Viewing

Hooper and Coury (1994) have described an intriguing and important HCI issue in the operation of periscopes in modern submarines. Traditionally, operators of standard submarine periscopes were required to physically rotate the periscope to view in different directions and, as a consequence, the operator adjusted his/her own orientation with respect to the submarine. Thus, in combination with visual cues from within the internal environment of the submarine, information pertaining to the spatial direction of the periscope was available to the operator. Hooper and Coury (1994) noted that the development of newer periscopes with advanced video systems and graphics workstations has led to a significant perceptual-motor translation issue. Specifically, the spatial information that was inherent in the physical operation of the periscope is no longer readily available in the computer displays that accompany new systems. As such, information such as periscope direction relative to the submarine and the direction of view now have to be provided within the graphic display (see Hooper & Coury, 1994, for examples). Hooper and Coury examined the best way to present this type of information. The results of their experiments indicated that, in some display situations, operators would adopt a strategy of mental rotation to determine orientation. This translation process would represent an additional cognitive demand on the operator.

An interesting aspect of this HCI scenario as described by Hooper and Coury (1994) is that it seems to be a case in which information pertaining to the global relation between perceptual and motor workspaces is incorporated into some form of image display and must therefore be extracted from that display. This rather indirect means of obtaining this information is then shown to have an impact on information processing demands on the operator. To date, our investigations involving the use of our relative organizational mapping model has been limited to the study of how different levels of spatial organization might interact to constrain perceptual-motor performance. The work by Hooper and Coury (1994) provides a new interesting avenue for us in which we might consider how information such as global relations can be used in perceptual–motor interactions, and the impact that removing or modifying this type of information might have on performance.

Automated Compensation

In the previous examples, the performance impact of the spatial relation between the perceptual and motor workspaces of an operator is readily apparent. Our own research efforts using the relative organizational mapping framework has been directed at the specific levels of spatial relations that have a bearing on perceptual–motor interactions. Although we have focused primarily on the interaction between levels of spatial organization and their influence on perceptual-motor translation, recent work by Macedo and colleagues (Macedo, Kaber, Endsley, Powanusorn, & Myung, 1998) has examined how a system might readily compensate for spatial misalignments between perceptual and motor workspaces. Specifically, Macedo and colleagues tested methods to automatically compensate for misalignments between display and operator orientation on a visual tracking task. Participants tracked an irregularly moving target on a display using a joystick-controlled cursor. Spatial alignment between the operator and the display was varied by orienting the display, operator, or both, at fixed angles of rotation. By monitoring joystick orientation and the operator's head rotation, Macedo et al. implemented a computer algorithm that served to rotate and align the axes of the joystick and the display, and automatically compensate for the misalignment. Their findings showed that tracking performance of the operator in situations of operator-display misalignment could be improved through automated compensation. These findings suggest that methods can be developed that decrease the burden of perceptual-motor translation on an operator in situations in which SR incongruencies might otherwise have a significant impact on performance. Thus, an engineering approach to facilitating the effective mapping of perceptual and motor workspaces would seem to hold promise. Whether algorithms can be developed to compensate for the impact of mapping at multiple levels of constraint will be an interesting empirical issue for the future.

SUMMARY AND IMPLICATIONS

The field of HCI offers a rich environment for the study of perceptual–motor interactions. The design of effective interfaces, and HCI in general has been and, continues to be, a significant challenge that demands an appreciation of human perceptual-motor behavior. The information-processing approach has provided a dominant theoretical and empirical framework for the study of perceptual-motor behavior in general, and for consideration of issues in HCI and human factors in particular. Texts in the area of human factors and HCI (including the present volume) are united in their inclusion of chapters or sections that pertain to the topic of human information-processing. Moreover, the design of effective interfaces reflects our knowledge of the perceptual (e.g., visual displays, use of sound, graphics), cognitive (e.g., conceptual models, desktop metaphors), and motoric constraints (e.g., physical design of input devices, ergonomic keyboards) of the human perceptual-motor system.

Technological advances have undoubtedly served to improve the HCI experience. For example, we have progressed beyond the use of computer punch cards and command–line interfaces to more complex tools, such as graphical user interfaces and speech recognition. As HCI has become not only more effective, but also by the same token more elaborate, the importance of the interaction between the various perceptual, cognitive, and motor constraints of the human system has also come to the forefront. In this chapter, we have presented an overview of some current topics of research in perceptual–motor interactions that we believe have relevance to HCI. Clearly, considerable research will be necessary to evaluate the applicability of these potentially relevant data to specific HCI design problems. Nevertheless, the experimental work to date, along with rational considerations, leads us to conclude this chapter with a few implications for early infusion to HCI.

First, we have outlined the benefits of integrating the frameworks and tools offered by the information-processing and movement dynamics approaches to capture the richness of perceptual–motor interactions, from processing and movement preparation through movement execution. For example, one area that we are pursuing in this regard involves a more detailed evaluation of principles of control-display movement. Current design principles (e.g., Warrick's principle, scale-side principle, clockwise-for-increase principle, etc.) are generally based on the expectations and preferences of the population of users. However, faced with a unique or novel display-control arrangement, such an approach can place principles in conflict, making it difficult to determine the optimal mapping (e.g., Brebner & Sandow, 1976; Petropoulos & Brebner, 1981). We hold that the integration of information processing and movement dynamics approaches could yield dependent measures and new metrics that would be effective in evaluating dynamic human performance.

Second, we have discussed the implications of recent research on attention and SR compatibility with respect to their impact on action selection and execution. Here, the implications are more straightforward. That the allocation of attention also carries an action-centered component means that an effective interface must also be sensitive to the specific action associated with a particular response location, the action relationship between that response and those around it, and the degree of translation required to map the perceptual-motor workspaces.

Finally, we presented our relative organizational mapping model to conceptualize and address empirically, potential multilevel constraints on perceptual–motor interactions. We hold that the examination of perceptual–motor interaction as a problem involving multiple levels of constraint acknowledges a range of potential interactions that can occur in a given HCI context. Indeed, this range extends from the specific arrangement of display and response aspects of an interface, to include the immediate workspace, the workspace within a room, that room within a building, and so on. As theoretical and empirical work continues, we believe the important implication of current research for HCI and human factors situations remains, to quote Norman (1988), "getting the mappings right."

References

Adams, J. A. (1971). A closed-loop theory of motor learning. *Journal of Motor Behavior, 3*, 111–150.

Adams, J. A., & Dijkstra, S. (1966). Short-term memory for motor responses. *Journal of Experimental Psychology, 71*, 314–318.

Allport, A. (1987). Selection for action: Some behavioural and neurophysiological considerations of attention and action. In H. Heuer & A. F. Sanders (Eds.), *Perspectives on perception and action* (pp. 395–419). Hillsdale, NJ: Erlbaum.

Bayerl, J., Millen, D., & Lewis, S. (1988). Consistent layout of function keys and screen labels speeds user responses. *Proceedings of the Human Factors Society 32nd Annual Meeting* (pp. 334–346). Santa Monica, CA: Human Factors Society.

Brebner, J., & Sandow, B. (1976). The effect of scale side on population stereotype. *Ergonomics, 19*, 571–580.

Broadbent, D. E. (1958). *Perception and communication*. New York: Pergamon.

Brooks, V. B. (1974). Some examples of programmed limb movements. *Brain Research, 1974, 71*, 299–308.

Card, S. K., English, W. K., & Burr, B. J. (1978). Evaluation of mouse, rate-controlled isometric joystick, step keys, and text keys for text selection on a CRT. *Ergonomics, 21*, 601–613.

Castillo, U. (1996). Grasping a fruit: Selection for action. *Journal of Experimental Psychology: Human Perception and Performance, 22*, 582–603.

Chapanis, A., & Lindenbaum, L. E. (1959). A reaction time study of four control-display linkages, *Human Factors, 1*, 1–7.

Cherry, E. C. (1953). Some experiments on the recognition speech, with one and with two ears. *Journal of the Acoustical Society of America, 25*, 975–979.

Chua, R. (1995). *Informational constraints in perception-action coupling*. Unpublished doctoral dissertation, Simon Fraser University, Burnaby, B.C., Canada.

Chua, R., & Elliott, D. (1993). Visual regulation of manual aiming. *Human Movement Science, 12*, 365–401.

Chua, R., & Weeks, D. J. (1997). Dynamical explorations of compatibility in perception-action coupling. In B. Hommel & W. Prinz (Eds.), *Theoretical issues in stimulus-response compatibility* (pp. 373–398). Amsterdam: Elsevier Science B.V.

Chua, R., Weeks, D. J., & Goodman, D. (1996). Compatibility in coordinative actions. Paper presented at the Annual Conference of the Psychonomic Society, Chicago.

Chua, R., Weeks, D. J., Ricker, K. L., & Poon, P. (2001). Influence of operator orientation on relative organizational mapping and spatial compatibility. *Ergonomics, 44*, 751–765.

Crossman, E. R. F. W., & Goodeve, P. J. (1983). Feedback control of hand movement and Fitts' Law. Paper presented at the meeting of the Experimental Psychology Society, Oxford, July 1963. Published in the *Quarterly Journal of Experimental Psychology, 35A*, 251–278.

Deutsch, J. A., & Deutsch, D. (1963). Attention: Some theoretical considerations. *Psychological Review, 70*, 80–90.

Elliott, D., Carson, R. G., Goodman, D., & Chua, R. (1991). Discrete vs continuous visual control of manual aiming. *Human Movement Science, 10*, 393–418.

English, W. K., Engelhart, D. C., & Berman, M. L. (1967). Display-selection techniques for text manipulation. *IEEE Transactions on Human Factors in Electronics, 8*, 5–15.

Fitts, P. M. (1954). The information capacity of the human motor system in controlling the amplitude of movement. *Journal of Experimental Psychology, 47*, 381–391.

Fitts, P. M. (1964). Perceptual-motor skills learning. In A. W. Melton (Ed.), *Categories of human learning* (pp. 243–285). New York: Academic Press.

Fitts, P. M., & Deninger, R. I. (1954). S-R compatibility: Correspondence among paired elements within stimulus and response codes. *Journal of Experimental Psychology, 48*, 483–491.

Fitts, P. M., & Posner, M. I. (1967). *Human performance*. Belmont, CA: Brooks-Cole.

Fitts, P. M., & Seeger, C. M. (1953). S-R compatibility: Spatial characteristics of stimulus and response codes. *Journal of Experimental Psychology, 46*, 199–210.

Hanna, G. B., Shimi, S. M., & Cuschieri, A. (1998). Task performance in endoscopic surgery is influenced by location of the image display. *Annals of Surgery, 4*, 481–484.

Hick, W. E. (1952). On the rate of gain of information. *The Quarterly Journal of Experimental Psychology, 4*, 11–26.

Hoffman, E. R., & Lim, J. T. A. (1997). Concurrent manual-decision tasks. *Ergonomics, 40*, 293–318.

Hooper, E. Y., & Coury, B. G. (1994). Graphical displays for orientation information. *Human Factors, 36*, 62–78.

Howard, L. A., & Tipper, S. P. (1997). Hand deviations away from visual cues: Indirect evidence for inhibition. *Experimental Brain Research, 113*, 144–152.

Hyman, R. (1953). Stimulus information as a determinant of reaction time. *Journal of Experimental Psychology, 45*, 188–196.

Ibbotson, J. A., Chua, R., & Weeks, D. J. (in preparation). Influence of action properties of target and distractor objects on response selection and execution. *Manuscript in preparation*.

Jeka, J. J., & Kelso, J. A. S. (1989). The dynamic pattern approach to coordinated behavior: A tutorial review. In S. A. Wallace (Ed.), *Perspectives on the coordination of movement* (pp. 3–45). Amsterdam: North-Holland.

Jervis, C., Bennett, K., Thomas, J., Lim, S., & Castiello, U. (1999). Semantic category interference effects upon the reach-to-grasp movement. *Neuropsychologia, 37*, 857–868.

Kantowitz, B. H., Triggs, T. J., & Barnes, V. E. (1990). Stimulus-response compatibility and human factors. In R. W. Proctor & T. G. Reeve (Eds.), *Stimulus-response compatibility: An integrated perspective* (pp. 365–388). Amsterdam: North-Holland.

Kelso, J. A. S. (1982). The process approach to understanding human motor behavior: An introduction. In J. A. S. Kelso (Ed.), *Human motor behavior: An introduction* (pp. 3–19). Hillsdale, NJ: Lawrence Erlbaum Associates.

Kelso, J. A. S. (1995). *Dynamic patterns: The self-organization of brain and behavior*. Cambridge, MA: MIT Press.

Kelso, J. A. S., Southard, D. L., & Goodman, D. (1979). On the coordination of two-handed movements. *Journal of Experimental Psychology: Human Perception and Performance, 5*, 229–238.

Kornblum, S., Hasbroucq, T., & Osman, A. (1990). Dimensional overlap: Cognitive basis for stimulus-response compatibility—A model and taxonomy. *Psychological Review, 97*, 253–270.

Lyons, J., Elliott, D., Ricker, K. L., Weeks, D. J., & Chua, R. (1999). Action-centred attention in virtual environments. *Canadian Journal of Experimental Psychology, 53*, 176–178.

Lyons, J., Weeks, D. J., & Chua, R. (2000a). Affordance and coding mechanisms in the facilitation of object identification. Paper presented at the Canadian Society for Psychomotor Learning and Sport Psychology Conference, Waterloo, Ontario, Canada.

Lyons, J., Weeks, D. J., & Chua, R. (2000b). The influence of object orientation on speed of object identification: Affordance facilitation or cognitive coding? *Journal of Sport & Exercise Psychology, 22* (Suppl.), S72.

Macedo, J. A., Kaber, D. B., Endsley, M. R., Powanusorn, P., & Myung, S. (1998). The effect of automated compensation for incongruent axes on teleoperator performance. *Human Factors, 40,* 541-553.

MacKenzie, C. L., Marteniuk, R. G., Dugas, C., Liske, D., & Eickmeier, B. (1987). Three dimensional movement trajectories in Fitts' task: Implications for control. *The Quarterly Journal of Experimental Psychology, 39A,* 629-647.

Mandryk, R. L., & MacKenzie, C. L. (1999). Superimposing display space on workspace in the context of endoscopic surgery. *ACM CHI Companion,* 284-285.

Marteniuk, R. G. (1976). *Information processing in motor skills.* New York: Holt, Rinehart and Winston.

Marteniuk, R. G., MacKenzie, C. L., Jeannerod, M., Athenes, S., & Dugas, C. (1987). Constraints on human arm movement trajectories. *Canadian Journal of Psychology, 41,* 365-378.

Marteniuk, R. G., MacKenzie, C. L., & Leavitt, J. L. (1988). Representational and physical accounts of motor control and learning: Can they account for the data? In A. M. Colley & J. R. Beech (Eds.), *Cognition and action in skilled behaviour* (pp. 173-190). Amsterdam: Elsevier Science Publishers.

Meegan, D. V., & Tipper, S. P. (1998). Reaching into cluttered visual environments: Spatial and temporal influences of distracting objects. *The Quarterly Journal of Experimental Psychology, 51A,* 225-249.

Meyer, D. E., Abrams, R. A., Kornblum, S., Wright, C. E., & Smith, J. E. K. (1988). Optimality in human motor performance: Ideal control of rapid aimed movements. *Psychological Review, 95,* 340-370.

Nicoletti, R., & Umiltà, C. (1989). Splitting visual space with attention. *Journal of Experimental Psychology: Human Perception and Performance, 15,* 164-169.

Nicoletti, R., & Umiltà, C. (1994). Attention shifts produce spatial stimulus codes. *Psychological Research, 56,* 144-150.

Norman, D. A. (1988). *The psychology of everyday things.* New York: Basic Books, Inc.

Petropoulos, H., & Brebner, J. (1981). Stereotypes for direction-of-movement of rotary controls associated with linear displays: The effects of scale presence and position, of pointer direction, and distances between the control and the display. *Ergonomics, 24,* 143-151.

Posner, M. I. (1982). Cumulative development of attentional theory. *American Psychologist, 37,* 168-179.

Posner, M. I., & Keele, S. W. (1969). Attentional demands of movement. *Proceedings of the 16th Congress of Applied Psychology.* Amsterdam: Swets and Zeitlinger.

Pratt, J., & Abrams, R. A. (1994). Action-centered inhibition: Effects of distractors in movement planning and execution. *Human Movement Science, 13,* 245-254.

Proctor, R. W., & Lu, C. H. (1994). Referential coding and attention-shifting accounts of the Simon effect. *Psychological Research, 56,* 185-195.

Proctor, R. W., & Van Zandt, T. (1994). *Human factors in simple and complex systems.* Boston: Allyn and Bacon.

Proctor, R. W., & Vu, K. L. (this volume). Human information processing.

Rubichi, S., Nicoletti, R., Iani, C., & Umiltà, C. (1997). The Simon effect occurs relative to the direction of an attention shift. *Journal of Experimental Psychology: Human Perception and Performance, 23,* 1353-1364.

Rutkowski, C. (1982). An introduction to the human applications standard computer interface, part I: Theory and principles. *Byte, 7,* 291-310.

Schöner, G., & Kelso, J. A. S. (1988). Dynamic pattern generation in behavioral and neural systems. *Science, 239,* 1513-1520.

Shneiderman, B. (1983). Direct manipulation: A step beyond programming languages. *IEEE Computer, 16,* 57-69.

Shneiderman, B. (1992). *Designing the user interface: Strategies for effective human-computer interaction.* Reading, MA: Addison-Wesley Publishing Company.

Simon, J. R. (1968). Effect of ear stimulated on reaction time and movement time. *Journal of Experimental Psychology, 78,* 344-346.

Simon, J. R., & Rudell, A. P. (1967). Auditory S-R compatibility: The effect of an irrelevant cue on information processing. *Journal of Applied Psychology, 51,* 300-304.

Stelmach, G. E. (1982). Information-processing framework for understanding human motor behavior. In J. A. S. Kelso (Ed.), *Human motor behavior: An introduction* (pp. 63-91). Hillsdale, NJ: Lawrence Erlbaum Associates.

Stoffer, T. H. (1991). Attentional focusing and spatial stimulus response compatibility. *Psychological Research, 53,* 127-135.

Stoffer, T. H., & Umiltà, C. (1997). Spatial stimulus coding and the focus of attention in S-R compatibility and the Simon effect. In B. Hommel & W. Prinz (Eds.), *Theoretical issues in stimulus-response compatibility* (pp. 373-398). Amsterdam: Elsevier Science B.V.

Tenner, E. (1996). *Why things bite back: Technology and the revenge of unintended consequences.* New York: Alfred A. Knopf.

Tipper, S. P., Howard, L. A., & Jackson, S. R. (1997). Selective reaching to grasp: Evidence for distractor interference effects. *Visual Cognition, 4,* 1-38.

Tipper, S. P., Lortie, C., & Baylis, G. C. (1992). Selective reaching evidence for action-centered attention. *Journal of Experimental Psychology: Human Perception and Performance, 18,* 891-905.

Treisman, A. M. (1964a). The effect of irrelevant material on the efficiency of selective listening. *American Journal of Psychology, 77,* 533-546.

Treisman, A. M. (1964b). Verbal cues, language, and meaning in selective attention. *American Journal of Psychology, 77,* 206-219.

Treisman, A. M. (1986). Features and objects in visual processing. *Scientific American, 255,* 114-125.

Treisman, A. M., & Gelade, G. (1980). A feature-integration theory of attention. *Cognitive Psychology, 12,* 97-136.

Tucker, M., & Ellis, R. (1998). On the relations between seen objects and components of potential actions. *Journal of Experimental Psychology: Human Perception and Performance, 24,* 830-846.

Turvey, M. T. (1990). Coordination. *American Psychologist, 45,* 938-953.

Umiltà, C., & Nicoletti, R. (1992). An integrated model of the Simon effect. In J. Alegria, D. Holender, J. Junca de Morais, & M. Radeau (Eds.), *Analytic approaches to human cognition* (pp. 331-350). Amsterdam: North-Holland.

Wallace, R. J. (1971). S-R compatibility and the idea of a response code. *Journal of Experimental Psychology, 88,* 354-360.

Weeks, D. J., & Proctor, R. W. (1990). Salient-features coding in the translation between orthogonal stimulus and response dimensions. *Journal of Experimental Psychology: General, 119,* 355-366.

Weir, P. L., Weeks, D. J., Welsh, T. N., Elliott, D., Chua, R., Roy, E. A., & Lyons, J. (2002). Influence of terminal action requirements on action-centered distractor effects. *Manuscript under review.*

Welford, A. T. (1968). *Fundamentals of skill.* London: Methuen.

Welsh, T. N., & Elliott, D. (2001). Hand deviations towards distracting stimuli reflect response competition in selective reaching. Paper presented at the Canadian Society for Psychomotor Learning and Sport Psychology Conference, Montreal, Quebec, Canada.

Worringham, C. J., & Beringer, D. B. (1989). Operator orientation and compatibility in visual-motor task performance. *Ergonomics, 32,* 387-399.

HUMAN INFORMATION PROCESSING:
AN OVERVIEW FOR HUMAN–COMPUTER
INTERACTION

Robert W. Proctor and Kim-Phuong L. Vu
Purdue University

It is natural for an applied psychology of human–computer interaction
to be based theoretically on information-processing psychology.
—Card, Moran, and Newell (1983, p. 13)

Human–computer interaction (HCI) is fundamentally an information processing task. In interacting with a computer, the human has specific goals and subgoals in mind. The user initiates the interaction by giving the computer commands that are directed toward accomplishing those goals. The commands may activate software programs designed to allow specific types of tasks, such as word processing or statistical analysis, to be performed. The resulting computer output, typically displayed on a screen, must provide adequate information for the user to complete the next step, or the user must enter another command to obtain the desired output from the computer. The sequence of interactions to accomplish the goals may be long and complex, and several alternative sequences, differing in efficiency, may be used to achieve these goals. During the interaction, the user is required to identify displayed information, select responses based on the displayed information, and execute those responses. The user must search the displayed information and attend to the appropriate aspects of it. She or he must also recall the commands and resulting consequences of those commands for different programs, remember information specific to the task that is being performed, and make decisions and solve problems during the process. For the interaction between the computer and user to be efficient, the interface must be designed in accordance with the user's information processing capabilities.

HUMAN INFORMATION PROCESSING APPROACH

The rise of the human information processing approach in psychology is closely coupled with the rise of the fields of cognitive psychology, human factors, and human engineering. Although research that can be classified as falling within these fields has been conducted since the last half of the 19th century, their formalization dates back to World War II (see Hoffman & Deffenbacher, 1992). As part of the war efforts, experimental psychologists worked along with engineers on applications associated with using the sophisticated equipment being developed. As a consequence, the psychologists were exposed not only to applied problems, but also to the techniques and views being developed in areas such as communications engineering. Many of the concepts from engineering, such as the notion of transmission of information through a limited capacity communications channel, were seen as applicable to analyses of human performance.

The human information processing approach is based on the idea that human performance, from displayed information to a response, is a function of several processing stages. The nature of these stages, how they are arranged, and the factors that influence how quickly and accurately a particular stage operates, can be discovered through appropriate research methods. It is often said that the central metaphor of the information processing approach is that a human is like a computer (e.g., Lachman, Lachman, & Butterfield, 1979). However, even more fundamental than the computer metaphor is the assumption that the human is a complex system that can be analyzed in terms of subsystems and their interrelation. This point is clearly evident in the work of researchers on attention and performance, such as Paul Fitts (1951) and Donald Broadbent (1958), who were among the first to adopt the information processing approach in the 1950s.

The systems perspective underlies not only human information processing but also human factors and HCI, providing a direct link between the basic and applied fields (Proctor & Van Zandt, 1994). Human factors in general, and HCI in particular, begin with the fundamental assumption that a human–machine system can be decomposed into machine and human subsystems, each of which can be analyzed further. The human information processing approach provides the concepts, methods, and theories for analyzing the processes involved in the human subsystem. Posner (1986) states, "Indeed, much of the impetus for the development of this kind of empirical study stemmed from the desire to integrate description of the human within overall systems" (p. V-6).

In the first half of the 20th century, the behaviorist approach predominated in psychology, particularly in the United States. Within this approach, numerous sophisticated theories of learning and behavior were developed that differed in many details (Bower & Hilgard, 1981). However, the research and theories of the behaviorist approach tended to minimize the emphasis on cognitive processes and were of limited value to the applied problems encountered in World War II. The information processing approach was adopted because it provided a way to examine topics of basic and applied concern such as attention that were relatively neglected during the behaviorist period. It continues to be the dominant approach in psychology, although contributions have been made from an alternative approach called the ecological perspective. The ecological perspective emphasizes the structure of the environment and the constraints of the environment on the interactions of the human with it. This perspective has had a recent impact on human factors through what is called ecological interface design (Flach, Hancock, Caird, & Vicente, 1995). We view the basic and applied research from the ecological perspective as valuable and complementary to that from the information processing perspective, but will not go into it in detail in this chapter.

Within HCI, human information processing analyses are used in two ways. First, empirical studies evaluate the information processing demands imposed by various tasks in which a human uses a computer. Second, computational models are developed to characterize human information processing when interacting with computers, and predict human performance with alternative interfaces. In this chapter, we survey methods used to study human information processing and summarize the major findings and the theoretical frameworks developed to explain them.

INFORMATION PROCESSING METHODS

Any theoretical approach makes certain presuppositions and tends to favor some methods and techniques over others. Information processing researchers have used behavioral and, to an increasing extent, psychophysiological measures, with an emphasis on chronometric (time-based) methods. There also has been a reliance on flow models that are often quantified through computer simulation or mathematical modeling.

Signal Detection Methods and Theory

One of the most useful methods for studying human information processing is that of signal detection (Macmillan & Creelman, 1991). In a signal detection task, some event is classified as a signal, and the subject's task is to detect whether the signal is present. Trials on which it is not present are called noise trials. The proportion of trials on which the signal is correctly identified as present is called the hit rate, and the proportion of trials on which the signal is incorrectly identified as present is called the false alarm rate. By using the hit rate and false alarm rate, it is possible to evaluate whether the effect of a variable is on discriminability or response bias.

Signal detection theory is often used as the basis for analyzing data from such tasks. This theory assumes that the response on each trial is a function of two discrete operations: encoding and decision. On the occurrence of the trial event, the subject collects the information presented and decides whether this information is sufficient to warrant a *signal present* response. The sample of information is assumed to provide a value along a continuum of evidence states regarding the likelihood of the signal being present. The noise trials form a probability distribution of states, as do the signal trials. The decision that must be made on a trial can be characterized as whether the event is from the signal or noise distribution. The subject is presumed to adopt a criterion value of evidence above which he or she responds *signal present* and below which he or she responds *signal absent*.

In the simplest form, the distributions are assumed to be normal and equal variance. In this case, a measure of detectability or discriminability, d', can be derived. This measure represents the difference in the means for the signal and noise distributions in standard deviation units. A measure of response bias, β, which represents the relative heights of the signal and noise distributions at the criterion, can also be calculated. This measure reflects the subject's overall willingness to say *signal present*, regardless of whether it actually is present. There are numerous alternative measures of detectability and bias based on different assumptions and theories, and many task variations to which they can be applied (see Macmillan & Creelman, 1991, for further discussion).

Signal detection analyses have been particularly useful because they can be applied to any task that can be characterized in terms of binary discriminations. For example, the proportion of words in a memory task correctly classified as old can be treated as a hit rate, and the proportion of new lures classified as old can be treated as a false alarm rate (Lockhart & Murdock, 1970). In cases such as these, the resulting analysis helps researchers determine whether variables are affecting discriminability or response bias.

An area of research in which signal detection methods have been widely used is that of vigilance (Parasuraman & Davies, 1977). In a typical vigilance task, a subject is asked to monitor a display for certain changes in it (e.g., the occurrence of a rare stimulus). The most common finding for vigilance tasks is called the vigilance decrement, in which the hit rate decreases as the time on the task increases. The classic example of this vigilance decrement is that, during World War II, British radar observers began to miss the enemy's radar signals after 30 minutes in a radar observation shift (Mackworth, 1948). Parasuraman and Davies concluded that, for many situations, signal detection analyses suggest that the primary cause of the vigilance decrement is an increasingly strict response criterion. That is, the false alarm rate as well as the hit rate decreases as a function of time on task. Perceptual sensitivity seems to be affected as well when the task requires the subject to compare rapidly presented events to information in memory to identify the events as a signals or nonsignals. Although signal detection theory can be used to help determine whether a variable affects encoding quality or decision, as in the vigilance example, it is important to keep in mind that the measures of discriminability and bias are based on certain theoretical assumptions. With regard to the vigilance decrement, Balakrishnan (1998) has argued, on the basis of an analysis that does not require the assumptions of signal detection theory, that the vigilance decrement is not a result of a biased placement of the response criterion, even when the signal occurs rarely and time on task increases.

Chronometric Methods

Chronometric methods, for which time is a factor, have been the most widely used methods for studying human information processing. Indeed, Lachman et al. (1979) portrayed reaction time as the main dependent measure of the information processing approach. Although many other measures have been used, reaction time still is widely used in part because of its sensitivity and in part because of the sophisticated techniques that have been developed for analyzing reaction time data.

A technique called the subtractive method, introduced by F. C. Donders (1868/1969) over a century ago, was revived in the 1950s and 1960s. This method provides a way to estimate the duration of a particular processing stage. The assumption of the subtractive method is that a series of discrete processing stages intervene between stimulus presentation and response execution. Through careful selection of tasks that differ in terms of a single stage, the reaction time for the easier task can be subtracted from that for the more difficult task to yield the time for the additional process. For example, Donders examined performance of two tasks, each of which used two stimuli. One was a go/no-go task in which a response was to be made to only one of the two stimuli; the other was a two-choice task in which one response was to be made to one stimulus and another response to the second stimulus. Donders reasoned that the choice task required a response-selection process that the go/no-go task did not and attributed the difference in reaction time for the two tasks to the response-selection stage.

The subtractive method has been used in recent years to estimate the durations of a variety of processes. One widely known application is Posner and Mitchell's (1967) use of the method to estimate the time to name letters. They had subjects classify pairs of letters as same or different. Reaction time to judge two physically identical letters (e.g., AA) as same was 70–100 ms shorter than that to judge two physically different letters (e.g., Aa) as same. Posner and Mitchell interpreted this difference in reaction time as the time to obtain name codes for the letters after the physical representations were formed. The subtractive method has also been used to estimate the rates of mental rotation (approximately 12–20 ms per degree of rotation; Shepard & Metzler, 1971) and memory search (approximately 40 ms per item; Sternberg, 1969). An application of the subtractive method to HCI would be, for example, to compare the time to find a target link on two web pages that are identical, except for the number of links displayed, and to attribute the extra time to the additional visual search required for the more complex web page.

The subtractive logic has several limitations (Pachella, 1974). First, it is only applicable when discrete, serial processing stages can be assumed. Second, the processing for the two tasks being compared must be equivalent except for the additional process that is being evaluated. This requires an assumption of pure insertion, which is that the additional process for the more complex of two tasks can be inserted without affecting the processes held in common by the two tasks. However, this assumption often is not justified.

Sternberg (1969) developed the additive factors method to allow determination of the processes involved in performing a task. The additive factors method avoids the problem of pure insertion because the crucial data are whether two variables affect reaction time for the same task in an additive or interactive manner. Sternberg assumed, as did Donders, that information processing occurs in a sequence of discrete stages, each of which produces a constant output that serves as input to the next stage in the sequence. With these assumptions, he showed that two variables that affect different stages should have additive effects on reaction time. In contrast, two variables that affect the same stage should have interactive effects on reaction time. Sternberg performed detailed analyses of memory search tasks in which a person holds a set of letters or digits in memory and responds to a target stimulus by indicating whether or not it is a member of the memory set. Based on the patterns of additive and interactive effects that he observed, Sternberg concluded that the processing in such tasks involves four stages: identification of the target, memory search, response selection, and response execution.

Both the subtractive and additive factors methods have been challenged on several grounds (Pachella, 1974). First, the assumption of discrete serial stages with constant output is difficult to justify in many situations. Second, both methods rely on analyses of reaction time only, without consideration of error rates. This can be problematic because performance is typically not error free, and, as described next, speed can be traded for accuracy. Despite these limitations, the methods have proved to be robust and useful (Sanders, 1998).

Speed-Accuracy Methods

The function relating response speed to accuracy is called the speed-accuracy operating characteristic (Pachella, 1974). The function, illustrated in Fig. 2.1, shows that very fast responses can be performed with chance accuracy, and accuracy will increase as responding slows down. Of importance is the fact that when accuracy is high, as in most reaction time studies, a small increase in errors can result in a large decrease in reaction time.

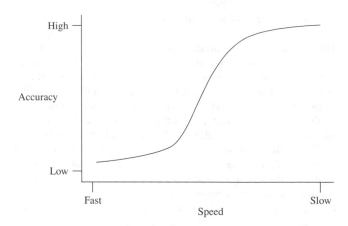

FIGURE 2.1. Speed-accuracy operating characteristic curve. Faster responding occurs at the cost of lower accuracy.

Several researchers have advocated that speed-accuracy studies be conducted instead of reaction time studies because they are potentially more informative, providing information about the intercept (time at which accuracy exceeds chance), asymptote (the maximal accuracy), and rate of ascension from the intercept to the asymptote, each of which may reflect different processes.

In speed-accuracy tradeoff studies, the speed-accuracy criterion is varied between blocks of trials by instructing subjects differently in different blocks regarding the relative importance of speed vs. accuracy, varying payoffs such that speed or accuracy is weighted more heavily, or imposing different reaction time deadlines (Wickelgren, 1977). Because the speed-accuracy criterion is manipulated in addition to any other variables of interest, many more trials must be conducted in a speed-accuracy study than in a reaction time study. Consequently, use of speed-accuracy methods has been restricted to situations in which the speed-accuracy relation is of major concern, rather than being widely adopted as the method of choice.

Psychophysiological Methods

In recent years, psychophysiological methods have come to be increasingly used in information processing research, as the apparatus and techniques for measuring physiological responses have become more sophisticated. Psychophysiological methods have the potential to provide details regarding the nature of processing by examining physiological activity as a task is being performed. The most widely used method currently involves measurement of electroencephalograms, which are recordings of changes in brain activity as a function of time measured by electrodes placed on the scalp (Rugg & Coles, 1995). Of most concern for information processing research are event-related potentials (ERPs), which are the changes in brain activity that are elicited by an event such as stimulus presentation or response initiation. ERPs are obtained by averaging across many trials of a task to remove background electroencephalogram noise and are thought to reflect postsynaptic potentials in the brain.

There are several features of the ERP that represent different aspects of processing. These features are labeled according to their polarity, positive (P) or negative (N), and their sequence or latency. The first positive (P1) and negative (N1) components are associated with early perceptual processes. They are called exogenous components because they occur in close temporal proximity to the stimulus event and have a stable latency with respect to it. Later components reflect cognitive processes and are called endogenous because they are a function of the task demands and have a more variable latency than the exogenous components. One such component that has been studied extensively is the P3 (or P300), which represents postperceptual processes. When an occasional target stimulus is interspersed in a stream of standards, the P3 is observed in response to targets, but not to standards. By comparing the effects of task manipulations on various ERP components, such as P3, their onset latencies, and their scalp distributions, relatively detailed inferences about the cognitive processes can be made.

One example of applying a P3 analysis to HCI is a study by Trimmel and Huber (1998). In their study, subjects performed three HCI tasks (text editing, programming, and playing Tetris) for 7 minutes each. They also performed comparable paper/pencil tasks in three other conditions. The P3 was measured after each experimental task by having subjects monitor a stream of high- and low-pitched tones, keeping count of each separately. The P3 varied as a function of type of task, as well as medium (computer vs. paper/pencil). The amplitude of the P3 was smaller following the HCI tasks than following the paper/pencil tasks, suggesting that the HCI tasks caused more fatigue or depletion of resources than the paper/pencil task. The P3 latency was shorter after the programming task than after the others, which the authors interpreted as an after-effect of highly focused attention.

Another measure that has been used extensively in studies of human information processing is the lateralized readiness potential (Eimer, 1998). This potential can be recorded in choice-reaction tasks that require a response with the left or right hand. It is a measure of differential activation of the lateral motor areas of the visual cortex that occurs shortly before and during execution of a response. The asymmetric activation favors the motor area contralateral to the hand making the response, because this is the area that controls the hand. Of importance, the lateralized readiness potential has been obtained in situations in which no overt response is ever executed, allowing it to be used as an index of covert, partial response activation. The lateralized readiness potential is thus a measure of the difference in activity from the two sides of the brain that can be used as an indicator of covert reaction tendencies, to determine whether a response has been prepared even when it is not actually executed. It can also be used to determine whether the effects of a variable are prior or subsequent to response preparation.

Electrophysiological measurements and recordings of magnetic fields do not have the spatial resolution needed to provide precise information about the brain structures that produce the recorded activity. Recently developed neuroimaging methods, including positron-emission tomography and functional magnetic resonance imaging, measure changes in blood flow associated with neuronal activity in different regions of the brain. These methods have poor temporal resolution, but much higher spatial resolution than the electrophysiological methods. Typically, both control and experimental tasks are performed, and the functional neuroanatomy of the cognitive processes is derived by subtracting the image during the control task from that during the experimental task. This subtractive method is based on many strong assumptions and has been questioned. Sartori and Umiltà (2000) have proposed to replace the subtractive method with the additive-factors method and what they call, the specific-effect method. For this latter method, the task that the subject is instructed to perform remains unchanged, but the processes required to perform it change. The logic of specific effects is to look for a crossover interaction between an independent variable and a brain area. Sartori and Umiltà argued that the specific-effect method should be used when investigating activations produced by different levels of a qualitative variable, whereas the additive-factor method should be used for quantitative variables. The use of these methods allows a researcher to distinguish between parallel and serial stages of processing and between local and distributed processing.

INFORMATION PROCESSING MODELS

Discrete and Continuous Stage Models

It is common to assume that the processing between stimuli and responses consists of a series of discrete stages for which the output for one stage serves as the input for the next, as Donders and Sternberg assumed. This assumption is made for the Model Human Processor (Card et al., 1983) and applications of the Executive-Process Interactive Control (EPIC) architecture (Meyer & Kieras, 1997), both of which have been applied to HCI. However, models can be developed that allow for successive processes to operate concurrently. A well-known model of this type is McClelland's (1979) cascade model, in which partial information at one subprocess, or stage, is transferred to the next. Each stage is continuously active, and its output is a continuous value that is always available to the next stage. The final stage results in selection of which of the possible alternative responses to execute.

In McClelland's (1979) cascade model, an independent variable may affect the rate of activation or the asymptotic level of activation within a particular stage. The activation rate determines the speed at which the final output is attained, and asymptotic level is analogous to the output in the discrete stage model. Even though the cascade model and discrete stage model make different assumptions about the relations among the processes, the patterns of interactions and additive effects in reaction time data can be interpreted in the same manner. That is, two variables that affect the rate parameter of the same stage will have interactive effects on reaction time. But, if the two variables affect the rate parameters of different stages, additive effects on reaction time will be observed. As long as the assumption is made that the final output of a stage does not vary as a function of the manipulations, then the additive factors logic can be used to interpret the reaction time patterns without the assumption of discrete stages.

According to J. Miller (1988), models of human information processing can be classified as discrete or continuous along three dimensions: Representation, transformation, and transmission. Representation refers to whether the input and output codes for the processing stage are continuous or discrete. Transformation refers to whether the operation performed by the processing stage (e.g., spatial transformation) is continuous or discrete. Transmission is classified as discrete if the processing of successive stages does not overlap temporally. The discrete stage model proposed by Sternberg (1969) has discrete representation and transmission, whereas the cascade model proposed by McClelland (1979) has continuous representation, transmission, and transformation.

Models can be intermediate to these two extremes. For example, J. Miller's (1988) asynchronous discrete coding model assumes that most stimuli are composed of features, and these features are identified separately. Discrete processing occurs for feature identification, but once a feature is identified, this information can be passed to response selection while the other features are still being identified.

Sequential Sampling Models

Sequential sampling models are able to account for both reaction time and accuracy, and consequently, the tradeoff between them (Van Zandt, Colonius, & Proctor, 2000). According to such models, information from the stimulus is sequentially sampled, resulting in a gradual accumulation of information on which selection of one of the alternative responses is based. A response is selected when the accumulated information exceeds a threshold amount required for that response. Factors that influence the quality of information processing have their effects on the rate at which the information accumulates, whereas factors that bias speed vs. accuracy or specific responses have their effects on the response thresholds.

Balakrishnan (1998) argued that sequential sampling may be a factor even when the experiment does not stress speed of responding. As described previously, he showed that an analysis of vigilance data that does not make the assumptions of signal detection theory suggests that attribution of the vigilance decrement to a change toward a more conservative response bias is incorrect. One reason why signal detection theory may lead to an incorrect conclusion is that the model assumes that the decision is based on a fixed sample of information, rather than information that is accumulating across time. Balakrishnan argued that even though there are no incentives to respond quickly in the typical vigilance task, subjects may choose not to wait until all of the stimulus information has been processed before responding. He proposed that a sequential sampling model, in which the subject continues to process the information until a stopping rule condition is satisfied, provides a better depiction. In this model, there are two potential sources of bias: the stopping rule and decision rule. Based on this model, Balakrishnan concluded that there is a response bias initially when the signal rate is low, and that the vigilance decrement is due to a gradual reduction of this response bias toward a more optimal decision during the timecourse of the vigil.

INFORMATION PROCESSING IN CHOICE-REACTION TASKS

In a typical choice-reaction task in which each stimulus is assigned to a unique response, it is common to distinguish between three stages of processing: stimulus identification, response selection, and response execution (Proctor & Van Zandt, 1994). The stimulus identification stage involves processes that are entirely dependent on properties of the stimuli. The response-selection stage concerns those processes involved in determining what responses are to be made to the stimuli. Response execution refers to the motor responses and their execution. Based on additive-factors logic, Sanders (1998) decomposes the stimulus identification stage into three subcategories, and the response execution stage into two subcategories, resulting in a total of six distinct stages (see Fig. 2.2).

Stimulus Identification

The preprocessing stage of stimulus identification refers to peripheral sensory processes involved in the conduction of the sensory signal along the afferent pathways to the sensory projection areas. These processes are affected by variables such as stimulus contrast and retinal location. As stimulus contrast, or intensity, increases, reaction time decreases until reaching and asymptote. For example, Bonin-Guillaume, Possamäi, Blin, and Hasbroucq (2000) had young and elderly subjects perform a two-choice reaction task, in which a left or right keypress was made to a bright or dim light positioned to the left or right. Stimulus intensity interacted with age group, with reaction times for the young adults being approximately 25 ms shorter to a bright stimulus than to a dim stimulus, and those for the older adults being approximately 50 ms shorter. The effect of stimulus intensity did not interact with variables that affect response selection and motor adjustment, suggesting that although the elderly subjects were slowed in sensory preprocessing, the deficiency in sensory preprocessing did not affect the efficiency of the other later processing stages.

Feature extraction involves lower level perceptual processing based in area V1 (the visual cortex) and other early visual cortical areas. Stimulus quality, word priming, and stimulus discriminability affect the feature extraction process. For example, manipulations of stimulus quality such as superimposing a grid, using dotted stimuli, with some dots shifted, etc., slow reaction time presumably by creating difficulty for the extraction of features. Identification itself is influenced by mental rotation and word frequency. Mental rotation refers to the finding that when a stimulus is rotated from the upright position, the time it takes to identify the stimulus increases as an approximately linear function of angular deviation from upright (Shepard & Metzler, 1971). This is presumed to affect a normalization process in which the image is mentally rotated in a continuous manner to the upright position.

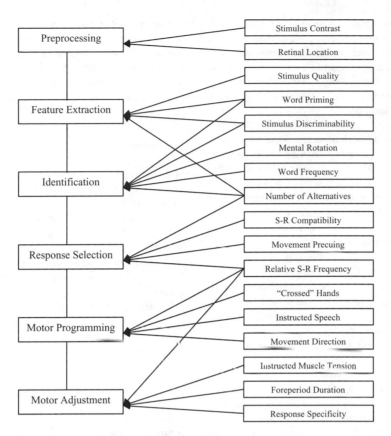

FIGURE 2.2. Information processing stages and variables that affect them, based on Sanders' (1998) taxonomy. S-R = stimulus-response.

Response Selection

Response selection refers to those processes involved in determining what response to make to a particular stimulus. It is affected by the variables of number of alternatives, stimulus-response compatibility, and precuing. As the number of stimulus-response alternatives increases, reaction time increases as a logarithmic function of the number of alternatives (Hick, 1952; Hyman, 1953). This relation is known as the Hick-Hyman law, which for N equally likely stimulus-response alternatives is:

$$RT = a + b \log_2 N, \qquad (1)$$

where a is the base processing time and b is the amount that RT increases with increases in N. The slope of the Hick-Hyman function is influenced by many factors. For example, the slope decreases as subjects become practiced at a task (Teichner & Krebs, 1974).

One of the variables that affects the slope of the Hick-Hyman function is that of stimulus-response compatibility, a variable that has considerable impact on response-selection efficiency. Compatibility effects refer to differences in speed and accuracy of responding as a function of how natural, or compatible, the relation between stimuli and responses is. Two types

of compatibility effects can be distinguished (Kornblum, Hasbroucq, & Osman, 1990). For one type, certain sets of stimuli are more compatible with certain sets of responses than with others. For example, the combinations of verbal-vocal and spatial-manual sets typically yield better performance than the combinations of verbal-manual and spatial-vocal sets (Wang & Proctor, 1996). For the other type, within a specific stimulus-response set, some mappings of individual stimuli to responses produce better performance than others. For example, if one stimulus has the meaning left, and the other the meaning right, performance is better if the left stimulus is mapped to the left response and the right stimulus to the right response, regardless of the stimulus and response modes.

Fitts and Seeger (1953) and Fitts and Deininger (1954) demonstrated both types of compatibility effects for spatially arranged display and response panels. However, it is not fully appreciated that compatibility effects occur for a wide variety of other stimulus-response sets. According to Kornblum et al. (1990), dimensional overlap (similarity) between the stimulus and response sets is the critical factor. When the sets have dimensional overlap, a stimulus will activate its corresponding response automatically. If this response is correct (compatible mapping), responding will be facilitated, but if it is not correct (incompatible mapping), responding will be inhibited. A second factor contributing to the advantage for the compatible mapping is that intentional translation of the stimulus into a response will occur quicker when the mapping is compatible than when it is not. Most contemporary models of stimulus-response compatibility incorporate both automatic and intentional response-selection routes (Hommel & Prinz, 1997), although they differ regarding the exact conditions under which each plays a role and the way in which they interact.

One reason why automatic activation is considered to contribute to compatibility effects is that such effects also occur when irrelevant stimulus information overlaps with the response set (Lu & Proctor, 1995). The most well-known phenomenon of this type is the Stroop color-naming effect, in which irrelevant color words that conflict with a to-be-named color produce considerable interference. A similar phenomenon, known as the Simon effect, occurs when stimulus location is irrelevant, but responses are spatial.

Compatibility effects are even more ubiquitous than the previous description would suggest. For completely unrelated stimulus and response sets that are structured, performance is better when structural correspondence is maintained (Reeve & Proctor, 1990). When stimuli and responses are ordered (e.g., a row of four stimulus locations and a row of four response locations), reaction time is faster when the stimulus-response mapping can be characterized by a rule (e.g., press the key at the mirror opposite location) than when the mapping is random (Duncan, 1977). Spatial compatibility effects also occur when display and response elements refer to orthogonal dimensions (Cho & Proctor, 2001). On the other hand, on occasion, stimulus-response compatibility effects sometimes do not occur under conditions in which one would expect them to occur. For example, when compatible and incompatible mappings are mixed within a single block, the typical compatibility effect is eliminated (Shaffer, 1965).

Because when and where compatibility effects are going to occur is not obvious, interface designers are likely to make poor decisions if they rely only on their intuitions. Payne (1995) had naïve subjects predict performance for four stimulus-response configurations that differed in terms of spatial mappings. For each configuration, a row of four stimulus locations was mapped to a row of four responses. In one condition, all four stimuli were mapped to the corresponding responses, and in a second condition, the stimuli were mapped to their mirror opposite response locations. In the remaining two conditions, mappings were mixed (i.e., two stimuli were mapped to their corresponding responses, and two to the opposite responses). Subjects correctly predicted that performance would be best for the compatible condition, but incorrectly predicted that performance would be better in the mixed conditions than in the mirror-opposite condition. Apparently, the subjects did not realize that there is a benefit when an "opposite" rule can be applied to all stimuli and that there are costs for the compatibly mapped pairs when mixed with incompatible pairs.

Vu and Proctor (2001) confirmed Payne's (1995) findings, but showed that after limited practice with the four mapping configurations, subjects were able to adjust their initial judgments of performance to more accurately match actual performance. The important point for HCI is that the designers need to be aware of the potential problems created by incompatibility

between display and response elements because their effects are not always obvious. The designer can get a better feel for the relative compatibility of alternative arrangements by using them himself or herself. However, after the designer selects a few arrangements that would seem to yield good performance, the alternatives should be tested on groups of users.

Response Execution

Motor programming refers to specification of the physical response that is to be made. This process is affected by variables such as relative stimulus-response frequency and movement direction. One factor that apparently influences this stage is movement complexity. The longer the sequence of movements, that is to be made on occurrence of a stimulus in a choice-reaction task, the longer the reaction time (Sternberg, Monsell, Knoll, & Wright, 1978). This effect is thought to be due to the time required to load the movement sequence into a buffer before initiating the movements.

One of the most widely known relations attributed to response execution is Fitts' law for the time to make aimed movements to a target location (Fitts, 1954). This law, as originally specified by Fitts, is:

$$\text{Movement Time} = a + b \log_2(2D/W), \quad (2)$$

where a and b are constants, D is distance to the target, and W is target width. However, there are slightly different derivations of the law. According to Fitts' law, movement time is a direct function of distance and an inverse function of target width. Fitts' law has been found to provide an accurate description of movement time in many situations, although alternatives have been proposed for certain situations. One of the factors that contributes to the increase in movement time as the index of difficulty increases is the need to make a corrective submovement based on feedback to hit the target location (Meyer, Abrams, Kornblum, Wright, & Smith, 1988).

Most HCI involves the users using text keys, step keys (arrows), a mouse, or a joystick to move a cursor to a target position. Consequently, the time to make these movements can be described in terms of Fitts' law. One implication of the law for interface design is that the slope of the function, b, may vary across different control devices, in which case, movement times will be faster for the devices that yield lower slopes. Card, English, and Burr (1978) conducted a study that evaluated how efficient text keys, step keys, a mouse, and a joystick are at a text selection task, in which users selected text by positioning the cursor on the desired area and pressing a button or key. They showed that the mouse was the most efficient and effective device for this task: Positioning time for the mouse and joystick could be accounted for by Fitts' law, with the slope of the function being less steep for the mouse; positioning time with the keys was proportional to the number of key strokes that had to be executed.

Another implication of Fitts' law is that anything that reduces the index of difficulty should decrease the time for motor movements. Walker, Smelcer, and Nilsen (1991) evaluated movement time and accuracy of menu selection for the mouse. Their results showed that reducing the distance to be traveled (which reduces the index of difficulty) by placing the initial cursor in the middle of the menu, rather than the top, improved movement time. In addition, placing a border around the menu item in which a click would still activate that item, and increasing the width of the border as the travel distance increases, also improved performance. The reduction in movement time by use of borders is predicted by Fitts' law because borders increase the size of the target area.

Gillan, Holden, Adam, Rudisill, and Magee (1992) noted that designers must be cautious when applying Fitts' law to HCI because factors other than distance and target size play a role when using a mouse. Specifically, they proposed that the critical factors in pointing and dragging are different than those in pointing and clicking [which was the main task in Card et al.'s (1978) study]. Gillan et al. showed that, for a text-selection task, both point-click and point-drag movement times can be accounted for by Fitts' law. For point-click sequences, the diagonal distance across the text object, rather than the horizontal distance, provided the best fit for pointing time. For point-drag, the vertical distance of the text provided the best fit. The reason why the horizontal distance is irrelevant is that the cursor must be positioned at the beginning of the string for the point-drag sequence. Thus, task requirements should be taken into account before applying Fitts' law to the interface design.

Motor adjustment is the last stage and deals with the transition from a central motor program to peripheral motor activity. Studies of motor adjustment have focused on the influence of foreperiod duration on motor preparation. In a typical study, a neutral warning signal is presented at various intervals prior to the onset of the imperative stimulus. Bertelson (1967) varied the duration of the warning foreperiod and found that reaction time reached a minimum at a warning interval of 150 ms and then increased slightly at the 200- and 300-ms foreperiods. However, the error rate increased to a maximum at the 150-ms foreperiod and decreased slightly at the longer foreperiods. This pattern, which is relatively typical, suggests that it takes time to attain a state of high motor preparation, and that this state reflects an increased readiness to respond quickly, but at the expense of accuracy.

MEMORY IN INFORMATION PROCESSING

Memory refers to effects of previous information on information processing. It may involve recollection of an immediately preceding event or one many years in the past, knowledge derived from everyday life experiences and education, or learned procedures to accomplish complex perceptual-motor tasks. Memory can be classified into several categories. Episodic memory refers to memory for a specific event, such as going to the movie last night, whereas semantic memory refers to general knowledge, such as what a movie is. Declarative memory is verbalizable knowledge, and procedural memory is knowledge that can be expressed nonverbally. In other words, declarative memory is knowing that something is the case, whereas procedural memory is knowing how to do something. For example, telling your friend your new phone number involves declarative memory, whereas riding a bicycle involves procedural knowledge. A memory test is regarded as explicit if a person is asked to judge whether a specific item or event has occurred before in a particular context; the test is implicit if the person is to make a judgment, such as whether a string of letters is a word or nonword, that can be made without reference to earlier priming events. In this section, we focus primarily on explicit episodic memory.

Three types of memory systems are commonly distinguished: Sensory stores, short-term memory (STM; or working memory), and long-term memory (LTM). Sensory stores, which we will not cover in detail, refer to brief modality-specific persistence of a sensory stimulus from which information can be retrieved for 1 or 2 s after offset of a stimulus (see Nairne, in press). STM and LTM are the main categories by which investigations of episodic memory are classified, and, as the terms imply, the distinction is based primarily on duration. The dominant view is that these are distinct systems that operate according to different principles, but there has been increasing debate over whether the processes involved in these two types of memories are the same or different.

Short-Term (Working) Memory

STM refers to representations that are currently being used or have recently been used and last for a short duration. A distinguishing characteristic of STM is that it is of limited capacity. This point was emphasized in G. A. Miller's (1956) classic article, "The Magical Number Seven Plus or Minus Two," in which he indicated that capacity is not simply a function of the number of items, but rather the number of chunks. For example, i, b, m are three letters, but most people can combine them to form one meaningful chunk of IBM. Consequently, memory span is similar for strings of unrelated letters and strings of meaningful acronyms or words. Researchers refer to the number of items that can be recalled correctly, in order, as memory span.

As most people are aware from personal experience, if distracted by another activity, information in STM can be forgotten quickly. Recall of a string of letters or single-syllable words that is within the memory span decreases to close to chance levels over a retention interval of 18 s when rehearsal is prevented by an unrelated distractor task (Brown, 1958; Murdock, 1961; Peterson & Peterson, 1959). This short-term forgetting was thought initially to be a consequence of decay of the memory trace due to prevention of rehearsal. However, Keppel and Underwood (1968) showed that proactive interference from items on previous lists is a significant contributor to forgetting. They found no forgetting at long retention intervals when only the first list in a series was examined, with the amount of forgetting being much larger for the second and third lists as proactive interference built up. Consistent with this interpretation, release from proactive inhibition (i.e., improved recall) occurs when the category of the to-be-remembered items on the current list differs from that of previous lists (D. D. Wickens, 1970).

The capacity limitation of STM noted by G. A. Miller (1956) is closely related to the need to rehearse the items. Research on determinants of the capacity limitation has shown that the memory span, the number of words that can be recalled correctly in order, varies as a function of word length. That is, the number of items that can be retained decreases as word length increases. Evidence has indicated that the capacity is the number of syllables that can be said in about 2 s (Baddeley, Thomson, & Buchanan, 1975; Schweickert & Boruff, 1980). That pronunciation rate is a critical factor suggests a time-based property of STM, which is consistent with a decay account. Consequently, the most widely accepted view is that both interference and decay contribute to short-term forgetting, with decay acting over the first few seconds of a retention interval and interference accounting for the largest part of the forgetting.

As the complexity of an HCI task increases, one consequence is to overload STM. Jacko (1997) and Jacko and Ward (1996) varied four different determinants of task complexity (multiple paths, multiple outcomes, conflicting interdependence among paths, or uncertain or probabilistic linkages) in a task that required use of a hierarchical menu to acquire specified information. When one determinant was present, performance was slowed by approximately 50%, and when two determinants were present in combination, performance was slowed further. That is, as the number of complexity determinants in the interface increased, performance decreased. Jacko and Ward attributed the decrease in performance for all four determinants to the increased STM load that they imposed.

The best-known model of STM is Baddeley and Hitch's (1974) working memory model, which partitions STM into three main parts: Central executive, phonological loop, and visuospatial sketchpad. The central executive is the least-defined component of the model, and is assumed to control and coordinate the actions of the phonological loop and visuospatial sketchpad. The phonological loop is composed of a phonological store that is responsible for storage of the to-be-remembered items, and an articulatory control process that is responsible for recoding verbal items to a phonological form and the rehearsal of those items. The items stored in the phonological store decay over a short interval and can be refreshed through rehearsal from the articulatory control process. The visuospatial sketchpad retains information regarding visual and spatial information, and it is involved in mental imagery.

The working memory model has been successful in explaining several phenomena of STM (Baddeley, 2000), for example, that the number of words that can be recalled is affected by word length. However, the model cannot explain why memory span for visually presented material is only slightly reduced when subjects engage in concurrent articulatory suppression (such as saying the words "the" aloud repeatedly). Articulatory suppression should monopolize the phonological loop, preventing any visual items from entering it. To account for findings of this type, Baddeley recently revised his working memory model to include an episodic buffer component. The revised model is shown in Fig. 2.3. The buffer is a limited capacity temporary store that can integrate information from the phonological loop, visuospatial sketchpad, and LTM. By attending to a given source of information in the episodic buffer, the central executive can create new cognitive representations that might be useful in problem-solving. Because the buffer was only recently introduced, there has not been much evidence in support or opposition to its assumed functions.

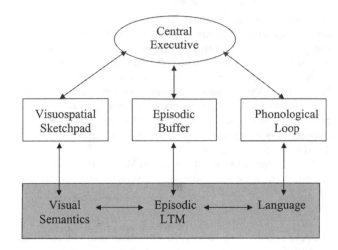

FIGURE 2.3. Baddeley's (2000) revised working memory model. LTM = long-term memory. From "The Episodic Buffer: A New Component of Working Memory?" by A. D. Baddeley, 2000, *Trends in Cognitive Sciences, 4,* p. 421. Copyright 2000 by Elsevier Science Ltd. Reprinted with permission.

Long-Term Memory

LTM refers to representations that can be remembered for durations longer than can be attributed to STM. LTM can involve information presented minutes ago or years ago. Initially, it was thought that the probability of an item being encoded into LTM was a direct function of the amount of time that it was in STM, or how much it was rehearsed. However, Craik and Watkins (1973) showed that rehearsal in itself is not sufficient, but rather that deep-level processing of the meaning of the material is the important factor in transferring items to LTM. They presented subjects with a list of words and instructed them that when the experimenter stopped the presentation of the words, they were to recall the last word starting with the letter *a*. The number of other words between instances of *a* words was varied, with the idea being that the amount of time a word was rehearsed would depend on the number of words before the next *a* word. At the end of the session, subjects were given a surprise recall test in which they were to recall all *a* words. The results showed that there was no effect of number of intervening words on recall, indicating that, although subjects rehearsed the words longer, their recall did not improve because they did not process the words deeply.

Craik and Watkins' (1973) results are consistent with the level of processing framework proposed by Craik and Lockhart (1972). According to this view, encoding proceeds in a series of analyses, from shallow perceptual features to deeper, semantic levels. The deeper the level of processing, the more strongly the item is encoded in memory. A key study supporting the levels of processing view is that of Hyde and Jenkins (1973). In their study, groups of subjects were presented a list of words for which they engaged in shallow processing (e.g., deciding whether each word contained a capital letter) or deep processing of it (e.g., identifying whether each word was a verb or a noun). Subjects were not told in advance that they would be asked to recall the words, but were given a surprise recall test at the end of the session. The results showed that the deep processing group recalled more words than the shallow processing group.

Another well-known principle for LTM is called encoding specificity. This principle states that the probability that a retrieval cue results in recollection of an earlier event is an increasing function of the match between the features encoded initially and those provided by the retrieval cue (Tulving & Thomson, 1973). An implication of this principle is that memory will be context dependent. Godden and Baddeley (1975) demonstrated a context-dependent memory effect by having divers learn a list of words either on land or under water, and recall the words on land or under water. Recall was higher for the group who learned on land when recall took place on land than under water. Similarly, recall was higher for the group who learned under water when recall took place under water than on land. A related principle is that of transfer appropriate processing, proposed initially by Morris, Bransford, and Franks (1977). They showed that, as typically found, deep-level, semantic judgments during study produced better performance than shallow rhyme judgments on a standard recognition memory test. However, when the memory test required decisions about whether the test words rhymed with studied words, the rhyme judgments led to better performance than the semantic judgments.

Recent research has confirmed that the levels of processing framework must be combined with the encoding specificity and transfer appropriate processing principles to explain the effects of processing performed during encoding. Although levels of processing has a strong effect on accuracy of explicit recall and recognition, Jacoby and Dallas (1981) found no effect on an implicit memory test. Later studies have shown a robust effect of levels of processing on implicit tests similar to that obtained for recall and recognition if the test is based on conceptual cues, rather than perceptual cues (see Challis, Velichkovsky, & Craik, 1996). Challis et al. constructed direct recognition tests, in which the words were graphemically, phonologically, or semantically similar to the studied words, that showed no levels of processing effect. They emphasized that to account for levels of processing results, it is necessary to specify the types of information produced by the levels of processing, the types of information required for the specific test, and how task instructions modify encoding and retrieval processes.

Other Factors Affecting Retrieval of Earlier Events

Memory researchers have studied many factors that influence long-term retention. Not surprisingly, episodic memory improves with repetition of items or events. Also, massed repetition (repeating the same item in a row) is less effective than spaced repetition (repeating the same item with one or more intervening items). This benefit for spaced repetition, called the spacing effect or lag effect, is widespread and occurs for both recall and recognition (Hintzman, 1974).

Another widely studied phenomenon is the generation effect, in which recall is better when subjects have to generate the to-be-remembered words rather than just studying the words as they are presented (Slamecka & Graf, 1978). In a generation effect experiment, subjects are divided into two groups: read

and generate. Each group receives a series of words, with each word spelled out completely for the read group and missing letters for the generate group. An example is as follows:

Read group: CAT; ELEPHANT; GRAPE; CAKE
Generate group: C_T; E_E_H_NT; G_APE; CAK_

The typical results show that subjects in the generate group can recall more words than those in the read group. One application of the generation effect to HCI is that when a user needs a password for an account, the system should allow the user to generate the password rather than providing the user with one, because the user would be more likely to recall it in the former than latter case.

Events that precede or follow an event of interest can interfere with recall of that event. The former is referred to as proactive interference, and was discussed in the section on STM, and the latter is referred to as retroactive interference. One area of research in which retroactive interference is of central concern is that of eyewitness testimony. Loftus and Palmer (1974) showed that subsequent events could distort a person's memory of an event that the person witnessed. Subjects were shown a sequence of events depicting a car accident. Subsequently, they were asked the question, "How fast were the cars going when they _____ each other." When the verb *contacted* was used, subjects estimated the speed to be 32 mph, and only one-tenth of them reported seeing broken glass. However, when the verb *smashed* was used, the estimated speed increased to 41 mph, and almost one-third of the subjects reported seeing broken glass. Demonstrations like these indicate not only that retroactive interference can cause forgetting of events, but that it also can cause the memory of events to be changed. More recent research has shown that completely false memories can be implanted (see Roediger & McDermott, 1995).

Another phenomenon that influences recall is part-set cuing, in which providing people with a subset of the to-be-remembered items during recall impairs recall performance instead of enhancing it. Slamecka (1968) presented subjects with a list of words twice and had the control group recall as many words as they could. In the other groups, subjects were provided a subset of the words that were on the list and were asked to recall the remaining ones. Recall of the remaining words was worse for the subset groups than for the control group. One explanation for the impairment of recall by part-set cuing is that providing the subset of items disrupts normal retrieval strategies. With respect to the password example in the previous paragraph, these results suggest that, if cues are given to help recall, they should not contain any part of the password itself.

Mnemonic techniques can also be used to improve recall. The basic idea behind mnemonics is to connect the to-be-remembered material with an established organizational structure that can be easily accessible later on. Two widely used mnemonic techniques are the pegword method (Wood & Pratt, 1987) and the method of loci (Verhaeghen & Marcoen, 1996). In the pegword method, a familiar rhyme provides the organizational structure. A visual image is formed between each pegword in the rhyme and the associated target item. At recall, the rhyme is generated, and the associated items come to mind. For example, if you wanted to remember to buy a computer mouse, a printer cable, and diskettes, the peg could be the rhyme "one is a bun, two is a shoe, three is a tree," and you could visualize a computer mouse in a bun, a shoe laced with a printer cable, and a tree full of diskettes. For the method of loci, locations from a well-known place, such as your house, are associated with the to-be-remembered items. Although specific mnemonic techniques are limited in their usefulness, the basic ideas behind them (utilizing imagery, forming meaningful associations, and using consistent encoding and retrieval strategies) are of broad value for improving memory performance.

ATTENTION IN INFORMATION PROCESSING

Attention is increased awareness directed at a particular event or action to select it for increased processing. This processing may result in enhanced understanding of the event, improved performance of an action, or better memory for the event. Attention allows us to filter out unnecessary information so that we can focus on a particular aspect that is relevant to our goals. Several influential information processing models of attention have been proposed.

Models of Attention

In an influential study, Cherry (1953) presented different messages to each ear through headphones. Subjects were to repeat aloud one of the two messages while ignoring the other. When subsequently asked questions about the two messages, subjects were able to accurately describe the message to which they were attending, but could not describe anything except physical characteristics, such as gender of the speaker, about the unattended message.

To account for findings such as these, Broadbent (1958) developed the filter theory, which assumes that the nervous system acts as a single-channel processor. According to filter theory, information is received in a preattentive temporary store and then is selectively filtered, based on physical features such as spatial location, to allow only one input to access the channel. Broadbent's filter theory implies that the meaning of unattended messages is not identified, but later studies showed that the unattended message could be processed beyond the physical level, in at least some cases (Treisman, 1964).

To accommodate the finding that meaning of an unattended message can influence performance, Treisman (1964) reformulated filter theory into what is called the filter-attenuation theory. According to attenuation theory, early selection by filtering still precedes stimulus identification, but the filter only attenuates the information on unattended channels. This attenuated signal may be sufficient to allow identification if the stimulus is one with a low identification threshold, such as a person's name or an expected event. Deutsch and Deutsch (1963) proposed that unattended stimuli are always identified, and the bottleneck occurs in later processing, a view called late-selection theory. The difference between attenuation theory and late-selection theory is that the latter assumes that meaning is fully analyzed, but the former does not. Based on a review of the literature, Pashler

(1998) recently concluded that the evidence supports early attenuation, but as an optional strategy rather than a structural bottleneck.

In divided attention tasks, a person must attend to multiple sources of information simultaneously. Kahneman (1973) proposed a unitary resource model that views attention as a single resource that can be divided up among different tasks in different amounts, based on task demands and voluntary allocation strategies. Unitary resource models provided the impetus for dual-task methodologies, such as performance operating characteristics, and mental workload analyses that are used widely in HCI (Eberts, 1994). The expectation is that multiple tasks should produce interference when their resource demands exceed the supply that is available.

Many studies have shown that it is easier to perform two tasks together when they use different stimulus or response modalities than when they use the same modalities. Performance is also better when one task is verbal and the other visuospatial than when they are the same type. These result patterns provide the basis for multiple resource models of attention such as that of C. D. Wickens (1984). According to multiple resource models, different attentional resources exist for different sensory-motor modalities and coding domains. Multiple resource theory captures the fact that multiple task performance typically is better when the tasks use different input–output modes than when they use the same modes. However, it is often criticized as being too flexible because new resources can be proposed arbitrarily to fit any finding of specificity of interference (Navon, 1984).

A widely used metaphor for visual attention is that of a spotlight that is presumed to direction attention to everything in its field (Posner & Cohen, 1984). Direction of attention is not necessarily the same as the direction of gaze, because the attentional spotlight can be directed independently of fixation. Studies show that when a location is cued as likely to contain a target stimulus, but then a probe stimulus is presented at another location, a spatial gradient surrounds the attended location such that items nearer to the focus of attention are processed more efficiently than those farther away from it (Yantis, 2000). The movement of the attentional spotlight to a location can be triggered by two types of cues: exogenous and endogenous. An exogenous cue is an external event such as the abrupt onset of a stimulus at a peripheral location that involuntarily draws the attentional spotlight to its location. Exogenous cues produce rapid performance benefits, which dissipate quickly, for stimuli presented at the cued location. An endogenous cue is typically a symbol such as a central arrowhead that must be identified before a voluntary shift in attention to the designated location can be made. The performance benefits for endogenous cues take longer to develop and are sustained for a longer period of time when the cues are relevant, indicating that their benefits are due to conscious control of the attentional spotlight (Klein & Shore, 2000).

In a visual search task, subjects are to detect whether a target is present among distractors. Treisman and Gelade (1980) developed Feature Integration Theory to explain the results from visual search studies. When the target is distinguished from the distractors by a basic feature such as color (feature search), reaction time, and error rate often show little increase as the number of distractors increases. However, when two or more features must be combined to distinguish the target from distractors (conjunctive search), reaction time and error rate typically increase sharply as the number of distractors increases. To account for these results, feature integration theory assumes that basic features of stimuli are encoded into feature maps in parallel across the visual field at a preattentive stage. Feature search can be based on this preattentive stage because a target-present response requires only detection of the feature. The second stage involves focusing attention on a specific location and combining features that occupy the location into objects. Attention is required for conjunctive search because responses cannot be based on detection of a single feature. According to feature integration theory, performance in conjunctive search tasks decreases as the number of distractors increases because attention must be moved sequentially across the search field until a target is detected or all items present have been searched. Feature integration theory served to generate a large amount of research on visual search that showed, as typically the case, that the situation is not as simple as depicted by the theory. This has resulted in modifications of the theory, as well as alternative theories. For example, Wolfe's (1994) Guided Search Theory maintains the distinction between an initial stage of feature maps and a second stage of attentional binding, but assumes that the second stage is guided by the initial feature analysis.

The role of attention in response selection has been investigated extensively using the psychological refractory period (PRP) paradigm (Pashler, 1998). In the PRP paradigm, a pair of choice-reaction tasks must be performed, and the stimulus-onset asynchrony (SOA) of the second stimulus is presented at different intervals. Reaction time for Task 2 is slowed at short SOAs, and this phenomenon is called the PRP effect. Experimental results have been interpreted with what is called locus of slack logic (Schweickert, 1978), which is an extension of additive factors logic to dual-task performance. The basic idea is that if a Task 2 variable has its effect prior to a bottleneck, that variable will have an underadditive interaction with SOA. This underadditivity occurs because, at short SOAs, the slack period during which post-bottleneck processing cannot begin can be used for continued processing for the more difficult condition. If a Task 2 variable has its effect after the bottleneck, the effect will be additive with SOA.

The most widely accepted account of the PRP effect is the response-selection bottleneck model (Pashler, 1998). The primary evidence for this model is that perceptual variables typically have underadditive interactions with SOA, implying that their effects are prior to the bottleneck. In contrast, post-perceptual variables typically have additive effects with SOA, implying that their effects are after the bottleneck. There has been dispute as to whether there is also a bottleneck at the later stage of response initiation (De Jong, 1993), and whether the apparent response-selection bottleneck is structural or simply a strategy adopted by subjects to comply with task instructions (Meyer & Kieras, 1997). This latter approach is consistent with a recent emphasis on the executive functions of attention in the coordination and control of cognitive processes (Monsell & Driver, 2000).

Automaticity and Practice

Attention demands are high when a person first performs a new task. However, these demands decrease and performance improves as the task is practiced. Because the quality of performance and attentional requirements change substantially as a function of practice, it is customary to describe performance as progressing from an initial cognitively demanding phase to a phase in which processing is automatic (Anderson, 1982; Fitts & Posner, 1967).

The time to perform virtually any task from choice-reaction time to solving geometry problems decreases with practice, with the largest benefits occurring early in practice. Newell and Rosenbloom (1981) proposed a power function to describe the changes in reaction time with practice:

$$RT = BN^{-\alpha}, \tag{3}$$

where N is the number of practice trials, B is reaction time (RT) on the first trial, and α is the learning rate. Although the power function has become widely accepted as a law that describes the changes in reaction time, Heathcote, Brown, and Mewhort (2000) indicated that it does not fit the functions for individual performers adequately. They showed that exponential functions provided better fits than power functions to 40 individual data sets and proposed a new exponential law of practice. The defining characteristic of the exponential function is that the relative learning rate is a constant at all levels of practice, whereas, for the power function, the relative learning rate is a hyperbolically decreasing function of practice trials.

PROBLEM SOLVING AND DECISION MAKING

Beginning with the work of Newell and Simon (1972), it has been customary to analyze problem solving in terms of a problem space. The problem space consists of the following: (1) an initial state, (2) a goal state that is to be achieved, (3) operators for transforming the problem from the initial state to the goal state in a sequence of steps, and (4) constraints on application of the operators that must be satisfied. The problem-solving process itself is conceived of as a search for a path that connects the initial and goal states.

Because the size of a problem space increases exponentially with the complexity of the problem, most problem spaces are well beyond the capacity of STM. Consequently, for problem solving to be effective, search must be constrained to a limited number of possible solutions. A common way to constrain search is through the use of heuristics. For example, people often use a means-ends heuristic for which at each step, an operator is chosen that will move the current state closer to the goal state (Atwood & Polson, 1976). Such heuristics are called weak methods because they do not require much knowledge about the exact problem domain. Strong methods, such as those used by experts, rely on prior domain-specific knowledge and do not require much search because they are based on established principles applicable only to certain tasks.

The problem space must be an appropriate representation of the problem, if the problem is to be solved. One important method for obtaining an appropriate problem space is to use analogy or metaphor. Analogy enables a shift from a problem space that is inadequate to one that may allow the goal state to be reached. There are several steps in using analogies (Holland, Holyoak, Nisbett, & Thagard, 1986), including detecting similarity between source and target problems, and mapping the corresponding elements of the problems. Humans are good at mapping the problems, but poor at detecting that one problem is an analog of another. An implication for HCI is that potential analogs should be provided to users for situations in which they are confronted by novel problems.

The concept of the mental model, which is closely related to that of the problem space, has become widely used in recent years (see chapter 3, this volume). The general idea of mental models with respect to HCI is that, as the user interacts with the computer, he or she receives feedback from the system that allows him/her to develop a representation of how the system is functioning for a given task. The mental model incorporates the goals of the user, the actions taken to complete the goals, and expectations of the system's output in response to the actions. A designer can increase the usability of an interface by using metaphors that allow transfer of an appropriate mental model (e.g., the desktop metaphor), designing the interface to be consistent with other interfaces with which the user is familiar (e.g., the standard Web interface), and conveying the system's functions to the user in a clear and accurate manner. Feedback to the user is perhaps the most effective way to communicate information to the user and can be used to guide the user's mental model about the system.

Humans often have to make choices for situations in which the outcome depends on events that are outside of their control. According to expected utility theory, a normative theory of decision making under uncertainty, the decision maker should determine the expected utility of a choice by multiplying the subjective utility of each outcome by the outcome's probability and summing the resulting values (see Proctor & Van Zandt, 1994). The expected utility should be computed for each choice, and the optimal decision is the choice with the highest expected utility. It should be clear from this description that, for all but the simplest of problems, a human decision maker cannot operate in this manner. To do so would require attending to multiple cues that exceed attentional capacity, accurate estimates of probabilities of various events, and maintenance of, and operation on, large amounts of information that exceeds STM capacity.

Research of Kahneman and Tversky (2000) and others has shown that what people do when the outcome associated with a choice is uncertain is to rely heavily on decision-making heuristics. These heuristics include representativeness, availability, and anchoring. The representativeness heuristic is that the probability of an instance being a member of a particular category is judged on the basis of how representative the instance is of the category. The major limitation of the representativeness heuristic is that it ignores base rate probabilities for the respective categories. The availability heuristic involves determining the probability of an event based on the ease with which instances of the event can be retrieved. The limitation is that availability is affected not only by relative frequency, but also by other factors. The anchoring heuristic involves making a judgment regarding probabilities of alternative states based on initial information, and then adjusting these probabilities from this initial anchor as additional information is received. The limitation of anchoring is that the initial judgment can produce a bias for the probabilities. Although heuristics are useful, they may not always lead to the most favorable decision. Consequently, designers need to make sure that the choice desired for the user in a particular situation is one that is consistent with the user's heuristic biases.

SUMMARY AND CONCLUSION

The methods, theories, and models in human-information processing are currently well developed. The knowledge in this area, of which we were only able to describe at a surface level in this chapter, pertains to a wide range of concerns in HCI, from visual display design to representation and communication of knowledge. For HCI to be effective, the interaction must be made compatible with the human information processing capabilities. Cognitive architectures that incorporate many of the facts about human information processing have been developed that can be applied to HCI. The Model Human Processor of Card et al. (1983) is the most widely known, but applications of other more recent architectures, including the ACT (Adaptive Control of Thought) model of Anderson and colleagues (Anderson, Matessa, & Lebiere, 1997), the Soar Model of Newell and colleagues (Howes & Young, 1997), and the EPIC Model of Kieras and Meyer (1997), hold considerable promise for the field, as demonstrated in chapter 5 of this volume.

References

Anderson, J. R. (1982). Acquisition of cognitive skill. *Psychological Review, 89*, 369–406.

Anderson, J. R., Matessa, M., & Lebiere, C. (1997). ACT-R: A theory of higher level cognition and its relation to visual attention. *Human-Computer Interaction, 12*, 439–462.

Atwood, M. E., & Polson, P. G. (1976). A process model for water jug problems. *Cognitive Psychology, 8*, 191–216.

Baddeley, A. D. (2000). The episodic buffer: A new component of working memory? *Trends in Cognitive Sciences, 4*, 417–423.

Baddeley, A. D., & Hitch, G. J. (1974). Working memory. In G. H. Bower (Ed.), *The psychology of learning and motivation* (Vol. 8, pp. 47–89). New York: Academic Press.

Baddeley, A. D., Thomson, N., & Buchanan, M. (1975). Word length and the structure of shortterm memory. *Journal of Verbal Learning and Behavior, 14*, 575–589.

Balakrishnan, J. D. (1998). Measures and interpretations of vigilance performance: Evidence against the detection criterion. *Human Factors, 40*, 601–623.

Bertelson, P. (1967). The time course of preparation. *Quarterly Journal of Experimental Psychology, 19*, 272–279.

Bonin-Guillaume, S., Possamäi, C.-A., Blin, O., & Hasbroucq, T. (2000). Stimulus preprocessing, response selection, and motor adjustment in the elderly: An additive factor analysis. *Cahiers de Psychologie Cognitive/Current Psychology of Cognition, 19*, 245–255.

Bower, G. H., & Hilgard, E. R. (1981). *Theories of learning.* Englewood Cliffs, NJ: Prentice-Hall.

Broadbent, D. E. (1958). *Perception and communication.* Oxford: Pergamon Press.

Brown, J. (1958). Some tests of the decay theory pf immediate memory. *Quarterly Journal of Experimental Psychology, 10*, 12–21.

Card, S. K., English, W. K., & Burr, B. J. (1978). Evaluation of the mouse, rate-controlled isometrick joystick, step keys, and text keys for text selection on a CRT. *Ergonomics, 21*, 601–613.

Card, S. K., Moran, T. P., & Newell, A. (1983). *The psychology of human-computer interaction.* Hillsdale, NJ: Erlbaum.

Challis, B. H., Velichkovsky, B. M., & Craik, F. I. M. (1996). Levels-of-processing effects on a variety of memory tasks: New findings and theoretical implications. *Consciousness and Cognition, 5*, 142–164.

Cherry, E. C. (1953). Some experiments on the recognition of speech, with one and with two ears. *Journal of the Acoustical Society of America, 25*, 975–979.

Cho, Y. S., & Proctor, R. W. (2001). Effect of initiating action on the up-right/down-left advantage for vertically arrayed stimuli and horizontally arrayed responses. *Journal of Experimental Psychology: Human Perception and Performance, 27*, 472–484.

Craik, F. I. M., & Lockhart, R. S. (1972). Levels of processing: A framework for memory research. *Journal of Verbal Learning and Verbal Behavior, 11*, 671–684.

Craik, F. I. M., & Watkins, M. J. (1973). The role of rehearsal in short-term memory. *Journal of Verbal Learning and Verbal Behavior, 12*, 599–607.

De Jong, R. (1993). Multiple bottlenecks in overlapping task performance. *Journal of Experimental Psychology: Human Perception and Performance, 19*, 965–980.

Deutsch, J. A., & Deutsch, D. (1963). Attention: Some theoretical considerations. *Psychological Review, 70*, 80–90.

Donders, F. C. (1868/1969). On the speed of mental processes. In W. G. Koster (Ed.), *Acta Psychologica, 30, Attention and Performance II* (pp. 412–431). Amsterdam: North-Holland.

Duncan, J. (1977). Response selection rules in spatial choice reaction tasks. In S. Dornic (Ed.), *Attention and performance VI* (pp. 49–71). Hillsdale, NJ: Erlbaum.

Eberts, R. E. (1994). *User interface design.* Englewood Cliffs, NJ: Prentice-Hall.

Eimer, M. (1998). The lateralized readiness potential as an on-line measure of central response activation processes. *Behavior Research Methods, Instruments, & Computers, 30*, 146–156.

Fitts, P. M. (1951). Engineering psychology and equipment design. In S. S. Stevens (Ed.), *Handbook of experimental psychology* (pp. 1287–1340). New York: Wiley.

Fitts, P. M. (1954). The information capacity of the human motor system in controlling the amplitude of movement. *Journal of Experimental Psychology, 47*, 381–391.

Fitts, P. M., & Deininger, R. L. (1954). S-R compatibility: Correspondence among paired elements within stimulus and response codes. *Journal of Experimental Psychology, 48*, 483–492.

Fitts, P. M., & Posner, M. I. (1967). *Human performance.* Belmont, CA: Brooks/Cole.

Fitts, P. M., & Seeger, C. M. (1953). S-R compatibility: Spatial characteristics of stimulus and response codes. *Journal of Experimental Psychology, 46*, 199–210.

Flach, J., Hancock, P., Caird, J., & Vicente, K. (1995). *Global perspectives on the ecology of human-machine systems.* Hillsdale, NJ: Erlbaum.

Gillan, D. J., Holden, K., Adam, S., Rudisill, M., & Magee, L. (1992). How should Fitts' law be applied to human-computer interaction? *Interacting with Computers, 4*, 291–313.

Godden, D. R., & Baddeley, A. D. (1975). Context-dependent memory in two natural environments: On land and underwater. *British Journal of Psychology, 66*, 325–331.

Heathcote, A., Brown, S., & Mewhort, D. J. K. (2000). The power law repealed: The case for an exponential law of practice. *Psychonomic Bulletin & Review, 7*, 185–207.

Hick, W. E. (1952). On the rate of gain of information. *Quarterly Journal of Experimental Psychology, 4*, 11–26.

Hintzman, D. L. (1974). Theoretical implications of the spacing effect. In R. L. Solso (Ed.), *Theories of cognitive psychology: The Loyola symposium* (pp. 77–99). Hillsdale, NJ: Erlbaum.

Hoffman, R. R., & Deffenbacher, K. A. (1992). A brief history of applied cognitive psychology. *Applied Cognitive Psychology, 6*, 1–48.

Holland, J. H., Holyoak, K. J., Nisbett, R. E., & Thagard, P. R. (1986). *Induction.* Cambridge: MIT Press.

Hommel, B., & Prinz, W. (Eds.). (1997). *Theoretical issues in stimulus-response compatibility.* Amsterdam: North-Holland.

Howes, A., & Young, R. M. (1997) The role of cognitive architecture in modeling the user: Soar's learning mechanism. *Human-Computer Interaction, 12*, 311–343.

Hyde, T. S., & Jenkins, J. J. (1973). Recall of words as a function of semantic, graphic, and syntactic orienting tasks. *Journal of Verbal Learning and Verbal Behavior, 12*, 471–480.

Hyman, R. (1953). Stimulus information as a determinant of reaction time. *Journal of Experimental Psychology, 45*, 188–196.

Jacko, J. A. (1997). An empirical assessment of task complexity for computerized menu systems. *International Journal of Cognitive Ergonomics, 1*, 137–148.

Jacko, J. A., & Ward, K. G. (1996). Toward establishing a link between psychomotor task complexity and human information processing. *19th International Conference on Computers and Industrial Engineering, 31*, 533–536.

Jacoby, L. L., & Dallas, M. (1981). On the relationship between autobiographical memory and perceptual learning. *Journal of Experimental Psychology: General, 110*, 306–340.

Kahneman, D. (1973). *Attention and effort.* Englewood Cliffs, NJ: Prentice Hall.

Kahneman, D., & Tversky, A. (Eds.). (2000). *Choices, values, and frames.* New York: Cambridge University Press.

Keppel, G., & Underwood, B. J. (1962). Proactive inhibition in short-term retention of single items. *Journal of Verbal Learning and Verbal Behavior, 1*, 153–161.

Kieras, D. E., & Meyer, D. E. (1997). An overview of the EPIC architecture for cognition and performance with application to human-computer interaction. *Human-Computer Interaction, 12*, 391–438.

Klein, R. M., & Shore, D. I. (2000). Relation among modes of visual orienting. In S. Monsell and J. Driver (Eds.), *Control of cognitive processes: Attention and performance XVIII* (pp. 195–208). Cambridge, MA: MIT Press.

Kornblum, S., Hasbroucq, T., & Osman, A. (1990). Dimensional overlap: Cognitive basis for stimulus-response compatibility—A model and taxonomy. *Psychological Review, 97*, 253–270.

Lachman, R., Lachman, J. L., & Butterfield, E. C. (1979). *Cognitive psychology and information processing: An introduction.* Hillsdale, NJ: Erlbaum.

Lockhart, R. S., & Murdock, B. B., Jr. (1970). Memory and the theory of signal detection. *Psychological Bulletin, 74*, 100–109.

Loftus, E. F., & Palmer, J. C. (1974). Reconstruction of automobile destruction: An example of the interaction between language and memory. *Journal of Experimental Psychology: Human Learning and Memory, 4*, 19–41.

Lu, C.-H., & Proctor, R. W. (1995). The influence of irrelevant location information on performance: A review of the Simon effect and spatial Stroop effects. *Psychonomic Bulletin & Review, 2*, 174–207.

Mackworth, N. H. (1948). The breakdown of vigilance during prolonged visual search. *Quarterly Journal of Experimental Psychology, 1*, 6–21.

Macmillan, N. A., & Creelman, C. D. (1991). *Detection theory: A user's guide*. New York: Cambridge University Press.

McClelland, J. L. (1979). On the time relations of mental processes: A framework for analyzing processes in cascade. *Psychological Review, 88*, 375–407.

Meyer, D. E., Abrams, R. A., Kornblum, S., Wright, C. E., & Smith, J. E. K. (1988). Optimality in human motor performance: Ideal control of rapid aimed movements. *Psychological Review, 86*, 340–370.

Meyer, D. E., & Kieras, D. E. (1997). A computational theory of executive cognitive processes and multiple-task performance: Part 2. Accounts of psychological refractory-period phenomena. *Psychological Review, 104*, 749–791.

Miller, G. A. (1956). The magical number seven plus or minus two: Some limits on our capacity for processing information. *Psychological Review, 63*, 81–97.

Miller, J. (1988). Discrete and continuous models of human information processing: Theoretical distinctions and empirical results. *Acta Psychologica, 67*, 191–257.

Monsell, S., & Driver, J. (Eds.). (2000). *Control of cognitive processes: Attention and performance XVIII*. Cambridge, MA: MIT Press.

Morris, C. D., Bransford, J. D., & Franks, J. J. (1977). Levels of processing versus transfer appropriate processing. *Journal of Verbal Learning and Verbal Behavior, 16*, 519–533.

Murdock, B. B., Jr. (1961). The retention of individual items. *Journal of Experimental Psychology, 62*, 618–625.

Nairne, J. S. (in press). Sensory and working memory. In A. F. Healy & R. W. Proctor (Eds.), *Experimental psychology*. Volume 4 of the *Handbook of psychology*, Editor-in-Chief: I. B. Weiner. New York: Wiley.

Navon, D. (1984). Resources—A theoretical soup stone? *Psychological Review, 91*, 216–234.

Newell, A., & Rosenbloom, P. S. (1981). Mechanisms of skill acquisition and the law of practice. In J. R. Anderson (Ed.), *Cognitive skills and their acquisition* (pp. 1–55). Hillsdale, NJ: Erlbaum.

Newell, A., & Simon, H. A. (1972). *Human problem solving*. Englewood Cliffs, NJ: Prentice-Hall.

Pachella, R. G. (1974). The interpretation of reaction time in information-processing research. In B. H. Kantowitz (Ed.), *Human information processing: Tutorials in performance and cognition* (pp. 41–82). Hillsdale, NJ: Erlbaum.

Parasuraman, R., & Daives, D. R. (1977). A taxonomic analysis of vigilance performance. In R. R. Mackie (Ed.), *Vigilance: Theory, operational performance, and physiological correlates* (pp. 559–574). New York: Plemum.

Pashler, H. (1998). *The psychology of attention*. Cambridge, MA: MIT Press.

Payne, S. J. (1995). Naïve judgments of stimulus-response compatibility. *Human Factors, 37*, 495–506.

Peterson, L. R., & Peterson, M. J. (1959). Short-term retention of individual verbal items. *Journal of Experimental Psychology, 58*, 193–198.

Posner, M. I. (1986). Overview. In K. R. Boff, L. Kaufman, & J. P. Thomas (Eds.), *Handbook of perception and human performance vol. II: Cognitive processes and performance* (pp. V3–V10). New York: Wiley.

Posner, M. I., & Cohen, Y. (1984). Components of visual orienting. In H. Bouma & D. G. Bouwhuis (Eds.) *Attention and performance X* (pp. 531–556). Hillsdale, NJ: Erlbaum.

Posner, M. I., & Mitchell, R. F. (1967). Chronometric analysis of classification. *Psychological Review, 74*, 392–409.

Proctor, R. W., & Van Zandt, T. (1994). *Human factors in simple and complex systems*. Boston, MA: Allyn & Bacon.

Reeve, T. G., & Proctor, R. W. (1990). The salient features coding principle for spatial- and symbolic-compatibility effects. In R. W. Proctor & T. G. Reeve (Eds.), *Stimulus-response compatibility: An integrated perspective* (pp. 163–180). Amsterdam: North-Holland.

Roediger, H. L., III, & McDermott, K. B. (1995). Creating false memories: Remembering words not presented in lists. *Journal of Experimental Psychology: Learning, Memory, and Cognition, 21*, 803–814.

Rugg, M. D., & Coles, M. G. H. (Eds.). (1995). *Electrophysiology of mind: Event-related brain potentials and cognition*. Oxford, UK: Oxford University Press.

Sanders, A. F. (1998). *Elements of human performance*. Mahwah, NJ: Erlbaum.

Sartori, G., & Umiltà, C. (2000). How to avoid the fallacies of cognitive subtraction in brain imaging. *Brain & Language, 74*, 191–212.

Schweickert, R. (1978). A critical path generalization of the additive factor method: Analysis of a Stroop task. *Journal of Mathematical Psychology, 18*, 105–139.

Schweickert, R., & Boruff, B. (1986). Short-term memory capacity: Magic number or magic spell? *Journal of Experimental Psychology: Learning, Memory, and Cognition, 12*, 419–425.

Shaffer, L. H. (1965). Choice reaction with variable S-R mapping. *Journal of Experimental Psychology, 70*, 284–288.

Shepard, R. N., & Metzler, J. (1971). Mental rotation of three-dimensional objects. *Science, 171*, 701–703.

Slamecka, N. J. (1968). An examination of trace storage in free recall. *Journal of Experimental Psychology, 76*, 504–513.

Slamecka, N. J., & Graf, P. (1978). The generation effect: Delineation of a phenomenon. *Journal of Experimental Psychology: Human Learning and Memory, 4*, 592–604.

Sternberg, S. (1969). The discovery of processing stages: Extensions of Donders' method. In W. G. Koster (Ed.), *Attention and Performance II. Acta Psychologica, 30*, 276–315.

Sternberg, S., Monsell, S., Knoll, R. L., & Wright, C. E. (1978). The latency and duration of rapid movement sequences. In G. E. Stelmach (Ed.), *Information processing in motor control and learning*. New York: Academic Press.

Teichner, W. H., & Krebs, M. J. (1974). Laws of visual choice reaction time. *Psychological Review, 81*, 75–98.

Treisman, A. M. (1964). Selective attention in man. *British Medical Bulletin, 20*, 12–16.

Treisman, A. M., & Gelade, G. (1980). A feature-integration theory of attention. *Cognitive Psychology, 12*, 97–136.

Trimmel, M., & Huber, R. (1998). After-effects of human-computer interaction indicated by P300 of the event-related brain potential. *Ergonomics, 41*, 649–655.

Tulving, E., & Thomson, D. M. (1973). Encoding specificity and retrieval processes in episodic memory. *Psychological Review, 80*, 359–380.

Van Zandt, T., Colonius, H., & Proctor, R. W. (2000). A comparison of two response time models applied to perceptual matching. *Psychonomic Bulletin & Review, 7*, 208–256.

Verhaeghen, P., & Marcoen, A. (1996). On the mechanisms of plasticity in young and older adults after instruction in the method of loci: Evidence for an amplification model. *Psychology & Aging, 11*, 164–178.

Vu, K.-P. L., & Proctor, R. W. (2001). Stimulus-response compatibility in interface design. In M. J. Smith, G. Salvendy, D. Harris, & R. J. Koubek (Eds.), *Usability evaluation and interface design: Cognitive engineering, intelligent agents, and virtual reality* (Vol. 1, pp. 1368-1372). Mahwah, NJ: Erlbaum.

Walker, N., Smelcer, J. B., & Nilsen, E. (1991). Optimizing speed and accuracy of menu selection: A comparison of walking and pull-down menus. *International Journal of Man-Machine Studies, 35*, 871-890.

Wang, H., & Proctor, R. W. (1996). Stimulus-response compatibility as a function of stimulus code and response modality. *Journal of Experimental Psychology: Human Perception and Performance, 22*, 1201-1217.

Wickelgren, W. A. (1977). Speed-accuracy tradeoff and information processing dynamics. *Acta Psychologica, 41*, 67-85.

Wickens, C. D. (1984). Processing resources in attention. In R. Parasuraman & D. R. Daives (Eds.), *Varieties of attention* (pp. 63-102). San Diego, CA: Academic Press.

Wickens, D. D. (1970). Encoding categories of words: An empirical approach to meaning. *Psychological Review, 77*, 1-15.

Wolfe, J. M. (1994). Guided Search 2.0: A revised model of visual search. *Psychonomic Bulletin & Review, 1*, 202-238.

Wood, L. E., & Pratt, J. D. (1987). Pegword mnemonic as an aid to memory in the elderly: A comparison of four age groups. *Educational Gerontology, 13*, 325-339.

Yantis, S. (2000). Goal-directed and stimulus-driven determinants of attentional control. In S. Monsell & J. Driver (Eds.), *Control of cognitive processes: Attention and performance XVIII* (pp. 195-208). Cambridge, MA: MIT Press.

·3·

MENTAL MODELS

Gerrit C. van der Veer and Maria del Carmen Puerta Melguizo
Vrije Universiteit

WHY BOTHER ABOUT MENTAL MODELS IN HUMAN–COMPUTER INTERACTION?

The group of people using information technology and complex interactive systems is getting broader and more diverse, both in work situations and for leisure activities. Information technology is embedded in telephones, TV sets, and home computers. Apart from this, public services are growing: electronic counters, ATM combined with electronic shop, etc. Users of these systems are not "professional" users and may frequently lack any education focused on the use of information technology. Additionally, the frequency of use will sometimes be low. As a consequence, users normally show low motivation for practice, for reading directions, and for formal training (that anyhow would not be feasible in most cases). Users, however, need to understand the functionality of the device, the relation of this to their task, and the dialogue for applying the device to be able to apply the system. As a conclusion, design methods for complex but sometimes infrequent human–machine interaction need to focus (among other things) on enabling the development of an "instant" mental model that allows useful interaction.

Carroll and Olson (1987) define mental model as the mental representation that reflects the user's understanding of the system. Although explicitly training a correct mental model has advantages, this approach is not generally the one taken for the design of *walk-up-and-use* systems and everyday household devices (Wickens & Hollands, 2000). As Norman (1988) stated, the consequences are that people often have erroneous mental models for this kind of system and have difficulties interacting correctly with them. The interest in mental models from human–computer interaction (HCI) is based on the idea that, by exploring what users can understand and how they reason about the systems, it is possible to design systems that support the acquisition of the appropriate mental model and to avoid errors while performing with them. In the following paragraphs, we will show some examples of "good" and "bad" designed systems and their consequences in the user's mental model and its performance with the system.

Examples of Problems

When things go wrong in design, one of the most striking problems is that the user does not understand the technology. The cheap answer in that case would be to blame the user for not being smart enough or not having had the right education. But, in fact, the designers could be blamed for not envisioning the problems that they caused to the future users. Users are human beings characterized by cognitive phenomena like planning, executing procedures, and evaluating states of the environment they live in. The users' actions are based on the users' needs in a certain moment, on the users' assessment of the situation (including the technology they perceive and the functionality they interpret to be available) and on the users' knowledge of procedures and expectations of the outcomes. In most situations, people will apply their previous knowledge, which means

they reuse their mental models (adjusted to the actual situation, of course). But, in case of unexpected situations with new elements, users will instantly develop a model of the current situation mainly based on their previous knowledge in similar situations. This means that the new situation will trigger the search for, and selection of, available knowledge.

Waern (1989) provides an early example of the problems that may occur when users of new systems are trying to reuse knowledge of existing systems. In an empirical study, she found that the transfer from a traditional electronic typewriter to a word processor suffered from the obvious analogies between the old and new task situation. Users expected the new system to work in the same way as the well-known old one (this was exactly what was advertised with the new system). Waern found various types of mismatch between the users' mental model and the actual working of the new system caused by confusing old and new system knowledge:

- Task mismatch: For example, replacing a word with another one of the same length. The old system required two subtasks—delete the old and type the new; the new system requires a single task—type the new word over the old one.
- Dialogue inconsistencies: For example, a space (the object manipulated by hitting the space bar) in the old system is a location on paper; in the new system, it is a character in the text string.
- Keystroke level confusion for moving the indicator of place of action: In the old system, the space bar and the new-line button are used; in the new system, arrow keys should be used.

The observations by Waern show that users develop a mental model of the new system that is analogous to the model of the old one at the levels just mentioned. Consequently, the users showed less than optimal performance (they performed more subtasks than needed to reach a specific goal), a consistent misunderstanding of the dialogue semantics, and made frequent typing errors. In general, the users imagined text objects like new line, space, as operators on the text object (identical to the function of the arrow keys), and this misunderstanding was a very stable characteristic of their mental model that prevented them from optimal performance for a considerable time. Prior knowledge of a device that is perceived to behave in a way analogue to a new device seems to be a strong determinant of the mental model applied to the new situation. Although on many occasions this can be considered an advantage, this is not always the case; one may wonder what would have happened if the new device would not have provided a space bar and a new line button, preventing the incorrect analogy.

If we systematically look for problems at the various levels of human-machine interaction, many other characteristic examples may be found of problems where users' knowledge and mental model do not match the knowledge needed to handle the situation:

Mismatch of Designers' Intention and the Users' Mental Model at the Level of Task Delegation. The Dutch railway company planned to remove the human-operated ticket desks at small railway stations. To gradually pursue this business goal, a first step was to develop ticket-selling machines for day trips, to be placed on the platforms. The early machines provided users with the possibility to buy tickets to all railway stations in the country. However, the journey had to start at the location where the ticket was issued. The machine did not mention this, and neither did the information campaign designed to change the travelers' behavior to buy tickets at the machine instead of at the desk. As a result, many travelers with a month pass for a certain trajectory were fighting the machines when they wanted to extend their journey to a destination not covered by their pass. Consequently, the intended buyer behavior did not develop according to business plans. The requirements defined for the first generation machines did not take into account actual traveler behavior patterns, even at the high level of planning and economic buying strategies. This is an example of a mismatch between the mental model users had of the machine and the conceptual model built into it.

Problems in Redesign of Task Semantics and Functionality. Sommerville Bentley, Rodden, & Sawyer (1994) discuss an air traffic control system redesign project. Part of their discussion concerns the question of what objects are relevant for the task space, and, hence, should be part of the system's task model. Traditionally, air traffic controllers in many countries use paper flight strips symbolizing an individual aircraft, with a number of attributes of that flight printed on the strip (see Fig. 3.1). While handling the flight, the strips get notes scribbled on them, and they are physically positioned and manipulated at the work desk area of the group managing the space the aircraft is in. All information on the strip, as well as its history, obviously can be handled by creating an electronic record and manipulating that record. At the abstract level (task level), the functionality is obvious, and its automation seems fine. Human users, though, in a situation of mental overload, use elements in the physical work situation to help them stay aware of both the level of workload in general and the individual subtasks that are currently running. The paper flight strips have precisely that function: they are visibly available as physical objects, they contain all of the relevant flight's attributes as well as the changes, and they can be grouped and manipulated, making the user literally feel the work in progress. The mental model of the task space in this case required that certain information be permanently perceptible and manipulatible by the users (e.g., assessing the workload according to the number of flight strips). The first version of the new system did not provide this functionality.

Issues Regarding Syntax Level of a Newly Designed Dialogue. The dialogue design should relate to human goal-driven behavior. For example, if a person's goal in using an ATM is to get money, users tend to stop the dialogue with the machine as soon as they receive that money. Identification of the user with the help of a bank card and PIN is only a lower level goal triggered by the need of the system to replace the (physical) identification and verification by bank employees during the transaction. Users will accept this subgoal only as far as, and as long as, it helps them reach their primary goal: getting money. The first generation of ATMs often returned the bank cards only after the whole transaction was finished, which resulted in a lot of users forgetting to take out their card. Psychological analysis of the users' goal structure in task situations will reveal what dialogue structure best matches expected behavior. In the ATM case, the aim should be to specify a design model that invites users to develop a mental model that includes card handling (as a subgoal and subtask).

Design Questions at the Representation Level. Specifying the system's image (the perceptible user interface) should take into account a whole series of aspects:

- Generic issues like human perception characteristics in relation to the actual work situation, the cultural meaning of symbols, etc.
- The task- and situation-dependent need for specific information, such as the actual system state, options for next user actions, ranges of values to be put in, etc.

An analysis of the system under development with the help of early usability tests will help the designer to make a good choice. First of all, the types of users should be taken into account in these studies. For example, if a system is designed to provide local information to be installed in a public place, some experimentation might show what questions are the first ones users would want to ask, as well as how best to state each individual question. In a menu-type dialogue these questions should be

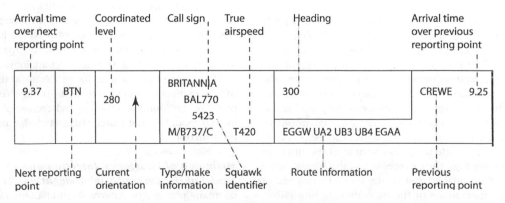

FIGURE 3.1. Example of a flight progress strip.

immediately recognizable at the user interface. Depending on the local situation, the options may be in one or more languages, or represented by pictograms that appropriately reflect the cultural symbol values of the types of users expected. If 95% of the people requesting information do not speak the local language, it does not make sense to use that language to indicate the option to choose another language. On-the-spot analysis of user behavior and user characteristics, with the help of an early prototype, may prevent the development of representations that are unusable, or otherwise do not fit the potential user population. Again, the design should refer to the users' natural mental models of/or in the current situation.

Examples of Valid Relations

Good design, from the point of the user's understanding, can be characterized as design in which care is taken to match the system to the user's potential knowledge. A system should be designed so that a user will be able to understand and hence to learn how to manage the system in an acceptable way (regarding investment of learning effort and time). Yet also, a system should be designed to counter the user's reasonable expectations. This means that the system should allow the user to apply human mechanisms of analogous thinking and of reuse of available knowledge. In actual situations, this means that the system either provides for quick and easy learning or shows clearly how already available knowledge (including skills) can be readily applied.

The next examples illustrate: how a new technology allows the transfer of available skills and knowledge (the mouse), how to help novice users to instantly develop a mental model for an infrequent interaction (the wizard), and how new knowledge can be presented exactly at the occasion that a user is motivated to learn and expand his or her repertoire for a certain functionality (hinting).

The Mouse. Douglas Englebart invented the mouse in 1963/1964 at Stanford Research Institute (Engelbart & English, 1968). The original implementation was a simple wooden box with wheels placed at right angles to track cursor movement when rolled over a flat surface (see Figs. 3.2 and 3.3). Later, it featured tiny wheels (mechanical rollers placed at right

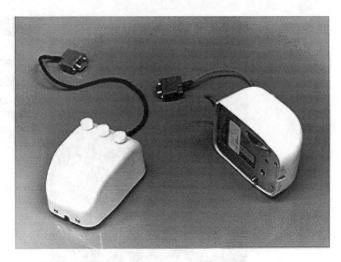

FIGURE 3.3. Early three-button mouse, 1968 (courtesy of the Bootstrap Institute).

angles; Ralston & Reilly, 1993), optical signals, or an upside down trackball. The mouse got its name because of its flexible two-dimensional movements and its tail (originally mice were not wireless).

Doug Englebart's invention started a wide variety of commercially successful designs. For example, it featured as a basis both the graphic user interface for the Apple Macintosh and the Microsoft Windows program.

The main reason for the success of the mouse, in whatever technical version, is the fact that the two-dimensional movements that the user performs with the mouse are directly translated into two-dimensional movements of the cursor on a screen. Even if the mouse is in fact spatially dislocated from the screen, the movement, in combination with the apparently immediate and isomorphic effect on the screen, makes users perceive that they are moving their point of attention (indicated by their hand) on the screen itself. No mental translation of the movement into the effect seems to be necessary. On the other hand, the readers can find out for themselves what kind of mental effort is needed to use the mouse when the tail is pointed to the wrong side: Even though one understands, the effort of running the mental model is considerable.

User Guidance. The main usability principle of wizards is visibility of goal, steps, and decisions (van Welie, 2001). Whenever a user needs to do a rather complex task (that consists of several subtasks) that is only infrequently performed, users may lack a mental model that is complete enough to guide them through execution. Wizards are solutions that help the users by providing them with an outline of the steps to be taken and guiding them through these. Successful wizards will state the goal of the operation at the onset, and show exactly where and when decisions have to be made. Examples like the Microsoft PowerPoint Pack and Go Wizard (see Fig. 3.4) or the Installshield installation program show the success of helping out if a mental model cannot be expected to be available in enough detail to leave it to the user to complete the job.

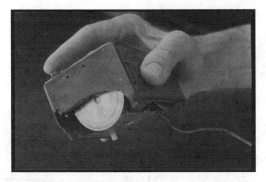

FIGURE 3.2. Engelbart's original mouse (courtesy of the Bootstrap Institute).

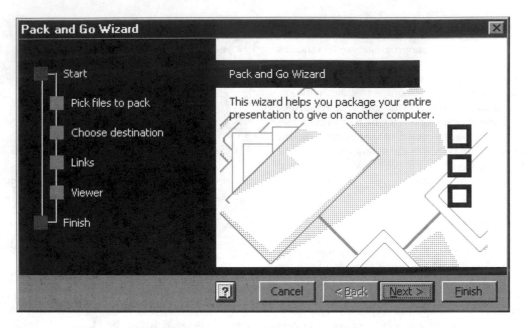

FIGURE 3.4. MS Pack and Go Wizard.

Recently, Van Welie analyzed this solution and developed it into a user interface design pattern on user guidance.

Hinting as a Base for Learning. Users frequently have the expectation (or even the knowledge) that certain functions are available in different ways. For example, many functions in Windows environments cannot only be reached by traversing the hierarchy of a menu, but additionally by using short cuts or choosing the right icon from a toolbar. At low levels of expertise with the system, users often do not remember the exact actions. Menu choices can be based on recognition, which is more readily available and more reliable early in learning. Consequently, for incidental use and early in the process of developing expertise, users still have to perform multiple menu actions to reach their goal. The user interface may help users to learn by providing hints for alternative, and often faster, actions. For example, Office 2000 menus, as well as the tool tips from the same package (see Fig. 3.5), provide hints to the user to reach the functionality they are currently considering, in different ways (e.g., the menu shows both the toolbar icons and the shortcuts, whenever available). Van Welie (2001) developed a generic design pattern for this type of facility, based on the visibility principle of incremental revealing. These types of hints may help the user to complete their mental model for future occasions.

Valid Mental Models Seem to Make a Difference

In this first section, we tried to explain the importance of designing systems that allow users to develop the correct mental model to work with complex interactive systems. For doing this, we showed a few examples of different designs and the consequences for the user's mental models and their performance with the system. In the next sections, we will try to understand what are mental models and how to explore whether the system to be designed is able to communicate knowledge to help develop the appropriate mental model to the user.

MENTAL MODELS IN PSYCHOLOGICAL THEORY

The question of how human beings represent information mentally and how they use it to interact with the world in adaptive ways is widely investigated by researchers in philosophy, cognitive psychology, and cognitive science (Sasse, 1997). As a

FIGURE 3.5. Examples of hinting from MS Office 2000.

result, different definitions of the concept have been made. In this review, we will refer to some of the more important ones.

Overview of Relevant Concepts and Theories

Craik (1943)—People Need Mental Models of Their World. The idea that people represent the world they interact with through mental models can be traced back to Kenneth Craik's suggestion in 1943. According to this author, mental models are representations in the mind of real or imaginary situations and can be constructed from perception, imagination, or from the comprehension of the discourse. Mental models underlie visual images, but they can also be abstract, representing situations that cannot be visualized (Johnson-Laird & Byrne, 2000).

The most important characteristic of these representations is that they are small-scale models of the situation and the possible actions. With this representation in mind, people are "able to try out various alternatives, conclude which is the best of them, react to future situations before they arise, utilize the knowledge of past events in dealing with the present and future, and in every way to react in a much fuller, safer, and more competent manner to emergencies which face it" (Craik, 1943).

Johnson-Laird (1983)—Human Reasoning Is About Meaning. Studying the interaction between humans and the world, one of the most influential theories is Johnson-Laird's (1983) theory of mental models. The theory seeks to provide a general explanation of human thought, reasoning processes, and language comprehension. Instead of the traditional symbolic logic theories, he proposed the theory of mental models to explain reasoning and emphasized the fact that people consider the semantic content of the problem and not only the syntactic structure. Although reasoning people do not always seem to apply the logical inference rules that represent only the syntax of a problem, reasoning people construct the mental model that represents the relevant (semantic) information of the problem. Another important aspect that he emphasizes is the fact that mental models are not exact representations of the perceived situation or problem, but simpler. This mental representation is built by associating the knowledge they already have with the incoming information (Johnson-Laird, Byrne, & Schaeken, 1992).

Early Definitions from Human–Computer Interaction— Mental Models of Computer Systems. The same year that Johnson-Laird published his book, another one with the same title appeared. The collection edited by Gentner and Stevens (1983) analyzes mental models in an interdisciplinary theoretical approach (cognitive psychology and artificial intelligence) and shows that mental models, and the mechanisms by which they are constructed, may differ according to the task or problem domain (Sasse, 1997). Among others, this book contains two of the earliest papers of mental models of computer systems by Norman (1983) and Young (1983). Both papers were an important beginning as a practical aid in the design of systems.

Norman (1983). Norman defines the mental model as the mental representation constructed through interaction with the target system and constantly modified throughout this interaction. From observations on a variety of tasks, Norman concluded that mental models are incomplete, have vague boundaries, are unstable over time, are unscientific, contain aspects of superstitions, are difficult, have restrictions to run, and are parsimonious.

- The fact that mental models are incomplete means that they are constrained by such things as the user's background, expertise, and the structure of the human information processing system.

- Because they have vague boundaries, operations and systems with certain relations or similarities can be mixed up. For example, the mental model of an operational system and the mental model of an application program used in the operational system can be mixed up.

- Mental models are unstable over time. This implies two things. First, unstable implies that people can forget the details of the system they are using, especially when those details (or the whole system) have not been used for some period. Second, as long as mental models are naturally evolving models, they change not only because people forget and mix up details, but also because new information is incorporated over time and through interaction with the system. So, learning and forgetting are two sides of the same coin.

- Mental models are unscientific and contain aspects of superstitions. For example, people maintain behavior patterns even when they know they are unneeded. According to Norman, people maintain this kind of behavior because normally they cost little in physical effort and save mental effort, and this is specially apt to be the case when a person has experience with a number of different systems, all very similar. We have to add that superstitious actions do not really seem to save mental effort and, in most of the cases, people know they are not necessary. People maintain superstitious behavior interacting with the computer because in doing so they feel more comfortable and more confident when using the system. For example, before shutting down the computer or logging out, some people first go back to their home directory. When asked, they explain that they know it is unnecessary, but they feel they are behaving nicely with the system and, as a consequence, the system will respond according to their expectancies to reach the goal.

- Mental models can only be run with restrictions because people experience gaps in their knowledge and insight in the process. In addition, mental models tend to be parsimonious, meaning that people prefer to know a limited set of elements about the reality they model. Mainly, this set is sufficient to understand the relevant parts of the system, even though people know perfectly well that in fact reality is more complex. The consequence is that people operate systems by using a restricted repertoire of actions, normally causing them to perform more steps in case of unfamiliar situations. Users tend to trade off extra physical actions for reduced mental complexity.

But even when mental models are not fully accurate, they are functional. Mental models denote the knowledge structure that users apply in planning actions, guiding the interaction with the system and execution of the planned task, evaluating the results according to the expectations the user has, and interpreting unexpected events when using the system.

In the study of the aspects that influence the interaction between human beings and systems, Norman maintains that it is necessary to consider five different concepts or, as he called them, *things*: the target system, the conceptual model, the mental model, the scientist's conceptualization of the mental model, and, finally, the system image.

- The target system refers to the system with which the user interacts. For information technology (IT) systems, this concerns hardware, software, and anything that the user is interacting with through the technology.
- The conceptual model is an accurate, consistent, and complete description of the system as far as relevant to the user. As Norman explains, this model is devised as a tool for the understanding or teaching of the system to the user. With this goal, the conceptual model must fulfill the criteria of learnability, functionality, and usability. The conceptual model can be also useful during the designing process of the system. As a consequence, conceptual models, even different conceptual models of the same system, can be developed by designers, teachers, scientists, etc.
- The mental model from which the user understands and predicts the behavior of the system is constructed through interaction with the target system and constantly modified throughout this interaction.
- Norman introduces the idea of system image that refers to the perceptible aspects of the system that are available to the user.
- The scientist's conceptualization of the mental model describes the content and structure of the user's mental model as understood by the scientist. It reflects the understanding and the knowledge the scientists have about the mental model of the user. Using the words of Johnson-Laird (1983): "Since cognitive scientists aim to understand the human mind, they, too, must construct a working model." Norman stresses the importance of using experimental psychology and observation techniques to figure out the mental models users actually have.

Later, Norman (1986) distinguished between the designer's and the researcher's conceptualizations. We think it is important to stress the idea that not in all cases of design it is possible to distinguish between the designer's and the researcher's conceptualizations. In some cases, in the design teams, there is not an expert scientist interested in the study of the mental model. The ideal situation of an expert in experimental and cognitive psychology studying the mental representation of users and collaborating directly in the design process often is not a real one. Instead, during the design process, designers pay very little attention to the way the future user understands and interacts with the system. We will show that it is really necessary to include experts in mental models and HCI in the design teams or, at least, to educate designers to understand the mental model of the user. Fortunately, technology companies, such as Apple Computer, Philips Design, Bell Labs, and Xerox's Palo Alto Research Center, seeking to design more usable products and services are beginning to tap the human behavior expertise of psychologists, anthropologists, and other social scientists in growing numbers (Bailey, 1996; Lear, 2000).

Young (1983). In the same volume and in parallel with Norman, instead of the mental model, Young (1983) uses the term user's conceptual model to refer to "... a more or less definite representation or metaphor that a user adopts to guide his actions and help him interpret the device's behavior" (p. 35).

Furthermore, Young (1983) introduces a distinction between different types of mental representations that users can have about the system (strong analogy, surrogate, mapping, coherence, vocabulary, problem space, psychological grammar, and commonality). Following the idea that it is possible to have different mental models about a system, representing different kinds of information, different classifications about the mental models were made (e.g., DiSessa, 1986; Nielsen, 1990).

Cañas and Antolí (1998): Mental Models Are Dynamic Representations.
Although here we are mainly focused with the common and compatible ideas authors have about what is a mental model, the reality is that there is not agreement about the definition of the concept, and even contradictory statements have been made. The result of this situation is that, nowadays, people are still trying to build a common, useful, and valid definition of the concept. According to Cañas and Antolí (1998), the fact that the concept "mental model" has been used by researchers who work in different tasks is the main reason for the disagreements. For example, Johnson-Laird (1983) worked in reasoning, whereas Norman (1983) used the concept in the field of HCI. Consequently, researchers focused on different aspects of mental models. Some give more importance to the information extracted by perceptual processes of the characteristics of the task (e.g., Johnson-Laird, 1983) and some pay more attention to the knowledge the person has stored in long-term memory (LTM) about the operations and the structure of the system (e.g., Norman, 1983). Differences between approaches are due to the way variables are represented (instantiation in specific symbols vs. direct representation), the relation of models with the memory structures [working memory (WM) vs. LTM], and the phenomena they intend to explain (syllogistic reasoning vs. causal reasoning). For example, a common debate in the literature is the question about if mental models are structures of knowledge in LTM or temporary creations in WM. Cañas and Antolí unify both ideas and, as we will see later, develop a model of mental model formation.

To understand Cañas and Antolí's ideas, it is important to refer to the concepts of WM and LTM. For a detailed explanation, see chapter 2. We will present just some of the most important characteristics:

- LTM is a relatively permanent storage and contains our autobiographical memories and the knowledge we have about facts.

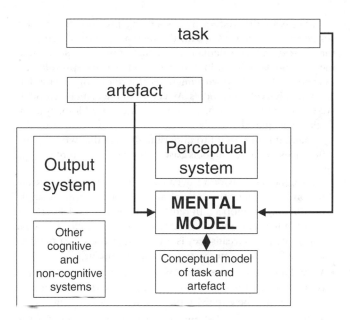

FIGURE 3.6. The role and place of mental models during the interaction with a physical system. From *Proceedings of the Ninth European Conference on Cognitive Ergonomics— Cognition and Cooperation* (p. 00), by J. J. Cañas and A. Antolí, (1998), European Association of Cognitive Ergonomics EACE, Unite de Recherche INRIA: Rocquencourt, Le Chesnay Cedex, France. Copyright 1998 by the authors. Adapted with permission.

- In LTM, it is possible to store very large amounts of information in a more or less permanent way.

- WM is a short-duration, limited-capacity memory system capable of simultaneously and actively storing and manipulating information needed to perform a task (Baddeley, 1986, 1995).

- Information can be maintained in WM for brief periods of time (around 30 s if rehearsal or other strategies are not used).

- WM holds relatively small amounts of information, approximately 7 ± 2 information items (Miller, 1956).[1]

- WM is the system where information is dynamically and actively manipulated to execute any complex cognitive task (e.g., learning, reasoning, and comprehension).

- The activated information in WM contains information retrieved from LTM as well as information from the environment and the task that the person perceives in a specific moment.

The basic hypothesis that Cañas and Antolí maintain is that a mental model is the dynamic representation created in WM by combining information stored in LTM and characteristics extracted from the environment (see Fig. 3.6). The knowledge from LTM is instantiated into a mental model in WM based on triggering events from the situation. This instantiated mental model is dynamic and functions as a runnable model in WM.

Classification

Some authors tried to make classifications of mental models (e.g., DiSessa, 1986; Nielsen, 1990). The classification presented later is partly derived from Sasse's (1997) analysis of those classifications.

Mental Models of Procedures vs. Semantics. DiSessa (1986) distinguishes between structural ("what is?") knowledge and functional ("how to?") knowledge about the system. According to this distinction, *structural models* provide the users with a detailed understanding of the system, whereas *functional models* represent the properties of the system needed to perform a specific task. The first kind of mental model contains information about the internal structure of the system, and it is independent of the specific task. However, the functional models contain information about how to use a selected set of functionality to perform a real-world task.

According to DiSessa, functional models seem to be ideal for nonexperts who want to use the system as a tool because they take less time and mental effort to acquire and maintain. But, although Sasse (1997) maintains that walk-up-and-use systems are the exception, we think that nowadays it is possible to perform more that one task even with this kind of system. This implies that users would have to acquire a multiplicity of overlapping functional models (one for every specific task that can be performed with the system). So, the advantage of less mental effort does not seem so evident any more. In conclusion, especially for complex systems that can perform different tasks, the neat division between structure and function is not possible. Finally, DiSessa introduces the idea of *distributed models*. They are not purely structural or functional, but accumulations of multiple partial explanation users hold of a system, tied in with their previous existing knowledge and experience.

A related distinction is the one made by Nielsen (1990) between general and specific models. Although general models describe users and systems in general terms, specific models are aimed at describing specific reasoning mechanisms used by a particular user group (e.g. novices vs. experts) or specific tasks that can be performed using the system. Related to the idea of specific models, later we will consider the user's virtual machine or UVM. Briefly, the UVM is a detailed description about the aspects of the system that are directly relevant for a specific task and/or a specific group of users.

Descriptive vs. Analytic Models. Normally, the mental representations that users and designers have are informal descriptions of the system. However, other descriptions of the system use more formal methods and show a more complex analysis of the system (e.g., the scientist's conceptualization and the conceptual model of Norman). Maintaining her distinction between researchers and designers, Sasse (1997) thinks that the analytic models are used more by researchers than by designers.

Internal vs. External Knowledge. An internalized model can reside inside the human mind or in the system. According

---

[1] It is possible to group together bits of information into meaningful items and increase in this way the WM capacity (chunking process).

...o Van der Veer, Kok, and Bajo (1999), internal models are for *execution*. The main idea is that there is an agent who uses the model to make decisions based on the behavior of the model and to make predictions on the behavior of the modeled reality. If the agent is human, we will call it mental model. But, if the agent is the system, we will call it program, database, etc. On the other hand, externalized models are mainly built for communication (Van der Veer et al., 1999) and are represented in some explicit form outside the human mind or the system. Examples of external models are the conceptual model and the scientist's conceptualization of Norman (1983).

Generic vs. Instantiated Models. The instantiated model is the idiosyncratic mental model the user develops in a specific situation. The generic model refers to the prototype mental model of the user envisaged by the designer and/or the researcher (the scientist's conceptualization of Norman).

Individual vs. Shared Mental Models. Until now, we spoke about the users mental model, defining it as the mental representation an individual has of the system through its interaction with the system. But, in group situations, individuals have shared mental models (Van Engers, 2001). The concept of shared mental models indicates that at least part of the mental model of several individuals is equivalent for all of them. Of course, this presupposes they are in equivalent situations. The main assumption is that shared mental models are based on common knowledge in LTM (Clark, 1996, refers to *common ground*).

Shared mental models and shared knowledge indicate sharing by individuals. This is different from the concept of group knowledge as used in computer-supported cooperative work (CSCW) approaches and in certain ethnographic investigations (Jordan & Henderson, 1995). Group knowledge is knowledge that a group of people share as a knowledge source (e.g., announcements on the note board of a department are group knowledge for all who know where the board is and what type of information may be posted on it). Other types of group knowledge include shared stories, which may point to a possible overlap and confusion between the concepts. Mental models, in our view, are mental (i.e., they are in the mind of a person). We speak of shared mental models when we may assume that the mental models of a group of people are, to a certain extend, at least equivalent in their effect on the individuals' behavior. Techniques for investigating shared mental models will be reviewed in "The Pathfinder Algorithm."

Summarizing Mental Models

As seen, a mental model is an internal representation that people form and use while interacting with the environment (problem, system, etc.). To a certain degree, this representation contains structural information about the properties of the system and functional knowledge about the task to perform. On the other hand, mental models are incomplete, unstable, and unscientific, but they can be used for planning, execution, evaluation, and interpretation of the system or problem the subject has to solve. Another important characteristic is that they are built while working with the system by associating previous knowledge stored in memory (LTM) with the incoming information from the context and tasks demands. As a matter of fact, it seems that mental models are built in WM. They can be run to produce expectations about the environment. Furthermore, mental models are naturally evolving models. As a consequence, mental models are changed and, most of the time, refined as additional information is acquired.

Finally, it is important to underline the fact that we may find differences in mental models based on several different causes:

- Mental models may differ because of the variations in the situation that trigger the mental model. For example, a single person may develop a different instant mental model depending on whether there is a need to explain to a colleague what a formula in a spreadsheet is, or whether there is a need to insert a formula in an existing real spreadsheet.
- Mental models change over time (e.g., because of new experiences) so that a single person may show different mental models when the expertise on a system is developing.
- Different people will show different mental models for the same situation, depending on all kinds of variables like details of individual knowledge, cognitive styles (visualizers vs. verbalizers; van der Veer, 1989).
- Different mental models of the systems may lead to differences in performance while interacting with the system (e.g., differences between experts and novices).

A PRAGMATIC APPROACH TOWARD APPLYING THE PSYCHOLOGICAL CONCEPT OF THE MENTAL MODEL

In the previous sections, we tried to expose briefly some of the most relevant definitions of mental models. As in this volume, we are focusing on HCI; we will focus on design and analysis of interactive systems. To this end, we need a pragmatic approach that allows us to apply the psychological insights in understanding the interaction between systems and their human users. A pragmatic approach may be derived from analyses like Norman (1983). The author considers the mental model any type of mental representation that develops during the interaction with the system and enables and facilitates the interaction with the system. Van der Veer (1990) found that several representations may be used by different subjects for identical interaction situations.

Our View on Mental Models

Following this pragmatic approach, our ideas about mental models can be described as follows. While working with the system, users activate a mental model constructed in WM from the previous knowledge (stored in LTM) they have with the situation and from the information they perceive from the environment (mainly the system and the task demands). When confronted with a new system, the only previous knowledge available is based on previous experiences with similar systems. Depending on the nature of the new system, this previous knowledge can be more or less appropriate in the new situation. This leads to the need of designing the new system in such a way that it is able to communicate the correct information about its use mainly via its interface or system image. The system image should be able to give the information needed to develop the appropriate user mental model. On the other hand, design should take into account the limitations and processes in WM to prevent information overload and confusion of the users, and to assist them with the knowledge available whenever adequate action in the situation requires this. The next question we need to solve is what kind of knowledge should be represented in the system's interface?

The "Appropriate" Mental Model

At this point, we think it is important to elaborate on the idea of the "appropriate" mental model. As stated earlier, a user's mental model is not an accurate description of the system and how to work with it, but a simpler version of the system. The user's mental model is neither complete nor correct in every detail. In agreement with Norman (1983), we take into account the limitations mental models have (see our explanation about Norman's ideas in previous sections. We consider a mental model to be appropriate as long as it is functionally applicable to plan and execute a task with the system, as well as functionally sufficient to evaluate results of our actions and to interpret any unexpected system action.

How to Design for the Appropriate Mental Model

Following the pragmatic approach, to understand the role that mental models play in improving the design process, it is necessary to analyze:

1. What users need to know to use the system for their tasks. Users need to understand (i.e., to have enough knowledge and receive enough information from the system to develop an adequate mental model whenever they need it). This means they need insight into the relevant functionality of the system for their role, and they need insight into the dialog needed to actually use the system. We will go into detail on the types of knowledge that users need in "What Type of Knowledge Do Users Need?"
2. What different users there are (whenever applicable) for the system to be designed. For example, in designing an ATM, there are different user roles to be considered:
   (a) The customer who wants to withdraw cash and needs to understand how to operate the machine with his bank card and PIN.
   (b) The bank employee who needs to resupply bank notes and printer paper, as well as perform maintenance.
   (c) Certain lawyers who need to analyze whether a transaction regarding an account of their client could be the result of fraudulent use of the ATM, for which they need

to understand the process and record content of keeping logs of machine transactions.

Of course, people with either role (b) or (c) may also, in some situations, take role (a). In "Modeling Users Task Knowledge for Complex Interactive Systems," we describe the analysis of roles as part of task analysis.

3. What are the requirements based on this for the engineers to build? Here, we need to explore what types of knowledge designers need to specify to make the new system usable. Designers need to develop their specifications from the point of view of the intended users. The first type of specification that should be developed is a focus on the knowledge that is needed by the future user. In "The User's Virtual Machine," we will elaborate on its elements. In complex interactive systems design, we consider the situation that various types of users may have different roles (defined by characteristic tasks), each of which, by definition, need different knowledge to apply the system for their own work. For each role, designers need to specify the knowledge a user in that role needs to handle the system for the role-specific tasks. Designers need to develop a system model based on the combination of UVMs for the different roles. Thus, specifying a single (i.e., complete) interactive system will amount to specifying various more or less overlapping UVMs or structures of knowledge needed by the various types of users. This last combined model is equivalent with what Norman (1983) labels the conceptual model. The conceptual model is still nothing more than the knowledge needed by the users as a whole to effectively use the system. Only based on this conceptual model should the actual design of hardware and software start.

What Type of Knowledge Do Users Need?

Whereas, for the human user, the system as such is a single more or less monolith thing; for the designer, it is crucial to find out what types of knowledge will be needed and, hence, have to be specified. In the first place, the system should speak a language that is clearly and immediately understandable for the users in relation to their intentions and tasks. But, in interactive systems, much more complex knowledge is needed. Users need to be able to instantiate a mental model of what is going on in the interaction based on knowledge available in human memory and information provided by the system at the relevant moment. To facilitate the design process, it is important to specify the knowledge needed, and the information to be provided, as an early phase in the design of the system.

In the following paragraphs, we explain the types of knowledge users should be aware of to interact with complex systems. Users need two types of knowledge:

1. What the system can do for them, i.e., what is the *functionality* of the system. The current section will go into detail about this.
2. How to communicate with the system? This type of knowledge is mostly labeled the *dialogue*.

Both types of knowledge should be considered separately by the designer, because both will lead to separate design specifications of the UVM. Both types of design specifications should be decided on in relation to each other (i.e., the dialogue should fit the functionality). For example, if the functionality is intended to provide synchronous communication between a user and another person, the dialogue style chosen should provide a natural way of human-to-human communication. Speech might be more natural than typed text, and typed text might be more natural than menu choices to indicate the type of speech acts. In Part IV of this volume, dialogues will be treated in detail. From the point of view of the user, there will not be a clear distinction between functionality and dialogue. The user will need an adequate mental model for performing a task with the system, which to the user mainly consists of applying a dialogue to acquire access to the functionality.

In the next paragraphs, we will show the different types of knowledge that are needed by users to build adequate mental models with respect to the functionality.

What Is Going on Behind the Screen? To a large extent, the user will never care about what's inside. For most current and almost all future IT users, operating systems, processors, memory size, speed of transfer, etc., are completely irrelevant as long as the performance of the system is acceptable from the point of view of human information processing capacities. If certain internal processes take more than a few seconds, or are unpredictable for the user, the system should make the user aware of exactly this, nothing else. Certain things going on behind the screen are needed to understand the interaction and perform the intended tasks successfully (e.g., "Is this object still on my clipboard?" "Is the original file untouched after I performed a 'safe as' operation?"). This type of knowledge is certainly needed for the intended users (text-composing MS Word users in the case of these examples).

What Is Going on Behind This Computer? Current interactive systems are seldom stand-alone. Even if the machine is connected to the outside world solely by exchanging diskettes, users need to know the reason for certain unexpected events, and users need to take measures for safe, reliable, and efficient interaction with external agents. In the case of more sophisticated connectivity, users (depending on their roles and tasks) need to understand aspects of local and worldwide connections and safety procedures (fire walls, signatures, authentication, etc.). In certain types of tasks, the actual location of certain processes should be known and understood ("Is my pin code transported through the network to the bank, or is it validated locally at the terminal?").

What Organizational Structure Is Behind My Computer? In the type of interactive systems that are increasingly being used, the user and the machine are not isolated or disconnected from other systems. A user needs to understand that other users, some with identical roles and others with different roles, are players in the same game, thus influencing the total process. Users need to know that others with a certain role may manage their workflow process or may monitor their actions.

Additionally, users need to understand that others, with various roles, will have different "rights" in relation to system objects ("Who may inspect or change or delete the e-mail I just sent?").

What Task Domain Is Available Through My Computer? Interactive systems are being intended for user tasks. But the fact that IT is used to support the task requires a clear insight for the user about what part of the task domain is in fact available for delegation to the machine, and what should/could best be performed in other (traditional?) ways. For example, if a group of fashion designers are applying an interactive system for their work, several tasks could be delegated with considerable added value. A beneficial change in the task domain could be the facility to search multimedia databases of video clips from cat walk shows, images of the people and the location the clothes are intended to serve, interviews with potential clients, works of art that are characteristic of the cultural ambiance of the intended buyers, etc. So far, possibly because the right media are not available yet, other parts of the task domain are not yet supported successfully. Examples of tasks in this domain are assessing the texture quality of fabric or impressionistic sketching of the process of moving clothes at a stage before the design of shape has developed enough detail for a mannequin to be actually "dressed."

What Are the Process and Time Aspects of Delegating Tasks to My Computer? One specific aspect of applying IT to support interactive systems is that this technology has a time scale that is completely different from that of human information processing. In many cases, this leads to trivial phenomena of systems providing reactions to user request amazingly fast or, alternatively, annoyingly slow and unpredictable. A more fundamental problem occurs if the computer is in fact a front end to a process where time characteristics are by definition, or intentionally, of a nonhuman scale (Hoc, 1989). In fact, the problem can go either way: (1) the process is slow in providing feedback (e.g., monitoring chemical plant processes), but the user is supposed to anticipate the outcome of previous actions to optimize the continuation of the process; and (2) the process is fast in performing irreversible tasks (e.g., delegating certain flight operations to the automatic pilot) and, consequently, the user is expected to asses the expected consequences in all details that effect safety and reliability. Both cases are requesting the same type of insight and knowledge from the user. Various recorded cases where pilots ended up fighting the automatic pilot are illustrations of designs where this type of knowledge was never considered. The same can be found in recorded disasters with nuclear power plants where potential consequences of task delegation were never modeled, and where, consequently, failure indications were completely inadequate and in fact did disturb human problem-solving processes.

What Type of Insight in Users' Knowledge Do Designers Need?

Taking into account the different types of knowledge users need to work with the system is the starting point, but designers need more. They need to develop an adequate dialogue style that fits

the functionality as well as the situation. Designing technology is thus translated into building prescriptive knowledge models, where the prescription is not intended for the users, but for the engineer who will have to implement the system. Of course, in prescribing this knowledge, the cognitive ergonomic insights in human information processing are of paramount importance. The system should be understandable, usable, and requiring the lowest level of effort for the user. Furthermore, for the designer, there is, apart from the functionality and the dialogue, still a 3rd component to take care of: the system's representation toward the user (see section in which we refer to Norman's concept of "systems image"). Sometimes users need extra information to create a mental model that fits the current situation. As far as the designer cannot expect that this information is immediately available for the user (not in WM) the interface could provide the user with it. Suppose the user is trying to find train information from the Dutch railway web site, and the user has to indicate the name of the town to travel to. Sometimes towns have several different names. For example, in the Netherlands, there is a city (in English called the Hague) that in Dutch can either be called Den Haag or s-Gravenhage. The designers of the previous version of the web site only implemented a single name for any destination. If the user has to indicate the name of the city he or she would be very grateful if he or she could choose instead of having to guess which name could be accepted by the machine. Representation is generally an important facility to provide users with knowledge that is needed for developing a valid mental model of the actual situation. For reasons illustrated here, representation is the 3rd issue that designers should consider in specifying the UVM.

The User's Virtual Machine. In specifying the technology to be developed, from the point of view of the user, or, rather, various types of intended users, Tauber (1988) introduced the concept of the UVM. The UVM is a complete specification of all aspects of the technology to be (re)designed *from the point of view of the prospective user*. Specifying the technology, in this case, is equal to defining all the relevant knowledge a user should need to use the new system. We will for now only mention the relevant subparts of the process of specifying the UVM. In "Modeling the User's Virtual Machine," we will go into detail of how to model the UVM.

The specification of the UVM can be structured into three subactivities that are strongly interrelated:

- Specifying the functionality of the system. This activity includes specifying the system details as far as relevant to the user (what the system provides to perform the task to a specific type of user).
- Specifying the dialogue between the users and the system. Modeling the language interface means to define the type(s) of language style in which the users express themselves to the system.
- Specifying the representation or the way the system is presented to the user. Modeling the system's actions and representation of relevant information for the user it is possible to influence the kind of control the user feels to have over the

system, as well as the amount of learning needed to operate the system.

The result of specifying the dialogue but mainly the representation corresponds to the "system image" of Norman (1983) as it represents the perceptible and the dynamic aspects of the system.

Different User's Virtual Machines. Specifying the details of technology from the point of view of designing for the user's mental model is equivalent to specifying all relevant knowledge needed for each type of users or UVMs (type being defined by role and characteristic tasks). Each distinguishable UVM consists of the functionality needed to perform the role-specific tasks, the dialogue needed to apply this functionality, and the representations needed to guide the user and provide adequate (role-specific) information about the system state. In this respect, we need to specify a different UVM for each role, but not for various users with the same task even if they vary in relevant cognitive psychological attributes, such as in the fields of cognitive style, perception, and motor action. In designing along those lines, adaptability still has to be taken care of. The adaptation (whether by the user or the system) will be clearly separated from specifying the knowledge needed to perform role-specific tasks. Special needs of a user (e.g., a reading handicap) or a situation (e.g. driving a car, which should not require a user to search on a screen) will ask for additional design measures, even if in many cases the functionality of the system should preferably be generic. For detailed analysis of adaptation issues, see chapters 15 and 22–26 of this volume.

Finally, specifying the previously described types of user knowledge with the aim to have the system developed by engineers requires a special type of knowledge modeling. We need a modeling language that is precise and unambiguous enough to be understood by engineers without deep insight in cognitive science. At the same time, this language should allow, or even better, enforce, the complete specification of what matters to the user and what is relevant from the point of view of human information processing. We will show in the next section how the specification of the UVM may benefit from a modeling approach and supporting tools. In "Designing for Users and Tasks," we will show how a concern for the user's mental model and knowledge fits in the overall landscape of design techniques that focus on designing interactive systems.

Modeling the User Virtual Machine

Many specification techniques have been developed for assisting (and provoking) designers to specify an adequate conceptual model (or set of UVMs) of a system. Most techniques are based at least partly on cognitive psychology [Goals-Operators-Methods-Selection rules (GOMS; Card, Moran & Newell, 1983), Cognitive Complexity Theory (CCT; Kieras & Polson, 1985), Tasks Action Grammar (TAG; Green, Schiele & Payne, 1988), and their offsprings, see chapter 5; de Haan, van der Veer, & van Vliet, 1991]. As an example, we consider User Action Notation (UAN; Hix & Hartson, 1993) for specifying and analyzing the details of technology. It seems an excellent candidate for formalizing all knowledge a user should need in using the system, hence, for specifying the UVM. One of the main special features of UAN is that it helps designers to consider separately the different aspects of the UVM: the functionality of the system is mainly specified in what UAN indicates as connection to computation. Different components of UAN specify the different aspects to take into account when considering the dialog. The user actions describe the part of dialogue performed by the user when the interface state and the interface actions describe the aspects of the dialogue performed by the system. The interface state refers to aspects of the dialogue that are not visible at a certain moment, but that are relevant for the user to know or understand. Aspects of the dialogue that are immediately visible to the user and are part of the representation aspect, in UAN are indicated as interface actions.

Originally, UAN (Hix & Hartson 1993) allows the specification of:

- All user actions at the level of keystrokes (mouse clicks, voice input, scribbles on drawing pads, etc). These are precisely the activities that a user needs to perform in any given situation, and, hence, the actions that should be guided by an adequate mental model.
- Interface feedback refers to meaningful elements to be presented to the user like cursor movements, highlighting, beeps, etc. This entry allows the designer to specify the location, sequence, and meaning of all system actions toward the user. Because nowadays the user is often confronted with system events that are initiated by actions of system agents or other users, the interface feedback needs to be exchanged for a Interface actions column. From the point of view of the user, the system acts either as a reaction to the users' behavior, or as the result of the actions of some other agent. The designer should keep in mind that it matters for the user to have a clear mental model for interpreting the system actions. The system could be just reacting (e.g., giving the user feedback about the status of a print command the user has just given). Alternatively, an unexpected event might occur that triggers users behavior (like an alarm from a security systems, the arrival of new e-mail, or an alert from the workflow system about a new task). Each of the situations from these examples will require the instantiation of a mental model for that specific occasion.
- Interface state information, intended to describe interface states and state changes that are relevant for the user to know (like "PIN being matched to info from credit card magnet strip"). In this slot should be specified all the information that the user needs to understand to develop a valid mental model of the interaction and its effects.
- Connection to computation as far as relevant for the user to know, such as "the amount of money X is now withdrawn from your account." Again, this slot allows the specification of all that is relevant for the user of what is going on behind the screen (see "What Type of Knowledge Do Users Need?").

Some additions to this type of specification of the UVM may be needed, especially when one needs to be precise and complete regarding the knowledge that is relevant for the user to have the best mental model at any given moment in interaction. In many cases of interaction, users know, or should know, that there are conditions to start an interaction, or, in other words, only in certain situations would the interaction make sense. A useful extension of specification tools or formalisms is to specify this user knowledge as an optional prestate descriptor or precondition as it is labeled in MAD (Scapin & Pierret-Golbreich, 1989). It triggers designers to describe relevant user knowledge about the system state preconditions for interactions to make sense. For example, when deleting a file, it needs to be selected first.

In some systems, there are repeated subtasks as part of the dialogue. Selecting an object before applying an operation to it (see the previous paragraph) is one example. Identification of the user before any transaction that results in a change of account is an example from banking systems. Formalisms like UAN can be adapted to trigger designers to specify separate modular interactions for reusable pieces of dialogue. Techniques like GOMS provide this as a standard feature. Such modeling helps to design more consistency throughout the interface, as well as making the designer aware of the knowledge structure (and the chunks of knowledge) users need to keep in mind as part of their mental models. Reuse of generic actions is desired because it can improve the learnability aspects of the design. With reuse, the user only needs to learn the interaction once and can apply it in other contexts.

To make timing understandable from the point of view of users and allowing them to develop a mental model that guides them during a process, any chosen formalism needs to indicate clearly how the temporal relation is between user actions and system actions:

- *Simultaneity:* In this case, system events are considered to occur at the same instant from the user point of view like pressing the mouse button and showing the matching animation movement of the icon on the screen. Designers should make this relation clear to the user in the system's representation. Even if the mouse button press is processed by the system resulting after some milliseconds in the resulting animation, for the user we want to suggest that the mouse operation is directly connected to the visible effect.
- *Consecutive events:* These should be understood by the user to occur in a temporal order. The sequence is relevant either because of causal relations between the events or because different processes happen to interfere and the exact nature of interference (the location in time of one process being interrupted by another) is relevant for the user understanding.

In some cases, it is very important to model the mental actions of the user, especially if users are supposed to make decisions based on comparison, retrieving knowledge, or reasoning with knowledge. GOMS (see chapter 5) was one of the first techniques to model this. A limited set of relevant mental operators seems to be needed. See Table 3.1 for a proposal of some useful mental operators. Designers should feel free to add more (and ask a psychologist) when needed. If mental actions are specified there may be interface actions at the same time. These

TABLE 3.1. Suggested List of Mental Actions to Be Specified in Relation to the UVM

Symbols	Meaning
wm	Working memory
ltm	Long-term memory
RECALL(x,y)	Retrieve x from memory type y
RETAIN(x,y)	Store x in memory type y
FORGET(x)	X need not be kept further in working memory
FIND(x,y)	Find x in the environment y (e.g., the interface)
CHOOSE(x,y)	Choose x from the set y

may be specified with the intention to assist the mental actions (e.g., providing the information needed in a decision, showing an animation to indicate the system is still waiting for a choice from the user). When used, the mental actions can give insight into the cognitive aspects that are needed, and the designer can be aware of whether the user needs to make a decision using elements visible on screen or from working memory.

There are two rules for the relation of mental action to other actions:

- Mental actions are not real (physical) user actions toward the outside world (including the interface) and therefore no interface action can ever be caused by them. This means that an interface action on the same line or below the mental action is never the effect of the mental action but to some user action or event above the mental action. It may well be that the mental action is facilitated by knowledge displayed through an interface action. Designers may in fact be triggered to consider this type of facilitation precisely by specifying the mental actions.
- The mental action itself can be related to some interface action or user action, depending on the context of the mental action. The sequential structure of interface actions and mental actions should be indicated. An example is a mental action being the first action of an interaction, such as reading the identity of the person who is trying to establish a telephone connection with me. In such cases, it is related to the interface prestate and should be explicitly specified as such.

One of the important assessment aspects of interactive systems is ease of learning, which is related to design issues like the number of different rules users need to apply. If a system needs less different rules for the same functionality, the system is easier to learn, and users will be able to apply rules even in situations that are new, by instantly developing an appropriate mental model based on analogy and on the expectation of consistency. This phenomenon has been used in various formal specification languages. For example, in Task-Action Grammar, a description of a device interface requires a dictionary of simple tasks. Simple tasks are defined as the application of a single rule. Each entry in the dictionary of simple tasks associates a concept name with a set of feature values (Green, Schiele, & Payne, 1988). For example, a simple task could be to move the cursor one character forward, and the rule for this (in case of the E_{max} editor) is move {direction = forward, unit = character}. For other directions or other units (line, paragraph), another feature value can be filled in the same rule. Applied to MS Office 2000 systems

the same rule could be applied for saving a word document, a spreadsheet, a database, a picture, etc.

Parallelism is an issue in complex interactive systems in which many users work together, synchronous or asynchronous. In Sage and Johnson (1998), this kind of cooperation is modeled by joining two UAN tables into one table. However, from the users' point of view, this is not relevant and all aspects can adequately be modeled by just including cooperation and communication processes as part of the UVM. The user is interacting with the system, and certain events are caused (and should be understood as such) by another user approaching the current user through the system. The users' mental model will contain the notion of other people interacting with them through the system. The users will normally not be bothered by the designer's problem of integrating different peoples' dialogues.

A final word regarding the interface action specification. In UAN, this column specifies the interface actions as elements of knowledge and their format. This is needed to specify what information is provided to the user, in what situation, at which time, and in which format. However, it is not enough to specify the total UVM. An additional specification should consist of sketches showing the layout of the output device (e.g., the screen layout), the graphics and colors, etc. In some cases three-dimensional sketches, such as cardboard mockups are needed, and in some cases animated video clips or sound tracks, etc.

In this section, we showed what aspects need to be specified for the UVM from the point of view of what a user should need to understand and be able to develop adequate mental models. In fact, all aspects mentioned have been incorporated into a recent version of UAN that has been developed by Venema (1999). Table 3.2 provides an example.

The example in Table 3.2 is about the arrival of new e-mail and the possibility of the user to read this at once or postpone this. In UAN, the formalism to specify the interaction has the shape of a template, which helps designers to consider all relevant elements of the UVM (of course, any part of the formalism that would not apply should simply be left empty).

- The top of the template allows for stating a prior condition for this dialogue whenever this condition should be known by the user (at this level of specification, we should not bother about technical conditions). In this case: The user should understand that the e-mail is running as a background process, although nothing is indicated on the screen. Additionally, there is no unread mail in the mailbox. This knowledge is relevant for initiating a mental model the moment that the interface provides information on new arriving e-mail.
- In addition, the top allows additional verbal information, intended to provide some key description of the template to describe any other user-relevant aspects that could not be put in the regular slots. In this case: The user should understand that another e-mail could arrive during handling the first new one and should be able to adjust the mental model accordingly if the dialogue starts to repeat.
- The left column indicates the actions that the user may perform. In this case, they are reactions to the systems initiative (column 2): The user has to choose to either accept or

TABLE 3.2. Example of an Interaction Specified in UAN Notation as Extended by Venema (1999)

Interaction: New E-mail			
About			**Interface Prestate**
This interaction describes the arrival of a new e-mail, in case all previous mails have been read. If another message arrives during this dialogue, the dialogue would restart immediately after the current one is finished.			1. E-mail running as a background process, although not represented at the interface. 2. All previous e-mails have been read (unread_mail = false until the new arrival).
User Actions	**Interface Actions**	**Interface State**	**Connection to Computation**
	! MESSAGE("New e-mail has arrived")		unread_mail ⇒ true
	! ASK("Read now", [Yes,No])	Previous dialogue postponed, not aborted	
CHOOSE (task to continue, from the set of previous tasks plus new_e-mail task)			
POINTERTO(<yesbutton>) CLICK(<yesbutton>) ‖ POINTERTO(<nobutton>) CLICK(<nobutton>)	MOVEPOINTER(<yesbutton>) SHOW_EMAIL([latest]) MOVEPOINTER(<nobutton>)		unread_mail ⇒ false unread_mail ⇒ true
	HIDE_MESSAGE( )	Previous dialogue enabled to continue	

postpone the proposal to read the new mail. The decision is in this case probably based on the user's mental model of the current task situation, including the urgency of competing activities. The decision of the user is in itself invisible and could be left out of the specification without changing the meaning for the design. But it could be useful for the designer to understand the fact of the choice and the possible requirements for knowledge. Based on this, the designer could decide to specify the representation on the screen of a reminder about the task to be postponed (either the previous running task or the handling of the new mail). After specification of the user's decision, the next time slot (below the horizontal line) shows the user's physical actions toward the interface: either presses the yes button or the no button.

- The second column shows the interface actions. In this case, the interface takes the initiative by telling the user that new mail has arrived and by providing the user information on the choice to be made. At the moment the user performs an action, the interface is mirroring this by showing the movement of the pointer and providing the new mail if the user has chosen this option. This column indicates in fact, either exactly all information that the interface provides at any moment to the user, or all information that changes (in the example: HIDE_MESSAGE( )). The template provokes preciseness at this point; but, if needed, sketches should be used to clarify the details of representation.

- The third column provides a specification of all knowledge of the interface state that is relevant for the user. In the case of our example, this is the information about the concurring processes in which the user may be engaged (which

one is postponed, which one is running in this phase of the dialogue). When the interface does not explicitly show the status of user-relevant processes, it should be specified that the user needs to keep them in mind (i.e., adjust the mental model about things to be done in the future and things to be done at the very moment).

- The last column shows any changes in the underlying technology, network, or other parts of the system that the user should understand. In our case, this concerns the status of the mailbox regarding unread mail.

The main conclusion of the current section is that well-chosen modeling formalisms may serve to urge designers to consider all aspects of the UVM and to specify these from the point of view of user-relevant knowledge and representations. The specification in Table 3.2 is exactly equivalent with the ideal mental model of an expert user for handling the announcement of new email.

Designing for Users and Tasks

In this section, we will show the relation between design techniques on one hand, and the role of users' mental models and knowledge on the other hand. There have been developed a variety of design approaches toward interactive systems. In Part V of this volume, the reader will find state-of-the-art accounts of the most relevant approaches. We will mainly point to some phases in design that have a strong relation to user knowledge. In the next paragraph, we will focus on the collection and modeling of knowledge on the task domain, as well as on the activities to develop and envision a future task world. The specific content of task models from the point of view of representing users' knowledge will be considered in "Modeling Users Tasks Knowledge for Complex Interactive Systems." The user knowledge and mental models aspects of detail specifications have in fact already been considered, because the concept of the UVM is a core of our concern about the future user's mental model. In "Specifying Details of Technology: The Users Virtual Machine." we will only briefly indicate the modeling of the UVM as a process of specifying future users' systems knowledge. Finally, in "Evaluating Design for Future Users' Mental Models," we investigate relations with early and late evaluation techniques (during task analysis and during specification and development of the UVM).

Task Analysis as a Process for Analyzing Users' Knowledge. The design process frequently starts by an extensive task analysis. The first task model we make is a descriptive task model and is used for analyzing the current task situation. A possible second task model is a prescriptive task model for the system that is to be designed.

All task analysis approaches start with one or more techniques for collecting, modeling, and analyzing task knowledge (in this respect especially chapters 48 through 50 are relevant). Task knowledge, in the case of complex interactive systems, may be collected both from individual people (task experts) and from the community of practice (by using ethnographic methods, document analysis, etc.). Task models, whether they describe an existing situation, or envision a future situation, are knowledge models. In the first case, they are descriptive, although they will be inspiring redesign by showing problems, failures, inconsistencies, or wishes. In the latter case, they are the bases for detail design in providing the knowledge that would be needed by the future community of users for the intended system to be successfully developed, implemented, and used. We prefer to split the two types of task models to make sure we do not mix reality and vision.

Analyzing the Current Task Situation (Task Model 1). In many cases, the design of a new system is triggered by an existing task situation. Either the current way of performing tasks is not considered optimal, or the availability of new technology is expected to allow improvement over current methods. A systematic analysis of the current situation may help to formulate design requirements, and at the same time may later allow evaluation of the design. In all cases where a current version of the task situation exists, it pays off to model this. In most cases, a combination of classical HCI techniques may be needed, such as structured interviews and other psychology-based techniques (Johnson & Johnson, 1991), as well as CSCW-related techniques such as ethnographic studies and interaction analysis (see chapter 50). The first type of techniques will help us to model knowledge (the basis for mental models in use) found in individual users, and the second type of techniques allows us to find group knowledge that may be related to, but is not identical to, the concept of shared mental models (see "Differences Between Mental Models of Experts and Novices in Some Other Knowledge Domains").

Envisioning the Future Task Situation (Task Model 2). Many design methods in HCI that start with task modeling are structured in a number of phases. After describing a current situation (task model 1) the method requires a redesign of the task structure to include technological solutions for problems and technological answers to requirements. Johnson and Johnson (1991) provide an example of a systematic approach where a second task model is explicitly defined in the course of design decisions. Task model 2 will, in general, be formulated and structured in the same way as the previous model, but, as we said, in this case it is not considered a descriptive model of users' knowledge. However, if the design finally results in implementation, task model 2 (in its final content, consistent with the ultimate design) may well be applied as a prescriptive model. It defines the knowledge an expert user of the new technology should possess to be able to develop adequate mental models in actual new work situations.

Modeling Users Task Knowledge for Complex Interactive Systems. For modeling any type of knowledge, one needs a conceptual framework to state clearly what types of entities are relevant for applying the modeled knowledge and what types of relations are relevant for analysis. The representation of task models, from our point of view, should allow the designer to understand completely all the relevant knowledge aspects of the domain for which a system is being designed. The task models will provide insight in the background knowledge that is (task model 1) or should be (task model 2) available, either in LTM or in the actual situation, including the interface, for users to develop an adequate mental model when interacting with the system. The various viewpoints on task knowledge as may be found in HCI (e.g., Johnson & Johnson, 1991) and CSCW (Jordan & Henderson, 1995) show us that the complex of knowledge needs to be analyzed from different viewpoints:

- Because current and future systems will frequently have multiple users with different roles, task models should cover the users' and the community of practice's knowledge of the agents involved and their organizational relations.
- Task models should feature the work for which the system is intended to be used, in all relevant details.
- Additionally, task models should model all aspects of the work situation that are relevant for the user to develop a mental model in actual situations.

Furthermore, each knowledge aspect relates to the others. In modeling task knowledge in this manner, designers are supported to analyze and specify from different angles, whereas design tools can be used to guard consistency and completeness.

The three viewpoints are an integration of the main focal points in the domains of HCI and CSCW. Both design fields consider agents (users vs. cooperating users or user groups) and work (activities or tasks, respectively, the objectives or the goals of interaction and the cooperative work). Moreover, especially CSCW stresses the situation in which technological support has to be incorporated. In HCI, this is only sometimes, and then mostly implicitly, considered. In this section, we will briefly mention our conceptual framework for task knowledge. This is the structure of knowledge elements that are relevant for understanding users' background knowledge for mental models when using interactive systems.

Agents. In human knowledge of task domains, agents often are people, either individuals or groups, but in general agents may also be systems, either relatively simple ones ("my answering machine") or complex ones ("the tax office"). In specifying user-relevant knowledge of the task domain, we need to make a distinction between actors, as acting individuals or systems, and the roles they play. Actors have to be described with task-relevant characteristics (e.g., for human actors the language they speak, the amount of typing skill or experience with MS Windows, because this is relevant for users when they consider delegating tasks to others). Roles indicate that certain subsets of tasks are allocated to an actor. Consequently, roles are generic for the task world. In general, actors know (or should know) that they are responsible for the tasks defined by their current role. More than one actor may perform the same role, and a single actor may have several roles at the same time. Each user should be aware of the role of other people (actors) that he or she is collaborating or communicating with. The user's relevant knowledge of the organization refers to the relation between actors and roles in respect to task allocation: Delegation and mandating responsibilities from one role to another is part of the organization.

Work. Users' knowledge of work concerns both the structural and the dynamic (process) aspect of work, so we take task as the basic concept. Users should know the relation between tasks, both in relation to other tasks (subtasks, supertasks) and in relation to temporal and causal aspects (tasks may trigger other tasks, and the completion of one task may be a precondition for performance of another one). Complex tasks may be split up between actors or roles. This means that a person can delegate a subtask to another person, probably in relation to an adequate role. For example, a manager who is organizing a major business process (a high level task) change will ask his secretary to organize meetings, collect documents, etc. (all of which are subtasks of the original one). Normally, people will perform a task because they want to reach a goal, and we may define a task as an activity performed by agents to reach the goal of that task. On the other hand, people may often choose various different tasks to reach a goal, depending on situational conditions. A special situation may arise when the start of a task is triggered by an event from the situation that is not normally related to the task domain (like power failure, interruption because of a calamity), especially if this event is communicated through the system to be designed. Care should be taken to allow the user to develop the needed mental model instantaneously to cope with the exceptional condition. Finally, we need to make a distinction between tasks and actions: People often perform meaningful actions that do not have a goal in itself. For example, when someone hits a return key on the machine keyboard, this is on purpose. But the meaning could be varied, such as: Go to the next cell in the same row (in a spreadsheet), insert a "return" character in the text string (input mode in a word processor), end of command (in defining a UNIX shell script), etc. We label these meaningful activities actions. Users should know and understand them, but they will not have a goal to perform them as such. In redesign, one might consider changing actions, omitting some, and, especially, choosing them to be consistent and easy to learn and use.

The unit level of tasks needs special attention. The lowest task level that people are able to consider in referring to their work is often labeled the *unit task* (Card, Moran, & Newell, 1983). Unit tasks are important elements in any user's task knowledge, being the smallest activities that users have in their repertoire to reach their goals. If someone is asked about his or her tasks, unit tasks are usually the smallest tasks mentioned. Card et al. (1983) state that unit tasks may also be identified in actual users' behavior by observing the time that people spend waiting to perform the next action in a dialogue. If this time exceeds a certain amount (1.35 s), this indicates that considerable thinking is going on between dialogue sequences; so, the transfer to a new unit task may be inferred. Unit tasks will often be role-related. For a manager who is organizing a meeting, the invitation of the participants may be a unit task (performed by delegating this to his secretary). For the secretary, the actual invitation is a complex task to be decomposed into writing letters, collecting addresses, putting letters in envelopes, and so on. For her, a unit task could be the task of putting the closed envelopes into the mailbox.

To apply a system, users need to understand (and sometimes learn) the relation between the unit tasks and the functionality that the system provides. If a user decides to delegate a unit task to the system, the mental model needs to provide knowledge of the user actions and understanding of the resulting system actions. The dialogue specified in Table 3.2 concerns a unit task that could be labeled "accept reading a new message or defer for later reading."

Situation. Users need knowledge of the situation to develop an optimal mental model to act. So, designers need to collect and model this relevant knowledge, which means detecting and describing relevant aspects of the environment (physical, conceptual, and social) and the objects in the environment. In this framework, objects are not defined in the sense of object-oriented methods. Each thing that is relevant for the user to the work in a certain situation is an object in the sense of task analysis; even the environment can be an object. Objects may be physical things, or conceptual (nonmaterial) things like messages, gestures, passwords, stories, or signatures. People in a task situation need knowledge of the things they work with. The objects may be changed (or destroyed or created) as result of a task. Also, objects may be relevant in conditions for tasks. For example, the objects "message" and "mailbox" are needed as elements in the users mental model of the mailer ("If there are already some messages in the mailbox that I did not read, I do not expect my machine to alert me when a new message arrives."). The history of past relevant events in the task situation is part of the actual environment if this features in conditions for task execution ("If I just handled new incoming mail, I expect the mailer to alert me of new messages.").

Specifying Details of Technology: The Users' Virtual Machine. After the task modeling activity, the next step in design is to specify the new system. Task model 2 gives the envisioned task world where the new system will be situated. From there, the details of the technology and the basic tasks that involve interaction with the intended new system need to be worked out. From our point of view (designing for the future users' understanding, knowledge, and mental models), the process of specifying the details of technology should result in a detailed description of the system as far as it is of direct relevance to the end-user. As indicated previously, the UVM includes both the technology semantics (matching to the functionality that the system offers the user for task delegation), the syntax (matching the dialogue for task delegation to the system), and the system's perceptible representations (Norman's system image).

When making the transition from task model 2 to designing the UVM, the tasks and the objects modeled in task model 2 determine the first sketch of the application. The task or object structure is used to create the main displays and navigational structure. From there on, the iterative refinement process takes off, guided by various evaluation techniques that will be discussed further.

Representations for Detailed Design. For representing the UVM, we already showed UAN or its offsprings to be good candidates. But, there are other families of techniques that can be used for modeling detail specifications of the UVM, such as variants of GOMS, TAG, etc. The reader should keep in mind that not any of the mentioned modeling approaches and formalisms is able to cover all relevant knowledge aspects of the representation. At least different types of sketching techniques need to be added (see "Modeling the User Virtual Machine"). From our point of view, the main reason for choosing any representation is to allow and provoke designers in this stage of design to consider and specify all relevant knowledge a user should have available when applying the intended system. Specifying this knowledge will help to consider how future users will be able to develop mental models in actual use situations, what and how they need to learn, and how the system can assist them in these activities.

Evaluating Design for Future Users' Mental Models. Evaluation during design of interactive systems is the topic of the third part of section V in this volume. Here, we will only consider evaluation in relation to questions regarding future users' knowledge and mental models. The main issue is twofold:

1. How to represent our early and later specifications in such a way that the meaning of the system for the future user can be made clear.
2. How to explore the mental model of future users in a relatively early stage (before the system design is complete and the system is implemented).

Early evaluation can be done by inspecting design specifications or by performing walkthrough sessions with designers and/or users. But independent of whether the evaluators are design specialists taking the stand for the potential users, or real (future) users, the main issue is to provide them with an optimal representation of the design decisions so far. An optimal representation, in our view, is a representation that helps the evaluator, the future user, or other stakeholders, to understand our envisioned design. In other words, the representation should allow the development of a mental model of the system to be, even if only parts, rough ideas, or vague sketches are specified yet.

Formal representations like UAN, GOMS, etc., however useful they are in being precise and unambiguous, and in documenting and analyzing the design decisions, are mostly not fit for making other people (let alone users) really understand the actual system or situation we are intending to develop. In all cases of early and later evaluations, we need to confront the evaluators with the design specifications so far, whether this is just part of the envisioned task world (task model 2) or whether it concerns details of the UVM like the dialogue or the systems presentation. We will need to represent our decisions in such a way that the evaluator is able to understand the design fragment we are evaluating. In other words, to allow the evaluator to develop a mental model of the intended task world or system details that we want to assess, we have to represent our design intentions or specifications in a way that both:

- shows our intentions as precisely as they are in the current state
- provides the evaluator with the feeling or understanding of freedom in interpretation where decisions have not yet been made or where intuitive interpretations might be needed to understand the impact of the design on the user.

It is precisely for this reason that early evaluation requires representations like scenarios, sketches, storyboards, and impressionistic video clips, where later evaluation should be based on detailed use cases combined with interactive prototypes (in many cases of only a part of the system, though). Each of these will help the user or expert evaluator to develop a mental model of the intended system, allowing reactions and eventually interactions that help the designer to assess the match with the intended virtual machine and mental model.

Assessing the early and later design ideas requires collecting and recording the users' (and other evaluators') mental models, knowledge, feelings, and behavior. There are many evaluation methods that will be dealt with in other parts of this book that we need not repeat here. From the point of user knowledge, learning, and mental models, most of them will provide relevant insight. In the next section of this chapter, we will focus on specific tools for assessing the mental models that are developed by providing the evaluators (or users) with early design specifications in the form of scenarios, etc.

HOW TO CONCEPTUALIZE AND MEASURE MENTAL MODELS

Theory Is Not Enough

In the past sections, we have shown that the design of interactive systems requires attention to the concept of mental models.

The basic psychological theories on mental models have been discussed, and a pragmatic approach toward applying the mental model concept in design has been elaborated. These issues are part of our general plea for designing for users and tasks: Usable systems require the availability of knowledge in such a way that users are able to instantiate a mental model for performing their intended task with a system in an actual situation. The design approach in this respect includes several phases related to user knowledge:

1. Understanding the user's task world knowledge (task analysis).
2. Envisioning a new task world (and as a model of the task world knowledge) when the technology to be designed would be implemented.
3. Specifying the UVM—the systems knowledge as far as it is relevant for the user or the different types of user roles.
4. Developing a representation of the intended task world and the intended system that can be presented to prospective users early in the design (e.g., scenarios, mock-ups, and early prototypes).
5. Analyzing the future user's mental model and feeding back the resulting conceptualization to the designer.

Steps 1 to 4 have so far been covered in the previous sections. Step 5 requires techniques to conceptualize the mental models provoked by the early scenarios.

In this section, we will show two complementary approaches and techniques toward the need to investigate real mental models. Mental models have been assessed using a variety of methods, including thinking-aloud and verbal protocols, online protocols, problem-solving performance, information retention over time, observations of system use, users explanations of systems, users predictions about system performance, cued association lists, etc. (Ferstl & Kintsch, 1999; Sasse, 1991). The main problem is that, because mental models are in the mind of people they cannot be accessed directly. It is necessary to create an intermediate representation that can be accessed and from which we can understand the mental model. The intermediate representation is the one we can visualize, analyze, and compare to study the mental models. Measuring the mental model therefore implies modeling (Van Engers, 2001). Assuming this fact, these methods maintain that mental models are linguistically mediated and can be represented as networks of concepts where the meanings for the concepts are embedded in their relationships to other concepts (Carley & Palmquist, 1992).

Two of the most frequently techniques used to explore mental models are pathfinder and teach-back protocols. For example, Bajo, Gonzalvo, Gómez-Ariza, and Puerta-Melguizo (1988) used the pathfinder to study the mental models of experts and novices in physics. On the other hand, Van der Veer (1990) used the teach-back protocols to study differences in mental models between users of computer systems. We think that a combination of methods can give compatible and complementary information about the contents of the user's mental model. As the reader can remember, the content of a mental model contains, in more or less degrees, and depending among other things of the user expertise with the system, semantic knowledge about the system and procedural knowledge about how to perform different tasks with this system (see "Mental Models of Procedures vs. Semantics"). As we will see, pathfinder is more likely to offer semantic information in a structured way, whereas teach-back is ideal to offer information about the procedural user's knowledge.

The Pathfinder Algorithm

Pathfinder is a graph theoretic technique that derives network structures from rated data (Schvaneveldt, 1990; Schvaneveldt et al., 1985). The main assumption of this method is that semantic proximity can be represented in terms of geometric space. Using this method, participants have to make pairwise estimations of the degree of relatedness of the relevant concepts in the considered domain. The concept's ratings are converted into proximities and analyzed with pathfinder. In the network, concepts are represented as nodes and relations between concepts are represented as links between the nodes (see Fig. 3.7). A weight corresponding to the strength of the relationship between two nodes is associated with each link. Concepts can be directly linked or not. The algorithm searches through the nodes to find the closest indirect path between concepts. A link remains in the network only if it is a minimum length path between the two concepts (Bajo, Cañas, Gonzalvo, & Gómez-Ariza, 1999).

Pathfinder provides a technique to find the average net of a group of users (function *ave*). This allows conceptualization of a group mental model or shared mental model. The underlying assumption is that this average structure represents the concept space as it is shared by a group of people. Of course this only makes sense if we assume the people in the group share common knowledge and use this knowledge in the same situations in a comparable way. We would not presume the people will have identical content in LTM, but that they have equivalent mental models when instantiated in the situation created by the pathfinder instructions. To do so, we should test the homogeneity of the group's PFNets.

Pathfinder allows comparison between different PFNets. One may want, for example, to compare mental models of a single group to test if this group has a certain degree of homogeneity. It is also possible to compare mental models of different groups of subjects or to compare mental models of different systems. Goldsmith and Davenport (1990) discuss two basic ways of measuring similarity between graphs with common node sets:

(a) Based on path lengths: The distances between all nodes in the networks are examined. If the distances between certain nodes in one net are similar to the distances between the same nodes in the other net, the networks are considered similar.

(b) Based on neighborhood similarities: In this case, the set of neighborhood's nodes of each node is considered. If the nodes in one net have similar neighborhoods as the same nodes in the other net, the networks are considered more similar. With this method, it is necessary to use the so-called C values. If we have two networks A and B with a common node set, let

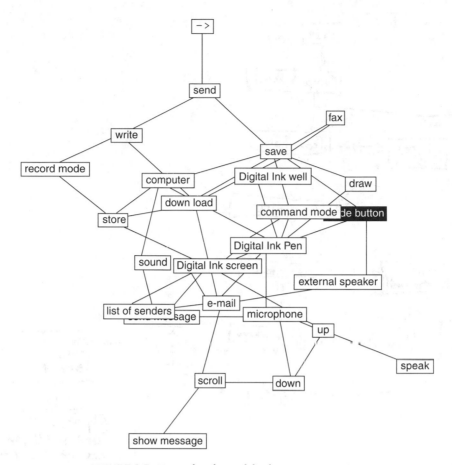

FIGURE 3.7. Example of a pathfinder representation.

E_A be the set of edges of network A and E_B the set of edges of network B. The similarity $C(A, B)$ is then defined as:

$$C(A, B) = |E_A \cap E_B|/|E_A \cup E_B|.$$

C is a value between 0 and 1. If C is 0, the networks do not have any links in common and therefore are very dissimilar. If C is 1, the networks are identical. Research has shown that, for empirically derived PFNets, the second way of measuring is most appropriate, because it can account for certain phenomena that the first cannot (Gonzalvo, Cañas, & Bajo, 1994).

Schvaneveldt claims that the mental maps presented by the pathfinder algorithm represent the concepts and relationships between concepts as they reside in LTM. The method, however, does not give guidelines for the selection of the concepts. Different ways have been used to grasp the central concepts in a certain domain. One possibility is to ask experts to express the relevant concepts; but, according to Van Engers (2001), this method may have the disadvantage that we miss the misconceptions that are typical for novices. Another approach is to analyze documents or interview material to find those concepts that are used in the domain of interest.

Another problem is the fact that pathfinder needs the subjects to make pairwise estimations of the degree of relatedness between concepts. Each concept is presented with all other concepts so the subjects has to make $[n * (n - 1)]/2$ similarity judgments. The problem is that the higher the number of relevant concepts, the higher the number of pairs of concepts that the subjects have to judge. For example, for 30 concepts, there are $(30 * 29)/2 = 435$ pairs of concepts, and if a person needs about 10 s per response, it takes around 1 hr and 20 min to perform the task. Finally, there is a lack of powerful statistical methods for the comparison of network structures (Ferstl & Kintsch, 1999).

Teach-back Protocols

The teach-back method is a hermeneutic method for measuring mental models and was developed in the framework of the conversation theory (Pask & Scott, 1972). Later, teach-back has been extended as a hermeneutic method to provoke the users to externalize their mental models (Van der Veer, 1990). Among other utilities, the teach-back method is suitable for detecting individual differences in mental representations (Van der Veer, 1990; Van der Veer et al., 1999).

Using this method, participants are asked individually to teach to an imaginary colleague how to solve the problem stated in the teach-back question. To respond, participants can write, make diagrams, create drawings, etc. (see Fig. 3.8 for an example

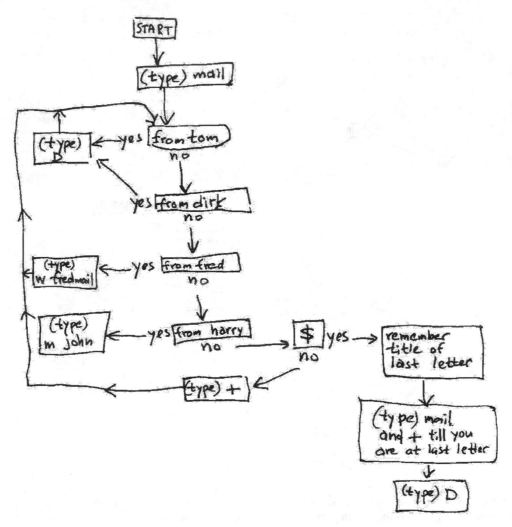

FIGURE 3.8. Example of a teach-back representation.

of a teach-back representation). Finally, the answered protocols are scored using categories. The function of the categories is to explore the mental models of the participants on characteristics that are relevant for the research questions of the specific study.

The most frequent questions used with this technique are the "what is?" type and the "how to?" type:

- "what is?": The goal of this kind of question was to explore the conceptual and semantic knowledge the participant has about the system. One example of a "what is?" question can be to "explain to an imaginary reader what Microsoft Outlook is."
- "how to?": The goal of this kind of question was to explore the procedural knowledge the user has about the system. For example, "explain to an imaginary reader how to send an e-mail with Microsoft Outlook."

One of the most important limitations of teach-back is that this descriptive method does not provide explanations. Another problem is that the method is only of use to obtain information on knowledge that the user is aware of and that can be represented in writing or drawing. Furthermore, teach-back requires more than a straightforward application of the technique. Both the definition of the relevant categories and the interpretation of the teach-back protocols require experience with the hermeneutic analysis.

EXAMPLES OF EMPIRICAL RESULTS ON MENTAL MODEL MEASUREMENT

After explaining briefly both methods and their problems, in the next sections we will show some of the results obtained in different domains using both techniques. We will also show our ideas about how to use both methods in an easy, complementary, and useful way to design complex systems.

Mental Models of Operating Systems

Comparing Operating Systems Using Teach-back. Van der Veer (1990) investigated mental models of different operating systems in a large project, including more than 700 Dutch school children between 11 and 18 years old. The two operating systems compared were a state-of-the-art commercially available direct manipulation interface (Mac OS) and a command-driven operating system (MS-DOS) both used at stand-alone PC systems. For all subjects, the mental model research was performed after exactly 10 hr of hands-on experience with one of the systems during school classes. None of the subjects had any hands-on experience with the other system. The investigation consisted of asking a "how to?" teach-back question. For data analysis, subject groups were selected that matched on verbal aptitude, computer literacy, level of education, and type of school. For half of the subjects that used the command interface, the teachers of the class showed the students a graphical representation of the systems semantics in a metaphorical way (using tubes between containers of objects, labeled with the command names, etc.).

The direct manipulation operating system of that time does (as do these interface today) show clearly all possible user actions, by presenting all objects that are immediately available for acting on. In addition, this type of interface shows immediately whether an operation is available at the current state or not. Also, it immediately reflects users' commands to the system by showing the movements of the mouse on the screen and indicating state changes of objects visibly. The command-driven operating system of that time does not show any availability of user actions or objects, and it does not show any effects of actions performed.

Several categories of representations were systematically found by hermeneutic analysis and could be reliably scored by different judges:

- Users of this age and level of experience mostly prefer verbal representations in their protocols, although 25% used visual-spatial images as well, and 28% used icons. Some used declarative descriptions of their "how to" knowledge, but more used procedural representations.
- Representations could be classified as knowledge at different levels: task level, semantics (system functionality), dialogue, and keystroke (see Fig. 3.9 for an example scored at keystroke level). In their teach-back protocols, most subjects represented several levels, nearly always adjacent ones.

Here, we show some of the most relevant relations found between system characteristics and teach-back protocols:

- Correctness and completeness of the mental model as judged from the teach-back protocols correlated positively with teachers' ratings of the subjects' competence in using the system.
- Correctness and completeness was not related to the operating system used by the subjects. Apparently, the students were able to learn both systems equally well.
- Those command system users that were not confronted with the additional graphical representation of the system semantics represented more keystroke level knowledge than the Mac users, as well as showed stronger preference for visual spatial representations. This indicates that they need a mental model that includes precisely those aspects of the system that

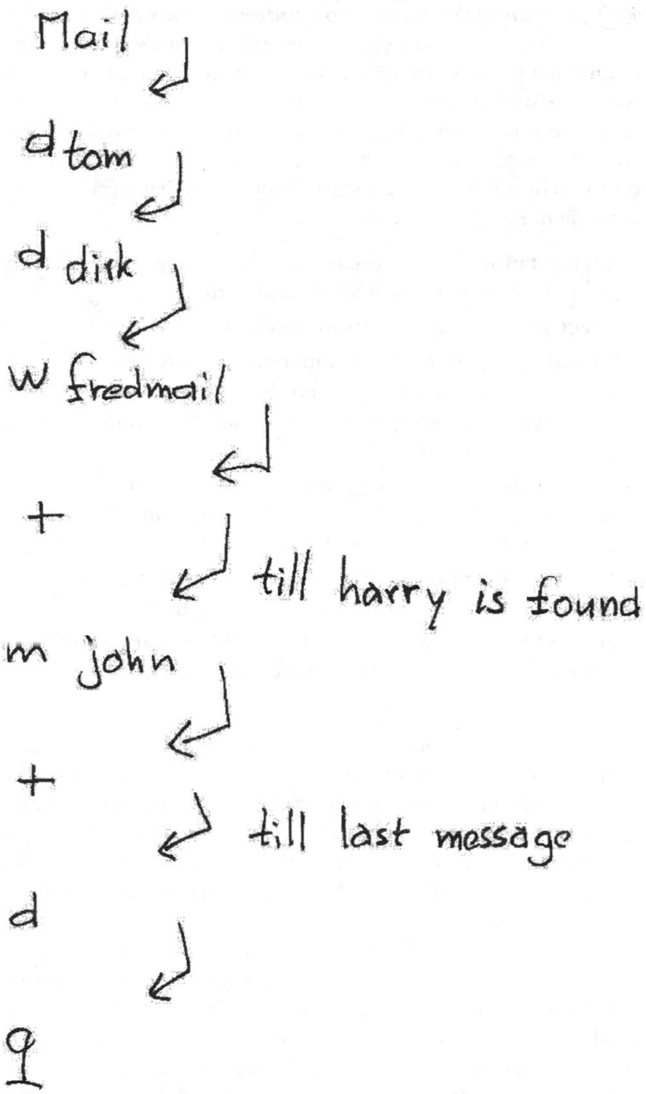

FIGURE 3.9. Example of a teach-back protocol for a "how to?" question on using e-mail at a command interface.

the system itself does not show: the available actions and any indication of the current workspace.

- Providing the command system users with a graphical representation of the systems semantics resulted in more representations on all system-related levels (functionality, dialogue, and keystroke). Moreover, the correctness and completeness of the teach-back protocols were superior to those of the command system users that did not receive this additional help.

Based on this study, we can conclude that teach-back helps to validly assess mental models of operating systems and that it is related to the system's presentation: Users tend to develop those aspects of system knowledge that are not represented at the interface. Moreover, the results show that providing well-chosen representations may increase the completeness and correctness of mental models.

An Example in Designing Computer Systems Using Teach-back. Mulder (2000) used the teach-back method to explore the mental models participants created after being confronted with envisioned new information systems. The early design ideas were represented by scenarios. He used "what is?" and "how to?" teach-back questions and from the teach-back protocols a set of suitable categories were obtained that were relevant for designers:

- Interpretation of functionality in relation to goals of the subject (e.g., "I would use it as a blackboard")
- Affective reactions like absurd, useless
- Lack of recognition of new functionality in which the design model and the scenario showed there was new functionality (e.g., "the same as a regular scale"; "the same things as with any other whiteboard")
- New functionality assumptions in which the mental model shows functionality that has not been presented in the scenario and is not part of the design model
- Dialogue assumptions (e.g., "I push the icon") in which the scenario did not mention any icon to be pushed
- Implementation assumptions in which hardware or software is represented in the mental model that was not presented in the scenario.

Using these categories, he scored the teach-back protocols of a group of 30 subjects. In the next phase of his study, the scored data were shown to expert designers. The most relevant result of his study was the fact that designers found that the information extracted from the future user, the mental model, could be very useful during the design process, especially in an early stage of the design.

Comparing Representation Effects in Information Retrieval Using Pathfinder. Klok (1998) studied the mental model that subjects develop from working with an information retrieval (IR) system. The search facility of the IR system was based on concepts and relations between them. Four groups of subjects learned to search among the IR system. The experimentally manipulated difference between groups was the representation used for the interaction between the subject and the system: formal, graphical, natural language, and a combination of the graphical and natural language representations. For each group, Klok calculated the average pathfinder net as an indication of the groups' mental model of the system (see Fig. 3.10). The comparison of the differences in performance and accuracy of the mental models between the four groups showed that a combination of natural language and a graphical representation yields better overall results than a single representation.

Effects of Users' Vertical Machine Consistency on Mental Models Using Pathfinder. Van Engers (2001) reports a study on the effect of consistency of the UVM on the resulting mental models. Two versions were specified and built of a menu-based interpersonal agenda system. The functionality of both was completely equivalent. For one version (i.e., in fact built for, and used by, a large government organization), the menu structure was developed after interaction guidelines. The menu structure of the other was experimentally manipulated to show inconsistency of the hierarchy (and, consequently, the dialogue) in relation to the functionality. Half of the subjects in the study had considerable knowledge of the consistent version; the other half had never used the system before.

For both versions, the conceptual model was developed and the respective designer's knowledge of this was modeled using pathfinder. The four user groups (with or without prior knowledge; using the consistent or inconsistent version) used their system for a standard amount of time and solved a set of problems with the system. After this, their mental model was investigated by applying pathfinder, and the similarity of their mental models with the designer's model was calculated. The results show:

- For the inconsistent version, prior knowledge has no differentiating effect; in all cases, the users' mental models differ considerable from the designer's model.
- For the consistent version, prior knowledge increases significantly the similarity of the users' mental model to the designer's model.
- The higher the similarity of the user's mental model to the designer's model, the shorter the solution time for the experimental tasks.

From these findings, we conclude that a consistently specified UVM enables users to develop a mental model that is more congruent with the design model. This congruency is an indication of the usability of the system.

Differences Between Mental Models of Experts and Novices in Some Other Knowledge Domains

An Example in Physics Using Pathfinder. Gonzalvo et al. (1994) provide empirical data to indicate that differences between experts and novices can be caused in part by the conceptual organization that people have in LTM and, consequently, the mental model they use when solving problems related to the specific domain. To describe the changes in mental models as a function of expertise, Bajo et al. (1998) used the pathfinder method in the domain of physics.

Comparisons between students and experts in physics showed important differences in the way they scattered the concepts around the network. Students seemed to organize kinematic concepts close in the network with velocity as a central concept, whereas experts scattered kinematic concepts around the network, close to concepts that define dynamic principles or energy states. Another important difference is the role of mass. Students seemed to relate mass directly to force, whereas experts related it to energy states. Both the importance of dynamic principles and the connection mass-energy imply a sophisticated understanding of the concept. Another important finding was that the organization was flexible and dependent on the context. When novices learned about the domain, their mental model became more similar to that of the experts. In another study, students solved problems that emphasized dynamic

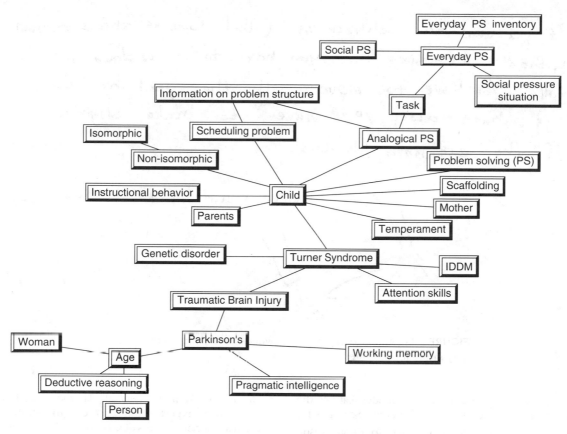

FIGURE 3.10. Group pathfinder net for the group that used a natural language information retrieval dialogue. IDDM = insulin-dependent diabetes mellitus.

aspects. After solving the problems, students' nets were similar to the ones of experts, especially showing the importance of dynamic principles.[2] The authors concluded that instructional practice can make use of this flexibility to elicit appropriate physic principles for solving problems. The results of this study indicate that if the context encourages students to think of sophisticated physics principles, they are able to do it.

An Example in Physics Using Teach-back. Van der Veer et al. (1999) used the teach-back method to study mental models in the domain of physics. They explored the individual differences in mental models and examined whether these differences could be systematically related with the level of expertise. Three teach-back questions that triggered different kind of answers were designed for the study:

- The first question, real life, did not refer to the physics as dealt with in education. For example, "Imagine you have to throw an egg to someone and that person has to catch it without breaking it. . . ." The goal of this question was to investigate if the subjects apply physics laws and principles in real life.

- The second teach-back question, general classroom, explicitly referred to the domain of physics as presented in education. Before this question, the subjects were confronted with five typical high school exercises from the domain. After this, they were asked to "explain to an imaginary fellow student how to solve a physics problem."

- The third teach-back question, actual exercise, implied solving a physics problem as dealt with in educational settings. "Use the following physics problem. . . ."

Scoring categories were developed based on literature about problem solving in the domain of physics, mainly in relation to the differences between novices and experts. The actual occurrence of the categories was confirmed by hermeneutic analysis, where the interjudge reliability was acceptable:

- Memory, indicating the use of past experiences to develop a mental model. Example: "You will know from handball that the main concern is the technique of how to catch."

- Theoretical explanation, referring to laws in the domain (even if sometimes applied incorrectly). For example, "If the mass is larger, the gravity increases."

---

[2]However, as cluster analysis showed, students still stress kinematics as an organizing principle, but experts did not.

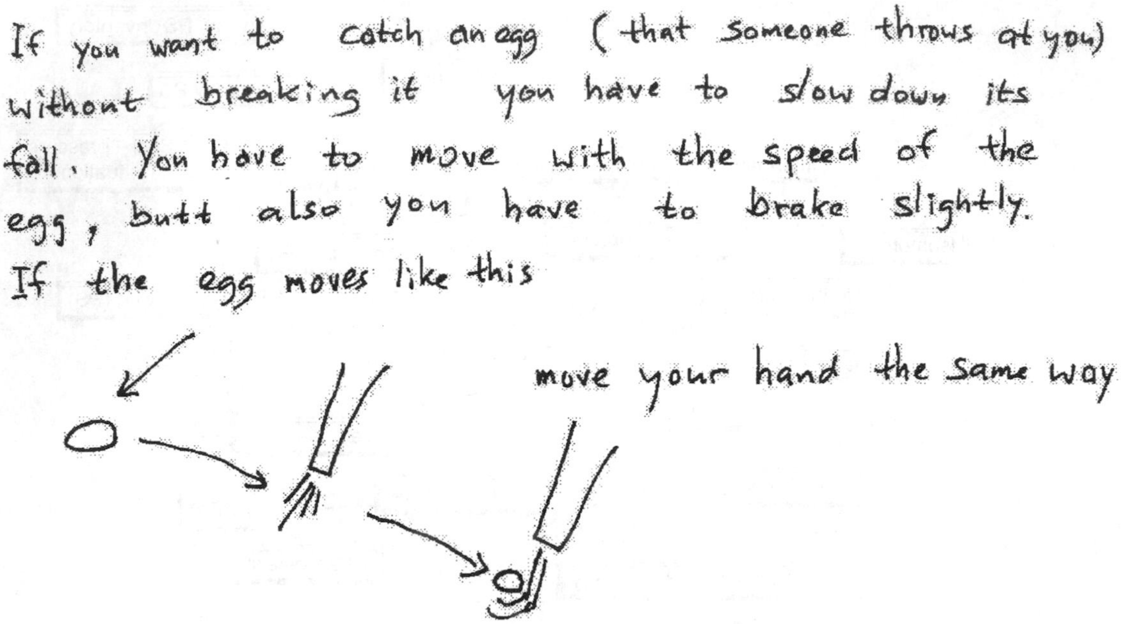

If you want to catch an egg (that someone throws at you) without breaking it you have to slow down its fall. You have to move with the speed of the egg, butt also you have to brake slightly. If the egg moves like this

move your hand the same way

FIGURE 3.11. A novice in physics showing a mental model of the category internal experiment.

- Internal experiment, where the mental model features the mental running of a process to find the outcome. See Fig. 3.11.
- Metaphor, where the mental model refers to a comparison from another (e.g., not the mechanics) domain, primarily indicated in the protocol with words such as like, same as.
- Concept, where a concept label is used as a chunk in WM without elaboration of its meaning. For example, "If the egg crashes, this is because of the force exerted on it."
- Procedure, where the mental model includes a structure of steps to be performed in a certain order to solve the problem. "First, I read the whole exercise. Then, I look at the relation between the questions. Next, I collect the data. Finally, I find the relevant formula and solve the problem."
- Formula, indicating that the mental model, contains the representation of a formula without an indication of the procedure to solve the equation. For example, km/hr = m∗f.

Sixty-eight novices (37 male and 31 female) and 18 experts (6 female and 12 male) participated in the study. The answered protocols were scored using the categories mentioned previously. To investigate expertise effects on the scored categories, an analysis of variance was conducted. The authors found significant differences between experts and novices in the categories theoretical explanation, internal experiment, and procedure. The results showed that novices scored lower in these categories than the experts. Experts' mental model can be characterized, in comparison with novices' mental models, as referring more to theory, containing procedural knowledge, and including the mental running of experiments.

Other relevant results were found when they analyzed separately the data obtained from the three teach-back categories:

- The results from the general classroom teach-back question showed experts did not score in the formula category, whereas a lot of students did. These results seem to show that novices often think in terms of a formula without an accompanying solution procedure when an educational question is asked.
- On the other hand, experts scored in all of the categories when performing the real-life question (against novices who stuck to internal experiments and some theory; see Fig. 3.11), showing that these type of problems stimulate a broad diversity of mental modeling to enable physic reasoning in experts.

In general, the authors interpreted their results as a probe of the usefulness of teach-back as an approach to investigate expert's and novice's mental models.

Applying Teach-back in Artistic Design Domains. Stork (1995) investigated the measurement of mental models of artistic design domains using the teach-back technique.

Hermeneutic analysis with acceptable interrater reliability showed that several scoring categories could be detected systematically over domains relating to music composition and graphic artistic domains. Mental model elements in working memory can be identified as:

- Reference to knowledge from episodic LTM contents: e.g., "painted like Jongkind or John Constable"; "sounds like Pastorius sound."
- The use of concept labels that seem to indicate chunks, without elaboration of its meaning: e.g., "I think in terms of dry or muddy sounds."

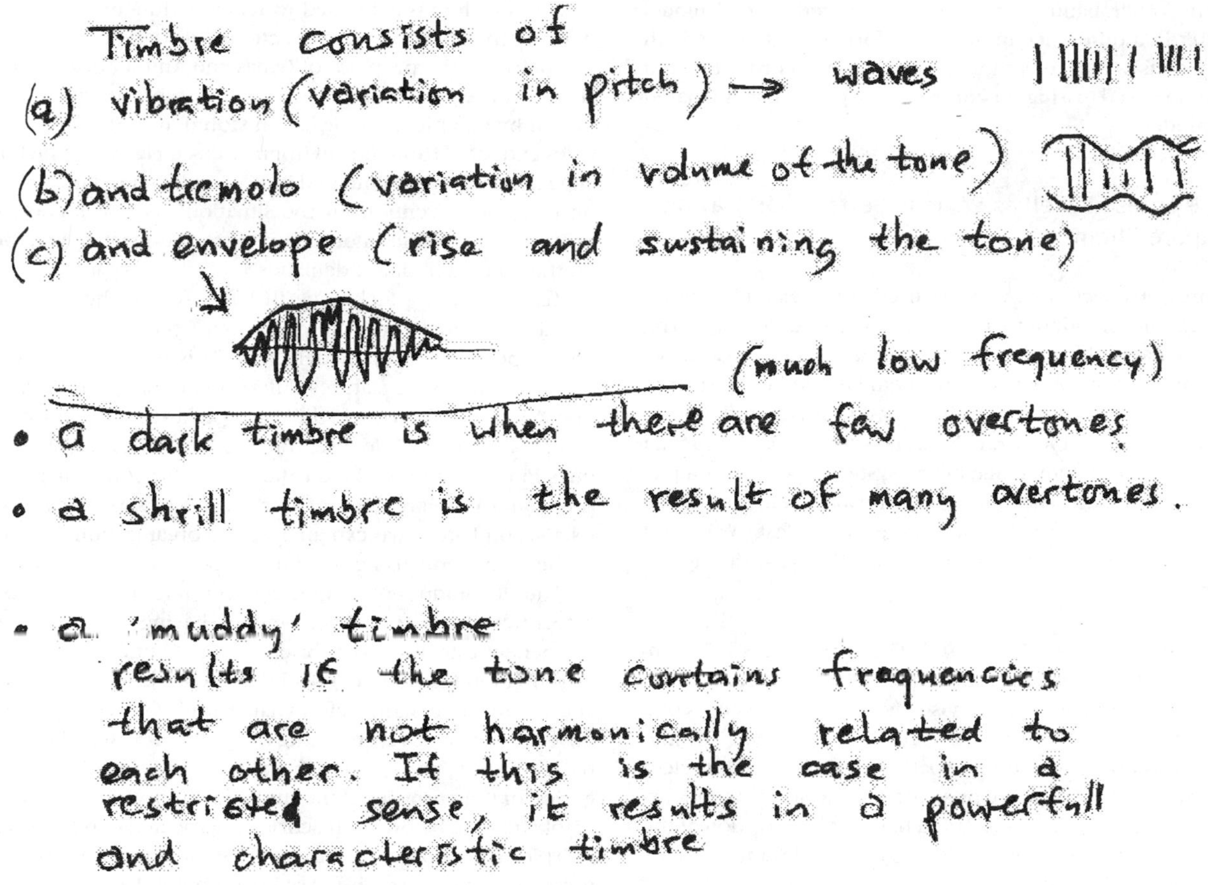

Timbre consists of

(a) vibration (variation in pitch) → waves

(b) and tremolo (variation in volume of the tone)

(c) and envelope (rise and sustaining the tone)

(much low frequency)

- a dark timbre is when there are few overtones.
- a shrill timbre is the result of many overtones.

- a 'muddy' timbre results if the tone contains frequencies that are not harmonically related to each other. If this is the case in a restricted sense, it results in a powerfull and characteristic timbre

FIGURE 3.12. A music composer referring to physical theory (the graphic representations of wave forms and sound envelope).

- Metaphors: e.g., "It should sound like bells far away"; "film story lines are imagined."
- A physical-theoretical model: e.g., "a harmonic overtone structure with formants in certain places...." (Some examples of this type are graphically presented in Fig. 3.12).
- Physical activity references: e.g., "the internal chaos that I throw on the cloth of the painting"; "by moving the finger placed on the string up and down, like violin players."
- Reference to an instrument technology: e.g., "I used a flat boy mouthpiece: thick, full"; "did not use a palette knife for painting."

Shared Mental Models, Relation to Team Problem Solving.
Van Engers, Vork, and Puerta-Melguizo (2000) report an experiment with a business management game in which they manipulated the prior development of a shared mental model by applying the qualitative modeling technique (Vennix, 1995). This technique applies guided group discussion and, in this case, the discussion focused on the domain of a business game.

After an explanation about the game, they worked with two different groups: groups in which the members have a shared mental model, and groups with an individual or nonshared mental model. For the groups that have a shared mental model, they used Vennix's shared conceptual modeling technique. To prevent the groups who did not build a shared mental model from having an incomparable situation, they let the members of these groups build their own individual mental models by using a questionnaire with teach-back questions on the domain of the business game. Additionally, they assigned predefined tasks to half of the groups and not predefined tasks to the other groups. In the case of predefined tasks, specific goals were assigned to specific group members. In the case of the other groups, all the tasks were assigned to the group as a whole.

The analysis of the game results showed an interaction effect of shared mental models and predefined tasks:

- If there was no predefined task allocation, groups without a shared mental model development phase developed better communication than groups with shared mental models.
- After the game, the mental model similarity (as measured with pathfinder) between these groups did not differ significantly.

We may conclude that a shared mental model to start with makes communication less needed. If needed, it helps to develop the mental model during actual cooperation.

On the other hand, starting with a shared mental model may result in a difference in group performance. Groups with predefined tasks differed in the game's target control: A shared mental model lead to a higher control in the game than no shared mental model.

Buying Apples as Well as Pears Is Better Than Having to Compare Them

In the previous sections, we presented some examples using teach-back and pathfinder to explore the mental models. The problem is that both techniques have limitations and, as a consequence, predictions obtained with them should be considered as partial descriptions of the user mental model. On the other hand, because they focus on different aspects of the mental representation and knowledge of the system, they are complementary techniques. The question is how to integrate those techniques in such a way that they are easy and fast to use, and to understand in such a way that they can be used during the design process.

Extracting Relevant Concepts. The advantages of pathfinder are that it is fast, the technique is easy to use, and the resulting representation shows visually how the concepts of a specific knowledge domain are organized (see Fig. 3.7). On the other hand, a serious limitation of pathfinder is how to preselect the relevant concepts that are presented to the subjects. Our hypothesis is that using teach-back and having a description of the conceptual model will allow us to extract the relevant concepts or categories of the system.

The conceptual model has to describe accurately and consistently the aspects of the functionality, dialog, and representation that are relevant to the different possible users (see "Modeling the User's Virtual Machine" for a detailed explanation). Consequently, all of the relevant concepts needed to understand and interact with a system are defined in the conceptual model. But, as long as it is accurate and complete, the conceptual model does not have to be organized conceptually in the same way as actual user's mental models.

Teach-back is a technique that focuses on user knowledge. Using this technique, we can explore the way the users understand the system. It is possible to compare the set of concepts users have with the conceptual model to explore the adequacy, correctness, and completeness of the user's mental model.

Knowledge About Functionality and Dialogue. To explore the user's mental representation means, among other things, to explore the way the components and functions of the system are organized according to the user. To explore the user's mental representation means, too, to explore the procedures the user needs to perform with the system to reach a specific goal. For example, if we want to know if the interface is able to create an appropriate mental model of the system, we need to explore if the semantic and procedural representations the user constructs when interacting with the system are adequate. In terms of design, the semantic knowledge refers to the functionality of the systems, whereas the procedural knowledge implies structuring a goal into the tasks needed to reach it (functionality) and to perform the appropriate user actions (dialogue).

Following the proposal of Cañas and Antolí (1998), we consider that the mental model is a dynamic representation created in WM by combining information stored in LTM and characteristics extracted from the environment (see Fig. 3.6). The knowledge from LTM is instantiated into a mental model in WM based on triggering events from the situation. As Cañas and Antolí proved, the specific information extracted from LTM depends on the context and task demands.

Teach-back is a technique that focuses on the user instantiated knowledge, and the contents of the teach-back protocols depend on the task demands. This means that the type of representation it features differs, depending on the instruction. "What is?" focuses on the conceptual-semantic knowledge, whereas "how to?" focuses on the procedures. Using these questions, we aim to extract the concepts (both semantic and procedural) that are relevant for the users in a specific situation. As a second step, we can analyze the organizational structure of these concepts using pathfinder.

The technique of extracting the relevant concepts (or categories) from teach-back implies having experience with hermeneutic analysis. Very briefly, hermeneutic analysis implies a group of judges that: (1) read the answer protocol as a whole and try to understand fully what is said; (2) try to formulate how the subject is representing the problem space of the teach-back question; and (3) classify the responses into relevant categories for the purposes of the study. To do so, raters need considerable training before their scoring is expected to be sufficiently reliable. This problem led us to the same conclusion that technology companies (such as Apple Computer, Microsoft, XEROX, Boeing, IBM, or Philips Design, among others) have reached: It is important to have experts in cognitive psychology (or other experts on mental models) on the design team.

CONCLUSIONS

Knowledge and insight of the concept of mental models, which are the actual instantiations of user knowledge, is important in the design of interactive systems. System design, as we showed, is starting with knowledge of users in an existing situation (task analysis). The next step is deriving problems and design triggers from this knowledge, and modeling a potential future world, again as a knowledge model, not a technology model. This future world, along with user-relevant knowledge aspects of technology (the UVM), should be matched with future users' mental models and knowledge again. Evaluation of all decisions in this trajectory means to a large extend to remodel the ideas and decisions in such a way that potential users and other evaluators will develop a mental model of the world to be or the system to be. Scenarios are well-established techniques to do this. Consequently, analysis of the evaluators reactions means insight in the knowledge and mental model aspects of our design ideas and specifications so far. Only after we have accepted the resulting specifications from the point of view of future users' understanding, will our specifications be given to the engineers to develop the real system.

As previously described, regarding conceptualizing actual mental models, applying the results of this to the improvement of design will frequently ask for experts like psychologists. Still, designers can handle some of the design knowledge, as has been collected from well-established successful examples. Design pattern collections frequently show the relation between issues of user knowledge and design solutions. We present three examples here taken from The Amsterdam Collection of Patterns in User Interface Design (2001):

- Unambiguous format: In many cases, the user is not familiar with the required syntax for a dialogue. In that case, the instantiation of an adequate mental model may fail. The solution proposed by this pattern is to present the user with fields for each data element and to label the fields with the data unit if there are doubts about the semantics of the field. Additionally, the pattern suggests providing a description of the format and sound defaults. For example, the time control panel in MS Windows helps solve the problem of inputting dates—a well-known example of cultural differences in conventions.
- Bread Crumbs: Users of complex systems easily get lost and need to know where in a hierarchical structure they are. This

will allow them to build a mental model of where they are and where they can go. Bread crumb is a way to show the path from the top level in a graphical way and to allow the user to go to any of the other higher level categories. An example is Sun's web site product pages.

- Progress: Whenever there is a silence in the dialogue, users need a mental model of whose turn it is and why nothing happens. Is the operation still being performed and how much longer will it take? An established pattern to allow the user to maintain a clear mental model of the situation is the progress pattern. The system provides feedback at a rate that gives the impression that work is in progress (e.g., by an animation every 2 s when downloading a file in Internet Explorer 5). Additionally, a valid indication of the progress may be provided.

In "User Guidance" and "Hinting as a Base for Learning," we showed some other examples. Many more guidelines and patterns may be found that help the designer make sure the user is able to develop relevant knowledge and a valid mental model. For more information, see chapter 51 or see Borchers (2001).

References

Amsterdam Collection of patterns in user interface design (2001). Available: http://www.cs.vu.nl/~martijn/patterns/index.html

Baddeley, A. D. (1986). *Working memory*. New York: Oxford University Press.

Baddeley, A. D. (1995). Working memory. In M. S. Gazzaniga (Ed.), *The cognitive neurosciences* (pp. 755–764). Cambridge, MA: The MIT Press.

Bailey, R. W. (1996). *Human performance engineering: Designing high quality, professional user interfaces for computer products, applications, and systems*. Englewood Cliffs, NJ: Prentice-Hall.

Bajo, M. T., Cañas, J. J., Gonzalvo, P., & Gómez-Ariza, C. (1999). Changes in categorization as a function of expertise and context in elementary mechanics. In D. Kayser & S. Vosniadou. *Modelling changes in understanding: Case studies in physical reasoning*. Elmsford, NY: Pergamon.

Bajo, M. T., Gonzalvo, P., Gómez-Ariza, C., & Puerta-Melguizo, M. C. (1998). Changes in categorization as a function of expertise and context in elementary mechanics. *X ESCOP Conference*. European Society for Cognitive Psychology. Jerusalem.

Borchers, J. (2001). *A pattern approach to interaction design*. New York: Wiley.

Cañas, J. J., & Antolí, A. (1998). The role of working memory in measuring mental models. In T. R. G. Green, L. Bannon, C. P. Warren, & J. Buckley (Eds.), *Proceedings of the Ninth European Conference on Cognitive Ergonomics—Cognition and Cooperation*. European Association of Cognitive Ergonomics (EACE): Institut National de Recherche en Informatique et en Automatique (INRIA), Rocquencourt, Le Chesnay Cedex, France. Rocquencourt.

Card, S. K., Moran, T. P., & Newell, A. (1983). *The psychology of human-computer interaction*. Hillsdale, NJ: Lawrence Erlbaum.

Carley, K., & Palmquist, M. (1992). Extracting, representing, and analyzing mental models. *Social Forces, 70*, 601–636.

Carroll, J. M., & Olson, J. (Eds.). (1987). *Mental models in human-computer interaction: Research issues about what the user software knows*. Washington, DC: National Academy Press.

Clark, H. (1996). *Using language*. Cambridge University Press.

Craik, K. J. W. (1943). *The nature of explanation*. Cambridge, UK: Cambridge University Press.

de Haan, G., van der Veer, G. C., & van Vliet, J. C. (1991). Formal modelling techniques in human-computer interaction. *Acta Psychologica, 78*, 26–76.

DiSessa, A. (1986). Models of computation. In D. A. Norman & S. W. Draper (Eds.), *User-centered system design: New perspectives in human-computer interaction*. Hillsdale, NJ: Lawrence Erlbaum Associates.

Engelbart, C., & English, W. K. (1968). A research center for augmenting human intellect. *AFIPS Conference Proceedings of the 1968 Fall Joint Computer Conference* (Vol. 33, pp. 395–410). San Francisco, December.

Ferstl, E. C., & Kintsch, W. (1999). Learning from text: Structural knowledge assessment in the study of discourse comprehension. In H. van Oostendorp & S. R. Goldman (Eds.), *The construction of mental representations during reading*. Hillsdale, NJ: Lawrence Erlbaum Associates.

Gentner, D. A., & Stevens, A. L. (Eds.). (1983). *Mental models*. Hillsdale, NJ: Lawrence Erlbaum Associates.

Goldsmith, T. E., & Davenport, D. M. (1990). Assessing structural similarity of graphs. In R. W. Schvaneveldt (Ed.), *Pathfinder associative networks*. Norwood, NJ: Ablex Publishing Corporation.

Gonzalvo, P., Cañas, J. J., & Bajo, M. T. (1994). Structural representation in knowledge acquisition. *Journal of Educational Psychology, 4*, 601–616.

Green, T. R. G., Schiele, F., & Payne, S. J. (1988). Formalisable models of user knowledge in human-computer interaction. In G. C. van der

Veer, T. R. G. Green, J. M. Hoc, & D. Murray (Eds.), *Working with computers: theory versus outcome* (pp. 3–44). London: Academic Press.

Hix, D., & Hartson, R. (1993). *Developing user interfaces: Ensuring usability through product & process*. New York: John Wiley & Sons, Inc.

Hoc, J. M. (1989). La conduite d'un processus continu á long délais de réponse: Une activité de diagnostic. *Le Travail Humain, 52,* 299–316.

Johnson, H., & Johnson, P. (1991). Task knowledge structures: Psychological basis and integration into system design. *Acta Psychologica, 78,* 3–26.

Johnson-Laird, P. N. (1983). *Mental models*. Cambridge, UK: Cambridge University Press.

Johnson-Laird, P. N., & Byrne, R. M. (2000). Mental models website: A gentle introduction. Available: http://www.tcd.ie/Psychology/Ruth_Byrne/mental_models/index.html

Johnson-Laird, P. N., Byrne, R. M., & Schaeken, W. (1992). Propositional reasoning by model. *Psychological Review, 99,* 418–439.

Jordan, B., & Henderson, A. (1995). Interaction analysis: Foundations and practice. *The Journal of the Learning Sciences, 4*(1), 39–103.

Kieras, D., & Polson, P. G. (1985). An approach to the formal analysis of user complexity. *International Journal of Man-Machine Studies, 22*(4), 365–394.

Klok, J. A. (1998). *A needle in a haystack*. Master's thesis. Enschede, The Netherlands: University Twente.

Lear, A. C. (2000). Uncovering technology's human side. *Computer: Innovative Technology for Computer Professionals, 7,* 24.

Miller, G. A. (1956). The magic number seven plus or minus two: Some limits on our capacity for processing information. *Psychological Review, 63,* 81–97.

Mulder, B. (2000). *The role of mental models in designing computer systems*. Masters Thesis, Vrije Universiteit, Amsterdam.

Nielsen, J. (1990). A meta-model for interacting with computers. *Interacting with Computers, 2,* 147–160.

Norman, D. A. (1983). Some observations on mental models. In D. A. Gentner & A. L. Stevens (Eds.), *Mental models*. Hillsdale, NJ: Erlbaum.

Norman, D. A. (1988). *The psychology of everyday things*. New York: Harper & Row.

Norman, D. A. (1986). Cognitive engineering. In D. A. Norman & S. W. Draper (Eds.), *User-centered design: New perspectives in human computer interaction*. Hillsdale, NJ: Lawrence Erlbaum Associates.

Pask, G., & Scott, B. C. E. (1972). Learning strategies and individual competence. *International Journal of Man-Machine Studies, 4,* 217–253.

Ralston, A., & Reilly, E. D. (Eds.). (1993). The mouse. *In Encyclopedia of computer science* (3rd ed., p. 900). International Thomson Computer Press. New York: Grove's Dictionaries, Inc.

Sage, M., & Johnson, C. (1998). *Pragmatic formal design: A case study in integrating formal models into the HCI development cycle* (pp. 134–154). 5th International Eurographics Workshop on Design Specifications and Verification of Interactive Systems DSV-IS98. Abingdon, UK.

Sasse, M. A. (1991). How to t(r)ap users' mental models. In M. J. Tauber & D. Ackermann (Eds.), *Mental models and human-computer interaction 2*. New York: Elsevier.

Sasse, M. A. (1997). *Eliciting and describing users' models of computer systems*. PhD Thesis. Birmingham, UK: University of Birmingham, Faculty of Science.

Scapin, D., & Pierret-Goldbreich, C. (1989). Towards a method for task description: MAD. *Work with display units, 89,* 371–380.

Schvaneveldt, R. W. (1990). *Pathfinder associative networks: Studies in knowledge organization*. Norwood, NJ: Ablex Publishing Corporation.

Schvaneveldt, R. W., Durso, F. T., Goldsmith, T. E., Breen, T. J., Cook, N. M., Tucker, R. G., & DeMaio, J. C. (1985). Measuring the structure of expertise. *International Journal of Man-Machine Studies, 23,* 699–728.

Sommerville, I., Bentley, R., Rodden, T., & Sawyer, P. (1994). Cooperative system design. *The Computer Journal, 37*(5), 357–366.

Stork, E. (1995). *Developing a generic method for measuring and evaluating relevant aspects of mental models on perceptual domains*. Master Thesis, Vrije Universiteit, Amsterdam.

Tauber, M. (1988). On mental models and the user interface. In G. C. van der Veer, T. R. G. Green, J. M. Hoc, & D. Murray (Eds.), *Working with computers: Theory versus outcome* (pp. 89–119). London: Academic Press.

Van der Veer, G. C. (1989). Individual differences and the user interface. *Ergonomics, 32,* 1431–1449.

Van der Veer, G. C. (1990). *Human-computer interaction: Learning, individual differences, and design recommendations*. PhD Thesis. Amsterdam, The Netherlands; Vrije Universiteit.

Van der Veer, G. C., Kok, E., & Bajo, T. (1999). Conceptualising mental representations of mechanics: A method to investigate representational change. In D. Kayser & S. Vosniadou. *Modelling changes in understanding: Case studies in physical reasoning*. Elmsford, NY: Pergamon.

Van Engers, T. (2001). *Knowledge management: The role of mental models in business systems design*. PhD Thesis, Department of Computer Science, Vrije Universiteit, Amsterdam.

Van Engers, T., Vork, L., & Puerta-Melguizo, M. C. (2000). Kennisproductiviteit in groepen. *Human resource development. Thema: Stimuleren van kennisproductiviteit, 2,* 76–82.

Van Welie, M. (2001). *Task-based user interface design*. PhD Thesis, Department of Computer Science, Vrije Universiteit, Amsterdam.

Venema, D. C. (1999). *The N-UAN: A new user action notation*. Masters Thesis, Department of Computer Science, Vrije Universiteit, Amsterdam.

Vennix, J. A. M. (1995). Building consensus in strategic decision making. *Group Decision and Negotiation, 4,* 335–355.

Waern, Y. (1989). *Cognitive aspects of computer supported tasks*. New York: John Wiley & Sons.

Wickens, C. D., & Hollands, J. G. (2000). *Engineering psychology and human performance*. Englewood Cliffs, NJ: Prentice-Hall.

Young, R. M. (1983). Surrogates and mappings: Two kinds of conceptual models for interactive devices. In D. A. Gentner & A. L. Stevens (Eds.), *Mental models*. Hillsdale, NJ: Erlbaum.

• 4 •

EMOTION IN HUMAN–COMPUTER INTERACTION

Scott Brave and Clifford Nass
Stanford University

Emotion is a fundamental component of being human. Joy, hate, anger, and pride, among the plethora of other emotions, motivate action and add meaning and richness to virtually all human experience. Traditionally, human–computer interaction (HCI) has been viewed as the ultimate exception: Users must discard their emotional selves to work efficiently and rationally with computers, the quintessentially unemotional artifact. Emotion seemed at best marginally relevant to HCI and at worst oxymoronic.

Recent research in psychology and technology suggests a very different view of the relationship between humans, computers, and emotion. After a long period of dormancy and confusion, there has been an explosion of research on the psychology of emotion (Gross, 1999). Emotion is no longer seen as limited to the occasional outburst of fury when a computer crashes inexplicably, excitement when a videogame character leaps past an obstacle, or frustration at an incomprehensible error message. It is now understood that a wide range of emotions plays a critical role in every computer-related, goal-directed activity, from developing a three-dimensional computer-aided design (CAD) model and running calculations on a spreadsheet, to searching the Web and sending an e-mail, to making an online purchase and playing solitare. Indeed, many psychologists now argue that it is impossible for a person to have a thought or perform an action without engaging, at least unconsciously, his or her emotional systems (Picard, 1997b).

The literature on emotions and computers has also grown dramatically in the past few years, driven primarily by advances in technology. Inexpensive and effective technologies that enable computers to assess the physiological correlates of emotion, combined with dramatic improvements in the speed and quality of signal processing, now allow even personal computers to make judgments about the user's emotional state in real time (Picard, 1997a). Multimodal interfaces that include voices, faces, and bodies can now manifest a much wider and more nuanced range of emotions than was possible in purely textual interfaces (Cassell, Sullivan, Prevost, & Churchill, 2000). Indeed, any interface that ignores a user's emotional state or fails to manifest the appropriate emotion can dramatically impede performance and risks being perceived as cold, socially inept, untrustworthy, and incompetent.

This chapter reviews the psychology and technology of emotion, with an eye toward identifying those discoveries and concepts that are most relevant to the design and assessment of interactive systems. The goal is to provide the reader with a more critical understanding of the role and influence of emotion, as well as the basic tools needed to create emotion-conscious and consciously emotional interface designs.

The seat of emotion is the brain; hence, we begin with a description of the psychophysiological systems that lie at the core of how emotion emerges from interaction with the environment. By understanding the fundamental basis of emotional responses, we can identify those emotions that are most readily manipulable and measurable. We then distinguish emotions from moods (longer term affective states that bias users' responses to any interface) and other related constructs. The following section discusses the cognitive, behavioral, and attitudinal effects of emotion and mood, focusing on attention and memory, performance, and user assessments of the interface. Designing interfaces that elicit desired affective states requires knowledge of the causes of emotions and mood; we turn to that issue in the following section. Finally, we discuss methods for measuring affect, ranging from neurological correlates to questionnaires, and describe how these indicators can be used both to assess users and to manifest emotion in interfaces.

UNDERSTANDING EMOTION

What is emotion? Although the research literature offers a plethora of definitions (Kleinginna & Kleinginna, 1981), two generally agreed-on aspects of emotion stand out: (1) emotion is a reaction to events deemed relevant to the needs, goals, or concerns of an individual; and (2) emotion encompasses physiological, affective, behavioral, and cognitive components. Fear, for example, is a reaction to a situation that threatens (or seems to threaten, as in a frightening picture) an individual's physical well-being, resulting in a strong negative affective state, as well as physiological and cognitive preparation for action. Joy, on the other hand, is a reaction to goals being fulfilled and gives rise to a more positive, approach-oriented state.

A useful model for understanding emotion, based on a simplified view of LeDoux's (1996) work in neuropsychology, is shown in Fig. 4.1. There are three key regions of the brain in this model: the thalamus, the limbic system, and the cortex. All sensory input from the external environment is first received by the thalamus, which functions as a basic signal processor. The thalamus then sends information simultaneously both to the cortex, for higher level processing, and directly to the limbic system (LeDoux, 1995). The limbic system,[1] often called the "seat of emotion," constantly evaluates the need/goal relevance of its inputs. If relevance is determined, the limbic system sends appropriate signals both to the body, coordinating the physiological response, and also to the cortex, biasing attention and other cognitive processes.

The direct thalamic-limbic pathway is the mechanism that accounts for the more primitive emotions, such as startle-based

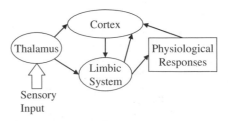

FIGURE 4.1. Neurological structure of emotion.

---

[1]The limbic system is often considered to include the hypothalamus, the hippocampus, and the amygdala. According to LeDoux, the amygdala is the only critical area (LeDoux & Phelps, 2000).

fear, as well as innate aversions and attractions. Damasio (1994) calls these the primary emotions. In an HCI context, onscreen objects and events have the potential to activate such primitive emotional responses (Reeves & Nass, 1996). For example, objects that appear or move unexpectedly (e.g., pop-up windows, sudden animations) and loud or sharp noises are likely to trigger startle-based fear. Visual stimuli that tend to be particularly arousing include images that fill a large fraction of the visual field either because the image or screen is large or because the eyes are close to the screen (Detenber & Reeves, 1996; Voelker, 1994), images that seem to approach the user (e.g., a rapidly expanding image on the screen, an image that appears to be flying out from the screen, or a character that walks toward the user), and images that move in peripheral vision (i.e., on the side of the screen; Reeves & Nass, 1996). Finally, certain images and sounds may be innately disturbing or pleasing due to their evolutionary significance (e.g., screeching or crying noises or explicit sexual or violent imagery; see, e.g., Lang, 1995; Malamuth, 1996).

Most of the emotions that we are concerned with in the design of human-computer interfaces—and the ones we will focus on in the remainder of this chapter—require more extensive cognitive (i.e., knowledge-based) processing. These secondary emotions, such as frustration, pride, and satisfaction, result from activation of the limbic system by processing in the cortex. Such cortical processing can occur at various levels of complexity, from simple object recognition (e.g., seeing the Microsoft Office Paperclip) to intricate rational deliberation (e.g., evaluating the consequences of erasing a seldom-used file), and may or may not be conscious. The cortex can even trigger emotion in reaction to internally generated stimuli (e.g., thinking about how difficult it will be to configure a newly purchased application).

Finally, an emotion can result from a combination of both the thalamic-limbic and the cortico-limbic mechanisms. For example, an event causing an initial startle/fear reaction can be later recognized as harmless by more extensive, rational evaluation (e.g., when you realize that the flash of your screen suddenly going blank is just the initiation of the screen saver). In other situations, higher level processing can reinforce an initial evaluation. Whatever the activation mechanism—thalamic or cortical, conscious or nonconscious—the cortex receives input from an activated limbic system, as well as feedback from the body, both contributing to the conscious experience of emotion.

The previous discussion provides a useful framework for considering one of the classic debates in emotion theory: Are emotions innate or learned? At one extreme, evolutionary theorists argue that all emotions (including complex emotions such as regret and relief) are innate, each evolved to address a specific environmental concern of our ancestors (Darwin, 1872/1998; Neese, 1990; Tooby & Cosmides, 1990; see also Ekman, 1994; Izard, 1992). These theories are consistent with a hypothesis of high differentiation within the limbic system, corresponding to each of the biologically determined emotions. From this perspective, it is also reasonable to speculate that each emotion is associated with a unique set of physiological and cognition-biasing responses.

At the other extreme, many emotion theorists argue that, with the exception of startle and innate affinity/disgust (which they would consider pre-emotional), emotions are almost entirely learned social constructs (Averill, 1980; Ortony & Turner, 1990; Shweder, 1994; Wierzbicka, 1992). Such theories emphasize the role of higher cortical processes in differentiating emotions and concede minimal, if any, specificity within the limbic system (and, consequently, within physiological responses). For example, the limbic system may operate in simply an on/off manner, or at most be differentiated along the dimensions of valence (positive/negative or approach/avoidance) and arousal (low/high) (Barrett & Russell, 1999; Lang, 1995). From this perspective, emotions are likely to vary considerably across cultures, with any consistency being based in common social structure, not biology.

Between these two extremes lie those who believe that there are basic emotions. Citing both cross-cultural universals and primate studies, these theorists contend that there is a small set of innate, basic emotions shared by all humans (Ekman, 1992; Oatley & Johnson-Laird, 1987; Panksepp, 1992). Which emotions qualify as basic is yet another debate, but the list typically includes fear, anger, sadness, joy, disgust, and sometimes also interest and surprise. Other emotions are seen either as combinations of these basic emotions or as socially learned differentiations within the basic categories (e.g., agony, grief, guilt, and loneliness are various constructions of sadness, Bowci, 1992). In this view, the limbic system is prewired to recognize the basic categories of emotion, but social learning and higher cortical processes still play a significant role in differentiation.

If the basic emotions view is correct, a number of implications for interaction design and evaluation emerge. First, the basic categories would likely be the most distinguishable, and therefore measurable, emotional states (both in emotion recognition systems as well as in postinteraction evaluations). Furthermore, the basic emotions would be less likely to vary significantly from culture to culture, facilitating the accurate translation and generalizability of questionnaires intended to assess such emotions. Lower variability also enables more reliable prediction of emotional reactions to interface content, both across cultures and across individuals. Finally, for users interacting with onscreen characters, depictions of the basic emotions would presumably be most immediately recognizable. If the social construction view of emotions is valid, then emotion measurement and assessment, prediction, and depictions are more challenging and nuanced.

DISTINGUISHING EMOTION FROM RELATED CONSTRUCTS

Mood

It is useful to distinguish among several terms often used ambiguously: emotion, mood, and sentiment. Emotion can be distinguished from mood by its object-directedness. As Frijda (1994) explains, emotions are *intentional*: They "imply and involve relationships with a particular object." We get scared *of* something, angry *at* someone, and excited *about* some event. Moods, on the other hand, although they may be indirectly caused by a particular object, are nonintentional; they are not directed at any object in particular and are thus experienced as more diffuse, global, and general. A person can be sad about something (an emotion) or generally depressed (a mood). Unfortunately, the English language often allows the same term to describe both emotion and mood (e.g., happy).

Another distinction between emotion and mood emerges from a functional perspective. As a reaction to a particular situation, emotions bias action—they prepare the body and the mind for an appropriate, immediate response. As such, emotions also tend to be relatively short-lived. Moods, in contrast, tend to bias cognitive strategies and processing over a longer term (Davidson, 1994). More generally, moods can be seen to serve as a background affective filter through which both internal and external events are appraised. A person in a good mood tends to view everything in a positive light, whereas a person in a bad mood does the opposite. The interaction between emotions and moods is also important. Moods tend to bias which emotions are experienced, lowering the activation thresholds for mood-related emotions. Emotions, on the other hand, often cause or contribute to moods.

When assessing user response to an interface, it is important to consider the biasing effects of user mood. Users entering a usability or experimental study in a good mood, for instance, are more likely to experience positive emotion during an interaction than users in a bad mood. Pretesting for mood and including it as a variable in analysis can, therefore, reduce noise and increase interpretive power. If pretesting users immediately before an interaction is inappropriate, there is a second noise-reducing option: assessment of temperament. Temperament reflects the tendency of certain individuals to exhibit particular moods with great frequency. Participants can be pretested for temperament at any point prior to the study, enabling the exclusion of extreme cases of depressive or excitable individuals (e.g., Bishop, Jacks, & Tandy, 1993). Finally, if user testing involves multiple stimuli, the order of presentation can also influence the results. For example, earlier stimuli may establish a mood that biases emotional reactions to subsequent stimuli. To combat this problem, the order of stimuli should be varied from participant to participant, when feasible.

Sentiment

Sentiment is also often confused with emotion. Unlike emotions (and moods), sentiments are not states of an individual, but assigned properties of an object. When a person says that they like an interface or find an interface to be frustrating, what they really mean is that that they associate the interface with a positive or frustrating emotional state; they expect interaction with the interface to lead to positive or frustrating emotions. The basis for this judgment often comes from direct experience and subsequent generalization, but may also arise from social learning (Frijda, 1994).

One reason for the confusion between emotions and sentiment is that many languages use the same words for both. For example, the word "like" can be used both to indicate prediction or opinion (sentiment), as well as a current emotional state (e.g., "I like receiving e-mail" vs. "I like the e-mail that just arrived").

Clore (1994, p. 108) offers an interesting explanation for this ambiguity, theorizing that sentiments are judged by bringing the object to mind and observing the affective reaction. But, whereas emotions and moods are fleeting—emotions lasting only seconds and moods lasting for hours or even days—sentiments can persist indefinitely and are thus responsible for guiding our propensities to seek out or avoid particular objects and situations. In this sense, sentiments are of critical importance for HCI because they motivate users to return to particular software products or web sites.

Although direct interaction with an object is the most accurate way for a user to create a sentiment—consider the colloquial phrase, "How do you know you don't like it unless you try it?"—sentiments can also be caused by assumptions based on the communicated properties of an object. A person may, for example, base a sentiment on someone else's description of their interaction with the object, or even immediately adopt the sentiment of someone they know or respect (e.g., consider the presumed influence of celebrities in software advertisements).

As a predictive construct, sentiments are often generalizations about a class of objects with a given recognizable property, i.e., stereotypes. Although some of these generalizations may be logical and accurate, others may not—in fact, they may not even be conscious. Negative experiences with a particular computer character, for example, may lead users to conclude that they dislike all character-based interfaces. However, using a character that people know and like already—Mickey Mouse, for example—may be able to leverage sentiment to an interface's advantage. Similarly, many people have well-established sentiments regarding certain types of applications, e.g., "I hate spreadsheet applications." For such users, interfaces that avoid triggering their negative stereotypes have the advantage. Positive stereotypes, on the other hand, should be encouraged whenever possible, such as when learning applications are framed as entertainment.

EFFECTS OF AFFECT

Attention

One of the most important effects of emotion lies in its ability to capture attention. Emotions have a way of being completely absorbing. Functionally, they direct and focus our attention on those objects and situations that have been appraised as important to our needs and goals so that we can deal with them appropriately. Emotion-relevant thoughts then tend to dominate conscious processing—the more important the situation, the higher the arousal, and the more forceful the focus (Clore & Gasper, 2000). In an HCI context, this attention-getting function can be used advantageously, as when a sudden beep is used to alert the user, or can be distracting, as when a struggling user is frustrated and can only think about his/her inability.

Emotion can further influence attention through a secondary process of emotion regulation (Gross, 1998). Once an emotion is triggered, higher cognitive processes may determine that the emotion is undesirable. In such cases, attention is often directed away from the emotion-eliciting stimulus for the purpose of distraction. For example, becoming angry with an onscreen agent may be seen as ineffectual (e.g., because it doesnot recognize your anger) or simply unreasonable. An angered user may then actively try to ignore the agent, focusing instead on other on- or offscreen stimuli, or even take the next step and completely remove the agent from the interaction (which could mean leaving an application or web site entirely). Positive emotions may likewise require regulation at times, such as when amusing stimuli lead to inappropriate laughter in a work environment. If the emotionally relevant stimulus is too arousing, however, regulation through selective attention is bound to fail (Wegner, 1994), because users will be unable to ignore the stimulus.

Mood can have a less profound but more enduring effect on attention. At the most basic level, people tend to pay more attention to thoughts and stimuli that have some relevance to their current mood state (Bower & Forgas, 2000). However, people also often consciously regulate mood, selecting and attending to stimuli that sustain desired moods or, alternatively, counteract undesired moods. An interface capable of detecting—or at least predicting—a user's emotional or mood state could similarly assume an affect regulation role, helping to guide attention away from negative and toward more positive stimuli. For example, a frustrated user could be encouraged to work on a different task, focus on a different aspect of the problem at hand, or simply take a break (perhaps by visiting a suggested online entertainment site).

Memory

Emotion's effect on attention also has implications for memory. Because emotion focuses thought on the evoking stimulus, emotional stimuli are generally remembered better than unemotional events (Thorson & Friestad, 1985). Negative events, which tend to be highly arousing, are typically remembered better than positive events (Newhagen & Reeves, 1991, 1992; Reeves & Nass, 1996, chap. 10; Reeves, Newhagen, Maibach, Basil, & Kurz, 1991). In addition, emotionality improves memory for central details while undermining memory for background details (see Heuer & Reisberg, 1992; Parrott & Spackman, 2000).

Mood also comes into play both in memory encoding and retrieval. Research has shown that people will remember "mood-congruent" emotional stimuli better than incongruent stimuli. Bower, Gilligan, and Monteiro (1981), for example, hypnotized subjects into either a happy or sad mood before having them read stories about various characters. The next day, subjects were found to remember more facts about characters whose mood had agreed with their own than about other characters. Similarly, on the retrieval end, people tend better to recall memories consistent with their current mood (Ellis & Moore, 1999). However, the reverse effect has also been shown to occur in certain situations: People will sometimes better recall mood-incongruent memories (e.g., happy memories while in a sad mood). Parrott and Spackman (2000) hypothesize that mood

regulation is responsible for this inverse effect: When a given mood is seen as inappropriate or distracting, people will often actively try to evoke memories or thoughts to modify that mood (see Forgas' (1995) Affect Infusion Model for insight into these contradictory findings (also see Erber & Erber, 2001). Finally, there is some evidence for mood-dependent recall: Memories encoded while in a particular mood are better recalled when in that same mood. This effect is independent of the emotional content of the memory itself (Ucros, 1989). It should be noted, however, that the effects of mood on memory are often unreliable and therefore remain controversial.

Performance

Mood has also been found to affect cognitive style and performance. The most striking finding is that even mildly positive affective states profoundly affect the flexibility and efficiency of thinking and problem solving (Hirt, Melton, McDonald, & Harackiewicz, 1996; Isen, 2000; Murray, Sujan, Hirt, & Sujan, 1990). In one of the best-known experiments, subjects were induced into a good or bad mood and then asked to solve Duncker's (1945) candle task. Given only a box of thumbtacks, the goal of this problem is to attach a lighted candle to the wall, such that no wax drips on the floor. The solution requires the creative insight to thumbtack the box itself to the wall and then tack the candle to the box. Subjects who were first put into a good mood were significantly more successful at solving this problem (Isen, Daubman, & Nowicki, 1987). In another study, medical students were asked to diagnose patients based on X-rays after first being put into a positive, negative, or neutral mood. Subjects in the positive-affect condition reached the correct conclusion faster than did subjects in other conditions (Isen, Rosenzweig, & Young, 1991). Positive affect has also been shown to increase heuristic processing, such as reliance on scripts and stereotypes. Although some have argued that such reliance is at the expense of systematic processing (Schwartz & Bless, 1991), more recent evidence suggest that heuristic processing and systematic processing are not mutually exclusive (Isen, 2000). Keeping a user happy may, therefore, not only affect satisfaction, but may also lead to efficiency and creativity.

Assessment

Mood has also been shown to influence judgment and decision making. As described earlier, mood tends to bias thoughts in a mood-consistent direction, while also lowering the thresholds of mood-consistent emotions. One important consequence of this is that stimuli—even those unrelated to the current affective state—are judged through the filter of mood (Clore et al., 2001; Erber & Erber, 2001; Niedenthal, Setterlund, & Jones, 1994). This suggests that a user in a good mood will likely judge both the interface and their work more positively, regardless of any direct emotional effects. It also suggests that a happy user at an e-commerce site would be more likely to evaluate the products or services positively.

Positive mood also decreases risk-taking, likely in an effort to preserve the positive mood. That is, although people in a positive mood are more risk-prone when making hypothetical decisions, when presented with an actual risk situation, they tend to be more cautious (Isen, 2000). In an e-commerce purchasing situation, then, one can predict that a low-risk purchase is more likely during a good mood, due to a biased judgment in favor of the product, whereas a high-risk purchase may be more likely in a less cautious, neutral, or negative mood (consistent with the adage that desperate people resort to desperate measures).

A mood's effect on judgment, combined with its effect on memory, can also influence the formation of sentiments. Sentiments are not necessarily determined during interaction with an object; they often are grounded in reflection. This is important to consider when conducting user tests, as the mood set by the interaction immediately prior to a questionnaire may bias like/dislike assessments of earlier interactions. Thus, varying order of presentation ensures both that later stimuli do not influence the *assessment* of earlier stimuli and that earlier stimuli do not influence the *experience* of later stimuli (as discussed earlier).

CAUSES OF EMOTION

What causes emotions? The answer to this question is critical for HCI because an understanding of emotions' antecedents will better enable us to design interfaces that encourage desired emotional states and understand interfaces that do not.

Needs and Goals

As we saw in the first section, emotions are reactions to situations deemed relevant to the needs and goals of the individual. Clearly, a user comes to a computer hoping to achieve certain application-specific goals—composing a document, sending an e-mail, finding a piece of information, etc. The degree to which an interface facilitates or hampers those goals has a direct effect on the emotional state of the user. An interface capable of detecting emotion could, therefore, use such information as feedback regarding whether the user's goals are being met, modifying its behavior as necessary. In an information-seeking context, for example, emotional reactions to displayed content could be used to improve the goal relevance of future retrievals. Similarly, if an interface detects frustration, desperation, or anger in a user, goals may be facilitated by trying a new approach or offering assistance (Klein, Moon, & Picard, 1999; Picard, 1997a). (If the particular goals implicated by an emotion are not clear, there can be advantages to an interface that empathizes with the user; Klein et al., 1999). More generally, user preferences can be automatically determined based on a user's emotional reactions to interface elements (Picard, 1997a).

There are also a host of more abstract needs underlying, and often adjacent to, application-specific goals. A user may have a strong need to feel capable and competent, maintain control, learn, or be entertained. A new user typically needs to feel comfortable and supported, whereas an expert is more focused on aesthetic concerns of efficiency and elegance. Acknowledging these more abstract goals in interface design can be as instrumental in determining a user's affective state as meeting or obstructing application-specific goals. Maslow's hierarchy (Maslow, 1968) presents a useful starting place for considering the structure of these more abstract user needs. In his later work, Maslow (1968) grouped an individual's basic needs into eight categories:

Physiological: hunger, thirst, bodily comforts, etc.
Safety/security: being out of danger
Social: affiliate with others, be accepted
Esteem: to achieve, be competent, gain approval and recognition
Cognitive: to know, to understand, and explore
Aesthetic: symmetry, order, and beauty
Self-actualization: to find self-fulfillment and realize one's potential
Transcendence: to help others find self-fulfillment and realize their potential.

When a particular situation or event is deemed as promoting these needs, positive emotion results. When someone or something hampers these needs, negative emotion results. The specific emotion experienced is due in part to the category of need implicated by the event. Fright, for example, is typically associated with threatened safety/security needs; love and embarrassment with social needs; pride with esteem needs; and curiosity with cognitive needs.

Within Maslow's (1968) framework, application-specific goals of a user can be seen as instruments ultimately serving these more basic needs. For example, a user who successfully enhances a digital family photograph may simultaneously be contributing to the fulfillment of social, esteem, cognitive, and aesthetic needs. However, interfaces can also directly address a user's basic needs. For example, a spell-checker interface that praises a user on his or her spelling ability, regardless of their actual performance, is a somewhat humorous, although illustrative, approach to acknowledging a user's esteem needs. Such interfaces, by enhancing the user's affective state, have been shown also to be viewed as more intelligent and likable (Reeves & Nass, 1996, chap. 4). As another example, an interface that takes care to establish a trusting and safe relationship with users may ultimately lead to more effective and cooperative interactions (Fogg, 1998). Educational software should address users' emotional needs, not only teaching the relevant content, but also ensuring users believe that they are learning. Optimized learning further requires a careful balance of esteem and self-actualization needs, offering appropriate levels of encouragement and challenge, as well as praise and criticism. Finally, one of the key arguments for social interfaces is that they meet the social needs of users (Reeves & Nass, 1996).

Although the type of need relevant in a situation offers some insight into emotional reaction, need category alone is not sufficient to differentiate fully among all emotions. Distinguishing frustration and anger, for example, cannot be achieved based solely on knowing the users' need; it also requires some notion of agency.

Appraisal Theories

"Appraisal" theories provide much greater predictive power than category or hierarchy-based schemes by specifying the critical properties of antecedent events that lead to particular emotions (Lazarus, 1991; Ortony, Clore, & Collins, 1988; Roseman, Antoniou, & Jose, 1996; Scherer, 1988). Ellsworth (1994), for example, describes a set of abstract elicitors of emotion. In addition to *novelty* and *valence*, Ellsworth contends that the level of *certainty/uncertainty* in an event has a significant impact on the emotion experienced. For instance, "uncertainty about probably positive events leads to interest and curiosity, or to hope," whereas "uncertainty about probably negative events leads to anxiety and fear" (Ellsworth, 1994, p. 152). Certainty, on the other hand, can lead to relief in the positive case and despair in the negative case.

Because slow, unclear, or unusual responses from an interface generally reflect a problem, one of the most common interface design mistakes—from an affective standpoint—is to leave the user in a state of uncertainty. Users tend to fear the worst when, for example, an application is at a standstill, the hourglass remains up longer than usual, or the hard drive simply starts grinding away unexpectedly. Such uncertainty leads to a state of anxiety that can be easily avoided with a well-placed, informative message or state indicator. Providing users with immediate feedback on their actions reduces uncertainty, promoting a more positive affective state (see Norman, 1990, on visibility and feedback). When an error has actually occurred, the best approach is to make the user aware of the problem and its possible consequences, but frame the uncertainty in as positive a light as possible (e.g., "this application has experienced a problem, but the document should be recoverable").

According to Ellsworth (1994), *obstacles and control* also play an important role in eliciting emotion. High control can lead to a sense of challenge in positive situations, but stress in negative situations. Lack of control, on the other hand, often results in frustration, which if sustained can lead to desperation and resignation. In an HCI context, providing an appropriate level of controllability, given a user's abilities and the task at hand, is thus critical for avoiding negative affective consequences. Control need not only be perceived to exist (Skinner, 1995; Wegner & Bargh, 1998), but must be understandable and visible, otherwise the interface itself is an obstacle (Norman, 1990).

Agency is yet another crucial factor determining emotional response (Ellsworth, 1994; Friedman & Kahn, 1997). When oneself is the cause of the situation, shame (negative) and pride (positive) are likely emotions. When another person or entity is the cause, anger (negative) and love (positive) are more likely. But, if fate is the agent, one is more likely to experience sorrow (negative) and joy (positive). An interface often has the opportunity to direct a user's perception of agency. In any anomalous situation, for example—be it an error in reading a file, inability to recognize speech input, or simply a crash—if the user is put in a position encouraging blame of oneself or fate, the negative emotional repercussions may be more difficult to diffuse than if the computer explicitly assumes blame (and is apologetic). For example, a voice interface encountering a recognition error can say, "This system failed to understand your command" (blaming itself), "The command was not understood" (blaming no one), or "You did not speak clearly enough for your command to be understood" (blaming the user).

Appraisal theories of emotion, such as Ellsworth's (1994), are useful not only in understanding the potential affective impacts of design decisions, but also in creating computer agents that exhibit emotion. Although in some cases scripted emotional responses are sufficient, in more dynamic or interactive contexts, an agent's affective state must be simulated to be believable. Ortony, Clore, and Collins' (1988) cognitive theory of emotion is currently the most commonly applied appraisal theory for such purposes (Bates, Loyall, & Reilly, 1994; Elliott & Brzezinski, 1998; for alternate approaches, see Ball & Breese, 2000; Bozinovski & Bozinovska, 2001; Scheutz, Sloman, & Logan, 2000). Appraisal theories can also be used to help model and predict a user's emotional state in real-time (Elliott & Brezezinski, 1998).

Contagion

Another cause of emotion that does not fit cleanly into the structure described previously is contagion (Hatfield, Cacioppo, & Rapson, 1994). People often catch other's emotions. Sometimes, this social phenomenon seems logical, such as when a person becomes afraid upon seeing another experience fear. At other times contagion seems illogical, such as when another person's laughter induces immediate, unexplainable amusement. Anticipatory excitement is another emotion that transfers readily from person to person.

Emotions *in* interfaces can also be contagious. For example, a character that exhibits excitement when an online product appears can make users feel more excited. Similarly, an attempt at light humor in a textual interface, even if unsuccessful, may increase positive affect (Morkes, Kernal, & Nass, 2000).

Moods and Sentiments

Mood and sentiment can also bias emotion. One of the fundamental properties of mood is that it lowers the activation threshold for mood-consistent emotions. Sentiment can act in a similar way. For example, interaction with an object, to which a sentiment is already attached, can evoke emotion either in memory of past interaction or in anticipation of the current interaction. Thus, an interface that proved frustrating in the past may elicit frustration before the user even begins working. In addition, sentiment can bias perception of an object, increasing the probability of eliciting sentiment-consistent emotions. For example, an application that users *like* can do no wrong, whereas one that users *dislike* does everything to anger them, regardless of the application's actual behavior. Of critical importance here is that sentiments need not be derived from direct experience; they may also be inferred from stereotypes or other generalizations.

Previous Emotional State

Finally, a user's previous emotional state can affect the experience of subsequent emotions. This occurs not only through the mechanism of mood—emotions can cause moods and moods then bias the activation thresholds of emotions—but also through the mechanisms of excitation transfer and habituation. Excitation transfer (Zillmann, 1991) is based on the fact that after an emotion-causing stimulus has come and gone, an activated autonomic nervous system takes some time to return to its deactivated state. If another emotion is triggered before that decay is complete, the residual activation (excitement) will be added to the current activation and be perceived as part of the current emotion. As Zillmann (1991) explains, "residues of excitation from a previous affective reaction will *combine* with excitation produced by subsequent affective stimulation and thereby cause an *overly intense* affective reaction to subsequent stimuli. . . . Residual arousal from anger, then, may intensify fear; residues from fear may intensify sexual behaviors; residual sexual arousal may intensify aggressive responses; and so forth" (p. 116). Thus, people who have just hit the purchase button associated with their Web shopping cart can become particularly angry when they are presented with multiple pages before they can complete their transaction: The arousal of buying increases the intensity of their frustration with the postpurchase process. Similarly, Reeves and Nass (1996) have argued that pictorial characters raise the volume knob on both positive and negative feelings about an interaction, because explicitly social interactions are more arousing than their nonsocial counterparts.

Habituation is, in some sense, the converse of excitation transfer. It posits that the intensity of an emotion decreases over time if the emotion is experienced repeatedly. One explanation for this effect relates back to appraisal theory: "Emotions are elicited not so much by the presence of favorable or unfavorable conditions, but by actual or expected changes in favorable or unfavorable conditions" (Frijda, 1988, p. 39). Repeated pleasurable affective states, therefore, become expected and thus gradually lose intensity. The same is true for negative affect; however, particularly extreme negative emotional states may never habituate (Frijda, 1988). This may be why negative experiences with frequently used interfaces (e.g., operating systems) are remembered more vividly than positive experiences.

CAUSES OF MOOD

Mood has a number of potential causes. The most obvious is emotion itself. Intense or repetitive emotional experiences tend to prolong themselves into moods. A user who is continually frustrated will likely be put in a frustrated mood, whereas a user who is repeatedly made happy will likely be put in a positive mood. Mood can also be influenced, however, by anticipated emotion, based on sentiment. For example, if users know that they must interact with an application that they dislike (i.e., they associate with negative emotion), they may be in a bad mood from the start.

Contagion

Similar to emotion, moods also exhibit a contagion effect (Neumann & Strack, 2000). For example, a depressed person will often make others feel depressed, and a happy person will often make others feel happy. Murphy and Zajonc (1993) have shown that even a mere smiling or frowning face, shown so quickly that the subject is not conscious of seeing the image, can affect a person's mood and subsequently bias judgment. From an interface standpoint, the implications for character-based agents are clear: Moods exhibited by onscreen characters may directly transfer to the user's mood. On-screen mood can also lead to perceived contagion effects: One smiling or frowning face on the screen can influence users' perceptions of other faces that they subsequently see on the screen, perhaps as a result of priming (Reeves, Biocca, Pan, Oshagan, & Richards, 1989; Reeves & Nass, 1996, chap. 22).

Color

Color can clearly be designed into an interface with its mood-influencing properties in mind. Warm colors, for example, generally provoke "active feelings," whereas cool colors are "much less likely to cause extreme reactions" (Levy, 1984). Gerard (1957, 1958), for example, found that red light projected onto a diffusing screen produces increased arousal in subjects, using a number of physiological measures (including cortical activation, blood pressure, and respiration), whereas blue light has essentially the opposite calming effect (see Walters, Apter, & Svebak, 1982). Subjective ratings of the correlations between specific colors and moods can be more complicated. As Gardano (1986) summarizes, "yellow (a warm color) has been found to be associated with both sadness (Peretti, 1974) and with cheerfulness (Wexner, 1954). Similarly, red (another warm color) is related to anger and violence (Schachtel, 1943) as well as to passionate love (Henry & Jacobs, 1978; Pecjak, 1970); and blue (a cool color), to tenderness (Schachtel, 1943) and sadness (Peretti, 1974). . . ." Nevertheless, as any artist will attest, carefully designed color schemes (combined with other design elements) can produce reliable and specific influences on mood.

Other Effects

A number of other factors can affect mood. For example, in music, minor scales are typically associated with negative emotion and mood, whereas major scales have more positive/happy connotations (Gregory, Worrall, & Sarge, 1996). Other possible influences on mood include weather, temperature, hormonal cycles, genetic temperament, sleep, food, medication, and lighting (Thayer, 1989).

MEASURING AFFECT

Measuring user affect can be valuable both as a component of usability testing and as an interface technique. When evaluating interfaces, affective information provides insight into what a user is feeling—the fundamental basis of liking and other sentiments. Within an interface, knowledge of a user's affect provides useful feedback regarding the degree to which a user's goals are being met, enabling dynamic and intelligent adaptation. In particular, social interfaces (including character-based interfaces) must have the ability to recognize and respond to emotion in users to execute effectively real-world interpersonal interaction strategies (Picard, 1997a).

Neurological Responses

The brain is the most fundamental source of emotion. The most common way to measure neurological changes is the electroencephalogram (EEG). In a relaxed state, the human brain exhibits an alpha rhythm, which can be detected by EEG recordings taken through sensors attached to the scalp. Disruption of this signal (alpha blocking) occurs in response to novelty, complexity, and unexpectedness, as well as during emotional excitement and anxiety (Frijda, 1986). EEG studies have further shown that positive/approach-related emotions lead to greater activation of the left anterior region of the brain, whereas negative/avoidance-related emotions lead to greater activation of the right anterior region (Davidson, 1992; see also Heller, 1990). Indeed, when one flashes a picture to either the left or the right of where a person is looking, the viewer can identify a smiling face more quickly when it is flashed to the left hemisphere, and a frowning face more quickly when it is flashed to the right hemisphere (Reuter-Lorenz & Davidson, 1981). Current EEG devices, however, are fairly clumsy and obstructive, rendering them impractical for most HCI applications. Recent advances in magnetoresonance imaging offer great promise for emotion monitoring, but are currently unrealistic for HCI because of their expense, complexity, and form factor.

Autonomic Activity

Autonomic activity has received considerable attention in studies of emotion, in part due to the relative ease in measuring certain components of the autonomic nervous system, including heart rate, blood pressure, blood pulse volume, respiration, temperature, pupil dilation, skin conductivity, and more recently, muscle tension (as measured by electromyography). However, the extent to which emotions can be distinguished on the basis of autonomic activity alone remains a hotly debated issue (see Ekman & Davidson, 1994, chap. 6; Levenson, 1988). On the one end are those, following in the Jamesian tradition (James, 1884), who believe that each emotion has a unique autonomic signature—technology is simply not advanced enough yet to fully detect these differentiators. On the other extreme, there are those, following Cannon (1927), who contend that all emotions are accompanied by the same state of nonspecific autonomic (sympathetic) arousal, which varies only in magnitude—most commonly measured by galvanic skin response, a measure of skin conductivity (Schachter & Singer, 1962). This controversy has clear connections to the nature-nurture debate in emotion, described earlier, because autonomic specificity seems more probable if each emotion has a distinct biological basis, whereas nonspecific autonomic

(sympathetic) arousal seems more likely if differentiation among emotions is based mostly on cognition and social learning.

Although the debate is far from resolved, certain measures have proven fairly reliable at distinguishing among "basic emotions." Heart rate, for example, increases most during fear, followed by anger, sadness, happiness, surprise, and finally disgust, which shows almost no change in heart rate (Cacioppo, Bernston, Klein, & Poehlmann, 1997; Ekman, Levenson, & Friesen, 1983; Levenson, Ekman, & Friesen, 1990). Heart rate also generally increases during excitement, mental concentration, and "upon the presentation of intense sensory stimuli" (Frijda, 1986). Decreases in heart rate typically accompany relaxation, attentive visual and audio observation, and the processing of pleasant stimuli (Frijda, 1986). As is now common knowledge, blood pressure increases during stress and decreases during relaxation. Cacioppo et al. (2000) further observe that anger increases diastolic blood pressure to the greatest degree, followed by fear, sadness, and happiness. Anger is further distinguished from fear by larger increases in blood pulse volume, more nonspecific skin conductance responses, smaller increases in cardiac output, and other measures, indicating that "anger appears to act more on the vasculature and less on the heart than does fear" (Cacioppo et al., 1997). Results using other autonomic measures are less reliable.

Combined measures of multiple autonomic signals show promise as components of an emotion recognition system. Picard, Vyzas, and Healey (2001), for example, achieved 81% recognition accuracy on eight emotions through combined measures of respiration, blood pressure volume, and skin conductance, as well as facial muscle tension (to be discussed in the next subsection). Many autonomic signals can also be measured in reasonably nonobstructive ways (e.g., through user contact with mice and keyboards; Picard, 1997a).

However, even assuming that we could distinguish among all emotions through autonomic measures, it is not clear that we should. In real-world social interactions, humans have at least partial control over what others can observe of their emotions. If another person, or a computer, is given direct access to users' internal states, they may feel overly vulnerable, leading to stress and distraction. Such personal access could also be seen as invasive, compromising trust. It may, therefore, be more appropriate to rely on measurement of the external signals of emotion (discussed below).

Facial Expression

Facial expression provides a fundamental means by which humans detect emotion. Table 4.1 describes characteristic facial features of six basic emotions (Ekman & Friesen, 1975; Rosenfeld, 1997). Endowing computers with the ability to recognize facial expressions, through pattern recognition of captured images, has proven to be a fertile area of research (Essa & Pentland, 1997; Lyons, Akamatsu, Kamachi, & Gyoba, 1998; Martinez, 2000; Yacoob & Davis, 1996); for recent reviews, see Cowie et al., 2001; Lisetti & Schiano, 2000; Tian, Kanade, & Cohn, 2001). Ekman and Friesen's (1977) Facial Action Coding System, which identifies a highly specific set of muscular

TABLE 4.1. Facial Cues and Emotion

Emotion	Observed Facial Cues
Surprise	Brows raised (curved and high)
	Skin below brow stretched
	Horizontal wrinkles across forehead
	Eyelids opened and more of the white of the eye is visible
	Jaw drops open without tension or stretching of the mouth
Fear	Brows raised and drawn together
	Forehead wrinkles drawn to the center
	Upper eyelid is raised and lower eyelid is drawn up
	Mouth is open
	Lips are slightly tense or stretched and drawn back
Disgust	Upper lip is raised
	Lower lip is raised and pushed up to upper lip or it is lowered
	Nose is wrinkled
	Cheeks are raised
	Lines below the lower lid, lid is pushed up but not tense
	Brows are lowered
Anger	Brows lowered and drawn together
	Vertical lines appear between brows
	Lower lid is tensed and may or may not be raised
	Upper lid is tense and may or may not be lowered due to brows' action
	Eyes have a hard stare and may have a bulging appearance
	Lips are either pressed firmly together with corners straight or down or open, tensed in a squarish shape
	Nostrils may be dilated (could occur in sadness too), unambiguous only if registered in all three facial areas
Happiness	Corners of lips are drawn back and up
	Mouth may or may not be parted with teeth exposed or not
	A wrinkle runs down from the nose to the outer edge beyond lip corners
	Cheeks are raised
	Lower eyelid shows wrinkles below it and may be raised but not tense
	Crow's-feet wrinkles go outward from the outer corners of the eyes
Sadness	Inner corners of eyebrows are drawn up
	Skin below the eyebrow is triangulated, with inner corner up
	Upper lid inner corner is raised
	Corners of the lips are drawn or lip is trembling

Based on Ekman & Friesen (1975). Adapted with permission.

movements for each emotion, is one of the most widely accepted foundations for facial recognition systems (Tian et al., 2001). In many systems, recognition accuracy can reach as high as 90–98% on a small set of basic emotions. However, current recognition systems are tested almost exclusively on produced expressions (i.e., subjects are asked to make specific facial movements or emotional expressions), rather than natural expressions resulting from actual emotions. The degree of accuracy that can be achieved on more natural expressions of emotion remains unclear. Further, "not all... emotions are accompanied by visually perceptible facial action" (Cacioppo et al., 1997).

An alternate method for facial expression recognition, capable of picking up both visible and extremely subtle movements of facial muscles, is facial elecromyography. Electromyography signals, recorded through small electrodes attached to the skin, have proven most successful at detecting positive vs. negative emotions, and show promise in distinguishing among basic emotions (Cacioppo, Bernston, Larsen, Poehlmann, & Ito, 2000). Although the universality (and biological basis) of facial expression is also debated, common experience tells us that, at least within a culture, facial expressions are reasonably consistent. Nonetheless, individual differences may also be important, requiring recognition systems to adapt to a specific user for greatest accuracy. Gestures can also be recognized with technologies similar to those for facial expression recognition, but the connection between gesture and emotional state is less distinct, in part due to the greater influence of personality (Cassell & Thorisson, 1999; Collier, 1985).

Voice

Voice presents yet another opportunity for emotion recognition (see Cowie et al., 2001, for an extensive review). Emotional arousal is the most readily discernible aspect of vocal communication, but voice can also provide indications of valence and specific emotions through acoustic properties such as pitch range, rhythm, and amplitude or duration changes (Ball & Breese, 2000; Scherer, 1989). A bored or sad user, for example, will typically exhibit slower, lower-pitched speech, with little high-frequency energy, whereas a user experiencing fear, anger, or joy will speak faster and louder, with strong high-frequency energy and more explicit enunciation (Picard, 1997a). Murray and Arnott (1993) provide a detailed account of the vocal effects associated with several basic emotions (see Table 4.2). Although few systems have been built for automatic emotion recognition through speech, Banse and Scherer (1996) have demonstrated the feasibility of such systems. Cowie and Douglas-Cowie's ACCESS system (Cowie & Douglas-Cowie, 1996) also presents promise (Cowie et al., 2001).

Self-Report Measures

A final method for measuring a user's affective state is to ask. Postinteraction questionnaires, in fact, currently serve as the primary method for ascertaining emotion, mood, and sentiment during an interaction. However, in addition to the standard complexities associated with self-report measures (such as the range of social desirability effects), measuring affect in this way presents added challenges. To begin with, questionnaires are capable of measuring only the conscious experience of emotion and mood. Much of affective processing, however, resides in

TABLE 4.2. Voice and Emotion

	Fear	Anger	Sadness	Happiness	Disgust
Speech rate	Much faster	Slightly faster	Slightly lower	Faster or slower	Very much slower
Pitch average	Very much higher	Very much higher	Slightly lower	Much higher	Very much lower
Pitch range	Much wider	Much wider	Slightly narrower	Much wider	Slightly wider
Intensity	Normal	Higher	Lower	Higher	Lower
Voice quality	Irregular voicing	Breathy chest tone	Resonant	Breathy blaring	Grumbled chest tone
Pitch changes	Normal	Abrupt on stressed syllables	Downward inflections	Smooth upward inflections	Wide downward terminal inflections
Articulation	Precise	Tense	Slurring	Normal	Normal

Based on Murray & Arnott (1993). Adapted with permission.

the limbic system and in nonconscious processes. Although it is debatable whether an emotion can exist without any conscious component at all, a mood surely can. Furthermore, questions about emotion, and often those about mood, refer to past affective states and thus rely on imperfect and potentially biased memory. Alternatively, asking a user to report on an emotion as it occurs requires interruption of the experience. In addition, emotions and moods are often difficult to describe in words. Finally, questions about sentiment, although the most straightforward given their predictive nature, are potentially affected by when they are asked (both because of current mood and memory degradation). Nevertheless, self-report measures are the most direct way to measure sentiment and a reasonable alternative to direct measures of emotion and mood (which currently remain in the early stages of development).

Several standard questionnaires exist for measuring affect (Plutchik & Kellerman, 1989, chaps. 1–3). The most common approach presents participants with a list of emotional adjectives and asks how well each describes their affective state. Izard's (1972) Differential Emotion Scale, for example, includes 24 emotional terms (such as delighted, scared, happy, and astonished) that participants rate on 7-point scales, indicating the degree to which they are feeling that emotion (from "not at all" to "extremely"). McNair, Lorr, and Droppleman's (1981) Profile of Mood States is a popular adjective-based measure of mood. Researchers have created numerous modifications of these standard scales (Desmet, Hekkert, & Jacobs, 2000, present a unique nonverbal adaptation), and many current usability questionnaires include at least some adjective-based affect assessment items (e.g., the Questionnaire for User Interface Satisfaction; Chin, Diehl, & Norman, 1988).

A second approach to questionnaire measurement of affect derives from dimensional theories of emotion and mood. Many researchers argue that two dimensions—arousal (activation) and valence (pleasant/unpleasant)—are nearly sufficient to describe the entire space of conscious emotional experience (Fledman Barrett & Russell, 1999). Lang (1995), for example, presents an interesting measurement scheme in which subjects rate the arousal and valence of their current affective state by selecting among pictorial representations (rather than the standard number/word representation of degree). Watson, Clark, and Tellegan's (1988) Positive and Negative Affect Schedule is a popular dimensional measure of mood. Finally,

to measure emotion as it occurs, with minimum interruption, some researchers have asked subjects to push one of a small number of buttons indicating their current emotional reaction during presentation of a stimulus (e.g., one button each for positive, negative, and neutral response; Breckler & Berman, 1991).

Affect Recognition by Users

Computers are not the only (potential) affect recognizers in human-computer interactions. When confronted with an interface—particularly a social or character-based interface—users constantly monitor cues to the affective state of their interaction partner, the computer (although often nonconsciously; see Reeves & Nass, 1996). Creating natural and efficient interfaces requires not only recognizing emotion in users, but also expressing emotion. Traditional media creators have known for a long time that portrayal of emotion is a fundamental key to creating the illusion of life (Jones, 1990; Thomas & Johnson, 1981; for discussions of believable agents and emotion, see, e.g., Bates, 1994; Maldonado, Picard, & Hayes-Roth, 1998).

Facial expression and gesture are the two most common ways to manifest emotion in screen-based characters (Cassell et al., 2000; Kurlander, Skelly, & Salesin, 1996). Although animated expressions lack much of the intricacy found in human expressions, users are nonetheless capable of distinguishing emotions in animated characters (Cassell et al., 2000; Schiano, Ehrlich, Rahardja, & Sheridan, 2000). As with emotion recognition, Ekman and Friesen's (1977) Facial Action Coding System is a commonly used and well-developed method for constructing affective expressions. One common strategy for improving the accuracy of communication with animated characters is to exaggerate expressions, but whether this leads to corresponding exaggerated assumptions about the underlying emotion has not been studied.

Characters that talk can also use voice to communicate emotion (Nass & Gong, 2000). Prerecorded utterances are easily infused with affective tone, but are fixed and inflexible. Cahn (1990) has successfully synthesized affect-laden speech using a text-to-speech system coupled with content-sensitive rules regarding appropriate acoustic qualities (including pitch, timing, and voice quality; see also Nass, Foehr, & Somoza, 2000). Users were able to distinguish among six different emotions with about 50% accuracy, which is impressive considering that people are generally only 60% accurate in recognizing affect in human speech (Scherer, 1981).

Finally, characters can indicate affective state verbally through word and topic choice, as well as explicit statements of affect (e.g., "I'm happy"). Characters whose nonverbal and verbal expressions are distinctly mismatched, however, may be seen as awkward or even untrustworthy. In less extreme mismatched cases, recent evidence suggests that users will give precedence to nonverbal cues in judgments about affect (Nass et al., 2000). This finding is critical for applications in which characters/agents mediate interpersonal communication (e.g., in virtual worlds or when characters read email to a user), because the affective tone of a message may be inappropriately masked by the character's affective state. Ideally, in such computer-mediated communication contexts, emotion would be encoded into the message itself, either through explicit tagging of the message with affect, through natural language processing of the message, or through direct recognition of the sender's affective state during message composition (e.g., using autonomic nervous system or facial expression measures). Mediator characters could then display the appropriate nonverbal cues to match the verbal content of the message.

OPEN QUESTIONS

Beyond the obvious need for advancements in affect recognition and manifestation technology, it is our opinion that there are five important and remarkably unexplored areas for research in emotion and HCI:

1. With which emotion should HCI designers be most concerned?

   Which emotion(s) should interface designers address first? The basic emotions, to the extent that they exist and can be identified, have the advantage of similarity across cultures and easy discriminability. Thus, designs that attempt to act on or manipulate these dimensions may be the simplest to implement. However, within these basic emotions, little is known about their relative manipulability or manifestability—particularly within the HCI context—or their relative impact on individuals' attitudes and behaviors. Once one moves beyond the basic emotions, cultural and individual differences introduce further problems and opportunities.

2. When and how should interfaces attempt to directly address users' emotions and basic needs (vs. application-specific goals)?

   If one views a computer or an interface merely as a tool, then interface design should solely focus on application-specific goals, assessed by such metrics as efficiency, learnability, and accuracy. However, if computers and interfaces are understood as a medium, then it becomes important to think about both uses and gratifications (Katz, Blumler, & Gurevitch, 1974; Rosengren, 1974; Rubin, 1986), that is, the more general emotional and basic needs that users bring to any interaction. Notions of "infotainment" or "edutainment" indicate one category of attempts to balance task and affect. However, there is little understanding of how aspects of interfaces that directly manipulate users' emotions compliment, undermine, or are orthogonal to aspects of interfaces that specifically address users' task needs.

3. How accurate must emotion recognition be to be useful as an interface technique?

   Although humans are not highly accurate emotion detectors—the problem of receiving accuracy (Picard, 1997a, p. 120)—they nonetheless benefit from deducing other's emotions and acting on those deductions (Goleman, 1995). Clearly, however, a minimum threshold of accuracy is required before behavior based on emotion induction is appropriate. Very little is known about the level of confidence necessary before an interface can effectively act on a user's emotional state.

4. When and how should users be informed that their affective states are being monitored and adapted to?

   When two people interact, there is an implicit assumption that each person is monitoring the other's emotional state and responding based on that emotional state. However, an explicit statement of this fact would be highly disturbing: "To facilitate our interaction, I will carefully and constantly monitor everything you say and do to discern your emotional state and respond based on that emotional state" or "I have determined that you are sad; I will now perform actions that will make you happier." However, when machines acquire and act on information about users without making that acquisition and adaptation explicit, there is often a feeling of surreptitiousness or manipulation. Furthermore, if emotion monitoring and adapting software are desired by consumers, there are clearly incentives for announcing and marketing these abilities. Because normal humans only exhibit implicit monitoring, the psychological literature is silent on the psychological and performance implications for awareness of emotional monitoring and adaptation.

5. How does emotion play out in computer-mediated communication?

   This chapter has focused on the direct relationship between the user and the interface. However, computers also are used to mediate interactions between people. In face-to-face encounters, affect not only creates richer interaction, but also helps to disambiguate meaning, allowing for more effective communication. Little is known, however, about the psychological effects of mediated affect, or the optimal strategies for encoding and displaying affective messages (see Maldonado & Picard, 1999; Rivera, Cooke, & Bauhs, 1996).

CONCLUSION

Although much progress has been made in the domain of affective computing (Picard, 1997a), more work is clearly necessary before interfaces that incorporate emotion recognition and manifestation can reach their full potential. Nevertheless, careful consideration of affect in interaction design and testing can be instrumental in creating interfaces that are both efficient and effective, as well as enjoyable and satisfying. Designers and theorists, for even the simplest interfaces, are well advised to thoughtfully address the intimate and far-reaching linkages between emotion and HCI.

ACKNOWLEDGMENTS

James Gross, Heidy Maldonado, Roz Picard, and Don Roberts provided extremely detailed and valuable insights. Sanjoy Banerjee, Daniel Bochner, Dolores Canamero, Pieter Desmet, Ken Fabian, Michael J. Lyons, George Marcus, Laurel Margulis, Byron Reeves, and Aaron Sloman provided useful suggestions.

References

Averill, J. R. (1980). A constructionist view of emotion. In R. Plutchik & H. Kellerman (Eds.), *Emotion: Theory, research, and experience* (Vol. 1, pp. 305–339). New York: Academic Press.

Ball, G., & Breese, J. (2000). Emotion and personality in conversational agents. In J. Cassell, J. Sullivan, S. Prevost, & E. Churchill (Eds.), *Embodied conversational agents* (pp. 189–219). Cambridge, MA: The MIT Press.

Banse, R., & Scherer, K. (1996). Acoustic profiles in vocal emotion expression. *Journal of Personality and Social Psychology, 70*(3), 614–636.

Barrett, L. F., & Russell, J. A. (1999). The structure of current affect: Controversies and emerging consensus. *Current Directions in Psychological Science, 8*(1), 10–14.

Bates, J. (1994). The role of emotions in believable agents. *Communications of the ACM, 37*(7), 122–125.

Bates, J., Loyall, A. B., & Reilly, W. S. (1994). An architecture for action, emotion, and social behavior, *Artificial social systems: Fourth European workshop on modeling autonomous agents in a multi-agent world*. Berlin: Springer-Verlag.

Bishop, D., Jacks, H., & Tandy, S. B. (1993). The Structure of Temperament Questionnaire (STQ): Results from a U.S. sample. *Personality & Individual Differences, 14*(3), 485–487.

Bower, G. H. (1992). How might emotions affect learning? In C. Sven-Åke (Ed.), *The handbook of emotion and memory: Research and theory* (pp. 3–31). Hillsdale, NJ: Lawrence Erlbaum Associates.

Bower, G. H., & Forgas, J. P. (2000). Affect, memory, and social cognition. In E. Eich, J. F. Kihlstrom, G. H. Bower, J. P. Forgas, & P. M. Niedenthal (Eds.), *Cognition and emotion* (pp. 87–168). Oxford: Oxford University Press.

Bower, G. H., Gilligan, S. G., & Monteiro, K. P. (1981). Selectivity of learning caused by affective states. *Journal of Experimental Psychology: General, 110*, 451–473.

Bozinovski, S., & Bozinovska, L. (2001). Self-learning agents: A connectionist theory of emotion based on crossbar value judgment. *Cybernetics and Systems, 32*, 5–6.

Breckler, S. T., & Berman, J. S. (1991). Affective responses to attitude objects: Measurement and validation. *Journal of Social Behavior and Personality, 6*(3), 529–544.

Cacioppo, J. T., Bernston, G. G., Klein, D. J., & Poehlmann, K. M. (1997). Psychophysiology of emotion across the life span. *Annual Review of Gerontology and Geriatrics, 17*, 27–74.

Cacioppo, J. T., Bernston, G. G., Larsen, J. T., Poehlmann, K. M., & Ito, T. A. (2000). The psychophysiology of emotion. In M. Lewis & J. M. Haviland-Jones (Eds.), *Handbook of emotions* (2nd ed., pp. 173–191). New York: The Guilford Press.

Cahn, J. E. (1990). The generation of affect in sythesized speech. *Journal of the American Voice I/O Society, 8*, 1–19.

Cannon, W. B. (1927). The James-Lange theory of emotions: A critical examination and an alternate theory. *American Journal of Psychology, 39*, 106–124.

Cassell, J., Sullivan, J., Prevost, S., & Churchill, E. (Eds.). (2000). *Embodied conversational agents*. Cambridge, MA: MIT Press.

Cassell, J., & Thorisson, K. (1999). The power of a nod and a glance: Envelope vs. emotional feedback in animated conversational agents. *Journal of Applied Artificial Intelligence, 13*, 519–538.

Chin, J. P., Diehl, V. A., & Norman, K. L. (1988). Development of an instrument measuring user satisfaction of the human-computer interface, *Proceedings of CHI '88 human factors in computing systems* (pp. 213–218). New York: ACM Press.

Clore, G. C. (1994). Why emotions are felt. In P. Ekman & R. J. Davidson (Eds.), *The nature of emotion: Fundamental questions* (pp. 103–111). New York: Oxford University Press.

Clore, G. C., & Gasper, K. (2000). Feeling is believing: Some affective influences on belief. In N. H. Frijda, A. S. R. Manstead, & S. Bem (Eds.), *Emotions and beliefs: How feelings influence thoughts* (pp. 10–44). Paris/Cambridge: Editions de la Maison des Sciences de l'Homme and Cambridge University Press (jointly published).

Clore, G. C., Wyer, R. S., Jr., Diened, B., Gasper, K., Gohm, C., & Isbell, L. (2001). Affective feelings as feedback: Some cognitive consequences. In L. L. Martin & G. C. Clore (Eds.), *Theories of mood and cognition: A user's handbook* (pp. 63–84). Mahwah, NJ: Lawrence Erlbaum Associates.

Collier, G. (1985). *Emotional expression*. Hillsdale, NJ: Lawrence Erlbaum Associates.

Cowie, R., & Douglas-Cowie, E. (1996). *Automatic statistical analysis of the signal and posidic signs of emotion in speech*. Paper presented at the Proceedings of the 4th international conference on spoken language processing (ICSLP-96), New Castle, DE.

Cowie, R., Douglas-Cowie, E., Tsapatsoulis, N., Votsis, G., Kollias, S., Fellenz, W., & Taylor, J. G. (2001). Emotion recognition in human-computer interaction. *IEEE Signal Processing Magazine, 18*(1), 32–80.

Damasio, A. R. (1994). *Descartes' error: Emotion, reason, and the human brain*. New York: Putnam Publishing Group.

Darwin, C. (1872/1998). *The expression of the emotions in man and animals*. London: HarperCollins.

Davidson, R. J. (1992). Anterior cerebral asymmetry and the nature of emotion. *Brain and Cognition, 20*, 125–151.

Davidson, R. J. (1994). On emotion, mood, and related affective constructs. In P. Ekman & R. J. Davidson (Eds.), *The nature of emotion* (pp. 51–55). New York: Oxford University Press.

Desmet, P. M. A., Hekkert, P., & Jacobs, J. J. (2000). When a car makes you smile: Development and application of an instrument to measure product emotions. In S. J. Hoch & R. J. Meyer (Eds.), *Advances in consumer research* (Vol. 27, pp. 111-117). Provo, UT: Association for Consumer Research.

Detenber, B. H., & Reeves, B. (1996). A bio-informational theory of emotion: Motion and image size effects on viewers. *Journal of Communication, 46*(3), 66-84.

Duncker, K. (1945). On problem-solving. *Psychological Monographs, 58*(Whole No. 5).

Ekman, P. (1992). An argument for basic emotions. *Cognition and Emotion, 6*(3/4), 169-200.

Ekman, P. (1994). All emotions are basic. In P. Ekman & R. J. Davidson (Eds.), *The nature of emotion: Fundamental questions* (pp. 7-19). New York: Oxford University Press.

Ekman, P., & Davidson, R. J. (Eds.). (1994). *The nature of emotion.* New York: Oxford University Press.

Ekman, P., & Friesen, W. V. (1975). *Unmasking the face.* Englewood Cliffs, NJ: Prentice-Hall.

Ekman, P., & Friesen, W. V. (1977). *Facial Action Coding System.* Consulting Psychologist Press.

Ekman, P., Levenson, R. W., & Friesen, W. V. (1983). Autonomic nervous system activity distinguishes among emotions. *Science, 221,* 1208-1210.

Elliott, C., & Brzezinski, J. (1998, Summer). Autonomous agents and synthetic characters. *AI Magazine, 19*(2), 13-30.

Ellis, H. C., & Moore, B. A. (1999). Mood and memory. In T. Dalgleish & M. J. Power (Eds.), *Handbook of cognition and emotion* (pp. 193-210). New York: John Wiley & Sons.

Ellsworth, P. C. (1994). Some reasons to expect universal antecedents of emotion. In P. Ekman & R. J. Davidson (Eds.), *The nature of emotion: Fundamental questions* (pp. 150-154). New York: Oxford University Press.

Erber, R., & Erber, M. W. (2001). Mood and processing: A view from a self-regulation perspective. In L. L. Martin & G. C. Clore (Eds.), *Theories of mood and cognition: A user's handbook* (pp. 63-84). Mahwah, NJ: Lawrence Erlbaum Associates.

Essa, I. A., & Pentland, A. P. (1997). Coding, analysis, interpretation, and recognition of facial expressions. *IEEE Transactions on Pattern Analysis and Machine Intelligence, 19*(7), 757-763.

Fledman Barrett, L., & Russell, J. A. (1999). The structure of current affect: Controversies and emerging consensus. *Current Directions in Psychological Science, 8*(1), 10-14.

Fogg, B. J. (1998). Charismatic computers: Creating more likable and persuasive interactive technologies by leveraging principles from social psychology. *Dissertation Abstracts International Section A: Humanities & Social Sciences, 58*(7-A), 2436.

Forgas, J. P. (1995). Mood and judgment: The Affect Infusion Model (AIM). *Psychological Bulletin, 117,* 39-66.

Friedman, B., & Kahn, P. H., Jr. (1997). Human agency and responsible computing: Implications for computer system design. In B. Friedman (Ed.), *Human values and the design of computer technology* (pp. 221-235). Stanford, CA: CSLI Publications.

Frijda, N. H. (1986). *The emotions.* Cambridge, New York: Cambridge University Press.

Frijda, N. H. (1988). The laws of emotion. *American Psychologist, 43*(5), 349-358.

Frijda, N. H. (1994). Varieties of affect: Emotions and episodes, moods, and sentiments. In P. Ekman & R. J. Davidson (Eds.), *The nature of emotion* (pp. 59-67). New York: Oxford University Press.

Gardano, A. C. (1986). Cultural influence on emotional response to color: A research study comparing Hispanics and non-Hispanics. *American Journal of Art Therapy, 23,* 119-124.

Gerard, R. (1957). *Differential effects of colored lights on psychophysiological functions.* Unpublished doctoral dissertation, University of California, Los Angeles.

Gerard, R. (1958, July). Color and emotional arousal [abstract]. *American Psychologist, 13,* 340.

Goleman, D. (1995). *Emotional intelligence.* New York: Bantam Books.

Gregory, A. H., Worrall, L., & Sarge, A. (1996). The development of emotional responses to music in young children. *Motivation and Emotion, 20*(4), 341-348.

Gross, J. J. (1998). Antecedent- and response-focused emotion regulation: Divergent consequences for experience, expression, and physiology. *Journal of Personality and Social Psychology, 74,* 224-237.

Gross, J. J. (1999). Emotion and emotion regulation. In L. A. Pervin & O. P. John (Eds.), *Handbook of personality: Theory and research* (2nd ed., pp. 525-552). New York: Guilford.

Hatfield, E., Cacioppo, J. T., & Rapson, R. L. (1994). *Emotional contagion.* Paris/Cambridge: Editions de la Maison des Sciences de l'Homme and Cambridge University Press (jointly published).

Heller, W. (1990). The neuropsychology of emotion: Developmental patterns and implications for psychpathology. In N. L. Stein, B. Leventhal, & T. Trabasso (Eds.), *Psychological and biological approaches to emotion* (pp. 167-211). Hillsdale, NJ: Lawrence Erlbaum Associates.

Henry, D. L., & Jacobs, K. W. (1978). Color eroticism and color preference. *Perceptual and Motor Skills, 47,* 106.

Heuer, F., & Reisberg, D. (1992). Emotion, arousal, and memory for detail. In C. Sven-Åke (Ed.), *The handbook of emotion and memory: Research and theory* (pp. 3-31). Hillsdale, NJ: Lawrence Erlbaum Associates.

Hirt, E. R., Melton, R. J., McDonald, H. E., & Harackiewicz, J. M. (1996). Processing goals, task interest, and the mood-performance relationship: A mediational analysis. *Journal of Personality and Social Psychology, 71,* 245-261.

Isen, A. M. (2000). Positive affect and decision making. In M. Lewis & J. M. Haviland-Jones (Eds.), *Handbook of emotions* (2nd ed., pp. 417-435). New York: The Guilford Press.

Isen, A. M., Daubman, K. A., & Nowicki, G. P. (1987). Positive affect facilitates creative problem solving. *Journal of Personality and Social Psychology, 52*(6), 1122-1131.

Isen, A. M., Rosenzweig, A. S., & Young, M. J. (1991). The influence of positive affect on clinical problem solving. *Medical Decision Making, 11*(3), 221-227.

Izard, C. E. (1972). *Patterns of emotions.* New York: Academic Press.

Izard, C. E. (1992). Basic emotions, relations among emotions, and emotion-cognition relations. *Psychological Review, 99*(3), 561-565.

James, W. (1884). What is an emotion? *Mind, 9,* 188-205.

Jones, C. (1990). *Chuck amuck: The life and times of an animated cartoonist.* New York: Avon Books.

Katz, E., Blumler, J. G., & Gurevitch, M. (1974). Utilization of mass communication by the individual. In J. G. Blumler & E. Katz (Eds.), *The uses of mass communications: Current perspectives on gratifications research* (pp. 19-32). Beverly Hills, CA: Sage.

Klein, J., Moon, Y., & Picard, R. W. (1999). *This computer responds to user frustration.* Paper presented at the human factors in computing systems: CHI'99 extended abstracts, New York.

Kleinginna, P. R., Jr., & Kleinginna, A. M. (1981). A categorized list of emotion definitions, with suggestions for a consensual definition. *Motivation and Emotion, 5*(4), 345-379.

Kurlander, D., Skelly, T., & Salesin, D. (1996). *Comic chat.* Paper presented at the Proceedings of SIGGRAPH'96: International conference on computer graphics and interactive techniques, New York.

Lang, P. J. (1995). The emotion probe. *American Psychologist, 50*(5), 372-385.

Lazarus, R. S. (1991). *Emotion and adaptation.* New York: Oxford University Press.

LeDoux, J. E. (1995). Emotion: Clues from the brain. *Annual Review of Psychology, 46*, 209–235.

LeDoux, J. E. (1996). *The emotional brain*. New York: Simon & Schuster.

LeDoux, J. E., & Phelps, E. A. (2000). Emotional networks in the brain. In M. Lewis & J. M. Haviland-Jones (Eds.), *Handbook of emotions* (pp. 157–172). New York: The Guilford Press.

Levenson, R. W. (1988). Emotion and the autonomic nervous system: A prospectus for research on autonomic specificity. In H. Wagner (Ed.), *Social psychophysiology: Perspectives on theory and clinical applications* (pp. 17–42). London: Wiley.

Levenson, R. W., Ekman, P., & Friesen, W. V. (1990). Voluntary facial action generates emotion-specific autonomic nervous system activity. *Psychophysiology, 27*, 363–384.

Levy, B. I. (1984). Research into the psychological meaning of color. *American Journal of Art Therapy, 23*, 58–62.

Lisetti, C. L., & Schiano, D. J. (2000). Automatic facial expression interpretation: Where human-computer interaction, artificial intelligence and cognitive science intersect. *Pragmatics and Cognition (Special Issue on Facial Information Processing: A Multidisciplinary Perspective), 8*(1), 185–235.

Lyons, M., Akamatsu, S., Kamachi, M., & Gyoba, J. (1998). Coding facial expressions with gabor wavelets. *Proceedings of the Third IEEE international conference on automatic face and gesture recognition* (pp. 200–205). New York: IEEE Press.

Malamuth, N. (1996). Sexually explicit media, gender differences, and evolutionary theory. *Journal of Communication, 46*, 8–31.

Maldonado, H., & Picard, A. (1999). The Funki Buniz playground: Facilitating multicultural affective collaborative play. *CHI99 extended abstracts conference on human factors in computing systems* (pp. 328–329). New York: ACM Press.

Maldonado, H., Picard, A., & Hayes-Roth, B. (1998). Tigrito: A high affect virtual toy. *CHI98 Summary conference on human factors in computing systems* (pp. 367–368). New York: ACM Press.

Martinez, A. M. (2000). *Recognition of partially occluded and/or imprecisely localized faces using a probabilistic approach*. Paper presented at the Proceedings of IEEE computer vision and pattern recognition (CVPR '2000).

Maslow, A. H. (1968). *Toward a psychology of being*. New York: D. Van Nostrand Company.

McNair, D. M., Lorr, M., & Droppleman, L. F. (1981). *Manual of the Profile of Mood States*. San Diego: Educational and Industrial Testing Services.

Morkes, J., Kernal, H. K., & Nass, C. (2000). Effects of humor in task-oriented human-computer interaction and computer-mediated communication: A direct test of SRCT theory. *Human-Computer Interaction, 14*(4), 395–435.

Murphy, S. T., & Zajonc, R. B. (1993). Affect, cognition, and awareness: Affective priming with suboptimal and optimal stimulus. *Journal of Personality and Social Psychology, 64*, 723–739.

Murray, I. R., & Arnott, J. L. (1993). Toward the simulation of emotion in synthetic speech: A review of the literature on human vocal emotion. *Journal Acoustical Society of America, 93*(2), 1097–1108.

Murray, N., Sujan, H., Hirt, E. R., & Sujan, M. (1990). The influence of mood on categorization: A cognitive flexibility interpretation. *Journal of Personality and Social Psychology, 59*, 411–425.

Nass, C., Foehr, U., & Somoza, M. (2000). *The effects of emotion of voice in synthesized and recorded speech*. Unpublished manuscript, Stanford, CA.

Nass, C., & Gong, L. (2000). Social aspects of speech interfaces from an evolutionary perspective: Experimental research and design implications. *Communications of the ACM, 43*(9), 36–43.

Neese, R. M. (1990). Evolutionary explanations of emotions. *Human Nature, 1*(3), 261–289.

Neumann, R., & Strack, F. (2000). "Mood contagion": The automatic transfer of mood between persons. *Journal of Personality and Social Psychology, 79*(2), 211–223.

Newhagen, J., & Reeves, B. (1991). Emotion and memory responses to negative political advertising. In F. Biocca (Ed.), *Televisons and political advertising: Psychological processes* (pp. 197–220). Hillsdale, NJ: Lawrence Erlbaum.

Newhagen, J., & Reeves, B. (1992). This evening's bad news: Effects of compelling negative television news images on memory. *Journal of Communication, 42*, 25–41.

Niedenthal, P. M., Setterlund, M. B., & Jones, D. E. (1994). Emotional organization of perceptual memory. In P. M. Niedenthal & S. Kitayama (Eds.), *The heart's eye* (pp. 87–113). San Diego: Academic Press, Inc.

Norman, D. (1990). *The design of everyday things*. Garden City, NJ: Doubleday.

Oatley, K., & Johnson-Laird, P. N. (1987). Towards a cognitive theory of emotions. *Cognition and Emotion, 1*(1), 29–50.

Ortony, A., Clore, G. C., & Collins, A. (1988). *The cognitive structure of emotions*. Cambridge, MA: Cambridge University Press.

Ortony, A., & Turner, T. J. (1990). What's basic about emotions. *Psychological Review, 97*(3), 315–331.

Panksepp, J. (1992). A critical role for "affective neuroscience" in resolving what is basic about basic emotions. *Psychological Review, 99*(3), 554–560.

Parrott, G. W., & Spackman, M. P. (2000). Emotion and memory. In M. Lewis & J. M. Haviland-Jones (Eds.), *Handbook of emotions* (2nd ed., pp. 476–490). New York: The Guilford Press.

Pecjak, V. (1970). Verbal synthesis of colors, emotions and days of the week. *Journal of Verbal Learning and Verbal Behavior, 9*, 623–626.

Peretti, P. O. (1974). Color model associations in young adults. *Perceptual and Motor Skills, 39*, 715–718.

Picard, R. W. (1997a). *Affective computing*. Cambridge, MA: The MIT Press.

Picard, R. W. (1997b). Does HAL cry digital tears? Emotions and computers. In D. G. Stork (Ed.), *Hal's Legacy: 2001's Computer as Dream and Reality* (pp. 279–303). Cambridge, MA: The MIT Press.

Picard, R. W., Vyzas, E., & Healey, J. (2001). Toward machine emotional intelligence: Analysis of affective physiological state. *IEEE Transactions on Pattern Analysis and Machine Intelligence, 23*(10), 1175–1191.

Plutchik, R., & Kellerman, H. (Eds.). (1989). *Emotion: Theory, research, and experience* (Vol. 4: The Measurement of Emotions). San Diego: Academic Press, Inc.

Reeves, B., Biocca, F., Pan, Z., Oshagan, H., & Richards, J. (1989). *Unconscious processing and priming with pictures: Effects on emotional attributions about people on television*. Unpublished manuscript, Stanford, CA.

Reeves, B., & Nass, C. (1996). *The media equation: How people treat computers, television, and new media like real people and places*. New York: Cambridge University Press.

Reeves, B., Newhagen, J., Maibach, E., Basil, M. D., & Kurz, K. (1991). Negative and positive television messages: Effects of message type and message content on attention and memory. *American Behavioral Scientist, 34*, 679–694.

Reuter-Lorenz, P., & Davidson, R. J. (1981). Differential contributions of the two cerebral hemispheres to the perception of happy and sad faces. *Neuropsychologia, 19*(4), 609–613.

Rivera, K., Cooke, N. J., & Bauhs, J. A. (1996). The effects of emotional icons on remote communication. *CHI '96 interactive posters*, 99–100.

Roseman, I. J., Antoniou, A. A., & Jose, P. E. (1996). Appraisal determinants of emotions: Constructing a more accurate and comprehensive theory. *Cognition and Emotion, 10*(3), 241–277.

Rosenfeld, A. (1997). Eyes for computers: How HAL could "see." In D. G. Stork (Ed.), *Hal's legacy: 2001's computer as dream and reality* (pp. 211–235). Cambridge, MA: The MIT Press.

Rosengren, K. E. (1974). Uses and gratifications: A paradigm outlined. In J. G. Blumler & E. Katz (Eds.), *The uses of mass communications: Current perspectives on gratifications research* (pp. 269–286). Beverly Hills, CA: Sage.

Rubin, A. M. (1986). Uses, gratifications, and media effects research. In J. Bryant & D. Zillman (Eds.), *Perspectives on media effects* (pp. 281–301). Hillsdale, NJ: Lawrence Erlbaum Associates.

Schachtel, E. J. (1943). On color and affect. *Psychiatry, 6,* 393–409.

Schachter, S., & Singer, J. E. (1962). Cognitive, social, and physiological determinants of emotional state. *Psychological Review, 69*(5), 379–399.

Scherer, K. (1981). Speech and emotional states. In J. K. Darby (Ed.), *Speech evaluation in psychiatry* (pp. 189–220): Grune and Stratton, Inc.

Scherer, K. R. (1988). Criteria for emotion-antecedent appraisal: A review. In V. Hamilton, G. H. Bower, & N. H. Frijda (Eds.), *Cognitive perspectives on emotion and motivation* (pp. 89–126). Dordrecht: Kluver Academic Publishers.

Scherer, K. R. (1989). Vocal measurement of emotion. In R. Plutchik & H. Kellerman (Eds.), *Emotion: Theory, research, and experience* (Vol. 4, pp. 233–259). San Diego: Academic Press, Inc.

Scheutz, M., Sloman, A., & Logan, B. (2000, November 11–12). *Emotional states and realistic agent behavior.* Paper presented at the GAME-ON 2000, Imperial College, London.

Schiano, D. J., Ehrlich, S. M., Rahardja, K., & Sheridan, K. (2000, April 1–6). *Face to interFace: Facial affect in (hu)man and machine.* Paper presented at the human factors in computing systems: CHI'00 conference proceedings, New York.

Schwartz, N., & Bless, H. (1991). Happy and mindless, but sad and smart?: The impact of affective states on analytic reasoning. In J. P. Forgas (Ed.), *Emotion and social judgment* (pp. 55–71). Oxford: Pergamon.

Shweder, R. A. (1994). "You're not sick, you're just in love": Emotions as an interpretive system. In P. Ekman & R. J. Davidson (Eds.), *The nature of emotions* (pp. 32–44). New York: Oxford University Press.

Skinner, E. A. (1995). *Perceived control, motivation, & coping.* Thousand Oaks: Sage Publications, Inc.

Thayer, R. E. (1989). *The biopsychology of mood and arousal.* New York: Oxford University Press.

Thomas, F., & Johnson, O. (1981). *The illusion of life: Disney animation.* New York: Hyperion.

Thorson, E., & Friestad, M. (1985). The effects on emotion on episodic memory for television commercials. In P. Cafferata & A. Tybout (Eds.), *Advances in consumer psychology* (pp. 131–136). Lexington, MA: Lexington.

Tian, Y.-L., Kanade, T., & Cohn, J. F. (2001). Recognizing action units for facial expression analysis. *IEEE Transactions on Pattern Analysis and Machine Intelligence, 23*(2), 1–19.

Tooby, J., & Cosmides, L. (1990). The past explains the present: Emotional adaptations and the structure of ancestral environments. *Ethology and Sociobiology, 11,* 407–424.

Ucros, C. G. (1989). Mood state-dependent memory: A meta-analysis. *Cognition and Emotion, 3,* 139–167.

Voelker, D. (1994). *The effects of image size and voice volume on the evaluation of represented faces.* Unpublished dissertation, Stanford University, Stanford, CA.

Walters, J., Apter, M. J., & Svebak, S. (1982). Color preference, arousal, and theory of psychological reversals. *Motivation and Emotion, 6*(3), 193–215.

Watson, D., Clark, L. A., & Tellegan, A. (1988). Development and validation of brief measures of positive and negative affect: The PANAS scales. *Journal of Personality and Social Psychology, 54,* 128–141.

Wegner, D. M. (1994). Ironic processes of mental control. *Psychological Review, 101,* 34–52.

Wegner, D. M., & Bargh, J. A. (1998). Control and automaticity in social life. In D. T. Gilbert, S. T. Fiske, & G. Lindzey (Ed.), *The handbook of social psychology* (4th ed., Vol. 1, pp. 446–496). Boston: McGraw-Hill.

Wexner, L. (1954). The degree to which colors (hues) are associated with mood-tones. *Journal of Applied Psychology, 38,* 432–435.

Wierzbicka, A. (1992). Talking about emotions: Semantics, culture, and cognition. *Cognition and Emotion, 6*(3/4), 285–319.

Yacoob, Y., & Davis, L. S. (1996). Recognizing human facial expresnions from long image sequences using optical flow. *IEEE Transactions on Pattern Analysis and Machine Intelligence, 18*(6), 636–642.

Zillmann, D. (1991). Television viewing and physiological arousal. In J. Bryant & D. Zillman (Eds.), *Responding to the screen: Reception and reaction processes* (pp. 103–133). Hillsdale, NJ: Lawrence Erlbaum Associates.

·5·

COGNITIVE ARCHITECTURE

Michael D. Byrne
Rice University

Designing interactive computer systems to be efficient and easy to use is important so that people in our society may realize the potential benefits of computer-based tools....Although modern cognitive psychology contains a wealth of knowledge of human behavior, it is not a simple matter to bring this knowledge to bear on the practical problems of design—to build an applied psychology that includes theory, data, and knowledge.

—Card, Moran, and Newell, 1983, p. vii

INTRODUCTION

Integrating theory, data, and knowledge about cognitive psychology and human performance in a way that is useful for guiding design in human–computer interaction (HCI) is still not a simple matter. However, there have been significant advances since Card et al. wrote the previous passage. One of the key advances is the development of cognitive architecture, the subject of this chapter. The chapter will first define "cognitive architecture" and then elucidate why cognitive architecture is relevant for HCI. To detail the present state of cognitive architectures in HCI, it is important to consider some of the past use of cognitive architectures in HCI research. Then, four architectures actively in use in the research community (LICAI/CoLiDeS, Soar, EPIC, and ACT-R/PM) and their applications to HCI will be examined. This chapter will conclude with a discussion of the future of cognitive architectures in HCI.

What are Cognitive Architectures?

Most any dictionary will list several different definitions for the word *architecture*. For example, dictionary.com lists among them "a style and method of design and construction," e.g., Byzantine architecture; "orderly arrangement of parts; structure," e.g., the architecture of a novel; and one from computer science: "the overall design or structure of a computer system." What, then would something have to be qualify as a cognitive architecture? It is something much more in the latter senses of the word architecture, an attempt to describe the overall structure and arrangement of a very particular thing—the human cognitive system. A cognitive architecture is a broad theory of human cognition based on a wide selection of human experimental data and implemented as a running computer simulation program. Young (Gray, Young, & Kirschenbaum, 1997; Ritter & Young, 2001) defines a cognitive architecture as an embodiment of "a scientific hypothesis about those aspects of human cognition that are relatively constant over time and relatively independent of task."

This idea has been a part of cognitive science since the early days of cognitive psychology and artificial intelligence, as manifested in the General Problem Solver (Newell & Simon, 1963), one of the first successful computational cognitive models. These theories have progressed a great deal since the General Problem Solver, and are gradually becoming more and more broad. One of the best descriptions of the vision for this area is presented in Newell's (1990) book *Unified Theories of Cognition*. In it, Newell argues that the time has come for cognitive psychology to stop collecting disconnected empirical phenomena and begin seriously considering theoretical unification in the form of computer simulation models. Cognitive architectures are attempts to do just this.

Cognitive architectures are distinct from engineering approaches to artificial intelligence, which strive to construct intelligent computer systems by whatever technologies best serve that purpose. Cognitive architectures are designed to simulate human intelligence in a humanlike way (Newell, 1990). For example, the chess program that defeated Kasparov, Deep Blue, would not qualify as a cognitive architecture, because it does not solve the problem (chess) in a humanlike way. Deep Blue uses massive search of the game space, whereas human experts generally look only a few moves ahead, but concentrate effectively on quality moves.

Cognitive architectures differ from traditional research in psychology in that work on cognitive architecture is integrative. That is, they include attention, memory, problem solving, decision making, learning, and so on. Most theorizing in psychology follows a divide-and-conquer strategy that tends to generate highly specific theories of a very limited range of phenomena; this has changed little since the 1970s (Newell, 1973). This limits the usefulness of such theories for an applied domain like HCI in which users employ a wide range of cognitive capabilities in even simple tasks. Instead of asking, "How can we describe this isolated phenomenon?," people working with cognitive architectures can ask, "How does this phenomenon fit in with what we already know about other aspects of cognition?"

Another important feature of cognitive architectures is that they specify only the human *virtual machine*, the fixed architecture. A cognitive architecture alone cannot do anything. Generally, the architecture has to be supplied with the knowledge needed to perform a particular task. The combination of an architecture and a particular set of knowledge is generally referred to as a model. In general, it is possible to construct more than one model for any particular task. The specific knowledge incorporated into a particular model is determined by the modeler. Because the relevant knowledge must be supplied to the architecture, the knowledge engineering task facing modelers attempting to model performance on complex tasks can be formidable.

Another centrally important feature of cognitive architectures is that they are software artifacts constructed by human programmers. This has a number of relevant ramifications. First, a model of a task constructed in a cognitive architecture is runnable and produces a sequence of behaviors. These behavior sequences can be compared with the sequences produced by human users to help assess the quality of a particular model. They may also provide insight into alternate ways to perform a task; that is, they may show possible strategies that are not actually utilized by the people performing the task. This can be useful in guiding interface design as well. Another feature of many architectures is that they enable the creation of quantitative models. For instance, the model may say more than just "click on button A and then menu B," but may include the time between the two clicks. Models based on cognitive architectures can produce execution times, error rates, and even learning curves. This is a major strength of cognitive architectures as an approach to certain kinds of HCI problems and will be discussed in more detail in the next section.

On the other hand, cognitive architectures are large software systems, which are often considered difficult to construct and maintain. Individual models are also essentially programs, written in the language of the cognitive architecture. Thus, individual modelers need to have solid programming skills.

Finally, cognitive architectures are not in wide use among HCI practitioners. Right now, they exist primarily in academic research laboratories. One of the barriers for practitioners is that learning and using most cognitive architectures is itself generally a difficult task. However, this is gradually, changing and some of the issues being addressed in this regard will be discussed in later. Furthermore, even if cognitive architectures are not in wide use by practitioners, this does not mean that they are irrelevant to practitioners. The next section highlights why cognitive architectures are relevant to a wide HCI audience.

Relevance to Human–Computer Interaction

For some readers, the relevance of models that produce quantitative predictions about human performance will be obvious. For others, this may be less immediately clear. Cognitive architectures are relevant to usability as an engineering discipline, have several HCI-relevant applications in computing systems, and serve an important role in HCI as a theoretical science.

At nearly all HCI-oriented conferences, and many online resources, there are areas where corporations recruit HCI professionals. A common job title in these forums is usability engineer. Implicit in this title is the view that usability is, at least in part, an engineering enterprise. Although people with this job title are certainly involved in product design, there is a sense in which most usability engineering would not be recognized as engineering by people trained in more traditional engineering disciplines, such as electrical or aerospace engineering. In traditional engineering disciplines, design is generally guided at least in part by quantitative theory. Engineers have at their disposal hard theories of the domain in which they work, and these theories allow them to derive quantitative predictions. Consider an aerospace engineer designing a wing. Like a usability engineer, the aerospace engineer will not start with nothing; a preexisting design often provides a starting point. But when the aerospace engineer decides to make a change in that design, there is usually quantitative guidance about how the performance of the wing will change as a result of the change in design. This guidance, although quantitative, is not infallible, thus the need for evaluation tools like wind tunnels. This is not unlike the usability engineer's usability test. However, unlike in usability testing, the aerospace engineer has some quantitative idea about what the outcome of the test will be, and this is not guided simply by intuition and experience, but by a quantitative theory of aerodynamics. In fact, this theory is now so advanced that few wind tunnels are being built anymore. Instead, they are being replaced by computer simulations, an outcome of the application of computational techniques to complex problems in aerodynamics called *computational fluid dynamics*. This has not entirely replaced wind tunnels, but the demand for wind tunnel time has clearly been affected by this development.

For the most part, the usability engineer lacks the quantitative tools available to the aerospace engineer. Every design must be subjected to its own wind tunnel (usability) test, and the engineer has little guidance about what to expect other than from intuition and experience with similar tests. Although intuition and experience can certainly be valuable guides, they often fall short of more hard quantitative methods. Perhaps the engineer can intuit that interface X will allow users to complete tasks faster than with interface Y, but how much faster? 10%? 20%? Even small savings in execution times can add up to large financial savings for organizations when one considers the scale of the activity. The paradigm example is the telephone operators studied Gray, John, and Atwood (1993), where even a second saved on an average call would save the telephone company millions of dollars.

Computational models based on cognitive architectures have the potential to provide detailed quantitative answers, and for more than just execution times. Error rates, transfer of knowledge, learning rates, and other kinds of performance measures are all metrics than can often be provided by architecture-based models. Even if such models are not always precisely accurate in an absolute sense, they may still be useful in a comparative sense. For example, if a usability engineer is comparing interface A with interface B and the model at his or her disposal does not accurately predict the absolute times to complete some set of benchmark tasks, it may still accurately capture the difference between the two interfaces, which may be more than enough.

Additionally, there are certain circumstances when usability tests are impractical or prohibitively costly, or both. For example, access to certain populations, such as physicians or astronauts, may be difficult or expensive, so bringing them in for repeated usability tests may not be feasible. Whereas developing a model of a pilot or an air traffic controller performing an expert task with specialized systems may be difficult at first, rerunning that model to assess a change made to the user interface should be much more straightforward than performing a new usability test for each iteration of the system. This is possible only with a quantitatively realistic model of the human in the loop, one that can produce things like execution times and error rates. Computational models can, in principle, act as surrogate users in usability testing, even for special populations.

Of course, some of these measures can be obtained through other methods such as GOMS[1] analysis (Card et al., 1983; John & Kieras, 1996) or cognitive predict walkthrough (Polson, Lewis, Reiman, & Wharton, 1992). However, these techniques were originally grounded in the same ideas as some prominent cognitive architectures and are essentially abstractions of the relevant architectures for particular HCI purposes. Also, architecture-based computational models provide things that GOMS models and cognitive walkthroughs do not. First, models are executable and generative. A GOMS analysis, on the other hand, is a description of the procedural knowledge the user has to have and the sequence of actions that must be performed to accomplish a specific task instance, whereas the equivalent computational model actually generates the behaviors, often in real time or faster. Equally important, computational models have the capacity to be reactive in real time. So, although it may be possible

---

[1]GOMS stands for goals, operators, methods, and selection rules. More will be said about GOMS later in this chapter.

to construct a GOMS model that describes the knowledge necessary and the time it will take an operator to classify a new object on an air traffic controller's screen, a paper-and-pencil GOMS model cannot actually execute the procedure in response to the appearance of such an object. A running computational model, on the other hand, can.

Because of this property, architecture-based computational models have some other important uses beyond acting as virtual users in usability tests. One such use is in intelligent tutoring systems (ITSs). Consider the Lisp tutor (Anderson, Conrad, & Corbett, 1989). This tutoring system contained an architecture-based running computational model of the knowledge necessary to implement the relevant Lisp functions and a module for assessing which pieces of this knowledge were mastered by the student. Because the model was executable, it could predict what action the student would take if the student had correct knowledge of how to solve the problem. When the student took a different action, this told the ITS that the student was missing one or more relevant pieces of knowledge. The student could then be given feedback about what knowledge is missing or incomplete, and problems that exercise this knowledge could be selected by the ITS. By identifying students' knowledge, and the gaps in that knowledge, it is possible to generate more effective educational experiences. Problems that contain knowledge the student has already mastered can be avoided, to not bore the student with things he or she already know. This frees up the student to concentrate on the material not yet mastered, resulting in improved learning (Anderson et al., 1989). Although the Lisp tutor is an old research system, ITSs based on the same underlying cognitive architecture with the same essential methodology have been developed for more pressing educational needs, such as algebra and geometry, and are now sold commercially (see www.carnegielearning.com).

There is another HCI-relevant application for high-fidelity cognitive models: populating simulated worlds or situations. For example, training an F-16 fighter pilot is expensive, even in a simulator, because that trainee needs to face realistic opposition. Realistic opposition consists of other trained pilots, so training one person requires taking several trained pilots away from their normal duties (i.e., flying airplanes on real missions). This is difficult and expensive. If, however, the other pilots could be simulated realistically, then the trainee could face opposition that would have useful training value, without having to remove already-trained pilots from their duties. There are many training situations like this, where the only way to train someone is to involve multiple human experts who must all be taken away from their regular jobs. However, the need for expensive experts can potentially be eliminated (or at least reduced) by using architecturally based cognitive models in place of the human experts. The U.S. military has already started to experiment with just such a scenario (Jones et al., 1999). There are other domains besides training where having realistic opponents is desirable, such as video games. Besides things like texture-mapped three-dimensional graphics, one of the features often used to sell games is network play. This enables players to engage opponents whose capabilities are more comparable with their own than typical computer-generated opponents. However, even with network play, it is not always possible for a gamer to find an appropriate opponent. If the computer-generated opponent was a more high-fidelity simulation of a human in terms of cognitive and perceptual-motor capabilities, then video game players would have no difficulty finding appropriate opponents without relying on network play. Although this might not be the most scientifically interesting use of cognitive architectures, it seems inevitable that cognitive architectures will be used in this way.

Cognitive architectures are also theoretically important to HCI as an interdisciplinary field. Many people (including some cognitive psychologists) find terms from cognitive psychology such as *working memory* (WM) or *mental model* vague and ill-defined. A harsh evaluation of explanations relying on such terms is found in Salthouse (1988, p. 3): "It is quite possible that interpretations relying on such nebulous constructs are only masquerading ignorance in what is essentially vacuous terminology." Computational cognitive architectures, on the other hand, require explicit specifications of the semantics of theoretical terms. Even if the architectures are imperfect descriptions of the human cognitive system, they are at a minimum well-specified and therefore clearer in what they predict than strictly verbal theories.

A second theoretical advantage of computational theories such as cognitive architectures is they provide a window into how the theory actually works. As theories grow in size and number of mechanisms, the interactions of those mechanisms become increasingly difficult to predict analytically. Computer simulations permit relatively rapid evaluations of complex mechanisms and their interactions. (For an excellent discussion of this topic, see Simon, 1996.) Another problem with theories based solely on verbal descriptions is that it can be very difficult to assess the internal coherence of such theories, whereas such assessment is much more straightforward with computational models. Verbal theories can easily hide subtle (and not so subtle) inconsistencies that make them poor scientific theories. Computational models, on the other hand, force explanations to have a high level of internal coherence; theories that are not internally consistent are typically impossible to implement on real machines.

Finally, HCI is an interdisciplinary field, and thus theories that are fundamentally interdisciplinary in nature are appropriate. Cognitive architectures are such theories, combining computational methods and knowledge from the artificial intelligence end of computer science with data and theories from cognitive psychology. Although cognitive psychology and computer science are certainly not the only disciplines that participate in HCI, they are two highly visible forces in the field. Psychological theories that are manifested as executable programs should be less alien to people with a computer science background than more traditional psychological theories.

Thus, cognitive architectures are clearly relevant to HCI at a number of levels. This fact has not gone unnoticed by the HCI research community. In fact, cognitive architectures have a long history in HCI, dating back to the original work of Card et al. (1983).

A BRIEF LOOK AT PAST SYSTEMS IN HUMAN–COMPUTER INTERACTION

The total history of cognitive architectures and HCI would be far too long to document in a single chapter; however, it is possible to touch on some highlights. Although not all of the systems described in this section qualify as complete cognitive architectures, they all share intellectual history with more current architectures and influenced their development and use in HCI. Finally, many of the concepts developed in these efforts are still central parts of the ongoing research on cognitive architecture.

The Model Human Processor and GOMS

The Psychology of Human–Computer Interaction (Card et al., 1983) is clearly a seminal work in HCI, one of the defining academic works in the early days of the field. Even though that work did not produce a running cognitive architecture, it was clearly in the spirit of cognitive architectures and was quite influential in the development of current cognitive architectures. Two particular pieces of that work are relevant here: the Model Human Processor (MHP) and GOMS.

The MHP represents a synthesis of the literature on cognitive psychology and human performance up to that time, and sketches the framework around which a cognitive architecture could be implemented. The MHP is a system with multiple memories and multiple processors, and many of the properties of those processors and memories is described in some detail (see Fig. 5.1). Card et al. also specified the interconnections of the processors and a number of general operating principles. In this system, there are three processors: one cognitive, one perceptual, and one motor. In some cases, the system behaves essentially serially. For instance, for the system to press a key in response to the appearance of a light, the perceptual processor must detect the appearance of the light and transmit this information to the cognitive processor. The cognitive processor's job is to decide what the appropriate response should be and then transmit that to the motor processor, which is responsible for actually executing the appropriate motor command. In this situation, the processors act serially, one after another. However, in more complex tasks, such as transcription typing, all three processors will often be working in parallel.

Besides specification of the timing for each processor and the connectivity of the processors, Card et al. laid out some general operating principles, ranging from very general and qualitative to detailed and quantitative. For example, Principle P9, the Problem Space Principle, states:

The rational activity in which people engage to solve a problem can be described in terms of (1) a set of states of knowledge, (2) operators for changing one state into another, (3) constraints on applying operators, and (4) control knowledge for deciding which operator to apply next. (Card et al., 1983, p. 27)

This is a particularly general and somewhat vague principle. In contrast, consider Principle P5, Fitts's law:

The time T_{pos} to move the hand to a target of size S which lies a distance D away is given by:

$$T_{pos} = I_M \log_2(D/S + .5)$$

where $I_M = 100 \, [70 \sim 120]$ ms/bit. (Card et al., 1983, p. 27)

This is a very specific principle that quantitatively describes hand movement behavior, which is highly relevant to, say, pointing with a mouse. Overall, the specification of the MHP is quite thorough and lays out a basis for a cognitive architecture able to do a wide variety of HCI-relevant tasks. However, Card et al. did not implement the MHP as a running cognitive architecture. This is likely for pedagogical reasons; it is not necessary to have a complete, running cognitive architecture for the general properties of that architecture to be useful for guiding HCI researchers and practitioners. At the time, computational modeling was the domain of a very specialized few in cognitive psychology.

Card et al. lay out another concept that has been highly influential throughout HCI and particularly in the community of computational modelers. This is GOMS, which stands for goals, operators, methods, and selection rules. GOMS is a framework for task analysis that describes routine cognitive skills in terms of the listed four components. Routine cognitive skills are those where the user knows what the task is and how to do the task without doing any problem solving. Text editing with a familiar editor is the prototypical case of this, but clearly a great many tasks in of interest in HCI could be classified as routine cognitive skills. Thus, the potential applicability of GOMS is quite broad. Indeed, GOMS has been applied to a variety of tasks; the web site www.gomsmodel.org lists 143 GOMS-related papers in its bibliography.

What does a GOMS analysis provide? Essentially, a GOMS analysis of a task describes the hierarchical procedural knowledge a person must have to successfully complete that task. Based on that, and the sequence of operators that must be executed, it is possible to make quantitative predictions about the execution time for a particular task. Other analyses, such as predictions of error, functionality coverage, and learning time are also sometimes possible. Since the original formulation presented in Card et al., a number of different forms of GOMS analysis have been developed, each with slightly different strengths and weaknesses (John & Kieras, 1996).

The core point as it relates to cognitive architectures is that GOMS analysis is originally based on a production rule analysis (Card, personal communication, 1999). Because this will come up several times, a brief introduction to production systems is warranted. Production rules are IF-THEN condition-action pairs, and a set of production rules (or simply productions or just rules) and a computational engine that interprets those productions is called a *production system*. In addition to productions, production systems contain some representation of the current state. This representation typically consists of a set of loosely structured data elements, such as propositions or attribute-value pairs. This set is called *working memory* or *declarative*

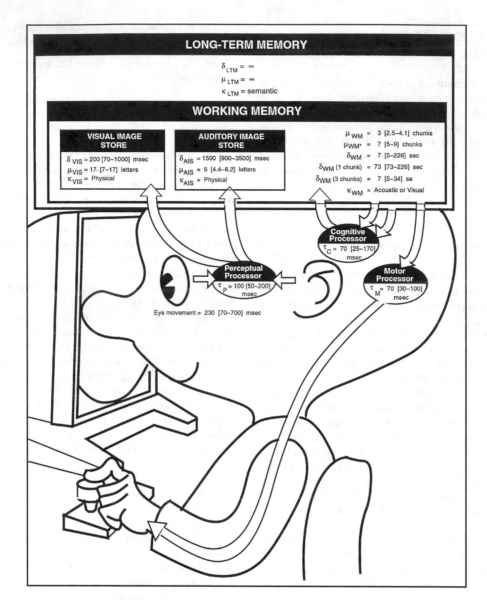

LONG-TERM MEMORY

$$\delta_{LTM} = \infty$$
$$\mu_{LTM} = \infty$$
$$\kappa_{LTM} = semantic$$

WORKING MEMORY

VISUAL IMAGE STORE

$$\delta_{VIS} = 200\ [70\sim1000]\ msec$$
$$\mu_{VIS} = 17\ [7\sim17]\ letters$$
$$\kappa_{VIS} = Physical$$

AUDITORY IMAGE STORE

$$\delta_{AIS} = 1500\ [900\sim3500]\ msec$$
$$\mu_{AIS} = 5\ [4.4\sim6.2]\ letters$$
$$\kappa_{AIS} = Physical$$

$$\mu_{WM} = 3\ [2.5\sim4.1]\ chunks$$
$$\mu_{WM}^* = 7\ [5\sim9]\ chunks$$
$$\delta_{WM} = 7\ [5\sim226]\ sec$$
$$\delta_{WM}\ (1\ chunk) = 73\ [73\sim226]\ sec$$
$$\delta_{WM}\ (3\ chunks) = 7\ [5\sim34]\ se$$
$$\kappa_{WM} = Acoustic\ or\ Visual$$

Cognitive Processor
$$\tau_C = 70\ [25\sim170]\ msec$$

Perceptual Processor
$$\tau_P = 100\ [50\sim200]\ msec$$

Motor Processor
$$\tau_M = 70\ [30\sim100]\ msec$$

Eye movement = 230 [70~700] msec

FIGURE 5.1. Model human processor. From *The Psychology of Human-Computer Interaction* (p. 26), by S. K. Card, T. P. Moran, & A. Newell, 1983, Hillsdale, NJ: Erlbaum. Copyright 1983 by Lawrence Erlbaum Associates. Adapted with permission.

memory. Because working memory is also a psychological term with somewhat different meaning in that literature, declarative memory will be used in all further discussions.

The operation of a production system is cyclic. On each cycle, the system first goes through a pattern-matching process. The IF side of each production tests for the presence of a particular pattern in declarative memory. When the IF conditions of a production are met, the production is said to fire, and the actions specified on the THEN side are executed. The actions can be things like pressing a button or even some higher level abstraction of action (e.g., "turn left"). Actions also include modifying the contents of declarative memory, which usually means that a different production or productions will match on the

next cycle. At this abstract and purely symbolic level, production systems are Turing complete and thus can compute anything that is computable (Newell, 1990), thus, they should be flexible enough to model the wide array of computations performed by the human cognitive system.

This is relevant to cognitive architectures because most cognitive architectures are (or contain) production systems. GOMS was actually abstracted from production rule analysis. Card et al. discovered that, for routine cognitive skills, the structure of the productions was quite similar across tasks and a more abstract representation was possible. This representation is the original GOMS formulation. Thus, translating a GOMS analysis into production rules, the language of most cognitive architectures, is generally straightforward. Similarly, for routine cognitive skills, it is often relatively simple to derive a GOMS analysis from the set of productions used to model the task. Models based on cognitive architectures can go well beyond routine cognitive skills, but this connection has certainly influenced the evolution of research on cognitive architecture and HCI. This connection has also fed back into research and development of GOMS techniques themselves, such as Natural GOMS Language (Kieras, 1988). This language allows the prediction of learning time for the knowledge described in a GOMS model based on a theory of transfer of training referred to as cognitive complexity theory (CCT), which will be described in more detail in the next section.

Cognitive Complexity Theory

When someone has learned to perform a task with a particular interface and must switch, doing the same task with a new interface, how much better off will that person be than someone just learning to do the task with the new interface? That is, how much is the knowledge gained from using the old interface transferred to using the new interface? This question has intrigued psychologists for at least a century, and having some answers to this question has implications for training programs, user interface design, and many other areas. Cognitive complexity theory (Bovair, Kieras, & Polson, 1990; Kieras & Polson, 1985) is a psychological theory of transfer of training applied to HCI. Most relevant to the current discussion, this theory is based on production rules. The major points of CCT are as follows:

• Knowledge of the procedures that people need to execute to perform routine tasks can be represented with production rules. The relevant production rules can be generated based on a GOMS analysis of the task to be modeled.

• The complexity of a task will be reflected in the number and content of the production rules. When certain conventions are adopted about the style of those production rules, complexity is reflected almost entirely in the number of rules.

• The time it takes to execute a procedure can be predicted with a production system that interprets those rules along with a set of typical operator times (e.g., the time it takes to type a three-letter command). The production interpreter used in this work was not intended to be a general cognitive architecture, but the production system framework is certainly consistent with current architectures.

• The time it takes to learn a task is a function of the number of new rules that the user must learn. "New" is clearly defined in this context. If the user already has a production, and a new task requires a rule that is similar (again, similarity is well-defined based on the production rule syntax), then the rule for the new task need not be learned.

• Some predictions about errors and speedup with practice can also be gleaned from the contents of the production rules.

Obviously, this was an ambitious agenda, and there are many subtleties. For example, the notion of a *task* as the term was used in the description of CCT actually includes more than just the task at an abstract level. Consider a simple instance of a text-editing task, deleting the word "redux" from the middle of a sentence. The actual commands needed to accomplish this task could be very different in different text editors, thus, modeling the "delete word" task would require two different sets of productions, one for each editor. That is, the necessary knowledge, and thus the production rules for representing it, are actually a function both of the task from the user point of view (e.g. "delete word") and the interface provided to accomplish the task. Transfer from one text editor to another therefore depends a great deal on the particulars of each interface. CCT thus predicts asymmetrical transfer: learning editor A after editor B should not be the same as learning editor B after editor A.

CCT models, like a GOMS analysis, omit modeling many details of user behavior. In general, anything that falls outside the domain of procedural knowledge (how-to-do-it knowledge) is not modeled. This means that the model does not model motor actions, such as keypresses, and instead has a DoKeystroke primitive operator. Nor do CCT models model things like natural language comprehension, clearly a requirement in text editing. CCT models also do not include any model of the perceptual processes required by users—the model was simply given information about the state of the display, and did not have to, for example, look to see if the cursor was over a particular character. This is the same scope as a typical GOMS model, although a CCT model is more formalized and quantitative than the GOMS models described by Card et al. (1983).

Despite these limitations (or perhaps in part because these limitations allowed the researchers to concentrate on the most central aspects of the phenomena), CCT fared very well. Numerous laboratory experiments provide empirical support for many of the claims of CCT (see, especially, Bovair et al., 1990). While the original development and validation of the CCT framework was focussed on the domain of command-line text editing, it has also been applied to menu-based systems (Polson, Muncher, & Engelbeck, 1986) and a control panel device (Kieras & Bovair, 1986). Singley and Anderson (1989) provide a strikingly similar analysis of transfer of training, as well as supporting empirical results, lending credence to the CCT analysis. CCT was certainly one of the most prominent early successes of computational modeling in HCI.

Collaborative Activation-Based Production System

CAPS (collaborative activation-based production system; Just & Carpenter, 1992) is a cognitive architecture designed to model individual differences in working memory (WM) capacity and the effects of working memory load. This speciality is applicable to a number of HCI situations. Certainly, some kinds of user interfaces can create excessive working memory demands, for example, phone-based interfaces. In phone-based interaction, options do not remain on a screen or in any kind of available storage; rather, users are forced to remember the options presented. This seems like a prime candidate for modeling with a system designed to capture the effects of working memory demand, and this is exactly what Huguenard, Lerch, Junker, Patz, and Kass (1997) did. Their data showed that, contrary to guideline advice and most people's intuition, restricting phone menus to only a few (three) items each does not reduce error rates. The CAPS-based model provided a clear theoretical account of this phenomenon. The model showed that short menus are not necessarily better in phone-based interactions because of two side-effects of designing menu hierarchies with few options at each level. First, for the same number of total items, this increases menu depth, which creates working memory demand. Second, with fewer items at each level, each individual item has to be more general and therefore more vague, especially at the top levels of the hierarchy. This forces users to spend WM resources on disambiguating menu items when they are in a situation where WM demands outstrip supply.

Another application of CAPS that is HCI-relevant is the account of postcompletion error provided by Byrne and Bovair (1997). What is a postcompletion error? Anecdotal evidence and intuition suggests that, when interacting with man-made artifacts, certain kinds of errors occur with greater frequency than others. In particular, there is an entire family of errors that seem intuitively common, these are errors that people make when there is some part of a task that occurs after the main goal of the task has been accomplished (hence "postcompletion"). Nearly everyone reports having made an error of this type at one time or another. Here are two prototypical examples:

- Leaving the original on the glass of a photocopier. The main goal one generally has when using a photocopier is "get copies," and this goal is satisfied before one removes the original document. This error is less common now that many photocopiers include document feeders; the more current equivalent is leaving a document on the glass in a flatbed scanner.
- Leaving one's bank card in an automated teller machine (ATM). Again, the main goal is something on the order of "get cash," and in many ATMs, card removal occurs after the cash is dispensed. This error was common enough in the first generation of ATMs that many ATMs are now designed in such a way that this error is now impossible to make.

There are many others, such as leaving the gas cap off after filling up the car's gas tank, leaving change in vending machines, and more—most readers can probably think of several others. Although numerous HCI researchers were aware of this class of error (e.g., Young, Barnard, Simon, & Whittington, 1989; Polson et al., 1992), no account had previously been developed that explained why this type of error is persistent, yet not so frequent that it occurs every time. The CAPS model provides just such an account and can serve as a useful example of the application of a cognitive architecture to an HCI problem.

Like most other production systems, CAPS contains two kinds of knowledge: declarative memory and productions. Declarative memory elements in CAPS also have associated with them an activation value, and elements below a threshold level of activation cannot be matched by productions' IF sides. Additionally, unlike most other production systems, the THEN side of a CAPS production may request that the activation of an element be incremented. For this to be truly useful in modeling working memory, there is a limit to the total amount of activation available across all elements. If the total activation exceeds this limit, then all elements lose some activation to bring the total back within the limit. This provides a mechanism for simulating human working memory limitations.

In Byrne and Bovair's postcompletion error model, there is a production that increments the activation of subgoals when the parent goal is active and unsatisfied. So, to use the photocopier example, the get copies subgoal supplies activation to all the unfulfilled subgoals throughout the task. However, when the get copies goal is satisfied, the activation supply to the subgoals stops. Because the goal to remove the original is a subgoal of that goal, it loses its activation supply. Thus, what the model predicts is that the postcompletion subgoals are especially vulnerable to working memory load, and lower capacity individuals are more at risk than higher capacity individuals. Byrne and Bovair conducted an experiment to test this prediction, and the data supported the model.

This is a nice demonstration of the power of cognitive architectures. Byrne and Bovair neither designed nor implemented the CAPS architecture, but were able to use the theory to construct a model that made empirically testable predictions, and those predictions were borne out. Whereas CAPS is unlikely to guide much future HCI work (its designers are no longer developing and supporting it because they have gone in a different direction), it provides an excellent example case.

CONTEMPORARY ARCHITECTURES

There are currently cognitive architectures that are being actively developed, updated, and applied to HCI-oriented tasks. Three of the four most prominent are production systems, or rather are centrally built around production systems. These three are Soar, EPIC, and ACT-R (particularly ACT-R/PM). Although all contain production rules, the level of granularity of an individual production rule varies considerably from architecture to architecture. Each one has a different history and each one has a unique focus. They all share a certain amount of intellectual history; in particular, they have all been influenced one way or another by the MHP and by each other. At some level, they may have more similarities than differences, whether this is because they borrow from one another or because the science is converging is still an open question. The fourth system is somewhat different than these three production system models and will be considered first.

LICAI/CoLiDeS

LICAI (Kitajima & Polson, 1997) is a good example of a nonproduction system architecture and has been used in an HCI context. All the work discussed up to this point more or less assumes that the users being modeled are relatively skilled with the specific interface being used; these approaches do a poor job of modeling relatively raw novices. One of the main goals of LICAI is addressing this concern. The paradigm question addressed by LICAI is, "How do users explore a new interface?"

Unlike the other architectures discussed, LICAI's central control mechanisms are not based on a production system. Instead, LICAI is built around an architecture originally designed to model human discourse comprehension, construction-integration (C-I; Kintsch, 1998). Like production systems, C-I's operation is cyclic. However, what happens on those cycles is somewhat different than what happens in a production system. Each cycle is divided into two phases: construction and integration (hence the name). In the construction phase, an initial input (e.g., the contents of the current display) is fed into a weakly constrained, rule-based process that generates a network of propositions. Items in the network are linked on the basis of their argument overlap. For example, the goal of graph data might be represented with the proposition (PERFORM GRAPH DATA). Any proposition containing GRAPH or DATA would thus be linked to that goal.

Once the construction phase completes, the system is left with a linked network of propositions. What follows is the integration phase, in which activation propagates through the network in a neural network-like fashion. Essentially, this phase is a constraint-satisfaction phase, which is used to select one of the propositions in the network as the preferred one. For example, the system may need to select the next action to perform while using an interface. Action representations will be added to the network during the construction phase, and an action will be selected during the integration phase. The action will be performed, and the next cycle initiated. Various C-I models have used this basic process to select things other than actions. The original C-I system used these cycles to select between different interpretations of sentences.

There are three main kinds of cycles in LICAI: one type selects actions, one generates goals, and one selects goals. This is in contrast to how most HCI tasks have been modeled in production system architectures; in such systems, the goals are usually included in the knowledge given to the system. This is not true in LICAI; in fact, the knowledge given to LICAI by the modelers is quite minimal. For the particular application of LICAI, which was modeling users who knew how to use a Macintosh for other tasks (e.g. word processing) and were now being asked to plot some data using a Macintosh program called CricketGraph (one group of users actually worked with Microsoft Excel), it included some very basic knowledge about the Macintosh GUI and some knowledge about graphing. Rather than supply the model with the goal hierarchy, Kitajima and Polson gave the model the same somewhat minimal instructions as the subjects. One of the major jobs of the LICAI model, then, was to generate the appropriate goals as they arose while attempting to carry out the instructions.

Again, this illustrates one of the strengths of using a cognitive architecture to model HCI tasks. Kitajima and Polson did not have to develop a theory of text comprehension for LICAI to be able to comprehend the instructions given to subjects, because LICAI is based on an architecture originally designed to do text comprehension, they essentially got that functionality gratis. Additionally, they did not include just any text comprehension engine, but one that makes empirically validated predictions about how people represent the text they read. Thus, the claim that the model started out with is roughly the same knowledge as the users is highly credible.

The actual behavior of the model is also revealing, because it exhibits many of the same exploratory behaviors as the users. First, the model pursues a general strategy that can be classified

as label-following (Polson et al., 1992). The model, like the users, had a strong tendency to examine anything on the screen that had a label matching, or nearly matching (i.e., a near synonym) a key word in the task instructions. When the particular subtask being pursued by the model contained steps that were well-labeled, the users were rapid, which the model predicted. Although this prediction is not counterintuitive, it is important to note that LICAI is not programmed with this strategy. This strategy naturally emerges through the normal operation of C-I through the linkages created by shared arguments. The perhaps less intuitive result—modeled successfully by LICAI—is the effect of the number of screen objects. During exploration in this task, users were slower to make choices if there were more objects on the screen, but only if those items all had what were classified as poor labels. In the presence of good labels (literal match or near synonym), the number of objects on the screen did not affect decision times, for the users or for LICAI.

The programmers who implemented the programs operated by the users put in several clever direct manipulation tricks. For example, to change the properties of a graph axis, one double-clicks on the axis and a dialog box specifying the properties of that axis appears. Microsoft Excel has some functionality that is most easily accessed by drag-and-drop. Fanzke (1994) found that, in a majority of first encounters with these kind of direct manipulations, users required hints from the experimenter to be able to continue, even after 2 min of exploration. LICAI also fails at these interactions because there are no appropriate links formed between any kind of task goal and these unlabeled, apparently static screen objects during the construction phase. Thus, these screen objects tend to receive little activation during the integration phase, and actions involving other objects are always selected.

Overall, LICAI does an excellent job of capturing many of the other empirical regularities in exploratory learning of a graphical user interface. This is an important issue for many interfaces, particularly any interface that is aimed at a walk-up-and-use audience. Whereas currently common walk-up-and-use interfaces, such at ATMs, provide simple enough functionality that this is not always enormously difficult, this is not the case for more sophisticated systems, such as many information kiosks.

More recently, LICAI has been updated (and renamed to CoLiDeS, for Comprehension-based Linked model of Deliberate Search; Kitajima, Blackmon, & Polson, 2000) to handle interaction with Web pages. This involves goals that are considerably less well-elucidated and interfaces with a much wider range of semantic content. To help deal with these complexities, LICAI has been updated with a more robust attentional mechanism and a much more sophisticated notion of semantic similarity based on Latent Semantic Analysis (Landauer & Dumais, 1997). CoLiDeS shows the effects of poor labels and poor hierarchical organization on Web navigation, getting lost in much the same way as real users. This is a promising tool for the analysis of semantically rich, but functionality-poor, domains such as the Web. Interestingly, Pirolli and Card (1999) have implemented a similar model in a modified version of ACT-R they term ACT-IF, where the IF stands for information forager. Whether these systems will ultimately converge, diverge, or simply complement one another is an open question.

Soar

The development of Soar is generally credited to Allan Newell (especially Newell, 1990), and Soar has been used to model a wide variety of human cognitive activity from syllogistic reasoning (Polk & Newell, 1995) to flying combat aircraft in simulated wargames (Jones et al., 1999). Soar was Newell's candidate for unified theory of cognition and was the first computational theory to be offered as such.

Whereas Soar is a production system, it is possible to think of Soar at a more abstract level. The guiding principle behind the design of Soar is Principle P9 from the MHP, the Problem Space Principle. Soar casts all cognitive activity as occurring in a problem space, which consists of a number of states. States are transformed through the application of operators. Consider Soar playing a simple game such as tic-tac-toe as player X. The problem space is the set of all the states of the tic-tac-toe board—not a very large space. The operators available at any given state of that space are placing an X at any of the available open spaces on the board. Obviously, this is a simplified example; the problem space and the available operators for flying an F-16 in a simulated wargame are radically more complex.

Soar's operation is also cyclic, but the central cycle in Soar's operation is called a decision cycle. Essentially, on each decision cycle, Soar answers the question, "What do I do next?" Soar does this in two phases. First, all productions that match the current contents of declarative memory fire. This usually causes changes in declarative memory, so other productions may now match. Those productions are allowed to fire, and this continues until no new productions fire. At this time, the decision procedure gets executed, in which Soar examines a special kind of declarative memory element, the preference. Preferences are statements about possible actions, for example, "operator o3 is better than o5 for the current operator" or "s10 rejected for supergoal state s7." Soar examines the available preferences and selects an action. Thus, each decision cycle may contain many production cycles. When modeling human performance, the convention is that each decision cycle lasts 50 ms, so productions in Soar are very low-level, encapsulating knowledge at a very small grain size. This distinguishes Soar productions from those found in other production systems.

Other than the ubiquitous application of the problem space principle, Soar's most defining characteristics come from two mechanisms developed specifically in Soar: universal subgoaling and a general-purpose learning mechanism. Because the latter depends on the former, universal subgoaling will be described first. One of the features of Soar's decision process is that it is not guaranteed to have an unambiguous set of preferences to work with. Alternately, there may be no preferences listing an acceptable action. Perhaps the system does not know any acceptable operators for the current state, or perhaps the system lacks the knowledge of how to apply the best operator. Whatever the reason, if the decision procedure is unable to select an action, an impasse is said to occur. Rather than halting or entering some kind of failure state, Soar sets up a new state in a new problem space with the goal of resolving the impasse. For example, if multiple operators were proposed, the goal of the new problem space is to choose between the proposed operators.

In the course of resolving one impasse, Soar may encounter another impasse and create another new problem space, and so on. As long as the system is provided with some fairly generic knowledge about resolving degenerate cases (e.g., if all else fails, choose randomly between the two good operators), this universal subgoaling allows Soar to continue even in cases where there is little knowledge.

Learning in Soar is a by-product of universal subgoaling. Whenever an impasse is resolved, Soar creates a new production rule. This rule summarizes the processing that went on in the substate. The resolution of an impasse makes a change to the superstate (the state in which the impasse originally occurred), this change is called a result. This result becomes the condition, or THEN, side of the new production. The condition, of IF, side of the production is generated through a dependency analysis by looking at any declarative memory item matched in the course of determining this result. When Soar learns, it learns only new production rules, and it only learns as the result of resolving an impasse. It is important to realize that an impasse is not equated with failure or an inability to proceed in the problem solving, but may arise simply because, for example, there are multiple good actions to take and Soar has to choose one of them. Soar impasses regularly when problem solving and thus learning is pervasive in Soar.

Not surprisingly, Soar has been applied to a number of learning-oriented HCI situations. One of the best examples is the recent work by Altmann (Altmann, 2001; Altmann & John, 1999). Altmann collected approximately 80 min of data from an experienced programmer while she worked at understanding and updating a large computer program by examining a trace. These data included verbal protocols (i.e., thinking aloud), as well as a log of the actions taken (keypresses and scrolling). About once every 3 min, the programmer scrolled back to find a piece of information that had previously been displayed. Altmann constructed a Soar model of her activity. This model is a kind of comprehension model that attempts to gather information about its environment; it is not a complete model of the complex knowledge of an experienced programmer. The model attends to various pieces of the display, attempting to comprehend what it sees, and issues commands. Comprehension in this context is not the same as in C-I–based models, but rather is manifested in this model as an attempt to retrieve information about the object being comprehended.

When an item is attended, this creates what Altmann termed an episodic trace, that is, a production that notes that the object was seen at a particular time. Because learning is pervasive, Soar creates many new rules like this. However, because of the dependency-based learning mechanism, these new productions are quite specific to the context in which the impasse originally occurred. Thus, the index into the model's (fairly extensive) episodic memory consists of very specific cues, usually found on the display. Seeing a particular variable name is likely to trigger a memory for having previously seen that variable name. Importantly, this memory is generated automatically, without need for the model to deliberately set goals to remember particular items.

Whereas Altmann and John (1999) is primarily a description of the model, Altmann (2001) discusses some of the HCI ramifications for this kind of always-on episodic memory trace, and discusses this in terms of display clutter. Although avoiding display clutter is hardly new advice, it is generally argued that it should be avoided for visual reasons (e.g., Tullis, 1983). However, what Altmann's model shows is that display clutter can also have serious implications for effective use of episodic memory. Clutter can create enormous demands for retrieval. Because more objects will generally be attended on a cluttered display, the episodic trace will be large, lowering the predictive validity for any single cue. Although this certainly seems like a reasonable account on the surface, it is unlikely that kind of analysis would have been generated if it had not been guided by a cognitive architecture that provided the omnipresent learning of Soar.

A second implication of Altmann's model is the surprising potential utility of browsing. Simple browsing creates an episodic trace of the objects encountered, regardless of intent, and thus browsing a complex interface may pay off later in the learning curve. Of course, for this to be most effective, the interface has to be structured correctly to best support retrieval of the episodic trace, and it is not clear exactly what the best organization would be to support such browsing.

Soar has also been used to implement models of exploratory learning, somewhat in the spirit of LICAI. There are two prominent models here, one called IDXL (Rieman, Young, & Howes, 1996) and a related model called Task-Action Learner (Howes & Young, 1996). These models both attempt to learn unfamiliar GUI interfaces. IDXL operates in the same graphing domain as LICAI, whereas the Task-Action Learner starts with even less knowledge and learns basic GUI operations, such as how to open a file. For brevity, only IDXL will be described in detail.

IDXL goes through many of the same scanning processes as LICAI, but must rely on very different mechanisms for evaluation, because Soar is fundamentally different than LICAI. IDXL models evaluation of various display elements as search through multiple problem spaces, one that is an internal search through Soar's internal knowledge and the other a search through the display. As items are evaluated in the search, Soar learns productions that summarize the products of each evaluation. At first, search is broad and shallow, with each item receiving a minimum of elaboration. However, that prior elaboration guides the next round of elaboration, gradually allowing IDXL to focus in on the "best" items. This model suggests a number of ways in which interface designers could thus help learners acquire the knowledge needed to utilize a new interface. Like the LICAI work, the IDXL work highlights the need for good labels to guide exploration. A more radical suggestion is based on one of the more subtle behavior of users and IDXL. When exploring and evaluating alternatives, long pauses often occur on particular menu items. During these long pauses, IDXL is attempting to determine the outcome of selecting the menu item being considered. Thus, one suggestion for speeding up learning of a new menu-driven GUI is to detect such pauses, and show (in some easily-undoable way) what the results of selecting that item would be. For instance, if choosing that item brings up a dialog box for specifying certain options, that dialog box could be shown in some grayed-out form, and would simply vanish if the user moved off that menu item. This would make the evaluation

of the item much more certain and would be an excellent guide for novice users. This is not unlike ToolTips for toolbar icons, but on a much larger scale.

A model that does an excellent job of highlighting the power of cognitive architectures is NTD-Soar (Nelson, Lehman, & John, 1994). NTD stands for "NASA Test Director," who

> . . . is responsible for coordinating many facets of the testing and preparation of the Space Shuttle before it is launched. He must complete a checklist of launch procedures that, in its current form, consists of 3000 pages of looseleaf manuals . . . as well as graphical timetables describing the critical timing of particular launch events. To accomplish this, the NTD talks extensively with other members of launch team over a two-way radio. . . . In addition to maintaining a good understanding of the status of the entire launch, the NTD is responsible for coordinating troubleshooting attempts by managing the communication between members of the launch team who have the necessary expertise. (p. 658)

Constructing a model that is even able to perform this task at all is a significant accomplishment. Nelson et al. were able to not only build such a model, but also this model was able to produce a timeline of behavior that closely matched the timeline produced by the actual NTD being modeled. That is, the ultimate result was a quantitative model of human performance and an accurate one at that.

It is unlikely that such an effort could have been accomplished without the use of an integrated cognitive architecture. This was a Soar model that made use of other Soar models. Nelson et al. did not have to generate and implement theory of natural language understanding to model the communication between the NTD and others, or the NTD reading the pages in the checklist, because one had already been constructed in Soar (Lewis, 1993). They did not have to construct a model of visual attention to manage the scanning and visual search of those 3,000 pages, because such a model already existed in Soar (Weismeyer, 1992). There was still a great deal of knowledge engineering that had to go on to understand and model this complex task, but using an integrated architecture greatly eased the task of the modelers.

Although this modeling effort was not aimed at a particular HCI problem, it is not difficult to see how it would be applicable to one. If one wanted to replace the 3,000-page checklist with something like a personal digital assistant (PDA), how could the PDA be evaluated? There are very few NTDs in the world, and it is unlikely that they would be able to devote much time to participatory design or usability testing. However, because an appropriate quantitative model of the NTD exists, it should be possible to give the model a simulated PDA and assess the impact of that change on the model's performance. Even if the model does not perfectly capture the effects of the change, it is likely that the model would identify problem areas and at least guide the developers in using any time they have with an actual NTD.

Soar has also been used as the basis for simulated agents in wargames (Jones et al., 1999). This model (TacAir-Soar) participates in a virtual battlespace in which humans also participate. TacAir-Soar models take on a variety of roles in this environment, from fighter pilots to helicopter crews to refueling planes. Because they are based on a cognitive architecture, they function well as agents in this environment. Their interactions are more complex than simple scripted agents, and they can interact with humans in the environment with English natural language. This is an ambitious model containing over 5,000 production rules. One of the major goals of the project is to make sure that TacAir-Soar produces humanlike behavior, because this is critical to their role, which is to serve as part of training scenarios for human soldiers. In large-scale simulations with many entities, it is much cheaper to use computer agents than to have humans fill every role in the simulation. Although agents (other than the ubiquitous and generally disliked paper clip) have not widely penetrated the common desktop interface, this remains an active HCI research area, and future agents in other roles could also be based on cognitive architecture rather than more engineering-oriented artificial intelligence models.

There have been many other applications of Soar to HCI-related problems. Ritter, Baxter, Jones, and Young (2000) report on a number of Soar models of GUI-based tasks. So the applicability of Soar to the development of a quantitative theory of HCI is clear. Soar has particular strengths in this regard relative to the other architectures covered in this chapter. In particular, Soar is more focused on learning than any of the other systems, and learnability is clearly an important property for many user interfaces. Second, it is known that Soar models scale up to very complex tasks, such as NTD-Soar and TacAir-Soar; tasks of this complexity have not been modeled in other architectures.

EPIC

With the possible exception of the NTD model, all of the models discussed up to this point have been almost purely cognitive models. That is, the perception and action parts of the models have been handled in an indirect, abstract way. These models focus on the cognition involved, which is not surprising given the generally cognitive background of these systems. However, even the original formulation of the MHP included processors for perception and motor control. In fact, user interfaces have also moved from having almost exclusively cognitive demands (e.g., one had to remember or problem solve to generate command names) to relying much more heavily on perceptual-motor capabilities. This is one of the hallmarks of the GUI, the shift to visual processing and direct manipulation rather than reliance on complex composition of commands.

However, providing accurate quantitative models for this kind of activity requires a system with detailed models of human perceptual and motor capabilities. This is one of the major foci and contributions of EPIC (for executive process interactive control). EPIC is the brainchild of Kieras and Meyer (see, especially, 1996, 1997; Kieras, Wood, & Meyer, 1997). The overall structure of the processors and memories in EPIC is shown in Fig. 5.2. This certainly bears some surface similarity to the MHP, but EPIC is substantially more detailed. EPIC was explicitly designed to pair high-fidelity models of perception and motor mechanisms with a production system. The perceptual-motor processors represent a new synthesis of the human performance literature, whereas the production system is the same one used in the CCT work discussed earlier.

Constructing a model in EPIC thus requires specification of both the knowledge needed in the form of production rules, as well as some relevant perceptual-motor parameters. Because there are a number of processors, there are quite a number of (mostly numeric) parameters in EPIC. There are two types of parameters in EPIC: standard, which are system parameters believed to be fixed across all tasks, and typical, which are free to vary across task situations, but have more-or-less conventional values. A standard parameter in EPIC is the duration of a production cycle in the Cognitive Processor, this is 50 ms. An example of a typical value is the time it takes the Visual Processor to recognize that a particular shape represents a right arrow, which is 250 ms.

All the processors in EPIC run in parallel with one another. So, while the Visual Processor is recognizing an object on the screen, the Cognitive Processor can be deciding what word should be spoken in response to some other input, while at the same time the Manual Motor Processor is pressing a key. The information flow is typical of traditional psychological models, with information coming in through the eyes and ears, and outputs coming from the mouth and hands. More specifically, what is modeled in each of the processors is primarily time course. EPIC's Visual Processor does not take raw pixels as input and compute that those pixels represent a letter A, instead, it determines whether the object on the screen can be seen and at what level of detail, and how long it will take for a representation of that object to be delivered to EPIC's declarative memory once the letter becomes available to the Visual Processor. The appearance of the letter can actually cause a number of different elements to be deposited into EPIC's declarative memory at different times (e.g., information about the letter's color will be delivered before information about the letter's identity).

Similarly, on the motor side, EPIC does not simulate the computation of the torques or forces needed to produce a particular hand movement. Instead, what EPIC computes is the time it will take for a particular motor output to be produced after the Cognitive Processor has requested it. This is complicated by the fact that movements happen in phases. Most importantly, each movement includes a preparation phase and an execution phase. The time to prepare a movement is dependent on the number of movement features that must be prepared for each movement and the features of the last movement prepared. Features that have been prepared for the previous movement can sometimes be reused, saving time. EPIC can make repeated identical movements rapidly because there is no feature preparation time necessary if the movements are identical. If they are not identical, the amount of savings is a function of how different the current movement is from the previous one. After being prepared, a movement is executed. The execution time for a movement corresponds roughly to the time it physically takes to execute the movement; the execution time for aimed movements of the hands or fingers are governed by Fitts's law, which was described previously. EPIC's motor processors can only prepare one movement at a time and can only execute one movement at a time, but may be preparing one movement while executing another. Thus, in some tasks, it may be possible to pipeline movements effectively to generate very rapid sequences of movements.

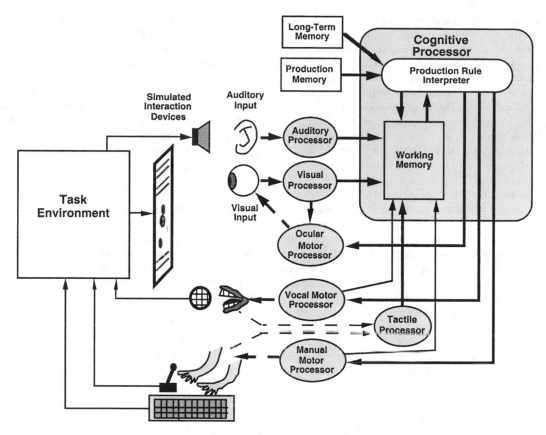

FIGURE 5.2. Overall structure of the EPIC architecture. From *The EPIC Architecture: Principles of Operation* [On-line], by D. E. Kieras & D. E. Meyer, 1996. Available: ftp://ftp.eecs.umich.edu/people/kieras/ EPICarch.ps. Adapted with permission.

EPIC's Cognitive Processor is a production system, the same one that was used for the earlier CCT work. One highly salient feature of this system is that multiple rules are allowed to fire on a production cycle. In fact, there is no upper bound on the number of productions that can fire on a cycle. Productions in this system are at a much higher grain size than productions in Soar, which gives EPIC a highly parallel quality at all levels. That is, all the processors work in parallel and EPIC's Cognitive Processor is itself capable of parallel processing.

This allows EPIC particular leverage in multiple-task situations. When more than one task is being performed, the tasks can execute in parallel. However, many of the perceptual-motor processors are effectively serial. People only have one set of eyes that can only be aimed at one place at a time; so, if multiple tasks are ongoing and they both require the eyes, there must be something that arbitrates. In EPIC, this additional knowledge about how to manage multiple tasks is termed *executive* knowledge, and the productions that implement this knowledge execute in parallel with the productions implementing the task knowledge.

Why is all this machinery and extra knowledge necessary? Because the world of HCI is changing. The GUI forced designers and analysts to consider more seriously the perceptual-motor constraints, and the propagation of computers with user interfaces away from the desktop and into mobile phones, kiosks, automobiles, and many, many other places create a huge demand on people's ability to multitask. Multiple-task issues have largely gone unmodeled and outside the theoretical scope of most psychological accounts in HCI, at least before EPIC.

Although LICAI and Soar have not been adequately equipped to deal with high-performance perception and action components of many tasks, EPIC is not equipped to handle some of the issues covered by other architectures. In particular, EPIC does not include any learning mechanisms, so it would be difficult to generate EPIC models for many of the domains Soar has approached successfully. However, this is not a fatal shortcoming, because there are a wide variety of domains in which learning is not an enormously key component and where high-fidelity modeling of perception and action, along with multiple-tasking, are central.

These are the kinds of domains to which EPIC has been applied. One of the first major applications of EPIC was to a deceptively simple dual-task paradigm known as the psychological refractory period (or PRP; see Meyer & Kieras, 1997). In this task, laboratory subjects are typically confronted with two choice reaction time tasks, something on the order of "either a red light or a green light will appear; if it's red, hit the 'L' key; if it's green, hit the 'J' key." This sounds simple, but the empirical literature is rich and shows a variety of subtle effects, for which EPIC provides the first unified account. Critically, what the EPIC models of these experiments show is that people's low-level strategies for scheduling the tasks play a large role in determining performance in this simple paradigm.

EPIC has also been used to model some of the classic psychological results from the short-term memory (or working memory) literature (Kieras, Meyer, Muller, & Seymour, 1999). This concerns the question of how much people can remember when they repeat a string of words (or numbers) to themselves. For example, when someone reads off a telephone number to someone else who has to dial that number a few moments later, the person who has to remember the number often speaks the number repeatedly. This is called *articulatory rehearsal* in the psychology literature, and although the phenomenon had been described in detail over the years, EPIC was the first serious quantitative model of this process. An accurate model of this process is clearly important in HCI applications, because many interfaces force people to remember many things over short periods, and rehearsal is a likely response to that demand. The field should have a better answer to understanding this problem than simply an admonition to reduce working memory demand. Although such an admonition is certainly good advice, more precise performance prediction is often warranted.

EPIC has been used to model several tasks with a more HCI-oriented flavor. One of those tasks is menu selection (Hornof & Kieras, 1997, 1999; Kieras & Meyer, 1997); but, for brevity, a detailed description of these models will be omitted. Another application of EPIC that definitely merits mention is the model of telephone assistance operators (TAOs), data originally presented in Gray et al. (1993). When a telephone customer dials "0," a TAO is the person who answers. The TAOs modeled here sat at a dumb terminal-style workstation and assisted customers in completing telephone calls. In particular, TAOs determine how calls should be billed, and this is done by speaking to the customer. The detailed EPIC models (Kieras et al., 1997) covered a subset of the possible billing types.

This provided a good test of EPIC because the task is performed under time pressure, and seconds—actually, milliseconds—counted in task performance. Second, this task is multimodal. The TAO must speak, type, listen, and look at a display. Third, very fine-grained performance data were available to help guide model construction. By now, it should come as no surprise to the reader that it was possible to construct an EPIC model that did an excellent job of modeling the time course of the TAO's interaction with the customer. However, this modeling effort went beyond just that and provided some insight into the knowledge engineering problem facing modelers using cognitive architectures as well.

Like other production system models, EPIC provides a certain amount of freedom to the modeler in model construction. Although the architecture used provides certain kinds of constraints, and these constraints are critical in doing good science and affecting the final form of the model (Howes & Young, 1997), the modeler does have some leeway in writing the production rules in the model. This is true even when the

production rule model is derived from another structured representation, such as a GOMS model, which was the case in the TAO model. In EPIC, it is possible to write a set of "aggressive" productions that maximize the system's ability to process things in parallel, although it is also possible to write any number of less aggressive sets representing more conservative strategies. EPIC will produce a quantitative performance prediction regardless of the strategy, but which kind of strategy should the modeler choose? There is generally no a priori basis for such a decision, and it is not clear that people can accurately self-report on such low-level decisions.

Kieras et al. (1997) generated an elegant approach to this problem, later termed *bracketing* (Kieras & Meyer, 2000). The idea is this: construct two models, one of which is the maximally aggressive version. At this end of the strategy spectrum, the models contain very little in the way of cognition. The Cognitive Processor does virtually no deliberation and spends most of its cycles simply reading off perceptual inputs and immediately generating the appropriate motor output. This represents the super-expert whose performance is limited almost entirely by the rate of information flow through the peripherals. At the other end of the spectrum, a model incorporating the slowest reasonable strategy is produced. The slowest reasonable strategy is one where the basic task requirements are met, but with no strategic effort made to optimize scheduling to produce rapid performance. The idea is that observed performance should fall somewhere in between these two extremes. Different users will tend to perform at different ends of this range for different tasks; so, this is an excellent way to accommodate some of the individual differences that are always observed in real users.

What was discovered by using this bracketing procedure to the TAO models was surprising. Despite the fact that the TAOs were under considerable time pressure and were extremely well-practiced experts, their performance rarely approached the fastest possible model. In fact, their performance most closely matched a version of the model termed the *hierarchical motor-parallel model*. In this version of the model, eye, hand, and vocal movements are executed in parallel with one another when possible; furthermore, the motor processor is used somewhat aggressively, preparing the next movement whereas the current movement was in progress. The primary place where EPIC could be faster, but the data indicated the TAOs were not, was in the description of the task knowledge. It is possible to represent the knowledge for this task as one single, flat GOMS method with no use of subgoals. On the other hand, the EPIC productions could represent the full subgoal structure or a more traditional GOMS model. Retaining the hierarchical representation—thus incurring time costs for goal management—provided the best fit to the TAOs performance. This provides solid evidence for the psychological reality of the hierarchical control structure inherent in GOMS analysis, because even well-practiced experts in fairly regular domains do not abandon it for some kind of faster knowledge structure.

The final EPIC-only model that will be considered is the model of the task first presented in Ballas, Heitmeyer, and Perez (1992). Again, this model first appeared in Kieras and Meyer (1997), but a richer version of the model is described in more detail later, in Kieras, Meyer, and Ballas (2001). The display used is a split screen, on the right half of the display, the user is confronted with a manual tracking task that is performed using a joystick. The left half of the display is a tactical task in which the user must classify objects as hostile or neutral based on their behavior. There were two versions of the interface to the tactical task: one a command-line style interface using a keypad and one a direct-manipulation-style interface using a touchscreen. The performance measure of interest in this task is the time taken to respond to events (such as the appearance of a new object or a change in state of an object) on the tactical display.

This is again a task well-suited to EPIC because the perceptual-motor demands are fairly extensive. This is not, however, what makes this task so interesting. What is most interesting is the human performance data: in some cases, the keypad interface was faster than the touchscreen interface, and in many cases the two yielded almost identical performance, and in some other cases the touchscreen was faster. Thus, general claims about the superiority of GUIs do not apply to this case, a more precise and detailed account is necessary.

EPIC provides just the tools necessary to do this. Two models were constructed for each interface, again using the bracketing approach. The results were revealing. In fact, the fastest possible models showed no performance advantage for either interface. The apparent direct-manipulation advantage of the touchscreen for initial target selection was almost perfectly offset by some type-ahead advantages for the keypad. The reason for the inconsistent results is that the users generally did not operate at the speed of the fastest possible model; they tended to work somewhere in between the brackets for both interfaces. However, they tended to work more toward the upper (slowest reasonable) bracket for the touchscreen interface. This suggests an advantage for the keypad interface, but the caveat is that the slowest reasonable performance bound for the touchscreen was faster than the slowest possible for the keypad. Thus, any strategy changes made by users in the course of doing the task, perhaps as a dynamic response to changes in workload, could affect which interface would be superior at any particular point in the task. Thus, results about which interface is faster are likely to be inconsistent—exactly what was found.

This kind of analysis would be impossible to conduct without a clear quantitative human performance model. Constructing and applying such a model also suggested an alternative interface that would almost certainly be faster than either, which is one using a speech-driven interface. One of the major performance bottlenecks in the task was the hands, and so voice-based interaction should, in this case, be faster. Again, this could only be clearly identified with the kind of precise quantitative modeling enabled by something like the EPIC architecture.

Despite this, the EPIC architecture is sometimes considered a less complete architecture than others because it does not include a learning mechanism, and thus would be unsuitable for the kinds of interfaces and tasks modeled with LICAI or Soar. However, LICAI and Soar might conversely be considered incomplete for their lack of detailed specification of perceptual-motor processors; the definition of a complete cognitive architecture is not entirely clear at this point. However, this incompleteness has been acknowledged by both the EPIC community and the Soar community, and some experimentation has

been done with a union of the two architectures, using Soar in place of EPIC's Cognitive Processor. This fusion is called EPIC-Soar and some of the initial results have been promising. For example, Chong and Laird (1997) demonstrated that Soar could indeed learn at least some of the complex executive knowledge needed to manage a dual-task situation (again, a combination of a tracking task and a simple decision-making task) in EPIC. Lallement and John (1998) looked at that same task with a slightly different version of EPIC-Soar that effectively negated the cognitive parallelism found in EPIC. One of the things they found was that the cognitive parallelism in EPIC was not actually necessary for performance in this particular task; whether this is true for other tasks is still an open question. Although the future of EPIC-Soar is uncertain at this point, the fact that the integration was not only possible but viable in that it has produced several running models is encouraging for a complete vision of cognitive architecture.

ACT-R/PM

ACT-R/PM (Byrne & Anderson, 1998, 2001; Byrne, 2001) represents another approach to a fully unified cognitive architecture, combining a very broad model of cognition with rich perceptual-motor capabilities. ACT-R/PM is an extension of the ACT-R cognitive architecture (Anderson, 1993; Anderson & Lebiere, 1998) with a set of perceptual-motor modules like those found in EPIC (hence ACT-R/PM). ACT-R has a long history within cognitive psychology, because various versions of the theory have been developed over the years. In general, the ACT family of theories have been concerned with modeling the results of psychology experiments, and this is certainly true of the current incarnation, ACT-R. Anderson and Lebiere (1998) show some of this range, covering areas as diverse as list memory (chapter 7), choice (chapter 8), and scientific reasoning (chapter 11).

However, ACT-R was not originally designed to model things like multimodal, multiple-task situations like those EPIC was designed to handle. Although ACT-R was certainly moving toward application to GUI-style interactions (Anderson, Matessa, & Lebiere, 1997), it was not as fully developed as EPIC. In fact, the standard version of ACT-R is incapable of showing any kind of time savings in even a simple dual-task situation because its operation is entirely serial. This issue is corrected with the inclusion of the PM portion of ACT-R/PM. In many ways, ACT-R/PM is a fusion of EPIC and ACT-R much in the spirit of EPIC-Soar. There are some differences, however, and the extent to which those differences result in serious differences in models in the HCI domain is not yet clear.

The ACT-R system at the heart of ACT-R/PM is, like EPIC and Soar, a production system with activity centered around the production cycle, which is also set at 50 ms in duration. However, there are many differences between ACT-R and the other architectures. First, ACT-R can only fire one production rule per cycle. When multiple production rules match on a cycle, an arbitration procedure called conflict resolution comes into play. Second, ACT-R has a well-developed theory of declarative memory. Unlike EPIC and Soar, declarative memory elements

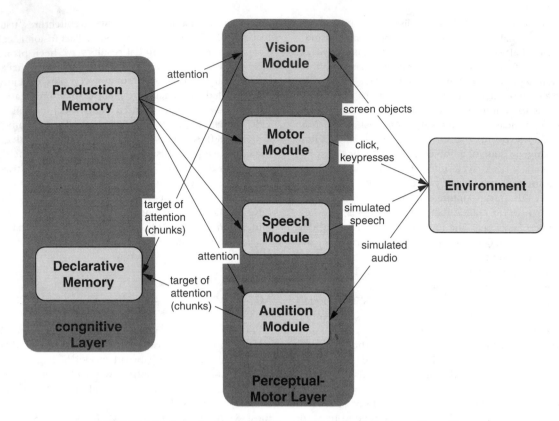

FIGURE 5.3. Overall structure of the ACT-R/PM architecture. From *The Atomic Components of Thought* (p. 173), by M. D. Byrne & J. R. Anderson, 1998, Hillsdale, NJ: Erlbaum. Copyright 1998 by Lawrence Erlbaum Associates. Adapted with permission.

in ACT-R are not simply symbols. Each declarative element in ACT-R also has associated with it an activation value, which determines whether and how rapidly it may be accessed. Third, ACT-R contains learning mechanisms, but is not a pervasive learning system in the same sense as Soar. These mechanisms are based on a rational analysis (Anderson, 1990) of the information needs of an adaptive cognitive system.

For example, consider conflict resolution. Each production in ACT-R has associated with it several numeric parameters, including numbers that represent the probability that if the production fires, the goal will be reached and the cost, in time, that will be incurred if the production fires. These values are combined according to a formula that trades off probability of success vs. cost and produces an expected gain for each production. The matching production with the highest expected gain is the one that gets to fire when conflict resolution is invoked. The expected gain values are noisy, so the system's behavior is somewhat stochastic, and the probability and cost values are learned over time so that ACT-R can adapt to changing environments.

Similarly, the activation of elements in declarative memory is based on a Bayesian analysis of the probability that a declarative memory element will be needed at a particular time. This is a function of the general utility of that element, reflected in what is termed its base-level activation, and that element's association with the current context. The more frequently and recently an

element has been accessed, the higher its base-level activation will be, and thus the easier it is to retrieve. This value changes over time according to the frequency and recency of use, thus, this value is learned. Associations between elements may also be learned, so that the activation an element receives based on its association with the current context can change as well. These mechanisms have helped enable ACT-R to successfully model a wide range of cognitive phenomena.

In ACT-R/PM, the basic ACT-R production system is augmented with four EPIC-like peripheral modules, as depicted in Fig. 5.3. Like EPIC, all of these modules run in parallel with one another, giving ACT-R the ability to overlap processing. The peripheral modules come from a variety of sources. ACT-R/PM's Vision Module is based on the ACT-R Visual Interface described in Anderson et al. (1997). This is a feature-based attentional visual system, but does not explicitly model eye movements. Recently, the Vision Module has been extended to include an eye movement model (Salvucci, 2001a) as well. The Motor Module is nearly identical to the Manual Motor Processor in EPIC and is based directly on the specification found in Kieras and Meyer (1996), and the Speech Module is similarly derived from EPIC. The Audition Module is a hybrid of the auditory system found in EPIC and the attentional system in ACT-R/PM's Vision Module.

One other important property of ACT-R/PM is that it is possible to have ACT-R/PM interact with the same software as the human users being modeled. There are some fairly restrictive conditions on how the software must be developed; but, if these conditions are met, then both the user and the model are forced to use the same software. This reduces the number of degrees of freedom available to the modeler in that it becomes impossible to force any unpleasant modeling details into the model of the user interface, because there is no model of the user interface. More will be said about this issue in the next section.

As described in the section on EPIC, multiple tasking is becoming increasingly critical to HCI endeavors. Thus, one of the first major modeling efforts with ACT-R/PM has been to dual-task phenomena; in fact, the same kinds of simple PRP dual tasks to which EPIC has been applied (Byrne & Anderson, 2001). However, as the most recent architecture of those described here, ACT-R/PM has not yet been as widely applied to HCI tasks as have the others. However, there have been some recent models that are more directly HCI endeavors than the PRP work.

The first example comes from the dissertation work of Ehret (1999). Among other things, Ehret developed an ACT-R/PM model of a fairly simple, but subtle, experiment. In that experiment, subjects were shown a target color and asked to click on a button that would yield that color. The buttons themselves had four types: blank, arbitrary icon, text label, and color. In the color condition, the task was simple: Users just found the color that matched the target, then clicked the button. In the text label condition, the task was only slightly more difficult: Users could read the labels on the buttons and select the correct one because the description matched the color. In the arbitrary icon condition, more or less random pictures appeared on each icon (e.g., a mailbox). Users had to either memorize the picture to color mapping, which they had to discover by trial and error, or memorize the location of each color, because the buttons did not change their function over time. In the hardest condition, the blank condition, users simply had to memorize the mapping between button location and color, which they had to discover through trial and error.

Clearly, the conditions will produce different average response times, and what Ehret found is that they also produced somewhat different learning curves over time. Ehret added an additional manipulation as well: After performing the task for some time, all the labeling was removed. Not surprisingly, the amount of disruption was different in the different conditions, reflecting the amount of incidental location learning that went on as subjects performed the task. The ACT-R/PM model that Ehret constructed did an excellent job of explaining the results. This model represented the screen with the built-in visual mechanisms from ACT-R/PM and learned the mappings between color and location via ACT-R's standard associative learning mechanisms. The initial difference between the various conditions was reproduced, as were the four learning curves. The model also suffered disruptions similar to those suffered by the human users when the labels were removed. This model is an excellent demonstration of the power of ACT-R/PM, exercising both the perceptual-motor capabilities of the system, as well as the graded learning in ACT-R's rational analysis-driven mechanisms.

Salvucci (2001b) describes an ACT-R/PM model that tests ACT-R/PM's ability to handle multimodal, high-performance situations in a very compelling task: this model drives an automobile driving simulator. This is not a robotics project; the ACT-R/PM model does not actually turn the steering wheel or manipulate the pedals, but rather it communicates with the automobile simulation software. The model's primary job is to maintain lane position as the car drives down the road. Salvucci (2001b) adds an additional task that makes it particularly interesting: The model dials telephone numbers on a variety of mobile phone interfaces. There were two factors that were crossed: Whether the telephone was dialed manually via keypad vs. dialed by voice, and whether the full telephone number needed to be dialed vs. a shortened speed-dial system. The model was also validated by comparison with data from human users.

What both the model and the human users showed is that dialing while not driving is faster than dialing while driving, and that steering performance can be disrupted by telephone dialing. Not surprisingly, the most disruptive interface was the full-manual interface, in which the full phone numbers were dialed on a keypad. This is due largely to the fact that dialing with the keypad requires visual guidance, causing the model (and the users) to take their eyes off the road. There was very little disruption associated with the voice interfaces, regardless of whether full numbers or speed-dial was used.

This is a nice illustration of the value of cognitive architectures for a number of reasons. First, the basic driving model could simply be reused for this task; it did not have to be reimplemented. Second, the model provides an excellent quantitative fit to the human data. Third, this is an excellent example of a situation where testing human users can be difficult. Testing human drivers with interfaces that degrade driving performance is dangerous, so simulators are generally used for this kind of evaluation. However, maintaining a driving simulator requires a great deal of space and is quite expensive. If someone wanted to test another variant of the telephone interface, it would be much faster and cheaper to give that interface to Salvucci's model than it would be to recruit and test human drivers.

There is other published and ongoing work applying ACT-R/PM to HCI problems (e.g., Byrne, 2001; Schoelles & Gray, 2000), but space considerations prohibit a more exhaustive review. The vitality of the research effort suggests that ACT-R/PM's combination of perceptual-motor modules and a strong theory of cognition will pay dividends as an HCI research tool.

In fact, this is not limited to ACT-R/PM; overall, cognitive architectures are an exciting and active area of HCI research. The four systems described here all take slightly different approaches and focus on slightly different aspects of various HCI problems, but there is clearly a great deal of cross-pollination. Lessons learned and advancements made in one architecture often affect other systems (e.g., the development of EPIC's peripheral systems clearly impacted both Soar and ACT-R).

Comparisons

An exhaustive and detailed comparison of the major cognitive architectures is beyond the scope of this chapter; however, an excellent comparison that includes a number of other architectures can be found in Pew and Mavor (1998). Certainly, the three production systems—Soar, EPIC, and ACT-R/PM—are related.

TABLE 5.1. Architecture Feature Comparison

	LICAI/CoLiDeS	Soar	EPIC	ACT-R/PM
Original focus	Text comprehension	Learning and problem solving	Multiple-task performance	Memory and problem solving
Basic cycle	Construction-integration	Decision cycle	Production cycle (parallel)	Production cycle (serial)
Symbolic or activation-based?	Both	Symbolic	Symbolic	Both
Architectural goal management	Special cycle types, supports vague goals	Universal subgoaling	None	Goal stack
Detailed perceptual-motor systems	No	No	Yes	Yes
Learning mechanisms	No	Yes, pervasive	No	Yes, but not pervasive
Large, integrated models	No	Yes	No	No
Extensive natural language	Yes	Yes	No	No
Support for users learning system	None	FAQs, some tutorial materials	None	Extensive tutorial materials, summer school
User community[a]	None	Some, primarily AI	None	Growing, primarily psychology

Note. FAQs = frequently asked questions.
[a] Outside of the researchers who have developed the system.

A major difference between them is their original focus; they were originally developed to model slightly different aspects of human cognition. However, as they develop, there appears to be more convergence than divergence. This is generally taken to be a good sign that the science is cumulating. Still, there are differences, and certainly between the production systems and LICAI/CoLiDeS. Many of the relevant comparisons are summarized in Table 5.1.

This table does not include the hybrid EPIC-Soar system because it is not clear what is in store for that system; however, this system would essentially contain most of the attributes of Soar, but would get a "yes" on the detailed perceptual-motor systems feature. Most of the entries on this table have been discussed previously, with the exception of the last two. Support for learning will be discussed in the next section. The presence and size of the user community have not been discussed, because it is not clear what role (if any) such a community plays in the veridicality of the predictions made by the system. However, it may be relevant to researchers for other reasons, particularly those trying to learn the system.

In addition, many of the values in this table are likely to change in the future. For example, a more pervasive learning mechanism for ACT-R is slated to be a part of the next revision of the theory. Whether this will result in ACT-R being as successful as Soar in modeling things like exploratory learning is not yet clear.

It is difficult to classify the value an architecture has on a particular attribute as an advantage or a disadvantage, because what constitutes an advantage for modeling one phenomenon may be a disadvantage for modeling others. For example, consider learning in Soar. Certainly, when attempting to model the improvement of users over time with a particular interface, Soar's

learning mechanism is critical. However, there are many applications for which modeling learning is not critical, and Soar's pervasive learning feature occasionally causes undesired side effects that can make model construction more difficult.

THE FUTURE OF COGNITIVE ARCHITECTURES IN HUMAN–COMPUTER INTERACTION

Beyond the incremental development and application of architectures like Soar and ACT-R/PM, what will the future hold for cognitive architectures in HCI? What are the challenges faced, and what is the ultimate promise? Currently, there are indeed a number of important limitations for cognitive architectures. There are questions they cannot yet address, and questions it is hard to see how they even would address. Other limitations are more pragmatic than in principle, but these are relevant as well.

First, there are a wide array of HCI problems that are simply outside the scope of current cognitive architectures. Right now, these architectures focus on cognition and performance, but there are other aspects of HCI, such as user preference, boredom, aesthetics, fun, and so on. Another important challenge, although one that might be overcome, is that these architectures have generally not been applied to social situations, such as those encountered in groupware or online communities (Olson & Olson, this volume; Preece & Maloney-Krichmar this volume). It is not in principle impossible to implement a model of social interaction in a cognitive architecture; however, the knowledge engineering problem here would certainly be a difficult one. How does one characterize and implement knowledge about social situations with enough precision to be implemented in a production system? It may ultimately be possible to do so, but it is unlikely that this will happen anytime soon.

One problem that will never entirely be resolved, no matter how diligent the modelers are, is the knowledge engineering problem. Every model constructed using a cognitive architecture still needs knowledge about how to use the interface and what the tasks are. By integrating across models, the knowledge engineering demands when entering a new domain may be reduced (the NTD is a nice example), but they will never be eliminated. This requirement will persist even if an architecture was to contain a perfect a theory of human learning—and there is still considerable work to be done to meet that goal.

Another barrier to the more widespread use of cognitive architectures in HCI is that the architectures themselves are large and complex pieces of software, and (ironically) little work has been done to make them usable or approachable for novices. For example: "EPIC is not packaged in a 'user-friendly' manner; full-fledged Lisp programming expertise is required to use the simulation package, and there is no introductory tutorial or user's manual." (Kieras & Meyer, 1997, p. 399). The situation is slightly better for Soar, which does have a frequently asked questions Web resource (http://ritter.ist.psu.edu/soar-faq/) and some tutorial materials (http://www.psychology.nottingham.ac.uk/staff/Frank.Ritter/pst/pst-tutorial.html). However, Soar has a reputation as being difficult to learn and use. Tutorial materials, documentation, and examples for ACT-R are available, and most years there is a 2-week "summer school" for those interested in learning ACT-R (see http://act.psy.cmu.edu/). However, the resources for ACT-R/PM are somewhat more limited, although there are some rudimentary examples and documentation (see http://chil.rice.edu/byrne/RPM/).

Another limiting factor is implementation. In order for a cognitive architecture to accurately model interaction with an interface, it must be able to communicate with that interface. Because most user interfaces are "closed" pieces software with no built-in support for supplying a cognitive model with the information it needs for perception (i.e., what is on the screen where) or accepting input from a model, this creates a technical problem. Somehow, the interface and the model must be connected. An excellent summary of this problem can be found in Ritter et al. (2000). A number of different approaches have been taken. In general, the EPIC solution to this problem has been to reimplement the interface to be modeled in Lisp, so the model and the interface can communicate via direction function calls. The ACT-R/PM solution is not entirely dissimilar. In general, ACT-R/PM has only been applied to relatively new experiments or interfaces that were initially implemented in Lisp, and thus ACT-R/PM and the interface can communicate via function calls. To facilitate the construction of models and reduce the modeler's degrees of freedom in implementing a custom interface strictly for use by a model, ACT-R/PM does provide some abilities to automatically manage this communication when the interface is built with the native GUI builder for Macintosh Common Lisp under MacOS and Allegro Common Lisp under Windows. If the interface is implemented this way, both human users and the models can interact with the same interface.

Despite these limitations, this is a particularly exciting time to be involved in research on cognitive architectures in HCI. There is a good synergy between the two areas, as cognitive architectures are certainly useful to HCI, so HCI is also useful for cognitive architectures. HCI is a complex and rich yet still tractable domain, which makes it an ideal candidate for testing cognitive architectures. HCI tasks are more realistic and require more integrated capabilities than typical cognitive psychology laboratory experiments, and thus cognitive architectures are the best theoretical tools available from psychology. Theories like EPIC-Soar and ACT-R/PM are the first complete psychological models that go from perception to cognition to action in detail. This is a significant advance and holds a great deal of promise.

Even some of the limitations have generated solid research in an attempt to overcome them. For example, Ritter et al. (2000) describes the implementation of a generic *sim-eye* and *sim-hand* as part of a more general program researching of what they term *cognitive model interface management systems*. For instance, they have implemented a virtual hand and eye in Tcl/Tk that can interact with either Soar or ACT-R via a socket-based interface. This is not a trivial technical accomplishment and may serve to help make it easier to develop computational cognitive models that interact with software systems.

The most intriguing development along this line, however, is recent work by St. Amant and Riedl (2001). They have implemented a system called VisMap, which directly parses the screen bitmap on Windows systems. That is, given a Windows display—any Windows display—VisMap can parse it and identify things like text, scroll bars, GUI widgets, and the like. It also has facilities for simulating mouse and keyboard events. This is an intriguing development, because it should in principle be possible to connect this to an architecture like EPIC or ACT-R, which would enable the architecture to potentially work with any Windows application in its native form.

Although there are still a number of technical details that would have to be worked out, this has the potential of fulfilling one of the visions held by many in the cognitive architecture community: a high-fidelity virtual user that could potentially use any application or even combination of applications. Besides providing a wide array of new domains to researchers, this could be of real interest to practitioners as well because this opens the door for at least some degree of automated usability testing. This idea is not a new one (e.g., Byrne, Wood, Sukaviriya, Foley, & Kieras, 1994; St. Amant, 2000), but technical and scientific issues have precluded its adoption on even a limited scale. This would not eliminate the need for human usability testing (see Ritter & Young, 2001, for a clear discussion of this point) for some of the reasons listed previously, but it could significantly change usability engineering practice in the long run.

The architectures themselves will continue to be updated and applied to more tasks and interfaces. There is a new version (version 5.0) of ACT-R currently under development, and this new version has definitely been influenced by issues raised by the PM system and numerous HCI concerns. New applications of EPIC result in new mechanisms (e.g., similarity-based decay in verbal working memory storage; Kieras et al., 1999) and new movement styles (e.g., click-and-point; Hornof & Kieras, 1999).

Applications like the World Wide Web are likely to drive these models into more semantically rich domain areas, and tasks that involve greater amounts of problem solving are also likely candidates for future modeling.

The need for truly quantitative engineering models will only grow as user interfaces propagate into more and more places and more and more tasks. Cognitive architectures, which already have a firmly established niche in HCI, seem the most promising road toward such models. Thus, as the architectures expand their range of application and their fidelity to the humans they are attempting to model, this niche is likely to expand. HCI is an excellent domain for testing cognitive architectures as well, so this has been, and will continue to be, a fruitful two-way street.

References

Altmann, E. M. (2001). Near-term memory in programming: A simulation-based analysis. *International Journal of Human-Computer Studies, 54*(2), 189-210.

Altmann, E. M., & John, B. E. (1999). Episodic indexing: A model of memory for attention events. *Cognitive Science, 23*(2), 117-156.

Anderson, J. R. (1990). *The adaptive character of thought.* Hillsdale, NJ: Erlbaum.

Anderson, J. R. (1993). *Rules of the mind.* Hillsdale, NJ: Erlbaum.

Anderson, J. R., Conrad, F. G., & Corbett, A. T. (1989). Skill acquisition and the LISP tutor. *Cognitive Science, 13*(4), 467-505.

Anderson, J. R., & Lebiere, C. (1998). *The atomic components of thought.* Mahwah, NJ: Erlbaum.

Anderson, J. R., Matessa, M., & Lebiere, C. (1997). ACT-R: A theory of higher level cognition and its relation to visual attention. *Human-Computer Interaction, 12*(4), 439-462.

Ballas, J. A., Heitmeyer, C. L., & Perez, M. A. (1992). Evaluating two aspects of direct manipulation in advanced cockpits. *Proceedings of ACM CHI 92 Conference on Human Factors in Computing Systems* (pp. 127-134). New York: ACM.

Bovair, S., Kieras, D. E., & Polson, P. G. (1990). The acquisition and performance of text-editing skill: A cognitive complexity analysis. *Human-Computer Interaction, 5*(1), 1-48.

Byrne, M. D. (2001). ACT-R/PM and menu selection: Applying a cognitive architecture to HCI. *International Journal of Human-Computer Studies, 55*, 41-84.

Byrne, M. D., & Anderson, J. R. (1998). Perception and action. In J. R. Anderson & C. Lebiere (Eds.), *The atomic components of thought* (pp. 167-200). Hillsdale, NJ: Erlbaum.

Byrne, M. D., & Anderson, J. R. (2001). Serial modules in parallel: The psychological refractory period and perfect time-sharing. *Psychological Review, 108*, 847-869.

Byrne, M. D., & Bovair, S. (1997). A working memory model of a common procedural error. *Cognitive Science, 21*(1), 31-61.

Byrne, M. D., Wood, S. D., Sukaviriya, P. N., Foley, J. D., & Kieras, D. E. (1994). Automating interface evaluation. *ACM CHI'94 Conference on Human Factors in Computing Systems* (pp. 232-237). New York: ACM Press.

Card, S. K., Moran, T. P., & Newell, A. (1983). *The psychology of human-computer interaction.* Hillsdale, NJ: Erlbaum.

Chong, R. S., & Laird, J. E. (1997). Identifying dual-task executive process knowledge using EPIC-Soar. In M. Shafto & P. Langley (Eds.), *Proceedings of the Nineteenth Annual Conference of the Cognitive Science Society* (pp. 107-112). Hillsdale, NJ: Erlbaum.

Ehret, B. D. (1999). *Learning where to look: The acquisition of location knowledge in display-based interaction.* Unpublished doctoral dissertation, George Mason University, Fairfax, VA.

Franzke, M. (1994). *Exploration, acquisition, and retention of skill with display-based systems.* Unpublished doctoral dissertation, University of Colorado, Boulder.

Gray, W. D., John, B. E., & Atwood, M. E. (1993). Project Ernestine: Validating a GOMS analysis for predicting and explaining real-world task performance. *Human-Computer Interaction, 8*(3), 237-309.

Gray, W. D., Young, R. M., & Kirschenbaum, S. S. (1997). Introduction to this special issue on cognitive architectures and human-computer interaction. *Human-Computer Interaction, 12*, 301-309.

Hornof, A., & Kieras, D. E. (1997). Cognitive modeling reveals menu search is both random and systematic. *Proceedings of ACM CHI 97 Conference on Human Factors in Computing Systems* (pp. 107-114). New York: ACM.

Hornof, A., & Kieras, D. (1999). Cognitive modeling demonstrates how people use anticipated location knowledge of menu items. *Proceedings of ACM CHI 99 Conference on Human Factors in Computing Systems* (pp. 410-417). New York: ACM.

Howes, A., & Young, R. M. (1996). Learning consistent, interactive, and meaningful task-action mappings: A computational model. *Cognitive Science, 20*(3), 301-356.

Howes, A., & Young, R. M. (1997). The role of cognitive architecture in modeling the user: Soar's learning mechanism. *Human-Computer Interaction, 12*(4), 311-343.

Huguenard, B. R., Lerch, F. J., Junker, B. W., Patz, R. J., & Kass, R. E. (1997). Working memory failure in phone-based interaction. *ACM Transactions on Computer-Human Interaction, 4*, 67-102.

John, B. E., & Kieras, D. E. (1996). The GOMS family of user interface analysis techniques: Comparison and contrast. *ACM Transactions on Computer-Human Interaction, 3*, 320-351.

Jones, R. M., Laird, J. E., Nielsen, P. E., Coulter, K. J., Kenny, P., & Koss, F. V. (1999). Automated intelligent pilots for combat flight simulation. *AI Magazine, 20*(1), 27-41.

Just, M. A., & Carpenter, P. A. (1992). A capacity theory of comprehension: Individual differences in working memory. *Psychological Review, 99*(1), 122-149.

Kieras, D. E., & Bovair, S. (1986). The acquisition of procedures from text: A production-system analysis of transfer of training. *Journal of Memory & Language, 25*(5), 507-524.

Kieras, D. E., & Meyer, D. E. (1996). The EPIC architecture: Principles of operation. Available: ftp://ftp.eecs.umich.edu/people/kieras/EPICarch.ps.

Kieras, D. E., & Meyer, D. E. (1997). An overview of the EPIC architecture for cognition and performance with application to human-computer interaction. *Human-Computer Interaction, 12*(4), 391-438.

Kieras, D. E., & Meyer, D. E. (2000). The role of cognitive task analysis in the application of predictive models of human performance. In J. M. Schraagen & S. F. Chipman (Eds.), *Cognitive task analysis* (pp. 237-260). Mahwah, NJ: Erlbaum.

Kieras, D. E., Meyer, D. E., & Ballas, J. A. (2001). Towards demystification of direct manipulation: cognitive modeling charts the gulf of execution, *Proceedings of ACM CHI 01 Conference on Human Factors in Computing Systems* (pp. 128-135). New York: ACM.

Kieras, D. E., Meyer, D. E., Mueller, S., & Seymour, T. (1999). Insights into working memory from the perspective of the EPIC architecture for modeling skilled perceptual-motor and cognitive human performance. In A. Miyake & P. Shah (Eds.), *Models of working memory: Mechanisms of active maintenance and executive control* (pp. 183–223). New York: Cambridge University Press.

Kieras, D., & Polson, P. G. (1985). An approach to the formal analysis of user complexity. *International Journal of Man-Machine Studies, 22*(4), 365–394.

Kieras, D. E., Wood, S. D., & Meyer, D. E. (1997). Predictive engineering models based on the EPIC architecture for multimodal high-performance human-computer interaction task. *Transactions on Computer-Human Interaction, 4*(3), 230–275.

Kieras, D. E. (1988). Towards a practical GOMS model methodology for user interface design. In M. Helander (Ed.), *Handbook of human-computer interaction* (pp. 135–157). New York: North-Holland.

Kintsch, W. (1998). *Comprehension: A paradigm for cognition.* New York: Cambridge University Press.

Kitajima, M., & Polson, P. G. (1997). A comprehension-based model of exploration. *Human-Computer Interaction, 12*(4), 345–389.

Kitajima, M., Blackmon, M. H., & Polson, P. G. (2000). A comprehension-based model of web navigation and its application to web usability analysis. In S. McDonald & Y. Waern & G. Cockton (Eds.), *People and computers XIV—Usability or else!* (Proceedings of HCI 2000) (pp. 357–373). New York: Springer.

Lallement, Y., & John, B. E. (1998). Cognitive architecture and modeling idiom: An examination of three models of the Wickens's task. In M. A. Gernsbacher & S. J. Derry (Eds.), *Proceedings of the Twentieth Annual Conference of the Cognitive Science Society* (pp. 597–602). Hillsdale, NJ: Erlbaum.

Landauer, T. K., & Dumais, S. T. (1997). A solution to Plato's problem: The latent semantic analysis theory of acquisition, induction, and representation of knowledge. *Psychological Review, 104*(2), 211–240.

Lewis, R. L. (1993). *An architecturally-based theory of human sentence comprehension.* Unpublished doctoral dissertation, University of Michigan, Ann Arbor.

Meyer, D. E., & Kieras, D. E. (1997). A computational theory of executive cognitive processes and multiple-task performance: I. Basic mechanisms. *Psychological Review, 104*(1), 3–65.

Nelson, G., Lehman, J. F., & John, B. E. (1994). Integrating cognitive capabilities in a real-time task. In A. Ram & K. Eiselt (Eds.), *Proceedings of the Sixteenth Annual Conference of the Cognitive Science Society* (pp. 353–358). Hillsdale, NJ: Erlbaum.

Newell, A. (1973). You can't play 20 questions with nature and win: Projective comments on the papers of this symposium. In W. G. Chase (Ed.), *Visual information processing.* New York: Academic Press.

Newell, A. (1990). *Unified theories of cognition.* Cambridge, MA: Harvard University Press.

Newell, A., & Simon, H. A. (1963). GPS, a program that simulates human thought. In E. A. Feigenbaum & J. Feldman (Eds.), *Computers and thought* (pp. 279–293). Cambridge, MA: MIT Press.

Olson, G. M., & Olson, J. S. (2002). Groupware and computer-supported cooperative work. In J. Jacko & A. Sears (Eds.), *Handbook of human-computer interaction.* Hillsdale, NJ: Erlbaum.

Pew, R. W., & Mavor, A. S. (Eds.). (1998). *Modeling human and organizational behavior: Application to military simulations.* Washington, DC: National Academy Press.

Pirolli, P., & Card, S. (1999). Information foraging. *Psychological Review, 106*(4), 643–675.

Polk, T. A., & Newell, A. (1995). Deduction as verbal reasoning. *Psychological Review, 102*(3), 533–566.

Polson, P. G., Lewis, C., Rieman, J., & Wharton, C. (1992). Cognitive walkthroughs: A method for theory-based evaluation of user interfaces. *International Journal of Man-Machine Studies, 36*(5), 741–773.

Polson, P. G., Muncher, E., & Engelbeck, G. (1986). A test of a common elements theory of transfer. *Proceedings of ACM CHI'86 Conference on Human Factors in Computing Systems* (pp. 78–83). New York: ACM.

Preece, J., & Maloney-Krichmar, D. (2002). Online communities: sociability and usability. In J. Jacko & A. Sears (Eds.), *Handbook of human-computer interaction.* Hillsdale, NJ: Erlbaum.

Rieman, J., Young, R. M., & Howes, A. (1996). A dual-space model of iteratively deepening exploratory learning. *International Journal of Human-Computer Studies, 44*(6), 743–775.

Ritter, F. E., Baxter, G. D., Jones, G., & Young, R. M. (2000). Cognitive models as users. *ACM Transactions on Computer-Human Interaction, 7*, 141–173.

Ritter, F. E., & Young, R. M. (2001). Embodied models as simulated users: Introduction to this special issue on using cognitive models to improve interface design. *International Journal of Human-Computer Studies, 55*, 1–14.

Salthouse, T. A. (1988). Initiating the formalization of theories of cognitive aging. *Psychology & Aging, 3*(1), 3–16.

Salvucci, D. D. (2001a). An integrated model of eye movements and visual encoding. *Cognitive Systems Research, 1*(4), 201–220.

Salvucci, D. D. (2001b). Predicting the effects of in-car interface use on driver performance: An integrated model approach. *International Journal of Human-Computer Studies, 55*, 85–107.

Schoelles, M. J., & Gray, W. D. (2000). Argus Prime: Modeling emergent microstrategies in a complex simulated task environment. In N. Taatgen & J. Aasman (Eds.), *Proceedings of the Third International Conference on Cognitive Modeling* (pp. 260–270). Veenendal, The Netherlands: Universal Press.

Simon, H. A. (1996). *The sciences of the artificial* (3rd ed.). Cambridge, MA: MIT Press.

Singley, M. K., & Anderson, J. R. (1989). *The transfer of cognitive skill.* Cambridge, MA: Harvard University Press.

St. Amant, R. (2000, Summer). Interface agents as surrogate users. *Intelligence magazine, 11*(2), 29–38.

St. Amant, R., & Riedl, M. O. (2001). A perception/action substrate for cognitive modeling in HCI. *International Journal of Human-Computer Studies, 55*, 15–39.

Tullis, T. S. (1983). The formatting of alphanumeric displays: A review and analysis. *Human Factors, 25*(6), 657–682.

Wiesmeyer, M. (1992). *An operator-based model of covert visual attention.* Unpublished doctoral dissertation, University of Michigan, Ann Arbor.

Young, R. M., Barnard, P., Simon, T., & Whittington, J. (1989). How would your favorite user model cope with these scenarios? *ACM SIGCHI Bulletin, 20*, 51–55.

·6·

MODELING HUMANS IN HUMAN–COMPUTER INTERACTION

Hidekazu Yoshikawa
Kyoto University

118

INTRODUCTION

The subject area of human modeling is extensive even if examined only in relation to human modeling within human–computer interaction (HCI), because there are at least three different approaches: from psychology, from physiology, or from sociology. In other words, there are the three aspects of cognitive, physical, and affective factors for consideration when modeling human behavior in HCI. However, the cognitive aspect is the most indispensable one from the viewpoint of applying the relevant knowledge during the various phases of design, analysis, and evaluation for HCI. In fact, many new input–output devices, as well as information processing technologies, are now available in the commercial market and the invention of new types of human interface for supporting our daily work by computer are expanding day by day. However, the cognitive ability of humans has not varied, but is almost at the same level as that of prehistoric man. Therefore, understanding generic features of human cognitive behavior in HCI is a fundamental requirement for all researchers and practitioners who are involved in the design, analysis, and evaluation of human–computer interfaces.

Because the basic knowledge on human perception, motor action, information processing, mental models, emotion, and cognitive architecture are already described in chapters 1 through 5, this chapter will provide a bird's eye view of modeling human cognitive behavior in HCI by summarizing these basic concepts. The discussion will then proceed on to making full use of these concepts for implementation into a computerized human cognitive simulator for various practical applications.

In the first section of this chapter, we will give you a comprehensive overview of human cognitive behavior characteristics when using human–computer interfaces. These characteristics are discussed in relation to modeling human information processing at the interface level and provide insight into the understanding of human error mechanisms. Two important HCI frameworks have been developed from examination of cognitive theories: a comprehensive scheme for human error analysis and a conceptual framework of modeling human behavior at the human–computer interface level.

Next, the discussion will proceed to the methodological exploitation of human modeling in computer simulation. We will start by describing the necessary constituents of computerized human modeling for engineering application, followed by an introduction of various implementation methods for the integration of human modeling into computers mainly within the area of process control. Two practical applications of human modeling simulation are presented. One application involves the simulation of human–machine interactions for process control, and the other describes introduction into the education and training environment.

This chapter will conclude by forecasting how and where further research is expected in the study of human modeling. The intent is to enable human modeling simulators to become personified interface agents that will partner with humans in virtual space through communication and collaboration to support human daily work over the Internet.

BASIC KNOWLEDGE REQUIRED FOR HUMAN MODELING IN HCI

Generic Characteristics of Human Cognitive Processing at the Interface Level

Basic Characteristics of the Human Memory System. Basic knowledge of the human memory system and character are briefly described from the viewpoint of cognitive psychology, by introducing some basic knowledge on human information processing.

Two Different Memory Systems. Human information processing is supported by two kinds of memory: short-term memory (STM) and long-term memory (LTM). LTM is the vast portion of memory in the human brain in which many experiences and knowledge that a person has learned in the past are stored and retrieved, as opposed to the limited capacity of STM in which humans conduct cognitive processing at a conscious level. The STM is alternatively called working memory (WM) because it is the work space for cognitive processing.

Human memory performs by memorizing past experiences and learned knowledge not in an exact form, but by trimming the real facts to a normalized form representing the meanings. Therefore, humans save memories as a theory or structured form called a *schema*. This was how human memory was described early in psychology. Then, with the emergence of cognitive psychology, the field of artificial intelligence (AI) describes forms of human memory with terms such as frame, script, semantic network, and so on.

Outside-world information obtained through the sensing organs of the eyes and ears is unconsciously filtered and codified and then stored as structured knowledge in the enormous storage of the LTM. Selected clues perceived from outside-world information are pattern-matched to stored knowledge in the LTM to drive a very prompt knowledge retrieval mechanism that differs from the keyword retrieval used in the computer database. Then retrieved knowledge from LTM is consciously held in the STM for further cognitive processing.

Model of Human Information Processor. With the analogy of a digital computer, the model of the human information processor was proposed by Card, Moran, and Newel (1983). The model was developed using various data assembled by experimental psychology researchers on the basic characteristics of human information processing, which involves sensing, perception, and cognition, as well as related human memory systems.

The complete picture of the human information processor model is shown in Fig. 6.1. As illustrated in the figure, the sensing organs that function as the input channels to the human information system are restricted to the eyes and ears. The perceived world information sensed by the eyes and ears is fed into the visual image storage and auditory image storage areas of the brain along with the original signal forms of the sensing organs. Those signals are then converted into either an auditory

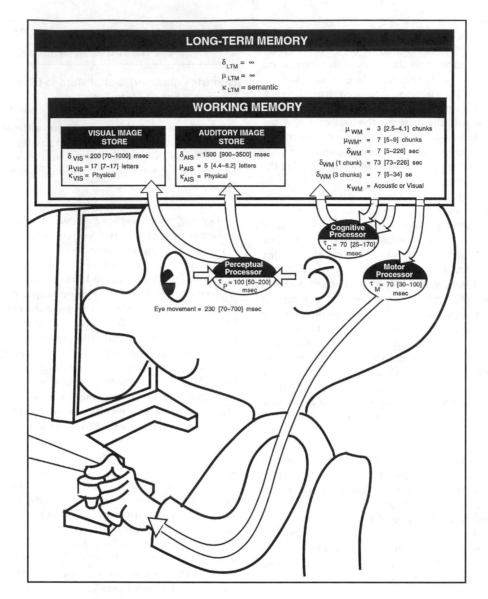

FIGURE 6.1. Model of human information processor by Card et al. (1983).

or visual image in the WM of the cognitive system. Interactive processing with semantic information stored in the LTM occurs periodically, followed by the trigger of a motor action.

The characteristic data of the human memory and processing systems are listed in Table 6.1. The retention time, retention capacity, and code type of the information are listed for the two memory systems: (1) visual and auditory perception and (2) cognitive processing by the WM and LTM, along with the periodic times for the three processing systems of perception, cognition, and motor action. The information given in Table 6.1 is often applied for desk-top estimation of human response time for various task executions with the HCI. We should remember at least two points about the special character of human information processing. (1) The retention capacity of human WM is limited to the short retention time of 7s, with the maximum number of memorized items reaching the magical number of 7 ± 2 chunks of information. (2) The capacity of memory retention is virtually limitless for LTM, but the memorized information is converted into semantic codes after the interpretation of cognitive processing within the WM. The word *chunk* as used here means a piece of information "structured as meaningful information." For example, if your birthday is November 30, 1950, it will be remembered in it's entirety as a chunk of information, because this is your unforgettable date, although it is actually made up of the three individual pieces of data: "November," "30," and "1950."

Model of Human Information Processing by Rasmussen. The model of the human information processor by Card does not take into account the different natures and functions of various cognitive processes, but a conceptual model

TABLE 6.1. Characteristics Values of Model Human Information Processor

Memory System	Retention Time (δ)	Retention Quantity (μ)	Code Type (κ)
Perception System			
Visual information storage	200 (70–1,000) ms	17 (7–17) characters	Physical
Auditory information storage	1,500 (900–3,500) ms	5 (4.4–6.2) characters	Physical
Cognition System			
Working memory	7 (5–226) s[a]	7 (5–9) characters	Auditory Visual
Long-term memory	∞	∞	Semantic
Processing System	Cycle Time (τ)		
Perception	100 (50–200) ms		
Cognition	70 (25–170) ms		
Motor action	70 (30–100) ms		

[a]Depends on the number of recalled items. 73 (73–226) s for 1 chunk and 7 (5–34) s for 3 chunks.

of human information processing proposed by Rasmussen (1986) does take these processes into consideration. The model is illustrated in Fig. 6.2, in which both higher and lower levels of psychological processing are considered as a set of block diagrams that depict various functional aspects of human cognitive information processing. According to Rasmussen's model, the sensed input from the outside world is perceived through the sensing organs, and then various cognitive processes work together in both conscious and unconscious modes to generate motor action to the outside world. The upper portion of the figure represents the conscious part of human information processing using both serial and symbolic operations, whereas the lower portion symbolizes the unconscious functions of parallel and analogue processing. Multiple cognitive processes, such as recall, association, inference, prediction, etc., are conducted in STM under the control mechanism of attention. This is where mutual connections are made between perceived outside-world information

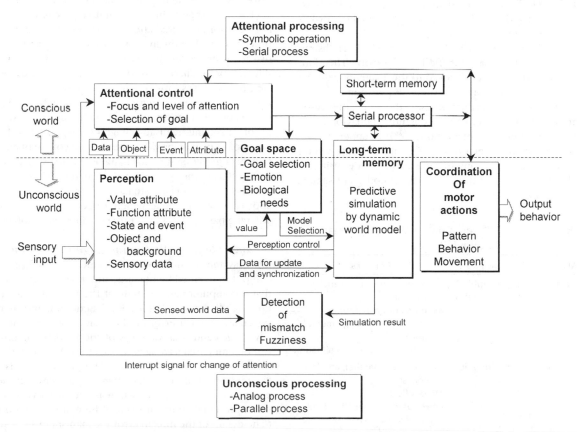

FIGURE 6.2. Model of human information system by J. Rasmussen.

from the sensing organs, and the stored memorized information being highlighted and activated in the enormous storage capacity of the LTM. This is a very complicated illustration, but it is a well thought out model from the viewpoint of a system engineer, because it includes whole essential elements of human cognitive information processing that can then be used to construct a workable human model for use in HCI work.

Control Model of Human Information Processing. The characteristics of the control model will be discussed based on human cognitive processing for problem-solving activities. The words *problem solving* are used to denote how humans behave when they have found a task to complete.

Two Different Control Modes. Humans will conduct problem solving by using knowledge bases recalled from LTM through pattern matching with incoming outside-world information. The rules of the knowledge bases humans use for problem solving are classified into two types: The symptomatic rule (S-rule) and the topographic rule (T-rule). The S-rule is a routinely used heuristics which depend on the specific situation, whereas the T-rule is a general rule of thumb or a fundamental principle that is independent from the specific situation. For example, driving a car. The S-rule is a simple rule of daily practice, such as "when a traffic signal turns red, step on the brake pedal to stop the car." On the other hand, the T-rule is basic knowledge of the car's mechanism, such as "the foot brake is connected to a brake pressure cylinder via piping, so the brake pressure cylinder activates the wheel disks to stop the car when you step hard on the brake pedal."

The S-rule is rapid reflex action, but the T-rule is abstract knowledge that is necessary for the brain to work. Therefore, humans prefer the S-rule as opposed to the T-rule for responding to a situation to act on something. When human beings can no longer cope with a situation using the S-rule, they resort to the more difficult T-rule. Human cognitive behavior is classified into two different control modes: the attentional mode of analysis by way of abstract thinking using the T-rule and the schematic mode of completing routine daily activities using the S-rule. The comparison of the two control modes are summarized in Table 6.2.

TABLE 6.2. Two Different Control Modes of Human Cognitive Process

Control Mode	Applied Rule	Remarks
Attentional mode	Topographic rule (T-rule)	Conscious, attentive, serial processing Resource intensive (time consuming and large mental effort) Problem solving in new situation
Schematic mode	Symptomatic rule	Unconscious, parallel, and fast processing Minor effort Powerful in familiar situation, but not effective in new, unexpected situations

Multiladder Model by Rasmussen. A descriptive model of human problem-solving behavior called the multiladder model was proposed by Rasmussen (1986) as illustrated in Fig. 6.3. In this model, the basic premise on human problem solving is stated as "human behavior is goal oriented such that a human will set local sub-goals and will attain those sub-goals to reach the final goal." The description of the meaning of the model depicted in Fig. 6.3 is rather lengthy. "When a human detects the necessity of action, the human will climb up the ladder of knowledge base analysis from activation, to observation, to identification, to interpretation and evaluation, then go down the ladder of knowledge base planning from selection of subgoals, to the definition of the task, to formation of a procedure, followed by the execution of the procedure, and then will proceed to subsequent cycles of a similar order. Shortcuts will occur between those processes at many levels in the decision making loop, because of familiarity with past experience." (The meaning and effects of short cut paths will be discussed later.)

When the previous models of human information processing as shown in Fig. 6.2 are seen as block diagrams within the whole control system of this multiladder model of human cognitive behavior, it becomes evident that there are two-layered structures of upper and lower level processors that perform specific aspects of the whole cognitive process. Those different processors will influence each other simultaneously, but each with different timeframes. The characteristics of the two-layer structure of human cognitive processing are summarized in Table 6.3, respectively, for the upper and lower levels of processors.

The advantage of human cognitive processing is that the laborious work by serial attentive processing in the upper level processors will gradually move to automatic parallel processing by the lower processors as the human becomes accustomed to the task or familiarized through the repetition of learning. In fact, the various short-cut paths that appear in Fig. 6.3 exhibit a variety of effects of "automatic processing by familiarization" in the course of moving through multiladder processes during problem solving. A disadvantage of this model is that various forms of human error will appear in the course of applying this human cognitive system to changing outside-world situations as a side-effect of the system's intrinsic nature. Therefore, the form of human error can be systematically predicted by this cognitive approach, by looking at which aspect of the cognitive system will take part in the whole process of problem solving.

Balance Sheet of the Human Cognitive System. The cognitive system of humans is fundamentally different from the computer invented by humans, and it has superior functioning that a computer cannot attain at the present time. Therefore, the cognitive system's superior functioning and its generated systematic human error are both sides of the same coin. The advantages and disadvantages of the human cognitive system can be compared as a balance sheet, as shown in Table 6.4. In this balance sheet, each item on the positive side is an advantageous aspect of the human cognitive system that the modern computer cannot realize, whereas the negative side explains the representative mechanism of human error by the system. (The details of the human error mechanism in Table 6.4 will be described in a later section.)

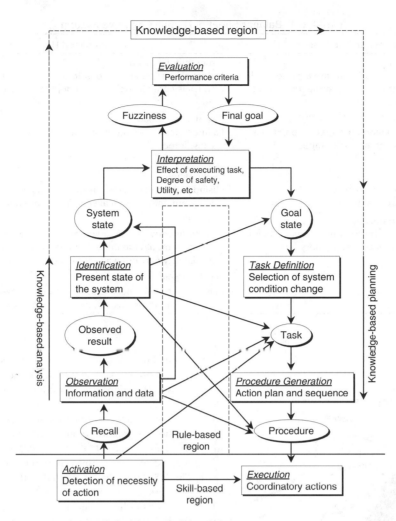

FIGURE 6.3. Multiladder model of human cognitive process by J. Rasmussen.

Modeling Human Information Processing Behavior With the Human–Computer Interface and Human Error

How a Human Views the Human–Computer Interface. Figure 6.4 is a famous picture you will find in some books on cognitive psychology. Probably you will see either an older woman or a young lady. When you see the older woman, you cannot see the other figure at the same time, because a certain constraining mechanism works unconsciously to maintain the visual consistency of what you see. When you see the older woman, she is recognized as the figure with the young lady hidden in the background and vice versa. The figure and background are not always fixed, and they reverse from time to time when you change the perspective of your visual attention. The

TABLE 6.3. Functional Characteristics of Two-Layered Structure of Human Cognitive System

	Processing Aspect	Psychological Aspect
Upper level processors	Operate in a longer time span Intermittent control for lower level processors Monitoring the progression state toward goal Offering means for attaining goal	Consciously attentive processing of planning and monitoring Selective processing by memory constraint of short-term memory (Selectivity)
Lower level processor	Automatic processing to perform individual functions of elementary cognitive process Operate in a short time by responding to a specific data set	Automatic execution of appropriate scheme or various knowledge structures of artificial intelligence, such as frame, script, semantic net, heuristics Prompt retrieval of memory by similarity matching or automatic activation by frequency gambling

TABLE 6.4. Balance Sheet of Human Cognitive System

Item	Advantage of Cognitive System	Predicted Human Error
1	Familiarization Workload of resource-intensive attentive process will be mitigated by control transfer to the lower level automatic processing	Strong-but-wrong error Failure of attention by unconscious execution of familiarized routine action, which is normally right but is not proper in a particular case
2	Selectivity Focus of attention is applied for a specific point in a whole world because of resource limitation of attention	Cognitive overload Failure to look over important data, because of too much data to be considered.
3	Effort after meaning Knowledge in long-term memory is stored as theorized or structured rather than the real fact	Confirmation bias Misjudgment will occur when applying the retrieved knowledge from long-term memory because of its distorted nature
4	Content-addressable retrieval Prompt memory retrieval by content or semantics from enormous storage of long-term memory	Availability More frequently experienced way of doing or recently conducted way is naturally brought to mind, but the unconscious application of this availability may bring about the performance failure in a particular case Matching bias Biases which were unconsciously stored in long-term memory are the source of performance failure

FIGURE 6.4. Fuzzy picture of elderly woman and young lady.

perspective of attention will constrain your brain's cognitive system on one side and regulate the context of your perceived world. Therefore, the direction of attention will control what you see.

The picture is a typical example of the effect of our visual perception. In relation to an interface, Rasmussen (1983) pointed out another effect of our perception system. A single gauge may be seen or cognitively perceived differently as depicted in Fig. 6.5. This effect is related to three distinct cognitive modes of users at the interface level, with the instrumentation perceived as either a signal, sign, or symbol, in accordance with the individual operator's cognitive mode.

What is the intention of Fig. 6.5? When the needle of the gauge is seen as a signal, as is the case in (a), the operator will manipulate the valve by hand to move the needle to the set point. In this instance, the operator works as a manual controller to maintain constant flow through the pipe. In condition (b), where the meter is viewed as a sign, the operator uses the position of needle as input for logical judgment. If the valve is open and the position of the needle is C, then the flow meter is functioning properly; but, if the needle position is D, then the flow meter should be calibrated. If the valve is closed and the needle is on position A, then the flow meter is working correctly; but, if the needle position is B, the flow meter should be calibrated. Lastly, in case (c), when the gauge is seen as a symbol, the operator has to think from the basic physical principle perspective to solve the problem. Because the valve is closed and the flow meter needle position is B, the flow meter is calibrated. Even after calibration, the meter needle is still on B. This combination of facts is perplexing to the operator and requires further investigation to determine where the problem is occurring. The user concludes that there must be a hole between the gauge and valve that is allowing water to leak from the system. The operator decides that the problem is solved.

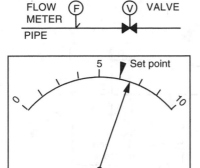

(a) In the case dial is seen as "signal"

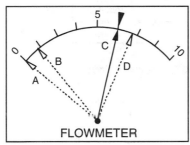

(b) In the case dial is seen as "sign"

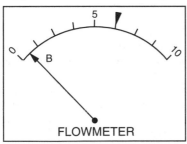

(c) In the case dial is seen as "symbol"

FIGURE 6.5. Three different ways of how the instruments are perceived by the operator.

In summary, the same visual input will change meaning based on the user's focus of attention or the context in which a task is performed.

Three Operator Cognitive Modes at the Interface Level.

In accordance with the different perception patterns of signal, sign, and symbol at the interface as described in the preceding section, the subsequent action of the operator is classified by Rasmussen (1983) into three modes: (1) skill-based action of unconscious, reflex behavior; (2) rule-based action of a conscious, pattern-matching response; and (3) the knowledge-based attentive action of logical and abstract thinking.

The three user cognitive modes of process control at the human–machine interface are illustrated in Fig. 6.6. When a user perceives the instrument input as signals of plant control, an operator will perform smoothly as if they are the automatic control of a multivariable system. If the operator behaves in a

rule-based mode, the instrument information will be perceived as signs, and the user will attempt to counteract the situation through the application of various rules of thumb for associative pattern matching. As a result, the user will try to recall the appropriate rules necessary to interpret the state, task, and procedure required to cope with the situation promptly. Both skill- and rule-based behaviors are effortless schematic modes, and a well-experienced operator can usually respond by either action mode.

On the other hand, in both of the action modes, the situation is abstractly recognized as symbol by operators within a knowledge-based mode. In a knowledge-based mode, the perceived world is analyzed by using internally formed *mental models* of the functional relationship or the cause–consequence relationship for the problem at hand. Based on such mental models, operators understand the reason or cause of the present state, predict future trends, and decide the appropriate action plan. This kind of knowledge-based behavior is an attention mode with a considerable amount of mental workload. This is, in a sense, the behavior of a novice person lacking the experience of an expert; but even experts would fall into this category when they encounter a new situation they are not accustomed to handling.

Human Error and the Generic Error Modeling System (GEMS) Dynamics Model

Human Error as Viewed from Cognitive Psychology. There are a variety of views of human error in the area of human factors research. Based on a behavioristic approach, Swain and Guttmann (1983) developed a technique for human error rate prediction called THERP. Human error is defined as human actions that deviate from the standard actions prescribed by a certain authority as the correct procedure. According to THERP, the human error is classified into two types: error of omission and error of commission. Error of omission is the failure to complete an appropriate human action, compared with the usually accepted standard action. Error of commission can be described as the completion of an extra untimely action or the reversal of an action, which deviates from the generally accepted standard. Hence, using this approach, it is possible to find human error by outside observation and to count the number of human errors which occurr.

As opposed to these approaches, the cognitive approach focuses on the mental mechanisms associated with why humans commit errors. An important point about the cognitive approach is that it is mainly concerned with determining if intent formation is accurate, as opposed to noticing whether or not humans actually commit an error. Accordingly, human error investigation based on the cognitive approach is classified into two types: the intent formation is accurate, but there are slips and lapses at the execution phase; and faulty intent formation. The distinction between slip and lapse is that a slip occurs with a failure of attention at the time of execution, whereas a lapse gives rise to memory failure. The subsequent slip and/or lapse after an original correct judgment would normally be detected by an attentive person during the execution stage and may be corrected before leading to serious failure.

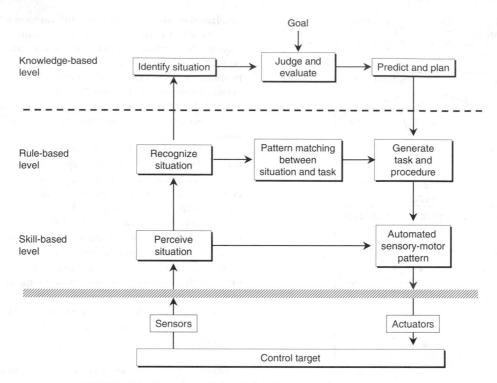

FIGURE 6.6. Operator's three behavior modes by J. Rasmussen.

On the other hand, if a slip and lapse occurs after a failure of judgment, such as the misdiagnosis of a disease by a medical doctor, the result may be a fatal mistake. Therefore, the consequences of the mistake are enormous even if neither a slip nor a lapse occurs after the misjudgment. Contrary to a slip and lapse, which is rather easy to recover from, once a person has made a judgment, it is very difficult to revise the conclusion by himself. The correction would most likely need to be revealed first by another person. Therefore, the mistake is also dangerous from the perspective of the rare possibility of recovery.

The previous statements are applicable to the daily life of an average person, but what kind of human error would be committed by a specialist such as a plant operator at the interface?

GEMS Dynamics Model. Embrey and Reason (1986) proposed a qualitative model of the plant operator's behavior at the interface level called GEMS dynamic model. It was developed as an extension of Rasmussen's model of the three cognitive modes of operators by considering the time transition during the operator's monitoring and control actions. The major features of the GEMS dynamic model is illustrated in Fig. 6.7, where the probable forms of human error committed by operators are distinguished into the following two categories: a failure of skill-based behavior at the monitoring phase causing a slip, and a failure of rule- and knowledge-based behaviors at the problem-solving phase causing a mistake. In the subsequent section, the detailed contents of the GEMS dynamic model are discussed in relation to the above detailed two types of operator failure (refer to Fig. 6.7).

Monitoring Failure (a Slip before Problem Detection). The routine work of skilled operators performing repetitive tasks in a familiar situation can be taken as a sequential set of preprogrammed actions. In this instance, most of the actions would be unconsciously executed with interruptions from time to time for attention checks in accordance with the progression of the situation. At the time of each intermittent check, upper level cognitive processors will interrupt the unconscious control loop to monitor whether or not the ongoing control action is proceeding as planned and whether or not the execution plan is enough to attain the goal.

The whole control plan in the operator's mind is assumed to be a chain of several elementary actions, as illustrated by the intended action sequence in the upper part of Fig. 6.7. A branching node would exist in between each elementary action to allow for several possible action routes afterward. The important route among them is the route of a slip sequence that has a high probability of being executed because of the unconscious activation of a wrong memory from past experience. The trigger mechanism for a wrong memory is called *availability*, as listed in the human error column of Table 6.4. This slip sequence is assumed to not only meet with the present intention, but would also lead to a fatal error. In this case, the ideal behavior is that the operator's attention check will just coincide with the branching node; but, if this check happens to shift from the branching node, the operator would deviate to the slip sequence unconsciously, which will result in a fatal error. In the field of cognitive psychology, this type of slip error is called a *strong-but-wrong* error or a double capture slip.

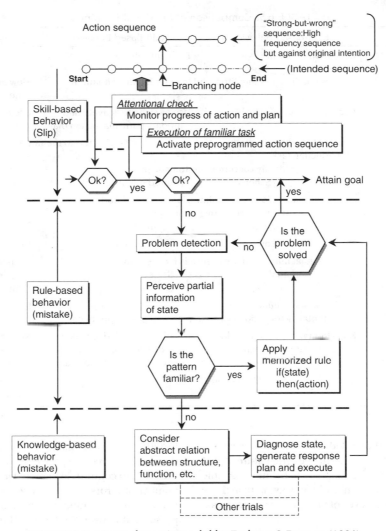

FIGURE 6.7. GEMS dynamic model by Embrey & Reason (1986).

Problem-Solving Failure (a Mistake in Problem Solving). When a certain problem is detected by an attention check, the operator's cognitive processing will proceed to the problem-solving stage as is indicated by the arrow after the branch block in Fig. 6.7. In terms of problem solving, the problem may be simple enough for the operator to cope with easily by his acquired heuristics, or it may be a very difficult problem requiring the operator to think about it using acquired in-depth knowledge. This knowledge may include information about the function and structure of the plant system or information about the first principle of physical law for the behavior of plant dynamics. Because of the generic nature of human beings in that humans prefer easy pattern-matching to laborious calculation and optimum search, the operator will tend to resort to the easier rule-based problem-solving method in the first stage. The operator will not move to the next problem-solving stage of the abstract knowledge-based mode until easy pattern matching and rule-based procedures are no longer successful in coping with the situation.

The dominant mechanism of the problem-solving stage is automatic initiation of lower level cognitive processors that have the knowledge structures, such as schema, frame, script, etc.—available either through automatic recall, by availability, or by conducting pattern matching. The perceived signs and symptoms are used as input data to assist in the recall of appropriate knowledge-base labels within the memory system. In this instance, the source of mistakes in problem solving lies in the superimposition of the following four factors: (1) availability, (2) defect of the knowledge base, (3) matching bias, and (4) cognitive overload (see Fig. 6.4). The availability effect is the major source of rule-based problem-solving errors for the same reason as slips during the monitoring phase, but the other three factors concern both problem-solving stages. Any defect of the knowledge base in LTM would occur at the time of memorization based on confirmation bias, and this becomes the reason for matching bias during the association phase. The last factor of cognitive overload will also occur during the association phase.

TABLE 6.5. Comparison of the Cognitive Modes

	Skill-Based Behavior	Rule-Based Behavior	Knowledge-Based Behavior
Action type	Execution of routine procedure	Problem solving	
	Smooth action with little workload for familiar situation		Slow and intermittent action for new, unfamiliar situation
Input information	Use as continuous signals	Use as signs to activate and update particular action plan	Use as symbols to drive particular mental model
Focus of attention	Things other than presently conducting task	Directed to the now tackling problem	
Control mode	Mainly concurrent, Automatic process (Schema)	(Stored rules)	Resource-intensive and attentional serial process
Predictability of error type	Mostly predictable intrusion of strong habit of Strong-but-wrong error (Action)	(Rule)	Variability, novice error
Sensitivity to stress	Low	Medium	High
Ratio of actual error occurrence to potential chances	Actual number of occurrence is high, but the ratio to potential error chances is small		Actual occurrence is rather low, but the ratio of potential chance is high
Influence of situation factors	Low or medium Mental factor is dominant (frequency of past experience)		External factor is dominant
Knowledge on external change of world as the trigger of error occurrence	Knowledge on external world change cannot be activated in good timing	Devoid of knowledge base to predict when and how external world would change	Neither knowledge on change of external world nor consideration
Self-detection	Mostly effective and rapid		Difficult, mostly detected by other person

Missing data in the associated knowledge base are unconsciously replaced by default values because of the generic nature of WM called *bounded rationality*. There appears to be more versatile error forms at this phase of human problem solving than those error forms found when the level of the problem becomes a more complicated one.

Comparison Between the Three Cognitive Modes of Operators. Taking into account the aforementioned discussion on human error by the GEMS dynamic system, various characteristics of the three modes of operator's cognitive behavior (i.e., skill-based, rule-based, and knowledge-based action at the interface level) are summarized in Table 6.5. The items compared in Table 6.5 cover various aspects of human cognitive behavior such as: (1) where each mode is applied, (2) what are the input data for action, (3) where the focus of attention is directed, (4) what is the control mode and what type of error is anticipated, (5) how does stress influence performance, (6) what is the probability of human error, (7) what is the influence of the situation, (8) how would the dynamic changes of the external world influence the error occurrence, and (9) whether or not the committed human error is detected and modified by the same person.

To summarize the operator's cognitive behavior from what has been mentioned overall, most of the cognitive behavior of a skilled operator is a skill-based and rule-based one. The dominant form of human error is the strong-but-wrong error. The actual number of skill- and rule-based errors is rather high, but

its ratio to the potential chance of error is rather low, and error detection and recovery are easy for the person who committed the error. In contrast, in knowledge-based errors, the ratio of the number of errors to the chances of error is high, and it is difficult to detect and modify such errors without the help of other persons. Moreover, the forms of human error from knowledge-based behavior are so multifaceted that it is very difficult to predict every case of human error beforehand.

To add a few discussion items on knowledge-based behavior, there are two situations in which a mismatch arises between the system response and operator behavior:

1. The operator cannot habituate with the system because of knowledge deficiency of system characteristics, no display of data, and too large a workload.
2. Although it is possible to adapt initially, the addition of improper operation of the system by trial and error worsens the system state so that operator is no longer able to cope with it.

Another factor of knowledge-based behavior is that problem-solving ability will deteriorate under high stress conditions by narrowing the view and the lack of consistency in the whole behavior.

Comprehensive Scheme for Human Error Analysis by the Cognitive Approach. As has been previously described, human error examined from the behaviorist approach is defined as the deviation of human action from the standard of reasonable

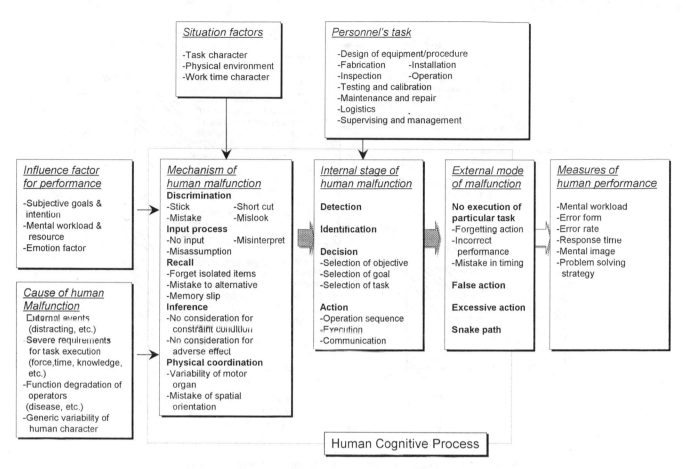

FIGURE 6.8. Comprehensive scheme of human error at human–machine interface.

action. The form of error is classified as either error of omission or error of commission. On the other hand, the cognitive psychology approach focuses on the mental mechanism of committing human errors and the factors affecting the occurrence of human error. How then should the whole framework of the human error be configured to integrate the two different approaches, and how should the framework be used to analyze human errors as observed in human performance data? To meet these requirements, a comprehensive scheme has been configured as shown in Fig. 6.8. This version is modified from the original scheme of human error mechanism as proposed by Rasmussen (1986) by the addition of various indexes to measure human performance. In Fig. 6.8, the block of human cognitive process is composed by the three elements: (1) the mechanism of human malfunction, (2) the internal stage of human malfunction, and (3) the external mode of malfunction. The direction of the arrows between the two adjacent boxes indicates the direction of influence that gives rise to human error from the internal to the external stages. The four boxes surrounding the block of the human cognitive process with influence arrows toward respective boxes within the block represents various factors affecting the internal process. Several indexes are designated in the box on the right-hand side of Fig. 6.8, which represent measures of human performance

and are used for human factors analysis and the evaluation of human interaction with an interface.

Conceptual Framework of Modeling Human Behavior at the Interface Level

The basic framework for modeling human cognitive behavior at the human–computer interface will be described in this section by extending the basic knowledge on the various characteristics of human cognitive behavior as introduced in the preceding sections. Concerning the premises for human modeling, Reason (1990) described the fallible machine, that is the following two conditions that are necessary for the development of a human model: (1) Not only right performance, but also the prediction of possible forms of human error based on theories of cognitive psychology, and (2) the framework of the human model can be extended for computer simulation via AI.

A basic framework for the human cognitive model at the human–machine interface level was created by Yoshikawa, Nakagawa, Nakatani, Furuta, and Hasegawa (1997), by combining the idea of the fallible machine by Reason (1990) and the first AI-based computer model of an operator called COSIMO, which was developed by Cacciabue, Mancini, and Bersini (1992). The

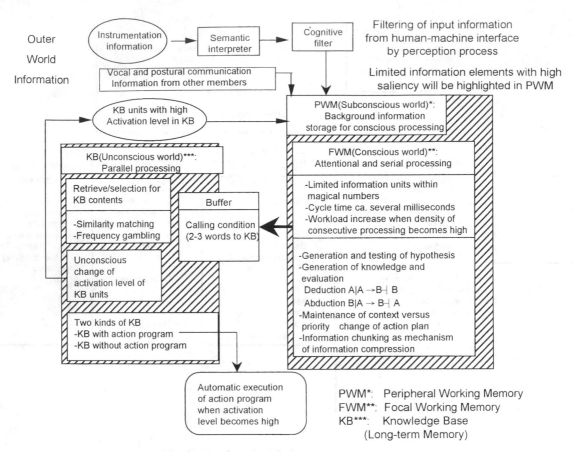

FIGURE 6.9. Conceptual framework of human cognitive model at human–machine interface.

framework is depicted in Fig. 6.9, and the background knowledge for this human model framework is itemized below with the highlighted major keywords as appears in Fig. 6.9.

1. The displayed information on the interface is originally quantitative numeric data of either digital or analog value. Semantic interpreter is the psychological function of human cognition that enables the quantitative data to be interpreted as meaningful information, e.g., a particular value is normal or high.

2. *Cognitive filter* is the selection filter that allows a user to choose only the more conspicuous or salient items from the many semantic information items coming from the semantic interpreter.

3. There are two kinds of human memory: LTM and WM. LTM is the storage of an extensive knowledge base, and the WM is divided into two kinds: focal working memory (FWM) and peripheral working memory (PWM). FWM is the major WM for conscious processing, with the capacity limitation of 7 ± 2 data items, but the information storage capacity in FWM can be increased by chunking or structuring the information. PWM is the working space between LTM and FWM, in which the elements of knowledge bases activated from LTM are ready for processing in the FWM.

4. The concept of activation of knowledge elements is introduced as the driving mechanism of the prior memory system.

There are two activation mechanisms: a specific activator and a general activator. Specific activators originate from the nature of FWM. The object attribute of FWM processing will correspond to some keywords in the FWM within a certain time span, and those keywords become the calling words or cues to put in the buffer situated between FWM and LTM. Then, the knowledge elements that have a high similarity to those calling words are highlighted or activated in PWM. With specific activators, there is a nondirectional activation mechanism called general activator that will activate the knowledge bases in the LTM naturally with no specific objective. For example, those knowledge elements that have been used many times in the past will be activated fairly extensively from the beginning, and they will be called on easily via the keywords in the buffer zone when the present situation resembles a past experience. Thus, knowledge retrieval from LTM via the buffer area occurs unconsciously through the use of similarity matching and frequency gambling.

5. There are two kinds of knowledge bases in the LTM: a knowledge base with motor programming and a knowledge base without motor programming. When a knowledge base with motor programming is activated above a certain level, it is automatically executed, and a motor action is the outcome.

6. Laborious cognitive actions, such as inference or hypothesis testing for problem solving, monitoring for confirmation of the context, etc., are conducted in the FWM by using both

the knowledge elements coming out of the LTM and perceived outside-world information. These two knowledge elements converge in the PWM of the background space of the conscious memory, for serial processing in FWM.

7. Creative activities by humans, such as generation of a new knowledge and its evaluation, are mainly conducted by the combination of *induction* and *abduction*. Induction is the logic that, "if A is true, then B" is true and "A is true," then "B is true" is true, whereas abduction is the logic that, "if A is true, then B" is true and "B is true," then we conjecture that "A is true" (but it is not always true).

METHODS OF HUMAN MODELING BY COMPUTER SIMULATION

Necessary Components of a Computerized Human Model for Engineering Application

The conceptual framework that depicts the application of cognitive psychology to various human factors issues related to the human–machine interface, (Fig. 6.9), is a comprehensive scheme of human modeling of cognitive behavior at the interface level. However, this is simply a human model that is qualitative in nature. For the purpose of an engineering application, it is necessary to refine the model so that it becomes more systematic, one that can actually work on a computer. The necessary modeling elements to configure a computerized human model applicable for engineering purposes are discussed by Furuta and Kondo (1993), as shown in Fig. 6.10. The three modeling

elements are named process model, knowledge model, and control model. The discussion of those three models will be detailed in the next section. The methodologies of how those modeling elements can be implemented using computer simulation techniques will be discussed in the subsequent section.

Process Model. It is suggested from knowledge of experimental psychology, as well as physiology of the human brain, that the human information processing system consists of various modules that govern various basic functions, and that those basic functional modules to some degree work independently from each other. The process models are models of various stages of such basic information processing. There are many assumptions about basic information processing for this model; but, if we refer to the multiladder model of human cognitive behavior by Rasmussen (1986) (see Fig. 6.3), there are at least five elements necessary for the model as a whole: (1) observation, (2) interpretation, (3) planning, (4) execution, and (5) memory.

Observation. Observation is the process of perceiving selectively the important attributes of those objects that exist in the environment surrounding a human. The process of qualitative interpretation of perceived information with its low-level semantic analysis and evaluation is also included in the process model. The situation recognition sign by Rasmussen (1983) falls in this class.

Interpretation. The interpretation process includes diagnosis of the objective system, identification and evaluation of the system state, and understanding and prediction of the system response. Interpretation performs higher level recognition and

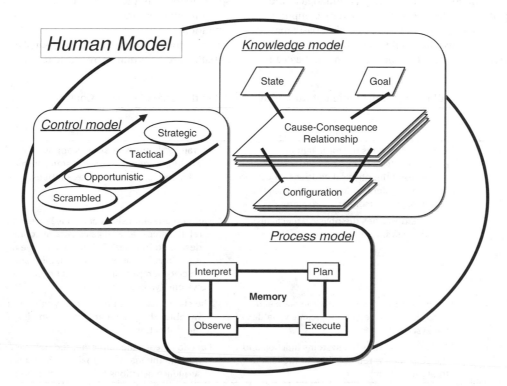

FIGURE 6.10. Three elements in the human model.

evaluation for perceived information than the previous observation process. Concretely, this process will be completed not only by the already mentioned simple methods of similarity matching and frequency gambling, but also by the three typical algorithms of topographical searching, symptomatic searching, and hypothesis generation and testing, which Rasmussen (1986) summarized as the operators' heuristics for daily problem solving.

Planning. Planning is the process of proposing the goal of action and then generating the procedure (i.e., the action sequence to fulfill the goal). The planning process includes the rule-based planning of simply associating a proper template for action from the LTM, the knowledge-based planning of configuration of an elaborate procedure from the elementary actions via a goal-means analysis, and the combination of both methods of rule-based and knowledge-based procedures.

Execution. The execution process performs the steps of the planned procedure one by one, and, as a result, the actions of operation, observation, and communication are executed in a cyclic manner. Execution also includes the monitoring action, whether or not the ongoing execution produces the expected effect to the system.

Memory. Retention and recall of the information in the STM and the association of knowledge elements from LTM are the processes included in the memory system.

Knowledge Model. The knowledge model relates to the semantic contents and the forms of the human information processing model. It is important to know how the original knowledge is described by the forms of rules, graphs, and logical and mathematical equations so that it can be handled by computer. Prior to going to this stage, it is necessary to define basic matters such as how humans would imagine the objective system, and what sort of words, notions, relations, etc., should be dealt

with in a human's behavior, such as thinking, action, and communication. The knowledge model deals with the conceptual structure of this kind of knowledge.

Correlation between the models of knowledge and processing is very multifaceted, but it is important to distinguish the knowledge model from the process model, because the knowledge on a particular content and its form does not correspond on a one-to-one basis to the particular process model. There are particular knowledge contents and forms suited for a certain process, and a certain type of knowledge would generally be used for a certain range of processes. For instance, the knowledge of the cause–consequence relationship can be used not only for prediction of system behavior, but also for the diagnosis of anomalies by observed symptoms and the reduction of a successful procedure to attain a certain goal.

Consider the whole span of the knowledge model by viewing process control as the example. The operator's knowledge of both the plant system itself and the tasks to control the plant can be constituted by the four different types of knowledge space: configuration space, cause-consequence space, state space, and goal space. The meaning of the four spaces is indicated in Table 6.6.

The cognitive process will work as a whole, although there exist different types of knowledge space. The different types of knowledge space and the different layers of the identical space are correlated with each other. For example, the following correlations should be considered for the knowledge spaces in Table 6.6:

1. The correlation paths in the cause-consequence space correlate with the various physical objects in the configuration space. In this way, the knowledge of the cause–consequence relationship is grounded by the knowledge of physical configuration.
2. The constraint conditions for goal satisfaction and for task initiation are determined by the state space. In this way, the

TABLE 6.6. Configuration of Knowledge Model in the Case of Process Control

Kind of Knowledge Space	Meaning	Remarks
Configuration Space	This space describes the form, figure, interconnection, layout, and the other static attributes of the physical entities that comprise the whole system, independently from the notion of any persons.	Hierarchical representation is applied for the description of whole-part relation of the system, such as whole system, subsystems, equipment, assemblies, and parts.
Cause–Consequence Relation Space	This space describes the physical mechanism of the whole system behavior.	Various co-relationships and cause–consequence relationships, as well as constraint conditions are described by a set of either qualitative or quantitative equations for the dynamic variables, such as position, velocity, temperature, etc., that determines the whole system dynamic behavior.
State Space	This space describes the possible state space made by the observable dynamic variables of the system.	The state space is represented by the set of symptom variables that are meaningful for the human to describe the distinction of different states.
Goal Space	This space describes the systems function and the goal–means relationship for the task to treat the system.	Function means the implemented behavior of the system to perform the intended goal; which task is a systematic sequence of actions to perform the intended goal.

TABLE 6.7. Four Different Modes of Control Model in COCOM

Control	Meaning
Scrambled Control	Human behavior in panic. No relation exists between the situation and the action taken, and the inexecutable action would be selected in a trial-and-error way, without evaluating the execution result. This is very extreme, and it would be a very rare occasion to appear in reality.
Opportunistic Control	Human response in a hurried situation, in which the next action is selected based on the immediate situation, but it lacks long-time prediction and planning. This would appear when human takes enough time to understand the situation. The behavior is more or less trial-and-error mode when the action is dependent on the eye-catching signs and routinely used heuristics.
Tactical Control	Conservative state of human behavior, in which human follows the prescribed plan, procedure, rule, and standard. This is when an appropriate template for action is available, and it is proper to follow that view of the situation; therefore, effective and minor errors in general.
Strategic Control	Proactive state of human behavior in which human grasps the whole situation and selects the action from a long-time viewpoint, including a future goal. The action selection is made nominally following the template, but with full consideration of the related situation; therefore, the most superior mode of control with effectiveness and robustness.

intimate relations are formed between the spaces of goal and state.

The essential point of the cognitive system being able to move freely from one knowledge space to another is that it becomes possible to model the generic nature of the flexibility in human thinking. The way humans view the system is versatile with respect to the amount of detail of system configuration ranging from the system as a whole, to subsystem level, to elementary parts, to the difference in knowledge levels of skill-based heuristics by a well-experienced person versus elementary basic principles by a novice. Therefore, there exists more or less a redundancy when we see human knowledge space as a whole, and it is possible to derive several different answers for one problem by using different types of knowledge. This is the source of the superior problem-solving ability of humans, and it brings about robustness that ensures that the human cognitive system will not die as suddenly as a machine, even if it encounters unexpected situations.

Control Model. In the human model, the control model is used for deciding the order of executing a sequence of various elementary cognitive processes to integrate into a whole cognitive system interaction to the perceived situation from the outside world. This explanation somewhat resembles the aforementioned process model, except the process model represents the whole picture of the human cognitive system from the view of experimental psychology in which human behavior is implicitly assumed as a one-directional process of stimulus input–response, selection–output response. However, the actual human behavior is goal-oriented in nature, in which the response starts from the consciousness of some goal to action, and the subsequent process is not the simple repetition of stimulus input–response, selection–output response. Here, the concern of the control model is the higher level control mode of complete human behavior rather than the lower level process model of stimulus input–response, selection–output response.

As an example that incorporates a control model into the human model, Hollnagel (1993) proposed the situation-dependent control model in his human model called COCOM. Hollnagel

defined four control modes as: scrambled mode, opportunistic mode, tactic mode, and strategic mode. The characteristics of those modes are as explained in Table 6.7. In COCOM, how the respective modes of human behavior would appear are described by the introduction of six control parameters: result of the immediate action, subjective available time, number of simultaneous goals, availability of plans, horizontal span of event, and execution mode. For example, the control mode would move to a more strategic mode if subjectively more time is available, while becoming more opportunistic if less time is available. If an extreme situation with no available time suddenly happens, a human would tend to fall into the scrambled mode from any initial mode.

Implementation of Human Modeling into Computers

In this section, the techniques necessary for implementing the modeling elements from the preceding section into the design of a human model simulator on a computer will be introduced.

Observation and State Recognition. This section introduces methods of the configuration of computerized models for the phase of human cognitive process that relates to how the human perceives the presented information at the interface level and then recognizes the situation.

Observation. Although there are many styles and methods for information presentation at the interface level, the entities of information are space-time patterns of multimedia information and are perceived by humans mainly through the sensing organs of the eyes and ears. When a human is in the skill-based mode, the perceived sensual information is recognized as a signal, and the amplitude of the response motor action is directly regulated by the quantitative space-time pattern of the signal. This kind of input–output process of human response by rule-based mode can be modeled in the form of transfer functions and can be implemented into computerized human models as a sort of multivariable digital controller. On the other hand, both the

(a) Stationary value

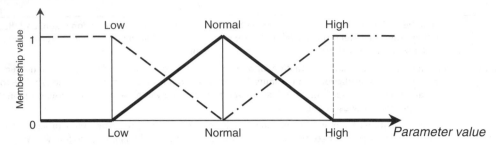

(b) Dynamic change

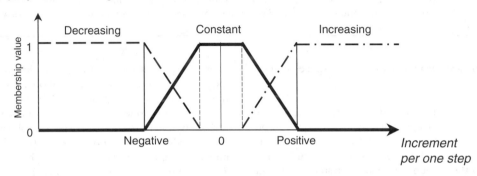

FIGURE 6.11. Membership functions for qualitative interpretation of observed information.

rule-based and knowledge-based behavior can be treated with the two-step approach: interpretation of quantitative space-time pattern information at the interface as the qualitative information of signs and symbols, followed by symbolic reasoning.

Conversion to Qualitative Information. The simplest method for the interpretation of the conversion of quantitative information to qualitative is to classify the information into several categories by comparing the original value with several threshold values. The threshold values can be reduced by conducting interviews with the target person for the modeling.

In case this method is insufficient, classification by fuzzy sets will be applied. In this method, the various membership functions are formed by threshold values as illustrated in Fig. 6.11(a), and the quantitative parameter is classified into a class of fuzzy sets (i.e., interpreted qualitative statement), which membership value exceeds 0.5. The obtained membership value is also treated as the degree of confidence of the relationship of the person to the observed parameter.

The case illustrated in Fig. 6.11(a) is an example of parameter interpretation at a steady state, whereas Fig. 6.11(b) illustrates interpretation in light of a time-varying parameter. In the latter case, the parameter value is monitored at multiple times during transition to calculate its difference from the initially set value. To interpret the degree of parameter change, the difference value is applied for the graph in Fig. 6.11(b), and when it is classified as either increase or decrease, the observation will finish with the reporting of either an increase or decrease.

When this observation process reaches the upper limit of the observation window, it will report as a constant state. (The upper limit value is determined by considering the time constant of the parameter change.)

Identification of State. The process of state identification by the observation process as described previously is mainly conducted by hypothesis generation through similarity matching and validation of the hypothesis. Here, the hypothesis is generated by "setting a particular system state and then reducing a set of symptom patterns to be observed at that system state." The measure of similarity is represented as "the degree of coincidence between the set of symptom patterns and the set of observations registered in the WM of the mental model after its qualitative observation process." Because there are important symptoms and not so important ones in view of hypothesis validation, it is necessary to take into account the importance of each symptom for evaluating the degree of similarity. Therefore, the set of hypotheses versus symptom patterns is represented by the assembly of three data sets described by Hypothesis Name (category of system state), Symptom, and Weight.

These sets of hypotheses versus symptom patterns are reduced by a simulation experiment or interview with the target persons for modeling and are prepared as the knowledge base in the human model.

Degree of Confidence for Hypothesis Formation. First, to differentiate the degree of coincidence of a particular hypothesis from the degree of confidence for the observation, the inner product value is calculated by the following equation:

$$\varphi = \sum_{i} \omega_i \cdot C_i$$

where ω_i and C_i are weights of the symptom i and the degree of confidence for the observation. Then, by using the obtained result, the normalized value S is calculated by the following Sigmoid function:

$$S = \frac{1}{1 + e^{-(a\phi + b)}}$$

where a and b are the parameters. The normalized value S is defined as the degree of similarity for a particular hypothesis on the basis of present observation. Those hypotheses of which the similarity value exceeds a certain threshold value are registered in the WM as the target of further consideration in the human model.

Validation of Hypotheses. When the degree of similarity is not very high or it is not possible to choose a single hypothesis of the highest similarity, then further observation will be conducted to look for a new symptom not yet observed to validate the hypothesis. When a new symptom is observed and registered in the WM, the value S will be recalculated, and when S reaches a certain standard by the process of hypothesis validation, then the single hypothesis is adopted so that the hypothesis will become a belief.

Implementation of Control Model. On introducing the control model into the human model, it is favorable to use a technique that cannot only alter the inference methods flexibly in accordance with changes of control mode, but can also simulate different behaviors of operators who have various personalities. For this purpose, the blackboard system developed by Engelmore and Morgan (1988) is a promising method for the

simulation of human cognitive behavior, because of its flexibility with a range of changing inference strategies for various applications in the field of AI.

Configuration of the Blackboard System and Execution Procedure. The general configuration scheme of the blackboard system is depicted in Fig. 6.12. It consists of a blackboard, knowledge source, and a blackboard control mechanism. The blackboard is a common database area for the whole system that is surrounded by various kinds of knowledge sources that work together as a problem-solving module for a specific purpose. The blackboard can be built either as a flat space with no structure or as a structured configuration that is geared toward a specific application. It is possible for all the knowledge sources to read and write to the blackboard, and the choice of whether or not to respond to the specific information on the blackboard is left to each knowledge source. Therefore, it is possible to realize various types of problem-solving strategies through the participation of many knowledge sources, strategies such as bottom up, top down, and the coordination of both. This is realized by the blackboard mechanism with proper execution procedures to control the knowledge sources as shown in Table 6.7.

Control of Cognitive Behavior Simulation by the Blackboard System. The simulation model of human cognitive behavior can be implemented by the blackboard system with the various constituent elements of process modeling from a choice of knowledge sources. In this case, the limited memory space of the WM corresponds to blackboard, and the function of attention is simulated by blackboard control mechanisms to imitate the attentive management of cognitive tasks under the restriction of cognitive resources. The allocation of cognitive resources is dependent on the priority of cognitive tasks, and the priority is determined by various control parameters used in the human model COCOM developed by Hollnagel (1993), as well as the saliency of the incoming information from the sensing organs. It is also possible to include a specific knowledge source that

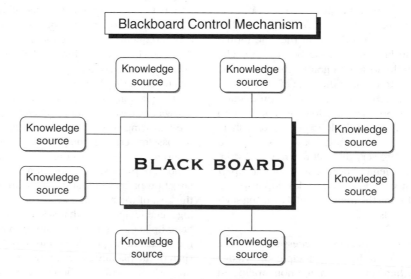

FIGURE 6.12. Configuration of blackboard system.

can modify the evaluation standard for prioritization in accordance with the situation change in the blackboard. Therefore, it is even possible to simulate the dynamic control of cognitive behavior depending on the contextual change of the problem-solving situation.

In addition to change of context, there are an assortment of characteristics observed in humans. For modeling human personality, Woods, Roth, and Pople (1987) proposed their human model simulator called CES. They suggested that various stereotypes, such as vagabond, Hamlet, garden path, inspector, and specialist can be modeled by altering the priority for the different evaluation standards. For example, if too high a priority is laid on newly discovered events, then the locus of thinking would swing around to and fro like vagabond, whereas too much priority on the cognitive tasks of interpretation and explanation of events would cause thinking to become too prudent, similar in nature to the character Hamlet, therefore allowing too much time to pass before the decision to take action. Through this mechanism of regulating the control model, it is possible to model the habits and biases of human thinking, as well as the various types of human error generated by the intrinsic nature of human variability.

Implementation Method of Interpretation and Planning.
In this section, three inference methods will be introduced that are applicable for both interpretation and planning, and then the issues of rule-based and knowledge-based planning will be discussed.

Use of Production Rules and Their Limitations. Application of a production rule like an if-then type statement is the simplest way for implementing the inference system in the human model. For example, the process of presenting the action goal for the identified system state is given by applying the following rule:

(Condition part)–State–Action goal–(Conclusion part)

When the state of the condition part appears in the WM, then the action goal in the conclusion part is written into the WM. It is not necessary to write all knowledge in the rule format and store the rules in the knowledge base beforehand. The knowledge represented in the form of a graph, etc., should be compiled in the rule format to use it at the time of association. Because most of the inference can be realized by production rules if you do not mind the programming load, it is possible to develop the system by using an available expert system shell. The higher level AI methods, such as model base reasoning and nonmonotonous inference, are very difficult to program by the production method or to implement using conventional expert system shells. Therefore, it would be wise to develop a special inference engine for developing the human model by using such higher level inference methods.

Model-Based Reasoning. It becomes necessary to use a deep inference method based on the first principle knowledge for the objective system when you encounter a nonanticipated difficult situation. The new approach called *ontology* in the area of AI will be required for realizing effective reasoning methods on such knowledge-based behavior at the human interface level. Lind (1994) proposed a reasoning model by the name of MFM (Multi-Level Flow Model), which is especially well suited to dealing with a plant operator's knowledge-based behavior at the human–machine interface. MFM is a descriptive model of the functional relationship among the various components in the whole plant system. MFM recognizes the three different kinds of flow (i.e., mass, energy, and control signal). An operator's behavior based on the first principle for knowledge-based diagnosis can be well simulated by constructing the MFM model for a particular system and devising a proper inference procedure for the functional relationships that are derived from the MFM model.

Nonmonotonous Inference. In human thinking, it is frequently noted that a person makes a tentative judgment based on the available information at a point in time even if some data are still missing, insufficient, or incorrect. Then, when new information is obtained, the person will change their mind to replace the previous judgment with a more proper and sensible one. In the field of AI, to implement some kind of change of mind into the human model simulation, Furuta and Kondo (1993) proposed the method of managing the truth value of belief by using the techniques of nonmonotonous inference. The essential issues to overcome when simulating this peculiar behavior of the human mind from this inference method are: whether or not the change of belief by new information can be effectively distinguished from the time change of state, and how to avoid the occurrence of contradiction that a person believes both a proposition and its counter-proposition at the same time.

Rule-Based and Knowledge-Based Planning. There are two types of planning in the planning process: rule-based planning of simply recalling a proper template and knowledge-based planning to refine the plan through means-ends analysis. As for the example of rule based, Yoshikawa et al. (1997) applied the Petri net model in their developed SEAMAID system. The system represents the operator procedure as a rule-based template of action sequences, because the operation procedure is normally described in the operation manual. Using their method, the execution sequence of the operator is controlled by the movement of a token in the Petri net.

The planning method of using AI is applicable for describing knowledge-based planning. The planning by AI is conducted in the following way. Various kinds of operators are assigned as the base materials in the planning. In this case, operator means a certain analysis procedure that deals with effect, selection condition, preposition condition, and concrete contents of the target problem. Generation of a plan is conducted by including the goal of the planning in the effect operator and then selecting suitable operators that satisfy selection condition. The generated plans from the previous step are then evaluated. When there exists a preposition condition that is not yet satisfied, the span of the planning is expanded so that the attainment of this preposition condition should be included as a subgoal. In this way, the recursive process of AI planning is repeated until all the preposition conditions are satisfied.

Human–Environment Interaction

Limitations of Traditional Artificial Intelligence. The framework of a human model in HCI or a human-machine system basically assumes that the human model would have knowledge of the objective system surrounding the human. It is true, and it should be when real humans make an introspection about our relationship with the surrounding environment; but, from the viewpoint of conducting a computer simulation, the issue is different.

There has been a philosophy in the traditional AI world that by conducting the symbol processing by the symbolic processors that have the complete symbols and knowledge about the world, it is possible to obtain the full understanding of the world. If we apply this notion for configuration of the human model, the information and knowledge about the situations that govern human behavior should be formed within the human model, and the whole cognitive process from the observation until the decision for action would basically proceed independently from the external environment. But there is a limitation to the understanding of human cognition based on this traditional AI thinking, and today the necessity for a new alternative approach has been widely recognized in the AI area.

Knowledge Existence in the Outer World. According to environmental psychology as noted by Gibson (1979) and "knowledge in outer world" by Norman (1988), human cognitive activity is not passively enclosed within the human brain, but will interact positively with the outer world to obtain the situational information and knowledge that exist in the actual world in a form that is understandable to a human and can be used by them when necessary for further interaction. Based on this notion, a human does not need to hold all situational information and knowledge in his/her brain; a human can perform more flexible thinking with less cognitive load than humans as viewed by traditional AI.

In human cognitive activity, there always arises the mutual interaction with the acting environment in the process of acquiring situational information and knowledge and the use of both. These processes determine a large part of human cognitive behavior. When dealing with the human-machine interaction at the interface level, it is not appropriate to implement all relevant knowledge into the human model, because it is thought that most of the attribute knowledge about a human-machine interface is acquired from the acting environment.

Simulation Scheme of Human-Environment Interaction. When considering the simulation of human-machine interaction from the viewpoint as described previously, it is reasonable to adopt the whole framework that the situational information and knowledge are implemented in the model of the plant, whereas the human model will obtain the information from the plant model when needed. When this actually occurs, there will be a flooding of communication between both models in addition to various low-level processing of the human model, such as calculation of body motion, head movement, range of visual sight, etc.

To avoid those problems, Furuta, Shimada, and Kondo (1999) proposed a different framework of simulating the human-machine interaction as illustrated in Fig. 6.13, where knowledge obtained from the environment and the interaction models on both the environment and human are separated from the human model and the plant model. In Fig. 6.13, the behavioral environment model is newly introduced to describe the human-machine interface from the viewpoint of both physical and cognitive aspects and through a set of submodules of mutual interaction with the environment. This combination model is used to process different kinds of human interactions with the surrounding environment, such as sensory-motor action patterns, semantic interpreter, cognitive filter, body motion and posture control of the human body, movement path calculation, eye direction control, sound/noise/voice recognition, utterance of sentences, etc. Moreover, the information among the three models of human-machine interaction are exchanged via message passing channels with a higher level of communication. In this way, the flexible computer simulation of the human-machine interaction can be realized without bringing complexity to the human model when describing human interactions with the environment.

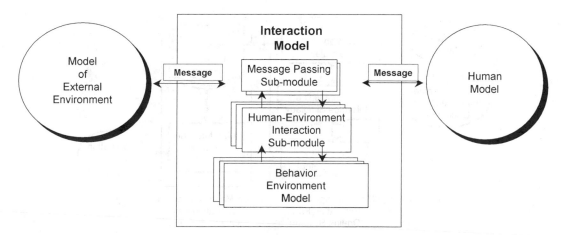

FIGURE 6.13. Simulation framework for human–environment interaction.

APPLICATION OF HUMAN MODELING FOR HCI

In the preceding section, the fundamental method of implementing the human model in the computer is described and also discussed are several issues on integrating the human model into the whole simulation of human–machine interaction. In this section, fulfillment of process control in the whole human–machine interaction simulation system will be presented with practical applications, followed by future trends and the prospect of human modeling in HCI.

Simulation of Human–Machine Interactions for Process Control

Objective of System Development. During the design of the human–machine interface, both hardware and software elements will become the direct target of multifaceted evaluation for human factors and safety aspects, etc. In this case, the hardware elements could be the control room, control board, control panel for plant systems; or the driver's seat and cockpit design for an automobile, train, or aircraft. The software elements include items such as operational procedures, an operations manual, display design, or the introduction of computerized aid. It is expected that the simulation of complex human–machine system interaction by using human modeling technology would be a useful engineering tool for evaluating the possibilities of various mismatches between the human and machine systems at the interface level, and for the prevention of inadvertent human errors, misjudgments, and the mitigation of cognitive load, so that the whole man–machine system will be assured as fault-tolerant and reliable.

Aimed at applying the previously stated objectives in the field of process control, there have been a lot of human model simulators developed thus far, such as COSIMO (Cacciabue et al., 1990), CES (Woods et al., 1987), COCOM (Hollnagel, 1990), and SEAMAID (Nakagawa et al., 2000), which were cited earlier in this chapter. A new possibility of human model application for human reliability analysis was first suggested by Kirwan (1996) as a tool for probabilistic safety assessment of nuclear power plant.

In the following, a detailed method for simulating human-machine interaction by computer will be described by taking the SEAMAID system as an example of human modeling applications for practical engineering purposes.

Overall Configuration of SEAMAID. SEAMAID has been developed with the purpose of practical application of human modeling for a variety of human–machine interface issues. It has been developed for performing real-time simulation of complex human-machine interaction in various emergency situations of a nuclear power plant. The whole system framework of SEAMAID is shown in Fig. 6.14. As seen in Fig. 6.14, it is divided into two parts: the Man–Machine System Simulation Part as an on-line system and the Man–Machine Interface Evaluation Part as an off-line system. The system functions of various elements of the SEMAID configuration are explained in Table 6.8. The most salient feature of SEAMAID is that the Man–Machine System Simulation Part is configured by three real-time simulators, namely: (1) the dynamic simulator of a nuclear power plant, (2) the human–machine interface simulator, and (3) the human model simulator (i.e., simulator for operator's cognitive behavior at the man–machine interface level). It is accomplished as a distributed computational system on a computer network. Because the SEAMAID is used for the evaluation of the man–machine system of the nuclear power plant, the plant simulator used in the SEAMAID is a real-time training simulator for a pressurized water reactor. But, if you would like to consider other machine systems, you can easily change to another machine simulator suitable for your problem via the distributed simulation scheme in use in the present SEAMAID. The important features of the simulators 2 and 3 are described in the subsequent sections.

Human–Machine Interface Simulator. The hardware equipment in the control room of a nuclear power plant as the physical environment can be understood as the interrelationship of structured sheets and icons as shown in Fig. 6.15, using the theoretical framework of AI. The human–machine interface simulator in the SEAMAID has a kind of on-line object-oriented

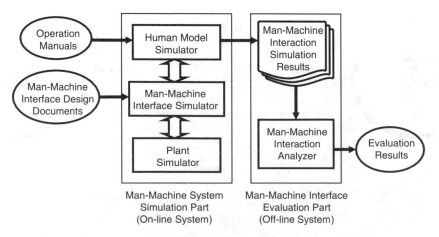

FIGURE 6.14. Overall configuration of SEAMAID system.

TABLE 6.8. System Functions of Various Elements in SEAMAID

System Parts	System Elements	Functions of System Parts and Elements
Man–Machine System Simulation Part	Human Model Simulator	Computer models of operators' cognitive behavior to monitor, control, and maintain various equipments at MMI and those in the plant.
	Man–Machine Interface Simulator	A model simulator of a real MMI with respect to its physical configurations, functions, and dynamical changes of values by use of AI.
	Plant Simulator	A model simulator to describe dynamic behavior of a machine system. In case of detailed simulation, engineering simulator is needed.
Man–Machine Interface Evaluation Part	Man–Machine Interaction Analyzer	Analysis and calculation of various indexes for interface evaluation, such as workload of operators, probability of human error, by using the results of man–machine interaction simulation.

Note. MMI = man–machine interface; AI = artificial intelligence.

database model that describes the hierarchical structure of man–machine interface equipment. The method of representing a distributed real-time simulation system by SEAMAID is illustrated in Fig. 6.16, with the man–machine interface simulator highlighted within the whole system. The dynamic changes of various sensors and actuators in the man–machine interface system can be simulated by rewriting the corresponding attributes of the instrumentation and controls equipment in the object-oriented database in accordance with the dynamic changes of both the plant simulator and the human model simulator. The on-line object-oriented database of the man–machine interface can be constructed easily by using a special graphic editor with icons.

The salient feature of the human–machine interface simulator in Fig. 6.16 is that it can realize the separation of the human–environment interaction model by introducing two shared memories for coupling the man–machine interface simulator with the human model simulator to avoid the bottleneck of data transfer between the two. On the shared memory 2, only the information on the operator's position and the actuator operation to the control board are given by the human model simulator, and the man–machine interface simulator will calculate the whole eye span within the operator's eyesight from his position. Then the man–machine interface simulator will calculate the responses of all the instrumentation within the operator's

eye span by taking into account the response characteristics of the instrumentation. The resultant responses from the instrumentation are then fed back to the shared memory 1 for the operator's use of alarm and meter reading within their eye span.

Human Model Simulator. The salient feature of the human model simulator in the SEAMAID is that it consists of two different models to describe three modes of the operator's cognitive behavior (Rasmussen, 1983) at the human–machine interface: the first human model is for skill- and rule-based behaviors to handle plant transitions in accordance with standard operating procedures, and the second model for knowledge-based behavior is used to diagnose plant anomalies by inference with the combination of monitored instrumentation signals and the preacquired knowledge bases for plant instrumentation and control systems behavior. The first model is realized by a hierarchical Petri net model simulator (Yoshikawa et al., 1997), whereas the second model by a knowledge engine is used to describe and control the operator's circular cognitive process of anomaly detection, symptom recall, hypothesis formation, future prediction, comparison with the monitored data, and hypothesis acceptance or rejection (Wu et al., 2000).

The human model simulator in the SEAMAID is illustrated in Fig. 6.17. To compare the framework of Fig. 6.17 with that of the

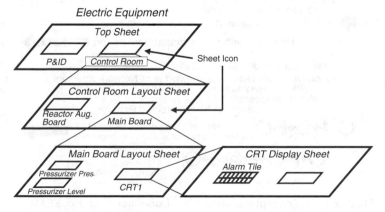

FIGURE 6.15. Configuration of man–machine interface simulator as an icon-based, object-oriented database. P&ID = piping and instrumentation diagram.

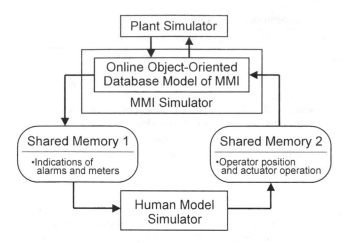

FIGURE 6.16. Computer implementation of human–environment interaction in SEAMAID. MMI = man–machine interface.

generic human model in Fig. 6.9, the fundamental features of the computational architecture of the human model simulator are itemized by the following five discussions:

1. There are two types of knowledge bases stored in the LTM in the human model simulator. One is a group of the piecewise knowledge bases of the network structure that describes the divided diagram of Piping and Instrumentation Diagrams of an nuclear power plant, whereas the other group of hierarchical Petri net models is especially developed for describing the plant operation procedure by modeling it as discrete, parallel events.

2. There are two types of Petri net models for the operation procedure. One is for physical motion at the interface level and the other is for the chain of cognitive processes without any physical motions.

3. In the case of the first type of Petri net model (with physical motion), basic physical motions, such as walk, bend, look at the meter, etc., are preprogrammed as the related subroutines for motion dynamics and stored in the motion library for common usage of motion calculations when the relevant Petri net is activated in the simulation.

4. The knowledge base of the second type (i.e., Petri net model), when activated in the PWM and selected by the high-priority index to rule-based process with consciousness in FWM, will be automatically processed in accordance with the processing sequence described in the Petri net model.

5. The knowledge base of the first type, when activated in PWM and used for knowledge-based attention processing in FWM, will be used by the inference engine to produce the chain of reasoning on cause–consequence relationships among the observed outside-world information (i.e., alarm signals and other selected instrumentation information).

Validation of the Human Model. Validation of the second model for knowledge-based behavior is especially difficult because the human cognitive process occurs mentally (i.e., it is a hidden process) and because it differs between persons and at different times because of intrapersonal and interpersonal differences. By using the knowledge engine for the second model, a Monte Carlo-type simulation experiment was conducted by changing various human cognitive parameters used in the model to explain the variation of diagnostic behaviors that were

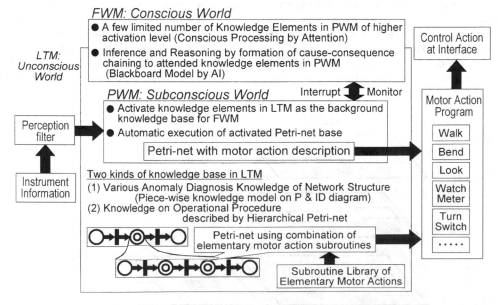

FIGURE 6.17. Conceptual framework of operator model in SEAMAID. PWM = peripheral working memory; LTM = long-term memory; FWM = focal working memory; AI = artificial intelligence; P&ID = piping and instrumentation diagram.

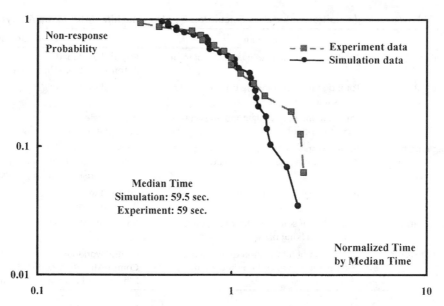

FIGURE 6.18 Intercomparison of time vs. reliability curves for LOCA event between simulation and experiment.

observed in the authors' conducted laboratory experiment using the Pressurized Water Reactor (PWR) plant simulator with the participation of several specialists in PWR plant operation. The comparison between the simulation and experiment is shown in Fig. 6.18, where two graphs of TRC (TRC = time vs. reliability curve) of plant diagnosis normalized by the median time are shown for the Loss-of-Coolant Accident (LOCA) case. As seen in Fig. 6.18, the results showed that the human model agreed with the experimental data with respect to the time trend of TRC of plant diagnosis and the median time (see details in Yoshikawa and Wu, 1999).

Industrial Application of SEAMAID. Thus far, the developed SEAMAID has been applied to the following three human factors issues: (1) estimation of the previously described TRCs both for anomaly detection and successful diagnosis of plant operators in an actual control room of a nuclear power plant (Wu et al., 2000; Yoshikawa & Wu, 1999), (2) virtual reality (VR)-based visualization of the simulated operator's motion by SEAMAID for monitoring and control of the control panel in the control room with verbalization of the thinking process by a synthesized voice to cope with plant transitions (Ishii, Wu, Shimoda, & Yoshikawa, 1999), and (3) comparative evaluation of different types of panel layout designs from the viewpoint of potential human error and workload in the real-size plant control room to cope with plant emergency situations (Nakagawa et al., 2000).

In case 1, the SEAMAID can be a useful tool for probabilistic safety assessment/human reliability analysis of nuclear power plants, whereas in case 2, it can be used to preview the design of the whole control room, showing where the operator will move in the event of plant emergency, and can be easily understood by virtue of the three-dimensional visualization of VR.

In case 3, the three-operator model version of SEAMAID (i.e., one shift supervisor, one reactor operator, and one turbine operator) was developed for checking the imbalance of each person's workload, the inconvenience of panel layout by two operators' collision during each individual person's movement, or hindered sight. For such an evaluation purpose, a man–machine interface evaluation system was separately developed on a personal computer, by which the calculated results of man–machine interaction by SEAMAID were used for calculating various measures of human factors to help designers' analysis based on assorted guidelines. This system on a personal computer corresponds to the Man–Machine Interface Evaluation Analysis Part of SEAMAID in Fig. 6.14. The objective measures given by the system are listed in Table 6.9, and are used for qualitative evaluation of the man–machine interface.

As seen from the application work of SEAMAID, the human model simulation will contribute to a variety of human factors analysis areas, such as the designing of man–machine interface and operational procedures, human error analysis, and training and education of operators.

Introduction to the Education and Training Environment

A promising application of the human model in HCI is expected for education and training for engineers, doctors, operators, and maintenance workers in many industrial fields, such as aviation, railroads, shipping, plant operations, or hospital operations. It is common sense that advanced computer-assisted instruction (CAI) will use the human model as the teacher model and student model in the system. It is probable that human modeling will be implemented for configuring new education and training systems that will actively use VR and multimedia information to create a common environment of virtual experience and information visualization. The framework of such advanced CAI is depicted in Fig. 6.19, in which three elements of the experiential

TABLE 6.9. List of Qualitative Indexes Calculated by Man–Machine Interaction Analyzer

Index	Meaning	Evaluation aspect
Moving length (horizontal)	Total moving distance in the control room	Physical workload Potential risk of selecting wrong meters and actuators
Moving length (up and down)	Total distance of standing up and sitting down in the control room	Physical workload
Eyesight moving length	Total distance of eyesight moving on the control panel	Work efficiency Potential risk of selecting wrong meters and actuators
Oblique eyesight moving length	Total distance of the eyesight for looking up and down at the control board	Physical workload Potential risk of selecting wrong meters and actuators Potential risk of reading wrong meters
Work time	Total time of finishing the task	Physical workload Work efficiency
Degree of selecting wrong control panel	Possibility of selecting the wrong panel during walking passage	Potential risk of selecting wrong panel
Information quantity of short-term memory	Integration of information quantity in the short-term over certain time span	Mental workload Complexity of procedure Efficiency of man–machine interface

CAI system are the human model, an intelligent tutoring system (ITS), and a VR interface.

Concerning the human model in Fig. 6.19, there are two types of potential human models: a Teacher model that corresponds to the teacher who is equipped with profound knowledge and skills to be represented as a standard model, or the Student model that corresponds to a student and should be constructed so that it can effectively grasp deviation of the student's knowledge and skill from the teacher's and delineate the missing knowledge and skills.

The real student in a CAI system learns the knowledge and skills represented in the VR interface as a place of virtual experience and information visualization, and uses it to master skills and knowledge. The ITS in the system plays the role of good teacher. It will educate the students in the way best suited for each student by comparing their performances with the teacher model.

Recently, there has been a lot of system developments, for example, the visual engineering system VIGOR developed by Doi, Kato, Umeki, Harashima, and Matsuda (1995), an operator training system with the combination of a VR control room and ITS that was developed by Matsubara, Toihara, Tsukinari, and Nagamachi (1997). The afforded functions by such VR-based information visualization will be introduced by showing the output information of the VENUS system (VENUS = VR-based education system for operating nuclear power plant) (Ishii et al., 1999).

The VENUS system is the expanded system based on the previously described SEAMAID, so that it is provided with the VR-based experiential training environment in which the animated behavior of a virtual operator in manipulating a virtual panel in a virtual control room can be observed by trainees, both visually and acoustically. The VENUS system is configured on a distributed computer network with the software system

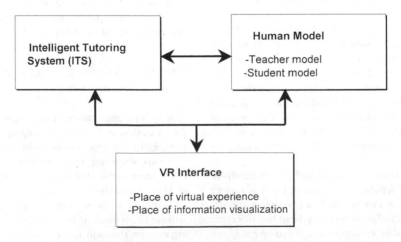

FIGURE 6.19. Configuration of experiential computed-assisted instruction system by virtual reality (VR).

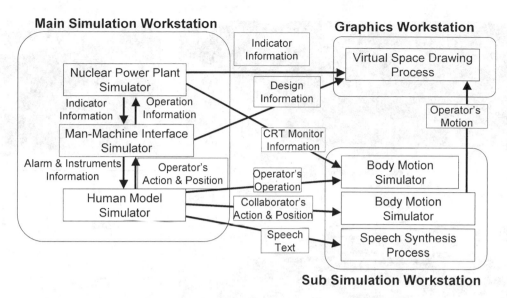

FIGURE 6.20. Software configuration of VENUS system.

configuration as illustrated in Fig. 6.20, in which various simulated data by SEAMAID on the main simulation workstation are transferred to both the subsimulation workstation and the graphic workstation to generate an assortment of visual-auditory information for the trainees. Some results of VR simulation by VENUS are shown in Fig. 6.21, which depicts the operator walking toward the control panel, the line of vision of the operator, the trajectory of the operator's focal point on the CRT display, and the snapshots of operator action. The behavior simulation of a virtual operator in a virtual control room by VENUS provides trainees with the following realistic information: (1) a VR model of a control room, (2) animated operator actions during a plant emergency, (3) voice narration to explain diagnostic processes by the operator, (4) changes of visual input from the operator's view, and (5) a trace of eye focal points when checking the plant status on a CRT display. This is a typical example of multiple information visualization by VR.

Personified Interface Agents and Future Trends

With the rapid progression of information technology, it is a prevailing trend in the world of the Internet that assorted animated characters generated by computer graphics are used as interface agents for humans. Because humans feel more friendly and affected when the characters have human-like shapes and behaviors, there are a lot of personified agents emerging these days. Therefore, it seems that the human model in HCI should be directed more toward attaining humanly behavior that reflects the human-to-human relationship in actual society so that agents can be considered social agents. A social agent is endowed with three qualities: cognitive, physical, and affective. In the last section of this chapter, the future trends of human modeling in HCI are projected mainly from the social image of process control gained from the previous discussion of social agents.

Virtual Collaborator: Constituents for the Personified Interface Agent. Virtual collaborator means an image of a virtual robot that is extended from the virtual operator in the virtual control room concept as configured by VENUS in the previous section. The virtual collaborator is a personified interface agent who has the following characteristics: (1) a kind of autonomous robotic, humanlike shape in three-dimensional virtual space; (2) intelligent, but obedient to humans; and (3) natural face-to-face communication with humans. However, it is necessary to achieve the following functions to configure such a personified interface that can communicate with humans naturally in virtual space: behave in the same form as a human; speak, make facial expressions and gestures like a human; hear and understand what humans speak; understand a human's thinking and feeling; and think and judge by itself and do so on behalf of humans. The image of such a virtual collaborator is depicted in Fig. 6.22, as an extension of the virtual operator by VENUS. The research framework toward VR simulation of human modeling in HCI is proposed by Yoshikawa et al. (2001), as depicted in Fig. 6.23, in which the research needed for fulfilling the whole scheme is organized by the following three subjects;

1. Extension of the human model simulator to deal with bidirectional conversation with humans, including the affective message.
2. Realization of various human-sensing methods, such as motion recognition, voice recognition, face recognition, and even the recognition of real humans' cognitive/emotional situation.
3. Development of flexible tools for constructing interactive virtual space and performing interactive simulation by integrating a human model simulator and human-sensing methods into virtual space.

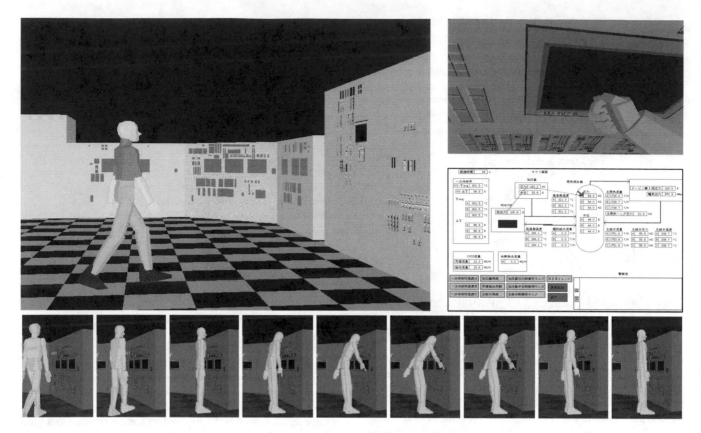

FIGURE 6.21. Simulation results of virtual operator's behavior in the virtual control room.

When virtual collaborators are realized in the future, the operators in process plants will no longer need to monitor and manipulate the control panels; instead, they will only have to direct the virtual collaborators to complete the desired tasks through voice and gesture guidance, as depicted in Fig. 22. Therefore, the human interface will be transformed from the traditional direct completion by mouth or keyboard to the third generation human interface of indirect agent controls. When the virtual collaborator is used in VR-based experiential CAI, it will either become a virtual teacher to the pupils or be a good partner to work with the human in virtual space.

Future Prospects of the Virtual Collaborator. When the technologies of autonomous-personified interface agents are superimposed on rapidly progressing Internet technologies, the future image of human–machine interaction in the age of information technology can be depicted as shown in Fig. 6.24. (This image is the chapter author whose background is mainly plant engineering.) From this figure, we can determine that the workers in the 21st century will no longer commute to a plant and work in the control room, but will work from home or at a nearby tele-work center functioning through tele-operation, tele-testing, and tele-maintenance—all by consulting with various virtual collaborators via the Internet.

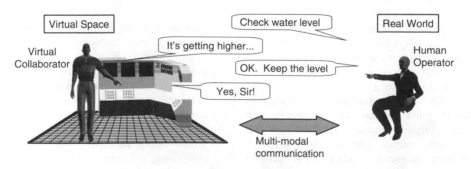

FIGURE 6.22. Image of mutual communication between virtual collaborator and real human operator.

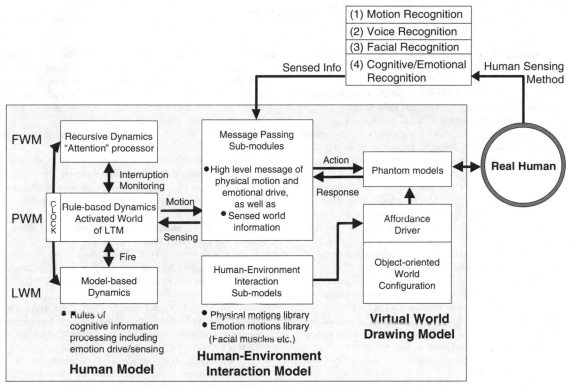

Virtual Collaborator in the Virtual World

FIGURE 6.23. Research framework toward virtual reality-based simulation of human–machine interaction. FWM = focal working memory; PWM = peripheral working memory; LWM = long-term working memory; LTM = long-term memory.

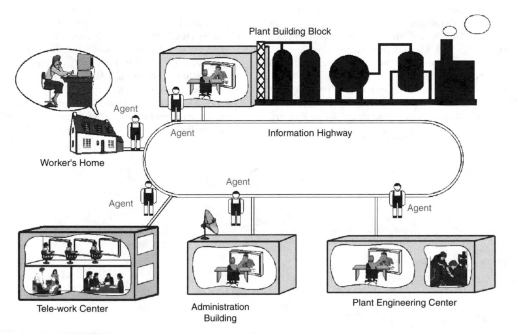

FIGURE 6.24. Image of mutual communication between human and machine in the age of information technology in the 21st century.

References

Cacciabue, P. C., Mancini, G., & Bersini, U. (1990). A model of operator behavior for man-machine system simulation. *Automatica, 26,* 1025-1034.

Card, S. K., Moran, T. P., & Newell, A. (1983). *The psychology of human-computer interaction.* Hillsdale, NJ: Erlbaum Associates.

Doi, M., Kato, N., Umeki, N., Harashima, T., & Matsuda, K. (1995). Visual engineering system—VIGOR: Virtual environment for visual engineering and operation. In Y. Anzai, K. Ogawa, & H. Mori (Eds.), *Symbiosis of human and artifact* (pp. 435-440). Amsterdam: Elsevier.

Embrey, D. E., & Reason, J. (1986). The application of cognitive models to the evaluation and prediction of human reliability. In *Proceedings of the international topical meeting on advances in human factors in nuclear power systems* (pp. 292-301). Knoxville, TN: American Nuclear Society.

Engelmore, R., & Morgan, T. (Eds.). (1988). *Blackboard systems.* Wokingham, UK: Addison & Wesley.

Furuta, K., & Kondo, S. (1993). An approach to assessment of plant man-machine systems by computer simulation of an operator's cognitive behavior. *Journal of Man-Machine Studies, 39-3,* 473-493.

Furuta, K., Shimada, T., & Kondo, S. (1999). Behavioral simulation of a nuclear power plant operator crew for human-machine system design. *Nuclear Engineering and Design, 188,* 97-109.

Gibson, J. J. (1979). *The ecological approach to visual perception.* Boston: Houghton Mifflin.

Hollnagel, E. (1993). *Human reliability analysis: context and control.* London: Academic Press.

Ishii, H., Wu, W., Li, D., Shimoda, H., & Yoshikawa, H. (1999). Development of a VR-based experienceable education system—A cyber world of virtual operator in virtual control room. In N. Callaos, N. Nada, A. Cherif, & M. Aveledo (Eds.), *Proceedings of the 3rd world multi-conference on SCI & ISAS* (pp. 473-478). Orlando: International Institute of Informatics and Systemics.

Kirwin, B. (1996). The requirements of cognitive simulations for human reliability and probabilistic safety assessment. In H. Yoshikawa & E. Hollnagel (Eds.), *CSEPC96 cognitive systems engineering in process control* (pp. 292-299). Kyoto, Japan: Kyoto University.

Lind, M. (1994). Modeling goals and functions of complex industrial plants. *Applied Artificial Intelligence, 8-2,* 259-283.

Matsubara, Y., Toihara, S., Tsukinari, Y., Nagamachi, M. (1997). Virtual learning environment for discovery learning and its application on operator training. *IEICE Transactions on Information and Systems, E80-D,* 176-188.

Nakagawa, T., Matsuo, S., Yoshikawa, H., Wu, W., Kameda, A., & Fumizawa, M. (2000). Development of effective tool for interactive design of human machine interfaces in nuclear power plant. In S. Kondo & K. Furuta (Eds.), *PSAM5—Probabilistic safety assessment and management* (pp. 2713-2718). Tokyo: Universal Academic Press, Inc.

Norman, D. A. (1988). *The psychology of everyday things.* New York: Basic Books.

Rasmussen, J. (1983). Skills, rules, and knowledge; signals, signs, and symbols, and other distinctions in human performance models. *IEEE Transactions on Systems, Man, and Cybernetics, SMC-13,* 257-266.

Rasmussen, J. (1986). *Information processing and human-machine interaction.* Amsterdam: North-Holland.

Reason, J. (1990). *Human error.* Cambridge, United Kingdom: Cambridge University Press.

Swain, A. D., & Guttmann, H. E. (1983). *Handbook of human reliability analysis with emphasis on nuclear power plant application.* USNRC Report, NUREG/CR-1278. Washington, DC: U.S. Nuclear Regulatory Commission.

Woods, D. D., Roth, E. M., & Pople, H. (1987). An artificial intelligence based cognitive environment simulation for human performance assessment. USNRC Report, NUREG/CR-4862. Washington, DC: U.S. Nuclear Regulatory Commission.

Wu, W., Yoshikawa, H., Nakagawa, T., Kameda, A., & Fumizawa, M. (2000). Human model simulation of plant anomaly diagnosis (HUMOS-PAD) to estimate time cognitive reliability curve for HRA/PSA practice. In S. Kondo & K. Furuta (Eds.), *PSAM5—Probabilistic safety assessment and management* (pp. 1001-1007). Tokyo: Universal Academic Press, Inc.

Yoshikawa, H., Nakagawa, T., Nakatani, Y., Furuta, T., & Hasegawa, A. (1997). Development of an analysis support system for man-machine design information. *Control Engineering Practice, 5,* 417-425.

Yoshikawa, H., Shimoda, H., Ishii, H., Nakagawa, T., Wu, W., Fumizawa, M., & Monta, K. (2001). Development of integrated simulation system SEAMAID for human-machine interaction in nuclear power plant: Its practical applications and future prospect. In *Preprints of the 8th IFAC/IFIP/IFORS/IEA Symposium on Analysis, Design, and Evaluation of Human-Machine Systems* (pp. 535-540). Kassel Germany: University of Kassel.

Yoshikawa, H., & Wu, W. (1999). An experimental study on estimating human error probability (HEP) parameters for PSA/HRA by using human model simulation. *Ergonomics, 42,* 1588-1595.

· II ·

COMPUTERS IN HUMAN–COMPUTER INTERACTION

Robert J. K. Jacob
Tufts University

The previous section of the handbook described the human in human–computer interaction; here we consider the computer. The problem of human–computer interaction can be viewed as two powerful information processors, the human and the computer, attempting to communicate with each other via a narrow-bandwidth, highly constrained interface (Tufte, 1989). Although the computer is evolving rapidly, the human is not. If we can predict anything about human–computer interaction 100 or even 1,000 years from now, it is that the humans will be quite similar to ourselves in size, shape, perceptual abilities, and mental capacities, whereas the computers might well be unrecognizable to us. As designers and builders of user interfaces, we must thus invent systems that fit the unchanging characteristics of humans, rather than try to bend the humans to fit our changing designs. Human–computer communication was once conducted via switches, knobs, jumper wires on patch boards, and status lights. These were convenient for the computer but awkward and difficult for humans; they required considerable training and adaptation, and computers were used by only a few, highly trained engineers. Now, as computer use becomes more widespread and computer power increases, we can spend more and more of that power on better communication with the user.

Today, the frontiers of research in human–computer communication seek new ways to increase the bandwidth and naturalness of this communication channel. The challenge is to design new interaction media that better fit and exploit the characteristics of humans that are relevant to communication. We can view the fundamental task of human–computer interaction as moving information between the brain of the user and the computer. Progress in this area attempts to increase the useful bandwidth across that interface by seeking faster, more natural, and more convenient communication means. On the user's side of the communication channel, the interface is constrained by the nature of human communication organs and abilities; on the computer side, it is constrained only by the devices and methods that we can invent.

The chapters in this section describe the state of the art and the frontiers of research in this area. They focus on the actual input and output devices that we use to connect human and computer but take us well beyond the familiar mouse, keyboard, and video display. All of our designs and plans for new user experiences and human–computer interfaces, from high-level conceptual and organizational issues to low-level mechanical ones, are eventually funneled through the actual input and output computer media covered here. They are the ultimate means by which the human and computer communicate information.

Hinckley surveys the world of input devices and lays out the field as a cornerstone of this section. We typically think of a mouse, a keyboard, and maybe a virtual reality glove or tracker, but Hinckley places these and other devices into a broader perspective and provides a taxonomy for understanding the range of existing and possible input devices.

Karat, Vergo, and Nahamoo cover speech both as an input and an output medium. They show how the problem of speech recognition goes beyond simply processing the input and requires understanding of natural language, dialogue, and the context of the application. They also discuss multimode dialogues, which combine speech with gestures and other actions to make a dialogue with a computer that is more like one with humans.

The principal means for humans to receive information is through vision. Unlike the earliest, text-based computer interfaces, our current interfaces reflect this and rely heavily on visual communication. Luczak, Roetting, and Oehme go beyond the familiar cathode ray tubes and liquid crystal displays to provide a systematic overview of the full range of visual displays. They cover the principles of display systems and survey the current technology in this field.

Another form of computer output is through the sense of touch—haptic output. Computers can generate visual images of highly complex and interactive objects and worlds, but we can typically only see them, not touch them. Iwata introduces the field of haptic output, in which computers synthesize physical sensations, much as they already do visual ones. He introduces the sensory mechanisms involved, both proprioceptive for the motion of the limbs and cutaneous for the direct sense of touch. He then describes a range of technologies for producing these sensations, some rather more exotic and less familiar than our visual display devices.

The final communication mode in our section is auditory output. We have already covered speech, but here we are interested in the use of other sounds in a user interface. Brewster surveys the field of nonspeech auditory output, ranging from realistic or representational sounds of physical objects to abstract, often more musical sounds. He shows how visual and auditory output have opposite characteristics in several respects so that combining the two can make a particularly effective interface.

Beyond these chapters, we might look toward the future of computer input and output media. It is easy to predict that computers both smaller and larger than today's desktop workstations will be widespread. This will be a force driving the design and adoption of future interface mechanisms. Small computers such as laptops, palmtops, personal digital assistants, and wearable computers are often intended to blend more closely into the user's other daily activities. They will certainly require smaller input–output mechanisms, and may also require more unobtrusive devices, if they are to be used in settings where the user is simultaneously engaged in other tasks, such as talking to people or repairing a piece of machinery. At the same time, computers will be getting larger. As display technology improves, as more of the tasks one does become computer-based, and as people working in groups use computers for collaborative work, an office-sized computer can be envisioned, with a display as large as a desk or wall (and resolution approaching that of a paper desk). Such a computer leaves considerable freedom for user interaction. If it is a large, fixed installation, then it could accommodate a special-purpose console or "cockpit" for high-performance interaction. Or the large display might be fixed, but the user or users move about the room, interacting with each other and with other objects in the room, using small, mobile interaction devices.

A broader trend is toward increased naturalness in user interfaces. Such interfaces seek to make the user's actions as close as possible to the user's thoughts that motivated those actions,

that is, to reduce the "Gulf of Execution" described by Hutchins, Hollan, and Norman (1986), the gap between the user's intentions and the actions necessary to communicate them into the computer. Such natural interfaces build on the equipment and skills humans have already acquired and exploit them for human–computer interaction. A prime reason for the success of direct manipulation interfaces (Shneiderman, 1983) is that they draw on analogies to existing human skills (pointing, grabbing, moving objects in space), rather than trained behaviors. Virtual reality interfaces, too, gain their strength by exploiting the user's preexisting abilities and expectations. Navigating through a conventional computer system requires a set of learned, unnatural commands, such as keywords to be typed in or function keys to be pressed. Navigating through a virtual reality system exploits the user's existing, natural "navigational commands," such as positioning his or her head and eyes, turning his or her body, or walking toward something of interest. Tangible user interfaces and ubiquitous computing similarly leverage real-world manipulation of real physical objects to provide a more natural interface (Ishii & Ullmer, 1997). The result in each case is to make interacting with the computer more "natural," more like interacting with the rest of the world and, incidentally, to make the computer more invisible.

Another trend in human–computer interaction is toward lightweight, noncommand, or passive interactions, which seek inputs from context and from physiological or behavioral measures. These provide information from a user without explicit action on his or her part. For example, behavioral measurements can be made from changes in the user's typing speed, general response speed, manner of moving the cursor, frequency of low-level errors, or other patterns of use. Passive measurements of the user's state may also be made with, for example, three-dimensional position tracking, eye tracking, computer vision, and physiological sensors. They can then subtly modify the computer's dialogue with its user to match the user's state (Picard, 1997). Blood pressure, heart-rate variability, respiration rate, eye movement and pupil diameter, galvanic skin response (the electrical resistance of the skin), and electroencephalograms can be used in this way. They are relatively easy and comfortable to measure, but interpreting them accurately and instantaneously within a user-computer dialogue is much more difficult. We might thus look into the future of interaction devices by examining a progression that begins with experimental devices used to measure some physical attribute of a person in laboratory studies. As such devices become more robust, they may be used as practical medical instruments outside the laboratory. As they become convenient, noninvasive, and inexpensive, they may find use as future computer input devices. The eye tracker is such an example (Jacob, 1993); other physiological monitoring devices may also follow this progression.

With the current state of the art, computer input and output are somewhat asymmetric. The bandwidth from computer to user is typically far greater than that from user to computer. Graphics, animations, audio, and other media can output large amounts of information rapidly, but there are hardly any means of inputting comparably large amounts of information from the user. Humans can receive visual images with very high bandwidth, but we are not very good at generating them. We can generate higher bandwidth with speech and gesture, but computers are not yet adept at interpreting these. User-computer dialogues are thus typically one-sided. Using lightweight, passive measurements as input media can help redress this imbalance by obtaining more data from the user conveniently and rapidly.

Perhaps the final frontier in user input and output devices will be to measure and stimulate neurons directly, rather than relying on the body's transducers (Moore, Kennedy, Mynatt, & Mankoff, 2001). Our goal is to move information between the brain of the user and the computer; current methods all require a physical object as an intermediary. Reducing or eliminating the physical intermediary could reduce the Gulf of Execution and improve the effectiveness of the human–computer communication, leading to a view of the computer as a sort of mental prosthesis, where the explicit human–computer communication actions vanish and the communication is direct, from brain to computer.

ACKNOWLEDGMENTS

I thank my students for their collaboration in this research, particularly Leonidas Deligiannidis, Stephen Morrison, Horn-Yeu Shiaw, and Vildan Tanriverdi. My recent work in these areas was supported by the National Science Foundation, National Endowment for the Humanities, Office of Naval Research, Naval Research Laboratory, Berger Family Fund, Tufts Selective Excellence Fund, and the MIT Media Lab.

References

Hutchins, E. L., Hollan, J. D., & Norman, D. A. (1986). Direct manipulation interfaces. In D. A. Norman & S. W. Draper (Eds.), *User centered system design: New perspectives on human computer interaction* (pp. 87–124). Hillsdale, NJ: Lawrence Erlbaum.

Ishii, H., & Ullmer, B. (1997). Tangible bits: Towards seamless interfaces between people, bits, and atoms. In *Proceedings of the Association for Computing Machinery Computer-Human Interaction CHI'97 Human Factors in Computing Systems Conference* (pp. 234–241). New York, NY: Addison-Wesley/ACM Press.

Jacob, R. J. K. (1993). Eye movement-based human-computer interaction techniques: Toward non-command interfaces. In H. R. Hartson & D. Hix (Eds.), *Advances in human-computer interaction* (Vol. 4, pp. 151–190). Norwood, NJ: Ablex. Retrieved from http://www.eecs.tufts.edu/~jacob/papers/hartson.txt [ASCII]; http://www.eecs.tufts.edu/~jacob/papers/hartson.pdf [PDF]

Moore, M., Kennedy, P., Mynatt, E., & Mankoff, J. (2001). Nudge and shove: Frequency thresholding for navigation in direct brain computer interfaces. In *Proceedings of the Association for Computing Machinery Computer-Human Interaction CHI 2001 Human Factors in Computing Systems Conference* (extended abstracts) (pp. 361–362). New York, NY: ACM Press.

Picard, R. W. (1997). Affective computing. Cambridge, MA: MIT Press.

Shneiderman, B. (1983). Direct manipulation: A step beyond programming languages. *IEEE Computer, 16*(8), 57–69.

Tufte, E. R. (1989). Visual design of the user interface. Armonk, NY: IBM.

·7·

INPUT TECHNOLOGIES AND TECHNIQUES

Ken Hinckley
Microsoft Research

INTRODUCTION: WHAT'S AN INPUT DEVICE ANYWAY?

Input to computers consists of sensed information about physical properties (such as position, velocity, temperature, or pressure) of people, places, or things. For example, the computer mouse senses the motion imparted by the user's hand. But using an "input device" is a multifaceted experience that encompasses all of the following:

- *The physical sensor:* On mechanical mice, this typically consists of a rolling ball and an optical encoder mechanism.
- *Feedback to the user:* For the mouse, feedback includes the visible cursor on the display, as well as the "clicking" sound and the tactile feel of the mouse buttons.
- *Ergonomic and industrial design:* This may include color, shape, texture, and the number and placement of buttons.
- *Interaction techniques:* Interaction techniques are the hardware and software elements that together provide a way for the user to accomplish a task. This includes details of how sensor data is filtered and interpreted, the use of buttons to support selection and dragging, and input–output constructs such as scroll bars.

A poor design in any of these areas can lead to usability problems, so it is important to consider all of these aspects when designing or evaluating input to a computer system.

This chapter emphasizes continuous input sensors and their use in applications but also includes a brief section on text entry. It discusses important questions to ask about input technologies, techniques to effectively use input signals in applications, and models and theories that can be used to evaluate interaction techniques as well as to reason about design options.

UNDERSTANDING INPUT TECHNOLOGIES

A designer who understands input technologies and the task requirements of users can choose input devices with properties that match all of the user's tasks as well as possible. But what are some of the important properties of input devices?

Pointing Device Properties

On the surface, the variety of pointing devices is bewildering. Fortunately, there is a limited set of properties that many devices share. These properties help a designer know what to ask about a device and how to anticipate potential problems. We first consider these device properties in general and then follow up with specific examples of common input devices and how these properties apply to them. See the reviews of Buxton (1995b), Jacob (1996), MacKenzie (1995), and Greenstein (1997) for further discussion.

Resolution and Accuracy. The resolution of a sensor describes how many unique units of measure can be addressed.

High resolution does not necessarily imply high accuracy, however, which is why you should always ask about both. For example, an 8-bit analog-to-digital converter can resolve a signal to one part in 256, but if noise is present, the measurement might only be accurate to 7 or fewer bits.

Sampling Rate and Latency. The sampling rate, measured in Hertz (Hz), indicates how often a sample is collected from a device. For pointing devices, an 80- to 100-Hz sampling rate is fairly typical. Latency (*lag*) is the delay between the onset of sampling and the resulting feedback to the user. It is impossible to completely eliminate latency from a system, and minimizing it can be difficult (Liang, Shaw, & Green, 1991). Latency of more than about 75 to 100 ms harms user performance for many interactive tasks (MacKenzie & Ware, 1993; Robertson, Card, & Mackinlay, 1989).

Noise, Aliasing, and Nonlinearity. Electrical or mechanical imperfections in a device can add noise to a signal. *Aliasing* resulting from an inadequate sampling rate may cause rapid signal changes to be missed, just as a spinning wheel appears to stop or rotate backward to the naked eye. Because of physical limitations, a sensor may exhibit an unequal response across its sensing range, known as nonlinearity. For example, a proximity range sensor that measures distance to a nearby object might be able to detect small changes when the object is nearby but only large changes if the object is far away (Hinckley, Pierce, Sinclair, & Horvitz, 2000).

Absolute versus Relative. Does the device sense only relative changes to its position, or is the absolute value known? With absolute devices, the nulling problem arises if the position of a physical intermediary, such as a slider on a mixing console, is not in agreement with a value set in the software. This can occur if the slider is used to control more than one variable. This problem cannot occur with relative devices, but time may be wasted clutching the device: If the mouse reaches the edge of the mouse pad, the user must pick up the mouse, move it, and put it back down in a comfortable position on the pad.

Control-to-Display Ratio. This is the ratio of the distance moved by an input device to the distance moved on the display, also known as C:D gain. For example, if moving a mouse 1 inch across the table results in the cursor moving 2 inches on the screen, the C:D gain is 2. Experts have criticized the concept of gain because it confounds what should be two measurements, device size and display size, in one arbitrary metric (Accot & Zhai, 2001; MacKenzie, 1995). The common belief that there is an optimal setting for the C:D gain is also controversial (Accot & Zhai, 2001) because gain often exhibits little or no effect in experiments (Jellinek & Card, 1990) and furthermore because faster performance may be offset by higher error rates (MacKenzie, 1995).

Physical Property Sensed. The type of physical property sensed by most pointing devices can be classified as position, motion, or force. For example, most tablets sense position, mice sense motion, and isometric joysticks (such as the IBM TrackPoint) sense force (Rutledge & Selker, 1990). The property sensed determines the type of transfer function that is most appropriate for the device (see Transfer Functions later in this chapter).

Number of Dimensions. A device that senses position, for example, might be a one-dimensional slider, a two-dimensional pointer, or even a three-dimensional position tracker.

Direct versus Indirect. On direct devices, the display surface is also the input surface. Examples include touchscreens, handheld devices with pen input, and light pens. All other devices are indirect. Indirect devices often involve a mechanical intermediary such as a stylus that can easily become lost or moving parts that are subject to damage, which is one reason that touchscreens are popular for high-volume applications such as shopping mall kiosks (Sears, Plaisant, & Shneiderman, 1992).

Metrics of Effectiveness. Various other criteria can distinguish devices, including pointing speed and accuracy, error rates, device acquisition time (time to pick up and put down the device), learning time, footprint (how much space the device occupies), user preference, and cost (Card, Mackinlay, & Robertson, 1990).

Taxonomies of Input Devices

When considering the relationships between various devices or when trying to devise or better understand new input devices, it can be useful to consider a "map" of all possible input devices. Researchers have proposed several such taxonomies that organize input devices according to the properties outlined above. However, there are so many such properties that the proposed taxonomies differ widely in complexity, level of description, and completeness. Buxton classifies continuous, manually operated input devices by the property sensed versus the number of dimensions (Buxton, 1983). This is perhaps the most useful high-level taxonomy for general interface design issues. Other device properties can be enumerated and organized into a decision tree that shows design choices along classification dimensions (Lipscomb & Pique, 1993), a structure that is particularly useful for understanding the confusing design space of joysticks. The taxonomy of Card, Mackinlay, and Robertson (1991) extends that of Buxton (1983) to include composition operators, allowing description of interaction techniques and combination devices such as a radio with multiple buttons and knobs. The completeness of the Card et al. taxonomy makes it appropriate for precise description of complex input devices or for distinguishing details of devices that are not captured by the Buxton taxonomy.

A Brief Tour of Pointing Devices

Often the term *mouse* is applied to any device that is capable of producing cursor motion, but what is the difference between a mouse and a touchscreen, for example? Indeed, most operating systems treat input devices uniformly as virtual devices, so one might be tempted to believe that pointing devices are completely interchangeable.

As suggested above, however, the details of what the input device senses, how it is held, the presence or absence of buttons, and many other properties can significantly impact the interaction techniques—and hence the end-user tasks—that a device can support effectively. The following tour discusses important properties to keep in mind for several common pointing devices. See also William Buxton's *Directory of Sources for Input Technologies* for a comprehensive list of devices on the market (Buxton, 2001).

Mice. Douglas Englebart and colleagues (English, Englebart, & Berman, 1967) invented the mouse in 1967 at the Stanford Research Institute. The long endurance of the mouse stands out in an era in which technologies have become obsolete like last year's fashions, but the mouse is still in use because its properties match the demands of desktop graphic interfaces (Balakrishnan, Baudel, Kurtenbach, & Fitzmaurice, 1997). For typical pointing tasks on a computer, one can point with the mouse about as well as with the hand itself (Card, English, & Burr, 1978). Furthermore, the mouse is stable: Unlike a stylus used on a tablet, the mouse does not fall over when released, saving the user the time of picking it up again later. The mouse also doesn't tend to move when you press a button, and the muscle tension required to press the button has minimal interference with cursor motion compared with other devices.[1] Finally, with mice, all of the muscle groups of the hand, wrist, arm, and shoulder contribute to pointing. This combination of muscle groups allows high performance for both rapid, coarse movements and slow, precise movements (Guiard, 1987; Zhai, Milgram, & Buxton, 1996).

Trackballs. A trackball is essentially a mechanical mouse that has been turned upside down. Because the trackball rolls in place, it stays at a fixed place and has a small working space (*footprint*). A trackball can also be mounted for use on an angled working surface. The buttons are located to the side of the ball, which can make them awkward to reach and hold while rolling the ball.

Isometric Joysticks. An isometric joystick is a force-sensing joystick that returns to center when released. Isometric joysticks require significant practice to achieve expert cursor control, but when integrated with a keyboard, it takes less time compared with the mouse for the user to acquire the joystick while typing or to return to typing after pointing (Rutledge & Selker, 1990). However, this reduction in acquisition time is usually not

---

[1] Some users, such as sufferers of repetitive strain injury or elderly persons, may find mouse buttons difficult to use. See Ergonomic Issues for Input Devices in this chapter and chapter 21, Designing Computer Systems for Older Adults.

enough to overcome the longer pointing time for the isometric joystick (Douglas & Mithal, 1994). Because isometric joysticks can have a tiny footprint, they are often used when space is at a premium.

Isotonic Joysticks. Isotonic joysticks sense the angle of deflection of the joystick, so most isotonic joysticks move from their center position. By contrast many isometric joysticks are stiff, with little or no "give" to provide the user feedback of how he or she is moving the joystick. Some hybrid designs blur the distinctions between the two types of joysticks, but the main questions to ask are as follows: Does the joystick sense force or angular deflection? Does the stick return to center (zero value) when released? and Does the stick move from the starting position? (See also Lipscomb & Pique, 1993).

Tablets. Tablets (known variously as touch tablets, graphics tablets, or digitizing tablets) sense the absolute position of a pointing device on the tablet. Tablets might be used with the bare finger, a stylus, or a puck.[2] Tablets can operate in absolute mode, in which there is a fixed C:D gain between the tablet surface and the display, or in relative mode, in which the tablet responds only to motion of the stylus. If the user touches the stylus to the tablet in relative mode, the cursor resumes motion from its previous position; in absolute mode, it would jump to the new position. Absolute mode is generally preferable for tasks such as drawing, handwriting, tracing, or digitizing, but relative mode may be preferable for typical desktop interaction tasks such as selecting graphic icons or navigating through menus. Indeed, tablets can operate in either mode, allowing coverage of a wide range of tasks (Buxton, Hill, & Rowley, 1985), whereas devices such as mice and trackballs can only operate in relative mode.

Touchpads. Touchpads are small touch-sensitive tablets commonly used on laptop computers. Touchpads typically respond in relative mode, because of the small size of the pad. Most touchpads support an absolute mode to allow Asian language input or signature acquisition, for example, but for these uses make sure the touchpad can sense contact from a stylus (and not just the bare finger). Touchpads often support clicking by recognizing tapping or double-tapping gestures, but accidental contact (or loss of contact) can erroneously trigger such gestures (MacKenzie & Oniszczak, 1998).

Touchscreens. Touchscreens are transparent touch-sensitive tablets mounted on a display. Parallax error is a mismatch between the sensed finger position and the apparent finger position due to viewing angle and displacement between the sensing and display surfaces; look for a small displacement to minimize this problem. Check the transmissivity of a touchscreen, because it may reduce the luminance and contrast ratio of the display. Depending on the mounting angle, touchscreens may result in arm fatigue. Also, when the user drags a finger across the screen, this must either move the cursor or drag an object on the screen; it cannot do both because there is no separation between the two actions (Buxton, 1990b). Finally, offset the selected position to be visible above the fingertip and present new information in an area of the screen unlikely to be occluded by the user's hand (see also Sears & Shneiderman, 1991).

Pen Input Devices. The issues noted above for touchscreens also apply to pen input on handheld devices. Furthermore, users often want to touch the screen using a bare finger as well as the stylus, so commonly used commands should be large enough to accommodate this. Keep in mind that there is no true equivalent of the mouse "hover" state for calling up tool tips, nor is there any extra button for context menus. However, pen dwell time on a target is sometimes an acceptable way to provide one of these functions.

Alternative Pointing Devices. For disabled use, or tasks in which the user's hands may be occupied, standard pointing devices may be unsatisfactory. Some alternatives, roughly ordered from most the practical solutions with current technology to the most preliminary solutions still in the research labs, include the following:

• *Software aids:* Because people are so resourceful and adaptable, modification of system software through screen magnification, sticky modifier keys, and other such techniques can often enable access to technology. Several such resources are listed on Microsoft's accessibility Web site (Microsoft, 2001).

• *Feet for input:* Foot-operated devices can provide effective pointing control (Pearson & Weiser, 1988); input using the knee is also possible (English et al., 1967). Foot switches and rocker pedals are useful for specifying modes or controlling secondary values (Balakrishnan, Fitzmaurice, Kurtenbach & Singh, 1999; Sellen, Kurtenbach, & Buxton, 1992).

• *Head tracking:* It is possible to track the position and orientation of the user's head, but unfortunately, the neck muscles offer low bandwidth cursor control compared with the hands. Head tracking is a natural choice for viewpoint control in virtual environments (Brooks, 1988; Sutherland, 1968).

• *Eye tracking:* Eye tracking has several human factors and technology limitations. The human eye fixates visual targets within the fovea, which fundamentally limits the accuracy of eye gaze tracking to 1° of the field of view (Zhai, Morimoto, & Ihde, 1999). The eye jumps around constantly, moving rapidly in saccades between brief fixation points, so a high sampling rate and intelligent filtering is necessary to make sense of eye-tracking data. If one uses eye gaze to execute commands, the so-called Midas touch problem results, because the user cannot glance at a command without activating it (Jacob, 1991). Combining manual input and eye tracking offers another approach (Zhai et al., 1999).

---

[2]A puck is a mouse that is used on a tablet. The primary difference is that a puck usually senses absolute position, whereas a mouse senses motion only.

• *Direct brain interfacing:* In some cases, direct access to the brain offers the only hope for communication. Electrodes have been surgically implanted in the motor cortex of human patients; signal processing of the data allows imagined hand movements to move a cursor on the screen (Moore, Kennedy, Mynatt, & Mankoff, 2001). Clinical trials of this approach are extremely limited at present.

See also Interaction Issues for Diverse Users (chapters 20–26), speech recognition technologies as discussed in chapters 8, 14, and 36, and alternative devices listed by Buxton (2001).

Input Device States

The integration of buttons with devices to support clicking and dragging may seem trivial, but failing to recognize these fundamental states of input devices can lead to design problems. The three-state model (Buxton, 1990b) enumerates the states recognized by commonly available input devices: Tracking (State 1), Dragging (State 2), and Out of Range (State 0), as shown in Table 7.1. The three-state model is useful to specify exactly what an input device can do in relation to the demands of interaction techniques.

TABLE 7.1. Summary of States in Buxton's Three-State Model (Buxton, 1990b)

State	Description
0	Out of range: The device is not in its physical tracking range
1	Tracking: Moving the device causes the tracking symbol to move
2	Dragging: Allows one to move objects in the interface

For example, Fig. 7.1 compares the states supported by a mouse to those supported by a touchpad. A mouse operates in the tracking state, or in the dragging state (when one moves the mouse while holding the primary mouse button), but it cannot sense the out of range state—it does not report an event if it is lifted. A touchpad also senses two states, but they are not the same two states as the mouse. A touchpad senses a tracking state (the cursor moves while one's finger is on the pad) and an out of range state (the pad senses when the finger is removed).

The continuous values that an input device senses also may depend on these states (Hinckley et al., 1998a). For example, the (x, y) position of one's finger is sensed while it contacts a touchpad, but nothing is sensed once the finger breaks contact. Thus, the touchpad has a single state in which it senses a position, whereas the mouse has two states in which it senses motion; any interaction technique that requires sensing motion in two different states will require special treatment on a touchpad.

WHAT'S AN INPUT DEVICE FOR? THE COMPOSITION OF USER TASKS

Input devices are used to complete elemental tasks on a computer. One way of reasoning about input devices and interaction techniques is to view the device or technique in light of the tasks that it can express. But what sort of tasks are there?

Elemental Tasks

Although computers can support many activities, at the input level some subtasks appear repeatedly in graphic user interfaces, such as pointing at a target on the screen or typing a character. Foley, Wallace, and Chan (1984) proposed that all user interface transactions are composed of the following six elemental tasks:

- *Select:* Indicating object(s) from a set of alternatives
- *Position:* Specifying a position within a range, such as a screen coordinate
- *Orient:* Specifying a rotation, such as an angle, or the three-dimensional orientation of an object in a virtual environment
- *Path:* Specifying a series of positions or orientations over time, such as drawing a freehand curve in a paint program
- *Quantify:* Specifying a single numeric value
- *Text:* Specifying symbolic data such as a sequence of characters

If a computer system allows the user to accomplish all six of these elemental tasks, then in principle the user can use the system to accomplish any computer-based task.

Compound Tasks and Chunking

Although elemental tasks are useful to find commonalities between input devices and different ways of accomplishing the same task on different systems, the level of analysis at which one specifies "elemental" tasks is not well defined. For example, a mouse can indicate an integral (x, y) position on the screen, but an Etch-a-Sketch separates positioning into two subtasks by providing a single knob for x and a single knob for y. If positioning is an elemental task, why do we find ourselves subdividing this task for some devices but not for others? What if we are working with three-dimensional computer graphics where a *Position* requires an (x, y, z) coordinate? One way to resolve this puzzle is to view all tasks as hierarchies of subtasks (Fig. 7.2). Observe that

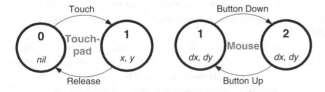

FIGURE 7.1. States models for a touchpad (left) and a standard mouse (right); adapted from Hinckley, Czerwinski, and Sinclair (1998a).

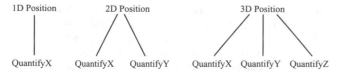

FIGURE 7.2. Task hierarchies for one-dimensional, two-dimensional, and three-dimensional position tasks.

whether a task is elemental depends on the input device being used: the Etch-a-Sketch supports separate *QuantifyX* and *QuantifyY* tasks, whereas the mouse supports a single compound *2D Position* task (Buxton, 1986).

From the user's perspective, a series of elemental tasks often seems like a single task. For example, scrolling a Web page to click on a link could be conceived as an elemental one-dimensional positioning task followed by a two-dimensional selection task, or it can be viewed as a compound navigation/selection task (Buxton & Myers, 1986). One can design the interaction so that it encourages the user to work at the higher level of the compound task, such as scrolling with one hand while pointing to the link with the other hand. This is known as chunking.

These examples show that the choice of device influences the level at which the user is required to think about the individual actions that must be performed to achieve a goal. This is an important point. The appropriate choice and design of input devices and interaction techniques can help to structure the interface such that there is a more direct match between the user's tasks and the low-level syntax of the individual actions that must be performed to achieve those tasks. In short, the design of the system has a direct influence on the steps of the workflow that must be exposed to the user (Buxton, 1986).

Multichannel Input Devices. Because many user tasks represent parts of larger compound tasks, it is sometimes desirable to provide a multichannel input device, which provides multiple controls on a single device (Zhai, Smith, & Selker, 1997). For example, many mice include dedicated scrolling devices such as wheels, isometric joysticks, or touchpads. One advantage of such devices is that they eliminate the need to remove visual attention from one's work, as is necessary when moving the cursor to interact with standard scroll bars.

Multiple Degree of Freedom Input Devices. Multiple degree of freedom (Multi-DOF) devices sense multiple dimensions of spatial position or orientation, unlike multichannel devices, which provide extra input dimensions as separate controls. Examples include mice that can sense when they are rotated or tilted (Balakrishnan et al., 1997a; Hinckley, Sinclair, Hanson, Szeliski, & Conway, 1999; Kurtenbach, Fitzmaurice, Baudel, & Buxton, 1997) and magnetic trackers that sense six DOFs (three-dimensional position and three-dimensional orientation). Numerous interaction techniques have been proposed to allow standard two-dimensional pointing devices to control three-dimensional positioning or orientation tasks (Conner et al., 1992). Such techniques may be less effective

for three-dimensional input tasks than well-designed Multi-DOF input techniques (Balakrishnan et al., 1997; Hinckley, Tullio, Pausch, Proffitt, & Kassell, 1997; Ware, 1990). However, three-dimensional input devices can be ineffective for many standard desktop tasks, so overall performance for all tasks must be considered. (See Hinckley, Pausch, Goble, and Kassell (1994b) and Zhai (1998) for further discussion of three-dimensional input techniques. See also chapter 31, Virtual Environments.)

Bimanual Input. People use both hands to accomplish most real-world tasks (Guiard, 1987), but with computers, there is little use of the nonpreferred hand for tasks other than typing. Bimanual input enables compound input tasks, such as navigation and selection, in which the user can scroll with the nonpreferred hand while using the mouse in the preferred hand (Buxton et al., 1986). This assignment of roles to the hands corresponds to a theory of bimanual action proposed by Guiard (1987), which suggests that the nonpreferred hand sets a frame of reference (scrolling to a location in the document) for the more precise actions of the preferred hand (selecting an item within the page using the mouse). Other applications for bimanual input include command selection with the nonpreferred hand (Bier, Stone, Pier, Buxton, & DeRose, 1993; Kabbash, Buxton, & Sellen, 1994), drawing programs (Kurtenbach et al., 1997), and virtual camera control and three-dimensional manipulation (Balakrishnan & Kurtenbach, 1999; Hinckley, Pausch, Proffitt, & Kassell, 1998b).

Gesture Input Techniques. With mouse input, or preferably pen input, users can draw simple gestures analogous to proofreader's marks to issue commands, such as scribbling over a word to delete it. Note that the gesture integrates the command (delete) and the selection of the object to delete. Another example is moving a paragraph by circling it and drawing a line to its new location. This integrates the verb, object, and indirect object of the command by selecting the command, selecting the extent of text to move, and selecting the location to move the text (Buxton, 1995b). A design trade-off is whether to treat a gesture as content ("ink") or recognize it immediately as a command. With careful design, the approaches can be combined (Kramer, 1994; Moran, Chiu, & van Melle, 1997). Marking menus use simple straight-line gestures to speed menu selection (Kurtenbach, Sellen, & Buxton, 1993). Multimodal Input (see chapter 14) often combines speech input with pen gestures. Future interfaces may use sensors to recognize whole-hand gestures (Baudel & Beaudouin-Lafon, 1993) or movements of physical tools, such as handheld computers (Bartlett, 2000; Harrison, Fishkin, Gujar, Mochon, & Want, 1998; Hinckley et al., 2000).

Gesture input has technical challenges and limitations. A pragmatic difficulty is to recognize the gesture, using techniques such as the Rubine classifier (Rubine, 1991). As the number of gestures increases, recognition becomes more difficult, and it is harder for the user to learn, remember, and correctly articulate each gesture, which together limit the number of commands that can be supported (Zeleznik, Herndon, & Hughes, 1996).

For a brief discussion of character recognition, see Character Recognition later in this chapter.

EVALUATION AND ANALYSIS OF INPUT DEVICES

In addition to general usability engineering techniques (see chapters 51–59), there are a number of techniques specifically tailored to the study of input devices.

Representative Tasks for Pointing Devices

To test an input device, it is useful to see how the device performs in a wide range of tasks that represent actions users may perform with it. Some examples are listed below (see also Buxton, 1995b). These tasks are appropriate for quick, informal studies or they can be formalized for quantitative experiments.

Target Acquisition. The user clicks back and forth between targets on the screen. Vary the distance between the targets and the size of the targets. This can be formalized using Fitts' Law (Fitts, 1954; MacKenzie, 1992a), as discussed later in this section.

Steering. The user moves the cursor through a tunnel. Typically, circular and straight-line tunnels are used; vary the length, width, and direction of travel. This task can be modeled using the Steering Law (Accot & Zhai, 1997); see Other Metrics and Models later in this section.

Pursuit Tracking. The user follows a randomly moving target with the cursor. Average root mean squared (RMS) error quantifies the performance. (See Zhai, Buxton, & Milgram, 1994, for an example of this type of study.)

Freehand Drawing. Try signing your name in different sizes (Buxton, 1995b) or rapidly drawing a series of XOXOs, for example.

Drawing Lines. Draw a box and see if it is easy to get the ends of the strokes to line up with one another. Also try a diamond and a box that is slightly off-axis (rotated 5°).

Tracing and Digitizing. Some tablets can digitize coordinates from paper drawings or maps. The accuracy and linearity of the device, particularly when the pointer is sensed through different types of materials that the user may need to digitize, should be tested.

Rapid or Slow Motion. Does the device respond quickly and accurately when rapidly moving back and forth or in circles? Can one hold the device still to bring up tool tips? Look for signs of jitter when slowing to select a tiny target.

Clicking, Double Clicking, and Dragging. Does the device support each of these tasks effectively? One should also consider the placement of the buttons. Are they awkward to reach or hold while moving the device? Does hitting the buttons cause unintended cursor motion? Are any of them easy to hit by mistake?

Ergonomic Issues for Input Devices

Many modern information workers suffer from repetitive strain injury (RSI). Researchers have identified many risk factors for such injuries, such as working under stress or taking inadequate rest breaks (Putz-Anderson, 1988). Some themes that encourage ergonomic device design (Pekelney & Chu, 1995) follow. (See also section Trends in Keyboard Design in this chapter for a brief discussion of ergonomic keyboards and chapter 19, Design of Computer Workstations, for a full treatment of this important topic.) Example ergonomic design themes include the following:

- *Reduce repetition* by using one-step instead of multiple-step operations or by providing alternative means (e.g., keyboard shortcuts) to accomplish tasks.
- *Minimize force* required for dynamic load to hold and move the device, as well as static load required to hold down a button. Also, avoid sharp edges that might put pressure on the soft tissues of the hand.
- *Encourage natural and neutral postures.* The device should accommodate a range of hand sizes and should not restrict hand motion. Design for the optimal direction of movement for the finger, hand, or arm used to operate the device. Devices should be held with a natural, slight flexion of fingers, straight or slight extension and ulnar deviation of the wrist, and slight pronation of the wrist (Keir, Back, & Rempel, 1999; Rempel, Bach, Gordon, & Tal, 1998).
- *Offer cues for use.* Devices should communicate a clear orientation for grip and motion to discourage inappropriate grips that may lead to problems.

Fitts' Law: A Design, Engineering, and Research Tool

Fitts' Law (Fitts, 1954) is an experimental paradigm that has been widely applied to the comparison and optimization of pointing devices. Fitts' Law is used to measure how effectively a pointing device can acquire targets on the screen, summarized as the bandwidth of the input device. Industry practitioners consider even a 5% difference in bandwidth to be important. (See MacKenzie, 1992a, and Douglas, Kirkpatrick, & MacKenzie, 1999, for details about conducting such studies.)

What Is Fitts' Law? Fitts' Law was first described by Paul Fitts in 1954. The law as he envisioned it had foundations in information theory, although psychomotor interpretations have also been suggested (Douglas & Mithal, 1997). Fitts conducted an experiment in which participants were asked to use a stylus to rapidly tap back and forth between two metal plates, missing the plates as infrequently as possible. Fitts measured

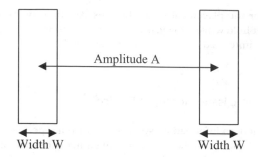

FIGURE 7.3. Fitts' task paradigm (see Fitts, 1954).

the movement time *MT* between the two plates, while varying the amplitude *A* of the movement, as well as the width *W* of the error tolerance (Fig. 7.3).

Fitts discovered that the average movement time is predicted by a logarithmic function of the ratio of *A* to *W*. The formulation of Fitts' Law typically used in device studies is:

$$MT = a + b \log_2(A/W + 1). \qquad (7.1)$$

Here *MT* is the movement time, *A* is the amplitude of the movement, and *W* is the width of the target that must be selected. The constants *a* and *b* are coefficients fit to the average of the observed data for *MT* for each combination of *A* and *W* tested in the experiment. One calculates *a* and *b* by the normal techniques of linear regression, using a statistical package or spreadsheet program. In typical rapid, aimed movement studies, it is not unusual to see Fitts' Law model the mean observed movement times with more than 85% of all variance explained by Equation 7.1. (For example experimental data, see MacKenzie, 1992a, 1992b.)

The Fitts Index of Difficulty (ID), measured in units of bits, is defined as

$$ID = \log_2(A/W + 1). \qquad (7.2)$$

The ID is the key concept of Fitts' Law because it describes moves in the task space using a single abstract metric of difficulty. The ID balances the intuitive concepts that far movements should take more time than short ones, whereas pointing to small targets should take longer than large ones. Note that the units of *A* and *W* cancel out, so the ID is dimensionless. It is arbitrarily assigned the unit of bits because of the base 2 logarithm. By substitution of the ID, Fitts' Law (Equation 7.1) often appears in the literature as

$$MT = a + b \ ID. \qquad (7.3)$$

The Index of Performance (IP), measured in bits/second, quantifies the bandwidth of a pointing device. In the information–theoretic view of Fitts' Law, the IP captures the rate of information transfer through the human–computer interface, defined as

$$IP = MT / ID. \qquad (7.4)$$

Dropping the constant offset *a* from Equation 7.3 and dividing

MT by *ID*, the *IP* can also be calculated as

$$IP = 1/b. \qquad (7.5)$$

When *bandwidth*, *throughput*, or *IP* is mentioned in experimental studies, this quantity is typically the inverse slope as calculated by Equation 7.5. Comparing the ratio of IP's is a good way to make comparisons across Fitts' Law studies (MacKenzie, 1992a). Although Equation 7.5 is normally reported as the *IP*, Equation 7.4 is sometimes used to calculate an IP for an individual experimental condition, or even for an individual pointing movement, where a "regression line" is nonsensical.

A final quantity known as the effective width is sometimes used to account for both speed and accuracy of performance in Fitts' Law studies. The effective width is the target width that would be required to statistically correct the subject's performance to a constant 4% error rate (MacKenzie, 1992a), using the standard deviation σ of endpoint coordinates around the target. For each combination of *A* and *W* tested in an experiment, the effective width W_e for that condition is calculated as

$$W_e = 4.133 \, \sigma. \qquad (7.6)$$

One then recalculates the ID by substituting W_e for *W* in Equation 7.2, yielding an effective ID (ID_e). This results in an effective IP_e in Equation 7.4. The IP_e thus incorporates both speed and accuracy of performance.

Applications of Fitts' Law. Fitts' Law was first applied to the study of input devices by Card et al. (1978), and since then it has evolved into a precise specification for device evaluation (Douglas et al., 1999) published by the International Standards Organization (ISO). Strictly speaking, Fitts' Law, as specified in Equation 7.1, only deals with one-dimensional motion. The law can be extended to two-dimensional targets by taking the target width *W* as the extent of the target along the direction of travel (MacKenzie & Buxton, 1992). The approach angle can also have some effect on target acquisition. The ISO standard uses circular targets and a movement pattern analogous to the spokes of a wheel to keep the width *W* constant while averaging performance over many approach angles (Douglas et al., 1999).

Fitts' Law has been found to apply across a remarkably diverse set of task conditions, including rate-controlled input devices (MacKenzie, 1992a), area cursors (Kabbash & Butxon, 1995), and even when pointing under a microscope (Langolf & Chaffin, 1976). A few important results for input devices are summarized below.

- *Tracking versus dragging states:* Experiments have shown that some devices exhibit much worse performance when used in the dragging state than in the tracking state (MacKenzie, Sellen, & Buxton, 1991). It is important to evaluate a device in both states.
- *Bandwidth of limb segments:* Different muscle groups exhibit different peak performance bandwidths. The general ordering is fingers > wrist > upper arm, although much depends on the exact combination of muscle groups used, the direction of movement, and the amplitude of movement (Balakrishnan &

MacKenzie, 1997; Langolf & Chaffin, 1976). For example, using the thumb and index finger together has higher bandwidth than the index finger alone (Balakrishnan & MacKenzie, 1997). A device that allows fine fingertip manipulations is said to afford a precision grasp, like holding a pencil, as opposed to a *power grasp*, such as the grip used to hold a suitcase handle (Mackenzie & Iberall, 1994; Zhai et al., 1996).

• *Effects of lag:* Fitts' Law can be adapted to predict performance in the presence of system latency (MacKenzie & Ware, 1993), yielding a form of the law as follows:

$$MT = a + (b + c \ LAG) \ ID. \qquad (7.7)$$

• *C:D gain:* Fitts' Law predicts that gain should have no effect on performance (Jellinek & Card, 1990) because for any constant gain g, the ratio $gA/gW = A/W$, which does not change the ID (9.2). However, this scale-invariant prediction is only valid within limits (Accot & Zhai, 2001; Guiard, 2001).

• *Scrolling and multiscale navigation:* The effectiveness of scrolling techniques varies depending on how far, and how precisely, one must scroll in a document, a problem which can be formulated as a Fitts' Law study (Hinckley & Cutrell, 2002). See Guiard, Buourgcois, Mottet, & Beaudouin-Lafon, 2001, for Fitts' Law in multiscale interfaces.

Other Metrics and Models of Input

Fitts' Law is emphasized here because it has found such wide influence and application. But Fitts' Law only deals with the elemental action of pointing. To go beyond pointing, other models are needed.

Steering Law. Steering a cursor through a narrow tunnel, as required to navigate a pull-down menu, is not a Fitts task because the cursor must stay within the tunnel at all times. For a straight line tunnel (see Fig. 7.4a) of width W and length A,

for example, the Steering Law predicts that the movement time is a linear function of A and W:

$$MT = a + b \ A/W. \qquad (7.8)$$

For a circular tunnel (Fig. 7.4b) of radius R, the Steering Law becomes

$$MT = a + b \ 2\pi R/W. \qquad (7.9)$$

See Accot & Zhai, 1997, for discussion of the Steering Law with arbitrary curved paths, as well as using the Steering Law to predict instantaneous velocity. Also note that the Steering Law only models successful completion of the task; errors are not considered.

Keystroke-Level Model (KLM). The KLM model predicts expert performance of computer tasks (Card, Moran, & Newell, 1980). One counts the elemental inputs required to complete the task, including keystrokes, homing times to acquire input devices, pauses for mental preparation, and pointing at targets. For each elemental input, a constant estimate of the average time required is substituted, yielding an overall time estimate. The model is only useful for predicting the performance of experts because it assumes error-free execution; it does not account for the searching and problem-solving behaviors of novices.

Numerous enhancements and extensions to the KLM have been devised, including GOMS (goals, objects, methods, and selection rules). The advantage of these kind of models is that they can predict the average expert completion time for short, repetitive tasks without requiring implementation of the tasks, training end users, and evaluating their performance (Olson & Olson, 1990). For further discussion of GOMS and related models, see chapter 58, Model-Based Evaluations. See also chapter 5, Cognitive Architecture, and chapter 6, Modeling Humans in HCI.

MAPPINGS: HOW TO GET THE MOST OUT OF AN INPUT SIGNAL

Once an input signal has been digitized and passed along to a host computer, the work of the interface designer has only begun. One still must determine how best to make use of that data to provide fast, accurate, comfortable, easy-to-learn, and satisfying computer interfaces.

Transfer Functions

A transfer function is a mathematical transformation that scales the data from an input device. Typically, the goal is to provide more stable and intuitive control, but one can easily design a poor transfer function that hinders performance. The choice of transfer function often depends on the type of input sensor used. A transfer function that matches the properties of an input device is known as an appropriate mapping. For force-sensing input devices, the transfer function should be a force-to-velocity function; for example, the force one exerts on the IBM

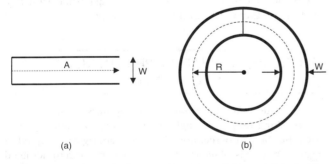

(a)   (b)

FIGURE 7.4. The Steering Law for (a) straight and (b) circular tunnels (Accot & Zhai, 1999). Steering through a straight tunnel is modeled by Equation 7.8. Steering through a circular tunnel of width W and Radius R is modeled by Equation 7.9. In both cases, the user starts at the vertical line and attempts to follow the dotted line without moving outside the tunnel. The dotted lines here are for illustration only and typically would not be shown to participants in an actual experiment.

Trackpoint isometric joystick controls the speed at which the cursor moves. Other appropriate mappings include position-to-position or velocity-to-velocity functions, used with tablets and mice, respectively.

A commonly seen example of an inappropriate mapping is calculating a velocity based on the position of the mouse cursor, such as to scroll a document. The resulting input is often difficult to learn to use and difficult to control; a better solution is to use the position of the cursor to control the resulting position within the document, as occurs when one grabs the handle of a scrollbar. Inappropriate mappings are sometimes the best design compromise given competing demands but should be avoided if possible. (For further discussion, see Zhai, 1993, and Zhai et al., 1997.)

Self-Centering Devices. Rate mappings are most appropriate for force-sensing devices or other devices that return to center when released. This property allows the user to stop quickly by releasing the device. The formula for a nonlinear rate mapping is

$$dx = Kx^{\alpha}, \qquad (7.10)$$

where x is the input signal, dx is the resulting rate, K is a gain factor, and α is the nonlinear parameter. The best values for K and α depend on the details of the device and application, and appropriate values must be identified by experimentation or optimal search (Zhai & Milgram, 1993). It may be necessary to ignore sensor values near zero so that noise in the sensor will not cause motion. The mappings used on commercial devices can be quite complex (Rutledge & Selker, 1990).

Motion-Sensing Devices. Most computer systems available today use an exponential transformation of the mouse velocity, known as an acceleration function, to modify the cursor response. Experimental testing of mouse acceleration functions suggest that they may not improve the overall pointing speed. Rather, they help to limit the footprint that is required to use the mouse (Jellinek & Card, 1990).

Absolute Devices. Sometimes one can temporarily break the 1:1 control-to-display mapping of absolute devices. Sears and Shneiderman (1991) described a technique for touchscreens that uses cursor feedback to allow high-resolution pointing. If the finger moves slightly, the movement of the cursor is dampened to allow fine adjustments. But if the finger continues to move, the cursor jumps to the absolute position of the user's finger again. In this way, fine adjustments are possible, but the absolute nature of the device is preserved. This technique uses the touchscreen in the cursor tracking state, so it cannot be applied if one uses the dragging state as the default (Buxton, 1990b).

Design Challenges for Real-Time Response

When designing the real-time response to a continuous input from the user, there are often several conflicting design challenges. From the author's experience, three common themes in such challenges are as follows.

1. *Never throw data from the user's input away:* This should be the mantra of the interaction designer. The data from the input device is the only direct evidence of the user's will. However, it is sometimes necessary to ignore artifacts of devices, such as noise or nonlinearity. Removing such irregularities without ruining the expressive bandwidth of a device can be challenging. For example, filtering noise by averaging input over several samples necessarily adds latency to the interaction (Liang et al., 1991). Another example is using a timeout (of perhaps 40 ms) to debounce switch closures; this introduces some lag, but otherwise the software might sense multiple false contacts.

2. *Immediate feedback:* It is usually important to provide a real-time response to the user's input with minimum latency. This is true even if the software is not yet sure what the user is trying to do. The user knows what he or she is trying to accomplish from the start of a movement or gesture; the computer must provide some kind of intermediate feedback so that any recognition delay does not annoy the user or interfere with performance of the task. For example, a first-in, first-out queue of recent inputs allows restoration of the cursor position if lifting one's finger from a touch tablet disturbs the cursor position, without delaying feedback of the current cursor position (Buxton et al., 1985). Speculative execution offers another approach to deal with ambiguous inputs (Bartlett, 2000).

3. *Make the response to the input readily perceptible.* For example, during direct manipulation of a zoom factor, providing an exponential transformation of the input makes the response uniformly controllable (Igarashi & Hinckley, 2000). Likewise, one may need to linearize the response of some sensors (e.g., Hinckley et al., 2000). It is also important to avoid sudden changes that may disorient the user. For example, instead of moving objects on the screen instantaneously, animated transitions can help the user to more readily perceive changes in response to an input (Robertson et al., 1989). Limiting the derivative of movement in response to an input (Igarashi & Hinckley, 2000) is another way to avoid sudden changes.

FEEDBACK: WHAT HAPPENS IN RESPONSE TO AN INPUT?

The literature often treats input to computers separately from the resulting output through visual, tactile, or auditory feedback. But human perception and action are tightly coupled in a single motor-visual feedback loop, so clearly both topics need to be considered in the design of interaction techniques. (For a discussion of audio feedback design and perception issues, refer to chapter 11, Nonspeech Auditory Output.)

From the technology perspective, one can consider feedback as passive or active. Passive feedback results from physical properties of the device, such as the shape, color, or feel of buttons when they are depressed. Active feedback is under computer control. For example, the SenseAble Phantom is a stylus connected to motors by an armature; computer control of the motors presents artificial forces to the user.

Passive Feedback

Several types of passive feedback are relevant to the design and use of input devices.

Passive Visual, Auditory, and Tactile Feedback. The industrial design of a device suggests the purpose and use of a device even before a user touches it. Mechanical sounds and vibrations that result from using the device provide confirming feedback of the user's action. The shape of the device and the presence of landmarks can help users orient a device without having to look at it (Hinckley, Pausch, Goble, & Kassell, 1994a; Hinckley et al., 1997).

Proprioceptive and Kinesthetic Feedback. These imprecise terms refer to sensations of body posture, motion, and muscle tension (Burdea, 1996; Gibson, 1962). These senses allow users to feel how they are moving an input device without looking at the device, and indeed without looking at the screen in some situations (Balakrishnan & Hinckley, 1999; Mine, Brooks, & Sequin, 1997). This may be important when the user's attention is divided between multiple tasks and devices (Fitzmaurice & Buxton, 1997). Sellen et al. (1992) report that muscular tension makes modes more salient to the user.

Input–Output Correspondence

Performance can be influenced by correspondences between input and output. For visual feedback, two types of input–output correspondence that should be considered are known as perceptual structure and kinesthetic correspondence.

Perceptual Structure. The input dimensions of a device, as well as the control dimensions of a task, can be classified as integral or separable. Jacob, Sibert, McFarlane, and Mullen (1994) explore two input devices, a three-dimensional position tracker with integral (x, y, z) input dimensions, and a standard two-dimensional mouse, with (x, y) input separated from (z) input by holding down a mouse button. For selecting the position and size of a rectangle, the position tracker is most effective. For selecting the position and gray-scale color of a rectangle, the mouse is most effective. The best performance results when the integrality or separability of the input matches that of the output.

Kinesthetic Correspondence. Graphic feedback on the screen should correspond to the direction that the user moves the input device (Britton, Lipscomb, & Pique, 1978). If the user moves a device to the left, then the object on the screen should likewise move left. However, users can easily adapt to certain kinds of noncorrespondences: When the user moves a mouse forward and back, the cursor actually moves up and down on the screen; if the user drags a scrollbar downward, the text on the screen scrolls upward.

Active Haptic Feedback

Haptic feedback research has sought to provide an additional channel of sensory feedback that might improve user interfaces. Haptic feedback includes force feedback (active presentation of forces to the user) and tactile feedback (active presentation of vibrotactile stimuli to the user). Here, I briefly discuss haptic feedback with an emphasis on pointing devices. Haptic feedback is popular for gaming devices, such as force feedback steering wheels and joysticks, but general-purpose pointing devices with force or tactile feedback are not yet common. (See also chapter 10, Haptic Interfaces, and Burdea, 1996.)

Adding force to a mouse or stylus may impose constraints on the mechanical design because a physical linkage is typically needed to reflect the forces. This may prevent a force feedback mouse from functioning like a traditional mouse because it may limit range of motion or preclude clutching by lifting the device. Some devices simulate forces by increasing resistance between the mouse and the pad. One can also use a vibrotactile stimulus, such as a vibrating pin under the mouse button or vibrating the shaft of an isometric joystick (Campbell, Zhai, May, & Maglio, 1999). Combination devices have also been explored (Akamatsu & MacKenzie, 1996).

Using force or tactile feedback to provide attractive forces that pull the user toward a target, or to provide additional feedback for the boundaries of the target, has been found to yield modest speed improvements in some target acquisition experiments, although error rates also may also increase (Akamatsu & MacKenzie, 1996; MacKenzie, 1995). When multiple targets are present, as on a computer screen with many icons and menus, haptic feedback for one target may interfere with the selection of another, unless one uses techniques such as reducing the haptic forces during rapid motion (Oakley, Brewster, & Gray, 2001). Finally, one should also consider whether software constraints, such as snap-to grids, are sufficient to support the user's tasks.

Another challenge for haptic feedback techniques results from the interaction between the haptic and visual channels. Visual dominance deals with phenomena resulting from the tendency for vision to dominate other modalities (Wickens, 1992). Campbell et al. (1999) showed that tactile feedback improves steering through a narrow tunnel, but only if the visual texture matches the tactile texture; otherwise tactile feedback harms performance.

KEYBOARDS AND TEXT ENTRY TECHNIQUES

Do Keyboards Have a Future?

Keyboards and typewriters have been in use for well over 100 years (Yamada, 1980). With the advent of speech recognition technology, it is tempting to believe that this is going to change and that keyboards will soon become irrelevant. But a recent study revealed that, when error correction is included, keyboard–mouse text entry for the English language is more than twice as fast as automatic recognition of dictated speech (Karat, Halverson, Horn, & Karat, 1999). Furthermore, speaking commands can interfere with short-term memory for words (Karl, Pettey, & Shneiderman, 1993), so even perfect speech recognition will be problematic for some text entry tasks. Thus, keyboards will continue to be widely used for text entry tasks for many years to come.[3]

Procedural Memory

Keyboards rely on the automation of skills in procedural memory, which refers to the human ability to perform complex sequences of practiced movements, seemingly without any cognitive effort (Anderson, 1980). Procedural memory enables touch typing on a keyboard with minimal attention when entering commonly used symbols. As a result, attention can be focused on mentally composing the words to type and on the text appearing on the screen. Hot keys (chorded key combinations for frequently used commands) can also be learned to allow rapid selection of commands.

The automation of skills in procedural memory takes lots of practice. This can be formalized as the power law of practice (Anderson, 1980):

$$T = aP^b, \qquad (7.11)$$

where T is the time to perform a task, P is the amount of practice, and the multiplier a and exponent b are fit to the observed data. (For a good example of applying the power law of practice, see MacKenzie, Kober, Smith, Jones, & Skepner, 2001, which discusses text entry on mobile phone keypads.) The power law of practice helps explain why keyboard designs are so resistant to change. Several keyboard layouts that are faster than qwerty have been discovered, but the performance advantage typically does not merit the upfront investment required to relearn the skill of touch-typing.

Trends in Keyboard Design

Many factors can influence typing performance, including the size, shape, activation force, key travel distance, and the tactile and auditory feedback provided by striking the keys (Lewis, Potosnak, & Magyar, 1997). Because these factors seem to be well understood, recent keyboard designs have stressed other means of improving overall input efficiency or user comfort. Some recent trends include the following.

Ergonomic Design. Split-angle keyboards may help maintain neutral posture of the wrist and thereby help avoid ulnar deviation (Honan, Serina, Tal, & Rempel, 1995; Marklin, Simoneau, & Monroe, 1997; Smutz, Serina, Bloom, & Rempel, 1994). Ulnar deviation is associated with increased pressure in the carpal tunnel, which can lead to repetitive stress injuries (Putz-Anderson, 1988; Rempel et al., 1998).

Integrated Pointing Devices. An integrated isometric joystick or touchpad can allow pointing without having to move one's hand away from the keyboard. Such devices are useful when space is limited or when there is no convenient flat surface for a mouse.

Miniature Keyboards. Small foldaway keyboards are becoming popular to allow rapid text entry on small handheld devices.

Wireless Keyboards. These devices allow using a keyboard in one's lap or at a distance from the display.

Extra Functionality. Many recent keyboards include extra function keys, such as Internet forward and back navigation. Keyboards can facilitate two-handed input by placing commonly used functions such as scrolling (Buxton et al., 1986) in a position convenient for the left hand[4] (MacKenzie & Guiard, 2001).

One-Handed Keyboards

Chording keyboards have a small number of keys that the user presses in tandem to enter text. Although chording keyboards take more time to learn, they can sometimes allow one to achieve higher peak performance because the fingers have minimal travel among the keys (Buxton, 1990a; Noyes, 1983). It is also possible to type using just half of a standard qwerty keyboard by using the space bar as a chord for the other half of the keyboard (Mathias, MacKenzie, & Buxton, 1996).

Soft Keyboards

Soft keyboards, which are especially popular on mobile devices, depict keys in a graphic user interface to allow typing with a touchscreen or stylus. The design issues for soft keyboards differ tremendously from normal keyboards. Soft keyboards require significant visual attention because *the user must look at the keyboard* to coordinate the motion of the pointing device. Only one key at a time can be touched, so much of the time is spent moving back and forth between keys (Zhai, Hunter, & Smith, 2000).

Character Recognition

Handwriting (even on paper, with no "recognition" involved) is much slower than skilled keyboard use, but its naturalness is appealing. Recognizing natural handwriting is difficult and error-prone for computers, although the technology continues to improve. To simplify the problem and make performance more predictable for the user, many handwriting recognizers

---

[3]For an extensive review, including technology and user requirements for successful use of speech recognition, see chapter 8, Conversational Interface Technologies.

[4]Approximately 95% of computer users hold the mouse in their right hand, since many left-handers learn to use the mouse this way. Therefore, for most users, placing functions on the left side reduces the demands on the mouse hand.

rely on single-stroke gestures, known as unistrokes (Goldberg & Richardson, 1993), such as graffiti for the 3COM PalmPilot. Unistroke alphabets are designed so that the stroke for each letter is easy for a computer to distinguish and recognize, while also being as simple as possible for users to learn and remember (MacKenzie & Zhang, 1997).

THE FUTURE OF INPUT

Researchers are looking to move beyond the current "WIMP" (windows, icons, menus, and pointer) interface, but there are many potential candidates for the post-WIMP interface. Nielsen (1993) suggested candidates such as virtual realities, sound and speech, pen input and gesture recognition, limited artificial intelligence, and highly portable computers. Weiser (1991) proposed the ubiquitous computing paradigm, which suggests that networked computers will increasingly become integrated with ordinary implements and that computers will be embedded everywhere in the user's environment.

Connecting the multiplicity of users, computers, and other digital artifacts through the Internet and wireless networking technologies represents the foundation of the ubiquitous computing vision; indeed, it seems likely that 100 years from now the phrase *wireless network* will seem every bit as antiquated as *horseless carriage* does today. Techniques that allow users to communicate and share information will become increasingly important. Biometric sensors or other convenient means for establishing identity will be important for security, but such technologies will also make services such as personalization of the interface and sharing data much simpler. As one early example of what is possible, the pick and drop technique (Rekimoto, 1997) supports cut-and-paste between multiple computers by using a stylus with a unique identifier code. From the user's perspective one just "picks up" data from one computer and "drops" it on another, but behind the curtain the computer senses the unique ID and uses it as a reference to an Internet location that stores the data. Other important input issues to effectively support multiple users working together include interaction on large work surfaces (Buxton, Fitzmaurice, Balakrishnan, & Kurtenbach, 2000; Moran et al., 1997), combining large and small displays (Myers, Stiel, & Gargiulo, 1998), and electronic tagging techniques for identifying objects (Want, Fishkin, Gujar, & Harrison, 1999).

Because of the diversity of locations, users, and task contexts, intelligent use of sensors to acquire contextual information will also be important to realize this vision of the future (Buxton, 1995a). For example, point-and-shoot cameras often have just one button, but a multitude of sensors detect lighting levels, the type of film loaded, and distance to the subject, thus simplifying the technical details of capturing a good exposure. This perspective may lead to an age of ubiquitous sensors where inexpensive, special-purpose, networked sensors collect information that allows greatly simplified interaction with many devices (Fraden, 1996; Saffo, 1997). Examples include mobile devices that sense location (Schilit, Adams, & Want, 1994; Want, Hopper, Falcao, & Gibbons, 1992) or that know how they are being held and moved (Bartlett, 2000; Harrison et al., 1998; Hinckley et al., 2000; Schmidt et al., 1999).

Current computers are inexpensive enough to be personal or even handheld. In the future, computing will be all but free, so more and more tools and everyday objects will include computing power and network connectivity. Hence many computers will be specialized, application-specific devices, rather than general-purpose tools. Application-specific input devices can be tailored to suit specific users and task contexts, such as allowing surgeons to view brain scans in three dimensions by rotating a miniature doll's head (Hinckley et al., 1994a). Tangible user interfaces may allow users to employ the world as its own interface by tightly coupling atoms and bits (Ishii & Ullmer, 1997). Cameras, microphones, and other sensors may imbue computers with perceptual apparatus that can help computers to perceive the world and the activity of users (Pentland, 1999).

Advances in technology will yield new "elemental" inputs. For example, scanners and digital cameras make the input of *images* a user—interface primitive. Tablets that sense multiple points of contact (Lee, Buxton, & Smith, 1985), sensors that detect the pressure exerted by a user's hand in a high resolution grid of samples (Sinclair, 1997), or digital "tape" that senses twist and curvature (Balakrishnan et al., 1999) all challenge the traditional concept of the pointing device because the interaction is no longer defined by a single point of contact. The difference is analogous to sculpting clay with the entirety of both hands rather than with the tip of a single pencil.

Another approach is to synthesize structure from low-level input (Fitzmaurice, Balakrisnan, & Kurtenbach, 1999). For example, a user might receive e-mail from a colleague saying, "Let's meet for lunch sometime next week." Pattern recognition techniques can automatically recognize that a calendar is needed, and select appropriate dates. The fields for a meeting request can be filled in with information gleaned from the message, thus eliminating the need for many input actions that otherwise might be required of the user (Horvitz, 1999).

We will continue to interact with computers using our hands and physical intermediaries, not necessarily because our technology requires us to do so, but because touching, holding, and moving physical objects is the foundation of the long evolution of tool use in the human species. The forms and capabilities of the technologies we use will continue to advance, but we are probably stuck with the basic human senses and cognitive skills. The examples enumerated above underscore the need to have a broad view of interaction. One must consider not only traditional "pointing devices," but also new sensors, high-dimensional input devices, and synthesis techniques that together will advance human interaction with computers.

ACKNOWLEDGMENTS

I thank Hugh McLoone for material on ergonomic device design, Gina Venolia for sharing many insights on input device design and evaluation, and Ed Cutrell and Steve Bathiche for commenting on early versions of this chapter.

References

Accot, J., & Zhai, S. (1997). Beyond fitts' law: models for trajectory-based hci tasks. *Proceedings of Computer-Human Interaction 1997: Association for Computing Machinery Conference on Human Factors in Computing Systems* (pp. 295-302). New York: Association for Computing Machinery. New York: Association for Computing Machinery.

Accot, J., & Zhai, S. (1999). Performance evaluation of input devices in trajectory-based tasks: an application of the steering law. *Proceedings of Computer-Human Interaction 1999: Association for Computing Machinery Conference on Human Factors in Computing Systems* (pp. 466-472). New York: Association for Computing Machinery.

Accot, J., & Zhai, S. (2001). Scale effects in steering law tasks. *Proceedings of Computer-Human Interaction 2001: Association for Computing Machinery Conference on Human Factors in Computing Systems* (pp. 1-8). New York: Association for Computing Machinery.

Akamatsu, M., & Mackenzie, I. S. (1996). Movement characteristics using a mouse with tactile and force feedback. *International Journal of Human-Computer Studies, 45*, 483-493.

Anderson, J. R. (1980). Cognitive skills. In *Cognitive Psychology and Its Implications* (pp. 222-254). San Francisco: W. H. Freeman.

Balakrishnan, R., Baudel, T., Kurtenbach, G., & Fitzmaurice, G. (1997a). The Rockin' Mouse: integral 3d manipulation on a plane. *Proceedings of Computer-Human Interaction 1997: Association for Computing Machinery Conference on Human Factors in Computing Systems* (pp. 311-318). New York: Association for Computing Machinery.

Balakrishnan, R., Fitzmaurice, G., Kurtenbach, G., & Singh, K. (1999a). Exploring interactive curve and surface manipulation using a bend and twist sensitive input strip. *Proceedings of Association for Computing Machinery 1999 Symposium on Interactive 3D Graphics* (pp. 111-118). New York: Association for Computing Machinery.

Balakrishnan, R., & Hinckley, K. (1999b). The role of kinesthetic reference frames in two-handed input performance. *Proceedings of Association for Computing Machinery 1999 Symposium on User Interface Software and Technology* (pp. 171-178). New York: Association for Computing Machinery.

Balakrishnan, R., & Kurtenbach, G. (1999c). Exploring bimanual camera control and object manipulation in 3d graphics interfaces. *Proceedings of Computer-Human Interaction 1999 Association for Computing Machinery Conference on Human Factors in Computing Systems* (pp. 56-63). New York: Association for Computing Machinery.

Balakrishnan, R., & Mackenzie, I. S. (1997b). Performance differences in the fingers, wrist, and forearm in computer input control. *Proceedings of Computer-Human Interaction 1997: Association for Computing Machinery Conference on Human Factors in Computing Systems* (pp. 303-310). New York: Association for Computing Machinery.

Bartlett, J. F. (2000, May/June). Rock 'n' scroll is here to stay. *Institute of Electrical and Electronics Engineers Computer Graphics and Applications (IEEE CG&A)*, 40-45.

Baudel, T., & Beaudouin-Lafon, M. (1993). Charade: remote control of objects using hand gestures. *Communications of the Association for Computing Machinery, 36*(7), 28-35.

Bier, E., Stone, M., Pier, K., Buxton, W., & DeRose, T. (1993). Toolglass and magic lenses: the see-through interface. *Proceedings of Association for Computing Machinery's Special Interest Group on Computer Graphics 1993* (pp. 73-80). New York: Association for Computing Machinery.

Britton, E., Lipscomb, J., & Pique, M. (1978). Making nested rotations convenient for the user. *Computer Graphics, 12*(3), 222-227.

Brooks, F. P. Jr. (1988). Grasping reality through illusion: interactive graphics serving science. *Proceedings of Computer-Human Interaction 1988: Association for Computing Machinery Conference on Human Factors in Computing Systems* (pp. 1-11). New York: Association for Computing Machinery.

Burdea, G. (1996). *Force and touch feedback for virtual reality*. New York: Wiley.

Buxton, W. (1983). Lexical and pragmatic considerations of input structure. *Computer Graphics, 17*(1), 31-37.

Buxton, W. (1986). Chunking and phrasing and the design of human-computer dialogues. *Information Processing 1986: Proceedings of the International Federation for Information Processing 10th World Computer Congress* (pp. 475-480). Amsterdam: North Holland Publishers.

Buxton, W. (1990a). The pragmatics of haptic input. *Proceedings of Computer-Human Interaction 1990: Association for Computing Machinery Conference on Human Factors in Computing Systems: Tutorial 26 Notes*. New York: Association for Computing Machinery.

Buxton, W. (1990b). A three-state model of graphical input. *Proceedings of Interact 1990* (pp. 449-456). Amsterdam: Elsevier Science.

Buxton, W. (1995b). Touch, gesture, and marking. In R. Baecker, J. Grudin, W. Buxton, & S. Greenberg (Eds.), *Readings in Human-Computer Interaction: Toward the Year 2000* (pp. 469-482). San Francisco: Morgan Kaufmann.

Buxton, W. (2002). A directory of sources for input technologies. Retrieved from the World Wide Web February 6, 2002, http://www.billbuxton.com/InputSources.html

Buxton, W., Fitzmaurice, G., Balakrishnan, R., & Kurtenbach, G. (2000, July/August). Large displays in automotive design. *Institute of Electrical and Electronics Engineers Computer Graphics and Applications (IEEE CG&A)*, 68-75.

Buxton, W., Hill, R., & Rowley, P. (1985). Issues and techniques in touch-sensitive tablet input. *Computer Graphics, 19*(3), 215-224.

Buxton, W., & Myers, B. (1986). A study in two-handed input. *Proceedings of Computer-Human Interaction 1986: Association for Computing Machinery Conference on Human Factors in Computing Systems* (pp. 321-326). New York: Association for Computing Machinery.

Campbell, C., Zhai, S., May, K., & Maglio, P. (1999). What you feel must be what you see: adding tactile feedback to the trackpoint. *Proceedings of Interact 1999: 7th International Federation for Information Processing conference on Human Computer Interaction* (pp. 383-390). Amsterdam: IOS Press.

Card, S., English, W., & Burr, B. (1978). Evaluation of mouse, rate-controlled isometric joystick, step keys, and text keys for text selection on a CRT. *Ergonomics, 21*, 601-613.

Card, S., Mackinlay, J., & Robertson, G. (1990). The design space of input devices. *Proceedings of Computer-Human Interaction 1990: Association for Computing Machinery Conference on Human Factors in Computing Systems* (pp. 117-124). New York: Association for Computing Machinery.

Card, S., Mackinlay, J., & Robertson, G. (1991). A morphological analysis of the design space of input devices. *Association for Computing Machinery Transactions on Information Systems, 9*(2), 99-122.

Card, S., Moran, T., & Newell, A. (1980). The keystroke-level model for user performance time with interactive systems. *Communications of the Association for Computing Machinery, 23*(7), 396–410.

Conner, D., Snibbe, S., Herndon, K., Robbins, D., Zeleznik, R., & van Dam, A. (1992). Three-dimensional widgets. *Proceedings of 1992 Symposium on Interactive 3D Graphics* (pp. 183–188, 230–231). New York: Association for Computing Machinery.

Douglas, S., Kirkpatrick, A., & Mackenzie, I. S. (1999). Testing pointing device performance and user assessment with the iso 9241, part 9 standard. *Proceedings of Computer-Human Interaction 1999: Association for Computing Machinery Conference on Human Factors in Computing Systems* (pp. 215–222). New York: Association for Computing Machinery.

Douglas, S., & Mithal, A. (1994). The effect of reducing homing time on the speed of a finger-controlled isometric pointing device. *Proceedings of Computer-Human Interaction 1994: Association for Computing Machinery Conference on Human Factors in Computing Systems* (pp. 411–416). New York: Association for Computing Machinery.

Douglas, S., & Mithal, A. (1997). *Ergonomics of computer pointing devices.* Heidelberg, Germany: Springer-Verlag.

English, W., Englebart, D., & Berman, M. (1967). Display-selection techniques for text manipulation. *Transactions on Human Factors in Electronics, 8*(1), 5–15.

Fitts, P. (1954). The information capacity of the human motor system in controlling the amplitude of movement. *Journal of Experimental Psychology, 47,* 381–391.

Fitzmaurice, G., & Buxton, W. (1997). An empirical evaluation of graspable user interfaces: towards specialized, space-multiplexed input. *Proceedings of Computer-Human Interaction 1997: Association for Computing Machinery Conference on Human Factors in Computing Systems* (pp. 43–50). New York: Association for Computing Machinery.

Fitzmaurice, G., Balakrisnan, R., & Kurtenbach, G. (1999). Sampling, synthesis, and input devices. *Communications of the Association for Computing Machinery, 42*(8), 54–63.

Foley, J., Wallace, V., & Chan, P. (1984, November). The human factors of computer graphics interaction techniques. *Institute of Electrical and Electronics Engineers Computer Graphics and Applications,* 13–48.

Fraden, J. (1996). *Handbook of modern sensors.* Heidelberg, Germany: Springer-Verlag.

Gibson, J. (1962). Observations on active touch. *Psychological Review, 69*(6), 477–491.

Goldberg, D., & Richardson, C. (1993). Touch-typing with a stylus. *Proceedings of Association of Computing Machinery INTERCHI 1993 Conference on Human Factors in Computing Systems* (pp. 80–87). New York: Association for Computing Machinery.

Greenstein, J. (1997). Pointing devices. In M. Helander, T. Landauer, & P. Prabhu (Eds.), *Handbook of Human-Computer Interaction* (pp. 1317–1348). Amsterdam: North-Holland.

Guiard, Y. (1987). Asymmetric division of labor in human skilled bimanual action: the kinematic chain as a model. *The Journal of Motor Behavior, 19*(4), 486–517.

Guiard, Y. (2001). Disentangling relative from absolute amplitude in fitts' law experiments. *Proceedings Computer-Human Interaction 2001: Association for Computing Machinery Conference on Human Factors in Computing Systems: Extended Abstracts* (pp. 315–316). New York: Association for Computing Machinery.

Guiard, Y., Buourgeois, F., Mottet, D., & Beaudouin-Lafon, M. (2001). Beyond the 10-bit barrier: fitts' law in multi-scale electronic worlds. *Proceedings of Joint Association Francophone d'Interaction*

Homme-Machine and British Human-Computer Interaction Group Conference (IHM-HCI) 2001 (pp. 573–587). Heidelberg, Germany: Springer-Verlag.

Harrison, B., Fishkin, K., Gujar, A., Mochon, C., & Want, R. (1998). Squeeze me, hold me, tilt me! an exploration of manipulative user interfaces. *Proceedings of Computer-Human Interaction 1998: Association for Computing Machinery Conference on Human Factors in Computing Systems* (pp. 17–24). New York: Association for Computing Machinery.

Hinckley, K., Cutrell, E., Bathiche, S., & Muss, T. (2002). Quantitative analysis of scrolling techniques. *Proceedings of Computer-Human Interaction 2002: Association for Computing Machinery Conference on Human Factors in Computing Systems* (pp. 65–72). New York: Association for Computing Machinery.

Hinckley, K., Czerwinski, M., & Sinclair, M. (1998a). Interaction and modeling techniques for desktop two-handed input. *Proceedings of the Association for Computing Machinery 1998 Symposium on User Interface Software and Technology* (pp. 49–58). New York: Association for Computing Machinery.

Hinckley, K., Pausch, R., Goble, J., & Kassell, N. (1994a). Passive real-world interface props for neurosurgical visualization. *Proceedings of Computer-Human Interaction 1994: Association for Computing Machinery Conference on Human Factors in Computing Systems* (pp. 452–458). New York: Association for Computing Machinery.

Hinckley, K., Pausch, R., Goble, J., & Kassell, N. (1994b). A survey of design issues in spatial input. *Proceedings of the Association for Computing Machinery 1994 Symposium on User Interface Software and Technology* (pp. 213–222). New York: Association for Computing Machinery.

Hinckley, K., Pausch, R., Proffitt, D., & Kassell, N. (1998b). Two-handed virtual manipulation. *Association for Computing Machinery Transactions on Computer-Human Interaction, 5*(3), 260–302.

Hinckley, K., Pierce, J., Sinclair, M., & Horvitz, E. (2000). Sensing techniques for mobile interaction. *Proceedings of Association for Computing Machinery 2000 Symposium on User Interface Software and Technology* (pp. 91–100). New York: Association for Computing Machinery.

Hinckley, K., Sinclair, M., Hanson, E., Szeliski, R., & Conway, M. (1999). The VideoMouse: a camera-based multi-degree-of-freedom input device. *Proceedings of Association for Computing Machinery 1999 Symposium on User Interface Software and Technology* (pp. 103–112). New York: Association for Computing Machinery.

Hinckley, K., Tullio, J., Pausch, R., Proffitt, D., & Kassell, N. (1997). Usability analysis of 3d rotation techniques. *Proceedings of Association for Computing Machinery 1997 Symposium on User Interface Software and Technology* (pp. 1–10). New York: Association for Computing Machinery.

Honan, M., Serina, E., Tal, R., & Rempel, D. (1995). Wrist postures while typing on a standard and split keyboard. *Proceedings of the Human Factors and Ergonomics Society 39th Annual Meeting* (pp. 366–368). Santa Monica, CA: Human Factors and Ergonomics Society (HFES).

Horvitz, E. (1999). Principles of mixed-initiative user interfaces. *Proceedings of Computer-Human Interaction 1999: Association for Computing Machinery Conference on Human Factors in Computing Systems* (pp. 159–166). New York: Association for Computing Machinery.

Igarashi, T., & Hinckley, K. (2000). Speed-dependent automatic zooming for browsing large documents. *Proceedings of Association for Computing Machinery 2000 Symposium on User Interface Software and Technology* (pp. 139–148). New York: Association for Computing Machinery.

Ishii, H., & Ullmer, B. (1997). Tangible bits: towards seamless interfaces between people, bits, and atoms. *Proceedings of Computer-Human Interaction 1997: Association for Computing Machinery Conference on Human Factors in Computing Systems* (pp. 234-241). New York: Association for Computing Machinery.

Jacob, R. (1991). The use of eye movements in human-computer interaction techniques: what you look at is what you get. *Association for Computing Machinery Transactions on Information Systems, 9*(3), 152-169.

Jacob, R., Sibert, L., McFarlane, D., & Mullen, M. Jr. (1994). Integrality and separability of input devices. *Association for Computing Machinery Transactions on Computer-Human Interaction, 1*(1), 3-26.

Jacob, R. (1996). Input devices and techniques. In A. B. Tucker (Ed.), *The Computer Science and Engineering Handbook* (pp. 1494-1511). Boca Raton, Florida: CRC Press.

Jellinek, H., & Card, S. (1990). Powermice and user performance. *Proceedings of Computer-Human Interaction 1990: Association for Computing Machinery Conference on Human Factors in Computing Systems* (pp. 213-220). New York: Association for Computing Machinery.

Kabbash, P., & Buxton, W. (1995). The prince technique: fitts' law and selection using area cursors. *Proceedings of Computer-Human Interaction 1995: Association for Computing Machinery Conference on Human Factors in Computing Systems* (pp. 273-279). New York: Association for Computing Machinery.

Kabbash, P., Buxton, W., & Sellen, A. (1994). Two-handed input in a compound task. *Proceedings of Computer-Human Interaction 1994: Association for Computing Machinery Conference on Human Factors in Computing Systems* (pp. 417-423). New York: Association for Computing Machinery.

Karat, C., Halverson, C., Horn, D., & Karat, J. (1999). Patterns of entry and correction in large vocabulary continuous speech recognition systems. *Proceedings of Computer-Human Interaction 1999: Association for Computing Machinery Conference on Human Factors in Computing Systems* (pp. 568-575). New York: Association for Computing Machinery.

Karl, L., Pettey, M., & Shneiderman, B. (1993). Speech-activated versus mouse-activated commands for word processing applications: An Empirical Evaluation. *International Journal of Man-Machine Studies, 39*(4), 667-687.

Keir, P., Back, J., & Rempel, D. (1999). Effects of computer mouse design and task on carpal tunnel pressure. *Ergonomics, 42*(10), 1350-1360.

Kramer, A. (1994). Translucent patches–dissolving windows. *Proceedings of Association for Computing Machinery 1994 Symposium on User Interface Software and Technology* (pp. 121-130). New York: Association for Computing Machinery.

Kurtenbach, G., Fitzmaurice, G., Baudel, T., & Buxton, W. (1997). The design of a gui paradigm based on tablets, two-hands, and transparency. *Proceedings of Computer-Human Interaction 1997: Association for Computing Machinery Conference on Human Factors in Computing Systems* (pp. 35-42). New York: Association for Computing Machinery.

Kurtenbach, G., Sellen, A., & Buxton, W. (1993). An empirical evaluation of some articulatory and cognitive aspects of 'marking menus'. *Journal of Human Computer Interaction, 8*(1), 1-23.

Langolf, G., & Chaffin, D. (1976). An investigation of Fitts' law using a wide range of movement amplitudes. *Journal of Motor Behavior, 8*, 113-128.

Lee, S., Buxton, W., & Smith, K. (1985). A multi-touch three dimensional touch-sensitive tablet. *Proceedings of Computer-Human Interaction 1985: Association for Computing Machinery Conference on Human Factors in Computing Systems* (pp. 21-25). New York: Association for Computing Machinery.

Lewis, J., Potosnak, K., & Magyar, R. (1997). Keys and keyboards. In M. Helander, T. Landauer, & P. Prabhu (Eds.), *Handbook of Human-Computer Interaction* (pp. 1285-1316). Amsterdam: North-Holland.

Liang, J., Shaw, C., & Green, M. (1991). On temporal-spatial realism in the virtual reality environment. *Proceedings of Association for Computing Machinery 1991 Symposium on User Interface Software and Technology* (pp. 19-25). New York: Association for Computing Machinery.

Lipscomb, J., & Pique, M. (1993). Analog input device physical characteristics. *Association for Computing Machinery's Special Interest Group on Computer-Human Interaction (SIGCHI) Bulletin, 25*(3), 40-45.

Mackenzie, C., & Iberall, T. (1994). *The grasping hand.* Amsterdam: North Holland.

MacKenzie, I. S. (1992a). Fitts' law as a research and design tool in human-computer interaction. *Human-Computer Interaction, 7*, 91-139.

MacKenzie, I. S. (1992b). Movement time prediction in human-computer interfaces. *Graphics Interface 1992* (pp. 140-150). San Francisco: Morgan Kaufmann.

MacKenzie, I. S. (1995). Input devices and interaction techniques for advanced computing. In W. Barfield & T. Furness (Eds.), *Virtual environments and advanced interface design* (pp. 437-470). Oxford, UK: Oxford University Press.

MacKenzie, I. S., & Buxton, W. (1992). Extending Fitts' law to two-dimensional tasks. *Proceedings of Computer-Human Interaction 1992: Association for Computing Machinery Conference on Human Factors in Computing Systems* (pp. 219-226). New York: Association for Computing Machinery.

MacKenzie, I. S., & Guiard, Y. (2001). The two-handed desktop interface: are we there yet? *Proceedings of Computer-Human Interaction 2001: Association for Computing Machinery Conference on Human Factors in Computing Systems: Extended Abstracts* (pp. 351-352). New York: Association for Computing Machinery.

MacKenzie, I. S., Kober, H., Smith, D., Jones, T., & Skepner, E. (2001). LetterWise: prefix-based disambiguation for mobile text input. *Proceedings of Association for Computing Machinery 2001 Symposium on User Interface Software and Technology* (pp. 111-120). New York: Association for Computing Machinery.

MacKenzie, I. S., & Oniszczak, A. (1998). A comparison of three selection techniques for touchpads. *Proceedings of Computer-Human Interaction 1998: Association for Computing Machinery Conference on Human Factors in Computing Systems* (pp. 336-343). New York: Association for Computing Machinery.

MacKenzie, I. S., Sellen, A., & Buxton, W. (1991). A comparison of input devices in elemental pointing and dragging tasks. *Proceedings of Computer-Human Interaction 1991: Association for Computing Machinery Conference on Human Factors in Computing Systems* (pp. 161-166). New York: Association for Computing Machinery.

MacKenzie, I. S., & Ware, C. (1993). Lag as a determinant of human performance in interactive systems. *Proceedings of Association for Computing Machinery INTERCHI 1993 Conference on Human Factors in Computing Systems* (pp. 488-493). New York: Association for Computing Machinery.

MacKenzie, I. S., & Zhang, S. (1997). The immediate usability of graffiti. *Proceedings of Graphics Interface 1997* (pp. 129-137). San Francisco: Morgan Kaufmann.

Marklin, R., Simoneau, G., & Monroe, J. (1997). The effect of split and vertically-inclined computer keyboards on wrist and forearm posture. *Proceedings of Human Factors and Ergonomics Society 41st Annual Meeting* (pp. 642-646). Santa Monica, CA: Human Factors and Ergonomics Society (HFES).

Mathias, E., Mackenzie, I. S., & Buxton, W. (1996). One-handed touch typing on a qwerty keyboard. *Human Computer Interaction, 11*(1), 1-27.

Microsoft (2001). Accessibility web site. Retrieved from http://www.microsoft.com/accessibility/

Mine, M., Brooks, F., & Sequin, C. (1997). Moving objects in space: expoiting proprioception in virtual-environment interaction. Proceedings of Association for Computing Machinery's Special Interest Group on Computer Graphics 1997. *Computer Graphics, 31*, 19-26.

Moore, M., Kennedy, P., Mynatt, E., & Mankoff, J. (2001). Nudge and shove: frequency thresholding for navigation in direct brain-computer interfaces. *Proceedings of Computer-Human Interaction 2001: Association for Computing Machinery Conference on Human Factors in Computing Systems: Extended Abstracts* (pp. 361-362). New York: Association for Computing Machinery.

Moran, T., Chiu, P., & van Melle, W. (1997). Pen-based interaction techniques for organizing material on an electronic whiteboard. *Proceedings of Association for Computing Machinery 1997 Symposium on User Interface Software and Technology* (pp. 45-54). New York: Association for Computing Machinery.

Myers, B., Stiel, H., & Gargiulo, R. (1998). Collaboration using multiple PDAs connected to a PC. *Proceedings of Association for Computing Machinery Computer Supported Cooperative Work (ISCW) 1998* (pp. 285-294). New York: Association for Computing Machinery.

Nielsen, J. (1993). Noncommand user interfaces. *Communications of the Association for Computing Machinery, 36*(4), 83-89.

Noyes, J. (1983). Chord keyboards. *Applied Ergonomics, 14*, 55-59.

Oakley, I., Brewster, S., & Gray, P. (2001). Solving multi-target haptic problems in menu interaction. *Proceedings of Computer-Human Interaction 2001: Association for Computing Machinery Conference on Human Factors in Computing Systems: Extended Abstracts* (pp. 357-358). New York: Association for Computing Machinery.

Olson, J., & Olson, G. (1990). The growth of cognitive modeling in human-computer interaction since GOMS. *Human-Computer Interaction, 5*, 221-266.

Pearson, G., & Weiser, M. (1988). Exploratory evaluation of a planar foot-operated cursor-positioning device. *Proceedings of Computer-Human Interaction 1988: Association for Computing Machinery Conference on Human Factors in Computing Systems* (pp. 13-18). New York: Association for Computing Machinery.

Pekelney, R., & Chu, R. (1995). Design criteria of an ergonomic mouse computer input device. *Proceedings of Human Factors and Ergonomics Society 39th Annual Meeting* (pp. 369-373).

Pentland, A. (1999). Perceptual intelligence. *Proceedings of Handheld and Ubiquitous Computing (HUC) 1999* (p. 74). Heidelberg, Germany: Springer-Verlag.

Putz-Anderson, V. (1988). *Cumulative trauma disorders: A manual for musculoskeletal diseases of the upper limbs*. Bristol, PA: Taylor & Francis.

Rekimoto, J. (1997). Pick-and-Drop: A direct manipulation technique for multiple computer environments. *Proceedings of Association for Computing Machinery 1997 Symposium on User Interface Software and Technology* (pp. 31-39). New York: Association for Computing Machinery.

Rempel, D., Bach, J., Gordon, L., & Tal, R. (1998). Effects of forearm pronation/supination on carpal tunnel pressure. *Journal of Hand Surgery, 23*(1), 38-42.

Robertson, G., Card, S., & Mackinlay, J. (1989). The cognitive coprocessor architecture for interactive user interfaces. *Proceedings of Association for Computing Machinery 1989 Symposium on User Interface Software and Technology* (pp. 10-18). New York: Association for Computing Machinery.

Rubine, D. (1991). Specifying gestures by example. *Computer Graphics, 25*(4), 329-337.

Rutledge, J., & Selker, T. (1990). Force-to-motion functions for pointing. *Proceedings of Interact 1990: The International Federation for Information Processing Conference on Human-Computer Interaction* (pp. 701-706) Amsterdam: North-Holland.

Saffo, P. (1997). Sensors: The next wave of infotech innovation. *Institute for the Future: 1997 Ten-Year Forecast* (pp. 115-122). Menlo Park, CA: Institute for the Future.

Schilit, B., Adams, N., & Want, R. (1994). Context-aware computing applications. *Proceedings of Institute of Electrical and Electronics Engineers Workshop on Mobile Computing Systems and Applications* (pp. 85-90). Santa Cruz, CA: Institute of Electrical and Electronics Engineers Computer Society.

Schmidt, A., Aidoo, K., Takaluoma, A., Tuomela, U., Van Laerhove, K., & Van de Velde, W. (1999). Advanced interaction in context. *Handheld and Ubiquitous Computing (HUC) 1999* (pp. 89-101). Heidelberg, Germany: Springer-Verlag.

Sears, A., Plaisant, C., & Shneiderman, B. (1992). A new era for high precision touchscreens. In Hartson & Hix (Eds.), *Advances in Human-Computer Interaction*, Vol. 3, pp. 1-33. Norwood, NJ: Ablex.

Sears, A., & Shneiderman, B. (1991). High precision touchscreens: design strategies and comparisons with a mouse. *International Journal of Man-Machine Studies, 34*(4), 593-613.

Sellen, A., Kurtenbach, G., & Buxton, W. (1992). The prevention of mode errors through sensory feedback. *Human Computer Interaction, 7*(2), 141-164.

Sinclair, M. (1997). The haptic lens. *Association for Computing Machinery's Special Interest Group on Computer Graphics 1997 Visual Proceedings* (p. 179). New York: Association for Computing Machinery.

Smutz, W., Serina, E., Bloom, T., & Rempel, D. (1994). A system for evaluating the effect of keyboard design on force, posture, comfort, and productivity. *Ergonomics, 37*(10), 1649-1660.

Sutherland, I. (1968). A head-mounted three dimensional display. *Proceedings of the Fall Joint Computer Conference* (pp. 757-764).

Want, R., Fishkin, K., Gujar, A., & Harrison, B. (1999). Bridging physical and virtual worlds with electronic tags. *Proceedings of Computer-Human Interaction 1999: Association for Computing Machinery Conference on Human Factors in Computing Systems* (pp. 370-377). New York: Association for Computing Machinery.

Want, R., Hopper, A., Falcao, V., & Gibbons, J. (1992). The active badge location system. *Association for Computing Machinery Transactions on Information Systems, 10*(1), 91-102.

Ware, C. (1990). Using hand position for virtual object placement. *Visual Computer, 6*(5), 245-253.

Weiser, M. (1991, September). The computer for the 21st century. *Scientific American*, 94-104.

Wickens, C. (1992). Chapter 3: attention in perception and display space. *Engineering Psychology and Human Performance* (pp. 74-115). New York: HarperCollins.

Yamada, H. (1980). A historical study of typewriters and typing methods: from the position of planning Japanese parallels. *Journal of Information Processing, 24*(4), 175-202.

Zeleznik, R., Herndon, K., & Hughes, J. (1996). SKETCH: an interface for sketching 3d scenes. *Proceedings of Association for Computing Machinery's Special Interest Group on Computer Graphics 1996* (pp. 163-170). New York: Association for Computing Machinery.

Zhai, S. (1993). Human performance evaluation of manipulation schemes in virtual environments. *Proceedings of 1993 Institute of Electrical and Electronics Engineers Virtual Reality Annual International Symposium (IEEE VRAIS'93)* (pp. 155-161). Santa Cruz, CA: Institute of Electrical and Electronics Engineers Computer Society.

Zhai, S. (1998). User performance in relation to 3d input device design. *Computer Graphics, 32*(4), 50-54.

Zhai, S., Buxton, W., & Milgram, P. (1994). The silk cursor: investigating transparency for 3d target acquisition. *Proceedings of Computer-Human Interaction 1994: Association for Computing Machinery Conference on Human Factors in Computing Systems* (pp. 459-464). New York: Association for Computing Machinery.

Zhai, S., Hunter, M., & Smith, B. A. (2000). The Metropolis keyboard—an exploration of quantitative techniques for virtual keyboard design. *Proceedings of Association for Computing Machinery 2000 Symposium on User Interface Software and Technology* (pp. 119-128). New York: Association for Computing Machinery.

Zhai, S., & Milgram, P. (1993). Human performance evaluation of isometric and elastic rate controllers in a 6d of tracking task. Proceedings of Society of Photo-Optical Instrumentation Engineers (SPIE) 2057: Telemanipulator Technology. Bellingham, WA: SPIE.

Zhai, S., Milgram, P., & Buxton, W. (1996). The influence of muscle groups on performance of multiple degree-of-freedom input. *Proceedings of Computer-Human Interaction 1996: Association for Computing Machinery Conference on Human Factors in Computing Systems* (pp. 308-315). New York: Association for Computing Machinery.

Zhai, S., Morimoto, C., & Ihde, S. (1999). Manual and gaze input cascaded (MAGIC) pointing. *Proceedings of Computer-Human Interaction 1999: Association for Computing Machinery Conference on Human Factors in Computing Systems* (pp. 246-253). New York: Association for Computing Machinery.

Zhai, S., Smith, B. A., & Selker, T. (1997). Improving browsing performance: A study of four input devices for scrolling and pointing tasks. *Proceedings of Interact 1997: The Sixth International Federation for Information Processing Conference on Human-Computer Interaction* (pp. 286-292). Boca Raton, FL: Chapman & Hall.

· 8 ·

CONVERSATIONAL INTERFACE TECHNOLOGIES

Clare-Marie Karat, John Vergo, and David Nahamoo
IBM T. J. Watson Research

A CONCEPTUAL FRAMEWORK FOR CONVERSATIONAL INTERFACE TECHNOLOGIES AND APPLICATION DEVELOPMENT

The goal of conversational technologies is to close the gap between human–computer interaction (HCI) and human–human interaction (HHI) by leveraging expertise in human-to-human conversational interaction. The basic tenets of human–computer interaction are to know your users, their tasks, and the context of use in which the users will complete their tasks. To provide a conceptual framework for this discussion of conversational interface technologies, we will describe the relevant characteristics of the users, conversational tasks, and context of use for applications that employ these technologies. To be successful, any solution that is developed must be designed with an understanding of these variables. An application can selectively mix and match elements of different interactive tasks and social and physical contexts of use to meet the needs of different users. After the discussion of the framework, the chapter examines current conversational interface technologies including automatic speech recognition, speech synthesis, natural language processing and understanding, and speaker recognition (verification, identification, and classification). Each section includes a discussion of how the technology works, its capabilities and limitations, user interface guidelines for application design, examples of successful applications, and commercially available tools, engines, and application protocol interfaces (APIs).

This chapter is related to several other chapters in the handbook because it is an overview, in this case of the state of conversational interface technologies. It describes how these technologies work and their capabilities and limitations, as well as key user, task, and physical and social context of use variables to consider in the design of applications employing these technologies. This chapter supplements many of the chapters in section III, part B, Interaction Issues for Diverse Users, by providing information related to conversational interfaces for children, the elderly, and users who have one or more impairments. This chapter also provides important information for the chapters in section III, part C, Interaction Issues for Special Applications, by describing the capabilities and limitations of the conversational interface technologies for special form factors and contexts of use. There is a particular tie to chapter 35, Conversational Speech Interfaces, which focuses on user interface design issues for telephony applications.

USER CHARACTERISTICS

Native and Nonnative Speakers

Conversational interface technologies employ engines that are developed based on user models of a particular language. Based on 30 years of research and development in academia and industry around the world, it is widely understood that native speakers of a particular language are significantly more successful in using a conversational technology application than are nonnative speakers, who may create sentences that do not conform to the standards of the language. Developers must identify the native language of the intended users of a system. If there are multiple native languages, the system may be developed to enable users to select their native language from a specified set of languages available to communicate with the system. Nonnative speakers will be able to use the system, but they generally will experience lower recognition accuracy. Alternatively, use of a constrained vocabulary or prescribed dialogue that severely restricts the possible responses users can make may increase the probability of successful task completion for nonnative speakers using a particular conversational system.

Casual Versus Expert Training

Users may employ a system that has conversational technologies embedded in it on a casual or infrequent basis or as a matter of daily, dedicated use. These users have very different skills and experience to draw on in using an application of this type. Experienced users have learned how to use the application. They may have trained the system on their voice models and may have learned over time how to communicate with the system most effectively. The users have a mental model of the system and remember from day to day the way the system works, what they need to do to complete their tasks, and how best to recover from errors. If users forget how to use the system or need to do something new, they can use recognition (visual or auditory prompts by the system) rather than recall to stimulate their actions or they can learn how to complete new tasks and enrich their user model of the system. Casual users also have a user model of the system, but it may be fairly primitive. They may expect to be able to "walk up and use" the system with no training or memory of previous use. Such users need to have very clear instructions and feedback from the system to successfully complete tasks and recover from errors.

Age

If a conversational system employs speech technology, be aware that current conversational technologies work most effectively with people who are over 14 years old. Currently, it is difficult to successfully employ these technologies with children because of the difficulty of building voice models that can accurately recognize the acoustic signals in the young people's speech. Children's language skills are continually improving and changing as grow. Their voices, and the pitch of their voices, also change as they are growing. Therefore, the recognition accuracy is lower, and it is more difficult, but not impossible, to create conversational interface applications for children. Some products are developed especially for children as young as 4 years old, and they have a constrained vocabulary, dialogue, and task set that enable the applications to achieve required levels of accuracy.

Physical Condition

The usability and usefulness of conversational technology applications varies depending on the user's physical state. In general, users who are not overly tired and do not have a bad cold

computer system, but a human (the unseen "Wizard") is actually formulating the responses to the user. WOz is an efficient technique for collecting realistic data before a system has been built.

The steps the user takes in the execution of the task will yield insight into what information needs to be represented as part of the user context. The tasks themselves will be represented as goals and functional capabilities in the system. Goals and user context are the essential elements in controlling the dialogue with the end user.

After an initial system has been built, it should be tested repeatedly with real users and modified based on the testing. The amount and type of training given to users before and during these sessions should be the same as the training end users will receive when the system is deployed. The testing takes place in the physical and social contexts of use (for example, in offices, homes, and cars; on the sidewalk of public streets; and in airports and train stations) where people want to use the solution individually or in teams to complete their tasks.

In summary, the conceptual framework identifies the critical user, task, and contextual variables for the application designer and developer to keep in mind while creating a conversational application. Development of a conversational application using UCD methods will give the product team the best chance of success. An application may employ one or more of the following conversational technologies to enable the target users to complete desired tasks in relevant contexts.

AUTOMATIC SPEECH RECOGNITION

How Does Automatic Speech Recognition (ASR) Work?

Continuous ASR systems work by analyzing the acoustic signal received through a microphone connected to the computer (see Fig. 8.1). The user dictates some text and the microphone captures the acoustic signal as digital data that is then analyzed by an acoustic model and a language model. The different speech recognition systems on the market differ in the building blocks (e.g., phonemes) that they employ to analyze the acoustic signal data. The analysis employs algorithms based on Hidden Markov

Models, a type of algorithm that uses stochastic modeling to decode a sequence of symbols to complete the computations (Rabiner, 1989; Roe & Wilpon, 1993). After the acoustic analysis is complete, the system analyzes the resulting strings of building-block data using a language model that contains the speaker voice model based on user training of the system, a base vocabulary, a personal vocabulary, and any specific domain topics (e.g., computer, medical) that were selected before the dictation. When the analysis is complete, the text appears on the computer screen.

Continuous speech recognition systems require computers with an approximate minimum of 200 MHz processor, 32 MB of RAM, 300 MB of available hard disk, and a 16-bit sound card with a microphone input jack and good recording. These requirements enable local decoding of the recognized speech. It is possible to have a system (e.g., small pervasive device) capture a user's speech and then decode it on a remote server and return the decoded text to the user, albeit with a short time delay. Multiple users can work with one installation of a speech recognition system because each user creates a user voice model and logs on with their individual user name so that the system uses their personalized speech files for processing. Each user can also create several user voice models to achieve the best recognition rates in environments with different levels of background noise (e.g., home, office, and mobile work locations).

Current Capabilities and Limitations of Speech Recognition Software

Before a user can begin to use speech recognition software, he or she must adjust the volume and position of the microphone with feedback from the system on the quality of the resulting audio. A good-quality microphone greatly improves recognition accuracy by improving the quality of the input data. Although speech recognition systems work in a speaker-independent mode—that is, without use of a user-specific voice model—the accuracy is much higher in a speaker- dependent mode in which the system can employ the user's voice model to analyze his or her speech. The user voice model is created during a process called enrollment. After adjusting the microphone, the user trains the system on the user's voice by reading words

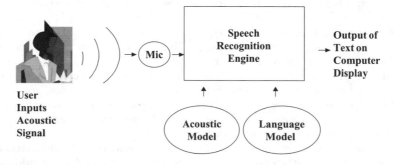

FIGURE 8.1. Overview of human–computer interaction model for speech recognition.

and stories presented through text on the screen. The system creates a voice model by analyzing how the user pronounces words in the prescribed text. With current speech recognition software, the enrollment process takes between 20 and 30 minutes. Recognition accuracy improves when the system can analyze a user's personal documents to determine the user's writing style and to add words to the user's personal vocabulary. Words can also be added to the system language model in everyday use of the software. Speech recognition systems allow the user to select and use specific language models for particular domains (e.g., computing, medicine, law) that significantly increase recognition accuracy.

Karat, Lai, Danis, and Wolf (1999) reported ASR system accuracy rates in the mid-90s for "in vocabulary" words (i.e., words that were in the 20,000-word vocabulary shipped with the system) in people's initial use of continuous speech recognition software. After frequent use, error rates of 2 to 5% are common. Karat, Horn, Halverson, and Karat (1999) tested three commercially available speech recognition systems in 1998 with users in initial and extended use and found initial use data of 13.6 corrected words per minute with an average of 1 incorrect, missing, or extra word for every 20 words and 1 formatting error every 75 words. Sears, Karat, Oseitutu, Kaimullah, and Feng (2000) found improved ASR system accuracy and higher user productivity when users employed IBM's ViaVoice Millennium speech recognition software (available in 1999 with a 64,000-word vocabulary) to complete similar transcription and composition tasks as the Karat, Horn, et al. study. For example, in the transcription tasks in Sears et al. (2000), both initial-use groups (traditional and disabled) produced about 13 words per minute (in the final document after corrections) with an average of 1 incorrect, missing, or extra word for every 39 words and 1 formatting error for every 81 words. The extended-use participants (those with 10 hours of experience with one of the three major speech systems on the market) in the Karat, Horn, et al. (1999) study achieved productivity approaching that for keyboard use. Extended-use participants produced about 28 corrected words per minute with speech and about 32 corrected words per minute with keyboard. The difference in performance of initial-use participants in the two studies appears to be due to the improvement in the user interface for error correction. Users can correct errors more easily and quickly with the new version of the speech recognition software. The product also provided a better quality microphone to reduce the number of recognition errors that occurred in the first place.

The usability of the error correction processes of ASR systems has improved significantly, and user practice with the system improves the performance, but more remains to be done. In general, there are three types of errors in ASR systems (Halverson, Horn, Karat, & Karat, 1999). Users can make direct errors where they misspeak, stutter, or press the wrong key. Second, users can make errors of intent, where they decide to restate something. The third type of error is an indirect error, where the speech recognition system misrecognizes what the user says. The indirect errors are difficult to detect during proofreading. Also, these indirect errors can lead to cascading errors where in the process of correcting one error, others occur. In the Karat, Horn, et al. (1999) study, we found that extended-use participants took about half of the steps to make a correction compared with novices (3.8 steps per correction compared with 7.3); novice users employed speech alone to make almost all corrections. They spent approximately 75% of their time correcting errors and 25% dictating (Halverson et al., 1999). Extended-use participants employed multimodal correction methods (corrections in which keyboard and mouse are used to either select the text to be corrected or to type in the correction or either keyboard- or mouse-only correction methods. For extended-use participants, about 25% of their time was spent correcting errors, and 75% dictating. In the Sears et al. (2000) study, we found that although overall error rates were lower for initial use participants, these participants spent 66% of their time on correction activities and 33% on dictation. One third of the correction time was spent simply navigating to the correction site. We found that the traditional users spent more time on navigation commands than the disabled users.

Conversational telephony systems (compared with menu-driven, touch-tone-based telephony systems) are examples of speaker-independent speech recognition systems. These systems have accuracy rates about 10% lower than the speaker-dependent speech recognition systems described above. Conversational telephony systems are server based compared with speaker-dependent speech recognition systems that are located on the individual's personal computer. The telephony systems must handle the signal degradation that occurs across the telephone lines. Telephony systems may be based on a combination of speech recognition and natural language processing (NLP) technologies. In telephony systems, a dialogue manager works with the speech recognition software to handle the course of the conversation with the user. The system will provide feedback to the user through the dialogue manager and recordings of human voice or text to speech. Telephony systems are more complex systems than the speaker-dependent conversational systems described above. They include a dialogue manager and have capabilities such as "barge in" and "talk ahead" that enable the user to redirect the action of the system and complete multiple requests before being prompted for additional information necessary to complete the task. Conversational telephony systems include interactive voice response (IVR) systems and new systems built using VoiceXML (please see section on NLP for a description of this technology).

User Interface Guidelines for ASR Application Design

The accuracy of speech recognition systems varies depending on the user's experience, age, articulation, native language, and physical state. In general, users who are over 14 years old, do not tend to mumble, can read at the fifth-grade level or above, are native speakers of the version of the software they are using (e.g., English, French, German), have some experience with speech recognition software, and are not overly tired or do not have a cold will have better success than others in using the speech recognition products currently available in the marketplace. Speaker-dependent speech recognition systems can effectively be used for composition, transcription, and transactions. A good-quality microphone will significantly improve performance. Factors in the physical environment such as background noise impact the accuracy rate. The form factor of the device must meet the needs of the user and task. It is difficult or impossible to compose a letter on a small, pocket-sized display. Working in a fairly quiet environment or one with a steady level of background noise will increase the effectiveness of the system. These systems work best when used in private settings.

If your application will have a constrained vocabulary, you can consider using speaker-independent ASR. ASR is used in the development of speech-enabled IVR systems. The HFES-200 standards committee is working on the development of human-factors standards for voice input–output and telephony. Most of the guidelines are focused on telephony applications that employ menu-driven, touch-tone-based IVR user interfaces (Gardner-Bonneau, 1999). There are a few guidelines related to IVR user interfaces employing speech recognition. The first states that the word *say* must be used to prompt the user to speak. Another guideline states that if enrollment is required for the application itself or for security reasons, the enrollment must take place in the same context as the application will be used. A third states that callers should be able to interrupt any prompt ("barge in") unless it is critical that the user hear the full message. A fourth guideline states that users should be able to say information that enables the application to handle a series of actions without interrupting the user with prompts (called "talk ahead").

If you are considering an automated telephony application, directed dialogues will be more successful than user-initiated or mixed-initiative conversations in the current environment because of technical and usability issues. There are issues with decreased accuracy of voice recognition over telephone lines. Telephony applications are best designed for users who are native speakers and who may be casual users of the system. Similar user characteristics regarding age, physical condition, and education mentioned above apply, with the exception that users will not need to have fifth-grade reading ability. Telephony systems are currently most applicable to simple transaction and collaboration tasks. Carefully consider the ability of the physical form factor for the device. Even simple transactions and collaborative tasks require both the system and the user to provide useful and usable feedback and both the system and the user must be able to gracefully recover from error situations. If there are high memory or cognitive load factors in the user's task, speech may not be the appropriate technology for the interface (Shneiderman, 2000). Telephony applications are best used in private or semipublic areas (phone banks) rather than in social or public areas where the user and the device degrade the experience of others. For assistance with the development of scenarios of use to guide these design decisions, please see a chapter written by John Karat on scenario use in the design of a speech recognition system (Karat, 1995). Please also see chapter 35 in this handbook on

conversational interfaces. There are helpful design guidelines for the limits of telephony systems available (Novick, Hansen, Sutton, & Marshall, 1999). More information available on human factors and voice interactive systems (Gardner-Bonneau, 1999) and designing effective speech interfaces (Weinshenk & Barker, 2000).

Examples of Successful Applications of Speech Recognition Technology

There are several successful applications of speaker-dependent speech recognition technology in domains such as medicine, law, finance, business, and computing in which users are working with a constrained vocabulary. An early success in this area was MedSpeak, an application for radiologists to use to dictate their reports (Lai & Vergo, 1997). In regard to the issue of universal access and speech recognition software, speech software is hands-free software and can be used by some individuals with disabilities. The software is useful for people with no uncorrected visual impairments or documented cognitive, hearing, or speech impairments. Speech recognition software has been the alternative of choice for many people with repetitive strain injuries (RSIs). For people with severe physical impairments who cannot use their hands, ASR systems for dictation and computer control can enable them to communicate with the outside world and be gainfully employed at home or in an office setting when they could not otherwise do so (Thomas, Basson, & Gardner-Bonneau, 1999). Efforts have been made to build environmental control units for "smart homes," which could offer the elderly and the disabled independence that they could not otherwise maintain. One example of good design of these types of systems is the personal home assistant developed in Europe (Burmeister, Machate, & Klein, 1997). We are currently researching improvements to speech user interfaces for individuals with spinal cord injuries at or above C6 with American Spinal Cord Injury Association (ASIA) scores of A or B (Sears et al., 2000). The research goal is to enable these individuals to more easily use the ASR technology and to be more effective with it. This research may make the ASR software more usable and useful for all users.

Telephony systems have been deployed in many industries (e.g., insurance, banking, finance, health care) to handle customer service inquiries regarding services available or order status, to make financial transactions, and to check account status. Currently, the most successful telephony systems are IVR systems with directed conversations with users. There are successful deployments of finance, airline reservation, and banking transaction systems using speech recognition in telephony applications (e.g., Schwab by Nuance, SpeechWorks by United Airlines, IBM/T. Rowe Price System; Wolf & Zadrozny, 1998). Other telephony systems enable users to manage their personal information such as e-mail, calendars, and to-do lists by providing a virtual secretary (e.g., Portico by General Magic). There are also telephony systems that allow users to speak a person's name to have that person's phone number dialed or that complete voice dialing.

Commercially Available ASR Tools, Engines, and APIs

Several companies offer development tools for speech recognition and telephony applications:

- IBM at www.ibm.com/software/speech/dev
- L&H at www.lhsl.com
- Microsoft at www.microsoft.com/iit
- Sun Microsystems at www.java.sun.com/products/java-media/speech

SPEECH SYNTHESIS

How Does Speech Synthesis Work?

Speech synthesis enables computers or other electronic systems such as telephones to output simulated human speech. Synthetic speech is based on the fields of text analysis, phonetics, phonology, syntax, acoustic phonetics, and signal processing (Spiegel & Streeter, 1997). There is a hierarchy of the quality and effectiveness of speech synthesis. The base level of achievement is to produce speech synthesis that is intelligible by human beings. The second level is to produce speech synthesis that simulates the natural quality of human speech. The third level of speech synthesis is to produce synthesized speech that is personalized to the user it is representing, that is, it has the intonation of the particular person's speech. The fourth and highest level of achievement in synthesized speech is to produce speech based on a person's own recordings so that the speech sounds just like the actual person's speech. Currently, speech synthesis technology has achieved the base level of quality and effectiveness, and concatenated synthesis can simulate the natural quality of human speech, although at great expense. The third and fourth levels of speech synthesis technology are the focus of research in laboratories around the world.

There are two types of speech synthesis commercially available today, concatenated synthesis and formant synthesis, the dominant form of the technology. Concatenated synthesis employs computers to assemble recorded voice sounds into speech output. It sounds fairly natural but can be prohibitively expensive for many applications because it requires large disk storage space for the units of recorded speech and significant computational power to assemble the speech units on demand. Concatenated synthesizers rely on databases of diphones and demisyllables to create the natural-sounding synthesized speech. Diphones are the transitions between phonemes, and demisyallables are the half-syllables recorded from the beginning of a sound to the center point or from the center point to the end of a sound (Weinschenk & Barker, 2000). After the voice units are recorded, the database of units is coded for changes in frequency, pitch, and prosody (intonation and duration). The coding process enables the database of voice units to be as efficient as possible.

Formant synthesis is a rule-based process that creates machine-generated speech in which a set of phonological rules is applied to an audio waveform that simulates human speech (see Fig. 8.2). Formant synthesis involves two complex steps. The first covers the conversion of the input text into a phonetic representation. The second encompasses the production of sound based on that phonetic representation. In the first step, the text is input from a database or file and is normalized so that any symbols or abbreviations are resolved as full alphabetic words. Then a pronunciation dictionary (for most words) and later a set of letter-to-sound rules (for exceptions) is employed to convert the words into phonemes. In the second step, the phonemes are analyzed using a sound inventory and intonation rules about pitch and duration, and the speech synthesis is the resulting output that is heard by the users through a speaker or headphone.

Current Capabilities and Limitations of Speech Synthesis Software

Formant synthesis produces speech that is highly intelligible but sounds unnatural. It has the power to produce nearly unlimited speech, however. The limitation of using formant synthesis is the complexity of the required linguistic rules to produce the accurate speech output. Using domain-specific information and assumptions produces a substantial improvement in the synthesizer's prosody. Users will be able to comprehend the speech at a higher rate and will perceive the voice to be more natural. In many cases, an application of speech synthesis will use a database that was never intended to be used as an input for speech synthesis. The database will have a variety of irregularities. To achieve higher comprehension and virtually 100% accuracy for users, these databases may be run against automatic preprocessors.

Some applications of concatenated synthesis attempt to reduce costs by basing the voice recordings on whole words. IVRs

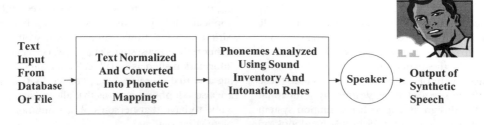

FIGURE 8.2. Model of text-to-speech synthesis.

and voicemail systems that use concatenated speech synthesis based on voice recordings of whole words sound unnatural and unevenly paced, and this makes the synthetic voice hard to understand or remember. An application of concatenated synthesis needs to be done correctly or not at all.

User Interface Guidelines for Application Design Using Speech Synthesis

Users of speech synthesis may be native or nonnative speakers and need little or no training to use these applications. There is no age restriction; however, the user must have an educational level that enables them to communicate effectively. Generally, users of speech synthesis systems need to have no auditory impairments and have the physical ability to access the system device. For applications for the disabled, the field of augmentative and alternative communication has found that the technology to provide nonspeaking individuals with a "voice" is too complex and cannot be effectively used by many individuals for whom it is intended. A specialist must access the potential user's language skills, symbol recognition skills, and physical ability to access the device (Thomas et al., 1999). Users of speech synthesis employ any of a variety of speakers or headphones for the output. They can the application to transact tasks in a variety of physical and social locations, given that the context of the task or communication is appropriate. For example, physical or social locations where distractions or noise would interfere with the comprehension of speech synthesis would create serious usability problems for the user. See Section IV, parts B and C of this handbook for information related to the developing conversational applications for the elderly and for users with impairments.

Examples of Successful Applications of Speech Synthesis

Successful applications of speech synthesis include information access, customer ordering, information for drivers, and interfaces for the disabled. Information access applications include systems for reverse directory service, railway timetables (Temem & Gitton, 1993), and voice access to e-mail (Yankelovich, Levow, & Marx, 1995). Customer ordering applications include catalog sales and telephone products and services. Information for travelers include driver information systems (e.g., Hertz "Never Lost" system) that provide a global positioning satellite, inertial guidance, a map database, and an in-car display with spoken-voice directions (Finney, 1995). Interfaces for the disabled include two types of solutions. The first is a set of solutions in which speech synthesis can be a "voice" for those with speech impairments. Speech synthesizers can match the user's gender, but additional work needs to be done to be able to match the user's age, pitch range, and personality characteristics. The second set of systems are "eyes" that read textual information to those individuals with vision impairments. Screen readers have been in use for many years (Thomas et al., 1999), and now there is e-mail for the blind. Speech synthesizers read books and other text for the visually impaired. The user places a book on an optical character reading system and hears the pages read. Designers of Web pages can access programs to ensure that sites are accessible to the blind. Newer applications of speech synthesis include navigating through digitally recorded speech segments and searching with text to speech.

Commercially Available Tools, Engines, and APIs

A number of text-to-speech tools and engines are available for developing speech synthesis applications:

- AT&T at www.att.com
- ELAN at www.elan.fr
- IBM at www.3dplanet.com
- L&H at www.lhsl.com
- Lucent at www.lucent.com
- Microsoft at www.hj.com
- Oki at www.oki.com

NATURAL LANGUAGE PROCESSING (NLP) AND UNDERSTANDING (NLU)

How Do NLP and NLU Work?

NLP refers to a wide range of processing techniques aimed at extracting, representing, responding to, and ultimately understanding the semantics of text. NLU is an area of NLP focused on the understanding of natural language text. Allen (1995) defines NLU as any system or technology that computes the representation of meaning, essentially restricting the discussion to the domain of computational linguistics. We are less concerned with the underlying algorithmic approach and therefore include any technology that allows a user to communicate with a system using a language that is not rigidly structured (i.e., a "formal" language). The focus of this chapter is on systems in which communication between the user and the system has constructs similar in syntax, grammar, and dialogue to the language of everyday human-to-human communication.

As a technology, NLU is independent from speech recognition, although the combination of the two yields a powerful human–computer interaction paradigm. When combined with NLU, speech recognition transcribes an acoustic signal into text, which then is interpreted by an understanding component to extract meaning. In a conversational system, a dialogue manager will then determine the appropriate response to give the user. Communication with the user can take place via a variety of modalities including speech in and out, text in and out, and handwriting. Figure 8.3 shows a block diagram of a prototypical multimodal conversational system that allows speech and keyboard natural language input and speech and graphical user interface (GUI) natural language output.

NLU is the process of analyzing text and taking some action based on the meaning of that text. The possible actions can vary considerably, depending on the task. A full discussion

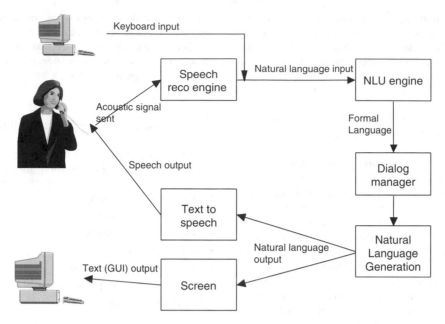

FIGURE 8.3. Diagram of a prototypical multimodal conversational system.

of NLU encompasses a wide range of applications and tasks well beyond those that are related to the topic of this chapter. Examples of NLU applications that we do not discuss include keyword search, document summarization, document categorization, language translation, and question answering. Summarization and categorization are rarely thought of in the context of conversational applications, so we eliminate them from our discussions.

It is tempting to include translation and question answering in a discussion of conversational technologies. NLU systems that translate language from one natural language to another will certainly be used in conversational applications in the future. For example, two people who do not share a common language will be able to converse with each other by having a computer translate, in real-time, from the language of the first speaker to the language of the second speaker, and vice-versa. We choose to exclude this technology because it is a fundamentally human-to-human interaction, not human-to-computer.

Question-answering systems take queries phrased in the form of a question and attempt to find the answer contained in a reference corpus. Generally speaking, most question-answering systems today, including research prototypes, allow the user to issue a question to the system, with the system subsequently returning its best guess at the passage in the corpus that contains the answer to the question. Unlike conversational systems, there is an assumption that the answer is absolute and factual, contained in a specific noun-phrase in the reference corpus, and that the answer does not depend on the context of the conversation. In fact, the interaction is essentially restricted to a single question–answer exchange between the user and the system. Question answering systems are designed to answer questions such as, "When did Columbus sail to America?" but are a long way from handling questions such as "Why is George Bush the president of the United States?" We therefore choose to exclude question answering from our discussion of conversational systems.

NLU has been an active area of research for many decades. The promise of NLU lies in the "naturalness" of the interaction. Because humans have deep expertise in interacting with each other through the use of language, it has been an implicit and explicit hypothesis in a wide array of research studies and technology development efforts that leveraging a user's ability to interact using language will result in systems with greater usability. Designing systems that use natural interaction techniques should mean the user is freed from learning the formal language of a system. We use the phrase *formal language* to include a broad array of interface technologies including languages such as SQL, UNIX command line commands, and programming and scripting languages and GUIs (including menu systems and other GUI widgets such as push buttons). Less user training, more rapid development of expertise, and better error recovery are all promising aspects of systems that use NLU.

Our description of NLU, as contrasted with speech recognition, leads us to include many types of applications as NLU applications. Generally speaking, if the system analyzes human-created text and attempts to transform the text into a formal representation for use by the system, it is an NLU application. The source of the text is not a determining factor in our definition of NLU, and we include systems in which the text originates from a speech recognizer, a GUI front end, the Internet, and so on.

Dialogue may be described as a series of related conversational interactions. It adds the richness of context and the knowledge of multiple interactions over time to the user interface, transforming natural language interfaces into conversational interfaces. Such interactions require the system to maintain a history of the interaction as well as the state of the interaction at all times (Chai, 2001). The IBM Natural Language engine and dialogue system (www.ibm.com) provides a feature called contextual inheritance, in which the context of a conversation can be inherited in later parts of the dialogue. For example, T. Rowe Price (www.trp.com) has a commercially available conversational system that allows users to trade mutual funds over the phone. A user may go through and entire dialogue to execute a trade, buying 2000 shares of fund A and then, say, "Do the same with fund B, but with 100 fewer shares." To accomplish this task, the dialogue manager must inherit the context of the conversation. Specifically, it understands "type of trade = buy," "amount = 2000," "unit = shares," "fund = A." Each parameter is inherited and then updated with new information from the conversation, which is subsequently acted upon.

Additional roles for the dialogue component may be to provide feedback to the user and ask the user to confirm critical operations. Two important components, which must be represented in any dialog system, are user goals and the instantaneous context of interaction. The history of a conversational interaction must be evaluated against long-term user goals, with prompts to the user designed to acquire the necessary information required to satisfy a goal. Dialogue technology may be embedded in an application to enable either user-initiated, system-initiated, or mixed- initiative conversations. These conversations are guided by dialogue management technology.

Mixed-Initiative Dialogue

Mixed initiative refers to dialogue in which the initiation of parts of the conversation changes from one participant in the dialogue to the other, possibly numerous times over the course of the conversation. Human-to-human conversation is clearly mixed initiative. Here is a simple example of mixed-initiative dialogue for an airline reservation system:

User: I'd like to fly to Paris.
System: Did you say Paris?
User: Yes, and I'd like to know what's available on Monday.

The user initiates the dialog in the first sentence, but the system takes control of the conversation with the second prompt. In the third utterance, the user responds to the system-initiated request in the second prompt, and initiates yet another request in the third sentence.

Walker, Fromer, Di Fabbrizio, Mestel, and Hindle (1998) compared two different designs for a spoken language interface (SLI), a system-initiated dialogue style and a mixed-initiated style in the context of an e-mail task. They measured the efficiency of the dialogue by counting the number of turns and elapsed time of the tasks. There results show that mixed-initiative dialogue is more efficient than system-initiative dialogue. User satisfaction is higher with the system-initiated dialogue, however. The authors speculate that this may be because system-initiated dialog is more predictable and easier to learn.

VoiceXML

VoiceXML is an emerging standard for distributed Web-based conversational interfaces (Lucas, 2000). Just as HTML gives Web designers a language for the development of the visual presentation of information, VoiceXML greatly simplifies the development of interactive voice applications. VoiceXML 1.0 was released in March 2000 and the specification can be found at www.voicexml.org. VoiceXML 1.0 provides a set of basic spoken dialogue capabilities that we characterize as directed dialogue. Directed dialogue is a structured form of dialogue in which the system directs the user to answer a series of specific questions or prompts the user for specific information. For example, a VoiceXML dialogue that handles a log-in scenario for an application might be structured as follows: System: "Please speak your name." User: Jose Jimenez." System: "Please say your password." User: "comedy." The highly structured nature of this dialogue makes it possible to achieve very high speech recognition accuracy.

VoiceXML also allows a more natural form of dialogue by providing for the use of input grammars. Although there is no limit to the complexity of the grammar a designer can create, the speech recognition accuracy will decrease as the grammar increases in complexity. Even with grammar-based input fields, VoiceXML still only supports a system-initiated dialogue model. Mixed- initiative dialogue is described in the emerging technologies section at the end of this chapter.

Current Capabilities and Limitations of NLU Systems

Ogden and Bernick (1997) cite Watt (1968), who defined the habitability of a language as the measure of how easily, naturally, and effectively people can use it. In particular, he focused on how well users could communicate while interacting in the course of trying to complete a task. For a natural language to be deemed habitable, users must stay within the conceptual, functional, syntactic, and lexical boundaries of the language, and the habitability of the language must match the capabilities of the system. For example, Chai (2001) described an NLU system that allows users to purchase notebook computers online. The conceptual domain of the system is limited to notebook computers. A user who attempts to use the system to buy groceries will have a frustrating experience.

The functional capabilities of the language can impact the structure of the conversational interaction. For example, rather than a user asking, "How much does the most expensive notebook cost?" a user might be forced to ask two questions: "What is the most expensive notebook?" The system replies: "The XYZ series." The user must then ask, "How much does the XYZ series cost?" The exchange must be done in two steps because of functional constraints of the system. The user must understand the functional capabilities and limitations of the system. One of the challenges of NLU systems is to communicate these capabilities and limitations to the user so that the system is usable and useful.

The syntactic coverage of the system is roughly equivalent to the number of paraphrases the system understands for a single command. The broader the syntactic coverage, the less user training required, and the more likely it is that a natural paraphrase will be understood by the system. Syntactic coverage is affected by the underlying NLU understanding technology. Word spotting and statistical pattern-matching technologies (Papineni & Roukos, 1997) provide wide latitude in how a command or sentence can be expressed and understood by a NLU system. Parsers such as the UNIX tool YACC and others based on BNF grammars are much more rigid. Good syntactic coverage can be achieved, but it is the responsibility of the conversational interface designer to ensure it happens. This is typically achieved through iterative user testing and development activities (Good, Whiteside, Wixon, & Jones, 1984; Kelly, 1984), with system performance leveling off as more data are collected and the system is updated with the depth and breadth of user responses to questions on the given topic.

The final domain, and probably the easiest to evaluate, is the lexical domain. Simply put, the lexical domain of a language is the set of all words in the language. Using the example derived from Chai (2001) above, a language that allows "I want to buy a notebook" and "want to buy a laptop" is more accommodating than one that understands only "notebook."

The syntactic and lexical domains combine to affect the parse rate of the system, the rate at which the system accurately translates a natural language phrase into a formal language statement, which is ultimately executed by the system. Tennant (1980) found that successful parse rates and task completion rates were not correlated, concluding that one cannot use parse rate as a cross-study–system comparison.

Walker, Litman, Kamm, and Abella (1997) developed the Paradise (PARAdigm for Dialog System Evaluation) evaluation framework for spoken language systems. Many aspects of the Paradise performance model apply to NLU. Paradise defines an overall user satisfaction metric that is a linear combination of task-based success metrics, dialogue efficiency, and dialogue quality. Paradise posits that dialogue efficiency should be maximized to maximize user satisfaction. Walker et al. did not make specific recommendations on which metrics should be used as part of a Paradise data-collection effort. Task-based success might be measured as task completion rate, average completion time, and so forth. Dialogue efficiency can be measured as the number of utterances in the dialogue, the duration of the dialogue, and so forth. Dialogue quality might be assessed via subjective questionnaires. Although the stated goal of Paradise is to inform the design of spoken language systems, it does not attempt to measure components of habitability, which have clear and direct impact on system design. The development of an evaluation framework based on habitability metrics is a potentially fruitful area of research.

One of the greatest and longest running challenges in the field of computer science is the Turing test (Turing, 1950).

The test consists of an interrogator, a second person, and a machine. The interrogator is connected to the second person and the machine via a terminal, therefore the interrogator cannot see his or her human and machine counterparts. His or her task is to determine which of the two candidates is the machine and which is the person by asking them questions. If the machine can convince the interrogator it is human, it is deemed to have intelligence. In 1990, Hugh Loebner agreed with the Cambridge Center for Behavioral Studies to underwrite a contest designed to implement the Turing Test. The $100,000 Loebner prize will go to the creator of the first computer with responses that are indistinguishable from a human's. This is probably the toughest evaluation criterion for NLU systems that exists.

Most NLU applications easily run on the desktop and notebook computers commonly available in 2002. In addition, research prototype systems are beginning to appear that move NLU functionality onto handheld systems. Cars are available that accept natural voice commands to control a limited range of features. NLU on the Internet is becoming more common (www.askjeeves.com, www.neuromedia.com) and illustrates client–server architecture for NLU. As computing power increases and digital technology pervades everyday life, we can expect to see a dramatic increase in the use of NLU in new environments, supporting more tasks.

Capturing and representing knowledge of a domain is a complex and labor-intensive process. As the scope of the application domain increases, it becomes increasingly difficult to build and maintain NLU applications. As a result, all successful examples of rich NLU interfaces have relatively narrow application domains. This has a direct impact on the definition of user profiles for any given application. Simply stated, the narrower the conceptual, functional, syntactic, and lexical domains of the target user population, the greater the chance of building an NLU application that satisfies its users. There is currently no way to quantify and measure these characteristics, and determining whether a particular application represents a tractable problem in NLU is largely an issue of experience and instinct on the part of designers and engineers.

User Interface Guidelines for NLU Applications

As with speech recognition technology, designers must consider many user characteristics when designing conversational systems based on NLU and dialog. The language users employ in the execution of any task will likely be impacted by whether they are native or nonnative speakers, their age, and their education. When assessing the user population of a particular application, limiting the range of each of these user characteristics will improve the performance of the system. The more homogenous the target user population is with respect to language use, the greater the likelihood of building a conversational application that can allow the users to complete their tasks. The more heterogeneous the user population, the more stress there will be on NLU parsing and pattern-matching engines to correctly translate a user's query into something the system understands correctly. Likewise, a heterogeneous user population makes it more difficult to design the output side of the dialogue so that all users understand the system-generated side of the dialogue with equal ease.

For many researchers and developers working on NLU, the promise of the technology is the elimination of the need to train users. By removing the constraints of formal language from human–computer interaction, we hope to take advantage of well-understood and frequently used modes of communication that our users employ every day. As of June 2001, there have been limited demonstrations of robust "walk up and use" conversational systems for serious commercial business applications. Even in the research world, conversational system technology, when it works well, is typically targeted at conceptually narrow application domains.

When humans communicate with each other, the language they use changes based on their understanding of the listener's linguistic and cognitive characteristics. The language adapts and converges based on feedback each participant in the conversation receives from the other (Fraser, 1993). Each conversational participant adopts features from the other's speech as a means to improve communication. This leads to the possibility that a conversational system, through the way it communicates with the user, can modify and even control the characteristics of the natural language issued by the user. Malhorta and Sheridan (1976) showed that users artificially constrain the syntax of their conversations when they interact with NLU systems. Ringle and Halstead-Nussloch (1989) specifically set out to understand if a user's natural language could be influenced to be more easily and accurately processed by the system while still feeling "natural" to the end user. Their strategy was to give feedback to users that presented an alternative form of sentences and commands used by participants. They reported success in their attempts to modify users' natural language to a more tractable form.

NLU and dialogue technologies span the entire range of conversational tasks, including composition, transcription, transaction, and collaboration. As NLU and dialogue technology continue to improve, virtually any nontrivial application will benefit from employing the technology in the interface. The human-centric word processor (Vergo, 1998) is an example of a compositional research prototype system. The T. Rowe Price mutual trading application is an example of a commercially available transactional system ("IBM/T. Rowe Price," 2001). Portico, a speech and natural language personal information management system (www.generalmagic.com), is a collaborative conversational system. IBM's Virtual Assistant (Wood, 2001) and DONALD systems are examples of collaborative conversational systems (Ramaswamy, Kleindienst, Coffman, Gopalakrishnan, & Neti, 1999).

Because of the wide range of applications that can and will employ conversational interfaces, there will be enormous range in both the physical and social context of use. The phone has already been identified as a ubiquitous input device for conversational applications. With the widespread adoption of wireless telephony, conversational interfaces are themselves becoming ubiquitous. As we have discussed, conversational interfaces are appearing on PCs, both as native applications and interfaces to World Wide Web applications. VoiceXML holds the promise of transforming the Web into a spoken language, conversational phenomena, transforming the workplace and many homes into locations of conversational interfaces. Cars are currently equipped with speech recognition technology, and it is natural to extend the interface to be conversational for long duration tasks such as navigation. The technology exists for PDAs with pen input capabilities to capture handwritten natural language input, send it via wireless local area network (LAN) to a server, which then recognizes the text and engages the user in a conversational interaction.

As the spread of NLU input devices increases, they will be found in virtually every social context imaginable. In many cases, the input modality (handwriting, speech recognition, keyboard, mouse, etc.) and output modalities (text to speech (TTS), GUI, etc.) will have to be chosen to ensure that social and cultural norms and constraints are maintained. It is possible that the impact of the social context of use for conversational applications will impact the interface to a degree greater than we have seen with traditional applications. To our knowledge, no one has studied this possibility.

Examples of Successful Applications of NLU

In recent years, there has been an explosion of commercial applications that use NLU. Some of the most popular speech recognition products such as IBM's ViaVoice, Dragon's Naturally Speaking, and L&H's Voice Express boast the ability to understand users when they issue commands in "natural language." In each of these cases, the "understanding" component is tightly coupled with the speech recognizer. In fact, there is no separate understanding engine— the speech recognized is simply set up to recognize a specific set of phrases. The application designer specifies a grammar (typically using a Backus Naur Form (BNF) notation or a variant thereof). The grammar is compiled and given to the recognizer. Such an environment requires the interface designer to completely describe the phrases that the system will understand.

Searching for information on the Internet has characteristics of NLU. The popular "Ask Jeeves" (www.askjeeves.com) uses question-answering techniques to determine the best URL to present to a user in response to a question. Users are encouraged to pose questions in plain English, although they can enter simple search terms. They in turn receive links to Web sites containing relevant information, services, and products. Ask Jeeves uses parsing technology to understand the query. Additionally, it uses data mining and a knowledge base to continually improve the performance of the search engine. Ask Jeeves does not employ dialog as part of the interface. Each question or query is handled independently without maintaining the context of the conversation.

One of the most promising and active areas of commercial development of NLU systems is in telephony. Call-center routing (Carpenter & Chou-Carroll, 1998) is an illustrative example of this type of NLU application. Call routing is the act of associating an incoming phone call with a particular destination. Although IVR systems provide this functionality, users frequently give them low scores for usability and overall user satisfaction. Carpenter and Chou-Carroll described a system that takes spoken natural language from users, prompts for additional information when required, and routes a phone call in the same manner as a human agent. They report a 94% routing accuracy when dealing with dozens of routing destinations. Zadrozny, Wolf, Kambhatla, and Ye (1998) described an integrated set of artificial intelligence systems called "conversation machines" that enable transactional telephony applications such as online banking and stock trading.

The human-centric word processor (Vergo, 1998) is a multimodal speech-and pen-based application that allows a user to simultaneously gesture with a pen and utter natural language commands to control a word-processing system. The system formalizes the gesture input from the pen and augments the spoken natural language utterance with the gesture information. The augmented natural language is then fed into the statistical NLU engine (Papineni, Roukos, & Ward, 1997), where it is converted to a formal language statement that is processed by the system's back-end logic. The NLU engine is a pattern matching engine, extracting features from the natural language input and matching it to one of a set of predefined formal language statements.

Other examples of NLU applications include e-mail routing (Walker, Fromer, & Narayanan, 1998) and information and database access (Androutsopoulos & Ritchie, 1995). Our list is far from complete. As NLU technology improves, researchers will experiment with it on an increasingly broader range of application domains.

Commercially Available NLU-Related Tools, Engines, and APIs

As of this writing, we are beginning to see the emergence of commercial NLU products targeting conversational interfaces.

- Ask Jeeves at www.askjeeves.com
- IBM at www.ibm.com
- VoiceXML at www.voicexml.org

SPEAKER RECOGNITION: VERIFICATION, IDENTIFICATION, AND CLASSIFICATION

What Is Speaker Recognition, and How Does It Work?

We divide the domain of speaker recognition into three classes: speaker verification, speaker identification, and speaker classification. Speaker verification is a binary decision-making process for accepting or rejecting the claimed identity of a speaker by processing a speech segment from the speaker. Speaker verification can be extended to include the option of nondecision where the system will ask for additional speech before making a

positive or negative decision. In speaker identification, the identity of a speaker is hypothesized from a list of enrolled speakers. Finally, speaker classification is the task of identifying a speech segment as belonging to one of the enrolled speakers or another speaker outside the available list. Speaker classification is an "open-set" speaker identification capability that handles speakers outside an enrolled list. When a speaker is identified as not being one of the existing speakers, it classifies the speech segment as being produced by a new speaker and also automatically creates a new cluster center for this speaker.

A speaker recognition task involves processing an utterance of a speaker to either verify the claimed identity in a speaker verification case or to assign an identity to the utterance in an identification or classification case. Similar to speech recognition processing, feature vectors are extracted from the spoken utterance. The most reliable and discriminative features today are based on short-term spectral processing of the utterance such as cepstral coefficients that capture the vocal tract characteristics. Fundamental frequency parameters such as pitch contour, representing voice source characteristics, are less reliable and are not commonly used. For practical reasons, the required length of a spoken utterance for analysis ranges between 2 and 5 seconds. The performance of the system improves by increasing the length of the utterance.

In a speaker verification task, a user utterance is produced in one of three ways: text independent, text dependent, and text prompted. Text independent provides the most flexible operation because the user has total freedom in composing the utterance. Text dependent is the most restrictive because the user must say a prescribed word sequence. Both text-independent and text-dependent verification approaches contain a security problem because an imposter can play back a recorded utterance from the claimed speaker. Text-prompted speaker verification provides a reliable mechanism to handle this problem by asking the speaker to repeat a new utterance every time access is requested. A text-dependent system provides the most accurate verification. Because the text for the training and test utterances are identical, the scoring based on template matching is an easier and more reliable task for the verification system because of the common contextual representation of the phoneme sequences for the test and training data.

In the speaker identification problem, the feature vector sequence of the input utterance is matched against the models for all the speakers in the identification database. The identity of the model with the closest score is assigned to the utterance. In the speaker verification problem, the acoustic model of the claimed identity is matched against the feature vectors of the spoken utterance. The acceptance or rejection decision is made based on this score compared with a prespecified threshold. In the classification problem, an extra step is necessary. In this case, usually the utterance is a long segment of speech with many speakers producing subsegments of the utterance. Therefore, there is a need to break the utterance into segments, each coming from a different speaker. These speech segments are then passed through an identification system that can assign an available identity to the segment or call it a new speaker if the closest score is larger than a chosen threshold.

There are three main approaches for voice-print modeling for speaker recognition problem. These are template matching, statistical modeling, and neural networks. Template matching is deployed in text-dependent processing. The voice print template is matched against the test utterance through dynamic time warping. This is to find the best time alignment between the test utterance and the template by minimizing the total Euclidean distance between the two. In text-independent or text-prompted cases, in which the spoken utterance varies from session to session, it is not practical to create templates for each possibility. In these cases, either structured or unstructured Guassian mixture models (GMM) are used to represent each speaker. The scores are calculated by finding the likelihood of the GMM producing the spoken utterance. A structured GMM is a multistate hidden Markov model (HMM), similar to the standard speech recognition HMMs, which are utterance dependent and model the phonetic structure of a sentence. The unstructured GMM is a single-state HMM with no phonetic structure.

To build the voice-print models for either verification or identification problem, each speaker has to provide speech samples. This is called the enrollment process. The length of the speech data necessary for building voice-print models ranges from 30 to 120 seconds. Similar to speech recognition, the performance of a speaker recognition system improves if each speaker provides more enrollment data.

Basic Capabilities and Requirements

Similar to all conversational technologies, central to the success of speaker recognition is the level of performance accuracy that they provide under variety of user and environment conditions as well as the computing resources and the system's response time. The performance of the speaker recognition technologies is affected by the health and the psychological and emotional state of the user. Although in many situations, these user conditions can be compensated for by users' focus and awareness of the requirement, more severe situations arise because of the audio channel variability. Microphone variation in different wireline and wireless telephony, speaker phones, being too close or far from the microphone, and array microphone introduces enough spectral modification in the recorded speech to potentially increase the error rate by a factor of two or more. Currently, the most effective approach for compensating the microphone variation impact is to ask the user to enroll from a variety of microphones so that voice-print models already take into account such differences. Similarly, interfering noise in the background and audio channel can have an adverse effect on the recognition. It is conceivable that, as with microphone variability, more data collection under a variety of noise and speaker conditions can improve the performance of the system. Speaker adaptation and channel normalization techniques such as cepstral mean normalization have been shown to improve the performance of the speaker recognition tasks.

Speaker recognition errors are classified as either false acceptance or false rejection. In a speaker verification task, false acceptance represents the case in which the system has erroneously given access to the wrong person. False rejection is the case where a valid user is wrongfully denied access to his or her account. In speaker verification, because the matching score is compared with a set threshold, it is possible to set this threshold to trade false acceptance rate for false rejection. In most cases, false acceptance is a critical and unacceptable error, whereas false rejection is simply an annoyance. In time-sensitive situations, (e.g., while trying to trade a stock whose price is falling rapidly) false rejection could also be critical. A popular practice today is to set the threshold for equal false acceptance and false rejection error rates.

Current state-of-the-art performance under normal telephony conditions and text-independent mode of operation is about 95% accuracy for speaker identification tasks with small populations (about 100 speakers) and around 70 to 90% accuracy for large populations (a few thousand speakers) using 3 to 5 seconds of telephone-quality speech. In the speaker verification task, the typical error rate is about 2 to 5% for 4 seconds of telephone-quality speech. The performance of the system improves by almost a factor of two with a text-dependent model. Three main issues arise with respect to the computing resources requirement. First, the storage requirement grows linearly with the size of the user population for any speaker recognition task. Second, although the central processing unit (CPU) requirements for a speaker verification task is independent of the population size, the CPU resources increase with the population size for the speaker identification task. A sublinear growth in CPU needs can be achieved by implementing fast-match techniques similar to those in speech recognition. Finally, to support global usage, duplicates of the same system need to be in different locations to handle network availability and channel performance issues.

User Interface Guidelines for Speaker Verification Applications

The user interface guidelines for speech recognition detailed in the earlier section of this chapter apply here as well. Some additional considerations for speaker identification and verification technologies are that these technologies are nonintrusive and have a good chance of being well accepted by users. Error handling and consideration of user's expertise represent the main interface design issues. In the future, when a combination of different biometrics such as face recognition and speaker recognition are combined, usability design needs to be further extended to identify the most beneficial combination of each approach. For example, face recognition does not intrude in a social setting such as a meeting while producing speech for the purpose of verification might be disruptive.

Examples of Successful Applications of Speech Recognition Verification Technology

Today's Caller-ID technology relies on the incoming phone number to inform the recipient of the identification of the calling party; however, Caller-ID's utility is diminished as the number of people having access to the same phone number increases. Speaker identification alleviates this shortcoming by informing the recipient of the exact identification of the calling party, assuming that the caller has completed enrollment in the system. Speaker identification technology can be used to announce the identification of a speaker anytime a change of speaker takes place during a conference call or in searching audio databases. When searching an audio recording of a radio or TV broadcast, a conference call, or a meeting, the user can find the segments spoken by a specific person, for example, a given actor, news anchorman, politician, or colleague.

Although speaker identification can enhance Caller-ID and audio search applications, the highest demand is for access control applications. A robust speaker verification technology can greatly improve the security of accessing critical Web and telephony services such as banking and shopping. This provides value to both users and service providers. As more services are offered through telephone and Web access, each service requires a user ID, a password or a personal identification number (PIN), and perhaps information such as one's mother's maiden name. These security data may be difficult for users to remember without writing them down, and once they are in writing, the documentation can be lost or stolen. A robust speaker verification system can effectively remove this problem by replacing the need for passwords and PINs.

There are other biometric technologies that share the main advantage of speaker recognition—the convenience for the user. These are fingerprint, face-recognition, signature-recognition, hand-recognition, and retinal-scanning technologies. The main advantage of speaker recognition is the pervasiveness of the availability of inexpensive microphones through either wired or wireless communication networks, as well as with any PCs, so that applications can be created with remote or local implementations of the speaker recognition engines. Although speaker recognition is nonintrusive, the accuracy is not as high as that for biometrics such as fingerprints.

Given the advantages of speaker verification over other biometric technologies (i.e., social acceptance and device availability), researchers have been working on many approaches to improve the performance of the technology. A recent success has been achieved by combining speaker verification and speech recognition technologies. This new approach is called conversational biometrics because it relies on a process of dialogue between the machine and the user in addition to the voice-print. Classical authentication is achieved through one of the following mechanisms: what we own, who we are, and what we know. Key- and card-based based mechanisms (such as a car key or magnetic badge for entering a building) are good examples of what we own. Passwords, PINs, mother's maiden name are good examples of what we know. Biometric technologies are based on who we are. Conversational biometrics combines text independent voice-print verification (i.e., who we are) and the content of answers to a small number of questions. The questions are addressed to the user through a spoken dialogue interaction, and the answers are recognized through a speech recognition system and matched against the answers that the user provided at the enrollment time. The questions addressed to the user can be randomly selected, follow a predefined sequence, or follow a business logic. Conversational biometrics overcomes the two main shortcomings of speaker verification, that is, the security problem of prerecorded speech and the high false acceptance error rate. It offers a very secure system with false acceptance rate as low as 0.00001% by insisting that both the voice-print and the content of the answers match the stored information. The questions are selected from a large set of information that are known only to the user and can be changed on regular basis (e.g., every 6 months). It is extremely difficult for any intruder to know the answers to all the questions, and even with prerecorded speech of another person, access to that person's account will be denied.

Commercially Available Speaker Verification Tools, Engines, and APIs

Currently, limited speaker verification technology offerings are available from Nuance; www.nuance.com, T-Netix; www.t-netix.com, Veritel; www. veritel.com, and VoiceRite; www.Voicerite.com. The established API for Speaker Verification is SVAPI.

ACKNOWLEDGMENTS

This chapter is based in part on work supported by the National Science Foundation (NSF) under grant no. 9910607. Any opinions, findings, and conclusions or recommendations expressed in this material are those of the authors and do not necessarily reflect the views of the NSF.

References

Allen, J. (1995). Natural Language Understanding, 2nd Ed., ISBN 0805303340. Addison-Wesley Pub. Co.

Burmeister, M., Machate, J., & Klein, J. (1997, March). Access for all: HEPHAISTOS—a personal home assistant. *Human Factors in Computing Systems—Computer Human Interaction (CHI) 97 Conference Proceedings* (Vol. 2, pp. 36–37). New York: Association of Computing Machinery.

Carpenter, B., & Chou-Carroll, J. (Nov. 30–Dec. 4, 1998). Natural language call routing: A robust, self-organizing approach. Paper presented at *the Fifth International Conference on Spoken Language Processing.* Sydney: Australia.

Chai, J. (2001, August). Natural language sales assistant—a Web-based dialog system for online sales. Presented at the Thirteenth Conference on Innovative Applications of Artificial Intelligence. WA: Seattle.

Finney, P. B. (1995, August 8). Business travel. New York Times, p. C4.

Fraser, N. M. (1993). Sublanguage, register and natural language interfaces. *Interacting with Computers, 5*, 441–444.

Gardner-Bonneau, D. (1999). Guidelines for speech-enabled IVR application design. In D. Gardner-Bonneau (Ed.), *Human factors and voice interactive systems* (pp. 147–162). Boston: Kluwer Academic.

Good, M. D., Whiteside, J. A., Wixon, D. R., & Jones, S. J. (1984). Building a User Derived Interface. In *Communications of the ACM*, Vol. 27, No. 10.

Halverson, C. A., Horn, D. A., Karat, C., & Karat, J. (1999). The beauty of errors: Patterns of error correction in desktop speech systems. In M. A. Sasse & C. Johnson (Eds.), *Human-computer interaction—INTERACT '99* (pp. 133–140). IBM/T. Rowe Price. (2001, June 5). *New York Times*, p. B2.

Kamm, C., & Helander, M. (1997). Design issues for interfaces using voice input. In M. Helander, T. K. Landauer, & P. Prabhu (Eds.), *Handbook of human-computer interaction* (pp. 1043–1059). Amsterdam: Elsevier.

Karat, J. (1995). Scenario use in the design of a speech recognition system. In J. Carroll (Ed.), *Scenario-Based Design* (pp. 109–133). New York: Wiley.

Karat, C., Halverson, C., Horn, D., & Karat, J. (1999). Patterns of entry and correction in large vocabulary continuous speech recognition systems. In M. Altom & M. Williams (Eds.), *Human factors in computing systems—CHI 99 Conference Proceedings* (pp. 568–575). New York: Association of Computing Machinery.

Karat, J., Horn, D., Halverson, C., & Karat, C. (2000). Overcoming unusability: Developing efficient strategies in speech recognition systems. *Human Factors in Computing Systems—Computer Human Interaction 2000 Conference Proceedings* (Vol. 2, pp. 141–142). New York: Association of Computing Machinery.

Karat, J., Lai, J., Danis, C., & Wolf, C. (1999). Speech user interface evolution. In D. Gardner-Bonneau (Ed.), *Human factors and voice interactive systems* (pp. 1–35). Boston: Kluwer Academic.

Kelly, J. F. (1984). An iterative design methodology for user-friendly natural-language office information applications. *ACM Transactions on Office Information Systems, 2*, 26–41.

Lai, J., & Vergo, J. (1997). MedSpeak: Report creation with continuous speech recognition. *Human Factors in Computing Systems—CHI 99 Conference Proceedings* (pp. 431–438). New York: Association of Computing Machinery.

Lucas, B. (2000). VoiceXML for Web-based distributed conversational applications. In *Communications of the ACM, 43*(9), 53–57.

Malhorta, A., & Sheridan, P. B. (1976). *Experimental determination of design requirements for a program explanation system* (IBM Research Report RC 5831). Armonk, NY: International Business Machines.

Marics, M. A., & Englebeck, G. (1997). Designing voice menu applications for telephones. In M. Helander, T. K. Landauer, & P. Prabhu (Eds.), *Handbook of human-computer interaction* (pp. 1085–1102). Amsterdam: Elsevier.

Ogden, W. C., & Bernick, P. (1997). Using Natural Language Interfaces. In M. G. Helender, T. K. Landauer, & P. V. Prabhu (Eds.), Handbook of Human-Computer Interaction (pp. 137–163). Elsevier, Amsterdam: North Holland.

Papineni, K. A., Roukos, S., & Ward, R. T. (1997). Feature-based language understanding. In *Proceedings of the 5th European Conference on Speech Communication and Technology, 3*, (pp. 1435–1438). Rhodes, Greece: European Speech Communication Association.

Novick, D. G., Hansen, B., Sutton, S., & Marshall, C. R. (1999). Limiting factors of automated telephone dialogues. In D. Gardner-Bonneau (Ed.), *Human factors and voice interactive systems* (pp. 163–186). Boston: Kluwer Academic.

Rabiner, L. R. (1989). A tutorial on hidden Markov models and selected applications in speech recognition. *Proceedings of Institute of Electrical & Electronics Engineers, 77*, 257–286.

Ramaswamy, G., Kleindienst, J., Coffman, D., Gopalakrishnan, P., & Neti, C. (1999, Sept.). A pervasive conversational interface for information interaction. *Eurospeech99*, Budapest, Hungary.

Ringle, M. D., & Halstead-Nussloch, R. (1989). Shaping user input: A strategy for natural language design. *Interacting with Computers, 1*, 227–244.

Roe, D. B., & Wilpon, J. G. (1993). Wither speech recognition: The next 25 years. *IEEE Communications Magazine, 11*, 54–62.

Sears, A., Karat, C., Oseitutu, K., Kaimullah, A., & Feng, J. (2000). Productivity, satisfaction, and interaction strategies of individuals with spinal cord injuries and traditional users interacting with speech recognition software. *Universal Access in the Information Society, 1*, 5–25.

Shneiderman, B. (2000). The limits of speech recognition. *Communications of the ACM, 43*(9), 63–65.

Temem, J., & Gitton, S. (1993). An experience with speech technologies applied to SNCF's Telephone Information Centres. *Applications of Speech Technology, 9*, 39–42.

Tennant, H. R. (1980). *Evaluation of natural language processors* (Report T-103). Urbana: University of Illinois, Coordinated Science Laboratory.

Thomas, J. C., Basson, S., & Gardner-Bonneau, D. (1999). Universal access and assistive technology. In D. Gardner-Bonneau (Ed.), *Human factors and voice interactive systems* (pp. 135–146). Boston: Kluwer Academic.

Turing, A. M. (1950). Computing machinery and intelligence. *Mind, 54*, 433–460.

Vergin, R., O'Shaughnessy, D., & Farhat, A. (1999, September). Generalized Mel-Frequency Cepstral Coefficients for Large Vocabulary Speaker Independent Continuous Speech Recognition, IEEE Trans. on Speech and Audio Processing (Vol. 7, pp. 525–532).

Vergo, J. (1998). A statistical approach to multimodal natural language interaction. *Proceedings of the AAAI'98 Workshop on Representations for Multimodal Human-Computer Interaction* (pp. 81–85). Madison WI: AAAI Press.

Walker, M., Fromer, J., & Narayanan, S. (1998). Learning Optimal Dialogue Strategies: A case study of spoken Dialogue Agent for E-Mail, 36th Meeting of the ACL, Montreal, Canada.

Walker, M. A., Litman, D. J., Kamm, C. A., & Abella, A. (1997, July). PARADISE: A Framework for Evaluating Spoken Dialogue Agents. *Proceedings of the Thirty-Fifth Annual Meeting of the Association for Computational Linguistics and Eighth Conference of the European Chapter of the Association for Computational Linguistics.* Madrid, Spain.

Walker, M. A., Fromer, J., Di Fabbrizio, G., Mestel, C., & Hindle, D. (1998, April). What can I say?: Evaluating a spoken language interface to email. *Proceedings of Association of Computing Machinery Computer Human Interaction 98 Conference on Human Factors in Computing Systems* (pp. 582–589). Los Angeles.

Watt, W. C. (1968, July). Habitability. *American Documentation*, (pp. 338–351).

Weinshenk, S., & Barker, D. T. (2000). *Designing effective speech user interfaces.* New York: Wiley.

Wilcox, L., Chen, F., Kimber, D., & Balasubramanian, V. (1994). Segmentation of speech using speaker identification. *Proceedings of ICASSP-94, I*, 161–164.

Wolf, C., & Zadrozny, W. (1998). Evolution of the conversation machine: A case study of bringing advanced technology to the marketplace. *Human Factors in Computing Systems—CHI 1998 Conference Proceedings* (pp. 488-495). New York: Association of Computing Machinery.

Wood D., Personal Communication.

Yankelovich, N., Levow, G., & Marx, M. (1995). Designing speech acts: Issues in speech user interfaces. In *Human factors in computing systems—CHI 95 Conference Proceedings* (pp. 369-376). New York: Association of Computing Machinery.

Zadrozny, W., Wolf, C., Kambhatla, N., & Ye, Y. (1998). Conversation machines for transaction processing. Presented at *the Fifteenth National Conference on Artificial Intelligence, American Association of Artificial Intelligence and Tenth Conference on Innovative Applications of Artificial Intelligence Conference (IAAI)*, Madison, Wisconsin.

·9·

VISUAL DISPLAYS

Holger Luczak
Aachen University of Technology

Matthias Roetting
Liberty Mutual Research Center for Safety and Health

Olaf Oehme
Aachen University of Technology

INTRODUCTION

Together with the general development of human–computer interaction, visual displays have evolved over the decades of computer development (cf. Baecker & Buxton, 1987; Nielsen, 1993). The first fully functional freely programmable computer, Konrad Zuse's Z3, used lamps to display four decimal digits with decimal points (Fig. 9.1). This and all the other early computers can be classified as batch systems. In addition to lamps, printers and teletypes were used as visual output generating devices. The next generation of computers, with line-oriented interfaces, used monochrome cathode ray tubes (CRTs) to display alphanumerical characters. With the advent of full-screen interfaces, CRTs capable of displaying gray-scale or color and graphics became common. With today's direct manipulation operating systems, displays capable of generating graphics with high resolution and millions of colors are the standard for the personal computer (PC).

The computer and its applications continue to develop, and some of these developments (e.g., virtual reality and augmented reality) require visual displays with characteristics very different from those found on standard PCs. On the other hand, technical devices with formerly restricted interaction capabilities are becoming more "intelligent" (ubiquitous computing). Hence, user interaction is more diverse, and one way of accomplishing this has been to implement visual interfaces with greater spatial resolution. Many of these devices show an evolution of the visual display similar to that of the computer: from simple lamps to graphic displays.

This chapter provides a systematic overview of current and future visual displays, their technical aspects, and the characteristics that are relevant to human–computer interaction.

FRAMEWORK

Visual displays can be distinguished by many criteria. The following list is helpful both to focus the discussion on certain display types commonly found in human–computer interaction and to structure the chapter, in which we focus on several of the following characteristics of visual displays:

- The variation of light or of a physical entity
- The nature of the data displayed, analog or digital

FIGURE 9.1. The output device of Konrad Zuse's Z3 computer with lamps for the decimal numbers (right) and arithmetic exception handling (left). Reprinted by permission of Horst Zuse. A low-resolution version of the photograph is available at http://irb.cs.tuberlin.de/~zuse/Konrad_Zuse/Z3-detail.htm (retrieved January 31, 2002).

- The spatial dimensionality of the image produced: one-, two-, or three-dimensional
- The set of displayable tokens: binary, fixed character set, or (practically) unlimited
- The physical principal of image generation: emission, transmission, light reflection, or a combination of these principles
- The number of colors the display can present: monochrome or polychrome
- The physical dimensions of the display

Applied to the field of human–computer interaction, only a small subset of these criteria is relevant.

Variation of Light or of a Physical Entity

Historically, many "displays" perceived by vision were based on the displacement of a physical entity: the hands of a clock, the arms of a railroad signal, flags used to communicate between ships using the semaphore alphabet, or smoke signals. Only since the introduction of artificial light did displays based on the variation of light become possible. The majority of visual displays used in human–computer interaction utilize light variation (e.g., indicator lights or light emitting diodes (LEDs), signaling that the power is on or the hard drive is read or creating the main visual display). There are still some visual indicators on common computers that rely on the displacement of a physical entity. Examples are the write-protect tab on a 3.5-inch floppy disk, the eject button on the floppy disk drive, or latches on notebook computers. Displays based on the displacement of a physical entity may have the advantage that they can be perceived by touch and therefore do not rely solely on the human visual system to convey their message, but the amount of information that can be displayed is restricted. Therefore, all displays discussed in this chapter are based on visual light, that is, electromagnetic radiation with a wavelength between about 380 nm and 780 nm.

Analog and Digital Data Displays

Analog displays translate a value of a variable into an angle or a length. Analog displays consist of a scale and an indicator or hand. Either the scale or the hand moves. There exist a number of guidelines for the design of analog displays (e.g., Baumann & Lanz, 1998; Burandt, 1986; Woodson, 1987). Because a computer is a digital device, all data displayed is of a digital nature, and therefore all commonly used displays are digital. Nonetheless, some programs mimic analog display devices, displaying them on the standard digital screen device. With a high spatial resolution of the digital display, no difference from an analog display can be perceived.

One-, Two-, or Three-Dimensional Displays

Visual displays can be distinguished by the spatial dimensionality of the image generated and hence the information displayed.

Most displays used in conjunction with a computer are two-dimensional, for example, the common CRT or flat panel displays or the displays found on handheld devices.

In recent years three-dimensional (3D) displays were developed and are applied in selected settings (see Displays for Selected Applications later in this chapter). They are either real 3D or virtual 3D. Most virtual 3D image displays require the user to wear glasses to perceive the three-dimensional image. Indicator lights, either lamps or LEDs, are referred to as one-dimensional (even if this classification is not correct in the strictest sense, because lamps and LEDs have spatial extension). They can be found as indicators that the power is on or that the hard drive is active on common computers.

Set of Displayable Tokens

Common PC displays have a sufficiently high resolution to display a virtually unlimited set of characters and graphics. A broader variety of displays with lower spatial resolution is found on other kinds of computerized technical devices. Here, the set of displayable tokens is often more restricted.

Basic displays are binary, being either on or off. A ternary display could be built from a two-color LED, in which either of the two colors is off or on. To transmit more information, single elements with even more discrete stages could be used (using, for example, different colors or brightness levels to distinguish stages). A more common way to increase the amount of displayable information is the grouping of a number of binary displays into a single unit. The classical seven-segment display (Fig. 9.2) and many custom displays are examples of this approach. Combining all these approaches, adding multiple binary displays, and enhancing them to display multiple colors and different levels of brightness results in the common CRT and flat panel displays.

Physical Principles

Different visual displays use different physical principles to generate an image; light can be emitted, transmitted, or reflected by the display. Examples from noncomputer displays might be helpful to explain the principle and to point out relevant advantages and restrictions. Writing on a piece of paper alters the reflective properties of the paper from a highly reflective white to a less reflective blue or black. Ambient light is needed to read what is written on the paper, but the contrast ratio between text and background remains the same at all lighting conditions. So, as long as the light is above a threshold and fairly evenly distributed, the text is equally readable. The same applies to reflectance-based displays such as liquid crystal

FIGURE 9.2. Seven-segment display.

displays (LCDs) without backlighting or electronic ink and paper (see Technologies in this chapter).

Transparency film used with an overhead projector and slides used with a projector are examples of transmission. Different parts of the transparency or slide transmit light of different wavelengths (i.e., color) in different amounts. Although transmission is used as a principle in displays, there are probably no displays in use in human–computer interaction that rely solely on transmission. Most displays combine transmission with other physical principles: a projection display consists of a light (emission), an LCD that generates the image and transmits the light, and finally the projection surface that reflects the light into the eye of the observer.

An example for emission is a lighthouse. Although its light can be easily seen at night, it is barely visible in bright daylight. Examples for emission-based displays are lamps, LEDs, CRTs, or electroluminescent displays (see Technologies). Similar to a lighthouse, these displays need to be brighter than the ambient light to be perceived.

Number of Colors the Display Can Present

Different displays can present differing numbers of colors. The simplest displays have two stages, either on or off, and therefore are monochrome. The greater the difference between these two states, expressed as luminance ratio and wavelength differences of the light falling in the observers eye, the more easily the viewer can distinguish them.

Other displays, called gray-scale displays, are able to adopt different luminance levels. More complex displays can reproduce color as well. The Commision International de l'Eclairage (CIE) chromaticity diagram is often used to represent a display's color capabilities. The diagram is based on the CIE standard XYZ system. X, Y, and Z are hypothetical primary colors. Figure 9.3 depicts the equal energy-matching functions of the standard XYZ system. The XYZ primaries were chosen so that whenever equal amounts of them are combined, they match

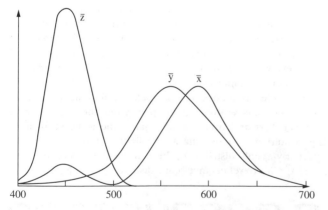

FIGURE 9.3. Equal energy-matching functions of the standard XYZ system. The curves $\bar{x}$, $\bar{y}$, and $\bar{z}$ show the relative amounts of the X, Y, and Z primary colors needed to match the color of the wavelength of light (after Kaufman, 1974).

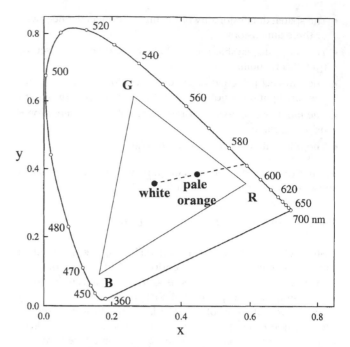

FIGURE 9.4. The CIE chromaticity diagram. The curved line, the locus of spectral colors, represents the x, y, and z values for all the spectral colors between 380 nm and 770 nm. The triangle depicts the colors that can be presented by a display that uses the three primary colors at the corners of the triangle. The hue of any color within the curved line can be determined by drawing a line from pure white (x = 0.33 and y = 0.33) through the color of interest to the locus line. A pale orange (x = 0.45 and y = 0.35) is marked by a black dot as an example. The saturation is given by the relative length of the dashed line between white and the color of interest, and white and the intersection of this line with the locus (cf. Kaufman, 1974).

white light. Independent of the absolute value of X, Y, and Z, relative amounts—denoted by lower case letters—can be used to describe a color. Because x, y, and z are relative amounts, x + y + z = 1. Consequently, when x and y are known, z is known as well, because z = 1 − (x + y). Therefore, a two-dimensional graph can be used to depict all the possible combinations of the XYZ primaries, shown here in Fig. 9.4. At the intersection of x = 0.33, and y = 0.33, pure white (i.e., achromatic light) can be found. The curved line, the locus of spectral colors, represents the x, y, and z values for all the spectral colors. The hue of any color within the curved line can be determined by drawing a line from pure white through the color of interest to the locus line. The saturation is given by the relative length of the line between white and the color of interest and white and the intersection of this line with the locus (see Fig. 9.4).

Because x and y are relative values, an additional parameter, not shown in Fig. 9.4, is needed to specify the luminance. It is denoted by Y, and hence this system of color characterization is referred to as the 'xyY system'.

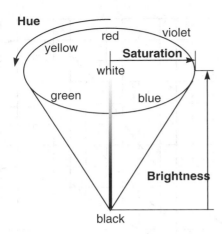

FIGURE 9.5. The hue, saturation and brightness system of specifying color.

To depict a great variety of different colors, common displays use the three primary colors, red, green, and blue. The actual scope of colors is restricted by the position of each of the three primaries in the CIE chromaticity diagram and lies within the triangle defined by these primaries (see Fig. 9.4 for an illustration).

Another restriction of the variety of colors is the representation of this information in the computer's hardware and software. Popular configurations devote 4 to 32 bits to the representation of color information, yielding 16 to 16,777,216 colors with different hues and saturations. Whereas the CIE chromaticity diagram is useful in depicting the color space, additional systems are used to describe color in the context of human–computer interaction. The specification of *hue, saturation, and brightness* (HSB; sometimes 'instead of brightness' the terms *luminance* or *lightness* are used) provides a device-independent way to describe color. For this model, the color space is an upside-down cone (see Fig. 9.5). On the edge of the cone base, the visible light spectrum is arranged in a circle by joining red and violet. Hue is the actual color and can be specified in angular degrees around the cone starting and ending at red $= 0°$ or $360°$, in percent or eight-bit values (0 to 255). Saturation is the purity of the color, measured in percent or eight-bit values from the center of the cone (min.) to the surface (max.). At 0% saturation, hue is meaningless. Brightness is measured in percent or eight-bit values from black (min.) to white (max.). At 0% brightness, both hue and saturation are meaningless.

RGB stands for the three basic colors—red, green, and blue—that are produced by the visual display. A number of other colors can be produced by *additive color mixing*. If any two of the color channels are mixed in equal proportions, new colors are created: blue and green create cyan (bright, light blue); red and blue, magenta (a bright pink); and red and green, yellow. If all three colors are mixed together equally at full power, the result is white light. Colors can be specified giving the relative power of each color in percent or as an 8-bit number (0 to 255 in decimal or 00 to FF in hexadecimal). So, for example, (0, 0, 0) specifies black; (255, 0, 0), a pure red, and (255, 255, 255), white.

CMYK stands for cyan, magenta, yellow, and black. This system is used to specify colors on the monitor for printing. In printing, colors are mixed subtractive and, using red, green, and blue, not many colors could be produced. By choosing cyan, magenta, and yellow as basic colors, many other colors, including red, green and blue, can be produced. Theoretically, when all three basic colors are printed over each other, the resulting color should be black. In practice, this is not the case, and a fourth printing process with black ink is used. Values for CMYK are often specified as percentages.

Physical Dimensions of the Display and Further Criteria

The physical dimensions of the display are relevant for the application and positioning of the display. Some other criteria, such as power consumption, emissions into the environment (e.g., heat, radiation), and susceptibility to influences of the environment are equally important when deciding on a display.

Current displays for PCs are either CRTs or flat panel LCD monitors, the latter of which is enjoying increasing market share (both are discussed later in this chapter; see Technologies). The main arguments for the flat panel displays are their smaller footprint, lower power consumption, and lower heat production. Price and a wider range of models favor the CRT. For group presentations in particular, projection displays are a common choice. Although their use requires a projection screen, the projectors have been shrinking in size and are highly transportable.

As part of augmented reality and virtual reality (both are discussed in Displays for Selected Applications) head-mounted displays (HMDs) and Computer Animated Virtual Environments (CAVEs) are the most common display systems. Head-up displays (HUDs) are used in military applications, research, and a few commercially available road vehicles. Issues of size and power consumption are of great importance in handheld and mobile computer applications, for which LCDs are currently the display of choice for most applications.

QUALITY CRITERIA FOR VISUAL DISPLAYS

Luminance

There are many concepts and terms that relate to the measurement of light (Sanders & McCormick, 1993). Common photometric quantities and units are shown in Table 9.1 (Cakir, Hart, & Stewart, 1979).

TABLE 9.1. Photometric Quantities and Units

Quantity	Symbol	Unit	Abbreviation
Luminous flux	Φ	Lumen	lm
Luminous intensity	I	Candela	cd = lm/sr
Illuminance	E	Lux	lx = lm/m^2
Luminance	L	Candela/m^2	cd/m^2

The fundamental photometric quantity is luminous flux (Φ), which describes the rate at which light energy is emitted from a source. Luminous flux is a concept similar to other flow rates, such as liters per minute (Sanders & McCormick, 1993). The luminous sensation caused by physical power is converted to a luminous power scale using the unit *lumen* instead of watts (Bosman, 1989).

Luminous intensity (I; also called radiant intensity) is the emitted flux per solid angle ω:

$$I = \frac{\Phi}{\omega} \tag{9.1}$$

It is measured in candela; the unit is lumen per steradian. The illuminance E is the luminous flux through a surface area of 1 m^2 (Bosman, 1989):

$$E = \frac{\partial \Phi}{\partial A} \tag{9.2}$$

It is measured in terms of luminous flux Φ per unit area A. One lumen per square meter is called a lux (lx). Often, the older unit, footcandle (fc), is used, which means lumen per square foot (lm/ft^2). One footcandle equals 10.76 lx (Sanders & McCormick, 1993). According to the German standard DIN 5035, the illumination in offices with computers should be no less than 500 lux and, for other offices, no less than 300 lux (LfAS, 2000).

Luminance (L) is a measure of intensity of light emitted from a light source per unit surface area normal to the direction of the light flux (Cakir et al., 1979). In this context, A is the area of the surface of the source and θ is the angle between the perpendicular from the surface and the direction of measurement (Boff & Lincoln, 1988):

$$L = \frac{I}{A} \cdot \cos\theta = \frac{\Phi}{\omega A}\cos\theta \tag{9.3}$$

The unit of luminance is lumen per steradian per square meter or candela per square meter (cd/m^2). Luminance is an important quantity. According to the German and European standard DIN EN 29241-3, the luminance of the display must be at least 35 cd/m^2, and brightness and contrast (see the following sections of this chapter) of a display must be adjustable. The background luminance of displays using dark symbols on a bright background should be at least 100 cd/m^2 (LfAS, 2000). Considering the difference sensitivity $L/\Delta L$, however, it makes no sense to increase the luminance above a certain level (see Fig. 9.6). At first, the difference sensitivity increases with an increasing viewing field luminance L_a. After an optimum level is reached, it declines with further increases in luminance (Hartmann, 1992) because of glare (see Glare).

Contrast

Strictly speaking, contrast is not a physiological unit but is nevertheless one of the most important photometry quantities insofar as it is related to many visual functions, such as visual acuity,

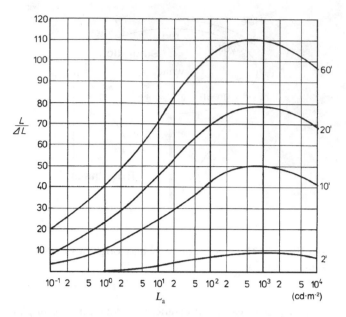

FIGURE 9.6. Difference sensitivity $L/\Delta L$ for four viewing objects of varying size over viewing field luminance L_a (Hartmann, 1992).

contrast sensitivity, speed of recognition, and so forth. There are different definitions of contrast. The *Weberscher Contrast* C_w (luminance contrast) after the International Lighting Commission is defined as follows:

$$C_w = \frac{L_{object} - L_{background}}{L_{background}} \tag{9.4}$$

The definition implies that there is a negative luminance contrast by presenting a dark object on a bright background and a positive contrast by presenting a bright object on a dark background (Ziefle, 1998b). This formula is not symmetric; if the luminance of the object is much greater than the background luminance, the numerical values of contrast are large and increase rapidly. If the background luminance is much greater than the object luminance, the luminance contrast tends to move asymptotically to the value $C_w = -1$ (Cakir et al., 1979).

The *Michelson Contrast* C_m (also called modulation contrast, depth of modulation, or relative contrast about the mean luminance) is generally used for periodic stimuli that deviate symmetrically above and below a mean luminance value (e.g., gratings or bar patterns) and is computed as follows (Boff & Lincoln, 1988):

$$C_m = \frac{L_{max} - L_{min}}{L_{max} + L_{min}} \tag{9.5}$$

L_{max} and L_{min} are maximum and minimum luminances in the pattern. The Michelson Contrast will take on a value between 0 and 1 (Flügge, Hartwig, & Weiershausen, 1985).

Sanders & McCormick (1993) specified another popular possibility of definition of the contrast (called *luminous contrast*) as the difference between the maximum and minimum luminance in relationship to the maximum luminance:

$$C_l = \frac{L_{\max} - L_{\min}}{L_{\max}} \qquad (9.6)$$

For a simple luminance increment or decrement relative to the background luminance (such as a single point), the *contrast ratio* is used (Boff & Lincoln, 1988). This simple ratio of two luminances is widely used in engineering (Bosman, 1989):

$$C_r = \frac{L_{object}}{L_{background}} \qquad (9.7)$$

These contrast measures can be converted from one to another. For example, given the contrast ratio C_r and knowledge of positive contrast conditions ($L_{\max} = L_{object}$; $L_{\min} = L_{background}$), the other visual contrasts can be calculated as follows:

Weberscher Contrast:

$$C_w = C_r - 1 \qquad (9.8)$$

Michelson Contrast:

$$C_m = \frac{C_r - 1}{C_r + 1} \qquad (9.9)$$

Luminous Contrast:

$$C_l = \frac{C_r - 1}{C_r} \qquad (9.10)$$

In a computer display, the contrast ratio of symbols to background has to be more than 3:1 respectively 1:3. The symbols must be represented sharply up to the edges of the screen. Usually, the contrast ratio for CRTs is 1:10 and for TFT LCDs, 1:50. Furthermore, the background luminance of the screen has to be evenly distributed. A maximum luminance ratio of 1.7:1 between the middle of the screen and any other part of the screen is tolerable (LfAS, 2000).

Glare

When the range of luminance in the visual field is too great, visual adaptation is disturbed and glare occurs. Glare caused by light sources in the field of view is called *direct glare*; glare caused by light being reflected by a surface in the field of view is called *reflected glare* (Fig. 9.7). Reflected glare can occur from (Sanders & McCormick, 1993) specular (smooth, polished, mirrorlike) surfaces, spread (brushed, etched, or pebbled) surfaces, diffuse (flat or matte) surfaces, or as a combination of the above three (compound). It increases when the luminance of the luminous surface or its size within the visual field is increasing (Cakir et al., 1979). Glare decreases when the glare source is more remote or when the areas surrounding them are made brighter. Experiments also show that visibility is decreased by glare, and the decrease is greatest when the source of the glare is in the line of vision (Boff & Lincoln, 1988). Therefore, no light sources should be placed behind the display, and displays should not be placed near the windows. It is advisable to position the display

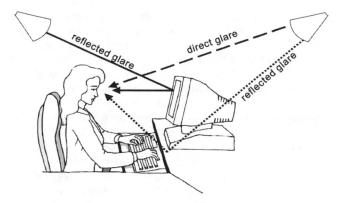

FIGURE 9.7. Direct and reflected glare.

at a right angle to the window (so that the line of vision is parallel to the window). The display also can be protected with curtains, blinds, or movable walls (LfAS, 2000). Lamps that can reflect in the monitor may not have a mean luminance of more than 200 cd/m² , and the maximum luminance must be less than 400 cd/m² according to German standard DIN 5035-7 (Fig. 9.8).

Glare also can be classified according to its effects on the observer. Discomfort glare produces visual discomfort but does not necessarily interfere with visual performance. For example, glare from interior lighting sources may often be a source of discomfort but is seldom a disability. Disability glare reduces visual performance and is often accompanied by discomfort. Both types of glare may, but need not, occur simultaneously (Cakir et al., 1979). Blinding glare is so intense that for an appreciable length of time after it has been removed, the viewer cannot see (Sanders & McCormick, 1993).

Readability

Reading a text from a visual display terminal (visual display unit or computer screen) is not the same as reading a text from a hardcopy (Sanders & McCormick, 1993). Various tests have shown that people proofread more slowly when the material is

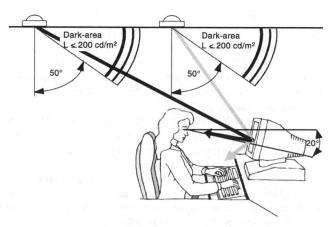

FIGURE 9.8. Glare can be avoided by limiting the lamp luminance and by correct positioning of the display.

presented on a CRT or LCD than when it is presented on hardcopy (Ziefle, 1998b). The reason for reading more slowly from a visual display than from hardcopy appears related to image quality (Gould et al., 1986, 1987; Harpster, Freivalds, Shulman, & Leibowitz, 1989). The characters or images on a visual display are formed when combinations of elements in a matrix (called pixel) are on or off (Sanders & McCormick, 1993). Today, typical resolutions of desktop computer monitors for word processing are 1,024 × 768 Extended Graphics Array (XGA), or 1,280 × 1,024 Super Extended Graphics Array (SXGA). The higher the resolution of the screen, the less is the difference between reading from a visual display and reading from hardcopy (Sanders & McCormick, 1993). A simple calculation helps to understand this concept. Using a 17-inch computer monitor with an XGA resolution and a viewing distance of 50 cm the visual angle of 1 pixel on the monitor is approximately 2 min of arc. The visual resolution of the human visual system (of normal-seeing people) is 1 min of arc. That means that the user of the display can see each pixel of the letters. In comparison, the printout from a modern laser printer has a resolution of 600 dpi; for the same viewing distance, the visual angle of 1 pixel is 0.3 min of arc. In this case, the visual system in not able to distinguish the separate pixels of the printed matrix.

Characters' readability also depends on polarity (positive or negative contrast condition). A bright background reduces glare and the visibility of reflections on the screen, but it increases the likelihood of perceiving flicker (Sanders & McCormick, 1993). Because the higher the background brightness the greater the sensitivity to flicker, displays with bright backgrounds must have a higher refresh rate than screens with dark backgrounds to avoid perceptions of flicker.

There are many more influences on readability, such as typography, reading distance, character size, hardware considerations (e.g., refresh rate), and color conditions. The interested reader is referred to Sanders & McCormick (1993).

TECHNOLOGIES

Lamps

Lamps can be classified into incandescent and gas discharge lamps, based on the physical principle of light generation. In an incandescent lamp a current is passed through a thin wire filament, which causes the wire to glow to white heat due to resistance. To prevent the filament from burning, it is sealed in an evacuated glass bulb. The spectrum of the emitted light is a function of the temperature of the filament, from red (1,500 K) over yellow and white to blue (6,500 K). Incandescent lamps are not very efficient; only about 5% of the energy is emitted as light, and the rest is emitted as heat.

In gas discharge lamps, the atoms or molecules of the gas inside a glass, quartz, or translucent ceramic tube are ionized by an electric current passing through the gas. Photons are emitted when electrons of the gas molecules, temporarily displaced from their respective energy levels, return to their old energy level. Most photons emitted are within the ultraviolet spectrum, so an additional transformation takes place in a coating inside the tube, transforming ultraviolet into visible light. The color of gas discharge lamps depends mainly on the mixture of gasses and the coating of the tube.

Lamps used as displays are often colored to make them more conspicuous or as a means of coding.

Light Emitting Diodes (LED)

LEDs date back to the beginning of the 19th century, when it was discovered that yellow light is emitted when a current passes through a silicon carbide detector (Brown University, 1999). In the 1960s and 1970s, the LEDs matured and became the standard display, first for the electronic calculator and shortly after for the watch. In both markets LEDs were eventually replaced by LCDs. Their role in human–computer interaction is currently restricted to serving as a simple status indicator for power or hard drive access. Recent technical breakthroughs, among them the availability of LEDs in almost any color, have led to the development of large public displays based on LEDs, and full-color miniature LED matrices for head-mounted and head-up displays are currently being developed.

Cathode Ray Tube (CRT)

The CRT is probably the most well-known visual display technology. It is the major component of television sets and most computer monitors. The CRT dates back to 1897, when Ferdinand von Braun invented the so-called Braun'sche Röhre (Oscilloscope). In 1998, 250 million displays for televisions and computers were produced worldwide, 90% of which were CRTs (Blankenbach, 1999). Figure 9.9 shows the major parts and components of a CRT (Chadha, 1995). A CRT consists of a glass bulb, a cathode, an electron gun, deflection coils, a mask, and a phosphor coating. The major component of the cathode ray tube is

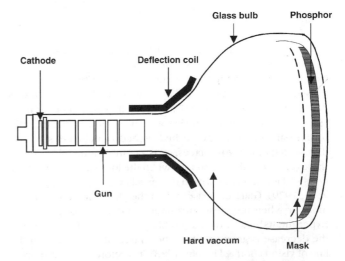

FIGURE 9.9. Major parts and components of a cathode ray tube.

an evacuated flask. At the back of the flask a heated cathode is charged with high voltage in the range of 10 kV to 30 kV. This cathode emits negatively charged electrons, which are attracted and accelerated by an anode that is located in front of the screen. The electron beam is diverted by an electromagnetic field, built up by the deflection coils, and thus directed toward the screen. A screen mask is attached in front of the ground glass plate, so that the electron beam is focused and then steered on the phosphorus layer, deposited on the front surface. As a consequence, the outside areas of the beam are prevented from mistakenly hitting adjoining phosphorus dots, which would lead to a blurred representation and chromatic distortion. Display screen masks are either dot mask screens, Trinitrons, or slot mask screens.

The electron beam is directed line-by-line across the screen and stimulates the phosphor coating to light up for a brief moment. The intensity of the light is modulated by the voltage applied at the cathode. The more voltage applied, the brighter the light shines.

Color representation is obtained both on the type of phosphor used and the electron beam. The phosphorus layer is made of a triad in the form of an array, consisting of a dot of each of the colors red, green, and blue. Color monitors therefore have three electron guns. The electron beam of each falls only on one colored dot in each triad (Bosman, 1989). Three dots, of one triad each, arc 1 pixel. With different light intensities, almost every mixture of colors is possible.

Display Screen Masks

Dot Mask Screen. As the name suggests, a thin metal or ceramic screen with multiple holes is used in a dot mask screen (Fig. 9.10). The shadow mask technique is applied, in which three electron beams pass through the holes and focus to a single point on the tube's phosphor surface. Thus, the other display screen dots are shaded. The electron guns are arranged in the

FIGURE 9.11. Slit mask arrangement.

form of a triangle (delta gun), which requires a high amount of adjustment, because insufficient adjustment will cause color defects. The measurement of the dot pitches of dot-type screens is taken diagonally between adjoining dots of the same color. Because of their horizontal measurement they cannot be easily compared with the distance between the dots of a Trinitron or a slot mask screen (Blankenbach, 1999).

Slit Mask (Trinitron Technology). Monitors based on the Trinitron technology use an aperture grill instead of a shadow mask (Fig. 9.11). The phosphorus surface does not consist of colored dots but a multitude of tiny vertical phosphorus wires. As with the dot mask, their arrangement is alternating. The electron guns are on a single line, easing the adjustment effort (Precht, Meier, & Kleinlein, 1997). Instead of a dot-type screen, the Trinitron has vertically taut wires. In contrast to the dot-type screen, the equivalent to the dot pitch, the stripe pitch (SP) of these monitors is measured by the horizontal distance between wires. Slit masks are relatively insensitive to the warmth that develops during use, because they do not bend, but merely change their length. To mechanical load (e.g., vibration) they are very sensitive. Depending on the size of the monitor, either one or two horizontal holding wires are used for stabilization. These wires are recognizable (e.g., on the Trinitron, they are visible as small gray lines in the top and down third).

Slot Mask Screen. This screen technology (Fig. 9.12) also has tiny phosphor wires, similar to monitors based on Triniton technology. The openings in the shadow mask are executed as slots (Precht et al., 1997).

Resolution. The resolution of a monitor mainly depends on its screen diagonal and its dot pitch or stripe pitch (SP). Because of the tubes' construction it is not possible to use the whole screen diagonal as viewable screen. Otherwise, the picture would not be rectangular. For these kinds of calculations, the effective screen diagonal D_{eff} is used. Typical values for the

FIGURE 9.10. Dot mask arrangement.

FIGURE 9.12. Slot mask arrangement.

effective screen diagonal depending on the screen diagonal, are shown in Table 9.2 (Matschulat, 1999).

The highest resolution ($x_{max} \times y_{max}$) that can be represented on the monitor can be calculated as follows:

$$x_{max} = \frac{D_{eff} \cdot \cos \alpha}{SP} \qquad (9.11)$$

The angle α (angle between the horizontal and the screen diagonal) can be calculated with the knowledge about the aspect ratio between the vertical (V) and horizontal (H):

$$\alpha = \arctan \left(\frac{V}{H} \right) \qquad (9.12)$$

As an example, a 17-inch monitor ($D_{eff} = 15.8$ inches) with a horizontal SP of $SP = 0.26$ mm and an aspect ratio $H{:}V = 3{:}4$ (common computer monitor) a maximum number of dots presentable in the horizontal of 1,235 respectively 926 dots in the vertical is possible. This means for our example, that with a resolution of 1,024 × 768 on the graphic card, every dot that has to be presented, and this is accomplished by at least one phosphorus wire. But with a resolution of 1,280 × 1,024, not all dots can be represented clearly.

Frequencies and Rates. The number of times that the image on the display is drawn each second is called the *display refresh rate* (also called *frame rate* or *vertical scanning rate*). The unit of measurement is Hertz (Hz). A high display refresh rate prevents a flickering image, because there is only a small amount of time between two successive stimulations of a single dot. A refresh rate of between 70 and 80 Hz is needed to ensure a flicker-free image on the display (Bosman, 1989). High display refresh rates reduce visual problems and improve the reading rate (Ziefle, 1998b).

The phosphorus dots are stimulated by the electron beam. Because they light up for only a fraction of a second, this stimulation has to happen several times a second. It passes line by line while the electron beam is writing from the left to the right and then returns to the beginning of the following line. Line building begins in the upper left-hand corner, line by line, until each dot has been stimulated one time. Then the electron beam goes back to the upper left-hand corner and begins to build the picture again.

The line frequency (or horizontal frequency, measured in kHz) is another characteristic of CRT displays. It measures the number of lines the electron beam can draw per second. In this case, line frequency, display refresh rate, and resolution are directly connected with each other. A monitor having a display refresh rate of 70 Hz and a resolution of 1,024 × 768 needs to have an electron beam that is capable of drawing 70 × 768 lines = 53,760 lines per second. This means that the monitor has to process a line frequency of at least 53.76 kHz. In case one wants to use a higher resolution, it is necessary to check whether the monitor is capable of processing such a line frequency so as not to damage it.

In addition to the frequencies described, the image representation mode can also lead to poor-quality representation. There are two image representation modes: interlaced and noninterlaced. With the interlaced-mode, the image formation is divided into two half images. Here, the electron beam first builds all uneven lines and then all even lines. This is the usual process for televisions (e.g., for Phase Alternation Line (PAL) 50 half images per second are normal, which means a vertical scanning frequency of 50 Hz, but a full screen frequency of only 25 Hz). Nowadays, monitors with noninterlaced modes are commonly used. In this case, the electron beam builds all lines one after another, without exception.

Representation Errors and Adjustment Possibilities. Almost every CRT monitor allows adjustment of brightness and contrast. Higher quality computer monitors are equipped with additional adjustments for horizontal and vertical image positioning and image geometry to prevent errors such as barrel distortion, pincushion distortion, trapezoid distortion, parallelogram distortion, or seagull distortion. Furthermore, diversion errors can be caused by an asymmetric deflection current.

Linearity Error. Geometric elements and letters should be represented at the same size everywhere on the screen. If displays are adjustable only vertically, problems can arise. An error of deflection can cause the marginal, equispaced raster elements to disperse. Therefore, the figures' widths change constantly as they move horizontally from the middle of the screen. This is called *symmetrical linearity error.* The pin-cushion correction can adjust this problem by over coming the asymmetrical error in the horizontal direction using a linearity coil. An adjustable

TABLE 9.2. Screen Diagonal of the Tube and Effective Screen Diagonal

Screen Diagonal	Effective Screen Diagonal D_{eff}
15 inches	12.78 inches, 32.5 cm
17 inches	15.76 inches, 40.0 cm
19 inches	18.20 inches, 46.2 cm
21 inches	19.98 inches, 50.7 cm

resistor placed in the vertical deflection control works against the asymmetrical error in the vertical direction.

Convergency. Color representation is obtained with three electron beams and three phosphorus layers, one each for the colors red, green, and blue. Each electron beam hits one layer. Congruent dots of the layers build a pixel that can be seen in any color, depending on the intensity of the beams. If the electron beams are not adjusted exactly, misconvergency is the consequence. "Ghosting" of the removed color arise, causing a blurred and distorted image. Misconvergency can also arise from environmental electromagnetic disturbances.

Moiré. Certain color combinations and pixel arrangements can cause interference called moiré. The interference can result when a mask is deposited imprecisely. The electrons therefore pass imprecisely through the mask, and streaks are seen on the screen. For this reason, some monitors are equipped with moiré-reduction, which allows the user to remove the streaks. High-grade monitors are able to maintain the focus at the same time.

Liquid Crystal Display (LCD)

LCDs have become increasingly viable in recent years and now tend to surpass CRTs. Early LCD commercial developments concentrated on small numeric and alphanumeric displays, which rapidly replaced LEDs and other technologies in applications such as digital watches and calculators (Bosman, 1989). Now there are even more complex displays for use in many applications, such as laptops, handheld computers, flat panel displays (FPDs), head-mounted displays (HMDs), and miniature televisions (e.g., those in seats on airplanes). Experiments have shown that people experience improved readability and less eye strain using LCDs compared with CRTs (Ziefle, 1998b). Other advantages of LCDs are as follows (Bosman, 1989):

- Power consumption is minimal.
- They operate at low voltages.
- In normal environments, their lifetimes can be long.
- Displays may be viewed either directly in transmission or reflection, or they may be projected onto large screens.
- The display does not emit light.

LCDs consist of two glass plates with microscopic lines or grooves on their inner surfaces and a liquid crystal layer between them (Fig 9.13). Liquid crystal materials do not emit light, so external or back illumination must be provided. The physical principle is based on the anisotropic material qualities of liquid crystals. When substances are in an odd state that is somewhat like a liquid and somewhat like a solid, their molecules tend to point in the same direction, like the molecules in a solid, but can also move around to different positions, like the molecules in a liquid. This means that liquid crystals are neither a solid nor a liquid but are closer to a liquid state. In an electric field, liquid crystals change their alignment and therefore their translucence. If no voltage is switched on, the light can pass through,

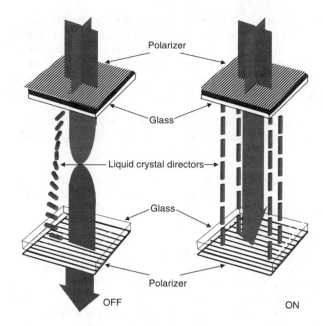

FIGURE 9.13. Principles of operation of a twisted nematic LCD.

and the pixels appear bright. When voltage is switched on, the pixels become dark (Matschulat, 1999).

The most common LCD is called the twisted nematic (TN) display. The microscopic lines of the glass plates are arranged orthogonally to each other, and the glass plates serve as polarizers (Precht et al., 1997). Their directions of translucence lie at right angles on top of one another, so that no light can pass through them. Because of the fine grooves on the inner surface of the two glass panels (arranged vertically on one panel and horizontally on the other), the liquid crystal is held between them and can be encouraged to form neat spiral chains. These chains can alter the polarity of light. In the so-called nematic phase, the major axes of the crystal's molecules tend to be parallel to each other.

Nonpolarized light from background illumination can pass the polarization filter with just one plane of polarization. It is twisted about 90° along the helix and can thus pass through the second polarization layer. The display appears to be bright when there is no electric current. By applying an electric current to twisted nematics, they untwist and straighten, changing the angle of the light passing through them so that it no longer matches the angle of the top polarizing filter. Consequently, no light passes through that portion of the LCD, which becomes darker than the surrounding areas, and the pixel appears black.

By applying different electric currents, gray scales can be produced with LCD technology. A disadvantage is the relatively low contrast, but this can be improved by applying a steep electro-optic characteristic line of the liquid crystals, and then low voltage is sufficient to change the translucence, causing the liquid crystals to twist the light by more than 90 degrees. Such displays are called super twisted nematic (STN), for example, double super twisted nematic (DSTN) or triple super twisted nematic (TSTN) (Schadt, 1996).

Traditionally, each pixel is addressed by horizontal and vertical electron stripes (passive matrix) which results in a poor contrast image. The electric field, however, expands over the pixel to be addressed to the entire horizontal and vertical electrodes. This results in disturbing stripes, or "ghosting". Other disadvantages are slow response times and prelighting of the pixels.

Colors can be produced by RGB-filters put in front of the pixels. A disadvantage is the relatively small translucence caused by the color filter and the polarizers.

Active Matrix LCD. By depositing an array of thin film transistors (TFTs) on the LCD, one per pixel, the problems mentioned above can be overcome. By turning on these transistors, the charge can be maintained, and the display contrast remains stable. To display a full range of colors, each element on an LCD is assigned to a primary color by a special filter placed on the face plate of the LCD. To guarantee high display contrast quality, brightness, and high quality of color representation, many transistors are needed. For a resolution of $1,024 \times 768$ pixel, 2.36 million transistors are required—one per primary color and subpixel. Dot pitch on LCDs is measured horizontally as the distance between two elements.

Scaling. The resolution for CRTs can be easily adjusted by changing the gating electronics, whereas the maximum resolution for TFTs is predetermined by the traditional pixel-by-pixel addressing. A smaller resolution has to be extrapolated with the scaling factor. Integer variables simply have to be doubled. Other scaling factors, however, are more complicated because then the scaling algorithm has to decide whether one or two pixels must be addressed, causing side effects. For this reason, only the necessary pixels are addressed if a lower resolution is desired. The pixels that are not required appear black, and the screen diagonal is reduced.

Representation Errors. The traditional errors caused by incorrect deflection do not occur because of the fixed matrix, but defective transistors can cause errors, leading to permanently bright or dark pixels. These pixels are distracting, particularly when all three transistors of an elementary cell are switched to bright, which causes a white pixel to appear on the screen. Furthermore, the representation of moving pictures is restricted. The response latency of liquid crystals is about 20 to 30 ms. Fast sequences (e.g., scrolling) are blurred.

LCD monitors with a standard Video Graphics Array (VGA) plug have to convert the analog VGA-signal back into a digital signal. By the use of a wrong A/D changer (e.g., an 18-bit A/D changer), the color representation can be affected, although the LCD monitor is actually capable of representing True Color (24 Bit).

CRTs easily represent black; the cathode simply does not emit electrons. LCDs, however, have to completely block out the light from the backlight. Technically, this is impossible, and as a consequence the contrast is reduced.

The technology of LCDs leads to another problem: When a display is viewed from an angled position, it appears darker and color representation is distorted. Recently developed technologies, such as in-plane-switching (IPS), multi-domain vertical alignment (MVA) and thin film transistors (TFT) have improved the width of the viewing angle. Worth mentioning is the welcome effect of the restricted viewing angle for privacy purposes, as in the case of automated teller machines.

Finally, the nominal screen diagonal of a LCD is equivalent to the effective screen diagonal. In contrast, a CRT's nominal screen diagonals are smaller throughout (see Table 9.2).

Plasma Display Panels (PDP)

The oldest electro-optical phenomenon able to produce light is an electrical discharge in gas (Bosman, 1989). Millions of years elapsed before this effect was identified, analyzed, and mastered by humans. The first attempts to produce a matrix display panel were made in 1954. Since then, research has continued, and a host of approaches have evolved (Bosman, 1989). The beauty of this technique is that, unlike front-view projection screens, one does not have to turn off the lights to see the image clearly. Therefore, plasmas are excellent for video conferencing and other presentation needs (Pioneer, 2001).

In plasma technology, two glass plates are laid with their parallel thin conducting paths at right angles to one another (Precht, 1997). The gap between the plates is evacuated and filled with a gas mixture (see Fig. 9.14). If sufficient high voltage is applied to the cross point of two orthogonal conducting paths, the gas ionizes and begins to shine (like many little gas discharge lamps). Color representation, however, is not easily obtained.

The inside of one glass plate is coated with a phosphorus layer, which is, according to the fundamental colors, composed of three different kinds of phosphorus. To trigger the PDP, there is a wired matrix below the phosphorus layer. The space between the glass plates is divided into gas-filled chambers. When voltage is applied to the wired matrix on the bottom of the display, the gas is transformed into a plasmatic state and emits ultraviolet radiation, causing the phosphorus to glow.

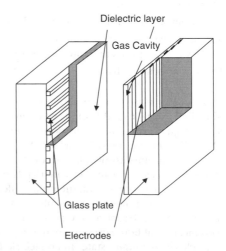

FIGURE 9.14. Major parts and components of a plasma display panel.

The advantage of this technology is the flat construction of large screens that perform extraordinarily well under most ambient light conditions. For example, even very bright light does not wash out the image on the screen. Another characteristic of a plasma panel is the extreme viewing angles both vertically and horizontally. With a 160° viewing angle, people sitting to the side of the screen will still be able to see the image without losing any information (Pioneer, 2001).

Plasma screens do have disadvantages. First, they consume a large quantity of electricity, making this technology unsuitable for battery-operated and portable computers. Furthermore, high voltage is required to ignite the plasma, and the phosphorus layers degrade over time. Finally, the pixels are oversized, so the user must be situated at a distance from the display.

Electroluminescent Displays (ELDs)

Similar to a CRT, ELDs rely on a phosphorous layer to emit visible light, but in the case of an ELD, a thin film of phosphorescent substance is sandwiched between two conducting electrodes. The top electrode is transparent. One of the electrodes is coated with vertical wires and the other with horizontal wires, forming a grid. When an electrical current is passed through a horizontal and a vertical wire, electrons are accelerated and pass through the phosphorous layer. The electrons excite the activator ions in the phosphorous layer as a result of direct impact. As in the CRT, the relaxation of the excited ions leads to the emission of visible light (Chadha, 1995; cf. Fig. 9.15).

Color ELDs can be manufactured in different ways, but the basic construction is similar to the monochrome displays. For a single-layer display, a phosphor is chosen that emits a large number of spectral components and that appears white. Additionally, a layer of RGB color filters is inserted between the transparent electrode layer and the glass substrate. So, depending on the color of the filter in front of it, pixels emit one of the RGB colors. Double-layer, full-color ELDs consist of two "displays" put in front of each other. The "display" at the back is similar to the color display described above, but with only blue and green filters used and without the front glass substrate. A layer of silicone oil insulates it from the second "display" that produces the red color. It consists of a transparent electrode, a dielectric, a red

light emitting phosphor, another dielectric, another transparent electrode, and the glass substrate (see Chadha, 1995).

The development of new, organic electroluminescent materials has helped to overcome some of the shortcomings of earlier ELDs, in particular the presentation of a wide range of colors with uniform levels of brightness and longevity. In addition, organic electroluminescent displays require considerably lower voltages and power than the older inorganic types (e.g., 15 V and 3 W instead of 100 V and 6 W; Ness Corporation, 2000).

Currently, ELDs are manufactured in sizes ranging from 1 to 18 inches with resolutions from 50 to 1,000 lines per inch (Rack, Naman, Holloway, Sun, & Tuenge, 1996). The brightness achieved with an (inorganic) ELD is reported to be 150 cd/m^2. At 170°, the viewing angle is significantly larger than the viewing angle of common LCDs, and the response time is 1 ms compared with the typical 20 ms for an LCD display (Grossman, 2000). The images are of high contrast (>7:1 at 500 lx; NN, 2001) and crisp quality (Chadha, 1995). Power consumption of the organic ELD is the lowest of all light-emitting flat panel displays but higher than for a LCD. ELDs operate well in harsh environments and over a wide temperature range (e.g., −40° to 85°C; Grossman, 2000), therefore ELDs are expected to be used in automobiles, personal digital assitants (PDAs), portable computers, and for televisions.

One type of ELD, the AC powder ELD, is used as backlight for LCD applications. The construction consists of a clear glass or plastic substrate, a transparent electrode, a layer of ZnS powder dispersed in a dielectric, an optional insulating layer, and an aluminum electrode. They emit monochrome light for which the wavelength (color) depends on the phosphor used.

Laser Display Technology

LASER is the acronym for light amplification by stimulated emission of radiation. Laser light is generated in the active medium of the laser. Energy is pumped into the active medium in an appropriate form and is partially transformed into radiation energy (Institut für Lasertechnik, 2001). In contrast to thermal emitters (normal lamps), a laser emits a concentrated and monochromatic light with high local and temporal coherence (see Fig. 9.16).

In contrast to conventional light sources, laser light (König, 1990) has rays that are nearly parallel with each other (has a small divergence), is focusable down to a wave length, has a high power density, has a monochromatic character (light of one wavelength), and has high local and temporal coherence

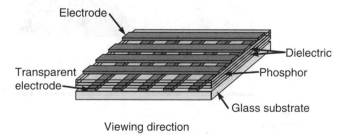

Viewing direction

FIGURE 9.15. Schematic drawing of the construction of a monochrome AC thin film electroluminescent display. The light is emitted from the phosphorous layer and passes through the transparent electrode and the glass substrate.

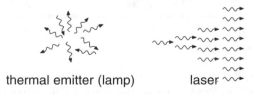

thermal emitter (lamp)   laser

FIGURE 9.16. Difference in dispersion of light between thermal emitters and laser light.

(the same phase). These characteristics can be used to form an image set up as in a conventional CRT system: the laser is directed line by line across the display area. The line diversion, horizontal and vertical, is achieved by the use of rotary or oscillating mirrors. Light output from the diode-pumped solid-state lasers is modulated according to the input signal, and red, green, and blue light is combined. This combined white light is then raster-scanned onto the screen to create an image. The laser projects the image either onto a flat area or directly to the retina (see the description of the virtual retinal display discussed later in Augmented Reality).

Electronic Ink and Paper

Currently two research groups are working on displays that promise to have paperlike properties: they can be viewed in reflective light, have a wide viewing angle, and are thin, flexible, and relatively inexpensive. Unlike paper, they are electrically writeable and erasable. One big advantage over other types of displays is their low power consumption, possibly extending the battery life of devices with such displays into months or even years. The approaches taken by the two research groups are somewhat different.

Based on developments of the Massachusetts Institute of Technology, E Ink Corporation uses an electronic ink technology. The principal components of electronic ink (see E Ink, 2001a) are tiny capsules, about 100 μm in diameter, filled with a clear fluid. Suspended in the fluid are positively charged white particles and negatively charged black particles. The particles will move toward an inversely charged electrode. So when a top transparent electrode is negatively charged, the white particles move to the top and become visible to the user. The black particles will at the same time move to a positively charged

TABLE 9.3. Comparison of Reflective Display Media (E Ink, 2001b)

Display Technology	White State Reflectance (%)	Contrast Ratio
Transflective Mono STN LCD (common PDA with touchscreen)	4.2	4.1
Transflective Mono TN LCD (common eBook with touchscreen)	4.0	4.6
E Ink (with touchscreen)	26.6	9.2
E Ink (no touchscreen)	38.1	10.0
Wall Street Journal Newspaper	61.3	5.3

bottom electrode, and the microcapsule appears white to the user. Similarly, the microcapsule will appear black when an electric field of the opposite direction is applied (see Fig. 9.17). The microcapsules are suspended in a 'carrier medium' that can be printed onto a sheet of plastic film (or onto many other surfaces). The film is laminated to a layer of circuitry that forms a pattern of pixels that can be controlled similarly to other displays.

A comparison of reflectance and contrast ratio of an electronic ink display prototype with different LCDs and paper is shown in Table 9.3 (E Ink, 2001b).

A current prototype of the display has a resolution of 80 dots per inch (dpi) (Deussen, 2001). E Ink Corporation predicts refresh rates of 1 to 10 Hz, allowing animation but not the display of video (Walker, 2000). Although currently only black-and-white versions of the display exist, work is underway for a color version.

An invention made 20 years ago at Xerox Palo Alto Research Center is the basis of Gyricon reusable paper (Fig. 9.18). It consists of an array of small balls of 0.03 to 0.1 mm in diameter embedded in oil-filled pockets in transparent silicone rubber. Every ball is half white and half black and can rotate freely. Each

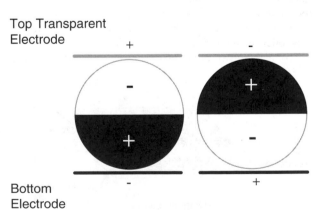

FIGURE 9.17. Operating principles of Electronic Ink (after E Ink, 2001a). Depending on the electric field applied to the transparent top electrode and the bottom electrode, either the positively charged white or the negatively charged black pigment chips will move to the top of the microcapsule.

FIGURE 9.18. Operating principles of Gyricon. Depending on the electric field applied to the transparent top electrode and the bottom electrode, either the positively charged black or the negatively charged white half of the little ball will move to the top of the microcapsule.

TABLE 9.4. A Selected List of National and International Standards, Regulations, and Seals of Approval Regulating Design and Use Aspects of Visual Displays

Ergonomics

Image quality	DIN EN 29241-3, ISO 13406-2, TCO'99, GS-Mark, Ergonomics Approved Mark
Reflection characteristics	DIN EN ISO 9241-7, ISO 13406-2, TCO'99, GS-Mark, Ergonomics Approved Mark
Color requirements	DIN EN ISO 9241-8, ISO 13406-2, TCO'99, GS-Mark, Ergonomics Approved Mark
Brightness and contrast adjustable	German ordinance for work with visual display units, European VDU directive 90/270 EEC, TCO'99, GS-Mark, Ergonomics Approved Mark
Tilt and swivel	German ordinance for work with visual display units, TCO'99, GS-Mark
Gloss of housing	German ordinance for work with visual display units, GS-Mark

Emissions

Noise	German ordinance for work with visual display units, European VDU directive 90/270 EEC, ISO 7779 (ISO 9296), TCO'99, GS-Mark
Electrostatic potential, electrical and magnetic fields	prEN 50279, TCO'99, Ergonomics Approved Mark
X-ray radiation	TCO'99

Energy Consumption

EPA Energy Star, TCO'99, VESA DPMS

Electrical Safety

TCO'99, GS-Mark

Documentation

Technical documentation and user manual	German Equipment Safety Law, GS-Mark

ball has an electric charge, and when an electric field is applied to the surface of the sheet, the balls rotate showing either their black or white side and attach to the wall of the sheet. They remain there displaying the intended image as long as no new electric field is applied (Gibbs, 1998).

A current prototype has a resolution of 200 dpi; future commercial versions are expected to have a resolution of 300 dpi (Walker, 2000). Bicolor versions other than black and white are conceivable, and full-color versions are in development.

Applications for electronic paper include products such as books and newspapers, rewriteable office paper, material for price tags and retail signs, wall-sized displays, foldable displays, and alternative displays for PDAs, mobile phones, and electronic readers.

STANDARDS

There exist a great number of national and international standards, regulations, and seals of approval that regulate diverse aspects of the design and use of visual displays. Among the areas covered are ergonomics, emissions, energy consumption, electrical safety, and documentation. Owed to constant technological advancement, only selected standards are listed in Table 9.4. Up-to-date information can be found on the Web sites listed in Table 9.5.

DISPLAYS FOR SELECTED APPLICATIONS

Augmented Reality

Augmented reality (AR) means the enrichment of the real world with virtual information (Azuma, 1997). For example, repair instructions for a machine tool or important installation tips can be placed directly into the worker's field of view (Plapper, Wenk, & Weck, 1999; see chapter 32, User-Centered Interdisciplinary Design of Wearable Computers, of this handbook for examples of such systems). One way to merge virtual information with the real world is to use a head-mounted display (HMD). An HMD can be worn like normal eye glasses. The information is superimposed onto the user's field of view via two integrated LCDs. Superimposition can happen in two ways: HMD with see-through mode and HMD with non-see-through mode, also called optical-see-through or video-see-through (feed-through), respectively (see Fig. 9.19).

Half-silvered mirrors must be installed to be able to see the real world and the virtual display information. Unfortunately, this reduces the see-through features of the glasses and increases size and the weight of the HMD (Azuma, 1997). Another problem is the time lag between the real-world information and the virtual information that is blended into the field of view (FOV), caused by the computing time necessary to generate the image. Another problem of currently available HMDs is the limited FOV of about 30°. The user cannot perceive actions taking place outside the FOV because of the construction of the HMD, making them unsafe and ineffective, especially in industrial areas (Oehme, Wiedenmaier, Schmidt, & Luczak, 2001) because the user has only a limited peripheral view of the real world.

Using an HMD with video-see-through, the viewer sees the real world and the virtual world through a monitor (one monitor for each eye). This means, that the viewer sees a quasi-real-time video film of the real world with superimposed virtual objects. This method allows a synchronization of the real world and the represented virtual information, but the viewer perceives both with a time lag. Another disadvantage is the loss of information caused by the fact that the real world's resolution is limited to the maximum resolution of the displays.

Another possibility to realize visual overlay is to use a virtual retinal display (VRD), also called retinal scanning display (RSD).

TABLE 9.5. Web Sites that Provide Information About National and International Standards, Regulations, and Seals of Approval for Visual Displays

European directives	European directives regarding health and safety are available via the Web site of the European Agency for Safety and Health at Work: http://europe.osha.eu.int/legislation/
U.S. Environmental Protection Agency (EPA)	The EPA promotes the manufacturing and marketing of energy-efficient office automation equipment with its Energy Star Program: http://www.energystar.gov/
National laws and ordinances	National laws, directives, and regulations regarding health and safety of many European and some other countries are available via the Web site of the European Agency for Safety and Health at Work: http://europe.osha.eu.int/legislation/
International Organization for Standardization (ISO)	The ISO is a worldwide federation of national standards bodies from 140 countries: http://www.iso.ch
European Agency for Safety and Health at Work	European standards are available via the Web site of the European Agency for Safety and Health at Work: http://europe.osha.eu.int/legislation/standards
DIN (Deutsches Institut für Normung)	DIN is the German Institute for Standardization http://www.din.de
TCO (The Swedish Confederation of Professional Employees)	TCO developed requirements for PCs. The TCO'99 label specifies ergonomic, ecological, energy consumption, and emission requirements: http://www.tco.se
Video Electronics Standards Association (VESA)	VESA creates standards for transmissions between computers and video monitors that signal inactivity: http://www.vesa.org
German Safety Mark	The *German safety approval mark* shows conformity with the German Equipment Safety Law and signals that the product, as well as the user manual and production process, has been tested by an authorized institution such as TÜV Rheinland: http://www.tuv.com
Ergonomics Approved Mark	The *Ergonomics Approved Mark* demonstrates that a visual display terminal complies with numerous ergonomic standards. The test mark is devised by TÜV Rheinland: http://www.tuv.com

In contrast to the HMD, the VRD reaches the retina directly with a single stream of pixels, thus guaranteeing a clear, sharp projection of different kinds of information (Microvision, 2000). Besides, an additional projection surface inside the glasses becomes unnecessary. Because of the higher light intensity of a laser, the half-silvered mirrors commonly used for see-through HMDs can be optimized to a maximum translucence. Thus, the see-through qualities of the eye glasses can be improved. Furthermore, the display's maximum resolution is no longer determined by the tolerances of manufacturing of the display pixels, but by the control logic and deflection mirror's quality.

Field of View. Apart from the resolution, the FOV is another important characteristic of augmented reality devices. It describes the maximum angle with which an HMD can represent information. Most producers only quote the diagonal FOV (similar to the screen diagonal for normal monitors) so that the horizontal and vertical components still need to be itemized. With HMDs, the FOV is constant, whereas the FOV on normal monitors can be influenced by changing the distance between the screen and the viewer.

To guarantee a comparable impression of size from one display to another, the representational size should be given as angle of view, because FOV can vary from display to display by a factor of two.

Monocular Versus Binocular. Binocular HMDs provide one display for each eye, whereas monocular HMDs provide only one display for either the left or right eye. By using binocular displays, stereoscopic images can be presented by providing slightly different images to the left and right eye. Monocular

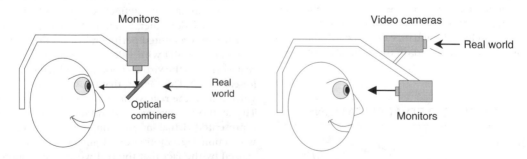

FIGURE 9.19. Optical-see-through (left) and video-feed-through (right) head-mounted displays.

displays should basically be used with the dominant eye. Unfortunately, this is not always possible because some HMDs do not allow one to install the displays on different sides of the device.

Virtual Reality

A virtual reality (VR) system "is a computer you can no longer see" (Williams, 1995). In contrast to AR systems, in a virtual reality system the viewer is immersed in an artificial "virtual" world, having no view of the real world. The left and the right eyes are provided with different image information, thus creating a three-dimensional view.

A commonly used output tool is the HMD. Two small displays (LCD or CRT) produce the image and send it to the eyes via a special optic. In this case, no half-silvered mirrors are used. Of course, a few problems arise because of not all users have the same distance between eyes (inter pupillary distance; IPD). Therefore, the HMD has to be adjustable to prevent differences in depth perception.

With current VR technology, only a limited FOV of 120° × 60° is possible, and the human FOV is much larger. A bigger FOV affords a higher amount of pixels per inch, but today this is limited by production capabilities.

Projector systems can reproduce large images of VR scenes on the front or back of one or more screens. Stereoscopic images for both eyes are projected either alternately or with different polarization at the same time. By using the first method, so-called shutter glasses alternately block each eye's view so that each perceives only the image intended for it. In this case, this is a disadvantage because either the refresh rate per eye or the vertical resolution is halved. Using the other method, both images, for the right and the left eye, are projected at the same time with different alignments (polarization). The images are separated by polarization glasses.

A further development in projectors is the CAVE (computer animated virtual environment). It consists of a cube with several panels onto which the images are projected from behind. Depending on the construction, there are C3 (two walls and the floor), C4, C5 or C6 respectively. A CAVE provides space for small groups, but the device can track and optimize the stereoscopic view for only one person. The others perceive distortions, especially on corners and edges. Current state-of-the-art in virtual environments technology, design, and implementation strategies, as well as health and safety concerns, are discussed in chapter 31, Virtual Environments, of this handbook.

Handheld and Mobile Devices

In recent years, there has been a movement away form desktop-based, general-purpose computers toward more task-specific information appliances. Mobile phones (cell phones) and personal digital assistants (PDAs) have increasingly become the focus of research as their dominance grows commercially (Schmidt et al., 1999). Chapter 32 of this handbook provides a good overview of "Designing Interfaces for Handheld Computers".

Display technologies are a significant component of PDAs and cell phones. People take their phones and PDAs everywhere, using them in various environments and situations to perform different tasks (Schmidt et al., 1999). Because of their mobility, such displays must have a small size and weight, as well as low energy consumption. This is why most of the displays for mobile units are monochrome LCDs. The displays also must be relatively resistant to impact and liquids.

The principal problems with such units are resolution and color (Weiland, Zachary, & Stokes, 2000). The resolution of the displays is restricted because of the small switch size and the manufacturing abilities of LCDs. Because of the higher energy consumption of color LCDs, color PDAs have a shorter operating time before recharge is required.

Newer technologies, such as HMD or VRD will undoubtedly improve to the level of desktop displays (Weiland et al., 2000). However, the currently available HMDs are either too big or too expensive (or both); it will take some time before such devices will provide a real alternative to LCDs. Another problem is the missing interaction concepts for HMDs. Usually, one interacts with a PDA by means of a touch screen. With an HMD, such an interaction in not possible because of the position of the HMD near the eyes.

VISUAL HAZARDS AND IMPAIRMENTS

Different areas of concern have been investigated in the last decades regarding the use of visual displays and human health. Among these are the following

Radiation and Fields

Almost the entire electromagnetic spectrum is emitted by visual display units (VDUs) especially CRTs, when the device is plugged in and operating. Visible light is emitted to form the intended image, but in addition, smaller amounts of long-wavelength ultra-violet (UV) and infrared (IR) radiation are generated. Electric and magnetic fields are emitted in three frequency ranges: Extremely low-frequency (ELF) fields of 50 or 60 Hz are generated by the power supply, the horizontal deflection coils emit fields between 15 and 35 kHz, and weak signals at higher radio frequencies (RF) are emitted from the VDU's electronic circuitry. Static electric fields are present and originate from the buildup of electric charge by electrons striking the front of the display screen. Very low-energy X-rays are produced inside the CRT, but the screen is made from leaded glass and thick enough to completely absorb them.

Exposure limits were set to protect the general population against biological effects of electric and magnetic fields. Since 1990, most video display terminal manufacturers have voluntarily complied with the limits set by the Swedish MPR2 recommendation and later with other even stricter standards. However, long-term health effects are believed to occur even at field strengths below these limits. A variety of health problems are attributed to weak electric and magnetic fields (e.g., "Hypersensitivity to Electricity" [EHS]; Müller, 2000).

Adverse Pregnancy Outcomes

In the late 1970s, it was suggested that working with a VDU could affect the outcome of a pregnancy. In Australia, Europe, and North America several adverse pregnancy outcomes (miscarriages or birth of malformed children) were noticed among women who worked with VDUs. This led to many epidemiologic and animal studies. Taken as a whole, these studies did not demonstrate any effects on pregnancy outcome due to EMF emitted from VDUs. Studies have suggested, however, that if there are effects on reproduction, they may be related to other work factors, such as job stress.

Eye Discomfort

A small number of reports describing dryness of the eyes following the use of VDUs have been documented. This can be caused by low relative humidity (20–30%) in the working area rather than by radiation emission from the VDU. In addition, eye fatigue can be caused by improper illumination. This confirms that the work environment should be within the recommended values, especially when working with VDUs.

Effects on the Skin

An excess of symptoms such as skin rashes or itching has been studied, particularly in Scandinavian countries. However, laboratory tests conducted on people with these symptoms showed their symptoms were not the result of EMF exposure.

Other Areas

Some individuals have experienced headaches or dizziness and musculo-skeletal discomfort. These are largely preventable if proper work environment and ergonomic measures are introduced for working with VDUs. Such measures include designing equipment, lighting, and other aspects of the environment to encourage proper posture and to reduce muscular and eye strain and other stress-producing tensions.

In summary, the work environment, including indoor air quality, job-related stress, and ergonomic issues such as posture and seating while using a VDU, are the determining factors of possible health effects associated with VDU use. These health concerns are discussed in greater detail in the proceedings of the Work With Display Units conference series (Berlinguet & Berthelette, 1990; Grieco, Molteni, Occhipinti, & Piccoli 1995; Knave & Widebäck, 1987; Luczak, Cakir, & Cakir, 1993; Miyamoto, Saito, Kajiyama, & Koizumi, 1997).

References

Azuma, R. (1997). A survey of augmented reality. *Presence, 6*, 355–385.

Baecker, R. M., & Buxton, W. A. S. (Eds.). (1987). *Readings in human-computer interaction, a multidisciplinary approach.* Los Altos, CA: Morgan Kaufman.

Baumann, K., & Lanz, H. (1998). *Mensch-Maschine-Schnittstellen elektronischer Geräte Leitfaden für Design und Schaltungstechnik [Human-Machine Interface for Electronic Appliances. Guideline for Design and Circuitry]* Berlin: Springer.

Berlinguet, L., & Berthelette, D. (Eds.). (1990). Work With Display Units 89. Amsterdam: North-Holland.

Blankenbach, K. (1999). Multimedia-Displays—Von der Physik zur Technik [Multimedia Displays—from Physics to Technology]. *Physikalische Blätter, 55*(5), 33–38.

Boff, K. R., & Lincoln, J. E. (1988). *Engineering data compendium—Human perception and performance.* Harry G. Armstrong Aerospace Medical Research Laboratory & Wright-Patterson Air Force Base, Ohio.

Bosman, D. (Ed.). (1989). *Display engineering: Conditioning, technologies, applications.* Amsterdam: Elesevier Science.

Brown University. (1999). Interactive information display tutorial. Brown University, Division of Engineering. Retrieved July 25, 2001 from http://display.engin.brown.edu/usdc/

Burandt, U. (1986). *Ergonomie für Design und Entwicklung [Ergonomics for Design and Development].* Köln: Schmidt.

Cakir, A., Hart, D. J., & Stewart, T. F. M. (1979). The VDT manual—Ergonomics, workplace design, health and safety, task organisation. Darmstadt: IFRA—Inca-Fiej Research Association.

Chadha, S. S. (1995). *An overview of electronic displays.* London: Applied Vision Association.

Deussen, N. (2001). E-Ink: Die ewige Zeitung. [The eternal newspaper] *Der Tagesspiegel.* Retrieved July 7, 2001, from http://www2.tagesspiegel.de/archiv/2001/07/24/akin-558100.html

E Ink. (2001a). *What is Electronic Ink?* Retrieved September 17, 2001, from http://www.eink.com/technology/index.htm

E Ink. (2001b). *Electronic Ink key performance benefits.* Internet document, http://www.eink.com/pdf/key_benefits.pdf, September 17th, 2001.

Flügge, J., Hartwig, G., & Weiershausen, W. (1985). *Studienbuch zur technischen Optik [Study book for technical optics].* 2. Aufl. Göttingen, Germany: Vanderhoeck und Ruprecht.

Gibbs, W. W. (1998). The reinvention of paper. *Scientific America Online.* Retrieved January 3, 2001, from http://www.sciam.com/1998/0998issue/0998techbus1.html

Gould, A., Alfaro, L., Finn, R., Haupt, B., Minuto, A., & Salaun, J. (1986). Why is reading slower from CRT displays than from paper? *Proceedings of the Human Factors Society 30th Annual Meeting* (pp. 834–836). Santa Monica, CA: Human Factors Society.

Gould, J., Alfaro, L., Varnes, V., Finn, R., Grischkowsky, N., & Minuto, A. (1987). Reading is slower from CRT displays than from paper: Attempts to isolate a single-variable explanatation. *Human Factors, 29*, 269–299.

Grieco, A., Molteni, G., Occhipinti, E., & Piccoli, B. (Eds.). (1995). Work With Display Units 94. Amsterdam: North Holland.

Grossman, S. (2000, May 1). Electroluminescent display triples brightness to 150 candelas/m². *Electronic Design.* Retrieved February 2, 2002, from www.planetee.com/planetee/servlet/DisplayDocument?ArticleID = 7642

Harpster, J., Freivalds, A., Shulman, G., & Leibowitz, H. (1989). Visual performance on CRT screens and hard-copy displays. *Human Factors, 31*, 247–257.

Henning, K., & Kutscha, S. (1993). *Informatik im Maschinenbau* [Computer Science in Mechanical Engineering]. 4. Aufl. Berlin: Springer, 1993.

Institut für Lasertechnik. (2001). Fraunhofer Institut für Lasertechnik Retrieved September 20, 2001, from http://www.ilt.fhg.de

Kaufman, L. (1974). *Sight and mind—An introduction to visual perception*. London: Oxford University Press.

Knave, B., & Widebäck, P.-G. (Eds.). (1997). *Work With Display Units 86*. Amsterdam: North-Holland.

König, W. (1990). Fertigungsverfahren Band 3—Abtragen [Manufacturing Processes]. 2. Neubearb. Aufl. Düsseldorf: VDI.

LfAS. (2000). Büro- und Bildschirmarbeitsplätze—ein Ratgeber für die Praxis [Office and VDU Workplaces—A Guidebook for the Practice]. Bayerisches Landesamt für Arbeitsschutz, Arbeitsmedizin und Sicherheitstechnik—herausgegeben im Auftrag des Bayerischen Ministeriums für Gesundheit, Ernährung und Verbrauchwerschutz. München.

Luczak, H., Cakir, A., & Cakir, G. (Eds.). (1993). *Work With Display Units 92*. Amsterdam: North-Holland.

Matschulat, H. (1999). Lexikon der Monitor-Technologie [Encyclopedia of Monitor Technology]. Aachen, Germany: Elektor-Verlag, 1999.

Microvision. (2000). Microvision Homepage. Retrieved August 3, 2001, from http://www.mvis.com/

Miyamoto, H., Saito, S., Kajiyama, M., & Koizumi, N. (Eds.). (1997, November). *Proceedings of WWDU'97*, Tokyo.

Müller, C. H. (2000). Projekt NEMESIS—Niederfrequente elektrische und magnetische Felder und Elektrosensibilität in der Schweiz—Abhandlung zur Erlangung des Titels Doktor der Naturwissenschaften der Eidgenössischen Technischen Hochschule Zürich [Project NEMESIS-Low-frequency electrical and magnetic fields and electrosensibility in Switzerland. Doctoral Dissertation]. Zürich: ETHZ

Ness Corporation. (2000). What is an organic electroluminescent display (OELD)? Retrieved September 17, 2001, from http://http://www.ness.co.kr/prod/ca2.htm

Nielsen, J. (1993). *Usability engineering*. San Diego, CA: Academic Press.

NN. (2001). Interactive information display tutorial—electroluminescent displays. Retrieved July 25, 2001, from http://display.engin.brown.edu/usdc/electroluminescent.html

Oehme, O., Wiedenmaier, S., Schmidt, L., & Luczak, H. (2001). Empirical studies on an augmented reality user interface for a head based virtual retinal display. In J. M. Smith, G. Salvendy, (Eds.), Systems, Social and Internationalization Design Aspects and Human-Computer Interaction. Volume 1, *Proceedings of the Human Computer Interaction International (HCII) 2001*.

Pioneer. (2001). Why choose Plasma. Retrieved September 6, 2001 from http://www.pioneerelectronics.com/Pioneer/CDA/Common/ArticleDetails/0,1484,1547,00.html

Plapper, V., Wenk, C., & Weck, M. (1999). Augmented Reality unterstützt den Teleservice [Augmented Reality Supports Teleservices]. *wt Werkstatttechnik* (1999), Nr. 89 H. 6, S. 293–294.

Precht, M., Meier, N., & Kleinlein, J. (1997). *EDV-Grundwissen: Eine Einführung in Theorie und Praxis der modernen EDV*. 4th ed. [Basic Computer Knowledge: An Introduction in Theory and Practice of Modern Computing]. Bonn, Germany: Addison-Wesley-Longman.

Rack, P. D., Naman, A., Holloway P. H., Sun, S.-S., & Tuenge R. T. (1996). Materials used in electroluminescent displays. *Materials Research Society Bulletin, 21*(3), 49–58.

Sanders, M. S., & McCormick, E. J. (1993). *Human factors in engineering and desing* (7th ed.) New York: McGraw-Hill.

Schadt, M. (1996). Optisch strukturierte Flüssigkeitskristall-Anzeigen mit großem Blickwinkelbereich [Optically structured LCDs with wide viewing angle]. Physikalische Blätter, *52*, 695–698.

Schmidt, A., Aidoo, K. A., Takaluoma, A., Tuomela, U., Laerhoven, K., & Velde, W. (1999, September). Advanced interaction in context. In: H. W. Gellersen (Ed.). *Handheld and Ubiquitous computing—First International Symposium*, HUC'99, Karlsruhe. Lecture notes in Computer Science Vol. 1707, Berlin: Springer (ISBN 3-540-66550-1).

Walker, A. (2000). *Don't say goodbye to paper just yet*. Retrieved September 17, 2001, from http://excite.quicken.ca/eng/soho/features/index.phtml?body = paperless.

Weiland, W., Zachary, W., & Stokes, J. (2000, July–August). Personal wearable computer systems. In: *Proceedings of IEA 2000/HFES 2000 Congress* (pp. 721–715). San Diego, CA, USA.

Williams, M. (1995). Auditory Virtual Environments. In K. Carr, R. England, (Eds.). Simulated and Virtual Realities (pp. 179–208). London: Taylor & Francis.

Woodson, W. E. (1987). Human factor reference guide for electronics and computer professionals. New York: McGraw-Hill.

Ziefle, M. (1998a). Effects of display resolution on visual performance. *Human Factors, 40*, 554–568.

Ziefle, M. (1998b). *Visuelle Faktoren bei der Informationsentnahme am Bildschirm* [Visual factors of information acquisition at VDUs]. Aachen, Germany: RWTH, Philosophische Fakultät, Habilitation.

·10·

HAPTIC INTERFACES

Hiroo Iwata
University of Tsukuba

INTRODUCTION

It is well known that sense of touch is vital to understanding the world. The use of force feedback to enhance human–computer interaction has often been a topic of discussion. A haptic interface is a feedback device that generates sensation to the skin and muscles, including a sense of touch, weight, and rigidity. Compared with ordinary visual and auditory sensations, haptics technology is difficult to synthesize. Visual and auditory sensations are gathered by specialized organs—the eyes and ears. However, a sensation of force can occur at any part of the human body and is therefore inseparable from actual physical contact. These characteristics lead to many difficulties in developing a haptic interface. Visual and auditory media are widely used in everyday life, although little application of haptic interface is used for information media.

In the field of virtual reality, haptic interface is one of the major areas of research. The last decade has seen significant advancements in the development of haptic interface. High-performance haptic devices have been developed, and some are commercially available. This chapter presents current methods and issues in haptic interface.

The first section describes mechanisms of haptic sensation and provides an overview of feedback technologies. The next three sections introduce examples of haptic interface technologies developed by the author. The last section presents applications of and prospects for haptic interface.

MECHANISM OF HAPTICS AND METHODS FOR HAPTIC FEEDBACK

Somatic Sensation

Haptic interface presents synthetic stimulation to somatic sensation, which comprises proprioception and skin sensation. Proprioception is achieved by mechanoreceptors of skeletal articulations and muscles. There are three types of joint position receptors: free nerve endings and Ruffini and Pacinian corpuscles. Ruffini corpuscles detect static force, whereas Pacinian corpuscles measure acceleration of the joint angle. Position and motion of the human body is perceived by these receptors. Force sensation is accomplished by mechanoreceptors of muscles—muscle spindles and Goldi tendons. These receptors detect contact forces applied by an obstacle in the environment.

Skin sensation is derived from mechanoreceptors and thermorecepters, and these receptors evoke on sense of touch. Mechanoreceptors of the skin are classified into four types: Merkel disks, Ruffini capsules, Meissner corpuscles, and Pacinian corpuscles. These receptors detect the edge of an object and stretching, velocity, and vibration of the skin, respectively.

Proprioception and Force Display

A force display is a mechanical device that generates a reaction force from virtual objects. Research activities into haptic

FIGURE 10.1. Desktop force display.

interfaces have recently increased, although the technology is still in a state of trial and error. There are several approaches to implementing haptic interfaces.

Exoskeleton Force Displays. An exoskeleton is a set of actuators attached to the hand or body. In the field of robotics research, exoskeletons have often been used as master manipulators for teleoperations. However, most master manipulators entail a great quantity of hardware and are therefore expensive, which restricts their applications. Compact hardware is needed to use them in human–computer interactions.

The first example of a compact exoskeleton suitable for desktop use was published in 1990 (Iwata, 1990a). The device applies force to the fingertips and the palm. Figure 10.1 shows an overall view of the system. Lightweight and portable exoskeletons also have been developed. Burdea and colleagues used small pneumatic cylinders to apply the force to the fingertips (Burdea, Zhuang, Roskos, Silver, & Langlana, 1992). Cyber Grasp (Fig. 10.2) is a commercially available exoskeleton, in which cables are used to transmit the force for more information, visit the following URL: ⟨http://www.vti.com⟩.

Tool-Handling Force Displays. Tool-handling force displays are the easiest way to realize force feedback. This configuration is similar to that of a joystick. Unlike the exoskeleton, tool-handling force displays do not need to be fitted to the user's hand. They cannot generate force between the fingers but do have practical advantages.

A typical example of this category is the pen-based force display (Iwata, 1993). A pen-shaped grip is supported by two three degrees of freedom (DOF) pantographs that enable a six

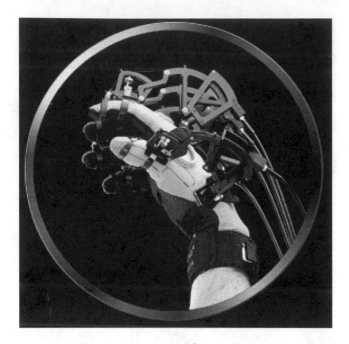

FIGURE 10.2. Cyber grasp.

FIGURE 10.3. PHANToM.

DOF force–torque feedback. Another example of this type is the Haptic Master, which was demonstrated at SIGGRAPH'94 (Special Interest Group on Computer Graphics). The device has a ball-shaped grip to which six DOF force–torque is fed back (Iwata, 1994). It employs a parallel mechanism in which top and base triangular platforms are connected by three sets of pantographs. This compact hardware has the ability to carry a large payload.

Massie and Salisbury developed the PHANToM, which has a 3 DOF pantograph (Massie & Salisbury, 1994). A thimble with a gimbal is connected to the end of the pantograph, which can then apply a three DOF force to the fingertips. The PHANToM is one of the most popular commercially available haptic interfaces (Fig. 10.3).

Object-Oriented Force Displays. Object-oriented force displays are a radical idea for the design of a haptic interface. These devices move or deform to simulate the shapes of virtual objects. Users can physically contact the surface of the virtual object. An example can be found in Tachi's work (Tachi, Maeda, Hirata, & Hoshino, 1994). Their device consists of a shape-approximation prop mounted on a manipulator. The position of the fingertip is measured, and the prop moves to provide a contact point for the virtual object. McNeely (1993) proposed an idea called Robotic Graphics, which is similar to Tachi's method. Hirose developed a surface display that creates a contact surface using a 4 by 4 linear actuator array (Hirota & Hirose, 1996). This device simulates an edge or a vertex of a virtual object.

Passive Props. A passive input device equipped with force sensors is a different approach to the haptic interface. Murakami and Nakajima (1994) used a flexible prop to manipulate a three-dimensional virtual object. The force applied by the user is measured, and the deformation of the virtual object is determined based on the applied force. Sinclair (1997) developed a force sensor array to measure pressure distribution. These passive devices allow the user to interact using their bare fingers; however, these devices have no actuators, so they cannot represent the shape of virtual objects.

Proprioception and Full-Body Haptics

One of the new frontiers of haptic interface is full-body haptics, including foot haptics. Force applied to the entire body plays important roles in locomotion, which is the most intuitive way to move about the real world. Locomotion interface is a device that provides the sensation of walking while the walker's body is in the real world. There are several approaches to realize locomotion interface.

Sliding Device. Controlling a steering bar or joystick is not as intuitive an action as locomotion. The project called Virtual Perambulator was aimed at developing a locomotion interface using a specialized sliding device (Iwata & Fujii, 1996). The primary object of the first stage was to allow for the walker's feet to change direction. The first prototype of the Virtual Perambulator was developed in 1989 (Iwata, 1990b). Figure 10.4 shows an overall view of the apparatus. A user wore a parachutelike harness and omni-directional roller skates. The user's trunk was attached to the framework of the system with harness. The omni-directional sliding device allows the feet to change direction by feet. Specialized roller skates equipped with four casters were developed, which enabled two-dimensional motion. The user could freely move his or her feet in any direction, and an ultrasonic range detector measured the motion of the feet. From the result of this measurement, an image of the virtual space,

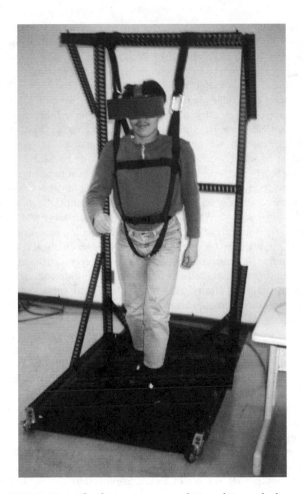

FIGURE 10.4. The first prototype of Virtual Perambulator.

FIGURE 10.5. TreadPort.

corresponding with the motion of the user, was displayed in a head-mounted display. The direction of locomotion in virtual space was determined according to the direction of the walker's step.

Treadmill. A simple device for virtual walking is a treadmill, ordinarily used for physical fitness. An application of this device to virtual building simulator was developed at the University of North Carolina (Brooks, 1986). This treadmill has a steering bar similar to that of a bicycle. A treadmill equipped with a series of linear actuators underneath the belt was developed at ATR (Noma, Sugihara, & Miyasato, 2000). The device is called a ground surface simulator, and it simulates the slope of virtual terrain. The TreadPort developed at University of Utah is a treadmill combined with a large manipulator connected to a walker (Christensen, 1998). The manipulator provides gravitational force while the walker is passing a slope. Figure 10.5 shows an overall view of the TreadPort.

The Omni-directional Treadmill employs two perpendicular treadmills, one inside the other. Each belt is consists of approximately 3,400 separate rollers, woven together into a mechanical fabric. Motion of the lower belt is transmitted to the user by the rollers, allowing omni-directional movement (Darken, Cockayne, & Carmein, 1997).

Foot Pad. A foot pad applied to each foot is an alternative of implementation of locomotion interface. Two large manipulators driven by hydraulic actuators were developed at the University of Utah and applied to a locomotion interface. The device is called BiPort (for more information, visit URL http://www.sarcos.com). These manipulators are attached to a user's feet. These manipulators can represent the viscosity of virtual ground. A similar device has been developed at the Cybernet Systems Corporation, which uses two 3 DOF motion platforms for the feet (Roston & Peurach, 1997). These devices, however, have not been evaluated or applied to Virtual Environment (VE).

Pedaling Device. In the battlefield simulator of Naval Post Graduate School Network (NPSNET) project, unicyclelike pedaling device is used for locomotion in the virtual battlefield (Pratt & Zyda, 1994). A user changes direction by twisting his or her waist.

The OSIRIS, simulator of night-vision battle, uses a stair stepper device (Lorenzo, 1995). A user changes direction by controlling the joystick or twisting the waist.

Gesture Recognition of Walking. Slaters et al. proposed locomotion in virtual environments by "walking in place." They recognized the gesture of walking using a position sensor and a neural network (Slater, Usoh, & Steed, 1994).

Skin Sensation and Tactile Display

The tactile display that stimulates skin sensation is a well-known technology. It has been applied to communication aids for blind persons as well as master system of teleoperation. A sense of vibration is relatively easy to produce, and a good deal of work has been done using vibration displays (Kontarinis & Howe, 1995; Minsky & Lederman, 1997). The micropin array is also used for tactile displays. Such a device has enabled the provision of a teltaction and communication aid for blind persons (Kawai & Tomita, 2000; Moy, Wagner, & Fearing, 2000). It has the ability to convey texture or two-dimensional geometry (Burdea, 1996).

The micropin array looks similar to object-oriented force displays, but it can only create the sensation of skin. The stroke distance of each pin is short, so the user cannot feel the three-dimensional-shape of a virtual object directly. The major role of tactile display is to convey sense of fine texture of object's surface. The latest research activities of tactile display focus on selective stimulation of mechanoreceptors of skin. As mentioned at the beginning of this section, there are four types of mechanoreceptors in the skin: Merkel disks, Ruffini capsules, Meissner corpuscles, and Pacinian corpuscles. Stimulating these receptors selectively, various tactile sensation such as roughness or slip can be presented. Micro air jet (Asanura et al., 1999) and micro electrode array (Kajimoto, Kawakami, Maeda, & Tachi, 1999) are used for selective stimulation.

TECHNOLOGIES IN FINGER–HAND HAPTICS

Exoskeleton

Exoskeleton is one of the typical forms of haptic interface. Figure 10.6 shows a detailed view of an exoskeleton that was introduced in the previous section (Iwata, 1990a).

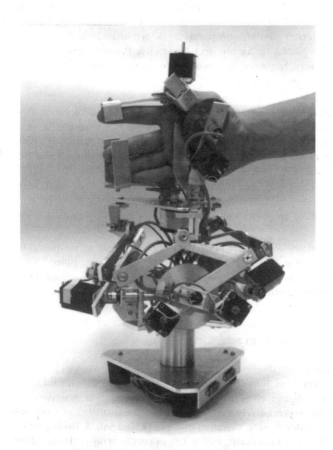

FIGURE 10.6. Desktop force display.

Force sensation contains six dimensions of information, three of dimensional force and three of dimensional torque. The core element of the force display is a six DOF parallel manipulator. The typical design feature of parallel manipulators is an octahedron called the Stewart platform. In this mechanism, two triangular platforms, one on the top and one on the base, are connected by six length-controllable cylinders. This compact hardware has the ability to carry a large payload. The structure does, however, have some practical disadvantages, including its small working volume and lack of backdrivability (reduction of friction). In our system, three sets of parallelogram linkages (pantograph) are employed instead of linear actuators. Each pantograph is driven by two Direct Current (DC) motors. Each motor is powered by a pulse width modulation (PWM) amplifier. The top end of the pantograph is connected with a vertex of the top platform by a spherical joint. This mechanical configuration has the same advantages as an octahedron mechanism. The pantograph mechanism improves the working volume and backdrivability of the parallel manipulator. The inertia of motion parts of the manipulator is so small that compensation is not needed.

The working space of the center of the top platform is a spherical volume with a diameter of approximately 30 cm. Each joint angle of the manipulator is measured by potentiometers, with linearity of 1%. The maximum payload of the manipulator is 2.3 Kg, which is more than a typical hand.

The top platform of the parallel manipulator is fixed to the palm of the operator by a U-shaped attachment, which enables the operator to move the hand and fingers independently. Three actuators are set coaxially with the first joint of the operator's thumb, index finger, and middle finger. The last three fingers work together. DC servo motors are employed for each actuator.

Tool-Handling Haptic Interface

Exoskeleton users feel uncomfortable when they put on these devices, which obstructs their practical use. Tool-handling devices provide a method of implementation for force display without the glovelike device. A pen-based force display has been proposed as another alternative (Iwata, 1993), and a six DOF force reflective master manipulator with a pen-shaped grip has been developed. Users are familiar with pens from everyday life. People also use spatulas and rakes to model solid objects, and these devices have stick-shaped grips that are similar to that of a pen. For this reason, the pen-based force display is easily applied to the design of three-dimensional shapes.

The human hand has the capability of six DOF motion in three-dimensional space. If a six DOF master manipulator is built using serial joints, each joint must support the weight of the upper joints. This characteristic requires large hardware. We use a parallel mechanism to the reduce size and weight of the manipulator. The pen-based force display employs two 3 DOF manipulators, and both ends of the pen are connected to these manipulators; the total DOF of the force display is six. Three DOF force and three DOF torque are applied at the pen. An overall view of the force display is shown in Fig. 10.7. Each three DOF manipulator is composed of a pantograph link.

FIGURE 10.7. Pen-based force display.

With this mechanism, the pen is free from the weight of the actuators

Figure 10.8 shows a diagram of the mechanical configuration of a force display. Joints MA1, MA2, MA3, MB1, MB2, and MB3 are equipped with DC motors and potentiometers; other joints move passively. The position of joints A and B are measured by potentiometers. A three-dimensional force vector is applied at joints A and B. Joint A determines the position of the pen's point, and joint B determines the pen's orientation. The working space

of the pen point is part of a spherical volume with a diameter of 44 cm. The rotational angle around the axis of the pen is determined by the distance between joints A and B. A screw motion mechanism converts rotational motion of the pen into transition of the distance between joints A and B.

Applied force and torque at the pen is generated by a combination of forces at points A and B. If these force are in the same direction, translational force is applied to the user's hand. If the forces are in the reverse direction, torque is generated around the yaw or the pitch axis. If the two forces are in opposition, torque around the roll axis is generated by the screw motion mechanism.

Object-Oriented Haptic Interface

The Basics of FEELEX. The author has demonstrated haptic interfaces to many people and found that some were unable to fully experience virtual objects through the medium of synthesized haptic sensation. There seem to be two reasons for this phenomenon. First, haptic interfaces only allow the user to touch the virtual object at a single point or at a group of points. These contact points are not spatially continuous because of the hardware configuration of the haptic interfaces; the user feels a reaction force thorough a grip or thimble. Exoskeletons provide more contact points, but these are achieved by using Velcro bands attached to specific part of the user's fingers, which are not continuous. Therefore, these devices cannot recreate a natural interaction sensation when compared with manual manipulation in the real world. The second reason users fail to perceive the sensation is related to the combination of visual and haptic displays. A visual image is usually combined with a haptic interface by using a conventional cathode ray tube or projection screen. Thus, the user receives visual and haptic sensation

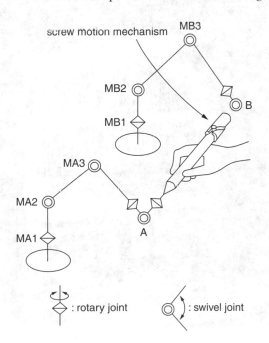

FIGURE 10.8. Mechanical configuration of a pen-based force display.

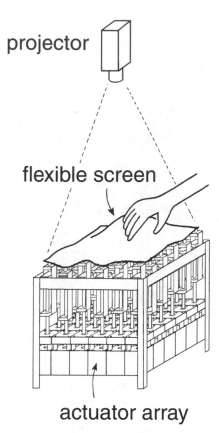

projector

flexible screen

actuator array

FIGURE 10.9. Basic design of FEELEX.

FEELEX 1. The FEELEX 1, developed in 1997, was designed to enable double-handed interaction using the entire palm. Therefore, the optimum size of the screen was determined to be 24 cm × 24 cm. The screen is connected to a linear actuator array that deforms its shape. Each linear actuator is composed of a screw mechanism driven by a DC motor. The screw mechanism converts the rotation of an axis of the motor to the linear motion of a rod. The motor must generate on the screen both motion and a reaction force. The diameter of the smallest motor that can drive the screen is 4 cm. Therefore, a 6 × 6 linear actuator array can be set under the screen. The deformable screen is made of a rubber plate and a white nylon cloth. The thickness of the rubber is 3 mm. Figure 10.10 shows an overall view of the device.

The screw mechanism of the linear actuator has a self-lock function that maintains its position while the motor is off. A hard virtual wall is difficult to simulate with tool-handling force displays. Considerable power is required to generate the reaction force of the virtual wall, which often leads to uncomfortable vibration. The screw mechanism does not present this problem. A soft wall can be represented by the computer-controlled motion of the linear actuators based on the data from the force sensors. Two strain gauges are used as force sensors, one of

through different displays and therefore has to integrate the visual and haptic images in his or her brain. Some users, especially elderly people, have difficulty with this integration process.

With these problems in mind, new interface devices have been developed. The project is named FEELEX (a conjunction of (*feel* and *flex*). The major goals of this project are to provide a spatially continuous surface that enables users to feel virtual objects with any part of the fingers or even the whole palm and to provide visual and haptic sensations simultaneously using a single device that doesn't require the user to wear an additional apparatus. FEELEX, A new visual–haptic display configuration, was designed to achieve these goals. Figure 10.9 illustrates the basic concept of FEELEX. The device is composed of a flexible screen, an array of actuators, and a projector. The flexible screen is deformed by the actuators to simulate the shape of virtual objects. An image of the virtual objects is projected onto the surface of the flexible screen. Deformation of the screen converts the two-dimensional image from the projector into a solid image. This configuration enables the user to touch the image directly using any part of their hand. The actuators are equipped with force sensors to measure the force applied by the user. The hardness of the virtual object is determined by the relationship between the measured force and its position of the screen. If the virtual object is soft, small applied force causes a large deformation.

FIGURE 10.10. Overall view of the FEELEX 1.

which is set at the top of each linear actuator. The strain gauge detects small displacements at the top of the linear actuator, caused by the force applied by the user. An optical encoder connected to the axis of the DC motor measures the position of the top of the linear actuator. The maximum stroke of the linear actuator is 80 mm, and the maximum speed is 100 mm/s.

The system is controlled via PC. The DC motors are interfaced by a parallel input–output unit, and the force sensors are interfaced by an Analog-Digital converter. The force sensors provide interaction with the graphics. The position and strength of the force applied by the user are detected by a 6 by 6 sensor array. The graphics projected onto the flexible screen are changed according to the measured force.

FEELEX 2. The FEELEX 2 was designed to improve the resolution of the haptic surface. To determine the resolution of the linear actuators, we considered the situation in which a medical doctor palpates a patient. Result of interviews with several physicians proved that they usually recognized a tumor using their index, middle, and third fingers. The size of a tumor is perceived by comparing it with the width of their fingers (i.e., two fingers large, three fingers large). Thus, the distance between the axis of the linear actuators should be smaller than the width of a finger. Considering this, the distance between the axis at linear actuators is set to 8 mm. This 8-mm resolution enables the user to hit at least one actuator when touching an arbitrary position on the screen. The size of the screen is 50 mm by 50 mm, which allows the user to touch the surface using three fingers.

To realize 8-mm resolution, a piston–crank mechanism is employed for the linear actuator. The size of the motor is much larger than 8 mm, so the motor should be placed at a position offset from the rod. The piston–crank mechanism can easily achieve this offset position. Figure 10.11 illustrates the mechanical configuration of the linear actuator. A servomotor from a radio-controlled car is selected as the actuator. The rotation of the axis of the servomotor is converted to the linear motion of the rod by a crankshaft and a linkage. The stroke of the rod is 18 mm, and the maximum speed is 250 mm/s. The maximum

FIGURE 10.12. Overall view of the FEELEX 2.

torque of the servomotor is 3.2 kg-cm, which applies a 1.1 Kgf force at the top of each rod. This force is sufficient for palpation with the fingers.

The flexible screen is supported by 23 rods, and the servomotors are set remotely from the rods. Figure 11.12 shows an overall view of the FEELEX 2. The 23 separate sets of piston-crank mechanisms can be seen in the picture.

Figure 10.13 shows the top end of the rods. The photo is taken while the flexible screen is off. The diameter of each rod is 6 mm. A strain gauge cannot be put on the top of the rod because of its small size. Thus, the electric current going to each servomotor is measured to sense the force. The servomotor generates a force to maintain the position of the crankshaft. When the user applies a force to the rod, the electric current on the motor increases to balance the force. The relationship between the applied force and the electric current is measured. The applied force at the top of the rods is calculated using data from the electric current sensor. The resolution of the force sensing capability is 40 gf.

Characteristics of the FEELEX. The performance of existing haptic interfaces is usually represented by the dynamic range of force, impedance, inertia, friction, and so on. However, these parameters are only crucial while the device is attached to the finger or hand. In the case of tool-handling haptic interface or the exoskeleton, the devices move with the hand

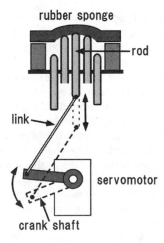

FIGURE 10.11. Piston-crank mechanism.

FIGURE 10.13. Top end of the rods (FEELEX 2).

even though the user doesn't touch the virtual objects. Therefore, inertia or friction degrades the usability, and the dynamic range of force determines the quality of the virtual surface. In contrast, FEELEX is entirely separate from the user's hand, so its performance is determined by the resolution and speed of the actuators. The resolution of the actuator corresponds to the smoothness of the surface, and the speed of the actuator determines the motion of the virtual object. FEELEX 2 has improved resolution and motion speed compared with FEELEX 1. Each actuator of the FEELEX 2 has a stroke rate of up to 7 Hz, which is much faster than the human pulse rate and can simulate the motion of a very fast virtual object. The rod pushes the rubber sponge so that the user feels as if the object is pulsating.

The major advantage of the FEELEX is that it allows natural interaction using only the bare hand. At Special Insterest Group on Computer Graphics (SIGGRAPH'98), nearly 2,000 subjects enjoyed the haptic experience. One of the contents of the FEELEX 1 system, known as Anomalocaris, was selected as a long-term exhibition at the Ars Electronica Center (Linz, Austria). Anomalocaris is a name of an animal that is belived to have lived during the Cambrian Era. The exhibition has been popular with visitors, especially children.

Another advantage of FEELEX is safety. The user doesn't wear special equipment while the interaction takes place. Exoskeleton and tool-handling force displays have control problems on their contact surfaces for the virtual objects. Vibration or unwanted forces can be generated back to the user, which can be dangerous. The contact surface of FEELEX is physically generated, so it is free from such control problems.

The major disadvantage of FEELEX is the degree of difficulty in its implementation. It requires a large number of actuators that must be controlled simultaneously. The drive mechanism of the actuator must be robust enough for rough manipulation. Because FEELEX provides the sensation of natural interaction, some users apply great force. Our exhibit at the Ars Electronica Center suffered from overload of the actuators.

Another disadvantage of the FEELEX is that it is limited in the shapes it can display. The current prototypes cannot present a sharp edge on a virtual object. Furthermore, the linear actuator array can only simulate the front surface of an object. Some participants at the Anomalocaris demonstration wanted to touch the back of the creature, but a separate mechanism would be required to achieve this.

TECHNOLOGIES IN FULL-BODY HAPTICS

Treadmill-Based Locomotion Interface

Basic Design of the Torus Treadmill. A key principle of treadmill-based locomotion interface is to make the floor move in a direction opposite to that of the walker. The motion of the floor cancels the displacement of the walker in the real world. The major problem of treadmill-based locomotion interface is devising a method that allows the user to change direction. An omni-directional active floor would enable a virtually infinite area. To realize an infinite walking area, geometric configuration of an active floor must be chosen. A closed surface driven by actuators has the ability to create this infinite floor. The following requirements for implementation of the closed surface must be considered: (1) The user and actuators must be put outside the surface, (2) the walking area must be a plane surface, and (3) the surface must be made of a material that stretches only minimally.

The shape of a closed surface, in general, is a surface with holes. If there are no holes, the surface is a sphere, which is the simplest infinite surface. However, the walking area of the sphere is not a plane surface. A very large diameter is required to make a plane surface on a sphere, which restricts implementation of the locomotion interface.

A closed surface with one hole like a doughnut is called torus. A torus can be implemented by a group of belts. These belts make a plane surface on which the user can walk. A closed surface with more than one hole cannot make a plane walking surface. Thus, the torus is the only form suitable for a locomotion interface.

Mechanism and Performance. The Torus Treadmill is a group of belts connected to each other. The Torus Treadmill is realized by these belts (Iwata, 1999). Figures 10.14 and 10.15 illustrate the basic structure of the Torus Treadmill. The Tours Treadmill employs twelve treadmills. These treadmills move the walker along an "X" direction. Twelve treadmills are connected side by side and driven in a perpendicular direction. This motion moves the walker along a "Y" direcion.

Figure 10.16 shows an overall view of the apparatus. Twelve treadmills are connected to four chains and mounted on four rails. The chain drives the walker in the Y direction. The rail supports the weight of the treadmills and the user. An AC motor is employed to drive the chains. The power of the motor is 200 W and is controlled by an inverter. The maximum speed of rotation is 1.2 m/s, and the maximum acceleration is 1.0 m/s². The deceleration caused by friction is 1.5 m/s². Frequency characteristics are limited by a circuit protector of the motor driver. The maximum switching frequency is 0.8 Hz.

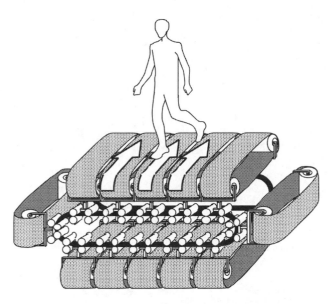

FIGURE 10.14. Structure of Torus Treadmill (X motion).

Each treadmill is equipped with an AC motor. To shorten the length of the treadmill, the motor is put underneath the belt. The power of each motor is 80 W and controlled by an inverter. The maximum speed of each treadmill is 1.2 m/s, and the maximum acceleration is 0.8 m/s². The deceleration caused by friction is 1.0 m/s². The width of each belt is 250 mm, and the overall walkable area is 1 m × 1 m.

A problem with this mechanical configuration is the gap between the belts in the walking area. To minimize this gap, we alternated the driver unit of each treadmill. The gap is only 2 mm wide in this design.

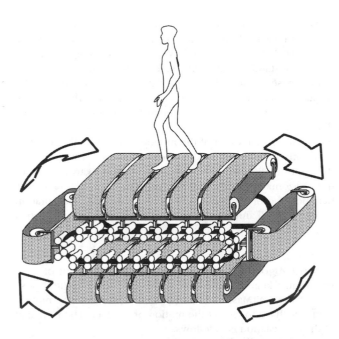

FIGURE 10.15. Structure of Torus Treadmill (Y motion).

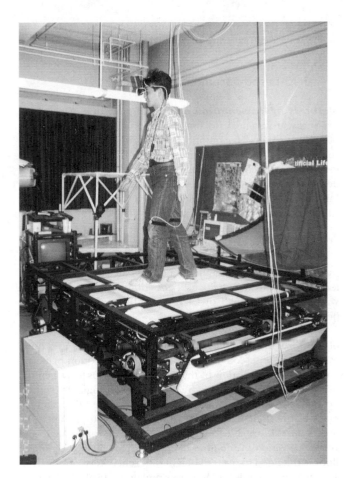

FIGURE 10.16. The Torus treadmill.

Control Algorithm of the Torus Treadmill. A scene of the virtual space is generated that corresponds to the results of motion tracking of the feet and head, which are measured by a Polhemus FASTRACK. The device measures six DOF motion. Sampling rate of each point is 20 Hz. A receiver is attached to each knee (we cannot put the sensors near the moving floor because a steel frame distorts magnetic field) and the length and direction of a step is calculated by the data from those sensors. The user's viewpoint in virtual space moves corresponding to the length and direction of the steps.

To keep the user in the center of the walking area, the Torus Treadmill must be driven in correspondence with the user. A control algorithm is required to achieve safe and natural walking. From our experience with the Virtual Perambulator project, the user should not be connected to a harness or mechanical devices because these restrict motion and inhibit natural walking. The control algorithm of the Torus Treadmill must be safe enough to allow the user to remove the harness. At the final stage of the Virtual Perambulator project, we succeeded in removing the harness using a hoop frame. The user can freely walk and turn around in the hoop, which supports the body while the feet slide. We simulated the function of the hoop in the control algorithm of the Torus Treadmill by putting circular deadzone in the center of the walking area. If the user steps out of the area, the floor moves in the opposite direction so that he or she is carried back into the deadzone.

Foot-Pad-Based Locomotion Interface

Methods of Presentation of Uneven Surface. One of the major research issues in locomotion interface is the presentation of uneven surface. Locomotion interfaces are often applied for simulation of buildings or urban spaces. Those spaces usually include stairs. A walker should have a sense of ascending or descending those stairs. In some applications of locomotion interface, such as training simulators or entertainment facilities, rough terrain needs to be presented.

Presentation of virtual staircase was tested at the early stage of the Virtual Perambulator project (Iwata & Fujii, 1996). A string is connected to the roller skate on each foot, and the string is pulled by a motor. When the user climbs the stairs, the forward foot is pulled up. When the user goes down the stairs, the backward foot is pulled up. However, this method was not successful because of instability.

Later, a six DOF motion platform was applied to final version of the Virtual Perambulator, in which a user walks in a hoop frame. The user stood on the top plate of the motion platform. Pitch and heave motion of the platform were used. When the user stepped forward to climb a stair, the pitch angle and vertical position of the floor increased. After completing the climbing motion, the floor returned to the neutral position. When the user stepped forward to go down a stair, the pitch angle and vertical position of the floor decreases. This inclination of the floor is intended to present height difference between the feet. The heave motion is intended to simulate vertical acceleration. This method failed in the simulation of stairs, however, because the floor was flat.

A possible method for creating height difference between the feet is the application of two large manipulators. The BiPort is a typical example, in which a four DOF manipulator driven by hydraulic actuators is connected to each foot. The major problem with this method is in how the manipulators trace a turning motion; when the user turns around, the two manipulators interfere each other.

The Torus Treadmill provides natural turning motion in that the user can physically turn on the active floor. Turning motion using the feet makes major contributions to human spatial recognition performance. Vestibular and proprioceptive feedback is essential to a sense of orientation (Iwata & Yoshida, 1999). The Torus Treadmill could theoretically modified for simulation of uneven surface. If we install an array of linear actuators on each treadmill, uneven floor could be realized by controlling the length of each linear actuator. This method is almost impossible to implement, however, because a very large number of linear actuators are required to cover the surface of the torus-shaped treadmills, and a control signal for each actuator must be transmitted wirelessly.

Basic Design of the GaitMaster. GaitMaster, a new locomotion interface that simulates an omni-directional, uneven

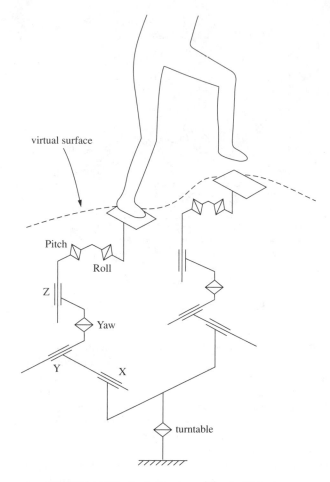

FIGURE 10.17. Basic Design of the GaitMaster.

surface, has recently been designed. The core elements of the device are two 6 DOF motion bases mounted on a turntable. Figure 10.17 illustrates the GaitMaster's basic configuration.

A user stands on the top plate of the motion base. Each motion base is controlled to trace the foot's position, and the turntable is controlled to trace the orientation of the user. The motion of the turntable removes interference between the two motion bases.

The X and Y motion of the motion base traces the horizontal position of the feet and cancels the motion by moving in the opposite direction. The rotation around the yaw axis traces the horizontal orientation of the feet. The Z motion traces vertical position of the feet and cancels its motion. The rotation around the roll and pitch axis simulates inclination of a virtual surface.

Control Algorithm of the GaitMaster. The control algorithm must keep the position of the user at the neutral position of the GaitMaster. To maintain the position, the motion platforms have to cancel the motion of the feet. The principal of the cancellation is as follows:

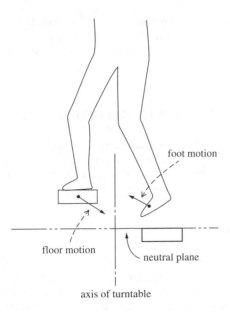

FIGURE 10.18. Canceling an ascending motion.

1. Assume that the right foot is in the forward position and the left foot is in the backward position while walking.
2. When the user steps forward with the left foot, his or her weight is displaced to the right foot.
3. The motion platform of the right foot goes backward in accordance with the displacement of the left foot, so that the user's central position is maintained.
4. The motion platform of the left foot follows the position of the left foot. When the user finishes stepping forward, the motion platform supports the left foot.

If the walker ascends or descends stairs, a similar procedure can be applied; the vertical motion of the feet is canceled using the same principal. Vertical displacement of the forward foot is canceled in accordance with the motion of the backward foot, so that the user's central position is maintained at the neutral height. Figure 10.18 illustrates the method of canceling an ascending motion.

The turntable rotates so that the two motion platforms can trace the user's rotational motion. If the user changes direction, the turntable rotates to trace his or her orientation. The turntable's orientation is determined according to direction of the feet; it rotates so that its orientation is at the center of the feet. The user can physically turn around on the GaitMaster using the control algorithm of the turntable.

Prototype GaitMaster. Figure 10.19 shows an overall view of the GaitMaster prototype. To simplify the mechanism of the motion platform, the surface of the virtual space was defined as sets of planar surfaces. Most buildings or urbane spaces can be simulated without inclining the floor. Thus, we can neglect the roll-and-pitch axis of the motion platforms. Each platform of the GaitMaster prototype is composed of three linear actuators; the top actuator has a yaw joint mounted on it. We disassembled

FIGURE 10.19. An overall view of the Gait Master.

a six DOF Stewart platform and made two XYZ stages. Three linear guides are applied to support the orientation of the top plate of the motion platform. The payload of each motion platform is approximately 150 Kg. A rotational joint around the yaw axis, equipped with a spring that moves the feet to the neutral direction, is mounted on each motion platform.

A turntable was developed using a large Direct Drive (DD) motor. The maximum angular velocity is 500°/s. A three DOF goniometer, which measures back-and-forth and up-and-down motion as well as yaw angle, is connected to each feet. The control algorithm mentioned in the previous section was implemented and succeeded in the presentation of virtual staircases.

APPLICATIONS FOR HAPTIC INTERFACE

Applications for Finger–Hand Haptics

Medicine. Medical applications for haptic interfaces are increasingly evident. Various surgical simulators have been developed using tool-handling force displays. We developed a simulator for laparoscopic surgery using the Haptic Master, and simulator software using PHANToM are commercially available.

Palpation is a common element of medical examinations. The FEELEX 2 is designed to be used as a palpation simulator. If we display a virtual tumor based on a computerized tomographic or magnetic resonance image, a doctor can palpate the internal organs before surgery, and this technique also can be applied to telemedicine. Connecting two FEELEXs via a communication line would allow a doctor to palpate a patient remotely.

Three-Dimensional Shape Modeling. The design of three-dimensional shapes definitely requires haptic feedback, and this is a typical application of tool-handling force displays. One of the most common applications of the PHANToM system is as a modeling tool. It allows a user to point contact, and this type of manipulation is best suited to precision modeling tasks. It is not effective in when the modeling task requires access to the whole shape, however. Designers use their palm or finger joints to deform a clay model. The FEELEX has the ability to support such natural manipulation.

Haptic User Interface. Today, touchscreens are widely used in automatic teller machines, ticketing machines, information kiosks, and so on. A touchscreen enables an intuitive user interface, although it lacks haptic feedback. Users can see virtual buttons but they cannot feel them—a serious problem for a blind person. The FEELEX provides a barrier-free solution to the touchscreen-based user interface. Figure 10.13 shows an example of a haptic touchscreen using the FEELEX 1.

Art. Interactive art may be one of the best applications of the FEELEX system. As discussed earlier, the Anomalocaris has been exhibited at a museum in Austria. It succeeded in evoking haptic interaction with many visitors. The FEELEX can be used for interactive sculptures. Visitors are usually prohibited from touching physical sculptures, but they can not only touch sculptures created with FEELEX, they can also deform them.

Applications for Locomotion Interfaces

As a serious application of our locomotion interface, we are working with the Ship Research Laboratory to develop an 'evacuation simulator (Yamao, Ishida, Ota, & Kaneko, 1996).' The Ship Research Laboratory is a national research institute that is part of Japan's ministry of transportation. Analysis of evacuation of passengers during maritime accidents is vital to ship safety, but it is impossible to carry out experiments with human subjects during an actual disaster. Therefore, this research laboratory introduced virtual reality tools to simulate disaster to analyze passenger evacuation. They built a virtual ship that models the generation of smoke and the inclination of the vessel. Evacuation experiments are carried out to construct a mathematical model of passengers' behavior in disaster. The Torus Treadmill will undoubtedly be effective in such experiments. Locomotion by walking is intuitive and inevitable in the study of human behavior in virtual environments. We are applying the system to research on human models of evacuation

in maritime accidents. In addition, the GaitMaster can be applied to areas that the Torus Treadmill cannot. Its application may include rehabilitation of walking or simulators for mountain climbing.

CONCLUSION AND FUTURE PROSPECTS

This chapter describes major topics of haptic interfaces. A number of methods have been proposed for implementing haptic interfaces. Future work of this research field will include two issues: the safety and psychology of haptics.

Safety Issues

Safety issues pose a serious problem in haptic interface because inadequate control of actuators may potentially injure the user. Exoskeleton and tool-handling force displays have control problems in their contact surface for the virtual objects. Vibration or unwanted forces can be generated back to the user, which can be dangerous. One of the major advantage of FEELEX is safety. The user of FEELEX does not wear special equipment while the interaction is taking place. The contact surface of the FEELEX is physically generated, so it is free from such control problems.

Locomotion interface have many more safety issues. The system supports the user's whole body, so that inadequate control can cause serious injury. Specialized hardware to ensure safety must be developed.

Psychology in Haptics

There have been many research findings regarding haptic sensation. Most of these are related to skin sensation, but and research activities that survey muscle sensation are few. Among these, Lederman and Klatzky's (1987) work is closely related to the design of the force display. Their latest work involves spatially distributed forces (Lederman & Klatzky, 1999). They performed an experiment involving palpation in which subjects were asked to find a steel ball placed underneath a foam-rubber cover. The results showed that steel balls smaller than 8 mm in diameter decreased the subjects' scores. This finding supports our specification for the FEELEX 2 in which the optimal distance between rods is 8 mm. This kind of psychological study will support future development of haptic interface.

Haptics technology is indispensable for human interaction in the real world. However, it is not commonly used in the field of human–computer interaction. Although there are several commercially avalable haptic interfaces, they are expensive and are limitated in their functionality. Image display has a history of more than 100 years. Today, image displays, such as television or motion pictures, are part of everyday life. In contrast, haptic interface has only a 10-year history. There are challenges to overcome to make haptic interface ready for popular use, but this new frontier of media technology will definitely make important contributions to our world.

References

Asanuma, N., Yokoyama, N., & Shinoda, H. (1999). A method of selective stimulation to epidermal skin receptors for realistic touch feedback (pp. 274-281). Proceedings of Institute of Electrical & Electronics Engineers Virtual Reality '99.

Brooks, F. P., Jr. (1986, October). A dynamic graphics system for simulating virtual buildings. *Proceedings of the 1986 Workshop on Interactive 3D Graphics* (pp. 9-21). New York: ACM.

Brooks, F. P., Jr., et al. (1990). Project GROPE—Haptic displays for scientific visualization. *Computer Graphics, 24,* 177-185.

Burdea, G. C. (1996). Force and touch feedback for virtual reality. New York: Wiley Interscience.

Burdea, G., Zhuang, J., Roskos, E., Silver, D., & Langlana, L. (1992). A portable dextrous master with force feedback. *Presence, 1,* 18-22.

Christensen, R., Hollerbach, J. M., Xu, Y., & Meek, S. (1998, November). Inertial force feedback for a locomotion interface. Proceedings of the American Society of Mechanical Engineers Dynamic Systems and Control Division, DSC-Vol. 64 (pp. 119-126). Anaheim.

Darken, R., Cockayne, W., & Carmein, D. (1997). The omni-directional treadmill: A locomotion device for virtual worlds. Proceedings of User Interface Software Technology '97 (pp. 213-221). Association for Computing Machinery.

Hirota, K., & Hirose, M. (1996). Simulation and presentation of curved surface in virtual reality environment through surface display. Proceedings of Institute of Electrical & Electronics Engineers, Virtual Reality Annual International Symposium '96.

Iwata, H. (1990a). Artificial reality with force-feedback: Development of desktop virtual space with compact master manipulator, ACM SIGGRAPH. *Computer Graphics, 24,* 165-170.

Iwata, H. (1990b). Artificial reality for walking about large scale virtual space [in Japanese]. *Human Interface News and Report, 5,* 49-52.

Iwata, H. (1993). Pen-based haptic virtual environment. Proceedings of the Institute of Electrical & Electronics Engineers, Virtual Reality Annual International Symposium '93.

Iwata, H. (1994, August). Desktop force display. Presented at Special Interest Group in Computer Graphics 94.

Iwata, H. (1999). Walking about virtual space on an infinite floor. Proceedings of Institute of Electrical & Electronics Engineers Virtual Reality '99 (pp. 236-293).

Iwata, H., & Fujii, T. (1996). Virtual Perambulator: A novel interface device for locomotion in virtual environment. *Proceedings of the Institute of Electrical & Electronics Engineers 1996 Virtual Reality Annual International Symposium* (pp. 60-65).

Iwata, H., & Yoshida, Y. (1999). Path reproduction tests using a torus treadmill. *Presence, 8,* 587-597.

Kajimoto, H., Kawakami, N., Maeda, T., & Tachi, S. (1999). Tactile feeling display using functional electrical stimulation. Proceedings of International Conference on Artificial Reality and Tele-existence '99 (pp. 107-114). Virtual Reality Society of Japan.

Kawai, Y., & Tomita, F. (2000). A support system for the visually impaired to recognize three-dimensional objects. *Technology and Disability, 12,* 13-20.

Kontarinis, D. A., & Howe, R. D. (1995). Tactile display of vibratory information in teleoperation and virtual environment. *Presence, 4,* 387-402.

Lederman, S. J., & Klatzky, R. L. (1987). Hand movements: A window into haptic object recognition. *Cognitive Psychology, 19,* 342-368.

Lederman, S. J., & Klatzky, R. L. (1999). Sensing and displaying spatially distributed fingertip forces in haptic interfaces for teleoperators and virtual environment system. *Presence 8.*

Lorenzo, M. (1995). *OSIRIS. Special Interest Group in Computer Graphics '95 Visual Proceedings* (p. 129). Association for Computing Machinery.

Massie, T., & Salisbury, K. (1994). The PHANToM haptic interface: A device for probing virtual objects. American Society of Mechanical Engineers Winter Anual Meeting. New York.

McNeely, W. (1993). Robotic Graphics: A new approach to force feedback for virtual reality. Proceedings of Institute of Electrical & Electronics Engineers, Virtual Reality Annual International Symposium '93.

Minsky, M., & Lederman, S. J. (1997). Simulated haptic textures: Roughness. Symposium on Haptic Interfaces for Virtual Environment and Teleoperator Systems. *Proceedings of the American Society of Mechanical Engineers Dynamic Systems and Control Division,* DSC-Vol. 58.

Moy, G. Wagner, C., & Fearing, R. S. (2000, April). A compliant tactile display for teletaction. Institute of Electrical & Electronics Engineers International Conference on Robotics and Automation. San Francisco.

Murakami, T., & Nakajima, N. (1994). Direct and intuitive input device for 3D shape deformation. Proceedings of Association for Computing Machinery Special Interest Group on Computer Human Interaction 1994, Conference on Human Factors in Computing Systems (pp. 465-470).

Noma, H., Sugihara, T., & Miyasato, T. (2000). *Development of ground surface simulator for Tel-E-Merge System.* Proceedings of Institute of Electrical & Electronics Engineers Virtual Reality 2000 (pp. 217-224).

Roston, G. P., & Peurach, T. (1997). A Whole body kinematic display for virtual reality applications. Proceedings of the Institute of Electrical & Electronics Engineers International Conference on Robotics and Automation (pp. 3006-3011).

Pratt, D. R., & Zyda, M. (1994). Insertion of an Articulated Human into a Networked Virtual Environment, Proc. of the 1994 AI, Simulation, and Planning in High Autonomy Systems Conference (pp. 7-9).

Sinclair, M. (1997). The haptic lens. Special Interest Group in Computer Graphics 97 Visual Proceedings (p. 179). Association for Computing Machinery.

Slater, M., Usoh, M., & Steed, A. (1995). Taking Steps: The Influence of a Walking Metaphor in Virtual Reality, ACM Transactions on Computer Human Interaction, Vol. 2, No. 3, pp. 201-219.

Tachi, S., Maeda, T., Hirata, R., & Hoshino, H. (1994). A construction method of virtual haptic space. Proceedings of International Conference on Artificial Reality and Tele-existence '94.

Yamao, T., Ishida, S., Ota, S., & Kaneko, F. (1996). Formal safety assesment—research project on quantification of risk on lives [MSC67/INF.9]. International Maritime Organization information paper.

·11·

NONSPEECH AUDITORY OUTPUT

Stephen Brewster
University of Glasgow

INTRODUCTION AND A BRIEF HISTORY OF NONSPEECH SOUND AND HUMAN–COMPUTER INTERACTION

The combination of visual and auditory information at the human–computer interface is a powerful tool for interaction. In everyday life, both senses combine to give complementary information about the world. Our visual system gives us detailed information about a small area of focus, whereas our auditory system provides general information from all around, alerting us to things outside our peripheral vision. The combination of these two senses gives much of the information we need about our everyday environment. Blattner and Dannenberg (1992) discussed some of the advantages of using this approach in multimedia/multimodal computer systems: "In our interaction with the world around us, we use many senses. Through each sense we interpret the external world using representations and organizations to accommodate that use. The senses enhance each other in various ways, adding synergies or further informational dimensions." They went on to say:

People communicate more effectively through multiple channels.... Music and other sound in film or drama can be used to communicate aspects of the plot or situation that are not verbalized by the actors. Ancient drama used a chorus and musicians to put the action into its proper setting without interfering with the plot. Similarly, non-speech audio messages can communicate to the computer user without interfering with an application. (pp. XVIII–XIX)

These advantages can be brought to the multimodal human–computer interface by the addition of nonspeech auditory output to standard graphical displays (see chapter 14 for more on multimodal interaction). While directing our visual attention to one task, such as editing a document, using sound we can still monitor the state of other tasks on our machine. Currently, almost all information presented by computers uses the visual sense. This means information can be missed because of visual overload or because the user is not looking in the right place at the right time. A multimodal interface that integrated information output to both senses could capitalize on the interdependence between them and present information in the most efficient way possible.

The classical uses of nonspeech sound can be found in the human factors literature (see McCormick & Sanders, 1982). Here it is used mainly for alarms and warnings or monitoring and status information. Buxton (1989) extended these ideas, suggesting that encoded messages could be used to present more complex information in sound. It is this type of auditory feedback that will be considered here.

The other main use of nonspeech sound is in music and sound effects for games and other multimedia applications. These kinds of sounds indicate to the user something about what is going on and try to create a mood for the piece (much as music does in film and radio). Work on auditory output for interaction takes this further and uses sound to present information—things that the user might not otherwise see or important events that the user might not notice.

The use of sound to convey information in computers is not new. In the early days of computing, programmers used to attach speakers to their computer's bus or program counter (Thimbleby, 1990). The speaker would click each time the program counter was changed. Programmers would get to know the patterns and rhythms of sound and could recognize what the machine was doing. Another everyday example is the sound of a hard disk. Users often can tell when a save or copy operation has finished by the noise their disk makes. This allows them to do other things while waiting for the copy to finish. Sound is therefore an important information provider, giving users information about things in their systems that they may not see.

Two important events kick-started the research area of nonspeech auditory output. The first was the special issue of the HCI journal in 1989 on nonspeech sound, edited by William Buxton (Butxton, 1989). This laid the foundations for some of the key work in the area; it included papers by Blattner on Earcons (Blattner, Sumikawa, & Greenberg, 1989), Gaver on Auditory Icons (Gaver, 1989), and Edwards on Soundtrack (Edwards, 1989). The second event was the First International Conference on Auditory Display held in Santa Fe in 1992 (Kramer, 1994a). For the first time, this meeting brought together the main researchers interested in the area to get discussion going, and the proceedings of this meeting (Kramer, 1994a) are still a valuable resource today (see the ICAD Web site at www.icad.org for the proceedings of the ICAD conferences). Resulting from these two events was a large growth in research in the area during the 1990s. One active theme of research is *sonification*—literally, visualization in sound (Kramer & Walker, 1999). A review of this area is beyond the scope of this chapter, but the interested reader is referred to the ICAD Web site for more details on this topic.

The rest of this chapter goes into detail on all aspects of auditory interface design. The next section presents some of the advantages and disadvantages of using sound at the interface. Then follows a brief introduction to psychoacoustics, or the study of the perception of sound. The fourth section gives information about the basic sampling and synthesis techniques needed for auditory interfaces. The fifth section describes the main techniques used for auditory information presentation, and the sixth goes through some of the main applications of sound in human–computer interaction, including sonic enhancement and graphic replacement in GUIs, sound for users with visual impairments, and sound for wearable and mobile computers. The chapter concludes with some conclusions about the state of research in this area.

WHY USE NONSPEECH SOUND IN HUMAN–COMPUTER INTERFACES?

Some Advantages Offered by Sound

It is important to consider why nonspeech sound should be used as an interaction technique. There are many reasons it is advantageous to use sound in user interfaces.

Vision and Hearing Are Interdependent. Our visual and auditory systems work well together. Our eyes provide high-resolution information around a small area of focus (with peripheral vision extending further). According to Perrott, Sadralobadi, Saberi, and Strybel (1991), humans view the world through a window of 80° laterally and 60° vertically. Within this visual field, focusing capacity is not uniform. The foveal area of the retina (the part with the greatest acuity) subtends an angle of only 2° around the point of fixation (Rayner & Pollatsek, 1989). Sounds, on the other hand, can be heard from all around the user—above, below, in front, or behind, but with a much lower resolution. Therefore, "our ears tell our eyes where to look"; if there is an interesting sound from outside our view, we turn to look at it to get more detailed information.

Superior Temporal Resolution. As Kramer (1994b) said, "Acute temporal resolution is one of the greatest strengths of the auditory system." (p. 8) In certain cases, reactions to auditory stimuli have been shown to be faster than reactions to visual stimuli (Bly, 1982).

Reduce the Load on the User's Visual System. Modern, large or multiple screen graphic interfaces use the human visual system intensively. This means that we may miss important information because our visual system is overloaded—we have just too much to look at. To stop this overload, information could be displayed in sound so that the load could be shared between senses.

Reduce the Amount of Information Needed on Screen. Related to the point above is the problem with information presentation on devices with small visual displays, such as mobile telephones, or personal digital assistants (PDAs). These have very small screens that can easily become cluttered. To solve this, some information could be presented in sound to release screen space.

Reduce Demands on Visual Attention. Another issue with mobile devices is that users who are using them on the move cannot devote all of their visual attention to the device; they must look where they are going to avoid traffic, pedestrians, and so forth. In this case visual information may be missed because the user is not looking at the device. If this information were played in sound, then the information would be delivered regardless of where the user was looking.

The Auditory Sense Is Underutilized. The auditory system is powerful and appears to be able to take on extra capacity (Bly, 1982). We can listen to (and some can compose) highly complex musical structures. As Alty (1995) noted, "The information contained in a large musical work (say a symphony) is very large. . . . The information is highly organised into complex structures and sub-structures. The potential therefore exists for using music to successfully transmit complex information to a user." (p. 410)

Sound Is Attention Grabbing. Users can choose not to look at something but it is more difficult to avoid hearing it. This makes sound useful for delivering important information.

Some Objects or Actions Within an Interface May Have a More Natural Representation in Sound. Bly (1982) suggested that: "perception of sound is different to visual perception, sound can offer a different intuitive view of the information it presents." (p. 14) Therefore, sound could allow us to understand information in different ways.

To Make Computers More Usable by Visually Disabled Users. With the development of graphic displays, user interfaces became much harder for visually impaired users to operate. A screen reader (see Sound for Users With Visual Impairments later in this chapter) cannot easily read this kind of graphic information. Providing information in an auditory form can help solve this problem and allow visually disabled persons to use the facilities available on modern computers.

Some Problems With Nonspeech Sound

The above reasons show that there are advantages to be gained from adding sound to human–computer interfaces. However, Kramer (1994b) suggested some general difficulties with sound:

Low Resolution. Many auditory parameters are not suitable for high-resolution display of quantitative information. Using volume, for example, only a few different values can be unambiguously presented (Buxton, Gaver, & Bly, 1991). Vision has a much higher resolution. The same also applies to spatial precision in sound. Differences of about 1° can be detected in front of the listener, and this falls to more than 30° to the side (see Three-Dimensional Sound in this chapter and Begault, 1994). In vision differences of an angle of 2 s can be detected in the area of greatest acuity in the central visual field.

Presenting Absolute Data Is Difficult. Many interfaces that use sound to present data do it in a relative way. A user hears the difference between two sounds so that he or she can tell if a value is going up or down. It is difficult to present absolute data unless the listener has perfect pitch (which is rare). With vision a user only has to look at a number or graph to get an absolute value.

Lack of Orthogonality. Changing one attribute of a sound may affect the others. For example, changing the pitch of a note may affect its perceived loudness and vice versa (see the next section, Perception of Sound).

Transience of Information. Sound disappears when it has been presented and can therefore cause problems. Users must remember the information that the sound contained, or some method of replaying must be provided. With vision the user can easily look back at the display and see the information again. (This is not always the case; think, for example, of an air conditioning system: Its sounds continue for long periods of time and become habituated. Sounds often continue in the background and only become apparent when they change in some way.)

Annoyance Due to Auditory Feedback. There is one problem with sound that has not yet been mentioned but it is the one most commonly brought up against the use of sound in user interfaces: *annoyance*. Because this is such an important topic, it will be discussed in detail later in this section.

Comparing Speech and Nonspeech Sounds for Interface Design

One obvious question is, Why not use speech for output? Why do we need to use nonspeech sounds? Many of the advantages presented above apply to speech as well as nonspeech sounds. There are, however, some advantages to nonspeech sounds. If we think of a visual analogy, speech output is like the text on a visual display and nonspeech sounds are like the icons. Presenting information in speech is slow because of its serial nature; to assimilate information, the user must typically hear it from beginning to end, and many words may have to be comprehended before a message can be understood. With nonspeech sounds, the messages are shorter and therefore more rapidly heard (although the user might have to learn the meaning of the nonspeech sound whereas the meaning is contained within the speech and so requires no learning, just as in the visual case). Speech suffers from many of the same problems as text in text-based computer systems, because this is also a serial medium. Barker and Manji (1989) claimed that an important limitation of text is its lack of expressive capability: It may take many words to describe something fairly simple. Graphical displays were introduced that speeded up interactions because users could see a picture of the application they wanted instead of having to read its name from a list (Barker & Manji, 1989). In the same way, an encoded sound message can communicate its information in fewer sounds. The user hears the sound then recalls its meaning rather than having the meaning described in words. Work has been done on increasing the presentation rate of synthetic speech (Aldrich & Parkin, 1989). This found that the accuracy of recognition decreased as the speech rate went up. Pictorial icons can also be more universal; they can mean the same thing in many different languages and cultures, and nonspeech sounds have similar universality (given some of the different musical cultures).

An important ability of the auditory system is *habituation* in which continuous sounds can fade into the "background" of consciousness after a short period of time. If the sound changes (or stops), then it would come to the foreground of attention because of the sensitivity of the auditory system to change (Buxton, 1989). Habituation is difficult to achieve with speech because of the large dynamic range it uses. According to Patterson (1982), "The vowels of speech are often 30 dB more intense than the consonants, and so, if a voice warning were attenuated to produce a background version with the correct vowel level the consonants would be near or below masked threshold." (p. 11) It is easier to habituate certain types of nonspeech sounds (think of an air conditioner, which you only

notice that it was on when it switches off) and sounds can be designed to facilitate habituation if required.

Baddeley (1990) provided evidence to show that background speech, even at low intensities, is much more disruptive than nonspeech sound when recalling information. He reported the "unattended speech effect." Unattended speech, that is, in the background, causes information to be knocked out of short-term memory, whereas noise or nonspeech sound does not. This problem is unaffected by the intensity of the speech, provided that it is audible. This uncovers a problem for speech at the interface, because it is likely to prove disruptive for other users in the same environment unless it is kept at a very low intensity, and, as noted previously, this can cause problems with the ability to hear consonants.

Nonspeech sounds are also good for presenting continuous information, for example, a graph of stock market data. With speech, particular values could be spoken out at particular times, but there would be no way to monitor the overall trend of the data. Methods for doing this in nonspeech sounds were developed more than 15 years ago (Mansur, Blattner, & Joy, 1985) and have proved to be effective (see the sound graphs later in this chapter for more on this).

This discussion has shown that there are many reasons to think of using nonspeech sounds in addition to speech in human–computer interfaces. There are as yet few interfaces that make good use of both. Speech in general is good for giving instructions and absolute values, nonspeech sounds are good at giving rapid feedback on actions, presenting continuous data and highly structured information quickly. Together, they are an effective means of presenting information nonvisually.

Avoiding Annoyance

The main concern potential users of auditory interfaces have is annoyance caused by sound pollution. There are two aspects to annoyance: A sound may be annoying to the user whose machine is making the noise (the primary user) and/or to others in the same environment who overhear it (secondary users). Buxton (1989) discussed some of the problems of sound and suggested that some sounds help us (information) and some impede us (noise). We therefore need to design sounds so that there are more informative ones and less noise. Of course, one person's informative sounds are another's noise, so it is important to make sure that the sounds on one computer are not annoying for a colleague working nearby.

There are few studies that particularly look at the problems of annoyance due to nonspeech sounds in computers. There are, however, many studies of annoyance from speech (e.g., Berglund, Harder, & Preis, 1994)—from the sounds of aircraft, traffic, or other environmental noise—and most of these suggest that the primary reason for the annoyance of sound is excessive volume. In a different context, Patterson (1989) investigated some of the problems with auditory warnings in aircraft cockpits. Many of the warnings were added in a "better safe than sorry" manner that lead to them being so loud, the pilot's first response was to try and turn them off rather than deal with the problem indicated. One of Patterson's main recommendations was that the volume of the warnings should be reduced.

A loud sound therefore grabs the attention of the primary user, even when the sound is communicating an unimportant event. Because the sound is loud, it travels from one machine to the ears of other people working nearby, increasing the noise in their environment.

How can annoyance be avoided? One key way is to avoid using intensity as a cue in sound design for auditory interfaces. Quiet sounds are less annoying. Listeners are also not good at making absolute intensity judgments (see the next section). Therefore, intensity is not a good cue for differentiating sounds.

Headphones could be used so that only the primary user hears sounds. This may be fine for users of mobile telephones and music players but is not always a good solution for desktop users who do not want to by physically tied to their desks or cut off from their colleagues. Manipulating sound parameters other than intensity can make sounds attention grabbing (but not annoying). Rhythm or pitch can be used to make sounds demanding because the human auditory system is good at detecting changing stimuli (for more see Edworthy, Loxley, & Dennis, 1991). Therefore, if care is taken with the design of sounds in an interface, specifically avoiding the use of volume changes to cue the user, then many of the problems of annoyance due to sound can be avoided, and the benefits of nonspeech sound can be taken advantage of.

PERCEPTION OF SOUND

This section will provide some basic information about the perception of sound that is applicable to nonspeech auditory output. The auditory interface designer must be conscious of the effects of psychoacoustics, or the perception of sound, when designing sounds for the interface. As Frysinger (1990) noted, "The characterization of human hearing is essential to auditory data representation because it defines the limits within which auditory display designs must operate if they are to be effective." (p. 31) Using sounds without regard for psychoacoustics may lead to the user being unable to differentiate one sound from another, unable to hear the sounds or unable to remember them. There is not enough space here to give great detail on this complex area; chapter 26 has more on the basics of human auditory perception. For more detail in general, see Moore (1997).

An important first question is, What is sound? Sounds are pressure variations that propagate in an elastic medium (in this case, the air). The pressure variations originate from the motion or vibration of objects. These pressure variations are what hit the listener's ear and start the process of perceiving the sound. The pressure variations plotted against time can be seen in Fig. 11.1. This shows the simplest form of sound: a sine wave with a single frequency component (as might be produced by a tuning fork). A sound is made up from three basic components: *Frequency* is the number of times per second the wave repeats itself (Fig. 11.1 shows three cycles); it is normally measured in

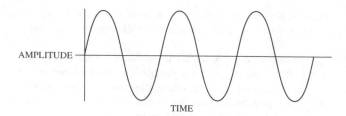

FIGURE 11.1. A sine wave.

Hertz (Hz). *Amplitude* is the deviation away from the mean pressure level, or force per unit area of a sound; it is normally measured in decibels (dB). *Phase* is the position of the start of the wave on the time axis (measured in milliseconds).

Sounds from the real world are normally much more complex than Fig. 11.1 and tend to be made up of many sine waves with different frequencies, amplitudes, and phases. Figure 11.2 shows a more complex sound made of three sine wave components (or *partials*) and the resulting waveform (in a similar way as a sound might be created in an additive synthesis system; see the section on sound synthesis and MIDI). *Fourier analysis* allows a sound to be broken down into its component sine waves (Gelfand, 1981).

The sounds in Figs. 11.1 and 11.2 are *periodic*—they repeat regularly over time. This is common for many types of musical instruments that might be used in an auditory interface. Many natural, everyday sounds (such as impact sounds) are not periodic and do not repeat. The sound in Fig. 11.2 is also *harmonic*—its partials are integer multiples of the lowest (or *fundamental*) frequency. This is again common for musical instruments but not for everyday sounds. Periodic harmonic sounds have a

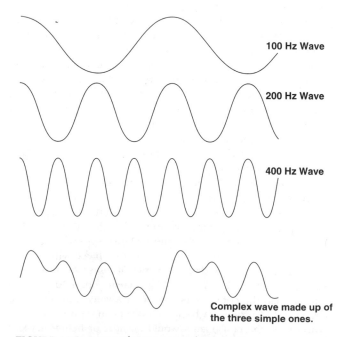

100 Hz Wave

200 Hz Wave

400 Hz Wave

Complex wave made up of the three simple ones.

FIGURE 11.2. A complex wave made up of three components with its fundamental at 100 Hz.

recognizable pitch, whereas nonperiodic, inharmonic sounds tend to have no clear pitch.

The attributes of sound described above are the physical aspects. There is a corresponding set of perceptual attributes (for more on each of these perceptual attributes, see chapter 26). *Pitch* is the perceived frequency of a sound. Pitch is roughly a logarithmic function of frequency. It can be defined as the attribute of auditory sensation in terms of which sounds may be ordered on a musical scale (Moore, 1997). In the Western musical system, there are 96 pitches arranged into 8 octaves of 12 notes. Tones separated by an octave have the frequency ratio 2:1. For example, middle C is 261.63 Hz, the octave above it is at 523.25 Hz, and the octave below is at 130.81 Hz. Pitch is one of the most useful and easily controlled aspects of sound and is useful for auditory interface designers. As Buxton et al. (1991) noted, however, "It is important to be aware of the myriad interactions between pitch and other attributes of sound." (p. 210) For example, pitch is affected by loudness; at less than 2 kHz an increase in intensity increases the perceived pitch, at 3 kHz and over an increase in intensity decreases the perceived pitch.

Humans can perceive a wide range of frequencies. The maximum range we can hear is from 20 Hz to 20 kHz. This decreases with age so that at age 70, a listener might only hear a maximum of 10 kHz. It is therefore important to make sure that the sounds in an auditory interface are perceivable by its users (also poor-quality loudspeakers may not be able to cope with the highest frequencies). Listeners are not good at making absolute judgments of pitch (Moore, 1997). Only 1% of the population has perfect pitch. Another important factor is *tone deafness*. Moore suggested that this is a misnomer and almost everyone is able to tell that two sounds are different; they are not always able to say which is higher or lower, however. This can often be overcome with practice, but it is important for the auditory interface designer to be aware of the problem. Mansur et al. (1985) gave evidence of one other important effect: "There appears to be a natural tendency, even in infants, to perceive a pitch that is higher in frequency to be coming from a source that is vertically higher in space when compared to some lower tone." (p. 171) This is important when creating an auditory interface because it could be used to give objects a spatial position. If only stereo position is available to provide spatial cues in the horizontal plane, then pitch could provide them in the vertical plane. Guidelines for the use of pitch (and the other parameters below) are described in later in this chapter; see Earcons.

Loudness is the perceived intensity of a sound. Loudness (L) is related to intensity (I) according to the *Power Law*: $L = k\ I^{0.3}$ (Gelfand, 1981). Therefore, a 10-dB increase in intensity doubles the perceived loudness of a sound. Loudness is again affected by the other parameters of sound. For example, sounds of between 1 kHz and 5 kHz sound louder at the same intensity level than those outside that frequency range. Humans can perceive a wide range of intensities: The most intense sound that a listener can hear is 120 dB louder than the quietest. This equates to a ratio of 1,000,000,000,000:1 (Moore, 1997). Buxton et al. (1991) also reported that listeners are "very bad at making absolute judgments about loudness." Also, "Our ability to make relative judgments of loudness are limited to a scale of about three different levels." (p. 210) It is also a primary cause

of annoyance so should be used sparingly by auditory interface designers.

Timbre is the "quality" of the sound. It is the attribute of auditory sensation in terms of which a listener can judge two sounds with the same loudness and pitch to be dissimilar. It is what makes a violin sound different from a piano even if both are playing the same pitch at the same loudness. Little is known about the structure or dimensions of timbre. It is known to be based partly on the spectrum and dynamics of a sound. Even though its structure is not well understood, it is one of the most important attributes of sound that an interface designer can use. As Blattner et al. (1989) said, "Even though timbre is difficult to describe and notate precisely, it is one of the most immediate and easily recognizable characteristics of sound" (p. 26) (both *auditory icons* and *earcons* use timbre as one of their fundamental design attributes; see Nonspeech Sound Presentation Techniques later in this chapter). Many of the synthesis techniques in the next section make it easy for a designer to create and use different timbres.

Duration is another important attribute of sound. Sounds of different durations are used to form rhythmic structures that are a fundamental part of music, for example. Duration can also affect the other parameters of sound. For example, for sounds of less than 1 s, loudness increases with duration. This is important in auditory interfaces because short sounds are often needed so that the auditory feedback can keep pace with the interactions taking place so they must be made loud enough for listeners to hear.

Direction is the position of the sound source. This is often overlooked but is an important aspect of sound in our everyday lives. As mentioned above, one of the key differences between sight and hearing is that sounds can be heard as coming from all around the listener. If a sound source is located to one side of the head, then the sound reaching the opposite ear will be reduced in intensity (*interaural intensity difference*, IID) and delayed in time (*interaural time difference*, ITD) (Begault, 1994). These two factors are key in allowing a listener to localize a sound in space. Humans can detect small changes in the position of a sound source. The minimum auditory angle (MAA) is the smallest separation between two sources that can be reliably detected. Strybel, Manligas, and Perrott (1992) reported that in the median plane sound sources only 1° apart can be detected. At 90° azimuth (directly opposite one ear), sources must be around 40° apart. This has important implications for auditory displays. It indicates that higher resolution sounds can be used when presented in front of the user (see Three-Dimensional Sound for more on sound positioning).

TECHNOLOGY AND SOUND PRODUCTION

Most modern computers have sophisticated sound hardware available for the auditory interface designer. This is normally for playing games but is sufficient to do most of the things required by an auditory interface. The aim of this section is to briefly describe some of the main technology that is important for designers to understand when creating interfaces. There are two main aspects to sound production: the first is sound synthesis and the second is sound sampling. A basic overview will be given focusing on aspects related to audio interfaces. For much more detail on sound synthesis and musical instrument digital interface (MIDI), see Roads (1996); for more on sampling, see Pohlmann (1995).

A Brief Introduction to Sound Synthesis and MIDI

The aim of sound synthesis is to generate a sound from a stored model, often a model of a musical instrument. For auditory interfaces, we need a wide range of high-quality sounds that we can generate in real time as the user interacts with the interface. Synthesizers come in three main forms: soundcards on personal computers (PCs), external hardware devices, and it is now becoming possible to do good-quality synthesis in software.

Most synthesizers are *polyphonic* (i.e., able to play multiple notes at the same time, as opposed to *monophonic*). This is important for auditory interface design as one might well want to play a chord made up of several notes. Most modern synthesizers are *multitimbral* (i.e., can play multiple different instruments at the same time). This is again important because in many situations a sound composed of two different instruments might be required. The main forms of synthesis will now be briefly reviewed.

Wavetable synthesis is one of the most common and low-cost synthesis techniques. Many of the most popular PC soundcards use it (such as the AuDigy™ series from Creative Labs and the Santa Cruz™ soundcards from Turtle Beach). The idea behind wavetable synthesis is to use existing sound recordings (which are often difficult to synthesize exactly) as the starting point and create convincing simulations of acoustical instruments based on them (Heckroth, 1994; Roads, 1996). A sample (recording) of a particular sound will be stored in the soundcard. It can then be played back to produce a sound. The sample memory in these systems contains a large number of sampled sound segments and can be thought of as a "table" of sound waveforms that may be looked up and used when needed. Wavetable synthesizers employ a variety of techniques, such as sample looping, pitch shifting, mathematical interpolation, and polyphonic digital filtering, to reduce the amount of memory required to store the sound samples or to get more types of sounds. More sophisticated synthesizers contain more wavetables (perhaps one or more for the initial *attack* part of a sound and then more for the *sustain* part of the sound and then more for the final *decay* and *release* parts). Generally, the more wavetables that are used, the better the quality of the synthesis, but this does require more storage. It is also possible to combine multiple separately controlled wavetables to create a new instrument.

Wavetable synthesis is not the best choice if you want to create new timbres because it lacks some of the flexibility of the other techniques below. Most wavetable synthesizers contain many sounds (often many hundreds), so there may not be a great need to create new ones. For most auditory interfaces, the sounds from a good-quality wavetable synthesizer will be perfectly acceptable.

Frequency modulation (FM) synthesis techniques generally use one periodic signal (the modulator) to modulate the frequency of another signal (the carrier; Chowning, 1975). If the modulating signal is in the audible range, then the result will be a significant change in the timbre of the carrier signal. Each FM voice requires a minimum of two signal generators. These generators are commonly referred to as *operators*, and different FM synthesis implementations have varying degrees of control over the operator parameters. Sophisticated FM systems may use four or six operators per voice, and the operators may have adjustable envelopes that allow adjustment of the attack and decay rates of the signal. FM synthesis is cheap and easy to implement and can be useful for creating expressive new synthesized sounds. If the goal is to recreate the sound of an existing instrument, however, then FM synthesis is not the best choice because it can generally be done more easily and accurately with wavetable-based techniques (Heckroth, 1994).

Additive (and subtractive) synthesis is the oldest form of synthesis (Roads, 1996). Basically multiple sine waves are added together to produce a more complex output sound (subtractive synthesis is basically the opposite: A complex sound has frequencies filtered out to create the sound required). Using this method it is theoretically possible to create any sound (because all complex sounds can be decomposed into sets of sine waves by Fourier analysis). However, it can be difficult to create particular sound. Computer musicians often use this technique because it is flexible and easy to create new and unusual sounds, but it may be less useful for auditory interface designers.

Physical modeling synthesis uses mathematical models of the physical acoustic properties of instruments and objects. Equations describe the mechanical and acoustic behavior of an instrument. The better the simulation of the instrument, the more realistic the sound produced. Mathematical models can be created for a wide range of instruments, and these can be manipulated in flexible ways. Nonexistent instruments can also be modeled and made to produce sounds. Physical modeling is an excellent choice for synthesis of many classical instruments, especially those of the woodwind and brass families. Its parameters directly reflect the ones of the real instrument, and excellent emulations can be produced. The downside is that it can require large amounts of processing power, which, in turn, can limit the polyphony of physical modeling synthesizers (Heckroth, 1994).

Musical Instrument Digital Interface (MIDI). MIDI allows real-time control of electronic instruments (such as synthesizers, samplers, drum machines, etc.) and is now widely used. It specifies a hardware interconnection scheme, a method for data communications, and a grammar for encoding musical performance information (Roads, 1996). For auditory interface designers the most important part of MIDI is the performance data, which is an efficient method for representing sounds. Most soundcards support MIDI with an internal synthesizer and also provide a MIDI interface to connect to external devices. Both Apple MacOS and Microsoft Windows provide good support for MIDI. Most programming languages now come with libraries supporting MIDI commands. For example, versions 1.3 and onwards of the Java programming language from Sun Microsystems has a built in software synthesizer (as part of the Java Sound package), which is directly accessible from Java code.

MIDI performance information is like a piano roll: Notes are set to turn on or off and play different instruments over time. A MIDI message is an instruction that controls some aspect of the performance of an instrument. A MIDI message is made up of a *status byte*, which indicates the type of the message, followed by up to two *data bytes* that give the parameters. For example the *Note On* command takes two parameters: one value giving the pitch of the note required and another the velocity. This makes it a compact form of presentation.

Performance data can be created dynamically from program code or by a sequencer. In an auditory interface, the designer might assign a particular note to a particular interface event, for example, a click on a button. When the user clicks on the button, a MIDI Note On event will be fired; when the user releases the button the corresponding Note Off event will be sent. This is a simple and straightforward way of adding sounds to a user interface. With a sequencer, data can be entered using classical music notation by dragging and dropping notes onto a stave or by using an external piano-style keyboard. This could then be saved to a file of MIDI for later playback (or could be recorded and played back as a sample; see the below).

A Brief Introduction to Sampling

In many ways sampling is simpler than synthesis. The aim is to make a digital recording of an analog sound and then to be able to play it back later with the played back sound matching the original as closely as possible. There are two important aspects: *sample rate* and *sample size*.

Sample Rate. This is the number of discrete "snapshots" of the sound that are taken, often measured per second. The higher the sampling rate, the higher the quality of the sound when it is played back. With a low sampling rate, few snapshots of the sound are taken, and the recording will not match well to the sound being recorded. The *sampling theorem* (Roads, 1996) states that "to be able to reconstruct a signal, the sampling frequency must be at least twice the frequency of the signal being sampled." As mentioned above, the limit of human hearing is around 20 kHz; therefore, a maximum rate of 40 kHz is required to be able to record any sound that a human can hear (Roads, 1996). The standard audio compact disk (CD) format uses a sample rate of 44.1 kHz, meaning that it can record all of the frequencies that a human can hear. If a lower sampling rate is used, then higher frequencies are lost. For example the .au sample format uses a sampling rate of 8 kHz, meaning that only frequencies of less than 4 kHz can be recorded. For more details on the huge range of sample formats, see Bagwell (1988).

Higher sampling rates generate much more data than lower ones and so may not always be suitable if storage is limited (for example, on a mobile computing device). As an auditory interface designer, it is important to think about the frequency range of the sounds needed in an interface, and this might allow the sample rate to be reduced. If the highest quality is required, then you must be prepared to deal with large audio files.

Sample Size. The larger the sample size, the better the quality of the recording because more information is stored at each snapshot of the sound. Sample size defines the volume (or dynamic) range of the sound. With an 8-bit sample, only 256 discrete amplitude (or quantization) levels can be represented. To fit an analog sound into one of these levels might cause it to be rounded up or down, and this can add noise to the recording. CD-quality sounds use 16-bit samples, giving 65,536 different levels, so the effects of quantization are reduced. Many high-quality samplers use 24-bit samples to reduce still further the problems of quantization.

The two main sizes used in most soundcards are 8 and 16 bits. As with sample rates, the main issue is size: 16-bit samples require a lot of storage, especially at high sample rates. Audio CD-quality sound generates around 10 MBytes of data per minute.

Comparing MIDI Synthesis to Sampling for Auditory Interface Design

MIDI is flexible because synthesizers can generate sound in real time as it is needed. If auditory interface designers do not know in advance all of the sounds they might want in their auditory interface, this can be effective: as the sound is needed it is just played by the synthesizer. A system that is based around samples can only play back samples that have been prerecorded and stored.

Another advantage of MIDI is that sounds can be changed. Once a sample has been stored, it is difficult to change it. For example, it is possible to change the speed and pitch of an audio steam independently with MIDI. If a sample is played back at a different speed, its pitch will change, which may cause undesirable effects.

MIDI commands are also very small; each command might only take up 2 or 3 bytes. Generating sounds from code in the auditory interface is straightforward. For instance, files containing high-quality stereo-sampled audio require about 10 MBytes of data per minute of sound, whereas a typical MIDI sequence might consume less than 10 KBytes of data per minute. This is because the MIDI file does not contain the sampled audio data; it contains only the instructions needed by a synthesizer to play the sounds.

Samples have the advantage that the interface designer knows exactly what the sound will be like. With MIDI, you are at the mercy of the synthesizer on the user's machine. It may be of poor quality, and therefore not play the sounds as they were designed. With samples, all of the information about the sound is stored so that one can guarantee it will sound like the recording made (within the limitations of the speakers on the user's machine).

It is also not possible to synthesize all sounds. Much of the work in the area of sound synthesis has focused on synthesizing musical instruments. There are few synthesizers that can do a good job of creating natural, everyday sounds. If a designer wishes to use natural sounds in an interface, then he or she is limited to using samples.

Three-Dimensional (3D) Sound

Much of the recorded sound we hear is in stereo. A stereo recording uses differences in intensity between the ears. From these differences, listeners can gain a sense of movement and position of a sound source in the stereo field. The perceived position is along a line between the two loudspeakers or inside the head between the listeners' ears if they are wearing headphones. This simple, inexpensive technique can give useful spatial cues at the auditory interface. This is being taken further to make sounds appear as if they come from around a user (in virtual 3D) when only a small number of loudspeakers (or even just a pair of headphones) are used (Begault, 1994). As well as the ITD and IID, in the real world we use our pinnae (the outer ear) to filter the sounds coming from different directions so that we know where they are coming from. To simulate sounds as coming from around the user and outside of the head when wearing headphones, sounds entering the ear are recorded by putting microphones into the listeners' ear canals. The differences between the sound at the sound source and at the eardrum are then calculated, and the differences, or head-related transfer functions (HRTFs), derived are used to create filters with which stimuli can be synthesized (Begault, 1994). This research is important because 3D auditory interfaces can be created that are more natural, with sounds presented around the user as they would be in real life.

Three-dimensional sounds over headphones can be generated by most current PC soundcards. These are often used for games but not commonly for everyday interactions; however, this is beginning to change (see the work on Nomadic Radio in the last section of this chapter).

The main problem with providing simulated 3D sound through headphones comes from the HRTFs used. If the user's ears are not like those of the head from which the HRTFs were generated, the sounds will seem to come from inside the user's head, and not outside. It is also easy to confuse front and back so that listeners cannot tell where a sound is. Vertical positioning is also difficult to do reliably. This means that many designers who use 3D sound in their interfaces limit the sounds to a plane cutting through the head horizontally at the level of the ears. This reduces the space in which sounds can be presented but avoids many of the problems of users not being able to localize the sounds properly.

NONSPEECH SOUND PRESENTATION TECHNIQUES

Two main types of information presentation techniques have emerged in the area of sound in human–computer interfaces: *auditory icons* and *earcons*. Substantial research has gone into developing both of these, and the main work is reviewed below.

Auditory Icons

Gaver (1989, 1997) developed the idea of auditory icons. These are natural, everyday sounds that can be used to represent actions and objects within an interface. He defined them as "everyday sounds mapped to computer events by analogy with everyday sound-producing events. Auditory icons are like sound effects for computers." (p. 69) Auditory icons rely on an analogy between the everyday world and the model world of the computer (Gaver, 1997; for more examples of the use of earcons see the work on Mercator and audio aura described later).

Gaver used sounds of events that are recorded from the natural environment, for example, tapping or smashing sounds. He used an "ecological listening" approach, suggesting that people do not listen to the pitch and timbre of sounds but to the sources that created them. When pouring liquid, a listener hears the fullness of the receptacle, not the increases in pitch. Another important property of everyday sounds is that they can convey multidimensional data. When a door slams, a listener may hear the size and material of the door, the force that was used, and the size of room on which it was slammed. This could be used within an interface so that selection of an object makes a tapping sound, the type of material could represent the type of object, and the size of the tapped object could represent the size of the object within the interface.

Gaver used these ideas to create auditory icons and from these built the SonicFinder (Gaver, 1989). This is an interface that ran on the Apple Macintosh and provided auditory representations of some objects and actions within the interface. Files were given a wooden sound, applications a metal sound, and folders a paper sound. The larger the object, the deeper the sound it made. Thus, selecting an application meant tapping it—it made a metal sound that confirmed that it was an application and the deepness of the sound indicated its size. Copying used the idea of pouring liquid into a receptacle. The rising of the pitch indicated that the receptacle was getting fuller and the copy progressing.

To demonstrate how the SonicFinder worked, a simple interaction is provided in Fig. 11.3 showing the deletion of a folder. In A), a folder is selected by tapping on it; this causes a "papery" sound, indicating that the target is a folder. In B), the folder is dragged toward the wastebasket, causing a scraping sound. In C), the wastebasket becomes highlighted, and a "clinking" sound occurs when the pointer reaches it. Finally, in D), the folder is dropped into the wastebasket, and a smashing sound occurs to indicate it has been deleted (the wastebasket becomes "fat" to indicate there is something in it).

Problems can occur with representational systems such as auditory icons because some abstract interface actions and objects have no obvious representation in everyday sound. Gaver used a pouring sound to indicate copying because there was no natural equivalent; this is more like a "sound effect." He suggested the use of movielike sound effects to create sounds for things with no easy representation. This may cause problems if the sounds are not chosen correctly because they will become more abstract than representational, and the advantages of auditory icons will be lost.

Gaver developed the ideas from the SonicFinder further in the *ARKola* system (Gaver, Smith, & O'Shea, 1991), which modeled a soft drinks factory. The simulation consisted of a set of nine machines split into two groups: those for input and those

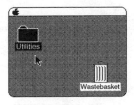

A) Papery tapping sound to show selection of folder.

B) Scraping sound to indicate dragging folder.

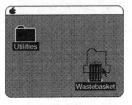

C) Clinking sound to show wastebasket selected.

D) Smashing sound to indicate folder deleted.

FIGURE 11.3. An interaction showing the deletion of a folder in the SonicFinder (from Gaver, 1989).

for output. The input machines supplied the raw materials; the output machines capped the bottles and sent them for shipping. Each machine had an on–off switch and a rate control. The aim of the simulation was to run the plant as efficiently as possible, avoid waste of raw materials, and make a profit shipping bottles. Two users controlled the factory, with each user able to see approximately one third of the whole plant. This form of plant was chosen because it allowed Gaver et al. to investigate how the sounds would effect the way users handled the given task and how they affected the way people collaborated. It was also an opportunity to investigate how different sounds would combine to form an auditory *ecology* (integrated set of sounds). Gaver et al. related the way the different sounds in the factory combined to the way a car engine is perceived. Although the sounds are generated by multiple distinct components, these combine to form what is perceived as a unified sound. If something goes wrong, the sound of the engine will change, alerting the listener to the problem, but in addition, to a trained ear the change in the sound would alert the listener to the nature of the problem. The sounds used to indicate the performance of the individual components of the factory were designed to reflect the semantics of the machine.

Each of the machines had a sound to indicate its status over time, for example, the bottle dispenser made the sound of clinking bottles. The rhythm of the sounds reflected the rate at which the machine was running. If a machine ran out of supplies or broke down its sound stopped. Sounds were also added to indicate that materials were being wasted. A splashing sound indicated that liquid was being spilled, the sound of smashing bottles indicated that bottles were being lost. The system was designed so that up to 14 sounds could be played at once. To reduce the chance that all sounds would be playing simultaneously, sounds were pulsed once a second rather than playing continuously.

An informal evaluation was undertaken in which pairs of users were observed controlling the plant, either with or without sound. These observations indicated that the sounds were effective in informing the users about the state of the plant and that the users were able to differentiate the different sounds and identify the problem when something went wrong. When the sounds were used, there was much more collaboration between the two users. This was because each could hear the whole plant and therefore help out if there were problems with machines that the other was controlling. In the visual condition, users were not as efficient at diagnosing what was wrong even if they knew there was a problem.

One of the biggest advantages of auditory icons is the ability to communicate meanings that listeners can easily learn and remember; other systems (for example earcons, see the next section) use abstract sounds, the meanings of which are more difficult to learn. Problems did, occur however, with some of the warning sounds used, as Gaver et al. indicated: "the breaking bottle sound was so compelling semantically and acoustically that partners sometimes rushed to stop the sound without understanding its underlying cause or at the expense of ignoring more serious problems." (p. 89) Another problem was that when a machine ran out of raw materials, its sound just stopped; users sometimes missed this and did not notice that something had gone wrong.

Design Guidelines for Auditory Icons. As yet few formal studies have been undertaken to investigate the best design for auditory icons, so there is little guidance for how to design them effectively. Mynatt (1994) has proposed the following basic methodology: (1) Choose short sounds that have a wide bandwidth, and for which length, intensity, and sound quality are roughly equal. (2) Evaluate the identifiability of the auditory cues using free-form answers. (3) Evaluate the learnability of the auditory cues that are not readily identified. (4) Test possible conceptual mappings for the auditory cues using a repeated-measures design in which the independent variable is the concept the cue will represent. (5) Evaluate possible sets of auditory icons for potential problems with masking, discriminability, and conflicting mappings. (6) Conduct usability experiments with interfaces using the auditory icons.

Earcons

Earcons were developed by Blattner et al. (1989). They use abstract, synthetic tones in structured combinations to create auditory messages. Blattner et al. (1989) defined earcons as "non-verbal audio messages that are used in the computer/user interface to provide information to the user about some computer object, operation or interaction." (p. 13) Unlike auditory icons, there is no intuitive link between the earcon and what it represents; the link must be learned. They use a more traditional musical approach than auditory icons.

Earcons are constructed from simple building blocks called *motives* (Blattner et al., 1989). These are short, rhythmic sequences of pitches that can be combined in different ways.

Blattner suggest the most important features of motives are as follows:

Rhythm: Changing the rhythm of a motive can make it sound very different. Blattner et al. (1989) described this as the most prominent characteristic of a motive.

Pitch: There are 96 pitches in the Western musical system, and these can be combined to produce a large number of different motives.

Timbre: Motives can be made to sound different by the use of different timbres, for example, playing one motive with the sound of a violin and the other with the sound of a piano.

Register: This is the position of the motive in the musical scale. A high register means a high pitched note and a low register a low note. The same motive in a different register can convey a different meaning.

Dynamics: This is the volume of the motive. It can be made to increase as the motive plays (crescendo) or decrease (decrescendo).

There are two basic ways in which earcons can be constructed. The first, and simplest, is by using *compound earcons*. These are simple motives that can be concatenated to create more complex earcons. For example, a set of simple, one element motives might represent various system elements such as *create*, *destroy*, *file*, and *string* (see Fig. 11.4A). These could then be concatenated to form earcons (Blattner et al., 1989). In the figure, the earcon for *create* is a high-pitched sound that gets louder; for *destroy*, it is a low-pitched sound that gets quieter. For *file*, there are two long notes that fall in pitch, and for *string* two short notes that rise. In Fig. 11.4B, the compound earcons can be seen. For the *create file* earcon, the *create* motive is simply followed by the *file* motive. This provides a simple and effective method for building up earcons.

Hierarchical earcons are more complex but can be used to represent more complex structures in sound. Each earcon is a node in a tree and inherits properties from the earcons above it. Figure 11.5 shows a hierarchy of earcons representing a family of errors. The top level of the tree is the family rhythm. This sound has a rhythm and no pitch; the sounds used are clicks. The rhythmic structure of Level 1 is inherited by Level 2, but

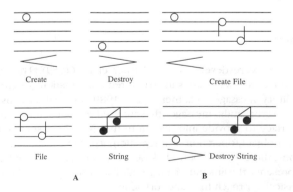

FIGURE 11.4. Compound earcons: **A** shows the four audio elements *create*, *destroy*, *file*, and *string*. **B** shows the compound earcons *create file* and *destroy string* (Blattner et al., 1989).

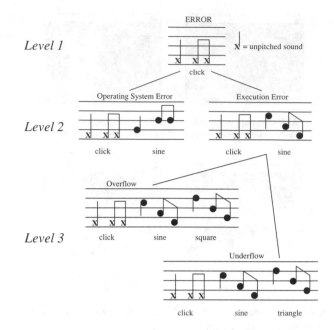

FIGURE 11.5. A hierarchy of earcons representing errors (from Blattner et al., 1989).

this time a second motive is added in which pitches are put to the rhythm. At this level, Blattner et al. suggested that the timbre should be a sine wave, which produces a "colorless" sound. This is done so that at Level 3 the timbre can be varied. At Level 3 the pitch is also raised by a semitone to make it easier to differentiate from the pitches inherited from Level 2. Other levels can be created where register and dynamics are varied.

Blattner et al. proposed the design of earcons but did not develop or test them. Brewster et al. (1994) carried out a detailed evaluation of compound and hierarchical earcons based on the design proposed by Blattner, simple system beeps and a richer design based on more complex musical timbres using psychoacoustical research (see Perception of Sound earlier in this chapter and chapter 26). In these experiments, participants were presented with earcons representing families of icons, menus, and combinations of both (examples can be heard at www.dcs.gla.ac.uk/~stephen/demos.shtml). They heard each sound three times and then had to identify them when played back. Results showed that the more complex musical earcons were significantly more effective than both the simple beeps and Blattner's proposed design, with more than 80% recalled correctly. Brewster et al. found that timbre was a much more important than previously suggested, whereas pitch on its own was difficult to differentiate. The main design features of the earcons used were formalized into a set of design guidelines, as follows.

Timbre: This is the most important grouping factor for earcons. Use musical instrument timbres with multiple harmonics because this helps perception and can avoid masking. These timbres are more recognizable and differentiable.

Pitch and register: If listeners are to make absolute judgments of earcons, then pitch and register should not be used as a cue on their own; a combination of register and another parameter gives better rates of recall. If register alone must be used, then there should be large differences (two or three octaves) between earcons. Much smaller differences can be used if relative judgments are to be made. The maximum pitch used should be no higher than 5 kHz and no lower than 125 Hz to 150 Hz so that the sounds are not easily masked and are within the hearing range of most listeners.

Rhythm, duration, and tempo: Make rhythms as different as possible. Putting different numbers of notes in each earcon is effective. Earcons are likely to be confused if the rhythms are similar even if there are large spectral differences. Very short note lengths might not be noticed, therefore do not use sounds less than 0.03 s. Earcons should be kept as short as possible so that they can keep up with interactions in the interface being sonified. Two earcons can be played in parallel to speed up presentation.

Intensity: This should not be used as a cue on its own because it is a major cause of annoyance. Earcons should be kept within a narrow intensity range so that annoyance can be avoided.

One aspect that Brewster also investigated was musical ability—because earcons are based on musical structures, is it only musicians who can use them? The results showed that the more complex earcons were recalled equally well by nonmusicians as by musicians, indicating that they are useful to a more general audience of users.

In a further series of experiments, Brewster (1998b) looked in detail at designing hierarchical earcons to represent larger structures (with over 30 earcons at four levels). These were designed building on the guidelines above. Users were given a short training period and then were presented with sounds, and they had to indicate where the sound was in the hierarchy.

Results were again good, with participants recalling more than 80% correctly, even with the larger hierarchy used. The study also looked at the learning and memorability of earcons over time. Results showed that even with small amounts of training, users could get good recall rates, and that the recall rates of the same earcons tested a week later was unchanged. More recent studies (Leplâtre & Brewster, 2000) have begun to investigate hierarchies of more than 150 nodes (for representing mobile telephone menus). For examples of earcons in use, see the sonically enhanced widgets, Audiograph, and Palm III work in The Applications of Auditory Output, later in this chapter.

Comparing Auditory Icons and Earcons

Earcons and auditory icons are both effective at communicating information in sound. There is more formal evidence of this for earcons as more basic research has looked at their design. There is less basic research into the design of auditory icons, but the systems that have used them in practice have been effective. More detailed research is needed into auditory icons to correct this problem and to provide designers with guidance on how to create effective sounds. It may be that each has advantages over the other in certain circumstances and that a combination of both is the best. In some situations, the intuitive nature of

auditory icons may make them favorable. In other situations, earcons might be best because of the powerful structure they contain, especially if there is no real-world equivalent of what the sounds are representing. Indeed, there may be some middle ground where the natural sounds of auditory icons can be manipulated to give the structure of earcons.

The advantage of auditory icons over earcons is that they are easier to learn and remember because they are based on natural sounds and the sounds contain a semantic link to the objects they represent. This may make their association to certain, more abstract actions or objects within an interface more difficult. Problems of ambiguity can also occur when natural sounds are taken out of the natural environment and context is lost (people also may have their own idiosyncratic mappings). If the meanings of auditory icons must be learned, they lose some of their advantages.

Earcons are abstract so their meaning must always be learned. This may be a problem, for example, in "walk up and use"-type applications. Research has shown that little training is needed if the sounds are well designed and structured. Leplâtre and Brewster (2000) have begun to show that it is possible to learn the meanings implicitly while using an interface that generates the sounds as it is being used. However, some form of learning must take place. According to Blattner et al. (1989), earcons may have an advantage when there are many highly structured sounds in an interface. With auditory icons, each must be remembered as a distinct entity because there is no structure linking them together. With earcons there is a strong structure linking them that can be manipulated easily. There is not yet any experimental evidence to support this.

"Pure" auditory icons and earcons make up the two ends of a presentation continuum from representational to abstract (see, Fig. 11.6). In reality, things are less clear. Objects or actions within an interface that do not have an auditory equivalent must have an abstract auditory icon made for them. The auditory icon then moves more toward the abstract end of the continuum. When hearing an earcon, the listener may hear and recognize a piano timbre, rhythm and pitch structure as a kind of "catch phrase" representing an object in the interface; he or she does not hear all the separate parts of the earcon and work out the meaning from them (listeners may also try and put their own representational meanings on earcons, even if the designer did not intend it as found by Brewster, 1998b). The earcon then moves more toward the representational side of the continuum. Therefore, earcons and icons are not necessarily as far apart as they might appear.

There are not yet any systems that use both types of sounds to their full extent, and this would be an interesting area to investigate. Some parts of a system may have natural analogs in sound, and therefore auditory icons could be used; other parts might

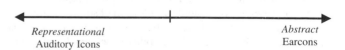

Representational   Auditory Icons   *Abstract*   Earcons

FIGURE 11.6. The presentation continuum of auditory icons and earcons.

be more abstract or structured, and earcons would be better. The combination of the two would be the most beneficial. This is an area ripe for further research.

THE APPLICATIONS OF AUDITORY OUTPUT

Auditory output has been used in a wide range of different situations and applications. This section will outline some of the main areas of use and will highlight some of the key papers in each area (for more uses of sound, see the ICAD or Association for Computing Machinery Computer-Human Interaction (ACM CHI) www.acm.org/sigchi series of conferences).

Sonic Enhancement and Graphic Replacement in Graphical User Interfaces

One long-running strand of research in the area of auditory output is in the addition of sound to standard graphical displays to improve usability. One reason for doing this is that users can become overloaded with visual information on large, high-resolution displays. In highly complex graphical displays, users must concentrate on one part of the display to perceive the visual feedback, so that feedback from another part may be missed. This is important for situations in which users must notice and deal with large amounts of dynamic data. For example, imagine working on your computer writing a report and are monitoring several ongoing tasks such as a compilation, a print job, and downloading files from the Internet. The word-processing task will take up your visual attention because you must concentrate on what you are writing. To check when your printout is done, the compilation has finished, or the files have downloaded, you must move your visual attention away from the report and look at these other tasks. This causes the interface to intrude into the task you are trying to perform. If information about these other tasks was presented in sound, you could continue looking at the report but hear information in the background about the other tasks. To find out how the file download was progressing, you could just listen to the download sound without moving your visual attention from the writing task.

There are two aspects to this work: *enhancement* and *replacement*. Research has gone into looking at how feedback from widgets can be improved by the addition of sound. Alternatively, feedback can be taken away from the visual sense altogether and be presented only in sound. Examples of both types of designs are described later.

One of the earliest pieces of work on sonic enhancement of an interface was Gaver's (1989) SonicFinder described earlier, which used auditory icons to present information about the Macintosh interface redundantly with the graphic display. Brewster (1998a) investigated the addition of sound to enhance graphic buttons. An analysis of the way buttons are used was undertaken, highlighting some usability problems. It was found that the existing, visual feedback did not indicate when mispresses of a button might have occurred. For example, the selection of a graphic button is shown in Fig. 11.7 (starting with

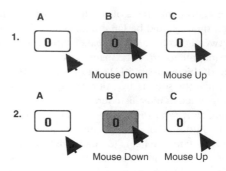

FIGURE 11.7. The visual feedback presented by a graphical button when selected. Figure **1** shows a correct selection and **2** shows a slip-off (from Brewster, 1998a).

1A and 2A). The button highlights when it is pressed down (Fig. 11.71B and 2B). There is no difference in feedback between a correct (Fig. 11.71C) and a misselection (Fig. 11.72C), in which the user moves the mouse off the graphic button before the selection is complete. The user could therefore "slip off" the button, fail to press it, and receive no feedback. This error can happen when the user is moving away from the button and on to some other task. For example, the user moves to a toolbar to press the "bold" button and then moves back to the text to position the cursor to start typing. The button press and the mouse move overlap, and the button is not pressed. It is difficult for the user to notice this because no feedback is given.

The problems could not be solved easily by adding more graphic feedback; the user is no longer looking at the button's location so any feedback given there will be missed. Feedback could be given at the mouse location, but we cannot be sure the user will be looking there either. Brewster designed a new button that used auditory feedback to indicate more about the state of the interaction. This was advantageous because sound is omni-directional, and the user does not need to focus attention on any part of the screen to perceive it.

Three earcons were used to improve the effectiveness of graphical buttons. An organ timbre was used for all of the sounds. When the user moved over a button, a continuous tone was played at 130 Hz at a volume just above the background sound level. This informed the user the cursor was over the target but could easily be habituated. When the mouse was pressed down over the graphic button, a continuous tone was played at 261 Hz. The third sound indicated that the graphic button had been successfully selected. This sound consisted of two short tones with a pitch of 1046 Hz and duration of 40 ms. This sound was not played if a slip-off error occurred. If the user pressed the button very quickly, then only the success sound was played to avoid unnecessary feedback.

An experimental evaluation of these sounds was undertaken. The participants were presented with a graphic number pad of buttons. Using this number pad, they had to enter as many five digit strings as possible within 15 minutes.

Results from the study showed that users recovered from slip-off errors significantly faster and with significantly fewer mouse clicks when sounds were present in the buttons. Users also significantly preferred the buttons with sound when asked to rate subjective preference. They also did not rate the buttons as more annoying than the standard graphic ones. An interesting point to note was the use of no sound when a sound was expected could be attention grabbing. The participants could easily recognize a slip-off because of the demanding nature of the success sound *not* being played. This is important because reducing the amount of feedback presented is one way to make sure that it is not annoying.

An example of a sonified widget that replaced a visual one is the auditory progress bar (Crease & Brewster, 1998). Progress bars are a common feature of most graphic interfaces indicating the current state of a task, such as downloading documents from the Web or copying files from one disk to another. If downloads are occurring over long periods of time then the progress bar is likely to get pushed behind the window in which the user is working, and completion or errors may not be noticed. A better presentation method would be to use sound so that the user could monitor the download while working in another window at the same time. In Crease and Brewster's progress bar four sounds were used:

Endpoint sound: This marked the end of the progress bar. This was a single bass guitar note played for 500 ms every second during the download and was of a fixed pitch (65 Hz). This sound was played as a discrete note every second rather than a continuous note to minimize any annoyance. A bass instrument was chosen because this sound is the root upon which the progress sound is based, in a similar way that a bass line is the root to a melody in a tune.

Progress sound: This was a single organ note played for 250 ms every second during the download, half a second after the endpoint sound started. The pitch of this note was used to indicate the percentage completed. The pitch of the note starts at 261 Hz, and as the task progresses it moved down toward 65 Hz (the pitch of the endpoint sound) in proportion to the amount of the task completed.

Rate of progress sound: This sound was used to indicate the rate at which the task was being completed. Each note used a piano instrument with a pitch of 65 Hz and played alongside the progress sound. Each note was 10 ms long. At least two notes would be played at regular intervals every second. As the rate of the task increased, more notes were played, up to a maximum of 12 per second.

Completion sound: Once the task was completed, three chords were played. Each chord consisted of two notes played for 250 ms with a pitch of 65 Hz except for the third chord that was played for 500 ms. Chords were played rather than individual notes to make this sound more demanding. The final chord was lengthened to indicate completion. If the task completed unsuccessfully, a similar but discordant sound alerted the user.

An experiment was conducted to test this purely auditory progress bar against a standard visual one. Users were required to type in text while simulated file downloads took place (represented either visually or sonically). The results showed that subjective workload was reduced with the audio progress

bar (people felt that using audio was much less demanding). People also preferred it to the standard one and completed their download tasks significantly faster compared with the standard visual one. These results showed that the visual progress bar could be removed from the display and users could perform better because they could share the tasks of typing and monitoring between their two senses.

Other widgets have been successfully sonified. Beaudouin-Lafon and Conversey (1996) showed that nonspeech sounds could improve usability of scrollbars, Maury, Athenes, and Chatty (1999) added sounds to improve menu selection times in drop-down menus. Brewster and colleagues have investigated a wide range of visual widgets, including scrollbars, menus, tool palettes, and drag and drop, and added sound to them. These widgets have also been included into a toolkit (Crease, Gray, & Brewster, 2000) that designers can use to add sound to their interfaces (for a demonstration of some of these widgets, see http://www.dcs.gla.ac.uk/research/audio_toolkit).

Sound for Users With Visual Impairments

One of the most important uses for nonspeech sound is in interfaces for people with visual disabilities (see chapter 26 for more on the causes and problems of visual disabilities). One of the main deprivations caused by blindness is decreased access to information. A blind person will typically use a screen reader and a voice synthesizer to use a computer (Edwards, 1995; Raman, 1996). The screen reader extracts textual information from the computer's video memory and sends it to the speech synthesizer to speak it. This works well for text but not for the graphical components of current user interfaces (chapter 26 discusses this in more detail). It is surprising to find that most of the commercial applications used by blind people make little use of nonspeech sound, concentrating on synthetic speech output. This is limiting (as discussed earlier) because speech is slow, can overload short-term memory, and is not good for presenting certain types of information; for example, it is not possible to render many types of images via speech, so these can become inaccessible to blind people. One reason sound hasn't been used more extensively is the question how to employ it effectively; as Edwards (1995) says, "Currently the greatest obstacle to the exploitation of the variety of communications channels now available is our lack of understanding of how to use them." (p. xvii) The combination of speech and nonspeech sounds can increase the amount of information presented to the user. As long as this is done in a way that does not overload the user, it can improve access to information. Some of the main research into the use of nonspeech auditory in interfaces for blind people is described here.

Soundtrack was an early attempt to create a word-processor designed to be used by blind persons and was developed by Edwards (1989). It used earcons and synthetic speech as output and was designed so that the objects a sighted user would see in an interface, for example, menus and dialogues, were replaced by auditory equivalents that were analogies of their visual counterparts. Its interface was constructed from auditory objects with which the user could interact. They were defined by a

File Menu	Edit Menu	Sound Menu	Format Menu
Alert	Dialog	Document 1	Document 2

FIGURE 11.8. Soundtrack's main screen (from Edwards, 1989).

location, a name, a sound, and an action. They were arranged into a grid of two layers (see Fig. 11.8), analogous to menus.

Each auditory object made a sound when the cursor entered it, and these could be used to rapidly navigate around the screen. Soundtrack used sine waves for its audio feedback. Chords were built up for each menu dependent on the number of menu items. For the edit menu, a chord of four notes was played because there were four menu items within it (cut, copy, paste, and find).

The base sounds increased in pitch from left to right, as in the normal representation of a musical scale (for example, on a piano), and the top layer used higher pitches than the bottom. Using these two pieces of information a user could quickly find his or her position on the screen. If any edge of the screen was reached, a warning sound was played. If at any point the user got lost or needed more precise information, he or she could click on an object, and it would speak its name.

Double-clicking on a menu object took the user to the second level of Soundtrack: the menu items associated with the chosen menu. Moving around is similar to moving on the main screen. Each item had a tone—when moving down the menu the pitch of the tone decreased, and when moving up the pitch increased.

Results from trials with blind users showed that it was very successful, and users could navigate around the interface and find menus and items easily. The main drawback of Soundtrack was that it was not a general solution to the problem of visual interfaces; it could only be applied to a word-processor, the same solution could not be used to help guide a user around the PC desktop, for example. Soundtrack did prove, however, that a full auditory representation of a visual interface could be created with all the interactions based in sound. It showed that the human auditory system was capable of controlling an interaction with a computer and therefore using purely auditory interfaces was possible.

The approach taken in Soundtrack was to take the visual interface to a word processor and translate it into an equivalent auditory form. The Mercator system (Mynatt & Edwards, 1995; Mynatt & Weber, 1994) took a broader approach. The designers' goal was to model and translate the graphic interfaces of X Windows applications into sound without modifying the applications (and thus create a more general solution than Soundtrack's). Their main motivation was to simulate many of the features of graphic interfaces to make graphic applications accessible to blind users and keep coherence between the audio and visual interfaces so that blind and sighted users could interact and work together on the same applications. This meant that the auditory version of the interface had to facilitate the same mental model as the visual one. This did not mean that they translated every pixel on the screen into an auditory form; instead, they modeled the interaction objects that were present. If there was a menu, dialogue box, or button on the screen, it was

displayed but in an auditory form. Modeling the pixels exactly in sound was ineffective because of the different nature of visual and auditory media and the fact that graphic interfaces had been optimized to work with the visual sense (for example, the authors claimed that an audio equivalent of overlapping windows was not needed because overlapping was just an artifact of a small visual display). Nonspeech sound was an important aspect of their design to make the iconic parts of a graphic interface usable.

Mercator used three levels of nonspeech auditory cues to convey symbolic information presented as icons in the visual interface. The first level addressed the question "what is this object?" In Mercator, the type of an interface object was conveyed with an auditory icon. For example, touching a window sounded like tapping a piece of glass, container objects sounded like a wooden box with a creaky hinge, and text fields used the sound of a manual typewriter. Although the mapping was easy for interface components such as trashcan icons, it was less straightforward for components that did not have simple referents in reality (e.g., menus or dialogue boxes, as discussed earlier). In Mercator, auditory icons were also parameterized to convey more detailed information about specific attributes such as menu length. Global attributes were also mapped into changes in the auditory icons. For example highlighting and graying-out are common to a wide range of different widgets. To represent these Mynatt et al. used sound filters. A low-pass filter was used to make the sound of a grayed-out object duller and more muffled.

The next two pieces of work look at how visual representations of data can be transferred into sound. Mansur et al. (1985) reported the first significant study of presenting data in sound. Their study, which laid out the research agenda for subsequent research in sound graphs, used sound patterns to represent two-dimensional line graphs. Their prototypical system provided blind people with a means of understanding line graphs similar to printed graphs for those with sight. Their study used auditory graphs that had a 3-s continuously varying pitch to present the graphed data. The value on the y-axis of the graph was mapped to pitch and the x-axis to time. This meant that a listener could hear the graph rise and fall over time in a way similar to how a sighted person could see the line rising and falling. In an experiment, the auditory graphs were compared with engraved plastic tactile representations of the same data.

In general, results were good, with blind participants able to extract much information about the graphs. They found that their approach was successful in allowing distinctions to be made between straight and exponential graphs, varying monotonicity in graphs, convergence, and symmetry. They did, however, find that there were difficulties in identifying secondary aspects of graphs such as the slope of the curves. They suggested that a full sound graph system should contain information for secondary aspects of the graph, such as the first derivative. Their suggestion was to encode this information by adding more overtones to the sound to change the timbre. They also suggested using special signal tones to indicate a graph's maxima or minima, inflection points, or discontinuities.

When comparing auditory representations with tactile ones, they discovered that mathematical concepts such as asymmetry, monotonicity, and the slopes of lines could be determined more quickly using sound. The relative accuracy of the subjects on five sets of test questions was virtually the same (sound = 84.6%, tactile = 88.9%), and they stated that better performance could be expected with greater training. Several further studies have been undertaken in this area to develop this presentation technique further, most notably Flowers and Hauer (1992; 1995).

Alty and Rigas developed the Audiograph system to present diagrams and drawings to blind people (Alty & Rigas, 1998; Rigas & Alty, 1997). Their main objective was to see if music alone could be used to convey meaningful information to users about the spatial layout of objects in a simple graphic drawing tool. The following information was communicated using music in the system: (1) the current position of the cursor or the position of a graphic object, (2) the nature of a graphic object (e.g., its type, size, and shape), and (3) the overall position of graphic objects using various scanning techniques. A coordinate point in the display was described using a musical mapping from coordinate value to pitch (high pitch = high coordinate value), and X and Y coordinates were distinguished by timbre (organ and piano, respectively) generated via a MIDI synthesizer. The earcon for a coordinate pair was the X coordinate motive followed by the Y coordinate motive. This was extended to present the coordinates of shapes such as lines, circles and squares. Rigas and Alty (1997) described a *note-sequence* technique that they used to do this. Taking a start point, play all of the notes between the start point and the position required. This communicates length through pitch and time. They found this method successful after experimental evaluation. For example, the horizontal line in Fig. 11.9 would be played as a series of notes with increasing pitch in the organ instrument but a static pitch in the piano. The rectangle would alternate between rising organ/static piano, static organ/falling piano, falling organ/static piano, and static organ/rising piano as the cursor moved round from the top left.

Users navigated the drawing space using the cursor keys. Rigas and Alty discovered that even before training, more than 80% of users recognized the shape of a circle and straight line presented with the method described. Their experiments also show that users could estimate the size of graphic objects as well as their overall shape to an accuracy of within 10%.

They also experimented with three different scanning techniques (using the note–sequence technique) to enable users to obtain an overall appreciation of the graphic space. These were (1) top-down scanning (presenting the drawing area starting at the top left-hand corner and scanning progressively down the area left to right playing any object encountered along the way), (2) center scanning (starting at the center of the screen and scan the area in increasing circles), and (3) ascending scanning (in which scanning of objects was done in ascending order of size).

Rigas and Alty revealed that participants could get a broad picture of the diagrams with these techniques. The diagrams reproduced by their blind subjects on raised grid paper showed that the number of objects perceived in the space and their distribution were broadly in agreement but that the perception of size was not as accurate. They stated that this may be partly explained by the difficulty of drawing tasks for their blind users.

These four investigations have shown that sophisticated nonspeech sounds can present a lot of complex information in compact form. They can make something as complex as a graphic interface usable by a blind or partially sighted person.

Sound for Wearable and Mobile Computers

One of the major growth areas in computing at the end of the 20th and the start of the 21st centuries has been in mobile computing. People no longer use computers only while sitting at a desk. Mobile telephones, personal digital assistants (PDAs), and handheld computers are now some of the most widely used devices (chapters 32 and 33 provide more details). One problem with these devices is that there is a limited amount of screen space on which to display information. The screens are small because the devices must be able to fit into the hand or pocket to be easily carried. Small screens can easily become cluttered with information and widgets, and this presents a difficult challenge for interface designers (Brewster & Cryer, 1999; Sawhney & Schmandt, 2000).

The graphic techniques for designing interfaces on desktop interfaces do not apply well to handheld devices. Screen resources are limited; often screens are black and white to reduce cost and power consumption. Memory and processing power are much reduced from desktop systems. In many cases, however, interface designs and interaction techniques have been taken straight from standard desktop graphic interfaces (where screen space and other resources are not a problem) and applied directly to mobile devices. This has resulted in devices that are difficult to use, with small text that is hard to read, cramped graphics, and little contextual information (Brewster & Murray, 2000; Hindus et al., 1995). Speech and nonspeech sounds provide an important way to solve these problems.

Another reason for using sound is that if users are performing tasks while walking or driving, they cannot devote all of their visual attention to the mobile device. Visual attentional resources must remain with the main task for safety. It is therefore difficult to design a visual interface that can work well under these circumstances. An alternative, sonically enhanced interface would require less visual attention and therefore potentially interfere less in the main activity in which the user is engaged.

Three main pieces of work are surveyed in this section, covering the main approaches taken in this area. The first adds sound to the existing interface of a mobile computer to improve

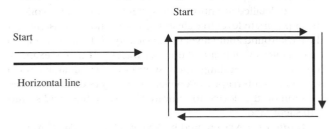

FIGURE 11.9. The musical description of graphic shapes from Rigas and Alty (1997).

usability; the second creates a purely auditory interface for a mobile, and the third an auditory environment that users move through to receive information.

Brewster developed the ideas of sonified buttons as described earlier and applied them to buttons on the 3Com Palm series of pen-based handheld computers (Brewster & Cryer, 1999). Many of the same feedback problems with buttons apply in handhelds as in desktops, but are worse because the screen is smaller (and may be difficult to see when the device is moving or the sun is shining). In addition, there is the problem of the stylus (or finger) obscuring the target on the display, which makes it difficult for users to know when they are pressing in the correct place. Simple earcons were used to overcome the problems. The same tasks were performed as described above, but two different button sizes were used: standard (16×16 pixels) and small (8×8 pixels). One aim of the experiment was to see if adding audio could reduce the size of the widgets so that screen space could be freed. Each condition had two 7-min treatments: standard visual-only buttons and visual plus sound buttons. Figure 11.10 shows the interface used in the experiment with the small buttons. Because of the limitations of the device on which the experiment was performed (a 3Com PalmIII handheld computer with stylus input), three simple earcons were played. The standard PalmIII key click sound was played when a button was successfully selected; a higher pitched version of this sound was played when the stylus was pressed on a graphic button (in this case helping the users know when they were on a target that could be difficult to see); and a lower pitched version of this sound was played when a button was misselected.

In general, the results confirmed those of the previous experiment. Subjective workload in the sonically enhanced buttons of both sizes was reduced, compared with their silent counterparts. In both conditions the addition of sound allowed the participants to enter significantly more five-digit strings than in the corresponding silent treatment. A further experiment (Brewster, 2002) tested the same interface in a mobile

FIGURE 11.10. Screenshot of the 3Com Palm III interface used by Brewster (2002).

environment with users entering five-digit strings as they walked. Results from this experiment showed that participants walked further when sound was added and that small buttons with sound allowed as many strings to be entered as the large, silent buttons. The suggested reason for this was that users did not have to concentrate so much of their visual attention on the device, because much of the feedback needed was in sound, and so could look where they were going. This would therefore allow the size of items on the display to be reduced without a corresponding drop in usability.

In a very detailed piece of work, Sawhney and Schmandt (1999, 2000) developed a wearable-computer-based personal messaging audio system called Nomadic Radio to deliver information and messages to users on the move. One of the aims of this system was to reduce the interruptions to a user caused by messages being delivered at the wrong time (for example mobile telephone calls being received in a meeting, a PDA beeping to indicate an appointment in the middle of a conversation). In the system, users wore a microphone and shoulder-mounted loudspeakers that provide a basic planar 3D audio environment (as described earlier) through which the audio was presented. A clock face metaphor was used with 12:00 in front of the user's nose, 3:00 by the right ear, 6:00 directly behind the head, and so forth. Messages were then presented in the position appropriate to the time that they arrived. The advantage of the 3D audio presentation (as described above) is that it allows users to listen to multiple sound streams at the same time and still be able to distinguish and separate each one (the "Cocktail party" effect).

The system used a context-based notification strategy that dynamically selected the appropriate notification method based on the user's attentional focus. Seven levels of auditory presentation were used from silent to full speech rendering. If the user was engaged in a task, the system was silent and no notification of an incoming call or message would be given (so as not to cause an interruption). The next level used ambient cues (based on auditory icons), with sounds such as running water indicating that the system was operational. These cues were designed to be easily habituated but to let the user know that the system was working. The next level was a more detailed form of auditory cue giving information on system events, task completions, and mode transitions. For example, a ringing telephone sound was used to indicate the arrival of voicemail. These were more attention grabbing than the ambient cues and would only be played if the user was not occupied. The next four levels of cue used speech, expanding from a simple message summary up to the full text of a voicemail message. These might be used if the person wearing Nomadic Radio was not involved in tasks that required detailed attention. The system attempted to work out the appropriate level to deliver the notifications by listening to the background audio level in the vicinity of the user (using the built-in microphone) and if the user was speaking. For example, if the user was speaking, the system might use an ambient cue so as not to interrupt the conversation. Users could also press a button on the device to indicate they were busy and so turn it to silent mode.

Audio Aura was created by Mynatt (Mynatt, Back, Want, & Fredrick, 1997; Mynatt, Back, Want, Baer, & Ellis, 1998) "to provide serendipitous information, via background auditory will have better success than other people in using these applications. The users need to be sufficiently rested to be able to speak clearly and not mumble. Some conversational technology applications are hands-free systems and can be used by many individuals with disabilities. These solutions are useful for people with no uncorrected visual or hearing impairments and no documented cognitive or speech impairments.

Education

Conversational technology applications assume a user has reading, comprehension and visual recognition skills approximately at the fifth-grade level. Some conversational applications (telephony) have much more limited reading requirements.

CONVERSATIONAL TASKS

Composition

Composition tasks have as their primary goal the creation of a document. We use the term *document* in the most general sense, including word processing documents, e-mails, instant messaging text, and so on. The composition task includes the dictation of text, along with the correction of any resultant errors in the text. We categorize composition applications as conversational in nature because the user speaks to the system to create the text and then engages in a dialogue to correct errors.

Transcription

Broadcast news, business meetings, call center calls, and conference calls are all examples of situations in which people engage in conversations and having a permanent, textual record of the conversation is valuable. A digital (textual) record of the conversation is searchable, is readable by deaf individuals, and supports advanced business intelligence applications. These are all examples of transcription tasks. Transcription is similar to composition in that a document is being created from speech, but it differs in that the primary user task has nothing to do with the creation of a document. In each of these scenarios, the user is engaged in an activity that is independent of the transcription process. They are not concerned with creation of a document that is a grammatically correct and well organized. In fact, they devote no cognitive processing to the creation of the document. They simply want an accurate record of the conversations.

Transaction

The third type of interaction, and a major focus of this chapter, is one in which users have as their goal the completion of one or more transactions, rather than the creation of a document or a permanent record of a conversation. Examples of transactional applications include financial account management, such as trading stocks and mutual fund and transferring money between accounts; e-commerce applications, such as the purchase

of computer equipment; searching for information on the Internet; controlling the environment (e.g., systems to control temperature and lights); and transacting a conversation (e.g., for an impaired user who cannot speak to have a verbal conversation with another person). In each case, a conversation between the user and the system results in an action that the system executes on behalf of the user.

Collaboration

The final type of conversational task we consider is collaboration. Collaborative conversational applications are characterized by tasks that result in human-to-human communication. This communication can be synchronous, in which speech recognition is an input modality to an instant-messaging application, or asynchronous, for the management of personal information such as e-mail, calendars, and to-do lists.

PHYSICAL AND SOCIAL CONTEXT OF USE

Conversational interfaces have to accommodate different user characteristics, support special features of each task, and adjust to the physical and social context of the interaction. The context of use deals with the audio channel and device characteristics, the physical location of the interaction, and the social situation of the interaction.

Audio Channel and Device Characteristics

Spoken language systems need a microphone, a speaker, a transmission channel, an audio subsystem for digitizing and reproducing speech, and system resources for running conversational technology engines such as speech recognition. The speech input–output capabilities of existing devices vary a lot. Personal computers (PCs), either desktop or laptop, act as clients and provide good-quality audio subsystems and speakers. There are a variety of headset, lavaliere, desktop, and handheld microphones that can be used in a wired or wireless connection to a PC and perform well under quiet conditions. Although headset microphones provide a good level of noise robustness, these microphones are not the top choice among users because of their head-mounting requirement.

Unfortunately, other microphone designs are susceptible to noise and interfering sounds that lead to higher error rates for the speech recognition technology. The microphone industry is currently working on a few alternatives such as microphone arrays and distance-sensitive microphones as well as noise cancellation algorithms to reduce the impact of interfering sounds on the recognition performance. Another promising approach is the development of wireless microphones that deploy BlueTooth technology to transmit the speech signal to the recognition system. The processor and memory resources available on the current PCs are more than sufficient to run most stand-alone applications such as speech dictation and navigation.

The audio channels for other system configurations and devices are not on par with PCs. Although most personal digital assistants (PDAs) have started to offer speech input capabilities, the quality of their microphones and audio subsystems degrade the speech signal, resulting in poor recognition performance for many applications. Nearly all PDAs do not have enough resources to enable large-vocabulary speech recognition processing for applications such as dictation. In telephony applications, the conversational technology engines are usually deployed on large servers. These servers have large resources for running all the engines necessary for a complete interaction, but the design challenge, driven by lowering the cost of deployment, has been about architectural and algorithmic advancements to accommodate running many conversational engines simultaneously on a given server.

The recognition performance of telephony systems for a given task is normally 10% lower than a similar PC configuration with a headset microphone in a quiet environment. However, noise, signal degradation due to packet loss in voice over internet protocol (VoIP), poor connection in cellular telephony, and the application of compression techniques can substantially reduce the performance of speech recognition systems and speaker-recognition tasks. Advances in new compression techniques suitable for both human–human communication as well as human–machine interaction hold strong promise for achieving high recognition accuracy. These new compression techniques are based on compressing mel filter ceptral coefficients (MFCC) parameters in which the coefficients are represented in frames with constant sampling intervals (Vergin, O'Shaughnessy, & Farhat, 1999).

Physical Context of Interaction

The nature of interfering noise varies as a function of the location of the conversational interaction. In a car and at home, a user can control the occurrence of background conversations. However, background conversations impose a serious problem on conversational systems in open offices and public places. Another important source of noise is the interfering sound generated by a radio, a television, or an audio system at home or in a car. In these cases, because the source of noise is known, an accurate measurement of the interfering signal is available, and we can expect a strong suppression of this kind of noise. Although this has been confirmed in a laboratory environment, commercial availability depends on providing a link between the source of the noise (e.g., a radio) and the speech recognition device. Finally, transient noise such as a dog barking at home, a cart passing by in the office, a public address system making an announcement in a public area, or a truck passing by a car with its windows down introduces difficult recognition challenges that are expected to be solved by better modeling and the collection of a large database of these types of interfering sounds.

Social Context

We change our speaking style depending on the social settings and protocols of our interaction. When interacting with a machine, users can be expected to speak slower and clearer, and this enhances recognition performance. On the other hand, in applications that create a transcript of a meeting or a conference call, expecting user cooperation for the benefit of the conversational technologies is unrealistic. In these situations, the speech that occurs suffers from many by-products of spontaneous speech, such as hesitation, change of mind, repetition, and other disfluencies. In addition, speech produced by more than one speaker talking at the same time creates challenging problems that are technically unsolved at this time. Another interesting situation arises when conversational technologies are deployed in places such as a public library where loud speech is inappropriate or in a public setting such as a plane where spoken utterances need to be kept confidential. In these situations, we use soft speech, which introduces additional challenges to recognition accuracy. Additional elements of context of interaction deal with multimodal and multilingual situations.

USER-CENTERED DESIGN APPROACH TO CONVERSATIONAL TECHNOLOGY APPLICATIONS

We strongly recommend applying iterative user-centered design (UCD) methods when designing and developing conversational technology applications. UCD starts with a careful definition of the target user profile(s) for the application under design. Because the syntax and lexicon of language are likely to vary tremendously among different user populations, it is especially important that this step be done thoroughly.

Following the definition of the user profile(s), the system designer performs a task analysis to understand how a user completes different tasks that the system will support. As part of the initial data-collection effort, the designer works with representative users to identify core user task scenarios and to define the steps, types of interaction, and words that users employ to complete a set of tasks. Data collection refers to the process of obtaining samples of conversations in which a user will engage over the course of completing a task with the system. The conversation is captured by the designer and used to define the lexicon, syntax, functionality, and conceptual range of the system under design. The actual user-generated sentences or commands are used to define the lexicon and syntax of the understanding engine. For word-spotting and parsing technologies, the sentences are converted to grammars that will be supported by the system. Pattern-matching engines require the sentences be "annotated" with formal language constructs. The annotated sentences are then used to "train" the natural language understanding (NLU) engine. For other types of conversational systems, these user-generated sentences are employed to build the vocabulary, command, or response structure of the particular type of conversational technology being employed in the solution.

It is common to use Wizard-of-Oz (WOz; Kelly, 1984) simulations in the early data-collection phase. WOz experiments use humans, in place of a computer system, to interpret a user's utterances. The user believes he or she is interacting with a cues, that is tied to people's physical actions in the workplace." (p. 566) In a similar way to Nomadic Radio, Audio Aura used auditory icons to provide background information that did not distract users.

The system used active badges so that the location of users could be identified and appropriate audio cues given, along with wireless headphones so that users could hear the sounds without distracting others. The location information from the active badges was combined with other data sources such as on-line calendars and email. Changes in this information triggered audio cues sent to the user through the headphones. Audio Aura is an audio equivalent to the ideas of ambient computing developed by Ishii and Ullmer (1997). Here, similar information is presented via visual cues in the physical environment.

Here are some examples of how the system might be used. First, the user might go the office coffee room and as he or she enters the room hear information about the number and type of e-mail messages currently waiting. This would give the user a cue as to whether to stay and talk to colleagues or go back to the office to answer the messages. In the second example, a user goes to a colleague's office but the occupant is not there. Audio Aura would play sounds indicating if the occupant has been in recently or has been away for a longer period of time. Members of work groups who are physically separated can find it difficult to get a sense of group activity levels. Data was collected about whether workgroup members were in the office, working on shared documents, or in meetings. Audio Aura could then use this to present a "group pulse" to all of the members of the group so that they knew what was going on.

The authors were keen to make sure the sounds were not distracting and attention grabbing; they were meant to give background information and not to be alarms. To this end, great care was taken with the cue design. They attempted to design "sonic ecologies"—groups of sounds that fitted together into a coherent whole. For example, one set of cues was based on a beach scene. The amount of new e-mail was mapped to seagull cries— the more mail, the more the gulls cried. Group activity levels were mapped to the sound of surf—the more activity going on within the group, the more active the waves became. These cues were subtle and did not grab users' attention, but some learning of the sounds would be needed as they are quite abstract.

One novel aspect of Audio Aura was the prototyping technique the designers used. They built initial systems using virtual reality modeling language (VRML) (a 3D modeling language which includes 3D audio; www.vrml.org). This allowed them to simulate their office building and then try out a range of audio cue designs before integrating all of the physical technology needed for the real system. For example, they could test the positioning of sensors to ensure that they were correctly placed before putting them into the real office environment.

CONCLUSIONS

Research into the use of nonspeech sounds for output at the human–computer interface began in the early 1990s, and there has been rapid growth since then. It has shown its benefits in a wide range of different applications, from systems for blind people to wearable computers. There are many good examples that designers can look at to see how sounds can be used effectively and design guidelines are now starting to appear. Two areas are likely to be key in its future growth. The first is in combining it with speech to make the most of the advantages of both. This is an area ripe for further investigation, and there are many interesting interaction problems that can be tackled when they are both used together. The second area in which nonspeech sound has a large part to play in the near future is in mobile computing devices. The number of these is increasing rapidly, but they are difficult to use because the screens are small. This is exactly the situation where sound has many advantages: It does not take up any precious screen space and users can hear it even if they cannot look at their device. This is an area that interface designers can bring about major usability improvements with nonspeech auditory output.

References

Aldrich, F. K., & Parkin, A. J. (1989). Listening at speed. *British Journal of Visual Impairment and Blindness, 7,* 16–18.

Alty, J., & Rigas, D. (1998, April). *Communicating graphical information to blind users using music: the role of context.* Paper presented at Association for Computing Machinery Computer–Human Interaction '98. Los Angeles, CA.

Alty, J. L. (1995, September). Can we use music in human-computer interaction? Paper presented at Human–Computer Interaction '95. Huddersfield, UK.

Baddeley, A. (1990). *Human memory: theory and practice.* London: Lawrence Erlbaum.

Bagwell, C. (1998, 14/11/1998). *Audio file formats FAQ.* Bagwell, C. Retrieved April 2001 from *http://home.sprynet.com/%7Ecbagwell/AudioFormats.html*

Barker, P. G., & Manji, K. A. (1989). Pictorial dialogue methods. *International Journal of Man-Machine Studies, 31,* 323–347.

Beaudouin-Lafon, M., & Conversy, S. (1996, April). *Auditory illusions for audio feedback.* Paper presented at Association for Computing Machinery Computer–Human Interaction '96. Conference Companion. Vancouver, Canada.

Begault, D. R. (1994). *3-D sound for virtual reality and multimedia.* Cambridge, MA: Academic Press.

Berglund, B., Harder, K., & Preis, A. (1994). Annoyance perception of sound and information extraction. *Journal of the Acoustical Society of America, 95,* 1501–1509.

Blattner, M., & Dannenberg, R. B. (Eds.). (1992). *Multimedia interface design.* New York: ACM Press, Addison-Wesley.

Blattner, M., Sumikawa, D., & Greenberg, R. (1989). Earcons and icons: Their structure and common design principles. *Human Computer Interaction, 4,* 11–44.

Bly, S. (1982). *Sound and computer information presentation* (unpublished doctoral dissertation). Livermore, CA: Lawrence Livermore National Laboratory.

Brewster, S. A. (1998a). The design of sonically-enhanced widgets. *Interacting with Computers, 11*, 211–235.

Brewster, S. A. (1998b). Using non-speech sounds to provide navigation cues. *ACM Transactions on Computer-Human Interaction, 5*, 224–259.

Brewster, S. A. (2002). Overcoming the lack of screen space on mobile computers. *Accepted for publication in Personal and Ubiquitous Technologies, 6*, 3.

Brewster, S. A., & Cryer, P. G. (1999, April). *Maximising screen-space on mobile computing devices.* Paper presented at the Summary Proceedings of Association for Computing Machinery Computer-Human Interaction '99. Pittsburgh, PA.

Brewster, S. A., & Murray, R. (2000). Presenting dynamic information on mobile computers. *Personal Technologies, 4*, 209–212.

Brewster, S. A., Wright, P. C., & Edwards, A. D. N. (1994). *A detailed investigation into the effectiveness of earcons.* Paper presented at the International Conference on Auditory Display '92. Santa Fe, New Mexico.

Buxton, W. (1989). Introduction to this special issue on nonspeech audio. *Human-Computer Interaction, 4*, 1–9.

Buxton, W., Gaver, W., & Bly, S. (1991, May). *Tutorial number 8: The use of non-speech audio at the interface.* Association for Computing Machinery Computer Human Interaction '91. New Orleans, Louisiana.

Chowning, J. (1975). Synthesis of complex audio spectra by means of frequency modulation. *Journal of the Audio Engineering Society, 21*, 526–534.

Crease, M., Gray, P., & Brewster, S. A. (2000, September). *Caring, sharing widgets.* Paper presented at British Computer Society Human Computer-Interaction 2000. Sunderland, UK.

Crease, M. C., & Brewster, S. A. (1998, November). *Making progress with sounds—The design and evaluation of an audio progress bar.* Paper presented at the Proceedings of the International Conference on Auditory Display '98. Glasgow, United Kingdom.

Edwards, A. D. N. (1989). Soundtrack: An auditory interface for blind users. *Human Computer Interaction, 4*, 45–66.

Edwards, A. D. N. (Ed.). (1995). *Extra-ordinary human-computer interaction.* Cambridge, UK: Cambridge University Press.

Edworthy, J., Loxley, S., & Dennis, I. (1991). Improving auditory warning design: Relationships between warning sound parameters and perceived urgency. *Human Factors, 33*, 205–231.

Flowers, J. H., & Hauer, T. A. (1992). The ear's versus the eye's potential to assess characteristics of numeric data: Are we too visuocentric? *Behavior Research Methods, Instruments, and Computers, 24*, 258–264.

Flowers, J. H., & Hauer, T. A. (1995). Musical versus visual graphs: Cross-modal equivalence in perception of time series data. *Human Factors, 37*, 553–569.

Frysinger, S. P. (1990). *Applied research in auditory data representation.* Extracting Meaning From Complex Data: Processing, Display, Interaction. The International Society for Optical Engineering (SPIE) symposium on electronic imaging. Springfield, VA.

Gaver, W. (1989). The SonicFinder: An interface that uses auditory icons. *Human Computer Interaction, 4*, 67–94.

Gaver, W. (1997). Auditory Interfaces. In M. Helander, T. Landauer, & P. Prabhu (Eds.), *Handbook of human-computer interaction* (2nd ed., pp. 1003–1042). Amsterdam: Elsevier.

Gaver, W., Smith, R., & O'Shea, T. (1991, May). *Effective sounds in complex systems: The ARKola simulation.* Paper presented at the Proceedings of Association for Computing Machinery Computer-Human Interaction '91. New Orleans, Louisiana.

Gelfand, S. A. (1981). *Hearing: An introduction to psychological and physiological acoustics.* New York: Marcel Dekker.

Heckroth, J. (1994). *A tutorial on MIDI and wavetable music synthesis: Crystal.* Retrieved April 2001 from the World Wide Web: *http://kingfisher.cms.shu.ac.uk/midi/main p.htm*

Hindus, D., Arons, B., Stifelman, L., Gaver, W., Mynatt, E., & Back, M. (1995). *Designing auditory interactions for PDAs.* Paper presented at the 8th Association for Computing Machinery symposium on user interface and software technology. Pittsburgh, PA.

Ishii, H., & Ullmer, B. (1997, April). *Tangible bits: Towards seamless interfaces between people, bits and atoms.* Paper presented at Association for Computing Machinery Computer-Human Interaction '97. Atlanta, Georgia.

Kramer, G. (Ed.). (1994a). *Auditory display.* Reading, MA: Addison-Wesley.

Kramer, G. (1994b). An introduction to auditory display. In G. Kramer (Ed.), *Auditory Display* (pp. 1–77). Reading, MA: Addison-Wesley.

Kramer, G., & Walker, B. (Eds.). (1999). *Sonification report: Status of the field and research agenda.* Santa Fe, NM: International Community for Auditory Display.

Leplâtre, G., & Brewster, S. A. (2000, April). *Designing non-speech sounds to support navigation in mobile phone menus.* Paper presented at International Conference on Auditory Display 2000, Atlanta, Georgia.

Mansur, D. L., Blattner, M., & Joy, K. (1985). Sound-Graphs: A numerical data analysis method for the blind. *Journal of Medical Systems, 9*, 163–174.

Maury, S., Athenes, S., & Chatty, S. (1999, April). *Rhythmic menus: Toward interaction based on rhythm.* Paper presented in the Extended Abstracts of Association for Computing Machinery Computer-Human Interaction '99. Pittsburgh, PA.

McCormick, E. J., & Sanders, M. S. (1982). *Human factors in engineering and design* (5th ed.). New York: McGraw-Hill.

Moore, B. C. (1997). *An introduction to the psychology of hearing* (4th ed.). London: Academic Press.

Mynatt, E., Back, M., Want, R., & Fredrick, R. (1997, November). *Audio aura: Light-weight audio augmented reality.* Paper presented at Association for Computing Machinery User Interface Software Technology '97. Banff, Canada.

Mynatt, E., & Edwards, K. (1995). Metaphors for non-visual computing. In A. D. N. Edwards (Ed.), *Extra-ordinary human-computer interaction* (pp. 201–220). Cambridge, UK: Cambridge University Press.

Mynatt, E. D. (1994, April). *Designing with auditory icons: How well do we identify auditory cues?* Paper presented at Computer-Human Interaction '94 conference. Boston, Massachusetts.

Mynatt, E. D., Back, M., Want, R., Baer, M., & Ellis, J. B. (1998, April). *Designing audio aura.* Paper presented at Association for Computing Machinery Computer-Human Interaction '98. Los Angeles, California.

Mynatt, E. D., & Weber, G. (1994, April). *Nonvisual presentation of graphical user interfaces: Contrasting two approaches.* Paper presented at Association for Computing Machinery Computer Human Interaction '94. Boston, Massachusetts.

Patterson, R. D. (1982). *Guidelines for auditory warning systems on civil aircraft* (CAA Paper 82017). London: Civil Aviation Authority.

Patterson, R. D. (1989). Guidelines for the design of auditory warning sounds. *Proceedings of the Institute of Acoustics, Spring Conference, 11*(5), 17–24.

Perrott, D., Sadralobadi, T., Saberi, K., & Strybel, T. (1991). Aurally aided visual search in the central visual field: Effects of visual load and visual enhancement of the target. *Human Factors, 33*, 389–400.

Pohlmann, K. (1995). *Principles of digital audio* (3rd ed.). New York: McGraw-Hill.

Raman, T. V. (1996, April). *Emacspeak—A speech interface*. Paper presented at Association for Computing Machinery Computer–Human Interaction '96. Vancouver, Canada.

Rayner, K., & Pollatsek, A. (1989). *The psychology of reading*. Englewood Cliffs, NJ: Prentice-Hall International.

Rigas, D. I., & Alty, J. L. (1997, July). *The use of music in a graphical interface for the visually impaired*. Paper presented at International Federation for Information Processing Interact '97. Sydney, Australia.

Roads, C. (1996). *The computer music tutorial*. Cambridge, MA: MIT Press.

Sawhney, N., & Schmandt, C. (1999). *Nomadic radio: Scalable and contextual notification for wearable messaging*. Paper presented at Association for Computing Machinery Computer–Human Interaction '99. Pittsburgh, Pennsylvania.

Sawhney, N., & Schmandt, C. (2000). Nomadic Radio: Speech and audio interaction for contextual messaging in nomadic environments. *ACM Transactions on Human-Computer Interaction, 7*, 353–383.

Strybel, T., Manligas, C., & Perrott, D. (1992). Minimum audible movement angle as a function of the azimuth and elevation of the source. *Human Factors, 34*, 267–275.

Thimbleby, H. (1990). *User interface design*. New York: ACM Press, Addison-Wesley.

Part

·III·

HUMAN–COMPUTER INTERACTION

Part

·IIIA·

INTERACTION
FUNDAMENTALS

Julie A. Jacko
Georgia Institute of Technology

It is a compelling exercise to examine a definition for the term *interaction* that existed before researchers, practitioners, designers, and educators were even interested in interactions of the human–computer variety. In 1942, *The Royal English Dictionary* defined interaction as "action of one body on another; mutual influence" (p. 304). It is this mutual influence that is still of interest when we speak of interaction in the domain of human–computer interaction, despite the fact that, in this day of sophisticated systems and complicated humans, we have become much more precise about the ways in which mutual influence is studied and understood. In contemporary, domain-relevant times, Marcus (1995, 1998) defines interaction as the means by which users communicate input to the system, as well as the feedback supplied by the system. The term implies all aspects of input devices (e.g., mice, joysticks, track pads, keyboards) and sensory feedback, which may be presented with, for example, visual displays, auditory cues, graphical buttons, and tactile surfaces, or some combination therein.

Thus, in this section, authors build on the knowledge presented in section I (Humans in Human–Computer Interaction) and Section II (Computers in Human–Computer Interaction) to focus on the mutual influence of humans and computers. This is accomplished with eight chapters, contributed by 10 different authors: Multimedia User Interface Design, by Alistair Sutcliffe; Visual Design Principles for Usable Interfaces, by Suzanne Watzman; Multimodal Interfaces, by Sharon Oviatt; Adaptive Interfaces and Agents, by Anthony Jameson; Network-Based Interaction, by Alan Dix; Motivating, Influencing, and Persuading Users, by B. J. Fogg; Human Error Identification in Human–Computer Interaction, by Neville A. Stanton; and Design of Computer Workstations, by Michael J. Smith, Pascale Carayon, and William J. Cohen.

Sutcliffe, in the first chapter of this section (Multimedia User Interface Design) examines the design issues associated with multimedia: the trade-offs between rich representation and confusing complexity, information overload, conceptual disorientation, distracting presentations, matching the message to the media, and making the theme stand out from the noise. In addition, cognitive implications of multimedia are presented, such as processing visual, auditory, haptic, and olfactory information; selective attention; arousal; and motivation. From this, he transitions to principles for multimedia design, media selection, content selection, layout, sequencing, navigation, interaction, attention, and integration. Sutcliffe concludes with the evaluation of the effectiveness of multimedia information delivery, as well as with a review of evaluation approaches.

Visual relationships are explored in chapter 13, in which Suzanne Watzman presents Visual Graphic Design Principles for Usable Interfaces. In this chapter, she introduces the reader to the process of good design by reviewing the three phases of an informed design process: the audit, design development, and implementation and monitoring. Visual design principles are reviewed in detail, including the five criteria for good design. She examines visual design principles at work in typography, variations in letterforms, and typographic guidelines. This information is compellingly coupled with specific design principles about building the design of a page, charts, diagrams, graphics and icons, and the use of color. The chapter then shifts to designing the experience and the challenges and opportunities of doing so.

The following chapter broadens the reader's perspective from primarily visual-focused designs to multimodal systems. In this chapter, the author, Sharon Oviatt, opens by defining multimodal systems as those that "process two or more combined user input modes—such as speech, pen, touch, manual gestures, gaze, and head and body movements—in a coordinated manner with multimedia system output." Thus, this chapter builds on Sutcliffe's depiction of multimedia, as well as Watzman's portrayal of visual design. Once the history and current state of the art are presented, Oviatt examines the cognitive science underpinnings of multimodal interfaces design, as well as methodological approaches to multimodal design and development. She also addresses the questions, what basic architectures and processing techniques have been used to design multimodal systems? What are the main future directions for multimodal interface design?

Anthony Jameson, in chapter 15, examines interaction from the perspective of adaptive interfaces and agents. He opens the chapter with discussions of supporting system use, supporting information acquisition, and usability challenges. Obtaining information about users is emphasized extensively, including self-reports and self-assessments, nonexplicit input, learning, inference and decision making, classification learning, collaborative filtering, and decision-theoretic methods. A number of empirical methods are also covered, such as Wizard-of-Oz studies, simulations using data from nonadaptive systems, controlled studies, and studies of actual system use. Jameson concludes with the future of user-adaptive systems.

Alan Dix contributes Network-Based Interaction in chapter 16 by first providing an overview of types of networks, and then examining networks as enablers, networks as mediators, networks as subjects, and networks as platforms. Within these focus areas, Dix provides information on remote resources, applications, network properties, user interface properties, media issues, network management, network awareness, locking, architectures for networked systems, and supporting infrastructure. He concludes with an examination of the future of networking.

The section then refocuses itself on motivating, influencing, and persuading users. In chapter 17, B. J. Fogg introduces The Functional Triad, which is a framework for persuasive technology that makes explicit three computer functions from the user's perspective: tools, media, and social actors. Within the framework, computers as persuasive tools, computers as persuasive media, and computers as persuasive social actors are examined in great detail. Fogg then transitions the chapter to a discussion of computers and credibility, as well as the ethics of computing systems designed to persuade. Within this context, several examples are provided of unique ethical issues in persuasive computing.

A section on interaction fundamentals would not be complete without a thorough examination of human errors in interaction. Neville Stanton contributes this material in chapter 18. The chapter opens with a comprehensive discussion of what constitutes error. Error research is then presented, which focuses on the approaches to collection and categorization of errors. Regulations and standards, as well as methods for predicting and analyzing errors, and designing for errors in simple and complex systems are covered. Finally, Stanton summarizes by providing a conclusion for designers that focuses on sorting out the implications of human error for design.

This section concludes with a chapter from Michael Smith, Pascale Carayon, and William Cohen titled Design of Computer Workstations. This chapter emphasizes the physical dimensions of interaction. For example, it reviews studies concerned with workstation issues related to computer use, including the ergonomic design of screens and displays, and accommodating keyboards, mice, and other input devices. The authors allocate additional attention to the visual environment, the auditory environment, and HVAC systems. The chapter aptly concludes with general principles of workstation design that are targeted on computer manufacturers, office furniture designers, and office designers.

This section sets the stage for the rest of the handbook volume. It integrates fundamental principles about humans, computers, and the mutual influence they can potentially have on each other.

References

Marcus, A. (1995). Principles of effective visual communication for graphical user interface design. In R. Baecker, J. Grudin, W. Buxton, & S. Greenberg (Eds.), *Readings in human-computer interaction* (2nd ed., pp. 425–441). Palo Alto, CA: Morgan-Kaufman.

Marcus, A. (1998). Metaphor design in user interfaces. *The Journal of Computer Documentation, 22*(2), 43–57.

The Royal English Dictionary (1942). Paris: Société Française D'Éditions.

·12·

MULTIMEDIA USER INTERFACE DESIGN

Alistair Sutcliffe

University of Manchester Institute of Science and Technology

INTRODUCTION

Design of multimedia interfaces currently leaves a lot to be desired. As with many emerging technologies, it is the fascination with new devices, functions, and forms of interaction that has motivated design rather than ease of use, or even utility of practical applications. Poor usability limits the effectiveness of multimedia products that might look good, but do not deliver effective use (Scaife, Rogers, Aldrich, & Davies, 1997). The multimedia market has progressed beyond the initial hype, and customers are looking for well-designed, effective, and mature products.

The distinguishing characteristics of multimedia are information-intensive applications that have a complex design space for presenting information to people. Design, therefore, has to start by modeling information requirements. This chapter describes a design process that starts with an information analysis, then progresses to deal with issues of media selection and integration. The background to the method and its evolution with experience can be found in several publications (Faraday & Sutcliffe, 1996, 1997b, 1998b; Sutcliffe & Faraday, 1994). A more detailed description is given in Sutcliffe (2002). The time-to-market pressure gives little incentive for design; so at first reading, a systematic approach may seem to be counter to the commercial drivers of development. However, I would argue that if multimedia design does not adopt a usability engineering approach, it will fail to deliver effective and usable products.

Multimedia applications have significant markets in education and training, although dialogue in many systems is restricted to drill-and-quiz interaction and simple navigation. This approach, however, is oversimplified: For training and education, interactive simulations, and microworlds are more effective (Rogers & Scaife, 1998). Multimedia has been used extensively in task-based applications in process control and safety critical systems (Alty, 1991; Hollan, Hutchins, & Weitzman, 1984); however, most transaction processing applications are currently treated as standard interfaces rather than multimedia-based designs. With the advent of the web and e-commerce, this view may change.

Design issues for multimedia user interfaces expand conventional definitions of usability (e.g., ISO 9241 part 11) into five components:

- *Operational usability* is the conventional sense of usability that concerns design of graphical user interface features such as menus, icons, metaphors, and navigation in hypermedia.
- *Information delivery* is a prime concern for multimedia or any information-intensive application, and raises issues of media selection, integration, and design for attention.
- *Learning*: Training and education are both important markets for multimedia, and hence learnability of the product and its content are key quality attributes. However, design of educational technology is a complex subject in its own right, and multimedia is only one part of the design problem (see chapter 42, Quintana et al., which deals with educational software design).
- *Utility*: In some applications, this will be the functionality that supports the user's task; in others, information delivery and learning will represent the value perceived by the user.
- *Aesthetic appeal*: The attractiveness of multimedia is now a key factor, especially for Web sites. Multimedia interfaces have to attract users and motivate them, as well as being easy to use and learn.

Multimedia design involves several specialisms that are technical subjects in their own right. For instance, design of text is the science (or art) of calligraphy that has developed new fonts over many years; visualization design encompasses the creation of images, either drawn or captured as photographs. Design of moving images, cartoons, video, and film are further specializations, as are musical composition and design of sound effects. Multimedia design lies on an interesting cultural boundary between the creative artistic community and science-based engineering. One implication of this cultural collision (or rather, one hopes, synthesis) is that space precludes "within media" design (i.e., guidelines for design of one particular medium) being dealt with in depth in this chapter. Successful multimedia design often requires teams of specialists who contribute from their own skill sets (Kristof & Satran, 1995; Mullet & Sano, 1995).

DEFINITIONS AND TERMINOLOGY

Multimedia essentially extends the graphical user interface paradigm by providing a richer means of representing information for the user by use of image, video, sound, and speech. Some views of what constitutes multimedia can be found in Bernsen (1994), who proposed a taxonomy of analogue versus discrete media, which he calls modalities, as well as visual, audio, and tactile dimensions. Heller and Martin (1995) take a more conventional view of classifying image, text, video, and graphics for educational purposes. The following definitions broadly follow those in the ISO standard 14915 on Multimedia User Interface Design (ISO, 1998). The starting point is to ask about the difference between what is perceived by someone and what is stored on a machine.

Communication concepts in multimedia can be separated into:

- *Message*: The content of communication between a sender and receiver.
- *Medium* (plural *media*): The means by which that content is delivered. Note that this is how the message is represented rather than the technology for storing or delivering a message. There is a distinction between perceived media and physical media, such as CD-ROM, hard disk, etc.
- *Modality*: The sense by which a message is sent or received by people or machines. This refers to the senses of vision, hearing, touch, smell, and taste.

A message is conveyed by a medium and received through a modality. A modality is the sensory channel that we use to send and receive messages to and from the world, essentially our senses. Two principal modalities are used in human–computer communication:

- *Vision*: All information received through our eyes, including text and image-based media.
- *Hearing*: All information received through our ears, as sound, music, and speech.

In the future, as multimedia converges with virtual reality, we will use other modalities more frequently: *haptic* (sense of touch), *kinaesthetic* (sense of body posture and balance), *gustation* (taste), and *olfaction* (smell). These issues are dealt with in chapter 14, Multimodal Interfaces (Oviatt), and chapter 31, Virtual Environments (Stanney).

Defining a medium is not simple because it depends on how it was captured in the first place, how it was designed, and how it has been stored. For example, a photograph can be taken on film, developed, and then scanned into a computer as a digitized image. The same image may have been captured directly by a digital camera and sent to a computer as an e-mail file. At the physical level, media may be stored by different techniques.

Physical media storage has usability implications for the quality of image and response time in networked multimedia. A screen image with 640 × 480 VGA resolution using 24 bits per pixel for good color coding gives 921,600 bytes; so, at 30 frames/s, 1 s needs around 25 megabytes of memory or disk space. Compression algorithms (e.g., MPEG [Moving Pictures Expert Group]) reduce this by a factor of 10. Even so, storing more than a few minutes of moving image consumes megabytes. The usability trade-off is between the size of the display footprint (i.e., window size), the resolution measured in dots per inch, and the frame rate. The ideal might be full screen high resolution (600 dpi) at 30 frames/s; with current technology, a 10-cm window at 300 dpi and 15 frames/s is more realistic. Physical image media constraints become more important on networks, when bandwidth will limit the desired display quality. Sound, in comparison, is less of a problem. Storage demands depend on the fidelity required for replay. Full stereo with a complete range of harmonic frequencies only consumes 100 kilobytes for 5 mins, so there are few technology constraints on delivery of high-quality audio.

COGNITIVE BACKGROUND

The purpose of this section is to give a brief overview of cognitive psychology as it affects multimedia design. More details can be found in section I, Humans in Human–Computer Interaction.

Perception and Comprehension

Our eyes scan images in a series of rapid jumps called saccades interleaved with fixations in which the eye dwells on a particular area. Fixations allow image detail to be inspected, so eye tracking gives some impression of the detail inspected in images. Generally, our eyes are drawn to moving shapes, then complex, different, and colorful objects. Visual comprehension can be summarized as "what you see depends on what you look at and what you know."

Multimedia designers can influence what users look at by controlling attention with display techniques, such as use of movement, highlighting, and salient icons. However, designers should be aware that the information people assimilate from an image also depends on their internal motivation, what they want to find, and how well they know the domain (Treisman, 1988). A novice will not see interesting plant species in a tropical jungle, whereas a trained botanist will. Selection of visual content therefore has to take the user's knowledge and task into account. Because the visual sense receives information continuously, it gets overwritten in working memory (Baddeley, 1986). This means that memorization of visually transmitted information is not always effective unless users are given time to view and comprehend images. Furthermore, users only extract very high-level or *gist* (general sense) information from moving images. Visual information has to be understood by using memory. In realistic images, this process is automatic; however, with nonrealistic images, we have to think carefully about the meaning, for example to interpret a diagram. Although extraction of information from images is rapid, it does vary according to the complexity of the image and how much we know about the domain. Sound is a transient medium, so unless it is processed quickly, the message can be lost. Even though people are remarkably effective at comprehending spoken language and can interpret other sounds quickly, the audio medium is prone to interference because other sounds can compete with the principal message. Because sound is transient, information in speech will not be assimilated in detail, and so only the gist will be memorized (Gardiner & Christie, 1987).

Selective Attention

We can only attend to a limited number of inputs at once. Although people are remarkably good at integrating information received by different senses (e.g., watching a film and listening to the sound track), there are limits determined by the psychology of human information processing (Wickens, Sandry, & Vidulich, 1983). Our attention is selective and closely related to perception; for instance, we can overhear a conversation in a room with many people speaking (the cocktail party effect). Furthermore, selective attention differs between individuals and can be improved by learning factors: for example, a conductor can distinguish the different instruments in an orchestra, whereas a typical listener cannot. However, all users have cognitive resource limitations, which means that information delivered on different modalities (e.g., by vision and sound) has to compete for the same resource. For instance, speech and printed text both require a language understanding resource, whereas video and a still image use image interpretation resources. Cognitive models of information processing architectures (e.g., Interacting Cognitive Subsystems: Barnard, 1985) can show that certain media combinations and media design will not result in effective comprehension, because they compete for the same cognitive resources, thus creating a processing bottleneck. We have two main perceptual channels for receiving

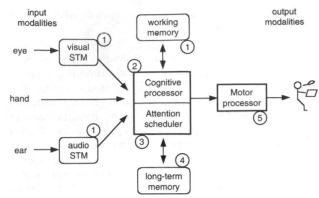

Bottlenecks
1. Capacity overflow: information overload
2. Integration: common message?
3. Contention: conflicting channels
4. Comprehension
5. Multi-tasking input/output

FIGURE 12.1. Approximate model of human information processing using a human as computer system analogy, based on the Model Human Processor (Card et al., 1983). For more on cognitive models, see chapter 2 (Proctor and Vu) and chapter 5 (Byrne). STM = short-term memory.

information: vision and hearing; information going into these channels has to be comprehended before it can be used. Information can be received in a language-based form either as speech or as written text viewed in an image. All such input competes for language understanding resources, hence making sense of speech and reading text concurrently is difficult (Barnard, 1985). Figure 12.1 shows the cognitive architecture of human information processing and resource limitations that lead to multimedia usability problems.

Capacity overflow (1) may happen when too much information is presented in a short period, swamping the user's limited working memory and cognitive processor's capability to comprehend, chunk, and then memorize or use the information. The connotation is to give users control over the pace of information delivery. Integration problems (2) arise when the message on two media is different, making integration in working memory difficult; this leads to the thematic congruence principle. Contention problems (3) are caused by conflicting attention between dynamic media, and when two inputs compete for the same cognitive resources (e.g., speech and text require language understanding). Comprehension (4) is related to congruence; we understand the world by making sense of it with our existing long-term memory. Consequently, if multimedia material is unfamiliar, we cannot make sense of it. Finally, multitasking (5) makes further demands on our cognitive processing, so we will experience difficulty in attending to multimedia input when performing output tasks.

Making a theme in a multimedia presentation clear involves directing the user's reading and viewing sequence across different media segments. Video and speech are processed in sequence, and text enforces a serial reading order by the syntactic convention of language; however, viewing image media is less predictable, because it depends on the size and complexity of the image, the user's knowledge of the contents, task and motivation (Norman & Shallice, 1986), and designed effects for salience. Attention-directing effects can increase the probability that the user will attend to an image component, although no guarantee can be given that a component will be perceived or understood.

Learning and Memorization

Learning is the prime objective in tutorial multimedia. However, the type of learning can be either skill training, in which case conducting an operational task efficiently and without errors is the aim, or a deeper understanding of the knowledge may be required. In both cases, the objective is to create a rich memory schema that can be accessed easily in the future. We learn more effectively by active problem solving or learning by doing. This approach is at the heart of constructivist learning theory (Papert, 1980), which has connotations for tutorial multimedia. Interactive microworlds in which users learn by interacting with simulations, or constructing and testing the simulation, give a more vivid experience that forms better memories (Rogers & Scaife, 1998). Multiple viewpoints help to develop rich schemata by presenting different aspects of the same problem, so the whole concept can be integrated from its parts. An example might be to explain the structure of an engine, then how it operates, and finally display a causal model of why it works. Schema integration during memorization fits the separate viewpoints together.

The implications from psychology are summarized in the form of multimedia design principles that amplify and extend those proposed for general UI design (e.g., ISO 9241 part 10 [ISO, 1997]). The principles are high-level concepts that are useful for general guidance, but they have to be interpreted in a context to give more specific advice.

• *Thematic congruence*: Messages presented in different media should be linked together to form a coherent whole. This helps comprehension as the different parts of the message make sense by fitting together. Congruence is partly a matter of designing the content so it follows a logical theme (e.g., the script or story line makes sense and does not assume too much about the user's domain knowledge) and partly a matter of attentional design to help the user follow the message thread across different media.

• *Manageable information loading*: Messages presented in multimedia should be delivered at a pace that is either under the user's control or at a rate that allows for effective assimilation of information without causing fatigue. The rate of information delivery depends on the quantity and complexity of information in the message, the effectiveness of the design in helping the user extract the message from the media, and the user's domain knowledge and motivation. Some ways of reducing information overload are to avoid excessive use of concurrent dynamic media and give the user time to assimilate complex messages.

• *Ensure compatibility with the user's understanding*: Media should be selected that convey the content in a manner compatible with the user's existing knowledge (e.g., the radiation symbol and road sign icons are used to convey hazards and dangers to users who have the appropriate knowledge and cultural background). The user's ability to understand the message is important for designed image media (diagrams, graphs) when interpretation is dependent on the user's knowledge and background.

• *Complementary viewpoints*: Similar aspects of the same subject matter should be presented on different media to create an integrated whole. Showing different aspects of the same object (e.g., picture and design diagram of a ship) can help memorization by developing richer schema and better memory cues.

• *Consistency* helps users learn an interface by making the controls, command names, and layout follow a familiar pattern. People recognize patterns automatically, so operating the interface becomes an automatic skill. Consistent use of media to deliver messages of a specific type can help by cueing users with what to expect.

• *Reinforce messages*: Redundant communication of the same message on different media can help learning. Presentation of the same or similar aspects of a message helps memorization by the frequency effect. Exposing users to the same thing in a different modality also promotes rich memory cues.

DESIGN PROCESS

Multimedia design has to address the problems inherent in the design of any user interface, viz. defining user requirements, tasks, and dialogue design; however, there are three issues that concern multimedia specifically:

• *Matching the media to the message*, by selecting and integrating media so the user comprehends the information content effectively.

• *Managing users' attention* so key items in the content are noticed and understood, and the user follows the message thread across several media.

• *Navigation and control* so the user can access, play, and interact with media in a flexible and predictable manner.

Figure 12.2 gives an overview of the design process that addresses these issues.

The method starts with requirements and information analysis to establish the necessary content and communication goals of the application. It then progresses to domain and user characteristic analysis to establish a profile of the user and the system environment. The output from these stages feeds into media selection and integration that match the logical specification of the content to available media resources. Design then progresses to thematic integration of the user's reading/viewing sequence and dialogue design. The method can be tailored to fit within different development approaches. For instance, in rapid applications development, storyboards, prototypes, and iterative build-and-evaluate cycles would be used. On the other hand, in a more systematic, software engineering approach, more detailed specifications and scripts will be produced before design

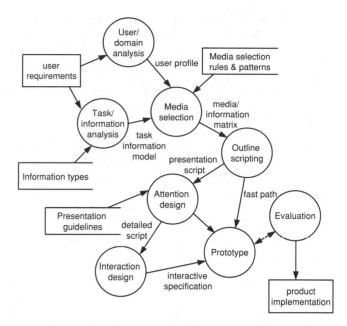

FIGURE 12.2. Overview of the multimedia design process expressed as a data flow diagram.

commences. Even though the process is described as a sequence, in practice the stages are interleaved and iterated; however, requirements, information modeling, and media selection should be conducted, even if they are not complete, before the media and attentional design stages commence.

Design approaches in multimedia tend to be interactive and user-centered. *Storyboards* are a well-known means of informal modeling in multimedia design (Nielsen, 1995; Sutcliffe, 1999). Originating from animation and cartoon design, storyboards are a set of images that represent key steps in a design. Translated to software, storyboards depict key stages in interaction and are used for conducting walkthroughs to explain what happens at each stage. Allowing the users to edit storyboards and giving them a construction kit to build their own encourages active participation. Storyboards are followed by building concept demonstrators using multimedia authoring tools (e.g., Macromedia Director, Toolbook) to rapidly develop early prototypes. *Concept demonstrators* are active simulations that follow a scenario script of interaction; departure from the preset sequence is not allowed. Several variations can be run to support comparison; however, the user experience is passive. In contrast, users can test *interactive prototypes* by running different commands or functions. The degree of interactivity depends on the implementation cost that increases as prototypes converge with a fully functional product.

Users, Requirements, and Domains

The starting point for multimedia, as in all applications, is requirements analysis. The difference in multimedia lies in the greater emphasis on information requirements. A variety of analytic approaches can be adopted, such as task analysis (see chapter 48, Redish and Wixon), contextual design (chapter 49, Holtzblatt), or scenario analysis (chapter 53, Rosson and Carroll). Requirements are listed and categorized into information, task-related, and nonfunctional classes. These will be expanded in subsequent analyses.

It is important to get a profile of the target user population to guide media selection. There are three motivations for user analysis:

- *Choice of modalities*: This is not only important for people with disabilities, but also for user preferences. Some people prefer verbal-linguistic material over image.
- *Tuning the content* presented to the level of users' existing knowledge. This is particularly important for training and educational applications.
- *Capturing the users' experience* of multimedia and other computer systems.

Acquiring information about the level of experience possessed by the potential user population is important for customization. User profiles are used to design training applications to ensure that the right level of tutorial support is provided, and to assess the users' domain knowledge so that appropriate media can be selected. This is particularly important when symbols, designed images, and diagrammatic notations may be involved. The role and background of users will have an important bearing on design. For example, marketing applications will need simple, focused content and more aesthetic design, whereas tutorial systems need to deliver detailed content. Information kiosk applications need to provide information, as do task-based applications, but decision support and persuasive systems (Fogg, 1998; see also, Fogg, chapter 17) also need to ensure users comprehend and are convinced by messages. Domain knowledge, including use of conventions, symbols, and terminology in the domain, is important because less experienced users will require more complete information to be presented.

The context and environment of a system will also have an important bearing on design. For example, tourist information systems in outdoor public areas will experience a wide range of lighting conditions, which can make image and text hard to read. High levels of ambient noise in public places or factory floors can make audio and speech useless. Hence, it is important to gather information on the location of use (office, factory floor, public/private space, hazardous locations), pertinent environmental variables (ambient light, noise levels, temperature), usage conditions (single user, shared use, broadcast), and expected range of locations (countries, languages, and cultures). Choice of language, icon conventions, interpretation of diagrams and choice of content all have a bearing on design of international user interfaces.

As well as gathering general information about the system's context of use, domain modeling can prove useful for creating the system metaphor. A sketch of the user's workplace—recording spatial layout of artefacts, documents, and information—can be translated into a virtual world to help users navigate to the information and services they need. Structural metaphors for organizing information and operational metaphors for controls and devices have their origins in domain

analysis. Domain models are recorded as sketches of the work environment showing the layout and location of significant objects and artefacts, accompanied by lists of environmental factors.

Information Analysis

Information types are amodal, conceptual descriptions of information components that elaborate the content definition. Information types specify the message to be delivered in a multimedia application and are operated on by mapping rules to select appropriate media resources. The following definitions are based on the Task-based Information Analysis Method (Sutcliffe, 1997) and ISO 14915, part 3 (ISO, 2000).

The information types are used in walkthroughs, in which the analyst progresses through the task/scenario/use case asking questions about information needs. This can be integrated with data modeling (or object/class modeling), so that the information in objects and their attributes can be categorized by the following types, using the decision tree in Fig. 12.3. The first question is whether information represents concrete facts about the real world or more abstract, conceptual information; this is followed by questions about the information that relates to change in the world or describes permanent states. Finally, the decision tree gives a set of ontological categories to classify information that expands on type definitions commonly found in software engineering specifications. More complex ontologies are available (Arens, Hovy, & VanMulken, 1993; Mann & Thompson, 1988), so the classification presented in Fig. 12.3 is a compromise between complexity and ease of use. A finer grained classification enables more finely tuned media selection decisions, but at the cost of more analysis effort.

Components are classified by walking through the decision tree using the definitions and the following questions:

- Is the information contained in the component physical or conceptual?
- Is the information static or dynamic (i.e., does it relate to change or not?).
- Which type in the terminal branch of the tree does the information component belong to?

It is important to note that one component may be classified with more than one type; for instance, instructions on how to get to the railway station may contain procedural information (the instructions <turn left, straight ahead, etc.>), and spatial or descriptive information (the station is in the corner of the square, painted blue). The information types are tools for thought, which can be used either to classify specifications of content or to consider what content may be necessary. To illustrate, for the task "navigate to the railway station," the content may be minimally specified as "instructions how to get there," in which case the information types prompt questions in the form "what sort of information does the user need to fulfil the task/user goal?" Alternatively, the content may be specified as a scenario narrative of directions, waymarks to recognize, and description of the target. In this case, the types classify

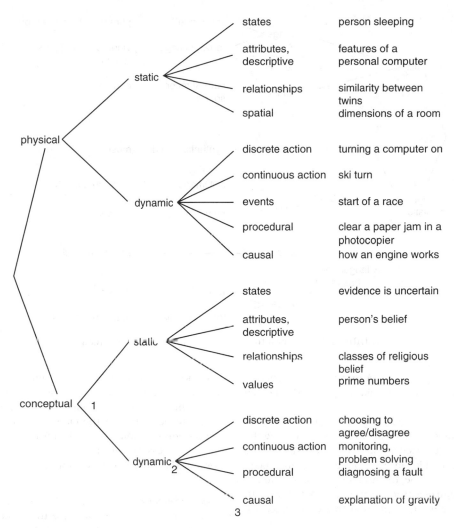

FIGURE 12.3. Decision tree for classifying information types. The first decision point reflects abstraction from the real world, the second points to change in time, and the third categorizes content.

components in the narrative to elucidate the deeper structure of the content. The granularity of components is a matter for the designer's choice and will depend on the level of detail demanded by the application. To illustrate the analysis:

Communication goal: Explain how to assemble a bookshelf from ready-made parts.

    Information component 1:
      Parts of the bookshelf, sides, back, shelves, connecting screws
      Mapping to information types:
        Physical-Static-Descriptive; parts of the bookshelf are tangible, don't change and need to be described
        Physical-Static-Spatial; dimensions of the parts, how they are organized
        Physical-Static-Relationship type could also be added to describe which parts fit together

Information component 2:
    How to assemble parts instructions
    Mapping to information types:
      Physical-Dynamic-Discrete action
      Physical-Dynamic-Procedure
      Physical-Static-State; to show final assembled bookshelf.

Media Selection and Combination

The information types are used to select the appropriate media resource(s). Media classifications have had many interpretations (Alty, 1997; Bernsen, 1994; Heller & Martin, 1995). The following classification focuses on the psychological properties of the representations rather than the physical nature of the medium (e.g., digital or analogue encoding in video). Note that these definitions are combined to describe any specific medium, so

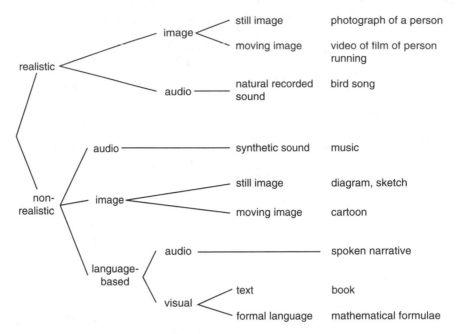

FIGURE 12.4. Decision tree for classifying media resources.

speech is classified as an audio, linguistic medium, whereas a cartoon is classified as a nonrealistic (designed) moving image.

The definitions may be usefully considered in two dimensions of abstraction: the designer's involvement in creating the medium and rate of change. Media resources are classified using the decision tree illustrated in Fig. 12.4. More fine-grained taxonomic distinctions can be made, for instance, between different signs and symbolic languages (see Bernsen, 1994); but, as with information types, richer taxonomies increase specification effort.

The approach to classifying media uses a walkthrough of the decision tree with the following questions that reflect the facets of the classification:

• Is the medium perceived to be realistic or not? Media resources captured directly from the real world will usually be realistic (e.g., photographs of landscapes, bird song sound recordings). In contrast, nonrealistic media are created by human action. However, the boundary case category that illustrates the dimension is a realistic painting of a landscape.

• Does the medium change over time or not? The boundary case here is the rate of change, particularly in animations where some people might judge 10 frames/s to be a video, but 5 slides in 1 min shown by a PowerPoint presentation to be a sequence of static images.

• Which modality does the resource belong to? In this case, the categories are orthogonal, although one resource may exhibit two modalities (e.g., a film with a sound track communicates in both visual and audio modalities).

Classification of media resources facilitates mapping of information types to media resources; however, the process of selection may also guide the acquisition or creation of appropriate media resources. So, if the selection process indicates the need for a resource that is not in the designer's media resource library, the classification guides the necessary acquisition/creation. Cost trade-offs will naturally be considered in this process. Finally, the classification provides a mechanism for indexing media resource libraries.

Media Selection. Recommendations for selecting and influencing the user's attention have to be interpreted according to the users' task and design goal. If information provision is the main design goal (e.g., a tourist kiosk information system), then persistence of information and drawing attention to specific items is not necessarily as critical as in tutorial applications. Task and user characteristics influence media choice; for instance, verbal media are more appropriate to language-based and logical reasoning tasks; and visual media are suitable for spatial tasks involving moving, positioning, and orienting objects. Some users may prefer visual media, whereas image is of little use for blind users. Media resources may be available for selection or have to be purchased from elsewhere. If existing media can be edited and reused, this is usually preferable to creating new media from scratch. Graphical images can be particularly expensive to draw, whereas capture of images by scanning is usually quick and cheap.

The mappings are used in multiple passes; for example, when a procedure for explaining a physical task is required, first a series of realistic images will be selected, followed by video and speech to integrate the steps, then text to summarize the key points. The guidelines that differentiate physical from abstract information are used first, followed by the other guidelines. The summary mappings in Table 12.1 are described in more detail in the following guidelines:

TABLE 12.1. Media Selection Example

Media Type	Information Type											
	Causation	Conceptual	Continuous Action	Descriptive	Discrete Action	Event	Physical	Procedure	Relationship	Spatial Information	State	Value
Realistic audio	Sound of rain and storms		Sound of skiing		Click of ON switch	Sound of the starting gun	*Noise of a tornado*			Echoes in a cave	Sound of snoring	Musical note encodes a value
Nonrealistic audio		Rising tone illustrates increasing magnetic force	Continuous tone signals progress of action	Morse code describes a ship	Tones signal open/close door	*Alarm siren*			Tones associate two objects	Sonar and Doppler effect	Continuous sound in a heart beat monitor	
Speech	*Tell someone why El Nino happens*	Tell someone about your religious beliefs	Tell someone what a ski turn looks like	Verbal description of a person	Tell someone how to turn computer on	*Tell someone race has started*	Tell someone how it feels to be in a storm	Speak instructions on engine assembly	Tell someone Jack and Jill are related	Tell someone pathway to and location of railway station	Tell someone "Jane's asleep"	Verbal report of numbers and figures
Realistic still image	*Photograph of El Nino storms and ocean currents*	Statue of Liberty photograph represents "freedom"	Set of photographs showing snap shots of action	*Overview and detail photographs of a car*	*Photograph of computer ON switch*	Photograph of the start of a race	*Photograph of a person's face*	*Photographs showing engine assembly*		*Photograph of a landscape*	Photograph of a person sleeping	
Nonrealistic still image	*Diagrams of ocean currents and sea temperature to explain El Nino*	*Hierarchy diagram of plant taxonomy*	Diagram with arrow depicting ski turn motion	Histogram of ageing population	Diagram showing where and how to press ON switch	Event symbol in a race sequence diagram		Explode parts diagram of engine with assembly numbers	*Graphs, histograms, Entity Relationship diagrams*	*Map of the landscape*	Waiting state symbol in race sequence diagram	*Charts, graphs, scatter plots*
Text	*Describe reasons for El Nino storms*	*Explain taxonomy of animals*	Describe ski turn action	*Describe a person's appearance*	Describe how to turn computer on	Report that the race has started	Report of the storm's properties	*Bullet point steps in assembling engine*	*Describe brother and sister relationship*	Describe dimensions of a room	*Report that the person is asleep*	*Written number one, two*
Realistic moving image	*Video of El Nino storms and ocean currents*		*Movie of person turning while skiing*	Aircraft flying		*Movie of the start of a race*	*Movie of a storm*	*Video of engine assembly sequence*		Fly through landscape	Video of a person sleeping	
Nonrealistic moving image	*Animation of ocean temperature change and current reversal*	Animated diagram of force of gravity	Animated mannequin doing ski turn		Animation showing operation of ON switch	Animation of start event symbol in diagram		*Animation of parts diagram in assembly sequence*	Animation of links on Entity Relationship diagram			
Language-based: formal, numeric	Equations, functions formalizing cause and effect	*Symbols denoting concepts, e.g., pt*			Finite state automata	Event-based notations		Procedural logics, process algebras	Functions, equations, grammars		State-based languages, e.g., Z	Numeric symbols

Note. Italics denote the preferred mappings for media and information types, and regular text shows other potential media uses for the information type.

- *Physical information*: For physical information, visual media (e.g., realistic still or moving image) are preferred, unless user or task characteristics override this choice (Alty, 1997; Baggett, 1989; Faraday & Sutcliffe, 1997a; e.g., a photograph is used to portray the landscape in a national park). When physical details need to be communicated precisely, such as the dimensions of a building, captions may be overlaid on an image. When a partial abstraction of physical information is desired, a nonrealistic image may be used (e.g., sketch or diagram).

- *Conceptual information*: Linguistic media (e.g., text or speech) are preferred for abstract or conceptual information (Booher, 1975; Faraday et al., 1998b; e.g., to convey sales objectives and commentary on the market strategy, choose text, bullet points or speech for the commentary). Abstract information with complex relationships may be shown by nonrealistic images (graphs, sketches, diagrams) or by graphical images with embedded text; for instance, a flowchart to portray the functions of a chemical process, with speech to describe the functions in detail.

- *Descriptive information*: Linguistic media (text, speech) are preferred for information describing the properties of objects, agents, or the domain (Booher, 1975; e.g., narrative text describes the properties of a chemical compound such as salt). When describing objects and agents with physical attributes, language may be combined with an image (Baggett, 1989).

- *Visual-spatial information*: A still image is effective for visual-spatial information (Bieger & Glock, 1984; May & Barnard, 1995; e.g., the location of cargo on a ship is shown by a diagram). Spatial, detailed information may be presented in a realistic image (e.g., photographs); whereas complex pathways may be conveyed by a moving image (e.g., animating a pathway).

- *Value information*: Numeric text and tables should be chosen for numeric values and quantitative information (Booher, 1975; Tufte, 1997; e.g., the height and weight of a person is given as 1.8 m, 75 kg). Graphs and charts are combined with captions and tables to summarize trends, differences, and categories in quantitative data. Speech is not effective for values because they usually need to be inspected during a task, so a persistent medium is advised.

- *Relationships in value information*: Nonrealistic images (e.g., charts, graphs) should be chosen to display relationships within and between sets of values or conceptual relationships between objects and agents (Bertin, 1983; Tufte, 1997; e.g., the values for rainfall in London for each month are displayed using a histogram).

- *Discrete action information*: For simple or discrete actions, still image media are effective (Hegarty & Just, 1993; Park & Hannafin, 1993; e.g., a series of still images of the coffee machine illustrates filling a coffee percolator with water [Andre & Rist, 1993]). Use of still image media for discrete actions allows the relationship between the action, the object acted on, and the agent performing the action to be inspected. Abstract actions (e.g., mental processes) may be described using speech or text.

- *Continuous action information*: For complex or continuous actions, moving image media are preferred (Sutcliffe & Faraday, 1994; e.g., turning while skiing is illustrated with a video). Complex physical action may be better illustrated with nonrealistic media (animated diagrams), so the coordination of motor actions can be inspected.

- *Event information*: For significant events and warnings, sound or speech should be used to alert the user (Pezdek & Maki, 1988; e.g., the outbreak of a fire is conveyed by sounding an alarm). Abstract events may have to be explained in language (Bernsen, 1994). Temporal information may be illustrated in sequence as lists or text or in graphical images as timelines (Ahlberg & Shneiderman, 1994). A still image may be used to deliver further information about the context of the event (e.g., a red marker on a diagram of the building to show the fire's location).

- *State information*: For states, still image or linguistic media are preferred (Faraday et al., 1997a; e.g., the state of the weather is shown by a photograph of a sunny day). Abstract states, such as a person's belief, may be explained in linguistic media or may be described in diagrams. If a sequence of discrete states is required, then a series of still images may be used as a slideshow.

- *Procedural information*: A series of images with text captions will be necessary for physical procedural information (Hegarty & Just, 1993; e.g., instructions for assembling a bookshelf from a kit are given as a set of images for each step, with text captions). To explain procedures, a combination of media may be necessary, such as a still image sequence with text, followed by an animation of the whole sequence. Nonphysical procedures may be displayed as formatted text (e.g., bullet points or numbered steps).

- *Causal information*: To explain causality, still and moving image media need to be combined with text (Narayanan & Hegarty, 1998; e.g., the cause of a flood is explained by text describing excessive rainfall with an animation of the river level rising and overflowing its banks. Causal explanations of physical phenomena may be given by introducing the topic using linguistic media, showing the cause and effect by a combination of still image and text with speech captions for commentary; integrate the message by moving image with voice commentary and provide a bullet point text summary.

The endpoint of media selection is media integration: One or more media will be selected for each information group to present complementary aspects of the topic. Some examples of media combination that amplify the basic selection guidelines are given in Table 12.1. The table summarizes the media selection and combinations for each information type; the first preference media choice(s) is shown in italics.

Aesthetics and Attractiveness. The previous guidelines were oriented to a task-driven view of media. Media selection, however, can also be motivated by aesthetic choice. These considerations may contradict some of the earlier guidelines, because the design objective is to please the user and capture their attention rather than deliver information effectively. First, a health warning should be noted: the old saying "beauty is in the eye of the beholder" has good foundation. Judgments of aesthetic quality suffer from considerable individual differences. A person's reaction to a design is a function of their motivation (see chapter 4, Brave and Nass), individual preferences, knowledge of the domain, and exposure to similar examples, to say nothing of peer opinion and fashion. Furthermore, attractiveness is often influenced more by content than the choice of media or presentation format. The following heuristics should therefore be interpreted with care and their design manifestations tested with users:

- *Dynamic media*, especially video, have an arousing effect and attract attention, hence video and animation are useful in improving the attractiveness of presentations. However, animation must be used with care, because gratuitous video that cannot be turned off quickly offends (Spool, Scanlon, Snyder, Schroeder, & De Angelo, 1999).
- *Speech* engages attention because we naturally listen to conversation. Choice of voice depends on the application: Female voices for more restful and information effects, male voices to suggest authority and respect (Reeves & Nass, 1996).
- *Image* needs to be selected with careful consideration of content. Images may be selected for aesthetic motivations, to provide a restful setting for more important foreground information (Mullet & Sano, 1995). Backgrounds in half shades and low saturation color provide more depth and interest in an image.
- *Music* has an important emotive appeal, but it needs to be used with care. Classical music may be counterproductive for a younger audience, whereas older listeners will not find heavy metal pop attractive.
- *Natural sounds*, such as running water, wind in trees, bird song, and waves on a seashore have restful properties.

Media integration rules may also be broken for aesthetic reasons. If information transfer is not at a premium, use of two concurrent video streams might be arousing for a younger audience, as MTV (music TV) and pop videos indicate. Multiple audio and speech tracks can give the impression of complex, busy, and interesting environments.

Image and Identity. Design of media for motivation is a complex area in its own right, and this topic is dealt with in more depth in chapter 17 (Fogg), so the treatment here will focus on media selection issues. Simple photographs or more complex interactive animations (talking heads or full-body mannequins) have an attractive effect. We appear to ascribe human properties to computers when interfaces give human-like visual cues (Reeves & Nass, 1996); however, the effectiveness of media representing people depends on the characters' appearance and voice (see Fig. 12.5). In human–human conversation, we modify our reactions according to our knowledge, or assumptions about, the other person's role, group identification, culture, and intention (Clark, 1996). For example, reactions to a military mannequin will be very different from those to the representation of a parson. Male voices tend to be treated as more authoritative than female voices.

Use of human-like forms is feasible with prerecorded video and photographs; however, the need depends on the application. Video representation of a lecturer can augment presentations, and video communication helps interactive dialogue. However, the rules of human conversation apply (Grice, 1975), so these need to be considered when sourcing or generating the necessary media.

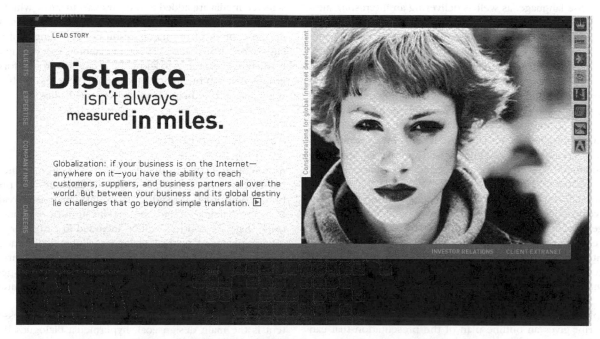

FIGURE 12.5. Effective use of human image for attraction. The picture attracts by the direction of gaze to the user, as well as by the appearance of the individual.

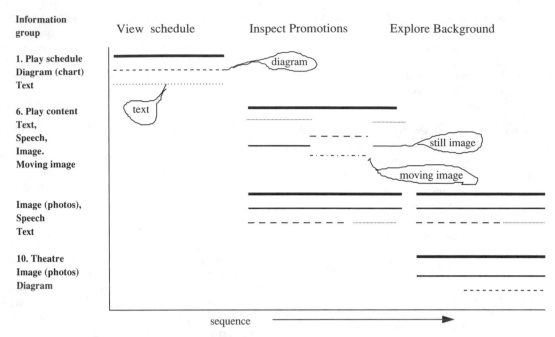

FIGURE 12.6. Bar chart representation of the first-cut presentation script after media selection. Information types are placed on the y axis with the task subgoals that determine the script order on the x axis with media resources represented by bars. For this browsing task, three separate sequences are planned for the view schedule, promotions, etc.

A good speaker holds our attention by a variety of tricks, such as maintaining eye contact, varying the voice tone, using simple and concise language, as well as delivering an interesting message. These general effects can be reinforced by projected personality. Friendly people are preferred over colder, more hostile individuals. TV announcers, who tend to be middle-aged, confident, but avuncular characters have been selected to optimize the attention-drawing power of a dominant yet friendly personality. Both sexes pay attention to extrovert, young personalities, whereas the male preference for beautiful young women is a particularly strong effect. These traits have been exploited by advertisers for a long time. There are lessons here for multimedia designers as the Web and interactive TV converge, and when we want media to convey a persuasive message (Reeves & Nass, 1996; chapter 17, Fogg).

Although the bounds of media selection are only set by the creative imagination of the designer, some fundamentals do not change: Design still needs to be motivated by users' goals, which may be work-related, learning, or having fun; nevertheless, testing with users is still essential.

The outcome of media selection is a first-cut script. The presentation sequence can be planned using bar charts to illustrate the presentation order organized according to the task, or application requirements (e.g., lesson plan); see Fig. 12.6. The script gives an outline plan of the presentation that can be annotated with navigational and dialogue controls. The information model defines the high-level presentation order. Key information items are annotated onto presentation bar charts when planning the sequence and duration of media delivery. The information types are ordered in a first-cut sequence. The selected media are added to the bar chart to show which media stream will be played over time. Decisions on timing depend on the content of the media resource (e.g., length of a video clip, frame display rate). Timeline diagrams can be augmented by hierarchy diagrams to show the classification of information groups that can be mapped to menu-style access paths or network diagrams to help planning hypermedia links.

Navigation and Control

Although discussion of the dialogue aspects of multimedia has been delayed until now, in practice, dialogue and presentation design proceed hand in hand. Task analysis provides the basis for dialogue design and specification of navigation controls. Dialogue controls will be included in early storyboards. Navigational and control dialogues allow flexible access to the multimedia content, and give users ability to control how media are played. Dialogue design may also involve specifying how users interact with tools, agents, and objects in interactive microworlds.

In information-intensive multimedia, in which access to content is the main design goal, hypermedia dialogues that link content segments will be appropriate. Good hypertext design is based on a sound information analysis that specifies the pathways between related items and use of cues to show the structure of the information space to the user. In document-based hypermedia (e.g., HTML and the Web), links can only access the whole media resource rather than point to components within it. The design issues are to plan the overall structure, segment complex structures into a hierarchy of subnetworks and then plan the implementation of each subnetwork. The access structure of most hypermedia will be hierarchical, organized according to the information model and categorization of content (e.g., information grouped by function, organization, task usage, or user preference). Navigation design transforms the user's conceptual model of an information space into a hypermedia structure. Unfortunately, individual users have different models, so this may not be an easy task. Implementing too many links to satisfy each user's view will make the system too complex and increase the chance of the user getting lost. Too few links will frustrate users who cannot find the associations they want. Unfortunately hypermedia systems assume a fixed link structure so the user is limited to the pathways provided by the designer. More open-ended hypermedia environments (e.g., Microcosm: Lowe & Hall, 1998) provide more flexibility via links with query facilities that allow access to databases. Dynamic links attached to hotspots in images or nodes in text documents provide access paths to a wider variety of data.

One problem with large hypermedia systems is that users get lost in them. Navigation cues, waymarks, and mini-map overviews can help to counter the effects of disorientation. *Mini-maps* give an overview of the hypertext area and a reference context for where users are in large networks. *Filters* help to reduce complexity by showing only a subset of nodes and links that the user is interested in. Having typed links helps filtering views, because the user can guess the information content from the link type (e.g., reference, example, source, related work, etc.). Other navigation facilities are *visit lists* containing a history of nodes traversed in a session and *bookmarks*, so users can tailor a hypermedia application with their own navigation aide memoires (Nielsen, 1995). Once the structure has been designed, link cues need to be located within media resources, so the appropriate cues need to be considered for each medium, e.g.,

- *Text media*: The Web convention is to underline and highlight text in a consistent color, e.g., blue or purple.

- *Images*: Link cues can be set as stand-alone icons or as active components in images. Icons need to be tested with users because the designer's assumed meaning can be ambiguous. Active components should signal the link's presence by captions or pop-up hover text, so the user can inspect a link before deciding whether to follow it.

- *Moving images*: Links from animation and film are difficult to design because the medium is dynamic; however, link buttons can be placed below a window. Active components within a moving image are technically more challenging, although overlaid buttons can provide the answer. Buttons may also be timed to pop up at appropriate times during the video.

- *Sound and speech* links are difficult for the same reason as with moving images. One solution is to use visual cues, possibly synchronized with the sound or speech track. If speech recognition is available, then voice commands can act as links, but these commands need to be explained to the user. Automatic links can also be used; but, if these are embedded within the speech/sound medium, then user controls to activate preset links before playing will need to be provided.

Navigation controls provide access to the logical content of multimedia resources; however, access to logical components may be constrained by limitations of physical media resources. For example, a movie may be logically composed of several scenes, but it can only be accessed by an approximate timer set to the beginning of the whole film. Worse still, a video clip may only be playable as a single segment, making implementation of navigation requirements impossible.

In many cases, controls will be provided by the media-rendering device (e.g., video player for .avi files or Quicktime movies). If controls have to be implemented from scratch, the following list should be considered for each media type:

- *Static media*: Size and scale controls to zoom and pan, page access if the medium has page segmentation as in text and diagrams, the ability to change attributes such as color and display resolution, font type, and size in text.

- *Dynamic media*: The familiar video controls of stop, start, play, pause, fast forward, and rewind; also, the ability to address a particular point or event in the media stream by a time marker or an index, e.g., "go to" component/marker, etc.

Navigation controls use standard user interface components (buttons, dialogue boxes, menus, icons, sliders) and techniques (form filling, dialogue boxes, and selection menus); for more guidance, see ISO 9241, parts 12, 14, and 17 (ISO, 1997) and ISO 14915, part 2 (ISO, 1998).

Media Integration and Design for Attention

Having selected the media resources, the designer must now ensure that the user will extract the appropriate information. An important consideration of multimedia design is to link the thread of a message across several different media. This section gives recommendations on planning the user's reading/viewing sequence, and guidelines for realizing these recommendations in presentation sequences, hypermedia dialogues, and navigation controls. The essential differences are timing and user control. In a presentation design, the reading/viewing sequence and timing are set by the designer, whereas the reading/viewing sequence in hypertext implementation and interactive dialogues is under user control.

Presentation techniques help to direct the user's attention to important information and specify the desired order of reading or viewing. The need for thematic links between information components are specified, and attention-directing techniques are selected to implement the desired effect.

The design issues are to:

- Plan the overall thematic thread of the message.
- Draw the user's attention to important information.

- Establish a clear reading/viewing sequence.
- Provide clear links when the theme crosses from one medium to another.

Design for attention is particularly important for images. User attention to time-varying media is determined by the medium itself (i.e., we have little choice but to listen to speech or to view animations in the order in which they are presented). The reading sequence is directed by the layout of text, although this is culturally dependent (e.g., western languages read left to right, Arabic languages in the opposite direction). However, viewing order in images is unpredictable unless the design specifically selects the user's attention.

In some cases, a common topic may be sufficient for directing the user's reading/viewing sequence; however, when the thread is important or hard to follow, designed effects for attention, or contact points, are necessary to aid the user's perception. The term contact point refers to a reference from one medium to another and comes from the research of Hegarty and Just (1993) and Narayanan and Hegarty (1998) that demonstrated comprehension is improved by reinforcing the links between information in different media. Two types of contact points are distinguished:

- *Direct contact points*: Attention-directing effects are implemented in both the source and destination medium (e.g., in the text, an instruction is given, "Look at the oblong component in Fig. 12.1," while the component is highlighted). Direct contact points create a strong cross reference between two media, but can become intrusive if overused. An example of a direct contact point is shown in Fig. 12.7, in which the speech track refers to an image component that is circled to draw the user's attention.

- *Indirect contact points* implement an attention-directing effect only in the source, or less frequently the destination, medium (e.g., "In Fig. 12.1, the assembly is shown" is spoken, with no highlighting being used in the image). Indirect contact points have less attention-directing force and work by temporal sequencing or spatial juxtapositoning. Indirect contact points are less intrusive, so they may be used more frequently without becoming disruptive.

In most cases, the attentional effect in a direct contact point will be actuated in sequence, although occasionally both effects may be presented concurrently if the order of the association is not important. In hypermedia implementations, direct contact points become a link cue in the source medium and a highlight anchor in the destination medium. Contact points are specified in the presentation bar chart illustrated in Fig. 12.8.

Multiple contact points may be organized in a logical order to follow the theme and connect a thread of topics (Faraday & Sutcliffe, 1998a, 1999). For instance, in a biology tutorial, explaining parts of a cell is organized with interleaved speech segments and a diagram describing the cell's components top to bottom, left to right. Highlighting techniques locate each component in turn, following the order of the spoken explanation. The attention-directing techniques described in the following section are used to implement contact points.

The design problem is how to direct the user's attention to the appropriate information at the correct level of detail.

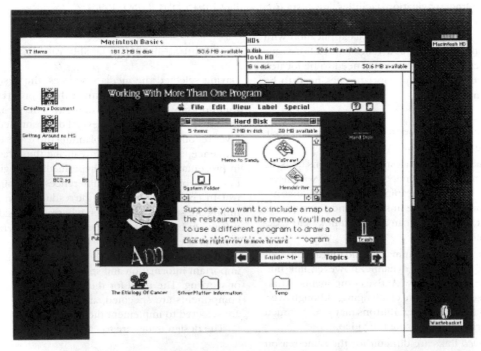

FIGURE 12.7. Direct contact point between text and a highlighted image, reinforced by speech.

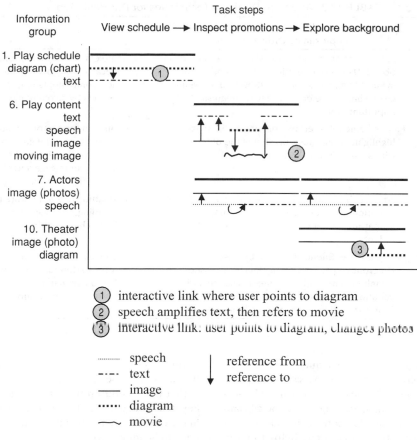

FIGURE 12.8. Contact points for the theater booking browse program task. Contact points are represented by vertical arrows between the presented media, shown as horizontal bars. The duration of media presentation is cross referenced to the task sequence, and the key information items to be made salient are specified.

Initially, users will tend to extract information from images at the scene level (i.e., major objects will be identified, but with very little descriptive detail [Treisman, 1988]). Regular layout grids help design composite images (Mullet & Sano, 1995) and encourage viewing sequences in image sets. Alternatively, the window frame can be set to control which parts of an image are viewed. Larger window frames will be attended to before smaller areas. A list of the key components that the user needs to focus on and the facts that should be extracted is checked against the initial presentation design to see if the key components will attract sufficient attention or whether the user is likely to be confused by extraneous detail.

Still Image Media. Highlighting techniques for designed and natural images, organized in approximate power of their effect, are summarized in Table 12.2. A common highlighting technique will pick out spatially distributed objects (e.g., change all the related objects to the same color); colocated objects can be grouped by using a common color or texture for their background or drawing a box around them. The highlighted area will set the granularity of the user's attention. Captions linked to objects in an image are another useful means of drawing attention and providing supplementary information (e.g., identity). Dynamic revealing of captions is particularly effective for directing the user's viewing sequence. Sequential highlighting is also useful for showing the pathways or navigational instructions.

Moving Image Media. Directing attention to components within moving images is difficult because of the dynamic nature of the medium. Design of film and video is an extensive subject in its own right, so treatment here will necessarily be brief. The following design advice is based on Hochberg (1986). The design objectives, as for other media, are how to draw the user's attention to key components within the video or animation.

First the content needs to be structured into scenes that correspond to the information script. To structure animation sequences and make scene boundaries obvious, use a cut, wipe, or dissolve to emphasize that a change in the content structure has taken place. However, cuts should be used with care and

TABLE 12.2. Attention-Directing Techniques for Different Media

	Highlight Techniques in Approximate Order of Power	Notes
Still image: designed and natural	Movement of or change in the shape/size/color of an object. Use of bold outline. Object marked with a symbol (e.g., arrow) or icon. Draw boundary; use color, shape, size or texture to distinguish important objects.	Some effects may compromise natural images because they overlay the background image with new components (e.g., arrow, arcs, icons). Group objects by a common technique.
Moving image	Freeze frame followed by applying a still image highlight. Zoom, close-up shot of the object. Cuts, wipes, and dissolve effects.	Change in topographic motion, in which an object moves across the ground of an image and is more effective than internal movement of an object's components. Size and shape may be less effective for highlighting a moving object.
Text	Bold, font size, type, color, or underlining. To direct attention to larger segments of text, use formatting, bullet points, subsections, and indentation.	Formatting techniques are paragraphs. Headings/titles are entry points. Indents show hierarchical nesting, with bullet points and lists.
Speech/sound	Familiar voice. Silence followed by onset of sound. Different voices, or a change in voice prosody (tonality), amplitude (loudness), change and variations in pitch (frequency), voice rate, change source direction, alarm sounds (police sirens).	Voices familiar to the user (e.g., close relatives) attract attention over nonfamiliar speech. Discourse markers "next," "because," "so," etc., draw attention to subsequent phrases.

continuity maintained between the two sequences if they are to be integrated. Continuity is manifest as the same viewpoint and subject matter in two contiguous shots. Change in background or action (e.g., an individual walking left in one clip and walking right in the next), are quickly noticed as a change. An establishing shot that shows the whole scene helps to introduce a new sequence and provide context. To provide detail of a newly introduced object or context, the object is shown filling the frame with a small amount of surrounding scene; whereas to imply a relationship or compare two objects, a tight two-shot with both objects together in the same frame is advisable. Attention-directing effects are summarized in Table 12.2.

Linguistic Media (Text and Speech). As with moving image, the literature is extensive, so the following heuristics are a brief summary (see Levie & Lentz, 1982 for more detail). Text may be structured to indicate subsections by indentation, formatting into paragraphs, columns, or segmented by background color. Bullet points or numbered sections indicate order more formally (e.g., for procedures). Different voices help to structure speech, while also attracting attention. If language is being used to set the context for accompanying media, it is important that the correct level of identification is set. For instance, a higher level concept, or the whole scene in an accompanying image, is described at the beginning of a script, then lower-level topics reset the user's focus. Discourse markers can make phrases and sentences more salient.

Adding contact points and attention-directing effects completes the design process; however, as with all user interfaces, there is no substitute for usability testing. Designs are constructed incrementally by iterations of design and evaluation that check for usability using standard methods, with additional memory and comprehension tests for multimedia. So, when

testing a design, ask the user to tell you what they understood the message to be. This can be done during the presentation with a think-aloud protocol to check that users did attend to key items and afterward by a memory test. If key components in the message are not being remembered, then the design may need to be improved.

CONCLUSIONS

Multimedia still poses many issues for further research. The design method described in this chapter, coupled with user-centered design, can improve quality; however, there is still a need for experts to create specific media resources (e.g., film/video, audio experts). Furthermore, considerable research is still necessary before we fully understand the psychology of multimedia interaction. Design for motivation and attractiveness is still poorly understood, and personality effects in media may not be robust when usability errors impede communication. The process by which people extract information from complex images still requires extensive research, although the increasing number of eyetracking studies is beginning to throw some light on this topic. In the future, language and multimodal communication will change our conception of multimedia from its current CD-ROM or Web-based form into interfaces that are conversational and multisensory. Multimedia will become part of wearable and ubiquitous user interfaces in which the media is part of our everyday environment. Design for multisensory communication will treat media and artefacts (e.g., our desks, clothes, walls in our homes) as a continuum, while managing the diverse inputs to multimedia from creative design, technology, and usability engineering will be one of the many interesting future challenges.

References

Ahlberg, C., & Shneiderman, B. (1994). Visual information seeking: Tight coupling of dynamic query filters with starfield displays. In B. Adelson, S. Dumais, & J. Olson (Eds.), *Proceedings CHI '94: Human factors in computing systems, celebrating independence, Boston, MA* (pp. 313-317). New York: ACM Press.

Alty, J. L. (1991). Multimedia: What is it and how do we exploit it? In D. Diaper, & N. V. Hammond (Eds.), *Proceedings of HCI '91: People and computers VI* (pp. 31-41). Cambridge, UK: Cambridge University Press.

Alty, J. L. (1997). Multimedia. In A. B. Tucker (Ed.), *Computer science and engineering handbook* (pp. 1551-1570). New York: CRC Press.

Andrc, E., & Rist, T. (1993). The design of illustrated documents as a planning task. In M. T. Maybury (Ed.), *Intelligent multimedia interfaces* (pp. 94-116). Cambridge, UK: AAAI/MIT Press.

Arens, Y., Hovy, E., & Van Mulken, S. (1993). Structure and rules in automated multimedia presentation planning. *Proceedings: IJCAI-93: Thirteenth International Joint Conference on Artificial Intelligence.*

Baddeley, A. D. (1986). *Working memory*. London: Oxford University Press.

Baggett, P. (1989). Understanding visual and verbal messages. In H. Mandl & J. R. Levin (Eds.), *Knowledge acquisition from text and pictures* (pp. 101-124). Amsterdam: Elsevier Science.

Barnard, P. (1985). Interacting cognitive subsystems: A psycholinguistic approach to short term memory. In A. Ellis (Ed.), *Progress in psychology of language* (Vol. 2, pp. 197-258). London: LEA.

Bernsen, N. O. (1994). Foundations of multimodal representations: A taxonomy of representational modalities. *Interacting with Computers, 6*(4), 347-371.

Bertin, J. (1983). *Semiology of graphics*. Madison: University of Wisconsin Press.

Bieger, G. R., & Glock, M. D. (1984). The information content of picture-text instructions. *Journal of Experimental Education, 53*, 68-76.

Booher, H. R. (1975). Relative comprehensibility of pictorial information and printed word in proceduralized instructions. *Human Factors, 17*(3), 266-277.

Card, S. K., Moran, T. P., & Newell, A. (1983). *The psychology of human computer interaction*. Hillsdale, NJ: Lawrence Erlbaum Associates.

Clark, H. H. (1996). *Using language*. Cambridge, UK: Cambridge University Press.

Faraday, P., & Sutcliffe, A. G. (1996). An empirical study of attending and comprehending multimedia presentations. In *Proceedings ACM Multimedia '96: 4th Multimedia Conference, Boston, MA 18-22 November 1996* (pp. 265-275). New York: ACM Press.

Faraday, P., & Sutcliffe, A. G. (1997a). Designing effective multimedia presentations. In *Proceedings: Human factors in computing systems CHI '97, Atlanta, GA* (pp. 272-279). New York: ACM Press.

Faraday, P., & Sutcliffe, A. G. (1997b). Multimedia: Design for the moment. *Proceedings: Fifth ACM International Multimedia Conference, Seattle, WA 9-13 November 1997* (pp. 183-192). New York: ACM Press.

Faraday, P., & Sutcliffe, A. G. (1998a). Providing advice for multimedia designers. In C. M. Karat, A. Lund, J. Coutaz, & J. Karat (Eds.), *CHI '98 Conference Proceedings: Human Factors in Computing Systems, Los Angeles, CA 18-23 April 1998* (pp. 124-131). New York: ACM Press.

Faraday, P., & Sutcliffe, A. G. (1998b). Using contact points for Web page design. *People and Computers XIII, Proceedings: BCS-HCI Conference,* Sheffield, UK.

Faraday, P., & Sutcliffe, A. G. (1999). Authoring animated Web pages using contact points. *Proceedings of CHI '99: Human Factors in Computing Systems, Pittsburgh, PA* (pp. 458-465). New York: ACM Press.

Fogg, B. J. (1998). Persuasive computer: Perspectives and research directions. *Proceedings: Human Factors in Computing Systems: CHI '98, Los Angeles CA 18-23 April 1998* (pp. 225-232). New York: ACM Press.

Gardiner, M., & Christie, B. (1987). *Applying cognitive psychology to user interface design*. Chichester, UK: Wiley.

Grice, H. P. (1975). Logic and conversation. *Syntax and Semantics, 3.*

Hegarty, M., & Just, M. A. (1993). Constructing mental models of text and diagrams. *Journal of Memory and Language, 32*, 717-742.

Heller, R. S., & Martin, C. (1995, Winter). A media taxonomy. *IEEE Multimedia*, 36-45.

Hochberg, J. (1986). Presentation of motion and space in video and cinematic displays. In K. R. Boff, L. Kaufman, & J. P. Thomas (Eds.), *Handbook of perception and human performance, 1: Sensory processes and perception*. New York: Wiley.

Hollan, J. D., Hutchins, E. L., & Weitzman, L. (1984). Steamer: An interactive inspectable simulation-based training system. *AI Magazine, 5*(2), 15-27.

ISO. (1997). *ISO 9241: Ergonomic requirements for office systems with visual display terminals (VDTs)*. International Standards Organisation.

ISO. (1998). *ISO 14915: Multimedia user interface design software ergonomic requirements, Part 1: Introduction and framework*. International Standards Organisation.

ISO. (2000). *ISO 14915-3: Software ergonomics for multimedia user interfaces. Part 3: Media selection and combination. Draft international standard*. International Standards Organisation.

Kristof, R., & Satran, A. (1995). *Interactivity by design: Creating and communicating with new media*. Mountain View, CA: Adobe Press.

Levie, W. H., & Lentz, R. (1982). Effects of text illustrations: A review of research. *Educational Computing and Technology Journal, 30*(4), 159-232.

Lowe, D., & Hall, W. (1998). *Hypermedia and the Web*. Chichester, UK: John Wiley.

Mann, W. C., & Thompson, S. A. (1988). Rhetorical structure theory: Toward a functional theory of text organisation. *Text, 8*(3), 243-281.

May, J., & Barnard, P. (1995). Cinematography and interface design. In K. Nordbyn, P. H. Helmersen, D. J. Gilmore, & S. A. Arnesen (Eds.), *Proceedings: Fifth IFIP TC 13 International Conference on Human-Computer Interaction, Lillehammer, Norway 27-29 June 1995* (pp. 26-31). London: Chapman & Hall.

Mullet, K., & Sano, D. (1995). *Designing visual interfaces: Communication oriented techniques*. Englewood Cliffs, NJ: SunSoft Press.

Narayanan, N. H., & Hegarty, M. (1998). On designing comprehensible interactive hypermedia manuals. *International Journal of Human-Computer Studies, 48*, 267-301.

Nielsen, J. (1995). *Multimedia and hypertext: The internet and beyond*. Boston, MA: AP Professional.

Norman, D. A., & Shallice, T. (1986). Attention to action: Willed and automatic control of behaviour. In G. E. Davidson & G. E. Schwartz (Eds.), *Consciousness and self-regulation* (Vol. 4, pp. 1-18). New York: Plenum.

Papert, S. (1980). *Mindstorms: Children, computers, and powerful ideas*. New York: Basic Books.

Park, I., & Hannafin, M. J. (1993). Empirically-based guidelines for the design of interactive multimedia. *Educational Technology Research and Development, 41*(3), 63–85.

Pezdek, K., & Maki, R. (1988). Picture memory: Recognizing added and deleted details. *Journal of Experimental Psychology: Learning, Memory and Cognition, 14*(3), 468–476.

Reeves, B., & Nass, C. (1996). *The media equation: How people treat computers, television and new media like real people and places*. Stanford, CA/Cambridge, UK: CLSI/Cambridge University Press.

Rogers, Y., & Scaife, M. (1998). How can interactive multimedia facilitate learning? In J. Lee (Ed.), *Intelligence and multimodality in multimedia interfaces: Research and applications*. Menlo Park, CA: AAAI Press.

Scaife, M., Rogers, Y., Aldrich, F., & Davies, M. (1997). Designing for or designing with? Informant design for interactive learning environments. In S. Pemberton (Ed.), *Proceedings: Human Factors in Computing Systems CHI '97, Atlanta GA 22–27 May 1997* (pp. 343–350). New York: ACM Press.

Spool, J. M., Scanlon, T., Snyder, C., Schroeder, W., & De Angelo, T. (1999). *Web site usability: A designer's guide*. San Francisco: Morgan Kaufmann.

Sutcliffe, A. G. (1997). Task-related information analysis. *International Journal of Human-Computer Studies, 47*(2), 223–257.

Sutcliffe, A. G. (1999). User-centred design for multimedia applications. *Proceedings. Vol. 1: IEEE Conference on Multimedia Computing and Systems, Florence,* Italy (pp. 116–123). Los Alamitos, CA: IEEE Computer Society Press.

Sutcliffe, A. G. (2002). *Multimedia and virtual reality: Designing multisensory user interfaces*. Mahwah, NJ: Lawrence Erlbaum Associates.

Sutcliffe, A. G., & Faraday, P. (1994). Designing presentation in multimedia interfaces. In B. Adelson, S. Dumais, & J. Olson (Eds.), *CHI '94 Conference Proceedings: Human Factors in Computing Systems 'Celebrating Interdependence', Boston, MA 24–28 April 1994* (pp. 92–98). New York: ACM Press.

Treisman, A. (1988). Features and objects: Fourteenth Bartlett memorial lecture. *Quarterly Journal of Experimental Psychology, 40A*(2), 201–237.

Tufte, E. R. (1997). *Visual explanations: Images and quantities, evidence and narrative*. Cheshire, CN: Graphics Press.

Wickens, C. D., Sandry, D., & Vidulich, M. (1983). Compatibility and resource competition between modalities of input, output and central processing. *Human Factors, 25*, 227–248.

·13·

VISUAL DESIGN PRINCIPLES FOR USABLE INTERFACES

Suzanne Watzman
Watzman Information Design

EVERYTHING IS DESIGNED: WHY WE SHOULD THINK BEFORE DOING

Take a moment and visualize.... Las Vegas at night.

What kind of image does this conjure up for you?

Flashing lights coming from all directions, one hotel's lighting display designed to outdo the one next door, as well as the casino signage down the street. At first glance, everything is exciting, colorful, and beautiful.

Now add a fireworks display to your picture. More color, more excitement. Where do you look first? There is so much going on; it is hard to take it all in, but you don't want to miss a thing. Your head turns in all directions. You look there; then, out of the corner of your eye, you see something else. Look over there!

Now the fireworks are at their peak, and the noise gets even louder. Any conversation with companions is impossible; it has become impossible to focus on any one thing for more than a split second. Overwhelmed, overloaded, everything screaming for your attention. Can you manage it? For how much longer? Do you begin to shake your head in despair and just give up? Do you wish you were somewhere else—NOW?

Making Things Easier to Use and Understand—Thinking About the User's Experience

The previous description is unfortunately an accurate analogy of many users' experiences as they attempt to work, play, and relax. New products, new services, and new technology, with new or less-than-familiar ways of interacting with them. These "users" are customers, electricians, grand parents, clerks, pilots, students. They are you and me. And, for a great majority, it's a jungle out there!

Las Vegas at night with fireworks. Or, monitors that are winking, blinking, distracting, disturbing, overwhelming and, after a short period of time, visually deafening. Now add the voices coming from boxes . . . ! Although this may seem like an exaggeration, for many this is exactly their experience.

It is our job as creators of these products, of our user's experience to be simple and useful; where technology and process are transparent as possible. To do our job well, we must play the role of user advocate, ensuring that the interfaces we design are not just merely exercising the technology, but assisting the user do a job, easily moving from task to task, getting work done, making life easier.

When we succeed, our products become effortless, even pleasurable to use. Good design is not noticed, it just works. That is the role of good design.

Defining Visual Design. Visual design is not merely a series of subjective choices based on favorite colors or trendy typefaces—at best a cosmetic afterthought considered if there is enough time and money. Good visual design is the tangible representation of product goals. It is concerned with the "look,"

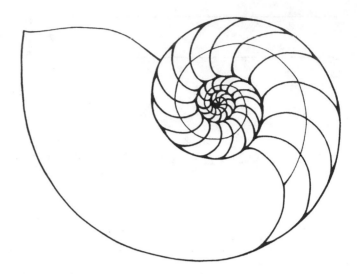

FIGURE 13.1. The nautilus shell is one of nature's many examples of a perfect synthesis of form and function. It is the result of years of evolution to survive in its current from. The work of evolution is transparent, as we admire its beauty. It is a perfect analogy for design and design process to create valuable user experiences and usable interfaces.

the method, and the style in which the information is presented. It should be the result of a thoughtful, well-considered process, not merely a decorative afterthought.

Applying the appropriate visual/experience design principles and tools while incorporating the user perspective (information design) enhances the value, perception, and usefulness of products. It is the best combination of project goals, the user perspective, and informed decision making.

The Role of the Designer. Visual design choices must be based on project goals, user perspective, and informed decision making. Although many aspects of design are quantifiable, there are visual principles that are less quantifiable, but equally important. Even though many individuals can learn the necessary skills, become visually literate, and able to make competent design decisions, it is a unique combination of creativity and talent that differentiates one design solution as more attractive and desirable than another.

One must not only have the talent, but also the understanding and skills to apply the principles required to present information in its most accessible, useful, and pleasing form. This is the role of the designer in the development of interfaces for interactive products: Understanding product goals and ensuring that interface is approachable, useful, and desirable. In an environment in which the interface is the only tangible representation of a product, and user perception determines product success, appropriate information presentation and visual design is the key. Designers understand visual principles within their usage context, and how to apply them appropriately to create attractive solutions that also solve problems.

The Process of Good Design—How Do We Get There from Here?

As interface designers, we are responsible for defining what the experience will be like when someone uses our product. We are defining, designing, then creating this experience to be one that is useful and meaningful, even pleasant and empowering. The designer must maintain an attitude of unbiased discovery and empathy for the user. In addition, one needs to create clearly defined goals, a good design, and evaluation process—that support and enhance these goals, and the flexibility to make changes as the process continues and products evolve.

An Information Design Process Is an Informed Design Process

An Information Design Process is a method of visually structuring and organizing information to develop effective communication. Information design is not superficial or decorative, rather a merging of functional, performance-based requirements with the most appropriate form for presentation of these requirements.

A thoughtful, well-designed solution will:

- Motivate users: It psychologically entices the audience, convincing them that they can successfully cope with the information and tasks at hand.
- *Increase ease of use and accessibility*: Decreases the effort needed to comprehend the information and provides a clear path through material that aids in skimming, quick reference and easy access.
- *Increase the accuracy and retention of information*: Users learn and retain information better when it is visually mapped and structured in obvious and intuitive ways.
- *Focus on needs of its users*: Different audiences have different requirements and styles of learning. This approach ensures solutions that provide ways for different types of users to access information.

An Information Design approach is part of a process that incorporates research, design, testing, and training to produce useful, cost-effective and desirable solutions.

Phase 1: The Audit. The goal of the Audit is to create a blueprint for the project, much like one would create an architectural blueprint before building before construction.

To do an Audit, one must begin the process by asking and answering a number of questions, acknowledging ongoing change and an ever-increasing palette of products and services. Asking questions occurs throughout the entire product life cycle, because the answers/design solutions reflect the user/use environment and affect the ongoing usefulness and value of the product.

To create a good design, ask and answer, on an ongoing basis, the following questions:

Audit Questions A

- Who are your users?
- How will they use this product?
- When will they use this product?
- Why will they use this product?
- Where will they use this product?
- How will your process evolve to support this product, as it evolves.

When the first set of questions are asked and answered, the next set of questions must be asked and answered as well:

Audit Questions B

- What is the most efficient, effective way for your user to accomplish their tasks and move on to the next of tasks?
- How can you most efficiently and effectively present information required for ease-of-use of this product?
- How can the design of this product be done to support ease-of-use and transition from task to task as a seamless, transparent, and even pleasurable experience?
- What are your technical and organizational parameters?

The Audit focuses on discovery. Many disciplines and organizational resources must be considered. Change is a given, because we begin with assumptions, and don't know what we don't know. The answers and their analysis in the context of organizational objectives provide the basis for the Audit Report, which will serve as the guide for Phase 2: Design and Development.

The Audit Report can be as simple as a two-page list or as complex as a comprehensive one-hundred page report. Since the goal is discovery, it include every aspect of the organization concerned with product the development cycle: project management, usability engineering, technical development, user support/documentation, visual communication and design, and content management. With these goals in mind, the result will be unbiased, accurate, comprehensive information to serve as a basis for design in the following Design Development phase.

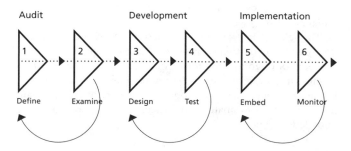

FIGURE 13.2. An Information Design Process is phased to ensure user and organizational needs are met. It is ongoing and iterative, throughout the life cycle of a product. Any change can trigger a recycling of this process to ensure solutions remain appropriate and useful.

Phase 2: Design Development. Using the Audit Report as a guideline, the process of Design Development begins. This is an ongoing, iterative process, with each iteration incorporating test results (from users) to make the product appropriate to the particular set of needs. In reality, the length of this process is often defined and limited by real-world deadlines, such as product release dates.

The Design Development phase includes design and testing. The designer/design team create a number of solutions based on results an objectives of the Audit Report, as well as other project specifics. The first, design ideas should be very broad, incorporating many ideas and options no matter how unrealistic or unusual. As ideas are tested and evaluated, other parameters defined and user feedback is incorporated, the solutions naturally become more refined. The surviving design ideas will then be based on solid information from user feedback, serving as a good basis for final design decisions.

In the beginning of this phase, the focus is on high-level concepts and navigation; how will the product work? What will it feel like to use? As these initial concepts are refined, design details become more specific. When the conceptual model and organizational framework are approved, the design of the look, or product package, begins. By the end of this phase, the design has been tested, approved, and specified, to be carried out in the Implementation and Monitoring phase.

Phase 3: Implementation and Monitoring. The Implementation phase focuses on delivering what has been defined, designed, and documented in the preceding phases. It is the final part of a *holistic* process by defining all of what is required to make a product succeed on an ongoing basis. This includes not only implementation of the design within the technology, but also product support, such as training creation of additional materials, and other support that enhances use and productivity. On going monitoring of solutions is key to continued product success, and must reflect the evolving technology and user needs.

This last phase is more consultative, and ongoing throughout the product lifecycle. This ensures that any changes, such as new technology and product developments, are reflected in the product itself. These may in fact trigger another Audit/Design/Testing cycle, although usually less extensive than the initial process.

Though the Implementation Phase is called the last phase, it reveals the evolutionary process of design and development. The goal of ongoing monitoring of solutions is to be aware of changes in user needs, technology and competition that impact user acceptance and satisfaction. Changes here often may result in the need to re-evaluate and re-design to incorporate this new knowledge gained.

VISUAL DESIGN PRINCIPLES

Many visual design principles can be easily explained and learned. Others, although easily defined, are more complex to explain and understand. They are often intuitive for those naturally skilled in aesthetics and have been formally trained in visual design/usability disciplines. Good visual interaction and experience design bridge many worlds: that of visual design, information presentation, and usability with that of aesthetics. The resulting solutions are not only usable, but also attractive, interesting, and pleasurable to use.

Universal Principles of Visual Communication and Organization

Every visual person, whether artist or graphic designer, understands the universal principles of visual organization. They are at work in everything we see and do. Though more conceptual in nature, they are the basis for every visual decision. To begin to understand these principles and how they work is to become visually literate.

There are three visual communication principles that are fundamental to all successful design solutions, related yet distinct in their meaning and application.

Harmony is the thoughtful combination of many and different parts into a pleasing, orderly whole. In interface design, this is achieved when all elements of a design appear to fit and work well together. Transitions place to place are effortless, and the techniques used to achieve harmony are transparent to the user.

Visual harmony has the same goal as musical harmony; complementing yet enhancing the basic piece. In the visual world, the golden rectangle of Greek architecture is one of the most widely known examples of this.

Balance is the pleasing harmony of various visual elements to achieve a sense of stability and comfort in design. Much like a clown balancing on a ball while juggling objects of different weights who must make adjustments for actions that are occurring, visual balance requires the same concerns and adjustments in the interaction world.

In design, all elements have visual weight, or heaviness. Depending on whether the design is symmetrical or asymmetrical, visual balance and a feeling of unity must be achieved for a solution to feel comfortable to the user.

There are a number of ways to achieve this. The simplest way is through *symmetry*, such as a page with centered type and illustrations. Though it is more likely to be successful, it is not as interesting and has the potential to be boring and static. *Asymmetrical* design is use of variation of elements, such as size, contrast, color, and placement to create visual tension and drama. Both are valid approaches and require skill and understanding of complex visual interaction to achieve a balanced and attractive design solution.

Simplicity in visual design is the embodiment of clarity, elegance, and economy. Although there are many ways to achieve the same result, a solution that is simple works, effortlessly devoid of unnecessary decoration. Simplicity in design appears easy, accessible, and approachable, even though it often requires more skill to achieve.

Achieving simple visual design solutions is no easy task, but two guidelines for creating simple design solutions are: "Less is more" (Mondrian) and "When in doubt, leave it out!" (Anonymous). The simplest, most refined design is direct and includes only the essential elements—as if by removing any of the remaining elements the composition would be rendered unintelligible or radically different.

Visual Design Tools and Techniques

New technologies are rapidly being created that go beyond simple automation of tasks and communication; they are revolutionizing processes and the resulting products. Before the revolution brought about by electronic publishing technology, many disciplines (e.g., writing, editing, design, publishing, programming) were discreet and clearly defined.

Today, new publishing and communication environments have brought to life the possbility of the *renaissance* publisher—one person who can create, design, publish, and distribute. Yet the process to arrive at successful solutions is very complex. One must remain focused on what factors determine success and with constant evaluation and adjustment of these factors in light of new developments.

The Five Criteria for Good Design

Before any work begins, participants in the process should have a clear understanding of the criteria for good design. These five questions are guidelines for evaluation of design solutions before, during, and after the process to ensure that all solutions remain valid as products, technology and user needs evolve.

- *Is it Appropriate?*
  Is the solution appropriate for the particular audience, environment, technology, culture?

- *Is it Durable?*
  Will the solution be useful over time? Can it be refined, transitioned, as the product evolves and is redefined?

- *Is it Verifiable?*
  Has the design been tested in the use environment by typical users? Has feedback been properly evaluated and used to improve the product?

- *Does it have Impact?*
  Does the design solution not only solve the problem, but also impact a look and feel, so that the user finds the product experience comfortable, useful, and desirable?

- *Is it Cost-Effective?*
  Can the solution be implemented and maintained? The cost of any design begins with the Audit and the Design phases, but continues after Implementation to ensure that it remains useful and cost-effective. The hard costs and soft costs of delivery the solution plus ongoing maintenance add up to the real cost of the design. Are there individuals with necessary skills and understanding to create, refine and maintain the design as time goes on?

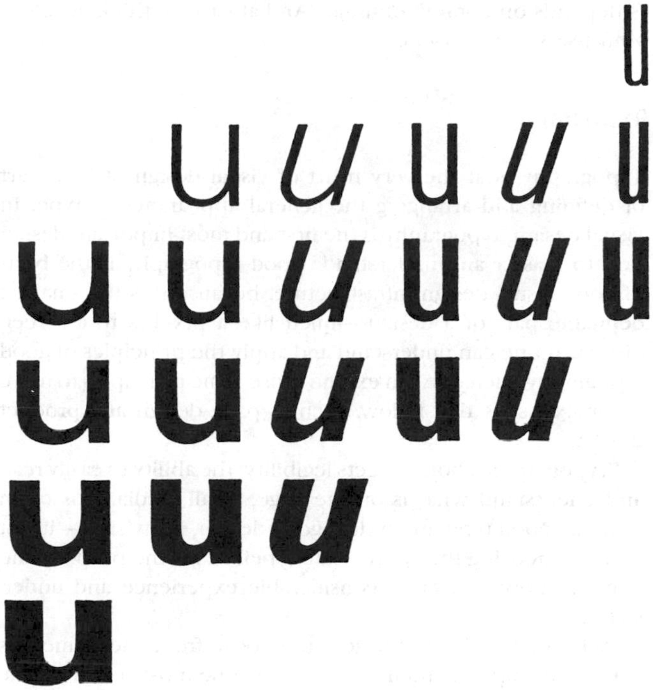

FIGURE 13.3. The Univers typeface was designed by Adrian Frutiger in 1957 to provide a set of compatible progressive variations easily identified by a series of numbers.

VISUAL DESIGN PRINCIPLES AT WORK

The following sections outline the various visual design disciplines and principles that are used when creating quality design solutions. Each topic can be a subject of extended study in its own right, because there is much to understand when presenting information most appropriately for every specific situation.

It is important to recognize that, as the design process evolves, new insights and information will be discovered that will have impact on the design solution. It is optimistic to base solutions on the early initial process, because the very nature of process means discovering what is unknown, yet critical. For that reason, all those involved in the design process must remain open and ready to incorporate new information that may impact or change the design, cause delays, but will more accurately reflect the users/customers.

For example, if a new feature was developed that would change a product's target audience from professional users to focus on executives, one would have to reconsider most critical interactions and content delivery. Executive users have less time and need different information. The design result might be a simpler interface, different content, perhaps larger typefaces, different visual "tone of voice," etc.

The most important principle to remember when thinking about design is: There are no rules, only guidelines. Everything is depends on context of usage. And always be thinking about your users, users, users.

Typography

Typography is at the very heart of visual design; it is the art of defining and arranging the general appearance of type. In visual design, typography is the first and most important design skill to master and understand. Good typography is the basis of good visual design infrastructure, because it is the smallest definable part of a design—much like a pixel is to a screen display. If one can understand and apply the principles of good typography, then one can extend those same principles to more complex issues that follow, such as page design and product design.

Typographic choice affects legibility; the ability to easily read and understand what is on the page, in all media. It is often said that good typography, like good design, is invisible—it just works. Choosing the appropriate typeface for the purpose and context, however, takes considerable experience and understanding.

With hundreds of typefaces to choose from and numerous ways to manipulate them—finding the one most suited for the intended audience is no easy task. Choosing the appropriate typeface is the difference between being able to read and understand something, or not. Given the current publishing environment with its lack of control, multiple media, and varied viewing contexts, and user needs it is an even more complex task.

The choice of typeface immediately impacts whether a communication is read and how it is perceived. A typeface can be used to set a mood. An old-fashioned typeface can make a newsletter look dated; a typeface with extreme thick and thin strokes in the letterform can look great in a brochure, but render a web page unreadable. A typeface specifically designed for online use can increase legibility as well as providing perceptual cues about approachability and quality of an interface. Thus, typographic choice ultimately impacts product acceptance.

A good choice makes the task of reading more enjoyable and effortless rather than frustrating and fatiguing. Though typographic choice might seem to be an insignificant issue, it is often the major factor affecting overall usability. The designer must have a clear understanding of the various concepts and principles that affect legibility when making choices about typography.

How the Eye Sees, Then Reads. The human eye does not read one letter at a time, or even one word at a time. It moves along a line of text, grouping the text to form comprehensible phrases of information. This motion of the eye during reading is known as *saccadic movement*. Typeface choice directly affects this process, making it easier or more difficult for the eye to group, read and understand information.

The following characteristics of type further illustrate principles that affect legibility of type and overall quality of the communication.

x-Height. This refers to the height of the main element of a lowercase letter and is equivalent to the height of a lowercase x. The x-height, not the point size, conveys the actual physical and psychological impression of the size of a letter.

Choosing a Typeface. The typefaces in Fig. 13.4 are the same point size, but appear different because of variations in x-heights of each typeface. Because of these variations, as well as other design elements of the letterforms themselves, some will be more or less readable and legible than others. This depends on resolution of the output/viewing devices, viewing environment, color, context, and a variety of other design issues. When choosing a typeface, it is critical to understand not only the characteristics of a typeface, but also the usage context and application environment as well.

Ascenders and Descenders. The ascender is the part of the lowercase letter that rises above the body (x-height) of the letter. The descender refers to the part of the lowercase letter that falls below the body (x-height) of the letter.

Serif and Sans Serif. Serif is the stroke that projects from the top or bottom of the main stroke of the letter. Some printed letters have no serifs at all; these letterforms are called sans serif (without serif).

hqx	hqx	hqx	hqx	hqx
PALATINO	BASKERVILLE	BODONI	SERIFA	HELVETICA

FIGURE 13.4. The x-height of a typeface (actual height of a lower-case x) is a key characteristic when deciding the visual size of a typeface, particularly where readability is the critical requirement. The above typefaces are the same point size. Some will seem larger (e.g., Helvetica) and easier to read than others (e.g., Serifa) though they are the same point size.

Readability Versus Legibility. Readability is ability to find what you need on the page; legibility is being able to read it when you get there. Effective page design makes a page readable; good use of typography makes it legible.

Legibility is determined by:

- typeface
- output/viewing device, resolution
- line length/column width
- letter spacing, word spacing, line spacing
- justified versus ragged columns
- movement
- color
- viewing environment

Contrast. Contrast can affect size, shape, color, and background color.

Variations. There are five ways to vary a typeface:

- lightface
- boldface
- condensed
- expanded
- italics

Note: In some typefaces, there may be additional increments of medium and extra bold, as well as combinations such as bold extended or bold italic.

Font. A font is made up of all of the characters of one size of one particular typeface. In addition to the alphabet and punctuation marks, certain fonts include symbols and special characters, such as the &.

Size. Type size is referred to in terms of point size. Because of differences in x-heights, 14-point Helvetica looks much larger than 14-point Times Roman.

This is 14-point Helvetica.

This is 14-point Times.

Kerning. Kerning is the adjustment of the spacing between letters to give the visual impression that they are all equidistant.

Families of Type. There are five families, or organizational groupings of type, based on their historical development. Although created over hundreds of years ago, all of these typefaces are in use today. But more important than knowing the date of creation is understanding how typefaces have evolved over time, and the resulting differences and similarities among them. The style of typeface is very much a reflection of trends, fashion, current events, and technical developments at the time the typeface was designed. See Fig. 13.5.

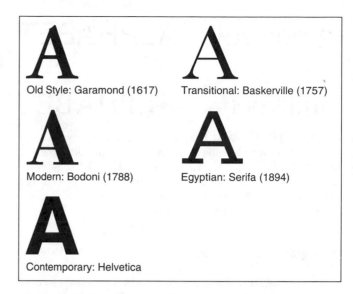

Old Style: Garamond (1617) Transitional: Baskerville (1757)

Modern: Bodoni (1788) Egyptian: Serifa (1894)

Contemporary: Helvetica

FIGURE 13.5. These five A's show typographic style from the 1600s through modern times, reflecting similar changes in tools, fashion, and current events. Ultimately, choice of output media should determine typeface choice, given details such as the thick and thin parts of the letterforms, negative space, viewing environment, output resolution, etc.

Variations in Letterforms

Variations in Stress. Early designers of type attempted to match the handwritten letterforms of the scribes as much as possible. The results were typefaces with a distribution of weight through the thinnest part of the letterform, creating a diagonal stress. A good example of this is Garamond. Over time, the stress became more vertical as in Baskerville, and later, completely vertical with Bodoni. With Century Expanded, there is a return to a slight diagonal stress. In Helvetica, you will find no noticeable stress at all.

Variations in Thicks and Thins. Typefaces also vary in the degree of contrast between the thick and thin strokes of the letters. Garamond illustrates the prominent characteristics of Old Style faces, little contrast between thick and thin strokes. The contrast is even greater in Transitional faces, and Modern faces such as Bodoni, which have the most extreme contrasts between thick and thin strokes. With Egyptian faces, such as Century Expanded, there is a return to less contrast. In Contemporary faces, such as Helvetica, there are no perceptible thick and thin strokes.

Variations in Serifs. Serifs also vary from one face to the next in their weight and in the way they are bracketed—the way in which the serif meets the vertical stroke of the letter. You can see this evolution in type from the Old Style heavy serif of Garamond through the Transitional serif of Baskerville to the refined Modern serif of Bodoni. This was followed by a return to the heavy serif in Century Expanded, an Egyptian face, and the elimination of the serif in Helvetica, a Contemporary face.

alphabet   ALPHABET

alphabet   ALPHABET

FIGURE 13.6. The serif typeface Century below versus the sans serif typeface Univers, above. Understanding the differences within the typeface as well as what the differences between a serif and a sans serif is important. Try setting a paragraph of each with exactly the same line length, size and spacing to compare the differences.

Typographic Guidelines

Serif Versus Sans Serif. Serif typefaces have a stronger baseline due to the "feet" created by serifs. This helps move the reader's eye horizontally across the line of type. Sans serif and serif typefaces can be effectively combined if one limits the number of typographic changes to prevent what could become visual chaos. The key is to ensure, that no matter what choices are made, they reinforce information hierarchy and overall design goals. See Figs. 13.6 and 13.7.

- Sans serif is often easier to read as online, though depending on the type size and monitor resolution, sans serif can be equally as legible if the appropriate size, style and color choice is made
- sans serif or serif can be effective for contrast, particularly when combined with size and weight changes.
- resolution and color impact choices

Combining Typefaces. When combining typefaces, it is important to decide whether the goal is harmony or contrast. As a general rule, it is wise not to use more than two different typefaces on one page. Excellent typography does not get in the way of the reader. Too many typefaces can jar and confuse

alphabet   ALPHABET

alphabet   ALPHABET

FIGURE 13.7. Serif typeface (Century) on the bottom versus sans serif typeface Univers, on the top, show the differences in each as well as where the information in each "is held" are the keys to optimum legibility and readability of text. Try covering up the lower half of lines of type with some other choices, to get an idea of where the information "is held" in each typeface. How easy or hard is it to understand the words or lines of text? Try this for each situation and you will begin to get an idea of what is critical for optimum readability and legibility of text.

the reader, create visual intrusions, and slow or the curtail reading.

Contrast in Weight (Boldness). Combining classic faces with a good differential factor can add useful contrast, such as extra bold Helvetica with regular-weight Times. Be wary of combining two faces that are both intricate, such as Gill Sans Bold and Souvenir. This can add too much contrast and visual complexity.

Output Device and Viewing Environment. The quality of publishing technologies and viewing environment vary greatly—laser printer versus video versus electronic media, etc. In choosing a typeface style, size, spacing, and leading, it is critical to consider the final output medium and its effect on legibility. Low-quality monitors and poor lighting have a major impact: serifs sometimes disappear, letters in small bold type fill in, and colored type may disappear altogether.

Letter Spacing. When letter spacing is too tight, the letters are hard to distinguish from each other, making them less legible. When letter spacing is too wide, the gaps between the letters do not allow the eye to recognize letter groups as easily. Optimal letter spacing is unnoticeable, the eye can skim across a line and quickly and easily understand.

Word Spacing. Too-tight word spacing makes words difficult to distinguish one from the next. When word spacing is too wide, gaps between words don't allow the eye to forms word groups as easily. When there is greater space between words than there is between lines, the reader's eye naturally falls to the closest word—which may be the word below instead across the line. This often occurs with low-resolution or poorly developed products.

Line Spacing/Leading. The space between lines of text, or leading, should increase in relation to type size. However, this adjustment must be done visually, not mathematically. You can also improve legibility by increasing the leading in relation to column width.

Line Length/Column Width. The correct line length is just long enough for the eye to easily move across the line without losing its place and easily drop down to continue reading the following lines.

Justified Versus Ragged Right (Flush Left). A justified column can leave uneven word spacing, creating rivers, or vertical white spaces, within in the paragraph. These rivers cause the eye to move vertically down the page, to naturally connect visually what is closest in proximity, instead of easily across the line of type. It is very difficult to prevent rivers in justified columns, unless much time and effort are applied. It is for this reason that, unless the type is manually set or adjusted, it is better to use a ragged right column. See Fig. 13.8.

Highlighting with Type. There are three basic ways to highlight type:

A justified column can leave uneven word spacing, creating rivers, or vertical white spaces within the paragraph. These rivers cause the eye to move vertically down the page, to naturally connect visually what is closest in proximity, instead of easily across the line of type. It is very difficult to prevent rivers in justified columns, unless much time and effort are applied. It is for this reason that, unless the type is manually set or adjusted, it is better use a ragged right column.

A justified column can leave uneven word spacing, creating rivers, or vertical white spaces within the paragraph. These rivers cause the eye to move vertically down the page, to naturally connect visually what is closest in proximity, instead of easily across the line of type. It is very difficult to prevent rivers in justified columns, unless much time and effort are applied. It is for this reason that, unless the type is manually set or adjusted, it is better use a ragged right column.

FIGURE 13.8. With current technology, the difference between a justified column and ragged right column can make a huge difference. In a poorly justified column, spaces within a justified line connect vertically down the page, distracting the eye from easily reading across a line of text.

- **Bold** or **Extra bold**
- *italics*—Italics are appropriate for short phrases versus long passages, because the slant of the italic appears lighter on is more difficult to read.
- UPPERCASE Because Uppercase letterforms are more angular, the eye attempts to connect lines and shapes that are in closest proximity and drawn in the same direction. In addition, uppercase letterforms take the full space from baseline to the top of the font space. The eye has limited shape and size cues to help differentiate between letters, words and sentences to create meaning.

It is only necessary to use one highlighting technique for emphasis.

Decorative Typefaces. These are of limited use for body text, because their irregular design make them less legible. They should be used for headlines with caution. Because they are essentially typographic fashion statements, decorative typefaces can either reinforce or distract from the overall message, or brand of a particular product or organization.

Black on White Versus White on Black and Dark on Light Background Versus Light on a Dark Background

Positive and Negative Type. White on black (or light on a dark background) is generally regarded as less legible and much more difficult to read over large areas. To the human eye, white letters on a black background appear smaller than their reversed equivalent. The amount of contrast between the color of type and the background is an especially important factor for online communication. Color adds exponential levels of complexity to these considerations since displays are inconsistent from one situation to another.

DESIGN PRINCIPLES: PAGE DESIGN

While typography deals with legibility, page design focuses on readability—the ability to read and comprehend information. Can the reader find what is needed on the page? The two important functions of page design are motivation and accessibility.

A well-designed page is inviting, drawing the eye into the information. Users are motivated to accept the invitation. A good page will ensure that the reader will continue by increasing the ease of understanding and accessibility of the information. (For purposes of simplicity, the term *page design* is used interchangeably to mean page, screen, and document design.)

Motivation and accessibility can be accomplished by providing the reader with ways to understand the information hierarchy quickly. At a glance, the page design should reveal easy navigation and clear, intuitive paths to discovering additional details and information. This is called visual mapping.

A page, site or product that is visually mapped and designed for easy navigation has:

- a clear underlying visual structure
- organizational landmarks
- graphic cues and other reader aids
- clearly differentiated information types, clearly structured examples, procedures, and reference tools
- well-captioned and annotated diagrams, matrices, charts, graphics, etc.

This kind of visual structuring will aid readers and

- provide a clear path through the information
- aid in skimming
- give a conceptual framework
- prevent a feeling of overload from too information

One could think of a table of contents as a simple visual map, because it quickly provides a simple overview, the order, and some details about the structure and content. What it does not reveal however, are priorities. Site maps or other diagrams will provide some of this information as well.

Building the Design of a Page

A good example of how visually mapping works effectively is visible in the following example of the evolution of a page from simple text. As more design elements are added, and the page becomes a combination of text, type, visual cues, and graphic elements. Information design techniques, drawn from cognitive science, can be used to improve communication effectiveness and performance.

Gray Page or Screen. "Raw text" interests few. When information is presented as a uniform undifferentiated block, people find it difficult and irritating to use and very easy to ignore. See Fig. 13.9.

Chunking. This involves structuring or simplifying the visual field by breaking like kinds of information into manageable chunks according to their subject matter. One sees things that are close together as related. Adding white space, rules, and

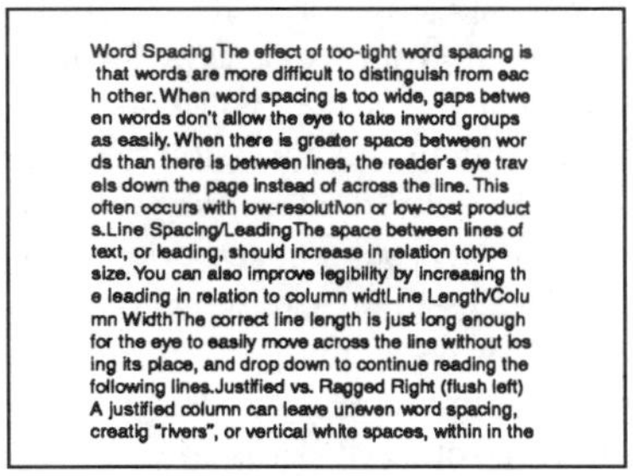

FIGURE 13.9. Gray page or screen.

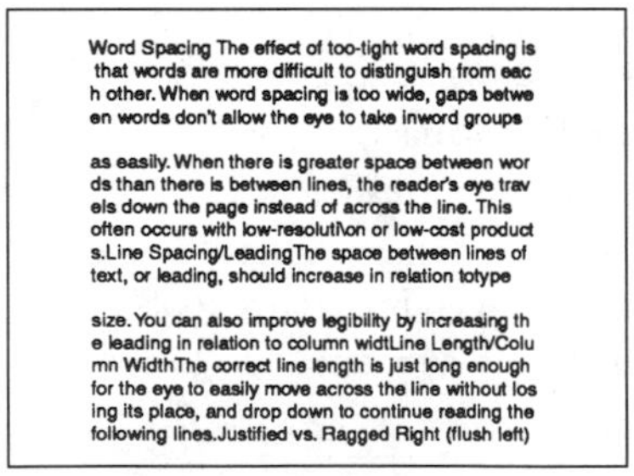

FIGURE 13.10. Chunking.

FIGURE 13.11. Queuing.

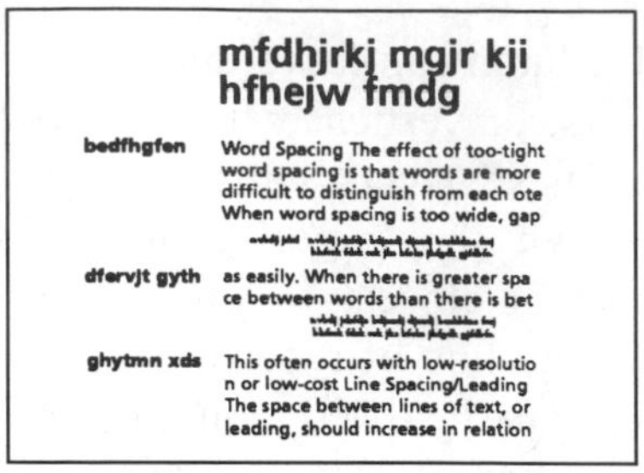

FIGURE 13.12. Filtering.

FIGURE 13.13. Mixing modes.

FIGURE 13.14. Abstracting.

other graphic devices can increase a grouping/chunk and to separate one chunk from another. See Fig. 13.10.

Queuing. This entails ordering chunks of information visually to reflect the content hierarchy. The design suits the user's requirements of subject matter, order, and importance. See Fig. 13.11.

Filtering. This step simplifies linguistic and visual order to filter out unnecessary background noise that interferes with the information being transmitted. Filtering builds a sense of layers of information through color, visual cues and symbols, bulleted lists and headers, making one page effective for a range of users and uses. See Fig. 13.12.

Mixing Modes. Different people learn by using different cognitive modes or styles. Some prefer text, others prefer illustrations, photos, diagrams, or formulas. To suit these naturally varied learning styles, information must be translated into several different modes, which are then carefully presented to avoid a confusing jumble. See Fig. 13.13.

Abstracting. The individual page or screen is a microcosm of the complete book, site product. The result is a complete codified system of graphic standards, effective for both the reader and the producer. Abstracting creates a system of standards that simplify text organization, create consistent approaches to preprocessing information, and establish a unique customized look for an organization's products. See Fig. 13.14.

Other Page Design Techniques

White Space. White space (or empty space) is one of the most underutilized tools of design, yet is extremely effective. It can be used to visually open up a page, focus attention, help group like kinds of information, as well as provide a rest for the reader's eye and create the perception of simplicity and ease of use.

The Grid. A grid is a system for distribution of visual elements in a clearly intelligible order. Grids, as part of a design system, determine the horizontal placement of columns, and the vertical placement of headlines, text, graphics, and images.

This visual organization or grid system is a series of consistent relationships, alignments, and spatial organization. The grid acts as a blueprint of the page that can be used again and again to create additional pages that appear related, but have different information. When the grid system is understood, it forms the basis for consistent application and extension of the design by others who also understand the intention of the system.

Every good design has an underlying structure or grid, as a basis to create a consistent look and feel to a program, web site, book, or sets of any of these. One could think of the grid as the visual analogy of the metal beams as a framework of a high-rise buildings. Each floor has the same underlying elements. Such as windows, elevators, plumbing, but depending on the use of each floor, will be built and look very different.

The grid is also a tool to improve usability. For example, if a user can anticipate a button to always appear in the same place, or help always available in the same way, this greatly improves the usefulness of the product or program and ultimately its success. Placement all visual elements such as buttons and help are specified on the grid.

Field of Vision. Field of vision refers to what a user can see on a page with little or no eye movement; it is the main area where the eye rests to view most of the page. A good design places key elements in the primary field of vision. It should reflect and reinforce the information hierarchy. Size, contrast, grouping, relationships, and movement are tools that create and reinforce field of vision.

The user will see first what is visually strongest, not necessarily what is largest or at the top of a page. This is particularly important for online information, because of limitations of page real estate and dense information environment.

One can easily experience these concepts, as well as the strength of peripheral vision, when looking at a page that has a banner advertisement or moving graphics. It is virtually impossible to ignore or focus attention on the primary field of vision when there is winking and blinking somewhere else on the page. Superfluous use of visual devices in fact reduces the value of the information by distracting and disturbing the user's desire and ability to focus, read, and understand.

Proximity. This concept applies to the placement of visual elements physically close, so they will be understood as related elements.

For example, if there were 3 images with captions on a page, it would be more useful to place each caption near the image if explains, though it might be more efficient to place the three captions together in one block of type on the page.

The Illusion of Depth. Though the online world exists on a two-dimensional space, various visual techniques can be used to create the illusion of depth, much like the painters of the Italian Renaissance period. Visual cues, such as layering, overlapping, perspective, size, contrast, and color can reinforce visual hierarchy by giving the illusion that one element appears on top of or in front of another.

Charts, Diagrams, Graphics, and Icons

The goal of any visual device is to provide the fastest, most efficient path to understanding ideas, as well as to make it clearer and more compelling. Useful, effective graphics can act much like visual shorthand, particularly important when the real estate of the page is limited. A good graphic can eliminate the need for text and communicate across cultures. However, a bad graphic that is unclear and must be reinforced by long captions can be worse than none at all.

The old cliché, a picture is worth a thousand words, is true only if it is efficient and effective. In stressful situations, people do have the time to read or the ability to focus on lengthy text or complex visuals. Though more difficult to achieve brevity and simplicity in such cases have greater value.

People prefer well-designed charts, diagrams, and illustrations that quickly and clearly communicate complex ideas and information such as comparisons or analysis. Studies show that images are retained long after the reader is finished reading. Done correctly, visual images can be used to make the information more memorable and effective. At a minimum, a good illustration or graphic can often improve performance simply because it increases user motivation.

Visuals are powerful communications tools. They can be used to:

- visualize data
- visualize new or abstract concepts
- visualize physical and technical concepts that are invisible to the eye
- communicate a large amount of information efficiently and effectively

Visuals can be used to explain and reinforce concepts, relationships, and data by making them tangible. They become

FIGURE 13.15. Zen calligraphy is an example of the historically close relationship between word and image. The great Zen master Hakuin of Kyoto, Japan (1768–1865) created this symbol to mean "dead," with additional notes saying, "Whenever anyone understands this, then he is out of danger."

thinking tools. Information is clarified, made easier to evaluate, and has greater impact. The use of visuals, whether they are photographs, charts, illustrations, icons, or diagrams, is a very effective way to communicate a message. Choosing the appropriate presentation of the concept is critical to the user's ability to effectively comprehend the message. In addition, a key to a successful visual is understanding the limitations of the display medium.

Tables, Charts, and Diagrams. These three types of graphics are discussed in order of complexity. Tables are the least difficult to create, charts the second most difficult, and diagrams the third. Illustrations, graphics, and other images and visuals are the most complex and require more conceptual and visual sophistication.

When is one more appropriate than the other? Determining which format is the most effective is illustrated in Fig. 13.16.

In addition to this list, it is important to remember that visual cues, such as color, shading, texture, lines and boxes, should be considered redundant cues and only used to provide additional emphasis to support the concept.

Icons and Graphic Cues. Icons and other graphic cues are another form of visual shorthand that help users locate and remember information. Choosing a style that is easily understood, and consistent with the overall style, is no easy task. It is important to choose style that is simple and consistently reinforced throughout a product.

More complex and unique symbols and icons can be used if usage takes place over a longer period, allowing familiarity and learning to take place. The Mastercard logo is basically two

intersecting circles, but after many years of reinforcement many will recognize it immediately without any text or other explanation. It is a very difficult task to create an icon that, without any explanation, communicates a concept across cultures.

For example, the use of a freestanding rectangular box with an open door flap indicates a mailbox, or in-box. This kind of mailbox is rarely used today, and was never in use in Europe (they have mail slots or upright boxes) or any other part of the world. Even the concept of mail delivery would be considered strange. This is a case where understanding had to be learned. Although simple ideas presented as icons are appropriate, a program with many complex icons using colloquial images would surely make a program agonizing for users from other cultures.

There is an important difference between an icon and an illustration, though often the two concepts are confused. If an icon has to be labeled, in fact it is really an illustration. The value of an icon as visual shorthand is lost, and it is better to use just the word or short phrase rather than both when screen real estate is at a minimum. Alternatively it might be more appropriate and useful as an illustration.

If an icon is memorable with minimal reinforcement, then it is successful. If after several times a user cannot remember the meaning of a particular icon, then it is of no value and should be eliminated. If a set of icons is being designed, there must be consistency of style (business-like vs. playful), light source (upper right or other), perspective and line style among the icons, as well as consistency with the product of which they are a part.

Illustrations and Photographs. As technology improves, the use of complex images will only be limited by the designer's

When to Use What Graphic

	If you want to show...	use a...
Groups	Group of related items, with a specific order	numbered list
Relationships	Relationships and steps involved in a process	flow chart
	Relationships between categories of ideas	table
	Relationships of tasks taking place over time	project plan table
Evaluate/Compare	Evaluate items against several criteria	rating table
	Evaluate items against one criteria	comparison table
	Compare more than one item to more than one variable	matrix diagram
	Compare several things in relation to one variable	bar chart
	Compare the relative parts that make up a whole	pie chart
Hierarchy	Hierarchical structure of an organization	organizational chart
Concepts	concept	Illustration and/or text icons, other graphics
	Abstract concept	complex images interactive components

FIGURE 13.16. When to use what graphic.

choice. Appropriateness is the most important component; one should not use cartoons for a company brochure, or use a low-resolution photograph of a control panel when a line illustration would be clearer.

Understanding meaning and implications of illustrations and photographs is no easy task, but there are some basic guidelines for making those choices.

Photographs represent existing objects easily, but issues of resolution and cross-media publishing can often make them unintelligible. If the photographs can be reproduced with proper resolution, cropping, and contrast, and the focus user on the required detail, then this is a good choice. They can also include cues for orientation and context that is more difficult to achieve in an illustration.

Often the reproduction quality of a photograph is unpredictable, no matter how it is simplified or cropped to focus attention. In this case, an illustration or line drawing is more effective. The obvious advantage of illustration is for visualization of concepts or objects that do not yet, or may never exist. Another advantage is the ability to focus by the design. For example, attention can be focused on a specific machine part by highlighting various lines and greying-out less important parts of the illustration. To do a similar thing in a photograph at best

would add time, complicate the image, and possibly never simplify the explanation.

No matter what method one uses for visual explanations, it must clarify and reinforce. If the goal of the image is to explain where to locate a piece of equipment, then an overview of the equipment in the environment is appropriate. If the goal is to show a particular aspect, such as a button, then the illustration should focus attention only on that. One can crop an image to focus attention on specifically what is being explained; it all depends on the goal of the photograph or illustration.

There are cases when the combination of photography and illustration are more effective than either alone. For example, a photograph of an object in its usage environment conveys information beyond the image of the object itself. If the goal was to show the location of a particular part of that object, then a detail, closely associated with the overall photograph or inset, would be even more useful than just a photo or illustration alone.

Guidelines

Visuals Should Reinforce the Message. Don't assume that the audience will understand how the visual reinforces the

FIGURE 13.17. Thirty centuries of development separate the Chinese ancient characters on the left from the modern writing on the right. The meaning of the characters is (from top to bottom): sun, mountain, tree, middle, field, frontier, and door.

FIGURE 13.18. It is obvious which of the above examples communicates a important message most quickly. The goal for the designer is to communicate the message in the most direct way, so that the user can understand and make decisions based on that information. Obviously, some situations are more critical than others, but it is no less important to begin design with consideration of the needs of users.

argument. A clear focused illustration with a concise caption will shorten comprehension and learning and cause the user to say "Aah, that is how it all fits together!"

Visuals should:

- Help clarify complex ideas
- Reinforce concepts
- Help the user understand relationships

Create a Consistent Visual Language. In creating graphics, it is important to establish a consistent visual language that works within the entire communication system. Graphics attract attention. When the user sees the screen, the eye automatically jumps to a visual, regardless of the fact that it may interrupt the flow of reading. A graphic should not create disharmony. This will only slow down the progress of comprehension and make it more difficult for the user to continue. It will also increase the effort needed to understand the relationship between the text and the visual.

Consider Both Function and Style. It is important to consider function versus decoration. Although it would be wonderful to see an artistic illustration tax forms or comic styled illustrations in annual reports, it would not be appropriate or reinforce a message or image of the communication and organization. A good graphic is appropriate to the context of the communication and reinforces and validates the message.

Focus on Quality Not Quantity. Graphics are only effective if they are carefully planned, well-executed, and used sparingly; like visual shorthand. One good diagram with a consice caption is more effective than several poorly thought out diagrams that require long explanations.

Work with a Professional. Most of can write a letter that clearly communicates the the message, but when it comes to writing the year-end sales report or the company brochure, we often turn to a professional writer to help us find the most effective, relevant, and interesting way to communicate our message. This holds true for the development of user interfaces, graphics, and other visuals that impact the look and feel and ultimately the overall success of a program.

Build Graphics Library to Create Visual Consistency, Organizational Identity, and a Streamlined Process. Because graphics often do require a professional, they can become time consuming and expensive to create. Once a visual language and style are established, start building a graphics library. If the same concepts are being illustrated repeatedly, this is an opportunity to streamline the development process by collecting them in one place and making them available for reuse. An organizational style can be created for these visual explanations. As time goes on, users will come to associate a particular style and method of explanation with the organization, which aids understanding as well as reinforcing an organization's product brand and identity.

Reinforce Shared Meaning (Common Visual Language).
A serious issue to consider when creating graphics, particularly conceptual diagrams, is shared meaning, whether it be across an organization or across the globe. The same diagram can be interpreted in entirely different ways by different people having different backgrounds and experience.

Truly effective graphics require extra time and effort, but the payoff can be tremendous. They are invaluable tools for promoting additional learning and action because they

- reinforce the message
- increase information retention
- shorten comprehension time

COLOR

Though color is reinforcing, or redundant visual cue, it is by far the most powerful element in visual communication. Color evokes immediate and forceful responses, both emotional and informational. Because color is a shared human experience, it is symbolic as well. And, like fashion, the perception and value of color changes over time.

Color can be used to trigger certain reactions or define a style. For example, in Western business culture, dark colors (such as navy blue) are generally considered to be conservative, whereas paler colors (such as pink) are regarded as feminine and not businesslike. However, in other cultures these color choices would have entirely different meaning.

The appropriate use of color can make it easier for users to absorb large amounts of information and differentiate information types and hierarchies. Research on the effects of color in advertising show that ads using one spot of color are noticed 200% more often than black and white ads, whereas full-color ads produce a 500% increase in interest.

Color is often used to add information to:

- show qualitative differences
- act as a guide through information
- attract attention/highlight key data
- indicate quantitative changes
- depict physical objects accurately

All in all, color is an immensely powerful tool. Like the tools of typography and page design, it can easily be misused. Research shows that whereas one color, well used, can increase communication effectiveness, speed, accuracy, and retention; multiple colors, poorly used, actually decrease effectiveness.

Because of its ready availability, it very tempting to apply it in superficial ways. For color to be effective, it should be used as an integral part of the design program, to reinforce meaning, not simply as decoration. The choice of color—while ultimately based on individual choice—should follow and reinforce content, as well as function.

Basic Principles of Color

Additive Primaries. The entire spectrum of light is made up of red, green, and blue light, each representing a third of the spectrum. These three colors are known as additive primaries, and all colors are made up of varying amounts of them. When all three are combined, they produce white light.

Subtractive Primaries. By adding and subtracting the three primaries, cyan, yellow, and magenta are produced. These are called subtractive primaries.

Green + Blue − Red = Cyan
Red + Blue − Green = Magenta
Red + Green − Blue = Yellow

Color on a computer display is created by using different combinations of red, green, and blue light. In print, colors are created with pigments rather than light. All pigments are made up of varying amounts of the subtractive primaries.

The three attributes of color are:

- *Hue*—the actual color
- *Saturation*—the intensity of the color
- *Value*—includes lightness and brightness:
  Lightness—how light or dark a color appears
  Brightness—this is often used interchangeably with lightness; the differences are as follows: lightness depends on the color of the object itself, and brightness depends on the amount of light illuminating the object.

How to Use Color

Less Is More . . . Useful and Understandable. Just as you can overload a page or screen with too many typefaces, you can have too many colors or make bad choices. Given the unpredictability of color displays, users, and viewing situations, the choice can get complicated. Color is often best used to highlight key information. As a general rule, use no more than three colors for primary information. An example is the use of black, red, and gray—black and red for contrasting information, gray for secondary. When thinking about color online, one must remember that each display will output color in a different way. Red and green should be used sparingly, since they spring forward. Blue is often used for backgrounds, since it recedes. Add to that the lighting situation and a variety of users. All these factors affect color choice.

Create a Color Logic, Use Color Coding. Use a color scheme that reinforces the hierarchy of information. Don't miscue the audience by using different colors for the same concept. Whenever possible, try to use colors that work with the product branding and identity or an established visual language. Create a color code that is easily understood by the user and reinforces meaning.

Create a Palette of Compatible Colors. Harmonious color can be created by using a monochromatic color scheme or differing intensities of the same hue. But make them different enough to be easily recognized and simple enough to be easily reproduced, no matter the medium.

Use Complementary Colors with Extreme Caution. These are colors that lie opposite each other on the color wheel. Let one dominate and use the other for accents. Never put them next to each other because the edges where they meet will vibrate. Though this was the goal of pop art in the 1960s, it makes pages impossible to read. One must check each particular display, because the calibration of monitors can unexpectedly cause this to happen.

Decisions Regarding Color in Typography Are Critical. Colored type appears smaller to the human eye than the same type in black. This is important to consider when designing user interfaces. One must also consider the smear effect on typography in displays, based on the color chosen, the strength of the projected light and interaction with colors surrounding it. Additionally, quality and calibration of displays impact characteristics of color online.

Consider the Viewing Medium. The same color looks different when produced by:

- a computer display
- an LCD projector
- color laser printer versus dot matrix output
- glossy versus dull paper

Context Is Everything. Though printed color is very familiar and more controllable, projected color is inconsistent and varies, depending on lighting, size of the color area, size and quantity of colored elements, lighting, output device. One must check all viewing possibilities to ensure that color choices are readable as well as legible, across all media used. What might look good on a laptop may not be readable when projected in a room of 500 people printed in the corporate brochure.

The amount of color will affect how it is viewed, as well as the best background choice. A blue headline is very readable on a white background, but if that background becomes a color, then readability can be reduced dramatically, depending on how it gets presented on each particular display.

Contrast Is Critical When Making Color Choices. Contrast is the range of tones between the darkest and the lightest elements. The desired contrast between what is being read (this includes graphics, photographs, etc.) must be clearly and easily differentiated from the background. If there is not enough contrast (of color, size, resolution, etc.), then differentiation and reading will be difficult or impossible. This is particularly a problem with online displays, because the designer has no control of display quality.

Quantity Affects Perception. Color used in a small area will be perceived differently by the eye than the same color used in a larger area. In the smaller area, the color will appear darker, in the larger area, lighter and brighter.

Use Color as a Redundant Cue When Possible. At least 9% of the population, mostly male, is color-deficient to some degree, so it is generally not a good idea to call out warning points only through color. With a combination of color and a different typeface, etc., you won't leave anyone in the dark.

We Live in a Global World, So When in Rome... Remember that different colors have different connotations in different cultures, religions, professions, etc.

- in the U.S., on February 14th, red means love
- in Korea, red means death
- in China, red is used in weddings and signals good luck and fortune
- in many countries, red means revolution
- to a competitor, red means first place
- to an accountant, red means a negative balance
- to a motorist, red means stop
- in emergencies, a red cross means medical help

Creating a System: Graphic Standards and Branding

With the explosion of new publishing media in a global marketplace, the need for guidelines for developing and producing consistent, quality communication has taken on a new urgency.

The new technology makes it easy to generate images, and offers a wealth of options. The danger lies in creating the visual chaos, with every element demanding attention beyond the point of sensory overload. With the new tools, it can happen faster, at a lower cost, and with greater distribution.

Graphic standards system provide guidelines and tools for structuring and organizing communications, and reinforce a brand across a corporation on the globe. Graphic standards are documented guidelines that explain the methodology behind the design. In addition, the guidelines and examples support those who wish to expand the system by explaining how to maintain a consistent brand and organizational look and feel as new products, features, and technology are introduced.

A graphic standards system will ensure:

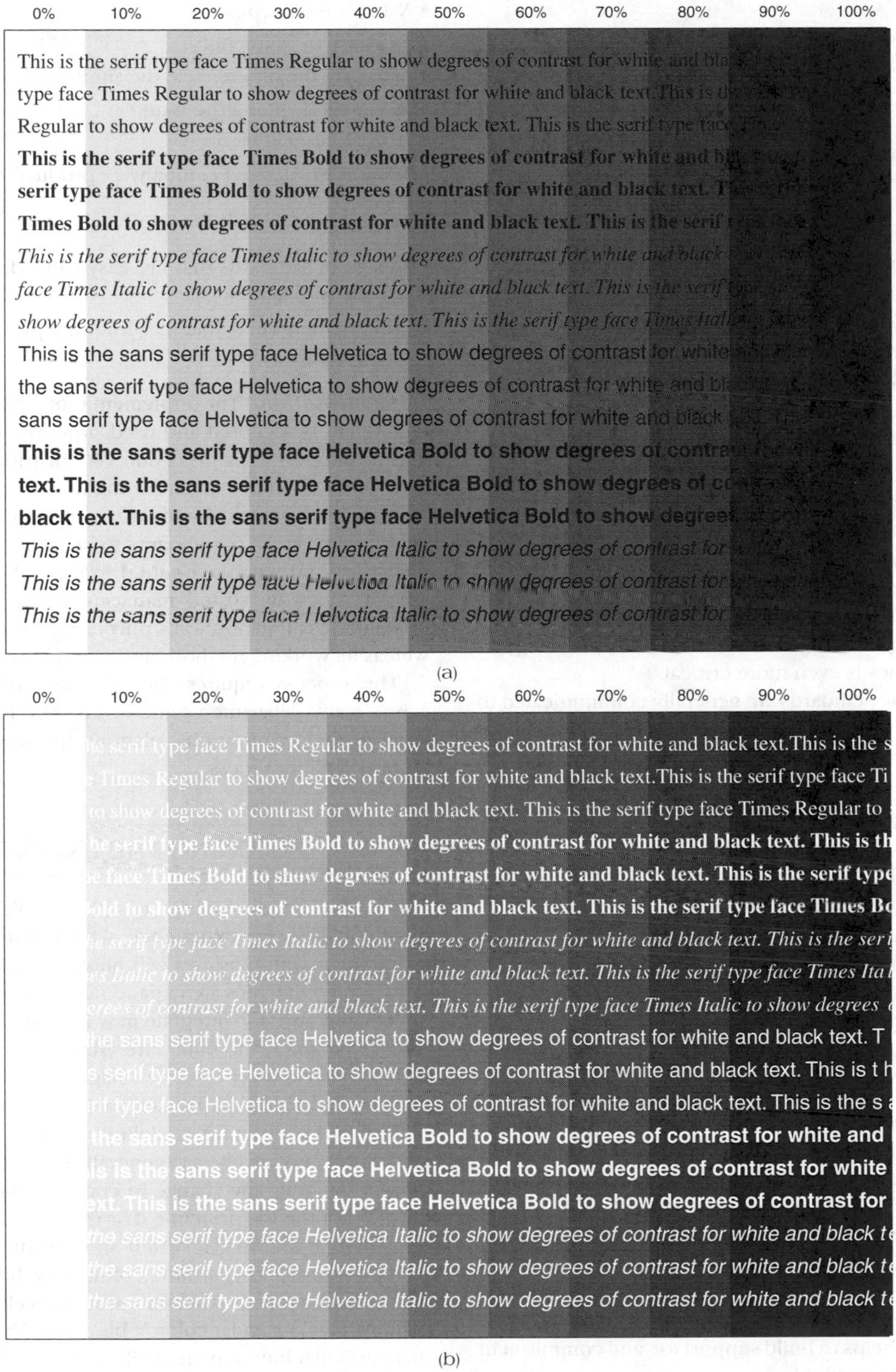

0%   10%   20%   30%   40%   50%   60%   70%   80%   90%   100%

This is the serif type face Times Regular to show degrees of contrast for white and black type face Times Regular to show degrees of contrast for white and black text.This is the Regular to show degrees of contrast for white and black text. This is the serif type face

This is the serif type face Times Bold to show degrees of contrast for white and bl serif type face Times Bold to show degrees of contrast for white and black text. T Times Bold to show degrees of contrast for white and black text. This is the serif

This is the serif type face Times Italic to show degrees of contrast for white and black face Times Italic to show degrees of contrast for white and black text. This is the serif ty show degrees of contrast for white and black text. This is the serif type face Times Itali

This is the sans serif type face Helvetica to show degrees of contrast for white the sans serif type face Helvetica to show degrees of contrast for white and bl sans serif type face Helvetica to show degrees of contrast for white and black

This is the sans serif type face Helvetica Bold to show degrees of contra text. This is the sans serif type face Helvetica Bold to show degrees of c black text. This is the sans serif type face Helvetica Bold to show degrees

This is the sans serif type face Helvetica Italic to show degrees of contrast for This is the sans serif type face Helvetica Italic to show degrees of contrast for This is the sans serif type face Helvetica Italic to show degrees of contrast for

(a)

0%   10%   20%   30%   40%   50%   60%   70%   80%   90%   100%

(b)

FIGURE 13.19. Trying examples in your context. Because color is not available in this particular edition, try your own experiment. Take a look at this illustration, and recreate a paragraph of text, with the background graded from 100% to 0%, choosing one color for the background. Then set lines of type in a variety of typefaces and sized, to see where it becomes legible or totally impossible to read. Remember to test your choices with each context and parameters. Such things as lighting, projection distance, and users' physiological constraints can make the difference whether something can be read or not.

- Built-in quality
  The system ensures that the company/product image is communicated to all audiences. Standards promote consistency in handling information across product lines, companies, projects, etc.

- Control over resources
  A system provides dramatic managerial control over resources of use of skills, time and materials. Well-developed standards build in flexibility. They can be adapted to new communications without having to go back to the original designer each time.

- A Streamlined development process
  They help structure thinking for content, design, and production by providing a guideline of predetermined solutions for particular communication problems. Typical problems are solved in advance or the first time they occur. Most importantly, graphics standards systems allow people to go on to higher level issues of communication effectiveness.

What Does a System Cover?. Corporate graphic standards historically had been applied to an organization's logo, stationery, business cards, and other printed materials. As the online portion of an organizations brand dominates, providing for cross-media guidelines is even more critical.

Corporate graphic standards are generally communicated to the organization in print and electronic form. Documentation often includes:

- Corporate Identity Manuals
  Style guides in both print and online illustrate the application of the standards across the company's publications and provide specifications for production and expansion.

- Templates and Guidelines
  These come in both print and electronic form.

- Editorial Style Guides
  Determine the use of product/service names, punctuation, and spelling, writing styles.

Developing the System. When developing an organizational graphics standards system, one must consider the global publishing needs, the resources available for producing products, and the skill level of those directing production.

To responsibly determine the overall corporate needs, a team effort is required. Personnel from areas such as information systems, graphic design, usability, and marketing, along with engineers, writers, and users should be involved in the process. This team approach helps to build support for, and commitment to, the organizational standards.

The development of a comprehensive system follows the Information Design Process of Audit, Development, and Implementation.

Audit. The Audit is a critical step in determining the scope and parameters of an organization's corporate graphic standards. Specific questions for the Audit phase include:

- What is the purpose?
- Who are the audiences?
- What are the differences; the similarities?
- Who will be doing the work?
- What tools will be used?
- What is the desired company or product image?

Development. Goals for the Development Phase include:

- Design of standards that are easy to read, use, and project a consistent quality image.
- Design of products that fit within the production parameters of the company.

Implementation. The Implementation phase must ensure that the system is accepted and used properly. This requires training and support, easy procedures for distributing and updating materials, and a manual explaining how to use the system.

The development of standards is in itself an educational process. It requires all participants to be aware of communication objectives and what is required to meet them. As alternatives are developed and tested, management has the opportunity to evaluate their organization's purpose, nature, and direction, as well as its working methods and communication procedures.

The process requires the commitment and involvement across many departments and levels. The result is an empowering of the organization—planting the seeds for growth and increased effectiveness.

DESIGNING THE EXPERIENCE

The heart of interface design is in the definition and creation of designing the user's experience; what is it really like for people facing the monitor, using a cell phone or an ATM. Though presentation possibilities are expanding day by day, our capacity to understand, use, and integrate new information and technology has not grown at the same rate. Making the most appropriate media choices, whether it be images, animation, or sound, to explain complex ideas to widely varied audiences is no easy task.

The most important guideline is: there are no rules, only guidelines. Though one can generally say visual principles work a particular way, any change in context would change the application of the the principle. For example, in the early days of the software industry, research showed that a particular blue worked well as a background color. Now, however, depending on the calibration of your monitor, as well as environmental lighting, that particular color of blue could be a disaster. In fact, that particular blue can often vibrate if type in particular colors is placed on this background. Of course, it would all depend on how much type, its size and boldness, viewing situation, etc. Sound complex? For this reason, one needs to understand the principles, test the ideas, and then test on every output device that will be used. Putting known guidelines together with experience continually gathered from the field will allow the designer to develop a clear understanding of what works well in a particular environment and user situations.

The next key guideline is to keep it simple. Although many tools are available, there is only one goal: to clearly communicate ideas. The designer must always ask: What is the most efficient and effective way to communicate this idea? A good illustration might work better (and take less bandwidth) than an animated sequence. A simple bold headline might allow the user to read the page than a banner moving across the page, constantly drawing the eye to the top. Animated icons are entertaining, but would they be appropriate or necessary to understand serious financial information?

There is a great temptation to use many new tools. The best tip is to use a tool only if it can explain an idea better than any other method or enhance an explanation or illustrate a point that otherwise could not be done as effectively or efficiently. The best design is not noticed, it just works. Products are used to get something done, not to notice the design. The best test of success is ease of understanding and completing the tasks, and moving on to the next.

Like all expensive real estate, online real estate has the same characteristic: location, location, location! With such a premium of space, and so much to accomplish in such a short time, being considerate and efficient with the screen real estate is the design goal. The appropriate use of all the design principles, graphics, icons, and illustrations make that goal possible if it is applied with understanding and consideration of each particular context of usage.

In the following sections are some of the issues and considerations when presenting interactive information. As one considers how the many elements impact the design of interfaces, the following principles must be considered very seriously.

Effective and Appropriate Use of the Medium

Transitioning a print document to an online environment requires a rethinking of how it must be presented. Viewing and navigating through online information require radically different design considerations and methods. Users do not necessarily view the information in a linear way, in a particular order, or time frame.

Interactive media viewed on computer screens have quite different characteristics and potential, particularly as information crosses platforms, resolutions, and environments.

Historically, we have come from the rich medium of print, where we can hold the entire product in our hand, view it, and choose what/when/how we wish to read. The mere physicality of a book provides many sensory cues that are not present on a two-dimensional monitor. As designers, we must find other ways to provide the same cues that allow people to use products comfortably and with confidence.

The Element of Time

This is the critical difference between static and interactive media. The sense of interaction with a product impacts the user's perception of usefulness and quality. In addition, animated cues (blinking cursors, etc.) and other implied structural elements (e.g., handles around selected areas) become powerful navigational tools if intuitively understood and predictably applied.

In addition, one must keep in mind how the product will be used. Will the user calmly sit down and use the product, will the tasks be interrupted over a period of hours, days, months, or years? Will the user be physically impaired, in a state of panic fumbling with a keypad. The element of time contributes to the design criteria and choices.

Consistent and Appropriate Visual Language

A major issue is the unpredictability and vastness of the products. Providing way-finding devices that are easy to recognize, understand, and remember, include:

- clear and obvious metaphors
- interface elements consistent with the visual style of other program parts, including consistent style for illustrations, icons, graphic elements, dingbats, shading, etc.
- guidelines for navigational aids, such as use of color, typography, page/screen structure, etc., consistent with other parts of product support.

Navigation Aids

When reading a book, there are many ways we can see our progress through it. We can use a bookmark, turn down the corner of a page, or use a pen to highlight information we want to remember. We can refer to the table of contents or index, and then flip directly to the desired page. We can use a finger as a placeholder and walk down the hall to show a colleague.

At no time can we ever see or touch the entire digital document (or program). If we cannot hold the entire document, how do we know where we are in relation to the whole? How do we get back to where we were? Or some where use haven't been?

Navigation aids provide readers with highways, maps, road signs, and landmarks as they move through the online landscape. They enhance discovering and communicating the underlying structure; providing a sense of place so that the user knows where they are, where they have been, how to move elsewhere, or return to the beginning.

Using familiar visual elements (e.g., from other products, releases, etc.) leverages existing knowledge. Graphic standards support this as well. When using a familiar page layout/grid structure, it is much easier to remember the zones in which like kinds of information appear. This ensures that whatever visual cues are applied can take advantage of the user's experience and save time for the designer.

Graphics/Icons

Graphic representations are very effective devices to orient users within a program. A visual map can offer an overall picture of the program's sections and interrelations. Graphics and icons can help support the function of the table of contents, index, and page numbers. In addition to the many new tools to highlight their functionality, they can be even more effective as guides through and around a product. The key here is to ensure that the intent and action have been clearly defined and designed.

Metaphor

We learn easiest when we have previous structure, or mental model with which to associate and expand information. If we have a basic understanding of the concepts we can easily add more information. The desktop metaphor for a software interface is easy to grasp as a way to organize data in a program, because the basic logic is similar to what we are familiar with in the real world. Using familiar visual analogies helps users understand and organize new information more easily.

Color

Once the monitor is paid for, color is free and a very seductive design tool. One must be sure to use it consistently. On the monitor, there is limited space to work with. When colors are assigned meanings, and those meanings are carried throughout the product, the colors can replace written explanations (e.g., the bars at the top of the screen are blue, so this must be the testing section or the yellow background always means an overview section).

Legibility

As discussed in typography, legibility is the ability to read the information on the page. The page can be a screen, and as such, has special considerations. Color, size, background, movement, viewing environment, lighting, resolution, all play a critical part in legibility.

Readability

Readable screens demand use of clear visual representations and concise, unambiguous text. A design can imply meaning by the placement of elements in particular areas, or zones alotted for certain types of information. This makes the screen easier to comprehend and more accessible. It also makes optimal use of a limited space.

Guidelines

Use the Analogy of a Poster As a Guide to Design. One analogy is to think of a home page as if it were a poster. A poster must grab one's attention quickly, in unpredictable and uncontrollable locations. Because of limited space, the viewer gets only hints of related information but no great amount of detail. Imagine if someone walking by could click on the speaker's name and get additional biographical information, or click on the location and get directions. This is kind of organization is hierarchical and a radical difference from the way information is presented in a brochure where order is fixed. There is a specific linear sequence; the chronology is implied by its binding or folding, though one may choose to read page 5 before page 1.

In online environments, the designer can rarely control how and in what order the user will access the product. This requires fundamental differences in presentation of information. We can make suggestions and best guesses but still must design with an awareness these major unknowns.

This idea goes hand-in-hand with using the laptop format as another design consideration. The home page, like a well-designed poster, should hint at all topics contained in the site, provide high level information about these topics, and suggest easy paths to access this a information. If information goes beyond the laptop format, the design must visually communicate to the user how they can know it is there by providing strong visual hints, so that they will investigate beyond what is immediately visible. One can imagine the changes required for smaller, hand-held, voice-activated devices.

Design for the Most Difficult Common Denominator. One must design the interface in anticipation of the worst-case scenario. If a majority of users will be using your product in a quiet room, with fast connections, perfect lighting, and large monitors, the requirements are different from a contractor accessing critical information on a laptop in the field. Often, the user profile is unknown, because new technology often defines new categories. But if, for example, the users will be on a variety of platforms and locations with constant interruption, then one has to design from this situation. It is critical to consider what the breadth of possibilities will be; and user testing, viewing, questioning can make the difference of product acceptance or not.

Avoid Overuse of Saturated Colors. Saturated colors, such as red, tend to jump out at the viewer, which is distracting and irritating. Thus, red is usually not a good choice for large areas of color on the screen. High impact is dependent on the contrast between background and foreground colors. For instance, when designing screens for a display with a black background, both yellow and white have a higher impact than red. What must also be considered is the variations in every viewing situation and how that affects contrast among the various elements on the page, as well as overall legibility and readability.

Consider Different Users' Levels of Skill. All navigation tools should be simple enough for the novice user, but must not slow down the expert. Detailed visual maps and other visual graphics/elements should be available for those users who need them, without getting in the way of the expert user who wants to bypass unnecessary explanation and jump immediately the desired section.

Be Aware of the Fatigue Factor. Although there is no definitive answer on fatigue caused by looking at a computer screen for long periods of time, it is a central factor to consider. According to H. John Durrett's book, *Color and the Computer*, looking at a well-designed computer screen should not cause any more fatigue than reading a book or writing a report. Though many would disagree with this statement, many people spend more time with their computer than a book and no doubt would have additional input on this subject.

As interactive media becomes a commodity, the focus will not be on what a product does, but how it does it; that will be the difference between product acceptance or product failure. Success or failure will be judged by the ease of use and understanding of its interface—the face of the product/program to the users.

Other Differences to Consider. There are a many differences that impact how and why we design our interfaces, and many of them have been discussed in other chapters in more detail. A designer should never forget differences such as vision and physical impairment (sight, motor skills, etc.) mental impairment and how that impacts ability to read, comprehend, and use the interfaces we design.

Use the Squint Test to Check the Design. A very simple self-test to check visual hierarchy is the squint test. Simply squint your eyes at the page you are evaluating, putting the details out of focus. As you look at the page, what is the first, most dominant element on the page? Is this what should be seen first? What should have primary, secondary, importance on the page? In cognitive psychology terms, this is called visual queuing. The visual ordering of what the user sees on the page is the goal of good interaction design.

CHALLENGES AND OPPORTUNITIES: CREATING YOUR OWN GUIDELINES

The challenges facing today's designers of interactive communications are great. How can we create products that are seen, read, understood, and acted on? How do we harness the power of the new technologies, given increasing variety and complexity? How can we make informed visual choices? and provide our users with a useful, usable and desireable experience?

WARNING: No book or seminar or technology alone will turn anyone into a professional designer! It requires years of extensive training and practical experience in a variety of disciplines. Much like a programmer or surgeon or fine cabinet maker, it is a lifelong endeavor.

The guidelines that follow are offered as starting points only, first steps in understanding how one makes informed design decisions—design that is the best, most thoughtful and appropriate integration of both form and function.

There Are No Universal Rules, Only Guidelines

If there were rules, everything would look the same and work perfectly according to those rules. Each situation is different with its own context and parameters.

Remember the Audience: Be a User Advocate

Throughout the process of development, audience needs are primary. Who are they? What requirements do they have? How and where are they using your product? Answers to these and other questions are the criteria to evaluate alternatives throughout the development process. As designers, we must understand and advocate from the point of view of the user.

Structure the Messages

Content must be analyzed to create a clear visual hierarchy (reflecting the information hierarchy) of major and minor elements. This visual layering of information helps the user focus on context and priorities.

Test the Reading Sequence. Apply the squint test. How does the eye travel across the page, screen, or publishing medium? What is seen first, second, third? Does this sequence support the objectives and priorities as defined in the Audit?

Form Follows Function

Be clear about the user and use environment first. The interface design should be its tangible representation and reinforce these goals.

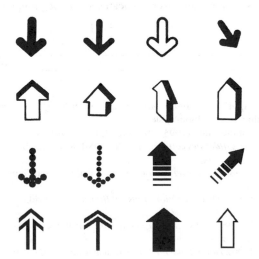

FIGURE 13.20. There are many ways to say the same thing. What is important is to create an appropriate visual style and design all visual elements in the same way. The goal is to create a consistent visual language throughout the entire product.

Keep Things Simple. The objective is to efficiently and effectively communicate a message, so that users can perform a task. Fewer words, type styles, and graphic elements generally mean less visual noise and greater comprehension. An obvious metaphor enhances intuitive understanding and use. The goal is to communicate ideas and information, not show off features or graphics.

People Don't Have Time to Read

In addition to writing clearly and concisely, it is important to design information in the most economical, accessible, intuitive format, enhanced by a combination of graphics and typography. Graphics are very powerful and can often be used to efficiently and effectively provide explanations while saving space on a page. Designer beware, as they can visually dominate a page, unless considered beforehand and designed as an integral part of the page.

Be Consistent

Consistent use of type, page structure, graphic, and navigational elements creates a visual language that reduces the effort needed read and understand the interface. The goal is to create a user experience that appears effortless and enjoyable throughout.

Start the Design Process Early

Don't wait until the last minute. Put together the development team of designers, usability professionals. engineers, researchers, writers, and user advocates at the beginning of the process. In particular, with interactive media, the traditional review and production process will change. The process is less of a hand-off and more of a team effort; it's more like making a film than writing a book.

No matter how varied an organization's products are, successfully applying the principles of good design will enable an organization to communicate more effectively with all its audiences and customers. This will have a direct impact on improving the value of the products and services, in addition to adding value to an organization's brand and identity.

Good Design Is Not About Good Luck

Good design is based on the principles of visual and interaction design applied appropriately and thoughtfully. Creating the most useful, successful design for an interactive product is difficult. By its nature, the design process is iterative, ongoing and experiential. There are usually several possible ways to solve a problem, with the final design decision dictated by the best choices based on requirements at a particular time.

Always and forever, remember the users, user, users. They are why we are here and have this work to do. They are everywhere, in places we have not yet imagined. As the world grows smaller and we are more connected, the opportunity lies in where and what we have not discovered.

I have been studying and practicing in this field of visual/interaction design in all its various flavors for over 20 years. Though contexts and technology change, the basic principles of visual design still apply. What does not change, however, is my focus on the user, user perspective, and use environment, with an ongoing goal to make things easier to use and understand.

This is only the very beginning. . . .

References

Apple Computer, Inc. (1992). *Macintosh human interface guidelines*. Reading, MA: Addison-Wesley.

Apple Computer, Inc. (1993). *Making it Macintosh. The Macintosh interface guidelines companion* [CD-ROM]. Reading, MA: Addison-Wesley.

Craig, J. (1998). *Designing with type*. New York: Watson-Guptil Publications.

Dondis, D. A. (1998). *A primer of visual literacy*. Cambridge, MA: MIT Press.

Howlett, V. (1996). *Visual interface design for Windows*. New York: Wiley Computer Publishing.

Krug, S. (2000). *Don't make me think*. Indianapolis, IN: Que Publishing.

Laurel, B. (Ed.). (1990). *The art of human-computer interface design*. Reading, MA: Addison-Wesley.

Laurel, B. (1991). *Computers as theater*. Reading, MA: Addison-Wesley.

Lynch, P. J., & Horton, S. (1999). *Web style guide: Basic design principles for creating web sites*. New Haven, CT: Yale University Press.

McCloud, S. (1993). *Understanding comics: The invisible art*. Northhampton, MA: Kitchen Sink Press.

Meggs, P. (1992). *A history of graphic design*. New York: Van Nostrand Reinhold.

Mullet, K., & Sano, D. (1995). *Designing visual interfaces: Communications oriented techniques*. Englewood Cliffs, NJ: Sunsoft/Prentice Hall.

Norman, D. A. (1990). *The design of everyday things*. New York: Doubleday Currency.

Olins, W. (1989). *Corporate identity: Making business strategy visible through design*. Boston, MA: Harvard Business School Press.

Rosenfeld, L., & Morville, P. (1998). *Information architecture for the World Wide Web*. Sebastopol, CA: O'Reilly.

Sano, D. (1996). *Designing large-scale web sites*. New York: John Wiley and Sons.

Schriver, K. A. (1997). *Dynamics in document design*. New York: Wiley Computer Publishing.

Shedroff, N. (2001). *Experience design*. Indianapolis, IN: New Riders.

Spiekermann, E., & Ginger, E. M. (1993). *Stop stealing sheep*. Mountain View, CA: Adobe Press.

Tufte, E. (1983). *The visual display of quantitative information*. Cheshire, CT: Graphics Press.

Tufte, E. (1990). *Envisioning information*. Cheshire, CT: Graphics Press.

Tufte, E. (1997). *Visual explanations*. Cheshire, CT: Graphics Press.

Weinman, L. (0000). *Coloring web graphics*.

Xerox Corporation (1988). *Xerox publishing standards*. New York: Watson-Guptil Publications.

Zapf, H. (1960). *About Alphabets*. Cambridge, MA: MIT Press.

·14·

MULTIMODAL INTERFACES

Sharon Oviatt
Oregon Health & Science University

WHAT ARE MULTIMODAL SYSTEMS, AND WHY ARE WE BUILDING THEM?

Multimodal systems process two or more combined user input modes—such as speech, pen, touch, manual gestures, gaze, and head and body movements—in a coordinated manner with multimedia system output. This class of systems represents a new direction for computing, and a paradigm shift away from conventional WIMP interfaces. Since the appearance of Bolt's (1980) "Put That There" demonstration system, which processed speech in parallel with touch-pad pointing, a variety of new multimodal systems has emerged. This new class of interfaces aims to recognize naturally occurring forms of human language and behavior, which incorporate at least one recognition-based technology (e.g., speech, pen, vision). The development of novel multimodal systems has been enabled by myriad input and output technologies currently becoming available, including new devices and improvements in recognition-based technologies. This chapter will review the main types of multimodal interfaces, their advantages and cognitive science underpinnings, primary features and architectural characteristics, and general research in the field of multimodal interaction and interface design.

The growing interest in multimodal interface design is inspired largely by the goal of supporting more transparent, flexible, efficient, and powerfully expressive means of human-computer interaction. Multimodal interfaces are expected to be easier to learn and use, and are preferred by users for many applications. They have the potential to expand computing to more challenging applications, to be used by a broader spectrum of everyday people, and to accommodate more adverse usage conditions than in the past. Such systems also have the potential to function in a more robust and stable manner than unimodal recognition systems involving a single recognition-based technology, such as speech, pen, or vision.

The advent of multimodal interfaces based on recognition of human speech, gaze, gesture, and other natural behavior represents only the beginning of a progression toward computational interfaces capable of relatively human-like sensory perception. Such interfaces eventually will interpret continuous input from a large number of different visual, auditory, and tactile input modes, which will be recognized as users engage in everyday activities. The same system will track and incorporate information from multiple sensors on the user's interface and surrounding physical environment to support intelligent adaptation to the user, task, and usage environment. Future adaptive multimodal-multisensor interfaces have the potential to support new functionality, to achieve unparalleled robustness, and to perform flexibly as a multifunctional and personalized mobile system.

WHAT TYPES OF MULTIMODAL INTERFACES EXIST, AND WHAT IS THEIR HISTORY AND CURRENT STATUS?

Multimodal systems have developed rapidly during the past decade, with steady progress toward building more general and robust systems, as well as more transparent human interfaces than ever before (Benoit, Martin, Pelachaud, Schomaker, & Suhm, 2000; Oviatt et al., 2000). Major developments have occurred in the hardware and software needed to support key component technologies incorporated within multimodal systems, as well as in techniques for integrating parallel input streams. Multimodal systems also have diversified to include new modality combinations, including speech and pen input, speech and lip movements, speech and manual gesturing, and gaze tracking and manual input (Benoit & Le Goff, 1998; Cohen et al., 1997; Stork & Hennecke, 1995; Turk & Robertson, 2000; Zhai, Morimoto, & Ihde, 1999). In addition, the array of multimodal applications has expanded rapidly, and presently ranges from map-based and virtual reality systems for simulation and training, to person identification/verification systems for security purposes, to medical and web-based transaction systems that eventually will transform our daily lives (Neti, Iyengar, Potamianos, & Senior, 2000; Oviatt et al., 2000; Pankanti, Bolle, & Jain, 2000).

In one of the earliest multimodal concept demonstrations, Bolt had users sit in front of a projection of "Dataland" in "the Media Room" (Negroponte, 1978). Using the Put That There interface (Bolt, 1980), they could use speech and pointing on an armrest-mounted touchpad to create and move objects on a two-dimensional (2-D) large-screen display. For example, the user could issue a command to "Create a blue square there," with the intended location of "there" indicated by a 2-D cursor mark on the screen. Semantic processing was based on the user's spoken input, and the meaning of the deictic "there" was resolved by processing the x, y coordinate indicated by the cursor at the time "there" was uttered. Since Bolt's early prototype, considerable strides have been made in developing a wide variety of different types of multimodal systems.

Among the earliest and most rudimentary multimodal systems were ones that supported speech input along with a standard keyboard and mouse interface. Conceptually, these multimodal interfaces represented the least departure from traditional graphical user interfaces (GUIs). Their initial focus was on supporting richer natural language processing to support greater expressive power for the user when manipulating complex visuals and engaging in information extraction. As speech recognition technology matured during the late 1980s and 1990s, these systems added spoken input as an alternative to text entry via the keyboard. As such, they represent early involvement of the natural language and speech communities in developing the technologies needed to support new multimodal interfaces. Among the many examples of this type of multimodal interface are CUBRICON, Georal, Galaxy, XTRA, Shoptalk, and Miltalk (Cohen et al., 1989; Kobsa, et al., 1986; Neal & Shapiro, 1991; Seneff, Goddeau, Pao, & Polifroni, 1996; Siroux, Guyomard, Multon, & Remondeau, 1995; Wahlster, 1991).

Several of these early systems were multimodal-multimedia map systems in which a user could speak or type and point with a mouse to extract tourist information or engage in military situation assessment (Cohen et al., 1989; Neal & Shapiro, 1991; Seneff et al., 1996; Siroux et al., 1995). For example, using the CUBRICON system, a user could point to an object on a map and ask: *"Is this <point> an air base?"* CUBRICON was an expert system with extensive domain knowledge, as well as natural language processing capabilities that included referent identification and dialogue tracking (Neal & Shapiro, 1991). With the Georal system, a user could query a tourist information system to plan travel routes using spoken input and pointing via a touch-sensitive screen (Siroux et al., 1995). In contrast, the Shoptalk system permitted users to interact with complex graphics representing factory production flow for chip manufacturing (Cohen et al., 1989). Using Shoptalk, a user could point to a specific machine in the production layout and issue the command: *"Show me all the times when this machine was down."* After the system delivered its answer as a list of time ranges, the user could click on one to ask the follow-up question, *"What chips were waiting in its queue then, and were any of them hot lots?"* Multimedia system feedback was available in the form of a text answer, or the user could click on the machine in question to view an exploded diagram of the machine queue's contents during that time interval.

More recent multimodal systems have moved away from processing simple mouse or touchpad pointing, and have begun designing systems based on two parallel input streams that each are capable of conveying rich semantic information. These multimodal systems recognize two natural forms of human language and behavior, for which two recognition-based technologies are incorporated within a more powerful bimodal user interface. To date, systems that combine either speech and pen input (Oviatt & Cohen, 2000) or speech and lip movements (Benoit et al., 2000; Stork & Hennecke, 1995; Rubin, Vatikiotis-Bateson, & Benoit, 1998) constitute the two most mature areas within the field of multimodal research. In both cases, the keyboard & mouse have been abandoned. For speech & pen systems, spoken language sometimes is processed along with complex pen-based gestural input involving hundreds of different symbolic interpretations beyond pointing[1] (Oviatt et al., 2000). For speech & lip movement systems, spoken language is processed along with corresponding human lip movements during the natural audiovisual experience of spoken interaction. In both of these subliteratures, considerable work has been directed toward quantitative modeling of the integration and synchronization characteristics of the two rich input modes being processed, and innovative time-sensitive architectures have been developed to process these patterns in a robust manner.

Multimodal systems that recognize speech and pen-based gestures first were designed and studied in the early 1990s (Oviatt, Cohen, Fong, & Frank, 1992), with the original QuickSet system prototype built in 1994. The QuickSet system is an agent-based collaborative multimodal system that runs on a hand-held PC (Cohen et al., 1997). With QuickSet, for example, a user can issue a multimodal command, such as "Airstrips . . . facing this way <draws arrow>, and facing this way <draws arrow>," using combined speech and pen input to place the correct number, length and orientation (e.g., SW, NE) of aircraft landing strips on a map. Other research-level systems of this type were built in the late 1990s. Examples include the Human-centric Word Processor, Portable Voice Assistant, QuickDoc, and MVIEWS (Bers, Miller, & Makhoul, 1998; Cheyer, 1998; Oviatt et al., 2000;

Waibel, Suhm, Vo, & Yang, 1997). These systems represent a variety of different system features, applications, and information fusion and linguistic processing techniques. For illustration purposes, a comparison of five different speech and gesture systems is summarized in Table 14.1. In most cases, these multimodal systems jointly interpreted speech and pen input based on a frame-based method of information fusion and a late semantic fusion approach, although QuickSet used a statistically ranked unification process and a hybrid symbolic/statistical architecture (Wu, Oviatt, & Cohen, 1999). Other very recent systems also have begun to adopt unification-based multimodal fusion and a hybrid architectural approach (Bangalore & Johnston, 2000; Denecke & Yang, 2000; Wahlster, 2001).

Although many of the issues discussed for multimodal systems incorporating speech and 2-D pen gestures also are relevant to those involving continuous three-dimensional (3-D) manual gesturing, the latter type of system presently is less mature (Sharma, Pavlovic, & Huang, 1998; Pavlovic, Sharma, & Huang, 1997). This primarily is because of the significant challenges associated with segmenting and interpreting continuous manual movements, compared with a stream of x, y ink coordinates. As a result of this difference, multimodal speech and pen systems have advanced more rapidly in their architectures and have progressed further toward commercialization of applications. However, a significant cognitive science literature is available for guiding the design of emerging speech and 3-D gesture prototypes (Condon, 1988; Kendon, 1980; McNeill, 1992), which will be discussed further in What Are the Primary Features of Multimodal Language? Existing systems that process manual pointing or 3-D gestures combined with speech have been developed by Koons and colleagues (Koons, Sparrell, & Thorisson, 1993), Sharma and colleagues (Sharma et al., 1996), Poddar and colleagues (Poddar, Sethi, Ozyildiz, & Sharma, 1998), and by Duncan and colleagues (Duncan, Brown, Esposito, Holmback, & Xue, 1999).

Historically, multimodal speech and lip movement research has been driven by cognitive science interest in intersensory audiovisual perception and the coordination of speech output with lip and facial movements (Benoit & Le Goff, 1998; Bernstein & Benoit, 1996; Cohen & Massaro, 1993; Massaro & Stork, 1998; McGrath & Summerfield, 1985; McGurk & MacDonald, 1976; MacLeod & Summerfield, 1987; Robert-Ribes, Schwartz, Lallouache, & Escudier, 1998; Sumby & Pollack, 1954; Summerfield, 1992; Vatikiotis-Bateson, Munhall, Hirayama, Lee, & Terzopoulos, 1996). Among the many contributions of this literature has been a detailed classification of human lip movements (visemes) and the viseme-phoneme mappings that occur during articulated speech. Existing systems that have processed combined speech and lip movements include the classic work by Petajan (1984), Brooke and Petajan (1986), and others (Adjoudani & Benoit, 1995; Bregler & Konig, 1994; Silsbee & Su, 1996; Tomlinson, Russell, & Brooke, 1996). Additional examples of speech and lip movement systems, applications, and relevant cognitive science research have been detailed elsewhere (Benoit et al., 2000). Although few existing multimodal interfaces

---

[1]However, other recent pen/voice multimodal systems that emphasize mobile processing, such as MiPad and the Field Medic Information System (Holzman, 1999; Huang et al., 2000), still limit pen input to pointing.

TABLE 14.1. Examples of Functionality, Architectural Features, and General Classification of Different Speech and Gesture Multimodal Applications

Multimodal System Characteristics	QuickSet	Human-Centric Word Processor	VR Aircraft Maintenance Training	Field Medic Information	Portable Voice Assistant
Recognition of simultaneous or alternative individual modes	Simultaneous and Individual modes	Simultaneous and Individual modes	Simultaneous and individual modes	Alternative Individual modes[a]	Simultaneous and individual modes
Type and size of gesture vocabulary	Pen input, multiple gestures, Large vocabulary	Pen input, deictic selection	3-D manual input, multiple gestures, small vocabulary	Pen input, deictic selection	Pen input deictic selection[b]
Size of speech vocabulary[c] and type of linguistic processing	Moderate vocabulary, grammar-based	Large vocabulary, statistical language processing	Small vocabulary, grammar-based	Moderate vocabulary, grammar-based	Small vocabulary, grammar-based
Type of signal fusion	Late semantic fusion, unification, hybrid symbolic/statistical MTC framework	Late semantic fusion, frame-based	Late semantic fusion, frame-based	No mode fusion	Late semantic fusion, frame-based
Type of platform and applications	Wireless handheld, varied map and VR applications	Desktop computer, word processing	Virtual reality system, aircraft maintenance training	Wireless handheld, medical field emergencies	Wireless handheld, catalog ordering
Evaluation status	Proactive user-centered design and iterative system evaluations	Proactive user-centered design	Planned for future	Proactive user-centered design and iterative system evaluations	Planned for future

[a]The FMA component recognizes speech only, and the FMC component recognizes gestural selections or speech. The FMC component also can transmit digital speech and ink data, and can read data from smart cards and physiological monitors.

[b]The PVA also performs handwriting recognition.

[c]A small speech vocabulary is up to 200 words, moderate 300–1,000 words, and large in excess of 1,000 words. For pen-based gestures, deictic selection is an individual gesture, a small vocabulary is 2–20 gestures, moderate 20–100, and large in excess of 100 gestures.

Note. VR = Virtual reality; 3-D = three-dimensional; MTC = Members-Teams-Committee; FMA = Field Medic Associate; FMC = Field Medic Coordinator; PVA = Portable Voice Assistant.

currently include adaptive processing, researchers in this area have begun to explore adaptive techniques for improving system robustness in noisy environmental contexts (Dupont & Luettin, 2000; Meier, Hürst, & Duchnowski, 1996; Rogozan & Deglise, 1998), which is an important future research direction. Although this literature has not emphasized the development of applications, nonetheless its quantitative modeling of synchronized phoneme/viseme patterns has been used to build animated characters that generate text-to-speech output and coordinated lip movements. These new animated characters are being used as an interface design vehicle for facilitating users' multimodal interaction with next-generation conversational interfaces (Cassell, Sullivan, Prevost, & Churchill, 2000; Cohen & Massaro, 1993).

Although the main multimodal literatures to date have focused on either speech and pen input or speech and lip movements, recognition of other modes also is maturing and beginning to be integrated into new kinds of multimodal systems. In particular, there is growing interest in designing multimodal interfaces that incorporate vision-based technologies, such as interpretation of gaze, facial expressions, and manual gesturing (Morimoto, Koons, Amir, Flickner, & Zhai, 1999; Pavlovic, Berry, & Huang, 1997; Turk & Robertson, 2000; Zhai et al., 1999). These technologies unobtrusively or *passively* monitor user behavior and need not require explicit user commands to a computer. This contrasts with *active input modes*, such as speech or pen, which the user deploys intentionally as a command issued to the system (see Fig. 14.1). Although passive modes may be attentive and less obtrusive, active modes generally are more reliable indicators of user intent.

As vision-based technologies mature, one important future direction will be the development of *blended* multimodal interfaces that combine both passive and active modes. These interfaces typically will be *temporally cascaded*, so one goal in designing new prototypes will be to determine optimal processing strategies for using advance information from the first mode (e.g., gaze) to constrain accurate interpretation of the following modes (e.g., gesture, speech). This kind of blended multimodal interface potentially can provide users with greater transparency and control, while also supporting improved robustness and broader application functionality (Oviatt & Cohen, 2000; Zhai et al., 1999). As this collection of technologies matures, there also is strong interest in designing new types of pervasive and mobile interfaces, including ones capable of adaptive processing to the user and environmental context.

As multimodal interfaces gradually evolve toward supporting more advanced recognition of users' natural activities in context, they will expand beyond rudimentary bimodal systems to ones that incorporate three or more input modes, qualitatively different modes, and more sophisticated models of multimodal interaction. This trend already has been initiated within biometrics research, which has combined recognition of multiple behavioral input modes (e.g., voice, handwriting) with physiological ones (e.g., retinal scans, fingerprints) to achieve reliable person identification and verification in challenging field conditions (Choudhury, Clarkson, Jebara, & Pentland, 1999; Pankanti et al., 2000).

WHAT ARE THE GOALS AND ADVANTAGES OF MULTIMODAL INTERFACE DESIGN?

Over the past decade, numerous advantages of multimodal interface design have been documented. Unlike a traditional keyboard and mouse interface or a unimodal recognition-based interface, multimodal interfaces permit flexible use of input modes. This includes the choice of which modality to use for conveying different types of information, to use combined input modes, or to alternate between modes at any time. Since individual input modalities are well suited in some situations, and less ideal or even inappropriate in others, modality choice is an important design issue in a multimodal system. As systems become more complex and multifunctional, a single modality simply does not permit all users to interact effectively across all tasks and environments.

Since there are large individual differences in ability and preference to use different modes of communication, a multimodal interface permits diverse user groups to exercise selection and control over how they interact with the computer (Fell et al., 1994; Karshmer & Blattner, 1998). In this respect, multimodal interfaces have the potential to accommodate a broader range of users than traditional interfaces—including users of different ages, skill levels, native language status, cognitive styles, sensory impairments, and other temporary illnesses or permanent handicaps. For example, a visually impaired user or one with repetitive stress injury may prefer speech input and text-to-speech output. In contrast, a user with a hearing impairment or accented speech may prefer touch, gesture or pen input. The natural alternation between modes that is permitted by a multimodal interface also can be effective in preventing overuse and physical damage to any single modality, especially during extended periods of computer use (R. Markinson,[2] 1993, personal communication).

Multimodal interfaces also provide the adaptability that is needed to accommodate the continuously changing conditions of mobile use. In particular, systems involving speech, pen, or touch input are suitable for mobile tasks and, when combined, users can shift among these modalities from moment to moment as environmental conditions change (Holzman, 1999; Oviatt, 2000b, 2000c). There is a sense in which mobility can induce a state of temporary disability, such that a person is unable to use a particular input mode for some period of time. For example, the user of an in-vehicle application may frequently be unable to use manual or gaze input, although speech is relatively more available. In this respect, a multimodal interface permits the modality choice and switching that is needed during the changing environmental circumstances of actual field and mobile use.

A large body of data documents that multimodal interfaces satisfy higher levels of user preference when interacting with simulated or real computer systems. Users have a strong

---

[2]R. Markinson, University of California at San Francisco Medical School, 1993.

Multimodal interfaces process two or more combined user input modes—such as speech, pen, touch, manual gestures, gaze, and head and body movements—in a coordinated manner with multimedia system output. They are a new class of interfaces that aim to recognize naturally occurring forms of human language and behavior, and that incorporate one or more recognition-based technologies (e.g., speech, pen, vision).

Active input modes are ones that are deployed by the user intentionally as an explicit command to a computer system (e.g., speech).

Passive input modes refer to naturally occurring user behavior or actions that are recognized by a computer (e.g., facial expressions, manual gestures). They involve user input that is unobtrusively and passively monitored, without requiring any explicit command to a computer.

Blended multimodal interfaces are ones that incorporate system recognition of at least one passive and one active input mode. (e.g., speech and lip movement systems).

Temporally cascaded multimodal interfaces are ones that process two or more user modalities that tend to be sequenced in a particular temporal order (e.g., gaze, gesture, speech), such that partial information supplied by recognition of an earlier mode (e.g., gaze) is available to constrain interpretation of a later mode (e.g., speech). Such interfaces may combine only active input modes, only passive ones, or they may be blended.

Mutual disambiguation involves disambiguation of signal or semantic-level information in one error-prone input mode from partial information supplied by another. Mutual disambiguation can occur in a multimodal architecture with two or more semantically rich, recognition-based input modes. It leads to recovery from unimodal recognition errors within a multimodal architecture, with the net effect of suppressing errors experienced by the user.

Visemes refer to the detailed classification of visible lip movements that correspond with consonants and vowels during articulated speech. A *viseme-phoneme mapping* refers to the correspondence between visible lip movements and audible phonemes during continuous speech.

Feature-level fusion is a method for fusing low-level feature information from parallel input signals within a multimodal architecture, which has been applied to processing closely synchronized input such as speech and lip movements.

Semantic-level fusion is a method for integrating semantic information derived from parallel input modes in a multimodal architecture, which has been used for processing speech and gesture input.

Frame-based integration is a pattern-matching technique for merging attribute-value data structures to fuse semantic information derived from two input modes into a common meaning representation during multimodal language processing.

Unification-based integration is a logic-based method for integrating partial meaning fragments derived from two input modes into a common meaning representation during multimodal language processing. Compared with frame-based integration, unification derives from logic programming, and has been more precisely analyzed and widely adopted within computational linguistics.

FIGURE 14.1. Multimodal interface terminology.

preference to interact multimodally, rather than unimodally, across a wide variety of different application domains, although this preference is most pronounced in spatial domains (Hauptmann, 1989; Oviatt, 1997). For example, 95% to 100% of users preferred to interact multimodally when they were free to use either speech or pen input in a map-based spatial domain (Oviatt, 1997). During pen/voice multimodal interaction, users preferred speech input for describing objects and events, sets and subsets of objects, out-of-view objects, conjoined information, past and future temporal states, and for issuing commands for actions or iterative actions (Cohen & Oviatt, 1995; Oviatt & Cohen, 1991). However, their preference for pen input increased when conveying digits, symbols, graphic content, and especially when conveying the location and form of spatially oriented information on a dense graphic display such as a map (Oviatt, 1997; Oviatt & Olsen, 1994; Suhm, 1998). Likewise, 71% of users combined speech and manual gestures multimodally, rather than using one input mode, when manipulating graphic objects on a CRT screen (Hauptmann, 1989).

During the early design of multimodal systems, it was assumed that efficiency gains would be the main advantage of designing an interface multimodally, and that this advantage would derive from the ability to process input modes in parallel. It is true that multimodal interfaces sometimes support improved efficiency, especially when manipulating graphical information. In simulation research comparing speech-only with multimodal pen/voice interaction, empirical work demonstrated that multimodal interaction yielded 10% faster task completion time during visual-spatial tasks, but no significant efficiency advantage in verbal or quantitative task domains (Oviatt, 1997; Oviatt, Cohen, & Wang, 1994). Likewise, users' efficiency improved when they combined speech and gestures multimodally to manipulate 3-D objects, compared with unimodal input (Hauptmann, 1989). In another early study, multimodal speech and mouse input improved efficiency in a line-art drawing task (Leatherby & Pausch, 1992). Finally, in a study that compared task completion times for a graphical interface versus a multimodal pen/voice interface, military domain experts averaged four times faster at setting up complex simulation scenarios on a map when they were able to interact multimodally (Cohen, McGee, & Clow, 2000). This latter study was based on testing of a fully functional multimodal system, and it included time required to correct recognition errors.

One particularly advantageous feature of multimodal interface design is its superior error handling, both in terms of error avoidance and graceful recovery from errors (Oviatt, 1999a; Oviatt, Bernard, & Levow, 1999; Oviatt & van Gent, 1996; Rudnicky & Hauptmann, 1992; Suhm, 1998; Tomlinson et al., 1996). There are user-centered and system-centered reasons why multimodal systems facilitate error recovery, when compared with unimodal recognition-based interfaces. For example, in a multimodal speech and pen-based gesture, interface users will select the input mode that they judge to be less error prone for particular lexical content, which tends to lead to error avoidance (Oviatt & van Gent, 1996). They may prefer speedy speech input, but will switch to pen input to communicate a foreign surname. Secondly, users' language often is simplified when interacting multimodally, which can substantially reduce the complexity of natural language processing and thereby reduce

recognition errors (Oviatt & Kuhn, 1998; see What Are the Primary Features of Multimodal Language? for discussion). In one study, users' multimodal utterances were documented to be briefer, to contain fewer complex locative descriptions, and 50% fewer spoken disfluencies, when compared with a speech-only interface. Thirdly, users have a strong tendency to switch modes after system recognition errors, which facilitates error recovery. This error resolution occurs because the confusion matrices differ for any given lexical content for the different recognition technologies involved in processing (Oviatt et al., 1999).

In addition to these user-centered reasons for better error avoidance and resolution, there also are system-centered reasons for superior error handling. A well-designed multimodal architecture with two semantically rich input modes can support *mutual disambiguation* of input signals. For example, if a user says "ditches," but the speech recognizer confirms the singular ditch as its best guess, then parallel recognition of several graphic marks can result in recovery of the correct plural interpretation. This recovery can occur in a multimodal architecture even though the speech recognizer initially ranks the plural interpretation ditches as a less preferred choice on its n-best list. Mutual disambiguation involves recovery from unimodal recognition errors within a multimodal architecture, because semantic information from each input mode supplies partial disambiguation of the other mode, thereby leading to more stable and robust overall system performance (Oviatt, 1999a, 2000a). Another example of mutual disambiguation is shown in Fig. 14.2. To achieve optimal error handling, a multimodal interface ideally should be designed to include complementary input modes, and so the alternative input modes provide duplicate

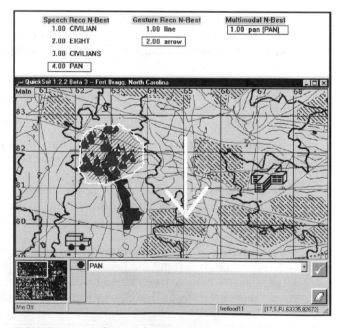

FIGURE 14.2. Multimodal command to "pan" the map, which illustrates mutual disambiguation occurring between incoming speech and gesture information, such that lexical hypotheses were pulled up on both n-best lists to produce a correct final multimodal interpretation.

functionality such that users can accomplish their goals using either mode.

In two recent studies involving more than 4,600 multimodal commands, a multimodal architecture was found to support mutual disambiguation and error suppression ranging between 19% and 41% (Oviatt, 1999a, 2000a). Improved robustness also was greater for challenging user groups (accented vs. native speakers) and usage contexts (mobile vs. stationary use). These results indicate that a well-designed multimodal system not only can perform more robustly than a unimodal system, but also in a more stable way across varied real-world users and usage contexts. Finally, during audiovisual perception of speech and lip movements, improved speech recognition also has been demonstrated for both human listeners (McLeod & Summerfield, 1987) and multimodal systems (Adjoudani & Benoit, 1995; Tomlinson et al., 1996).

WHAT METHODS AND INFORMATION HAVE BEEN USED TO DESIGN NOVEL MULTIMODAL INTERFACES?

The design of new multimodal systems has been inspired and organized largely by two things. First, the cognitive science literature on intersensory perception and intermodal coordination during production has provided a foundation of information for user modeling, as well as information on what systems must recognize and how multimodal architectures should be organized. For example, the cognitive science literature has provided knowledge of the natural integration patterns that typify people's lip and facial movements with speech output (Benoit, Guiard-Marigny, Le Goff, & Adjoudani, 1996; Ekman, 1992; Ekman & Friesen, 1978; Fridlund, 1994; Hadar, Steiner, Grant, & Rose, 1983; Massaro & Cohen, 1990; Stork & Hennecke, 1995; Vatikiotis-Bateson et al., 1996), and their coordinated use of manual or pen-based gestures with speech (Kendon, 1980; McNeill, 1992; Oviatt, DeAngeli, & Kuhn, 1997). Given the complex nature of users' multimodal interaction, cognitive science has and will continue to play an essential role in guiding the design of robust multimodal systems. In this respect, a multidisciplinary perspective will be more central to successful multimodal system design than it has been for traditional GUI design. The cognitive science underpinnings of multimodal system design are described later.

Secondly, high-fidelity automatic simulations also have played a critical role in prototyping new types of multimodal systems (Dahlbäck, Jëonsson, & Ahrenberg, 1992; Oviatt et al., 1992). When a new multimodal system is in the planning stages, design sketches and low-fidelity mock-ups may initially be used to visualize the new system and plan the sequential flow of human–computer interaction. These tentative design plans then are rapidly transitioned into a higher fidelity simulation of the multimodal system, which is available for proactive and situated data collection with the intended user population. High-fidelity simulations have been the preferred method for designing and evaluating new multimodal systems, and extensive data collection with such tools preferably is completed before a fully functional system ever is built.

During high-fidelity simulation testing, a user interacts with what she believes is a fully functional multimodal system, although the interface is actually a simulated front-end designed to appear and respond as the fully functional system would. During the interaction, a programmer assistant at a remote location provides the simulated system responses. As the user interacts with the front end, the programmer tracks her multimodal input and provides system responses as quickly and accurately as possible. To support this role, the programmer makes use of automated simulation software that is designed to support interactive speed, realism with respect to the targeted system, and other important characteristics. For example, with these automated tools, the programmer may be able to make a single selection on a workstation field to rapidly send simulated system responses to the user during a data collection session.

High-fidelity simulations have been the preferred method for prototyping multimodal systems for several reasons. Simulations are relatively easy and inexpensive to adapt, compared with building and iterating a complete system. They also permit researchers to alter a planned system's characteristics in major ways (e.g., input and output modes available), and to study the impact of different interface features in a systematic and scientific manner (e.g., type and base rate of system errors). In comparison, a particular system with its fixed characteristics is a less flexible and suitable research tool, and the assessment of any single system basically amounts to an individual case study. Using simulation techniques, rapid adaptation and investigation of planned system features permits researchers to gain a broader and more principled perspective on the potential of newly emerging technologies. In a practical sense, simulation research can assist in the evaluation of critical performance trade-offs and in making decisions about alternative system designs, which designers must do as they strive to create more usable multimodal systems.

To support the further development and commercialization of multimodal systems, additional infrastructure that will be needed in the future includes: (1) simulation tools for rapidly building and reconfiguring multimodal interfaces, (2) automated tools for collecting and analyzing multimodal corpora, and (3) automated tools for iterating new multimodal systems to improve their performance (see Oviatt et al., 2000, for further discussion).

WHAT ARE THE COGNITIVE SCIENCE UNDERPINNINGS OF MULTIMODAL INTERFACE DESIGN?

This section discusses the growing cognitive science literature that provides the empirical underpinnings needed to design next-generation multimodal interfaces. The ability to develop multimodal systems depends on knowledge of the natural integration patterns that typify people's combined use of different input modes. In particular, the design of new multimodal systems depends on intimate knowledge of the properties of different modes and the information content they carry, the unique characteristics of multimodal language and its processability, and the integration and synchronization characteristics

---

Ten Myths of Multimodal Interaction

Myth #1: If you build a multimodal system, users will interact multimodally.

Myth #2: Speech and pointing are the dominant multimodal integration pattern.

Myth #3: Multimodal input involves simultaneous signals.

Myth #4: Speech is the primary input mode in any multimodal system that includes it.

Myth #5: Multimodal language does not differ linguistically from unimodal language.

Myth #6: Multimodal integration involves redundancy of content between modes.

Myth #7: Individual error-prone recognition technologies combine multimodally to produce even greater unreliability.

Myth #8: All users' multimodal commands are integrated in a uniform way.

Myth #9: Different input modes are capable of transmitting comparable content.

Myth #10: Enhanced efficiency is the main advantage of multimodal systems.

(from Oviatt, 1999b)

FIGURE 14.3. Ten myths of multimodal interaction: separating myth from empirical reality.

of users' multimodal interaction. It also relies on accurate prediction of when users are likely to interact multimodally, and how alike different users are in their specific integration patterns. The relevant cognitive science literature on these topics is very extensive, especially when consideration is given to all of the underlying sensory perception and production capabilities involved in different input modes currently being incorporated in new multimodal interfaces. As a result, this section will be limited to introducing the main cognitive science themes and findings that are relevant to the more common types of multimodal systems.

This cognitive science foundation also has played a key role in identifying computational myths about multimodal interaction and replacing these misconceptions with contrary empirical evidence. Figure 14.3 summarizes 10 common myths about multimodal interaction, which are addressed and discussed in more detail elsewhere (Oviatt, 1999b). As such, the literature summarized in this section aims to provide a more accurate foundation for guiding the design of next-generation multimodal systems.

When Do Users Interact Multimodally?

During natural interpersonal communication, people are always interacting multimodally. Of course, in this case, the number of information sources or modalities that an interlocutor has

available to monitor is essentially unlimited. However, all multimodal systems are constrained in the number and type of input modes they can recognize. Also, a user can compose active input during human–computer interaction that either is delivered multimodally or that is delivered entirely using just one mode. That is, although users in general may have a strong preference to interact multimodally rather than unimodally, this is no guarantee that they will issue every command to a system multimodally, given the particular type of multimodal interface available. Therefore, the first nontrivial question that arises during system processing is whether a user is communicating unimodally or multimodally.

In the case of speech and pen-based multimodal systems, users typically intermix unimodal and multimodal expressions. In a recent study involving a visual-spatial domain, users' commands were expressed multimodally 20% of the time, with others just spoken or written (Oviatt et al., 1997). Predicting whether a user will express a command multimodally also depends on the type of action he or she is performing. In particular, users almost always express commands multimodally when describing spatial information about the location, number, size, orientation, or shape of an object. In one study, users issued multimodal commands 86% of the time when they had to add, move, modify, or calculate the distance between objects on a map in a way that required specifying spatial locations (Oviatt et al., 1997). They also were moderately likely to interact multimodally when selecting an object from a larger array, for example, when deleting a particular object from the map. However, when performing general actions without any spatial component, such as printing a map, users expressed themselves multimodally less than 1% of the time. These data emphasize that future multimodal systems will need to distinguish between instances when users are and are not communicating multimodally, so that accurate decisions can be made about when parallel input streams should be interpreted jointly versus individually. They also suggest that knowledge of the type of actions to be included in an application, such as whether the application entails manipulating spatial information, should influence the basic decision of whether to build a multimodal interface at all.

In a multimodal interface that processes passive or blended input modes, there always is at least one passively tracked input source providing continuous information (e.g., gaze tracking, head position). In these cases, all user input would by definition be classified as multimodal, and the primary problem would become segmentation and interpretation of each continuous input stream into meaningful actions of significance to the application. In the case of blended multimodal interfaces (e.g., gaze tracking and mouse input), it still may be opportune to distinguish active forms of user input that might be more accurately or expeditiously handled as unimodal events.

What Are the Integration and Synchronization Characteristics of Users' Multimodal Input?

The past literature on multimodal systems has focused largely on simple selection of objects or locations in a display, rather than considering the broader range of multimodal integration patterns. Since the development of Bolt's (1980) Put That There system, speak-and-point has been viewed as the prototypical form of multimodal integration. In Bolt's system, semantic processing was based on spoken input, but the meaning of a deictic term such as "that" was resolved by processing the x,y coordinate indicated by pointing at an object. Since that time, other multimodal systems also have attempted to resolve deictic expressions using a similar approach, for example, using gaze location instead of manual pointing (Koons et al., 1993).

Unfortunately, this concept of multimodal interaction as point-and-speak makes only limited use of new input modes for *selection* of objects—just as the mouse does. In this respect, it represents the persistence of an old mouse-oriented metaphor. In contrast, modes that transmit written input, manual gesturing, and facial expressions are capable of generating symbolic information that is much more richly expressive than simple pointing or selection. In fact, studies of users' integrated pen/voice input indicate that a speak-and-point pattern only comprises 14% of all spontaneous multimodal utterances (Oviatt et al., 1997). Instead, pen input more often is used to create graphics, symbols and signs, gestural marks, digits, and lexical content. During interpersonal multimodal communication, linguistic analysis of spontaneous manual gesturing also indicates that simple pointing accounts for less than 20% of all gestures (McNeill, 1992). Together, these cognitive science and user-modeling data highlight the fact that any multimodal system designed exclusively to process speak-and-point will fail to

provide users with much useful functionality. For this reason, specialized algorithms for processing deictic-point relations will have only limited practical use in the design of future multimodal systems. It is clear that a broader set of multimodal integration issues needs to be addressed in future work. Future research also should explore typical integration patterns between other promising modality combinations, such as speech and gaze.

It also is commonly assumed that any signals involved in a multimodal construction will co-occur temporally. The presumption is that this temporal overlap then determines which signals to combine during system processing. In the case of speech and manual gestures, successful processing of the deictic term "that square" in Bolt's original system relied on interpretation of pointing when the word that was spoken to extract the intended referent. However, one empirical study indicated that users often do not speak deictic terms at all and, when they do, the deictic frequently is not overlapped in time with their pointing. In fact, it has been estimated that as few as 25% of users' commands actually contain a spoken deictic that overlaps with the pointing needed to disambiguate its meaning (Oviatt et al., 1997).

Beyond the issue of deixis, users' input frequently does not overlap at all during multimodal commands to a computer. During spoken and pen-based input, for example, users' multimodal input is sequentially integrated about half the time, with pen input preceding speech 99% of the time, and a brief lag between signals of 1 or 2 (Oviatt et al., 1997). This finding is consistent with linguistics data revealing that both spontaneous gesturing and signed language often precede their spoken lexical analogues during human communication (Kendon, 1980; Naughton, 1996). The degree to which gesturing precedes speech is greater in topic-prominent languages, such as Chinese, than it is in subject-prominent ones like Spanish or English (McNeill, 1992). Even in the speech and lip movement literature, close but not perfect temporal synchrony is typical, with lip movements occurring a fraction of a second before the corresponding auditory signal (Abry, Lallouache, & Cathiard, 1996; Benoit, 2000).

In short, although two input modes may be highly interdependent and synchronized during multimodal interaction, synchrony does not imply simultaneity. The empirical evidence reveals that multimodal signals often do not co-occur temporally at all during human-computer or natural human communication. Therefore, multimodal system designers cannot necessarily count on conveniently overlapped signals to achieve successful processing in the multimodal architectures they build. Future research needs to explore the integration patterns and temporal cascading that can occur among three or more input modes—such as gaze, gesture, and speech—so that more advanced multimodal systems can be designed and prototyped.

In the design of new multimodal architectures, it is important to note that data on the order of input modes and average time lags between input modes has been used to determine the likelihood that an utterance is multimodal versus unimodal, and to establish temporal thresholds for fusion of input. In the future, weighted likelihoods associated with different utterance segmentations, for example, that an input stream containing speech, writing, speech should be segmented into [S / W S] rather than [S W / S], and with intermodal time lag distributions, will be used to optimize correct recognition of multimodal user input (Oviatt, 1999b). In the design of future time-critical multimodal architectures, data on users' integration and synchronization patterns will need to be collected for other mode combinations during realistic interactive tasks, so that temporal thresholds can be established for performing multimodal fusion.

What Individual Differences Exist in Multimodal Interaction, and What Are the Implications for Designing Systems for Universal Access?

When users interact multimodally, there actually can be large individual differences in integration patterns. For example, previous empirical work on multimodal pen/voice integration has revealed two main types of users—ones who habitually deliver speech and pen signals in an overlapped or simultaneous manner, and others who synchronize signals sequentially with pen input preceding speech by up to 4 s (Oviatt, 1999b). These users' dominant integration pattern could be identified when they first began interacting with the system and then persisted throughout their session (Oviatt, 1999b). That is, each user's integration pattern was established early and remained consistent, although two distinct integration patterns were observed among different users. As previously described, substantial differences also have been observed in the degree to which manual gestures precede speech for members of different linguistic groups, such as Chinese, Spanish, and Americans. All of these findings imply that future multimodal systems capable of adapting temporal thresholds for different user groups potentially could achieve greater recognition accuracy and interactive speed.

Both individual and cultural differences also have been documented between users in modality integration patterns. For example, substantial individual differences also have been reported in the temporal synchrony between speech and lip movements (Kricos, 1996). In addition, lip movements during speech production are less exaggerated among Japanese speakers than Americans (Sekiyama & Tohkura, 1991). In fact, extensive interlanguage differences have been observed in the information available from lip movements during audiovisual speech (Fuster-Duran, 1996). These findings have implications for the degree to which disambiguation of speech can be achieved through lip movement information in noisy environments or for different user populations. Finally, nonnative speakers, the hearing impaired, and elderly listeners all are more influenced by visual lip movement than auditory cues when processing speech (Fuster-Duran, 1996; Massaro, 1996). These results have implications for the design and expected value of audiovisual multimedia output for different user groups in animated character interfaces.

Finally, gender, age, and other individual differences are common in gaze patterns, as well as speech and gaze integration (Argyle, 1972). As multimodal interfaces incorporating gaze become more mature, further research will need to explore these gender and age-specific patterns, and to build appropriately adapted processing strategies. In summary, considerably more research is needed on multimodal integration and synchronization patterns for new mode combinations, as well as for diverse and disabled users for whom multimodal interfaces may be especially suitable for ensuring universal access.

Is Complementarity or Redundancy the Main Organizational Theme That Guides Multimodal Integration?

It frequently is claimed that the propositional content conveyed by different modes during multimodal communication contains a high degree of redundancy. However, the dominant theme in users' natural organization of multimodal input actually is complementarity of content, not redundancy. For example, speech and pen input consistently contribute different and complementary semantic information—with the subject, verb, and object of a sentence typically spoken, and locative information written (Oviatt et al., 1997). In fact, a major complementarity between speech and manually oriented pen input involves visual-spatial semantic content, which is one reason these modes are an opportune combination for visual-spatial applications. Whereas spatial information is uniquely and clearly indicated via pen input, the strong descriptive capabilities of speech are better suited for specifying temporal and other nonspatial information. Even during multimodal correction of system errors, when users are highly motivated to clarify and reinforce their information delivery, speech and pen input express redundant information less than 1% of the time. Finally, during interpersonal communication, linguists also have documented that spontaneous speech and manual gesturing involve complementary rather than duplicate information between modes (McNeill, 1992).

Other examples of primary multimodal complementarities during interpersonal and human-computer communication have been described in past research (McGurk & MacDonald, 1976; Oviatt & Olsen, 1994; Wickens, Sandry, & Vidulich, 1983). For example, in the literature on multimodal speech and lip movements, natural feature-level complementarities have been identified between visemes and phonemes for vowel articulation, with vowel rounding better conveyed visually, and vowel height and backness better revealed auditorily (Massaro & Stork, 1998; Robert-Ribes et al., 1998).

In short, actual data highlight the importance of complementarity as a major organizational theme during multimodal communication. The designers of next-generation multimodal systems therefore should not expect to rely on duplicated information when processing multimodal language. In multimodal systems involving both speech and pen-based gestures and speech and lip movements, one explicit goal has been to integrate complementary modalities in a manner that yields a synergistic blend, such that each mode can be capitalized on and used to overcome weaknesses in the other mode (Cohen et al.,

1989). This approach to system design has promoted the philosophy of using modes and component technologies to their natural advantage, and of combining them in a manner that permits mutual disambiguation. One advantage of achieving such a blend is that the resulting multimodal architecture can function more robustly than an individual recognition-based technology or a multimodal system based on input modes lacking natural complementarities.

What Are the Primary Features of Multimodal Language?

Communication channels can be tremendously influential in shaping the language transmitted within them. From past research, there now is cumulative evidence that many linguistic features of multimodal language are qualitatively very different from that of spoken or formal textual language. In fact, it can differ in features as basic as brevity, semantic content, syntactic complexity, word order, disfluency rate, degree of ambiguity, referring expressions, specification of determiners, anaphora, deixis, and linguistic indirectness. In many respects, multimodal language is simpler linguistically than spoken language. In particular, comparisons have revealed that the same user completing the same map-based task communicates fewer words, briefer sentences, and fewer complex spatial descriptions and disfluencies when interacting multimodally, compared with using speech alone (Oviatt, 1997). One implication of these findings is that multimodal interface design has the potential to support more robust future systems than a unimodal design approach. The following is an example of a typical user's spoken input while attempting to designate an open space using a map system: *"Add an open space on the north lake to b-include the north lake part of the road and north."* In contrast, the same user accomplished the same task multimodally by encircling a specific area and saying: *"Open space."*

In previous research, hard-to-process disfluent language has been observed to decrease by 50% during multimodal interaction with a map, compared with a more restricted speech-only interaction (Oviatt, 1997). This drop occurs mainly because people have difficulty speaking spatial information, which precipitates disfluencies. In a flexible multimodal interface, they instead use pen input to convey spatial information, thereby avoiding the need to speak it. Further research is needed to establish whether other forms of flexible multimodal communication generally ease users' cognitive load, which may be reflected in a reduced rate of disfluencies.

During multimodal pen/voice communication, the linguistic indirection that is typical of spoken language frequently is replaced with more direct commands (Oviatt & Kuhn, 1998). In the following example, a study participant made a disfluent indirect request using speech input while requesting a map-based distance calculation: *"What is the distance between the Victorian Museum and the, uh, the house on the east side of Woodpecker Lane?"* When requesting distance information multimodally, the same user encircled the house and museum while speaking the following brief direct command: *"Show*

distance between here and here." In this research, the briefer and more direct multimodal pen/voice language also contained substantially fewer referring expressions, with a selective reduction in co-referring expressions that instead were transformed into deictic expressions. This latter reduction in co-reference would simplify natural language processing by easing the need for anaphoric tracking and resolution in a multimodal interface. Also consistent with fewer referring expressions, explicit specification of definite and indefinite reference is less common in multimodal language (Oviatt & Kuhn, 1998). Current natural language processing algorithms typically rely heavily on the specification of determiners in definite and indefinite references to represent and resolve noun phrase reference. One unfortunate by-product of the lack of such specifications is that current language processing algorithms are unprepared for the frequent occurrence of elision and deixis in multimodal human–computer interaction.

In other respects, multimodal language clearly is different than spoken language, although not necessarily simpler. For example, users' multimodal pen/voice language departs from the canonical English word order of S-V-O-LOC (i.e., Subject-Verb-Object-Locative constituent), which is observed in spoken language and also formal textual language. Instead, users' multimodal constituents shift to a LOC-S-V-O word order. A recent study reported that 95% of locative constituents were in sentence-initial position during multimodal interaction. However, for the same users completing the same tasks while speaking, 96% of locatives were in sentence-final position (Oviatt et al., 1997). It is likely that broader analyses of multimodal communication patterns, which could involve gaze and manual gesturing to indicate location rather than pen-based pointing, would reveal a similar reversal in word order.

One implication of these many differences is that new multimodal corpora, statistical language models, and natural language processing algorithms will need to be established before multimodal language can be processed optimally. Future research and corpus collection efforts also will be needed on different types of multimodal communication and in other application domains, so that the generality of previously identified multimodal language differences can be explored.

WHAT ARE THE BASIC WAYS IN WHICH MULTIMODAL INTERFACES DIFFER FROM GRAPHICAL USER INTERFACES?

Multimodal research groups currently are rethinking and redesigning basic user interface architectures, because a whole new range of architectural requirements has been posed. First, GUIs typically assume that there is a single event stream that controls the underlying event loop, with any processing sequential in nature. For example, most GUIs ignore typed input when a mouse button is depressed. In contrast, multimodal interfaces typically can process continuous and simultaneous input from parallel incoming streams. Secondly, GUIs assume that the basic interface actions, such as selection of an item, are atomic and unambiguous events. In contrast, multimodal systems process input modes using recognition-based technologies, which are designed to handle uncertainty and entail probabilistic methods of processing. Thirdly, GUIs often are built to be separable from the application software that they control, although the interface components usually reside centrally on one machine. In contrast, recognition-based user interfaces typically have larger computational and memory requirements, which often makes it desirable to distribute the interface over a network so that separate machines can handle different recognizers or databases. For example, cell phones and networked personal digital assistants (PDAs) may extract features from speech input, but transmit them to a recognizer that resides on a server. Finally, multimodal interfaces that process two or more recognition-based input streams require time-stamping of input, and the development of temporal constraints on mode fusion operations. In this regard, they involve uniquely time-sensitive architectures.

WHAT BASIC ARCHITECTURES AND PROCESSING TECHNIQUES HAVE BEEN USED TO DESIGN MULTIMODAL SYSTEMS?

Many early multimodal interfaces that handled combined speech and gesture, such as Bolt's Put That There system (Bolt, 1980), have been based on a control structure in which multimodal integration occurs during the process of parsing spoken language. As discussed earlier, when the user speaks a deictic expression, such as "here" or "this," the system searches for a synchronized gestural act that designates the spoken referent. Although such an approach is viable for processing a point-and-speak multimodal integration pattern, as discussed earlier, multimodal systems must be able to process richer input than just pointing, including gestures, symbols, graphic marks, lip movements, meaningful facial expressions, and so forth. To support more broadly functional multimodal systems, general processing architectures have been developed since Bolt's time. Some of these recent architectures handle a variety of multimodal integration patterns, as well as the interpretation of both unimodal and combined multimodal input. This kind of architecture can support the development of multimodal systems in which modalities are processed individually as input alternatives to one another, or those in which two or more modes are processed as combined multimodal input.

For multimodal systems designed to handle joint processing of input signals, there are two main subtypes of multimodal architecture. First, there are ones that integrate signals at the *feature level* (i.e., early fusion) and others that integrate information at a *semantic level* (i.e., late fusion). Examples of systems based on an early feature-fusion processing approach include those developed by Bregler and colleagues (Bregler, Manke, Hild, & Waibel, 1993), Vo and colleagues (Vo et al., 1995), and Pavlovic and colleagues (Pavlovic et al., 1997, 1998). In a feature-fusion architecture, the signal-level recognition process in one mode influences the course of recognition in the other. Feature fusion is considered more appropriate for closely temporally synchronized input modalities, such as speech and lip movements (Rubin et al., 1998; Stork & Hennecke, 1995).

In contrast, multimodal systems using the late semantic fusion approach have been applied to processing multimodal speech and pen input or manual gesturing, for which the input modes are less coupled temporally. These input modes provide different but complementary information that typically is integrated at the utterance level. Late semantic integration systems use individual recognizers that can be trained using unimodal data, which are easier to collect and already are publicly available for speech and handwriting. In this respect, systems based on semantic fusion can be scaled up easier in number of input modes or vocabulary size. Examples of systems based on semantic fusion include Put That There (Bolt, 1980); ShopTalk (Cohen et al., 1989); QuickSet (Cohen et al., 1997); CUBRICON (Neal & Shapiro, 1991); Virtual World (Codella et al., 1992); Finger-Pointer (Fukumoto, Suenaga, & Mase, 1994); VisualMan (Wang, 1995); Human-Centric Word Processor, Portable Voice Assistant (Bers et al., 1998); the VR Aircraft Maintenance Training System (Duncan et al., 1999); and Jeanie (Vo & Wood, 1996).

As an example of multimodal information processing flow in a late-stage semantic architecture, Figure 14.4 illustrates two input modes (e.g., speech and manual or pen-based gestures) recognized in parallel and processed by an understanding component. The results involve partial meaning representations that are fused by the multimodal integration component, which also is influenced by the system's dialogue management and interpretation of current context. During the integration process, alternative lexical candidates for the final multimodal interpretation are ranked according to their probability estimates on an n-best list. The best-ranked multimodal interpretation then is sent to the application invocation and control component, which transforms this information into a series of commands to one or more back-end application systems. System feedback typically includes multimedia output, which may incorporate text-to-speech and nonspeech audio, graphics and animation, and so forth. For examples of feature-based multimodal processing flow and architectures, especially as applied to multimodal speech and lip movement systems, see Benoit et al. (2000).

There are many ways to realize this information processing flow as an architecture. One common infrastructure that has been adopted by the multimodal research community involves *multiagent architectures*, such as the Open Agent Architecture (Cohen, Cheyer, Wang, & Baeg, 1994; Martin, Cheyer, & Moran, 1999) and Adaptive Agent Architecture (Kumar & Cohen, 2000). In a multiagent architecture, the many components needed to support the multimodal system (e.g., speech recognition, gesture recognition, natural language processing, and multimodal integration) may be written in different programming languages, on different machines, and with different operating systems. Agent communication languages are being developed that can handle asynchronous delivery, triggered responses, multicasting and other concepts from distributed systems, and that are fault-tolerant (Kumar & Cohen, 2000). Using a multiagent architecture, for example, speech and gestures can arrive in parallel or

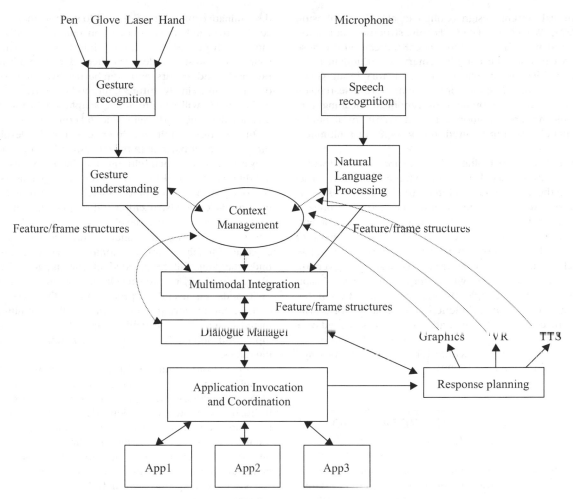

FIGURE 14.4. Typical information processing flow in a multimodal architecture designed for speech and gesture. (VR = virtual reality; TTS = text-to-speech; App = application)

asynchronously via individual modality agents, with the results recognized and passed to a facilitator. These results, typically an n-best list of conjectured lexical items and related time-stamp information, then are routed to appropriate agents for further language processing. Next, sets of meaning fragments derived from the speech and pen signals arrive at the multimodal integrator. This agent decides whether and how long to wait for recognition results from other modalities, based on the system's temporal thresholds. It fuses the meaning fragments into a semantically and temporally compatible whole interpretation before passing the results back to the facilitator. At this point, the system's final multimodal interpretation is confirmed by the interface, delivered as multimedia feedback to the user, and executed by any relevant applications. In summary, multiagent architectures provide essential infrastructure for coordinating the many complex modules needed to implement multimodal system processing, and they permit doing so in a distributed manner that is compatible with the trend toward mobile computing.

The core of multimodal systems based on semantic fusion involves algorithms that integrate common meaning representations derived from speech, gesture, and other modalities into a combined final interpretation. The semantic fusion operation requires a common meaning representation framework for all modalities, and a well-defined operation for combining partial meanings that arrive from different signals. To fuse information from different modalities, various research groups have independently converged on a strategy of recursively matching and merging attribute/value data structures, although using a variety of different algorithms (Cheyer & Julia, 1995; Pavlovic & Huang, 1998; Shaikh et al., 1997; Vo & Wood, 1996). This approach is considered a *frame-based integration* technique. An alternative logic-based approach derived from computational linguistics (Calder, 1987; Carpenter, 1990, 1992) involves the use of *typed feature structures* and *unification-based integration*, which is a more general and well understood approach. Unification-based integration techniques also have been applied to multimodal system design (Cohen et al., 1997; Johnston et al., 1997; Wu et al., 1999). Feature-structure unification is considered well suited to multimodal integration, because unification can combine complementary or redundant input from both modes, but it rules out contradictory input. Given this foundation for multimodal integration, more research still is needed on the development of canonical meaning representations that are common among different input modes that will need to be represented in new types of multimodal systems.

When statistical processing techniques are combined with a symbolic unification-based approach that merges feature structures, then the multimodal architecture that results is a *hybrid symbolic/statistical* one. Hybrid architectures represent one major new direction for multimodal system development. Multimodal architectures also can be hybrids in the sense of combining Hidden Markov Models and neural networks. New hybrid architectures potentially are capable of achieving very robust functioning, compared with either an early- or late-fusion approach alone. For example, the Members-Teams-Committee hierarchical recognition technique, which is a a hybrid symbolic/statistical multimodal integration framework trained over a labeled multimodal corpus, recently achieved 95.26% correct recognition performance, or within 1.4% of the theoretical system upper bound (Wu et al., 1999).

WHAT ARE THE MAIN FUTURE DIRECTIONS FOR MULTIMODAL INTERFACE DESIGN?

The computer science community is just beginning to understand how to design innovative, well-integrated, and robust multimodal systems. To date, most multimodal systems remain bimodal, and recognition technologies related to several human senses (e.g., haptics, smell, taste) have yet to be well represented or included at all within multimodal interfaces. The design and development of new types of systems that include such modes will not be achievable through intuition. Rather, it will depend on knowledge of the natural integration patterns that typify people's combined use of various input modes. This means that the successful design of multimodal systems will continue to require guidance from cognitive science on the coordinated human perception and production of natural modalities. In this respect, multimodal systems only can flourish through multidisciplinary cooperation, as well as teamwork among those representing expertise in the component technologies.

Most of the systems outlined in this chapter have been built during the past decade, and they are research-level systems. However, in some cases they have developed well beyond the prototype stage, and are being integrated with other software at academic and federal sites, or appearing as newly shipped products. To achieve commercialization and widespread dissemination of multimodal interfaces, more general, robust, and scalable multimodal architectures will be needed, which just now are beginning to emerge. Future multimodal interfaces also must begin incorporating input from multiple heterogeneous information sources, such that when combined, they contain the discriminative information needed to support more robust processing than individual recognition technologies. To support increasingly pervasive multimodal interfaces, these information sources also must include data collected from a wide array of input modes and sensors, and from both active and passive forms of user input. Finally, future multimodal interfaces, especially mobile ones, will require active adaptation to the user, task, ongoing dialogue, and environmental context.

In the future, multimodal interfaces could be developed that provide a better balance between system input and output, so they are better matched with one another in expressive power. Multimodal interfaces have the potential to give users more expressive power and flexibility, as well as better tools for controlling sophisticated visualization and multimedia output capabilities. As these interfaces develop, research will be needed on how to design whole multimodal-multimedia systems that are capable of highly robust functioning. To achieve this, a better understanding will be needed of the impact of visual displays, animation, text-to-speech, and audio output on users' multimodal input and its processability. One fertile research domain for exploring these topics will be multimedia animated character design and its impact on users' multimodal input and interaction with next-generation conversational interfaces.

In conclusion, multimodal interfaces are just beginning to model human-like sensory perception. They are recognizing and identifying actions, language, and people that have been seen, heard, or in other ways experienced in the past. They literally reflect and acknowledge the existence of human users, empower them in new ways, and create for them a voice. They also can be playful and self-reflective interfaces that suggest new forms of human identity as we interact face to face with animated personas representing our own kind. In all of these ways, novel multimodal interfaces, as primitive as their early bimodal instantiations may be, represent a new multidisciplinary science, a new art form, and a sociopolitical statement about our collective desire to humanize the technology we create.

ACKNOWLEDGMENTS

The author thanks the National Science Foundation for their support over the past decade, which has enabled her to pursue basic exploratory research on many aspects of multimodal interaction, interface design, and system development. The preparation of this chapter has been supported by NSF Grants IRI-9530666, IIS-0117868, and by NSF Special Extension for Creativity (SEC) Grant IIS-9530666. This work also has been supported by contracts DABT63-95-C-007 and N66001-99-D-8503 from DARPA's Information Technology and Information Systems Office, and by Grant N00014-99-1-0377 from ONR. Thanks also to Phil Cohen and others in the Center for Human-Computer Communication for many insightful discussions, and to Dana Director and Rachel Coulston for expert assistance with manuscript preparation. LEA, Inc. is acknowledged for granting permission to reprint Table 14.1 and Fig. 14.4, and ACM for allowing the reprint of Figs. 14.2 and 14.3.

References

Abry, C., Lallouache, M.-T., & Cathiard, M.-A. (1996). How can coarticulation models account for speech sensitivity to audio-visual desynchronization? In D. G. Stork & M. E. Hennecke (Eds.), *Speechreading by humans and machines: Models, systems and applications* (pp. 247–255). New York: Springer Verlag.

Adjoudani, A., & Benoit, C. (1995, September 18–25). Audio-visual speech recognition compared across two architectures. *Proceedings of the Eurospeech Conference* (Vol. 2, pp. 1563–1566). Madrid, Spain.

Argyle, M. (1972). Nonverbal communication in human social interaction. In R. Hinde (Ed.), *Nonverbal communication* (pp. 243–267). Cambridge, MA: Cambridge University Press.

Bangalore, S., & Johnston, M. (2000). Integrating multimodal language processing with speech recognition. In B. Yuan, T. Huang, & X. Tang (Eds.), *Proceedings of the International Conference on Spoken Language Processing (ICSLP 2000)* (Vol. 2, pp. 126–129). Beijing, China: Chinese Friendship Publishers.

Benoit, C. (2000). The intrinsic bimodality of speech communication and the synthesis of talking faces. In M. Taylor, F. Neel, & D. Bouwhuis (Eds.), *The Structure of Multimodal Dialogue II* (pp. 485–502). Amsterdam: John Benjamins.

Benoit, C., Guiard-Marigny, T., Le Goff, B., & Adjoudani, A. (1996). Which components of the face do humans and machines best speechread? In D. G. Stork & M. E. Hennecke (Eds.), *Speechreading by humans and machines: Models, systems, and applications: Vol. 150 of NATO ASI Series. Series F: Computer and Systems Sciences* (pp. 315–325). Berlin, Germany: Springler-Verlag.

Benoit, C., & Le Goff, B. (1998). Audio-visual speech synthesis from French text: Eight years of models, designs and evaluation at the ICP. *Speech Communication, 26*, 117–129.

Benoit, C., Martin, J.-C., Pelachaud, C., Schomaker, L., & Suhm, B. (2000). Audio-visual and multimodal speech-based systems. In D. Gibbon, I. Mertins, & R. Moore (Eds.), *Handbook of multimodal and spoken dialogue systems: Resources, terminology and product evaluation* (pp. 102–203). Dordrecht: Kluwer Academic Publishers.

Bernstein, L., & Benoit, C. (1996). For speech perception by humans or machines, three senses are better than one. *Proceedings of the International Conference on Spoken Language Processing (ICSLP '96)* (Vol. 3, pp. 1477–1480). New York: IEEE Press.

Bers, J., Miller, S., & Makhoul, J. (1998, February 8–11). "Designing conversational interfaces with multimodal interaction," *Proc. DARPA Broadcast News Transcription and Understanding Workshop* (pp. 319–321), Lansdowne, VA: Morgan Kaufmann Publishers.

Bolt, R. A. (1980). Put-that-there: Voice and gesture at the graphics interface. *Computer Graphics, 14*(3), 262–270.

Bregler, C., & Konig, Y. (1994). Eigenlips for robust speech recognition. *Proceedings of the International Conference on Acoustics Speech and Signal Processing (IEEE-ICASSP)* (Vol. 2, pp. 669–672). New York: IEEE Press.

Bregler, C., Manke, S., Hild, H., & Waibel, A. (1993). Improving connected letter recognition by lipreading. *Proceedings of the International Conference on Acoustics, Speech and Signal Processing (IEEE-ICASSP)* (Vol. 1, pp. 557–560). Minneapolis, MN: IEEE Press.

Brooke, N. M., & Petajan, E. D. (1986). Seeing speech: Investigations into the synthesis and recognition of visible speech movements using automatic image processing and computer graphics. *Proceedings of the International Conference Speech Input and Output: Techniques and Applications, 258*, 104–109.

Calder, J. (1987). Typed unification for natural language processing. In E. Klein & J. van Benthem (Eds.), *Categories, polymorphisms, and unification* (pp. 65–72). Edinburgh, Scotland: Center for Cognitive Science, University of Edinburgh.

Carpenter, R. (1990). Typed feature structures: Inheritance, (in)equality, and extensionality. *Proceedings of the ITK Workshop: Inheritance in Natural Language Processing* (pp. 9–18). Tilburg: Institute for Language Technology and Artificial Intelligence, Tilburg University, The Netherlands.

Carpenter, R. (1992). *The logic of typed feature structures.* Cambridge, UK: Cambridge University Press.

Cassell, J., Sullivan, J., Prevost, S., & Churchill, E. (Eds.). (2000). *Embodied conversational agents.* Cambridge, MA: MIT Press.

Cheyer, A. (1998, January). MVIEWS: Multimodal tools for the video analyst. *International Conference on Intelligent User Interfaces (IUI '98)* (pp. 55–62). New York: ACM Press.

Cheyer, A., & Julia, L. (1995, May). Multimodal maps: An agent-based approach. *International Conference on Cooperative Multimodal Communication (CMC' 95)* (pp. 103–113). Eindhoven, The Netherlands.

Choudhury, T., Clarkson, B., Jebara, T., & Pentland, S. (1999). Multimodal person recognition using unconstrained audio and video. *Proceedings of the 2nd International Conference on Audio-and-Video-based Biometric Person Authentication* (pp. 176–181). Washington, DC.

Codella, C., Jalili, R., Koved, L., Lewis, J., Ling, D., Lipscomb, J., Rabenhorst, D., Wang, C., Norton, A., Sweeney, P., & Turk C. (1992). Interactive simulation in a multi-person virtual world. *Proceedings of the Conference on Human Factors in Computing Systems (CHI '92)* (pp. 329–334). New York: ACM Press.

Cohen, M. M., & Massaro, D. W. (1993). Modeling coarticulation in synthetic visual speech. In M. Magnenat-Thalmann & D. Thalmann (Eds.), *Models and techniques in computer animation.* Tokyo: Springer-Verlag.

Cohen, P. R., Cheyer, A., Wang, M., & Baeg, S. C. (1994). An open agent architecture. *AAAI '94 Spring Symposium Series on Software Agents* (pp. 1–8). Menlo Park, CA: AAAI Press. [Reprinted in Huhns and Singh (Eds.). (1997). *Readings in Agents* (pp. 197–204). San Francisco: Morgan Kaufmann.]

Cohen, P. R., Dalrymple, M., Moran, D. B., Pereira, F. C. N., Sullivan, J. W., Gargan, R. A., Schlossberg, J. L., & Tyler, S. W. (1989). Synergistic use of direct manipulation and natural language. *Proceedings of the Conference on Human Factors in Computing Systems (CHI '89)* (pp. 227–234). New York: ACM Press. (Reprinted in Maybury & Wahlster (Eds.), (1998). *Readings in Intelligent User Interfaces* (pp. 29–37). San Francisco: Morgan Kaufmann.)

Cohen, P. R., Johnston, M., McGee, D., Oviatt, S., Pittman, J., Smith, I., Chen, L., & Clow, J. (1997). Quickset: Multimodal interaction for distributed applications. *Proceedings of the Fifth ACM International Multimedia Conference* (pp. 31–40). New York: ACM Press.

Cohen, P. R., McGee, D. R., & Clow, J. (2000). The efficiency of multimodal interaction for a map-based task. *Proceedings of the Language Technology Joint Conference (ANLP-NAACL 2000)* (pp. 331–338). Seattle, WA: Association for Computational Linguistics Press.

Cohen, P. R., & Oviatt, S. L. (1995). The role of voice input for human-machine communication. *Proceedings of the National Academy of Sciences, 92*(22), 9921–9927. Washington, DC: National Academy of Sciences Press.

Condon, W. S. (1988). An analysis of behavioral organization. *Sign Language Studies, 58*, 55–88.

Dahlbäck, N., Jönsson, A., & Ahrenberg, L. (1992, January). Wizard of Oz studies—Why and how. In Wayne D. Gray, William E. Hefley, &

Dianne Murray (Eds.), *Proceedings of the International Workshop on Intelligent User Interfaces* (pp. 193-200). New York: ACM Press.

Denecke, M., & Yang, J. (2000). Partial information in multimodal dialogue. *Proceedings of the International Conference on Multimodal Interaction* (pp. 624-633). Beijing, China.

Duncan, L., Brown, W., Esposito, C., Holmback, H., & Xue, P. (1999). Enhancing virtual maintenance environments with speech understanding. Boeing M&CT TechNet, Intranet publication.

Dupont, S., & Luettin, J. (2000, September). Audio-visual speech modeling for continuous speech recognition. *IEEE Transactions on Multimedia, 2*(3), 141-151. Piscataway, NJ: Institute of Electrical and Electronics Engineers.

Ekman, P. (1992, January). Facial expressions of emotion: New findings, new questions. *American Psychological Society, 3*(1), 34-38.

Ekman, P., & Friesen, W. (1978). *Facial action coding system.* Palo Alto, CA: Consulting Psychologists Press.

Fell, H., Delta, H., Peterson, R., Ferrier, L., Mooraj, Z., & Valleau, M. (1994). Using the baby-babble-blanket for infants with motor problems. *Proceedings of the Conference on Assistive Technologies (ASSETS '94)* (pp. 77-84). Marina del Rey, CA.

Fridlund, A. (1994). *Human facial expression: An evolutionary view.* New York: Academic Press.

Fukumoto, M., Suenaga, Y., & Mase, K. (1994). Finger-pointer: Pointing interface by image processing. *Computer Graphics, 18*(5), 633-642.

Fuster-Duran, A. (1996). Perception of conflicting audio-visual speech: an examination across Spanish and German, In D. G. Stork & M. E. Hennecke (Eds.), *Speechreading by humans and machines: Models, systems and applications* (pp. 135-143). New York: Springer Verlag.

Hadar, U., Steiner, T. J., Grant, E. C., & Clifford Rose, F. (1983). Kinematics of head movements accompanying speech during conversation. *Human Movement Science, 2*, 35-46.

Hauptmann, A. G. (1989). Speech and gestures for graphic image manipulation. *Proceedings of the Conference on Human Factors in Computing Systems (CHI '89)* (Vol. 1, pp. 241-245). New York: ACM Press.

Holzman, T. G. (1999). Computer-human interface solutions for emergency medical care. *Interactions, 6*(3), 13-24.

Huang, X., Acero, A., Chelba, C., Deng, L., Duchene, D., Goodman, J., Hon, H., Jacoby, D., Jiang, L., Loynd, R., Mahajan, M., Mau, P., Meredith, S., Mughal, S., Neto, S., Plumpe, M., Wang, K., & Wang, Y. (2000). MiPad: A next-generation PDA prototype. *Proceedings of the International Conference on Spoken Language Processing (ICSLP 2000)* (Vol. 3, pp. 33-36). Beijing, China: Chinese Military Friendship Publishers.

Johnston, M., Cohen, P. R., McGee, D., Oviatt, S. L., Pittman, J. A., & Smith, I. (1997). Unification-based multimodal integration. *Proceedings of the 35th Annual Meeting of the Association for Computational Linguistics* (pp. 281-288). San Francisco: Morgan Kaufmann.

Karshmer, A. I., & Blattner, M. (organizers). (1998). *Proceedings of the 3rd International ACM Proceedings of the Conference on Assistive Technologies (ASSETS '98).* Marina del Rey, CA. Available: http://www.acm.org/sigcaph/assets/assets98/assets98index.html

Kendon, A. (1980). Gesticulation and speech: Two aspects of the process of utterance. In M. Key (Ed.), *The relationship of verbal and nonverbal communication* (pp 207-227). The Hague: Mouton.

Kobsa, A., Allgayer, J., Reddig, C., Reithinger, N., Schmauks, D., Harbusch, K., & Wahlster, W. (1986). Combining deictic gestures and natural language for referent identification. *Proceedings of the 11th International Conference on Computational Linguistics* (pp. 356-361). Bonn, Germany.

Koons, D., Sparrell, C., & Thorisson, K. (1993). Integrating simultaneous input from speech, gaze, and hand gestures. In M. Maybury (Ed.),

Intelligent multimedia interfaces (pp. 257-276). Cambridge, MA: MIT Press.

Kricos, P. B. (1996). Differences in visual intelligibility across talkers, In D. G. Stork & M. E. Hennecke (Eds.), *Speechreading by humans and machines: Models, systems and applications* (pp. 43-53). New York: Springer Verlag.

Kumar, S., & Cohen, P. R. (2000). Towards a fault-tolerant multiagent system architecture. *Fourth International Conference on Autonomous Agents 2000* (pp. 459-466). Barcelona, Spain: ACM Press.

Leatherby, J. H., & Pausch, R. (1992, July). Voice input as a replacement for keyboard accelerators in a mouse-based graphical editor: An empirical study. *Journal of the American Voice Input/Output Society, 11*(2), 69-76.

MacLeod, A., & Summerfield, Q. (1987). Quantifying the contribution of vision to speech perception in noise. *British Journal of Audiology, 21*, 131-141.

Martin, D. L., Cheyer, A. J., & Moran, D. B. (1999). The open agent architecture: A framework for building distributed software systems. *Applied Artificial Intelligence, 13*, 91-128.

Massaro, D. W. (1996). Bimodal speech perception: A progress report, In D. G. Stork & M. E. Hennecke (Eds.), *Speechreading by humans and machines: Models, systems and applications* (pp. 79-101). New York: Springer Verlag.

Massaro, D. W., & Cohen, M. M. (1990, January). Perception of synthesized audible and visible speech. *Psychological Science, 1*(1), 55-63.

Massaro, D. W., & Stork, D. G. (1998). Sensory integration and speechreading by humans and machines. *American Scientist, 86*, 236-244.

McGrath, M., & Summerfield, Q. (1985). Intermodal timing relations and audio-visual speech recognition by normal-hearing adults. *Journal of the Acoustical Society of America, 77*(2), 678-685.

McGurk, H., & MacDonald, J. (1976). Hearing lips and seeing voices, *Nature, 264*, 746-748.

McNeill, D. (1992). *Hand and mind: What gestures reveal about thought.* Chicago: University of Chicago Press.

Meier, U., Hürst, W., & Duchnowski, P. (1996). Adaptive bimodal sensor fusion for automatic speechreading. *Proceedings of the International Conference on Acoustics, Speech and Signal Processing (IEEE-ICASSP)* (pp. 833-836). New York: IEEE Press.

Morimoto, C., Koons, D., Amir, A., Flickner, M., & Zhai, S. (1999). Keeping an eye for HCI. *Proceedings of SIBGRAPI '99, XII Brazilian Symposium on Computer Graphics and Image Processing* (pp. 171-176). State University of Campinas, São Paulo, Brazil.

Naughton, K. (1996). Spontaneous gesture and sign: A study of ASL signs co-occurring with speech. In L. Messing (Ed.), *Proceedings of the Workshop on the Integration of Gesture in Language & Speech* (pp. 125-134). Newark, DE: University of Delaware.

Neal, J. G., & Shapiro, S. C. (1991). Intelligent multimedia interface technology. In J. Sullivan & S. Tyler (Eds.), *Intelligent user interfaces* (pp. 11-43). New York: ACM Press.

Negroponte, N. (1978, December). *The media room.* Report for ONR and DARPA. Cambridge, MA: MIT, Architecture Machine Group.

Neti, C., Iyengar, G., Potamianos, G., & Senior, A. (2000). Perceptual interfaces for information interaction: Joint processing of audio and visual information for human-computer interaction. In B. Yuan, T. Huang, & X. Tang (Eds.), *Proceedings of the International Conference on Spoken Language Processing (ICSLP '2000)* (Vol. 3, pp. 11-14). Beijing, China: Chinese Friendship Publishers.

Oviatt, S. L. (1997). Multimodal interactive maps: Designing for human performance [Special issue]. *Human-Computer Interaction, 12*, 93-129.

Oviatt, S. L. (1999a). Mutual disambiguation of recognition errors in a multimodal architecture. *Proceedings of the Conference on Human*

Factors in Computing Systems (CHI '99) (pp. 576–583). New York: ACM Press.

Oviatt, S. L. (1999b) Ten myths of multimodal interaction. *Communications of the ACM, 42*(11), 74–81. New York: ACM Press. (Translated into Chinese by Jing Qin and published in the Chinese journal *Computer Application.*)

Oviatt, S. L. (2000a). Multimodal system processing in mobile environments. *Proceedings of the Thirteenth Annual ACM Symposium on User Interface Software Technology (UIST 2000)* (pp. 21–30). New York: ACM Press.

Oviatt, S. L. (2000b). Taming recognition errors with a multimodal architecture. *Communications of the ACM, 43*(9), 45–51.

Oviatt, S. L. (2000c). Multimodal signal processing in naturalistic noisy environments. In B. Yuan, T. Huang, & X. Tang (Eds.), *Proceedings of the International Conference on Spoken Language Processing (ICSLP 2000)* (Vol. 2, pp. 696–699). Beijing, China: Chinese Friendship Publishers.

Oviatt, S. L., Bernard, J., & Levow, G. (1999). Linguistic adaptation during error resolution with spoken and multimodal systems [Special issue]. *Language and Speech, 41*(3–4), 415–438.

Oviatt, S. L., & Cohen, P. R. (1991). Discourse structure and performance efficiency in interactive and noninteractive spoken modalities. *Computer Speech and Language, 5*(4), 297–326.

Oviatt, S. L., & Cohen, P. R. (2000 March). Multimodal systems that process what comes naturally. *Communications of the ACM, 43*(3), 45–53.

Oviatt, S. L., Cohen, P. R., Fong, M. W., & Frank, M. P. (1992). A rapid semi-automatic simulation technique for investigating interactive speech and handwriting. *Proceedings of the International Conference on Spoken Language Processing, 2,* 1351–1354.

Oviatt, S. L., Cohen, P. R., & Wang, M. Q. (1994). Toward interface design for human language technology: Modality and structure as determinants of linguistic complexity. *Speech Communication, 15,* 283–300.

Oviatt, S. L., Cohen, P. R., Wu, L.,Vergo, J., Duncan, L., Suhm, B., Bers, J., Holzman, T., Winograd, T., Landay, J., Larson, J., & Ferro, D. (2000). Designing the user interface for multimodal speech and gesture applications: State-of-the-art systems and research directions. *Human Computer Interaction, 15*(4), 263–322. Reprinted in J. Carroll (Ed.), *Human-Computer Interaction in the New Millennium* (ch. 19, 421–456.). Boston: Addison-Wesley Press.

Oviatt, S. L., DeAngeli, A., & Kuhn, K. (1997). Integration and synchronization of input modes during multimodal human-computer interaction. *Proceedings of the Conference on Human Factors in Computing Systems (CHI '97)* (pp. 415–422). New York: ACM Press.

Oviatt, S. L., & Kuhn, K. (1998). Referential features and linguistic indirection in multimodal language. *Proceedings of the International Conference on Spoken Language Processing, 6,* 2339–2342.

Oviatt, S. L., & Olsen, E. (1994). Integration themes in multimodal human-computer interaction. In K., Shirai, S., Furui, & K. Kakehi, (Eds.), *Proceedings of the International Conference on Spoken Language Processing, 2,* 551–554.

Oviatt, S. L., & van Gent, R. (1996). Error resolution during multimodal human-computer interaction. *Proceedings of the International Conference on Spoken Language Processing, 2,* 204–207.

Pankanti, S., Bolle, R. M., & Jain, A. (Eds.). (2000). Biometrics: The future of identification. *Computer, 33*(2), 46–80.

Pavlovic, V., Berry, G., & Huang, T. S. (1997). Integration of audio/visual information for use in human-computer intelligent interaction. *Proceedings of the IEEE International Conference on Image Processing* (pp. 121–124). New York: IEEE Press.

Pavlovic, V., & Huang, T. S. (1998). Multimodal prediction and classification on audio-visual features. *AAAI '98 Workshop on Representations for Multi-modal Human-Computer Interaction* (pp. 55–59). Menlo Park, CA: AAAI Press.

Pavlovic, V., Sharma, R., & Huang, T. (1997). Visual interpretation of hand gestures for human-computer interaction: A review. *IEEE Transactions on Pattern Analysis and Machine Intelligence, 19*(7), 677–695.

Petajan, E. D. (1984). *Automatic lipreading to enhance speech recognition,* Ph.D. thesis, University of Illinois at Urbana-Champaign.

Poddar, I., Sethi, Y., Ozyildiz, E., & Sharma, R. (1998, November). Toward natural gesture/speech HCI: A case study of weather narration. In M. Turk, (Ed.), *Proceedings of the 1998 Workshop on Perceptual User Interfaces (PUI '98)* (pp. 1–6). San Francisco, CA.

Robert-Ribes, J., Schwartz, J.-L., Lallouache, T., & Escudier, P. (1998). Complementarity and synergy in bimodal speech: Auditory, visual, and auditory-visual identification of French oral vowels in noise. *Journal of the Acoustical Society of America, 103*(6), 3677–3689.

Rogozan, A., & Deglise, P. (1998). Adaptive fusion of acoustic and visual sources for automatic speech recognition. *Speech Communication, 26*(1–2), 149–161.

Rubin, P., Vatikiotis-Bateson, E., & Benoit, C. (Eds.). (1998). Audio-visual speech processing [Special issue]. *Speech Communication, 26,* 1–2.

Rudnicky, A., & Hauptman, A. (1992). Multimodal interactions in speech systems. In M. Blattner & R. Dannenberg (Eds.), *Multimedia interface design* (pp. 147–172). New York: ACM Press.

Sekiyama, K., & Tohkura, Y. (1991). McGurk effect in non-English listeners: Few visual effects for Japanese subjects hearing Japanese syllables of high auditory intelligibility. *Journal of the Acoustical Society of America, 90,* 1797–1805.

Seneff, S., Goddeau, D., Pao, C., & Polifroni, J. (1996). Multimodal discourse modelling in a multi-user multi-domain environment. In T. Bunnell & W. Idsardi (Eds.), *Proceedings of the International Conference on Spoken Language Processing* (Vol. 1, pp. 192–195). Newark, DE: University of Delaware & A. I. duPont Institute.

Shaikh, A., Juth, S., Medl, A., Marsic, I., Kulikowski, C., & Flanagan, J. (1997). An architecture for multimodal information fusion. *Proceedings of the Workshop on Perceptual User Interfaces (PUI '97)* (pp. 91–93). Banff, Alberta, Canada.

Sharma, R., Huang, T. S., Pavlovic, V. I., Schulten, K., Dalke, A., Phillips, J., Zeller, M., Humphrey, W., Zhao, Y., Lo, Z., & Chu, S. (1996, August). Speech/gesture interface to a visual computing environment for molecular biologists. *Proceedings of 13th International Conference on Pattern Recognition (ICPR '96)* (Vol. 3, pp. 964–968). Vienna, Austria.

Sharma, R., Pavlovic, V. I., & Huang, T. S. (1998). Toward multimodal human-computer interface [Special issue]. *Proceedings of the IEEE, 86*(5), 853–860.

Silsbee, P. L., & Su, Q. (1996). Audiovisual sensory intergration using Hidden Markov Models. In D. G. Stork & M. E. Hennecke (Eds.), *Speechreading by humans and machines: Models, systems and applications* (pp. 489–504). New York: Springer Verlag.

Siroux, J., Guyomard, M., Multon, F., & Remondeau, C. (1995, May). Modeling and processing of the oral and tactile activities in the Georal tactile system. *International Conference on Cooperative Multimodal Communication, Theory & Applications,* Eindhoven, Netherlands: Springer.

Stork, D. G., & Hennecke, M. E. (Eds.). (1995). *Speechreading by humans and machines.* New York: Springer Verlag.

Suhm, B. (1998). *Multimodal interactive error recovery for non-conversational speech user interfaces,* Ph.D. thesis, Fredericiana University, Germany: Shaker Verlag.

Sumby, W. H., & Pollack, I. (1954). Visual contribution to speech intelligibility in noise. *Journal of the Acoustical Society of America, 26,* 212-215.

Summerfield, A. Q. (1992). Lipreading and audio-visual speech perception, *Philosophical Transactions of the Royal Society of London, Series B, 335,* 71-78.

Tomlinson, M. J., Russell, M. J., & Brooke, N. M. (1996). Integrating audio and visual information to provide highly robust speech recognition. *Proceedings of the International Conference on Acoustics, Speech and Signal Processing (IEEE-ICASSP)* (Vol. 2, pp. 821-824). New York: IEEE Press.

Turk, M., & Robertson, G. (Eds.). (2000). Perceptual user interfaces [Special issue]. *Communications of the ACM, 43*(3), 32-70.

Vatikiotis-Bateson, E., Munhall, K. G., Hirayama, M., Lee, Y. V., & Terzopoulos, D. (1996). The dynamics of audiovisual behavior of speech. In D. G. Stork & M. E. Hennecke (Eds.), *Speechreading by Humans and Machines: Models, Systems, and Applications, Vol. 150 of NATO ASI Series. Series F: Computer and Systems Sciences* (pp. 221-232). Berlin, Germany: Springler-Verlag.

Vo, M. T., Houghton, R., Yang, J., Bub, U., Meier, U., Waibel, A., & Duchnowski, P. (1995, January 22-25). Multimodal learning interfaces. *Proceedings of the DARPA Spoken Language Technology Workshop.* Austin, Texas: Morgan Kauffman Publishers (San Francisco, CA.)

Vo, M. T., & Wood, C. (1996). Building an application framework for speech and pen input integration in multimodal learning interfaces.

Proceedings of the International Conference on Acoustics Speech and Signal Processing (IEEE-ICASSP) (Vol. 6, pp. 3545-3548). New York: IEEE Press.

Wahlster, W. (1991). User and discourse models for multimodal communciation. In J. W. Sullivan & S. W. Tyler (Eds.), *Intelligent user interfaces* (chap. 3, pp. 45-67). New York: ACM Press.

Wahlster, W. (2001, March). SmartKom: Multimodal dialogs with mobile web users. *Proceedings of the Cyber Assist International Symposium* (pp. 33-34). Tokyo: Tokyo International Forum.

Waibel, A., Suhm, B., Vo, M. T., & Yang, J. (1997, April). Multimodal interfaces for multimedia information agents. *Proceedings of the International Conference on Acoustics Speech and Signal Processing (IEEE-ICASSP)* (Vol. 1, pp. 167-170). New York: IEEE Press.

Wang, J. (1995). Integration of eye-gaze, voice and manual response in multimodal user interfaces. *Proceedings of the IEEE International Conference on Systems, Man and Cybernetics* (pp. 3938-3942). New York: IEEE Press.

Wickens, C. D., Sandry, D. L., & Vidulich, M. (1983). Compatibility and resource competition between modalities of input, central processing, and output. *Human Factors, 25,* 227-248.

Wu, L., Oviatt, S., &. Cohen, P. (1999). Multimodal integration—A statistical view. *IEEE Transactions on Multimedia, 1*(4), 334-341.

Zhai, S., Morimoto, C., & Ihde, S. (1999). Manual and gaze input cascaded (MAGIC) pointing. *Proceedings of the Conference on Human Factors in Computing Systems (CHI '99)* (pp. 246-253). New York: ACM Press.

ADAPTIVE INTERFACES AND AGENTS

Anthony Jameson
DFKI, German Research Center for Artificial Intelligence, and International University in Germany

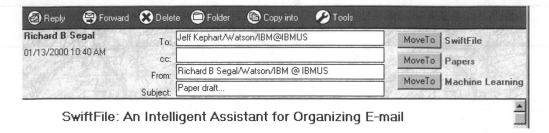

SwiftFile: An Intelligent Assistant for Organizing E-mail

FIGURE 15.1. Screen shot from SWIFTFILE, showing its three shortcut buttons for the filing of the current e-mail message. Note. From "Incremental Learning in SwiftFile," by R. B. Segal and J. O. Kephart, 2000, in P. Langley (Ed.), *Machine Learning: Proceedings of the 2000 International Conference*, San Francisco: Morgan Kaufmann. Copyright 2000 by the authors. Adapted with permission.

INTRODUCTION

As its title suggests, this chapter covers a broad range of interactive systems. But they all have one idea in common: that it can be worthwhile for a system to learn something about each individual user and adapt its behavior to them in some nontrivial way.

The perception and discussion of systems based on this idea has tended to be dominated by extreme examples, both real and hypothetical. So we should start with a look at a relevant system that is more typical of current trends (Fig. 15.1). SWIFTFILE (Segal & Kephart, 2000) is designed to expedite the tedious task of filing incoming e-mail messages into folders. By observing and analyzing the way an individual user files messages, the system learns to predict the three most likely folders for any new message. It enhances the usual e-mail interface with three buttons that name the most likely folders. If the user notices that one of these guesses is correct, she can click on the corresponding button, saving herself the mental and physical effort of selecting the correct folder via the usual methods. If all of the guesses are wrong (a relatively rare event; cf. Simulations Using Data From a Nonadaptive System)—or if the user simply does not wish to pay any attention to the buttons—she can file the message in the usual way.

Concepts. The key idea embodied in SWIFTFILE and the other systems discussed in this chapter is that of *adaptation to the individual user*. Depending on their function and form, systems that adapt to their users have been given labels ranging from *adaptive interfaces* through *user modeling systems* to *software agents* or *intelligent agents*. Starting in the late 1990s, the broader term *personalization* became popular, especially in connection with commercially deployed systems. To be able to discuss the common issues that all of these systems raise, we will refer to them as *user-adaptive systems* (UASs). To simplify exposition, we will use the symbol S to refer to an interactive computing system or device and U to refer to its user. Figure 15.2 introduces some concepts that can be applied to any UAS; Fig. 15.3 shows the form that they take in SWIFTFILE.

A UAS makes use of some type of information about the current individual user U, such as the choices U has made when filing messages into folders. In the process of *user model acquisition*, S performs some type of learning and/or inference on the basis of the information about U to arrive at some sort of *user model*, which in general concerns only limited aspects of U (such as her mail-filing habits). In the process of *user-model application*, S applies the user model to the relevant features of the current situation to determine how to adapt its behavior to U.

A UAS can be defined as: An interactive system that adapts its behavior to individual users on the basis of processes of user model acquisition and application that involve some form of learning, inference, or decision making. This definition distinguishes UASs from *adaptable* systems: Ones that the individual user can explicitly tailor to her own preferences (e.g., by choosing options that determine the appearance of the user interface). The relationship between adaptivity and adaptability will be discussed later.

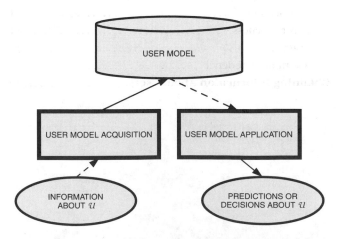

FIGURE 15.2. General schema for the processing in a user-adaptive system. Ovals: input or output; rectangles: processing methods; cylinder: stored information; dotted arrows: use of information; solid arrows: production of results.

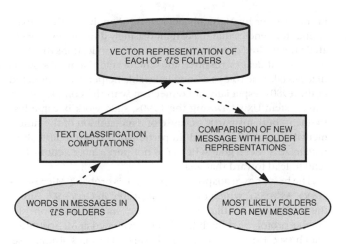

FIGURE 15.3. Application of the schema of Fig. 15.2 to the example of SWIFTFILE.

Chapter Preview. The next two sections in this chapter address the question, "What can user-adaptivity be good for?" They examine in turn nine different functions that can be served by user-adaptivity, giving examples ranging from familiar commercially deployed systems to research prototypes. The following section discusses some usability challenges that are especially important in connection with UASs, challenges that have stimulated most of the controversy that has surrounded these systems. The two following sections consider two key design decisions: What types of information about each user should be collected, and what techniques should be used for the processes of learning, inference, and decision making that are involved in user model acquisition and application? The next-to-last section looks at several approaches to the empirical study of UASs, and the concluding section comments on the reasons why their importance is likely to continue to grow.

FUNCTIONS: SUPPORTING SYSTEM USE

Some of the ways in which user-adaptivity can be helpful involve support for a user's efforts to operate a system successfully and effectively. This section considers four types of support.

Taking Over Parts of Routine Tasks

The first function of adaptation was illustrated by SWIFTFILE. Systems in this category take over some of the work that $\mathcal{U}$ would normally have to perform herself—routine tasks that may place heavy demands on a user's time, although typically not on her intelligence or knowledge. Typical tasks of this sort include e-mail management (see, e.g., Maes, 1994) and appointment scheduling (see, e.g., Mitchell, Caruana, Freitag, McDermott, & Zabowski, 1994).

The primary benefits of this form of adaptation are savings of time and effort for $\mathcal{U}$. The potential benefits are greatest where $\mathcal{S}$ can perform the entire task without input from $\mathcal{U}$. In most cases, however, $\mathcal{U}$ is kept in the loop (as with SWIFTFILE), because $\mathcal{S}$'s ability to predict what $\mathcal{U}$ would want done is limited.

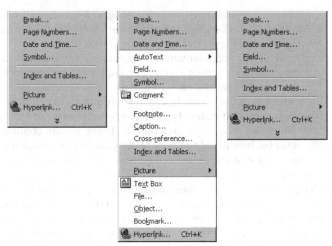

FIGURE 15.4. Example of adaptation in SMART MENUS. $\mathcal{U}$ accesses the "Insert" menu. Not finding the desired option, $\mathcal{U}$ clicks on the extension arrows and selects the "Field" option. When $\mathcal{U}$ later accesses the same menu, "Field" now appears in the main section.

Adapting the Interface

A different way of helping a person to use a system more effectively is to adapt the user interface so that it fits better with $\mathcal{U}$'s way of working with the system. Interface elements that have been adapted in this way include menus, icons, and the system's processing of signals from input devices such as keyboards.

A recent example is provided by the SMART MENUS feature that Microsoft introduced in Windows 2000. Figure 15.4 illustrates the basic mechanism (cf. Fig. 15.5): An infrequently used menu option is initially hidden from view; it appears in the main part of a menu only after $\mathcal{U}$ has selected it for the first time. (It will be removed later if $\mathcal{U}$ does not select it often enough.) The idea is that, in the long run, the menus should contain just the items that $\mathcal{U}$ accesses regularly, so that $\mathcal{U}$ needs to spend less time searching within menus.

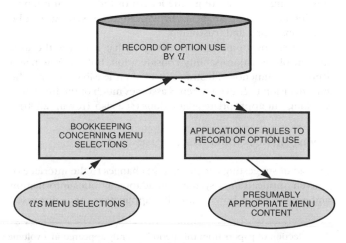

FIGURE 15.5. Overview of adaptation in SMART MENUS.

Although no published evaluation of SMART MENUS is available at the time of this writing, some relevant earlier research had been conducted before the appearance of SMART MENUS. For example, Sears and Shneiderman (1994) examined *split menus*, a different way of separating a menu into high- and low-frequency regions, with the several most frequently used options appearing at the top of the menu. In a field study and an experiment, split menus yielded generally positive results in terms of selection times and subjective preferences. These results confirm the potential benefits of this type of organization; but since the menus in these studies remained constant for each user, the results do not reflect the problems that can arise when menus change during use. Frequent changes in the arrangement of menu items can produce a strong variant of the general usability problem of *predictability*. As will be discussed later, this usability problem is especially acute for very frequently used interface elements, sometimes to the point of making virtually any form of adaptation undesirable.

One important goal of interface adaptation is to take into account special perceptual or physical impairments of individual users so as to allow them to use a system more efficiently, with minimal errors and frustration (cf. the chapters by Jacko et al., by Sears, and by Stephanidis in this handbook). Trewin and Pain (1998) have developed a method for recommending adjustments to the parameters of a computer keyboard to compensate for several types of physical disability. For example, some users often inadvertently press the same key twice in succession. Trewin and Pain's system includes a mechanism for recognizing this tendency on the basis of $\mathcal{U}$'s normal typing behavior and computing an optimal "bounce key" interval, during which a given key cannot be reactivated. $\mathcal{U}$ is given the option of having the computed interval applied to her keyboard.

Many user interfaces, although not *adaptive*, are *adaptable* (cf. the discussion of this distinction in the chapter by Stephanidis in this handbook). They offer $\mathcal{U}$ the opportunity to specify desired properties of the user interface explicitly—e.g., the aspects of the processing of keyboard input that Trewin and Pain (1998) address. Although *adaptability* is often an attractive alternative to adaptation, the keyboard example illustrates several typical limitations: $\mathcal{U}$ may not know what options exist or how she can set them. She may have no idea what the best setting is for her (e.g., the length of the optimal bounce-key interval), and trial and error with different settings can be time-consuming and frustrating.

One strong point of adaptability is that it leaves the user in control—a major usability consideration. But, as Trewin and Pain's recommender illustrates, it is often feasible to leave the final decision to $\mathcal{U}$ even when $\mathcal{S}$ assumes much of the burden of working out appropriate adaptations (see also Trewin, 2000).

Giving Advice About System Use

Instead of suggesting (or executing) changes to the interface of a given application, a system can adaptively offer information and advice about how to use that application. As is discussed in the chapter by Mehlenbacher in this handbook, there exist various tendencies that make it increasingly difficult for users to attain the desired degree of mastery of the applications that they use. A good deal of research into the development of systems that can take the role of a knowledgeable helper was conducted in the 1980s, especially in connection with the complex operating system UNIX.[1] During the 1990s, such work became less frequent, perhaps partly because of a recognition of the fundamental difficulties involved. In particular, it is often difficult to recognize a user's goal when $\mathcal{U}$ is not performing actions that tend to lead toward that goal.

The best-known adaptive help system is probably Microsoft's OFFICE ASSISTANT, which originally appeared in OFFICE 97. Part of the technology for the OFFICE ASSISTANT was derived from the LUMIÈRE prototype, which had been developed at Microsoft Research (see Horvitz, Breese, Heckerman, Hovel, & Rommelse, 1998). In Fig. 15.6 (cf. Fig. 15.7), LUMIÈRE is proposing help topics on the basis of $\mathcal{U}$'s recent actions—a source of information that can be especially valuable when $\mathcal{U}$ requires some advice but does not know which concepts she could use to find it in either the system itself or a nonadaptive help system. The type of spontaneous intervention shown in Fig. 15.6 can also help $\mathcal{U}$ to expand her knowledge of the system's functionality, even when she is able to perform her tasks with the functions that she is already familiar with.

When adaptive help is given spontaneously, and not just on demand, it can endanger the usability goals of *unobtrusiveness* and *controllability*. Figure 15.6 illustrates one way in which the original LUMIÈRE prototype dealt with this problem, using decision-theoretic methods: The assistance window in Fig. 15.6 appears only if the "Likelihood that help is needed" by $\mathcal{U}$ exceeds a given threshold. The window is initially small, and it includes a "volume control" with which $\mathcal{U}$ can raise or lower the threshold for $\mathcal{S}$'s future interventions. If $\mathcal{U}$ does not hover over the assistance window or interact with it, the window spontaneously vanishes after displaying a brief apology in its title bar.

In the deployed OFFICE ASSISTANT, the decision-theoretic methods for deciding when to offer help or were replaced with a relatively simple rule-based system which, for example, did not take into account $\mathcal{U}$'s level of competence or her willingness to be interrupted. The resulting tendency of the OFFICE ASSISTANT to pop up in "distracting" ways (see, e.g., Schaumburg, 2001) has created a rather distorted impression of the potential value of adaptive help systems. More recent research has continued to explore ways in which UASs can take into account users' cognitive resource limitations with decision-theoretic methods (see, e.g., Horvitz, 1999; Horvitz, Jacobs, & Hovel, 1999; Jameson et al., 2001; Wolfman, Lau, Domingos, & Weld, 2001).

Controlling a Dialogue

Much of the early research on UASs concerned systems that conducted natural language dialogues with their users (see, e.g., Kobsa & Wahlster, 1989). During the 1990s, attention shifted to interaction modalities that were more widely available and that

---

[1] A collection of papers from this period recently appeared in a volume edited by Hegner, McKevitt, Norvig, and Wilensky (2001).

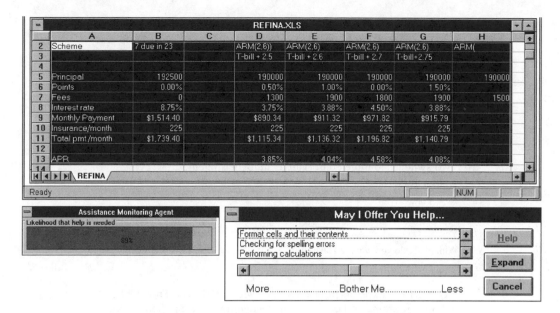

FIGURE 15.6. Example of assistance offered by the LUMIÈRE prototype. *U* has just searched through several menus, selected the entire spread-sheet, and paused. *Note.* From "The Lumière Project: Bayesian User Modeling for Inferring the Goals and Needs of Software Users," by E. Horvitz, J. Breese, D. Heckerman, D. Hovel, and K. Rommelse, 1998, in G. F. Cooper and S. Moral (Eds.), *Uncertainty in Artificial Intelligence: Proceedings of the Fourteenth Conference* (pp. 256–265), San Francisco: Morgan Kaufmann. Copyright 1998 by Morgan Kaufmann Publishers. Adapted with permission.

made it possible in many cases to implement adaptation straightforwardly. Advances in the technology of natural language and speech processing (cf. the chapters by Karat et al. and by Lai & Yankelovich in this handbook) have led to a recent reawakening of interest in user-adaptive dialogue systems (see, e.g., Haller, McRoy, & Kobsa, 1999; Zukerman & Litman, 2001).

Natural language dialogue has served as an interaction modality in connection with most of the functions of user-adaptivity

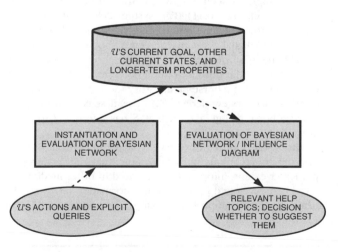

FIGURE 15.7. Overview of adaptation in LUMIÈRE.

discussed in this and the following sections, such as the provision of help and the recommendation of products. But there is also a type of adaptivity that is largely characteristic of natural language dialogue: Adaptation of the system's *dialogue strategy*—a policy for determining when and how S should provide information, acquire information from U, and perform other dialogue acts.

An informative example is TOOT (Litman & Pan, 2000), a prototype spoken dialogue system for retrieving online train schedules. The basic idea underlying the adaptation is that different dialogue strategies are appropriate in different situations. For example, if a given U's speech is understood poorly by S, S may need to adopt a conservative dialogue strategy, acquiring just one piece of information at a time and frequently requesting confirmation. If, on the other hand, U's speech is understood well by S, more efficient dialogue strategies can be used. Figure 15.8 (cf. Fig. 15.9) shows two excerpts from an example dialogue in which S starts with an ambitious, potentially efficient strategy and then, noticing that the dialogue is not progressing well, shifts downward twice until it finally achieves success with a very conservative strategy.

In a controlled empirical evaluation of TOOT's adaptation (discussed later in Controlled Studies), it was shown that TOOT adapted to different extents in different dialogues and that this adaptation led to a much higher rate of successful dialogue completion (65% vs. 23%) than the use of a single dialogue strategy (the first one shown in Fig. 15.8). User-adaptivity also led to a

Beginning with ambitious dialogue strategy:

S: Hi, this is AT&T Amtrak Schedule System. This is TOOT. How may I help you?

U: I need to find a train going to New York City.
(Interpretation by speech recognizer: "DC I don't care on Thursday".)

S: Which city do you want to leave from?

U: Baltimore.

S: Do you want me to find the trains from Baltimore to DC on Thursday at any time now?

U: No

S: ...

After two switches to more conservative dialogue strategies:

S: Which city do you want to go to?

U: New York City.

S: Do you want to go to New York City?

U: Yes.

S: Which city do you want to leave from?

U: ...

FIGURE 15.8. Excerpts from a dialogue with TOOT. *Note.* From "Predicting and Adapting to Poor Speech Recognition in a Spoken Dialogue System," by D. J. Litman and S. Pan, 2000, in *Proceedings of the Seventeenth National Conference on Artificial Intelligence*, Austin, TX (pp. 722–728). Copyright 2000 by the American Association for Artificial Intelligence. Adapted with permission.

slight increase in subjective satisfaction. Some issues that arise in the design and interpretation of controlled studies of this type will be discussed later.

FUNCTIONS: SUPPORTING INFORMATION ACQUISITION

We are constantly hearing that information overload is a typical problem of our age, especially because of the explosive growth of the internet and, in particular, the World Wide Web.

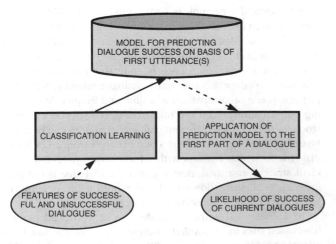

FIGURE 15.9. Overview of adaptation in TOOT.

In addition to the vast number of electronic documents of various sorts, users now have access to a vast number of products available for sale, people that they can get in touch with, and systems that can teach them about some topic. The second major type of function of UASs is to help people to find what they need in a form that they can deal with.

Helping Users to Find Information

We will first look at the broad class of systems that help *U* to find relevant electronic documents, which may range from brief news stories to complex multimedia objects. A relatively novel example of such a system is the ADAPTIVE NEWS SERVER (Fig. 15.10; cf. Fig. 10.11): This system delivers news stories to small, portable computing devices, such as mobile phones and personal digital assistants. Screen A shows an overview of available news stories; the first two are about American football games and the third is about a horse race. After *U* has selected and read the horse racing story (Screen B), *S* generates an adapted overview screen (C): The first two stories now concern horse racing, a sport that *S* has inferred *U* to be interested in.

The systems in this category[2] typically draw from the vast repertoire of techniques for analyzing textual information (and to a lesser extent, information presented in other media) that have been developed in the field of information retrieval. The forms of adaptive support are in part different in three different situations.

[2]Surveys of parts of this large area are provided by, among others, Brusilovsky (1996, 2001), Hanani, Shapira, and Shoval (2001), and Mladenic (1999).

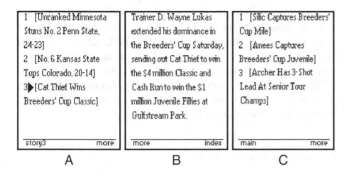

1 [Unranked Minnesota Stuns No. 2 Penn State, 24-23] 2 [No. 6 Kansas State Tops Colorado, 20-14] 3 ▶[Cat Thief Wins Breeders' Cup Classic] story3 more	Trainer D. Wayne Lukas extended his dominance in the Breeders' Cup Saturday, sending out Cat Thief to win the $4 million Classic and Cash Run to win the $1 million Juvenile Fillies at Gulfstream Park. more index	1 [Silic Captures Breeders' Cup Mile] 2 [Anees Captures Breeders' Cup Juvenile] 3 [Archer Has 3-Shot Lead At Senior Tour Champs] main more
A	B	C

FIGURE 15.10. Sequence of three screens presented by the ADAPTIVE NEWS SERVER. Note. From a slide supplied by Michael J. Pazzani. Copyright 2000 by Michael J. Pazzani. Adapted with permission.

Support for Browsing. $\mathcal{U}$ may actively search for desired information by examining information items and pursuing cross-references among them, as in hypermedia systems such as the World Wide Web or in the ADAPTIVE NEWS SERVER. A UAS can help focus $\mathcal{U}$'s browsing activity by recommending or selecting promising items or directions of search, on the basis of what S has been able to infer about $\mathcal{U}$'s information needs. Instead of selecting some documents at the expense of others, as in the ADAPTIVE NEWS SERVER, systems with more communication bandwidth typically highlight recommended hyperlinks and/or provide separate lists of recommendations. (Both of these methods are used, for example, in the WEBWATCHER system of Joachims, Freitag, & Mitchell, 1997.)

Support for Query-Based Search or Filtering. Many systems provide some sort of query mechanism, such as a search engine or a keyword-based filtering mechanism. But explicit queries often only roughly reflect $\mathcal{U}$'s actual information need. A user model constructed on the basis of other sources of information can help to improve the selection of the documents presented to $\mathcal{U}$ and/or the appropriateness of the way in which they are presented (e.g., their sorting in terms of relevance).

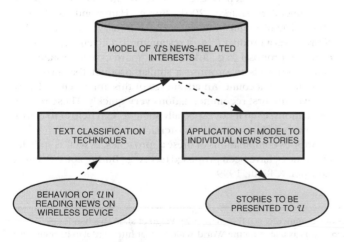

FIGURE 15.11. Overview of adaptation in ADAPTIVE NEWS SERVER.

In one straightforward approach, which is used in the ADAPTIVE NEWS SERVER, S first processes a query in a nonadaptive way and then consults its model of $\mathcal{U}$'s interests when deciding about the further filtering and/or presentation of the results.

Spontaneous Provision of Information. A number of systems present information that may be useful to $\mathcal{U}$ even while $\mathcal{U}$ is simply working on some task, making no effort to find information. For example, WATSON (Budzik, Hammond, & Birnbaum, 2001) monitors the text that a user is typing into a word processor and presents links to relevant documents in a Web browser. (Other systems in this category are described by Maglio, Barrett, Campbell, & Selker, 2000, and by Rhodes, 2000.) The usability challenge of maintaining unobtrusiveness is especially important here, because $\mathcal{U}$ is not actively searching for information (see Maglio & Campbell, 2000, for an experimental study of how best to achieve this goal).

Tailoring Information Presentation

Even in cases where it is clear which document a system should present to $\mathcal{U}$, the best specific way of presenting it may vary from one user to the next. Figure 15.12 shows part of a screen from the tourist information system AVANTI that describes a particular

Personalize the table below.	
Services offered	
Facilities:	Credit cards accepted: Yes Groups allowed: Yes Animals allowed: Yes
Laundry service:	Yes, accessible
Car service to station:	No
Car park:	None
Cash-register service:	Yes, accessible
Lift:	Connects all floors
Shared bathrooms:	Yes
Access to public areas:	Presents no obstacles; 3 stairs into the bar

💡 If you are no longer interested in specific information for wheelchair-bound or dystrophic people, please click here.

FIGURE 15.12. Part of a screen from the AVANTI tourist information system. Note. From "Adaptable and Adaptive Information Provision for All Users, Including Disabled and Elderly People," by J. Fink, A. Kobsa, and A. Nill, 1998, New Review of Hypermedia and Multimedia (Vol. 4, pp. 163–188). Copyright 1998 by Taylor Graham Publishers. Adapted with permission.

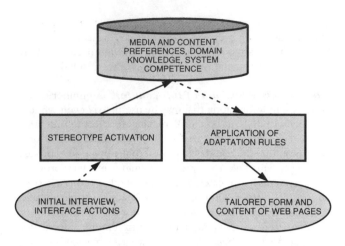

FIGURE 15.13. Overview of adaptation in AVANTI.

hotel in Siena, Italy (Fink, Kobsa, & Nill, 1998). S's current model of U states that U is interested in information for wheelchair-bound or dystrophic visitors; accordingly, information about the accessibility of several parts of the hotel is included. For other users, this information would be omitted on the grounds that it would only clutter the display. This type of variation in presentation can be realized to some extent through explicit adaptation by U. Indeed, in the example U is given the opportunity to "Personalize the table below" by specifying which attributes are to be shown. But the limitations of pure adaptability that were discussed in connection with interface adaptation require this approach to be combined with some spontaneous system adaptivity (cf. also the chapter by Stephanidis in this handbook). In the example in Fig. 15.12 (cf. Fig. 15.13), S has actually inferred from U's previous actions that U may no longer be interested in information for users with mobility limitations; S therefore gives U the option of allowing S to perform further adaptations to eliminate such information.

Evaluations of AVANTI confirmed that mobility-impaired users appreciated the additional accessibility information that S made available and that user-controlled adaptation was taken advantage of mainly by experienced users of the system.

Another class of systems in which the tailoring of information to individual users can be especially beneficial comprises systems that present medical information to patients (see, e.g., Hirst, DiMarco, Hovy, & Parsons, 1997; Jones et al., 1999).

Properties of users that may be taken into account in the tailoring of documents include: U's degree of interest in particular topics; U's knowledge about particular concepts or topics; U's preference or need for particular forms of information presentation (e.g., AVANTI generates textual descriptions in lieu of maps for visually impaired users); and the display capabilities of U's computing device (e.g., web browser vs. cell phone).

Even in cases where it is straightforward to determine the relevant properties of U, the automatic creation of adapted presentations can require sophisticated techniques of natural language generation (see, e.g., Hirst et al., 1997) and/or multimedia presentation generation (see, e.g., André & Rist, 1995). Various less complex ways of adapting hypermedia documents to individual users have also been developed (see Brusilovsky, 1996, Section 6).

Recommending Products

One of the most practically important categories of UAS today comprises the product recommenders that are found in many commercial Web sites and also, increasingly, in mobile information servers.[3] A screen from a typical film recommender system, the MOVIECENTRAL Web site, is shown in Fig. 15.14. The user has already rated 10 films, as is required by S before any recommendation can be made. On the basis of these ratings, S has identified a set of *neighbors* for U—other users with tastes similar to U's. For the movie *2001, A Space Odyssey*, S has generated a "predicted rating" for U by examining the ratings of the subset of U's neighbors who have rated this movie. U, who has seen this movie, has provided feedback by giving her actual rating of it. Another common form of adaptation, also shown in the figure, is the presentation of reviews that have been supplied by similar users. The general approach to adaptation summarized in Fig. 15.15 is called *collaborative filtering*. Many studies have shown that collaborative filtering techniques produce recommendations whose accuracy for an individual user is usefully high.

Some product recommenders require U to specify her evaluation criteria explicitly instead of simply rating individual items. For example, if U is looking for a suitable dog with the help of PERSONALOGIC, the system will ask how important it is to U that the dog should be easy to train (cf., Fig. 15.22). This method offers a natural alternative to collaborative filtering when relatively complex and important decisions are involved for which it is worthwhile for U to think carefully about the attributes of the products in question.

A third approach is exemplified by the FINDME family of recommenders (Burke, 2001; Burke, Hammond, & Young, 1997). The distinguishing feature is an iterative cycle in which S proposes a product (e.g., a restaurant in a given city), U criticizes the proposal (e.g., asking for a "more casual" restaurant), and S proceeds to propose a similar product that takes the critique into account. An advantage of this approach is that U receives the first recommendations very quickly. These recommendations, even if not especially suitable, can help U to clarify her own product evaluation criteria.[4]

Various combinations of these approaches, as well as specific other ideas, have been proposed (see, e.g., Burke, 2001; Schafer, Konstan, & Riedl, 1999).

---

[3]For a more general treatment of the human–computer interaction aspects of e-commerce, see the chapter by Vergo et al. in this handbook.
[4]At the time of this writing, a restaurant recommender from the FINDME family was available on the World Wide Web at http://infolab.ils.nwu.edu/entree/pub/

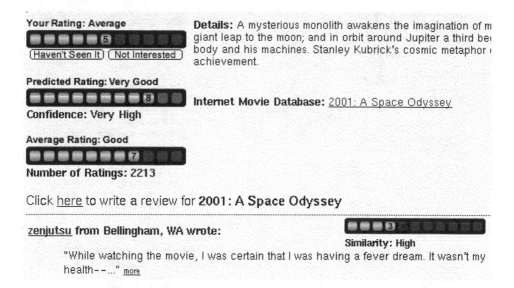

Your Rating: Average

⬛⬛⬛⬛⬛5⬛⬛⬛⬛
(Haven't Seen It) (Not Interested)

Details: A mysterious monolith awakens the imagination of m
giant leap to the moon; and in orbit around Jupiter a third bei
body and his machines. Stanley Kubrick's cosmic metaphor
achievement.

Predicted Rating: Very Good

⬛⬛⬛⬛⬛⬛⬛8⬛⬛
Confidence: Very High

Internet Movie Database: 2001: A Space Odyssey

Average Rating: Good

⬛⬛⬛⬛⬛⬛7⬛⬛⬛
Number of Ratings: 2213

Click here to write a review for **2001: A Space Odyssey**

zenjutsu **from Bellingham, WA wrote:**

⬛⬛⬛3⬛⬛⬛⬛⬛⬛
Similarity: High

"While watching the movie, I was certain that I was having a fever dream. It wasn't my health‒‒..." more

FIGURE 15.14. Part of a screen from the MOVIECENTRAL film recommendation Web site describing the movie *2001, A Space Odyssey*. The screen shot was made from http://www.qrate.com/ in January 2001 and edited for compactness. This Web site is no longer in operation.

Product recommenders address several problems that computer users typically experience when they search for products:

1. U may not know what aspects of the products to attend to or what criteria should determine her decision. Some recommenders either make it less necessary for U to be explicitly aware of her evaluation criteria (as when collaborative filtering is used) or help U to learn about her own criteria during the course of the interaction with S.

2. If U is unfamiliar with the concepts used to characterize the products, she may be unable to make effective use of any search or selection mechanisms that may be provided. Product recommenders generally reduce this communication gap by allowing U to specify her criteria (if this is necessary at all) in terms that are more natural to her.

3. U may have to read numerous product descriptions in various parts of the site, integrating the information found to arrive at a decision. Once a product recommender has acquired an adequate user model, S can take over a large part of this work, often examining the internal descriptions of a much larger number of products than U could deal with herself.

From the point of view of the vendors of the products concerned, the most obvious potential benefit is that users will find one or more products that they consider worth buying, instead of joining the notoriously large percentage of browsers who never become buyers (cf. Schafer et al., 1999). A related benefit is the prospect of cross-selling: S's model of U can be used for the recommendation of further products that U might not have considered herself. Finally, some vendors aim to build up customer loyalty with recommenders that acquire long-term models of individual customers: If U believes that S has acquired an adequate model of her, U will tend to prefer to use S again rather than starting from scratch with some other system.

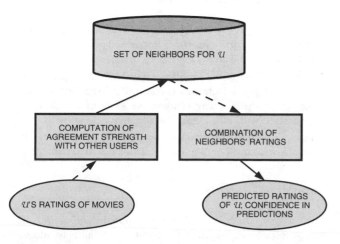

FIGURE 15.15. Overview of adaptation in MOVIECENTRAL.

Supporting Collaboration

The increasing tendency for computer users to be linked via networks has made it increasingly feasible for users to collaborate, even in a spontaneous way and without prior acquaintance. A system that has models of a large number of users can facilitate such collaboration by taking into account the ways in which users match or complement each other.

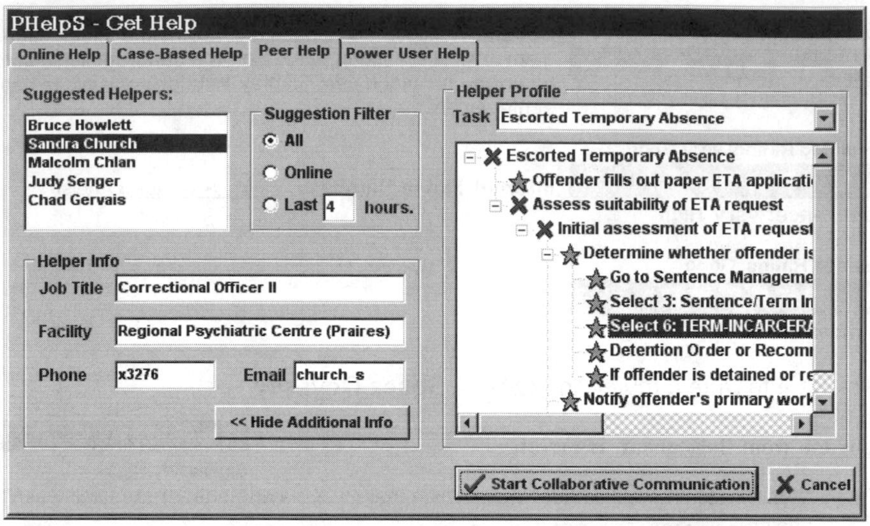

FIGURE 15.16. Screen shot from the PHELPS system. $\mathcal{U}$ is having difficulty arranging an escorted temporary absence for a prisoner. $\mathcal{S}$ offers information on a number of possible helpers at various places in Canada. The window in the right-hand side of the screen shows a profile of a potential helper's knowledge of the task in question. *Note.* From "Supporting Peer Help and Collaboration in Distributed Workplace Environments," by J. E. Greer, G. I. McCalla, J. A. Collins, V. S. Kumar, P. Meagher, and J. Vassileva, 1998, *International Journal of AI and Education* (Vol. 9, pp. 159–177). Copyright 1998 by The International Artificial Intelligence in Education Society. Adapted with permission.

A well-known example of a system of this sort is PHELPS (see, e.g., Collins et al., 1997; Greer et al., 1998). PHELPS is part of the Offender Management System used in the Correctional Services of Canada. It is designed to support relatively inexperienced workers (e.g., trainees) who are not sure how to handle a particular task that they are working on (see, e.g., the screen in Fig. 15.16 and the schema in Fig. 15.17). The system suggests helpers who (a) have the relevant specific knowledge, (b) are available to provide help in the time frame required, (c) have not been overburdened with other help requests in the recent past; and (d) have other relevant positive characteristics (e.g., speaking the same language as $\mathcal{U}$). PHELPS was found to work as expected and to be accepted by representative potential users in a small-scale study with four trainees (Greer et al., 1998).

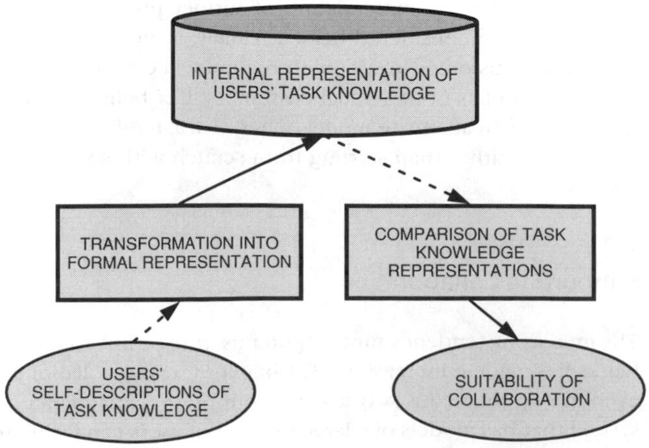

User modeling has been applied in connection with several (partially overlapping) types of collaboration:

- In computer-supported learning environments, in which the idea of *collaborative learning* has gained popularity in recent years (see, e.g., Paiva, 1997).
- As a way of providing "intelligent help" for complex tasks (see, e.g., Vivacqua & Lieberman, 2000). Putting a human expert into the loop is a way of avoiding some of the difficulties associated with fully automatic adaptive help systems.
- In environments for computer-supported cooperative work within organizations (see, e.g., McDonald & Ackerman, 2000).

Supporting Learning

Research on *student modeling*—or *learner modeling*, as it has been called more often in recent years—aims to add user-adaptivity to computer-based tutoring systems and learning

FIGURE 15.17. Overview of adaptation in PHELPS.

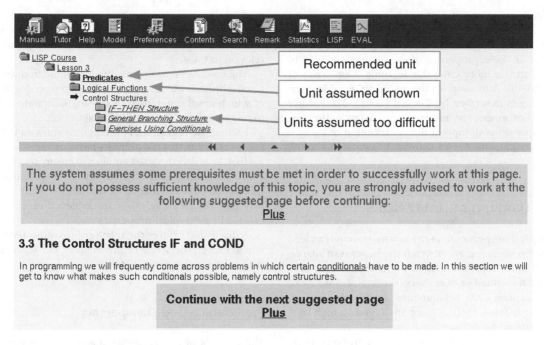

FIGURE 15.18. Example screen from ELM-ART showing the system's assessment of the suitability of particular learning units for the current user. Screen shot made from http://www.psychologie. uni–trier.de:8000/elmart in December 2000. In the actual system, the different colors of the folder icons are clearly distinguishable. Adapted with the permission of Gerhard Weber.

environments (cf. Corbett & Koedinger, 1997, and the chapter by Emurian & Durham in this handbook).[5]

Increasingly, learning environments are being made available on the World Wide web. An example is ELM-ART (Weber & Specht, 1997), which teaches users the programming language LISP. Figure 15.18 (see also Fig. 15.19) illustrates just one of the system's adaptive functions: the way in which it guides learners to parts of a course that it would be appropriate for them to study at a given time, given the skills that they have been observed to possess so far. S signals the suitability of learning units to U in two different ways, without restricting U's freedom to explore the learning environment on her own. Link annotations (realized here as color-coded folders) indicate the extent to which a visit to a given unit is recommended. The button at the bottom of the screen recommends a single unit that it would be especially appropriate for U to visit next.

Interaction in *intelligent tutoring systems* and *intelligent learning environments* can take many forms, ranging from tightly system-controlled tutoring to largely free exploration by the learner. In addition to navigation support (illustrated here by ELM-ART), aspects of the system that can be adapted to the individual user include: (a) the selection and form of the instructional information presented; (b) the content of problems and tests; and (c) the content and timing of hints and feedback.

Learner modeling systems may adapt their behavior to any of a broad variety of aspects of the user, such as: (a) U's knowledge of the domain of instruction, including knowledge acquired prior to and during the use of S; (b) U's learning style, motivation, and general way of looking at the domain

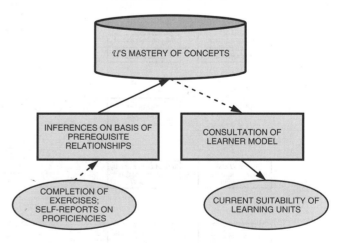

FIGURE 15.19. Overview of adaptation in ELM-ART.

[5]Good sources of literature include the *International Journal of Artificial Intelligence in Education* and the proceedings of the biennial Conferences on Artificial Intelligence in Education (see, e.g., Lajoie & Vivet, 1999).

in question; and (c) the details of $\mathcal{U}$'s current processing of a problem.

The underlying assumption is that the adaptation of $\mathcal{S}$'s behavior to some of these properties of the learner can lead to more effective and/or more enjoyable learning. One series of studies that directly demonstrates the added value of learner-adaptive tutoring is described by Corbett (2001). Many other evaluation studies assess the effectiveness of an entire system and therefore do not pinpoint the contribution of learner-adaptivity (see, e.g., Corbett, McLaughlin, & Scarpinatto, 2000, p. 91).

USABILITY CHALLENGES

Some of the typical properties of user-adaptive systems can lead to usability problems that may outweigh the benefits of adaptation to the individual user. Discussions of these problems have been presented by a number of authors (see, e.g., Höök, 2000; Lanier, 1995; Norman, 1994; Schaumburg, 2001; Shneiderman, 1995; Wexelblat & Maes, 1997). Figure 15.20 gives a high-level summary of many of the relevant ideas.

The USABILITY GOALS shown in the third column correspond to several generally desirable properties of interactive systems. Those listed in the top three boxes (PREDICTABILITY AND TRANSPARENCY, CONTROLLABILITY, and UNOBTRUSIVENESS) correspond to general usability principles (see, e.g., the chapters by Stewart

& Travis, by Cockton et al., and by van der Veer & Puerta Melguizo in this handbook). The remaining two goals, maintenance of PRIVACY and of BREADTH OF EXPERIENCE, are especially relevant to UASs.

The column TYPICAL PROPERTIES lists some frequently encountered (although not always necessary) properties of UASs, each of which has the potential of causing difficulties with respect to one or more of the usability goals.

Each of the remaining two columns shows a different strategy for ensuring that the usability goals are nonetheless realized: Each of the PREVENTIVE MEASURES aims to ensure that a typical property is not present in such a way that it would cause problems. Each of the COMPENSATORY MEASURES aims to ensure in some other way that one or more goals are achieved despite the threats created by the typical properties.

A discussion of all of the relationships indicated in Fig. 15.20 would exceed the scope of this chapter, but some remarks will help to clarify the main ideas.

Predictability and Transparency

The concept of *predictability* refers to the extent to which a user can predict the effects of her actions. *Transparency* is the extent to which she can understand system actions and/or has a clear picture of how the system works (cf. the chapter by van der Veer & Puerta Melguizo in this handbook). These properties are

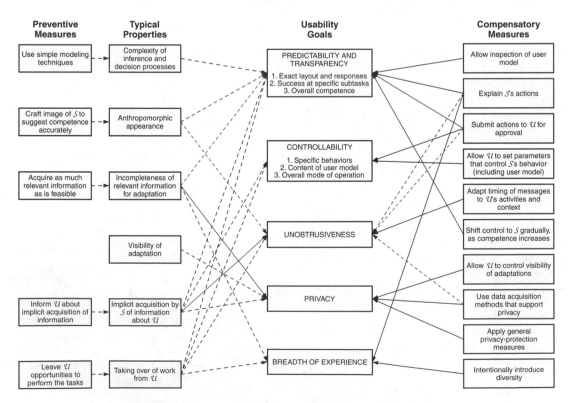

FIGURE 15.20. Overview of usability challenges for user-adaptive systems. Solid and dashed arrows denote positive and negative causal influences, respectively; further explanation is given in the text.

grouped together here because they are associated with largely the same set of other variables.

As the numbered items in the box PREDICTABILITY AND TRANSPARENCY indicate, users can try to predict and understand a system on several different levels of detail.

1. *Exact layout and responses:* Especially detailed predictability is important when interface elements are involved that are accessed frequently by skilled users, for example, icons in control panels or options in menus. If the layout and behavior of the system is highly predictable—in fact, essentially identical—over time, skilled users can engage in *automatic processing* (see, e.g., Hammond, 1987): They can use the parts of the interface quickly, accurately, and with little or no attention. In this situation, even minor deviations from complete predictability on a fine-grained level can have the serious consequence of making automatic processing impossible or error-prone.

2. *Success at specific subtasks:* Users may desire only more global predictability and transparency when S is performing some more or less complex task on U's behalf (e.g., searching for suitable products on the web): In the extreme case, S may want only to predict (or evaluate) the quality of the result of a complex system action.

3. *Overall competence:* The most global form of predictability and transparency concerns U's ability to assess S's overall level of competence—the degree to which S tends in general to perform its tasks successfully. With many types of system, high overall competence can be taken for granted; but as we have seen, the processes of acquiring and applying user models do not in general ensure a high degree of accuracy. If U seriously overestimates S's competence, she may rely on S excessively; if she underestimates S, she will not derive the potential benefits that S can provide. A factor that is especially important with regard to this global level is the way in which the adaptive part of S is presented to U. Some UASs (such as the OFFICE ASSISTANT) have used lifelike characters, for various reasons. As has often been pointed out, such anthropomorphic representations can invoke unrealistically high expectations concerning system competence, not only with regard to capabilities like natural language understanding, but also with regard to S's ability to understand and adapt to U.

In general, the levels and degrees of predictability and transparency that are necessary or desirable in a given case can depend on many factors, including the function that is being served by the adaptation and U's level of skill and experience. The same is true of the choice of the measures that are most appropriate for the achievement of predictability and transparency.

Controllability

Controllability refers to the extent to which U can bring about or prevent particular actions or states of S if she has the goal of doing so. Although controllability tends to be enhanced by transparency and predictability, these properties are not perfectly correlated. For example, when U clicks on a previously unused option in SMART MENUS, she can predict with certainty that it will be moved to the main part of its menu; but U has no control over whether this change will be made—except through the drastic step of deactivating the entire SMART MENUS mechanism.

A typical measure for ensuring some degree of control is to have S submit any action with significant consequences to U for approval. This measure may have negative impact on the usability goal of *unobtrusiveness*; so it is an important interface design challenge to find ways of making recommendations in an unobtrusive fashion that still makes it easy for U to notice and follow up on them (cf. Figs. 15.1, 15.6, and 15.12).

Like predictability and transparency, controllability can be achieved on various levels of granularity (see the items in the box labeled CONTROLLABILITY in Fig. 15.20). Especially since the enhancement of controllability can come at a price, it is important to consider what kinds of control will really be desired. For example, there may be little point in submitting individual actions to U for approval if U lacks the knowledge or interest required to make the decisions. Wexelblat and Maes (1997) recommend making available several alternative types of control for users to choose from.

Unobtrusiveness

We will use the term *obtrusiveness* to refer to the extent to which S places demands on the user's attention that reduce U's ability to concentrate on her primary tasks. This term, and the related words *distracting* and *irritating*, are often heard in connection with UASs. Figure 15.20 shows that (a) there are several different reasons why UASs can easily turn out to be obtrusive and (b) there are equally many corresponding strategies for minimizing obtrusiveness. Some of these measures can lead straightforwardly to significant improvements—for example, when it is recognized that distracting lifelike behaviors of an animated character are not really a necessary part of the system.

Privacy

UASs typically gather data about individual users and use these data to make decisions that may have more or less serious consequences. Users may accordingly become concerned about the possibility that their data will be put to inappropriate use. Privacy concerns tend to be especially acute in e-commerce contexts (Ghosh & Swaminatha, 2001) and with some forms of support for collaboration, because in these cases (a) data about U are typically stored on computers other than the user's own; (b) the data often include personally identifying information; and (c) there may be strong incentives to use the data in ways that are not dictated by U's own interests. As will be discussed in the section on Obtaining Information About Users, different means of acquiring information about users can have different consequences with regard to privacy. On the other hand, many of the measures that can be taken to protect privacy—for example, a policy of storing as little personally identifying data as possible—are not specific to UASs (see the chapters by Diller & Masten and by Friedman & Kahn in this handbook).

Breadth of Experience

When a UAS helps U with some form of information acquisition, much of the work of examining the individual documents, products, and/or people involved is typically taken over by S. A consequence can be that U ends up learning less about the domain in question than she would with a nonadaptive system (cf. Lanier, 1995). For example, if S recommends apartments in a given region to U (see, e.g., Burke et al., 1997; Shearin & Lieberman, 2001) and U simply accepts one of the recommendations, U may learn less about the real estate market in that region than she would if she took the trouble to search systematically through a real estate Web site. One point of view here (see, e.g., Shneiderman & Maes, 1997, p. 53) is that it should be up to the user to decide whether she prefers to learn about a given domain or to save time by delegating work to a system. It may be worthwhile to give U a continuous spectrum of possibilities between complete control over a task and complete delegation of it. For example, many product recommendation systems allow users to alternate freely between pursuing S's recommendations and browsing through product descriptions in the normal way.

A second way in which adaptivity can narrow the user's experience is through excessive reliance on an incomplete user model. For example, suppose that an apartment seeker has so far shown interest only in apartments in the Bellvue area: If S accordingly supplies information only about other apartments in this area, S may never discover that U is willing to consider apartments in other areas. Some systems mitigate this problem by systematically proposing solutions that are *not* dictated by the current user model (see, e.g., Linden, Hanks, & Lesh, 1997; Shearin & Lieberman, 2001).

OBTAINING INFORMATION ABOUT USERS

Some of the usability challenges discussed in the previous section are closely connected with the ways in which information about individual users is acquired—a consideration that also largely determines the success of a system's adaptation. The next two subsections will look, respectively, at (a) information that U supplies to S explicitly for the purpose of allowing S to adapt; and (b) information that S obtains in some other way.

Explicit Self-Reports and Self-Assessments

Self-Reports About Objective Personal Characteristics. Information about objective properties of the user (such as age, profession, and place of residence) often has implications that are relevant for system adaptation—for example, concerning the topics that U is likely to be knowledgeable about or interested in. This type of information also has the advantage of changing relatively infrequently. Many UASs request information of this type from users, but the following caveats apply:

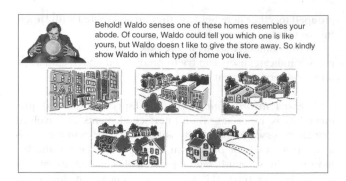

FIGURE 15.21. Example of a screen with which the LifeStyle Finder elicits demographic information. *Note.* From "Lifestyle Finder: Intelligent User Profiling Using Large-Scale Demographic Data," by B. Krulwich, 1997, AI M*agazine*, 18(2), pp. 37–45. Copyright 1997 by the American Association for Artificial Intelligence. Adapted with permission.

1. Specifying information such as profession and place of residence may require a fair amount of tedious menu selection and/or typing.
2. Because information of this sort can often be used to determine the user's identity, U may justifiably be concerned about privacy issues. Even in cases where such concerns are unfounded, they may discourage U from entering the requested information.

Some systems (e.g., the adaptive tour guide AMPRES, described by Rössel, 2000) address these problems by (a) restricting requests for personal data to the few pieces of information (if any) that S really requires; and (b) explaining the uses to which the data will be put. A novel approach was tried in the Web-based LifeStyle Finder prototype (Fig. 15.21; Krulwich, 1997), which was characterized by a playful style and an absence of requests for personally identifying information. Of the users surveyed, 93% agreed that the LifeStyle Finder's questions did not invade their privacy.

It is sometimes possible to avoid requests for explicit input about personal characteristics by accessing sources where similar information has already been stored.

Self-Assessments With Respect to General Dimensions. It is sometimes helpful for S to have an assessment of a property of U that can be expressed naturally as a position on a particular general dimension: the level of U's interest in a particular topic, the level of her knowledge about it, or the importance that U attaches to a particular evaluation criterion. Often an assessment is arrived at through inference on the basis of other evidence. But it may be quicker, if feasible, to ask U for an explicit assessment. Figure 15.22 shows a typical rating scale. Some systems use checkboxes, which make it possible for U to give quick "yes/no" ratings for a larger number of dimensions (see, e.g., Pazzani & Billsus, 1999).

This assessment method raises basically the same problems that have long been familiar in fields, such as psychology and

How important to you is a dog that's easy to train?

Some dogs are more stubborn than others. If you want a dog that will learn and obey your commands, place more emphasis here. If you don't plan to teach your dog more than a few tricks, this won't matter much.

FIGURE 15.22. Rating scale from the PERSONALOGIC decision guide for dog seekers. Part of a screen shot made from http://www.purina. personalogic.com in July 2001. Before mid-2001, many similar recommenders had been available via http://www.personalogic.com

marketing research, in which questionnaires are regularly used. For example, users may not know the exact meaning of the various points on the scale, and they may be inclined to answer in a way that seems socially desirable (see, e.g., King & Bruner, 2000). Too often, the design of this type of scale for a UAS does not adequately take into account the methodological knowledge that has been built up with regard to such problems (see, e.g., the chapters by Blomberg et al. and by Karat in this handbook).

The effort involved in this type of self-assessment is more cognitive than physical, but it can still be enough to discourage users. So as with requests concerning personal characteristics, it is in general worthwhile to consider ways of minimizing such requests, making responses optional, and ensuring that the purpose is clear.

Self-Reports on Specific Evaluations. Instead of asking U to assess her position on a dimension explicitly, some systems try to infer U's position on the basis of her explicitly evaluative responses to specific items. For this purpose, S may present icons (e.g., "thumbs-up" and "thumbs-down"), checkboxes, or rating scales. The items that U evaluates can be (a) items that U is currently experiencing directly (e.g., the current web page), (b) actions that S has just performed (see, e.g., Billsus & Pazzani, 2000; Wolfman et al., 2001), (c) items that U must judge on the basis of a description (e.g., the abstract of a talk; a table listing the attributes of a physical product), or (d) the mere name of an item (e.g., a movie) that U may have had some experience with in the past (see, e.g., Fig. 15.14). The physical effort required is usually that of a simple action such as a mouse click. The cognitive effort depends in part on how directly available the item is: In the third and fourth cases just listed, U may need to perform memory retrieval and/or inference to arrive at an evaluation.

Even when the effort is minimal, users often do not like to bother with explicit evaluations that do not constitute a necessary part of the task they are performing. One way of dealing with this problem is to develop inference methods that can make use of even a small number of explicit assessments of individual items. Another approach is to have S interpret naturally occurring actions of U (e.g., selection vs. skipping of news

stories whose headlines are displayed on a screen) as implicit evaluations. These two approaches were evaluated by Billsus and Pazzani (2000) in two variants of the ADAPTIVE NEWS SERVER.

Responses to Test Items. In systems that support learning, it is often natural to administer tests of knowledge or skill. In addition to serving their normal educational functions, these tests can yield valuable information for S's adaptation to U. An advantage of tests is that they can be constructed, administered, and interpreted with the help of a large body of theory, methodology, and practical experience (see, e.g., Wainer, 2000).

Outside of a learning context, users will in general be reluctant to invest time in tests of knowledge or skill, unless these can be presented in an enjoyable form (see, e.g., the color discrimination test used by Gutkauf, Thies, & Domik, 1997, to identify perceptual limitations relevant to the automatic generation of graphs). But it is sometimes possible to design a part of a system so that it functions as a concealed test, in addition to serving its primary function (e.g., of providing information). For example, the AMPRES hypertext system (Rössel, 2000) includes an introductory sequence in which U can request explanations of various concepts that are used within the system. This sequence is designed so that the pattern of U's requests can be interpreted as if they were answers to items on a knowledge test.

Nonexplicit Input

Naturally Occurring Actions. The broadest and most important category of information about users includes all of the actions that U performs with S that do not have the express purpose of revealing information about U to S. These actions may range from major actions like purchasing an expensive product to minor ones like scrolling down a web page. The more significant actions tend to be specific to the particular type of system that is involved (e.g., e-commerce sites vs. learning environments). Within some domains, there has been considerable research on ways of interpreting particular types of naturally occurring user actions. For example, researchers interested in adaptive hypertext navigation support have developed a variety of ways of analyzing U's navigation actions to infer U's interests and/or to propose navigation shortcuts (see, e.g., Goecks & Shavlik, 2000).

In their purest form, naturally occurring actions require no additional investment by the user, because they are actions that U would perform anyway. The main limitation is that they are hard to interpret; for example, the fact that a given web page has been displayed in U's browser for 4 min does not reveal with certainty which (if any) of the text displayed on that page U has actually read. Some designers have tried to deal with this tradeoff by designing the user interface in such a way that the naturally occurring actions are especially easy to interpret. For example, a web-based system might display just one news story on each page, even if displaying several stories on each page would normally be more desirable.

The interpretation of naturally occurring actions by S can raise privacy and transparency issues (cf. Fig. 15.20) that do not arise in the same way with explicit self-reports and self-assessments: Whereas the latter way of obtaining information about the user can be compared with interviewing, the former way is more like eavesdropping—unless U is informed about the nature of the data that are being collected and the ways in which they will be used (cf. Martin, Smith, Brittain, Fetch, & Wu, 2001).

Previously Stored Information. Sometimes a system can access relevant information about U that has been acquired and stored independently of S's interaction with U:

1. If U has some relationship (e.g., patient, customer) with the organization that operates S, this organization may have information about U that it has stored for reasons unrelated to any adaptation (e.g., U's medical record or address).

2. Relevant information about U may be stored in publicly available sources such as electronic directories or web homepages. For example, Pazzani (1999) explores the idea of using a user's web homepage as a source of information for a restaurant recommending system.

3. If there is some other system that has already built up a model of U, S may be able to access the results of that modeling effort and try to apply them to its own modeling task. There is a line of research that deals with *user modeling servers* (see, e.g., Kobsa, 2001): systems that store information about users centrally and supply such information to a number of different applications. Some of the major commercial personalization software is based on this conception (see Fink & Kobsa, 2000, for an overview).

Relative to all of the other types of information about users, previously stored information has the advantage that it can in principle be applied right from the start of the first interaction of a given user with a given system. To be sure, the interpretability and usefulness of the information in the context of the current application may be limited. Moreover, questions concerning privacy and transparency may be even more important than with the interpretation of naturally occurring actions.

Low-Level Indices of Psychological States. The next two categories of information about U have become practically feasible only in recent years, with advances in the miniaturization of sensing devices (cf. the chapter by Hinckley in this handbook).

The first category of sensor-based information (discussed at length by Picard, 1997) comprises data that reflect aspects of a user's psychological state, such as: (a) anger and frustration, which have especially clear relevance in the context of automated spoken dialogs with customers; (b) attraction to particular items, especially relevant for recommender systems; and (c) stress and cognitive load, which can be important factors when U is performing a challenging task (or several tasks at once), such as driving.

Two categories of sensing devices have been employed: (a) devices attached to U's body (or to the computing device itself) that transmit physiological data, such as electromyogram signals, the galvanic skin response, blood volume pressure, and the pattern of respiration; and (b) video cameras and microphones that transmit psychologically relevant information about U, such as features of her facial expressions (Picard, 1997) or her speech (e.g., pitch, intensity, and quality of articulation; or more linguistic features, such as the length of utterances and the occurrence of pauses, Müller et al., 2001).

With both categories of sensors, the extraction of meaningful features from the low-level data stream requires the application of pattern recognition techniques. These typically make use of the results of machine learning studies in which the relationships between low-level data and meaningful features have been learned.

Although it is sometimes possible to recognize a psychological state (such as anger) on the basis of sensor data alone, often this type of information needs to be combined with other types before reliable recognition is possible.

One advantage of sensors is that they supply a continuous stream of data, the cost to U being limited to the physical and social discomfort that may be associated with the carrying or wearing of the devices. These factors are significant now, but further advances in miniaturization—and perhaps changing attitudes as well—seem likely to reduce their importance.

Signals Concerning the Current Surroundings. As computing devices become more portable, it is becoming increasingly important for a UAS to have information about U's current surroundings (cf. the chapter by Stephanidis in this handbook). Here, again, two broad categories of input devices can be distinguished:

1. Devices that receive explicit signals about U's surroundings from specialized transmitters. Some mobile systems that are used outdoors (see, e.g., Dey, Abowd, & Wood, 1998) employ Global Positioning System (GPS) technology. More specialized transmitters and receivers are required, for example, if a portable museum guide system is to be able to determine which exhibit U is looking at.

2. More general sensing or input devices. For example, Schiele, Starner, Rhodes, Clarkson, and Pentland (2000) describe the use of a miniature video camera and microphone (each roughly the size of a coin) that enable a wearable computer to discriminate among different types of surroundings

(e.g., a supermarket vs. a street). The use of general-purpose sensors eliminates the dependence on specialized transmitters. On the other hand, the interpretation of the signals requires the use of sophisticated machine learning and pattern recognition techniques.

LEARNING, INFERENCE, AND DECISION MAKING

A distinguishing feature of UASs is the central role of techniques for user model acquisition and application (Fig. 15.2). These techniques enable S to learn about individual users and make inferences and decisions about them. The present section describes the most important properties of several commonly used computational paradigms, which differ in terms of the contributions that they can make and the conditions under which they can be applied effectively.

Classification Learning

Many UASs use learning methods from a broad category of machine learning techniques called *classification learning*. A great variety of methods have been developed within this paradigm, including: decision trees, probabilistic classifiers, neural networks, case-based reasoning, and specialized text-classification methods.[6]

From a broad perspective, the differences among these methods are less important than the basic nature of a classification learning problem, which will be illustrated with examples below: The learning procedure starts with a set of *training examples*, each of which is characterized in terms of its *features*. Each training example has been *classified*, that is, assigned to one of a set of two or more categories. On the basis of these examples, the procedure learns a *classifier*: a model that is capable of assigning a new item to one of the same set of categories. Usually the assignment for a new item cannot be made with certainty; accordingly, some methods yield a set of possible assignments for each item, each assignment being associated with some index of S's confidence.

Example Systems. These concepts are illustrated clearly by the SWIFTFILE system (Fig. 15.1), which learns how to classify a user's e-mail messages in (more or less) the same way as U would herself. At any given time, the training examples are the messages that U has filed so far, and the categories correspond to U's e-mail folders. Once S has learned how to classify like U, S can *predict* how U will classify any given message—though of course not with perfect accuracy.

The particular type of model that SWIFTFILE learns takes advantage of the fact that each item to be classified (i.e., each e-mail message) contains a large number of words that serve as features. SWIFTFILE uses well-established text classification methods from the information retrieval field to arrive at a representation of each e-mail folder as a *weighted word-frequency vector*. To classify a new message, SWIFTFILE essentially compares the distribution of the words in its text with the current representations of U's folders (Segal & Kephart, 1999, 2000).

Here are some further examples of the use of classification learning for UASs:

1. Many systems that recommend or select documents (e.g., the ADAPTIVE NEWS SERVER, use some form of classification learning to learn to predict which documents a given U will like. The relevant features of the documents may include, in addition to the words in the text, attributes like length or date of appearance. Recommenders of this type are sometimes called *content-based* recommenders, because their predictions are based mainly on the content of the documents in question, as opposed, say, to the ratings of those items by other users.

2. CASPER (Bradley, Rafter, & Smyth, 2000) filters the search results returned by a job-finding Web site by learning to predict which job offers the current U will like or dislike, on the basis of U's ratings of previous job offers. It makes use of the *nearest neighbor* classification method, and it employs a relatively sophisticated scheme for analyzing the features of job offers.

3. By observing how U enters appointments into a calendar system, the CALENDAR APPRENTICE (Mitchell et al., 1994) learns how to predict the properties of appointments that U makes, such as their location and duration. It uses the method of *decision tree induction*, which in effect yields rules for predicting the remaining features of an appointment on the basis of features that have already been specified.

Requirements. Two of the prerequisites for the application of classification learning methods may be hard to fulfill in some cases:

1. It may not be straightforward to characterize the items in question in terms of features. For example, when the items are images or music clips, it may be a challenging task to extract useful features (e.g., concerning content or style) from the digital representations of the items. When the items are nonelectronic objects to which the computer has no direct access (e.g., products offered for sale in an e-commerce site), information about their adaptation-relevant features must be obtained from some existing information source or entered specifically for the purpose of enabling adaptation.

2. It may not be possible for S to process an adequate number of training examples before it has to begin classifying new items. Depending on the nature of the system in question, users may or may not be willing to grant S ample training time before they expect useful adaptation to occur. There is currently a good deal of research into classification methods for UASs that can learn on the basis of a minimal number of examples. SWIFTFILE's text classification method rates well in this respect, and sometimes the

---

[6]Han and Kamber (2001), Langley (1996), and Mitchell (1997) offer broad overviews of statistical and machine learning techniques, some of which are related to those discussed in this section. The role of machine learning in UASs is reviewed by Webb, Pazzani, and Billsus (2001).

To identify neighbors for $\mathcal{U}$:

1. Store $\mathcal{U}$'s ratings of items.
2. For each user $\mathcal{U}^*$ in a sample of other users, compute an *agreement strength* with $\mathcal{U}$ on the basis of:
   - the difference between the ratings of $\mathcal{U}^*$ and $\mathcal{U}$;
   - the number of items rated by both $\mathcal{U}^*$ and $\mathcal{U}$.
3. Decide whether to add $\mathcal{U}^*$ as a neighbor on the basis of:
   - the agreement strength of $\mathcal{U}^*$ with $\mathcal{U}$;
   - the total number of items rated by $\mathcal{U}^*$;
   - the degree to which the total proportion of items rated by $\mathcal{U}$'s neighbors would be increased if $\mathcal{U}^*$ were added.

To make recommendations for $\mathcal{U}$:

1. For each item, use the ratings of $\mathcal{U}$'s neighbors to compute:
   - a predicted rating for $\mathcal{U}$;
   - a degree of confidence in this prediction;
   - the extent of disagreement among $\mathcal{U}$'s neighbors with regard to that item.
2. Base recommendations on these factors (among others).

FIGURE 15.23. Summary of a typical algorithm for generating recommendations through collaborative filtering.

nearest neighbor method (exemplified by the CASPER system) does so as well (see, e.g., Billsus & Pazzani, 2000). A different approach is to allow $\mathcal{U}$ to give $\mathcal{S}$ hints, in the form of explicit self-reports, that allow $\mathcal{S}$ to create an initial, partially accurate model that serves as a basis for further learning (see, e.g., Section 4 of Pazzani & Billsus, 1997).

Collaborative Filtering

The paradigm of *collaborative filtering* was illustrated with the movie recommender system MovieCentral. More generally, the approach is used for the prediction of the *responses* (e.g., ratings or purchases) of a user to *items* (e.g., documents or products) to which other users have previously responded. The distinguishing property of the paradigm is the fact that each item is characterized in terms of the previous responses of other users, not in terms of its intrinsic features.[7] Figure 15.23 summarizes a typical computational procedure.

Requirements. As with classification learning, certain prerequisites of the pure collaborative filtering paradigm are in some cases hard to fulfill:

1. For some of the items about which a prediction is to be made for $\mathcal{U}$, there may not be a sufficiently large number of responses available in the database that have been made to these items by users similar to $\mathcal{U}$ (or indeed by any users). For example, when new items continually enter the database and remain interesting for only a short time (e.g., news stories that are being fed in by a news service), they may have lost their importance by the time enough responses to them have been accumulated.

2. $\mathcal{U}$ may not be willing to give a sufficient number of responses to items before receiving useful recommendations. In particular, $\mathcal{U}$ may want to be able to specify a general preference

explicitly, such as "I like science fiction movies." The only way for $\mathcal{U}$ to convey a preference like this in MovieCentral would be by requesting the opportunity to rate a number of science fiction movies and giving all of them high ratings.

Combinations With Other Paradigms. Some ingenious schemes have been devised for combining the basic strategy of collaborative filtering with other methods, so as to overcome some of the limitations of the pure form (see, e.g., Good et al., 1999, for an empirical comparison of a number of hybrid methods). Some researchers have explored ways of identifying suitable neighbors who have not necessarily responded to many of the same items as $\mathcal{U}$; instead, $\mathcal{S}$ can take into account their similarity to $\mathcal{U}$ in terms of personal characteristics (Pazzani, 1999) or interest profiles (Balabanović & Shoham, 1997). Good et al. (1999) go a step further by introducing artificial "neighbors" who serve to fill in the gaps left by a user's human neighbors: Each such neighbor is an agent that implements a content-based prediction method, such as one that uses text classification methods.

Decision-Theoretic Methods

Basic Characteristics. A remarkable fact about the two paradigms discussed so far is that they are almost entirely *data-based*: They make use of virtually no general knowledge about users, their goals, or the items that they are dealing with. By contrast, the next paradigm represents a class of more *theory-based* methods: The system designers build into their models a good deal of knowledge about the variables that are relevant in a given interaction situation.

For example, the partial *Bayesian network* in Fig. 15.24 shows a few of the assumptions made by the designers of the Lumière adaptive help prototype.[8] Each rectangle represents a

---

[7]The less frequently used term *social recommendation* is on the whole more apt than *collaborative filtering*, because neither active collaboration nor filtering are essential aspects of the approach.

[8]A good introductory overview of Bayesian networks and other decision-theoretic techniques is given by Jensen (2001); the classic work is by Pearl (1988). An early survey of applications within UASs is given by Jameson (1996), and recent developments are discussed by Zukerman and Albrecht (2001).

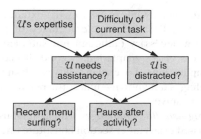

FIGURE 15.24. Part of a Bayesian network used in the LUMIÈRE prototype for inferring the likelihood that U requires assistance. *Note.* From "The Lumière Project: Bayesian User Modeling for Inferring the Goals and Needs of Software Users," by E. Horvitz, J. Breese, D. Heckerman, D. Hovel, and K. Rommelse, 1998, in G. F. Cooper and S. Moral (Eds.), *Uncertainty in Artificial Intelligence: Proceedings of the Fourteenth Conference* (pp. 256–265), San Francisco: Morgan Kaufmann. Copyright 1998 by the authors. Adapted with permission.

variable about which S in general has only an uncertain belief. For example, the probability that the value of U NEEDS ASSISTANCE? is *true* at a given moment might be estimated as 89%, as in the situation illustrated in Fig. 15.6. Arrows represent probabilistic relationships among the variables, which can usually be interpreted as causal influences. For example, this network states that U's need for assistance from S will depend largely on U's expertise in using S and on the difficulty of the task that U is currently dealing with. In turn, U's need for assistance will influence the occurrence of behaviors of U such as surfing through menus and/or pausing after performing some actions.

Sometimes S obtains information that leads to certainty about the value of a particular variable. This new certain belief is then propagated through the Bayesian network according to an applicable algorithm, and S's beliefs concerning the other variables in the network are adjusted accordingly. For example, when a "pause after activity" is observed, the probabilities associated with U NEEDS ASSISTANCE? and U IS DISTRACTED? will in general increase.[9]

Potential Advantages. Relative to the simpler data-based models discussed in the previous two sections, theory-based Bayesian networks offer several potential advantages, including the following:

1. S can make useful inferences about a user U without first acquiring any long-term model of U on the basis of data. For example, S might offer the advice shown in Fig. 15.6 after observing just a few of U's actions within a spreadsheet application.

2. With the help of other closely related decision-theoretic techniques, the probabilistically expressed beliefs generated by a Bayesian network can be used to make adaptation *decisions*. Such a decision-making process quantitatively takes into

account the possible consequences of S's actions and the overall *utility* of these consequences. For example, S can quantitatively weigh the expected benefits of offering advice against the possible costs in terms of time and distraction. Note that the two paradigms discussed previously—classification learning and collaborative filtering—yield predictions about a user (e.g., about the films that U is most likely to enjoy); they do not yield decisions as to what S should do (e.g., how many films should be recommended to U and in what form?). With these paradigms, therefore, the system designer must specify in advance how S's predictions are to be translated into S's actions. By contrast, decision-theoretic methods allow S to select actions more flexibly, taking into account U's perceived priorities and the details of the current situation (cf. Jameson et al., 2001). To be sure, this greater flexibility may sometimes diminish the predictability and transparency of S's actions.

Requirements. The major challenge in the development of decision-theoretic UASs concerns the construction of suitable general models, which are almost always much more complex than the partial model shown in Fig. 15.24. If the construction is purely theory-based, one or more persons with relevant knowledge must specify the qualitative and quantitative relationships among the variables of the model, some of which are typically unobservable. It is in general difficult to make well-founded judgments about all aspects of such a model. In recent years, many techniques have been developed for learning decision-theoretic models at least partly on the basis of data (see, e.g., Heckerman, 1998). In connection with UASs, these methods have been applied successfully with relatively simple models (see, e.g., Müller et al., 2001) that are similar to the types of data-based model that were discussed in the previous two subsections. The question of how to learn more complex, theoretically interpretable decision-theoretic models at least partly on the basis of data is a challenge that is being addressed in current research (see, e.g., Wittig & Jameson, 2000).

Other Approaches

Two other largely theory-based paradigms are worth mentioning briefly, although they are at present somewhat less widely used than the three paradigms discussed so far.

Techniques for Plan Recognition. In the field of artificial intelligence, many approaches have been developed to the problem of recognizing a person's plans on the basis of her observed actions. For UASs, these techniques offer the possibility that a system can interpret a user's actions as steps in the execution of a plan that is intended to achieve some goal; S may then be able to assist U in various ways, which depend in part on the function that the adaptation is intended to serve:

---

[9]More complex, application-specific Bayesian networks are used to estimate what particular help topics are most likely to be relevant to U; see Fig. 3 of Horvitz et al. (1998).

1. Some systems that aim to take over routine actions from $\mathcal{U}$ do so by recognizing a plan that $\mathcal{U}$ needs to execute repeatedly and offering to execute the plan for $\mathcal{U}$ in future situations. This category includes some systems for *programming by example* (see, e.g., Lieberman, 2001).
2. Help systems and tutoring systems may point out problems with $\mathcal{U}$'s plan or remind $\mathcal{U}$ of steps that need to be taken.
3. Systems that conduct dialogs can choose their own dialog contributions in accordance with $\mathcal{U}$'s perceived plan.

An overview of uses of plan recognition in interactive systems is given by Carberry (2001).

The Stereotype Approach. A stereotype-based system (see, e.g., Rich, 1989) distinguishes a set of categories, called *stereotypes*, that a given user may belong to. For example, in the AVANTI tourist information system, each stereotype corresponds to a group of visitors that have a certain set of capabilities and information needs. The system provides a set of rules for assigning each user to one or more stereotypes on the basis of $\mathcal{U}$'s observed behavior or other information about $\mathcal{U}$ (such as self-reports). Once $\mathcal{U}$ has been categorized in this way, $\mathcal{S}$ can ascribe to $\mathcal{U}$ properties and/or take actions that are associated with the stereotype(s) in question.

The stereotype approach was the first inference paradigm to be widely used for UASs (see, e.g., Rich, 1979). It is currently employed in some commercial personalization servers (cf. Fink & Kobsa, 2000), among other systems.

The inference processes in stereotype-based systems can be realized with a variety of computational techniques. The emphasis is less on sophisticated computation than on realistic specification of the content of the stereotypes and the rules for activating them.

EMPIRICAL METHODS

Like any other type of interactive computing system or device, a UAS cannot be designed on the basis of first principles alone. No matter how sophisticated the techniques that are used to realize the system, what ultimately counts is how well the larger "system" that includes the user(s) works. So, at various points in the design process, some sort of empirical work will need to be done to ensure that the design is in touch with reality.

The full repertoire of empirical methods in human-computer interaction (cf. Section V of this handbook) is in principle applicable to UASs. This section will focus on four categories of empirical study that are especially relevant and/or raise some important general issues when they are applied to UASs.[10] With each type of study, we will consider what it can tell us about each of the following two questions:

1. *Accuracy of modeling:* One important difference between UASs and other interactive systems is that a UAS typically derives testable hypotheses about each individual user $\mathcal{U}$. It is therefore often worthwhile to ask to what extent $\mathcal{S}$'s modeling of $\mathcal{U}$ is accurate. First, reasonable accuracy of $\mathcal{S}$'s modeling is in general a necessary (although not sufficient) condition for the success of $\mathcal{S}$'s adaptation. Second, it can be hard to know the implications of overall usability results if the accuracy of the modeling is not known—unless the results turn out to be conclusively positive.

2. *Meeting usability challenges:* The general usability challenges discussed in an earlier section constitute one reason why it is important not to restrict empirical studies to the question of modeling accuracy.

Wizard-of-Oz Studies

Systems that adapt to their users are in one methodological respect similar to systems that make use of speech (cf. the chapter by Lai & Yankelovich in this handbook): They attempt to realize a capability that is so far possessed to the highest degree by humans. Consequently, as with speech interfaces, valuable information can sometimes be obtained from a *Wizard-of-Oz study*: In a specially created setting, a human takes over a part of the processing of the to-be-developed system $\mathcal{S}$ for which humans are especially well suited (cf. the chapters by Lai & Yankelovich, by Beaudouin-Lafon & Mackay, and by Pew in this handbook).

The left-hand side of Fig. 15.25 summarizes a typical study of this type that was conducted in an early phase of the development of the LUMIÈRE prototype. Whereas the users believed they were interacting with an adaptive help system, the help was actually provided by usability experts.

Assessing Accuracy. A Wizard-of-Oz study can yield an *upper bound* estimate of the highest level of modeling accuracy that might be attainable given the available information—as long as one can assume that the human "wizards" are more competent at the type of assessment in question than a fully automatic system is likely to be in the foreseeable future. For example, if the expert advisors in this particular study had shown no ability at all to recognize the users' goals, perhaps the entire project would have been reconsidered.

Assessing Usability. The example study brought to light a subtle problem: the tendency of the advisors' recommendations to become self-fulfilling prophecies. This problem is related to the general usability challenge concerning *breadth of experience*. Here, however, the limited accuracy of the experts' "user models" did not lead to a narrowing of the users' experience; instead, it caused users to be led into new areas that were presumably of no real interest to them. Note that the experts' behavior in this study also suggested a way of avoiding this problem. Of course any study in which the system is simulated by a human will reveal little about usability issues that involve details of the appearance and behavior of the user interface, such as the behavior of the animated characters that personify the OFFICE ASSISTANT.

---

[10]More extended discussions of empirical methods for UASs are provided by Langley and Fehling (1998), Höök (2000), and Chin (2001).

Method

Subjects

- Were told an experimental help system would track their activity and make guesses about how to help them.
- Received the advice via a computer monitor.

Experts

- Worked in a separate room.
- Viewed subjects' activity via a monitor.
- Conveyed advice by typing.
- Were not informed about the assigned spreadsheet tasks.

Results

Difficulty of experts' task

- Experts showed some ability to identify $\mathcal{U}$'s goals and needs.
- They were often uncertain about:
  1. $\mathcal{U}$'s goals – sometimes recognized with an "Aha!" reaction after a period of confusion;
  2. the value of providing different kinds of assistance.

Consequences of poor advice

- Users typically examined advice carefully.
- Even when advice was off the mark, subjects would often become distracted by it and begin to experiment with the features described.
- This behavior gave experts false confirmation of successful goal recognition.
- Experts then gave further advice along the same lines.

How experts improved

- Experts became more skillful in offering advice in this situation.
- For example, they learned to give conditional advice:
  "If you are trying to do X, then ...."

FIGURE 15.25. Summary of a Wizard-of-Oz study conducted in the LUMIÈRE research project. Summarized on the basis of p. 258 of Horvitz et al. (1998).

Simulations Using Data From a Nonadaptive System

This second type of empirical study is uniquely applicable to UASs. It focuses entirely on the issue of accurate modeling of $\mathcal{U}$. A typical basic procedure is as follows.

Given a database of behavioral data on how a number of users have used a *nonadaptive* version of a system S, simulate a use situation in which S receives parts of these data incrementally as information about $\mathcal{U}$, checking how fast and how well S can acquire and apply a model of $\mathcal{U}$.

A clear example of a study of this type was conducted with the SWIFTFILE system (Segal & Kephart, 2000). The training examples with respect to each $\mathcal{U}$ were $\mathcal{U}$'s previously filed messages. SWIFTFILE's learning method was applied to these messages just as if the system had done the learning while $\mathcal{U}$ was originally filing them. The overall result of the study was that S quickly and consistently attained high accuracy in predicting the folder that $\mathcal{U}$ would choose, including the correct folder about 90% of the time in its three guesses for each message. To be sure, this result applies only if the messages for which $\mathcal{U}$ decided to create a new folder are not counted, which is a reasonable policy.

Moreover, it is possible to evaluate many alternative variants of S on the basis of the same data, whereas each user could work with only one variant at any given time. Figure 15.26 shows the results of an analysis of this type. The results indicate how often SWIFTFILE would have suggested the correct folder, on the average, using each possible number of suggestion buttons between 1 and 5. On the basis of these results, Segal and Kephart

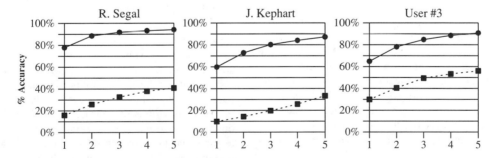

FIGURE 15.26. Comparison of alternative variants of SWIFTFILE on the basis of data concerning the use of a nonadaptive system. Upper curves: Accuracy of SWIFTFILE's predictions with different numbers N of suggestion buttons, shown on the x-axis. Lower curves: Accuracy of the naive strategy that simply predicts the N most frequently used folders. *Note.* From "MailCat: An Intelligent Assistant for Organizing E-mail," by R. B. Segal and J. O. Kephart, 1999, in *Proceedings of the Third International Conference on Autonomous Agents* (pp. 276–282). Copyright 1999 by the Association for Computing Machinery, Inc. Adapted with permission.

(1999) argue that the optimal number of buttons is 3. The results also allow us to reject the hypothesis that S could do just as well by always suggesting the N most commonly used folders.

As these examples show, precise, thorough analyses can be performed without any investment of users' time, and without waiting for enough data from each user to accumulate, which could take months in the case of an e-mail system.

Assessing Accuracy. In some respects, the accuracy estimates that can be obtained with data from a nonadaptive system are actually more realistic than those that could be obtained in studies involving interaction with an adaptive system: When using the real SWIFTFILE system, users might sometimes be inclined to accept one of S's folder suggestions even when they would normally have chosen some other folder: U can save time and mental effort by relying uncritically on S's classifications. In this case, the predictions of S would be self-fulfilling prophecies to some extent, like some of the advice of the experts in the Wizard-of-Oz study discussed previously; and the resulting accuracy estimates would be inflated to some unknown extent. Simulation studies can avoid the problem of self-fulfilling prophecies. Still, they should not be relied on exclusively, because the self-fulfilling prophecy phenomenon is one that can arise with real system use. The most complete picture is given by a combination of simulation studies and studies of real use, a procedure followed by Segal and Kephart, who also report on a study of actual system use (Segal & Kephart, 1999, Section 6).

Assessing Usability. Simulations with data from a nonadaptive system cannot yield information about how users actually interact with a corresponding adaptive system. Therefore, they do not directly address the typical usability challenges discussed earlier. At best, results about S's modeling accuracy can form a basis for speculation about particular aspects of S's usability, such as S's predictability.

Controlled Studies

The third category of study is one that is familiar from many evaluation studies in the human–computer interaction field: Two or more variants of a given system are compared within a controlled setting in which users perform more or less predefined tasks. When a UAS is involved, typically an adaptive and a nonadaptive variant of the system are compared. With the current state of the art, the key question is often whether the adaptive version shows any advantages. In a few years, we may see more studies in which the main focus is on the comparative evaluation of different adaptive variants.

The evaluation of the dialogue system TOOT (Litman & Pan, 2000) provides examples of the types of information that can be obtained from a controlled comparison. Six novice users conducted dialogues with an adaptive version of TOOT that was able to change its dialogue strategy in the way illustrated in Fig. 15.8, starting with the most ambitious strategy. Six other users conducted dialogues with a nonadaptive version that always used the most ambitious strategy.

The central result of the study concerns the likelihood that a given dialogue will end successfully, with U hearing the schedule information for the desired train. The likelihoods were 65% and 23% for the adaptive and nonadaptive versions, respectively, the difference being statistically significant. Taken by itself, this result would simply confirm that a system that often shifts to a more long-winded, conservative dialogue strategy will have a higher likelihood of successful dialogue completion. But the dialogues conducted with the adaptive version were actually somewhat shorter than those with the nonadaptive version (although the difference in length was not statistically significant).

These results show that the adaptive version of TOOT was clearly preferable to the nonadaptive version used in this study. But as Litman and Pan acknowledge, it is still quite possible that some other nonadaptive version could outperform the adaptive version by consistently applying a single, more conservative dialogue strategy. The general point is that it can be difficult to demonstrate that user-adaptivity yields the best results in a given situation. It is not enough to show that a particular nonadaptive version of the system performs less well than an adaptive version. The real question is whether *any* single nonadaptive system could feasibly be identified in advance that could do as well as an adaptive one.

This point is reminiscent of the lessons learned from attempts to evaluate the relative merits of competing user interface paradigms through empirical comparisons: No matter how flawless the methodology may be, it is hard to generalize the results with confidence beyond the particular system variants, tasks, and user groups that figured in the studies.

Assessing Accuracy. One typical obstacle to the assessment of modeling accuracy on the basis of data concerning system use was already discussed in the previous subsection: the problem of potentially self-fulfilling prophecies. The adaptation in TOOT illustrates a still more fundamental obstacle: TOOT switches to a new dialogue strategy when it predicts that the current strategy will not lead to success; but once the switch has occurred, there is no way of determining what would actually have happened with the original strategy.

Assessing Usability. Controlled studies offer an opportunity to compare users' subjective ratings of two or more system variants. In addition to overall satisfaction, these ratings may include assessments of the key usability variables predictability, transparency, controllability, and unobtrusiveness. Generalizing such ratings to situations of real use is problematic, however. Various factors that may strongly influence U's desire for adaptive features may be absent in the controlled setting, such as: time pressure, personal significance of the tasks performed with S, and distractions due to competing tasks and events. These contextual factors can be taken into account in studies of actual use.

Studies of Actual System Use

The final major category of user studies is likewise familiar from its frequent use with nonadaptive systems: Some more or less complete version of S is employed by users in a more or less realistic setting, and various objective and/or subjective variables are assessed.

Observation and Interviewing. The typical field study of a UAS involves a prototype that is used by several people in their normal everyday setting. Sawhney and Schmandt (2000, pp. 377–380) used this method to evaluate their NOMADIC RADIO, a wearable system for the context-sensitive transmission of audio messages. They observed and interviewed two experienced users of mobile phones and pagers who tried out a prototype of the NOMADIC RADIO during a 3-day period. Although this type of study cannot rigorously test general hypotheses, the study did yield numerous suggestions that led to design improvements. Most of these concerned the way in which the system fit into the users' everyday patterns of work and social interaction. One result concerned S's assignment of priorities to messages, which it used to determine the nature and timing of notification: The users found that explicit, distinct auditory cues to a message's priority were less useful than simply having a message presented in a manner appropriate to its priority. (Note that what is involved here is a tradeoff between transparency and unobtrusiveness; cf. Fig. 15.20.) More generally, one user found the original scheme of auditory cues to be too complex to be used effectively during everyday activities and interactions. As was noted previously, it would be more difficult to obtain useful results concerning questions like this from a controlled study.

Use of Questionnaires. More extensive data on actual use—including use over a long period of time—can be obtained through questionnaires. For example, Schaumburg (2001) obtained data from 105 largely experienced users of the OFFICE ASSISTANT, each of whom spent 10–15 min filling out a questionnaire. Subjects reported both on their actual use of the system (e.g., whether they tended to use it when confronted with a problem) and on their subjective reactions to it (in particular, evaluations concerning key usability variables).

Assessing Accuracy. Field studies can yield information about modeling accuracy if it is possible after the fact to check S's predictions against U's actual behavior—for example, when S makes recommendations that U can either accept or reject (see, e.g., the interesting analysis of this type conducted by Mitchell et al., 1994). But, as was discussed previously, it may be at least as effective to employ data concerning the use of a nonadaptive version of the system for this purpose.

Assessing Usability. As the examples given have indicated, studies of actual use yield especially valuable information concerning the bottom-line question of the overall usefulness and usability of a UAS in real situations. If the study is designed accordingly, comparisons between competing system versions can be made (see, e.g., the large-scale field trial of a personalized medical information system described by Jones et al., 1999). As always with empirical studies, care must be taken in generalizing

beyond the particular system versions, user groups, and tasks employed in the study.

THE FUTURE OF USER-ADAPTIVE SYSTEMS

This chapter has shown that adaptive interfaces, agents, and other user-adaptive systems do not represent a smooth and easy road to more successful human–computer interaction: They present a complex set of usability challenges; they require carefully designed methods of acquiring information about users, as well as relatively sophisticated computational techniques that are not needed in other types of interactive system. And even when all of these requirements have been dealt with, it is often tricky to prove empirically that user-adaptivity has actually added any value. It is no wonder that some experts believe that the interests of computer users are better served by continued progress within more familiar paradigms of user-centered system design (see, e.g., Shneiderman & Maes, 1997).

On the other hand, our understanding of the complex challenges raised by UASs has been growing steadily, and they are now familiar and valued elements in a number of types of system.

Growing Need for User-Adaptivity

Increases in the following variables suggest that the functions served by user-adaptivity will continue to grow in importance.

Diversity of Users and Contexts of Use. As several chapters in this handbook make clear, computing devices are being used by an ever-increasing variety of users in an increasing variety of contexts. (See Section B, *Interaction Issues for Diverse Users*, as well as the chapter by Stephanidis & Savidis.) It is therefore becoming harder to design a system that will be suitable for all users and contexts without some sort of user-adaptivity or user-controlled adaptability. As has been discussed at several points in this chapter, adaptability has its limitations.

Number and Complexity of Interactive Systems. The functions of user-adaptivity discussed in the second section partly involve helping users to deal effectively with interactive systems and tasks even when they are not able or willing to gain complete understanding and control in each individual case. This goal becomes increasingly important as the number—and in some cases the complexity—of the systems that people must deal with continues to increase, because of factors ranging from the growth of the World Wide Web to the proliferation of miniature interactive computing devices.

Scope of Information to Be Dealt With. Even when using a single, relatively simple system, users today can often access a much larger and more diverse set of objects of interest than they could a few years ago—be they documents, products, or potential collaborators. It is therefore becoming relatively more attractive to delegate some of the work of dealing with these objects, even to a system that has an imperfect model of the user's requirements.

Increasing Feasibility of Successful Adaptation

As the need for user-adaptivity increases, so—fortunately—does its feasibility, largely because of advances in the following areas:

Ways of Acquiring Information About Users. Most of the methods for acquiring information about users are becoming more powerful with advances in technology and research. They therefore offer the prospect of substantial increases in the quality of adaptation, although methods for ensuring users' privacy call for equal attention.

Advances in Techniques for Learning, Inference, and Decision. In addition to the more general progress in the fields of machine learning and artificial intelligence, communities of researchers have been focusing on the specific requirements of computational techniques that support user-adaptivity. Consequently, noticeable progress is being made every year in this area.

Attention to Empirical Methods. The special empirical issues and methods that are involved in the design and evaluation of user-adaptive systems have been receiving increasing attention from researchers, as emphasis has shifted from high technical sophistication to ensuring that the systems enhance the users' experience.

Concluding Remarks

The future role of user-adaptive systems will not be the result of a sudden paradigm shift motivated by a desire to emulate interaction among humans. It will be shaped by continuing technical progress and increases in understanding along the many frontiers reviewed in this chapter.

ACKNOWLEDGMENTS

Preparation of this chapter was supported by the German Ministry of Education and Research under Grant 01 IW 001 as part of the MIAU project. Valuable comments on an earlier version were supplied by Eric Horvitz, Rabia Aziz, Irfan Jaffry, Abdul Khan, Olga Lamonova, Shivaz Mehta, and Marie Norlien.

References

André, E., & Rist, T. (1995). Generating coherent presentations employing textual and visual material. *Artificial Intelligence Review, 9,* 147–165.

Balabanović, M., & Shoham, Y. (1997). Fab: Content-based, collaborative recommendation. *Communications of the ACM, 40*(3), 66–72.

Billsus, D., & Pazzani, M. J. (2000). User modeling for adaptive news access. *User Modeling and User-Adapted Interaction, 10,* 147–180.

Bradley, K., Rafter, R., & Smyth, B. (2000). Case-based user profiling for content personalisation. In P. Brusilovsky, O. Stock, & C. Strapparava (Eds.), *Adaptive hypermedia and adaptive web-based systems: Proceedings of AH 2000* (pp. 62–72). Berlin: Springer.

Brusilovsky, P. (1996). Methods and techniques of adaptive hypermedia. *User Modeling and User-Adapted Interaction, 6,* 87–129.

Brusilovsky, P. (2001). Adaptive hypermedia. *User Modeling and User-Adapted Interaction, 11,* 87–110.

Budzik, J., Hammond, K., & Birnbaum, L. (2001). Information access in context. *Knowledge-Based Systems, 14,* 37–53.

Burke, R. D. (2001). Knowledge-based recommender systems. *Encyclopedia of Library and Information Science, 69.*

Burke, R. D., Hammond, K. J., & Young, B. C. (1997). The FindMe approach to assisted browsing. *IEEE Expert, 12*(4), 32–40.

Carberry, S. (2001). Techniques for plan recognition. *User Modeling and User-Adapted Interaction, 11,* 31–48.

Chin, D. N. (2001). Empirical evaluation of user models and user-adapted systems. *User Modeling and User-Adapted Interaction, 11,* 181–194.

Collins, J. A., Greer, J. E., Kumar, V. S., McCalla, G. I., Meagher, P., & Tkatch, R. (1997). Inspectable user models for just-in-time workplace training. In A. Jameson, C. Paris, & C. Tasso (Eds.), *User modeling: Proceedings of the Sixth International Conference, UM97* (pp. 327–337). Vienna: Springer Wien New York.

Corbett, A. (2001). Cognitive computer tutors: Solving the two-sigma problem. In M. Bauer, P. Gmytrasiewicz, & J. Vassileva (Eds.), *UM2001, User modeling: Proceedings of the Eighth International Conference* (pp. 137–147). Berlin: Springer.

Corbett, A., McLaughlin, M., & Scarpinatto, K. C. (2000). Modeling student knowledge: Cognitive tutors in high school and college. *User Modeling and User-Adapted Interaction, 10,* 81–108.

Corbett, A. T., & Koedinger, K. R. (1997). Intelligent tutoring systems. In M. Helander, T. K. Landauer, & P. V. Prabhu (Eds.), *Handbook of human-computer interaction* (pp. 849–874). Amsterdam: North-Holland.

Dey, A. K., Abowd, G. D., & Wood, A. (1998). Cyberdesk: A framework for providing self-integrating context-aware services. *Knowledge-Based Systems, 11,* 3–13.

Fink, J., & Kobsa, A. (2000). A review and analysis of commercial user modeling servers for personalization on the world wide web. *User Modeling and User-Adapted Interaction, 10,* 209–249.

Fink, J., Kobsa, A., & Nill, A. (1998). Adaptable and adaptive information provision for all users, including disabled and elderly people. *New Review of Hypermedia and Multimedia, 4,* 163–188.

Ghosh, A. K., & Swaminatha, T. M. (2001). Software security and privacy risks in mobile e-commerce. *Communications of the ACM, 44*(2), 51–57.

Goecks, J., & Shavlik, J. (2000). Learning users' interests by unobtrusively observing their normal behavior. In H. Lieberman (Ed.), *IUI 2000: International Conference on Intelligent User Interfaces* (pp. 129–132). New York: ACM.

Good, N., Schafer, J. B., Konstan, J. A., Borchers, A., Sarwar, B., Herlocker, J., & Riedl, J. (1999). Combining collaborative filtering with personal agents for better recommendations. In *Proceedings of the Sixteenth National Conference on Artificial Intelligence* (pp. 439-446). Orlando, FL.

Greer, J. E., McCalla, G. I., Collins, J. A., Kumar, V. S., Meagher, P., & Vassileva, J. (1998). Supporting peer help and collaboration in distributed workplace environments. *International Journal of AI and Education, 9*, 159-177.

Gutkauf, B., Thies, S., & Domik, G. (1997). A user-adaptive chart editing system based on user modeling and critiquing. In A. Jameson, C. Paris, & C. Tasso (Eds.), *User modeling: Proceedings of the Sixth International Conference, UM97* (pp. 159-170). Vienna: Springer Wien New York.

Haller, S., McRoy, S., & Kobsa, A. (Eds.). (1999). *Computational models of mixed-initiative interaction.* Dordrecht, The Netherlands: Kluwer.

Hammond, N. (1987). Principles from the psychology of skill acquisition. In M. M. Gardiner & B. Christie (Eds.), *Applying cognitive psychology to user-interface design* (pp. 163-188). Chichester, England: Wiley.

Han, J., & Kamber, M. (2001). *Data mining: Concepts and techniques.* San Francisco: Morgan Kaufmann.

Hanani, U., Shapira, B., & Shoval, P. (2001). Information filtering: Overview of issues, research and systems. *User Modeling and User-Adapted Interaction, 11*, 203-259.

Heckerman, D. (1998). A tutorial on learning with Bayesian networks. In M. I. Jordan (Ed.), *Learning in graphical models.* Dordrecht: Kluwer.

Hegner, S. J., McKevitt, P., Norvig, P., & Wilensky, R. L. (Eds.). (2001). *Intelligent help systems for UNIX.* Dordrecht, The Netherlands: Kluwer.

Hirst, G., DiMarco, C., Hovy, E., & Parsons, K. (1997). Authoring and generating health-education documents that are tailored to the needs of the individual patient. In A. Jameson, C. Paris, & C. Tasso (Eds.), *User modeling: Proceedings of the Sixth International Conference, UM97* (pp. 107-118). Vienna: Springer Wien New York.

Höök, K. (2000). Steps to take before IUIs become real. *Interacting with Computers, 12*, 409-426.

Horvitz, E. (1999). Principles of mixed-initiative user interfaces. In M. G. Williams, M. W. Altom, K. Ehrlich, & W. Newman (Eds.), *Human factors in computing systems: CHI '99 conference proceedings* (pp. 159-166). New York: ACM.

Horvitz, E., Breese, J., Heckerman, D., Hovel, D., & Rommelse, K. (1998). The Lumière project: Bayesian user modeling for inferring the goals and needs of software users. In G. F. Cooper & S. Moral (Eds.), *Uncertainty in artificial intelligence: Proceedings of the Fourteenth Conference* (pp. 256-265). San Francisco: Morgan Kaufmann.

Horvitz, E., Jacobs, A., & Hovel, D. (1999). Attention-sensitive alerting. In K. B. Laskey & H. Prade (Eds.), *Uncertainty in artificial intelligence: Proceedings of the Fifteenth Conference* (pp. 305-313). San Francisco: Morgan Kaufmann.

Jameson, A. (1996). Numerical uncertainty management in user and student modeling: An overview of systems and issues. *User Modeling and User-Adapted Interaction, 5*, 193-251.

Jameson, A., Großmann-Hutter, B., March, L., Rummer, R., Bohnenberger, T., & Wittig, F. (2001). When actions have consequences: Empirically based decision making for intelligent user interfaces. *Knowledge-Based Systems, 14*, 75-92.

Jensen, F. V. (2001). *Bayesian networks and decision graphs.* New York: Springer.

Joachims, T., Freitag, D., & Mitchell, T. (1997). WebWatcher: A tour guide for the World Wide Web. In M. E. Pollack (Ed.), *Proceedings of the Fifteenth International Joint Conference on Artificial Intelligence* (pp. 770-777). San Francisco, CA: Morgan Kaufmann.

Jones, R., Pearson, J., McGregor, S., Cawsey, A. J., Barrett, A., Craig, N., Atkinson, J. M., Gilmour, W. H., & McEwen, J. (1999). Randomised trial of personalised computer based information for cancer patients. *British Medical Journal, 319*, 1241-1247.

King, M. F., & Bruner, G. C. (2000). Social desirability bias: A neglected aspect of validity testing. *Psychology & Marketing, 17*, 79-103.

Kobsa, A. (2001). Generic user modeling systems. *User Modeling and User-Adapted Interaction, 11*, 49-63.

Kobsa, A., & Wahlster, W. (Eds.). (1989). *User models in dialog systems.* Berlin: Springer.

Krulwich, B. (1997). Lifestyle Finder: Intelligent user profiling using large-scale demographic data. *AI Magazine, 18*(2), 37-45.

Lajoie, S. P., & Vivet, M. (Eds.). (1999). *Artificial intelligence in education: Open learning environments: New computational technologies to support learning, exploration, and collaboration.* Amsterdam: IOI Press.

Langley, P. (1996). *Elements of machine learning.* San Francisco: Morgan Kaufmann.

Langley, P., & Fehling, M. (1998). *The experimental study of adaptive user interfaces* (Technical Report 98-3). Institute for the Study of Learning and Expertise, Palo Alto, CA.

Lanier, J. (1995). Agents of alienation. *interactions, 2*(3), 66-72.

Lieberman, H. (Ed.). (2001). *Your wish is my command: Programming by example.* San Francisco: Morgan Kaufmann.

Linden, G., Hanks, S., & Lesh, N. (1997). Interactive assessment of user preference models: The automated travel assistant. In A. Jameson, C. Paris, & C. Tasso (Eds.), *User modeling: Proceedings of the Sixth International Conference, UM97* (pp. 67-78). Vienna: Springer Wien New York.

Litman, D. J., & Pan, S. (2000). Predicting and adapting to poor speech recognition in a spoken dialogue system. In *Proceedings of the Seventeenth National Conference on Artificial Intelligence* (pp. 722-728). Austin, TX.

Maes, P. (1994). Agents that reduce work and information overload. *Communications of the ACM, 37*(7), 30-40.

Maglio, P. P., Barrett, R., Campbell, C. S., & Selker, T. (2000). SUITOR: An attentive information system. In H. Lieberman (Ed.), *IUI 2000: International Conference on Intelligent User Interfaces* (pp. 169-176). New York: ACM.

Maglio, P. P., & Campbell, C. S. (2000). Tradeoffs in displaying peripheral information. In T. Turner, G. Szwillus, M. Czerwinski, & F. Paternò (Eds.), *Human factors in computing systems: CHI 2000 conference proceedings* (pp. 241-248). New York: ACM.

Martin, D. M., Smith, R. M., Brittain, M., Fetch, I., & Wu, H. (2001). The privacy practices of web browser extensions. *Communications of the ACM, 44*(2), 45-50.

McDonald, D. W., & Ackerman, M. S. (2000). Expertise Recommender: A flexible recommendation system and architecture. In P. Dourish & S. Kiesler (Eds.), *Proceedings of the 2000 Conference on Computer-Supported Cooperative Work* (pp. 231-240). New York: ACM.

Mitchell, T., Caruana, R., Freitag, D., McDermott, J., & Zabowski, D. (1994). Experience with a learning personal assistant. *Communications of the ACM, 37*(7), 81-91.

Mitchell, T. M. (1997). *Machine learning.* Boston: McGraw-Hill.

Mladenic, D. (1999). Text-learning and related intelligent agents: A survey. *IEEE Intelligent Systems, 14*(4), 44-54.

Müller, C., Großmann-Hutter, B., Jameson, A., Rummer, R., & Wittig, F. (2001). Recognizing time pressure and cognitive load on the basis of speech: An experimental study. In M. Bauer, P. Gmytrasiewicz, & J. Vassileva (Eds.), *Proceedings of the Eighth International Conference on User Modeling.* Berlin: Springer.

Norman, D. A. (1994). How might people interact with agents. *Communications of the ACM, 37*(7), 68-71.

Paiva, A. (1997). Learner modelling for collaborative learning environments. In B. du Boulay & R. Mizoguchi (Eds.), *Artificial intelligence in education: Knowledge and media in learning systems* (pp. 215-222). Amsterdam: IOI Press.

Pazzani, M., & Billsus, D. (1997). Learning and revising user profiles: The identification of interesting web sites. *Machine Learning, 27*, 313-331.

Pazzani, M. J. (1999). A framework for collaborative, content-based and demographic filtering. *Artificial Intelligence Review, 13*, 393-408.

Pazzani, M. J., & Billsus, D. (1999). Evaluating adaptive web site agents. Presented at the Workshop on Recommender Systems: Algorithms and Evaluation, 22nd International Conference on Research and Development in Information Retrieval, Berkeley, CA.

Pearl, J. (1988). *Probabilistic reasoning in intelligent systems: Networks of plausible inference.* San Mateo, CA: Morgan Kaufmann.

Picard, R. W. (1997). *Affective computing.* Cambridge, MA: MIT Press.

Rhodes, B. J. (2000). *Just-in-time information retrieval.* Unpublished doctoral dissertation, School of Architecture and Planning, Massachusetts Institute of Technology. Cambridge, MA. (Available: http://rhodes.www.media.mit.edu/people/rhodes/research/).

Rich, E. (1979). User modeling via stereotypes. *Cognitive Science, 3*, 329-354.

Rich, E. (1989). Stereotypes and user modeling. In A. Kobsa & W. Wahlster (Eds.), *User models in dialog systems* (pp. 35-51). Berlin: Springer.

Rössel, M. (2000). *Ein System zur individualisierten Informationsvermittlung—dargestellt am Beispiel eines multimedialen Branchenkatalogs der Technischen Keramik [A system for individualized information presentation—as exemplified by a multimedia product catalog for technical ceramics].* Unpublished doctoral dissertation, Bereich Wirtschaftsinformatik I, Universität Erlangen-Nürnberg.

Sawhney, N., & Schmandt, C. (2000). Nomadic Radio: Speech and audio interaction for contextual messaging in nomadic environments. *ACM Transactions on Computer-Human Interaction, 7*, 353-383.

Schafer, J. B., Konstan, J., & Riedl, J. (1999). Recommender systems in e-commerce. In *Proceedings of the ACM Conference on Electronic Commerce*, Minneapolis, MN.

Schaumburg, H. (2001). Computers as tools or as social actors?—The users' perspective on anthropomorphic agents. *International Journal on Intelligent Cooperative Information Systems, 10*, 217-234.

Schiele, B., Starner, T., Rhodes, B., Clarkson, B., & Pentland, A. (2000). Situation aware computing with wearable computers. In W. Barfield & T. Caudell (Eds.), *Augmented reality and wearable computers.* Mahwah, NJ: Erlbaum.

Sears, A., & Shneiderman, B. (1994). Split menus: Effectively using selection frequency to organize menus. *ACM Transactions on Computer-Human Interaction, 1*, 27-51.

Segal, R. B., & Kephart, J. O. (1999). MailCat: An intelligent assistant for organizing e-mail. In *Proceedings of the Third International Conference on Autonomous Agents* (pp. 276-282).

Segal, R. B., & Kephart, J. O. (2000). Incremental learning in SwiftFile. In P. Langley (Ed.), *Machine Learning: Proceedings of the 2000 International Conference.* San Francisco: Morgan Kaufmann.

Shearin, S., & Lieberman, H. (2001). Intelligent profiling by example. In J. Lester (Ed.), *IUI 2001: International Conference on Intelligent User Interfaces* (pp. 145-151). New York: ACM.

Shneiderman, B. (1995). Looking for the bright side of user interface agents. *interactions, 2*(1), 13-15.

Shneiderman, B., & Maes, P. (1997). Direct manipulation vs. interface agents. *interactions, 4*(6), 42-61.

Trewin, S. (2000). Configuration agents, control, and privacy. *Proceedings of the ACM Conference on Universal Usability* (pp. 9-16). Arlington, VA.

Trewin, S., & Pain, H. (1998). A model of keyboard configuration requirements. In *Proceedings of the Third International ACM Conference on Assistive Technologies* (pp. 173-181). New York: ACM.

Vivacqua, A., & Lieberman, H. (2000). Agents to assist in finding help. In T. Turner, G. Szwillus, M. Czerwinski, & F. Paternò (Eds.), *Human factors in computing systems: CHI 2000 conference proceedings* (pp. 65-72). New York: ACM.

Wainer, H. (Ed.). (2000). *Computerized adaptive testing: A primer* (2nd ed.). Hillsdale, NJ: Erlbaum.

Webb, G., Pazzani, M. J., & Billsus, D. (2001). Machine learning for user modeling. *User Modeling and User-Adapted Interaction, 11*, 19-29.

Weber, G., & Specht, M. (1997). User modeling and adaptive navigation support in WWW-based tutoring systems. In A. Jameson, C. Paris, & C. Tasso (Eds.), *User modeling: Proceedings of the Sixth International Conference, UM97* (pp. 289-300). Vienna: Springer Wien New York.

Wexelblat, A., & Maes, P. (1997). *Issues for software agent UI.* Unpublished manuscript, available from http://wex.www.media.mit.edu/people/wex/.

Wittig, F., & Jameson, A. (2000). Exploiting qualitative knowledge in the learning of conditional probabilities of Bayesian networks. In C. Boutilier & M. Goldszmidt (Eds.), *Uncertainty in artificial intelligence: Proceedings of the Sixteenth Conference* (pp. 644-652). San Francisco: Morgan Kaufmann.

Wolfman, S. A., Lau, T., Domingos, P., & Weld, D. S. (2001). Mixed initiative interfaces for learning tasks: SMARTedit talks back. In J. Lester (Ed.), *IUI 2001: International Conference on Intelligent User Interfaces* (pp. 167-174). New York: ACM.

Zukerman, I., & Albrecht, D. W. (2001). Predictive statistical models for user modeling. *User Modeling and User-Adapted Interaction, 11*, 5-18.

Zukerman, I., & Litman, D. (2001). Natural language processing and user modeling: Synergies and limitations. *User Modeling and User-Adapted Interaction, 11*, 129-158.

·16·

NETWORK-BASED INTERACTION

Alan Dix

Lancaster University and vfridge limited

INTRODUCTION

In some ways, this chapter could be seen as redundant in a human–computer interaction book; surely networks are just an implementation mechanism, a detail below the surface—all that matters are the interfaces that are built on them. On the other hand, networked interfaces, especially the web, but increasingly also mobile devices, have changed the way we view the world and society. Even those bastions of conservatism, the financial institutions, have found themselves in sea-change and a complete restructuring of the fundamentals of businesses . . . just an implementation detail.

Structure

This chapter begins with a brief overview of types of networks and then deals with network-based interaction under four main headings:

- networks as enablers: things that are only possible with networks
- networks as mediators: issues and problems because of networks
- networks as subjects: understanding and managing networks
- networks as platforms: algorithms and architectures for distributed interfaces

In addition, there will be a section taking a broader view of the history and future of network interaction and the societal effects and paradigm changes engendered, especially by more recent developments in global and wireless networking.

ABOUT NETWORKS

The word network will probably make many think of accessing the Internet and the web. Others may think of a jumble of ethernet wires between the PCs in their office. In fact, the range of networking standards, including physical cabling (or lack of cabling), and the protocols that computers use to talk down those cables is extensive. Although most of the wire-based networks have been around for some time, they are in a state of flux because of increases in scale and the demands of continuous media. In the wireless world, things are changing even more rapidly with two new generations of data service being introduced over the next 2 years.

As an aid to seeing the broader issues surrounding these changing (and, in some cases potentially ephemeral) technologies, we can use the following two dimensions to classify them:

- global vs. local
  How spatially distant are the points connected—ranging from machines in the same room (IrDa, Bluetooth), through those in a building/site (LAN) to global networks (Internet, mobile phone networks)

	fixed	flexible
local	LAN	PAN IrDa bluetooth wireless LAN
	— WAN —	
global	Internet	GSM, GPRS, etc.
	mobile	

- fixed vs. flexible
  How permanent are the links between points of the network, from physically fixed machines, to self reconfiguring devices that recognize other devices in their vicinity.

The fixed vs. flexible dimension is almost, but not quite, terrestrial vs. wireless. The not quite is because fixed networks increasingly involve wireless links. Also, it is often possible, when visiting another organization, to plug a portable computer into a (wired) ethernet network and find you have access to the local printers, Internet connections, etc.—flexible wire-based networking.

Let's look at a few network technologies against these dimensions. Traditional office LANs (local area networks) are squarely in the local–fixed category, whereas the Internet is largely in the global–fixed category. Corporate WANs (wide area networks), connecting offices within the same national or international company, sit somewhere between.

Mobile phones have been placed within the global–fixed category as well. This may seem strange—the phone can go anywhere. However, the interconnections between phones are fixed and location-independent. If two mobile phones are in the same room, it is no easier to connect between them than if they are at opposite ends of the earth (bar a shorter lag time perhaps). Similarly, the Internet although increasingly accessible through mobile devices and phones, is largely based on fixed domain names, Internet protocol (IP) numbers, and uniform resource locators (URLs).

Given the placement of mobile phones is a little ambiguous, and it is possible to detect the location of phones and thus deliver location-based content, some of the phone technologies have been listed in the global–flexible category (the global system for mobile communications, or GSM, and the general packet radio service, or GPRS, are the names of the first- and second-generation data services).

There is obviously a steadily increasing data rate, and third-generation services are looking toward being able to cope with heavy media content (a Hewlett-Packard advert has two climbers at the top of a mountain watching a television soap over a mobile connection). However, the most significant differences are the charging and connectivity model. With GSM, you connect when required to the Internet, and this is treated like any other telephone call, usually meaning pay per minute while connected. In contrast, GPRS is based on sending small packets of data (the P in the acronym). The connection to the Internet is treated as always on, and packets of data are sent to or from the phone as required. Charging is also typically by data use.

In the local–flexible category, there is a host of existing and emerging technologies. At the most mundane are wireless ethernet networks (so-called 802.11 devices, such as Apple airport or PC WaveLAN cards;) IEEE 802.11. These merely treat the machine the same as if plugged into the local–fixed network. At a more local scale, infrared (IrDa) enabled devices can talk to one another if their infrared sensors are within line of sight. These are also more flexible as a personal digital assistant (PDA) placed near a mobile phone will be able to use the phone's modem with little intervention (except perhaps to carefully align the devices).

Bluetooth has been much hyped and uses radio to connect close devices in a similar fashion to IrDa (but faster and more intelligently) (Bluetooth, 2001). For example, a Bluetooth hands-free headset can connect to a Bluetooth phone without having to plug in with a piece of wire.

Finally, research in wearable computers has suggested using the body itself as the connection between worn devices in a personal area network (PAN) (Zimmerman, 1996). The future is networked and we will become the network!

On the whole, we have seen in the last 10 years the main focus of network-based interaction has moved anticlockwise in this picture from fixed/local networks (mainly LAN), through fixed global networks (the Internet and web explosion), through global mobile networks (mostly phone-based, but including the wireless application protocol (WAP), i-mode, etc.) and moving toward flexible local connections between devices.

NETWORKS AS ENABLERS: THINGS THAT ARE ONLY POSSIBLE WITH NETWORKS

It can be the case that the network is no more than an implementation detail—for example, using a networked disk rather than a local one. However, there are also many applications that are only possible because the network is there, for example, video-conferencing. The key feature of networks is the access to remote resources of some kind or other.

Remote Resources

Four kinds of remote things are made accessible by networks:

- People
- Physical things
- Data
- Computation

These may be remote because they are far away from where you normally are, or because you are yourself on the move and hence away from one's own resources (colleagues, databases, etc.). Thus, mobility can create a need for any or all the above.

People. Networks mean we can communicate and work with others in distant places. This is often a direct action, e-mailing someone, engaging in a video-conference. These are all the normal study of computer-supported cooperative work (CSCW) and groupware (see chapter 29).

Interaction with remote people may also be indirect. Recommender systems gather information about people's preferences and use this to suggest further information, services, or goods based on your own preferences and those of others who have similar tastes (Resnick & Varian, 1997). Because the people making recommendations are in different locations from each other, the data on who selected what must be stored centrally.

Collaborative virtual environments also offer the ability for remote people to interact, but by embedding them within an apparently local virtual reality world. Although the people you are dealing with may be half a world away, their avatar (a virtual presence, perhaps a cartoon character, photo, or robotlike creature) may seem only a few yards or meters away in the virtual world.

Physical Things. We can also view and control remote things at a distance. For example, live web cams in public places allow us to see things (and people) there. Similarly, the cameras mounted around rockets as they prepare to take off (and then usually destroyed during the launch) allow the mission controllers to monitor critical aspects of the physical system, as do the numerous telemetry sensors that will also be related via some sort of closed network. Of course the launch command itself will be relayed to the rocket by the same closed network, as will the ongoing mission, perhaps the Mars robots, via wireless links.

In the rocket example, it would be dangerous to be in the actual location; in other circumstances, it is merely expensive or inconvenient. Telescopes are frequently mounted in distant parts of the world where skies are clearer than those above the laboratories to which they belong. To avoid long international trips to remote places, some of these now have some form of remote control and monitoring using the Internet (Lavery, Kilgourz, & Sykeso, 1994).

At a more personal level, the systems within certain high-end cars are controlled using a within-car network (called CAN). Even an adjustable heated seat may require dozens of control wires; but, with a network, only one power and one control cable are needed. The engine management system, lighting assemblies, radio, CD player, wind screen wipers, each have a small controller that talks through the network to the driver's console (although critical engine systems will usually have a separate circuit).

Many household appliances are now being made Internet-ready. In some cases, this may mean an actual interface—for example, an Internet fridge that can scan the bar codes of items as you put them in and out, and then warn you when items are getting out of date, generate a shopping list of items for you, and even order from your favorite store (Electrolux, 1999). Others have instead or in addition connectivity for maintenance purposes, sending usage and diagnostic data back to the manufacturer so that they can organize service or repair visits before the appliance fails in some way.

Data. Anyone using the web is accessing remote data. Sometimes, data is stored remotely purely for convenience, but often data is necessarily stored remotely because:

- it is shared by many remote people
- central storage helps maintain control, security, or privacy
- it is used by a single user at different locations (web e-mail)
- it is too extensive to be stored locally (e.g., large databases and fat client/server).

In the case of the web, the data is remote because it is accessed by different people at different locations, the author(s) of the material, and all those who want to read it.

Even the web is quite complex; we may perceive a web page as a single entity, but in fact it exists in many forms. The author of the page will typically have created it offline on their own PC. They then upload the page (which effectively means copying it) onto the web server. Any changes the author makes after uploading the page will not be visible to the world until it is next uploaded. When a user wants to see the page and enters a URL or clicks on a link, their browser asks the web server for the file, which is then copied into the browser's memory and displayed to the user. You can tell the browser has a copy because you can disconnect from the Internet and still scroll within the file. If you access the same page again quite soon, your browser may choose to use the copy it holds rather than going back to the web server; again, potentially meaning you see a slightly out-of-date copy of the page. Various other things may keep their own cached copies, including web proxies and firewalls (Fig. 16.1).

This story of copied data in various places is not just about the web, but true to some extent or other of all shared networked data. With people or physical things, we do not expect to have the actual person or thing locally, just a representation. This is equally true for shared data, except that the representation is so much like the real thing, it is far less obvious to the user.

For shared networked data, even the real thing may be problematic, there may be no single golden copy; but, instead, many variants, all with equal right to be called the real data.

You don't even escape networking issues if you only access data locally on your own PC. Networking issues may still arise if your data is backed up over the network.

Computation. Sometimes it is remote computational resources that are accessed over the network. The most obvious example of this are large supercomputers. These have enormous computational power, and scientists wishing to use them will often prebook time slots to perform particularly intensive calculations, such as global weather simulations, analysis of chemical structure, stress calculations, etc. Because these machines are so expensive, programs for them are typically developed on other less powerful computers and then uploaded over the network when the supercomputer is available.

If the data required as input or output for the calculation are not too great, fairly simple means can be used to upload the programs and data. However, some calculations work on large volumes of data; for example, data from microwave readings of the upper atmosphere to probe the ozone hole generate terabytes (millions of millions of bytes) of data per second. High-capacity networks are being created in many countries to enable both high-volume data for this sort of application and also the expected data required for rich media (Foster, 2000; Foster & Kesselman, 1999, 2000; GRID, 2001).

Sometimes calculations need to be performed centrally, not because the central computer is powerful, but because the local device is a computational lightweight. For example, one may want to create a remote analysis package in which engineers in the field enter data into a PDA or phone interface, but where complex stress calculations are conducted on a small server back in the office. The data on materials and calculations involved may not be extensive by supercomputer standards, but may still be too much for a hand-held device.

Because transporting large volumes of data is not always practical, calculations are often performed where the data is. (In performing any computation program, data and computational engine must all be in the same place. If they are not together, then one or the other must move or be copied to bring them together; Ramduny & Dix, 1997.) For example, when you perform a database access, the request for the data is usually transmitted to the database server as an SQL query, for example,

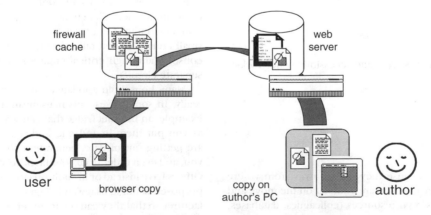

FIGURE 16.1. Copies of a web page in many places.

"SELECT name, salary FROM payroll WHERE salary > 70,000." In principle, the complete contents of the payroll database could be downloaded to your PC and the selection of appropriate records conducted locally; however, it would be more costly to transmit the data, hence the calculation is effectively transmitted to the database server.

Even when the volume of data is not large or the frequency of access would make it cost-effective to transmit it, security or privacy reasons may prevent the download of data. For example, some data sets are available to search to a limited degree on the web, but charge for a download or CD of the complete data set. My own hcibook.com site (Dix, Finlay, Abowd, & Beale, 1998) is rather like this, allowing searching of the book's contents online and displaying portions of the text, but not allowing a full download because readers are expected to buy the book.

Security considerations may also prohibit the distribution of programs themselves if they contain proprietary algorithms. Also, if the source of the program is not fully trusted, one may not want to run these programs locally. The latter is the reason that Java applets are run in a software sandbox confining the ability of the applet to access local files and other potentially vulnerable resources.

SETI (Aug. 2001) is an interesting example of remote computation (SETI@home). Normally, remote computation involves a device of low-computational power asking a central computer to do work for it. In the case of SETI, large calculations are split up and distributed over large numbers of not particularly powerful computers.

The same technique is used in PC farms, in which large numbers of PCs are networked together to act as a form of supercomputer. For example, in CERN (the home of the web), data from high-energy collisions may consist of many megabytes of data for each event, with perhaps hundreds of significant events per second (CERN, Aug. 2001). Data from each event are passed to a different PC, which then performs calculations on the data. When the PC finishes, it stores the results and then adds itself back to a pool of available machines.

In the coming years, we are likely to see both forms of remote computation. As devices become smaller and more numerous, many will become simply sensors or actuators communicating with central computational and data servers (although central here may mean one per room, or even one per body). On the other hand, inspired by SETI, several companies are pursuing commercial ways of harnessing the spare, and usually wasted, computational power of the millions of home and office PCs across the world.

Applications

The existence of networks, particularly the global networks offered by the Internet and mobile phone networks, has made many new applications possible and changed others. Several of the more major application areas made possible by networks are covered in their own chapters: groupware (chapter 29), on-line communities (chapter 30), mobile systems (chapter 32), e-commerce (chapter 39), telecommunications (chapter 40) and, of course, the web (chapter 37).

In addition, networking impinges on many other areas. Handheld devices (chapter 32) can operate alone, but are increasingly able to interact with one another and with fixed networks via wireless networking. Similarly, wearable computing (chapter 33) is expected to be interacting with one another via short-range networks, possibly carried through our own bodies (makes mobile phones seem positively safe) and information appliances (chapter 38) will be Internet connected to allow remote control and maintenance. In the area of government and citizenship (chapter 41), terms such as e-democracy and e-government are used to denote not just the technological ability to vote or access traditional government publications online, but a broader agenda whereby citizens feel a more intimate connection to the democratic process. Of course, education, entertainment, and game playing are also making use of networks.

Throughout this chapter, we will also encounter broader issues of human abilities, especially concerned with time and delays, involving aspects of virtually all of part II (human perception, cognition, motor skills, etc.). Also, we will find that networking raises issues of trust and ethics (chapters 65 and 62) and, of course, the global network increases the importance of culturally and linguistically accessible information and interfaces (chapter 23).

Networking has already transformed many people's working lives, allowing telecommuting, improving access to corporate information while on the move, and enabling the formation of virtual organizations. Networks are also allowing whole new business areas to develop, not just the obvious applications in e-shopping and those concerned with web design.

The Internet has forced many organizations to create parallel structures to handle the more direct connections between primary supplier and consumer (disintermediation). This paradoxically is allowing more personalized (if not personal) services and often a focus on customer–supplier and customer–customer communication (Light, 2001; Siegal, 1999). This restructuring may also allow the more flexible businesses to revolutionize their high street (or mall) presence, allowing you to buy shoes in different sizes or next-day fitting services for clothes (Dix, 2001b).

The complexity of installing software and the need to have data available anywhere at any time has driven the nascent application service provider (ASP) sector. You do not install software yourself, but use software hosted remotely by providers who charge on a usage rather than once-off basis. By storing the data

---

For those who have not come across it, the SETI (the Search for Extra-Terrestrial Intelligence) project is analyzing radio signals from outer space looking for patterns or regularities that may indicate transmissions from an alien civilization. You can download a SETI screensaver that performs calculations for SETI when you are not using your machine. Each SETI screensaver periodically gets bits of data to analyze from the central SETI servers and then returns results. This means that the SETI project ends up with the combined computational resources of many hundreds of thousands of PCs.

with third parties, an organization can offload the majority of its backup and disaster management requirements.

NETWORKS AS MEDIATORS: ISSUES AND PROBLEMS BECAUSE OF NETWORKS

This section takes as a starting point that an application is networked and looks at the implications this has for the user interface. This is most apparent in terms of timing problems of various kinds. This section is really about when the network is largely not apparent, except for the unintended effects it has on the user.

We will begin with a technical introduction to basic properties of networks and then see how these affect the user interface and media delivery.

Network Properties

Bandwidth and Compression. The most commonly cited network property is bandwidth—how much data can be sent per second. Those who have used dial-up connections will be familiar with 56K modems, and those with longer memories or using mobile phone modems may recall 9.6K modems or less. The K in all of these refers to thousands of bits (0/1 value) per second (strictly Kbps) rather than bytes (single character) that are more commonly seen in disk and other memory sizes. A byte takes 8 bits; taking into account a small amount for overhead, you can divide the bits per second by 10 to get bytes per second.

Faster networks between machines in offices are more typically measured in megabits per second (again strictly Mbps, but often just written M)—for example, the small telephone cable ethernet is rated at either 10 Mbps or 100 Mbps.

As numbers, these do not mean much; but, if we think about them in relation to real data, the implications for users become apparent.

A small word processor document may be 30 Kb (kilo*bytes*). Down a 9.6K modem, this will take approximately half a minute; on a 56K modem, this is reduced to 5 s; for a 10 Mb ethernet, this is 30 ms. A full screen web-quality graphic may be 300 Kb,

Note that I am using the formula:

$$\text{download time } T = \frac{F \times 10}{M}$$

where:
F = size of file in bytes
10 is the number of raw bits per byte
M = modem speed in bits per second

taking 5 min of 9.6K modem, less than 1 min on a 56K modem, or $\frac{1}{3}$ s on a 10 Mb ethernet. (*Note.* These are theoretical minimum times if there is nothing else using the network.)

Rich media, such as sound or video, put a greater load again. Raw, uncompressed HI-FI quality sound needs more than 200 kilo bits per second and video tens of mega bits per second. Happily, there are ways to reduce this; otherwise, digital audiovisual would be impossible over normal networks.

Real media data has a lot of redundant information—areas of similar color in a picture, successive frames in a video are similar, sustained notes in music, etc. Compression techniques use this similarity to reduce the actual amount of data that needs to be sent (e.g., rather than sending a whole new frame of video, just send the differences from the last frame). Also, some forms of compression make use of human perceptual limits: for example, MP3 stores certain pitch ranges with greater fidelity than others, because the human ear's sensitivity is different at different pitches (MPEG, 2001); also, JPEG images give less emphasis to accurate color hue than the darkness/lightness (JPEG, 2001). Between them, these techniques can reduce the amount of information that needs to be transferred significantly, especially for richer media, such as video. Thus, the actual bandwidth and the effective bandwidth, in terms of the sorts of data that are transmitted, may be very different.

Latency and Start-up. Bandwidth measures how much data can be transferred; latency is how long each bit takes (Fig. 16.2). In terms of a highway, bandwidth would be how many lanes and latency is the time it takes to travel the length of the highway. The latency is because of two factors. The first is the speed of transmission of electricity through wires or light through optical networks. This may seem insignificant, but for a beam of light to travel across the Atlantic, it would take 20 ms and in practice

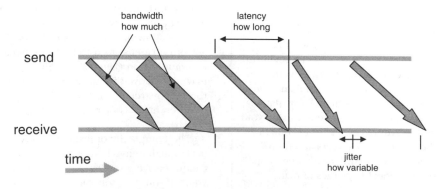

FIGURE 16.2. Bandwidth, latency, and jitter.

this hop takes more like 70 ms. For satellite-based communications, the return trip to and from a geostationary satellite takes nearly a second. Think about the typical delay you can hear on a transcontinental telephone call. The second factor contributing to latency is that every electronic switch or computer routing signal will have to temporarily store, then decide what to do with the signal before passing it on to the next along the chain. Typically, this is a more major influence and, in practice, trans-Atlantic Internet traffic will take nearer 250 ms from source to final destination, most of which is in various computer centers at one end or other.

Latency is made worse by set-up time. Every time you establish an Internet connection, a conversation is established between your computer and the machine hosting the web server: "Hello, are you there?" "Yes, I'm here. What do you want?" "I'd like to send you some data." "Great, I'm waiting." "OK, here it is then." (This is called handshaking.) Each turn in this conversation involves a round trip, network latency on both outward and return paths and processing by both computers. This is before the web server proper even gets to look at your request. Similar patterns happen as you dial a telephone number.

Latency and setup time are critical, because they often dominate the delay for the user except for very large files or streaming audiovisual media. Early web design advice (by those concerned about people with slow connections, but who clearly had never used one) used to suggest having only as much text as would fit on a single screen. This was intended to minimize the download time. However, this ignores set-up times. A long text page does not take long to load even on a slow connection, once the

connection to the web server has been established. Then, it is far faster to scroll in the browser than to click and wait for another small page to load. A similar problem is the practice of breaking large images up into a jigsaw of small pieces. There are valid reasons for this—allowing rollover interaction or where parts of the image are of different kinds (picture/text)—however, it is also used without such reasons; and each small image requires a separate interaction with the server encountering latency and set-up delays.

Jitter and Buffering. Suppose you send letters to a friend every 3 days, and the postal service typically takes 2 days to deliver letters (the average latency in network terms). Your friend will receive letters every 3 days, just delayed from when you sent them. Now, imagine that the postal system is a little variable, sometimes letters take 2 days, but occasionally they are faster and arrive the next day and sometimes slower and take 3 days. You continue to send letters every 3 days, but if a slow letter is followed by a fast one, your friend will receive them only 1 day apart. If, on the other hand, a fast letter is followed by a slow one, the gap becomes 5 days. This variability in the delay is called *jitter*. (*Note.* The fast letters are just as problematic as the slow ones; a fast letter followed by a normal speed one still gives a 4-day gap.)

Jitter does not really matter when sending large amounts of data, or when sending one-off messages. However, it is critical for continuous media. If you just played video frames or sound when it arrived, jitter would mean that the recording would keep accelerating and slowing down (see Figs. 16.3a and 16.3b).

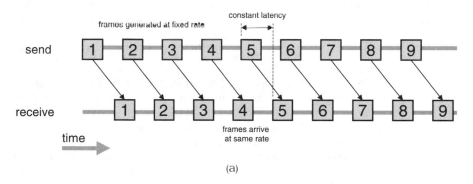

FIGURE 16.3a. No jitter, no problem.

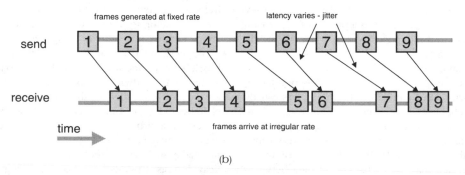

FIGURE 16.3b. Jitter causes irregular reception.

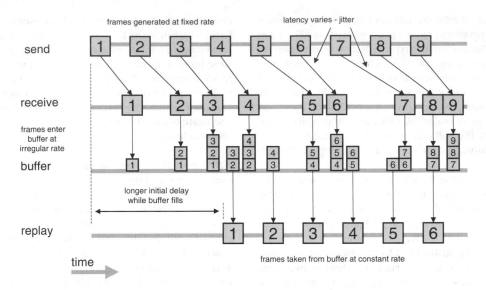

frames generated at fixed rate

latency varies - jitter

send

receive

frames enter buffer at irregular rate

buffer

longer initial delay while buffer fills

replay

time

frames taken from buffer at constant rate

FIGURE 16.4. Buffering smooths jitter, but adds delay.

Jitter can be partially alleviated by buffering (Fig. 16.4). Imagine your friend's postman holds back one letter for 3 days and then starts giving letters to your friend one every third day. If your mail always arrives in exactly 2 days, the postman will always hold exactly one letter, because mail will arrive as fast as he passes it on. If, however, a letter arrives quickly, he will simply hold two letters for a few days; and, if it is slow, he will have a spare letter to give. Your friend's mail is now arriving at a regular rate, but the delay has increased to (a predictable) 5 days. Buffering in network multimedia behaves exactly the same, holding back a few seconds audio/video data and then releasing it at a constant rate.

Reliability and Loss, Datagram, and Connection-Based Services. Virtually all networks are designed on the principle that there will be some loss or damage to data en route. This arises for various reasons. Sometimes there is electrical interference in a wire, sometimes the internal computers and routers in the network may have too much traffic to cope with, etc. This is normal, and network software is built to detect damaged data and cope with lost data.

Because of this, the lowest layers of a network are assumed to be lossy. Any data damaged in transit are discarded, and computers and hardware en route can choose to discard data if they get busy. So, when one computer sends a packet of data to another, it can assume that, if the packet of data arrives, it will be intact, but it may never arrive at all.

Some network data, in particular certain forms of real-time multimedia data, are deliberately sent in this unreliable, message-at-a-time form called datagrams. However, it is usually easier to deal with reliable channels and so higher levels of the network create what are called connection-based services on top of the unreliable lower level service. Internet users may have come across the term TCP/IP. IP is the name of an unreliable low-level service that routes packets of data between computers. TCP is a higher level connection-based service built

on top of IP. The way TCP works is that the computer wanting to make a connection contacts the other, and they exchange a few (unreliable IP) messages to establish the link. Once the link is established, the sending computer tags messages it sends with sequence data. Whenever the receiving computer has all the data up to a certain point, it sends an acknowledgment. If the sending computer does not get an acknowledgment after a certain time, it resends the data (Stevens, 1998, 1999).

With TCP, the receiving computer cannot send a message back; when it notices a gap, it has to wait for the sending computer to resend after the timeout. While it is awaiting the resend, it cannot process any of the later data. Notice what this means: Reliability is bought at the price of potential delays.

Quality of Service and Reservation. The previous properties are not just determined by a raw network's characteristics, such as the length of wires, types of routers, modems, etc. They are also affected by other traffic—its volume and nature. If 10 PCs are connected to a single 10 Mbps network connection and require high-volume data transfers (perhaps streaming video), then there is only, on average, 1 Mbps available for each. If you are accessing a network service that requires trans-Atlantic connections during peak hours, then intermediate routers and hubs in the network are likely to become occasionally overloaded, leading to intermittent packet loss, longer average delays, and more variability in delay, hence jitter.

For certain types of activity, in particular real-time or streaming rich-media, one would like to be able to predict or guarantee a minimum bandwidth, maximum delay and jitter, etc. These are collectively called quality of service issues (Campbell & Nahrstedt, 1997). Some network protocols allow applications to reserve a virtual channel with guaranteed properties; however, the most common large-scale network, the Internet, does not have such guarantees, it operates solely on a best endeavor basis. Upgrades to the underlying protocol that have been under development for some years (called by the catchy name IPv6) will allow some differentiation of different types of traffic. This may allow routers to make decisions to favor time-critical data, but it will still not be able to reserve guaranteed capacity.

Encryption, Authentication, and Digital Signatures. Some networks, such as closed office networks, offer no greater worries about security of information than talking together (both are capable of being bugged, but with similar levels of difficulty). However, more open networks, such as the Internet, or phone networks, mean that data is traveling through a third party and public infrastructure to get to its recipients. Increasing use of wireless devices also means that the data sent between devices is more easily able to be monitored or interfered with by third parties. One option is to only use physically secure networks; but, for economic reasons, this is often not an option. Furthermore, solutions that do not rely on the network itself being secure are more robust. If you rely on, for example, a private dedicated line between two offices and assume it is secure, then if someone does manage to tap into it, all your interoffice communication is at risk.

The more common approach now is to assume the networks are insecure and live with it. This gives rise to two problems:

secrecy—how to stop others from seeing your data
security—how to make sure data is not tampered with

The first problem is managed largely by encryption methods, thus ensuring that even if someone reads all your communications, they cannot understand them (Schneier, 1996). The "https" in some URLs is an example of this, denoting that the communication to the web server is encrypted.

The second problem, security, has various manifestations. Given communications are via a network, how do you know that you are talking to the right person/machine. Authentication mechanisms deal with this. In various ways, they allow one machine to verify (usually by secret information that can only be known by the true intended party) that it is talking to the right party.

Even if you know that you are talking to the right person/machine, how do you know that the data you receive has not been changed? This is like receiving a signed letter, but unbeknownst to you, someone has added some lines of text above the signature; although it really comes from the person you think, the message is not as was sent. If data is being encrypted, then this may often implicitly solve this problem, because any tampered data is uninterpretable by a third party, who therefore cannot alter it in a meaningful way.

If secrecy is not an issue, however, encryption is an unnecessary overhead and instead digital signatures generate a small data block that depends on the whole of the message and secret information known to the sender. It is possible for the recipient to verify that the signature block corresponds to the data sent and the person who is supposed to have sent it. One example of this are signed applets, in which the Java code is digitally signed so that you can choose to only run Java programs from trusted parties.

User Interface Properties

Network Transparency. One of the goals of many low-level network systems is to achieve transparency, that is to make it invisible to the user where on the network a particular resource lies. When you access the web, you use the same kind of URL and same kind of interface whether the web server is in Arizona, Australia, or Armenia. I know that when, at home, I send an e-mail between two machines less than 2 meters apart, the message actually goes all the way across the Atlantic and back; but this is only because I have quite a detailed understanding of the computers involved. As a user, I press send mail on one machine, and it arrives near instantaneously on the other.

Although network transparency has many advantages to the user (you do not care about routes through the network, etc.), there are limits to its effectiveness and desirability. Some years ago, I was at a Xerox lab in Welwyn Garden City in the UK. Randy Trigg was demonstrating some new features of Notecards (an early hypertext system; Halasz, Moran, & Trigg, 1987). The version was still under development and every so often would hit a problem and a LISP debugger window would appear. After using it for a while, it suddenly froze—no debugger window, no error message, just froze. After a few embarrassed seconds, he hit a control key and launched the debugger. A few minutes of frantic scanning through stack dumps, program traces, etc., and the reason became clear to him. He had demonstrated a feature that he had last used on his workstation at Palo Alto. The feature itself was not at fault, but required an obscure font that he had on his own workstation, but not on the machine there in Welwyn. When Notecards had requested the font, the system might have thrown up an error window or substituted a similar font. However, in the spirit of true network transparency, the location of the font should not matter. Having failed to find it on the local machine, it proceeded to interrogate machines on the local network to see if they had it; it then proceeded to scan the Xerox UK network and world network. Eventually, if we had waited long enough, it would have been found on Randy's machine in Palo Alto. Network transparency rarely extends to timing!

Transparency has also been critiqued for CSCW purposes (Mariani & Rodden, 1991). It may well be very important to users where resources and people are. For mobile computing also, an executive takes a laptop on the plane only to discover that the files needed are residing on a network file server rather than on the machine itself. If the interface hides location, how can one predict when and where resources will be available.

Delays and Time. As evident, one of the issues that arises again and again when considering networks is time: How long are the delays, how long to transfer data, how variable, etc.? Networking is not the only reason for delays in applications, but is probably one of the most noticeable; the web has often been renamed the "World Wide Wait." There is a long-standing literature on time and delays in user interfaces. This is not as extensive as one might think, largely because for a long time, the prevailing perception in the human–computer interaction community was that temporal problems would go away (with some exceptions), leading to what I called the "myth of the infinitely fast machine" (Dix, 1987).

100 ms	Feedback for hand-eye coordination tasks needs to be less than 100–200 ms to feel fluid. This is probably related to the fact that there are delays of this length in our motor-sensory system anyway. For aural feedback, the timescales are slightly tighter again.
1 s	Timescale for apparent cause–effect links, such as popping a window after pressing a button. If the response is faster than this, the effect seems "immediate."
5–10 s	Waits longer than this engender annoyance and make it hard to maintain task focus. This may be related to short-term memory decay.

One of the earlier influential papers was Ben Shneiderman's review of research findings on delays (Shneiderman, 1984), mainly based on command line interfaces. More recently, there have been a series of workshops and special journal issues on issues of time, sparked largely by web delays (Clarke, Dix, Ramduny, & Trepess, 1997; Howard & Fabre, 1998; Johnson & Gray, 1996).

There are three main timescales that are problematic for networked user interfaces: The 100 ms time is hard to achieve if the interaction involves even local network traffic. The 1-s time is usually achievable for local networks (as are assumed by X-Windows systems), but more problematic for long haul networks. The 5–10 ms time is, in principle, achievable for even the longest transcontinental connections; but, when combined with bandwidth limitations or overload of remote resources, it may become problematic. This is especially evident on web-based services in which the delay between hitting a link and retrieving a page (especially a generated page) may well exceed these limits, even for the page to begin to draw.

The lesson for user interface designers is to understand the sort of interaction required and to ensure that parts of the user interface are located appropriately. For example, if close hand-eye coordination is required, it must run locally on the user's own machine: In the case of the web, in an applet, or JavaScript code, etc. If the nature of the application is such that parts of the application cannot reside close enough to the user for the type of interaction required, then, of course, one should not simply have a slow version of (say) dragging an icon around, but instead change the overall interaction style to reflect the available resources.

Two of the factors that alleviate the effects of longer delays are predictability of the delay and progress indicators. Both give the user some sense of control or understanding over the process, especially if users have some indication of expected delays before initiating an action (Johnson, 1997). The many variable factors in networked systems make predicting delays very difficult, increasing the importance of giving users some sense of progress. The psychological effect of progress indicators is exploited (cynically) by those web browsers that have progress bars that effectively lie to the user, moving irrespective of any real activity (try unplugging a computer from the network and

attempting to access a web page; some browsers will hit 70% on the progress bar before reporting a problem). Other network applications use recent network activity to predict remaining time for long operations (such as large file downloads). Other solutions include generating some sort of intermediate low-quality or partial information while the full information is being generated or downloaded (e.g., progressive image formats or web pages designed to partially display).

For virtual reality using head-mounted displays, as well as hand-eye coordination tasks, we also have issues of the coordination between head movements and corresponding generated images. The timescales here are even tighter as the sensory paths are faster within our bodies, hence less tolerant of external delays. The brain receives various indications of movement: The position and changes of neck and related muscles, the balance sensors in the inner ear, and visual feedback. Delays between the movement of the generated environment and head movement lead to dissonance between these different senses, and have an effect rather like being at sea, with corresponding disorientation and nausea. Also, any delays reduce the sense of immersion—being there within the virtual environment. Early studies of virtual reality showed that users' sense of immersion was far better when they were given very responsive wire frame images than if they were given fully rendered images at a delayed and lower frame rate (Pausch, 1991).

Coping Strategies. People are very adaptable. When faced with unacceptable delays (or other user interface problems), users develop ways to workaround or ameliorate the problem—coping strategies. For example, web users may open multiple windows so that they can view one page while reading another (McManus, 1997) and users of Telnet for remote command line interfaces may type test characters to see whether the system has any outstanding input (Dix, 1994).

Coping strategies may hide real problems, so it is important not to assume that just because users do not seem to be complaining or failing that everything is all right. We can also use the fact that users are bright and resourceful by building interface features that allow users to adopt coping strategies where it would be impossible or impractical to produce the interface response we would like. For example, where we expect delays, we can ensure that continual interaction is not required (by perhaps amassing issues requiring user attention in a batch fashion), thus allowing users to more easily multitask. Unfortunately, this latter behavior is not frequently seen—the 'myth' lives on, and most networked programs still stop activity and await user interaction whenever problems are encountered.

Timeliness of Feedback/Feedthrough, Pace. Although feedback is one of the most heavily used terms in human–computer interaction, we may often ignore the complex levels of feedback when dealing with near instantaneous responses of graphical user interfaces. In networked systems with potentially long delays, we need to unpack the concept. We have already discussed some of the critical timescales for feedback. For hand-eye coordination, getting feedback below the 100 ms threshold is far more important than fidelity; quickly moving wire frames or simple representations are better than dragging a exact image with drop shadow.

For longer feedback cycles, such as pressing a button, we need to distinguish:

- syntactic feedback—that the system has recognized your action
- intermediate feedback—that the system is dealing with the request implied by your action (and if possible progress toward that request)
- semantic feedback—that the system has responded and the results obtained

The direct manipulation metaphor has led to an identification between these levels, and many systems provide little in the way of syntactic or intermediate feedback, relying solely on semantic feedback. In networked systems where the semantic feedback includes some sort of remote resource, it is crucial to introduce specific mechanisms to supply syntactic and intermediate feedback; otherwise, the system may simply appear to have ignored the user's action (leading to repeated actions with potentially unforeseen consequences) or even frozen or crashed.

This also reminds us of a crucial design rule for slow systems: Wherever possible, make actions idempotent—that is invoking the same action twice should, where possible, have the same effect as a single action. This means that the try again response to a slow system does not lead to strange results.

For collaborative systems or those involving external or autonomous resources (remote-controlled objects, environmental sensors, software agents), we must also consider feedthrough. Feedback is experiencing the effect of one's own actions; feedthrough is the effect of one's own actions on other people and things, and experiencing the effects of their actions yourself. For example, in an online chat system, you type a short message and press send, and your message appears in your transcript (feedback); then, sometime later, it also appears in the transcript of the other chat participants (feedthrough).

Feedback is needed to enable us to work out whether the actions we have performed are appropriate, hence (typically) need to be much quicker than feedthrough responses. This is fortunate, because feedthrough by its very nature usually requires network transmission and ensuing delays. The exception to the rule that feedthrough can afford to be slower is where the users are attempting to perform some close collaborative task (e.g., positioning some items using direct manipulation) or

where there is a second fast communication channel (e.g., on the telephone, user A says to user B, "see the red box," but the relevant item has not appeared yet on B's screen).

Potentially more important to users of collaborative systems than bandwidth or even raw delays is pace—the rate at which it is possible to interact with a remote resource or person. This is partly determined by lower level timings, but is also heavily influenced by interface design. For example, you know that someone is sitting at their desk and send them an urgent e-mail. The time that it takes to get a response will be hardly affected by the raw speeds between your machine and your colleague, but more determined by factors such as how often the e-mail client checks the server for new e-mail and whether it sounds an alert when new e-mail arrives, or simply waits there until your colleague chooses to check the in-box.

Race Conditions and Inconsistent Interface States. Alison and Brian are using an online chat program.

Alison writes, "It's a beautiful day. Let's go out after work." and then begins to think about it.

Brian writes, "I agree totally." and then has to leave to go to a meeting.

At almost the same time, Alison writes, "Perhaps not. I look awful after the late party."

Unfortunately, the messages are so close to simultaneous that both Alison and Brian's machines put their own contribution first, so Alison sees the chat window as in Figure 16.5a and Brian sees it as in Fig. 16.5b. Brian thinks for a few moments and then writes, "No, you look lovely as ever." But, unfortunately, Alison never sees this, because she takes one look at Brian's previous remark and shuts down the chat program.

This type of incident, where two events happen so close together that their effects overlap, is called a race condition. Race conditions may lead to inconsistent states for users as in this example, or may even lead to the software crashing. Although, in principle, race conditions are possible however fast the underlying network, the likelihood of races occurring gets greater as the network (and other) delays get longer.

Even some of the earliest studies in collaborative systems have shown the disorienting effects of users seeing different views of their shared information space, even when this is simply a matter of seeing different parts of the same space (Stefik, Bobrow, Foster, Lanning, & Tatar, 1987).

Consistency becomes an even greater problem in mobile systems where wireless connections may be temporarily lost,

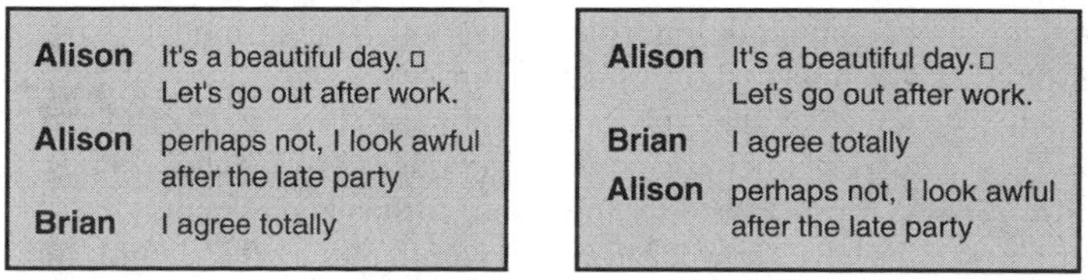

Alison	It's a beautiful day. □ Let's go out after work.
Alison	perhaps not, I look awful after the late party
Brian	I agree totally

(a) Alison's chat window

Alison	It's a beautiful day.□ Let's go out after work.
Brian	I agree totally
Alison	perhaps not, I look awful after the late party

(b) Brian's chat window

FIGURE 16.5. Consistency breakdown.

or devices may be unplugged from fixed networks while on the move. During these periods of disconnection, it is easy for several people to be updating the same information, leading to potential problems when their devices next become network connected.

In a later section, we will discuss mechanisms and algorithms that can be used to maintain consistency, even when delays are long and race conditions likely to occur.

Awareness. Returning to Alison and Brian. After Brian has typed his response, he may not know that Alison has not seen his second contribution.

Awareness of who is around and what they are doing is a major issue in CSCW (e.g., Dourish & Bellotti, 1992; McDaniel & Brinck, 1997). It has various forms:

- being able to tell easily, when you want to know, what other people are doing
- being made aware (via alerts, very salient visual cues, etc.) when significant events occur (e.g., new user arrives, someone makes a contribution)
- having a peripheral awareness of who is around and what they are up to

Awareness is not just about other people. In any circumstance where the environment may change, but not through your own direct action, you may need to know what the current state is and what is happening. This is not confined to networked applications, but applies to any hidden or invisible phenomena; for example, background indexing of your hard disk contents. In networked applications, anything distant is invisible unless it is made visible (audible) in the interface.

One of the earlier influential experiments to demonstrate the importance of peripheral awareness was ArKola (Gaver, Smith, & O'Shea, 1991). This was a simulated bottling factory where two people worked together to maintain the factory, supplying, maintaining, etc., the process. The participants could not see all of the factory at once and so relied on the sounds produced to be aware of its smooth running or if there were any problems. For example, the sound of breaking glass might suggest that the end of the production line has run out of crates; but, if it immediately stopped, one would assume that the other participant had sorted the problem out.

The numerous forms of shared video and audio spaces are another example of this; several people, usually in distant offices establish long-term, always-on audio, video, or audio/video links between their offices (Buxton & Moran, 1990; Olson & Bly, 1991). Sometimes these are used for direct communication, but most of the time just give a peripheral awareness that the other person is there and the sort of activity they are doing. This can be used for functional purposes (e.g., knowing when the other person is interruptable), but also for social purposes—feeling part of a larger virtual office. Other systems have allowed larger numbers of, usually deliberately low quality and so less intrusive, web-cam views of colleagues' offices and shared areas (Roussel, 1999, 2001). The aims are the same: To build social cohesion, to allow at-a-glance reading of one another's situation, and to promote accidental encounters.

A form of awareness mechanism is now common on the web with buddy lists that tell you when friends are online (ICQ, Aug. 2001). Currently, I know of no examples of rich media experiments in a domestic environment; for example, a virtual kitchen share with your elderly mother in Minnesota (although the Casablanca project at Interval (Hindus, Mainwaring, Leduc, Hagström, & Bayley, 2001) has worked with shared electronic sketch pads in the home). However, the increasing prevalence of continuously connected households and information appliances will surely make this common in the future (see chapter 38).

Trying to capture all this information within a computer display can be distracting, uses up valuable screen space, and of course assumes that the computer is there and on. For this reason, several projects have looked at ambient interfaces, which in various ways make the physical environment reflect the virtual. These interfaces monitor various events in the electronic worlds and then change things in the physical environment: Lights on the wall, moving strings hung from the ceiling, even a shaking pot plant (Lock, Allanson, & Phillips, 2000). Again, this is not fundamentally limited to networked environments, but is of course not very useful when the relevant activity is close at hand anyway.

The other side of this is finding out what people are doing to signal this to others. For computer activity—are you logged on, have you been typing recently, what web page are you viewing—this is, in principle, available, although the various layers of software may make it hard for an awareness service to discover. For noncomputer aspects, this is more problematic—are you in the room, busy, with other people—and may require a range of sensors in the environment, ultrasound, video, etc., with corresponding privacy issues (see, e.g., Bellotti, 1993). Monitoring of everyday objects is another way to achieve this; for example, one experiment used electronic coffee cups with sensors to tell when they were picked up and moved around (Gellersen, Beigl, & Krull, 1999). As more and more devices become networked, it may be that we do not need special sensors, just use the combined information from those available, although the privacy issues remain.

In collaborative virtual reality environments, knowing that other people are around (as avatars) is as important as in a physical world, but harder because of limited senses (usually just vision and sound) and limited field of view. Furthermore, there are computational costs in passing information, such as audio or even detailed positional information, around the network when there are tens, hundreds, or thousands of users. Various spatial models have been developed to analyze and implement the idea of proximity in virtual space (Benford, Bowers, Fahlen, Mariani, & Rodden, 1994; Dix et al., 2000; Rodden, 1996; Sandor, Bogdan, & Bowers, 1997). These seek to formalize concepts of: (a) where your focus of attention is within the virtual world and thus whether you require full-quality audio and visual representation of others; (b) broader areas where you would expect some peripheral awareness where potentially degraded information can be used; and (c) those parts of the space for which you need no awareness information.

Media Issues

When describing the intrinsic network properties, issues for continuous media were mentioned several times. This is because, with the possible exception of close hand-eye coordination tasks, continuous media put some of the tightest requirements on the underlying networks.

Interactive Conversation and Action. Most demanding of all are audiovisual requirements of interactive conversation. Anyone who has had a transcontinental telephone conversation will have some feeling for the problems a second or two delay can cause. While actually speaking, the delays are less significant; however, turntaking becomes very problematic. This is because the speaker in a conversation periodically (and subconsciously) leaves short (200–300 ms) gaps in the flow of speech. These moments of silence act as entry points for the other participant who is expected to either acknowledge with a "go on" sound, such as "uhm" or perhaps a small nod of the head, or can use to break in with their own conversation. Entries at other points would be seen as butting in and rude, and lack of feedback responses can leave the speaker uncertain as to the listener's understanding. The 200–300 ms is again almost certainly related to the time it takes for the listener's sensory system to get the relevant aural information to the brain, and for it to signal the relevant nod, acknowledgment, or start to speak. Clearly, our conversational system is finely tuned to the expected intrinsic delays of face-to-face conversation.

When network delays are added, it is no longer possible to respond within the expected 200–300 ms window. The speaker therefore gets no responses at the appropriate points, and it is very hard for the listener to break into the flow of speech without appearing rude (by the time they hear the gap and speak, the speaker has already restarted). Some telephone systems are half-duplex (i.e., only allow conversation in one direction at a time); this means that the various vocalizations ("uhu," "hmm," etc.) that give the speaker feedback will be lost entirely while the speaker is actually talking. It is not uncommon for the speaker to have to resort to saying, "Are you there?" because of a loss of sense of presence.

These effects are similar whether one is dealing with pure audiostream (as with the telephone), videostreams (as with desktop conferencing), or distributed virtual environments. One virtual reality project in the UK conducted all its meetings using a virtual environment in which the participants were represented by cuboid robotlike avatars (called blockies) (Greenhalgh, 1997). The project ended with an online virtual party. As the music played, the participants (and their avatars) danced. Although clearly enjoying themselves, the video of the party showed an interesting phenomenon. Everyone danced alone. There are various reasons for this; for example, it was hard to determine the gender of a potential dancing partner. However, one relates directly to the network delays. Although everyone hears the same music, they all were hearing it at slightly different time; furthermore, the avatars for other people will be slightly delayed from their actual movements. Given popular music rhythms operate at several beats per second, even modest delays mean that your partner appears to dance completely out of time.

Reliability. As well as delays, we have noted that network connections may not always be reliable (i.e., information may be lost). Videostreams and audiostreams behave very differently in the presence of dropped information. Imagine you are watching a film on a long airflight. The break in the sound when the pilot makes an announcement is much more difficult than losing sight of the screen for a moment or two as the passenger in front stands up. On a smaller scale, a fraction of a second loss of a few frames of video just makes the movie seem a little jerky; a smaller loss of even a few tens of milliseconds of audio signal would make an intrusive click or distortion. In general, reliability is more important for audiostreams than videostreams, and where resources are limited, it is typically most important to reserve the quality of service for the audiostream.

Sound and Vision. Why is it that audio is more sensitive than video? Vision works (largely) by looking at a single snapshot. Try walking around the room with your eyes shut, but opening them for glances once or twice a second. Apart from the moment or two as your eyes refocus, you can cope remarkably well. Now turn a radio on with the sound turned very low and every second turn the sound up for a moment and back to silent—potentially an interesting remixing sound, but not at all meaningful. Sound, more than vision, is about change in time. Even the most basic sounds, pure tones, are measured in frequencies—how long between peaks and troughs of air pressure? For more complex sounds, the shape of the sound through time—how its volume and frequency mix changes—are critical. For musical instruments, it is hard to hear the difference between instruments that are playing a continuous note, but instantly differentiable by their attack or how the note starts. (To get some idea of the complexity of sound, see Mitsopoulos' thesis and web pages (Mitsopoulos, 2000) and for an insight into the way different senses affect interaction, see my own AVI'96 paper (Dix, 1996.)

Compression. As we discussed previously, it is possible to produce reliable network connections, but this introduces additional delays. Compression can also help by reducing the overall amount of audiovisual data that needs to be transmitted, but again may introduce additional delays. Furthermore, simple compression algorithms require reliable channels (both kinds of delays). Special algorithms can be designed to cope with dropped data, making sure that the most important parts of the signal are replicated or spread out so that dropped data leads to a loss in quality rather than interruption.

Jitter. As noted previously, jitter is particularly problematic for continuous media. Small variations in delay can lead to jerky video play back, but is again even worse for audiostreams. First of all, a longer than normal gap between successive bits of audio data would lead to a gap in the sound just like dropped data.

Perhaps even more problematic, what do you do when subsequent data arrives closer together—play it faster? Changing the rate of playing audio data does not just make it jerky, but changes the frequency of sound rendering it meaningless.

(*Note.* Actually, there are some quite clever things you can do by digitally speeding up sound, but not changing its frequency. These are not useful for dealing with jitter, but can be used to quickly overview audio recordings, or catch up on missed audiostreams (Arons, 1997; Stifelman, Arons, & Schmandt, 2001).

For real-time audiostreams, such as video-conferencing or voice over the Internet, it is hard to do anything about this. The best one can do is drop late data and do some processing of the audiostream to smooth out the clicks this would otherwise generate.

Broadcast and Prerecorded Media. Where media is prerecorded or being broadcast, but where a few seconds delay are acceptable, it is possible to do far better. Recall that in several places we saw that better quality can be obtained if we are prepared to introduce additional delays.

If you have used real-audio or real-video broadcasts, you will know that the quality is quite acceptable and does not have many of the problems described. This is partly because of efficient compression, meaning that video is compressed to a fraction of a percent of its raw bandwidth and so can fit down even a modem line. However, this would not solve the problems of jitter. To deal with this, the player at the receiving end buffers several seconds of audiovisual data before playing it back. The buffering irons out the jitter, giving continuous quality.

Try it out for yourself. Tune onto a radio channel and simultaneously listen to the same broadcast over the Internet with real-audio. You'll clearly hear up to a minute delay between the two.

Public Perception: Ownership, Privacy, and Trust

One of the barriers to consumer e-commerce has been distrust of the transaction mechanisms, especially giving credit card details over the web. Arguments that web transactions are more secure than phone-based credit cards transactions or even using a credit card at your local restaurant (which gets both card number and signature) did little to alleviate this fear. This was never a major barrier in the United States as it was, for example, in Europe; but across the world it has been a concern, slowing down the growth of e-shopping (or really e-buying), but that is another story (Dix, 2001a).

It certainly is the case that transactions via secure channels can be far more secure than physical transactions, where various documents can be stolen or copied en route and are in a format much more easy to exploit for fraud. However, knowing that a transaction is secure is more than the mechanisms involved; it is about human trust. Do I understand the mechanisms well enough, and the people involved well enough, to trust my money to it?

In fact, with a wider perspective, this distrust is very well founded. Encryption and authentication mechanisms can ensure that I am talking to a particular person or company and that no one else can overhear. But how do I know to trust that

person? Being distant means I have few of the normal means available to assess the trustworthiness of my virtual contact. In the real world, I may use the location and appearance of a shop to decide whether I believe it will give good service. For mail order goods, I may use the size, glossiness, and publication containing an advert to assess its expense and again use this to give a sense of quality of the organization. It is not that these physical indicators are foolproof, but that we are more familiar with them. In contrast, virtual space offers few intrinsic affordances. It is easy and quite cheap to produce a very professional web presence, but it may be little more than a facade. This is problematic in all kinds of electronic materials, but perhaps most obvious when money is involved.

Even if I trust the person at the other end, how do I know whether the network channel I am using is of a secure kind? Again, the affordances of the physical world are clear: In a closed office vs. in the open street vs. in a bar frequented by staff of a rival firm. We will say different things, depending on the perceived privacy of the location. In the electronic world, we rely on "https" at the beginning of a URL (how many ordinary consumers know what that means?), or an icon inserted by the e-mail program to say a message has been encrypted or signed. We need to trust not only the mechanisms themselves, but also the indicators that tell us what mechanisms are being used (Fogg et al., 2001; Millett, Friedman, & Felten, 2001).

In nonfinancial transactions, issues of privacy are also critical. We have already seen several examples where privacy issues occur. As more devices become networked, especially via wireless links and our environment and even our own bodies become filled with interlinked sensors, issues about who can access information about you become more significant. This poses problems at a technical level (ensuring security between devices), at an interface level (being able to know and control what or who can see specific information) and at a perception level (believing in the privacy and security of the systems). There are also legal implications. For example, in the UK, currently (2001) it is illegal for mobile telecomm operators to give location information to third parties (Sangha, 2001).

The issue of perception is not just a minor point, but perhaps the dominant one. Networks, and indeed computer systems in general, are by their nature hidden. We do not see the bits traveling down the wires or through the air from device to device, but have to trust the system at even the most basic level. As human–computer interaction specialists, we believe ourselves a little above the mundane software engineers who merely construct computer systems, because we take a wider view and understand that the interaction between human and electronic systems has additional emergent properties. It is this complete sociotechnical unit that achieves real goals. For networked systems, this view is still far too parochial.

Imagine if the personal e-mail of millions of people was being sucked into the databanks of a transnational computer company and only being released when accessed through the multinational's own web interface. The public outcry! Imagine hotmail, yahoo mail, etc. How is it that, although stored on distant computers, perhaps half the world away, millions of people feel that it is their mailbox and trust the privacy of web mail more than perhaps their organization's own mail system. This feeling of

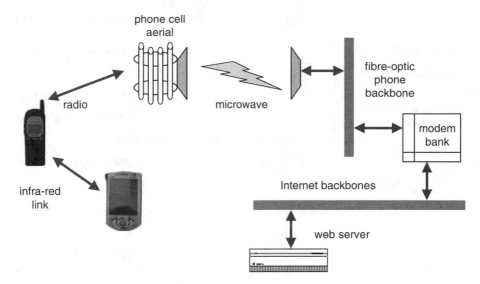

FIGURE 16.6. The long path from PDA to web.

ownership of remote resources is more than the technology that protects the security of such systems; it is a cultural phenomenon and a marketing phenomenon. The web mail product is not just technology, or interface, but formed by every word that is written about the product in advertisements, press releases, and media interviews (Dix, 2001a).

NETWORKS AS SUBJECTS: UNDERSTANDING AND MANAGING NETWORKS

When using a networked application, you do not really care what kind of network is being used, whether the data is sent over copper wires, fiber optic, microwave, or satellite. All you care is that the two ends manage to communicate, and the effects any of the above have on the end-to-end network properties, such as bandwidth, discussed in the previous section. However, there are times when the network's internals cannot be ignored.

Those involved in installing or managing networks need to understand the internal workings of the network to optimize performance and find faults. For ordinary users, when things go wrong in a networked application, they effectively become a network manager; thus understanding something of the network can help them to deal with the problem. Even when things are working, having some awareness of the current state of the network may help one predict potential delays, avoid problems, and minimize costs.

We will start this section by looking at some of the technical issues that are important in understanding networks. This parallels the section on User Interface Properties, but is focused on the internal properties of the network. We will then look at the interface issues for those managing networks and the ways in which interfaces can make users aware of critical network state. Finally, we will look briefly at the way models of networks can be used as a metaphor for some of the motor and cognitive behaviors of humans.

Network Models

Layers. Networking is dominated by the idea of layers; lower levels of the network offer standard interfaces to higher levels so that it is possible to change the details of the lower level without changing the higher levels. For example, imagine you are using your PDA to access the Internet while on a train (see Fig. 16.6). The web browser establishes a TCP/IP connection to the web server, requests a web page, and then displays it. However, between your PDA and the web server, the message may have traveled through an infrared link to your mobile phone, which then used a cell-based radio to send it to a mobile-phone station, then via a microwave link to a larger base station onto a fiber optic telephone backbone and via various copper wires to your Internet service provider's (ISP's) modem bank. Your ISP is then connected into another fiber optic Internet backbone and eventually via more fiber optic, copper, and microwave links to the web server. To complicate things even further, it may even be that the telephone and Internet backbones may share the same physical cabling at various point. Imagine if your poor PDA had to know about all of this.

In fact, even your PDA will know about at least 5 layers:

- Infrared—how the PDA talks to the phone
- Modem—how the PDA uses the phone to talk to your ISP
- IP—how your PDA talks to the web-server computer
- TCP—how data is passed as a reliable connection-based channel between the right program on your PDA and the web-server computer
- HTTP—how the browser talks to the web server

Each of these hides most of the lower levels, so your browser needs to know nothing about IP, the modem, or the infrared connection while accessing the web page.

The nature of these layers differs between different types of networks; e.g., WAP, for sending data over mobile phones and

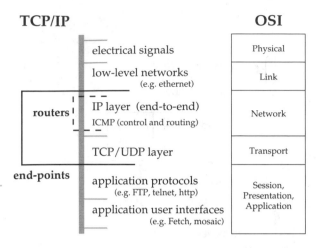

FIGURE 16.7. OSI seven layers and TCP/IP.

220 mail.server.net ESMTP

HELO mypc.mydomain.com

250 mail.server.net Hello mypc.mydomain.com[100.0.1.7],

pleased to meet you

MAIL From:<myself@mydomain.com>

250 <myself@mydomain.com>... Sender ok

RCPT To: <a-friend@theirdomain.com>

250 <a-friend@theirdomain.com>... Recipient ok

DATA

354 Enter mail, end with "." on a line by itself

first line of message

..dotty 2nd line

last line

.

250 KAA24082 Message accepted for delivery

QUIT

221 mail.server.net closing connection

FIGURE 16.8. Protocol to send e-mail via SMTP.

devices, has five defined layers (Arehart et al., 2000), and the ISO OSI reference model has seven layers (ISO/IEC 7498, 1994). (see Fig. 16.7).

Protocols. Systems at the same layer typically require some standard language to communicate. This is called a protocol. For higher levels, this may be quite readable. For example, to send an Internet e-mail message, your mail program connects to an SMTP server (a system that relays messages) using TCP/IP and has the exchange shown in Fig. 16.8.

At lower levels, data is usually sent in small packets, which contain a small amount of data plus header information saying where the data is coming from, where it is going, and other bookkeeping information, such as sequence numbers, data length, etc.

Even telephone conversations, except those using predigital exchanges, are sent by chopping up your speech into short segments at your local telephone exchange, sending each segment as a packet, and then reassembling the packets back into a continuous stream at the other end (Stevens, 1998, 1999).

Internetworking and Tunneling. This layering does not just operate within a particular network standard, but between different kinds of networks, too. The Internet is an example of an internet (notice little "i") that is a network which links together different kinds of low-level network. For example, many PCs are connected to the Internet via an ethernet cable. Ethernet sends its own data in packets like those in Fig. 16.9. The IP also has packets of a form like Fig. 16.9. When you make an Internet connection via ethernet, the IP packets are placed in the data portion of the ethernet packet, so you get something a bit like Fig. 16.10.

This placing of one kind of network packet inside the data portion of another kind of network is also used in virtual private networks in a process called tunneling. These are used to allow a secure network to be implemented using a public network like the Internet. Imagine a company has just two offices: one in Australia and the other in Canada. When a computer in the Sydney office sends data to a computer in the Toronto office, the network packet is encrypted, put in the data portion of an Internet packet, and sent via the Internet to a special computer in the Toronto office. When it gets there, the computer at the Toronto office detects its virtual private network data, extracts the encrypted data packet, decrypts it, and puts it onto its own local network where the target computer picks it up. As far as both ends are concerned, it looks as if both offices are on the same LAN, and any data on the Internet is fully encrypted and secure.

Routing. If two computers are on the same piece of physical network, each can simply listen out for packets that are destined for them, so sending messages between them is easy. If, however, messages need to be sent between distant machines (e.g., if you are dialed into an ISP in the UK and are accessing a web server in the United States), the message cannot simply be broadcast to every machine on the Internet (Fig. 16.11). Instead, at each stage, it needs to be passed in the right direction between different parts of the network. Routers perform

header				body
to address	from address	info	data length	data .....

FIGURE 16.9. Typical network packet format (simplified).

ethernet header   ethernet body

ethernet to addr.	ethernet from addr.	IP flag data len. etc.	IP to addr	IP from addr	other header IP info	Internet data .....
			IP header			IP body

FIGURE 16.10. Internet protocol (IP) packet inside ethernet packet.

The 32-bit IP number space allows for 4 billion addresses. This sounds like quite a lot, However, these have been running out because of the explosive growth in the number of Internet devices and wasted IP numbers because of the way ranges of numbers get allocated to subnetworks. The new version of TCP/IP, IPv6, which is beginning to be deployed, will have 128-bit addresses, which will require 16 numbers (IPng, 2001). This will allow unique IP numbers for every phone, PDA, Internet-enabled domestic appliance, or even an electronic paper clip.

this task. They look at the address of each packet and decide where to pass it. If it is a local machine, this might be to simply put it onto the relevant local network; but, if not, it may need to pass it on to another intermediate machine. Routers may be stand-alone boxes in network centers or may be a normal computer. Often, a file server acts as a router between a LAN and the global network.

As well as routers networks are also linked by hubs and switches that make several different pieces of physical network behave as if they were one local network, they are also linked by gateways that link different kinds of networks. The details of these are not important, but they add more to the sheer complexity of even small networks.

Addresses. To send messages on the Internet or any other network, you need to have the address of where they are to go (or at least your computer needs it). In a phone network, this is the telephone number and on the Internet it is an IP number. The IP number is a 32-bit number, normally represented as a group of 4 numbers between 0 and 255 (e.g., 212.35.74.132), which you will have probably seen at some stage when using a web browser or other Internet tool. It is these IP numbers that are used by routers to send Internet data to the right place.

However, with any network, there is a problem of how you get to know the address. With phone numbers, you simply

look up the person's name in a telephone directory or phone the operator. Similarly, most networks have a naming scheme and some way to translate these into addresses. In the case of the Internet, domain names (e.g., acm.org, www.hcibook.com, magisoft.co.uk) are the naming system. There are so many of these and they are changed relatively rapidly, so there is no equivalent of a telephone directory, but instead special computers called domain name servers (DNS) act as the equivalent of telephone directory enquiries operators. Every time an application needs to access a network resource using a domain name (e.g., to lookup a URL), the computer has to ask a DNS what IP address corresponds to that domain name. Only then can it use the IP address to contact the target computer.

The DNS system is an example of a white pages system. You have an exact name and want to find the address for that name. In addition there are so-called yellow pages systems in which you request, for example, a color postscript printer and are told of addresses of systems supplying the service. Sometimes this may be mediated by brokers that may attempt to find the closest matching resource (e.g., a non-postscript color printer) or even perform translations (e.g., the Java JINI framework; Edwards & Rodden, 2001).

This latter form of resource discovery system is most important in mobile systems and ubiquitous computing in which we are particularly interested in establishing connections with other geographically close devices.

A final piece in the puzzle is how one gets to know the address of the name server, directory service, or brokering service. In some types of network, this may be managed by sending broadcast requests to a network, "Is there a name server out there?" In the case of the Internet, this is normally explicitly set for each machine as part of its network settings.

All Together... If you are in your web browser and try to access the URL http://www.meandeviation.com/qbb/, the following stages happen:

1. Send IP-level request to DNS asking for www.meandeviation.com
2. Wait for reply
3. DNS sends reply 64.39.13.108
4. Establish TCP-level connection with 64.39.13.108
5. Send HTTP request ("GET /qbb/ HTTP/1.1")
6. Web server sends the page back in reply
7. Close TCP connection

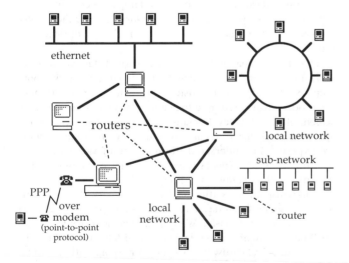

FIGURE 16.11. Routers send messages in the right direction through complex networks.

Most of these stages are themselves simplified; all will involve layering on top of lower level networks, and most stages involve several substages (e.g., establishing a TCP-level connection requires several IP-level messages). The basic message is that network internals are multilevel, multistage, and pretty complicated.

Network Management

Those most obviously exposed to this complexity are the engineers managing large national and international networks, both data networks, such as the main Internet backbones and telecoms networks. The technical issues outlined are compounded by the fact that the different levels of network hardware and network management software are typically supplied by different manufacturers. Furthermore, parts of the network may be owned and managed by a third party and shared with other networks. For example, a trans-Atlantic fiber optic cable may carry telecoms and data traffic from many different carriers.

When a fault occurs in such a network, it is hard to know whether it is a software fault or a hardware fault, where it is happening, and who is responsible. If you send engineers out to the wrong location, it will cost both their time and also increase the time the service is unavailable. Typically, the penalties for inoperative or reduced quality services are high—you need to get it right fast.

This is a specialized and complex area, but clearly of increasing importance. It poses many fascinating user interface challenges: How to visualize complex, multilayered structures, and how to help operators trace faults in these. Although I know that it is a topic being addressed by individual telecom companies, published material in the human–computer interaction literature is minimal. The exception is visualization of networks, both physical and logical, which is quite extensive. (Dodge and Kitchin's *Atlas of Cyberspace* and associated Web site (Dodge & Kitchin, 2001) is the comprehensive text in this area.)

Ordinary system administrators in organizations face similar problems, albeit on a smaller scale. A near universal experience of misconfigured e-mail systems, continual network failures and performance problems certainly suggests this is an area ripe for effective interface solutions, but again there is very little in the current human–computer interaction literature.

Finally, it appears that everyone is now a network administrator, even a first-time home PC user must manage modem settings, name server addresses, SMTP and POP servers, and more. It is interesting that for most such users the interface they use is identical to that supplied for full systems administrators. Arguably, this may ease the path for those who graduate from single machines to administering an office or organization, perhaps less than 5% of users. Unfortunately, it makes life intolerable for the other 95%. The only thing that makes this possible at all is that the welcome disks for many ISPs offer step-by step instructions or may automatically configure the system. These complications are compounded if the user wishes to allow access through more than one ISP or connect into a fixed network. Because many home users now have several PCs and other devices that need to be networked, this is not a minor issue.

MacOS Location Manager

To edit a location:

- Select it as the edit location (different from the current location)
- Select the configuration item you want to change in the edit interface
- Change and confirm the corresponding control panel for the current location
- Confirm the change in the edit interface
- Remember to change everything back for the current configuration

---

If we look at the current state of the two most popular PC systems, Microsoft Windows and MacOS, the picture is not rosy.

On Windows at least most of the settings for an ISP are concentrated within one configuration file and so dialing into an ISP means selecting the right file and double clicking. Unfortunately, swopping between fixed networks is far more complicated involving recording half a dozen IP numbers and other settings, and changing these in different control panels every time one changes network. For those on portables, moving between different sites, this is very difficult.

On MacOS, setting up a new ISP involves at least three control panels (a side effect of organizing the control panels to correspond to logical network layers). On the positive side, the MacOS Location Manager uses the same interface for switching between ISPs or between different fixed networks, and both are equally easy (once configured). Unfortunately, the interface model for editing configurations for a particular location is perhaps one of the most incomprehensible I've encountered, involving a confusion between the location being edited and current location. Given the slickness of most other aspects of MacOS and the ease of networking Macs with AppleTalk, this appears to indicate a lack of perceived importance somewhere within the design process, made more surprising by the emphasis placed on Internet readiness in the marketing of iMacs and iBooks.

I would hope that both these Windows and MacOS interfaces will improve over the lifetime of this book. The problems in both emphasize partly the intrinsic complexity of networking— yes it does involve multiple logically distinct settings, many of which relate to low-level details. However, it also exposes the apparent view that those involved in network administration are experts who understand the meaning of various internal networking terms. This is not the case even for most office networks and certainly not at home.

This is just initially setting up the system to use. For the home user, debugging faults has many of the same problems as large networks. You try to visit a Web site and get an error box.... Is the Web site down? Are there problems with the phone line, the modem, the ISP's hardware? Are all your configuration settings right? Has a thunderstorm 3,000 miles away knocked out a vital network connection? Trying to understand a multilayered, nonlocalized and, when things work, largely hidden system is intrinsically difficult; and where diagnostic tools for this are provided, they assume an even greater degree of expertise.

The much heralded promise of devices that connect up to one another within our homes and about our bodies is going to throw up many of the same problems. Some old ones may ease as explicit configuration becomes automated by self-discovery between components, but this adds further to the hiddenness and thus difficulty in managing faults, security, etc. You can imagine the scenario: The sound on my portable DVD stops working and produces a continuous noise—why? There are no cables to check of course (wireless networking), but hours of checking and randomly turning devices on and off narrow the problem down to a fault in the washing machine that is sending continuous "I finished the clothes" alerts to all devices in the vicinity.

Network Awareness

One of the problems noted previously is the hiddenness of networks. This causes problems when things go wrong, because one has not an appropriate model of what is going on, but also sometimes even when things are working okay.

We discussed previously some of the network properties that may affect usability—bandwidth, delay, jitter, etc. These are all affected to some extent by other network load, the quality of current network connections, etc. So predicting performance (and knowing whether or not to panic if things appear to go slow) needs some awareness of the current state of the network.

PCs using wireless networks usually offer some indication of signal strength (if one knows where to look and what it means), although this is less common for line quality for modems. As wireless devices and sensors become smaller, they will not have suitable displays for this and explicitly making users aware of the low-level signal strength of an intelligent paper clip may not be appropriate. However, as interface designers, we do need to think how users will be able to cope and problem solve in such networks.

Only more sophisticated network management software allows one to probe the current load on a network. This does matter. Consider a small home network with several PCs connected through a single modem. If one person starts a large download, they will be aware that this will affect the performance of the rest of their web browsing. Other members of the household will just experience a slowing down of everything and not understand why. In experiments at MIT, the level of network traffic has been used to jiggle a string hanging from the ceiling so that heavy traffic leads to a lot of movement (Wisneski et al., 1998). The movements are not intrusive, so give a general background awareness of network activity. Although supplying a ceiling mounted string may not be the ideal solution for every home, other more prosaic interface features are possible.

Cost awareness is also very important. In the UK, current generation mobile data services are charged by connection time. So knowing how long you have been connected and how long things are likely to take becomes critical. If these charges differ at peak hours of the day, calculating whether to read your e-mail now, or do a quick check and download the big attachments later can become a complex decision. With moves to data volume-based charging, the calculation will be replaced by some attempt to estimate the volume of data that is likely to be involved in initiating an action vs. the value of that data. Do you want to click on that link if the page it links to includes large graphics, perhaps an applet or two?

Network Confusion. If the preceding does not sound confusing enough, the multilayered nature of networked applications means that it is hard to predict the possible patterns of interference between things implemented at different levels or even at the same level. Again, this is often most obvious when things go wrong, but also because unforeseen interactions may mean that two features that work perfectly well in isolation may fail when used together.

This problem, feature interaction, has been studied particularly in standard telecoms (although certainly not confined to it). Let us look at an example of feature interaction. Telephone systems universally apply the principle that the caller, who has control over whether and when the call is made, is the person who pays. In the exceptions (free phone numbers, reverse charges), special efforts are made to ensure that subscribers understand the costs involved. Some telephone systems also have a feature whereby a caller who encounters a busy line can request a call-back when the line becomes free. Unfortunately, at least one company implemented this call-back feature so that the charging system saw the call-back as originating from the person who had originally been called. Each feature seemed to be clear on its own, but together meant you could be charged for calls you did not want to make.

With N features, there are $N(N-1)/2$ possible pairs of interactions to consider, $N(N-1)(N-2)/6$ triples, etc. This is a well-recognized (but not solved) problem with considerable efforts being made using, for example, formal analysis of interactions to automatically detect potential problems. It is worth noting that this is not simply a technical issue. The charging example shows that it is not just who pays that matters, but the perceptions of who pays. This particular interaction would have been less of a problem if the interface of the phone system had, for example, said (in generated speech) "You have had a call from XXX. Press call back. You will be charged for this call." The hybrid feature interaction research group at Glasgow are attempting to build a comprehensive list of such problems in telecoms (HFIG, 2001), but these sorts of problems are likely to be found increasingly in related areas, such as ubiquitous computing and resource discovery.

The Network Within

So far, the story is pretty bleak from a user interface viewpoint—a complex problem, of rapidly growing importance, with relatively little published work in many areas. One good thing as a human–computer interaction practitioner about understanding the complexity of networks, is that they help us understand better the workings of the human cognitive and motor systems.

For at least five decades, computational models have been used to inspire cognitive. Also, of course, cognitive and neurological models have been used to inspire computational models in artificial intelligence and neural networks. However, our bodies are not like a single computer, but in various ways more like a networked system.

First because several things can happen at once. The interacting cognitive subsystems (ICS) model from APU Cambridge (Barnard & May, 1995) takes this into account looking at various parts of the cognitive system, the conversions between representations between these parts, and the conflicts that arise if the same part is used to perform different tasks simultaneously. Similarly, the very successful PERT-style GOMS analysis used on the NYNEX telephone operators interface used the fact that the operator could be doing several things simultaneously with no interfering parts of their bodies and brains (Gray, John, & Atwood, 1992; John, 1990).

We are also like a networked system in that signals take an appreciable time to get from our senses to our brains and from our brains to our muscles. The famous homunculus from Card, Moran, and Newell's Model Human processor makes this very clear with timings attached to various paths and types of mental processing (Card, Moran, & Newell, 1983). In fact, the sorts of delays within our bodies (from 50 to 200 ms on different paths) are very similar to those found on international networks.

In industrial control, one distinguishes between open-loop and closed-loop control systems. Open-loop control is where you give the machine an instruction and assume it does it correctly (like a treasure map; "Ten steps North, turn left, three steps forward, and dig"). This assumes the machine is well calibrated and predictable. In contrast, closed-loop control uses sensors to constantly feedback and modify future actions based on the outcomes of previous ones (e.g., "Follow the yellow brick road until you come to the Emerald City"). Closed-loop control systems tend to be far more robust, especially in uncontrolled environments, like the real world. (see Fig. 16.12).

Not surprisingly, our bodies are full of closed-loop control systems (e.g., the level of carbon dioxide in your lungs triggers the breathing reflex). However, closed-loop control can become difficult if there are delays—you have not received feedback from the previous action when starting the next one. Delays either mean one has to slow down the task or use some level of prediction to work out what to do next based on feedback of actions before the last one. This breakdown of closed-loop control in the face of (especially unexpected) delays is one of the reasons hand-eye coordination tasks, such as mouse movement, breakdown if delays exceed a couple of hundred milliseconds. The feedback loops in our bodies for these tasks assume normal delays of around 200 ms and are robust to variations around this figure, but adding delays beyond this starts to cause breakdown.

The delays inside our bodies cause other problems, too. The path from our visual cortex into our brain is far faster (by a 100 ms or so) than that from our touch and muscle tension sensors around our bodies. If we were designing a computer system to use this information, we might consider having a short 100-ms tape loop, so that we could store the video input until we had the appropriate information from all senses. However,

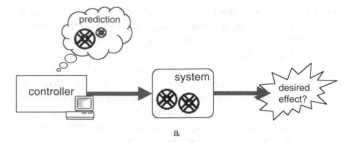

FIGURE 16.12a. Open-loop control.

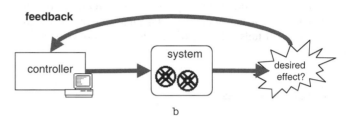

FIGURE 16.12b. Closed-loop control.

the sheer volume of visual information means that our brains do not attempt to do this. Instead, there is a part of our brains that predicts where it thinks our bodies are and what it is feeling based on previous nerve feedback and what it knows the muscles have been asked to do. The same bit of the brain then monitors what actually did happen (when the nerve signals have made their way up the spinal column to the brain) and gives an uncomfortable, or shocked, sensation when a mismatch occurs. For example, if you go to pick something up, but because of poor light or a strange-shaped object, you touch it earlier or later than you would expect. Tickling is also connected with this lack of ability to predict the sensations (why it is difficult to tickle yourself).

Race conditions also occur within this networked system of our bodies; for example, getting letter inversions while typing in which signals to the two hands get processed in the wrong order. Steve Brewster and I also used race conditions to understand what goes wrong in certain kinds of mis-hits of onscreen buttons (Dix & Brewster, 1994). In certain circumstances, two almost simultaneous commands from our brain to our hand to release the mouse button and to our arm to move to a new mouse location can get out of order, meaning the mouse moves out of the target before it is released. This analysis allowed us to design an experiment that forced this very infrequent error to occur much more frequently and therefore make it easier to assess potential solutions.

NETWORKS AS PLATFORMS: ALGORITHMS AND ARCHITECTURES FOR DISTRIBUTED INTERFACES

User interfaces are hard enough to construct on a single machine; concurrent access by users on networked machines is a nightmare!

Happily, appropriate algorithms, architectures, toolkits, and frameworks can help . . . a bit.

Accessing Shared Objects

We saw in the section on User Interface Properties how race conditions within networked systems can lead to inconsistencies within the user interface and within the underlying data structures. Fortunately, there are a range of techniques for dealing with this.

Locking. The standard technique, used in databases and file systems, for dealing with multiple accesses and the same data object is locking. When a user's application wants to update a particular database record, it asks the database manager for a lock on the record. It can then get a copy of the record and send back the update knowing that nothing else can access it in the mean time. Users are typically unaware that locking is being performed; the act of opening a file or opening a edit form for a database record establishes a lock and later, when the file or edit form is completed and closed, the lock is released.

Although this is acceptable for more structured domains, there are problems in more dynamic domains, such as shared editing. Locking a file when one user is editing it is no good, because we want several people to edit the same file at the same time. In these cases, more lightweight forms of locking can be used at finer granularities—at paragraph, sentence, or even per character level. For example, the act of clicking over a paragraph to set a text entry position may implicitly request a paragraph lock, which is released when you go on to edit another paragraph. However, implicit and informal locks, because they are not apparent, can lead to new problems. For example, a user may click on a paragraph, do some changes, but before moving on to another part of the document get interrupted. No one else can then edit the paragraph. To avoid this, the more informal locks are often time limited or can be forcibly broken by the server if another user requests a lock on the same object.

Replication. In systems such as Lotus Notes, users do not lock central copies of data, but instead each user (or possibly each site) has its own replica of the complete Notes database. Periodically, theses replicas are synchronized with central copies or with each other. Updates can happen anywhere by anyone with a replica. Conflicts may of course arise if two people edit the same note between synchronizations. Instead of preventing such conflicts, the system (and software written using it) accepts that such conflicts will occur. When the replicas synchronize, conflicts are detected and various (configurable) actions may occur: Flagging the conflicts to users, adding conflicting copies as versions, etc.

This view of replicate and worry later is essential in many mobile applications because attempts to lock a file while disconnected would first of all require waiting until a network connection could be made and, worse, if the network connection is lost while the lock is still in operation could lead to files being locked for very long periods. Other examples of replication in research environments include the CODA at CMU, which allows replication of a standard UNIX file system (Kistler & Satyanarayanan, 1992) and Liveware, a contact information system that replicates and synchronizes when people meet in a manner modeled after the spread of computer viruses (Witten, Thimbleby, Coulouris, & Greenberg, 1991). Although Liveware is now quite a few years old and was spread using synchronizing floppy disks, the same principle is now being suggested for PDAs and other devices to exchange information via IrDA or Bluetooth.

Optimistic Concurrency for Synchronous Editing. The "do it now and see if there are conflicts later" approach is also called optimistic concurrency, especially in more synchronous settings. For example, in a shared editing system, the likelihood of two users editing the same sentence at the same time is very low. An optimistic algorithm does not bother to lock or otherwise check things when the users start to edit in an area, but in the midst or at the end of their edits checks to see if there are any conflicts and attempts to fix them.

There are three main types of data that may be shared:

- orthogonal data—in which the data consists of attributes of individual objects/records that can all be independently updated
- sequential data—particularly text, but any form of list in which the order is not determined by an attribute property
- complex structural data—such as directory trees, taxonomic categories, etc.

In terms of complexity for shared data, these are in increasing difficulty.

Orthogonal data, although by no means trivial, is the simplest case. There is quite a literature on shared graphical editors, which all have this model—independent shapes and objects with independent attributes such as color, size, and position. When merging updates from two users, all one has to do is look at each attribute in turn and see whether it has been changed by only one user, in which case the updated value is used, or if it has been changed by both, in which case either the last update is used or the conflict is flagged.

Structured data is most complicated: What do you do if someone has created a new file in directory D, but at the same time someone else has deleted the directory? I know of no optimistic algorithms for dealing with this in the CSCW literature. CODA deals with directory structures (normal UNIX file system), but takes a very simple view of this as it only flags inconsistencies and does not attempt to fix them.

Algorithms for shared text editing sit somewhere between the two and have two slightly different problems, both relating to race conditions when two or more users are updating the same text:

- dynamic pointers—If user A is updating an area of text in front of user B, then the text user B is editing will effectively move in the document.
- deep conflict—What happens if user A and user B's cursors are at the very same location and they perform insertions/deletions?

Figure 16.13 shows an example of the first of these problems. The deeper conflict occurs when both cursors are at the same point, say after the "Y" in "XYZ." Adonis types "A" and at the same

Imagine two users, Adonis and Beatrice.

They are working using a shared editor and their current document reads:

> Adonis is⌶ and Beatrice is⌶.
> 　　　　Ⓐ　　　　　　　　Ⓑ

The sentence is partial and both users are about to type in their prime personal characteristic in order to complete it.

Adonis' insertion point is denoted by the boxed Ⓐ and Beatrice's insertion point is the boxed Ⓑ.

Beatrice types first yielding:

> Adonis is⌶ and Beatrice is beautiful⌶.
> 　　　　Ⓐ　　　　　　　　　　　　　Ⓑ

Adonis then types 'adorable', but unfortunately the implementor of the group editor was not very expert and the resulting display was:

> Adonis is adorable⌶ and Beatrice is ⌶eautiful.
> 　　　　　　　　　Ⓐ　　　　　　　　　Ⓑ

Beatrice's insertion point followed the 36th character before Adonis' insertion, and followed the 36th character after.

Reasonable but wrong! The actual text should clearly read:

> Adonis is adorabl⌶ and Beatrice is beautiful⌶.
> 　　　　　　　　Ⓐ　　　　　　　　　　　　　Ⓑ

This correct behavior is called a dynamic pointer as opposed to the static pointer "character position 36".

FIGURE 16.13. Dynamic pointers from (Dix, 1995).

time Beatrice types "B" should we have "XYABZ" or "XYBAZ" or even lose one or other character? Or, if Adonis types "A" and Beatrice presses delete, should we have "XYZ" or "XAZ"?

A number of algorithms exist for dealing with this (Mauve, 2000; Sun & Ellis, 1998; Vidot, Cart, Ferriz, & Suleiman, 2000), with most stemming from the dOPT algorithm used in the Grove editor (Ellis & Gibbs, 1989). These algorithms work by having various transformations that allow you to reorder operations. For example, if we have two insertions (labeled a and b) performed at the same time at different locations:

(a) Insert text a at location n
(b) Insert text b at location m

but decide to give insert a preference, then we have to transform b to b′ as follows:

(b′)　if　(m < n), insert text b at location m　　　　(i)
　　　if　(m = n), insert text b at location m　　　　(ii)
　　　if　(m > n), insert text b at location m +
　　　　　length (text a)　　　　　　　　　　　　(iii)

Case (i) says that if the location of insert b is before insert a, you do not have to worry. Case (iii) says that if it is after insert a, you have to shift your insert along accordingly. Case (ii) is the difficult one where a conflict occurs and has to be dealt with carefully to ensure that the algorithm generates the same results no matter where it is. The version described previously would mean that B's cursor gets left behind by A's edit. The alternative would be to make case (ii) the same as case (iii), which would mean B's cursor would be pushed ahead of A's.

In early work in this area, I proposed regarding dynamic pointers as first-class objects and using these in all representations of actions (Dix, 1991, 1995). This means that rules like case (i) and case (iii) happen for free, but the deep conflict case still needs to be dealt with specially.

Groupware Undo. The reason that undo is complicated in groupware is similar to the problems of race conditions in optimistic concurrency.

In the case of optimistic concurrency, user A has performed action a and user B has performed action b, both on the same initial state (Fig. 16.14). The problem is to transform user B's action into one b′ that can be applied to the state after action a yet still means the same things as the original action b.

In the case of groupware undo, we may have the situation where user A has performed action a, followed by user B performing action b, then user A decides to undo action a. How do we transform action b so that the transformed b′ means the same before action a as b did after.

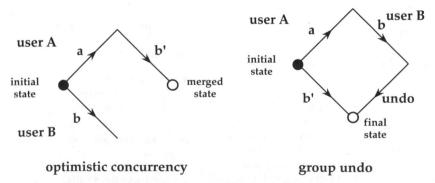

optimistic concurrency　　　　　group undo

FIGURE 16.14. Multiuser transformations.

Similar, but slightly different, transformation rules can be produced for the case of undo and also dynamic pointers can be used for most cases.

As with optimistic concurrency, there is slightly more work on group undo in shared graphical editors where the orthogonal data makes conflicts easier (Berlage & Spenke, 1992).

Real Solutions? Although these various algorithms can ensure there is no internal inconsistency and that all participants see the same thing, they do not necessarily solve all problems. Look again at the case of Alison and Brian's chat in the section on User Interface Properties. Certainly, in the case of group undo, when Gregory Abowd and I published the first paper on the topic (Abowd & Dix, 1992), we proposed various solutions, but also recommended that as well as an explicit undo button systems ought to provide sufficient history to allow users to recreate what they want to without using the undo button. This is not because it is impossible to find a reasonable meaning for the undo button, but because in the case of group undo, there are several reasonable meanings. Choosing the meaning a user intends is impossible, and so it may sometimes be better not to guess.

Architectures for Networked Systems

Software architecture is about choosing what (in terms of code and functionality) goes where. For applications on a single machine, these are often logical distinctions between parts of the code. For networked systems 'where' includes physical location and this choice of location makes an enormous difference to the responsiveness of the system.

The simplest systems are almost always centralized client-server architectures in which the majority of computation and data storage happens in the central server. Many of the problems of race conditions and potential inconsistencies disappear. However, this means that every interaction requires a network interaction with the server meaning that feedback may be very slow. At the opposite extreme are replicated peer–peer architectures in which all the code is running on users' own PCs and the PCs communicate directly with one another. Feedback can now be instantaneous, but the complexity of algorithms to maintain consistency, catch-up late joiners, etc., can be very complex. Most systems operate somewhere between these two extremes, with a central "golden copy" of shared data, but with some portion of the data on individual PCs, PDAs, or other devices to allow rapid feedback.

In web applications, options are constrained by the features allowed in HTML and web browsers. Note that even a web form is allowing some local interaction (filling in the form), as well as some centralized interaction (submitting it). Applets allow more interaction, but the security limitations mean that they can only talk back to the server where they originated. Thus, true peer–peer architectures are impossible on the web, but can be emulated by chat servers and similar programs that relay messages between clients. For Intranets, it is easier to configure browsers so that they accept applets as trusted and thus with greater network privileges, or to include special plug-in components to perform more complicated actions.

Mobile systems have yet more issues, because they need to be capable of managing disconnected operation (when they have no physical or wireless connection to the network). This certainly means keeping cached copies of central data. For example, the AvantGo browser for PDAs downloads copies of web pages onto a PDA and synchronizes these with their latest versions every time the PDA is docked into a desktop PC (AvantGo, 2001). Although this is now beginning to change with higher bandwidth mobile data services, it is still the case that access while mobile is slower and more expensive than while using a fixed connection. This pushes one toward replicated architectures with major resynchronization when connected via cheap/fast connections and more on-demand or priority-driven synchronization when on the move.

Supporting Infrastructure

To help manage networked applications, various types of supporting infrastructure are being developed.

Awareness Servers. These servers keep track of which users are accessing particular resources (e.g., visiting a particular Web page) so that you can be kept informed as to whether others are near you in virtual space, or whether friends are online and active (ICQ, 2001; Palfreyman & Rodden, 1996; SUN Microsystems, 2001).

Notification Servers. These servers serve a similar role for data, allowing client programs to register an interest in particular pieces of shared data. When the data is modified or accessed, the interested parties are notified. For example, you may be told when a web page has changed or when new items have been added to a bulletin board. Some notification servers also manage the shared data (Patterson, Day, & Kucan, 1996), where others are 'pure' just managing the job of notification (Ramduny, Dix, & Rodden, 1998).

Event/Messaging Systems. These systems allow different objects in a networked environment to send messages to one another in ways that are more convenient than that allowed by the raw network. For example, they may allow messages to be sent to objects based on location independent names, so that objects do not have to know where each other are.

Resource Discovery. Systems such as the Java JINI framework and Universal Plug-and-Play allow devices to find out about other devices close to them (e.g., the local printer) and configure themselves to work with one another. As ubiquitous and mobile devices multiply, this will become increasingly important.

HISTORY, FUTURES, AND PARADIGM SHIFTS

History

Given the anarchic image of the web, it is strange that the Internet began its development as a U.S. military project. The suitability of the Internet for distributed management and

Timeline—Key events for the Internet	
1968	First proposal for ARPANET; military & government research contracted to Bolt, Beranek, & Newman
1971	ARPANET enters regular use
1973/4	Redesign of lower level protocols leads to TCP/IP
1983	Berkeley TCP/IP implementation for 4.2 BSD—public domain code
1980s	rapid growth of NSFNET—broad academic use
1990s	WWW and widespread public access to the Internet
2000	WAP on mobile phones; web transcends the Internet

independent growth stems not from an egalitarian or anticentralist political agenda, but from the need to make the network resilient to nuclear attack with no single point of failure.

In the 1970s when the Internet was first developing, it and other networks were mainly targeted at connecting together large computers. It was during the 1980s, with the rise of personal computing, that local networks began to become popular.

However, even before that point, very local networks at the laboratory-bench level had been developed to link laboratory equipment, for example, the IEEE 488 designed originally to link Hewlett-Packard's proprietary equipment and then becoming an international standard. Ethernet, too, began life in commercial development at Xerox before becoming the de facto standard for local networking.

Although it is technical features of the Internet (decentralized, resilient to failures, hardware independent) that have made it possible for it to grow, it is the web that has made it become part of popular consciousness. Just as strange as the Internet's metamorphosis from military standard to anarchic cult is the web's development: from medium of exchange of large, high-energy physics data sets to e-commerce, home of alternative web pages and online sex!

Paradigm Shift

During the 1970s through to the mid-1990s, networks were a technical phenomenon, enabling many aspects of business and academic life, but with very little public impact. However, this has changed dramatically over the last 5 years and now we think of a networked society. The Internet and other network technologies, such as SMS text messages, are not only transforming society, but at a popular and cultural level defining an era.

International transport, telecommunications, and broadcasting had long before given rise to the term global village. However, it seems this was more a phrase waiting for a meaning. Until recently, the global village was either parochial—telephoning those you already know, or sanitized—views of distant cultures through the eyes of the travel agent or television camera. It is only now that we see chat rooms and web home pages allowing new contacts and friendships around the world, or at least among those affluent enough to get Internet access.

Markets, too, have changed because of global networks. It is not only the transnationals who can trade across the world; even my father-in-law runs a thriving business selling antiques through eBay.

Marketing has also had to face a different form of cross-cultural issue. Although selling the same product, a hoarding in Karachi may well be different from an advertisement in a magazine in Kentucky, reflecting the different cultural concerns. Global availability of web pages changes all that. You have to create a message that appeals to all cultures—a tall order. Those who try to replicate the targeting of traditional media by having several country-specific Web sites may face new problems. The global access to even these country-specific pages means that the residents of Kentucky and Karachi can compare the different advertisements prepared for them and, in so doing, see how the company views them and their cultures.

Economics drives so much of popular as well as business development of networked society. One of the most significant changes in the UK in recent years was from changes in charging models. In the United States, local calls have long been free and hence so were Internet connections to local points of presence. The costs of Internet access in the UK (and even more important the perception of the cost) held back widespread use. The rise of free or fixed-charge unmetered access changed nearly overnight the acceptability and style of use. Internet access used to be like a lightning guerrilla attack into the web territory, quick in and quick out before the costs mounted up, but is now a full-scale occupation.

The need for telecom companies across the world to recover large investments in wireless band franchises, combined with use rather than connection-based charging made possible by GPRS and third-generation mobile services, makes it likely that we will see a similar growth in mobile access to global-networked information and services.

In an article in 1998, I used the term PopuNET to refer to a change in society that was not yet there, but would come. PopuNET is characterized by network access:

everywhere, everywhen by everyone

This pervasive, permanent, popular access is similar to the so-called Martini principle applied more recently to mobile networking—anytime, anyplace, anywhere. Of course, Martini never pretends to be anything but exclusive, so not surprisingly these differ on the popular dimension. "Anyplace, anywhere" does correspond to the pervasive "everywhere" and "anytime" to the permanent "everywhen." However, there is a subtle difference, especially between anytime and everywhen. Anytime means that at any time you choose you can connect. Everywhen means that at all places and all times you are connected. When this happens, one ceases to think of connectedness and it simply becomes part of the backdrop of life.

The changes in charging models have brought the UK closer to everywhen, and the United States and many other parts of the world are already well down the path. Always-on mobile connectivity will reinforce these changes.

PopuNET will demand new interfaces and products, not just putting web pages on TV screens or spreadsheets on fridge doors. What these new interfaces will be is still uncertain.

Futures (Near)

It is dangerous to predict far into the future in an area as volatile as this. One development that is already underway, which will make a major impact on user interfaces, is short-range networking that will enable various forms of wearable and ubiquitous networks. Another is the introduction of network appliances, which will make the home 'alive' on the network.

We have considered network aspects of continuous media at length. The fact that the existing Internet TCP/IP protocols do not enable guaranteed quality-of-service will put severe limits on its ability to act as an infrastructure for services such as video-on-demand or video sharing between homes. The update to TCP/IP, which has been under development for several years, IPv6, will allow prioritized traffic; it falls short of real guaranteed quality-of-service (IPng, 2001).

It seems that this is impasse. One of the reasons that IPv6 has taken so long is not the technical difficulty, but backward compatibility and the problems of uptake on the existing worldwide infrastructure. So, evolutionary change is hard. However, revolutionary change is also hard; one cannot easily establish a new parallel international infrastructure overnight. Or, perhaps one can.

Mobile phone services have started with an infrastructure designed for continuous voice and are, through a series of quite dramatic changes, moving this toward a fully mixed media/data service that is also a global network. Furthermore, more and more non–web-based Internet services are using HTTP, the web protocol, to talk to one another to be firewall friendly. This means that mobile phones and PDAs can be web connected without being Internet connected. So, perhaps the future is not an evolution of Internet protocols, but a gradual replacement of the Internet by nth-generation mobile networks. Instead of the Internet on phones, we may even see mobile phone networking standards being used over wires.

References

Abowd, G. D., & Dix, A. J. (1992). Giving undo attention. *Interacting with Computers, 4*(3), 317–342.

Arehart, C., Chidambaram, N., Guruprasad, S., Homer, A., Howell, R., Kasippillai, S., Machin, R., Myers, T., Nakhimovsky, A., Passani, L., Pedley, C., Taylor, R., & Toschi, M. (2000). *Professional WAP.* Birmingham, UK: Wrox Press.

Arons, B. (1997). SpeechSkimmer: A system for interactively skimming recorded speech. *ACM Transactions on Computer Human Interaction, 4*(1), 3–38.

AvantGo. (2001). Hayward, CA: AvantGo, Inc. [On-line]. http://www.avantgo.com/

Barnard, P., & May, J. (1995). Interactions with advanced graphical interfaces and the deployment of latent human knowledge. In F. Paterno (Ed.), *Eurographics Workshop on the Design, Specification and Verification of Interactive Systems.* Berlin: Springer Verlag.

Bellotti, V. (1993). *Design for privacy in ubiquitous computing environments. Proceedings of CSCW'93* (pp. 77–92). New York: ACM Press.

Benford, S., Bowers, J., Fahlen, L., Mariani, J., & Rodden, T. (1994). Supporting cooperative work in virtual environments. *The Computer Journal, 37*(8), 635–668.

Berlage, T., & Spenke, M. (1992, August 10–14). The GINA interaction recorder. *Proceedings of the IFIP WG2.7 Working Conference on Engineering for Human–Computer Interaction,* Ellivuori, Finland.

Bluetooth (2001). Official Bluetooth SIG Website [On-line]. Available http://www.bluetooth.com/

Buxton, W., & Moran, T. (1990, September). EuroPARC's integrated interactive intermedia facility (IIIF): Early experiences. In S. Gibbs & A. A. Verrijn-Stuart (Eds.), Multi-user interfaces and applications (pp. 11–34). New York: North-Holland.

Campbell, A., & Nahrstedt, K. (Eds.). (1997). *Building QoS into distributed systems.* Kluwer, Boston.

Card, S. K., Moran, T. P., & Newell, A. (1983). *The psychology of human computer interaction.* Hillsdale, NJ: Lawrence Erlbaum Associates.

CERN. (0000). European Organisation for Nuclear Research [On-line]. Available: http://public.web.cern.ch/Public/

Clarke, D., Dix, A., Ramduny, D., & Trepess, D. (Eds.). (1997, June). Workshop on Time and the Web, Staffordshire University. [Online] available as: http://www.hiraeth.com/conf/web97/papers/

Dix, A. J. (1987). The myth of the infinitely fast machine. In *People and computers, Volume III—Proceedings of HCI'87* (pp. 215–228). New York: Cambridge University Press.

Dix, A. J. (1991). *Formal methods for interactive systems.* New York: Academic Press.

Dix, A. J. (1994). *Seven years on, the myth continues* (Contract No. (RR9405). University of Huddersfield, Huddersfield, UK.

Dix, A. J. (1995). Dynamic pointers and threads. *Collaborative Computing, 1*(3), 191–216.

Dix, A. J. (1996). Closing the loop: Modelling action, perception and information (Keynote). In T. Catarci, M. F. Costabile, S. Levialdi, & G. Santucci (Eds.), *AVI'96—Advanced Visual Interfaces* (pp. 20–28), Gubbio, Italy. New York: ACM Press.

Dix, A. (2001a, Autumn). artefact + marketing = product. *Interfaces, 48,* 20–21.

Dix, A. (2001b, September). Cyber-economies and the real world. Keynote—South African Institute of Computer Scientists and Information Technologists Annual Conference, SAICSIT 2001. Pretoria, South Africa.

Dix, A., & Brewster, S. A. (1994). Causing trouble with buttons. In D. England (Ed.), *Ancilliary Proceedings of HCI'94.* Glasgow, Scotland. London, UK: BCS HCI Group.

Dix, A., Finlay, J., Abowd, G., & Beale, R. (1998). *Human–computer interaction,* 2nd ed. Englewood, Cliffs, NJ: Prentice Hall.

Dix, A., Rodden, T., Davies, N., Trevor, J., Friday, A., & Palfreyman, K. (2000, September). Exploiting space and location as a design framework for interactive mobile systems. *ACM Transactions on Computer-Human Interaction, 7*(3), 285–321.

Dodge, M., & Kitchin, R. (2001). *Atlas of cyberspace.* Reading, MA: Addison Wesley.

Dourish, P., & Bellotti, V. (1992). Awareness and coordination in shared workspaces. *Proceedings of CSCW'92* (pp. 107-114).

Edwards, W. K., & Rodden, T. (2001). *Jini, example by example*. Englewood Cliffs, NJ: SUN Microsystems Press, Prentice-Hall PTR.

Electrolux. (1999). Screenfridge [On-line]. Available http://www.electrolux.com/screenfridge/

Ellis, C. A., & Gibbs, S. J. (1989). Concurrency control in groupware systems. *SIGMOD Record, 18*(2), 399-407.

Fogg, B. J., Marshall, J., Laraki, O., Osipovich, A., Varma, C., Fang, N., Paul, J., Rangnekar, A., Shon, J., Swani, P., & Treinen, M. (2001). What makes web sites credible? A report on a large quantitative study. *Proceedings of CHI2001*, Seattle, 2001. Also *CHI Letters, 3*(1), 61-68.

Foster, I. (2000, December 7). Internet computing and the emerging grid. Nature, webmatters [On-line]. Available: http://www.nature.com/nature/webmatters/grid/grid.html

Foster, I., & Kesselman, C. (Eds.). (1999). *The Grid: Blueprint for a new computing infrastructure*. San Francisco, CA: Morgan-Kaufmann.

Gaver, W. W., Smith, R. B., & O'Shea, T. (1991, April). Effective sounds in complex situations: The ARKola simulation. In S. P. Robertson, G. M. Olson, & J. S. Olson (Eds.), *Reaching Through Technology—CHI'91 Conference Proceedings* (pp. 85-90). New York: ACM Press.

Gellersen, H.-W., Beigl, M., & Krull, H. (1999). The MediaCup: Awareness technology embedded in an everyday object. *Handheld & Ubiquitous Computing, Lecture Notes in Computer Science, 1707*, 308-310.

Gray, W. D., John, B. E., & Atwood, M. E. (1992). The precis of project ernestine or an overview of a validation of goms. In P. Bauersfeld, J. Bennett, & G. Lynch (Eds.), *Striking a Balance, Proceedings of the CHI'92 Conference on Human Factors in Computing Systems* (pp. 307-312). New York: ACM Press.

Greenhalgh, C. (1997). Analysing movement and world transitions in virtual reality tele-conferencing. In J. A. Hughes, W. Prinz, T. Rodden, & K. Schmidt (Eds.), *Proceeding of the ECSCW 97* (pp. 313-328). Dordrecht, The Netherlands: Kluwer Academic Publishers.

GRID. (2001). GRID Forum home page [On-line]. Available: http://www.gridforum.org/

Halasz, F., Moran, T., & Trigg, R. (1987). NoteCards in a nutshell. In *Proceedings of the CHI + GI* (pp. 45-52). New York: ACM.

HFIG. (2001). Human feature interaction group home page, University of Glasgow, Scotland. http://www.dcs.gla.ac.uk/research/hfig/

Hindus, D., Mainwaring, S. D., Leduc, N., Hagström, A. E., & Bayley, O. (2001). Casablanca: designing social communication devices for the home. *Proceedings of CHI 2001* (pp. 325-332). New York: ACM Press.

Howard, S., & Fabre, J. (Eds.). (1998). Temporal aspects of usability. (special issue), *Interacting with Computers, 11*(1), 1-105.

ICQ. (2001). [On-line]. Available: http://www.icq.com/products/whatisicq.html

IEEE 802.11. (2001). Working Group [On-line]. Available: http://www.ieee802.org/11/

IPng. (2001). IP Next Generation (IPng) Working Group home page [On-line]. Available: http://playground.sun.com/pub/ipng/html

ISO/IEC 7498. (1994). Information technology—Open Systems Interconnection—Basic Reference Model: The Basic Model. International Standards Organisation [On-line]. Available: http://www.iso.org

John, B. E. (1990). Extensions of GOMS analyses to expert performance requiring perception of dynamic visual and auditory information. In J. C. Chew & J. Whiteside (Eds.), *Empowering People—Proceedings of CHI'90 Human Factors in Computer Systems* (pp. 107-115). New York: ACM Press.

Johnson, C. (1997). What's the web worth? The impact of retrieval delays on the value of distributed information [On-line].

In Clarke, 1997. Available: http://www.hiraeth.com/conf/web97/papers/johnson.html

Johnson, C., & Gray, P. (Eds.). (1996, April). Temporal aspects of usability (report of workshop in Glasgow, June 1995). *SIGCHI Bulletin, ACM, 28*(2), 32-61.

JPEG. (2001). Joint Photographic Experts Group home page [On-line]. Available: http://www.jpeg.org/public/jpeghomepage.htm

Kistler, J. J., & Satyanarayanan, M. (1992, February). Disconnected operation in the CODA file system. *ACM Transactions on Computer Systems, 10*(1), 3-25.

Lavery, D., Kilgourz, A., & Sykeso, P. (1994). Collaborative Use of X-windows applications in observational astronomy. In G. Cockton, S. Draper, & G. Wier (Eds.), *People and Computers IX* (pp. 383-396). New York: Cambridge University Press.

Light, A. (2001). *Interaction at the producer-user interface: An interdisciplinary analysis of communication and relationships through interactive components on websites for the purpose of improving design*. DPhil Thesis, University of Sussex, UK.

Light, A., & Wakeman, I. (2001, February). Beyond the interface: Users' perceptions of interaction and audience on websites. In D. Clarke & A. Dix (Eds.), *Special issue interfaces for the active web (Part 1). Interacting with Computers* (Vol. 13, pp. 401-426). Amsterdam, The Netherlands: Elsevier.

Lock, S., Allanson, J., & Phillips, P. (2000). User-driven design of a tangible awareness landscape. *Symposium on Designing Interactive Systems 2000* (pp. 434-440). [On-line]. Available: http://www.acm.org/pubs/citations/proceedings/chi/347642/p434-lock/

Mariani, J. A., & Rodden, T. (1991). The impact of CSCW on database technology. In *Proc. ACM Conference on Computer Supported Cooperative Work* (includes critique of 'transparency' in a CSCW setting). New York: ACM Press.

Mauve, M. (2000). Consistency in replicated continuous interactive media. *Proceedings of CSCW'2000* (pp. 181-190). New York: ACM Press.

McDaniel, S. E., & Brinck, T. (1997, October). Awareness in collaborative systems: A CHI 97 workshop (report). *SIGCHI Bulletin, 29*(4).

McManus, B. (1997). Compensatory actions for time delays. In Clarke, 1997 [On-line]. Available: http://www.hiraeth.com/conf/web97/papers/barbara.html

Millett, L. I., Friedman, B., & Felten, E. (2001). Cookies and web browser design: Toward informed consent online. *Proceedings of CHI2001*, Seattle; WA, 2001. Also *CHI Letters, 3*(1), 46-52.

Mitsopoulos, E. (2000). *A principled approach to the design of auditory interaction in the non-visual user interface*. DPhil Thesis, University of York, UK.

MPEG. (2001). Moving Picture Experts Group home page [On-line]. Available: http://www.cselt.it/mpeg/

Olson, M., & Bly, S. (1991). The Portland experience: A report on a distributed research group. *International Journal of Man-Machine Studies, 34*, 211-228.

Palfreyman, K., & Rodden, T. (1996, November). A protocol for user awareness on the world wide web. *Proceedings of CSCW'96* (pp. 130-139). New York: ACM Press.

Patterson, J. F., Day, M., & Kucan, J. (1996). Notification servers for synchronous groupware. *Proceedings of CSCW'96* (pp. 122-129). New York: ACM Press.

Pausch, R. (1991). Virtual reality on five dollars a day. In S. P. Robertson, G. M. Olson & J. S. Olson (Eds.), *CHI'91 Conference Proceedings*, (pp. 265-270). Reading, MA: Addison Wesley.

Ramduny, D., & Dix, A. (1997). Why, what, where, when: Architectures for co-operative work on the WWW. In H. Thimbleby, B. O'Connaill, & P. Thomas, *Proceedings of HCI'97* (pp. 283-301). Bristol, UK: Springer.

Ramduny, D., Dix, A., & Rodden, T. (1998). Getting to know: The design space for notification servers. *Proceedings of CSCW'98* (pp. 227-235). Available: http://www.hcibook.com/alan/papers/GtK98/

Resnick, P., & Varian, H. R. (Guest Eds.). (1997). Special issue on recommender systems. *CACM, 40*(3), 56-89.

Rodden, T. (1996). Populating the application: A model of awareness for cooperative applications. In M. S. Ackerman (Ed.), *Proceedings of the 1996 ACM Conference on Computer-Supported Cooperative Work (CSCW'96)* (pp. 87-96). New York: ACM Press.

Roussel, N. (1999, January). Beyond webcams and videoconferencing: Informal video communication on the web. In *Proceedings of The Active Web* (pp. 65-69). Stafford, UK.

Roussel, N. (2001, May). Exploring new uses of video with videoSpace. In *Proceeding of EHCI'01, the 8th IFIP Working Conference on Engineering for Human–Computer Interaction, Lecture Notes in Computer Science*. New York: Springer-Verlag. [On-line] Available: http://www.hiraeth.com/conf/activeweb/

Sandor, O., Bogdan, C., & Bowers, J. (1997). Aether: An awareness engine for CSCW. In J. Hughes (Ed.), *Proceedings of the Fifth European Conference on Computer Supported Cooperative Work (ECSCW'97)* (pp. 221-236). Dordrecht, The Netherlands: Kluwer Academic.

Sangha, A. (2001, June). Legal implications of location based advertising. Interview for the WAP Group [On-line]. Available: http://www.thewapgroup.com/53762_1.DOC

Schneier, B. (1996). *Applied cryptography*, 2nd ed. New York: Wilcy.

SETI. (0000). SETI@home—The search for extra-terrestrial intelligence [On-line]. Available: http://setiathome.ssl.berkeley.edu/

Shneiderman, B. (1984). Response time and display rate in human performance with computers. *ACM Computing Surveys, 16*(3), 265-286.

Siegal, D. (1999). *Futurize your enterprise*. New York: Wiley.

Stefik, M., Bobrow, D. G., Foster, G., Lanning, S., & Tatar, D. (1987). WYSIWIS revisited: Early experiences with multiuser interfaces. *ACM Transactions on Office Information Systems, 5*(2), 147-167.

Stevens, R. (1998). *UNIX network programming, Volume 1, 2nd ed.: Networking APIs: Sockets and XTI*. Englewood Cliffs: Prentice Hall.

Stevens, R. (1999). *UNIX network programming, Volume 2, 2nd ed.: Interprocess communications*. Englewood Cliffs: Prentice Hall.

Stifelman, L., Arons, B., & Schmandt, C. (2001). The Audio Notebook: Paper and pen interaction with structured speech. *Proceedings of CHI2001*, Seattle, WA, 2001. Also *CHI Letters, 3*(1), 182-189.

Sun, C., & Ellis, C. (1998). Operational transformation in real-time group editors: issues, algorithms, and achievements. *Proceedings of CSCW'98* (pp. 59-68). New York: ACM Press.

SUN Microsystems. (2001). Awarenex [On-line]. Available: http://www.sun.com/research/features/awarenex/

Vidot, N., Cart, M., Ferriz, J., & Suleiman, M. (2000). Copies convergence in a distributed real-time collaborative environment. *Proceedings of CSCW'2000* (pp. 171-180). New York: ACM Press.

Wisneski, G., Ishii, H., Dahley, A., Gorbet, M., Brave, S., Ullmer, B., & Yarin, P. (1998). Ambient display: Turning architectural Space into an interface between people and digital information. In *Proceedings of the First International Workshop on Cooperative Buildings (CoBuild'98), Darmstadt, Germany (February 25-26, 1998). Lecture Notes in Computer Science* (Vol. 1370). Heidelberg, Germany: Springer-Verlag.

Witten, I. H., Thimbleby, H. W., Coulouris, G., & Greenberg, S. (1991). Liveware: A new approach to sharing data in social networks. *International Journal of Man-Machine Studies, 34*, 337-348.

Zimmerman, T. G. (1996). Personal area networks: Near-field intrabody communication. *IBM Systems Journal, 35*(3&4).

All web links and other related links and material are available from: http://www.hiraeth.com/alan/hbhci/network/

·17·

MOTIVATING, INFLUENCING, AND PERSUADING USERS

B.J. Fogg
Stanford University

INTRODUCTION

Since the advent of modern computing in 1946, the uses of computing technology have expanded far beyond their initial role of performing complex calculations (Denning & Metcalfe, 1997). Computers are not just for scientists any more; they are an integral part of workplaces and homes. The diffusion of computers has led to new uses for interactive technology, including one that computing pioneers of the 1940s never imagined: Using computers to change people's attitudes and behavior—in a word, *persuasion*.

Today, creating successful human–computer interactions (HCIs) requires skills in motivating and persuading people. However, interaction designers do not often view themselves as agents of influence. They should. The work they perform often includes crafting experiences that change people—change the way people feel, change what they believe, and change the way people behave. Consider these common challenges: How can designers motivate people to register their software? How can they get people to persist in learning an online application? How can they create experiences that build product loyalty? Often, the success of today's interactive product hinges on changing people's attitudes or behaviors.

Sometimes the influence elements in HCI are small, almost imperceptible, such as creating a feeling of confidence or trust in what the computing product says or does. Other times, the influence element is large, even life altering, such as motivating someone to quit smoking. Small or large, elements of influence are increasingly present in Web sites, productivity tools, video games, wearable devices, and other types of interactive computing products. Because of the growing use of computing products and the unparalleled ability of software to scale, interaction designers may well become leading change agents of the future. Are we ready?

The study of computers as persuasive technologies, referred to as "captology," is a relatively new endeavor when compared with other areas of HCI (Fogg, 1997, 1998). Fortunately, understanding in this area is growing. HCI professionals have established a foundation that outlines the domains of applications, useful frameworks, methods for research, design guidelines, best-in-class examples, as well as ethical issues (Berdichevsky & Neuenschwander, 1999; Fogg, 1999; Khaslavsky & Shedroff, 1999; King & Tester, 1999; Tseng & Fogg, 1999). This chapter will not address all these areas in depth, but it will share some key perspectives, frameworks, and design guidelines relating to captology.

DEFINING PERSUASION AND GIVING HIGH-TECH EXAMPLES

What is "persuasion"? As one might predict, scholars do not agree on the precise definition. For the sake of this chapter, *persuasion is a noncoercive attempt to change attitudes or behaviors*. There are some important things to note about this definition. First of all, persuasion is *noncoercive*. Coercion—the use of force—is not persuasion; neither is manipulation or deceit. These methods are shortcuts to changing how people believe or behave, and for interaction designers these methods are rarely justifiable.

Next, persuasion requires an *attempt* to change another person. The word *attempt* implies intentionality. If a person changes someone else's attitude or behavior without intent to do so, it's an accident or a side effect; it's not persuasion. This point about intentionality may seem subtle, but it is not trivial. Intentionality distinguishes between a *side effect* and a *planned effect* of a technology. At its essence, captology focuses on the planned persuasive effects of computer technologies.

Finally, persuasion deals with *attitude changes or behavior changes* or both. Although some scholars contend persuasion pertains only to attitude change, other scholars would concur with my view, including behavior change as a target outcome of persuasion. Indeed, it's these two outcomes—attitude change and behavior change—that are fundamental in the study of computers as persuasive technologies.

Note how attitude and behavior changes are central in two examples of persuasive technology products. First, consider the CD-ROM product "5 A Day Adventures" (www.dole5aday.com). Created by Dole Foods, this computer application was designed to persuade kids to eat more fruits and vegetables. Using 5 A Day Adventures, children enter a virtual world with characters like "Bobby Banana" and "Pamela Pineapple," who teach kids about nutrition and coach them to make healthy food choices. The program also offers areas where children can practice making meals using fresh produce, and the virtual characters offer feedback and praise. This product clearly aims to change the attitudes children have about eating fruits and vegetables. But even more important, the product sets out to change their eating behaviors.

Next, consider another example: Amazon.com. The goal of this Web site is to persuade people to buy products again and again from Amazon. Almost everything on the Web site contributes to this end result: User registration, tailored information, limited-time offers, third-party product reviews, one-click shopping, confirmation messages, and more. Dozens of persuasion strategies are integrated into the overall experience. Although the Amazon online experience may appear to be focused on providing mere information and seamless service, it's really about persuasion—buy things now and come back for more.

THE 5TH WAVE: PERSUASIVE INTERACTIVE TECHNOLOGY

Computing systems did not always contain elements of influence. It's only been in recent years that interactive computing became mature enough to spawn applications with explicit elements of influence. The dramatic growth of technologies designed to persuade and motivate represents the fifth wave of focus in end-user computing. The fifth wave leverages advances from the four previous waves (see Fig. 17.1).

The first wave of computing began over 50 years ago and continues today. It's the focus on function. The energy and attention of computer professionals focused on just getting computing devices to work properly, and then to make them more and more capable. In short, the first wave is function.

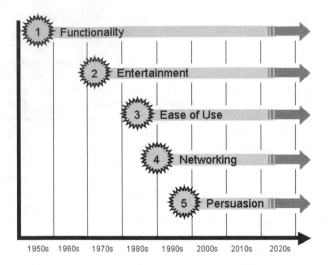

FIGURE 17.1. Major waves in computing.

TABLE 17.1. Persuasive Technologies Can Be Used in Various Domains

Domains for Persuasive Technologies	Example
Commerce—Buying and Branding	To buy a certain product
Education, Learning, & Training	To engage in activities that promote learning
Safety	To drive more safely
Environmental Conservation	To re-use shopping bags
Occupational Productivity	To set and achieve goals at work
Preventative Health Care	To quit smoking
Fitness	To exercise with optimal intensity and frequency
Disease Management	To manage diabetes better
Personal Finance	To create and adhere to a personal budget
Community Involvement/Activism	To volunteer time at a community center
Personal Relationships	To keep in touch with their aging parents
Personal Management & Improvement	To avoid procrastination

The second wave of computing began to rise in the 1970s with the emergence of digital gaming, first represented by companies like Atari and with products like Pong. This wave is entertainment, and it continues to swell because of continued attention and energy devoted to computer-based fun.

The third wave of computing came in the 1980s when human factors specialists, designers, and psychologists sought to create computers for ordinary people. This third wave is ease of use. Although new developments like the computer mouse and the graphical user interface came before 1980, it was a consumer product—the Apple Macintosh—that generated widespread attention and energy to making computers easier to use. Like the previous two waves, the third wave keeps rolling today. It provides the foundation for most work in HCI arenas.

The fourth major wave in computing is networking. Beginning as early as the 1960s, networked computers proliferated rapidly during the 1980s and became a major wave in the early 1990s with the invention and release of the World Wide Web. Networking computers via the internet and other technology infrastructures will continue to be a major wave in computing.

And, this brings us to the fifth wave: Computers designed to persuade. Early signs of this wave appeared in the 1970s and 1980s with a handful of computing systems designed to motivate health behaviors and work productivity. But it was not until the late-1990s—specifically the commercialization of the World Wide Web—that more than a handful of people began to devote attention and energy to making interactive systems capable of motivating and influencing users. This fifth wave—persuasion—is new and could be as significant as the four waves that have come before.

DOMAINS WHERE PERSUASION AND MOTIVATION MATTERS

Captology is relevant to systems designed for many facets of human life. The most obvious domain is in promoting commerce—buying and branding, especially via the web.

Although promoting commerce is perhaps the most obvious and lucrative application, at least 11 other domains are potential areas for persuasive technology products. The various domains, along with a sample target behavior change, are summarized in Table 17.1.

The domains described reflect how much persuasion is part of ordinary human experience, from personal relationships to environmental conservation. Interactive technologies have been, and will continue to be, created to influence people in these 12 domains, as well as in others that are less apparent. The way various computing products incorporate persuasion and motivation principles will evolve as computing technology advances and as people adopt a wider array of interactive systems for a wider range of human activities. The influence elements in these systems can be readily apparent, or they can be woven into the fabric of an interactive experience, a distinction explored in the next section.

PERSUASION AND INTERACTIVE TECHNOLOGY: TWO LEVELS OF ANALYSIS

One key insight in captology is to see that persuasion in computing products takes place on two levels: macro and micro. On the macro level, one finds products designed for an overall persuasive outcome. For example, the Dole "5 A Day Adventures" CD ROM and Amazon.com Web site are designed specifically for persuasion. For these and other products, persuasion and motivation are the sole reason these products exist. I use the word *macrosuasion* to describe this type of big-picture target outcome.

On the other hand, one finds computing products with what I call *microsuasion*. These products could be word processing programs or spreadsheets; they do not necessarily have a persuasive outcome as the overall goal of the product. But they will incorporate smaller elements of influence to achieve other goals. Microsuasion can be designed into dialogue boxes, visual elements, interactions sequences, and more. In productivity software, microsuasion can lead to increased productivity or stronger brand loyalty. The following examples will help clarify the distinction between macrosuasion and microsuasion.

Examples of Macrosuasion

One notable example of macrosuasion is a product named "Baby Think It Over." A U.S. company (www.btio.com) designed this computerized doll to simulate the time and energy required to care for a baby, with the overall purpose of persuading teens to avoid becoming parents prematurely. Used as part of many school programs in the United States, the Baby Think It Over infant simulator looks, weighs, and cries something like a real baby. The computer embedded inside the doll triggers a crying sound at random intervals; in order to stop the crying sound, the teen caregiver must pay immediate attention to the doll. If the caregiver fails to respond appropriately, the computer rembedded inside the doll records the neglect. After a few days of caring for the simulated infant, teenagers generally report less interest in becoming a parent in the near future (see www.btio.com), which—along with reduced teen pregnancy rates—is the intended outcome of the device.

Next, consider Scorecard.org as another example of macrosuasion. Created by the Environmental Defense Foundation, this Web site helps people find information about pollution threats in their neighborhoods. When users enter their zip code, the site lists names of the polluting institutions in their area, gives data on chemicals being released, and outlines the possible health consequences. But that's not all. Scorecard.org then encourages users to take action against the polluting organizations and makes it easy to contact policy makers to express concerns. This Web site aims to increase community activism to pressure officials and offending institutions into cleaning up the environment. The entire point of this Web site is to get people to take action against polluting institutions in their neighborhoods. That's macrosuasion.

Examples of Microsuasion

Most computing products were not created with persuasion as the main focus. Larger software categories include applications for productivity, entertainment, and creativity. Yet, these same products often use influence elements as part of the overall experience. Examples of interactive products using microsuasion are plentiful and sometimes subtle. A word processing program may encourage users to spell check text, or a Web site devoted to high school reunions may reward alumni for posting a current photograph online. This is persuasion on a micro level.

For a deeper look at microsuasion, consider the personal finance application Quicken, created by Intuit (www.intuit.com). Quicken is a productivity software product. Its overall goal is to simplify the process of managing personal finances. Quicken uses microsuasion to accomplish this overall goal. For example, the application reminds users to take financial responsibility,

such as paying bills on time. Also, the software tracks personal spending habits and shows results in graphs, allowing projections into future financial scenarios. In addition, the software praises users for doing necessary but menial tasks, like balancing their online check registry. These microsuasion elements—reminders, visualizations, and praise—are influence strategies embedded in the Quicken experience to change what users think and how they act. Ideally, when these microsuasion elements succeed, users benefit from Quicken's approach to managing personal finances.

Like Quicken, educational software often uses microsuasion. The overall point of most educational applications and interactive experiences is to teach facts and skills, not to persuade. But to get users to stick with the program or to believe what is presented, many products will incorporate motivational elements, as well as building credibility perceptions of the content. The product may seek to persuade the learner that the content is important, that the learner is able to successfully master it, and that following the guidelines of the program will lead to the greatest success. Note how these smaller elements of the program—the microsuasions—contribute to the overall goal: Learning. Furthermore, interactive educational products will often incorporate elements of games, which leads to a large area related to microsuasion: computer-based gaming.

Video games are typically rich in microsuasion elements. The overall goal of most games is to provide entertainment, not to persuade. But during the entertainment experience, players can be bombarded with microsuasion elements, sometimes continuously. Video games can leverage the seven basic intrinsic motivators: Challenge, curiosity, fantasy, control, competition, cooperation, and recognition (Malone & Lepper, 1987). Video games can also incorporate other categories of microsuasion, such as social influence dynamics.

Captology is relevant to computing products designed with macrosuasion in mind—like Baby Think It Over—and to those that simply use microsuasion to make the product more successful, like Quicken. In both cases, designers must understand how to create interactive experiences that change the way people think and behave, whether it be for a single overall outcome or for near-term outcomes that are the building blocks of a larger experience.

NO UNIVERSAL THEORY OF PERSUASION

Creating interactive technology experiences that motivate and influence users would be easy if persuasion were fully understood. It's not. Our understanding is limited, despite the fact that the study of persuasion extends back at least 2,000 years. The fields of psychology, marketing, advertising, public information campaigns, and others have developed theories and perspectives on how to influence and motivate people, but all approaches have limitations. The reality is this: We have no universal theory or framework for persuasion. In other words, no single set of principles fully explains what motivates people, what causes them to adopt certain attitudes, and what leads them to perform certain behaviors (Ford, 1992). In some ways, this is not a surprise. Human psychology is complex, and persuasion is a large domain, often with fuzzy boundaries. Without a universal theory of persuasion, we must draw from a set of theories and models that describe influence, motivation, or behavior change in specific situations and for specific types of people. This limitation creates an additional challenge for designers of persuasive technology products.

Because computing technology creates new possibilities for influencing people, work in captology can lead to new frameworks which, although not perfect, enhance the knowledge and practice in HCI. One such framework is the "Functional Triad."

THE FUNCTIONAL TRIAD: A FRAMEWORK FOR PERSUASIVE TECHNOLOGY

Computers play many roles, some of which go unseen and unnoticed. From a user's perspective, computers function in three basic ways: as tools, as media, and as social actors. In the last two decades, researchers and designers have discussed variants of these functions, usually as metaphors for computer use (e.g., Kay, 1984; Verplank, Fulton, Black, & Moggridge, 1993). However, these three categories are more than metaphors; they are basic ways that people view or respond to computing technologies. These categories also represent three basic types of experiences that motivate and influence people.

Described in more detail elsewhere (Fogg, 1999, 2002), the Functional Triad is a framework that makes explicit these three computer functions—tools, media, and social actors. First, as this framework suggests, computer applications or systems function as tools, providing users with new abilities or powers. Using computers as tools, people can do things they could not do before, or they can do things more easily.

The Functional Triad also suggests that computers function as media, a role that has grown dramatically during the 1990s as computers became increasingly powerful in displaying graphics and in exchanging information over a network such as the Internet. As a medium, a computer can convey either symbolic content (e.g., text, data graphs, icons) or sensory content (e.g., real-time video, virtual worlds, simulations).

Finally, computers also function as social actors. Empirical research demonstrates that people form social relationships with technologies (Reeves & Nass, 1996). The precise causal factors for these social responses have yet to be outlined in detail, but I propose that users respond socially when computers do at least one of the following: (1) adopt animate characteristics (e.g., physical features, emotions, voice communication), (2) play animate roles (e.g., coach, pet, assistant, opponent), or (3) follow social rules or dynamics (e.g., greetings, apologies, turn taking).

The Functional Triad is not a theory; it is a framework for analysis and design. In all but the most extreme cases, a single interactive technology is a mix of these three functions, combining them to create an overall user experience.

In captology, the Functional Triad is useful because it helps show how computer technologies can use different techniques for changing attitudes and behaviors. For example, computers as tools persuade differently than computers as social actors. The strategies and theories apply to each function differently. The paragraphs that follow use the Functional Triad to highlight aspects of persuasive technology, including general design strategies and approaches for creating computing products that persuade and motivate.

Computers as Persuasive Tools

In general, computers as *persuasive tools* change attitude and behavior by increasing a person's abilities or making something easier to do (e.g., Tombari, Fitzpatrick, & Childress, 1985). Although one could propose numerous possibilities for persuasion in this manner, below are four general ways in which computers persuade as tools: by (1) increasing self-efficacy, (2) providing tailored information, (3) triggering decision-making, and (4) simplifying or guiding people through a process.

Computers That Increase Self-efficacy. Computers can increase self-efficacy (Lieberman, 1992), an important contributor to attitude and behavior change processes. Self-efficacy describes individuals' beliefs in their ability to take successful action in specific domains (Bandura, 1997; Bandura, Georgas, & Manthouli, 1996). When people perceive high self-efficacy in a given domain, they are more likely to take action (Bandura, 1997; Bandura et al., 1996). Self-efficacy is a perceived quality. So if individuals believe that their actions are likely to be more effective and productive because they are using a specific computing technology, they are more likely to perform a particular behavior. As a result, functioning as tools, computing technologies can make individuals feel more efficient, productive, in control, and generally more effective (DeCharms, 1968; Kernal, 1999; Pancer, George, & Gebotys, 1992). For example, a heart rate monitor may help people feel more effective in meeting their exercise goals when it provides ongoing information on heart rate and calories burned. Without the heart rate monitor, people could still take their pulse and calculate calories, but the computer device—whether it be worn or part of the exercise machinery—makes these tasks easier. The ease of tracking heart rate and calories burned likely increases self-efficacy in fitness behavior, making it more likely the individual will continue to exercise (Brehm, 1997; Strecher, DeVellis, Becker, & Rosenstock, 1986; Thompson, 1992).

Computers That Provide Tailored Information. Next, computers act as tools when they tailor information, offering people content that is pertinent to their needs and contexts. Compared with general information, tailored information increases the potential for attitude and behavior change (Beniger, 1987; Dijkstra, Liebrand, & Timminga, 1998; Jimison et al., 1997; Nowak et al., 1999; Strecher, 1999; Strecher et al., 1994).

One notable example of a tailoring technology is the Web site discussed earlier, Scorecard.org, which generates information according to an individual's geographical location to achieve a persuasive outcome. After people enter their zip code into this Web site, the web technology reports on chemical hazards in their neighborhood, identifies companies who create those hazards, and describes the potential health risks. Although no published studies document the persuasive effects of this particular technology, outside research and analysis suggests that making information relevant to individuals increases their attention and arousal, which can ultimately lead to increased attitude and behavior change (Beniger, 1987; MacInnis & Jaworski, 1989; MacInnis, Moorman, & Jaworski, 1991; Strecher, 1999).

Computers That Trigger Decision Making. Technology can also influence people by triggering or cueing a decision-making process. For example, today's web browsers launch a new window to alert people before they send information over insecure network connections. The message window serves as a signal to consumers to rethink their planned actions. A similar example exists in a very different context. Cities concerned with automobile speeding in neighborhoods can use a stand-alone radar trailer that senses the velocity of an oncoming automobile and displays that speed on a large screen. This technology is designed to trigger a decision-making process regarding driving speed.

Computers That Simplify or Guide People Through a Process. By facilitating or simplifying a process for users, technology can minimize barriers that may impede a target behavior. For example, in the context of web commerce, technology can simplify a multistep process down to a few mouse clicks. Typically, to purchase something online, a consumer needs to select an item, place it in a virtual shopping cart, proceed to checkout, enter personal and billing information, and verify an order confirmation. Amazon.com and other e-commerce companies have simplified this process by storing customer information so that consumers need not reenter information every transaction. By lowering the time commitment and reducing the steps to accomplish a goal, these companies have reduced the barriers for purchasing products from their sites. The principle used by web and other computer technology (Todd & Benbasat, 1994) is similar to the dynamic Ross and Nisbett (1991) discuss on facilitating behaviors through modifying the situation.

In addition to reducing barriers for a target behavior, computers can also lead people through processes to help them change attitudes and behaviors (Muehlenhard, Baldwin, Bourg, & Piper, 1988; Tombari et al., 1985). For example, a computer nutritionist can guide individuals through a month of healthy eating by providing recipes for each day and grocery lists for each week. In general, by following a computer-led process, users are exposed to information they may not have seen otherwise and are engaged in activities they may not have done otherwise (Fogg, 2002).

Computers as Persuasive Media

The next area of the Functional Triad deals with computers as *persuasive media*. Although "media" can mean many things, here the focus is on the power of computer simulations. In this role, computer technology provides people with experiences, either first hand or vicarious. By providing simulated experiences, computers can change people's attitudes and behaviors. Outside the world of computing, experiences have a powerful impact on people's attitudes, behaviors, and thoughts

(Reed, 1996). Experiences offered via interactive technology have similar effects (Bullinger, Roessler, & Mueller-Spahn, 1998; Fogg, 2000).

Three types of computer simulations that are relevant to persuasive technologies:

- simulated cause-and-effect scenarios
- simulated environments
- simulated objects

The paragraphs that follow discuss each simulation type in turn. (*Note.* Other taxonomies for simulations exist. For example, see Alessi, 1991; de Jong, 1996; Gredler, 1986.)

Computers That Simulate Cause and Effect. One type of computer simulation allows users to vary the inputs and observe the effects (Hennessy & O'Shea, 1993)—what one could call "cause-and-effect simulators." The key to effective cause-and-effect simulators is their ability to demonstrate the consequence of actions immediately and credibly (Alessi, 1991; Balci, 1998; Balci, Henrikson, & Roberts, 1986; Crosbie & Hay, 1978; de Jong, 1991; Hennessy & O'Shea, 1993; Zietsman & Hewson, 1986). These computer simulations give people first-hand insight into how inputs (such as putting money in a savings account) affect an output (such as accrued retirement savings). By allowing people to explore causes and effects of situations, these computer simulations can shape attitudes and behaviors.

Computers That Simulate Environments. A second type of computer simulation is the "environment simulator." These simulators are designed to provide users with new surroundings, usually through images and sound. In these simulated environments, users have experiences that can lead to attitude and behavior change (Bullinger et al., 1998), including experiences that are designed as games or explorations (Lieberman, 1992; Schlosser & Kanifer, 1999; Schneider, 1985; Woodward, Carnine, & Davis, 1986).

The efficacy of this approach is demonstrated by research on the Tectrix Virtual Reality Bike (an exercise bike that includes a computer and monitor that show a simulated world). Pocari, Zedaker, and Maldari (1998) found that people using an exercise device with computer simulation of a passing landscape exercised harder than those who used an exercise device without simulation. Both groups, however, felt that they had exerted themselves a similar amount. This outcome caused by simulating an outdoor experience mirrors findings from other research: People exercise harder when outside rather than inside a gym (Ceci & Hassmen, 1991).

Environmental simulators can also change attitudes. Using a virtual reality environment in which the people saw and felt a simulated spider, Carlin, Hoffman, and Weghorst (1997) were able to decrease the fear of spiders in his participants. In this research, participants wore a head-mounted display that immersed them into a virtual room, and they were able to control both the number of spiders and their proximity. In this case study, Carlin found that the virtual reality treatment reduced the fear of spiders in the real world. Similar therapies have been used for fear of flying (Klein, 1999; Wiederhold,

TABLE 17.2. Computer-based Simulations Offer Different Ways to Persuade

Simulation Type	Key Advantages
Cause-and-effect simulators	• Allow users to explore and experiment • Show cause-and-effect relationships clearly and quickly • Persuade without being overly didactic
Environment simulators	• Can create situations that reward and motivate people for a target behavior • Allow rehearsal: practicing a target behavior • Can control exposure to new or frightening situations • Facilitate role playing: adopting another person's perspective
Object simulators	• Fit into the context of a person's normal life • Are less dependent on imagination or suspension of disbelief • Make clear the impact on normal life

Davis, Wiederhold, & Riva, 1998), agoraphobia (Ghosh & Marks, 1987), claustrophobia (Bullinger et al., 1998), and fear of heights (Bullinger et al., 1998), among others (Kirby, 1996).

Computers That Simulate Objects. The third type of computer simulations are "object simulators." These are computerized devices that simulate an object (as opposed to an environment). The Baby Think It Over infant simulator described earlier in this chapter is one such device. Another example is a specially equipped car created by Chrysler Corporation, designed to help teens experience the effect of alcohol on their driving. Used as part of high school programs, teen drivers first navigate the special car under normal conditions. Then the operator activates an onboard computer system, which simulates how an inebriated person would drive—sluggish brakes, inaccurate steering, and so on. This computer-enhanced care provides teens with an experience designed to change attitudes and behaviors about drinking and driving. Although the sponsors of this car do not measure the impact of this intervention, the anecdotal evidence is compelling (e.g., see Machrone, 1998).

Table 17.2 lists the three types of simulations and outlines what advantage each type of simulation offers as far as persuasion and motivation are concerned.

Computers as Persuasive Social Actors

The final corner of the Functional Triad focuses on computers as *persuasive social actors*, a view of computers that has only recently become widely recognized (Reeves & Nass, 1996). Past empirical research has shown that individuals form social relationships with technology, even when the stimulus is rather impoverished (Fogg, 1997; Marshall & Maguire, 1971; Moon & Nass, 1996; Muller, 1974; Nass, Fogg, & Moon, 1996;

Nass, Moon, Fogg, Reeves, & Dryer, 1995; Nass & Steuer, 1993; Nass, Moon, Morkes, Eun-Young, & Fogg, 1997; Parise, Kiesler, Sproull, & Waters, 1999; Quintanar & Crowell, 1982; Reeves & Nass, 1996). For example, individuals share reciprocal relationships with computers (Fogg & Nass, 1997a; Parise et al., 1999), can be flattered by computers (Fogg & Nass, 1997b), and are polite to computers (Nass, Moon, & Carney, 1999).

In general, I propose that computers as social actors can persuade people to change their attitudes and behaviors by (1) providing social support, (2) modeling attitudes or behaviors, and (3) leveraging social rules and dynamics.

Computers That Provide Social Support. Computers can provide a form of social support to persuade, a dynamic that has long been observed in human–human interactions (Jones, 1990). Although the potential for effective social support from computer technology has yet to be fully explored, a small set of empirical studies provide evidence for this phenomenon (Fogg, 1997; Fogg & Nass, 1997b; Nass et al., 1996; Reeves & Nass, 1996). For example, computing technology can influence individuals by providing praise or criticism, thus manipulating levels of social support (Fogg & Nass, 1997b; Muehlenhard et al., 1988).

Outside the research context, various technology products use the power of praise to influence users. For example, the Dole 5 A Day Adventures CD-ROM, discussed earlier, uses a cast of over 30 onscreen characters to provide social support to users who perform various activities. Characters such as "Bobby Banana" and "Pamela Pineapple" praise individuals for checking labels on virtual frozen foods, for following guidelines from the food pyramid, and for creating a nutritious virtual salad.

Computers That Model Attitudes and Behaviors. In addition to providing social support, computer systems can persuade by modeling target attitudes and behaviors. In the natural world, people learn directly through first-hand experience and indirectly through observation (Bandura, 1997). When a behavior is modeled by an attractive individual or is shown to result in positive consequences, people are more likely to enact that behavior (Bandura, 1997). Lieberman's research (1997) on a computer game designed to model health-maintenance behaviors shows the positive effects that an onscreen cartoon model had on those who played the game. In a similar way, the product "Alcohol 101" (www.centurycouncil.org/underage/education/a101.cfm) uses navigable onscreen video clips of human actors dealing with problematic situations that arise during college drinking parties. The initial studies on the Alcohol 101 intervention shows positive outcomes (Reis, 1998). As we look to the future of persuasive technologies computer-based characters, whether artistically rendered or video images, are increasingly likely to serve as models for attitudes and behaviors.

Computers That Leverage Social Rules and Dynamics. Computers have also been shown to be effective persuasive social actors when they leverage social rules and dynamics (Fogg, 1997; Friedman & Grudin, 1998; Marshall & Maguire, 1971; Parise et al., 1999). These rules include turntaking, politeness norms, and sources of praise (Reeves & Nass, 1996). The rule of reciprocity—that we must return favors to others—is among the most powerful of social rules (Gouldner, 1960) and is one that has been shown to also have force when people interact with computers. Fogg and Nass (1997a) showed that people performed more work and better work for a computer that assisted them on a previous task. In essence, users reciprocated help to a computer. On the retaliation side, the inverse of reciprocity, the research showed that people performed lower quality work for a computer that had served them poorly in a previous task. In a related vein, Moon (1998) found that individuals followed rules of impression management when interacting with a computer. Specifically, when individuals believed that computer interviewing them was in the same room, they provided more honest answers, compared to interacting with a computer believed to be a few miles away. In addition, subjects were more persuaded by the proximate computer.

The previous paragraphs outline some of the early demonstrations of computers as social actors that motivate and influence people in predetermined ways, often paralleling research from long-standing human–human research.

Functional Triad Summary

Table 17.3 summarizes the Functional Triad and the persuasive affordances that each element offers.

In summary, the Functional Triad can be a useful framework in captology, the study of computers as persuasive technologies. It makes explicit how a technology can change attitudes and behaviors—either by increasing a person's capability, or by providing users with an experience, or by leveraging the power of social relationships. Each of these paths suggests related persuasion strategies, dynamics, and theories. One element that is common to all three functions is the role of credibility. Credible tools, credible media, and credible social actors will all lead to increased power to persuade. This is the focus of the next section.

TABLE 17.3. A Summary of the Functional Triad

Function	Essence	Persuasive affordances
Computer as a *tool* or *instrument*	Increases capabilities	• reduces barriers (time, effort, cost) • increases self-efficacy • provides information for better decision making • changes mental models
Computer as *medium*	Provides experiences	• provides first-hand learning, insight, visualization, resolve • promotes understanding of cause-and-effect relationships • motivates through experience, sensation
Computer as *social actor*	Creates relationship	• establishes social norms • invokes social rules and dynamics • provides social support or sanction

COMPUTERS AND CREDIBILITY

One key issue in captology is computer credibility, a topic that suggests questions such as "Do people find computers to be credible sources?" "What aspects of computers boost credibility?" and "How do computers gain and lose credibility?" Understanding the elements of computer credibility promotes a deeper understanding of how computers can change attitudes and behaviors, because credibility is a key element in many persuasion processes (Gahm, 1986; Lerch & Prietula, 1989; Lerch, Prietula, & Kulik, 1997).

Credibility has been a topic of social science research since the 1930s (for reviews, see Petty & Cacioppo, 1981; Self, 1996). Virtually all credibility researchers have described credibility as a perceived quality made up of multiple dimensions (e.g., Buller & Burgoon, 1996; Gatignon & Robertson, 1991; Petty & Cacioppo, 1981; Self, 1996; Stiff, 1994). This description has two key components germane to computer credibility. First, credibility is a perceived quality; it does not reside in an object, a person, or a piece of information. Therefore, in discussing the credibility of a computer product, one is always discussing the *perception* of credibility for the computer product.

Next, researchers generally agree that credibility perceptions result from evaluating multiple dimensions simultaneously. Although the literature varies on exactly how many dimensions contribute to the credibility construct, the majority of researchers identify trustworthiness and expertise as the two key components of credibility (Self, 1996). Trustworthiness, a key element in the credibility calculus, is described by the terms *well-intentioned, truthful, unbiased,* and so on. The trustworthiness dimension of credibility captures the perceived goodness or morality of the source. Expertise, the other dimension of credibility, is described by terms such as *knowledgeable, experienced, competent,* and so on. The expertise dimension of credibility captures the perceived knowledge and skill of the source.

Extending research on credibility to the domain of computers, I have proposed that *highly credible computer products will be perceived to have high levels of both trustworthiness and expertise* (Fogg & Tseng, 1999). In evaluating credibility, a computer user will make an assessment of the computer product's trustworthiness and expertise to arrive at an overall credibility assessment.

When Does Credibility Matter?

Credibility is a key component in bringing about attitude change. Just as credible people can influence other people, credible computing products also have the power to persuade. Computer credibility is not an issue when there is no awareness of the computer itself or when the dimensions of computer credibility—trustworthiness and expertise—are not at stake. In these cases, computer credibility does not matter to the user. But, in many cases, credibility is key. Seven categories outline when credibility matters in human–computer interactions (Tseng & Fogg, 1999).

1. *When computers act as a knowledge repository:* Credibility matters when computers provide data or knowledge to users. The information can be static information, such as simple web pages or an encyclopedia on CD-ROM. But computer information can also be dynamic. Computers can tailor information in real-time for users, such as providing information that matches interests, personality, or goals. In such cases, users may question the credibility of the information provided.

2. *When computers instruct or tutor users:* Computer credibility also matters when computers give advice or provide instructions to users. Sometimes it's obvious why computers give advice. For example, auto navigation systems give advice about which route to take, and online help systems advise users on solving a problem. These are clear instances of computers giving advice. However, at times, the advice from a computing system is subtle. For example, interface layout and menu options can be a form of advice. Consider a default button on a dialogue box. The fact that one option is automatically selected as the default option suggests that certain paths are more likely or profitable for most users. One can imagine that if the default options are poorly chosen, the computer program could lose some credibility.

3. *When computers report measurements:* Computer credibility is also at stake when computing devices act as measuring instruments. These can include engineering measurements (e.g., an oscilloscope), medical measurements (e.g., a glucose monitor), geographical measurements (e.g., devices with GPS technology), and others. In this area, I observed an interesting phenomemon in the 1990s when digital test and measurement equipment was created to replace traditional analogue devices. Many engineers, usually those with senior status, did not trust the information from the digital devices. As a result, some engineers rejected the convenience and power of the new technology because their old analog equipment gave information they found more credible.

4. *When computers report on work performed:* Computers also need credibility when they report to users on work the computer has performed. For example, computers report the success of a software installation or the eradication of viruses. In these cases and others, the credibility of the computer is at issue if the work the computer reports does not match what actually happened. For example, suppose a user runs a spell check and the computer reports no misspelled words. If the user later finds a misspelled word, then the credibility of the program will suffer.

5. *When computers report about their own state:* Computers also report their own state, and these reports have credibility implications. For example, computers may report how much disk space they have left, how long their batteries will last, how long a process will take, and so on. A computer reporting about its own state raises issues about its competence in conveying accurate information about itself, which is likely to affect user perceptions of credibility.

6. *When computers run simulations:* Credibility is also important when computers run simulations. This includes simulations of aircraft navigation, chemical processes, social dynamics, nuclear disasters, and so on. Simulations can show cause-and-effect relationships, such as the progress of a disease in a population or the effects of global warming. Similarly, simulations can replicate the dynamics of an experience, such as piloting an aircraft or caring for a baby. Based on rules that humans provide, computer simulations can be flawed or biased. Even if the bias is not intentional, when users perceive that the computer simulation lacks veridicality, the computer application will lose credibility.

7. *When computers render virtual environments:* Related to simulations is the computer's ability to create virtual environments for users. Credibility is important in making these environments believable, useful, and engaging. However, virtual environments do not always need to match the physical world; they simply need to model what they propose to model. For example, like good fiction or art, a virtual world for a fanciful arcade game can be highly credible if the world is internally consistent.

Web Credibility Research and Guidelines for Design

When it comes to credibility, the web is unusual. The web can be the most credible source of information, and the web can be one of the least credible sources. There's a direct connection between web credibility and persuasion via the web: When a Web site gains credibility, it also gains the power to change attitudes, and, at times, behaviors; when a Web site lacks credibility, it will not be effective in persuading or motivating users.

Some Web sites today offer users low-quality—or outright misleading—information. As a result, credibility has become a major concern for those seeking or posting information on the web (Caruso, 1999; Johnson & Kaye, 1998; Kilgore, 1998; McDonald, Schumann, & Thorson, 1999; Nielsen, 1997; Sullivan, 1999). Web users are likely becoming more skeptical of what they find online and may be wary of web-based experiences in general.

Interaction designers face increasing challenges to design web experiences that motivate web users to adopt specific behaviors, such as the following:

- spend time on the site
- register personal information
- purchase things online
- fill out a survey
- click on the ads
- set up a virtual community
- download software
- bookmark the site and return often

If web designers can influence people to perform these actions, they have been successful. These are key behavioral outcomes. But what makes a Web site credible?

Because quantitative research was lacking on web credibility, I led a research team in a series of studies on how different elements of Web sites affect people's perception of credibility. Of these, one study in particular, the "Overview Study," investigated a wide range of web credibility elements (Fogg et al., 2000; Fogg et al., 2001). Over 1,400 people participated in the Overview Study, both from the United States and Europe, evaluating 51 different Web site elements. Data showed which elements boost and which elements hurt perceptions of web credibility. We found these 51 elements fell into 1 of 7 factors. In order of impact, the five types of Web site elements that increased credibility perceptions were "real-world feel," "ease of use," "expertise," "trustworthiness," and "tailoring." The two types of elements that hurt credibility were "commercial implications" and "amateurism."

Whereas the specific results of the Overview Study have been reported elsewhere (Fogg et al., 2001), the following paragraphs outline the key design implications from this research. Each guideline comes from 1 of the 7 scales that emerged from the study data and suggest to designers ways to create more credible Web sites. (For more research along these lines, see www.webcredibility.org)

• *Guideline 1: Design Web sites to convey the "real world" aspect of the organization:* According to the study data, one effective way to enhance the credibility of a Web site is to include elements that highlight the brick-and-mortar nature of the organization it represents. In this study, we included web elements such as a listing a physical address and showing employee photographs. Many other possibilities exist that were not examined. The overall implication seems clear: To create a site with maximum credibility, designers should highlight features that communicate the legitimacy and accessibility of the organization.

• *Guideline 2: Make Web sites easy to use:* In the HCI community, we have long emphasized ease of use, so a guideline advocating ease of use is not new. However, the Overview Study adds another important reason for making Web sites usable: It will enhance the site's credibility. In the Overview Study, people awarded a Web site credibility points for being usable (e.g., "The site is arranged in a way that makes sense to you"), and they deducted credibility points for ease-of-use problems (e.g., "the site is difficult to navigate"). Although this information should not change how we, as HCI professionals, design user experiences for the web, it does add a compelling new reason for investing time and money in usable design—it makes a site more credible. Going beyond the data, one could reasonably conclude that a simple, usable Web site would be perceived as more credible than a site that has extravagant features but is lacking in usability.

• *Guideline 3: Include markers of expertise:* Expertise is a key component in credibility, and the data in the Overview Study support the idea that web sites that convey expertise can gain credibility in users' eyes. Important "expertise elements" in this study included listing an author's credentials and including citations and references. It's likely that many other elements also exist. Many Web sites today miss opportunities to legitimately convey expertise to their users.

• *Guideline 4: Include markers of trustworthiness:* Trustworthiness is another key component in credibility. As with expertise, the Overview Study suggests that Web site elements that convey trustworthiness will lead to increased perceptions of credibility. In this research, we tested how people assessed specific "trustworthiness" elements: linking to outside materials

and sources, stating a policy on content, and so on. Of course, other markers of trustworthiness exist. We propose that Web site designers who concentrate on conveying the honest, unbiased nature of their Web site will end up with a more credible— and therefore more effective—Web site.

• *Guideline 5: Tailor the user experience:* Although not as vital as the previous suggestions, tailoring does make a difference. Our study shows that tailoring the user experience on a Web site leads to increased perceptions of Web credibility. For example, people think a site is more credible when it acknowledges that the individual has visited it before. To be sure, tailoring and personalization can take place in many ways. Tailoring extends even to the type of ads shown on the page: Ads that match what the user is seeking seem to increase the perception of Web site credibility.

• *Guideline 6: Avoid overly commercial elements on a Web site:* Although many Web sites, especially large Web sites, exist for commercial purposes, our study suggests that users penalize sites that have an aggressively commercial flavor. For example, web pages that mix ads with content to the point of confusing readers will be perceived as not credible. In the Overview Study, mixing ads and content received the most negative response of all. But it is important to note that ads do not always reduce credibility. In this study and elsewhere (Kim, 1999), quantitative research shows that banner ads done well can enhance the perceived credibility of a site. It seems reasonable that, as with other elements of people's lives, we accept commercialization to an extent but become wary when it is overdone.

• *Guideline 7: Avoid the pitfalls of amateurism:* Most web designers seek a professional outcome in their work. The Overview Study suggests organizations that care about credibility should be ever vigilant—and perhaps obsessive—to avoid small glitches in their Web sites. These "small" glitches seem to have a large impact on web credibility perceptions. Even one typographical error or a single broken link is damaging. Although designers may face pressures to create dazzling technical features on Web sites, failing to correct small errors undermines that work.

Despite the growing body of research, much remains to be discovered about web credibility. The study of web credibility needs to be an ongoing concern because three things continue to evolve: (1) web technology, (2) the type of people using the web, and (3) people's experiences with the web. Fortunately, what researchers learn about designing for web credibility can translate into credible experiences in other high-tech devices that share information, from mobile phones to gas pumps.

THE ETHICS OF COMPUTING SYSTEMS DESIGNED TO PERSUADE

In addition to research and design issues, captology addresses the ethical issues that arise from design or distributing persuasive interactive technologies. Persuasion is a value-laden activity. By extension, creating or distributing an interactive technology that attempts to persuade is also value laden. Ethical problems arise when the values, goals, and interests of the creators do not match with those of the people who use the technology. HCI professionals can ask a few key questions to get insight into possible ethical problem areas:

- Does the persuasive technology advocate what's good and fair?
- Is the technology inclusive, allowing access to all, regardless of social standing?
- Does it promote self-determination?
- Does it represent what's thought to be true and accurate?

Answering no to any of these questions suggests the persuasive technology at hand could be ethically questionable, and perhaps downright objectionable (For a longer discussion on ethics, see Friedman & Kahn, later in this volume).

Although it's clear that deception and coercion are unethical in computing products, some behavior change strategies, such as conditioning, surveillance, and punishment, are less cut and dry. For example, operant conditioning—a system of rewards—can powerfully shape behaviors. By providing rewards, a computer product could get people to perform new behaviors without their clear consent or without them noticing the forces of influence at work.

Surveillance is another common and effective way to change behavior. People who know they are being watched behave differently. Today, computer technologies allow surveillance in ways that were never before possible, giving institutions remarkable new powers. Although advocates of computer-based employee surveillance (e.g., DeTienne, 1993) say that monitoring can "inspire employees to achieve excellence," opponents point out that such approaches can hurt morale or create a more stressful workplace. When every keystroke and every restroom break is monitored and recorded, employees may feel they are part of an electronic sweatshop.

Another area of concern is when technologies use punishment—or threats of punishment—to shape behaviors. Although punishment is an effective way to change outward behaviors in the short term, punishment has limited outcomes beyond changing observable behavior, and many behavior change experts frown on using it. The problems with punishment increase when a computer product punishes people. The punishment may be excessive or inappropriate to the situation. Also, the long-term effects of punishment are likely to be negative. In these cases, who bears responsibility for the outcome?

Discussed elsewhere in more detail (Berdichevsky, 1999; Fogg, 1998), those who create or distribute persuasive technologies have a responsibility to examine the moral issues involved.

PERSUASIVE TECHNOLOGY: POTENTIAL AND RESPONSIBILITY

Computer systems are now becoming a common part of everyday life. Whatever the form of the system, from a desktop computer to a smart car interior to a mobile phone, these interactive experiences can be designed to influence our attitudes and affect our behaviors. They can motivate and persuade by merging the power of computing with the psychology of persuasion.

We humans are still the supreme agents of influence—and this will not change any time soon. Computers are not yet as effective as skilled human persuaders, but at times computing technology can go beyond what humans can do. They never forget, they do not need to sleep, and they can be programmed to never stop trying. For better or worse, computers provide us with a new avenue for changing how people think and how they act.

To a large extent, we as a community of HCI professionals will help create the next generation of technology products, including those products designed to change people's attitudes and behaviors. If we take the right steps—raising awareness of persuasive technology in the general public and encouraging technology creators to follow guidelines for ethical interactive technologies—we may well see persuasive technology reach its potential, enhancing the quality of life for individuals, communities, and society.

References

Alessi, S. M. (1991). Fidelity in the design of instructional simulations. *Journal of Computer-Based Instruction, 15*, 40–47.

Balci, O. (1998, December 13–16). *Verification, validation, and accreditation*. Paper presented at the Winter Simulation Conference. Washington, DC.

Balci, O., Henrikson, J. O., & Roberts, S. D. (1986, December 8–10). Credibility assessment of simulation results. In J. R. Wilson, J. O. Henriksen, & S. D. Roberts (Eds.), *1986 Winter Simulation Conference Proceedings*. Washington, DC.

Bandura, A. (1997). *Self-efficacy: The exercise of control*. New York: W. H. Freeman.

Bandura, A., Georgas, J., & Manthouli, M. (1996). Reflections on human agency. In *Contemporary psychology in Europe: Theory, research, and applications*. Seattle, WA: Hogrefe & Huber.

Beniger, J. R. (1987). Personalization of mass media and the growth of pseudo-community. *Communication Research, 14*(3), 352–371.

Berdichevsky, D., & Neunschwander, E. (1999). Towards an ethics of persuasive technology. *Communications of the ACM, 42*(5), 51–58.

Brehm, B. (1997, December). Self-confidence and exercise success. *Fitness Management, December*, 22–23.

Buller, D. B., & Burgoon, J. K. (1996). Interpersonal deception theory. *Communication Theory, 6*(3), 203–242.

Bullinger, A. H., Roessler, A., & Mueller-Spahn, F. (1998). From toy to tool: The development of immersive virtual reality environments for psychotherapy of specific phobias. In *Studies in health technology and informatics* (Vol. 58, pp. 103–111). Amsterdam: IOS Press.

Carlin, A. S., Hoffman, H. G., & Weghorst, S. (1997). Virtual reality and tactile augmentation in the treatment of spider phobia: A case report. *Behaviour Research & Therapy, 35*(2), 153-158.

Caruso, D. (1999, November 22). Digital commerce: Self indulgence in the Internet industry. *The New York Times.*

Ceci, R., & Hassmen, P. (1991). Self-monitored exercise at three different PE intensities in treadmill vs. field running. *Medicine and Science in Sports and Exercise, 6*(23), 732-738.

Crosbie, R. E., & Hay, J. L. (1978). The credibility of computerised models. In R. E. Crosbie (Ed.), *Toward real-time simulation. Languages, models and systems.* La Jolla, CA: Society of Computer Simulation.

de Jong, T. (1991). Learning and instruction with computer simulations. *Education & Computing, 6,* 217-229.

DeCharms, R. (1968). *Personal causation: The internal affective determinants of behavior.* New York: Academic.

Denning, P., & Metcalfe, R. (1997). *Beyond calculation: The next fifty years of computing.* New York: Springer-Verlag.

DeTienne, Kristen Bell (1993, September-October). Big brother or friendly coach? Computer monitoring in the 21st century. *The Futurist, 27*(5), 33-37.

Dijkstra, J. J., Liebrand, W. B. G., & Timminga, E. (1998). Persuasiveness of expert systems. *Behaviour and Information Technology, 17*(3), 155-163.

Fogg, B. J. (1997). *Charismatic computers: Creating more likable and persuasive interactive technologies by leveraging principles from social psychology.* Doctoral thesis, Stanford University, Stanford, CA.

Fogg, B. J. (1998). Persuasive computers: Perspectives and research directions. *Proceedings of the Conference on Human Factors in Computing Systems, CHI 98.* Los Angeles, CA. New York: ACM Press.

Fogg, B. J. (1999). Persuasive technologies. *Communications of the ACM, 42*(5), 26-29.

Fogg, B. J. (2002). *Persuasive technologies: Using computer power to change attitudes and behaviors.* San Francisco, CA: Morgan Kaufmann.

Fogg, B. J., Marshall, J., Laraki, O., Osipovich, A., Varma, C., Fang, N., Paul, J., Rangnekar, A., Shon, J., Swani, P., & Treinen, M. (2000). Elements that affect web credibility: Early results from a self-report study. *Proceedings of the ACM CHI 2000 Conference on Human Factors in Computing Systems.* New York: ACM Press.

Fogg, B. J., Marshall, J., Laraki, O., Osipovich, A., Varma, C., Fang, N., Paul, J., Rangnekar, A., Shon, J., Swani, P., & Treinen, M. (2001). What makes a web site credible? A report on a large quantitative study. *Proceedings of the ACM CHI 2001 Conference on Human Factors in Computing Systems.* New York: ACM Press.

Fogg, B. J., & Nass, C. (1997a). How users reciprocate to computers: An experiment that demonstrates behavior change. *Proceedings of the Conference on Human Factors in Computing Systems, CHI 97.* New York: ACM Press.

Fogg, B. J., & Nass, C. (1997b). Silicon sycophants: The effects of computers that flatter. *International Journal of Human-Computer Studies, 46*(5), 551-561.

Fogg, B. J., & Tseng, H. (1999). The elements of computer credibility. *Proceedings of the Conference on Human Factors and Computing Systems, CHI 99.* Pittsburgh, PA. New York: ACM Press.

Ford, M. E. (1992). *Motivating humans: Goals, emotions, personal agency beliefs.* Newbury Park, CA: Sage.

Friedman, B., & Grudin, J. (1998). Trust and accountability: preserving human values in interactional experience. *Proceedings of the Conference on Human Factors in Computing Systems, CHI 98.* Los Angeles, CA. New York: ACM Press.

Gahm, G. A. (1986). *The effects of computers, source salience and credibility on persuasion.* Doctoral thesis. State University of New York at Stony Brook.

Gatignon, H., & Robertson, T. S. (1991). *Innovative decision processes.* Englewood Cliffs, NJ: Prentice-Hall.

Ghosh, A., & Marks, I. M. (1987). Self-treatment of agoraphobia by exposure. *Behavior Therapy, 18*(1), 3-16.

Gouldner, A. W. (1960). The norm of reciprocity: A preliminary statement. *American Sociological Review, 25,* 161-178.

Gredler, M. B. (1986). A taxonomy of computer simulations. *Educational Technology, 26,* 7-12.

Hennessy, S., & O'Shea, T. (1993). Learner perceptions of realism and magic in computer simulations. *British Journal of Educational Technology, 24*(2), 125-38.

Jimison, H. B., Street, R. L., Jr., & Gold, W. R. (1997). Patient-specific interfaces to health and decision-making information. *LEA's communication series.* Mahwah, NJ: Lawrence Erlbaum.

Johnson, T. J., & Kaye, B. K. (1998). Cruising is believing? Comparing Internet and traditional sources on media credibility measures. *Journalism and Mass Communication Quarterly, 75*(2), 325-340.

Jones, E. E. (1990). *Interpersonal perception.* New York: W. H. Freeman.

Kay, A. (1984). Computer software. *Scientific American, 251,* 53-59.

Kernal, H. K. (1999). *Effects of design characteristics on evaluation of a home control system: A comparison of two research methodologies.* Paper presented to SIGGRAPH Annual Conference 2000. New Orleans, LA.

Khaslavsky, J., & Shedroff, N. (1999). Understanding the seductive experience. *Communications of the ACM, 42*(5), 45-49.

Kilgore, R. (1998). Publishers must set rules to preserve credibility. *Advertising Age, 69*(48), 31.

Kim, N. (1999). *World Wide Web credibility: What effects do advertisements and typos have on the perceived credibility of web page information?* Unpublished honors thesis, Stanford University. Stanford, CA.

King, P., & Tester, J. (1999). The landscape of persuasive technologies. *Communications of the ACM, 42*(5), 31-38.

Kirby, K. C. (1996). Computer-assisted treatment of phobias. *Psychiatric Services, 4*(2), 139-140, 142.

Klein, R. A. (1999). Treating fear of flying with virtual reality exposure therapy. *Innovations in Clinical Practice: A Source Handbook, 17,* 449-465.

Lerch, F. J., & Prietula, M. J. (1989). How do we trust machine advice? Designing and using human-computer interfaces and knowledge based systems. In G. Salvendy & M. J. Smith (Eds.), *Designing and using human-computer-interface and knowledge-based systems* (pp. 411-419). Amsterdam: Elsevier.

Lerch, F. J., Prietula, M. J., & Kulik, C. T. (1997). The Turing effect: The nature of trust in expert system advice. In *Expertise in context: Human and machine.* Cambridge, MA: MIT Press.

Lieberman, D. (1992). The computer's potential role in health education. *Health Communication, 4,* 211-225.

Lieberman, D. (1997). Interactive video games for health promotion. In W. G. R. Street & T. Mannin (Eds.), *Health promotion and interactive technology.* Mahwah, NJ: Lawrence Earlbaum.

Machrone, B. (1998, July 1). Driving drunk. *PC Magazine.*

MacInnis, D. J., & Jaworski, B. J. (1989). Information processing from advertisements: Toward an integrative framework. *Journal of Marketing, 53*(4), 1-23.

MacInnis, D. J., Moorman, C., & Jaworski, B. J. (1991). Enhancing and measuring consumers' motivation, opportunity, and ability to process brand information from ads. *Journal of Marketing, 55*(4), 32-53.

Malone, T., & Lepper, M. (1987). Making learning fun: A taxonomy of intrinsic motivation for learning. In R. E. Snow & M. J. Farr

(Eds.), *Aptitude, learning, and instruction*. Hillsdale, NJ: Lawrence Earlbaum.

Marshall, C., & Maguire, T. O. (1971). The computer as social pressure to produce conformity in a simple perceptual task. *AV Communication Review, 19*(1), 19-28.

McDonald, M., Schumann, D. W., & Thorson, E. (1999). Cyberhate: Extending persuasive techniques of low credibility sources to the World Wide Web. In D. W. Schumann & E. Thorson (Eds.), *Advertising and the World Wide Web*. Mahwah, NJ: Lawrence Erlbaum.

Moon, Y. (1998). The effects of distance in local versus remote human-computer interaction. *Proceedings of the Conference on Human Factors in Computing Systems, CHI 98* (pp. 103-108). New York: ACM.

Moon, Y., & Nass, C. (1996). How "real" are computer personalities? Psychological responses to personality types in human-computer interaction. *Communication Research, 23*(6), 651-674.

Muehlenhard, C. L., Baldwin, L. E., Bourg, W. J., & Piper, A. M. (1988). Helping women "break the ice": A computer program to help shy women start and maintain conversations with men. *Journal of Computer-Based Instruction, 15*(1), 7-13.

Muller, R. L. (1974). *Conforming to the computer: Social influence in computer-human interaction*. Doctoral thesis. Syracuse University.

Nass, C., Fogg, B. J., & Moon, Y. (1996). Can computers be teammates? *International Journal of Human-Computer Studies, 45*(6), 669-678.

Nass, C., Moon, Y., & Carney, P. (1999). Are people polite to computers? Responses to computer-based interviewing systems. *Journal of Applied Social Psychology, 29*(5), 1093-1110.

Nass, C., Moon, Y., Fogg, B. J., Reeves, B., & Dryer, D. C. (1995). Can computer personalities be human personalities? *International Journal of Human-Computer Studies, 43*(2), 223-239.

Nass, C., & Steuer, J. (1993). Voices, boxes, and sources of messages: Computers and social actors. *Human Communication Research, 19*(4), 504-527.

Nass, C. I., Moon, Y., Morkes, J., Eun-Young, K., & Fogg, B. J. (1997). Computers are social actors: A review of current research. In B. Friedman (Ed.), *Human values and the design of computer technology*. Stanford, CA: CSLI & Cambridge Press.

Nielsen, J. (1997). *How users read on the web*. Available: www.useit.com/alertbox/9710a.html

Nowak, G. J., Shamp, S., Hollander, B., Cameron, G. T., Schumann, D. W., & Thorson, E. (1999). Interactive media: A means for more meaningful advertising? *Advertising and consumer psychology*. Mahwah, NJ: Lawrence Erlbaum.

Pancer, S. M., George, M., & Gebotys, R. J. (1992). Understanding and predicting attitudes toward computers. *Computers in Human Behavior, 8*, 211-222.

Parise, S., Kiesler, S., Sproull, L., & Waters, K. (1999). Cooperating with life-like interface agents. *Computers in Human Behavior, 15*(2), 123-142.

Petty, R. E., & Cacioppo, J. T. (1981). *Attitudes and persuasion—Classic and contemporary approaches*. Dubuque, IA: W. C. Brown.

Porcari, J. P., Zedaker, M.S., & Maldari, M. S. (1998). Virtual motivation. *Fitness Management, 14*(13), 48-51.

Quintanar, L., Crowell, C., & Pryor, J. (1982). Human-computer interaction: A preliminary social psychological analysis. *Behavior Research Methods and Instrumentation, 14*(2), 210-220.

Reed, E. (1996). *The necessity of experience*. New Haven: Yale University Press.

Reeves, B., & Nass, C. I. (1996). *The media equation: How people treat computers, television, and new media like real people and places*. New York: Cambridge University Press Center for the Study of Language and Information.

Reis, J. (1998). *Research results: National data analysis*. The Century Council. Available: www.centurycouncil.org/alcohol101/dem_nat.cfm

Ross, L., & Nisbett, R. E. (1991). *The person and the situation: Perspectives of social psychology*. New York: McGraw-Hill.

Schlosser, A. E., & Kanifer, A. (1999). Current advertising on the Internet: The benefits and usage of mixed-media advertising strategies. In D. W. Schumann & E. Thorson (Eds.), *Advertising and the World Wide Web* (pp. 41-62). Mahwah, NJ: Lawrence Erlbaum.

Schneider, S. J. (1985). Computer technology and persuasion: The case of smoking cessation. *Proceedings of COMPINT 85: Computer Aided Technologies*. Washington, DC: IEEE.

Self, C. S. (1996). Credibility. In M. Salwen & D. Stacks (Eds.), *An integrated approach to communication theory and research*. Mahwah, NJ: Erlbaum.

Stiff, J. B. (1994). *Persuasive communication*. New York: The Guilford Press.

Strecher, V. J. (1999). Computer-tailored smoking cessation materials: A review and discussion. Special Issue: Computer-tailored education. *Patient Education & Counseling, 36*(2), 107-117.

Strecher, V. J., DeVellis, B. M., Becker, M. H., & Rosenstock, I. M. (1986). The role of self-efficacy in achieving health behavior change. *Health Education Quarterly, 13*(1), 73-92.

Strecher, V. J., Kreuter, M., Den Boer, D.- J., Kobrin, S., Hospers, H. J., & Skinner, C. S. (1994). The effects of computer-tailored smoking cessation messages in family practice settings. *Journal of Family Practice, 39*(3), 262-270.

Sullivan, C. (1999). Newspapers must retain credibility. *Editor & Publisher, 1*, 4-5.

Thompson, C. A. (1992). *Exercise adherence and performance: Effects on self-efficacy and outcome expectations*. Doctoral thesis. Illinois Institute of Technology.

Todd, P. A., & Benbasat, I. (1994). The influence of decision aids on choice strategies under conditions of high cognitive load. *IEEE Transactions on Systems, Man, & Cybernetics, 24*(4), 537-547.

Tombari, M. L., Fitzpatrick, S. J., & Childress, W. (1985). Using computers as contingency managers in self-monitoring interventions: A case study. *Computers in Human Behavior, 1*(1), 75-82.

Tseng, S., & Fogg, B. J. (1999). Credibility and computing technology. *Communications of the ACM, 42*(5), 39-44.

Verplank, B., Fulton, J., Black, A., & Moggridge, B. (1993). *Observation and invention: Uses of scenarios in interaction design*. Handout for tutorial at INTERCHI '93, Amsterdam.

Wiederhold, B. K., Davis, R., Wiederhold, M. D., & Riva, G. (1998). The effects of immersiveness on physiology. In *Studies in health technology and informatics* (Vol. 58, pp. 52-60). Amsterdam: IOS.

Woodward, J. P., Carnine, D., & Davis, L. G. (1986). Health ways: A computer simulation for problem solving in personal health management. Special Issue: Technological advances in community health. *Family & Community Health, 9*(2), 60-63.

Zietsman, A. I., & Hewson, P. W. (1986). Effect of instruction using microcomputer simulations and conceptual change strategies on science learning. *Journal of Research in Science Teaching, 23*(1), 27-39.

HUMAN ERROR IDENTIFICATION
IN HUMAN–COMPUTER INTERACTION

Neville A. Stanton
Brunel University

HUMAN ERROR

We are all familiar with the annoyance of errors we make with everyday devices, such as switching on an empty kettle, or making mistakes in the programming sequence with video cassette recorder. People have a tendency to blame themselves for human error. However, the use and abuse of the term has led some to question the very notion of human error (Wagenaar & Groeneweg, 1988). Human error is often invoked in the absence of technological explanations. Chapanis (1999) wrote that, back in the 1940s, he noted that pilot error was really designer error. This was a challenge to contemporary thinking, and shows that design is all important in human error reduction. He became interested in why pilots often retracted the landing gear instead of the landing flaps after landing the aircraft. He identified the problems as designer error rather than pilot error, because the designer had put two identical toggle switches side-by-side: one for the landing gear, the other for the flaps. Chapanis proposed that the controls were separated and coded. Separation and coding of controls are now standard human factors practice. Half a century after Chapanis's original observations, the idea that one can design error-tolerant devices is beginning to gain credence (Baber & Stanton, 1994). One can argue that human error is not a simple matter of one individual making one mistake, so much as the product of a design that has permitted the existence and continuation of specific activities that could lead to errors (Reason, 1990).

Human error is an emotive topic and psychologists have been investigating its origins and causes since the dawn of the discipline (Reason, 1990). Traditional approaches suggested that errors were attributable to individuals. Indeed, so-called Freudian slips were treated as the unwitting revelation of intention: Errors revealed what a person was really thinking but did not wish to disclose. More recently, cognitive psychologists have considered the issues of error classification and explanation (Senders & Moray, 1991). The taxonomic approaches of Norman (1988) and Reason (1990) have enabled the development and formal definition of several categories of human error (such as capture errors, description errors, data-driven errors, associated activation errors, and loss of activation errors), whereas the work of Reason (1990) and Wickens (1992) attempts to understand the psychological mechanisms that combine to cause errors (such as failure of memory, poor perception, errors of decision making, and problems of motor execution). Reason (1990), in particular, has argued that we need to consider the activities of the individual if we are to be able to identify what may go wrong. Rather than viewing errors as unpredictable events, this approach regards them to be wholly predictable occurrences based on an analysis of an individual's activities. Reason's definition proposes that errors are:

. . . those occasions in which a planned sequence of mental or physical activities fail to achieve its intended outcome, [and] when these failures cannot be attributed to the intervention of some chance agency. (Reason, 1990, p. 9)

If errors are no longer to be considered as random occurrences, then it follows that we should be able to identify them and predict their likelihood. The impetus to achieve this has been fueled in the wake of several recent and significant incidents, most notably in the nuclear industry, and there now exist several human error identification (HEI) techniques. The aims of this chapter are to:

1. Consider human error classifications
2. Look at systems approaches to human error
3. Consider how human error can be predicted
4. Examine the validation evidence
5. Look at human error in the context of design

HUMAN ERROR CLASSIFICATION

The development of formal human error classification schemes has assisted in the anticipation and analysis of error. The anticipation of error has come about through the development of formal techniques for predicting error, which is dealt with in the Predicting Human Error section. The analysis of error is assisted by taxonomic systems and interpretation of underlying psychological mechanisms. Three contemporary systems are presented in the work of Norman (1981), Reason (1990), and Wickens (1992).

Norman (1981) reported research on the categorization of action slips, in which he presented the analysis from 1,000 incidents. Underpinning the analysis was a psychological theory of schema activation. He argued that action sequences are triggered by knowledge structures (organized as memory units and called schemas). The mind comprises a hierarchy of schemas that are invoked (or triggered) if particular conditions satisfied or events occur. The theory seems particularly pertinent as a description of skilled behavior. The classification scheme is presented in Table 18.1.

In Neisser's (1976) seminal work on Cognition and Reality, he puts forward a view of how human thought is closely coupled with a person's interaction with the world. He argued that knowledge of how the world works (e.g., mental models) leads to the anticipation of certain kinds of information, which in turn directs behavior to seek out certain kinds of information and provide a ready means of interpretation. During the course of events, as the environment is sampled, the information serves to up date and modify the internal, cognitive schema of the world that will again direct further search. An illustration of the perceptual cycle is shown in Fig. 18.1.

The perceptual cycle can be used to explain human information processing in programming a videocassette recorder. For example (assuming that the individual has the correct knowledge of the videocassette recorder they are programming), their mental model will enable them to anticipate events (such as the menu items they expect to see), search for confirmatory evidence (look at the panel on the video machine), direct a course of action (select channel, day of the week, start time, end time, etc.) and continually check that the outcome is as expected (menu item and data field respond as anticipated). If they uncover some data they do not expect (such as a menu item not previously encountered or the data field does not accept their input), they are required to source a wider knowledge of the

TABLE 18.1. Taxonomy of Slips With Examples

Taxonomy of Slips	Examples of Error Types
Slips that result from errors in the formation of intention	Mode errors: erroneous classification of the situation Description errors: ambiguous or incomplete specification of intention
Slips that result from faulty activation of schemas	Capture errors: similar sequences of action, where stronger sequence takes control Data-driven activation errors: external events that cause the activation of schemas Association-activation errors: currently active schemas that activate other schemas with which they are associated Loss-of-activation errors: schemas that lose activation after they have been activated
Slips that result from faulty triggering of active schemas	Blend errors: combination of components from competing schemas Premature activation errors: schemas that are activated too early Failure to activate errors: failure of the trigger condition or event to activate the schema

world to consider possible explanations that will direct future search activities. The completeness of the model is in the description of process (the cyclical nature of sampling the world) and product (the updating of the world model at any point in time).

This interactive schema model work well for explaining how we act in the world. As Norman's (1981) research shows, it may also explain why errors occur as they do. If, as schema theory predicts, action is directed by schema, then faulty schemas or faulty activation of schemas will lead to erroneous performance. As Table 18.1 shows, this can occur in at least three ways. First, we can select the wrong schema from misinterpretation of the situation. Second, we can activate the wrong schema because of similarities in the trigger conditions. Third, we can activate schemas too early or too late. Examples of these types of errors are presented in Table 18.1.

Of particular interest is the problem of mode errors. Norman (1981) singled this error type out as requiring special attention in design of computing systems. He pointed out that the

misclassification of the mode that the computing system was in could lead to input errors, which may have serious effect. In word processors, this may mean the loss of documents, in video recorders this may mean the loss of recordings, and in flight decks this may mean damage to aircraft.

Casey (1993) describes the case of how an apparently simple mode error by a radiotherapy technician working in a cancer care center led to the death of a patient. The Therac-25 she was operating was a state-of-the-art, million dollar machine that could be used as both a high-power X-ray machine (25 million electron volts delivered by typing "x" on the keyboard) and low-power electron beam machine (200 rads delivered by typing "e" on the keyboard). After preparing the patient for radiation therapy, the radiotherapist went to her isolated control room. She accidentally pressed the "x" key instead of the "e" key, but quickly realized this and selected the edit menu so that she could change the setting from X-ray to electron beam. Then she returned to the main screen to wait for the "beam ready" prompt. All of this occurred within 8 s. Unknown to her, the

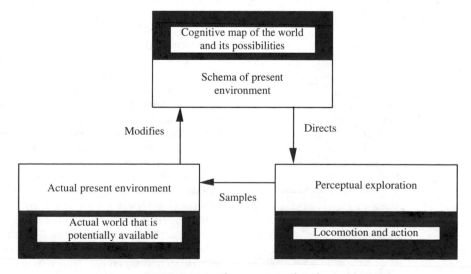

FIGURE 18.1. The perceptual cycle.

rapid sequence of inputs had never been tested on the machine before, and it had actually entered a hybrid mode, delivering blasts of 25,000 rads, which was more than 125 times the prescribed dose. When the "beam ready" prompt was displayed, the radiotherapist pressed the "b" key to fire the beam. The high-energy beam was delivered to the patient and the computer screen displayed the prompt "Malfunction 54." Unaware that the machined had already fired, she reset it and pressed "b" again. This happened for a third time until the patient ran out of the room reporting painful electric shocks. On investigation, the problem with the machine modes was found, but not before other overdoses had been given. This case study demonstrates the need to consider the way in which design of a system can induce errors in users. A thorough understanding of human error is required by the design team. Error classifications schemes can certainly help, but they need to be supported by formal error prediction techniques within a user-centerd design approach.

Reason (1990) developed a higher level error classification system, incorporating slips, lapses, and mistakes. Slips and lapses are defined by attentional failures and memory failures, respectively. Both slips and lapses are examples of where the action was unintended, whereas mistakes are associated with intended action. The taxonomy is presented in Table 18.2.

Wickens (1992) takes the information processing framework to consider the implication of psychological mechanisms in error formation. He argues with mistakes in which the situation assessment and/or planning are poor, whereas the retrieval action execution is good. With slips, the action execution is poor, whereas the situation assessment and planning are good. Finally, with lapses, the situation assessment and action execution are good, but memory is poor. A summary of these distinctions is shown in Table 18.3.

Wickens (1992) is also concerned with mode errors, with particular reference to technological domains. He suggests that a pilot raising the landing gear while the aircraft is still on the runway is an example of a mode error. Wickens proposed that mode errors are a result of poorly conceived system design that allows the mode confusion to occur and allow the operation in an inappropriate mode. Chapanis (1999) argued that the landing gear switch could be rendered inoperable if the landing

TABLE 18.2. Basis Error Types With Examples

Basic Error Type	Example of Error Type
Slip	Action intrusion
	Omission of action
	Reversal of action
	Misordering of action
	Mistiming of action
Lapse	Omitting of planned actions
	Losing place in action sequence
	Forgetting intended actions
Mistake	Misapplication of good procedure
	Application of a bad procedure
	Misperception
	Poor decision making
	Failure to consider alternatives
	Overconfidence

TABLE 18.3. Error Types and Associated Psychological Mechanisms

Error Type	Associated Psychological Mechanism
Slip	Action execution
Lapse and mode errors	Memory
Mistake	Planning and intention of action
Mistake and mode errors	Interpretation and situation assessment

gear detects weight on the wheels, because the aircraft would be on the ground.

The taxonomies of errors can be used to anticipate what might go wrong in any tasks. Potentially, every task or activity could be subject to a slip, lapse, or mistake. The two approaches represented within the taxonomies are a schema-based approach and an error list-based approach. Examples of these two approaches will be presented next in the form of formal human error identification techniques.

PREDICTING HUMAN ERROR

An abundance of methods for identifying human error exist, some of which may be appropriate for the analysis of consumer products. In general, most of the existing techniques have two key problems. The first of these problems relates to the lack of representation of the external environment or objects. Typically, human error analysis techniques do not represent the activity of the device and material that the human interacts with, in more than a passing manner. Hollnagel (1993) emphasizes that Human Reliability Analysis often fails to take adequate account of the context in which performance occurs. Second, there tends to be a good deal of dependence made on the judgment of the analyst. Different analysts, with different experience, may make different predictions regarding the same problem (called intraanalyst reliability). Similarly, the same analyst may make different judgments on different occasions (interanalyst reliability). This subjectivity of analysis may weaken the confidence that can be placed in any predictions made. The analyst is required to be an expert in the technique, as well as the operation of the device being analyzed if the analysis has a hope of being realistic.

Two techniques are considered here because of the inherent differences in the way the methods work. SHERPA (Systematic Human Error Reduction and Prediction Approach) is a divergent error prediction method: It works by associating up to 10 error modes with each action. In the hands of a novice, it is typical for there to be an overinclusive strategy for selecting error modes. The novice user would rather play safe than be sorry and tend to predict many more errors than actually occur. This might be problematic; crying wolf too many times might ruin the credibility of the approach. TAFEI (Task Analysis For Error Identification), by contrast, is a convergent error prediction technique: It works by identifying the possible transitions between the different states of a device and uses the normative description of behavior (provided by hierarchical task analysis; HTA) to identify potentially erroneous actions. Even in the hands of a novice, the technique seems to prevent the individual generating too many false alarms, certainly no more than they do using heuristics. In fact, by constraining the user of TAFEI to the problem space surrounding the transitions between device states, it should exclude extraneous error prediction. Indeed, this was one of the original aims for the technique when it was originally developed (Baber & Stanton, 1994).

Systematic Human Error Reduction and Prediction Approach (SHERPA)

SHERPA represents the error-list approach. At its core is an error taxonomy that is not unlike the classification schemes presented in the previous section. The idea is that each task can be classified into one of five basic types. SHERPA (Embrey, 1986) uses HTA (Annett, Duncan, Stammers, & Gray, 1971), together with an error taxonomy to identify credible errors associated with a sequence of human activity. In essence, the SHERPA technique works by indicating which error modes are credible for each task step in turn, based on an analysis of work activity. This indication is based on the judgment of the analyst and requires input from a subject matter expert to be realistic. A summary of the procedure is shown in Fig. 18.2.

The process begins with the analysis of work activities, using HTA. HTA (Annett et al., 1971) is based on the notion that task performance can be expressed in terms of a hierarchy of goals (what the person is seeking to achieve), operations (the activities executed to achieve the goals), and plans (the sequence in which the operations are executed). An example of HTA for the programming of a video cassette recorder is shown in Fig. 18.3.

For the application of SHERPA, each task step from the bottom level of the analysis is taken in turn. First, each task step is classified into a type from the taxonomy, into one of the following types:

- Action (e.g., pressing a button, pulling a switch, opening a door)
- Retrieval (e.g., getting information from a screen or manual)
- Checking (e.g., conducting a procedural check)
- Information communication (e.g., talking to another party)
- Selection (e.g., choosing one alternative over another)

This classification of the task step then leads the analyst to consider credible error modes associated with that activity, as shown in Table 18.4.

For each credible error (i.e., those judged by a subject matter expert to be possible), a description of the form that the error would take is given as illustrated in Table 18.5. The consequence of the error on system needs to be determined next, because this has implications for the criticality of the error. The last four steps consider the possibility for error recovery, the ordinal probability of the error (high, medium of low), its criticality (either critical or not critical), and potential remedies. Again, these are shown in Table 18.5.

As Table 18.5 shows, there are six basic error types associated with the activities of programming a VCR. These are:

TABLE 18.4. Error Modes and Their Description

Error Mode	Error Description
Action	
A1	Operation too long/short
A2	Operation mistimed
A3	Operation in wrong direction
A4	Operation too much/little
A5	Misalign
A6	Right operation on wrong object
A7	Wrong operation on right object
A8	Operation omitted
A9	Operation incomplete
A10	Wrong operation on wrong object
Information retrieval	
R1	Information not obtained
R2	Wrong information obtained
R3	Information retrieval incomplete
Checking	
C1	Check omitted
C2	Check incomplete
C3	Right check on wrong object
C4	Wrong check on right object
C5	Check mistimed
C6	Wrong check on wrong object
Information communication	
I1	Information not communicated
I2	Wrong information communicated
I3	Information communication incomplete
Selection	
S1	Selection omitted
S2	Wrong selection made

A. Failing to check that the VCR clock is correct.
B. Failing to insert a cassette.
C. Failing to select the program number.
D. Failing to wait.
E. Failing to enter programming information correctly.
F. Failing to press the confirmatory buttons.

The purpose of SHERPA is not only to identify potential errors with the current design, but to guide future design considerations. The structured nature of the analysis can help to focus the design remedies on solving problems, as shown in the remedial strategies column. As this analysis shows, quite a lot of improvements could be made. It is important to note, however, that the improvements are constrained by the analysis. This does not address radically different design solutions that may remove the need to program at all.

Task Analysis For Error Identification (TAFEI)

TAFEI represents the schema-based approach. It is developed on the notion of interaction between person and machine. TAFEI (Baber & Stanton, 1994; Stanton & Baber, 1996) explicitly analyzes the *interaction* between people and machines. TAFEI analysis is concerned with task-based scenarios. This is done by mapping human activity onto machine states. An overview of

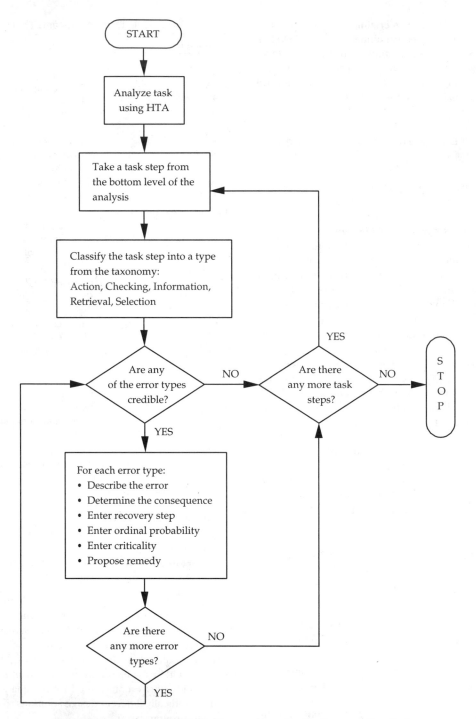

FIGURE 18.2. The Systematic Human Error Reduction and Prediction Approach (SHERPA). HTA = hierarchical task analysis.

the procedure is shown in Fig 18.4. TAFEI analysis consists of three principal components: HTA, state-space diagrams (SSDs; that are loosely based on finite state machines; Angel & Bekey, 1968), and transition matrices. HTA provides a description of human activity, SSD provides a description of machine activity, and a transition matrix provides a mechanism for determining potential erroneous activity through the interaction of the human and the device. In a similar manner to Newell and Simon (1972), legal and illegal operators (called *transitions* in the TAFEI methodology) are identified.

In brief, the TAFEI methodology is as follows. First, the system to be addressed needs to be defined. Next, the human activities and machine states are described in separate analyses. The basic building blocks are HTA (describing human activity;

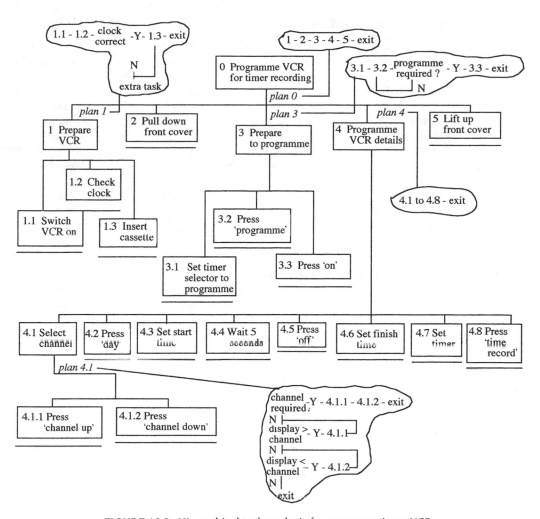

FIGURE 18.3. Hierarchical task analysis for programming a VCR.

see Fig. 18.3) and state-space diagrams (describing machine activity). These two types of analysis are then combined to produce the TAFEI description of human–machine interaction, as shown in Fig 18.5.

From the TAFEI diagram, a transition matrix is compiled, and each transition is scrutinized. In Table 18.5, each transition has been classified as impossible (i.e., the transition cannot be performed), illegal (the transition can be performed, but it does not lead to the desired outcome), or legal (the transition can be performed and is consistent with the description of error-free activity provided by the HTA), until all transitions have been analyzed. Finally, illegal transitions are addressed in turn as potential errors, to consider changes that may be introduced (Table 18.6).

Thirteen of the transitions defined as illegal can be reduced to a subset of six basic error types:

A. Switch VCR off inadvertently.
B. Insert cassette into machine when switched off.
C. Program without cassette inserted.
D. Fail to select program number.
E. Fail to wait for ON light.
F. Fail to enter programming information.

In addition, one legal transition has been highlighted because it requires a recursive activity to be performed. These activities seem to be particularly prone to errors of omission. These predictions then serve as a basis for the designer to address the redesign of the VCR. A number of illegal transitions could be dealt with fairly easily by considering the use of modes in the operation of the device, such as switching off the VCR without stopping the tape and pressing play without inserting the tape. As with the SHERPA example, the point of the analysis is to help guide design effort to make the product error-tolerant.

VALIDATION OF HUMAN ERROR IDENTIFICATION

There have been a few attempts to validate HEI techniques (Baber & Stanton, 1996; Kennedy, 1995; Kirwan, 1992a, 1992b; Whalley & Kirwan, 1989; Williams, 1989). For instance, Whalley and Kirwan (1989) evaluated six HEI techniques (Heuristics, Potential Human Error Cause Analysis (PHECA), Skill–Rules–Knowledge model (SRK), SHERPA, Technique for Human Error

TABLE 18.5. The Systematic Human Error Reduction and Prediction Approach (SHERPA) Description

Task Step	Error Mode	Error Description	Consequence	Recovery	P	C	Remedial Strategy
1.1	A8	Fail to switch VCR on	Cannot proceed	Immediate	L		Press any button to switch VCR on
1.2	C1 C2	Omit to check clock Incomplete check	VCR clock time may be incorrect	None	L	!	Automatic clock setting and adjust via radiotransmitter
1.3	A3	Insert cassette wrong way around	Damage to VCR	Immediate	L	!	Strengthen mechanism
	A8	Fail to insert cassette	Cannot record	Task 3	L		On-screen prompt
2	A8	Fail to pull down front cover	Cannot proceed	Immediate	L		Remove cover to programming
3.1	S1	Fail to move timer selector	Cannot proceed	Immediate	L		Separate timer selector from programming function
3.2	A8	Fail to press PROGRAM	Cannot proceed	Immediate	L		Remove this task step from sequence
3.3	A8	Fail to press ON button	Cannot proceed	Immediate	L		Label button START TIME
4.1.1	A8	Fail to press UP button	Wrong channel selected	None	M	!	Enter channel number directly from keypad
4.1.2	A8	Fail to press DOWN button	Wrong channel selected	None	M	!	Enter channel number directly from keypad
4.2	A8	Fail to press DAY button	Wrong day selected	None	M	!	Present day via a calendar
4.3	I1 I2	No time entered Wrong time entered	No program recorded Wrong program recorded	None None	L L	! !	Dial time in via analogue clock Dial time in via analogue clock
4.4	A1	Fail to wait	Start time not set	Task 4.5	L		Remove need to wait
4.5	A8	Fail to press OFF button	Cannot set finish time				Label button FINISH TIME
4.6	I1 I2	No time entered Wrong time entered	No program recorded Wrong program recorded	None None	L L	! !	Dial time in via analogue clock Dial time in via analogue clock
4.7	A8	Fail to set timer	No program recorded	None	L	!	Separate timer selector from programming function
4.8	A8	Fail to press TIME RECORD button	No program recorded	None	L	!	Remove this task step from sequence
5	A8	Fail to lift up front cover	Cover left down	Immediate	L		Remove cover to programming

Note. P = ordinal probability; C = criticality; L = Low; M = Medium; H = High; ! = error critical for system performance.

Rate Prediction (THERP), and HAZard and OPerability study (HAZOP)) for their ability to account for the errors known to have contributed to four genuine incidents within the nuclear industry. More recently, Kirwan (1992b) has developed a comprehensive list of eight criteria to evaluate the acceptability of these techniques at a more qualitative level. In an unpublished study, Kennedy (1995) has included Kirwan's criteria when examining the ability of the techniques to predict 10 actual incidents retrospectively. Although these studies failed to identify a clear favorite from among these HEI techniques, all three studies indicated impressive general performance using the SHERPA method. SHERPA achieved the highest overall rankings, and Kirwan (1992b) recommends a combination of expert judgment, together with the SHERPA technique as the most valid approach.

The strength of these studies lies in the high level of ecological or face validity that they achieve. The methodologies make use of the opinions of expert assessors for the prediction of errors contributing to real-world events. However, these studies do raise several methodological concerns. Specifically, the number of assessors using each technique is small (typically 1 to 3) and the equivalence of assessment across techniques is brought into question because different people are assessing each HEI technique. A second methodological concern centers on the use of subjective rating scales. It is doubtful that the assessors will share the same standards when rating the acceptability or usefulness of an HEI technique. This factor combined with the small number of assessors for each technique means that these data should be accepted with some degree of caution.

In light of these criticisms, Baber and Stanton (1996) aimed to provide a more rigorous test of the predictive validity of SHERPA and TAFEI. Predictive validity was tested by comparing the errors identified by an expert analyst with those observed during 300 transactions with a ticket machine on the London Underground. Baber and Stanton (1996) suggest that SHERPA and TAFEI provided an acceptable level of sensitivity based on the data from two expert analysts (0.8). The strength of this latter study over Kirwan's is that it reports the use of the method in detail as well as the error predictions made using SHERPA and TAFEI. Stanton and Baber (2002) report a study on the

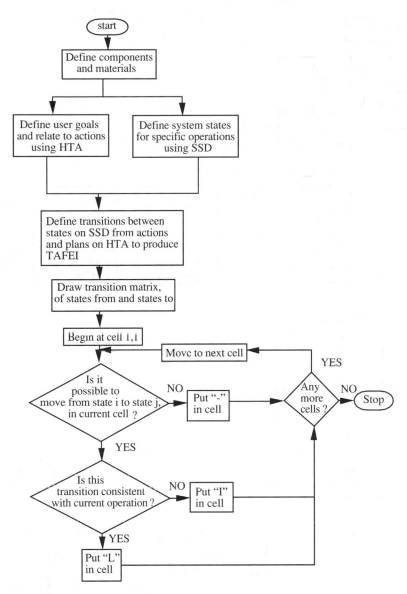

FIGURE 18.4. The Task Analysis For Error Identification (TAFEI) procedure. HTA = hierarchical task analysis; SSD = state-space diagram.

performance of SHERPA and TAFEI using a larger pool of analysts and using novice analysts to examine the important issue
of ease of acquiring the method. They report reliability values
of between 0.4 and 0.6, and sensitivity values of between 0.7
and 0.8. This compares favorably with Hollnagel, Kaarstad, and
Lee's (1999) analysis of the Cognitive Reliability and Error Analysis Method, for which they claim a 68.6% match between predicted outcomes and actual outcomes.

The research into HEI techniques suggests that the methods
enable analysts to structure their judgment. However, the results
run counter to the literature in some areas (such as usability evaluation), which suggest the superiority of heuristic approaches
(Neilsen, 1993). The views of Lansdale and Ormerod (1994)
may help us to reconcile these findings. They suggest that, to
be applied successfully, an heuristic approach needs the support of an explicit methodology to "ensure that the evaluation

is structured and thorough" (p. 257). Essentially, SHERPA and
TAFEI provide a semistructured approach that forms a framework for the judgment of the analyst without constraining it.
It seems to succeed precisely because of its semistructured nature that alleviates the burden otherwise placed on the analyst's
memory while allowing them room to use their own heuristic
judgment.

APPLYING TAFEI TO INTERFACE DESIGN

A study by Baber and Stanton (1999) shows how TAFEI can
be used as part of the interface design process for a computer
workstation. Their case study is based on a design project for
a medical imaging software company. The software was used
by cytogeneticists in research and hospital environments. The

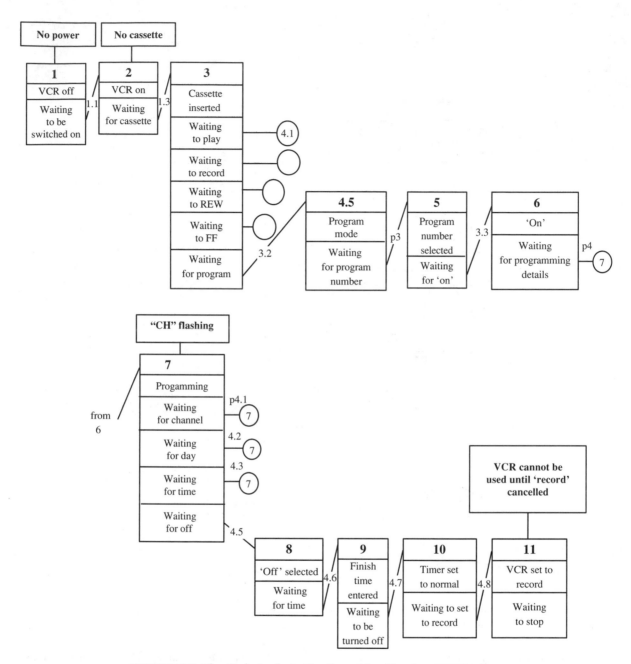

FIGURE 18.5. The Task Analysis For Error Identification (TAFEI) description.

existing software comprised a menu-driven, text-based interface. The task of metaphase finding has six main subtasks:

1. The set-up task (where the computer is fired up, the microscope is calibrated, the scan is defined, and the cells are prepared)
2. The capture task (where the drug is applied to the cells on the slide using a pipette and the images are captured)
3. The processing task (where the background is subtracted from the cell image and the image processing is performed)
4. The analysis task (where the data are graphed and tabulated)

5. The storage task (where images are assessed and selected images and data are saved)
6. The shut-down task (where the computer and microscope are shut down)

Using TAFEI to model the existing system, Baber and Stanton (1999) found that the sequence of activities required to conduct the metaphase finding tasks were not supported logically by the computer interface. The two main problems were the confusing range of choices offered to users in each system state and the number of recursions in the task sequence required to

TABLE 18.6. The Transition Matrix

To state:

From state:	1	2	3	4.5	5	6	7	8	9	10	11
1	—	L	I	—	—	—	—	—	—	—	—
2	L	—	L	I	—	—	—	—	—	—	—
3	L	—	—	L	—	—	—	—	—	—	—
4.5	I	—	—	—	L	I	—	—	—	—	—
5	I	—	—	—	—	L	I	I	—	—	—
6	I	—	—	—	—	—	L	I	—	—	—
7	I	—	—	—	—	—	L	L	—	—	—
8	I	—	—	—	—	—	—	—	L	—	—
9	I	—	—	—	—	—	—	—	—	L	—
10	I	—	—	—	—	—	—	—	—	—	L
11	—	—	—	—	—	—	—	—	—	—	—

L = legal; I = illegal; — = impossible.

perform even the simplest of tasks. Interviews with the cytogeneticists revealed some fundamental problems with the existing software, such as the existing system was not error tolerant, there was a lack of consistency between commands, the error and feedback messages were not meaningful to users, and the same data had to be entered several times in the procedure. The lack of trust that the users had with the system meant that they also kept a paper-based log of the information they entered into the computer, meaning even greater duplication of effort.

From their analysis of the tasks sequences, Baber and Stanton (1999) developed a TAFEI diagram of the ideal system states that would lead the user through the metaphase finding task in a logical manner. One of the software engineers described the analysis as "a sort of video playback of someone actually doing things with the equipment." The TAFEI analysis was a revelation to the company, who had some vague idea that all was not well with the interface, but had not conceived of the problem as defining the task sequence and user actions. As a modeling approach, TAFEI does this rather well. The final step in the analysis was to define and refine the interface screen. This was based on the task sequence and TAFEI description, to produce a prototype layout for comment with the users and software engineers. Specific task scenarios were then tried out to see if the new interface would support user activity. Following some minor modification, the company produced a functional prototype for user performance trials.

Baber and Stanton (1999) report successful completion of the interface design project using TAFEI. They argue that the method supports analytical prototyping activity. TAFEI enables designers to focus attention on the task sequence, user activity, and interface design. It also highlights potential problem points in the interaction, where errors are likely to occur. Although it is accepted that these considerations might also form part of other methods (e.g., storyboards), TAFEI does this in a structured and rigorous manner.

APPLYING SHERPA TO SAFETY CRITICAL SYSTEMS

Studies reported on SHERPA show how it can be applied to the evaluation of energy distribution (Glendon & Stanton, 2000) and oil extraction (Stanton & Wilson, 2000) in the safety critical industries. Both of these represent multiperson systems. The study by Glendon and Stanton (2000) compares the impact of the implementation of a new safety management system on error potential in an energy distribution company. The first assessment of human error potential in electrical switching operations was undertaken in 1994. Following this, the company undertook major organization restructuring activities. This included a major safety management initiative, which was to separate operational control of the energy distribution system from safety management. In the original assessment in 1994, it became obvious that safety was often secondary to operational performance of the system. To remove this conflict, a separate safety management center was established. This enabled the operational center to pass control of part of the energy distribution system over to the safety management center, when safety issues arose. Thus, operational control and safety management were not in conflict. When the safety issues were resolved, control of that

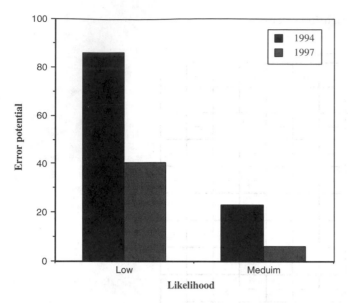

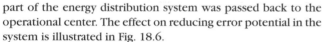

FIGURE 18.6. Systematic Human Error Reduction and Prediction Approach (SHERPA) analysis of an energy distribution system.

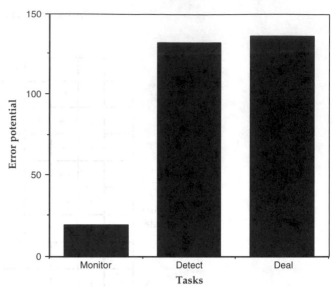

FIGURE 18.7. Systematic Human Error Reduction and Prediction Approach (SHERPA) analysis of drilling for oil tasks.

part of the energy distribution system was passed back to the operational center. The effect on reducing error potential in the system is illustrated in Fig. 18.6.

Statistical analyses, using the Binomial test, were conducted to determine the difference between the error rates in 1994 with those in 1997. All comparisons between critical (low and medium likelihood) errors proved statistically significant ($p < .001$). It should be noted that there were no high likelihood errors in the analyses of this system, all of which have probably been designed out of the system as it has evolved. Figure 18.6 shows a reduction in the error potential by more than 50% for critical errors from 1994 to 1997. This suggests that the system is likely to be safer. The SHERPA technique assisted in the redesign process after the 1994 analysis, by proposing design reduction strategies. The results of these proposals are seen in the 1997 analysis.

Stanton and Wilson (2000) describe the assessment of oil drilling task undertaken offshore using the SHERPA technique. Surprising, the oil industry lacked a method for assessing the robustness of its working practices. SHERPA scrutinizes the minutiae of human action and provides some basis for assessing the magnitude of risk rising from human activity, which can be used in turn to inform which intervention strategies are likely to make the working environment safer. Drilling is a highly integrated team activity comprising the Site Controller, Drilling Supervisor, Geologist, Toolpusher, Directional Driller, Assistant Driller, Mudlogger, Mud Engineer, Derrickman, Crane Operator, Roughnecks, and Roustabouts. It is also a fairly dangerous occupation as the incidents at Piper Alpha and Ocean Odyssey show. It is argued that the risks arising from human operations can be controlled in the same way as engineering risk. Stanton and Wilson (2000) conducted a detailed task analysis of drilling operations covering monitoring the status of the well (monitor),

detecting abnormalities (detect), and dealing with kicks (deal). The results of the SHERPA analysis are shown in Fig. 18.7.

As Fig. 18.7 shows, the error potential in the detection of abnormalities tasks and the dealing with kicks task are considerably greater than the monitoring the well task. This should be some cause for concern and certainly explains some of the problems that oil companies have observed. The proposed remedies from the SHERPA analysis included strategies such as computer-prompted checks of drilling variables, placement of drilling parameters on separate displays, redesign of trend displays, computer-prompted alarm levels, automatic transmission of information, electronic links between the drilling team members, computer-generated tables of mud loss, computer-based procedures, online weather systems, and automated shutdown procedures. All of these proposals are really about the design of the human–computer interface. Stanton and Wilson (2000) argued that adopting these approaches should help organizations design safer systems and help prevent catastrophic disasters in the future.

CONCLUSIONS

The case studies in this chapter show that human error identification techniques can be used both in the design of new systems as well as the evaluation of existing systems. Stanton and Young (1999) argue that the methods may have the greatest impact at the prototyping stage, particularly considering one of the key design stages: *analytic prototyping*. Although in the past, it may have been costly to alter design at structural prototyping, and perhaps even impossible, with the advent of computer-aided design, it is made much simpler. It may even be possible to compare alternative designs at this stage with such technology. In terms of the analytical prototyping of human interfaces, Baber and Stanton (1999) proposed three main forms: functional analysis (i.e., consideration of the range of functions the device supports), scenario analysis (i.e., consideration of the device with regard to a particular sequence of activities), and structural analysis (i.e., nondestructive testing of the interface from a user-centered perspective). The case studies of how TAFEI and SHERPA were applied on interface design projects show how system improvements may be made. Validation evidence shows that both SHERPA and TAFEI can be used to predict most of the errors that people make (Stanton & Steverage, 1998; Baber & Stanton, 1996). Stanton and Young (1999) have extended this analysis to other ergonomics methods. There are three take-home messages from this chapter:

1. Most technology-induced errors are entirely predictable.
2. Structured methods, such as SHERPA and TAFEI, produce reliable and valid error data.
3. Ergonomics methods should be used as part of the formative design process to improve design and reduce errors.

References

Angel, E. S., & Bekey, G. A. (1968, March). Adaptive finite state models of manual control systems. *IEEE Transactions on Man-Machine Systems*, 15–29.

Annett, J., Duncan, K. D., Stammers, R. B., & Gray, M. J. (1971). *Task analysis. Training information no. 6.* London: HMSO.

Baber, C., & Stanton, N. A. (1994). Task analysis for error identification: A methodology for designing error tolerant consumer products. *Ergonomics, 37*(11), 1923–1941.

Baber, C., & Stanton, N. A. (1996). Human error identification techniques applied to public technology: Predictions compared with observed use. *Applied Ergonomics, 27*(2), 119–131.

Baber, C., & Stanton, N. A. (1999). Analytical prototyping, In J. M. Noyes & M. Cook (Eds.), *Interface technology: The leading edge* (pp. 174–194). Baldock: Research Studies Press.

Casey, S. (1993). *Set phasers on stun-and other true tales of design, technology and human error.* Santa Barbara, CA: Aegean Publishing Company.

Chapanis, A. (1999). *The Chapanis chronicles: 50 years of human factors research, education, and design.* Santa Barbara, CA: Aegean Publishing Company.

Embrey, D. E. (1986). SHERPA: A systematic human error reduction and prediction approach. Paper presented at the International Meeting on Advances in Nuclear Power Systems. Knoxville, TN.

Glendon, A. I., & Stanton, N. A. (1999). Safety culture in organisations. *Safety Science, 34*, 193–214.

Hollnagel, E. (1993). *Human reliability analysis: Context and control.* London: Academic Press.

Hollnagel, E., Kaarstad, M., & Lee, H.-C. (1998). Error mode prediction. *Ergonomics, 42*, 1457–1471.

Kennedy, R. J. (1995, March 22). Can human reliability assessment (HRA) predict real accidents? A case study analysis of HRA. In A. I. Glendon & N. A. Stanton (Eds.), *Proceedings of the risk assessment and risk reduction conference.* Birmingham, UK: Aston University.

Kirwan, B. (1992a). Human error identification in human reliability assessment. Part 1: Overview of approaches. *Applied Ergonomics, 23*, 299–318.

Kirwan, B. (1992b). Human error identification in human reliability assessment. Part 2: Detailed comparison of techniques. *Applied Ergonomics, 23*, 371–381.

Lansdale, M. W., & Ormerod, T. C. (1994). *Understanding interfaces.* London: Academic Press.

Neisser, U. (1976). *Cognition and reality: Principles and implications of cognitive psychology.* San Francisco: Freeman.

Newell, A., & Simon, H. A. (1972). *Human problem solving.* Englewood Cliffs, NJ: Prentice Hall.

Nielsen, J. (1993). *Usability engineering.* Boston: Academic Press.

Norman, D. A. (1981). Categorisation of action slips. *Psychological Review, 88* (1), 1–15.

Norman, D. A. (1988). *The psychology of everyday things.* New York: Basic Books.

Reason, J. (1990). *Human error.* Cambridge: Cambridge University Press.

Senders, J. W., & Moray, N. P. (1991). *Human error.* Hillsdale, NJ: LEA.

Stanton, N. A., & Baber, C. (1996). A systems approach to human error identification. *Safety Science, 22*, 215–228.

Stanton, N. A., & Baber, C. (2002). Error by design. Manuscript submitted for publication.

Stanton, N. A., & Steverage, S. (1998). Learning to predict human error. *Ergonomics, 41*(11), 1737–1756.

Stanton, N. A., & Wilson, J. (2000, January/February). Human factors: Step change improvements in effectiveness and safety. *Drilling Contractor*, 36–41.

Stanton, N. A., & Young, M. S. (1999). *A guide to methodology in ergonomics.* London: Taylor & Francis.

Wagenaar, W. A., & Groeneweg, J. (1988). Accidents at sea: Multiple causes, impossible consequences. *International Journal of Man-Machine Studies, 27*, 587–598.

Whalley, S. J., & Kirwan, B. (1989, June). An evaluation of five human error identification techniques. Paper presented at the 5th International Loss Prevention Symposium. Oslo.

Wickens, C. D. (1992). *Engineering psychology and human performance.* New York: Harper Collins.

Williams, J. C. (1989). Validation of human reliability assessment techniques. *Reliability Engineering, 11*, 149–162.

·19·

DESIGN OF COMPUTER WORKSTATIONS

Michael J. Smith, Pascale Carayon, and William J. Cohen
University of Wisconsin–Madison

INTRODUCTION

Millions of people use personal computers (PCs), laptop computers (LPCs), and related information technology (IT) daily to conduct work, to shop, to communicate, and to have fun. The technological devices that connect people to PCs, LPCs, and IT have grown in number and have shrunk in size. Now it is possible to carry computing and IT on your person and use them while you walk, talk, sit, run, or swim. The boundaries of fixed use environments and limited activities for computing have fallen to be replaced with highly mobile computing and communications. Portability, universal access, enhanced usability and expanded communications capabilities have led to the use of computers in almost any conceivable activity. This introduces a host of ergonomic concerns related to the design of work areas (and activities) in which computing is used. Although there has been decades of research and applications that have defined important considerations in the ergonomic design of fixed computer work areas (Cakir, Hart, & Stewart, 1979; Grandjean, 1979, 1984; Smith & Cohen, 1997; Stammerjohn, Smith, & Cohen, 1981), very little has been done to define the design of work areas for mobile computing and communications.

Ergonomics is the science of fitting the environment and activities to the capabilities, dimensions, and needs of people. Ergonomic knowledge and principles are applied to adapt working conditions to the physical, psychological, and social nature of the person. The goal of ergonomics is to improve performance while enhancing comfort, health, and safety. Computer workstation design is more than just making the computer interfaces easier to use or making furniture adjustable in various dimensions. It also involves integrating design considerations with the work environment, the task requirements, social aspects of work, and job design.

A fundamental perspective in this chapter is that work area (workstation) design influences employee comfort, health, motivation, and performance. We will examine basic ergonomic considerations (principles, practices, concerns) that can be used to develop guidance for the design of work areas (workstations) for the use of computing and related IT.

HISTORICAL PERSPECTIVE

Forty years ago, a person would interact with several different types of information sources and technologies when engaging in activities such as work. The person might look at hard copy documents, take notes with a pen and a paper tablet, use a fixed location telephone, talk face-to-face with colleagues, type on a typewriter, and perform many other tasks during the course of the working day. The diversity of activities led people to actively move around during the day. Then, 30 years ago, many people started to spend most of their workday in sedentary work sitting in front of a computer terminal. This type of human–machine system led to restricted physical movement, and employee attention directed toward the computer monitor.

Since the initial work of Hultgren and Knave (1973), Ostberg (1975), and Gunnarsson and Ostberg (1977) in Sweden, there have been hundreds (maybe thousands) of research studies from every corner of the globe examining the working conditions of computer users and their associated health complaints. There have been several international conferences devoted to these issues starting in 1980, and there are conferences on these issues already programmed through the decade of the 2000s. The findings from this research and the meetings have generally indicated that poor ergonomic and job design conditions are associated with large numbers of computer users complaining about visual discomfort, musculoskeletal discomfort and pain, and psychological distress (Berlinguet & Berthelette, 1990; Bergqvist, 1984; Bergqvist, Knave, Voss, & Wibom, 1992; Bullinger, 1991; Bullinger & Ziegler, 1999; Cakir et al., 1979; Carayon, 1994; Carayon, Swanson, & Smith, 1987; Cohen et al., 1995; Cohen, Piotrkowski, & Coray, 1987; Conway et al., 1996; Grandjean, 1979, 1984; Grandjean & Vigliani, 1980; Grieco, Molteni, Occhipinti, & Piccoli, 1995; Gunnarsson & Ostberg, 1977; Hales, Sauter, Peterson, & Fine, 1994; Hultgren & Knave, 1973; Knave & Wideback, 1987; Knave, Wibom, Voss, Hedstrom, & Bergqvist, 1985; Lim & Carayon, 1993, 1995; Lim, Rogers, Smith, & Sainfort, 1989; Luczak, Cakir, & Cakir, 1993; Ostberg, 1975; Piotrkowski, Cohen, & Coray, 1992; Saito et al., 1997; Sauter, Schleifer, & Knutson, 1991; Smith, 1984, 1987, 1997; Smith & Carayon, 1995; Smith, Carayon, Eberts, & Salvendy, 1992; Smith & Cohen, 1997; Smith, Cohen, Stammerjohn, & Happ, 1981; Smith et al., 1996; Smith & Salvendy, 1989a,b, 1993, 1997, 2001; Smith, Salvendy, Harris, & Koubek, 2001; Stammerjohn et al., 1981).

In fact, if these adverse ergonomic conditions are repeated daily over a long period, more or less chronic aches and pains of the upper extremities and back can occur, and may involve not only muscles, but also other soft tissues, such as tendons and nerves. Long-lasting, adverse ergonomic conditions may lead to a deterioration of joints, ligaments, and tendons. Reviews of field studies (Bergquist, 1984; Bernard, 1997; Carayon, Smith, & Haims, 1999; Grandjean, 1979, 1987; Hagberg et al., 1995; Putz-Anderson, 1987; Smith, 1984, 1987, 1997), as well as general experience, have shown that these conditions may be associated with a higher risk of:

1. Inflammation of the joints
2. Inflammation of the tendon sheaths
3. Inflammation of the attachment points of tendons
4. Symptoms of degeneration of the joints in the form of chronic arthroses
5. Painful induration of the muscles
6. Disc troubles
7. Peripheral nerve disorders

The relationship between the user and technology is reciprocal. It is a system in which the constraints of one element affect the performance of the other. Various aspects of the system, such as the task requirements, the work demands, the environment, and the workstation, also influence how effectively and comfortably the technology can be used by the person. One of the consequences of this is that the design of the workspace limits the nature and effectiveness of the interaction between the person and the technology. For instance, inadequate space for carrying out physical activities can lead to constrained postures, which together with long-lasting static loading produces discomfort in the back and shoulders. This leads to symptoms of tiredness, muscle cramps, and pain. In addition, heavy workload, chronic repetition, and other biomechanical strains can cause similar problems. These adverse postural, repetition, and workload exposures lead to reduced performance and productivity, and, in the long run, they may also affect employee well-being and health.

Today, people carry their computing and communications with them and engage in activities in any available work area. The potential for serious ergonomic risks is high for these new modes of working. We will first examine problems and design considerations for fixed workstations where PCs are used. Then, we will look at mobile computing and IT.

STUDIES ON THE DESIGN OF FIXED WORKSTATIONS

Workstation design is a major element in ergonomic strategies for improving user comfort and particularly for reducing musculoskeletal problems. Often, the task requirements will have a role in defining the layout and dimensional characteristics of the workstation. The relative importance of the screen, input devices, and hard copy (e.g., source documents) depends primarily on the task, and this defines the design considerations necessary to improve operator performance, comfort, and health. Data entry jobs, for example, are typically hard copy oriented. The operator spends little time looking at the screen, and tasks are characterized by high rates of input (keying/talking). For this type of task, it is logical for the layout to emphasize the input devices and the source documents, because these are the primary items used in the task, whereas the screen is of lesser importance. On the other hand, when conducting a search on the Internet, the greatest amount of time is spent looking at the screen; and often there is no hard copy. For this type of task, the screen should be emphasized, with some important consideration also given to the input devices.

Grandjean and his colleagues conducted a series of workstation design studies from the late 1970s through the mid-1980s (Grandjean et al., 1982a, 1982b; Grandjean, Hünting, & Nishiyama, 1984; Grandjean, Hünting, & Pidermann, 1983; Hünting, Grandjean, & Maeda, 1980a; Hünting, Läubli, & Grandjean, 1980b; Hünting, Läubli, & Grandjean, 1981; Läubli, Hünting, & Grandjean, 1981). In the field studies of Hünting et al. (1981) and Läubli et al. (1981), several significant relationships were discovered between the design of workstations, employee postures, and the incidence of health complaints and medical findings. These results can be summarized as follows: Physical discomfort and/or the number of medical findings in the neck-shoulder-arm-hand area were likely to increase when:

1. The keyboard level above the floor was too high.
2. Forearms and wrists could not rest on an adequate support.
3. The keyboard level above the desk was too high.
4. Operators had a marked head inclination.
5. Operators adopted a slanting position of the thighs under the table from insufficient space for their legs.

6. Operators engaged in marked sideward twisting of their hands while operating a keyboard.

Grandjean (1984) proposed the consideration of the following features of workstation design:

1. The furniture should be as flexible as possible with adjustment ranges to accommodate the anthropometric diversity of the users.
2. Controls for workstation adjustment should be easy to use.
3. There should be sufficient knee space for seated operators.
4. The chair should have an elongated backrest with an adjustable inclination and a lumbar support.
5. The keyboard should be moveable on the desk surface.

Cohen et al. (1995) used a case study method to determine the effects of task demands, customer needs, and organizational environment on the recommendations for ergonomic redesign in a large pension and insurance organization. The purposes of the study were to provide recommendations regarding future purchases of system furniture, to suggest retrofitting options for those offices already using newer workstations, and to complete a thorough job analysis to ensure that workstation recommendations were appropriate for job requirements. A specific issue was the improvement of workstation design to reduce employee musculoskeletal discomfort. At the time of the study, the organization was planning a major renovation of 40,000 workstations and chairs.

Cohen et al. (1995) identified the following health risk factors that led to awkward postures and undue loads on the musculoskeletal system:

1. Static postures of the trunk, neck and arms.
2. Awkward twisting and reaching motions.
3. Poor lighting and glare.
4. The placement of the keyboard on uneven working surfaces.
5. Insufficient work surface space.
6. Insufficient knee and toe space.
7. The inability for the chair armrests to fit under the working surfaces.

For some workstations, the computer and keyboard were placed on a swivel base that could be turned 180 degrees so that adjacent employees could use one computer. This caused extended arm reaching and back twisting when using the computer. Frequently, the base became stuck or was blocked by desk clutter. The swivel base was unstable and required continuous adjustment while typing to keep it aligned with the employee. If the keyboard was removed from the swivel to bring it closer to the user, there was no room for wrist placement in front of the keyboard.

The chairs had some common deficiencies, such as poor back and shoulder support, inadequate padding in the backrest and seat pan, arm rests that did not fit under working surfaces, and a lack of appropriate seat pan height adjustment. Recommendations were made for chairs with full back rests, better padding, removable arm rests, and seat pan height adjustment.

Derjani-Bayeh and Smith (1999) conducted a prospective intervention study to examine the benefits of ergonomic redesign for computer users at fixed workstations. The study took place in a consumer products call center where shoppers could order products from a catalog using a telephone or an Internet site. There were three ergonomic interventions studied. In the first condition, expert ergonomists provided modifications to current workstation configurations to maximize their fit with the incumbent employee. In the second condition, new workstation accessories (keyboard tray, monitor holder, document holder, wrist rest, foot rest, and/or task lighting) were added as needed to improve the employees' fit with their workstations and general environment. In the third condition, the same factors as the second condition were added, but in addition a new chair with multiple adjustments was also added.

There were 80 volunteer subjects who participated. They were drawn from a larger pool of volunteers. The participants for the third condition were randomly selected from the larger pool, and then subjects for conditions 1 and 2 were matched to these selections based on type of job, age, gender, and length of experience with the company. Baseline measurements of self-reported health status were collected using a questionnaire survey. Follow-up measurements were taken directly after implementation of the ergonomic improvements and then 12 months later. The results indicated that subjects working under conditions 2 and 3 showed reductions in the extent and intensity of musculoskeletal health complaints, but not the subjects in condition 1.

RECOMMENDATIONS FOR FIXED WORKSTATION DESIGN FEATURES

Based on the findings from field studies and standards dealing with computer workstation design (ANSI/HFS-100, 1988; CSA, 1989), the following general guidance is proposed for the design of fixed computer workstations.

The recommended size of the work surface is dependent on the task(s), the documents, and the technology. The primary working surface (e.g., those supporting the keyboard, the display, and documents) should be sufficient to:

1. Allow the screen to be moved forward or backward for a comfortable viewing distance.
2. Allow a detachable keyboard to be placed in several locations on the working surface.
3. Allow source documents to be positioned for easy viewing and proper musculoskeletal alignment.
4. Add working surfaces (e.g., secondary working surfaces) as required to store, lay out, read, and/or write on documents or materials.
5. Provide adequate knee and leg room for repositioning movements while working.
6. Use a table top that is as thin as possible for better thigh and knee clearance.
7. Establish a comfortable table height that provides the necessary thigh/knee clearance and that allows input devices (keyboard, mouse) to be at comfortable heights.

Sometimes computer workstations are configured so that multiple pieces of equipment and source materials can be equally accessible to the user. In this case, additional working surfaces are necessary to support these additional tools and should be arranged to allow for easy movement from one surface to another. Proper clearances under each working surface should be maintained, as well as a comfortable height.

It is important to provide unobstructed room under the working surface for the feet and legs so that operators can easily shift their posture. Kneespace height and width and toe depth are the three key factors for the design of clearance space under the working surfaces. The recommended minimum width for leg clearance is 51 cm, whereas the preferred minimum width is 61 cm (ANSI/HFS-100, 1988). The minimum depth under the work surface from the operator edge of the work surface should be 38 cm for clearance at the knee level and 59 cm at the toe level (ANSI/HFS-100, 1988). A good workstation design accounts for individual body sizes and often exceeds minimum clearances to allow for free postural movement.

Table height has been shown to be an important contributor to computer user musculoskeletal problems (Grandjean et al., 1983; Hünting et al., 1981). In particular, tables that are too high cause the keyboard and other input devices to be positioned too high. This puts undue pressure on the joints, hands, wrists, arms, shoulders, and/or neck. Normal desk height of 30 inches (76 cm) is often too high for keyboard and mouse use by most people. It is desirable for table heights to vary with the height of the user, particularly if the chair is not height adjustable. Height-adjustable working surfaces are effective for this. Adjustable multisurface tables encourage good posture by allowing the keyboard and screen to be independently adjusted to appropriate keying and viewing heights for each individual and each task. Tables that cannot be adjusted easily are a problem when used by multiple individuals of differing sizes, especially if the chair is not height adjustable. When adjustable tables are used, the ease of adjustment is essential. Adjustments should be easy to make and operators should be instructed how to adjust the workstation to be comfortable.

Specifications for seated working surfaces' heights vary with whether the table is adjustable or at one fixed height, and with a single working surface or multiple working surfaces. The proper height for a nonadjustable working surface is about 70 cm (27.5 inches) from the floor to top of the working surface. Adjustable tables allow vertical adjustments of the keyboard and display. Some allow for independent adjustment of the keyboard and display. For a single adjustable working surface, the working surface height adjustment range should be between 60–80 cm (23.5–31.5 inches). In the case of independently adjustable working surfaces for the keyboard and the screen, the appropriate height ranges are 59–71 cm (23–28 inches) for the keyboard surface (ANSI/HFS-100, 1988) and 90–115 cm (35.5–45 inches) for the screen surface (Grandjean, 1987).

THE DESIGN OF WORKSTATIONS FOR USE WITH PORTABLE TECHNOLOGY

Imagine yourself sitting at the airport and your flight has been delayed for 2 hours. Your have your laptop with you, so you decide to get some work done while you wait. You could rent a cubicle or kiosk at the airport that would provide you with a high-speed internet connection, a stationary telephone, a working surface (desk or table), a height adjustable chair, and some privacy (noise control, personal space). The characteristics of these work areas do not often conform to the best principles of ergonomic design. It is likely that the cubicle will provide some improvement over sitting with the laptop on your lap, but the characteristics may not meet the recommendations presented in this chapter. Such situations are acceptable for short exposures of up to 60 min, but longer exposures may lead to musculoskeletal discomfort, pain, and injury (if chronic). Now imagine that you have been told to stay in the boarding area because it is possible that the departure may be sooner than 2 hours. You get out your laptop, connect it to your cell phone, and place them on your lap. (That's why they are called laptops). You are sitting in a nonadjustable chair with poor back support.

This scenario is all too common. You can walk through O'Hare International Airport on any given day and see hundreds of people sitting at the boarding gate working on their laptop that is sitting on their laps. Now, imagine a palm-held device that allows you to access your e-mail or to connect to the Internet. This device can be operated while you are standing in line at the airport to check in or sitting at the boarding gate like the laptop users. You can stand or sit punching at miniature buttons (sometimes with a stylus because they are so small) and interact with the interconnected world. Again, this scene is all too familiar in almost any venue (airport, restaurant, street, office). So, what is wrong with all of this portable capability to connect and work and play? What is wrong is the revolution in IT devices for portable use has regressed ergonomic conditions for the users from the vast improvements made with fixed PCs over the last 2 decades to adverse conditions that would horrify any safety and health practitioner. This has happened with many types of technology in the name of portability, continuous connectivity, and productivity.

Let's start with the laptop as a prime example of the influence of portability and efficiency. Recently the Human–Computer Interaction Committee of the International Ergonomics Association (IEA) produced a guideline for the use of LPCs to improve ergonomic conditions (Saito et al., 2000). This was prompted by the ever-increasing sales of laptops and the replacement of fixed PCs by portable laptops.

The primary advantage of the laptop is easy portability so the user can take the computer anywhere to do work. S/he can use it at the office, then take it home to finish work, or take it with her/him to make a presentation at a meeting. The convenience and effectiveness of easy, lightweight portability are very high. In addition, all of the files are with the laptop, so nothing is mistakenly left behind at the office (or home). However, the comfort and health factors are often very low, because the person uses the laptop in all manner of environments, workstations, and tasks that diminish the consistent application of good ergonomic principles. An important feature of the IEA laptop guideline (Saito et al., 2000) is to encourage situations of use that mirror the best practices of ergonomic conditions for fixed PCs relating to technology, workstation, environment, and task design. The following material (*in italics*) has been extracted directly (with some minor editing) from the Saito et al. (2000) article entitled "Ergonomic Guidelines for Using Notebook

Personal Computers," which appeared in *Industrial Health, 38*, 421–434, 2000.

Work Environment and Workstation Layout: "Create an environment that fits your work."

1. *Use your laptop in a proper environment (lighting, temperature, noise, and so on). In particular, make sure the work area is neither too bright nor too dark.*
2. *Allocate enough space on your desk when placing a laptop.*

Chair and Desk: "Adjust chair height to match your physique."

1. *Adjust your chair height based on the height of the keyboard, such that your forearm is parallel to the surface of the keyboard.*
2. *If your feet do not lie flat on the floor, provide a footrest.*
3. *Provide enough space underneath the desk.*

Keyboard: "Set the keyboard to a desirable angle, and use a palm rest if necessary."

1. *Adjust the angle of the keyboard based on your posture and preferences.*
2. *Make sure there is space in front of the keyboard for you to comfortably rest your wrists (this space can be on the desktop surface itself if the keyboard is thin).*
3. *If the keyboard seems difficult to use, use an external keyboard.*

Working Posture: "Avoid unnatural postures, and change your posture occasionally."

1. *Avoid staying in postures where you are bent too far forward or backward, or twisted, for an extended duration.*
2. *Laptop users tend to view the display from too close, so make sure you maintain a distance of at least 40–50 cm between the display and your eyes.*
3. *Alternate near vision with far vision (i.e., observe object located at least 6 m far) as frequent as possible.*
4. *Make sure your wrists are not at an unnatural angle.*

Non-keyboard Input Devices: "Use a mouse as your pointing device if at all possible."

1. *If a mouse can be connected to your laptop, then do so as often as possible. Use a mouse pad whenever you use a mouse.*
2. *When you cannot connect a mouse, make sure you understand the built-in pointing device, and use the pointing device appropriately.*

These laptop guidelines provided by Saito et al. (2000) are useful when the laptop is used as a fixed PC at a docking station or at a desk (worktable). However, they do not provide as much help in the situations that were described earlier when the laptop is used at the airport, or when using a hand-held personal digital assistant (PDA). In situations where there is not a fixed workstation, the device is typically positioned wherever is convenient. Very often, such positioning creates bad postures for either the legs, back, shoulders, arms, wrists/hands, or neck. In addition, the smaller dimensions of the manual input devices (touch pad, buttons, keyboard, joy stick, roller ball) make motions much more difficult, and these are often posturally constrained. If the devices are used continuously for a prolonged period (such as 1 hour or more), muscle tension builds up and discomfort in joints, muscles, ligaments, tendons, and nerves can occur.

To reduce the undesirable effects of the poor workstation characteristics that lead to the discomfort, the following recommendations are given:

1. If you are using a laptop on your lap, find a work area where you can put the laptop on a table (rather than on your lap). Then arrange the work area as closely as possible with the recommendations presented in the IEA laptop guidelines (Saito et al., 2000).

2. If you are using a hand-held PDA, you should position yourself so that your back is supported. It is preferable to use the device sitting down. Of course, if you are using the PDA as you are walking, then this is not possible. If the PDA has a voice interface, then use an ear piece and a microphone so that you do not have to hold it with your hand.

3. Never work in poor postural conditions for more than 30 minutes continuously. Take at least a 5-min break (preferably 10 min) away from the laptop/PDA use, put the device down (away), get up and stretch for 1 min or more, and then walk for 2–3 minutes. If you are using a hand-held PDA in a standing position, then during your break, put it away. Do 1 min of stretching, and then sit down for 4 min. That may mean sitting on the floor, but preferably you will sit where you can support your back.

4. Buy equipment that provides the best possible input interfaces and displays (screens, headphones, typing pads). Because these devices are small, the perceptual motor requirements for their use are much more difficult (sensory requirements, motion patterns, skill requirements, postural demands). Therefore, screens should provide easily readable characters (large, understandable), and input buttons should be easy to operate (large, properly spaced, easily accessible).

5. Only use these devices when you do not have access to fixed workstations that have better ergonomic characteristics. Do not use these devices continuously for more than 30 min.

THE CHAIR AS AN ELEMENT OF THE WORKSTATION

It was not until the last 40 years that sitting posture and chairs (seats) became topics for scientific research, especially for ergonomics and orthopedics. Studies have revealed that the sitting position, as compared with the standing position, reduces static muscular efforts in the legs and hips, but increases the physical load on the intervertebral discs in the lumbar region of the spine.

The debate over what constitutes proper seated posture is not yet fully resolved. Is an upright seated posture most healthy, or is a relaxed posture with a backward-leaning trunk more healthy? Interesting experiments by the Swedish surgeons

Nachemson and Elfstrom (1970) and Andersson and Ortengreen (1974) offer some guidance about this. These authors measured the pressure inside the intervertebral discs, as well as the electrical activity of the back muscles in relation to different sitting postures. When the backrest angle of the seat was increased from 90 to 120 degrees, subjects exhibited an important decrease of the intervertebral disc pressure and the electromyographic activity of the back. Because heightened pressure inside intervertebral discs means that they have more stress, it was concluded that a sitting posture with reduced disc pressure is more healthy and desirable.

The results of the Swedish studies indicated that leaning the back against an inclined backrest transfers some of the weight of the upper part of the body to the backrest. This reduces considerably the physical load on the intervertebral discs and the static strain of the back and shoulder muscles. Thus, some computer users seem to instinctively get into the proper posture when they lean backward.

Most ergonomic standards for computer workstations are based on a more traditional view about a healthy sitting posture. Mandal (1982) reports that the "correct seated position" goes back to 1884 when the German surgeon Staffel recommended the well-known upright position. Mandal (1982, p. 520) stated:

But no normal person has ever been able to sit in this peculiar position (upright trunk, inward curve of the spine in the lumbar region and thighs in a right angle to the trunk) for more than 1–2 minutes, and one can hardly do any work as the axis of vision is horizontal. Staffel never gave any real explanation why this particular posture should be better than any other posture. Nevertheless, this posture has been accepted ever since quite uncritically by all experts all over the world as the only correct one.

It is indeed a fact that the sitting posture of students in the lecture hall or of any other audience is very seldom a correct upright position of the trunk. On the contrary, most people lean backward (even with unsuitable chairs) or in some cases lean forward with elbows resting on the desk. It is most probable that these two preferred trunk positions are associated with a substantial decrease of intervertebral disc pressure, as well as lessened tension of muscles and other tissues in the lumbar and thoracic spine (Andersson & Ortengreen, 1974; Nachemson & Elfstrom, 1970). Thus, sitting in general and particularly when using a computer is probably most comfortable and healthy when leaning back, or when the arms are supported when leaning forward.

Poorly designed chairs can contribute to computer user discomfort. Chair adjustability in terms of height, seat angle, and lumbar support helps to provide trunk, shoulder, neck, and leg postures that reduce strain on the muscles, tendons, and discs. The postural support and action of the chair help maintain proper seated posture and encourage good movement patterns. A chair that provides swivel action encourages movement, whereas backward tilting increases the number of postures that can be assumed.

The chair height should be adjustable so that the computer operator's feet can rest firmly on the floor with minimal pressure beneath the thighs. The minimum range of adjustment for seat pan height should be between 38–52 cm (15–20.5 inches) to accommodate a wide range of stature.

To enable short users to sit with their feet on the floor without compressing their thighs, it may be necessary to add a footrest. A well-designed footrest has the following features:

1. It is inclined upwards slightly (about 5–15 degrees).
2. It has a nonskid surface.
3. It is heavy enough that it does not slide easily across the floor.
4. It is large enough for the feet to be firmly planted.
5. It accommodates persons of different stature.

The seat pan is where the person sits on the chair. It is the part of the chair that directly supports the weight of the buttocks. The seat pan should be wide enough to permit operators to make slight shifts in posture from side to side. This not only helps to avoid static postures, but also accommodates a large range of individual buttock sizes. The seat pan should not be overly U-shaped because this can lead to static sitting postures. The minimum seat pan width should be 45 cm (18 inches), and the depth between 38–43 cm (15–17 inches) (ANSI/HFES-100, 1988). The front edge of the seat pan should be well-rounded downward to reduce pressure on the underside of the thighs that can affect blood flow to the legs and feet. This feature is often referred to as a "waterfall" design. The seat needs to be padded to the proper firmness that ensures an even distribution of pressure on the thighs and buttocks. A properly padded seat should compress about one-half to one-inch when a person sits on it.

Some experts feel that the seat front should be elevated slightly (up to 7 degrees), whereas others feel it should be lowered slightly (about 5 degrees). There is some disagreement among the experts about the correct answer. Because of this disagreement, many chairs allow for both front and backward angling of the front edge of the seat pan. The operator can then angle the chair's front edge to a comfortable position. The seat pan height and angle adjustments should be accessible and easy to use from a seated position.

The tension and tilt angle of the chair's backrest should be adjustable. Inclination of the chair backrest is important for operators to be able to lean forward or back in a comfortable manner while maintaining a correct relationship between the seat pan angle and the backrest inclination. A backrest inclination of about 110 degrees is considered an appropriate posture by many experts. However, studies have shown that operators may incline backward as much as 125 degrees, which also is an appropriate posture. Backrests that tilt to allow an inclination of up to 125 degrees are therefore a good idea. The backrest tilt adjustments should be accessible and easy to use. An advantage of having an independent tilt angle adjustment is that the backrest tilt will then have little or no effect on the front seat height or angle. This also allows operators to shift postures readily.

Chairs with high backrests are preferred because they provide support to both the lower back and the upper back (shoulder). This allows employees to lean backward or forward, adopting a relaxed posture and resting the back and shoulder muscles. A full backrest with a height around 45–51 cm (18–20 inches) is recommended. To prevent back strain, it is also recommended that chairs have lumbar (midback) support, because the lumbar region is one of the most highly strained parts of the spine when sitting.

For most computer workstations, chairs with rolling castors or wheels are desirable. They are easy to move and facilitate postural adjustment, particularly when the operator has to reach for equipment or materials that are on the secondary working surfaces. Chairs should have five supporting legs.

Another important chair feature is armrests. Both pros and cons to the use of armrests at computer workstations have been advanced. On the one hand, some chair armrests can present problems of restricted arm movement, interference with the operation of input devices, pinching of fingers between the armrest and table, restriction of chair movement (such as under the work table), irritation of the arm or elbows from tissue compression when resting on the armrest, and adoption of awkward postures. Properly designed armrests can overcome these problems. Armrests can provide support for resting the arms to prevent or reduce arm, shoulder, and neck fatigue. Removable armrests are an advantage, because they provide greater flexibility for individual operator preference. For specific tasks, such as using a numeric keypad, a full armrest can be beneficial in supporting the arms. Many chairs have height-adjustable armrests that are helpful for operator comfort, and some allow for adjusting the angle of the armrests as well.

ADDITIONAL WORKSTATION CONSIDERATIONS

Providing the capability for the screen to swivel and tilt up/down gives the user the ability to better position the screen for easier viewing. Reorientation of the screen around its vertical and horizontal axes can help to position a screen to reduce screen reflections and glare. Reflections can be reduced by simply tilting the display slightly back or down, or to the left or right away from the source of glare. The perception of screen reflections depends not only on screen tilt, but also on the operator's line of sight.

An important component of the workstation that can help reduce musculoskeletal loading is a document holder. When properly designed, proportioned, and placed, document holders reduce awkward inclinations of the head and neck and frequent movements of the head up and down and back and forth. They permit source documents to be placed in a central location at the same viewing distance as the computer screen. This eliminates needless head and neck movements and reduces eyestrain. In practice, some flexibility about the location, adjustment, and position of the document holder should be maintained to accommodate both task requirements and operator preferences. Dainoff (1982) showed the effectiveness of an inline document holder. The document holder should have a matte finish so that it does not reflect light.

Privacy requirements include both visual and acoustical control of the workplace. Visual control prevents physical intrusions, contributes to confidential/private conversations, and prevents the individual from feeling constantly watched. Acoustical control prevents distracting and unwanted noise (from machine or conversation) and permits speech privacy. Although certain acoustical methods and materials, such as free-standing panels, are used to control general office noise level, they can also be used for privacy. Planning for privacy should not be made at the expense of visual interest or spatial clarity. For instance, providing wide visual views can prevent the individual from feeling isolated. Thus, a balance between privacy and openness enhances user comfort, work effectiveness, and office communications. Involving the employee in decisions of privacy can help in deciding the compromises between privacy and openness.

The use of a wrist rest when keying can help to minimize extension (backward bending) of the hand/wrist, but the use of a wrist rest for operator comfort and health has generated some debate because there are trade-offs for comfort and health. When the hand or wrist is resting on the wrist rest, there is compression of the tissue that may create increased carpal canal pressure or local tissue ischemia. On the other hand, the wrist rest allows the hands and shoulders to be supported with less muscular tension that is beneficial to computer operator comfort. At this time, there is no scientific evidence that the use of a wrist rest either causes or prevents serious musculoskeletal disorders of the hands, wrists, or shoulders. Thus, the choice to use a wrist rest should be based on employee comfort and performance considerations until scientific evidence suggests otherwise.

If used, the wrist rest should have a fairly broad surface (5 cm minimum) with a rounded front edge to prevent cutting pressure on the wrist and hand. Padding further minimizes skin compression and irritation. Height adjustability is important so that the wrist rest can be set to a preferred level in concert with the keyboard height and slope.

Arm holders are also available to provide support for the hands, wrists, and arms while keyboarding and have shown to be useful for shoulder comfort. The placement of the arm holder should not induce awkward postures in its use. The device should be placed within easy reach of the operator, especially when it will be used frequently during work.

When keyboard trays are used, they should allow for the placement of other input devices directly on the tray instead of on other working surfaces.

THE VISUAL ENVIRONMENT

Visual displays tend to be of two types: the cathode ray tube (CRT) and the diode matrix (LED/LCD) flat panels. Fixed workstations with personal computers tend to use CRTs because they are cheap, whereas LPCs and hand-held devices are almost exclusively flat panel because of size and weight limitations. Both types of displays have characteristics that lead to problems from environmental influences. For instance, luminance sources in the environment that fall on the screen wash out characters on the screen, and the accumulation of dust particles on the screen may distort images. These conditions affect the ability to read the screen and can lead to visual fatigue and dysfunction. Specific characteristics of the environment, such as illumination and glare, have been related to computer operator vision strain problems. The main visual functions involved in computer work are accommodation, convergence, and adaptation.

The alignment of lighting in relation to the computer workstation, as well as levels of illumination in the area surrounding a computer workstation, have been shown to influence the ability of the computer operator to read hard copy and the computer screen (Cakir et al., 1979; Dainoff, 1983; Grandjean, 1987; Stammerjohn et al., 1981). Readability is also affected by the differences in luminance contrast in the work area. The level of illumination affects the extent of reflections from working surfaces and from the screen surface. Mismatches in these characteristics and the nature of the job tasks have been postulated to cause the visual system to overwork and lead to visual fatigue and discomfort (Cakir et al., 1979; NAS, 1983).

Generally, it has been shown that excessive illumination leads to increased screen and environmental glare, and poorer luminance contrast (Ghiringhelli, 1980; Gunnarsson & Ostberg, 1977; Läubli et al., 1981). Several studies have shown that screen and/or working surface glare are problematic for visual disturbances (Cakir et al., 1979; Gunnarsson & Ostberg, 1977; Läubli et al., 1981; Stammerjohn et al., 1981). Research by van der Heiden et al. (1984) has shown that computer users spend a considerable amount of their viewing time looking at objects other than the screen. Bright luminance sources in the environment can produce reflections and/or excessive luminance contrasts that may create excessive pupillary response that leads to visual fatigue.

All surfaces within the visual field of an operator should be of a similar order of brightness as much as practical to achieve uniformity. The temporal uniformity of the surface luminance is as important as the static spatial uniformity. Rhythmically fluctuating surface luminances in the visual field are distracting and reduce visual performance. Such unfavorable conditions prevail if the work requires the operator to alternately glance at a bright and then a dark surface, or if the light source generates an oscillating light. Sometimes fluorescent lights can have a noticeable stroboscopic effect on moving reflective objects, and sometimes when reflected from a computer CRT screen. When fluorescent tubes wear out or are defective, they develop a slow, easily perceptible flicker, especially at the visual periphery. Flickering light is extremely annoying and causes visual discomfort.

Lighting is an important aspect of the visual environment that influences computer screen and hard copy readability, glare on the screen, and viewing in the general environment. The intensity of illumination or the illuminance being measured is the amount of light falling on a surface. In practice, this level depends on both, the direction of flow of the light and on the spatial position of the surface being illuminated in relation to the light flow. Illuminance is measured in both the horizontal and vertical planes. At computer workplaces, both the horizontal and vertical illuminances are important. A document lying on a desk is illuminated by the horizontal illuminance, whereas the computer screen is illuminated by the vertical illuminance. In an office that is illuminated from overhead luminaries, the ratio between the horizontal and vertical illuminances is usually between 0.3 and 0.5. So, if the illuminance in a room is said to be 500 lux, this implies that the horizontal illuminance is 500 lux, whereas the vertical illuminance is between 150 and 250 lux (0.3–0.5 of the horizontal illuminance).

The illumination required for a particular task is determined by the visual requirements of the task and the visual ability of the employees concerned. The illuminance in workplaces that use computer screens should not be as high as in workplaces that exclusively use hard copy. Lower levels of illumination will provide better computer screen image quality and reduced screen glare. Illuminance in the range of 300–700 lux measured on the horizontal working surface (not the computer screen) is normally preferable. The lighting level should be set up according to the visual demands of the tasks performed. For instance, higher illumination levels are necessary to read hard copy and lower illumination levels are better for work that just uses the computer screen. Thus, a job in which a hard copy and a computer screen are both used should have a general work area illumination level of about 500–700 lux; and a job that only requires reading the computer screen would have a general work area illumination of 300–500 lux.

Conflicts can arise when both hard copy and computer screens are used by different employees who having differing job task requirements or differing visual capabilities and are working in the same room. As a compromise, room lighting can be set at the lower level (300 lux) or intermediate level (500 lux) and additional task lighting for hard copy tasks can be provided at each workstation as needed. Such additional lighting must be carefully shielded and properly placed to avoid glare and reflections on the computer screens and adjacent working surfaces of other employees. Furthermore, task lighting should not be too bright in comparison with the general work area lighting, because the contrast between these two different light levels may produce eyestrain.

The surface of the computer screen reflects light and images. The luminance of the reflections decreases character contrast and disturbs legibility; it can be so strong that it produces a glare. Image reflections are annoying, especially because they also interfere with focusing mechanisms; the eye is induced into focusing between the text and the reflected image. Thus, reflections are also a source of distraction. Stammerjohn et al. (1981), as well as Elias and Cail (1983), observed that bright reflections on the screen are often the principal complaint of operators.

Luminance is a measure of the brightness of a surface, the amount of light leaving the surface of an object, either reflected by the surface (as from a wall or ceiling), emitted by the surface (as from the CRT characters), or transmitted (as light from the sun that passes through translucent curtains). High-intensity luminance sources (such as windows) in the peripheral field of view should be avoided. In addition, a balance among luminance levels within the computer user's field of view should be maintained. To reduce environmental glare, the luminance ratio within the user's near field of vision should be approximately 1:3 and approximately 1:10 within the far field of vision. For luminance on the screen itself, the character-to-screen background luminance contrast ratio should be at least 7:1. To give the best readability for each operator, it is important to provide screens with adjustments for character contrast and brightness. These adjustments should have controls that are obvious and easily accessible from the normal working position (e.g., located at the front of the screen).

Experts have traditionally recommended a viewing distance between the screen and the operator's eye of 45–50 cm, but no more than 70 cm. However, experience in field studies has shown that users may adopt a viewing distance greater than 70 cm and still be able to work efficiently and comfortably. Thus, viewing distance should be determined in context with other considerations. It will vary depending on the task requirements, computer screen characteristics, and individual visual capabilities. For instance, with poor screen or hard copy quality, it may be necessary to reduce viewing distance for easier character recognition. Typically, the viewing distance should be 50 cm or less because of the small size of the characters on the computer or IT device screen.

ERGONOMIC IMPLEMENTATION ISSUES

Implementing a workplace change, such as improving workstation design or work methods is a complex process because it impacts many elements of the work system (Derjani-Bayeh & Smith, 1999; Hagberg et al., 1995; Smith & Carayon, 1995; Smith & Sainfort, 1989). Managers, designers, and engineers often like to believe that technological enhancements are easy to make, and that performance and health improvements will be immediate and substantial. Proper implementation involves changes in more than the workstation; for instance, there needs to be consideration of the work organization, job content, task improvements, job demands, and socialization issues. Planning for change can help the success of implementation and reduce the stress generated by the change. But the success of implementing change depends heavily on the involvement and commitment of the concerned parties, in particular management, technical staff and support staff, first line supervision, and the employees.

There is universal agreement among change management experts that the most successful strategies for workplace improvements involve all elements (subsystems) of the work system that will be affected by the change (Hendrick, 1986; Lawler, 1986; Smith & Carayon, 1995). Involvement assumes that there is an active role in the change process, not just providing strategic information. Active participation generates greater motivation and better acceptance of solutions than passively providing information and taking orders. Active participation is achieved by soliciting opinions and sharing authority to make decisions about solutions. However, one drawback of active participation is the need to develop consensus among participants who have differing opinions and motives. This usually takes more time than traditional decision making and can bring about conflict among subsystems. Another drawback is that line employees often do not always have the technical expertise necessary to form effective solutions.

Participative ergonomics can take various forms, such as design decision groups, quality circles, and worker-management committees. Some of the common characteristics of these various programs are: Employee involvement in developing and implementing ergonomic solutions, dissemination and exchange of information, pushing ergonomics expertise down to lower levels, and cooperation between experts and nonexperts. One of the characteristics of participatory ergonomics is the dissemination of information (Noro, 1991). Participative ergonomics can be beneficial to reduce or prevent resistance to change because of the information provided to the various members of the organization concerned with the new technology. Uncertainty and lack of information are two major causes of resistance to change and have been linked to increased employee stress. If employees are informed about potential ergonomics changes in advance, they are less likely to actively resist the change.

Training computer users about how the new workstation functions and operations is important especially if the adjustment controls are neither obvious nor intuitive. Hagberg et al. (1995) have indicated that employee training is a necessary component to any ergonomic program for reducing work-related musculoskeletal disorders. Green and Briggs (1989) found that adjustable workstations are not always effective without appropriate information about benefits of adjustments and training in how to use the equipment. Hagberg et al. (1995) suggested the following considerations for ergonomics training programs:

1. Have employees involved in the development and process of training. Using employee work experiences can be helpful in illustrating principles to be learned during training. In addition, using employees as instructors can be motivational for the instructors and learners.

2. Use active learning processes in which learners participate in the process and apply hands-on methods of knowledge and skill acquisition. This approach to learning enhances acquisition of inputs and motivation to participate.

3. Apply technology to illustrate principles such as audiovisual equipment and computers. Much like active processes, technology provides opportunities for learners to visualize the course materials and test their knowledge dynamically and immediately.

4. Use of on-the-job training is preferred over classroom training. Both can be effective when used in combination.

SUMMARY OF RECOMMENDATIONS

A computer workstation (work area) is comprised of the computer, input and output interfaces, the furniture where the computer is used, and the physical environment in which the computer is used. The design of these elements and how they fit together play a crucial role in user performance and in minimizing potential adverse discomfort and health consequences. The recommendations presented in this chapter address the physical environment and implementation issues. Important consideration should also be given to organizational factors and task-related factors as they affect and/or depend on the individual. Unique situational factors need to be considered when an ergonomic intervention is being implemented. Generalizing these recommendations to all situations in which computers (or IT devices) are used is a mistake. The approach presented in this chapter emphasizes the adaptation of the work area to a user's needs so that sensory and musculoskeletal loads are minimized.

Positioning of computers in the workplace is important to provide a more productive work environment. Computer workstations should be placed at right angles to the windows, and windows should not be behind or in front of the operator. This will reduce the possibility of reflections on the screen, which can otherwise reduce legibility. In addition, bright reflections coming from light sources can be reduced by placing these light sources on either side of and parallel to the line of vision of the operator. Moreover, illumination should be adapted to the quality of the source documents and the task required. This is done by having a high enough illumination to enhance legibility, yet low enough to avoid excessive luminance contrasts. Recommended levels range between 500–700 lux.

The keyboard and other input devices should be movable on the work surface, but stable when in use. A wrist rest or support surface of 15 cm in depth is useful to rest the wrist and forearms. Computer workstations should allow for adjustments that promote good posture. A computer workstation without adjustable keyboard (input device) height and without adjustable height and distance of the screen is not reasonable for continuous work of more than 30 min. The controls for adjusting the dimensions of a workstation and chair should be easy to use. Such adjustability is particularly important at workstations and chairs used by more than one employee. Furthermore, sufficient space for the user's thighs, knees, and legs should be provided to allow for comfort and to avoid unnatural or constrained postures.

With regard to chair design, a backward-leaning posture allows for relaxation of the back muscles and decreases the load on the intervertebral discs. Chair seat pan height should be easily adjustable and fit a wide range of statures. The chair should have a full backrest that can incline backward of up to 125 degrees. In addition, the backrest should have a lumbar support and a slightly concave form at the thoracic level.

These recommendations are not exhaustively inclusive of all aspects of computer workstation design. The chapter described other important workstation criteria in various sections.

References

Andersson, B. J. G., & Ortengreen, R. (1974). Lumbar disc pressure and myoeletric back muscle activity. *Scandinavian Journal of Rehabilitation Medicine, 3*, 115–121.

ANSI. (1988). *American national standard for human factors engineering of visual display terminal workstations* (ANSI/HFS Standard No. 100-1988). Santa Monica, CA: The Human Factors Society.

Bergqvist, U., Knave, B., Voss, M., & Wibom, R. (1992). A longitudinal study of VDT work and health. *International Journal of Human-Computer Interaction, 4*(2), 197–219.

Bergqvist, U. O. (1984). Video display terminals and health: A technical and medical appraisal of the state of the art. *Scandinavian Journal of Work, Environment and Health, 10*(Suppl. 2), 87.

Berlinguet, L., & Berthelette, D. (Eds.). (1990). *Work with display units 89.* Amsterdam: Elsevier Science Publishers.

Bullinger, H. J. (1991). *Human aspects in computing: Design and use of interactive systems and information management* (Vol. 18B). Amsterdam: Elsevier Science Publishers.

Bullinger, H.-J. (1991). *Human aspects in computing: Design and use of interactive systems and work with terminals* (Vol. 18A). Amsterdam: Elsevier Science Publishers.

Bullinger, H.-J., & Ziegler, J. (1999). *Human-computer interaction: Ergonomics and user interfaces* (Vol. 1). Mahwah, NJ: Lawrence Erlbaum Associates.

Cakir, A., Hart, D. J., & Stewart, T. F. M. (1979). *The VDT manual.* Darmstadt, Germany: Inca-Fiej Research Association.

Carayon, P. (1994). A longitudinal study of quality of working life among computer users: Preliminary results. In A. Grieco, G. Molteni, E. Occhipinti, & B. Piccoli (Eds.), *Proceedings of the Fourth International Scientific Conference on Work with Display Units* (pp. 39–44). Milan, Italy: Elsevier Science Publishers.

Carayon, P., Swanson, N., & Smith, M. J. (1987). Objective and subjective ergonomic evaluations of automated offices. In J. M. Flach (Ed.), *Proceedings of the Fourth Midcentral Ergonomic/Human Factors Conference* (pp. 358–366). Urbana, IL: University of Illinois.

Cohen, B. G. F., Piotrkowski, C. S., & Coray, K. E. (1987). Working conditions and health complaints of women office workers. In G. Salvendy, S. L. Sauter, & J. J. Hurrel (Eds.), *Social, ergonomic and stress aspects of work with computers* (Vol. 10A, pp. 365–372). Amsterdam: Elsevier Science Publishers.

Cohen, W. J., James, C. A., Taveira, A. D., Karsh, B., Scholz, J., & Smith, M. J. (1995). Analysis and design recommendations for workstations: A case study in an insurance company. In *Proceedings of the Human Factors and Ergonomics Society 39th Annual Meeting* (Vol. 1, pp. 412–416). San Diego, CA: Human Factors and Ergonomics Society.

Conway, F. T., Smith, M. J., Cahill, J., & Legrande, D. (1996). Psychological Mood State of Tension and Musculoskeletal Pain. In O. Brown (Ed.), *Human Factors in Organizational Design and Management* (pp. 303–308). Amsterdam: Elsevier.

CSA. (1989). *A guideline on office ergonomics: A National Standard of Canada* (Document No. CAN/CSA-Z412-m89). Rexdale, Ontario: Canadian Standards Association.

Dainoff, M. J. (1982). Occupational stress factors in visual display terminal (VDT) operation: A review of empirical research. *Behaviour and Information Technology, 1*(2), 141–176.

Dainoff, M. J. (1983, December). Video display terminals: The relationship between ergonomic design, health complaints and operator performance. *Occupational Health Nursing,* 29–33.

Derjani-Bayeh, & Smith, M. J. (1999). Effect of physical ergonomics on VDT worker's health: A longitudinal intervention field study in a service organization. *Inter. J. of Human-Computer Interaction, 11*(2), 109–135.

Elias, R., & Cail, F. (1983). *Constraints et astreints devant les terminaux a ecran cathodique* (p. 1109). Paris: Institut National de Recherche et de Securite.

Ghiringhelli, L. (1980). Collection of subjective opinions on use of VDUs. In E. Grandjean & E. Vigliani (Eds.), *Ergonomic aspects of visual display terminals* (pp. 227–232). London, England: Taylor & Francis, Ltd.

Grandjean, E. (1979). *Ergonomical and medical aspects of cathode ray tube displays.* Zurich, Switzerland: Federal Institute of Technology.

Grandjean, E. (1984). Postural problems at office machine work stations. In E. Grandjean (Ed.), *Ergonomics and health in modern offices* (pp. 445–455). London, England: Taylor & Francis, Ltd.

Grandjean, E. (1987). Design of VDT workstations. In G. Salvendy (Ed.), *Handbook of human factors* (pp. 1359–1397). New York: John Wiley and Sons.

Grandjean, E., Hünting, W., & Nishiyama, K. (1982a). Preferred VDT workstation settings, body posture and physical impairments. *Journal of Human Ergology, 11*(1), 45–53.

Grandjean, E., Hünting, W., & Nishiyama, K. (1984). Preferred VDT workstation settings, body posture and physical impairments. *Applied Ergonomics, 15*(2), 99–104.

Grandjean, E., Hünting, W., & Pidermann, M. (1983). VDT workstation design: Preferred settings and their effects. *Human Factors, 25*(2), 161–175.

Grandjean, E., Nishiyama, K., Hünting, W., & Pidermann, M. (1982b). A laboratory study on preferred and imposed settings of a VDT workstation. *Behaviour and Information Technology, 1,* 289–304.

Grandjean, E., & Vigliani, E. (Eds.). (1980). *Ergonomic aspects of visual display terminals.* London, England: Taylor & Francis, Ltd.

Green, R. A., & Briggs, C. A. (1989). Effect of overuse injury and the importance of training on the use of adjustable workstations by keyboard operators. *Journal of Occupational Medicine, 31,* 557–562.

Grieco, A., Molteni, G., Occhipinti, E., & Piccoli, B. (1995). *Work with display units 94.* Amsterdam: Elsevier Science Publishers.

Gunnarsson, E., & Ostberg, O. (1977). *Physical and emotional job environment in a terminal-based data system* (p. 35). Stockholm: Department of Occupational Safety, Occupational Medical Division, Section for Physical Occupational Hygiene.

Hagberg, M., Silverstein, B., Wells, R., Smith, M. J., Hendrick, H., Carayon, P., & Peruse, M. (1995). *Work related musculoskeletal disorders (WRMSDs): A reference book for prevention.* London, England: Taylor & Francis, Ltd.

Hales, T. R., Sauter, S. L., Peterson, M. R., & Fine, L. J. (1994). Musculoskeletal disorders among visual display terminal users in a telecommunications company. Special Issue: Telecommunications. *Ergonomics, 37*(10), 1603–1621.

Hendrick, H. (1986). Macroergonomics: A conceptual model for integrating human factors with organizational design. In O. Brown & H. Hendrick (Eds.), *Human factors in organizational design and management* (pp. 467–477). Amsterdam: Elsevier Science Publishers.

Hultgren, G., & Knave, B. (1973). Contrast blinding and reflection disturbances in the office environment with display terminals. *Arbete Och Halsa.*

Hünting, W., Grandjean, E., & Maeda, K. (1980a). Constrained postures in accounting machine operators. *Applied Ergonomics, 11*(3), 145–149.

Hünting, W., Läubli, T., & Grandjean, E. (1980b). Constrained postures of VDU operators. In E. Grandjean & E. Vigliani (Eds.) *Ergonomic Aspects of Visual Display Terminals* (pp. 175–184). London: Taylor & Francis, Ltd.

Hünting, W., Läubli, T., & Grandjean, E. (1981). Postural and visual loads at VDT workplaces: I. Constrained postures. *Ergonomics, 24*(12), 917–931.

Knave, B., & Wideback, P. G. (1987). *Work with display units.* Amsterdam: Elsevier Science Publishers.

Knave, B. G., Wibom, R. I., Voss, M., Hedstrom, L. D., & Bergqvist, O. V. (1985). Work with video display terminals among office employees: I. Subjective symptoms and discomfort. *Scandinavian Journal of Work, Environment and Health, 11*(6), 457–466.

Läubli, T., Hünting, W., & Grandjean, E. (1981). Postural and visual loads at VDT workplaces: II. Lighting conditions and visual impairments. *Ergonomics, 24*(12), 933–944.

Lawler, E. E. (1986). *High-involvement management.* San Francisco, CA: Jossey-Bass Publishers.

Lim, S. Y., & Carayon, P. (1993). An integrated approach to cumulative trauma disorders in computerized offices: The role of psychosocial factors, psychological stress and ergonomic risk factors. In M. J. Smith & G. Salvendy (Eds.), *Proceedings of the Fifth International Conference on Human-Computer Interaction* (Vol. 19A, pp. 880–885). Orlando, FL: Elsevier Science Publishers.

Lim, S. Y., & Carayon, P. (1995). Psychosocial and work stress perspectives on musculoskeletal discomfort. In *Proceedings of PREMUS 95*. Montreal, Canada.

Lim, S. Y., Rogers, K. J. S., Smith, M. J., & Sainfort, P. C. (1989). A study of the direct and indirect effects of office ergonomics on psychological stress outcomes. In M. J. Smith & G. Salvendy (Eds.), *Work with computers: Organizational, management, stress and health aspects* (pp. 248–255). Amsterdam: Elsevier Science Publishers.

Luczak, H., Cakir, A., & Cakir, G. (1993). *Work with display units 92*. Amsterdam: Elsevier Science Publishers.

Mandal, A. C. (1982). The seated man: Theories and realities. In *Proceedings of Human Factors Society* (pp. 520–524). Santa Monica, California: Human Factors and Ergonomics Society.

Nachemson, A., & Elfstrom, G. (1970). Intravital dynamic pressure measurements in lumbar discs. *Scandinavian Journal of Rehabilitation Medicine* (Suppl. 1), pp. 1–40.

National Academy of Sciences (NAS). (1983). *Video terminals, work and vision*. Washington, DC: National Academy Press.

Noro, K. (1991). Concepts, methods and people. In K. Noro & A. Imada (Eds.), *Participatory ergonomics* (pp. 3–29). London, England: Taylor & Francis, Ltd.

Ostberg, O. (1975, November/December). Health problems for operators working with CRT displays. *International Journal of Occupational Health and Safety*, 24–52.

Piotrkowski, C. S., Cohen, B. G., & Coray, K. E. (1992). Working conditions and well-being among women office workers. Special Issue: Occupational stress in human-computer interaction: II. *International Journal of Human Computer Interaction, 4*(3), 263–281.

Saito, S., Piccoli, B., Smith, M. J., Sotoyama, M., Sweitzer, G., Villanueva, M. B. G., Yoshitake, R. (2000). Ergonomic guidelines for using notebook personal computers. *Industrial Health, 38*, 421–434.

Sauter, S. L., Schleifer, L. M., & Knutson, S. J. (1991). Work posture, workstation design, and musculoskeletal discomfort in a VDT data entry task. *Human Factors, 33*(2), 151–167.

Smith, M. J. (1984). Health issues in VDT work. In J. Bennet, D. Case, J. Sandlin, & M. J. Smith (Eds.), *Visual display terminals* (pp. 193–228). Englewood Cliffs, NJ: Prentice Hall.

Smith, M. J. (1987). Mental and physical strain at VDT workstations. *Behaviour and Information Technology, 6*(3), 243–255.

Smith, M. J., Carayon, P. C. (1995). New technology, automation and work organization: Stress problems and improved technology implementation strategies. *International Journal of Human Factors in Manufacturing, 5*, 99–116.

Smith, M. J., & Carayon, P., Eberts, R., & Salvendy, G. (1992). Human-computer interaction. In G. Salvendy (Ed.), *Handbook of industrial engineering* (pp. 1107–1144). New York: John Wiley and Sons, Inc.

Smith, M. J., & Cohen, W. J. (1997). Design of computer terminal workstations. In G. Salvendy (Ed.), *Handbook of human factors and ergonomics* (2nd ed., pp. 1637–1638). New York: John Wiley & Sons, Inc.

Smith, M. J., Cohen, B. G., Stammerjohn, L. W., & Happ, A. (1981). An investigation of health complaints and job stress in video display operations. *Human Factors, 23*(4), 387–400.

Smith, M. J., & Sainfort, P. C. (1989). A balance theory of job design for stress reduction. *International Journal of Industrial Ergonomics, 4*, 67–79.

Smith, M. J., & Salvendy, G. (1989a). *Designing and using human-computer interfaces and knowledge based systems* (Vol. 12B). Amsterdam: Elsevier Science Publishers.

Smith, M. J., & Salvendy, G. (1989b). *Work with computers: Organizational, management, stress and health aspects* (Vol. 12A). Amsterdam: Elsevier Science Publishers.

Smith, M. J., & Salvendy, G. (1993a). *Human computer interaction: Applications and case studies* (Vol. 19A). Amsterdam: Elsevier Science Publishers.

Smith, M. J., & Salvendy, G. (1993b). *Human-computer interaction: Software and hardware interfaces* (Vol. 19B). Amsterdam: Elsevier Science Publishers.

Smith, M. J., & Salvendy, G. (2001). *Systems, social and internationalization design aspects of human-computer interaction*. Mahwah, NJ: Lawrence Erlbaum Associates.

Smith, M. J., Salvendy, G., Harris, D., & Koubek, R. J. (2001). *Usability evaluation and interface design*. Mahwah, NJ: Lawrence Erlbaum Associates.

Smith, M. J., Stammerjohn, L., Cohen, B., & Lalich, N. (1980). Video display operator stress. In E. Grandjean & E. Vigliani (Eds.), *Ergonomic aspects of visual display terminals* (pp. 201–210). London, England: Taylor & Francis, Ltd.

Stammerjohn, L. W., Smith, M. J., & Cohen, B. G. F. (1981). Evaluation of work station design factors in VDT operations. *Human Factors, 23*(4), 401–412.

van der Heiden, G. H., Braeuninger, U., & Grandjean, E. (1984). Ergonomic studies on computer aided design. In E. Grandjean (Ed.), *Ergonomic and health aspects in modern offices*. London, England: Taylor & Francis, Ltd.

·IIIB·

INTERACTION FOR DIVERSE USERS

Gregg Vanderheiden
University of Wisconsin–Madison

DIMENSIONS OF DIVERSITY

The phrase *design for diversity* usually brings to mind special or targeted populations (disabilities, gender, language, culture, etc.). It is, however, important to remember that almost all users find themselves in diverse environments and situations that can affect design requirements. Within a single day, as a person drives, walks, shops, and recreates, at home, in the office, and in other venues, the differing demands and constraints of the different environments or activities will require different interfaces, often on the same device. In addition, people may switch preferences even within a single environment and activity. For example, individuals sitting at their workstation may want to stand up and walk around the room while continuing to compose. People using the keyboard to carry out one task may prefer to use speech or pointing for another, or the user may simply want to take a break to spread the demands on their body.

Thus, diversity applies not only to specific users but to environments and other situational constraints that any user might encounter. The interesting thing to note is the number of times different user needs parallel different situational needs (common to all users), and the commonality of solutions for both. This section is about diversity among users, environments and tasks, and the resulting diversity in interface needs.

ACCOMMODATING DIVERSITY WITHIN A PRODUCT

In the past, technology limited our ability to accommodate individuals with such a wide variety of interface requirements. Interfaces were implemented in hardware, and adjustability was mechanical and difficult or expensive. Today, it is hard to find any product where the interface is not controlled by a microprocessor and software instructions. As a result, the products can exhibit very different behaviors at different times in order to fit the needs or whim of the user. With software, even the appearance, language, and language level used on the buttons can change from one user to the next.

This introduction explores a number of concepts relating to the development of products with alternate or flexible interfaces. It begins with a discussion about accommodating diversity from the point of view of people who have disabilities. This point of departure was chosen because many of the concepts are most easily introduced within this context. The discussion then shifts to a discussion of evolving interfaces as they apply to all users (with and without disabilities), including people of different age, gender, cultures, as well as users in different environments or pursuing different activities, or simply users in different moods. Finally, the implications of these approaches to human augmentation and operation of devices by machine intelligence is introduced.

IF A PERSON CANNOT OPERATE THE WORLD AS IT IS CURRENTLY DESIGNED . . .

If someone is unable to use the environments and devices he or she encounters in daily life effectively, there are three things that can be done:

1. Change the individual.
2. Change the world.
3. Create a bridge between the two.

For individuals who have disabilities, Approach 1 includes such things as surgery, rehabilitation, training, and personal assistive technologies. Glasses, hearing aids, wheelchairs, prosthetics, and so forth are examples of personal assistive technologies. In the future, individuals may also carry specialized interface technologies or devices that can act as "universal remote consoles" that can be used to control other devices as they are encountered in daily life (see below).

Approach 2 (changing the world) is commonly called "universal design," "accessible design," or "barrier-free design" in the United States. In other countries, this general concept goes by the terms "design for all" or "inclusive design." Approach 2 involves designing standard environments and products so that they are usable by people with as wide a range of abilities (or constraints) as is commercially possible. The objective is to make the world directly usable by as many people (both with and without disabilities) as is possible.

Approach 3 (Bridge technologies) refers to technologies used specifically to adapt a particular device or environment to a particular individual. These bridges can take the form of adaptive assistive technologies such as screen readers or custom modifications to environments or products. Such technologies generally do not move around with the individual and are not seen as an extension of the individual.

Two Types of Assistive Technology

This model divides assistive technology (AT) into two types: personal assistive technology (P-AT) and adaptive assistive technology (A-AT). The term personal assistive technology is used here to refer to technologies that are an extension of the person, rather than an adaptation to a particular device or environment. P-AT is with the person whenever he or she encounters devices or environments where it is needed and is generally used for a variety of tasks.

Adaptive assistive technology (A-AT) is associated with a particular location, device, or function. If P-AT is an extension of the individual, A-AT is a bridge to or adaptation of a device or environment. This distinction becomes important when considering the accessibility or usability of a product. In general, a building is considered accessible to people in wheelchairs if they can use it with their wheelchair (we do not require that the building be usable if the wheelchair is left at the door) because they will always have the AT (wheelchair) with them when they "encounter" the building. A card catalog system in a library that is screen-reader compatible but has no screen reader installed would not be considered accessible to users who are blind because they would not have a screen reader with them when they encounter the computer and would not be allowed to load foreign software into the system if they did.

This suggests a definition of *accessibility* that includes compatibility with AT, but only when AT will be present and usable when it is needed.

Accessible: Able to be used effectively by Individuals either directly or with the assistive technologies that they will have with them and can use when they encounter the environment, device, or system.

This essentially means "usable by individuals including their personal assistive technologies."

Network-Based Personal Assistive Technology

Interestingly, it may be possible in the future for individuals to have personal assistive technologies which they do not actually carry on their person. We are rapidly approaching a time when it is practical to have AT that resides "in the ether," and that could be pulled down and be used whenever and wherever an individual needed them. This is already true for some Internet-based interactions. People with disabilities can use the Internet to invoke "virtual" assistive technologies from any location on the Internet. This "virtual AT" capability is incomplete today, however, because when using the Internet from a library or other public place, users generally must first be able to use the (Library's) terminal before they can invoke or call up Internet-based AT services. There are also some locations where full Internet access may not be available and the AT service cannot be reached, or where rerouting of Internet content through the virtual AT or downloading of the virtual AT may not be permitted. For this "virtual personal assistive technology" model to be fully effective, a means to automatically fetch and install the virtual AT software must be provided as the person approaches the device, and the devices need to be designed to allow users to substitute their alternate interface or interface modification in place of or as a part of the standard interface on the product. Such virtual AT would be considered P-AT because users would have it with them or available to them whenever they encountered the products with which they needed to use it.

Human Augmentation As A Personal Assistive Technology

Another type of personal assistive technology that is evolving is personal augmentation. Hearing aids are probably the earliest type of electronic augmentation in that they allow individuals to hear better than they would without such devices. We also have prosthetic limbs and cochlear implants and are at the early stages of developing artificial vision.

At present, these human augmentations fall into the category of artificial partial restorations of ability. The term *partial* is used because they do not restore the full function of the healthy human organ. The term *artificial* is used because the restored functionality is not necessarily the same exact type of functionality. For example, a cochlear implant does not provide the same functionality as the healthy human ear; it does not even provide the same functionality or characteristics as is experienced by someone who is hard of hearing. Traditional design guidelines for users with low vision or who are hard of hearing may not help people using artificial vision or hearing, or these guidelines may not help them as much as guidelines that take the specific strengths and weaknesses of the artificial restorations or augments into account.

For example, people practicing universal design may create a product that can be used both by people with full hearing or by people who are hard of hearing. However, to think of people who use cochlear implants as being "hard of hearing" would result in a mischaracterization of their hearing abilities. For example, most individuals who are hard of hearing have an ever-increasing dropoff in the high frequencies while retaining better mid- to low-frequency hearing. In contrast, people with a cochlear implants may be able to "hear" equally well at high and low frequencies depending on how the cochlear implant presents information to the user. However the type of hearing they have at any frequency may be quite different from that of someone with a hearing impairment. Guidelines that seek to provide guidance for addressing individuals with a full range of hearing abilities need to take into account this different type of hearing. Similarly, as we become able to provide individuals with artificial vision, we will find that it, too, differs from traditional low vision. Such differences are important to product designers. A better understanding of partial artificial abilities can lead to better design of both the partial artificial technologies themselves and of methods for designing products and environments to work with them.

NON-DISABILITY-RELATED HUMAN ENHANCEMENTS

If things continue to progress in the direction we are currently headed, it is likely that even people without functional impairments will be augmenting their abilities. Electronic or electromechanical enhancements are increasingly realistic as advances in biocompatibility continue. These advances and the personal enforcements that appear as mainstream performance enhancement strategies may bring new meaning to the term "human diversity."

Nondisability Parallels to Functional Limitation

The same considerations that make products more usable by people with functional limitations can make products more usable by individuals without disabilities who happen to be operating under constrained conditions. For example, cell phones that are easy to operate without sight are easier for individuals to operate while driving a car. Devices that are used by individuals who are deaf can also be used by individuals in very loud environments or in environments where there is enforced silence, such as a library or a meeting. At the World Wide Web (WWW) Conference in 1999, a panel consisting of mobile computing experts and disability access experts was convened. An observation was made that almost anything that improved access for people with disabilities could also improve mobile computing, and vice versa. In a follow-up to the comment, one of the panel members asked if there was anything incompatible

between the two sets. The panel and audience spent 15 minutes trying to identify factors that were beneficial to disability Internet access that would not also improve mobile Internet computing and vice versa; they were unable to identify anything.

A year later at the next WWW Conference, wireless telephone representatives stated that the best guidelines available for the design of Web content for individuals using telephone-based browsers was to follow the Web Content Accessibility Guidelines (for disabilities).

ARTIFICIAL AGENT ACCESS AS DIVERSITY ACCOMMODATION?

Another trend in human-computer interfaces is the increased use of artificial intelligence. The next-next-generation interface for the Internet will likely be artificial-agent based (in addition to direct-access). With these interfaces, people would not interact directly with the content on the Web when searching and carrying out other activities; rather they could pose their questions (or requests) to artificial agents that would then interact with the Internet.

A number of advances both in language comprehension and processing, as well as in the semantic markup of the Internet, would be required before such systems would be functional. Key to this would be the creation of Internet content that is *machine perceivable*, *machine operable*, and *machine comprehendible*.

Machine perceivable refers to the ability of a machine to access the information given the limitations on machine vision and hearing at any time. Today, graphics are accompanied by Alt-Text so that a machine would not need to interpret them but could perceive their meaning through the Alt-Text. In the future, text that is presented graphically might be scanned and read, and some elementary charts or diagrams might also be machine perceivable. Similarly, clear, articulate speech may soon be considered machine perceivable. It is likely to be some time, however, before complex scenes and images or complex auditory information would be machine perceivable unless it were accompanied by text equivalents (Alt-Text or text descriptions or transcripts).

Machine operable means that the interface elements would be operable given the device's perceptual limitations. Interface elements that do not appear until users carry out a visual task (such as pointing) and that would not be evident by examining the content code would not be machine operable, nor would server-side image maps for which the machine would have to literally click on every pixel to discover the links because it is unable to process the visual image. Client-side image maps with proper semantic markup, however, could be easily deciphered, analyzed, and operated.

Machine comprehendible is a more difficult factor. Using clear and simple language can facilitate the ability of machines to process text and understand it. Software that can create summaries and generate indexes by reading clearly written material already exists. Actually understanding the content of most Web pages will follow as machine intelligence advances.

Machine Access = Disability Access?

This raises an interesting proposition:

> Does making Web content (or any technology) interface with both human (without disabilities) and machine capabilities result in content (technology) that is accessible to people with disabilities (directly or via assistive technology)?

Such content (or technology) would likely have an interface that was either directly operable by users with disabilities or operable via machine that could translate or transform the interface and present it in a format appropriate to the abilities of the user.

Machine Access = Preference Accommodation?

Machines will also eventually be able to translate content and interfaces to match people with different skills, abilities, languages, cultures, genders, backgrounds, preferences, situations, and attitudes. Finally, remember that many who have no disabilities or limitations that prevent the use of standard interfaces may prefer to employ artificial agents and personal augmentative technologies to access and use a wide variety of technologies in our environments. They may approach vending machines and simply speak or think "Coca-Cola" and leave it up to their personal digital assistant (PDA) or built-in implants to select the right soda and then handle arrangements for payment. They may even approach a sandwich machine and have the agent compare the offerings with the known preferences (or dietary or medical requirements) so that users are only presented with a subset of choices that relates to them (with occasional variety mixed in as users' desires and safety allow).

CONCLUSION

The need to accommodate the needs of individuals with diverse abilities, constraints, and preferences is already causing designers to think differently about interface technologies. We need to think about how to create interfaces that can adapt to a wider variety of users and situations. Mobile computing, for example, requires us to create interface strategies that can accommodate users as they operate under given constraints (while walking or driving, when it is noisy or silent) and yet still be able to take advantage of the increased efficiency that is possible when the individuals are not under constraints (e.g., at workstation).

The chapters in this section explore different dimensions of user diversity. If one is designing products for a single type of user, these chapters can help target product design for that group. If the task is to design a product for general use, the chapters can help to identify the range of factors that must be accommodated directly or indirectly to create flexible, adaptable designs. Either way, interface designers have a daunting task as we set out to create an egalitarian electronic environment. It is ironic that advances in technology may actually give us the tools to build these new flexible interfaces faster than we can understand the problems and figure out how the interfaces should be designed to be maximally effective; however, the collected knowledge represented here is a good start in that direction.

·20·

GENDERIZING HUMAN–COMPUTER INTERACTION

Justine Cassell
MIT Media Lab

INTRODUCTION

In this chapter, I discuss video games as an example of how a technology has been designed for a particular gender and how this design process illustrates a particular relationship between gender and the design of human–computer interfaces (HCI)—and technology in general. I label that relationship a *genderization* of information technology (IT) and return to that sarcastic wording a little later. The goal of this chapter is to problematize the endeavor of trying to figure out solutions to designing ITs for girls and women. I make this look even more difficult than we already thought it was by showing some of the unintended consequences of well-meaning solutions. I argue that girls were, for a long time, not taken into account in the design of computer games; however, designing games "specially for girls" risks ghettoizing them as a population that needs "special help" in their relation to technology. In contrast to this stance, my own design philosophy, which I call "underdetermined design," encourages both boys and girls to express aspects of self-identity that transcend stereotyped gender categories. In the course of the chapter, I conclude by questioning the following a priori: In designing technology, do we even want to talk about how and whether girls are different from boys? Who cares about those differences—about whether their *wetwear* is different? I believe that there *are* differences, and important ones at that. But until we can get away from a kind of a *deficit model* of girls and technology, we may need to watch our step in designing explicitly with one gender in mind.

In fact, gender as an analytic category only emerged in the late 20th century. Earlier theorists referred to primary oppositions between men and women, or to the "woman question," but they did not employ gender as a way of talking about systems of sexual or social relations (Scott, 1986). Today, however, the binary opposition between the sexes carries much weight and leads us to speculate about "masculine" and "feminine" qualities, likes and dislikes, and activities. We have become used to seeing "masculine" and "feminine" as natural dichotomies—a classification system that mirrors the natural world. The binary opposition between masculine and feminine is a purely cultural construct, however, a construct that is conceived of differently in different cultures, historical periods, and contexts. Thus, in some cultures, fishing is women's work, and in others it is exclusively the province of men. In medieval times, women were considered to be sexually insatiable; the Victorians considered them naturally frigid (Scott, 1986). The Malagasy of Madagascar attribute indirect, ornate, and respectful speech that avoids confrontation to men; women are held to be overly direct and incapable of repressing their excitability and anger (Keenan, 1974, cited in Gal, 1991). In the United States, however, men's speech is described as "aggressive," "forceful," "blunt," and "authoritarian," whereas women's speech is characterized as "gentle," "trivial," "correct," and "polite" (Kramarae, 1980, cited in Gal, 1991).

Much of the dialogue about gender and technology mirrors this essentialist trend—girls, boys, can't, can—and that's something we don't want to take for granted because it may not be serving us well. The structure of this chapter is as follows:

- First, I offer statistics about who uses a computer, what she or he does with the computer, and what she or he believes about the computer.
- Then I provide an extended example from video games, illustrating how these statistics were used to launch an entrepreneurial movement to build video games for girls.
- I continue by detailing the course of the girls' game movement and what some of its consequences were.
- I outline three design strategies within the girls' game movement.
- Finally, I offer conclusions and applications of that work to other information technology design domains, for example, women on the web.

BACKGROUND

In this section, I discuss the background that led to the girls' game movement, that is, the statistics about differential use of the computer and how those statistics led to different design philosophies.

Boys' and Girls' Differential Use of Computers

First, let's consider some data on who uses the computer and to whom the computer is perceived to belong. Both girls and boys in kindergarten judge the computer to be a boy's toy (Wilder, Mackie, & Cooper, 1985), and children of that age already demonstrate a gender gap in use, with boys spending more time than girls at the computer, a gap that increases between the ages of 2 and 7 years (Huston, Wright, Marquis, & Green, 1999). The magnitude of this difference between boys and girls continues to increase with age; thus, among fourth- through sixth-grade students "heavy users" of computers are overwhelmingly boys—the ratio of boys to girls is 4 to 1 (Sakamoto, 1994). Among secondary-school-age children (11 to 18 years), boys are at least 3 times more likely than girls to use a computer at home, participate in computer-related clubs or activities at school, or attend a computer camp. In 1982, only 5% of high school girls, as opposed to 60% of boys, enrolled in computer classes or used the computer outside of class (Lockheed, 1982). More recent statistics come from a 1998 survey demonstrating that high school boys predominate in all kinds of computer classes (design and technology, programming desktop publishing, artificial intelligence) except for word processing, where girls predominate (American Association of University Women (AAUW) Educational Foundation Commission on Technology and Teacher Education, 2000). Teachers appear to depend on similar gender stereotypes in their assessment of students. Culley (1993) found that teachers attributed secondary-school girls' high computer exam scores to hard work and diligence, whereas boys, even those who did less well on exams, were thought to have intuitive interest and a "flair" for computers. Despite the increasing prevalence of computers in schools and homes, these statistics have not changed significantly since the early 1980s (AAUW Educational Foundation Commission on

Technology, 2000). Although the majority of studies have examined the state of affairs in North America, the same situation is found internationally (Janssen Reinen & Tjeed, 1993; Makrakis, 1993).

What are children, both boys and girls, using the computer for? Giacquintta, Bauer, and Levin (1993) found that by third grade, boys conceptualize computers differently from girls. Boys are more likely to play games, to program, and to see the computer as a recreational toy. Girls tend to view the computer as a tool, a means to accomplish a task, such as word processing or other clerical duties (Culley, 1993; Ogletree & Williams, 1990). In an informal study of an inner-city after-school computer program, I asked boys and girls why they were there. The boys tended to find the question ridiculous. One said, for example, "It's fun. I mean, there are all these computers for me to play with." The girls tended to be far more serious in their answers. One said, "Well, I really think this is a good opportunity for me to better my situation in life, and I believe that I can get a better job if I know how to use a computer." Adult women are also more likely than men to report that they see the computer as a tool rather than as an interesting artifact in its own right (Dennett, Branger, & Honey 1998). That difference in how the girls and boys see their involvement with computers turns out to be mirrored by how designers see boys and girls. When educators with software design experience were asked to design software specifically for boys or for girls, they tended to design learning tools for the girls and games for the boys. When they were asked to design software for generic "students," they again designed games—the type of software that they had designed for boys (Huff & Cooper, 1987). If this seems difficult to comprehend, an illuminating parallel can be drawn from cooking. Before James Beard began to host a television cooking show in 1946, home cooking in the United States was a woman's domain, and it was thought to be unmanly to cook at home.[1] James Beard explicitly addressed the notion that men could cook, and he said it was fun. What was the result? Once men could cook and it was fun, domestic devices for the kitchen were no longer called "appliances" but "gadgets." Think of the electric rotisserie, the bread machine, the coffee maker—when men took over kitchen technology, cooking began to resemble a game.

Mens' and Women's Differential Involvement in Information Technology

Let's look at what happens when these kids grow up. Men report more interest in computers than women do (Giacquinta et al., 1993; Morlock, Yando, & Nigolean, 1985), and men are more likely to work in computer-related fields. In 1990, approximately 70% of all employed computer specialists were men, a figure that had not changed throughout the 1980s despite the fact that the computer fields were growing rapidly. In addition, the 30% of women in these fields were concentrated in lower paid, less prestigious jobs (Kramer & Lehman, 1990). Although the computer industry continues to grow and to

diversify, the statistics are still dismally weighted toward men. According to the most recent Computing Research Association Taulbee Survey (*2000 CRA Taulbee Survey Results*, 2001), only 19% of the bachelor's degrees in computer science or computer engineering were awarded to women. Women received 26% of the master's degrees, 15% of the PhDs, and constituted 18% of currently enrolled PhD students. In addition, 14% of assistant professors, 13% of associate professors, and 8% of full professors in computer science were women among the 214 universities surveyed. Unpleasantly enough, those statistics are getting worse, not better, and in fact, over the last decade there has been an overall 24% decrease in the number of women getting degrees in computer science.

But why worry, one might ask? What's the big deal? Who wants children to spend their time face to face with a computer? Some parents say, "So, my little girl doesn't want to play with a video game? Shouldn't I be happy? I mean, shouldn't I say, I've got a daughter who is getting outside? So, maybe it's disposition, maybe it's just a different way of seeing the world, a different thing that girls and boys like. Like, girls don't play with lawnmowers as much as boys do either, and no one's getting all upset about it." There are a couple of reasons to worry. One is that in today's job market, computer literacy is important and it's only getting more important. In fact, the National Science Foundation has predicted that by 2010, one in four jobs in the United States will require computational literacy—Not just computational literacy, but the higher paying the job is, the more technical fluency that will be required. Computer games constitute the most frequent use of computers for children aged 2 to 18 years (Roberts, Foehr, Rideout, & Brodie, 1999), and computer and video games bootstrap computer literacy (Greenfield, 1996; Kiesler, Sproull, & Eccles, 1985), so this means that if girls are playing fewer computer games, they may be getting less technical fluency than boys; boys are then getting higher paid jobs requiring that fluency, and girls don't have access to those jobs.

We should also worry because there is evidence that the problem is not one of inherent interest or ability but of access (Kinnear, 1995). Kiesler et al. (1985) reported the following:

Even in preschool, males dominate the school computers. In one preschool, the boys literally took over the computer, creating a computer club and refusing to let the girls either join the computer club or have access to the computer. As a result, the girls spent very little time on the computer. When the teachers intervened and set up a time schedule for sharing computer access, the girls spent as much time on the computer as the boys.... Apparently, girls can enjoy the computer and do like to use it, but not if they have to fight with boys in order to get a turn. (p. 254)

In another study, first-grade girls working on the computer in mixed-gender groups were more likely to be laughed at, to be criticized, and to have their competence questioned than when they were working alone or in all-girl groups. In addition, the girls were frequently interrupted by male students, whereas the reverse was not true (Nicholson, Gelpi, Young, & Sulzby, 1998).

---

[1] It was unmanly even to *enjoy* food too much. As a 1937 self-help book advised, "When the waitress puts the dinner on the table, the old men look at the dinner. The young men look at the waitress" (Burgess, 1937, cited in Stern & Stern, 1991).

In fact, continued exposure to computer games decreases preexisting gender differences (Greenfield, 1996), and when educators really make an effort to ensure that girls have equal time to spend on the computer, girls show equal ability in programming (Linn, 1985) and in technology-enhanced science classes (Mayer-Smith, Pedretti, & Woodrow, 2000). Woodrow (1994) found that boys' greater experience and more positive attitude toward computers did not actually result in higher performance with computers.

A final reason to worry about these findings comes from girls' own perception of the correlation between gender and computer use. Many girls do not believe that they are good at math or at computer science (Busch, 1996); those who *are* good at computers may not believe that they are good at being girls. An example comes from responses that I received several years ago to an advertisement looking for research assistants to work on the topic of gender and computer games. I received many, many responses, and many of the responses from the young women went along the lines of "please, please, please hire me. All my life I've been waiting for this, I really have to look at this issue, this is so important to me. It's not important to me personally because I wasn't raised like a girl, but my younger sister isn't good at the computer." Similar comments were reported by Huber and Schofield (1998). What those girls were saying is that they are technically fluent, but that's because they are not "real girls," and that to me is the saddest thing of all: that one should find one's gender to be incompatible with one's abilities and that one should have to deny one's gender to accept one's abilities.

HEY! LET'S DESIGN COMPUTER GAMES FOR GIRLS!

In the mid-1990s, unprecedented numbers of women were becoming entrepreneurs, starting new businesses at twice the rate of men, and many of these women were thinking that these businesses could do some kind of good in the world (Moore & Buttner, 1997). Many saw their entrepreneurial enterprises as explicit sites to undermine conventional and stereotypical notions about "a woman's place" and to explicitly help the next generation of women to feel empowered (Goffee & Scase, 1985). Having looked at the same statistics discussed earlier, some of these entrepreneurs started companies to equalize the technology playing field for girls. At this point in time, until around 1996, fewer than 25% of game purchasers were girls. In terms of use (counting the fact that girls sometimes went to boys' houses to play with games) statistics were still in the 20% range. When girls did play video games, they found little to welcome them. In a study of 100 arcade games (cited in Provenzo, 1991), 92% contained no female roles whatsoever. Of the remaining 8%, the majority (6%) had female characters playing the "damsel in distress," and 2% had them playing active roles. More recently, in 1998, *Next Generation* magazine concluded that despite dramatic increases in the number of female game characters, "they all seem to be constructed around very simple aesthetic stereotypes. In the East, it's all giggling schoolgirls and sailor uniforms, but in the West the recipe appears to be bee-sting lips, a micro-thin waist, and voluminous, pneumatic breasts" ("Girl Trouble," 1998).

Of course, not all of the companies reading statistics about how few girls played video games had feminist goals. Thus, some of these entrepreneurs, and some other established companies, regardless of the good that they could do in the world, looked at these statistics and realized that there was a whole unexploited market out there to develop. The net result was that, in the mid-1990s, a number of companies began to build computer and video games for girls. In what follows, the terms *computer game* and *video game* will be used interchangeably. Of course, they aren't strictly interchangeable; originally video games were played on a console, such as Nintendo or Sega, but today most of these games have been translated into home computer use, and many of the consoles—for example, the new Sony PlayStation with its Internet capability—are starting to look more like computers.

Barbie Fashion Designer: Designing for "the Girl"

Thus, these companies began to build video games for girls. So many small software companies opened, and so many of the larger companies joined the bandwagon, that the phenomenon acquired the name of the "girls' game movement." But the game that initiated the trend, and that led the other companies to think that there was a niche, was Barbie Fashion Designer (Fig. 20.1).

Barbie Fashioner Designer allows players to design clothes for a Barbie doll and then put them on a virtual Barbie doll and have that Barbie doll walk down the runway; likewise, the designs can be printed on special fabric that comes with the kit, cut out and sewn into clothes for a physical Barbie doll. One of the interesting things about Barbie Fashioner Designer, as Subrahmanyarn and Greenfield (1998) argued, is that it's not a game *qua* game, it's a game as accessory for doll play. This means that as software it still fits into girls' existent Barbie play. Barbie Fashion Designer did extraordinarily well on the market: 500,000 copies sold during its first 2 months, which is more than the industry expected—and more than any other

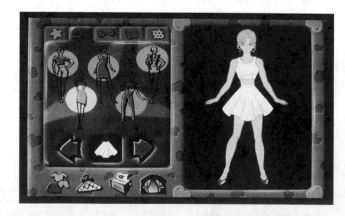

FIGURE 20.1. Barbie Fashion Designer software.

FIGURE 20.2. Characters in the Rockett software.

children's software title in history. From the angle of genderizing IT, one of the interesting aspects of this Mattel game is that although relying on stereotypical doll play, it confounds stereotypes by inviting girls to build, to construct, to imagine. In fact, when I invited some MIT undergraduates to play with a whole series of girl games, many of the students said that the game they preferred was Barbie Fashion Designer. In fact, one young man, more honest than many in the class, said, "Yeah, this was like totally the kind of game that when I was a kid I would have said, eeww, that's for girls . . . make the skirt red!"

Other Games for "the Girl"

Many of the girls games that arrived on the market around the same time as Barbie Fashion Designer also involved stereotypical girl play. Purple Moon's first game was called "Rockett's First Day at School" (Fig. 20.2). A press release for Purple Moon describes its products as "guided by the complete and unique understanding of girls and girls' play motivations" that emerged from "thousands of hours" of research. Purple Moon's Brenda Laurel told *Wired* (Beato, 1997), "I agreed that whatever solution the research suggested, I'd go along with. Even if it meant shipping products in pink boxes." Rockett games involve social relationships and are described as "friendship adventures for girls." Players can decide what Rockett's next step will be during the day. Thus, for example, on Rockett's first day, she arrives at the front door of her school to find another girl wearing exactly the same outfit. Players can decide: Does she make friends with the person, does she go home and change, or does she stomp into school?

Another game from the same era that also relied on research about girls' stereotypical interests was called Talk About Me (Fig. 20.3). The heroine of this game is the player herself, who can look up horoscopes, answer quizzes about romance, read interviews with potential role models who talk about how they mix family and work, and test out clothing mixes and matches. The software also has a space for the girl to keep a diary.

These three games, representative of others in the genre, demonstrate that the software put out during this initial phase of the girls' game movement was designed for *the* girl. In the toy industry, boys' toys are divided by *type* of boy; there are toys designed for jock boys, geek boys, action-oriented boys, and the simulation-game boy. In computer games, likewise, software game companies market action games, adventure games, fighting games, physical skill games, sports, role playing, simulation, and each of these targets a different segment of the population.

FIGURE 20.3. Let's Talk About Me software.

The software we've just reviewed, however, demonstrates a rather unidimensional and stereotypical view of girlhood. And software for boys, although more varied, still conveys stereotypical notions of what it means to be a boy. The situation can be compared with 1960s and 1970s feminism, which sought to break down the fixed ascription of gender roles, promoting a ideal where everyone was free to choose identities and activities they found most comfortable. Marlo Thomas's *Free to Be . . . You and Me* (1974) for example, as a book, record, and television special, encouraged boys to explore their feelings and to play with dolls and sought to encourage more competitive attitudes in girls. As Sherry Turkle explained during a *Nightline* (Nightline, 1997) discussion of the girls' games movement, "If you market to girls and boys according to just the old stereotypes and don't try to create a computer culture that's really more inclusive for everyone, you're going to just reinforce the old stereotypes. . . . We have an opportunity here to use this technology, which is so powerful, to make of ourselves something different and better."

Isn't Traditional Femininity a Viable Option?

The games for girls depended on traditional and stereotypical visions of girlhood. But what's wrong with fostering a space that's "girl only"? Can't traditional femininity be good, a viable choice for girls and women? And what's so great about these other "boy games" anyway? They can encourage violence and certain undesirable kinds of competition. Why not foster different social and cultural values than those that dominate the boys' game market? So highlighting traditional femininity and demonstrating that it's okay is one good thing about these girls' games. Girl Games' Laura Groppe argued, "I want girls to know that it's OK to be a girl!" (Russo, 1997). Seiter (1993) called on us to value girls' cultural tastes and interests, even as we push toward more empowering fantasies, because there are so many other forces in society that belittle and demean girls:

Something was gained and lost when marketers and video producers began exploiting little girls as a separate market. Little girls found themselves in a ghettoized culture that no self-respecting boy would take an interest in, but for once, girls were not required to cross over, to take on an ambiguous identification with a group of male characters. . . . The choice is not made out of identification with an insipid and powerless femininity but out of identification with the limited sources of power and fantasy that are available in the commercial culture of femininity. (pp. 157–158, 171)

NO! LET'S APPROPRIATE GAMES FOR BOYS

Historically, gender was an unexploited category in video-game design, with male designers developing games based on their own tastes and cultural assumptions without considering how these approaches might be anything other than gender-neutral. *Nightline* (1997) quoted ID's Todd Hollenshead, "What we try to do is make games that we think are fun, and they're not targeted to any specific gender." Yet we live in a culture where the male or the masculine remains the invisible norm. Women may dress in male clothing, but not vice versa. Girls play with boys' toys, but boys are ridiculed for playing with girls' toys. As long as the male choice is the norm, unself-conscious efforts are likely to simply perpetuate male dominance. This seems to have been the case with video-game design; remember Huff and Cooper's finding that when game designers designed for "children," they designed products identical to those they designed for boys (and different from those they designed for girls).

When faced with an unexplored market, then, some software companies dealt with the issue of designing for girls quite differently; rather than shifting the content, they added new characters to existing games and gave those characters female bodies (in some parallel universe of wasp-waisted females; see, for example, the character in Fig. 20.4). This character was not always a draw for girls, but it was a nod in the direction of the female audience. As one 12-year-old girl said after switching from the single female character in the game "Odyssey" to one of the male characters, "I don't like the way she dies. The male characters scream when they're slaughtered. The female character whimpers." An example of this strategy comes from Sega, who introduced female protagonists into many of its fighting games, giving them strengths and capabilities that are attractive to both male and female players and then giving them supermodel bodies. As Lee McEnany Caraher, corporate spokesperson for Sega in North America, said:

FIGURE 20.4. Female fighter from Soul Caliber (released September 1999).

The girls are babes in our game; they're babes. You know, they have big breasts and they wear scanty clothes. But the clothes don't fly off and that kind of stuff. I don't have a problem with representing women as babes, because when I go to the gym that's what they look like. They're not "gorgeous-gorgeous" but they're built. And if you were a real martial arts contender, you'd be built too (Glos & Goldin, 1998).

There are women and girls who are passionate about playing these computer games that have traditionally been geared toward men. Many of these women, who call themselves "game grrrls," see video games as highly competitive and see that competitiveness as being key for the acquiring the kind of skills needed in the real world. They see combat games as essential places for men and women to compete in an arena in which physical strength doesn't play a role. They reject what they see as a traditional and old-fashioned view of femininity from the girls' games movement, all the while criticizing the characters they find in boys' games as being the product of male erotic fantasies.

Isn't Appropriation a Viable Option? The Case of Lara Croft

Let's look at Lara Croft (Fig. 20.5) as an example of differing views on the appropriation of traditionally male games. Lara Croft Tomb Raider was introduced in 1997, around the same time as Barbie Fashion Designer. In the marketing of the game, Core Designs said that this was a game targeted equally toward men and women, that Lara Croft was a strong independent woman. One of the designers, Toby Gard, said, "Lara was designed to be a tough self-reliant, intelligent woman." Which is great, because it's good to have strong role models for girls in computer games, right? Then he continued, "She confounds all

FIGURE 20.5. Lara Croft, Tomb Raider.

the sexist clichés, apart from the fact that she's got an unbelievable figure, strong independent women are the perfect fantasy girls, the untouchable is always the most desirable" (Whitta, 1997). This is no longer a comment about girl players. Who are we designing for here? As Cal Jones said in *PC Gaming World*, "The problem with Lara is that she was designed by men for men. How do I know this? Because if you genetically engineered a Lara shaped woman, she would die within around 15 seconds, since there's no way her abdomen could house all her vital organs" (Jones, 1997).

Female gamers have also objected to many of the company's efforts to promote the game to male players, including their hiring of a scantily clad female model to impersonate Croft at computer trade shows, or the development of an ad campaign based on the theme "Where the Boys Are," showing lusty boys abandoning strip clubs in search of Lara (Brown, 1997). An underground industry in home-developed nude shots of Lara Croft, including a Nude Raider Web site, suggest the dangers in linking female empowerment to images couched in terms of traditional sex appeal (Whitta, 1997). Game magazine coverage of Lara Croft and the attempts of other game companies to imitate the success of Tomb Raider explain the phenomenon almost entirely in terms of her erotic appeal to young male players. Corrosive Software's Kate Roberts asks, "Would Tomb Raider have sold as many copies if Lara had been wearing a nice warm sweater and sweatpants?" ("Girl Trouble," 1998).

This is the other side of the coin: To use the computer, do girls have to be girly-girls or tomboys? Aren't other options open? That is, many of these games are leading to—undoubtedly unintentionally—increased gender stereotypicity. Thus, increased gender stereotypicity may be one unintended

outcome of this attempt to give girls increased access to technology. When one reads designers' comments, it becomes clear how this happens: "this is what girls want," says Sarina Simons of Philips Media (Newsweek, Oct. 28, 1996)." They want Barbie, we give them Barbie. "One girl game designer has suggested she wanted to give her product "Cooties" so that boys would stay away," "boys, in general, like competitive win-lose situations, high scores and body counts. It's almost the opposite for girls" (Cassell & Jenkins, 1998, pp. 24, 299).

WHAT ARE "BOYS" AND "GIRLS" ANYWAY?

These statements may be true for many girls and many boys, but they are not true for all girls and boys. They certainly are not accurate depictions of "girlness" and "boyness," which are quite context dependent, because gender is a context-dependent notion. Meanings for the opposition between male and female differ by historical period and by culture, as described earlier. They also differ in particular contexts within one historical period and one culture. Kafai asked children to build video games to teach math. It turned out that there were real differences between the video games the girls built and the video games the boys built. The girls' games were often about relationships among people, a drama, or a plot line, and the boys games were often about achieving intermediate goals on the way to a final goal (Kafai, 1996). A couple of years later, Kafai ran a similar study: She once again asked boys and girls to build video games, but this time to teach science. She found no differences between boys and girls (Kafai, 1998). Teach math, big differences; teach science, no differences. We don't know why the context of science elicits fewer differences between boys and girls than the context of math, but it does. Likewise, Hurtig (Hurtig, Kail, & Rouch, 1991) showed that when asked to categorize the people in a photograph of "successful executives," viewers named the photo as being of "men and women." When different viewers were asked to categorize the people in the same photograph, this time called a photograph of "a group of friends," the categories of male and female did not come into play. Hurtig concluded that gender is only a variable when it is at issue, that is, when socially constructed categories having to do with what we expect of men and women are evoked. The context that we believe we're seeing affects our understanding of gender.

A Short Note About Methodology

At this point a methodological remark is in order. A chief executive officer (CEO) of one of the games-for-girls companies is fond of describing her research strategy as follows: "I hire women who look like they could still be teenagers and I send them around America to throw slumber parties, and everybody wears feet-y pajamas and they eat popcorn, and they watch videos and they tell scary stories, and they talk about what they'd like from technology." In 1997, I shared a panel with this CEO, and she described her company's new product, which happened to be a kind of technologically enhanced nail polish. At the same panel, I was talking about a project I was directing that included 3,062 children in 139 countries. I invited the children to think through how to use technology to make the world a better place for children and then invited them to implement those ideas (Cassell, in press). We flew around the world, handed out computers and Internet connections, and hooked up kids; we built an online Web site that allowed the kids to communicate with one another in five languages. Interestingly, although we hadn't planned it this way, roughly 60% of the participants were girls, and when the children themselves voted to send 100 children to Cambridge in November 1998 for a summit with world leaders, 60% of those elected to participate were girls. I spent a lot of time listening to these children talk about how they wanted to use technology; amazingly, not a single one referred to nail polish. Instead, they described things suchas eradicating poverty. Methods are important, and the context is as well.

Willis (1991) concluded that "It matters little that many nursery schools now mix the dolls and trucks on their play-area shelves if everyone—children in particular—perceives toys as originating in a boy-versus-girl context." The color-coding of products, the narrow casting of children's programs, and the targeting of advertisements for specific genders results in a culture that gives children clear signals about gender-appropriate fantasies and desires (Fleming, 1996). It is perhaps not surprisingly, then, that the market research supporting the growth of the girls' game movement has located fairly stereotypical conceptions of feminine taste. Desires are manufactured by the toy industry itself long before researchers get a chance to talk with the girls and find out "what girls really want from technology." Appeals to such empirical research as a justification for design and development decisions run the risk of reinforcing (and naturalizing) this gender-polarized play culture rather than offering girls an escape from its limitations on their choices.

ANOTHER OPTION: UNDERDETERMINED DESIGN FOR ME (WHOEVER I Am)

So far, we've seen a split between designing for *the girl* and challenging the benefits of designing for any girls at all. In my own work, I've tried to find a third position. My students and I have depended on theories that see gender as dynamic, performative, and context dependent (Butler, 1990). We didn't see that it was our place to design a game for girls or a game for boys. We didn't see that it was our place to claim to know what "girl" or "boy" was, because there's too much diversity. So we decided to design computer games that in their very use would allow children to decide who they were and to discover who they were in the richest way that we could. I call this design philosophy *undetermined design* (Cassell, 1998), that is, design that allows users to engender themselves, to attribute to themselves a gendered identity of any one of a number of sorts, to create or perform themselves through technology use. We chose to express these notions through narrative games because, as Ochs and Taylor wrote (1995), "[G]ender identities are constituted through actions and demeanors . . . among other routes, children come to understand family and gender roles through differential modes of acting and expressing feelings in narrative activity." (p. 98)

Many of the insights on which we relied came from feminist pedagogy, which had radically changed educational practice years before. For example, value subjective experience, value how people see themselves and their own experiential knowledge. Transfer authority from the front of the classroom to the whole room. In our case, transfer authority from the software designer to the user. Allow a multiplicity of viewpoints and privilege *voice*—speaking out—with everything that *voice* entails (Cassell & Ryokai, 2001). The term *voice* in narrative theory has referred to whether an author speaks through a narrator or a character or speaks as him- or herself; it is the taking of different perspectives on a story. But popular books on adolescence, and much feminist theory, use the terms *voice*, *words*, and *language* metaphorically "to denote the public expression of a particular perspective on self and social life, the effort to represent one's own experience, rather than accepting the representations of more powerful others" (Gal, 1991, p. 172).

In this perspective, we understand the kinds of activities that have been described as "what girls really do," not as neutral or isolated acts but instead as involving the person becoming and acting in the world as part of the construction of a complex identity. In this case, designing "games for girls" misses the point. We should, rather, expand the range of activities we can perform on a computer so as to encourage identity formation as a part of the game.

One example is Rosebud (Glos & Cassell, 1997), a system that makes stuffed animals into children's allies and partners, facilitating the use of technology with which children may not be familiar and making the computer not a tool but one voice in a multiparty conversation. The stuffed animal is unique in a number of ways that are important for the different kinds of narratives that children (and adults) tell: (a) because it represents a sentient being, the child can attribute social goals to the stuffed animal, thus giving the child an imaginary partner to share in his or her experiences; (b) the stuffed animal plays an early role in the child's narrative life: a listener for the child's early stories to him- or herself, the subject of other stories, and the hero of plays put on by groups of children; (c) stuffed animals are solidly gender-neutral toys until preadolescence (at which point boys deny liking them anymore but often refuse to throw them out); (d) stuffed animals become keepsake objects that continue to play a role in the people's memories of their lives throughout the life span.

In the Rosebud system (Fig. 20.6) the computer recognizes children's stuffed animals (via an infrared transmitter in the toy and a receiver in the computer) and asks the child to tell about the stuffed animal or, in a subsequent interaction, calls the stuffed animal by name and recalls what it has heard. The child is asked to tell a story about the stuffed animal, any story at all, with prompts along the way. The computer is an encouraging listener, as well as a teacher, pushing the child to write, write more, edit, and improve. The child is in charge of the interaction, deciding which stuffed animal(s) to play with and what story to tell.

The collaboration between child, computer, and stuffed animal ends with the child recording the story in her or his own voice. The story is saved into the stuffed animal, and the child can then ask the stuffed animal to repeat the story. Rosebud

FIGURE 20.6. The Rosebud system.

supports storytelling by one child and one stuffed animal but also by multiple children, each with his or her own stuffed animal, working together. In the literal and metaphoric sense of "voice," Rosebud encourages the establishing of voice through an open-ended storytelling framework for the child. It promotes collaborative learning not only among several users and through peer review, but through presenting the computer as a supportive learning partner rather than an authoritative viewpoint. Rosebud focuses on collaboration by allowing multiple-toy use and multiple-author storybooks, so that several children can write a story together about all of their stuffed animals. Likewise, because the toy serves as a storage device, children can trade their stories by lending their stuffed animals to a friend.

Underdetermined design allows users to create or perform themselves through using technology. In this way, the very use of technology is a kind of design of that technology, as users appropriate the technology to fit their needs, and use technology to re-envision whom they might become. There's no explicit mention of gender in any of these toys. In our testing with children, we have discovered that both boys and girls are equally likely to play with these toys, and the children tell whatever story they have in their heads. The computer doesn't correct, edit, or encourage in a particular direction. The computer listens and stores, and the very act of having their story heard is extremely important for all children, but perhaps particularly so for girls who often feel unheard.

WHAT ABOUT "JUST GOOD DESIGN"?

Underdetermined design is in many ways similar to user-centered or participatory design (Muller, in press). Participatory design is not a single theory or technique for accomplishing software design. Rather, it is a set of perspectives that share concern for a "more humane, creative and effective relationship between those involved in technology's design and its use" (Suchman, 1993, p. viii). The goal of the participatory design movement is to encourage active participation in the design process by users of computer systems and to make this participation

empowering (Greenbaum & Kyng, 1991). In practical terms, this stance translates into conceiving of users as an essential part of the design team, and therefore bringing them in early during the design phase of new technology. The points of contact between participatory design and underdetermined design are not surprising given their political commonalities; both raise questions about democracy, power, and control in the workplace (Balka, 1997).

Nonetheless, although advocates of participatory design do bring users into the lab early in the product development cycle, the product itself is still static, constructed in the absence of the users, and no commitment is expressed to making a product that allows different kinds of use by different users at different moments. Underdetermined design, on the other hand, makes the system about design, so that the design and construction cycle continues into the use of the system itself.

Also similar to the philosophy of underdetermined design is what has been called "gender fair." One mathematical game that has received attention for its gender fairness is the Logical Journey of the Zoombinis (published by Broderbund software), which features small blue creatures that can be personalized with one of five kinds of hairstyles, eyes or eyewear, nose colors, and feet or footwear. Rubin, Murray, O'Neil, and Ashley (1997) showed that both boys and girls were equally engaged with this software. Some children (in general, the girls) were more interested in building the characters, and some (in general, the boys) were more interested in solving the logical puzzles that led the Zoombinis to their destination, but all the children were eager to play the game. Unfortunately, as pointed out by Castell and Bryson (1998), engagement is not sufficient. Because it is the logical puzzles that carry educational value, those children that become caught up in character development may be missing the point of the software. Underdetermined design, on the other hand, integrates the design and construction cycle into the goals of the software.

Some of the goals of underdetermined design are also similar to the principles of "learning with understanding" (Bransford, Brown, & Cocking, 1999). This educational philosophy relies on four features: that learning environments be learner-centered (anchor learning in meaningful, authentic problems to help learners make connections), knowledge-centered (foster symbolization and abstraction of underlying principles), assessment-centered (opportunities for feedback, reflection, and revision), and community-centered (enable guided inquiry in a collaborative community). Interestingly, learning environments that meet these criteria appear to demonstrate fewer gender differences than nonsituated, traditional textbook-based learning (Boaler, 1994). For girls to profit, however, these features must be implemented for all learners, not just for girls. In this context, Boaler highlighted an interesting tension in the discourse around how to change mathematics education, a tension that also arises in speaking of designing technology for women. It has been posited that girls need more real-life contexts for their learning; that their learning needs to involve intuition, creativity, emotion, and relativism. Some have interpreted this to mean that girls are less able to think abstractly, and this has led to programs to train girls to think in ways similar to boys. Or, alternatively, special mathematics units have been devised for girls to link math to girls' interests: calculating the geometry of dress hems, counting Beanie Babies. However, as was described earlier for computer games, setting girls up as a "problem-space" in mathematics risks attributing to girls special (less capable) needs that need to be specially helped. One cannot simply graft "girls' interests" onto an existent lesson. A design philosophy such as underdetermined design makes the interests of the user integral to the working of the system. In fact, in educational applications, students participate in the design of their own learning environments, with no weakening of underlying learning potential.

LESSONS LEARNED: DESIGNING FOR WOMEN

McIntosh (1983) posited five interactive phases of change that occur when new perspectives on gender are brought to the attention of curriculum designers. In her example, the field of history traverses the following five stages:

1. womanless history
2. women in history
3. women as a problem, anomaly, or absence in history
4. women as history
5. history redefined or reconstructed to include us all

As we move in this chapter from the example of video games for girls to other aspects of designing technology for women, it is instructive to apply McIntosh's model to the design of technology. We have left Phase 1 behind: No longer is it possible to build womanless technology. Currently, there is widespread recognition of the importance of taking gender into account in interface design (witness the presence of this chapter in a handbook on HCI). We have passed through Phase 2: Public perception of the role of women in technology has changed radically, due to the efforts of activist computer scientists and historians who have highlighted, among others, Ada Lovelace's seminal role in the birth of the computer and Grace Hopper's essential contribution to computing. Now, however, we find ourselves at a stage where women seem to pose a problem for the technology design. Tech companies pay consultants to help them figure out how to design for women. One gender and technology consulting firm refers to its ability to help companies succeed at "the notoriously selective and lucrative demographic of teenage girls." A consultant for online businesses advertises its knowledge of "what makes women click": a six-step program from initiating the relationship through subtle tactics of banner and home-page design, through deepening the relationship by asking motivating survey questions. The goal is "the inside track to get inside women's minds and keep them inside" the Web site. In fact, many Web sites for women have sprung up, but the majority treat the same topics as women's magazines, which have been around for hundreds of years (the banner on one women's Web site invites readers to learn about "making your home a haven for your family").

How do we progress, then, from Phase 3 to Phase 4, from the *problem* of women in the interface to women as central

to the design of the interface? Perhaps paradoxically, I believe that we will reach this phase when we consider the diversity of men as well as women in our design of the human-computer interface. That is, the users of technology must come to have many faces: men and women, young and old, American and Bangladeshi. Women are central to technology when all users are central to technology, when all users are diverse, and when all kinds of users design technology. Women are central when technology is designed for human needs, by all kinds of humans. No longer will women be conceived of as having special needs any more than any other group. "Different folks, different strokes" might be the motto of Phase 4. When only one population is seen as needing specially designed technology, it is almost inevitably seen as external to the normal practice of technology. The marked is almost always perceived as the lesser. When only one population is special, one risks ghettoizing it. I asked a young boy what he thought of a new video game. He responded, "That is so stupid, not even a girl would like that."

Phase 4 is a net advance over what we see today, but can we imagine a world where Phase 5 is possible, where the notion of interface design is radically reconstructed so as to include, and because of including, us all? This follows from Phase 4: It is not just users of technology that benefit when the preferences and needs of all are taken into account; technology itself changes for the better as well. When we attempt to put ourselves in multiple perspectives during the design process, we are more likely to solve intractable problems. It is in Phase 5 that I believe underdetermined design becomes most important. Whereas Phase 4 may rely on a multiplicity of viewpoints and a multiplication of interfaces, Phase 5 relies on no viewpoint and only one barely designed interface. Such a development in the field of human-computer interface would truly change the nature of the field. Interfaces might be more intrinsic to the internal working of the system, rather than applied afterward as a "skin." Interfaces might be constructed on the fly according to user specifications or even constructed by users themselves. Note that these interfaces are not "gender free." If the example of video games for girls has taught us anything, it is that there is no such thing as "gender-free" software. Because this is the case, we can only integrate the dynamic nature of gender construction—of self-construction—into the software itself.

CONCLUSION: FOCUS GROUPS ARE OKAY, BUT IMPLEMENTING IS POWER

Currently, there's a lot of gendering going on in information technology, but that gendering takes a narrow notion of what it means to be a particular gender. To conclude, evoke the concept of the "common divide" (Snitow, 1990). Ann Snitow described this feminist form of the reassertion of sexual difference as one side of "a common divide [that] keeps forming in both feminist thought and action between the need to build the identity 'woman' and give it solid political meaning and the need to tear down the very category 'woman' and dismantle its all-too-solid history." I believe it's become as common in interface design as elsewhere. What that means is that when we try and do something to equalize opportunities for women, there have always been two approaches: to value traditional femininity or to deny differences between men and women. We're seeing exactly the same thing in technology. Do we encourage girls to beat boys at their own game, or do we construct a girl-only space? The problem is that both sides ultimately start from the assumption that computers are boys' toys, and thus both scenarios can result in the pejorativization of girls' interests. What can we do? We can ensure that there is an abundance of applications, games, and Web sites because this ensures innovation. The proliferation of technologies ensures a proliferation of designers to design those technologies—both men and women—to be attractive to girls and women of all sorts. Finally, to be designed for can be

dangerous. To be stuck in a room for a focus group is okay, but implementing things yourself is power.

ACKNOWLEDGMENTS

Some of the arguments in this chapter were first explored in Cassell and Jenkins (1998). Thanks to coauthor Henry Jenkins for illuminating my understanding of the issues and allowing me to expand on our joint text. Parts of the chapter were also introduced to audiences at IBM, Mattel, the Tech Museum of San Jose, the UMBC, and Xerox PARC. I am grateful for the participants' feedback. Finally, thanks to Anita Borg, Penny Eckert, and Ivy Ross for inspiration.

References

2000 CRA Taulbee Survey Results. (2001). Retrieved August 15, 2001, from http://www.cra.org/statistics/survey/00/

American Association of University Women Educational Foundation Commission on Technology and Teacher Education. (2000). *Tech-Savvy: Educating girls in the new computer age*. Washington, DC: Author.

Balka, E. (1997). Participatory design in women's organizations: The social world of organizational structure and the gendered nature of expertise. *Gender, Work and Organizations, 4*, 99–115.

Beato, G. (1997, April). Girl games: Computer games for girls is no longer an oxymoron. *Wired*.

Bennett, D., Brunner, C., & Honey, M. (1998). Girl Games and technological desire. In J. Cassell & H. Jenkins (Eds.), *From Barbie to Mortal Kombat: Gender and computer games* (pp. 72–88). Cambridge, MA: MIT Press.

Boaler, J. (1994). When do girls prefer football to fashion? An analysis of female underachievement in relation to "realistic" mathematics context. *British Educational Research Journal, 20,* 551–564.

Bransford, J., Brown, A., & Cocking, R. (1999). *How people learn: Brain, mind, experience, and school*. Washington, DC: National Academy Press.

Brown, J. (1997, November 11). GameGirls turn on to female gamers. *Wired*.

Busch, T. (1996). Gender, group composition, cooperation, and self-efficacy in computer studies. *Journal of Educational Computing Research, 15*, 125–135.

Butler, J. (1990). *Gender trouble*. New York: Routledge.

Cassell, J. (1998). Storytelling as a nexus of change in the relationship between gender and technology: A feminist approach to software design. In J. Cassell & H. Jenkins (Eds.), *From Barbie to Mortal Kombat: Gender and computer games* (pp. 298–326). Cambridge, MA: MIT Press.

Cassell, J. (in press). "We have these rules inside": The effects of exercising voice in a children's online forum. In S. Calvert, R. Cocking, & A. Jordan (Eds.), *Children in the digital age*. New York: Praeger Press.

Cassell, J., & Jenkins, H. (Eds.). (1998). *From Barbie to Mortal Kombat: Gender and computer games* (pp. 24, 299). Cambridge, MA: MIT Press.

Cassell, J., & Ryokai, K. (2001). Making space for voice technologies to support children's fantasy and storytelling. *Personal Technologies, 5*, 203–224.

Castell, S. D., & Bryson, M. (1998). Retooling play: Dystopia, dysphoria, and difference. In J. Cassell & H. Jenkins (Eds.), *From Barbie to Mortal Kombat* (pp. 232–261). Cambridge, MA: MIT Press.

Culley, L. (1993). Gender equity and computing in secondary schools: Issues and strategies for teachers. In J. Beynon & H. Mackay (Eds.), *Computers into classrooms: More questions than answers* (pp. 147–158). London: Falmer.

Fleming, D. (1996). *Powerplay: Toys as popular culture*. Manchester, UK: Manchester University Press.

Gal, S. (1991). Between speech and silence. In M. di Leonardo (Ed.), *Gender at the crossroads of knowledge: Feminist anthropology in the postmodern era* (pp. 175–203). Berkeley: University of California Press.

Giacquinta, J. B., Bauer, J. A., & Levin, J. E. (1993). *Beyond technology's promise*. Cambridge, UK: Cambridge University Press.

Girl Trouble. (1998, January). *Next Generation*, pp. 98–102.

Glos, J., & Cassell, J. (1997, March). *Rosebud: Technological Toys for Storytelling*. Paper presented at the CHI '97, Atlanta, Georgia.

Glos, J., & Goldin, S. (1998). An interview with Lee McEnany Caraher (Sega). In J. Cassell & H. Jenkins (Eds.), *From Barbie to Mortal Kombat: Gender and computer games* (pp. 192–213). Cambridge: MIT Press.

Goffee, R., & Scase, R. (1985). *Women in charge: The experience of female entrepreneurs*. London: Allen and Unwin.

Greenbaum, J., & Kyng, M. (Eds.). (1991). *Design at work: Cooperative design of computer systems*. Hillsdale, NJ: Lawrence Erlbaum Associates.

Greenfield, P. M. (Ed.). (1996). *Video Games as Cultural Artifacts*. Norwood, NJ: Ablex.

Huber, B. R., & Schofield, J. W. (1998). "I like computers, but many girls don't": Gender and the sociocultural context of computing. In H. Bromley & M. Apple (Eds.), *Education/Technology/Power* (pp. 103–132). Albany: SUNY Press.

Huff, C., & Cooper, J. (1987). Sex bias in educational software: The effects of designers' stereotypes on the software they design. *Journal of Applied Social Psychology, 17*, 519–532.

Hurtig, M.-C., Kail, M., & Rouch, H. (Eds.). (1991). *Sexe et genre: de la hiérarchie entre les sexes*. Paris: Editions du Centre National de la Recherchie Scientifique.

Huston, A., Wright, J. C., Marquis, J., & Green, S. B. (1999). How young children spend their time: Television and other activities. *Developmental Psychology, 35*, 912–925.

Janssen Reinen, I., & Tjeed, D. (1993). Some gender issues in educational computer use: Results of an international comparative study. *Computers and Education, 20,* 353-365.

Jones, C. (1997, December 30). Lara Croft, female enemy number one? *The Mining Company Guide.*

Kafai, Y. (1998). Video game designs by boys and girls: Variability and consistency of gender differences. In J. Cassell & H. Jenkins (Eds.), *From Barbie to Mortal Kombat: Gender and computer games* (pp. 90-114). Cambridge, MA: MIT Press.

Kafai, Y. B. (1996). Electronic playworlds: Gender differences in children's constructions of video games. In Y. Kafai & M. Resnick (Eds.), *Contructionism in practice: Designing, learning, and thinking in a digital world* (pp. 97-123). Mahwah, NJ: Lawrence Erlbaum Associates.

Kiesler, S., Sproull, L., & Eccles, J. (1985). Poolhalls, chips and war games: Women in the culture of computing. *Psychology of Women Quarterly, 4,* 451-462.

Kinnear, A. (1995). Introduction of microcomputers—a case-study of patterns of use and childrens perceptions. *Journal of Educational Computing Research, 13,* 27-40.

Kramer, P. E., & Lehman, S. (1990). Mismeasuring women—a critique of research on computer ability and avoidance. *Signs, 16,* 158-172.

Linn, M. C. (1985). Fostering equitable consequences from computer learning environments. *Sex Roles, 13,* 229-240.

Lockheed, M. (1982). *Evaluation of computer literacy at the high school level* (Evaluation of Computer Services). Princeton, NJ: Elsevier Science, Ltd.

Makrakis, V. (1993). Gender and computers in schools in Japan: The "we can, I can't paradox." *Computers and Education, 20,* 191-198.

Mayer-Smith, J., Pedretti, E., & Woodrow, J. (2000). Closing of the gender gap in technology enriched science education: a case study. *Computers & Education, 35,* 51-63.

McIntosh, P. (1983). *Interactive phases of curricular re-vision: A feminist perspective* (Working Paper No. 124). Wellesley, MA: Center for Research on Women.

Moore, D. P., & Buttner, E. H. (1997). *Women entrepreneurs: Moving beyond the glass ceiling.* London: Sage.

Morlock, H., Yando, T., & Nigolean, K. (1985). Motivation of video game players. *Psychological Reports, 57,* 247-250.

Muller, M. J. (in press). Participatory design: The third space in HCI. In J. A. Jacko & A. Sears (Eds.), *Handbook for Human-Computer Interaction.* New York: Lawrence Erlbaum Associates.

Nicholson, J., Gelpi, A., Young, S., & Sulzby, E. (1998). Influences of gender and open-ended software on first graders' collaborative composing activities on computers. *Journal of Computing in Childhood Education, 9,* 3-42.

Nightline, A. (1997). Revolution in a box, part 12 [Television broadcast]. *Nightline.* New York: American Broadcasting Company.

Ochs, E., & Taylor, C. (1995). The "Father Knows Best" dynamic in dinnertime narratives. In K. Hall & M. Bucholtz (Eds.), *Gender articulated: Language and the socially constructed self.* New York: Routledge & Kegan Paul.

Ogletree, S. M., & Williams, S. W. (1990). Sex and sex-typing effects on computer attitudes and aptitude. *Sex Roles, 23,* 703-712.

Provenzo, E. (1991). *Video kids, making sense of Nintendo.* Cambridge, MA: Harvard University Press.

Roberts, D., Foehr, U., Rideout, V., & Brodie, M. (1999). *Kids & media @ the new millennium.* Menlo Park, CA: Kaiser Family Foundation.

Rubin, A., Murray, M., O'Neil, K., & Ashley, J. (1997). *What Kind of educational computer games would girls like?* Paper presented at the American Educational Research Association March 24-28, 1997, Chicago, IL.

Russo, M. (1997). *Software for girls: A mother's perspective.* Retrieved Dec. 2001 from http://www.superkids.com/aweb/pages/features/girls/jrcl.shtml

Sakamoto, A. (1994). Video game use and the development of the socio-cognitive abilities in children: Three surveys of elementary school students. *Journal of Applied Social Psychology, 24,* 21-24.

Scott, J. (1986). Gender: A useful category of historical analysis. *American Historical Review, 91,* 1053-1075.

Seiter, E. (1993). *Sold separately: Children and parents in consumer culture.* New Brunswick, NJ: Rutgers University Press.

Snitow, A. (1990). A gender diary. In M. Hirsch & E. F. Keller (Eds.), *Conflicts in feminism* (pp. 9-43). New York: Routledge & Kegan.

Stern, J., & Stern, M. (1991). *American gourmet: Classic recipes, deluxe delights, flamboyant favorites, and swank "company" food from the '50s and '60s.* New York: HarperCollins.

Subrahmanyarn, K., & Greenfield, P. (1998). Computer games for girls: What makes them play. In J. Cassell & H. Jenkins (Eds.), *From Barbie to Mortal Kombat: Gender and computer games* (pp. 46-71). Cambridge, MA: MIT Press.

Suchman, L. (1993). Forward. In D. Scholar & A. Namioka (Eds.), *Participatory design principles and practices* (pp. vii-x). Hillsdale, NJ: Lawrence Erlbaum Associates.

Thomas, M. (1974). *Free to Be . . . You and Me.* New York: McGraw-Hill.

Whitta, G. (1997, August). If looks could kill. *PC Gamer.*

Wilder, S., Mackie, D., & Cooper, J. (1985). Gender and computers: Two surveys of computer-related attitudes. *Sex Roles, 13,* 215-228.

Willis, S. (1991). *A primer for daily life.* London: Routledge, Chapman and Hill.

Woodrow, J. (1994). The development of computer-related attitudes of secondary students. *Journal of Educational Computing Research, 11,* 307-338.

· 21 ·

DESIGNING COMPUTER SYSTEMS
FOR OLDER ADULTS

Sara J. Czaja and Chin Chin Lee
University of Miami School of Medicine

INTRODUCTION

The increased number of older people in the population and the increased reliance on technology in most societal contexts has created a need for the systematic study of technology and older adults. In 2000, persons aged 65 years or older represented 13% of the U.S. population, and it is estimated that people in this age group will represent 20% of the population by 2030 (Fig. 21.1). In addition, the older population itself is getting older; currently, there are about 15 million people over age 75 years (Administration on Aging, 2000).

At the same time that the population is aging, information technology (IT) is becoming integral to work, education, and daily life. In 2000, about 51% of households in the United States owned a personal computer (PC) and approximately, 41% of these households had access to the Internet (U.S. Department of Commerce, 2000). Other forms of technology such as automatic teller machines, telephone-based menu systems, and information kiosks at shopping malls or in public buildings are also common. Simple tasks such as fueling an automobile, banking, or locating a book in the library increasingly involve interaction with some form of technology. Technology has also become more diverse, and the varieties of computer tools that are available have increased at an unprecedented rate. In the near future, television, telephone, and other communication media will become integrated with computer network resources.

In essence, to function independently and interact successfully with the environment, people of all ages must learn how to assimilate technology into their lives. In this regard, the National Research Council (1997) recently issued a report that stressed the importance of making the Nation's Information Infrastructure (NII) accessible to as many people as possible, including people of all ages and varying abilities. The report also pointed out that although the usability of systems has improved substantially, current interfaces still exclude many people, such as those who are older or who have disabilities, from effective NII access. Given that information technologies are becoming integral to education, work, and daily living, not being able to successfully interact with these technologies will have increasingly negative ramifications for individuals. People must learn new methods of information seeking and to use new systems such as computers. In fact, Nickerson and Landauer (1997) suggested that making information networks easier to use and providing quality information to people as opposed to sheer quantity may be the single greatest challenge facing the discipline of human–computer interaction in the near future.

Given that older people represent an increasing large proportion of the population and are likely to be active users of technology, issues surrounding aging and computer technology are of critical importance within the domain of human–computer interaction. We need to understand the implications of age-related changes in functional abilities for the design and implementation of computer systems. We also need to understand the preferences of older people with respect to the design of computer interfaces and software applications. Although this topic has received increased attention within the research community, there are still many unanswered questions.

The intent of this chapter is to summarize the current state of knowledge regarding computer technology and older adults. Topics discussed include acceptance of technology by the elderly, training, and hardware and software design. A detailed discussion of the aging process is not provided (there are many excellent sources of this material, e.g., Birren & Schaie, 2001; Charness, 1985; Fisk & Rogers, 1997; Rogers & Fisk, 2001). Instead, a brief review brief review of age-related changes in abilities that have relevance to the design of computer systems is presented. Finally, suggestions for design guidelines and areas of needed research are summarized. It is hoped that this chapter will serve to motivate researchers and system designers to consider older adults as an important component of the community of computer users.

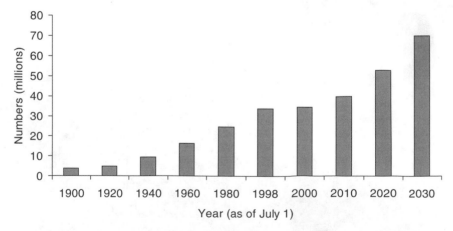

FIGURE 21.1. Number of persons aged 65+, 1900–2030. Source: U.S. Department of Health and Human Services Administration on Aging (2000).

USE OF COMPUTER TECHNOLOGY
BY OLDER ADULTS

There are a number of settings in which older people are likely to use computers, including the workplace, the home, and health care and service settings. Despite a trend toward increased computer use among people over the age of 55, however, use of computers among this age group is still low compared with other age groups. As shown in Fig. 21.2, about 26% of households of older people in the United States had a computer compared with about 55% of households of persons aged 35 to 54 years. Use of the Internet among older people is also lower than that of younger age groups (Fig. 21.3). Only 30% of people aged 50+ were Internet users in 2000, and although the number of Internet users in this age group is increasing at the same rate as the overall population, Internet users aged 50+ are still less than half that of users aged 16 to 40 years (U.S. Department of Commerce, 2000). Furthermore, people with a disability such as impaired vision or problems with manual dexterity such as arthritis are only half as likely to access the Internet as those without a disability (National Telecommunications and Information Administration, 2000). As shown in Fig. 21.4, the incidence of visual impairments and chronic conditions such as arthritis increases with age.

There are a number of ways that computers and the Internet may be beneficial to older people. For example, computer networks can facilitate linkages between older adults and health care providers and communication with family members and friends, especially those who are live faraway. It is common within the United States for family members to be dispersed among different geographic regions. In fact, nearly 7 million Americans are long-distance caregivers for older relatives (Family Caregiving Alliance, 1997). Clearly, network linkages can make it easier for family members to communicate, especially among those who live in different time zones. Computers may also be used to help older people communicate with health care providers or other older people and may help older

people become involved in continuing education. For example, telemedicine applications allow direct communication between health care providers and patients. There are also many opportunities to enroll in distance-learning courses online. These opportunities will be enhanced with future developments in technology such as video conferencing. Computers and the Internet can also help older people access information about health care and community services and resources. Finally, the Internet may also be used to facilitate the performance of routine tasks such as financial management or shopping. Access to these resources and services may be particularly beneficial for older people who have mobility restrictions or lack of transportation. In fact, recent data (U.S. Department of Commerce, 2000) indicate that the most common reasons people use the Internet are for e-mail, online shopping, and bill paying.

As will be demonstrated in this chapter, older adults find technologies such as computers to be valuable, and older people are receptive to using this type of technology. Available data (e.g., Czaja & Sharit, 1998; Mead, Spaulding, Sit, Meyer, & Walker, 1997) also indicate, however, that although older people are generally willing and able to use computers, they typically have more problems with them than younger adults do. Data also suggest that some of the primary reasons older people do not use computers and the Internet are lack of access to the technology, lack of knowledge, and cost (Morrell, Mayhorn, & Bennett, 2000). Before the full benefits of computer technology can be realized for older people, the usefulness and usability of these technologies for this population must be maximized. The following section reviews, in more detail, potential use of computer technology by older adults.

Work Environments

One setting where older people are likely to encounter computer technology is the workplace. The rapid introduction of computers and other forms of automated technology into occupational settings implies that most workers need to interact with

Percent of U.S. Households with a Computer and Internet Access, Selected Years

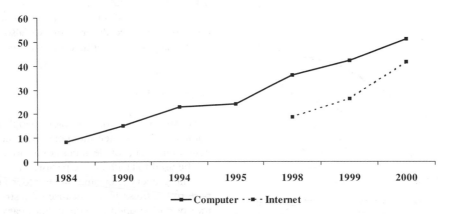

FIGURE 21.2. Trends of computer and internet access. Source: U.S. Census Bureau (1999).

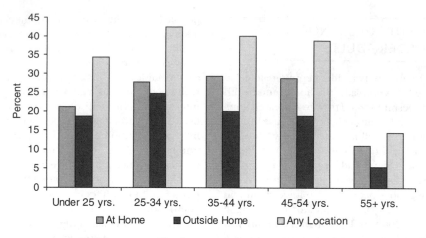

FIGURE 21.3. Precent of U.S. persons using the Internet. Source: U.S. Department of Health and Human Services Administration on Aging (2000).

computers simply to perform their jobs. Computer-interactive tasks are becoming prevalent within the service sector, office environments, and manufacturing industries. For example, office personnel need to use word-processing, e-mail, and database management packages to perform routine office tasks. Cashiers, sales staff, and bank clerks also use computers on a routine basis, and computer tasks are becoming more prevalent within manufacturing and process control industries. In 1997, half of the employed adults in the United States used a computer on the job compared with 45% in 1993 and 25% in 1984 (U.S. Census Bureau, 1999). In 1995, at least 3 million Americans were telecommuting for purposes of work, and this number is expected to increase by 20% per year over the next decade (Nickerson & Landauer, 1997). Telecommuting may be particularly appropriate for older adults because they are more likely than younger people to have mobility restrictions. Telecommuting would allow them to work from home; it also allows for more flexible work schedules.

Given that the workforce is aging, the implications for older workers of the explosion of technology in the workplace needs to be carefully evaluated. For example, issues such as worker retraining and skill obsolescence are especially important for

older workers because they are unlikely to had have experience with computers. Furthermore, although computers reduce the physical demands of tasks, they generally increase the mental demands. These changes in job demands need to be carefully evaluated for older workers because there are age-related changes in cognition. It may be that computer tasks are more stressful for older people or that some types of computer tasks are more appropriate than others for older adults. Czaja and Sharit (1993) examined age differences in the performance of three simulated real-world computer tasks commonly performed in work environments—a data entry task, a file modification task, and an inventory management task. They found age differences in task performance and subjective assessments of task difficulty and stress. For all three tasks, older people exhibited longer times and a greater number of errors than younger people. They also found the tasks to be more stressful and found the tasks, especially the inventory management task (the most demanding of the three tasks), to be more difficult and mentally challenging for the older people. Furthermore, the older participants reported more fatigue following performance of all three tasks.

To ensure that older adults are able to be successfully integrated into today's work environments, we need to understand if there are age differences in the performance of computer-based tasks and, if so, we need to understand the nature of these performance differences. We also need information on the efficacy of design interventions.

Home Environments

There are a number of ways older people can use computers at home to enhance their independence and quality of life. Home computers can provide access to information and services and can be used to facilitate the performance of tasks such as banking and grocery shopping. Many older people have problems performing these tasks because of restricted mobility, lack of transportation, inconvenience, and fear of crime (Nair, 1989). Data from the National Health Interview Survey (Dawson,

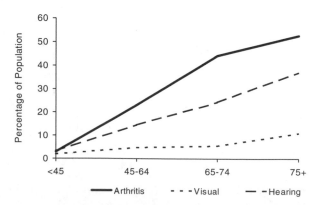

FIGURE 21.4. Perceptual and motor impairments.

Hendershot, & Fulton, 1987) indicate that about 27% of people aged 65+ require help with activities such as shopping. Home computers can also be used to expand educational, recreational, and communication opportunities.

Several studies (e.g., Czaja, Guerrier, Nair, & Landauer, 1993; Eilers, 1989; Furlong, 1989) have shown that older adults are receptive to using e-mail as a form of communication and that e-mail is effective in increasing social interaction among the elderly. Increased social connectivity can so be beneficial for older people, especially those who are isolated or who live alone. In 1998, about 31% of all older Americans lived alone, including 41% of older women and 17% of older men (Administration on Aging, 2000). A recent study (Cody, Dunn, Hoppin, & Wendt, 1999) found that older adults who learned to use the Internet had more positive attitudes toward aging, higher levels of perceived social support, and higher levels of connectivity with friends and relatives. The success of SeniorNet also points to the receptivity of older people to using computers for activities such as communication and continuing education. SeniorNet is a non-profit organization whose mission is to provide people over the age of 50 with access to computer technology. Members learn to use computers for communicating, continuing education, financial management, and other activities such as desktop publishing. Currently, the organization has more than 39,000 members and 210 learning centers throughout the United States. Senior-Net has taught more than 100,000 older adults to use computer technology.

Computers may also be used to augment older people's memory by providing reminders of appointments, important dates, and medication schedules. Chute and Bliss (1994) showed that computers can be used for cognitive rehabilitation and for the training and retraining of cognitive skills such as memory and attention. Schwartz and Plude (1995) found that compact disc–interactive systems are an effective medium for memory training with the elderly. Computers can also enhance the safety and security of older people living at home. As noted, a large proportion of older women live alone. Systems can be programmed to monitor home appliances, electrical, and ventilation systems and can be linked to emergency services.

Health Care

Computer technology also holds the promise of improving health care for older people. Electronic links can be established between health care professionals and older clients, providing caregivers with easy access to their patients and allowing them to conduct daily status checks or to remind patients of home health care regimes. For example, Leirer, Morrow, Tanke, and Pariante (1991) demonstrated that voice mail was effective in reducing medication noncompliance among a sample of elderly people. Similarly, Tanke and Leirer (1993) found that voice mail increased the compliance of a sample of elderly patients to tuberculosis medication appointments. Conversely, the elderly patient may use the computer to communicate with their caregivers. They may ask health-related questions or indicate that they are experiencing some type of problem.

Computers may also be used for health care assessment and monitoring. Ellis, Joo, and Gross (1991) demonstrated that older people could successfully use a microcomputer-based health risk assessment. In the future, computers may be used to monitor a patient's physical functioning, such as measuring blood pressure, pulse rate, temperature, and so forth. The use of computers in health care management offers the potential of allowing many people who are at risk for institutionalization to remain at home. There are vast opportunities for research within this area.

Computer technology may also be beneficial for family caregivers who are providing care for an older person with a chronic illness or disease such as dementia. Generally, the prevalence of chronic conditions or illnesses such as dementia, diabetes, heart disease, or stroke increases with age, and consequently older adults (especially the "oldest old") are more likely to need some form of care or assistance. Approximately 7 million people aged 65+ years have mobility or self-care limitations, and about 4 million Americans suffer from Alzheimer's disease (AD). Family members are the primary and preferred source of help for elders. Currently, at least one in four households in the United States has at least one caregiver (National Alliance for Caregiving and the Alzheimer's Association, 1999).

Current information technologies offer the potential of providing support and delivering services to caregivers and other family members. Computer networks can link caregivers to each other, to health care professionals, and to community service and educational programs. Information technology can also enhance a caregiver's ability to access health-related information or information regarding community resources. Gallienne, Moore, and Brennan (1993) found that access to a computer network, ComputerLink, increased the amount of psychological support provided by nurses to a group of homebound caregivers of Alzheimer's patients and enabled caregivers to access a support network that allowed them to share experiences, foster new friendships, and gather information on the symptoms of the disease. Technology can also aid caregivers' ability to manage their own health care needs and those of the patient by providing access to information about medical problems, treatments, and prevention strategies. Software is available on several health-related topics such as stress management, caregiving strategies, and nutrition. For example, the Alzheimer's Association has a home page on the World Wide Web: www.alzheimers.com.

Telephone-Linked Care for Alzheimer's Disease (TLC-AD) is an automated telecommunications system for AD caregivers that are being evaluated at the Boston site of the REACH (Resources for Enhancing Alzheimer's Caregiver Health) program. REACH is a multisite program, funded by the National Institute on Aging and the National Institute of Nursing Research, which is evaluating the efficacy of a variety of interventions with respect to reducing burden and enhancing the quality of life of family caregivers of Alzheimer's patients. The interventions are broad based and multifaceted, consisting of psychosocial and educational services, behavioral interventions, environmental modifications, and technology interventions such as TLC-AD, which monitors the primary caregiver's stress and health status. It also provides a voice-mail caregiver support network, an "ask the expert" call option, and a respite function to provide caregivers with a break from their caregiving duties. Telecommunications technology, including the telephone and video conferencing, is also being used for counseling or psychotherapy. Telephone-based therapy offers several advantages over face-to-face therapy. For example, it eliminates the need for both clients and therapists to travel and allows for more flexible appointment scheduling.

The Miami site of the REACH program is evaluating family-therapy intervention augmented by a computer–telephone system (CTIS) for family caregivers. The intent of the CTIS system is to enhance a family-therapy intervention by facilitating the caregiver's ability to access formal and informal support services. The system enables caregivers to communicate with therapists, family, and friends; to participate in "online" support groups; to send and receive messages; and to access information databases such as the Alzheimer's Association Resource Guide. A respite function is also provided. In addition, the CTIS system provides the therapist with enhanced access to both the caregivers and their family members. To date, experience with the system has been positive, with high acceptance of the system by caregivers. Preliminary data indicates that the majority of caregivers like the system and find it valuable and easy to use. The most common reason that caregivers use the system is to communicate with other family members, especially those who do not live nearby. The data also indicate that the system facilitates communication with other caregivers. Most caregivers reported that they found the participation in the online discussion groups to be valuable (Czaja & Rubert, in press).

Clearly, computer technology holds the promise of improving the quality of life for older adults and their families. However, for the full potential of technology to be realized for these populations the needs and abilities of older adults must be considered in system design. The following section provides a brief review of age-related changes in abilities that have implications for the design of computer systems.

UNDERSTANDING THE OLDER COMPUTER USER

There are several age-related changes in functional abilities that have relevance to the design of computer systems. These include changes in sensory–perceptual processes, motor abilities, response speed, and cognitive processes. In this chapter, a brief review of aging is provided as a framework for understanding the potential implications of the aging process for system design. Of course, there are substantial individual differences in rate and degree of functional change. Within any age group, young or old, there is significant variability in range of abilities, and this variability tends to increase with age.

Sensory Processes

There are a number of changes in visual abilities that have relevance to the design of computer systems. Currently, about 14 million people in the United States suffer from some type of visual impairment, and, as shown in Fig. 21.4, the incidence of visual impairment increases with age. Generally with increased age there is a loss of static and dynamic visual acuity. Older adults also experience a reduction in the range of accommodation, a loss a contrast sensitivity, decreases in dark adaptation, declines in color sensitivity (especially in the blue region), and heightened susceptibility to problems with glare. Visual search skills and the ability to detect targets against a background also decline with age (Kline & Schieber, 1985).

Although most older people will not experience severe visual impairments, they may experience declines in eyesight sufficient to make it more difficult to perceive and comprehend visual information. This has vast implications for the design of computer systems given that computer communications is primarily based on visually presented text. Charness, Schumann, and Boritz (1992) reported that the majority of older participants, in their study of word processing, experienced some difficulty reading the screen and that these difficulties may have contributed to the lower performance of older people. In a more recent study, Charness and Holley (2001) found that older adults had more difficulty than younger adults selecting small targets on a computer screen, especially when they were using a mouse as opposed to a light pen. Visual decrements may make it more difficult for older people to perceive small icons on toolbars, to read e-mail, or to locate information on complex screens or Web sites. Age-related changes in vision also have implications for the design of written instructions and computer manuals.

Aging is also associated with declines in auditory acuity. Most older adults experience some decline in auditory function. Age-associated losses in hearing include a loss of sensitivity for pure tones, especially high-frequency tones; difficulty understanding speech, especially if the speech is distorted; problems localizing sounds; problems in binaural listening; and increased sensitivity to loudness (Schieber, Fozard, Gordon-Salant, & Weiffenbach, 1991). These changes in audition are also relevant to design of computer systems. Older people may also find it difficult to understand synthetic speech because it is typically characterized by some degree of distortion. Multimedia systems may also be problematic for older adults. High-frequency alerting sounds such as beeps or pings may also be difficult for older adults to detect.

Motor Skills

Aging is also associated with changes in motor skills including slower response times, declines in ability to maintain continuous movements, disruptions in coordination, loss of flexibility, and greater variability in movement (Rogers & Fisk, 2000). The incidence of chronic conditions such as arthritis also increases with age (Fig. 21.4). These changes in motor skills have direct relevance to the ability of older people to use current input devices such as a mouse or keyboard. For example, various aspects of mouse control, such as moving, clicking, fine-positioning, and dragging, are likely to be difficult for older people.

A study by Smith, Sharit, and Czaja (1999) also examined age differences in the performance of basic computer mouse control techniques (pointing, clicking, double-clicking, and dragging). The data indicated that the older participants had more difficulty performing the tasks than younger participants, especially the complex tasks such as double-clicking. Furthermore, age-related changes in psychomotor abilities such as manual dexterity were related to performance differences.

Other studies that have examined age differences in mouse performance have found similar results. For example, Riviere and Thakor (1996) found that older adults were less successful in performing a tracking task with a mouse than were younger adults. Increased task difficulty resulted in greater age differences. Walker, Millians, and Worden (1996) compared older and younger people on a basic target acquisition task and found that older people had more difficulty, especially with smaller and more distant targets. Charness, Bosman, and Elliott (1995) compared people of different ages on a target acquisition task using mice and keyboard control in one study and target acquisition and scroll-bar dragging tasks using mice and light pens in another study. Although the performance of the older people improved with practice, they performed more slowly in all cases, especially with the mouse. Findings from these studies suggest that alternative input devices might be beneficial for older people. In fact, recent data (Charness & Holley, 2001) suggest that for target selection tasks age differences in performance may be minimized with the use of alternative input devices such as a light pen.

Cognitive Abilities

Age-related changes in cognition also have relevance to the performance of computer-based tasks. It is well established that component behaviors comprising cognition decline with age. Processes that decline include attentional processes, working memory, discourse comprehension, problem solving and reasoning, inference formation and interpretation, and encoding and retrieval in memory (Park, 1992). Aging is also associated with declines in information processing speed (Salthouse, 1985). Older people tend to take longer to process incoming information and typically require more time to respond.

Generally, computer tasks are characterized by cognitive demands. For example, component abilities such as psychomotor speed and attention are important predictors of performance of data-entry tasks (Czaja & Sharit, 1998), and skills such as spatial memory and motor skills are important to word-processing performance. Similarly searching for information on the Internet is a complex cognitive task and involves cognitive skills such as memory, reasoning, attention, learning, and problem solving. Given that there are age-related declines in these component abilities use of current information technologies such as the Internet is likely to be challenging for older adults. In fact, Morrell and Echt (1996) posited that age-related declines in cognitive abilities such as working memory, processing speed, and text comprehension are influential in age-related differences in the acquisition of computer skills. They recommend that software, instructional materials, and training protocols be designed to reduce demands on these cognitive mechanisms.

Although the current technologies may present challenges for older adults, the literature generally suggests that older people are receptive to using new technologies. The following section summarizes the literature on aging and attitudes toward computers.

OLDER ADULTS' ACCEPTANCE OF COMPUTER TECHNOLOGY

One issue that needs to be considered when discussing aging and computers is the willingness of older people to use computer technology. Clearly, the success of this technology in improving the lives of the elderly is highly dependent on the degree to which older people are willing to interact with computer systems. A commonly held belief is that older people are resistant to change and unwilling to interact with "high-tech" products; available data dispute this stereotype, however. The majority of studies that have examined the attitudes of older people toward computer technology indicate that older people are receptive to using computers. They may experience more computer anxiety and less computer efficacy; however, attitudes toward technology and comfort using technology is largely influenced by experience and the nature of interactions with computer systems.

For example, a study (Dyck & Smither, 1994) that examined the relationship among computer anxiety, computer experience, gender, and education found that older adults had more positive attitudes toward computers than younger adults; however, the older people expressed less computer confidence than the younger people. The results also indicated that people who had experience with computers had more positive attitudes and greater computer confidence. Jay and Willis (1992) also found that experience with computers has an influence on computer attitudes. They examined the attitudes toward computers of a sample of older adults before and after participation in a 2-week computer training course. They found that the attitudes of the training participants became more positive following training. Specifically, the participants expressed greater computer comfort and computer efficacy.

Charness et al. (1992) did not find age differences in computer anxiety in a study examining the effects of age, training technique, and computer anxiety on learning to use a word processor. They did, however, find that computer experience had an impact on computer anxiety such that the computer-related anxiety was reduced for all participants following computer training. In a study examining age differences in acquisition of word-processing skills (Czaja, Hammond, Blascovich, & Swede, 1989), the results showed that posttraining attitudes were related to the training experience such that people who rated the training experience and their own performance positively also had more positive posttraining attitudes toward computers. Furthermore, there were no age effects for either pre- or posttraining computer attitudes. Danowski and Sacks (1980) also found that a positive experience with computer-mediated communication resulted in positive attitudes toward computers among a sample of older people. A more recent study (Czaja & Sharit, 1998) examined age differences in computers as a function of computer experience among a community sample of 384 adults ranging in age from 20 to 75 years. The results indicated that in general older people perceived less comfort and efficacy using computers than younger people; however, experience with computers resulted in more positive attitudes for all participants irrespective of age. Kalasky, Sharit, and Czaja (1999) also found that experience with computer systems increased the comfort of older people toward using computers.

Generally, user satisfaction with a system, which is determined by how pleasant the system is to use, is considered an important aspect of usability. User satisfaction is an especially important usability aspect for systems that are used on a discretionary basis such as home computers (Nielsen, 1993). Kelley and colleagues (Kelley, Morrell, Park, & Mayhorn, 1999) found that the most important predictor of continued use of a bulletin board system, ELDERCOMM, was success at initial training. They also found that the most positive attitudes toward computers were found among persons who used the system most often. Cody and colleagues (Cody et al., 1999) in their study of Internet use among older adults also found that people who spent the most time online had the least computer anxiety and high computer efficacy. Finally, in our study of e-mail (Czaja et al., 1993), all participants found it valuable to have a computer in their home; however, the perceived usefulness of the technology was an important factor with respect to use of the technology. When the participants were asked what type of computer applications they would like available, the most common requests included emergency response, continuing education, health information, and banking/shopping.

The available data suggest that older people are not "technophobic" and are willing to use computers; however, the nature of their experience with computers and available computer applications are important determinants of attitudes and willingness to use computer systems.

OLDER ADULTS AND THE ACQUISITION OF COMPUTER SKILLS

As shown in Table 21.1, a number of studies have examined the ability of older adults to learn to use computer technology. These studies span a variety of computer applications and also vary with respect to training strategies such as conceptual versus procedural training (Morrell, Park, Mayhorn, & Echt, 1995) or computer-based or instructor-based versus manual-based training (Czaja, Hammond, Blascovich, & Swede, 1989). In addition, the influence of variables, such as attitude toward computers and computer anxiety on learning have been examined. Overall, the results of these studies indicate that older adults are, in fact, able to use computers for a variety of tasks; however, they often have more difficulty acquiring computer skills than younger people and require more training and more help during training. Also, when compared with younger adults on performance measures such as speed, they often achieve lower levels of performance.

For example, Egan and Gomez (1985) conducted a series of experiments in an attempt to identify individual difference

TABLE 21.1. Computer Training and Older Adults

Study	Age Range	Application	Findings
Caplan & Schooler (1990)	18–60	Drawing software	Conceptual model detrimental for older adults; older adults had lower performance
Charness, Schumann, & Boritz (1992)	25–81	Word processing	Older adults took longer to complete training and required more help
Czaja, Hammond, Blascovich, & Swede (1989)	25–70	Word processing	Older adults took longer to complete task problems and made more errors
Czaja, Hammond, & Joyce (1989)	25–70	Word processing	Older adults took longer to complete task problems and made more errors; goal-oriented training improved performance
Egan & Gomez (1985)	28–62	Word processing	Age and spatial memory were significant predictors of performance
Elias, Elias, Robbins, & Gage (1987)	18–67	Word processing	Older adults had longer learning times and required more help
Garfein, Schaie, & Willis (1988)	49–67	Spreadsheet	No age differences
Gist, Rosen, & Schowoerer (1988)	Mean = 40	Spreadsheet	Older adults performed more poorly on a posttraining test; modeling improved performance
Hartley, Hartley, & Johnson (1984)	18–75	Line editor	No age differences in performance efficiency; older adults took longer to complete training and required more help
Morrell, Park, Mayhorn, & Echt (1995)	Young-old (X = 68.6) Old-old (X = 79.9)	Electronic bulletin board	Procedural training superior to conceptual; age effects
Zandri & Charness (1989)	20–84	Calendar and notepad	Advanced organizer not beneficial; age effects for training time and needed more help

variables that predict ability to learn text editing. They found that age and spatial memory were significant predictors of learning difficulty. Elias, Elias, Robbins, and Gage (1987) conducted a study to examine age differences in the acquisition of text-editing skills and to identify sources of difficulty encountered by older adults. The training program included an audio tape and a training manual. The results indicated that all participants were able to learn the fundamentals of word processing; however, the older adults required more time to complete the training program and required more help. The older people also performed more poorly on a review examination. Garfein, Schaie, and Willis (1988) examined the ability of older adults to learn a spreadsheet package. They also attempted to identify component abilities that are predictive for computer novice to acquire computer skills. The results of their study indicated that all participants were able to operate a computer and use the spreadsheet package after only two brief 90-min sessions of training. There were no significant age effects for the performance measures; however, this may be because the age range of the participant was restricted and only included people ranging in age from 49 to 67 years. In terms of other factors affecting computer proficiency, they found that fluid intelligence was an important predictor of performance. Gist, Rosen, and Schwoerer (1988) also examined the influence of age and training method on the acquisition of a spreadsheet program. The training program consisted of two approaches: tutorial or behavioral modeling. The tutorial approach involved a computer-based step-by-step interactive instructional package. The behavioral modeling involved watching a videotape of a middle-aged male demonstrating the use of the software and then practicing the procedure. The results indicated that the modeling approach was superior to the tutorial approach for both younger and older participants. They also found that older adults performed more poorly on a posttraining test.

Zandri and Charness (1989) investigated the influence of training method on the ability of older people to use a calendar and notepad system. Specifically, they examined if providing an advanced organizer would impact on the acquisition of computer skills for younger and older adults. They also examined if learning with a partner would have an influence on learning. The results indicated that for the older adults who received training without a partner, the advanced organizer resulted in better performance. For the other group of older people, there was no performance effect. For the younger subjects, having the advanced organizer resulted in worse performance if they learned alone, but it made no difference if they learned with partners. These results suggest that the provision of an advanced organizer may be differentially effective for older people under some learning conditions. Furthermore, the older people were about 2.5 times slower than the younger people in the training sessions and they required about three times as much as help.

In a follow-up study, Charness et al. (1992) examined the impact of training techniques and computer anxiety on the acquisition of word-processing skills in a sample of younger and older adults. In the first study, 16 computer novices ranging in age from 25 to 81 years learned word-processing skills under a self-paced training program. Half the participants received the organizer prior to training. Overall, the provision of the organizer did not improve performance. The results also indicated

that older adults took about 1.2 times more than younger adults to complete training, and they required more help.

Czaja, Hammond, Blascovich, and Swede (1989) evaluated the efficacy of three training strategies for novice adults in leaning to use a word-processing program: instructor-based, manual-based, and online training. The results indicated that younger adults were more successful in learning the word-processing program. The results also indicated that the manual and instructor-based training were superior to online training for all participants.

In a follow-up study, Czaja, Hammond, and Joyce (1989) attempted to identify a training strategy that would minimize age differences in learning text editing. Two training programs were evaluated, a goal-oriented program and a traditional approach that included a manual and lecture. The goal-oriented approach introduced the elements of text editing in an incremental fashion, moving from a simpler to a more complex task. The training sessions included problem-solving tasks with objectives of discovering and achieving methods for completing the tasks. The manual was written as a series of goal-oriented units. It used simple language, similarities were drawn between computer and familiar concepts, and the amount of necessary reading was minimized. The results indicated that posttraining performance was better for participants who were trained using the goal-oriented approach. These participants took less time to complete the tasks and made fewer mistakes. Despite of training manipulation, the performance was still lower for older than for younger adults.

Caplan and Schooler (1990) evaluated whether providing the participants with a conceptual model of the software would improve their ability to learn a painting software program. They provided half of the participants with an analogical model of a painting software program before training. The results indicated that the model was beneficial for younger adults but detrimental for older adults. Similar results were found in Morrell, Park, Mayhorn, and Echt (1995) study. The examined the ability of young-old (aged 60–74 years) and old-old (aged 75–89) adults to perform tasks on ELDERCOMM, a bulletin board system. The participants were presented with procedural instructional materials or a combination of conceptual information and procedural instructions. The results indicated that all participants performed better with the procedural instruction material. They also found that the young-old adults had better performance than the old-old adults; the old-old adults made more performance errors. The investigators concluded that conceptual training may not be beneficial for older adults because they need to translate the model into actions that may increase working-memory demands.

More recently, Rogers, Fisk, Mead, Walker, and Cabrera (1996) assessed the efficacy of several instructional methods in teaching older adults to use automatic teller machines. The results indicated that training method did have an influence on performance such that an online tutorial, which provided specific practice on the task components, was superior to written instructions and written instructions accompanied by graphics. The authors discussed the importance of providing older adults with actual training on technologies such as automatic teller machines: Sole reliance on instructional materials or self-discovery may not be optimal for this population, especially for more complex technological applications such as the Internet. Identification of training strategies that are efficacious for older adults is especially important given that continual developments in technology will require lifelong learning for people of all ages. In this regard, Mead and Fisk (1998) examined the impact of the type of information presented during training on the initial and retention performance of younger and older adults learning to use automaticteller technology. Specifically, they compared two types of training: concept and action. The concept training presented factual information, whereas the action training was procedural in nature. The action training was found to be superior for older adults. They showed superior speed and accuracy immediately after training and superior speed following the retention interval. They concluded that presenting procedural information to older adults during training was more important than presenting conceptual information. Mead et al. (1997) examined the effects of type of training on efficiency in a World Wide Web search activity. The participants were trained with a "hands-on" Web navigation tutorial or a verbal description of available navigation tools. The hands-on training was found to be superior, especially for older adults. Older adults who received hands-on training increased the use of efficient navigation tools. These findings suggest that type of training strategy has an impact on the ability of older people to acquire skills. Generally, the data suggest that procedural "hands-on" training with an action component is superior for older adults. However, more research is needed to identify training strategies that facilitate the ability of older people to acquire computer-based skills.

Generally, the literature indicates that older adults are able to use computers for routine tasks and that they are able to learn a wide variety of computer applications. However, they are typically slower to acquire computer skills than younger adults and generally require more help and "hands-on" practice. Furthermore, they typically need training on basic computer concepts, such as mouse and windows management, in addition to training on the application area of interest. They may also require information of the types of technologies that are available, the potential benefits associated with using these technologies, and where and how to access these technologies. Finally, greater attention needs to be given to the design of training and instructional materials to accommodate age-related changes in perceptual and cognitive abilities. On the basis of these findings, Table 21.2 presents some recommendations for the development of training programs for older people.

Aging and Computer Task Performance

Only a handful of studies have examined the ability of older people to perform computer-based tasks. In an early study, Czaja and Sharit (1993) examined age differences in the performance of three simulated real-world computer tasks; a data-entry task: a file modification task, and an inventory management task. They also evaluated differences in subjective assessments of task difficulty and stress. For all three tasks, the older people exhibited longer times and a greater number of errors than the younger people.

TABLE 21.2. Training Guidelines

1. Allow extra time for training; self-paced learning schedules appear to be optimal.
2. Ensure that help is available and easy to access; create a supportive learning environment.
3. Ensure that the training environment is free from distractions.
4. Training materials should be well organized, and important information should be highlighted.
5. Make use of illustrations in training manuals when possible.
6. Training manuals should be designed so that they are concise and easy to read. "How to" information should be presented in a procedural, step-by-step format.
7. Allow the learner to make errors, but provide immediate feedback regarding how to correct mistakes.
8. Provide sufficient practice on task components.
9. Provide an active learning situation; allow the learner to "discover" ways of accomplishing tasks.
10. Structure the learning situation so that the learner proceeds from the simple to the more complex.
11. Minimize demands on spatial abilities and working memory.
12. Familiarize the learner with basic concepts regarding hardware and software and use of the equipment; address any concerns the learner has about use of the equipment (e.g., "Will I break the computer if I do this?").
13. Emphasize distinctions between computers and typewriters.

In a follow-up study, Czaja and Sharit (1998, 1999, in press) examined age differences in a simulated data-entry task, a customer service representative task (information search and retrieval), and an accounts balancing task (commonly performed in the banking industry). They also examined the influence of task experience on performance and the relationship between component cognitive abilities and performance. For all three tasks, older adults performed at lower levels than younger adults with respect to output, and these differences remained task experience. However, there were no age differences in accuracy for the data-entry task after controlling for differences in output. Furthermore, interventions such as redesigning the screen and reconfiguring the timing of the computer mouse improved the performance of all participants. The results also indicated that prior computer experience and component cognitive abilities such as working memory and psychomotor speed were important predictors of performance for all three tasks. These findings have important implications for training and interface design.

Other investigators have examined age as a potential factor impacting on ability to use the computer for information search and retrieval. This is an important area of investigation given that this is one of the most common reasons people use technology such as computers. For example, this type of activity is also central to use of the Internet. Also, in many work settings such as department stores, airlines, hotels, utility and health insurance companies, and education, institutional workers are required to search through computer databases and access information to respond to customer requests. Generally, the findings from these studies indicate that although older adults are capable of performing these types of tasks, there are age-related differences in performance. Furthermore, these differences appear to be related to age differences in cognitive abilities.

For example, Westerman, Davies, Glendon, Stammers, and Matthews (1995) examined the relationship between spatial ability, spatial memory, vocabulary skills, and age and the ability to retrieve information from a computer database, which varied according to how the database was structured (e.g., hierarchical vs. linear). In general, they found that the older subjects were slower in retrieving the information than the younger adults; however, there were no age-related differences in accuracy. The learning rates also differed for the two groups such that the older people were slower than the younger people. They found that the slower response on the part of the older adults was dependent on general processing speed.

Freudenthal (1997) examined the degree to which latencies on an information retrieval task were predicted by movement speed and other cognitive variables in a group of younger and of older adults. The participants were required to search for answers to questions in a hierarchical menu structure. Results indicated that the older subjects were slower than the younger subjects on overall latencies for information retrieval and that this slowing increased with each consecutive step in the menu. Similar to Westerman et al. (1995), that movement speed was a significant predictor of overall latency. Other cognitive abilities, such as reasoning speed, spatial ability, and memory, were also predictive of response latencies; however, memory and spatial abilities only predicted latency on steps further into the menu structure. Freudenthal (1997) suggested that deep menu structures may not be appropriate for older adults because navigation through these types of structures is dependent on spatial skills that tend to decline with age. Vicente, Hayes, and Williges (1987) also found that age, spatial ability, and vocabulary were highly predictive of variance in search latency for a computer-based information retrieval task. They postulated that people with low spatial ability tend to "get lost" in the database. Mead, Jamieson, Rousseau, Sit, and Rogers (1996) examined the ability of younger and older adults to use an online library database. Overall, the younger adults achieved more success in performing the searches than the older adults. They also used more efficient search strategies. The older adults also made more errors when formulating search queries and had more difficulty recovering from these errors. Czaja and Sharit (in press), in their investigation of this type of task, which simulated a customer service representative for a health insurance company, also found that spatial skills were important predictors of performance as were response speed and prior computer experience. Kubeck, Miller-Albrecht, and Murphy (1999) investigated age differences in finding information using the Internet in a naturalistic setting. They found that older people tended to use less efficient search strategies than younger people and were also less likely to find "correct answers." However, they also found that with training older adults were successful in their searches and had positive reactions to their "Internet experiences."

As discussed, information seeking is a complex process and places demands on cognitive abilities such as working memory, spatial memory, reasoning, and problem solving. Information seeking within electronic environment also requires special skills such as knowledge related to the search system. Given that older adults typically experience declines in cognitive abilities such as working memory and are less likely than younger people to have knowledge of the structure and organization of search systems, a relevant question is the degree to which they will experience difficulty searching for information in electronic environments. Generally, the available literature suggests that older adults are able to search and retrieve information within "electronic environments." They appear to have more difficulty than do younger adults, however, and tend to use less efficient navigation strategies. The also appear to have problems remembering where and what they searched. To maximize the ability of older people to successfully interact with electronic information systems such as the Internet and have access to the "information highway," we need to have an understanding of the source of age-related difficulties. This type of information will allow us to develop interface design and training strategies to accommodate individual differences in performance. Currently, there is little information on problems experienced by older people when attempting to learn and navigate the Internet, especially in real-world contexts.

DESIGNING COMPUTER SYSTEMS TO ACCOMMODATE OLDER ADULTS

Hardware Considerations

As discussed, there are of age-related changes in functioning that have implications for the design of computer systems. For example, careful attention needs to be paid to the design of the display screen, labeling of the keyboard, and design and layout of instructional materials and manuals. Design features such as character size and contrast are especially important for older computer users. Generally, larger characters and high-contrast displays are beneficial for older people. This may not be a major problem with most computers used in the home or the workplace because it is relatively easy to enlarge screen characters; however, it may be an issue for computers in public places such as information kiosks and automated teller machines. In addition to character size and contrast, it is also important to minimize the presence of screen glare.

The organization and amount of information on a screen is also important because there are declines in visual search skills and selective attention with age. Only necessary information should be presented on a screen, and important information should be highlighted. Further principles of perceptual organization, such as grouping, should be applied. Caution must also be exercised with respect to the use of color coding because there are declines in color discrimination abilities with age.

Design and labeling of the keyboard also needs special consideration. In our study of e-mail (Czaja et al., 1993) people commonly confused the "send" and the "cancel" keys. Even though these keys were labeled, they were identical in size and shape and close in proximity. These findings underscore the importance of clearly differentiating keys so that labeling is easy to read and keys are easily identified.

Input Devices

Although there is a growing body of research examining the relative merits and disadvantages of various input devices, there are only a few studies that have examined age effects. Generally, these studies suggest that commonly used input devices such as keyboards and the mouse may be problematic for older people. More research needs to be directed toward identifying the efficacy of alternative input devices for older people.

For example, Charness and Holley (2001) suggested that older adults may find a light pen may be easy to use because it is a direct addressing device and eliminates the need translate the target selection device onto the display; however, light pens may be difficult to use for people with arthritis or tremors because it may be difficult for them to grasp and point the pen. Ellis et al. (1991), in their study of a computerized health care system, concluded that a touchscreen version of the appraisal system may eliminate the interface problems experienced by older users. Casali (1992) evaluated the ability of persons with impaired hand and arm function (age unknown) and nondisabled persons to perform a target acquisition task with five cursor control devices: a mouse, a trackball, cursor keys, a joystick, and a tablet. She found that even persons with profound disabilities were able to operate each device by using minor modifications and unique operating strategies. The mouse, track ball, and tablet resulted in better performance than the joystick and cursor key for all participants. Consistent with Charness, she found that small targets were problematic for physically impaired users as was the task of dragging. In a follow-up study (Casali & Chase, 1993) with a sample of persons with arm and hand disabilities, the data indicated that although the mouse, trackball, and tablet tended to result in quicker performance, these devices were more error prone than the keyboard and joystick. In addition, performance improved with practice.

Clearly much work needs to be done to evaluate which types of input devices are optimal for older people, especially for those who have restrictions in hand function. It appears that input devices such as mice and trackballs might not be appropriate for older adults, or at least that they will require more practice to effectively use these types of devices. It may be that speech-recognition devices will eliminate many of the problems (visual and movement) associated with manual input devices. In this regard, Ogozalek and Praag (1986) found that although using a simulated listening typewriter compared with a keyboard editor made no difference in performance of a composition task for either younger and older people, but the voice input was strongly preferred by all participants. More recently, Kalasky, Czaja, Sharit, and Nair (1999) found that a commercially available speech-recognition program was robust in accepting the speech input of both younger and older adults. The technology was also acceptable to both user groups, and the data suggest that the older participants found the system more useful than their younger counterparts. These results are encouraging and suggest that speech interfaces may be especially beneficial for older people.

Software Considerations

There has been little research examining the impact of the design of the software interface on the performance of older computer users. Given that there are age-related changes in cognitive

processes, such as working memory and selective attention, it is likely that interface style (e.g., function keys vs. menus) will have a significant influence on the performance of older adults. The limited data that are available support this conclusion.

Joyce (1990) evaluated the ability of older people to learn a word-processing program as a function of interface style. The participants interacted with the program using one of three interface styles: on-screen menu, functions keys, and pull-down menus. She found that people using the pull-down menus performed better on the word-processing tasks. Specifically, they performed the tasks more quickly and made fewer errors. They also executed a greater number of successful editorial changes. Joyce hypothesized that the pull-down menu reduced the memory demands of the task because the users did not have to remember editing procedures. Although the on-screen menu provided memory cues, the names of the menu items were not reflective of menu contents, thus requiring the user to remember which items were contained in which menu. The names of the pull-down menus were indicative of menu contents.

Egan and Gomez (1985) found that a display editor was less difficult for older adults than a line editor. The line editor was command based and required the user to remember command language and produce complicated command syntax. Using the display editor, changes were made by positioning a cursor at the location of change and using labeled function keys rather than a command language, thus there were fewer memory demands associated with the display editor. Furthermore, the display editor was less complex than the line editor, and in accordance with the age–complexity hypothesis, we would anticipate smaller age effects (Cerella, Poon, & Williams, 1980).

In our study of e-mail (Czaja et al., 1989a), we found that study participants sometimes had difficulty remembering the address name required to access a particular application. We also had to provide on-screen reminders of basic commands (e.g., "Press enter after entering the address") even though the system was simple and there were no complex command procedures. Similarly, in our research on text-editing (Czaja et al., 1989b), we found that the older participants had more difficulty remembering editing procedures than younger adults.

Charness and Holley (2001) examined the differential effect of a keystroke-based command, menu, and menu + icon interface on word-processing skill acquisition among a sample of younger and older adults. They found that the menu and the menu + icon interface yielded better performance for all participants. The menu conditions provided more environmental support and were associated with fewer memory demands.

Consistent with the cognitive aging literature, these results suggest that systems that place minimal demands on working memory would be suitable for older people. In addition, minimal demands should be placed on spatial abilities. In this regard, interface styles such as Windows should prove beneficial; windows may be problematic for older adults, however, because it is spatially organized, and it is sometimes easy to "get lost" within a Windows environment. Also, Windows may initially increase cognitive demand because users are required to learn window operations. On-screen aids such as maps

TABLE 21.3. Summary of Interface Design Guidelines
for Older Adults

1. Maximize the contrast between characters and screen background.
2. Minimize screen glare.
3. Avoid small targets and characters (fonts <12).
4. Minimize irrelevant screen information.
5. Present screen information in consistent locations (e.g., error messages).
6. Adhere to principles of perceptual organization (e.g., grouping).
7. Highlight important screen information.
8. Avoid color discriminations among colors of the same hue or in the blue-green range.
9. Clearly label keys.
10. Maximize size of icons.
11. Use icons that are easily discriminated and meaningful; label icons if possible.
12. Provide sufficient practice on the use of input devices such as a mouse or trackball.
13. Provide sufficient practice on window operations.
14. Minimize demands on spatial memory.
15. Provide information on screen location.
16. Minimize demands on working memory.
17. Avoid complex command languages.
18. Use operating procedures that are consistent within and across applications.
19. Provide easy-to-use, online aiding and support documentation.

and history markers may also prove to be beneficial for older people.

Clearly, there are a number of interface issues that need to be investigated. At the present time, we can only offer general guidelines with respect to designing interfaces to accommodate the needs of older users. Table 21.3 presents a summary of these guidelines. An abundance of research needs to be carried out within this area.

CONCLUSIONS

There are many areas where older people are likely to interact with computer technology, including the workplace, the home, and service and health care settings. Current data indicate that older adults are generally receptive to using this type of technology but often have more difficulty than younger people acquiring computer skills and using current computer systems. This presents a challenge for human–computer interaction community. For many older people, especially those who are frail, isolated, or who have some type of mobility restrictions, access to computers holds the promise of enhancing independence by providing linkages to goods and resources, facilitating communication, and enhancing the ability to perform routine tasks such as banking and shopping. Computer technology can also help older people access information on health and other topics and manage personal finances. As developments in computer technology continue to emerge, users will have faster access to more channels of information in a myriad of formats.

Although research in this area has grown, there are many unanswered questions. For example, issues of screen design, input devices, and interface style are largely unexplored, and our knowledge within these areas is limited. To maximize the potential benefits of computer technology for the older population and optimize the interactions of older people with these types of systems, we must perceived them as active technology users and include them in system design and evaluation efforts. In essence, to design interfaces for information systems so that they are useful and usable for older people, it is important to understand (a) why technology may be difficult to use, (b) how to design technology for easier and more effective use, and (c) how to effectively teach people to take advantage of available technologies. Answers to these questions will not only serve to benefit older adults, but all potential computer users.

References

Administration on Aging, U.S. Department of Health and Human Services. (2000). *A profile of older Americans: 2000.* Washington, DC: U.S. Government Printing Office.

Birren, J. E., & Schaie, K. W. (2001). *Handbook of the psychology and aging.* San Diego, CA: Academic Press.

Caplan, L. J., & Schooler, C. (1990). The effects of analogical training models and age on problem-solving in a new domain. *Experimental Aging Research, 16,* 151–154.

Casali, S. P. (1992). Cursor control use by persons with physical disabilities: Implications for hardware and software design. *Proceedings of the 36th Annual Meeting of the Human Factors Society* (pp. 311–315). Atlanta, Georgia.

Casali, S. P., & Chase, J. (1993). The effects of physical attributes of computer interface design on novice and experienced performance of users with physical disabilities. *Proceedings of the 37th Annual Meeting of Human Factors and Ergonmics Society* (pp. 849–853). Seattle, Washington.

Cerella, J., Poon, L. W., & Williams, D. (1980). Age and the complexity hypothesis. In L. W. Poon (Ed.), *Aging in the 1980's* (pp. 332–340). Washington, DC: American Psychological Association.

Charness, N. (1985). *Aging and human performance.* New York: John Wiley & Sons.

Charness, N., Bosman, E. A., & Elliot, R. G. (1995, August). Senior-friendly input devices: Is the pen mightier than the mouse? *Paper presented at the 103rd Annual Convention of the American Psychological Association Meeting.* New York, New York.

Charness, N., & Holley, P. (2001, June). Minimizing computer performance deficits via input devices and training. *Presentation prepared for the Workshop on Aging and Disabilities in the Information Age.* Baltimore, Maryland, John Hopkins University.

Charness, N., Schumann, C. E., & Boritz, G. A. (1992). Training older adults in word processing: Effects of age, training technique and computer anxiety. *International Journal of Aging and Technology, 5,* 79–106.

Chute, D. L., & Bliss, M. E. (1994). ProsthesisWare: Concepts and caveats for microcomputer-based aids to everyday living. *Experimental Aging Research, 20,* 229–238.

Cody, M. J., Dunn, D., Hoppin, S., & Wendt, P. (1999). Silver surfers: Training and evaluating Internet use among older adult learners. *Communication Education, 48,* 269–286.

Czaja, S. J., Guerrier, J. H., Nair, S. N., & Laudauer, T. K. (1993). Computer communication as an aid to independence for older adults. *Behavior & Information Technology, 12*, 197-207.

Czaja, S. J., Hammond, K., Blascovich, J., & Swede, H. (1989b). Age-related differences in learning to use a text-editing system. *Behavior and Information Technology, 8*, 309-319.

Czaja, S. J., Hammond, K., & Joyce, J. B. (1989a). *Word processing training for older adults*. Final report submitted to the National Institute on Aging (Grant No. 5 R4 AGO4647-03).

Czaja, S. J., & Rubert, M. (in press). Telecommunications technology as an aid to family caregivers of persons with dementia. *Psychosomatic Medicine.*

Czaja, S. J., & Sharit, J. (1993). Age differences in the performance of computer based work as a function of pacing and task complexity, *Psychology and Aging, 8*, 59-67.

Czaja, S. J., & Sharit, J. (in press). Examining age differences in performance of a complex information search and retrieval task. *Psychology and Aging.*

Czaja, S. J., & Sharit, J. (1998). Ability-performance relationships as a function of age and task experience for a data entry task. *Journal of Experimental Psychology: Applied, 4*, 332-351.

Czaja, S. J., & Sharit, J. (1999). Performance of a complex computer-based trouble shooting task in the bank industry. *International Journal of Cognitive Ergonomics and Human Factors, 3*, 1-22.

Czaja, S. J., Sharit, J., Ownby, R., Roth, D. L., & Nair, S. (2001). Examining age differences in performance of a complex information search and retrieval task. *Psychology and Aging, 16*, 564-579.

Danowski, J. A., & Sacks, W. (1980). Computer communication and the elderly. *Experimental Aging Research, 6*, 125-135.

Dawson, D., Hendershot, G., & Fulton, J. (1987). Aging in the eighties: Functional limitations of individuals aged 65 years and older. *National Center for Health Statistics Advance Data 1987, 133*, 1-11.

Dyck, J. L., & Smither, J. A. (1994). Age differences in computer anxiety: The role of computer experience, gender and education. *Journal of Education Computing Research, 10*, 239-248.

Egan, D. E., & Gomez, L. M. (1985). Assaying, isolating, and accommodating individual differences in learning a complex skill. *Individual Differences in Cognition, 2*, 174-217.

Eilers, M. L. (1989). Older adults and computer education: "Not to have the world a closed door." *International Journal of Technology and Aging, 2*, 56-76.

Elias, P. K., Elias, M. F., Robbins, M. A., & Gage, P. (1987). Acquisition of word-processing skills by younger, middle-aged, and older adults. *Psychology and Aging, 2*, 340-348.

Ellis, L. B. M., Joo, H., & Gross, C. R. (1991). Use of a computer-based health risk appraisal by older adults. *Journal of Family Practice, 33*, 390-394.

Family Caregiving Alliance. (1997). *Annual report: California's caregiver resource center system fiscal year 1996-1997*. San Francisco: Family Caregiver Alliance.

Fisk, A. D., & Rogers, W. (1997). *The handbook of human factors and the older adults*. San Diego, CA: Academic Press.

Freudenthal, D. (1997). *Learning to use interactive devices; age differences in the reasoning process*. Unpublished master's thesis, Eindhoven University of Technology, The Netherlands.

Furlong, M. S. (1989). An electronic community for older adults: The SeniorNet network. *Journal of Communication, 39*, 145-153.

Gallienne, R. L., Moore, S. M., & Brenna, P. F. (1993). Alzheimer's caregivers: Psychosocial support via computer networks. *Journal of Gerontological Nursing, 12*, 1-22.

Garfein, A. J., Schaie, K. W., & Willis, S. L. (1988). Microcomputer proficiency in later-middle-aged adults and older adults: Teaching old dogs new tricks. *Social Behavior, 3*, 131-148.

Gist, M., Rosen, B., & Schwoerer, C. (1988). The influence of training method and trainee age on the acquisition of computer skills. *Personal Psychology, 41*, 255-265.

Hartley, A. A., Harley, J. T., & Johnson, S. A. (1984). The older adult as a computer user. In P. K. Robinson, J. Livingston, & J. E. Birren (Eds.), *Aging and technological advances* (pp. 347-348). New York: Plenum Press.

Jay, G. M., & Willis, S. L. (1992). Influence of direct computer experience on older adults attitude towards computer. *Journal of Gerontology: Psychological Sciences, 47*, 250-257.

Joyce, B. J. (1990). *Identifying differences in learning to use a text-editor: The role of menu structure and learner characteristics*. Unpublished master's thesis, State University of New York at Buffalo.

Kalasky, M. A., Czaja, S. J., Sharit, J., & Nair, S. (1999). Is speech technology robust for older populations? *Proceedings of the 43rd annual Meeting of the HFES, 43*, 123-128.

Kelly, C. L., Morrell, R. W., Park, D. C., & Mayhorn, C. B. (1999). Predictors of electronic bulletin board system use in older adults. *Educational Gerontology, 25*, 19-35.

Kline, D. W., & Schieber, F. J. (1985). Vision and aging. In J. E. Birren & K. W. Schaie (Eds.), *Handbook of the psychology and aging* (pp. 296-331). New York: Van Nostrand Reinhold.

Kubeck, J. W., Miller-Albrecht, S. A., & Murphy, M. D. (1999). Finding information on the World Wide Web: Exploring older adults' exploration. *Educational Gerontology, 25*, 167-183.

Leirer, V. O., Morrow, D. G., Tanke, E. D., & Pariante, G. M. (1991). Elders nonadherence: Its assessment and medication reminding by voice mail. *The Gerontologist, 31*, 514-520.

Mead, S. E., & Fisk, A. D. (1998). Measuring skill acquisition and retention with an ATM simulator: The need for age-specific training. *Human Factors, 40*, 516-523.

Mead, S. E., Sit, R. A., Jamieson, B. A., Rousseau, G. K., & Rogers, W. A. (1996, August). Online library catalog: Age-related differences in performance for novice users. *Paper presented at the Annual Meeting of the American Psychological Association*. Toronto, Canada.

Mead, S. E., Spaulding, V. A., Sit, R. A., Meyer, B., & Walker, N. (1997). Effects of age and training on World Wide Web navigation strategies. *Proceedings of the Human Factors and Ergonomics Society 41st Annual Meeting* (pp. 152-156). Human Factors and Ergonomics Society.

Morrell, R. W., & Echt, K. V. (1996). Instructional design for older computer users: The influence of cognitive factors. In W. A. Rogers, A. D. Fisk, & N. Walker (Eds.), *Aging and skilled performance: Advances in theory and application* (pp. 241-265). Mahwah, NJ: Lawrence Erlbaum Associates.

Morrell, R. W., Mayhorn, C. B., & Bennett, J. (2000). A survey of World Wide Web in middle-aged and older adults. *Human Factors, 42*, 175-185.

Morrell, R. W., Park, D. C., Mayhorn, C. B., & Echt, K. V. (1995, August). Older adults and electronic communication networks: Learning to use ELDERCOMM. *Paper presented at the 103rd Annual Convention of the American Psychological Association*. New York, New York.

Nair, S. (1989). *A capability-demand analysis of grocery shopping problems encountered by older adults*. Thesis design, submitted to the department of Industrial Engineering, State University of New York at Buffalo, in partial fulfillment for the requirements for Master of Science.

National Alliance for Caregiving and the Alzheimer's Association. (1999). *Caregiving's heavy toll on family demands a response: New study challenges private, public forces to do their part*. Bethesda, MD.

National Research Council. (1997). *More than screen deep: Toward every-citizen interfaces to the nation's information infrastructure.* Washington, DC: National Academy Press.

National Telecommunications and Information Administration. (2000). *Falling through the Net: Toward digital inclusion.* Washington, DC: U.S. Census Bureau.

Nickerson, R. S., & Landauer, T. K. (1997). Human-computer interaction: Background and issues. In M. G. Helander, T. K., Landauer, & P. V. Prabhu. *Handbook of human-computer interaction* (2nd ed., pp. 3-32). Amsterdam, The Netherlands: Elsevier.

Nielsen, J. (1993). *Useability engineering.* New York: Academic Press.

Ogozalek, V. Z., & Praag, J. V. (1986). Comparison of elderly and younger users on keyboard and voice input computer-based composition tasks. *Proceedings of CHI '86 Human Factors in Computing Systems* (pp. 205-211). New York: ACM.

Park, D. (1992). Applied cognitive aging research. In F. I. M. Craik & T. A. Salthouse (Eds.), *The handbook of aging and cognition* (pp. 44-494). Mahwah, NJ: Lawrence Erlbaum Associates.

Riviere, C. N., & Thakor, N. V. (1996). Effects of age and disability on tracking tasks with a computer mouse: Accuracy and linearity. *Journal of Rehabilitation Research and Development, 33,* 6-15.

Rogers, W., & Fisk, A. (2000). Human factors, applied cognition, and aging. In F. I. M. Craik & T. A. Salthouse (Eds.), *The handbook of aging and cognition.* Mahwah, NJ: Lawrence Erlbaum Associates.

Rogers, W. A., & Fisk, A. D. (2001). *Human Factors Interventions for the Health Care of Older Adults.* Mahwah, NJ: Lawrence Erlbaum Associates.

Rogers, W. A., Fisk, A., Mead, S. E., Walker, N., & Cabrera, E. F. (1996). Training older adults to use automatic teller machines. *Human Factors, 38,* 425-433.

Salthouse, T. A. (1985). Speed of behavior and its implication for cognition. In J. E. Birren & K. W. Schaie (Eds.), *Handbook of the psychology of aging* (pp. 400-426). New York: Van Nostrand Reinhold.

Schieber, F., Fozard, J. L., Gordon-Salant, S., & Weiffenbach, J. W. (1991). Optimizing sensation and perception in older adults. *International Journal of Industrial Ergonomics, 7,* 133-162.

Schmidt, F. L., Hunter, J. E., Outerbridge, A. N., & Goff, S. (1988). Joint relation of experience and ability with job performance: Tests of three hypotheses. *Journal of Applied Psychology, 73,* 46-57.

Schwartz, L. K., & Plude, D. (1995, August). Compact disk-interactive memory training with the elderly. *Paper presented at the 103rd Annual Convention of the American Psychological Association.* New York: New York.

Smith, N. W., Sharit, J., Czaja, S. J. (1999). Aging, motor control, and performance of computer mouse tasks. *Human Factors, 41,* 389-396.

Tanke, E. D., & Leirer, V. O. (1993). Use of automated telephone reminders to increase elderly patient's adherence to tuberculosis medication appointment. *Proceedings of the Human Factors and Ergonomics Society, 37th Annual Meeting* (pp. 193-196). Human Factors and Ergonomics Society.

U.S. Census Bureau. (1999). *Computer use in the United States.* Washington, DC: Author.

U.S. Department of Health and Human Services Administration on Aging (2000). *Falling through the net: Toward digital inclusion.* Washington, D.C.: U.S. Department of Commerce.

Vicente, K. J., Hayes, B. C., & Williges, R. C. (1987). Assaying and isolating individual differences in searching a hierarchical file system. *Human Factors, 29,* 349-359.

Walker, N., Millians, J., Worden, A. (1996). Mouse accelerations and performance of older computer users. *Proceedings of Human Factors and Ergonomics Society 40th Annual Meeting* (pp. 151-154). Santa Monica, CA: Human Factors and Ergonomics Society.

Westerman, S. J., Davies, D. R., Glendon, A. I., Stammer, R. B., & Matthews, G. (1995). Age and cognitive ability as predictors of computerized information retrieval. *Behaviour and Information Technology, 14,* 313-326.

Zandri, E., & Charness, N. (1989). Training older and younger adults to use software. *Educational Gerontology, 15,* 615-631.

·22·

HUMAN–COMPUTER INTERACTION FOR KIDS

Amy Bruckman and Alisa Bandlow
Georgia Institute of Technology

DESIGNING FOR CHILDREN

How is designing computer software and hardware for kids different from designing for adults? At the time of this writing, little formal research has been done on this topic, and most that has been done has focused on designing educational software. Evaluation of these products is primarily related to learning outcomes, not usability, but usability is a prerequisite for learning. For student projects in a Georgia Tech graduate class, Educational Technology: Design and Evaluation, many student designers are never able to show whether the educational design of their software is successful. What they find instead is that usability problems intervene, and they are unable even to begin exploring pedagogical efficacy. If children can't use educational technology effectively, they certainly can't learn through the process of using it. Usability is similarly important for entertainment, communications, and other applications.

In designing for children, people tend to assume that they are creative, intelligent, and capable of great things if given support and the right tools. If children can't use technologies we've designed, it is our failure as designers. These assumptions are constructive, because users generally rise to designers' expectations. In fact, the same assumptions are useful in designing for adults. Designers of software for children start out at an advantage, because they tend to believe in their users. However, they may be at a disadvantage, because they no longer remember the physical and cognitive differences of being a child.

In this chapter, we

- describe how children's abilities change with age, as it relates to HCI;
- discuss how children differ from adults cognitively and physically, for those characteristics most relevant for HCI;
- review recommendations on laboratory-based testing with kids;
- discuss participatory design with children as design partners; and
- review genres of computer technology for kids and design recommendations for each genre.

HOW ARE CHILDREN DIFFERENT?

As people develop from infants to adults, their physical and cognitive abilities increase over time (Kail, 1991; Miller & Vernon, 1997; Thomas, 1980). The Swiss psychologist Jean Piaget was a leading figure in analyzing how children's cognition evolves (Piaget, 1970). He showed that children do not simply lack knowledge and experience, but they also fundamentally experience and understand the world differently than adults do. He divided children's development into a series of stages (Piaget, 1970, pp. 29–33): sensorimotor (birth–2 years), preoperational (ages 2–7), concrete operational (ages 7–11), and formal operational (ages 11 and up).

Contemporary research recognizes that all children develop differently, and individuals may differ substantially from this typical picture (Schneider, 1996). However, this general characterization remains useful.

In the sensorimotor stage, children's cognition is heavily dependent on what their senses immediately perceive. Software for children this young is difficult to design. Little interaction can be expected from the child. Obviously, all instruction must be given in audio, video, or animation because babies can't read. Furthermore, babies generally cannot be expected to use the mouse effectively, even with large targets. Jumpstart Baby by Knowledge Adventure (http://www.knowledgeadventure.com) is recommended for ages 9 to 24 months. The child is presented with a mobile of spinning icons, each representing a different activity. One icon spins at a time. The child can either click on the icon for the activity he or she wishes to play or hit any key while the icon for that activity is spinning. This eliminates the need for the child to click on a specific mouse target. Within each activity, the infant simply hits any key to advance the animation through fixed patterns. For example, in the jigsaw puzzle activity, hitting a key puts a puzzle piece in place. The user can't chose which piece or where it goes. The puzzle is finished one piece at a time by simply having the child hit the keyboard periodically.

Another title aimed at this age group is Play with the Teletubbies. Also by Knowledge Adventure, it is aimed at ages 1 to 4 years. In this program, an animated world runs semi-autonomously. To give the child greater feedback, mouse movement is accentuated by surrounding the cursor with a shower of sparkles. The sparkles intensify when the mouse moves and when it touches an active part of the scene. The animation proceeds on its own most of the time, but the child's actions add additional sounds and may modify the animation slightly. For example, click on a Teletubby, and it waves hello and then continues what it was doing. Clicking on a specific part of the scene is occasionally necessary to move the story forward. Repetition is used extensively.

Jumpstart Baby is designed in accordance with adult expectations of what a baby should like. The narrator, a teddy bear, addresses the child and invites him or her to play. A motherly voice helps the child play hide-and-seek to find where teddy is hiding. The design is in strong conformance with adult stereotypes of what babies like. In contrast, Teletubbies is out of harmony with those stereotypes. Many adults find the television show and software bizarre and grating, but it is wildly popular with toddlers. The designers of the original British Broadcasting Company (BBC) television series, Anne Wood and Andy Davenport, used detailed observations of young children's play and speech in their design. Wood comments,

Our ideas always come from children. If you make something for children, the first question you must ask yourself is, "What does the world look like to children?" Their perception of the world is very different to that of grown-ups. We spend a lot of time watching very young children: how they play; how they react to the world around them; what they say." (Davenport & Wood, 1997)

Focus groups also played an important role (BBC, 1997). Young children are so radically different from adults that innovative design requires careful fieldwork.

FIGURE 22.1. Children playing with Music Blocks. (Photo Courtesy of Amy Williams of Neurosmith.)

Whereas toddlers' interaction with software on a standard desktop computer affords limited possibilities, specialized hardware can expand the richness and complexity of interactions. For example, Music Blocks by Neurosmith is recommend for ages 2 and up. Five blocks fit in slots in the top of a device rather like a "boom box" portable music player. Each block represents a phrase of music. Each side of the block is a different instrumentation of that musical phrase. Rearranging the blocks changes the music (http://www.neurosmith.com). Interaction of this complexity would be impossible for 2-year-olds using a screen-based interface, but it is quite easy with specialized hardware.

In the preoperational stage (ages 2–7), children's attention span is brief. They can only hold one thing in memory at a time. They have difficulty with abstractions. They can't understand situations from another person's point of view. Although some children may begin to read at a young age, designs for this age group generally assume the children are still preliterate. It is reasonable to expect that children at this age can click on specific mouse targets, but they must be relatively large. Most designers generally avoid making it necessary for the child to use the keyboard.

In the concrete operational stage (ages 7–11), "we see children maturing on the brink of adult cognitive abilities. Though they cannot formulate hypothesis, and though abstract concepts such as ranges of numbers are often still difficult, they are able to group like items and categorize" (Schneider, 1996, p. 69). Concrete operational children are old enough to use relatively sophisticated software, but still young enough to appreciate a playful approach. It is reasonable to expect simple keyboard use; children's ability to learn to type increases throughout this age group. It is also reasonable to expect relatively fine control of the mouse.

Finally, by the time a child reaches the formal operational stage (ages 12 and up), designers can assume his or her thinking is generally similar to that of an adult, but their interests and tastes remain different, of course. Designing for this age group is much less challenging, because adult designers can at least partially rely on their own intuitions.

In the next sections, we'll focus on several characteristics of children most relevant for HCI research: dexterity, speech, reading, background knowledge, and interaction style.

Dexterity

Young children's fine motor control is not equal to that of adults (Thomas, 1980), and they are physically smaller. Devices designed for adults may be difficult for children to use. Joiner, Messer, Light, and Littleton (1998) noted that "the limited amount of research on children has mainly assessed the performance of children at different ages and with different input devices." Numerous studies confirm that children's performance with mice and other input devices increases with age (Joiner et al., 1998). Compared with adults, children have difficulty holding down the mouse button for extended periods and have difficulty performing a dragging motion (Strommen, 1994). Kids have difficulty with marquee selection. Marquee selection is a technique for selecting several objects at once using a dynamic selection shape. In traditional marquee selection, the first click on the screen is the initial, static corner of the selection shape (typically a rectangle). Dragging the mouse controls the diagonally opposite corner of the shape, allowing the user to change the dimensions of the selected area to encapsulate the necessary objects. Dragging the mouse away from the initial static corner increases the size of the selection rectangle, whereas dragging the mouse toward the initial static corner decreases the size of the selection rectangle. A badly placed initial corner can make it difficult and sometimes impossible to select and encapsulate all of the objects. Berkovitz (1994) experimented with a new encirclement technique: The initial area of selection is specified with an encircling gesture, and moving the mouse outside of the area enlarges it.

Kids may have trouble double-clicking, and their small hands may have trouble using a three-button mouse (Bederson, Hollan, Druin, Stewart, Rogers, & Proft, 1996). As with adults, point-and-click interfaces are easier to use than drag-and-drop (Inkpen, 2001; Joiner et al., 1998). Inkpen (2001, p. 30) noted that "Despite this knowledge, children's software is often implemented to utilize a drag-and-drop interaction style. Bringing solid research and strong results . . . to the forefront may help make designers of children's software think more about the implications of their design choices."

Strommen (1998) noted that because young children can't reliably tell their left from their right, interfaces for kids should not rely on that distinction. In his Actimates interactive plush toy designs, the toys' left and right legs, hands, and eyes always perform identical functions.

Speech

Speech recognition has intriguing potential for a wide variety of applications for children. O'Hare and McTear (1999) studied use of a dictation program by 12-year-olds and found that using speech recognition allowed them to generate text more quickly and accurately than typing did. They noted that dictation automatically avoids some of the errors children would otherwise make, because the recognizer generates correct spelling and capitalization. This is desirable in applications in which generating correct text is the goal. If instead the goal is to teach children to write correctly (and, for example, to capitalize their sentences), then dictation software may be counterproductive.

Whereas O'Hare and McTear (1999) were able to use a standard dictation program with 12-year-olds, Nix, Fairweather, and Adams (1998) noted that speech recognition developed for adults will not work with very young children. In their research on a reading tutor for children aged 5 to 7 years old, they first tried a speech recognition program designed for adults. The recognition rate was only 75%, resulting in a frustrating experience for their subjects. Creating a new acoustic model from the speech of children in the target age range, they were able to achieve an error rate of less than 5%. Further gains were possible by explicitly accounting for common mispronunciations and children's tendency to respond to questions with multiple words where adults would typically provide a one-word answer. Even with the improved acoustic model, the recognizer still made mistakes. To avoid frustrating the children with incorrect feedback, they chose to have the system never tell the child they were wrong. When the system detects what it believes to be a wrong answer, it simply gives the child an easier problem to attempt.

Reading

The written word is the main vehicle for most communication between humans and computers. Consequently, designing computer technology for children with developing reading skills presents a challenge. Words must be chosen that are at an appropriate reading level for the target population. Larger font sizes are generally preferred. Bernard, Mills, Frank, and McKnown (2001) found that kids aged 9 to 11 years old prefer 14-point over 12-point fonts. Surprisingly, at the time of this writing, this is the only known empirical study in this area. Most designers follow the rule of thumb that the younger the child, the larger the font should be.

Designing for preliterate children presents a special challenge. Audio, graphics, and animation must substitute for all functions that would otherwise be communicated in writing. The higher production values required can add significantly to development time and cost.

Background Knowledge

Many user interfaces are based on metaphors (Erickson, 1990) from the adult world. Jones (1992) noted that children are less likely to be familiar with office concepts such as file folders and in-out boxes. In designing an animation system for kids, Halgren, Fernandes, and Thomas (1995) found many kids to be unfamiliar with both the metaphor of a frame-based film strip and that of a video cassette recorder. It is helpful to choose metaphors that are familiar to kids, although kids often have success in learning interfaces based on unfamiliar metaphors if they are clear and consistent (Schneider, 1996).

Interaction Style

Children's patterns of attention and interaction are quite different from those of adults. Children are easily distractible. Hanna, Risden, and Alexander (1997) used a funny noise as an error message and found that the children repeatedly generated the error to hear the noise. Similarly, Halgren and colleagues (1995) found that children would click on any readily visible feature just to see what would happen, and they might click on it repeatedly if it generated sound or motion in feedback. They chose to redesign their interface to hide advanced functionality in drawers. Children found the drawer metaphor familiar. "By hiding the advanced tools, the novice users would not stumble onto them and get lost in their functionality. Rather, only the advanced users who might want the advanced tools would go looking for more options. This redesign allows the product to be engaging to and usable by a wider range of ages and abilities" (Halgren et al., 1995, p. 521).

Children are more likely than adults to work with more than one person at a single computer. They enjoy doing so to play games (Inkpen, 1997) and may be forced to do so because of limited resources in school (Stewart, Raybourn, Bederson, & Druin, 1998). Teachers may also create a shared-computer setup to promote collaborative learning. When multiple children work at one machine simultaneously, they need to negotiate sharing control of input devices. Giving students multiple input devices increases their productivity and their satisfaction (Inkpen, 1997; Inkpen et al., 1995; Stewart et al., 1998). Inkpen et al. compared two protocols for transferring control between multiple input devices: give and take. In a give protocol, the user with control clicks the right mouse button to cede it to the other user; in a take protocol, the idle user clicks to take control. In one study with 12-year-olds and another with 9- to 13-year-olds, they found that girls solve more puzzles with a "give" protocol, but boys are more productive with a "take" protocol (Inkpen, 1997; Inkpen et al., 1995). (For more on issues of gender and HCI, see chapter 21 by Justine Cassell in this volume.)

CHILDREN AND USABILITY TESTING

Several usability guidelines developed for work with adults become more important when applied to children. For example, it is important to emphasize that it is the software that is being tested, not the participant (Rubin, 1994). Children might become anxious at the thought of taking a test, and test taking may conjure up thoughts of school. The researcher can emphasize that even though the child is participating in a test, the child is not the one being tested (Hanna et al., 1997). Rubin recommended that you show the participant where the video cameras are located, let them know what is behind the one-way mirror, and tell them whether people will be watching. With children, showing them what is behind the one-way mirrors and taking them around the lab gives them "a better sense of control and trust in you" (Hanna et al., 1997, p. 12).

Hanna et al. (1997) developed the following set of guidelines for laboratory-based usability testing with children:

- The lab should be made more child-friendly by adding colorful posters, but avoid going overboard because too many extra decorations may become distracting to the child.
- Try to arrange furniture so that children are not directly facing the video camera and one-way mirror because the children may choose to interact with the camera and mirror rather than to do the task at hand.
- Children should be scheduled for 1 hour of lab time. Preschoolers will generally only be able to work for 30 min but will need extra time to play and explore. Older children will become tired after an hour of concentrated computer use, so if the test will last longer than 45 min, children should be asked if they would like to take a short break at some point during the session.
- Hanna and colleagues suggested that researchers explain confidentiality agreements to children by telling them that designs are "top-secret." Parents should also sign the agreements because they will inevitably also see and hear about the designs.
- Children up to 7 or 8 years old will need a tester in the room with them for reassurance and encouragement. They may become agitated by being alone or following directions from a loudspeaker. If a parent will be present in the room with the child, it is important to explain to the parent that he or she should interact with the child as little as possible during the test. Older siblings should stay in the observation area or a separate room during the test because they may eventually be unable to contain themselves and start to shout out directions.
- Hanna et al. suggested that researchers not ask children if they want to play the game or do a task because this gives them the option to say no. Instead, researchers should use phrases such as, "Now I need you to . . . " or "Let's do this . . . " or "It's time to . . . ."

Disagreement over Use of Video

There is some disagreement over the use of video cameras in research. Druin (1999) and her design team prefer not to use video cameras during observations of children. They found that children tended to "freeze" or "perform" when they saw a video camera in the room. There are also technical difficulties to deal with. Her research team found that even with smaller cameras, it was difficult to capture data in small bedrooms and large public spaces. The sound and speech captured in public spaces was difficult to understand or even inaudible. Finally, it was difficult to know where to place cameras because they did not know where children would sit, stand, or move within the environment. Druin did, however, encourage her design team to use video cameras (along with journal writing, team discussion, and adult debriefing) as a way to record their brainstorming sessions and other design activities.

Goldman-Segall (1996) argued that digital video data is an important part of ethnographic interviews and observations. When using video, the researcher doesn't have to worry about remembering or writing down every detail: "She can concentrate fully on the person and on the subtleties of the conversation." The researcher also has access to "a plethora of visual stimuli which can never be 'translated' into words in text," such as body language, gestures, and facial expressions. It is especially important to be able to review the body language of children as they interact with software. Hanna et al. (1997, pp. 13–14) stated that children's "behavioral signs are much more reliable than children's responses to questions about whether or not they like something, particularly for younger children. Children are eager to please adults, and may tell you they like your program just to make you happy." Video is extremely useful in being able to study these behavioral signs because researchers may miss some important signs and gestures during the actual observation or interview.

Children as Design Partners

Participatory design is an "approach towards computer systems design in which the people destined to *use* the system play a critical role in *designing* it" (Schuler & Namioka, 1993). With children, this idea is even more important: Because they are physically and cognitively different from adults, their participation in the design process may offer significant insights. Schuler and Namioka wrote:

[Participatory design] assumes that the workers themselves are in the best position to determine how to improve their work and their work life . . . It views the users' perceptions of technology as being at least as important to success as fact, and their feelings about technology as at least as important as what they can do with it." (1993, p. xi)

Empowering children in this way and including them in the design process can be difficult because of the traditionally unequal power relationships between young people and adults.

Cooperative Inquiry

Druin (1999) developed new research methods that include children in various stages of the design process, developing new technologies for children with children. This approach, called *cooperative inquiry*, is a combination of participatory design, contextual inquiry, and technology immersion. Children and adults work together on a team as research and design partners. She and her colleagues reiterated the idea that "each team member has experiences and skills that are unique and important, no matter what the age or discipline" (Alborzi et al., 2000).

In this model, the research team frequently observes children interacting with software, prototypes, or other devices to gain insight into how child users will interact with and use these tools. When undertaking these observations, adult and child researchers both observe, take notes, and interact with the child users. During these observations, there are always at least two note takers and one interactor, and these roles can be filled by either an adult or child team member. The interactor is the researcher who initiates discussion with the child user and asks questions concerning the activity. If there is no interactor or if the interactor takes notes, the child being observed may feel uncomfortable, as if he or she is "on stage" (Druin, 1999). Other researchers have found that the role of interactor can be useful for members of the design team. Scaife and Rogers (1999) successfully involved children as informants in the development of ECOi, a program that teaches children about ecology. They wanted children to codesign certain animations in ECOi. Rather than simply having the software designer observe the children as they played with and made comments about the ECOi prototypes, the software designer took on the role of interactor to elicit suggestions directly. Through these on-the-fly, high-tech prototyping sessions, they learned that "it was possible to get the software designer to work more closely with the kids and to take on board some of their more imaginative and kid-appealing ideas" (Scaife & Rogers, 1999, p. 43).

When working as design partners, children are included from the beginning. The adults do not develop all the initial ideas and then later see how the children react to them. The children participate from the start in brainstorming and developing the initial ideas. The adult team members need to learn to be flexible and learn to break away from carefully following their session plans, which is too much like school. Children can perform well in this more improvisational design setting, but the extent to which they can participate as a design partner depends on his/her age. Children younger than 7 years have difficulty expressing themselves verbally and being self-reflective. These younger children also have difficulty in working with adults to develop new design ideas. Children older than age 10 are typically beginning to become preoccupied with preconceived ideas of the way "things are supposed to be." In general, it has been found that children aged 7 to 10 years are the most effective prototyping partners. They are "verbal and self-reflective enough to discuss what they are thinking" and understand the abstract idea that their low-tech prototypes and designs are going to be turned into technology in the future. They also don't get bogged down with the notion that their designs must be similar to preexisting designs and products.

Through her work with children as design partners, Druin (1999; Druin et al., 2001) has discovered that there are stumbling blocks on the way to integrating children into the design process and to help adults and children work together as equals. One set of problems deals with the ability of children to express their ideas and thoughts. When the adult and children researchers are doing observations, it is best to allow each group to develop its own style of note taking. Adults tend to take detailed notes, and children tend to prefer to draw cartoons with short, explanatory notes. It is often difficult to create one style of note taking that will suit both groups. Because children may have a difficult time communicating their thoughts to adults, low-tech prototyping is an easy and concrete way for them to create and discuss their ideas. Art supplies such as paper, crayons, clay, and string allow adults and children to work on an equal footing. A problem that arises in practice is that because these tools are childlike, adults may believe that only the child needs to do such prototyping. It is important to encourage adults to participate in these low-tech prototyping sessions.

The second set of problems emerge from the traditionally unequal power relationships between adults and children. In what sense can children be treated as peers? When adults and children are discussing ideas, making decisions, or conducting research, traditional "power structures" may emerge. In conducting a

usability study, the adult researcher might lead child users through the experiment rather than allowing them to explore freely on their own. In a team discussion, the children may act as if they are in a school setting by raising their hands to speak. Adults may even inadvertently take control of discussions. Is it sensible to set up design teams where children are given equal responsibilities to those of adult designers? Getting adults and children to work together as a team of equals is often the most difficult part of the design process. It is to be expected that it may take a while for a group to become comfortable and efficient when working together. It can take up to 6 months for an "intergenerational design team to truly develop the ability to build upon each other's ideas" (Druin et al., 2001, p. 400). To help diffuse such traditional adult–child relationships, adults are encouraged to dress casually, and there always should be more than one adult and more than one child on a team. A single child may feel outnumbered by the adults, and a single adult might create the feeling of a school environment in which the adult takes on the role of teacher. Alborzi et al. (2000) starts each design session with 15 min of snack time, where adults and children can informally discuss anything. This helps both adults and children to get to know each other better as "people with lives outside of the lab" (Alborzi et al., 2000, p. 97) and to improve communication within the group.

Although there have been many successes in having children participate as design and research partners in the development of software, there are still many questions to be answered about the effectiveness of this approach. Scaife and Rogers (1999) attempted to address many of the questions and problems faced when working with children. The first question deals with the multitude of ideas and suggestions produced by children. Children say outrageous things. How do you decide which ideas are worthwhile? When do you stop listening? The problem of selection is difficult because in the end it is the adult who will decide which ideas to use and which ideas to ignore. Scaife and Rogers suggest creating a set of criteria to "determine what to accept and what not to accept with respect to the goals of the system. . . . You need to ask what the trade-offs will be if an idea or set of ideas are implemented in terms of critical 'kid' learning factors: that is, how do fun and motivation interact with better understanding?" (Scaife & Rogers, 1999, p. 47).

In addition to deciding which of the children's ideas to use, there is also the problem of understanding the meaning behind what the child is trying to say. Adults tend to assume that they can understand what kids are getting at, but kid talk is not adult talk. It is important to remember that children have "a different conceptual framework and terminology than adults" (Scaife & Rogers, 1999).

Another problem with involving children, particularly with the design of educational software, is that "children can't discuss learning goals that they have not yet reached themselves" (Scaife & Rogers, 1999). Can children make effective contributions about the content and the way they should be taught, something which adults have always been responsible for? Adults have assumptions about what is an effective way to teach children. Kids tend to focus on the fun aspects of the software rather than the educational agenda. There may exist a mismatch of expectations if kids are using components of the software in unanticipated ways. Involving children in the design and evaluation process may help detect where these mismatches occur in the software.

GENRES OF TECHNOLOGY FOR KIDS

Technology for kids falls into two broad categories: education and entertainment. When game companies try to mix these genres, they may use the term "edutainment." New products for kids increasingly include specialized hardware as well as software.

Entertainment

Designers of games and other entertainment software rarely write about how they accomplish their job. Talks are presented each year at the Game Developer's Conference (http://www.gdconf.com), and some informal reflections are gathered as conference proceedings. Attending the conference is recommended for people who wish to learn more about current issues in game design. The magazine *Game Developer* is the leading publication with reflective articles on the game design process.

Most game designers are men, and they work by simply designing games that they themselves would like to play. This simple design technique is easy and requires little if any background research with users. With this approach, they are able to appeal quite effectively to the core gaming audience: young men and teenage boys. Gaming companies are, however, increasingly recognizing that people outside that group represent a large potential market for their products. Designing for teenagers is relatively easy; designing for very young children presents substantial challenges. The younger the target audience, the more it is necessary to use sound design methodology, consulting with target users at every stage of the design process. (For more on interactive entertainment, see chapter 43 by Jesse Schell in this volume.)

Brenda Laurel pioneered the use of careful design methods for nontraditional game audiences in her work with the company Purple Moon in the mid-1990s. Laurel aimed to develop games that appeal to preteen girls both to tap this market segment and also to give girls an opportunity to become fluent with technology. Many people believe that use of computer games leads to skills that offer later advantages at school and work. (See chapter 21 on gender and HCI by Justine Cassell, this volume.) Through extensive interviews with girls in their target age range, Purple Moon was able to create successful characters and game designs. However, the process was so time-consuming and expensive that the company failed to achieve profitability fast enough to please its investors. The company was closed in 1999, and its characters and games were sold to Mattel. Purple Moon perhaps did more research than was strictly necessary, particularly because their area was so new. The broader lesson is that the game industry typically does not budget for needs analysis and iterative design early in the design process. "Play testing" and "quality assurance" typically take place relatively late in the

design cycle. Designers contemplating incorporating research early in their design process must consider the financial cost. (For more on game design and evaluation, see chapter 46 by Randy Pagulayan et al. in this volume.)

Game designer Carolyn Miller (1998) highlighted seven mistakes ("kisses of death") commonly made by people trying to design games for kids:

- "Death kiss #1: Kids love anything sweet."
  Miller wrote that "sweetness is an adult concept of what kids should enjoy." (p. 423) Only very young children will tolerate it. Humor and good character development are important ingredients. Don't be afraid to use off-color humor or to make something scary.
- "Death kiss #2: Give 'em what's good for 'em."
  Miller advised, "don't preach, don't lecture, and don't talk down—nothing turns kids off faster." (p. 425)
- "Death kiss #3: You just gotta amuse 'em."
  "Don't assume that just because they are little, they aren't able to comprehend serious themes." (p. 426)
- "Death kiss #4: Always play it safe!"
  Adult games often rely on violence to maintain dramatic tension. Because designers probably wouldn't want to include this in games for children, they need to find other ways to maintain dramatic tension. Don't let your game become bland.
- "Death kiss #5: All kids are created equal."
  Target a specific age group, and take into consideration humor, vocabulary, skill level, and interests. If you try to design for everyone, your game may appeal to no one.
- "Death kiss #6: Explain everything."
  In an eagerness to be clear, some people overexplain things to kids. Kids are good at figuring things out. Use as few words as possible, and make sure to use spoken and visual communication as much as possible.
- "Death kiss #7: Be sure your characters are wholesome!"
  Miller warns that if every character is wholesome, the results are predictable and boring. Characters need flaws to have depth. Miller identifies a number of common pitfalls in assembling groups of characters. It's not a good idea to take a "white bread" approach, in which everyone is white and middle class. On the other end of the spectrum, it's also undesirable to take a "lifesaver approach" with one character for each ethnicity. Finally, you also need to avoid an "off-the-shelf" approach, in which each character represents a stereotype: "You've got your beefy kid with bad teeth; he's the bully. You've got the little kids with glasses; he's the smart one." (p. 429) Create original characters that have depth and flaws that they can struggle to overcome (Miller, 1998).

Education

To design educational software, we must expand the concept of user-centered design (UCD) to one of learner-centered design (LCD; Soloway, Guzdial, & Hay, 1994). There are several added steps in the process:

- Needs analysis
  - For learners
  - For teachers
- Select pedagogy
- Select media/technology
- Prototype
  - Core application
  - Supporting curricula
  - Assessment strategies
- Formative evaluation
  - Usability
  - Learning outcomes
- Iterative design
- Summative evaluation
  - Usability
  - Learning outcomes

In our initial needs analysis, for software to be used in a school setting, we need to understand not just learners but also teachers. Teachers have heavy demands on their time and are held accountable for their performance in ways that vary between districts and between election years.

Once we understand our learner and teacher needs, we need to select an appropriate *pedagogy*—an approach to teaching and learning. For example, behaviorism views learning as a process of stimulus and reinforcement (Skinner, 1968). Constructivism sees learning as a process of active construction of knowledge through experience. A social-constructivist perspective emphasizes learning as a social process (Newman, Griffin, & Cole, 1989). (A full review of approaches to pedagogy is beyond the scope of this chapter.)

Next, we're ready to select the media we will be working with, matching their affordances to our learning objectives and pedagogical approach. Once the prototyping process has begun, we need to develop not just software or hardware, but (for applications to be used in schools) also supporting curricular materials and assessment strategies.

"Assessment" should not be confused with "evaluation." The goal of assessment is to judge an individual student's performance. The goal of evaluation is to understand to what extent our learning technology design is successful. An approach to assessing student achievement is an essential component of any school-based learning technology. For both school and free-time use, we need to design feedback mechanisms so that learners can be aware of their own progress. It is also important to note whether learners find the environment motivating. Does it appeal to all learners, or more to specific gender, learning style, or interest groups?

As in any HCI research, educational technology designers use formative evaluation to understand informally what needs improvement in their learning environment and to guide the process of iterative design. Formative evaluation must pay attention first to usability and second to learning outcomes. If students can't use the learning hardware or software, they certainly won't learn through its use. Once it's clear that usability has met a minimum threshold, designers then need to evaluate whether learning outcomes are being met. After formative evaluation and iterative design are complete, a final summative evaluation serves to document the effectiveness of the design and justify its use by learners and teachers. Summative evaluation must similarly pay attention to both usability and learning outcomes.

A variety of quantitative and qualitative techniques are commonly used for evaluation of learning outcomes (Gay & Airasian, 2000). Most researchers use a complementary set of both quantitative and qualitative approaches. Demonstrating educational value is challenging, and research methods are an ongoing subject of research.

This represents an idealized learner-centered design (LCD) process. Just as many software design projects do not in reality follow a comprehensive user-centered design (UCD) process, many educational technology projects do not follow a full LCD process. Generally, LCD is substantially more time-consuming than UCD. Although it may in some cases be possible to collect valid usability data in a single session, learning typically takes place over longer periods of time. To get meaningful data, most classroom trials take place over weeks or months. Furthermore, classroom research needs to fit into the school year at the proper time. If you are using Biologica (Hickey, Kindfield, Horwitz, & Christie, 2000) to teach about genetics, a designer must wait until genetics is covered in the curriculum during the school year. There may be only one or two chances per year to test this educational technology. It frequently takes many years to complete the LCD process. In the research community, one team may study and evolve one piece of educational technology over many years. Yet in a commercial setting, educational products need to get to market rapidly, and this formal design process is rarely used. (For more on the development of educational software, see chapter 42 by Chris Quintana et al. in this volume.)

Genres of Educational Technology

In 1980, Taylor divided educational technology into three genres:

1. Computer as tutor
2. Computer as tool
3. Computer as tutee

Suppose that we are learning about acid rain. If the computer is serving as *tutor*, it might present information about acid rain and ask the child questions to verify the material was understood. If the computer is a *tool*, the child might collect data about local acid rain and input that data into an ecological model to analyze its significance. If the computer is a *tutee*, the child might program his or her own ecological model of acid rain.

With the advent of the Internet, we must add a fourth genre:

4. Computer-supported collaborative learning (CSCL)

In a CSCL study of acid rain, kids from around the country might collect local acid rain data, enter it into a shared database,

TABLE 22.1. Genres of Children's Software

Genre	Description
Entertainment	Games created solely for fun and pleasure
Educational	Software created to help children learn about a topic using some type of pedagogy (an approach to teaching and learning)
Edutainment	A mix of the entertainment and educational genres.
Computer as tutor	Often referred to as "drill and practice" or "computer-aided instruction" (CAI), this approach is grounded in behaviorism. Children are presented with information and then quizzed on their knowledge.
Computer as tool	The learner directs the learning process, rather than being directed by the computer. This approach is grounded in constructivism, which sees learning as an active process of constructing knowledge through experience.
Computer as tutee	Typically, the learner uses construction kits to help reflect on what he or she learned through the process of creation. This approach is grounded in constructivism and constructionism.
Computer-supported collaborative learning (CSCL)	Children use the Internet to learn from and communicate with knowledgeable members of the adult community. Children can also become involved in educational online communities with children from different geographic regions. This approach is grounded in social constructivism.

analyze the aggregate data, and talk online with adult scientists who study acid rain. This is, in fact, the case in the NGS-TERC Acid Rain Project (Tinker, 1993).

Computer as Tutor. In most off-the-shelf educational products, the computer acts as tutor. Children are presented with information and then quizzed on their knowledge. This approach to education is grounded in behaviorism (Skinner, 1968). It is often referred to as "drill and practice" or "computer-aided instruction." The computer tracks student progress and repeats exercises as necessary.

Researchers with a background in artificial intelligence have extended the drill and practice approach to create "intelligent tutoring systems." Such systems try to model what the user knows and tailor the problems presented to an individual's needs. Many systems explicitly look for typical mistakes and provide specially prepared corrective feedback. For example, suppose a child adds 17 and 18 and gets an answer of 25 instead of 35. The system might infer that the child needs help learning to carry from the 1s to the 10s column and present a lesson on that topic. One challenge in the design of intelligent tutors is in accurately modeling what the student knows and what their errors might mean.

Byrne, Anderson, Douglass, and Matessa (1999) experimented with using eye tracking to improve the performance

of intelligent tutors. Using an eye tracker, the system can tell whether the student has paid attention to all elements necessary to solve the problem. In early trials with the eye tracker, they found that some of the helpful hints the system was providing to the user were never actually read by most students. This helped guide their design process. They were previously focusing on how to improve the quality of hints provided; however, this is irrelevant if the hints are not even being read (Byrne et al., 1999).

An interesting variation on the traditional "computer as tutor" paradigm for very young children is the Actimates line interactive plush toys. Actimates Barney and other characters lead children in simple games with educational value, such as counting exercises. The "tutor" is animated and anthropomorphized. The embodied form lets young children use the skills they have in interacting with people to learn to interact with the system, enhancing both motivation and ease of use (Strommen, 1998; Strommen & Alexander, 1999). (For more on computer-based tutoring systems, see chapter 34 by Emurian and Durham in this volume.)

Computer as Tool. When the computer is used as a tool, agency shifts from the computer to the learner. The learner is directing the process, rather than being directed. This approach is preferred by constructivist pedagogy, which sees learning as an active process of constructing knowledge through experience. The popular drawing program Kid Pix is an excellent example of a tool customized for kids' interests and needs. Winograd comments that Kid Pix's designer Craig Hickman "made a fundamental shift when he recognized that the essential functionality of the program lay not in the drawings that it produced, but in the experience for the children as they used it" (Winograd, 1996, p. 60). For example, Kid Pix provides several ways to erase the screen, including having your drawing explode or be sucked down a drain.

Simulation programs let learners try different possibilities that would be difficult or impossible in real life. For example, Biologica (an early version was called Genscope) allows students to learn about genetics by experimenting with breeding cartoon dragons with different inherited characteristics such as whether they breathe fire or have horns (Hickey et al., 2000). Model-It lets students try different hypotheses about water pollution and other environmental factors in a simulated ecosystem (Soloway et al., 1996).

The goal of such programs is to engage students in scientific thinking. The challenge in their design is how to get students to think systematically and not simply try options at random. Programs like Model-It provide the student with "scaffolding." Initially, students are given lots of support and guidance. As their knowledge evolves, the scaffolding is "faded," allowing the learner to work more independently (Guzdial, 1994; Soloway et al., 1994).

Computer as Tutee. Seymour Papert commented that much computer-aided instruction is "using the computer to program the child" (Papert, 1992, p. 163). Instead, he argued that children should learn to program the computer and through this process gain access to new ways of thinking and understanding the world. Early research argued that programming would improve children's general cognitive skills, but empirical trials produced mixed results (Clements, 1986; Clements & Gullo, 1984; Pea, 1984). Some researchers argue that the methods of these studies are fundamentally flawed, because the complexity of human experience cannot be reduced to pre- and posttests (Papert, 1987). The counterargument is that researchers arguing that technology has a transformative power need to back up their claims with evidence of some form, whether quantitative or qualitative (Pea, 1987; Walker, 1987). More recently, the debate has shifted to the topic of technological fluency. As technology increasingly surrounds our everyday lives, the ability to use it effectively as a tool becomes important for children's success in school and later in the workplace (Resnick & Rusk, 1996).

In the late 1960s, Feurzeig and colleagues (W. Feurzeig, personal communication, 1996) at BBN invented Logo, the first programming language for kids. Papert extended Logo to include "turtle graphics," in which kids learn geometric concepts by moving a "turtle" around the screen (Papert, 1980). A variety of programming languages for kids have been developed over subsequent years, including Starlogo (Resnick, 1994), Boxer (diSessa & Abelson, 1986), Stagecast (Cypher & Smith, 1995), Agentsheets (Repenning & Fahlen, 1993), MOOSE (Bruckman, 1997), and Squeak (Guzdial & Rose, 2001). Lego Mindstorms (originally "Lego/Logo") is a programmable construction kit with physical as well as software components (Martin & Resnick, 1993). Another programmable tool bridging the gap between physical constructions and representations on the screen is Hypergami, a computer-aided design tool for origami developed by Michael Eisenberg and Ann Nishioka Eisenberg at the University of Colorado at Boulder. Students working with Hypergami learn about both geometry and art (Eisenberg, Nishioka, & Schreiner, 1997).

In most design tools, the goal is to facilitate the creation of a product. In educational construction kits, the goal instead is what is learned through the process of creation. So what makes a good construction kit? In a 1996 *Interactions* article entitled "Pianos, Not Stereos: Creating Computational Construction Kits," Resnick, Bruckman, and Martin (1996) discussed the art of designing construction kits for learning ("constructional design"):

FIGURE 22.2. Penguins created using Hypergami. (Photo Courtesy of Mike and Ann Eisenberg.)

The concept of learning-by-doing has been around for a long time. But the literature on the subject tends to describe specific activities and gives little attention to the general principles governing what kinds of "doing" are most conducive to learning. From our experiences, we have developed two general principles to guide the design of new construction kits and activities. These constructional-design principles involve two different types of connections:

- Personal connections. Construction kits and activities should connect to users' interests, passions, and experiences. The point is not simply to make the activities more "motivating" (though that, of course, is important). When activities involve objects and actions that are familiar, users can leverage their previous knowledge, connecting new ideas to their pre-existing intuitions.

- Epistemological connections. "Construction kits and activities should connect to important domains of knowledge—more significantly, encourage new ways of thinking (and even new ways of thinking about thinking). A well-designed construction kit makes certain ideas and ways of thinking particularly salient, so that users are likely to connect with those ideas in a very natural way, in the process of designing and creating." (Resnick et al., 1996, p. 42)

Bruckman (2000, p. 370) adds a third design principle:

- Situated support. "Support for learning should be from a source (either human or computational) with whom the learner has a positive personal relationship, ubiquitously available, richly connected to other sources of support, and richly connected to every-day activities."

Computer-Supported Collaborative Learning. Most tools for learning have traditionally been designed for one child working at the computer alone; however, learning is generally recognized to be a social process (Newman et al., 1989). With the advent of the Internet came new opportunities for children to learn from one another and from knowledgeable members of the adult community. This field is called computer-supported collaborative learning (CSCL; Koschmann, 1996).

CSCL research can be divided into four categories:

- Distance education: Attempts to organize online something like a traditional classroom.

- Information retrieval: Research projects in which students use the Internet to find information.

- Information sharing: Students debate issues with one another. One of the first such tools was the computer-supported intentional learning environment, a networked discussion tool designed to help students engage in thoughtful debate as a community of scientists does (Scardamalia & Bereiter, 1994). They may also collect scientific data and share it with others online. In the One Sky, Many Voices project, students learn about extreme weather phenomena by sharing meteorological data they collect with other kids from around the world and also by talking online with adult meteorologists (Songer, n.d.). In the Palaver Tree Online project, kids learn about history by talking online with adults who lived through that period of history (Ellis & Bruckman, 2001). A key challenge in the design of information sharing environments is how to promote serious reflection on the part of students (Guzdial, 1994; Kolodner & Guzdial, 1996).

- Technological samba schools: In *Mindstorms*, Seymour Papert has a vision of a "technological samba school." At samba schools in Brazil, a community of people of all ages gather together to prepare a presentation for carnival. "Members of the school range in age from children to grandparents and in ability from novice to professional. But they dance together and as they dance everyone is learning and teaching as well as dancing. Even the stars are there to learn their difficult parts" (Papert, 1980, p. 178). People go to samba schools not just to work on their presentations, but also to socialize and be with one another. Learning is spontaneous, self-motivated, and richly connected to popular culture. Papert imagines a kind of technological samba school where people of all ages gather together to work on creative projects using computers. The Computer Clubhouse is an example of such a school in a face-to-face setting (Resnick & Rusk, 1996). MOOSE Crossing is an Internet-based example (Bruckman, 1998). A key challenge in the design of such environments is how to grapple with the problem of uneven achievement among participants. When kids are allowed to work or not work in a self-motivated fashion, typically some excel but others do little (Elliott, Bruckman, Edwards, & Jensen, 2000). (For more on the design of online communities for kids, see the Introduction to Section IIIc by Jennifer Preece in this volume.)

Child Safety Online

One challenge in the design of Internet-based environments for kids is the question of safety. The Internet contains information that is sexually explicit, violent, and racist. Typically, such information does not appear unless one is looking for it; however, it is unusual but possible to stumble across it accidentally. Filtering software blocks access to useful information as well as harmful (Schneider, 1997). Furthermore, companies that make filtering software often fail to adequately describe how they determine what to block, and they may have unacknowledged political agendas that not all parents will agree with. Resolving this issue requires a delicate balance of the rights of parents, teachers, school districts, and children (Electronic Privacy Information Center, 2001). Another danger for kids online is the presence of sexual predators and others who wish to harm children. Although such incidents are rare, it is important to teach kids not to give out personal information online such as their last name, address, or phone number. Kids who wish to meet an online friend face-to-face should do so by each bringing a parent and meeting in a well-populated public place like a fast-food restaurant. A useful practical guide "Child Safety on the Information Superhighway" is available from the Center for Missing and Exploited Children (http://www.missingkids.org). Educating kids, parents, and teachers about online safety issues is an important part of the design of any online software for kids.

CONCLUSION

To design for kids, we must have a model of what kids are and what we would like them to become. Adults were once kids. Many are parents. Some are teachers. We tend to think that we know kids—who they are, what they are interested in, what they like—but we do not have as much access to our former selves as many would like to believe. Furthermore, our fundamental notions of childhood are in fact culturally constructed and change over time. Karin Calvert wrote about the changing notion of childhood in the United States and the impact it has had on artifacts designed for children and child rearing:

In the two centuries following European settlement, the common perception in America of children changed profoundly, having first held to an exaggerated fear of their inborn deficiencies, then expecting considerable self-sufficiency, and then, after 1830, endowing young people with an almost celestial goodness. In each era, children's artifacts mediated between social expectations concerning the nature of childhood and the realities of child-rearing: before 1730, they pushed children rapidly beyond the perceived perils of infancy, and by the nineteenth century they protected and prolonged the perceived joys and innocence of childhood." (Calvert, 1992, p. 8)

Although Calvert was reflecting on the design of swaddling clothes and walking stools, the same role is played by new technologies for kids such as programmable Legos and drill and practice arithmetic programs: These artifacts mediate between our social expectations of children and the reality of their lives. If you believe that children are unruly and benefit from strong discipline, then you are likely to design computer-aided instruction. If you believe that children are creative and shouldn't be stifled by adult discipline, then you might design an open-ended construction kit like Logo or Squeak. In designing for kids, it is crucial to become aware of one's own assumptions about the nature of childhood. Designers should be able to articulate their assumptions and be ready to revise them based on empirical evidence.

References

Alborzi, H., Druin, A., Montemayor, J., Platner, M., Porteous, J., Sherman, L., Boltman, A., Taxén, G., Best, J., Hammer, J., Kruskal, A., Lal, A., Schwenn, T. P., Sumida, L., Wagner, R., & Hendler, J. (2000). *Designing Story Rooms: Interactive storytelling spaces for children*. Paper presented at the Proceedings of the Symposium on Designing interactive systems: Processes, practices, methods, and techniques, Brooklyn, NY.

Bederson, B., Hollan, J., Druin, A., Stewart, J., Rogers, D., & Proft, D. (1996). *Local tools: An alternative to tool palettes*. Paper presented at the Proceedings of the Association for Computing Machinery Symposium on User Interface Software and Technology, Seattle, WA.

Berkovitz, J. (1994). *Graphical interfaces for young children in a software-based mathematics curriculum*. Paper presented at the Proceedings of the Association for Computing Machinery Conference on Human Factors in Computings Systems: Celebrating Interdependence, Boston, MA.

Bernard, M., Mills, M., Frank, T., & McKnown, J. (2001, Winter). *Which fonts do children prefer to read online?* Internet newsletter,

Software Usability Research Laboratory (SURL). Retrieved March 2002 from http://wsupsy.psy.twsu.edu/surl/usabilitynews/41/onlinetext.htm

British Broadcasting Company. (1997). *Teletubbies* press release [Web site]. BBC Education. Retrieved March 2002 from http://www.bbc.co.uk/education/teletubbies/information/pressrelease/

Bruckman, A. (1997). *MOOSE Crossing: Construction, community, and learning in a networked virtual world for kids.* Unpublished Ph.D., dissertation, Massachusetts Institute of Technology, Cambridge, MA.

Bruckman, A. (1998). Community support for constructionist learning. *Computer Supported Cooperative Work, 7,* 47-86.

Bruckman, A. (2000). Situated support for learning: Storm's weekend with Rachael. *Journal of the Learning Sciences, 9,* 329-372.

Byrne, M. D., Anderson, J. R., Douglass, S., & Matessa, M. (1999, May). *Eye tracking the visual search of click-down menus.* Paper presented at the Proceedings of the Association for Computing Machinery Conference on Human Factors in Computings Systems: The Computer Human Interaction is the limit, Pittsburgh, PA.

Calvert, K. (1992). *Children in the house: The material culture of early childhood, 1600-1900.* Boston: Northeastern University Press.

Clements, D. H. (1986). Effects of Logo and CAI environments on cognition and creativity. *Journal of Educational Psychology, 78,* 309-318.

Clements, D. H., & Gullo, D. F. (1984). Effects of computer programming on young children's cognition. *Journal of Educational Psychology, 76,* 1051-1058.

Cypher, A., & Smith, D. C. (1995, May). *End user programming of simulations.* Paper presented at the Proceedings of the Association for Computing Machinery Conference on Human Factors in Computing Systems, Denver, CO.

Davenport, A., & Wood, A. (1997). *TeleTubbies FAQ.* BBC education Web site. Retrieved from http://www.bbc.co.uk/education/teletubbies/information/faq/q18.shtml

diSessa, A. A., & Abelson, H. (1986). Boxer: A reconstructible computational medium. *Communications of the ACM, 29,* 859-868.

Druin, A. (1999, May). *Cooperative inquiry: Developing new technologies for children with children.* Paper presented at the Proceedings of the Association for Computing Machinery Conference on Human Factors in Computings Systems: The Computer Human Interaction is the limit, Pittsburgh, PA.

Druin, A., Bederson, B., Hourcade, J. P., Sherman, L., Revelle, G., Platner, M., & Weng, S. (2001, June). *Designing a digital library for young children: An intergenerational partnership.* Paper presented at the Proceedings of the Joint Conference on Digital Libraries, Roanoke, VA.

Eisenberg, M., Nishioka, A., & Schreiner, M. E. (1997, January). *Helping users think in three dimensions: Steps toward incorporating spatial cognition in user modelling.* Paper presented at the Proceedings of the International Conference on Intelligent User Interfaces, Orlando, FL.

Electronic Privacy Information Center. (2001). *Filters and Freedom 2.0: Free speech perspectives on Internet content control.* Washington, DC: Author.

Elliott, J., Bruckman, A., Edwards, E., & Jensen, C. (2000, June). *Uneven achievement in a constructionist learning environment.* Paper presented at the Proceedings of the International Conference on the Learning Sciences, Ann Arbor, MI.

Ellis, J. B., & Bruckman, A. S. (2001, June). *Designing palaver tree online: Supporting social roles in a community of oral history.* Paper presented at the Proceedings of the SIG-CHI on Human Factors in Computing Systems, Seattle, WA.

Erickson, T. (1990). Working with interface metaphors. In B. Laurel (Ed.), *The art of human-computer interface design* (pp. 65-73). Reading, MA: Addison Wesley.

Gay, L. R., & Airasian, P. (2000). *Education research: Competencies for analysis and application* (6th ed.). Upper Saddle River, NJ: Merrill.

Goldman-Segall, R. (1996). Looking through layers: Reflecting upon digital video ethnography. *JCT: An Interdisciplinary Journal For Curriculum Studies, 13.*

Guzdial, M. (1994). Software-realized scaffolding to facilitate programming for science learning. *Interactive Learning Environments, 4,* 1-44.

Guzdial, M., & Rose, K. (Eds.). (2001). *Squeak: Open personal computing and multimedia:* Prentice Hall.

Halgren, S., Fernandes, T., & Thomas, D. (1995, May). *Amazing Animation™: Movie making for kids design briefing.* Paper presented at the Proceedings of the Association for Computing Machinery Conference on Human Factors in Computings Systems, Denver, CO.

Hanna, L., Risden, K., & Alexander, K. (1997). Guidelines for usability testing with children. *Interactions, 4,* 9-14.

Hickey, D. T., Kindfield, A. C. H., Horwitz, P., & Christie, M. A. (2000, June). *Integrating instruction, assessment, and evaluation in a technology-based genetics environment: The GenScope follow-up study.* Paper presented at the Proceedings of the International Conference of the Learning Sciences, Ann Arbor, MI.

Inkpen, K. (1997, June). *Three important research agendas for educational multimedia: Learning, children and gender.* Paper presented at the Proceedings of Graphics Interface, Calgary, Alberta.

Inkpen, K. (2001). Drag-and-drop versus point-and-click: Mouse interaction styles for children. *ACM Transactions Computer-Human Interaction, 8,* 1-33.

Inkpen, K., Gribble, S., Booth, K. S., & Klawe, M. (1995, May). *Give and take: Children collaborating on one computer.* Paper presented at the Proceedings of the Association for Computing Machinery Conference on Human Factors in Computings Systems, Denver, CO.

Joiner, R., Messer, D., Light, P., & Littleton, K. (1998). It is best to point for young children: A comparison of children's pointing and dragging. *Computers in Human Behavior, 14,* 513-529.

Jones, T. (1992). Recognition of animated icons by elementary-aged children. *Association for Learning Technology Journal, 1,* 40-46.

Kail, R. (1991). Developmental changes in speed of processing during childhood and adolescence. *Psychological Bulletin, 109,* 490-501.

Kolodner, J., & Guzdial, M. (1996). Effects with and of CSCL: Tracking learning in a new paradigm. In T. Koschmann (Ed.), *CSCL: Theory and practice.* Mahwah, NJ: Lawrence Erlbaum Associates.

Koschmann, T. (Ed.). (1996). *CSCL: Theory and practice.* Mahwah, NJ: Lawrence Erlbaum Associates.

Martin, F., & Resnick, M. (1993). LEGO/Logo and electronic bricks: Creating a scienceland for children. In D. L. Ferguson (Ed.), *Advanced educational technologies for mathematics and science* (pp. 61-90). Berlin: Springer-Verlag.

Miller, C. (1998, March). *Designing for kids: Infusions of life, kisses of death.* Paper presented at the Proceedings of the Game Developers Conference, Long Beach, CA.

Miller, L. T., & Vernon, P. A. (1997). Developmental changes in speed of information processing in young children. *Developmental Psychology, 33,* 549-554.

Newman, D., Griffin, P., & Cole, M. (1989). *The construction zone: Working for cognitive change in school.* Cambridge, England: Cambridge University Press.

Nix, D., Fairweather, P., & Adams, B. (1998, April). *Speech recognition, children, and reading.* Paper presented at the Proceedings of the Association for Computing Machinery Conference on Human Factors in Computings Systems, Los Angeles, CA.

O'Hare, E. A., & McTear, M. F. (1999). Speech recognition in the secondary school classroom: An exploratory study. *Computers & Education, 3*(8), 27–45.

Papert, S. (1980). *Mindstorms: Children, computers, and powerful ideas*. New York: Basic Books.

Papert, S. (1987, January–February). Computer criticism vs. technocentric thinking. *Educational Researcher*, 22–30.

Papert, S. (1992). *The children's machine*: Basic Books.

Pea, R. (1984). On the cognitive effects of learning computer programming. *New Ideas in Psychology, 2*, 137–168.

Pea, R. (1987, June–July). The aims of software criticism: Reply to Professor Papert. *Educational Researcher*, 4–8.

Piaget, J. (1970). *Science of education and the psychology of the child*. New York: Orion Press.

Repenning, A., & Fahlen, L. E. (1993, May). *Agentsheets: A tool for building domain-oriented visual programming environments*. Paper presented at the Proceedings of the Association for Computing Machinery Conference on Human Factors in Computing Systems, Amsterdam, The Netherlands.

Resnick, M. (1994). *Turtles, termites, and traffic jams: Explorations in massively parallel microworlds*. Cambridge, MA: MIT Press.

Resnick, M., Bruckman, A., & Martin, F. (1996). Pianos not stereos: Creating computational construction kits. *Interactions, 3*(5), 40–50.

Resnick, M., & Rusk, N. (1996). The Computer Clubhouse: Preparing for life in a digital world. *IBM Systems Journal, 35*, 431–440.

Rubin, J. (1994). *Handbook of usability testing*. New York: Wiley.

Scaife, M., & Rogers, Y. (1999). Kids as informants: Telling us what we didn't know or confirming what we knew already? In A. Druin (Ed.), *The design of children's technology* (pp. 27–50). San Francisco: Morgan Kaufmann.

Scardamalia, M., & Bereiter, C. (1994). Computer support for knowledge-building communities. *Journal of the Learning Sciences, 3*, 265–283.

Schneider, K. G. (1996). Children and information visualization technologies. *Interactions, 3*(5), 68–73.

Schneider, K. G. (1997). *The Internet filter assessment project (TIFAP)*. Retrieved March 2002 from http://www.bluehighways.com/tifap/learn.htm

Schuler, D., & Namioka, A. (Eds.). (1993). *Participatory design: Principles and practices*. Hillsdale, NJ: Lawrence Erlbaum Associates.

Skinner, B. F. (1968). *The technology of teaching*. New York: Appleton-Century-Crofts.

Soloway, E., Guzdial, M., & Hay, K. E. (1994). Learner-centered design: The challenge for HCI in the 21st century. *Interactions, 1*, 36–48.

Soloway, E., Jackson, S. L., Klein, J., Quintana, C., Reed, J., Spitulnik, J., Stratford, S. J., Studer, S., Jul, S., Eng, J., & Scala, N. (1996, April). *Learning theory in practice: Case studies of learner-centered design*. Paper presented at the Proceedings of the Association for Computing Machinery Conference on Human Factors in Computings Systems, Vancouver, Canada.

Songer (n.d.), N. Kids as global scientists. Retrieved March 2002 from http://onesky.engin.umich.edu/

Stewart, J., Raybourn, E. M., Bederson, B., & Druin, A. (1998, April). *When two hands are better than one: Enhancing collaboration using single display groupware*. Paper presented at the Proceedings of the Association for Computing Machinery Conference on Human Factors in Computings Systems, Los Angeles, CA.

Strommen, E. (1994, April). *Children's use of mouse-based interfaces to control virtual travel*. Paper presented at the Proceedings of the Association for Computing Machinery Conference on Human Factors in Computings Systems: Celebrating Interdependence, Boston, MA.

Strommen, E. (1998, April). *When the interface is a talking dinosaur: Learning across media with ActiMates Barney*. Paper presented at the Proceedings of the Association for Computing Machinery Conference on Human Factors in Computings Systems, Los Angeles, CA.

Strommen, E., & Alexander, K. (1999, May). *Emotional interfaces for interactive aardvarks: Designing affect into social interfaces for children*. Paper presented at the Proceedings of the Association for Computing Machinery Conference on Human Factors in Computings Systems: The CHI is the limit, Pittsburgh, PA.

Taylor, R. P. (Ed.). (1980). *The computer in the school, tutor, tool, tutee*. New York: Teachers College Press.

Thomas, J. R. (1980). Acquisition of motor skills: Information processing differences between children and adults. *Research Quarterly for Exercise and Sport, 51*, 158–173.

Tinker, R. (1993). *Thinking about science*. Concord, MA: Concord Consortium.

Walker, D. F. (1987). Logo needs research: A response to Professor Papert's paper. *Educational Researcher*, June/July 9–11.

Winograd, T. (1996). Profile: Kid Pix. In T. Winograd (Ed.), *Bringing design to software* (pp. 58–61). New York: ACM Press.

· 23 ·

GLOBAL AND INTERCULTURAL
USER-INTERFACE DESIGN

Aaron Marcus
Aaron Marcus and Associates, Inc.

INTRODUCTION

The concept of "user interfaces for all," as set forth by Stephanidis (2000) implies the availability of and easy access to computer-based products and services among all peoples in all countries worldwide. Successful computer-based products and services developed for users in different countries and among different cultures consist of partially universal, general solutions and partially unique, local solutions to the design of user interfaces (UIs). Global enterprises seek to mass distribute products and services with minimal changes to achieve cost-efficient production, maintenance, distribution, and user support. Nevertheless, it is becoming increasingly important, technically viable, and economically necessary to produce localized versions for certain markets. UIs must be designed for specific user groups, not merely given a superficial "local" appearance for quick export to foreign markets.

Insufficient attention to localization can lead to embarrassing or sometimes critical miscommunication. For example, in Chinese, Coca-Cola means "bite the wax tadpole" or "female horse stuffed with wax," depending on the dialect, which caused the company to change its name to a phonetic equivalent that means "happiness in the mouth." Similarly, Pepsi's slogan "Come alive with Pepsi" becomes "Pepsi brings your ancestors back from the grave" (Hendrix, 2001). Differences of culture can lead to significant business implications, as Saudi Arabia's Higher Committee for Scientific Research and Islamic Law banned Pokemon video games because they "possessed the minds" of Saudi children, thus closing off one of the Middle East's largest markets to the Japanese Nintendo's multibillion-dollar enterprise (Saudi Arabia Issues Edict, 2001).

By contrast, attention to localization of language leads to greater comprehension, which can lead to a drop in customer-service costs when instructions are displayed in users' native languages. Moreover, localization can lead to greater attention and retention on the part of viewers or customers. This implication is especially significant for Web-based communication, in which Forrester Research (1998) reported that visitors remain twice as long reviewing local-language sites as they do English-only sites, and business users are three times more likely to buy when communication is in their own language.

By managing the user's experience with familiar structures and processes, surprise at novel approaches, and preferences and expectations, the UI designer can achieve compelling forms that enable the UI to be more usable and acceptable. Globalization of product distribution requires a strategy for the design process that enables efficient product development, marketing, distribution, and maintenance. Globalization of UI design, the content and form of which is so much dependent on visible languages and effective communication, improves the likelihood that users will be more productive and satisfied with computer-based products and services in many different global locations.

From the designer's perspective, two primary objectives should be (a) provide a consistent UI and balance user experience across all appropriate products and services and (b) design products and services, with their necessary support systems, that are appropriately internationalized (prepared for localization) and localized (i.e., designed for specific markets). Before discussing globalization, localization, and culture issues, we review briefly essential concepts of UI design.

Demographics, experience, education, and roles in organizations of work or leisure characterize users. Their individual needs and wants, hence their goals as well as their group roles, define their tasks. User-centered, task-oriented design methods account for these aspects and facilitate the attainment of effective UI design.

UIs conceptually consist of metaphors, mental models, navigation, interaction, and appearance, which may be defined as follows (Marcus, 1995, 1998):

- *Metaphors:* essential concepts conveyed through words and images or through acoustic or tactile means. Metaphors concern both overarching concepts that characterize interaction and individual items, such as the "trashcan" standing for "deletion" within the "desktop" metaphor.

- *Mental models:* organization of data, functions, tasks, roles, and people in groups at work or play. The term (similar to but distinct from cognitive models, task models, user models, and so forth), is intended to convey the organization observed in the UI itself, which is presumably learned and understood by users and which reflects the content to be conveyed as well as the available user tasks.

- *Navigation:* movement through mental models, afforded by windows, menus, dialogue areas, control panels, and other methods of navigation. The term implies dialogue and process, as opposed to structure (i.e., sequences of content potentially accessed by users, as opposed to the static structure of that content).

- *Interaction:* the means by which users communicate input to the system and the feedback supplied by the system. The term implies all aspects of command and control devices (e.g., keyboards, mice, joysticks, microphones) as well as sensory feedback (e.g., changes of state of virtual graphical buttons, auditory displays, and tactile surfaces).

- *Appearance:* verbal, visual, acoustic, and tactile perceptual characteristics of displays. The term implies all aspects of visible, acoustic, and haptic languages (e.g., typography or color, musical timbre or cultural accent within a spoken language, and surface texture or resistance to force).

Localization concerns go well beyond language translation. They may affect each component of a UI—from choices of metaphorical references, to hierarchies in the mental model, to colors and graphics.

This chapter discusses the development of UIs intended for users in many different countries with different cultures and languages, presenting a survey of important issues, as well as recommended steps in the development of UIs for an international and intercultural user population. With the rise of the World Wide Web and application-oriented Web sites, the challenge of designing good UIs becomes not only a theoretical issue, but an immediate, practical matter. This topic is discussed from a user perspective rather than a technology and code perspective. The chapter (a) introduces fundamental definitions of globalization

TABLE 23.1. Examples of Differing Displays for Currency, Time, and Physical Measurements

Item	U.S. Examples	European Examples	Asian Examples
Currency	$1,234.00 (U.S. Dollars)	€1.234 (Euros, Germany)	¥1,234 (Japanese yen)
Time measures	8:00 PM, August 24, 1999 8:00 PM, 8/24/99	20:00, 24 August 1999 (England) 20:00, 24.08.99 (Germany, traditional) 20:00, 1999-08-24 (ISO 8601 Euro standard)	20:00, 1999.08.24, or Imperial Heisei 11, or H11 (Japan)
Physical measures	3 lb, 14 oz 3' 10", 3 feet and 10 inches	3.54 kg, 8.32 m (England) 3,54 kg, 8,32 m (Euro standard)	3.54 kg, 8.32 m in Roman or Katakana chars

in UI design, (b) demonstrates why globalization is vital to the success of computer-based communication products, (c) introduces cultural dimensions, and (4) shows their effect on Web designs.

GLOBALIZATION

Definitions: Globalization, Internationalization, and Localization

Globalization refers to the entire process of preparing products or services for worldwide production and consumption and includes issues at international, intercultural, and local scales. In an information-oriented society, globalization affects most computer-mediated communication, which in turn affects UI design. The discussion that follows refers particularly to UI design.

Internationalization refers to the process of preparing code that separates the localizable data and resources (that is, items that pertain to language and culture needed for input and output) from the primary functionality of the software. Software created in this way does not need to be rewritten or recompiled for each local market. International issues refer to geographic, political, linguistic, and typographic issues of nations, or groups of nations. An example of efforts to establish international standards for some parts of UIs is the International Organization of Standardization's (ISO) draft human factors standards in Europe for color legibility standards of cathode-ray tube devices (ISO, 1989). Another example is the legal requirement for bilingual English and French displays in Canada or the quasi-legal denominations for currency, time, and physical measurements, which differ from country to country (Tables 23.1 and 23.2).

Intercultural issues refer to the religious, historical, linguistic, aesthetic, and other, more humanistic issues of particular groups or peoples, sometimes crossing national boundaries. Examples (see Table 23.3) include calendars that acknowledge various religious time cycles; terminology for color, type, and signs reflecting various popular cultures; and organization of content in Web search criteria reflecting cultural preferences.

Localization refers to the process of customizing (including language translation but potentially other changes, also, e.g., content hierarchies, graphics, colors, icons and symbols, etc.) the data and resources of code that are needed for a specific market. Translation is accomplished manually by in-house staff or by one or more vendors performing that service. Translation can also be accomplished semiautomatically using software provided by third-party firms (e.g., Systran). Localization can take place for specific, small-scale communities, often with

TABLE 23.2. Examples of Differing Displays for Other Data Formats

Item	U.S. Examples	European Examples	Asian Examples
Numerics	1,234.56 (also Can., China, UK)	1 234,56 (Finland, Fr, Lux., Portugal., Sweden) 1.234,56 (Albania, Denmark, Greece, Netherlands) 1'234.56 (Switz: German, Italian) 1'234,56 (Switz: French)	1,234.56
Telephone numbers	1-234-567-8901, ext. 23 1.234.567.8901 (123) 456-7890	1234 56 78 90 (Austria) (123) 4 5 6 78 90 (Germany) (12) 3456 789 (Italy) +46(0)12 345 67 +49 (1234) 5678-9 (Switzerland)	+81-53-478-1481 (Japan) 82 2 3142 1100 (Korea) +82-(0)2-535-3893 (Korea) 86 12 34567890 (China)
Address formats	Title, first name, middle initial, last name Department Company Number, street, City, state, Zip Code Country	Paternal name, maternal name, first name Company, department Street, number City, district/region Postal code, country (Order may vary from country to country)	Family name, first name Department Company Number, street, neighborhood, district Postal code, city (Japan)

Note. Adapted in part from Aykin (2000).

TABLE 23.3. Examples of Differing Cultural References

Item	North America/Europe Example	Middle-Eastern Example	Asian Example
Sacred colors	White, blue, gold, scarlet (Judeo-Christian)	Green, light blue (Islam)	Saffron yellow (Buddhism)
Reading direction	Left to right	Right to left	Top to bottom

Item	USA	France, Germany	Japan
Web search	"Culture" doesn't imply political discussions	"Culture" implies political discussions	"Culture" implies tea ceremony discussions
Sports references	Baseball, football, basketball; golf is a sport	Soccer	Sumo wrestling baseball; golf is a religion

(*Note*. Derived in part from Aykin (2000).)

unified language and culture, and, usually, at a scale smaller than countries or significant cross-national ethnic "regions." Examples include affinity groups (e.g., French "twenty-somethings," or U.S. Saturn automobile owners), business or social organizations (e.g., German staff of DaimlerChrysler or Japanese golf club members), and specific intranational groups (e.g., India's untouchables or Japanese housewives). With the spread of Web access, "localization" may come to refer to groups of shared interests that may also be geographically dispersed. Note that this broad definition of "culture" is not accepted by all theorists. For example, see Clausen, as reported by Yardley (2000). For the purposes of this chapter, however, this broad definition is used.

Localization changes may need to consider any or all of the items shown in Table 23.4.

Table 23.5 demonstrates the complexities, even within English-language users. Preparing texts in local languages may require use of additional or different characters. The ASCII system, which uses 7 or 8 bits to represent characters, supports English, and the ISO 8859-1 character set supports Western European languages, such as Spanish, French, and German. Other character encoding systems include EBCDIC, Shift-JIS, UTF-8, and UTF-16. ISO has established specific character sets for languages such as Cyrillic, Modern Greek, Hebrew, Japanese, and so on. The new Unicode system (ISO 10646; Graham, 2000) uses 16 bits to represent 65,536 characters, which is sufficient to display Asian languages such as Japanese and Korean and permits easier translation and presentation of character sets.

Advantages and Disadvantages of Globalization

The business justification for globalization of UIs is complex but compelling. Clear business reasons can drive decisions to localize content on the Web. If the content (functions and data) is likely to be of value to target population outside of the original market, it is usually worthwhile to plan for international and intercultural factors in developing a product, so that it may be efficiently customized, for example, having separate text files that more easily can be translated. Rarely can a product achieve global acceptance with a "one-size-fits-all" solution. Developing a product for international, intercultural audiences usually involves more than merely a translation of verbal language, however. Visible or otherwise perceptual (e.g., auditory) language also must be revised, and other UI characteristics may need to be altered.

Developing products ready for global use, although increasing initial development costs, gives rise to potential for increased international sales. For some countries, however, monolithic domestic markets may inhibit awareness of and incentives for globalization. For example, because the United States has in the

TABLE 23.4. Examples of Localization Changes

Address formats	Electrical and electronic plug formats and nomenclature	Keyboard formats	Punctuation symbols and usage
Alphabetic sequence and nomenclature	Energy formats	Language differences	Reading and writing direction
Arithmetic operations symbolism	Environmental standards (green-compliancy, low energy, low pollution, etc.)	Licensing standards	Sorting sequences
Business standards (quotes, tariffs, contracts, agreement terms, etc.)	File formats	Measurement units (length, volume, weight, electricity, energy, temperature, etc.)	Style formats
Calendar	Font nomenclature, sizes, faces, and byte formats	Monetary or currency formats	Telephone, fax, temperature formats
Character handling	Frequency (e.g., gigahertz)	Multilingual usage	Text length
Colors	Hyphenation and syllabification	Name formats	Video recording and playback formats
Content categories	Icons and symbols	Negative formats	Voltage units and formats
Date and time formats	Intellectual property (protection via patents, copyrights, trademarks)	Numeric formats and number symbols	Weight formats
Documentation nomenclature and formats		Packaging	
		Paper formats	

TABLE 23.5. Comparison of English-Language User Community Conventions

	United States	United Kingdom
Dates	Month/Day/Year: March 17, 2001, 3/17/01	Day/Month/Year: 17 March 2001, 17/03/01
Time	12-hour clock, AM/PM No leading zero (8:32 AM)	24-hour clock Leading zero (08:32)
Currency	$189.56, 56¢	GB£189.56, £189.56, 56p
Spelling	Center Color	Centre Colour
Terminology	Truck Bathroom	Lorrie Toilet
Book spine title	Top-down	Bottom-up

past been such a large producer and consumer of software, it is not surprising that some U.S. manufacturers have targeted only domestic needs. To penetrate some markets (e.g., France), the local language may be a nearly absolute requirement. Recent reports show that European Economic Community countries must provide local variations to gain user acceptance; English-only portals are a barrier to business success (Vickers, 2000).

Some software products are initiated with international versions (but usually released in sequence because of limited development resources). Other products are "retrofitted" to suit the needs of a particular country, language, or culture as needs or opportunities arise. In some cases, the later, ad hoc solution may suffer because of the lack of original planning for globalization.

Globalization Development Process

The "globalized" UI development process is a sequence of partially overlapping steps, some of which are partially or completely iterative.

• *Plan:* Define the challenges or opportunities for globalization; establish objectives and tactics; determine budget, schedule, tasks, development team, and other resources. Globalization must be specifically accounted for in each item of project planning; otherwise, cost overruns, delays in schedule, and lack of resources are likely to occur.

• *Research:* Investigate dimensions of global variables and techniques for all subsequent steps (e.g., techniques for analysis, criteria for evaluation, media for documentation, etc.). In particular, identify items among data and functions that should be targets for change and identify sources of national, cultural, or local reference. Globalized user-centered design stresses the need to research adequately users' wants and needs according to a sufficiently varied spectrum of potential users across specific dimensions of differentiation.

• *Analyze:* Examine results of challenges or opportunities in the prospective markets, refine criteria for success in solving problems or exploiting opportunities (write marketing or technical requirements), determine key usability criteria, and define the design brief—the primary statement of the design's goals. At this stage, globalization targets should be itemized.

• *Design:* Visualize alternative ways to satisfy criteria using alternative prototypes; based on prior or current evaluations, select the design that best satisfies criteria for both general good UI design and globalization requirements; prepare documents that enable consistent, efficient, precise, accurate implementation.

• *Implement:* Build the design to complete the final product (e.g., write code using appropriate, efficient tools identified in planning and research steps).

• *Evaluate:* At any stage, review or test results in the marketplace against defined criteria for success (e.g., conduct focus groups, test usability on specific functions, gather sales and user feedback). Identify and evaluate matches and mismatches, then revise the designs. Test prototypes or final products with international, intercultural, or specific localized user groups to achieve globalized UI designs.

• *Document:* Record development history, issues, and decisions in specifications, guidelines, and recommendation documents. Honold (1999), for example, noted that German and Chinese cell phone users require different strategies for documentation and training that are related to cultural differences predicted by classical culture models.

• *Maintain:* Determine which documents, customer-response services, and other processes will require specialized multiple languages and changes in media or delivery techniques. Prepare appropriate guidelines and templates.

• *Train:* Determine which documents and processes will require multiple languages, different graphics, pacing, media, or distribution methods. Prepare appropriate guidelines and templates.

Critical Aspects for Globalization: General Guidelines

Beyond the UI development process steps identified in the previous section, the following guidelines can assist developers in preparing a "checklist" for specific tasks. Recommendations below are grouped under UI design terms referred to earlier:

• *User demographics*
— Identify national and cultural target user populations and segments within those populations, then identify possible needs for differentiation of UI components and the probable cost of delivering them.

— Identify potential savings in development time through the reuse of UI components, based on common attributes among user groups. For example, certain primary (or top-level) controls in a Web-based, data-retrieval application might be designed for specific user groups, so as to aid comprehension and to improve appeal. Lower level controls, on the other hand, might be more standardized, unvarying formlike elements.

— Consider legal issues in target communities; these may involve issues of privacy, intellectual property, spamming, defamation, pornography and obscenity, vandalism (e.g., viruses), hate speech, fraud, theft, exploitation and abuse (children, environment, elderly, etc.), legal jurisdiction, seller–buyer protection, and so on.

- *Technology*
  — Determine the appropriate media for the appropriate target user categories.
  — Account for international differences to support platform, population and software needs, including languages, scripts, fonts, colors, file formats, and so forth.
  — Research and provide appropriate software for code development and content management.
- *Metaphors*
  — Determine optimum minimum number of concepts, terms, and primary images to meet target user needs.
  — Check for hidden miscommunication and misunderstanding.
  — Adjust metaphorical images or text to account for national or cultural differences. For example, in relation to metaphors for operating systems, Chavan (1994) states that Indians relate more naturally to bookshelves, books, chapters, sections, and pages, rather than the desktop, file folders, and files.
- *Mental models*
  — Determine optimum minimum varieties of content organization.
  — Consider how hierarchies may need to change in detail and overall in terms of breadth and depth. Choong and Salvendy (1999) noted that Chinese and North American users tended to organize the contents of a house in different ways and that if one group were given the hierarchies of the other, the group had more difficulty navigating the hierarchy.
- *Navigation*
  — Determine need for navigation variations to meet target user requirements, determine cost–benefit, and revise as feasible.
- *Interaction*
  — Determine optimum minimum variations of input and feedback. For example, because of Web-access-speed differences for users in countries with very slow access, it is usually important to provide text-only versions, without extensive graphics, as well as alternative text labels to avoid graphics that take considerable time to appear.
- *Appearance*
  — Determine optimum minimum variations of visual and verbal attributes. Visual attributes include layout, icons and symbols, typography, color, and general aesthetics. Verbal attributes include language, formats, and ordering sequences. For example, many written Asian languages, such as Chinese and Japanese, contain symbols with many small strokes. This factor seems to lead to an acceptance of higher visual density of marks in complex public information displays than is typical in Western countries.

An Example of Specific Guidelines: Appearance

Guidelines for visual and verbal appearance are described in this section. Additional details can be found in DelGaldo and Nielsen (1996), Fernandes (1995), Marcus et al. (1999), and Nielsen (1990).

- *Layout and orientation*
  — Adjust layout of menus, tables, dialogue boxes, and windows to account for varying reading directions and size of text. Roman languages read only left to right, but Asian languages may read in several directions (e.g., in Japanese, to the right and down, or down and to the right; Arabic/Hebrew may include right-reading Roman text within left-reading lines of text).
  — If dialogue areas use sentencelike structure with embedded data fields or controls, these areas require special restructuring to account for language changes that significantly alter sentence format. For example, German sentences often have verbs at the ends of sentences, whereas English and French place them in the middle.
  — As appropriate, change layout of imagery that implies or requires a specific reading direction. Left-to-right sequencing may be inappropriate or confusing for use with right-to-left reading scripts and languages.
  — Check for misleading arrangements of images that lead the viewer's eye in directions inconsistent with language reading directions.
  — For references to paper and printing, use appropriate printing formats and sizes. For example, in the United States, standard office letterhead paper size is 8.5 × 11 inches; Europeans use A4 paper size, which is 210 × 297 mm.
- *Icons, symbols, and graphics*
  — Avoid use of text elements and punctuation within icons and symbols to minimize the need for versions to account for varying languages and scripts.
  — Adjust the appearance and orientation to account for national or cultural differences. For example, using a postal mailbox as an icon for e-mail may require different images for different countries.
  — Consider using signs or derivatives from international signage systems developed for safety, mass transit, and communication (see American Institute of Graphic Arts, Olgyay, and Pierce). They require little or no translation and may require minimal culture-specific support information.
  — Avoid puns and local, unique references that will not transfer well. Note that many "universal" signs may be covered by international trademark and copyright use (e.g., Mickey Mouse and the "Smiley" smiling face).
  — Check for appropriateness and use with caution the following:
  — Animals, people, body parts and positions, puns or plays on words, colors, national emblems, signs, hand gestures, and religious, mythological signs.
  — Consider whether selection symbols, such as the X or check marks, convey correct distinctions of selected versus unselected items. Some users may interpret an X as crossing out what is not desired, not selection.
  — Be aware that office equipment (e.g., telephones, mailboxes, folders, and storage devices) clothing, and people differ significantly from nation to nation.
- *Typography*
  — Consider character coding schemes, which differ dramatically for different languages. ASCII (American Standard Code for Information Interchange) is used primarily for

English, but single-byte schemes are used for European languages and double-byte schemes are used for Asian languages. These differences, as well as bidirectional fonts for Hebrew and Arabic display, make it more challenging to support multilingual UIs. Without accounting for character-coding schemes, it is more difficult for users to access content easily.

— Use fonts available for the range of languages required.

— Consider whether special font characters are required for currency, physical measurements, and so forth.

— Use appropriate alphabetic sequence and nomenclature (e.g., U.S. "zee" versus Canadian/English "zed").

— Ensure appropriate decimal, ordinal, and currency number usage. Formats and positioning of special symbols vary from language to language.

— Consider appropriate numeric formats for decimal numbers and their separators (Aykin, 2000):

1,234.56	Canada, China, United Kingdom, United States
1.234,56	Albania, Argentina, Denmark, Greece, Netherlands
1 234,56	Finland, France, Luxembourg, Portugal, Sweden
1'234.56	Switzerland (German, Italian)
1'234,56	Switzerland (French)

Other numeric issues include names of characters, standards for display of negative and positive numbers, percent indication use of leading zeros for decimal values (e.g., 0.1 or .1), list separators, and lucky and unlucky numbers (e.g., lucky telephone numbers in Asian countries sometimes sell for higher prices).

— Use appropriate temperature formats (e.g., Fahrenheit, Centigrade, Kelvin).

— Use appropriate typography and language for calendar, time zone, and telephone and fax number references.

— Consider the following date and time issues, among others:

Calendars (e.g., Gregorian, Moslem, Jewish, Indian, Chinese, Japanese, etc.)

Character representation (Hindu-Arabic, Arabic, Chinese, Roman, etc.)

Clock of 12 or 24 hours

Capitalization rules

Days considered for start of week and for weekend

Format field separators

Maximum and minimum lengths of date and time

Names and abbreviations for days of week and months (two-, three-, and multicharacter standards)

Short and long date formats for dates and times

Time zone(s) appropriate for a country and their names

Use of AM and PM character strings

Use of daylight savings time

Use of leading zeros

— Consider monetary format issues such as the following:

Credit/debit card formats and usage conventions

Currency names and denominations

Currency symbols (local versus international versions)

Currency conversion rates

Monetary formats, symbols, and names

Rules for combining different monetary formats (e.g., required multiple currency postings of international dollars plus local national currency)

Validating monetary input

— Consider name and address formats:

Address elements and number of lines

Address line order

Address: street numbers first or last, punctuation (e.g., 4, route de Monastère versus 1504 South 58th Street versus Motza Illit 11)

Address: zip/postal codes—alphanumerics

Address: zip/postal codes—sequence with city, province/state, country

Character sets

Datafield labels (family name versus last name versus surname; first name versus given name versus Christian name)

Field labels (last name versus surname, city/town/district/province, etc.)

Location and location order: neighborhood, district, city/town, state/province)

Name formats (e.g., family name first for Asian names, family names in all caps in Asia); number of names, number of last names or surnames (maternal versus paternal), prefixes/titles (e.g., German double title: Dr. Ing.), suffixes (e.g., Jr.)

Name formats: double family names (Note, for example, that even within Spanish-speaking countries, some list double family names with maternal first, others with paternal first.) Number of last names/surnames (maternal, paternal)

Number of names (first, middle, initials, etc.)

Prefixes and suffixes

Zip and postal codes (numeric versus alphanumeric, typography, order in relation to city, state/province, or country)

— Consider telephone, fax, and mobile phone number formats:

Grouping of digits varies from country to country

Internal dialing (initial area-code zeros) versus external (without)

Numeric versus alphanumeric (e.g., +1-510-601-0994 versus 510-POP-CORN)

Number grouping, separators (e.g., (,), +, -, period, space, etc.)

Use of plus sign for country codes

Use of parentheses for area codes

Format for multiple sequential numbers of businesses (slash, commas, etc.)

• *Color*

— Follow perceptual guidelines for good color usage. For example, use warm colors for advancing elements and cool colors for receding elements; avoid requiring users to recall in short-term memory more than 5 ± 2 different coded colors.

— Respect national and cultural variations in colors, where feasible, for the target users.

— Follow appropriate professional/popular usage of colors, color names, denotation, and connotation.

- *Aesthetics*
  — Respect, where feasible, different aesthetic values among target users. For example, some cultures have significant attachment to wooded natural scenes, textures, patterns, and imagery (e.g., the Finnish and the Japanese) that might be viewed as exotic or inappropriate by other cultures.

  — Consider specific culture-dependent attitudes. For example, Japanese viewers find disembodied body parts, such as eyes and mouths, unappealing in visual imagery.

- *Language and verbal style*
  — Consider which languages are appropriate for the target users, including the possibility of multiple national languages within one country. For example, English and French within Canada; French, German, and Italian within Switzerland; French or Dutch in Belgium; and in Israel, Hebrew, Arabic, French (official), and English (unofficial). India has more than 20; South Africa has 7, with English third among native languages. Note also, that some languages have different dialects of forms (e.g., Mexican, Argentinian, and Castillian (Spain) Spanish; Parisian, Swiss, and Canadian French).

  — Consider which dialects are appropriate within language groupings and check vocabulary carefully (e.g., for British versus American terms in English, Mexican versus Spanish terms in Spanish, or Mainland China versus Taiwanese terms in Chinese).

  — Consider the impact of varying languages on the length and layout of text. For example, German, French, and English versions of text generally have increasingly shorter lengths. Some Asian texts are 50–80% shorter than English; some non-English roman-character prose texts can be 50 to 200% longer. Some labels can be even longer.

  Example (from Aykin (2000)):

English	Undo	Dutch	Ongedaan maken
English	Autoscroll	Swedish	Automatisk rullning
English	Preferences	German	Bildschirmeinstellungen

  — Consider the different alphabetic sorting or ordering sequences for the varied languages and scripts that may be necessary and prepare variations that correspond to the alphabets. Note that different languages may place the same letters in different locations, for example, Å comes after A in French but after Z in Finnish.

  — Consider differences of hyphenation, insertion point location, and emphasis (i.e., use of bold, italic, quotes, double quotes, brackets, etc.).

  — Use appropriate abbreviations for such typical items as dates, time, and physical measurements. Remember that different countries have different periods of time for "weekends" and the date on which the week begins.

GLOBALIZATION CASE STUDY

Planet Sabre

An example of globalization is the UI design for Sabre's Planet Sabre, one of the world's largest private extranets used by one third of the all travel agents. Sabre contained approximately

42 terabytes of data about airline flights, hotels, and automobile rentals that enabled almost $2 billion of bookings annually, receiving up to one billion "hits" per day. The author's firm worked closely with Sabre over a period of 5 years to develop the Planet Sabre UI (Marcus, 2001).

The UI development process emphasized achieving global solutions from the beginning of the project. For example, requirements mentioned allowing for the space needs of multiple languages for labels in windows and dialogue boxes. Besides supporting English, Spanish, German, French, Italian, and Portuguese, the UI design proposed switching icons for primary application modules so they would be more gender, culture, and nation appropriate (see Figs. 23.1 and 23.2).

Figure 23.1 shows the initial screen of Planet Sabre, with icons representing the primary applications or modules within the system conveyed through the metaphor of objects on the surface of a planet. The postal box representing the electronic mail functions depicts an object that users in the United States would recognize immediately; however, users in many other countries would have significant difficulty recognizing this object, because postal boxes come in different physical forms.

Figure 23.2 shows a prototype version of a Customizer dialogue box, in which the user can change preferences (e.g., icons, so they appear throughout the UI with more recognizable images such as the depiction of the passenger).

At every major stage of prototyping, designs developed in the United States were taken to users in international locations for evaluation. User feedback was relayed to the development team and affected later decisions about all aspects of the UI design.

CULTURAL DIMENSIONS

Localization includes considerations of the target market cultures. The Web in particular enables global distribution of products and services through Internet Web sites, intranets, and extranets. Professional analysts and designers generally agree that well-designed UIs improve the performance and appeal of the Web, helping to convert "tourists" or "browsers" to "residents" and "customers." In a global economy, user differences may reflect worldwide cultures. This section analyzes some of the needs, wants, preferences, and expectations of different cultures through reference to a cross-cultural theory developed by Geert Hofstede (1997).

For example, consider the order in which one prefers to find information. If one is planning a trip by train, is it preferable to see the schedule information first or read about the organization and assess its credibility? Different cultures look for different data to make decisions.

Culture: An Additional Issue for Globally Oriented UI Designers

Cultures, even within some countries, are very different. As noted earlier, sacred colors in the Judeo-Christian West (e.g., red, blue, white, gold) are different from Buddhist saffron yellow or Islamic green. Subdued Finnish designs for

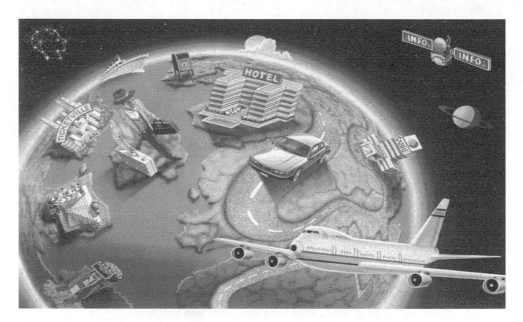

FIGURE 23.1. Example of Planet Sabre home screen showing typical icons for passenger, airline booking, hotel rental, car rental, and e mail (post box).

background screen patterns (see Fig. 23.3) might not be equally suitable in Mediterranean climates. These differences reflect strong cultural values. Batchelor, as reported by Boxer (2001), analyzed the history of Western culture's "fear" of color in his apppropriately titled *Chromophobia*. How might these cultural differences be understood?

Many analysts in organizational communication have studied cultures thoroughly and published classic theories; other authors have applied these theories to analyze the impact of culture on business relations and commerce (e.g., Elashmawi & Harris, 1998; Harris & Moran, 1993; Lewis, 1991). Many of these works are not well known to the UI design community. Anthropologically-oriented theorists include Hofstede (1997), Hall (1969), Victor (1992), and Trompenaars & Turner (1998), each of whom provide valuable insight into illuminating the challenges of cross-cultural communication on the Web.

Recent publications add to the depth of this research, but also follow some of the controversy regarding the permanence

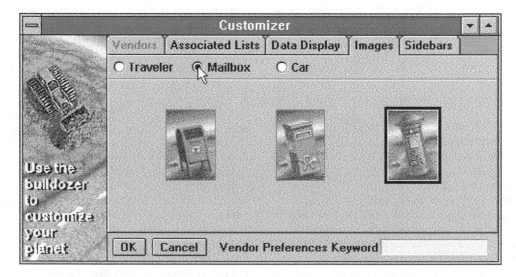

FIGURE 23.2. Example of dialogue box in the Customizer application, by which users can change the icons to become more culturally relevant.

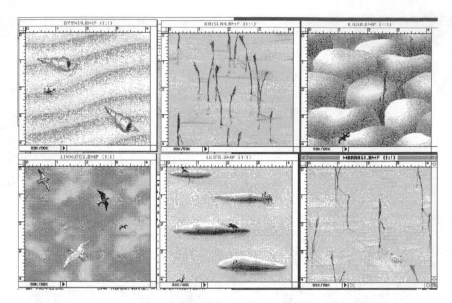

FIGURE 23.3. TeamWare Finnish screen patterns.

and identity of some characterizations of cultures. For example, Nisbett and colleagues as reported by Goode (2000), asserted that essential thinking patterns seem to be culturally influenced and are not universal. They point to Western culture's devotion to logical reasoning, categorization, and desire to understand situations and events as linear, cause-and-effect sequences, whereas Eastern minds seem to accept contradiction more readily.

Clausen, as reported by Yardley (2000) argued that the term anthropological term *culture* refers to "the [essentially inescapable] total structure of life of a particular society," whereas many today use *culture* to refer to (optional) "shared values." As decribed by Yardley, Clausen argued that "the 'culture' of the Internet has none of the characteristics of a real culture. It is not a total way of life; it did not evolve among a distinct people; nobody inherited it or was raised in it; it makes no moral demands, has no religion at its center, and produces no art." Although complex, its rules are procedural. Some of Clausen's statements are, from the perspective of Internet "dwellers," likely to stir debate, as are his assertions that in terms of the strict definition, culture no longer exists in the United States.

The permanence of cultural attributes is questioned further in articles by Ona and Spindle (2000) and Herskovitz (2000), who noted the rise of individualism in classically collectivist Japan and the acceptance, as well as the influence of Japanese pop-cultural artifacts (e.g., music, movies, and television) in Asian nations that were recently mortal enemies of Japan.

The relation of culture to economic success, discussed by some theorists including Hofstede (1997), have made differing assertions about the relation of culture to economic success. Hofstede believes it is a complex mixture of culture plus geography and idiosyncratic drifts of technology (i.e., both culture and creativity). The subject is again in the news, as some theorists argue for strong cultural determinism, as reported by Stille (2001).

Although many cultural anthropology theorists are potentially valuable for UI designers, the application of Hofstede's ideas will demonstrate the value of this body of research for the fields of UI design and specifically Web design. The following section introduces Hofstede's concept of cultural dimensions and applies them to Web UIs.

Hofstede's Dimensions of Culture

During the period from 1978–1983, the Dutch social psychologist Geert Hofstede conducted detailed interviews with hundreds of IBM employees in 53 countries. Through standard statistical analysis of large data sets, he was able to determine patterns of similarities and differences among the replies. From this analysis, he formulated his theory that world cultures vary along consistent, fundamental dimensions. Because his subjects were constrained to one multinational corporation's worldwide employees, and thus to one company culture, he ascribed their differences to the effects of their national cultures. (One weakness of his theory is that he maintained that each country has a single dominant culture.) In 1997, Hofstede published a more accessible version of his research publication in *Cultures and Organizations: Software of the Mind.* His focus was not on defining culture as refinement of the mind but rather on essential patterns of thinking, feeling, and acting that are well established by late childhood. These cultural differences manifest themselves in a culture's choices of symbols, heroes and heroines, rituals, and values.

Hofstede rated 53 countries on indices for each of five dimensions, normalized to values (usually) of 0 to 100. His five dimensions (indices) of culture are: power distance, collectivism/individualism, femininity/masculinity, uncertainty avoidance, long- and short-term time orientation.

Each of Hofstede's terms are examined in this section, focusing on their implications for UI and Web design. Illustrations of characteristic Web sites are also provided. Hofstede's complete data for all countries are shown in Table 23.6.

Power Distance

Power-distance (PD) refers to the extent to which less powerful members expect and accept unequal power distribution within a culture. High PD countries tend to have centralized political power and exhibit tall hierarchies in organizations with large differences in salary and status. Subordinates may view the boss as a benevolent dictator and are expected to do as they are told. Parents teach obedience and expect respect. Teachers possess wisdom and are esteemed. Inequalities are expected and even may be desired. Low PD countries tend to view subordinates and supervisors as closer together and more interchangeable, with flatter hierarchies in organizations and less difference in salaries and status. Parents and children, and teachers and students, may view themselves more as equals. Equality is expected and generally desired.

Hofstede noted that these differences are hundreds or even thousands of years old. He does not believe they will disappear quickly from traditional cultures, even with powerful global telecommunication systems. Based on this definition, power distance may influence the following aspects of UI design:

- Access to information: highly (high PD) versus less highly (low PD) structured
- Hierarchies in mental models: tall versus shallow
- Emphasis on the social and moral order (e.g., nationalism or religion) and its symbols: significant and frequent versus minor and infrequent use
- Focus on expertise, authority, certifications, official logos: strong versus weak
- Prominence given to leaders versus citizens, customers, or employees
- Importance of security, restrictions, or barriers to access: explicit, enforced, and frequent restrictions on users versus transparent, integrated, and implicit freedom to roam
- Social roles used to organize information (e.g., a manager's section obvious to all but inaccessible to nonmanagers): frequent versus infrequent

These PD differences are illustrated by university Web sites from two countries with very different power distance indices (PDIs) (Figs. 23.4, and 23.5a,b): the Universiti Utara Malaysia (www.uum.edu.my) in Malaysia, with a PDI of 104, the highest in Hofstede's analysis; and the Ichthus Hogeschool (www.ichthus-rdam.nl) and the Technische Universiteit Eindhoven (www.tue.nl) in The Netherlands, with a PDI of 38.

The Malaysian Web site features strong axial symmetry, a focus on the official seal of the university, photographs of faculty or administration leaders conferring degrees, and monumental buildings in which people play a small role. A top-level menu selection provides a detailed explanation of the symbolism of the official seal and information about the leaders of the university.

The Dutch Web sites feature a visual emphasis on students (not leaders), a stronger use of asymmetric layout, and photos of both genders in illustrations. These Web sites emphasize the power of students as consumers and equals. Students even have the opportunity to operate a WebCam and take their own tour of the Ichthus Hogeschool.

Individualism versus Collectivism

Individualism in cultures implies loose ties; people are expected to look after themselves or their immediate family but no one else. Collectivism implies that people are integrated from birth into strong, cohesive groups that protect them in exchange for unquestioning loyalty.

Hofstede found individualistic cultures value personal time, freedom, challenge, and such extrinsic motivators as material rewards at work. In family relations, they value honesty/truth, talking things out, using guilt to achieve behavioral goals, and maintaining self-respect. Their societies and governments place individual socioeconomic interests over the group, maintain strong rights to privacy, nurture strong private opinions (expected from everyone), restrain the power of the state in the economy, emphasize the political power of voters, maintain strong freedom of the press, and profess the ideologies of self-actualization, self-realization, self-government, and freedom.

At work, collectivist cultures value training, physical conditions, skills, and the intrinsic rewards of mastery. In family relations, they value harmony more than honesty/truth (and silence more than speech), use shame to achieve behavioral goals, and strive to maintain face. Their societies and governments place collective socioeconomic interests over the individual, may invade private life and regulate opinions, favor laws and rights for groups over individuals, dominate the economy, control the press, and profess the ideologies of harmony, consensus, and equality.

Individualism and collectivism may influence the following Web UI aspects:

- Motivation based on personal achievement: maximized (expect the extraordinary) for individualist cultures versus underplayed (in favor of group achievement) for collectivist cultures
- Images of success: demonstrated through materialism and consumerism versus achievement of sociopolitical agendas
- Rhetorical style: controversial/argumentative speech and tolerance or encouragement of extreme claims versus official slogans and subdued hyperbole and controversy
- Prominence given youth and action versus aged, experienced, wise leaders and states of being
- Importance of individuals versus products shown by themselves or with groups
- Underlying sense of social morality: emphasis on truth versus relationships
- Emphasis on change: what is new and unique versus tradition and history.
- Willingness to provide personal information versus protection of personal data differentiating the individual from the group

TABLE 23.6. Indices from Hofstede (1997), *Cultures and Organizations: Software of the Mind*

	PDI		IDV		MAS		UAI		LTO	
	Rank	Score	Rank	Score	Rank	Score	Rank	Score	Rank	Score
Arab countries	7	80	26/27	38	23	53	27	68		
Argentina	35/36	49	22/23	46	20/21	56	10/15	86		
Australia	41	36	2	90	16	61	37	51	15	31
Austria	53	11	18	55	2	79	24/25	70		
Bangladesh									11	40
Belgium	20	65	8	75	22	54	5/6	94		
Brazil	14	69	26/27	38	27	49	21/22	76	6	65
Canada	39	39	4/5	80	24	52	41/42	48	20	23
Chile	24/25	63	38	23	46	28	10/15	86		
China									1	118
Columbia	17	67	49	13	11/12	64	20	80		
Costa Rica	42/44	35	46	15	48/49	21	10/15	86		
Denmark	51	18	9	74	50	16	51	23		
East Africa	21/23	64	33/35	27	39	41	36	52		
Equador	8/9	78	52	8	13/14	63	28	67		
Finland	46	33	17	63	47	26	31/32	59		
France	15/16	68	10/11	71	35/36	43	10/15	86		
Germany FR	42/44	35	15	67	9/10	66	29	65	14	31
Great Britain	42/44	35	3	89	9/10	66	47/48	35	18	25
Greece	27/28	60	30	35	18/19	57	1	112		
Guatemala	2/3	95	53	6	43	37	3	101		
Hong Kong	15/16	68	37	25	18/19	57	49/50	29	2	96
India	10/11	77	21	48	20/21	56	45	40	7	61
Indonesia	8/9	78	47/48	14	30/31	46	41/42	48		
Iran	29/30	58	24	41	35/36	43	31/32	59		
Ireland (Republic of)	49	28	12	70	7/8	68	47/48	35		
Israel	52	13	19	54	29	47	19	81		
Italy	34	50	7	76	4/5	70	23	75		
Jamaica	37	45	25	39	7/8	68	52	13		
Japan	33	54	22/23	46	1	95	7	92	4	80
Malaysia	1	104	36	26	25/26	50	46	36		
Mexico	5/6	81	32	30	6	69	18	82		
Netherlands	40	38	4/5	80	51	14	35	53	10	44
New Zealand	50	22	6	79	17	58	39/40	49	16	30
Nigeria									22	16
Norway	47/48	31	13	69	52	8	38	50		
Pakistan	32	55	47/48	14	25/26	50	24/25	70	23	0
Panama	2/3	95	51	11	34	44	10/15	86		
Peru	21/23	64	45	16	37/38	42	9	87		
Philippines	4	94	31	32	11/12	64	44	44	21	19
Poland									13	32
Portugal	24/25	63	33/35	27	45	31	2	104		
Salvador	18/19	66	42	19	40	40	5/6	94		
Singapore	13	74	39/41	20	28	48	53	8	9	48
South Africa	35/36	49	16	65	13/14	63	39/40	49		
South Korea	27/28	60	43	18	41	39	16/17	85	5	75
Spain	31	57	20	51	37/38	42	10/15	86		
Sweden	47/48	31	10/11	71	53	5	49/50	29	12	33
Switzerland	45	34	14	68	4/5	70	33	58		
Taiwan	29/30	58	44	17	32/33	45	26	69	3	87
Thailand	21/23	64	39/41	20	44	34	30	64	8	56
Turkey	18/19	66	28	37	32/3	45	16/17	85		
Uruguay	26	61	29	36	42	38	4	100		
USA	38	40	1	91	15	62	43	46	17	29
Venezuela	5/6	81	50	12	3	73	21/22	76		
West Africa	10/11	77	39/41	20	30/31	46	34	54		
Yugoslavia	12	76	33/35	27	48/49	21	8	88		
Zimbabwe									19	25

Note. PD = power-distance; IDV = collectivism/individualism; MAS = femininity/masculinity; UAI = uncertainty avoidance; LTO = long- and short-term time orientation.

FIGURE 23.4. High power distance: Malaysian university Web site.

The effects of these differences can be illustrated on the Web by examining national park Web sites from two countries with very different IC indices (Figs. 23.6, 23.7, and 23.8). Glacier Bay National Park (www.nps.gov/glba/evc.htm) is located in the United States, which has the highest IC index rating (91). The Web site from the National Parks of Costa Rica (www.tourism costarica.com; Fig. 23.7) is located in a country with an IC index rating of 15. The third image (Fig. 23.8) shows a lower level of the Costa Rican Web site.

Note the differences in the two groups of Web sites. The U.S. Web site features an emphasis on the visitor, his or her goals, and possible actions in coming to the park. The Costa Rican Web site features an emphasis on nature, downplays the individual tourist, and uses a slogan to emphasize a national agenda. An even more startling difference lies below the What's Cool menu link. Instead of a typical Western display of new technology or experience to consume, the screen is filled with a massive political announcement that the Costa Rican government has signed an international agreement against the sexual exploitation of children and adolescents.

Masculinity versus Femininity

Masculinity and femininity refer to gender roles, not physical characteristics. Hofstede (1997) focused on the traditional assignment to masculine roles of assertiveness, competition, and toughness and to feminine roles of orientation to home and children, people, and tenderness. He acknowledged that in different cultures different professions are dominated by different genders. (For example, women dominate the medical profession in Russia, whereas men dominate in the United States.) However, in masculine cultures, the traditional distinctions are strongly maintained, whereas feminine cultures tend to collapse the distinctions and overlap gender roles (both men and women can exhibit modesty, tenderness, and a concern with both quality of life and material success). Traditional masculine work goals include earnings, recognition, advancement, and challenge. Traditional feminine work goals include good relations with supervisors, peers, and subordinates; good living and working conditions; and employment security.

The following list shows some typical masculinity (MAS) index values, where a high value implies a strongly masculine culture:

95 Japan
79 Austria
63 South Africa
62 USA
53 Arab countries
47 Israel
43 France
39 South Korea
05 Sweden

Because Hofstede's definition focuses on the balance between roles and relationships, masculinity and femininity may logically be expressed on the Web through different emphases.

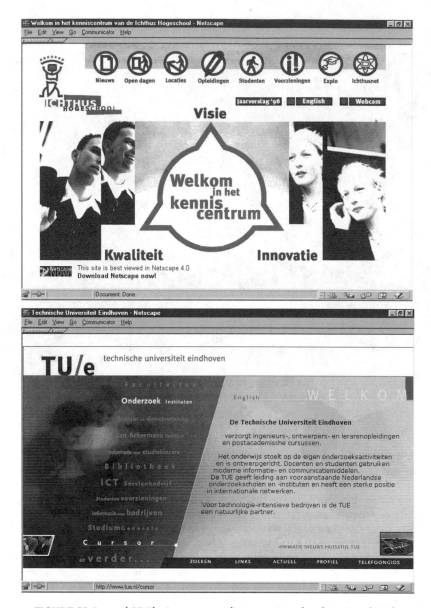

FIGURE 23.5a and 23.5b. Low power distance: Dutch educational Web sites.

High masculinity cultures would focus on the following UI and design elements:

- Traditional gender/family/age distinctions
- Work tasks, roles, and mastery, with quick results for limited tasks
- Navigation oriented to exploration and control
- Attention gained through games and competitions
- Graphics, sound, and animation used for utilitarian purposes

Feminine cultures would emphasize the following UI elements:

- Blurring of gender roles
- Mutual cooperation, exchange, and support (versus mastery and winning)

- Attention gained through poetry, visual aesthetics, appeals to unifying values

Examples of MAS differences on the Web can be illustrated by examining Web sites from countries with very different MAS indices (Figs. 23.9, 23.10, and 23.11). The Woman.Excite Web site (woman.excite.co.jp) is located in Japan, which has the highest MAS value 95. The Web site narrowly orients its search portal toward a specific gender, which this company does not do in other countries.

The ChickClick (www.chickclick.com) U.S. Web site (MAS = 52) consciously promotes the autonomy of young women (leaving out later stages in a woman's life.)

The Excite Web site (www.excite.com.se) from Sweden, with the lowest MAS value of 5, makes no distinction in gender

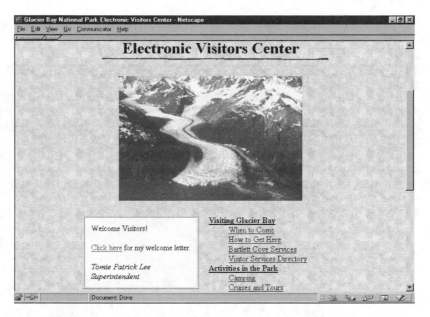

FIGURE 23.6. High individualist value: U.S. National Park Web site.

or age. (With the exception of The Netherlands, another low MAS country, all other European Web sites provide more preselected information.)

Uncertainty Avoidance

People vary in the extent to which they feel anxious about uncertain or unknown matters, as opposed to the more specific feeling of fear caused by known or understood threats. Cultures vary in their avoidance of uncertainty, creating different rituals and having different values regarding formality; punctuality; legal, religious, and social requirements; and tolerance for ambiguity.

Hofstede notes that cultures with high uncertainty avoidance (UA) tend to have high rates of suicide, alcoholism, and accidental deaths, and high numbers of prisoners per capita. Businesses may have more formal rules, require longer career commitments, and focus on tactical operations rather than strategy. These cultures tend to be expressive; people talk with their hands, raise their voices, and show emotions. People seem active, emotional, even aggressive; shun ambiguous situations; and

FIGURE 23.7. Low individualist value: Costa Rican National Park Web site.

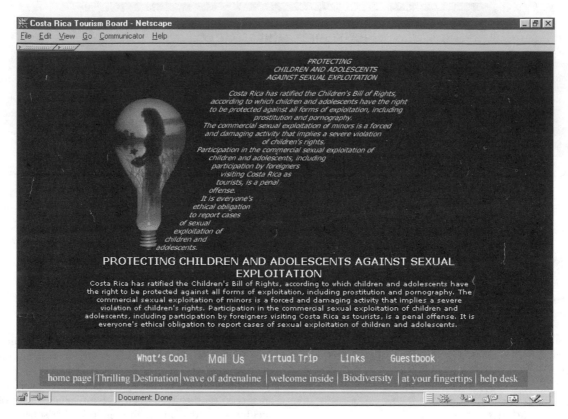

FIGURE 23.8. "What's Cool": Political message about exploitation of children.

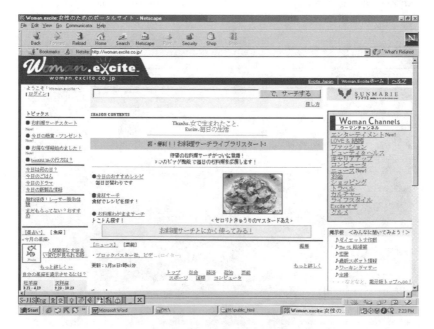

FIGURE 23.9. High masculinity Web site: Excite.com for women in Japan.

FIGURE 23.10. Medium masculinity Web site: ChickClick.com in the United States.

expect structure in organizations, institutions, and relationships to help make events clearly interpretable and predictable. Teachers are expected to be experts who know the answers and may speak in cryptic language that excludes novices. In high UA cultures, what is different may be viewed as a threat, and what is "dirty" (unconventional) is often equated with what is dangerous.

By contrast, low UA cultures tend to be less expressive and less openly anxious; people behave quietly without showing aggression or strong emotions (although their caffeine consumption may be intended to combat depression from their inability to express their feelings). People seem easy-going, even relaxed. Teachers may not know all the answers (or there may be more than one correct answer), run more open-ended classes, and are expected to speak in plain language. In these cultures, what is different may be viewed as simply curious or perhaps ridiculous.

Based on this definition, uncertainty avoidance may influence contrary aspects of UI and Web design. High UA cultures would emphasize the following:

FIGURE 23.11. Low masculinity Web site: Swedish Excite.com.

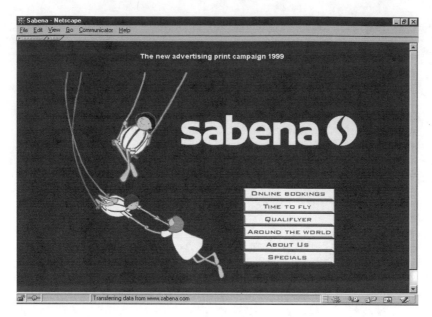

FIGURE 23.12. High uncertainty avoidance: Sabena Airlines Web site from Belgium.

- Simplicity, with clear metaphors, limited choices, and restricted data
- Attempts to reveal or forecast results of actions before users act
- Navigation schemes intended to prevent users from becoming lost
- Mental models and help systems that focus on reducing "user errors"
- Redundant cues (color, typography, sound, etc.) to reduce ambiguity

Low UA cultures would emphasize the reverse:

- Complexity with maximal content and choices
- Acceptance (even encouragement) of wandering and risk, with a stigma on "overprotection"
- Less control of navigation; for example, links might open new windows leading away from the original location
- Mental models and help systems might focus on understanding underlying concepts rather than narrow tasks
- Coding of color, typography, and sound maximize information

Examples of UA differences can be illustrated on the Web by examining airline Web sites from two countries with very different UA indices (Figs. 23.12 and 23.13). The Sabena Airlines Web site (www.sabena.com) is located in Belgium, a country with a UA of 94, the highest of the cultures studied. This Web site shows a home page with very simple, clear imagery and limited choices.

The British Airways Web site (www.britishairways.com) from the United Kingdom (UA = 35) shows more complexity of content and choices with pop-up windows, multiple types of controls, and "hidden" content that must be displayed by scrolling.

Long- versus Short-Term Time Orientation

In the early 1980s, shortly after Hofstede first formulated his cultural dimensions, work by Michael Bond convinced him that a fifth dimension needed to be defined. Long-term orientation (LTO) seemed to play an important role in Asian countries influenced by Confucian philosophy over thousands of years that shared the following beliefs:

- A stable society requires unequal relations.
- The family is the prototype of all social organizations; consequently, older people (parents) have more authority than younger (and men more than women).
- Virtuous behavior to others means not treating them as one would not like to be treated.
- Virtuous behavior in work means trying to acquire skills and education; working hard; and being frugal, patient, and persevering.

Western countries, by contrast, were more likely to promote equal relationships, emphasize individualism, to focus on treating others as you would like to be treated, and to find fulfillment through creativity and self-actualization. When Hofstede and Bond developed a survey specifically for Asia and reevaluated earlier data, they found that long-term orientation cancelled out some of the effects of masculinity/femininity and uncertainty avoidance. They concluded that Asian countries are oriented to practice and the search for virtuous behavior, whereas Western countries are oriented to belief and the search for truth. Of the

FIGURE 23.13. Low uncertainty avoidance: The British Airways Web site.

23 countries compared, the following showed the most extreme values:

  118 China (ranked 1)   29 USA (17)
   80 Japan (4)   0 Pakistan (23)

High LT countries would emphasize the following aspects of UI design:

- Content focused on practice and practical value
- Relationships as a source of information and credibility
- Patience in achieving results and goals

Low LT countries would emphasize the contrary:

- Content focused on truth and certainty of beliefs
- Rules as a source of information and credibility
- Desire for immediate results and achievement of goals

Examples of LTO differences on the Web can be illustrated by examining versions of the same company's Web site from two countries with different LT values (Figs. 23.14 and 23.15). The Siemens Web site (www.siemens.co.de) from Germany (LT of 31) shows a typical Western corporate layout that emphasizes crisp, clean, functional design aimed at achieving goals quickly.

The Chinese version from Beijing requires more patience to achieve navigational and functional goals.

Design Issues

Hofstede noted that some cultural relativism is necessary: It is difficult to establish absolute criteria for what is noble and what is disgusting. There is no escaping bias; all people develop cultural values based on their environment and early training as children. Not everyone in a society fits the cultural pattern precisely, but there is enough statistical regularity to identify trends and tendencies. These trends and tendencies should not be treated as defective or used to create negative stereotypes, but recognized as different patterns of values and thought. In a multicultural world, it is necessary to cooperate to achieve practical goals without requiring everyone to think, act, and believe identically.

This review of cultural dimensions raises many issues about UI design, especially for the Web.

- How formal or rewarding should interaction be?
- What will motivate different people: money, fame, honor, achievement?
- How much conflict can people tolerate in content or style of argumentation?
- Should sincerity, harmony, or honesty be used to make appeals?
- What role exists for personal opinion versus group opinion?
- How well are ambiguity and uncertainty avoidance received?
- Will shame or guilt constrain negative behavior?
- What role do community values play in individualist versus collectivist cultures?
- Does the objective of distance learning change what can be learned in individualist versus collectivist cultures? Should these sites focus on tradition? On skills? On expertise? On earning power?
- How should online teachers or trainers act: as friends or as gurus?
- Would job sites differ for individualist versus collectivist cultures?

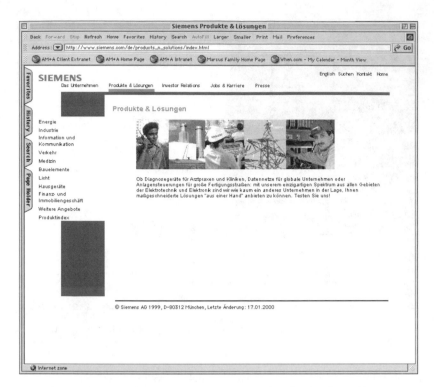

FIGURE 23.14. Low Long-term orientation: Web site form Siemens Germany.

- Should there be different sites for men and women in different cultures?
- Would personal WebCams be acceptable?
- How much advertising hyperbole can be tolerated?
- Would an emphasis on truth as opposed to practice and virtue require different types of procedural Web sites for Western or Asian audiences?

If cross-cultural theory becomes an accepted element of UI design, then we may need to change our current practices and develop new tools. We need to make it feasible to

FIGURE 23.15. High long-term orientation: Web site from Siemens in China.

develop multiple versions of Web sites in a cost-effective manner, perhaps through templates or through specific versioning tools. As the Web continues to develop globally, answering these questions, and exploring and then exploiting these dimensions of culture, will become a necessity for successful theory and practice.

CONCLUSIONS AND FUTURE RESEARCH ISSUES

To achieve culturally sensitive, successful global access to UIs provides many design challenges in the UI development process. Progress in technology increases the number and kinds of functions, data, platforms, and users of computer-based communication media. The challenge of enabling more people and more kinds of people to use this content and these tools effectively will depend increasingly upon global solutions. By recognizing the need for and benefit to users of UI designs intended for international and intercultural markets, developers will achieve greater success and increased profitability through the global distribution and acceptance of their products.

The recommendations provided in this chapter are an initial set of heuristics that will assist developers in achieving global solutions to their product development. Design methodologies must support globalization throughout the development process. In addition, it is likely that international and intercultural references will change rapidly, requiring frequent updating of designs. Future work on global UI design may address the following issues:

1. How might global UIs be designed to account for different kinds of intelligence? Gardner (1985) identified the following dimensions of intelligence: verbal/image comprehension, word/image fluency, numerical/graphical fluency, spatial visualization, associative memory, perceptual speed, reasoning, interpersonal awareness, and self-awareness. These dimensions suggest users might have varying strengths of conceptual competence with regard to using UIs on an individual basis, but these might also vary internationally, or inter-culturally, due to influences of language, history, or other factors.

2. How might metaphors, mental models, and navigation be designed precisely for different cultures that differ by such dimensions as age, gender, national or regional group, or profession? The author posed this as a question to the UI analysis/design community (Marcus, 1993). The topic is discussed broadly in (DelGaldo & Nielson, 1996). Furthermore, what means can be developed to enable these variations to be produced in a cost-effective manner using templates?

The taxonomic analyses of global issues for UIs, the theoretical basis for their component selection, the criteria for their evaluation, and their design methodology are emerging in the UI development field. Nevertheless, designers should be aware of the scope of the activity, know sources of insight, and incorporate professional techniques in their development process to improve the value and success of their international and intercultural products.

ACKNOWLEDGMENTS

The author acknowledges (Marcus et al., 1999) and (Marcus & Gould, 2000a, 2000b), which were edited to create this chapter. Mr. Marcus thanks his staff at Aaron Marcus and Associates, Inc.; Sabre; Ms. Emile Gould, Rensellaer Polytechnic Institute; Dr. Andrew Sears and Prof. Julie Jacko, the editors of this book; and Dr. Constantine Stephanidis, editor of *User Interfaces for All*, for their assistance in preparing this chapter. The author also acknowledges DelGaldo and Niclsen (1996), Fernandes (1995), and Nielsen (1990), which provided a basis for points raised in this chapter, and the advice of Peter Siemlinger, Vienna, Austria, and Professor Andreas Schneider, Tama Art University, Tokyo, Japan; and Nuray Aykin, AT&T.

References

Alvarez, G. M., Kasday, L. R., & Todd, S. (1998). How we made the Web site international and accessible: A case study. Proc. 4th Hum. Fact. and the Web Conf. Holmdel, NJ.

American Institute of Graphic Arts (AIGA). (1981). *Symbol Signs*. Visual Communication Books, Hastings House, New York.

Aykin, N. Internationali(z?)(s?)ation and Locali(z?)(s?)ation of Web Sites. (2000). AT&T Labs, private communication.

Boxer, S. (2001, April 28). Vivid color in a world of black and white. *New York Times*, p. A15ff.

Chavan, A. L. (1994). *A design solution project on alternative interface for MS Windows*. Unpublished master's thesis, New Delhi, India.

Choong Y., & Salvendy, G. (1999). Implications for Design of Computer Interfaces for Chinese Users in Mainland China. In *International Journal of Human-Computer Interaction, 11*(1), 29–46.

Coriolis Group. (1998). *How to build a successful international Web site*. Scottsdale, AZ: Author.

Cox, T., Jr. (1994). *Cultural diversity in organizations*. San Francisco: Berrett-Koehler.

Crystal, D. (1987). *The Cambridge encyclopedia of language*. Cambridge, England: Cambridge University Press.

Daniels, P. T., & Bright, W. (Eds.). (2000). *The world's writing systems*. Sandpoint, ID: MultiLingual Computing.

Day, D. L., del Galdo, E. M., & Prabhu, G. V. (Eds.). (2000, July). *Designing for global markets 2*, 2nd International Workshop on Internationalisation of Products and Systems. Rochester, NY: Backhouse Press.

Day, Donald. (2000, 13–15 July). "Gauging the Extent of Internationalization Activities," Proc., 2nd International Workshop on Internationalization of Products and Sysems (pp. 124–136). Baltimore, MD: Backhouse Press, Rochester, NY, ISBN 0-9656691-4-9.

Day, Donald L., & Lynne M. Dunckley. (2001, 12–14 July). Designing for Global Markets 3, Third International Workshop on Internationalisation of Products and Systems (IWIPS 2001). Milton Keynes, UK, ISBN 0-7492-53258.

DelGaldo, E., & Nielsen, J. (Eds.). (1996). *International user interfaces*. New York: Wiley.

Doi, Takeo, *The Anatomy of Dependence*. Kodansha International, New York, 1973.

Doi, Takeo, *The Anatomy of Self: The Individual versus Society*. Kodansha Internat., N.Y., 1986.

Dreyfuss, Henry, *Symbol Sourcebook*. Van Nostrand Rhinehold, New York, 1966.

Elashmawi, F., & Harris, P. R. (1998). *Multicultural management 2000: Essential cultural insights for global business success*. Houston, TX: Gulf.

Fernandes, T. (1995). *Global interface design: A guide to designing international user interfaces*. Boston: AP Professional.

Forrester Research, Inc. (2001, March) "The Global User Experience," Forrester Research, Inc. Cambridge, MA: www.Forrester.com.

Forrester Research, Inc. (1998). JIT Web localization. Retrieved July 1998 from www.Forrester.com. Cambridge, MA: Author.

French, Tim, & Smith, Andy. (2000, 13–15 July). "Semiotically Enhanced Web Interfaces: Can Semiotics Help Meet the Challenge of Cross-Cultural Design?", Proc., 2nd Internat. Wkshop. on Internat. of Products and Systems (pp. 23–38). Baltimore, MD: Backhouse Press, Rochester, NY, ISBN 0-9656691-4-9.

Gardner, H. (1985). *Frames of mind, the theory of multiple intelligences*. New York: Basic Books.

Goode, E. (2000, August 8). How culture molds habits of thought. *New York Times*, p. D1ff.

Gould, Emilie, W., Zakaria, Norhayati, & Shafiz Affendi Mohd. Yusof. (2000, 25 September). "Applying Culture to Website Design: A comparison of Malaysian and US Websites," *Proc.*, IPCC/SIGDOC, Boston.

Graham, T. (2000). *Unicode: A primer*. Sandpoint, ID: MultiLingual Computing and Technology.

Hall, E. T. (1969). *The hidden dimension*. New York: Doubleday.

Harel, Dan, & Girish Prabhu. (1999, 22–23 May). "Global User Experience (GLUE), Design for Cultural Diversity: Japan, China, and India," *Designing for Global Markets*, Proc. of First International Workshop on Internationalization of Products and Systems (IWIPS-99) (pp. 205–216). Rochester, NY, ISBN 0-9656691, Backhouse Press, Rochester, NY.

Harris, J., & McCormack, R. (2000). "Tanslation is Not Enough." Sapient, www.Sapient.com.

Harris, P. R., & Moran, R. T. (1993). *Managing cultural differences*. Houston, TX: Gulf.

Hendrix, A. (2001, April 15). The nuance of language. *San Francisco Chronicle*, p. A10.

Herskovitz, J. (2000, December 26). J-Pop takes off: Japanese music, movies, TV shows enthrall Asian nations. *San Francisco Chronicle*, p. C2.

Hofstede, G. (1997). *Cultures and organizations: Software of the mind, intercultural cooperation and its importance for survival*. New York: McGraw-Hill.

Hoft, Nancy, L. (1995). *International Technical Communication: How to Export Information About High Technology*. John Wiley and Sons, Inc., New York, ISBN: 0-471-03743-5.

Honold, P. (1999). Learning how to use a cellular phone: Comparison between German and Chinese users. *Journal of Society for Technical Communication, 46*, 196–205.

International Organization for Standardization. (1989). *Computer display color* (Draft Standard Document 9241-8). Geneva, Switzerland: Author.

International Standards Organization. (1990). *ISO 7001: Public Information Symbols*. Geneva, Switzerland: The American National Standards Institute (ANSI).

International Standards Organization. (1993). *ISO 7001: Public Information Symbols: Amendment 1*. Geneva, Switzerland: The American National Standards Institute (ANSI).

Kimura, Doreen, "Sex Differences in the Brain," *Sci. Amer., 267*(3), 118–125.

Konkka, Katja, & Koppinen, Anne. (13–15 July). "Mobile Devices: Exploring Cultural Differnces in Separating Professional and Personal Time," Proc., Second International Workshop on Internationalization of Products and Sysems (pp. 89–104). Baltimore, MD: Backhouse Press, Rochester, NY, ISBN 0-9656691-4-9.

Kurosu, M. (1997, August 24–29). Dilemma of usability engineering. In G. Salvendy, M. Smith, & R. Koubek (Eds.), *Design of Computing Systems: Social and Ergonomics Considerations: Vol. 2: Proc. 7th Int. Conference on Human-Computer Interaction* (HCI International '97 (pp. 555–558). San Francisco, USA: Amsterdam: Elsevier.

Leventhal, L. et al. "Assessing User Interfaces for Diverse User Groups: Evaluation Strategies and Defining Characteristics," *Behaviour & Information Technology*, 1996, 10 pp., in preparation. Discusses cultural diversity in UI design.

Lewis, R. (1991). *When cultures collide*. London: Nicholas Brealey.

Lingo Systems. (1999). *The Guide to Translation and Localization*. IEEE Computer Society, Los Alamitos, CA, ISBN 0-7695-0022-6.

LISA, The Localization Industry Primer. The Localization Industry Standards Association, LISA Administration, 7, rute du Monastère, 1173 Féchy, Switz., www.lisa.org, 1999, 35 pp.

Marcus, A. (1992). *Graphic Design for Electronic Documents and User Interfaces*, Addison-Wesley, Reading.

Marcus, A. (1993, April). "Human Communication Issues in Advanced UIs," *Communications of the ACM, 36*(4), 101–109.

Marcus, A. (1993, October). Designing for diversity. *Proceedings of the 37th Human Factors and Ergonomics Society* (pp. 258–261). Seattle, WA.

Marcus, A. (1995). Principles of effective visual communication for graphical user-interface design. In R. Baecker et al. (Eds.), *Readings in human–computer interaction* (2nd ed., pp. 425–441). Palo Alto, CA: Morgan Kaufman.

Marcus, A. (1998). Metaphor design in user interfaces. *Journal of Computer Documentation, 22*, 43–57.

Marcus, A. (2000). "International and Intercultural User Interfaces." In Dr. C. Stephanidis (Ed.), *User Interfaces for All*. (pp. 47–63). Lawrence Erlbaum Associates, New York.

Marcus, A., & Gould E. W. (2000). Crosscurrents: Cultural dimensions and global web user-interface design. *Interactions, 7*, 32–46.

Marcus, A., et al. (1999, June). Globalization of user-interface design for the Web. *Proc.*, 5th Hum. Fact. and Web Conf. Gathersburg, MD.

Marcus, A. (2001). UI design of Sabre: A case study. *Information Design Jounal, 10*, 188–206.

Neustupmy, J. V. (1987). *Communicating with the Japanese*. The Japan Times, Tokyo.

Nielsen, J. (Ed.). (1990). *Designing user interfaces for international use. Vol. 13: Advances in Human Factors/Ergonomics*. Amsterdam: Elsevier Science.

Olgyay, Nova. (1995). *Safety Symbols Art*. Van Nostrand Reinhold, New York.

Ona, Y., & Spindle, B. (2000, December 30). Japan's long decline makes one thing rise: Individualism. *Wall Street Journal*, p. A1.

Ota, Yukio. (1973). *Locos: Lovers Communications System* (in Japanese), Pictorial Institute, Tokyo. The author presents the system of universal writing that he invented.

Ota, Yukio. (1987). *Pictogram Design*, Kashiwashobo, Tokyo, ISBN 4-7601-0300-7. The author presents a world-wide collection of case studies in visible language signage systems.

Peírce, Todd. (1996). *The International Pictograms Standard*. ST Publications, Cincinatti, OH.

Peng, Kaiping. (2000). *Readings in Cultural Psychology*: Theoretical, Methodolgical and Empirical Developments During the Past

Decade (1989–1999). John Wiley and Sons, Custom Services, ISBN 0-471-38803-3.

Perlman, Gary. (13–15 July). "ACM SIGCHI Intercultural Issues," Proc., Second International Workshop on Internationalization of Products and Systems (pp. 183–195). Baltimore, MD: Backhouse Press, Rochester, NY, ISBN 0-9656691-4-9.

Pierce, T. (1996). *The International Pictograms Standard.* Cincinatti, OH: ST Publications. (Available also on CD-ROM).

Prabhu, G. V., Chen, B., Bubie, W., & Koch, C. (1997, 24–29 August). Internationalization and localization for cultural diversity. In Salvendy et al. (Eds.), *Design of Computing Systems: Cognitive Considerations. Vol. 1: Proc. 7th Int. Conf. on Human-Computer Interaction* (HCI International '97, pp. 149–152). San Francisco, Amsterdam: Elsevier.

Prabhu, Girish V., & delGaldo, Elisa M. (Ed.). (1999, 20–22 May). Designing for Global Markets 1, Proc., First International Workshop on Internationalization of Products and Systems (p. 226). Rochester, NY: Backhouse Press, New York, ISBN 0-9656691-2-2.

Prabhu, Girish, & Dan Harel. (1999, 22–26 August). "GUI Design Preference Validation for Japan and China: A Case for KANSEI Engineering?," Proc., 8th Int. Conf. on Human-Computer Interaction (HCI International '99). Munich, Germany: Amsterdam, Elsevier.

Saudi Arabia issues edict against Pokemon. (2001, March 27). *San Francisco Chronicle*, p. F2.

Shahar, Lucy, & Kurz, David. (1995). *Border Crossings: American Interactions with Israelis.* Intercultural Press, Yarmouth, Maine, ISBN 1-977864-31-5.

Stephanidis, C. (Ed.). (2000). *User interfaces for all.* New York: Lawrence Erlbaum Associates.

Stille, A. (2001, 13 January). An old key to why countries get rich: It's the culture that matters, some argue anew. *New York Times*.

Storti, Craig. (1994). *Cross-Cultural Dialogues: 74 Brief Encounters with Cultural Difference* (p. 140). Intercultural Press, Yarmouth, ME, ISBN 1-1877864-28-5.

Tannen, Deborah. (1990). *You Just Don't Understand: Women and Men in Conversation.* William Morrow and Company, Inc., New York, ISBN 0-688-07822-2.

Trompenaars, F., & Turner, C. H. (1998). *Riding the waves of culture.* New York: McGraw-Hill.

Vickers, B. (2000, November 22). Firms push to get multilingual on the Web. *Wall Street Journal*, p. B11A.

Victor, D. A. (1992). *International business communication.* New York: HarperCollins.

Yardley, J. (2000, 7 August). *Faded Mosaic* nixes idea of "cultures" in U.S. *San Francisco Examiner*, p. B3.

Yeo, Alvin Wee. (2001, 31 March–5 April). "Global-Software Development Lifecycle: An Exploratory Study," Proc., CHI 2001 (pp. 104–111). Seattle, WA, ISBN 1-58113-327-8.

URLS *and Other* Information Resources

American National Standards Institute (ANSI): www.ansi.org

Bibliography of Intercultural publications: www.HCIBib.org//SIGCHI/Intercultural

China National Standards; China Commission for Conformity of Elect. Equip. (CCEE) Secretariat; 2 Shoudu Tiyuguan, NanLu, 100044, P. R. China; Tel: +86-1-8320088, ext. 2659, Fax: +86-1-832 0825

Cultural comparisons: www.culturebank.com, www.webofculture.com, www.iir-ny.com

Digital divide: www.digitaldivide.gov, www.digitaldivide.org, www.digitaldividenetwork.org

Globalization and Internet language statistics: language: www.euromktg.com/globstats/, www.sapient.com, www. world-ready.com/biblio.htm

Glossary, six languages: www.bowneglobal.com/bowne.asp?page=9&language=1

International Organization for Standardization (ISO): http://www.iso.ch/

Internationalization providers: www.basistech.com, www.cij.com, www.Logisoft.com

Internationalization resources: www.world-ready.com/r_intl.htm, www.worldready.com/biblio.htm

Japan Info. Processing Society; Kikai Shinko Bldg., No. 3-5-8 Shiba-Koen, Minato-ku, Tokyo 105, Japan; Tel: +81-3-3431-2808, Fax: +81-3-3431-6493

Japanese Industrial Standards Committee (JISC); Min. of Internat. Trade and Industry; 1-3-1, Kasumigaseki, Chiyoda-ku, Tokyo 100, Japan; Tel: +81-3-3501-9295/6, Fax: +81-3-3580-1418

Java Internationalization: http://java.sun.com/docs/books/tutori

Localization Industry Standards Organization (LISA): www.lisa. org

Localization providers: www.Alpnet.com, www.Berlitz.com, www.globalsight.com, www.lhsl.com, www.Lionbridge.com, www.Logisoft.com, www.Logos-usa.com, www.translations.com, www.Uniscape.com

Machine translation providers: www.babelfish.altavista.com, www.IDC.com, www.e-Lingo.com, Lernout & Hauspie <www.lhsl.com>, www.Systransoft.com

Microsoft global development: www.eu.microsoft.com/ globaldev/

Simplified English: userlab.com/SE.html

Internet users survey, Nua: www.nua.ie/surveys/how_many_ online

Unicode: www.unicode.org/, www-4ibm.com/software/developer/library/glossaries/unicode.html

World-Wide Web Consortium: www.w3.org/International, www.w3.org/WAI

·24·

INFORMATION TECHNOLOGY
FOR COGNITIVE SUPPORT

Alan F. Newell, Alex Carmichael, Peter Gregor, and Norman Alm
University of Dundee

INTRODUCTION

Well-designed communication and information technology (IT) systems have great and unrealized potential to enhance quality of life and independence for those with cognitive dysfunction, including elderly people by

- allowing them to retain a high level of independence and control over their lives,
- providing appropriate levels of monitoring and supervision of "at-risk" people, without violating privacy,
- keeping people intellectually and physically active, and
- providing communication methods to reduce social isolation.

Using technology to augment human cognitive capacity is not a new idea, taking the term *technology* in its widest sense to include tools that humans have developed to help them cope with their environment. The first cognitive function to be augmented was probably memory. Mnemonic methods help with this task, and they are still in wide use today. One of the commonest is to link what is to be remembered with a structure that is easier to recall, such as the layout of a city or a narrative that has already been memorized. Most people rely to a large extent on the extended memory that the written word affords us. When written language was widely introduced, it is said that many worried this would be a backward step, because it would make learned people reliant on it, and the ability to memorize would wither away. Indeed, few people today, other than professional entertainers, could perform the memory feats that were required of learned people in the past. All most of us can hope to do is to know how to find out what we need to know on any occasion.

The increasing power and decreasing size of computer technology, along with its capacity to provide communication as well as computation, offers the possibility of sophisticated help for cognitive problems, if we can develop the appropriate software. Computers have the potential to act as a kind of scaffolding for cognitive tasks, taking over functions affected by illness, accident, or aging. They could also provide prompts for daily living, if they were able to track successfully through the user's sequence of tasks and actions.

Particular strengths of computers as assistants for people with cognitive impairments include computers being consistent and tireless and their not becoming emotionally involved in a shared task. In addition, multimedia and multimodal systems can provide rich interaction that may be particularly advantageous for users with cognitive dysfunction. For example, such systems have great potential to address the problems of memory loss and the other problems presented by dementia. In addition, communication systems that use synthetic speech, predictive programs that can facilitate writing, and a range of nonlinguistic methods of communication, can be used by those with speech and language dysfunction caused by cognitive impairments.

A great deal of work will be required to realize this potential, both in uncovering new knowledge about how cognition works and in developing assistive systems based on this knowledge.

Because of the wide range of skills and knowledge needed to understand the problems people with cognitive impairment face, it is essential that research work in this field be multidisciplinary, including psychologists, members of the health and therapeutic professions, and engineers. It is also vital to involve potential users of the technology at all stages of development as partners in the research.

In addition, it is essential that interface designers be more aware of the range of abilities among users. For example, access to the Internet for disabled people is often thought to be synonymous with access for blind people, but in fact blind people form only a small percentage of the disabled population. Ogozalec (1997) pointed out that if current trends continue in the United States, by 2030 one fifth of the population will be over 65 years of age and commented that "It is difficult to categorise and draw conclusions about 'the elderly,' since they comprise such a diverse and heterogeneous population." This diversity, particularly of cognitive function, ought to be taken into account if we are to make software and the Internet available to as large a percentage of the population as possible.

We describe the major types of cognitive impairment and include within this discussion the effects of aging. We then illustrate the development of systems to support people with cognitive impairment with specific projects in which we have been involved. They will not cover all aspects of cognitive impairment, but they do illustrate a methodology and an approach to developing assistive technology that may have a wider relevance.

In the concluding part of this chapter, we address the development of methodologies that we believe will be valuable for designers of systems to support people with cognitive impairment, and also assist designers of general systems to take into account the needs of people with cognitive impairments.

Cognitive Impairment

The use of the term *cognitive impairment* implies that two categories of human cognitive systems exist—impaired and unimpaired. This, however is not the case, although it can reasonably be stated that there are "normal" or "average" cognitive systems. It is this "normal" system to which the vast majority of the experimental cognitive psychology literature relates. In many contexts, this level of explanation is suitable for indicating what most people are capable of. It should always be borne in mind, however, that in "real-world" situations, there is no marked distinction between that which is "normal" and that which is not. In other words, everyone has some limits to their cognitive ability. Some have a highly specific impairment, some more diffuse problems, and there are also some that experience interrelated constellations of impairments. In addition the cognitive abilities of any one human being will change with time. Significant changes can occur over a period of minutes, days, and weeks. Aging can have substantial effects on cognitive ability, which is particularly marked in some age-related conditions such as dementia.

For ease of exposition, the forms of cognitive impairment identified and described here in the main refer to categories of impairment. It should be noted, however, that all these categories lie somewhere on a continuum and, while they are delineated on the basis of educational or clinical/medical criteria, such "cut-off points," are relatively arbitrary in the context of the wide variability of cognitive ability across the population. It is also worth noting that within the context of "normal" cognitive systems, there is significant diversity among people in regard to differential preferences for types of material and ways of approaching and processing information. For example, some people may be considered primarily "verbal" and will tend to excel in language-based tasks, relative to those considered "visuospatial" (see, for example, Lohman, 2000). Thus, many of the types of impairment addressed below can to some extent be construed as the extremities of "normal" diversity.

Intelligence Quotient

The most widely known dimension of general cognitive ability is probably "intelligence." Scientific investigation of this dimension has had a controversial past, and many aspects of this are beyond the scope of this chapter (for a more comprehensive account, see Gould, 1997). One underlying reason for such controversy is that the word *intelligence* has a rather nebulous definition. In day-to-day usage, this is rarely problematic, but the differences between scientific and lay definitions can cause misunderstanding (e.g., Sternberg, 2000). Such misunderstanding can lead to controversy because most definitions of intelligence include connotations that are considered socially important and can thus often be highly emotive. Despite these difficulties, the investigation of intelligence has provided many insights into a wide range of more particular cognitive abilities and has also developed methods for quantifying general intellectual ability such as the various forms of IQ (intelligence quotient) test. Again, some controversy has surrounded the use of these tests over the years (see Gould, 1997; Kaufman, 2000), but such tests have been widely used and are commonly accepted as a general benchmark of a person's intellectual capability.

An IQ score of 100 is, by definition "normal," with about 50% of the population scoring above and 50% scoring below, but it should be noted that elderly people are not generally included in the standardization of these scores. Approximately 50% of the population are considered to be within the bounds of "normality" and deviate to either side of 100 by no more than 10 points. The "not-normal" 50% are about evenly distributed above and below this band. Thus, about a quarter of the (nonelderly) population fall below the level of what is considered "normal." Although the terminology varies across cultures and over time, around 20% of the population have IQ scores between 75 and 90 and would generally be classified as "slow learners." The final 5% will generally have special needs that on the whole are best addressed on an individual basis (Kaluger & Kolson, 1987). Furthermore and as an example of the emotive connotations associated with the issue of intelligence, it is worth noting that the first official (American Association on Mental Retardation (AAMR), USA, 1921) classifications scheme associated with IQ tests further broke down this latter 5% (Detterman, Gabriel, & Ruthsatz, 2000). These classifications were "moron" (IQ 50–75), "imbecile" (IQ 25–50), and "idiot" (IQ < 25), terms

that are wholly unacceptable today as a description of anyone with a cognitive impairment.

An IQ score reflects a person's intellectual ability as a whole. A low score may be due to the whole system functioning at a suboptimal level, but a similar result may also be due to impairment of one or more component abilities. There are many tests of IQ, some of which will give some indication of this, whereas others do not. Some IQ tests are broken down into subtests that reflect the relative levels of ability in the component cognitive abilities, such as the verbal and visuospatial abilities mentioned above. The more common forms of cognitive impairment are described below, but these areas are covered in more detail in chapters 1 and 2 of this book.

For any information in the outside world to enter the cognitive system, it must first be detected and transmitted by the sensory apparatus. In an important sense, this is not simply the "start" of the process because aspects of attention will influence what is and is not detected or transmitted, and, to a certain extent, how. Basic perceptual processing creates a sensory-specific representation of the stimulus event. Streams of such stimulus events are summated into meaningful cognitive entities (e.g., strokes on a page recognized as letters and numerals are summated into a name and telephone number). These will then be either passed immediately to short-term memory, further processed by working memory, "rehearsed" for maintenance in short-term memory or for encoding into longer term storage. Rehearsal, for example, could refer to rote rehearsal of a telephone number between reading and dialing it or to more elaborate processing to associate it with relevant extant memories to improve the chance of subsequent recall.

Output from the cognitive system will generally be initiated in response to some form of external stimulus, or probe, by accessing extant memories relevant to the probe using executive processes to organize them in a task-relevant way and then producing a response. Output of this kind has been most commonly studied with the use of memory tests, minimizing the influence of "intellectual processing" (problem solving etc.) per se and emphasizing the registration, rehearsal, and encoding of information, the effects of decay, interference, and other forms of forgetting and the effectiveness of different cues (probes) in eliciting specific memories (e.g., recall vs. recognition).

Virtually all aspects of the processing touched on in the previous paragraph are shaped by attention, and it is important to note that, regardless of impairment, although we all have some control over attention, it can also be the case that attention can have some control over us. This is to say, we can utilize attention to focus on searching a list for a particular telephone number while ignoring the chatter of people around us. Having read the number, however, our attention can exert its own control if someone calls our name and asks if we have made that call yet. Despite our best efforts at rehearsal, it is likely that our attention will be "grabbed" by our name, and the ensuing question and this brief distraction can be enough to lose the information from temporary storage.

In general terms, mild to moderate global cognitive impairment will be associated with decrements in efficiency across most of the processing stages outlined above, in aspects of the utilization of attention, or both. Thus, in many ways the user requirements of "slow learners" are equivalent to those of many elderly people. The following section describes some of the main decrements in cognitive ability related to interacting with computer-type systems with notable differences between older and younger people addressed as appropriate.

INTERFACE DESIGN TO SUPPORT PEOPLE WITH COGNITIVE IMPAIRMENT

It is important to develop special technology to provide support for people with various types of cognitive impairment, and some such projects are described later. It is also important, however, to address the challenge of providing access to mainstream technology, both software and the Internet, for people with cognitive impairments.

When designing or specifying mainstream technology for such users, designers must focus on the characteristics of the potential users and be fully aware of the range of cognitive diversity, even among those without clinical dysfunction. This is rarely mentioned in human-interface design, for which the cognitive diversity of the human race has not been the focus of a large body of research. It is also important to consider the effects of age on cognitive function. As Worden, Walker, Bharat, and Hudson (1997, p. 266) commented, "It is known that, as people age, their cognitive perceptual and motor abilities decline with negative effects on their ability to perform many tasks. Computers can play an increasingly important role in helping older adults function well in society. Despite this, little research has focused on computer use of older adults."

A key aspect of any intellectual task with regard to interactive technology for people with mild or moderate global cognitive impairment is speed (Salthouse, 1991). Whatever level of performance a person can achieve in a given situation, it will be made worse if the task must be done under externally imposed time constraints, whether these are actual or simply inferred by the user. Thus the design of any interaction should, wherever possible, allow every step to be carried out at the user's own pace. This issue also raises the first distinction between older and younger people. A relatively greater proportion of the extra time needed by older people is due to age-related declines in their sensory systems, particularly in hearing and vision rather than cognitive impairment per se. For example, given comparable levels of cognitive impairment, an older person will need relatively more time to perceive the relevant stimulus before cognitive processes can be brought to bear on it. Thus, for many older people the requirement for extra time can be reduced, although rarely removed, if care is taken to present text and other aspects of on-screen layout in a suitably clear way (see, for example, Carmichael, 1999; Charness & Bosman, 1994). Beyond this, clear text and presentation layout will always be worth considering carefully because such aspects have been found to benefit those who do not specifically need it, albeit to a less marked extent (Freudenthal, 1999; Pirkl, 1994).

Another key concept related to interface design for people with cognitive impairment is complexity, and how to avoid it. Complexity in interfaces can manifest itself in many different ways and at many different levels. A truly comprehensive coverage of this is beyond the scope of this chapter, but some illustrative examples will be given to elucidate this idea.

The use of language in an interactive system should be given careful consideration, and the syntax and vocabulary should be kept as straightforward and "everyday" as the context allows. This is particularly pertinent for any form of instruction. If the requirements of a particular stage of an interaction cannot be captured in a few simple concrete statements, then serious consideration should be given to redesigning the interaction itself. Similarly, any on-screen display should be kept as uncluttered as practicable and wherever possible should present the user with only a single issue (menu, subject, decision, etc.) at any particular point in time. Similarly, but on a larger scale, progression through an interaction should be kept, again wherever practicable, as linear as possible. That is, the user should only need to consider one thing at a time. Any requirement to deal with different things in parallel will markedly increase the possibility of errors and general user dissatisfaction (Detterman et al., 2000; Salthouse, 1985).

Unfortunately, because designers of a system have a comprehensive understanding of the functions of that system, they are unlikely to be able to assess issues of complexity from the users' point of view, particularly that of a novice user. In addition, prescriptive checklists for avoiding complexity ultimately will be inadequate because the optimum approach always depends on the specifics of the task the interactive system is intended to support (Carmichael, 1999). This is one of the main issues that highlights the importance of early and rigorous user involvement in the design of interactive systems.

This is particularly important in the case of young designers developing interfaces for older users. Research into cognitive changes in later life (aged 50+) has indicated the heuristic value of the concepts of *fluid* and *crystallized* abilities (Horn & Cattell, 1967). In general terms, fluid abilities (e.g., novel problem solving) are those that decline with age, and crystallized (e.g., existing world knowledge) are those that do not. Research that has looked at the very old (aged 80+) tends to find that ultimately everything declines but that the crystallized abilities follow a markedly slower trajectory (Bäckman, Small, Wahlin, & Larsson, 2000). This is another distinction between younger and older people, because younger people tend to have much less accumulated knowledge. There are advantages and disadvantages on both sides of this distinction. Rabbitt (1993) showed that in many circumstances, relevant accumulated knowledge can ameliorate decline in fluid ability (e.g., well-learned strategies). In other circumstances, however, the opposite can be the case, wherein a well-learned, but essentially inappropriate, strategy can put a greater burden on the associated fluid abilities.

The Effects of Attention

Many of the constraints imposed by cognitive impairment can be further shaped by decrements in various aspects of attention. One major aspect of this is generally referred to as selective attention, which allows us to focus on salient aspects of a task and at the same time helps us to actively ignore irrelevant aspects. The efficiency of selective attention is markedly diminished in most forms of cognitive impairment. This factor further supports the recommendation to present the user with just one thing at a time, which will avoid the user erroneously devoting time and cognitive resources to processing irrelevant information. Similarly, if the nature of the interaction requires the user to attend to some critical information at a particular time or location, appropriately obvious highlighting should be employed to grab the users' attention. These issues become emphasized in situations in which selective attention must be maintained over periods of more than just a few minutes.

Another aspect of attention known to be less efficient in cognitive impairment is referred to as divided attention. In general this refers to the ability to allocate cognitive resources appropriately when trying to do two or more distinct cognitive tasks, or distinct portions of the same task, at the same time. Many scenarios in which the user is required to do more than one thing at a time can simply demand more cognitive resources than are available. Declines in the efficiency of divided attention, however, can mean that, even if the tasks involved demand no more than the available resources, they may not be allocated appropriately. Generally speaking the interactive system should be designed to relieve the user of this kind of burden. It is difficult to be prescriptive about suitable solutions to this problem because the appropriate approach will depend on the specifics of the interaction involved, but some general ideas may be of use. For example the provision of some form of notepad function may be helpful for temporarily recording information for subsequent use, although great care is required to ensure that the instantiation of such a function, and its utilization, does not in itself put further cognitive load on the user. Another possibility would be the provision of an overview of the task at hand that could show, or remind, users "where they are" and what they have and have not done.

Memory Loss and Dementia

Limitations in memory affect people with age- and non-age-related cognitive impairment. Thus, wherever practicable, interactive systems should be designed to take the burden of memory off the user, for example, by judicious use of prompts and reminders. Also careful consideration of the steps in an interaction and the way they are presented to the user can help mitigate the most common problem of deficient short-term and working memory. Even with the best design efforts, however, such problems are likely to make users with cognitive impairment relatively error prone. It is thus important to ensure the interactive system allows for error correction in an easy-to-use form. Also, to ensure that the user spots such errors, the system should provide feedback regarding user actions and, when appropriate, elicit active confirmation from them.

Various forms of dementia can exaggerate the relatively mild effects of normal aging on the cognitive system. At the age of 60 years, about 1% of the population is diagnosed with dementia. This percentage approximately doubles for every subsequent 5 year age band (e.g., 4% at 70 years, 16% at 80 years;

Bäckman et al., 2000). Alzheimer's disease accounts for about 60% of the elderly population with dementia, depending on the diagnostic criteria used, and a further 25% is vascular dementia (i.e., related to circulatory problems). Most of the vascular dementias are referred to as multiinfarct dementias and tend to be caused by series of ministrokes and thus tend to have more diffuse and less predictable effects on ability than a major stroke. The remaining 15% of dementias are made up of various relatively rare conditions (Bäckman et al., 2000).

Regardless of the various causes and effects, all forms of dementia involve damage to the brain, such damage being more or less widespread, affecting cortical or subcortical areas (or both). In general terms, damage to the cortex results in cognitive and perceptual impairment, whereas damage to the subcortical areas is related more to physical impairment. There are, however, a number of well-known problems related to the diagnosis of dementia. Two of these are relevant to interface design. The first involves the "grey area" between the worst effects of normal aging and the initial effects of pathological aging at the onset of dementia. The second is that the effects of depression in later life can closely mimic those of dementia. A significant decline in cognitive ability caused by depression may in time be reversed or, if due to a pathological condition, may continue at a more or less rapid pace. These additional complexities further expand the overall diversity of the impact of cognitive impairment in relation to human interface design, both with regard to the general level of ability and in the variation of the level over periods of days, weeks, months, and even years. The convolution of this situation is further augmented by the effects of a relatively greater probability of ill health among older people. It is estimated that around 80% of those aged 65 and over have at least one chronic illness and many have more than one. In addition to the effects of health per se, there is also potential for cognitive ability to be affected by medication and interactions between different medicines.

Despite the above, there are some fairly systematic changes associated with extreme old age and dementia that are relevant to human-interface design. In general, the first ability to deteriorate in dementia, particularly with Alzheimer's disease, is episodic memory (as distinct from semantic memory). Episodic memory is memory for events, usually from the viewpoint of personal experience, rather than for facts. Remembering that "X is the capital of Y" is the product of semantic memory, whereas remembering when and where you were and who you were with while you were reading that fascinating book that told you that "X is the capital of Y" is the product of episodic memory. This generalized decrement in episodic memory may be related to findings in normal aging research, such as disproportionate decrements in source memory (i.e., specifically remembering *where* an item was rather than *what* it was) and prospective memory (i.e., remembering to do something in the future). These changes have important implications for successful navigation of interactive systems. For example, keeping track of where one has just been is often an important prompt to where one is going now; the kind of scenario in which a user might ask themselves, "How did I get it to do this last time?" causes problems for people with poor memory.

Visuospatial, Iconic, and Verbal Abilities

In addition to memory problems, there are, at least at the level of the population, marked deterioration of visuospatial and verbal abilities in older people. Decline in visuospatial abilities can cause difficulty with decoding layouts and utilizing any inherent organization. Related to this is a deterioration in iconic memory that, given the graphic nature of many interfaces, can be problematic in its own right. Among the limitations in verbal ability is the diminution of vocabulary. Particular difficulties have also been found in the comprehension of abstract and metaphorical phrases, with the tendency being to take them literally. Such conditions can develop into more global aphasia (e.g., Broca's aphasia, related to the production of speech, and Wernicke's aphasia, related to comprehension). There is also a likelihood that the general difficulty with recall of proper nouns, found in normal aging, can develop into more profound anomia. The depth of such problems may not be apparent to an outside observer because the ability to read aloud may be well preserved, regardless of the extent to which the content is properly understood and subsequently remembered.

Another distinct form of global cognitive impairment is autism, including a set of rarer but related syndromes (Kaluger & Kolson, 1987). The precise causes of autism are not clearly understood. Briefly stated, it is a general neurological disorder that impacts the normal development of the brain particularly in relation to social interaction and communication skills. Its effects will usually become apparent within the first 3 years of life. People with autism typically have difficulties in verbal and nonverbal communication and social interactions. The disorder makes it difficult for them to communicate with others and relate to the outside world, they also tend to have relatively low IQ scores. Closely related to autism is Asperger's syndrome. People with Asperger's experience similar social communication difficulties but generally demonstrate a normal IQ. There are several generally similar conditions, some of which have varying physical and behavioral elements associated with them. These come under the collective heading of pervasive developmental disorders, and all tend to produce difficulties with communication. An important element of these social communication difficulties in the context of this chapter is an inability to grasp the implications of metaphorical or idiomatic language. This is similar to that mentioned above for dementia, but in autism tends to be more profound. There is, however, some evidence to show that people with autism or Asperger's syndrome are more able to communicate with computers than with people, or with people via computers, rather than face-to-face, and thus properly designed computer systems may have potential for assisting such user groups.

COGNITIVE PROSTHESES

In addition to the need to provide better interfaces to everyday software for people with impairments, specially designed computer systems have great potential to support people with a wide variety of cognitive impairment. We consider some specific examples of such software so that the reader is able to see

how such development can proceed. Within the Applied Computing Department at Dundee University, a number of projects have developed computer-based systems to support people with cognitive dysfunction, and examples from the work of this group can illustrate the range of areas where this technology has particular potential. These examples will also clearly show the need to use an appropriate methodology for research in this area and also illustrate the need for research into developing more sensitive and effective methodologies.

Software As a Cognitive Scaffolding and a Prompt for Communication

One sequence of projects took as its starting point the improvement of communication systems for physically impaired nonspeaking people. It became apparent that this could be done effectively by developing models of the cognitive tasks involved in communication. This research has now spawned a new area of development, which is cognitive support for people with dementia, in which communicative impairment is only part of their range of difficulties.

For severely physically impaired nonspeaking people, even with current speech output technology, speaking rates of 2 to 10 words per minute are common, whereas unimpaired speech proceeds at 150 to 200 words per minute. In an attempt to improve this, a certain amount of progress has been made in the area of using computers to replace or augment some of the cognitive aspects of communication. Although the cognitive processes underlying language use are incompletely understood, a number of theories that attempt to explain language use have been used to improve the functionality of communication systems for nonspeaking people. This approach to the problem usually involves taking a sociolinguistic view of language. Instead of focusing on the building blocks, taking a "bottom-up" approach, the interaction as a whole is analyzed, paying attention to its goals, taking a "top-down" approach to the communication. This may well be a realistic simulation of the natural process because the production of speech by an unimpaired speaker occurs at such a rate that conscious processing and controlling of the speech at a microlevel is not possible. In common with other learned skills, speech is produced, to some extent, automatically, with the speaker being aware of giving high level instructions to the speech production system but leaving the details of its implementation to the system.

The nonconscious control of much of speech production has been modeled in the CHAT prototype (Alm, Arnott, & Newell, 1992). This produced quick greeting, farewell, and feedback remarks by giving the user semiautomatic control of exactly what form the remarks would take, within parameters which the user had previously selected. This mimicked the phenomenon of a speaker responding automatically to greetings and other commonly occurring speech routines, without giving the process any detailed thought.

The CHAT-like conversation described illustrates an attempt to achieve a particular communicative goal, achieving social closeness by observing social etiquette. Some recent research efforts have been directed at finding ways to incorporate large chunks of text into an augmented conversation to help users carry out topic discussion. This has been driven by the observation that a great deal of everyday discourse is reusable in multiple contexts.

Much of this type of discourse takes the form of conversational narratives. Research into the conversational narrative at a sociolinguistic level indicates several interesting characteristics. These include the way in which narratives are told and to whom they are told. For example, a recent event is told repeatedly for a limited time to most people with whom the speaker has contact. As the event recedes in history, the narrative is retold when it is relevant to the topic of conversation. The length of the narrative depends on its age—the older a story is, the more embellished it becomes—and the time available within the conversation. The version of the story (although the sequence of events remain the same, the details or embellishments of a story differ) depends on the conversational partner (some stories are not relevant to specific people).

One of the ways to make the retrieval of text chunks easier is to anticipate the chunks of text that the user may want to use. This has been achieved by modeling the way in which conversational narratives are used using techniques from the fields of artificial intelligence and computational linguistics (Waller, 1992). The prototypes developed constantly adapt to the users' language, thus mirroring his or her perception of where conversation items are stored. In this way, the system adapts to the way the user thinks instead of having the user learn a new retrieval (coding) system.

One of the arguments against using prediction such as this was raised when word-prediction systems were first developed in the early 1980s. Therapists and teachers were concerned that nonspeaking people, especially children, would select what was offered on the screen rather than what they originally wanted to say. Although this may happen, research into predictive systems applied to writing suggests that they may carry over the help they offer and have a wider effect on the user's ability development. Some of this research reports an increase in written output by reluctant writers and people with spelling problems (Newell, Arnott, Booth, & Beattie, 1992). A general improvement in spelling has also been noted. Children with language dysfunction and/or learning disabilities have shown improvement in text composition (Newell, Booth, & Beattie, 1991). This research is in the writing domain, but its results suggest that predictive systems can offer assistance without becoming mere substitutes for creative expression.

Also, it is true of completely unimpaired conversation that speakers often change direction in their communication depending on chance occurrences or on the sudden recollection of a point they would like to include. Thus, there is a degree of opportunism in all conversations. Another argument in favor of offering predicted phrases and sequences is that the current situation for most augmented communicators is that their conversations tend to be quite sparse, with the control of the interaction residing with unaided speaking partners. If it is not possible to go boating on the lake, easily going off in any direction you please, is it not preferable to build a boardwalk out over the water than to stay on the shore?

One of the motivations to improve communication systems for nonspeaking people is the fact that they are commonly perceived by people who do not know them as intellectually below the level at which they in fact operate. It is often reported by nonspeaking people that they are considered unintelligent or immature by strangers. It may be that the issue of "perceived communicative competence" is one that needs increased attention (McKinlay, 1991).

Related to this, an interesting finding emerged from work in which one of the authors of this chapter was involved. Here, a prototype communication system was used to evaluate listeners' impressions of the content of computer-aided communication based on prestored texts, as compared with naturally occurring dialogues. The user was able to use only prestored texts to conduct the conversations. Most of the text was material about one subject (holidays). A number of rapidly accessible comments and quick feedback remarks were also available. The unaided conversations were between pairs of volunteers who were asked to converse together on the topic of holidays. Transcripts of randomly sampled sections of the conversations and audio recordings of reenactments of the samples with pauses removed were rated for social competence on a six-item scale (coefficient alpha $= 0.83$) by 24 judges. The content of the computer-aided conversations was rated significantly higher than that of the unaided samples ($p < .001$). The judges also rated the individual contributions of the computer-aided communicator and the unaided partners on how "socially worthwhile and involving" these appeared. There was no significant difference between the ratings of their respective contributions ($p > .05$; Todman, Elder, & Alm, 1995).

This finding came as something of a surprise to the researchers because the original purpose had been to establish whether conversations using prestored material would simply be able to equal naturally occurring conversations in quality of content. Of course, the pauses in actual computer-aided conversation do have an effect on listeners' impressions of the quality of the communication, but this finding is still of interest because it suggests that in some ways augmented communication could have an edge over naturally occurring talk. A plausible explanation for this finding is that naturally occurring talk is full of high-speed dysfluencies, mistakes, substitutions, and other messy features that listeners tend to discount with their ability to infer what the speaker is intending to say. Prestored material is by its nature selected because it may be of particular interest, and it is expressed more carefully than quick-flowing talk and thus may appear more orderly and dense with meaning than natural talk.

An addition to conversational narratives, another common structure in everyday communication is the script, particularly where the speaker is undertaking some sort of transaction. Scripts may be a good method of organising prestored utterances to attempt to overcome the problem of memory load when operating a complex communication system based on a large amount of prestored material. Users' memory load can be reduced by making use of their existing long-term memory to help them locate and select appropriate utterances from the communication system. Schank and Abelson (1977) proposed a theory that people remember frequently encountered situations in structures in long-term memory that they termed *Scripts*. A script captures the essence of a stereotypical situation and allows people to make sense of what is happening in a particular situation and to predict what will happen next. Other research (Vanderheyden, Demasco, McCoy, & Pennington, 1996) has shown the potential that similar schema-based techniques offer to this field.

An initial experiment was devised (Alm, Morrison, & Arnott, 1995) to investigate the potential of a script-based approach to transactional interactions with a communication system, and a prototype system was developed to facilitate this experiment. The aim was to ascertain whether a transactional interaction could be conducted using a script-based communication system. It was decided to simulate a particular transactional interaction that could be reasonably expected to follow a predictable sequence of events (i.e., one that would be amenable to the script approach) to find out whether a computer-based script could enable a successful interaction to happen.

The transaction chosen for the experiment was that of arranging the repair of a household appliance over the telephone. Although the script interface was a relatively simple one devised for the purpose of this experiment, it was successful in facilitating the interaction and produced a significant saving in the amount of physical effort required.

To take this work further, a large-scale project was undertaken to incorporate scripts into a more widely usable device. The user interface of this system is made up of three main components: scripts, rapidly produced speech acts, and a unique text facility. The scripts component is used in the discussion phase of a conversation and consists of a set of scripts with which the user can interact. The rapid speech act component contains high-frequency utterances used in the opening and closing portions of a conversation and in giving feedback and consists of groups of speech-act buttons. This facility is based on previous work with CHAT. The unique utterance component is used when there are no appropriate prestored utterances available and consists of a virtual keyboard, a word-prediction mechanism, and a notebook facility.

To provide access to a set of scripts, an interface was devised that involved a pictorial representation of the scenes in the script. The pictorial approach was taken to attempt to give users easier access to the stored material and to assist them in varying levels of literacy skills. In this interface, scripts are presented to the user as a sequence of cartoon-style scenes. The scenes give users an indication of the subject matter and purpose of the script and assists them in assessing quickly whether the script is appropriate for current needs. Each scene is populated with realistic objects chosen to represent the conversation tasks that can be performed. The user thus receives a pictorial overview of the script, what happens in it, and what options are available. This assists the user to see quickly what the script will be able to do in the context of the current conversation. An example of the interface for the system can be seen in Fig. 24.1, which shows a scene within the doctor script.

Research into picture recognition and memory structures has demonstrated that groups of objects organized into realistic scenes corresponding to stereotypical situations better assist recognition and memory compared with groups of arbitrarily placed objects (Mandler, 1984; Mandler & Parker, 1976). The

scene-based interface using a realistic arrangement of objects within a scene was therefore chosen to facilitate recognition and remembering by the user and thus reduce the cognitive load required to locate suitable objects during a conversation.

Because it would be impractical to provide scripts for every conceivable situation, it was decided to provide users with a limited number of scripts together with an authoring package with which they can develop their own custom scripts with help from their therapists.

As a start, it was decided to develop six complete scripts. These were chosen after discussions with a user advisory group about situations in which they found difficulty communicating. The scripts developed were "at the doctor," "at the restaurant," "going shopping," "activities of daily living," "on the telephone," "meeting someone new," and "talking about emotions."

The system uses the script to guide the user through a dialogue. There is a prediction mechanism that predicts the next most probable stage in the dialogue that the user will need (based on the script), so the user can usually follow a predicted path through a conversation. This prediction mechanism monitors the sequences of objects selected and uses this information to modify future predictions.

Help for Aphasia

Communication systems for nonspeaking people have been described that in some way model the cognitive processes underlying communication. In the case of most physically disabled nonspeaking people, this is necessary to speed up the communication process. However, in the case of speech problems caused by a stroke or other trauma (aphasia), the person trying to communicate will also have cognitive problems. Interestingly, it was the objections that conversation modeling might provide an active prompt to communication that suggested a way of possibly helping people who might need such a prompt to initiate communication at all.

In a research project investigating the possibility of prompting people with aphasia in their communication, a predictive communication system was developed with a simple interface (Waller, Dennis, Brodie, & Cairns, 1998). The system held personal sentences and stories that were entered with the help of a carer. The user could then retrieve the prestored conversational items, with the system offering probable items based on previous use of the system. The interface was designed to be as simple as possible so as to be usable by people who were unfamiliar with technology and with cognitive difficulties.

To access the sentences and stories, the user is led through a sequence of choices on the screen. First they are offered a choice of conversational partners, then a list of topics most likely to be appropriate for the chosen partner, then a list of the four most common sentences for that partner and that topic. The user can choose to speak one of the sentences through a speech synthesizer, or have the system look for more suggestions. The sentences and topic categories are personal to each user, and the order in which topic words or sentences are presented depends on the past use of the system. Thus the system is specific to users, both in the information content and how it adapts to individual ways of communicating.

I have an appointment, my name is James Smith.

FIGURE 24.1. The script system user interface showing a scene from the doctor script. A text preview and display box appears at the top of the user interface. The main interface area (bottom right) contains the scene image. The function buttons on the left side of the interface are, from top to bottom, "I'm listening" rapid speech act button; button to access the main rapid speech act interface; scene navigation backtrack to previous scenes button; scene navigation overview button; tool button to access the notepad, and additional system control facilities.

The system was evaluated with five adults with nonfluent aphasia who were able to recognize, but not produce, familiar written sentences. There was little change in the underlying comprehension and expressive abilities of the participants while not using the system. When making use of the system, the results showed that some adults with nonfluent aphasia were able to initiate and retain control of the conversation to a greater extent when familiar sentences and narratives were predicted. In other words, users' existing and residual abilities (e.g., small sight vocabulary, pragmatic knowledge of conversation) were to a degree augmented by the computer functioning as a cognitive prosthesis. This project indicated that a communication system based on prompting could be of help to people with cognitive and communication difficulties.

Support for Dementia

Dementia, which involves the loss of short-term memory in elderly people, is a serious problem for the person and for their family and carers. It can rule out most social activities and interactions because these depend on a working short-term memory for effective participation. This includes even the essential ability to communicate.

As well as being valuable with all older people, reminiscence is an important tool used to help elderly people who have dementia (Feil, 1993; Sheridan, 1992) because, although their short-term memory may be impaired, their long-term memory is often more or less intact (Rau, 1993). The difficulty is accessing these long-term memories without the capability of keeping a conversation going, which depends on short-term memory. What can be a help are activities that do not require patients to maintain the structure of a conversation, for instance, looking at and commenting on a series of photographs that evoke reminiscences. This can provide a framework for meaningful interactions.

The tools used in such reminiscence work can also include videos, sound, music, and written materials. These are all in separate media, however, and it can be time-consuming searching for a particular photo, sound, or film clip. Bringing all these media together into a multimedia scrapbook could mean a more lively content for reminiscence sessions that could be easily

used. The intention of this project was to begin to develop a system that can act as a "conversation prosthesis," giving the user the support needed to carry out a satisfying conversation about the past.

An investigation was first undertaken to determine which aspects of multimedia would be most helpful for such a reminiscence experience and the best way to present them. A number of prototype interfaces for a multimedia reminiscence experience were developed. These included text, photographs, videos, and songs from the past life of the city. The materials were collected with the assistance of the University and Dundee City archives and library and the Dundee Heritage Project. The prototypes were demonstrated for people with dementia and staff at a day center run by Alzheimer Scotland Action on Dementia. The following conclusions were drawn from these evaluation sessions:

1. *Is it better for the display to use the metaphor of a real-life scrapbook or a standard computer screen?* Six of the staff members preferred the book presentation, three preferred the screen, and two had no preference. Interestingly, this was almost the reverse of the preference shown by the group of people who had dementia, the majority preferring the screen presentation. The preference shown for the screen presentation could be due to reduced cognitive ability. Having a book presentation may give the person with dementia more information to process than they are comfortably able to. They would first have to see the book and recognize it as such before moving on to seeing the picture.

2. *How should the scrapbook material be organized, by subject or by media?* The majority of the staff evaluators preferred the arrangement by subject, saying it was more logical. Some were unsure which was preferable; however no one showed a preference for the arrangement by media. The clients with dementia reflected these findings. Despite preferring the arrangement being by subject, the majority of evaluators could see benefits from having access to both arrangements. It was concluded that for basic reminiscence sessions, the arrangement by subject is preferable. But access to the arrangement by media should be an option to make the software available for use in other ways.

3. *How does each individual medium add to the reminiscence process, and what effects are produced by the various media (e.g., sounds, pictures, videos, music)?* It was found that with videos the clients were only able to strongly identify with them when they triggered specific personal memories, whereas songs and photographs were more generally appreciated. Most of the videos and photographs and all the songs were able to spur conversations, however. Attention stayed longest on the songs, which were particularly enjoyed when played repeatedly with everyone singing along. The staff did, however, feel that some individual clients had enjoyed the videos most.

One general finding was that the multimedia presentation produced a great deal of interest and motivation from the people with dementia with whom we worked. Staff also were keen to see the idea developed further. A new project is now beginning that will fully develop the ideas produced by this preliminary

work and produce a fully functional multimedia reminiscence system. The project will be carried out by a multidisciplinary team consisting of a software engineer, a designer, and a psychologist specializing in aging.

In addition to its benefits for the person with dementia, participation in reminiscence activities has been shown to have a positive outcome for carers who take part. Thus, help for reminiscence is not merely a tool to stimulate interaction, but also contributes to improved quality of life for the person with dementia and their family.

Recent work using videos to present life histories for people with dementia has shown that new technologies, when sensitively and appropriately applied, can bring a substantial added impact to supportive and therapeutic activities for people with cognitive problems (Cohen, 2000). The reminiscence system developed in this project will, we hope, have an immediate application and will also serve as exploratory efforts in developing a range of computer-based entertainment systems for people with dementia.

SYSTEMS TO SUPPORT PEOPLE WITH DYSLEXIA

Dyslexia is a language disorder that is still in the process of being investigated and thoroughly understood. A word-processing environment has been developed to alleviate some of the visual problems encountered by some people with dyslexia when they read and produce text. In the absence of a full understanding of the nature of dyslexia, however, the researchers' approach was to identify some of the most commonly noted problems that dyslexics encounter when reading and producing text. On the basis of these common difficulties, they identified ways in which each individual might be able to minimize the consequences of their own particular problems by manipulating the appearance of their word-processing environment and the text presented within it. This stage of the development process involved the start of an iterative cycle of prototype development and evaluation with dyslexic computer users. The work in progress ultimately led to a software system that provides a highly (and easily) configurable environment for dyslexic people, with a wide range of differing preferences and problems, to use for reading and producing text. The researchers' approach has been to examine the parameters of the situation, including cost, cost–benefit, existing software, and user demand, with a view to finding an optimal path to the production of a software system that is of real use in practical situations. It should be noted that although users were extensively involved throughout the development process, the researchers did not always rely heavily on their input to further development; rather, they ensured that at all times the process was sensitive to user need and opinion by ensuring their full participation but also was informed by any relevant sources, ranging from documented wisdom about the subject to hunches about what might work.

This development of a word processor to assist dyslexics is thus a particularly illuminating example of how the needs of people with a particular type of cognitive impairment can be effectively factored into the design and development process. This research and development will thus be described in some

depth as an example and to assist the reader in appreciating the generic importance of this type of approach, which requires both a knowledge of the underlying syndrome but also a methodology that encourages an innovative approach to development.

Dyslexia, as has been noted, is still being investigated as a language disorder; at present it has a number of definitions. The British Dyslexia Association offers this description:

Dyslexia is best described as a combination of abilities and difficulties which affect the learning process in one or more of reading, spelling, writing and sometimes numeracy/language. Accompanying weaknesses may be identified in areas of speed of processing, short-term memory, sequencing, auditory and/or visual perception, spoken language and motor skills.

"Some dyslexics have outstanding creative skills. Others have strong oral skills. Whilst others have no outstanding talents, they all have strengths. Dyslexia occurs despite normal intellectual ability and conventional teaching. It is independent of socio-economic or language background. (British Dyslexia Association, http://www.bda-dyslexia.org.uk)

The symptoms of dyslexia vary greatly and the reasons for the existence of these symptoms are also varied. However, two distinct types of dyslexia are recognised: acquired dyslexia and developmental dyslexia.

Acquired dyslexia is associated with those people who have difficulties caused by damage to the brain; before the occurrence of brain damage, no difficulties would have been identified. This area can be subdivided into disorders in which the visual analysis system is damaged (peripheral dyslexia) and disorders in which processes beyond the visual analysis system are damaged, resulting in difficulties affecting the comprehension or pronunciation of written words (central dyslexia).

Peripheral dyslexia is associated with difficulties such as misreading letters within words and migrating letters between words. Central dyslexia concerns issues such as poor comprehension due to an impaired semantic system and the inability to read unfamiliar words, whereas familiar words are read easily. Examples would include *monkey* being read as *ape* (a semantic error), and *patient* being read as *parent* (a visual error).

Developmental dyslexia, which is sometimes known as congenital dyslexia, exists from birth. It has been described as follows (Critchley, 1964):

Developmental dyslexia is a learning disability which initially shows itself by difficulty in learning to read, and later by erratic spelling and by lack of facility in manipulating written as opposed to spoken words. The condition is cognitive in essence, and usually genetically determined. It is not due to intellectual inadequacy or to lack of sociocultural opportunity, or to emotional factors, or to any known structural brain defect.

One of the main features of dyslexia is the individual nature of the disorder. The condition is not typically characterized by a single difficulty but by a range of difficulties that vary in combination and in intensity between individuals, giving rise to an enormous variation between individuals in the problems encountered. Each dyslexic person thus has a range of difficulties that need to be addressed differently from other dyslexic people. Dyslexia is an example of the need to design for dynamic diversity.

The wide-ranging characteristics of dyslexia provide a challenge for technological assistance because a single approach will not be appropriate for the range of problems presented by a group of dyslexic people. Computer technology offers the opportunity to provide reading and writing systems that are highly configurable for each individual user, but they need to be based on an understanding of the problems which dyslexics have in reading and writing and some of the visual problems that can affect them. The approach adopted in this research was to offer dyslexic users a range of appropriate visual settings for the display of a word processor, together with the opportunity to easily configure the way in which text is displayed to them. The user can select, by experimentation, the settings that best suit them. These settings are then saved and later recalled each time that person uses the word processor. This approach affords the potential to make computer-based text significantly easier to read than printed text, as well as to improve the usability of computer word-processing systems for a wide range of users with dyslexia.

Some Common Problems of Dyslexia

It was first necessary to determine the parameters of the computer display that should be offered for configuration by the user. An initial investigation revealed that some of the most commonly encountered problems are as follows (adapted from Willows, Kruk, & Corcos, 1993, and D. Shaw, personal communication, March, 1994):

1. *Number and letter recognition:* One of the fundamental problems people with dyslexia face is the recognition of alphanumeric symbols. This is often seen when letters that are similar in shape, such as *n* and *h* or *f* and *t*, are confused. The problem is exacerbated with the introduction of uppercase letters. In addition, many adults with dyslexia who are capable of reading printed letters have difficulty reading cursive writing.

2. *Letter reversals:* Many people with dyslexia are prone to reversing letters, which results in a particular letter being interpreted as another letter. Examples of these characters would be *b*, *d* and *p* and *q*. This problem can result in poor word recognition with words containing reversal characters being substituted for other words such as *bad* for *dad*.

3. *Word recognition:* In addition to the substitution effect caused as a result of letter reversals, words that have similar outline shapes can be substituted by people with dyslexia. A typical example of this problem are the words *either* and *enter*. Both words have the same start and finishing characters, and this, allied with the fact that both words also have the same overall shape, make them candidates for being substituted for each other when they occur in the text.

4. *Number, letter and word recollection:* Even once the ability to recognize numbers and letters is mastered, it can still prove difficult for a person with dyslexia to recall the actual form and shape of a character. Many have so much difficulty recalling

upper- and lowercase characters that they continue to print later in life. Similarly, poor visual memory means that people with dyslexia have little ability to distinguish whether a word "looks right."

5. *Spelling problems:* Because of the problems discussed in 1–4, people with dyslexia can have great difficulty with spelling. Their spelling appears to reflect a phonic strategy with words such as *of* and *all* being spelled *ov* and *olb*.

6. *Punctuation recognition:* People with dyslexia appear to have difficulty recognizing punctuation.

7. *Fixation problems:* Another problem found in many people with dyslexia is their lack of ability to scan text without losing their place. Many find it difficult to move from the end of one line to the beginning of the next and also find themselves getting lost in the text.

8. *Word additions and omissions:* People with dyslexia may add or remove words from a passage of text, apparently at random. This is manifested by words being duplicated, extra words being added, or word order being reversed.

9. *Poor comprehension:* With the variety of errors caused by the factors described above, a person with dyslexia may perceive a totally different passage of text from the one that is actually in front of them. Thus, these individuals display poor comprehension skills because the text they perceive is significantly different from the actual text. A minor version of this effect can also occur in normal readers, who can completely miss typographic errors. In this case, however, reading comprehension is usually improved.

Computer Aid for the Problems of Dyslexia

In an attempt to alleviate some of the problems discussed above, people with dyslexia, particularly within the education system, are encouraged to use computers for text manipulation. The use of a computer keyboard has the potential to alleviate the problems of character recollection, but this only helps with the recollection of characters, not the recognition of them once they are on the screen.

There is, however, strong evidence to suggest that the use of lexical and spelling aids can greatly assist with spelling problems exhibited by dyslexics (e.g., Newell & Booth, 1991). Merely highlighting an incorrect word and offering a replacement, however, may not be enough because one of the other problems some people with dyslexia face is an inability to tell if a word "looks right." Grammar-checking facilities may also be of use, but merely picking out these errors may not be enough because the dyslexic may be incapable of selecting the appropriate corrections.

Some people with dyslexia are sensitive to color, and colored acetate screen, tinted glasses, or lighting conditions can improve their ability to read text. Many also report interference from peripheral vision, indicating that anything that can be done to reduce screen clutter outside the main screen window, such as making the document page fill the whole screen, may be of benefit (D. Shaw, personal communication, May 1996).

The approach taken in our research was to investigate ways in which reading and document production could be improved

for people with dyslexia by making visual changes to the environment in which the person is working and by providing an easy-to-use interface for those with the disorder to configure this environment to their own particular preferences. This approach recognizes the diversity within the dyslexic community. A diagnostic system could have been developed for use with this population, in which a variety of tests were performed and conclusions drawn, leading to a third-party setting up an optimized word-processing system for each individual. The researchers believed, however, that presenting to the individual a configuration interface that was easy to manipulate and that enabled the user to see the consequences of various changes before accepting them, would produce a system that not only left the user in control, but that also could give rise to a substantial body of research data on which to build future system improvements.

From the above discussion of the difficulties encountered by people with dyslexia, the researchers considered ways that the screen image of the text could be manipulated. These included foreground and background color; character typeface; font and spacing; making letters or words with similar shapes distinguishable by using different typeface, font, or color; and presenting text in narrower columns or with different spacing between words or lines to reduce fixation problems. The researchers investigated potentially promising ideas by implementing prototypes and evaluating their utility with users who have dyslexia, with an overall view to developing a configuration system that will enable them to set up their own optimized environment

An Experimental Text Reader for People With Dyslexia

The first stage of the research was to develop an experimental text reader. This prototype presented the user with an easily configurable interface that allowed for a number of display variables to be altered. Initially, these were background, foreground, and text colors; font size and style; and the spacing between paragraphs, words, and characters. The interface was designed in such a way that it gave visual feedback on selections before they were confirmed and made minimal use of text instruction.

This was evaluated using 12 computer-literate dyslexic students from higher education by using "think-aloud" techniques as well as the use of questionnaires and interviews. At various development stages, the students were asked to try the system with a view to seeing if it was possible for them to put together a display that improved their ability to read text from the screen. All the users were able to find a setting that was subjectively superior for them to standard black text on a white background with Times Roman 10- or 12-point text, but the screen layouts developed by the subjects were extremely varied. This highlighted the individual nature of the disorder and the diverse characteristics of a interface that would be appropriate for this group.

Each person appeared to have his or her own favorite color combination, although all the testers liked brown text on a green background. Subjects were in greatest agreement about the selection of a typeface: Almost all the testers rated the sans-serif Arial the best. All reported that increasing the spacing between the characters, words, and lines was beneficial. The most interesting point that arose during the testing, however, was the fact that at the beginning of the evaluation period, the subjects did not appear to be aware that altering these variables might be of use.

A second prototype was developed based on Microsoft Word for Windows macros to provide the required configuration interfaces. This was based on the concept of an evolutionary system, rather than a fixed prototype. It was clear that there would be a substantial advantage to developing a "dyslexic configuration," but this design decision raised an interesting deviation from the received wisdom of the desirability of WYSIWYG ("What You See Is What You Get"). In the case of a dyslexic user, what one sees should be whatever he or she can read best, and print previewing facilities would have to be used to show how the layout will appear when printed.

This prototype provided a facility to enhance character reversals (e.g., *b* for *d*) by using color, font, and size. This idea of coloring reversal characters provided interesting and unanticipated results, which are described later. Fixation problems were tackled by reducing the page width, and a speech synthesizer that could read the text on the screen was also included.

There were two distinct parts to the overall solution, a preference program and a reading–editing program. The first allowed users to experiment with the various parameters, and the second made use of these preferences within a reading and editing environment. The preference program menu presents the user with various options and variables, together with a preview facility, to enable the user to experiment with and finally store their data in a preferences file.

The fact that a unique user environment, tailored to the need of each individual is provided, means that the document is (deliberately) not WYSIWYG. A print option thus allows the user to print the document as it appeared with special formatting applied to it or as it would appear without any special formatting.

This second prototype was developed as an add-on module to Microsoft Word 6/7 and evaluated in a similar fashion to the earlier prototype by 7 users with dyslexia (age range of 15 to 30 years). The users found the system easy and intuitive to use, reporting that each of the options had an effect on their ability to read. The options that allowed the user to change the color scheme of the document appeared to be the most helpful, but font size and spacing, column width, and indications of reversals were also reported to assist reading by some or all of the users.

The reversals option provided the most interesting results. The reason for the improvement was not always that the reversal characters were clearly distinguished and easy to read. Instead, subjects claimed that the sporadic coloring "broke up the text" and resulted in their being less likely to "get lost," that is, the system was reducing fixation problems rather than recognition problems.

As the testing progressed, the testers appeared to be surprised by the effect some of the changes had on their ability to read the document. Comments included the following: "I would never have thought of doing that," or "I don't think that will do me much good" before finding that a feature did indeed help. None of the test subjects were aware of how simple changes could dramatically affect their ability to read.

The prototypes were developed from the perspective that the user population was diverse and that the design process must accommodate potential changes in preferences over time. The fact that dyslexia is an individual disorder and that the users were unaware of how easy it was to improve their reading potential by changing visual aspects of the reading environment was evident during the testing sessions. This development has shown important generic issues in developing systems for people with cognitive dysfunction and illustrates how a standard user-centered design methodology is not appropriate for these user groups.

RESEARCH METHODOLOGIES

The research described above gives a flavor of successful approaches to developing human interfaces and software to support people with various types of cognitive impairment. Much of the methodology used in these developments, however, had to be developed *ab initio*. Traditional user-centered design does not have the flexibility for these user groups, and most research and development in the field of communication and IT to support people with disabilities has, to date, concentrated on the development of special systems and on accessibility features for younger, mainly physically or sensorially disabled people. Similarly the human interfaces to most computer systems for general use have been designed, either deliberately or by default, for a "typical," younger user (Newell & Cairns, 1993; Newell, 1995; Newell & Gregor, 1997). Knowledge from these fields does not necessarily transfer comfortably to the challenges encompassed in universal design (Beirmann, 1997; Hypponen, 1999; Sleeman, 1998; Stephanidis, 2001) and, in particular, the widely varying and often declining abilities associated with the range of cognitive impairments.

This section addresses the particular issues for the design process that accompany cognitive impairment and suggests a paradigm and methodology to support the process of designing software that is as near to the universal accessibility ideal as is possible, derived from the approach to specific projects described above.

Software systems, which are not of a very specific "prosthetic" nature, need to address the wide variations in the nature and severity of cognitive impairment among individuals. This demand is further complicated by the fact that, in general, as people grow older their abilities change. This process of change includes a decline over time in the cognitive, physical, and sensory functions, and each of these will decline at different rates relative to one another for each individual. This pattern of capabilities varies widely between individuals, and as people grow older, the variability between people increases. In addition, any given individual's capabilities vary in the short-term because of, for example, a temporary decrease in or loss of function due to a variety of causes, such as the effects of drugs, illness, blood sugar levels, and states of arousal.

This collection of phenomena presents a fundamental problem for the designers of computing systems, whether they be generic systems for use by all ages or specific systems designed to compensate for loss of function. Systems tend to be developed for a typical user and either by design or by default, this user is young, fit, male, and, crucially, has abilities that are static over time. These abilities are assumed to be broadly similar for everybody. Not only is this attitude wrong, in that is does not take account of the wide diversity of abilities among users, but it also ignores the fact that these abilities are dynamic over time.

Current software design also typically produces an artifact that is static and which has no, or very limited, means of adapting to the changing needs of users as their abilities change. Even the user-centered paradigm (e.g., International Organization of Standardization, 1999; Nielsen, 1993; Preece, 1994, Shneiderman, 1992) looks typically at concerns such as representative user groups, without regard for the fact that the user is not a static entity. It is thus important not only to be aware of the diverse characteristics of people with cognitive dysfunction, but also the dynamic aspects of their abilities.

It is clear that people with cognitive impairments, whatever their cause, can have characteristics very different from most human-interface and software designers. It is also clear that in these circumstances user-centered design principles need to be employed if appropriate technology is to be developed for this user group (Gregor & Newell, 1999). These methodologies, however, have been developed for user groups with relatively homogonous characteristics. People with dementia, for example, are a diverse group, and even small subsets of this group tend to have a greater diversity of functionality than is found in groups of able young people.

An additional complication is that there can be serious ethical issues related to the use of such people as "subjects" in the software development process. Many of these are medically related but also include, for example, the ability to obtain informed consent. It is thus suggested that the standard methodology of user-centered design is not appropriate for designing for this user group.

The importance of research and development taking into account the full diversity of the potential user population, including cognitive diversity, was addressed by Newell in his keynote address to InterCHI '93, where the concept of "ordinary and extraordinary human computer interaction" was developed (Newell, 1993; Newell & Cairns, 1993; Newell & Gregor, 1997). Market share is clearly an important consideration, and it has given impetus, not only by demographic trends, but also by recent legislation in the United States and other countries, to accessibility of computer systems for people with disabilities. In terms of the workplace, both the Americans with Disabilities Act and the (UK) Disability Discrimination Act put significant requirements on employers to ensure that people with disabilities are able to be employed within companies and to provide appropriate technology so that such employees have full access to the equipment and information necessary for their employment. Increasingly there is political pressure to increase this access, and more and more requirements for improved access

by disabled people are being enshrined in legislation. As commented earlier, however, *access* does not only mean that people with wheelchairs can maneuver buildings; it also means that there must be provision for people with cognitive disabilities to be able to operate computers and other equipment essential to the workplace.

An important additional factor in the "value for money" equation is that design that takes into account the needs of those with slight or moderate cognitive dysfunction can produce better design for everyone. A situation in which this has not occurred is illustrated by the problems the majority of users have had with videotape recorders. If the designers had considered those with cognitive impairments within their user group, it is possible that they may have been able to design more usable systems. Another example is an e-mail system that was designed to be simple to use by older people with reduced cognitive functioning, which was found to be preferred by executives to the standard e-mail system they were used to using.

Some people are impaired from birth, but people also may become temporarily or permanently disabled by accident or even through normal functioning within their employment. This is particularly noticeable in cognitive functioning. Short-term changes in cognitive ability occur with everyone. These can be caused by fatigue, noise levels, blood sugar fluctuations, lapses in concentration, stress, or a combination of such factors; they can produce significant changes over minutes, hours, or days. In addition, alcohol and drugs can also induce serious changes in cognitive functioning, which is recognized in driving legislation but not in terms of how easy it is to use computer-based systems.

Most people at one time or another, will exhibit cognitive functional characteristics that are significantly outside the normal range. Although they would not consider themselves disabled, nor would their peers, their ability to operate standard equipment may well be significantly reduced.

The questions that designers need to consider include the following:

- Does the equipment that I provide comply with the legislation concerning use by employees who may be cognitively disabled?

- To what extent do I need to take into account the needs of employees who are not considered "disabled" but who have significant temporary or permanent cognitive dysfunction?

- Should I make specific accommodation for the known reductions in cognitive abilities that occur as employees get older (e.g., less requirement for short-term memory or the need to learn new operating procedures)?

- What are the specific obligations designers and employers have to provide systems that can be operated by employees whose cognitive ability has been reduced due to the stress, noise, or other characteristics of the workplace?

The argument is that it would be unusual for anyone to go through their working life without, at some or many stages, being significantly cognitively disabled. If equipment designers took this into account, it is probable that the effectiveness and efficiency of the workforce could be maintained at a higher level than would be the case if the design of the equipment was based on an idealistic model of the characteristics of the user.

The Disabling Environment

In addition to the user having characteristics that can be considered disabled, it is also possible for them to be disabled by the environments within which they have to operate. Newell and Cairns (1993) made the point that the human–machine interaction problems of an able-bodied (ordinary) person operating in an high workload, high stress, or environmentally extreme (i.e., extraordinary) environment has close parallels to a disabled (extraordinary) person, operating in an ordinary situation (e.g., an office).

High workloads and the stress levels to which this can lead often reduce the cognitive performance of the human operator. For example, a very noisy environment cannot only create a similar situation to hearing or speech impairment but can also lead to reduced cognitive performance. The stress level in the dealing room of financial houses can be high and is often accompanied by high noise levels. A significant advance could be achieved by designing software for this environment that assumes users would be hearing impaired and have a relatively low cognitive performance. It is interesting to speculate as to whether such systems would produce higher productivity and better decision making, as well as less stress for the operators. Other examples of extreme environments in which people have to operate are the battlefield or space. The stress and fatigue caused by working within such environments means that their performance is similar to that which could be achieved by a very disabled person operating in a more normal environment. It is not always clear that the equipment such people need to operate has been designed with this view of the user.

It is important to describe the users of technology in terms of their functional ability related to technology rather than generic definitions of either medical conditions or primarily medical descriptions of their disabilities. Unfortunately, most statistical data is presented as generic and medically categorizations of disability. Gill and Shipley (1999), however, did define disabled user groups in terms of their functional ability, with specific emphasis on telephone use. They estimated that within the European Union, which has a population of 385 million, there were 9 million people with cognitive impairment that could lead to problems using the telephone. These figures do not take into account multiple impairments, and the authors pointed out that, in the elderly population in particular, there may be a tendency toward cognitive, hearing, vision, and mobility impairments all being present to a varying extent, and these may interact when considering the use of technological systems. It is this multiple minor reduction in function, often together with a major disability, which means that the challenges to technological support for older people have significantly different characteristics from those of younger disabled people and of the nondisabled young population.

There has been some movement in mainstream research and development in technology, both in academia and industry, away from a technology-led focus of user-led approach, and this has led to the development of user-centered design principles and practices in many industries. In addition, a number of initiatives have been launched to promote a consideration of people with disabilities within the user group in mainstream product development teams with titles including Universal Design, Design for All, Accessible Design, and Inclusive Design. The Design for All/Universal Design movement has been valuable in raising the profile of disabled users of products and has laid down some important principles. This approach, however, has tended not to place much significance on cognitive impairment, and it becomes more difficult to use a traditional user-centered design approach, particularly if use by people with disabilities is included as a factor in the design process.

Newell and Gregor (2000) suggested that a new design approach should be developed that would be based on the already accepted user-centered design methodology. There are some important distinctions between traditional user-centered design with able-bodied users and the approach needed when the user group either contains or is exclusively made up of people with cognitive dysfunction. These include the following:

- Much greater variety of user characteristics and functionality
- The difficulty in finding and recruiting "representative users"
- Situations in which "design for all" is not appropriate (e.g., if a task requires a high level of cognitive ability)
- The need to specify exactly the characteristics and functionality of the user group
- Conflicts of interest between user groups, including temporarily able bodied
- Tailored, personalizable, and adaptive interfaces
- Provision for accessibility using additional components (hardware and software)

The balance in the design process also needs to shift from a focus on user needs, rather than focusing on the users themselves. There will be additional problems when considering people with cognitive dysfunction, which will include the following:

- The lack of a truly representative user group
- Difficulties of communication with users
- Ethical issues (Alm, 1994; Balandin & Raghavendra, 1999)
- A different attitude is required from the designer
- Obtaining informed consent from some users
- Inability of users to communicate their thoughts

There can thus be particularly difficult ethical problems when involving users with cognitive disabilities in the design process. In addition, it is often necessary to involve clinicians when users with cognitive disabilities are involved, so some of the user-centered design actually focuses on professional advice about the user, rather than direct user involvement. Even with these problems, however, it is not impossible to include users with cognitive dysfunction in the design process.

The Inclusion of Users With Disabilities Within Research Groups

In Dundee, users with disabilities have a substantial involvement in the research, and they have made a significant contribution both to the research and to the commercial products that have grown from this research. There are two major ways in which users are involved: as disabled consultants on the research team, when they act essentially as "test pilots" for prototype systems, and by the traditional user-centered design methodology of having user panels, formal case studies, and individual users who assess and evaluate the prototypes produced as part of the research.

The contribution made by clinicians is also vital to the research, and these are full members of the research team. Dundee's Applied Computing Department is also one of the few computing departments that has employed speech therapists, nurses, special-education teachers, linguists, and psychologists.

User-Sensitive Inclusive Design

Some significant differences must be introduced into the user-centered design paradigm if users with disabilities are to be included, and this is particularly important if the users have cognitive dysfunction. To ensure that these differences are fully recognized by the field, the title *user-sensitive inclusive design* has been suggested. The use of the term *inclusive* rather than *universal* reflects the view that *inclusivity* is a more achievable, and in many situations more appropriate, goal than universal design or design for all. *Sensitive* replaces *centered* to underline the extra levels of difficulty involved when the range of functionality and characteristics of the user groups can be so great that it is impossible in any meaningful way to produce a small representative sample of the user group, or often to design a product that truly is accessible by all potential users.

Design for Dynamic Diversity

In addition to the aspects of user-sensitive inclusive design described above, it is necessary to make designers fully aware of the range of diversity that can be expected with people with cognitive impairment and also of the changing nature of the cognitive functioning of people. It has thus been suggested by Gregor and Newell (2001) that this be drawn particularly to the attention of designers by introducing the concept of designing for dynamic diversity. This process, described earlier, entails a recognition that peoples' abilities are diverse at any given age and that as they grow older the diversity grows dynamically; it also involves a recognition that even any given individual's abilities will vary according to factors such as mood, fatigue,

sugar levels, and so on. Only by taking on board the factors associated with designing for dynamic diversity will software design produce artifacts that are not static and that have no (or limited) means of adapting to the changing needs of users as their abilities change.

As has been seen above, metaphors and processes in use at present are ineffective in meeting the needs of this design paradigm or in addressing the dynamic nature of diversity. New processes and practices are needed to address design issues: awareness must be raised among the design, economic, and political communities, and research is needed to find methods to pin down this moving target.

A Storytelling Metaphor

Researchers need to consider how best to promulgate the concepts behind universal usability and the results of user-sensitive inclusive research. User-sensitive inclusive design needs to be an attitude rather than simply a mechanistic application of a set of "design for all" guidelines. This offers a further challenge to the community. The dangers of using such studies to produce more extensive guidelines was referred to earlier, but it is important that the results of user-sensitive inclusive design are made available to other designers and researchers. It is, however, too early to lay down principles and practices that must be followed by designers, and it may even be impossible to do this for some of the unconstrained environments in which designers have to work. It is thus suggested that we follow a storytelling approach, in which information about accessibility issues and design methods that focus on accessibility is presented in narrative form, with particular examples to illustrate generic principles. This is, in some sense, an extension of the single case study methodology, but without the constraints academics sometimes apply to these studies. This methodology could provide useful information to designers in a form that they will find easy to assimilate and act upon. This will thus assist in their education and help them to design more accessible and better products for everyone.

CONCLUSION

Although it is not necessary for human-interface designers whose systems may be used by people with cognition impairment to be fully versed in all aspects of cognition, they should have general background knowledge of the subject. They should also be in contact with experts in other disciplines, such as psychology, and have access to appropriate clinical knowledge. In addition, the development of the concept of and a methodology for user-sensitive inclusive design, design for dynamic diversity, and storytelling methods for communicating results will facilitate researchers in the specialized field and also provide mainstream engineers with an effective and efficient way of including people with disabilities within the potential user groups for their projects. If both of these can be achieved, it will go some way toward providing appropriate technological support for people with cognitive impairment.

References[1]

Alm, N. (1994). Ethical issues in AAC research. In J. Brodin & E. Björk-Åkesson (Eds.), *Methodological issues in research in augmentative and alternative communication. Proceedings of the Third International Society of Augmentative and Alternative Communication Research Symposium* (pp. 98–104). Jönköping, Sweden: Jönköping University Press.

Alm, N., Arnott, J. L., & Newell, A. F. (1992). Prediction and conversational momentum in an augmentative communication system. *Communications of the ACM, 35*, 46–57.

Alm, N., Morrison, A., & Arnott, J. L. (1995, October). A communication system based on scripts, plans and goals for enabling non-speaking people to conduct telephone conversations. In *Proceedings of the IEEE Conference on Systems, Man & Cybernetics*. Vancouver, Canada.

Bäckman, L., Small, B. J., Wahlin, Å., & Larsson, M. (2000). Cognitive functioning in very old age. In F. I. M. Craik & T. A. Salthouse (Eds.), *The handbook of aging and cognition*. Mahwah, NJ: Lawrence Erlbaum Associates.

Balandin, S., & Raghavendra, P. (1999). Challenging oppression: Augmented communicators' involvement in AAC research. In F. T. Loncke, J. Clibbens, H. H. Arvidson, & L. L. Lloyd (Eds.), *Augmentative and alternative communication, new directions in research and practice* (pp. 262–277). London: Whurr.

Beirmann, A. W. (1997). *More than screen deep—towards an every-citizen interface to the national information infrastructure*. Washington, DC: Computer Science and Telecommunications Board, National Research Council, National Academy Press.

Carmichael, A. R. (1999). *Style guide for the design of interactive television services for elderly viewers*. Winchester, UK: Independent Television Commission.

Charness, N., & Bosman, E. A. (1994). Age-related changes in perceptual and psychomotor performance: Implications for engineering design. *Experimental Aging Research, 20*, 45–61.

Cohen, G. (2000). Two new intergenerational interventions for Alzheimer's disease patients and their families. *American Journal of Alzheimer's Disease, 15*, 137–142.

Critchley, M. (1964). *Developmental dyslexia*. London: Heinemann.

Detterman, D. K., Gabriel, L. T., & Ruthsatz, J. M. (2000). Intelligence and mental retardation. In R. J. Sternberg (Ed.), *Handbook of intelligence*. Cambridge, England: Cambridge University Press.

Feil, N. (1993). *The validation breakthrough*. Baltimore, MD: Health Professions Press.

Freudenthal, A. (1999). *The design of home appliances for young and old consumers*. Delft, The Netherlands: Delft University Press.

Gill, J., & Shipley, T. (1999). *Telephones, what features do people need*. London: Royal National Institute for the Deaf.

Gould, S. J. (1997). *The mismeasure of man* (2nd ed.). Harmondsworth, England: Penguin.

Gregor, P., & Newell, A. F. (1999). The application of computing technology to interpersonal communication at the University of Dundee's Department of Applied Computing. *Technology and Disability, 10*, 107–113.

Gregor, P., & Newell, A. F. (2001, May, Portugal). "Designing for dynamic diversity—making accessible interfaces for older people." In *Proceedings of the European Community/National Science Foundation Workshop on Universal Accessibility and Ubiquitous Computing (WUAUC)*. New York: ACM (in press).

Horn, J. L., & Cattell, R. B. (1967). Age differences in fluid and crystallised intelligence. *Acta Psychologica, 26*, 107–129.

Hypponen, H. (1999). *The handbook on inclusive design for telematics applications*. Helsinki: EU INCLUDE Project.

International Organization for Standardization. (1999). ISO 13407. *Human-centered design processes for interactive systems*. Geneva, Switzerland: Author.

Kaluger, G., & Kolson, C. L. (1987). *Reading and learning disabilities* (2nd ed.). Columbus, OH: Bell & Howell.

Kaufman, A. S. (2000). Tests of intelligence. In R. J. Sternberg (Ed.), *Handbook of intelligence* (pp. 445–476). Cambridge, England: Cambridge University Press.

Lohman, D. F. (2000). Complex information processing and intelligence. In R. J. Sternberg (Ed.), *Handbook of intelligence* (pp. 285–340). Cambridge, England: Cambridge University Press.

Mandler, J. M. (1984). *Stories, scripts and scenes: Aspects of schema theory*. Mahwah, NJ: Lawrence Erlbaum Associates.

Mandler, J. M., & Parker, R. E. (1976). Memory for descriptive and spatial information in complex pictures. *Journal of Experimental Psychology: Human Learning and Memory, 2*, 38–48.

McKinlay, A. (1991). Using a social approach in the development of a communication aid to achieve perceived communicative competence. In J. Presperin (Ed.), *Proceedings of the 14th Annual Conference of the Rehabilitation Engineers Society of North America* (pp. 204–206). Washington, DC: The RESNA Press.

Newell, A. F. (1993, September). Interfaces for the ordinary and beyond. *IEE Software, 10*(5), 76–78.

Newell, A. F. (1995). Extra-ordinary human computer operation. In A. D. N. Edwards (Ed.), *Extra-ordinary human-computer interaction*. Cambridge, England: Cambridge University Press.

Newell, A. F., Arnott, J. L., Booth, L., & Beattie, W. (1992). Effect of the PAL word prediction system on the quality and quantity of text generation. *Augmentative and Alternative Communication, 8*, 304–311.

Newell, A. F., & Booth, L. (1991). The use of lexical and spelling aids with dyslexics. In C. Singleton (Ed.), *Computers & literacy skills* (pp. 35–44). Hull, UK: University of Hull.

Newell, A. F., Booth, L., & Beattie, W. (1991). Predictive text entry with PAL and children with learning difficulties. *British Journal of Educational Technology, 22*, 23–40.

Newell, A. F., & Cairns, A. Y. (1993, October). Designing for extra-ordinary users. *Ergonomics in Design*, 10–16.

Newell, A. F., & Gregor, P. (1997). Human computer interfaces for people with disabilities. In M. Helander, T. K. Landauer, & P. Prabhu (Eds.), *Handbook of human-computer interaction* (pp. 813–824). Amsterdam: Elsevier.

Newell, A. F., & Gregor, P. (2000). User sensitive inclusive design—in search of a new paradigm. *Proceedings of the ACM Conference on Universal Usability* (pp. 39–44). New York: ACM.

Nielsen, J. (1993). *Usability engineering*. London: Academic Press.

Ogozalec, V. Z. (1997). A comparison of the use of text and multimedia interfaces to provide information to the elderly. *Proceedings of Computer Human Interaction '97* (pp. 65–71). New York: ACM Press.

---

[1] Web sites focused on universal design include http://www.design.ncsu.edu/cud/ud/ud.html; http://www.stakes.fi/include; http://www.trace.wisc.edu, and http://www.w3.org/WAI.

Pirkl, J. J. (1994). *Transgenerational design, products for an aging population.* New York: Van Nostrand Reinhold.

Preece, J. (1994). *A guide to usability—human factors in computing.* London: Addison Wesley and Open University.

Rabbitt, P. M. A. (1993). Does it all go together when it goes? The Nineteenth Bartlett Memorial Lecture. *The Quarterly Journal of Experimental Psychology, 46A,* 385-434.

Rau, M. T. (1993). *Coping with communication challenges in Alzheimer's disease.* San Diego, CA: Singular.

Salthouse, T. (1985). *A theory of cognitive aging.* Amsterdam: North Holland.

Salthouse, T. A. (1991). *Theoretical perspectives on cognitive aging.* Mahwah, NJ: Lawrence Erlbaum Associates.

Schank, R., & Abelson, R. (1977). *Scripts, plans, goals, and understanding.* Mahwah, NJ: Lawrence Erlbaum Associates.

Sheridan, C. (1992). *Failure-free activities for the Alzheimer's patient.* London: Macmillan Press.

Shneiderman, B. (1992). *Designing the user interface: Strategies for effective human-computer interaction.* Reading, MA: Addison-Wesley.

Sleeman, K. D. (1998). Disability's new paradigm, implications for assistive technology and universal design. In I. Placencia Porrero & E. Ballabio (Eds.), *Improving the quality of life for the european citizen* (Assistive Technology Research Series Vol. 4; xx-xxiv). Amsterdam: IOS Press.

Stephanidis, C. (Ed.). (2001). *User interfaces for all.* Mahwah, NJ: Lawrence Erlbaum Associates.

Sternberg, R. J. (2000). The concept of intelligence. In R. J. Sternberg (Ed.), *Handbook of intelligence* (pp. 3-15). Cambridge, England: Cambridge University Press.

Todman, J., Elder, L., & Alm, N. (1995). Evaluation of the content of computer-aided conversations. *Augmentative and Alternative Communication, 11,* 229-234.

Vanderheyden, P. B., Demasco, P. W., McCoy, K. F., & Pennington, C. A. (1996). A preliminary study into schema-based access and organization of re-usable text in AAC. *Proceedings of the RESNA '96 Conference* (pp. 59-61). Arlington, VA: RESMA Press.

Waller, A. (1992). Providing narratives in an augmentative communication system. Unpublished PhD dissertation, University of Dundee, Scotland.

Waller, A., Dennis, F., Brodie, J., & Cairns, A. (1998). Evaluating the use of TalksBac, a predictive communication device for nonfluent adults with aphasia. *International Journal of Language and Communication Disorders, 33,* 45-70.

Willows, D. M., Kruk, R. S., & Corcos, E. (1993). *Visual processes in reading and reading disabilities.* Mahwah, NJ: Lawrence Erlbaum Associates.

Worden, A., Walker, N., Bharat, K., & Hudson, S. (1997). Making computers easier for older adults to use. *Proceedings of CHI '97* (pp. 266-271). New York: ACM Press.

· 25 ·

PHYSICAL DISABILITIES AND COMPUTING TECHNOLOGIES: AN ANALYSIS OF IMPAIRMENTS

Andrew Sears
UMBC

Mark Young
Maryland Rehabilitation Center

INTRODUCTION

Computing devices are becoming smaller, more mobile, more powerful, and less expensive. As a result, they are making their way into every aspect of our lives. Although computing devices can be convenient tools for traditional computer users, they can also serve as barriers for individuals with impairments. A design process that considers the impairments of potential users can turn these barriers into powerful tools that increase employment opportunities, provide enhanced communication capabilities, and enable increased independence (Young, Tumanon, & Sokal, 2000).

Cognitive, perceptual, and physical impairments (PIs) can all hinder the use of computing technologies, but the focus of this chapter is on specific physical impairments that contribute to disability. More specifically, we focus on PIs that may hinder an individual's ability to physically interact with computing technologies (e.g., PIs affecting the upper body). Therefore, we do not address PIs that affect the lower body or hinder the production of speech. For additional information regarding cognitive or perceptual impairments, see chapters 24 and 26, respectively. Chapters 21 and 22 may also provide useful insights because they address the issues involved in designing technologies for elderly users and children, two groups whose cognitive, perceptual, and physical capabilities may require special attention.

The objectives of this chapter are to

- provide an introduction to specific PIs that can hinder the use of traditional computing devices,
- highlight critical characteristics of these PIs that must be considered when designing computing systems,
- discuss the relationship between PIs that result from health conditions and those that result from environmental or contextual factors,
- summarize the results of existing human–computer interaction (HCI) research involving individuals with PIs, and
- offer observations based on the literature that may provide guidance for future research and development efforts.

We believe that by beginning with an understanding of the PI involved, it is possible to design computing technologies that lessen or even eliminate the associated disabilities. We begin by presenting a set of definitions for *impairment, disability*, and *handicap* that are offered by the World Health Organization (WHO). By formally defining these terms and using the terms in a way that is consistent with these definitions, we hope to eliminate potential ambiguity and confusion. Subsequently, we define the subset of PIs that are of primary concern when addressing an individual's ability to physically interact with computing technologies. Through this definition, we further clarify the scope of this chapter.

Given these definitions, we proceed to discuss the relationship between health conditions and PI. Understanding the underlying health condition is critical because this often provides valuable insights into the nature of the resulting impairment. In this context, we describe common health conditions (e.g., cerebral palsy, spinal cord injuries) associated with PIs that can hinder an individual's ability to interact with computing devices. We identify the associated PIs, their important characteristics, and any additional impairments that may prove critical when designing computing technologies. Although health conditions are most often associated with PIs, both contextual and environmental factors can also hinder interactions with computing devices. As a result, we briefly discuss the relationship between the environment, context, physical impairments, and disabilities. Through this discussion, we highlight critical similarities and differences between those PI that result from health conditions and the difficulties individuals may experience as a result of the environment or activities in which they are engaged.

We conclude by reviewing recent research that has focused on addressing the needs of individuals with PI, discuss existing technologies in the context of various PI, and offering directions for additional research.

DEFINING IMPAIRMENT

WHO first published the International Classification of Impairments, Disabilities, and Handicaps in 1980. The final draft of the second version of this document (ICIDH-2) provides a set of definitions that serve a foundation for this our discussion (WHO, 2000). The ICIDH-2 model acknowledges the complex relationships that exist among health conditions, impairments, disabilities, and handicaps. This model also highlights the potentially important role of both the context and environment in which activities are taking place. Figure 25.1 highlights the relationships between impairments, disabilities, and handicaps as well as the influence of both health conditions and context. Both health conditions and context can directly result in impairments, disabilities, or handicaps. Contextual factors (e.g., environment and tasks) can also affect the relationship among impairments, health conditions, and handicaps. For example, a specific impairment may or may not result in a disability, depending on the environment in which the individual is located and the tasks in which the individual is engaged.

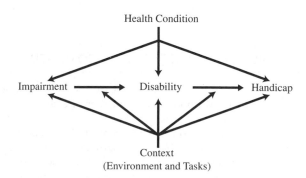

FIG. 25.1. Model illustrating the relationship between health conditions, disabilities, impairments, handicaps, and contextual factors.

The following definitions are adapted from the ICIDH-2:

- *Health condition:* A disease, disorder, injury, or trauma. Examples include spinal cord injuries, arthritis, cerebral palsy, stroke, multiple sclerosis, and amyotropic lateral sclerosis.
- *Impairment:* A loss or abnormality of body structure or function. Examples of physical impairments include a loss of muscle power, reduced mobility of a joint, uncontrolled muscle activity, and absence of a limb. Impairments can be caused by health conditions. For example, an individual with arthritis (health condition) may experience weakness, stiffness, and a reduced range of motion (impairments).
- *Disability (activity limitations):* Difficulties an individual may have in executing a task or action. Examples include difficulty communicating, learning, performing tasks, or using standard computing technologies (e.g., keyboard, mouse). Disabilities are usually caused by impairments; an individual has a disability; disabilities are activity-specific. For example, a loss of muscle power in both arms (impairment) may interfere with an individual's ability to use a standard keyboard and mouse (disability).
- *Handicap (participation restrictions):* Problems an individual may experience in involvement in life situations. Handicaps can be caused by disabilities; an individual experiences a handicap; handicaps occur at a social level. For example, an individual may have difficulty communicating (disability) that places restrictions on the individual's ability to participate in educational and work activities, to participate in the exchange of information, and to maintain social relationships (handicaps).

Given these definitions, the distinction between a disability and a handicap is, in many ways, dependent on the individual. If an individual has an impairment (e.g., bending the fingers may be difficult) that makes it difficult to use a standard keyboard, then this individual has a disability with respect to the operation of this device. However, an individual only experiences a handicap if the inability to use the keyboard (e.g., disability) affects the individual's normal life activities. Our focus is on designing computing technology to lessen or even eliminate disabilities that may result from PI, the environment, or the context in which an individual is interacting with this technology. As a result, the potentially subtle differences between disabilities and handicaps are not important.

PHYSICAL IMPAIRMENTS AND COMPUTING

Impairments, as described by the WHO, can affect every structure and function of the body. As a result, there are impairments that interfere with the internal workings of the body as well as how we interact with our environment. In this chapter, we focus on the subset of PIs that are of particular interest when discussing the use of computers. More specifically, we focus on PIs that interfere with physical interactions with computing devices (e.g., interacting with a keyboard or mouse). Impairments, even PIs, that interfere with vision, hearing, and other activities involved in the use of computing devices are not discussed in

this chapter (see chapters 24 and 26 for coverage of some of these issues). Four categories of PI are particularly relevant for the discussion that follows:

- *Structural deviations:* Situations in which there is significant deviation or loss with regard to anatomic parts of the body. Examples include the partial or total absence of a body part (e.g., missing finger, hand, arm) as well as situations in which a body part deviates from the norm in either position or dimension.
- *Mobility (of bone and joint) functions:* Mobility functions address an individual's ability to move a joint or bone. This includes both the range and ease of movement. For example, an individual may have limited range of motion or experience difficulty when bending his or her fingers (e.g., arthritis).
- *Muscle power functions:* Muscle power functions address an individual's ability to generate force by contracting a muscle or muscle group. Paresis (i.e., weakness) refers to the partial loss of power and can be caused by a variety of conditions (e.g., brain injury). Paralysis refers to the complete loss of power and can also be caused by various conditions (e.g., spinal cord injuries). Muscle tone and endurance functions may also be important in certain circumstances but tend to be less critical for most interactions with computing systems.
- *Movement functions:* Although additional movement functions exist, we focus on an individual's ability to control voluntary and involuntary movements. Difficulty controlling movements that involve a rapid change in direction (e.g., dysdiadochokinesia) would fit in the voluntary movement category. Uncontrolled shaking or trembling of the hands (e.g., essential tremors or tremors associated with Parkinson's disease) would fall under the involuntary movement category.

Although there are many other types of PI, including stability of the joints, motor reflex functions, and gait pattern functions, the four categories identified above account for the majority of PIs that hinder the use of computing technologies. As we discuss various health conditions, it is useful not only to identify specific PIs that may occur, but important characteristics of the PIs that may influence the design of alternative strategies or technologies. PIs can vary in many ways, but the following four dimensions are perhaps the most critical:

- PIs can be *permanent* or *temporary*.
- PIs may be *continuous* or *intermittent*. For example, individuals with multiple sclerosis experience relapses and remissions.
- PIs may be *progressive* (get more severe with time), *regressive* (get less severe with time), or *static* (no change with time).
- The severity of a PI can range from *mild* to *severe*. In other cases, severity is *variable*.

Temporary PI can be problematic because individuals may be hesitant to invest significant time or effort in adapting to new interaction techniques. If a PI is permanent, individuals will develop accommodation strategies, and learning new interaction techniques may be acceptable. Intermittent PI create difficulties because any alternative interaction technique that is provided must be accessible when needed without hindering interactions when the PI is not present. Progressive and regressive PIs require accommodations that can adapt as the impairment becomes more severe. Adaptation may be automatic or user-directed. When the severity of the PI is variable, the accommodation must also be able to adapt. However, the required adaptation can be more complex than that required for intermittent, progressive, and regressive PI because severity may increase or decrease at any given time, or the PI could completely disappear temporarily. PIs that are permanent, continuous, static, and of a fixed severity are perhaps the easiest to address. Under these conditions, the individuals are likely to adapt to their impairment, develop their own accommodation strategies, and be more willing to learn new interaction techniques. When PIs change with time (i.e., intermittent, progressive, regressive, or variable), providing effective solutions becomes more difficult.

HEALTH-CONDITION-INDUCED IMPAIRMENTS

It may be tempting to view all individuals with a particular PI as benefiting equally from a given accommodation. Unfortunately, this perspective can result in solutions that fail to adequately address the needs of the individual. Understanding the underlying health condition provides valuable insight into the nature of the associated PIs, which in turn provides insight into the characteristics of the accommodations that will prove beneficial.

We briefly describe some of the more common diseases, disorders, and injuries that result in PI that hinder the use of computing systems. For each health condition, the PIs that may result are identified and described. For additional information regarding any of the underlying health conditions, we refer the reader to the Merck Manual (Berkow, 1997) or the references included throughout the text that follows. Our descriptions assume that appropriate treatments (e.g., medication, surgery, therapy) have already taken place. Although we are not addressing the issues involved in perceptual, speech, or cognitive impairments in this chapter, we do highlight situations in which these impairments may be present. We do this for two reasons: to provide additional insights into difficulties individuals with specific health conditions may face (e.g., vision, hearing, or cognition difficulties). and to highlight situations in which specific alternative interaction strategies or technologies may not be appropriate (e.g., speech recognition).

Amyotrophic Lateral Sclerosis (ALS)

Amyotrophic lateral sclerosis (ALS) is also known as Lou Gehrig's disease (Taylor & Lieberman, 1988). ALS is a progressive disease that begins with weakness in the hands (less often, it begins in the feet). ALS progresses at a variable rate, but 50% of those affected die within 3 years of the initial symptoms, and only 10% live beyond 10 years. As the disease progresses, the weakness spreads to additional muscles and becomes more severe (eventually leading to paralysis that spreads throughout muscles of the body). With time, spasticity occurs resulting in muscles becoming tight. Spasms and tremors can also occur. ALS only affects those nerves that stimulate muscle action. Cognitive functions

and sensation remain intact. ALS is characterized primarily by muscle power PI (weakness progressing to paralysis), but movement PI also occurs (stiff muscles, spasms, and tremors). The PIs are permanent, continuous, and progressive. Initially, they are mild, but with time they become severe. Note that with time the muscles involved in producing speech are affected.

Arthritis

Arthritis refers to the inflammation of a joint (Hicks, 1988; Schumacher, 1993). The two most common forms of arthritis are osteoarthritis and rheumatoid arthritis. Osteoarthritis (OA) is a chronic joint disorder that can cause joint pain and stiffness. It is characterized by degeneration of joint cartilage and adjacent bone, is the most common joint disorder, and is one of the most common causes of physical impairments (Berkow, 1997). OA is not an inevitable part of aging, but it does occur more often in the elderly. OA typically progresses slowly after symptoms first appear. OA can result in significant mobility PI (reduced range of motion, weakness, and difficulty with repetitive motions). The PIs tend to be permanent, intermittent or continuous, progressive, and can range from mild to severe.

Rheumatoid arthritis (RA) is an autoimmune disease that can cause swelling, pain, and often the eventual destruction of the joint's interior. RA is characterized by symmetric inflammation of the joints. RA typically appears first between 25 and 50 years of age, affecting 1% of the population. Most often, RA starts subtly, progressing at a highly variable rate. In some rare cases, RA spontaneously disappears and treatment is often successful, but RA can result in significant impairment of mobility (reduced range of motion, weakness, and difficulty with repetitive motions). The PIs tend to be permanent, continuous, and progressive. The severity of the PI is typically mild initially but in 10% of the cases will progress to severe.

Brain Injury

Brain injury (BI) is a term used to described a collection of injuries (Horn & Zasler, 1996; Rosenthal, Griffith, Bond, & Miller, 1990). Technically, there is a disconnect between the cause of the injury (e.g., penetration of the skull by a foreign object or the rapid acceleration or deceleration resulting from a forceful blow to the head or vehicular accident) and the injury itself. Common head injuries include skull fractures, cerebral contusions (bruises on the brain), cerebral lacerations (torn brain tissue), concussions (brief loss of consciousness after an injury), and intracranial hematomas (collection of blood in the brain or between the brain and skull). Head injuries are the most common cause of death and disability for individuals under 50 years of age. BI are not always the result of trauma. For example, anoxia can also result in BI. The consequences of a brain injury depend on the area of the brain affected and the severity of the injury. Possible consequences include death; language, vision, and motor control difficulties; and periodic headaches. As with a stroke, other parts of the brain can often assume the responsibilities of the damaged portion, lessening the severity of the resulting impairments. This ability is more prevalent in children, because their brains are more adept at shifting functions to different parts of the brain. The most common PIs are movement oriented (e.g., difficulty controlling muscles). After rehabilitation, the PI are considered permanent, continuous, and static. Severity may range from mild to severe.

Cerebral Palsy

Cerebral palsy (CP) is a condition that results from brain injury that typically occurs before, during, or shortly after birth. CP is not a disease and is not progressive (Molnar, 1992). Over 90% of those affected with CP survive to adulthood. There are four main types of CP: spastic, choreoathetoid, ataxic, and mixed. Spastic CP occurs in approximately 70% of those individuals with this condition and is characterized by both movement and muscle power PIs (stiff muscles and weakness). Choreoathetoid CP is characterized by movement PIs (spontaneous, slow, uncontrolled muscle movements; abrupt and jerky movements) and occurs in 20% of individuals with CP. Ataxic CP occurs in approximately 10% of those individuals with CP and is characterized by movement and muscle power PIs (poor coordination, weakness, trembling, difficulty with rapid or fine movements). Mixed CP occurs in many individuals and is characterized by a combination of two of the previously mentioned forms of CP. Seizures are also possible (most often with ataxic CP). In all four forms, the PIs are permanent, continuous, static, and may range from mild to severe. It is important to note that in all forms of CP, speech and intelligence can be affected.

Missing Limbs or Digits (Amelia or Amputation)

Technically, missing limbs or digits is an impairment, not a health condition (Banerjee, 1982). However, as with tremors, it is possible for limbs or digits to be missing without any associated health condition (e.g., congenital absence of a limb or digit is not considered a health condition). The PI associated with missing limbs or digits is permanent, continuous, static, and can vary from mild to severe depending on the use and effectiveness of a prosthetic device.

Multiple Sclerosis

Multiple sclerosis (MS) is a disorder in which the nerve fibers associated with the eye, brain, and spinal cord loose patches of myelin (a protective, insulating sheath; Adams & Victor, 1993). MS is progressive, but there is not single pattern that typifies its progression. It results in numerous symptoms, including a variety of movement and muscle power PIs (clumsiness, tremors, stiff muscles, weakness). It often results in periods of flare-ups, alternating with periods of relatively good health; in other individuals, the symptoms become more severe within weeks or months. As flare-ups become more frequent, the symptoms become more severe and may become permanent. The PIs are permanent, intermittent, and progressive; they may be mild initially but often progress to become severe. In addition to multiple PIs, MS can result in visual impairments.

Muscular Dystrophy

Muscular dystrophy (MD) is a group of inherited muscle disorders (Pidcock & Christensen, 1997). The two most common forms which only affect males, are Duchenne's MD (affecting 20 to 30 of every 100,000 boys) and Becker's MD (affecting 3 of every 100,000 boys). Duchenne's MD is more severe, appearing between the ages of 3 and 7; most children must use a wheelchair by 10 or 12, and death is common by age 20. Becker's MD is less severe, with symptoms appearing around age 10; few children are confined to a wheelchair by age 16, and more than 90% live beyond age 20. Both forms begin with muscle power PIs (weakness) that spreads and becomes more severe. This is often followed by mobility PI (joints that cannot be fully extended due to contracted muscles). Some PIs may not be present early (e.g., inability to fully extend the elbows). In both forms, the PIs are permanent, continuous, and progressive. Typically, the PIs are mild initially and progressively worsen.

Parkinson's Disease

Parkinson's disease (PD) is a degenerative disorder of the nervous system (Duvaisin, 1991). PD is characterized by degeneration of the nerve cells in the basal ganglia, which results in reduced production of dopamine (the main neurotransmitter of the basal ganglia). PD affects approximately 1 in 250 individuals over the age of 40 and 1 in 100 individuals over the age of 65. PD typically begins with mild symptoms and progresses gradually. It often begins with resting tremors (movement PIs). As the disease progresses, initiating a movement can become difficult, muscles become stiff, and bending the elbow can become difficult or uncomfortable (movement and mobility PI). The tremors vary dramatically. Some individuals never develop tremors. For others, they may be permanent or temporary; continuous or intermittent; progressive, regressive, or static; and can vary from mild to severe. Other PIs, including difficulty initiating movements and stiff muscles are permanent, continuous, and progressive. They are mild initially but progress to become severe. Still other PIs, such as difficulty bending the elbow, may never occur. Individuals with PD often speak softly in a monotone and may stutter.

Repetitive Stress Injury

Repetitive stress injuries (RSI) are known by a variety of names including cummulative trauma disorders and repetitive trauma syndromes (Moore & Garg, 1992). RSI refers to a collection of disorders or injuries that are believed to be associated with repetitive activities, but the precise cause is still a subject of debate. RSI are most commonly associated with the wrists but can affect other parts of the body. RSI most often result in mobility PIs (reduced range of motion, weakness, and difficulty with repetitive motions). The PIs are permanent, continuous or intermittent, progressive, and can vary from mild to severe.

Stroke

A stroke is defined as the death of brain tissue resulting from a lack of blood flow and insufficient oxygen to the brain (Wade, Langton Hewer, Skilbeck, & David, 1985). A stroke typically occurs when a blood vessel becomes blocked or bursts. The symptoms depend on the area of the brain that is affected but include vision, speech, cognitive, and physical impairments. Because other parts of the brain can assume the responsibilities of the damaged portion, it is possible to lessen the severity of the impairments that are apparent immediately following a stroke. Rehabilitation services are critical in this respect. The most common PIs are movement-oriented (e.g., difficulty controlling muscles). After rehabilitation, the PI can be defined as permanent, continuous, and static. The severity of the PI may be range from mild to severe.

Spinal Cord Injury

Spinal cord injuries (SCI) occur when the spinal cord (a collection of nerves extending from the base of the brain through the spinal column) is compressed, cut, damaged, or affected by disease (Stiens, Goldstein, Hammond, & Little, 1997). The spinal cord contains motor nerves (controlling movement) and sensory nerves (providing information about temperature, pain, position, and touch). The consequences of an SCI depend on the level and completeness of the injury.

The level of an injury is based on the nerves that are affected. The spinal column is divided into four areas: cervical (neck), thoracic (chest), lumbar (lower back), and sacral (tail bone). There are seven cervical vertebrae, numbered C1 through C7 from top to bottom. There are 12 vertebrae in the thoracic region (T1–T12), 5 in the lumbar region (L1–L5), and 5 in the sacral region. The level of an SCI refers to the location of the damaged nerves. In the cervical region, injuries are labeled based on the vertebrae immediately below the damaged nerves (e.g., level C1 refers to damage to the nerves just above vertebrae C1). Damage to nerves just below C7 and above T1 are referred to as level C8. In all other regions, injuries are labeled based on the vertebrae immediately above the damaged nerve (level T3 refers to injuries to the nerves between T3 and T4). In general, injuries higher on the spinal cord will result in greater impairment. Because PIs that affect the upper body have the most significant impact on the use of computing devices, Table 25.1 summarizes the possible consequences of high-level SCI.

The completeness of an injury is often assessed using a scale defined by the American Spinal Injury Association (ASIA, 2001). Although there are additional details and assessments must be completed by trained professionals, the following summaries provide sufficient detail for our purposes. Table 25.2 provides a brief description of the ASIA scores.

SCIs result in muscle power PIs. The level of the injury determines which muscles are affected. The completeness of the

TABLE 25.1. Relationship Between the Level of a Spinal Cord Injury and the Resulting Effects on Muscles

Level of Injury	Effect
C1 to C5	Paralysis of muscles used for breathing, controlling the arms, and controlling the legs
C5 or C6	Paralysis of the legs; some ability to flex the arms remains
C6 or C7	Paralysis of the legs; paralysis of part of the wrists and hands; much of the ability to move the shoulders and to bend the elbows remains
C8 or T1	Legs and trunk paralyzed; hands paralyzed; arms remain relatively normal

injury determines how much muscle power is lost. Injuries with ASIA scores of A or B result in a complete loss of muscle power (i.e., paralysis) in the affected muscles. Injuries that are classified as ASIA C or D are more difficult to describe because the amount of muscle power that remains can be highly variable (i.e., weakness or paresis). PI associated with SCI are considered permanent, continuous, and static, with the severity of the PI ranging from mild to severe depending on the completeness of the injury.

Tremors

A tremor is an involuntary movement (Hallett, 1991). Tremors are produced by involuntary muscle activity and are often described as rhythmic shaking movements. Some tremors occur when muscles are in use (action tremors), whereas others occur when the muscles are resting (resting tremors). Intention tremors occur when an individual makes a purposeful motion.

TABLE 25.2. Residual Motor and Sensory Function Associated With Each American Spinal Injury Association (ASIA) Score

ASIA Score/ Completeness	Description of Motor and Sensory Function
A	Complete; no residual motor or sensory function below the level of the injury
B	Motor complete, sensory incomplete; there is no residual motor function, but some sensory capabilities remain in tact
C	Motor and sensory incomplete; most muscles below the level of injury are below three-fifths of normal strength
D	Motor and sensory incomplete; most muscles below the level of injury are at or above three-fifths of normal strength
E	Normal motor and sensory function

Tremors are a common consequence of various health conditions (e.g., MS or stroke). In these situations, the tremors are described under the appropriate health condition (e.g., see MS or stroke).

Essential tremors usually begin in early adulthood and slowly become more obvious (essential tremors that begin in the elderly are referred to as senile tremors). Essential tremors typically stop when the arms (or legs) are at rest, but become more obvious when the arms are held away from the body or in awkward positions. Essential tremors are relatively fast but result in little movement. The vocal cords can be affected by essential tremors, resulting in inconsistent speech. Essential tremors are permanent, intermittent, and can be progressive. Essential tremors are typically mild initially but may progress to be severe and can be variable. Essential tremors have no known cause, and for this reason we include tremors (which are technically a movement PI) in this list of health conditions.

Summary

Although the examples listed are some of the most common health conditions that result in PI, this is by no means a complete list. Table 25.3 summarizes the PIs that are frequently associated with each health condition, important characteristics of each PI, and any additional impairments that may be associated with the health condition that may influence the design of alternative strategies or technologies.

SITUATIONALLY-INDUCED IMPAIRMENTS AND DISABILITIES

As Fig. 25.1 indicates, both the environment in which an individual is working and the current context (e.g., the activities in which the person is engaged) can contribute to the existence of impairments, disabilities, and handicaps. Many environmental characteristics, including lighting, noise, vibration, and temperature, can affect an individual's ability to interact with computing technologies. Similarly, the context in which the activities are occurring can also hinder an individual's ability to interact with computing technologies. For example, the user may be engaged in other multiple activities that place demands on cognitive, perceptual, or physical capabilities (e.g., eyes-busy, hands-busy, or computationally intensive tasks). Stress can also affect an individual's ability to interact with computers.

In some of these situations, the environment or context creates temporary impairments. In other situations, there is no impairment, but a temporary disability does exist. For example, extended exposure to cold temperatures can make it difficult for an individual to bend one's fingers, creating a temporary mobility impairment (similar to arthritis). In contrast, an environment that is vibrating (e.g., a moving vehicle) does not alter any of the bodies structures or functions but can still make it difficult to use a stylus to enter text using gesture recognition. When the individual's physical capabilities are altered, the environment or context create a temporary PI. When the individual's physical capabilities are not altered but are instead rendered inadequate for the task the environment or context create a temporary disability. In both situations, the designer's goal is to minimize the negative impact of the environment and context on the computing activity.

Although extensive research has studied dual-task scenarios, in which participants were required to attend to two tasks at the same time, these studies tend to highlight decreases in cognitive performance rather than physical performance. At the same time, individuals are finding themselves in situations in which they are engaged in a secondary computing task with increasing frequency. For example, paramedics' hands may be busy providing medical care while they engage in a secondary computing activity to take notes and complete required forms. Extreme temperatures can affect both mental and physical activities with performance decreasing with extended exposure. An environment that is vibrating, such as a moving vehicle, can interfere with the performance of physical activities. Walking can also make more complex the physical activities required to interact with a mobile computing device.

Many factors, including temperature, movement, and multiple tasks, can adversely affect an individual's ability to perform the physical actions required to interact with computing technologies. Although some studies have been conducted (e.g., by the military), these relationships have yet to be systematically investigated and reported within the HCI literature. At the same time, the fundamental characteristics of these situationally-induced impairments and disabilities (SIID) are clear. Unlike impairments associated with health conditions that tend to be permanent, SIIDs are temporary. As a result, individuals experiencing them are less likely to develop their own accommodation strategies and are therefore more likely to benefit from technology-based accommodations. Because SIIDs are temporary, they are also intermittent. SIIDs may be progressive, regressive, or static depending on the conditions. Finally, SIIDs may range from mild to severe or could be variable. Effectively addressing the highly variable nature of SIID is an interesting and challenging area in need of additional research.

HCI RESEARCH AND PI

Researchers have approached the issue of designing computing technologies for individuals with PIs from a variety of perspectives. Many projects were clearly motivated by desire to address the needs of individuals with disabilities. A subset of these projects focuses on specific health conditions or PI. For example, researchers have investigated the development of a new interaction alternative for individuals with cerebral palsy (Roy, Panayi, Erenshteyn, Foulds, & Fawcus, 1994), cursor control for individuals with spinal cord injuries (Casali, 1992), and text entry for individuals with spinal cord injuries (Sears, Karat, Oseitutu, Karimullah, & Feng, 2001). Other projects are motivated by specific impairments or disabilities regardless of the underlying cause. For example, researchers have investigated the effect neck range of motion has on the use of head controls (LoPresti, Brienza, Angelo, Gilbertson, and Sakai, 2000) and the use of force feedback for individuals with various movement and muscle power PI that affect the hands (Keates, Langdon,

TABLE 25.3. Common Health Conditions That Can Result in Physical Impairments That Hinder Interactions
With Standard Computing Technologies

Health Condition	Common PI and Characteristics	Additional Impairments
Amyotrophic lateral sclerosis (ALS) or Lou Gehrig's Disease	Muscle power: Weakness often begins in hands or feet. With time, loss of muscle power spreads and becomes more severe. Permanent, continuous, progressive, mild progressing to severe (paralysis).	In time, muscles throughout the body can be affected, including those involved in producing speech.
	Movement: Muscles can become stiff. Spasms and tremors can occur. Permanent, continuous, progressive, mild to severe.	
Arthritis	Osteoarthritis Mobility: Reduced range of motion, weakness, difficulty with repetitive motions. Permanent, intermittent or continuous, progressive, mild to severe.	None.
	Rheumatoid arthritis Mobility: Reduced range of motion, weakness, difficulty with repetitive motions. Permanent, continuous, progressive, mild to severe.	None.
Brain injury	Movement: A wide range of movement-oriented difficulties are possible depending on the location and severity of the injury. Permanent, continuous, static, mild to severe.	Speech, vision, and cognitive impairments are possible.

Cerebral palsy	Spastic
	  Muscle power: Weakness. Permanent, continuous, static, mild to severe. Paralysis does not occur.
	  Movement: Stiff muscles. Permanent, continuous, static, mild to severe.
		Speech and intelligence can be affected.

	Choreoathetoid Movement. Spontaneous, slow, uncontrolled muscle movements as well as jerky, abrupt movements. Permanent, continuous, static, mild to severe.	Speech and intelligence can be affected.

	Ataxic
	  Muscle power: Weakness. Permanent, continuous, static, mild to severe. Paralysis does not occur.
	  Movement: Poor coordination, trembling, difficulty with rapid or fine movements. Permanent, continuous, static, mild to severe.
		Speech and intelligence can be affected.

	Mixed Any combination of two forms listed above.	
Missing limbs or digits	Mobility. Permanent, continuous, static, mild to severe.	None.
Multiple sclerosis	Muscle power: Weakness. Permanent, intermittent, progressive, mild to severe. Paralysis does not occur. Movement: Clumsiness, tremors, stiff muscles. Permanent, intermittent, progressive, mild to severe.	Vision can be affected.
Muscular dystrophy	Muscle power: Weakness that spreads and becomes more severe. Permanent, continuous, progressive, mild to severe. Paralysis does not occur. Mobility: Joints that cannot be fully extended. Permanent, continuous, progressive, mild to severe.	None.
Parkinson's disease	Movement: Tremors (pill rolling) of one side are often the first sign. Permanent or temporary; continuous or intermittent; progressive, regressive, or static; mild to severe. Difficulty initiating movements and stiff muscles. Permanent, continuous, progressive, mild to severe. Mobility: Difficulty bending the elbow. Permanent, continuous, progressive, nonexistent to severe.	Speech can be affected. Cognitive impairments are possible.
Repetitive stress injuries	Mobility: Reduced range of motion, weakness, difficulty with repetitive motions. Permanent, continuous or intermittent, progressive, mild to severe.	None.
Stroke	Movement: A wide range of movement-oriented difficulties are possible depending on the location (in the brain) and severity of the injury. Stroke does not always result in movement disorders. Permanent, continuous, static, mild to severe.	Speech, vision, and cognitive impairments are possible.
Spinal cord injuries	Muscle power: The specific muscles affected are determined by the level of the injury. How much power is lost is determined by the completeness of the injury. Permanent, continuous, static, mild to severe.	None.
Tremors	Movement: Tremors often affect the hands. Permanent, intermittent, progressive or static, mild to severe or variable.	Speech can be affected.

Clarkson, and Robinson, 2000). Both of these articles described studies that included individuals with a variety of underlying health conditions. Many articles that focus on the issues involved in addressing the needs of individuals with a specific health condition or PI are discussed below.

Another group of projects address issues associated with PI but do not identify any specific PI or health conditions as the motivation for the research. For example, several articles discuss communications issues for individuals "with severe speech and motor impairments" (e.g., McCoy, Demasco, Jones, Pennington, Vanderheyden, and Zickus, 1994), whereas others explore speech-based interfaces for "motor-control challenged computer users" (Manaris and Harkreader, 1998), new techniques for eye tracking that "could be useful to some people with physical disabilities" (Patmore and Knapp, 1998), and the use of word-prediction for text entry by "people with severe physical disabilities" (Garay-Vitoria and González-Abascal, 1997). Many articles that fit into this category are also discussed.

We have three goals for this literature review: (a) to summarize the current state of our knowledge with respect to designing computing technologies for individuals with PIs, (b) to highlight numerous unsolved problems in need of additional research, and (c) to provide pointers to numerous related articles that are not discussed. Although there are many ways to organize the articles we review, we choose to group articles by health condition or PI whenever possible. More focused articles are discussed first (e.g., articles that address specific health conditions or PI). Next, we present those articles that address less precisely defined PI (e.g., various impairments that affect the hands, regardless of the underlying health condition). The last set of articles have a clear association with PIs, but do not discuss any specific PI (e.g., articles that present a technology specifically in the context of individuals with PI but do not identify the PI or health condition). Occasionally, articles that may focus on a specific health condition are included in other sections because of the PI or technology discussed (e.g., Doherty, Cockton, Bloor, & Benigno, 2000). Throughout this section, we use terminology that is consistent with the definitions presented earlier. As a result, our terminology may differ from that used in the original articles.

Before discussing specific articles, we must acknowledge that there are several related bodies of literature that we do not discuss. First, we do not discuss the numerous articles that simply mention individuals with disabilities as one of the possible groups of users that may benefit from the results because this is beyond the scope of this chapter. For example, many articles discussing multimodal interfaces mention potential benefits for individuals with disabilities at some point, but few are designed specifically for this group of users (see chapter 14 for an introduction to multimodal interfaces and pointers to the relevant research). Second, we do not discuss those articles that focus on the more general problem of designing accessible computer systems without a significant focus specifically on PI. For articles of this nature, the reader is referred to the proceedings of the Association for Computing Machinery's ASSETS (2000) and Universal Usability (2000) conferences and the European Research Consortium for Informatics and Mathematics (2001)

Workshops on User Interfaces for All. Stephanidis (2001) also discussed these issues. We encourage the reader to review this literature because it often provides valuable insights into the methods that can be used to more effectively integrate accessibility concerns into the overall development process. Finally, we do not discuss the extensive literature available in the rehabilitation and assistive technology communities. As with the literature that discusses accessibility from a more general perspective, we encourage the reader to explore the rehabilitation and assistive technology literature. More specifically, we refer the reader to the proceedings of the annual Closing the Gap conferences (Closing the Gap, 2001), Rehabilitation Engineering and Assistive Technology Society of North America conferences (RESNA, 2001), and the California State University—Northridge (2001) conferences. The *IEEE Transactions on Rehabilitation Engineering* often includes articles discussing the technical details of new devices and various books can provide valuable insights (e.g., Gray, Quatrano, and Liberman, 1998; King, 1999). Several books and Internet sites also can serve as useful starting points when exploring these issues (e.g., Edwards, 1995; Stephanidis, 2001; Trace Center, 2001).

Spinal Cord Injuries

SCI are one of the most frequently studied health conditions in the HCI community. Given the relationship between the level of injury and the resulting impairments, studies typically include individuals with injuries that are no lower than C7. The following articles discuss various topics including cursor control, validating keystroke-level models, a brain-controlled interface, and the use of speech recognition for communications-oriented activities.

Casali (1992) presented a study investigating the efficacy of five cursor control devices when used by individuals with SCI. Twenty individuals with SCI and 10 individuals with no PIs participated in this study. Participants with SCI were divided into two groups using a custom assessment test (see Casili, 1995; Casali & Chase, 1993) that evaluates upper extremity motor skills. As a result, three groups of participants were discussed: low motor skills, high motor skills, and participants with no PIs. Each participant used a trackball, mouse, tablet, and joystick in addition to the cursor keys to complete a series of tasks that required either selecting or dragging a target. Both target size and distance were varied systematically. Several configurations were made available for the mouse, tablet, and joystick to accommodate the abilities of the participants.

Numerous statistics are reported detailing significant differences due to group, device, target size, distance, and mode (selection vs. dragging). Numerous significant interactions were also identified. The reader is referred to the original article for the complete details, but several particularly interesting results are summarized here. First, participants in the low motor skill group took longer to complete the tasks than the individuals with no PI. In contrast, there was no significant difference between the participants in the high motor skill group and the participants with no PI. Second, the rank ordering of the devices (with respect to target acquisition time) was the same for all three groups. For all three groups, the mouse, trackball, and tablet resulted in the shortest task completion times, the cursor keys took longer, and the joystick was the slowest of the five devices. Third, target size becomes more important as motor skills decrease. Interestingly, the authors noted that all participants, even those in the low motor skill group, were able to complete the tasks with all five devices with minimal customization. However, they also noted that individuals in the low and high motor skills groups experienced substantial difficulty when using the mouse. More specifically, the majority of these individuals found holding the mouse button while moving the mouse (e.g., dragging an object) difficult and ended up relying on a toggle switch customization that was made available.

Koester and Levine (1994) investigated the ability of keystroke-level models to predict performance improvements that individuals with SCI would experience when using word-prediction software to enter text. Word-prediction software can reduce the number of keystrokes required to enter text but also increases the cognitive and perceptual demands placed on users. This tradeoff should prove useful for individuals with PIs, but empirical studies confirming these benefits are lacking. In this article, the authors described a study that provides insights into (a) the benefits word-prediction software can provide and (b) the effectiveness of keystroke-level models for predicting performance by individuals with PIs.

Eight individuals with no PIs participated in the study. All of these participants used a mouthstick to interact with the keyboard. None had previous experience with a mouthstick. Six individuals with high-level SCI also participated. These participants used their normal method of interacting with a keyboard (two used a mouthstick, four used hand splints). They completed the tasks with and without word prediction to allow performance improvements to be evaluated. Two interaction strategies were explored when using word prediction: (a) searching the list of possible words before every keystroke and (b) entering the first two letters and then searching the list of words before every subsequent keystroke. Possible keystroke savings ranged from approximately 21 to 41% for the first strategy and 15 to 30% for the second strategy, depending on the task. Finally, two models were explored: (a) a generic model based on parameters derived from the existing literature combined with new data from individuals with no PI and (b) a user-driven model based on parameters derived from the data collected in this study.

Extensive statistics are reported describing the accuracy of the generic and user-driven models for predicting differences in data-entry rates that would be observed as a result of the two interaction strategies. As may be expected, the user-driven models were more accurate than the generic models. User-driven models resulted in errors of approximately 6 percentage points across all participants, whereas the generic model resulted in errors of 53 and 11 percentage points for the participants with and without PIs, respectively. Clearly, using data that more accurately reflects the performance of the individuals being modeled results in more accurate models. For example, the user-driven models incorporated data that highlighted several important differences that were missed by the generic model: participants

with PIs who were experienced with the input devices were faster at typing individual characters than the novice nonimpaired participants, actual keypress times are longer when using word prediction (especially for the participants with PIs), participants with PIs took longer to search the list of alternative words than nonimpaired participants. Perhaps more interesting than the model-accuracy results were the changes in data-entry rates observed for the two groups of participants. Overall, the models predicted modest improvements in performance (ranging from slight decreases to improvements of over 40%). Results for nonimpaired participants followed this same pattern, but the participants with PIs showed a consistent, and large, decrease in data-entry rates when using word prediction (ranging from approximately 20–50% reductions in data-entry rates). Also of interest are differences between the two groups of participants in terms of the effectiveness of the two interaction strategies used when using word prediction. Nonimpaired users were, on average, 4% slower with Strategy 2 (enter two characters, then search the word list), whereas users with PIs were almost 11% faster with Strategy 2.

Mason, Bozorgzadeh, and Birch (2000) described a brain-controlled switch that can be activated by imagining movement. Earlier articles described studies in which this technology was used by individuals with no PIs (Lisogurski & Birch, 1998; Bozorgzadeh, Birch, & Mason, 2000). These studies demonstrated that this technology could effectively identify both real and imagined finger movements. In this article, the authors described a study that involved two individuals with high-level SCI. Neither participant had any residual motor or sensory function in their hands. Their results indicated false-positive rates of less than 1% (i.e., the system detected activity when none was intended) and hit rates of 35 to 48% (i.e., the system detected activity when it was intended). Although higher hit rates are desirable, these results were viewed as positive because they confirmed that individuals with no residual motor or sensory function in their hands could activate the system by imagining finger movements. Furthermore, these results were obtained with relatively little training and technology that was not customized to the individual participants. As described later in Section 6.6, significant research is underway that explores the possibility of using electrophysiological data (as was done in this study) as input to computing systems.

Sears et al. (2001) discussed the results of an experiment designed to explore the effectiveness of speech recognition for communications-oriented activities when used by individuals with SCI as well as individuals with no PIs. The individuals with SCI all had injuries at or above C6 with ASIA scores of A or B, resulting in either no use or limited use of their arms and in their hands being paralyzed. After training and practice, participants completed several composition and transcription tasks. An analysis of data-entry rates, error rates, and the quality of the resulting documents provided interesting and encouraging results. Overall, there were no significant differences between the two groups of users. On average, the participants generated text at approximately 13 words per minute (wpm) with few errors. An analysis of subjective satisfaction ratings revealed several differences, with the participants with PIs exhibiting significantly more positive attitudes with respect to the time required, the effort involved in correcting errors, and the overall ease of use of the software. A more detailed analysis revealed significant differences with respect to the processes employed by the two groups of participants. For example, the participants with PIs spent more of their time dictating and interrupted their dictation more often to correct errors. As a result, they also spent less of their time navigating from one location to another within the document. Oseitutu, Feng, Sears, and Karat (2001) provided additional details that highlight the different strategies adopted by these two groups of participants.

Cerebral Palsy

CP can result in a variety of impairments due to reduced muscle power or a loss of control over voluntary or involuntary movements. In addition, CP can result in both impaired speech and intelligence. As a result, designing technologies for individuals with CP can be particularly challenging. In this section, we discuss the results of a single study that was motivated explicitly by CP. Additional studies discussed later (see Cursor-Control Technologies) may have been motivated by CP. For example, Alm, Todman, Elder, and Newell (1993) describe a system designed for individuals with severe speech and motor impairments including an evaluation that included one individual with CP.

Roy, Panayi, Erenshteyn, Foulds, and Fawcus (1994) discussed the design of a gestural interface designed for individuals with speech and motor impairments due to CP. Fourteen students with CP, aged 5 to 17, were observed during their regular school schedules. They observed numerous communicative acts that combined facial expressions, eye gaze, vocalization, dysarthcic speech (slurred, slow, difficult to produce, and difficult to understand), and upper extremity gestures using the head, arms, hands, and upper torso. Subsequently, four students participated in sessions in which data were collected for a predefined set of gestures (using a three-dimensional magnetic tracker and electromyographic electrodes). Computer recognition was tested for two similar gestures, providing encouraging recognition rates of approximately 96% and 93%. An evaluation of the complete set of gestures, or the subsequent use of such gestures by individuals with CP, was not reported in this article; however, the preliminary data provides encouragement with respect to the use of computer-based gesture recognition as a communications tool for individuals with CP.

Impairments of the Hands and Arms

Several studies have focused on the difficulties individuals experience because of movement or muscle power PIs that affect their hands and arms. These studies often include participants with a range of underlying health conditions (e.g., CP, SCI, stroke) as well as varying PIs. The unifying theme for these studies is that they all focus on the ability of the study participants to interact with computers using the arms and hands.

Trewin and Pain (1999) discussed the errors that occur when individuals with PI that affect their hands and arms interact with a keyboard and mouse. The authors began by acknowledging the difficulties individuals with PIs may experience using a keyboard and mouse but proceeded to highlight the lack of empirical data regarding the nature and frequency of these difficulties. Twenty individuals with various PIs and six individuals with no PIs participated in this study. The difficulties each individual experiences are associated with either a health condition (e.g., stroke, CP, spina bifida) or a PI (e.g., wrist stiffness, coordination loss, spasms), and the method each individual normally used to type was listed (e.g., right hand only, several fingers on both hands, thumb and first two fingers of left hand). Each participant completed three tasks: one used the keyboard, one used the mouse, and one used both the keyboard and mouse. Several important categories of typing errors were identified, with each type of error occurring more often for the participants with PI. The following six categories account for the majority of performance errors (in decreasing order of frequency):

- *Long key presses:* A key was pressed longer than the default key repeat delay and would have resulted in multiple characters being entered if the default delay were used. Participants with PIs pressed keys longer and the duration of keypresses was more variable. This was the most common problem, with 2,610 keypresses lasting longer than the default key repeat delay; however, this does not represent the number of keypresses that resulted in multiple keys because key repeat was disabled for many users and the key repeat delay was extended for the remaining participants.

- *Additional keys (local):* These are errors where a key near the desired key was accidentally activated by the body part being used to type. Of the 265 errors that fall in this category, almost 98% involved an adjacent key being pressed.

- *Missing keys:* A movement intended to generate a character failed. This could be because some other key was pressed (additional key error) or because insufficient force was applied to the key. This error occurred 179 times.

- *Simultaneous keys:* The participant failed to press two keys simultaneously when necessary (e.g., using the shift key). Participants with PI found it difficult to press two keys simultaneously, resulting in 56 errors. Nine participants used the caps-lock key as an alternative to pressing two keys simultaneously, and one participant simply omitted capital letters and punctuation. Another participant indicated that he normally avoided capital letters unless absolutely necessary, but did enter them during this study.

- *Bounces:* A key was accidentally pressed more than one time. Forty-four bounce errors occurred, but reasons for these errors are not identified.

- *Additional keys (remote):* These are errors in which a key that was not near the desired key was accidentally activated by a body part that was not being used to type. Thirty-seven instances of remote additional key presses were identified. Most often this involved pressing an extra key on the bottom row of the keyboard.

Several interesting results were also reported with regard to the mouse-based tasks:

• *Pointing:* Simply pointing at targets was difficult. Participants with PIs took almost 3 times longer than those with no PIs when pointing at targets of various sizes. Although participants with no PI did not make any pointing errors, those participants with PIs made many errors. Seventy percent of the participants with PIs had error rates greater than 10%, and 40% of the participants had error rates in excess of 20%.

• *Clicking on a target:* The mouse down and mouse up positions were not identical for only 6% of the mouse clicks by the participants with no. More than 28% of the mouse clicks by participants with PIs involved some kind of movement.

• *Attempts to click the mouse button more than one time:* Participants with PIs failed at this task more than 39% of the time, whereas those participants with no PIs failed 9% of the time. Interestingly, the reasons for the failures differed between the two groups of users. For participants with no PIs, most failures (66%) were caused by clicking the wrong number of times, with the remainder (33%) resulting from missing the target. In contrast, for participants with PI, 33% occurred when the participant moved the mouse during the clicks, 27% resulted from missing the target, 25% involved the wrong number of clicks, and 14% happened because the user clicked too slowly.

• Participants with PIs took longer to complete dragging tasks than participants with no PIs. Participants with no PIs experienced a failure rate of 5%, with all failures occurring because they missed the target. Participants with PIs were more error prone, with 55% of their attempts failing; 40% of the failures involved missing the target when releasing the mouse button, and almost 34% involved releasing the mouse button accidentally while dragging the object, 14% involved getting stuck (e.g., moving the mouse to a location where they were no longer physically capable of completing the dragging task) and having to give up, and 11% involved situations in which the user accidentally caused the window to scroll by dragging the object out of the current window.

The difficulties identified in this study provide valuable insights into the experiences of individuals with PIs. Numerous additional details are provided in the original article that could guide future research aimed at making the standard keyboard and mouse more effective for individuals with PIs. For additional related work, see Trewin (1996) and Trewin and Pain (1998a, 1998b).

Keates et al. (2000) discussed the use of haptic feedback for individuals with PIs. More specifically, they investigated the potential of force feedback to facilitate point-and-click activities. They report on two pilot studies that focus on individuals with movement and muscle power PIs but unaffected sensitivity of touch. Six individuals with various health conditions that result in PIs participated in the studies. Four participants had choreoathetoid CP, one had Friedrich's ataxia, and one had Kalman-Lamming's syndrome. Participants completed a series of point-and-click activities. In the first study, the efficacy of several alternative forms of feedback were explored: a pointer

trail, changing the color of the target when the cursor is over it, force feedback gravity wells that draw the cursor toward targets, vibrating the mouse when the cursor is over the target, or a combination of all of these forms of feedback. Their results indicated that gravity may reduce target selection times by 10 to 20%. Similar benefits were also seen for pointer trails, a standard accessibility option in the Windows operating systems. Vibrating the cursor increased target selection times dramatically, with all participants expressing a dislike for this form of feedback. Interesting, these results are similar to those reported for participants with no PIs using the Phantom input device (Oakley, McGee, Brewster, & Gray, 2000).

In Keates et al.'s (2000) second study, the relationship between force feedback and target size was explored. Force feedback reduced the time required to complete the target selection tasks by 30 to 50% and error rates by approximately 80%. Without force feedback the time required increased as target size decreased. This pattern was greatly reduced when force feedback was provided. Interestingly, at least one participant was unable to complete the tasks without force feedback but could perform the required activities when force feedback was provided. These results suggest that force feedback could prove useful for individuals with PI. It should be noted, however, that these results may be unique to the specific form of force feedback used in this study. Careful evaluation of other forms of force feedback is still necessary.

Keates, Clarkson, and Robinson (2000) discussed the use of existing user models for describing the behavior of computer users with PI. The concern is that these models tend to be calibrated based on performance of computer users with no PIs. The article investigates the use of the model human processor (Card, Moran, & Newell, 1983) for describing the behavior of computer users with PIs. This model combines perceptual, cognitive, and motor activities to predict the time required to complete a task. Tasks were developed to allow the time required for fundamental perceptual, cognitive, and motor activities to be determined for two groups of users. The first group consisted of six individuals with PIs caused by various health conditions (i.e., quadraplegia, MD, spastic CP, choreoathetoid CP, and Friedrich's ataxia). A second group of three individuals with no PI also participated. Their results are summarized in Table 25.4.

Participants with PIs appear to take longer to complete fundamental perceptual and cognitive activities. These results are consistent with those discussed above (Koester & Levine, 1994) but should be interpreted carefully given varied health conditions of the participants. In particular, the fact that individuals with CP participated and that CP can affect more than just physical activities suggests that a more detailed review of the results is necessary. Fortunately, the authors provided the results for

TABLE 25.4. Average Time Required by Individuals With and Without Physical Impairments (PIs) to Complete Fundamental Perceptual, Cognitive, and Motor Tasks

	Perceptual	Cognitive	Motor
No PI	80 ms	93 ms	70 ms
PI	100 ms	110 ms	110 ms, 210 ms, 300 ms

each participant, and a careful review suggests that the additional time required for perceptual and cognitive activities is not due to the participation of individuals with CP. Less surprising is the additional time required for motor activities. It is also not surprising that the time required by participants with PI varied dramatically, with the two slowest participants requiring almost 3 times as long as the two fastest participants. These results confirm the need to develop user models that are specific to the individuals who may use the system.

Through observation and an analysis of the different times recorded for motor activities, the authors concluded that additional cognitive or perceptual activities are taking place when individuals with PIs complete basic motor tasks (as compared with individuals with no PIs). Based on additional data, the authors suggested that additional cognitive activities are the likely explanation. Although their results are not definitive, they confirm the need for additional research that focuses on the time required to complete basic perceptual or cognitive activities. These results also demonstrate the need to study the activities that occur when individuals with PIs perform basic motor activities.

Impairments in Infants

Infants with severe PIs often develop to be passive, with limited or nonexistent speech. This is true even when cognitive skills are normal. Fell, Delta, Peterson, Ferrier, Mooraj, and Valleau (1994) discussed the development and evaluation of the Baby-Babble-Blanket as a technology to enable infants with severe PIs to control their environment and communicate. One goal was for this technology to help these infants improve their motor skills. The key component of the system is a pad, containing 12 equally spaced switches, that can be placed on the floor. When a switch is pressed, a computer provides feedback. The feedback varies but can include digitized speech, music, and sounds. The authors reported the results of a single-subject experiment involving a 5-month-old infant with poor muscle tone, hydrocephaly, and club feet. This infant was able to activate switches, became more active when sounds were played in response to switch activations, associated switches with a desired effect (i.e., hearing his mother's voice), and was able to modify these associations when the feedback was moved from one switch to another.

Significant Speech and Physical Impairments (SSPI)

Although the terminology may vary, several studies discuss technologies for individuals with significant speech and physical impairments. The article by Roy et al. (1994) discussed earlier could have been included here but was listed separately because of its explicit focus on impairments caused by CP. Of the three articles summarized below, one presents the results of an evaluation including a single participant with CP, whereas the other two discuss technologies without explicitly mentioning any health conditions. An individual's ability to speak can be impaired when PIs affect the muscles involved in producing speech or when the cognitive processes involved in producing speech are impaired.

Alm et al. (1993) described the development of a computer-aided communication technology for individuals with SSPI. This article focuses on situations in which the speech impairment results from physical rather than cognitive impairments. Under these conditions, the individual can understand incoming communications but has difficulty expressing thoughts. Although most existing systems store words (and require the user to construct sentences) or a limited number of prerecorded phrases or sentences, the authors described a system that tracks the conversation and helps the user select the next thing to say. The system allows users to specify the type of utterance they wish to generate using a limited number of parameters. For instance, the user can select greetings or wrap-up comments. In a prototype system designed to support conversations about vacations, the user specified the person (i.e., me or you), time (i.e., past, present, future), and topic (e.g., who, where, when), and a set of possible utterances were displayed.

The system was evaluated with two participants. The evaluation system was loaded with 1,600 utterances developed to support conversations about vacations. For comparison purposes, three person-to-person conversations were recorded. The mean rate of speech was 144.4 wpm. One participant with no PIs took part in eight conversations with student volunteers. The mean rate of speech was 88.2 wpm. The participant speaking with the assistance of the computer spoke 67.4 wpm, whereas the conversation partners spoke 132.9 wpm. Although selecting an utterance could require one to four clicks, only 2 of 422 selections actually involved more than two clicks. This suggests that the desired utterances were easily accessed given the current design of the system. Informal evaluations of the conversations yielded positive results, suggesting that further study is appropriate.

The second participant had CP. His primary method for communicating was a word board with 400 words that he could point to. He augmented this with some gestures, limited vocalizations, and a word and phrase storage device that could be used for names of people, places, and other special words. Because of his difficulties reading, the interface was simplified by reducing the number of text windows and reducing the options available to change perspectives (e.g., person, time, topic). The system was used to augment his normal communication techniques. The results indicate that the system allowed the participant to express himself more fully, increasing his vocabulary from 143 to 534 words and the number of times he took control of the conversations from 10 to 27.

Many important questions must still be addressed for systems such as the one described in this article. One critical issue is the effectiveness of such a system as the number of conversation topics increases. As more topics are discussed, the number of phrases increases, and the task of managing this information will become more complex. Another critical issue is how the user can efficiently enter and organize these phrases as they customize the system for use in new situations.

Demasco, Newell, and Arnott (1994) discussed the development of system to support communications activities that is based on principles of visual information seeking. Building on five key principles derived from the literature, a computerized word board application was designed. Key features include the separation of navigation and selection, which allows these activities to be supported by different devices and expanding the collection of words that are accessible using word associations (e.g., is–like, is–a, has–a, goes–with). At the time the article was written, the system was under development and a variety of directions for research were presented. The authors also provide an interesting discussion of the limitations of existing computer-based communications systems, including those that use word prediction, coding systems, and level-based systems.

Albacete, Chang, Polese, and Baker (1994) also discussed the design of iconic languages for use by individuals with SSPI. This approach is based on semantic compaction in which ambiguous icons are combined to represent concepts. For example, "APPLE VERB" could mean *eat* while "APPLE NOUN" means *food*. An icon algebra is defined, including operators to combine, mark, provide context for, enhance, or invert icons. The approach described in this system can allow a vocabulary of several thousand words to be represented with 50 to 120 icons. This approach served as the foundation for the Minspeak system.

Input Using Electrophysiological Data

Traditional input techniques involved some kind of physical activity. Most often, the hands and arms are involved, but other technologies allow users to provide input by moving their eyes, heads, tongue, or eyelids (see Cursor-Control Technologies). Recently, a number of researchers have been investigating input techniques that do not require physical activities. The article by Mason et al. (2000) discussed earlier provides one example in which electrophysiological data were used to generate input to a computer. In this section, several additional articles that discuss the use of electrophysiological data are discussed. This is a new, exciting, and promising form of input for individuals with significant PI.

Patmore and Knapp (1998) discussed a new approach to eye tracking that uses the electrooculogram (EOG) and visual evoked potentials (VEP) as input. Although a number of EOG-based eye-tracking systems have been discussed, additional research is needed to make such systems reliable (see LaCourse & Hludik, 1990; Patmore & Knapp, 1995). The VEP is a response to a flash of light that is highly sensitive to the stimulus' distance from the center of the field of vision. The authors presented the technical details of system that combines the EOG, the VEP, and fuzzy-logic to improve accuracy. Although promising, significant additional research is necessary before such a system will be sufficiently reliable for use under realistic conditions.

Allanson, Rodden, and Mariani (1999) described a toolkit designed to explore the use of electrophysiological data in interactions with computers. They discussed the variety of electrophysiological data available including the electroencephalograph (EEG), electromyograph (EMG), and galvanic skin resistance (GSR). The EEG is an electrical trace of brain activity, the EMG can detect the state (i.e., complete relaxation, partial contraction, complete contraction) of a muscle, and the GSR measures changes in the resistance of the skin. The authors provided several interesting references describing the use of the EEG to turn switches on and off (Kirkup, Searle, Craig, McIsaac, and Moses, 1997) and differentiated between five tasks (Keirn & Aunon, 1990) and the EMG to control computer games (Bowman, 1997) or manipulate objects on a computer (Lusted & Knapp, 1996). Their toolkit integrates these signals using predefined widgets that allow the signals to be used to drive applications.

Kübler et al. (1999) discussed the use of slow cortical potentials in a 2-s rhythm to control cursor movements on a computer screen. Three individuals with advanced ALS participated in extensive practice sessions to learn to control this signal. Ultimately, two of the three reached accuracy rates between 70 and 80%, which allowed these individuals to select letters and words displayed on a computer screen.

Doherty et al. (2000) discussed the use of formative experiments and contextual design to assess the potential of applications controlled by eye movements, the EOG signal, and the EEG signal (i.e., the Cyberlink interface). An initial study involving 44 participants with various PIs demonstrated some expected limitations (e.g., the mouse was not effective when arm and hand control was limited; eye tracking was not effective when peripheral vision was impaired) as well as unexpected difficulties (e.g., some quadriplegic participants could not produce signals below the neck and therefore could not operate the GSR device). Interestingly, the Cyberlink interface was the only alternative that all 44 participants could use to navigate a maze. Additional studies were used to address several practical and technical issues identified in the earlier phases of the project. Overall, their results confirm that interfaces built upon electrophysiological data can be useful for individuals with PI but that substantial additional research is necessary before the full potential of these technologies will be realized.

Kennedy, Bakay, Moore, Adams, and Goldwaithe (2000) discussed an experimental interface that requires the implantation of special electrodes into specific areas of the brain. At present, three individuals have had the electrode implanted. The first individual died from her underlying disease, and the third had had the surgery too recently to provide useful results. Therefore, results from the second individual are described. For this individual, EMG signals from various muscles are used to supplement the neural signals detected by the electrode. With 5 months of practice, the individual learned to control cursor movements. Results for three tasks are reported. These tasks involved moving the cursor across the screen, placing the cursor on an icon or button, and selecting the icon or button by holding the cursor still for a predefined period of time (or activating a predefined EMG signal). Through these tasks, this individual was able to invoke synthesized speech for predefined messages, spell names, and answer questions. As with the other articles summarized in this section, this article demonstrates the potential of electrophysiological data while confirming the need for additional research.

Cursor-Control Technologies

Cursor control and text generation are perhaps the two most common activities when interacting with computers. In fact, cursor control can form the foundation for text generation and therefore may be the most fundamental task in which individuals with PIs engage. Numerous technologies have been explored for controlling the location of the cursor on a computer screen and many of the articles summarized earlier focus on cursor control tasks. Although new eye-tracking technologies are being explored, commercial systems are readily available and can prove useful for individuals with PI. The issues involved in using these technologies are not discussed in this chapter, but Sibert and Jacob (2000) provided useful pointers to the recent literature on this topic. In the following sections, technologies are discussed that allow individuals to control the cursor using their head, tongue, eyelids, or speech.

Head. For individuals who have limited use of their hands and arms, other options must be found for interacting with computers. One option is to control the cursor location using head movements. In this section, we discuss several articles that explore the issues involved in interacting with computers via head movements.

Radwin, Vanderheiden, and Lin (1990) discussed the results of a study comparing a standard mouse to a lightweight ultrasonic head-controlled pointing device. Ten participants with no PIs completed a variety of tasks that required the selection of circular targets. The size of the targets and the distance to the targets were systematically varied. Radwin et al. reported movement times, cursor path distance, and deviation of the cursor from the optimal path. Distance, target size, and device all significantly affected movement times. Interestingly, target size had a greater effect for the head-controlled device, with movement times decreasing more dramatically than when the mouse was used. For example, moving from the small to medium sized targets reduced movement times by 249 ms when using the mouse and 525 ms when using the head-controlled device. The direction of the movement also affected movement times, with the horizontal movements resulting in the shortest times for the mouse and vertical movements resulting in the shortest times for the head-controlled device. Interesting results are also reported for the cursor path distance and amount the cursor deviated from the optimal path when the head-controlled device was used: Vertical and horizontal movements resulted in shorter paths and smaller errors than diagonal movements.

Results are also reported for two individuals with CP with details provided for one of these individuals. The results for this individual highlight the importance of carefully assessing the abilities of individuals with PI when designing computer systems for their use. For participants with no PI, vertical and horizontal movements were the most efficient with diagonal movements creating more difficulty. For this participant with CP, movements to the right resulted in more difficulty than any other direction. Interestingly, providing torso support allowed this individual to complete the tasks more efficiently, with the greatest improvement being for those tasks that involved moving to the right.

Lin, Radwin, and Vanderheiden (1992) reported on a similar study that focused on the relationship between control-display gain and performance with both mouse and head-controlled device. Ten participants with no PIs participated in this study.

Movement times and deviation of the cursor from the optimal path were reported. Extensive analyses of the results are provided, highlighting the importance of target size, movement distance, and gain. Gain had an effect on movement times, but this effect was smaller than that of target size and movement distance. A U-shaped curve was observed for both the mouse and head-controlled device. The optimal mouse gain was between 1.0 and 2.0, whereas the optimal gain for the head-controlled device was between 0.3 and 0.6. The reader is referred to the original article, as well as to Schaab, Radwin, Vanderheiden, and Hansen (1996), for additional information regarding the relationship between control-display gain and performance with head-controlled devices.

Malkewitz (1998) described a system that combines head movements to the control cursor location with speech to click the mouse buttons, activate hotkeys, and emulate the keyboard. The goal was to provide access to standard applications by emulating standard input devices (i.e., keyboard and mouse). Speech recognition is restricted to a predefined set of commands and the ability to enter individual letters. Full dictation is not supported for various reasons. Although preliminary results were encouraging, the authors noted the need for additional research.

Evans, Drew, and Blenkhorn (2000) described the design of a head-operated joystick. Unlike many other projects that investigated the use of head-controlled devices that effectively emulate a mouse, this article explores a device that emulates a joystick. Also unlike most of the articles summarized in this chapter, this article focuses on the technical details of the device and provides little information about its use. The authors briefly summarize the results of an evaluation involving 40 participants (9 with undocumented impairments). It is reported that all 40 individuals used the device successfully, but detailed measures were not provided. The nine individuals with impairments used not only this device, but also a commercially available device that emulated a mouse. Interestingly, all nine preferred the joystick emulation provided by this new device over the mouse emulation provided by the commercial device. It is speculated that this preference is due, at least in part, to the fact that users can rest their head in a neutral position once the cursor is placed in the correct location when using a device that emulates a joystick.

LoPresti et al. (2000) conducted a study that investigated the relationship between neck range of motion and the use of head-controlled pointing devices. Fifteen individuals with no PIs participated in the study. Ten individuals with various health conditions that resulted in PIs affecting neck movements also participated. The individuals with no PIs were confirmed to have greater neck range of motion than those participants with PIs. All participants completed both target tracking and icon selection tasks. A head-mounted display was used as opposed to a traditional desktop monitor. The results confirmed that the individuals with PI were less accurate and took longer when completing the icon selection tasks. They were also less accurate when performing the target tracking task. Furthermore, their performance was much more variable than that of the participants with no PIs. As with several other studies, the authors found that models developed based on individuals with no PIs do not accurately represent the performance of individuals with PIs. A more detailed review of the results indicated that vertical movements were faster than horizontal movements and that horizontal movements were faster than diagonal movements. LoPresti (2001) reported similar results.

Tongue. Salem and Zhai (1997) provided a brief description of a system that allows the cursor location to be controlled using a tongue-operated isometric joystick like those available on many portable computers. A Trackpoint joystick was mounted in a mouthpiece, similar to those used by athletes that is fitted to the individual's upper teeth. Two individuals participated in a pilot study. Neither had experience with an isometric pointing device. Both used the tongue-controlled joystick as well as a standard finger-controlled joystick. Participants completed multiple trials involving 15 tasks. For each task, participants selected 1 of 20 buttons as quickly as possible. Initial performance was substantially better when using their fingers, but with practice the gap between the two devices narrowed. During the last trial, the tongue-based device was 5% slower than the finger-based device for one participant and 57% slower for the other. The authors suggested that with additional research performance with the tongue-based joystick could improve further.

Eye Lid. Shaw, Loomis, and Crisman (1995) presented a system designed to allow individuals to control various devices, including computers, by opening and closing their eyes. Both eyes are monitored and the duration of each eye closure is identified as blink, short, long, or super long. Audio feedback is provided to help users with timing their input. The application described in this article is the control of a powered wheelchair, but this technology could also be used to interact directly with computers. In fact, computer control is demonstrated through a simulation program that users interact with before controlling a wheelchair. Two individuals used the system. The first (Participant A) had a high-level SCI resulting in no residual use of his arms or hands. The second (Participant B) had a stroke and could only control a single eyelid. With 6 hours of practice, Participant A was able to move from interacting with the simulation software to using the powered wheelchair. He found this device to be less obtrusive than the chin control he normally used to control his wheelchair. After 12 hours of practice, Participant B was able to navigate the maze presented by the simulation software and made progress toward controlling the powered wheelchair. Given the minimal motor control this individual exhibited, this is a noteworthy accomplishment.

Speech. Manaris and Harkreader (1998) investigated a speech-based interface for accessing all of the functionality provided by a computer. Their system, SUITEKeys, allows users to access the functionality of a standard keyboard and mouse using speech recognition. Users can access the keyboard by saying *press <key name>*, *release <keyname>*, or simply *<key name>* to initiate a complete keystroke. Saying *repeat <key name> <number of repetitions>* produces multiple keystrokes. The cursor can be moved by saying *move <direction>* to move at a slow speed until another command is issued, *stop* to stop the cursor where it is, *move <direction> <distance>* to move a specified distance, or *position <area>* to move the cursor to the center of five predefined regions of the screen. Clicking mouse buttons can be accomplished by saying *press <button name>*, *release <button name>*, *click <button name>*, or *double-click <button name>*. A pilot study involving three participants with PI that interfered with the use of a standard keyboard and mouse is described. Each participant completed a data-entry task using their normal input technique (i.e., a keyboard and mouse, directing an assistant to perform the necessary actions, using the keyboard via a mouthstick) and using SUITEKeys. The results indicate that even with minimal practice, these three individuals were able to achieve reasonable data-entry and accuracy rates with the SUITEKeys system. Additional study is required to evaluate the effectiveness of this system with additional practice.

Text Entry Technologies

Computers are frequently used to generate text. Although the keyboard is the most frequently used alternative, other technologies can also be used to generate text. Some individuals with PI select letters from on-screen keyboards, others use speech recognition, and still others interact with the standard keyboard. When using an on-screen keyboard, text entry is effectively turned into a cursor-control task. As a result, all of the cursor-control technologies discussed earlier can also be used for text entry. The article by Sears et al. (2001) discussed the use of speech recognition for text entry, but was included in the section on spinal cord injuries because of its primary focus. All three articles in the section on SSPI could be included here but are discussed separately because of their focus on individuals with a specific collection of impairments. The articles included below discuss technologies that can be used for text entry without focusing on any health condition or PI.

Keyboard-Based Text Entry. Matias, MacKenzie, and Buxton (1993) described a one-handed keyboard based on the traditional QWERTY design. The basic design allows users to access all of the keys on one side of the keyboard at a time, switching between sides by pressing and holding the spacebar. A study with 10 participants with no PIs is reported. Each participant used their nondominant hand when typing with one hand. Participants completed 10 sessions, with no more than one session per day. Each session included a two-handed pretest, several blocks of one-handed typing, and a two-handed posttest. Speed and errors improved significantly over the 10 sessions. By the end of 10 sessions, participants were typing 34.7 wpm with an error rate of 7.44% with the one-handed keyboard compared with 64.9 wpm and 4.20% errors with the two-handed keyboard. This design was intended to leverage existing knowledge and skill associated with the QWERTY keyboard. While this system could prove useful for individuals with PIs that hinder the use of one hand, the efficacy of this design for individuals with limited experience using the QWERTY keyboard is uncertain.

Many researchers have investigated techniques that allow users to generate text using fewer keystrokes than is normally required. These techniques typically predict words given just a couple of letters using statistics that describe how frequently various words are used. Some systems also use statistics regarding how frequently various words follow words that have already been entered. Other techniques present multilevel interfaces where users repeatedly select categories until the desired word is available. Once the word is selected, an appropriate ending (e.g., "s" or "ing") can be added. Demasco and McCoy (1992) presented a theoretical discussion of a word-based virtual keyboard that uses a level-based interface. McCoy et al. (1994) discussed a word-based text entry technique in which users specify uninflected content words and the system adds appropriate endings to the words as well as additional words (e.g., articles, prepositions) to complete the sentence. Garay-Vitoria and González-Abascal (1997) discussed the potential of syntactic analysis of previously entered words to enhance word-prediction results. Boissiere and Dours (2000) described a system that predicts word endings without assistance from the user. Unlike many of the articles discussed earlier, these take either a theoretical or systems approach to the problem. Consequently, some of the systems have not been implemented. Systems that are implemented are described, but user-based evaluations are not reported.

Speech Recognition–based Text Entry. Goette (1998) conducted a field study to identify factors that influence the successful adoption of speech-recognition software for both dictation-oriented and environmental control tasks. Individuals with various health conditions (e.g., MS, MD, SP, arthritis) participated. Most individuals who stopped using the speech-recognition software (53%) did so within 3 months. Interestingly, those individuals who successfully adopted the software had higher expectations for the system and expected greater benefits but also believed the system would be easy to use compared with those individuals who stopped using the software. When SR was successfully adopted, it was used for a wide variety of computer-based tasks rather than a few isolated activities. Four guidelines were proposed for successful outcomes: managing expectations by understanding the potential benefits and limitations, selecting the correct system for the tasks to be accomplished, obtaining thorough training, and trying the system for an extended period of time before purchasing it.

CONCLUSIONS

The most obvious conclusion is that additional research in designing computer technologies for people with physical impairments is necessary. PIs vary dramatically in terms of severity, temporal variability, and the body parts that are affected. Numerous technologies can be used in a variety of configurations. Given this variability and the limited number of studies reported in the literature, it is clear that only a fraction of the important questions have been investigated. For example, speech recognition is often recommended for use by individuals with PI, but little is known about how these potential users actually interact with this technology. At present, a single study as been reported that evaluated the use the speech recognition by individuals with PI for dictation-oriented activities (Sears et al., 2001). The opportunities for additional research are unlimited.

Unfortunately, conducting informative studies including individuals with PIs can be difficult. Unlike traditional computer users with no PIs, the pool of potential participants is limited. Even when appropriate participants can be found, additional factors can make such studies difficult. For example, individuals with PIs may need more frequent breaks, may be able to participate for less time, and may find it more difficult to arrange transportation to the site where the study is occurring. Finally, carefully documenting health conditions and the associated PIs often requires adding a physician, an occupational therapist, or both to the research team.

Although additional research is needed, the existing literature does provide insights that can prove useful to both practitioners and researchers. Individually, the articles highlight the potential benefits and limitations of various technologies for specific groups of individuals as well as new technologies that may prove useful for specific tasks or individuals in the future. When viewed as a whole, several lessons that extend beyond the technology or user groups studied become clear. We conclude by highlighting four such lessons, emphasizing the potential of new technologies that will provide alternative methods of interacting with computers, and stressing the need for additional research investigating the potential benefits as well as the limitations of existing technologies.

PI Does Not Imply Disability

Impairments can, but do not always, result in disabilities (see Fig. 25.1). Several of the studies discussed earlier demonstrate this fact. Casali (1992) included three groups of participants including individuals with PIs resulting in low motor skills, PIs resulting in high motor skills, and no PIs. In this study, the high motor skills group was able to complete cursor manipulation tasks just as quickly as the group with no PIs. This suggests that the high motor skills group did not experience any disability as a result of their PI. The authors did not provide detailed descriptions of the PI experienced by the individuals in this group or their residual motor skills, but these results clearly indicate that it is possible for PIs to be sufficiently mild such that they will not interfere with certain computing activities.

Similarly, Sears et al. (2001) described a study where individuals with high-level spinal cord injuries were able to compose text just as quickly as individuals with no PIs. These results also demonstrate that PI do not automatically translate into disabilities. Casali (1992) showed that the severity of the PI can influence whether an individual experiences a disability. In contrast, these results show that even severe PI do not result in disabilities if interfaces are designed appropriately. Several other studies provided additional evidence that technology can reduce or eliminate disabilities resulting from PIs (e.g., Keates, Langdon, et al., 2000).

PIs Affect Cognitive, Perceptual, and Motor Activities

PIs are expected to affect an individual's ability to complete basic motor activities but are not necessarily expected to affect performance on cognitive and perceptual activities. Many of the articles discussed above confirm differences in performance for motor activities (e.g., LoPresti et al., 2000; Radwin et al., 1990), More important, the unexpected affect of PIs on cognitive and perceptual activities become apparent when analyzing both fundamental interactions and the high-level strategies adopted by individuals with PIs. For example, Keates, Clarkson, and Robinson (2000) provided data showing that individuals with PIs required 18% longer to complete basic cognitive activities and 25% longer for basic perceptual activities.

Koester and Levine (1994) provided an interesting and practical example of the potential consequences of focusing on motor activities without simultaneously addressing cognitive and perceptual activities. Word prediction is often recommended for individuals with PI because it reduces the number of keystrokes they must enter. Interestingly, Koester and Levine found that each keystroke took longer when using word prediction software, especially for the individuals with PI. Since the physical actions did not change, they attribute this increase in keystroke time to unspecified cognitive factors. They also found that individuals with PI spent longer than individuals with no PI when scanning the list of possible words. As a result, word prediction actually resulted in substantial decreases in data-entry rates when compared with data entry that was not assisted by word prediction.

Differences in high-level strategies are also evident. Sears et al. (2001) found that individuals with and without PIs responded differently when using the same speech recognition system to accomplish the a variety of tasks. Individuals with PIs interrupted their dictation more often, spent a greater percentage of their time dictating as opposed to issuing commands, and navigated shorter distances. The results reported by Koester and Levine (1994) also highlight how differences in performance for basic cognitive, perceptual, and motor activities translates into differences for higher level strategies. In their study, individuals with and without PIs interacted with word-prediction software using two alternative interaction strategies. Individuals with PIs were faster when using the first strategy, whereas individuals with no PIs were faster using the second strategy. These differences can be traced directly to the differences reported earlier for basic activities such as entering keystrokes and scanning lists of alternative words.

These studies confirm the importance of integrating individuals with PIs into the process when designing new computer systems. Although existing models of computer user behavior tend to be based on results from traditional computer users with no PI, it is clear that new models based on the performance of individuals with PIs will prove more accurate (Koester and Levine, 1994). These new models must address differences not only for motor activities, but for cognitive and perceptual activities as well.

Basic Actions Can Be Difficult

A number of articles highlight the difficulty individuals with PIs may experience with even the most basic interaction activities.

Trewin and Pain (1999) highlighted a variety of difficulties with activities as fundamental as pressing keys on a keyboard. Their results also highlighted numerous difficulties using a mouse. Dragging objects was difficult, as were activities that required multiple mouse clicks, but these actions can be difficult for individuals without PI as well (Franzke, 1995; MacKenzie, Sellen, & Buxton, 1991). More important, other actions that are normally relatively easy, such as pointing and clicking on objects also resulted in difficulty. Casali (1992) also illustrated the difficulty individuals with PIs can have with activities that require the mouse button to be held down while dragging the mouse. This same article highlighted a simple accommodation that helped reduce these difficulties: A simple toggle switch was provided that allowed users to avoid holding the mouse button during dragging operations. Similarly, Evans et al. (2000) provides insights that may help reduce the difficulties that individuals with PIs experience with some of these basic actions. Although many head-pointing devices emulate a mouse, their results indicate that an implementation that emulates a joystick may prove more effective because it provides a neutral resting position that allows the user to rest while the cursor does not move. This may prove particularly useful given the difficulty individuals with PI experience holding the cursor still while clicking the mouse button (Trewin & Pain, 1999).

Unfortunately, these difficulties become even more significant as target of the action gets smaller (Casali, 1992). Radwin et al. (1990) showed that performance with head-controlled devices was affected by target size even more than the mouse. As a result, individuals with PIs that use head-controlled devices would be expected to experience substantial difficulties as target sizes decrease. At the same time, careful use of force feedback in the form of gravity may help reduce these difficulties (Keates, Langdon, et al., 2000).

Standardized Descriptions of Health Conditions and PIs

Perhaps the greatest hindrance for both practitioners and researchers is the lack of accepted standards for describing the capabilities and limitations of study participants. For studies including traditional computer users with no PI, we typically provide basic demographics such as age, gender, and computer experience.

For these users, we typically assume "normal" perceptual, cognitive, and motor abilities. Occasionally, researchers even may report that participants all had "normal" or "corrected to normal" vision or hearing if these abilities are particularly relevant for a given study. For individuals with PI (or cognitive or perceptual impairments for that matter), we can no longer assume that the individual's perceptual, cognitive, and motor abilities are "normal."

Unfortunately, existing studies rarely provide sufficient details regarding health conditions or associated PIs. Many articles simply list a health condition for each participant (e.g., CP, choreoathetoid CP, incomplete SCI at C6, Friedrich's ataxia). This provides a general sense as to the PIs that may exist but does not provide a detailed understanding of the participants' abilities and limitations. Trewin and Pain (1999) do provide a brief description of each participant. A typical description included a health condition, an informal description of the associated impairment, and, for many participants, a description of their normal interactions with computers. Sears et al. (2001) described the health condition of their study participants as well as the resulting PI.

When a health condition results in a well-defined set of PIs, a precise description of the underlying health condition may be sufficient. For example, an SCI at or above C5 with an ASIA score of A or B results in paralysis of the muscles used to control the arms and legs. Given this health condition, we know that an individual would not be able to use his or her hands or arms when interacting with a computer. In contrast, a SCI at or above C5 with an ASIA score of C or D results in much more ambiguous PIs. Similarly, indicating that an individual has CP provides general insights, but the resulting PIs can vary dramatically, as noted earlier. PIs can also change with time. Some PIs are not present initially, others are progressive or regressive, and in still other situations the severity of the PI is variable. When a health condition does not map to a precise set of PIs, it is critical to describe both the health condition and the resulting PIs (and any other impairments that may affect interactions with computers).

Health conditions are best described by physicians. Physical examinations can often provide the necessary information, but access to medical records may be sufficient. Ideally, a standardized scale will be used to describe the health condition. For example, the ASIA scale can be used to describe the completeness of a spinal cord injury. Such assessments focus on the status of the individuals health, not necessarily on their physical abilities. Although these assessments focus on an individual's current status, they also provide insights into changes that may occur in the future. Most of the time, health conditions do not map directly to a well-defined set of PI. In these situations, it is recommended that an occupational therapist become involved in assessing each participant's PI. Occupational therapists can evaluate specific skills or abilities, including sensory motor, cognitive, and motor skills. The primary focus should be on those skills and abilities that may affect the individual's ability to interact with computers. Numerous standardized tests exist. Assessments by occupational therapists focus on the impairments that may affect an individual's ability to perform certain activities without regard to the underlying health condition. These assessments focus on an individual's current status and therefore do not address changes that may occur because of the progression of a disease.

Future Directions

The existing literature highlight the potential benefits and limitations of various technologies for specific groups of individuals, provide examples of situations in which PIs do not result in disabilities, illustrate the potential impact of PIs on cognitive, perceptual, and motor activities, confirm the difficulties that individuals with PIs may experience with basic actions (e.g., pressing keys and clicking the mouse button) and with more complex actions, and reenforce the need for standardized descriptions of health conditions and PIs.

This same literature confirmed the potential of various new technologies, whereas highlighting the need for additional research. Several studies illustrated the use of body parts that are normally ignored when designing computer interfaces (e.g., Shaw et al., 1995; Salem & Zhai, 1997). Others explored the potential of using electrophysiological data to control computers (e.g., Doherty et al., 2000, Kennedy et al., 2000; Kübler et al., 1999). These studies were exploratory, but they highlight the potential of these new technologies, especially for individuals with severe PIs.

We must continue to explore technologies that support alternative methods of interacting with computers, but we also need a better understanding of both the potential and limitations of existing technologies. Existing studies provide examples in which individuals with PIs were just as fast and accurate as individuals with no PIs, but important differences were apparent for satisfaction ratings as well as the strategies that the individuals with PI employed. Other studies highlighted situations in which PIs resulted in longer task completion times, higher error rates, or more subtle differences that only became apparent when performance was analyzed in detail. Future studies should attend to changes in cognitive, perceptual, and motor activities. These changes may reveal themselves through task completion times, error rates, satisfaction ratings, or through changes in the strategies that individuals adopt when completing a task. To allow both researchers and practitioners to more effectively interpret and generalize the results of such studies, we must carefully document both the health condition and the PI of study participants.

Although research into the efficacy of various technologies for individuals with PI is important, studies that assess the outcomes when assistive technologies are actually used is also critical (Fuhrer, 2001). Research studies highlight the potential of a technology but do not guarantee its success when used in the field. Outcomes research will ultimately determine the success or failure of the technologies developed by the HCI community for individuals with PIs. As a result, it is important for HCI researchers and system developers to work closely with those rehabilitation professionals who evaluate the needs of individuals with PIs, determine which technologies they should be using, and assess the success for failure of these technologies.

ACKNOWLEDGMENTS

This material is based on work supported by the National Science Foundation (NSF) under Grants IIS-9910607 and IIS-0121570. Any opinions, findings, and conclusions or recommendations expressed in this material are those of the authors and do not necessarily reflect the views of the NSF.

References

Adams, R. D., & Victor, M. (1993). *Principles of neurology* (5th ed.). New York: McGraw Hill.

Albacete, P. L., Chang, S. K., Polese, G., & Baker, B. (1994). Iconic language design for people with significant speech and multiple impairments. *Proceedings of ASSETS 94* (pp. 23–30). New York: Association of Computing Machinery.

Allanson, J., Rodden, T., & Mariani, J. (1999). A toolkit for exploring electro-physiological human-computer interaction. *Proceedings of INTERACT '99* (pp. 231–237). Amsterdam: IOS Press.

Alm, N., Todman, J., Elder, L., & Newell, A. F. (1993). Computer aided conversation for severely physically impaired non-speaking people. *Proceedings of InterCHI 93* (pp. 236–241). New York: ACM.

American Spinal Injury Association. (2001). ASIA Impairment Scale. Retrieved June 27, 2001, from http://www.asia-spinalinjury.org/publications/2001_Classif_worksheet.pdf

ASSETS. (2000). ACM Special Interest Group on Computers and the Physically Handicapped. Retrieved June 28, 2001, from http://www.acm.org/sigcaph/

Banerjee, S. (1982). *Rehabilitation management of amputees.* Baltimore: Williams & Wilkins.

Berkow, R. (1997). *The Merck manual of medical information home edition.* Whitehouse Station, NJ: Merck Research Laboratories.

Boissiere, P., & Dours, D. (2000). VITIPI: A universal writing interface for all. *Proceedings of the 6th ERCIM Workshop: User Interfaces for All.* ERCIM EEIG Sophia-Antipolis Cedex, France.

Bowman, T. (1997). VR meets physical therapy. *Communications of the ACM, 40,* 59–60.

Bozorgzadeh, Z., Birch, G. E., & Mason, S. G. (2000). The LF-ASD BCI: On-line identification of imagined finger movements in spontaneous EEG with able-bodied subjects. *Proceedings of ICASSP 2000.* IEEE.

Card, S., Moran, T. P., & Newell, A. (1983). *The psychology of human-computer interaction.* Mahwah, NJ: Lawrence Erlbaum Associates.

Casali, S. P. (1992). Cursor control device used by persons with physical disabilities: Implications for hardware and software design. *Proceedings of the Human Factors and Ergonomics Society 36th Annual Meeting* (pp. 311–315). Santa Monica, CA: Human Factors and Ergonomics Society.

Casali, S. P. (1995). A physical skills based strategy for choosing an appropriate interface method (pp. 315–341). In A. D. N. Edwards (Ed.), *Extra-ordinary human-computer interaction.* Cambridge, England: Cambridge University Press.

Casali, S. P., & Chase, J. D. (1993). The effects of physical attributes of computer interface design on novice and experienced performance of users with physical disabilities. *Proceedings of the Human Factors and Ergonomics Society 37th Annual Meeting* (pp. 849–853). Santa Monica, CA: Human Factors and Ergonomics Society.

California State University, Northridge. (2001). CSUN Center on Disabilities Annual Conference. Retrieved June 26, 2001, from http://www.csun.edu/cod/conf2002/index.html

Closing the Gap. (2001). Closing the Gap: Computer technology in special education and rehabilitation. Retrieved June 26, 2001, from http://www.closingthegap.com/

Conference on Universal Usability. (2000). Association for computing Machinery's Special Interest Group on Computer-Human Interaction. Retrieved June 28, 2001, from http://www.acm.org/sigchi/cuu

Demasco, P. W., & McCoy, K. F. (1992). Generating text from compressed input: An intelligent interface for people with severe motor impairments. *Communications of the ACM, 35*(5), 68–78.

Demasco, P., Newell, A. F., & Arnott, J. L. (1994). The application of spatialization and spatial metaphor to augmentative and alternative communication. *Proceedings of ASSETS 94* (pp. 31–38). New York: Association for Computing Machinery.

Doherty, E., Cockton, G., Bloor, C., & Benigno, D. (2000). Mixing oil and water: Transcending method boundaries in assistive technology for traumatic brain injury. *Proceedings of ASSETS 2000* (pp. 110–117). New York: Association for Computing Machinery.

Duvaisin, R. (1991). *Parkinson's disease: A guide for patients and families.* New York: Raven Press.

Edwards, A. D. N. (1995). *Extra-ordinary human-computer interaction.* Cambridge, England: Cambridge University Press.

European Research Consortium for Informatics and Mathematics. (2001). ERCIM Working Group: User interface for all. Retrieved June 27, 2001, from http://ui4all.ics.forth.gr/html/workshop.html

Evans, D. G., Drew, R., & Blenkhorn, P. (2000). Controlling mouse pointer position using an infrared head-operated joystick. *IEEE Transactions on Rehabilitation Engineering, 8,* 107–117.

Fell, H. J., Delta, H., Peterson, R., Ferrier, L. J., Mooraj, Z., & Valleau, M. (1994). Using the baby-babble-blanket for infants with motor problems: An empirical study. *Proceedings of ASSETS 94* (pp. 77–84). New York: Association of Computing Machinery.

Franzke, M. (1995). Turning research into practice: Characteristics of display-based interaction. *Proceedings of CHI '95* (pp. 421–428). New York: Association of Computing Machinery.

Fuhrer, M. J. (2001). Assistive technology outcomes research: Challenges met and yet unmet. *American Journal of Physical Medicine and Rehabilitation, 80,* 528–535.

Garay-Vitoria, N., & González-Abascal, J. (1997). Intelligent word-prediction to enhance text input rate. *Proceedings of Intelligent User Interfaces 97* (pp. 241–244). New York: Association of Computing Machinery.

Goette, T. (1998). Factors leading to the successful use of voice recognition technology. *Proceedings of ASSETS 98* (pp. 189–196). New York: Association of Computing Machinery.

Gray, D. B., Quatrano, L. A., & Liberman, M. L. (1998). Designing and using assistive technology. Baltimore: Paul H. Brooks.

Hallett, M. (1991). Classification and treatment of tremor. *Journal of the American Medical Association, 266;* 1115.

Hicks, J. E. (1988). Approach to diagnosis of rheumatoid disease. *Archives of Physical Medical Rehabilitation, 69,* S79.

Horn, L. J., & Zasler, N. D. (Eds.). (1996). *Medical rehabilitation of traumatic brain injury.* Philadelphia: Hanley & Belfus.

Karat, C.-M., Halverson, C., Karat, J., & Horn, D. (1999). Patterns of entry and correction in large vocabulary continuous speech recognition systems. *Proceedings of CHI '99* (pp. 568–575). New York: Association of Computing Machinery.

Keates, S., Clarkson, J., & Robinson, P. (2000). Investigating the application of user models for motion-impaired users. *Proceedings of ASSETS 2000* (pp. 129–136). New York: Association of Computing Machinery.

Keates, S., Langdon, P., Clarkson, J., & Robinson, P. (2000). Investigating the use of force feedback for motor-impaired users. *Proceedings of the 6th ERCIM Workshop: User Interfaces for All.* Sophia-Antipolis, France: ERCIM EEIG.

Keirn, Z. A., & Aunon, J. I. (1990). Man-machine communications through brain-wave processing. *IEEE Engineering in Medicine and Biology Magazine, 9,* 55–57.

Kennedy, P. R., Bakay, R. A. E., Moore, M. M., Adams, K., & Goldwaithe, J. (2000). Direct control of a computer from the human central nervous system. *IEEE Transactions on Rehabilitation Engineering, 8*, 198-202.

King, T. W. (1999). Assistive technology: Essential Human Factors. Boston: Allyn and Bacon.

Kirkup, L., Searle, A., Craig, A., McIsaac, P., & Moses, P. (1997). EEG-based system for rapid on-off switching without prior learning. *Medical and Biological Engineering and Computing, 35*, 504-509.

Koester, H. H., & Levine, S. P. (1994). Validation of a keystroke-level model for a text entry system used by people with disabilities. *Proceedings of ASSETS 94* (pp. 115-122). New York: Association of Computing Machinery.

Kübler, A., Kochoubey, B., Hinterberger, T., Ghanayim, N., Perelmouter, J., Schauer, M., Fritsch, C., Taub, E., & Birbaumer, N. (1999). The thought translation device: A neurophysiological approach to communication in total motor paralysis. *Experimental Brain Research, 124*, 223-232.

LaCourse, J. R., & Hludik, F. C. J. (1990). An eye movement communication-control system for the disabled. *IEEE Transactions on Biomedical Engineering, 37*, 1215-1220.

Lin, M. L., Radwin, R. G., & Vanderheiden, G. C. (1992). Gain effects on performance using a head-controlled computer input device. *Ergonomics, 35*, 159-175.

Lisogurski, D., & Birch, G. E. (1998). Identification of finger flexions from continuous EEG as a brain computer interface. *Proceedings of IEEE Engineering in Medicine and Biology Society 20th Annual International Conference.* IEEE.

LoPresti, E. F. (2001). Effect of neck range of motion limitations on the use of head controls. *CHI 2001 Extended Abstracts* (pp. 75-76). New York: Association of Computing Machinery.

LoPresti, E., Brienza, D. M., Angelo, J., Gilbertson, L., & Sakai, J. (2000). Neck range of motion and use of computer head controls. *Proceedings of ASSETS 2000* (pp. 121-128). New York: Association of Computing Machinery.

Lusted, H. S., & Knapp, R. B. (1996). Controlling computers with neural signals. *Scientific American, 275*, 58-63.

MacKenzie, S., Sellen, A., & Buxton, W. (1991). A comparison of input devices in elemental pointing and dragging tasks. *Proceedings of CHI 91* (pp. 161-166). New York: Association of Computing Machinery.

Malkewitz, R. (1998). Head pointing and speech control as a hands-free interface to desktop computing. *Proceedings of ASSETS 98* (pp. 182-188). New York: Association of Computing Machinery.

Manaris, B., & Harkreader, A. (1998). SUITEKeys: A speech understanding interface for the motor-control challenged. *Proceedings of ASSETS 98* (pp. 108-115). New York: Association of Computing Machinery.

Mason, S. G., Bozorgzadeh, Z., & Birch, G. E. (2000). The LG-ASD brain computer interface: On-line identification of imagined finger flexions in subjects with spinal cord injuries. *Proceedings of ASSETS 2000* (pp. 109-113). New York: Association of Computing Machinery.

Matias, E., MacKenzie, I. S., & Buxton, W. (1993). Half-QWERTY: A one-handed keyboard facilitating skill transfer from QWERTY. *Proceedings of InterCHI 93* (pp. 88-94). New York: Association of Computing Machinery.

McCoy, K. F., Demasco, P. W., Jones, M. A., Pennington, C. A., Vanderheyden, P. B., & Zickus, W. M. (1994). A communication tool for people with disabilities: Lexical semantics for filling in the pieces. *Proceedings of ASSETS 94* (pp. 107-114). New York: Association of Computing Machinery.

Molnar, G. E. (1992). Cerebral palsy. In G. E. Molnar (Ed.), *Pediatric rehabilitation* (pp. 481-533). Baltimore: Williams & Wilkins.

Moore, J. S., & Garg, A. (Eds.). (1992). Ergonomics: Low-back pain, carpal tunnel syndrome, and upper extremity disorders in the workplace. *Occupational Medicine: State of the Art Reviews, 7*, 593-790.

Oakley, I., McGee, M. R., Brewster, S., & Gray, P. (2000). Putting the feel in "Look and Feel." *Proceedings of CHI 2000* (pp. 415-422). New York: Association of Computing Machinery.

Oseitutu, K., Feng, J., Sears, A., & Karat, C.-M. (2001). Speech recognition for data entry by individuals with spinal cord injuries. *Proceedings of the 1st International Conference on Universal Access in Human-Computer Interaction.* Mahwah, NJ: Lawrence Erlbaum Associates.

Patmore, D. W., & Knapp, R. B. (1995). A cursor controller using evoked potentials and EOG. *Proceedings of RESNA 95 Annual Conference* (pp. 702-704). Arlington, VA: Rehabilitation Engineering and Assistive Technology Society of North America.

Patmore, D. W., & Knapp, R. B. (1998). Towards an EOG-based eye tracker for computer control. *Proceedings of ASSETS 98* (pp. 197-203). New York: Association of Computing Machinery.

Pidcock, F. S., & Christensen, J. R. (1997). General and neuromuscular rehabiliation in children. In B. O'Young, M. Young, & S. Stiens (Eds.), *PM & R secrets* (pp. 402-406). Philadelphia: Hanley & Belfus.

Radwin, R. G., Vanderheiden, G. C., & Lin, M. L. (1990). A method for evaluating head-controlled computer input devices using Fitts' Law. *Human Factors, 32*, 423-438.

Rehabilitation Engineering and Assistive Technology Society of North America. (2001). [Website]. Retrieved June 26, 2001, from http://www.resna.org/

Rosenthal, M., Griffith, E. R., Bond, M. R., & Miller, J. D. (Eds.). (1990). *Rehabilitation of the adult and child with traumatic brain injury* (2nd ed.). Philadelphia: F. A. Davis.

Roy, D. M., Panayi, M., Erenshteyn, R., Foulds, R., & Fawcus, R. (1994). Gestural human-machine interaction for people with severe speech and motor impairment due to cerebral palsy. *CHI 94 Conference Companion* (pp. 313-314). New York: Association of Computing Machinery.

Salem, C., & Zhai, S. (1997). An isometric tongue pointing device. *Proceedings of CHI 97* (pp. 538-539). New York: Association of Computing Machinery.

Schaab, J. A., Radwin, R. G., Vanderheiden, G. C., & Hansen, P. K. (1996). A comparison of two control-display gain measures for head-controlled computer input devices. *Human Factors, 38*, 390-403.

Schumacher, H. R., Jr. (Ed.). (1993). *Primer on the rheumatic diseases.* Atlanta, GA: Arthritis Foundation.

Sears, A., Karat, C.-M., Oseitutu, K., Karimullah, A., & Feng, J. (2001). Productivity, satisfaction, and interaction strategies of individual with spinal cord injuries and traditional users interacting with speech recognition software. *Universal Access in the Information Society, 1*, 1-12.

Shaw, R., Loomis, A., & Crisman, E. (1995). Input and integration: Enabling technologies for disabled users (pp. 263-277). In A. D. N. Edwards (Ed.), *Extra-ordinary human-computer interaction.* Cambridge, England: Cambridge University Press.

Sibert, L. E., & Jacob, R. J. K. (2000). Evaluation of eye gaze interaction. *Proceedings of CHI 2000* (pp. 281-288). New York: Association of Computing Machinery.

Stephanidis, C. (2001). *User interfaces for all: Concepts, methods, and tools.* Mahwah: NJ: Lawrence Erlbaum Associates.

Stiens, S., Goldstein, B., Hammond, M., & Little, J. (1997). Spinal cord injuries. In B. O'Young, M. Young, & S. Stiens (Eds.), *PM & R secrets* (pp. 253-261). Philadelphia: Hanley & Belfus.

Taylor, R. G., & Lieberman, J. S. (1988). Rehabilitation of patient with diseases affecting the motor unit. In J. A. DeLisa (Ed.), *Rehabilitation medicine: Principles and practice* (pp. 811–820). Philadelphia: J. B. Lippincott.

Trace Center. (2001). [Web site]. College of Engineering, University of Wisconsin—Madison. Retrieved June 26, 2001, from http://www. trace.wisc.edu

Trewin, S. (1996). A study of input device manipulation difficulties. *Proceedings of ASSETS 96* (pp. 15–22). New York: Association of Computing Machinery.

Trewin, S., & Pain, H. (1998a). A study of two keyboard aids to accessibility. *Proceedings of the HCI '98 Conference on People and Computers XIII* (pp. 83–97). Heidelberg: Springer-Verlag.

Trewin, S., & Pain, H. (1998b). A model of keyboard configuration requirements. *Proceedings of ASSETS 98* (pp. 173–181). New York: Association of Computing Machinery.

Trewin, S., & Pain, H. (1999). Keyboard and mouse errors due to motor disabilities. *International Journal of Human-Computer Studies, 50,* 109–144.

Wade, D. T., Langton Hewer, R., Skilbeck, C. E., & David, R. M. (1985). *Stroke: A critical approach to diagnosis, treatment, and management.* Chicago: Year Book.

World Health Organization. (2000). *International classification of functioning, disability, and health.* Geneva: Author.

Young, M. A., Tumanon, R. C., & Sokal, J. O. (2000, Summer). Independence for people with disabilities: A physician's primer on assistive technology. *Maryland Medicine,* 28–32.

·26·

PERCEPTUAL IMPAIRMENTS AND
COMPUTING TECHNOLOGIES

Julie A. Jacko
Georgia Institute of Technology

Holly S. Vitense
University of Wisconsin-Madison

Ingrid U. Scott
Bascom Palmer Eye Institute, University of Miami

INTRODUCTION

The human perceptual systems are critical to successful interaction with computing technologies. In fact, the visual system dominates in most traditional interaction scenarios, with speech and audition becoming more and more important in interactions involving contemporary interfaces and systems. This chapter serves to introduce these perceptual systems, at a fundamental level, and then highlights recent technological advancements in human–computer interaction (HCI) research that strive to enhance the perceptual experience for those who possess visual, auditory, or speech deficits.

The framework for the structure of this chapter developed as a result of a research project that was conducted with research funding awarded to the first author by the Intel Corporation. Part of this project involved a comprehensive review of the literature so that a categorization scheme could be developed that represents the major categories of impairment. Five major categories emerged from the literature: hearing impairments, mental impairments, physical impairments, speech impairments, and

visual impairments. Figure 26.1 illustrates that each of the five, overarching categories comprises a collection of clinical diagnoses unique to that category. The diagnoses are depicted in Fig. 26.1 by $A_1, \ldots, A_n, B_1, \ldots, B_n, C_1, \ldots, C_n, D_1, \ldots, D_n$, and $E_1, \ldots, E_n$. Each diagnosis, in turn, influences certain functional capabilities that are critical to the access of information technologies (depicted in Fig. 26.1 by $(Y_1, \ldots, Y_n)$. A subset of these functional capabilities can be directly linked to specific classes of technologies. Consider, for example, a person who has been diagnosed with a specific type of hearing impairment (represented as B_3 in Fig. 26.1). This hearing impairment results in a reduction in the functional capability Y_2. However, for example, the functional capability represented as Y_2 does not impede a person's access to any of the classes of technology represented in Fig. 26.1. In contrast, consider a person who has received a diagnosis of a mental impairment represented as C_2 in Fig. 26.1. This diagnosis results in a reduction in the functional capability Y_4, which in turn impedes this person's ability to successfully access information using a personal computer and communication devices. From this conceptual representation it is apparent that

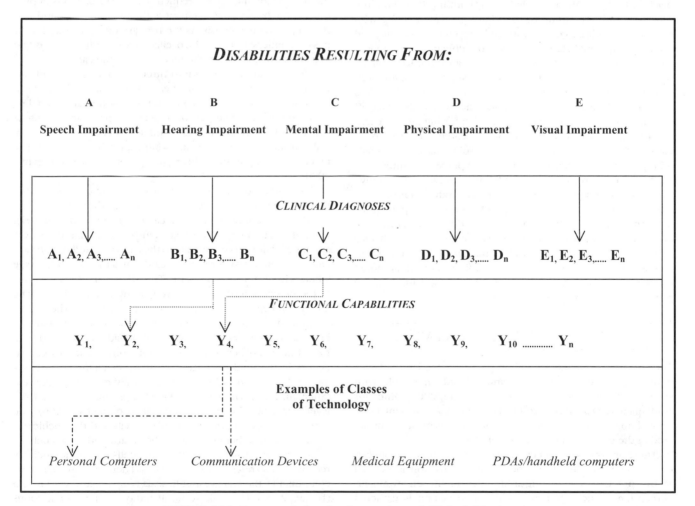

FIGURE 26.1. Framework for the Integration of Clinical Diagnoses, Functional Capabilities, and Access to Classes of Technologies.

much more knowledge is needed for researchers to possess an accurate depiction of the empirical relationships that exist between diagnoses, functional capabilities and access to specific classes of technologies. More specifically, emerging from this conceptual framework are several key research areas in need of investigation:

1. It is critical to establish empirical links between clinical diagnoses and sets of functional capabilities.
2. It is necessary to define the set of functional capabilities required to access information technologies.
3. It is essential to establish empirical bases for the influence of specific functional capabilities on access to specific classes of technologies.

Although the framework illustrates five categories of impairment, this chapter focuses on only three: hearing, speech, and visual. Mental (cognitive) impairments and physical (motor) impairments are covered in chapters 24 and 25 of this volume, respectively.

Thus, the chapter is organized into five major sections. First, auditory and speech impairments are discussed in tandem because they are often closely linked. Within this section, the sensation and perception of sound are presented, along with common pathologies of auditory and speech impairments, and then the functional classifications of the effects of hearing loss and speech problems. The chapter then transitions to visual impairments, in which visual function is first addressed, followed by a discussion of the three most common causes of vision loss in the United States. The section concludes with a discussion of specific visual functions. Finally, the chapter highlights recent technological advancements in HCI research for a variety of classes of technology, aimed at seeking solutions for individuals with perceptual impairments, including perceptual interfaces, multimedia interfaces, multimodal interfaces, and adaptive interfaces. Although these technological advancements are discussed briefly in this chapter within the context of perceptual impairments, it should be noted that additional information on perceptual–motor interactions may be found in the chapter authored by Chua, Weeks, and Goodman in this volume. In addition, chapter 11 on nonspeech auditory output, authored by Brewster, further focuses attention on audition and auditory processes as they relate to nonspeech output.

AUDITORY AND SPEECH IMPAIRMENTS

There are an estimated 28 million people in the United States with a significant hearing loss and 2.1 million people with speech impairments (National Ear Care Plan, 2002, Office for Students with Disabilities, 2000). These groups represent a substantial portion of the population whose needs must be considered in the design of emerging technologies.

There are two key aspects to studying auditory and speech impairments: sensation and perception. Sensation is defined as the process of detecting a stimulus in the environment; perception is the manner in which information is gathered and processed by the senses and then interpreted. Perception involves the development of internal models that represent the world, which typically requires considerable cognitive processing (Levine, 2000). Perception is based on sensations, so they are only as accurate as the information provided by the sensory system. As a result, impairments to hearing and speech have widespread ramifications on the formulation of perceptions, which can influence a person's ability to interact with computing technologies.

Because auditory and speech impairments are often closely associated, the first step in studying impairments in these sensory systems is to understand the sensation of hearing. As expected, the sensation of hearing is closely tied to the structure of sound waves and the mechanics of the ear.

Auditory and Speech Systems

Sensation of Hearing. The sensory receptors of the auditory system are sensitive to sound waves, which can be described as constantly fluctuating air pressure. The ear captures this fluctuating air pressure and converts it into perceptions of pitch, timbre, and loudness. The frequency of the oscillations of pure tones is the primary determinant of the pitch of the sound, and the amplitude or intensity is the primary determinant of loudness. To understand how the oscillations are collected, it is first important to understand the mechanics of the ear.

The ear can be divided into three distinct components according to function: the outer ear, middle ear, and inner ear. The outer ear is responsible for capturing sound and funneling it to the eardrum. The part of the ear just past the eardrum is considered the middle ear. The middle ear acts as the mechanical transformer and transmits sound into the inner ear. The inner ear is where the auditory receptors are located (Levine, 2000).

The part of the ear that is visible, the ear canal up to the tympanic membrane (eardrum), constitutes the outer ear. The pinna is the peripheral and noticeable portion of the outer ear. Once the sound waves are collected by the pinna, the waves are directed down the ear canal to the eardrum (Stillman, 1980). Once the sound waves reach the eardrum, a three-layered tissue vibrates in response to pressure. This vibration ultimately leads to the perception of sound (Coren, Ward, & Enns, 1999). The middle ear begins on the other side of the eardrum. The middle ear consists of an air-filled chamber that allows air pressure to be equalized between the inside and outside of the head. The role of the middle ear is to compensate for the energy loss that occurs in the fluid of the inner ear by mechanically increasing the force of the vibrations before they reach the inner ear (Stillman, 1980). Three interlocking bones, the ossicles, act as a lever to facilitate this transfer of energy (Coren et al., 1999). The inner ear, consisting of semicircular canals and the cochlea, is a fluid-filled chamber encased in a bony labyrinth. The cochlea, a spiral-shaped structure, contains the cells that act as auditory receptors, also referred to as hair cells. There are three cochlear components, the scala vestibuli, scala tympani, and scala media (Levine, 2000). The scala vestibuli and tympani are continuous with each other at the opening of the cochlear. The structure separating scala media from the scala tympani is the basilar membrane. When motion occurs in the fluid within the inner ear, the basilar membrane vibrates, which causes the cilia on the hair cells to move. This bending of the hair cells initiates a neural signal (Proctor & Proctor, 1997). The neural signal then, in turn, is used to form a conscious experience or perception. The cognitive processes involved with perceiving sound are complex. According to the information-processing approach, there are multiple levels of processing that occur at each stage of sensory processing. These stages range from registering the stimulus on the receptor to a final conscious representation in memory (Coren et al., 1999).

Perception of Hearing. The design of computing technologies for individuals with auditory and speech impairments must consider elements of the perception of hearing. Elements such as pitch, timbre and loudness cannot be ignored in the design of systems that rely on the auditory system. Pitch and timbre are two psychological qualities that are primarily affected by the frequency of an auditory stimulus. Loudness is another psychological quality related to the frequency and intensity of the auditory stimulus.

Pitch. Pitch is a subjective quality of an auditory stimulus that cannot be measured directly (Moore, 1982). As a result, pitch is often closely related to the frequency of a pure tone or the fundamental frequency of a complex tone containing many harmonics. Variations in pitch contribute to the melodies found in music, where the higher the frequency, the higher the pitch. Pitch is not identical to the frequency of a tone (Levine, 2000), however; rather, it is also influenced by the pattern of harmonics, and not just the fundamental frequency (Proctor & Proctor, 1997). This is because the fundamental frequency can be removed, yet the pitch remains the same. These and other factors are considered in the two predominate theories of auditory pitch perception: the frequency theory and the place theory.

The frequency theory, suggested by Wever and Bray (1930), attributes a human's well-developed abilities to discriminate between nearby frequencies, to the vibration of the basilar membrane. The basilar membrane vibrates in unison in response to pure tones. These vibrations then in turn reproduce the firing of the auditory nerve (Levine, 2000). As a result, the frequency of the firing is the neural code for pitch (Proctor & Proctor, 1997). Critics of this theory state that because the maximum firing rate of neurons is limited to 1,000 action potentials per second, the firing rate of the neurons could not match the frequencies over the entire range of human hearing (Levine, 2000). In response to this criticism, Wever and Bray (1937) modified their theory to incorporate the idea of phase locking that allowed high-frequency sounds to be replicated.

The second theory of auditory pitch perception, Helmholtz's (1977, 1954) place theory, suggests that different places on the basilar membrane are affected by different frequencies (as cited in Levine, 2000). This is a result of the belief that neighboring regions of the basilar membranes are not connected and have varying amounts of tension (Levine, 2000). Thus, the neural

code for frequency correspond to the particular neurons that are being stimulated (Proctor & Proctor, 1997). More recent work conducted by Von Békésy (1960) provides evidence that the basilar membrane operates in a manner consistent with both the frequency and place theories (as cited in Proctor & Proctor, 1997). Von Békésy demonstrated that waves travel down the basilar membrane from the base of the opening at a frequency corresponding to the tone (Proctor & Proctor, 1997).

Pitch is related to 'tone deafness,' which is actually a misnomer because nearly everyone is able to discriminate between the pitches of two tones when their frequencies differ over time (Moore, 1982). However, the ability to recognize and define a pitch of a musical tone without reference to a comparison tone is quite rare. Only about 1% of the population has this ability, which is referred to as absolute pitch (Moore, 1982).

Timbre. Timbre is the sensation closely related to the quality of sound. Because typical sounds encountered in everyday life are complex, containing multiple frequencies with relative phases, the discrimination between these frequencies conveys a lot about the quality of the sound, or timbre. Timbre is defined as the attribute of sound that allows a listener to judge that two sounds, with the same loudness and pitch, are dissimilar (Moore, 1982). This psychological property, allows people to differentiate between sounds generated by different instruments even when playing the same note. Timbre represents the perceived quality that accounts for differences in sound not related to pitch. These differences can be described as the harmonic of a sound (Levine, 2000). Timbre allows one sound to be discriminated from another when both have the same pitch, loudness, and duration. For example, timbre enables a person to distinguish between two vowels spoken from the same speaker at the same loudness for the same duration (Yost & Sheft, 1993).

Unlike pitch and loudness, timbre is multidimensional in the sense that there is no single scale along which timbres of different sounds can be compared or ordered. As a result, techniques similar to factor analysis can be used to determine the principle dimensions of variation within a set of stimuli (Moore, 1982). This type of procedure reduces the number of dimensions required to determine differences between amplitudes of frequency bands (Moore, 1982).

Loudness. Similar to pitch and timbre, loudness is a subjective quality of an auditory stimulus. Loudness is often described in terms of intensity, usually specified in decibels (dB). Phons and sone scales can be used when describing and measuring loudness. A phon is the level in dB of an equally loud 1000-Hz tone. Phon scales exist that represent equal loudness contours. The sone scale is another way to measure loudness. A sone is the loudness of a 1000-Hz tone presented at 40 dB. As a result, 1 sone equals 40 phons (Yost, 1994). Because the loudness of sound cannot be measured directly, these scales allow subjective descriptions of loudness to be related to physical descriptions of frequency and intensity.

The loudest sound humans can hear without damaging their ears is 120 dB (Moore, 1982). Humans are most sensitive to frequencies in the range of 1000 to 5000 Hz. The threshold increases rapidly at the very high and low frequencies. The highest audible frequency varies considerably with age. The loss of sensitivity with increasing age (presbycusis) is much greater at high than at low frequencies. At high frequencies, variability between different observers is greater (Moore, 1982); however, the ultimate ability to detect faint sounds is not limited by absolute sensitivity but rather the level of ambient noise. As a result, special consideration must be paid to the environment in which systems relying on auditory stimuli are used. Adaptation and fatigue in the auditory system are much less marked than in the visual system, and the effects have different time courses (Moore, 1982). This makes the auditory system an attractive modality for tasks that place great demands on the visual system.

Pitch, timbre, and loudness represent elements of the perception of hearing that are vital to the design of current and emerging computing technologies. They represent qualities affected by the frequency and intensity of an auditory stimulus. With an understanding of the sensation and perception of hearing, the focus of the discussion can now shift to the perception of speech, which is strongly tied to hearing.

Perception of Speech. Just as with the perception of hearing, the perception of speech is also an important aspect to consider in the design process. An understanding of the perception of speech allows systems relying on speech input to be developed that are sensitive to individuals with speech impairments. (See chapter 7, Input Technologies and Techniques, and chapter 8, Conversational Interface Technologies, in this handbook for additional information on these topics.) To fully understand speech perception, it is first important to have knowledge of the mechanics of speech. The most familiar units of speech are words. Words can be further dissected into syllables and then into phonemes. Phonemes represent speech sounds (Moore, 1982). In studying speech perception, speech can be thought of as being composed of a series of sounds that correspond to particular phoneme. Speech sounds have a one-to-one correspondence with phonemes (Moore, 1982).

Speech sounds are not organized in a linear manner. Rather, the ability to perceive speech is a complex process (Miller & Eimas, 1995). Speech is delivered in a time-varying continuous signal, yet listeners perceive speech as discrete linguistic units (phonemes, syllables, and words), whereas in actuality these linguistic units overlap. As a result, it is difficult to segment discrete linguistic units and correspond them to unique acoustic features (Miller & Eimas, 1995). Despite this difficulty, the study of speech perception primarily centers on the transfer of complex acoustic speech patterns into linguistic units at the perceptual level (Moore, 1982). More specifically, speech perception focuses on the listener's ability to identify or discriminate phonetic contrasts (Goodman, Lee, & DeGroot, 1994). This process is not fully understood (Yost, 1994). Empirical evidence is mixed regarding which linguistic unit (phonemes, syllables, or words) listeners use to analyze speech. Regardless of what unit is used, it is known that listeners extract abstract properties of the speech signal comparable to representations stored in long-term memory (Miller & Eimas, 1995).

Speech perception does not proceed via simple transformation from physical cues directly available in the sound waveform; rather, linguistic context is also important (Moore, 1982). A given linguistic unit is not represented by a given, fixed acoustic pattern in the speech wave. Instead the pattern varies according to the preceding and following sounds (Moore, 1982). The acoustic form of any given word typically also varies substantially when spoken by different speakers at different rates of speech or with different emotional force (Goodman et al., 1994; Miller & Eimas, 1995).

Because of the complexity of the perceptual mechanics of speech perception, there are a number of stages of analysis and representation. Theories need to consider not only what linguistic unit to analyze, but also how many and what kind of analyses are involved in the linguistic content (Miller & Eimas, 1995). There are two basic approaches to how speech perception is processed: autonomous and interactive. Fodor's (as cited in Miller & Eimas, 1995) modularity thesis is one autonomous approach that assumes a speech signal is processed in a strictly bottom-up fashion. This approach prescribes a process in which the end product of one stage is assumed to provide the input to the next level. According to Connine (as cited in Miller & Eimas, 1995), the interactive approach, on the other hand, assumes that the construction of linguistic representation is determined by information gained from the acoustic signal and from information from higher level processing. These two sources of information interact to contribute to speech perception. There is a heated debate over these two approaches of how speech perception is processed. The study of speech perception processing is not the only field within speech perception in which theories are debated. From a general perspective, there are several theoretical approaches to speech perception.

Three general, theoretical approaches to speech perception are briefly discussed here. A first approach is based on invariant features or cues. This theoretical approach assumes that invariant acoustic properties corresponding to individual features could be uncovered if the speech signal is examined in a proper way (Miller & Eimas, 1995). The models adhering to this approach provide knowledge regarding the acoustic structure of the linguistic content (Miller & Eimas, 1995). A second approach focuses on processes occurring within the listener. This motor theory of speech perception takes into account context sensitivity and variability when analyzing the speech signal. Rather than looking at invariant features in the waveform, the listeners are thought to reconstruct the phonetic intent of the speaker from the waveform (Miller & Eimas, 1995). Lastly, a third approach is a direct-realist approach to speech perception. This approach shares the motor theory's basic assumption that the objects of perception in speech are articulated gestures. However, the direct-realist approach assumes that cognitive mediation is not necessary to support perception (Miller & Eimas, 1995). These theoretical approaches to speech perception represent only a subset of approaches available.

The sensation and perception of hearing and the perception of speech provide a foundation for the design of accessible information technologies. To further understand the challenges imposed as a result of an auditory or speech impairment, the discussion now concentrates on the common pathologies of auditory and speech impairments.

Common Pathologies of Auditory and Speech Impairments

Auditory Impairments. Types of auditory function are described with respect to auditory threshold, which defines both amplitude and frequency of sound. Amplitude is the loudness of the sound measured in decibels (dB). Frequency is the pitch of sound measured in hertz (Hz) (Cook & Hussey, 1995). The degree of hearing loss can be categorized into two types, conductive or sensorineural. Conductive hearing loss describes a disorder involving the outer or middle ear. Typically, this disorder can reduce an individual's sensitivity to hear sounds below 60 dB (Stein, 1988). Sensorineural hearing loss is a condition, which stems from damage to the cochlea or auditory (eighth) cranial nerve. Of these two types, sensorineural is more likely to be irreversible and can result in hearing loss ranging from the reduction in thresholds for certain frequencies, to the total loss of sensitivity (Stein, 1988). To further understand these two types of hearing loss, a distinction is made as to when the loss occurred. If the loss is present at birth or occurs within the first 3 years of life, it is categorized as congenital or prelingual. Prelingual implies that the child has not developed the capability to speak. On the contrary, a loss that is acquired or postlingual occurs after speech development (Stein, 1988). One of the most common types of acquired hearing loss occurring in the elderly is presbycusis. An estimated 23% of the elderly between the ages of 65 to 74 years and 32% over 70 years of age are affected (national ear care plan, 2002). Presbycusis is the gradual loss of high-frequency hearing, usually beginning at age 65 and older (Stein, 1988). Regardless of the type of hearing loss, the degree of impairment is defined and labeled in accordance with the ability to hear pure tones at 500, 1000 and 2000 Hz. The average level of hearing intensity measured in decibels (dB), for an individual, corresponds to a degree of hearing impairment. Associated with these predefined hearing averages are the presumptions of sounds heard without amplification, the degree of handicap, and probable needs of an individual to resume activities of daily living, as shown in Table 26.1 (Stewart & Downs, 1984).

In general, a hearing impairment is considered slight if the loss is between 20 and 30 dB, mild from 30 to 45 dB, moderate from 60 to 75 dB, profound from 75 to 90 dB, and extreme from 90 to 110 dB (Cook & Hussey, 1995). When an individual's primary disability is auditory, other modalities, such as visual and tactile are used to convey information.

Speech Impairments. Speech is defined as the oral expression of language. Language can be defined as a system of symbols (i.g., letters) that are organized according to a set of rules (i.g., grammar) within the context of social interaction (Cook & Hussey, 1995). Language permits the communication of complex information between individuals (Kaiser, Alpert, & Warren, 1988). In general, language and communication disorders, such as speech impairments, are often closely linked to other types of disability. For instance, nearly all individuals who have a developmental disability require some form of language intervention to learn functional communication skills (Kaiser et al., 1988).

The identification and classification of clinical speech impairments are clearly defined. However, a precise prediction of specific language and communication impairments based on disability classifications is very difficult to define (Kaiser et al., 1988). Speech impairments can be evaluated clinically in terms of audibility, intelligibility and functional efficiency. Audibility, as the name implies, is a person's ability to speak at a level sufficient to be heard. Intelligibility relates to the ability to articulate and link phonetic units of speech. Functional efficiency is the ability to speak at a sufficient rate so that speech can be comprehended (American Medical Association, 1990). Judgments as to the level of impairment should be made with reference to classification ratings and percentages, as shown in Table 26.2 (American Medical Association, 1990). Table 26.1 shows the classification breakdown with respect to audibility, intelligibility, and functional efficiency (American Medical Association, 1990).

In addition to the inherent ability to produce speech, the environment contributes considerably to communication development and speech disorders. Elements of social interaction are critical to the normal development of language. Language is a social behavior that is learned in social contexts. These environmental factors can either facilitate or retard the affects of communication disorders (Kaiser et al., 1988).

Discussion of these common pathologies of auditory and speech impairments along with the descriptions of the auditory and speech systems provides valuable information for designers. However, knowledge of common pathologies is fundamentally important to the classification of functional abilities. Links between these pathologies and a person's functional capabilities inform design.

Functional Classifications of the Effects of Hearing Loss and Speech Problems

Effects of Hearing Loss. The effects of hearing loss cascade down through many senses, as is evident in communication and cognitive functional limitations. Speech and language development are significantly retarded as a result of congenital or prelingual hearing loss (Rampp, 1979). Reading ability is also affected by auditory processing disorders. There has been research to suggest a possible link between speech representation at a phonetic level and reading. Ultimately, these deficiencies in speech, language, and reading can result in learning disabilities (Rampp, 1979). The severity and duration of hearing loss at a young age greatly affects the level at which speech, language, and reading development are impaired. Deafness (profound hearing loss of 95 dB or higher) occurring perlingually strongly impedes the ability to communicate. At an older age (postlingual) the effects of deafness are not as severe because an individual would be able to acquire the ability to read lips and continue to speak (Rampp, 1979). Research has also explored the relationship between language disorders and auditory skill, and speech and reading acquisition. There may be a

TABLE 26.1. Effects of Hearing Loss

Average Hearing 500–2000 Hz (ANSI)	Description of Hearing Impairment	Sounds Heard Without Amplification	Degree of Handicap[a]	Probable Needs
0–15 dB	Normal range	All speech sounds	None	None
15–25 dB	Slight hearing loss	Vowel sounds heard clearly; may miss unvoiced consonant sounds	Mild auditory dysfunction in language learning	Consideration of need for hearing aid; lip-reading; auditory training; speech therapy, preferential seating
25–40 dB	Mild hearing loss	Hears only some louder voiced speech sounds	Auditory learning dysfunction, mild language retardation, mild speech problems, inattention	Hearing aid, lip-reading, auditory training, speech therapy
40–65 dB	Moderate hearing loss	Misses most speech sounds at normal conversational level	Speech problems, language retardation, learning dysfunction, inattention	All the above, plus consideration of special classroom situation
65–95 dB	Severe hearing loss	Hears no speech sounds of normal conversation	Severe speech problems, language retardation, learning dysfunction, inattention	All the above, plus probable assignment to special classes
More than 95 dB	Profound hearing loss	Hears no speech or other sounds	Severe speech problems, language retardation, learning dysfunction, inattention	All the above, plus probable assignment to special classes

Note. Adapted from Stewart and Downs (1984).
[a]If not treated in first year of life.

possible link between hearing the phonetic levels in words and relating the phonetic segments of printed words (Rampp, 1979). As a result, speaking, reading, and language development are all skills likely to be affected by a measurable hearing loss.

The ramifications at any level of uncorrected hearing loss can be great. The ability to hear plays an ever increasing role in emerging systems. For instance, the detection of notification sounds used as prompts is becoming increasingly more important. The use of sound to convey messages is not new. Items such as fire alarms, doorbells, phones, alarm clocks have been prompting users to action for years. Drawing on this universal acceptance, computers have adopted this form of auditory prompting and warning. All too often in complex environments, a user's visual attention is overloaded, and thus a visual warning could go undetected. The auditory channel provides another modality for communication between a user and computers. Auditory notifications have certain advantages as well as disadvantages and need to be carefully applied. Computer systems are constantly expanding their notification and warning schemes. In some cases this leads to more specific and unique auditory signals. The meaning of an auditory notification may depend on the tone of the signal or when the signal is given. In such cases, the ability to absolutely identify a signal is crucial. Some systems have even gone a step further by using the full capability of the auditory channel. These systems use localized three-dimensional signals to convey meaning. Because of the uniqueness of the growing user population, much consideration in the design process must be paid to making technology accessible to everyone, including individuals with hearing impairments.

Effects of Speech Problems. The purpose of language is to mediate the behavior of others in the context of social interaction. As a result, the nature of social interaction is integral to the development of communication (Kaiser et al., 1988). Individuals who have difficulty communicating are limited in their ability to achieve their goals. Even relatively minor speech impairments may seriously impair social conversation and information exchange (Kaiser et al., 1988). In this era of ubiquitous computing, speech impairments can limit an individual's use of emerging technologies such as a speech recognition system. For cases in which speech is degraded, speech recognition systems must be sensitive enough to adapt to impaired speech. Regardless of the cause of the speech impairment, devices and techniques are available to enhance the ability to communicate among individuals who experience difficulty speaking in an understandable manner. Hearing loss and many other types of impairments are associated to some degree with speech degeneration. For instance, nearly all individuals with developmental disabilities require some form of language intervention to learn functional communication skills (Kaiser et al., 1988). As a result, the prevalence of speech impairments has the potential to span a large proportion of the population and must be taken into consideration in the design of emerging technologies.

Summary

Beginning with descriptions of sensation and perception and of related common pathologies, we laid the foundation for a

TABLE 26.2. Speech Classification Chart [Adapted from American Medical Association, 1990]

	Audibility	Intelligibility	Functional Efficiency
Class 1: 0–10% speech impairment	Can produce speech loud enough for most of the needs of everyday speech communication, although this sometimes may require effort and occasionally may be beyond the individual's capacity.	Can perform most of the articulatory acts necessary for everyday speech communication, although listeners occasionally ask the individual to repeat and the individual may find it difficult or even impossible to produce a few phonetic units.	Can meet most the demands of articulation and phonation for everyday speech communication with adequate speed and ease, although occasionally the individual may hesitate or speak slowly.
Class 2: 15–35% speech impairment	Can produce speech loud enough for many of the needs of everyday speech communication; is usually heard under average conditions; however, may have difficulty in automobiles, buses, trains, stations, restaurants, etc.	Can perform many of the necessary articulatory acts for everyday speech communication; can speak name, address, etc. and be understood by a stranger but may have numerous inaccuracies; sometimes appears to have difficulty articulating.	Can meet many of the demands of articulation and phonation for everyday speech communication with adequate speech and ease but sometimes gives impression of difficulty, and speech may sometimes be discontinuous, interrupted, hesitant, or slow.
Class 3: 40–60% speech impairment	Can produce speech loud enough for some of the needs of everyday speech communication, such as close conversation but has considerable difficulty in such noisy places as listed for Class 2; the voice tires rapidly and tends to become inaudible after a few seconds.	Can perform some to the necessary articulatory acts for everyday speech communication; can usually converse with family and friends; however, strangers may find it difficult to understand the individual; may often be asked to repeat.	Can meet some of the demands of articulation and phonation for everyday speech communication with adequate speed and ease but often can only sustain consecutive speech for brief periods; may give the impression of being rapidly fatigued.
Class 4: 65–85% speech impairment	Can produce speech loud enough for a few of the needs of everyday speech communication; can barely be heard by a close listener or over the telephone, perhaps may be able to whisper audibly but has no voice	Can perform a few of the necessary articulatory acts for everyday speech communication; can produce some phonetic units; may have approximations for a few words such as names of family; however, these are unintelligible out of context.	Can meet a few of the demands of articulation and phonation for everyday speech communication with adequate speed and ease, such as single words or short phrases, but cannot maintain uninterrupted speech flow; speech labored, rate is impractically slow.
Class 5: 90–100% speech impairment	Can not produce speech loud enough for the needs of everyday speech communication.	Can perform none of the articulatory acts necessary for everyday speech communication.	Can meet none of the demands of articulation and phonation for everyday speech communication with adequate speed or ease.

Note. Adapted from the American Medical Association (1990).

discussion of the functional classifications of the effects of auditory and speech impairments. These classifications of functional effects exemplify the primary abilities individuals with auditory and speech impairments require to access information technology. Because these abilities are directly related to how individuals interact with information technologies, the more that is understood about the impairment, the better the design can accommodate. This section along with the visual impairments section provides the clinical and psychophysical background necessary for the design of accessible technologies.

VISUAL FUNCTION

Definitions

When considering the impact of visual loss on an individual's ability to use a computer effectively, it is first necessary to understand various dimensions of visual performance. Many

terms have been used to refer to abnormalities in visual function, including *disorder, impairment, disability*, and *handicap* (Colenbrander, 1977). Although often used as synonyms, there are distinct differences. For instance, whereas *disorder* and *impairment* describe aspects of an organ's condition, *disability* and *handicap* describe aspects of a patient's condition.

A *disorder* refers to an abnormality in the anatomy or physiology of an organ and, in the case of a visual *disorder*, may occur anywhere in the visual system. Examples of visual disorders include corneal scar, cataract, macular degeneration, optic atrophy, or occipital stroke. It is important to recognize that knowing the specific visual disorder provides no information concerning the functional capacity of the eye.

An *impairment* refers to a functional abnormality in the organ. Thus, varying degrees of visual impairment can be measured in terms of specific visual functions, such as visual acuity, contrast sensitivity, visual field, or color vision. Although such impairment measures demonstrate how well the eye functions, they do not reveal the impact of the visual disorder on the patient's ability to perform everyday activities. For example, a physician may state that the patient's visual acuity has dropped by four lines on the eye chart, while the patient reports an inability to see well enough to use a computer.

A *disability* refers to the ability of a patient (rather than an organ) to perform tasks, such as daily living skills, vocational skills, reading, writing, mobility skills, and so on. Because disability implies a broader perspective (the focus is on the person as a whole rather than on a specific organ), it is no longer entirely vision specific. For instance, although computer skills may be reduced by vision loss, they may also suffer because of conditions such as arthritis. It is the combination of visual and nonvisual skills that determines the abilities or diabilities of an individual. Vision substitution techniques (such as the use of a white cane and increased reliance on memory and hearing) may be helpful in improving the ability of an individual to perform specific tasks.

A *handicap* refers to an even broader perspective, that is, the socioeconomic consequences of a disability. The degree of handicap is determined by the difference between an individual's abilities and what is expected of the individual, both in terms of societal demands and self-image. Handicap may also be understood as the extra effort patients have to make to achieve the same goals as nonhandicapped individuals with respect to such factors as economic independence, independent mobility, and independent living.

Although a disorder may cause an impairment, an impairment may cause a disability, and a disability may result in a handicap, these links are not rigid. An analysis of these various dimensions of vision loss permits identification of interventions at each link that may improve the functional status and quality of life of an individual with visual loss (Fig. 26.1; Fletcher, 1999). For example, if one were interested in improving the ability of an individual with vision loss to use a computer effectively, possible interventions include medical and surgical intervention to impact the visual disorder and impairment; visual aids and adaptive devices to impact visual impairment and disability; and social interventions, training, counseling, and education to impact visual disability and handicap. The design of a graphic user interface that increases the ability of an individual with vision loss to perceive graphic and textual information would have a beneficial effect on the impairment, disability, and handicap of that individual.

Epidemiology

Low vision has been defined as a permanent visual impairment that is not correctable with glasses, contact lenses, or surgical intervention and that interferes with normal everyday functioning (Mehr, 1975). It is estimated that approximately 1% of the population of Western countries suffers from low vision (Dowie, 1988; Strong, Pace, & Plotkin, 1988), 14 million Americans (1 in every 20) have visual impairment (Kupfer, 2000), and that visual impairment affects nearly 5 million Americans to the extent that they are unable to read newsprint, even with the aid of full refractive correction (Nelson & Dimitrova, 1993). Among Americans 65 years of age and older, nearly 8% have low vision, and among Americans 85 years and older, this percentage increases to 25% (Nelson, 1987).

Because the number of Americans over the age of 65 years is projected to more than double between 1995 and 2030 (National Advisory Eye Council, 1993), the burden of low vision is expected to increase markedly. Vision loss has been ranked third, behind arthritis and heart disease, among conditions that cause persons older than 70 years to need assistance in activities of daily living (LaPlante, 1988).

Selected Visual Disorders

In the United States, the most common causes of decreased vision are age-related macular degeneration (AMD), diabetic retinopathy, and glaucoma.

Age-Related Macular Degeneration. AMD is the leading cause of irreversible visual loss in the Western world in individuals over 60 years of age. The *macula* is that part of the retina that is responsible for central vision. The prevalence of severe visual loss due to AMD increases with age. In the United States, at least 10% of persons between the ages of 65 and 75 years have some central vision loss due to AMD; among individuals over the age of 75 years, 30% have vision loss due to AMD.

Risk factors for this disease and its progression include age, sunlight exposure, smoking, light ocular and skin pigmentation, and elevated serum cholesterol levels. The role of nutrition has not been fully identified as a risk factor, but a diet low in antioxidants and lutein may be a contributing factor.

AMD is a bilateral disease in which visual loss in the first eye usually occurs at about 65 years of age; the second eye becomes involved at the rate of approximately 10% per year. The two main types of AMD are atrophic and exudative. The atrophic ("dry") form of the disease is generally a slowly progressive disease that accounts for approximately 90% of cases. It is characterized by the deposition of abnormal material beneath the retina (drusen), and degeneration and atrophy of the central retina (also known as the macula); patients typically note slowly progressive central visual loss. Although much less common, the exudative ("wet") form of the disease is responsible for about 88% of legal blindness attributed to AMD. This form of the disease, which often occurs in association with atrophic AMD, is characterized by the growth of abnormal blood vessels beneath the central retina (macula); these abnormal blood vessels elevate and distort the retina and may leak fluid and blood beneath or into the retina. Vision loss may be sudden onset and rapidly progressive (in contrast to the atrophic form of the disease, in which vision loss generally occurs progressively over several months or years). AMD can cause profound loss of central vision, but the disease generally does not affect peripheral vision, and therefore patients typically retain their ability to ambulate independently.

There currently is no effective treatment to reverse the retinal damage that has already occurred due to AMD. To try to prevent further vision loss, recommendations made to patients include eye protection against sunlight exposure (sunglasses with ultraviolet light protection), no smoking, optimal control of serum cholesterol level, and a diet rich in dark green leafy and orange vegetables (antioxidants are believed to reduce the damaging effects of light on the retina through their reducing and free-radical scavenging actions; lutein is a macular pigment). The potential benefit of vitamin and mineral supplementation is currently being investigated in a randomized, controlled clinical trial. Laser photocoagulation and photodynamic treatment of the abnormal blood vessels found in patients with exudative macular degeneration may help to prevent severe vision loss in some cases. Surgical rotation of the retina away from the area of abnormal blood vessels has also been effective in some cases. Other treatment modalities currently under investigation include radiotherapy and such drugs as corticosteroids and thalidomide.

Diabetic Retinopathy. Approximately 16 million Americans suffer from diabetes mellitus, most of whom will develop diabetic retinopathy within 20 years of their diagnosis. In fact, after 20 years of diabetes, nearly 99% of those with insulin-dependent diabetes mellitus and 60% with non-insulin-dependent diabetes mellitus have some degree of diabetic retinopathy. Diabetic retinopathy is the leading cause of legal blindness in Americans aged 20 to 65 years, with 10,000 new cases of blindness annually. One million Americans have proliferative diabetic retinopathy, and 500,000 have macular edema.

The major risk factor for diabetic retinopathy is duration of diabetes; it is estimated that at 15 years, 80% of diabetics will have background retinopathy and that of these 5 to 10% will progress to proliferative changes. Other risk factors include long-term diabetic control (as reflected in serum levels of glycosylated hemoglobin), hypertension, smoking, and elevated serum cholesterol.

There are two main types of diabetic retinopathy: nonproliferative and proliferative. Nonproliferative diabetic retinopathy refers to retinal microvascular changes that are limited to the confines of the retina and include such findings as microaneurysms, dot and blot intraretinal hemorrhages, retinal edema, hard exudates, dilation and bleeding of retinal veins, intraretinal microvascular abnormalities, nerve fiber layer infarcts, arteriolar abnormalities, and focal areas of capillary nonperfusion. Nonproliferative diabetic retinopathy can affect visual function through two mechanisms: intraretinal capillary closure resulting in macular ischemia, and increased retinal vascular permeability resulting in macular edema. Clinically significant macular edema is defined as any one of the following: (a) retinal edema located at or within 500 μm of the center of the macula, (b) hard exudates at or within 500 μm of the center if associated with thickening of adjacent retina, and (c) a zone of retinal thickening larger than one optic disc area if located within one disc diameter of the center of the macula.

Proliferative diabetic retinopathy is characterized by extraretinal fibrovascular proliferation, that is, fibrovascular changes that extend beyond the confines of the retina and into the vitreous cavity. Fibrovascular proliferation in proliferative diabetic retinopathy may lead to tractional retinal detachment and vitreous hemorrhage. High-risk proliferative diabetic retinopathy is defined by any combination of three of the four following retinopathy risk Factors: (a) presence of vitreous or preretinal hemorrhage, (b) presence of new vessels, (c) location of new vessels on or near the optic disc, and (d) moderate to severe extent of new vessels (Diabetic Retinopathy Study Research Group, 1979).

Management of diabetic retinopathy includes referring the patient to an internist for optimal glucose and blood pressure control. In the Early Treatment Diabetic Retinopathy Study, focal or grid laser photocoagulation treatment for clinically significant macular edema reduced the risk of moderate visual loss, increased the chance of visual improvement, and was associated with only mild loss of visual field (Early Treatment Diabetic Retinopathy Study Research Group, 1985, 1995). Panretinal laser photocoagulation treatment of eyes with high-risk proliferative diabetic retinopathy reduced the risk of severe visual loss by 50% compared with untreated control eyes (Diabetic Retinopathy Study Research Group, 1981). Surgery is often indicated for nonclearing vitreous hemorrhage and for tractional retinal detachment involving or threatening the macula.

Glaucoma. Primary open angle glaucoma (POAG) is the most prevalent type of glaucoma, affecting 1.3 to 2.1% of the general population over the age of 40 years in the United States, where it is the leading cause of irreversible blindness among Blacks and the third leading cause among Whites (following AMD and diabetic retinopathy). It is also responsible for 12% of legal blindness. Risk factors for the disease include increasing age (especially greater than 40 years), African ethnicity, positive family history of glaucoma, diabetes mellitus, and myopia (nearsightedness).

POAG is a chronic, slowly progressive optic neuropathy characterized by atrophy of the optic nerve and loss of peripheral vision. Central vision is typically not affected until late in the disease. Because central vision is relatively unaffected early in the cause of illness, visual loss generally progresses without symptoms and may remain undiagnosed for some time. Although usually bilateral, the disease may be asymmetrical. POAG is associated with increased intraocular pressure, but normal-tension glaucoma may cause glaucomatous vision loss in patients with normal intraocular pressure. Thus, "normal" eye pressure does not rule out the presence of glaucoma.

Treatment of POAG includes topical or systemic medications, laser surgery, or surgery to lower the intraocular pressure to a level at which optic nerve damage no longer occurs. Visual field testing is performed regularly to evaluate for progressive loss of peripheral vision, and the optic nerve is examined regularly to evaluate for evidence of progressive optic atrophy (clinical signs of glaucoma in the optic disc include asymmetry of the neuroretinal rim, focal thinning of the neuroretinal rim, optic disc hemorrhage, and any acquired change in the disc rim appearance or the surrounding retinal nerve fiber layer).

Specific Visual Functions

Visual Acuity. Visual acuity is the most common measure of central visual function and refers to the smallest object resolvable by the eye at a given distance. It is defined as the reciprocal of the smallest object size that can be recognized. Visual acuity is expressed as a fraction in which the numerator is the distance at which the patient recognizes the object and the denominator is the distance at which a standard eye recognizes the object. For instance, a visual acuity of 20/60 means that the patient needs an object 3 times larger or 3 times closer than a standard eye requires. The traditional visual acuity chart presents symbols of decreasing size with fixed high contrast. The visual acuity chart used most often in the clinical setting is the Snellen acuity chart, which is comprised of certain letters of the alphabet; the size of the letters are constant on a given line of the eye chart and decrease in size the lower the line on the chart. In accurate Snellen notation, the numerator indicates the test distance and the denominator indicates the letter size seen by the patient.

Contrast Sensitivity. Contrast sensitivity refers to the ability of the patient to detect differences in contrast and is defined as the reciprocal of the lowest contrast that can be detected. This may be measured with the Pelli-Robson chart, in which letters decrease in contrast rather than size, or the Bailey-Lovie chart, in which letters of a fixed low contrast are varied in size. Contrast sensitivity is considered a more sensitive indicator of visual function than Snellen acuity and may provide earlier detection of such pathology as retinal and optic nerve disease.

Visual Field. Visual field is classically defined as a three-dimensional graphic representation of differential light sensitivity at different positions in space. Perimetry refers to the clinical assessment of the visual field. Typically, visual field is assessed with kinetic or static perimetry. During kinetic perimetry, a test object of fixed intensity is moved along several meridians toward fixation, and points where the object is first perceived are plotted in a circle. During static perimetry, a stationary test object is increased in intensity from below threshold until perceived by the patient, and threshold values yield a graphic profile section. Although peripheral visual field loss often produces difficulty for patients in orientation and mobility functions, macular field loss (either centrally or paracentrally) often causes difficulties with reading. For instance, the presence of central or paracentral visual field loss is a more powerful predictor of reading speed than is visual acuity (Fletcher, Schuchard, Livingstone, Crane, & Hu, 1994).

Color Vision. Evaluation of color vision may be performed using pseudoisochromatic color plates, which are quick and commonly available; they consist of circles in various colors such that a person with normal color vision function can distinguish a number from the background pattern of circles. Ishihara or Hardy-Rand-Rittler pseudoisochromatic color plates are designed to screen for congenital red and green color deficiencies, whereas Lanthony tritan plates may be used to detect blue and yellow defects, which are frequently present in acquired disease. With the Farnsworth-Munsell 100-hue test, the patient must order 84 colored disks; the time-consuming nature of this test limits its clinical use. The Farnsworth Panel D-15 is a shorter and more practical version (using 15 disks) but is less sensitive. Most color vision defects are nonspecific.

Summary

Studies have demonstrated that ophthalmic patients are at high risk for decreased functional status and quality of life (Parrish, et al., 1997; Scott et al., 1994; Scott, Schein, Feuer, Folstein, & Bandeen-Roche, 2001). Patients' functional status and quality of life may be improved by interventions that increase visual function, such as surgery to repair retinal detachment or remove epiretinal membrane (Scott, Smiddy, Feuer, & Merikansky, 1998) and surgery to remove cataract (Applegate et al., 1987; Brenner, Curbow, Javitt, Legro, & Sommer, 1993; Donderi & Murphy, 1983; Javitt, Brenner, Curbow, Legro, & Street, 1993; Steinberg et al., 1994). In addition, functional status and quality of life may be improved by interventions, such as low-vision devices and services, which permit patients to use their remaining vision more effectively (Scott, Smiddy, Schiffman, Feuer, & Pappas, 1999). Prior studies have demonstrated the effect of low-vision interventions on objective task-specific measures of functional abilities such as reading speed, reading duration, and ability to read a certain print size (Nilsson, 1990; Nilsson & Nilsson, 1986; Rosenberg, Faye, Fischer, & Budicks,1989; Sloan, 1968). However, few data exists concerning the ability of the visually impaired to use computers and how modifications of a graphic user interface may increase accessibility of computer usage to patients with visual impairment (Jacko, Barreto, Marmet, et al., 2000; Jacko, Barreto, Scott, Rosa, & Pappas, 2000; Jacko, Dixon, Rosa, Scott, & Pappas, 1999; Jacko, Rosa, Scott, Pappas, & Dixon, 1999; Jacko, Rosa, Scott, Pappas, & Dixon, 2000; Jacko & Sears, 1998). The rapid propagation of visual displays beyond the desktop to such devices as handheld computers and mobile telephones, as well as the increasing population of persons with visual impairment, increase the importance of graphic user interface modifications that facilitate usage by visually impaired individuals.

HIGHLIGHTING TECHNOLOGICAL ADVANCEMENTS IN HCI RESEARCH

With nearly every aspect of today's society involving some type of computer technology, there is an ever-growing need to understand how individuals with perpetual impairments can access technology. Without special modifications, the typical personal computer currently poses several challenges to users with hearing, speech, or visual impairments. As a result, the HCI research community is placing great emphasis on the design of universally acceptable technologies. According to Stephanidis et al. (1998, p. 6), "universal access in the Information Society signifies the right of all citizens to obtain equitable access to, and maintain effective interaction with, a community-wide pool of information resources and artifacts." Accessibility has been a term traditionally associated with elderly individuals, individuals with disabilities, and others with special needs (Stephanidis et al., 1999); however, because of the current influx of new technologies into the market, the population of users who may possess special needs is growing. As a result, accessibility has taken on a more comprehensive connotation. This connotation denotes that all individuals with varying levels of abilities, skills, requirements, and preferences be able to access information technologies (Stephanidis et al., 1999). Universal access also implies more than just adding on features to existing technologies. Rather, the concept of universal access emphasizes that accessibility be incorporated directly into the design (Stephanidis et al., 1998). Perceptual and adaptive interfaces are two ideal examples of how universal accessibility can be achieved.

Perceptual Interfaces

The concept of perceptual design describes a perspective of design that defines interactions in terms of human perceptual capabilities. In a sense, it strives to humanize interaction. The design of perceptual interfaces adheres to the idea that lessons learned from psychological research about perception can be applied to interface design (Reeves & Nass, 2000). In adopting the concept of perceptual design, several opportunities surface for the creation of innovative perceptual user interfaces. Interactions with these interfaces can be described in terms of three particular human perceptual capabilities: chemical senses (i.e., taste and olfaction), cutaneous senses (i.e., skin and receptors), and vision and hearing (Reeves & Nass, 2000). Although commonly used computer technology limits the effectiveness of chemical senses, the technology can be extended to incorporate the cutaneous, visual, and hearing senses.

There exists extensive literature regarding vision and hearing. In terms of vision, research has focused on topics such as visual mechanics, color, brightness and contrast, objects and forms, depth, size, and movement. Hearing research includes psychophysical factors such as loudness, pitch, timbre, and sound localization; physiological mechanisms such as the auditory components of the ear, and the neural activity associated with hearing; and the perception of speech such as units of speech and the mechanics of word recognition (Reeves & Nass, 2000). With respect to the cutaneous senses, augmented graphical user interfaces (GUIs) with haptic feedback have been around since the early 1990s. Akamatsu and Sato (1994) conducted the first research with a haptic mouse that produced haptic feedback via fingertips and force feedback via controlled friction. Engel, Goossens, and Haakma (1994) found that directional two-degrees of freedom force feedback improved speed and error rates in a targeting task.

The strength of perceptual user interfaces comes from the ability of designers to combine an understanding of natural human capabilities with computer input–output devices, and machine perception and reasoning (Turk & Robertson, 2000). General examples of how capabilities can be combined with technology include speech and sound recognition and generation, computer vision, graphic animation, touch-based sensing and feedback, and user modeling (Turk & Robertson, 2000).

From an applied research standpoint, the concepts of perceptual interfaces are housed within multimedia and multimodal interfaces. Both multimedia and multimodal interfaces offer increased accessibility to technologies for individuals with perceptual impairments. Distinctions can be drawn between perceptual, multimedia, and multimodal interfaces, shown in Fig. 26.2. Perceptual interfaces prescribe humanlike perceptual capabilities to the computer. Multimedia and multimodal

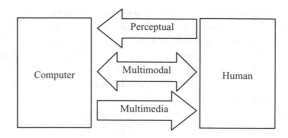

FIGURE 26.2. Perceptual, multimodal, and multimedia interfaces (flow of information).

interfaces can be considered applied extensions of this concept. Multimedia interfaces elicit perceptual and cognitive skills to interpret information presented to the user, whereas multimodal interfaces use multiple modalities for human–computer interaction. Multimedia interfaces focus on the media, whereas multimodal interfaces focus on the human perceptual channels (Turk & Robertson, 2000). The strength and capabilities of multimedia and multimodal interfaces with respect to individuals with perceptual impairments are described in more depth in the next two sections.

Multimedia Interfaces. Multimedia interfaces have grown from the need to display diverse forms of information in a flexible and interactive way. Multimedia can be simply defined as computer-controlled interactive presentations (Chignell & Waterworth, 1997). The broadness of this definition directly corresponds to the broadness of the field of multimedia research. Chapter 12 in this handbook contains a comprehensive discussion of multimedia, including topics dealing with the cognitive implications related to multimedia, selecting media for the message, navigation, and interaction and evaluation of multimedia. For the purpose of this chapter, the discussion of multimedia is limited to a brief overview of types and potential strengths of multimedia as they relate to enhancing accessibility to information technologies for individuals with perceptual impairments.

There are three approaches to multimedia: performance, presentation, and document (Chignell & Waterworth, 1997). In the performance approach, multimedia is a kind of theatrical play that is conveyed through "actors." The timing of the actors' performances is orchestrated in an effort to entertain and educate (Waterworth & Chignell, 1997). Presentation multimedia is a modern version of slide shows in which video clips and animation enhance a sequence of slides. The goal of the presentation approach is to convey ideas to the user (Chignell & Waterworth, 1997). Lastly, the document approach focuses on text and ideas. It can be thought of as an enhanced document that elaborates ideas in the text. All of these approaches provide additional opportunities to convey perceptual information to the user.

Multimedia allows communication between users and computers in a sensory manner. As such, the essential aspect of designing multimedia interfaces is selecting the media and modalities. Information can be taken from one modality and presented in another (Chignell & Waterworth, 1997). For instance, data presented visually could be converted and displayed audibly; multimedia could be used to enhance a GUI for a user with a hearing impairment, and information that would be commonly conveyed via auditory feedback could be provided visually. An example of this would be to have an icon display, rather than the traditional beep sounding, when an error has been committed. Other examples of how multiple modalities can enhance accessibility to information technologies are discussed in the multimodal section.

The potential strength of multimedia interfaces comes from its ability to use images, text, and animation to connect with users. To facilitate that type of exchange, however, much consideration must be paid to users' sensations and perceptions via the auditory, speech, and visual channels. For instance, when designing graphic images, basic knowledge of color vision is needed to ensure that colors can be discriminated, foreground can be separated from background, highlights will attract attention, and grouping of objects is apparent (Gillan, 1998). An understanding of human sensation and perception is necessary if perpetual interfaces are to reach their full potential. When perceptual interfaces are used, however, they provide alternative modalities to access technology for individuals with perpetual impairments. Multimodal interfaces are a second type of perceptual interface that provide this same benefit.

Multimodal Interfaces. Multimodal interfaces, as they are discussed in this chapter, are interfaces that support a wide range of perceptual capabilities (i.e., auditory, speech, and visual) as a means to facilitate human interaction with computers. Multimodal interfaces are discussed in a much broader sense (e.g., development of multimodal processing) in chapter 14.

With the growing complexity of technology and applications, a single modality no longer permits users to interact effectively across all tasks and environments (Oviatt et al., 2002). The strength of a multimodal design is its ability to allow users the freedom to use a combination of modalities or the best modality for their needs. These interfaces make the most effective use of the variety of human sensory channels, alone and in combination. Ultimately, multimodal interfaces offer expanded accessibility of computing and promote new forms of computing that were not previously available to individuals with perceptual impairments.

The development and application of multimodal interfaces for the purpose of making technologies accessible to users with impairments, and perceptual impairments in particular, is growing. Because most information presented on a GUI is visual, there is great interest in the research community to find alternative ways of displaying this information. One of the common approaches to conveying visual information in a nonvisual way is through the use of the tactile modality. Specific research has been conducted in the realm of tactile displays with respect to visual impairments. Not only can tactile displays provide information regarding a graphic's identity, but also the depth, location, and perception of its purpose. The use of tactile systems can also provide navigational information. Research has shown that tactile output of directional information offers support to the blind as they explore images (Kurze, 1998). Research has also been conducted in the area of movable dynamic tactile displays that present information to one or several fingertips in a

Braille-type manner. The Braille dots move in a wave of lifted and lowered series of pins (Fricke & Baehring, 1994). Along the same line of research, the use of a bidimensional single-cell Braille display combined with a standard Braille cell has been evaluated. Although initial research found this new combined device not to be an improvement over a standard stand-alone Braille display, continued research is yielding improvement (Ramstein, 1996). Tactile output via force feedback has been considered as a means of conveying numerical information. Yu, Ramloll, and Brewster (2000) have worked on a system that converts data typically displayed visually into haptic and auditory output. Because data visualization techniques are not typically appropriate for blind people or people with visual impairments, this system translates graphs into friction and textured surfaces along with auditory feedback.

Another common approach to converting visual information in a nonvisual way is through the use of the speech modality. Speech-recognition systems serve as an alternative modality for users and computers to interact. These systems recognize human speech and translate it into commands or words understood by the computer. Chapter 35 in this handbook offers an introduction to the technology of speech recognition systems and design issues associated with incorporating speech recognition into applications. This chapter's discussion of speech recognition systems concentrates instead on its use related to individuals with perceptual impairments. The integration of speech input and output into applications offers an alternative to purely graphic environments (Yankelovich & Lai, 1998). This type of technology-driven design allows for applications to be suited to a wide variety of individuals with disabilities (Danis & Karat, 1995). For instance, individuals with visual impairments can use a computer solely by voice activation.

Speech recognition systems are traditionally associated with the concept of dictation. Products such as Dragon Naturally Speaking (Scansoft) offers speech recognition products for dictation. Some specific packages are geared toward particular professions, such as medical and legal (Cunningham & Coombs, 1997). Other common dictation systems are IBM ViaVoice, and SRI International Eduspeak offer continuous speech recognition technology. Microsoft's Voicenet VRS provides speaker-independent speech recognition with online adaptation, noise robustness, and dynamic vocabularies and grammars (Huang et al., 1995). Nuance Communications Nuance 8 is another popular natural speech interface software that facilitates access to information, transactions, and services over the telephone (Nuance, 1999). Speech-driven menu navigation systems, such as IN CUBE Voice Command (Command Corp.) for window navigation, have also been developed (Karshmer, Brawner, & Reiswig, 1994; Karshmer, Ogden, Brawner, Kaugars, & Reiswig, 1994; Wilcox, 1999).

Because hearing loss and many other types of disabilities are associated to some degree with speech degeneration, speech recognition systems may not be the sole answer for replacing or augmenting GUIs (Rampp, 1979). For cases in which speech is degraded, speech recognition systems must be sensitive enough to adapt to impaired speech. Regardless of what caused the speech impairment, devices and techniques can be applied to augment the communicative abilities of individuals who experience difficulty speaking in an understandable manner. Augmentative and alternative communication (AAC) is a field of study concerned with providing such devices and techniques. McCoy, Demasco, Pennington, and Badman (1997) described a prototype system aimed at users with cognitive impairments. This prototype is designed to aid communication and provide language intervention benefits across several user populations. An iconic language approach has been applied to aid individuals with significant speech and multiple impairments (SSMI). Research has been conducted on the use of icon language design, based on the theory of icon algebra (Chang, 1990) and the theory of conceptual dependency (Schank, 1972). These methodologies are then used in interactive interface design (Albacete, Chang, & Polese, 1994). Individuals with SSMI also commonly have difficulty with word processing. Through the use of an animated graphic display, phoneme probabilities of speech can be more easily isolated and recognized. This allows users the opportunity to interpret their speech rather than forcing the computer to do it automatically (Roy & Pentland, 1998). The more feedback and modalities users are provided, the more efficiently they can interact with speech recognition by assisting the computer to correctly interpret their speech. Speech recognition systems must also take into account the possible extent of cognitive burden placed on the user. If voice is used to navigate through a GUI, certain prosodic features (e.g., pauses) of the user's speech, resulting from a high cognitive load, may affect performance (Baca, 1998). Additional challenges associated with the implementation of speech recognition systems are discussed in chapter 35 as well. Overall, these systems need to account for the users' varying levels of hearing, speech, and cognitive abilities.

The auditory channel serves not only to convey information, but also to receive information. The use of auditory feedback is extremely useful to many computer users. Ongoing research has looked at the use of bidirectional sound as a standard element of an interface environment. A prototype named the Voice Enabled Reading Assistant (VERA), written using an aural-oral user interface (A-OUI) model, was developed to provide bidirectional sound. The A-OUI model captures qualities and functions of plain text files needed for user interfaces to present information to the auditory, visual, and tactile senses. The use of the VERA prototype can be applied to many types of office and Internet text and data (Ryder & Ghose, 1998).

Much work with respect to audio feedback has focused on conveying information from a GUI. For instance, research conducted by Darvishi, Guggiana, Munteanu, and Schauer (1994) and Darvishi, Munteanu et al. (1994) looked at mapping GUIs into auditory domains through 'impact sounds' based on physical modeling. Ultimately, these sounds were used to convey information to the user regarding the objects in a GUI environment. Audio feedback has also been used to provide visually impaired users with a sense of depth perception by varying the location of the sound sources in a three-dimensional environment. Because depth perception is a function of vision, any cues that can be conveyed though other modalities are vital. This is particularly crucial when working with

three-dimensional computer applications, such as a CAD package (Mereu & Kazman, 1996). The use of auditory feedback has also been studied with regard to enhancing synthesis speech output. Through the development of spatial audio processing systems, a greater benefit of synthesis speech was achieved (Crispien, Wurz, & Weber, 1994). IBM's Websphere Voice Server uses synthesis speech to make GUIs accessible to those who are visually impaired or blind by converting screen information to speech or Braille. The users are continuously kept informed of screen activity and cursor movement. Reading typed characters, words, and sentences are features that can be made automatic. This software aids in the use of windows, menus, dialog boxes, and other controls (Thatcher, 1994). Scansoft's Realspeak a trainable, text-to-speech system that produces synthetic speech, sounds natural by reproducing the characteristics of original speech. WRITE:OUTLOUD is an additional text-to-speech product commonly used (Friedlander, 1997). Regardless of what product is used, text-to-speech software enables information to be collected and utilized in a more rapid manner.

In addition to the benefits already discussed that enable users with perceptual impairments to access technology, multimodal interfaces also provide superior support of error handling. More specifically, multimodal interfaces have been shown to have the ability to avoid errors and recover more efficiently from errors when they do occur. This is an important element of universal accessibility. Enabling users to use a technology not only includes being able to access the technology but also the ability to use it in an efficient manner. Reducing or avoiding errors is a major key to improving efficiency. There are several reasons a multimodal interface provides better error handling than a unimodal interface. (Oviatt, 1999):

- Users can select the input mode considered to be less error prone for a particular context, which is assumed to lead to error avoidance.
- The ability to use several modalities permits users the flexibility to leverage their strength by using the most appropriate modality. It is possible that the more comfortable the user, the less likely he or she is to make errors.
- Users tend to switch modes after systems errors. For instance, users are likely to switch input modes when they encounter a system recognition error.
- Users are less frustrated with error when interacting within a multimodal interface, even when errors are as frequent as in a unimodal interface. The reduction in frustration may be the result of the user feeling as thought they have more control over the system because they can switch modes.

Ultimately, multimodal systems aid in reducing errors (error handling) because they offer parallel or duplicated functionality that allow users to accomplish the task using one of several modalities (Oviatt & Cohen, 2000). These benefits stem from the user-centered and system-centered design perspectives from which multimodal interfaces are built (Oviatt et al., 2002; Oviatt & Cohen, 2000). Like multimodal interfaces, adaptive interfaces also benefit from a user-centered perspective.

Adaptive Interfaces

Adaptive interfaces have great potential for accommodating a wide range of users in a variety of work contexts. As a result, much research has been conducted on the design and implementation of adaptive interfaces. An example of early research in the field of adaptive GUI is illustrated by the work of Mynatt and Weber (1994) with the MERCATOR project and by the GUIB (Textual and Graphical User Interfaces for Blind People) project (Petrie, Morley, & Weber, 1995). Both of these projects focused on making environment-level adaptations to GUIs to make them more accessible (Stephanidis, 2001a). Mercator interfaces model graphical objects and their hierarchical relationships. The model serves to predict a user's interaction (Edwards & Mynatt, 1994; Edwards, Mynatt, & Stockton, 1994). The goal is to provide visually impaired users with an interface that is more accessible. By better understanding how users with low vision interact with a computer, interfaces can be designed more effectively. Outputs such as synthetic and digitized speech, refreshable Braille, and nonspeech sounds make an interface easier to use. Auditory icons and earcons are two predominate areas of nonspeech sound research. Auditory icons, developed by Gaver (1989), are everyday sounds that occur in the world mapped to the computer world. Graver first applied the concept of auditory icons to SonicFinder. Jacko (1996) and Jacko and Rosenthal (1997) investigated the uses of auditory icons in educational software for children. For more information regarding SonicFinder and other applications of auditory icons, please refer to chapter 11 of this handbook. Earcons are a second method for employing nonspeech audio in GUIs. Earcons, developed by Blattner, Sumikawa, and Greenberg (1989), are audio messages that provide the user with information about objects or operations of the computer. Earcons are explained in more depth in chapter 11 as well. The first version of GUIB adapted the GUI by combining Braille speech and nonspeech audio together to construct a nonvisual interface so that users who are blind can access the GUI (Emiliani, 2001).

Configurable interface designs based on user models have been heavily researched. The user models have been defined with respect to visual, cognitive, motor, and other abilities. Through the use of these models, custom computer systems, including both hardware and software, can be created (McMillan & Wisniewski, 1994). Semantic abstraction of user interaction, named abstract widgets, is another modeling approach that provides great flexibility. Abstract widgets separate the user interface from the application functionality. This allows users to interact with interfaces as they choose, independently from their environment (Kawai, Aida, & Saito, 1996). The use of adaptation determinates, constituents, goals, and rules are yet another approach of an adaptation strategy. This strategy is based on the fact that these important attributes, which categorize adaptation, can be used to formulate adaptation rules. These adaptation rules, in turn, assist the development

of intelligent user interfaces. This approach can be customized to the requirements of different application domains and user groups (Stephanidis, Karagiannidis, & Koumpis, 1997). Approaches such as described by pervasive accessible technology (PAT) allow individuals with disabilities to use standard interface devices that adapt to the user to communicate with information technology infrastructures (Paciello, 1996). Based on an individual's disability, the implementation of a user interface management system (UIMS) model provides the versatility needed to adapt interfaces to individuals. The selection of input devices, presentation of information on the screen, and choice in selection and activation method can all be adapted to fit specific user needs (Buhler, Heck, & Wallbruch, 1994). Current research focuses on operationally reliable software infrastructures that support alternative physical realizations through the abstractions of objects. More specifically, systems such as Active X by Microsoft and JavaBeans by SunSoft represent componentware technology (Stephanidis, 2001b). These systems represent a mainstream effort to provide technological structures that provide more adequate support for accessibility. They are also two examples of currently available tools that can be used to generate code for the adaptation of various interface components.

Some hypermedia systems research has focused on adaptive applications that keep track of the user's evolving aspects, such as preferences and domain knowledge. This information is stored to create a user model, which in turn is the basis for user interface adaptation (De Bra, Houben, & Wu, 1999). Adaptive hypermedia can also be designed based on task models, as the basis from which hypermedia systems are developed. These task models support the design and development of hypermedia. Different task models are associated with different types of users (Paterno & Mancini, 1999). Task models reflect the user's view of the activities to be performed. In addition to personalizing the content of hypermedia systems, adaptive navigation support has also been researched. Prototype systems have been developed to demonstrate how different navigational possibilities can be made available based on a user model (Pilar da Silva, Van Durm, Duval, & Olivie, 1998).

Another example of adapting interfaces to users is through the use of the EZ Access protocols developed by the Trace Center at the University of Wisconsin—Madison. EZ Access protocols are a set of techniques that modify an interface to fit a specific user's needs. These protocols work across disabilities and with a range of products. The most common use of EZ Access protocols is in touchscreen kiosks (Vanderheiden, 1998). For more information regarding adaptive interfaces (e.g., the functions of adaptation, user property modeling, and approaches to inference and decision making), refer to chapter 15 in this handbook.

Summary

Through technological advancements in HCI research, the concepts of perceptual and adaptive interfaces have emerged. These two categories of technologies provide vast opportunities for individuals with perceptual impairments to fully access electronic information. More specifically, multimedia and multimodal systems have been shown to greatly accommodate users of varying abilities.

CONCLUSIONS

To provide a context for the topic of perceptual impairments and computing technologies, Fig. 26.1 was presented, demonstrating that to achieve universal access across classes of computing technologies, researchers must be prepared to address several challenging issues: We must establish empirical links between clinical diagnoses and sets of functional capabilities, define the set of functional capabilities required to access information technologies, and establish empirical bases for the influence of specific functional capabilities on accessing to specific classes of technologies.

This chapter aimed to establish a basis for addressing such issues by examining three types of impairment (hearing, speech, and visual), specific diagnoses within each type, and their resulting functional abilities. Finally, advancements in perceptual interfaces, multimodal interfaces, multimedia interfaces, and adaptive interfaces was discussed, which can be applied across a variety of classes of technology to enhance the perceptual experience of individuals who possess such impairments.

ACKNOWLEDGMENTS

This chapter was made possible through funding provided to the first author by the National Science Foundation (BES-9896304) and by the Intel Corporation.

References

Akamatsu, M., & Sato, S. (1994). A multi modal mouse with tactile and force feedback. *International Journal of Human-Computer Studies, 40*, 443–453.

Albacete, P. L., Chang, S. K., & Polese, G. (1994). Iconic language design for people with significant speech and multiple impairments. *Proceedings for the ASSETS '94, the First Annual International ACM/SIGCAPH Conference on Assistive Technologies, Marina Del Rey, CA* (pp. 23–30). New York: Association for Computing Machinery.

American Medical Association. (1990). *Guides to the evaluation of permanent impairment* (3rd ed.). Milwaukee, WI: Author.

Applegate, W. B., Miller, S. T., Elam, J. T., Freeman, J. M., Wood, T. O., & Gettlefinger, T. C. (1987). Impact of cataract surgery with lens implantation on vision and physical function in elderly patients. *JAMA, 257*, 1064–1066.

Baca, J. (1998). Comparing effects of navigational interface modalities on speaker prosodics. *Proceedings for the ASSETS '98, the Third Annual International ACM/SIGCAPH Conference on Assistive Technologies, Marina Del Rey, CA* (pp. 3–10). New York: Association for Computing Machinery.

Blattner, M., Sumikawa, D., & Greenberg, R. (1989). Earcons and icons: Their structure and common design principles. *Human Computer Interaction, 4*, 11–44.

Brenner, M. H., Curbow, B., Javitt, J. C., Legro, M. W., & Sommer, A. (1993). Vision change and quality of life in the elderly. Response to cataract surgery and treatment of other chronic ocular conditions. *Archives of Ophthalmology, 111*, 680–685.

Buhler, C., Heck, H., & Wallbruch, R. (1994). A uniform control interface for various electronic aids. In W. L. Zagler, G. Busby, & R. R. Wagner (Eds.), *Computers for handicapped persons. Proceedings for the 4th International Conference* (pp. 51–58). Vienna: Springer-Verlag.

Chang, S. K. (Ed.). (1990). *Principles of visual programming systems.* New York: Prentice-Hall.

Chignell, M., & Waterworth, J. (1997). Multimedia. In G. Salvendy (Ed.), *Handbook of human factors and ergonomic* (pp. 1808–1861). New York: Wiley.

Colenbrander A. (1977). Dimensions of visual performance. *Transactions of the American Academy of Ophthalmology and Otolaryngology, 83*, 332–337.

Cook, M., & Hussey, S. M. (1995). *Assistive technologies: Principles and practice.* St. Louis, MO: Mosby-Year Book.

Coren, S., Ward, L. M., & Enns, J. T. (1999). Sensation and perception (5th ed.). New York: Harcourt Brace College Publishers.

Crispien, K., Wurz, W., & Weber, G. (1994). Using spatial audio for the enhanced presentation of synthesized speech within screen readers for blind computer user. In W. L. Zagler, G. Busby, & R. R. Wagner (Eds.), *Computers for handicapped persons. Proceedings for the 4th International Conference* (pp. 144–153). Vienna: Springer-Verlag.

Cunningham, C., & Coombs, N. (1997). *Information access and adaptive technology.* Phoenix, AZ: American Council on Education and The Oryx Press.

Danis, D., & Karat, J. (1995). Technology-driven design of speech recognition systems. *Conference Proceedings on designing Interactive Systems: Processes, practices, methods, & techniques, Ann Arbor, MI* (pp. 17–24). New York: Association for Computing Machinery.

Darvishi, A., Guggiana, V., Munteanu, E., & Schauer, H. (1994). Synthesizing non-speech sound to support blind and visually impaired computer users. In W. L. Zagler, G. Busby, & R. R. Wagner (Eds.), *Computers for Handicapped Persons. Proceedings for the 4th International Conference* (pp. 385–393). Vienna: Springer-Verlag.

Darvishi, A., Munteanu, E., Guggiana, V., Schauer, H., Motavalli, M., & Rauterberg, M. (1994). Automatic impact sound generation for using in non-visual interfaces. *Proceedings for the ASSETS '94, the First Annual International ACM/SIGCAPH Conference on Assistive Technologies, Marina Del Rey, CA* (pp. 100–106). New York: Association for Computing Machinery.

De Bra, P., Houben, G., & Wu, H. (1999). AHAM: A Dexter-based reference model for adaptive hypermedia. *Proceedings of the 10th ACM Conference on Hypertext and Hypermedia: Returning to Our Diverse Roots, Darmstadt Germany* (pp. 147–156). New York: Association for Computing Machinery.

Diabetic Retinopathy Study Research Group. (1979). Four risk factors for severe visual loss in diabetic retinopathy: DRS Report 3. *Archives of Ophthalmology, 97*, 654–655.

Diabetic Retinopathy Study Research Group. (1981). Photocoagulation treatment of proliferative diabetic retinopathy: Clinical application of diabetic retinopathy study (DRS) findings. DRS Report 8. *Ophthalmology, 88*, 583–600.

Donderi, D. C., & Murphy, S. B. (1983). Predicting activity and satisfaction following cataract surgery. *Journal of Behavioral Medicine, 6,* 313-328.

Dowie A. T. (1988). *Management and practice of low visual acuity.* London: The Eastern Press.

Early Treatment Diabetic Retinopathy Study Research Group. (1985). Photocoagulation for diabetic macular edema. *Archives of Ophthalmology, 103,* 1796-1806.

Early Treatment Diabetic Retinopathy Study Research Group. (1995). Focal photocoagulation treatment of diabetic macular edema. ET-DRS Report 19. *Archives of Ophthalmology, 113,* 1144-1155.

Edwards, W. K., & Mynatt, E. D. (1994). An architecture for transforming graphical interfaces. *Proceedings of ACM Conference on User Interface Software and Technology (UIST),* Marina Del Rey, CA (pp. 39-47). New York: Association for Computing Machinery.

Edwards, W. K., Mynatt, E. D., & Stockton, K. (1994). Providing access to graphical user interfaces—not graphical screens. *Proceedings for the ASSETS '94, the First Annual International ACM/SIGCAPH Conference on Assistive Technologies,* Marina Del Rey, CA (pp. 47-54). New York: Association for Computing Machinery.

Emiliani, P. L. (2001). Special needs and enabling technologies: An evolving approach to accessibility. In C. Stephanidis (Ed.), *User interfaces for all—concepts, methods and tools* (pp. 97-113). Mahwah, NJ: Lawrence Erlbaum Associates.

Engel, F. L., Goossens, P., & Haakma, R. (1994). Improved efficiency through I and E feedback: A trackball with contextual force feedback. *International Journal of Human-Computer Studies, 41*(6), 949-974.

Fletcher, D. C. (Ed.). (1999). *Low vision rehabilitation.* Hong Kong: American Academy of Ophthalmology.

Fletcher, D. C., Schuchard, R. A., Livingstone, C. L., Crane, W. G., & Hu, S. (1994). Scanning laser ophthalmoscope macular perimetry and applications for low vision rehabilitation clinicians. *Ophthalmology Clinics of North America, 7,* 257-265.

Fricke, J., & Baehring, H. (1994). Displaying laterally moving tactile information. In W. L. Zagler, G. Busby, & R. R. Wagner (Eds.), *Computers for handicapped persons. Proceedings for the 4th International Conference (ICCHP'94)* (pp. 461-468). Vienna: Springer-Verlag.

Friedlander, C. (1997). Speech facilities for the reading disabled. *Communications of the ACM, 40,* 24-25.

Gaver, W. (1989). The SonicFinder: An interface that uses auditory icons. *Human Computer Interaction, 4,* 67-94.

Gillan, D. J. (1998). The psychology of multimedia: Principles of perception and cognition. *Proceedings of the 1998 ACM Conference on Human Factors in Computing Systems (CHI'98),* Los Angeles, CA (pp. 143-144). New York: Association for Computing Machinery.

Goodman, J. C., Lee, L., & DeGroot, J. (1994). Developing theories of speech perception: Constraints from developmental data. In J. C. Goodman & H. C. Nusbaum (Eds.), *The development of speech perception: The transition from speech sounds to spoken words* (pp. 3-33). London: MIT Press.

Huang, X., Acero, A., Adcock, J., Hon, H., Goldsmith, J., Liu, J., & Plumpe, M. (1996). Whistler: A trainable text-to-speech system. *International Conference of Spoken Language Processing,* Philadelphia, PA (pp. 2387-2390). Piscataway: IEEE.

Jacko (1996). The identifiability of auditory icons for use in educational software for children. *Interacting with Computers, 8*(3), 121-133.

Jacko, J. A., & Rosenthal, D. J. (1997). Age-related differences in the mapping of auditory icons of visual icons in computer interfaces for children. *Perceptual and Motor Skills, 84,* 1223-1233.

Jacko, J. A., Barreto, A. B., Marmet, G. J., Chu, J. Y. M., Bautsch, H. S., Scott, I. U., & Rosa, R. H. (2000). Low vision: the role of visual acuity in the efficiency of cursor movement. *Proceedings of the Fourth International ACM Conference on Assistive Technologies (ASSETS 2000),* Arlington, VA (pp. 26-28). New York: Association for Computing Machinery.

Jacko, J. A., Barreto, A. B., Scott, I. U., Rosa, R. H., & Pappas, C. J. (2000). Using electroencephalogram to investigate stages of visual search in visually impaired computer users: preattention and focal attention. *International Journal of Human-Computer Interaction, 12,* 135-150.

Jacko, J. A., Dixon, M. A., Rosa, R. H., Scott, I. U., & Pappas, C. J. (1999). Visual profiles: A critical component of universal access. *Proceedings of the Association for Computing Machinery Conference on Human Factors in Computing Systems (ACM CHI 99)* (pp. 330-337). New York: Association of Computing Machinery.

Jacko, J. A., Rosa, R. H., Scott, I. U., Pappas, C. J., & Dixon, M. A. (1999). Linking visual capabilities of partially sighted computer users to psychomotor task performance. In H.-J. Bullinger, J. Ziegler, & N. J. Mahwah (Eds.), *Human-computer interaction: Communication, cooperation, and application design; proceedings of the 8th international Conference on Human-Computer Interaction* (pp. 975-979). Mahwah, NJ: Lawrence Erlbaum.

Jacko, J. A., Rosa, R. H., Scott, I. U., Pappas, C. J., & Dixon, M. A. (2000). Visual impairment: the use of visual profiles in evaluations of icon use in computer-based tasks. *International Journal of Human-Computer Interaction, 12,* 151-165.

Jacko, J. A., & Sears, A. (1998). Designing interfaces for an overlooked user group: considering the visual profiles of partially sighted users. *The 3rd ACM/SIGCAPH Conference on Assistive Technologies (ASSETS 98),* Marina del Rey, California (pp. 75-77). New York: Association for Computing Machinery.

Javitt, J. C., Brenner, M. H., Curbow, B., Legro, M. W., & Street, D. A. (1993). Outcomes of cataract surgery. Improvement in visual acuity and subjective visual function after surgery in the first, second, and both eyes. *Archives of Ophthalmology, 111,* 686-691.

Kaiser, A. P., Alpert, C. L., & Warren, S. F. (1988). Language and communication disorders. In V. B. Van Hassek (Ed.), *Handbook of developmental and physical disabilities* (pp. 395-422). New York: Pergamon Press.

Karshmer, A. I., Brawner, P., & Reiswig, G. (1994). An experimental sound-based hierarchical menu navigation system for visually handicapped use of graphical user interfaces. *Proceedings for the ASSETS '94, the First Annual International ACM/SIGCAPH Conference on Assistive Technologies,* Marina Del Rey, CA (pp. 123-128). New York: Association for Computing Machinery.

Karshmer, A. I., Ogden, B., Brawner, P., Kaugars, K., & Reiswig, G. (1994). Adapting graphical user interfaces for use by visually handicapped computer users: Current results and continuing research. In W. L. Zagler, G. Busby, & R. R. Wagner (Eds.), *Computers for Handicapped Persons. Proceedings for the 4th International Conference* (pp. 16-24). Vienna: Springer-Verlag.

Kawai, S., Aida, H., & Saito, T. (1996). Designing interface toolkit with dynamic selectable modality. *Proceedings of the second annual ACM conference on Assistive technologies,* Vancouver, Canada (p. 72). ACM: Association for Computing Machinery.

Kupfer, C. (2000). The National Eye Institute's low vision education program: Improving quality of life. *Ophthalmology, 107,* 229-230.

Kurze, M. (1998). TGuide: A guidance system for tactile image exploration. *Proceedings for the ASSETS '98, the Third Annual International ACM/SIGCAPH Conference on Assistive Technologies,* Marina Del Rey, CA (pp. 85-91). New York: Association for Computing Machinery.

LaPlante, M. P. (1988). Prevalence of conditions causing need for assistance in activities of daily living. In M. P. LaPlante (Ed.), *Data on disability from the National Health Interview Survey, 1983-1985* (p. 3). Washington, DC: National Institute on Disability and Rehabilitattion Research.

Levine, M. W. (2000). *Fundamentals of sensation and perception* (3rd ed.). New York: Oxford University Press.

McCoy, K. F., Demasco, P., Pennington, C. A., & Badman, A. L. (1997). Some interface issues in developing intelligent communication aids for people with disabilities. *Proceedings of the 1997 International Conference on Intelligent user interfaces (IUI'97)*, Orlando, FL (pp. 163-170). New York: Association for Computing Machinery.

McMillan, W. W., & Wisniewski, L. (1994). A rule-based system that suggests computer adaptations for users with special needs. *Proceedings for the ASSETS '94, the First Annual International ACM/SIGCAPH Conference on Assistive Technologies*, Marina Del Rey, CA (pp. 129-135). New York: Association for Computing Machinery.

Mehr, E. B. (1975). *Low vision care* (p. 254). Chicago: Professional Press.

Mereu, S. W., & Kazman, R. (1996). Audio enhanced 3D interfaces for visually impaired users. *Proceedings of the 1996 ACM Conference on Human Factors in Computing Systems (CHI'96)*, Vancouver Canada (pp. 72-78). New York: Association for Computing Machinery.

Miller, J. L., & Eimas, P. D. (1995). Speech, Language, and Communication. New York: Academic Press.

Moore, B. C. J. (1982). *An introduction to the psychology of hearing* (2nd ed.). New York: Academic Press.

Mynatt, E. D., & Weber, G. (1994). Nonvisual presentation of graphical user interfaces: contrasting two approaches. *Proceedings of the 1994 ACM Conference on Human Factors in Computing Systems (CHI'94)*, Boston, MA (pp. 166-172). New York: Association for Computing Machinery.

National Advisory Eye Council. (1993). *Vision research: A national plan 1994-1998*. Bethesda, MD: National Eye Institute, National Institute of Health.

National Ear Care Plan. (2002). Prevalence of Hearing Loss. Retrieved from http://www.necp.com/001/welcome.htm on march 20, 2002.

Nelson, K. A. (1987). Visual impairment among elderly Americans: statistics in transition. *Journal of Visual Impairment and Blindness, 81*, 331-334.

Nelson, K., & Dimitrova, G. (1993). Statistical brief #36: "Severe visual impairment" in the U.S. and the states. *Journal of Visual Impairment and Blindness, 87*, 80-85.

Nilsson, U. (1990). Visual rehabilitation of patients with and without educational training in the use of optical aids and residual vision: A prospective study of patients with advanced age-related macular degeneration. *Clinical Vision Science, 6*, 3-10.

Nilsson, U. L., & Nilsson, S. E. G. (1986). Rehabilitation of the visually handicapped with advanced macular degeneration. *Documenta Opthalmologica, 62*, 345-367.

Nuance. (1999). *Nuance 6*. Retrieved (March 20, 2002) from http://www.nuance.com/index.htma?SCREEN=nuance6

Office for Students with Disabilities. (2000). *People with speech impairments*. Retrieved (March 20, 2002) from http://spot.pcc.edu/osd/acspeech.htm

Oviatt, S. (1999). Mutual disambiguation of recognition errors in a multimodal architecture. *Proceedings of the 1999 ACM Conference companion on Human Factors in Computing Systems: Common Ground*, Pittsburgh, PA (pp. 576-583). New York: Association for Computing Machinery.

Oviatt, S., & Cohen, P. (2000). Multimodal interfaces that process: What comes naturally. *Communications of the ACM, 43*(3), 45-53.

Oviatt, S., Cohen, P., Suhm, B., Bers, J., Wu, L., Holzman, T., Winograd, T., Vergo, J., Duncan, L., Landay, J., Larson, J., & Ferro, D. (2002). Designing the user interface for multimodal speech and gesture applications: State-of-the-art systems and research directions from 2000 and beyond. *Human-Computer Interaction in the New Millennium*. New York: ACM Press.

Paciello, M. G. (1996). Designing for people with disabilities. *Interactions, 3*, 15-16.

Parrish, R. K. II, Gedde, S. J., Scott, I. U., Feuer, W. J., Schiffman, J. C., Mangione, C. M., & Montenegro-Piniella, A. (1997). Visual function and quality of life among patients with glaucoma. *Archives of Ophthalmology, 115*, 1447-1455.

Paterno, F., & Mancini, C. (1999). Developing adaptable hypermedia. *Proceedings of the 1999 ACM Conference on International Conference on Intelligent User Interfaces (IUI'99)*, Redondo Beach, CA (pp. 163-170). New York: Association for Computing Machinery.

Petrie, H., Morley, S., & Weber, G. (1995). Tactile-based direct manipulation in GUIs for blind users. *Proceedings of the 1995 ACM Conference on Human Factors in Computing Systems (CHI'95), Conference Companion*, Denver, CO (pp. 428-429). New York: Association for Computing Machinery.

Pilar da Silva, D., Van Durm, R., Duval, E., & Olivie, H. (1998). Adaptive navigational facilities in educational hypermedia. *Proceedings of the ninth ACM Conference on Hypertext and hypermedia: links, objects, time and space—structure in hypermedia systems (HyperText'98)*, Pittsburgh, PA (pp. 291-292). New York: Association for Computing Machinery.

Proctor, R. W., & Proctor, J. D. (1997). Sensation and perception. In G. Salvendy (Ed.), *Handbook of human factors and ergonomics* (pp. 43-88). New York: Wiley.

Rampp, D. L. (1979). Hearing and learning disabilities. In L. J. Bradford & W. G. Hardy (Eds.), *Hearing and hearing impairment* (pp. 381-389). New York: Grune & Stratton.

Ramstein, C. (1996). Combining hepatic and Braille technologies: Design issues and pilot study. *Proceedings for the ASSETS '96, the Second Annual International ACM/SIGCAPH Conference on Assistive Technologies*, Vancouver Canada (pp. 37-44). New York: Association for Computing Machinery.

Reeves, B., & Nass, C. (2000). Perceptual Bandwidth. *Communications of the ACM, 43*(3), 65-70.

Rosenberg, R., Faye, E., Fischer, M., & Budicks, D. (1989). Role of prism relocation in improving visual performance of patients with macular dysfunction. *Optometry and Vision Science, 66*, 747-750.

Roy, D., & Pentland, A. (1998). A phoneme probability display for individuals with hearing disabilities. *Proceedings for the ASSETS '98, the Third Annual International ACM/SIGCAPH Conference on Assistive Technologies*, Marina Del Rey, CA (pp. 165-168). New York: Association for Computing Machinery.

Ryder, J. W., & Ghose, K. (1998). Multi-sensory browser and editor model. *Proceedings of the 1999 ACM symposium on applied computing*, Langdale, Cumbria, United Kingdom (pp. 443-449). New York: Association for Computing Machinery.

Schank, R. C. (1972). Conceptual dependency: A theory of natural language understanding. *Cognitive Psychology, 3*, 532-631.

Scott, I. U., Schein, O. D., West, S., Bandeen-Roche, K., Enger, C., & Folstein, M. F. (1994). Functional status and quality of life measurement among ophthalmic patients. *Archives of Ophthalmology, 112*, 329-335.

Scott, I. U., Smiddy, W. E., Feuer, W., & Merikansky, A. (1998). Vitreoretinal surgery outcomes. Results of a patient satisfaction/functional status survey. *Ophthalmology, 105*, 795-803.

Scott, I. U., Smiddy, W. E., Schiffman, J., Feuer, W. J., & Pappas, C. J. (1999). Quality of life of low-vision patients and the impact of low-vision services. *American Journal of Ophthalmology, 128*, 54-62.

Scott, I. U., Schein, O. D., Feuer, W. J., Folstein, M. F., & Bandeen-Roche, K. (2001). Emotional distress in patients with retinal disease. *American Journal of Ophthalmology, 131*, 584–589.

Sloan, L. L. (1968). Reading aids for the partially-sighted: factors which determine success or failure. *Archives of Ophthalmology, 80*, 35–38.

Stein, L. K. (1988). Hearing impairment. In V. B. Van Hassek (Ed.), *Handbook of developmental and physical disabilities* (pp. 271–294). New York: Pergamon Press.

Steinberg, E. P., Tielsch, J. M., Schein, O. D., Javitt, J. C., Sharkey, P., Cassard, S. D., Legro, M. W., Diener-West, M., Bass, E. B., Damiano, A. M. (1994). National study of cataract surgery outcomes. Variation in 4-month postoperative outcomes as reflected in multiple outcome measures. *Ophthalmology, 101*, 1131–1140.

Stephanidis, C. (2001a). The concept of unified user interfaces. In C. Stephanidis (Ed.), *User interfaces for all* (pp. 371–388). Mahwah, NJ: Lawrence Erlbaum Associates.

Stephanidis, C. (2001b). User interfaces for all: New perspectives into human-computer interaction. In C. Stephanidis (Ed.), *User interfaces for all—concepts, methods and tools* (pp. 3–17). Mahwah, NJ: Lawrence Erlbaum Associates.

Stephanidis, C., Karagiannidis, C., & Koumpis, A. (1997). Decision making in intelligent user interfaces. *Proceedings of the 1997 international conference on Intelligent user interfaces IUI'97*, Orlando, FL (pp. 195–202). New York: Association for Computing Machinery.

Stephanidis, C., Salvendy, G., Akoumianakis, D., Arnold, A., Bevan, N., Dardailler, D., Emiliani, P. L., Iakovidis, I., Jenkins, P., Karshmer, A., Korn, P., Marcus, A., Murphy, H., Oppermann, C., Stary, C., Tamura, H., Tscheligi, M., Ueda, H., Weber, G., & Ziegler, J. (1999). Toward an information society for all: HCI challenges and R&D recommendations. *International Journal of Human-Computer Interaction, 11*, 1–28.

Stephanidis, C., Salvendy, G., Akoumianakis, D., Bevan, N., Brewer, J., Emiliani, P. L., Galetsas, A., Haataja, S., Iakovidis, I., Jacko, J., Jenkins, P., Karshmer, A., Korn, P., Marcus, A., Murphy, H., Stary, C., Vanderheiden, G., Weber, G., & Ziegler, J. (1998). Toward an information society for all: An international R&D agenda. *International Journal of Human-Computer Interaction, 10*, 107–134.

Stillman, R. (1980). Auditory Brain Mechanisms. In P. Levinson & C.

Sloan (Eds.), Auditory Processing and Language (pp. 1–18). London: Grune & Stratton, Inc.

Stewart, J. M., & Downs, M. P. (1984). Medical management of the hearing-handicapped child. In J. L. Northern (Ed.), *Hearing disorders* (2nd ed., pp. 267–278). Boston: Little, Brown.

Strong, J. G., Pace, R. J., & Plotkin, A. D. (1988). Low vision services: A model for sequential intervention and rehabilitation. *Canadian Journal of Public Health, 79*, 50–54.

Thatcher, J. (1994). Screen reader/2: Access to OS/2 and the graphical user interface. *Proceedings for the ASSETS '94, the First Annual International ACM/SIGCAPH Conference on Assistive Technologies*, Marina Del Rey, CA (pp. 39–46). New York: Association for Computing Machinery.

Turk, M., & Robertson, G. (2000). Perceptual user interfaces. *Communications of the ACM, 43*(3), 32–34.

Vanderheiden, G. C. (1998). Universal design and assistive technology in communication and information technologies: Alternatives or complements? *Assistive Technology, 10*, 29–36.

Waterworth, J. A., & Chignell, M. H. (1997). Multimedia interaction. In M. Helander, T. K. Landauer, & P. Prabhu (Eds.), *Handbook of human-computer interaction* (pp. 915–946). New York: Elsevier Science.

Wever, E. G., & Bray, C. W. (1930). Present possibilities for auditory theory. *Psychological Review, 37*, 365–380.

Wever, E. G., & Bray, C. W. (1937). The perception of low tones and the resonance-volley theory. *Journal of Psychology, 3*, 101–114.

Wilcox, R. (1999). *Commercial speech recognition*. Retrieved (March 14, 2002) http://www.tiac.net/users/rwilcox/speech.html

Yankelovich, N., & Lai, J. (1998). Designing speech user interfaces. *Proceedings of the 1998 ACM Conference on Human Factors in Computing Systems (CHI'98)*, Los Angeles, CA (pp. 131–133). New York: Association for Computing Machinery.

Yost, W. A. (1994). *Fundamentals of hearing: An introduction* (3rd ed.). New York: Academic Press Inc.

Yost, W. A., & Sheft, S. (1993). Auditory perception. In W. A. Yost, A. N. Popper, & R. R. Fay (Eds.), *Human psychophysics* (pp. 193–236). New York: Springer-Verlag.

Yu, W., Ramloll, R., & Brewster, S. (2000). Haptic graphs for blind computer users. *Proceedings of the First Workshop on Haptic Human-Computer Interaction*, Glasgow Scotland (pp. 102–107). UK: British Computer Society.

·IIIC·

INTERACTION ISSUES FOR SPECIAL APPLICATIONS

Jenny Preece
UMBC

During the last 10 years, the Internet's growth and the decreasing cost of hardware and software has enabled a wider range of people to use and own personal computers and computer-controlled devices. Well and less well-educated, young and old, males and females, and people who are physically challenged in various ways now have access to computers at home, at work, and in public places. This need to cater for a broad range of users, coupled with the low cost of processing power, has encouraged designers to create a wide spectrum of products for a huge range of tasks. In turn, these trends have led to the development of a variety of interaction styles. For example, there are hand-held devices that users interact with via touchscreens, a stylus, keys or buttons, or that have to be directed at sensors, or that respond to sound or voice input. It is also becoming increasing common for sensors linked to computing devices to be embedded in clothing (i.e., wearables) or placed in human bodies to help monitor and control vital organs. We are seeing the graphical user interfaces (GUIs) of the 1990s with WIMP-style interaction (i.e., windows, icons, menus, and pointing) being supplemented with a greater range of multimodal forms of interaction.

Whatever the type of application, it goes without saying that good usability is essential. Successful products are easy to learn and to use, and memorable so that users do not have to invest heavily in relearning operations each time they use them. Products should also be engineered so that users avoid making errors, and they must be safe and satisfying to use too. Well-established guidelines (e.g., Shneiderman, 1998) and evaluation heuristics (e.g., Nielsen & Mack, 1994) have helped designers to create products with these important characteristics. However, more and more products now also have to support other kinds of user experience if

they are to be successful in the marketplace. For example, communications devices must be socially acceptable, aesthetically pleasing, and comfortable to carry around, as well as being functional and easy to use in a variety of environments and contexts. Computerized toys must be fun, engaging, and entertaining. Learning environments must motivate, reward, challenge, and entice students to want to interact with them and want to learn. Designing products that are functional, usable, and also provide these other user experiences (Preece, Rogers, & Sharp, 2002) is challenging, especially as the speed of technological development is outpacing the development of guidelines and heuristics to guide product design. We do not have established guidelines and heuristics tailored for developing hand-held devices, online communities, or computerized toys, and reliable web guidelines and heuristics are only just emerging.

The chapters that follow in this section address many of these issues. They also illustrate how the concept of usability is being broadened to include a wider range of user experiences that go beyond being satisfying to use.

Documentation: Not Yet Implemented But Coming Soon!, by Mehlenbacher, should be essential reading for everyone, because documentation, whether online or not, is still the last resort for a frustrated user who just does not know what to do next. This chapter starts with a discussion about three myths: nobody reads documentation; humans use documentation poorly; and transparent interfaces will eventually eliminate the need for documentation. Mehlenbacher then makes a plea for a user-centered, task-centered, and a goal-directed approach to designing documentation, rather than for a feature or system-oriented approach that tends to result in poor documentation design. As well as discussing characteristics of users, their tasks, and environments of use, Mehlenbacher stresses the need to support universal usability—the concept of designing for a wide range of users who may differ physically, cognitively, culturally, emotionally, and in terms of gender, education, and income.

Stu Card's chapter on information visualization starts by introducing some well-known examples of visualization software. This provides a foundation for asking: "what is information visualization and why does it work?" The next part of the chapter discusses Card's reference model in which he analyzes and describes, via notation, different types of visualizations. The chapter ends with a discussion of the value of visualization for crystallizing knowledge and for sensemaking.

Groupware and Computer-Supported Cooperative Work, by Olson and Olson, starts by acknowledging the origins of the term groupware as "software . . . that is designed to run over a network in support of activities of a group or organization." The authors point out that, whereas early groupware tended to support only one cell in the well-known 'same and different time, same and different place' matrix (Ellis et al., 1991), more recent software is designed to support more combinations and the transitions between them. Not only has the design of many stand-alone applications improved, the World Wide Web has also made greater integration possible and extended the reach of many applications to a wide range of users. Different kinds of groupware systems are reviewed in terms of social, task, environmental and communication issues, technical challenges, and implications for security, privacy, trust, and sociality.

Online Communities: Focusing on Sociability and Usability, by Preece and Maloney-Krichmar, picks up some of the themes discussed in the chapter on groupware, but this time in relation to online communities. This chapter primarily focuses on text and graphical community environments; virtual environments are discussed elsewhere by Stanney. The authors start by discussing the multidisciplinary nature of this research field and proceed to discuss the sociability and usability needs of different types of communities. Three sets of issues are discussed that have a particularly strong influence on sociability: the community's purpose, the characteristics of people and groups in the community, and the policies that both developers and participants create to guide behavior as the community develops. For usability, the main topics of concern are navigation, access, and information design. The chapter also briefly discusses online communities for patient support, education, and commerce, and ends with suggestions for future research.

Stanney's account of *Virtual Environments* discusses how these environments "extend the realm of computer interaction . . . to multi-modal interactions that more closely parallel natural world exchanges. [They] allow users to interact with concrete and familiar artifacts using known perceptual and cognitive skills." This chapter reviews the current state of the art in virtual environments technology, suggests design and implementation strategies, and discusses health and safety concerns. The review of hardware and discussion of advantages and disadvantages of various types of head-mounted displays, haptic and other forms of responses, software and modeling capabilities, etc., will be particularly informative for newcomers to this topic. The discussion of the Internet2 provides a glimpse into what future developments may be on the horizon. The problems of nausea, visual disturbances, and drowsiness when using

these products and their potential to compromise users' safety are also discussed. As Stanney comments, "technology is currently driving VE use," so her focus on usability is particularly welcome. Stanney also reminds us that "users who engage in what seems like harmless violence in the virtual world may become desensitized to violence and mimic this behavior in the look-alike real world"—an alarming thought.

The chapter by Siewiorek and Smailagic on *User-Centered Interdisciplinary Design of Wearable Computers* discusses how the authors' methodology (a variant of user-centered design) has been used by students to design over two dozen wearable computers for a diverse range of applications. There is also a review of how technology has advanced to take advantage of high-performance computing and a discussion of how the web can be used to support interdisciplinary teamwork. The chapter ends with a brief description of challenges for future research.

A Cognitive Systems Engineering Approach to the Design of Decision Support Systems, authored by Smith and Geddes, has three clear goals: to discuss issues relevant to the design of decision support systems; to address practical design questions; and to illustrate these issues via case study examples. The authors start by reviewing literature about human performance, such as work on errors, cognitive biases, and human expertise, and then proceed to broaden the discussion to include research on how well humans perform monitoring tasks, the need to balance the type and amount of tasks that are allocated to humans and machines, the effects of excessive mental workloads, trust, user acceptance, and other issues. Four case studies help to show how theory can be put into practice.

Computer-Based Tutoring Systems: A Behavioral Approach, by Emurian and Ashley, presents a detailed review of previous research that shows how thinking among researchers in this field has evolved. The authors then describe how they iteratively integrated theoretical concepts into the design and evaluation of a Java tutor illustrating how theory and practice can go hand-in-hand. Research students who read this chapter will see an example of how empirical research can inform instructional design, and practitioners will be brought up to date on current theory and practice in this field.

Conversational Speech Interfaces, authored by Lai and Yankeloviche, points out that speech is natural for people, but it is anything but natural for machines, which is the crux of the controversy about the efficacy of speech interfaces. Although some researchers and developers look to speech as a panacea for usable interfaces, others argue that computers will never be able to simulate, far less understand, human speech with its nuances and need for contextualization. So, an important question to ask is, "Which kinds of tasks are well supported by speech and under what conditions and which are not?" With an air of caution, the authors outline the challenges associated with designing speech interfaces and suggest ways of dealing with them. They also discuss how speech can be integrated with other modalities, such as graphical displays, buttons, and controls.

Lazar's chapter, *Special Applications: The World Wide Web*, emphasizes that users' experiences on the web are influenced by the browsers they use and the response time of the Internet, which can vary greatly depending on the time of day, the day of the week, and what is happening in the world. Designers therefore do not have control over users' experiences. Furthermore, many web designers do not know who their users will be or the type of equipment they will use. Therefore, because the majority of users do not have state-of-the-art equipment with high-speed Internet access, web designers need to develop versions that will work on older, slower machines with modems. Designing for universal usability is particularly important in web design. As well as reviewing the dos and don'ts of web design, the author points out that portable, hand-held devices now offer web access via their small screens. The constraint of a small screen size creates additional challenges for designers.

The last chapter in this section is on *Information Appliances*, authored by Sharpe and Stenton. The authors define information appliances as: portable devices, that may be pervasive and that are designed for satisfying individuals' information needs, such as videorecorders, calendars, camera, toasters, PCs, etc. The general characteristics of such appliances are discussed and illustrated with non-tech examples, as well as computer-based examples (e.g., a Palm Pilot for capturing and organizing

notes). The authors point out that "information appliances are used in a rich context of individual and social activity that exerts powerful influences on appliance design." For example, people's moods and emotions vary according to where they are, what else is happening, and what they have to do next, which must be addressed in the design of successful appliances.

This eclectic group of chapters provides insights into the diversity of challenges for which today's interaction designers must find solutions. Some unifying themes running through many chapters in this section are: the need for designers to pay attention to users both as individuals and groups; and to acknowledge that users have emotional and social needs, as well as cognitive, physical, task, and work needs. Understanding the social and physical contexts in which systems will be used is, therefore, essential. Several authors also recognize the need to design for universal usability so that products are usable by the widest possible range of users (Shneiderman, 2000). Designing for universal usability not only increases a product's potential market share, but it also helps to reduce inequalities between people, cultures, and nations.

References

Ellis, C. A., Gibbs, S. J., & Rein, G. L. (1991). Groupware: Some issues and experiences. *Communications of the ACM, 34*(1), 38–58.

Nielsen, J., & Mack, R. L. (Eds.). (1994). *Usability inspection methods.* New York: John Wiley & Sons.

Preece, J., Rogers, Y., & Sharp, H. (2002). *Interaction design: Beyond human-computer interaction.* New York: John Wiley & Sons.

Shneiderman, B. (1998). *Designing the user interface. Strategies for effective human-computer interaction.* Reading, MA: Addison-Wesley Longman.

Shneiderman, B. (2000). Universal usability. *Communications of the ACM, 43*(5), 84–91.

· 27 ·

DOCUMENTATION: NOT YET IMPLEMENTED, BUT COMING SOON!

Brad Mehlenbacher
North Carolina State University

MYTHS ABOUT DOCUMENTATION

Nobody reads documentation. If they did, the developers of hardware and software systems would surely strive to create support materials that meet the high standards they have set for their primary systems. Still, a common goal shared by the developers of support information, whether hard copy or online, is to create documentation that is usable, aesthetically motivating, functionally approachable, consistent, task oriented, and that provides comprehensive coverage of the system that it is designed to support.

Interestingly, however, documentation has only recently received resources in industry and attention in the research literature. As Schriver (1997) points out, though "Documents play a role in almost everyone's daily activities . . . surprisingly, knowledge about creating documents for audiences is not yet well developed" (p. 3). Much of the resistance or lack of interest in documentation development and evaluation has come as a result of three prevalent myths subscribed to by technical specialists and the general public (Mehlenbacher, 1993). Briefly, the myths are as follows.

Myth 1: Nobody Reads Documentation

We read that sentence and we chuckle because we all agree that, not only does nobody read documentation (Rettig, 1991), but few design teams bother to factor documentation into their product design efforts. However, the argument that nobody reads documentation is also an oversimplification. Rather, nobody reads documentation unless they think they need it and, when they do read documentation, they satisfice, skip, scan, and skim; as Redish (1993) reminds us, "You can't assume that, just because you wrote a document, people will read it" (p. 15). Moreover, when users do read documentation, they still tend to fail to accomplish their tasks.

The most commonly cited shortcoming of documentation is that it is too complex or that it focuses too much on the features of the software or hardware that it supports (Sullivan & Flower, 1986). As Carroll and his colleagues (Carroll, 1990; Carroll, Smith-Kerker, Ford, & Mazur-Rimetz, 1987; Van der Meij & Carroll, 1995) have asserted since the early 1980s, most manuals would benefit dramatically from a considerable reduction in size.

Documents designed for learning adults should contain only procedural, task-supporting information. Brockmann (1990) reminds us that task-oriented documentation appeals to adults who, most of us agree:

- Are impatient learners and want to get started quickly on something productive
- Skip around in manuals and online documents and rarely read them fully
- Make mistakes but learn most often from correcting such mistakes
- Are best motivated by self-initiated exploration
- Are discouraged, not empowered, by large manuals with each task decomposed into its subtask minutiae (p. 113)

Myth 2: Humans Use Documentation Poorly

Importantly, the human characteristics described previously do not translate into documentation users are idiots but, rather, they teach us a great deal about the widespread and inappropriate design of manuals themselves. Thus, Odescalchi's (1986) comparison of how users employ a task-oriented manual versus a manual organized around software features revealed that users expressed higher dissatisfaction with the software-oriented manual, committed more errors, and were less productive than when using the task-oriented manual.

Notably, poor documentation is not produced in isolation from the software or hardware that it supports. Most often, documentation is produced much too late in the development cycle and, worse, is designed to describe only the features of the system or to compensate for shortcomings in the primary system. As Johnson-Eilola (1996) has observed, ". . . corporations [tend] to view technical communication as something to be added on to a primary product. Because the value is located in a discrete, technological product such as a piece of software, support becomes easily devalued, added at the end of the project (with too little time or too few staff members), or perhaps omitted entirely" (p. 248). But if the primary application is poorly designed, how well can any documentation support it? If hard copy documentation is poorly designed, how can we expect that moving it online will improve readers' contexts for using it: we are, after all, asking them to move from a well-learned medium (books) to an unknown hypertextual environment or, worse, to a resolution-reduced Portable Document Format (PDF) facsimile of the hard copy documentation.

Myth 3: Transparent Interfaces Will Eventually Eliminate the Need for Documentation

This is a common and widespread myth, even among human–computer interaction and usability researchers who, I would maintain, ought to know better. The popular joke goes: If the system is poorly designed, documentation will fix it; if the documentation is poorly designed, training will fix it; if the training is poorly designed, the help desk will fix it; and so on.

My colleagues and I (Duffy, Mehlenbacher, & Palmer, 1989; Duffy, Palmer, & Mehlenbacher, 1993) and others (Ong, 1982; Schriver, 1993) have argued vehemently against the myth of technical transparency. Even technologies that are traditionally assumed to be transparent—e.g., pencils, typewriters, and writing—are deeply embedded in our cultural upbringing and educational system. Because computer systems are constantly evolving symbol-making and exchanging systems, it is only natural that they will continue to transform the way human beings carry out their tasks (Penzias, 1989). Thus, transparency will always be an illusive and unobtainable goal, and new technologies will always require support systems—hard copy or online—that help users understand their capabilities.

It is also erroneous to assume that moving documentation online will better serve user needs. Poor hard copy documentation is going to produce poor online documentation and, I would even argue, well-designed hard copy documentation will not necessarily produce usable online documentation (Selber, Johnson-Eilola, & Mehlenbacher, 1997; Tomasi & Mehlenbacher, 1999).

Finally, the argument that, in time, designers and researchers will develop a comprehensive list of guidelines for designing usable systems and documentation is problematic as well. Guidelines and principles for design can be useful, but they also frequently ignore the designer's rhetorical situation and are either too rigid or too generic to be applied easily to the situation at hand (Marshall, Nelson, & Gardiner, 1987; Mehlenbacher, 1992; Mosier & Smith, 1986; Schriver, 1989; Wright, 1979). Thus, any guideline for writing or for designing requires that we ask, "How was the guideline created?" and "How does the guideline apply to our specific design situation?"

DOCUMENTATION IS DIFFICULT

Audiences are changing, growing, and collapsing in ways that are difficult to anticipate (Hill & Mehlenbacher, 1998). Computer expertise is difficult to locate. That is, there are few individuals we would define as experts whom we can find at any single location—are we looking for an expert on the same word processor, a similar statistical analysis application, or on a version of the particular programming language that we have been using for several months? Although the number of software and hardware alternatives has expanded exponentially, the number of individuals claiming total familiarity with a wide variety of systems has decreased. The most common form of user assistance—"Excuse me, do you know how I . . . ?—is no longer a useful strategy for approaching problems encountered during our computer interactions (Duffy et al., 1993).

In addition, most users are not developing a level of familiarity with their computers that computer use once demanded. Thus, we no longer feel that our interactions with computers necessitate the in-depth, comprehensive knowledge of the machine that they once did. In the early days of computing, conceptions of the computer user differed significantly from today's conception. Whereas early users of computers were primarily programmers, engineers, and technical specialists, frequent users of computers today often employ computers with specific tasks in mind, infrequently, and for limited periods of time. We create a slide presentation for a talk we are giving tomorrow morning, or update our bank balance occasionally, or check our e-mail for new messages daily. Our knowledge, goals, and strategies for computing do not necessarily transfer from one computing situation to another or from one application to another. When a colleagues asks us, "How do I . . . ?," we usually respond, "Well, on my machine you do the following. . . ."

So while access to human expertise has decreased, access to written expertise or documentation has increased, exponentially when one accounts for the influence of the World Wide Web. In this respect, our purposes for reading are growing as well. Users read to do (to perform a task), read to learn (to learn about something), read to assess (to figure out a document's contents or usefulness), read to learn to do (to acquire knowledge for completing tasks later), and occasionally read purely for pleasure with information goals in check.

Our applications are changing. Users are suffering the unpleasant effects of creeping featurism and deceptive intuitiveness. Norman (1990) coined the term "creeping featurism," and Karen Schriver (1997) ultimately defined it as "manufacturers' seeming obsession with adding newfangled features to their products" (p. 227). The designer's desire to provide as many functions and features as possible in each new release of a product was very popular in the 1980s, but is losing some momentum these days, thankfully, as usability issues become more important. This tendency helps explain why MS Word 4.01 required less than 1 Meg and, depending on how many add-on's you choose, the most recent version requires at least 100 times that.

When users fail to be able to accomplish even simple tasks with technologies, it is common to hear designers and engineers exclaim a great many things about human error. In fact, Schriver (1997) points out that almost 94% of the users she observed blamed themselves for the problems they encountered while using software applications. Butler (1996), moreover, has traced the history of usability engineering to studies during World War Two ". . . when equipment complexity began to exceed the human limits of safe operation" (p. 61), a characterization that surely stresses human error rather than excessive complexity. Whatever the case, understanding human behavior is critical to understanding whether or not interactions with technology will be successful or not.

Because technical transparency is a myth, whenever human beings interact with technology, learning is almost always part of the process. In terms of computer applications, then, an interface is perceived, interpreted, understood, and used depending on the knowledge that users bring to the situation and the tasks they are trying to accomplish. But user knowledge is a thorny construct because it includes factual and conceptual knowledge about *computing* (the particular application and its functionality) and about the *task domain* (experience, complexity, frequency of exposure) and about the mapping of the tasks to the tool at hand.

Previously, I suggested that even the interface of the pencil would mean little to a first-time user without life-long instruction, significant assistance and training, hours of practice, templates, trial-and-error, and without special situations for use (types, surfaces, calligraphic goals, grapholectic knowledge).

Figure 27.1 captures the contemporary version of the pencil. The interface is that of a popular commercial word processor and, although I have opened several more tool bars than usual to make my point, I would argue that the interface is far from intuitive, even though it has been with us in various forms for almost 20 years. Many computer users spend the majority of their time using word processors to create memos, reports, proposals, articles, books, presentations, and so on. However, few of them receive formal training in how to use word processors. Even experienced word processor users rarely choose the "double click-on-word + delete" in favor of backspacing (one letter at a time) until they have deleted a word, although the speed of the former is slightly greater than the speed of the latter set of actions.

Creeping featurism is not the only technological culprit that causes unfriendly user technology–documentation interactions. Some hardware and software demands that documentation be

FIGURE 27.1. Screen capture of popular commercial word processing interface—functionality vs. user friendliness?

consulted *before* we interact with it (e.g., modems, installation situations, complicated build-it-yourself devices). Other technologies require user–documentation interactions so that users can customize their settings, depending on the particulars of their bandwidth and peripheral setups (and these documentation situations grow in number as the high-tech industry presses forward inventing new variations of multimodal interaction and introducing converging incompatibilities; see Oviatt's chapter, Multimodal Interfaces, this volume). Finally, the development of new versions of hardware and software—and the legacy challenges that result—creates a neverending need for more creative and integrated documentation solutions.

DEFINING DOCUMENTATION

Documentation includes both hard copy and online support materials that help users achieve goals and accomplish tasks within the contexts of their primary work (Duffy et al., 1993; Selber et al., 1997). Documentation then can include everything from error messages (for reparative assistance) to user's manuals and online help (for task-oriented assistance) to intelligent online tutorials (for instructional and conceptual assistance). Rather than focus exclusively on the various genres of documentation or on whether the documentation is printed or online, however, it is more fruitful to view documentation strategically from a rhetorical perspective.[1] Documentation, then, is both about the production of documents and about documentation as a process that is intimately connected to the development of the primary software or hardware systems.

Fawcett, Ferdinand, and Rockley (1993) and numerous other writing researchers (Draper, 1998; Hayes, 1989) describe documentation design as a five-step process that includes the following:

1. Determine the purpose of the information by
   - Determining what you are writing about
   - Determining who you are writing for
   - Determining the objectives of the information
2. Select the medium of communication (such as print or video)
3. Organize the information by
   - Selecting appropriate organizational patterns
   - Organizing the information using one or more organizational technique
4. Formalize the organization
5. Evaluate the organization (in Barnum & Carliner, 1993, pp. 44–45)

Importantly, defining the documentation that you plan to design is much messier than that: it is a recursive rather than linear (step-by-step) process. However, as ill-structured as documentation processes can be, developers should always attempt to develop a documentation plan that anticipates five critical dimensions of all support documentation:

1. The knowledge and attributes of the intended *audience*.
2. The *task types* or activities the audience will be expected to accomplish.
3. The *information goals* that the audience will bring to the problem situation.

---

[1]My perspective is influenced by the work of several contemporary rhetorical theorists (e.g., Bitzer, 1968; Burke, 1969; Kinneavy, 1971). See also C. R. Miller's (1985) comprehensive survey of the relationship between rhetorical theory and technical communication research.

4. The *physical and rhetorical differences* presented by different media.
5. The *genre* or information type being developed.

Audience Knowledge and Attributes

Since the 1980s, information developers have been developing models of documentation audiences that they hope will help them anticipate the diverse information needs of different user types. The most common model of documentation users has been based on their experience and finds its theoretical roots in psychology (Chi, Feltovich, & Glaser, 1981, 1982; Gentner & Stevens, 1983; Simon, 1979). Thus, Kearsley (1988) describes three levels (or dimensions) of user experience in his conceptual models of help. Users can be defined in terms of their expertise with the computer, with the particular task domain, and with the particular application software. However, if we add that users can either be expert, intermediate, or novice users in each of these domains (Shneiderman, 1998), we generate nine user types: novice with the application, experienced with the computer, and intermediate with the task area being just one type of user (see Table 27.1).

Another method of describing the audiences for documentation is in terms of their demographic characteristics, for example, their level of education, economic standing, geographic placement, subcultural values and expectations, or their age (see Bruckman's chapter Designing for Children and Czaja's chapter, Designing for the Elderly, this volume). This perspective toward users finds its roots ultimately in the North American Classification System (see http://www.ntis.gov/product/naics.htm), a system that influences most contemporary marketing analyses of audiences, their tastes, and preferences. Thus, a demographic characteristic such as economic standing has been shown repeatedly to influence Internet and computer usage (Carver, 2000; Shneiderman, 2001; U.S. Department of Commerce, 2000) and this might, in turn, influence how one approaches the audience for a online chat environment being developed.

Learning styles as well influence the way users access information (Kolb, 1984; Sadler-Smith & Riding, 1999; Schmeck, 1988). Felder (1993) suggests that students (or, for our purposes, documentation users) can be characterized broadly according to five questions:

1. What type of information does the student preferentially perceive: sensory—sights, sounds, physical sensations, or intuitive—memories, ideas, insights?
2. Through which modality is sensory information most effectively perceived: visual—pictures, diagrams, graphs, demonstrations, or verbal—sounds, written and spoken words, and formulas?

3. With which organization of information is the student most comfortable: inductive—facts and observations are given, underlying principles are inferred, or deductive—principles are given, consequences and applications are deduced?
4. How does the student prefer to process information: actively—through engagement in physical activity or discussion, or reflectively—through introspection?
5. How does the student progress toward understanding: sequentially—in a logical progression of small incremental steps, or globally—in large jumps, holistically?

Mehlenbacher, Miller, Covington, & Larsen (2000), for example, found that reflective, global learners tended to perform better than hands on, sequential learners in web-based courses on communication for engineering and technology. One might extrapolate that learning styles would equally influence the way readers respond to task-oriented documentation developed for the web.

Audiences also bring various literacies to their problem-solving situations. Kintsch and Vipond (1979) stress that text processing is improved when information developers establish a hierarchy of topics, chunk information into meaningful units, and maintain a consistency of topics so that readers can establish thematic continuity (cf. Kintsch, 1986). Shneiderman (1998) explicitly distinguishes between domain experience ("I am an expert in accounting") and computer experience ("I am somewhat familiar with Excel"), and other researchers (Bernhardt, 1993; Hodes, 1998; Kress, Leeuwen, & Gress, 1995; Tufte, 1983) have distinguished between textual and visual literacies as well. Kostelnick and Roberts (1998) collapse these two approaches to information literacy under a single rhetorical perspective, but research on learning styles (e.g., Felder, 1993) suggests that user orientations will differ significantly, depending on how information is presented. Presentation and content are interdependent, and information developers should attempt to link user preferences to information display.

Importantly, cognitive and physical abilities factor strongly into the way that users interact with documentation, especially when the documentation has been placed online. Kolatch (2000) suggests that it is critical for designers to understand the cognitive abilities of their users if designers aim to anticipate the information needs that those users bring to interactions with documentation or systems (see Newell, Carmichael, Gregor, & Alm's chapter, Cognitive Impairments, this volume). To this end, researchers have created interfaces that account for various cognitive abilities, including users who have suffered brain injuries (Cole & Dehdashti, 1990), and institutions such as the Trace Research and Development Center at the University of Wisconsin-Madison (http://trace.wisc.edu) have amassed hard copy and online resources for designers interested in various audience types.

Anticipating the physical abilities of different users is, in some ways, much more complex, because it involves user abilities that include ambulatory, haptic, visual, and auditory considerations (see Brewster's chapter, Non-speech Auditory Output; Iwata's chapter, Haptic Output; Jacko, Vitense, & Scott, Perceptual Impairments; Luczak, Roetting, & Oehme's chapter, Visual Displays; and Sears, Motor Impairments, this volume). Certainly, as documentation moves online, issues of access for

TABLE 27.1. Nine-Part Matrix of User Experience Across Domains (Computer, Task, and Application)

	Expertise with Computer	Expertise with Task Domain	Expertise with Application
Novice			X
Transfer			
Expert	X	X	

diverse audiences become more and more important. For example, Loomis and Lederman (1995) have explored notions of tactual perception and Mynatt (1997) has described systems that support blind users. Recently, Shneiderman (2001) has heightened our sensitivity to issues in universal usability by holding the first ACM Conference on Universal Usability in Arlington, VA, in 2000 (see the conference's accompanying Web site, http://universalusability.org).

Finally, it is important to note that the audiences of documentation are subject to a host of particularized emotional, motivational, and affective attributes (see Nass & Brave's chapter, Human Emotions, and Fogg's chapter Motivating, Influencing, and Persuading Users, this volume). Research reminds us that most users of documentation and instructions are rushed, frustrated, task-oriented, and frequently hostile because their primary goal has been *interrupted* by the technology they are reading about (Duffy et al., 1993; Shneiderman, 1998). Because of this complex mix of emotions, users almost never read a manual from front-to-back (Redish, 1993). Designers, therefore, need to be aware of principles of effective information chunking and the importance of meaningful subject heading and chapter divisions (thus anticipating user scanning and skipping). Indeed, establishing when a task begins and when a task ends is critical for producing support materials that address users where their problems exist (e.g., when users read the error message "Printer not found" and access the online support, they do not want to read about printing in general). They want to read *only* about finding a missing printer.

Finally, users often hold mistaken models of the systems they are working with and these errors in intentionality (Kay & Thomas, 1995) or user misconceptions (Mirel, 1998). Indeed, these misconceptions can produce problem tangles that result from users' initial misunderstandings and lead to increasingly confusing mismatches between the system state and the users' representation of the original problem (Mack & Nielsen, 1994). All these problematic user situations can generate serious errors unless designers and information developers are able to anticipate them in advance and attempt to account for them in both the interface and the support materials.

Task Types and Activities

Another way of managing documentation is to focus entirely on the tasks and activities that users are likely to engage in while using the primary system. Prior to task-oriented models for documentation (Barker, Coe, 1996; Duffy et al., 1993), manuals and online systems were frequently organized around system features and topics rather than human needs and goals (Goodall, 1991). When users encounter an impasse in their performance, they redefine their tasks or subgoals to resolve their impasse with the primary system. Users in general move recursively

through six subgoals as they attempt to remove the impasse:

1. *Representing problems:* A term from cognitive psychology, this stage involves users attempting to understand their situation or problem as they work through it. How users characterize their problem situation is critical to developing a representation of their state and the possible solutions that are available given the constraints of the situation. The prior experience of users and their ability to apply that experience to interpret their particular situation greatly influences the form of their representation.

2. *Accessing information:* With detection and representation of the impasse accomplished, users must access a source of information. By access, I mean users must identify how to get into the manual, help system, or online documentation for the primary system. For some products, this is a significant challenge for users.

3. *Navigating information:* In selecting a topic, users are forced to begin navigating the document or online space. Many of the tools for navigating online are different from navigating hard copy texts (despite the emulation of familiar touchstones, such as tables of contents and indexes).

4. *Scanning information:* Few users read all the support materials; rather, they search for particular headings, information items, or procedures, assuming that the entire text is not relevant to their problem representation. This is a critical design opportunity and Keyes, Sykes, and Lewis (1989) recommend that information developers chunk information, provide queues about hierarchy, filter out irrelevant material, mix modes to appeal to various audience types, and abstract or simplify complex concepts for users.

5. *Understanding information:* Having located relevant information, users must be able to comprehend the text and graphics. This goal is easier to summarize than it is to produce: usable, readable document design is the topic of entire books (e.g., Barker, 1998; Schriver, 1997).

6. *Transferring information:* Ultimately, users must take what they have learned back to the primary system. Finding a solution does not necessarily make it easy for them to apply it (calculate the amount of steps they will go through between the documentation and the resolution of their problem).[2]

Importantly, these six subgoals of the general goal "How do I . . ." occur outside of the primary workspace and, so, affective issues should not be underestimated. Studies continually reveal that, even when hardware and software are well-designed (and they rarely are), humans are prone to committing errors and require meaningful feedback that allows them to undo or redo their actions accurately; this is a general principle of usability as well. Humans learn through trial and error, and making errors is roughly 50% of a human's natural learning curve (see Stanton's Designing for Errors chapter, this volume).[3]

---

[2]The six subgoals described here have been influenced by Kieras and Polson's (1985) cognitive-complexity theory, Moran's (1983) internal-external task-mapping, Norman's (1984, 1990) stages and levels of interaction, Shneiderman's (1998) semantic-syntactic model, and Suchman's (1987) and Frohlich and Luff's (1989) social and collaborative models of user behavior.

[3]Additionally, Panko's (1997–2000) Spreadsheet Research Web site is devoted to an analysis of incidents and causes of human error using spreadsheet software (http://panko.cba.hawaii.edu/ssr/).

Information Goals and Documentation

When producing information or support materials for a given technology, one of the earliest challenges is to explicitly establish the parameters of the technology that the documentation needs to support (e.g., to help users send e-mail, should designers explain how to turn the computer on or how TCP/IP works or can they always assume that the users' problems will remain confined to the e-mail program?). Ummelen (1997) distinguishes between information types, arguing that users either want procedural information (how-to-do-it, instructions, syntactical elaboration) or declarative information (how-it-works, explanatory, conceptual elaboration). Carroll and colleagues (Carroll, 1990; Carroll et al., 1987; Van der Meij & Carroll, 1995) assert that adult learners are only interested in procedural, task-supporting information; however, other researchers (Charney, Reder, & Kusbit, 1991) extend minimalism to account for learning situations that involve complex technologies.

Indeed, arguments about information goals inevitably return researchers to challenges of audience definition. Thus, numerous researchers (Brockman, 1992; Redish, 1988; Sticht, 1985) make strong distinctions between reading to learn and reading to do, arguing that people generally read documentation to do. For example, Redish, Battison, and Gold (1985) assert that

> If . . . owners read the [manual] at all, they are likely to skim through it when they first get it. After that, they will probably only go back to it when they need a specific piece of information. That's how people use manuals (p. 135).

Mehlenbacher et al. (2000) speculate that many other reading goals occur, including, for example, reading to learn to do, reading to analyze, reading to compare, confirm, or correct, and reading to summarize.

Tomasi & Mehlenbacher (1999) explicate six information goals connected to online support systems designed around examples. The information goals are also useful for understanding documentation in general, especially when we connect documentation information types with the reader questions that the types are designed to address (see Table 27.2).

Most information goals map conveniently onto existing documentation genres (e.g., "What can I accomplish with this application?" is usually found in the introductory tutorial guide), but many support documents attempt to be all things to all users. Careful consideration of the information goals of a document's audience can dramatically reduce user confusion and misinterpretation.

Ultimately, Ummelen (1997) argues that users prefer procedural information over declarative information, but she also notes that declarative information is used to a fairly high degree (as much as 40%) and has positive effects on delayed task performance, reasoning, and knowledge. Draper (1998) as well speculates that a tension exists between providing documentation that supports tasks (task-oriented and procedural information) vs. documentation that maximizes learning (explanatory and conceptual information). Designers should consider carefully the benefits of building conceptual information

TABLE 27.2. Six Information Types and the Reader Questions That the Types Are Designed to Address

Information Type	Reader Questions Addressed
Goal exploration	What can I accomplish with this application?
Definitional and descriptive	What is this particular feature? What is it used for?
Procedural and immediate	How do I do this? What steps does this require?
Diagnostic and state explication	What just happened? Huh? Where am I?
Example-based and medium term	How does this example work? How do I copy it step-by-step? What are the various parts?
Conceptual and long term	How do all these application features work together? What would I like to learn how to do with this application if I only had the time?

into their documentation, especially if it acts to scaffold user understanding and practice over extended periods of time and use.

Physical and Rhetorical Differences in Media

Print and online documentation are created to help users achieve goals as they negotiate the very real constraints of various time/space frames (Selber et al., 1997). If, for example, users are motivated to quickly complete a task (time) and not to understand its broad implications or varied uses (frame), then a user's guide or online help system may be the best solution (see Table 27.3).

Because of the interplay between user, task, primary workspace, and secondary materials, differences between print and online support can play a critical role in user success. Selber et al. (1997) outline the physical differences between hard copy pages and computer displays in terms of four parameters:

Resolution: Computer screens are much harder to read because their resolution is typically much lower than hard copy materials and because the area of the physical display is reduced.

Display area: Computer screens rarely allow readers to view more than one page at once, and also limit readers' abilities to mark their position in the documentation via marginal notes, dog-tagging, the physical chunking of a group of pages and then flipping back-and-forth between two relevant information types, or via highlighting or underscoring reminders.

TABLE 27.3. Rhetorical Frameworks for Online Support

	Help	Documentation	Tutorials
Users	Expert	Intermediate	Novice
Goals	Narrow/ short-term	Medium/ short-term	Broad/long-term
Time/Space Frames	Parasitic/ internal	Parallel/ internal	Encompassing/ external

TABLE 27.4. Physical Differences Between Print and Online Support Materials

	Pages	Screens
Resolution	70–1,200 dots/inch	50–100 dots/inch
Display area	Generally larger	Generally smaller
Aspect ratio	Generally taller than wide	Generally wider than tall
Presence	Physical Static Immutable	Virtual Static Dynamic Interactive Mutable

TABLE 27.5. Rhetorical Differences Between Print and Online Support

	Pages	Screens
Organizational	Linear Familiar Hierarchical Logical/deductive Fixed	Linear and nonlinear Familiar and unfamiliar Hierarchical and nonhierarchical Logical/deductive Associative and dynamic
Navigational	Familiar Limited Static	Familiar and unfamiliar Robust Static and dynamic
Contextual	Generally rich	Generally poor

Aspect ratio: The horizontal/vertical ratio of most computer screens (4 × 3) realigns traditional designer and reader expectations about the use of white space and, so, unfortunately, most computer screens invite a reduction in white space and therefore a reduction in readability.

Presence: The static, immutable nature of hard copy documentation invites readers to assume that it is less likely to disappear or to change from one user situation to another; ironically, the dynamic, interactive nature of online documentation provides critical opportunities for user customization or information tailoring (see Table 27.4).

Rhetorical differences between print and online information are more difficult to establish, in part because research conflicts on the strengths and weaknesses of the alternative media. For example, online libraries of documents allow users to find online books faster, although information *within* printed manuals is frequently easier to locate (Barnett, 1998; Landauer, 1995), and print manuals use conventions that are more familiar to users and that rarely require the system knowledge or special skills that navigation with online documents routinely requires (Reece & Scheiber, 1993; Zimmerman, Tipton, Bilsing, & Green, 1993). Still, Tomasi and Mehlenbacher (1999) assert that differences between the two media do exist and that emulating print documentation online is problematic. Selber et al. (1997) stress that the three major differences between print and online information are organizational, navigational, and contextual:

• *Organizational:* Users read print and online documentation in both linear and nonlinear ways (Chignell & Valdez, 1992), although online information frequently uses dynamic and associative structures that invite users to organize and reorganize information in various ways.

• *Navigational:* The spatial comfort that holding a printed book can provide is often lost to users, and they find themselves having trouble getting a sense of text (Haas, 1989) that includes an accurate, global understanding of the online information they are accessing.

• *Contextual:* User actions rely heavily on contingent, situated, recursive actions rather than on contextual plans (Beabes & Flanders, 1995; Boy, 1992; Suchman, 1987; Winograd & Flores, 1987), and therefore documentation must begin with a focus on the contexts and purposes of use rather than on the formal characteristics of different documentation types (see Table 27.5).

Documentation Genres and Information Types

Many beginning information developers and technical writers are hired to write a manual for this system or that, although few system developers are able to specify what the manual should contain. Unfortunately, too many system developers assume that the primary purpose of a manual is to describe what the system does. An approach based on Kinneavy's *Theory of Discourse* (1971) allows us to reframe the question from what type of documentation is being written to one of classifying the aims of users in specific situations or contexts: What do they want or expect from the document? What questions should the documentation be able to answer? The function of information should always be connected to the context of its use, and not to the writer's intentions or a static description of its form (Bethke, Dean, Kaiser, Ort, & Pessin, 1981).

Barker (1998) provides a table of Sample Titles for Software Manuals and Help Systems (p. 158) that information developers should find useful. The table lists common documentation genres or forms, although he stresses that user analyses and documentation objectives are critical to producing any of the following document types (see Table 27.6):

DEVELOPING DOCUMENTATION

Designing usable documentation is as messy as developing usable software and hardware. Donald Schön (1987) has characterized the mess that developers must confront as follows:

Designers put things together and bring new things into being, dealing in the process with many variables and constraints, some initially known and some discovered through designing.... Designers juggle variables, reconcile conflicting values, and maneuver around constraints—a process in which, although some design products may be superior to others, there are no unique right answers (p. 42).

But the complexity of design is well known to developers of human–computer interfaces and information designers who are

TABLE 27.6. Documentation Types and Brief Description of Typical Contents (Extension of Barker, 1998, p. 158)

Documentation Titles or Genres[a]	Description
Product Packaging and Labeling*	Product highlights, marketing reviews, screenshots, and sometimes brief installation instructions
Read-me file*	Last-minute bug fixes, feature updates, and information about errors in documentation (often a negative introduction, unfortunately, to the system)
Getting Started Guide*	Brief overview of system and introductory features and a walkthrough of some basis system features
User's Guide	Procedures for most system functions
Reference Manual or Reference Guide	Reference or support-level information; designed primarily for experienced users
Quick Reference Card or Pocket Reference	Brief overview of commands, menus, tools, and essential system information
Documentation Titles or Genres, Continued	Description, Continued
Manual	Parts of traditional user's guide, installation information, and reference section (all-inclusive title)
Error Messages*	Feedback that indicates a user or system error has occurred (frequently poorly written and unhelpful)
Tutorial	Series of lessons that introduce basic (and sometimes advanced) features of the system
Context-Sensitive Help Messages*	Rollover, mouse over, or bubble information that supplements icons or system features
Help or Online Help System	Help on using help, table of contents, alphabetic index, search engine, and procedural information
Online Guide, Coach, or Wizard*	Embedded user assistance aimed at taking users step-by-step through tasks (sometimes called "intelligent" help)
Online Documentation*	Parts of traditional user's guide, installation information, and reference section (all-inclusive title)

[a]Documentation titles or genres that are followed by an asterisk are items that I added to Barker's (1998) list; although items such as Error Messages are often viewed as outside the documentation domain, I would argue that information developers ought to be more involved in producing usable feedback at all levels of the system development.

both increasingly being expected to work with limited development cycles and with numerous design constraints. These include, for example, system portability issues, international audience considerations, and development tool inflexibility.

Because the challenges facing product developers are in many ways equivalent to the challenges facing information developers, especially given the documented advantages of integrating information products with technical products (Tomasi & Mehlenbacher, 1999), this section will be brief (readers should refer to Section V, this volume, for additional information on the development process in general).

As with human–computer interface design, information designers need to perform the activities described in Table 27.7.

The four processes—needs assessment, product specification, design goals, and evaluation and usability testing—encapsulate all the activities involved in the production of support materials for a given primary system. The final process stage, evaluation and usability testing, should ideally drive the entire design process and, although outside the scope of this chapter, I urge both system developers and information developers to acquaint themselves with the numerous resources available on the subject (Alder & Winograd, 1992; Duin, 1993; Dumas & Redish, 1993; Hackos & Redish, 1998; Lee, 1999; Lindgaard, 1994; Mayhew, 1999; Mehlenbacher, 1993; Nielsen, 1997, 1999; Nielsen & Mack, 1994; Rubin, 1994; Schell, 1986; Skelton, 1992; Trenner & Bawa, 1998).

CHALLENGES IN DOCUMENTATION: INFORMATION FUTURES

Maes, Goutier, and van der Linden (1992) assert that, "It can confidently be assumed that in the coming decades a growing number of readers will be confronted with a growing number and a growing variety of reading situations in which information is offered online" (p. 175). Indeed, the NSF Indicators Report on Science and Technology, Public Attitudes, and Public Understanding (1998) indicates that the existing and potential audience for online information has grown dramatically: In 1997, 57% of Americans and almost 90% of all college graduates reported using computers at work, at home, or both; 43% of Americans lived in a household with one or more computers, and 16% reported having access to the web from their home computer.

In this chapter, however, with my emphasis on taking a rhetorical perspective toward documentation, I have only minimally addressed the dramatic movement online and its implications for documentation design and evaluation (Hill & Mehlenbacher, 1998). Instead, I have described documentation in the most general terms as print or online support materials designed to assist different user types in their attempt to understand and accomplish tasks using primary systems. But the accelerated movement of information online has important social-organizational implications for documentation producers and users. At the very least, as Dicks' (1994) warns, the movement of documentation online will have profound implications for information developers because, "Some developers of traditional computer documentation (typically technical writers) and

TABLE 27.7. Documentation Development Process and Related Activities
(see Mehlenbacher & Dicks, in preparation; Selber et al., 1997)

Documentation Development Process	Description/Related Activities
Needs Assessment	Address broad-based questions: For example, What should our support materials accomplish? What are the support materials of our competitors accomplishing? What kinds of support can we realistically develop and maintain? What features are most useful and least useful? What documents are most and least difficult to produce (Beabes & Flanders, 1995; Holtzblatt's Contextual Design chapter; Blomberg, Burrell, & Guest's The Ethnographic Approach chapter, this volume)
Product Specification	Develop audience profiles and perform task analyses (see Redish & Wixon's Task Analysis chapter, this volume); generate topic outlines and style/design specifications (Stewart & Travis' Guidelines, Standards, and Style Guides chapter, this volume); assign project responsibilities and review protocols; generate schedules, cost estimates, dependencies, and risks; establish sign-off goals (Beaudouin-Lafon & Mackay's Prototype Development and Tools chapter, this volume; Wood, 1998)
Design Goals	Establish design goals and criteria that can be applied during the evaluation and usability testing phase of the development process; make sure that the goals emphasize usability of the product and its support materials and that they include both global and local design principles; Appendix A outlines 17 usability principles for print and, especially, for online document design; principles here are viewed more as heuristics than as guidelines for design; design guidelines have numerous well-documented shortcomings; they frequently ignore the designer's rhetorical situation, and are often too rigid or generic to be applied easily to the situation at hand (Marshall et al., 1987; Mehlenbacher, 1992; Mosier & Smith, 1986; Schriver, 1989; Wright, 1979, 1980)
Evaluation and Usability Testing	Forming an evaluation team; identifying evaluation goals; selecting evaluation methods; developing realistic scenarios; enlisting real users; implementing the evaluation; analyzing the data; implementing changes

classroom training (typically instructional designers and stand-up trainers) will resist moving to an approach that meliorates the need for their services as currently defined" (p. 117).

Fusing System and Documentation Development

The movement of documentation online is surely disturbing traditional dichotomies between product developers and information developers as well (see Lee & Mehlenbacher, 2000, for a discussion of some of the tensions experienced between technical specialists and technical communicators). They are both, after all, using complex technological tools to produce artifacts designed to help humans do things. Open-source movements are accelerating this blurring of roles by turning system-producing organizations into support-providing organizations that develop shared systems.

But what about the complex technological tools used to produce artifacts designed to help humans do things? Some might argue that I should have addressed them more than ill-structured concepts such as audience and task orientation in this chapter. My experience teaching undergraduate courses on software documentation design and graduate courses in online information design and evaluation for the past decade has taught me that describing how to work with existing tools designed to help developers create documentation (e.g., FrameMaker, RoboHelp, WinHelp, ForeHelp, HTML, XML, and so on) is problematic.[4]

Tools are not the problem with documentation today, and they are unlikely to solve the problem any time soon. From the documentation developer's perspective, the problem is the relationship we take to the role of documentation and to the human beings using it. Johnson-Eilola (1997a) argues that, "Computer documentation is most frequently valued for its technically efficient qualities: speed and accuracy in use. Although such measures are obviously critical in the overall success of documentation, the emphasis on these criteria tends to discourage documentation designers from thinking beyond very narrow, functional forms of content analysis" (p. 120).

The future of documentation, that is, a future that transcends task-oriented, minimalist support materials (Carroll, 1998), is most likely a future in which input–output model of documentation development from primary system developer to secondary documentation writer to end user becomes a triangle of interrelated technology users (in the broadest sense of the term).

Bringing Information to the Digital Surface

The future of documentation is inextricably bound to the (often romantically conceived) notion of the future of information. The traditional response to the relationship between information technologies and information has been that our proto-electronic era will ultimately result in language erosion, a flattening of historical perspectives, the waning of the private self, and the

---

[4]Readers interested in tools issues should see the archived articles of the Online Information Special Interest Group of the Society for Technical Communication's online journal, *Hyperviews*, Welinski (2000), or the host of technology magazines currently devoted to the topic (e.g., *PC World* or *Mac User*).

decentering of what is real with what is simulated (Birkerts, 1994; Doheny-Farina, 1996; Kirschenbaum, 1995). Indeed, Johnson-Eilola (1997b) has asserted that "many adults are terrified of this place" and "do not understand or relate to these networked spaces in the same ways that our children and students do; we tend to criticize them unfairly" (p. 186). Hill and Mehlenbacher (1998), however, have noted that the flattening that Birkerts (1994) worries about may in fact indicate that our children interpret reading activities as simultaneous, object-oriented, parallel, and contingent. Our children are just as likely to consult directions as they are to e-mail a friend, enter a MOO (Multi-User Domain, Object-Oriented), search an archived list-serv, select an item that has no label, or join an online forum to seek answers to questions they are motivated to solve.

Hobart and Schiffman (1998) posit that, although "Computers are enormously powerful information-processing machines, they are also, culturally speaking, toys," adding that "When we 'play' with [computers] we follow freely the 'absolute binding,' logical rules of information processing, which epitomizes play activity. Moreover, as play, this rule-governed activity justifies itself; it needs no further rationale" (p. 236). The implications of this perspective toward documentation production are significant for they undermine traditional notions of productivity in the workplace and existing approaches to the support of technological systems. Play invites a definition of productivity and support that incorporates creativity, uncertainty, business-to-business collaboration, and rationality bounded by the particular rules of the game. This definition of production, then, integrates data and information profusion and simulation, leading to what Hobart and Schiffman (1998) define as an "Indeterminacy in the sciences of complexity" that ultimately limits our ability to find closure in most information processing situations:

Closure . . . allows us to circumscribe our natural, human drive for order, which might otherwise de-generate into compulsive, Rabelaisian list making. It prevents us from being overwhelmed by the information we perceive. Both the classical and modern information idioms have provided closure through classifying and analyzing information. The contemporary idiom serves no such function. . . .

Loss of closure neither symbolizes nor sanctions the "anything goes" of cultural relativism. Even less does it directly imply our being overwhelmed by experience, although we may well feel our age's information overload. Rather, it simply means the displacement and supplanting of well-delineated information structures with the open-ended, unlimited activity of information processing. Just as closure through analysis converted information to knowledge in the modern age, so too does computer simulation convert it to play (pp. 258–259).

Acknowledging the lack of closure that contemporary technologies invite shakes traditional notions of work, organizational hierarchy, and the information user. Gray (April 24–30, 2001) stresses that emerging entrepreneurial approaches to working life are inevitable: "It's a kind of radicalization of the notion of autonomy, in which an autonomous life is seen as a succession of different episodes, activities or projects one after the other so that the value of that working life is not its consistency or continuity, the way it might have been in the past, but rather its variety, its spontaneity, its responsiveness to the moment."

Davenport and Beck (April 22–28, 2001) suggest that one outcome of this spontaneity is what they call the Symptoms of Organizational ADD (Attention Deficit Disorder) as:

1. An increased likelihood of missing key information when making decisions
2. Diminished time for reflection or anything but simple information transactions such as e-mail or voice mail
3. Difficulty holding other's attention (e.g., having to increase the glitziness of presentations and the number of messages to get and keep attention)
4. Decreased ability to focus when necessary

Working with these emerging work-related and organizational opportunities and constraints are individual users or, as Goldbach (August 1–7, 2000) describes them: The People formally known as Users. Thus, his first rule in response to the literatures on design, technology, and users is, "We cherish the fact that people are innately curious, playful, and creative. We therefore suspect that technology is not going to go away: it is too much fun."

Developers, Users, and Information Flow

So, from the documentation developer's perspective, human information processing and attention is the problem, but it is not an entirely novel problem either: in 1945, after all, Bush (1945) warned

There is a growing mountain of research. But there is increased evidence that we are being bogged down today as specialization extends. The investigator is staggered by the findings and conclusions of thousands of other workers—conclusions which he cannot find time to grasp, much less to remember, as they appear. Yet specialization becomes increasingly necessary for progress, and the effort to bridge between disciplines is correspondingly superficial.

Large, integrated documents are complex facts of life. Working with information technologies requires flexibility, creativity, and active knowledge construction. Consider several of the user rules for behaving described by contributors to *Mondo 2000*: "Information wants to be free, . . . Access to computers and anything which may teach you something about how the world works should be unlimited and total, . . . Do It Yourself, and . . . Surf the edge" (cited in Sobchack, 1996, pp. 84–85). Interpreted in the light of information technology and notions of play, these become rules for users motivated to take personal responsibility for their information needs. As Pasmore (1994) confesses, describing a typical user-manual scenario

. . . each time I learn to do something with one keystroke on my computer that used to take four or five. I have a sort of inner calculus that tells me when my frustration has built up to a point that it's time to read my manual. If I'm in the middle of something important, I'll wait a bit longer; if I have free time, I may look at the manual just out of curiosity. I'm sure that my learning would be faster if I took a concentrated course, many of which are available at the university for a nominal fee. But that much concentrated time is hard for me to find. So I waste time by seconds and minutes instead, which I'll bet in the span of a year add up to a lot more than the time I would spend in the course (p. 77).

This observation is certainly what providers of online alternatives to lifelong learning hope to capitalize on. Michaels (May 15, 2000) reports that, "An estimated 90,000 courses at U.S. colleges and universities are delivered by some form of distance learning" (up from 26,000 courses in 1994-1995, according to the National Center for Education Statistics; NCES 98-062, October 1997) and the corporate desire for prestige educational content is growing. It would seem that users of all types are pursuing lifelong learning, and this trend has powerful implications for product developers, information developers, and for users motivated to integrate technologies into their everyday work and leisure practices.

CHALLENGES IN DOCUMENTATION: BACK FROM THE FUTURE

When I step back into most high-tech organizations today, I am struck by the datedness of the challenges that their documentation developers face: establishing equity of pay and status; gaining access to the same technologies as their product developers; building systematic training programs that establish modular, minimalist, consistent, and task-oriented information designs; overcoming basic translation problems (see Marcus' chapter, Internationalization: Designing for Diversity in Culture and Language, this volume), and finding ways to integrate documentation testing into the production process.

Contextualized by these challenges, my conviction that all user assistance should be integrated seamlessly with the primary system it supports seems unrealistic (Mirel, 1999; Tomasi & Mehlenbacher, 1999). My fascination with the implications of audio and animation, three-dimensional data visualization, device portability, video and simulated learning environments, natural-language processing systems, and alternative tools for collaboration and documentation annotation seems esoteric and elitist. But these developments are occurring all around me, if I am interpreting contemporary research in computer science, engineering, and human–computer interaction research correctly.

Turning to Csikszentmihalyi (1990) and Schön (1983), I am provided with the confidence to set out a small number of huge challenges. Csikszentmihalyi (1990) provides me with a powerful citation from which to conclude:

Activity and reflection should ideally complement and support each other. Action by itself is blind, reflection impotent. Before investing great amounts of energy in a goal, it pays to raise the fundamental questions: Is this something I really want to do? Is this something I enjoy doing? Am I likely to enjoy it in the foreseeable future? Is the price that I—and others—will have to pay worth it? Will I be able to live with myself if I accomplish it? (p. 226).

I ask technical specialists and interface design researchers and practitioners to give the documentation problem a little thought, because I think it can teach you something important about the limitations of your current products: Should you produce artifacts that require so many resources to help humans use them? Is there a way that you could build these helpful resources directly into your products rather than forcing users to wander off looking for them? Are there ways of creating a community of users who can help you identify and strengthen the user response to your product over time?

I ask information developers to question the motivations of product developers. Better yet, learn how to solve technical problems for the technical specialists, so that you will not have to write documentation that nobody reads about products that were built requiring documentation. Ask yourself if you should document poorly designed products? Should you document products that were created without your involvement or feedback? Should you create documentation for users you have never met, talked to, or watched using your documents?

I ask users to demand a great deal from product and information developers, perhaps even to refuse to purchase their boxes, CDs, and books if they do not anticipate the complex problem solving most users are forced to endure. How many products do you use that allow you to perform all your desired tasks without difficulties? When you decide that a product is worth learning about, do you actively engage in developing both a procedural and conceptual understanding of the system? Finally, have you ever considered finding and cultivating solutions, audiences, and purposes for doing things that the developers were unable to anticipate?

Have you ever accessed documentation, read it, and discovered solutions to your difficulties, or is it true that nobody reads documentation?

References

Adler, P. S., & Winograd, T. A. (Eds.). (1992). *Usability: Turning technologies into tools.* New York: Oxford University Press.

Barker, T. T. (1998). *Writing software documentation: A task-oriented approach.* Needham Heights, MA: Allyn & Bacon.

Barnett, M. (1998). Testing a digital library of technical manuals. *IEEE Transactions on Professional Communication, 41*(2), 116–122.

Beabes, M. A., & Flanders, A. (1995). Experiences with using contextual inquiry to design information. *Technical Communication, 42*(3), 409–420.

Bernhardt, S. A. (1993). The shape of text to come: The texture of print on screens. *College Composition and Communication, 44,* 151–175.

Bethke, F., Dean, P., Kaiser, E., Ort, E., & Pessin, F. (1981). Improving the usability of programming publications. *IBM Systems Journal, 20*(3), 306–320.

Bevan, N. (1998). Usability issues in web site design. *Proceedings of Usability Professionals' Association (UPA).* Washington, DC.

Birkerts, S. (1994). *The Gutenberg elegies: The fate of reading in an electronic age.* Boston, MA: Faber & Faber.

Bitzer, L. F. (1968). The rhetorical situation. *Philosophy and Rhetoric, 1*, 1-14.

Boy, G. (1992). Computer integrated documentation. In E. Barrett (Ed.), *Sociomedia* (pp. 507-532). Cambridge, MA: MIT Press.

Brockmann, J. (1990). The why, where and how of minimalism. *SIGDOC '90: The 9th Annual International Conference Proceedings* (pp. 111-119). New York: ACM.

Brockmann, R. (1992). *Writing better computer user documentation: From paper to online* (2nd ed). New York: John Wiley & Sons.

Burke, K. (1969). *A grammar of motives*. Berkeley, CA: University of California Press.

Bush, V. (1945). As we may think. *Atlantic Monthly, 176*(1), 101-108 [on-line]. Available: http://www.theatlantic.com/unbound/flashbks/computer/bushf.htm

Butler, K. A. (1996, January). Usability engineering turns 10. *Interactions, 3*(1), 59-75.

Carroll, J. M. (1990). *The Nurnberg funnel: Designing minimalist instruction for practical computer skill*. Cambridge, MA: MIT Press.

Carroll, J. M. (Ed.). (1998). *Minimalism beyond the Nurnberg funnel*. Cambridge, MA: MIT Press.

Carroll, J. M., Smith-Kerker, P. L., Ford, J. R., & Mazur-Rimetz, S. A. (1987). The minimal manual. *Human-Computer Interaction, 3*(2), 123-153.

Carver, J. (2000). *Internet and computer usage by low-income groups*. College Park, MD: Department of Computer Science White Paper [on-line]. Available: http://www.otal.umd.edu/UUGuide/carver/

Charney, D. H., Reder, L. E., & Kusbit, G. W. (1991). Improving documentation with hands-on problem-solving. *Proceedings of the First Conference on Quality in Documentation* (pp. 134-153). Waterloo, Ontario: Centre for Professional Writing, University of Waterloo.

Chi, M. T., Feltovich, P. J., & Glaser, R. (1981). Categorization and representation of physics problems by experts and novices. *Cognitive Science, 5*, 121-152.

Chi, M. T., Glaser, R., & Rees, E. (1982). Expertise in problem solving. In R. J. Sternberg (Ed.), *Advances in the Psychology of Human Intelligence* (Vol. 1, pp. 7-75). Hillsdale, NJ: Lawrence Erlbaum Associates.

Chignell, M. H., & Valdez, J. F. (1992). Methods for assessing the usage and usability of documentation. *Proceedings of the Third Conference on Quality in Documentation* (pp. 5-27). Waterloo, Ontario: Centre for Professional Writing, University of Waterloo.

Coe, M. (1996). *Human factors for technical communicators*. New York: John Wiley & Sons.

Cole, E., & Dehdashti, P. (1990). Interface design as a prosthesis for individuals with brain injuries. *SIGCHI Bulletin, 22*(1), 28-32.

Csikszentmihalyi, M. (1990). *Flow: The psychology of optimal experience*. New York: HarpersCollins.

Davenport, T. H., & Beck, J. C. (2001, April 22-28). The attention economy. *Ubiquity, 2*(14) [on-line]. Available: http://www.acm.org/ubiquity/book/t_davenport_2.html

Dicks, R. S. (1994). Integrating online help, documentation, and training. *SIGDOC '94: The 12th Annual International Conference Proceedings* (pp. 115-118). New York: ACM.

Doheny-Farina, S. (1996). *The wired neighborhood*. New Haven, CT: Yale University Press.

Duffy, T. M., Mehlenbacher, B., & Palmer, J. E. (1989). The evaluation of online help systems: A conceptual model. In E. Barrett (Ed.), *The society of text: Hypertext, hypermedia, and the social construction of information* (pp. 362-387). Cambridge, MA: MIT Press.

Duffy, T. M., Palmer, J. E., & Mehlenbacher, B. (1993). *Online help: Design and evaluation*. Human-Computer Interaction Series. Norwood, NJ: Ablex.

Duin, A. H. (1993). Test drive—Techniques for evaluating the usability of documents. In C. M. Barnum & S. Carliner (Eds.), *Techniques for technical communicators* (pp. 306-335). New York: Macmillan.

Draper, S. W. (1998). Practical problems and proposed solutions in designing action-centered documentation. In J. M. Carroll (Ed.), *Minimalism beyond the Nurnberg funnel* (pp. 349-374). Cambridge, MA: MIT Press.

Dumas, J. S., & Redish, J. C. (1993). *A practical guide to usability testing*. Greenwich, CT: Ablex.

Fawcett, H., Ferdinand, S., & Rockley, A. (1993). The design draft—Organizing information. In C. M. Barnum & S. Carliner (Eds.), *Techniques for technical communication* (pp. 43-78). New York: Macmillan.

Felder, R. M. (1993). Reaching the second tier: Learning and teaching styles in college science education. *Journal of College Science Teaching, 23*(5), 286-290 [on-line]. Available: http://www2.ncsu.edu/unity/lockers/users/f/felder/public/Papers/Secondtier.html

Frohlich, D. M., & Luff, P. (1989). Conversational resources for situated action. *CHI '89 Conference on Human Factors in Computing Systems* (pp. 253-258). Austin, TX: ACM.

Gentner, D., & Stevens, A. (1983). *Mental models*. Hillsdale, NJ: Lawrence Erlbaum.

Goldbach, B. (2000, August 1-7). Just turn me off. *Ubiquity, 1*(23) [on-line]. Available: http://www.acm.org/ubiquity/views/b_goldbach_1.html

Goodall, S. D. (1991). Online help in the real world. *SIGDOC '91: The 9th Annual International Conference Proceedings* (pp. 21-29). New York: The Association for Computing Machinery (ACM), Special Interest Group on Documentation (SIGDOC).

Gray, J. (2001, April 24-30). Work in the coming age. *Ubiquity, 2*(10) [on-line]. Available: http://www.acm.org/ubiquity/views/j_gray_1.html

Haas, C. (1989). Seeing it on the screen isn't really seeing it: Computer writers' reading problems. In G. Hawisher & C. Selfe (Eds.), *Critical perspectives on computers and composition instruction* (pp. 16-29). New York: Teacher's College Press.

Hackos, J. T., & Redish, J. C. (1998). *User and task analysis for interface design*. New York: Wiley & Sons.

Hayes, J. R. (1989). Writing research: The analysis of a very complex task. In D. Klahr & K. Kotovsky (Eds.), *Complex information processing: The impact of Herbert A. Simon* (pp. 209-234). Hillsdale, NJ: Lawrence Erlbaum.

Hill, C. A., & Mehlenbacher, B. (1998). Transitional generations and World Wide Web reading and writing: Implications of a hypertextual interface for the masses. *TEXT Technology, 8*(4), 29-47.

Hobart, M. E., & Schiffman, Z. S. (1998). *Information ages: Literacy, numeracy, and the computer revolution*. Baltimore, MD: The Johns Hopkins University Press.

Hodes, C. L. (1998). Understanding visual literacy through visual information processing. *Journal of Visual Literacy, 18*(2), 131-136.

Hyperviews: Online archives. Online Information Special Interest Group of the Society for Technical Communication. Available: http://www.stcsig.org/oi/hyperviews/archive/archive.htm

Johnson-Eilola, J. (1996). Relocating the value of work: Technical communication in a post-industrial age. *Technical Communication Quarterly, 5*, 245-270.

Johnson-Eilola, J. (1997a). Wild technologies: Computer use and social possibility. In S. A. Selber (Ed.), *Computers and technical communication: Pedagogical and programmatic perspectives* (pp. 97-128). Greenwich, CT: Ablex.

Johnson-Eilola, J. (1997b). Living on the surface: Learning in the age of global communication networks. In I. Snyder (Ed.), *Page to screen:*

Taking literacy into the electronic era (pp. 185–210). New York: Allen & Unwin.

Kay, J., & Thomas, R. C. (1995). Studying long-term system use. *Communications of the ACM, 38*(7), 61–69.

Kearsley, G. (1988). *Online help Systems: Design and implementation*. Norwood, NJ: Ablex.

Keyes, E., Sykes, D., & Lewis, E. (1989). Technology + design = Information Design. In E. Barrett (Ed.), *Text, context, and hypertext* (pp. 251–264). Cambridge, MA: MIT Press.

Kieras, D. E., & Polson, P. G. (1985). An approach to the formal analysis of user complexity. *International Journal of Man-Machine Studies, 22*, 365–394.

Kinneavy, J. L. (1971). *A theory of discourse: The aims of discourse*. Englewood Cliffs, NJ: Prentice-Hall.

Kintsch, W. (1986). Learning from text. *Cognition and Instruction, 3*(2), 87–108.

Kintsch, W., & Vipond, D. (1979). Reading comprehension and readability in educational practice and psychological theory. In L. G. Nilsson (Ed.), *Perspectives on memory research: Essays in honor of Uppsala University's 500th anniversary* (pp. 329–365). Hillsdale, NJ: Lawrence Erlbaum.

Kirschenbaum, M. G. (1995). The cult of print: Review of Birkerts's "The Gutenberg Elegies." *Postmodern Culture, 6*(1) [on-line]. Available: http://jefferson.village.virginia.edu/pmc/text-only/issue.995/review-7.9957

Kolatch, E. (2000). Designing for users with cognitive disabilities. College Park, MD: Department of Computer Science White Paper [on-line]. Available: http://www.otal.umd.edu/UUGuide/erica/

Kolb, D. A. (1984). *Experiential learning: Experience as the source of learning and development*. Englewood Cliffs, NJ: Prentice-Hall.

Kostelnick, C., & Roberts, D. D. (1998). *Designing visual language: Strategies for professional communicators*. Needham Heights, MA: Allyn & Bacon.

Kress, G. R., Leeuwen, T. V., & Gress, G. R. (1995). *Reading images: The grammar of visual design*. New York: Routledge.

Landauer, T. K. (1995). *The trouble with computers: Usefulness, usability and productivity*. Cambridge, MA: MIT Press.

Lee, M. F., & Mehlenbacher, B. (2000). Technical writer/subject-matter expert interaction: The writer's perspective, the organizational challenge. *Technical Communication, 47*(4), 544–552.

Lee, S. H. (1999). Usability testing for developing effective interactive multimedia software: Concepts, dimensions, and procedures. *Educational Technology & Society, 2*(2) [on-line]. Available: http://ifets.gmd.de/periodical/vol_2_99/sung_heum_lee.html

Lindgaard, G. (1994). *Usability testing and system evaluation: A guide for designing useful computer systems*. London, UK: Chapman & Hall Computing.

Loomis, J. M., & Lederman, S. J. (1986). Tactual perception. In K. R. Boff, L. Kaufman, & J. P. Thomas (Eds.), *Handbook of perception and human performance* (pp. 31.31–31.41). New York: John Wiley & Sons/Interscience.

Mack, R. L., & Nielsen, J. (1994). Executive summary. In J. Nielsen & R. L. Mack (Eds.), *Usability inspection methods* (pp. 1–23). New York: John Wiley & Sons.

Maes, A., Goutier, S., & van der Linden, E.-J. (1992). Online reading and offline tradition: Adapting online help facilities to offline reading strategies. *SIGDOC '92: The 10th Annual International Conference Proceedings* (pp. 175–182). New York: The Association for Computing Machinery (ACM), Special Interest Group on Documentation (SIGDOC).

Marshall, C., Nelson, C., & Gardiner, M. M. (1987). Design guidelines. In M. M. Gardiner & B. Christie (Eds.), *Applying cognitive psychology to user-interface design* (pp. 221–278). New York: John Wiley & Sons.

Mayhew, D. J. (1999). *The usability engineering lifecycle: A practitioner's handbook for user interface design*. San Francisco, CA: Morgan Kaufmann.

Mehlenbacher, B. (1992). Navigating online information: A characterization of extralinguistic factors that influence user behavior. *SIGDOC '92: The 10th Annual International Conference Proceedings* (pp. 35–46). New York: The Association for Computing Machinery (ACM), Special Interest Group on Documentation (SIGDOC).

Mehlenbacher, B. (1993). Software usability: Choosing appropriate methods for evaluating online systems and documentation. *SIGDOC '93: The 11th Annual International Conference Proceedings* (pp. 209–222). New York: ACM.

Mehlenbacher, B., & Dicks, R. S. (in preparation). A pedagogical framework for faculty-student research and public service in technical communication. In D. Selfe, K. Kitalong, & T. Bridgeford (Eds.), *Innovative approaches to technical communication: Teaching, administration, and curriculum*. Raleigh, NC: North Carolina State.

Mehlenbacher, B., Miller, C. R., Covington, D., & Larsen, J. (2000). Active and interactive learning online: A comparison of web-based and conventional writing classes. *IEEE Transactions on Professional Communication, 43*(2), 166–184.

Michaels, J. W. (2000, May 15). Perspectives: Webucation. *Forbes Magazine* [on-line]. Available: http://www.forbes.com/forbes/2000/0515/6511092a.html

Miller, C. R. (1985). Invention in technical and scientific discourse: A prospective survey. In M. G. Moran & D. Journet (Eds.), *Research in technical communication: A bibliographic sourcebook* (pp. 117–162). Westport, CT: Greenwood Press.

Mirel, B. (1998). Minimalism for complex tasks. In J. M. Carroll (Ed.), *Minimalism beyond the Nurnberg funnel* (pp. 179–218). Cambridge, MA: MIT Press.

Mirel, B. (1999). Complex queries in information visualizations: Distributing instruction across documentation and interfaces. *SIGDOC '99: The 17th Annual International Conference Proceedings* (pp. 1–8). New York: ACM.

Moran, T. P. (1983). Getting into a system: External-internal task mapping analysis. *CHI '83 Conference on Human Factors in Computing Systems* (pp. 45–49). New York: ACM.

Mosier, J. N. & Smith, S. L. (1986). Application of guidelines for designing user interface software. *Behaviour and Information Technology, 5*, 39–46.

Mynatt, E. D. (1997). Transforming graphical interfaces into auditory interfaces for blind users multimodal interfaces. *Human-Computer Interaction, 12*(1/2), 7–45.

Nielsen, J. (1994). Heuristic evaluation. In J. Nielsen & R. L. Mack (Eds.), *Usability inspection methods* (pp. 25–62). New York: John Wiley & Sons.

Nielsen, J. (1997). Usability engineering. In A. B. Tucker, Jr. (Ed.), *The computer science and engineering handbook* (pp. 1440–1460). Boca Raton, FL: CRC Press.

Nielsen, J. (1999). *Designing web usability: The practice of simplicity*. Indianapolis, IN: New Riders Publishing.

Nielsen, J., & Mack, R. L. (Eds.). (1994). *Usability inspection methods*. New York: John Wiley & Sons.

Norman, D. A. (1984). Stages and levels in human-computer interaction. *International Journal of Man-Machine Studies, 21*, 364–375.

Norman, D. A. (1990). *The design of everyday things*. New York: Basic.

NSF. (1998, December 15). *NSF indicators report on science and technology, public attitudes and public understanding. National Science Foundation* [on-line]. Available: http://www.nsf.gov/sbe/srs/seind98/c7/c7s4.htm#c7s412

Odescalchi, E. K. (1986). Productivity gain attained by tasked oriented information. In *33rd International Technical Communication Conference Proceedings* (pp. 359-362). Washington, DC: Society for Technical Communication.

Ong, W. J. (1982). *Orality and literacy: The technologizing of the word*. New York: Methuen.

Pasmore, W. A. (1994). *Creating strategic change: Designing the flexible, high-performing organization*. New York: John Wiley & Sons.

Penzias, A. (1989). *Ideas and information: Managing in a high-tech world*. New York: Simon & Schuster.

Redish, J. C. (1988). Reading to learn to do. *The Technical Writing Teacher, 15*(3), 223-233.

Redish, J. C. (1993). Understanding readers. In C. M. Barnum & S. Carliner (Eds.), *Techniques for technical communication* (pp. 14-41). New York: Macmillan.

Redish, J. C., Battison, R., & Gold, E. (1985). Making written information accessible to readers. In L. Odell & D. Goswami (Eds.), *Writing in nonacademic settings* (pp. 129-153). New York: Guilford Press.

Reece, G., & Scheiber, H. J. (1993). Designing for dual delivery: Online and paper. *Proceedings of the International Professional Communication Conference* (pp. 85-87). New York: Institute of Electrical and Electronics Engineers.

Rettig, M. (1991). Nobody reads documentation. *Communications of the ACM, 34*(7), 19-24.

Rubin, J. (1994). *Handbook of usability testing: How to plan, design, and conduct effective Tests*. New York: John Wiley & Sons.

Sadler-Smith, E., & Riding, R. (1999). Cognitive style and instruction preferences. *Instructional Science, 27*, 355-371.

Schell, D. A. (1986). Testing online and print user documentation. *IEEE Transactions on Professional Communication, 29*(4), 87-92.

Schmeck, R. R. (Ed.). (1988). *Learning strategies and learning styles*. New York: Plenum Press.

Schön, D. (1983). *The reflective practitioner: How professionals think in action*. New York: Basic Books/HarperCollins.

Schön, D. (1987). *Educating the reflective practitioner*. San Francisco, CA: Jossey-Bass.

Schriver, K. A. (1989). Evaluating text quality: The continuum from text-focused to reader-focused methods. *IEEE Transactions on Professional Communication, 32*(4), 238-255.

Schriver, K. A. (1993). Quality in document design: Issues and controversies. *Technical Communication, 40*(2), 239-257.

Schriver, K. A. (1997). *Dynamics in document design: Creating texts for readers*. New York: John Wiley & Sons.

Selber, S. A., Johnson-Eilola, J. D., & Mehlenbacher, B. (1997). Online support systems: Tutorials, documentation, and help. In A. B. Tucker, Jr. (Ed.), *The computer science and engineering handbook* (pp. 1619-1643). Boca Raton, FL: CRC Press.

Shneiderman, B. (1998). *Designing the user interface: Strategies for effective human-computer interaction* (3rd ed.). Reading, MA: Addison-Wesley.

Shneiderman, B. (2001, March/April). CUU: Bridging the digital divide with universal usability (pp. 11-15). *Interactions* [on-line]. Available: http://sigchi.org/cuu/interactions.html

Simon, H. A. (1979). *Models of thought*. New Haven, CT: Yale University Press.

Skelton, T. M. (1992). Testing the usability of usability testing. *Technical Communication, 39*(3), 343-359.

Sobchack, V. (1996). Democratic franchise and the electronic frontier. In Z. Sardar & J. R. Ravetz (Eds.), *Cyberfutures: Culture and politics on the information superhighway* (pp. 77-89). New York: New York University Press.

Suchman, L. A. (1987). *Plans and situated actions: The problem of human machine communication*. Cambridge, UK: Cambridge University Press.

Sticht, T. (1985). Understanding readers and their uses of text. In T. M. Duffy & R. Waller (Eds.), *Designing usable texts* (pp. 315-340). Orlando, FL: Academic Press.

Sullivan, P., & Flower, L. (1986). How do users read computer manuals? Some protocol contributions to writers' knowledge. In B. T. Petersen (Ed.), *Convergences: Transactions in reading and writing* (pp. 163-178). Urbana, IL: National Council of Teachers of English.

Tomasi, M. D., & Mehlenbacher, B. (1999). Re-engineering online documentation: Designing examples-based online support systems. *Technical Communication, 46*(1), 55-66.

Trenner, L., & Bawa, J. (Eds.). (1998). *The politics of usability: A practical guide to designing usable systems in industry*. Berlin, Germany: Springer-Verlag.

Tufte, E. (1983). *The visual display of quantitative information*. Cheshire, CT: Graphics Press.

Ummelen, N. (1997). *Procedural and declarative information in software manuals: Effects on information use, task performance and knowledge*. Amsterdam: Rodopi.

U.S. Department of Commerce (2000). Americans in the information age: Falling through the net. *National Telecommunications Information Administration (NTIA) Report* [on-line]. Available: http://www.ntia.doc.gov/ntiahome/digitaldivide/index.html

Van der Maij, H., & Carroll, J. M. (1995). Principles and heuristics for designing minimalist instruction. *Technical Communication, 42*(2), 243-262.

Welinske, J. (2000, January). The future of online help development. *Los Angeles STC Chapter, WinWriters Group*. Los Angeles, CA [on-line]. Available: http://www.winwriters.com/talks/lastc/lastc.htm

Winograd, T., & Flores, F. (1987). *Understanding computers and cognition: A new foundation for design*. Reading, MA: Addison-Wesley.

Wood, L. E. (Ed.). (1998). *User interface design: Bridging the gap from user requirements to design*. Boca Raton, FL: CRC Press.

Wright, P. (1979). Quality control aspects of document design. *Information Design Journal, 1*, 33-42.

Wright, P. (1980). Usability: The criterion for designing written information. In P. A. Koler, M. E. Wrolstad, & H. Bouma (Eds.), *Processing visible language* (Vol. 2, pp. 183-206). New York: Plenum Press.

Zimmerman, D. E., Tipton, M., Bilsing, L., & Green, D. (1993). Hype, hypertext, and reality: What research tells us about hypertext. *Proceedings of the International Professional Communication Conference* (pp. 71-85). New York: Institute of Electrical and Electronics Engineers.

APPENDIX: USABILITY PRINCIPLES FOR

DOCUMENTATION DESIGN

TABLE A27.1. Usability Principles for Documentation Design (Mehlenbacher, 2001)

Accessibility	Has the online documentation been viewed on different platforms, browsers, modem speeds?
	Have you considered alternatives to traditional paper stock in terms of size?
	Is the documentation ADA compliant (e.g., colors such as red and yellow are problematic for visually challenged users)?
	Have ISO-9000 standards been considered?
Aesthetic appeal	Does the design appear minimalist (uncluttered, readable, memorable)?
	Are graphics or colors used aesthetically?
	Are distractions minimized (e.g., movement, blinking, scrolling, animation, etc.)?
Authority and authenticity	Does the documentation establish a serious tone or presence?
	Is humor or anthropomophic expressions used minimally?
	Is direction given for further assistance if necessary?
Completeness	Are levels clear and explicit about the end or parameters of the documentation?
	Are there different levels of use and, if so, are they clearly distinguishable?
Consistency and layout	Does every page or screen display begin with a title/subject heading that describes contents?
	Is there a consistent icon design and graphic display across pages or screens?
	Is layout, font choice, terminology use, color, and positioning of items the same throughout the documentation (<4 of any of the above is usually recommended)?
Customizability and maintainability	Does printing of the online documentation require special configuration to optimize presentation and, if so, is this indicated in the documentation?
	Are individual preferences/sections clearly distinguishable from one another?
	Is manipulation of the documentation possible and easy to achieve?
Error support and feedback	Is an interface design solution possible that prevents a problem from occurring or requiring documentation in the first place?
	When users scan or select something, does it differentiate itself from other information chunks or unselected items?
	Do cross-references, menu instructions, prompts, and error messages appear in the same place on each page or screen?
Examples and case studies	Are examples, demonstrations, or case studies of user experiences available to facilitate product learning?
	Are the examples divided into meaningful sections (e.g., overview, demonstration, explanation, and so on)?
Genre representation	Does the documentation contain task-oriented help, tutorials, reference material and, especially, troubleshooting information?
	Is help easy to locate and access online?
	Is the help table of contents or menu organized functionally, according to user tasks and not according to product features or generic topics?
Intimacy and presence	Is an overall tone that is present, active, and engaging established?
	Does the documentation act as a learning environment for users and not simply as a warehouse of unrelated topics or links?
Metaphors and maps	Does the documentation establish an easily recognizable metaphor that helps users identify additional support materials in relation to each other, their state in the system, and options available to them?

Note. Cf. Bevan, 1998; Nielsen, 1994, 1997; Selber, Johnson-Eilola, & Mehlenbacher, 1997.

TABLE A27.2. Usability Principles for Documentation Design (Mehlenbacher, 2001)

Navigability and user movement	Does the online documentation clearly separate navigation from content?
	How many levels down can users traverse and, if more than three, is it clear that returning to their initial state is possibly easy to do? (is a clear hierarchy established?)
	Can users see where they are in the overall documentation at all times?
	Do the locations of navigational elements remain consistent (e.g., tables or menus)?
	Is the need to scroll or traverse multiple pages for a single topic minimized across screens or pages?
Organization and information relevance	Is a site map or index available?
	Is the overall organization of the documentation clear from the majority of pages or screens?
	Are primary options emphasized in favor of secondary and tertiary ones?
Readability and quality of writing	Is the text in active voice and concisely written (> 4 < 15 words/sentence)?
	Are terms consistently plural, verb + object or noun + verb, etc., avoiding unnecessarily redundant words?
	Does white space highlight a modular text design that separates information chunks from each other?
	Are bold and color texts used sparingly to identify important text (limiting use of all capitals and italics to improve readability)?
	Can users understand the content of the information presented easily?
Relationship with real-world tasks	Is terminology and labeling meaningful, concrete, and familiar to the target audience?
	Do related and interdependent functions appear on the same screen or page?
	Is sequencing used naturally, if sequences of common events are expected?
	Does the documentation allow users to easily complete their transactions or tasks?
Reliability and functionality	Do all the titles, menus, icons, links, and opening windows work predictably across the online information or mean the same thing across the print documentation?
Typographic cues and structuring	Does the text use meaningful discourse cues, modularization, or chunking?
	Is information structured by meaningful labeling, bulleted lists, or iconic markers?
	Are legible fonts and colors used?
	Is the principle of left-to-right placement linked to most-important to least-important information?

Note. Cf. Bevan, 1998; Nielsen, 1994, 1997; Selber, Johnson-Eilola, & Mehlenbacher, 1997.

INFORMATION VISUALIZATION

Stuart Card
Xerox PARC

INTRODUCTION

The working mind is greatly leveraged by interaction with the world outside it. A conversation to share information, a grocery list to aid memory, a pocket calculator to compute square roots—all effectively augment a cognitive ability otherwise severely constrained by limited knowledge, by limited attention, and by limitations on reasoning. But the most profound leverage on cognitive ability is the cognitive ability to invent new representations, procedures, or devices that augment cognition far beyond its unaided biological endowment, and to bootstrap these into even more potent inventions.

This chapter is about one class of inventions for augmenting cognition, collectively called *information visualization*. Other senses could be used in this pursuit—audition, for example, or a multimodal combination of senses—the broader topic is really *information perceptualization*. But, in this chapter, we restrict ourselves to visualization. Visualization uses the sense with the most information capacity; recent advances in graphically agile computers have opened opportunities to exploit this capacity, and many visualization techniques have now been developed. A few examples suggest the possibilities.

Example 1: Finding Videos With the FilmFinder

The use of information visualization for search is illustrated by the FilmFinder (Ahlberg & Shneiderman, 1994a, 1994b). Unlike typical movie finder systems, the FilmFinder is organized not around searching with keywords, but rather around rapid browsing and reacting to collections of films in the database. Figure 28.1 shows a scattergraph of 2000 movies, plotting rated quality of the movie as a function of the year when it was released. Color differentiates type of movies, comedy from drama and the like. The display provides an overview, the entire universe of all the movies and also some general features of the collection.

It is visually apparent, for example, that a good share of the movies in the collection were released after 1965, but also that there are movies going back as far as the 1920s. Now, the viewer "drills down" into the collection by using the sliders in the interface so as to show only movies with Sean Connery that are between 1 and 4½ hr in length (Fig. 28.2). As the sliders are moved, the display zooms in to show about 20 movies. It can be seen that these movies were made between 1960 and 1995, and all have a quality rating higher than 4. Because there is now room on the display, titles of the movies appear. Experimentation with the slider shows that restricting maximum length to 2 hr cuts out few interesting movies. The viewer chooses the highly rated movie, "Murder on the Orient Express" by double-clicking on its marker. Up pop details in a box (Fig. 28.3) giving names of other actors in the movie and more information. The viewer is interested in whether two of these actors, Anthony Perkins and Ingrid Bergman, have appeared together in any other movies. She selects their names in the box, then requests another search (Fig. 28.4). The result is a new display of two movies. In addition to the movie she knew about, there is one other movie, a drama entitled "Goodbye, Again," made around

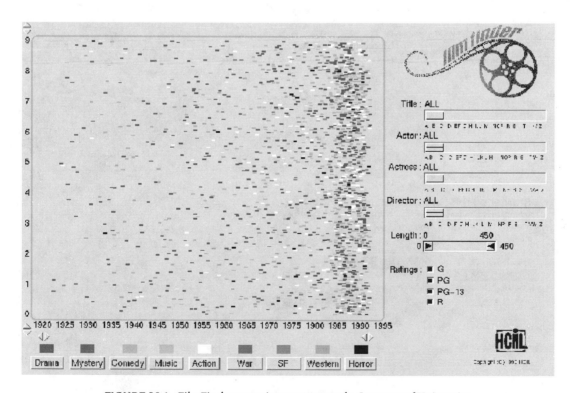

FIGURE 28.1. FilmFinder overview scattergraph. Courtesy of University of Maryland.

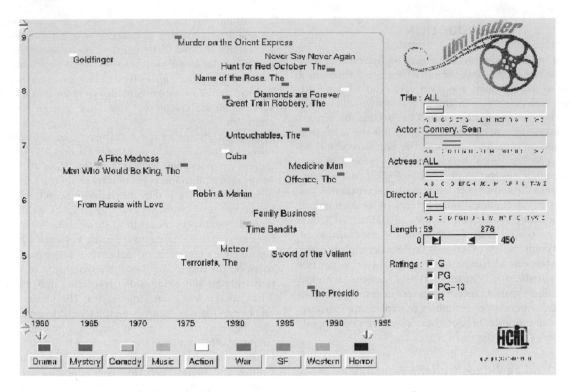

FIGURE 28.2. FilmFinder scattergraph zoom-in. Courtesy of University of Maryland.

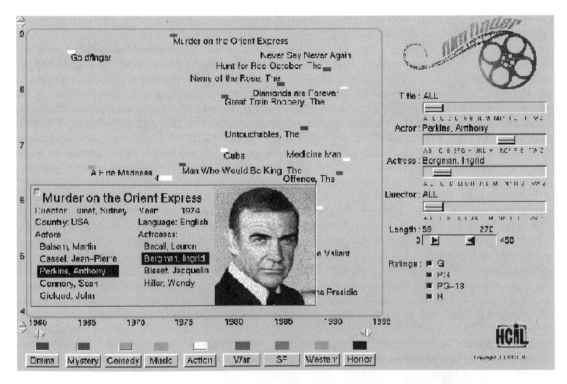

FIGURE 28.3. FilmFinder details on demand. Courtesy of University of Maryland.

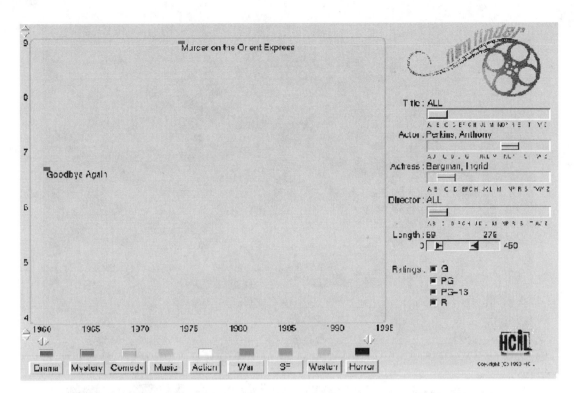

FIGURE 28.4. FilmFinder retrieval by example. Courtesy of University of Maryland.

1960. The viewer is curious about this movie and decides to watch it.

Information visualization has allowed a movie viewer in a matter of seconds to find a movie she could not have specified at the outset. To do this, the FilmFinder used several techniques from information visualization: an *overview* of the collection showing its structure, *dynamic queries* in which the visualization seems to change instantaneously with control manipulations, *zooming in* by adding restrictions to the set of interest, *details on demand* in which the user can display temporarily detail about an individual object, and *retrieval by example* in which selected attributes of an individual object are used to specify a new retrieval set.

Example 2: Monitoring Stocks With TreeMaps

Another example of information visualization, this time for monitoring, is the TreeMap visualization on the SmartMoney.com Web site[1] that is shown in Fig. 28.5a. Using this visualization, an investor can monitor more than 500 stocks at once, with data updated every 15 min. Each colored rectangle in the figure is a company. The size of the rectangle is proportional to its market capitalization. Color of the rectangle shows movement in the stock price. Bright yellow corresponds to about a 6% increase in price, bright blue to about a 6% decrease in price. Each business sector is identified with a label like "Communication."

Those items marked with a letter N have an associated news item.

In this example, the investor's task is to monitor the day's market and notice interesting developments. In Fig. 28.5a, the investor has moved the mouse over one of the bright yellow rectangles, and a box identifying it as Erickson, with a +9.28% gain for the day, has popped up together with other information. Clicking on a rectangle in the TreeMap gives the investor a popup menu for selecting even more detail. The investor can either click to go to World Wide Web links on news or financials, or drill down, for example, to the technology sector (Fig. 28.5b) or down further to individual companies in the software part of the technology sector (Fig. 28.5c). When she does so, the investor immediately notes interesting relationships. The software industry is now larger than the hardware industry, for example, and despite a recent battering at the time of this figure, the Internet industry is also relatively large. Microsoft is larger than all the other companies in its industry combined. Selecting a menu item to look at year-to-date gains (Fig. 28.6), the investor immediately notes interesting patterns: Microsoft stock shows substantial gains, whereas Oracle is down. Dell is up, but Compaq is down. Tiny Advanced Micro is up, whereas giant Intel is neutral. Having noticed these relationships, the investor drills down to put up charts or analysts positions for companies whose gains in themselves, or in relation to a competitor, are interesting. For example, the investor is preparing a report on the computer industry for colleagues and notices how AMD is making gains

---

[1] www.smartmoney.com

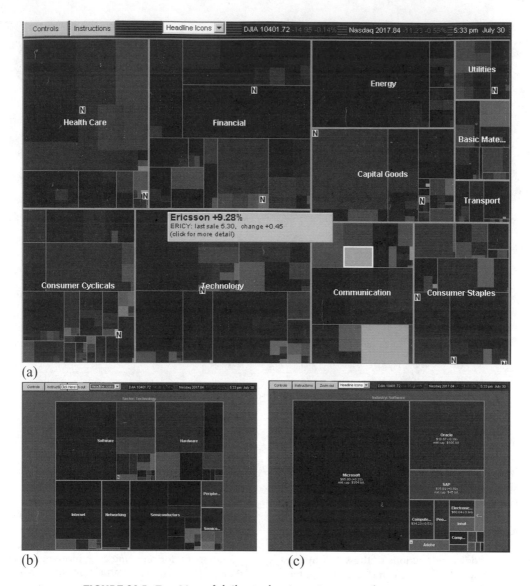

FIGURE 28.5. TreeMap of daily stock prices. Courtesy of SmartMoney.com.

against Intel or how competition for the Internet is turning into a battle between Microsoft and AOL/Time Warner.

Example 3: Sensemaking With Permutation Matrices

As a final information visualization example, consider the case proposed by Bertin (1977/1981) of a hotel manager analyzing hotel occupancy data (Table 28.1) in order to increase her return. To search for meaningful patterns in her data, she represents it as a permutation matrix (Fig. 28.7a). A permutation matrix is a graphic rendition of a cases × variables display. In Fig. 28.7a, each cell of Table 28.1 is rendered as a small bar of a bar chart. The bars for cells below the mean are white, those above the bar are black. By permuting rows and columns, patterns emerge that lead to making sense of the data.

In Fig. 28.7a, the set of months, which form the cases, are repeated so as to be able to see periodic patterns across the end of the cycle. By comparing the pairs of rows visually, rows are found that are similar. These are reordered and grouped (Fig. 28.7b). By this means, it is discovered that there seem to be two patterns of yearly variation. One pattern in Fig. 28.7b, is semiannual, dividing the year into the cold months of October through April and the warm months of May through September. The other pattern breaks the year down into four distinct regions. These patterns form beginnings of a *schema*, that is, a framework in terms of which the raw data we can be encoded and described it in a more compact language. Instead of talking about the events of the year in terms of individual months, the hotel manager can now talk in terms of two series of periods, the semiannual one and the four distinct periods. There is a *residue* of information not included as part of the new

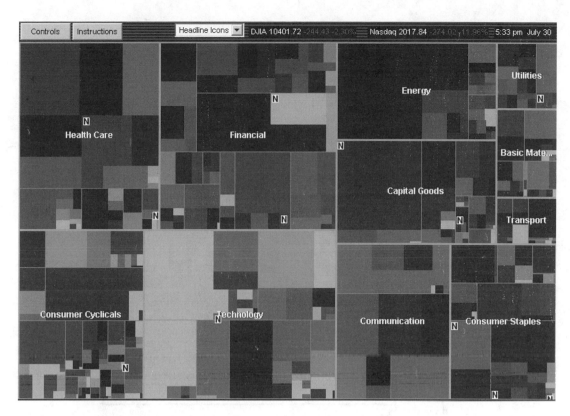

FIGURE 28.6. TreeMap of year-to-date stock prices. Courtesy of SmartMoney.com.

description. Sensemaking proceeds by the omission and re-coding of information into more compact form (see Resnikoff, 1989). This residue of information may be reduced by finding a better or more articulated schema, or it may be left as noise. Beyond finding the basic patterns in the data, the hotel manager wants to make sense of the data relative to a purpose: She wants

to increase the occupancy of the hotel. Therefore, she has also permuted general indicators of activity in Fig. 28.7b, such as %Occupancy and Length of Stay, to the top of the diagram and put the rows that correlate with these below them. This reveals that Conventions, Businessmen, and Agency Reservations—all of which generally have to do with convention business—are

TABLE 28.1. Data for Hotel Occupancy (Based on Bertin, 1977/1981)

ID	Variable	Jan	Feb	Mar	Apr	May	June	July	Aug	Sept	Oct	Nov	Dec
1	% Female	26	21	26	28	20	20	20	20	20	40	15	40
2	% Local	69	70	77	71	37	36	39	39	55	60	68	72
3	% USA	7	6	3	6	23	14	19	14	9	6	8	8
4	% South America	0	0	0	0	8	6	6	4	2	12	0	0
5	% Europe	20	15	14	15	23	27	22	30	27	19	19	17
6	% Middle East/Africa	1	0	0	8	6	4	6	4	2	1	0	1
7	% Asia	3	10	6	0	3	13	8	9	5	2	5	2
8	% Businessmen	78	80	85	86	85	87	70	76	87	85	87	80
9	% Tourists	22	20	15	14	15	13	30	24	13	15	13	20
10	% Direct reservations	70	70	75	74	69	68	74	75	68	68	64	75
11	% Agency reservations	20	18	19	17	27	27	19	19	26	27	21	15
12	% Air crews	10	12	6	9	4	5	7	6	6	5	15	10
13	% Under 20	2	2	4	2	2	1	1	2	2	4	2	5
14	% 20–35	25	27	37	35	25	25	27	28	24	30	24	30
15	% 35–55	48	49	42	48	54	55	53	51	55	46	55	43
16	% Over 55	25	22	17	15	19	19	19	19	19	20	19	22
17	Price of rooms	163	167	166	174	152	155	145	170	157	174	165	156
18	Length of stay	1.7	1.7	1.7	1.91	1.9	2	1.54	1.6	1.73	1.82	1.66	1.44
19	% Occupancy	67	82	70	83	74	77	56	62	90	92	78	55
20	Conventions	0	0	0	1	1	1	0	0	1	1	1	1

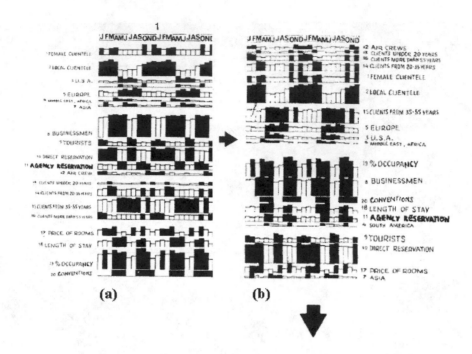

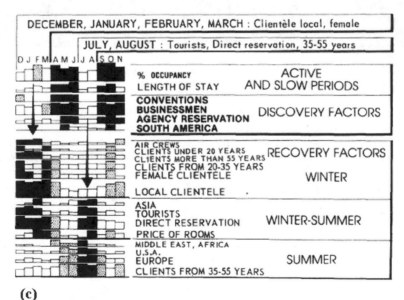

FIGURE 28.7. Hand drawn permutation matrix representation of hotel data. (a) Initial matrix of variables. (b) Permuted matrix to group like patterns together. (c) Permutation matrix in simplified form for presentation. *Note.* From *Graphics Constructions and Graphic Information-Processing* (pp. 24–31), by J. Bertin, 1977/1981, Berlin: Walter De Gruyter. Copyright 1977/1981 by Walter De Gruyter. Reprinted with permission.

associated with higher occupancy. This insight comes from the match in patterns internal to the visualization; it also comes from noting why these variables might correlate as a consequence of factors external to the visualization. She also discovers that there are marked differences between winter and summer guests during the slow periods. In winter, there are more local guests,

women, and age differences. In summer, there are more foreign tourists and less variation in age.

This visualization was useful for sensemaking on hotel occupancy data, but it is too complicated to communicate the high points. The hotel manager therefore creates a simplified diagram (Fig. 28.7c). By graying some of the bars, the main points are more readily graspable, while still preserving the data relations. A December convention, for example, does not seem to have the effect of the other conventions in bringing in guests. It is shown in gray as residue in the pattern. The hotel manager suggests moving it to another month where it might have more effect.

What Is Information Visualization?

The FilmFinder, the TreeMap, and the permutation matrix hotel analysis are all examples of the use of information visualization. We can define information visualization as follows:

Information visualization is the use of computer-supported, interactive, visual representations of abstract data in order to amplify cognition. (Card, Mackinlay, & Shneiderman, 1999)

Information visualization needs to be distinguished from related areas: Scientific visualization is like information visualization, but applied to scientific data, typically physically based. The starting point of a natural geometrical substrate for the data, whether the human body or earth geography, tends to emphasize finding a way to make visible the invisible (say, velocity of air flow) within an existing spatial framework. The chief problem for information visualization, in contrast, is often finding an effective mapping between abstract entities and a spatial representation. Both information visualization and scientific visualization belong to the broader field of data graphics, which is the use of abstract, nonrepresentational visual representations to amplify cognition. Data graphics, in turn, is part of information design, which concerns itself with external representations for amplifying cognition. At the highest level, we could consider information design a part of external cognition, the uses of the external world to accomplish some cognitive process. Characterizing the purpose of information visualization as amplifying cognition is purposely broad. Cognition can be the process of writing a scientific paper or shopping on the Internet for a cell phone. Generally, it refers to the intellectual processes in which information is obtained, transformed, stored, retrieved, and used. All of these can be advanced by means of external cognition generally and information visualization in particular.

Why Does Visualization Work?

Visualization aids cognition not because of some mystical superiority of pictures over other forms of thought and communication, but rather because visualization helps the user by making the world outside the mind a resource for thought in fairly specific ways. We list six groups of these in Table 28.2 (Card et al., 1999): Visualization amplifies cognition by (1) increasing the memory and processing resources available to the users, (2) reducing search for information, (3) using visual representations to enhance the detection of patterns, (4) enabling perceptual inference operations, (5) using perceptual attention mechanisms for monitoring, and (6) encoding information in a manipulable medium. The FilmFinder, for example, allows

the representation of a large amount of data in a small space in a way that allows patterns to be perceived visually in the data. Most important, the method of instantly responding in the display to the dynamic movement of the sliders allowed users rapidly to explore the multidimensional space of films. The TreeMap of the stock market allows monitoring and exploration of many equities. Again, there is the representation of much data in little space. In this case, the display manages the user's attention, drawing it to those equities with unusually large changes, and supplying the means to drill down into the data to understand why these movements may be happening. In the hotel management case, the visual representation makes it easier to notice similarities of behavior in a multidimensional attribute space, then to cluster and re-represent these. The final product is a compact (and simplified) representation of the original data that supports a set of forward decisions. In all these cases, visualization allows the user to examine a large amount of information, to keep an overview of the whole while pursuing details, to keep track of (by using the display as an external working memory) many things, and to produce an abstract representation of a situation through the omission and recoding of information.

Historical Origins

Drawn visual representations have a long history. Maps go back millennia. Diagrams were an important part of Euclid's books on geometry. Science, from earliest times, used diagrams to record observations, induct relationships, explicate methodology of experiments, and classify and conceptualize phenomena (for a discussion, see Robin, 1992). For example, Fig. 28.8 is a hand-drawn illustration, in Newton's first scientific publication, illustrating how white light is really composed of many colors. Sunlight enters from the window at right and is refracted into many colors by a prism. One of these colors can be selected (by an aperture in a screen) and further refracted by another prism, but the light stays the same color showing that it has already been reduced to its elementary components. As in Newton's illustration, early scientific and mathematical diagrams generally had a spatial, physical basis and were used to reveal the hidden, underlying order in that world.

Surprisingly, diagrams of abstract, nonphysical information are apparently rather recent. Tufte (1983) dates abstract diagrams to Playfair (1786) in the eighteenth century. Figure 28.9 is one of Playfair's earliest diagrams. The purpose was to convince readers that English imports were catching up with imports. Starting with Playfair, the classical methods of plotting data were developed—graphs, bar charts, and the rest.

Recent advances in the visual representation of abstract information derive from several strands that became intertwined. In 1967, Bertin, a French cartographer published his theory of *The Semiology of Graphics* (Bertin, 1967/1983, 1977/1981). This theory identified the basic elements of diagrams and their combination. Tufte, from the fields of visual design and data graphics, published a series of seminal books setting forth principles for the design of data graphics (Tufte, 1983, 1990, 1997) that emphasized maximizing the density of useful information.

TABLE 28.2. How Information Visualization Amplifies Cognition (Card, Mackinlay, & Shneiderman, 1999)

1. Increased Resources	
High-bandwidth hierarchical interaction	Human moving gaze system partitions limited channel capacity so that it combines high spatial resolution and wide aperture in sensing the visual environments (Restnikoff, 1989).
Parallel perceptual processing	Some attributes of visualizations can be processed in parallel compared to text, which is serial.
Offload work from cognitive to perceptual system	Some cognitive inferences done symbolically can be recoded into inferences done with simple perceptual operations (Larkin & Simon, 1987).
Expanded working memory	Visualizations can expand the working memory available for solving a problem (Norman, 1993).
Expanded storage of information	Visualizations can be used to store massive amounts of information in a quickly accessible form (e.g., maps).
2. Reduced Search	
Locality of processing	Visualizations group information used together reducing search (Larkin & Simon, 1987).
High data density	Visualizations can often represent a large amount of data in a small space (Tufte, 1983).
Spatially indexed addressing	By grouping data about an object, visualizations can avoid symbolic labels (Larkin & Simon, 1987).
3. Enhanced Recognition of Patterns	
Recognition instead of recall	Recognizing information generated by a visualization is easier than recalling that information by the user.
Abstraction and aggregation	Visualizations simplify and organize information, supplying higher centers with aggregated forms of information through abstraction and selective omission (Card et al., 1991; Resnikoff, 1989).
Visual schemata for organization	Visually organizing data by structural relationships (e.g., by time) enhances patterns.
Value, relationship, trend	Visualizations can be constructed to enhance patterns at all three levels (Bertin, 1967/1983).
4. Perceptual Inference	
Visual representations make some problems obvious	Visualizations can support a large number of perceptual inferences that are very easy for humans (Larkin & Simon, 1987).
Graphical computations	Visualizations can enable complex specialized graphical computations (Hutchins, 1996).
5. Perceptual Monitoring	Visualizations can allow for the monitoring of a large number of potential events if the display is organized so that these stand out by appearance or motion.
6. Manipulable medium	Unlike static diagrams, visualizations can allow exploration of a space of parameter values and can amplify user operations.

Both Bertin's and Tufte's theories became well known and influential. Meanwhile, within statistics, Tukey (1977) began a movement on exploratory data analysis. His emphasis was not on the quality of graphical presentation, but on the use of pictures to give rapid, statistical insight into data relations. For example, "box and whisker plots" allowed an analyst to get a rapid characterization of data distributions. Cleveland and McGill (1988) wrote an influential book, *Dynamic Graphics for Statistics*, explicating new visualizations of data with particular emphasis on the visualization of multidimensional data.

In 1985, the National Science Foundation launched an initiative on scientific visualization (McCormick & DeFanti, 1987). The purpose of this initiative was to use advances in computer graphics to create a new class of analytical instruments for scientific analysis, especially as a tool for comprehending the large data sets being produced in the geophysical and biological sciences. Meanwhile, the computer graphics and artificial intelligence communities were interested in the automatic design of visual presentations of data. Mackinlay's thesis APT (Mackinlay, 1986a, 1986b) formalized Bertin's design theory, added psychophysical data, and used these to build a system for automatically generating diagrams of data, tailored for some purpose. Roth and Mattis (1990) built a system to do more complex visualizations, such as some of those from Tufte. Casner (1991) added a representation of tasks. This community was interested not so much in the quality of the graphics as in the automation of the match between data characteristics, presentational purpose,

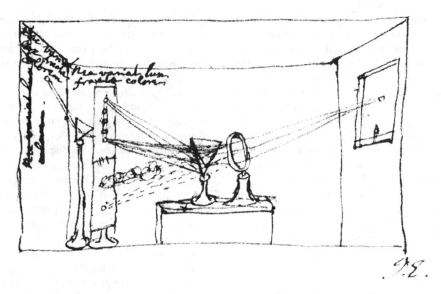

FIGURE 28.8. Newton's optics illustration. From his first scientific publication. *Note.* From *The Scientific Image: From Cave to Computer*, by H. Robin, 1992, New York: H. N. Abrams, Inc. Copyright 1992 by H. N. Abrams, Inc. Reprinted with permission.

and graphical presentation. Finally, the user interface community saw advances in graphics hardware opening the possibility of a new generation of user interfaces. The first use of the term "information visualization" was probably in Robertson, Card, and Mackinlay (1989). Early studies in this community focused on user interaction with large amounts of information: Feiner and Beshers (1990) presented a method, worlds within worlds, for showing six-dimensional financial data in an immersive virtual reality. Shneiderman (1992) developed a technique called dynamic queries for interactively selecting subsets of data items

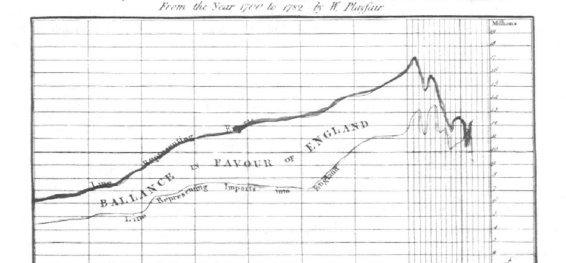

FIGURE 28.9. Playfair's chart of English imports and exports. *Note.* From *The Visual Display of Quantitative Information*, by E. R. Tufte, 1983, Cheshire, CT: Graphics Press. Copyright 1983 by Graphics Press. Reprinted with permission.

and TreeMaps, a space-filling representation for trees. Card, Robertson, and Mackinlay (1993) presented ways of using animation and distortion to interact with large data sets in a system called the Information Visualizer that used focus + context displays to present large amounts of information nonuniformly. The emphasis for these studies was on the means for cognitive amplification rather than on the quality of the graphics presentations.

The remainder of this chapter will concentrate on the techniques that have been developed for mapping abstract information to interactive visual form to aid some intellectual task. The perceptual foundations of this effort are beyond the scope of this chapter, but are covered in Ware (2000). Further details on information visualization techniques are treated in a text by Spence (2000). The classic papers in information visualization are collected in Card, Mackinlay, and Shneiderman (1999).

THE VISUALIZATION REFERENCE MODEL

Mapping Data to Visual Form

Despite their seeming variability, information visualizations can be analyzed systematically. Visualizations can be thought of as adjustable mappings from data to visual form to the human perceiver. In fact, we can draw a simple Visualization Reference Model of these mappings (Fig. 28.10). Arrows follow from *Raw Data* (data in some idiosyncratic format) on the left, through a set of *Data Transformations* into *Data Tables* (canonical descriptions of data in a variables × cases format extended to include metadata). The most important mapping is the arrow from Data Tables to *Visual Structures* (structures that express variable values through a vocabulary of visual elements—spatial substrates, marks, and graphical properties). Visual Structures

can be further transformed by *View Transformations*, such as visual distortion or three-dimensional (3D) viewing angle, until they finally form a *View* that can be perceived by human users. Thus, Raw Data might start out as text represented as indexed strings or arrays. These might be transformed into document vectors, normalized vectors in a space with dimensionality as large as the number of words. Document vectors, in turn, might be reduced by multidimensional scaling to create the analytic abstraction to be visualized, expressed as a Data Table of x, y, z coordinates that could be displayed. These coordinates might be mapped into a Visual Structure that is a surface on an information landscape, which is then viewed at a certain angle.

Transformations at different places in the model can achieve similar final effects. When a point is deleted from the visualization, has the point been deleted from the data set? Or is it still in the data merely not displayed? Chi and Riedl (1998) calls this the view-value separation, and it is an example of just one issue where identifying the locus of a transformation using the Visualization Reference Model helps to avoid confusion.

But information visualization is not only about creation of visual images, but also the interaction with those images in the service of some problem. In the Visualization Reference Model, another set of arrows flow back from the human at the right into the transformations themselves, indicating the adjustment of these transformations by user-operated controls. It is the rapid reciprocal reaction between the generation of images by machine and the selection and parametric adjustment of those images, giving rise to new images, that gives rise to the attractive power of interactive information visualization.

Data Structures

It is convenient to express Data Tables as tables of objects and their attributes, as in Table 28.3. For example, in the FilmFinder, the basic objects (or "cases") are films. Each film is associated

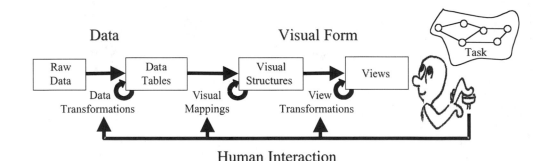

FIGURE 28.10. Reference model for visualization (Card et al., 1999). Visualization can be described as the mapping of data to visual form that supports human interaction in a workspace for visual sense making.

TABLE 28.3. A Data Table About Films

FilmID	230	105	540	...
Title	Goldfinger	Ben Hur	Ben Hur	...
Director	Hamilton	Wyler	Niblo	...
Actor	Connery	Heston	Novarro	...
Actress	Blackman	Harareet	McAvoy	...
Year	1964	1959	1926	...
Length	112	212	133	...
Popularity	7.7	8.2	7.4	...
Rating	PG	G	G	...
FilmType	Action	Action	Drama	...

Note. From *Information Visualization: Using Vision to Think*, by S. K. Card, J. D. Mackinlay, and B. Shneiderman, 1999, San Francisco: Morgan Kaufmann Publishers. Copyright 1999 by Morgan Kaufmann Publishing. Reprinted with permission.

with a number of attributes or variables, such as title, stars, year of release, genre type, and so forth. The heavy vertical double black line in the Data Table separates data in the table to the right of the line from the metadata, expressed as variable names, to the left of the line. The heavy horizontal black line across the table separates input variables from output variables. That is, the table can be thought of as a function:

$$f(\text{input variables}) = \text{output variables}.$$

So

$$\text{Year (FilmID} = 105) = 1959.$$

Variables imply a scale of measurement, and it is important to keep these straight. The most important to distinguish are:

N = Nominal (are only = or $\neq$ to other values),
O = Ordinal (obeys a < relation), or are
Q = Quantitative (can do arithmetic on them).

A nominal variable N is an unordered set, such as film titles {Goldfinger, Ben Hur, Star Wars}. An ordinal variable O is a tuple (ordered set), such as film ratings ⟨G, PG, PG-13, R⟩. A quantitative variable Q is a numeric range, such as film length [0, 360].

In addition to the three basic types of variables, there are subtypes that represent important properties of the world associated with specialized visual conventions. We sometimes distinguish the subtype *Quantitative Spatial* (Q_s) for intrinsically spatial variables common in scientific visualization and the subtype *Quantitative Geographical* (Q_g) or spatial variables that are specifically geophysical coordinates. Other important subtypes are similarity metrics *Quantitative Similarity* (Q_m), the temporal variables *Quantitative Time* (Q_t) and *Ordinal Time* (O_t). We can also distinguish Interval Scales (I) (like Quantitative scales; but, since there is not a natural zero point, it is not meaningful to take ratios). An example would be dates. It is meaningful to subtract two dates (June 5, 2002 − June 3, 2002 = 2 days), but it does not make sense to divide them (June 5, 2002 ÷ June 3, 2002 = Undefined). Finally, we can define an *Unstructured* scale (U), whose only value is present or absent. An example is an error flag. The scales are summarized in Table 28.4.

Scale types can be altered by transformations and this is sometimes convenient. For example, quantitative variables can

TABLE 28.4. Classes of Data and Visual Elements

| Class | Data Classes | | Visual Classes | |
	Description	Example	Description	Example
U	*Unstructured* (can only distinguish presence or absence)	ErrorFlag	*Unstructured* (no axis, indicated merely whether something is present or absent)	Dot
N	*Nominal* (can only distinguish whether two values are equal)	{Goldfinger, Ben Hur, Star Wars}	*Nominal Grid* (a region is divided into subregions, in which something can be present or absent)	Colored circle
O	*Ordinal* (can distinguish whether one value is less or greater, but not difference or ratio)	⟨Small, Medium, Large⟩	*Ordinal Grid* (order of the subregions is meaningful)	Alphaslider
I	*Interval* (can do subtraction on values, but no natural zero and cannot compute ratios)	[10 Dec. 1978–4 June 1982]	*Interval Grid* (region has a metric but no distinguished origin)	Year axis
Q	*Quantitative* (can do arithmetic on values)	[0–100] kg	*Quantitative Grid* (a region has a metric)	Time slider
Q_s	—Spatial variables	[0–20] m	—Spatial grid	
Q_m	—Similarity	[0–1]	—Similarity space	
Q_g	—Geographical coordinates	[30° N–50° N] Latitude	—Geographical coordinates	
Q_t	—Time variable	[10–20] μsec	—Time grid	

be mapped by data transformations into ordinal variables

$$Q \rightarrow O$$

by dividing them into ranges. For example, film lengths [0, 360] minutes (type Q) can be broken into the ranges (type O),

$$[0, 360] \text{ minutes} \rightarrow \langle \text{Short, Medium, Long} \rangle.$$

This common transformation is called *classing*, because it maps values onto classes of values. It creates an accessible summary of the data, although it loses information. In the other direction, nominal variables can be transformed to ordinal values

$$N \rightarrow O$$

based on their name. For example, film titles {Goldfinger, Ben Hur, Star Wars} can be sorted lexicographically

$$\{\text{Goldfinger, Ben Hur, Star Wars}\}$$
$$\rightarrow \langle \text{Ben Hur, Goldfinger, Star Wars} \rangle.$$

Strictly speaking, we have not transformed their values; but, in many uses (e.g., building alphabetically arranged dictionaries of words or sliders in the FilmFinder), we can act as if we had.

Variable scale types form an important class of metadata that, as we shall see, is important for proper information visualization. We can add scale type to our Data Table in Table 28.3 together with cardinality or range of the data to give us essentially a codebook of variables as in Table 28.5.

Visual Structures

Information visualization maps data relations into visual form. At first, there might seem to be a hopelessly open set of visual forms that could result. Careful reflection, however, reveals what every artist knows, that visual form is subject to strong constraints. Visual form that reflects the systematic mapping of data relations

TABLE 28.5. Data Table With Metadata Describing the Types of Variables

FilmID	N[a]		230	105	
Title	N		Goldfinger	Ben Hur	...
Director	N		Hamilton	Wyler	...
Actor	N		Connery	Heston	...
Actress	N		Blackman	Harareet	...
Year	Q_t		1964	1959	...
Length	Q		112	212	...
Popularity	Q		7.7	8.2	...
Rating	O		PG	G	...
FilmType	N		Action	Action	...

[a] N = nominal; Q = quantitative; O = ordinal.

Note. From *Information Visualization: Using Vision to Think*, by S. K. Card, J. D. Mackinlay, and B. Shneiderman, 1999, San Francisco: Morgan Kaufmann Publishing. Copyright 1999 by Morgan Kaufmann Publishing. Reprinted with permission.

onto visual form, as in information visualization or data graphics, is subject to even more constraints. It is a genuinely surprising fact, therefore, that most information visualization involves the mapping data relations onto only a half dozen components of visual encoding:

- Spatial substrate
- Marks
- Connection
- Enclosure
- Retinal properties
- Temporal encoding.

Of these mappings, the most powerful is how data are mapped onto the spatial substrate, that is, how data are mapped into spatial position. In fact, one might say that the design of an information visualization consists first in deciding which variables are going to get spatial mappings and then how the rest of the variables are going to make do with the coding mappings that are left.

Spatial Substrate. As we mentioned, the choice of which variables are going to map onto spatial position is the most important choice in designing an information visualization. This decision gives importance to spatially encoded variables at the expense of variables encoding using other mappings. Space is perceptually dominant (MacEachren, 1995). It is good for discriminating values and picking out patterns. It is easier, for example, to identify the difference between a sine and a tangent curve when encoded as a sequence of spatial positions than as a sequence of color hues.

Empty space itself, as a container, can be treated as if it had metric structure. Just as we classified variables according to their scale type, we can think of the properties of space in terms of the scale type of an axis of space (cf. Engelhardt, Bruin, Janssen, & Scha, 1996). There are axis scale types corresponding to the variable scale types (see Table 28.4). The most important axes are:

- U = Unstructured (no axis, indicated merely whether something is present or absent)
- N = Nominal Grid (a region is divided into subregions, in which something can be present or absent)
- O = Ordinal Grid (the ordering of these subregions is meaningful)
- Q = Quantitative Grid (a region has a metric).

Besides these, it is convenient to make additional distinctions for frequently used subtypes, such as Spatial axes (Qs).

Axes can be linear or radial; essentially, they can involve any of the various coordinate systems for describing space. Axes are an important building block for developing Visual Structures. Based on the Data Table for the FilmFinder in Table 28.5, we represent the scatterplot of Fig. 1 as composed of two orthogonal quantitative axes:

$$\text{Year} \rightarrow Q_x,$$
$$\text{Popularity} \rightarrow Q_y.$$

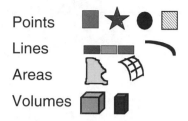

FIGURE 28.11. Types of marks.

	Spatial		Object	
Extent	(Position)	—|—|—|	Gray Scale	■ ■ ■ □
	Size	● ● · ·		
Differential	Orientation	— / | \	Color	■ ■ ■ □
			Texture	■ ▨ ▨ ▨
			Shape	■ ★ ● ◆

FIGURE 28.12. Retinal properties (Card et al., 1999). The six retinal properties can be grouped by whether they form a scale with a natural zero point (extend) and whether they deal with spatial distance or orientation (spatial).

The notation states that the Year variable is mapped to a quantitative X-axis and the Popularity variable is mapped to a quantitative Y-axis. Other axes are used for the FilmFinder query widgets. For example, an ordinal axis is used in the radio buttons for film ratings,

$$\text{Ratings} \rightarrow O_y.$$

A nominal axis is used in the radio buttons for film type,

$$\text{FilmType} \rightarrow N_x.$$

Marks. Marks are the visible things that occur in space. There are four elementary types of marks (Fig. 28.11):

P = Points (0D)
L = Lines (1D)
A = Areas (2D)
V = Volumes (3D).

Area marks include surfaces in three dimensions as well as two-dimensional (2D) bounded regions. Unlike their mathematical counterpart, point and line marks actually take up space (otherwise they would be invisible) and may have properties like shape.

Connection and Enclosure. Point marks and line marks can be used to signify other sorts of topological structure: graphs and trees. These allow showing relations among objects without the geometrical constraints implicit in mapping variables onto spatial axes. Instead, we draw explicit lines. Hierarchies and other relationships can also be encoded using enclosure. Enclosing lines can be drawn around subsets of items. Enclosure can be used for trees, contour maps, and Venn diagrams.

Retinal Properties. Other graphical properties were called retinal properties by Bertin (1967/1983), because the retina of the eye is sensitive to them independent of position. For example, the FilmFinder in Fig. 28.1 uses color to encode information in the scatterplot:

$$\text{FilmID(FilmType)} \rightarrow P(\text{Color}).$$

This notation says that the FilmType attribute for any FilmID case is visually mapped onto the color of a point.

Figure 28.12 shows Bertin's six "retinal variables" separated into spatial properties and object properties according to which area of the brain they are believed to be processed (Kosslyn, 1994). They are sorted according to whether the property is good for expressing the extent of a scale (has a natural zero point) or whether its principal use is for differentiating marks (Bertin, 1977/1981). Spatial position, discussed earlier as basic visual substrate, is shown in the position it would occupy in this classification.

Other graphical properties have also been proposed for encoding information. MacEachren (1995) has proposed crispness (the inverse of the amount of distance used to blend two areas or a line into an area), resolution (grain with raster or vector data will be displayed), transparency, and arrangement (e.g., different ways of configuring dots). He further proposes dividing color into value (essentially the gray level of Fig. 28.12), hue, and saturation. Graphical properties from the perception literature that can support preattentive processing have been suggested candidates for coding variables, for example, curvature, lighting direction, or direction of motion (see Healey, Booth, & Enns, 1995). All of these suggestions require further research.

Temporal Encoding. Visual Structures can also encode information temporally: Human perception is very sensitive to changes in mark position and the mark's retinal properties. We need to distinguish between temporal data variables to be visualized:

$$Q_t \rightarrow \text{some visual representation}$$

and animation, that is, mapping a variable into time,

$$\text{some variable} \rightarrow \text{Time}.$$

Time as animation could encode any type of data (whether it would be an effective encoding is another matter). Time as animation, of course, can be used to visualize time as data.

$$Q_t \rightarrow \text{Time}.$$

This is natural, but not always the most effective encoding. Mapping time data into space allows comparisons between two points in time. For example, if we map time and a function

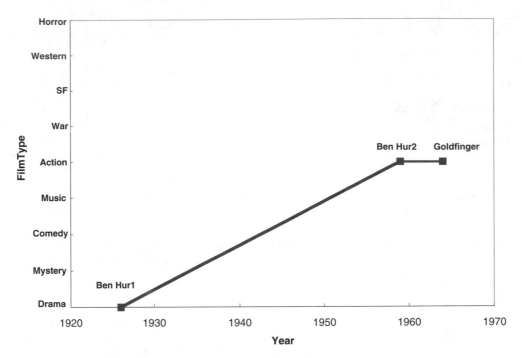

FIGURE 28.13. Mapping from data to visual form that violates expressiveness criterion.

of time into space (e.g., time and accumulated rainfall),

$$Q_t \rightarrow Qx \qquad \text{[make time be the X-axis]}$$
$$f(Q_t) \rightarrow Qy, \qquad \text{[make accumulated rainfall}$$
$$\text{be the Y-axis,}$$

then we can directly experience rates as visual linear slope, and we can experience changes in rates as curves. This encoding of time into space allows us to make much more precise judgments about rates than would be possible from encoding time as time. Animation can be used to enhance the ability of the user to track changes of view. If the user clicks on some structure causing it to enlarge and other structures to become smaller, animation can effectively convey the change and the identity of objects across the change, whereas simply viewing the two end states is confusing. Another use is to enhance a visual effect. Rotating a complicated object, for example, will induce 3D effects (hence allow better reading of some visual mappings).

Expressiveness and Effectiveness

Visual mappings transform Data Tables into Visual Structure and then into a visual image. This image is not just an arbitrary image. It is an image that has a particular meaning it must express. That meaning is the data relation of which it is the visual transformation. We can think of the image as a sentence in a visual language (Mackinlay, 1986b) that expresses the relations in the Data Table. To be a good information visualization, the mappings must satisfy some constraints. The first constraint is that the

mapping must be expressive. A visualization is said to be *expressive* if and only if it encodes all the data relations intended and no other data relations. The first part of expressiveness turns out to be easier than the second. Suppose we plot FilmType against Year using the data-to-visual mapping in Fig. 28.13. The problem of this mapping is that the nominal movie rating data are expressed by a quantitative axis. That is, we have tried to map:

$$\text{FilmType}(N) \rightarrow \text{Position}(Q).$$

In so doing, we have expressed all the data relation visually, but the visualization also implies relationships that do not exist. For example, the 1959 version of Ben Hur does not have a film type that is five times greater than the 1926 version of Ben Hur, as implied in the figure. Wisely, the authors of the FilmFinder chose the mapping:

$$\text{FilmType}(N) \rightarrow \text{Color}(N).$$

Of course, there are circumstances in which color could be read as ordinal or even possibly quantitative, but the miscellaneous order of the buttons in Fig. 28.1 discourages such an interpretation and the relatively low effectiveness of color for this purpose also discourages this interpretation (see Table 28.7).

Table 28.6 shows the mappings chosen by authors of the FilmFinder. The table shows the Data Table's metadata and data, and how they are mapped onto the Visual Structure. Note that the nominal data of the PG ratings is mapped onto a nominal

TABLE 28.6. Metadata and Mappings of Data Onto Visual Structure in the FilmFinder

Metadata			Data				Visual Structure			
Variable	Type[a]	Range					Type	Visual Mark	Control	Transformation Affected
FilmID			230	105	540	...				
Title	N	All titles	Goldfinger	Ben Hur	Ben Hur	...	O	—	Alphaslider	Select item
Director	N	All directors	Hamilton	Wyler	Niblo	...	O	—	Alphaslider	Select item
Actor	N	All actors	Connery	Heston	Novarro	...	O	—	Alphaslider	Select item
Actress	N	All actresses	Blackman	Harareet	McAvoy	...	O	—	Alphaslider	Select item
Year	Q	[1926, 1989]	1964	1959	1926	...	Q	Point (X)	—	
Length	Q	[0, 450]	112	212	133		Q	—	2-Sided slider	Clip range
Popularity	Q	[1, 9]	7.7	8.2	7.4	...	Q	Point (Y)	2-Sided slider	Clip range
Rating	O	{G, PG, PG-13, R}	PG	G	G	...	O	—	4 Radio buttons	Select item
FilmType	N	{Drama, Mystery, Comedy, Music, Action, War, SF, Western, Horror}	Action	Action	Drama		N	Color (X, Y)	9 Radio buttons	Select item

[a] N = nominal; Q = quantitative; O = ordinal.

Note. From *Information Visualization: Using Vision to Think*, by S. K. Card, J. D. Mackinlay, and B. Sheiderman, 1999, San Francisco: Morgan Kaufmann Publishing. Copyright 1999 by Morgan Kaufmann Publishing. Reprinted with permission.

visualization technique (colors). Note also that names of directors and stars (nominal variables) are raised to ordinal variables (through alphabetization) and then mapped onto an ordinal axis. This is, of course, a common way to handle searching among a large number of nominal items.

Some properties are more effective than others for encoding information. Position is by far the most effective all-around representation. Many properties are more effective for some types of data than for others. Table 28.7 gives an approximate evaluation for the relative effectiveness of some encoding techniques based on MacEachren (1995). We note that spatial position is effective for all scale types of data. Shape, on the other hand, is only effective for nominal data. Gray scale is most effective for ordinal data. Such a chart can suggest representations to a visualization designer.

Taxonomy of Information Visualizations

We have shown that the properties of data and visual representation generally constrain the set of mappings that form the basis

TABLE 28.7. Relative Effectiveness of Position and Retinal Encodings

		Spatial	Q	O	N	Object	Q	O	N
Extent	(Position)		●	●	●	Gray scale	◑	●	○
	Size		●	●	●				
Differential	Orientation		◑	◑	●	Color	◑	◑	●
						Texture	◑	◑	●
						Shape	○	○	●

Note. From *Information Visualization: Using Vision to Think*, by S. K. Card, J. D. Mackinaly, & B. Shneiderman, 1999, San Francisco: Morgan Kaufmann Publishing. Copyright 1999 by Morgan Kaufmann Publishing. Reprinted with permission.

Q = quantitative; O = ordinal; N = nominal.

for information visualizations. Taken together, these constraints provide the basis of a taxonomy of information visualizations. Such a taxonomy is given in Table 28.8. Visualizations are grouped into four categories. First are *Simple Visual Structures*, the static mapping depicted in Fig. 28.10 of data onto multiple spatial dimensions, trees, or networks plus retinal variables. Here it is worth distinguishing two cases. There is a perceptual barrier at three (or, in special cases, four) variables, a limit of the amount of data that can be perceived as an immediate whole. Bertin (1977/1981) calls this elementary unit of visual data perception the "image." Although this limit has not been definitively established in information visualization by empirical research, there must be a limit somewhere or else people could simultaneously comprehend a thousand variables. We therefore divide visualizations into those that can be comprehended in an elementary perceptual grasp (three or, in special cases, four variables)—let us call these *direct reading visualizations*—and those more complex than that barrier—which we call *articulated reading visualizations* in which multiple actions are required.

Beyond the perceptual barrier, direct composition of data relationships in terms of 1, 2, or 3 spatial dimensions plus remaining retinal variables is still possible, but rapidly diminishes in effectiveness. In fact, the main problem of information visualization as a discipline can be seen as devising techniques for accelerating the comprehension of these more complex *n*-variable data relations. Several classes of techniques for *n*-variable visualization, which we call *Composed Visual Structures*, are based on composing Simple Visual Structures together by reusing their spatial axes.

A third class of Visual Structures—*Interactive Visual Structures*—comes from using the rapid interaction capabilities of the computer. These visualizations invoke the parameter-controlling arrows of Fig. 28.10. Finally, a fourth class of visualizations—*Attention-Reactive Visual Structuress*—comes from interactive displays, in which the system reacts to user

TABLE 28.8. Taxonomy of Information Visualization Techniques

I. Simple Visual Structures

Direct Reading

One-Variable [X]
  Lists
  1D object charts
  1D scatterplots
  Pie charts
  Folded dimensions
  Distributions
  Box plots
Two-Variable [XY]
  2D object charts
  2D scatterplots
Three-Variable
  [XYR]
    Retinal scatterplot
    Kahonen diagrams
    Retinal topographies
  [(XY)Z]
    Information landscapes
    Information surfaces
  [XYZ]
    3D scatterplots
Four-Variable
  [XYZR]
    3D retinal scatterplots
    3D topographies

—Barrier of Perception—
Articulated Reading
  n-Variable
    $[XYR^{n-2}]$
      2D retinal scatterplots
    $[XYZR^{n-3}]$
      2D retinal scatterplots

Trees
  Node and link trees
  Enclosure trees
  TreeMaps
  Cone trees
Networks
Time

II. Composed Visual Structures

Single Axis
Composition $[XY^n]$
  Permutation matrices
  Parallel coordinates
Double Axis
Composition $[(XY)^n]$
  Graphs
Recursive Composition
  2D in 2D $[(XY)^{XY}]$
    Scatterplot matrices
    Prosection matrices
    Hierarchical axes
  Marks in 2D $[(XY)^R]$
    Stick figures
    Color icons
    Shape coding
    Keim spirals
  3D in 3D $[(XYZ)^{XYZ}]$
    Worlds within worlds

III. Interactive Visual Structures

Dynamic queries
Magic lens
Overview + detail
Linking and brushing
Extraction and comparison
Attribute explorer

IV. Focus + Context Attention-Reactive Visual Abstraction

Data-based Methods
  Filtering
  Selective aggregation
View-based Methods
  Micro-macro readings
  Highlighting
  Visual transfer functions
  Perspective distortion
  Alternate geometries

Note. 1D = one-dimensional; 2D = two-dimensional; 3D = three-dimensional.

actions by changing the display, even anticipating new displays, so as to lower cost of information access and sensemaking to the user. To summarize:

- Simple Visual Structures
  Direct Reading
  Articulated Reading
- Composed Visual Structures
  Single-Axis Composition
  Double-Axis Composition
  Recursive Composition
- Interactive Visual Structure
- Attention-Reactive Visual Structure

These classes of techniques may be combined with each other to get visualizations that are more complex. To help us keep track of the variable mapping into visual structure, we will use a simple shorthand notation for listing the element of the Visual Structure that the Data Table has mapped into. We will write, for example, $[XYR^2]$ to note that variables map onto the X-axis, the Y-axis, and two retinal encodings. [OX] will

indicate that the variables map onto one spatial axis used to arrange the objects (i.e., the cases), whereas another was used to encode the objects' values. Examples of this notation appear in Table 28.8 and Fig. 28.21.

SIMPLE VISUAL STRUCTURES

The design of information visualizations begins with mappings from variables of the Data Table into the Visual Structure. The basic strategy for the visualization designer could be described as follows:

1. Determine which variables of the Analytic Abstraction to map into spatial position in the Visual Structure.
2. Combine these mappings to increase dimensionality (e.g., by folding).
3. Use retinal variables as an overlay to add more dimensions.
4. Add controls for interaction.
5. Consider attention-reactive features to expand space and manage attention.

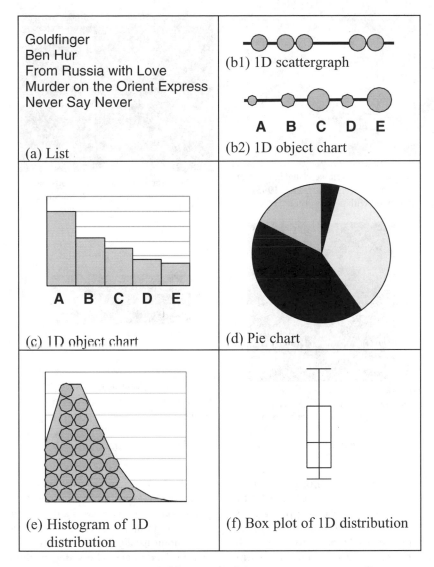

FIGURE 28.14. One-variable visual abstractions. 1D = one-dimensional.

We start by considering some of the ways in which variables can be mapped into space.

One-Variable Visual Displays

One-variable visual displays may actually use more than one visual dimension. This is because the data variable or attribute is displayed against some set of objects using some mark and the mark itself takes space. Or, more subtly, it may be because one of the dimensions is used for arranging the objects and another for encoding via position the variable. A simple example would be when the data are just visually mapped into a simple text list as in Fig. 28.14a. The objects form a sequence on the Y-dimension and the width of the marks (the text descriptor) takes space in the X-dimension. By contrast, a one-dimensional scattergraph

(Fig. 28.14b) does not use a dimension for the objects. Here, the Y-axis is used to display the attribute variable (suppose these are distances from home of gas stations); the objects are encoded in the mark (which takes a little bit of the X-dimension).

More generally, many single-variable visualizations are in the form $v = f(o)$, where v is a variable attribute and o is the object. Figure 28.14c is of this form and uses the Y-axis to encode the variable and the X-axis for the objects. Note that if the objects are, as usual, nominal, then they are reorderable: Sorting the objects on the variable produces easily perceivable visual patterns.

For convenience, we have used rectangular coordinates, but any other orthogonal coordinates could be used as the basis of decomposing space. Figure 28.14d uses θ from polar coordinates to encode, say, percentage voting for different presidential candidates. In Fig. 28.14e, a transformation on the data side has

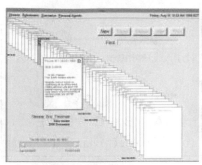

(a) Off-axis 1-variable visual abstraction: LifeLines (Freeman & Fertig, 1995). Reprinted with permission.

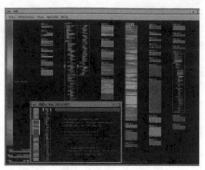

(b) Folded long 1-variable visual abstraction: SeeSoft (Eick, Steffen, & Sumner, 1992). Reprinted with permission.

(c) Spiraled long 1-variable visual abstraction: VisDB (Keim & Kriegel, 1994). Reprinted with permission.

(d). 1-variable visual abstraction used as a control. (Eick, 1993). Reprinted with permission.

FIGURE 28.15. Uses of one-variable visual abstractions.

transformed variable *o* into a variable representing the distribution, then mapped that onto points on the Y-axis. In Fig. 28.14f, another transformation on the data side has mapped this distribution into 2nd quartiles, 3rd quartiles, and outlier points, which is then mapped on the visual side into a box plot on the Y-axis. Simple as they are, these techniques can be very useful, especially in combination with other techniques.

One special, but common, problem is how to visualize dimensions that are very large. This problem occurs for single-variable visualizations, but also may also occur for one dimension of a multivariable visualization. Figure 28.15 shows several techniques for handing the problem. In Fig. 28.15a (Freeman & Fertig, 1995), the visual dimension is laid out in perspective. Even though each object may take only one or a few pixels on the axis, the objects are actually fairly large and selectable in the diagram. In Fig. 28.15b (Eick, Steffen, & Sumner, 1992), the objects (representing lines of code) are laid out on a *folded* Y-axis. When the Y-axis reaches the bottom of the page, it continues offset at the top. In Fig. 28.15c (Keim & Kriegel, 1994), the axis is wrapped in a square spiral. Each object is a single pixel, and its value is coded as the retinal variable color hue. The objects have been sorted on another variable; hence, the rings show the correlation of this attribute with that of the sorting attribute.

One-variable visualizations are also good parts of controls. Controls, in the form of slides, consume considerable space on the display (e.g., the controls in Fig. 28.1) that could be used for additional information communication. Figure 28.15d shows a slider on whose surface is a distribution representation of the number of objects for each value of the input variable, thereby communicating information about the slider's sensitivity in different data ranges. The slider on the left of Fig. 28.15b has a one-variable visualization that serves as a legend for the main visualization: It associates color hues with dates and allows the selection of date ranges.

Two-Variable Visual Displays

As we increase the number of variables, it is apparent that their mappings form a combinatorial design space. Figure 28.16 schematically plots the structure of this space (leaving out the use of multiple lower-variable diagrams to plot higher variable combinations). Two-variable visualizations can be thought of as a composition of two elementary axes (Bertin, 1977/1981; Mackinlay, 1986b), which uses a single mark to encode the position on both those axes. Mackinlay calls this *mark composition*, and it results in a 2D scattergraph (Fig. 28.16g). Note that, instead of mapping onto two positional visual encodings, one positional axis could be used for the objects and the data variables could be mapped onto a position encoding and a retinal encoding (size) as in Fig. 28.16f.

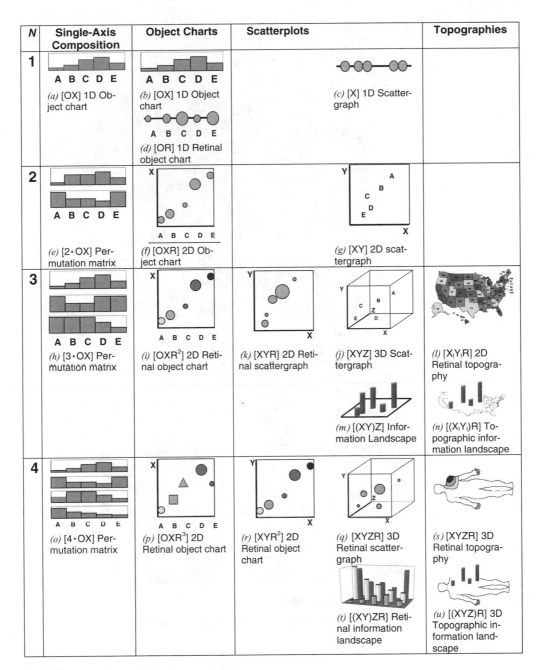

N	Single-Axis Composition	Object Charts	Scatterplots		Topographies
1	*(a)* [OX] 1D Object chart	*(b)* [OX] 1D Object chart *(d)* [OR] 1D Retinal object chart	*(c)* [X] 1D Scattergraph		
2	*(e)* [2·OX] Permutation matrix	*(f)* [OXR] 2D Object chart	*(g)* [XY] 2D scattergraph		
3	*(h)* [3·OX] Permutation matrix	*(i)* [OXR²] 2D Retinal object chart	*(k)* [XYR] 2D Retinal scattergraph	*(j)* [XYZ] 3D Scattergraph *(m)* [(XY)Z] Information Landscape	*(l)* [X₁Y₁R] 2D Retinal topography *(n)* [(X₁Y₁)R] Topographic information landscape
4	*(o)* [4·OX] Permutation matrix	*(p)* [OXR³] 2D Retinal object chart	*(r)* [XYR²] 2D Retinal object chart	*(q)* [XYZR] 3D Retinal scattergraph *(t)* [(XY)ZR] Retinal information landscape	*(s)* [XYZR] 3D Retinal topography *(u)* [(XYZ)R] 3D Topographic information landscape

FIGURE 28.16. Simple Visual Structures. 1D = one-dimensional; 2D = two-dimensional; 3D = three-dimensional.

Three-Variables and Information Landscapes

By the time we get to three data variables, there are several ways to produce a visualization. We can use three separate visual dimensions to encode the three data variables in a *3D scattergraph* (Fig. 28.16j). We could also use two spatial dimensions and one retinal variable in a *2D retinal scattergraph* (Fig. 28.16k). Or, we could use one spatial dimension as an object dimension, one as a data attribute dimension, and two retinal encodings for the other variables, as in an *object chart* (Fig. 28.16i). Because Fig. 28.16i uses multiple retinal encodings, it may not be as effective as other techniques. Note that because they all encode three data variables, we have classified 2D and 3D displays together. In fact, one popular three-variable information visualization that lies between 2D and 3D is the *information landscape* (Fig. 28.16m). This is essentially a 2D scattergraph with one data variable extruded into the third spatial dimension. Its essence is that two of the spatial dimensions are more tightly coupled and often relate to a 2D visualization. For example, the two dimensions might form a map

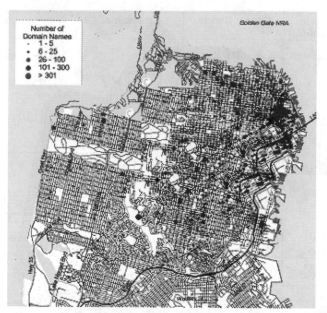

(a) X_lY_lR Retinal topography. Courtesy Matthew Zook.

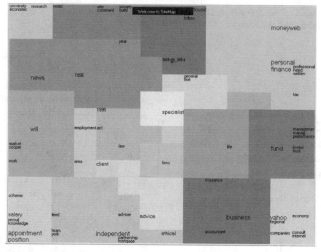

(b) X_sY_sR Retinal similarity topography. Courtesy Xia Lin.

FIGURE 28.17. Retinal information topographies.

with the bars showing the Gross Domestic Product of each region.

Another special type of three-variable information visualization is a 2D *information topography*. In an information typography, space is partly defined by reference to external structure. For example, the topography of Fig. 28.17a is a map of San Francisco, requiring two spatial variables. The size of blue dots indexes the number of domain names registered to San Francisco street addresses. Looking at the patterns in the visualization shows that Internet addresses have especially concentrated in the Mission and South of Mission districts. Figure 28.17a uses a topography derived from real geographical space. Various techniques, such as multidimensional scaling, factor analysis, or connectionist self-organizing algorithms, can create abstract spaces based on the similarities among collections of documents or other objects. These abstract similarity spaces can function like a topography. An example is Fig. 28.17b, where the pages in a Web site are depicted as regions in a similarity space. To create this diagram,[2] a web crawler crawls the site and indexes all the words and pages on the site. Each page is then turned into a document vector to represent the semantic content of that page. The regions are created using a neural network learning algorithm (see Lin, Soergel, & Marchionini, 1991). This algorithm organizes the set of web pages into regions. A visualization

<hr />

[2]This figure is produced by a program called SiteMap by Xia Lin and associates. See http://faculty.cis.drexel.edu/sitemap/index.html

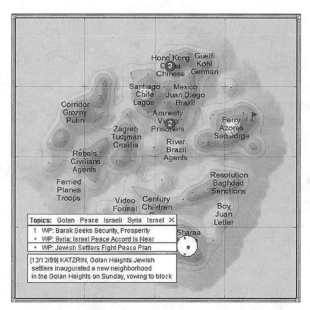

(a) News stories based on ThemeScapes(Wise et al., 1995). Courtesy of NewsMaps.com.

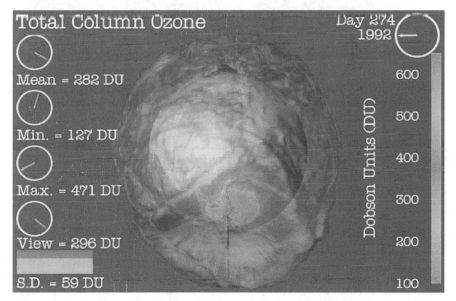

(b) Ozone layer surrounding earth. L. Treinish. Courtesy of IBM.

FIGURE 28.18. Three-dimensional information surface topographies.

algorithm then draws boundaries around the regions, colors them, and names them. The result, called a *Kahonen diagram* after its original inventor, is a type of retinal similarity topography.

Information landscapes can also use marks that are surfaces. In Fig. 28.18a, topics are clustered on a similarity surface, and the strength of each topic is indicated by a 3D contour. A more extreme case is Fig. 28.18b, where an information landscape is established in spherical coordinates and the amount of ozone is plotted as a semitransparent overlay on the ρ-axis.

n-Variables

Beyond three variables, direct extensions of the methods we have discussed become less effective. It is possible, of course to make plots using two spatial variables and $n-2$ retinal variables, and the possibilities for four variables are shown in Fig. 28.16. These diagrams can be understood, but at the cost of progressively more effort as the number of variables increases. It would be very difficult to understand an $[XYR^{20}]$ retinal scattergraph, for example.

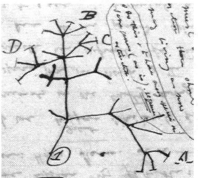

(a) Tree from Darwin's notes, from (Robin, 1992). Courtesy of Syndics of Cambridge University Library, Cambridge, UK.

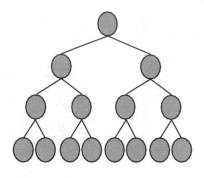

(b) Typical link and node tree layout.

(c) Circular tree of evolution of life. Courtesy of David Hillis, University of Texas.

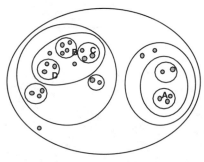

(d) Tree in (a) drawn using enclosure.

FIGURE 28.19. Trees.

Trees

An interesting alternative to showing variable values by spatial positioning is to use explicitly drawn linkages of some kind. Trees are the simplest form of these (see Fig. 28.19). Trees map cases into subcases. One of the data variables in a Data Table (e.g., the variable ReportsTo in an organization chart) is used to define the tree. There are two basic methods for visualizing a tree: Connection and Enclosures.

Connection. Connection uses lines to connect marks signifying the nodes of the tree. Logically, a tree can be drawn merely by drawing lines between objects located randomly positioned on the plane, but such a tree would be visually unreadable. Positioning in space is important. Figure 28.19a is a tree from Charles Darwin's notebook (Robin, 1992) drawn to help him work out

the theory of evolution. Lines proceed from ancestor species to new species. Note that, even in this informal setting intended for personal use, the tree uses space systematically (and opportunistically). There are no crossed lines. A common way of laying out trees is to have the depth in the tree map onto one ordinal access as in Fig. 28.19b, whereas the other axis is nominal and used to separate nodes. Of course, trees could also be mapped into other coordinate systems: For example, there can be circular trees in which the r-axis represents depth and the θ-axis is used to separate nodes as in the representation of the evolution species in Fig. 28.19c[3]. It is because trees have no cycles that one of the spatial dimensions can be used to encode tree depth. This partial correlation of tree structure and space makes trees relatively easy to lay out and interpret, compared with generalized networks. Hierarchical displays are important not only because many interesting collections of information, such as

---

[3]The figure is from David Hillis, University of Texas.

organization charts or taxonomies, are hierarchical data, but also because important collections of information, such as Web sites, are approximately hierarchical. Whereas practical methods exist for displaying trees up to several thousand nodes, no good methods exist for displaying general graphs of this size. If a visualization problem involves the displaying of network data, a practical design heuristic is to see whether the data might not be forced into a display as a modified tree, such as a tree with a few nontree links. A significant disadvantage of trees is that as they get large, they acquire an extreme aspect ratio, because the nodes expand exponentially as a function of depth. Consequently, any sufficiently large tree (say >1,000 nodes) resembles a straight line. Circular trees, such as Fig. 28.19c are one way of trying to buy more space to mitigate this problem. Another disadvantage of trees is the significant empty space between nodes to make their organization easily readable. Various tricks can be used to wrap parts of the tree into this empty space, but at the expense of the tree's virtues of readability.

Enclosure. Enclosure uses lines to enclose hierarchically nested subsets of the tree. Figure 28.19d is an enclosure encoding of Darwin's tree in Fig. 28.19a. We have already seen one use of tree enclosure, TreeMaps (Fig. 28.5). TreeMaps make use of all the space and stay within prescribed space boundaries, but they do not represent the nonterminal nodes of the tree very well and similar leaves can have wildly different aspect ratios. Recent variations on TreeMaps found ways to "squarify" nodes (Shneiderman & Wattenberg, 2001), mitigating this problem.

Networks

Networks are more general than trees and may contain cycles. Networks may have directional links. They are useful for describing communication relationships among people, traffic in a telephone network, and the organization of the Internet. Containment is difficult to use as a visual encoding for network relationships; so, most networks are laid out as node and link diagrams. Unfortunately, straightforward layouts of large node and link diagrams tend to resemble a large wad of tangled string.

We can distinguish the same types of nodes and links in network Visual Structures that we did for spatial axes: Unstructured (unlabeled), Nominal (labeled), Ordinal (labeled with an ordinal quantity), or Quantitative (weighted links). Retinal properties, such as size or color, can be used to encode information about links and nodes.

As in the case of trees, spatial positioning of the nodes is extremely important. Network visualizations escape from the strong spatial constraints of simple Visual Structures only to encounter another set of strong spatial constraints of node links crossing and routing. Networks and trees are not so much an alternative to the direct graphical mappings we have discussed so far as they are another set of techniques that can be overlaid on these mappings. Small node and link diagrams can be laid out opportunistically by hand or by using graph drawing algorithms that have been developed (Battista, Eades, Tamassia, &

Tollis, 1994; Cruz & Tamassia, 1998; Tamassia, 1996) to optimize minimal link crossing, symmetry, and other aesthetic principles.

For very large node and link diagrams, additional organizing principles are needed. If there is an external topographic structure, it is sometimes possible to use the spatial variables associated with the nodes. Figure 28.20a shows a network based on call traffic between cities in the United States (Becker, Eick, & Wilks, 1995). The geographical location of the cities is used to lay out the nodes of the network. Another way to position nodes is by associating nodes with positions in a similarity space, such that nodes that have the strongest linkages to each other are closest together. There are several methods for computing node nearness in this way. One is to use multidimensional scaling (Fairchild, Poltrock, & Furnas, 1988). Another is to use a "spring" technique, in which each link is associated with a Hooke's Law spring weighted by strength of association, and the system of springs is solved to obtain node position. Eick and Willis (1993) have argued that the multidimensional scaling technique places too much emphasis on smaller links. They have derived an alternative that gives clumpier (and hence more visually structured) clusters of nodes. If positioning of nodes corresponds perfectly with linkage information, then the links do not add more visual information. If positioning does not correspond at all with linkage information, then the diagram is random and obscure. In large graphs, node positions must have a partially correlated relationship to linkage to allow the emergence of visual structure. Note that this is what happens in the telephone traffic diagram Fig. 28.20a. Cities are positioned by geographical location. Communication might be expected to be higher among closer cities, so the fact that communications is heavy between coasts stands out.

A major problem in a network such as Fig. 28.20a is that links may obscure the structure of the graph. One solution is to route the links so that they do not obscure each other. The links could even be drawn outside the plane in the third dimension. There are limits to the effectiveness of this technique, however. Another solution is to use *thresholding* as in Fig. 28.20b. Only those links representing traffic greater than a certain threshold are included; the others are elided, allowing us to see the most important structure. Another technique is *line shortening* as in Fig. 28.20c. Only the portion of the line near the nodes is drawn. At the cost of giving up the precise linkage, it is possible to read the density of linkages for the different nodes. Figure 28.20d (Cox, Eick, & Wills, 1997) is a technique used to find patterns in an extremely large network. Telephone subscribers are represented as nodes on a hexagonal array. Frequent pairs are located near each other on the array. Suspicious patterns are visible because of the sparseness of the network.

The insightful display of large networks is difficult enough that many information visualization techniques depend on interactivity. One important technique, for example, is node aggregation. Nodes can be aggregated to reduce the number of links that have to be drawn on the screen. Which nodes are aggregated can depend on the portion of the network on which the user is drilling down. Similarly, the sets of nodes can be interactively restricted (e.g., telephone calls greater than a certain volume) to reduce the visualization problem to one within the capability of current techniques.

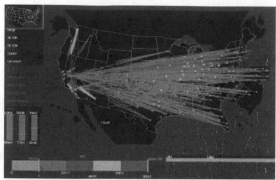

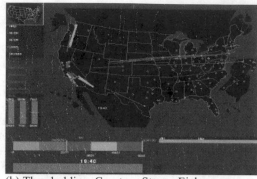

(a) Telephone traffic after California earthquake
(Becker, Eick, & Wilks, 1995). Reprinted with permission.

(b) Thresholding. Courtesy Steven Eick.

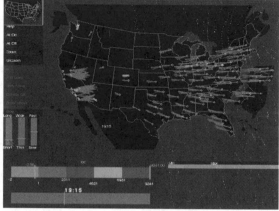

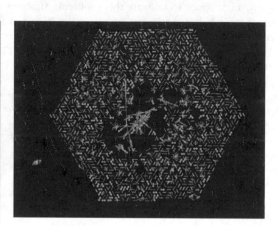

(c) Line shortening. Courtesy Steven Eick.

(d) Visualization to detect telephone fraud
(Cox, Eick, & Wills, 1997). Reprinted with permission.

FIGURE 28.20. Network methods.

COMPOSED VISUAL STRUCTURES

So far, we have discussed simple mappings from data into spatial position axes, connections and enclosures, and retinal variables. These methods begin to run into a barrier around three variables as the spatial dimensions are used up and as multiple of the less efficient retinal variables needed. Most interesting problems involve many variables. We shall therefore look at a class of methods that reuse precious spatial axes to encode variables. This is done by composing a compound Visual Structure out of several simple Visual Structures. We will consider five subclasses of such composition: mark composition, case composition, single-axis composition, double-axis composition, and recursive composition. Schematically, we illustrate these possibilities in Fig. 28.21.

Single-Axis Composition

In single-axis composition, multiple variables that share a single axis are aligned using that axis as illustrated in Fig. 28.21a. An example of single-axis composition is a method due to Bertin

(1977/1981) called *permutation matrices*. In a permutation matrix (Fig. 28.16o, for example) one of the spatial axes is used to represent the cases and the other a series of bar charts (or rows of circles of different size or some other depiction of the value of each variable) to represent the values. In addition, bars for values below average may be given a different color, as in Fig. 28.7 to enhance the visual patterns. The order of the objects and the order of the variables may both be permuted until patterns come into play. Permutation matrices were used in our hotel analysis example. They give up direct reading of the data space to handle a larger number of variables. Of course, as the number of variables (or objects) increases, manipulation of the matrices becomes more time-consuming and visual interpretation more complex. Still, permutation matrices or their variants are one of the most practical ways of representing multivariable data.

If we superimpose the bar charts of the permutation matrix atop each other, then replace the bar chart with a line linking together the tops of the bars, we get another method for handling multiple variables by single-axis composition—*parallel coordinates* (Inselberg, 1997; Inselberg & Dimsdale, 1990) as shown in Fig. 28.22. A problem is analyzed in parallel coordinates by interactively restricting the objects displayed (the lines)

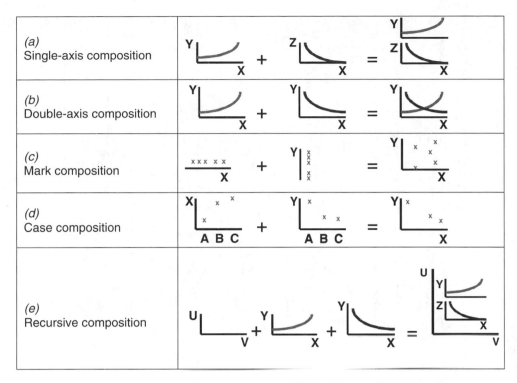

FIGURE 28.21. Composition types (Partially based on a slide by Jock Mackinlay)

to look at cases with common characteristics. In Fig. 28.22, parallel coordinates are being used to analyze the problem of yield from a certain processor chip. X1 is chip yield, X2 is quality, X3 through X12 are defects, and the rest of the variables are physical parameters. The analysis, looking at those subsets of data with high yield and noticing the distribution of lines on the other parameters, was able to solve a significant problem in chip processing.

Both permutation matrices and parallel coordinates allow analyses in multidimensional space, because they are efficient in the use (and reuse) of spatial position and the plane. Actually, they also derive part of their power from being interactive. In the case of permutation matrices, interactivity comes in reordering the matrices. In the case of parallel coordinates, interactivity comes in selecting subsets of cases to display.

Double-Axis Composition

In double-axis composition, two visual axes must be in correspondence, in which case the cases are plotted on the same axes as a multivariable graph (Fig. 28.21b). Care must be taken that the variables are plotted on a comparable scale. For this reason, the separate scales of the variables are often transformed to a common proportion change scale. An example would be change in price for various stocks. The cases would be the years and the variables would be the different stocks.

Mark Composition and Case Composition

Composition can also fuse diagrams. We discussed that each dimension of visual space can be said to have properties as summarized in Table 28.4. The visual space of a diagram is composed from the properties of its axis. In *mark composition* (Fig. 28.21c), the mark on one axis can fuse with the corresponding mark on another axis to form a single mark in the space formed by the two axes. Similarly, two object charts can be fused into a single diagram by having a single mark for each case. We call this latter form *case composition* (Fig. 28.21d).

Recursive Composition

Recursive composition divides the plane (or 3D space) into regions, placing a subvisualization in each region (Fig. 28.21e). We use the term somewhat loosely, because regions have different types of subvisualizations. The FilmFinder in Fig. 28.1 is a good example of a recursive visualization. The screen breaks down into a series of simple Visual Structures and controls: a three-variable retinal scattergraph (Year, Rating, FilmType) + a one-variable slider (Title) + a one-variable slider (Actors) + a one-variable slider (Actresses) + a one-variable slider (Director) + a one-variable slider (FilmLength) + a one-variable radio button control (Rating) + a one-variable button-set (FilmType).

Three types of recursive composition deserve special mention: 2D-in-2D, marks-in-2D, and 3D-in-3D. An example of

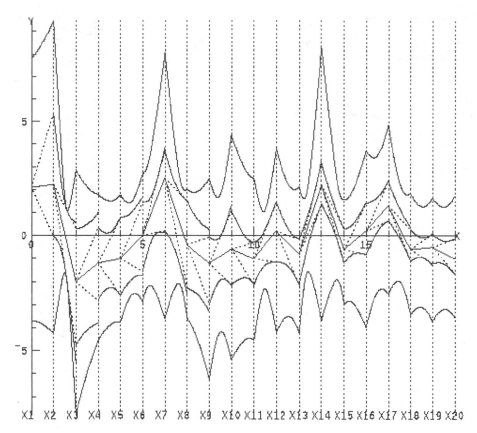

FIGURE 28.22. Single-axis composition: parallel coordinates (Insel-berg, 1997). Reprinted with permission.

2D-in-2D composition is the "prosection matrix" (Tweedie, Spence, Dawkes, & Su, 1996) shown in Fig. 28.23a. Each smaller square in the prosection matrix represents a pair of parameters plotted against each other. The coloring shows which values of the plotted pair give excellent (red region) or partly good (gray regions) performance for the design of some device. The arrangement of the individual matrices into a supermatrix re-defines the spatial dimensions (i.e., associates it with different variables) within each of the cells, and the cells themselves are arranged in an overall scheme that systematically uses space. In this way, the precious spatial dimension is effectively expanded to where all the variables can reuse it. An important property of techniques similar to this one is that space is defined at more than one grain size, and these levels of grain become the basis for a macro-micro reading (Tufte, 1990).

An example of marks-in-2D composition is the use of "stick figure" displays. This is an unusual type of visualization in which the recursion is within the mark instead of within the use of space. Figure 28.23b shows a mark that is itself composed of submarks. The mark is a line segment with four smaller line segments protruding from the ends. Four variables are mapped onto the angle of these smaller line segments and a fifth onto the angle of the main line segment. Two additional variables are mapped onto the position of this mark in a 2D display. A

typical result is the visualization in Fig. 28.23c, which shows five weather variables around Lake Ontario, the outline of which clearly appears in the figure.

Feiner and Beshers (1990) provide an example of the third re-cursive composition technique, 3D-in-3D composition. Suppose a dependent variable is a function of six continuous variables, $y = f(x, y, z, w, r, s)$. Three of these variables are mapped onto a 3D coordinate system. A position is chosen in that space, say $x1$, $y1$, $z1$. At that position, a new 3D coordinate system is presented with a surface defined by the other three variables (Fig. 28.23d). The user can thus view $y = f(x1, y1, z1, w, r, s)$. The user can also slide the second-order coordinate system to any location in the first, causing the surface to change appropriately. Note that this technique combines a composed visual interaction with interactivity on the composition. Multiple second-order coordi-nate systems can be displayed at the space simultaneously, as long as they do not overlap by much.

INTERACTIVE VISUAL STRUCTURES

In the examples we have considered so far, we have often seen that information visualization techniques were enhanced by being interactive. Interactivity is what makes visualization a

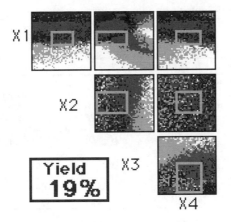

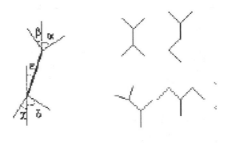

(a) 2D-in-2D: Attribute Explorer (Tweedie, Spence, Dawkes, & Su, 1996). Reprinted with permission.

(b) Marks-in-2D. Composition of a stick figure mark (Pickett & Grinstein, 1988). Reprinted with permission.

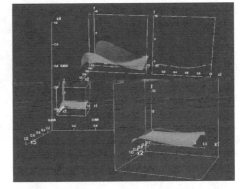

(c) Visualization of stick figures showing weather around Lake Ontario (Pickett & Grinstein, 1988). Reprinted with permission.

(d) 3D-in-3D: Worlds-within-worlds (Feiner & Beshers, 1990). Reprinted with permission.

FIGURE 28.23. Recursive composition.

new medium, separating it from generations of excellent work on scientific diagrams and data graphics. Interactivity means controlling the parameters in the visualization reference model (Fig. 28.10). This naturally means that there are different types of interactivity, because the user could control the parameters to data transformations, to visual mappings, or to view transformations. It also means that there are different forms of interactivity based on the response cycle of the interaction. As an approximation, we can think of there as being three time constants that govern interactivity, which we nominally take to be 0.1 s, 1 s, and 10 s (Card, Moran, & Newell, 1986), although the ideal value of these may be somewhat less, say 0.07 s, 0.7 s, and 7 s). The first time constant is the time in which a system response must be made, if the user is to feel that there is a direct physical manipulation of the visualization. If the user clicks on a button or moves a slider, the system needs to update the display in less than 0.1 s. Animation frames need to take less than 0.1 s. The second time constant, 1 s, is the time to complete an immediate action, for example, an animated sequence such as zooming into the data or rotating a tree branch.

The third time constant, 10 s (meaning somewhere in the 5- to 30-s interval), is the time for completing some cognitive action, for example, deleting an element from the display. Let us consider a few well-known techniques for interactive information visualizations.

Dynamic Queries

A general paradigm for visualization interaction is dynamic queries, the interaction technique used by the FilmFinder in Fig. 28.1. The user has a visualization of the data and a set of controls, such as sliders, by which subsets of the Data Table can be selected. For example, Table 28.9 shows the mappings of the Data Table and controls for the FilmFinder. The sliders and other controls will select which subset of the data is going to be displayed. In the FilmFinder, the control for Length is a two-sided slider. Setting one end to 90 min and the other end to 120 min will select for display only those cases of the Data Table whose year variable lies between these limits.

TABLE 28.9. Visual Marks and Controls for FilmFinder

Metadata			Data				Visual Form				
Variable	Type[a]	Range	$Case_i$	$Case_j$	$Case_k$		Type	Visual Structure	Control	Transformation Affected	
FilmID	N	All IDs	230	105	540	...	→	N	Points	Button	All (details)
Title	N	All titles	Goldfinger	Ben Hur	Ben Hur	...	→Sort	O		Alphasider	Select cases
Director	N	All directors	Hamilton	Wyler	Niblo	...	→Sort	O		Alphasider	Select cases
Actor	N	All actors	Connery	Heston	Novarro	...	→Sort	O		Alphasider	Select cases
Actress	N	All adresses	Blackman	Harareet	McAvoy	...	→Sort	O		Alphasider	Select cases
Year	Q	[1926, 1989]	1964	1959	1926	...	→	Q	X-axis	Axis	Clip range
Length	Q	[0, 450]	112	212	133		→	Q		Two-sided slider	Clip range
Popularity	Q	[1, 9]	7.7	8.2	7.4	...	→	Q	Y-axis	Axis	Clip range
Rating	Q	{G, PG, PG-13, R}	PG	G	G	...	→	O		Radio buttons	Select cases
Film Type	N	{Drama, Mystery, Comedy, Music, Action, War, SF, Western, Horror}	Action	Action	Drama	...	→	N	Color	Radio buttons	Select cases

[a]N = nominal; Q = quantitative; O = ordinal.

Note: From Card, Mackinlay, & Shneiderman (1999). Reprinted with permission.

The display needs to change within the 0.1 s of changing the slider.

Magic Lens (Movable Filter)

Dynamic queries is the name of one type of interactive filter. Another type is a movable filter that can be moved across the display, as in Fig. 28.24a. These *magic lenses* are useful when it is desired to filter only some of the display. For example, a magic lens could be used with a map that showed the population of any city it was moved over. Multiple magic lenses can be used to cascade filters.

Overview + Detail

We can think of an overview + detail display (Fig. 28.24b) as a particular type of magic lens, one that magnifies the display and has the magnified region off to the side so as not to occlude the region. Displays have information at different grain sizes. A geographical information system map may have information at the level of a continent, as well as at the level of a city. If the shape of the continent can be seen, the display will be too coarse to see the roadways of a city. Overview + detail displays show that data at more than one level, but they also show where the finer grain display fits into the larger grain display. In Fig. 28.24b, from SeeSoft (Eick et al., 1992), a system for visualizing large software systems, the amount of magnification in the detail view is large enough that two concatenated overview + detail displays are required. Overview + detail displays are thus very helpful for data navigation. Their main disadvantage is that they require coordination of two visual domains.

Linking and Brushing

Overview + detail is an example of coordinating dual representations of the same data. These can be coordinated interactively with *linking* and *brushing*. Suppose, for example, we wish to show power consumption on an airplane, both in terms of the physical representation of the airplane and a logical circuit diagram. The two views could be shown and linked by using the same color for the same component types. In brushing, running the cursor over a part of one of the views causes highlighting both in that view and in the other view.

Extraction and Comparison

We can also use interaction to extract a subset of the data to compare with another subset. An example of this is in the SDM (Spatial Data Management) system (Chuah, Roth, Mattis, & Kolojejchick, 1995) in Fig. 28.24c. The data are displayed in a 3D information landscape, but the perspective interferes with the ability to compare it. Information is therefore extracted from the display (leaving ghosts behind) and placed in an orthogonal viewing position where it can be compared using 2D. It could also be dropped into another display. Interactivity makes possible these manipulations, while keeping them coordinated with the original representations.

Attribute Explorer

Several of these interactive techniques are combined in the Attribute Explorer (Tweedie, Spence, Dawkes, & Su, 1996). Figure 28.24d shows information on four attributes of houses. Each attribute is displayed by a histogram, where each square making up the histogram represents an individual house. The user selects a range of some attribute, say price. Those pixels making up the histogram on price have their corresponding pixels linked representing houses highlighted on the other attributes. Those houses meeting all the criteria are highlighted in one color; those houses meeting, say, all but one are highlighted in another color. In this way, the user can tell about the "near misses." If the users were to relax one of the criteria only a little (say, reducing price by $100), then the user might be able to gain more on another criterion (say, reducing a commute by 20 miles).

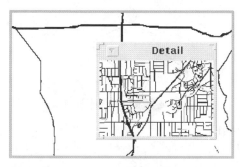

(a) Magic Lens (Bier, Stone, Pier, Buxton, & DeRose, 1993): Detail of map. Courtesy of Xerox Corp.

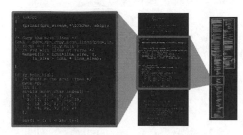

(b) Cascading overview + detail: SeeSoft (Eick et al., 1992). Reprinted with permission.

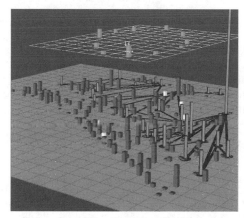

(c) Extract and compare SDM (Roth, Chuah, & Mattis, 1995). Reprinted with permission.

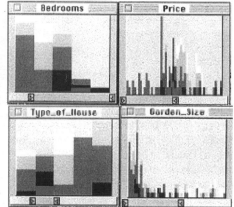

(d) Attribute Explorer (Tweedie et al., 1996). Courtesy of Robert Spence. Reprinted with permission.

FIGURE 28.24. Interaction techniques.

FOCUS + CONTEXT ATTENTION-REACTIVE ABSTRACTIONS

So far, we have considered visualizations that are static mappings from Data Table to Visual Structure and those where the mappings Data Table to Visual Structure are interactively controlled by the user. We now consider visualizations in which the machine is no longer passive, but its mappings from Visual Structure to View are altered by the computer according to the its model of the user's degree of interest. We call these visualizations *attention-reactive*, because the machine dynamically reacts to changes in the locus of attention of the user.

In principle, we can associate a cost of access with every element in the Data Table. Take the FilmFinder in Fig. 28.3. Details about the movie "Murder on the Orient Express" are accessible at low cost in terms of time, because they are presently visible on the screen. Details of "Goldfinger," a movie with only a mark on the display, take more time to find. Details of "Last Year at Marienbad," a movie with no mark on the display, would take much more time. The idea is that with a model for predicting

users' changes in interest, the system can adjust its displays to make costs lower for information access. For example, if the user wants some detail about a movie, such as the director, the system can anticipate that the user is more likely to want other details about the movie as well and therefore display them all at the same time: The user does not have to execute a separate command; the cost is therefore reduced.

Focus + context views are based on several premises. First, the user needs both overview (context) and detail information (focus) during information access, and providing these in separate screens or separate displays is likely to cost more in user time. Second, information needed in the overview may be different from that needed in the detail. The information of the overview needs to provide enough information for the user to decide where to examine next or to give a context to the detailed information rather than the detailed information itself. As Furnas (1981) has argued, the user's interest in detail seems to fall away in a systematic way with distance as information objects become farther from current interest. Third, these two types of information can be combined within a single dynamic display, much as human vision uses a two-level focus and context strategy. Information broken into multiple displays (separate legends for a

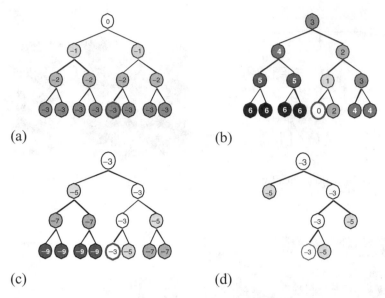

FIGURE 28.25. Degree-of-interest calculation for fish-eye visualization. (a) Intrinsic DOI. (b) Distance DOI. (c) DOI = Intrinsic DOI − Distance DOI. (d) After eliding nodes in (c) with a threshold of DOI < −5.

graph, for example) seem to degrade performance because of reasons of visual search and working memory.

Furnas (1981) was the first to articulate these ideas systematically in his theory of *fisheye views*. The essence of focus + context displays is that the average cost of accessing information is reduced by placing the most likely needed information for navigation and detail where it is fastest to access. This can be accomplished by working on either the data side or the visual side of the visual reference model (Fig. 28.10). We now consider these techniques in more detail.

Data-based Methods

Filtering. On the data side, focus + context effects can be achieved by filtering out which items from the Data Table are actually displayed on the screen. Suppose we have a tree of categories taken from *Roget's Thesaurus*, and we are interacting with one of these, "Hardness."

Matter
  ORGANIC
    Vitality
      Vitality in general
      Specific vitality
    Sensation
      Sensation in general
      Specific sensation
  INORGANIC
    Solid
      Hardness
      Softness
    Fluid

Fluids in general
Specific fluids

Of course, this is a small example for illustration. A tree representing a program listing or a computer directory or a taxonomy could easily have thousands of lines, a number that would vastly exceed what could fit on the display and hence would have a high cost of accessing. We calculate a degree of interest (DOI) for each item of the tree, given that the focus is on the node Hardness. To do this, we split the DOI into an intrinsic importance part and a part that varies with distance from the current center of interest and use a formula from Furnas (1981).

$$DOI = \text{Intrinsic Importance DOI} - \text{Distance DOI}.$$

Figure 28.25 shows schematically how to perform this computation for our example. We assume that the intrinsic DOI of a node is just its distance of the root (Fig. 28.25a). The distance part of the DOI is just the traversal distance to a node from the current focus node (Fig. 28.25b). It turns out to be convenient to use negative numbers for the intrinsic DOI, so that the maximum amount of interest is bounded, but not the minimum amount of interest. We subtract these two numbers (Fig. 28.25c) to get the DOI of each node in the tree. Then, we apply a minimum threshold of interest (−5 in this case) and only show nodes more interesting than that threshold. The result is the reduced tree:

Matter
  INORGANIC
  ORGANIC
    Solid
      Hardness
      Softness
    Fluid

The reduced tree gives local context around the focus node and progressively less detail farther away. But it does seem to give the important context.

Selective Aggregation. Another focus + context technique from the data side is selective aggregation. Selective aggregation creates new cases in the Data Table that are aggregates of other cases. For example, in a visualization of voting behavior in a presidential election, voters could be broken down by sex, precinct, income, and party affiliation. As the user drills down on, say, male Democrats earning between $25,000 and $50,000, other categories could be aggregated, providing screen space and contextual reference for the categories of immediate interest.

View-based Methods

Micro-macro Readings. Micro-macro readings are diagrams in which "detail cumulates into larger coherent structures" (Tufte, 1990). The diagram can be graphically read at the level of larger contextual structure or at the detail level. An example is Fig. 28.26. The micro reading of this diagram shows three million observations of the sleep (lines), wake (spaces), and feeding (dots) activity of a newborn infant. Each day's activity is repeated three times on a line to make the cyclical aspect of the activity more clearly visible. The macro reading of the diagram, emphasized by the thick lines, shows the infant transitioning from the natural human 25-hr cycle at birth to the 24-hr solar day. The macro reading serves as context and index into the micro reading.

Highlighting. Highlighting is a special form of micro-macro reading in which focal items are made visually distinctive in some way. The overall set of items provides a context for the changing focal elements.

Visual Transfer Functions. We can also warp the view with viewing transformations. An example is a visualization called the *bifocal lens* (Spence & Apperley, 1982). Figure 28.27a shows a set of documents the user would like to view, but which is too large to fit on the screen. In a bifocal lens, documents not in a central focal region are compressed down to a smaller size. This could be a strict visual compression. It could also involve a change in representation. We can talk about the visual compression in terms of a visual transfer function Fig. 28.27b, sometimes conveniently represented in terms of its first derivative in Fig. 28.27c. This function shows how many units of an axis in the original display are mapped into how many units in the resultant display. The result could be compression or enlargement of a section of the display. As a result of applying this visual transfer function to Fig. 28.27a, the display is compressed to Fig. 28.27d. Actually, the documents in the compressed region have been further altered by using a semantic zooming function to give them a simplified visual form. The form of Fig. 28.27c shows that this is essentially a step function of two different slopes. An example of a 2D step function is the Table Lens (Fig. 28.28a). The Table Lens is a spreadsheet in which the columns of selected cells are expanded to full size in X and the rows of selected cells are expanded to full size in Y. All other cells are compressed and their content represented only by a graphic. As a consequence, spreadsheets up to a couple orders of magnitude larger can be represented.

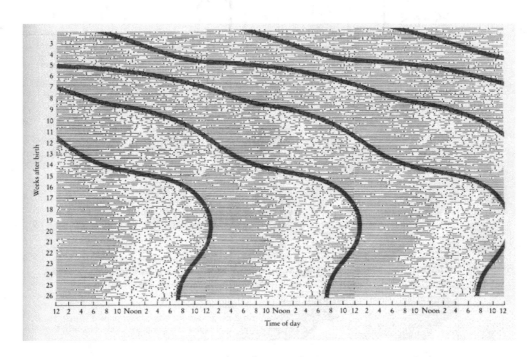

FIGURE 28.26. Micro-macro reading (Winfree, 1987). Courtesy of Scientific American Library.

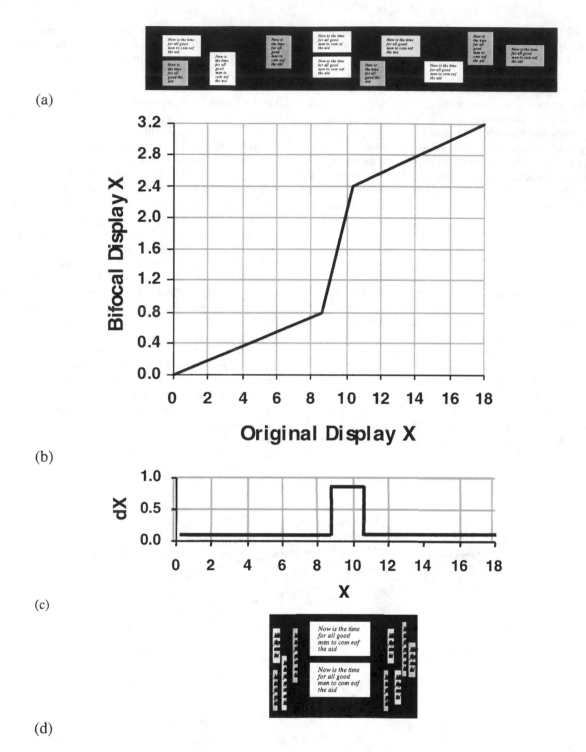

(a)

(b)

(c)

(d)

FIGURE 28.27. Bifocal + transfer function. (a) View of document space. (b) Visual transfer function. (c) First derivative of visual transfer function region. (d) Resulting bifocal view of documents.

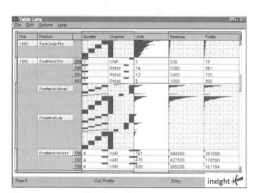

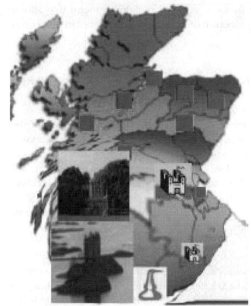

(a) Table Lens. Courtesy of Inxight Software.

(b) Nonlinear distortion of UK. Courtesy of Alan Keahey.

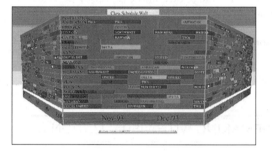

(c) Fisheye menus (Bederson, 2000). Reprinted with permission.

(d) Perspective Wall (Mackinlay, Robertson, & Card, 1991). Reprinted with permission.

FIGURE 28.28. Attention-reactive visualizations.

By varying the visual transfer function (see, e.g., the review by Leung and Apperley (1994)), a wide variety of distorted views can be generated. Figure 28.28b shows an application in which a visual transfer function is used to expand a bubble around a local region on a map. The expanded space in the region is used to show additional information about that region. Distorted views must be designed carefully so as not to damage important visual relationships. Bubble distortions of maps may

change whether roads appear parallel to each other. However, distorted views can be designed with flat and transition regions to address this problem. Figure 28.27a does not have curvilinear distortions. Focus + context visualizations can be used as part of compact user controls. Keahey (2001) has created an interactive scheme in which the bubble is used to preview a region. When the user releases a button over the region, the system zooms in far enough to flatten out the bubble. Bederson (2000) has developed a focus + context pull-down menu that allows the viewing and selection of large lists of typefaces in text editor Fig. 28.28c.

Perspective Distortion. One interesting form of distorting visual transfer functions is 3D perspective. Although it can be described with a 2D distorting visual transfer function, it is usually not experienced as distorting by users because of the special perceptual mechanisms humans have for processing 3D. Figure 28.28d shows the Perspective Wall (Mackinlay, Robertson, & Card, 1991). Touching any place on the walls animates its transition into the central focal area. The user perceives the context area of the wall as an undistorted 2D image in a 3D space, rather than as a distorted 2D image. However, the same sort of compression is still achieved in the nonfocus area.

Alternate Geometries. Instead of altering the size of components, focus + context effects can also be achieved by changing the geometry of the spatial substrate itself. One example is the hyperbolic tree (Lamping & Rao, 1994). A visualization such as a tree is laid out in hyperbolic space (which itself expands exponentially, just like the tree does) and then projected on to the Euclidean plane. The result is that the tree seems to expand around the focal nodes and to be compressed elsewhere. Selecting another node in the tree animates that portion to the focal area. Munzner and Burchard (1995) have extended this notion to 3D hyperbolic trees and used them to visualize portions of the Internet.

SENSEMAKING WITH VISUALIZATION

Knowledge Crystallization

The purpose of information visualization is to amplify cognitive performance, not just to create interesting pictures. Information visualizations should do for the mind what automobiles do for the feet. So here we return to the higher level cognitive operations of which information visualization is a means and a component. A recurrent pattern of cognitive activity to which information visualization would be useful (although not the only one!) is "knowledge crystallization." In knowledge crystallization tasks, there is a (sometimes ill-structured) goal that requires the acquisition and making sense of a body of information, as well as the creative formulation of a knowledge product, decision, or action. Examples would be writing a scientific paper, business or military intelligence, weather forecasting, or buying a laptop computer. For these tasks, there is usually a concrete outcome of the task—the submitted manuscript of a paper, a delivered briefing, or a purchase. Knowledge crystallization does have characteristic processes, however, and it is by amplifying these that information visualization seeks to intervene and amplify the user's cognitive powers. Understanding

TABLE 28.10. Knowledge Crystallization Operators

Acquire information	Monitor
	Search
	Capture (make tacit knowledge explicit)
Make sense of it	Extract information
	Fuse different sources
	Find schema
	Recode information into schema
Create something new	Organize for creation
	Author
Act on it	Distribute
	Apply
	Act

of this process is still tentative, but the basic parts can be outlined:

> Acquire information.   Make sense of it.   Create something new.   Act on it.

In Table 28.10, we have listed some of the more detailed activities these entail. We can see examples of these in our initial examples.

Acquire Information. The FilmFinder is concentrated largely on acquiring information about films. *Search* is one of the methods of acquiring information in Table 28.10, and the FilmFinder is an instance of the use of information visualization in search. In fact, Shneiderman (Card et al., 1999) has identified a heuristic for designing such systems:

> Overview first, zoom and filter, then details-on-demand

The user starts with an overview of the films and then uses sliders to filter the movies, causing the overview to zoom in on the remaining films. Popping up a box gives details on the particular films. The user could use this system as part of a knowledge crystallization process, but the other activities would take place outside the system. The SmartMoney system also uses the TreeMap visualization for acquiring information, but this time the system is oriented toward monitoring, another of the methods in Table 28.10. A glance at the sort of chart in Fig. 28.5 allows an experienced user to notice interesting trends among the hundreds of stocks and industries monitored. Another method of acquiring information, capture, refers to acquiring information that is tacit or implicit. For example, when users browse the World Wide Web, their paths contain information about their goals. This information can be captured in logs, analyzed, and visualized (Chi & Card, 1999). It is worth making the point that acquiring information is not something that the user must necessarily do explicitly. Search, monitoring, and capture can be implicitly triggered by the system.

Make Sense of It. The heart of knowledge crystallization is sensemaking. This process is by no means as mysterious as it

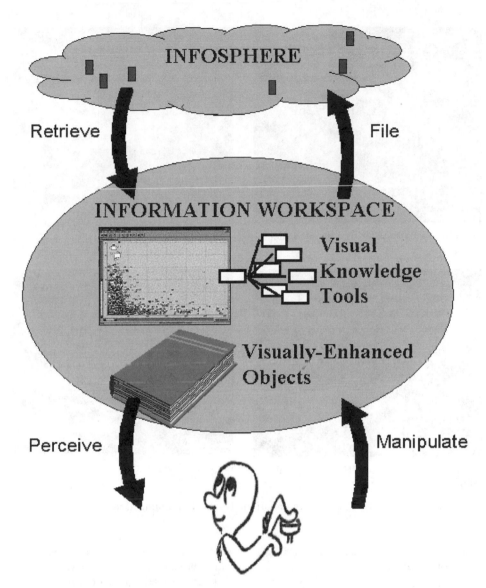

FIGURE 28.29. Levels of use for information visualization (Card et al., 1999). Reprinted with permission.

might appear. Because sensemaking involving large amounts of information must be externalized, the costs of finding, organizing, and moving information around has a major impact on its effectiveness. The actions of sensemaking itself can be analyzed. One process is extraction. Information must be removed from its sources. In our hotel example, the hotel manager extracted information from hotel records. A more subtle issue is that information from different sources must be fused, that is, registered in some common correspondence. If there are six called-in reports of traffic accidents, does this mean six different accidents, one accident called in six times, or two accidents reported by multiple callers? If one report merely gives the county, while another just gives the highway, it may not be easy to tell. Sensemaking involves finding some schema, that is,

some descriptive language, in terms of which information can be compactly expressed (Russell, Stefik, Pirolli, & Card, 1993). In our hotel example, permuting the matrices brought patterns to the attention of the manager. These patterns formed a schema she used to organize and represent hotel stays compactly. In the case of buying a laptop computer, the schema may be a table of features by models. Having a common schema then permits compact description. Instances are recoded into the schema. Residual information that does not fit the schema is noted and can be used to adjust the schema.

Create Something New. Using the schema, information can be reorganized to create something new. It must be organized into a form suitable for the output product and that product

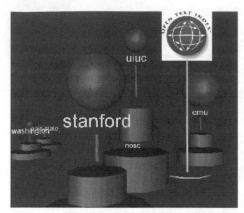

(a) Infosphere: (Bray, 1996). Reprinted with permission.

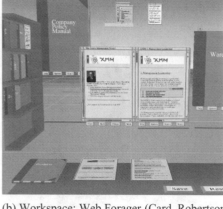

(b) Workspace: Web Forager (Card, Robertson, & York, 1996). Reprinted with permission.

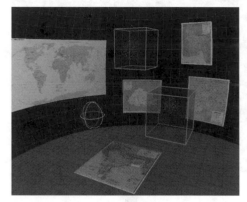

(c) Workspace: STARLIGHT: (Risch et al., 1997). Reprinted with permission.

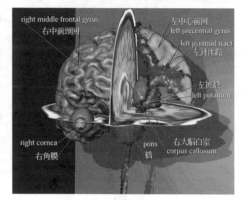

(d) Visually-enhanced object: Voxel-Man. Courtesy of University of Hamburg.

FIGURE 28.30. Information visualization applications.

must be authored. In the case of the hotel example, the manager created the presentation of Fig. 28.7c.

Act On It. Finally, there is some consequential output of the knowledge crystallization task. That action may be to distribute a report or give a briefing, to act directly in some way, such as setting up a new promotion program for the hotel or buying a laptop on the basis of the analysis, or by giving directives to an organization.

Levels for Applying Information Visualization

Information visualization can be applied to facilitate the various subprocesses of knowledge crystallization just described. It can also be applied at different architectural levels in a system. These have been depicted in Fig. 28.29. At one level is the use of visualization to help users access information outside the immediate environment—the *infosphere*, such as information on the Internet or from corporate digital libraries. Figure 28.30a shows such a visualization of the Internet (Bray, 1996). Web

sites are laid out in a space such that sites closer to each other in the visualization tend to have more traffic. The size of the disk represents the number of pages in the site. The globe size represents the number of out-links. The globe height shows the number of in-links.

The second level is the *information workspace*. The information workspace is like a desk or workbench. It is a staging area for the integration of information from different sources. An information workspace might contain several visualizations related to one or several tasks. Part of the purpose of an information workspace is to make the cost of access low for information in active use. Figure 28.30b shows a 3D workspace for the Internet, the Web Forager (Card, Robertson, & York, 1996). Pages from the World Wide Web that users access through clicking on URLs or searches appear in the space. These can be organized into piles or books related to different topics. Figure 28.30c shows another document workspace, STARLIGHT (Risch et al., 1997). Documents are represented as galaxies of points in space such that similar documents are near each other. In the workspace, various tools allow linking the documents to maps and other information and analytical resources.

The third level is *visual knowledge tools*. These are the sort of tools that allow schema forming and re-representation of information. The permutation matrices in Fig. 28.7, the SeeSoft system for analyzing software in Fig. 28.15b and the Table Lens in Fig. 28.28a are examples of visual knowledge tools. The focus is on determining and extracting the relationships.

The final level is *visually enhanced objects*, coherent information objects enhanced by the addition of information visualization techniques. An example is Fig. 28.30d, in which voxel data of the brain have been enhanced through automatic surface rendition, coloring, slicing, and labeling. Abstract data structures representing neural projects and anatomical labels have been integrated into a display of the data. Visually enhanced objects focus on revealing more information from some object of intrinsic visual form.

Information visualization is a set of technologies that use visual computing to amplify human cognition with abstract information. The future of this field will depend on the uses to which it is put and how much advantage it gives to these. Information visualization promises to help us speed our understanding and action in a world of increasing information volumes. It is a core part of a new technology of human interfaces to networks of devices, data, and documents.

ACKNOWLEDGMENTS

This chapter benefited from generous discussion with Jock Mackinlay and tutorial collaborations with Ed Chi of PARC.

References

Ahlberg, C., & Shneiderman, B. (1994a). *Visual information seeking using the filmfinder.* Paper presented at the Conference Companion of CHI'94, ACM Conference on Human Factors in Computing Systems, New York.

Ahlberg, C., & Shneiderman, B. (1994b). *Visual information seeking: Tight coupling of dynamic query filters with starfield displays.* Paper presented at the Proceedings of CHI'94, ACM Conference on Human Factors in Computing Systems, New York.

Battista, G. D., Eades, P., Tamassia, R., & Tollis, I. G. (1994). Annotated bibliography on graph drawing. *Computational Geometry: Theory and Applications, 4*(5), 235-282.

Becker, R. A., Eick, S. G., & Wilks, A. R. (1995, March 1). Visualizing network data. *IEEE Transactions on Visualization and Computer Graphics, 1,* 16-28.

Bederson, B. B. (2000). *Fisheye menus.* Paper presented at the UIST 2000, ACM Symposium on User Interface Software and Technology, New York.

Bertin, J. (1967/1983). *Semiology of graphics: Diagrams, networks, maps* (W. J. Berg, Trans.). Madison, WI: University of Wisconsin Press.

Bertin, J. (1977/1981). Graphic constructions (P. Scott, Trans.). In *Graphics constructions and graphic information-processing* (pp. 24-31). Berlin: Walter De Gruyter.

Bray, T. (1996, May 7-11). Measuring the web. *Computer Networks and ISDN Systems, 28,* 992.

Card, S. K., Mackinlay, J. D., & Shneiderman, B. (1999). *Information visualization: Using vision to think.* San Francisco: Morgan Kaufmann Publishers.

Card, S. K., Moran, T. P., & Newell, A. (1986). The model human processor: An engineering model of human performance. In J. Thomas (Ed.), *Handbook of perception and human performance* (chapter 45, pp. 41-35). New York: John Wiley & Sons.

Card, S. K., Robertson, G. G., & York, W. (1996). *The WebBook and the Web Forager: An information workspace for the world-wide web.* Paper presented at the Proceedings of CHI'96, ACM Conference on Human Factors in Computing Systems, New York.

Casner, S. (1991, April 2). Task-analytic approach to the automated design of graphic presentations. *ACM Transactions on Graphics, 10,* 111-151.

Chi, E. H., & Card, S. K. (1999). *Sensemaking of evolving web sites using visualization spreadsheets.* Paper presented at the Infovis 1999, IEEE Conference on Information Visualization 1999, San Francisco.

Chi, E. H., & Riedl, J. T. (1998). *An operator interaction framework for visualization spreadsheets.* Paper presented at the Proceedings of InfoVis'98, IEEE Symposium on Information Visualization Research Triangle Park, NC.

Chuah, M. C., Roth, S. F., Mattis, J., & Kolojejchick, J. A. (1995). *SDM: Malleable information graphics.* Paper presented at the Proceedings of InfoVis'95, IEEE Symposium on Information Visualization, New York.

Cleveland, W. S., & McGill, M. E. (1988). *Dynamic graphics for statistics.* Pacific Grove, CA: Wadsworth and Brooks/Cole.

Cruz, I. F., & Tamassia, R. (1998). Graph drawing tutorial [On-line]. Available: http://www.cs.brown.edu/people/rt/papers/gd-tutorial/gd-constraints.pdf

Eick, S. G., Steffen, J. L., & Sumner, E. E. (1992). Seesoft—a tool for visualizing software. *IEEE Transactions on Software Engineering, 18*(11-Nov.), 957-968.

Eick, S. G., & Wills, G. J. (1993, October 25-29). *Navigating large networks with hierarchies.* Paper presented at the Proceedings of IEEE Visualization'93 Conference, San Jose, CA.

Engelhardt, Y., Bruin, J. D., Janssen, T., & Scha, R. (1996). The visual grammar of information graphics. In S. University (Ed.), *Artificial intelligence in design workshop notes* (pp. 24-27).

Fairchild, K. M., Poltrock, S. E., & Furnas, G. W. (1988). Semnet: Three-dimensional representations of large knowledge bases. In R. Guindon (Ed.), *Cognitive science and its applications for human-computer interaction* (pp. 201-233). Hillsdale, NJ: Lawrence Erlbaum Associates.

Feiner, S., & Beshers, C. (1990). Worlds within Worlds: Metaphors for exploring n-dimensional virtual worlds. In *ACM Symposium on User Interface Software* and technology, Snowbird, Utah (pp. 76-83).

Freeman, E., & Fertig, S. (1995). *Lifestreams: Organizing your electronic life.* Paper presented at the Proceedings of AAAI Fall Symposium on AI Applications in Knowledge Navigation, Cambridge, MA.

Furnas, G. W. (1981). The fisheye view: A new look at structured files. In S. K. Card, J. D. Mackinlay, & B. Shneiderman (Ed.), *Readings in information visualization: Using vision to think* (pp. 312-330). San Francisco: Morgan Kaufmann Publishers, Inc.

Healey, C. G., Booth, K. S., & Enns, J. T. (1995). High-speed visual estimation using preattentive processing. *ACM Transactions on Computer-Human Interaction, 3*(2), 107-135.

Inselberg, A. (1997). *Multidimensional detective*. Paper presented at the Proceedings of InfoVis'97, IEEE Symposium on Information Visualization, IEEE Information Visualization, Phoenix, AZ.

Inselberg, A., & Dimsdale, B. (1990). *Parallel coordinates: A tool for visualizing multi-dimensional geometry*. Paper presented at the Proceedings of IEEE Visualization'90 Conference, Los Alamitos, CA.

Keahey, T. A. (2001, October 22-23). *Getting along: Composition of visualization paradigms*. Paper presented at the Infovis 2001. IEEE Information Visualization 2001, San Diego, CA.

Keim, D. A., & Kriegel, H.-P. (1994, September). VisDB: Database exploration using multidimensional visualization. *IEEE Computer Graphics and Applications* (pp. 40-49).

Kosslyn, S. M. (1994). *Image and brain: The resolution of the imagery debate*. Cambridge, MA: The MIT Press.

Lamping, J., & Rao, R. (1994). *Laying out and visualizing large trees using a hyperbolic space*. Paper presented at the Proceedings of UIST'94, ACM Symposium on User Interface Software and Technology, Marina del Rey, CA.

Larkin, J., & Simon, H. A. (1987). Why a diagram is (sometimes) worth ten thousand words. *Cognitive Science, 11*, 65-99.

Leung, Y. K., & Apperley, M. D. (1994, June). A review and taxonomy of distortion-orientation presentation techniques. *ACM Transactions on Computer-Human Interaction, 1*(2), 126-160.

Lin, X., Soergel, D., & Marchionini, G. (1991). *A self-organizing semantic map for information retrieval*. Paper presented at the Proceedings of SIGIR'91, ACM Conference on Research and Development in Information Retrieval, Chicago, IL.

MacEachren, A. M. (1995). *How maps work*. New York: The Guilford Press.

Mackinlay, J. D. (1986a). *Automatic design of graphical presentations*. Unpublished PhD, Stanford University, Stanford, CA.

Mackinlay, J. D. (1986b). Automating the design of graphical presentations of relational information. *ACM Transactions on Graphics, 5*(2), 110-141.

Mackinlay, J. D., Robertson, G. G., & Card, S. K. (1991). *The perspective wall: Detail and context smoothly integrated*. Paper presented at the Proceedings of CHI'91, ACM Conference on Human Factors in Computing Systems, New York.

McCormick, B. H., & DeFanti, T. A. (1987, November 6). Visualization is scientific computing. *Computer Graphics, 21*, 247-305.

Munzner, T., & Burchard, P. (1995, December 14-15, 1995). *Visualizing the structure of the world wide web in 3D hyperbolic space*. Paper presented at the Proceedings of VRML '95 (pp. 33-38). San Diego, CA.

Norman, D. A. (1993). *Things that make us smart*. Reading, Massachusetts: Addison-Wesley.

Playfair, W. (1786). *The commercial and political atlas*. London, England.

Pickett, R. M., & Grinstein, G. G. (1998). Iconographic displays for visualizing multidimensional data. Paper presented at Proceedings of IEEE Conference on Systems, Man, and Cybernetics. Piscataway, NJ: IEEE Press.

Resnikoff, H. L. (1989). *The illusion of reality*. New York: Springer-Verlag.

Risch, J. S., Rex, D. B., Dowson, S. T., Walters, T. B., May, R. A., & Moon, B. D. (1997). *The STARLIGHT information visualization system*. Paper presented at the Proceedings of IEEE International Conference on Information Visualization, London, England.

Robertson, G. G., Card, S. K., & Mackinlay, J. D. (1989). *The cognitive co-processor for interactive user interfaces*. Paper presented at the Proceedings of UIST'89, ACM Symposium on User Interface Software and Technology, Williamsburg, VA.

Robertson, G. G., Card, S. K., & Mackinlay, J. D. (1993). Information visualization using 3D interactive animation. *Communications of the ACM, 36*(4), 57-71.

Robin, H. (1992). *The scientific image: From cave to computer*. New York: H. N. Abrams, Inc.

Roth, S. F., & Mattis, J. (1990). *Data characterization for intelligent graphics presentation*. Paper presented at the Proceedings of CHI'90, ACM Conference on Human Factors in Computing Systems, New York.

Russell, D. M., Stefik, M. J., Pirolli, P., & Card, S. K. (1993). *The cost structure of sensemaking*. Paper presented at the Proceedings of INTERCHI'93, ACM Conference on Human Factors in Computing Systems, Amsterdam.

Shneiderman, B. (1992). Tree visualization with Tree-Maps: A 2-dimensional space filling approach. *ACM Transactions on Graphics, 11*(1), 92-99.

Shneiderman, B., & Wattenberg, M. (2001). *Ordered tree layouts*. Paper presented at the IEEE Symposium on Information Visualization, San Diego, CA.

Spence, R. (2000). *Information visualization*. Harlow, England: Addison-Wesley.

Spence, R., & Apperley, M. (1982). Data base navigation: An office environment for the professional. *Behavior and Information Technology, 1*(1), 43-54.

Tamassia, R. (1996, December 6). Strategic directions in computational geometry working group report. *ACM Computing Surveys, 28*, 591-606.

Tufte, E. R. (1983). *The visual display of quantitative information*. Cheshire, CT: Graphics Press.

Tufte, E. R. (1990). *Envisioning information*. Cheshire, CT: Graphics Press.

Tufte, E. R. (1997). *Visual explanations: Images and quantities, evidence and narrative*. Cheshire, CT: Graphics Press.

Tukey, J. W. (1977). *Exploratory data analysis*. Reading, MA: Addison-Wesley.

Tweedie, L. A., Spence, R., Dawkes, H., & Su, H. (1996). *Externalising abstract mathematical models*. Paper presented at the Proceedings of CHI'96, ACM Conference on Human Factors in Computing Systems, Boston.

Ware, C. (2000). *Information visualization: Perception for design*. San Francisco: Morgan Kaufmann Publishers.

Winfree, A. T. (1987). *The timing of biological clocks*. New York: Scientific American Books, Inc.

Wise, J. A., Thomas, J. J., Pennock, K., Lantrip, D., Pottier, M., & Schur, A. (1995). *Visualizing the non-visual: Spatial analysis and interaction with information from text documents*. Paper presented at the *Proceedings of Infovis '95, IEEE Symposium on Information Visualization*.

Winograd, 1988). For example, it is hard to specify that all junk mail should be deleted; the system requires an enumeration of all the sources of junk mail. Those who are successful in dealing with this overload use primitive manual routines (Whittaker & Sidner, 1996).

Kraut et al. (1998) reported that greater Internet use, which in their sample was mostly e-mail, led to declines in social interactions with family members and an increase in depression and loneliness. Not surprising, these results triggered widespread discussion and debate, both over the substance of the results and the methods used to obtain them. Recently, Kraut et al. (in press) have reported new results that suggest these initial effects may not persist. Interpersonal communication is one of the principal uses of the Internet, and the possible implications of this kind of communication for social life is important to understand. Indeed, Putnam (2000) has wondered whether the Internet can be a source of social cohesiveness. These kinds of questions need to be addressed by additional large-scale studies of the kind conducted by Kraut and his colleagues.

Conferencing Tools—Voice and Video

Real-time meetings are among the more difficult situations to support at a distance. The most familiar technology to support remote meetings is video conferencing, where special facilities are linked by high bandwidth connections, and compatible camera/projection systems are required at all points to be connected. Although there is good new technology to support video and audio over Internet protocol (see VoIP, 2001, for details), it is still time consuming and confusing to set up a connection, and multipoint conferences require expensive multipoint conference units. For point-to-point conferencing, however, the units and connect time are inexpensive. The rise of the use of standards for conferencing, such as H323 (see OpenH323 Project, 2001, and H323 Information Site, 2001, for more details about this standard), is leading to important interoperability, much like that in e-mail and voice telephony.

Although the new systems allow less expensive connectivity, they are not all of the same quality, and some aspects of quality strongly affect the experience. High-quality audio is essential. Video is important for some cues about who is speaking and whose turn it is next (Veinott, Olson, Olson, & Fu, 1999). But, without immediate audio (less than a half second), conversational flow is severely disrupted. People with experience with these technologies will mute the audio channel that accompanies the videoconferencing (which typically has a delay of over a second) and make a second telephone call to connect the audio over more standard high-quality channels.

The other key feature of successful remote meetings is the ability to share the objects they are talking about: The agenda, the to-do list, the latest draft of a proposal, etc. More traditional videoconferencing technologies often offer an object camera, onto which the participants can put a paper agenda, Powerpoint slides, or a manufactured part. For digital objects, there are now a number of products that will allow meeting participants to share the screen or, in some cases, the operation of the actual application. NetMeeting is a current popular application for allowing remote participants to see and manipulate digital material at a remote location. Some companies are using electronic whiteboards, both in a collocated meeting and in remote meetings to mimic the choreography of people using a physical whiteboard. Furthermore, in some "collaboratories" (see later section), scientists can even operate remote physical instruments from a distance and jointly discuss the results.

A number of distributed organizations have adopted remote conferencing to coordinate their work, but not without consequences (Finn, Sellen, & Wilbur, 1997). A controlled comparison of high-quality videoconferencing with shared digital objects against collocated meetings with the good support for digital objects showed the quality of the work was the same. This is good news on the face of it. However, the remote groups are less satisfied with the process and outcome, perhaps because it requires more effort to organize the flow of work and more attention to understand the subtleties of the interactions (Olson, Olson, & Meader, 1995). Meetings with video, however, are much more preferred to those with audio only, even though the quality of the work is indistinguishable from meetings with video. In a related study, however, it was shown video was much more important (a much greater value over audio) for people who did not know each other or were from different cultures—a situation that is common in long-distance work (Veinott et al., 1999).

There is a strong tendency for video conferences to be quite formal and stilted. The easy flow of interaction that one finds in face-to-face or even telephone interactions are readily disrupted by any delays. Poor image quality makes it difficult to pick up expressions and gestures. Eye contact is especially difficult to make, because cameras are usually far from where the remote people are projected on the screen. In short, there are many details of how to configure video conferences to allow for more spontaneous, free-flowing interaction. Research is needed on which details really matter for a remote meeting's success.

Instant Messaging, Chat, and MUDs

Chat systems are like instant e-mail; people type typically short comments or questions to each other, and with the "send" button, the message is instantly shown on the other person's window. There is a trail of the conversation so far in both participants' windows so that people can keep track of the thread easily by scrolling back through recent history. A good chat system can feel like a conversation. Because most chat systems allow a number of participants, one might suspect that the conversation would get chaotic and be hard to follow. But people in chats have no more trouble following than those in a face-to-face meeting of the same number of participants, because the conversation is typically purposeful, the participants add a few more conversational cues, like naming the person or topic to which they are referring, and the recent history allows people to review the conversational threads (McDaniel, Olson, & Magee, 1996). An example of a chat session is shown later in Fig. 29.3.

·29·

GROUPWARE AND COMPUTER-SUPPORTED COOPERATIVE WORK

Gary M. Olson and Judith S. Olson
University of Michigan

The networking of information technology devices has enabled a broad new class of tools that go under the generic name of groupware.[1] The broad study of the development and use of groupware is known quite widely as Computer-Supported Cooperative Work (CSCW).[2] These tools have emerged as new social and organizational challenges have arisen. Television and radio long ago broadened our awareness of and interest in activities all over the world. The telegraph and telephone enabled new forms of organizing to emerge. Groupware is allowing greater geographical and temporal flexibility in conducting a wide range of intellectual work.

Groupware is software designed to run over a network in support of the activities of a group or organization. These activities can occupy any of several combinations of same/different place and same/different time. Groupware has been designed for all of these combinations. Early groupware applications each tended to focus on only one of these cells, but more recently groupware that supports several cells and the transitions among them has emerged.

Whereas CSCW emerged as a formal field of study in the mid-1980s, there were a number of important antecedents. The earliest efforts to create groupware used time-shared systems, but were closely linked to the development of key ideas that propelled the personal computer revolution. Vannever Bush described a vision of something similar to today's World Wide Web in an influential essay published shortly at the end of World War II (Bush, 1945). Doug Engelbart's famous demonstration at the 1968 International Federation for Information Processing (IFIP) meeting in San Francisco included a number of key groupware components (see the report by Engelbart & English, 1968). These components included support for real-time, face-to-face meetings, audio and video conferencing, discussion databases, information repositories, and workflow support. Group decision support systems and computer-supported meeting rooms were explored in a number of business schools (see reviews by Kraemer & Pinsonneault, 1990; McLeod, 1992). Work on office automation included many groupware elements, such as group workflow management, calendaring, e-mail, and document sharing (Ellis & Nutt, 1980). A good summary of early historical trends, as well as reprints of key early articles, appear in Grief's (1988) important anthology of readings.

Today support for collaboration at a distance is included in many commercial products. There are a large number of specific groupware-based commercial products, like a host of e-mail applications, Lotus Notes, and NetMeeting. In addition, groupware functions are increasingly appearing as options in operating systems or specific applications. These trends suggest that groupware functionality will become widespread and familiar. However, there are still many research issues about how to design such systems and what effects they have on the individuals, groups, and organizations that use them.

ADOPTING GROUPWARE IN CONTEXT

Groupware systems are intended to support groups, who are usually embedded in an organization. As a result, there are a number of issues that bear on groupware success. In a justly famous set of papers, Grudin (1988, 1994) pointed out a number of problems that groupware systems have (see also Markus & Connolly, 1990). In brief, he pointed out that developers of groupware systems need to be concerned about the following issues (Grudin, 1994, p. 97):

1. *Disparity in work and benefit:* Groupware applications often require additional work from individuals who do not perceive a direct benefit from the use of the application.
2. *Critical mass and Prisoner's dilemma problems:* Groupware may not enlist the critical mass of users required to be useful, or can fail because it is never in any one individual's advantage to use it.
3. *Disruption of social processes:* Groupware can lead to activity that violates social taboos, threatens existing political structures, or otherwise demotivates users crucial to its success.
4. *Exception handling:* Groupware may not accommodate the wide range of exception handling and improvisation that characterizes much group activity.
5. *Unobtrusive accessibility:* Features that support group processes are used relatively infrequently, requiring unobtrusive accessibility and integration with more heavily used features.
6. *Difficulty of evaluation:* The almost insurmountable obstacles to meaningful, generalizable analysis and evaluation of groupware prevent us from learning from experience.
7. *Failure of intuition:* Intuitions in product development environments are especially poor for multiuser applications, resulting in bad management decisions and error-prone design processes.
8. *The adoption process:* Groupware requires more careful implementation (introduction) in the workplace than product developers have confronted.

However, there are reasons for optimism. In a recent survey of the successful adoption of group calendaring in several organizations, Palen and Grudin (in press) observed that organizational conditions in the 1990s were favorable for the adoption of group tools than they were in the 1980s. Furthermore, the tools themselves had improved in reliability, functionality, and usability. There is increased collaboration readiness and collaboration technology readiness. But there are still significant challenges in supporting group work at a distance (Olson & Olson, 2000).

[1] Barry Wellman (2001) laments the fact that such tools are not called "netware," a name that is now a trademarked company name.
[2] Groupware and CSCW have a variety of denotations, some of them quite narrow. We choose to use these terms quite broadly, as do many others.

TECHNICAL INFRASTRUCTURE

Groupware requires networks, and network infrastructure is a key enabler as well as a constraint on groupware. The technical possibilities are very different for advanced networks like Abilene (www.internet2.edu/abilene/) when contrasted with the commodity Internet with slow links to the desktop. Heterogeneity in network conditions across both space and time still remains a major technical challenge.

The World Wide Web and its associated tools and standards have had a major impact on the possibilities for groupware (Berners-Lee, 1999; Schatz & Hardin, 1994). Early groupware mostly consisted of stand-alone applications that had to be downloaded and run on each client machine. Increasingly, group tools are being written for the web, requiring only a web browser and perhaps some plug-ins. This makes it much easier for the user, and also helps with matters such as version control. It also allows for better interoperability across hardware and operating systems.

However, security on the Internet is a major challenge for groupware. In some sense, the design of Internet protocols are to blame, because the Internet grew up in a culture of openness and sharing (Abbate, 1999; Longstaff et al., 1997). E-commerce and sensitive application domains like medicine have been a driver for advances in security, but there is still much progress to be made (Camp, 2000; Longstaff et al., 1997).

Personal computing was a great enabler of collaborative applications. However, we are now undergoing a liberation of computing from the desktop. Laptops, personal digital assistants, wearables, and even multifunction cell phones provide a wide range of platforms for the individuals involved in group work. More and more applications are being written to operate across these diverse environments (e.g., Tang et al., 2001). These devices vary in computational power, display size and characteristics, network bandwidth, and connection reliability, providing interesting technical challenges to make them all interoperate smoothly.

Additional flexibility is being provided by the development of infrastructure that lies between the network itself and the applications that run on client workstations, called middleware. This infrastructure makes it easier to link together diverse resources to accomplish collaborative goals. For instance, the emerging Grid technologies allow the marshalling of powerful, scattered computational resources (Foster & Kesselman, 1999). Middleware provides such services as identification, authentication, authorization, directories, and security in uniform ways that facilitate the interoperability of diverse applications (see Internet2, 2001, for pointers to a variety of projects in this area).

COMMUNICATION TOOLS

We now turn to a review of specific kinds of groupware, highlighting their various properties and uses. We have grouped this review under several broad headings. We do not aim to be exhaustive, but rather seek to illustrate the variety of kinds of tools that have emerged to support human collaborative activities over networked systems. We also highlight various research issues pertaining to these tools.

E-mail

E-mail is almost as ubiquitous as the telephone. Because messages can be exchanged across networks and different base machines and software applications, it is nearly the universal service that the telephone is (Anderson, Bikson, Law, & Mitchell, 1995). Because of this widespread use, it is often called the first and perhaps the only successful groupware application (Anderson et al., 1995; O'Hara-Devereaux & Johnson, 1994; Satzinger & Olfman, 1992; Sproull & Kiesler, 1991). More recently, with the use of MIME as the standard for attachments and Postscript and others for representation, it is easier and easier to send spreadsheets, complexly formatted text documents, graphics, etc. Because of the speed and ease of use, people use e-mail to attempt to span distance and time, and to easily disseminate information to broad communities (Garton & Wellman, 1995).

Researchers have shown that this widespread use has had a number of effects on how people behave. It has had large effects on communication in organizations: It changes the social network of who talks to whom (Sproull & Kiesler, 1991), the power of people who formerly had little voice in decisions (Finholt, Sproull, & Kiesler, 1990), and the tone of what is said and how it is interpreted (Sproull & Kiesler, 1991). For example, with e-mail, people who were shy found a voice; they could overcome their reluctance to speak to other people by composing text, not speech to another face. This invisibility, however, also has a more general effect: Without the social cues in the recipient's face being visible to the sender, people will flame, send harsh or extremely emotive (usually negative) messages. Although people adapt their styles over time, it is still a concern for the steady stream of young users new to the medium (Arrow et al., 1996; Hollingshead, McGrath, & O'Connor, 1993).

As with a number of other designed technologies, people use e-mail for things other than the original intent. People use it for managing time, reminding them of things to do, and keeping track of steps in a workflow (Carley & Wendt, 1991; Mackay, 1989; Whittaker & Sidner, 1996). But because e-mail was not designed to support these tasks, it does not do it very well; people struggle with reading signals about whether they have replied or not (and to whom it was cc'd); they manage folders poorly for reminding them to do things, etc.

In addition, because e-mail is so widespread, and it is easy and free to distribute a single message to many people, people experience information overload. Many people get hundreds of e-mail messages each day, many of them mere broadcasts of things for sale or events about to happen, much like classifieds in the newspaper. Efforts to use artificial intelligence techniques to block and/or sort incoming e-mail are of little success, mainly because it is difficult to specify exactly what people do and do not want to receive (Malone, Grant, Lai, Rao, & Rosenblitt, 1987;

GROUPWARE AND COMPUTER-SUPPORTED COOPERATIVE WORK

Gary M. Olson and Judith S. Olson
University of Michigan

The networking of information technology devices has enabled a broad new class of tools that go under the generic name of groupware.[1] The broad study of the development and use of groupware is known quite widely as Computer-Supported Co-operative Work (CSCW).[2] These tools have emerged as new social and organizational challenges have arisen. Television and radio long ago broadened our awareness of and interest in activities all over the world. The telegraph and telephone enabled new forms of organizing to emerge. Groupware is allowing greater geographical and temporal flexibility in conducting a wide range of intellectual work.

Groupware is software designed to run over a network in support of the activities of a group or organization. These activities can occupy any of several combinations of same/different place and same/different time. Groupware has been designed for all of these combinations. Early groupware applications each tended to focus on only one of these cells, but more recently groupware that supports several cells and the transitions among them has emerged.

Whereas CSCW emerged as a formal field of study in the mid-1980s, there were a number of important antecedents. The earliest efforts to create groupware used time-shared systems, but were closely linked to the development of key ideas that propelled the personal computer revolution. Vannever Bush described a vision of something similar to today's World Wide Web in an influential essay published shortly at the end of World War II (Bush, 1945). Doug Engelbart's famous demonstration at the 1968 International Federation for Information Processing (IFIP) meeting in San Francisco included a number of key groupware components (see the report by Engelbart & English, 1968). These components included support for real-time, face-to-face meetings, audio and video conferencing, discussion databases, information repositories, and workflow support. Group decision support systems and computer-supported meeting rooms were explored in a number of business schools (see reviews by Kraemer & Pinsonneault, 1990; McLeod, 1992). Work on office automation included many groupware elements, such as group workflow management, calendaring, e-mail, and document sharing (Ellis & Nutt, 1980). A good summary of early historical trends, as well as reprints of key early articles, appear in Grief's (1988) important anthology of readings.

Today support for collaboration at a distance is included in many commercial products. There are a large number of specific groupware-based commercial products, like a host of e-mail applications, Lotus Notes, and NetMeeting. In addition, groupware functions are increasingly appearing as options in operating systems or specific applications. These trends suggest that groupware functionality will become widespread and familiar. However, there are still many research issues about how to design such systems and what effects they have on the individuals, groups, and organizations that use them.

ADOPTING GROUPWARE IN CONTEXT

Groupware systems are intended to support groups, who are usually embedded in an organization. As a result, there are a number of issues that bear on groupware success. In a justly famous set of papers, Grudin (1988, 1994) pointed out a number of problems that groupware systems have (see also Markus & Connolly, 1990). In brief, he pointed out that developers of groupware systems need to be concerned about the following issues (Grudin, 1994, p. 97):

1. *Disparity in work and benefit:* Groupware applications often require additional work from individuals who do not perceive a direct benefit from the use of the application.
2. *Critical mass and Prisoner's dilemma problems:* Groupware may not enlist the critical mass of users required to be useful, or can fail because it is never in any one individual's advantage to use it.
3. *Disruption of social processes:* Groupware can lead to activity that violates social taboos, threatens existing political structures, or otherwise demotivates users crucial to its success.
4. *Exception handling:* Groupware may not accommodate the wide range of exception handling and improvisation that characterizes much group activity.
5. *Unobtrusive accessibility:* Features that support group processes are used relatively infrequently, requiring unobtrusive accessibility and integration with more heavily used features.
6. *Difficulty of evaluation:* The almost insurmountable obstacles to meaningful, generalizable analysis and evaluation of groupware prevent us from learning from experience.
7. *Failure of intuition:* Intuitions in product development environments are especially poor for multiuser applications, resulting in bad management decisions and error-prone design processes.
8. *The adoption process:* Groupware requires more careful implementation (introduction) in the workplace than product developers have confronted.

However, there are reasons for optimism. In a recent survey of the successful adoption of group calendaring in several organizations, Palen and Grudin (in press) observed that organizational conditions in the 1990s were favorable for the adoption of group tools than they were in the 1980s. Furthermore, the tools themselves had improved in reliability, functionality, and usability. There is increased collaboration readiness and collaboration technology readiness. But there are still significant challenges in supporting group work at a distance (Olson & Olson, 2000).

---

[1] Barry Wellman (2001) laments the fact that such tools are not called "netware," a name that is now a trademarked company name.
[2] Groupware and CSCW have a variety of denotations, some of them quite narrow. We choose to use these terms quite broadly, as do many others.

TECHNICAL INFRASTRUCTURE

Groupware requires networks, and network infrastructure is a key enabler as well as a constraint on groupware. The technical possibilities are very different for advanced networks like Abilene (www.internet2.edu/abilene/) when contrasted with the commodity Internet with slow links to the desktop. Heterogeneity in network conditions across both space and time still remains a major technical challenge.

The World Wide Web and its associated tools and standards have had a major impact on the possibilities for groupware (Berners-Lee, 1999; Schatz & Hardin, 1994). Early groupware mostly consisted of stand-alone applications that had to be downloaded and run on each client machine. Increasingly, group tools are being written for the web, requiring only a web browser and perhaps some plug-ins. This makes it much easier for the user, and also helps with matters such as version control. It also allows for better interoperability across hardware and operating systems.

However, security on the Internet is a major challenge for groupware. In some sense, the design of Internet protocols are to blame, because the Internet grew up in a culture of openness and sharing (Abbate, 1999; Longstaff et al., 1997). E-commerce and sensitive application domains like medicine have been a driver for advances in security, but there is still much progress to be made (Camp, 2000; Longstaff et al., 1997).

Personal computing was a great enabler of collaborative applications. However, we are now undergoing a liberation of computing from the desktop. Laptops, personal digital assistants, wearables, and even multifunction cell phones provide a wide range of platforms for the individuals involved in group work. More and more applications are being written to operate across these diverse environments (e.g., Tang et al., 2001). These devices vary in computational power, display size and characteristics, network bandwidth, and connection reliability, providing interesting technical challenges to make them all interoperate smoothly.

Additional flexibility is being provided by the development of infrastructure that lies between the network itself and the applications that run on client workstations, called middleware. This infrastructure makes it easier to link together diverse resources to accomplish collaborative goals. For instance, the emerging Grid technologies allow the marshalling of powerful, scattered computational resources (Foster & Kesselman, 1999). Middleware provides such services as identification, authentication, authorization, directories, and security in uniform ways that facilitate the interoperability of diverse applications (see Internet2, 2001, for pointers to a variety of projects in this area).

COMMUNICATION TOOLS

We now turn to a review of specific kinds of groupware, highlighting their various properties and uses. We have grouped this review under several broad headings. We do not aim to be exhaustive, but rather seek to illustrate the variety of kinds of tools that have emerged to support human collaborative activities over networked systems. We also highlight various research issues pertaining to these tools.

E-mail

E-mail is almost as ubiquitous as the telephone. Because messages can be exchanged across networks and different base machines and software applications, it is nearly the universal service that the telephone is (Anderson, Bikson, Law, & Mitchell, 1995). Because of this widespread use, it is often called the first and perhaps the only successful groupware application (Anderson et al., 1995; O'Hara-Devereaux & Johnson, 1994; Satzinger & Olfman, 1992; Sproull & Kiesler, 1991). More recently, with the use of MIME as the standard for attachments and Postscript and others for representation, it is easier and easier to send spreadsheets, complexly formatted text documents, graphics, etc. Because of the speed and ease of use, people use e-mail to attempt to span distance and time, and to easily disseminate information to broad communities (Garton & Wellman, 1995).

Researchers have shown that this widespread use has had a number of effects on how people behave. It has had large effects on communication in organizations: It changes the social network of who talks to whom (Sproull & Kiesler, 1991), the power of people who formerly had little voice in decisions (Finholt, Sproull, & Kiesler, 1990), and the tone of what is said and how it is interpreted (Sproull & Kiesler, 1991). For example, with e-mail, people who were shy found a voice; they could overcome their reluctance to speak to other people by composing text, not speech to another face. This invisibility, however, also has a more general effect: Without the social cues in the recipient's face being visible to the sender, people will flame, send harsh or extremely emotive (usually negative) messages. Although people adapt their styles over time, it is still a concern for the steady stream of young users new to the medium (Arrow et al., 1996; Hollingshead, McGrath, & O'Connor, 1993).

As with a number of other designed technologies, people use e-mail for things other than the original intent. People use it for managing time, reminding them of things to do, and keeping track of steps in a workflow (Carley & Wendt, 1991; Mackay, 1989; Whittaker & Sidner, 1996). But because e-mail was not designed to support these tasks, it does not do it very well; people struggle with reading signals about whether they have replied or not (and to whom it was cc'd); they manage folders poorly for reminding them to do things, etc.

In addition, because e-mail is so widespread, and it is easy and free to distribute a single message to many people, people experience information overload. Many people get hundreds of e-mail messages each day, many of them mere broadcasts of things for sale or events about to happen, much like classifieds in the newspaper. Efforts to use artificial intelligence techniques to block and/or sort incoming e-mail are of little success, mainly because it is difficult to specify exactly what people do and do not want to receive (Malone, Grant, Lai, Rao, & Rosenblitt, 1987;

Winograd, 1988). For example, it is hard to specify that all junk mail should be deleted; the system requires an enumeration of all the sources of junk mail. Those who are successful in dealing with this overload use primitive manual routines (Whittaker & Sidner, 1996).

Kraut et al. (1998) reported that greater Internet use, which in their sample was mostly e-mail, led to declines in social interactions with family members and an increase in depression and loneliness. Not surprising, these results triggered widespread discussion and debate, both over the substance of the results and the methods used to obtain them. Recently, Kraut et al. (in press) have reported new results that suggest these initial effects may not persist. Interpersonal communication is one of the principal uses of the Internet, and the possible implications of this kind of communication for social life is important to understand. Indeed, Putnam (2000) has wondered whether the Internet can be a source of social cohesiveness. These kinds of questions need to be addressed by additional large-scale studies of the kind conducted by Kraut and his colleagues.

Conferencing Tools—Voice and Video

Real-time meetings are among the more difficult situations to support at a distance. The most familiar technology to support remote meetings is video conferencing, where special facilities are linked by high bandwidth connections, and compatible camera/projection systems are required at all points to be connected. Although there is good new technology to support video and audio over Internet protocol (see VoIP, 2001, for details), it is still time consuming and confusing to set up a connection, and multipoint conferences require expensive multipoint conference units. For point-to-point conferencing, however, the units and connect time are inexpensive. The rise of the use of standards for conferencing, such as H323 (see OpenH323 Project, 2001, and H323 Information Site, 2001, for more details about this standard), is leading to important interoperability, much like that in e-mail and voice telephony.

Although the new systems allow less expensive connectivity, they are not all of the same quality, and some aspects of quality strongly affect the experience. High-quality audio is essential. Video is important for some cues about who is speaking and whose turn it is next (Veinott, Olson, Olson, & Fu, 1999). But, without immediate audio (less than a half second), conversational flow is severely disrupted. People with experience with these technologies will mute the audio channel that accompanies the videoconferencing (which typically has a delay of over a second) and make a second telephone call to connect the audio over more standard high-quality channels.

The other key feature of successful remote meetings is the ability to share the objects they are talking about: The agenda, the to-do list, the latest draft of a proposal, etc. More traditional videoconferencing technologies often offer an object camera, onto which the participants can put a paper agenda, Powerpoint slides, or a manufactured part. For digital objects, there are now a number of products that will allow meeting participants to share the screen or, in some cases, the operation of the actual application. NetMeeting is a current popular application for allowing remote participants to see and manipulate digital material at a remote location. Some companies are using electronic whiteboards, both in a collocated meeting and in remote meetings to mimic the choreography of people using a physical whiteboard. Furthermore, in some "collaboratories" (see later section), scientists can even operate remote physical instruments from a distance and jointly discuss the results.

A number of distributed organizations have adopted remote conferencing to coordinate their work, but not without consequences (Finn, Sellen, & Wilbur, 1997). A controlled comparison of high-quality videoconferencing with shared digital objects against collocated meetings with the good support for digital objects showed the quality of the work was the same. This is good news on the face of it. However, the remote groups are less satisfied with the process and outcome, perhaps because it requires more effort to organize the flow of work and more attention to understand the subtleties of the interactions (Olson, Olson, & Meader, 1995). Meetings with video, however, are much more preferred to those with audio only, even though the quality of the work is indistinguishable from meetings with video. In a related study, however, it was shown video was much more important (a much greater value over audio) for people who did not know each other or were from different cultures—a situation that is common in long-distance work (Veinott et al., 1999).

There is a strong tendency for video conferences to be quite formal and stilted. The easy flow of interaction that one finds in face-to-face or even telephone interactions are readily disrupted by any delays. Poor image quality makes it difficult to pick up expressions and gestures. Eye contact is especially difficult to make, because cameras are usually far from where the remote people are projected on the screen. In short, there are many details of how to configure video conferences to allow for more spontaneous, free-flowing interaction. Research is needed on which details really matter for a remote meeting's success.

Instant Messaging, Chat, and MUDs

Chat systems are like instant e-mail; people type typically short comments or questions to each other, and with the "send" button, the message is instantly shown on the other person's window. There is a trail of the conversation so far in both participants' windows so that people can keep track of the thread easily by scrolling back through recent history. A good chat system can feel like a conversation. Because most chat systems allow a number of participants, one might suspect that the conversation would get chaotic and be hard to follow. But people in chats have no more trouble following than those in a face-to-face meeting of the same number of participants, because the conversation is typically purposeful, the participants add a few more conversational cues, like naming the person or topic to which they are referring, and the recent history allows people to review the conversational threads (McDaniel, Olson, & Magee, 1996). An example of a chat session is shown later in Fig. 29.3.

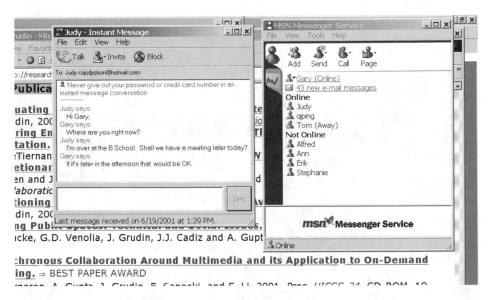

FIGURE 29.1. Example of instant messaging in action.

A more recent outgrowth of chat systems is a set of features collectively called "instant messaging." America On Line's Instant Messenger, ICQ ("I seek you"), Microsoft's Messaging System, and Jabber are all examples of popular current system. Once a participant registers on the system, they can explicitly permit specific others ("buddies") to see if they are on the system or not (or various states like "be right back"), and receive and send messages. Figure 29.1 shows an example of an instant messaging session. Nardi, Whittaker, and Bradner (2000) provide an account of how people in one organization used instant messaging in their work.

A different constellation of features surrounding the basic chat appears in MUDs and MOOs. The word MUD comes from the initial incarnation of this type if interaction in the game world, Multi-user Dungeons and Dragons. MOOs are object-oriented MUDs. The chats in MUDs are enhanced by various descriptions of places and things, some of which can be programmed to react to various actions of another participant. For example, instead of merely saying something, the participant can state, "Judy smiles." Judy could then leave behind a magic book that when picked up could open and state its secret message. Some people have picked up the room metaphor and object actions to be used in a work setting, with automatic announcements when people enter a meeting, agenda items being ticked off, etc. One MUD, Waterfall Glen, is one such work-related example in use at the Argonne National Lab (Churchill & Bly, 1999). Babble is a MUDlike system that can be useful for discussions and awareness of others' activities, in use by some groups at IBM (Erickson et al., 1999).

Instant messaging is growing very rapidly, and as of yet there is not much research on its spread, its use, and its effects. Much of the growth is driven by informal socializing, and in light of some early reports about the isolating tendencies of some Internet usage (Kraut et al., 1998), it would be especially important to look at the role that instant messaging might be playing in establishing and sustaining social relations.

COORDINATION SUPPORT

Meeting Support

In addition to video and audio conferencing to support the conversations of meetings, technologists have also attempted to use various groupware tools to help in supporting the process and the objects used in meetings.

Some meeting-support software imposes structure on the process of the meeting, embodying various brainstorming and voting procedures. Group Decision Support Systems (GDSSs) arose from a number of business schools, focusing on large meetings of stakeholders intent on going through a set series of decisions, such as prioritizing projects for future funding (Nunamaker, Dennis, Valacich, Vogel, & George, 1991). With the help of a facilitator and some technical support, the group was led through a series of stages: brainstorming without evaluating, evaluating alternatives from a variety of positions, prioritizing alternatives, etc. These meetings were held in specialized rooms in which individual computers were embedded in the tables, networked to central services, and summary displays shown center stage. Figure 29.2 shows an example of such a computer-supported meeting room. A typical scenario involved individuals silently entering ideas into a central repository, and after a certain amount of time, they were shown ideas one at a time from others and asked to respond with a new idea triggered by that one. Later, these same ideas were presented to the individuals who were then asked to rank or rate them according to some fixed criterion, like cost. Aggregates of individuals' opinions were computed, discussed further, and presented for vote. The system applied computational power (for voting and rating mechanisms), and networking control (for parallel input) to support typically weak aspects of meetings. These systems were intended to gather more ideas from participants, because one did not have to wait for another to stop

FIGURE 29.2. A computer-supported meeting room.

speaking to get a turn. Anonymous voting and rating were intended to ensure equal participation, not dominated by those in power.

Evaluations of these GDSSs have been reviewed producing some generalizations about their value (Hollingshead et al., 1993; Kraemer & Pinsoneault, 1990; McLeod, 1992). The systems indeed fulfill their intent in producing more ideas in brainstorming and having more evaluative comments because of anonymity. Decisions are rated as higher in quality, but the meetings take longer and the participants are less satisfied than those in traditional meetings.

A second class of technologies to support real-time meetings are less structured, more similar to individual workstation support. In these systems, groups are allowed access to a single document or drawing, and can enter and edit into them simultaneously at will. Different systems enforce different locking mechanisms (e.g., paragraph or selection locking) so that one person does not enter while another deletes the same thing (Ellis, Gibbs, & Rein, 1991). Some also allow parallel individual work, in which participants view and edit different parts of the same document, but can also view and discuss the same part as well. This kind of unstructured shared editor has been shown to be very effective for certain kinds of free-flowing meetings, like design or requirements meetings (Olson, Olson, Storrøsten, & Carter, 1993). The rated quality of the meeting products (e.g., a requirements document or plan) was higher when using these technologies than with traditional whiteboard or paper-and-pencil support, but like working in GDSSs, people were slightly less satisfied. The lower satisfaction here and with GDSSs may reflect the newness of the technologies; people may not have yet learned how to persuade, negotiate, or influence

each other in comfortable ways—to harness the powers inherent in the new technologies.

These new technologies did indeed change the way in which people worked. They talked less and wrote more, building on each other's ideas instead of generating far-reaching others. The tool seemed to focus the groups on the core ideas, and keep them from going off on tangents. Many participants reported really liking doing work in the meetings rather than spending time only talking about the work.

A third class of meeting room support appears in electronic whiteboards. For example, the LiveBoard (Elrod et al., 1992), SoftBoard, and SmartBoard are approximately $4' \times 6'$ rear-projection surfaces that allow pen input, much the way a whiteboard or flipchart does. People at Xerox PARC and Boeing have evaluated the use of these boards in meetings in extended case studies. In both cases, the board was highly valued because of its computational power and the fact that all could see the changes as they were made. At both sites, successful use required a facilitator who was familiar with the applications running to support the meeting. At Xerox, suggestions made in the meeting about additional functionality were built into the system so that it eventually was finely tuned support for their particular needs (Moran et al., 1996). For example, they did a lot of list making of freehand text items. Eventually, the board software recognized the nature of a list and an outline, with simple gestures changing things sensibly. For example, if a freehand text item was moved higher in a list, the other items adjusted their positions to make room for it. The endproduct was not only a set of useful meeting tools, but also a toolkit to allow people to build new meeting widgets to support their particular tasks.

Meetings are important, although often despised, organizational activities. Laboratory research of the kind just reviewed has shown quite clearly that well-designed tools can improve both work outcomes and participant satisfaction. However, meetings in organizations seldom use such tools. Inexpensive mobile computing and projection equipment combined with many commercial products mean that such tools are within reach of most organizations. It is a puzzle for researchers to figure out why the adoption of effective tools has been so slow.

Workflow

Workflow systems lend technology support to coordinated asynchronous (usually sequential) steps of activities among team members working on a particular task. For example, a workflow system might route a travel reimbursement voucher from the traveler to the approving party to the accounts payable to the bank. The electronic form would be edited and sent to the various parties, their individual to-do lists updated as they received and/or completed the tasks, and permissions and approval granted automatically as appropriate (e.g., allowing small charges to an account if the charges had been budgeted previously or, simply if there was enough money in the account). Not only is the transaction flow supported, but also often records kept about who did what and when they did it. It is this later feature that has potentially large consequences for the people involved and will be discussed later.

These workflow systems were often the result of work re-engineering efforts, focusing on making the task take less time and to eliminate the work that could be automated. Not only do workflow systems therefore have a bad reputation in that they often are part of workforce reduction plans, but also for those left, their work is able to be monitored much more closely. The systems are often very rigid, requiring, for example, all of a form to be filled in before it can be handed off to the next in the chain. They often require a great deal of rework because of this inflexibility. It is because of the inflexibility and the potential monitoring that the systems fall into disuse (e.g., Abbott & Sarin, 1994).

The fact that workflow can be monitored is a major source of user resistance. In Europe, such monitoring is illegal, negotiated out of workflow by powerful organized workers (Prinz & Kolvenbach, 1996). In the United States, it is not illegal, but many employees complain about its inappropriate use. For example, in one software engineering team where workflow had just been introduced to track bug reports and fixes, people in the chain were sloppy about noting who they had handed a piece of work off to. When it was discovered that the manager had been monitoring the timing of the handoffs to assign praise or blame, the team members were justifiably upset (Olson & Teasley, 1996). In general, managerial monitoring is a feature that is not well received by people being monitored (Markus, 1983). If such monitoring is mandated, workers' behavior will conform to the specifics of what is being monitored (e.g., time to pass an item off to the next in the chain) rather than perhaps to what the real goal is (e.g., quality as well as timely completion of the whole process).

Group Calendars

A number of organizations have now adopted online calendars, mainly to view people's schedules to arrange meetings. The calendars also allow a form of awareness, allowing people to see if a person who is not present is expected back soon. Individuals benefit only insofar as they offload scheduling meetings to others, like to an administrative assistant, who can write as well as read the calendar. In some systems, the individual can schedule private time, blocking the time but not revealing to others their whereabouts. By this description, online calendaring is a classic case of what Grudin (1988) warns against, a misalignment of costs and benefits: The individual puts in the effort to record his/her appointments so that another, in this case a manager or coworker, can benefit from ease of scheduling. However, since the early introduction of electronic calendaring systems, many organizations have found successful adoption (Grudin & Palen, 1995; Mosier & Tammaro, 1997; Palen & Grudin, in press). Apparently such success requires a culture of sharing and accessibility, something that exists in some organizations and not others (Ehrlich, 1987; Lange, 1992).

Awareness

In normal work, there are numerous occasions where people find out casually whether others are in and in some cases what they are doing. A simple walk down the hall to a printer offers numerous glances into people's offices, noting where their coat is, whether others are in talking, whether there is intense work at a computer, etc. This kind of awareness is unavailable to workers who are remote. Some researchers have offered various technology solutions: Some have allowed one to visually walk down the hall at the remote location, taking a 5 s-glance into each passing office (Bellotti & Dourish, 1997; Fish, Kraut, Root, & Rice, 1993). Another similar system, called Portholes, provides periodic snapshots instead of full-motion video (Dourish & Bly, 1992). Because of privacy implications, these systems have had mixed success. The places in which this succeeds are those in which the individuals seem to have a reciprocal need to be aware of each other's presence, and a sense of cooperation and coordination. A contrasting case is the instant messaging system, in which the user has control as to what state they wish to advertise to their partners about their availability. The video systems are much more lightweight to the user but more intrusive; the IM ones give the user more control, but require intention in action.

Another approach to signaling what one is doing occurs at the more micro level. And again, one captures what is easy to capture. When people are closely aligned in their work, there are applications that allow each to see exactly where in the shared document the other is working and what they are doing (Gutwin & Greenberg, 1999). If one is working nearby the other, this signals perhaps a need to converse about the directions each is taking. Empirical evaluations have shown that such workspace awareness can facilitate task performance (Gutwin & Greenberg, 1999).

As mentioned previously, instant messaging systems provide an awareness capability. Most systems display a list of buddies and whether they are currently on line or not (see Fig. 29.1). Nardi and her colleagues (2000) found that people liked this aspect of instant messaging. Because wireless has allowed constant connectivity of mobile devices like personal digital assistants, this use of tracking others is likely to grow. But, again, there are issues of monitoring for useful or insidious purposes, and the issues of trust and privacy loom large (see Godefroid, Herbsleb, Jagadeesan, & Li, 2000).

Studies of attempts to conduct difficult intellectual work within geographically distributed organizations show that one of the larger costs of geographical distribution is the lack of awareness of what others are doing or whether they are even around (Herbsleb, Mockus, Finholt, & Grinter, 2000). Thus, useful and usable awareness tools that mesh well with trust and privacy concerns could be of enormous organizational importance. This is a rich research area for CSCW.

INFORMATION REPOSITORIES

Repositories of Shared Knowledge

In addition to sharing information generally on the web, in both public and intranet settings, there are applications, like Lotus Notes, that are explicitly built for knowledge-sharing. The goal in most systems is to capture knowledge that can be reused by others, like instruction manuals, office procedures, training, and boilerplate or templates of commonly constructed genres, like proposals or bids. Experience shows, however, that these systems are not easy wins. Again, similar to the case of the online calendaring systems described above the person entering information into the system is not necessarily the one benefiting from it. In a large consulting firm, where consultants were quite competitive in their bid for advancement, there was indeed negative incentive for giving away one's best secrets and insights (Orlikowski & Gash, 1994). An implementation of Lotus Notes failed here. In other organizations where such a repository has been successful, the incentives were better aligned.

Sometimes subtle design features are at work in the incentive structure. In another adoption of Lotus Notes, in this case to track open issues in software engineering, the engineers slowly lost interest in the system because they assumed that their manager was not paying attention to their contributions and use of the system. The system design, unfortunately, made the manager's actual use invisible to the team. Had they known that he was reading daily what they wrote (although he never wrote anything himself), they would have continued to use the system (Olson & Teasley, 1996). A simple design change that would make the manager's reading activity visible to the team would have significantly altered their adoption. In another similar case, sales people recorded their contacts with major clients in a repository. The incentive to share in this way was aligned with their goal to not embarrass themselves in overcontacting the client. Better yet would be a shared commission in cooperative sales.

The web of course provides marvelous infrastructure for the creation and sharing of information repositories. Lotus Notes itself is now widely used in its web version, Domino. Environments for sharing like Worktools (worktools.si.umich.edu) are built on top of Lotus Notes. Document management tools like WebEx make it easy to share in a web environment. Systematic research on the use of such improved tools is needed.

Capture and Replay

Tools that support collaborative activity can create traces of that activity that later can be replayed and reflected on. The Upper Atmospheric Research Collaboratory (UARC) explored the replay of earlier scientific campaign sessions (Olson et al., 2001), so that scientists could reflect on their reactions to real-time observations of earlier phenomena. Using a VCR metaphor, they could pause where needed, and fast forward past uninteresting parts. This reflective activity could also engage new players who had not been part of the original session. Abowd (1999) has explored such capture phenomena in an educational experiment called Classroom 2000. Initial experiments focused on reusing educational sessions during the term in college courses. We do not yet fully understand the impact of such promising ideas.

SOCIALITY

Social Filtering

We often find the information we want by contacting others. Social networks embody rich repositories of useful information on a variety of topics. A number of investigators have looked at whether the process of finding information through others can be automated. The kinds of recommender systems that we find on Web sites like Amazon.com are examples of the result of such research. The basic principle of such systems is that an individual will tend to like or prefer the kinds of things (e.g., movies, books) that someone who is similar to him/her likes. They find similar people by matching their previous choices. Such systems use a variety of algorithms to match preferences with those of others and then recommend new items. Resnick and Varian (1997) edited a special issue of the *Communication of the ACM* on recommender systems that included a representative set of examples. Herlocker, Konstan, and Riedl (2001) used empirical methods to explicate the factors that lead users to accept the advice of recommender systems. In short, providing access to explanations for why items were recommended seems to be the key. Recommender systems are emerging as a key element of e-commerce (Schafer, Konstan, & Riedl, 2001). Accepting the output of recommender systems is an example of how people come to trust technical systems. This is a complex topic and relates to issues like security that we briefly described earlier.

Trust of People Via the Technology

It has been said that trust needs touch, and indeed in survey studies, coworkers report that they trust those who are collocated more than those who are remote (Rocco, Finholt, Hofer, & Herbsleb, 2000). Interestingly, those who spend the most time on the phone chatting about non–work-related things with their remote coworkers show higher trust than those they communicate with using only fax and e-mail. But laboratory studies show that telephone interaction is not as good as face-to-face. People using just the telephone behave in more self-serving, less trusting ways than they do when they meet face-to-face (Drolet & Morris, 2000).

What can be done to counteract the mistrust that comes from the impoverished media? Rocco (1998) had people meet and do a team-building exercise the day before they engaged in the social dilemma game, with only e-mail to communicate with. These people, happily, showed as much cooperation and trust as those who discussed things face-to-face during the game. This is important. It suggests that, if remote teams can do some face-to-face teambuilding before launching on their project, they will act in a trusting/trustworthy manner. Zheng, Bos, Olson, Gergle, and Olson (2001) found that using chat for socializing and sharing pictures of each other also led to trustful relations. Merely sharing a resume did not.

Since it is not always possible to have everyone on a project meet face-to-face before they launch into the work, what else will work? Researchers have tried some things, but with mixed success. When the text is translated into voice, it has no effect on trust, and when it is translated into voice and presented in a moving human-like face, it is even worse than text-chat (Jensen, Farnham, Drucker, & Kollock, 2000; Kiesler, Sproull, & Waters, 1996). However, Bos, Gergle, Olson, and Olson (2001) found that interactions over video and audio led to trust, albeit of a seemingly more fragile form.

If we can find a way to establish trust without expensive travel, we are likely to see important productivity gains. Clearly, the story is not over. However, we must not be too optimistic. In other tasks, video does not produce "being there." There is an overhead to the conversation through video; it requires more effort than working face-to-face (Olson et al., 1995). Today's video over the Internet is both delayed and choppy, producing cues that people often associate with lying. One does not trust someone who appears to be lying. Trust is a delicate emotion; today's video might not just do it in a robust enough fashion.

INTEGRATED SYSTEMS

Media Spaces

As an extension of video conferencing and awareness systems, some people have experimented with open, continuous audio and video connections between remote locations. In a number of cases, these experiments have been called Media Spaces. For example, at Xerox, two labs were linked with an open video link between two commons areas (Olson & Bly, 1991), the two locations being Palo Alto, CA, and Portland, OR. Evaluation of these experiments showed that maintaining organizational cohesiveness at a distance was much more difficult than when members are collocated (Finn et al., 1997). However, some connectedness was maintained. Where many of these early systems were plagued with technical difficulties, human factors limitations, or very large communication costs, in today's situation, it might actually be possible to overcome these difficulties, making media a possibility for connecting global organizations. A new round of experimental deployments with new tools is needed.

Collaborative Virtual Environments

Collaborative virtual environments are three-dimensional embodiments of MUDs. The space in which people interact is an analog of physical space, with dimensions, directions, rooms, and objects of various kinds. People are represented as avatars, simplified, geometric, digital representations of people, who move about in the three-dimensional space (Singhal, 1999). Similar to MUDs, the users in a meeting situation might interact over some object that is digitally represented, like a mock-up of a real thing (an automobile engine, an airplane hinge, a piece of industrial equipment) or with visualizations of abstract data (e.g., a three-dimensional visualization of atmospheric data). In these spaces, one can have a sense as to where others are and what they are doing, similar to the simplified awareness systems described previously. In use, it is difficult to establish mutual awareness or orientation in such spaces (Hindmarsh, Fraser, Heath, Benford, & Greenhalgh, 1998; Park, Kapoor, & Leigh, 2000). There have even been some attempts to merge collaborative virtual environments with real ones, although with limited success so far (Benford, Greenhalgh, Reynard, Brown, & Koleva, 1998).

What people seem to want is more like the Holodek in Star Trek. These environments are complicated technically, and perhaps even more complicated socially. In real life, we have developed interesting schemes that trigger behavior and interpretation of others' behavior as a function of real distance, a field called "Proxemics" (Hall, 1982). Only when these subtle behaviors are incorporated into the virtual environment will we have a chance of simulating appropriate interhuman behavior in the virtual three-dimensional world.

Collaboratories

A collaboratory is a laboratory without walls (Finholt & Olson, 1997). From a National Research Council report, a collaboratory is supposed to allow "... the nation's researchers [to] perform their research without regard to geographical location—interacting with colleagues, accessing instrumentation, sharing data and computational resources [and] accessing information in digital libraries" (National Research Council, 1993, p. 7). Starting in the early 1990s, these capabilities have been configured into support packages for a number of specific sciences (see review in Finholt, in press). Figure 29.3 shows a screen dump from the Upper Atmospheric Research Collaboratory (Olson et al., 2001), in which space scientists have access to geographically remote instruments, as well as each other through a simple chat facility.

A number of companies have also experimented with similar concepts, calling them virtual collocation. The goal there is to support geographically dispersed teams as they carry out product design, software engineering, financial reporting, and

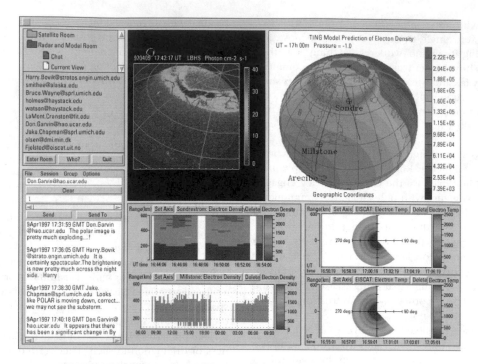

FIGURE 29.3. Screen dump from a collaboratory in upper atmospheric physics.

almost any business function. In these cases, suites of off-the-shelf groupware tools have been particularly important and have been used to support round-the-clock software development among overlapping teams of engineers in time zones around the world. (Carmel, 1999). There have been a number of such efforts, and it is still unclear as to their success or what features make their success more likely (Olson & Olson, 2000).

CONCLUSIONS

Groupware functionality is steadily becoming more routine in commercial applications. Similarly, suites of groupware functions are being written into operating systems. The prospect is that many of the functions we have described in this article will be ordinary elements of infrastructure in future-networked computing systems. There are still lots of issues for researchers to resolve.

Prognosticators looking at the emergence of groupware and the convergence of computing and communication media have forecast that distance will diminish as a factor in human interactions (e.g., Cairncross, 1997). However, to paraphrase Mark Twain, the reports of distance's death are greatly exaggerated. Even with all our emerging information and communications technologies, distance and its associated attributes of culture, time zones, geography, and language will continue to affect how humans interact with each other. Emerging distance technologies will allow greater flexibility for those whose work must be done at a distance, but we believe (see details in Olson & Olson, 2000) that distance will continue to be a factor in understanding these work relationships.

ACKNOWLEDGMENTS

Preparation of this chapter was facilitated by several grants from the National Science Foundation (Grants IIS-9320543, IIS-9977923, ATM-9873025, IIS-0085951, and cooperative agreement IRI-9216848). We are also grateful to several anonymous reviewers for helpful comments on an earlier draft.

References

Abbate, J. (1999). *Inventing the internet*. Cambridge, MA: MIT Press.

Abbott, K. R., & Sarin, S. K. (1994). Experiences with workflow management: Issues for the next generation. *Proceedings of the Conference on Computer Supported Cooperative Work* (pp. 113–120). New York, NY: The Association for Computing Machinery.

Abowd, G. D. (1999). Classroom 2000: An experiment with the instrumentation of a living educational environment. *IBM Systems Journal, 38,* 508–530.

Anderson, R. H., Bikson, T. K., Law, S. A., & Mitchell, B. M. (1995). *Universal access to e-mail: Feasibility and societal implications.* Santa Monica, CA: Rand.

Arrow, H., Berdahl, J. L., Bouas, K. S., Craig, K. M., Cummings, A., Lebei, L., McGrath, J. E., O'Connor, K. M., Rhoades, J. A., & Schlosser, A. (1996). Time, technology, and groups: An integration. *Computer Supported Cooperative Work, 4,* 253-261.

Bellotti, V., & Dourish, P. (1997). Rant and RAVE: Experimental and experiential accounts of a media space. In K. E. Finn, A. J. Sellen, & S. B. Wilbur (Eds.), *Video-mediated communication* (pp. 245-272). Mahwah, NJ: Lawrence Erlbaum Associates.

Benford, S., Greenhalgh, C., Reynard, G., Brown, C., & Koleva, B. (1998). Understanding and constructing shared spaces with mixed-reality boundaries. *ACM Transactions on Computer-Human Interaction, 5,* 185-223.

Berners-Lee, T. (1999). *Weaving the web.* New York: HarperSan-Francisco.

Bos, N., Gergle, D., Olson, J. S., & Olson, G. M. (2001). Being there vs. seeing there: Trust via video. *Short Paper at the Conference on Human Factors in Computing Systems: CHI-2001.* New York, NY: The Association for Computing Machinery.

Bush, V. (1945). As we may think. *Atlantic Monthly, 176*(1), 101-108.

Cairncross, F. (1997). *The death of distance: How the communications revolution will change our lives.* Boston, MA: Harvard Business School Press.

Camp, L. J. (2000). *Trust and risk in Internet commerce.* Cambridge, MA: MIT Press.

Carley, K., & Wendt, K. (1991). Electronic mail and scientific communication: A study of the Soar extended research group. *Knowledge: Creation, Diffusion, Utilization, 12,* 406-440.

Carmel, E. (1999). *Global software teams.* Upper Saddle River, NJ: Prentice-Hall.

Churchill, E., & Bly, S. (1999). It's all in the words: Supporting work activities with lightweight tools. *Proceedings of GROUP '99* (pp. 40-49). New York, NY: The Association for Computing Machinery.

Dourish, P., & Bly, S. (1992). Portholes: Supporting awareness in a distributed work group. *Proceedings of CHI 92* (pp. 541-547). New York, NY: The Association for Computing Machinery.

Drolet, A. L., & Morris, M. W. (2000). Rapport in conflict resolution: Accounting for how nonverbal exchange fosters coordination on mutually beneficial settlements to mixed motive conflicts. *Journal of Experimental Social Psychology, 36,* 1, 26-50.

Ehrlich, S. F. (1987). Strategies for encouraging successful adoption of office communication systems. *ACM Transactions on Office Information Systems, 5,* 340-357.

Ellis, C., & Nutt, G. (1980). Office information systems and computer science. *Computing Surveys, 12*(1), 27-60.

Ellis, C. A., Gibbs, S. J., & Rein, G. L. (1991). Groupware: Some issues and experiences. *CACM, 34*(1), 38-58.

Elrod, S., Bruce, R., Gold, R., Goldberg, D., Halasz, F., Janssen, W., Lee, D., McCall, K., Pedersen, E., Pier, K., Tang, J., & Welch, B. (1992). LiveBoard: A large interactive display supporting group meetings, presentations, and remote collaboration. *Proceedings of CHI '92* (pp. 599-607). New York, NY: The Association for Computing Machinery.

Engelbart, D., & English, W. (1968). A research center for augmenting human intellect. *Proceedings of FJCC, 33,* 395-410.

Erickson, T., Smith, D. N., Kellogg, W. A., Laff, M. R., Richards, J. T., & Bradner, E. (1999). Socially translucent systems: Social proxies, persistent conversation, and the design of 'Babble.' *Proceedings of CHI '99* (pp. 72-79).

Finholt, T. A. (in press). Collaboratories. In B. Cronin (Ed.), *Annual Review of Information Science and Technology.*

Finholt, T. A., & Olson, G. M. (1997). From laboratories to collaboratories: A new organizational form for scientific collaboration. *Psychological Science, 8,* 28-36.

Finholt, T., Sproull, L., & Kiesler, S. (1990). Communication and performance in ad hoc task groups. In J. Galegher, R. Kraut, & C. Egido (Eds.), *Intellectual Teamwork: Social and Technological Foundations of Cooperative Work* (pp. 291-325). Hillsdale, NJ: Lawrence Erlbaum Associates.

Finn, K., Sellen, A., & Wilbur, S. (Eds.). (1997). *Video-mediated communication.* Hillsdale, NJ: Lawrence Erlbaum Associates.

Fish, R. S., Kraut, R. E., Root, R. W., & Rice, R. E. (1993). Video as a technology for informal communication. *Communications of the ACM, 36*(1), 48-61.

Foster, I., & Kesselman, C. (1999). *The grid: Blueprint for a new computing infrastructure.* San Francisco, CA: Morgan Kaufmann.

Garton, L., & Wellman, B. (1995). Social impacts of electronic mail in organizations: A review of the research literature. In B. R. Burleson (Ed.), *Communication Yearbook* (pp. 434-453). Thousand Oaks, CA: Sage.

Godefroid, P., Herbsleb, J. D., Jagadeesan, L. J., & Li, D. (2000). Ensuring privacy in presence awareness systems: An automated verification approach. *Proceedings of CSCW 2000* (pp. 59-68). New York: ACM.

Greif, I. (Ed.). (1988). *Computer-supported cooperative work: A book of readings.* San Mateo, CA: Morgan Kaufmann.

Grudin, J. (1988). Why CSCW applications fail: Problems in the design and evaluation of organizational interfaces. *Proceedings of the Conference on Computer Supported Cooperative Work* (pp. 85-93). New York, NY: The Association for Computing Machinery.

Grudin, J. (1994). Groupware and social dynamics: Eight challenges for developers. *Communications of the ACM, 37*(1), 92-105.

Grudin, J., & Palen, L. (1995). Why groupware succeeds: Discretion or mandate? *Proceedings of the European Computer Supported Cooperative Work* (pp. 263-278). Amsterdam, The Netherlands: Kluwer Publishing.

Gutwin, C., & Greenberg, S. (1999). The effects of workspace awareness support on the usability of real-time distributed groupware. *ACM Transactions on Computer-Human Interaction, 6,* 243-281.

H323 Information Site. (2001). Web site last accessed July 28, 2001 [online]. Available: http://www.packetizer.com/iptel/h323/

Hall, E. T. (1982). *The hidden dimension.* New York: Anchor Doubleday Books.

Herbsleb, J. D., Mockus, A., Finholt, T. A., & Grinter, R. E. (2000). Distance, dependencies, and delay in a global collaboration. *Proceedings of CSCW 2000* (pp. 319-328). New York: ACM.

Herlocker, J. L., Konstan, J. A., & Riedl, J. (2000). Explaining collaborative filtering recommendations. *Proceedings of CSCW 2000* (pp. 241-250). New York: ACM.

Hindmarsh, J., Fraser, M., Heath, C., Benford, S., & Greenhalgh, C. (1998). Fragmented interaction: Establishing mutual orientation in virtual environments. *Proceedings of Conference on Computer-Supported Cooperative Work* (pp. 217-226). New York, NY: The Association for Computing Machinery.

Hollingshead, A. B., McGrath, J. E., & O'Connor, K. M. (1993). Group performance and communication technology: A longitudinal study of computer-mediated versus face-to-face work. *Small Group Research, 24,* 307-333.

Internet2 Middleware Project. (2001). Site last accessed on July 29, 2001 [on-line]. Available: middleware.internet2.edu

Jensen, C., Farnham, S. D., Drucker, S. M., & Kollock, P. (2000). The effect of communication modality on cooperation in on-line environments. *Proceedings of CHI '2000* (pp. 470-477). New York: ACM Press.

Kiesler, S., Sproull, L., & Waters, K. (1996). Prisoner's dilemma experiment on cooperation with people and human-like computers. *Journal of Personality and Social Psychology, 70*(1), 47-65.

Kraemer, K. L., & Pinsonneault, A. (1990). Technology and groups: Assessments of empirical research. In J. Galegher, R. Kraut, & C. Egido

(Eds.), *Intellectual teamwork: Social and technological foundations of cooperative work* (pp. 373–405). Hillsdale, NJ: Lawrence Erlbaum Associates.

Kraut, R., Kiesler, S., Boneva, B., Cummings, J., Helgeson, V., & Crawford, A. (in press). Internet paradox revisited. *Journal of Social Issues*.

Kraut, R., Patterson, M., Lundmark, V., Kiesler, S., Mukopadhyay, T., & Scherlis, W. (1998). Internet paradox: A social technology that reduces social involvement and psychological well-being. *American Psychologist, 53*, 1017–1031.

Lange, B. M. (1992). Electronic group calendaring: Experiences and expectations. In D. Coleman (Ed.), *Groupware* (pp. 428–432). San Mateo, CA: Morgan Kaufmann.

Longstaff, T. A., Ellis, J. T., Hernan, S. V., Lipson, H. F., McMillan, R. D., Pesanti, L. H., & Simmel, D. (1997). Security on the Internet. In *The Froehlich/Kent Encyclopedia of Telecommunications* (Vol. 15, pp. 231–255). New York: Marcel Dekker.

Mackay, W. E. (1989). Diversity in the use of electronic mail: A preliminary inquiry. *ACM Transactions on Office Information Systems, 6*, 380–397.

Malone, T. W., Grant, K. R., Lai, K. Y., Rao, R., & Rosenblitt, D. A. (1989). The information lens: An intelligent system for information sharing and coordination. In M. H. Olson (Ed.), *Technological support for work group collaboration* (pp. 65–88). Hillsdale, NJ: Lawrence Erlbaum Associates.

Markus, M. L. (1983). *Systems in organization: Bugs and features.* San Jose, CA: Pitman.

Markus, M. L., & Connolly, T. (1990). Why CSCW applications fail: Problems in the adoption of interdependent work tools. *Proceedings of the Conference on Computer Supported Cooperative Work* (pp. 371–380). New York, NY: The Association for Computing Machinery.

McDaniel, S. E., Olson, G. M., & Magee, J. S. (1996). Identifying and analyzing multiple threads in computer-mediated and face-to-face conversations. *Proceeding of the ACM Conference on Computer Supported Cooperative Work* (pp. 39–47). New York, NY: The Association for Computing Machinery.

McLeod, P. L. (1992). An assessment of the experimental literature on electronic support of group work: Results of a meta-analysis. *Human-Computer Interaction, 7*, 257–280.

Moran, T. P., Chiu, P., Harrison, S., Kurtenbach, G., Minneman, S., & van Melle, W. (1996). Evolutionary engagement in an ongoing collaborative work process: A case study. *Proceeding of the ACM Conference on Computer Supported Cooperative Work* (pp. 150–159). New York, NY: The Association for Computing Machinery.

Mosier, J. N., & Tammaro, S. G. (1997). When are group scheduling tools useful? *Computer Supported Cooperative Work, 6*, 53–70.

Nardi, B. A., Whittaker, S., & Bradner, E. (2000). Interaction and outeraction: Instant messaging in action. *Proceedings of the ACM Conference on Computer Supported Cooperative Work* (pp. 79–88). New York, NY: The Association for Computing Machinery.

National Research Council. (1993). *National collaboratories: Applying information technology for scientific research.* Washington, DC: National Academy Press.

Nunamaker, J. F., Dennis, A. R., Valacich, J. S., Vogel, D. R., & George, J. F. (1991). Electronic meeting systems to support group work. *Communications of the ACM, 34*(7), 40–61.

O'Hara-Devereaux, M., & Johansen, R. (1994). *Global work: Bridging distance, culture & time.* San Francisco, CA: Jossey-Bass.

Olson, G. M., Atkins, D., Clauer, R., Weymouth, T., Prakash, A., Finholt, T., Jahanian, F., & Rasmussen, C. (2001). Technology to support distributed team science: The first phase of the Upper Atmospheric Research Collaboratory (UARC). In G. M. Olson, T. Malone, &

J. Smith (Eds.), *Coordination theory and collaboration technology* (pp. 761–783). Hillsdale, NJ: Lawrence Erlbaum Associates.

Olson, G. M., & Olson, J. S. (2000). Distance matters. *Human-Computer Interaction, 15*, 139–179.

Olson, J. S., Olson, G. M., & Meader, D. K. (1995). What mix of video and audio is useful for remote real-time work? *Proceedings of CHI '95* (pp. 362–368). New York, NY: The Association for Computing Machinery.

Olson, J. S., Olson, G. M., Storrøsten, M., & Carter, M. (1993). Group work close up: A comparison of the group design process with and without a simple group editor. *ACM Transactions on Information Systems, 11*, 321–348.

Olson, J. S., & Teasley, S. (1996). Groupware in the wild: Lessons learned from a year of virtual collocation. *Proceeding of the ACM Conference on Computer Supported Cooperative Work* (pp. 419–427). New York, NY: The Association for Computing Machinery.

Olson, M. H., & Bly, S. A. (1991). The Portland experience: A report on a distributed research group. *International Journal of Man-Machine Studies, 34*, 211–228.

OpenH323 Project (2001). Web site last accessed on July 28, 2001 [online]. Available: http://www.openh323.org/

Orlikowski, W. J., & Gash, D. C. (1994). Technological frames: Making sense of information technology in organizations. *ACM Transactions on Information Systems, 12*, 174–207.

Palen, L., & Grudin, J. (in press). Discretionary adoption of group support software. In B. E. Munkvold (Ed.), *Organizational implementation of collaboration technology.*

Park, K. S., Kapoor, A., & Leigh, J. (2000). Lessons learned from employing multiple perspective in a collaborative virtual environment for visualizing scientific data. *Proceedings of ACM CVE '2000 Conference on Collaborative Virtual Environments* (pp. 73–82). New York, NY: The Association for Computing Machinery.

Prinz, W., & Kolvenbach, S. (1996). Support for workflows in a ministerial environment. *Proceedings of the Conference on Computer Supported Cooperative Work* (pp. 199–208). New York, NY: The Association for Computing Machinery.

Putnam, R. D. (2000). *Bowling alone: The collapse and revival of American community.* New York: Simon & Schuster.

Resnick, P., & Varian, H. R. (Eds.). (1997). Special section: Recommender systems. *Communications of the ACM, 40*(3), 56–89.

Rocco, E. (1998). Trust breaks down in electronic contexts but can be repaired by some initial face-to-face contact. *Proceedings of CHI '98* (pp. 496–502). New York: ACM Press.

Rocco, E., Finholt, T., Hofer, E. C., & Herbsleb, J. (2000). Designing as if trust mattered. CREW Technical Report, University of Michigan, Ann Arbor, MI.

Satzinger, J., & Olfman, L. (1992). A research program to assess user perceptions of group work support. *Proceeding of CHI '92* (pp. 99–106). New York, NY: The Association for Computing Machinery.

Schafer, J. B., Konstan, J., & Riedl, J. (2001). Electronic commerce recommender applications. *Journal of Data Mining and Knowledge Discovery, 5*(1/2), 115–152.

Schatz, B. R., & Hardin, J. B. (1994). NCSA mosaic and the World Wide Web: Global hypermedia protocols for the internet. *Science, 265*, 895–901.

Singhal, S. (1999). *Networked virtual environments: Design and implementation.* New York: Addison-Wesley.

Sproull, L., & Kiesler, S. (1991). *Connections: New ways of working in the networked organization.* Cambridge, MA: MIT Press.

Tang, J. C., Yankelovich, N., Begole, J., van Kleek, M., Li, F., & Bhalodia, J. (2001). ConNexus to Awarenex: Extending awareness to mobile users. *Proceedings of CHI 2001* (pp. 221–228). New York: ACM.

Veinott, E., Olson, J. S., Olson, G. M., & Fu, X. (1999). Video helps remote work: Speakers who need to negotiate common ground benefit from seeing each other. *Proceedings of the Conference on Computer-Human Interaction, CHI '99* (pp. 302–309). New York, NY: ACM.

VoIP. (2001). Hot links about voice over IP. Site last accessed on July 28, 2001 [on-line]. Available: http://www.protocols.com/voip.htm

Wellman, B. (2001). Design considerations for social networkware: Little boxes, glocalization, and networked individualism. Unpublished manuscript.

Whittaker, S., & Sidner, C. (1996). Email overload: Exploring personal information management of email. *Proceeding of CHI '96* (pp. 276–283). New York, NY: ACM.

Winograd, T. (1988). A language/action perspective on the design of cooperative work. *Human Computer Interaction, 3,* 3–30.

Zheng, J., Bos, N., Olson, J. S., Gergle, D., & Olson, G. M. (2001). Trust without touch: Jump-start trust with social chat. *Short Paper at the Conference on Human Factors in Computing Systems CHI-01.* New York, NY: ACM.

·30·

ONLINE COMMUNITIES: FOCUSING ON SOCIABILITY AND USABILITY

Jenny Preece and Diane Maloney-Krichmar
UMBC

BACKGROUND

Millions of people meet online to chat, to find like-minded people, to debate topical issues, to play games, to give or ask for information, to find support, to shop, or just to hang-out with others. They go to chat rooms, bulletin boards, join discussion groups, or they create their group using instant messaging software. Short messaging (also known as texting) is also gaining popularity in some parts of the world.

These online social gatherings are known by a variety of names, including *online community*, a name coined by early pioneers like Howard Rheingold, who describes these online communities as "cultural aggregations that emerge when enough people bump into each other often enough in cyberspace" (Rheingold, 1994, p. 57).

The Scope of This Chapter

There is no accepted definition of online community. The term means different things to different people (Preece, 2000), so this chapter starts by examining definitions and descriptions of online community from different disciplines, and briefly traces how the topic has emerged. The second section outlines research from social psychology, sociology, communications studies, computer-supported cooperative work (CSCW), and human-computer interaction (HCI) that informs our understanding of why people interact they way they do in online communities. The third section brings many of these ideas together in the context of design and evaluation of online communities, outlines a design methodology, and proposes a framework for supporting social interaction (i.e., sociability) and designing usability. The fourth section returns to research and briefly reviews key techniques that are being used to research online communities, and discusses the challenges of doing online communities research. The fifth section provides a brief summary of the chapter and proposes two agendas: one for practitioners wanting to create successful new online communities and the other for researchers looking to break new ground.

Our aim in this chapter is to promote better understanding of social interaction online and how this contributes to developing better sociability and usability, and to promote research in this new field. Throughout this chapter, we address the following questions. How do people interact in online communities? What is a successful online community? How can we improve sociability and usability for the millions of people participating in online communities? This chapter, therefore, focuses on web-based online communities supported by text and graphical user interfaces (GUIs), although much of the discussion is also relevant to three-dimensional (3-D) computer virtual environments, the topic of Chapter 31.

What Is an Online Community?

In 1996, a multidisciplinary group of academics held a workshop at which they identified the following core characteristics of online communities (Whittaker, Issacs, & O'Day, 1997, p. 137):

- Members have a shared goal, interest, need, or activity that provides the primary reason for belonging to the community.
- Members engage in repeated, active participation and there are often intense interactions, strong emotional ties, and shared activities occurring between participants.
- Members have access to shared resources, and there are policies for determining access to those resources.
- Reciprocity of information, support, and services between members is important.
- There is a shared context of social conventions, language, and protocols.

In addition, they also agreed that the following characteristics, although not as essential, could significantly impact interactions online: evidence of people having different roles; people's reputations; awareness of membership boundaries and group identity; initiation criteria for joining the community; history and existence over a period of time; notable events or rituals, shared physical environments; and voluntary membership. Not surprisingly many of these characteristics appear in other definitions, too. Several speak of continuing relationships cemented by rituals and history that create a sense of belonging. Depending on one's perspective and academic discipline, the different characteristics take on different levels of importance. Hence, there are several views about what an online community is.

Sociology is an obvious discipline to look for a definition, but it is worth remembering that for more than 50 years, sociologists have defined and redefined the concept of community (Wellman, 1982). Finding a suitable definition that everyone can agree with is therefore not an easy task. Furthermore, definitions change over time. Until the advent of telecommunications technology, definitions of community focused on close-knit groups in a single locale. Things such as birth and physical location determined belonging to a community. Social relationships were with a stable and limited set of individuals and interaction was primarily face to face. Because it was difficult to maintain relationships over long distances because of the slowness and cost of communicating, physical separation from the community often reduced not only contact, but also the strength of a person's membership in the community (Gergen, 1997; Jones, 1997; Rheingold, 1993). However, modern transportations, increased personal mobility, and the development of modern telecommunications systems made these concepts less useful for defining communities. Researchers therefore consider the strength and nature of relationships between individuals to be a more useful basis for defining community (Hamman, 1999; Haythornthwaite & Wellman, 1998; Wellman, 1997; Wellman & Gulia, 1999b).

Particularly potent indicators of community that have been adopted by many online community researchers include the concepts of people with shared interests, experiences, and/or needs engaged in supportive and sociable relations, where they obtain important resources, develop strong interpersonal feelings of belonging and being wanted, and forge a sense of shared identity (Jones, 1997; Rheingold, 1993; Wellman, 2000).

The notion of strong and weak ties is useful in further defining relationships (Granovetter, 1973, 1982). Granovetter's work suggests that the strength of an interpersonal tie can be measured by assessing the amount of time invested in maintaining the tie, the emotional intensity and degree of intimacy of the tie, and the level of reciprocal services that characterize the relationship (Granovetter, 1973). A parent–child relationship is an example of a strong tie. Typically, each of us has only a few strong ties, compared with many weak ties. For example, special interest groups and work-related groups with mailing lists (paper or electronic), telephone trees, theater groups, international organizations (e.g., Green Peace, Amnesty International, ACM, etc.) facilitate hundreds and thousands of weak-tie relationships between members. These weak-tie groups contain people that share some common interests, but do not rely on each other for strong emotional support, or regular, daily, or weekly help. Although the Internet helps to support strong ties, such as those between family members, it is particularly good for weak-tie relationships (Wellman & Gulia, 1999a). Because weak tie relationships are more numerous and diverse than strong tie ones, they provide a larger social network for obtaining and disseminating information and resources than strong-tie relationships (Walther & Boyd, 2002). They are important, therefore, for information exchange, making new contacts, and raising awareness about new ideas (Granovetter, 1973; Kling, 1996; Walther & Boyd, 2002; Wellman, 2000). These networks enable people to discuss topics and contact others with whom they would otherwise not communicate. Some of these relationships would probably flounder without their online component because of geographical distance. Furthermore, the longer such relationships last, the strong the ties tend to become (Walther, Anderson, & Park, 1994).

In contrast, technology-oriented definitions describe online communities by the software that supports them. It is common to hear talk of chat, bulletin board, listserver, UseNet News, MUDs (Multi-user dungeons), MOOs (Object-oriented MUDs), and web-based communities. Such descriptions are concise and meaningful to those who know about software; and although they indicate what conversation protocols are like, they say little about social interaction in the community. For example, two defining characteristics are whether software is synchronous or asynchronous (Ellis, Gibbs, & Rein, 1991). Synchronous technologies require all participants to be available (although not in the same place) at the same time, and communication usually involves short comments, as occurs in chats, for example. Asynchronous technologies (such as bulletin boards or e-mail) do not require participants to be available at the same time. Correspondence via asynchronous technologies, therefore, tends to take longer because it more closely resembles written notes in which one person raises or debates issues and others respond days, weeks, or even months later. Because they are geared to different communication tasks, it is becoming increasingly common to find both synchronous and asynchronous technologies together on community sites.

Enthusiasts of gaming and 3-D immersive environments focus on spatial relations in representations in which participants move around in the form of avatars. These environments are based on spatial metaphors that encourage assumptions about participants' behavior and relationships according, at least partly, to their spatial relationships. For example, can it be assumed avatars that are next to each in the same room are talking to each other, whereas avatars in different rooms are not. Perhaps avatars that are frequently seen together represent friends or at least collaborators.

In contrast to issues identified by both sociologists and technologists, e-business entrepreneurs take a pragmatic view of community (Hagel & Armstrong, 1997; McWilliam, 2000; Williams & Cothrel, 2000). For them, any chat or bulletin board on a Web site is a potential community because it can draw customers to the site—a concept known as "stickiness." Consequently, before the dot.com crash online communities were spawning everywhere. Whereas this market-driven approach is in keeping with the drive to promote commerce on the web (Hagel & Armstrong, 1997; Jones, 1999), it pays little attention to the complexity of interaction online and the need to support and guide it. This may explain why many are ghost towns.

Communities for professionals and others who share knowledge and resources are often referred to as "communities of practice" (Wenger, 1998) to distinguish them from special interest communities and support communities. Their members often have a shared task and well-defined roles (Feenberg, 1993), and they offer professionals emotional support as well as information and discussion (Moon & Sproull, 2000; Sproull & Faraj, 1997; Williams & Cothrel, 2000).

Another kind of community is physical communities that are supported by an online network, known as community networks to distinguish them from communities that primarily exist online. Early examples include Seattle Community Network (Schuler, 1996) and Blacksburg Electronic Village (Cohill & Kavanaugh, 1997), but there are now hundreds of community networks (Carroll & Rosson, 2001). These community networks usually focus on neighborhood issues, and the online communication supplements face-to-face meetings. Increasingly, many people are meeting online and physically, and the distinction between the two is becoming blurred; but, there are also people scattered across the globe who can only interact virtually (Lazar, Tsoa, & Preece, 1999).

Instant messaging and telephone texting communities (particularly in Europe) are also gaining popularity, especially with teenagers who like to keep contact with friends while moving from location to location. Many teenagers switch effortlessly between media, texting, e-mailing, and chatting. Judging online activity by what is seen in a single medium is, therefore, likely to give a distorted picture (Brown et al., 1999).

This variety of definitions and descriptions has led some researchers to seek new terms. For example, "online social space" avoids the sociologically inaccurate usage of the term community (Farnham, Smith, Preece, Bruckman, & Schuler, 2001). However, because online community is still the most widely used term, we will use it in this chapter to refer to social activity that involves groups of people interacting online. Such communities may be long or short–term, large or small, national or international, and completely or only partially virtual.

Emergence of Online Communities

The listserver, bulletin board, and chat technology that supports many of today's online communities changed comparatively little during the last 20 years, until the web became widespread in the mid-1990s. Since then, there has been a steady flow of new versions and new technologies; but the biggest changes are in how the technology is being used, and who is using it.

Increasingly combinations of different types of synchronous and asynchronous technology are embedded in Web sites supported with information, links to other sites, and search facilities. Linking into online communities via small mobile devices, such as telephones and PalmPilots, is also becoming popular, and no doubt we will see more access via other small devices during the next few years.

Early online communities for education (Hiltz, 1985), networked communities (Hiltz & Turoff, 1993; Rheingold, 1993; Schuler, 1994), and office communities (Sproull & Kiesler, 1991) were developed for known groups of users whose characteristics, needs, and skills were known and who had the same or similar communications software. Since then, the number of computer users has increased dramatically. In addition, the demographic composition of the user population has also changed to include people of all ages, different cultures, educational backgrounds, experience, and technical skills. A recent survey by the Pew Foundation confirms this trend. Pew reports that more than one hundred million Americans had Internet access in 2001 (Rainie & Packel, 2001). The range of people participating in various kinds of online communities has also changed. Although some communities require members to have particular skills or qualifications, there are millions of open communities in which anyone with Internet and web access can participate. Consequently, the majority of users in these open communities and many others are not technical people or skilled office workers. Today's online community participants come from all walks of life.

Early descriptions of online communities were anecdotal and tended to make comparisons with face-to-face communication, but chatting and sending messages online is becoming a normal part of many people's lives, particularly for young people. Online people do almost everything that people do when they get together, but they do it with words on screens, leaving their bodies behind independent of local time or location (Rheingold, 1994).

Sherry Turkle, an early researcher, reported that those who lack confidence in face-to-face situations often become more confident online and lose their inhibitions. She documented many cases of this phenomenon and, using her knowledge of psychotherapy, she explained how people explore new personas online in which they act out facets of their personalities that are problematic in face-to-face situations (Turkle, 1995). For example, people who are shy and find making relationships hard become bolder online because they do not have to face the person with whom they interact, and if the going gets tough, they can switch their computer off.

As well as having advantages for individuals, whole communities can benefit from becoming networked as Rheingold and Schuler have described. Rheingold told the story of life in the WELL (Rheingold, 1993, 1994), one of the first and most famous networked communities situated in the San Francisco Bay area. Schuler focused on design and development issues associated with creating the Seattle Community Network (Schuler, 1996). This experience led him to propose a noble set of core values

to guide future online community development. These values included: conviviality and culture, education, strong democracy, health and human services, economic equity, opportunity and sustainability, and information and communication.

From the late 1990s, the combination of less expensive computing power, the web, and several successful service providers enticed tens of thousands of people into online communities, which has aroused strong interest among researchers in HCI, sociology, anthropology, psychology, linguistics, communications studies, and information systems. This multidisciplinary group is interested in all aspects of social interaction online. The input from this broad range of fields, each with its own literature, theory, and research paradigms, makes studying online communities an intellectually rich research area.

Social scientists seek to answer questions about how the Internet is changing our lives (e.g., Kraut et al., 1998), how communities form and function (e.g., Smith & Kollock, 1999; Wellman & Gulia, 1999b; Wellman et al., 1996), and the policy issues concerned with privacy, security, etc. (Kahin & Keller, 1995). Linguists and psychologists try to understand how conversation, discourse (e.g., Herring, 1999), interaction, and social relationship development is different online from offline. Technology-oriented research address questions about design for sociability and usability (e.g., Erickson et al., 1999), supporting and visualizing interaction online (e.g., Sack, 2000a; Smith & Fiore, 2001; Viegas & Donath, 1999).

During this time period, several edited books appeared that document some of this research and made it more widely available (e.g., Jones, 1998; Kiesler, 1997; Smith & Kollock, 1999). Multiple perspectives and different skills provide many benefits, but one of the drawbacks is that material is scattered across many journals and conferences, which is a problem for future research that seeks to build on previous studies. Other books cover e-business, which somewhat mistakenly heralded online communities as a panacea for drawing customers to online sites (e.g., Hagel & Armstrong, 1997) and provide guidance for practitioners on how to develop successful online communities for business (e.g, Figallo, 1998; Kim, 2000). Specialist graduate courses have been developed that add to curricula offerings in information systems (e.g., www.umbc.edu/onlinecommunities), computer science (e.g., http://www.cc.gatech.edu/~asb/past-classes.html), and sociology (e.g., http://www.sscnet.ucla.edu/soc/faculty/kollock/classes/cyberspace/index.htm), and texts are appearing that attempt to distill the field for students and practitioners (e.g., Preece, 2000).

There has also been a strong research thrust into 3-D immersive environments, which helps to cast light on interaction, relationship development, identity, etc., in such worlds. Other exciting research challenges involve developing GUI communities to support large numbers of people with standard equipment well.

SOCIAL INTERACTION

The theory and research that informs our understanding of online communities is drawn from a broad range of disciplines. Consequently, there is a large body of potentially relevant research; so, we have had to be selective and focus on ideas that we consider most central for understanding the basics of this field. The first part of this section discusses communication between pairs and small groups, whereas the second part examines research that addresses community issues.

Communicating Online

In online textual environments, people represent themselves through their words, and both syntax and semantics convey meaning. However, nonverbal cues that help us understand each other (e.g., body language, facial expressions, and voice tone) are missing when people communicate via narrow bandwidth media, such as text. Developing shared understanding (i.e., establishing common ground), a sense of social presence, empathy, and trust are therefore usually harder, which in turn makes developing social relationships slower and more difficult.

All technologies have strengths and weaknesses, which developers need to understand. For example, video-conferencing conveys some nonverbal communication; but, because of limitations of communications bandwidth, screen size and resolution, subtle body language, and important contextual information about participants' moods, the context in which they are participating and their environment is lost (Olson & Olson, 2000). The developers' job is to select or develop technology that matches the communication tasks of the community, and their social and practical needs. The researchers' job is to elucidate fundamental knowledge that supports that process. The following discussion outlines some of that research.

Common Ground. Common ground theory is a linguistic theory that has been applied extensively in CSCW research to explain how the properties of different media effect communication (Olson & Olson, 1997). Common ground theory provides a framework for understanding how two people or a small group develop shared understanding in a conversation (Clark & Brennan, 1991). For instance, if person A speaks to person B about my daughter, the two of them must understand that she is referring to the child playing in the living room and not to girls playing in the street three blocks away. The process of acquiring this common understanding is grounding, which varies from situation to situation. Grounding takes one form in face-to-face conversation and other forms in computer-mediated communication supported by different types of software, and yet other forms when calling directory assistance, chatting with a friend, or participating in a debate. Grounding is, therefore, influenced both by the communication medium and the communication task.

Grounding occurs through several rounds of checking that a conversation partner has heard and correctly understood what is being said. This sounds cumbersome, but conversations usually follow an identifiable pattern. For example, by noticing how much attention a partner is paying to a comment, the speaker can judge whether there is shared understanding. Utterances, gazes, nodding, and facial expressions indicate that the person is paying attention and understands. People generally do this unconsciously with as little effort as possible, checking and then repeating or repairing incomplete comments when in doubt.

The amount and type of effort required for establishing common ground varies between different communication media. Techniques that work in one medium may not work so well in another. For example, a nod works in a face-to-face conversation, but is useless in a bulletin board or chat discussion. Similarly, an agreed short-hand communication language used by a group of friends for texting in England may not be understood outside the community; so, establishing common ground will be difficult. Furthermore, people who are unfamiliar with a particular medium will not have had time to develop their own ways of supporting grounding.

Factors that affect the ease with which common ground is established include:

- sharing the same physical space, i.e., *co-presence*
- being able to see each other, i.e., *visibility*
- being able to hear each other and detect voice tone, i.e., *audibility*
- both partners experiencing the conversation at roughly the same time, i.e., *co-temporality*
- sending and receiving more or less simultaneously, i.e., *simultaneity*
- keeping turns in order, i.e., *sequentiality*
- being able to review messages, i.e., *reviewability*
- being able to revise messages; i.e., *revisability*

Surprisingly, face to face is not necessarily the best for all types of communication, nor are high-bandwidth synchronous environments; it depends on the communication task. For example, video images do not contribute much in information transfer tasks; voice alone is adequate, although participants may prefer video (Sellen, 1994). Text-only environments can be preferable when the content of the conversation is potentially embarrassing, as in a discussion about a rape incident (Newell & Gregor, 1997). Asynchronous textual communication is preferable when having time to reflect is useful or when participants cannot be co-present. In a study of recovering alcoholics communicating via a bulletin board, participants reported that they liked being able to send messages any time of the day or night and having time to reflect before replying (King, 1994). People with poor typing skills or those who like to reflect may also prefer asynchronous textual media.

However, because face-to-face is the default we are used to, it has become the standard for judging other media. There are also times when no matter which media is available face-to-face communication is preferable because there is no substitute for the commitment of being there, sharing a hug, and getting a broad understanding of the context in which the conversation is occurring (Olson & Olson, 2000).

Social Presence, Identity, and Relationships. Social presence theory (Short, Williams, & Christie, 1976) speaks about how successfully media convey a sense of participants being physically co-present. Although it focuses on some of the same issues as common ground, its origins are in communications studies and social psychology rather than linguistics (Rice, 1987, 1993). Consequently, social presence theory takes a different perspective. It helps to explain how social behavior is affected by characteristics of different media, whereas common ground focuses on conversation. Media richness theory is similar to social presence, but it has a media-oriented perspective and was developed 10 years later, with, apparently, little knowledge of earlier work on social presence (Daft & Lengel, 1986; Rice, 1993).

Like common ground, social presence depends not only on the words people speak, but also on nonverbal cues, body language, and information about the speakers' context (Rice, 1993; Rice & Love, 1987). Reduced social cues (i.e., gestures, body language, facial expression, appearance, voice tone, etc.) are caused by not having sufficient bandwidth to carry this information (Culnan & Markus, 1987; Walther, 1993). In textual systems, for example, both task information and social information are carried in the same single verbal/linguistic channel which, although adequate for much task information, does not carry nonverbal information, which may be needed for interaction (Walther, 1994; Walther et al., 1994). Consequently, many clues about the communicators' emotional states are filtered out. Gaze and tonal information, for example, are missing.

When people meet each other for the first time, they develop mental models of each other and the content of their discussion (Norman, 1986). Their opinions are influenced partly by such things as age, gender, physical appearance, the context of the meeting, etc. Furthermore, they tend to be developed very quickly, but can be remarkably powerful and resistant to change, even when evidence suggests they are not completely correct (Wallace, 1999). So, another feature of reduced social presence, particularly in low bandwidth environments, is that the way people form impressions of each other is different, which can have both positive and negative affects, depending on the situation.

Because people communicate without knowing the circumstances and broader context in which comments are made, misunderstandings can occur, especially if the comment was abrupt, poorly explained, out of context, and so on. Annoyed, the person receiving the message may respond in an angry tone, possibly escalating the problem and causing an argument. Misunderstandings are particularly common among people who are not used to using the media, because they have not had time to get used to it and to develop ways of getting around this problem. People may also make unwarranted, angry attacks, known as flaming, encouraged by the fact that they do not have to face the person who they are attacking or take responsibility for their behavior (Hiltz, Johnson, & Turoff, 1986; Spears & Lea, 1992; Sproull & Keisler, 1986).

Conversely, there are times when not being able to see the person with whom you converse and knowing you may never meet them can be a positive feature of these environments, because people are encouraged to disclose more about themselves (Lea, O'Shea, Fung, & Spears, 1992; Spears, Russell, & Lee, 1990; Walther, 1996). This is why remarkably candid comments are sometimes made online about personal health problems, emotional relationships, and feelings. Furthermore, when people discover they have similar problems, opinions, or experiences,

they may feel closer, more trusting, and be prepared to reveal even more. When conversations are limited to just a few topics, a false sense of feeling similar and shared identify can develop. This has a snowball effect, in that the more people discover that they are similar to each other, the more they tend to like each other and the more they will disclose about themselves. This is known as self-disclosure reciprocity, and it is powerful online (Wallace, 1999). It works by, "If you tell me something about yourself, I'll tell you something about me."

Another phenomenon that has been noticed in research on people using low-bandwidth systems is that these users tend to send fewer messages during the same time period as those communicating face to face or via video-conferencing (Hiltz et al., 1986; Ogan, 1993; Walther, 1993). Some online relationships may, therefore, be slower to develop; but, given sufficient time, strong relationships can form that are comparable with those formed face-to-face (Walther, 1993). Furthermore, online relationships may be extremely rich (Spears & Lea, 1992). Encouraging participants to be particularly careful about what they say and how they say it early in relationships can be helpful until they become experienced with the medium and find ways to deal with the lack of visual cues (Rice & Barnett, 1986). For example, phrasing a comment tentatively to avoid appearing aggressive (Wallace, 1999), or prefacing it with IMHO ("in my humble opinion") can achieve this goal. Emoticons (also known as "smilies") are also used as softeners (Lehnert, 1998, provides a list). Placing additional personal material (e.g., pictures, personal stories) on web pages associated with the community can also help people to get to know each other online.

The way people choose to portray themselves online is of considerable research interest. An aspect that has received particular attention is how gender is portrayed and revealed. Whether done intentionally or unintentionally, many online participants have discovered that there can be consequences from revealing one's gender online. For example, women may get unwanted attention (Bruckman, 1993; Herring, 1992; Turkle, 1995, 1999); so, some avoid harassment by switching or disguising their gender. This behavior may fool other participants effectively, but linguists and those sensitive to gendered differences in conversational style can usually detect semantic and syntactic differences between the way women and men express themselves (Herring, 1992; Reid, 1993). For example, women tend to be more self-deprecating, apologetic, and include more adjectives in their speech (Tannen, 1990, 1994). Women also tend to avoid criticism by phrasing their questions in defensive ways (Herring, 1992).

What this research says to online community developers is that they need to look for ways of educating participants about how their online behavior may be perceived and help them to find ways of preventing misunderstanding that can damage online relationships. For example, taking the time to check that you have understood what the other person is really saying can be important (Zimmer & Alexander, 1996). Simulating physical presence via avatars is a frequently used technique in graphic environments, particularly gaming environments and chats, such as ActiveWorlds.com. By representing themselves as an avatar, participants can disguise their real identities and influence how others perceive them.

There is, however, a cost in screen real estate for using avatars. If too many are present at once, the screen becomes cluttered. Another problem is that avatars may move across the screen and out of view very quickly. Small, more abstract graphical representations that avoid this problem but give visual feedback about the number of people present in an environment, what they are doing, and who is speaking are being developed to support social presence online and also contribute to representing individuals' identities. One of the first environments to show this idea was chat circles, a chat environment in which participants are represented as small circles (Viegas & Donath, 1999). A variation on this theme is used in Babble, another chat environment (Erickson et al., 1999) that supports a community of practice for IBM researchers. In this application, small colored circles represent different participants. The relative position of these circles also indicates who are talking to whom, and who the most active participants are. However, as with any innovation that discloses information, there may be a downside for some people. For examples, people who read but do not send messages (i.e., silent participants, also known as lurkers) will also be shown. For people whose intention is not to be seen, such representations therefore pose a threat and may stop them from participating. Whether lurking should be encouraged or not is debatable, and opinions vary. In part, such judgments need to be related to the community's purpose, and we return to this topic later (Nonnecke & Preece, 2001).

Empathy and Trust. Additional support for these ideas comes from research on empathy, which is defined as "knowing what another person is feeling, feeling what another person is feeling, and responding compassionately to another person" (Levenson & Ruef, 1992). Research shows that empathy is strongest between similar people and people who share similar experiences, such as people in the same profession or siblings (Eisenberg & Strayer, 1987; Etchegoyen, 1991; Ickes, 1993, 1997). In fact, the more similar people are, the easier it is for them to understand each other (Hodges & Wegner, 1997). This phenomenon is particularly noticeable in patient support communities, where participants experience similar problems, discomfort, and treatment. Comments such as: "we're all in this together" are frequently seen (Preece, 1998, 1999a; Schoch & White, 1997). However, empathy, like common ground and social presence, depends heavily on nonverbal communication, such as gaze and body language (Eisenberg & Strayer, 1987; Etchegoyen, 1991; Lanzetta & Englis, 1989); so it, too, is influenced by the properties of different communication media (Preece, 2000).

Because trust seems to be similar to empathy, it is likely to be influenced by the properties of the media in a similar way. Trust can be defined as: "the expectation that arises within a community of regular, honest and cooperative behavior, based on commonly shared norms, on the part of the members of the community" (Fukuyama, 1995). Revealing personal information about one's health, agreeing to cooperate on a project, or making a purchase require trust. The more that is risked, the more trust in needed. Considerable research effort is being focused on understanding how trust develops online stimulated by the needs of e-business. Procedures and mechanisms are being sought to support trust online. These should involve evidence of good past performance and truthful promises and guarantees of similar future behavior (Shneiderman, 2000). Ways of supporting and managing trust in online communities is also acknowledged to be important (Kollock & Smith, 1999).

One example of successful online trust management is e-bay's reputation management system (Kollock, 1999). In this system, ratings of customers' satisfaction of transactions with a particular vendor are compiled to provide a history that can be examined by potential customers. Furthermore, knowing that a vendor has a good reputation encourages co-operation when things do not go quite as expected, because there is a basis for trusting that the problem will be put right in a timely way. However, it is hard to see how reputation systems could be used more widely in online communities without damaging some participants' confidence to participate.

A related trust issue concerns the persistence of conversations online (Erickson, 1999). Savvy online community participants who understand technology are reluctant to enter into online conversations that involve disclosing personal information, because they know that it can be retrieved, even after they themselves have deleted the text. They realize that their information could be dredged up, even years later, and they could be damaged. For example, revealing details about a health or personal problem could affect the cost of their health insurance. An unfavorable comment about a manager could prevent them gaining a much-deserved promotion.

Of course, what influences individuals, pairs, and small groups also impacts on the community, but communities also have a character and dynamics of their own.

Group Dynamics Online

Just as theory from psychology and linguistics have been adapted and applied to understand how people communicate online, theories from social psychology, sociology, and other branches of the social sciences are being drawn on to help explain how communities form and change.

Social Network Theory. Social network theory is a branch of sociology that examines the patterns and characteristics of social connections and their relationship to individual's lives and societal organization. This theory is used as a framework to study how people relate to each other through computer-mediated networks (Wellman, 1997; Wellman & Frank, 2001; Wellman et al., 1996). Wellman and Frank (2001) believe that a multilevel approach is required to understand the interactive effects of characteristics of computer-mediated networks. These characteristics include the composition of networks, the network size, the range of the network, the frequency of contact between people, the density of interpersonal ties, the characteristics of members, the history of the network, and the resource available in the network (Wellman & Frank, 2001).

Critical Mass. It is well known that if there are too few people contributing to an online discussion, it will die because there will be insufficient new messages to hold the interest of existing members. The number of people needed to make an online community viable and to attract others is known as its critical mass (Markus, 1987, 1990; Morris & Ogan, 1996). However, although critical mass is a useful concept for explaining success and failure (Rice, Grant, Schmitz, & Torobin, 1990) and interactions online (Ackerman & Starr, 1995), it is of limited practical value because it is so hard to quantify. What may be enough people in one community may not be in another because members of different kinds of communities have different expectations. Further research is therefore needed to quantify critical mass for different kinds of communities and situations.

Reciprocity and Social Dilemma. Reciprocity means giving back to the community, as well as taking from it. It is a central concept for explaining the success and failure of communities. In communities that function well, "whatever is given ought to be repaid, if only to ensure that more is available when needed. Repayment of support and social resources might be in the form of exchanges of the same kind of aid, reciprocating in another way or helping a mutual friend in the network" (Wellman & Gulia, 1999b). Even if reciprocity does not happen immediately, it can happen months or years later, possibly to another person in the community (Constant, Sproull, & Kiesler, 1996; Wellman & Gulia, 1999b). In healthy communities, reciprocity is a general and accepted norm among members.

The problem is that often behavior that benefits an individual can damage the group. Furthermore, in certain situations, individuals can gain benefit without it being obvious to the community that they are not contributing and are therefore damaging the community effort. For example, if a community agrees that each of its members should donate a certain amount of money or time to achieve a community goal, and then some people do not contribute, then they benefit, particularly if no one knows about their selfish act, and the community loses. Similarly, if participants in a small community decide to read messages in a topic discussion, but not to post because they do not want to spend time contributing, the community as a whole will suffer because there will not be sufficient critical mass for it to be viable. This tension between what is best for an individual and for the group is a social dilemma (Axelrod, 1984; Kollock, 1998), and it is at the heart of most social interactions (Kollock & Smith, 1999).

Furthermore, online it may be particularly tempting for people to take and not to give back because the chance of meeting people from the online community in person is likely to be extremely low, so there are no serious implications for future interactions (Walther, 1994).

Roles, Rituals, Norms, and Policies. Governance covers many issues, from registration to moderation to democracy online and is also strongly influenced by the cultural norms of the community. Communities that have a strong cultural basis, such as church groups, environmentalists, alcoholics anonymous, etc., that already have rules and norms in operating in their offline versions, can import them online. New communities that only exist online will have to develop their own governance procedures from scratch and gradually develop norms as members get to know each other and start to debate and

agree what is acceptable and what is not. Baym's research provides insight into how an online audience community devoted to soap operas start to do this. As fans discuss the story lines and characters of their favorite soap opera, they share their views and values, learn from the rich network of relationships, and develop shared norms (Baym, 2000).

Old issues have to be addressed online. What type of governance should there be? Should it be democratic or not? How democratic should it be? If so, what kinds of policies and social procedures are needed? Diversity University, for example, has a sophisticated democratic process for calling votes. Another example concerns freedom of speech. Should freedom of speech be limited if it is racist, obscene, blasphemous, or aggressive? A short clearly worded statement saying what is acceptable may be useful. Early in the existence of the WELL, for example, its members decided that complete freedom was important (Rheingold, 1993). Other communities develop guiding policies. For example, the Down Syndrome Online Advocacy Group (http://www.dsoag.com) simply requests: "Do not communicate to someone else that which you would not want communicated to you" (Lazar, Hanst, Buchwater, & Preece, 2000).

Having rules is fine, but how should they be enforced? There is no point making rules if they are not enforced. Moderators perform one of the best-known roles in online communities, but the extent of their roles may not be so well known. Moderators performed many different tasks (Berge, 1992; Collins & Berge, 1997; Salmon, 2000), including:

- Facilitating so that the group is kept focused and on topic.
- Managing the list, e.g., archiving, deleting, and adding subscribers.
- Filtering messages and deciding which ones to post; typically, this involves removing flames, libelous posts, spam, inappropriate or distracting jokes, and generally keeping the ratio of relevant messages high, which is often described as the signal/noise ratio.
- Being the expert, which involves answering frequently asked questions (FAQs) or directing people to online FAQs and understanding the topics of discussion.
- Editing text, digests, or formatting messages.
- Promoter of questions that generate discussion.
- Marketing the list to others so that they join, which generally involves providing information about it.
- Helping people with general needs.
- Being a fireman by ensuring that flaming and ad hominem attacks are done offline.

Levels of activity vary between moderators, from reading, making judgments, and taking action on every single message and updating FAQs regularly to stepping in just occasionally with a remark to deter a future transgression. Most moderators are self-taught or learn by observing others on the job (Feenberg, 1989). Knowing when to push discussions back on topic and when not to can be difficult, as the following quote from an experienced moderator illustrates: "Hmmm. How inviolable should the original purpose be? I manage a list that now only rarely touches on [the] topic it was originally supposed to talk about. So? The conversation is shaped by the community's current and compelling interests. The original topic re-emerges when someone needs to talk about it, when it has some kind of immediate relevance to someone's life. Fine with me" (Berge, 1992).

To protect themselves from unwarranted criticism, moderators often follow accepted policies, which are made public. Having clearly defined policies is also useful for coordinating two or more moderators. Helping roles, norms, and rules get developed is often done by community leaders or managers who work with the community. Skill is needed to make sure that there is enough structure to protect and guide the community's evolution, but not so much that it is stifled.

There are usually two sides to creating rules. The rule can stop unwanted behavior, but it can also deter people from joining and inhibit contributions to the community, particularly if there are too many rules and people feel stifled by them. For example, registering deters casual visitors intent on disrupting the community, but may discourage others too. Some communities get around this problem by allowing anyone to visit for a limited period with limited privileges. Others have a light registration procedure, but newcomers go through a probationary period in which their behavior is observed.

Participants in online communities often carve out roles for themselves just as they do in physical communities. For example, there are protagonists, experts, people who befriend others, people who always try to respond, witty people, sarcastic people, lurkers who watch silently, etc. Roles vary according to the type of the community, but can be extremely important in the early days of developing a community (Kim, 2000). Dynamic or charismatic characters help to draw others to the community.

Support for Social Interaction

How research informs design and management of online communities depends on many factors, including the purpose of the community, the needs of participants, and the policies that develop. For example, emotional and health support communities are quite different from scholarly communities. Table 30.1 summarizes some ways the key concepts just discussed and proposes some ways for supporting social interaction online.

Knowledge from research can be fed into design, development, and management of online communities to inform those processes.

TABLE 30.1. Suggestions for Supporting Social Interaction in Online Communities

Support For	Issues and Potential Solutions
Grounding	Support communication by encouraging participants to check that they share a common understanding. Different types of software provide different support. For example, turn-taking can be a problem in busy chats, whereas turn-taking is clearer in threaded bulletin boards. Helping to make the identity of individuals clear in synchronous environments and providing short-hand versions of common words and phrases can help. Chatters and texters also tend to develop their own short-hand language. Encouraging participants to check for common ground is helpful. Supporting social presence also helps.
Social presence	Avatars simulate being there and provide more identity for individuals Thumbnail pictures can also be used (Zimmer & Alexander, 1996). Other techniques include links to personal home pages and graphical representations (Donath, Lee, Boyd, & Goler, 2001; Erickson et al., 1999). Participants also need to be aware that it can take longer to develop relationships online (Walther, 1996).
Discouraging misunderstanding & aggression	Encourage participants to explain themselves clearly and to check each other's intentions and look for common ground (Zimmer & Alexander, 1996). Appoint moderators to check messages. Keep discussions on topic (Collins & Berge, 1997; Salmon, 2000).
Prevent flames	Registration helps to deter *ad hoc* flamers. Support moderators with tools to identify flames and spam (Seabrook, 2001).
Relationship formation	Supporting social presence, empathy, and trust helps. Pay particular attention to early interactions and encourage long-term communication (Wallace, 1999; Walther, 1993). Moderators and mentors can also help.
Encouraging empathy	Support social presence (Preece, 1999b). Provide a clear statement of the community's purpose (Preece, 2000). Allow participants to explore their similarities by facilitating private communication and providing space to tell stories (Preece, 1999a).
Encouraging trust	Support formation of long-term relationships. Provide a record of past behavior (e.g., reputation management; Kollock, 1999).
Encouraging critical mass	Provide a clear statement of purpose so people know what to expect and support the purpose (e.g., by keeping discussions on topic, etc.; Preece, 2000). Stage events (Kim, 2000) and make sure there is always new content.
Discouraging social dilemma	Encourage reciprocity with rewards (e.g., acknowledge helpful responses). Encourage good community norms and values.

DEVELOPING AND EVALUATING ONLINE COMMUNITIES

Involving participants in software design helps to ensure that their social and political needs are taken into account (Eason, 1988; Greenbaum & Kyng, 1991; Muller, 1992; Mumford, 1983; Schuler, 1994; Schuler & Namioka, 1993). What makes online communities different from most other software development is that communities evolve continuously, because community is a process not an entity (Fernback, 1999). The role of community developers and managers is therefore to start this evolution by providing suitably designed software and to help guide the community's social evolution. Schuler, advocates participatory design with a focus on core social values (Schuler, 1994). Cliff Figallo, one of the developers of the WELL, focuses on building relationships and increasing customer loyalty through online community to maintain competitive business edge (Figallo, 1998). Others with an interest in building online communities for e-business promote various business models of community (Hagel & Armstrong, 1997; Williams & Cothrel, 2000b). Kim documents best practices and proposes nine design strategies that are based on three sound principles: design for growth and change; create and maintain feedback loops; and empower members over time (Kim, 2000). Preece advocates a process of *participatory community-centered development* (PCCD) composed of two key components: software design, particularly designing usability, and guiding social development, that is, supporting sociability (Preece, 2000). PCCD borrows concepts from user-centered design (Norman, 1986), contextual inquiry (Beyer & Holtzblatt, 1998), and participatory design (Greenbaum & Kyng, 1991; Muller, 1992; Mumford, 1983; Schuler & Namioka, 1993), and has been deployed successfully in a number of online community development projects (Lazar et al., 1999; Lazar, Hanst, Buchwater, & Preece, 2000; Lazar & Preece, 1999a; Preece, 2000).

The first stage of PCCD is the community needs assessment and user task analysis requircs, which involves understanding the community's social needs, individual's communication task needs, and any technical constraints that must be considered. The second stage involves developing a conceptual model of the community space and then either building or selecting software with suitable usability, and starting to plan the sociability support that will be needed. The third stage is refining sociability and usability. The fourth and final stage involves seeding the community with participants, publicizing it, and creating events so others will come, and welcoming, nurturing, and guiding the community as it grows until it becomes self-sufficient. PCCD is iterative and benefits from multidisciplinary input and extensive participant involvement in which potential community members review and inform the process through different evaluation processes.

Although the PCCD approach can be used for any community development there are significant differences between the various technologies (listserv, bulletin board, chat, Usenet, 3-D environments, etc.) available for supporting online communities. Their relative strengths and weaknesses are described in Table 30.2.

The web makes it possible to integrate synchronous and asynchronous technologies so that users can benefit from both. For example, messages are left on boards or sent via e-mail to coordinate and schedule chat or virtual world sessions. Instant messages are used to signal that a document has been posted for review and so on. These combinations of technologies and the Web site on which they reside provide a richer basis for community than any single technology could on its own; they are the community. Furthermore, it is becoming increasingly difficult to distinguish between technologies, for example, instant messaging systems, chats, and virtual worlds often share common features. Wireless access from hand-held mobile devices will also become increasingly common, which incorporates "texting" and web browsing. Texting and short messaging systems (SMS) are already extensively used in some parts of Europe. Technology choices must also ensure that all users will be able to participate with the equipment they own, and that software is intuitive, straightforward, and pleasant to use (Preece, 2000). There are three design issues that are key to the success of online communities: supporting sociability, designing usability, and criteria for evaluating online communities.

Supporting Sociability and Designing Usability

Sociability is concerned with planning and developing social policies and supporting social interaction. Usability has been defined by many authors and operationalized over the last 20 years (e.g., Bennett, 1984; Dumas & Redish, 1999; Nielsen, 1993; Nielsen & Mack, 1994; Preece, Rogers, & Sharp, 2002; Shackel, 1990; Shneiderman, 1986). Sociability is a newer concept that still needs to be operationalized (Preecc, 2000). Because online communities are evolving continuously, developers must accommodate changes by regularly revisiting sociability and usability decisions. Developers of traditional systems office systems, record systems, and air traffic control do not need to deal with this type of continuous evolutionary change. This is a challenge for many software developers who are not used to working on a continuously moving target.

Key components of usability, often described as principles, guidelines, or heuristics, depending on their role in design and evaluation (Preece et al., 2002) are by now well understood and can be used as a framework to guide development. But accepted frameworks for sociability have not yet been established, because sociability is a new concept with many components as discussed in the previous section. Despite there being gaps in our fundamental knowledge of social interaction in online communities, a framework is needed to guide designers' thinking and to help them focus on key issues so that they do not become bogged down in details. Preece's pillars of participatory community-centered development aims to provide such a framework (Preece, 2001). The key components of sociability in this framework are the community's purpose, its people, and the policies that help to guide online behavior. The key components of usability are dialogue and social support, information design, navigation, and access. Applying this framework and showing how the components that make it up are related is a step toward systematically incorporating sociability

TABLE 30.2. Characteristics, Advantages, and Disadvantages of Various Online Community Technologies

Technology	Traits
Mailing lists/Listserver	**Characteristics** Asynchronous, available 24/7, may be moderated or unmoderated. Can be used for mass broadcast messages come to you—push technology. The list may be hosted by a company/institution or individually purchased and supported. Listservers deliver messages in two forms—either they trickle through as they are sent or a moderator collects them into a digest. Members have to register. **Advantages** Easy to use/good for newbies. No special equipment required beyond e-mail capability. Good for sending announcements and newsletters. Good for broadcasting messages and discussions. Participants may take time to reflect, compose, and edit items posted to the list. Visitors have to register—may help to create a feeling of community. **Disadvantages** Members have to register—may discourage participation. Lists with a large number of postings may be overwhelming to readers. Everything posted to the list comes to each member. Contexts for responses have to be provided by including parts of previous messages. If a digest is sent, it can be difficult to respond to a particular message because messages are linked and they are not threaded or ordered.
UseNet News newsgroups	**Characteristics** Asynchronous 24/7. Collection of discussions on various topics hosted on the Internet, cross posting between UseNet News groups is common and spamming is frequent. Users have to go to UseNet to read messages (pull technology). Open communities, no registration required to post. Usually not moderated. **Advantages** Open communities, no registration required to post—may encourage wider participation. No special equipment beyond Internet access. A large number of newsgroups exist on the Internet with a wide range of topics. It is easy to find an existing group to match your interests. Participants may take time to reflect, compose, and edit items posted to the list. **Disadvantages** Open communities, no registration required to post—may create a sense of anonymity that can lead to inappropriate messages and hostile postings (flaming). Spamming is frequent. The volume of messages in some groups may be overwhelming.
Message Boards, Bulletin Boards, Discussion, or Forum	**Characteristics** Asynchronous 24/7. Users have to go to a site to read messages (pull technology). May be moderated or nonmoderated. Usually require registration, but may be open. Discussions are threaded or linear. Many bulletin board services are set up to send an e-mail to signal new messages, responses, and/or topics of interest. **Advantages** No special equipment beyond Internet access. Participants may take time to reflect, compose, and edit items posted to the list. It is easy to find an existing group to match your interests. Discussion threads provide historical context. Linear organization provides separate topics for each conversation and is good for in-depth discussion. Participants may take time to reflect, compose, and edit items posted to the list. Many bulletin boards provide good search facilities that enable participants to search on topics, or people, or messages sent on or between particular dates, etc. Emoticons and other icons are also becoming increasingly common so participants can signal the content of their message and their mood. **Disadvantages** Newcomers may find it hard to break into the conversations. Following threads may become confusing. May be difficult and time-consuming to moderate a large board. Group norms may develop that stifle new points of view and participation.
Real-timer, Text-based Chats	**Characteristics** Synchronous, text environments. Messages are short and conversation moves on quickly. Real-time auditoriums may be structured to accommodate a large number of persons in a public chat. Instant messaging provides real-time chats for private groups. Participants register, pull technology—you have to go to the site.

(Continued)

TABLE 30.2. (Continued)

Technology	Traits
	Advantages Provides a sense of immediacy. Allows people to communicant in real-time. Good for quick exchanges, holding meetings, conducting interviews, and to hang out and relax. Newcomers can learn to participate in chats easily. Participation is fast paced. Disadvantages Must be online at a specific time to participate. No time to reflect, compose, and edit postings. Several conversations may appear at the same time and be confusing for participants. Conversations may get intertwined because messages appear on a first-come, first-displayed basis. Some types of real-time chat may require special download and configuration.
Immersive Graphic Environments	Characteristics Synchronous, interactive, navigable environments with 3-D graphics, sound, animation, and customizable characters (avatars). Highly versatile gaming e-business, learning, and entertainment environments. May be moderated or nonmoderated, open or public. Pull technology—you have to go to the site. Advantages Interactive, visual, and aural environments allow individuals creative freedom to express themselves. Provide highly collaborative environments. May provide a broader experience by generating a stronger sense of presence and engagement for some types of interaction. Disadvantages Many types of immersive environments require high-memory computers with audio ports, headsets, microphones, and fast Internet access. May require downloading programs or plug-ins that work with specific browsers. The space can become crowded with avatars that limit interaction. Unclear how much value is added by these environments.

Note. Figallo, 1998; Kim, 2000; Preece, 2000.

and usability into design and development of online communities (Preece, 2000). The components of sociability—purpose, people, and policies—will now be described briefly.

Purpose: Defining the community's purpose is important so that potential participants can immediately find out about the communities' goals (Kim, 1998; Preece, 1999c, 2000). Giving the community a meaningful name, and providing a clear, readable definition of its purpose helps to discourage people from joining who are not committed and encourages empathy by bringing like-minded people together. This in turn may encourage common ground to be established more easily, and may mitigate any effects of pour social presence online, and may foster trust. These effects can discourage off-topic discussions and can help to reduce frustration.

People: The sociability and usability needs of participants are central in community development. As in other kinds of software development, individual differences must be taken seriously, but so must the collective needs of the community. Communities for children will have different characteristics from those for adults. Support communities are different from religious, ethnic, and political discussion groups. Knowing who the members of the community will be enables developers to cater to their needs. Some communities deliberately try to restrict access to make achieving this purpose easier; others achieve the same thing by defining themselves narrowly. If the community is intended for a wide range of users, different versions of the interface may be needed. For example, basic information such as help and governance policies could be provided in different languages. There could be different versions for people with disabilities, limited experience, children, and seniors?

Policies: Supporting development of governance is often better than letting serendipity take its course. Every community will have its own culture and as it develops, agreed sets of values, norms, and other governance procedures will develop. Deciding which policies are needed, particularly early in a community's life and working with participants to develop them, and then making sure they are enforced, is an important task. Policies must be strong enough to guide community behavior, but flexible enough to change as the community evolves.

The basic requirements for the usability are similar to those for other software. Software should be consistent (e.g., have a consistent look and feel); users should be in control of what the software does, not controlled by it; and the way the software responds should be predictable (Shneiderman, 1998). Other definitions state, for example, that software should be: effective to use, efficient to use, safe to use, have good utility, be easy to learn, and easy to remember how to use (Preece et al., 2002). Although different aspects of usability for online community software is discussed (e.g., Erickson et al., 1999), coherent usability guidelines are not available. The components of usability—dialogue and social support, information display, navigation, and access—will now be described.

Dialogue and social support: These usability issues include how long it takes to learn the dialogue protocol, how difficult it is to send or read messages, or perform other actions. Users should also be satisfied with the nature of the dialogue and social support, make few errors, and be able to remember what to do when they return to the community on future occasions. Increasingly textual systems are appearing on the market with more advanced features, including ability to include in messages and ways of signaling message content and participants moods.

Information display: These usability issues include how easy it is to find information (e.g., Help) and to perform tasks with information-oriented goals with few or no errors. Other issues include whether users are satisfied with and like the information design and how it is structured.

Navigation: Navigation is a key usability issue for any web application, including online communities; particularly communities of practice that involve a large amount of information exchange. Key issues include the length of time it takes to learn to navigate through the community and its associated information resources, and the time and ease with which particular information can be found or a part of the community can be reached. How memorable and intuitive the navigation system is depends on a number of things, including the metaphor it is based on, the breadth vs. depth of the menu system, how intuitive the icons and menu names are, etc. The number of errors or dead-ends that users go down and their satisfaction with the navigation system are also key considerations. Threading and improved search facilities are making it easier to navigate many systems.

Access: This is an increasingly important usability feature for online communities. Developers have to ask themselves whether users can access the community with the equipment that they have available and whether they can read and send messages and whether response times are reasonable. If software has to be downloaded, users must be able to do this with comparative ease and in a timely manner. Whereas research on 3-D GUIs may suggest how to solve problems associated with low social presence in textual environments, these systems require high-bandwidth communications technology to use them satisfactorily. The majority of the world's users will not have access to such systems for many years, so attention to access is important for bringing greater equality to the Internet and ensuring that those from poorer regions of the world can participate. Alternatives may also be found that may include wireless telephone and other hand-held devices.

Relating Sociability With Usability

Sociability and usability are closely related and often influence each other. (In many respects, sociability is a new component of usability.) Consider, for example, taking a decision on whether community members should register to join a community. The decision to have registration, what the policy says, what information is requested from registrants, what promises are made about privacy and security, involves sociability issues. The mechanics of registering has to be designed in the software and involves usability decisions. The registration form (if a form is used) should have a clear, consistent design that reduces frustrating errors. The way terms are used should be consistent and meaningful and so should the typography. The form should also be engineered to reduce the possibility of users making frustrating errors. Table 30.3 contains nine questions that online community participants frequently ask and discusses some of the possible solutions (Preece, 2001) for improving sociability

and usability. It informs online community development by providing the users' perspective.

Determinants of Success

Excellent evaluations are published in accounts of novel systems. For example, Lili Cheng and her colleagues discuss a series of tests to evaluate prototypes of HutchWorld, a 3-D graphical chat environment for cancer patients. As well as finding ways of improving and fine-tuning their design, these researchers learned that patients wanted asynchronous communication so that they could plan to meet online to chat synchronously (Cheng, Stone, Farnham, Clark, & Zaner-Godsey, 2000). Erickson and his colleagues evaluated IBM's Babble system, to test the efficacy of a graphical representation of users online behavior and to ascertain how well it was liked (Erickson & Kellog, 2000). But, despite an increasing interest in online community design, there has been little attempt to identify criteria that indicate whether a particular community is successful or even what these criteria might be and how they could be assessed and measured?

Roxanne Hiltz discusses possible determinants of success for educational online communities (Hiltz, 1994), and a recent paper by Preece provides a more general initial set of possible determinants for sociability and usability. Some indicators of good sociability could include: the number of participants in the community (high in successful communities); the number of lurkers (the ideal number depends on critical mass of the community; Nonnecke, 2000; Nonnecke & Preece, 2000); the number of messages (high in successful communities); the number of messages per participant (high); how much reciprocity there is as indicated by, for example, the number of responses per participant (high); the amount of on-topic discussion (high); how empathic the interaction is (high in support groups, but it would vary according to the type of community); the level of trust (high); participants' satisfaction with social interaction in the community (high); the number and type of incidents that produce uncivil behavior (low in successful communities); average duration of membership (high); the percentage of people who are still members after a certain period of time (high); etc. (Preece, 2001).

Some determinants of good usability might be: speed of learning to use the interface (should be high in successful communities); retention, i.e., how much a user remembers about the mechanics of interacting with the online community software (should be high in successful communities); productivity, i.e., how long it takes to do standard tasks, such as reading or sending, searching, etc. (should be high); the number of errors that occur when doing communication tasks (should be low); users' satisfaction using the software (should be high); etc. (Preece, 2001).

Table 30.4 provides a summary of some possible determinants of success for online communities and relates them to the sociability and usability framework. In most cases, the determinants do not speak directly about purpose and policy, but they provide evidence that is indirectly indicative. To gain an overall impression of the success of the community,

TABLE 30.3. Nine Questions That Users Ask Some Sociability Implications, and Usability Solutions

Users' Questions	Sociability Implications	Usability Solutions
1. Why should I join this? community? (Purpose)	Consider what the title and content should communicate about the community's purpose. What information is needed and how should it be presented?	Provide a clear title and statement of purpose that is concise and consistent. Graphics should not detract from the main message.
2. How do I join or leave? (Policy)	Should the community be open or closed? This will depend on the sensitivity of topics discussed and whether participation needs to be controlled, etc.	Consider requiring registration. If there is registration, provide clear instructions, make the procedure short, and give reassurance that personal details are private and will not be revealed to third parties.
3. What are the rules? (Policy)	What kind of policies will support the community's purpose? Is a moderator needed to enforce rules or arbitrate in disputes? Are disclaimers, copyright regulations, etc., needed?	Provide clearly, concisely worded policies and appropriately position them. If moderation is needed, provide tools and policies to support the moderators.
4. How do I communicate with others in the community? (Policy)	Consider what newcomers will need to enable them to feel part of the community and communicate with others. What do the old-timers need? How might participants' needs change over time? Is private communication important?	Determine what kind of usability support is needed for different groups in the community. Consider providing templates, emoticons, FAQs, single messages, or digests for listservers, search facilities, ability to send private messages (i.e., back channel), etc.
5. Can I do what I want easily and get what I want? (Purpose)	Consider the social needs of the community. What's the community's purpose and who is it for? For example, is broadcast, private communication, long-term information, and synchronous and asynchronous communication needed?	Decide how to support different communication tasks (e.g., synchronous and asynchronous media), FAQs, enable users to express content and feelings and search, provide help at the right level, allow private communication, etc.
6. If I give, will I get back?	How can reciprocity be encouraged?	Acknowledge responses to questions, offers of help and support.
7. Is the community safe? (Policy)	Consider whether a moderator and stronger rules are needed to ensure appropriate behavior and support the community's purpose. Is confidentiality, security, & privacy important? How will trust be encouraged?	Find ways to: protect personal information; secure transaction processing; support private discussion; protect people from aggression, support trust by providing evidence of past behavior.
8. Can I express myself as I wish? (Purpose)	Determine the kind of communication a community with this purpose wants. How should it be supported?	Provide emoticons, content icons, consider whether avatars, personal pages, seamless links to private e-mail, etc., are needed.
9. Why should I come back? (Purpose & Policy)	Decide how to keep people interested and entice them to keep coming back. The question being asked is what's in it for me?	Provide changing content: e.g., news broadcasts, real-time discussions, encourage provocateurs and leaders to stimulate social interaction, focus on purpose, etc.

Note. From *Online Communities: Designing Usability, Supporting Sociability*, by J. Preece, 2000, Chichester, England: John Wiley & Sons. Copyright 2000 by John Wiley & Sons. Adapted with permission.

several measures of sociability and usability are needed. Furthermore, evidence from interviews and ethnographic studies will also be useful. (See later section for more about these methods.)

DIFFERENT TYPES OF COMMUNITIES

The use of the Internet to link individuals with others sharing common interests provides the scaffolding for building communities that offer support, solidarity, information, and social capital (Wellman & Frank, 2001). "The human need for affiliation is at least as strong as the need for information" (Kahin & Keller, 1995). Even though the Internet provides exposure to diverse groups and ideas, people are most strongly drawn to online groups that share their interests and concerns (Preece, 2000; Wellman, 2000; Wellman & Frank, 2001).

Online communities of interest or practice have developed to support all kinds of interests. For example, there are communities for expatriates, gardeners, genealogists, hobbyists, professionals, gamers, and also senior citizens, who have become one of the largest demographic groups on the Internet (Rainie & Packel, 2001). Spiritual groups create online communities to promote their beliefs, and there are interfaith health groups and bible study groups. Communities of practice create new products, processes, and services online. The darker side of the Internet is also represented by groups of Neo-Nazis, child pornographers, and the Klu Klux Klan, who have established online communities to recruit new members and support their organizations (Breeze, 1997; Church, 1996; Furlong, 1996;

TABLE 30.4. Some Examples of Determinants of Success for Sociability and Usability of Online Communities

Framework	Design Criteria	Examples of Determinants of Success
Sociability	Purpose	How many and what kinds of messages or comments (or comments per member) are being sent? How on-topic is the discussion? How much interactivity is occurring? How much and what kind of reciprocity occurs? What is the quality of the peoples' contributions and interactions?
	People	How many and what kinds of people are participating in the community? What do they do and what roles are they taking? How experienced are they? What are their ages, gender, and special needs, etc.?
	Policy	What policies are in place? For example, registration and moderation policies to deter uncivil behavior. How effective are the policies? How is relationship development being encouraged? For example, what kinds of policies encourage trustworthiness and how effective are these policies?
Usability	Dialogue and social support	How long does it takes to learn about dialogue and social support? How long does it actually take to send or read a message, or perform some other action, etc.? Are users satisfied? How much do users remember about dialogue and social support, and how many errors do they make?
	Information design	How long does it take to learn to find information (e.g., Help)? How long does it takes to achieve a particular information-oriented goal? How satisfied are users? How much do users remember after using the system? Can users access the information they need without errors?
	Navigation	How long does it take to learn to navigate through the communication software and Web site or to find something? Can users get where they want to go in a reasonable time? How much do users remember about navigation? How satisfied are they? How many and what kinds of errors do they make.
	Access	Can users get access to all the software components that they need? Can they download them and run them in reasonable time? Are response times reasonable? What problems do they encounter when trying to download and run software?

Note. From *Online Communities: Designing Usability, Supporting Sociability*, by J. Preece, 2000, Chichester, England: John Wiley & Sons. Copyright 2000 by John Wiley & Sons. Adapted with permission.

Gunderson, 1997; Moon & Sproull, 2000; Wenger & Snyder, 2000).

Technology user communities include the customers of corporations. Microsoft, for example, has a gateway to information and services that invites their users to join Microsoft Communities http://communities.microsoft.com/: ".... launching pad for communicating online with others about Microsoft products, technologies, and services. Converse with peers and experts in open forums." One of the best-known online technology user groups is the Linux developers. They are creating a collaborative open-source, PC-based operating system (Moon & Sproull, 2000), and the community is an informally bound group of people who share their expertise and passion for this joint project. This vigorous community comprises more than 3,000 developers, living in more than 90 countries on five continents. Through their interactions the community has developed their own ways interacting and norms for behaving (Baym, 2000; Wenger & Snyder, 2000).

There are many different kinds of online communities, and we cannot describe them all; so, in this section, we discuss patient support, education, and e-business communities. Although each community is unique and has its own characteristics, communities that share a common purpose generally share some characteristics.

Patient Support

We are witnessing doctor–patient relationships being transformed by the Internet (Rice, 2001). Patients are learning about their own problems and going to doctors empowered to discuss them on a more equal basis. Some doctors embrace this change; others feel that their expertise is challenged (Kahin & Keller, 1995). Patients who come online want to learn about their diseases; find information; get support; help fellow-suffers; and be less afraid. They can get information from Web sites, but online communities are more personal. Talking to other patients can be comforting and reassuring in ways that talking to even the most skillful and communicative physician may not be. Furthermore, getting enough face-to-face interaction with doctors is a problem everywhere in the world, and attending face-to-face support groups may not be convenient. Online communities enable patients to share experiences and relate to each other's problems (Davidson, Dickerson, & Dickerson, 2000). Other patients have been there (Preece, 1998) and can respond empathetically (Ickes, 1997), which may encourage strong relationships to develop, thus making these communities some of the most important on the Internet. The benefits that an online health community can provide for its members are especially valuable for people who lack mobility, or are socially or geographically isolated (Cummings, Sproull, & Kiesler, 2001; Davidson et al., 2000; Sproull & Keisler, 1986).

People access online health communities through the web pages, bulletin boards, listservs, and chat sites in which they create a sort of group narrative that is also typical of face-to-face self-help groups (Rappaport, 1996). Typically, an individual starts a thread by posting a question or comment to which others reply; all the threads can be read by anyone on the site. Many online health communities also provide opportunities for members to communicate privately by sending private e-mail or having side conversations (e.g., whispering in a chat room; Cummings et al., 2001; Preece, 1998, 1999a, 2000; Preece & Ghozati, 2000). It is well documented that many people choose not to post messages, but do spend a lot of time reading the conversational threads—i.e., lurking. In a study of lurking behavior in 77 listserver patient support communities, 45% of the members did not post during the 3-month period of the study (Nonnecke & Preece, 2000). Lurking in the 21 technical support groups that were also studied was much higher, at around 82%, which suggests that different categories of communities may indeed exhibit different characteristics.

According to the *Pew Internet & American Life: Online Life Report*, sixty million American adults, or 55% of those with Internet access, have researched a disease or medical condition on the Internet, and the number continues to grow (Rainie & Packel, 2001). A large proportion of those researching a disease or medical condition online go to bulletin board, UseNet News, or listserver communities (Rice, 2001). Forty-eight percent of those who sought health information online reported that the advice they found improved the way they take care of themselves and 55% said that access to the Internet improved the way they get medical and health information (Fox & Rainie, 2000). A recent edited volume by Ron Rice and James Katz is crammed full of facts and figures (Rice & Katz, 2001).

Online health communities also have a dark side. Physicians are rightly concerned about patients getting incorrect information. A University of Michigan 1999 survey of 400 health-related sites found that 6% provided incorrect information, and the Federal Trade Commission estimates that only about half of the content on health and medical Web sites is reviewed by doctors (Fox & Rainie, 2000). An increasing number of online communities now support question and answer sessions with real doctors (e.g., drkoop.com, drweil.com), but some doctors do not like this practice either, because online doctors do not see patients or know their backgrounds. These are real dangers, and patients need encouragement to become discerning consumers of medical information. The American Medical Association has launched a campaign to inform consumers that they must check the quality of health information they get online (Fox & Rainie, 2000). In addition, members of online health communities may experience negative, hostile, or malicious exchanges that come about in the online environment because people have a lack of fear of social sanctions and feelings of depersonalization (Fox, 1996; Fox & Rainie, 2000; Sproull & Keisler, 1986).

Privacy is also a big concern for those people accessing health sites on the Internet (Katz & Aspden, 2001). Eighty-nine percent (89%) of those who use the Internet to get health information express concern about a health site selling or giving away information about their online activity. Eighty-five percent fear that their insurance company might raise their rates or deny them coverage if they find out what health sites they visit, and 52% express concern that their employer could find out what online health sites they had visited (Fox & Rainie, 2000).

Supporting people coping with illness who may lack both physical and emotional stamina also requires special features. For example, as we mentioned, the developers of the Hutch-World 3-D synchronous chat environment for cancer patients discovered that they needed to include asynchronous communication so that patients did not have to be present at a particular time. Treatment regimes, days when patients felt unwell, and doctors' visits frequently prohibited their involvement in synchronous chats, which was frustrating. Also, they liked being able to leave messages when they felt like it and the ability to organize synchronous chats with other patients, family, friends, and caregivers (Cheng et al., 2000). Ninety-three percent of those people who got health information online say that being able to access help and information 24 hours a day is very important (Fox & Rainie, 2000). A recent study of an online self-help community for hearing impaired people brings new evidence that participants reported above average benefits when family and friends also participated in the online support group (Cummings et al., 2001). The study's findings also supported previous research that people who lack social support are more likely to actively participate online. So, apart from privacy and security, key sociability and usability concerns involve supporting communication and personal relationship development among people whose disease may limit their access.

Education

Distance education in which students learn from materials on the web is becoming widespread. Consequently, some students may not interact with classmates face to face, which is a concern because learning is an intrinsically social process (Hiltz, 1998; Vygotsky, 1978, 1986). Online communities, therefore, have a role in bringing social interaction to learning and supporting the learning process. Used creatively, they help to prevent digital diploma mills from developing (Noble, 1998), in which the students' learning experiences are limited to reading and absorbing facts from the web (Winner, 1995). Technology can be used to create learning communities that foster collaborative learning so that students can learn together and benefit from sharing ideas and resources supported by skillful moderators and mentors (Hiltz, 1998; Salmon, 2000). Supported by both physical and virtual communities, students are succeeding with ambitious projects that they could not have done without the Internet (Lazar et al., 1999; Lazar & Preece, 1999b). They can communicate with others in the same region, country, or across the globe, and find state-of-the-art research on the web that their professors do not know about. This adds a new dimension to learning that can be threatening for professors. More and more professors are having to accept that their role is to guide students to meaningful learning activities in a learner-centered process rather than to be the teacher in a traditional teacher-centered one (Berge & Collins, 1995; Hiltz, 1998).

Amy Bruckman describes two types of online educational communities: knowledge-building communities and "technological samba school" (Bruckman, 1999). Knowledge-based educational communities focus on knowledge sharing and collaborative learning through projects in which, for example, children from around the world collect and share data to build an understanding of environmental issues (See the TERC/National Geographic Acid Rain project at http://globalab.terc.edu/) Projects such as the Jason project (www.jasonproject.org/) allow students to participate in scientific research through interactive video-conferencing, and remote control of instruments provide rich collaborative learning environments.

Seymour Papert introduced a term technological samba schools to describe a process in which a community of people of all ages engage in a creative project using computers (Papert, 1980). He got the idea from watching a community of Brazilians—children to grandparents—learning to samba; everyone was teaching and learning. The MOOSE Crossing Project (www.cc.gatech.edu/elc/moose-crossing) is an example of a technological samba school. A MUD, it provides children with programming languages that are easy to learn so they can build virtual places and objects. As they work, they are learning creative writing and computer programming in a peer-supported environment (Bruckman, 1999).

Online professional groups that serve educators have also evolved into communities with large numbers of people seeking information and support. For example, WMST-L, a listserv created in 1991, provides a forum for women from around the world to share women studies teaching materials and ideas, network, and provide emotional, social, and professional support (Korenman, 1999). MediaMOO, a MUD (www.cc.gatech.edu/~asb/MediaMoo) for media researchers, functions like an "endless reception for a conference on media studies" (Bruckman, 1999, p. 13).

Sociability and usability considerations for educational applications depend on the purpose of the community and whether it is a closed, class-based community or an open community. Small, class-based communities have a small number of participants so critical mass can be a problem unless there are goals to motivate student involvement. Some professors therefore set tasks to be graded, but intrinsic motivation for participation is obviously more desirable. Support for discussion, collaborative project work, and access to resources (i.e., information, tools, etc.) are needed and sufficient moderation to protect students against inappropriate behavior and guide discussion. Better tools to support moderators would also be welcomed by educators. Privacy is another concern, both to protect students' grades and also comments made in discussions.

E-business

Driven by surveys such as the *Pew Internet Project: Internet Tracking Report, More online, doing more* (Pew, 2001), it has been reported that 52% of the 104 million American adults who have Internet access have bought a product online. E-business companies view building online brand communities as a marketing strategy. These companies seek to build a new kind of relationship with their customers through online communities (Hagel & Armstrong, 1997). Many companies began using the Internet by developing Web sites that provided product information, direct sales, and customer service; but, increasingly, they are hosting interactive consumer-to-consumer online community sites organized around their brand, products and services to create reinforcing, and competitively distinctive, and long-lasting relationships with consumers (McWilliams, 2000). Tightly interwoven with sales details and product information, companies use stickiness technologies like e-mail, chat rooms, affinity groups, and bulletin boards to encourage customers to stay at the site longer interacting with each other, the company,

and buying products and services (de Figueiredo, 2000; Preece, 2000). For example, the REI site (www.rei.com) reinforces the company's image as a high-quality retailer of outdoor gear that cares about their customers by embedding online communities in its Web site. These communities encourage consumers to communicate with each other about interesting hiking, biking, boating, and skiing locations and even helps them match up with each other for trips and activities. The Kodak Company's Web site, www.kodak.com, also has an embedded discussion board that serves as a gathering place for discussion of photography and visual storytelling. Companies like these hope to gain marketing edge and build product loyalty by fostering genuine relationships with and between their customers (McWilliams, 2000). Service providers, such as Yahoo and MicroSoft, host large numbers of online communities to encourage traffic to their sites where they carry advertisements.

E-business companies want to expand their markets by reaching customers worldwide via the Internet, market their products directly to the consumer, and accumulate detailed customer profiles for target marketing new and existing products and services. In addition, communications between consumers at these sites provide the companies with valuable feedback about the needs, likes, and dislikes of their consumers (de Figueiredo, 2000; Hagel & Armstrong, 1997; McWilliams, 2000; Preece, 2000; Venkatraman, 2000). Companies use customer communities to test new product ideas, involve customers in product development, monitor customers' purchase patterns, and gauge early demand for products. This type of information is used to make their brick and mortar stores, as well as their e-business sites, more responsive and efficient (Tedeschi, 2001; Venkatraman, 2000). The online arm of Toys'R Us, ToysRUs.com, noticed customers' preference for indigo game consoles and certain game titles through advanced promotion, sweepstakes promotions, and advanced sales at their online site of Nintendo's Game Boy Advance hand-held console and the software that went with it. The company had not planned to carry the most popular title sold online, Hot Potato, in their retail stores and the manufacturer of the consoles had produced consoles in equal quantities of three colors. However, as a result of the information gained from their online site, the company changed their in-store marketing plans and worked with the manufacturer to produce more indigo consoles in future production runs (Tedeschi, 2001).

People who buy from web-based retailers want value for money, their personal details to be secure and private, and to receive goods and services in a timely manner. Trust and privacy are key issues for customers. In a recent study at Brigham Young University that surveyed 4,000 adult Internet users, researchers discovered that credit card fear is the single most important factor that distinguished people who shop online from those who do not. Nonshoppers are afraid that their credit card will be stolen and that merchandise will not be delivered (Stellin, 2001). e-Bay's reputation system and Amazon review process help to alleviate some of these fears. The infrastructure for e-business sites must be designed to make it easy for customers to use the site without "sacrificing their trust about reliability, security, and privacy" (Venkatraman, 2000).

Designers of e-business communities must also consider the companies' business model and brand strategy and realize that online markets may be better served by focusing on customers rather than on products (Hagel & Armstrong, 1997; McWilliams, 2000; Tedeschi, 2001). Four Smart Ways to Run Online Communities (Williams & Cothrel, 2000) point out that member development, asset management, and community relations involve issues related to technological choices, social policies, and practices. Designs have to be both socially and technically feasible (Figallo, 1998; Preece et al., 2002) and Kim stresses practices for actively growing online communities, including staging events and ensuring that there is always fresh content (Kim, 2000).

RESEARCH TECHNIQUES

There is a large catalog of research techniques that can be drawn on from the social science, psychology, HCI, and CSCW. Which are actually chosen at any time depends on the question(s) to be addressed, and the training and skills of those doing the research. However, there are some special challenges associated with researching into online communities. Intervening in how a community functions changes the fundamental nature of the community and would also be unethical. Conducting surveys can also be tricky because a community's population may change from day to day or be unknown, which makes unbiased sampling impossible.

At a workshop in the early 1990s, Stu Card characterized the growth of new disciplines in four stages (Card, 1991; Olson & Olson, 1997). At the time he was considering HCI, but his model is widely applicable and can be used to analyze the emergence of interest in online communities. The first stage in the model involves starting to build, observe, and evaluate communities, which continues to intensify during the second stage. In the third stage, dimensions of success are identified that lead to the development of theories and laws in the fourth stage, which characterize a mature discipline.

When applied to online communities, this model helps to explain how this new field of research is developing. There are examples of case studies, rich ethnographic descriptions, and anecdotes about experiences in online communities. Surveys, interviews, and data logging are also starting to be widely used. Models and theory from other fields are imported to support and help explain the phenomena observed. Some of these may have particular techniques associated with them as in social network theory that uses sociograms to show social relationships.

In this section, we briefly review the most commonly used research techniques and explain why and when they are used.

Ethnography and Associated Techniques

Ethnography is a popular approach for understanding the dynamics of online communities, particularly early in the study of a community. This research tool, borrowed from anthropology and sociology, is a qualitative research method for understanding how technology is used in situ. The purpose of ethnographic research is to build a rich understanding of a group or situation from the point of view of its members/participants (Fetterman, 1998). Ethnographic research is becoming an increasingly popular method for studying the Internet because of the

unique way it contributes to understanding technology, and "the culture that enables it and is enabled by it" (Hine, 2000). Ethnography is especially useful for studying online communities because it causes little disturbance to the community. It is also useful because research questions are refined throughout the study as valuable details become known.

A variety of data collection techniques are used in ethnographic research, including participant observation in which researchers participate in the community. This involves observing what is happening, doing in-depth interviews, taking notes, collecting artifacts, and participating in the activities of the community to gain a better understanding about how the community functions (Preece et al., 2002; Walcott, 1999).

There are important ethical considerations. How much and how often should inform the community about their study and how much information should be revealed about the data sources (Herring, 1996). Brief descriptions of two examples help to illustrate how ethnography is used to understand online communities and how the researchers dealt with the ethical issues.

The first is a longitudinal study by Nancy Baym in which she joined an online community interested in soap operas as a participant observer for over a year to understand how the community functions (Baym, 1997, 2000). Baym comments: "As a longtime fan of soap operas, I was thrilled to discover this group. It was only after I had been reading daily and participating regularly for a year that I began to write about it. As the work evolved, I have shared its progress with the group members and found them exceedingly supportive and helpful" (Baym, 1997). By adopting this honest approach, she gained the trust of the community, who offered support and helpful comments. Because the researcher's presence can be hidden so easily and people's privacy abused, sensitivity to these ethical issues is needed (Markham, 1998). As Dr. Baym participated, she learned who the key characters were, how people interacted, their values, and the types of discussions in which they engaged. She also adapted interviewing and survey techniques to support her observations and to enrich her account of the community (Baym, 2000).

In a study lasting several years, David Silver observed and compared the day-to-day activities in two networked communities: the Blacksburg Electronic Village and the Seattle Community Network (Silver, 1999). These participant observations set the stage for face-to-face and online interviews that led to a deeper understanding of the differences between the two communities. He realized that the more market-driven nature of the Blacksburg Electron Village, which was originally set-up with funding from commercial organizations (Cohill & Kavanaugh, 1997), compared with the grass-roots development of the Seattle Community by community reactivists (Schuler, 1994) had far-reaching consequences for the character of the two communities. Silver asked the communities' permission to study them and frequently shared his findings with them. This also enabled him to check that he had understood particular events from the community's and participant's perspective. On some occasions, Silver also did face-to-face interviews.

Techniques that are frequently used with ethnography for data analysis include content analysis, discourse analysis, and various types of linguistic analysis. For example, content analysis was used to examine how much of the communication in a patient support community was empathic and how much was factual (Preece, 1999a) and to compare the type of communication that occurred in different kinds of communities (Preece & Ghozati, 1998, 2000).

In discourse and other types of linguistic analysis, the researchers focus more strongly on the intentions of the communicators. For example, Susan Herring did a study in which she investigated why textual computer-mediated communication is so popular, despite the inherent incoherence caused by repetition of messages, fragmented discussions, and breaks in turn-taking, etc. Herring suggests that a possible explanation is "the ability of users to adapt to the medium, and to except incoherence in exchange for greater interactivity" and communication (Herring, 1999). In other words, if users get enough from the technology they will put up with problems or find creative ways to work around them—a concept that has been called adaptive structuration by some researchers (DeSanctis & Gallupe, 1987; Hiltz & Turoff, 1993).

Erickson (1997) proposes using the concept of genre to analyze online discourse. He believes that analysis of the purpose of the communication, its regularity of form and substance, and the institutional (for those that are institutionally based), social, and technological forces that affect communication is more important to understanding online communication than the relationships between community members. This method may also be useful when looking at online communities like The Palace™, which are supported by a graphical environment for meeting and chatting, where participants do not form lasting relationships, share few values, and do not count on each other for help or provide information.

Not only does ethnography fit well within Card's model of how early studies are conducted in new disciplines, it is also a fundamental approach for understanding community, having developed from anthropology. The technique causes minimal disruption to the community and provides rich descriptions laced with persuasive anecdotes. Although quantitative comparisons are sometimes made (Fetterman, 1998; e.g., using content or other analysis techniques), they are often missing, which is seen as a limitation of this approach by some researchers.

Data Logging

Data logging can be used to examine mass interaction without disturbing the community. Examples include studies of the demography of UseNet News (Smith, 1999) and lurking in list-server communities (Nonnecke & Preece, 2000). Smith's (1999) research mapping the social structure of the Usenet provides a general topology that shows the amount of activity and relationships within this huge and geographically diverse network. The study looked at the technical and social components of the UseNet examining variation across the whole system in hierarchies, newsgroups, posts, posters, and cross posting (Smith & Kollock, 1999).

Nonnecke and Preece conducted a demographic study of lurking (those who read, but do not post) on e-mail discussion lists in health and software support groups (Nonnecke &

Preece, 2000). Lurkers are of interest to researchers because estimates generally assume that lurkers make up more than 90% of the population of online communities. However, as previously described, the results of this study showed that there were considerably fewer in these communities. Different communities, supported by different kinds of software, are likely to vary on this as on most other characteristics.

Visualization tools are starting to be used by online community researchers to explore trends in large data sets. MIT Media Lab's Sociable Media Research Group, http://smg.media.mit.edu, engages in research projects to develop intuitive visual representations of social information that provide a vivid sense of the abstract space in online groups (Donath et al., 2001; Xiong & Donath, 1999).

A paper by Mark Smith and Andrew Fiore (2001) provides a review of some of these techniques and an example of the application of the treemap visualization technique (Shneiderman, 1992) for analyzing UseNet News groups. It also discusses a tool for visualizing discourse in very large conversations (e.g., Sack, 2000b).

Together, data logging and ethnography provide a broad picture of online community activity in which both qualitative and quantitative aspects are represented.

Questionnaires

Questionnaires are useful for collecting demographic information and have the advantage that they can be distributed by hand to local participants or posted via e-mail or on the web (Harper, Slaughter, & Norman, 1997; Lazar & Preece, 1999a). In a study to identify the defining characteristics of online community, Terry Roberts e-mailed questionnaires to a selection of UseNet News Groups (Roberts, 1998). Three dimensions were used to select the groups: topic area, traffic, and the gender balance in the groups. Using analysis of variance primarily, Roberts identified six dimensions that "add up to a factor that one might call community." Another study used online questionnaires to assess the resistance of different demographic groups to participating in an online community for career changers (Andrews, Preece, & Turoff, 2001). This study emphasizes the importance of developing a thorough understanding of a demographic group's distinctive characteristic to build sustainable online communities for the target audience.

Although the Internet is a powerful and inexpensive distribution mechanism, there are major sampling problems because Internet populations are often unknown. For this reason, national census records are being used to obtain unbiased samples (see www.webuse.umd.edu).

Experiments and Quasi-experiments

Laboratory studies are valuable for testing the usability of the interface and users' reactions to new user interface features. For example, trust is a key factor for developing relationships in e-business, which can be investigated using laboratory experiments (Bos, Gergle, Olson, & Olson, 2001; Zheng, Bos, Olson, & Olson, 2001) to examine the effect of providing customers with different information about trust policies. However, it cannot usually be assumed that the results apply directly to online communities in the wild. However, some researchers are working to develop quasi-experimental techniques with better ecological validity. For example, B. J. Fogg and his colleagues (2001) worked with two companies on an experiment in which they intervened to change banner ads to study users' perception of the reputability of the web. Such approaches could conceivably be used to investigate the impact of change in software design on online communities. Roxanne Hiltz has also developed powerful quasi-experimental approaches for comparing performance of students learning online with similar groups learning in classrooms (Hiltz et al., 1986; Hiltz, Turoff, & Johnson, 1989).

As with many new areas of research, researching online communities poses new challenges. Gradually, as this field matures, we can expect to see other techniques imported from other disciplines and adapted to provide both qualitative and quantitative information about life online.

BRIEF SUMMARY AND AGENDA FOR FUTURE WORK

In this chapter, we pointed out that there is no single definition of online community; different researchers focus on different issues (first section). In the second section, we reviewed research that has contributed to our understanding of online community, starting with research about communication between pairs and small groups. Then we discussed group dynamics and interaction. Many of these concepts were then drawn together in the third section, which focused on how knowledge of sociability and usability can be used in online community development. The participatory community-centered development method and a framework for sociability and usability were proposed.

The fourth section examined three groups of online communities: patient support communities, education communities, and e-business communities. Many other groups could have been discussed, such as religious, sports, and entertainment communities, which each have particular needs and characteristics; but, because of space limitations, we focused on these three popular groups.

In the fifth section, we returned to the issue of research and discussed key techniques used to research online communities. Currently, ethnography tends to dominate, but data logging is becoming increasingly popular. Because these methods do not attract attention, it is important for researchers to be cognizant of participants' rights to privacy and other ethical concerns.

The previous sections of this chapter now set the scene for proposing an agenda for future work for researchers and practitioners.

Researchers

Detailed agendas for future interdisciplinary research online has been proposed (Brown et al., 1999a, 1999b, 1999c). There are many topics that could be mentioned; some that are specifically relevant to this chapter include:

- Application of fundamental community concepts from the social sciences to understand online communities (e.g., development of trust, social dilemma, reciprocity, weak and strong ties, etc.) and development of new theories that explain social interaction online.
- Techniques for showing and supporting social interaction.
- Comparative studies of communities that look for similarities and differences.
- Case studies of the relationship between physical–virtual relationships, particularly the roles that online communities play in people's lives.
- Scalability, which includes taking account of universal usability and sociability in large communities of tens of thousands or millions of people.
- Interactive styles between small hand-held devices and between these devices and the web.
- Development methods, frameworks to support sociability and usability in online community development, and techniques and measures for assessing the success of online communities.

Practitioners

Similarly, there are many topics for practitioners. Some of the most important ones include:

- Creating development processes that take account of sociability and usability. Every community is different, so it is essential to pay attention to the details of its purpose and the needs of the members.
- Focusing on designing for universal sociability and usability when appropriate. Versions for low bandwidth are particularly important.
- Pay attention to different stages of development of online communities and be sure to provide moderator support early in the community's life. Support moderators and managers well, so that they can support the community.
- Provide access from PCs, mobiles, web-tops, phones, hand-held machines, and wearables with well-designed interfaces and interaction.
- Continue to find ways to integrate asynchronous and synchronous software, so that users are not shocked by suddenly interacting with a new interface.
- Develop ways of scaling online communities to support large numbers of people from different cultures, with different kinds of experience using a variety of equipment for a variety of purposes (e.g., political communities, health communities, cultural communities, etc.).
- Develop ways of evaluating and measuring success that go beyond membership and participation metrics to reveal how well the sociability support and usability design supports the community.

Future research and development must focus on better understanding and developing theories to explain social interaction in textual and graphical communities accessed via a range of devices. Already existing usability methods need to be adapted for online communities. Most important, researchers and practitioners must pay attention to universal sociability and usability, so that the millions of people who do not have state-of-the-art broadband communications and who do not speak English can participate. To achieve this will also require the cooperation of hardware and software manufacturers, telecoms, service providers, and input from multidisciplinary research teams.

ACKNOWLEDGMENTS

Portions of the text (parts of the first and second sections) were adapted from *Online Communities: Designing Usability, Supporting Sociability*, by J. Preece, 2000, Chichester, England: John Wiley & Sons. Copyright 2000 by John Wiley & Sons. Adapted with permission. The authors thank John Wiley & Sons for allowing us to use this material.

References

Ackerman, M. S., & Starr, B. (1995). *Social activity indicators: Interface components for CSCW systems.* Paper presented at the ACM Symposium on User Interface: Interface components for CSCW systems.

Andrews, D., Preece, J., & Turoff, M. (2001, January 3–6, 2001). *A conceptual framework for demographic groups resistant to online community interaction.* Paper presented at the IEEE Hawaiian International Conference on System Science (HICSS).

Axelrod, R. (1984). *The evolution of cooperation.* New York: Basic Books.

Baym, N. (1997). Interpreting soap operas and creating community: Inside an electronic fan culture. In S. Kiesler (Ed.), *Culture of the Internet* (pp. 103–119). Mahwah, NJ: Lawrence Erlbaum Associates.

Baym, N. K. (2000). *Tune in, log on: Soaps, fandom, and online community.* Thousand Oaks, CA: Sage Publications.

Bennett, J. (1984). Managing to meet usability requirements. In J. Bennett, D. Case, J. Sandelin, & M. Smith (Eds.), *Visual display terminals: Usability issues and health concerns.* Englewood Cliffs, NJ: Prentice-Hall.

Berge, Z. L. (1992). *The role of the moderator in a Scholarly Discussion Group (SDG)* [On-line]. Available: http://star.ucc.nau.edu/star.ucc.nau.edu/~mauri/moderate/zlbmod.html

Berge, Z. L., & Collins, M. P. (1995). Computer-mediated communication and the online classroom: Overview and perspectives. In Z. Berge & M. Collins (Eds.), *Computer mediated communication and the online classroom* (Vol. I, pp. 1–10). Cresskill, NJ: Hampton Press, Inc.

Beyer, H., & Holtzblatt, K. (1998). *Contextual design. defining customer-centered systems.* San Francisco, CA: Morgan Kaufmann Publishers, Inc.

Bos, N., Gergle, D., Olson, J. S., & Olson, G. M. (2001). *Being there versus seeing there: Trust via video.* Paper presented at the CHI 2001, Seattle, WA.

Breeze, M.-A. (1997). Quake-ing in my boots: <Examining> clan: Community<construction in an online gamer population. *Cybersociology Magazine*, 1–5.

Brown, J., van Dam, A., Earnshaw, R., Encarnacao, J., Guedj, R., Preece, J., Shneiderman, B., & Vince, J. (1999a). Human-centered computing, online communities and virtual environments. *ACM SIGCHI Interactions, 6*(5), 9–16.

Brown, J., van Dam, A., Earnshaw, R., Encarnacao, J., Guedj, R., Preece, J., Shneiderman, B., & Vince, J. (1999b). Human-centered computing, online communities and virtual environments. *IEEE Computer Graphics, 19*(6), 70–74.

Brown, J., van Dam, A., Earnshaw, R., Encarnacao, J. G. R., Preece, J., Shneiderman, B., & Vince, J. (1999c). Special report on human-centered computing, online communities and virtual environments. *ACM SIGGRAPH Computer Graphics, 33*(3), 42–62.

Bruckman, A. (1993). *Gender swapping on the internet.* Paper presented at the The Internet Society (INET '93) Conference, San Francisco, CA.

Bruckman, A. (1999). *The day after Net day: Approaches to educational use of the Internet* [On-line]. Available: asb@cc.gatech.edu

Card, S. (1991). *Presentation on the theories of HCI at the NSF workshop on human computer interaction.* Washington, DC: National Science Foundation.

Carroll, J. M., & Rosson, M. B. (2001). *Better home shopping of new democracy? Evaluating community network outcomes.* Paper presented at the CHI 2001, Seattle, WA.

Cheng, L., Stone, L., Farnham, S., Clark, A. M., & Zaner-Godsey, M. (2000). *Hutchworld: Lessons learned. A collaborative project: Fred Hutchsinson.* Cancer Research Center & Microsoft Research. Paper presented at the Virtual Worlds Conference 2000, Paris, France.

Church, S. (1996, August). Irish expatriate keeps in touch via the Web. *CMC Magazine, 3*(8), 1–3.

Clark, H. H., & Brennan, S. E. (1991). Grounding the communication. In L. Resnick, J. M. Levine, & S. D. Teasley (Eds.), *Perspectives on socially shared cognition* (pp. 127–149). Washington, DC: APA.

Cohill, A. M., & Kavanaugh, A. L. (1997). *Community networks: Lessons from Blacksburg, Virginia.* Norwood, MA: Artech House.

Collins, M. P., & Berge, Z. L. (1997, March 24–28). *Moderating online electronic discussion groups.* Paper presented at the 1997 American Educational Research Association (AREA) Meeting, Chicago, IL.

Constant, D., Sproull, L., & Kiesler, S. (1996). The kindness of strangers: The usefulness of electronic weak ties for technical advice. *Organization Science, 7*(2), 119–135.

Culnan, M. J., & Markus, M. L. (1987). Information technologies. In F. M. Jablin, L. L. Putnam, K. H. Roberts, & L. W. Porter (Eds.), *Handbook of organizational communication: An interdisciplinary perspective* (pp. 420–443). Newbury Park, CA: Sage.

Cummings, J. N., Kiesler, S. B., & Sproull, L. (2002). Beyond hearing: Where real world and online support meet. *Group Dynamics: Theory, Research, and Practice, 6*(1), 78–88.

Daft, R. L., & Lengel, R. H. (1986). Organizational information requirements, media richness and structural design. *Management Science, 32*, 554–571.

Davidson, K., Dickerson, J., & Dickerson, S. (2000). Who talks: The social psychology of illness support groups. *American Psychologists, 55*(2), 205–217.

de Figueiredo, J. M. (2000). Finding sustainable profitability in electronic commerce. *Slorn Management Review, 41*(4), 41–52.

DeSanctis, G., & Gallupe, R. B. (1987). A foundation for the study of group decisions support systems. *Management Science, 33*(5), 589–609.

Donath, J., Lee, H.-Y., Boyd, D., & Goler, J. (2001, June 10). *Loom2-intuitively visualizing Usenet* [On-line]. Available: http://smg.media.mit.edu/projects/loom2/

Dumas, J. S., & Redish, J. C. (1999). *A practical guide to usability testing* (rev. ed.). Exeter, England: Intellect.

Eason, K. D. (1988). *Information technology and organisational change*. London: Taylor Francis.

Eisenberg, N., & Strayer, J. (1987). Critical issues in the study of empathy. In N. Eisenberg & J. Strayer (Eds.), *Empathy and its development* (pp. 3-13). Cambridge, UK: Cambridge University Press.

Ellis, C. A., Gibbs, S. J., & Rein, G. L. (1991). Groupware: Some issues and experiences. *Communications of the ACM, 34*(1), 38-58.

Erickson, T. (1997, January 6-10). *Social interaction on the the net: Virtual community as participatory genre*. Paper presented at the Thirtieth Hawaii International Conference on System Sciences, Maui, HI.

Erickson, T. (1999). *Persistent conversation: Discourse as document*. Paper presented at the Thirtieth Hawaii International Conference on Systems Science, Maui, HI.

Erickson, T., & Kellog, W. A. (2000). Social translucence: An approach to designing systems that support social processes. *ACM Transactions on Computer-Human Interaction, 7*(1), 59-83.

Erickson, T., Smith, D. N., Kellog, W. A., Laff, M., Richards, J. T., & Bradner, E. (1999). *Socially translucent systems: Social proxies, persistent conversation, and the design of Babble*. Paper presented at the CHI 99 Human Factors in Computing Systems, Philadelphia, PA.

Etchegoyen, R. H. (1991). *The fundamentals of psychoanalytic technique*. New York: Karnac Books.

Farnham, S., Smith, M. A., Preece, J., Bruckman, A., & Schuler, D. (2001, March 31-April 5). *Integrating diverse research and development. Approaches to the construction of social cyberspaces*. Paper presented at the Chi 2001, Seattle, WA.

Feenberg, A. (1989). The written word: On the theory and practice of computer conferencing. In R. Mason & A. Kaye (Eds.), *Mindweave: Communication, Computers and Distance Education*. New York: Pergamon Press.

Feenberg, A. (1993). Building a global network: The WBSI experience. In L. M. Harrisim (Ed.), *Global networks: Computers and international communication* (pp. 185-220). Cambridge, MA: MIT Press.

Fernback, J. (1999). There is a there there. Notes toward a definition of cybercommunity. In S. Jones (Ed.), *Doing internet resarch. Critical issues and methods for examining the net* (pp. 203-220). Thousand Oaks, CA: Sage Publications.

Fetterman, D. M. (1998). *Ethnography: Step by step* (2nd ed.; Vol. 17). Thousand Oaks, CA: Sage Publications.

Figallo, C. (1998). *Hosting web communities*. New York: John Wiley & Sons, Inc.

Fogg, B. J., Marshall, J., Kameda, T., Solomon, J., Rangnekar, A., Boyd, J., & Brown, B. (2001). *Web credibility research: A method for online experiments and early study results*. Paper presented at the CHI 2001, Seattle, WA.

Fox, S. (1996). *Listserv as a virtual community: A preliminary analysis of disability-related electronic listservs*. Paper presented at the Speech Communication Association. San Diego, CA.

Fox, S., & Rainie, L. (2000). *The online health care revolution: How the Web helps Americans take better care of themselves*. Washington, DC: The Pew Internet & American Life Project.

Fukuyama, F. (1995). *Trust*. New York: Free Press Paperbacks, Simon & Schuster.

Furlong, M. (1996). *Transcript of Mary Furlong Interview*. Web Networks, Inc. [On-line]. Available: www.transmitmedia.com/svr/vault/furlong/maryf_transcript.html (July 12, 2001).

Gergen, K. (1997). Social saturation and the populated self. In G. E. H. C. L. Selfe (Ed.), *Literacy, technology, and society: Confronting the issues* (pp. 12-36). Upper Saddle River, NJ: Prentice Hall.

Granovetter, M. (1973). The strength of weak ties. *American Journal of Sociology, 78*, 1360-1380.

Granovetter, M. (1982). The strength of weak ties: A network theory revisited. In P. M. N. Lin (Ed.), *Social Structure and Network Analysis* (pp. 105-130). Beverly Hills, CA: Sage.

Greenbaum, J., & Kyng, M. E. (1991). *Design at work: Cooperative design of computer systems*. Hillsdale, NJ: Lawrence Erlbaum Associates.

Gunderson, G. (1997, Fall). Spirituality, community, and technology: An interfaith health program goes online. *Generations, 21*, 42-45.

Hagel, J. I., & Armstrong, A. G. (1997). *NetGain: Expanding markets through virtual communities*. Boston, MA: Harvard Business School Press.

Hamman, R. B. (1999). *Computer networks linking network communities: Effects of AOL use upon pre-existing communities* [On-line]. Available: http://www.socio.demon.co.uk/cybersociety/

Harper, B., Slaughter, L., & Norman, K. (1997). *Questionnaire administration via the WWW: A validation and reliability study for a user satisfaction questionnaire*. Paper presented at the Proceedings of WebNet 97: International Conference on the WWW, Internet and Intranet.

Haythornthwaite, C., & Wellman, B. (1998). Work, friendship, and media use for information exchange in a networked organization. *Journal of the American Society for Information Science, 49*(12), 1101-1114.

Herring, S. (1992, October). *Gender and participation in computer-mediated linguistic discourse*. ERIC Clearinghouse on Languages and Linguistics.

Herring, S. (1999). *Interactional coherence in CMC*. Paper presented at the Thirty-Second Annual Hawaii International Conference on Systems Sciences, Maui, HI.

Herring, S. C. (Ed.). (1996). *Computer-mediated communication: Linguistic, social and cross-cultural perspectives*. Philadelphia, PA: John Benjamins Publishing Company.

Hiltz, S. R. (1985). *Online communities: A case study of the office of the future*. Norwood, NJ: Ablex Publishing Corp.

Hiltz, S. R. (1994). *The virtual classroom. Learning without limits via computer networks*. Norwood, NJ: Ablex Publishing Corporation.

Hiltz, S. R. (1998). *Collaborative learning in asynchronous learning networks: Building learning communities*. Paper presented at the WEB98, Orlando, FL.

Hiltz, S. R., Johnson, K., & Turoff, M. (1986). Experiments in group decision making; communication process and outcome in face to face versus computerized conferencing. *Human Communication Research, 13*, 225-252.

Hiltz, S. R., & Turoff, M. (1993). *The network nation: Human communication via computer* (Rev. ed.). Cambridge, MA: MIT Press.

Hiltz, S. R., Turoff, M., & Johnson, K. (1989). Experiments in group decision making. 3. Disinhibition, deindividuation, and group process in pen name and real name computer conferences. *Decision Support Systems, 5*, 217-232.

Hine, C. (2000). *Virtual ethnography*. Thousand Oaks, CA: Sage Publications.

Hodges, S. D., & Wegner, D. M. (1997). Automatic and controlled empathy. In W. Ickes (Ed.), *Empathic accuracy* (pp. 311-339). New York: The Guilford Press.

Ickes, W. (1993). Empathic accuracy. *Journal of Personality, 61*, 587-610.

Ickes, W. (1997). *Empathic accuracy*. New York: The Guilford Press.

Jones, Q. (1997). Virtual-communities, virtual-settlements and cyber-archaeology: A theoretical outline. *Journal of Computer Mediated Communication, 3*, 3.

Jones, S. (1999). Studying the net: Intricacies and issues. In S. Jones (Ed.), *Doing Internet research. Critical issues and methods for examining the net* (pp. 1–27). Thousand Oaks, CA: Sage Publications.

Jones, S. E. (1998). *Virtual culture: Identity & communication in cybersociety*. London: Sage.

Kahin, B., & Keller, J. (1995). *Public access to the internet*. Cambridge, MA: MIT Press.

Katz, J. E., & Aspden, P. (2001). Networked communication practices and security and privacy of electronic health care records. In R. E. Rice & J. E. Katz (Eds.), *The Internet and health communication* (pp. 393–415). Thousand Oaks, CA: Sage Publications.

Kiesler, S. E. (1997). *Culture of the Internet*. Mahwah, NJ: Lawrence Erlbaum Associates.

Kim, A. J. (1998). *Secrets of successful web communities*. Amy Jo Kim (January 8, 1999).

Kim, A. J. (2000). *Community building on the Web: Secret strategies for successful online communities*. Berkeley, CA: Peachpit Press.

King, S. (1994). Analysis of electronic support groups for recovering addicts. *Interpersonal Computing and Technology: An Electronic Journal for the 21st Century (IPCT), 2*(3), 47–56.

Kling, R. (1996). Social relationships in electronic forums: Hangouts, salons, workplaces and communities. In R. Kling (Ed.), *Computerization and controversy: Value conflicts and social choices* (2nd ed.). San Diego, CA: Academic Press.

Kollock, P. (1998). The economies of online cooperation: Gifts and public goods in cyberspace. In M. Smith & P. Kollock (Eds.), *Communities in cyberspace* (pp. 220–239). London: Routledge.

Kollock, P. (1999). The production of trust in online markets. In E. J. Lawlwer, M. Macy, S. Thyne, & H. A. Walker (Eds.), *Advances in group processes* (Vol. 16). Greenwich, CT: JAI Press.

Kollock, P., & Smith, M. (1999). Chapter 1. Communities in cyberspace. In M. A. S. P. Kollock (Ed.), *Communities in cyberspace* (pp. 3–25). London: Routledge.

Korenman, J. (1999). Email forums and women's studies: The example of WMST-L. In B. Pattanaik (Ed.), *CyberFeminism* (pp. 80–97). Melbourne, Australia: Spinifex Press.

Kraut, R., Patterson, M., Lundmark, V., Kiesler, S., Mukopadhyay, T., & Scherdis, W. (1998). Internet paradox: A social technology that reduces social involvement and psychological well-being? *American Psychologist, 53*(9), 1017–1031.

Lanzetta, J. T., & Englis, B. G. (1989). Expectations of cooperation and competition and their effects on observer's vicarious emotional responses. *Journal of Personality and Social Psychology, 56*, 543–544.

Lazar, J., Hanst, E., Buchwater, J., & Preece, J. (2000). Collecting user requirements in a virtual population: A case study. *WebNet Journal: Internet Technologies, Applications and Issues, 2*(4), 20–27.

Lazar, J., & Preece, J. (1999a). Designing and implementing web-based surveys. *Journal of Computer Information Systems, xxxix*(4), 63–67.

Lazar, J., & Preece, J. (1999b). *Implementing service learning in an online communities course*. Paper presented at the Proceedings of the 1999 Conference of the International Association for Information Management. Charlotte, NC, 22–27.

Lazar, J., Tsoa, R., & Preece, J. (1999). One foot in Cyberspace and the other on the ground: A case study of analysis and design issues in a hybrid virtual and physical community. *WebNet Journal: Internet Technologies, Applications and Issues, 1*(3), 49–57.

Lea, M., O'Shea, T., Fung, P., & Spears, R. (1992). 'Flamming' in computer-mediated communication: Observations, explanations, and implications. In M. Lea (Ed.), *Contexts of computer mediated communication*. London: Harvester-Wheatsheaf.

Lehnert, W. G. (1998). *Internet 101. A beginners guide to the Internet and the World Wide Web*. Reading, MA: Addison-Wesley Longman, Inc.

Levenson, R. W., & Ruef, A. M. (1992). Empathy: A physiological substrate. *Journal of Personality and Social Psychology, 63*, 234–246.

Markham, A. N. (1998). *Life online: Researching real experience in virtual space*. Walnut Creek, CA: Altamira.

Markus, M. L. (1987). Toward a critical mass theory of interactive media: Universal access, interdependence and diffusion. *Communication Research, 14*, 491–511.

Markus, M. L. (1990). Toward a critical mass theory of interactive media: Universal access, interdependence and diffusion. In J. Funk & C. Steinfield (Eds.), *Organizations and communication technology* (pp. 194–218). Newbury Park, CA: Sage.

McWilliams, G. (2000). Building stronger brands through online communities. *Sloan Management Review, 41*(3), 43–54.

Moon, J. Y., & Sproull, L. (2000). Essence of distributed work: The case of the Linus kernal. *firstmonday, 5*(11), November 5th, 2000. [Online: www.firstmonday.org/issues/issue5_11/index.html]

Morris, M., & Ogan, C. (1996). The internet as mass medium. *Journal of Communication, 46*(1).

Muller, M. J. (1992). *Retrospective on a year of participatory design using the PICTIVE technique*. Paper presented at the Proc. CHI'92 Human Factors in Computing Systems. Monterey, CA.

Mumford, E. (1983). *Designing participatively*. Manchester, UK: Manchester Business School.

Newell, A. F., & Gregor, P. (1997). Human computer interaction for people with disabilities (chapter 35). In M. G. Helander, T. K. Landauer, & P. V. Prabhu (Eds.), *Handbook of human computer interaction* (pp. 813–824). Amsterdam, Holland: Elsevier.

Nielsen, J. (1993). *Usability engineering*. Boston: AP Professional.

Nielsen, J., & Mack, R. L. (1994). *Usability inspection methods*. New York: John Wiley & Sons, Inc.

Noble, D. (1998). Digital diploma mills: The automation of higher education. *firstmonday, 3*(1), January 5th, 1998. [Online: www.firstmonday.dk/issues/issue3_1/index.html]

Nonnecke, B. (2000). *Why people lurk online*. London: South Bank University.

Nonnecke, B., & Preece, J. (2000). *Lurker demographics: Counting the silent*. Paper presented at the CHI 2000 Conference on Human Factors in Computing Systems, The Hague, The Netherlands.

Nonnecke, B., & Preece, J. (2001). *Why lurkers lurk*. Paper presented at the AMCIS Conference, Boston.

Norman, D. A. (1986). Cognitive engineering. In D. Norman & S. Draper (Eds.), *User-centered systems design*. Hillsdale, NJ: Lawrence Erlbaum Associates.

Ogan, C. (1993). Listserver communication during the Gulf War: What kind of medium is the electronic bulletin board. *Spring Journal of Broadcasting and Electronic Media, Spring*, 177–195.

Olson, G. M., & Olson, J. S. (1997). Research on computer supported cooperative work (chapter 59). In M. Helander, T. K. Landauer, & P. Prabhu (Eds.), *Handbook of human computer interaction* (2nd ed.) (pp. 1433–1456). Amsterdam: Elsevier.

Olson, G. M., & Olson, J. S. (2000). Distance matters. *TOCHI*.

Papert, S. (1980). Computer-based microworlds as incubators for powerful ideas. In R. P. Taylor (Ed.), *The computer in the school: Tutor, tool, tutee*. New York: Teacher College Press.

Preece, J. (1998). Empathic communities: Reaching out across the Web. *Interactions Magazine, 2*(2), 32–43.

Preece, J. (1999a). Empathic communities: Balancing emotional and factual communication. *Interacting with Computers, 12*, 63–77.

Preece, J. (1999b). Empathy online. *Virtual Reality, 4*, 1–11.

Preece, J. (1999c, December). What happens after you get online? Usability and sociability. *Information Impacts Magazine*.

Preece, J. (2000). *Online communities: Designing usability, supporting sociability*. Chichester, England: John Wiley & Sons.

Preece, J. (2001). *Designing usability, supporting sociability: Questions participants ask about online communities*. Paper presented at the Human-Computer Interaction INTERACT'01 Tokyo.

Preece, J. (2001). Sociability and usability in online communities: Determining and measuring success. *Behaviour & Information Technology, 20*(5), 347–356.

Preece, J., & Ghozati, K. (1998). *In search of empathy online: A review of 100 online communities*. Paper presented at the Proceedings of the 1998 Association for Information Systems Americas Conference, Baltimore, MD.

Preece, J., & Ghozati, K. (2000). Experiencing empathy online. In R. E. R. A. J. E. Katz (Ed.), *The Internet and health communication: Experience and expectations*. Thousand Oaks, CA: Sage.

Preece, J., Rogers, Y., & Sharp, H. (2002). *Interaction design: Beyond human-computer interaction*. New York: John Wiley & Sons.

Rainie, L., & Packel, D. (2001). *More online, doing more*. Washington, DC: Pew Internet & American Life Project.

Rappaport, S. H. (1996). *Supporting the 'clinic without walls' with an event-directed messaging system integrated into an electronic medical record*. Paper presented at the Proceedings of the American Medical Informatics Association.

Reid, E. (1993). Electronic chat: Social issues on internet relay chat. *Media Information Australia, 67*, 62–70.

Rheingold, H. (1993). *The virtual community. Homesteading on the electronic frontier*. Reading, MA: Addison-Wesley Publishing.

Rheingold, H. (1994). A slice of life in my virtual community. In L. M. Harasim (Ed.), *Global networks: Computers and international communication* (pp. 57–80). Cambridge, MA: MIT Press.

Rice, R., & Love, G. (1987). Electronic emotion: Socioemotional content in a computer-mediated communication network. *Communication Research, 14*(1), 85–108.

Rice, R. E. (1987). Computer mediated communication and organizational innovations. *Journal of Communication, 37*, 85–108.

Rice, R. E. (1993). Media appropriateness. Using social presence theory to compare traditional and new organizational media. *Human Communication Research, 19*(4), 451–484.

Rice, R. E. (2001). The Internet and health communication: A framework of experiences. In R. E. Rice & J. E. Katz (Eds.), *The Internet and health communication: Experiences and expectations* (pp. 5–46). Thousand Oaks, CA: Sage Publications.

Rice, R. E., & Barnett, G. (1986). Group communication networks in electronic space: Applying metric multidimensional scaling. In M. McLaughlin (Ed.), *Communication yearbook 9* (pp. 315–326). Newbury Park, CA: Sage.

Rice, R. E., Grant, A. E., Schmitz, J., & Torobin, J. (1990). Individual and network influences on the adoption and perceived outcomes of electronic messaging. *Social Networks, 12*, 17–55.

Rice, R. E., & Katz, J. E. (2001). *The Internet and health communication: Experience and expectations*. Thousand Oaks, CA: Sage Publications.

Roberts, T. L. (1998). *Are newsgroups virtual communities?* Paper presented at the CHI 98 Human Factors in Computing Systems, Los Angeles, CA.

Sack, W. (2000a). *Conversation map: A content-based UseNet newsgroup browser*. Paper presented at the International Conference on Intelligent User Interfaces, New Orleans, LA.

Sack, W. (2000b). *Discourse diagrams: Interface design for very large-scale conversations*. Paper presented at the 33rd Annual Hawaii International Conference on System Sciences, Maui, HI.

Salmon, G. (2000). *E-moderating: The key to teaching and learning online*. London: Kogan-Page.

Schoch, N. A., & White, M. D. (1997). *A study of the communication patterns of particpants in consumer health electronic discussion groups*. Paper presented at the Proceedings of the 60th ASIS Annual Meeting, Washington, DC.

Schuler, D. (1994). Community networks: Building a new participatory medium. *CACM, 37*(1), 39–51.

Schuler, D. (1996). *New community networks: Wired for change*. Reading, MA: ACM Press & Addison-Wesley Publishing Co.

Schuler, D., & Namioka, A. E. (1993). *Participatory design: Principles and practices*. Mahwah, NJ: Lawrence Erlbaum Associates.

Seabrook, R. (2001). *Electronic discussion group moderators' experiences with flame messages and implications for the desig of a multi-modal adaptive content-sensitive filtering too*. Paper presented at the CHI'2001, Seattle, WA.

Sellen, A. (1994). Remote conversations: The effects of mediating talk with technology. *Human-Computer Interaction, 10*(4), 401–444.

Shackel, B. (1990). Human factors and usability. In J. Preece & L. Keller (Eds.), *Human-computer interaction: Selected readings* (pp. 27–41). Hemel Hempstead, UK: Prentice-Hall.

Shneiderman, B. (1986). *Designing the user interface: Strategies for effective human-computer interaction* (1st ed.). Reading, MA: Addison-Wesley.

Shneiderman, B. (1992). Tree visualization with treemaps: A 2-d space filling approach. *ACM Transactions on Graphics, 11*(1), 92–99.

Shneiderman, B. (1998). *Designing the user interface: Strategies for effective human-computer interaction* (3rd ed.). Reading, MA: Addison-Wesley.

Shneiderman, B. (2000). Designing trust into online experiences. *Communications of the ACM, 43*(12), 57–59.

Short, J., Williams, E., & Christie, B. (1976). *The social psychology of telecommunications*. London: John Wiley and Sons.

Silver, D. (1999). Localizing the global village: Lessons from the Blacksburg electronic village. In R. B. Browne & M. W. Fishwick (Eds.), *The global village: Dead or alive?* (pp. 79–92). Bowling Green, OH: Popular Press.

Smith, M. A. (1999). Invisible crowds in cyberspace: Mapping the social structure of the internet. In M. A. Smith & P. Kollock (Eds.), *Communities in Cyberspace* (pp. 195–219). London: Routledge.

Smith, M. A., & Fiore, A. T. (2001). *Visualization components for persistent conversations*. Paper presented at the CHI 2001, Seattle, WA.

Smith, M. A., & Kollock, P. (1999). *Communities in Cyberspace*. London: Routledge.

Spears, M., Russell, L., & Lee, S. (1990). De-individuation and group polarization in computer-mediated communication. *British Journal of Social Psychology, 29*, 121–134.

Spears, R., & Lea, M. (1992). Social influence and the influence of 'social' in computer mediated communication. In M. Lea (Ed.), *Contexts of computer mediated communication*. Hemel Hempstead, UK: Harvest Wheatsheaf.

Sproull, L., & Faraj, S. (1997). The network as social technology. In S. Keisleer (Ed). *Culture of the Internet* (pp. 35–51). Mahwah, NJ: Lawrence Erlbaum.

Sproull, L., & Kiesler, S. (1991). *Connections: New ways of working in the networked organization*. Cambridge, MA: MIT Press.

Sproull, L., & Keisler, S. (1986). Reducing social context cues; Electronic mail in organizational communication. *Management Science, 32*, 1492–1512.

Stellin, S. (2001, July 9). Painting some pictures of the online shipper. *The New York Times*, p. C4.

Tannen, D. (1990). *You just don't understand. Men and women in conversation*. New York: William Morrow and Co., Ltd.

Tannen, D. (1994). *Talking from 9 to 5*. New York: William Morrow and Company, Inc.

Tedeschi, B. (2001, July 9). E-commerce report. *The New York Times*, p. C7.

Turkle, S. (1995). *Life on the screen. Identify in the age of the Internet*. New York: Simon & Schuster.

Turkle, S. (1999, Winter 1999/2000). Tinysex and gender trouble. *IEEE Technology and Society Magazine, 4*, 8-20.

Venkatraman, N. (2000). Five steps to a dot-com strategy: How to find your footing on the Web. *Sloan Management Review, 41*(3), 15-27.

Viegas, F. B., & Donath, J. S. (1999). *Chat circles*. Paper presented at the CHI 99 Human Factors in Computing Systems, Pittsburgh, PA.

Vygotsky, L. (1978). *Mind in society*. Cambridge, UK: Cambridge University Press.

Vygotsky, L. (1986). Thought and language. Cambridge, MA: MIT Press.

Walcott, H. F. (1999). *Ethnography: A way of seeing*. Walnut Creek, CA: AltaMira Press.

Wallace, P. (1999). *The psychology of the Internet*. Cambridge, UK: Cambridge University Press.

Walther, J. B. (1993). Impression development in computer-mediated interaction. *Western Journal of Communications, 57*, 381-398.

Walther, J. B. (1994). Anticipated ongoing interaction versus channel effects on relational communication in computer-mediated interaction. *Human Communication Research, 20*(4), 473-501.

Walther, J. B. (1996). Computer-mediated communication: Impersonal, interpersonal, and hyperpersonal interaction. *Communications Research, 23*(1), 3-43.

Walther, J. B., Anderson, J. F., & Park, D. W. (1994). Interpersonal effects in computer-mediated interaction: A meta-analysis of social and antisocial communication. *Communication Research, 21*, 460-487.

Walther, J. B., & Boyd, S. (2002). *Attraction to computer-mediated social support*. In C. A. Lin & D. Atkin (Eds.). *Communication technology and society: Audience adoption and uses* (pp. 153-188). Cresskill, NJ: Hampton Press.

Wellman, B. (1982). Studying personal communities. In P. M. N. Lin (Ed.), *Social structure and network analysis*. Beverly Hills, CA: Sage.

Wellman, B. (1997). An electronic group is virtually a social network. In S. Kiesler (Ed.), *Culture of the Internet* (pp. 179-205). Mahwah, NJ: Lawrence Erlbaum Associates.

Wellman, B. (2000). Changing connectivity: A future history of Y2.03K'. *Sociological Research Online, 4*(4). [Online: www.socresonline. org.UK/4/4/ wellman.html]

Wellman, B., & Frank, K. (2001). Getting support from personal comunities. In N. Lin, R. Burt, & K. Cook (Eds.), *Social capital: Theory and research*. Chicago, IL: Aldine De Gruytere.

Wellman, B., & Gulia, M. (1999a). Net surfers don't ride alone. In B. Wellman (Ed.), *Networks in the global village* (pp. 331-366). Boulder, CO: Westview Press.

Wellman, B., & Gulia, M. (1999b). Virtual communities as communities: Net surfers don't ride alone. In P. Kollock & M. Smith (Eds.), *Communities in Cyberspace*. Berkeley, CA: Routledge.

Wellman, B., Salaff, J., Dimitrova, D., Garton, L., Gulia, M., & Haythornthwaite, C. (1996). Computer networks as social networks: Collaborative work, telework and virtual community. *Annual Review of Sociology, 22*, 213-238.

Wenger, E. (1998). *Communities of practice*. Cambridge, UK: Cambridge University Press.

Wenger, E. C., & Snyder, W. M. (2000). Communities of practice: The organizational frontier. *Harvard Business Review, 78*(1), 139-145.

Whittaker, S., Issacs, E., & O'Day, V. (1997). Widening the net. Workshop report on the theory and practice of physical and network communities. *SIGCHI Bulletin, 29*(3), 27-30.

Williams, R. L., & Cothrel, J. (2000). Four smart ways to run online communities. *Sloan Management Review, 41*(4), 81-91.

Winner, L. (1995). Peter Pan in cyberspace. *Educom Review, 30*(3).

Xiong, R., & Donath, J. (1999). *PeopleGarden: Creating data portraits for users* [On-line]. Available: http://graphics. lcs.mit.edu/ ~becca/papers/pgarden [2001, 6/12]

Zheng, J., Bos, N., Olson, J. S., & Olson, G. M. (2001). *Trust without touch: Jump-start trust with social chat*. Paper presented at the CHI 2001, Seattle, WA.

Zimmer, B., & Alexander, G. (1996). The Rogerian interface: For open, warm empathy in computer-mediated collaborative learning. *Innovations in Education and Training International, 33*(1), 13-21.

·31·

VIRTUAL ENVIRONMENTS

Kay M. Stanney
University of Central Florida

INTRODUCTION

Virtual environments (VEs) offer a unique medium for human–computer interaction (HCI), one that allows users egocentric perspectives on three-dimensional (3D) digital worlds, surrounding them with tangible objects to be manipulated and venues to be traversed. They extend the realm of computer interaction from the purely visual to multimodal interactions that more closely parallel natural world exchanges. VEs move beyond the contrived conventions of traditional HCI, allowing users to interact with concrete and familiar artifacts using known perceptual and cognitive skills. Users can also interact with artificial autonomous agents or collaborate with other users who also have representations within the virtual world (see chapter 15).

Although VE technology has tremendous potential and applications using this technology have grown tremendously over the past decade, users of such systems often encounter difficult multimodal interaction techniques, navigational challenges, and adverse effects, such as sickness and aftereffects. Considerable HCI research and development are required if such systems are to be widely adopted into practice. This chapter reviews the current state-of-the-art in VE technology, provides design and implementation strategies, discusses health and safety concerns and potential countermeasures, and presents the latest in VE usability engineering approaches. Current efforts in a number of application domains are reviewed. The chapter should enable readers to better specify design and implementation requirements for VE applications and prepare them to use this advancing technology in a manner that minimizes health and safety concerns.

SYSTEM REQUIREMENTS

VEs provide multimodal computer-generated experiences, which are driven by the hardware and software used to generate the virtual world (see Fig. 31.1). The hardware interfaces consist primarily of the:

- interface devices used to present multimodal information and sense the VE
- tracking devices used to identify head and limb position and orientation
- interaction techniques that allow users to navigate through and interact with the virtual world

The software interfaces include the:

- modeling software used to generate VEs
- autonomous agents that inhabit VEs
- communication networks used to support multiuser VEs

Hardware Requirements

Remarkable advances in hardware technologies have been realized in the past half-decade, which will better support computer generation of VEs. VEs require very large physical memories, high-speed processors, high-bandwidth mass storage capacity, and high-speed interface ports for interaction devices (Durlach & Mavor, 1995). The memory bandwidth problem is being assuaged by 100 MHz (peak bandwidth 800 Mbyte/s) and 133 MHz (peak bandwidth 1.1 Gbyte/s) SDRAM, with promises of 400 MHz and higher speed memory in the near term (Stanney & Zyda, 2002). The gigahertz barrier has been surpassed by both AMD's Athlon and Intel's Pentium III and Willamette, and more advances will soon be realized with Intel's Itanium and McKinley, Sun's Ultrasparc III, Hewlett-Packard's 8700, and IBM's Power4, the latter of which can build an eight-processor system in a hand-sized module. The future looks even brighter, with promises of massive parallelism in computing via the realization of molecular feature size limits in integrated circuits (Appenzeller, 2000). With the rapidly advancing ability to generate complex and large-scale virtual worlds, hardware advances in multimodal input–output (I/O) devices, tracking systems, and interaction techniques are

Hardware Requirements   Software Requirements

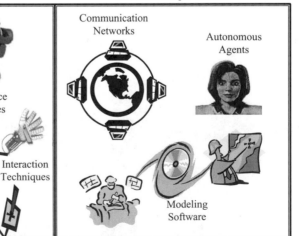

Interface
Devices

Communication
Networks

Autonomous
Agents

Interaction
Techniques

Tracking
Devices

Modeling
Software

FIGURE 31.1. Hardware and software requirements for virtual environment generation.

needed to support generation of increasingly engaging virtual worlds.

Multimodal I/Os. To present a multimodal VE (see chapter 14), multiple devices are used to present information to VE users. Stanney and Zyda (2002) suggest that the single largest challenge in presenting multimodal VE applications is the advances required in peripheral computer connections. Currently, the serial ports used to connect position trackers and other peripheral devices are typically those designed for character input rather than high-speed data transfer. This creates an input port speed problem that must be resolved if multimodal VE systems are to effectively present information to users.

In terms of peripheral devices, the one that has received the greatest attention, both in hype and disdain, is almost certainly the head-mounted display (HMD). HMDs can be used to present 3D visual scenes that are updated as users move their head about the virtual world (see chapter 9). Although this often provides an enticing and engaging experience, because of poor optics, sensorial mismatches, and slow update rates, these devices are also often associated with adverse effects, such as eyestrain and nausea (May & Badcock, 2002). In addition, while HMDs have come down substantially in weight, rendering them more suitable for extended wear, they are still hindered by cumbersome designs, obstructive tethers, suboptimal resolution, and insufficient field of view. These adverse effects may be the reason behind why, in a recent review of HMD devices, more than 40% had been discontinued by their manufacturers (Bungert, 2001). Of the HMDs available, there are several low- to mid-cost models that are relatively lightweight (approximately 39–1000 g) and provide a horizontal field of view (30–35 degrees per eye) and resolution (180 K to 2.4 M pixels/liquid crystal display) exceeding predecessor systems. The way of the future seems to be wearable computer displays (e.g., Microvision, MicroOptical), which can incorporate miniature LCDs directly into conventional eyeglasses (Lieberman, 1999). If designed effectively, these devices should eliminate the tethers and awkwardness of current designs, while enlarging the field of view and enhancing resolution. Stanney and Zyda (2002, p. 3) suggest that, "With advances in wireless and laser technologies and miniaturization of LCDs, during the next decade visual display technology should realize the substantial gains necessary to provide high fidelity virtual imagery in a lightweight non-cumbersome manner."

Spatially immersive displays (SIDs) resolve some of the problems of resolution and weight with cumbersome HMDs. SIDs physically surround viewers, often in a room-sized display, with a panorama of imagery generally projected via fixed front or rear projection display units (Majumder, 1999). Examples of this type of display include the Cave Automated Virtual Environment (CAVE) (Cruz-Neira, Sandin, & DeFanti, 1993), ImmersaDesk, PowerWall, Infinity Wall, and VisionDome (Majumder, 1999). In addition, current efforts are focused on developing desktop/office-sized stereo displays, one of the earliest of which was the Responsive Workbench (Agarwala et al., 1997). More recent efforts include the Totally Active WorkSpace (TAWS), which is a CAVE-like structure where the user works on a glass desk surface (Johnson, Leigh, DeFanti, Sandin, Brown, & Dawe, 1999). Another is the Personal Penta Panel (P3), which is an open box made out of five flat panels into which users place their tracked head and hands and are presented with a surround stereo view. The Personal Augmented Reality Immersive System (PARIS) is a desktop augmented reality device. Such desktop display systems have advantages over SIDs because they are smaller, easier to configure in terms of mounting cameras and microphones, easier to integrate with gesture and haptic devices, and more readily provide access to conventional interaction devices, such as mice, joysticks, and keyboards. Given the many options for a VE visual display (e.g., HMD, SID, desktop stereo display), VE developers must determine which is most appropriate for their application context.

VEs often provide 3D audio, which can greatly enhance the interactive experience. Currently, the most common platforms used for implementing spatialized audio include the AuSim Convolvotron, Lake Huron, and Tucker Davis Technologies PD-1, as well as a software solution from Sound Lab (Shilling & Shinn-Cunningham, 2002). With true 3D audio, a sound can be placed in any location, right or left, up or down, near or far (Begault, 1994). These systems use a head-related transfer function (HRTF) to represent the manner in which sound sources change as a listener moves his/her head that can be specified with knowledge of the source position, as well as the position and orientation of the listener's head (Butler, 1987; Cohen, 1992). The HRTF is dependent on the physiological makeup of the listener's ear (i.e., the pinna does a nonlinear fitting job in the HRTF). Recent advances have allowed for the development of personalized HRTF functions (Crystal Rivers Engineering, 1995). These functions have yet to include reverberation or echoes, which would cause auditory cues to seem more realistic and provide robust relative source distance information (Shilling & Shinn-Cunningham, 2002). In addition, current personalized functions still require a significant amount of calibration time and fail to provide adequate cues for front-to-back or up-down localization. Ideally, a more generalized HRTF could be developed that would be applicable to a multitude of users. This may be possible because the transfer functions of the external ear have been found to be similar across different individuals, although there tends to be a downward shift in spectra frequency with increasing physical size (Middlebrooks, Makous, & Green, 1989).

Although not as commonly incorporated into VEs as visual and auditory interfaces, current haptic devices (see chapter 10) can be used to provide net force and torque feedback to users, such as that experienced during tool usage (Basdogan & Srinivasan, 2002; Biggs & Srinivasan, 2002). Existing haptic devices are less effective at presenting tactile feedback, such as simulating skin contact or dynamic flexibility of a body segment. Three mechanical stimuli can provide such tactile sensation, a displacement of the skin over an extended period of time; a transitory (few milliseconds) displacement of the skin; and a transitory displacement of the skin that is repeated at a constant or variable frequency (Geldard, 1972). The attributes of the skin are difficult to characterize in a quantitative fashion because the skin has variable thresholds for touch (vibrotactile thresholds) and can perform complex spatial and temporal summations, which are a function of the type and position of the mechanical stimuli (Hill, 1967). As the stimulus changes, so too does the sensation of touch, thus creating a challenge for those attempting to model synthetic haptic feedback. To complicate the matter further, the sensations of the skin adapt with exposure to a stimulus, decreasing in sensitivity to a continued stimulus and, at times, even disappearing when a stimulus is still present. To communicate the sensation of synthetic remote touch, it is thus essential to have an understanding of: the mechanical stimuli that produce the sensation of touch; the vibrotactile thresholds; the effect of a sensation; the dynamic range of the touch receptors; and the adaptation of these receptors to certain types of stimuli (Stanney, Mourant, & Kennedy, 1998). The human haptic system needs to be more fully characterized, potentially through a computational model of the physical properties of the skin to generate such synthesized haptic responses. From the technology side, Biggs and Srinivasan (2002) suggest that, for the foreseeable future, haptic devices will be limited by development of new actuator devices. Advances in this area are worth pursuing, because incorporating haptic feedback in VEs can substantially enhance performance (Burdea, 1996).

Although other multimodal interaction is possible (e.g., gustatory, olfactory, vestibular), there has been limited research and development beyond the primary three interaction modes (i.e., visual, auditory, haptic) (Stanney & Zyda, 2002).

Tracking Systems. Tracking systems allow determination of user's head or limb position and orientation, or the location of hand-held devices to allow interaction with virtual objects and traversal through 3D computer-generated worlds (Foxlin, 2002). Tracking is what allows the visual scene in a VE to coincide with a user's point of view, thereby providing an egocentric real-time perspective. Tracking systems must be carefully coupled with the visual scene, however, to avoid unacceptable lags (Kalawsky, 1993). Advances in tracking technology have been realized in terms of drift-corrected gyroscopic orientation trackers, outside-in optical tracking for motion capture, and laser scanners (Foxlin, 2002). Foxlin suggests the future of tracking technology is likely hybrid tracking systems, with an acoustic-inertial hybrid on the market (see http://www.isense.com/products/) and several others in research labs (e.g., magnetic-inertial, optical-inertial, and optical-magnetic). In addition, ultra-wideband radio technology holds promise for an improved method of omni-directional point-to-point ranging.

Tracking technology also allows for gesture recognition, in which human position and movement are tracked and interpreted to recognize semantically meaningful gestures (Turk, 2002). Tracking devices that are worn (e.g., gloves, bodysuits) are currently more advanced than passive techniques (e.g., camera, sensors), yet the latter hold much promise for the future, because they are more powerful and less obtrusive than those that must be worn.

Interaction Techniques. Although one may think of joysticks and gloves when considering VE interaction devices, there are many techniques that can be used to support interaction with and traversal through a VE. Interaction devices support traversal, pointing and selection of virtual objects, tool usage (e.g., through force and torque feedback), tactile interaction (e.g., through haptic devices), and environmental stimuli (e.g., temperature, humidity) (Bullinger, Breining, & Braun, 2001).

Supporting traversal throughout a VE, via motion interfaces, is of primary importance (Hollerbach, 2002). Motion interfaces are categorized as either active (i.e., locomotion) or passive (i.e., transportation). Active motion interfaces require self-propulsion to move about a VE (e.g., treadmill, pedaling device, foot platforms). Passive motion interfaces transport users within a VE without significant user exertion (e.g., inertial motion, as in a flight simulator or noninertial motion, such as in the use of a joystick or gloves). The utility, functionality, cost, and safety of locomotion interfaces beyond traditional options (e.g., joysticks) have yet to be proven. In addition, beyond physical training, concrete applications for active motion interfaces have yet to be clearly delineated.

Another interaction option is speech control (see chapters 8 and 36). Speaker-independent continuous speech recognition systems are currently commercially available (Huang, 1998). For these systems to provide effective interaction, however, additional advances are needed in acoustic and language modeling algorithms to improve the accuracy, usability, and efficiency of spoken language understanding.

Gesture interaction allows users to interact through nonverbal commands conveyed via physical movement of the fingers, hands, arms, head, face, or other body limbs (Turk, 2002). Gestures can be used to specify and control objects of interest, direct navigation, manipulate the environment, and issue meaningful commands.

A variety of interaction techniques can be coupled to support natural and intuitive interaction. Combining speech interaction with nonverbal gestures and motion interfaces can provide a means of interaction that closely captures real-world communications.

Software Requirements

Software development of VE systems has progressed tremendously, from proprietary and arcane systems, to development kits that run on multiple platforms (i.e., general-purpose operating systems to workstations; Pountain, 1996). VE system components are becoming modular and distributed, thereby allowing VE databases (i.e., editors used to design, build, and maintain virtual worlds) to run independently of visualizer and other multimodal interfaces via network links. Standard Application Program Interfaces (APIs; e.g., OpenGL, Direct-3D, Mesa) allow multimodal components to be hardware-independent. VE programming languages are advancing, with APIs, libraries, and particularly scripting languages, thus allowing nonprogrammers to develop virtual worlds (Stanney & Zyda, 2002). Advances are also being made in modeling of autonomous agents and communication networks used to support multiuser VEs.

Modeling. A VE consists of a set of geometry, the spatial relationships between the geometry and the user, and the change in geometry invoked by user actions or the passage of time (Kessler, 2002). Generally, modeling starts with building the geometry components (e.g., graphical objects, sensors, viewpoints, animation sequences; Kalawsky, 1993). These are often converted from CAD data. These components then get imported into the VE modeling environment and rendered when appropriate sensors are triggered. Color, surface textures, and behaviors are applied during rendering. Programmers control the events in a VE by writing task functions, which become associated with the imported components.

A number of 3D modeling languages and toolkits are available that provide intuitive interfaces and run on multiple platforms and renderers (e.g., AC3D Modeler, Clayworks, MR Toolkit, MultiGen Creator and Vega, RealiMation, Renderware, VRML, WorldToolKit; Stanney & Zyda, 2002). In addition, there are scene management engines that allow programmers to work at a higher level, defining characteristics and behaviors for more holistic concepts (RealiMation, 2000). There have also been advances in photorealistic rendering tools that are evolving toward full-featured, physics-based global illumination rendering systems (Heirich & Arvo, 1997; Merritt & Bacon, 1997). Taken together, these advances in software modeling allow for generation of complex and realistic VEs that can run on a variety of platforms, permitting access to VE applications by both small- and large-scale application development budgets.

Autonomous Agents. Autonomous agents are synthetic or virtual human entities that possess some degree of autonomy, social ability, reactivity, and proactiveness (Allbeck & Badler, 2002; also see chapter 15). They can have many forms (e.g., human, animal), which are rendered at various levels of detail and style, from cartoonish to physiologically accurate models. Such agents are a key component of many VE applications involving interaction with other entities, such as adversaries, instructors, or partners (Stanney & Zyda, 2002).

There has been significant research and development in modeling embodied autonomous agents. As with object geometry, agents are generally modeled offline and then rendered during real-time interaction. Although the required level of detail varies, modeling of hair and skin adds realism to an agent's appearance (Allbeck & Badler, 2002). There are a few toolkits available to support agent development, with one of the most notable offered by Boston Dynamics, Inc. (http://www.bdi.com/), a spin-off from the MIT Artificial Intelligence Laboratory. Boston Dynamic's products allow VE developers to work directly in a 3D database, interactively specifying agent behaviors, such as paths to traverse and sensor regions. The resulting agents move realistically, respond to simple commands, and travel about a VE as directed.

Networks. Distributed networks allow multiple users at diverse locations to interact within the same VE (see chapters 16 and 30). Improvements in communication networks are required to allow realization of such shared experiences in which users, objects, processes, and autonomous agents from diverse locations interactively collaborate (Durlach & Mavor, 1995). Yet the foundation for such collaboration has been built within the Next Generation Internet (NGI) and Internet2. The NGI initiative (http://www.ngi.gov/) is connecting a number of universities and national labs at speeds 100 to 1,000 times faster than the 1996 Internet to experiment with collaborative-networking technologies, such as high-quality video-conferencing and audio and video streams. In addition,

Internet2 is using existing networks (e.g., NSF's vBNS–Very high-speed Backbone Network Service) to determine the transport designs necessary to carry real-time multimedia data at high speed (http://apps.internet2.edu/). Distributed VE applications can leverage the special capabilities (i.e., high bandwidth, low latency, low jitter) of these advancing network technologies to provide shared virtual worlds (Singhal & Zyda, 1999).

DESIGN AND IMPLEMENTATION STRATEGIES

Although many conventional HCI techniques can be used to design and implement VE systems, there are unique cognitive, content, products liability, and usage protocol considerations that must be addressed (see Fig. 31.2).

Cognitive Aspects

The fundamental objective of VE systems is to provide multimodal interaction or, when sensory modalities are missing, perceptual illusions that support human information processing in pursuit of a VE application's goals, which could range from training to entertainment. Ancillary, yet fundamental, to this goal is to minimize cognitive obstacles, such as navigational difficulties, that could render a VE application's goals inaccessible.

Multimodal Interaction Design. VEs are designed to provide users with direct manipulative and intuitive interaction with multisensory stimulation (Bullinger et al., 2001). The number of sensory modalities stimulated and the quality of this multisensory interaction is key to the realism and potential effectiveness of VE systems (Popescu, Burdea, & Trefftz, 2002). Yet, there is currently a limited understanding on how to effectively provide such sensorial parallelism (Burdea, 1996). When multimodal sensory information is provided to users, it is essential to consider the design of the integration of these multiple sources of feedback. This can be achieved through consideration of the coordination between sensing and user command and the transposition of senses in the feedback loop. Command coordination considers user input as primarily monomodal and feedback to the user as multimodal. Designers need to consider which input modality is most appropriate for a given task, if there is any need for redundant user input, and whether or not users can effectively handle such parallel input (Stanney, Mourant, & Kennedy, 1998; also see chapter 7). Sensorial transposition occurs when a user receives feedback through other senses than those expected, which may occur because a command coordination scheme has substituted available sensory feedback for those that cannot be generated within a VE. Sensorial substitution schemes may be one for one (e.g., visual for force) or more complex (e.g., visual for force and auditory; visual and auditory for force). If designed effectively, command coordination and sensory substitution schemes should provide multimodal interaction that allows for better user control of the VE.

Perceptual Illusions. When sensorial transpositions are used, there is an opportunity for perceptual illusions to occur.

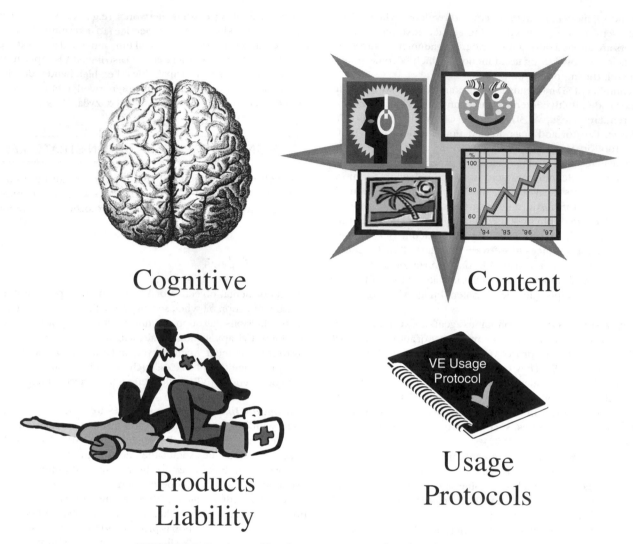

Cognitive

Content

Products
Liability

Usage
Protocols

FIGURE 31.2. Design and implementation considerations for virtual environments.

With perceptual illusions, certain perceptual qualities perceived by one sensory system are influenced by another sensory system (e.g., "feel" a squeeze when you *see* your hand "grabbing" a virtual object). Such illusions could simplify and reduce the cost of VE development efforts (Storms, 2002). For example, when attending to a high-quality visual image coupled with a *low*-quality auditory display, auditory-visual cross-modal perception allows for an increase in the perceived quality of the visual image. Thus, in this case, if the visual image is the focus of the task, there may be no need to use a high-quality auditory display. Unfortunately, little is known about how to leverage these phenomena to reduce development costs while enhancing one's experience in a VE. Perhaps the one exception is vection (i.e., a compelling illusion of self-motion throughout a virtual world), which is known to be enhanced via a number of display factors, including a wide field of view and high-spatial frequency content (Hettinger, 2002). Other such

illusions exist (e.g., visual dominance) and could likewise be leveraged if perceptual and cognitive design principles are identified that can be used to trigger and capitalize on these illusory phenomena.

Navigation and Wayfinding. Effective multimodal interaction design and use of perceptual illusions can be impeded if navigational complexities arise. Navigation is the aggregate of wayfinding (i.e., cognitive planning of one's route) and the physical movement that allows travel throughout a VE (Darken & Peterson, 2002). A number of tools and techniques have been developed to aid wayfinding in virtual worlds, including maps, landmarks, trails, and direction finding. These tools can be used to display current position, current orientation (e.g., compass), log movements (e.g., "breadcrumb" trails), demonstrate or access the surround (e.g., maps, binoculars), or provide guided movement (e.g., signs, landmarks; Chen & Stanney, 1999).

Darken and Peterson (2002) provide a number of principles concerning how best to use these tools. If effectively applied to VEs, these principles should lead to reduced disorientation and enhanced wayfinding in large-scale VEs.

Content Development

Content development is concerned with the design and construction of the virtual objects and synthetic environment that support a VE experience (Isdale, Fencott, Heim, & Daly, 2002). Whereas this medium can leverage existing HCI design principles, it has unique design challenges that arise from the demands of real-time, multimodal, collaborative interaction. In fact, content designers are just starting to appreciate and determine what it means to create a full sensory experience with user control of both point of view and narrative development. Aesthetics is thought to be a product of agency (e.g., pleasure of being), narrative potential, presence and co-presence (e.g., existing in and sharing the virtual experience), as well as transformation (e.g., assuming another persona; Church, 1999; Murray, 1998). Content development should be about stimulating perceptions (e.g., sureties, surprises), as well as contemplation over the nature of being (Fencott, 1999; Isdale et al., 2002).

Existing design techniques—for example, from entertainment, computer games, and theme parks development—can be used to support VE content development (see chapters 44 and 47). Game development techniques that can be leveraged in VE content development include, but are not limited to, providing a clear sense of purpose, emotional objectives, perceptual realism, intuitive interfaces, multiple solution paths, challenges, a balance of anxiety and reward, as well as an almost unconscious flow of interaction (Isdale et al., 2002). From theme park design, content development suggestions include: having a story that provides the all-encompassing theme of the VE and thus the rules that guide design; providing location and purpose; using cause and effect to lead users to their own conclusions; and anchoring users in the familiar (Carson, 2000a, 2000b). Although these suggestions provide guidelines for VE content development, considerable creativity is still an essential component of the process. Isdale et al. (2002) suggest that the challenges of VE content development highlight the need for art to compliment technology.

Products Liability

Those who implement VE systems must be cognizant of potential products liability concerns. Exposure to a VE system often produces unwanted side effects that could warrant users incapable of functioning effectively upon return to the real world. These adverse effects may include nausea and vomiting, postural instability, visual disturbances, and profound drowsiness (Stanney et al., 1998). As users subsequently take on their normal routines, unaware of these lingering effects, their safety and well-being may be compromised. If a VE product occasions such problems, liability of VE developers or system administrators could range from simple accountability (i.e., reporting what happened) to full legal liability (i.e., paying compensation for damages; Kennedy & Stanney, 1996a; Kennedy, Kennedy, & Bartlett, 2002). To minimize their liability, manufacturers and corporate users should design systems and provide usage protocols to minimize risks, warn users about potential aftereffects, monitor users during exposure, assess users' risk, and debrief users upon postexposure.

Usage Protocols

To minimize products liability concerns, VE usage protocols should be carefully designed. Adverse responses to VE exposure vary directly with the stimulus intensity of the VE and the susceptibility of the individual exposed (Stanney, Kennedy, & Kingdon, 2002). To minimize VE stimulus intensity, the following should be considered:

- Are system lags/latencies minimized and stable?
- Are frame rates optimized?
- Is the interpupillary distance of the visual display adjustable?
- Is a large field of view causing excessive vection, such that the spatial frequency content of the visual scene should be reduced?
- Is multimodal feedback well integrated such that sensory conflicts are minimized?

Self-report has been found useful in gauging individual susceptibility. In particular, a modification of the Motion History Questionnaire (MHQ) (Kennedy & Graybiel, 1965) has been found effective in determining susceptibility to motion sickness associated with VE exposure (Kennedy et al., 2001). The MHQ assesses susceptibility based on past occurrences of sickness in inertial environments. Those individuals determined to be susceptible to motion sickness can be expected to experience more than twice the level of adverse effects to VE exposure as compared with nonsusceptible individuals (Stanney, Kingdon, Nahmens, & Kennedy, 2002). These individuals should thus be carefully monitored during and after VE exposure.

Regardless of the strength of the stimulus or the susceptibility of the user, following a systematic usage protocol can minimize the adverse effects associated with VE exposure. During VE exposure, the room should be arranged such that there is adequate airflow and comfortable thermal conditions, because sweating often precedes an emetic response (Stanney, Kennedy, & Kingdon, 2002). All individuals, regardless of their susceptibility, should be educated about the potential risks of VE exposure, including the potential for nausea, malaise, disorientation, headache, dizziness, vertigo, eyestrain, drowsiness, fatigue, pallor, sweating, increased salivation, and vomiting. Users should be prepared for the transition into the HMD by informing them that there will be a perceptual adjustment period. To minimize fatigue, which can exacerbate adverse effects, all equipment should be adjusted to ensure comfortable fit and unobstructed movement. Particularly for strong VE stimuli, initial exposure should be short (e.g., 10 min) and re-exposure should be prohibited for 2–5 days. All users should be monitored during exposure, and red flags should be prudently attended to (e.g., profuse sweating, burping, verbal frustration, restricted head or body movement). Users demonstrating any of these behaviors should be observed closely, because they may experience an emetic response, and extra care should be taken with these individuals postexposure. Some individuals may be unsteady upon postexposure. These individuals may need assistance when initially standing up after exposure. After exposure, the well-being of users should be assessed, for which a derivative of the field sobriety test can be used (Kennedy & Stanney, 1996b). Depending on the strength of the VE stimulus, the amount of time after exposure that users must remain on the premises before driving or participating in other such high-risk activities should be determined. Individuals should be informed that upon postexposure, they may experience disturbed visual functioning, visual flashbacks, as well as unstable locomotor and postural control for prolonged periods after exposure.

HEALTH AND SAFETY ISSUES

The health and safety risks associated with VE exposure complicate usage protocols and lead to products liability concerns. It is thus essential to understand these issues if one is going to utilize VE technology. There are both physiological and psychological risks associated with VE exposure, the former being related primarily to sickness and aftereffects and the latter primarily being concerned with the social impact (see Fig. 31.3).

Cybersickness, Adaptation, and Aftereffects

Motion sickness-like symptoms and other aftereffects (i.e., balance disturbances, visual stress, altered hand-eye coordination) are unwanted by-products of VE exposure. The sickness related to VE systems is commonly referred to as "cybersickness" (McCauley & Sharkey, 1992). Some of the most common symptoms exhibited include dizziness, drowsiness, headache, nausea, fatigue, and general malaise (Kennedy, Lane, Berbaum, & Lilienthal, 1993). More than 80% of users will experience some level of disturbance, with approximately 12% ceasing exposure prematurely from this adversity (Stanney, Kingdon et al., 2002). Of those who dropout, approximately 10% can be expected to have an emetic response (i.e., vomit); however, only 1–2% of all users will have such a response. These adverse effects are known to increase in incidence and intensity with prolonged exposure duration (Kennedy, Stanney, & Dunlap, 2000). Although most users will experience some level of adverse effects, symptoms vary substantially from one individual

Cybersickness

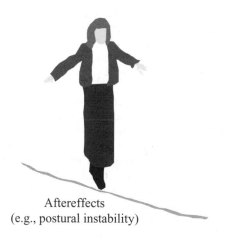

Aftereffects
(e.g., postural instability)

Social Impact

FIGURE 31.3. Health and safety issues for virtual environments.

to another, as well as from one system to another (Kennedy & Fowlkes, 1992). These effects can be assessed via the Simulator Sickness Questionnaire (Kennedy et al., 1993), with values above 20 requiring due caution (i.e., warn and observe users; Stanney, Kennedy, & Kingdon, 2002).

To overcome such adverse effects, individuals generally undergo physiological adaptation during VE exposure. This adaptation is the natural and automatic response to an intersensorily imperfect VE and is elicited because of the plasticity of the human nervous system (Welch, 1978). Because of technological flaws (e.g., slow update rate, sluggish trackers), users of VE systems may be confronted with one or more intersensory discordances (e.g., visual lag, a disparity between seen and felt limb position). They must compensate for these discordances by adapting their psychomotor behavior or visual functioning to perform effectively in the VE. Once interaction with a VE is discontinued, these compensations persist for some time after exposure, leading to aftereffects.

Once VE exposure ceases and users return to their natural environment, they are likely unaware that interaction with the VE has potentially changed their ability to effectively interact with their normal physical environment (Stanney & Kennedy, 1998). Several different kinds of aftereffects may persist for prolonged periods after VE exposure (Welch, 1997). For example, hand-eye coordination can be degraded via perceptual-motor disturbances (Kennedy, Stanney, Ordy, & Dunlap, 1997; Rolland, Biocca, Barlow, & Kancherla, 1995), postural sway can arise (Kennedy & Stanney, 1996), as can changes in the vestibulo-ocular reflex, or one's ability to stabilize an image on the retina (Draper, Prothero, & Viirre, 1997). The implications of these aftereffects are that:

1. VE exposure duration may need to be minimized.
2. Highly susceptible individuals or those from clinical populations (e.g., those prone to seizures) may need to avoid or be banned from exposure.
3. Users should be closely monitored during VE exposure.
4. Users' activities should be closely monitored for a considerable period of time postexposure to avoid personal injury or harm.

Social Impact

VE technology, like its ancestors (e.g., television, computers), has the potential for negative social implications through misuse and abuse (Kallman, 1993; also see chapter 62). Yet violence in VE is nearly inevitable, as evidenced by the violent content of popular video games. Such animated violence is a known favorite over the portrayal of more benign emotions, such as cooperation, friendship, or love (Sheridan, 1993; also see chapter 4). The concern is that users who engage in what seems like harmless violence in the virtual world may become desensitized to violence and mimic this behavior in the look-alike real world.

Currently, it is not clear whether or not such violent behavior will result from VE exposure; early research, however, is not reassuring. Calvert and Tan (1994) found VE exposure to significantly increase the physiological arousal and aggressive thoughts of young adults (Calvert & Tan, 1994). Perhaps more disconcerting was that neither aggressive thoughts nor hostile feelings were found to decrease from VE exposure, thus providing no support for catharsis. Such increased negative stimulation may then subsequently be channeled into real-world activities. The ultimate concern is that VE immersion may potentially be a more powerful perceptual experience than past, less interactive technologies, thereby increasing the negative social impact of this technology (Calvert, 2002). A proactive approach is needed, which weighs the risks and potential consequences associated with VE exposure against the benefits. Waiting for the onset of harmful social consequences should not be tolerated.

VIRTUAL ENVIRONMENT USABILITY ENGINEERING

Most VE user interfaces are fundamentally different from traditional graphical user interfaces, with unique I/O devices, perspectives, and physiological interactions. Thus, when developers and usability practitioners attempt to apply traditional usability engineering methods to the evaluation of VE systems, they find few if any that are particularly well suited to these environments (for notable exceptions, see Gabbard & Hix, 1997; Hix & Gabbard, 2002; Stanney, Mollaghasemi, & Reeves, 2000). There is a need to modify and optimize available techniques to meet the needs of VE usability engineering, as well as to better characterize factors unique to VE usability, including sense of presence and VE ergonomics.

Usability Techniques for VE Systems

Assessment of usability for VE systems must go beyond traditional approaches, which are concerned with the determination of effectiveness, efficiency, and user satisfaction (see chapter 56). Evaluators must consider whether multimodal I/O is optimally presented and integrated, navigation is supported to allow the VE to be readily traversed, object manipulation is intuitive and simple, content is immersive and engaging, and the system design optimizes comfort while minimizing sickness and aftereffects. The affective elements of interaction become important when evaluating VE systems (see chapter 59). It is an impressive task to ensure all of these criteria are met.

Gabbard and Hix (1997) have developed a taxonomy of VE usability characteristics that can serve as a foundation for identifying and evaluating usability criteria particularly relevant to VE systems. Stanney et al. (2000) used this taxonomy as the foundation on which to develop an automated system, MAUVE (Multi-Criteria Assessment of Usability for Virtual Environments), which assesses VE usability in terms of how effectively each of the following are designed: navigation, user movement, object selection and manipulation, visual output, auditory output, haptic output, presence, immersion, comfort, sickness, and aftereffects. MAUVE can be used to support expert evaluations of VE systems, similar to the manner in which traditional heuristic evaluations are conducted. Because of such issues as cybersickness and aftereffects, it is essential to use these or other techniques to ensure the usability of VE systems, not only to avoid rendering them ineffective, but also to ensure that they are not hazardous to users.

Sense of Presence

One of the usability criteria unique to VE systems is sense of presence. VEs have the unique advantage of leveraging the imaginative ability of individuals to psychologically transport themselves to another place, one that may not exist in reality (Sadowski & Stanney, 2002). To support such transportation, VEs provide physical separation from the real world by immersing users in the virtual world via an HMD, CAVE, SID or other visual display then impart sensorial sensations via multimodal feedback that would naturally be present in the alternate environment. Focus on generating such presence is one of the primary characteristics distinguishing VEs from other means of displaying information.

Presence has been defined as the subjective perception of being immersed in and surrounded by a virtual world rather than the physical world one is currently situated in (Stanney, Salvendy, et al., 1998). VEs that engender a high degree of presence are thought to be more enjoyable, effective, and well received by users (Sadowski & Stanney, 2002). To enhance presence, designers of VE systems should spread detail around a scene; let user interaction determine when to reveal important aspects; maintain a natural and realistic, yet simple appearance; and utilize textures, colors, shapes, sounds, and other features to enhance realism (Kaur, 1999). To generate the feeling of immersion within the environment, designers should isolate users from the physical environment (use of an HMD may be sufficient), provide content that involves users in an enticing situation supported by an encompassing stimulus stream, provide natural modes of interaction and movement control, and utilize design features that enhance vection (Stanney et al., 2000). Presence can be assessed via Witmer and Singer's (1998) Presence Questionnaire or techniques used by Slater and Steed (2000), as well as a number of other means (Sadowski & Stanney, 2002).

Virtual Environment Ergonomics

Ergonomics, which focuses on fitting a product or system to the anthropometric, musculoskeletal, cardiovascular, and psychomotor properties of users, is an essential element of VE system design (McCauley-Bell, 2002). Supporting user comfort, while donning cumbersome HMDs or unwieldy peripheral devices, is an ergonomics challenge of paramount importance because discomfort could supersede any other sensations (e.g., presence, immersion). If a VE produces discomfort, participants may limit their exposure time or possibly avoid repeat exposure. Overall, physical discomfort should thus be minimized and user safety maximized.

Ergonomic concerns affecting comfort include visual discomfort resulting from visual displays with improper depth cues, poor contrast and illumination, or improperly set interpupillary distances (Stanney et al., 2000). Physical discomfort can be driven by restrictive tethers; awkward interaction devices; or heavy, awkward, and constraining visual displays. To enhance the ergonomics of VE systems, several factors should be considered, including (McCauley-Bell, 2002):

- Is operator movement inhibited by the location, weight, or window of reach of interaction devices or visual displays?
- Does layout and arrangement of interaction devices and visual displays support efficient and comfortable movement?
- Is any limb overburdened by heavy interaction devices or visual displays?
- Do interaction devices require awkward and prolonged postures?
- If a seat is provided, does it support user movement and is it of the right height with adequate back support?
- If active motion interfaces are provided (e.g., treadmills), are they adjustable to ensure fit to the anthropometrics of users?
- Are the noise and sound levels within ergonomic guidelines and do they support user immersion?

APPLICATION DOMAINS

Boman, Piantanida, and Schlager (1993) anticipated that applications driven by VE technology would become commonplace within a decade. Similarly, Durlach and Mavor (1995) suggested that serious commercial applications would not be realized for 5–10 years. These insights appear well founded, because there are currently a number of viable applications (see Fig. 31.4). Yet, while practical applications exist today (Stone, 2002), most are still in the developmental stages. This may be because there is a lack of understanding of which applications are appropriate for VE technology (Stanney, Salvendy, et al., 1998). Technology is currently driving VE use rather than specifying where this technology can best be used to meet specific training/performance requirements. Stanney, Salvendy, et al. (1998, pp. 172–174) suggest that "identification of application areas particularly suited to VE technology was of essential importance to the success of this technology."

Yet, there are several challenges impeding widespread commercial use of VE technology. Stanney, Salvendy, et al. (1998, p. 171) suggest that

Currently, there are poor data links for industrial use, a lack of quality of experience (e.g., malaise and low presence), and a lack of structured publicly available case study evaluations that specify the added value provided by the use of VE applications over existing systems (Wilson, D'Cruz, Cobb, & Eastgate, 1996). Thus, VE technology does not fare well in the evaluation of cost versus performance and utility versus safety trade-offs. Second, there are the often unrecognized or ignored adverse aftereffects, which will almost certainly lead to products liability lawsuits engendered by incidence and accidents following prolonged VE exposure. Third, usability issues (e.g., visual discomfort and cybersickness) currently make the use of VE systems less than desirable.

Despite these shortcomings, forerunners in the field have developed a number of applications, of which Stone (2002)

FIGURE 31.4. Example of virtual environment application domains.

has provided an impressive overview. In engineering, VE technology supports the ability to convert CAD data into real-time interactive models that can inform business and product development processes. Such models are allowing system and product designs to be created in a more intuitive and natural manner, with multidisciplinary design teams communicating their ideas via the VE medium (Davies, 2002). In the military, VE applications are seen as a means of enhancing preparedness for a wide range of activities at reduced operating costs (Knerr, Breaux, Goldberg, & Thurman, 2002).

In education, the potential for enhanced learning via participatory first-person VE experiences is often lauded (Cobb, Neal, Crosier, & Wilson, 2002), yet early projects have failed to demonstrate any significant gains (Moshell & Hughes, 2002). This is likely because of the current complexity and novelty of the technology, as well as the need for instructional design techniques particularly well suited to educational VE applications. Early efforts, however, will likely lead the way to new and dynamic approaches in experiential learning.

In this information age, VE provides an impressive forum for facilitating the exploration of complex data sets (Bryson, 1996, 2002), thus information visualization is a natural application for VE technology (see chapter 28). Dynamic 3D visualization can help an individual to identify and understand abstract concepts through interaction with dynamic parameter interrelationships and experiencing trends over time. Impressive applications,

from the visualization of the stock market (Siebert, 1994) to the National Library of Medicine's Visible Human Project (Spitzer, Ackerman, Scherzinger, & Whitlock, 1996) currently exist.

VE technology has the potential to revolutionize medical care, via enhanced medical records, interactive diagnostics, support of preoperative planning, and augmented surgery, as well as support of patient education and physician training (Satava & Jones, 2002). Some of the earliest VE successes came in the area of psychotherapy, where VE technology has been used to treat acrophobia, fear of flying, arachnophobia, and other psychological disorders (Strickland, Hodges, North, & Weghorst, 1997; North, North, & Coble, 2002).

Probably the most exciting and highly publicized applications are those in entertainment (Badiqué et al., 2002). Current VE applications include games, online communities, location-based entertainment, theme parks, and other venues. From interactive arcades to cybercafes, the entertainment industry has leveraged the unique characteristics of the VE medium, providing dynamic and exciting experiences. Such applications tend to set the pace, both technological and in terms of content, for advances in VE technology, in general.

Stone (2002) suggests that, although there is currently no single "killer" VE market, the technological revolution over the close of the preceding millennium has set a foundation that should ensure the presence of VE applications for at least the next two decades.

CONCLUSIONS

VE technology has made substantial gains over the past decade, which promise to revolutionize the fields of engineering, system design, education, training, medicine, and entertainment. These applications bring unique levels of interaction, immersion, and collaboration not seen by more conventional media. Applications of this technology can be greatly enhanced through the consideration of a number of factors (e.g., hardware and software design specification, implementation strategies, health and safety issues, evaluation techniques), with which HCI practitioners have considerable experience. By integrating these factors into the VE application development life cycle, a more effective system can result that is readily adopted by users.

ACKNOWLEDGMENTS

The author thanks Branka Wedell for her contribution to Fig. 31.1. This material is based on work supported in part by the National Science Foundation (NSF) under Grant No. IRI-9624968, the Office of Naval Research (ONR) under Grant No. N00014-98-1-0642, the Naval Air Warfare Center Training Systems Division (NAWC TSD) under Contract No. N61339-99-C-0098, and the National Aeronautics and Space Administration (NASA) under Grant No. NAS9-19453. Any opinions, findings, conclusions, or recommendations expressed in this material are those of the author and do not necessarily reflect the views or endorsement of the NSF, ONR, NAWC TSD, or NASA.

References

Agrawala, M., Beers, A. C., McDowall, I., Fröhlich, B., Bolas, M., & Hanrahan, P. (1997). The two-user Responisve Workbench: Support for collaboration through individual views of a shared space (pp. 327–332). *SIGGRAPH '97*. New York: ACM Press.

Allbeck, J. M., & Badler, N. I. (2002). Embodied autonomous agents. In K. M. Stanney (Ed.), *Handbook of virtual environments: Design, implementation, and applications* (pp. 313–332). Mahwah, NJ: Lawrence Erlbaum Associates, Inc.

Appenzeller, T. (2000, May 1). The chemistry of computing: Computers made of molecule-size parts could build themselves. *US News and World Report* [On-line]. Available: http://www.usnews.com/usnews/issue/000501/chips.htm

Badiqué, E., Cavazza, M., Klinker, G., Mair, G., Sweeney, T., Thalmann, D., & Thalmann, N. M. (2002). Entertainment applications of virtual environments. In K. M. Stanney (Ed.), *Handbook of virtual environments: Design, implementation, and applications* (pp. 1143–1166). Mahwah, NJ: Lawrence Erlbaum Associates, Inc.

Basdogan, C., & Srinivasan, M. A. (2002). Haptic rendering in virtual environments. In K. M. Stanney (Ed.), *Handbook of virtual environments: Design, implementation, and applications* (pp. 117–134). Mahwah, NJ: Lawrence Erlbaum Associates, Inc.

Begault, D. (1994). *3-D sound for virtual reality and multimedia.* Boston, MA: Academic Press.

Biggs, S. J., & Srinivasan, M. A. (2002). Haptic interfaces. In K. M. Stanney (Ed.), *Handbook of virtual environments: Design, implementation, and applications* (pp. 93–116). Mahwah, NJ: Lawrence Erlbaum Associates, Inc.

Boman, D. K., Piantanida, T. P., & Schlager, M. S. (1993). *Virtual environment systems for maintenance training. Vol. 4: Recommendations for government-sponsored R&D* (Technical Report No. 8216). Menlo Park, CA: SRI International.

Bryson, S. (1996). Virtual environments in scientific visualization. *Communications of the ACM, 39*(5), 62–71.

Bryson, S. (2002). Information visualization in virtual environments. In K. M. Stanney (Ed.), *Handbook of virtual environments: Design, implementation, and applications* (pp. 1101–1118). Mahwah, NJ: Lawrence Erlbaum Associates, Inc.

Bullinger, H.-J., Breining, R., & Braun, M. (2001). Virtual reality for industrial engineering: Applications for immersive virtual environments. In G. Salvendy (Ed.), *Handbook of industrial engineering: Technology and operations management* (3rd ed., pp. 2496–2520). New York: Wiley.

Bungert, C. (2001). *HMD/headset/VR-helmet comparison chart* [On-line]. Available: http://www.stereo3d.com/hmd.htm

Burdea, G. (1996). *Force and touch feedback for virtual reality.* New York: Wiley.

Butler, R. A. (1987). An analysis of the monaural displacement of sound in space. *Perception & Psychophysics, 41*, 1–7.

Calvert, S. L. (2002). The social impact of virtual environment technology. In K. M. Stanney (Ed.), *Handbook of virtual environments: Design, implementation, and applications* (pp. 663–680). Mahwah, NJ: Lawrence Erlbaum Associates, Inc.

Calvert, S. L., & Tan, S. L. (1994). Impact of virtual reality on young adult's physiological arousal and aggressive thoughts: Interaction versus observation. *Journal of Applied Developmental Psychology, 15*, 125–139.

Carson, D. (2000a). Environmental storytelling, Part 1: Creating immersive 3D worlds using lessons learned from the theme park industry [On-line]. Available: http://www.gamasutra.com/features/20000301/carson_01.htm

Carson, D. (2000b). Environmental storytelling, Part 2: Bringing theme park environment design techniques lessons to the virtual world [On-line]. Available: http://www.gamasutra.com/20000405/carson_01.htm

Chen, J. L., & Stanney, K. M. (1999). A theoretical model of wayfinding in virtual environments: Proposed strategies for navigational aiding. *Presence: Teleoperators and Virtual Environments, 8*(6), 671–685.

Church, D. (1999, August). Formal abstract design tools. *Games Developer Magazine*, 44–50 [On-line]. Available: http://www.gamasutra.com/features/19990716/design_tools_02.htm

Cobb, S., Neal, H., Crosier, J., & Wilson, J. R. (2002). Development and evaluation of virtual environments for education. In K. M. Stanney (Ed.), *Handbook of virtual environments: Design, implementation, and applications* (pp. 911–936). Mahwah, NJ: Lawrence Erlbaum Associates, Inc.

Cohen, M. (1992). Integrating graphic and audio windows. *Presence: Teleoperators and Virtual Environments, 1*(4), 468–481.

Cruz-Neira, C., Sandin, D. J., & DeFanti, T. A. (1993). Surround-screen projection-based virtual reality: The design and implementation of the CAVE. *ACM Computer Graphics, 27*(2), pp 135–142.

Crystal Rivers Engineering. (1995). *Snapshot: HRTF measurement system.* Groveland, CA.

Darken, R. P., & Peterson, B. (2002). Spatial orientation, wayfinding, and representation. In K. M. Stanney (Ed.), *Handbook of virtual environments: Design, implementation, and applications* (pp. 493-518). Mahwah, NJ: Lawrence Erlbaum Associates, Inc.

Davies, R. C. (2002). Applications of system design using virtual environments. In K. M. Stanney (Ed.), *Handbook of virtual environments: Design, implementation, and applications* (pp. 1079-1100). Mahwah, NJ: Lawrence Erlbaum Associates, Inc.

Draper, M. H., Prothero, J. D., & Viirre, E. S. (1997). Physiological adaptations to virtual interfaces: Results of initial explorations. *Proceedings of the Human Factors & Ergonomics Society 41st Annual Meeting* (pp. 1393). Santa Monica, CA: Human Factors & Ergonomics Society.

Durlach, B. N. I., & Mavor, A. S. (1995). *Virtual reality: Scientific and technological challenges*. Washington, DC: National Academy Press.

Fencott, C. (1999). Content and creativity in virtual environment design. In *Proceedings of Virtual Systems and Multimedia '99* (pp. 308-317). University of Abertay Dundee, Scotland [On-line]. Available: http://www-scm.tees.ac.uk/users/p.c.fencott/vsmm99

Foxlin, E. (2002). Motion tracking requirements and technologies. In K. M. Stanney (Ed.), *Handbook of virtual environments: Design, implementation, and applications* (pp. 163-210). Mahwah, NJ: Lawrence Erlbaum Associates, Inc.

Gabbard, J. L., & Hix, D. (1997). *A taxonomy of usability characteristics in virtual environments* [On-line]. Available: http://csgrad.cs.vt.edu/~jgabbard/ve/taxonomy/.

Geldard, F. A. (1972). *The human senses* (2nd ed.). New York: Wiley.

Heirich, A., & Arvo, J. (1997). Scalable Monte Carlo image synthesis. *Parallel Computing* (Special issue on Parallel Graphics & Visualization), *23*(7), 845-859.

Hettinger, L. J. (2002). Illusory self-motion in virtual environments. In K. M. Stanney (Ed.), *Handbook of virtual environments: Design, implementation, and applications* (pp. 471-492). Mahwah, NJ: Lawrence Erlbaum Associates, Inc.

Hill, J. W. (1967). *The perception of multiple tactile stimuli* (Report No. 4823-1). Palo Alto, CA: Stanford University, Stanford Electronics Laboratory.

Hix, D., & Gabbard, J. L. (2002). Usability engineering of virtual environments. In K. M. Stanney (Ed.), *Handbook of virtual environments: Design, implementation, and applications* (pp. 681-699). Mahwah, NJ: Lawrence Erlbaum Associates, Inc.

Hollerbach, J. (2002). Locomotion interfaces. In K. M. Stanney (Ed.), *Handbook of virtual environments: Design, implementation, and applications* (pp. 239-254). Mahwah, NJ: Lawrence Erlbaum Associates, Inc.

Huang, X. D. (1998). Spoken language technology research at Microsoft. In 16th ICA and 135th ASA '98, Seattle, WA.

Isdale, J., Fencott, C., Heim, M., & Daly, L. (2002). Content design for virtual environments. In K. M. Stanney (Ed.), *Handbook of virtual environments: Design, implementation, and applications* (pp. 519-532). Mahwah, NJ: Lawrence Erlbaum Associates, Inc.

Johnson, A., Leigh, J., DeFanti, T., Sandin, D., Brown, M., & Dawe, G. (1999). Next-generation tele-immersive devices for desktop transoceanic collaboration. *Proceedings of IS&T/SPIE Conference on Visual Communications and Image Processing '99* (pp. 1420-1429). San Jose, CA.

Kalawsky, R. S. (1993). *The science of virtual reality and virtual environments*. Wokingham, England: Addison-Wesley.

Kallman, E. A. (1993). Ethical evaluation: A necessary element in virtual environment research. *Presence: Teleoperators and Virtual Environments, 2*(2), 143-146.

Kaur, K. (1999). *Designing virtual environments for usability*. Unpublished doctoral dissertation, City University, London.

Kessler, G. D. (2002). Virtual environment models. In K. M. Stanney (Ed.), *Handbook of virtual environments: Design, implementation, and applications* (pp. 255-276). Mahwah, NJ: Lawrence Erlbaum Associates, Inc.

Kennedy, R. S., & Fowlkes, J. E. (1992). Simulator sickness is polygenic and polysymptomatic: Implications for research. *International Journal of Aviation Psychology, 2*(1), 23-38.

Kennedy, R. S., & Graybiel, A. (1965). *The Dial test: A standardized procedure for the experimental production of canal sickness symptomatology in a rotating environment* (Report No. 113, NSAM 930). Pensacola, FL: Naval School of Aerospace Medicine.

Kennedy, R. S., Kennedy, K. E., & Bartlett, K. M. (2002). Virtual environments and products liability. In K. M. Stanney (Ed.), *Handbook of virtual environments: Design, implementation, and applications* (pp. 543-554). Mahwah, NJ: Lawrence Erlbaum Associates, Inc.

Kennedy, R. S., Lane, N. E., Berbaum, K. S., & Lilienthal, M. G. (1993). Simulator sickness questionnaire: An enhanced method for quantifying simulator sickness. *International Journal of Aviation Psychology, 3*(3), 203-220.

Kennedy, R. S., Lane, N. E., Grizzard, M. C., Stanney, K. M., Kingdon, K., Lanham, S., & Harm, D. L. (2001, September 5-7). Use of a motion history questionnaire to predict simulator sickness. Driving Simulation Conference 2001, Sophia-Antipolis (Nice), France.

Kennedy, R. S., & Stanney, K. M. (1996a). Virtual reality systems and products liability. *Journal of Medicine and Virtual Reality, 1*(2), 60-64.

Kennedy, R. S., & Stanney, K. M. (1996b). Postural instability induced by virtual reality exposure: Development of a certification protocol. *International Journal of Human-Computer Interaction, 8*(1), 25-47.

Kennedy, R. S., Stanney, K. M., & Dunlap, W. P. (2000). Duration and exposure to virtual environments: Sickness curves during and across sessions. *Presence: Teleoperators and Virtual Environments, 9*(5), 466-475.

Kennedy, R. S., Stanney, K. M., Ordy, J. M., & Dunlap, W. P. (1997). Virtual reality effects produced by head-mounted display (HMD) on human eye-hand coordination, postural equilibrium, and symptoms of cybersickness. *Society for Neuroscience Abstracts, 23*, 772.

Knerr, B. W., Breaux, R., Goldberg, S. L., & Thurman, R. A. (2002). National defense. In K. M. Stanney (Ed.), *Handbook of virtual environments: Design, implementation, and applications* (pp. 857-872). Mahwah, NJ: Lawrence Erlbaum Associates, Inc.

Lieberman, D. (1999). Computer display clips onto eyeglasses. *Technology News* [On-line]. Available: http://www.techweb.com/wire/story/TWB19990422S0003

May, J. G., & Badcock, D. R. (2002). Vision and virtual environments. In K. M. Stanney (Ed.), *Handbook of virtual environments: Design, implementation, and applications* (pp. 29-64). Mahwah, NJ: Lawrence Erlbaum Associates, Inc.

Majumder, A. (1999). *Intensity seamlessness in multiprojector multisurface displays* (Technical Report). Chapel Hill, NC: Department of Computer Science, University of North Carolina. Available: http://www.cs.unc.edu/~majumder/survey/paper.html.

McCauley, M. E., & Sharkey, T. J. (1992). Cybersickness: Perception of self-motion in virtual environments. *Presence: Teleoperators and Virtual Environments, 1*(3), 311-318.

McCauley-Bell, P. R. (2002). Ergonomics in virtual environments. In K. M. Stanney (Ed.), *Handbook of virtual environments: Design, implementation, and applications* (pp. 807-826). Mahwah, NJ: Lawrence Erlbaum Associates, Inc.

Merritt, E. A., & Bacon, D. J. (1997). Raster3D: Photorealistic molecular graphics. *Methods in Enzymology, 277*, 505-524.

Middlebrooks, J. C., Makous, J. C., & Green, D. M. (1989). Directional sensitivity of sound-pressure levels in the human ear canal. *Journal of the Acoustical Society of America, 86,* 89-108.

Moshell, J. M., & Hughes, C. E. (2002). Virtual environments as a tool for academic learning. In K. M. Stanney (Ed.), *Handbook of virtual environments: Design, implementation, and applications* (pp. 893-910). Mahwah, NJ: Lawrence Erlbaum Associates, Inc.

Murray, J. H. (1998). Hamlet on the holodeck: The future of narrative in cyberspace. MIT Press [Outline On-line]. Available: http://web.mit.edu/jhmurray/www/HOH.html

North, M. M., North, S. M., & Coble, J. R. (2002). Virtual reality therapy: An effective treatment for psychological disorders. In K. M. Stanney (Ed.), *Handbook of virtual environments: Design, implementation, and applications* (pp. 1065-1078). Mahwah, NJ: Lawrence Erlbaum Associates, Inc.

Popescu, G. V., Burdea, G. C., & Trefftz, H. (2002). Multimodal interaction modeling. In K. M. Stanney (Ed.), *Handbook of virtual environments: Design, implementation, and applications* (pp. 435-454). Mahwah, NJ: Lawrence Erlbaum Associates, Inc.

Pountain, D. (1996, July). VR meets reality: Virtual reality strengthens the link between people and computers in mainstream applications. *Byte Magazine* [On-line]. Available: http://www.byte.com/art/9607/sec7/art5.htm

RealiMation. (2000). A scene data management approach to real time 3d software applications. Technical Report TP001 [On-line]. Available: http://www.realimation.com/overview/technical/technical_papers.htm

Rolland, J. P., Biocca, F. A., Barlow, T., & Kancherla, A. (1995). Quantification of adaptation to virtual-eye location in see-thru head-mounted displays. *IEEE Virtual Reality Annual International Symposium '95* (pp. 56-66). Los Alimitos, CA: IEEE Computer Society Press.

Sadowski, W., & Stanney, K. (2002). Presence in virtual environments. In K. M. Stanney (Ed.), *Handbook of virtual environments: Design, implementation, and applications* (pp. 791-806). Mahwah, NJ: Lawrence Erlbaum Associates, Inc.

Satava, R. M., & Jones, S. B. (2002). Medical applications of virtual environments. In K. M. Stanney (Ed.), *Handbook of virtual environments: Design, implementation, and applications* (pp. 937-958). Mahwah, NJ: Lawrence Erlbaum Associates, Inc.

Sheridan, T. B. (1993). My anxieties about virtual environments. *Presence: Teleoperators and Virtual Environments, 2*(2), 141-142.

Shilling, R. D., & Shinn-Cunningham, B. (2002). Virtual auditory displays. In K. M. Stanney (Ed.), *Handbook of virtual environments: Design, implementation, and applications* (pp. 65-92). Mahwah, NJ: Lawrence Erlbaum Associates, Inc.

Siebert, L. (1994). *Visual stockbroking.* M.Sc Thesis, Department of Computer Science. The University of Manchester, UK [On-line]. Available: http://www.cs.man.ac.uk/aig/students/Loren.html

Singhal, S. (1999). *Networked virtual environments—Design and implementation.* New York: ACM Press Books, SIGGRAPH Series.

Slater, M., & Steed, A. (2000). A virtual presence counter. *Presence: Teleoperators and Virtual Environments, 9*(5), 413-434.

Spitzer, V., Ackerman, M. L., Scherzinger, A. L., & Whitlock, D. (1996). The visible human male: A technical report. *Journal of the American Medical Information Association, 3*(2), 118-130.

Stanney, K. M., & Kennedy, R. S. (1998, October 5-9). Aftereffects from virtual environment exposure: How long do they last? *Proceedings of the 42nd Annual Human Factors and Ergonomics Society Meeting* (pp. 1476-1480). Chicago, IL.

Stanney, K. M., Kennedy, R. S., & Kingdon, K. (2002). Virtual environments usage protocols. In K. M. Stanney (Ed.), *Handbook of virtual environments: Design, implementation, and applications* (pp. 721-730). Mahwah: NJ: Lawrence Erlbaum Associates, Inc.

Stanney, K. M., Mollaghasemi, M., & Reeves, L. (2000, August). *Development of MAUVE, the multi-criteria assessment of usability for virtual environments system* (Final Report, Contract No. N61339-99-C-0098). Orlando, FL: Naval Air Warfare Center–Training Systems Division.

Stanney, K. M., Mourant, R., & Kennedy, R. S. (1998). Human factors issues in virtual environments: A review of the literature. *Presence: Teleoperators and Virtual Environments, 7*(4), 327-351.

Stanney, K. M., Kingdon, K., Nahmens, I., & Kennedy, R. S. (2002). *What to expect from immersive virtual environment exposure: Influences of age, gender, body mass index, and past experience.* Manuscript submitted for publication.

Stanney, K. M., Salvendy, G., Deisigner, J., DiZio, P., Ellis, S., Ellison, E., Fogleman, G., Gallimore, J., Hettinger, L., Kennedy, R., Lackner, J., Lawson, B., Maida, J., Mead, A., Mon-Williams, M., Newman, D., Piantanida, T., Reeves, L., Riedel, O., Singer, M., Stoffregen, T., Wann, J., Welch, R., Wilson, J., & Witmer, B. (1998). Aftereffects and sense of presence in virtual environments: Formulation of a research and development agenda. Report sponsored by the Life Sciences Division at NASA Headquarters. *International Journal of Human-Computer Interaction, 10*(2), 135-187.

Stanney, K. M., & Zyda, M. (2002). Virtual environments in the 21st century. In K. M. Stanney (Ed.), *Handbook of virtual environments: Design, implementation, and applications* (pp. 1-14). Mahwah, NJ: Lawrence Erlbaum Associates, Inc.

Stone, R. J. (2002). Applications of virtual environments: An overview. In K. M. Stanney (Ed.), *Handbook of virtual environments: Design, implementation, and applications* (pp. 827-856). Mahwah, NJ: Lawrence Erlbaum Associates, Inc.

Storms, R. L. (2002). Auditory-visual cross-modality interaction and illusions. In K. M. Stanney (Ed.), *Handbook of virtual environments: Design, implementation, and applications* (pp. 455-470). Mahwah, NJ: Lawrence Erlbaum Associates, Inc.

Strickland, D., Hodges, L., North, M., & Weghorst, S. (1997). Overcoming phobias by virtual exposure. *Communications of the ACM, 40*(8), 34-39.

Turk, M. (2002). Gesture recognition. In K. M. Stanney (Ed.), *Handbook of virtual environments: Design, implementation, and applications* (pp. 223-238). Mahwah, NJ: Lawrence Erlbaum Associates, Inc.

Welch, R. B. (1978). *Perceptual modification: Adapting to altered sensory environments.* New York: Academic Press.

Welch, R. B. (1997). The presence of aftereffects. In G. Salvendy, M. Smith, & R. Koubek (Eds.), *Design of computing systems: Cognitive considerations* (pp. 273-276).

Wilson, J. R., D'Cruz, M.D., Cobb, S. V., & Eastgate, R. M. (1996). *Virtual reality for industrial applications: Needs and opportunities.* Nottingham, UK: Nottingham University Press.

Witmer, B., & Singer, M. (1998). Measuring presence in virtual environments: A Presence Questionnaire. *Presence: Teleoperators and Virtual Environments, 7*(3), 225-240.

USER-CENTERED INTERDISCIPLINARY DESIGN OF WEARABLE COMPUTERS

Daniel P. Siewiorek and Asim Smailagic
Carnegie Mellon University

INTRODUCTION TO WEARABLE COMPUTING

The convergence of a variety of technologies makes possible the current paradigm shift in information processing. Continued advances in semiconductor technology makes possible high-performance microprocessor requiring less power and less space. Decades of research in computer science have provided the technology for hands-off computing using speech and gesturing for input. Miniature heads-up displays weighing less than a few ounces have been introduced. Combined with mobile communication technology, it is possible for users to access information anywhere.

Sensors make the computing system an active part of the environment. As the user modifies the environment, the information can be automatically accumulated by the system, thereby eliminating the costly and error-prone process of information acquisition. Much like personal computers allow accountants and bookkeepers to merge their information space with their workspace (i.e., a sheet of paper), mobile computers allow mobile processing and the superposition of information on the user's workspace. Mobile computing deals in information rather than programs, becoming tools in the user's environment much like a pencil or a reference book.

Carnegie Mellon University's (CMU's) Wearable Computers Project is helping to define the future for not only computing technologies, but also for the use of computers in daily activities. The goal is to develop a new class of computing systems, with a small footprint that can be carried or worn by a human and able to interact with computer-augmented environments. By rapid prototyping of new artifacts and concepts, CMU has established a new paradigm of wearable computers (Smailagic & Siewiorek, 1994). More than two dozen generations of wearable computers have been designed and built over the last decade, with most tested in the field. The user-centered, interdisciplinary, concurrent system design methodology (Siewiorek, Smailagic, Bass, Siegel, Martin, 1998; Siewiorek, Smailagic, & Lee, 1994; Smailagic et al., 1995) has lead to a factor of more than 200 increase in the complexity of the artifacts while essentially holding design effort constant. The application domains range from inspection, maintenance, manufacturing, and navigation to on-the-move collaboration, position sensing, global communication, real-time speech recognition, and language translation. In the course of developing wearable systems to support these applications, we have identified or refined several conceptual frameworks regarding personal computing. At the core of these ideas is the notion that wearable computers should seek to merge the user's information space with his or her workspace. Information tools, such as wearable computers, must blend seamlessly with existing work environments, providing as little distraction as possible. This requirement often leads researchers to investigate replacements for the traditional console interfaces, such as a keyboard or mouse, which generally require a fixed physical relationship between the user and device. Identifying effective interaction modalities for wearable computers, as well as accurately modeling user tasks in the supporting software, are among the most significant challenges faced by wearable system designers. Because wearable computers represent a new paradigm in computing, there is no consensus on the mechanical/software human–computer interface or the capabilities of the electronics. Thus, iterative design and user evaluation made possible by the rapid design/prototyping methodology is essential for quick definition of this new class of computers.

Other representative examples of wearable computers are described in Dey, Futakawa, Salber, and Abowd (1999), Feiner, MacIntyre, and Höllerer (1997), Healey and Picard (1998), Mann (1997, 1998), Najjar, Thompson, and Ockerman (1997), and Starner, Weaver, and Pentland (1997). The applications described in these papers highlight the functionality supported by the input–output modalities and mobile information access provided by wearable computing.

In this chapter, we describe the CMU wearable computers and design methodology, and illustrate its effectiveness by describing three wearable computers and summarizing their design activities.

EVOLUTION OF WEARABLE COMPUTERS

The family tree of CMU wearable computers, as shown in Fig. 32.1, classifies wearable computers into several application categories and presents their development over the last decade. Each mark represents a distinct generation of CMU wearable computer and is placed under its corresponding application domain. The four wearable computer designs marked with a star have been awarded prestigious international design awards (VuMan 3, MoCCA or Mobile Communication and Computing Architecture, Digital Ink, and Promera). The following are major application areas:

- Plant operation
- Manufacturing
- Language Translation
- Maintenance
- Medical
- Invisible Computing/Smart Rooms
- Mobile Worker Concepts
- Navigation

Figure 32.2 depicts the increase in microprocessor performance (measured in millions of instructions per second or MIPS) as a function of time. In the early 1960s, Gordon Moore of Intel made the observation/prediction that the capacity of semiconductor chips was doubling every year. Similar trends have been noted for microprocessor speed, magnetic disk storage capacity, and network bandwidth. The points depicted in Fig. 32.2 are the performance thresholds necessary for each of the user interface types. Thus, a textual interface requires 1 MIPS, a graphical user interface (GUI) 10 MIPS, a handwriting interface 30 MIPS, a speech recognition interface 100 MIPS, natural language understanding 1,000 MIPS, and vision understanding 10,000 MIPS. The thresholds represent the time when microprocessors have

(by Operational Delivery Dates and Application Areas)

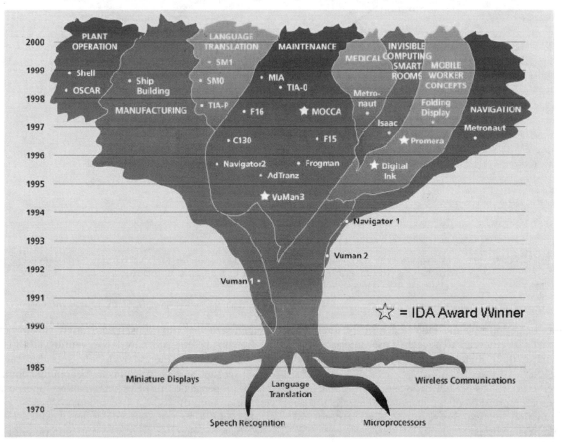

FIGURE 32.1. Family tree of CMU wearable computers. CMU = Carnegie Mellon University. IDA = Industrial Design Association.

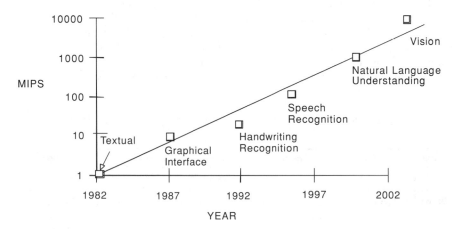

FIGURE 32.2. Performance thresholds necessary for each of the user interface types. MIPS = millions of instructions per second.

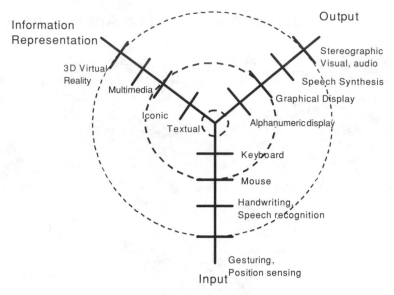

FIGURE 32.3. Kiviat graph representation of wearable computer use modalities. 3D = three-dimensional.

the capability of supporting the indicated form of interface. However, it takes several years to develop a user interface style that often emerges quite a while after the technology threshold has been passed.

Because ease of use is so closely associated with human reaction, it is much more difficult to quantify. There are at least three basic functions related to ease of use: input, output, and information representation.

Figure 32.3 summarizes several points for each of these basic functions. Note that, unlike the continuous variables for capacity and performance, the ease of use metrics are discrete.

Just as the performance of microprocessors has increased over time as shown in Fig. 32.2, the characteristics of the user interface shown in Fig. 32.3 are also moving out with time. For example, the keyboard with an alphanumeric display using textual information is representative of timesharing systems of the early 1970s. The keyboard and mouse, graphical output, and iconic desktop are representative of personal computers of the early 1980s. The addition of handwriting recognition input, speech synthesis output, and multimedia information began

emerging in the early 1990s. It takes approximately one decade to broadly disseminate new input, output, and informational representations. In the 2000s decade, speech recognition, position sensing, and eye tracking are becoming common inputs. Heads-up projection displays should allow superposition of information onto the user's environment.

Because of the requirements of lightweight and endurance (i.e., battery operating life), wearable computers have lagged top-of-the-line desktop computers in performance and capacity by about 5 years. Tables 32.1 through 32.4 illustrate the evolution of research wearable computer systems with respect to four of their key attributes: display, processor, user interface, and wireless communication. The evolution of displays has witnessed improvements in their resolution, wearability, and color palettes, as shown in Table 32.1. These improvements enabled the use of head-mounted displays in a broader variety of applications. In particular, the users required nonobtrusive displays that would neither block their field of vision nor be too bulky to wear. Recent displays fit these characteristics well. Processors have evolved to meet the demands of more complex wearable

TABLE 32.1. Research-Wearable Computer Display Technology Evolution

| | | | Display/Output | | |
Year	Computer	Creator	Display Type	Resolution	Color
1991	VuMan 1	CMU	Private Eye	320 × 240	Monochrome
1993	Herald 1	MIT	Private Eye	320 × 240	Monochrome
1994	Fast I	Georgia Tech	Virtual Vision	320 × 240	Color NTSC
1994	VuMan 3	CMU	Private Eye	320 × 240	Monochrome
1995	Navigator 2	CMU	Virtual Vision	640 × 480	Monochrome
1995	Fast II	Georgia Tech	Seattle Sight	640 × 480	Monochrome
1996	Wearcomp6	MIT	Kopin	640 × 480	Monochrome
2001	Spot	CMU	IBM Watson	640 × 480	24-Bit Color

Note. CMU = Carnegie Mellon University; MIT = Massachusetts Institute of Technology; NTSC = National Television Standards Committee.

TABLE 32.2. Performance Evolution of Research-Wearable Computers

Year	Computer	Creator	Processor
			Processor
1991	VuMan 1	CMU	8 MHz 80188
1992	VuMan 2	CMU	13 MHz 80C188
1993	Navigator 1	CMU	25 MHz 386
1993	Herald 1	MIT	12 MHz 286
1994	VuMan 3	CMU	20 MHz 386
1994	Fast I	Georgia Tech	33 MHz 486
1995	Navigator 2	CMU	33 MHz 486
1995	Fast II	Georgia Tech	100 MHz 486
1995	Herald 2	MIT	50 MHz 486
1996	C-130/TIA-P	CMU	100 MHz 486
1996	Lizzy 2	MIT	100 MHz 486
1997	MoCCA	CMU	133 MHz 586
1999	Itsy/Cue	Compaq/CMU	133 MHz StrongARM
2001	Spot	CMU	203 MHz StrongARM SA-1110

Note. CMU = Carnegie Mellon University; MIT = Massachusetts Institute of Technology.

computer applications, as shown in Table 32.2. Some inspection applications required only 20 MHz processors to run, whereas speech translation or augmented manufacturing applications require at least one order of magnitude improvement over that level of performance. Table 32.3 illustrates the use of different user interface modes in representative research examples of wearable computers over time, from textual and GUI interfaces, to multimedia interfaces. Wireless communication, described in Table 32.4, enable access to remote information, as well as real-time collaboration. The evolution of wearable computers and their applications has increased demand on processor performance, to be able to run applications using audio and video modalities, as well as communications bandwidth to enhance remote access/collaboration.

USER-CENTERED INTERDISCIPLINARY CONCURRENT SYSTEM DESIGN METHODOLOGY

A User-Centered Interdisciplinary Concurrent System Design Methodology (UICSM; Siewiorek et al., 1994), based on user-centered design and rapid prototyping, has been applied to the design of wearable computers. Based on user interviews, and observation of their operations, baseline scenarios are created for current practice. A visionary scenario is created to indicate how technology could improve the current practice and identify opportunities for technology injection. The visionary scenario forms the basis from which the requirements for the design are derived as well as evaluating design alternatives. Both scenarios are reviewed with the end user. A technology search generates candidates for meeting the design requirements. Several architectures, each appropriate to the various disciplines, are generated next: hardware, software, mechanical, shapes/materials, and human interaction modes. User feedback on scenarios and storyboards become input to the conceptual design phase.

TABLE 32.3. User Interface Evolution of Representative Research-Wearable Computers

Year	Computer	Creator	Interface Modes
			User Interface
1980	WearComp 0	MIT	Text/flash bulbs
1991	VuMan 1	CMU	GUI, blueprints, buttons
1992	VuMan 2	CMU	Text, maps, pictures, buttons
1993	Navigator 1	CMU	Phrase-based speech recognition
1993	Herald 1	MIT	Text, Twiddler
1994	VuMan 3	CMU	Text, schematics, dial, buttons
1995	Fast II	Georgia Tech	Speech recognition
1995	ADTranz	CMU	Internet phone, QuickCam, pen
1996	Isaac	CMU	Audio control of smart room
1996	C-130/TIA-P	CMU	Speech recognition, audio language translation, pen
1996	Wearcomp6	MIT	Video
1997	MoCCA	CMU	Audio bulletin boards, video, teleconferencing, pen
1997	Touring Machine	Columbia University	Trackpad, orientation tracker, video
1997	ITI-ALC	CMU	Integrate technical information for Air Logistic Centres, pen
1998	Smart Modules	CMU	Speech-to-text, text-to-text language translation, text-to-speech
1998	MIA	CMU	Multimedia—sketch, photographs, speech-to-text, pen
1999	Itsy/Cue	Compaq/CMU	Speech recognition, text-to-speech, pen
2001	Spot	CMU	Multimedia—video, speech recognition, text-to-speech, dial

Note. MIT = Massachusetts Institute of Technology; CMU = Carnegie Mellon University; GUI = graphical user interface.

Designers alternate between the abstract and the concrete; preliminary sketches are evaluated, new ideas emerge, and more precise drawings are generated. This iterative process continues with soft mock-ups, appearance sketches, and computer and machine shop prototypes until finally the product is fabricated.

As a result of UICSM, we have achieved a 4-month design cycle for each new generation of wearable computers. The

TABLE 32.4. Evolution of Wireless Communication in Selected Research-Wearable Computers

Year	Computer	Creator	Medium
			Wireless
1993	Navigator 1	CMU	Cell phone/modem
1994	Fast I	Georgia Tech	Wireless network
1995	ADtranz	CMU	WaveLAN 1.6 Mbps, CDPD
1996	Metronaut	CMU	Two-way pager
1996	Wearcomp6	MIT	Amateur radio modems
1996	Lizzy 2	MIT	CDPD
1997	MoCCA	CMU	CDPD
1999	Itsy/Cue	Compaq/CMU	WaveLAN 802.11b 2 Mbps
2001	Spot	CMU	WaveLAN 802.11b 11 Mbps

Note. CMU = Carnegie Mellon University; MIT = Massachusetts Institute of Technology; CDPD = Cellular Digital Packaged Data.

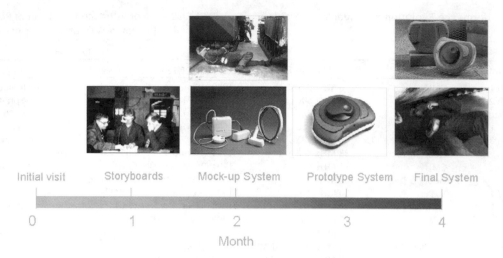

Initial visit   Storyboards   Mock-up System   Prototype System   Final System

0   1   2   3   4

Month

FIGURE 32.4. Four-month design cycle.

cycle time of the new products is ideally suited to the academic semester. As depicted in Fig. 32.4, student designers initially visit the user site for a walkthrough of the intended application. A second visit after a month of design, ending the conceptual phase, elicits responses to storyboards of the use of the artifact and the information content on the computer screen. After the second month, a software mock-up of the system running on a previous generation wearable computer is evaluated in the end-user's application, representing the results of the detail design phase. During the third month, implementation takes place, and a prototype of the system receives a further user critique. The final system is delivered after the fourth month for field trial evaluation.

The goal of the UICSM is to allow as much concurrency as possible in the design process. The design cycle includes monthly "builds," an evolving system integration demonstration that solicits end-user feedback each month. The System Build phases and the iterative nature of user-centered design in a 4-month design cycle are also represented in Fig. 32.5, illustrating the concurrent activities in four disciplines for the system depicted in Fig. 32.4. The application supported by VuMan 3 is Vehicle Inspection. The final design of VuMan 3 is described later.

Stage 0 System Build

The first visit to the users site generated the major attributes and design requirements imposed by the heavy military vehicle maintenance application (Fig. 32.5). The first step in the inspection process, the Limited Technical Inspection (LTI), was chosen for the application. The LTI is a checklist of more than 600 items, in a 50-page paper document. The selection of maintenance checklist application represents the primary design decision. We observed the maintenance process, the operating environment, the tools, and supporting activities. Maintenance workers wore gloves and operated in confined, environmentally harsh spaces in both subdued lighting and bright sunlight. The mechanics worked in dust, fuel, and corrosive chemicals.

Mechanics reviewed procedures in manuals stored at a central repository before returning to the vehicle, occasionally with copies of critical manual pages. Further task analysis yielded the first design constraints—the environmental sealed housing, an input device that could be operated with gloves, a single integrated housing whose interface could be operated independent of orientation.

Stage 1 System Build

During the second phase, we conceptualized the new product. A second visit after a month of design elicited responses to storyboards of the use of the artifact and the information content on the computer screen. The storyboard mock-up of the application includes a slide show of the application on a previous generation wearable computer and represents the first build phase in the evolution. Estimates of the processing and data storage requirements indicated the type of processor and memory cards. The data collected during a vehicle inspection would be uploaded to a logistic computer. Another goal was less than 5 min training time. The user interface adopted uniform, on-screen control icons to simplify screen manipulation.

Stage 2 System Build

This stage involves several physical mock-ups, as well as a software prototype on an existing platform; in this case, VuMan 2. On the third visit to the maintenance base, we demonstrated the prototype application software and displayed various forms for the input dial (Fig. 32.6). Based on user feedback, design proceeded with making the wooden mock-up to visualize and configure the housing. After exploring the implications on the housing configuration, it was decided to decrease the footprint of the design by increasing its thickness with a two-board design, the main processor board (Fig. 32.7), and a PCMCIA controller board. The PCMCIA Card Controller chip (82365SL) minimized chip count of the interface between the processor ISA bus and the PC Card socket.

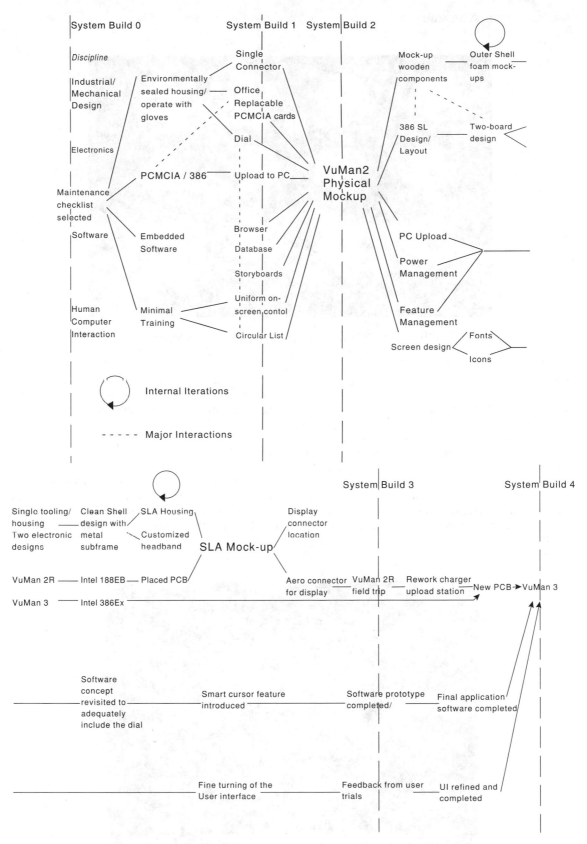

FIGURE 32.5. Concurrent design activities and system builds. SLA = stereolithography Appartus; PCB = Printed Circuit Board; UI = user interface; PCMCIA = Personnel Computer Memory Card International Association.

641

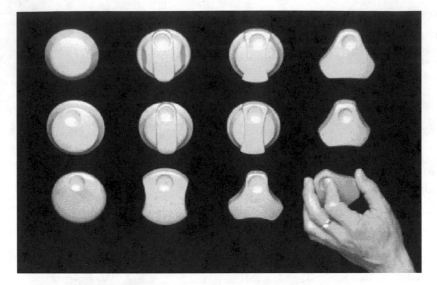

FIGURE 32.6. Different forms for the input dial for VuMan 3.

Stage 3 System Build

To reduce the risk of switching to a more complex electronic component set from one less complex, a two-phase approach was adopted. The first system, VuMan 2R, would use an existing processor from VuMan 2, the Intel 188EB; the second system, VuMan 3, would incorporate an Intel 386 processor. The 386 processor board would cover the batteries, whereas the 188 would have cut-outs for the batteries resulting in a thinner housing. The housing for both designs would be the same; the only difference was that the 386 housing would have an increased vertical height. Only one set of housing tools was required. The SLA (StereoLithography Appartus) mock-up model of the housing was created at the end of this stage.

Stage 4 System Build

The final wearable computer (Fig. 32.8) was a $5'' \times 6.25'' \times 2''$ unit weighing less than 2 pounds, including a rotary dial input device integrated with an environmental sealed housing, Private Eye display with a customized headband, and a smart docking

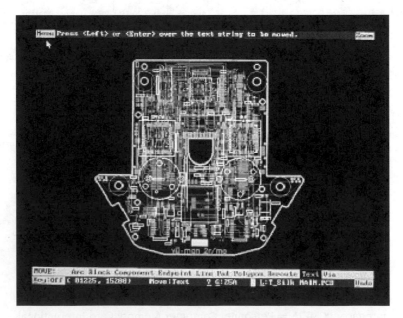

FIGURE 32.7. Main processor board design for VuMan 2R.

FIGURE 32.8. Final implementation of VuMan 3 wearable computer.

station that monitors the use of the NiCd rechargeable batteries and also acts as a communication link to a host computer system.

WEB-BASED DESIGN METHODOLOGY

The design methodology described in this chapter is web-based and defines intermediary design products that document the evolution of the design. These products are posted on the web so that even remote designers and end users can participate in the design activities. Table 32.5 is the Project Matrix used in the course. The design methodology proceeds through three phases: conceptual design, detailed design, and implementation culminating in System Builds 1, 2/3, and 4, respectively. Activities with the same number are conducted concurrently in the various disciplines. The underlined activities have outputs that are placed on the web that document design decisions so that other groups and our remote partners can follow and participate in the design. At different times, different disciplines initiate activities with other disciplines contributing to the results. For example, the Human–Computer Interaction group performs a field evaluation and produces the problem scenario that is reviewed and refined by the other groups (step 1 in Table 32.5). Individual disciplines are responsible for generating technology-specific product/feature matrices (Step 4 in Table 32.5) for the target technologies identified (Step 2) to support the visionary scenario (Step 3). Table 32.6 is an example product feature matrix of speech recognition systems, identifying features that were studied and evaluated in Step 4. The Conceptual Design Phase concludes with discipline specific architecture definitions (Step 6) The other two phases proceed in a similar manner. Cross functional teams ensure consistency between disciplines. Group leaders form a Project Management Team responsible for execution of the methodology. Each phase culminates in web products, a written report, and an oral presentation produced by the entire group. These activities are represented at the bottom of Table 32.5.

TAXONOMY OF WEARABLE COMPUTERS

Wearable/mobile computers can engage the senses of an individual to enhance the performance of that individual, as well as a team that the individual is associated with. We treat

collaboration as an enhancement of the human senses. In that respect, three examples of CMU wearable computers can be classified by the taxonomy shown in Table 32.7. VuMan 3 (Smailagic, Siewiorek, Stivoric, & Martin, 1998) represents a stand-alone wearable computer performing data recording functions. The C-130 uses one-to-one collaboration in the form of a help desk. MoCCA involves a group in collaboration. The following subsections summarize the final design for these three examples.

Stand-Alone Computer: VuMan 3

VuMan 3 design represents an experience in electronics/mechanical/software co-design, in particular software/mechanical integration, by coupling a novel customized input device, the rotary dial, and the application software. VuMan 3 functions as a data recorder and facilitates the filling out of forms. It was designed for the LTI of amphibious vehicles for the U.S. Marines at Camp Pendleton, CA. The result was a new user interface paradigm: circular input, circular visualization.

The LTI checklist consists of a number of sections, with about 100 items in each section. The users can manually progress from item to item by using the dial to select "next item" or "next field." Selectable items are represented as boldface as shown in Fig. 32.9. During the Second System Build Stage, users indicated that dial motion could be reduced by taking advantage of the fact that, on average, more than 80% of the items on an LTI are serviceable. The Smart Cursor feature represents built-in intelligence in the user interface and was designed to help automate navigation of the checklist. The Smart Cursor includes:

• An input pattern recognizer, which keeps track of what fields the user selects on a given screen, forming a "working set." If the working set remains the same over two or more screens, the Smart Cursor starts moving the cursor automatically to the fields in the working set. In essence, this is a macro recorder that runs continuously during the work session and employs a heuristic about when to repeat recorded inputs.

• A domain-specific heuristic, developed through studies of how users usually navigate through LTI hypertext documents (e.g., their behavior in the presence of multiple options). A high-level navigation pattern was found, which the input pattern recognizer could not identify. The knowledge about this high-level pattern was encoded in a navigation template. The system then uses a heuristic to decide when to apply this navigation template. This results in skipping to the next status box (top of right screen, Fig. 32.9b), when selecting the serviceable option.

A dial is simpler and inherently less expensive than a trackball, because it has only one degree of freedom (and hence one transducer, making it less expensive to manufacture). A trackball is a continuous two-dimensional input device, more complex to build and operate (manual dexterity). A dial requires a rethinking of the user interface to map a one-dimensional input device to a two-dimensional selection surface. Information displaying software requires a tight integration between input device and

TABLE 32.5. Project Matrix

Discipline	Product Development Phases		
	Conceptual Design	Detailed Design	Implementation
Hardware	1 Review field data/refine HCI Problem scenario 2 Select/refine HCI Target technologies 3 Review/refine HCI Visionary scenario 4 HW Product/feature matrices 4 HW Feasiblity studies 5 Input to PM Design Decision tracking form (HW selection criteria and choices) 6 HW architecture	9 Input to HCI Design scenario 10 Add to Task/Issue Tracking form 10 Add Resolutions to Task/Issue Tracking form 11) Status list HW tasks and issues 12 Provide input to schedule 13 Resolve issues 13 Perform unit HW implementation 14 HW design phase summary 14 User evaluation and feedback plan	17 Updates to status list of HW tasks and issues 19 Input to HCI Demo Script 20 Inputs to PM Integration Tree 21 Integrate HW components 22 Dry run of demo and testing
Software	1 Review field data/refine HCI Problem scenario 2 Select/refine HCI Target technologies 3 Review/refine HCI Visionary scenario 4 SW Product/feature matrices 4 SW Feasibility studies 5 Input to PM Design Decision tracking form (SW selection criteria and choices) 6 SW architecture	9 Input to HCI Design scenario 10 Add to Task/Issue Tracking form 10 Add resolutions to Task/Issue Tracking form 11 Status list SW design tasks/issues 12 Input to schedule 13 Resolve issues 13 Perform unit SW implementation 14 SW design phase summary 14 User evaluation and feedback plan	18 Updates to status list of SW tasks and issues 19 Input to HCI Demo Script 20 Input to PM Integration Tree 21 Integrate SW components 22 Dry run of demo and testing 23 Archive and document source and object code
Mechanical/ Industrial	1 Review field data/refine HCI Problem scenario 2 Select/refine HCI Target technologies 3 Review/refine HCI Visionary scenario 4 MEI Product/feature matrices 4 MEI Feasibility studies 5 Input to PM Design Decision tracking form (MEI selection criteria and choices) 6 MEI forms, mockups, drawings, analyses	9 Input to HCI Design scenario 10 Add to Task/Issue Tracking form 10 Add resolutions to Task/Issue Tracking form 11 Status list MEI design tasks/issues 12 Input to schedule 13 Resolve issues 13 Perform unit MEI implementation 14 Design phase summary 14 User evaluation and feedback plan	18 Updates to status list of MEI tasks and issues 19 Input to HCI Demo Script 20 Input to PM Integration Tree 21 Integrate MEI components 22 Dry run of demo and testing
Human–Computer Interface	1 Field evaluation reports and data 1 Problem scenario 2 Target technologies 3 Visonary scenario 4 HCI Feasibility studies 5 Review product/feature matrices 5 Review feasibility studies 5 Refined solution scenario 6 Initial user interface concepts	9 Input to HCI Design scenario 10 Add to Task/Issue Tracking form 10 Add resolutions to Task/Issue Tracking form 11 Status list of UI design tasks/issues 12 Input to schedule 13 Resolve issues 13 Perform unit UI implementation 14 Design phase summary 14 User evaluation and feedback plan 15 Coordinate user evaluation and prepare feedback report	18 Updates to status list of HW tasks and issues 19 Produce Demo Script 20 Inputs to PM Integration Tree 21 Integrate user interfaces 22 Dry run of demo and testing 23 Archive and document source and object code
Cross-Functional Groups	1 Review field data/refine HCI Problem scenario 2 Review/refine HCI Target technologies 3 Review/refine HCI Visionary scenario	9 Subsystem interface specifications 10 Add to task/Issue Tracking form 10 Add resolutions to Task/Issue Tracking form 11 Status list subsystem tasks/issues 12 Input to schedule	18 Updates to status list of subsystem tasks and issues 19 Input to HCI Demo Script 20 Input to PM Integration Tree 21 Integrate subsystem 22 Dry run of demo and testing

(Continued)

TABLE 32.5. (Continued)

Discipline	Product Development Phases		
	Conceptual Design	Detailed Design	Implementation
	4 Product/feature matrices	13 Resolve issues	
	4 Feasiblity studies	13 Perform unit subsystem	
	5 Input to PM Design Decision tracking	implementation	
	form (selection criteria and choices)	14 Design phase summary	
	6 High-level design	14 User evaluation and feedback plan	
Team	2 Requirement Table	15 Product design specification	21 Product design specification
Products	6 Requirement—Feature table	16 Presetnation slides	22 Presentation Slides
Link to group	7 Product design specification	16 Detailed Design Phase Report	23 Final Report
membership	8 Presentation slides		
and meeting	8 Conceptual Design Phase Report		
pages			

Note. HW = hardware; HCI = human–computer interface; PM = project matrix; SW = software; MEI = mechanical/industrial; UI = user interface.

software. Options to be selected can be logically arranged in a circular list. An option is highlighted. Rotation of the dial 1 position clockwise changes the highlighted option to the next one clockwise in the circular list. The same applies for counterclockwise. There may be more, less, or equal number of positions on the dial than in the circular list. Depression of one or more buttons performs the function specified by the highlighted option, which may include entering the highlighted information into a database, providing auxiliary information, selecting a hypertext link, selecting an option on a World Wide Web (WWW) page, etc. From a visual perspective, selectable items can be placed around the outside of the screen and can be navigated

TABLE 32.6. Product Feature Matrix for Speech Recognition Systems Indicating Functions
and Hardware/Software Requirements (Circa 1999)

Company	Product	OS	Processor	Memory	Space	Price	Features
Dragon Systems	Naturally Speaking (personal, deluxe versions)	Win95, WinNT	Pentium 133 MHz; also requires sound card and CD ROM	48 MB	100 MB	$129 (personal) $695 (deluxe)	230,000 Vocabulary (personal version); additional features in deluxe version; dictate into any application; voice control of mouse pointer; multiple users; ability to write macros; playback of recorded speech
Dragon Systems	Dragon Dictate 3.0	Win95, WinNT	Pentium 133 MHz; also requires sound card and CD ROM	32 MB	36 MB	$149	Discrete speech; dictation playback; text-to-speech conversion
IBM	Via Voice Gold	Win95, WinNT	Pentium 150 MHz; also requires Sound Blaster and CD ROM drive	32MB (Win95) 42MB (WinNT)	125 MB	$99	Dictate into any application; Macros; 22,000 vocabulary (can expand to 64,000); includes a noise cancellation headset microphone
IBM	Simply Speaking Gold	Win95 WinNT	Pentium 100 MHz; also requires Sound Blaster and CD ROM drive	32 MB	36 MB	$99	Discrete speech; dictation playback; text-to-speech conversion
Phillips	SpeechMagic	Win3.1	486 66 MHz; also requires Philips LFH 6210 Accelerator Board	16 MB	>500 MB	$???	64,000 word vocabulary; speaker adaptation; provides a correction editor, editing, and playback of recordings; vocabulary manager for adding new words; Windows DDE support and native API provided for integration.

Note. OS = operating system; DDE = dynamic data exchange; API = application program interface.

TABLE 32.7. Taxonomy of Example CMU Wearble Computers With Respect to collaboration

Number of Collaborators	Taxonomy	Wearable Computer
0	Stand-alone (data recorder)	VuMan 3
1	One-to-One Help Desk	C-130
N	Group Collaboration	MoCCA

Note. CMU = Carnegie Mellon University; MoCCA = mobile communication and computing architecture.

in a circular fashion. Thus, the screen perception corresponds to the input device and the end user feels as if he/she is dealing with a single, unified device. Application software was modified to reflect the dial design decisions, and the final result was a new user interface paradigm: circular input, circular visualization. The rotational input provides discrete, tactilely discernible

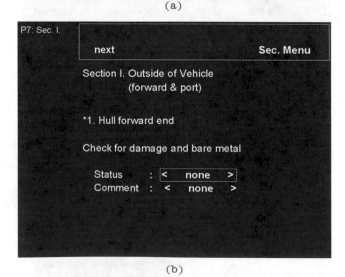

(a)

(b)

FIGURE 32.9. LTI information screens: (a) selecting an inspection region and (b) item requiring filling in the status box. LTI = limited technical inspection.

positions. The dial can be operated while wearing gloves, and the interface operation is unambiguous and the same no matter where the dial is worn on the body.

One-to-One Help Desk: C-130

A multimedia collaboration system with head-mounted display, video camera, and wireless communications provides access to electronic maintenance manuals and remote access to a human help desk expert on the C-130 flight line. Use of wearable computers for collaboration has been reported in Bauer, Heiber, Kortuem, & Segall (1998), Billinghurst, Weghorst, & Furness (1997), and Rekimoto (1996). The C-130 project was designed to use collaboration to facilitate training and to increase the number of trainees per trainer. Inexperienced users were being trained to perform a cockpit inspection and the trainers were remotely located. The trainee loads the inspection procedures and performs the inspection. The C-130 system also used the dial as an input device. A desktop system managed the normal job order process and was used by the instructors in the C-130 project to observe the trainee's behavior. In collaboration, the instructor looks over the shoulder (through a small video camera attached to the top of the trainee's head mounted display) and advises the trainee. In addition to a two-way audio channel, the instructor can provide advice via a cursor for indicating areas on a captured video image that is being shared through a whiteboard. The instructor manages the sharing session and whiteboard. The trainee's use of the whiteboard is limited to observation. This arrangement allows the instructor to "reach over the shoulder and point."

We observed the functions performed by the technicians to generate the functionality provided by the user interface. The wearable user has two main functions: navigating through the checklist and initiating collaboration. All navigational links are either through the natural sequence of the checklist or through a simple menu. The trainee uses a dial to sequence through active regions of the screen. The interface is organized as a hierarchy with a sequential list of inspection steps embedded in the hierarchy. Figure 32.10 illustrates an example of the user interface that supports a checklist application. The left side of the screen displays procedures and the right side presents diagrams. Each screen gives the ability to go to the next or previous step of the inspection or return to the index of all steps. The trainee can indicate when an action described by the instructions fails to produce the correct result by checking a box near that action. The trainee with a camera on top of the head-mounted display is shown in Fig. 32.11.

The C-130 application software was constructed using WWW browsers and took advantage of the hypertext linking facility supported by that software. Special helper applications were written in Java to enable collaborative design sessions using a shared whiteboard. The database component consisted of a collection of databases both newly created and pre-existing. The middleware included a WWW server that provided interfaces to the databases and generation of HTML with the information from the databases included. The middleware also included the communication aspects of the collaboration software. The

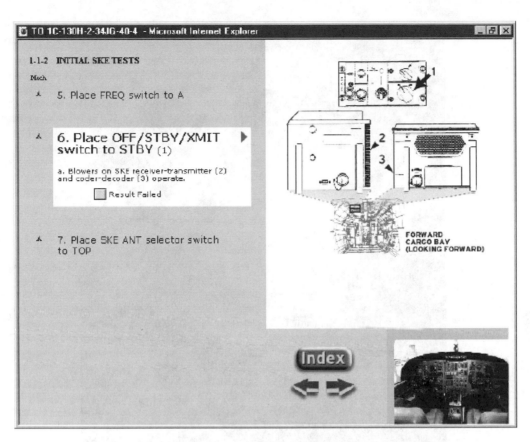

FIGURE 32.10. User interface screen for C-130 troubleshooting.

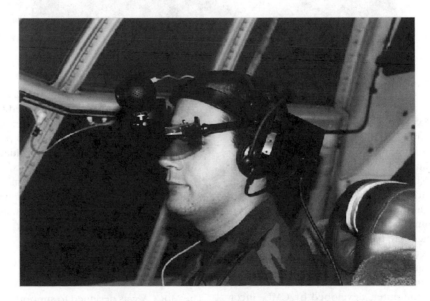

FIGURE 32.11. C-130 trainee with a head-mounted display and camera.

FIGURE 32.12. C-130 wearable computer.

database and middleware components of the software were resident on the desktop.

Microsoft's Internet Explorer was used for the C-130 project, because it allowed easy integration with the dial input device. Internet Explorer allows Tab and Shift-Tab as mechanisms for navigating through the links on a page. A clockwise rotation of the dial was mapped into the Tab and counterclockwise into the Shift-Tab, and consequently, the browser could be used without modification. The software design was chosen primarily to enable the utilization of commercially available components both for display and communication. In this case, it was the software that constrained the electronics to be sufficiently powerful to support a fully functional operating system that in turn, supported the WWW software.

The C-130 wearable computer, developed by CMU, incorporates a 133 MHz 586 processor, 32 MB DRAM, 2 GB IDE (integrated drive electronics) disk, full-duplex sound chip, and spread spectrum radio (2 Mbps, 2.4 GHz) in a ruggedized, hand-held, pen-based system designed to support speech translation applications. The C-130 wearable computer, shown in Fig. 32.12 was adapted so that a dial provided all of the necessary input capability except for the video camera and associated video capture card.

The C-130 wearable computer has also been demonstrated with the Dragon speech translation system in several foreign countries and supported experimentation with the use of electronic manuals for F-16 maintenance.

Group Collaboration: MoCCA

The MoCCA was designed to support a group of geographically distributed field service engineers (FSEs). The FSEs spend up to 30–40% of their time in a car driving to customer sites. Half of what they service is third-party equipment for which they may not have written documentation. The challenge was to provide a system that allowed the FSEs to access information and advice from other FSEs while on customer sites and while commuting between sites. Synchronous and asynchronous collaboration are supported for both voice and digitized information (Smailagic, 1998).

Conflicting Design Requirements. An additional challenge arose from user interviews, which suggested that the FSEs desired all of the functionality of a laptop computer, including a larger color display with an operational cycle of at least 8 hr. In addition, the system should be very light, preferably less than 1 pound in weight, and require access to several Legacy databases that existed on different corporate computing systems. Further discussions with the FSEs indicated that the most frequently used databases were textually oriented. Only on rare occasions was access to graphical databases required. However, when required, they were absolutely necessary. To address this set of conflicting requirements, a novel architecture was created that combined a lightweight alphanumeric satellite computer with the high functionality of a base unit that is included in the FSEs toolkit. The base unit can be carried into any customer site providing instant access to the global infrastucture.

Architecture. MoCCA consists of the following units (Fig. 32.13):

- A base unit, about the size of a small laptop computer, which is connected to a remote server wirelessly through a Cellular

Digital Packaged Data (CDPD) wireless connection. Located in the FSE's toolkit, the base unit has a large color display for viewing schematics.

- A cellular phone is tethered to the base unit and communicates wirelessly with the local cellular provider.

- A hand-carried satellite unit, which is in communications with the stationary base unit. The satellite unit displays the contents of the base unit screen. The satellite unit weighs less than 0.75 pounds.

- A microphone and headset combination that is wireless linked to the cellular phone.

Figure 32.14 is a view of the functional system architecture. The software architecture uses a thin client approach to minimize the amount of software on the system by exploiting web-browsing technology and wireless CDPD Internet connection to communicate with a server. The satellite unit is not running the browser; it is merely displaying whatever is currently on the base unit display.

The legacy databases and access software reside on the server machine located at the home office. It contains a database of field service information that was designed and customized to the specific needs of FSEs. When the FSE presses one of the user interface buttons or link on one of the browser web pages, a request is sent to software developed for the MoCCA unit to initiate a query on the Field Service Database. The results from the query are dynamically formatted into a web page. Then, all the appropriate links are made to enable related queries, and the page is passed across the CDPD network to the client machine

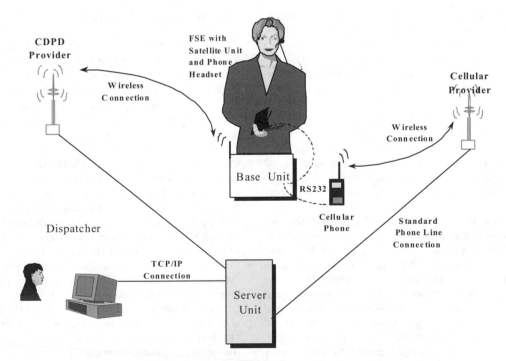

FIGURE 32.13. MoCCA system architecture. MoCCA = mobile communication and computing architecture; CDPD = cellular digital packaged data; FSE = field service engineer.

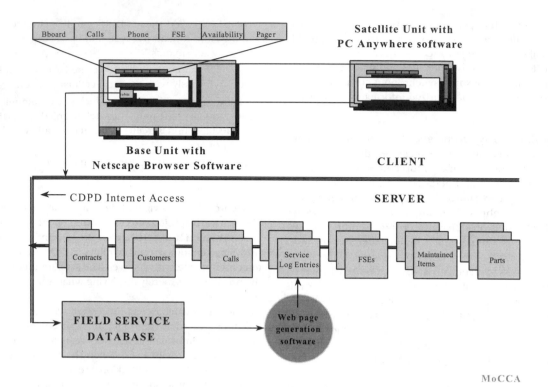

FIGURE 32.14. MoCCA data architecture. MoCCA = mobile communication and computing architecture; Bboard = bulletin board; FSE = field service engineer; CDPD = cellular digital packaged data.

that displays the web page. Monitor software residing on the server can watch certain directories and when data changes, notify the FSEs browser to change the color of one of the user interface buttons to alert them to rerun the query.

A summary of the integrated user interface software is presented in Fig. 32.15. The FSE Call List is the central screen, and all other screens can be accessed following hypertext links or the screen buttons. From the Call List screen, the user can select a primary function to be performed, such as database query (Call List) or messages via a pager. Depending on the primary function selected, subsequent secondary screens will display more detailed information. In the case of pager messages, the secondary screens allow the user to enter new information.

There are six buttons that appear on all screens of the user interface. The Bulletin Board button provides access to a phone-based voice bulletin board in which FSEs may asynchronously collaborate to solve problems. The Calls button accesses the summary of active field service calls for the engineer. The Phone button invokes an autodialer keypad. The FSE button brings up a directory of FSEs, and the Availability button shows their current status. The Pager button accesses the list of current Pager messages. The FSE Directory, Pager Message List, Calls List, and Availability Form are all web pages, generated automatically from various field service databases.

Consider the following typical usage scenario of MoCCA. The FSE checks the current status and the status of other FSEs

by clicking on the FSE Information button. This leads to a web page (Fig. 32.16a), generated from the MoCCA database, listing the names of all the FSEs, their cell phone numbers, their e-mail addresses, and their current status. The FSE changes his status from Off Duty to Available to let others know that he can be reached by cell phone. This is done by clicking on the Availability button at the top of the screen leading to the Availability Selection page (Fig. 32.15). Clicking on the Available button leads back to the FSE Information page. To check the list of calls assigned to the FSE, he clicks on the Calls button at the top of the screen, and brings up the Calls List screen (Fig. 32.16b). The page is assembled from a database query of the calls assigned to the FSE. It filters the list of all calls to only display those with an Open status. The table lists each call with the customer name, the time the call was received, the service contract, the contact person, their phone number, and a short description of the problem. Each underlined entry leads to further information about that field. If the FSE wants to review the assigned calls, and the previous problems at a site, he clicks on the Call Query button, which brings up the Call List Queries page. To get more details on a particular call, the FSE clicks on the description of the problem, showing the Detailed Call Information List. By clicking on the Service Log button, the FSE can review the Service Log entries associated with that call, leading to the Call Logging list. If there is a Tip recorded with information, the FSE can examine its contents. After reviewing the call history, he can look at the current Calls List again.

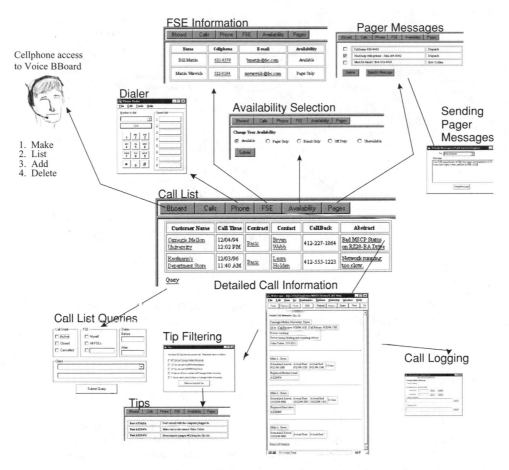

FIGURE 32.15. Summary of the integrated user interface software. FSE — field service engineer; BBoard = bulletin board.

Bboard	Calls	Phone	FSE	Availability	Pages

Name	Cellphone	E-mail	Availability
Bill Martin	621-8579	bmartin@dec.com	Available
Martin Warwick	522-0164	mwarwick@dec.com	Pager Only

(a)

Bboard	Calls	Phone	FSE	Availability	Pages

Customer Name	Call Time	Contract	Contact	CallBack	Abstract
Carnegie Mellon University	12/04/94 12:02 PM	Basic	Bryan Webb	412-227-1864	Bad MSCP Status on RZ28-BA Drive
Kaufmann's Department Store	12/03/96 11:40 AM	Basic	Laura Holden	412-555-1223	Network running too slow.

Query

(b)

FIGURE 32.16. Sample screen images for (a) Field Service Engineer (FSE) information and (b) call list. Bboard = bulletin board.

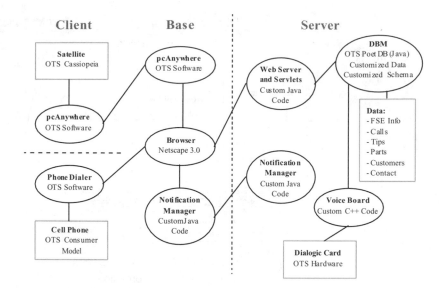

FIGURE 32.17. MoCCA software architecture. MoCCA = mobile communication and computing architecture; OTS = off-the-shelf; DBM = data base manager.

The satellite unit enabled FSEs to wear a very small, lightweight computer with display. An important design decision was to use the same user interface as the base station. The Cassiopeia satellite unit has a screen size of 2.5″ × 4.75″ and a resolution of 480 × 240 pixels, whereas the base station's resolution is 640 × 480 pixels. Only a 480 × 240 pixel window was used for the screens shown in Fig. 32.16.

The architecture of the MoCCA software is depicted in Fig. 32.17. The architecture can be broken into three different subsystems: Client, Base, and Server. The communication among the subsystems uses different mechanisms. The Client communicates with the Base unit using a serial connection. The Base communicates with the server by TCP/IP over a variety of media. The software was divided into functions to facilitate parallel development. The Client appears as a window into the Base. This functionality was achieved by using an off-the-shelf product called PC Anywhere. Netscape Navigator browser runs on the Base unit and can view the web pages generated by the Server. The browser retrieves the information from the Notification Manager and displays it as the first page. The Notification Manager informs the FSE of a newly arrived event and is represented by a button. The Server contains two data types, the first contains FSE and customer data, the second contains the voice bulletin board data. The Server contains a database manager and a voice database manager, which are accessed via servlet calls. A Java database called Poet was chosen. Poet is a *cross-platform* database supporting both C and Java schemas, and allows for easy integration of different software components. The Notification Manager server informs the Notification Manager on the Base of changes to the databases. Both Notification Managers are implemented as custom Java applets. The server also contains a hardware component, the Dialogic board, with associated software that controls the voice bulletin board system and dialing in.

Figure 32.18 depicts the base unit components and the satellite unit hardware. The base unit includes a 586 133 MHz processor, running the Windows 95 operating system. There are two PCMCIA slots on the base unit: one is occupied by the AirCard CDPD/Modem, and the other is occupied by the modem dialing to the cell phone. Radio frequency (RF) transceivers link the cell phone and the headset.

USER EVALUATION

This section summarizes some of the user evaluation studies performed with these systems.

VuMan 3

The VuMan 3 user trials were performed at the U.S. Marine base at Camp Pendleton, CA. The user trial results indicated potential savings by reducing maintenance crews from two to one, a decrease of up to 40% in inspection time because of reduced motion of climbing in and out of tight spaces to read/write clipboards, a further 30% of not having to type in handwritten notes (Fig. 32.19), and savings of over two orders of magnitude in the weight of maintenance documentation.

C-130

The C-130 application was evaluated at the 911th Air National Guard, Pittsburgh, PA. The collaboration function of the C-130 application allows for immediate collaboration when problems occur. This results in a significant savings of time, compared with a situation without the wearable system.

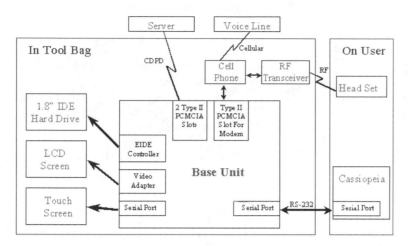

FIGURE 32.18. MoCCA base and satellite unit block diagram. MoCCA = mobile communication and computing architecture; CDPD = cellular digital packaged data; RF = radio frequency; IDE = integrated drive electronics; LCD = liquid crystal display, EIDE = enhanced integrated drive electronics; PCMCIA = Personnel Computer Memory Card International Association.

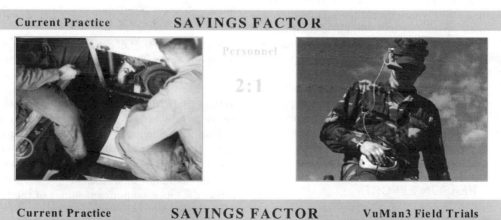

FIGURE 32.19. Saving factors from usage of VuMan 3.

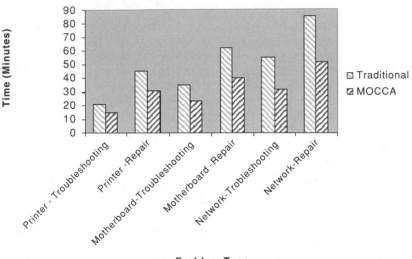

FIGURE 32.20. MoCCA performance experiments. MoCCA = Mobile Communication and Computing Architecture.

MoCCA

The MoCCA field tests were performed at Digital's facility in Forest Hills, Pittsburgh, PA. Five FSEs participated in tests that included performing a set of typical operations related to troubleshooting and repair operations on computing equipment. Each of the FSEs performed all of these operations. The subject systems included printers, motherboards, and networks. The use of MoCCA contributed to a significant savings of time (35–40%; Fig. 32.20). During these field tests, the FSEs used the system for the first time. A larger savings in time is expected with continued use. In addition, MoCCA allowed the FSEs to fix some problems immediately, which otherwise would have required return trips to find and bring back manuals.

PERSON EFFORT

The person effort represents the total amount of effort required to complete all phases of the design methodology. Records were kept throughout the projects to evaluate the design methodology. Table 32.8 summarizes the effort among phases in the design. Because of the custom nature of VuMan 3, the design phase required relatively more resources.

TABLE 32.8. Design Effort

Artifact	Conceptual Design (%)	Detailed Design (%)	Implementation/ Integration/ Evaluation (%)	Total Effort (person days)
VuMan 3	19	48	33	690
MoCCA	29	39	36	305
C-130	24	38	38	290

Note. MoCCA = mobile communication and computing architecture.

FUTURE RESEARCH

There are several challenges that research must address to make mobile access to information effective. Following is a partial list of those research challenges.

• User interface models: What is the appropriate set of metaphors for providing mobile access to information (i.e., what is the next desktop or spreadsheet)? These metaphors typically take over a decade to develop (i.e., the desktop metaphor started in the early 1970s at Xerox PARC and required over a decade before it was widely available to consumers). Extensive experimentation working with end-user applications will be required. Furthermore, there may be a set of metaphors, each tailored to a specific application or a specific information type.

• Input–output modalities: While several modalities mimicking the input–output capabilities of the human brain have been the subject of computer science research for decades, the accuracy and ease of use (i.e., many current modalities require extensive training periods) are not yet acceptable. Inaccuracies produce user frustrations. In addition, most of these modalities require extensive computing resources that will not be available in low-weight, low-energy wearable computers. There is room for new, easy-to-use input devices, such as the dial developed at CMU for list-oriented applications.

• Quick interface evaluation methodology: Current approaches to evaluate a human computer interface requires elaborate procedures with scores of subjects. Such an evaluation may take months and is not appropriate for use during interface design. These evaluation techniques should especially focus on decreasing human errors and frustration.

• Matched capability with applications: The current thought is that technology should provide the highest performance capability. However, this capability is often unnecessary to complete an application, and enhancements such as full-color graphics require substantial resources and may actually decrease ease of use by generating information overload for the user. For example, one informal survey of display requirements for military planning estimates 85% of the applications can be performed with an alphanumeric display, 10% with simple graphics, and only 5% require full bitmap graphics. Interface design and evaluation should focus on the most effective means for information access and resist the temptation to provide extra capabilities simply because they are available.

CONCLUSIONS

In this chapter, we have described a UICSM, as applied to design and implementation of over two dozen novel generations of wearable computers at CMU. The methodology is web-based, defines three phases of a design and implementation cycle, and documentation of the design evolution. We have defined a taxonomy of wearable computers with respect to collaboration and presented some evaluation results. Although the complexity of the prototype artifacts has increased by over two orders of magnitude, the total design effort has remained nearly constant. Field studies have been conducted and significant savings have been observed. The results of this research should allow us to set the design direction and make appropriate decisions in the future design of wearable computer systems.

ACKNOWLEDGMENTS

This material is based on work supported by the National Science Foundation under Grant No. 9901321. We also acknowledge the support provided by the Defense Advanced Research Project Agency.

References

Bauer, M., Heiber, T., Kortuem, G., & Segall, Z. (1998). A collaborative wearable system with remote sensing. *Proceedings of the Second International Symposium on Wearable Computers* (pp. 10–17). IEEE Computer Society Press, Los Alamitos, CA.

Billinghurst, M., Weghorst, S., & Furness T. III. (1997). Wearable computers for three-dimensional CSCW. *Proceedings of the First International Symposium on Wearable Computers* (pp. 39–47). IEEE Computer Society Press, Los Alamitos, CA.

Dey, A., Futakawa, M., Salber, D., & Abowd, G. (1999). The conference assistant: Combining context-awareness with wearable computing. *Proceedings of the Third International Symposium on Wearable Computers* (pp. 21–28). IEEE Computer Society Press, Los Alamitos, CA.

Feiner, S., MacIntyre, B., & Höllerer, T. (1997). A touring machine: Prototyping 3D mobile augmented reality systems for exploring the urban environment. *Proceedings of the First International Symposium on Wearable Computers* (pp. 74–81). IEEE Computer Society Press, Los Alamitos, CA.

Healey, J., & Picard, R (1998). StartleCam: A cybernetic wearable camera. *Proceedings of the Second International Symposium on Wearable Computers* (pp. 42–49). IEEE Computer Society Press, Los Alamitos, CA.

Mann, S. (1997). A historical account of the "WearComp" and "WearCam" inventions developed for applications. In Personal imaging, *Proceedings of the First International Symposium on Wearable Computers* (pp. 66–73). IEEE Computer Society Press, Los Alamitos, CA.

Mann, S. (1998). Humanistic intelligence: "WearComp" as a new framework and application for intelligent signal processing. *Proc. IEEE*, 2123–2151.

Najjar, L., Thompson, J. C., & Ockerman, J. J. (1997). A wearable computer for quality assurance in a food processing plant. *Proceedings of the First International Symposium on Wearable Computers* (pp. 163–164). IEEE Computer Society Press, Los Alamitos, CA.

Rekimoto, J. (1996). Transvision: A hand-held augmented reality system for collaborative design. *Proceedings of Virtual Systems and Multimedia*.

Siewiorek, D. P., Smailagic, A., Bass, L., Siegel, J., & Martin, D. (1998). Adtranz: A mobile computing system for maintenance and collaboration. *Proceedings of the Second IEEE International Symposium on Wearable Computers* (pp. 25–32). IEEE Computer Society Press, Los Alamitos, CA.

Siewiorek, D. P., Smailagic, A., & Lee, J. C. (1994). An interdisciplinary concurrent design methodology as applied to the Navigator wearable computer system. *Journal of Computer and Software Engineering, 2*(3), 259–292.

Smailagic, A. (1998). An evaluation of audio-centric CMU wearable computers. *ACM Journal on Special Topics in Mobile Networking, 6*, 59–68.

Smailagic, A., & Siewiorek, D. P. (1994). The CMU mobile computers: A new generation of computer systems. *Proceedings of the IEEE COMPCON 94* (pp. 467–473).

Smailagic, A., Siewiorek, D. P., Anderson, D., Martin, T., & Stivoric, J. (1995). Benchmarking an interdisciplinary concurrent design methodology for electronic/mechanical design. *Proceedings of the ACM/IEEE Design Automation Conference* (pp. 514–519).

Smailagic, A., Siewiorek, D. P., Stivoric, J., & Martin, R. (1998). Very rapid prototyping of wearable computers: A case study of custom versus off-the-shelf design methodologies. *Journal on Design Automation for Embedded Systems, 3*, 217–230.

Starner, T., Weaver, J., & Pentland, A. (1997). A wearable computer based American sign language recognizer. *Proceedings of the Second IEEE International Symposium on Wearable Computers* (pp. 130–137). IEEE Computer Society Press, Los Alamitos, CA.

·33·

A COGNITIVE SYSTEMS ENGINEERING APPROACH TO THE DESIGN OF DECISION SUPPORT SYSTEMS

Philip J. Smith
The Ohio State University

Norman D. Geddes
Applied Systems Intelligence, Incorporated

One of the most important uses of computers has been in assisting humans in decision making. There are many ways to use computer technology to improve performance on decision-making tasks. This includes providing:

- Improved training to the human decision makers.
- Improved forms of communication and coordination.
- External memory or perceptual aids.
- Enhanced access to relevant data and information.
- Active decision support systems where the computer is an active participant in the problem-solving and decision-making process.

This chapter focuses on interaction design issues (Preece, Rogers, Sharp, & Benyon, 1994) in the development and use of active decision support systems (DSSs), recognizing that such systems may incorporate these other approaches to enhancing performance as well, such as embedded just-in-time training or improved memory and perceptual aids.

Three points are emphasized in this discussion. The first is that the success of a DSS depends on its interactions with its human users, not just at the surface level of its user interface, but at the semantic level, and that these interactions need to be considered in the broader task context (Amalberti, 1999; Endsley & Kaber, 1999; Miller, Pelican, & Goldman, 2000; Parasuraman, 2000; Rasmussen, Pejtersen, & Goldstein, 1994). In that sense, the design of a DSS needs to be viewed from the perspective of cooperative problem solving, where the computer and the person interact with and influence each other (Hoc, 2000; Jones & Mitchell, 1995; Smith, McCoy, & Layton, 1997). From this viewpoint, the underlying functionality is just as important as the surface level representation or interface, as it is the overall interaction that determines ultimate performance. This perspective helps emphasize the importance of considering what functions the DSS performs and how it performs them, how this functionality and its results are reflected in the surface interface, how the user communicates with the DSS, and how the user develops and makes use of his mental model of the software.

A second area of emphasis is that the use of technology to provide decision aiding can be conceptualized as cooperative work between several individuals (Hutchins, 1990, 1995; Olson & Olson, 1997; Orasanu & Salas, 1991; Rasmussen, Brehner, & Leplat, 1991; Smith et al., 2000), with the computer as the medium through which they cooperate. This teamwork could involve system users who are physically or temporally distributed. In addition, it may be a useful conceptual approach to think of the design team as working cooperatively with the users, trying to communicate with them and extend or augment their capabilities, and is doing so through the software and other artifacts (such as paper documents) that they have developed. This perspective is useful as a reminder that we need to consider the psychology of designers, as well as the psychology of the users of systems, applying that understanding to help us take advantage of the strengths of both groups. In addition, it needs to be recognized that the cognitive processes and performances of both designers and users of systems are often very

dependent on the specific operational context in which they are working. Thus, we need to consider models of human performance that are situated, rather than just referring to abstract models of human cognition.

The third emphasis is on a human-centered approached to design (Billings, 1996). Although complete automation is one approach for trying to improve performance, this chapter focuses on applications where the underlying tasks and task environments are sufficiently complex that complete automation (i.e., total reliance on the design team to anticipate and deal adequately with all possible scenarios through its design), is implausible as a solution. Furthermore, it should be noted that, even in designs that are thought of as complete automation, this is rarely the case. Even in such cases, people still generally wind up interacting with the software either to overcome or recover from its limitations in unanticipated situations, or to perform maintenance and upgrades of the software.

The goal of this chapter is therefore threefold. The first is to provide a discussion of the considerations relevant to the design of a DSS (Helander, Landauer, & Prabhu, 1997). The second is to provide very practical lists of questions that need to be asked as part of the design process. The third is to use case studies to provide concrete, detailed illustrations of how these considerations can be addressed in specific application contexts.

HUMAN PERFORMANCE ON DECISION-MAKING TASKS

There are several reasons why an understanding of human performance is important to the designer of a DSS (Salvendy, 1977). The first is that the motivation for developing such systems is to increase efficiency (reduce production costs or time) and/or to improve the quality of performance. Thus, it is important for the design team to be able to efficiently and effectively complete the initial problem definition and knowledge engineering stages of a project, identifying areas where improvements are needed. An understanding of human performance makes it possible to, in part, take a top-down approach to this, looking to see whether certain classic human performance limitations are influencing outcomes in a particular application. A second motivation for understanding human performance is that a human-centered approach to the design of a DSS requires the incorporation and support of the user's skills. To do this effectively, knowledge of human performance (perception, learning and memory, problem-solving, decision making, etc.) is essential. A number of these aspects of human performance are covered in chapters 2, 3, 5, and 18, so this chapter will just highlight some of the factors most relevant to the design of a DSS and discuss how such factors are relevant to this design task.

Errors and Cognitive Biases

In very broad terms, human errors can be classified as slips and as mistakes (Norman, 1981). Slips arise through a variety of cognitive processes, but are defined as behaviors where the person's actions do not match his intentions. Generally, this refers to cases where the person has the correct knowledge to achieve some goal but, because of some underlying perceptual, cognitive, or motor process, fails to correctly apply this knowledge. As Norman describes it: "Form an appropriate goal but mess up in the performance, and you've made a slip" (1988, p. 106). Mistakes, on the other hand, refer to errors resulting from the accurate application of a person's knowledge to achieve some goal, but where that knowledge is incomplete or wrong.

DSSs are potentially useful for dealing with either of these sources of errors. If slips or mistakes can be predicted by the design team, then tools can be developed to either help prevent them, recover from them, or reduce their impacts.

Slips. Norman (1988) discusses six categories of slips. Knowing something about these different causes can help the designer look for possible manifestations in a particular application area. These six categories as defined in Norman include:

- Capture errors, "in which a frequently done activity suddenly takes charge instead of (captures) the one intended" (p. 107).
- Description errors, in which "the intended action has much in common with others that are possible" and "the internal description of the intention was not sufficient . . . often resulting in performing the correct action on the wrong object" (p. 108).
- Data-driven errors, in which an automatic response is triggered by some external stimulus that triggers the behavior at an inappropriate time.
- Associative activation errors in which, similar to a data-driven error, something triggers a behavior at an inappropriate time; but, in this case, the trigger is some internal thought or process.
- Loss-of-activation errors, or "forgetting to do something" (p. 110).
- Mode errors, or performing an action that would have been appropriate for one mode of operation for a system, but is inappropriate for the actual mode or state that the system is in.

Mistakes. As defined previously, mistakes are caused by incorrect knowledge (the rule, fact, or procedure that the person believes to be true is incorrect, resulting in an error) or incomplete (missing) knowledge.

Cognitive Biases. The literature on human error also provides other useful ways to classify errors in terms of surface-level behavior or the underlying cognitive process. Many of these are discussed under the label of cognitive biases (Poulton, 1989), including:

- Gambler's fallacy (Perceiving patterns in a series of events that are in reality random sequences. Gilovich, Vallone, and Tversky (1985) for instance, found that individuals believe they see streaks in basketball shooting even when the data show that the sequences are essentially random.)
- Insensitivity to sample size (Failing to understand that the law of large numbers implies that the probability of observing an extreme result in an average decreases as the size of the

sample increases). Tversky and Kahneman (1974), for example, found that subjects believed that "the probability of obtaining an average height greater than 6 feet was assigned the same value for samples of 1000, 100 and 10 men."

- Incorrect revision of probabilities (Failure to revise probabilities sufficiently when data are processed simultaneously, or conversely, revising probabilities too much when processing the data sequentially.)
- Ignoring base rates (Failure to adequately consider prior probabilities when revising beliefs based on new data.)
- Use of the availability heuristic (Tversky (1982) suggests that the probability of some type of event is in part judged based on the ability of the person to recall events of that type from memory, thus suggesting that factors like recency may incorrectly influence judgments of probability.)
- Attribution errors (Jones & Nisbett (1971) describe this by noting that "there is a pervasive tendency for actors to attribute their actions to situational requirements, whereas observers tend to attribute the same actions to stable personal dispositions.")
- Memory distortions caused by the reconstructive nature of memory (Loftus, 1975) describes processes that distort memories based on the activation of a schema as part of the perception of an event, and the use of that schema to reconstruct the memory of the event based on what the schema indicates should have happened rather than what was actually perceived originally. Smith, Giffin, Rockwell, and Thomas (1986) and Pennington and Hastie (1992) provide descriptions of similar phenomena in decision-making tasks.

In recent years, there has been considerable controversy regarding the nature and validity of many of these explanations as "cognitive biases" (Fraser, Smith, & Smith, 1992; Koehler, 1996), suggesting that it is important to carefully understand the specific context to generate predictions as to whether a particular behavior will be exhibited. Regarding the incorrect revision of probabilities as an illustration, Navon (1978) suggested that, in many real-world settings, data are not independent, and that what appears to be conservatism in revising probabilities in a laboratory setting may be the result of the subjects applying a heuristic or cognitive process that is effective in real-world situations in which data are correlated.

A more detailed example is provided by the popular use of the label "hypothesis fixation." This behavior refers to some process that leads the person to form an incorrect hypothesis and to stick with that hypothesis, failing to collect critical data to assess its validity or to revise it in the face of conflicting data. A variety of cognitive processes have been hypothesized to cause this behavior. One example is called biased assimilation, where the person judges a new piece of data as supportive of his hypothesis (increasing the level of confidence in the hypothesis) based simply on the consideration of whether that outcome could have been produced under the hypothesis. This contrasts with inferential processes based on a normative model that suggest that beliefs should be revised based on the relative likelihood of an outcome under the possible competing hypotheses, and that would, in the same circumstance, lead to a reduction in the level of confidence in that hypothesis.

Another example of a cognitive process discussed in the literature as relevant to hypothesis fixation is the so-called confirmation bias. This phenomenon, described in Wason (1960) and in Mynatt et al. (1977), is concerned with the person's data collection strategy. As an example, in a study asking subjects to discover the rules of particle motion in a simulated world, Mynatt et al. described performance by concluding that there was "almost no indication whatsoever that they intentionally sought disconfirmation." Later studies, however, have suggested that this bias is really an adaptive positive test strategy that is effective in many real-world settings, and that it is the unusual nature of the task selected by Mynatt et al. that makes it look like an undesirable bias (Klayman & Ha, 1987). Smith et al. (1986) further suggest that, in real-world task settings in which such a strategy might lead to confirmation of the wrong conclusion, experts have domain-specific knowledge or rules that help them to avoid this strategy and to ensure that they collect the critical disconfirming evidence.

Designer Error. Many of the error-producing processes discussed previously appear to focus on system operators. It is important to recognize, however, that the introduction of a DSS into a system is a form of cooperative work between the users and the design and implementation team and that, like the users, the design team is susceptible to errors. These errors may be caused by slips or mistakes. In the case of mistakes, it may be caused by inadequate knowledge about the application area (the designers may fail to anticipate all of the important scenarios) or incorrect knowledge. It may also be caused by a failure to adequately understand or predict how the users will actually apply the DSS, or how its introduction will influence their cognitive processes and performances. In addition, sources of errors associated with group dynamics may be introduced. Whether such errors are caused by a lack of sufficient coordination, where one team member assumes that another is handling a particular issue, or caused by the influence of group processes on design decisions (Janis, 1982; McCauley, 1989; Tetlock, 1998; Tetlock, Peterson, McQuire, Chang, & Feld, 1992), the result can be some inadequacy in the design of the DSS, or in the way the user and the DSS work together. (Note that such group processes are also important potential sources of errors when system operators work together as part of teams, perhaps with technological support.)

Systems Approaches to Error. Reason (1991) cautions designers against fixating on the immediately preceding causes of an accident when trying to prevent future occurrences. He reminds us that, in many system failures, a number of conditions must coincide for the failure to occur. This presents a variety of leverage points for preventing future occurrences, many of which may seem very remote from the actual accident or system failure. Many of these changes focus on preventing the conditions that could precipitate the system failure, rather than improving performance once the hazardous situation has been encountered.

Errors and Cognitive Biases—Implications for Design. As described at the beginning of this section, the value of this literature to the designer is in guiding initial knowledge engineering studies to identify opportunities for improvement in the decision-making processes to be supported by the DSS. It is also critical to the designer in making sure that the DSS does not cause new design-induced errors. This literature also serves to emphasize two additional considerations:

- The occurrence of errors are often caused by the co-occurrence of a particular problem-solving strategy with a given task environment. Strategies that were adaptive in one setting may no longer be adaptive in the newly designed environment. People are more likely to revert to general problem-solving strategies when they encounter a newly introduced system, and thus may be more susceptible to cognitive biases.

- For routine situations, people tend to use knowledge-rich problem-solving strategies (Newell, 1990) that overcome the errors that might be induced by particular general problem-solving strategies. Thus, people may be particularly susceptible to the potential errors associated with these general problem-solving strategies in exactly those situations where they are supposed to provide the critical safety net, namely during rare, idiosyncratic events that the design team has failed to anticipate.

In addition, this review emphasizes the need to take a broad systems perspective, recognizing that the design and implementation team are a potential source of error, and understanding that, if the goal is to prevent or reduce the impact of errors, then the solution is not always to change performance at the immediate point where an error was made. It may be more effective to change other aspects of the system so that the potentially hazardous situation never even arises.

Finally, the emphasis on designer error implies that the design process for a DSS needs to be viewed as iterative and evolutionary. Just because a tool has been fielded, it is not safe to assume that no further changes will be needed. As Horton and Lewis (1991) discuss, this has organizational implications (ensuring adequate communications regarding the actual use of the DSS, budgeting resources to make future revisions), as well as architectural implications (developing a system architecture that enables revisions in a cost-effective manner).

Human Expertise

As discussed in previous chapters, people have a variety of constraints that influence how effectively they process certain kinds of information. These include memory, perceptual, and information processing constraints (Anderson, 1993; Chi, Glaser, & Farr, 1988). As a result, there are certain decision tasks where the computational complexity or knowledge requirements limit the effectiveness of unaided individual human performance.

From a design perspective, these information processing constraints offer an opportunity. If important aspects of the application are amenable to computational modeling, then a DSS may provide a significant enhancement to performance, either in terms of efficiency or the quality of the solution. (Note that, even if development of an adequate computational model is not feasible, there may be other technological improvements, such as more effective communications environments, that could enhance performance. However, this chapter is focusing on DSSs that incorporate active information processing functions by the software.)

A good example of this is flight planning for commercial aircraft. Models of aircraft performance considering payload, winds, distance, and aircraft performance characteristics (for the specific aircraft as well as the general type of aircraft) are sufficiently accurate to merit the application of DSSs that use optimization techniques to generate alternative routes and altitude profiles to meet different goals in terms of time and fuel consumption (Smith, McCoy, & Layton, 1997). This example also provides a reminder, however, that it is not just human limitations that are relevant to design. It is equally important to consider human strengths, and to design systems that complement and are compatible with users' capabilities. In flight planning, for instance, this includes designing a system that allows the person to incorporate his judgments into the generation and evaluation of alternative flight plans, considering the implications of uncertainty in the weather, air traffic congestion, and other factors that may not be incorporated into the model underlying the DSS.

This cooperative systems perspective has several implications. First, the designer needs to have some understanding of when and how the person needs to be involved in the alternative generation and selection processes. This requires insights into how the person makes decisions, in terms of problem-solving strategies as well as in terms of access to the relevant knowledge and data, and how the introduction of a DSS can influence these problem-solving processes. It also implies that the strengths underlying human perceptual processes need to be considered through display and representation aiding strategies (Jones & Schkade, 1995; Kleinmuntz & Schkade, 1993; Larkin, 1989; Tufte, 1997) to enhance the person's contributions to the decision-making process by making important relationships more perspicuous.

Descriptive Models. The literature on human problem solving and decision making provides some very useful general considerations for modeling human performance within a specific application. This literature, which is reviewed in more detail in chapters 2 and 6, describes a variety of descriptive models of problem solving. This includes the use of heuristic search methods (Clancey, 1985; Michie, 1986; Russell & Norvig, 1995). This modeling approach (Newell & Simon, 1972) conceptualizes problem solving as a search through a space of problem states and problem-solving operations to modify and evaluate the new states produced by applying these operations. The crucial insight from modeling problem solving as search is an emphasis on the enormous size of the search space resulting from the application of all possible operations in all possible orders. In complex problems, the size of the space precludes exhaustive search of the possibilities to select an optimal solution. Thus, some heuristic approach, such as satisficing or elimination by aspects (Tversky, 1972), is needed to guide selective search of this space, resulting in the identification of an acceptable (if not optimal) solution.

Another aspect of problem solving that Newell and Simon noted when they formulated the search paradigm was the importance of problem representation. Problem representation issues emphasize that a task environment is not inherently, objectively meaningful but requires interpretation for problem solving to proceed. Some interpretations or representations are more likely to lead to successful solutions than others. In fact, an important component of expertise is the set of features used to describe a domain (Lesgold et al., 1988), sometimes referred to as a domain ontology.

Task-specific problem spaces allow problem solvers to incorporate task- or environment-induced constraints, focusing attention on just the operations of relevance to preselected goals (Sewell & Geddes, 1990; Waltz, 1975). For such task-specific problem-solving, domain-specific knowledge (represented in computational models as production rules, frames, or some other form) may also be incorporated to increase search efficiency or effectiveness (Laird, Newell, & Rosenbloom, 1987; Shalin, Geddes, Bertram, Szczepkowski, & DuBois, 1997).

These models based on symbolic reasoning have been specialized in a number of powerful ways for specific generic tasks, such as diagnosis or planning (Chandrasekaran, 1988; Miller, Galanter, & Pribram, 1960). For example, computational models of planning focus on the use of abstraction hierarchies to improve search efficiency (Sacerdoti, 1974), deal with competing and complementary goals (Wilenski, 1983), and mixed top-down/bottom-up processing (Hayes-Roth & Hayes-Roth, 1979) to opportunistically take advantage of data as it arises.

In contrast to such sequential search models, models based on case-based reasoning, recognition primed decision making, or analogical reasoning (Hammond, 1989; Klein, 1993; Kolodner, 1993; Riesbeck & Schank, 1989) focus on prestructured and indexed solutions based on previously experienced situations. These modeling approaches suggest that human experts in complex operational settings rarely describe their cognitions as a sequential search to construct alternative solution states, but rather as recognition processes that match features of the situation to prototypical, preformulated response plans (Rasmussen, 1983; Zuboff, 1988). In complex dynamic domains like aviation or fire-fighting, these preformulated plans incorporate a great deal of implicit knowledge, reflecting the constraints of the equipment, other participants in the work system, and previous experience. Over time, solutions sensitive to these constraints result in familiar accepted methods that are then triggered by situational cues. Although these accepted methods are not guaranteed to result in a successful outcome, their familiarity assists in the timely management of distributed cognition. However, it is important to recognize that such recognition-based processes can cause familiar task features to invoke only a single interpretation and hinder creative departure from the norm. This limitation can be critical when the task environment contains a new, unexpected feature that requires a new approach.

Normative Optimal Models. In addition to these descriptive models, there is a large literature that characterizes human decision making relative to various normative models based on optimal processes, which help to emphasize important factors within the task and task environment that should be considered in making a decision to arrive at a better outcome. These normative models are based on engineering models, such as statistical decision and utility theory, information theory, and control theory (Rouse, 1980; Sheridan & Ferrel, 1974). By contrasting human performance with optimal performance on certain tasks and emphasizing the factors that should influence decision making to achieve a high level of performance, this literature helps the designer to look for areas where human performance may benefit from some type of DSS.

Finally, the literature on human expertise emphasizes the ability of people to learn and adapt to novel situations, and to develop skeletal plans that guide initial decisions or plans, but that are open to adaptation as a situation unfolds (Geddes, 1989; Suchman, 1987). The literature also emphasizes variability, both in terms of individual differences (Jennings, Benyon, & Murray, 1991), and in terms of the ability for a single individual to use a variety of decision-making strategies in some hybrid fashion to be responsive to the idiosyncratic features of a particular scenario.

Human Expertise—Implications for Design. An understanding of the literature on human expertise is of value to the designer in a number of different ways. First, in terms of the initial problem-definition and knowledge engineering stages, familiarity with models of human problem solving and decision making can help guide the designer in looking for important features of performance within that application. Second, in an effort to provide cognitive compatibility with the users, many of the technologies underlying DSSs are based on these same computational models of human performance. Third, even if the underlying technology is not in some sense similar to the methods used by human experts in the application, the designer needs to consider how the functioning of the DSS system should be integrated within the user's decision-making processes.

In terms of some specific emphases, the previous discussion suggests:

• Don't fixate on active DSS technologies as the only way to improve system performance. Enhancing human performance through changes in procedures, improvements in passive communication tools, better external memory aids, etc., may be more cost-effective in some cases. In addition, such changes may be needed to complement an active DSS to make its use more effective.

• Look for ways in which tacit consideration of ecological constraints allow people to perform expertly (Flach, Hancock, Caird, & Vincente, 1995). These constraints may make what seems like a very difficult information processing task much less demanding.

• Related to the consideration of ecological constraints was the earlier suggestion that problem representation has a strong influence on how easily a problem solver can find a good solution. This issue of perspicuity, however, is dependent not only on the characteristics of the task and task environment, but also on the nature of the problem solver. Thus, to enhance human performance as part of the decision-making process, it is important to consider ways of representing the process that enable human perceptual and cognitive processes to work effectively. Similarly, the problem representation used within the DSS needs to consider the characteristics of the computational processes embedded in the software. In addition, problem representation can affect the performance of the designer. Consequently, alternative problem representations need to be generated as part of the design process to help the designer think more effectively.

• Consider the applicability of normative optimal models of performance for the task to focus attention on factors that should be influencing current performance, and to understand the strategies that people are using to be responsive to these factors. Aspects of these normative models may also be appropriate for more direct inclusion in the DSS itself, even though the strategies used by people do not fully reflect the optimal processes highlighted by these normative models. In addition to considering the task structure highlighted by normative optimal models, use knowledge of the variety of different models of human problem solving and decision making to guide knowledge engineering efforts, making it easier and more efficient to understand how people are currently performing the tasks. Thus, all of these models of decision making represent conceptual tools that help the designer to more effectively and efficiently understand performance in the existing system, and to develop new tools and procedures to improve performance.

• Whether the DSS is designed to process information "like" the user at some level, or whether it uses some very different processing strategy, the user needs to have an appropriate and effective mental model of what the system is and is not doing (Kotovsky, Hayes, & Simon, 1985; Lehner & Zirk, 1987; Nickerson, 1988; Zhang, 1997; Zhang & Norman, 1994). To help ensure such cognitive compatibility, the designer therefore needs to understand how people are performing the task.

• Consider design as a prediction task. One role of the designer is as psychologist, trying to predict how users will use the DSS system, and how it will influence their cognitive processes and performances.

Additional Cognitive Engineering Considerations

The preceding sections focused on our understanding of human performance on decision-making tasks and the implications of that literature for design. Below, some additional considerations based on studies of the use of DSSs are presented.

The Human Operator as Monitor. Numerous studies make it clear that, given the designs of DSSs for complex tasks must be assumed to be brittle, there is a problem with designs that naively assume that human expertise can be incorporated as a safety net by asking a person to simply monitor the output of the DSS, with the responsibility for overriding the software if a problem is detected. One problem with this role is that, in terms of maintaining a high level of attentiveness, people do not perform well on such sustained attention tasks. As Bainbridge (1983) notes: "We know from many 'vigilance' studies (Mackworth, 1950) that it is impossible for even a highly motivated human being to maintain effective visual attention towards a source of information on which very little happens, for more than about half an hour. This means that it is humanly impossible to carry out the basic function of monitoring for unlikely abnormalities."

A related issue is the problem of loss of skill. As Bainbridge further notes, if the operator has been assigned a passive monitoring role, he "will not be able to take over if he has not been reviewing his relevant knowledge, or practicing a crucial manual skill." Thus, a major challenge for retaining human expertise within the system is "how to maintain the effectiveness of the human operator by supporting his skills and motivation."

Complacency. Studies reported by Parasuraman and Riley (1997) introduce further concerns about assigning the person the role of critiquing the computer's recommendations before acting. These studies discuss how the introduction of a DSS system can lead to overreliance by the human user when the software is generating the initial recommendations (Skitka, Mosier, & Burdick, 1999).

Excessive Mental Workload. Designs that relegate the person to the role of passive monitor run the risk of a vigilance decrement because of insufficient engagement and mental workload. At the other extreme, designers must recognize that "clumsy automation" that leaves the person with responsibility for difficult parts of the task (such as coping with an emergency), but that adds additional workload from the awkward interactions now required to access information and functions embedded in the DSS (such as navigating through complex menus to view certain information), and can actually impair performance because of the added mental workload of interacting with the DSS (Wiener & Nagel, 1988).

Lack of Awareness or Understanding. Even if the person is sufficiently engaged with the DSS and the underlying task, designers need to consider how to ensure that the user has an accurate mental model of the situation and the functioning and state of the DSS (Billings, 1996; Larkin & Simon, 1987; Mitchell & Miller, 1986; Roth, Bennett, & Woods, 1987; Sarter & Woods, 1993). If the person does not have such an understanding, then it may be difficult for him to intervene at appropriate times or to integrate the computer's inputs into his own thinking appropriately (Geddes, 1997). This concern has implications for selection of the underlying technology and conceptual model for a system, as well as for the design of the visual or verbal displays intended to represent the state of the world and the state of the software for the user, including explanations of how the DSS has arrived at its recommendations (Clancey, 1983; Hasling, Clancey, & Rennels, 1984).

Lack of Trust and User Acceptance. As outlined previously, overreliance can be a problem with certain assignments of roles to the person and the computer. At the other extreme, lack of trust or acceptance of the technology can eliminate or reduce its value (Hollnagel, 1990; Muir, 1987). This lack of acceptance can result in outright rejection of the DSS (in which case it either is not purchased or is not installed or used), or a tendency to underutilize it for certain functions. It is important to note that this lack of acceptance can be because of resistance to change (Cartwright & Zander, 1960), even if there is no intrinsic weakness in the DSS; because of general beliefs held by the operators (rightly or wrongly) about how the software will influence their lives; or because of beliefs about how well such a DSS can be expected to perform (Andes & Rouse, 1992).

Active Biasing of the User's Cognitive Processes. Complacency is one way in which a DSS can influence the person's cognitive processing. Studies have also shown, however, that the use of a DSS can also actively alter the user's cognitive processes. The displays and recommendations presented by the software have the potential to induce powerful cognitive biases, including biased situation assessment and failures by the user to activate and apply his expertise because normally available cues in the environment are no longer present. The net result is that the person fails to exhibit the expertise that he would normally contribute to the decision-making task if working independently, not because he lacks that expertise, but because the computer has influenced his cognitive processes in such a way that this knowledge is never appropriately activated. Studies have shown that these biasing effects can induce practitioners to be 31% more likely to arrive at an incorrect diagnosis on a medical decision-making task (Guerlain et al., 1996) and 31% more likely to select a very poor plan on a flight planning task (Smith, McCoy, & Layton, 1997).

Distributed Work and Alternative Roles. Much of the literature focuses on the interactions between a single user and the DSS. Increasingly, however, researchers and system developers are recognizing that one approach to effective performance enhancement is to think in terms of a distributed work paradigm, in which the software may be one "agent" in this distributed system (Geddes & Lizza, 1999) or in which the software may be viewed primarily as a mediator to support human–human interactions (Hutchins, 1990, 1995; Olson & Olson, 1997; Orasanu & Salas, 1991; Rasmussen et al., 1991; Smith et al., 1997). One of the benefits of such a distributed approach to system performance is that it opens up the potential to consider a variety of different architectures for distributing the work in terms of the locus of control or responsibility, and the distribution of knowledge, data, and information processing capabilities (Sheridan, 1997; Smith et al., 2000).

Organizational Failures. The previous discussions focus on design-induced errors made by the system operators. It is equally important for the design team to recognize that part of the design process is ensuring that the management of the organization into which the DSS is introduced is functioning effectively as a safety net. This means that the design team needs to ensure that the organization has established effective procedures to detect significant problems associated with the introduction and use of the DSS, and that such problems are communicated to the levels of management in which responsibility can and will be taken to respond effectively to these problems (Horton & Lewis, 1991).

CASE STUDIES

Four case studies are presented focusing on the designs of several different DSSs. Within these case studies, the issues raised earlier are discussed in more detail within the context of real-world applications, along with presentations of design solutions that provide concrete illustrations of how to deal with those issues.

Case Study A. Interactive Critiquing as a Form of Decision Support

This case study emphasizes the challenge for interaction designers in dealing with the use of decision support tools that have the potential for brittle performance from known or unknown limitations. Although there are no perfect solutions to this problem, there are a number of approaches to help reduce its impact.

This case study looks at a tool that incorporates three complementary approaches to the design of an expert system to improve overall performance, while reducing the potential impact of brittle performance by the expert system. The first approach to deal with the impact of brittle performance by an expert system is to design a role that encourages the user to be fully engaged in the problem solving, and to apply his knowledge independently without being first influenced by the software. In this case study, the approach explored to achieve this is the use of the computer as a critic, rather than as the initial problem solver (Fischer, Lemke, Mastaglio, & Morch, 1991; Miller, 1986; Silverman, 1992). Thus, instead of asking the person to critique the performance of the software, the computer is assigned the role of watching over the person's shoulder. Note that this is more accurately described as having the design team try to anticipate all of the scenarios that can arise, and then, for all of those scenarios, trying to incorporate the knowledge necessary to detect possible slips or mistakes on the part user and to provide alerts and assistance in recovering.

The second approach to reduce sensitivity to brittleness in the computer's performance is to incorporate metaknowledge into the expert system that can help it to recognize situations in which it may not be fully competent. By doing so, the software may be able to alert the person to be especially careful because the computer recognizes that this is an unusual or difficult case.

The third approach to deal with brittleness is to develop problem-solving strategies that reduce susceptibility to the impact of slips or mistakes. In the software discussed in this case study, this is accomplished by incorporating a problem-solving strategy that includes the collection of converging evidence using multiple independent problem-solving strategies and sources of data to arrive at a final conclusion.

What follows, then, is a discussion of a specific expert system that incorporates these three strategies for dealing with brittleness. Empirical testing of the system suggests that this approach can significantly enhance performance, even on cases where the software is not fully competent.

The Application Area. The specific problem area considered in this case study is the design of a decision support tool to assist blood bankers in the identification of compatible blood for a transfusion. One of the difficult tasks that blood bankers must complete as part of this process is the determination of whether the patient has any alloantibodies present in his blood serum, and, if so, what particular antibodies are present.

Diagnosis as a Generic Task. Abstractly, this is a classic example of the generic task of abduction or diagnosis (Josephson & Josephson, 1994). It involves deciding what tests to run (what

data to collect), collection and interpretation of those data, and forming hypotheses, as well as deciding what overall combination of problem-solving strategies to use. Characteristics of this generic task that make it difficult for people include:

- The occurrence of multiple solution problems, in which more than one "primitive" problem is present at the same time.
- The occurrence of cases where two or more "primitive" problems are present, in which one problem masks the data indicative of the presence of the other.
- The existence of "noisy" data.
- The existence of a large "data space," in which data must be collected sequentially, so that the person must decide what data to collect next and when to stop.
- The presence of time stress, in which an answer must be determined quickly (Elstein, Shulman, & Sprafka, 1978; Smith et al., 1998).

In this application, the "primitive" problems are the individual antibodies that may be present in the patient's blood serum. Time stress can arise when the patient is in an emergency situation and needs a transfusion quickly.

Sample Problem. To illustrate the nature of this diagnosis task, consider the following partial description of an interaction with the decision support tool AIDA (the Antibody IDentification Assistant) that is the focus of this case study (Guerlain et al., 1999). Initially, the medical technologist needs to determine whether the patient is type A, B, AB, or O, and whether the patient is Rh-positive. Then, the technologist determines whether the patient shows evidence of any autoantibodies or alloantibodies. As part of this determination, displays like that shown in Fig. 33.1 are provided by AIDA.

To make visual scanning easier on this data display, some of the rows have been highlighted by the technologist in yellow (shown in light gray on this black-and-white version). In addition, to reduce memory load, the technologist has marked a number of intermediate conclusions, indicating that the patient is likely to have antibodies against the C antigen (marked by the technologist in orange), that the f, V, Cw, Lua, Kpa, and Jsa antibodies are unlikely (marked in blue), and that the other antibodies shown as the labels across the top row can be ruled out (marked in green). (These additional markings are all shown in darker gray on this black-and-white version.) These color-coded intermediate answers are transferred to all other data displays to reduce the memory load for the technologist. Figure 33.1 also provides an example of a critique that AIDA has provided in response to an error made by the technologist in marking anti-S as ruled out. This critique was generated by the expert model (rule-based system) underlying AIDA, which monitors all of the test selections and markings made by the technologist as he solves the case using the interface provided by AIDA.

Thus, what this figure serves to illustrate is that:

- The technologist is responsible for completing the analysis, deciding what tests to run and what intermediate and final conclusions to reach, and is thus very engaged in the task.

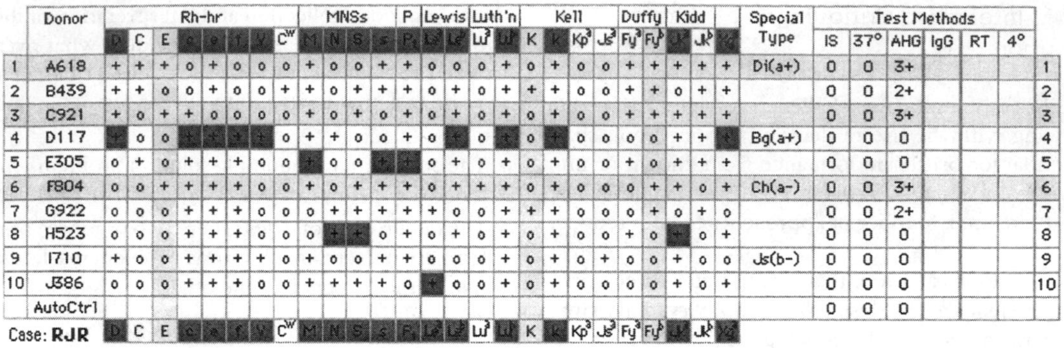

	Donor	Rh-hr								MNSs				P	Lewis		Luth'n		Kell				Duffy		Kidd			Special Type	Test Methods						
		D	C	E	c	e	f	V	C^w	M	N	S	s	P_1	Le^a	Le^b	Lu^a	Lu^b	K	k	Kp^a	Js^a	Fy^a	Fy^b	Jk^a	Jk^b	Xg^a		IS	37°	AHG	IgG	RT	4°	
1	A618	+	+	+	o	+	o	o	o	+	o	o	+	+	o	o	o	+	o	+	o	o	+	+	+	+	+	Di(a+)	0	0	3+				1
2	B439	+	+	o	o	+	o	o	+	+	o	+	+	+	o	+	o	+	+	+	o	o	+	+	o	+	+		0	0	2+				2
3	C921	+	o	+	+	o	o	o	o	+	o	+	o	+	o	+	o	+	o	+	o	o	+	+	+	+	+		0	0	3+				3
4	D117	+	o	o	+	+	+	+	o	+	+	o	o	+	o	+	o	+	o	+	o	o	o	o	+	+	+	Bg(a+)	0	0	0				4
5	E305	o	+	o	+	+	+	o	o	+	o	o	+	+	o	+	o	+	o	+	o	o	+	o	+	+	+		0	0	0				5
6	F804	o	o	+	+	+	+	o	o	+	o	+	+	+	o	+	+	+	o	+	o	o	o	+	+	o	+	Ch(a-)	0	0	3+				6
7	G922	o	o	o	+	+	+	o	o	o	+	+	+	+	+	o	o	+	+	+	o	o	o	+	o	+	o		0	0	2+				7
8	H523	o	o	o	+	+	+	o	o	o	+	+	o	+	o	+	o	+	o	+	o	o	+	o	+	o	+		0	0	0				8
9	I710	+	o	o	+	+	+	+	o	+	o	o	+	+	o	o	o	+	o	+	o	+	o	o	+	o	o	Js(b-)	0	0	0				9
10	J386	o	o	o	+	+	+	o	o	+	+	+	+	o	+	o	o	+	o	+	o	o	o	o	+	o	+		0	0	0				10
	AutoCtrl																												0	0	0				
Case: **RJR**		D	C	E	c	e	f	V	C^w	M	N	S	s	P_1	Le^a	Le^b	Lu^a	Lu^b	K	k	Kp^a	Js^a	Fy^a	Fy^b	Jk^a	Jk^b	Xg^a								

(a)

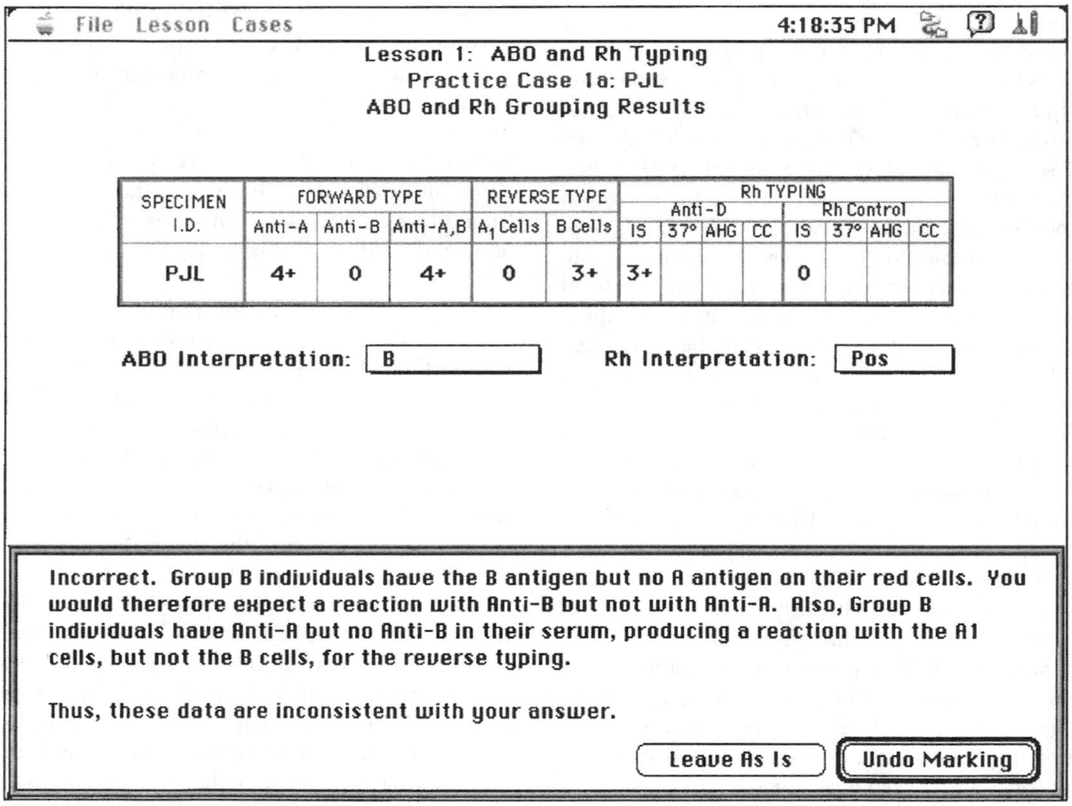

File Lesson Cases
4:18:35 PM
Lesson 1: ABO and Rh Typing
Practice Case 1a: PJL
ABO and Rh Grouping Results

SPECIMEN I.D.	FORWARD TYPE			REVERSE TYPE		Rh TYPING							
						Anti-D				Rh Control			
	Anti-A	Anti-B	Anti-A,B	A_1 Cells	B Cells	IS	37°	AHG	CC	IS	37°	AHG	CC
PJL	4+	0	4+	0	3+	3+				0			

ABO Interpretation: B
Rh Interpretation: Pos
Incorrect. Group B individuals have the B antigen but no A antigen on their red cells. You would therefore expect a reaction with Anti-B but not with Anti-A. Also, Group B individuals have Anti-A but no Anti-B in their serum, producing a reaction with the A1 cells, but not the B cells, for the reverse typing.
Thus, these data are inconsistent with your answer.
Leave As Is
Undo Marking

(b)

Figure 33.1. (a) Full test panel with intermediate results marked using color-coded markers provided by AIDA (shown here in black and white). For the panel shown above, at this point in the analysis it looks like anti-Fyb is present if the reaction is being caused by a single antibody (Fyb is on all the reacting cells—the 2+ and 3+ cells—and only on those cells). However, it looks like anti-E and anti-K are present if the reaction is being caused by 2 antibodies. AIDA = Antibody Identification Assistant. (b) Sample ABO and Rh panel with feedback provided to the user regarding a possible slip or mistake.

• The computer provides an interface that makes it easy for the technologist to select the tests to run and to view the resultant data (using the highlighting to make visual scanning easier), as well as to remember intermediate conclusions (using color-coded markings to indicate these conclusions).

• Although the primary motivation for the technologist in marking the data forms is to make his task easier, when he does so the computer is provided with a great deal of data regarding what the user is thinking, and can use that data to make inferences based on its expert model about when to provide a critique cautioning the user that he may have made a slip or mistake.

The Need for Decision Aiding. Initial laboratory and field studies indicated that the task of determining the alloantibodies present in a patient's blood is a difficult one for many technologists. A variety of causes of errors were observed, including slips, perceptual distortions, incorrect or incomplete knowledge (Guerlain et al., 1999), and cognitive biases (Fraser et al., 1992).

The Design Solution. It was the judgment of the design team that this task was sufficiently complex that, given the available development resources, it was unlikely that all of the possible scenarios could be anticipated and dealt with by the design team. It was also noted that, for this task, the cost of an erroneous diagnosis was potentially high. Thus, it was decided that, instead of automating the task, a decision support system should be developed that kept the person very much engaged in the task, and that provided other safety nets to reduce the chances of error for those cases where the computer's knowledge was incomplete. This conclusion was reinforced by a preliminary study regarding the impact of role on diagnostic accuracy in antibody identification. Guerlain et al. (1996) showed that, when the user was asked to critique the computer's performance instead of having the computer critique the user's performance, the final answer was wrong 31% more often in cases where the computer's knowledge was incomplete when, in that case, the person was assigned the role of critic rather than having the computer critique the person.

Critiquing—Additional Design Considerations. The literature provided additional guidance in deciding whether and how to develop this software to play the role of a critic. For example, Miller (1986) developed a prototype system called ATTENDING that focused on anesthesiology. Based on studies of its use, Miller suggested that critiquing systems are most effective in applications in which the user has a task that is frequently performed, but that requires the person to remember and apply a great deal of information to complete a case. Miller's conclusion was that, on such tasks, the person is more susceptible to slips and mistakes and would therefore benefit significantly from the decision support system.

A second consideration was how intrusive the interactions with the knowledge-based system would be for the user. A number of researchers have suggested that an interface that requires that the user spend significant time entering data and informing the computer about what he has concluded is too cumbersome, and will therefore be unlikely to be adopted in actual practice (Berner, Brooks, Miller, Masarie, & Jackson, 1989; Harris & Owen, 1986; Miller, 1986; Shortliffe, 1990).

A third consideration was concern over the potential for complacency if the person played the role of critic, letting the computer complete an initial assessment and then having the person decide whether to accept this assessment. Parasuraman and Riley (1997) have shown that, in such a role, there is a risk that the person will become overreliant on the computer and will not adequately apply his knowledge in completing the critique. (Note, however, that a person could become overreliant even with the roles reversed, because the person might start to get careless and assume the computer will always catch his slips. Administrative controls, based on regular monitoring of the person's performance, might help reduce such complacency, but this is an as yet unexplored aspect regarding the use of critiquing systems.)

A final consideration was the timeliness of critiques. Waiting for the user to reach a final answer before providing a critique, as has been the case for many critiquing systems, is potentially inefficient and objectionable, because the user may have invested considerable time and effort in arriving at a mistaken answer that the computer could have headed off earlier in the person's problem-solving process. Furthermore, if the critique is given well after the user has made a slip or mistake, it may be difficult for him to remember exactly why he arrived at that intermediate conclusion (thus making it harder to decide whether to accept the computer's critique). This consideration therefore suggests the need for an interface that provides the computer with data about the user's intermediate conclusions rather than just the user's final answer, so that critiquing can be more interactive.

Based on these considerations, AIDA was developed as a critiquing system that supported the user as he completed the antibody identification task. To provide immediate, context-sensitive critiques, an interface was developed that encouraged the user to mark intermediate conclusions on the screen. As suggested earlier, these markings in fact reduced the perceptual and memory loads for the user, thus encouraging this form of communication and allowing the computer to detect and respond immediately with context-sensitive critiques to potential slips and errors made by the person.

Complementary Strategies to Reduce Susceptibility to Brittleness. The considerations outlined previously focused on how to keep the person engaged in the diagnosis task and how to avoid triggering the cognitive biases that can arise if the computer suggests an answer before the person has explored the data himself (Smith et al., 1997). AIDA also incorporated two additional design strategies to reduce the potential for error. One was the incorporation of metaknowledge into the knowledge-based system. The other was to include a problem-solving strategy that was robust, even in the face of slips or mistakes by either the person or the computer (i.e., the design team).

Metaknowledge was included to help the computer identify cases in which its knowledge might be incomplete. Such rules were developed by identifying the potential weak points in the computer's problem-solving process. An example was the recognition that AIDA uses a thresholding technique when applying rules such as:

If a test cell's reactions are 0 on a test panel for all three of the testing conditions (IS, LISS and IgG), as they are for the first test cell (Donor A478) shown in Fig. 33.1, and if e is present (shown by a + in the column labeled e in the row corresponding to the first test cell) on that test cell but E is not (shown by a 0 in the column labeled E in the row corresponding to the first test cell), then mark e as ruled out.

This heuristic usually produces the correct inference. Because it does not directly reason with some form of probabilistic reasoning (Pearl, 1988; Shafer, 1996; Shafer & Pearl, 1990); however, it can sometimes lead to ruling out an antibody that is actually present in the patient's serum. This is most likely to happen when the reaction strengths are weak. Thus, AIDA was provided with a rule that monitored for weak reactions on cells, and when this was detected and the user was detected trying to complete rule-outs without first enhancing the reactions with some alternative test phase, the system cautioned the user that errors can result when rule-outs are made without first using some technique to strengthen the reactions. In this way, AIDA put the user on alert to be especially careful in applying his normal strategies and rules.

A second strategy incorporated into AIDA as protection against brittleness was to monitor for the collection and consideration of converging evidence by the technologist. This problem-solving strategy was noted in one of the experts involved in the development of AIDA. She used this strategy to catch her own errors and those of the technologists working under her supervision. The basic strategy was based on the assumption that any one data source or line of reasoning could be susceptible to error, and that it is therefore wise to only accept a conclusion if independent types of data (test results that are not based on the same underlying data or process) and independent sets of heuristics or problem-solving strategies have been used to test that conclusion. Based on this expert strategy, AIDA monitored the user's problem-solving process to see whether such converging evidence had been collected before reaching a final answer. If not, AIDA cautioned the user and suggested types of converging evidence that could be collected.

Evaluation of the Implemented Solution. From the standpoint of human–computer interaction, the key question is how effectively the person and system perform in this cooperative problem-solving paradigm. To gain insights into this, an empirical study was conducted using AIDA.

This study of 37 practitioners at 7 different hospitals found that those blood bankers using AIDA with its critiquing functions turned on made significantly fewer errors ($p < .01$) than those that used AIDA as a passive interface. Errors in the final diagnoses (answers) were reduced by 31–63%.

In those cases in which AIDA was fully competent (Posttest Cases 1, 2, and 4), errors (% of technologists getting the wrong final answer) were reduced to 0% when the critiquing functions were on (Treatment Group). In the case where AIDA was not fully competent (Posttest Case 2), AIDA still helped reduce errors by 31% (Table 33.1).

These empirical results support the potential value of interactive critiquing, supplemented with appropriate metaknowledge and problem-solving strategies that were embedded in the

software, as a design strategy for applications where concerns regarding the potential brittleness of the technology are deemed significant.

Case Study A—Conclusions. The goal of this case study was to highlight the issues associated with trying to design a system that is less susceptible to the errors induced by the brittleness of the underlying technology. In very practical terms, it suggests that the following questions should be asked during the design process:

- Is critiquing an appropriate role for the decision-support system? (This involves consideration of the potential brittleness of the technology, the reduced operational cost savings likely to result from having the computer play the role of critic instead of actively completing the problem solving itself as the initial problem-solver, and the costs of errors that could be induced if the computer played the role of primary or initial problem solver instead of critic.)

- Will this assignment of roles reduce the potential for design-induced error from the computer suggesting a solution early in the user's problem-solving process, thus potentially inducing cognitive biases that inhibit effective use of the person's knowledge during the problem solving?

- Will this assignment of roles help to ensure that the user is not complacent? Are additional administrative controls needed to help reduce the potential for complacency?

- Can training be developed to further reduce the potential for errors that could be induced by brittle performance by the computer?

- Has an interface and supporting functionality been developed that is unobtrusive, so that the user provides the computer with sufficient data to provide timely, context-sensitive critiques in a fashion that is acceptable to the user in terms of workload and ease of use?

- Have potential sources of brittleness been identified as part of the design process, and has metaknowledge been incorporated into the software to alert the user to be especially careful in such situations?

- Have critiquing functions been incorporated that evaluate the user's overall problem-solving process to determine whether converging evidence has been collected as protection against slips or mistakes made by the user or the design team?

- Have the factors that make diagnosis difficult for people been explicitly considered in determining the scenarios that need to be considered in developing the software's critiquing functions?

- Have management practices been established to monitor for evidence of any loss of skill or overreliance on the software to catch slips that the user is making? (And, have any concerns over such a "Big Brother" role been considered and addressed?)

Case Study B. Software Associates: Decisions in Real-Time, High-Stakes Settings

With the rapidly increasing technical complexity of many aspects of modern society, human decision makers are more and

TABLE 33.1. Percentage of Blood Bankers Arriving at the Wrong Final Answer on Four Test Cases

Posttest Case	Control Group (%)	Treatment Group (%)
1	33	0
2	50	19
3	38	0
4	63	0

more challenged by the need to make important decisions with large consequences in quick order. As we have seen throughout this handbook, humans are limited in their cognitive capabilities in ways that can compromise their decision-making behaviors in such real-time, complex tasks. Associate systems are a form of real-time, intelligent DSS specifically designed to help human decision makers in complex and high-stakes task environments while keeping the humans responsible for the outcome. As decision aids, associates are more flexible than conventional automation, while enabling very rapid planning and execution in dynamic environments. This case study describes the features found in two aviation associate systems that help keep humans both involved in the tasks and effective in their performance. The internal completeness and complexity of an associate illustrates the extent to which models of human behavior and human knowledge can be embedded inside computer processes for aiding humans.

The Application Area. Flight is a foreign environment for humankind. As the demand for performance has increased, aviation systems have become increasingly complex. This is particularly true in military aviation. Present military systems frequently exceed the ability of the human operator to sense his situation, organize his response, and execute that response in a correct and timely way. The deficits of human ability, compared with the demands of aviation, are felt across virtually all aspects of human behavior in aviation systems.

Modern aviation systems exist in a complex, high-energy environment in which changes occur quickly. In this environment, the projection of future events is uncertain and unexpected events are not infrequent. Decision time is often short, driven by traffic density, speeds, and weather phenomena. The constraints posed by terrain and weather, propulsion, and control systems are inviolate in nature. Aviation is also no longer the domain of the lone eagle. In both peacetime and war, the sheer number of aircraft operating in a confined airspace requires coordination. Each participant must consider the actions and intentions of the others within the airspace in determining his own course of action. Failure to consider and resolve all interactions and constraints can result in a dangerous, if not fatal, course of action.

Although conventional automation has often been designed to replace human control and decision making, associate systems seek to augment and enhance human judgment and responsibility. In aviation, automation of many low-level functions has been attempted, sometimes with excellent results. Recent experience with highly automated aircraft systems, however, has led to a growing realization that conventional automation can lead to dangerous isolation of the humans responsible for operation of the aviation systems. Some of the risks of overautomation in aviation are discussed in chapter 46 of this handbook.

The Need for Decision Aiding. The presence of the human in aviation task domains raises concern for human cognitive capabilities and limitations. Noticing and interpreting all of the relevant features and choosing an appropriate response within the tight temporal constraints of the domain is a challenge to any intelligent agent—human or machine. Moreover, as discussed earlier, human intelligence suffers from a well-established limitation of serial information processing and the tendency to fixate on a small number of problems and solution options. Adding to the challenge is the high stakes of aviation: even a small error in judgment or execution can result in substantial damage and the loss of many lives.

The Design Solution. The design of software associates represents an alternative paradigm to the design solution discussed in Case Study A, which focused on an interactive critiquing system. The decision as to the design approach to take depends on a number of factors, including the maturity of the software for that application and practical constraints such as the need for very fast responses. The first associate system to achieve comprehensive, real-time, intelligent behavior was the Pilot's Associate, a U.S. defense project sponsored by the Defense Advanced Research Projects Agency (DARPA) and the U.S. Air Force (Lizza & Banks, 1991; Rouse, Geddes, & Hammer, 1990). The Pilot's Associate development program was conducted in parallel by Lockheed Martin Corporation and McDonnell Douglas Corporation (now part of The Boeing Company) between 1986 and 1992, concluding with the successful evaluation of a real-time, intelligent support system for combat aircraft pilots that was hosted in a flight-worthy avionics computer system. This system was immediately followed by the U.S. Army Rotorcraft Pilot's Associate (RPA), which was developed by The Boeing Company from 1993 to 1998 and flight tested on an AH-64D Apache Longbow combat helicopter in 1998. By the end of the year 2000, a number of additional associate systems were in development for a variety of military and high-stakes commercial environments.

The defining characteristics of an associate system are its behavioral completeness and its relationship with its human operator. An associate system attempts to provide complete task coverage within a domain and address all areas of human physical and cognitive limitations that are encountered. As a result, functions are not allocated exclusively to the human operator *or* to the associate, but rather to both the human operator *and* the associate (Geddes, 1997; Jones & Mitchell, 1995; Sewell, Geddes, & Rouse, 1987). This completeness of task coverage allows the associate to perform any task that the human could perform, but only when so authorized by its user.

Because the associate is in principle capable of fully autonomous behavior, its relationship with its user is of paramount importance. By design, the user is always in charge of the associate. The associate must adapt its behavior to meet the desires of the user, not the other way around. The empowerment of the associate is distinct from its capability, extending only to those tasks and functions that have been permitted by the user. It is the associate's responsibility to follow the human's lead.

The architecture and processing within an associate system is very different from conventional software designs. The hallmark of an intelligent system is its ability to reorganize its primitive behaviors to adapt to a large set of situations. Its gross behaviors emerge from the interactions of its primitive behaviors. With appropriate design, a fairly small number of primitive processes, combined with representations of knowledge, can produce a vast set of intelligent behaviors. The Pilot's Associate and the RPA both were composed of more than 100 small processes, each performing very specific tasks. The tasks were not executed repetitively in a fixed sequence, but were instead executed only when appropriate. Internal to each associate was a control strategy that permitted the associate system to change its focus as the situation itself changed. The tasks inside each associate can be grouped into four main task areas: situation assessment, planning, acting and coordinating.

Situation Assessment. Situation Assessment (SA) is the primary interface to the sensory aspects of the aviation system. It is responsible for receiving sensor data at a variety of levels and from sources with varying reliability and credibility, and forming a coherent view of the world at many levels of abstraction and aggregation. As the incoming data changes, SA must update its world state model and advise interested processes of the changes. These data are maintained for the benefit of all other processes within the associate system, which use these data as client processes.

Planning. Based on the situation as determined by SA, plans are formulated by the associate system. The plans may be of many kinds and cover different periods of time at different levels of abstraction. In both the Pilot's Associate and the RPA, the planning capability could independently formulate and propose plans to the human users, and could complete the details of partial plans provided by the users.

Mission planning is a case of proactive or deliberative planning, in which the goals to be achieved and the plans to be used are determined well in advance of the performance of the plans. Response planning has a much shorter time frame than the proactive mission planning and attempts to deal with a specific deviation or perturbation to the mission plan. As a result, the state of the environment is known with greater certainty and the scope of the response planning is narrower than a full mission replanning task. Response plans in the associate system are normally intended to be performed while they are being planned and, therefore, the plans must be prepared within the time constraints of the situation.

Although a case can be made for integrated planning that unifies both mission and tactical planning, neither the Pilot's Associate nor the RPA attempted to perform unified planning. For the Pilot's Associate, there were three separate planners, one for mission routing, one for emergency responses to system malfunctions, and one for planning and performing air combat tactics. The RPA had a larger suite of special purpose planners whose outputs were loosely combined. Both associates treated planning as a partial ordering of abstract activities, rather than a fully detailed sequence of primitive actions.

Acting. An associate system is not simply a passive system, but also has the capability to act on behalf of its human users. Given a set of plans and an evolving situation, the associate system may issue commands directly to the active elements of the aircraft systems, such as sensors, communications, propulsion, flight controls, and secondary support systems. For any agent, human, or computer to perform a plan, the activity in the plan must be expressed in terms of primitive actions that are fully defined. Whereas the plan itself may not detail the specific sequence of actions, the agent responsible for plan execution

must have knowledge that relates the content of the plan to the detailed actions to be performed. For humans, translating the contents of a plan into specific actions is a matter of training and understanding the cause-and-effect relationships in the domain. An untrained human would have little success performing plans intended for a trained pilot or air traffic controller—he simply would not know how. To enable a computer-based plan execution agent to perform plans, the associate has access to knowledge about the sequence of actions and state triggers that provide nominal execution of the plan. This knowledge is represented directly as a generalized sequence of actions and state triggers.

Action and plan knowledge is used by the associate system to monitor the performance of activities by the plan execution agents, both human and machine. This process is known as intent interpretation (Geddes, 1989; Rubin, Jones, & Mitchell, 1988) and allows the associate to track the human users' activities and conform its behavior to their intentions. The most efficient knowledge representation is to use the plan generation structure and the execution knowledge as the means for interpreting actions. This knowledge is used to explain each action that is issued in terms of the activities being executed or the potentially useful plans that were not previously known to be active.

A third process within the associate used to support aiding of actions provides classifications for errors and methods for error recovery and remediation. If an action is issued that was not expected, that is, the interpretation of the action did not result in it being associated with current activity or plan or even any potentially useful plan, it is analyzed as a potential error. Because humans are capable of innovative behaviors that produce unanticipated benefits, unexpected actions are not considered an error unless the action fits into a pattern of error typical for a human operator or the consequences of the action in its circumstances are predicted to be adverse. The error classification knowledge, consequence knowledge, and error recovery knowledge support an error tolerant operational system that advises the pilot of errors and assists him in recovery.

The execution of actions with the required precision and timeliness is based on information about the world state. As a result, the fourth area of knowledge to support action execution is information requirements. For each plan that is to be performed, the execution agent needs to have access to the information that is referenced in the state triggers for the actions of the plan. In addition, information requirements must address conditions that warrant pausing and halting the plan. The knowledge about information requirements allows an information management process within the associate system to configure information presentations as needed to support the execution of the actions of a plan in a timely and precise fashion (Hammer & Small, 1995).

The user interface to an associate system is surprising in its simplicity as a result of the information management process. When the associate is aiding the user, there are few overt outward signs of the behavior of the associate. Instead, the user finds that his situation, options, and potential outcomes are organized and displayed for him without the need to perform frequent interactions with the user interface.

Coordination of Behaviors. To support coordination of multiple humans and associate systems, an associate system has knowledge and data that can be shared across participants. To enable a shared assessment of the situation, the participating associate systems can directly exchange concepts from their own local situation assessment processes. All participants also use a common framework for representing planning, permitting the direct exchange of plan and goal information. A shared common representation of the situation and each other's plans and goals permits the early detection of potential conflicts in activities.

Although the associate system will normally remain within the bounds of the behaviors that it has been authorized to perform by its human counterparts, it is also capable of recognizing the need to alter its permissions under extreme situations. At times, the level of task demand experienced by the pilot may warrant additional aiding beyond what might have been requested. Based on workload and performance models, an associate system can determine that the operator needs additional assistance and automatically begin to provide it if so doing will prevent serious adverse consequences (Rouse, Geddes, & Curry, 1987).

Associate systems define a very comprehensive DSS capability. As with the introduction of other far-reaching technologies, associate systems have opened new potential and new challenges in the relationship between human and computer.

Results. Both the Pilot's Associate and the RPA were heavily tested and evaluated over the course of their development. Although high-fidelity, man-in-the-loop mission simulations were used by both projects, the RPA also used large-scale purely digital, constructive simulations and flight test demonstrations to validate the behavioral results found in manned simulations.

Technical Correctness. A major concern for the Pilot's Associate and the RPA was the correctness of their behaviors. As a result, both programs performed extensive testing at both the algorithm level and the mission/task level using high-fidelity manned simulations. Because a behavioral anomaly might be the result of an error in the algorithms within the associate, or an error in the knowledge in the associate, detailed analysis of all behavioral anomalies was attempted.

Assessing correctness proved to be extremely problematic, except in the most obvious cases of programming errors. The fidelity of the simulation environment was critical to creating all of the data and interactions that defined the complexity of combat aviation, but this high level of fidelity also resulted in great difficulty in repeating suspect behaviors. Furthermore, only the eventual outcome of a series of decisions could be used to assess the correctness of a decision, because an associate system is empowered to reason beyond the boundaries of fixed doctrine. Both the Pilot's Associate and the RPA were capable of surprising levels of tactical innovation. Clearly, violations of doctrine were not necessarily incorrect behaviors when the results of the decisions were successful.

A major source of correctness assessment for the Pilot's Associate was the expert judgments of a group of about 25 experienced combat pilots, who observed the behaviors of the Pilot's Associate frequently over a period of several years.

A similar approach was used by the RPA to analyze suspect behaviors.

Confirming test results as correct or incorrect for associate systems can be expected to be a lengthy process, because the outcome space of the associate can be large. By carefully sampling the test parameter space during development, the ability to demonstrate the behaviors of the system in a consistent way is enhanced. In addition to confirming that the system behaves as defined, it is also necessary to show that the system does not create behaviors that are undesirable. Proving that a behavior does not exist generally requires exhaustive enumeration of the outcomes. This is rapidly becoming intractable even for conventional systems design. To support the concerns of potential certifying authorities, for example in medicine or in aviation, a tractable process for estimating undesirable outcomes still needs to be developed for intelligent systems such as associates.

Utility. The utility of associate systems as DSS depends on their ability to help the humans reach better decisions than they could without the associate systems in the actual task environment. Utility assessments have been pursued in three ways.

First, both the Pilot's Associate and the RPA were extensively evaluated for performance improvements using planned experiments with high-fidelity manned simulations and neutral, objective pilots supplied by the military. These settings are also very problematic for evaluation, because adequate performance measures were difficult to define and even more difficult to collect. Many very obvious effects of the associate systems decision support were averaged over mission and task periods that obscured the effect. Despite this, many important operational measures were improved by the associate systems interactions. This positive effect was borne out by universal subjective praise by the pilots who operated these systems in simulation tests.

A second approach to utility assessment was the use of unmanned, discrete event simulations to represent the long-term, campaign-level effects of small changes in mission performance by individual aircraft. Although not undertaken for the Pilot's Associate, the RPA made important use of this approach and was able to show important long-term gains from small improvements in mission task performance resulting from use of associate systems.

The third approach has been smaller, independent studies of specific aspects of associate systems, rather than trying to evaluate their more global effects. These studies have included experiments in information management for aviation cockpits, evaluation of the conflict detection approach in a multi-cockpit manned air traffic management simulator, and evaluations of correctness and trust in lower fidelity task simulations.

Acceptance. The Pilot's Associate and RPA had many significant issues of user trust and commitment to overcome. At the outset, pilots were extremely skeptical of automation as a direct result of the experiences with its brittleness in combat situations (Sewell, Geddes, & Rouse, 1987). Associates were assumed to be similar in behavior to automation. Furthermore, military pilots must be extremely self-assured to take the risks inherent in their jobs. It was widely felt by the pilots that no machine could produce anything other than weak or outright silly advice.

The results of using the Pilot's Associate and the RPA were dramatically different, however. Whereas most pilots maintained an opinion of guarded trust, all pilots universally embraced the associates as immediately and deeply helpful. Many of the test subject pilots acknowledged that, despite their years of operational experience in combat aircraft, they actually learned new and powerful strategies and tactics from their interactions with the associate system.

Beyond the acceptance by the operators of a DSS, it is also important to evaluate the acceptance of the owning organizations that must bear the costs of fielding, maintaining, and training for the use of a DSS. Both the Pilot's Associate and the RPA were expensive products to develop, each costing in excess of $50 million. Since that time, dedicated efforts have been made to create the tools for knowledge engineering, systems integration, testing, and training for associate systems. As a result, an order of magnitude reduction in nonrecurring engineering cost for deploying an associate system has been achieved.

Aviation, as a domain, is an important challenge to our present ability to produce intelligent decision-aiding systems. The potential benefits to successful intelligent aiding are high: a reduction in accidents and incidents, greater economic efficiency and reliability of aviation systems and operations, and greater combat effectiveness of military air operations. The introduction of this technology into aviation has been slow, however. At the writing of this chapter, there are no operational aviation associate systems.

Case Study C. Decisions and Problem Representations

One of the early, highly successful areas of decision support systems was the field of operations research. Emerging as a formal discipline during World War II, operations research pioneered the use of the mathematics of constrained optimization to address many problems in industry, transportation, and system design. Modern Enterprise Resource Planning (ERP) systems provide an example of a computationally intensive decision support system most often implemented using numerical optimization methods.

This case study illustrates the issues that can arise when the problem representation and decision processes within the DSS are markedly different from the thought processes of its human users. Although the numerical representations used in ERP systems permit efficient computing of solutions to large problems, users can sometimes suffer from serious misunderstandings of both the input assumptions and the solution results.

The Application Area. The modern business enterprise is a complex social and economic organization, operating in a world of uncertainty. ERP is an approach to business planning that attempts to allocate business resources in quantity and time to activities in an effort to produce the best return on value for the business stakeholders.

Decisions must be made at many levels throughout a business enterprise. Although some of the decisions are clearly responsive to changes in the internal and external state of affairs for the business, other decisions are more deliberative or strategic.

The two major decisions facing a business enterprise are how to increase revenues and how to reduce costs. These two decisions interact at many levels of a large enterprise. Decisions that increase revenue may also increase cost, and decisions to reduce costs can lead to negative effects on revenues. Often, these interactions are between business activities that are widely separated in the enterprise in terms of time, space, and lines of authority. A decision made by a production engineering manager may conflict with the decision of a marketing manager, undermining a promotional campaign; a shipping decision by a product manager may be undermined by a training decision made by a human resources specialist. The larger an enterprise, the more difficult it becomes to detect and avoid resource allocation and scheduling conflicts. These conflicts create inefficiencies in the business enterprise that can reach a toll of tens of millions of U.S. dollars per year for the larger international corporations.

One approach to reducing the inefficiencies that result from undetected and unresolved conflicts within an organization's activities is to centralize the decision making process. Whereas the complexities of an enterprise-wide, centralized decision process are far beyond the cognitive ability of even a group of expert humans, suitable computer algorithms and computing platforms to achieve this became available in the 1980s. This led first to the development of Manufacturing Resource Planning systems, then to modern ERP systems.

The Design Solution. Although different approaches to formulating the allocation of resources and schedules for activities exist within the overall product offerings of ERP, the dominant approaches are based on optimization (Callaway, 1999; O'Leary, 2000). For illustration in this case study, we will describe a typical ERP system without reference to a particular product design.

The output of a typical ERP system is similar to a schedule of activities for the enterprise for a period of time, typically on the order of weeks to months. The time scale of the schedule is often to the hour or smaller units of time. Each activity in the schedule will consume or occupy enterprise resources, such as cash, machines, inventory, and people's time. At least some of the activities also generate revenue, which is in turn translated into cash, machines, inventory, and people. The overall value of the enterprise changes with changes in its resources, and the ERP system determines the schedule that appears to provide the best value over the time period for the enterprise.

In addition to the schedule of activities, the ERP system normally determines the amount of unused or slack resources resulting from the scheduled activities, and those resources that are at (or sometimes above) full utilization. Resources that are underutilized are candidates for reduction in available quantity and resulting savings in costs. Resources at full availability may be causing a lost revenue opportunity, and hence may be candidates for increased investment.

The inputs to an ERP system are intended to represent the comprehensive set of resources and activities that the enterprise can control. As a result, ERP systems are normally connected to, or contain, financial accounting systems, marketing demand forecasting systems, inventory systems, shipping and receiving systems, personnel data systems, production control systems, and quality management systems. Data from all of these systems are needed to determine the quantity of each resource that is available and the financial impact of using or creating an increment of a resource.

The users of an ERP system must also supply the system with the intended or desired set of activities to be performed within the time frame of the schedule. Most ERP systems contain a means for defining the business processes of the enterprise as sequences of activities, which may in turn be connected hierarchically to lower level sequences of activities until the level of the primitive activities is reached. One such formalism for specifying process sequences is the IDEF notation. Each primitive activity is mapped to its input resource utilization and it output resource production, including such things as inventory consumption and creation, labor hours, machine time, energy use, floor space, and other financial factors. Each primitive activity also has precedence relationships with other activities in its sequence that require that the predecessor and successor activities have a specified start and end time relationship.

Finally, the users of an ERP system must define the value function that is used to evaluate the benefits of any particular scheduled allocation of activities and resources. This value function is normally related to the costs, assets, and revenues of the enterprise. It may also include many other less definitive aspects of the business, such as the value of its skilled labor force, its corporate intellectual property in the form of managed knowledge, as well as patents and trademarks, or the value of its brands in terms of customer perceptions of service, quality, and loyalty.

The ERP schedule of activities and resource allocation is created by allocating activities and resources to times in accordance with the precedence relationships necessary to execute the business processes. The schedule is feasible if it does not violate any of the resource availabilities or the time limits for the schedule; otherwise, it is infeasible and could not be actually executed. The algorithms and heuristics used by different vendors to accomplish the creation of a feasible schedule vary greatly and are the subject of numerous patents and trade secrets. Typically, the algorithms are run in an iterative fashion in which the solution (the schedule) is systematically improved to increase its value function or to gain feasibility. Because an ERP system must deal with a very large combinatoric problem, the time to execute an ERP solution may be lengthy even on expensive high-end computing platforms. ERP systems operate in a manner that is compatible with deliberative planning, but they may have difficulty with responsive or tactical planning over short time frames.

Results. The early successes of Manufacturing Resource Planning systems led to enthusiastic efforts by large industry to install the larger, more complex ERP systems. By the mid-1990s, numerous ERP products were available on the market and many large installations were underway. Often, these large installations were undertaken without in-depth evaluation of the product benefits because partial trial implementations or business simulations were not adequate to characterize the final expected benefits.

Technical Correctness. The correctness of the underlying algorithms of most ERP systems was established by mathematical theorem and proof. Given the correctness of the mathematics used by most ERP software, it would seem an obvious result

that the business enterprise must perform better and achieve greater value using the planned allocations determined by the ERP system than it would any other way. These expectations were confirmed in a limited way by business simulations.

The technical correctness of the algorithms, however, is not the only issue in understanding the notion of correctness within an ERP. It is also important to examine the correctness of the underlying representations used in the DSS. Because each enterprise is different in many respects from any other, a significant amount of customization is necessary to complete the installation of an ERP system. This includes the specification of resources and the values that they generate, among many other data items needed by the ERP system. The correctness of this underlying data may be extremely difficult to evaluate, and the subject of much speculation and unproven opinion. The results achieved by other companies in using any specific ERP implementation may not be indicative of the results for another company.

Utility. In actual use, the utility of ERP systems is very difficult to evaluate. The financial results of using an ERP system are influenced by the acceptance and utility of the system, but they are also strongly influenced by changes in the marketplace of the business enterprise completely beyond its direct control. Weather disasters, labor unrest, supplier or customer business failures, new competitive technologies, or general economic downturns may make a very successful use of ERP still look bad to the bottom line. Similarly, strong economic times may serve to mask an otherwise very poor ERP installation.

The makers and installers of ERP systems are quick to cite many successful installations. Quantitative results defining the levels of success achieved are harder to document.

The use of ERP has not been without major disappointments to some in the commercial world, however. Whereas the mathematics of most optimization engines is compelling in its correctness, the decisions that result are not always useful (Radding, 2000; Wheatley, 2001). A recent example is the serious loss of value experienced by a major international footwear manufacturer as a result of use of a major brand of ERP system. (Wilson, 2001). Although the footwear company accused the ERP system supplier of a faulty product, the ERP system supplier argued that the manufacturer had configured the system with faulty data. Regardless of the cause, the footwear company estimated its losses at more than $400 million as a direct result of the ERP system production plan, and its stock lost more than 15% value as a result of the problems.

Acceptance. Although initial market acceptance of the anticipated benefits of ERP was high, and some companies have obtained real benefits, it is also clear that expectations of ERP may have been overly optimistic. Currently, ERP systems are under criticism for their high cost and low return on investment across many industries. A significant number of major ERP installations remain incomplete and several have been abandoned altogether (Wheatley, 2001).

The uneven results of ERP provide an important conclusion about the organization and underlying representations used in DSS. Although computational efficiency is always a concern for large-scale, complex systems, representational issues may be even more important. These representational issues have implications for the internal computations as well as for training and communication of the processes and outcomes of the computations to the user.

If representations are used that are not sufficiently expressive and that cannot be readily validated, the correctness of the algorithms that manipulate the representations may be nearly irrelevant. It is well known that good decisions rarely come from bad data. It is even less likely that good decisions can come from weak representations of the problem characteristics. In addition, if the system design is based on a philosophy that does not support adequate interactions with the user, then it will be difficult for the person to evaluate the recommendations of the software and integrate his own knowledge and judgments.

Case Study D. Reducing the Burden of Training

Although the need for DSSs is often seen in complex settings, other applications can also achieve large benefits through decision aiding. One use of DSS that has had dramatic economic benefits has been the use of electronic performance support systems (EPSS) to enhance performance and reduce training requirements for many common tasks.

This case study describes the Retail Performance Support System (RPSS), an award-winning implementation of an EPSS system for retail sales associates of a major family footwear retail company. The RPSS illustrates how a human-centered DSS can enhance both the economic outcome and the job quality of ordinary task environments by reducing training time and enhancing customer service.

The Application Area. Retailing is a major sector of the world economy. Success in retailing depends on providing the retail customer with the desired products at an agreeable price. Although simple in concept, delivering this service with consistently high quality and at a profit is a difficult undertaking, depending on many factors, such as store location, stocking, layout and merchandising, pricing, and quality of customer service. Complicating the efforts to provide high quality of service is the need to provide substantial training to the retail sales force. Customers expect to interact with a knowledgeable sales

associate who is well informed about the product offerings, can provide helpful suggestions about product features, and can execute transactions in a correct and efficient manner. However, many sales associate job opportunities are at the entry level and may be filled by people who have little or no sales experience. In addition, the mobility of the workforce in commercially active communities can be high, leading to large turnover of retail sales personnel.

Recognizing the challenge of efficiently training its workforce, Payless ShoeSource, Inc., undertook the development and fielding of the RPSS with design and implementation support from a number of outside companies. Payless ShoeSource, Inc., is the largest family footwear retailer in North America, selling more than 200 million pairs of shoes to approximately 150 million customers each year. The Payless ShoeSource chain operates almost 5,000 stores across North America, featuring quality footwear in a self-service shopping format.

Payless retail sales personnel are in three categories, with the characteristics shown in Table 33.2. Each year, Payless estimated it was expending 33 person-years of labor and $3 million for the training of its retail sales personnel. One approach to improving the performance of the retail sales staff is to provide decision and performance support. Payless undertook the development of the RPSS as a means to reduce its training requirements while improving customer service.

The Design Solution. The vision of the RPSS within Payless ShoeSource, Inc., was as a tool for re-engineering the way work is done in its stores. The concept is a holistic approach to ensure that Payless' business objectives are accomplished through a combination of decision and training support systems that drive business processes, reduce workload, or provide critical learning to promote the behaviors Payless believes will increase sales and delight customers. Payless' desire is to become a company that has a sustainable competitive advantage by learning faster and better than its competitors.

The initial concept was to build a system within the sales associates' work context—i.e., the system should reflect their daily work environment (Table 33.2). The key to achieving this goal was to design RPSS so that learning is an integral part of doing. The RPSS was intended to support learning and provide knowledge and information at the moment of need, extending

TABLE 33.2. Characteristics of the Retail Sales Force

	Managers	Full-Time Sales	Part-Time Sales
Typical age	28–38	18–24	18–24
Male:female ratio	50:50	30:70	10:90
Education	AA degree typical	High School	High School
Ethnicity	Many multilingual	Many multilingual	Many multilingual
Time to train (Pre-RPSS)	72 hours	30 hours	20 hours
Computer literacy	70%	50%	50%
Retail experience	Almost always	Some	Some
Shoe retail experience	Not necessarily	Not necessarily	Not necessarily
Annual turn rate	15%	60%	96%
Average tenure	5–6 years	1 year	7–8 months

Used with permission of Payless ShoeSource, Inc., and Ariel Performance Centered Systems, Inc.

human sales associate capabilities and increasing Payless' competitive advantage.

One of the success criteria for Payless was to design the RPSS so that it provided day 1 support for a diverse audience. Some of the concerns included the large number of people with no or little computer skills and potentially lower education level. Another significant consideration based on the needs of the end-users was the ability to support multiple languages. In many store locations, English is not the primary language. As such, the RPSS had to be designed in a way to facilitate translating content, as well as a way to deliver multiple language versions of the system.

The system, which is aimed at store managers and sales associates, provides just-in-time support for most in-store processes. The system uses a graphical user interface that features a simulated storefront with objects familiar to store associates. The user can access underlying information and guidance by selecting the objects appearing on the store graphic or by using a text-based menu system. The RPSS also provides an interactive agent, a sock puppet named SeeMore. SeeMore engages associates through the use of guided tours of the system, search assistance, and, ultimately, proactive context sensitive support.

The RPSS organizes its information and knowledge around a detailed task analysis, known as a performance support map. The performance support map outlined all of the tasks, the type of learning/performance needed to support that task, and the knowledge requirements for the task. In all, almost 1,400 tasks were identified by the development team.

The RPSS system consists of several major content component types to support the sales associate. These include application programs for specific tasks, such as staff scheduling and inventory tracking; job aids to provide procedural instructions for tasks; references to provide in-depth discussions of topics; and computer-based learning modules for self-paced instruction.

Results. Payless ShoeSource, Inc., has historically held a high commitment to defining and instituting best business practices across its retail operations. In the past, Payless relied on traditional delivery systems, such as instructor-led, on the job training, and business practice manuals to train its staff. Although some of the business practices were available as on-line documentation, access to these sources was not efficient. The distribution of current operational practices across all 5,000 operating locations was a cumbersome and expensive process. In addition, use of such source material was interruptive of the personal interactions between customers and sales associates necessary to provide high-quality service. Further degradation in operational practices was occurring because of the need to rely on word-of-mouth advice within the store staff.

The RPSS is partially deployed across Payless ShoeSource as of the summer of 2001. It is showing the potential to significantly improve the cost and performance of the retail sales force within Payless ShoeSource, Inc. The company reports better-than-expected results from the initial deployed version, but plans future significant extensions to gain an even larger positive impact. RPSS is currently successful in bringing just the appropriate information to the associate at the precise moment of need. Changes in operational practices can now be distributed across the entire retail network by updating the information within the RPSS. The result is updated content that is immediately accessible in a familiar way and expressed in familiar terms.

Assessment. As a part of its decision to field the RPSS across all of its stores, Payless conducted a controlled field evaluation of the system. Results from stores using a partial implementation of RPSS were compared with those without it, balanced for issues including seasonal variations, market size, geographic location, store size, and time in the marketplace. The RPSS implementation contained approximately 10% of the content planned for the full RPSS implementation. The results showed an increase in sales of 4.8% using the pilot RPSS system. Although these results show meaningful gains in effectiveness, Payless expects even greater returns when the full RPSS content is in place. The company is currently reporting a good trend in sales related to the use of RPSS, adding sales to each store each week. These gains were achieved with no dedicated training to support the deployment of the RPSS, further suggesting that the reduction in training sought from the RPSS will be achievable. Payless reports that new sales associates are productive on their first day of work using the RPSS.

Acceptance. After the field trials, Payless management endorsed the RPSS as a strategic business thrust for the company. As of year-end 2000, about 12% of Payless' 5,000 stores were using the RPSS. Currently, the RPSS is being deployed in 108 new stores each week. The company expects to be fully deployed with RPSS in 2001. The system provides true "day 1" performance, therefore significantly reducing the cost of sales associate training in the Payless' annual training budget. Moreover, it provides Payless with added commercial advantage, helping its sales force outperform its competition. Although Payless ShoeSource has not released the latest estimate of their annual savings, the company expects to exceed their earlier estimates of the savings.

The RPSS system was awarded the design prize for electronic support systems at the Online Learning/Performance Support 2000 Conference. The award was sponsored by Lakewood Publications, the Performance Support Leadership Council, and the EPSS InfoSite. The RPSS was co-developed by Payless Shoe-Source, Inc., and a number of supporting companies, especially Ariel Performance Centered Systems, Inc., which participated in the design and implementation phases.

The success of the RPSS illustrates that significant economic advantage can be created by providing DSS in everyday domains in which human knowledge is taxed by constant change and high workforce mobility. The RPSS directly applied many principles of human computer interaction discussed in this chapter within its architecture, allowing rapid utility which high acceptance by both the users and the owners of the system.

CONCLUSIONS

The goal of this chapter has been to highlight the large number of cognitive factors that are relevant to the design of a DSS, in terms of the design of the underlying technology and its role, in terms of the interaction of the user or users and this technology, and in terms of the factors influencing designer performance. The discussions of specific issues, along with the case studies, are intended to make concrete the questions regarding human–computer interaction that need to be considered as part of the design process and to illustrate how the answers to these questions can lead to the design of more effective decision support systems.

References

Amalberti, R. (1999). Automation in aviation: A human factors perspective. In D. Garland, J. Wise, & V. Hopkins (Eds.), *Handbook of aviation* (pp.173–192). Mahwah, NJ: Lawrence Erlbaum and Associates.

Anderson, J. R. (1993). *Rules of the mind*. Hillsdale, NJ: Lawrence Erlbaum and Associates.

Andes, R. C., & Rouse, W. B. (1992). Specification of adaptive aiding systems. *Information and Decision Technology, 18*, 195–207.

Bainbridge, L. (1983). Ironies of automation. *Automatica, 19*, 775–779.

Berner, E., Brooks, C., Miller, R., Masarie, F., & Jackson, J. (1989). Evaluation issues in the development of expert systems in medicine. *Evaluation and the Health Professions, 12*, 270–281.

Billings, C. E. (1996). *Aviation automation: The search for the human-centered approach*. Hillsdale, NJ: Lawrence Erlbaum Associates.

Callaway, E. (1999). *Enterprise resource planning: Integrating applications and business processes across the enterprise*. Computer Technology Research Corporation.

Cartwright, D., & Zander, A. (Eds.). (1960). *Group dynamics: Research and theory* (2nd ed). Evanston, IL: Row & Peterson.

Chandrasekaran, B. (1988). Generic tasks as building blocks for knowledge based systems: The diagnosis and routine design examples. *Knowledge Engineering Review, 3*(3), 183–210.

Chi, M., Glaser, R., & Farr, M. (Eds.). (1988). *The nature of expertise*. Hillsdale, NJ: Lawrence Erlbaum and Associates, Publishers.

Clancey, W. J. (1983). The epistemology of a rule-based expert system—A framework for explanation. *Artificial Intelligence, 20*, 215–251.

Clancey, W. J. (1985). Heuristic classification. *Artificial Intelligence, 27*, 289–350.

Elstein, A. S., Shulman, L. S., & Sprafka, S. A. (1978). *Medical problem solving: An analysis of clinical reasoning*. Cambridge, MA: Harvard University Press.

Endsley, M., & Kaber, D. (1999). Level of automation effects on performance, situation awareness and workload in a dynamic control task. *Ergonomics, 42*(3), 462–492.

Flach, J., Hancock, P., Caird, J., & Vincente, K. (Eds.). (1995). *Global perspectives on the ecology of human-machine systems*. Hillsdale, NJ: Lawrence Erlbaum and Associates, Publishers.

Fischer, G., Lemke, A. C., Mastaglio, T., & Morch, A. I. (1991). The role of critiquing in cooperative problem solving. *ACM Transactions on Information Systems, 9*(3), 123–151.

Fraser, J. M., Smith, P. J., & Smith, J. W. (1992). A catalog of errors. *International Journal of Man-Machine Systems, 37*, 265–307.

Geddes, N. D. (1989). *Understanding human operator's intentions in complex systems*. Doctoral Dissertation, Georgia Institute of Technology, Atlanta, GA.

Geddes, N. D. (1997). Associate systems: A framework for human-machine cooperation. In M. J. Smith, G. Slavendy, & R. J. Koubek (Eds.), *Design of computing systems: Social and ergonomic coniderations. Advances in human factors/ergonomics* (Vol. 21B, pp. 237–242). New York: Elsevier.

Geddes, N. D., & Lizza, C. S. (1999). Shared plans and situations as a basis for collaborative decision making in air operations. SAE World Aeronautics Conference, SAE Paper 1999-01-5538.

Gilovich, T., Vallone, R., & Tversky, A. (1985). The hot hand in basketball: On the misperception of random sequences. *Cognitive Psychology, 17*, 295–314.

Guerlain, S., Smith, P. J., Obradovich, J. H., Rudmann, S., Strohm, P., Smith, J. W., & Svirbely, J. (1996). Dealing with brittleness in the design of expert systems for immunohematology. *Immunohematology, 12*, 101–107.

Guerlain, S., Smith, P. J., Obradovich, J. H., Rudmann, S., Strohm, P., Smith, J. W., Svirbely, J., & Sachs, L. (1999). Interactive critiquing as a form of decision support: An empirical evaluation. *Human Factors, 41*, 72–89.

Hammer, J. M., & Small, R. L. (1995). An intelligent interface in an associate system. In W. B. Rouse (Ed.), *Human/technology interaction in complex systems* (Vol. 7, pp. 1–44). Greenwich, CT: JAI Press.

Hammond, K. J. (1989). *Case-based planning: Viewing planning as a memory task*. Boston, MA: Academic Press, Inc.

Harris, S., & Owens, J. (1986). Some critical factors that limit the effectiveness of machine intelligence technology in military systems applications. *Journal of Computer-Based Instruction, 13*, 30–34.

Hasling, D. W., Clancey, W. J., & Rennels, G. (1984). Strategic explanations for a diagnostic consultation system. *International Journal of Man-Machine Systems, 20*, 3–19.

Hayes-Roth, B., & Hayes-Roth, F. (1979). A cognitive model of planning. *Cognitive Science, 3*(4), 275–310.

Helander, M., Landauer, T., & Prabhu, P. (Eds.). (1997). *Handbook of human-computer interaction*. Amsterdam: Elsevier.

Hoc, J.-M. (2000). From human-machine interaction to human-machine cooperation. *Ergonomics, 43*, 833–843.

Hollnagel, E. (1990). Responsibility issues in intelligent decision support systems. In D. Berry & A. Hart (Eds.), *Expert systems: Human issues*. Cambridge, MA: The MIT Press.

Horton, F., & Lewis, D. (Eds.). (1991). *Great information disasters*. London: Association for Information Management.

Hutchins, E. (1990). The technology of team navigation. In J. Galegher, R. Kraut, & C. Egido (Eds.), *Intellectual teamwork: Social and technical bases of collaborative work*. Hillsdale, NJ: Lawrence Erlbaum.

Hutchins, E. (1995). *Cognition in the wild*. Cambridge, MA: MIT Press.

Janis, I. (1982). *Groupthink* (2nd ed.). Boston: Houghton Mifflin.

Jennings, F., Benyon, D., & Murray, D. (1991). *Acta Psychologica, 78*, 243–256.

Jones, D. R., & Schkade, D. A. (1995). Choosing and translating between problem representations. *Organizational Behavior and Human Decision Processes, 61*(2), 214–223.

Jones, E., & Nisbett, R. (1971). The actor and the observer: Divergent perceptions of the causes of behavior. In E. Jones, D. Kanouse, H. Kelley, R. Nisbett, S. Valins, & B. Weiner (Eds.), *Attributions: Perceiving the causes of behavior* (pp. 79–94). Morristown, NJ: General Learning Press.

Jones, P. M., & Mitchell, C. M. (1995). Human-computer cooperative problem solving: Theory, design, and evaluation of an intelligent associate system. *IEEE Transactions on Systems, Man, and Cybernetics, 25*, 1039–1053.

Josephson, J., & Josephson, S. (1994). *Abductive inference* (pp. 31-38). New York: Cambridge University Press.

Klayman, J., & Ha, Y. (1987). Confirmation, disconfirmation, and information in hypothesis testing. *Psychological Review, 94*, 211-228.

Klein, G. A. (1993). A recognition-primed decision (RPD) model of rapid decision making. In G. A. Klein, J. Oransanu, R. Calderwood, & C. E. Zsambok (Eds.), *Decision making in action: Models and methods* (pp. 138-147). Norwood, NJ: Ablex.

Kleinmuntz, D. N., & Schkade, D. A. (1993). Information displays and decision processes. *Psychological Science, 4*(4), 221-227.

Koehler, J. J. (1996). The base rate fallacy reconsidered: Descriptive, normative, and methodological challenges. *Behavioral and Brain Sciences, 19*(1), 1-53.

Kolodner, J. (1993). *Case-based reasoning* (Chaps. 1-3). San Mateo, CA: Morgan Kaufman Publishers.

Kotovsky, K., Hayes, J. R., & Simon, H. A. (1985). Why are some problems hard? Evidence from Tower of Hanoi. *Cognitive Psychology, 17*, 248-294.

Laird, J. E., Newell, A., & Rosenbloom, P. S. (1987). SOAR: An architecture for general intelligence. *Artificial Intelligence, 33*, 1-64.

Larkin, J. H. (1989). Display-based problem solving. In D. Klahr & K. Kotovsky (Eds.), *Complex information processing: The Impact of Herbert A. Simon*. Hillsdale, NJ. Erlbaum.

Larkin, J. H., & Simon, H. A. (1987). Why a diagram is (sometimes) worth ten thousand words. *Cognitive Science, 11*, 65-99.

Lehner, P. E., & Zirk, D. A. (1987). Cognitive factors in user/expert-system interaction. *Human Factors, 29*(1), 97-109.

Lesgold, A., Glaser, R., Rubinson, H., Klopfer, D., Feltovich, P., & Wang, Y. (1988). Expertise in a complex skill: Diagnosing X-ray pictures. In M. Chi, R. Glaser, & M. Farr (Eds.), *The nature of expertise*. Hillsdale, NJ: Lawrence Erlbaum Associates, Publishers.

Lizza, C. S., & Banks, S. B. (1991, June). Pilot's associate: A cooperative knowledge based system application. *IEEE Expert, 6*, 18-29.

Loftus, E. (1975). Leading questions and the eyewitness report. *Cognitive Psychology, 7*, 560-572.

Mackworth, N. (1950). Researches on the measurement of human performance. Reprinted in H. W. Sinaiko (Ed.), *Selected papers on human factors in the design and use of control systems*, 1961. New York: Dover Publications.

McCauley, C. (1989). The nature of social influence in groupthink: Compliance and internalization. *Journal of Personality and Social Psychology, 22*, 250-260.

Michie, D. (1986). *On machine intelligence* (2nd ed.). Chicester, UK: Ellis Horwood Limited.

Miller, C., Pelican, M., & Goldman, R. (2000). "Tasking" interfaces to keep the operator in control. In M. Benedict (Ed.), *Proceedings of the 5th International Conference on Human Interaction with Complex Systems*. Urbana, IL.

Miller, G. A., Galanter, E., & Pribram, K. H. (1960). *Plans and the structure of behavior*. NY: Henry Holt and Company.

Miller, P. (1986). *Expert critiquing systems: Practice-based medical consultation by computer*. New York: Springer-Verlag.

Mitchell, C. M., & Miller, R. A. (1986). A discrete control model of operator function: A methodology for information display design. *IEEE Transactions on Systems, Man and Cybernetics, 16*, 343-357.

Muir, B. (1987). Trust between humans and machines. *International Journal of Man-Machine Studies, 27*, 527-539.

Mynatt, C., Doherty, M., & Tweney, R. (1977). Confirmation bias in a simulated research environment: An experimental study of scientific inference. *Quarterly Journal of Experimental Psychology, 30*, 85-95.

Navon, D. (1978). The importance of being conservative. *British Journal of Mathematical and Statistical Psychology, 31*, 33-48.

Newell, A. (1990). *Unified theories of cognition*. Cambridge, MA: Harvard University Press.

Newell, A., & Simon, H. (1972). *Human problem solving*. Englewood Cliffs, NJ: Prentice Hall, Inc.

Nickerson, R. (1988). Counting, computing, and the representation of numbers. *Human Factors, 30*, 181-199.

Norman, D. A. (1981). Categorization of action slips. *Psychological Review, 88*(1), 1-15.

Norman, D. A. (1988). *The design of everyday things*. New York: Doubleday.

O'Leary, D. E. (2000). *Enterprise resource planning systems: Systems, life cycle, electronic commerce and risk*. Cambridge, UK: Cambridge University Press.

Olson, G. M., & Olson, J. S. (1997). Research on computer supported cooperative work. In M. Helander, T. Landauer, & P. Prabhu (Eds.), *Handbook of human-computer interaction* (pp. 1433-1456). Amsterdam: Elsevier.

Orasanu, J., & Salas, E. (1991). Team decision making in complex environments. In J. O. G. Klein, R. Calderwood, & C. Zsambok (Eds.), *Decision making in action: Models and methods*. New Jersey: Ablex.

Parasuraman, R. (2000). Designing automation for human use: Empirical studies and quantitative models. *Ergonomics, 43*, 931-951.

Parasuraman, R., & Riley, V. (1997). Humans and automation: Use, misuse, disuse and abuse. *Human Factors, 39*, 230-253.

Pearl, J. (1988). *Probabilistic reasoning in intelligent systems: Networks of plausible inference*. San Mateo, CA: Morgan Kaufman.

Pennington, N., & Hastie, R. (1992). Explaining the evidence: Tests of the story model for juror decision making. *Journal of Personality and Social Psychology, 62*, 189-206.

Poulton, E. C. (1989). *Bias in quantifying judgements*. Hillsdale, NJ: Lawrence Erlbaum.

Preece, J., Rogers, Y., Sharp, H., & Benyon, D. (1994). *Human-computer interaction*. Reading, MA: Addison-Wesley.

Radding, A. (2000, February 22). Knowledge management appears on the ERP radar. *IT Management*. Earthweb. Available: www.itmanagement.earthweb.com/erp/archivcs

Rasmussen, J. (1983). Skills, rules and knowledge: Signals signs symbols and other distinctions in human performance models. *IEEE Transactions on Systems Man and Cybernetics, SMC-13*(3), 257-266.

Rasmussen, J., Brehner, B., & Leplat, J. (Eds.). (1991). *Distributed decision making: Cognitive models for cooperative work*. New York: John Wiley and Sons.

Rasmussen, J., Pejtersen, A., & Goldstein, L. (1994). *Cognitive systems engineering*. New York: John Wiley & Sons, Inc.

Reason, J. (1991). *Human error*. Cambridge, UK: Cambridge Press.

Riesbeck, C. K., & Schank, R. C. (1989). *Inside case-based reasoning*. Hillsdale, NJ: Lawrence Erlbaum.

Roth, E. M., Bennett, K. B., & Woods, D. D. (1987). Human interaction with an 'intelligent' machine. *International Journal of Man-Machine Studies, 27*, 479-525.

Rouse, W. B. (1980). *Systems engineering models of human machine interaction*. New York: Elsevier

Rouse, W. B., Geddes, N. D., & Curry, R. E. (1987). An architecture for intelligent interfaces: Outline of an approach to supporting operators of complex systems. *Human-Computer Interaction, 3*, 87-122.

Rouse, W. B., Geddes, N. D., & Hammer, J. M. (1990, March). Computer-aided fighter pilots. *IEEE Spectrum, 27*, 38-41.

Rubin, K. S., Jones, P. M., & Mitchell, C. M. (1988). OFMspert: Inference of operator intentions in supervisory control using a blackboard structure. *IEEE Transactions on Systems Man and Cybernetics, SMC-18*(4), 618-637.

Russell, S., & Norvig, P. (1995). *Artificial intelligence: A modern approach*. Englewood Cliffs, NJ: Prentice-Hall, Inc.

Sacerdoti, E. D. (1974). Planning in a hierarchy of abstraction spaces. *Artificial Intelligence, 5*(2), 115-135.

Salvendy, G. (Ed.). (1997). *Handbook of human factors and ergonomics* (2nd ed.). New York: John Wiley and Sons, Inc.

Sarter, N., & Woods, D. (1993). *Cognitive engineering in aerospace applications: Pilot interaction with cockpit automation* (NASA Contractor Report 177617). Moffett Field, CA: NASA Ames Research Center.

Sewell, D. R., & Geddes, N. D. (1990). A plan and goal based method for computer-human system design. *Human computer interaction: INTERACT 90* (pp. 283-288). New York: North Holland.

Sewell, D. R., Geddes, N. D., & Rouse, W. B. (1987). Initial evaluation of an intelligent interface for operators of complex systems. In G. Salvendy (Eds.), *Cognitive engineering in the design of human-computer interaction and expert systems* (pp. 551-558). New York: Elsevier.

Shafer, G. (1996). *Probabilistic expert systems*. Philadelphia: Society for Industrial and Applied Mathematics.

Shafer, G., & Pearl, J. (Ed.). (1990). *Readings in uncertain reasoning*. San Mateo, CA: Morgan Kaufmann.

Shalin V. L., Geddes, N. D. Bertram, D., Szczepkowski, M. A., & DuBois, D. (1997). Expertise in dynamic physical task domains. In P. Feltovich, K. Ford, & R. Hoffman (Eds.), *Expertise in context: Human and machine* (pp. 194-217). Cambridge, MA: MIT Press.

Sheridan, T. B. (1997). Supervisory control. In G. Salvendy (Ed.), *Handbook of human factors* (2nd ed., pp. 1295-1327). New York: John Wiley and Sons.

Sheridan, T. B., & Ferrell, W. R. (1974). *Man-machine studies: Information, control, and decision models of human performance*. Cambridge, MA: MIT Press.

Shortliffe, E. (1990). Clinical decision support systems. In E. Shortliffe & L. Perreault (Eds.), *Medical informatics: Computer applications in health care* (pp. 466-500). New York: Addison Wesley.

Silverman, B. G. (1992). Survey of expert critiquing systems: Practical and theoretical frontiers. *Communications of the ACM, 35*(4), 106-128.

Skitka, L., Mosier, K., & Burdick, M. (1999). Does automation bias decision making? *International Journal of Human-Computer Systems, 51*, 991-1006.

Smith, P. J., Billings, C., Chapman, R., Obradovich, J. H., McCoy, E., & Orasanu, J. (2000). Alternative architectures for distributed cooperative problem-solving in the national airspace system. In M. Benedict (Ed.), *Proceedings of the 5th International Conference on Human Interaction with Complex Systems*. Urbana, IL.

Smith, P. J., Giffin, W., Rockwell, T., & Thomas, M. (1986). Modeling fault diagnosis as the activation and use of a frame system. *Human Factors, 28*(6), 703-716.

Smith, P. J., McCoy, E., & Layton, C. (1997). Brittleness in the design of cooperative problem-solving systems: The effects on user performance. *IEEE Transactions on Systems, Man, and Cybernetics, 27*(3), 360-371.

Smith, P. J., McCoy, E., Orasanu, J., Billings, C., Denning, R., Rodvold, M., Gee, T., & VanHorn, A. (1997). Control by permission: A case study of cooperative problem-solving in the interactions of airline dispatchers and ATCSCC. *Air Traffic Control Quarterly, 4*, 229-247.

Smith, P. J., Obradovich, J. H., Guerlain, S., Rudmann, S., Strohm, P., Smith, J., Svirbely, J., & Sachs, L. (1998). Successful use of an expert system to teach diagnostic reasoning for antibody identification. *Proceedings of the 4th International Conference on Intelligent Tutoring Systems* (pp. 354-363).

Suchman, L. (1987). *Plans and situated actions: The problem of human-machine communication*. New York: Cambridge University Press.

Tetlock, P. (1998). Social psychology and world politics. In D. Gilbert, S. Fiske, & G. Lindzey (Eds.), *The handbook of social psychology* (4th ed., Vol. 2, pp. 868-912). New York: McGraw Hill.

Tetlock, P., Peterson, R., McQuire, M., Chang, S., & Feld, P. (1992). Assessing political group dynamics: A test of the groupthink model. *Journal of Personality and Social Psychology, 63*, 403-425.

Tufte, E. R. (1997). *Visual explanations*. Chesire, CT: Graphics Press.

Tversky, A. (1972). Elimination by aspects: A theory of choice. *Psychological Review, 79*, 281-299.

Tversky, A. (1982). *Judgment under uncertainty: Heuristics and biases*. Cambridge, UK: Cambridge University Press.

Tversky, A., & Kahneman, D. (1974). Judgment under uncertainty: Heuristics and biases. *Science, 185*, 1124-1131.

Wallace, T. F., & Kremzar, M. H. (2001). ERP: Making it happen: The implementers guide to success with enterprise resource planning (3rd ed.). New York: John Wiley & Sons.

Waltz, D. (1975). Understanding line drawings of scenes with shadows. In P. H. Winston (Ed.), *Psychology of computer vision*. Cambridge, MA: MIT Press.

Wason, P. (1960). On the failure to eliminate hypotheses in a conceptual task. *Quarterly Journal of Experimental Psychology, 12*, 129-140.

Wheatley, M. (2000, June 1). ERP training stinks. *CIO Magazine*. Available: www.cio.com/archive

Wiener, E. L., & Nagel, D. C. (1988). *Human factors in aviation*. New York: Academic Press.

Wilensky, R. (1983). *Planning and understanding: A computational approach to human reasoning*. Reading, MA: Addison-Wesley.

Wilson, T. (2001, March 1). Supply chain debacle. *Internet Week*. Available: www.internetweek.com

Zhang, J. (1997). The nature of external representations in problem solving. *Cognitive Science, 21*(2), 179-217.

Zhang, J., & Norman, D. A. (1994). Representations in distributed cognitive tasks. *Cognitive Science, 18*, 87-122.

Zuboff, S. (1988). *In the age of smart machines*. New York: Basic Books.

·34·

COMPUTER-BASED TUTORING SYSTEMS:
A BEHAVIORAL APPROACH

Henry H. Emurian
UMBC

Ashley G. Durham
Centers for Medicare and Medicaid Services

INTRODUCTION

Students are not equally advantaged to learn something new. They differ in such fundamental factors as preparation, motivation, maturity, study skills, and available energy, to name just a few. An optimal learning environment that takes such individual differences into consideration as factors that influence learning and the achievement of excellence may require a human teacher to interact with the individual student (Bloom, 1984). A teacher can continuously adjust a learning encounter based on the student's responses, and this feature is the essence of tutoring (Skinner, 1968, p. 37). The obvious assumption is that the teacher knows how to do that optimally, making such a human teacher a scarce resource, indeed. To the extent that the art and science of effective teaching can be codified as a set of enumerated principles and procedures that collectively determine an optimal learning environment for the individual student, it is reasonable to attempt to make those conditions available to all learners. This would enable learners to achieve a specific educational outcome without regard to the availability of a human teacher to monitor and manage the moment-by-moment process of learning by an individual student.

A computer-based tutoring system, as a generic teacher, has the objective of changing the behavior of a learner by structuring a series of interactive experiences with the system to achieve a criterion of mastery at the level of the individual student (Bostow, Kritch, & Tompkins, 1995). For the purposes of this chapter, a tutoring system is characterized by the requirement that a learner actively constructs responses and/or selects assessment options during an interactive dialog with the system. Those responses are evaluated for accuracy, and the outcome of that evaluation determines the sequence and content of the dialog events that follow. These events systematically change the behavior of the learner until the final educational objective has been achieved. A computer-based tutoring system is an information system that manages those interactive events.

This approach to operationalizing the essential features of a computer-based tutoring system allows the inclusion of programs that encompass a broad range of interdisciplinary applications and objectives, as long as those systems exhibit the above features. Such programs, then, might include drill and practice programs for acquiring domain knowledge and specific skills in such areas as mathematics, foreign language, and music (Alessi & Trollip, 1991; Merrill et al., 1996). Also included would be programs that foster generalized creative problem-solving skills (Ray, 1995) and those that teach introductory statistics (Shute, Gawlick, & Gluck, 1998). The potential range of programs is represented by the *looking at technology in context* framework for organizing learning theory and educational practice research from information transmission models (e.g., information presentation) studied in the research laboratory to constructivist models (e.g., information generation) applied to interconnected schools (Cognition and Technology Group at Vanderbilt, 1996).

The proposed operational definition of a computer-based tutoring system is to provide individual instruction to achieve specific task mastery by the user. As such, it would not include a consideration of the important stream of computer-based instructional delivery systems and authoring languages that commenced in the early 1960s with the use of the IBM 1500 at Penn State University and the University of Alberta (Buck & Hunka, 1992; Szabo & Montgomerie, 1992). It would also not include studies that show improvements in cognitive performance (e.g., Raven's Colored Matrices) that have been reported following a learner's use of such computer-delivered programs as Logo (Klein & Darom, 2000) or studies that show improvements in solving algebra word problems following a learner's use of a computer-based Word Problem-Solving tutor (Wheeler & Regian, 1999). These types of computer-based instructional delivery systems and tutorials, while demonstrably useful for groups of learners, may not always include the features that are critical to a focus on the individual learner's documented attainment of a criterion of mastery in a knowledge domain. These features are discussed later.

It is worthwhile, moreover, to acknowledge that since the invention of the computer and the diffusion of information technology into our lives, nothing has changed about the ways that people learn (Bransford, Brown, & Cocking, 2000). Learning is a process that includes the actions of study and practice (Swezey & Llaneras, 1997), sometimes for years (Ericsson & Lehmann, 1996), and the assessment of effectiveness as a change in the learner (Skinner, 1953, 1954), a change that might be observed and documented by other people or even by the learner as a self-evaluating authority. In the case of a computer-based tutoring system, a computer program documents that change in the learner's behavior, hence determining whether learning competency has occurred.

Theoretical and applied models of learning, as they relate to academic performance throughout the traditional school years from kindergarten through college, have been proposed and evaluated for centuries. Models of record include the pedagogy of Comenius (1657), Cicero's commentary on the importance of drill and practice (cited in Dale, 1967), cognitive science foundations of instruction (Rabinowitz, 1993), constructivism and the technology of instruction (Duffy & Jonassen, 1992), commentary on science and mathematics education (Bransford, Brown, & Rodney, 1999), and neurophysiological models of learning and memory (Rolls, 2000). To the extent that controversy continues to exist regarding pedagogy applied to formal educational settings, that controversy will not be resolved by the introduction of computer-based tutoring systems. It would be anticipated that the latter systems, however skillfully developed, would nonetheless reflect an incomplete understanding of basic issues in student learning (Tennyson & Elmore, 1997) and instructional design (Tennyson & Schott, 1997).

Nevertheless, providing opportunities to achieve a designated learning outcome at the level of the individual learner and across a wide range of applications has motivated the proliferation of computer-based tutoring systems as evidenced by the availability of commercial products and by the ever-increasing volume of research articles addressing the design, implementation, and effectiveness of these systems. The range of available systems, from tutoring systems to teach the alphabet to children[1] to simulations to support flying an aircraft,[2] makes it a daunting task to categorize computer-based tutoring systems based on common features of learning theory and instructional design.

Recent reviewers of this developing literature, however, identify historical landmarks in this stream of work (Brock, 1997; LeLouche, 1998; Shute & Psotka, 1996; Szabo & Montgomerie, 1992). Such organizing perspectives can be valuable to illuminate the sources and consequences of progress in this field of investigation and applications.

One class of landmarks is attributable to naming the available computing hardware power and software complexity to handle a particular knowledge domain in terms of recording learner input, providing corrective interventions for input errors, managing knowledge about the learner, and implementing alternative instructional strategies. The focal points of such landmarks are often identified as (1) programmed instruction, (2) computer-assisted instruction, (3) and intelligent computer-assisted instruction. The boundaries are arbitrary, however, and there is a continuum of programmatic complexity whereby improvements in the quality and quantity of interactive events are sometimes anthropomorphized as possessing human-like qualities (e.g., intelligence).

A second class of landmarks is attributable to tutoring system designs that craft a set of interactions based on theoretical models of cognitive functioning. The latter approach is exemplified by the Advanced Computer Tutoring Project at Carnegie Mellon University (Anderson, Corbett, Koedinger, & Pelletier, 1995). The interactive events programmed in the *cognitive tutors* are based on a production-rule theory of cognitive skill development, the ACT-R theory (Anderson, 1993), whereby the learner acquires cumulative units of goal-related knowledge. In this stream of work is included the research by Sohn and Doane (1997) that adopted a cognitive theoretical approach to understand how knowledge and processing demands influenced a user's acquisition of UNIX™ commands in a computer-based system. For the most part, however, the evolution of such computer-based systems to manage increasingly complex knowledge domains and instructional strategies reflects improvements in computer processing speed and memory, rather than a change in our understanding of the principles of learning (Szabo & Montgomerie, 1992).

Against that background, this chapter will focus upon an *atheoretical* approach to tutoring system design. This approach is based on the behavioral tradition of crafting a series of learner–environment interactions to support the progressive development of a complex repertoire. It is *atheoretical* in that it will focus on the interactions themselves as the explanation of the antecedents to knowledge and skill acquisition, and it will rely on those antecedents, rather than external explanatory metaphors, such as cognitive models, to explain the process and outcome of learning.

PROGRAMMED INSTRUCTION

Programmed instruction presents information to a learner in a planned sequence of steps that are interspersed with direct questions or other prompts (e.g., blanks) requiring a student's constructed response to demonstrate competency in the material presented. A *frame* consists of a unit of information together with the assessments to document learning. The essential features of programmed instruction are as follows: (1) comprehensibility of each unit; (2) tested effectiveness of a set of frames; (3) skip-proof frames; (4) self-correcting tests; (5) automatic encouragement for learning; (6) diagnosis of misunderstandings; (7) adaptations to errors by hints, prompts, and suggestions; (8) learner-constructed responses based on recall; (9) immediate feedback; (10) successive approximations to a terminal objective; and (11) student-paced progress (Scriven, 1969; Skinner, 1958; Vargas & Vargas, 1991). A program of study consists of many frames designed to promote the achievement of a demonstrable set of competencies, for the individual learner, at the conclusion of study.

A frame-based program of study is intended to promote generalized understanding, rather than rote memorization. Table 34.1 presents an example of a series of frames that leads a learner to understand the concept of a conductor (Deterline, 1962, p. 17). Several frames in this example, to include the final frame, do not lend themselves to automated assessment, and the importance of a personal interaction between a student and a teacher, at some point in a frame-based instructional system, is addressed in the personalized system of instruction discussed later.

There is nothing novel about this instructional technology. In the *Meno*, Socrates taught a slave boy the proof of the Pythagorean theorem by using simple diagrams to lead the learner, by small incremental steps, to significant generalizations (Corey, 1967). The anticipated use of computers in implementing this individualized instructional technology was stated over 30 years ago (Scriven, 1969, p. 15):

The 'far-out' entry in the teaching stakes (from the viewpoint of the 1960s) is the super teaching machine, the teaching computer. The idea is straightforward. The computer begins with certain background data about the student, and feeds him a few appropriate . . . frames of instruction. On the basis of his responses to these frames, plus the background data (previous courses and grades, I.Q., etc.), the computer chooses the most appropriate instructional sequences for the next section of the program—and so on.

This latter description is prescient of adaptive systems (Benyon & Murray, 1993) and intelligent tutoring systems (Shute & Psotka, 1996).

The intellectual force behind the development of programmed instruction is attributable to Harvard psychologist B. F. Skinner. In an early landmark paper entitled *The Science of Learning and the Art of Teaching* (Skinner, 1954), Skinner

---

[1] JumpStart Toddlers by Knowledge Adventure, Inc.
[2] F-22 Air Dominance Fighter by Infogames Entertainment, Inc.

TABLE 34.1. A Series of Frames

1. A conductor will carry electric current. A wire or any substance that will carry or conduct an electric current is called a _____.	conductor
2. A copper wire will conduct or carry an electric current because copper wire is a good _____.	conductor
3. A conductor is a substance that will carry or _____ an electric current. Rubber is not a conductor, so rubber will not _____ an _____.	conduct conduct electric current
4. An insulator will not conduct an electric _____. Rubber is a good _____ because it will _____. (complete)	current insulator not conduct an electric current (or) not conduct current
5. Electric current can flow or travel along a _____, but cannot flow along an _____.	conductor insulator
6. You could receive a "shock" from a copper wire unless the copper wire is surrounded by an _____.	insulator
7. An insulator is a substance or material that will _____. (complete)	not conduct electric current (or) not let current flow (or) stop current
8. A conductor will _____. (complete)	conduct an electric current (or) carry current

Note. From W. A. Deterline, An Introduction to Programed Instruction, by W. A. Deterline, 1962, New York: Allyn & Bacon. Copyright 1962 by Allyn & Bacon. Reprinted with permission.

argued how principles derived from the laboratory experimental analysis of behavior could be directly applied to education. The contribution was the argument that the intended changes in a student's behavior—the goal of education—are a function of lawful behavioral processes that follow directly from the skillful presentation of contingencies of reinforcement that promote such changes and their maintenance over repetition and time at the level of the individual student. The focus was the application of experimentally derived principles that would foster the development of increasingly complex repertoires of behaviors, at the level of the individual learner, that are otherwise taken to be evidence of *understanding*, *thinking*, *problem solving*, *cognitive functioning*, and so on. The challenge, of course, was to design a course of study whose outcome would provide convincing evidence that such objectives had been achieved. But the objective of this teaching-oriented strategy was not to demonstrate that one teaching method was superior to another teaching method, as evidenced by comparisons between group averages. Rather, the objective was to determine ways to design an educational environment so that each learner can attain a specific level of competency in a subject matter (Bijou, 1970).

Skinner referred to the need for *mechanical help* (Skinner, 1954, p. 29) in the management of the many thousands of contingencies of reinforcement that would be required to apply these principles to the individual learner, without regard to the knowledge domain. In that latter regard, the history of published attempts to provide automated and mechanized support for student learning can be traced at least as far back as the work of Pressey (1927), who reported the design of a machine to administer and score tests, and most importantly, to teach. The purpose of the machine was to free the teacher from overseeing and managing a student's moment-by-moment progress of essential drill, thereby freeing the teacher to meet the needs of the individual learner as required. The continuation of this approach to individualized instruction is exemplified by Skinner (1958). In this latter paper, the application of behavior principles to the design of frames of material was elaborated as an extension of the early contribution of Pressey (1927).

The emphasis on a teaching technology that focused on the individual learner generated a stream of research and scholarly books on programmed instruction (e.g., Calvin, 1969; Green, 1967; Holland, 1960; Lange, 1967; Margulies & Eigen, 1962). One of the first presentations of a programmed instruction text was authored by Holland and Skinner (1961). The book itself was a frame-based presentation of introductory psychology material. This text was followed by journal articles that began to focus on the elements of programmed instruction, such as the frequency of overt responses (Kemp & Holland, 1966) and techniques for instructional design of material (Holland, 1967). The early use of a linear progression of frames was extended to branching approaches in which branching paths to a learning objective were determined by a history of prior interactions with the material, not just an immediately prior frame (Crowder, 1962). This latter approach also adopted multiple-choice tests of competency to determine succeeding paths in learning.

Meta-analyses of the effectiveness of programmed instruction in higher education were generally supportive (Kulik, Cohen, & Ebeling, 1980), but research and corresponding journal activity declined during the 1970s. Moreover, some scholarly sources broadened their focus. For example, *Programmed Learning and Educational Technology*, a journal published by the Association for Educational and Training Technology since 1967, changed its title to *Education and Training Technology International* in 1989 and again to *Innovations in Education and Training International* in 1995. There were several factors influencing this trend.

Although the design of programmed instruction material to foster such intellectual skills as "thinking" had been discussed (Peel, 1967), it became increasingly apparent that it was difficult to operationalize many important objectives of education, such as mastering concepts, principles, rules, and cognitive strategies, as well as mastering the intellectual skills of reading, writing, and mathematical problem solving (Case & Bereiter, 1984). The inability of the behavioral approach to provide guidance on formulating the size of incremental steps leading to a terminal performance motivated learning theorists such as Gagne (1968) to propose an instructional technology based on hierarchical task analysis for identifying and sequencing intellectual skill development. Furthermore, the readiness of a learner to acquire new skills was recognized to require a consideration of developmental factors, not just learning technology factors (Case, 1978). These important contributions to an ontology of learning reflect an attempt to organize the stages in learning where specification of the exact learner–environment interactions taking place within those stages has yet to be realized. This is evidenced by the use of labels applied to learners, such as *intelligent, exceptional, average, novice*, and *expert*, to explain the outcome of a process of learning rather than the instructional events that account for the outcome itself. Operationalizing the instructional events, in terms of learner–media interactions, is a requirement for programmed instruction that is often problematic.

These latter developmental orientations are not necessarily incompatible with programmed instruction approaches, and the two differ primarily in terms of the methods used to group and classify the many functional components and interrelationships of a developing complex repertoire. In the history of educational technology, it is often the case that "educated person" is a term applied to a complex repertoire that is a by-product of pedagogical customs in which the specification of all the steps leading to an intended educational outcome is problematic. The importance of this is to be understood in terms of its impact on the expectations of the development of computer-based programmed instruction tutoring systems, which require an algorithm of interaction and knowledge acquisition that may not exist for very advanced areas of intellectual functioning. It is this combinatoric problem related to student–system interactions that motivated many early proponents of automated instructional systems—programmed instruction—to embrace a contextual orientation that included developmental readiness

for learning (Case, 1978) and the achievement of superordinate objectives whereby the learner is taught to apply self-regulation strategies to his or her own knowledge acquisition processes (Schunk, 2000).

PERSONALIZED SYSTEM OF INSTRUCTION

Programmed instruction material, by itself, provides an incomplete set of pedagogical tools. Some reports question its assumptions about the effectiveness of stepwise progress as sufficient motivation to engage a learner's attention (Tennyson & Elmore, 1997), and other studies suggest that students are not enthusiastic about the material's format (Day & Payne, 1987). Much earlier, however, Keller (1968) had proposed a comprehensive learning environment that included programmed instruction as a potential, but not necessary, component. This learning environment came to be known as the personalized system of instruction (PSI), and its design may overcome at least some of the objections about programmed instruction material.

The essential features of PSI are similar to programmed instruction, and they are as follows: (1) self-paced progression, (2) unit perfection across success stages in learning, (3) lectures as vehicles of motivation rather than as sources of critical information, (4) the emphasis on text to transmit information, and (5) the use of proctors in an interpersonal interaction for testing. It is this last feature that best distinguishes PSI from programmed instruction, although none of the components of PSI is necessarily frame-based after the design of a programmed instruction system. The social interaction between student and proctor takes place after study of material by the student. The interaction is intended to broaden the student's repertoire into the ". . . understanding of a principle, a formula, or a concept, or the ability to use an experimental technique" (Keller, 1968, p. 84). It is this assessment of *understanding* as a shared history between a speaker and a listener (Skinner, 1957, p. 280) that has led to a consideration of the importance of synchronous and asynchronous collaborative learning environments in which learners seek an equilibrium of understanding where the latter is based on overlapping agreement by members of a verbal community (e.g., Jehng & Chan, 1998).

Because the combinatoric challenges of programmed instruction approaches may falter in this later regard, an informed dialog between a student and a teacher is a test of mutual understanding and a method of remediation of higher order intellectual skills whose component parts are difficult to operationalize at a level of rigor necessary for automated instructional systems (Emurian, 2001). Ferster and Perrott (1968) adopted the PSI in a comprehensive learning environment, which included frequent assessment of a learner's competency by a proctor. Finally, the effectiveness of the PSI was demonstrated in a series of meta-analyses (Kulik, Kulik, & Bangert-Drowns, 1990; Kulik, Kulik, & Cohen, 1979) and integrative commentary (Buskist, Cush, & DeGrandpre, 1991).

A modified PSI approach is used in conjunction with a Java-programmed instruction tutoring system that is discussed later. The approach is modified in the sense that the interpersonal interactions among students and among instructors and students at the conclusion of the tutoring experience are more collaborative than proctoring. A modified approach to PSI was also reported by Crosbie and Kelly (1993) in a course in which computers were used primarily to administer and score objective tests automatically, and proctors were used in that course to confirm the test outcomes and to discuss incorrect student responses that were challenged by the student.

COMPUTER-BASED SYSTEMS

As early as 1961, researchers began to focus on the use of computers to automate programmed instruction systems (Coulson, 1962a). The CLASS system was a Philco 2000-based programmed instruction system supporting up to 20 learners (Coulson, 1962b). It provided a branching approach to instructional design, and it was intended to be one component in the entire complex of educational support, such as films, television, lectures, and textbooks. The PLATO II system was an ILLIAC-based programmed instruction system initially supporting two concurrent learners (Bitzer, Braunfelf, & Lichtenberger, 1962). Both the CLASS and PLATO II systems were based on the principles of programmed instruction, and both systems were able to support the multimedia information delivery that was available at that time. An example of a frame from the PLATO II system is presented here.

In the PLATO II system, textual information for both learning and testing was presented to the student by a television display, and a student's responses were input by a keyset. Figure 34.1 presents an example "text" slide displaying domain knowledge, and Fig. 34.2 presents an example of an "answer" slide. Correct input occasioned progression in the sequence, and incorrect input provided the opportunity to review the previous text slide and to branch to a series of smaller step remedial text slides. The authors anticipated the *student model* by suggesting

3 –

EACH POSITIVE INTEGER IS REPRESENTED IN DECIMAL NOTATION BY COMBINING THE TEN DIGITS:

0, 1, 2, 3, 4, 5, 6, 7, 8, 9.

THUS THE SYMBOL '3,549' IS INTERPRETED TO MEAN:

$3 \times 10^3 + 5 \times 10^2 + 4 \times 10 + 9$

i.e., $3 \times 1000 + 5 \times 100 + 4 \times 10 + 9$.

FIGURE 34.1. An example "text" slide displaying domain knowledge from the PLATO II system. PLATO II: A multiple-student, computer-controlled, automatic teaching device. From *Programmed Learning and Computer-Based Instruction* (p. 210), by J. E. Coulson (Ed.), 1962, Santa Monica, CA: System Development Corporation. Copyright 1962 by System Development Corporation. Reprinted with permission of John Wiley & Sons, Inc.

QUESTION: GIVE THE POSITIVE, NONTRIVIAL DIVISORS OF 51 IN INCREASING ORDER.

$d_1 =$ [ ]

$d_2 =$ [ ]

FIGURE 34.2. An example "answer" slide from the PLATO II system. PLATO II: A multiple-student, computer-controlled, automatic teaching device. From *Programmed Learning and Computer-Based Instruction* (p. 211), by J. E. Coulson (Ed.), 1962, Santa Monica, CA: System Development Corporation. Copyright 1962 by System Development Corporation. Reprinted with permission of John Wiley & Sons, Inc.

the presentation of information based on a diagnosis of student errors, together with the history of successful prior interactions with the system.

The decline in journal activity related to programmed instruction was offset by the emergence of computer-assisted instruction, as evidenced by the CLASS and PLATO II systems. One of the first reports of a computer-based instructional program to appear in the general scientific literature was published in the journal *Science* in 1969 (Suppes & Morningstar, 1969). The fact that such a program was published in this prestigious scientific journal was evidence of the recognition of the potential impact of these pedagogical approaches to formal education.

Suppes and Morningstar (1969) reported the results of the use of computer-assisted instruction that provided drill and practice in the skills of arithmetic computation for students in grades one through six. The results of a multiyear project over several states generally showed greater improvement, determined by changes in scores on a test administered before and after the 189 days of the study, by students who had used the computer-assisted instruction in comparison with control students who received conventional classroom instruction, which often included drill and practice at a group level. The computer was a PDP-1, and the student interacted with the system by way of a model-33 teletype.

> The teletype printed each individual problem and then positioned itself in readiness to accept the answer in the appropriate place. The student typed in the answer. If his answer was correct, he proceeded to the next problem. If he gave the wrong answer, the teletype printed 'No, try again,' and presented the problem once more. If the student gave the wrong answer for the third time, he was given the correct answer and the teletype automatically proceeded to the next problem (Suppes & Morningstar, 1969, p. 344).

The system exhibited *branching intelligence* and *adaptivity*, in that successive daily drill sessions were adjusted in difficulty depending on the percentage of problems that were solved correctly in a given session. The system also selected a review drill based on the student's performance over several prior successive sessions. Computer-assisted drill and practice followed an introduction of concepts by a teacher in the classroom.

Although the presentation by Suppes and Morningstar (1969) did not cite any behavioral principles or literature, procedurally the computer-based tutoring system exhibited many of the features of programmed instruction to include self-paced progression, immediate feedback for performance, and the *successive approximation* of skill in arithmetic computation by the presentation of increasingly more difficult problems across successive practice sessions. The one notable difference was that a learner was advanced in the program after three attempts at entering the correct answer, even if the third attempt was incorrect. A strict behavioral approach might have required a correct solution prior to the program's advancement to another problem. However, the software, even though sophisticated for 1965, did not have the capability to diagnose the source of learner errors and to put the student on a remedial path. That capability was to appear later in the development of computer-assisted instruction.

Despite these promising background developments in computer-delivered programmed instructional material, the trend in research and applications over the next two decades or more gave witness to the broadening interdisciplinary scope of computer-based instructional systems as evidenced by the reviews cited earlier (Brock, 1997; LeLouche, 1998; Shute & Psotka, 1996; Szabo & Montgomerie, 1992). Studies that focused on programmed instruction, per se, and computer-based technology began to re-emerge in the behavioral literature only in the early 1990s. One of the first such studies was conducted by Tudor and Bostow (1991) who investigated the presentation of a 315-frame instructional program, controlled by an IBM PC Junior® microcomputer, to teach college students how to construct effective programmed instruction. Comparisons on a competency test administered before and after the students' use of the tutoring system showed gains in knowledge for an experimental group whose members were required to recall and type each tested item for a frame of information. Importantly, an integrative posttutor assessment that required the construction of a sample frame, based on all principles taught, showed the superiority of the instructional condition that required overt constructed responses during knowledge acquisition that occurred over the completion of the 315 separate frames in the tutoring system.

These results were confirmed in a subsequent study (Kritch & Bostow, 1998) that showed improved posttutor performance when learners emitted a relatively high density of constructed responses across 176 successive frames, in comparison with a low-density condition. Figure 34.3 presents a summary of the experimental conditions and an example of three frames, three posttutor test items, and three questionnaire items. The first frame shows the use of *prompting* and *matching*, because the correct response is embedded in the frame and because the answer is partially presented. The third frame shows the requirement for recall under conditions of *prompting*. The first sample test item shows the requirement for *response generality*, which occurs as a function of completion of all frames. These features together reflect the method of *successive approximations* to

the objectives of learning (Sulzer-Azaroff & Mayer, 1991, p. 324). Finally, the importance of constructed responses, as they enhance the accuracy of posttutor assessments, was also demonstrated when the frames were prepared on videodisc that presented textual, graphics, and audio information (Kritch, Bostow, & Dedrick, 1995).

Another stream of research in the behavioral literature studied effects of delay intervals between constructed responses and progression in a computer-based tutoring system on response accuracy. Crosbie and Kelly (1994) presented 1,711 programmed instruction frames of the textbook by Holland and Skinner (1961) on an IBM-compatible PC to learners over 15 consecutive daily sessions. It was found that the presentation of a 10-sec postresponse delay interval, during which the frame material, the learner's constructed response, and the correct responses were simultaneously displayed, improved response accuracy overall. The researchers speculated that this effect was attributable to a diminution of *racing*, which was associated with a learner's insufficient study of the presented frames and with careless errors that might be easily correctable when incorrectly answered frames were repeated. These effects were confirmed in a systematic replication (Kelly & Crosbie, 1997) in which test performance of learning was superior for the imposed delay condition even after 1 month following completion of the computer-based programmed instruction tutoring system.

These background studies show, for the most part, that applications of information technology to the implementation of programmed instruction approaches have emphasized the student's learning of textual information in relatively simple knowledge domains. Recent studies have manipulated the parameters of the human–computer interaction, with the intention to improve instructional design.

THE LEARN UNIT

The contingencies of reinforcement underlying the steps in a programmed instruction approach have been conceptualized as a *learn unit* (Greer & McDonough, 1999). This unit consists of (1) discriminative stimulus for a response, the response itself, and a reinforcer consequence presented for response accuracy; and (2) branching or corrective remediation presented for response inaccuracy. A learn unit is specified with sufficient operational rigor to be countable. The size of the required response may change as a learner progresses through a programmed instruction tutor. Several elementary units, then, may combine into a higher order unit in which reinforcement is provided for the production of several prior and smaller units. This approach provides a mechanism for documenting and quantifying the total "learning force" required to progress from simple to complex units of performance. It is this approach that was adopted in the case studies to follow.

CASE STUDIES

Over the past several years, our work has focused on the investigation of factors related to the acquisition and retention of UNIX™ and Java™ by undergraduate and graduate students

High Density (HD) Overt responses to 176 frames	Low Density (LD) Overt responses to every other frame	Zero Density (ZD) Passive reading, key tapping to advance	Control for Time (CT) Passive reading, advance when HD advanced
1. A player piano is told what notes to play from a long scroll of paper with tiny holes punched through it. The paper scroll is like a script of commands that tells the piano what n---s to play.	1. A player piano is told what notes to play from a long scroll of paper with tiny holes punched through it. The paper scroll is like a script of commands that tells the piano what n---s to play.	1. A player piano is told what notes to play from a long scroll of paper with tiny holes punched through it. The paper scroll is like a script of commands that tells the piano what notes to play.	1. A player piano is told what notes to play from a long scroll of paper with tiny holes punched through it. The paper scroll is like a script of commands that tells the piano what notes to play.
2. With a player piano, the music is programmed. The scroll of paper is a script of commands that tells the player ----- what notes to play.	2. With a player piano, the music is programmed. The scroll of paper is a script of commands that tells the player piano what notes to play.	2. With a player piano, the music is programmed. The scroll of paper is a script of commands that tells the player piano what notes to play.	2. With a player piano, the music is programmed. The scroll of paper is a script of commands that tells the player piano what notes to play.
3. Like a player piano, a computer can be programmed. A computer program is like a sc---pt of commands that tells the c------r what to do.	3. Like a player piano, a computer can be programmed. A computer program is like a sc---pt of commands that tells the c------r what to do.	3. Like a player piano, a computer can be programmed. A computer program is like a script of commands that tells the computer what to do.	3. Like a player piano, a computer can be programmed. A computer program is like a script of commands that tells the computer what to do.

↑ ————————————————— Yoked ————————————————— ↑

Sample Test Items
1. The statements that cause a computer program to take actions are called -----.
2. The command that erases any previous material from the screen is the ----- command.
3. The command that tells the program to start a new frame is the ----- ----- command.

Questionnaire Items (1=very much dislike; 2=dislike; 3=neutral; 4=like; 5=very much like)
How would you describe your attitude about the instructional program that you experienced today?
How would you describe your attitude about computer assisted instructional programs in general?
How would you describe your attitude about computer assisted instructional programs that specifically teach program commands like those taught in the instructional program you just experienced?

FIGURE 34.3. A summary of the experimental conditions and an example of three frames, three posttutor test items, and three questionnaire items. The first row gives the experimental density conditions, which reflect how much interaction was required by the learner to progress from one frame to the next. From "Degree of Constructed-Response Interaction in Computer-Based Programmed Instruction," by K. M. Kritch and D. E. Barstow, 1998, *Journal of Applied Behavior Analysis, 31,* 387–398. Copyright 1998 by the Society for the Experimental Analysis of Behavior. Reprinted with permission.

who are information systems majors. The design of computer-based tutoring systems to accomplish the objectives follows the principles of programmed instruction. What follows is a presentation of a series of studies showing the transition of this instructional technology from the research laboratory to the classroom.

Unix Tutor

The first study compared the effectiveness of a learner's acquisition and retention of sequences of UNIX commands under conditions of recognition, recall, and a combination of recognition and recall. The details of the procedure and the theoretical

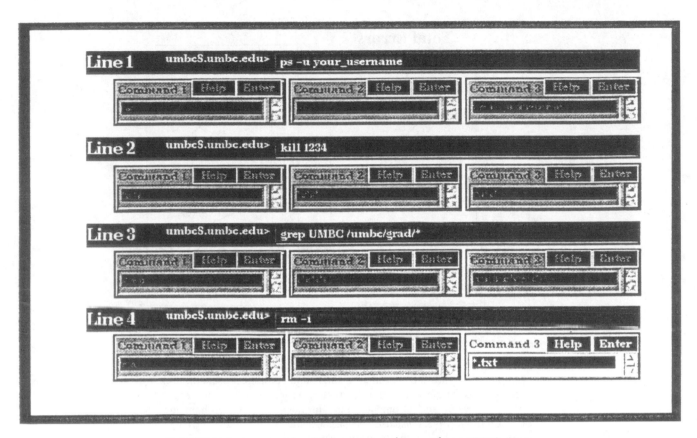

FIGURE 34.4. An example of the menu interface. umbc = UMBC. From "Learning and Retention With a Menu and a Command Line Interface," by A. G. Durham and H. H. Emurian, 1998, *Computers in Human Behavior*, 14, 597–620. Copyright 1998 by Elsevier Science. Reprinted with permission.

rationale for investigating these conditions are presented in Durham and Emurian (1998a). Within the context of a computer-based tutoring system, 36 student participants initially learned a series of up to 12 UNIX commands that were required to satisfy a common file, directory, or process manipulation objective. Initial learning occurred with a menu interface (recognition), a command line interface (recall), and a hybrid menu-command line interface (recognition-plus-recall). After a 4-week interval, participants repeated the task.

Figure 34.4 presents an example of the menu interface. Four three-item lines of UNIX commands and arguments were selected to achieve an objective involving common file, directory, and process manipulations. Within each Command Box was a scroll window with all the possible 51 commands and arguments used in the study. Each Command Box contained the identical list of possible commands and arguments, and each box was a separate *knowledge unit*, designed to teach the input for that particular field.

Each line of acceptable input contained a maximum of three fields, but no fewer than two fields: three columns of field values within a row of items that could be entered as one continuous line, including spaces, at the UNIX system prompt. The participant was presented with a description of the objectives of each series of commands, e.g., remove the read/write protection

from a-file.txt; print a-file.txt; delete a-file.txt; return to home directory. The correct value for each field in a row was obtained by clicking the "Help..." button within the Command Box. When help was selected, a pop-up panel appeared describing and displaying the UNIX command or argument to be selected.

The second interface condition required the acquisition to occur under conditions of item recall by substituting a command line interface for the selection list interface. An error was recorded whenever the subject entered an incorrect UNIX command or left the field blank. Spelling errors or strings not within the list of acceptable commands in the tutor cleared the input field, but they did not count as an error. The features of the command line interface best reflect the programmed instruction approach.

The third interface condition replicated the previous two procedures by synthesizing the menu and command line interfaces. The task consisted of successive iterations through the selection-based menu interface and the keyin command line interface, presented sequentially.

For the three interface conditions, trials for each successive line of input were repeated until the subject entered all three fields correctly without committing an error or requesting help. A criterion of task mastery was reached when three successive error-free and help-free passes for each of the three objectives

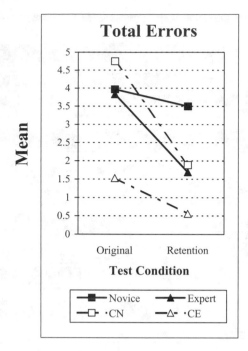

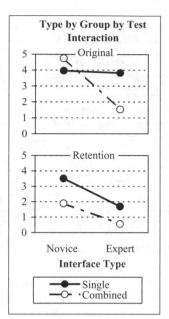

FIGURE 34.5. Mean total errors for all participants for the menu and command line interfaces within the single and combined presentation groups across original and retention testing. From "Learning and Retention With a Menu and a Command Line Interface," by A. G. Durham and H. H. Emurian, 1998, *Computers in Human Behavior, 14,* 597–620. Copyright 1998 by Elsevier Science. Reprinted with permission.

were achieved. After a 4-week interval, the sequence of events was replicated, with exactly the same tasks and mastery criteria.

The terminal performance was the learner's correct production of the 12 UNIX commands, as a serial stream, with no error and no help selection. Each command was first mastered as a single item in context, and each command box was a single learn unit. The learner repeated the interaction with a single command box until the command had been entered accurately. The next level of learn unit was the line consisting of the three successive items in that row. The learner repeated the line until it was entered with no error and no help selection on any Command Box in the row. The highest level of learn unit was the production of all 12 commands with no error and no help selection for three successive iterations through the interface. Thus, there was a maximum of 17 learn units in this programmed instruction tutoring system.

Figure 34.5 presents mean total errors for all participants for the menu and command line interfaces within the single and combined presentation groups across original learning and retention testing. Errors declined between original learning and retention testing, and a three-way interaction was observed. Without regard to the presence of a *transfer-appropriate processing effect* (Roediger & Guynn, 1996), which predicts facilitation from recognition to recall, the importance of acquisition under conditions of recall was revealed during retention testing.

Recall learning was robust, and this outcome, which was observed in the research laboratory, influenced the design of our computer-based tutoring system for Java.

The importance of this background study is to be understood in terms of providing baseline knowledge for the instructional designer. As a programmed instruction tutoring system, the outcome was achieved by symbol constructions whose accurate productions led to progress through the tutor itself. Thus, the contingencies were artificial, and the reinforcers were not programmed for the learners to produce consequences that would otherwise be anticipated to occur when one uses these commands in a real-life setting to achieve important objectives (Ferster, 1967). Thus, the meaning of the symbols to the learner was related to a synthetic context, constraining the generality of the outcomes (Sidman, 1960). Although the artificial symbolic manipulations supported the value of active recall, it was essential to broaden the scope of these research findings to a realistic task environment. That was the motivation for the next development in this work.

Java Tutoring System

The next series of studies extended this background approach to the design and implementation of a programmed instruction tutoring system for a Java Applet.[3] This system was intended to

---

[3]The tutoring system is freely accessible on the World Wide Web: http://webct.umbc.edu/public/JavaTutor/index.html

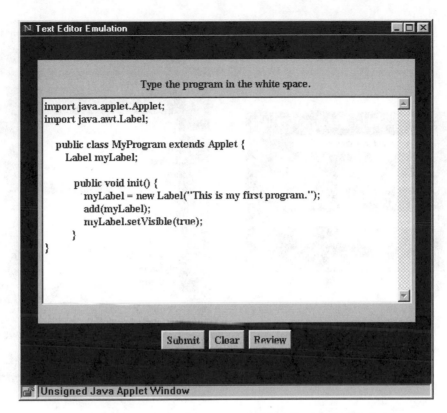

FIGURE 34.6. The code for the Java Applet that displays a Label object in a browser window. The code is displayed in an interface explained below. From *Managing Information Technology in a Global Environment* (pp. 155–160), by H. H. Emurian and A. G. Durham, 2001, Hershey, PA: Idea Group Publishing. Copyright 2001 by Idea Group Publishing. Reprinted with permission.

be used in the classroom as a component of a laboratory course in graphical user interface design. The rationale for using this tutoring system was to ensure a common and documented history in symbol manipulation in a group of students. This background set of experiences provided a competency reference point from which to determine the readiness of students to benefit from the instructional approaches that came later in the course. Details regarding this tutoring system have been published elsewhere (Emurian, Hu, Wang, & Durham, 2000).

Figure 34.6 presents the code for the Java Applet that displays a Label object in a browser window. The code is displayed in one of the learning interfaces described here. The production of the displayed stream of symbols is the terminal performance, the objective of learning. The program consists of 24 atomic items of code concatenated in a stream of 36 total items. The tutoring system design was intended to foster the learning of the individual items, the meaning of each item, the relationship of one item to another, the relationship of one item to the entire program, and the errorless production of the final stream of items.

The several stages in the tutoring system were based on a functional classification of verbal behavior that is assumed to underlie the acquisition of the form and meaning of textual information in context (Skinner, 1957), and the acquired

functional interrelationships among the information items are fostered after the techniques in verbal memory studies (Li & Lewandowsky, 1995). The progression from general context through details and synthesis follows the elaboration theory of instruction (Reigeluth & Darwexeh, 1982). Repetitions of components of the interfaces to be described are grounded in the power function of learning (Lane, 1987). Classroom applications are based on the personalized system of instruction (Keller, 1968). Although initial explorations of this tutoring system were undertaken with undergraduate and graduate students in information systems courses, the learner was assumed to have no prior knowledge of Java and no differentiated motivation for undertaking mastery of the program (Coffin & MacIntyre, 1999; Ryan, 1999).

Figure 34.7 presents the first two interfaces. The left view shows the symbol familiarity interface that requires the learner to copy the displayed symbol into the keyin field. The purpose of this interface is to generate a minimal transcription repertoire (Skinner, 1957) with the symbol set prior to the acquisition of a symbol's meaning. The successive symbols are the atomic units in the program, and this sequence affords prior nondifferential exposure to the code, which itself generates associative learning by contiguous temporal and sequential pairing. The

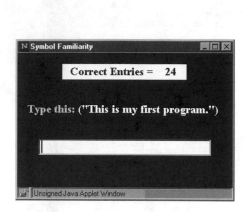

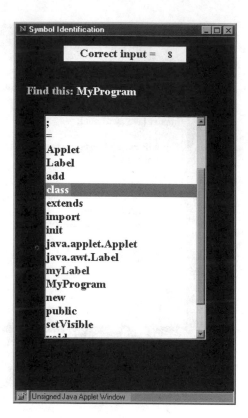

FIGURE 34.7. The first two interfaces. The left view shows the symbol familiarity interface that requires the learner to copy the displayed symbol into the keyin field. The right view shows the symbol identification interface that requires the learner to select the displayed symbol from a list of all symbols. From "Learning Java: A Programmed Instruction Approach using Applets," by H. H. Emurian, X. Hu, J. Wang, and A. G. Durham, 2000, *Computers in Human Behavior*, 16, 395–422. Copyright 2000 by Elsevier Science. Reprinted with permission.

right view shows the symbol identification interface that requires the learner to select the displayed symbol from among the 24 items presented in a list. The purpose of this interface is to sharpen a learner's discrimination of the formal properties among the symbols (Catania, 1998a). The sequence of selections followed the atomic units in the program, and this experience also contributes to associative development of the stream of symbols in the program.

Figure 34.8 presents the item interface (top view) and the serial stream interface (bottom view). The item interface teaches up to three items of code, and each of the three keyin fields is a separate knowledge unit. The purpose of this interface is to require accurate contextual response constructions by recall and to assess the understanding of an item's meaning by multiple-choice assessment. Each knowledge unit provides the occasion to observe the correct Java item prior to typing the item into the keyin field by recall. The explanation or meaning of the item is observed, and a multiple-choice test on the item's meaning must be passed to progress from one knowledge unit to the next. All entries for each unit must be completed correctly to progress to a succeeding unit. As a learner enters a unit

correctly in the keyin field, the unit is displayed in formatted and cumulative sequence in the white text area. Upon completion of the three individual items, the serial stream interface requires entering those items as a single unit. This increases the size of the response required in the serial stream learn unit. If the learner is not able to enter the stream of three items correctly, the tutor branches back to the item interface, and that cycle continues until the serial stream is entered correctly. Then the next item interface is presented.

In this version of the tutor, there were 14 item interfaces, and the 36 total items learned were distributed across 10 lines of code. The item and serial stream interfaces reflected 86 learn units (36 items, 36 multiple-choice tests, and 14 serial streams).

The next interface advanced the response requirement to a row of items, and there were 10 rows. The progression in the complexity of the response and in the format for its entry reflects another application of *successive approximations* (Sulzer-Azaroff & Mayer, 1991, p. 394) to the final performance. Figure 34.9 presents a representation of the first row across the six iterations (i.e., passes) that were required through this interface. On each successive iteration, the graphic label prompts

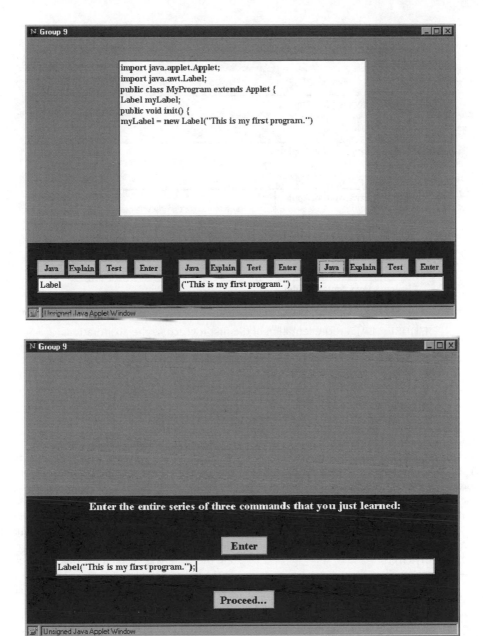

FIGURE 34.8. The item interface (top view) and the serial stream interface (bottom view). From "Learning Java: A Programmed Instruction Approach using Applets," by H. H. Emurian, X. Hu, J. Wang, and A. G. Durham, 2000, *Computers in Human Behavior*, 16, 395–422. Copyright 2000 by Elsevier Science. Reprinted with permission.

and background colors are gradually withdrawn as the stimulus control is transferred to the open area available in a text editor window. This shift in stimulus control reflects the process of *fading* (Sulzer-Azaroff & Mayer, 1991, p. 311). During the first iteration, the learner is required to pass a multiple-choice test on the objectives of each row immediately after entering the row correctly. If the learner is not able to enter the row, the tutor recycles through the item and serial stream interfaces that

contain the code for that row. In this version of the tutor, there were 70 learn units in this interface, 20 on the first iteration (10 rows and 10 multiple-choice tests) and 10 on each of the following five iterations.

Figure 34.10 presents total number of learn units encountered by a test learner across the six successive iterations of this interface. The figure shows the minimum number of learn units available on each iteration, together with the number of units

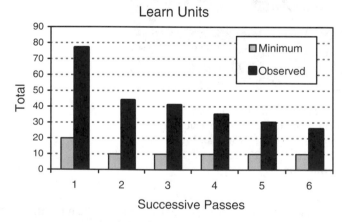

Row 1	import java.applet.Applet;	Test!
Row 1	import java.applet.Applet;	Hint?
Row 1	import java.Applet.Applet;	Review?
Row 1	import java.Applet.Applet;	Review?
Row 1	import java.applet.Applet;	
	import java.applet.Applet;	

FIGURE 34.9. A representation of the first row across the six iterations (i.e., passes) that were required through this interface. From "Learning Java: A Programmed Instruction Approach using Applets," by H. H. Emurian, X. Hu, J. Wang, and A. G. Durham, 2000, *Computers in Human Behavior*, 16, 395–422. Copyright 2000 by Elsevier Science. Reprinted with permission.

encountered by the learner to complete each iteration. The total observed units reflect the additional learn units that were encountered whenever the learner selected a review of the item interfaces that supported the code in a given row. The figure shows a marked reduction in observed learn units between iterations 1 and 2, and this was followed by a gradual reduction

Learn Units

FIGURE 34.10. Total number of learn units encountered by a test learner across the six successive iterations of the row interface. From "Learning Java: A Programmed Instruction Approach using Applets," by H. H. Emurian, X. Hu, J. Wang, and A. G. Durham, 2000, *Computers in Human Behavior*, 16, 395–422. Copyright 2000 by Elsevier Science. Reprinted with permission.

in observed learn units across iterations 2 through 6. Even after six iterations, however, this learner required learn unit support in excess of the minimum presented during the sixth iteration.

Figure 34.11 presents two versions of accurate code displayed in the final interface, which emulated a text editor window. This interface advances the requirement for a correct response to be the stream of 36 Java items. These features reflect *successive approximation* and *fading*. The format for writing the code was relaxed for this interface, and the code was evaluated as a stream of characters. If the learner is not able to enter the code correctly, a review is available that recycles the learner back to the sixth iteration of the row interface. From there, the learner could cycle back to the item interfaces as needed. Once the code is entered correctly, the learner gains access to additional information that is required to compile the code and to run the Applet on the World Wide Web.

Classroom Applications

The programmed instruction tutoring system is used in the classroom as one component in the personalized system of instruction. During the first class period of a 14-period course in graphical user interfaces, based on the Java Abstract Windowing Toolkit, students are given a 2.5-hr interval in which to work on the tutor. Before engaging the tutor, each student completes a series of likert-type questionnaires (Critchfield, Tucker, & Vuchinich, 1998) that reflect the student's current confidence in being able to use each of the 24 atomic units to construct a Java program. The scale anchors are *1 = No confidence, a novice*

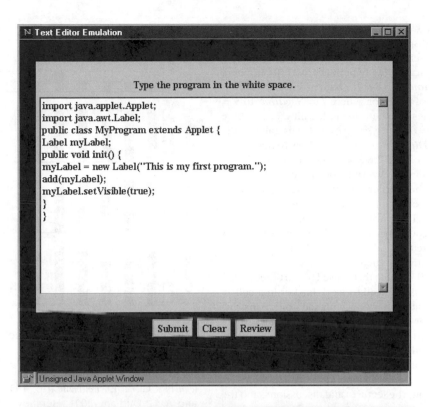

FIGURE 34.11. Two versions of accurate code displayed in the final interface, which emulated a text editor window. From "Learning Java: A Programmed Instruction Approach using Applets," by H. H. Emurian, X. Hu, J. Wang, and A. G. Durham, 2000, *Computers in Human Behavior*, 16, 395–422. Copyright 2000 by Elsevier Science. Reprinted with permission.

to 5 = *High confidence, an expert.* Each student also attempts to write the code to display a text string as a Label object in the browser window. Self-report rating data are also collected on the following four scales: (1) experience with Java, where *1 = No experience, a novice* to *5 = High experience, an expert*; (2) overall impression of the tutor, where *1 = Negative*, to *5 = Positive*; (3) effectiveness of the tutor in learning Java, where *1 = Not effective* to *5 = Highly effective*; and (4) usability of the interfaces, where *1 = Difficult to use* to *5 = Easy to use.* These data are collected online with the assessment tools available in the WebCT™ instructional delivery platform. After the 2.5-hr interval or whenever the learner completes the tutor, the assessments for confidence and for writing the code are repeated.

During the second class period, the authors use a lecture format to review the code presented in the tutor. During this time, the students write the code in a text editor as it is being discussed. The compilation of the Java source code is then discussed, along with the HTML file that is used to start the Applet. The students are then encouraged to work further in a collaborative context with each other and with the course instructors and assistants. After all students have successfully run the Applet in the browser on the World Wide Web, the assessments for confidence and writing the code are repeated. These latter assessments are also administered again during the very last class of the semester. The combination of the programmed instruction tutoring system, the lectures and discussions, and the collaborative learning environment completes the personalized system of instruction for this introductory exercise.

Figure 34.12 presents self-report data on the four scales by a class of 12 graduate students (Emurian & Durham, 2001). Data are partitioned into two groups. Six students completed all interfaces in the tutor, and six students were working on the last interface when the 2.5-hr period expired. These self-report data show that inexperienced learners generally were positive in their work with the programmed instruction tutor interfaces. Although the measurements for the six students who did not complete the tutor are graphically less than the other students, significant differences were not supported between the two groups in their self-report ratings.

Figure 34.13 presents median self-report ratings of confidence for all 12 learners across the four assessment occasions. These data show that the group of learners showed a pronounced increase in confidence immediately after using the tutor. Confidence ratings thereafter were observed to increase over the remaining assessment occasions. The importance of these data are to be understood as a descriptive autoclitic verbal performance (Catania, 1998b, p. 407; Skinner, 1957, p. 313), which is based on a learner's ability to anticipate future behavior to use a symbol effectively. The performance indicates that a learner is able to describe his or her own competency. The impact of a supportive affective context in automated instructional systems has been increasingly recognized by educational scholars (e.g., Tennyson, 1999), and these descriptive autoclitic responses show the positive impact on learners of the programmed instruction tutoring system and the remaining course delivery as a personalized system of instruction.

Figure 34.14 presents cumulative total correct Java programs written by these 12 learners on the last three assessment

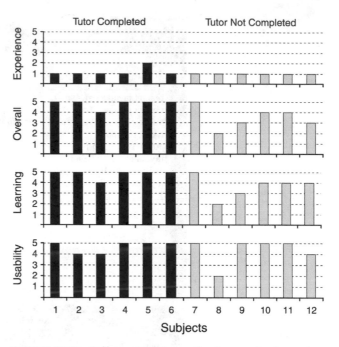

FIGURE 34.12. Self-report data on the four scales by a class of 12 graduate students. From *Managing Information Technology in a Global Environment* (pp. 155–160), by H. H. Emurian and A. G. Durham, 2001, Hershey, PA: Idea Group Publishing. Copyright 2001 by Idea Group Publishing. Reprinted with permission.

occasions. The instructions were to write the Java code that was taught by the tutor to display a text string, as a Label object, in a browser window. The code was written into a window on the WebCT assessment, and the accuracy of the code was determined by an error-free compilation. These data show that immediately after using the tutor, only two learners were able to write the code correctly in the assessment window. This outcome was observed even though six learners had entered the code correctly in the final interface of the tutor. After the students had run the Applet on the web, however, eight learners wrote the code correctly in the assessment window. But, during the final class period, only 1 of the 12 students was able to write the code correctly. This outcome presents a challenge for interpretation, especially in light of the reported beneficial effects of the tutoring system on learner confidence. An explanation for this latter finding requires consideration of several factors.

First, the overall course project objectives included only a single Applet subclass (. . . **My Program extends Applet** . . .) along with many other custom and built-in Java classes that controlled a user's interactions with a web-based information system. After the first two sessions of instruction, the properties of the Applet class were not discussed in detail further. The strength of the originally learned response, then, may have declined over time, and such changes have been attributable to *passive decay with disuse* (Cowan, Saults, & Nugent, 2001) and to inadequate *overlearning* (Postman, 1962) in the verbal memory literature. Moreover, competing sources of influence

Median Confidence Ratings

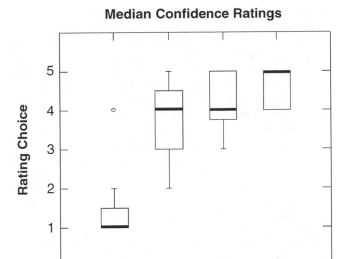

FIGURE 34.13. Box plots showing median confidence ratings for all 12 learners for the 24 distinct items that were used to compose the program. PRE = Pre-Tutor, POST = Post-Tutor, APPLET = Post-Applet, and FINAL = Final Class. From *Managing Information Technology in a Global Environment* (pp. 155–160), by H. H. Emurian and A. G. Durham, 2001, Hershey, PA: Idea Group Publishing. Copyright 2001 by Idea Group Publishing. Reprinted with permission.

from other subsequently learned Java items and class files suggest the involvement of symbol *interference* (Rehder, 2001; Underwood, 1957). The disuse interpretation is supported by the retention results with the UNIX tutor in which errors were observed on the second occasion of using the tutor, despite the more efficient learning that was observed. In the present situation, then, a more sensitive index of retention might have included performance data during reacquisition of the program taught by the Java-programmed instruction tutoring system.

Second, examination of the students' Java code that was submitted during the final assessment showed many instances of the students' attempts to use advanced, perhaps interfering, techniques to accomplish the requested outcome. Students attempted to write more sophisticated Java code at the end of the course than immediately after using the tutor, simply because they knew more object-oriented techniques by the end of the course. Still, they often produced programs that would not compile. Given the semester's experience in interpreting error messages, it is conceivable that a student might have discovered the source of an error and corrected it, if that opportunity had been a component of the final assessment.

These observations show that the effectiveness of repetition of a knowledge domain, presented within the context of a programmed instruction series of successive approximations to a terminal performance, requires documentation under more than a single occasion of rehearsal. Previous research, not directly related to programmed instruction, also emphasizes

Correct Programs Written

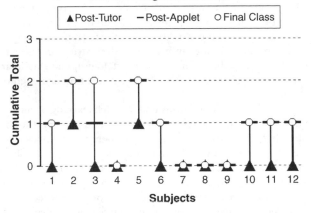

FIGURE 34.14. Cumulative total correct Java programs written by the 12 learners on the last three assessment occasions. From *Managing Information Technology in a Global Environment* (pp. 155–160), by H. H. Emurian and A. G. Durham, 2001, Hershey, PA: Idea Group Publishing. Copyright 2001 by Idea Group Publishing. Reprinted with permission.

the importance of assessing retention, as an index of learning strength, of a newly acquired skill after a delay interval (Davis & Bostrom, 1993; Healy et al., 1995; Shute & Gawlick, 1995; Simon, Grover, Teng, & Whitcomb, 1996). Finally, studies using computer-based programmed instruction show retention effectiveness after a 5-week delay (Kritch et al., 1995) and after a 1-month delay (Kelly & Crosbie, 1997) following completion of a tutoring system.

Despite these observations, the personalized system of instruction (Keller, 1968), with the Java-programmed instruction tutor as a component, has been adopted by the authors to good advantage in the classroom because it generates a history of symbol use and confidence in each individual student. It combines both teaching and testing within a single conceptual framework: programmed instruction. It allows the needs of the individual student to be met because it frees the teacher from relying exclusively on traditional approaches, such as lecturing and writing on the board, to deliver technical information to a group of students. Most important, perhaps, the present tutoring system combines knowledge delivery with learning, assessment, and documentation of competence.

CONCLUSIONS

Learning occurs when there is a documented change in behavior that results from interactions with one's environment. To be informed, an instructional history is required. When a knowledge domain lends itself to enumeration of its components at atomic levels, the instructional history may be formulated by its representation within the context of programmed instruction tutoring systems, which emerged from the scholarly discipline of the experimental analysis of behavior. Such approaches to the design of an instructional system have the advantage of overseeing and managing the moment-by-moment process of learning at the level of the individual student. Although the initial promise of the multimedia CLASS and PLATO systems has yet to be realized, perhaps, the application of computer-based information systems in this area offers the full range of computer hardware and software innovations to the instructional designer. It is understood that information technology applied in this area is but one tool to assist the development of human cognition (Mayer, 2000).

Identifying the conditions under which learning occurs continues to be the source of discourse and inquiry (Koubek, Benysh, & Tang, 1997), as does a consideration of how learning mechanisms should be implemented in training technologies (Swezey & Llaneras, 1997). How the parameters of these interactions should be structured to achieve specific educational objectives, in relationship to the status of the learner and the knowledge domain, has been the subject of inquiry for centuries, and the power function of learning is an effective model when applied to account for the acquisition and retention of a broad range of learning outcomes, to include intellectual skills (e.g., Anderson, 1987; Carlson & Yaure, 1990; Lane, 1987). The process of learning, then, is independent of a supportive instructional technology, which may be offered by computer-based systems or achieved by a learner who is skilled in self-regulation.

In that latter regard, Young (1996) proposed an instance of a student exhibiting a self-regulating style of learning that followed a self-directing and self-monitoring rehearsal strategy: "When I study for a test, I practice saying the important facts over and over to myself" (p. 18). It is, perhaps, a truism that effective learners already know how to regulate their learning environments and motivational status to achieve a criterion of mastery that is self-governed (Schunk, 2000; Skinner, 1968; Zimmerman, 1994). Effective self-management is itself a skill requiring a history whose components can be identified and made public in a scientific account of learning (Skinner, 1953). Programmed instruction approaches may be best suited for students who have not mastered the art of studying, and one important benefit of completing a programmed instruction tutoring system is that it teaches a learner how to acquire knowledge independent of the domain. The ultimate objective, then, of a programmed instruction tutoring system is to free the learner for advanced study undertaken with traditional forms of knowledge codification, such as a book (Brock, 1997).

Analyses of computer programming have been approached in the literature as a complex problem-solving activity (e.g., Campbell, Brown, & DiBello, 1992) and the acquisition of programming skills (e.g., Soloway, 1985; Van Merrienboer & Paas, 1990) For the most part, however, studies in the behavior of computer programming and program comprehension, ranging from early evaluations of conditional constructions (Sime, Green, & Guest, 1973) to recent simulations of memory representations by expert programmers (Altmann, 2001), provide observations and explanations of pre-existing human–computer interactions. Although identifying and classifying the strength of prior learning may cast light on the design effectiveness of alternative programming languages (Pane, Ratanamahatana, & Myers, 2001), educators seek to understand the user's history that leads to group membership as a *novice* or *expert* (Durham & Emurian, 1998b). Understanding that history suggests an instructional technology, and if the component steps can be identified, that summative history lends itself to implementation with computer-based programmed instruction.

A programmed instruction approach to learning a Java Applet was presented as an exemplar of this latter instructional technology that was implemented as a web-based tutoring system. The Applet program was operationalized as a serial stream, and a succession of learn units was programmed to concatenate a series of atomic units into a unitary serial stream of textual items. In the most elementary form, the learn unit required an item of Java code as the constructed response. In its terminal form, the learn unit required the serial stream of Java code as the constructed response. The utility of this approach for novice students to acquire competency in symbol manipulation was documented in the classroom where the programmed instruction tutoring system was applied within the context of a modified personalized system of instruction. The resulting repertoire, together with the positive affective responses by the learners, set the occasion for the continued mastery of advanced details and concepts of the Java programming language and the general properties of software design with the object-oriented model.

References

Alessi, S. M., & Trollip, S. R. (1991). *Computer-based instruction: Methods and development* (pp. 91–118). Englewood Cliffs, NJ: Prentice-Hall, Inc.

Altmann, E. M. (2001). Near-term memory in programming: A simulation-based analysis. *International Journal of Human-Computer Studies, 54,* 189–210.

Anderson, J. R. (1987). Skill acquisition: Compilation of weak-method problem solutions. *Psychological Review, 94,* 192–210.

Anderson, J. R. (1993). *Rules of the mind.* Hillsdale, NJ: Lawrence Erlbaum Associates, Inc.

Anderson, J. R., Corbett, A. T., Koedinger, K. R., & Pelletier, R. (1995). Cognitive tutors: Lessons learned. *Journal of Learning Science, 4,* 167–207.

Benyon, D., & Murray, D. (1993). Adaptive systems: From intelligent tutoring to autonomous agents. *Knowledge-Based Systems, 6,* 197–219.

Bijou, S. W. (1970). What psychology has to offer education—Now. *Journal of Applied Behavior Analysis, 3,* 65–71.

Bitzer, D. L., Braunfield, W. W., & Lichtenberger, W. W. (1962). PLATO II: A multiple-student, computer-controlled, automatic teaching device. In J. E. Coulson (Ed.), *Programmed learning and computer-based instruction* (pp. 205–216). New York: Wiley & Sons.

Bloom, B. S. (1984). The 2 sigma problem: The search for methods of group instruction as effective as one-to-one tutoring. *Educational Researcher, 13,* 4–16.

Bostow, D. E., Kritch, K. M., & Tompkins, B. F. (1995).

Computers and pedagogy: Replacing telling with interactive computer-programmed instruction. *Behavior Research Methods, Instruments, & Computers, 27*, 297–300.

Bransford, J. D., Brown, A. L., & Cocking, R. R. (2000). *How people learn: Brain, mind, experience, and school* (Expanded Edition). Washington, DC: National Academy Press.

Bransford, J. D., Brown, A. L., & Rodney, R. C. (1999). *How people learn: Brain, mind, experience, and school*. Washington, DC: National Academy Press.

Brock, J. F. (1997). Computer-based instruction. In G. Salvendy (Ed.), *Handbook of human factors and ergonomics* (pp. 578–593). New York: Wiley.

Buck, G. H., & Hunka, S. M. (1992). Development of the IBM 1500 Computer-Assisted Instruction System. *IEEE Annals of the History of Computing, 17*, 19–31.

Buskist, W. B., Cush, D., & DeGrandpre, R. J. (1991). The life and times of PSI. *Journal of Behavioral Education, 1*, 215–234.

Calvin, A. D. (1969). *Programmed instruction: Bold new adventure*. Bloomington, IN: Indiana University Press.

Campbell, K. C., Brown, N. R., & DiBello, L. A. (1992). The programmer's burden: Developing expertise in programming. In R. R. Hoffman (Ed.), *The psychology of expertise* (pp. 269–294). New York: Springer-Verlag.

Carlson, R. A., & Yaure, R. G. (1990). Practice schedules and the use of component skills in problem solving. *Journal of Experimental Psychology: Learning, Memory, and Cognition, 16*, 484–496.

Case, R. (1978). A developmentally based theory and technology of instruction. *Review of Educational Research, 48*, 439–463.

Case, R., & Bereiter, C. (1984). From behaviourism to cognitive behaviorism to cognitive development: Steps in the evolution of instructional design. *Instructional Science, 13*, 141–158.

Catania, A. C. (1998a). *Learning*. Upper Saddle River, NJ: Prentice-Hall, Inc.

Catania, A. C. (1998b). The taxonomy of verbal behavior. In K. A. Lattal & M. Perone (Eds.), *Handbook of research methods in human operant behavior* (pp. 405–433). New York: Plenum Press.

Coffin, R. J., & MacIntyre, P. D. (1999). Motivational influence on computer-related affective states. *Computers in Human Behavior, 15*, 549–569.

Cognition and Technology Group at Vanderbilt. (1996). Looking at technology in context: A framework for understanding technology and education research. In D. C. Berliner & R. C. Calfee (Eds.), *Handbook of educational psychology* (pp. 807–840). New York: Macmillan Publishing.

Comenius (1657). *The great didactic*. "education, history of" Encyclopædia Britannica [On-line]. Available: http://search.eb.com/bol/topic?eu=108334&sctn=5&pm=1 [Accessed April 5, 2001].

Corey, S. M. (1967). The nature of instruction. In P. C. Lange (Ed.), *Programmed instruction: The sixty-sixth yearbook of the National Society of the Study of Education* (pp. 5–27). Chicago, IL: The University of Chicago Press.

Coulson, J. E. (1962a). *Programmed learning and computer-based instruction*. New York: Wiley & Sons.

Coulson, J. E. (1962b). A computer-based laboratory for research and development in education. In J. E. Coulson (Ed.), *Programmed learning and computer-based instruction* (pp. 191–204). New York: Wiley & Sons.

Cowan, N., Saults, S., & Nugent L. (2001). The ravages of absolute and relative amounts of time on memory. In H. L. Roediger III, J. S. Nairne, I. Neath, & A. M. Surprenant (Eds.), *The nature of remembering* (pp. 315–330). Washington, DC: American Psychological Association.

Critchfield, T. S., Tucker, J. A., & Vuchinich, R. E. (1998). Self-report methods. In K. A. Lattal & M. Perone (Eds.), *Handbook of research*

methods in human operant behavior (pp. 435–470). New York: Plenum Press.

Crosbie, J., & Kelly, G. (1993). A computer-based personalized system of instruction course in applied behavior analysis. *Behavior Research Methods, Instruments, & Computers, 25*, 366–370.

Crosbie, J., & Kelly, G. (1994). Effects of imposed postfeedback delays in programmed instruction. *Journal of Applied Behavior Analysis, 27*, 483–491.

Crowder, N. A. (1962). Intrinsic and extrinsic programming. In J. E. Coulson (Ed.), *Programmed learning and computer-based instruction* (pp. 58–66). New York: Wiley & Sons.

Dale, E. (1967). Historical setting of programmed instruction. In P. C. Lange (Ed.), *Programmed instruction: The Sixty-sixth yearbook of the National Society of the Study of Education* (pp. 28–54). Chicago, IL: The University of Chicago Press.

Davis, S. I., & Bostrom, R. P. (1993). Training end users: An experimental investigation of the roles of the computer interface and training methods. *Management Information Systems Quarterly, 17*, 61–85.

Day, R., & Payne, L. (1987, January). Computer managed instruction: An alternative teaching strategy. *Journal of Nursing Education, 26*, 30–36.

Deterline, W. A. (1962). *An introduction to programmed instruction*. Englewood Cliffs, NJ: Prentice-Hall, Inc.

Duffy, T. M., & Jonassen, D. H. (1992). *Constructivism and the technology of instruction: A conversation*. Hillsdale, NJ: Lawrence Erlbaum Associates.

Durham, A. G., & Emurian, H. H. (1998a). Learning and retention with a menu and a command line interface. *Computers in Human Behavior, 14*, 597–620.

Durham, A. G., & Emurian, H. H. (1998b). Factors affecting skill acquisition and retention with a computer-based tutorial system. *Proceedings of the 4th World Congress on Expert Systems* (pp. 738–744). New York: Cognizant Communication Corporation.

Emurian, H. H. (2001, April–June). The consequences of e-Learning. *Information Resources Management Journal, 14*, 3–5.

Emurian, H. H., & Durham, A. G. (2001). A personalized system of instruction for teaching Java. In M. Khosrowpour (Ed.), *Managing information technology in a global environment* (pp. 155–160). Hershey, PA: Idea Group Publishing.

Emurian, H. H., Hu, X., Wang, J., & Durham, A. G. (2000). Learning Java: A programmed instruction approach using Applets. *Computers in Human Behavior, 16*, 395–422.

Ericsson, K. A., & Lehmann, A. C. (1996). Expert and exceptional performance: Evidence of maximal adaptation to task constraints. *Annual Review of Psychology, 37*, 273–305.

Ferster, C. B. (1967). Arbitrary and natural reinforcement. *The Psychological Record, 17*, 341–347.

Ferster, C. B., & Perrott, M. C. (1968). *Behavior principles*. New York: Appleton-Century-Crofts.

Gagne, R. M. (1968). *The conditions of learning*. New York: Holt, Rinehart, & Winston.

Green, E. J. (1967). *The learning process and programmed instruction*. New York: Holt, Rinehart, & Winston, Inc.

Greer, R. D., & McDonough, S. H. (1999). Is the learn unit a fundamental measure of pedagogy? *The Behavior Analyst, 22*, 5–16.

Healy, A. F., King, C. L., Clawson, D. M., Sinclair, G. P., Rickard, T. C., Crutcher, R. J., Ericsson, K. A., & Bourne, L. E. (1995). Optimizing the long-term retention of skills. In A. F. Healy & L. E. Bourne (Eds.), *Learning and memory of knowledge and skills: Durability and specificity* (pp. 1–65). Thousand Oaks, CA: SAGE Publications.

Holland, J. G. (1960). Teaching machines: An application of principles from the laboratory. *Journal of the Experimental Analysis of Behavior, 3*, 275–287.

Holland, J. G. (1967). A quantitative measure for programmed instruction. *American Educational Research Journal, 4,* 87-101.

Holland, J. G., & Skinner, B. F. (1961). *The analysis of behavior: A program for self-instruction.* New York: McGraw-Hill Co.

Jehng, J. J., & Chan, T. (1998). Designing computer support for collaborative visual learning in the domain of computer programming. *Computers in Human Behavior, 14,* 429-448.

Keller, F. S. (1968). Goodbye teacher. *Journal of Applied Behavior Analysis, 1,* 79-89.

Kelly, G., & Crosbie, J. (1997). Immediate and delayed effects of imposed postfeedback delays in computerized programmed instruction. *The Psychological Record, 47,* 687-698.

Kemp, F. D., & Holland, J. G. (1966). Blackout ratio and overt responses in programmed instruction: Resolution of disparate results. *Journal of Educational Psychology, 57,* 109-114.

Klein, P. S., & Darom, O. N. E. (2000). The use of computers in kindergarten, with or without adult mediation: Effects on children's cognitive performance and behavior. *Computers in Human Behavior, 16,* 591-608.

Koubek, R. J., Benysh, S. A. H., & Tang, E. (1997). Learning. In G. Salvendy (Ed.), *Handbook of human factors and ergonomics* (pp. 130-149). New York: Wiley.

Kritch, K. M., & Bostow, D. E. (1998). Degree of constructed-response interaction in computer-based programmed instruction. *Journal of Applied Behavior Analysis, 31,* 387-398.

Kritch, K. M., Bostow, D. E., & Dedrick, R. F. (1995). Level of interactivity of videodisc instruction on college students' recall of AIDS information. *Journal of Applied Behavior Analysis, 28,* 85-86.

Kulik, J. A., Cohen, P. A., & Ebeling, B. J. (1980). Effectiveness of programmed instruction in higher education: A meta-analysis of findings. *Educational Evaluation and Policy Analysis, 2,* 51-64.

Kulik, C. C., Kulik, J. A., & Bangert-Drowns, R. L. (1990). Effectiveness of mastery learning programs: A meta-analysis. *Review of Educational Research, 60,* 265-299.

Kulik, J. A., Kulik, C. C., & Cohen, P. A. (1979). A meta-analysis of outcome studies of Keller's personalized system of instruction. *American Psychologist, 34,* 307-318.

Lane, N. E. (1987). *Skill acquisition rates and patterns: Issues and training implications.* New York: Springer-Verlag.

Lange, P. C. (1967). *Programed instruction: The Sixty-sixth yearbook of the National Society for the Study of Education.* Chicago: The University of Chicago Press.

LeLouche, R. (1998). The successive contributions of computers to education: A survey. *European Journal of Engineering Education, 23,* 297-308.

Li, S., & Lewandowsky, S. (1995). Forward and backward recall: Different retrieval processes. *Journal of Experimental Psychology: Learning, Memory, and Cognition, 21,* 837-847.

Margulies, S., & Eigen, L. D. (1962). *Applied programmed instruction.* New York: John Wiley & Sons, Inc.

Mayer, R. E. (2000). An interview with Richard E. Mayer: About technology. *Educational Psychology Review, 12,* 477-483.

Merrill, P. F., Hammons, K., Vincent, B. R., Reynolds, P. L., Christensen, L., & Tolman, M. N. (1996). *Computers in education* (pp. 65-86). Boston: Allyn & Bacon.

Pane, J. F., Ratanamahatana, C. A., & Myers, B. A. (2001). Studying the language and structure in non-programmers' solutions to programming problems. *International Journal of Human-Computer Studies, 54,* 237-264.

Peel, E. A. (1967). Programmed thinking. *Programmed Learning & Educational Technology, 4,* 151-157.

Postman, L. (1962). Retention as a function of degree of overlearning. *Science, 135,* 666-667.

Pressey, S. L. (1927). A machine for automatic teaching of drill material. *School and Society, 23,* 549-552.

Rabinowitz, M. (1993). *Cognitive science foundations of instruction.* Hillsdale, NJ: Lawrence Erlbaum Associates.

Ray, R. D. (1995). A behavioral systems approach to adaptive computerized instructional design. *Behavior Research Methods, Instruments, & Computers, 27,* 293-296.

Rehder, B. (2001). Interference between cognitive skills. *Journal of Experimental Psychology: Learning, Memory, and Cognition, 27,* 451-469.

Reigeluth, C. M., & Darwexeh, A. N. (1982). The elaboration theory's procedures for designing instruction: A conceptual approach. *Journal of Instructional Development, 5,* 22-32.

Roediger, H. L., & Guynn, M. J. (1996). Retrieval processes. In E. L. Bjork & R. A. Bjork (Eds.), *Memory* (pp. 197-236). New York: Academic Press.

Rolls, E. T. (2000). Memory systems in the brain. *Annual Review of Psychology, 41,* 599-631.

Ryan, S. D. (1999). A model for the motivation for IT retraining. *Information Resources Management Journal, 12,* 24-32.

Schunk, D. H. (2000). *Learning theories: An educational perspective* (pp. 355-401). Upper Saddle River, NJ: Prentice-Hall, Inc.

Scriven, M. (1969). The case for and use of programmed texts. In A. D. Calvin (Ed.), *Programmed Instruction: Bold New Adventure* (pp. 3-36). Bloomington: Indiana University Press.

Shute, V. J., & Gawlick, L. A. (1995). Practice effects of skill acquisition, learning outcome, retention, and sensitivity to relearning. *Human Factors, 37,* 781-803.

Shute, V. J., Gawlick, L. A., & Gluck, K. A. (1998). Effects of practice and learner control on short- and long-term gain and efficiency. *Human Factors, 40,* 296-310.

Shute, V. J., & Psotka, J. (1996). Intelligent tutoring systems: Past, present, and future. In D. Jonassen (Ed.), *Handbook of research on educational communications and technology* (pp. 570-600). New York: Macmillan Publishing.

Sidman, M. (1960). *Tactics of scientific research.* New York: Basic Books.

Sime, M. E., Green, T. R. G., & Guest, D. J. (1973). Psychological evaluation of two conditional constructions used in computer languages. *International Journal of Man-Machine Studies, 5,* 105-113.

Simon, J. S., Grover, V., Teng, J. T. C., & Whitcomb, K. (1996). The relationship of information system training methods and cognitive ability to end-user satisfaction, comprehension, and skill transfer: A longitudinal field study. *Information Systems Research, 7,* 466-490.

Skinner, B. F. (1968). *The technology of teaching.* Englewood Cliffs, NJ: Prentice-Hall, Inc.

Skinner, B. F. (1953). *Science and human behavior.* New York: The Free Press.

Skinner, B. F. (1954). The science of learning and the art of teaching. *Harvard Educational Review, 24,* 86-97.

Skinner, B. F. (1957). *Verbal behavior.* New York: Appleton-Century-Crofts.

Skinner, B. F. (1958). Teaching machines. *Science, 128,* 969-977.

Sohn, Y. W., & Doane, S. M. (1997). Cognitive constraints on computer problem-solving skills. *Journal of Experimental Psychology: Applied, 3,* 288-312.

Soloway, E. (1985). From problems to programs via plans: The content and structure of knowledge for introductory LISP programming. *Journal of Educational Computing Research, 1,* 157-172.

Sulzer-Azaroff, B., & Mayer, G. R. (1991). *Behavior analysis for lasting change.* Orlando, FL: Holt, Rinehart, & Winston, Inc.

Suppes, P., & Morningstar, M. (1969). Computer-assisted instruction. *Science, 166,* 343-350.

Swezey, R. W., & Llaneras, R. E. (1997). Models in training and instruction. In G. Salvendy (Ed.), *Handbook of human factors and ergonomics* (pp. 514–577). New York: Wiley.

Szabo, M., & Montgomerie, T. C. (1992). Two decades of research on computer-managed instruction. *Journal of Research on Computing in Education, 25*, 113–134.

Tennyson, R. D. (1999). Goals for automated instructional systems. *Journal of Structural Learning and Intelligent Systems, 13*, 215–226.

Tennyson, R. D., & Elmore, R. L. (1997). Learning theory foundations for instructional design. In R. D. Tennyson, F. Schott, N. M. Seel, & S. Dijkstra (Eds.), *Instructional design: International perspectives* (pp. 55–78). Mahwah, NJ: Lawrence Erlbaum Associates.

Tennyson, R. D., & Schott, F. (1997). Instructional design theory, research, and models. In R. D. Tennyson, F. Schott, N. M. Seel, & S. Dijkstra (Eds.), *Instructional design: International perspectives* (pp. 1–16). Mahwah, NJ: Lawrence Erlbaum Associates.

Tudor, R. M, & Bostow, D. E. (1991). Computer-programmed instruction: The relation of required interaction to practical application. *Journal of Applied Behavior Analysis, 24*, 361–368.

Underwood, B. J. (1957). Interference and forgetting. *Psychological Review, 64*, 49–60.

Van Merrienboer, J. J. G., & Paas, F. G. W. C. (1990). Automation and schema acquisition in learning elementary computer programming: Implications for the design of practice. *Computers in Human Behavior, 6*, 273–298.

Vargas, E. A., & Vargas, J. S. (1991). Programmed instruction: What it is and how to do it. *Journal of Behavioral Education, 1*, 235–251.

Wheeler, J. L., & Regian, J. W. (1999). The use of a cognitive tutoring system in the improvement of the abstract reasoning component of word problem solving. *Computers in Human Behavior, 15*, 243–254.

Young, J. D. (1996). The effect of self-regulated learning strategies on performance in learner controlled computer-based instruction. *Educational Technology Research and Development, 44*, 17–27.

Zimmerman, B. (1994). Dimensions of academic self-regulation: A conceptual framework for education. In D. H. Schunk & B. J. Zimmerman (Eds.), *Self-regulation of learning and performance* (pp. 3–21). Hillsdale, NJ: Erlbaum.

·35·

CONVERSATIONAL SPEECH INTERFACES

Jennifer Lai
IBM T. J. Watson Research Center

Nicole Yankelovich
Sun Microsystems

WHAT IS A CONVERSATIONAL SPEECH INTERFACE?

Conversation—a dialogue between two or more people—is a skill mastered early in life and practiced frequently. A conversational speech interface is a dialogue between a person and a computer. A major benefit of incorporating speech into an application is that speech is natural for people. Most of us find speaking easy. Although speech is natural for humans, it certainly is not for computers. Speech technology, like the other recognition technologies, lacks 100% accuracy. Speech designers need to design to the strengths and weaknesses of the technology to optimize the overall user experience.

The goal in creating speech applications is to simulate the roles of a human speaker and listener convincingly enough to produce successful communication between a computer and the human collaborator. Successful communication occurs when the sender and receiver of the message achieve a shared understanding. Human-to-human conversations are characterized by turn taking, shifts in initiative, as well as verbal and nonverbal feedback to indicate understanding. Herb Clark (1993) says, "Speaking and listening are two parts of a collective activity." When designing dialogues for speech systems, it is important to work toward establishing what Clark calls a common ground or shared context. The common ground is built on past conversations, the immediate surroundings and context of the conversation, as well as the cultural background of both parties. Because language use is deeply ingrained in human behavior, successful speech interfaces should be based on an understanding of the different ways that people use language to communicate. Speech applications should adopt language conventions that help people know what they should say next and that avoid conversational patterns that violate standards of polite, cooperative behavior.

At first, it might seem strange to think about designing an interaction with a machine that emulates human-to-human conversational conventions. After all, the goal of the user is to complete a task, and to do so in the simplest and most efficient way possible. Users rarely state their goals in terms of needing the system to be a conversational partner, or ensuring that the computer is cooperative and avoids behaviors that are considered rude. However, studies published in The Media Equation (Reeves & Nass, 1996) have shown that people react to new media with their "old brains," and that we are hard-wired to respond automatically and unconsciously to speaking computers as if they were other people. "Individuals behave toward and make attributions about voice systems using the same rules and heuristics they would normally apply to other humans" (Nass & Gong, 2000).

The earlier chapter on Conversational Speech Technologies describes the types of systems that can be used to create conversational speech interfaces. This chapter focuses on design. We first examine the importance of designing with speech in mind from the onset of a project. After that, we examine a range of issues a designer must consider in creating an effective speech user interface. These issues include choosing a style for the dialogue, dealing with latency, crafting the system prompts, and using strategies for error correction. The chapter ends with a discussion of the various ways that users can be involved in the life cycle of a speech application.

DESIGN FOR SPEECH

When to Use Speech

A crucial factor in determining the success of a speech application is whether or not there is a clear benefit to using speech. Because speech is such a natural medium for communication, users' expectations of a speech application tend to be extremely high. This means speech is best used when the need is clear—for example, when the user's hands and eyes are busy—or when speech enables something that cannot otherwise be done, such as accessing electronic mail or an online calendar over the telephone when a computer is not available.

Speech is well suited to some tasks, but not for others. Tables 35.1 and 35.2 list characteristics that can help you determine when speech input and output are appropriate choices.

User motivation is a critical factor in the success of a speech application. A motivated user is willing to do such things as spend time training the system to the sound of his voice, or make an effort to enunciate when speaking. A clear way to motivate users is to increase their productivity. For example, consider a speech application that will replace keyboard entry. Although users who do not type well will be motivated to use speech because they see a clear benefit, others who are skilled typists could see their throughput go down and be less inclined to try to make speech work for them.

The telephone company's use of speech recognition to automate collect calling provides an example of how motivated users contribute to the success of a speech based service. People making a collect call want their call to go through, so they answer prompts carefully. People accepting collect calls are also motivated to cooperate, because they do not want to pay for unwanted calls or miss important calls from their friends and family. Automated collect calling systems save the company

TABLE 35.1. When Is Speech Input Appropriate?

Use When...	Avoid When...
No keyboard is available (e.g., over the telephone, at a kiosk, or on a portable device).	Task requires users to talk to other people while using the application.
Task requires the user's hands to be occupied so they cannot use a keyboard or mouse (e.g., maintenance and repair, graphics editing).	Users work in a very noisy environment.
Commands are embedded in a deep menu structure.	Task can be accomplished more easily using a mouse and keyboard.
Users are unable to type or are not comfortable with typing.	Privacy is an issue.
Users have a physical disability (e.g., limited use of hands).	

TABLE 35.2. When Is Speech Output Appropriate?

Use When . . .	Avoid When . . .
Task requires the user's eyes to be looking at something other than the screen (e.g., driving, maintenance, and repair).	Large quantities of information must be presented.
Situation requires grabbing users' attention.	Task requires user to compare data items.
Users have a physical disability (e.g., visual impairment).	Information is personal or confidential.
Interface is trying to embody a personality.	

Sender	Subject	Date & Time	Size
1. Arlene Rexford	Learn about Java	Mon Oct 28 11:23	2 K
2. Shari Jackson	Re: Boston rumors	Fri Jul 18 09:32	3 K
3. Hilary Binda	Change of address	Wed Jul 16 12:59	1 K
4. Arlene Rexford	Class Openings	Tue Jul 21 12:46	8 K
5. George Fitz	Re: Boston rumors	Tue Jul 21 12:46	1 K

FIGURE 35.1. E-mail message information.

money and benefit users. Telephone companies report that callers prefer talking to the computer because they are sometimes embarrassed by their need to call collect and they feel that the computer makes the transaction more private.

In addition to the users' motivation to make speech work for them, it is important for the designer to look at the environment that the users are working in. Do they work in an environment that is very loud, such as a factory? Or do they work in a cubicle where their speech would disturb co-workers or where privacy would be an issue?

Including speech in an application because it is a novelty means it probably will not get used. Including it because there is some compelling reason increases the likelihood for success.

Don't Translate Graphics into Speech

After you determine that speech is an appropriate interface technique, consider how speech will be integrated into the application. Generally, a successful speech application is designed with speech in mind from the beginning. It is rarely effective to add speech to an existing graphical application or to translate a graphical application directly into a speech-only one. Doing so is akin to translating a command-line program directly into a graphical user interface. The program may work, but the most effective graphical programs are designed with the graphical environment in mind from the outset.

Graphical applications do not translate well into speech for several reasons. First, graphical applications do not always reflect the vocabulary, or even the basic concepts, that people use when talking to one another in the domain of the application. Consider a calendar application, for example. Most graphical calendar programs use an explicit visual representation of days, months, and years. There is no concept of relative dates (e.g., "the day after Labor Day" or "a week from tomorrow") built into the interface. When people speak to each other about scheduling, however, they make extensive use of relative dates. A speech interface to a calendar, whether speech-only or multimodal, is therefore more likely to be effective if it allows users to speak about dates in both relative and absolute terms. By basing the speech interface design exactly on the graphical interface design, relative dates would not be included in the design, and the usability of the calendar application would be compromised.

Optimize for Spoken Communication

Information organization is another important consideration. Presentations that work well in the graphical environment can fail completely in the speech environment. Reading exactly what is displayed on the screen is rarely effective. Likewise, users find it awkward to speak exactly what is printed on the display.

Consider the way in which many e-mail applications present message headers. An inbox usually consists of a chronological, sometimes numbered, list of headers containing information such as sender, subject, date, time, and size (see Fig. 35.1). You can scan this list and find a subject of interest or identify a message from a particular person. Imagine if someone read this information out loud to you, exactly as printed. It would take a long time! The day, date, time, and size information, which you can easily ignore in the graphical representation, becomes quite prominent. It does not sound very natural, either. By the time you hear the fifth header, you may also have forgotten that there was an earlier message with the same subject.

An effective speech interface for an e-mail application would probably not read the date, time, and size information from the message header unless the user requested it. Better still would be an alternate organization scheme that groups messages into categories, perhaps by subject or sender (e.g., "You have two messages about 'Boston rumors'" or "You have two messages from Arlene Rexford"), so that the header list contains fewer individual items. Reading the items in a more natural spoken form would also be helpful. For example, instead of "Three. Hilary Binda. Change of address," the system might say "Message 3 from Hilary Binda is about Change of address."

On the speech input side, speaking menu commands can also be awkward and unnatural. In one e-mail program, a menu called "Move" contains a list of mailbox names. Translating this interface to speech would force the user to say something like "Move. Weekly Reports." A more natural interface would allow the user to say "Move this to my Weekly Reports folder" or the shorter "Move to Weekly Reports." The natural versions are a little longer, but are probably something the user can remember to say without looking at the screen.

Challenges

Even if you design an application with speech in mind from the outset, you face substantial challenges before your application is robust and easy to use. Understanding these challenges and assessing the various trade-offs that must be made during the design process will help to produce the most effective interface.

Transience: What Did You Say? Speech is *transient*. Once you hear it or say it, it's gone. By contrast, graphics are *persistent*. A graphical interface typically stays on the screen until the user performs some action.

Listening to speech taxes users' short-term memory. Because speech is transient, users can remember only a limited number of items in a list, and they may forget important information provided at the beginning of a long sentence. Likewise, while speaking to a dictation system, users often forget the exact words they have just spoken.

Users' limited ability to remember transient information has substantial implications for the speech interface design. In general, transience means that speech is not a good medium for delivering large amounts of information.

The transient nature of speech can also provide benefits. Because people can look and listen at the same time, speech is ideal for grabbing attention or for providing an alternate mechanism for feedback. Imagine receiving a notification about the arrival of an e-mail message while working on a spreadsheet. Speech might give the user the opportunity to ask for the sender or the subject of the message. The information can be delivered without forcing the user to switch contexts.

Invisibility: What Can I Say? Speech is *invisible*. The lack of visibility makes it challenging to communicate the functional boundaries of an application to the user (Yankelovich, 1996). In a graphical application, menus and other screen elements make most or all of the functionality of an application visible to a user. By contrast, in a speech application, it is much more difficult to indicate to the user what actions they may perform, and what words and phrases they must say to perform those actions.

To illustrate the problem of invisible functional boundaries, consider the speech-only calendar application described previously. Here's what an interaction might sound like:

Computer: Currently, you have "lunch with Julie." Later, at 2:30, you have "Staff Meeting."
User: What's on my calendar the Monday after Easter?
Computer: On Monday, April 8, you have "Call Fidelity Monitor," you have "Barbara Munch's birthday."
User: What does Paul have?
Computer: On Monday, April 8, Paul has "Customer visit." From 10 o'clock to 10:30, "Meeting with Bob."
User: What about the next day?
Computer: On Tuesday, April 9, Paul has no appointments.

The dialogue has progressed smoothly so far, but now the user gets into trouble:

User: What do I have tomorrow morning?
Computer: Sorry?

Although this query about tomorrow morning sounds perfectly reasonable given the prior dialogue, the user has just crossed the functional boundary of what the application supports. This calendar application supports browsing calendars only in units of whole days. The decision to constrain the grammar was driven primarily by the desire to maintain reasonable recognition performance. Perhaps the decision seems arbitrary, but it's a slippery slope. Once you allow users to talk about "morning," "afternoon," and "evening," they'll expect to be able to say things like "early afternoon," "after 6 p.m.," "around noon," and so forth. At some point, the constraints of the speech recognition technology dictate that only a limited set of utterances be allowed. Designers must draw the line someplace.

This example illustrates one side of the problem: Users assume they can say things that the application does not support. The flip side of the problem is that users do not know how to say things that the application does support. When the calendar application is first accessed, users hear what they have on their calendar from now until the end of the current day:

Today at 4 p.m. you have "HI Reading Group." At 6:30 p.m., you have "Dinner with Jane."

At that point, the user can hang up, switch to some other application, or do a calendar lookup, which includes queries such as:

What do I have tomorrow?
How about a week from Monday?
What does Paul have?
What's on Bob's calendar for April 5th?
What did he have yesterday?

The functionality is powerful, but hidden. One solution, of course, is to create a much more detailed initial prompt that spells out the range of options available to the user. The disadvantage of this approach is that speech output is slow and temporal. Not only are long prompts costly in terms of the time they take, users often only remember the end, or the beginning, of what was said. It can be hard to actually explain the sort of input that is acceptable in a short instruction. Another solution is to offer help that includes a set of example queries. A lower tech solution, and one that is often effective, is to provide printed wallet-sized cards with the samples. Cards with samples work well when they can be taped to the monitor where the user always works, but if they are needed for a telephony-based solution, they can sometimes be forgotten.

Asymmetry. Speech is *asymmetric*. People can produce speech easily and quickly, but they cannot listen nearly as easily and quickly. This asymmetry means people can speak faster than they can type, but listen much more slowly than they can read. Consider the following input and output rates (Schmandt, 1994):

Speaking rate:	175–225 words per minute (wpm)
Typing rate:	80–100 wpm (good typist)
Listening rate:	175–225 wpm
Reading rate:	350–500 wpm.

The asymmetry has design implications for what information to speak and how much to speak. A speech interface designer must balance the need to convey lots of instructions to users with users' limited ability to absorb spoken information.

Speech Synthesis Quality. Synthetic speech, also known as text-to-speech (TTS), is speech produced by a computer. Given that today's synthesizers still do not sound entirely natural, the choice whether to use synthesized output, recorded output, or no speech output is often a difficult one. Although recorded speech is much easier and more pleasant for users to listen to, it presents a technical challenge when the information being presented is dynamic. For example, recorded speech cannot be used to read people their e-mail messages over the telephone. This is because the text is not known in advance and thus cannot be prerecorded. Most systems on the market today use recorded speech for system prompts that do not change and synthesized speech for dynamic text. This mixing of voice types is found within sentences (e.g., e-mail headers are read as a combination of recorded speech and synthetic speech), as well as from sentence to sentence. In addition to mixing types of speech, many commercial systems also mix the gender of the voices, commonly using a female voice talent for the natural speech and a male voice for the text-to-speech.

Mixing recorded and synthesized speech, however, is not generally a good idea. Although users often report not liking the sound of text-to-speech, they are, in fact, able to adapt to the synthesizer better when it is not mixed with recorded human speech. Listening is considerably easier when the voice is consistent. In a study conducted by Gong and Lai (2001), 24 users interacted with a virtual assistant to manage both e-mail and calendar tasks. Half of the participants were in a mixed-voice condition (TTS and human speech) and the other half heard only TTS. Their task performance, self-perception of performance, and attitudinal responses were measured. Users interacting with the TTS-only interface performed the task significantly better, whereas users interacting with the mixed-voices interface *thought* they did better and had more positive attitudinal responses.

Deciding which text-to-speech engine to use can sometimes seem like a daunting experience, because everybody seems to have an opinion about which engine is easiest to understand and most pleasant to listen to. The quality of the synthetic speech should be evaluated along the lines of its acceptability, naturalness, and intelligibility. It is important to ask users to evaluate different speech engines against each other because these qualities are always relative (Francis & Nusbaum, 1999). Although these subjective differences in opinion will probably always exist, a study by Lai et al. (2000) showed that there were no significant differences in comprehension levels for longer messages (i.e., with a word length ranging from 100 to 500 words) among five major commercial text-to-speech engines. The engines examined were DECtalk, AcuVoice, IBM, L&H, and Lucent. The study looked at the comprehensibility of synthetic speech for various tasks ranging from short, simple e-mail messages to longer news articles on mostly obscure topics. Comprehension accuracy was measured for both TTS and for recorded human speech. Participants who did not take notes while listening performed significantly worse for all synthetic voice tasks when compared with recorded speech tasks. Performance for synthetic speech degraded as the message got longer and more detailed.

As a rule of thumb, use recorded speech for systems when all the text to be spoken is known in advance, or when it is important to convey a particular personality to the user. Use synthesized speech when some or all of the text to be spoken is not known in advance, or when storage space is limited. Recorded audio requires substantially more disk space than synthesized speech. Another rule of thumb is to set the output rate for synthetic speech at around 180 words per minute. This is often the default speed setting for text-to-speech engines. This rate may need to be adjusted depending on your user population. For example, people who use TTS on a regular basis, such as those who are visually impaired, will be comfortable at speaking rates almost three times the default rate (500 wpm; Raman, 1997).

Speech Recognition Performance. Speech recognizers are not perfect listeners. They make mistakes. A big challenge in designing speech applications, therefore, is working with imperfect speech recognition technology. Although this technology improves constantly, it is unlikely that, in the foreseeable future, it will approach the robustness of computers in science fiction movies.

An application designer should understand the types of errors that speech recognizers make and the common causes of these errors. Table 35.3 lists common errors and their causes.

Sometimes making an out-of-grammar request results in a recognition failure (also called a rejection). The other major pitfall of out-of-grammar requests is that they often get mapped to some other legal utterance in the grammar. If the user says "repeat the message" and repeat is a function that has not yet been implemented, the system might just as easily recognize

TABLE 35.3. Causes for Speech Recognition Errors

Problem	Cause
Rejection/ misrecognition	Voice is substantially different from stored voice models (e.g., kids' voices).
	User speaks one or more words not in the vocabulary.
	User's sentence does not match any active grammar.
	User speaks before system is ready to listen.
	Words in active vocabulary sound alike and are confused (e.g., "too," two").
	User pauses too long in the middle of a sentence.
	User speaks with a disfluency (e.g., restarts sentence, stutters, "ummm").
	User has an accent or cold.
	Computer's audio is not configured properly.
	User's microphone is not properly adjusted.
Misfire	Background speech triggers recognition.
	User is talking with another person.
	Nonspeech sound (e.g., cough, laugh).

"delete the message." Because this type of error is common, it is important to use specific confirmation messages. Imagine a system where this recognition error (delete instead of repeat) was followed-up with an unspecific confirmation (e.g., "Are you sure you want to do this? Please say yes or no."). The user would say yes, thinking that he had just agreed to repeating the message, and the deletion would proceed without the user ever knowing.

Even worse than the wrong thing being recognized, recognition errors can cause the user to form an incorrect model of how the system works. For example, if the user says "Read the next message," and the recognizer hears "Repeat the message," the application will repeat the current message, leading the user to believe that "Read the next message" is not a valid way to ask for the next message. If the user then says "Next," and the recognizer returns a rejection error, the user now eliminates "Next" as a valid option for moving forward. Unless there is a display that lists all the valid commands, users cannot know if the words they have spoken should work; therefore, if they do not work, users often assume they are invalid.

Some recognition systems adapt to users over time, but good recognition performance still requires cooperative users who are willing and able to adapt their speaking patterns to the needs of the recognition system. This is why providing users with a clear motivation to make speech work for them is essential.

Recognition: Flexibility Vs. Accuracy.

A flexible system allows users to speak the same commands in many different ways. The more flexibility an application provides for user input, the more likely errors are to occur (Makhoul & Schwartz, 1994). In designing a command-and-control style interface, therefore, the application designer must find a balance between flexibility and recognition accuracy.

For example, a calendar application may allow the user to ask about tomorrow's appointments in ways such as:

"What about tomorrow?"
"What do I have tomorrow?"
"What's on my calendar for tomorrow?"
"Read me tomorrow's schedule."
"Tell me about the appointments I have on my calendar tomorrow."

This may be quite natural in theory, but, if recognition performance is poor, users will not accept the application. On the other hand, applications that provide a small, fixed set of commands also may not be accepted, even if the command phrases are designed to sound natural (e.g., "Lookup tomorrow"). Users tend to forget the exact wording of fixed commands. What seems natural for one user may feel awkward for another. The section on Natural Dialogue Studies describes a technique for collecting data from users to determine the most common ways that people talk about a subject. In this way, applications can offer some flexibility without causing recognition performance to degrade dramatically.

SPEECH USER INTERFACE DESIGN ISSUES

Understanding the characteristics of speech input and output systems, as well as people's cognitive abilities as they relate to speaking and listening, is an important prerequisite to design. In this section, we examine some more specific design issues, starting with telephony applications that use speech input and output. After that, we look at issues involved when mixing speech with other interface modalities, such as graphical displays and physical buttons and controls.

Design Issues for Telephony Applications

Speech technologies are a good fit for using with telephony applications. Designing an application without a display takes a different set of skills and sensitivities than designing one with a strong visual component.

Dialogue Styles. The first consideration is what style of interaction you should provide.

Directed Dialogue (System Initiated). This is the most commonly used style of interaction in speech-based telephony systems on the market today. With a directed dialogue, the user is instructed or "directed" what to say at each prompt (Kamm, 1994). Systems that use this style can often be recognized by the use of the key word "say" in the prompts. "Welcome to ABC Bank. You can check an account balance, transfer funds, or pay a bill. Say Balance, Transfer, or Pay." The reason most systems use this dialogue style is to increase the accuracy of the speech recognition. By telling the user what to say, the chances of the user speaking an out-of-vocabulary utterance are much lower. This style is sometimes referred to as system-initiated, because the user plays a more passive role, waiting for the system to prompt him for what is needed. Directed dialogue systems come in two major flavors. The first is similar to form filling in which the user is prompted for pieces of information to fill in the "blanks" on the form.

Let's look at an example of a form-filling application:

Computer: Welcome to the Phone Travel Company. What city are you traveling to?
User: San Francisco.
Computer: What city are you traveling from?
User: New York.
Computer: On what day will you be traveling?
User: Saturday, May sixth.
Computer: Around what time of day would you like to leave?
User: Around 5 p.m.

It would have been a lot easier for the user to say "I'm interested in flights between New York and San Francisco leaving the evening of May 6th."

A second type of directed dialogue systems rely heavily on the use of menus, presenting the user with successive lists of valid choices. These systems are sometimes a carryover from the telephone interactive voice response systems that they are replacing. With the prompt, "Please select one of the following options: checking, savings, or bills," one can see a direct mapping from the old system that probably said: "press 1 for checking, 2 for savings, or 3 for bills." However, because a directed dialogue accepts spoken text as input, the type of information requested can be more varied than with a voice response unit. For example, with the prompt, "What city please?," it is much easier to say Boston than to try to map the characters to keys on the telephone pad. These systems are often very list-oriented and rely on the use of barge-in technology so the user can speak his selection as soon has he has heard it. Barge-in is the ability for the user to "interrupt" the system while it is speaking. The drawback with this type of interaction is that if the user is inexperienced with the application, he will not know about the barge-in feature, and will most likely wait quietly until the entire list is presented before speaking. This may cause him to forget the suggested wording or even what all the choices are. Let's look at an example of an interface that relies on list presentation combined with barge-in.

Computer: Choose one of the following options: stock quotes, travel, weather, news reports, ...
User: (User interrupts, barging-in) News.
Computer: Pick one of the following categories: top stories, traffic, sports, ski reports. . . .
User: Sports.
Computer: Please say a team name or say what are my choices for a list of all the teams we track for you.
User: What are my choices?
Computer: You can select the New York Knicks, Los Angeles Lakers. . . .

With this type of directed dialogue, accuracy can be very high even under a variety of acoustic environments, because the active vocabulary is constrained by the choices that were most recently presented. In some of these systems, an attempt is made to support the expert user by accepting selections from lists that have not yet been presented.

User-initiated. Unlike a directed dialogues, which work well for novice users, user-initiated systems are intended for use by expert users. With this type of interaction style, the system is passive and does not prompt the user for specific information. However, these types of dialogues do not work well with new users who do not know what to say, and tend to have lower accuracy rates.

Computer: How can I help you?
User: I would like to buy 100 shares of Amazon.com at the market price with funds from my current account number 345198.

Conversational Systems (Mixed Initiative). The type of interaction that is easiest for users and feels the most natural, is a mixed initiative system. In this style of interaction, the system and the user take turns initiating the communication depending on the flow of the conversation and the status of the task. With a mixed initiative system, the prompts can fluctuate between open-ended prompts and more directed prompts depending on the context. Here's a portion of dialogue:

Computer: What would you like to do? (open ended)
User: What is the price of Fidelity Magellan?
Computer: Price of Fidelity Magellan is 36.05 per share. What else?
User: I want to buy 500 shares. (which shares to buy is implied in the previous interaction)
Computer: Confirming purchase of 500 shares of Fidelity Magellan fund, please say yes or no. (directed)
User: No, skip the purchase. I'd like to transfer $1,000 from my index fund.
Computer: Purchase canceled. To which fund would you like to transfer? (prompt for missing information)

In this example, we see the system functioning as a cooperative partner in the conversation.

Along with the decision of what dialogue style to use, several important decisions need to be made about the wording of the prompts. One of these decisions is whether to use personal pronouns or not in the prompts. Compare the two prompts below:

"Confirming purchase of 500 shares of Fidelity Magellan fund, please say yes or no."

vs.

"Are you sure you want to buy 500 shares of Fidelity Magellan fund, please say yes or no."

or

"This system can be used to check account balances, transfer funds, or pay bills."

vs.

"I can check your account balance, transfer funds, or pay bills."

The decision whether to use personal pronouns or not, along with other prompt wording decisions, should be driven by the tone of the conversation that you want to set, and the "personality" that you want to establish for the system. These decisions can be aided by a natural dialogue study discussed later in this chapter.

Feedback and Latency. In conversations, timing is critical. People read meaning into pauses. Unfortunately, processing delays in speech applications often cause pauses in places where they do not naturally belong. For example, users may interpret a system delay, as they would in a human–human interaction, to mean that they were not heard or not understood. This would cause them to repeat themselves, and more than likely, the user will speak at the same time the system prompt is finally played, causing a collision. The collision generates a recognition error and often the whole scenario will be played over again, leading to increased user frustration. A good way to deal with this problem is to play an audio file to indicate that the system is working prior to the playing of the system prompt. This way, it is clear to the user that the system is busy dealing with his request and he should not speak again yet. Another way to facilitate the turn-taking process in a spoken interaction is to use tones. Auditory tones signal to the user that it is his turn to speak (Ballentine & Morgan, 1999). This is especially true if the system is implemented in such a way that does not support barge-in by the user.

Giving users adequate feedback is especially important in speech-only interfaces. Processing delays, coupled with the lack of peripheral cues to help the user determine the state of the application, make consistent feedback a key factor in achieving user satisfaction. When designing feedback, recall that speech is a slow output channel. This issue must be balanced with a user's need to know several vital facts:

- Is the recognizer processing or waiting for input?
- Has the recognizer heard the user's speech?
- If heard, was the user's speech correctly interpreted?

Verification should be commensurate with the cost of performing an action (Weinschenk, 2000). *Implicitly verify* commands that present data and *explicitly verify* commands that destroy data or trigger actions. For example, it would be important to give the user plenty of feedback before authorizing a large payment, whereas it would not be as vital to ensure that a date is correct before checking a weather forecast. In the case of the payment, the feedback should be explicit (e.g., "Do you want to make a payment of $1,000 to Boston Electric? Say yes or no."). The feedback for the forecast query can be implicit (e.g., "Tomorrow's weather forecast for Boston is...."). In this case, the word "tomorrow" serves as feedback that the date was correctly (or incorrectly) recognized. If correct, the interaction moves forward with minimal wasted time.

Prompting. Well-designed prompts lead users smoothly through a successful interaction with a speech-only application. Many factors must be considered when designing prompts, but the most important is assessing the trade-off between flexibility and performance. The more you constrain what the user can say to an application, the less likely they are to encounter recognition errors. On the other hand, allowing users to enter information flexibly can often speed the interaction (if recognition succeeds), feel more natural, and avoid forcing users to memorize commands. Here are some tips for creating useful prompts.

Use *explicit prompts* when the user input must be tightly constrained. For example, after recording a message, the prompt might be "Say cancel, send, or review." This sort of prompt directs the user to say just one of those three keywords. Directed dialogue systems use explicit prompts. More conversational systems, however, will occasionally fall back on this type of prompt when it is critical that the input be correctly recognized, as with transactions that involve booking travel. "Are you sure you want to book this flight from JFK to LAX. Please say Yes or No."

Use *implicit prompts* when the application is able to accept more flexible input. These prompts rely on conversational conventions to constrain the user input. For example, if the user says "Send mail to Bill," and "Bill" is ambiguous, the system prompt might be "Did you mean Bill Smith or Bill Jones?" Users are likely to respond with input such as "Smith" or "I meant Bill Jones." While possible, conversational convention makes it less likely that they would say "Bill Jones is the one I want."

Using *variable prompts* is a good way to try to simulate a human–human conversation. Given a certain condition or state of the system (e.g., the ready state, or a system response to silence), it is preferable not to play the exact same system prompt every time. Subtle variations in the wording impart a much more natural feel to the interaction. Note the following possibilities for the ready state in a natural language understanding system: "What now?," "I'm ready to help," "What's next?"

Another interaction that we can model on human speech is the use of *tapered prompts*. Tapering can be accomplished in one of two ways. If an application is presenting a set of data, such as current quotes for a stock portfolio, drop out unnecessary words once a pattern is established.

For example:

"As of 15 minutes ago, Sun Microsystems	was trading	at 45 up $\frac{1}{2}$,
	Motorola	was	at 83 up $\frac{1}{8}$, and
	IBM	was	at 106 down $\frac{1}{4}$."

Tapering can also happen over time. That is, if you need to tell the user the same information more than once, make it shorter each time. For example, you may wish to remind users about the correct way to record a message. The first time they record a message in a session, the instructions might be lengthy. The next time shorter and the third time just a quick reminder. For example:

Start recording after the tone. Pause for several seconds when done.
Record after the tone, then pause.
Record then pause.

Removable hints are a variation on tapering. Each prompt in the system includes an implicit, conversational question followed by an explicit hint:

"What banking service would you like? I'd recommend saying "check balance," "transfer funds," or pay bills."

The key to hints is that they are designed to be removed. After the user has successfully negotiated a prompt three or four times, the hint can be removed, leaving only the implicit question that the user has become accustomed to hearing: "What banking service would you like?" This personalized approach provides a means of offering help for commands that are infrequently used and tapering prompts that are frequently encountered.

Use *incremental prompts* to speed interaction for expert users and provide help for less experienced users. This technique involves starting with a short prompt. If the user does not respond within a time-out period, the application prompts again with more detailed instructions. For example, the initial prompt might be: "Which service?" If the user says nothing, then the prompt could be expanded to: "Say banking, address book, or yellow pages." Incremental prompts have been used successfully in a number of systems, but suffer from several problems. A first-time user is just as likely to say something to the first prompt, as they are to say nothing. This often results in a recognition error. Another common pitfall with incremental prompts is that they tend to cause collision errors where both the system and the user speak at the same time. When presented with the prompt "Which service?" if the user is confused, he is likely to pause for a moment thinking about what he should say. Just as he gets around to speaking, the system will have hit the end of the time-out period and play the next prompt. It is not uncommon for timing problems to happen repeatedly once started, increasing the user's frustration along the way.

Providing Help. There are two basic types of help. In the first case, the user initiates the playing of a help message by requesting help. Common queries for help in speech systems are "Help," "What can I say?," or "What are my choices?," and it is good to support several variations of this request in the grammar. In the second type of help, the system detects that the user is in trouble and presents the user with help without his having asked for it. Although not all forms of user problems are detectable by the system, one can presume that if a series of rejection errors (explained in the next section) or long silences occur, that the user probably needs some assistance.

Help, when given, needs to be specific to the user's current task. There are few things as frustrating as being stuck and asking for help only to find that the information presented, while long and detailed, has no bearing on the current situation. The other major problem with help functions is that users rarely resort to requesting help. This is probably a combination of previous negative experiences with similar functions and feeling somewhat self-conscious about asking a machine for help.

A good approach for Help is to provide the user with an example of a successful utterance just as he starts the downward spiral of repeated errors. If several rejection errors have occurred in a row, the system should present the user with sample wording that is expected to work, based on the context that he is in. The difficult part is determining the context. With simple transaction-based applications, context can be established through the use of keyword spotting. For example, imagine a stock trading application that allows the user to either check the price of a stock, or buy and sell stocks. If the user is trying to sell stock and encountering problems, the system could prompt: "To sell a stock, you could say for example: Sell 100 shares of IBM at the market price." If the context it not known, it is simple to present the user with a three-way branch to determine which function the user is interested in. Let's look at a sample interaction that starts with the user speaking an out-of-grammar request:

"I want to liquidate my portfolio today. The market is high, get rid of everything I've got."

Because this utterance does not map to any existing structure in the grammar, the system would return a rejection error and usually a repair prompt such as "Sorry?" would be played. The problem with playing a repair prompt is that it does not help the user to construct a successful utterance. Alternatively, the system could fall back to a directed dialogue to determine the function and then present the user with an example of a valid utterance.

Computer: Would you like to buy, sell, or check stock prices?
User: I would like to sell.
Computer: To sell a stock, you could say, for example: Sell 100 shares of Sun Microsystems at the market price.
User: Sell 35 shares of Microsoft at the market.

In a more complex system, with a much broader function set than the sample application we just looked at, context is harder to establish. If the exact function can be determined (e.g., forwarding e-mail) then the system can present an explicit model for moving ahead. For example: "To forward a message, say the first and last name of another subscriber. For example say: forward this message to Jennifer Lai."

However, the error might be such that the system can determine that a user is working with his e-mail messages, but cannot establish which function is being asked for. In this case, one option is to list the functions that are available at that branch. Combining this tactic with a more constrained vocabulary is sure to get the interaction moving forward again.

Here's an example of how that dialogue might go:

Computer: Let's start over. Here are the e-mail functions I know about: forward, reply, delete, or create a new message. Which one do you want?
User: I want to forward a message.
Computer: To forward a message, say the first and last name of another subscriber. For example say: forward this message to Nicole Yankelovich.
User: Forward this message to Alishia Grant.

If name recognition was the problem in the first place, the system might still not be able to correctly process the request, but at least we now know the requested function and the user feels that the system is working cooperatively with him. If the system returns a recognition error, it is easy to establish that the name is at fault because we already have the function. This dialogue could continue as follows:

Computer: Oh dear. I didn't catch that name. Say just the first name.
User: Alishia.
Computer: I think you said Alice, is that correct?
User: No, that is not correct.
Computer: Thanks for your patience with this. Please spell the first name.
User: A L I S H I A

System looks up all subscribers with a first name of Alishia.

Computer: OK, great job! You must want your note to go to Alishia Grant.

User: (relieved) yes.
Computer: Forwarding message to Alishia Grant.

Handling Errors. The way a system handles recognition errors can dramatically affect the quality of a user's experience. If either the application or the user detects an error, an effective speech user interface should provide one or more mechanisms for correcting the error. Although this seems obvious, correcting a speech input error is not always easy! If the user speaks a word or phrase again, the same error could reoccur depending on the cause of the error (Yankelovich, Levow, & Marx, 1995).

Recognition errors can be divided into three categories: rejection, substitution, and insertion (Schmandt, 1994; Ballentine, 1999). With multimodal systems, such as dictation applications, there is an additional category of errors that is called *deletions*, in which a word spoken by the user is dropped out. A rejection error is said to occur when the recognizer has no hypothesis about what the user said. A substitution error involves the recognizer mistaking the user's utterance for a different legal utterance, as when "send a message" is interpreted as "seventh message." With an insertion error, the recognizer interprets noise as a legal utterance—perhaps others in the room were talking or the user inadvertently tapped the telephone.

Rejection Errors. In handling rejection errors, you want to avoid the "brick wall" effect—that every rejection is met with the same "I didn't understand" response. Users get frustrated very quickly when faced with repetitive error messages. Instead, give progressive assistance: a short error message the first couple of times, and if errors persist, offer more detailed assistance. For example, here is one progression of error messages that a user might encounter:

Sorry?
What did you say?
Sorry. Please rephrase.
I didn't understand. Speak clearly, but don't overemphasize.
Still no luck. Wait for the prompt tone before speaking.

Because background noise and early starts are common causes of misrecognition (Weinschenk & Barker, 2000), simply repeating the command can sometimes solve the problem and is the simplest thing to ask the user to do. Persistent errors are often a sign of out-of-vocabulary utterances; so, asking the user to rephrase the request can often result in a correct recognition. Another common problem is that users respond to repeated rejection errors by exaggerating their speech; thus, it can be helpful to remind them to speak normally and clearly.

Progressive assistance does more than bring the error to the user's attention; the user is guided toward speaking a legal utterance by successively more informative error messages that consider the probable context of the misunderstanding. Repetitiveness and frustration are reduced. In a usability study of an application that used progressive assistance, one participant commented: "When you've made your request three times,

it's actually nice that you don't have the exact same response. It gave me the perception that it's trying to understand what I'm saying." Progressive assistance is one way to make your application seem cooperative rather than rude or combative.

Substitution Errors. Although rejection errors are frustrating, substitution errors can be damaging. If the user asks a weather application for "Kuai," but the recognizer hears "Goodbye" and then hangs up, the interaction could be completely terminated. In situations like this, you should explicitly verify that the user's utterance was correctly understood.

As described in the section on Feedback and Latency, verification should be commensurate with the cost of the action that would be effected by the recognized utterance. Reading the wrong stock quote or calendar entry will make the user wait a few seconds, but sending a confidential message to the wrong person by mistake could have serious consequences.

A component built into your application that converts phrases meaning the same thing into a canonical form can also help to compensate for minor substitution errors. For example, the following calendar queries could all be interpreted the same way:

What does Nicole have May sixth?
What do Nicole have on May six?
What is on Nicole's schedule May sixth?

This means that some substitution errors (e.g., "Switch to weather," misrecognized as "Please weather") will still result in the correct action.

Insertion Errors. Spurious recognition typically occurs because of background noise. The illusory utterance will either be rejected or mistaken for an actual command; in either case, the previous methods can be applied. The real challenge is to prevent insertion errors. Provide users with a keypad command to turn off the speech recognizer to talk to someone, sneeze, or simply gather their thoughts. Pressing the keypad command again can restart the recognizer with a simple prompt, such as "What now?" to indicate that the recognizer is listening again.

Whatever the type of error, a general technique for avoiding errors in the first place is to filter recognition results for unlikely user input. For example, a scheduling application might assume that an error has occurred if the user appears to want to schedule a meeting for 3 a.m.

Correction Strategies. If errors do occur, it is important to provide a means for users to correct the error (assuming they notice it). Flexible correction mechanisms that allow a user to correct a portion of the input are helpful. For example, if the user asks for a weather forecast for Boston for Tuesday, the system might respond, "Tomorrow's weather for Boston is. . . ." A flexible correction mechanism would allow the user to just correct the day: "No, I said Tuesday."

When possible, using an alternate form of input, such as the telephone keypad, can alleviate the user's frustration. If the user is at a prompt where three choices are available and the user has encountered several rejection errors, the user could be instructed: "Press any key when you hear the option you want." The telephone keypad works well when the requested input is numeric (e.g., telephone numbers, account or social security numbers). Getting users to type alphabetic text using a telephone keypad is not a good idea, and it is to avoid this type of input that speech systems are usually recommended in the first place.

In an earlier example of dialogue we saw, the system prompted the user to spell his input to help with the recognition of a first name. Spelling is a good alternate form of input to help resolve a recognition problem. It works best when the system has requested the user to spell the input so that it can load the appropriate recognition models. There are a couple of speech recognition challenges involved with spelling. The first is if the user spells something and intermingles the spelled input with the regular input (e.g., "Please forward this e-mail message to Mukund M U K U N D Aribiaten A R I B I A T E N."). The system will not find a match in the grammar and will most likely try to make words out of each spelled letter. The second challenge with spelling is the "e set," which has a high degree of confusability because all the letters sound alike: b c d e g p t v z.

Some systems use a keypad shortcut (referred to as DTMF input or touchtone) as a fallback method for experienced users when certain functions are not being recognized. For example, if navigating through a unified messaging system (e-mail, voice-mail, fax) under difficult acoustic conditions (e.g., low signal area for cell phones), the system could support the use of the pound key (#) for playing the next message, 7 to delete, and 4 to repeat the message.

Another strategy is to have the system take its best guess at the requested function. This is a good tactic to take when the number of functions enabled at that particular branch in the dialogue is too large to list for the user. A reasonable prompt is: "I think you are trying to create a message, is that correct?" If the user answers in the affirmative, the conversation moves forward and the system can present the user with a sample valid utterance for that function. However, if the best guess is wrong, it is not a good idea to keep iterating through the N-best choices, because this only leads to user frustration. If the response to the best guess is negative, a better solution is to reprompt with a restricted set of choices. Be sure to eliminate the choice that is definitely wrong (e.g., create a message in the previous example). The goal is to move away from a very general repair prompt such as, "I'm sorry I don't understand, please try again" towards a directed prompt that will increase the likelihood of success. A series of errors in a row is an indication that simply having the user repeat the utterance, or rephrase it, is not working.

The best guess tactic can be combined with another correction strategy, switching to more constrained grammar, to increase its likelihood of success. For example, in the prior prompt, a directive of what utterances are available to the user can be added: "I think you are trying to create a message, is that correct? Please say yes or no." The problem with this strategy, is that the user cannot always be relied on to reply either yes or no when prompted to do so, as seen in the example below.

Computer: Do you want to book this flight? Please say yes or no.
User: Do have anything closer to 11 o'clock rather than noon?
Computer: Was that a yes?
User: No.

To help the user understand the boundaries of the system, one strategy is to include in the grammar the most commonly asked for functions that have not been implemented. For example, in a system that gives driving directions within the city limits of Miami, the system designer could consider including requests to commonly asked for areas that are outside of Miami, so that the system can respond appropriately. For example:

Computer: Welcome to Directions on Demand, the in-car system for driving directions to any location in Miami.
User: I want to go to Key West.
Computer: I think you said Key West, is that correct?
User: Yes.
Computer: I'm sorry, Key West is a well-known tourist attraction that is well outside the city limits of Miami. Is there someplace in Miami that you would like to go to?
User: How about the Little Havana?

This same type of problem can occur with a statistical-based Natural Language Understanding system, where the recognition of the speech is not tied to a grammar. In this case, the spoken utterance gets decoded correctly, but does not map to an existing function. For example in an e-mail system, the user could ask to open his Address Book, or to perform an e-mail–related function that has not yet implemented (e.g., "File this message in my Current Projects folder."). These requests should not be met with a failure repair prompt, but with an explanation that the user is requesting a function that has not yet been implemented.

Using Speech in Multimodal or Embedded Applications

Multimodal applications include other input and output modalities along with speech. For example, speech integrated with a desktop application would be multimodal, as would speech augmenting the controls of a personal note taker or a radio. Although many of the design issues for a multimodal application are the same as for a speech-only one, some specific issues are unique to applications that provide users with multiple input mechanisms, particularly graphical interfaces driven by the keyboard and mouse.

Feedback and Latency. As in speech-only systems, performance delays can cause confusion for users. Fortunately, a graphic display can show the user the state of the recognizer (processing or waiting for input) that a speech-only interface cannot. If a screen is available, displaying the results of the recognizer makes it obvious if the recognizer has heard and if the results were accurate.

As described previously, the transient nature of speech sometimes causes people to forget what they just said. When dictating, particularly when dictating large amounts of text, this problem is compounded by recognition errors. When a user looks at dictated text and sees it is different from what they recall saying, making a correction is not always easy because they will not necessarily remember what they said or even what they were thinking. Access to a recording of the original speech is extremely helpful in aiding users in the correction of dictated text.

The decision of whether or not to show infirm (not yet finalized) results is a problem in continuous dictation applications. Infirm results are words that the recognizer is hypothesizing that the user has said, but for which it has not yet committed a decision. As the user says more, these words may change. Infirm text can be hidden from the user, displayed in the text stream in reverse video (or some other highlighted fashion), or shown in a separate window. Eventually, the recognizer makes its best guess and finalizes the words. An application designer makes a trade-off between showing users words that may change and having a delay before the recognizer is able to provide the finalized results. Showing the infirm results can be confusing, but not showing any words can lead the user to believe that the system has not heard them.

Dealing with Microphone Issues.
An important issue to consider when designing multimodal applications is the impact of recognition errors from microphone problems. What type and quality of microphone will most users have? It is good to understand if users will be using a head-mounted microphone (most common because they are fairly inexpensive) or a higher quality hand-held microphone. With both of these, distance and angle of the microphone in relation to the mouth can cause fluctuations in the error rate. Usually, the users are unaware of these variations and fail to understand why sometimes they are understood really well, and at other times, nothing they say seems to be transcribed correctly. Additional problems can occur if using a noise-canceling microphone. These are good to use in noisy environments, but they are also very sensitive to directional issues. If the user turns his head away, or angles the microphone the wrong way, the recognition accuracy will suddenly drop off. It is no surprise that users, who are involved in their task, do not take notice of the position of the microphone and the distance from the mouth. Another problem is that users can find themselves speaking too softly or loudly.

It is a good idea to include in the interface a strong visual indicator of the current audio input level. The problem is that users often get involved in the task at hand and disregard the visual feedback. As such, designers might consider adding some auditory cue that input levels have dropped below a certain minimum or exceeded a maximum. This solution has the potential to annoy the user if not designed carefully, because we have all experienced how disruptive system sounds (e.g., chimes) can be when trying to get a task accomplished.

Prompting.
Prompts in multimodal systems can be spoken or printed. Deciding on an appropriate strategy depends greatly on the content and context of the application. If privacy is an issue, it is probably better not to have the computer speak out loud.

On the other hand, even a little bit of spoken output can enable eyes-free interaction and can provide the user with the sense of having a conversational partner rather than speaking to an inanimate object.

With a screen available, explicit prompts usually involve providing the user with a list of valid spoken commands. These lists can become cumbersome even when they are organized hierarchically.

Another strategy is to let users speak any text they see on the screen, whether it is menu text or button text or field names. In applications that support more than simple spoken commands, one strategy is to list examples of what the user can say next, rather than a complete laundry list of every possible utterance.

Handling Errors.
Multimodal speech systems that display recognition results make it easier for users to detect errors. If a rejection error occurs, no text will appear in the area where recognition results are displayed. If the recognizer makes a misrecognition or misfire error, the user can see what the recognizer thinks was said and correct any errors.

Even with feedback displayed, an application should not assume that users will always catch errors. Filtering for unexpected input is still helpful, as is allowing the user to switch to a different input modality if recognition is not working reliably.

INVOLVING USERS

Involving users in the design process throughout the life cycle of a speech application is crucial. A natural, effective interface can only be achieved by understanding how, where, and why target users will interact with the application.

Natural Dialogue Studies

At the very early stages of design, users can help to define application functionality and, critical to speech interface design, provide input on how humans carry out conversations in the domain of the application. This information can be collected by performing a *natural dialogue study*, which involves asking target users to talk with each other while working through a scenario. For example, if you are designing a telephone-based e-mail program, you might work with pairs of study participants. Put the participants in two separate rooms. Give one participant a telephone and a computer with an e-mail program. Give the other only a telephone. Have the participant with only the telephone call the participant with the computer and ask to have his or her mail read aloud. Leave the task open-ended, but add a few guidelines such as "be sure to answer all messages that require a response."

In general, natural dialogue studies are quick and inexpensive. It is not necessary to include large numbers of participants. In some natural dialogue studies, it is advantageous to include a subject matter expert. For example, if you wish to automate a telephone-based financial service, study participants might call up and speak with an expert customer service representative from the financial service company.

Natural dialogue studies are an effective technique for refining application requirements, collecting appropriate vocabulary, determining commonly used grammatical constructs, discovering effective interaction patters, helping with prompt and feedback design, and getting a feeling for the tone of the conversation.

Refining Application Requirements and Functionality. The natural dialogue studies can help to uncover unanticipated behaviors. As mentioned in the section on Design for Speech, users in a calendar study made extensive use of relative dates. For example, study participants said:

What's on his calendar for Monday?
Any appointments tomorrow?
And the day after that?
Is anything happening on Veteran's Day?
So, either the 11th or the 12th?
I'll get back to you at the beginning of the week.

This functionality was completely absent from the graphical calendar application the users were familiar with, and never came up in design discussions. The study pointed out how central relative dates were in calendar-related conversations and resulted in a speech user interface design based around their use.

Collecting Appropriate Vocabulary. Being able to select appropriate vocabulary is essential since out-of-vocabulary utterances are a common cause of recognition errors. As a designer, having to invent the vocabulary at the outset of the design process is quite difficult and error prone. Especially if a graphical application already exists, the temptation is to use the same vocabulary as in the graphical application. In the same calendar natural dialogue study, the study participants almost never used any of the words from the graphical calendar application, even though the study design involved an administrative assistant who had the graphical calendar open on her screen while she was talking on the phone. Speech user interfaces seem much more effective when they select the same vocabulary that humans use when talking about the domain.

Determining Commonly Used Grammatical Constructs. Grammatical constructs, both for prompts and for user input, are perhaps even more important than vocabulary. In a natural dialogue study for a multimodal drawing program, the study pointed out a command grouping that had not been anticipated—combined shape and color selection:

"I want a slate blue polygon."

The same study also suggested a new feature idea—allowing users to refer to objects that they created by name:

"Bring the tree to the front."

In a study of a system designed to take messages, there were many instances of users answering a question and proceeding with the next conversational move without a pause. For example:

Assistant: Hi. Michelle's in Abe Woll's office. Would you like me to take a message?
Visitor: Sure. Tell her I'm looking for a video card and a video camera for a system.

These studies made it clear that, although technically challenging to implement, this sort of functionality would be important to the success of the final system.

Discovering Effective Interaction Patterns. Predesign studies are also extremely helpful in pointing to effective interaction techniques. In a study prior to the design of a multimodal online catalog, a number of different navigational strategies emerged. Some people browsed the catalog hierarchically:

Women's clothing
Sweaters
V-neck

But others were interested in clothing items with a certain characteristic ("What do you have in cashmere?") or knew exactly what they wanted ("I want a men's cotton mock turtleneck in black."). The natural dialogue study led to a system design that supported the different browsing strategies. In addition, the study brought to light the importance of maintaining context to streamline the ordering dialog. Users often indicated information such as gender, size, and color before explicitly asking to place an order. The human customer service representative certainly never forgot any of these details. It was important that the computer not forget them either.

The multimodal drawing study helped provide insights into the sort of proactive verbal help that might be appropriate to integrate into a graphics editor. Observations revealed times when help conversations took place and also when they did not. When participants began to use the graphics editor, they asked their more knowledgeable partner quite a few questions. But once they started focusing on creating a drawing, these help-related interactions rarely occurred. It is just as important to know when *not* to talk as it is to know when to talk.

Helping with Prompt and Feedback Design. Prompts can be much easier to design after a natural dialogue study, particularly if you have the benefit of studying interactions with an expert—someone accustomed to performing the exact task you want the computer to perform. For example, the calendar study involved an administrative assistant who routinely managed others' calendars. Here's what a dialogue involving calendar lookups sounded like:

Ben: I would like to figure out what Tom's calendar looks like for next week.... Specifically, the late afternoon of the 23rd and the 24th.
Asst: OK. 23rd. He has a meeting at 9 o'clock.
Ben: No, How about later that afternoon?
Asst: Ok, it's open.

Ben: How about the 24th?
Asst: 24th. What time?

Notice the snappiness of the assistant's answers and how she combines feedback ("24th") with the next conversational move ("What time?").

In the online catalog study, an experienced customer service representative who was highly regarded by her employers for her ability to help customers over the telephone was enlisted to help with the study. Here's what a typical interaction sounded like:

Rep: (displays first-level directory) And Mary which of the product lines would you like to see today?
Mary: In the womens.
Rep: (displays women's clothing directory) These are our categories. Any particular categories that you're looking for there?
Mary: Sweaters.
Rep: (displays sweaters directory) Sweaters. OK, in the sweater line we carry the four different categories
Mary: How about the drifter.

Not only did the study point out the routine conversations, such as this, but it also made it possible to observe how the representative handled difficult questions about color, size, and fit. In some cases, such as when describing colors, we decided we could codify her knowledge and offer color descriptions to users of the automated system:

Customer: It's tough to see the difference between dark mahogany and the espresso.
Rep: Yes.
Customer: Can you describe those for me?
Rep: Dark mahogany has almost a burgundy cast, where the espresso is definitely a brown.

In other cases, however, such as in the discussions about size and fit, we came to the conclusion that the caller would have to be transferred to a live operator.

Customer: I guess the large would be the size.
Rep: OK. So the large actually fits like a 14/16.
Customer: Oh, it's that big! No, medium. It's for an 11-year-old child.
Rep: OK, medium. 10/12.

These conversations had too much complexity in terms of real-world knowledge, grammatical variance, and unpredictable vocabulary to attempt to design a speech interface to support those interactions. Sometimes there is no substitute for human intervention. Natural dialogue studies can help you understand the limitations of your design.

Getting a Feeling for the Tone of the Conversations.
Although conversational tone is not something we can quantify, it is an important aspect of human–human communication. In the online catalog project for example, the client's management said it was a requirement for the computer to treat customers in a friendly, unhurried, low-pressure manner. Tone also turned out to be key in the message-taking study. In their natural setting, office visitors exhibited "abbreviated" politeness:

Visitor: Is Sam here?
Tester: He's around but I don't know where he is. Do you want to leave a message?
Visitor: Yes. Tell him Anthony stopped by.

The interactions were very brief (10 to 20 seconds). Although visitors used few words, their tone (and facial expressions) indicated that this brevity was not a sign of rudeness or displeasure; rather, it was simply appropriate given the social context.

In sum, natural dialogues can serve as an effective starting point for a speech user interface design. Not only do they help in the design of grammars, feedback, and prompts, but they also point out instances in which speech technology cannot be effectively applied.

With some creative thinking about a study design, it is usually easy to construct a setting for interaction that approximates the interaction that will occur in the software system. Although it is useful to think through the ways a tester will handle various situations (when a completely natural setting is not being studied), it is also important not to script the dialogues. Let the tester's natural social inclinations drive the interaction, or set up a situation where the conversational partners are pairs of study participants. Avoiding scripts not only results in more natural dialogues, it also allows for the opportunity of unexpected interactions. In the final analysis, it is these unexpected events that often give the software its sparkle.

Wizard-of-Oz Studies

Once a preliminary application design is complete, but before the speech application is implemented, a wizard-of-oz study can help test and refine the interface. The speech data collected from this study can be used to help develop the grammar (Rudnicky, 1995) or, in the case of a natural language understanding system, to statistically train the engine on potential utterances. In these studies, a human wizard—usually using software tools—simulates the speech interface. Major usability problems are often uncovered with these types of simulations. (The term Wizard of Oz comes from the classic movie in which the "wizard" is actually a human hidden behind a curtain, controlling an impressive display.)

Given a speech-only e-mail project, a wizard-of-oz study could involve bringing participants into a lab and telling them that they will be interacting with computer that will read their e-mail to them. When they call a telephone number, the human wizard picks up the phone and manipulates the computer so that recordings of synthesized speech speak to the participant. As the participant asks to navigate through the mailbox, listen or reply to messages, the wizard carries out the operations and has the computer speak the responses. The experimenter can play the role of the wizard in a control room adjacent to the usability lab. It is preferable that the participants be visible to the wizard through a one-way mirror and heard through an audio system connecting the usability lab to the control room.

The Wizard plays the correct prompt based on the input spoken by the participant. Often, none of the participants suspect that they are not interacting with a real speech system (Kelley, 1984; Dahlbäch, 1993).

Although a system that accepts typed text and generates synthetic speech dynamically gives the wizard much more flexibility, it has two major drawbacks. The first is that typing text and generating synthesized speech introduce more latency than a system really should have. The second is that it introduces a lot of variability and makes a comparison of findings across participants harder to do. We prefer to prerecord (with synthetic speech) all the possible responses that the system can make before the start of the study. Each recording is grouped according to function, and represented by a push-button that the wizard presses to have the prompt played. One must be sure in a wizard-of-oz study to introduce some simulations of speech recognition errors. It is not hard for experimenters who have been in the speech field for a while to detect an utterance that would cause a failure in a real speech system (e.g., stuttering, false starts, or prolonged pauses). If the experimenter is less experienced with speech, it is sufficient to play a repair prompt sporadically, approximately 1 of 10 utterances. The problem with randomly playing repair prompts, however, is that it can sometimes baffle the user if it is in response to a well-formulated utterance that was accepted in the past (e.g., "Next message please.").

Because computer tools are usually necessary to carry out a convincing simulation, wizard-of-oz studies are more time-consuming and complicated to run than natural dialogue studies. If a prototype of the final application can be built quickly, it may be more cost-effective to move directly to a usability study.

Usability Studies

A usability study assesses how well users are able to carry out the primary tasks that an application is designed to support. Conducting such a study requires at least a preliminary software implementation. The application need not be complete, but some of the core functionality must be working. Usability studies can be conducted either in a laboratory or in the field. Study participants are typically presented with one or more tasks that they must figure out how to accomplish using the application.

With speech applications, usability studies are particularly important for uncovering problems because of recognition errors, which are difficult to simulate effectively in a wizard-of-oz study, but are a leading cause of usability problems. The effectiveness of an application's error recovery functionality must be tested in the environments in which real users will use the application.

Conducting usability tests of speech applications can be a bit tricky. Two standard techniques used in tests of graphical applications—facilitated discussions and speak-aloud protocols—cannot be used effectively for speech applications. A facilitated discussion involves having a facilitator in the room with the study participant. Any human–human conversation, however, can interfere with the human–computer conversation, causing recognition errors. Speak-aloud protocols involve asking the study participant to verbalize their thoughts as they work with the software. Obviously, this is not desirable when dealing with a speech recognizer. It is best, therefore, to have study participants work in isolation, speaking only into a telephone or microphone. A tester should not intervene unless the participant becomes completely stuck. A follow-up interview can be used to collect the participant's comments and reactions.

SUMMARY

An effective speech application is one that uses speech to enhance a user's performance of a task or enable an activity that cannot be done without it. Designing an application with speech in mind from the outset is a key success factor. Incorporating speech into an interface involves dealing with challenges such as transience, invisibility, and asymmetry as well as having a clear understanding of the current capabilities of synthesized output and speech recognition. Speech designers must decide on a dialogue style for their application, ensure that system latency does not create pauses in unnatural places, and provide adequate feedback so that the conversation can proceed smoothly. The key to facilitating successful speech-only interactions is in crafting the prompts so they cue the user what to say, and are not so detailed as to be burdensome for the user. Also, the way the system is designed to handle recognition errors and provide help can make a dramatic difference in the quality of a user's experience.

Involving users in the design process throughout the life cycle of a speech application is critical. Basing the dialogue design on a natural dialogue study ensures that the input grammar will match the phrasing actually used by people when speaking in the domain of the application. A natural dialogue study also ensures that prompts and feedback follow conversational conventions that users expect in a cooperative interaction. Once an application is designed, wizard-of-oz and usability studies provide opportunities to test interaction techniques and refine application behavior based on feedback from prototypical users.

References

Ballentine, B., & Morgan, D. (1999). *How to Build a Speech Recognition Application: A Style Guide for Telephony Dialogues.* Published by Enterprise Integration Group, Inc., San Ramon, California.

Clark, H. (1993). *Arenas of language use.* Chicago: University of Chicago Press.

Dahlbäch, N., Jonsson, A., & Ahrenberg, L. (1993). Wizard-of-oz studies—why and how. *IUI '93 Conference Proceedings,* 193-199.

Francis, A., & Nusbaum, H. (1999). Evaluating the quality of synthetic speech. In D. Gardner-Bonneau (Ed.), *Human factors and voice interactive systems.* Boston, MA: Kluwer Academic Publishers.

Gong, L., & Lai, J. (2001). Shall we mix synthetic speech and human speech? Impact on users' performance, perception, and attitude. *CHI '2001 Conference on Human Factors in Computing Systems*, Seattle, WA.

Kamm, C. (1994). User interfaces for voice applications. In D. B. Roe & J. G. Wilpon (Eds.), *Voice communication between humans and machines*. Washington, DC: National Academy Press.

Kelley, J. F. (1984, March). An iterative design methodology for user-friendly natural language office information applications. *ACM Transactions on Office Information Systems, 2*(1), 26–41.

Lai, J., Wood, D., & Considine, M. (2000). The effect of task conditions on the comprehensibility of speech. *CHI '2000 Conference on Human Factors in Computing Systems*. The Hague, Netherlands.

Makhoul, J., & Schwartz, R. (1994). State of the art in continuous speech recognition. In D. B. Roe & J. G. Wilpon (Eds.), *Voice communication between humans and machines*. Washington, DC: National Academy Press.

Nass, C., & Gong, L. (2000, September). Speech interfaces from an evolutionary perspective. *Communications of the ACM, 43*(9).

Raman, T. V. (1997). *Auditory user interfaces: Towards the speaking computer*. Boston, MA: Kluwer Academic Publishers.

Reeves, B., & Nass, C. (1996). *The media equation: How people treat computers, television and new media like real people and places*. New York: Cambridge University Press/CSLI.

Rudnicky, A. (1995). The design of spoken language interfaces. In A. Syrdal, R. Bennett, & S. Greenspan (Eds.), *Applied speech technology*. Boca Raton, FL: CRC Press.

Schmandt, C. (1994). *Voice communication with computers: Conversational systems*. New York: Van Nostrand Reinhold.

Weinschenk, S., & Barker, D. (2000). *Designing effective speech interfaces*. New York: John Wiley & Sons.

Yankelovich, N. (1996, November/December). How do users know what to say? *ACM Interactions, III*(6).

Yankelovich, N., Levow, G. A., & Marx, M. (1995, May 7–11). Designing speech acts: Issues in speech user interfaces. *'95 Conference on Human Factors in Computing Systems*. Denver, CO.

·36·

THE WORLD WIDE WEB

Jonathan Lazar
Towson University

INTRODUCTION

In the decade of the 1990s, the World Wide Web grew from an experimental application used by researchers, to an integral part of work, play, shopping, and education. But the World Wide Web is not like traditional information system design. There are a number of quirks of the web environment, which pose a challenge for human–computer interaction researchers and professionals. These quirks include a distributed, unreliable network; a diverse, distributed population of users; and various different technologies, browsers, platforms, connection speeds, and plug-ins.

Along with these challenges, the web has "sped up the clock" of information systems development. Organizations want web sites created, and they want them yesterday! The more traditional creation of systems and applications provides a relatively structured development process. As development time is steadily decreased because of client demand, structured planning, user analysis, usability testing, and other good habits of system design are frequently diminished. The result is a Web site product that is less user-friendly.

The empirical research on users and user behavior is just beginning to emerge, which hopefully will impact on Web site design, and improve the functionality and usability of Web sites. The research related to the web is not limited only to the usability of Web sites. The research includes a full range of issues in the development life cycle: requirements gathering, usability testing, training, documentation, user behavior, user demographics, and evaluation. This chapter will present the latest research on a number of different topics related to the World Wide Web. The World Wide Web certainly intersects with a number of other topics related to human–computer interaction that are discussed in this handbook. It is expected that those who are interested in the web environment will also be especially interested in chapter 16: Network-Based Interaction, chapter 23: Global/Intercultural User-Interface Design, chapter 30: Online Communities, chapter 32: User-Centered Interdisciplinary Design of Wearable Computers, chapter 38: E-Commerce Interface Design, and chapter 40: Government Roles in Human–Computer Interaction.

CHALLENGES OF DESIGNING FOR THIS ENVIRONMENT

The World Wide Web is essentially an application domain that runs on the network of the Internet, where people share documents and files, usually in the form of documents called web pages. The World Wide Web uses a series of languages and protocols, such as HTML and HTTP. Because of the inherent design of the Internet (as a distributed network), there are a number of issues that influence the interaction experience of the World Wide Web.

An Unpredictable, Distributed Network

The web is a distributed, packet-switched network, and the response times are inherently unpredictable. This is related to a number of factors, such as network load, distance, connection speed, and server speed (Sears & Jacko, 2000). Most of these factors are outside of the control of the web developer. Not only can responses to web page requests become slow, but some web page requests may not be able to be fulfilled because of factors outside of the user's control (Lazar & Norcio, 2000). For instance, a remote Web site may have failed, which is not caused by the actions of the user, and at the same time, the user cannot do anything about it (Johnson, 1998). Therefore, web browsing is an inherently unpredictable experience.

Browser Incompatibility

Another major challenge to web design is the issue of browser incompatibility. Users are accessing the web using many different browsers (Internet Explorer, Netscape Navigator, Lynx, Opera, Neoplanet, etc.), different versions of those browsers (3.0, 4.0, 5.0), and different platforms (Windows, Mac, Linux, etc.). Although there are programming standards specified by the World Wide Web Consortium <http://www.w3.org>, these standards are rarely followed exactly. Instead, the individual web browsers are designed to support many HTML standards, but not support some other standards, and many browsers add some of their own tags (Lazar, 2001). Not only is it important to support different browser applications, but also to support older versions of browsers, because it may take users up to 2 years to upgrade browsers one version (e.g., from version 4.0 to version 5.0; Lynch & Horton, 1999; Nielsen, 2000).

Numerous browser incompatibility examples exist. For instance, the <marquee> tag is only supported in Internet Explorer and is not a part of the HTML standards. The <colgroup> tag is a part of the HTML standards, but is not supported by the Netscape browser. The browser incompatibilities only get worse when dealing with cascading style sheets and JavaScript. The result can truly be called chaos. The same web page may appear differently in each of five different browsers. In addition, if code is written sloppily (with missing end tags and incorrectly nested HTML tags), most versions of Internet Explorer will "assume" the missing tags, whereas most versions of Netscape Navigator will not assume the missing tags and will just not present the web content at all (Lazar, 2001). Although it would be nice if the browsers all followed accepted standards, in reality, it is the web designer who must face up to the challenge and design a site that appears appropriately and is easy to use in a number of different browser applications/versions/platforms.

Another challenge in designing for the web is the absence of training. In mainstream software applications (such as WordPerfect, MS-Excel, and MS-Access) or organizationally specific software applications (such as records applications, shared network applications, Intranets, etc.), users receive training and documentation on how to use the application (Hoffer, George, & Valacich, 1999). Users of web browsers rarely receive training, and even if training is received, the training relates to the use of the browser (Lazar, 2001; Lazar & Norcio, 2000). It is impossible to provide adequate training on how to navigate through the many thousands of Web sites that the user may visit. Therefore, Web sites should be self-explanatory, with a minimum learning

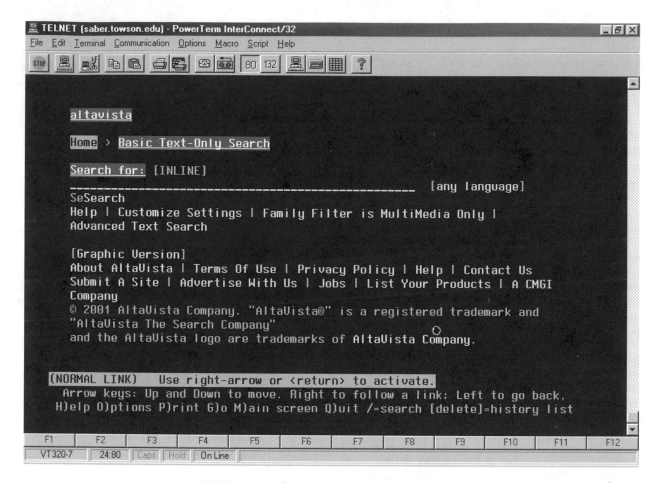

FIGURE 36.1. A first-generation text browser, Lynx.

Generations of Web Browsers

As the web environment has grown, users have been moving through three generations of web browsers. The first web browsers were text-based. In fact, the original standards for sharing documents were based on text standards. Users could share documents that were strictly text-based, using applications such as Gopher and Lynx. Lynx (see Fig. 36.1) is still used today, primarily by users who want to avoid graphics because they have slow connection speeds or they want fast downloads. In addition, Lynx can be a useful tool for testing a Web site, because if a Web site is usable with a text-only browser, such as Lynx, chances are good that it will meet basic requirements for accessibility for users with disabilities (Lazar, 2001).

The second generation of web browsers could be considered to be graphical, desktop browsers. These are the graphical browsers such as Netscape Navigator and Internet Explorer that are commonly used today by a majority of users. Multimedia is a major component of these web browsers. Web content can include graphics, animation, audio, and video. One of the

main challenges of this environment is the fact that the users access many different Web sites, most of which probably have a very different layout, design, and navigation scheme. What this means is that the user is, in effect, dealing with two different interfaces. One interface is the *browser interface* (see Fig. 36.2). This interface consists of the scrollbars, pull-down menus, title bar, buttons, and other traditional user interface widgets. This interface remains consistent in daily use. The other interface is the *site interface*, the interface of the Web site that is being viewed. This interface lacks consistency and is inherently unpredictable. The site interface changes from Web site to Web site. Although the user can be provided training on how to use the browser interface, it is impossible to provide training on how to use the site interface, because the site interface changes based on what Web site the user is currently viewing (Lazar, 2001). Users cannot develop a mental model of how a Web site works because it changes from site to site (with the exception of organizational Intranets).

Because users spend a small amount of time on a large number of Web sites, consistency is an important design goal for Web sites (Nielsen, 2000). Designers should try to follow conventional web style, for example, that hyperlinks are underlined and unvisited links appear as blue text, visited links appear as purple or red links (Nielsen, 2000). Although graphic designers

FIGURE 36.2. A second-generation graphical web browser.

may want their site to appear to be different, to be unique, in reality, the user needs a Web site that is consistent with other Web sites, and even similar in terms of the design and layout (Lazar, 2001). For instance, the Microsoft Office software would be impossible to use if the interface was unpredictable and changed every day (Lazar, 2001). Yet, Web sites that are designed with unique interfaces that are incomprehensible to the user accomplish the same thing. Because most users spend a small amount of time on a large number of Web sites, users do not learn from visit to visit (Lazar, 2001; Nielsen, 2000). Instead of different designs, the content is what should differentiate between Web sites.

The third generation of web browsers are those that are now included in mobile devices such as PalmPilots and mobile phones, using protocols such as WAP (the Wireless Access Protocol) and WML (the Wireless Markup Language). Users can now access some web content through portable, hand-held devices. Major design challenges for both the devices themselves, and the web content designed for the devices, are the screen size and the input design (Marcus & Chen, 2002). For instance, in designing the PalmPilot, one of the design goals was the get the most value out of every screen pixel (Bergman & Haitani, 2000). On such a small screen, no pixels can be wasted. In terms of designing the web content, the challenge is to "shrink" the web page so that the same content is available and usable on the smallest screens. At this point, these browsers are primarily text-based, because graphics and animation would greatly decrease usability with such small screens and slow wireless connections. However, this is beginning to change with the newest generation of wireless networks and devices (Sacher & Loudon, 2002). In addition, because each screen of data is so small, users frequently are required to scroll and navigate within screens and between screens (Metz, 2001). Another problem is the input design. WAP-enabled phones may use traditional phone keypads, which are insufficient input devices for the tasks being performed, because they essentially double the

number of keys that must be pressed. Other portable devices, such as the Blackberry, have full keyboards, but the keys are very small. Although WAP accesses web content, there are major challenges to developing effective web content for portable devices, and this is beyond the scope of this chapter. For more information on mobile and portable devices, see chapter 32: User-Centered Interdisciplinary Design of Wearable Computers.

USABILITY OF WEB SITES

When Web sites are created, the usability of the Web site should be an important concern. Web sites should be designed so that they are easy to use. The users should be able to easily find the content that interests them, and users should not frequently be confused or get "lost in cyberspace." There are a number of separate areas relating to the usability of Web sites. These include navigation design, accessibility, internationalization, and information architecture.

Navigation Design

In web navigation, users need information on where they have been, where they are, and where they can go. There are two different types of web navigation: the navigation mechanisms built into the web browser and the site navigation built into the Web site. Some mechanisms for users to determine where they have been are already built into the browser (Cockburn & Jones, 1996). The browser records URLs that were visited previously and allows users to go back through previously visited pages using the "back" button. However, the navigation pattern recorded by the browser and accessible by the back button can usually be only one path in a hierarchy, and the back button cannot be used to access other previously visited paths within the same hierarchy (Cockburn & Jones, 1996). In addition, the

FIGURE 36.3. An example of sectional navigation (on the left side of the page).

links in the web page appear in a different color when they have previously been visited (usually red or purple) than when they have not been visited (usually blue). It is interesting to note that the navigation information offered as a part of the most popular browsers has not really changed since much earlier versions of the browsers.

It is extremely important to provide navigation throughout a Web site. Web site navigation should be like the map at a shopping mall—it should let the users know where they are and where they can go (Lazar, 2001). Frequently, users are directed to the home page of a Web site, the so-called "front door" to the site, either through the URL or through a link from another site. However, a Web site is a collection of web pages, and the user must therefore be provided with a way of navigating through the many web pages on the Web site. Users may have accessed a search engine to find a Web site; however, once a user enters a specific Web site, navigation, not a search engine, is the main method for finding information on all but the largest Web sites (Lee, 1999). Additionally, once users enter a Web site, many users do not use the navigation mechanisms available in their browser (such as the "back" button) to retrace their path, but instead, navigate forward, using the navigation provided by the web page (Spool, Scanlon, Schroeder, Snyder, & DeAngelo, 1999).

Navigation is a major part of the user experience on the Web site, therefore, careful attention needs to be paid to designing appropriate and usable navigation that allows users to get where they want to go. For a small Web site, it might be feasible to provide a link to every page on the site, but once there are more than 10–15 web pages, this approach is no longer feasible. Therefore, links are usually provided to the different content sections of a Web site (also called sectional navigation). See Fig. 36.3 for an example of sectional navigation on the Seattle Times Web site. From each of those "content" pages (equivalent to sections of the newspaper), links are provided for specific content pages (equivalent to news articles). If there are a number of different, clearly defined user populations for a large Web site, it is also possible to provide separate navigation through a technique called audience-splitting, where the home page provides separate links to different web pages. Each of these web pages then have appropriate navigation maximized for a specific user population (Lazar, 2001; Lynch & Horton, 1999). Furthermore, all pages on a Web site should provide some level of navigation beyond just a link to the home page, because the user should not be required to return to the home page every time that they want to begin a new navigation path through the site structure (Newfield, Sethi, & Ryall, 1998). Web site navigation that provides an overview to the users of what is available on the site can also reduce user disorientation (Shneiderman, 1997). An additional method of navigation is through the use of path navigation, also called "breadcrumbs navigation" (Nielsen, 2000). In path navigation, users are presented with hierarchical information on the path that they took to get to the current web page, which can help in providing the user an overview of the structure of the Web site. Path navigation and traditional sectional navigation are not mutually exclusive. Some web pages (such as Yahoo!) provide both types of navigation.

In addition to providing users with site navigation that allows them to reach their task goals, the navigation on the web page must itself be easy to use. For instance, navigational links for the Web site should be only text-based. If site navigation is provided using only graphical objects (with no alternative text), anyone who is using a text browser or a graphical browser with the graphics turned off will not be able to navigate through the site. If navigation is provided using graphical objects, the download time will also be increased (see the section on download time). Another reason to provide text-based navigation is that it allows the users to utilize the navigation as soon as the Web site starts to download, without waiting for all graphical objects to load (Fleming, 1998; Lazar, 2001). See Fig. 36.4 for an example of this problem. Whatever approach for navigation is chosen should remain constant throughout the entire

Web site, so as not to disorient the user, who may wonder whether they are still on the same Web site (Fleming, 1998; Shubin & Meehan, 1997). Not only should the navigation remain constant, but users should also be informed as to where on the site they are (Nielsen, 2000). This can be done in a number of different ways. If there is a navigational link to the page that the user is currently viewing make that link a deactivated link, a different color, or some indication placed next to the link (Lazar, 2001). The name of the current web page can also be presented, right next to the navigational link under which the web page can be found.

Another consideration for site navigation is where on the web page the navigation will be located. It is strongly advised not to place navigation on either the bottom of the web page or the left side of the screen, because it is questionable whether the users will actually see the navigation. Rather, navigation should appear on the top of the web page and/or the left side of the web page, because these areas will generally be the first that the users look at (Lazar, 2001). Figure 36.5 presents navigation that is not easily visible to the user. The site navigation will only appear if the user scrolls down, which assumes that the user is aware of the navigation. Further information on Web site navigation is available in Jennifer Fleming's book, *Web Navigation: Designing the User Experience*.

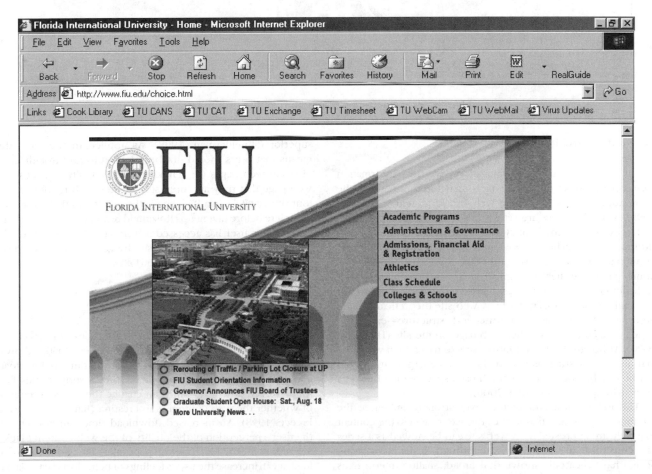

FIGURE 36.4. Graphical navigation that has not loaded yet (below "Colleges & Schools").

FIGURE 36.5. Navigation that is hard for the user to find.

Information Architecture

Closely related to the navigation is the information architecture—how the Web site is structured, and how the users can navigate through the Web site (Rosenfeld & Morville, 1998). Most Web sites are set up in some form of a top-down hierarchy, similar to a family tree or an organizational chart (Rosenfeld & Morville, 1998). Usually, there is a "top page" (the home page, where the user enters the Web site), "middlemen home pages" (either for topical areas, or for different user audiences), and "content pages" (where the content of interest is actually located; Lazar, 2001). A Web site theoretically could be designed so that it is a full network structure—each web page is linked to every other web page on the site (Huizingh, 2000). Although this would allow users to navigate in whatever manner they wished, as a Web site became larger, this would not be feasible, because too much page space would be taken up providing links (Huizingh, 2000).

The information architecture will strongly influence the number of web pages that it takes the user to access the content that they are interested in. A Web site can be viewed as a series of menus that a user must navigate through. Research on menu design has repeatedly shown that broad, shallow menu trees (where users go through few menus with many choices) are superior (Shneiderman, 1998). As applied to the web, this means that users should not be required to go through many different web pages (or "clicks") on a Web site to access the web page that they are interested in. Research has found that four or five clicks in a Web site is the maximum that a user will complete before giving up (Rosenfeld & Morville, 1998). Therefore, once a user has accessed a Web site, the user should be able to access all pages on the site by going through no more than four or five clicks (Lazar, 2001).

Download Time

Download time is one of the major frustrations of users (Lazar, 2001; Lightner, Bose, & Salvendy, 1996; Nielsen, 2000; Pitkow & Kehoe, 1996). When web pages take a long time to download, it can concern users. A number of studies support this fact. An increased download time can change the user's perception of whether the web content is interesting (Ramsay, Barbesi, & Preece, 1998). An increased download time can also change the user's perception of the quality of the web content (Jacko, Sears, & Borella, 2000). If a web page takes a long time to download, it can increase the user's feelings of being lost (Sears, Jacko, & Dubach, 2000). The longer a user must wait for a web page to

download, the harder it is for the user to remember information about the context of the task and the structure of the Web site (Shubin & Meehan, 1997). In addition, a long download time may cause users to believe that an error has occurred (Lazar & Norcio, 1999, 2000, 2002). General human–computer interaction guidelines by Shneiderman suggest a computer response time of no more than 10–15 seconds from the point at which the user has entered their commands (Shneiderman, 1998). Nielsen suggests that a 10-second response time really is the limit for the web (Nielsen, 2000). Although the actual download time of a web page can vary because of a number of factors (server load, network traffic, etc.), only one of these factors can actually be designed as a part of the Web site itself: the file size (Lazar, 2001; Sears & Jacko, 2000). Designers can control the size of the web documents themselves, by limiting the use of graphics, animation, and applets, and by "cleaning up" the code by eliminating any unneeded HTML tags. Many web development applications can add extra and unneeded HTML tags, and the web designer should try and eliminate these. It is also helpful to create web pages that "appear" to download faster than they actually do. For instance, all navigational links and as much content as possible should be provided with text, because text is the first thing to appear, before any graphics. As soon as the text appears, the user can start interacting with the web page. If all navigational links are provided with graphical pictures, this means that the user cannot interact with the navigational links until all pictures have downloaded. In addition, if large graphics must be included as a part of a web page, it might be helpful to have a low-resolution (and low-file size) version of the graphic appear first, to be replaced by the full image, when the page completely loads. This approach can be used by providing two different source files within the *<img>* tag for the graphics, one, using *src = "filename"* for the large graphic, and one using *lowsrc = "filename"* for the smaller graphic that will download first.

Accessibility to Users With Disabilities

Another important usability consideration is designing a Web site that is accessible to many different populations, including users with disabilities. For instance, users with visual impairment might be viewing a Web site using adaptive devices, such as screen readers (e.g., IBM Home Page Reader, JAWS, or HAL) or braille printers (Paciello, 2000). Users with hearing impairment may be unable to utilize streaming audio or other multimedia (Paciello, 2000). Web sites should be made accessible to these populations. The World Wide Web Consortium has created a set of guidelines for accessibility, which is available at <http://www.w3.org/WAI>. A very basic measure to improve accessibility is to provide textual equivalents for any multimedia content (graphics, sound, video, etc.; Sullivan & Matson, 2000). That way, any users who are not able to utilize graphical or audio content could still use adaptive devices to present the text (Laux, 1998). Frames and clickable bitmaps also decrease accessibility (Laux, 1998; Paciello, 2000). Unfortunately, a large portion of the most popular Web sites still are not accessible (Sullivan & Matson, 2000). The U.S. Federal Government

requires (as of June 2001) their Web sites to be accessible, and it is predicted that, at some point, similar rules will apply to large American corporations (Lazar et al., 2001). Other countries are also moving toward requiring certain categories of Web sites to be accessible (Paciello, 2000).

Designing for accessibility is part of a larger trend called universal usability, in which information systems are designed to be easy to use for all populations, including users with disabilities, older users, economically disadvantaged users, and users with different technology, download speed, and browsers (Shneiderman, 2000). Hochheiser and Shneiderman (2001a) suggest the use of universal usability statements on Web sites, similar to privacy statements, which inform the user about usability concerns, such as the web browser requirements, display requirements, accessibility, and suggested connection speed. Much as packaged software provides information on the minimum system requirements, such as processor speed, disk storage, and RAM, a Web site should include information such as the necessary browser requirements, suggested connection speed, and any plug-ins that are required for a successful user interaction with the Web site (Hochheiser & Shneiderman, 2001a). Universal usability is closely related to the concepts of transgenerational design and universal design. Transgenerational design attempts to design products that do not have accessibility problems related to age, whereas universal design focuses on accessibility and usability to maximize the number of people who can use a product (Keates, Clarkson, Harrison, & Robinson, 2000). Although universal design focuses on designing objects such as homes, furniture, public spaces, and transportation, universal usability focuses on computer equipment, informational systems, and Web sites. An example of a Web site that is not universally usable is in Fig. 36.6. This Web site requires the use of the Flash plug-in, and if a user is accessing the Web site using a Netscape browser, nothing but a background graphic appears, and no information is presented to the user on how to proceed. The Web site in Fig. 36.6 is also completely unusable with a text browser. The University of Maryland has a comprehensive Web site on universal usability at <http://www.otal.umd.edu/UUPractice/>. More information on accessibility is available in chapter 24: Information Technology for Cognitive Support, chapter 25: Physical Disabilities and Computing Technologies, and chapter 26: Perceptual Impairments and Computer Technologies.

Internationalization

A Web site can be accessed by anyone connected to the Internet, including users outside of the Web site's country of origin. These international users should be considered in the Web site design. To make a Web site more internationally usable, globalization is one option (Alvarez, Kasday, & Todd, 1998). Globalization is when country-specific cultural references, symbols, and abbreviations are removed. For instance, the Macintosh "trash can" is frequently interpreted as a postal box by British users (Sears et al., 2000). Another common problem is numerical formats such as date, time, and currency (Sears et al., 2000). For instance, date formats can be percieved differently, so that

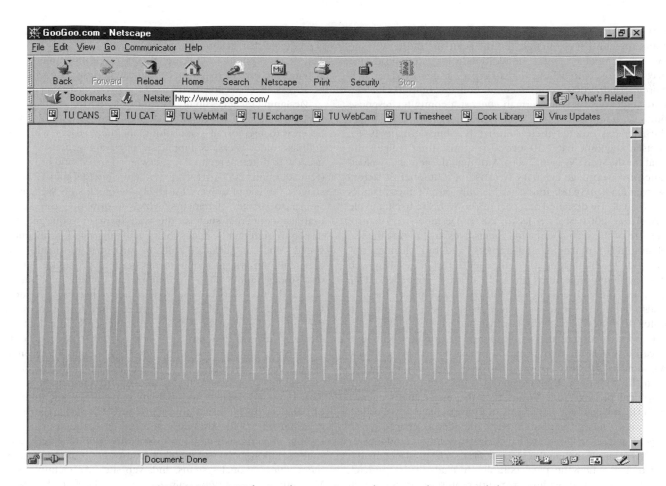

FIGURE 36.6. A Web site that requires a plug-in application and does not provide any alternatives or instructions to the user, and is therefore not universally usable.

03-08-01 would represent March 8, 2001 in the United States, but it would represent August 3, 2001 in Europe (Sears et al., 2000). Writing the date out as 8-March-2001 can clear up the confusion.

If it is expected that a Web site will be accessed by a large number of people from a specific culture, the Web site should be localized. Localization is when the Web site is added to and modified to meet the specific needs of a different population (Alvarez et al., 1998). An example of this would be to present the Web site in multiple languages. Many larger scale companies (such as Disney; see Fig. 36.7 and Amazon.com) simply have multiple Web sites for different countries, with each Web site created to meet the language needs, as well as the cultural needs, of a specific population. After redesigning a Web site for a local population, usability testing with users from the specific culture is especially important (Nielsen, 1996). Usability problems that exist for users of a different culture might not be obvious (or even known) to usability testers from the United States.

In addition to multilingual Web sites, some search engines, such as Google <http://www.google.com> are starting to provide language translation services built in. If a Web site

in a different language is returned as a part of a query, Google will attempt to translate the Web site into the local language. Although these translation services are certainly not perfect, this is an important step toward the internationalization of Web sites. More information on internationalization of user interfaces is available in chapter 23: Global/Intercultural User-Interface Design.

Use of Graphics and Animation

The use of graphics and animation can be a highly controversial topic in web development circles. Some designers feel that it is always better to add more graphics and images to a page to attract the user's attention. On the other hand, human–computer interaction experts point out that all of these images can be distracting to the users, who are trying to complete a task. A number of important findings come out of a study by Sears, Jacko, and Dubach (2000). A large number of graphics on a Web site may improve user perception of ease of use in a Web site, but this perception might be related to the user's connection speed. In their experiment, Sears et al. found that web

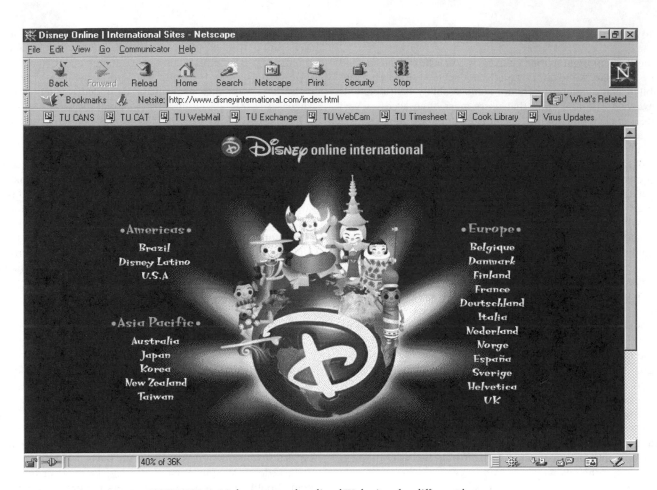

FIGURE 36.7. Links to many localized Web sites for different languages and cultures.

users from Switzerland, who tend to have lower connection speeds, were not as impressed by high-end graphics as users from the United States (Sears et al., 2000). Furthermore, as the download time for web pages increased, users found that they felt lost (Sears et al., 2000). There is certainly a direct correlation between an increased number of graphics and an increased download time. However, the download time is magnified by a slow dial-up connection. How users perceive a large number of multimedia objects on a Web site may therefore be affected by the local connection speed of the user. If a web page contains mainly text, and it takes a long time to download, users may feel more positive about the experience than if the same web page takes a long time to download, and it contains many graphics. The users are likely to perceive that the long download time is due to the graphics themselves (Jacko et al., 2000). Ideally, the user would be aware that he can control their environment and simply turn off all graphics (Lazar, 2001). However, for this to be effective, the site must be designed to be just as usable when graphics are turned off as when they are turned on, meaning that all multimedia objects have textual equivalents (for more information, see the section on accessibility issues in this chapter). Unfortunately, many Web sites are not designed with textual browsing usability in mind (Rowan, Gregor, Sloan,

& Booth, 2000; Sullivan & Matson, 2000). In addition, many users simply are not aware of the control that they have over their browsing environment (Ellis & Kurniawan, 2000; Lightner et al., 1996).

Animation is another type of multimedia enhancement with a debatable value. Animation on a web page is frequently used as a marketing technique to grab the user's attention (Lazar, 2001). This has the same effect of distracting users from their tasks. Experimental research by Zhang (2000) has determined that animation does decrease the user performance on information-seeking tasks. This is not to say that all animation is bad; certainly, animation might be appropriate as content on entertainment sites. However, animation does distract users and lowers performance on information-seeking tasks.

Another example of a high-end enhancement is the mouseover, which is commonly seen on Web sites. A mouseover is where an image changes when the user's mouse is placed over the image. Nielsen points out that most mouseovers have absolutely no value to the user, because they do not provide any additional content or useful information to the user (Nielsen, 2000). Instead, mouseovers increase the time that it takes for the web page to download, because of the additional code and graphics required to perform the mouseover.

How Users Find A Web Site

Another important usability consideration is how the users find the Web site. Although this might not seem at first to be a usability consideration, there are some usability considerations involved. The four ways that users most often access a Web site are: (1) typing in the domain name (either through knowledge or guessing), (2) using a search engine, (3) following a link from a different Web site, and (4) selecting a Web site saved as a bookmark in their browser (Cockburn & Jones, 1996). Because a bookmarked Web site implies that the user has previously visited a Web site, the following sections will discuss the other three ways that users find a Web site.

When a URL is chosen for a Web site, the domain name should be one that is short, easy to remember, and not commonly misspelled. The URL should then be "marketed" to the user population, so that they are familiar with the location of the Web site and will know how to find it (Lazar, 2001). However, users frequently attempt to find Web sites by "guessing" at the URL, based on their knowledge of the organization that sponsors the Web site. Therefore, a domain name should be one that matches the user's mental model and is easily guessable by the user. This means that a domain name should not add additional underscores or dashes. If an organization has a short name, this might be appropriate for the domain name. If an organization has a long name, if any acronyms are used, these acronyms might be appropriate as a domain name. For example, the Web site for Omicron Delta Kappa, the National Leadership Honor Society, is <http://www.odk.org>, and the organization is commonly known as "ODK." This would be easier to guess than any of the following:

<http://www.omicrondeltakappa.org>—long and easily misspelled.

<http://www.omicron-delta-kappa.org>—how do users know that dashes are needed?

<http://www.odknet.org>—how do users know that "net" is a part of the domain name?

<http://www.odkweb.org>—how do users know that "web" is a part of the domain name?

Many users find Web sites using search tools such as Hotbot, Google, or Altavista. With the exception of very experienced users, most users tend to have trouble forming complex queries to search electronic information in general (Marchionini, 1995) and search the web specifically (Jansen & Pooch, 2001; Shneiderman, 1997). Because a majority of web users tend not to be experienced with complex, boolean-based searches, most web queries submitted to search engines tend to be simple and use few terms (Jansen & Pooch, 2001). The different search engines also have different interfaces, offer different types of searching, and return results using different approaches, further confusing users (Shneiderman, 1997). In addition, users will not view all web pages returned as part of the query, and, in fact, typically view around 10 of the Web sites found in the search query (Jansen & Pooch, 2001). Therefore, it is important for the web designer to "go the extra mile" to make sure that users can easily find the Web site when using

a search engine. A number of actions can be taken to help users find a Web site through a search engine. For instance, <meta> tags can be placed in the HTML document, supplying appropriate keywords for the Web site. A number of search tools allow Web sites to manually be added to their search databases (Lazar, 2001). Information about the Web site, such as URL, owner, description, and keywords, can be supplied to the search engine, allowing for more effective searches.

When users follow a link from one Web site to another, it should be clear to them what Web site they are about to access. Clear information should be provided as to what the link is. The link should clearly describe the Web site and should not say something such as "click here" (Nielsen, 2000). In addition, link titles (using the *title* attribute of the <a> tag in HTML) can help inform the user more about the link (Nielsen, 2000). If a link title is placed in the HTML, when the user moves their mouse over a link, a bubble will pop up, describing the link in more detail.

DEVELOPMENT METHODOLOGIES FOR THE WEB

It is accepted that user involvement is necessary for building successful informational systems. With user involvement, appropriate attention can be paid to the system functionality and usability needed by the user population. These two aspects of informational systems, functionality and usability, are predictors of user acceptance of both traditional informational systems and Web sites (Abels, White, & Hahn, 1997; Davis, 1989; Goodwin, 1987; Lederer, Maupin, Sena, & Zhuang, 1998). User involvement should be more than simply users performing usability testing. Not only must a Web site be easy to use, it must also provide some content that the users are interested in (the functionality!). Unfortunately, with the short development times expected for Web site development, frequently users are not involved in any way with the development of the Web site. The result of this is that many Web sites (1) do not offer content that users are interested in and/or (2) are not considered by users to be easy to use. Although some might think that Web sites do not have a targeted set of users, in reality, a majority of Web sites are targeted toward a specific population (Lazar, 2001).

Before creating or redesigning a Web site, it is important to determine what the mission of the Web site is and who the target user population is. Although some Web sites (such as search engines and news agencies) are geared toward the entire user population of the web, most Web sites are in fact geared toward a targeted population of users (Lazar, 2001). It is important to determine in advance who the targeted users are. Requirements gathering should take place with those targeted users to determine what their content needs are and why they would return to a Web site. A parallel of the target user population is the mission of the Web site. Why is the site being brought into existence? What do the site owners hope that the site will achieve? Web sites generally fall into one of three categories: informational, e-commerce, or entertainment (Lazar, 2001), and a large portion of Web sites are informational (Huizingh, 2000). If there is a mismatch between what content the users want,

and the content offered by the Web site, then users may not return to the Web site. User involvement in the web development process can ensure that the users have a Web site with appropriate functionality and usability. There are a number of different levels of user involvement in the web development process.

Users can be involved in web development at a very basic level by performing usability testing. In this model, users are not involved with the creation (or redesign) of the Web site and have no say in the content development. Users are only brought in after the Web site has been developed, but before it "goes live." Users perform usability testing on the Web site, with the hope that the users' feedback can improve the usability of the Web site before it is released to the public.

A more active level of user involvement is the user-centered web development model by Lazar (2001). In this model, users are involved at two stages of web development: requirements gathering and usability testing. Determining appropriate content for a Web site is especially important. One study found that the web content offered was the best predictor of the overall user experience (Sinha, Hearst, Ivory, & Draisin, 2001). Another study found that, over time, as the users get acquainted with the interface of a Web site, that the importance of appropriate content increases, to encourage users to return to the Web site (Davern, Te'eni, & Moon, 2000). Users help to determine what content the Web site should have, and what usability standards would be appropriate for the Web site. After a working model of the Web site has been created, the users then usability test the Web site to look for any usability problems (Lazar, 2001). This web development model was used to design and redesign the Web sites at many different organizations. Three well-documented examples are Indiana University (Corry, Frick, & Hansen, 1997), PlayFootball.com (sponsored by the National Football League; Clarke, 2001), and the Eastman Kodak Company (Yu, Prabhu, & Neale, 1998). Users were at the center of the web design process for these Web sites. The user-centered development model has also been used in schools, governmental agencies, religious groups, and nonprofit organizations. This two-stage model of user involvement (requirements gathering and testing) is similar to the user involvement described in many systems analysis and design projects (Hoffer et al., 1999; Whitten & Bentley, 1997).

The ultimate level of user involvement in Web site development is the participatory design model. In participatory design, users actually become members of the design team and take part in all phases of development. Ellis and Kurniawan (2000) describe the use of participatory design to redesign a Web site for senior citizens. Participatory design was a useful methodology in this case because it allowed a deeper understanding of the usability challenges faced by older adults (Ellis & Kurniawan, 2000). The limitation to participatory design for Web site development is that participatory design requires that the users make a great commitment of time. This is only feasible when users can provide the time for being a part of the design team, and when users will be using a Web site frequently enough that they will make the commitment to join the participatory design team. More information on participatory design is available in chapter 54: Participatory Design: The Third Space in Human–Computer Interaction.

USABILITY TESTING AND EVALUATION OF WEB SITES

With any type of informational system, usability testing and other forms of evaluation are important steps in the development process (Nielsen, 1994). In the environment of the web, where there are so many potential usability problems, getting user feedback on ease of use is especially important. There are a number of new developments that relate specifically to the usability testing of Web sites and the use of the web for usability testing. More general information on usability testing is available in chapter 56: User-Based Evaluations and chapter 57: Inspection-Based Evaluations.

Usability Testing Over the Web

Usability testing for a Web site can take place in a traditional usability laboratory or in a user's natural task environment (Nielsen & Mack, 1994). But, for Web sites, usability tests can also take place over the web, where the usability test subject and the human factors team are separated by time and/or space (Hammontree, Weiler, & Nayak, 1994; Hartson, Castillo, Kelso, & Neale, 1996). Usability data can be collected through remote videoconferencing or automatic data logging (Hammontree et al., 1994; Hartson et al., 1996). At a more basic level, usability data could be collected simply by asking users to note problems with usability and other ways in which the system does not meet their needs (Hartson et al., 1996). If videoconferencing equipment is not available, even simple phone conversations can be used (Hammontree et al., 1994). There are a number of advantages to performing usability testing over the web. The network itself is part of the user's environment and usage patterns, so usability testing performed over the web will more realistically match the user's task environment (Hartson et al., 1996). The costs associated with renting usability equipment or travel costs for either users or developers to visit usability test labs can be lessened (Hartson et al., 1996). A wider range of users might be available to take part in remote usability testing (Hammontree et al., 1994). Realistic aspects of the user experience on the web—such as high server transaction loads, slow download time, and web server failure—may not be present in the usability lab, but are apparent when performing usability testing over the web (Lazar, Tsao, & Preece, 1999). In addition, remote usability testing might be helpful in continuing to improve usability after an informational system has been implemented, when usability testing in a laboratory frequently does not take place (Hartson & Castillo, 1998).

Usability Guidelines for the Web

Usability testing can be performed by representative users (user-based testing) or interface developers/experts (usability inspections; Nielsen, 1994; Nielsen & Mack, 1994). Usability inspections include the heuristic review, cognitive walkthrough, consistency inspection, and guidelines review (Nielsen & Mack,

1994). The heuristic review, where interface experts compare a new interface with a short set (7–10) of design rules, is one of the most popular types of usability inspections. New sets of heuristics have been developed, specifically for Web sites. Levi and Conrad (1996), based on their usability work at the Bureau of Labor Statistics, developed a set of nine usability heuristics. These heuristics focus on the layout of commonly accessed information, clear navigation, and consistent and organized information (Levi & Conrad, 1996). Another set of heuristics for web usability is Jakob Nielsen's "Top 10 Mistakes in Web Design," which is available at: <http://www.useit.com/alertbox/9605.html>. These heuristics focus on use of frames, page design, download time, and use of high-end graphics. Barnum adapted Nielsen's 1994 heuristics (Nielsen & Mack, 1994) for the web (Barnum, 2001). The 10 Nielsen/Barnum heuristics focus on navigation, clear language, browser compatibility, and consistency (Barnum, 2001). Fogg et al. (2001) recently published a set of heuristics for improving the credibility of Web sites. Web site credibility is the perception of trustworthiness and expertise (Fogg et al., 2001). These heuristics for trustworthiness focus on making Web sites that convey the physical (building-based) nature of the organization, have clear policies, and convey the quality and expertise of the organization (Fogg et al., 2001).

Another expert evaluation that is frequently used is the guidelines review. A guidelines review is when an expert compares an interface with a large set (possibly in the 100s) of design guidelines (Shneiderman, 1998). One set of guidelines is the *Web Style Guide*, from Lynch and Horton (1999). This is also sometimes known as the *Yale Style Guide*. Another set of web-specific usability guidelines is the *Web Content Accessibility Guidelines*, created by the World Wide Web Consortium. These guidelines focus on designing Web sites that are easy to use for users with disabilities. These guidelines are available at http://www.w3.org/WAI.

Automated Usability Testing

Automated usability testing is an emerging trend for Web sites. In automated usability testing, a web page is compared by a computer with a human-created list of usability guidelines. WEBSAT, the Web Static Analyzer Tool, was created at the National Institute of Standards and Technology (Scholtz, Laskowski, & Downey, 1998). WEBSAT compares a web page to two different sets of usability guidelines. WEBSAT is available for download at <http://zing.ncsl.nist.gov/WebTools/WebSAT/overview.html>. LIFT is a tool that finds usability and accessibility problems. More information, and a trial version, is available at: http://liftonline.usablenet.com. Another example of an automated usability testing tool is BOBBY. BOBBY focuses on Web site usability problems related to accessibility. When one enters the URL of a web page, the BOBBY service immediately scans the web page and reports problems related to accessibility. This service also ranks the accessibility problems according to severity. BOBBY is available at <http://www.cast.org/bobby/> and can be seen in Fig. 36.8. Both WEBSAT and BOBBY are available free-of-charge, but a number of companies sell more advanced automated usability testing applications. An excellent summary of usability testing applications that are available on a fee-paying basis can be found in an article by Brajnik (2000).

Evaluation Surveys for Web Sites

After a Web site has been implemented, web-based surveys are frequently placed on Web sites to assist in evaluating whether the Web site is meeting the needs of the users (Lazar & Preece, 2001). For instance, surveys have been used extensively in evaluating the Human–Computer Interaction/Ease of Use Web site at IBM (Dong & Martin, 2000; Lisle, Dong, & Isensee, 1998). There are a number of important considerations in designing web-based surveys. Surveys should first be written on paper, and the survey questions should be pretested for clarity (Lazar & Preece, 1999). Once the web version of the survey has been coded, it should be usability tested, and any opportunities for error (such as a check box instead of a radio button when a user should only select one choice) should be noted and eliminated (Lazar & Preece, 1999). An appropriate sampling methodology should be developed, so that it can be determined whether the web surveys received are actually representative of the user population, or if the responses are biased (Lazar & Preece, 2001).

Although organizations can certainly create and use their own surveys for evaluating Web sites, a number of Web site evaluation surveys have already been created and tested. These evaluation surveys look at different aspects of a Web site. For instance, the WEBMAC (WEB Motivational Analysis Checklist) suite of surveys have been developed by Small and Arnone (2000) and focus on the motivational quality of a Web site. Motivational quality is the concept of why users explore a Web site and return at a later time (Small & Arnone, 2000). Different WEBMAC surveys exist for business Web sites, e-commerce sites, and educational sites, and are available at <http://istweb.syr.edu/~digital/CDC/Resources/resources.htm>. Another survey tool that has been developed is the WAMMI (Web Analysis and MeasureMent Inventory). The WAMMI focuses on measuring user satisfaction with a Web site (Kirakowski, Claridge, & Whitehand, 1998). More information about the WAMMI is available at <http://www.wammi.com>. A third survey under development is the Information Quality Survey. The information quality survey, by Zhang, Keeling, and Pavur (2000), focuses on the user's perception of information quality of the Web site. Although WebMAC and WAMMI are commercially available products, the Information Quality survey is still under development.

Other Methods of Web Site Evaluation

Evaluation of currently existing Web sites can also be performed without direct user involvement in the evaluation, but through analysis of existing data on user behavior. Web site logs (on the web server) record data such as the web pages and related

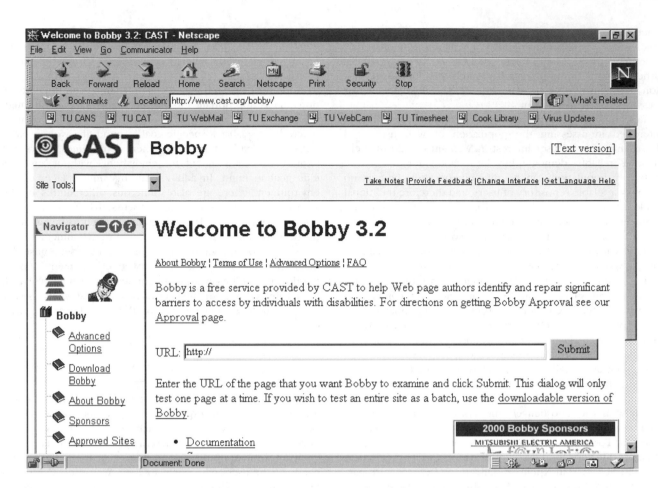

FIGURE 36.8. BOBBY, an automated usability testing tool.

objects that were requested, the time of the request, and the browsers used (Bertot, McClure, Moen, & Rubin, 1997; Stout, 1997). In addition, data about what other Web sites are driving traffic to the web server, as well as pages requested in error, might be available (Bertot et al., 1997; Stout, 1997). These logs can be useful in evaluation, because they provide a record of what web pages (and related files) are most popular and which web pages are least popular. Data about browser usage can help ascertain what browsers are important benchmarks for compatibility testing (Lazar, 2001). Data on how users find the Web site, and pages requested in error, can assist with the marketing of the Web site (Lazar, 2001). Numerous software packages are available to assist in statistical analysis.

This log data can also be analyzed using interactive visualization, which can assist in determining patterns of requests, sources of requests, and patterns of site navigation across time (Hochheiser & Shneiderman, 2001b). Information scent is another emerging topic, in which user navigation paths can be analyzed to determine the information goals of users (Chi, Pirolli, & Pitkow, 2000). Using this feedback on information scent, designers can redesign Web sites to better meet the information needs of users on a specific site (Chi et al., 2000).

CURRENT RESEARCH

The World Wide Web has realistically only been in existence for approximately a decade and has only been heavily used for perhaps 6 or 7 years. Whereas some areas of human–computer interaction have been researched for more than two decades, research related to the World Wide Web is still emerging. This section will offer a sampling of some of the newest areas of human–computer interaction research related to the World Wide Web.

One area of increasing research is user behavior on the web. Tauscher and Greenberg (1997) report on how frequently users revisit the same web pages. In their study, more than half of the web pages visited by an individual had been previously viewed by that same individual (Tauscher & Greenberg, 1997). Byrne, John, Wehrle, and Crow (1999) developed a "taskonomy," reporting what web-based tasks are most common for users to perform. These tasks include using information, locating information on a page, providing information, and configuring a browser application (Byrne et al., 1999). The tasks performed closely relate to the design of the web browser, because the browser application should closely support the user's tasks.

Since the user interacts with the web browser application, this seems to be a promising area of research to improve the user experience (Chen, Wang, Proctor, & Salvendy, 1997). Other areas of browser-related research may include navigation mechanisms (Newfield et al., 1998), bookmarks, error messages, and screen layout.

Researchers, policymakers, and marketers are certainly interested in assessing the population of web users and determining overall user interest in Web sites. Unfortunately, creating reliable demographic information for the entire web user population can be difficult, because there is no centrally available repository of users, and therefore traditional random sampling is hard to do (Lazar & Preece, 2001). The most reliable statistics have come out of Georgia Tech, with their WWW User Surveys. These surveys were performed twice a year from 1994 to 1998, and the results are available at <http://www.gvu.gatech.edu/user_surveys>. The Georgia Tech Web site indicates that they are about to start another major survey. Some surveys (mostly from marketing companies) attempt to determine which Web sites are most popular among the population of web users. For instance, Media Metrix <http://www.mediametrix.com> reports on the popularity of Web sites. Other population studies look at more specific web tasks, such as overall online purchasing habits of web users (Bellman, Lohse, & Johnson, 1999).

Much of the research on electronic commerce has dealt mainly with sales and delivery models. However, research on the human part of the e-commerce transaction is finally starting to appear. For a successful task completion, an e-commerce site must provide very good usability, as well as quality products at a good price, fast, inexpensive delivery, and easy, secure payment (Lazar & Norcio, 2001). Miles, Howes, and Davies (2000) provide a good framework for examining the human–computer interaction issues in electronic commerce. Many of the problems in e-commerce sites may relate to usability. For instance, in research reported by Tilson, Dong, Martin, and Kieke (1998), users were unhappy at the number of clicks to view an item, disliked the pull-down menus, and had trouble identifying which graphics were links. In a New York Times article, it was reported that when IBM improved the usability of their e-commerce site

to better meet the needs of the users, sales increased nearly 400% (Tedeschi, 1999). For more information on e-commerce, it is suggested that the reader look at chapter 38 on E-Commerce Interface Design.

Additional research may focus on training and documentation that is appropriate for the web environment. Although many traditional information systems tasks are goal-oriented, where the user has a specific goal in mind, the web in many ways is an exploratory environment. Sometimes, users just explore the environment for enjoyment, without any specific goal in mind. In addition to this unique facet of the environment, there are also numerous other quirks of the web environment (which were discussed in an earlier portion of this chapter). Training methods need to be modified to meet the different needs of the web environment. In addition, the documentation that is provided to users must also meet the unique aspects of the web environment. Providing documentation on using an application such as MS-Word, with a limited, well-defined set of possible task goals, is very different from documentation for those exploring the web environment.

SUMMARY

As more organizations build Web sites, as more people become web users, and as more organizational applications are deployed through web browsers, the human–computer interaction aspect of the web becomes increasingly important. For-profit companies and nonprofit organizations alike cannot ignore the needs of the users who come to their Web sites. Careful attention must be paid to user concerns throughout the entire development process for Web sites. Although the research on human–computer interaction issues in the web environment is starting to appear, more empirical studies must still be performed. More case studies of web development projects based around user concerns need to be published. The area of web development changes rapidly, and the research on user issues in Web sites must keep up with it to inform Web site design for improved user interactions.

References

Abels, E., White, M., & Hahn, K. (1997). Identifying user-based criteria for web pages. *Internet Research: Electronic Networking Applications and Policy, 7*(4), 252–262.

Alvarez, M., Kasday, L., & Todd, S. (1998). How we made the Web site international and accessible: A case study. *Proceedings of the 1998 Human Factors and the Web Conference* [On-line]. Available: www.research.att.com/conf/hfweb

Barnum, C. (2001). Usability testing and research. New York: Longman.

Bellman, S., Lohse, G., & Johnson, E. (1999). Predictors of online buying behavior. *Communications of the ACM, 42*(12), 32–38.

Bergman, E., & Haitani, R. (2000). Designing the Palm Pilot: A conversation with Robert Haitani. In E. Bergman (Ed.), *Information appliances and beyond*. San Francisco, CA: Morgan Kaufmann Publishers.

Bertot, J., McClure, C., Moen, W., & Rubin, J. (1997). Web usage statistics: Measurement issues and analytical techniques. *Government Information Quarterly, 14*(4), 373–395.

Brajnik, G. (2000). Automatic web usability evaluation: What needs to be done? *Proceedings of the Human Factors and the Web* [On-line]. Available: http://www.tri.sbc.com/hfweb/

Byrne, M., John, B., Wehrle, N., & Crow, D. (1999). The tangled web we wove: A taskonomy of www use. *Proceedings of the CHI 99: Human Factors in Computing* (pp. 544–551).

Chen, B., Wang, H., Proctor, R., & Salvendy, G. (1997). A human-centered approach for designing world wide web browsers. *Behavior Research Methods, Instruments, & Computers, 29*(2), 172–179.

Chi, E., Pirolli, P., & Pitkow, J. (2000). The scent of a site: A system for analyzing and predicting information scent, usage, and usability of

a web site. *Proceedings of the CHI 2000: Human Factors in Computing* (pp. 161-168).

Clarke, J. (2001). Key factors in developing a positive user experience for children on the web: A case study. *Proceedings of the 2001 Human Factors and the Web Conference* [On-line]. Available: http://www.optavia.com/hfweb/index.htm

Cockburn, A., & Jones, S. (1996). Which way now? Analysing and easing inadequacies in WWW navigation. *International Journal of Human-Computer Studies, 45*(1), 105-129.

Corry, M., Frick, T., & Hansen, L. (1997). User-centered design and usability testing of a web site: An illustrative case study. *Educational Technology Research and Development, 45*(4), 65-76.

Davern, M., Te'eni, D., & Moon, J. (2000). Content versus structure in information environments: A longitudinal analysis of website preferences. *Proceedings of the International Conference on Information Systems* (pp. 564-570).

Davis, F. (1989). Perceived usefulness, perceived ease of use, and user acceptance of information technology. *MIS Quarterly, 13*(3), 319-340.

Dong, J., & Martin, S. (2000). Iterative usage of customer satisfaction surveys to assess an evolving web site. *Proceedings of the Human Factors and the Web* [On-line]. Available: http://www.tri.sbc.com/hfweb/

Ellis, R. D., & Kurniawan, S. (2000). Increasing the usability of online information for older users: A case study in participatory design. *International Journal of Human-Computer Interaction, 12*(2), 263-276.

Fleming, J. (1998). *Web navigation: Designing the user experience.* Sebastopol, CA: O'Reilly & Associates.

Fogg, B., Marshall, J., Laraki, O., Osipovich, A., Varma, C., Fang, N., Paul, J., Rangnekar, A., Shon, J., Swani, P., & Treinen, M. (2001). What makes Web sites credible? A report on a large quantitative study. *Proceedings of the CHI 2001: Human Factors in Computing* (pp. 61-68).

Goodwin, N. (1987). Functionality and usability. *Communications of the ACM, 30*(3), 229-233.

Hammontree, M., Weiler, P., & Nayak, N. (1994). Remote usability testing. *Interactions, 1*(3), 21-25.

Hartson, H., & Castillo, J. (1998). Remote evaluation for post-deployment usability improvement. *Proceedings of the Conference on Advanced Visual Interfaces* (pp. 22-29).

Hartson, R., Castillo, J., Kelso, J., & Neale, W. (1996). Remote evaluation: the network as an extension of the usability laboratory. *Proceedings of the CHI: Human Factors in Computing* (pp. 228-235).

Hochheiser, H., & Shneiderman, B. (2001a). Universal usability statements: Marking the trail for all users. *Interactions, 8*(2), 16-18.

Hochheiser, H., & Shneiderman, B. (2001b). Using interactive visualizations of WWW log data to characterize access patterns and inform site design. *Journal of the American Society for Information Science and Technology, 52*(4), 331-343.

Hoffer, J., George, J., & Valacich, J. (1999). *Modern systems analysis and design.* Reading, MA: Addison-Wesley.

Huizingh, E. (2000). The content and design of Web sites: an empirical study. *Information & Management, 37,* 123-134.

Jacko, J., Sears, A., & Borella, M. (2000). The effect of network delay and media on user perceptions of web resources. *Behaviour and Information Technology, 19*(6), 427-439.

Jansen, B., & Pooch, U. (2001). A review of web searching studies and a framework for future research. *Journal of the American Society for Information Science and Technology, 52*(3), 235-246.

Johnson, C. (1998). Electronic gridlock, information saturation, and the unpredictability of information retrieval over the world wide web. In P. Palanque & F. Paterno (Eds.), *Formal methods in human-computer interaction* (pp. 261-282). London: Springer.

Keates, S., Clarkson, P., Harrison, L., & Robinson, P. (2000). Towards a practical inclusive design approach. *Proceedings of the ACM Conference on Universal Usability* (pp. 45-52).

Kirakowski, J., Claridge, N., & Whitehand, R. (1998). Human centered measures of success in web site design. *Proceedings of the Human Factors and the Web* [On-line]. Available: http://www.research.att.com/conf/hfweb/

Laux, L. (1998). Designing web pages and applications for people with disabilities. In C. Forsythe, E. Grose, & J. Ratner (Eds.), *Human factors and web development.* Mahwah, NJ: Lawrence Erlbaum Associates.

Lazar, J. (2001). *User-centered web development.* Sudbury, MA: Jones and Bartlett Publishers.

Lazar, J., Kumin, L., & Wolsey, S. (2001). Universal usability for web sites: Current trends in the U.S. law. *Proceedings of the Universal Access in Human-Computer Interaction 2001 Conference* (pp. 1083-1087).

Lazar, J., & Norcio, A. (1999). To err or not to err, that is the question: Novice user perception of errors while surfing the web. *Proceedings of the Information Resource Management Association 1999 International Conference* (pp. 321-325).

Lazar, J., & Norcio, A. (2000). System and training design for end-user error. In S. Clarke & B. Lehaney (Eds.), *Human-centered methods in information systems: Current research and practice* (pp. 76-90). Hershey, PA: Idea Group Publishing.

Lazar, J., & Norcio, A. (2001). User considerations in E-commerce transactions. In Q. Chen (Ed.), *Human-computer interaction: Issues and challenges* (pp. 185-195). Hershey, PA: Idea Group Publishing.

Lazar, J., & Norcio, A. (2002). Novice user perception of error on the web: Experimental findings. Paper under review.

Lazar, J., & Preece, J. (1999). Designing and implementing web-based surveys. *Journal of Computer Information Systems, 39*(4), 63-67.

Lazar, J., & Preece, J. (2001). Using electronic surveys to evaluate networked resources: From idea to implementation. In C. McClure & J. Bertot (Eds.), *Evaluating networked information services: Techniques, policy, and issues* (pp. 137-154). Medford, NJ: Information Today.

Lazar, J., Tsao, R., & Preece, J. (1999). One foot in cyberspace and the other on the ground: A case study of analysis and design issues in a hybrid virtual and physical community. *WebNet Journal: Internet Technologies, Applications, and Issues, 1*(3), 49-57.

Lederer, A., Maupin, D., Sena, M., & Zhuang, Y. (1998). The role of ease of use, usefulness, and attitude in the prediction of world wide web usage. *Proceedings of the Conference on Computer Personnel Research* (pp. 195-204).

Lee, A. (1999). Web usability: A review of the research. *SIGCHI Bulletin, 31*(1), 38-40.

Levi, M., & Conrad, F. (1996). A heuristic evaluation of a world wide web prototype. *Interactions, 3*(4), 50-61.

Lightner, N., Bose, I., & Salvendy, G. (1996). What is wrong with the world wide web? A diagnosis of some problems and prescription of some remedies. *Ergonomics, 39*(8), 995-1004.

Lisle, L., Dong, J., & Isensee, S. (1998). Case study of development of an ease of use web site. *Proceedings of the 1998 Human Factors and the Web Conference* [On-line]. Available: http://www.research.att.com/conf/hfweb/

Lynch, P., & Horton, S. (1999). *Web style guide: Basic design principles for creating web sites.* New Haven: Yale University Press.

Marchionini, G. (1995). *Information seeking in electronic environments.* Cambridge, England: Cambridge University Press.

Marcus, A., & Chen, E. (2002). Designing the PDA of the future. *Interactions, 9*(1), 34–44.

Metz, C. (2001, March 29). WAP usability: What's holding it back? *PC Magazine.*

Miles, G., Howes, A., & Davies, A. (2000). A framework for understanding human factors in web-based electronic commerce. *International Journal of Human-Computer Studies, 52*, 131–163.

Newfield, D., Sethi, B., & Ryall, K. (1998). Scratchpad: Mechanisms for better navigation in directed web searching. *Proceedings of the ACM UIST 1998 Conference* (pp. 1–8).

Nielsen, J. (1994). *Usability engineering.* Boston: Academic Press.

Nielsen, J. (1996). International usability engineering. In E. DelGaldo & J. Nielsen (Eds.), *International user interfaces.* New York: John Wiley & Sons.

Nielsen, J. (2000). *Designing web usability: The practice of simplicity.* Indianapolis: New Riders Publishing.

Nielsen, J., & Mack, R. (Eds.). (1994). *Usability inspection methods.* New York: John Wiley & Sons.

Paciello, M. (2000). Web accessibility for people with disabilities. Lawrence, KS: CMP Books.

Pitkow, J., & Kehoe, C. (1996). Emerging trends in the WWW population. *Communications of the ACM, 39*(6), 106–110.

Ramsay, J., Barbesi, A., & Preece, J. (1998). A psychological investigation of long retrieval times on the World Wide Web. *Interacting with Computers, 10*, 77–86.

Rosenfeld, L., & Morville, P. (1998). *Information architecture for the world wide web.* Sebastopol, CA: O'Reilly and Associates.

Rowan, M., Gregor, P., Sloan, D., & Booth, P. (2000). Evaluating web resources for disability access. *Proceedings of the ACM ASSETS Conference* (pp. 80–84).

Sacher, H., & Loudon, G. (2002). Uncovering the new wireless interaction paradigm. *Interactions, 9*(1), 17–23.

Scholtz, J., Laskowski, S., & Downey, L. (1998). Developing usability tools and technique for designing and testing web sites. *Proceedings of the Human Factors and the Web* [On-line]. Available: http://www.research.att.com/conf/hfweb/

Sears, A., & Jacko, J. (2000). Understanding the relation between network quality of service and the usability of distributed multimedia documents. *Human-Computer Interaction, 15*(1), 43–68.

Sears, A., Jacko, J., & Dubach, E. (2000). International aspects of WWW usability and the role of high-end graphical enhancements. *International Journal of Human-Computer Interaction, 12*(2), 243–263.

Shneiderman, B. (1997). Designing information-abundant web sites: Issues and recommendations. *International Journal of Human-Computer Studies, 47*(1), 5–29.

Shneiderman, B. (1998). *Designing the user interface: Strategies for effective human-computer interaction* (3rd ed.). Reading, MA: Addison-Wesley.

Shneiderman, B. (2000). Universal usability: Pushing human-computer interaction research to empower every citizen. *Communications of the ACM, 43*(5), 84–91.

Shubin, H., & Meehan, M. (1997). Navigation in web applications. *Interactions, 4*(6), 13–17.

Sinha, R., Hearst, M., Ivory, M., & Draisin, M. (2001). Content or graphics? An empirical analysis of criteria for award-winning websites. *Proceedings of the Human Factors and the Web 2001 Conference* [On-line]. Available: http://www.optavia.com/hfweb/index.htm/

Small, R., & Arnone, M. (2000). Evaluating the effectiveness of web sites. In B. Clarke & S. Lehaney (Eds.), *Human-centered methods in information systems: Current research and practice* (pp. 91–101). Hershey, PA: Idea Group Publishing.

Spool, J., Scanlon, T., Schroeder, W., Snyder, C., & DeAngelo, T. (1999). *Web site usability: A designer's guide.* San Francisco: Morgan Kaufmann Publishers.

Stout, R. (1997). *Web site stats.* Berkeley, CA: Osborne McGraw Hill.

Sullivan, T., & Matson, R. (2000). Barriers to use: Usability and content accessibility on the web's most popular sites. *Proceedings of the ACM Conference on Universal Usability* (pp. 139–144).

Tauscher, L., & Greenberg, S. (1997). Revisitation patterns in world wide web navigation. *Proceedings of the CHI 97: Human Factors in Computing* (pp. 399–406).

Tedeschi, B. (1999, August 30). Good web site design can lead to healthy sales. *The New York Times.*

Tilson, R., Dong, J., Martin, S., & Kieke, E. (1998). A comparison of two current e-commerce sites. *Proceedings of the ACM Conference on Computer Documentation* (pp. 87–92).

Whitten, I., & Bentley, L. (1997). *Systems analysis and design methods.* Boston: Irwin McGraw-Hill.

Yu, J., Prabhu, P., & Neale, W. (1998). A user-centered approach to designing a new top-level structure for a large and diverse corporate web site. *Proceedings of the 1998 Human Factors and the Web Conference* [On-line]. Available: http://www.research.att.com/conf/hfweb/

Zhang, P. (2000). The effects of animation on information seeking performance on the world wide web: Securing attention or interfering with primary tasks? *Journal of the Association for Information Systems, 1*(1) [Online]. Available: http://jais.aisnet.org/

Zhang, X., Keeling, K., & Pavur, R. (2000). Information quality of commercial web site home pages: an explorative analysis. *Proceedings of the International Conference on Information Systems* (pp. 164–175).

·37·

INFORMATION APPLIANCES

W. P. *Sharpe*
Appliance Studio Ltd.

S. P. *Stenton*
Hewlett-Packard Research Labs

INTRODUCTION

The notion of information appliances emerged two decades ago. In their early conception, they evolved as electronic diaries and phone lists. Visionaries (Kay & Goldberg, 1977) foretold of personal information devices in book format that would provide all the functions of personal information management, access to reference material and personal communication. Initial market entrants ranged from pocket calculators plus personal information management features to mini-portable PCs with stripped-down capabilities to sustain portability, battery life, and mobile use. Apart from some of the personal organizers, most of these early attempts failed because the technology was not advanced enough to build appliances at a price more than a few corporate globetrotters (BIS Strategic Decisions, 1993) would be willing to pay and because the concept of information appliance was not well understood.

While many product failures were littering the fire sale pages of the technical magazines of the early adopters, a wide range of information products were building successful positions in the markets where the question of whether or not they were an information appliance did not matter. Sales of mobile, cordless, and corded phones and calculators continued to grow throughout the 1980s and 1990s. Cameras, camcorders, dictaphones, video recorders, televisions, and radios all sold in numbers envied by the digital pretenders to the information appliance category. Despite the academic ridicule of the interface-challenged video recorder (Thimbleby, 1990), by 1992 this appliance had reached 82% of U.K. family households (Central Statistical Office, 1992). In this chapter, we take this broad category of information products as all belonging to the field of information appliances, and we look at the properties that can ensure their successful evolution as they gain new information capabilities. Rather than see information appliances as derived purely from the evolution of PCs, we propose it is more useful to look at the appliance space as emerging from all these different products evolving alongside the PC.

The evolution of PCs toward ideas of information appliances followed the success of the PC in the early 1990s. The PC-web-peripheral infrastructure has created its own route toward applianceness by extending and specializing the PC's functions through additional interface modalities. Such PC-web-peripherals have been in constant tension with appliance evolution because of the inherent contradiction between the generality of the PC model and restricted functionality that is the defining notion of appliances. Despite this contradiction, the PC continues to play a dominant role in the evolution of the information appliance space. Although retaining their generality, PCs are acquiring a growing set of devices that act as companion products, or as extensions to the interface for specific applications. Personal Digital Assistants (PDAs), e-books, and MP3 players emerged first as media slaves to PC masters, each one sold with a PC CD-ROM and connecting cable. Another evolutionary branch of the PC is seen in the new Internet appliance products that several companies are trying to establish, which are essentially tablet PCs dedicated to web applications of browsing and e-mail.

Whereas the PC and its surrounding devices will continue to play a major role in the evolution of appliances, we take the point of view that too much attention on its central role is misleading for those wanting to understand what appliances are for and how to design them. Because of the central role it has played so far in mediating our use of other devices, there has been insufficient attention in the world of human computer interface design to how to give appliances their own direct interface to the user appropriate to the task. In the coming age of appliances, we expect them all to interoperate as peers, with the PC just playing its part as a particularly powerful and flexible one. We have been constrained to use the PC for many purposes for which its general nature is ill-suited, solely because technology has been too expensive to be widely distributed into many specialized devices, and the processing demands of natural interfaces such as speech and pen have put them beyond reach. These restrictions are rapidly disappearing at the same time as communications infrastructure is rapidly expanding, with the result that we can anticipate an explosion of appliance innovation over the next few years.

Concern with whether a particular product is, or is not, an information appliance, and debates about the role of PCs distract from the central question of how to guide products into widespread, everyday use as they absorb information processing capabilities. In this chapter, our aim is to describe these attributes under the heading of "applianceness," with some examples from the work of Hewlett-Packard (HP) Laboratories in Bristol, the Appliance Studio Ltd., and others.

APPLIANCES AND "APPLIANCENESS"

The usual first port of call for an appliance definition is the dictionary. Norman (1998) quotes the following from the American Heritage Dictionary (3rd edition):

• **Appliance** *n*. A device or instrument *designed to perform a specific function*, especially an electrical device, such as a toaster, for household use. Synonyms: tool, instrument, implement, utensil. From which he derives the definition:

• **Information appliance** *n*. An appliance specializing in information: knowledge, facts, graphics, images, video, or sound. An information appliance is designed to perform a specific activity, such as music, photography, or writing. A distinguishing feature of information appliances is the ability to share information among themselves.

Of course the true meaning of words is in their everyday usage. Dictionary definitions are merely snapshots taken for reference as modern languages evolve. Many people in the electronics industry are currently using "appliance" to mean primarily hand-held PCs, organizers, and specialized Internet browsers or e-mailers (i.e., specialized or reduced function PCs). Which devices we come to call appliances may well depend on their penetration into our everyday lives.

The perspective we put forward in this chapter starts from the notion of an appliance in the everyday world as a *device of*

specialized and widespread use (e.g., television, dishwasher, toaster, vacuum cleaner). If you accept this usage, then there are already many *information* appliances: digital cameras, telephones, printers, scanners, etc., that do one specific thing to information of a certain type. So we prefer to think of appliance-ness as the features that together take a product category into pervasive everyday use and to look at the way these features apply to the evolving landscape of information products.

Applianceness, as we use it here, is therefore a set of properties that guide the design process toward simple, helpful devices that exploit the potential of embedded Information Technology (IT) in everyday things. This has become an important opportunity in the electronics industry because of the explosion of technologies crossing size, price and performance thresholds, and the advent of a supporting and almost pervasive wired and wireless communication infrastructure. We now need to shift the focus of our attention from the one computer that does everything to many more specialized devices that need to be simple in themselves and in the ways they work together.

Applianceness is nothing to do with technical sophistication. It is true, technology sophisticates like information appliances, and are the early adopters of them, just as they are the early (and only) adopters of complicated devices. Conversely, appliance-ness *is* critical to pervasive technology adoption by nontechno enthusiasts.

Information Appliance Characteristics

Information appliances as we conceive them have the following characteristics:

- Everyday things
- Focused function
- Work together as peers
- Portable
- Personal
- Support task closure
- Direct user interface

Everyday Things. Successful appliances become everyday things, requiring everyday skills to use them. These everyday skills may be ones we and our culture have invested time to acquire (e.g., writing and, over the last 10 years, typing). It is a point commonly made that appliances should be easy to use, and to become an everyday thing certainly requires that ease of use is attainable. This does not necessarily mean instant or walk-up ease of use. Riding a bicycle is an acquired skill, which once mastered makes this conveyance easy to use. Using a knife and fork is a similarly acquired skill anyone over the age of three takes for granted. Neither of these examples could be said to be easy to use on first encounter. Some everyday appliances in the arts take years to master. Playing a musical instrument is an interface skill that many amateur musicians make a hobby of mastering, and a small but significant number of professionals make a living out of the ease of use they have attained (Fig. 37.1).

FIGURE 37.1. The use of some appliances is an acquired skill.

Focused Function. Successful appliances do one thing well. They support a single use or activity, rather than a single function. Appliances are seen as more useful if they do have a single function that can be used through a *consistent* simple user model for a *wide variety* of purposes (e.g., telephone). However, the singularity of the activity supported and its generality of purpose is the key to an appliance's utility and acceptance. A telephone can be useful in a wide variety of situations or contexts for a wide variety of purposes. The activity it supports (communication between remote parties) is the same in each one.

Work Together as Peers. This feature appears third in the list to help the flow of the characterization being developed, but it is the most important aspect of appliances in our view. The combination of single, simple function with a generality of purpose benefits from what we are calling a "network effect" if devices can work together as peers, without configuration and with a consistent user model. The ability to work together as peers exchanging what has been variously called content, data, media, information, or "stuff" (Taylor, 1999) unlocks the potential for each appliance to support a single activity. If a digital camera captures images and can point and shoot (transfer) those images to a printer for printing, there is no requirement for the camera to support printing or the printer to capture images. The camera does not need to know about the working of the printer and vice versa.

Portable. It is not a necessity that all appliances should be portable. However, many information appliances become more

useful when they are portable, allowing the user to take the tool to the task, instead of the task to the tool. Portability increases the number of situations in which a device's use can be employed. A singularity of function and the ability to work as a peer with other appliances makes it easier to create portable devices by reducing the component count and the power requirements.

Personal. To say that the preferred usage of an appliance is a personal one is not necessarily to imply ownership (i.e., that the user must be the owner). What we mean here is that the device should be stateless and totally available to an individual at the time of use. A company-owned photocopier is a stateless device. It does not carry over context from one user to the next. When it is in use, it is supporting the personal task of the current user. This is the notion of personal use to which we refer. Portable examples from everyday life include a pen, a tape measure, a torch, and a newspaper. Even though owned, they can easily be loaned to another person for casual use.

Task Closure. There is a sense of psychological task closure directly associated with the use of the device. Using a knife to cut up an apple before eating it creates a sense of closure even though the apple is still to be eaten. The subtask of creating bite-sized pieces is complete. Using a telescope to identify a far-off galleon as friend or foe may be the first step in a prolonged engagement, but the telescope can be said to have served its purpose by enabling the identification process to be completed. Similarly, looking up an address in a digital organizer provides psychological task closure even though the complete task may be to send a letter or e-mail. The question to ask is whether the activity supported by the appliance fulfils a meaningful task partitioning. A camera captures images. Capturing an image has a sense of task closure. However, in most analogue cameras, the results of the image capture are only available after further steps. Viewing the image captured can only be done when photographs are received back from the developers. It is interesting to note that, in this example, digital cameras with displays for reviewing captured images are delaying the point at which psychological closure is achieved (i.e., after the image is displayed rather than after pressing the shutter switch).

Direct User Interface. A direct interface with the user is clearly associated with the activity supported. Much has been published about the need for simple and direct interfaces (Buxton, 2001; Norman, 1998). The goal for appliance design as well as for any other IT device is to attain a level of transparency that allows the user to concentrate on the task at hand rather than operating the device. This need not be an immediate experience but should be attainable with reasonable experience, where reasonable can be measured by the commitment required against the benefits of transparency. While learning to drive a car, the learner's experience shifts from an acute awareness of the coordination of pressing pedals, pulling levers, and turning a wheel to the task-focused experience of driving down a series of roads to a destination. For many information appliances, the benefits offered are not as high as the mobility and independence offered by a car. The commitment required to reach transparency must therefore be low, and so appliance functions must

be offered through compellingly simple physical affordances. Gaver (1991) describes affordances as "properties of the world that are compatible with and relevant for people's interactions." Affordances offer a direct link between perception and action. For example, the presence of a steel plate rather than a handle on a door suggests pushing as a means of opening. Affordances reflect the intimate relationship between human physical abilities and perceptions, and the physical and implied operational characteristics of objects and devices. If the perceived affordances of an interface match the intended operation of the device, then it will be easy to use.

CONCEPTUAL APPROACH TO APPLIANCE DESIGN

This section outlines the design implications of the appliance characteristics described and looks at some examples by way of illustration. We have grouped the features of applianceness into two main foci: operational context and direct interaction. In some respects, these two groups represent a macro and micro view of appliance design. The chapter concludes with a description of the *network effect*, which covers the requirements for appliances to work as peers and is, we believe, the most important driver for the future success of appliances.

Operational Context

Operational context
Whole system
Right thing here & now
Attention
Ambience, Affect & Attitude
Single function Open purpose
Closure

Human-machine interface design is as old as the design of the earliest tools when skilled craftsmen manufactured artifacts tailored to the intended user. After the industrial revolution, manufacturing processes enabled mass production of the same design for use by many individuals. This shift to one-size-fits-many led to an increase in the number of potential uses for any one design. User interface design changed from personal intimacy to a general understanding of human capabilities and contexts. The PC today is an extreme example. The standard design of the device allows mass manufacture, while supporting the different requirements of its millions of users. More than 50 million households in the United States alone have at least one PC. It is not that every PC user uses the same set of functions, but that there is a portfolio of functions offering something to every user. The problem of understanding the context of use is solved by a shotgun approach to functionality. The casualty of this approach is the interface, which becomes a one-size-fits-all function—all people and all contexts. The desktop style of the interface assumes the user is sitting up and paying attention, in an environment where they have ample attention available for

selecting and tuning the application they need for the task in hand, and will then continue interacting with that application for some time.

The interface to the PC has been the subject of thoughtful critique for some years (Norman, 1998). How much of an achievement you believe the PC's user interface to be depends on the success criteria you choose. In terms of penetration and financial success, there is little dispute. However, outside the context of the sit-up, pay attention desktop, the number of contexts within which the current PC format is useful is extremely limited.

The PC has some of the appliance features we have talked about. Transparency in some tasks, such as word processing and web browsing, is attainable. PCs we would argue have become everyday things, and typing is almost an acquired everyday skill. However, a PC is not an appliance. It does not interact with other devices as a peer; it dominates and controls other devices. It does not have a singularity of use, or offer personal, stateless use. It currently offers no physical affordances, although the adoption of the technology in schools and almost all office-based workplaces will lead to the acculturation of its interface for future generations in the same way that we all know how to turn on a television and surf the channels.

For appliances to support a single activity and offer a direct and transparent interface to their operation, the shotgun approach to context will not work. The resulting multiplicity of functions will defeat the simplicity of use and obscure the quality of applianceness. To build successful appliances, designers must pay attention to the two-way relationships between the device's use, the physical environment, and the personal and social cognitive processes at play. We call this the whole system approach.

Whole System Approach. For information appliances to pervade our everyday lives, they have to break free of the context of sit-up, focused attention, and long-duration applications that is typical of PC usage. They have to project and utilize successful affordances for instant use and to work seamlessly as peers providing task closure in their appropriate context. To achieve this, we need to understand the relationships between the use of appliances and the environments we inhabit in which they will be used. It is important to understand that there are *two-way* causal relationships between the environments in which we operate, our cognitive and cultural processes and the way we use materials to distribute and extend our cognitive capabilities over time and space. Hollan, Hutchins, and Kirsch (2001) provide an extended treatment of how our cultures accumulate material means of thought, and Clark (1997) illustrates how cognition is embodied (i.e., it is shaped by our physical and cultural attributes).

The approach we find useful is that cognitive systems are the culmination of the interplay between language, thought, and environment, and that our perceptions and actions both determine and are determined by the materials with which we interact and think. The calendar on the wall is an ambient reminder of commitments in the month ahead in the same way that a clock is a gentle reminder of the time. The knot in your handkerchief, like the calendar, is an aid to prospective memory for a more

immediate obligation. We humans build up very rich physical environments that help reduce the load on our cognitive functions. Look around your office and count the number of paper appliances: notepad, sticky pad, papers, books, posters, etc. We fail to notice the sheer diversity of one common medium, because its many uses have become intimately adapted to the way we as human beings work in our environment. There are many things we can learn by looking at this example in more detail and observing the rich use of space and physical forms of information that people use to help them in their work.

At a top level, studies of individual office workers have found that people maintain a high-level categorization of their work into:

- The *active* focus of attention, the current task in hand.
- The *anticipated* set of work that is to be done, or is in process but pending other events.
- The *stored* and *archived* materials that belong to completed work, or infrequently needed resources.

(Frohlich, 1992; Frohlich & Perry, 1994; Walker & Stenton, 1991). The first thing to note is that physical space is used to encode and support this distinction in the way that work is arranged for access. Active work needs to command full attention, and so is given pride of place, with the appropriate tools occupying the precious desk space. The work in progress is very often left visible in ways that help people remember and order the things they have to do—a physical to-do list; the use of piles of documents is one of the most common ways of achieving this (e.g., Hollan et al., 2001; Kirsh, 1995). These need to be placed so that they can serve as reminders without dominating the available space. Archived material can be removed to less accessible places with more formal means of indexing and access.

Moving down to a finer level of analysis reveals extraordinary richness in the physical means by which work is done and why dreams of the paperless office are just that. Consider the following examples:

- Building up a cash flow forecast spreadsheet from several source materials on the business.
- Picking up a pile of marketing reports, turning to previously selected parts marked with stickies, laying out important diagrams, and gradually building up a proposal on pieces of paper, whiteboard, and the word processor.

Studies such as O'Hara et al. (2001) have shown how, in instances such as these, key aspects of the external material context are closely connected with the internal cognitive processes, and are critical to accomplishing the task.

We will look in more detail at the direct interaction model this leads to for the individual appliances; here, we want to draw attention to the overall systemic nature of people's interaction with the world, where each device should fit into the overall flow of activity. The standard configuration PC provides strong support only for a subset of what we need for active tasks. In the case of office work, the prevailing desktop metaphor gives us a modest form of the breakdown between the three categories of information: you can fill the screen with the active task; folders

on the desktop can signal work in progress; and the file system holds the archive. But the effectiveness of the PC drops off very quickly as you move to the tasks needing richer material interaction. The PC is evolving to address this problem through the increasingly common use of multiple displays that allow people to maintain anticipated items, messaging windows, etc., in a display that only requires peripheral attention.

As yet, there are few examples of appliances deployed within the workspace that approach the richness supported by paper. This reflects both the paperless and cyber preoccupations of mainstream computing, but also technology limitations. If we had very cheap dynamic surfaces of the sort promised by emerging technologies, then there is no doubt there would be an explosion of market innovation. One category that has achieved some success is electronic or printing whiteboards that let you print off a copy of what you have done. At HP Labs in Bristol, we extended these into more of an appliance model. In addition to printing, a copy could be directed back over the network to the user's PC. This connection into a wider appliance world through printing and PC communication brings some important benefits:

- Immediate sharing of meeting results can strengthen the sense of commitment.
- The lack of an intermediate step of writing up notes gives the record greater authenticity.
- Direct copy of the information can preserve the full messiness of the way information was created—all the squiggles or scaffolding (Whittaker, Geelhoed, & Robinson, 1993) that were used to convey meaning.
- Persistence of the information back at the desk on the PC screen can be supported through thumbnail versions that help the transition and translation to action, and so provide further elements of an environment of distributed cognition.

This example shows the value of appliances that are well-designed to augment physical interaction, and are well-positioned in the surrounding physical and social system.

Right Thing Here and Now. The rhetoric of the computer industry is "anything, anytime, anywhere," and the implementation strategy is one of wireless communications infrastructure, network servers, and mobile PC and PC companions. The typical advertising image of someone in the wild beyond still able to run their corporate empire through their trusty laptop celebrates the value of removing barriers between where we happen to be and the power of global information systems. From our point of view, this is only half the story of the potential of information appliances, and we need a corrective emphasis on using the same infrastructure to support the appropriate use in a particular context—"the right thing, here, now."

Look at the way we use maps and signs to solve the problems of finding our way about when we are walking or driving vehicles (Fig. 37.2). Maps are an anywhere, anytime form of information resource and very useful ones, too. Wherever we are, they provide the same information and are appropriate for planning our route at several different levels of abstraction. We

FIGURE 37.2. Road signs demonstrate situated instruction.

can see the big picture: for example, "I'll take the motorway to Bristol, and then the local road out to Thornbury." They will also make us aware of where the main intersections will be, and give us a rough idea of their layout and how we will navigate them. By contrast, signage is a right thing here and now form of information; it is designed to enable us to make the right decisions at the critical moment as we approach each decision point. It references the information we find in a map—place names—and connects it to relevant action in the moment—turn left. It is positioned so that we encounter it only in the place where it is relevant to action, and its form and position are intended to be within the information processing capabilities we have available while continuing to control a vehicle or cross a road. At a finer grain, road signage often uses simple maplike forms as we approach decision points to help us frame our decision (be ready for the third exit from the junction) and then simpler action-oriented forms at the key decision point. At an even finer grain, signs will give us very local, specific information—Do Not Enter, Keep Left—that has no connection to our ultimate goal, but is a form of simple interaction language for the physical space. Generally, good space design will force such decisions rather than require us to read or interpret symbolic instructions. In general, wayfinding is a very rich domain for understanding how different levels of representation can help people move back and forth between high-level intentional planning and fine-grained situated action.

Turning to appliance design, we can see how an appliance that is tuned to simple action in a specific time and place should be thought of with respect to a wider context of activity. Key points for approaching appliance design are:

- Think about how activity is distributed across *all* the times and places relevant to the application.
- Look for how appliance opportunities can provide action-oriented information in the here and now.

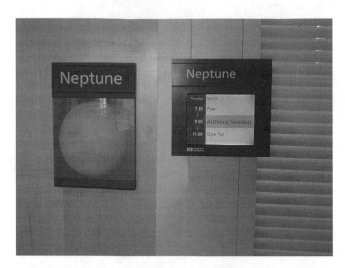

FIGURE 37.3. Room sign outside the Neptune meeting room.

FIGURE 37.4. The industrial design model of the Piglet proto-type.

• Conversely, look at how the appliance solution might be simplified by leaving more complex tasks to other parts of the system.

In moving into a new building at HP Labs in Bristol, we took this approach to creating a number of situated appliances (the whiteboard example in Fig. 37.3 was one). In particular, we explored the potential of signage appliances, specifically for signaling the status of meeting rooms. Most organizations have a number of meeting rooms of various shapes and sizes that are shared among all the building occupants. Usually, there is some sort of process for booking them, and usually there is heavy competition for them, with resulting problems from bookings not cancelled, meetings running over into other people's time, and so on. The system implemented at HP comprised two primary components. First there was room-booking software that was accessible by anyone over the Intranet from their PC. With this, anyone could see the current bookings on all the rooms and make their own. Secondly, room signs were built and installed outside every room. These were kept very simple and just displayed the bookings for the day—the times and who had made the booking. This provided the information people needed as they were arriving at their meeting or searching for an empty room: "Am I in the right place at the right time?" "This room was booked, but it's gone time for the meeting and is empty—I can probably use it anyway." In this way, people were getting the right information in the right place and at the right time.

Another example comes from field trials of a prototype appliance called "Piglet." The device predated the PalmPilot's release, but was similar in appearance (see Fig. 37.4). The functionality of the Piglet was limited to scribbling notes using the pen interface that could then be compiled into notebook sections, to-do lists, calendar entries, or sent via an in-built digital radio to other Piglets. The field trials explored its use among a team of support engineers, management, and administrative staff. The device was very popular and quickly established itself as a handy tool for frequent use through the day.

The reported success of the appliance was attributed to the transparency of the interface and the immediacy of the capture and communicate functions. The pen interface offered a socially acceptable way of discretely replying to urgent messages during meetings. The instant-on scribble pad lowered the cost threshold for sending a message and so more updating messages were sent, keeping the team informed of work as it progressed. More problems were discussed in a timely fashion, allowing the team's distributed expertise to be shared when and where it was needed. The Piglet's organizing functions were almost always used back at the desk leading the research team to develop the notion of a Personal Information Lifecycle (Bedford-Roberts et al., 1995), the bumper sticker of which was "capture now—organize later." The value demonstrated by Piglet can now be seen in the explosive uptake of text messaging on mobile phones in Europe, which allows people to keep multiple short interactions going with work and social colleagues throughout the day.

An example of capture now—organize later comes from the PalmPilot for which an excellent account of the design approach is given by Haitani (Bergman & Haitani, 2000). Applications share the load with the PC and focus on what aspects you really need with you on the road (e.g., for expense claims, the application focuses on recording the details of each event and assumes that report generation will be done back at the PC). This contrasts with the notion of the hand-held device as a small version of the desktop PC, trying to offer the full functionality through a restricted interface. In another example of here and now usage, Haitani describes the use of the schedule application via the external button: ".... you are in a hurry and you pull out your Palm and you want to know where you're supposed to be. You press one button and it shows you your schedule. No struggle, no effort, no confusion; it just does what you want it to do. Your reaction is a contented sigh, and it's very calming." You accessed the information you wanted without having to go through the cognitive effort of operating a device and waiting for the application to boot up.

Attention. The bulk of experience in interaction design for information products has been developed in the context of a human facing a computer. In this situation, you have the luxury of the person's full attention, in a comfortable environment, with distractions at a (mostly) manageable level. But an information appliance will typically have to work in a situation where the person's attention is only available for a few moments or needs to be directed *through* the appliance to the task in hand. For example, you are at your desk, working on your PC preparing a report. The phone rings, you answer, and start talking to a client. You need to take some notes. Do you open up a new blank document in your word processor and start typing, or reach for your pen and jot down the notes? Most people reach for their pen. Attention needs to be directed toward the other person in the conversation and maintaining the dialogue with them. Note-taking has to be achieved within the cognitive processing left over. Some people can use a word processor with enough flexibility for real-time note-taking; very few can sustain that flexibility and keep up a conversation at the same time. Pen and paper are available for use immediately, and because of our deep familiarity with them, we are able to take notes with them while sustaining spoken interaction (Harper & Sellen, 1995; Luff, Heath, & Greatbatch, 1992).

So, an appliance designer must ask, what is the dominant focus of attention of the person in the instant they will use the appliance, and how will this condition their capacity to interact with it? Many things that make our environment work for us make use of our peripheral awareness and attention, and things that can be taken in at a glance. Road signs have to perform their wayfinding function without commanding the attention that is needed for control of the vehicle. We can walk down the street, avoiding people without conscious attention, carrying on a conversation. Is my colleague off the phone yet and available for an interruption. I can tell just by overhearing his voice. John in marketing has a suit on today; oh, yes, there's a client coming in.

In our work, we have explored this notion of peripheral awareness in the design of appliances that support office communication. A number of studies have looked at the role of cameras to enhance remote telephone communication (see Finn, Sellen, & Wilbur, 1997, for an overview). Most have focused on the value of presenting images of the people participating in the conversation through video conferencing. O'Conaill, Whittaker, and Wilbur (1993) studied telephone usage in an office environment and discovered a further opportunity for the use of images. They reported that the first hurdle to a successful conversation was actually getting hold of the person; some 64% of attempted calls failed because the intended recipient was away from their phone. Further studies of remote collaboration had shown that talking heads (live video pictures of the conversational partners' heads) added nothing to the effectiveness or satisfaction of the communication (Whittaker, Geelhoed, & Robinson, 1993). Camera images were therefore used to help initiate phone conversations rather than enhance them.

Glance cameras were used between colleagues to maintain peripheral awareness of presence and availability. In this way, these low-resolution images were used to reproduce many of the properties of sharing a physical space, by extending the lightweight visual interactions. People used them to extend the opportunistic synchronization that goes on all the time between people working in the same space (e.g., using a bit of attention to look out for a good moment to talk, while getting on with the work).

There were two very important aspects of the design. First, they were mounted solely within a *personal* space, so it was a purely personal decision to use them, and they did not intrude on anyone else's privacy. Secondly, if you wanted to look at another person's workspace, then you could only do so by putting your own camera online. This reciprocity meant that you could create a community of colleagues who were sharing views of their respective spaces, in just the way that co-located colleagues can look over to each other's desks (cf. Gaver, Moran, MacLean, Lovstrand, Dourish, Carter, & Buxton, 1992).

As mobile phones become more pervasive, there is a growing need for similar approaches to be developed for gauging the receiver's availability and the acceptability of making or receiving a call in different situations (Pedersen, 2001; Schmidt, Takaluama, & Mantyjarvi, 2000). Unless the whole world is wired with cameras, it is difficult to envisage a glance camera solution to mobile phones. In fact, two things have happened: the control over whether to receive a call has stayed firmly with the receiver; and text messaging has become popular as a less intrusive way of contacting someone. This latter behavior is the same as observed by the Piglet study (Bedford-Roberts, Nelson, Stenton, Stott, & Lacohee, 1992).

The mobile phone pushes the boundaries of attention distribution in many situations. In certain situations, the attention deficit caused by mobile phone use is a safety hazard. Too much attention is diverted when dialing to be safely used while driving. This has created an opportunity for hands-free number look-up through simple speech recognition systems.

In many situations, it is the mobile phone's attention-destroying capabilities that have become unacceptable. Politeness constraints are being breached and remedial steps taken. Certain railway carriages are being designated mobile phone free; schools are banning their use on the premises. People are asked to switch off their mobile phones in cinemas and theatres. In these situations, it is the obtrusive nature of phone use that causes society to act to constrain their operational context (cf. Churchill & Wakeford, 2001; Perry, O'Hara, Sellen, Brown, & Harper, 2001).

Rather than falling foul of these issues, appliance design needs to turn outward to all the surrounding influences that govern appropriate use.

Ambience, Affect, and Attitude. The central point we have been making throughout this section is the simple observation that appliances are used in a rich surrounding context of individual and social activity that exerts powerful influences on appliance design. In understanding this context, we need to take into account not only the presenting nature of the task and its place in an overall activity flow, such as sending a quick message while walking along, but also how that task is governed by influences on our emotions—are we relaxed, in a hurry, in a safe or threatening environment. We group these influences under the general heading of ambience, meaning the way our surroundings signal different possibilities—all the way from precise

physical things, such as entrances and stairways, to more diffuse social possibilities for interaction, such as the way a restaurant or hotel lobby is arranged.

The ambience governs our affective, or emotional, response to the situation, and this will underpin our attitude—the role and approach we take toward the task, and what we will perceive as the possibilities presented by particular appliances in the situation. For example, is this a place where I feel it is safe and appropriate to stand and give my attention to an information kiosk that requires concentrated interaction.

The role of an appliance is to satisfy a purpose; the role of the appliance interface is to facilitate its manipulation toward satisfying that purpose. Emotion and attitude are key to how people will perceive the interface and its possibilities, and so must be given explicit attention both in the design of the device and in consideration of the ambience in which it will operate. This is quite different from the conventional human–computer interaction design premise, that physical affordances are used to guide the largely symbolic interaction and that affective aspects are largely side effects. We are not accustomed in computer application design to making the emotional response and the way it governs interaction a central concern. The exception, of course, is in computer games, and now the emerging possibilities of interactive toys. Excellent accounts of what can be learned from these areas are in Bergman and Haitani (2000).

In an earlier chapter of this book, Brave and Nass (2001) give comprehensive coverage of the role of emotion in our engagement with artifacts. Picard (1997) also describes the role of emotion in decision making. She points to research that concludes that, if there is no emotion attached to experience, then the ability to make good decisions is severely impaired. The observation comes from correlation between the decision-making ability and emotional deficit in patients with frontal lobe disorders (Damasio, 1994). Emotion is usually thought of at best as a distraction to effective decision making and action and at worst a weakness or flaw clouding thought, impeding logic and reliability. From the analyses of Brave and Nass and Picard, it seems fair to conclude that emotion is a necessary component in cognition and everyday action. It predetermines our reactions and drives us on. The thrill of scientific discovery and artistic creativity are emotions that have fueled our technical and cultural development.

If emotion and affect have such a close binding to the way we think and perform, then it is important that in appliance design we understand not only the situational antecedents, but also the interface implications for affordance and perception setting. The technogadget design aimed at the early adopting technophile or the home hi-fi design for audio-buffs might not create the right emotions for an extended family photo-sharing appliance. The exclamation, "I couldn't use one of those!" is in many cases born out of fear rather than inexperience.

BANG! Did that make you jump? Of course it didn't. Affect is created first and foremost by prelinguistic, sensory inputs. Fight or flight responses are triggered by primitive sensory perceptions: a loud noise, a bright flash, the rapid movement of an object in the visual periphery. More subtle reactions can be set by the way an object looks. Just as a facial expression can foretell the kind of interaction you are likely to have with someone, the physical appearance of an appliance can set the scene for the role it plays and the operating characteristics it affords. Disney does this to great effect in the way that the smells, lighting, and sounds surround the people in the pre-ride lines of their theme parks. The attitude of the guests is set before they step into the carriage. In this respect, the effect of an appliance's appearance is more than a fashion statement.

In addition to the attitude set by the physical appearance of an appliance, ambience can also determine successful integration into everyday life. The ambience of most family rooms is one of sit back and relax. The browsing and passive consumption origins of the TV fit well with this. The PC, on the other hand, demands that users sit up and deliver focused attention, yet the PC and the digital TV are set on the road to competition. The ultimate outcome of this competition may be decided by the conflicting attitudes of productivity and relaxation that surround these devices as they both attempt to take on the role of the other. The sit-up posture of the PC has meant it has made the transition easily into game applications, but only poorly into watching movies. Likewise, doing e-mail or home banking on a TV is never going to be very easy or rewarding. For appliance design to unlock the door to pervasiveness, the prevailing ambience of the device's operational context must be supported.

To complicate matters, the ambience of the family room may itself be affected by the predominance of the devices it hosts. In pre-TV days, the eyes-free entertainment of the radio afforded additional activities while listening to Flash Gordon's inevitable rush to the next cliff-hanging punctuation. Board games, playing cards, and creative pastimes were pursued, encouraging a more sit-up style of activity around a table. The advent of the TV moved the couch and easy chairs into a respectful theater position facing the screen. Tables became shorter, providing a surface on which to place drinks and snacks, and the dominant style became sit back. The phrase *couch potato* was coined.

A survey of 2,000 U.S. Internet users by Scarborough Research (2001) reported that 50% of home Internet users have a PC in the same room as the TV; of those, 91% watch the TV while surfing the Internet. It is unclear whether it is an indictment of the quality of U.S. TV or a tribute to its attraction that new pastimes are pursued while it is on (as with the radio in earlier days) rather than substituting for it. Perhaps the repetitive nature of the schedules enables TV viewing to be a mostly eyes-free activity, with occasional glances prompted by the sound track to catch moments of interest. It is interesting to note that the convergence of TV and web functions into PCs and digital TVs has not been a market success. This may say something about the conflict between the social aspect of TV viewing and the solitary nature of web browsing, which is best supported by a communal TV and personal browser.

Situational dynamics can also affect roles constructed for the use of appliances. Just as the arrangement of the tables in a meeting room can affect the style of the meeting, so control of the writing surface in the room has social effects. It orchestrates contribution and imparts dominance to the current holder of the pen. A shift to personal slates transmitting to a shared surface changes the dynamic of interaction from orchestrated dominance to more participation and negotiated control.

More socially rooted attitudes surfaced during the glance camera experiments we described previously. The cameras allowed colleagues to share views of cubicles in office buildings over the network. This provided the opportunity to determine one another's availability by glancing, just as if looking through the door or over the cubicle partition. For some people, this was a very positive development; for others, it signaled an intrusion of privacy and the feeling of being watched. This example shows how powerfully prevailing values can determine the way the purpose of an appliance is viewed.

Understanding the prevailing values, the ambience and the resulting coordination of social roles will determine what functions are appropriate to the appliance. It is worth recognizing that a wide boundary may need to be drawn around the relevant social context to determine how an appliance's purpose will be constructed and maintained.

Single in Function, Open in Purpose. To be clear what we mean here, it is worth taking time to distinguish between purpose and function. By purpose, we mean the use to which the function is being employed. When a knife is used to cut the peel off an apple, the purpose is to peel the apple. The knife's function is to cut. If the knife is used to cut a hole in a sack, the purpose may be to access the sack's contents. The function is still to cut. From these examples, it can be seen that purpose is constructed by a human interacting with the appliance in a specific context. People can also appropriate an appliance to any purpose they like, not just the ones for which it is meant. If the same knife is tied to the end of a piece of string and dangled adjacent to a wall, its purpose is to pull the string taut (and so create a plumb line). Its function is to provide weight. In this respect, people are very flexible and adaptable in finding something that will do what they need. In the absence of pen and paper or reliable short-term memory, a calculator can be used to record and store a telephone number in a "capture now—organize later" situation.

Fewer, simpler, functions result in flexibility of purpose, and so unlock more uses in more situations. For example, a kitchen sink has the simple function of holding liquids (usually water). This function can be used to wash dishes, clothes, bicycle wheels, small children, pets, anything small enough to fit in or over the sink. Any detergents or solvents can be used that can legally and safely be disposed of down the local drain. Phones and cameras are similar in the way that a simple function provides great flexibility of purpose. Contrast this with a dishwasher or a washing machine, both of which offer rinsing and drying functions. The added functions of these devices reduce the number of purposes for which they can be appropriated i.e., cleaning only items of a certain size and material with specifically designed detergent. This does not make them bad appliances, but illustrates that, in focusing the functions of an appliance, it is worth striving to do so through simple functions that preserve open purposes.

Even a simple single function can become specialized and so reduced in application. A knife has one function: to cut. A potato peeler has one and the same function, but is specialized to cut the peel off vegetables and fruit. In this way, the knife is more open in purpose than the potato peeler.

The Swiss Army penknife might be thought of as being an exception to the less-is-more rule we are proposing here. In our view, the Swiss Army penknife is a collection of appliances held together on a common clip. Each has one, or at most, two functions and differing degrees of openness of purpose. Buxton (2001) suggests that the increased bulk incurred by the multiple functions makes the penknife less portable and so reduces its potential contexts of use. This trade-off between functionality and portability has parallels in appliance design where size and complexity of use can increase with the number of functions supported. One measure of usefulness can be said to be the variety of things it is possible to do with an appliance rather than the number of functions it has.

The way knowledge workers use paper is a powerful illustration of this principle of maximum usefulness being derived from simple functions. In a study of the requirements of potential appliance users, Kidd (1994) usefully extends the notion of the knowledge worker (Druker, 1973) and identifies the information needs of this class of office work. The capture, presentation, and arrangement of notes is reported to be far more useful than database filing and retrieval, the traditional target of IT support and PDA design. For knowledge workers, the important interaction is between the information they gather and the changing world model inside their head.

Paper provides the ideal medium for capturing, reviewing, and organizing information for the purpose of this assimilating and informing by the knowledge worker. The common representation used is scribbled and annotated notes on papers spread across a workspace. Systems that impose a structure for storing information destroy the flexibility of purpose offered by paperbound scribblings. Once informed, the knowledge worker may file away notes but rarely, if ever, retrieves them for future use. Thus, the notion of a portable database as a structured information repository extending mental capacity and performance is misplaced for this category of work. Instead, devices should support quick capture of information and ideas, and allow these notes to be spread physically across a workspace and metaphorically across business contexts. The key function is displaying the marks as made to inform the knowledge worker. The open purpose is supplied by the use of the knowledge worker's expertise across the range of its application.

Closure. Mailing a letter in a mailbox provides a sense of completion for the sending of a letter. Until that point, there is a sense of unresolved tension while the addressed and stamped envelope sits on the desk waiting to commence its journey. Hitting the send button to send an e-mail creates the same feeling, whereas "transmission successful" only gets you part of the way there when you send a fax. In this situation, completion only seems to come when the receiver acknowledges receipt. The same need for closure was reported in the Piglet study. As part of the list of improvements, many users asked for a way to check the intended recipient had viewed a message. Messages in the Piglet experiment are best thought of as wireless Post-its.

But, what is the difference between mailing a letter or sending an e-mail and sending a fax or transmitting a scribbled message? For the role of the devices, there is no difference. The mail

box, the e-mailer, the fax machine, and the Piglet had all fulfilled their purpose in providing a portal to the delivery system, the other side of which the sender has no more part to play. The sense of task completion for the use of the device should then be fulfilled. We would argue that, for the mailbox, the PC, the fax, and the Piglet it is, and that sending is a meaningful stage in the larger process of corresponding. The unresolved feeling with a fax machine and the Piglet, we believe, has its roots in the reliability of timely reception. The shared nature of most office fax machines and the short expiry date of the average fax content leave senders nervous about timely receipt. Similarly, although the Piglets were personal devices, they could be out of range or switched off, and like faxes the content of messages sent had a short expiry date. On the other hand, physical mailboxes and intrays are part of the always-on mail infrastructure, which provides a personalized service. Most office PCs are personal devices in daily use. It could be argued that e-mail message content covers a range of expiry timescales from imminent to Christmas letters. We would note that most popular e-mail systems allow flags to be set to trigger notification that a message has at least been opened.

The point to be made here is that task closure should be thought of in terms of the simple use of a tool fulfilling its purpose. The broader context of the overall process that the user is engaged in may complicate the picture and suggest requirements for greater functionality (such as acknowledged receipt), but these should be considered cautiously, and reduced to activities that achieve the right sense of closure in the sense of completion of a well-defined task, eliminating the need for the user to remember further steps to be carried out in the future. Saving state is not in itself bad, but it should only be used where keeping information is intrinsic to the content of the task—such as storing a phone number or a downloaded piece of music— and then must eliminate all aspects of state that are side effects of the underlying implementation.

Direct User Interface

> **Direct User Interface**
>
> Sensory motor interaction
> Immediacy
> Transparency
> Mediation vs. Automation

How a function is supported matters. It will determine the different purposes for which the function is used. It is easy (and common) in designing applications to think that applications that achieve the same function are therefore equivalent and redundant in purpose. For example, when we first implemented a simple electronic version of Post-it sticky notes for the PC, we often got the reaction, "But I can send messages with e-mail, why would I want these as well?" The observable evidence though is that people will use different channels of communication in significantly different ways. The fact that a note pops up on a colleague's screen presumes a certain sort of familiarity and permission to occupy that space, just as it would if you stuck a

physical note on their monitor screen. This makes it appropriate for the sort of quick "Time for lunch?" query that e-mail was inappropriate for. The rapid adoption of instant messaging in offices is now demonstrating the value of this sort of lightweight communication. This example shows that the precise embodiment of an application concept has a very big effect on just what it can and will be used for, and this should be a central concern of appliance designers.

Direct manipulation is a powerful way to change the perceived usefulness of a device. The landmark contributions were introduced to the PC interface, via the Apple Macintosh, building on the insights created at Xerox PARC, to simplify the user's interactions with PC (Levy, 1994). For all the criticisms of the PC's complexity, the direct manipulation interface has played a central role in keeping that complexity down to a manageable level for millions of users. We believe that the appliance world also faces the danger of presenting people with an overwhelming complexity of possibilities and that direct manipulation is a central concept in producing simple and effective solutions.

For the successful adoption of appliances, action needs to be timely and capable of execution with complete focus on the task requirements (cf. Perry et al., in submission). The need for timeliness may seem like a truism. It is, but it is worth keeping in mind because in many everyday situations it is hard to achieve and failure can be fatal for everyday acceptance. Designers of PDAs have striven for over a decade to get close to the timely and direct manipulation of a pocket diary. Early devices relied on tiny keyboards for entering diary appointments or contact details. The Apple Newton used handwriting recognition to approximate an everyday interface. The device had a dedicated following among the technophiles and early adopters, but did not cross the chasm into popular use (to use Moore's model of adoption—Moore, 1991).

The more recent palm form factor PDAs (Palm PCs and Pilots) with scribble or shorthand interfaces have gone some way to closing the gap further. Scribble interfaces support the "capture now, organize later" principle, whereas shorthand systems like Graffiti (Palm) and Unistroke (Xerox) allow quick and direct entry of information at the cost of time spent learning the shorthand.

This section of our conceptual approach to appliance design will focus on the interface requirements of direct interaction and will consider the need for *sensory motor interaction*, *transparency*, and *immediacy*. The section will close with a look at the decisions surrounding the choice between *mediation* and *automation*.

Sensory Motor Interaction. The human mind is embodied and a large part of what it does to enable us to function as physical beings in a physical world goes on without any conscious thought (Lakoff & Johnson, 1999), but computer application design with its roots in symbol manipulation has been slow to take account of our physical capabilities for interaction.

Until very recent history, the design of artifacts was rooted in long traditions of understanding the way people work as embodied beings, because, in large part, there was a deep relationship between the physical form of a thing and its purpose. In the domain of symbolic and linguistic artifacts, such as books,

or signs, for example, a slow process of experimentation has adapted them to fit in with our physical natures: in the case of signs, the information is embodied in such a way as to support our physical task of getting around; in the case of books, the physical interaction of page turning and riffling supports the reading task.

Computers entered the world disconnected from these design traditions, concerned only with the abstractions of symbol manipulation, for tasks where the human's responsibility was to provide input and output to the calculation at a symbolic level. There was no connection between physical form and the content or purpose of the calculations. The explosive growth of new applications merely increased the disconnection, with only the advent of the desktop metaphor and the mouse to recover a modicum of physical interaction. Slowly, computers have found their way back into things that fit our physical natures, either through the enormous deployment of processing power to make everything in our lives smart or through the steady emergence of the appliance design approach we are exploring here—designing things around *human* inputs and outputs rather than computer ones. Thus, all the sensing and acting devices such as cameras, printers, speakers, and microphones that are considered peripheral in the conventional view of computer systems, because they are outside the central concern of automation, must become central in the design process, because all interaction is mediated by our physical senses and actions—humans are embodied.

We have an enormous repertoire of highly effective sensory motor ways of dealing with our environment: pushing, pulling, turning, poking, seeing, hearing, speaking, kicking, leaning, holding, shaking, plucking, stroking, blowing, striking, pumping, licking, sliding, marking, writing, riffling, swinging, pedaling, and climbing are all candidates for the appliance designer. Symbolic or conscious cognitive activity may also be necessary for situated action, but it should be a side effect of appliance interaction and not the focus of it. The understanding of how our embodied nature guides our thinking, and how conscious thought can be engaged and entrained by physical interaction is a research frontier (Lakoff & Johnson, 1999; Maturana & Varela, 1987; Wilson, 1998), so appliance designers find themselves pioneering in this area, generating good design practice that will feed this emerging understanding of our embodied cognition. In this chapter, we can only give some examples to suggest the approach. Previous work has been developed under the general concept of affordances (Gaver, 1991; Gibson, 1979; Norman, 1990). This is useful work, but there is a danger that affordance becomes no more than a shorthand for "whatever you can make obvious to the user," and is therefore just an after-the-fact descriptive term. We believe that, for the idea of affordances to continue to inform appliance design, it needs to be further developed and derived from deep and detailed investigations of the relationship between sensory motor and conscious symbolic interaction.

An example of such a study is Kirsh (1995). Kirsh studied a range of everyday activities, such as cooking, assembly, and packing in supermarkets, workshops, and playrooms. His analysis is concerned with the way we use space as a resource for problem solving and planning to relieve ourselves of any need

for conscious analytical processing—deliberation as he calls it. His study reveals three general categories of spatial resources:

- Spatial arrangements that simplify choice.
- Spatial arrangements that simplify perception.
- Spatial dynamics that simplify internal computation.

For example, the way a grocery packer in a supermarket uses a buffer zone to assist in the packing activity, sorting things by size into categories, picking from the categories.

In a study of office work, Kidd (1994) reported that knowledge workers found the physical experience of flipping and browsing through books, magazines, and newspapers and spatially organizing paper notes into clusters and piles more effective for organizing their active work than the more symbolic approach of indexing and searching information that is so effective for searching through large online sources. The PC makes use of a desktop metaphor, but a metaphor is all it is and its interface limitations are exposed when it comes to assimilating information from a number of sources. Hence, knowledge workers prefer the clutter of their real desktop to the limited real estate of its PC equivalent. The sensory motor actions of holding, sliding, and riffling are less obtrusive and more conducive to mentally organizing and making sense of information than dragging, dropping, selecting, and opening within the narrowly selective salience mechanism of windows. At a more detailed level, O'Hara, Taylor, Sellen, and Newman (2001) have looked at the detailed material interactions that are involved when someone is producing a written document using multiple sources—as it might be compiling a market research report. They found that concurrent views, spatial organization of documents, and annotations were all coordinated and played an essential part in accomplishing the task; there was a close relationship between the material context and the thinking processes involved.

An extensive study of the affordances of paper in Sellen and Harper (2001) significantly deepens this understanding of the direct and material aspects of paper that are central to its usefulness in supporting individual, team, and organizational activities.

We are not suggesting here that appliances should mimic paper in all respects—otherwise why invent them—but rather that an appreciation for the detailed interaction that goes on in the everyday use of paper-based information provides insights into the sorts of ways that appliances need to work to make full use of human sensory motor capabilities.

Immediacy. Emotion and impulses to act arise in the moment and quickly dissipate if there is no ability to act on them quickly and easily. The impulse buy at supermarket checkouts aims to exploit this phenomenon, even if you thought about picking up that magazine as you came into the store, you were not going to make a detour to find it. But it's there, you pick it up, and drop it in the basket.

Whereas satisfying purchase impulses is two-edged, with its flip side of succumbing to the persuasion of people who want our money, there are many cases where we convert an intent into a future plan because it cannot be accomplished in the moment. Placing another item on an already overcrowded

to-do list adds to the background stress levels of our lives. We found from our Piglet field trials that the just-do-it-now attitude fostered by the scribble messaging capability resulted in more messages and fewer things to remember. Many of the voices in the focus groups conducted after the trial echoed an appreciation of this feature.

Appliance users demonstrate a high sensitivity to lack of immediacy caused by system delays, interface moding, or simply too many interaction steps. Our experience in appliance design in field trials and lab experiments has shown that usage of any appliance drops off very quickly as steps are added to the interaction beyond two because:

- It just takes too long to accomplish in the moment.
- People cannot remember what is required for anything they do not do regularly. There is no sensory-motor script available, therefore it creates a focus of attention on the interface that competes with the focus on the task.
- Task attention also causes the emotional energy to drain. It becomes too much trouble, and the moment is lost.

The PalmPilot (Bergman & Haitani, 2000) is a good example of this sort of immediacy. The number of buttons on the front was kept very small so they can be taken in with a single glance and memorized by their physical position. This contrasts with a PDA that presents 20 or more application icons in which your eyes have to scan to find the right one, which requires more explicit attention. On the Palm, buttons also allowed the reduction of two steps to one for accessing key functions—the button turns on the power as well as selecting function. In addition, only the most frequently used features are kept instantly visible at the interface to reduce clutter. As Haitani points out, this requires that you really do know what are the most important features, and this can lead to asymmetries that are counterintuitive from an engineering perspective (e.g., 'Add' is frequently needed for phone numbers, but 'Delete' is less frequent, so it can be put behind a menu). Haitani's analogy is the stapler that is kept on top of the desk for frequent use, whereas the staple remover is in the drawer, where the extra step to access it is not a problem for the less frequent use.

The always-on Internet of DSL, cable modems, and GPRS mobile phones will dramatically increase the number of things for which the Internet is used. In the dial-up days of V90 modems, it is hardly worth waiting for modems to squeal at each other to find out when the next bus will arrive or what the latest football score is. Low emotion desires that are not worth a lot of effort, or that can be easily satisfied by other means, are brought above the cost threshold as the always-connected feature takes out a lengthy step. The result is a similar one to the effect of instant-on provided by PDAs. Without this feature, who would wait for a Palm PC to boot up to check a diary appointment or jot down a telephone number? Frohlich, Dray, and Silverman (2001), in a study of family perspectives on the future of home PCs, report a "strong latent desire" for "brief, intermittent, and casual" use. The rapid uptake of always-on connections to the Internet in the United States is confirmation that making an action instant dramatically increases its value.

Transparency. A pencil is a transparent tool; once the skill of writing is acquired, it feels just like an extension of the hand—its control surfaces have become extensions of our own, so that we can focus *through* them on the task in hand—writing, doodling, taking notes, or whatever.

The goal of transparency is to create such a focus *through* the tool *on* the task. This requires that the attention needed to operate the device itself is very low, either intrinsically, or because of deep familiarity. Following road signs in a car is only possible because the controls of the car have become deeply familiar through extended use. Beating a computer-controlled Brazilian soccer team in the World Cup final of FIFA 2001 is only possible because the gamepad controls have become extensions of our fingers. Some interaction styles have to be learned and so become transparent; others have natural affordance that has evolved over thousands of years of cultural development.

Moding of an appliance interface should be avoided as it forces the interaction up to a conscious, plan-making level to guide the choice, instead of allowing the interaction to proceed directly from affordance to action. For many people, the use of a keyboard has reached a high level of transparency with respect to text entry for word processing, e-mail, etc., but then transparency is lost when special keys and commands are needed to change the mapping of the keys or to execute application specific functions.

The evolution of cameras from the time when the user set all parameters, through moding for different weather types and distances, to just point and shoot is a good example of design focusing in on transparency. A simple camera lets us just pay attention to the scene through the viewfinder and it does the rest. Thus, the camera's interaction model is not only immediate enough to operate in the moment, but is also transparent enough to also capture the moment. The phone is similarly transparent; once we have made the connection, its operation sinks beneath the level of explicit attention and we can talk happily for hours, unless the transmission quality deteriorates and forces too much effort to be made on the basic understanding. This problem is a signal detection problem and is not the same as losing transparency. Cognitive load can be increased in a number of ways during interaction, and not all of these are through a loss of transparency, although the effect is very similar. In the phone example, the appliance did not become any less transparent with the loss of sound quality. The extra effort was required to extract words from a degraded signal. Listening to a speaker speak in a language that you are not fluent in has a similar effect, but is because of translation not detection. Listening to a technical speaker in an unfamiliar field of research can increase load through lexical access and the need for logical inference. None of these is the same as having to explicitly translate interface elements to work out a desired action, but they make the demands on transparency more severe.

Musical instruments, skis, and woodworking tools are all examples of interfaces that are transparent for those who are practiced in their use and not for the rest of population. The attention required when operating such appliances decreases over time, and transparency is achieved as the student moves from conscious incompetence through conscious competence to unconscious competence. Thus, if transparency is not immediate,

all is not lost. If mastery of the appliance has its own intrinsic rewards, many people will find deep pleasure in the process of acquiring transparent competence. Similarly, for the devices we own, we become familiar with their quirks, and if the ramp is gentle enough some learning can be tolerated (e.g., in the controls for your wristwatch). Designers of games have to give considerable attention to managing the learning ramp, to ensure that the user is engaged but not overwhelmed by the initial interaction, and is drawn on to learn more capabilities. Similar approaches could be applied to appliance interfaces.

Because you cannot in general expect people to become deeply familiar with the interface of your device, transparency can present quite a challenge. The place to start is with basic media affordances. Good examples today of acculturated affordances can be found in all the consumer products, like radios, televisions, audio and video tape players, cameras, and wristwatches that have found their way into widespread use. Good interface examples are the familiar triangle and square symbols for *play*, *rewind*, and *stop* controls of a wide variety of audiovisual appliances that have become standard serial search devices on both hardware and software media appliances. The downward action of a camera shutter button and the milled winding arm of an analogue wristwatch would each elicit the same affordance mime.

Mediation Vs. Automation. What do appliances *do*? And what do people *do* when they are using them? Appliances stand at the intersection of two radically different outlooks on how we put technology to work for us. On the one hand, we have the history of media, from papyrus manuscripts to movies, and, on the other hand, we have computational devices from hourglasses to PCs and online automatic shopping agents. The approach we have developed to appliance design puts ideas of media ahead of those of computation, in contrast to conventional application design. When we think of a medium, our intuitive sense of what it does is that it makes something available to our senses, and what it means is up to us. It is this perspective that underpins many of the ideas we have already discussed on the interaction of the user with an appliance:

- Media are single in function, but can be used for a wide variety of purposes.
- Sensory motor interaction is organized around physical marking and manipulation of media.
- Immediacy and transparency can often be achieved by keeping to intuitive levels of media interaction, such as jotting down a note.

Here, we want to continue to expand this idea by looking at appliances as *mediating communication between people*. When we use some form of computational device, like a spreadsheet, then we always need in some way to think about the device sharing our interpretation of what is presented through the medium. For example, you are doing a forecast for your business, you put a column of figures in a spreadsheet, and set up a cell for the total to be automatically calculated. You print out the spreadsheet and give a copy to a colleague for discussion. The two of you get together and work on it, annotate it, and then

you go back to your computer and redo the numbers. Look at each of these steps. When the computer filled in the total, you relied on it to manipulate symbols for amounts and for the calculation you wanted, in ways that match their mathematical meaning to you, and to do meaning-preserving transformations. When you printed out the spreadsheet to use the paper as a medium for sharing, the printer converted the displayed numbers in the spreadsheet to marks on paper without interpretation. The printer mediated the discussion and the spreadsheet automated the calculation.

In the late 1980s, a number of researches described the ways in which conversation is *co-produced* by speaker and listener through very fine-grained interaction (Grosz & Sidner, 1986; Levin & Moore, 1977; Reichman, 1985; Stenton, 1987; Whittaker & Stenton, 1988). A naïve understanding of conversation would describe it as well-constructed, complete utterances that are exchanged in a fairly orderly way, being interpreted and replied to by each speaker in turn. The reality is that, even when one speaker is delivering an extended utterance, there is a lot of interaction going on between the participants to reinforce or shape what is said, indicate a desire for turn-taking, and so on. We have found that close respect for this underlying reality of communication is the core of a general principle. Whenever you can increase the fluency, granularity, physical richness, and general ease of communication between people across time and space then they can invent all sorts of ways to get their work done. Attempting to automate all or parts of this communication constrains the types of conversation possible and so narrows the purpose. By mediating interaction between people, or people and information, it is often possible to open up the appliance to a much broader use. As a general rule, there is more value in mediating communication between people than in automating tasks for individuals.

Thus, the general design stance is, whenever possible, to think of an appliance as supporting shared interaction around a *medium of communication*, rather than as implementing an *application*. If this can be done, then it will allow a simple function to be kept very open in purpose, in the way explored previously. This is not to argue against any specialization, but to suggest that the communicative and interactive media aspects are kept as the design center.

What are the inputs and outputs of a phone call? That is a question best answered by the participants once the call is over. It is not something that the designer of the phone system has any involvement in. The phone system mediates the interaction. That is its function. The people create its social purpose, such as to chat, negotiate, argue, or whatever. For the phone to interject, "You appear to be having an argument. Would you like some help?" would be absurd. In this context, designers do not design conversations. It would be far too limiting if they did. On the other hand, appliance designers can look at the dynamics of conversation, and how material means support them, and design appliances that create new conversational possibilities. The electronic whiteboard described previously was one such example. Another example (Omnishare) is given in the following section.

The calculator is an exception that illustrates the mediate rule. It automates numeric transformations through its

understanding of the semantics of numbers and mathematical functions. We would argue that numbers and the grammar of mathematics are a special case of a formalized unambiguous content form unlike any other. In fact, we spend many years developing the mental skills that allow this special language to be used consistently across all situations in our daily lives. Thus, the formality of mathematics allows its functions to be automated, enabling calculators to be focused in function yet open in purpose. The same cannot be argued for Natural Language (NL) translation systems. Four decades of research has grappled with the ambiguity and context dependency of language use. At the dawn of the 21st century, NL systems still have to make the trade-off of coverage against accuracy. NL applications are a good example of where successful automation is achievable, but at the expense of openness of purpose. Successful small vocabulary NL systems rely on limited conversation capabilities and a rigid semantic model.

In the gray area between automation and mediation are text retrieval engines, Thesauri, and single-word translation systems. On the one hand, there are no semantic transformations being carried out. Words are retrieved along with the other words around them, an associated block of words, or a paired word from another language. On the other hand, browsing and look-up functions can be said to have been automated. We would argue that these low-level functions are similar to the ones that convert the output of an image scan to a format for communication with a rendering device or the conversion of a file to ink on paper. Such transformations are context free requiring no intelligent intervention by the device.

Context-aware appliances, such as in-car navigation systems, use Global Positioning System (GPS) technology to determine where they are and mediate navigation in the same way that a map does, but with the added value of a "you are here" marker. Built-in route-finding technology can tell you when to turn to take an optimum route. Navigation is mediated and not automated because the system stops short of controlling the car. Suggestions can be ignored by the driver, thereby causing the system to recalculate the route. Speech output provides 'in the moment' sensory motor instructions, in a manner sensitive to the eyes-busy driving situation. The mediation approach allows navigation appliances to be open in purpose functioning on many road systems and in any driving conditions. The converse is true for in-flight navigation systems and autopilots; automation is possible here because the number and complexity of routes are many times smaller and the traffic conditions far less variable.

In the case of the glance cameras described previously, the mediate approach made ambient social context available across the network and made it possible for social norms to operate. This made it easy for an individual to decide on a wide range of possibilities: they are out—leave a voicemail; they are on the phone—call them and park a call-back; they are in a meeting, but it looks casual; wander by and see if you can catch their eye, etc.

Omnishare. Having covered the key features of direct user interface design, we would like to finish off this section with a

FIGURE 37.5. The Omnishare product released in 1994.

description of an appliance, Omnishare[1] (Fig. 37.5). We include it here because it is a good example of an appliance designed with all the features of a direct user interface in mind.

Working together at a distance is a problem that has attracted huge amounts of investigation over the years. Our own work on this was done in the early 1990s. We were interested in the possibilities for desktop PC conferencing, and in particular the potential of the emerging capabilities of PC-based video technology. We conducted systematic studies of how people could use the different modalities of communication:

- Voice for conversation.
- Video images to see each other.
- Shared whiteboard for illustration.
- Keyboard/mouse/pen+tablet for writing on the whiteboard.

We found a video view of talking heads had no measurable impact on task effectiveness, measured by time to completion, and

---

[1] Omnishare became a product, but was hosted in a rapidly growing PC business for which it became a distraction, so unfortunately was withdrawn soon after launch. In the meantime, it won a *Business Week* industrial design award.

various subjective measures of quality of the output produced and the interaction that produced it. What *was* key to effective interaction was being able to *see what you are talking about*, and this is most simply achieved by voice conversation with a way to *point at and mark* the item under discussion.

These ideas were developed into the Omnishare—a pen- and tablet-based appliance that supported simultaneous interactive voice and data in the form of interactive fax. The two interactive data forms supported were a blank shared whiteboard and a shared image. The software interface was designed by Hunter (1995) in the "software as hardware" style of Mathis (1991). Key concepts of Omnishare were:

- The primary focus was to *mediate* the conversation through a *transparent* interface that could be picked up and used during a phone call.
- The design center was a *shared interactive surface*.
- The primary user model was *pointing to* and *marking* materials that were brought into the telephone conversation.

The model of shared surfaces—a whiteboard and shared document image—were deliberately chosen as the simplest everyday interaction models available, with the widest applicability. An alternative design center would be shared applications, but the demands of transparency and immediacy in the context of a phone call would have been difficult to satisfy.

When a document image was being shared, either person could turn pages with the two views kept synchronized. Each user could manipulate their own cursor and see the other person's. In this way, it was possible to direct the other person's attention with such as utterances as "You see *this* here....," "What about *this* milestone, how would it be affected if we outsourced *this part*." This is what linguists call deixis and is an important part of everyday speech. In addition to pointing with the cursor, each user could write with the pen to annotate the document or image being viewed. Both people could also erase the marks, just as you could on a whiteboard.

Connectivity was provided by a shared voice and data modem for simplicity and the potential for widest deployment on a single standard telephone line. Documents were sent to it from a PC, a scanner, or a fax machine.

Prototypes received very high scores on all user trials and were especially valuable to people trying to reach closure on complex materials under time pressure (e.g., marketing materials, advertising copy, contracts, or schedules). We found the fluency of interacting with a pen was critical to the user experience and a powerful illustration of the sensitivity of the core value to the right embodiment of interaction.

THE NETWORK EFFECT AND NAÏVE MEDIA

> The Network Effect
>
> Work together as peers
> Naïve media
> Conversations
> Surfaces
> Metaphor

So far, we have been looking at what makes an appliance work at an immediate and intuitive level within a wider context of activity. Central to those properties, and the defining notion of an appliance, is that it is simple and focused in what it does. A key enabler of this simplicity is that appliances should be able to work together in unplanned and flexible ways to get a job done that cannot be achieved by one of them alone. To the extent that things just work together with the same simplicity that they work individually, it becomes possible for them each to do one thing really well. The goal of appliance design should always be products that are individually great and together even better.

The foundation of a world of appliances that work together will be standards for interoperability that:

- Enable direct, peer-to-peer interaction of any two devices without recourse to online updates and downloads.
- Allow new devices to appear in the environment without everything else having to be updated.

If we were in the appliance age and could assume this pervasive, peer-to-peer interoperability, then all appliances would benefit from the network effect—anyone buying a new product would immediately find all their existing appliances would become a bit more useful, and they would all contribute their value to the new product (Fig. 37.6).

The notion of the network effect was coined by Bob Metcalfe to express the idea that the value of a network increases according to the square of the number of devices attached to it. When there are only three phones on a network, there are only three different ways to connect them. Add a fourth, and it is available to all three existing subscribers, doubling the number of possible connections to six, and so on. Of course, in practice, not all the new connections are of any interest or value to the subscribers, and this can mean that networks grow slowly until a threshold is reached where the effect takes off. At this point, a new logic of participation takes over; instead of needing a reason to join, you need an excuse for not joining because the network has become the way things are done.

Look at the evolution of the web. Before the invention of the World Wide Web, the Internet was growing steadily and quite rapidly, with a range of services: e-mail, newsgroups, and so on. However, in the realm of access to information services, progress was relatively slow; services such as Compuserve and AOL could not gain any general momentum because they failed to reach a critical mass in which it was worth putting up all the specialized information pages for the people who would want

FIGURE 37.6. Appliances should be human-centered, individually useful, and just work together.

them. These services had their own standards for the protocols that operated between the information services and the client computers, so the information world was split into incompatible user populations. There was a very high churn rate among subscribers because the range of services on offer did not add up to enough to retain them. There was little incentive on the content side to put up information for the small proportion of the total user population who would want it; and users could rarely find the specific information they really needed.

The value of the World Wide Web was that by standardizing the model of information interaction between servers and clients, it made the network effect take off between online pages of information and users. It broke down the walls between incompatible groups of content and users so any client could talk to any information on the network. This unlocked an explosion in specialized services that could now find enough users out on the network to justify the effort of putting them up, and users who found more and more information relevant to their specialized interests. At this point, the logic of participation took over—no business, political party, club, etc., that expects to stay in business can realistically avoid having a Web site, any more than it can avoid having a phone. So people sign up to the Web expecting to find access to the things relevant to them and further reinforce the shift.

The situation in the world of appliances is now very similar to what it was in the world of the Internet before the web. There are many islands of connectivity between different devices, and the world's standards bodies are ever busy adding to the family of interconnection standards, but there is no universal model to which all devices conform. What we need is a standard that will allow for direct and immediate peer-to-peer interaction between any two devices that have a chance encounter and need to communicate. Just as in the world of the web, you want something sensible to happen, even if you lack the latest content plug-in for the superanimated graphics, or whatever, that the site would

like to offer you. Imagine, you have received a fax on your new Internet-enabled smart phone, but it is 10 pages long and you want to print it out. You want to walk over to a nearby printer, send it over a high-speed infrared IrDA or Bluetooth link, and have it printed. Or you are at a friend's house with your digital camera and want to send the pictures to the big TV screen they have, so that you can chat about your holiday.

There is a lot of unnecessary confusion concerning the standards that are needed for this to come about. Three things are required:

- Universal peer-to-peer networking that works for all wired and wireless links, down to opportunistic brief encounters of very low-end devices.
- Universal peer-to-peer content exchange that respects the unpredictable combinations of capabilities of devices, and degrades gracefully for low-end products.
- A set of stable, consistent, and predictable metaphors by which people can understand how all these devices should interact without having to read an instruction manual each time.

At the time of writing, there is a hype and excitement being generated around Bluetooth—a standard for short-range wireless communication between appliances. However, this standard is both too much and too little. The World Wide Web took off because there was already a universal peer-to-peer *network* standard in place—TCP/IP—to which a new layer of content exchange standards—http and html—could be added. What we need is for the notion of universal, peer-to-peer networking to be extended uniformly across all devices and all sorts of connections; what we do not want is further proliferation of standards that combine the network communication technology with the application and content, which is what has happened so far with infrared, wireless, and some new wired technologies. Each new networking technology is driven by a section of the IT industry that looks to serve its own particular interests by pushing a family of application-level protocols for connecting the devices it happens to be interested in. The standards for the physical links such as infrared (IrDA), very local wireless (Bluetooth) are good for what they do. Now we need them to be brought into a model of universal peer-to-peer interaction.

For universal content exchange to work between appliances, there needs to be a similar shift in the industry similar to what happened when http and html became the standard for information exchange among computer clients and servers. This is more than a matter of a particular technical standard winning out against competitors; it requires that a particular *metaphor of interaction* also becomes established in the minds of users.

The World Wide Web is not just a protocol, it is a consistent interaction model that people can use for accessing, presenting, and navigating information. In many important respects, as other writers have commented (Nielsen Norman Group, 2000), the roots of html in a graphical layout language represent a setback of many years in producing simple and usable interfaces because the interaction supported by updating pages is so clunky—all the more significant, therefore, that it has been as successful as it has. The key to this is that those same roots

mean that the user has a very simple intuitive physical metaphor of *pages* at the center of the interaction. Pages with names and addresses build very strongly on familiar ideas of information, and lead new users by a very gentle learning ramp into more levels of interaction such as forms. Point and click is not hard to master as a way of moving between pages, and nothing more is needed for huge amounts of value to be extracted from the Web.

The purpose of this chapter is to guide designers in producing appliance solutions. If the world of universal interaction had come about, then we would all know what had emerged as the universal, stable set that should be taken as a given by all appliance designers. We confidently predict that such an emergent order will appear in the industry over the next few years, but until then we need to get on with building useful products. Each designer will find himself severely limited by the number of products he can impact in addition to his own and will therefore be unable to take a system wide view of the interaction model. The guidance we give here represents our view on the key aspects such a universal solution is likely to have, and so should be applicable both to limited solutions that can be built now and also relevant to a future of universal interaction.

The basic idea we need is that *all appliances are understood by the user as acting on common, user comprehensible, everyday media*, not as *interacting with each other*. This is a critical change of perspective. It is only hard to make because most information product designers have grown up in the computer world where the PC, or some other computer, controls all the other devices that are conceived of as peripherals, whose job is to get information in and out of the place where all the real work happens—the computer. In this model, interaction with each information tool is mediated by another one—the computer. By now, it should be clear that the world of appliances is the precise inverse of this: peripherals are the tools by which a physical, embodied human picks up, browses, arranges, shares, sends, and general messes around with information. Computation, when it happens, supports this interaction—a web browser renders the information I want to look at, a printer does the million and one things necessary to get a set of marks into the right place on a sheet of paper, and a camera handles focus and light intensity to capture an image.

This change of perspective becomes very obvious when we turn to the world of everyday things that already work in the way we propose. Think of the tools and appliances in the kitchen; they all work on food, and they work together in the sense that they do particular things, with respect to particular food, in ways that can be helpfully combined by a human being without specialist knowledge. By contrast, except in a few specialized instances, the tools do not have specialized interfaces to *each other*. A knife is used on a chopping board, the food is transferred to a pan, which is put on the hob. Food is served up onto plates, and we use our knives and forks to eat. Look at what is happening in each of these interactions. It is the food, and the human's intentions around it, that organizes each of the activities and brings different tools together: the chopping board to *support* it while the knife *cuts* it; the pan *contains* it, while the hob *heats* it. The exceptions are in the areas where one tool controls another, such as a bottle opener extracting a cork, or the hob heating via the pan in ways of which we need a reasonable understanding to get good results. In these cases, the model of combination is still very simple: the two tools combine around a single action such as opening or heating.

Turning to a closer example, look at the way that paper, documents, and related physical media provide us with a consistent model to interact with all the information appliances that we use in the office. I can take a piece of paper from the printer to the copier, fax machine, notice board, ring binder, filing cabinet, or shredder, and each one will do one thing with each—each one has a clear *verb* attached to it. When they combine, for instance in putting a row of box files onto a shelf, or pinning a notice to a notice board, they again do so by extending a basic function of one of them—displaying, storing.

When devices have interfaces by which they *control each other*, then generally we are in the realm of expert knowledge and specialist interests concerned with the internal structure of artifacts, such as maintaining cars or central heating systems. Home heating systems are an interesting case because of the way that a number of devices—furnace, radiators, hot water system, room thermostat—are all related together to provide the overall function of warmth and hot water. In everyday life, this is probably one of the closest systems of appliances we encounter where the devices talk to each other in the sorts of ways so familiar to computer system designers. Anecdotal observation and reporting suggest that most people have difficulty mastering the interactions between the different devices, because this requires them to build a model of the interaction for which there is no everyday analogue.

Returning then to information appliances, we want to coin the term *naïve media* for the common material around which all appliance interaction should be organized.[2] Our experience strongly supports the position that people have a clear intuitive sense of information and how it can be handled at the level of everyday physical actions: display a picture, copy it, send it, store it, and clip it to a letter. They do not confuse (much) the information itself with the different ways it can be handled, any more than they confuse food with all the different kitchen implements, or information that can be represented on paper with documents, slides, and files. As we have stressed throughout this chapter, everyday action with information appliances needs to be rooted in the basic sensory motor behaviors that get us around the world. So, the pragmatic way to arrive at naïve media is by asking whether you can apply simple information verbs to them. By this criterion, we start to assemble a preliminary list:

- Human conversation
- Handwriting
- Text
- Drawings

---

[2] Oxford English Dictionary. **Naïve a.** Natural, unaffected, simple, artless. **b. Naïve realism** The belief, attributed to nonphilosophers, that the world is directly perceived as it really is.

- Music
- Pictures
- Movies

If your appliance has a clear relationship, expressed as a simple verb, with respect to one or more media, then you are on the road to a simple interaction model, and a consistent relationship with other appliances. People can easily combine devices that do simple things to naïve media—my camera captured an image, I want to print it on this printer, or display it on this picture frame—how hard is that? Quite hard technically to make it just work, but from a user perspective, it is obvious. Even more complex devices, such as PDAs, should be built so that their interactions with others always obey naïve expectations; more sophisticated levels should be built on top.

With this level as the foundation, we suggest that there are three stable concepts for intuitive, systemwide, user interaction:

- Conversations
- Surfaces
- Application metaphors

Communication happens between people through naïve media that are manipulated at a physical level by one or both parties to the communication. This book you are reading has carried marks made first on the screen of my computer through to a piece of paper in your hand. The phone conversation you will have later connects the air in two distant places to vibrate in sympathy with each other as a medium for voice and hearing. At the most fundamental level, an information appliance is *always* providing a medium of communication between people. At one end, this is the highly asymmetric communication of a book or a movie; at the other, it is the intimate exchange of initiative of face-to-face dialogue. We have highly developed social models for dealing with all these different conversations, and good appliances will aim to provide media that support, not automate, these interactions. This is the foundation for the mediation vs. automation approach we outlined earlier. Presented with a medium of interaction, humans will rapidly develop social protocols for their conversations that are far more sophisticated than any application could ever embody.

Building on the idea of conversations, the design concept of a *surface* embodies the shared medium and can be provided with the appropriate affordances to match the nature of the conversation. Returning to our wayfinding example, guidebooks, maps, and road signs all come together in a user's mind as facets of a single type and source of information: publicly available geographic information. Each one chooses a different physical surface tuned to a different setting and style of physical interaction for the conversation. A simple way to think about appliances is in the way they extend these surfaces over time and space, breaking the limitations of nondigital media.

Why did we just say nondigital and not physical media? There is an unfortunate tendency to talk of digital media, and being digital, as if we humans had suddenly acquired extrasensory abilities to interact with disembodied information. In fact, of course, we *only* interact with information once it has entered the embodied, analogue, physical world. The power of digital *technology* is to extend over space and time the reach of *physical media.*

A PC-shared whiteboard application, for example, takes the commonplace office whiteboard notion of an information surface, but makes it available for marking by people at a distance, projecting the surface across distance and synchronizing the marking, mediated by their computer screens. By studying the appropriate affordances for this sort of interaction, the Omnishare project at HP was able to produce a product that was very closely matched to the physical interaction appropriate to using a whiteboard in conversation. A key observation was the need to support gesture as well as marking, so that it was easy to point to information under discussion without necessarily marking it—just as can be done in a real conversation around a whiteboard (O'Conaill, Geelhoed, & Toft, 1994).

The third concept required is the *application metaphor.* An application metaphor confines and reduces an information medium to something that we can manage. Thus, PCs use a desktop metaphor, and the web uses pages, forms, chat rooms, and so on. Books themselves reduce the medium of text to a rich set of culturally stable interactions—pages, chapters, indices, etc. One approach to appliances is of course to borrow the existing organizing ideas for information—books, pages, desktops and so on—and then extend them with digitally enabled capabilities—hyperlinks, searching, or whatever. It is here that particular care is needed to preserve systemwide consistency between appliances so that they can still support the unplanned and opportunistic interaction that is so desirable. Because an application metaphor specializes naïve media in some way, there is even less chance of getting systemwide agreement on it among different appliance vendors. We suggest here that, unless you can guarantee interaction at the level of the application metaphor, you at least fall back on a consistent behavior at the level of naïve media.

Consider, for example, the idea of the electronic business card. Business cards are a stable cultural way of sharing important personal information. Because sharing is so important, we have made the copying function redundant and established a social protocol of exchanging the physical embodiment of the information. What happens after the exchange is up to the receiver, who may file it, enter it into a handwritten book, PDA, or whatever. The important point is that those later steps in no way interfere with the basic function of allowing the human to get a copy of the information. It does not seem to have occurred to the makers of all the PDAs and phones that, even in the absence of a shared encoding for all the fields of a business card, it would be a step forward if they could exchange information at the level where a *human* could interpret it, even if the device could not.

The emergence of XML is an important enabler for the emergence of stable application metaphors, but by itself will not guarantee interoperability. That only happens when there are culturally stable information forms that XML can encode.

So far, there is little progress toward a universal interaction model for appliances that meets all the criteria we have covered. HP's Jetsend (Williams, 1998) was one, introduced in 1997 and promoted in the industry for the interaction of image

information among printers, scanners, projectors, cameras, and the like. Unfortunately, HP appears to be no longer actively promoting this, and a substitute has yet to appear. In the meantime, we believe that the naïve media approach outlined here provides a guide for practical approaches to appliance interworking and simplicity of function for individual appliances.

CONCLUSIONS

In this guide to the design of information appliances, our aim has been to explore the concept of applianceness as a range of features that can take a product category into pervasive everyday use as it takes full advantage of emerging digital technology; we think this is more productive than worrying whether a product is or is not an appliance. These features arc based in understanding how humans interact at a simple sensory motor level with the everyday things around them and letting these shape the information-handling capabilities. The core of the ideas is to make devices do one thing well and work together. These twin goals are best served by attention to simple functions that are open in purpose and that can be combined together through the notion of naïve media. Life will get a lot easier for designers when a pervasive standard emerges for interworking; until then, designers must do their best to manage the overall system context of their own appliance.

ACKNOWLEDGMENTS

The foundations for this chapter were laid by our many colleagues in Hewlett-Packard Laboratories who over the last 10 years pioneered a wide range of appliance concepts. Thanks are also due to Kenton O'Hara for his review and comments on earlier drafts, and to Ruth Sharpe for patient proofreading of a changing manuscript.

References

Bedford-Roberts, J., Nelson, A. L., Stenton, S. P., Stott, C., & Lacohee, H. (1995). Piglet: A pen-based, personal communicator for light touch messaging. In P. Thomas (Ed.), *Developments in personal systems* (p. 140).

Bergman, E., & Haitani, R. (2000). Designing the Palm Pilot: A conversation with Rob Haitani. In E. Bergman (Ed.), *Information appliances and beyond* (pp. 81–103). San Francisco: Morgan Kaufmann.

BIS Strategic Decisions. (1993). *Mobile professional market segmentation study.* MA: BIS Strategic Decisions.

Buxton, W. (2001). Less is more (more or less)· Some thoughts on the design of computers and the future. In P. Denning (Ed.), *When cyberspace becomes our world.* New York: ACM Press.

Central Statistical Office. (1992). *Social trends.* (Vol. 22, p. 177). London: HMSO.

Churchill, E. F., & Wakeford, N. (2001). Framing mobile collaboration and mobile technologies. In B. Brown, N. Green, & R. Harper (Eds.), *Wireless world: Social and interactional implications of wireless technology.* New York: Springer Verlag.

Clark, A. (1997). *Being there: Putting brain, body and world together again.* Cambridge, MA: MIT Press.

Damasio, A. R. (1994). *Descartes' error: Emotion, reason and the human brain.* New York: Gosset/Putnam Press.

Drucker, P. F. (1973). *Management: Tasks, responsibilities and practices.* New York: Harper & Row.

Finn, K., Sellen, A. J., & Wilbur, S. (Eds.). (1997). *Video-mediated communication.* Mahwah, NJ: Lawrence Erlbaum.

Frohlich, D. M. (1992). *A survey of office work practice* (Hewlett-Packard Laboratories Bristol Technical Report HPBL-92-121). Bristol, UK: HP Laboratories.

Frohlich, D. M., Dray, S., & Silverman, A. (2001). Breaking up is hard to do: Family perspectives on the future of the home PC. *International Journal of Human-Computer Studies, 54,* 701–724.

Frohlich, D. M., & Perry, M. (1994). *The paperful office paradox.* (Hewlett-Packard Laboratories Bristol Technical Report HPBL-94-20). Bristol, UK: HP Laboratories.

Gaver, W., Moran, T., MacLean, A., Lovstrand, L., Dourish, P., Carter, K., & Buxton, W. (1992). Realizing a video environment: EuroPARC's RAVE system. In *Proceedings of CHI '92 Conference on Human Factors in Computing Systems* (pp. 27–35).

Gaver, W. W. (1991). Technology affordances. *CHI 91 Conference Proceedings* (pp. 79–84). New York: ACM Press.

Gibson, J. J. (1979). *The ecological approach to visual perception.* New York: Houghton Mifflin.

Grosz, B. J., & Sidner, C. L. (1986). Attentions, intentions and the structure of discourse. *Computational Linguistics, 12,* 175–204.

Harper, R., & Sellen, A. (1995). Collaborative tools and practicalities of professional work at the International Monetary Fund. In *Proceedings of CHI '95 Conference on Human Factors in Computing Systems* (pp. 122–129). Vancouver, New York: ACM Press.

Hunter, A. (1995). Pen-based interaction for a shared telephone workspace. In J. Nielsen (Ed.), *Advances in human-computer interaction* (pp. 115–158). Norwood, NJ: Ablex.

Hollan, J., Hutchins, E., & Kirsh, D. (2001). Distributed cognition: A new theoretical foundation for human-computer interaction research. *ACM Transactions on Human-Computer Interaction.*

Kay, A., & Goldberg, A. (1977). Personal dynamic media. *Computer, 10*(3), 31–41.

Kidd, A. L. (1994, April 24–28). The marks are on the knowledge worker. *Proceedings of the ACM CHI '94: Human Factors in Computing Systems.* Boston, MA. New York: ACM Press.

Kirsh, D. (1995). The intelligent use of space. *Artificial Intelligence, 73,* 31–68.

Lakoff, G., & Johnson, M. (1999). *Philosophy in the flesh.* New York: Basic Books.

Levin, J. A., & Moore, J. A. (1977). Dialogue games: Metacommunication structures for natural language interaction. *Cognitive Science, 4,* 395–421.

Levy, S. (1994). *Insanely great: The life and times of Macintosh the computer that changed everything.* New York: Viking.

Luff, P., Heath, C., & Greatbatch, D. (1992). Tasks in interaction: Paper and screen based documentation in collaborative activity. In *Proceedings of CSCW '92, Conference on Computer-Supported Cooperative Work* (pp. 163–170). Toronto, Canada. New York: ACM Press.

Mathis, B. (1991). Why can't a computer be more like a car? VUE 2.0 Visual User Environment. *Innovation, 10*(1), 32-34.

Maturana, H. R., & Varela, F. J. (1987). *The tree of knowledge*. Boston: Shambhala Publications.

Moore, G. A. (1991). *Crossing the chasm*. New York: Harper Business.

Nielsen Norman Group. (2000, December). *WAP usability report*.

Norman, D. A. (1990). *The design of everyday things*. New York: Doubleday.

Norman, D. A. (1998). *The invisible computer: Why good products can fail, the personal computer is so complex and information appliances are the solution*. Cambridge, MA: MIT Press.

O'Conaill, B., Geelhoed, E., & Toft, P. (1994). DeskSlate: A shared workspace for telephone partners. *Proceedings of CHI '94 Conference Companion* (pp. 303-304). New York: ACM.

O'Conaill, B., Whittaker, S., & Wilbur, S. (1993). Conversations over video conferences: An evaluation of the spoken aspects of video-mediated communication. *Human-Computer Interaction, 8*, 389-428.

O'Hara, K., Taylor, A., Sellen, A., & Newman, W. (in press). Understanding the materiality of writing while reading from multiple sources. *International Journal of Human-Computer Studies*.

Pedersen, E. R. (2001). Calls.calm: Enabling caller and Callee to collaborate. In *Extended Abstracts of CHI '01 Conference on Human Factors in Computing Systems* (pp. 235-236). Seattle, WA. New York: ACM Press.

Perry, M., O'Hara, K., Sellen, A., Brown, B., & Harper, R. (2001). Dealing with mobility: Understanding access anytime, anywhere. *Transactions on Human-Computer Interaction, 8*(4), 323-347.

Picard, R. (1997). *Affective computing*. Cambridge, MA: MIT Press.

Reichman, R. (1985). *Getting computers to talk like you and me*. Cambridge, MA: MIT Press.

Scarborough Research. (2001, May 9). *The First Scarborough National Internet Study*. New York: Scarborough Research.

Schmidt, A., Takaluama, A., & Mantyjarvi, J. (2000). Context aware telephony over WAP. *Personal Technologies, 4*, 225-229.

Sellen, A. J., & Harper, R. H. R. (2001). *The myth of the paperless office*. Cambridge, MA: MIT Press.

Stenton, S. P. (1987). Dialogue management for co-operative knowledge-based systems. *The Knowledge Engineering Review, 2*(2), 99-121.

Taylor, J. T. (1999). Engineering the information age. *IEE President's Inaugural Address*. London: Institute of Electrical Engineers.

Thimbleby, H. (1990). *User interface design*. ACM Press Frontier Series. New York: ACM Press.

Walker, M. A., & Stenton, S. P. (1992). A personal information management survey (Hewlett-Packard Laboratories Bristol Technical Report HPBL-11-92). Bristol, UK: HP Laboratories.

Whittaker, S. J., Geelhoed, E., & Robinson, E. (1993). Shared workspaces: How do they work and when are they useful? *International Journal of Man Machine Studies, 39*, 813-842.

Whittaker, S. J., & Stenton, S. P. (1988). Cues and control in expert-client dialogues. *Proceedings of the 26th ACL* (pp. 123-130).

Williams, P. (1998). JetSend: An appliance communication protocol. *In Proceedings of the IEEE Workshop on Networked Appliances* (pp. S1-S3).

Wilson, F. R. (1998). *The hand*. New York: Pantheon Books.

Part

·IV·

APPLICATION DOMAINS

Arnold M. Lund
Sapient

Human–Computer Interaction (HCI), according to the Association of Computing Machinery SIGCHI (1992), is "a discipline concerned with the design, evaluation and implementation of interactive computing systems for human use and with the study of major phenomena surrounding them." The emphasis in this definition is the traditional one within HCI on the direct interaction of the user with an application. The driver for the rich diversity of these computing systems, however, is business and other institutions; these institutions keep most HCI professionals employed. To get a feeling for the phenomena that will serve as the basis of future designs, it is valuable to look at application areas driving the technology producing the phenomena. An important aspect of the work in these application areas is that design is not just about helping the user accomplish their goals, it is also about design to enable institutions to accomplish their goals through shaping user's goals, expectations, behavior, and even emotional reactions. In many cases design can even be thought of as being about creating and supporting relationships between institutions and users.

The following chapters provide a sample of important application domains where HCI professionals are active. These range from government (chapter 40, Government Roles in HCI by Jean Scholtz) to important public service areas such as health care (chapter 41, Human–Computer Interaction in Health Care by François Sainfort) and education (chapter 42, A Framework for Understanding the Development of Educational Software by Chris Quintana, Joseph Krajcik, Elliot Soloway, and Cathleen Norris). The chapters include work that touches the infrastructure that shapes our society such as telecommunications (chapter 39, The Evolution of HCI Work During the Telecommunications Revolution by Edmond Israelski and Arnold Lund), e-commerce (chapter 38, E-Commerce, Interface Design by John Vergo, Sunil Noronha, Joseph Kramer, Jon Lenchner, and Thomas A. Cofino), and transportation (chapter 45, Human–Computer Interface in Aerospace by Amy R. Pritchett; chapter 44, Motor Vehicle Driver Interfaces by Paul Green).

There are also discussions of HCI in areas such as entertainment that add joy to life and that may stretch the more traditional principles of HCI (e.g., chapter 43, Understanding Entertainment: Story and Gameplay Are One by Jesse Schell and chapter 46, User-Centered Design in Game, by Randy J. Pagulayan, Kevin Keeker, Dennis Wixon, Ramon L. Romero, and Thomas Fuller).

The chapters and the domains they present have several elements in common. In general, grounding in the principles, methodologies, research, and theories discussed elsewhere in the *HCI Handbook* is assumed. A practitioner entering one of these areas must learn to adapt what has been learned previously to the unique properties of the domain and the organization in which they will be practicing. The chapters each describe a unique set of needs that the HCI professional is attempting to meet through design. The specific needs within and between domains differ based on what the users hope to accomplish and the context in which they hope to accomplish it. This highlights the importance in HCI of understanding the whole user, not just inside the skin of the user but also the user in the context of and interaction with the world in which they live. The needs described also suggest that the practitioner is not just designing to help users accomplish their goals. The chapters talk about how designers attempt to design to support the goals of the institutions providing the applications for the users. Scanning the domains also reveals the impact of organizational constraints that shape the day-to-day practice of the professional and what it could mean for how HCI as a field can help the practitioner demonstrate the value of user-centered design.

For most applications, one way to classify users is in terms of whether they are required to use an application (as in a business application that is required as part of a user's job) versus whether they have ready access to competing options (as in an application aimed at the consumer market). Obviously this is a classification with a fuzzy boundary. In practice, it is common to see management require the use of an application only to see employees find ways to retain the previous way of doing things

or to see them use the application in unexpected and perhaps less effective ways. Similarly, a new consumer application may be bundled with a product that is chosen for other reasons, and so the consumer gets the application whether they like it or not.

In general, however, this distinction highlights the institutional imperatives behind the applications and design goals faced by HCI professionals. In situations in which users are being required to use an application, traditional usability considerations of ease of use and usefulness are important. From the institution's perspective, the goal is typically to help users become more productive and more effective. When consumer products are being designed, users need to be attracted to the product and be sufficiently satisfied that they continue to use it. Often this means that in addition to the more traditional dimensions of usability, the application also needs to be compelling. It may need to be aesthetically appealing on first exposure (e.g., to attract purchasers) and fun and enjoyable when experienced.

Figure IV.1 illustrates the properties of this broader view of the interface. The applications discussed in these chapters often are composed of some kind of content, functions for interacting with and deriving value from the content, and a presentation environment that provides a first impression, guides interaction, and offers much of the feeling of an experience. The functions for interacting with the content and the presentation itself can serve as content for the user, as in the case of many entertainment applications. All three "layers" interact to create the experience that users have as they try to accomplish their goals. The chapters, however, also illustrate that as in the bricks and morter world, the provider of the experience also has goals that they want to accomplish. These goals might be to educate, to

communicate, or to guide behavior in some way. The goals are defined to deliver the value that caused the development effort to be initiated. The experience that the HCI professional is shaping, therefore, must be done with both ends in mind. It must enable the user to meet their needs while accomplishing the institution's goals. When done effectively, a kind of implicit relationship is created between the institution and the user, and the value generated becomes the return on investment that provides the cost–benefit justification for HCI activity.

The user research that has characterized most HCI work in companies has tended to focus on removing barriers that keep users from taking advantage of the functionality in the application. It has focused, in other words, on ease of use. Often, of course, much is learned about the match between the functions of the application and the requirements of the work, but ease of use tends to be the main goal and usefulness a secondary issue. More recently, however, HCI professionals have realized that understanding the total user—the user in context—is critical for delivering efficiency, effectiveness, and satisfaction. Ethnographic techniques (Millen, 2000; Blomberg, Giacomi, Mosher, & Swenton-Wall, 1993; Beyer & Holtzblatt, 1997) have been proving increasingly valuable in providing the kinds of user data that make a design not just easy to use but also useful, and both are needed to deliver applications that will be successful when deployed. Modeling experience increases the value of the data by creating a model of changes in the user experience as a function of contextual variables and using that model to drive the identification of opportunities to meet needs in new ways and to drive the user profiles and scenarios that serve to shape subsequent design and usability testing activities.

Interestingly, these layers and the usability and ethnographic research techniques that shape them lend themselves to skills specialization within companies. HCI professionals entering the field may find they qualify for a variety of jobs with unfamiliar names. In Fig. IV.2 Mok (2000) provided a visual description of job titles within experience design, and there are titles that touch each of the layers illustrated in Fig. IV.1. Mok's representation also shows the research specializations that address ensuring that users find applications easy to use and useful. In the future, there may be HCI specializations designed to study organizational and user behavior to shape interfaces more effectively to deliver the value that institutions seek. When institutions begin to appreciate the direct link between designed experiences and the bottom line, the demand for HCI grows.

One of the effects of working within an organization is that organizational constraints impact the nature of the work. Although there often has been a greater appreciation of the importance of designing for the user in accomplishing institutional goals, there has also been an increasing demand to do more with less, to become more productive. This has required practitioners to seek out methodologies that will allow them to obtain as much information about users as quickly as possible to shape the most important aspects of a design. It has also led to an interest in tools that can help people design more efficiently. Sometimes lost, however, are the improvements in productivity and design effectiveness that should come from designing based on sound theory and lessons gained across projects. Short project time frames often drive either a "there is only time to invent it here"

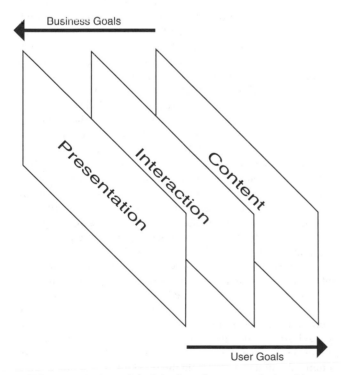

FIGURE IV.1. A model of the interface as a relationship.

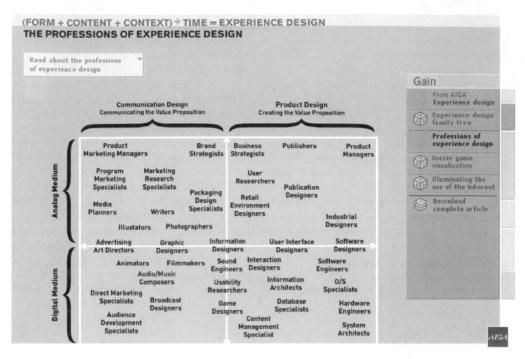

FIGURE IV.2. Job titles within experience design. Reprinted with permission of Clement Mok.

or "copy but don't bother to improve" attitude. An approach that seems to have been encouraged during the recent focus on the Web is to get something out even if it isn't perfect, use user feedback (e.g., complaints) to identify areas that need to be fixed, and then launch a new release.

It is the nature of these application domains that they will persist and evolve. As long as there are business and user needs in the area and meeting those needs more and more effectively is required (e.g., in the case of government or education) or provides business opportunities (e.g., in the case of e-commerce,

telecommunications, entertainment, aerospace, and transportation), there will be an opportunity for HCI professionals to practice. The HCI professional's influence on users will depend on whether they can deliver value to the institutions they support, within the organizational constraints within which they work. The professional's ability to deliver, in turn, will depend on a rich understanding of the fundamentals of HCI, a growing and vibrant field; a deep understanding of the whole user; and the professional's ability to design productively and effectively easy-to-use, useful, and compelling experiences.

References

ACM SIGCHI Curricula for Human Computer Interaction. (1992). New York: ACM Press.

Beyer, H., & Holtzblatt, K. (1997). *Contextual design: A customer-centered approach to systems design.* San Francisco, CA: Morgan Kaufmann.

Blomberg, J., Giacomi, J., Mosher, A., & Swenton-Wall, P. (1993). Ethnographic field methods and their relation to design. In D. Dehuler &

A. Namioka (Eds.), *Participatory design: Principles and practices.* Mahwah, NJ: Lawrence Erlbaum Associates.

Millen, D. R. (2000). Rapid ethnography: Time deepening strategies for HCI field research, In *Conference Proceedings on Designing Interactive Systems: Processes, Practices, Methods, and Techniques* (pp. 280–286). New York: ACM Press .

Mok, C. (2000). The professions of experience design. *GAIN, 1,* 7.

·38·

E-COMMERCE INTERFACE DESIGN

John Vergo, Sunil Noronha, Joseph Kramer,
Jon Lechner, and Thomas Cofino
IBM

This chapter is meant to serve as a resource to human–computer interaction (HCI) practitioners and researchers who wish to understand the existing body of research in the area of e-commerce systems as it applies to HCI. In particular, we report on research that will be useful to designers of interactive e-commerce systems. We identify issues that are unique to e-commerce systems design, which HCI practitioners need to understand. We attempt to answer the question, What do HCI practitioners need to do differently for e-commerce systems design?

We look at the design of these systems as a two-actor system, with the role of the customer and the marketer being the two primary roles. The chapter is not a primer on marketing, Internet marketing, e-commerce systems, or HCI methods. Our goal is to identify issues in the design of e-commerce systems that are important for HCI practitioners to understand and to provide valuable references to existing research that will inform the activities of e-commerce systems interface designers. Additionally, we identify interesting research opportunities for HCI practitioners in the area of e-commerce interface design. Please see chapter 36, The World Wide Web, by Jonathan Lazar for more general, Web-related HCI issues.

INTRODUCTION

E-commerce systems have forever changed the way that sellers and buyers interact with each other. The World Wide Web routinely brings marketers and their customers in contact with each other through interactive systems. For success in this channel, the use of appropriate HCI development methods is critical for shaping the design of these interactive systems. Thus, HCI practitioners must take into account the needs of both customers and marketers when designing Web user experiences, a new twist on existing methods, to optimize across business results for the firm and user experience for the customer.

Although traditional models of buyer behavior well served the needs of marketing professionals in the past, conducting e-commerce successfully via the Internet today requires modification of traditional models to address the unique characteristics of this new channel. As the World Wide Web rapidly brought the ultimate consumer into direct contact with the original manufacturer of goods or provider of services, businesses learned that their ability to apply traditional marketing techniques effectively to the online world was disappointing. Businesses responded to the promise of the Internet by rapidly deploying Web sites, but often didn't provide adequate transactional and informational capabilities to facilitate generating the sales they desired (Cramer, 2001). Those sites, such as Amazon.com, that supported their customers' needs well quickly moved into leadership positions in the marketplace. Even in hotly contested areas such as book retailing, early efforts by giants such as Barnes & Noble and Borders were disappointing. However, by reformulating their business strategy and business goals, and improving their Web site, barnesandnoble.com is now taking back market share from Amazon (Helft, 2001).

As HCI practitioners, we understand that we must design applications that facilitate user task accomplishment and achieve our firm's business goals. For example, in a recent Forrester study (Forrester Research, 2000), 42% of U.S. consumers bought online again because of a previous good online purchase experience with the site. Yet the marketers we support are usually not able to effectively translate their marketing prowess from the bricks-and-mortar domain to the Internet domain. Does greater interactivity translate into increased user goal achievement? Do more ad impressions yield greater sales? Do flashy ad graphics spur consumers to new heights of purchasing? Our challenge, then, is to set the appropriate context in which HCI practitioners can leverage marketers' expertise in traditional channels to their advantage over the Internet channel, while at the same time serving the needs of the customer. Often, customer needs and marketing needs conflict. HCI practitioners must balance the two to maximize business results.

In the first years of the 21st century, the stock market is making it painfully obvious to investors that too many firms are competing on the Web for too few customer dollars. Take the case of Petopia.com, a promising startup launched in August 1999 with financing of $79 million at launch and $45 million later in 2000 (Enos, 2000), which failed in a crowded field (Petopia.com Cuts Staff, 2000) against competitors PetSmart.com, Amazon-backed Pets.com, and PetStore.com. Even though Petopia.com was very well financed, a key indicator of future success for a startup, the market wasn't large enough to support the three top competitors. Exceptional financing did not guarantee success. Yet for the firm that understands how to serve their customers over the Web better than its competitors, success is attainable, even in a crowded field. That success is affected by many variables—a sound business model, a strong value proposition for the customer, an efficient and effective Web site, and first-mover advantage, among others—but for HCI practitioners, success depends on how well we use the medium of the Web to serve the needs of the customer and the marketer. Should we use lots of high-quality images or fancy graphics simply because we can? Should we assume that all customers know how to use common Web interaction techniques? The answer to these questions is "not necessarily." What, then, should we do? How should we apply the capabilities of technology to serve both the customer and the marketer?

The importance of good usability engineering is not lost on business executives. For instance, usability was rated as the most important contributor to success for a bank or brokerage Web site by financial services executives (Forrester Research, 2001b). Even though this data comes from commercial research and is perhaps not as rigorous as practitioners in the field would want, it demonstrates recognition of the importance of focusing on usability to achieve business goals. Gerald Lohse, in a useful survey of the field, was even stronger on this point, arguing that "the quality of the customer experience drives the success or failure of a site (Lohse, 2000)." In a study by Lohse and Spiller, interface design features having the largest influence on Web store traffic and sales were identified. They found that product list navigation features that reduced the time it took to purchase online were the largest contributor to higher sales (Lohse & Spiller, 2000). Other important product list features that had the greatest impact on sales included price, small product images, and more descriptive product naming.

We believe that knowing the customer is crucial for the HCI practitioner. User interface (UI) designers need to build in-depth a priori knowledge of the target customer population to inform their UI designs. Through our work we have accumulated some understanding of users on the Internet who come to a Web site to buy a computer. We know, for example, that customers like to view pricing information very early in the buying process; we know that they rely on images of the computers to make a purchase decision; and we know that they tend to buy additional computers made by the same firm that made their first computer. Shouldn't designers leverage this knowledge to better inform their designs? Yet showing pricing information may not be enough, if it isn't dynamically changing as the customer changes the configuration. Showing images may not be enough if the images don't provide the specific details in which the customer is interested, such as a view of the front, sides, and back of the computer. There exist proven methods that allow us to determine, through detailed knowledge of our customers, the interactions that should be used to facilitate the customer's buying decision. Only through such knowledge of the customer are UI designers able to select the most appropriate information model, interaction methods, and transactional capabilities for the Web site they must design. But understanding only customer needs isn't enough for success on the Internet. We must also take into account the marketer's needs to support the firm's business goals to maximize the return for the firm. Thus, we must extend existing HCI methods to accommodate both the needs of the customer and the firm's marketing needs.

FRAMEWORK

We start with the presentation of a framework of HCI activities, which will be familiar to most HCI practitioners. We progress by examining each step of the HCI user experience design process in more detail. As we examine the design process, we identify issues that are specific to e-commerce interface design and marketing's role in the design process. We conclude with a summary of key insights and point to interesting directions for future research.

Eight Phases of User Experience Design

In Fig. 38.1, view the process of e-commerce user experience design as a sequence of eight major phases that contain crucial marketing and design choices. The remainder of this chapter examines each of the phases listed above, in the context of designing e-commerce user experiences. We present a definition of each phase, as well as a series of relevant questions typically posed by an interface designer. In particular, we focus on variants of questions that are commonly found in the e-commerce design space. Because of the two-actor nature of e-commerce interface design, HCI practitioners are likely to be confronted with many unfamiliar questions that originate from marketing. These choices must be resolved through activities that integrate principles and objectives from both marketing and HCI.

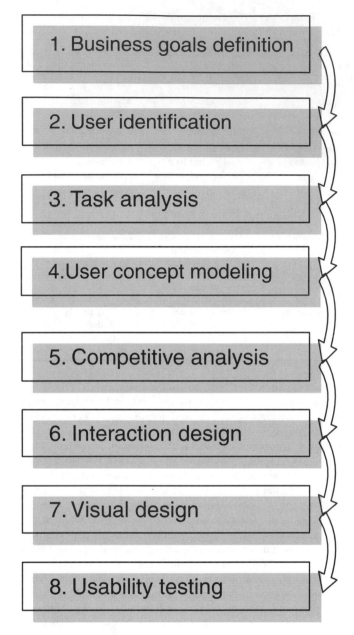

FIGURE 38.1. The eight phases of user experience design.

In this handbook, please see chapter 48, Task Analysis, by Janice (Ginny) Redish and Dennis Wixon, and chapter 54, Participatory Design: The Third Space in HCI, by Michael Muller. Three chapters provide more detail on the topic of evaluation: User-Based Evaluations by Joseph S. Dumas (chapter 56), Inspection-Based Evaluations by Gilbert Cockton, Darryn Lavery, and Alan Woolrych (chapter 57), and Model-Based Evaluations by David Kieras (chapter 58).

For instance, terminology differences can cause confusion. The terminology of HCI is often not the same as the terminology of marketing. Although concepts may be shared across these two disciplines, terminology describing the concepts may not. The HCI practitioner must be aware of this, and be able to bridge

TABLE 38.1. The Language Barrier: A Comparison of Human–Computer Interaction (HCI) and Marketing Concepts and Terminology

Concept	HCI Perspective	Marketing Perspective
Individual	End user	Customer
Artifact	User experience	Whole-product, brand image
Values	Ease of use	Share of wallet
Measures of success	Task completion rate, error free rate, time on task	Profit, revenue, site stickiness, conversion ratio
Major phase: understanding the customer	User modeling	Customer segmentation
Key decisions	What tasks should be supported?	What needs should the product satisfy?
Contributing disciplines	Systems analysis	Market research
Dominant paradigms	User-centered design (UCD)	One-to-one marketing, utility theory (Kleindorfer, Kunreuther, & Schoemaker, 1993)

both disciplines. Table 38.1 above shows examples of these differences in terminology. For instance, where HCI practitioners refer to measures of success to determine the effectiveness of their work, they often think about this concept as it involves the end user. Questions such as the following are answered: What is the task completion rate for the typical user? What percentage of users are satisfied with their experience on the Web site? What percentage of users had great difficulty in achieving their goals on the Web site? Marketers, on the other hand, when discussing measures of success, often think about how much the latest marketing campaign increased revenue on the Web site, or how much more often a customer purchased a product after seeing it on the Web site. Although each group is rightly concerned with measuring the effectiveness of their efforts, what is measured is often different. Without each side understanding the perspective of the other, it is difficult to avoid confusion and frustration when working together. Worse still is that the HCI practitioners may be thinking they are satisfying the needs of the marketers, when in fact from the marketers' perspective they are not doing so at all.

THE HCI DESIGN FRAMEWORK

Business Goals Definition

Business goals are specific outcomes that a business seeks to obtain. These goals are set in the first phase of developing the e-commerce user experience, although they may be revised somewhat in later phases.

The key questions that must be answered include the following: How should we create differentiation from our competitors? Should we use technology, price, a novel business model, brand image, product design, product availability, marketing, service, customer support, or other factors as the primary differentiators of our business? Should we seek market share before profitability? How fast should we grow? Should we seek to be the market leader, or be second or third behind a single dominant firm? What are our competitors doing? Whatever the business goals, the key question for HCI practitioners is: How are these goals supported in the design of the user experience?

As more and more firms fought for market dominance on the Internet, numerous experiments were conducted to determine how best to reach the customer. Some of the most successful experiments succeeded because the firms identified new e-business models to attract customers. For the business owner seeking success via the Web, a sound business model is even more essential than it is in the physical world. Barriers to entry are extraordinarily low on the Web, allowing many companies to spring up virtually overnight to compete with each other and the established firms for the customer's business, in ways not possible in the physical world.

A good example of such successful e-business models is e-Bay (Enos, 2001). By providing a safe, easily understood marketplace that matched sellers with buyers for all sorts of items, and charging each side of the transaction a fee along the way, e-Bay was able to dominate the virtual auction marketplace for consumers. This was so enticing to consumers that tens of thousands of people joined e-Bay, first as buyers, then many as sellers. e-Bay's business model demanded that their audience needed to be many thousands of consumers, and there should be a balance between buyers and sellers. They actively worked to grow their user base as quickly as possible, as well as balancing the number of buyers with the number of sellers. Envisioning success, they designed their Web site to make it easy for many simultaneous buyers to find items to buy, and for sellers to offer items for sale. They used proactive email messages to alert buyers that another buyer had outbid them, and to notify prospective buyers that a desired item had come up for auction. They provided sellers with a safe, reliable means to collect what they were owed, while also providing the buyer with a similar safe, reliable means to ensure that what they bought was just as the seller had represented it—e-Bay claims that only 1 auction in 40,000 is fraudulent among the 6,000,000 auctions posted each day (According to e-Bay spokeman Kevin Pursglove; see Savino, 2001). Using Web technology, they were able to do business on a worldwide scale, even when they were first starting out.

Setting appropriate business goals is essential to managing the rapid growth that many firms using the Web experience. Should a firm first seek to maximize revenue, profit, or share? The answer for startups is different than for established firms. E-Bay needed to maximize share, in the hope that doing so would lead to owning the customers forever afterward. Amazon's business model was similar to e-Bay's with regard to the need for a very large customer base. It adopted a business goal similar to e-Bay's with respect to garnering market share, to the point where they routinely incurred expenses not long endurable in the physical world, such as using more expensive shipping means to ensure the fastest delivery possible and shipping

partial orders immediately at no extra cost; offering to meet any competitor's pricing; and carrying the most extensive product line in their market area. Both firms supported their business strategies and goals with the newest technology delivering pleasing user experiences.

This new function required new models of user interaction. For e-Bay, suddenly, people with little or no computer expertise had to be able to quickly and easily engage in a complex transaction with a stranger to buy or sell something with a strong assurance that they would be happy with the result of the interaction. For e-Bay's business model to succeed, many people had to engage in these transactions successfully. This required the designers to develop methods of educating potential customers in a nonintrusive and nonthreatening way and to seek out new technology to support the functions upon which the business model depended, such as efficiently processing thousands of small payments promptly. Amazon developed correlations of purchases with classes of buyers, effectively personalizing the shopping experience for each customer by offering suggestions about additional products popular with similar buyers.

What can we, as designers, take from these examples and others like them? First, that the business goals of the Web site, with respect to the target audience, be well defined: It must be clear who the intended audience is, what their needs are and how they will be reached; the value proposition of the Web site must be readily apparent to the target audience and able to satisfy its needs; and the proposed business transaction must be compatible with the channel. We then must understand the customers' mental model for the tasks we expect them to perform as they visit our Web site. We must support those tasks with appropriate information design, interaction techniques, and visual design such that the customers are able to complete the tasks easily and efficiently, using a Web site that is consistent with their mental model.

When attempting to improve an existing site, it is even more essential to set specific, measurable business goals, such as increasing conversion rates from 2 to 4%, increasing user goal achievement on key tasks from 50 to 70%, reducing contact center call volumes on site navigation help by 30%, or increasing completed transactions by 25%. Be certain to take a thorough set of baseline measurements at the start of the work. These measurements set the frame of reference against which measurable business goals are stated. Too often, business goals are vague or ill defined, such as improving the look and feel of the site, or just "refreshing" the site, making it impossible to measure improvements or determine the return on investment for the work undertaken. Even if goals are clearly stated and accurately measured, it is critical to understand that interface design and usability will not guarantee the ability to meet the stated goals. The total user experience includes factors that are frequently not controllable by interface designers, such as price, product quality, sacrifice gap (how close the actual product is to the customer's "ideal" product), etc.

It is instructive to use personalization as a way to illustrate how setting, understanding, and using business goals impact the user experience. Personalization means many things to many people. If the design team is given the option of personalizing the site, how do they move forward? Well-crafted business goals can give the team important direction. If a business goal is to increase revenue, then designing an effective up-sell/cross-sell capability based on customer profiling might be the right design to investigate. If improving customer loyalty is the business goal, then creating a loyalty program based on prior site purchases or visits might be the answer. If the business goal is to create a superior user experience, measured by greater user goal achievement, then it may be significantly easier to guide the user in finding a product or service if we design the user experience based on our knowledge of the history of the customer's visits to our site, then dynamically alter the site organization and navigation on a customer-by-customer basis. All these decisions, made early in the design process, are based on satisfying specific, measurable business goals.

User Identification

A clear specification of the target audience is a key prerequisite to the design of any user experience. From an HCI practitioner's perspective, this phase requires the identification of users' behavioral or psychological characteristics that have the potential to significantly affect design parameters. From a marketer's perspective, this step, commonly referred to as "customer segmentation," requires the identification of the groups of customers that are likely to be the most profitable consumers of the products and services that are being offered. A vague or ambiguous specification of the target customer population ultimately results in business failure through lost sales, either by wasting sales and design effort on less profitable customers or by turning away the more profitable customers through e-commerce experiences that are poorly designed for their needs.

The key questions that must be answered during the user identification phase are the following: Who are the intended users of the system? Which subgroups (segments) must be given preference when there are unavoidable design tradeoffs? What are the characteristics of the preferred (e.g., most profitable) customers? Given an arbitrary customer, can any of these characteristics actually be used to classify the customer as a preferred customer? Can detection of the customer's characteristics be done dynamically (e.g., in real-time while the customer browses the Web site)? Given different kinds of identifying information about the customer, what corresponding design actions can be taken to improve the user experience? Or conversely, what is the minimum amount of customer information needed to offer a significant improvement in user experience? What are the implications from the viewpoint of protecting the customer's privacy?

Both marketers and HCI practitioners have traditionally developed techniques to answer some of these questions, with differing objectives and levels of detail. Unfortunately, a straightforward union of their respective techniques and bodies of knowledge does not appear to be adequate for designing e-commerce user experiences. The discussion in the following paragraphs will point out where existing techniques are useful and where there is a need to innovate.

Identification of the complete target user population is usually the responsibility of the business owner, because this is integrally linked to the mission, business goals, and the core

capabilities of the organization. The range of products and services that the organization is capable of delivering through the Internet channel implicitly defines the space of possible customers. Thus, the early project definition interviews conducted by HCI practitioners with the business owner and key stakeholders will reveal the range of possible customers, as defined by their needs and wants, that the business is capable of satisfying.

However, on closer inspection many of these customers will be found to be unprofitable. It is common knowledge that every Web site has many more "browsers" than "buyers"; customers differ both in terms of their purchasing power as well as in terms of the demands they place on a business, e.g., sales effort needed and after-sales support costs. Marketing wisdom has it that attempting to uniformly service everybody is a sure route to failure; it is better to focus business resources on the best customers. It is therefore a marketing truism that customers must be segmented into tiers according to potential profitability, based either on their short-term needs, or, as propounded by the current one-to-one marketing movement, based on their lifetime value to the business (Peppers & Rogers, 1997). Traditional marketing segmentation techniques have been widely applied to the former problem; addressing lifetime value has proved to be more difficult because of the difficulty in assessing loyalty and long-term customer behavior. Of the segments that are produced by these techniques, marketing managers select a few of the top segments as the primary targets, and a few as secondary targets. The remaining segments do not receive explicit marketing attention. As with traditional channels, e-commerce UI designers must also keep the same selected target segments in focus, as a minimum requirement for business success.

However, these target segment definitions are often unusable from a design standpoint. Traditional marketing techniques are typically based on telephone survey data and employ demographic, psychographic, and firmographic data about customers along with questions about purchase intent. Some models use information about past purchases as well (Colombo & Jiang, 1999; Levin & Zahavi, 2001). None of the traditional techniques uses fine-grained information about the interactive behavior of the customer. This is not surprising, because traditional channels have either been less interactive than the Internet (e.g., mail order), or have relied heavily on a human's interpersonal skills to deal with the customer (e.g., salesperson in a retail store). The Internet is the first channel in which marketers (and really, HCI designers) have needed to classify and choreograph customer interactions on a massive scale, while simultaneously optimizing profitability and ease of use. There is a vast body of literature on consumer behavior in traditional channels (Lilien, Kotler, & Moorthy, 1992), and some of it is at least partially applicable to interactive system design (e.g., the classification of human choice processes; Bettman, Luce, & Payne, 1998). However, there is a need to extend marketing segmentation techniques by basing them on behavioral data relating to Web interactions rather than static customer attributes. Because such segmentations would directly connect specific shopping behaviors to the desirable target segments, they would provide a useful basis for making design choices.

The second factor that makes traditional marketing segmentations unusable is the heightened need for "diagnosability" on the Internet (i.e., the need to detect the relevant customer characteristics in real-time during the experience to provide the right user experience). For example, if one of the most desirable segments is identified through characteristics such as "young males aged 17 through 28," today it is nearly impossible to detect that a person who has just visited one's Web site falls into that category. In general, privacy concerns preclude asking the customer for such information, although this may be possible on occasion (e.g., clicking on a link to "Products for Small Businesses" may be taken to imply that the customer owns or works for a small business). Considerable ingenuity may be required to obtain the relevant information, and this is usually not just the HCI practitioner's problem, but the marketer's as well, because the best solutions involve business plans. For example, the relevant information may have been revealed by the customer during a past purchase and may be present in the company databases; it could be used if the customer's identity could be determined as soon as he or she accessed the Web site. Identifying the customer could be possible if they were offered something of value in return for identifying themselves (e.g., simplified access to product support information). The mechanics of identification could be simplified if, for example, the customer were sent a promotional e-mail containing a unique link into the Web site. Thus, many of the design challenges involved in making effective use of target customer characteristics hinge on the ability of the marketer and the HCI practitioner to jointly foresee and resolve "diagnosability" issues.

The target customer set can be segmented in more than one way, and in general, multiple segmentations are needed to design rich e-commerce experiences. Every segment, when diagnosable and based on interaction needs, corresponds to one or more branch points somewhere in the overall interaction sequence. For example, if customers are segmented into "domain experts who know exactly what they want to purchase" and "novices who need to learn about the different purchasing options available," two different interaction subsequences may be implied, corresponding to a fast path for the former segment and a detailed presentation of the product range for the latter. Conversely, each branch point should be viewed as implicitly defining a customer segment, and the choice made by the customer at that point as an implicit revelation of one of their characteristics. Thus, every set of choices that could be presented during the experience has implications of interest to both designers and marketers and bears careful design.

Marketers most commonly identify and segment customers according to their purchasing needs (or by products offered, which is usually equivalent), which is useful to designers because they can often simplify the experience by organizing content into mutually exclusive areas of interest to customers. However, designers also need to classify customers according to goals and triggers (see the task analysis section), because the context associated with each goal or trigger often dictates a different user experience. For example, a customer purchasing a new computer for the first time is different from one seeking to replace an existing computer; the latter will think of their prospective purchase in terms of comparison to their older machine and will potentially need assurances of compatibility and assistance in migrating their existing data. Such differences

between customers are usually revealed during the task analysis phase. The designer must analyze the conversation that occurs during a naturalistic purchasing activity for variability in interaction requirements (detailed use case scenarios) and knowledge (conceptual user model). Systematic classification of these patterns of variability reveals the appropriate segmentation.

User identification is an absolute prerequisite for the design of personalized user experiences. Although there has been considerable discussion of personalization in the literature (Riecken, 2000), by and large it has not been deployed successfully because of many design and business challenges, the most important of which is delivering a nontrivial amount of value to the customer (Bayers, 1998; Fournier, Dobscha, & Mick, 1998; Nunes & Kambil, 2001). As mentioned above, the flip side of personalization and user identification is the loss of privacy. In general, the solution to the privacy problem is primarily not technical (e.g., cookie management or automated exchange of privacy policies), but psychological: ensuring that enough value is offered, directly in context of the use of identifying information, to reassure the customer that loss of their anonymity is justified. An example is product support and maintenance: the ability to automatically upgrade device drivers or receive safety recall notices or consumer alerts is so convenient that many customers are willing to identify themselves when they access a Web site.

Task Analysis

Task analysis deals with eliciting goals, values, triggers, and activities, and finally incorporating these into task flows, the sequence of activities leading to goal completion. The HCI notions of goals, values, and triggers are closely related to the traditional marketing notions of needs and wants. A detailed understanding of these user attributes allows the designer to be confident that an assembled list of tasks to be supported by an application is correct and complete. Correctness means that the tasks one is supporting are appropriate ones, satisfying both business and user goals. Completeness means the list of supported tasks is exhaustive.

HCI practitioners are used to customercentric task analysis, but in the two-actor setting of e-commerce one should also include some representation from marketing. Indeed, as designers, marketing is one of *our* customers. Exactly what part should marketing play in task analysis? How are the marketing notions of needs and wants related to the traditional HCI notions of goals, values, and triggers that one tries to elicit as part of task analysis? As a designer or information architect, how can one translate an understanding of these attributes into more effective requirements and designs? Lastly, how can a detailed understanding of these user attributes be used to measure the success of a design?

These questions will be handled in turn, but it is best to first give definitions of the terms needs, wants, goals, values, and triggers. Needs are defined in the marketing literature to be basic human requirements, such as the need for food, air, clothing, and shelter (Kotler, 2000, p. 11). Needs give rise to wants when the need is directed to specific objects that might

satisfy it. For example the need for food might give rise to the desire (or "want") to eat watermelon on a hot day. People also have strong needs for such things as recreation, education, entertainment, self-esteem, transportation, and so forth. Needs are often interrelated.

For product marketers a key part of understanding a want is discovering the associated needs. Web sites sometimes fail because they fail to address some of their customers' needs, some of which the customer may have a hard time even articulating. Kotler (2000, p. 21) distinguishes five types of needs:

- Stated needs (the customer wants an expensive car)
- Real needs (the customer wants a car whose operating cost, not initial price, is low)
- Unstated needs (the customer expects good service from the dealer)
- Delight needs (the customer would like the dealer to include a gift of a U.S. road atlas)
- Secret needs (the customer wants to be seen by friends as a savvy customer)

The Web is unique in its ability to cater to some of these types of needs. Buyers are apt to be less secretive about their secret needs when shopping on the Web because the shopping experience is more private. Also, the Web has the potential to be more effective at speaking to a customer's real needs. For example, for certain customers it is easier to do a convincing job of establishing total cost of ownership over the Web than in person-to-person selling. On the Web, assertions regarding cost of ownership can readily be analyzed and compared with competitors' claims, while in person-to-person selling the details of such an assertion would rarely be completely absorbed by the customer, and a certain degree of salesmanship would necessarily pervade the conversation.

The notion of value is one defined primarily in the marketing literature. In marketing, the value of a product is defined to be its benefit divided by its cost (Kotler, 2000, p. 11). Alternatively, a marketer may talk about values, as in what details in the buying decision the user finds valuable. Values are thus tied to the marketing concept of a product's value. By understanding what a customer finds valuable about a product, one can translate that into a value statement or a set of "value propositions" for a product. In our inquiry, we are interested not just in a product's value propositions, but also in the value propositions of the product in the context of electronic commerce. In other words, in the customer's mind, what gives the electronic transaction value as compared with the transaction conducted over traditional channels, and how should one tailor the electronic experience to cater to associated customer expectations?

These value notions, which have their origins in marketing, are of great importance to the HCI practitioner. When talking of the value propositions for a product, one is led to the further notions of relative and competitive value propositions. Relative value propositions refer to the attributes that distinguish a given product in a company's catalog from others in the same category. Competitive value propositions refer to the attributes that distinguish a given product of one firm from the analogous

products of competitors. In designing an e-commerce site, one needs to be mindful of when it is important to call out a product's relative value proposition and when to call out a product's competitive value proposition. In other words, when is it appropriate to bring up competitors, and the competitive advantages one may have over them, and when is it best to ignore the competition and focus on comparing products within one's own product line? This result should fall out as a natural finding of the task analysis phase.

Finally, one ought to be cognizant of the special values to the customer of the electronic transaction. That is, what is special about the Internet versus person-to-person selling, or other channels, that is inducing the customer to use this form of shopping? What are the customer's expectations? The more precise one can be about these values, the better they can later be measured.

Next let us move on to goals. Goals are multifaceted and may refer to the subgoals going into making a purchase, the greater goal that this purchase may be part of, or simply the goal of completing the purchase decision. Understanding a user's goal structure (Keeney, 1992) is at the heart of the ultimate output of task analysis, the construction of task flows. Triggers refer to the initiating events that give rise to the tasks described in the task flows.

Marketing must have representation in all aspects of task analysis so that marketing tasks, triggers, and goals are fully understood. Later in the design process, one cannot effectively make optimal tradeoffs necessary to support all users without understanding the tasks of all users. For example, a customer may come to an e-commerce site with a promotional code obtained from a newspaper ad. It thus may be important to have a clear path to the associated promotion from the very first page of the shopping experience. Without eliciting this trigger from the marketing user, designers might not have a motivation to provide a path to promotions directly on the home page. Just like the other users of the system, marketing has needs, wants, values, and goals. The process of unearthing these attributes parallels the associated process for standard customers.

Traditional task analysis asks users to enumerate their current or envisioned tasks, and the associated events that trigger the tasks. A problem with this approach is that a user's tasks are only undertaken to service needs, wants, and goals and are informed by values. Thus there may be more effective ways of satisfying user needs then piecing together known tasks.[1] Furthermore, if users don't have an existing method of shopping online for a company's products, today's tasks at the bricks and mortar store only offer a glimpse at what tomorrow's tasks might be at the online store. It is better to start with a deep understanding of needs, wants, values, and goals with a keen eye to those that pertain specifically to the e-commerce purchasing process, and from that, create a suitable set of tasks and task flows. Additionally, needs, wants, and values can be quantified, and goals placed into a hierarchy, to inform the inevitable design tradeoffs that must be made when trying to support multiple task flows

using a limited set of Web pages. To sum up, the surest path to gathering thorough requirements is to deeply understand all of the users' needs and wants, and the most effective way to make design decisions with confidence is to have a gauge on these various attributes.

Finally, the same gauge that one develops to measure the relative importance of the various user attributes can be incorporated into a final measurement tool. The measurement tool can assess how well one has satisfied the various needs, wants, values, and goals that have been identified. The assessment of importance of these factors, given by prospective users, were somewhat a priori in the sense that they were given before the user had any experience with the online store. Thus, one should not assume that these initial assessments are either constant across all users or valid for the same users, following a first experience with the online store. Hence the importance of these various factors should be requeried as part of the testing instrument.

Modeling of User Concepts in the Product Domain

User concept modeling is an activity that attempts to capture and formally represent the customer's knowledge of the product domain It is variously known as information architecture, content modeling, or navigation architecture, with the name usually reflecting the designer's bias and approach to executing the activity. The key questions that must be answered are as follows: What is the total set of content that must be made available to the customer? What is the best way to classify and organize it? What must be done to ensure that the customer will find the content meaningful and easy to understand? What is the level of knowledge that can be expected of the customer? How does the designer compensate for gaps in the customer's knowledge? What are the most effective ways in which the customer can manipulate the content? Ultimately, how does the designer guarantee that customers can find the information they need, when they need it?

There is considerable variation, among both marketers and HCI designers, with respect to the modeling constructs and the model elicitation and content creation processes used. Marketers view content from the perspective of product-line design and product presentation. They also tend to place heavy emphasis on promoting brand awareness and organizing content according to various types of promotional campaigns. Over the last few years, the dominant approach to marketing on the Internet has been to treat it as yet another channel, akin to traditional mail order. Consequently, much of current practice of Internet marketing tends to expose its origins in database marketing, with an emphasis on crafting static messages and advertisements (in effect, electronic brochures) and failing to adequately use the highly interactive nature of the channel. One of the most common consequences is that the organization of the site is determined by first breaking up each Web page into regions allocated to specific tasks (e.g., a rectangular region at the

---

[1] An alternative approach is to take a group of users and have them engage in generating "blue-sky task flows" where they try to brainstorm desirable, but not necessarily feasible task flows. See Dayton, McFarland, and Kramer (1998).

top for an advertisement, a central region to highlight two of the most popular products, the right and bottom sides for varying promotions), and subsequently squeezing product information into each region. The result is a content structure that makes perfect sense to a marketer, but is not necessarily meaningful to the customer. The worst form of this type of "site organization through negotiation" occurs in large companies with complex organizational structures and multiple business units. Space on the site is allocated according to demands for visibility of specific pieces of content; the resulting structure more closely reflects the business's organizational structure than the intrinsic nature of the product and service offerings. Avoiding this situation is one of the reasons for the emphasis placed by some HCI practitioners on deriving the Web site's structure from a conceptual user model that is created through conversations with customers rather than content providers (Kramer, Noronha, & Vergo, 2000).

Even when internal organizational boundaries are not reflected in the structure of a Web site, the "marketer's mental model" may become the basis for structuring content, to the detriment of the user experience. Marketers usually carry a mental model of the product domain that reflects the way they run their business; for example, each business unit may offer a set of product lines, which may be organized into brands, which in turn comprise different product series, each of which contains several product models, which can have multiple configurations, and so on. The customer segments being served also appear within this hierarchical structure, at a varying level. The marketers model may differ drastically from the user model, resulting is severe usability problems.

One of the concepts that marketers (and HCI practitioners as well) have been surprisingly slow to grasp is that customers on the Web have a completely different perspective and mental model from that of the marketer. Many of the structuring concepts that are of paramount importance to marketers (e.g., brand names) have a much lower significance from a customer's perspective. The particular combinations of features that happen to appear together in a given product model due to the intricacies of the product development process may not make sense to a customer who is concerned with the value offered by the model. Furthermore, marketers tend to create a number of variations on the same product model to address specific customer microsegments. This would be fine in many traditional channels in which the customers can be reached in each channel in different ways with just the models that are right for them, but fails on the Internet when all the products from all the channels are simultaneously offered on a single Web site. The result is a confused customer facing multiple products that look essentially the same from their perspective, although the products are quite different from a marketer's perspective.

The important lesson here is that traditional product offerings may not be able to be dropped into an e-commerce experience in a straightforward manner; marketers may need to redesign their product line's breadth and depth to support easy comparisons between the choices at each level in the space of offerings. The problem is not the traditional one of explaining a single product to millions of customers (as in a broadcast advertisement), but that of explaining the entire range of product offerings to a single customer with a specific need.

Among HCI practitioners, the two dominant themes used to determine the organization of content are the task-oriented and the object-oriented approaches. The former begins with an enumeration of the tasks that the customer may wish to perform and presents this list as the first choice encountered by the customer. When the customer selects a task, he or she is then presented with the next set of (sub)activities that can be performed to proceed with the task. The interaction sequence determines the basic structure of the experience, and enough information to perform the subactivity is provided at each step of the sequence. The advantage of structuring the content this way is that the designer can easily ensure that the user is provided with everything needed to complete the most important use cases. The disadvantage is a loss of flexibility: It is difficult to anticipate all possible use cases, and new use cases require new task flows to be created. Furthermore, customers do not have control over their interaction sequences. In addition, the intrinsic relationships between the different pieces of content tend to be obscured.

The object-oriented approach starts with the premise that customers carry mental models of objects or "things" from their ordinary experience with the physical world, and reflecting these objects in the user experience gains the major advantage of familiarity. As realized in participatory design methods such as Bridge (Dayton et al., 1998), this approach begins with an enumeration of the objects and actions via protocol analysis, a conversation with the customer that is analyzed to yield the nouns and verbs that constitute their understanding of the world (Ericsson & Simon, 1993). The objects are placed in relationship to each other via techniques such as CARD, PICTIVE, and PANDA (Muller et al., 1995). These methods use low-tech materials such as 4×6 index cards to represent objects, which customers manipulate while describing the activities that they would perform during the e-commerce experience. In particular, customers step through the details of their interaction sequences while using the cards, and document the information needed about each object by attaching sticky notes to the cards. One of the interobject relationships that receives the most attention is the containment relationship: determining which object is contained in ("is part of" or "is inside") another object. Containment is important because it closely relates to the concept of location (i.e., the place where an object can be found during the user experience) and thus is the key to good navigation. Indeed, in pure object-oriented methods such as Bridge, access to an object is entirely determined via the containment relationship. To move from one object to another, the customer must either access a contained object or perform some action on the object. In the latter case, the actions (also known as "methods") are documented via sticky notes on the object.

Constructing the object model in this manner by eliciting the mental maps of the target customers ensures that any design that is derived via instantiation of these objects as concrete views (see the section on visual design) feels familiar (i.e., "intuitive") even to customers encountering it for the first time.

Furthermore, using these object views to step through all the interaction sequences ensures that all the information required to complete the major use cases will be available to the customer. However, unlike the task-oriented approach, the object-oriented approach tends to be passive in terms of prompting the customer to perform the next step of the interaction; there is no explicitly defined "next" step, and the customer has to select one of the methods available on the visible objects to proceed with their task. This is both an advantage and a disadvantage: It increases flexibility but can make novices feel uncertain about what they should do. Incorporating a process orientation during any part of the interaction requires the reification of the flow using a metaphor such as a "wizard" or a "shopping assistant."

Because the goal of user concept modeling is to enhance communication by ensuring the design's fidelity with respect to the customer's knowledge of the product domain, this step relies primarily on participatory design activities with several customers. However, it is sometimes necessary to involve content providers as well. This occurs when the product domain is quite new and has not yet entered the mainstream of customer consciousness. Furthermore, marketing objectives such as promoting brand awareness may demand the use of brand names in place of generic object names (e.g., "IBM NetVista" instead of "Desktop Computer"). The designer is then faced with the problem of resolving the tension between the customer's real conceptual model and the marketer-driven conceptual model. In object-oriented approaches, such dissonance is easily detected, and compromises created (e.g., "NetVista Desktop"); the object model is never abandoned. In other approaches such as task-oriented design, the dissonance is evidenced by excessive tweaking of terminology, and often manifests itself as a high level of user dissatisfaction with technical jargon, or "marketing hype." In both approaches, one helpful step toward resolution is to establish explicit criteria and a scoring system for evaluating the goodness of the model (e.g., familiarity with each phrase), specifying criteria for both ease of use and marketing.

Competitive Analysis

Competitive analysis is a well-known and vital activity in the marketing world. It is less well known, but also a useful activity, in e-commerce design. In marketing, competitive analysis is used to hone product positioning and set performance benchmarks for product development and for marketing initiatives. In HCI and e-commerce design, competitive analysis is useful to compare different online selling approaches and benchmark the effectiveness of an e-commerce design against that of the competition.

What kinds of insight can e-commerce site designers get from the competitive analyses done by marketers? After all, the e-commerce Web site is a product/service, albeit a free one, just like any other product a company may offer. What are the main questions one should seek to answer when doing a competitive analysis of Web sites, and what are the preferred methods of doing such a competitive analysis?

The first step in a marketing competitive analysis is to identify the competition. Next, one identifies the major strategies and strategy groups that the competition falls into. Depth of product line, price points, and service levels are some of the characteristics that help stratify strategy groups. The market position (i.e., the relative market dominance) of each competitor is then assessed, in terms of three important variables: market share, mind share, and "share of heart." The distinction between mind share and heart share is that mind share is correlated with the question "What is the first company that comes to mind in this market segment?" and heart share is correlated with the question "What company would you prefer to buy the product from?" (Kotler, 2000, pp. 225–226).

A parallel, effort in any marketing competitive analysis is the benchmarking of best practices across the industry. A central problem when benchmarking best practices is identifying best practice companies. Good sources of information in this regard include customers, suppliers, distributors, and major consulting companies, the latter of which often maintain sizeable files on best practices across industries.[2] Unlike in traditional marketing channels, for the Web one may not get good benchmarking leads from suppliers and distributors unless the Web site is a business-to-business site. Still, customers are a great source of information about which companies they consider to have best Web practices, as are the prominent management consulting companies. There is also online information, available through Forrester (Forrester Research, 2001a), Gomez Advisors (Gomez Advisors, 2001), and others.[3] Lastly, of course, the designer ought to stay in touch with the latest innovations in Web-based commerce and understand what is working and what is not working on the Web across industries.

Oddly, a competitive analysis is not often considered a part of the formal methodologies of HCI. Our thesis is that because the e-commerce Web site (or any of a company's other public Web sites) is a product in its own right, the techniques and lessons drawn from marketing competitive analyses should carry over to it.

Competitive analysis allows a company to position its product offerings to fill voids and hone its marketing messages to grab mind and heart share. Attention to marketing's competitive analyses can, however, lead the Web site designer to the so-called usability versus branding dilemma. On the one hand, the marketing team wants to build up a brand image and call out differentiating features of the company's products. On the other hand, the Web site designer wants to make navigation as smooth as possible and so does not want to introduce brand names when simple terms from the customer's conceptual model of the product space are available. For example, a number of computer manufactures sell systems that run a version of the UNIX operating system. Most of these companies have a brand name

---

[2]For a more extensive discussion of benchmarking best practices, including a breakdown of this activity into detailed steps, see (Kotler, 2000): 227.
[3]Ziff Davis (2001) has a reasonable compendium of best practices.

for their version of UNIX (e.g., Sun, Solaris; Hewlett Packard, HPUX; IBM, AIX). In what circumstances should the designer use the branded name such as Solaris, and in what circumstances the more familiar terminology UNIX?

Just as results from the competitive analyses of competitors and their products inform the design of a Web site, a competitive analysis of the competitors' Web sites should do the same. This also provides a benchmark to measure the success of new designs. If one is redesigning a Web site, a competitive analysis of the existing Web site provides another such benchmark. It is a good idea to test with customers to see what the competition is and is not doing effectively.

Often, scrutinizing a competitor's Web presence can tell something about its attributes that cannot be gleaned readily from looking at its presence in the market in other ways. In other words, an HCI Web-based competitive analysis can inform the marketing competitive analysis.

Some questions that an HCI practitioner might ask when doing an HCI Web-based competitive analysis include the following: How are the firm and its competitors filling the needs of the target customers? How could one design a Web site to fill customer needs more effectively? How are the firm and its products being positioned? What are the differentiators? Are the differentiators called out sufficiently on the Web site?

Interaction Design

Interaction design is the practice of conceiving, planning, and constructing the interactive artifacts of an interface that a user will manipulate to control the interface. Interaction design for e-commerce applications focuses on tasks related to e-commerce such as learning about, comparing, and buying products. The tasks, and the individual steps a user will take in the execution of those tasks, are identified and documented in the aforementioned task analysis phase. Interaction design, along with the next step, visual design, is the realization of the interactive elements of an interface. The realization of the interaction design yields an abstract wire frame, sometimes referred to as the logical design, although this design lacks most of the visual elements created during the layout of Web pages. These abstract views do include the objects and actions needed for each step of the task flows.

The key questions that must be answered during the interaction design phase are as follows: What is the best way to organize data and interactive elements on the Web site? In what order should information be laid out (on a particular page)? What information should be visible at various points of the experience? What is the model of interaction? How should the presentation of information be altered based on the specific use case in which the customer is engaged? What are the relationships between the various elements of the conceptual model? Is the conceptual model realized in the interface congruent with the customer's conceptual mental model? Does the interactive design facilitate choice between products? Can all the high-priority customer goals be accomplished by manipulating the interface? What will make this a satisfying interactive experience for the customer?

The Bridge methodology was among the first that dealt with the entire user interface design process from start to completed design. Bridge employs a three-stage process. First, user needs (requirements) are represented as task flows. Task flows are then mapped into task objects, using the task object design method. Finally, task objects are mapped into graphical user interface objects, completing the design process. OVID (Berry, Isensee, & Roberts, 1997) uses similar interaction design techniques. Both methods advocate the use of interaction diagrams to explicitly document the steps necessary to complete every task, and the user concepts that need to be available for use by the user at each point in the process. These techniques are partially supported by unified modeling language tools, such as Rational Rose, through the use of sequence and collaboration diagrams (isomorphic forms of interaction diagrams).

In both the Bridge and OVID methods, concepts in the user conceptual model are identified as necessary to the user to complete one or more tasks. Based on the specific requirements of each task, one or more views are created. The decision on how many views to create, and the information contained in each view, is best determined through iterative user testing (UCD and participatory design). For e-commerce systems, many products are likely to need at least two views. One view will typically give the user a complete, detailed set of specifications for the product. This may be important for the use case in which the customer is comparing the product on a Web site with the product on another Web site (addressing the competitive value proposition). A second view of the same product might contain a significantly reduced amount of information in order to facilitate the comparison of the product with other similar products on the same Web site (addressing the relative value proposition). The number of views needed for a product will depend on the complexity of the product and the number of use cases that involve the product. Of course, views of user concepts need to be created for all the concepts that are presented to the customer in the course of interacting with the Web site. Some obvious examples of concepts are a "store," a "shopping cart," "products," "catalog," and a "confirmation of purchase."

Marketing requirements in the interaction design area have the potential to be in conflict with the interaction design that is generated strictly from user centered design activities. In a large company, the corporate Web site may have hundreds of organizations within the company vying for some of the user traffic on the Web site. This desire frequently manifests itself as requests for "links" on the organization's pages. The golden ring is to receive a link on the coveted corporate home page. When requests of this type are made and decisions levied, they directly impact the Web site organization and interaction design. What is most problematic in this scenario is that the requests and decisions are typically made without the benefits of the entire UCD process we have illustrated in our framework and thus have the likely potential to negatively impact the usability of the Web site.

Another problematic issue related to interaction design of e-commerce applications is the common assertion that the fewer clicks it takes to put something in the shopping cart, the better the design. Some people have advocated that all items must be capable of being added to a shopping cart in three clicks. If the

underlying rationale for this requirement is "things should be made as simple as possible—but no simpler"[4] then we agree with the motivation, but we take exception with the form of the requirement. It is likely that some e-commerce applications will require more than three clicks because of the nature and complexity of the task. The most important factor in determining the number of steps needed to execute a task is how the customer approaches the problem. If it is natural and necessary for a typical customer to take 12 steps in buying a particular type of product, then the Web site design should support the customer in the execution of the 12 steps. Using reductio ad absurdum, if a maximum of three clicks is good, a maximum of one click is better, but having everything on the home page for any nontrivial e-commerce site will yield an unusable design.

Visual Design

Visual design is the process of applying visual treatments to the logical interaction design. It is important to note that a single interaction design can be given many visual design treatments, each equally valid but perhaps not equally usable. The views created during interaction design are used to define the precise information that must be rendered during visual design.

The key questions that must be answered during the visual design process are the following: Does the visual design reinforce the user conceptual model? Does the visual presentation of information contribute to the ease of use of the Web site? Does the visual design create an aesthetically pleasing experience? Is the underlying data that is presented on the Web site rendered in a way that makes the data valuable and meaningful to the customer at each step in the task flows? Does the visual design help establish and maintain the corporate brand image?

There are few methods that explicitly support the creation of visual designs from abstract designs. Bridge was developed for a GUI windows environment (Microsoft, Motif, etc.), and was recently extended to include the realization of interaction and visual design for the World Wide Web environment (Kramer et al., 2001). These methods are fundamentally usercentric in their approach to design. We have observed a conflict between requirements generated through UCD activities, and requirements that flow from marketing and business interests in the company.

Marketing has traditionally been intimately involved in visual design through its print and television advertising activities. The visual design of advertisements has a completely different set of goals than does visual design of traditional interactive systems. Marketers seldom if ever worried about user tasks or goal achievement in the design of advertisements. They were more often concerned with being consistent with the firm's brand image as they crafted their messages. Advertisements fall into two broad categories (Strauss & Frost, 2000), brand advertising (impression advertising) and direct response advertising. Brand advertising attempts to build or maintain brand awareness by simply getting a message in front of a customer a given number of times. Direct response advertising has as its goal to make the customer take action, such as purchase a product or request product information. Neither scenario has facilitation of complex, HCI as its main goal. To our knowledge, no one has examined and proposed a method for integrating branding and interaction requirements into a visual design method.

HCI practitioners should ensure that the user experience they design supports the firm's brand image. If a major focus of the company is producing products with leading-edge industrial design, that focus should be represented in the design of the user experience. Of course, this could be taken too literally. Suppose a company were strongly associated in the customers' minds with the color red. If the visual designers were instructed to develop their visual design in varying shades of the firm's characteristic red, it is certainly possible that such extensive use of red-on-red would negatively impact the readability of the Web site. Because this, in turn, negatively impacts the user experience on the Web site, the visual designers need to find another way to be consistent with the firm's brand image that doesn't hurt the user experience, yet leverages the firm's strong customer association with the particular shade of red for which it is known.

Another important consideration for visual design is to be sure it properly supports the logical design of the Web site. During the interaction design phase, abstract views were developed with the characteristic that each view presented the customer with all the data needed, but no more, to complete a particular task in the task flow. The visual treatment of this data needs to be consistent with the user's conceptual model. An important property of product data, for instance, is containment, that is, some data elements are contained as components of other data elements. A specific computer system model, for example, might be part of a series of systems, which in turn might be part of a category of computer systems—the model X98 is part of the QED series, which is part of the laptop computer category. This containment relationship—models are contained in series, which are contained in categories—is developed during construction of the user's conceptual model. Visual designs that violate the containment relationship cause confusion in the customer's mind, because the customer expects something different than what is presented. Anything that confuses the customer is detrimental to a satisfying user experience for the customer. Thus, the HCI practitioner needs to work closely with the visual designer to ensure that all characteristics of the data being rendered be consistent with the customer's conceptual model.

Usability Testing

As with any product or service, understanding whether a product meets its intended target user population both in a "usable" and "useful" manner is critical to a successful product launch and continued success in the marketplace. In fact marketing and HCI each primarily concentrates on one branch of these dimensions. The "useful" strand in this equation is the focus of marketing and it occurs during market testing and launch. In this phase, marketing explicitly tests assumptions

---

[4]A famous quote attributed to Albert Einstein.

made during concept development, such as target values of attributes and core product concepts. On the other hand the HCI perspective focuses on the "usable" portion of the equation. HCI specifically attempts to measure the consumers' ability to use the product for its intended purpose, whether that purpose is of any intrinsic value to that consumer or not.

In this respect, e-commerce experiences do not differ; they represent an experiential type of product that for analysis needs to be decoupled from the product or service being offered within the e-commerce experience. "Useful" in an e-commerce experience has to answer a channel value question, and "usable" has to answer a channel experience question.

Thus, the key questions that marketing asks are different from the key questions that HCI asks. Marketing asks the following: Are the customers' values met by the channel? Are there customer needs, including latent needs, that can be especially well satisfied by this channel? Can the channel's value proposition stimulate an unidentified need? On the other hand, HCI asks the following: Is it clear how the customer will be able to use this e-commerce experience? Can we create new features or improve on existing features to meet the requirements of e-commerce customers? Does the user experience continue to be expedient and intuitive for the different customer populations?

Even in areas where marketers and HCI practitioners observe similar phenomena, the inherent focus of the discipline distinguishes the approach and the outcome. For example, both marketing and HCI are interested in doing competitive analysis of the Web channel. Marketing wishes to understand the competition to maintain a marketplace differentiation of the brand on the Web. The method that they use to achieve this goal is ordinarily focus groups. HCI practitioners wish to establish competitive baselines for usability metrics on the Web site. They typically engage in controlled usability evaluations to understand the effectiveness of the user interface, at a detailed level sufficient to directly inform design.

Both marketing and HCI practitioners have developed methods that assist them in answering the questions posed above. Table 38.2 examines the methods and the challenges in applying them to e-commerce experiences.

TABLE 38.2. Summary of Usability Assessment Methods

Usability Assessment Method	Description	Stage of Development Process Where Applied	Shortcoming for E-Commerce Applications
Observation	Observation in the field of real usage of e-commerce interface with an observer recording the user's thoughts or observations	After release of the interface into the field	Often difficult to gain access to consumers actively using the interface; purchasing decisions may not occur in as rapid a time frame as the observer desires
Heuristic evaluation	Involves a small set of expert evaluators examining an interface's compliance with recognized usability principles	On a fairly robust version of a high-fidelity prototype or postdevelopment	Current sets of heuristics focus largely on task-dominated environments and are less suitable for experiential situations
Cognitive walkthroughs	Evaluator simulates the user's decision-making process with the interface on an explicit task (Usability Inspection Methods, J. Nielsen and R. L. Mack, eds. New York: Wiley, p. 125)	After the specification of a fairly detailed design of the user interface, could be performed on a paper simulation	Explicitly focused on the heuristic of ease of learning, which is secondary in most e-commerce experiences
Questionnaires and interviews	Set of questions administered by an evaluator, often with Likert scale subjective questions	On a fairly robust version of a prototype, but could be performed on a paper simulation	Existing standardized questionnaires such as QUIS, WAMMI, etc. (see Proceedings of SIGCHI '88, pp. 213–218), do not reflect the experiential nature of e-commerce experiences.
Focus groups	Planned discussions with recruited target subjects following a script	Could be carried out at any stage of the development process but used primarily at the outset of the project	Tends to be suitable to obtain aggregate feelings but suffers from group-influenced biases and does not reflect individual's true feelings
Web log analysis	Mechanized path reconstruction and user visit duration recording	On a fairly robust version of a high-fidelity prototype or postdevelopment	Accurately reports the "real world" but does not provide input into "context"
User feedback	Comments garnished from users ad hoc or in response to predetermined questions	Could be carried out at any stage of the development process but used primarily when a prototype exists	Highly subjective or biased by experimenter questions

Aside from the shortcomings mentioned in Table 38.2 above, all usability assessment methods suffer the additional shortcoming in that they don't address a fundamental e-commerce concern. The focus of an e-commerce experience assessment should include measurement of customer propensity to complete transactions, that is, a simulated transaction may not be an accurate estimator of actual customer buying propensity because real money is not at stake in the simulation.

The premise of this chapter is that existing methods used by HCI practitioners to evaluate products and services fall short of meeting the requirements demanded by e-commerce experiences. Ordinarily when a designer submits a product or service for evaluation, the customer information that they require is evaluating the fit of the product to the customer's need. However, in an e-commerce experience, the designer has an explicit second set of requirements that emanate from the marketing organization that demand that the interface not only satisfy the customer's task need but also satisfy the marketing need to sell the product. It therefore is important for the practitioner to uncover the customer's latent unexpressed preferences so that they can identify the characteristics of the user experience that increase the likelihood that the customer's visit results in a sale.

The HCI practitioner has to be prepared to address not only the customer's ability to successfully complete the tasks of finding the desired product and checkout, but also determine what design principles could have been used to encourage the customer to upgrade their purchase. The customer did not express this upgrade need, yet the optimal interface might have increased customer satisfaction by actually doing exactly that. Observation, Web log analysis, customer feedback—none of these would uncover this marketing requirement need, let alone provide design suggestions.

The standardized questionnaires used to evaluate user experiences leave similar gaps in our understanding. QUIS lists the primary factors being evaluated as screen factors, terminology and system feedback, learning factors, and system capabilities. These can be directly attributed to software product satisfaction criteria but fail to give any indication on how to maximize the user satisfaction in an e-commerce experience. Other tests often focus on performance-related measurements such as speed and accuracy that again fail to answer the marketing requirements being placed on e-commerce experience designers. For example, it is intuitively obvious that a user might evaluate an interface satisfactorily on performance-related measures yet never use the interface again because of dissatisfaction with other factors (e.g., ease of navigation).

A significant set of papers have been written suggesting criteria that should be used in evaluating Web sites; however, few of the evaluations provide clear theoretical basis for evaluation (Novak, Hoffman, & Yung, 2000). suggests that creating compelling commercial Web sites depending on facilitating the sense of flow. A successful e-commerce Web site must provide enough challenge to arouse the consumer but not too much that they become frustrated.

Limayem et al., 2000 indicated that perceived consequences was a strong influencer of likelihood for purchasing online. Customers had expectations that a company would reflect the lowered expense of selling a product online into cheaper prices online. Perceived negative consequences such as security also strongly influence the likelihood of consumers from purchasing online. Other important perceived consequences found were "possibility of saving time," "expectation of improved customer service," and "comparative shopping." Although these measures may all be valid additions to a heuristic evaluation or might even be uncovered during a "talk aloud protocol" usability evaluation, none of these are present in the standardized questionnaires such as QUIS and WAMMI.

Several studies have been conducted over the last few years investigating consumers' perceived obstacles to shopping online. Limayem et al., 2000 bring an excellent overview of a number of previous studies, and the reader is advised to review that paper for a comprehensive review. Some of the obstacles reported include product variety and customer service, lack of security, and network reliability. The authors of the paper developed a questionnaire based on constructs that were gathered through a belief elicitation process. The authors found that the main antecedents that accounted for more than 36% of the variance in the rationale for persons shopping online were intentions and behavioral control. Intentions were explained by perceived consequences, attitude, personal innovativeness, subjective norms, and behavioral control, which accounted for 53% of the variance. Another important paper on this topic is Novak et al. (2000), who studied a conceptualization of flow theory, defined as the cognitive state achieved during navigation, determined by high levels of skill and control, high levels of challenge and arousal, focused attention and interactivity/telepresence. Their study showed that measuring the aforementioned constructs could lead to a better understanding of the requirements for creating compelling sites.

What one should take away from this discussion is that a combination of existing methods, such as heuristic or talk-aloud protocol evaluations, uncovers only some of the obstacles that prevent users from completing their explicitly articulated goals. The factors that can influence customer satisfaction and customer behavior that will directly satisfy marketing requirements cannot, at this time, be systematically uncovered using existing usability evaluation techniques.

CONCLUDING REMARKS

There is growing evidence to support the intuitive notion that improving the user experience of an e-commerce Web site will likely improve sales on the Web site, all other things being equal. Business executives are becoming much more aware of this connection, and therefore focusing more attention on improving user experience on their websites. The importance of setting specific goals and associated quantifiable metrics to guide the design process was stressed.

There is also evidence that today's HCI methods, as used for e-commerce Web site design, do not adequately take into account needs other than user (customer) needs, such as the needs of the marketing function of a firm. We argued that taking these needs explicitly into account improves business results for the firm and may also improve the user experience of the firm's Web site. To do so, extensions to existing methodologies need

to be developed. These extensions represent promising areas of research for HCI practitioners.

The eight-phase e-commerce interface design process was discussed in some detail. Highlighted were key questions needing resolution during each phase. A discussion of what user experience designers need to do differently for e-commerce systems was presented. Also highlighted were needed extensions to the activities of some of the phases, specifically to address the differences between e-commerce systems and traditional interactive systems.

References

Bayers, C. (1998, May). The promise of one to one (a love story). *Wired, 6*, 130-135, 185-187.

Berry, D., Isensee, S., & Roberts, D. (1997). OVID: Object View and Interaction Design (Tutorial). *Proceedings of ACM CHI 97 Conference on Human Factors in Computing Systems, 2*, 186-187. New York: Association of Computing Machinery.

Bettman, J. R., Luce, M. F., & Payne, J. W. (1998). Constructive consumer choice processes. *Journal of Consumer Research, 25*, 187-217.

Chin, J. P., Diehl, V. A., & Norman, K. L. (1998). Development of an instrument measuring user satisfaction of the human–computer interface. *Proceedings of SIGCHI '88* (pp. 213-218). New York: ACM/SIGCHI.

Colombo, R., & Jiang, W. (1999). A stochastic RFM model. *Journal of Interactive Marketing, 13*, 2-12.

Cramer, J. J. (2001, January 8). State of the Web: The cost of being committed. *The Industry Standard*. Available: http://thestandard.net/article/0,1902,21348,00.html

Dayton, T., McFarland, A., & Kramer, J. (1998). Bridging user needs to object oriented GUI prototype via task object design. In L. E. Wood (Ed.), *User interface design: Bridging the gap from user requirements to design* (pp. 15-56). Boca Raton, FL: CRC Press.

Enos, L. (2000, December 6). Petco gobbles up Petopia's assets. *E-Commerce Times*. Available: http://www.ecommercetimes.com/perl/printer/5880/

Enos, L. (2001, June 19). Can anyone catch e-Bay? *E-Commerce Times*. Available: http://www.ecommercetimes.com/perl/story/11349.html

Ericsson, K. A., & Simon, H. A. (1993). *Protocol analysis verbal reports as data* (rev. ed.). Cambridge, MA: MIT Press.

Forrester Research. (2000). *Retail & media data overview*. Cambridge, MA: Forrester Research.

Forrester Research. (2001a). Forrester Research. Retrieved from http://www.forrester.com

Forrester Research. (2001b). *Next-generation financial sites*. Cambridge, MA: Forrester Research.

Fournier, S. M., Dobscha, S., & Mick, D. G. (1998). Preventing the premature death of relationship marketing. *Harvard Business Review*, 42-51.

Gomez Advisors. (2001, Jan./Feb.). Retrieved from http://www.gomez.com

Helft, M. (2001, April 17). barnesandnoble.com's Rosy News. *The Industry Standard*. Available: http://www.thestandard.com/article/0,1902,23761,00.html

Jurek Kirakowski and Bozena Cierlik (1998). Measuring the Usability of Web Sites. Paper presented at the HFES Annual Conference, Chicago.

Keeney, R. L. (1992). *Value-focused thinking: A path to creative decision making*. Cambridge, MA: Harvard University Press.

Kleindorfer, P. R., Kunreuther, H., & Schoemaker, P. J. H. (1993). *Decision sciences: An integrative perspective*. Cambridge, England: Cambridge University Press.

Kotler, P. (2000). *Marketing management*. New York: Prentice Hall.

Kramer, J., Noronha, S., & Bardon, D. (2001). Designing object oriented E-commerce Web sites. Retrieved from http://www-3.ibm.com/ibm/easy/eou_ext.nsf/Publish/1826

Kramer, J., Noronha, S., & Vergo, J. (2000). A user centered design approach to personalization. *Communications of the ACM, 43*, 44-48.

Levin, N., & Zahavi, J. (2001). Predictive modeling using segmentation. *Journal of Interactive Marketing, 15*, 2-22.

Lilien, G. L., Kotler, P., & Moorthy, K. S. (1992). *Marketing models*. Englewood Cliffs, NJ: Prentice-Hall.

Lohse, G. L. (2000). Usability and profits in the digital economy. In *Proceedings of People and Computers XIV Usability or Else*. London: Springer-Verlag London.

Lohse, G. L., & Spiller, P. (2000). Internet retail store design: How the user interface influences traffic and sales. *Journal of Computer Mediated Communication, 5*. [On-line]. Available: http://www.ascusc.org/jcmc/vol5/issue2/lohse.htm

Muller, M. J., Tudor, L. G., Wildman, D. M., White, E. A., Root, R. W., Datyon, T., Carr, B., Diekmann, B., & Dykstra-Erickson, E. (1995). Bifocal tools for scenarios and representations in participatory activities with users. In J. M. Carroll (Ed.), *Scenario-based design for human computer interaction* (pp. 135-163). New York: Wiley.

Novak, T. P., Hoffman, D. L., & Yung, Y.-F. (2000). Measuring the customer experience in online environments: A structural modeling approach. *Marketing Science, 19*, 22-42.

Nunes, P. F., & Kambil, A. (2001). Personalization? No thanks. *Harvard Business Review, 79*, 32.

Peppers, D., & Rogers, M. (1997). *Enterprise one to one tools for competing in the interactive age* (1st ed.). New York: Currency Doubleday.

Petopia.com Cuts Staff, Loses Heart. (2000, October 27). *San Francisco Business Times*.

Riecken, D. E. (2000). Special issue on personalization. *Communications of the ACM, 43*, 89-91.

Savino, L. (2001, May 23). FBI charges dozens in Internet auction scam. *The Kansas City Star*.

Strauss, J., & Frost, R. (2000). *E-marketing* (2nd ed.). New York: Prentice Hall.

Ziff Davis. (2001). Ziff Davis e-commerce best practices. Retrieved from http://www.zdnet.com/ecommerce/filters/sublanding/0,10385,6006111,00.html

·39·

THE EVOLUTION OF HUMAN–COMPUTER INTERACTION DURING THE TELECOMMUNICATIONS REVOLUTION

Edmond Israelski
Abbott Laboratories

Arnold M. Lund
Sapient

INTRODUCTION

This chapter describes the role of Human–Computer Interaction (HCI) in the design and evaluation of computer-based user interfaces within the telecommunications industry. The telecommunications industry has been a fertile ground for applying user-centered design to HCI and was the home for many pioneers in the area. Communications is a vital part of the economies of all nations, shaping the cultures and the quality of lives of citizens. This chapter provides an overview of the range of design problems faced within telecommunications, a sense of where new issues are emerging, and practical guidance for practitioners working in the area. The chapter first covers the history of human factors in telecommunications from the perspective of the Bell System, where HCI in telecom was born in 1948. Next, design and evaluation issues are addressed by particular applications areas within telecommunications. Methodological trends and learnings are covered followed by discussions of HCI Standards impacting telecommunications and the impact of government regulations.

PRACTICING HCI WITHIN TELECOMMUNICATIONS

A Brief History

Alexander Graham Bell was committed to enhancing the lives of those with hearing disabilities, and it was out of this commitment that the invention of the telephone emerged. The patent was issued March 7, 1876. In 1934, the principle of universal service was established as a national goal—providing each American with access to affordable basic telephone service—and was built into the corporate culture. Universal access to empowering technologies continues to motivate much HCI work in the telecommunications industry to this day.

As AT&T grew, some of the earliest human factors work began. In the 1920s and 1930s, for example, landmark studies of worker productivity were undertaken at the Western Electric Hawthorne plant. The focus on designing systems that involve people really took off during and following World War II, as it did in most industries. The evolution of HCI work in the Bell System took two general paths that are nicely described by Hanson (1983). One path addressed the needs of AT&T's customers, and the other addressed the needs and skills of the employees of what was eventually to become "the biggest company on earth" (Klienfield, 1981, p. 1).

The earliest concerns addressed by behavioral scientists at AT&T were the quality and intelligibility of transmitted speech. The switches in these early days connected calls by mechanically switching relays (e.g., the No. 5 Crossbar) and can be thought of as early computers. Improvements in the design of the telephone (e.g., the invention of the switchhook and the design of the handset based on the anthropometric properties of the users) can be thought of as early interface design activities. In 1945, John E. Karlin was hired out of S. S. Stevens'

psychoacoustics lab at Harvard to start a lab within Bell Laboratories. According to Edmund Klemmer (1989), many consider Karlin to be the parent of human factors in private industry. Karlin's department initially focused on user preferences, but its work grew to include telephone design and a variety of new network applications. One of the early observations that continues to support the importance of user research today is that customers were not particularly good at predicting their own preferences without having actual experience with the alternatives.

In the 1950s, Karlin's department conducted many studies that evaluated customer's abilities to use all-number dialing. There were also early demonstrations of the feasibility of speech recognition from a customer's perspective, and the impact on user perceptions of sampled speech. In the late 1950's, Dick Hatch's department developed SIBYL, a system that could be programmed to simulate a wide range of network attributes. One of its first uses was to compare push-button versus rotary dialing and to demonstrate the superiority of the current arrangement of numbers on the telephone pad compared to the typical calculator arrangement (see, for example, Lund, 1999, for a Goals, Operators, Methods, and Selection rules (GOMS) analysis of dialing and references to many of these studies). The early 1960s saw a unique focus on a broader view of customer satisfaction as much of this work was incorporated into the Trimline and Princess phones. Phones would never again come only in black.

In 1965, the first ESS (electronic switching system) was introduced into the network, a change that was comparable to the change from mechanical computing devices to all electronic computing. From the late 1960s and through the 1970s, work continued on improving transmission quality and the phones used to communicate over the network. In addition, however, the first work on multimodal communication was undertaken. The Picturephone visual telephone service was designed, and service was actually offered in limited areas, Chicago and Pittsburgh, in 1970. Many of the challenges of desktop video conferencing that are still being studied were first uncovered at this time. Early studies were conducted to determine how to create an environment where nonprogrammers, that is, business customers, could program their own services (or if it was even possible).

Human factors work quickly grew until there was work not only in consumer services, but also groups devoted to military systems, operations support systems, business systems, systems training, personnel subsystems and analysis, and research. Research included Tom Landauer's HCI group, Mike Wish's work on social psychology, Saul Sternberg's work on information processing, Bela Julesz groundbreaking work on sensory and perceptual processing, and O. Fujimura's linguistic research. By the early 1980s, HCI efforts devoted to customer interfaces had grown to cover a wide variety of areas. The vision of an "automated office" served as the focus of many studies. Video conferencing and team activity were being studied. Teletext trials were underway, and a "home of the future" was built and provided a context for user testing. Teletext is the retrieval of text and preloaded graphics over telephone lines and was the precursor to the current graphic Internet browser. There were

a variety of business systems being designed in which communications and data capabilities were combined. Speech recognition services were being explored, along with environments for customer "programming" of their own services using workstations and telephones. Desktop information systems were being tested. The quality of packetized voice services (as used now in IP telephony services) was being tested. The cognitive processing of text during problem solving was being studied to deliver more effective customer instruction, and artificial-intelligence-based systems were being developed to aid in designing new customer interfaces. Many of these areas were reflected in a review of behavioral science activity in the Bell System and a subsequent review of work that continued in AT&T postdivestiture (Turner, 1983).

The behavioral science community numbered in the hundreds toward the end of the Bell System era but was geographically distributed. The AT&T Behavioral Science Coordinating Committee worked to create a community and to work on issues of importance to the community and to Bell System customers. The committee was composed of volunteers from the community itself, supported by executive sponsors. Among the activities of the committee was an internal conference at which members of the community could gather to share research and design experiences and knowledge. These activities served to create a network of relationships that persisted even as the Bell System ended and has continued to influence HCI work throughout the telecommunications industry.

On January 1, 1984, the Bell System was broken apart. Seven regional operating companies (RBOCs) were formed to provide local service, each with roughly equal numbers of customers (although widely differing in geographic area.) GTE was not a part of the Bell System but shared territory that overlapped with most of the RBOCs. AT&T's major telecommunications competitors were MCI and Sprint, but there was a sense that AT&T's real competition was IBM. One of the concepts behind divestiture was the realization that telecommunications and computing were converging industries. IBM realized that switches were really just large computers, and computers could be switches. IBM acquired communications capabilities (including the acquisition of Rohm, a Private Branch Exchange (PBX) company), and AT&T began to move into the computing industry. AT&T developed a partnership with Sun Microsystems and for a brief time offered its own line of personal computers, and later acquiring NCR. AT&T Bell Labs also split, and Bellcore was formed to support the RBOCs. With divestiture and the creation of Bellcore, the behavioral science community also split. AT&T's Behavioral Science Coordinating Committee continued, as did internal conferences and other activities. Bellcore also initiated a coordination effort and internal conferences to support the increasingly distributed HCI community. At divestiture, the same kinds of activities that were taking place within Bell Labs (Fetz, Pilc, Rubinstein, & Wish, 1989) took place within Bellcore (Burkhart & Root, 1994).

The RBOCs, however, quickly felt the need to focus on the unique characteristics of their own customers and business goals. NYNEX and US WEST formed early human factors groups working on HCI issues and research, with additional groups being formed in PacTel, SBC, Ameritech (Lund, 1994b), and finally a small group in Bell South. GTE also started a human factors group to develop new product concepts and to explore emerging HCI issues for the company (Virzi & Sorce, 1994). As the industry has gone through a variety of mergers and splits, HCI work has continued in the various companies formed. Internationally, many telecoms have significant human factors organizations including British Telecom, France Telcom, Telenor, Telia, Australian Telstra, and NTT, among others. Data Networking companies have also added human factors groups including Cisco, 3 Com, and others.

More recently, Sprint PCS has developed a strong human factors program, especially supporting its wireless division. When Lucent was split from AT&T, and AVAYA subsequently split from Lucent, human factors HCI work was maintained in each of the companies. Nortel and BNR, and the companies spinning out of them, have long had strong HCI programs, as have companies providing telecommunications equipment (e.g., Nokia, Ericsson, Siemens, and Phillips). Increasingly, however, companies normally associated with the computing industry (e.g., Microsoft, Intel, Sun, and of course IBM) have offered communications capabilities as part of new applications, and HCI work has been critical in these efforts. Most recently, companies such as Sapient have found it important to evolve new forms of HCI work (e.g., information architecture) to support telecommunications design problems.

The lesson appears to be that HCI work is critical to the success of communications and information products, and once its impact is realized the lesson is learned. An observation is that many behavioral scientists from the old Bell System are now spread throughout many of these companies and have helped maintain a professional community that spans competitive borders. Although there has been a steady increase in competitive pressure and the need to keep information proprietary, there has also been a sharing of nonproprietary lessons and insights that has served to benefit users and that probably has helped move the entire industry forward.

Application Trends

The basic telephone network supporting "plain old telephone service" (a.k.a. POTS) can be thought of as consisting of a network of computerized switches that connect a caller to the location or locations being called (see Fig. 39.1). The switches of one company will typically work with the switches of another to make this happen if the call is a long distance call. The switches have user interfaces that monitor status and can be used to update the software. In addition to the switches, there is a signaling infrastructure to help ensure calls are routed properly. The switches and the signaling infrastructure can be combined with a database and a system to program the database to create an environment that allows new services to be created relatively quickly. These services are based on a logic that consists of making routing decisions based on time of day and day of week, triggers from touch-tone entries, destination numbers, and other events. Applications can also be attached to the switches and programmed independently. Such an application might be a speech recognition application for

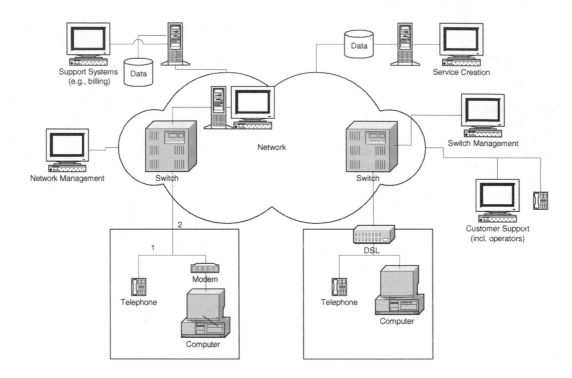

1 Inside Wiring
2 Loop
3 Switch Interface
4 Network Management

FIGURE 39.1. HCI interfaces in a prototypical network.

voice dialing. This switching network is supported by a variety of operations systems that collect information about activity on the network and that generate bills, monitor status, and track resources. Operations personnel, service representatives, and, increasingly, field personnel (e.g., installers) access these systems to run the network. Field personnel may access the systems using a variety of wireless devices. Local and long distance companies typically own the infrastructure and sell access, transport, and services. Equipment companies, however, may also sell switches (PBXs) and private network products directly to businesses who will connect them to the public network.

The POTS network interfaces with wireless networks (and may include wireless components itself). Although the particular technology used, or the signaling protocol, may vary, and although the particular functions that are used to operate the network will certainly be different, at the highest level a wireless network is similar to the POTS network in the diversity of user interfaces that need to be designed. Packet-switched data networks (such as the Internet), using routers instead of switches, bill in a different way and are used by Applications Service Providers (ASPs) to offer services. As with the POTS network, there typically are network management, billing, provisioning, application server, and other functions that need to be designed.

The general direction of networking is to make the network ubiquitous (Lund, 2000). Microprocessors are being embedded in more and more of the artifacts that surround us. Devices with the ability to link to the network with wireless and wireline interfaces are becoming increasingly diverse. The interfaces may use small screens and handwriting input, they may use speech input and output, or they may involve interactive televisions with integrated communications capabilities. Perhaps more important, these various channels of access to information and communications functionality will need to provide support to a common experience. More and more applications will have multiple interfaces, and users will move from one to the other as the context of use dictates. Companies that used to be thought of as telecommunications companies are selling a variety of consumer devices and are offering communications, television, information, and other services, as well as providing a way for content companies to offer their services to customers. Although design for operations and customer relationship management functions will continue to be needed and to evolve, new kinds of applications will also need to be designed.

HCI professionals in telecommunications will need to master not just the design of interfaces enabling people to derive meaning from content, they will also contribute an understanding of the role communications can play and how information can enhance communications. Increasingly they will be faced with designing for novel environments in which the design principles have not yet been well studied and will need to derive design principles from more distant domains. In addition, they will need to design not just with the focus on ease of use and

usefulness that characterized earlier business applications, applications will need to have the "color"—the excitement and compelling attributes—needed for successful consumer applications in a competitive market.

Roles for HCI Professionals

Many user interface design and evaluation opportunities exist in telecommunications for HCI professionals. The convergence of computers, Internet technologies, and communications products and services has created a vast domain. Microprocessors have become common in telecommunication devices. This pervasiveness broadens the reach of HCI design issues to include the following applications:

- Operations and support systems (OSS) software
- Web applications for end users and system administrators
- Intranets for telecommunications vendors and suppliers
- Call centers for customer relationship management
- Network administration client applications
- Handheld devices: Cellular phones with Internet access, personal digital assistants (PDAs), telephone craft devices
- Telephone terminal devices
- Switch administration user interfaces
- Telephone operator systems

An overview of the types of HCI applications in telecommunications—is a large and complex domain—given in the section titled Issues by Application Area—Overview Model.

Roles for HCI professionals are similar to those discussed elsewhere in this handbook and include (a) the development of performance support systems (e.g., training and documentation) and (b) all aspects of user-centered design (UCD). Some UCD methods often used include the following:

- contextual Inquiry (e.g., user profiles, user environment, task analysis, reengineering the task flow, process modeling [e.g., (service representatives, repair, yellow pages)];
- prototyping (e.g., Graphical User Interfaces (GUIs), and Web, microdesign for cellular phones and PDAs [Schmidt, 2000], interactive voice response—and Automatic speech recognition—enabled devices, interactive television);
- usability testing;
- field trials;
- focus groups and other market research methodologies; and
- information architecture roles for Web-based applications.

ORGANIZATIONS AND LABS

Because divestiture and the reduction in the size of individual telecommunications companies and the human factors staff that support them, there are typically three and sometimes four general types of HCI groups within telecommunications companies. There is usually a group within the information technology (IT) organization. This group tends to focus on traditional workstation-based applications used by internal personnel. The applications are often for operator and directory services and service representatives supporting customers and also for the many administrative and operational interfaces required to keep a network running. The goal of this type of group is primarily productivity improvements and user effectiveness. To ensure development efficiency, interoperability and the reliability required of network systems, the group in the IT organization has to be well integrated into the standardized IT processes. They often develop design standards and guidelines to aid developers when there are insufficient HCI professionals to support the work and to help resolve differences. The challenge these groups face is that resources for ensuring ease of use for employees are often in relatively short supply. Discretionary corporate resources tend to go toward capital investment and revenue generation (i.e., new products). In designing new technologies, the IT HCI group therefore often must depend more on accumulated knowledge of what users want than on iterative prototyping and design processes.

The second kind of HCI group focuses on products. Their core function is ensuring that new products meet the customers' needs. They may get involved early in the process and have the opportunity to help invent or shape new product concepts based on a deep understanding of customers. Or, as occurs in many companies, all too often they may be pulled in late in the product development process to test or screen a product before it is delivered to customers. The goal is typically to be involved as early as possible in defining the requirements of what will be built, whether built within the company or by a vendor. Early involvement clearly is the most cost-effective use of the resources and results in the greatest impact on revenue. Although the ideal process involves an iterative prototyping and design process with increasing fidelity over the course of the development cycle, most projects tend to provide for one iteration in about the middle of the cycle. When the HCI professionals are able to be an integral part of a product line, supporting each new product as it is conceived, they do accumulate a knowledge of the users of that type of a product that allows them to build on a history of customer understanding and to target new testing.

The product and IT groups occasionally come into conflict. The area of conflict tends to be around customer support. The product groups are responsive to marketing, who approach their products ideally with an eye toward the sales channel and customer support. But customer support and e-commerce sales channels are functions housed typically in IT-developed systems. It is therefore important for HCI groups within companies to work together with the common goal of user satisfaction, efficiency, and effectiveness when the goals of the groups converge.

A third function that exists in at least some of the companies is often housed in the safety or legal area. This is more often a single person than a group, but they take the perspective of user health and well-being as they work with users interacting with computing technologies throughout the company. They may also carry responsibilities in the area of helping ensure compliance with regulatory and legal requirements. Often this role is more a balance of serving as an advocate for users while also meeting corporate goals than a design role.

When Bell Labs was part of the old Bell System, in Bellcore, and in many of the larger telecommunications companies around the world, there have been groups that focused on more forward-looking HCI research. Generally this work was envisioned as yielding new product ideas. Shortly after divestiture, groups doing this kind of research also existed within US WEST and NYNEX, and until recently work of this type happened within GTE Labs. In U.S. telecommunications companies this forward-looking work is virtually all gone, although some more focused product ideation activities continue. The current thinking is often that the speed with which technology is evolving and the need to grow revenue quickly requires that the R&D to drive new products be carried out by the companies supplying new services to the telecommunications providers. Competition is not yet sufficiently great that the lack of diversity in offerings across companies has become a problem. A need that virtually all the companies have is to predict more effectively which products are likely to be successful in the market before significant development resources are invested.

Almost every way of organizing these functions has been tried at one time or another and in one company or another. At Ameritech and in the early PacTel, all the functions were centralized in one group. At companies such as US WEST and NYNEX, there were different groups handling the different functions. At QWEST, there is an IT group, but there is no longer a human factors product group; rather, there are individual user-experience people scattered throughout the lines of business. In the old Bell System, there were several studies of ways to organize user interface support. A cycle was recognized that proceeded roughly as follows. An advocate would become convinced that HCI support was needed and would start to build a group. The group would support the organization, would be highly valued, and would continue to grow. Eventually other organizations without HCI support would request it from this group, and it would continue to grow. Over time, however, they would begin to demand support within their own organizations. The centralized group might then be broken up and distributed to be part of the various teams. In some cases, the individuals would be used in novel ways, for example, as systems engineers, and would lose their professional identity. In other cases, the work would be sufficiently valuable that more people would be hired and added to the organization to start a group, and the cycle would repeat. In most of the companies, it has been recognized that a critical mass of colocated HCI professionals is important, and ideally they should have a manager from the profession who can coach, evaluate the work, and define strategic direction and serve as an advocate.

Most of the usability labs within the telecommunications industry have relied on the default arrangement of a test room and an observation room, with audio and video recordings of the users' faces, their interaction with the input device, and the output (Blatt, Jacobson, & Miller, 1994; Dayton, Tudor, & Root, 1994; de Vries, van Gelderen, & Brigham, 1994; Fowler, Stuart, Lo, & Tate, 1994; Lund, 1994a). This has been appropriate because many of the applications being tested have been for internal systems or for businesses. Laboratory testing has also been used (probably more so in the early human factors studies within the Bell System) to do parametric research to establish requirements and to do head-to-head comparisons of alternative designs. The laboratory also is an environment that lends itself to unique instrumentation assembled to test emerging devices such as handheld information appliances, remote controls, and so on. In addition, the basic functions of recording user expressions, comments, and behaviors are also supported in "portable labs" taken into the field. For example, in a service center a camera might be set up on a tripod with a wireless microphone to capture the user, a scan-converter might be attached to a workstation to make a copy of what is being displayed on the screen, and a camera might be set up and aimed to capture keystrokes. The three video sources might be time-stamped so they can be combined on editing back in the lab.

Most labs are equipped with rapid prototyping tools for building GUIs and browser-based applications. Most companies also have the ability to prototype a speech or interactive voice response application and to tap a telephone line in the lab to collect response information. Interactive television applications have often been simulated by converting personal computer (PC) output and delivering it over a television. Touchscreen-based PC simulations have been used for complex digital telephones and feature-rich cellular phones. The tactile feel is not simulated and is sacrificed to get early user-performance data on interactions between displays and simulated controls.

In addition to the more officelike testing arrangements and the portable labs, a variety of other laboratory arrangements have been used in the past as well. AT&T Bell Labs and more recently Bell South have built "homes of the future," and AT&T participated in the Disney experiment with a community of the future. Both Ameritech and US WEST Advanced Technologies had spaces designed to represent key areas within homes in a natural way (e.g., including a multimedia center, a home office, and so on). AT&T Bell Labs, Bellcore (e.g., the video wall), and Ameritech instrumented teleconferencing rooms to for HCI work in collaborative technologies, and Ameritech instrumented a classroom of the future to study new educational technologies. At one point, Ameritech even built the exterior corner of a home inside one of their labs to demonstrate and test the user interface for a new security system that members of the group had invented and constructed. Before Lucent and AT&T split, Bell Labs housed a consumer products lab where consumers could quickly screen new product concepts.

The debate, of course, within telecommunications reflects the perennial discussion in the larger HCI community. Is it better, in some sense, to evaluate user interfaces in the lab where there is more control over variables that might influence behavior, or is it better to evaluate them in the field that is more natural? Experience suggests that the best prediction of how something will be used is when real people use something that is virtually identical to the real thing in the real environment of use to accomplish real goals. This of course does not happen until the project is done, and changes are often prohibitively expensive. But the closer designers can come, for example, with field testing of a high-fidelity prototype, the better the prediction. Simulations of the real world should be the next best thing, although there is little systematic comparison of simulations versus controlled laboratory situations. It is early in the process, however, when larger changes can be made least expensively.

Low-fidelity prototyping in laboratory situations seems capable of capturing those larger issues. Control is also important by definition when one is trying to isolate a variable and can also be important when deciding which of two approaches to a design is significantly better. The solution adopted in many organizations is to argue that beginning-to-end involvement in the process and iterations between design and testing through the process is what is most effective. Unfortunately, at most companies this tends to be rare, and the various techniques are employed based on opportunities and the goals at any given point in the process.

One novel testing environment that telecommunications companies have uniquely been positioned to deliver has been called the Test Town. Because the services are offered over a telecommunications network, and most telecommunications science and technology labs connect to those networks, services may be offered in the field and user interactions captured for later analysis. Bellcore conducted a trial known internally as the "Red Bank" trial, in which local homes were offered services out of the Bellcore Labs in New Jersey. As people used the services, the patterns of use were analyzed and changes to the service were made. At Ameritech, a town in Illinois was wired and services were offered to the town. The town and others like it where application were tested and design changes were tried became the focus of a corporate advertising campaign (Lund, 1996) including prominent television commercials featuring actors portraying human factors professionals. The Nortel Meridian screen phone was designed based on the results of the Red Bank trial, and the Home Receptionist services offered by US WEST for the telephone were in turn designed based on field research conducted by Ameritech and US WEST.

Another increasingly common approach is to apply ethnographic techniques derived from anthropology to product and system design. Contextual analysis was an early step in this direction and worked well for specific applications. But several of the HCI groups in telecommunications have been going further and integrating anthropologists into their HCI organizations for enhanced cross-fertilization. At Media One Labs (now AT&T Broadband Labs) a study was conducted by anthropologists to understand what having broadband in the home meant to people. The changes that were observed over time informed initiatives and product designs. At US WEST Advanced Technologies an anthropological study was conducted to understand what customers were attempting to do, why and when they called service centers, and what the service centers thought was happening. The resulting insights improved the collaborative problem-solving activity that took place when service representatives and customers interacted.

ISSUES BY APPLICATION AREA

Overview Model

As described in the first sections of this chapter, the area of telecommunications is a rich ground for HCI professionals. There are many existing and emerging user interface domains that require the involvement of human factors professionals.

Table 39.1 shows a sampling of the types of users (e.g., end users, expert users, operators, administrators, servicing personnel, etc). Many parts of the UCD toolkit are needed to address this variety of user types, each with different user profiles, environments, and task flows.

Network Management

HCI design is critical here. Element management systems (EMSs) are the mainstay. EMSs cover the following functions for telecommunications network management:

- *Configuration.* This is the area in which new local and long-haul lines, loops, and trunks and the transmission equipment are configured. The complexities of this domain have previously required technician training of 6 months. In the Bell System era, this type of personnel investment was feasible, but today telecommunication service providers cannot afford to have long training cycles because of tighter labor markets and higher turnover. HCI professionals have substantially contributed to the design of intuitive user interfaces that require far less initial and follow-up training. Much of this work is now in the process of migrating from client/server configuration applications to Web applications. This is even more reason to use the methods of UCD.

- *Provisioning.* This is a function requiring database look ups and sequential modal tasks that result in a circuit or service being installed for a customer. For example, when you order a custom calling feature such as Caller ID (which shows the name and number of a caller on a telephone with a display) many different provisioning elements are triggered from updating software on the serving telephone switch in a central office to setting parameters in databases residing on adjunct server platforms that are part of the telephone network. CRUD (create, review, update, and delete) kinds of functions are the GUI page type usually employed. Graphic display of the provisioning elements may also be required in the HCI driven design.

- *Performance monitoring.* Here the technical performance of a switch or other network element is monitored. Attempts have been made to use newer technologies such as virtual reality applications that let technicians see the big picture of a network element (e.g., capacity utilization, transmission quality [line noise, echo, distortion, etc.]) and then zoom into an element. In zooming, the technician can actually go to screen simulations of the actual switch and drill down to look at the front panels of circuit cards or the back plane to look for faults via signaling lights. The use of virtual reality is probably more of a hit on the trade-show floor than a real enhancement to user performance and success. An interesting job in telecommunications monitoring is communications satellite controller. The telecommunications satellite networks are controlled by technicians in ground stations, who send signals to the orbiting satellites to fire small rocket thrusters. These thrust adjustments keep the satellites in their proper orbits.

- *Alarms and troubleshooting (network operations centers).* All of the above types of interactions can also be performed in a network operations center. This is usually a large

TABLE 39.1. User Types and Products

User Type	Example Products and Services	Considerations
Casual end users	• Voice terminals • Personal-computer based telephones • Handheld devices • Network based features (e.g., Call Waiting) • Voicemail • Interactive voice response systems • Multimedia systems • Self-installed digital server lines	• Little opportunity for training • Low propensity to read long instructions • Only a few features get used regularly • Complex extra features are mostly ignored • Includes residential and business end users
Expert Users		
Executive assistants	• Multiline telephone systems • Fax machines, teleconferencing	• Use many special features (e.g., transfer, conference, hold, and automatic call back) • Sophisticated users exploiting system capabilities
Service center agents	• Telemarketing Applications • Automatic call distribution	• Selling • Billing inquiries, • Help desk troubleshooting
Operators (attendants)	• Long distance operators • Directory assistance	• High-volume applications • Seconds saved translate to large dollar savings
System administrators	• Telecom switch administration • Network management systems	• Complex user interfaces • Long training periods • Configuration, maintenance, and system performance
Installation and maintenance technicians	• Graphical User Interfaces/Web applications that are operations support systems • Enterprise systems (e.g., Private Branch Exchanges, small business switches) • Handheld devices for field use • Specialized testing systems	• Many complex systems for testing and installation • Requires extensive training • Multiple vendors with little user interface consistency • Difficult cryptic command line systems are still being designed for some enterprise systems (customer premises systems)
Telecommunications Managers	• Operations support systems including billing and service provisioning	• User interfaces for telecom managers need to quickly give the big picture of the status and performance of networks and equipment
Sales personnel	• Special software for sales configurations	• Sales needs to be able to access and demonstrate telecom applications

mission-control-type room with large displays and multiple monitors at workstations. The large displays create excellent opportunities for critical HCI design. Data visualization of complex and rapidly changing data is a real challenge in design. Thorough contextual inquiry including detailed task analysis is critical for design success.

• *Security.* These systems allow administrators to control who else can configure, monitor, and troubleshoot the network and its elements. Usually these are also CRUD-type screens. Interestingly, newer technologies are being used for the input of security data including speaker recognition and biometrics input such as fingerprint recognition and retinal scans.

• *Billing systems.* They are the revenue collection engines that drive the telecommunications machine. Traditionally GUI client server applications with input from stored program control network switches these systems are also migrating to web technologies.

In a few years, it is likely that all network management will be done using Web technologies. In this way, client applications are always up to date and the most difficult problems will be updating plug-ins for the system administrators browsers. Most data will be stored as Extensible Markup Language (XML)

databases and Extensible Stylesheet Language (XSL) scripts will translate to the platforms needed by telecommunications administrators users. These users will be able to configure, troubleshoot, and monitor their systems from PC workstations, PDAs, cellular phones, and dedicated information appliances. All of these innovations should increase the productivity of the technical caretakers of modern telecommunications systems. Thanks to computers and HCI telecommunications craftworker productivity has skyrocketed in the last 30 years. A typical Class 5 central office switch needed a technical crew of 20 people to operate in the early years. Today, a Class 5 switch of equivalent processing power is usually unmanned and is maintained by remote technicians who are get dispatched to a central office site only as a last resort to replace something like a power supply or a circuit pack card. A remote center might monitor and control 10 or more end office switches. (The telephone network has five classes of switches in a network hierarchy starting with the local end office telephone switch, a Class 5, and continuing up the switching hierarchy to a Class 1 switch that would serve as an international gateway. The technical jobs associated with operations, maintenance, and administration (OA&M) have been thoroughly analyzed by human factors specialists and are summarized by Israelski & Reilly, 1988).

Fieldwork

Computers are also pervasive for the outside fieldwork domains in telecommunications. In the old days, field technicians such as cable pullers, cable splicers, and lineman (lineworkers) used traditional handtools, pots of molten lead for sealing cable splices, and simple electronic measuring devices for testing pairs of wires to establish and repair circuits. Today, microprocessors have found their way into the outside parts of the network, which is referred to as the outside Plant. Human factors specialists have done much work to apply human engineering principles to these outside plant designs as described by Hanson and Israelski (1977). A fairly typical application that companies are currently considering involves a wireless handheld terminal (e.g., a hardened PDA) for access to legacy system data. Supervisory instructions may be downloaded to people in the field over this same device. Managers would like some positioning information to be available as well, so they can tell where people are and how best to redirect them depending on job priorities. (Unions, of course, tend to resist both the real-time direction and the positioning information because it is seen as taking away the ability of the field technicians to organize their own work and time.)

Examples of modern devices that are currently in use include (a) handheld field devices (testing units such as PDAs, specialized microprocessor-controlled testing appliances such as testing fiber-optic splices and alignment of microwave transmitters) and (b) work-tracking time reporting, inventory, and documentation systems. In addition, some exploration of speech interfaces for field technicians has taken place, including investigation of speech-based browsing of remote legacy data while viewing a head-mounted virtual display.

Service Centers and Operators

While often the least "exciting" areas from the perspective of the HCI designer, productivity improvements for service centers and operators is often among the most important for telecommunications companies. The larger companies have thousands and thousands of people in these roles; they take months to become fully proficient in their jobs, and their tenure on the job is relatively short. Furthermore, as perhaps the most important interface to a company's customers, their performance directly impacts customer satisfaction and often revenue. In the Bell System, it is was estimated that saving 1 s of an average operator's time would save $10 million each year.

HCI is critical to the success of CRM, or customer relationship management. Call centers are either outbound or inbound. Outbound call centers usually place calls to customers and include the much-maligned telemarketers. Some of these outbound systems will use random dialing or filtered calling lists to place calls with a recorded introduction. A live telemarketing agent will come on the line if live voice energy is detected. Some sophisticated techniques can distinguish answering machine greetings from live voice to avoid connecting a live agent to give a sales pitch to an answering machine. As described in the earlier section Ideation and Forward Looking Work, the Privacy

Manager feature developed with the help of the Ameritech Human Factor's group was one weapon used to ward off unsolicited telemarketing calls. Interestingly, the local telephone companies have been compared to arms dealers providing "weapons" for both sides (i.e., telemarketing organizations and the consumer on the receiving end). HCI professionals have done studies that provide key data in the timing and delivery of such messages. Israelski et al. (1989) studied critical user timing for many telephony features and presented data that show a mean time from of 1,143 ms from picking up a telephone receiver until speaking occurs. This particular piece of data is an important input in the design of the previously described outbound call distribution systems. Human performance-based timing parameters such as voice onset time are used by telecommunications developers to optimize system performance with user performance.

Inbound calling is another important area for HCI. Call center agents will use a variety of workstation applications that are increasing Web based to handle many functions, including billing, collections, sales, maintenance, and service configuration.

The design of the call center workflow has been a subject of study by telecommunications human factors specialists as well. The Ameritech human factors group did extensive studies using the MicroSaint mathematical modeling tool to optimize sales agent workflows. The SBC human factors group (Bushcy, Mauney, & Deelman, 1998) did large-scale studies looking at sales agents cognitive and selling styles to model and optimize sales performance.

The systems that distribute inbound calls to call centers are servers that are usually called automatic call distribution systems (ACDs). Human factors plays an important role in the design of these systems as well in design topics covering the following:

- *Interactive Voice Response Systems (IVR/VRU).* These systems require touch-tone input from telephone keypads and output voice prompts. Today, IVRs seem to be everywhere. Poorly designed systems that do not follow good UCD practices are notoriously frustrating and the butt of many jokes about being stuck in "voicemail jail" with no chance to escape to a live agent. Schumacher, Hardzinski, and Schwartz (1995) provide a useful set of user interface design guidelines for IVR systems.
- *Web and GUI system administrator applications.* These are used by call center managers to optimize the call center's performance and customer satisfaction against parameters such as average work time of a call (in large call centers, HCI work that shaves off seconds from call handling can result in multimillion-dollar cost savings), calls in cue, calls on hold, average hold time, and abandonments (callers who hang up).

Operator Services

Some of the earliest work done by human factors professionals in telecommunications centered on telephone operators. At John Karlin's Bell Labs in the 1960s and 1970s, much work was done to optimize the telephone operator work flow and system interactions. Included in that work were human factors studies that were integral parts of the design process for services such

as directory or information services (411), network services, calling cards, collect calling, and third-party billing.

Hanson (1983) described Bell Labs studies that allowed credit-card calls to be introduced to the world in the 1970s. Many thought that typical telephone users would not be able to reliably enter calling-card (telephone credit card) numbers. Previously, operators needed to handle such calls. Requiring live operators would have severely hampered growth of these important revenue-generating calls. But extensive lab and field HCI studies that investigated different voice prompts, digit-entry timeouts, and feedback tones resulted in the system that is in use today. Speech control of calls previously handled by operators have made for further advances in automated operator services.

An interesting sidelight is that computer-controlled call handling introduced in the 1960s allowed telephone calling volumes to increase dramatically without the need to add legions of switchboard operators. Landauer (1995) noted this exception to his past challenge to demonstrate a computer-based system that really does enhance productivity. If not for software-controlled central office switches, an astronomical number of telephone operators would have been needed to handle call volumes. At its peak, the Bell System employed more than 1.1 million people, and about 100,000 were telephone operators (Klienfield, 1981.)

PRODUCT IDEATION AND DEVELOPMENT

The standard development philosophy involves beginning to end involvement and iterative prototyping and testing (see, for example, Lund & Tschirgi, 1991). In most HCI groups, design and testing are combined in the same group. Before divestiture, there often was more time to implement the ideal procedure, and even to supplement it with experimental tests of design alternatives. The ideal is rarely achieved, of course. Today, when testing takes place earlier in a project life cycle, it tends to take place using lower fidelity prototypes and what has been called "discount usability." Often, the experience is that new projects show up having recently heard about their need for HCI expertise and ask for someone to look at what has been developed before it is released. The experience, however, is generally a positive one and because many projects tend to span multiple releases, HCI personnel are often included on project teams earlier as each iteration takes place.

Research conducted at Ameritech and US WEST Advanced Technologies suggested that usability is made up of ease of use and usefulness, and these drive user satisfaction (e.g., Schwartz & Seifert, 1996). User satisfaction, in turn, results in usage and reduces product churn. Usage is what delivers business value (e.g., more revenue or improved productivity). With an increased focus on consumer applications, there has also been some suggestion that making applications compelling (e.g., aesthetically pleasing) can be important in stimulating the initial purchase or usage (which then allows people to assess ease of use and usefulness). User research early in the development process when change is less expensive often informs an understanding of what is useful, whereas user research with higher fidelity prototypes reveals more about the ease of use of an interface.

Carefully shaping a series of low- to high-fidelity prototypes can focus user attention on different aspects of a design, while controlling incidental design elements either by leaving them out or establishing user reactions through the base of previous testing.

When a usability test takes place during product development, the default test is similar to that in other industries. Users representing the target market are brought in one at a time. They are asked to go through a series of likely scenarios of usage using a prototype. The environment may attempt to approximate the context of usage, although most often an "office" environment is provided. The study may be conducted in such a way that the experimenter observes from behind a one-way mirror and only minimally interacts with the user, or the experimenter may sit next to the user and probe for detail about thoughts, feelings, and expectations as the user goes through the scenarios. As the user thinks out loud while interacting with prototype and responds to questions, his or her expressions and comments are captured. What the user sees and hears and the user's input are also captured. Supplemental questions may be asked before or after the testing as appropriate. Because often only one of these tests is possible during a tight development cycle, it typically takes place late in design or early in development. To get some of the benefits of an iterative process, heuristic and expert reviews of design concepts may take place earlier in the design process. The details of these tests vary somewhat depending on the specific type of applications being tested. The following sections describe some of these applications and the HCI work that has been done.

Interactive Voice Response Systems

The heart of the design problem for most communications services is that they have audio output and touch-tone input. The ability of users to take advantage of the audio output is constrained by their working memory, and that memory may be further limited if the touch-tone input requires special processing (e.g., translating the combinations of touch-tones into letters). At Ameritech in the early 1990s, it was concluded that the complexity of applications such as banking may be such that either the interface is too complex to use (no matter how well designed) or takes too long to use. Two technologies that help reduce this complexity for users are speech recognition and display screens. The speech recognition interface works well when the model embedded in the design matches the model that users have in their longer term memory, and the display bypasses the problem by simply reducing memory requirements.

Telecommunications services, like many workstation applications, build on well-learned conventions. Once the industry began to converge on a standard way of interacting with messaging systems (see ISO/IEC 13714), the key assignments could be used as the basis of many new services without having to educate the users about those assignments. Before these functional key assignments became standard, most interfaces depended on menus. The issues in the design of the menus, because of the serial nature of the audio interface, included the tradeoff of depth versus breadth (and the maximum length of menus that should be allowed) and the menu syntax (verb–object versus

object–verb). Menu items and prompts need to be short, and the critical information needs to be presented in just the right place. A hierarchy of menus, prompts, and information entry fields is referred to as an interactive voice response (IVR) application. A typical example of an IVR is the interface to a service center, where users can get some information automatically and after a series of choices may be connected with a service representative.

The other key to communications services is that the most successful ones have been—perhaps not too surprisingly—services that help people communicate or that manage communications. Voicemail and Caller ID have been two of the most successful, and services to screen unwanted callers (e.g., telemarketers) have been more recent successes. Many years of creating audio-based information services and transactions (e.g., banking) have historically been less successful, although the emergence of screens and accurate speech recognition appear to be suggesting attractive services might be possible. Perhaps as important are the cultural changes that have prepared people to expect the services and to use them to enhance their lives. The success of many telecommunications services appear to build on the success of previous services that have trained users on how to use them and have helped them articulate their desires for new services.

IVR prototypes have been created using systems originally designed to create personal voicemail systems, using computer simulations, and with other techniques. To test them in a lab, the line connecting the phone being used by the person being tested to the simulator is usually tapped, and the audio exchange and the touch tones are captured. Sometimes the touch tones are time stamped and saved directory in a database, along with codes representing prompts.

Automatic Speech Recognition Systems

Speech recognition creates new possibilities. When it works well, users often expect speech recognition to require less effort than an IVR interface. When the speech recognition interface delivers on the expectation depends on its design and the quality of the speech recognizer. A speech recognizer recognizing 85% of utterances may require a very different interface than one that recognizes more than 95% of entries (a rate that approaches touch-tone accuracy). In general, experience has shown that the best use of speech recognition is not to simply replace the touch-tone numbers in an IVR with speech entry. Rather, the best speech interface requires an interface that takes advantage of the properties of human-to-human dialogue. This is not to say the interface needs to be a natural language understanding interface (which will eventually be possible), but rather just as one person can guide the responses of another by the way they phrase their statements and listen, so can the system.

Usability testing of speech recognition interfaces still requires a great deal of fine tuning of the interface. The prompts need to be crafted to stimulate the right behaviors. For example, "Will you accept a call from X?" has a very different effect than "Will you accept a call from X, yes or no?" Like the touch-tone input, speech input and the output prompts are typically

captured by tapping the line between the phone being used in the test and the system delivering the prototype. One way in which the data are used that is important, however, is as captured speech samples. The captured speech samples are used to shape templates to ensure they can handle a variety of accents, vocabulary, and so on. For most speech recognition applications, testing in realistic field conditions has also been important to create successful templates and interface designs. For example, if the person using the application is going to be using it from a pay phone or a cellular phone in a roadside coffee shop, testing should take place in places where that kind of noise is typical. If the application is only tested in a laboratory, it may not work nearly as well in the field, and so the error correction protocol (which is so important for speech recognition applications) may be wrong for the level of accuracy.

Visual Interfaces

Another technology that has proven to be important is the appearance of visual interfaces for output. Bellcore was able to demonstrate that screen phones made many services that would otherwise be unusable into easy-to-use and valuable services. For much of the 1990s, these services were supported using a protocol called the analog display services interface (ADSI) protocol. This was a text-based interface that could be used to support communications services and to support information services. In the late 1990s, personal communications system (PCS) and business phones began to come with text-based interfaces. The PCS phones even started to support wireless application protocols (WAP) interfaces that supported access to Web-based information. Screen phones also began to appear that were really Internet phones, integrating Internet access and visual interfaces to services. If telecommunications and data networks support more internet protocol (IP)-telephony services, a richer set of communications interfaces should be possible. The next step will be to support speech input and visual output, with integrated communications capabilities and touch-tone backup.

Communications and information services are now being accessed using telephones, screen and Internet telephones, workstations, interactive televisions and set top boxes, PCS phones with screens, wireless PDAs, and other devices. Each has its own design challenges based on the size and properties of the visual output, and the functions available for input and how they map to software functions. For many of these hardware–software combinations design standards and conventions have yet to be defined. Because of the input and output properties, and because of network access bandwidth that may range from narrow to broadband, different kinds of content experiences are supported. The devices may also have somewhat different contexts of use. Communications on an interactive television, for example, is being used in an environment that typically has low interactivity, that is largely passive (and in fact parallel activities may be taking place), and that may be social. Communication over a PCS phone is typically taking place in a mobile environment with a high likelihood of interruptions. Communications over an Internet phone may be taking place

in a kitchen. The design challenge is further complicated, however, when an application may need to be accessed from each of these interfaces. In essence, applications increasingly will need to have multiple interfaces designed rather than just one, and skills will need to transfer and a consistent experience delivered despite the idiosyncratic properties of the device and the context of use.

Although the general approach to testing doesn't necessarily vary across the various devices, it is important that they be tested in environments that simulate the context of use. Testing a wireless PDA service in a lab may not reveal the information that is critical to usability if the device eventually will be used by a technician hanging from the side of a telephone pole (and trying to use the application with only one hand). When testing in the field, the challenge of capturing input, output, and user thought and expectations can be challenging. The tools that have been developed for testing applications delivered to a telephone or a workstation (tools usually developed for other purposes and adapted for usability testing use) haven't been needed yet. Small cameras and other accessories are beginning to emerge, however, and are likely to be the basis for usability tools for these devices in the future. In the meantime, much usability testing of these devices involves some combination of testing using emulators on a workstation and field observations and video.

Business Applications

Business applications are tested using similar processes to residential applications, although with an eye toward different contexts of use. Small businesses can range from those that purchase applications that are basically residential applications to those that have expectations and may be in the process of becoming large businesses. Small businesses typically are highly focused on their businesses, however, and so are seeking services that will provide value. They are typically not interested in the technology per se. High-end residential users may be more sophisticated in the knowledge they bring to an interaction than many small business users.

Large businesses tend to be similar to small telephone companies in the services that are being used and how they are managed. Their services may be purchased by one person, managed by another (often a former telephone company employee), and used by still others.

Group Communication Systems (Teleconferencing and Multimedia Collaboration)

The application area of group communications systems combines the methods and principles of HCI, its subset of computer-supported cooperative work (CSCW), social psychology, and group behavior. The technologies that apply to group communications include audio conference bridges, video conferencing systems (PC- and Room-based systems), and multimedia conferencing systems including application sharing, white boards, and video.

Video telecommunications has been around since 1927, when then Secretary of Commerce Herbert Hoover made a one-way, full-motion video call from Washington, DC, to New York City. At the turn of the millennium the promise of widespread use of video telephony is still unmet. Sight does add communication value to voice alone as described by Angiolillo and colleagues (Angiolillo, Blanchard & Israelski, 1993; Angiolillo, Blanchard, Israelski, & Mane, 1997). Video accomplishes all of the following:

- It improves nonvebral information, such as head-nodding gestures or pointing.
- It adds emphasis or phrasing to a verbal message.
- It creates a sense of presence to make the listener more attentive.
- It improves turn taking in a conversation by using visual cues.
- It enhances one's memory of a conversation.
- It relays emotional states to add clarity to a communication.
- It provides feedback to the communication participants.

It should be noted that there are still many behavioral barriers and usability issues that need to be resolved for video to achieve widespread usage (Noll, 1992). It is true that costs continue to come down as bandwidth availability increases and people are becoming more comfortable with personal use of video devices, but adoption continues to be weak. Some usability issues outlined by Angiolillo et al. (1997) include problems with video call setup, ensuring spontaneous communications, privacy and control of video modes, self-view alternatives such as mirror view and preview modes, multiple-view choice problems, camera control, and delay and lip and sound synchronization.

Multimedia communications takes audio and video and extends the richness of the media by including Software application sharing, such as a spreadsheet or presentation; a white-board for collaborative text and graphic creation and annotation; and rapid file sharing. Current commercial examples of such multimedia systems include: Microsoft's Netmeeting, Intel's Pro Share and Avaya's MMCX. Fussell and Benimoff (1995) summarized many of the usability issues surrounding multimedia collaborative systems. Some of these issues requiring additional research include nonverbal sources of information (e.g., eye gaze and good eye contact, gestures) and perspective taking (e.g., creating a shared physical context, communicators' spatial relationships, the salience of a participant, and creating a shared social context). Fussell and Benimoff (1995) also outlined from their literature review implications for multimedia product design. They pose the following questions for researchers to answer:

- What information should be included in the visual field?
- What size should the video window be?
- How can spatial relationships between participants be established and maintained?
- How can mismatches be reduced between the assumptions individuals make about what others are viewing and what they are actually viewing?

- How should systems deal with workspaces that are fully or only partially shared?
- How can construction of a shared social context be facilitated?
- Can multimedia conferencing improve face-to-face communication?

Technology will advance and make strides in addressing these issues. Mulbach et al. (1995) among others have looked systematically at the effect of variables that can enrich the video aspects of multimedia conferencing. They found that having a three-dimensional stereoscopy effect increases telepresence or the degree to which participants of a telemeeting get the impression of sharing space with the remote site. Similarly, they found that an enhanced video setup that eliminates horizontal and vertical eye contact angles (makes it look like participants truly have eye contact) seems to be advantageous in conferences with more than two persons per site. An excellent reference on the use of video in communications is the volume edited by Finn, Snellen, and Wilbur (1997) covering video-mediated communications.

NEW PRODUCT IDEATION AND FORWARD-LOOKING WORK

HCI involvement in new product ideation has been relatively rare. In general, HCI professionals have been engaged once the project team is formed around a concept, and user research has been employed to shape the concept and subsequently the design. When it takes place, it usually has come from two sources. Companies such as Bell Labs, Bellcore, GTE Research, and more recently AT&T Broadband Labs had (and in some cases still have) HCI groups in organizations charged with either coming up with new technologies and finding applications for them or with coming up with new product ideas. This is a typically either a behavior- or technical-research-driven process (or both).

The other approach is when HCI professionals have a relationship with marketing in which marketing engages the professionals as part of a more traditional brainstorming exercise. The assumption is that the HCI professionals bring knowledge of the user to the exercise, almost as if knowledgeable users themselves were participating in the brainstorming. Sometimes, HCI professionals will in fact bring users in on behalf of marketing and conduct product ideation sessions. At other times, again in service of this marketing-driven effort, they will go out to the field explicitly looking for new product ideas. Before the Ameritech human factors group was absorbed into the SBC group, it was a part of the Ameritech Product Marketing division. During this time, it enjoyed a position of involvement in many new product designs at the earliest conceptual stages. Examples include the following:

- *Privacy Manager (Product of the Year in 1998 by In-teractive Week).* This is an interactive service by which calls without Caller ID were redirected to a network server for the caller to record their name and missing identity for delivery to the subscriber. The subscriber had several choices to accept the call or manage their privacy. Data showed 70% of unknown callers refused to identify themselves and hung up and thus never bothered the subscriber. Clear, concise, and self-evident user interfaces were designed by the human factors group using lab studies and large field trials with friendly users.
- *Talking Call Waiting (Product of the Year in 1999 by Interactive Week).* This service uses a highly optimized text-to-speech platform in the network to deliver the name of the calling party after the usual Call Waiting beep. Call Waiting is one of the many customized smart telephone features that allows a call to be delivered to a busy telephone line if the subscriber takes appropriate actions. Human factors professionals did enormous amounts of user research to optimize the pronunciation of residential and business names using both lab and field studies.

One study that was conducted at Ameritech involved studying all the kinds of messages that people exchange in their lives and the roles those messages play. The goal was to look for new kinds of products that could meet the needs otherwise fulfilled by the messages. It was noted, for example, that many messages are exchanged around significant events such as weddings. These include messages between the couple and their family and friends and between the couple and vendors supporting the wedding. The messages may just be details, or they may contain content that is filled with or stimulates strong emotions. Memories of the details and the emotions typically make up much of what is cherished in subsequent years about the activity. A service suggested by this activity is one in which all the voice messages are saved at the request of the couple and recorded later on a CD-ROM as an audio memory book.

A different approach was taken at US WEST around electronic shopping. It didn't begin with a need for a product, but the work was undertaken with the plan to derive product ideas from it. High-tech power shoppers were the focus of an ethnographic study trying to determine what the relationship between digital and bricks-and-mortar commerce would be in the future. The study included storytelling sessions around collages that participants created, instrumenting computers, shadowing online and bricks-and-mortar shopping trips, field observations, diaries, and telephone interviews. It also included brainstorming by the participants around solutions to needs that seemed to be emerging from the research. In this case, a location-sensitive shopping support application that would make coupons available, provide information about a product, and simplify price comparisons (and that supporting customized bundling options by stores) was suggested. An analogous study was undertaken at Ameritech, where E-Labs (a company using anthropological techniques to derive user-centered business plans and product designs, now part of Sapient) was engaged to conduct an ethnography of communications in the home.

In general, given the size of the companies, products are expensive to develop and deploy (and often, the deployment costs are the biggest element in the costs). Products need to capture a sizable market share to be worth the development effort. A rule of thumb has been that 1 out of 10 ideas is worth pursuing, and 1 out of 10 of those is worth deploying, and only 1 out of 10 of those will have a sizable impact on the bottom line. HCI professionals have helped to screen these concepts to find the

winners by various kinds of user testing. Prototypes have been built and tested. Movies have been made in which it appears that users are interacting with the product, and customers are then able to identify with the characters and imagine and evaluate the value of the applications. Another approach to screening that has been tried in various forms is to provide a scenario of use of the concept and then to have customers rate the concept depicted in the scenario along a variety of user value dimensions. These can be compared with a database of past product concepts that have been tried in the market to make guesses about which ones will be effective or not. For the most promising concepts, prototyping and field testing (e.g., in a "Test Town") provide powerful technical and marketing data about potential costs and revenue opportunities. HCI professionals are typically involved in these tests, often in collaboration with market research. Emerging technologies will create many opportunities for the application of UCD. We are starting to see increasing usage of the following technologies:

- Packet-based networks (IP networks, e.g., voice over IP, video over IP, disappearance of circuit-switched networks)
- Wireless access protocol and related technologies (e.g., Internet access for handhelds, such as cellphones and PDAs).
- VXML (e.g., voice control over the Internet using a XML language customized for voice commands)
- Broadband interactive television (DSL and cable modem; e.g., Internet and communications functions available through the television and other devices in the home)

Research Issues and the Future of HCI in Telecommunications

HCI research within telecommunications, in particular, research that is not devoted to design questions associated with specific product interfaces, has taken many forms, including pure, applied, design, and product ideation research. As noted previously, pure HCI research has largely disappeared from telecommunications in the United States, with the changes in the Bell System and the increasingly intense focus of Bell Labs and Bellcore (now Telcordia) on near-term revenue generation. Some vestiges remain at Verizon Labs (formerly GTE Labs) and TRI Technology Resources (the R&D subsidiary of SBC). Such research also continues in Europe, where basic research continues at BT Labs in the UK, Telenor R&D in Norway, the German-government-sponsored Heinrich Hertz Institute and the many multination consortium research projects funded by the European Commission.

Bell Labs and Bellcore, however, for many years contributed to more applied questions. This work included work by Landauer and Dumais (1997) on latent semantic analysis, Kraut's (1987) work on teamwork, Nielsen's work on hypertext and "discount" usability methods (Nielsen, 1989; Nielsen & Molich, 1990), Dennis Egan on individual differences and new interfaces to electronic information such as Superbook (Egan et al., 1989, 1991). Landauer (1995) summarized lessons from much of this work in *The Trouble With Computers: Usefulness, usability and productivity*. US WEST Advanced Technologies, NYNEX,

GTE Labs, and Ameritech also studied a variety of areas that could conceivably influence new product concepts or product design but that were pursued in a product-independent context. At Ameritech, for example, a series of studies were conducted to develop a way of assessing the usability (including ease of use, usefulness, and satisfaction) of various applications independent of interface type. This work was used to drive more efficient techniques for improving the usability of designs and to better predict product success. It also suggested an approach to modeling the impact of user interface design changes to prioritize development activities.

Design research refers to generalizable research emerging out of specific design problems. At Ameritech, for example, work was conducted on various aspects of how people interact with video and multimedia content and with the Internet. At SBC, US WEST, and Ameritech, work was conducted on how people use the touch-tone pad as an input device. NYNEX demonstrate the value of a goals, operators, methods and selection rules (GOMS) analysis for systems with highly practiced users.

Product ideation activities typically have been less published because they tend to be most closely tied to proprietary activities within the business. HCI professionals, however, have played many roles across the industry in this area. At many companies they have participated in more traditional ideation brainstorming activities, bringing to bear their experience with users in the domain of interest. In some cases, the ideas generated were taken through prototyping and user testing. In others, they were presented to groups of users either in focus groups or in sessions where the concepts were screened along a variety of dimensions important for market success. In the Bell System there were early studies of people in the context of use to understand fundamental user needs that could be used to inform product design. This included studies of office workers and the use of information services in real homes. As noted earlier, AT&T Bell Labs even built an early version of a home of the future in which people could be studied in context.

One particularly effective approach for concepts that were too difficult to prototype was to simulate the products and create videotapes simulating users interacting with the concepts. This used the ability of video to engage viewers to give potential users a sense of how they would or would not use the application in their own lives. At Ameritech, another approach involved an advanced inventing technique previously used at Motorola to create intellectual property. In this technique, a domain was defined, brainstorming was conducted within the domain, the ideas were categorized and compared with currently protected intellectual property, and then new intellectual property was generated.

TECHNOLOGY TRENDS AND LEARNINGS

The general trend in technology is smaller, faster, cheaper, and ubiquitous. Moore's law predictions seem to be extensible to cover processing power and memory, and bandwidth is expanding at a similarly rapid pace. IP packet speeds have gone from

1.5 Kbits to 10 Gbits and beyond. The practical impact is that it is possible to imagine almost anything empowered with processing power. When this is coupled with emerging wireless technologies, each device is not even limited by its own capacity. Although almost anything seems possible, not everything will of course be useful.

This evolution in the technology is happening even as markets become more competitive. With competition, the network bandwidth becomes a commodity, and prices are driven down. As prices drop, profits typically drop. One way to respond is to capture new markets, and this has driven the convergence of networking technologies, wired and wireless telephony (local and long distance), Internet (running largely over the telecommunications backbone), and cable, as well as emerging business strategies. The split of the US WEST MediaOne cable network and US WEST Communications, the move to offering cable over VDSL and to provide data services to wireless devices that are linked to the wired network, and the subsequent acquisition of MediaOne by AT&T (which had already acquired McCaw Wireless) represent this trend. This has in turn shaped the work that many HCI professionals in telecommunications have been performing. They have had to develop skills in areas outside of traditional telephony (e.g., learning from the entertainment industry), but also have brought insights from telecommunications to these new businesses. Over time, network management of these new integrated works, field services, and customer relationship management services should experience a variety of innovative improvements.

Capturing new markets, however, will only be effective through this period of deregulation, mergers, and divestitures. Eventually, the revenue will need to come from new services that take advantage of these new networks. HCI professionals will need to be able to design for an increasingly diverse range of technologies and contexts. As noted, professionals now are being called to design Web applications and GUIs, PDAs, and PCS phone interfaces; speech recognition dialogues; and interfaces for interactive television. In the future, the interface may include interfaces that are provided over new kinds of hardware (e.g., refrigerators, cars, jewelry, and so on). More challenging, perhaps, will be the need to design applications that provide an integrated experience across multiple devices. Applications are being discussed where various subsets of the application functions may be accessed over a wireless device, a workstation, and an interactive television. More about some of these design challenges and the implications for preparation can be found in Lund (2000).

HCI STANDARDS IN TELECOMMUNICATIONS

The standards activities that are important for HCI professionals in telecommunications can be thought of in roughly three categories, although the boundaries are often fuzzy. There are standards that address traditional telecommunications issues. ITU-T is perhaps the best example of a standards body that takes ownership of this area. There are standards that address human–computer interactions, which is now probably best thought of as a superset of the area that includes telecommunications. In

the past these standards tended to focus on workstations such as those used to manage networks, but increasingly they apply to IP telephony, wireless PDAs, Internet phones, and other applications that represent the convergence of communications and information technologies that characterize telecommunications today. ISO 9241 and HFES 200 are examples of these standards. The third category of standards would be those standards governing the physical conditions in which computing technologies are used. The HFES 100 standard for visual display terminals would be an example of this category. These issues, however, are more often handled through regulation than through standards.

HCI in telecommunications is so diverse that it is difficult to determine which standards activities are not potentially relevant to professionals working in the area. Some companies, of course, are less concerned about standards before they are finalized, and many are not focusing on the requirements of international markets. Nevertheless, standards in HCI ideally represent best practices in the field and can create precedents that serve as the learning that users bring to new applications. The *SIGCHI Bulletin* has had a regular column covering HCI standards activities over the last few years (see, for example, Blanchard, 1998, for a summary relevant to telecommunications and a recent article by Reed et al., 2000, which summarized many current issues in the standards area).

Committees sponsored by Human Factors and Ergonomics Society (HFES) for telecommunications professionals, HFES 100, HFES 200, and HFES 300, are currently relevant. They are producing standards that will be submitted to American National Standard Institute (ANSI) for standardization. HFES 100 defines standards for visual display terminals, and HFES 300 is addressing repetitive stress syndrome. HFES 200 is the HCI standard; it nationalizes software design sections of ISO (International Organization of Standardization) 9241 and is adding new material on accessibility, the use of color in user interfaces, and designing for speech input–output (including interactive voice response and other interfaces used for telephony applications). The HFES 200 committee produced much that became ISO 9241, the international standard covering workstation ergonomics issues. The ISO 9241 standard contains standards ranging from the physical design of computers and terminals to user interface design guidance. ISO 9241 Part 11 is becoming increasingly influential in defining a standard perspective on what usability is and how to measure it. It is defined from the perspective of the effectiveness, efficiency, and satisfaction for the user. Another ISO standard that is relevant is ISO 13407, "Human-centered design processes for interactive systems." It is being developed by a working group of ISO TC159 SC4 (an ISO committee working on human–system interaction issues).

In 1993 ANSI published T1.232, a user interface design recommendation for computer systems used to control telephone and communications networks. Over the last few years, ANSI T1M1.5 has been working on standards for network management applications. Several years ago, the voice messaging industry met to agree on standards for messaging interfaces. This work was submitted to ANSI, and ANSI X3V1.9 (who also are responsible for the ANSI QWERTY standard and who have published standards on symbols and icons) worked with ISO to create

an international messaging standard. The result was ISO/IEC 13714, "Information Technology—Document Processing and Related Communications—User Interface to Telephone-Based Services—Voice Messaging Applications." This standard has in turn influenced design guidelines for other telephony interfaces, including those created for IP–telephony environments.

More recent efforts that interest professionals working on telecommunications applications include the work of the World Wide Web Consortium (W3C), including XML. XML promises to provide part of what is needed to create easy-to-use multichannel interfaces. IEEE has a program called the VoiceXML Forum (which is currently meeting as part of AVIOS). The VoiceXML forum is attempting to create a standard that will help in building speech interfaces to information services. Finally, ANSI has created the Information Infrastructure Standards Panel (ISSP) to facilitate the development of standards critical to the Global Information Infrastructure. Several of the working items approved by the ISSP are relevant to HCI professionals in telecommunications, although at this point the ISSP still is not exerting a great deal of influence. Of emerging interest will be standards for interactive broadband applications (including Interactive television). Current standards activities tend to focus on technical standards, although these standards are such that the flexibility designers will have when designing applications will be influenced by the standards.

Regulation

The telecommunications industry is a critical element in ensuring the productivity of society and the welfare of its people. Not too surprisingly, then, it is heavily regulated, and many of these regulations have an impact on the interfaces between people and technology. In 2000 and 2001, there was much discussion around the U.S. Occupational Safety and Health Administration (OSHA) requirements as they impact conditions that can cause repetitive stress injuries (see OSHA Releases Final ERGO Standard, 2000). These injuries can be experienced by some of the hundreds of thousands of telecommunications workers sitting in front of computer terminals executing the same tasks over and over again under pressure and by technical people in the field interacting with computers under physically demanding conditions. OSHA has long been actively involved in monitoring working conditions within the telecommunications industry.

In 1990, the U.S. Congress passed the Americans With Disabilities Act (ADA). It is intended to extend the Civil Rights protections to people with disabilities. The general goal is to remove the barriers that keep people with disabilities from participating in mainstream society. One area relevant to telecommunications is in employment and the need to provide accommodation for disabilities. Computer technologies have been both a challenge and an opportunity (see Fig. 39.2). They have been a challenge for people with visual limitations, for example, when more graphic user interfaces were first introduced. On the other hand, the introduction of new speech interfaces enabled others to take on jobs that would otherwise be difficult if not impossible to perform. The ADA also required that telephone communications services be made accessible to individuals who

FIGURE 39.2. Testing to ensure universal access (courtesy of Qwest).

have impaired hearing or speech. This resulted in the requirement that all common carriers provide nationwide telecommunications relay services (TRS). These services use operators to serve as an interface between people who use TDDs (nonvoice telecommunications devices for the deaf) and those who use the general voice telephone network. Interestingly, this service is being explored as a way to overcome language barriers as well; and the technologies are expected to evolve over time as speech recognizers and synthesizers improve. A more recent development is represented by hearings in Congress in 1999. These hearing made it clear that e-commerce applications (e.g., online catalogs) are in the same category as physical stores and need to accommodate people with disabilities. Web sites, for example, will need to have interfaces for people irrespective of their disabilities and for each potential channel of access. Many telecommunications companies have not yet met this requirement in their customer interfaces nor in the interfaces they supply for their business customers.

The Federal Communication Commission's (FCC) Telecommunications Act of 1996 was the first major overhaul of telecommunications law in more than a half a century. The primary goal of the law was to open up competition; however, one of the provisions in it has had particular meaning to the FCC leadership. Section 255 of the act was designed to ensure that new services provided in this competitive environment would be usable by anyone, whether or not they had a disability. The provision was viewed as the equivalent of a "curb cut." On September 29, 1999, the commission released a Report and Order establishing the rules and policies to implement Section 255. These rules and policies were based on the work of a variety of advocacy groups for people with disabilities, and experts in disabilities and telecommunications (several from the industry itself). Section 255 requires providers of telecommunications service to ensure that equipment and services are accessible to persons

with disabilities, if readily achievable; that networks be designed to support universal design; and that voicemail and interactive voice response systems be of special focus initially. Although IP-telephony, unified messaging, and other emerging services have not yet been explicitly discussed in the rules, many experts believe that given the original intention of this section of the Act, they will be covered. Furthermore, although the initial focus is on the regulated carriers, it would be reasonable over time to expect these regulations to also apply to new and perhaps nontraditional communication service providers if their communications services are viewed as in some sense "basic" and required for full participation in society.

We have focused here on the primary federal regulations of interest, but HCI professionals in the field may find themselves faced with other regulations as well. Section 508 of the Rehabilitation Act (Final Rule, issued by U.S. Access Board, for "Electronic and Information Technology Accessibility Standards," 36 CFR Part 1194) requires that when federal departments or agencies develop, procure, maintain, or use electronic and information technology, the technology shall be accessible. This is relevant when the applications HCI professionals are developing will be sold to federal department. Often state regulators had strict requirements for providing accessible technologies, and PacTel, NYNEX, and Ameritech were early leaders in building relationships with their local regulators through special attention to people with disabilities. Many countries, especially in Europe, also have a strong regulatory tradition protecting the needs of consumers of telecommunications products.

RESOURCES

Professional Organizations

HCI professionals in telecommunications participate in as many different organizations as there are HCI applications. They tend to be members of Association for Computing Machinery Special Interest Group for Computer Human Interaction (ACM SIGCHI), HFES, and Usability Professional Organization (UPA). Within HFES, they may be members of the Telecommunications Technical Group, Computer Systems Technical Group, or, more recently, the Internet Technical Group. These groups typically meet and present their work at the annual HFES conference. Within SIGCHI, they tend to participate in the annual computer human interaction (CHI) conference, but HCI professionals are also often seen at conferences focused on special topics of interest (e.g., CSCW, the conference on computer supported cooperative work; DIS, designing interactive systems; and CUU, the conference on universal usability). UPA's annual conference tends to draw people who are fairly new to practicing HCI and provides a focus for sharing experiences of what works and what doesn't work. AVIOS is probably the leading conference for talking about speech applications. A conference that serves as an international gathering specifically for professionals working in telecommunications is the International Symposium on Human Factors in Telecommunications HFT, which tends to meet once every 3 years. Its proceedings provide one of the best overviews in one place of HCI work in telecommunications around the world.

Web Sites

Here is a short listing of HCI/telecommunications–related Web sites:

- International Human Factors in Telecommunications: http://www.hft.org/
- Communications Technical Group (Human Factors and Ergonomics Society): http://www.hfes.org/Escapes/HFES-TechnicalGroups.html
- Computer Systems Technical Group (Human Factors and Ergonomics Society): http://cstg.hfes.org
- Internet Technical Group (Human Factors and Ergonomics Society): http://www.InternetTG.org/
- International Telecommunications Union: http://www.itu.ch/
- International Organization for Standardization home page: http://www.iso.ch/index.html
- European Telecommunications Standards Organization—Human Factors Technical Committee: http://www.etsi.org/hf/

Listservers

Professionals in telecommunications now frequent a variety of discussion groups. These include the following:

- Usability topics: UTEST, http://hubcap.clemson.edu/~tharon/utest/
- Broadband topics: Usableitv@yahoogroups.com
- Web design topics: http://www.sigchi.org/web/
- Wireless topics: wirelessui@egroups.com
- Wireless topics: usablemobile@smartgroups.com

References

Angiolillo, J. S., Blanchard, H. E., & Israelski, E. W. (1993, May/June). Video telephony. *AT&T Technical Journal, 72*(3), 7–20.

Angiolillo, J. S., Blanchard, H. E., Israelski, E. W., & Mane, A. (1997). A Technology constraints of video mediated communications. In K. E. Finn, A. J. Sellen, & S. B. Wilbur (Eds.), *Video mediated communications*. Mahwah, NJ: Lawrence Erlbaum Associates.

Blanchard, H. E. (1998). Update on recent HCI and usability standards. *ACM SIGCHI Bulletin, 30*(2), 14–15.

Blatt, L., Jacobson, M., & Miller, S. (1994). Designing and equipping a usability laboratory. *Behavior and Information Technology, 13,* 81–93.

Burkhart, B. J., & Root, R. W. (1994). 'Bellcore's user-centered design approach.' In M. E. Wiklund (Ed.), *Usability in practice* (pp. 489–515). Cambridge, MA: AP Professional.

Bushey, R., Mauney, M. J., & Deelman, T. (1998), The application of a behavioral categorization technique in the development of user models. *Proceedings of the Human Factors and Ergonomics Society 42nd Annual Meeting,* 434–438.

Dayton, T., Tudor, L. G., & Root, R. W. (1994). Bellcore's user-centered design support center. *Behavior and Information Technology, 13,* 57–66.

de Vries, G., van Gelderen, T., & Brigham, F. (1994). Usability laboratories at Philips: Supporting research, development, and design for consumer and professional products. *Behavior and Information Technology, 13,* 119–127.

Egan, D. E., Lesk, M., Ketchum, R. D., Lochbaum, C. C., Remde, J. R., Littman, M. L., & Landauer, T. K. (1991). Hypertext for the electronic library? CORE sample results. *Hypertext 1991* (pp. 299–312). New York: ACM Press.

Egan, D. E., Remde, J. R., Gomez, L. M., Landauer, T. K., Eberhardt, J., & Lochbaum, C. C. (1989). Formative design-evaluation of SuperBook. *TOIS, 7,* 30–57.

Fetz, B. H., Pilc, R. J., Rubinstein, C., & Wish, M. (Eds). (1989). *AT&T Technical Journal* (special issue on designing the human interface), *68*(5).

Finn, K. E., Sellen, A. J., & Wilbur, S. B. (Eds). (1997). *Video mediated communications.* Mahwah, NJ: Lawrence Erlbaum Associates.

Fowler, C., Stuart, J., Lo, T., & Tate, M. (1994). Using the usability laboratory: BT's experiences. *Behavior and Information Technology, 13,* 146–153.

Fussell, S. R., & Benimoff, N. J. (1995). Social and cognitive processes in interpersonal communications: Implications for advanced telecommunications technologies. *Human Factors, 37,* 228–250.

Hanson, B. L. (1983). A brief history of applied behavioral science at bell laboratories. *The Bell System Technical Journal* (special issue on human factors and behavioral science), *62,* 1571–1590.

Hanson, B. L., & Israelski, E. W. (1977, June). Human factors engineering in the outside plant: Bringing out the best. *Bell Labs Record,* 30–35.

Israelski, E. W., & Reilly, R. R. (1988). Telecommunications craftworkers. In S. Gael (Ed.), *The job analysis handbook for business, industry and government.* New York: Wiley.

Israelski, E. W., Angiolillo-Bent, J. S., Brems, D. J., Hoag, L. L., Roberts, L. A., & Wells, R. S. (1989, September). Generalizable user interface research, *AT&T Technical Journal, 68*(5), 31–43.

ISO/IEC 13714—Voice Messaging Applications. ISO International Standards Organization.

Klemmer, E. T. (1989). An interview with John E. Karlin. In E. T. Klemmer (Ed.), *Ergonomics: Harness the power of human factors in your business* (pp. 197–201). Norwood, NJ: Ablex.

Klienfield, S. (1981). *The biggest company on earth: A profile of AT&T.* New York: Holt Rinehart Winston.

Kraut, R. E. (Ed.). (1987). *Technology and the transformation of white-collar work.* Hillsdale, NJ: Lawrence Erlbaum Associates.

Landauer, T. K. (1995). *The trouble with computers: Usefulness, usability and productivity.* Cambridge, MA: MIT Press.

Landauer, T. K., & Dumais, S. T. (1997). A solution to Plato's problem: The latent semantic analysis theory of acquisition, induction and representation of knowledge. *Psychological Review, 104,* 211–240.

Lund, A. M. (1994a). Ameritech's usability laboratory: From prototype to final design. *Behavior and Information Technology, 13,* 67–80.

Lund, A. M. (1994b). The evolution of broadband work in Ameritech's customer interface systems and human factors department. In M. E. Wiklund (Ed.), *Usability in Practice* (pp. 457–488). Cambridge, MA: AP Professional.

Lund, A. M. (1996). A case study of an ad campaign. *Ergonomics and Design, 4*(4), 5–11.

Lund, A. M. (1999). Why do people dial wrong numbers? In D. J. Bonneau (Ed.), *Human factors and voice interactive systems* (pp. 187–203). Boston, MA: Kluwer Academic.

Lund, A. M. (2000). Convergence of telephony, TV, and computing in the home. In W. Karwowski (Ed.), *International encyclopedia of ergonomics and human factors.* London: Taylor and Francis.

Lund, A. M., & Tschirgi, J. E. (1991). Designing for people: Integrating human factors into the product realization process. *IEEE Journal on Selected Areas in Communications, 9,* 496–500.

Mulbach, L., Boecher, M., & Suwita, A. (1995). Telepresence in Video communications: A Study in Stereoscopy and Individual Eye Contact. *Human Factors, 37*(2), 290–305.

Nielsen, J. (1989). The matters that really matter for hypertext usability, *Proceedings of ACM Hypertext '89 Conference* (pp. 239–248). New York: ACM Press.

Nielsen, J., & Molich, R. (1990). Heuristic evaluation of user interfaces. *Proceedings of ACM CHI'90* (pp. 249–256). New York: ACM Press.

Noll, A. M. (1992, January 12). VideoPhone: A flop that won't die. *The New York Times,* 13.

OSHA releases final ergo standard. (2000). *Human Factors and Ergonomics Society Bulletin, 43*(12), 1.

Reed, P., Holdaway, K., Isensee, S., Buie, E., Fox, J., Williams, J., & Lund, A. (1999). User interface guidelines and standards: Progress, issues, and prospects. *Interacting With Computers, 12*(2), 119–142.

Schumacher, R., Hardzinski, M., & Schwartz, A. (1995). Increasing the usability of interactive voice response systems: Research and guidelines for phone based systems. *Human Factors, 37,* 251–264.

Schwartz, A. L., & Seifert, C. (1996). Can a usable product flash 12:00?: Perceived usability as a function of usefulness. In *Proceedings of the Human Factors and Ergonomics Society's 40th Annual Meeting,* (pp. 313–317). Santa Monica, CA: Human Factors and Ergonomics Society.

Turner, P. A. (Ed.). (1983). *The Bell System Technical Journal* (special issue on human factors and behavioral science), *62*(6).

Virzi, R. A., & Sorce, J. F. (1994). GTE Laboratories Incorporated: The evolution of usability within new service design. In M. E. Wiklund (Ed.), *Usability in practice* (pp. 559–590). Cambridge, MA: AP Professional.

·40·

GOVERNMENT ROLES IN HUMAN–COMPUTER INTERACTION

Jean C. Scholtz
National Institute of Standards and Technology

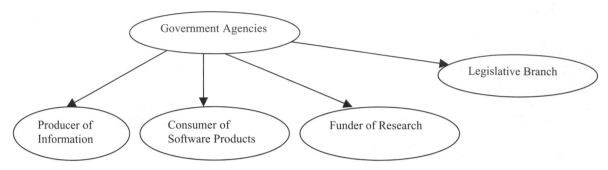

FIGURE 40.1. Government roles in software research and development.

INTRODUCTION

In the current human–computer interaction (HCI) literature, case studies frequently address such domains as office work, health care, e-business, and education. Case studies looking at HCI issues in some aspect of government are less abundant. Why is this? What are the HCI issues in government aspects of information technology, and why are so few researchers investigating this domain?

Governments do not tend to be early adopters of technology. Also, the technology they purchase tends to have a long life once it is in place. Therefore, it is prudent for government agencies to purchase mature technology. Because of the necessity to serve the public, it is disruptive to have unreliable technology. Moore (1991) described the market penetration of new technology in a bell curve with innovators and early adopters constituting the front end, the early majority and the late majority equally consuming the majority of the curve, and the laggards bringing up the last of the curve. Moore summarized these groups as follows:

- Innovators aggressively seek out new technology and are most generally technologists.
- Early adopters are not technologists but buy into new technology early on, relying mostly on their own instincts and judgments.
- Early majority users are practically oriented and want to see a reference in their own community before buying into the technology. It's important for them to see and understand how a new technology will work in their business.
- Late majority users wait until the technology has become an accepted practice before they adopt it.
- Laggards are groups who only buy into the technology when it is no longer possible to avoid it.

Given this categorization, the majority of government agencies fit into the late majority grouping. This may account in part for the small number of case studies of HCI issues in the government. By the time many government agencies adopt a technology, the product has matured and many issues of concern to the HCI community have been resolved or at least assessed.

Government covers a large territory. Governments perform many functions from the day-to-day services provided to citizens to highly specialized services provided by government agencies such as air traffic control and the military.

The government, like many large corporations, also has a number of roles with respect to software. These roles are illustrated in Fig. 40.1. It purchases software for employees to use; it produces information products for use by citizens; it passes legislation concerning standards for the safety and accessibility of products used in its jurisdiction. Additionally, some government agencies fund research in information technology. In each of these roles, there is a need for involvement on the part of HCI researchers and practitioners. All four of these aspects are discussed in this chapter and illustrated with case studies or specific examples. Figure 40.2 shows the HCI involvement for each of the roles the government plays in software research and development and shows the influence of the different agencies on each other.

The information in this chapter has been based primarily on knowledge about the workings of the U.S. government. However, the background research for this chapter has turned up similar issues and initiatives in many other countries. Searches for "e-government" on the Web turned up pages from numerous countries, including Russia and Estonia (e.g., Estonia Government Home Page, 2001). Estonia conducts all government business via a secure HTTPS server, with proposed laws, commenting procedures, and voting all available online (Perera, 2000).

THE GOVERNMENT AS A PRODUCER OF INFORMATION: E-GOVERNMENT

Just as more and more businesses are turning to e-commerce, more and more governments are turning to e-government. One aspect of e-government is providing services to the general public using electronic means. Many citizens and public officials view the Internet as an efficient and effective way to get more information to people and to conduct services. Today, those with access to the Internet can use a variety of services such as registering to vote, renewing automobile registrations, obtaining tax forms, getting medical information, getting information about passports and visas, reviewing legislative proceedings, and making reservations for accommodations at state parks. The vision

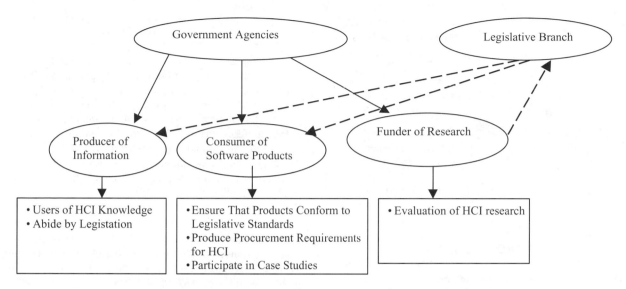

FIGURE 40.2. Government influence on HCI with respect to roles.

for e-government in New Zealand states: "New Zealanders will be able to gain access to government information and services, and participate in our democracy, using the Internet, telephones and other technologies as they emerge." (E-government Vision for New Zealand, 2001). In the United Kingdom, the Office of the e-Envoy (Office of the e-Envoy, 2001) is leading the drive to get the country online, to ensure that the country, its citizens, and its businesses derive maximum benefit from the knowledge economy (UK Online, 2001). The goal of the UK online project is as follows: "UK online is for everyone. It is a drive to enable everyone in the UK to gain access to the Internet by 2005 and to make the UK one of the world's leading knowledge economies. This partnership between government, industry, the voluntary sector, trades unions and consumer groups aims to make technology clear and easy to use. Ukonline.gov.uk is a key part of the UK online initiative."

The International City/County Management Association (ICMA) and Public Technology, Inc. (PTI) conducted a survey of local United States governments in the fall of 2000 on electronic government (PTI/ICMA, 2000). Responses from more than 1,500 local governments were received. More than 80% of the governments responding had a Web presence, and another 10% planned to create one within the next year. All jurisdictions with populations over 500,000 surveyed had a Web site. These Web sites are advertised to citizens through local government publications and through newspaper and television announcements. More than 80% of the governments report trying to overcome digital divide issues by providing public access terminals in city and county facilities. Few local governments currently offer financial transactions through the Web, but many are planning to offer some transactions in the near future. Lack of technology and Web expertise along with the financial resources to support development are top barriers encountered in trying to establish government Web services. Security issues are also a priority. Government officials feel that Web sites are a public service, and only a small percentage of those surveyed allow paid advertising on their sites. A somewhat larger percentage had a formal policy

about paid advertising, but most jurisdictions currently had no policy in place. Government officials felt that e-government has changed the role of the staff. More demands are being placed on employees, and agency work practices and processes must be reengineered to accommodate these changes.

The PTI Web site (Public Technology, 2001) contains links to a number of municipal Web sites. Typical services available on these Web sites include tax payments, payments for parking tickets, business tax filing, applications for permits, and payment of utility bills. Interactive services are also available in some municipalities including access to property information, crime reporting, requesting street light repair, filing complaints, and commenting on municipal budgets. However, interactive services are rarer than the more basic one-way information services. Few sites provide search engines to facilitate site navigation. Some municipalities use the Internet to involve more citizens by offering streaming video of local council meetings. Portal Web sites are also available in some communities to facilitate access to government information and services. For federal government information in the United States, the First Gov Web site (First Gov, 2001) provides this service. This portal site contains featured subjects and groups of governmental links organized into categories such as agriculture and food, arts and culture, health, benefits and grants, consumer safety, money and taxes, and library and references.

HCI Requirements for E-Government Services

The Hart-Teeter Report (1999) prepared for The Council for Excellence in Government surveyed the general public, government officials, and business and nonprofit customers of government to determine their current experience with e-government and their future expectations. All three groups viewed e-government as having the potential to greatly improve current government operations. Only half of all adults in the general population were familiar with the concept of e-government,

but more than 90% of the government officials were familiar with it. More than two thirds of the American online population had visited a government Web site and rated them highly. More than 50% of government customers agreed that conducting business with the federal government had become easier because of e-government services. E-government services viewed as most useful by the general public were access to medical information from the National Institutes of Health, viewing of candidates' voting records, access to Social Security benefit information, online motor vehicle registration, and online student loan applications. More than two thirds of the general public felt that tax dollars should be invested in providing more e-government services. However, the public also was concerned about privacy and security as more information becomes accessible on-line.

Every-Citizen Interface. Is there a difference between services offered by e-government and by e-commerce? Although the groups offering e-commerce would like a consumer to think that their site is the only place to make a certain purchase, this is usually not the case. If a user cannot navigate a particular commercial site, another site selling similar products is most likely available. This is not true in the case of the government. For example, if a person needs information about census data in the United States, the most expedient way to obtain this is to visit the U.S. Census Bureau site. Because government Web sites are made possible by tax dollars, citizens expect to be able to access these sites easily and locate the information they need. In addition, there might be a necessity to complete the task. It may not be essential that a person purchase a particular book, but completing the renewal of a driver's license is a necessity.

It is not effective nor in the interest of the majority of citizens if electronic information provided by the government can only be accessed by a few, privileged individuals. *More Than Screen Deep* (National Research Council, 1997) identified the research necessary to produce every-citizen interfaces to the nation's information infrastructure. This report was the combined work from a workshop attended by a number of prominent HCI researchers. The researchers identified a set of characteristics that they felt were necessary for every-citizen interfaces, which should be:

- Easy to understand
- Easy to learn
- Error tolerant
- Flexible and adaptable
- Appropriate and effective for the task
- Powerful and efficient
- Inexpensive
- Portable
- Compatible
- Intelligent
- Supportive of social and group interactions
- Trustworthy (secure, private, safe, and reliable)
- Information centered
- Pleasant to use

For more detailed discussions on interfaces to support diverse groups of users, see section IIIB of this handbook.

Section 508: Accessibility Legislation. In the United States, Section 508 of the Federal Rehabilitation Act (2001) went into effect June 25, 2001. It legislates that federal agencies' electronic and information technology is accessible to people with disabilities. This applies to software, hardware, and telecommunications products purchased for use within U.S. governmental agencies as well to Web sites that are used by these agencies to provide information to the general public. This section summarizes the guidelines that have been established by the Department of Justice for a Web site to be accessible (Department of Justice, 2001). More discussion of Section 508 for hardware and software accessibility appears in the following section, The Government As a Consumer of Software.

The following recommendations are given in a Department of Justice Survey available from the Department of Justice Web site:

- Provide alt-text equivalents for all nontext objects on a Web page.
- Text captioning should be provided for audio and video on a Web page and synchronized with the video.
- The Web page should be viewable and understandable in the absence of color.
- Web pages should be readable whether or not a style sheet is imposed.
- Text links should be provided for each active region of a server-based image map.
- Columns and rows of tables should be identified within each cell to make tables understandable when a screen reader is used.
- Frames should be titled to facilitate navigation.
- Web pages should not use techniques that cause them to flicker at rates between 2 Hz and 55 Hz.
- Dynamic Web pages generated by scripts need to identify the context that is generated
- If plug-ins are needed to view documents or objects obtainable from a Web page, links to those plug-ins should be available on that page.
- Electronic forms need to make all labels and fill-ins accessible to users.
- Timed responses must accommodate users with slower response times.
- Conduct user testing of the Web site by individuals with disabilities.

Private organizations are also establishing guidelines for Web accessibility including the Web Accessibility Forum (WAI; Web Accessibility Forum, 2001) and the National Committee for Information Technology Standards (NCITS; National Committee for Information Technology Standards Home Page, 2001). Many of these guidelines have already been incorporated in those in the Section 508 implementation.

Case Studies and Ongoing Research. In addition to accessibility, government Web sites are often difficult to access and use because of the nature of the data they contain. Much government information exists in large databases and is statistical in nature. Even textual information is difficult to use because of the terminology used. The following examples give some insights into design requirements for governmental agency Web sites and user interfaces for accessing government information.

Case Study: Security and Exchange Commission. In 1996, the Electronic Data Gathering, Analysis, and Retrieval system (EDGAR) was fully installed at the Securities and Exchange Commission (Kambil & Ginsburg, 1996). This commission regulates financial issuers and market makers in the United States and mandates certain financial disclosures that are made available to the public. These disclosure documents are read by the investors to make decisions about financial investments. Before the use of the EDGAR system, it was expensive and inconvenient for market makers to gain access to the financial disclosures. Data feeds for this information ran from $75,000 to $150,000 per year. Paper copies of individual disclosures can be ordered by mail for $20 to $30 per filing. Public access reference rooms hooked to the data feeds are available but only in four cities, and there are limited terminals for access. Moreover, there is a tremendous amount of data processed. The disclosure documents amount to 60 GB of text data per year.

Although making these documents available for access with decreased costs is valuable, the researchers also found the need for more value-added services. Searching the documents is difficult but can be facilitated if the disclosure documents are delivered in a marked up or tagged format suitable for electronic searching. Electronic searching also pointed out cases of inconsistencies with names for the companies; in some cases the company name was used, and in other cases the name listed on the stock exchange was used. Value-added services can be delivered, but this will necessitate a procedural change in the way documents are prepared and delivered. Standardized tags for marking up the documents would facilitate the addition of value-added services such as searching. Other value added services could facilitate customization of the user interface so that users or groups of users could view only those types of financial disclosures they wish to see.

Case Study: U.S. Census Bureau. The development of user interfaces for the U.S. Census Bureau (University of Maryland Human Computer Interaction Laboratory, Census Project, 2001) by researchers from the Human–Computer Interaction Laboratory (HCIL) at the University of Maryland focused on navigation through databases, user interfaces, and the use of data visualization tools to facilitate understanding the data. The technique of query previews was used to allow the user to pose a query by adjusting slider bars corresponding to attributes of the data. The visual representation of the results shows the user the amount of information in the database that matches the selected set of attributes. This gives users feedback on the underlying data set, thus allowing them to assess whether to generalize the query or to make it more specific. This type of technique is useful because it is difficult for users who are not well versed in Census data to gain an understanding of the type and amount of data available from standard query feedback. Moreover, queries on large data sets take time to complete, making successive queries slow and frustrating for the user. The query preview technique allows the user to compose an acceptable query in a reasonable amount of time.

Research Study: Searching Unfamiliar Data. Not only is much of the information from government agencies stored in large databases, it is often accessible only if the correct terminology is available to query that database. Slaughter (2001) researched the use of visual exploration and browsing mechanisms to supplement searching of the database when the user is not familiar with the terminology. Another approach to this problem is the use of ontologies for converting a user query in normal everyday terms to the specific vocabulary of the database (MetaData Research Program, 2001). Researchers at the University of California at Berkeley have developed entry vocabulary modules that facilitate users performing searches on databases with unfamiliar meta-data. The researchers have used this technique to assist in searches of the US Patent and Library of Congress databases.

Case Study: Library of Congress. Libraries are a category of government agencies interested in making their services available electronically. Another HCIL project developed a user interface for the Library of Congress National Digital Library. (Plaisant, Marchionini, Bruns, Komlodi, & Campbell, 1997). As in many governmental agency interfaces, the interface at the Library of Congress needed to accommodate a wide range of users. At the time the research was conducted, only a small portion of the library's collection was digital, so it was necessary to represent this fact the users. The issue of accommodating different types of users was addressed by setting up specific introductory tours for five general types of users, where the type was determined by the goal of the visit. Filters were proposed to facilitate user navigation through the collection. Filters included such things as geographic location, chronological period, and format of data. However, objects in the collection had different amounts of meta-data, making the type of data difficult to classify.

Case Study: Public Library Design. A design project undertaken by Philips Corporate Design (Jorna et al., 1997) also looked at incorporating technology into the Public Library of Eindhoven. Here the emphasis was on facilitating the typical user scenario: coming into the public library, locating something of interest, and leaving the library with a selected item. An operational process within the library was developed through user interviews. Additionally, a number of user requirements were collected including the following:

- Provide interest-group-related information in specific areas of the library.
- Integrate technology gradually.
- Provide computer technology that is nonthreatening.
- Provide browsing applications for all users.
- Promote what the library has available.
- Develop more supporting and advisory services.

One of the findings of the research team was that users performed different activities in different parts of the library. There was a need for different tools to support different uses of the library, but those tools had to be part of a coherent set. The design team developed the concepts for a number of tools that were mapped onto the current operation process within the library. Navigation and utilization of electronic services in this case were facilitated by leveraging off the knowledge library patrons already held about the process in a library.

THE GOVERNMENT AS A CONSUMER OF SOFTWARE

Much government business occurs in an office environment. Software purchased for this environment has the same requirements as software in any typical office environment. The problem is how to ensure that a usability engineering cycle has been employed during product development and that the resulting product meets quantifiable usability standards. Other government agencies are dealing with more specialized tasks and could benefit from improved user interfaces. In this section, several methods for ensuring usability are discussed. These are appropriate for general-purpose software as well as for software for more specialized tasks. Case studies present examples of HCI concepts and methodologies applied to software developed for specialized tasks.

The Industry Usability Reporting project (IUSR) at the National Institute of Standards and Technology is an attempt to raise the level of visibility of usability so that this quality can be considered during procurement procedures (IUSR, 2001). Although this project was developed for general propose commercial software, the work is certainly applicable for software acquired by the government. There are two parts to the IUSR project. First is the development of the common industry format (CIF) for reporting on usability results. The second part of the project involves pilot studies performed by a number of industrial affiliates to identify and measure internal effects and dependencies on the usability metrics. The CIF was developed by a team of usability professionals and specifies not what should be tested, but what should be reported when a summative usability test is performed on a software product. The CIF Web page (http://www.nist.gov/iusr) specifies the exact descriptions for reporting, but the general items to be included are demographic information of participants in the user test; tasks given to the participants; the equipment setup used in the test; and metrics for effectiveness, efficiency, and user satisfaction of the product as used to perform the given tasks. The recommended use of the CIF is for an interested purchaser to notify a vendor of the intent to purchase a product with a request for a CIF to be provided. This request must be negotiated with the vendor as to nondisclosure agreements as well as time and costs needed to produce the CIF. Once the CIF is delivered, a usability professional on the staff of the consumer company would interpret the results. The usability professional would examine the user demographics and tasks to determine the relevance for the consumer company. Also, the reported metrics would be examined to determine if they fall within the acceptable range for the consumer company. If either the user demographics or user tasks are not relevant, the consumer company has several options. They can either negotiate with the vendor to have another CIF produced using different participants and different tasks for the summative evaluation, or the consumer company can undertake the testing themselves. Once the tasks and participants are judged to be similar to those of the consumer company, then the usability professional must work with others in the company to determine if the usability metrics of the product under consideration will have a positive or adverse affect on the total cost of ownership of the software. If the reported metrics are unsatisfactory, the consumer can either look for other software or can work with the vendor to uncover usability problems as well as suggestions for improvement.

In the second part of the project, consumer organizations are examining different costs of software ownership and ways to measure and track these costs. Examples include cost of internal support staff, staff productivity, training, and staff turnover. Once baseline measures of some of these costs have been established, then consumer companies will be able to determine if and how changes in the usability metrics for new software products affect the total cost of ownership. This should help consumer companies develop acceptable ranges of metrics for different categories of software.

On the surface, it would seem that consumer companies receive most of the benefits in the IUSR project. However, in the ongoing pilot work, vendors are finding the value of the CIF as a document for discussing tasks, demographics, and usability requirements with large consumer organizations.

Lewis, Henry, and Mack (1990) presented a benchmark technique to evaluate integrated office software. In their case study, the researchers developed scenarios and benchmark measures and conducted both quantitative and qualitative analyses. Two systems were compared, one that was loosely integrated and one that was more tightly integrated. Significant differences were found in performance and error rates, but there were no differences attributable to satisfaction ratings. This suggests that testing measures other than user satisfaction are important if performance is a concern. It also suggests that benchmark techniques could be used to decide whether a new system under consideration for procurement will produce a significant advantage. Both private companies and government agencies should consider this type of comparison when investing a significant amount in commercial software.

Government software for more specialized user populations is acquired through government contracts. At the 1995 Conference on Human Factors in Computer Systems (CHI'95) a workshop conducted by Winkler and Buie (1995) outlined the major challenges of incorporating HCI and usability engineering into government contracted software. The major challenges the workshop identified were as follows:

• *Rigid specification for contracts.* It is difficult to include the language needed for usability into the current requests for proposals. Furthermore, it is difficult for contracting teams to assemble a multidisciplinary team because many contracts require contractors to have degrees in computer science or engineering.

Human factors and usability professionals may have degrees in psychology or technical communications.

• *Insufficient access to users.* A major challenge is the inability of bidders for contracts to talk to users during the proposal phase. Therefore, it is completely unrealistic to be able to assess user requirements and to adequately include those costs and processes in the submitted proposal.

• *Long development cycles.* The contracting and development cycles of specialized software can take years. Processes, users, and users' knowledge can change considerably during that period of time.

• *Entrenched development methodologies.* In many government agencies, a waterfall method of software development is mandated to facilitate the ability of the agency to monitor and evaluate interim progress on the contract deliverables.

• *Cost-benefit considerations.* Software systems are built for the government to satisfy perceived needs of citizens. Financial considerations are heavily weighted in making procurement decisions, so any type of activity, such as usability, that may not be well understood by government procurement officials will be viewed as costly and unnecessary.

• *Sparse communications between the government and the HCI community.* HCI professionals working in the government agencies are practitioners. They need help from the HCI research community, but there is little dialogue that occurs. Understanding the needs of various government agencies with respect to HCI is complex. These needs should be of interest to the HCI community, but a way to connect the two communities must be developed.

The proposed approaches suggested by the CHI'95 workshop participants include influencing policymakers and procurement processes in the government, integrating usability engineering into the government-recognized software engineering process and using standards as leverage. Although this will take time, some progress has been made in these areas. A later section describes how usability language can be incorporated into the contracting request. Access to users is a barrier that must be overcome to produce more intelligent bids based on a better understanding of user requirements. This will require a major change to the current contracting process. To achieve this, procurement officials and policymakers must be able to see the benefits of incorporating usability engineering processes into product acquisition.

Incorporating Usability Into Software Procurement Requirements

Government agencies issue a procurement document when there is a need for new software to be developed. In the United States this document is referred to as a request for proposal (RFP). Contractors prepare bids based on the work that is requested in the RFP. One way to address usability issues is to specifically request the necessary processes, deliverables, and acceptance criteria using the appropriate sections of the RFP (Buie & Kreitzberg, 1998).

Typically, an RFP contains the following sections: supplies and costs, description of work, packaging, inspection and acceptance, deliverables or performance, contract administration, special contract requirements, contract clauses, statements to the offeror, instructions, and evaluation for award. Usability requirements should be included in the sections on description of work, deliverables or performance, statements to the offeror, and evaluation factors for award. Table 40.1 shows these sections, describes what they usually contain, and describes the usability information that should be incorporated into each section.

By specifying usability requirements upfront in the RFP, contractors know that usability expertise will be required and can develop their proposal to include the required processes and to incorporate this into the costs of the projects. On the agencies' part, usability expertise is also required to be able to generate the specifications in the RFP and to be able to adequately evaluate the deliverables associated with usability during the contract period.

Standards and the Procurement Process

The U.S. Government has passed legislation effective June 25, 2001, stating that all federal agencies electronic and information

TABLE 40.1. Sections of an RFP and Suggested Usability Insertions

Section Label	Current Contents	Usability Additions
Description of work	• Overview of the work • Needs of the agency • Work to be performed by contractor	• Usability requirements of the agency; specific usability objectives • Usability engineering activities expected to be performed by the contractor
Deliverables or performance	Specifies the software to be delivered, the documentation training modules, etc.	Requests for documentation of usability engineering process, including tak analysis reports, user interface requirements, usability test plans, Common Industry Format (CIF) reports
Instructions to offerors	Any special instructions	Statements to the effect that usability engineering must be incorporated into the software engineering process used in development
Evaluation	Criteria the agency will use to evaluate the proposals and make awards	Emphasis on the importance of a well-thought-out usability engineering process and deliverables

technology must be accessible to people with disabilities (Section 508, 2001). In accordance with this legislation, the U.S. Access Board is required to develop the standards, and the General Services Administration (GSA) is to provide technical assistance. The Federal Information Technology Accessibility Initiative (FITAI) is an interagency initiative created to assist in this effort. The standards have been arrived at by voluntary consensus through user and vendor forums. The standards fall into several categories:

- *Software and operating systems.* These standards deal mainly with vision impairments, keyboards, animated displays, color, navigation, and forms.
- *Web-based.* These standards deal mostly with vision impairments and compatibility of the user interface with screen readers (see The Government As a Producer of Information).
- *Telecommunications.* These standards ensure compatibility with hearing aids, cochlear implants, assistive listening devices, TeleCommunications device for the deaf (also TDD) (acoustic coupler with a keyboard and message display), an adjustable volume for output.
- *Video or multimedia products.* These standards describe captioning and audio descriptions on training and informational multimedia.
- *Embedded, closed products (information kiosks, copiers, printers, calculators).* These standards deal with access features that must be built in and do not require the attachment of special devices.
- *Computers.* These standards apply to biometric identification and mechanical controls.

The standards specified in Section 508 apply to documentation and any support of information technology products as well as to the product itself. Detailed technical specifications can be viewed at the FITAI (2001) Web site. These standards can be specified as requirements in the contract request. In addition, requests for compliance with other standards, such as those developed by the International Organization for Standardization (ISO) can be written into the requirements as well. ISO standards are discussed in the following section, HCI Standards and the Government's Role.

Case Studies

The case studies described in this section have been grouped into three categories: software used by specialized government agencies; special purpose software with highly complex interfaces; and military systems. These three categories are definitely not mutually exclusive but the case studies have been placed in one of the categories based on the focus of the software system being developed. For example, case studies of air traffic control are grouped under special purpose software with highly complex interfaces while software being developed for economists at the International Monetary Fund is grouped under software used by specialized government agencies.

Software Used by Specialized Government Agencies. The case studies in this section illustrate the need for close examination of the processes currently in place and the need to consider how to best support those processes when developing electronic tools.

Case Study: International Monetary Fund. An interesting ethnography study by Harper and Sellen (1995) was conducted at the International Monetary Fund (IMF) to determine design criteria for collaborative tools. IMF membership consists of most of the world's countries. Members in the IMF contribute to monetary resources used to provide low interest, multicurrency loans to members when needed. Interestingly, the IMF was chosen as a domain to examine professional work because of the high number of professionals on the staff, the complexity of the work carried out, and the willingness of the IMF to allow researchers access to all facets of their work. A network had been installed at the IMF before the study. It had been thought that this network would enable more sharing of data and reports between the economists than had actually occurred.

Individual economists at the IMF are responsible for maintaining data on a member country. When a particular need arises for a country, the economist responsible for that country is joined by a team of four or five others to produce a mission report. During the production of this mission report, the work is divided so that each economist is responsible for a particular aspect of the report. Checks and balances are put in place as the construction of these reports involves much judgment on the part of the economists. The group working on the mission report evolves a set of interpretations to be used through the report generation. With respect to collaboration, the researchers found that information involving a high degree of judgment in its production is difficult to share and is best interpreted by the producer of the report. Statistical data, on the other hand, is sharable by the nature of the information itself. Social interaction is crucial to the sharing of interpreted data but not to objective information.

This study also looked at the reuse of reports compiled at the IMF. Effective reuse of documents for other purposes is achieved only through interpretation of the initial reports by the authors. The study also found that paper documents supported a redesign procedure more effectively than current electronic documents. In this study paper, documents were more easily woven into the collaboration activities and supported a higher level of design as opposed to electronic documents.

Case Study: Collaboration Tools for the German Ministry. A different approach was followed in a study of collaboration tool design for a German Ministry (Pankoke-Babatz, Mark, & Klockner, 1997). In this study, the team at the German Ministry was given a collaborative tool and trained in its use. User advocates made site visits on a regular basis to help with problems and to understand how the tool fit with the work practices of the office. They found that the users of the system did not experiment with the system but rather started using it immediately in their work process. Thus, although many inconsistencies and other problems were found, the team was able to identify early on design features that were incompatible with their work

practices. The disadvantage to using the system immediately in a real-world work setting is that the system has to be a product, not a prototype. Hence, a considerable amount of time is needed to modify the software as it is necessary to ensure that any changes made are robust and reliable.

Case Study: Italian Public Administration. Catarci, Matarazzo, and Raiss (2000) worked on the Usability Working Group created by the Italian authority for distribution of software in the public administration (Autorita per l'Informatica nella Pubblica Amministrazione; AIPA). They found a number of problems in applying usability practices to software developed for the public administration. Although customers regarded usability as a chief source of dissatisfaction with current products, neither the customers nor the software suppliers were well informed about usability practices and principles. Therefore, usability information collected from customers was not accomplished in a systematic fashion. Technological constraints and backward compatibility are driving factors in software development that compete with usability efforts.

Case Study: Maryland Department of Juvenile Justice. The Human–Computer Interaction Laboratory (HCIL) and the Maryland Department of Juvenile Justice (DJJ) worked jointly on developing a tool for finding specialized programs for troubled juveniles (Ellis, Rose, & Plaisant, 1997). Twenty-two site visits were conducted to understand the current workflow of the Department of Juvenile Justice and to propose a system that could be placed effectively into the current workflow. During the design of the prototype, the researchers noted that the management team had a different view of what was needed than did the actual users of the system. The users were concerned with matching the new system to the way they currently conducted their work. A change in work practices through a new software system is certainly feasible, but designers must proceed with extreme caution to ensure that all parties affected buy into this change and its implications. This also shows the importance of having both users and management involved in the design efforts.

Case Study: The Chicago Mercantile Exchange. Another study (Parikh & Lohse, 1995) examined trading activities in the Chicago Mercantile Exchange. Commodity trading of items such as crude oil, gold, silver, agricultural goods, and currencies is currently conducted by brokers using voice and hand signals to indicate goods to sell or a desire to buy. These brokers are extremely efficient and complete on the average of 60 trades per hour. A first attempt to build an electronic system to support these activities omitted crucial actions that can be easily and quickly accomplished by humans. Human brokers are capable of representing both buyers and sellers, completing trades by combining several buyers or sellers, and rapidly adjusting to the market price. Current interfaces supporting commodities markets have been less successful than hoped as trading was actually slowed down using the sequential processes supported electronically.

Training Issues for Government Agencies. Clichelli (1998) pointed out that training and performance support are needed to reduce downtime, minimize disruptions, and improve user satisfaction when new software is introduced into government agencies. There is a clear need to identify the current processes and to match the knowledge and capabilities of the new system to what is currently being done. She discussed the concept of a knowledge map that diagrams the information and knowledge that workers need to perform successfully.

Special Purpose Software With Highly Complex Interfaces. Human factors researchers who are interested in complex displays and problem solving often study air traffic control displays. Air traffic control is a good example of an area in which there is an abundant amount of information available. To facilitate the air traffic controller's job, relevant information has to be readily accessible from the user interface so that quick actions and plans can be undertaken. There are many other possibilities in aviation and in military operations for studying humans interacting with complex systems such as logistics support, decision support, and operation planning. There have been studies conducted on displays for situational awareness and assessment in both the aviation and military domains. These systems involve either collaboration among a number of participants or a synthesis of a number of different data sources. Many of these systems place heavy cognitive loads on users. Other external factors may be present that contribute to overall user stress and confusion.

The majority of cockpit display studies were done in the 1970s and 1980s. They are now out of date because of the ability to bring more information with greater accuracy into the cockpit. The importance of reexamining cockpit displays of traffic information with the advent of an increase the information available in the cockpit is discussed in Johnson (1998). In particular, he looked at the Traffic Information System (TIS), which provides low-resolution radar data on nearby aircraft to the flight deck and the Automatic Dependent Surveillance Broadcast System (ADSB), which provides accurate ground track, speed, range, and bearing. For a thorough discussion of issues in air traffic control, the reader is referred to Pritchett (chapter 45, this volume).

New user interfaces and information may be required when new work practices are put into place. The U.S. Federal Aviation Administration (FAA) proposed a Free Flight Initiative (Free Flight Initiative Home Page, 2001) that would require additional situational awareness challenges for cockpit displays. Currently, air traffic controllers monitor the views of various sectors, responding to requests from pilots, making adjustments for weather conditions, and instructing pilots on the aircraft speed, flight levels, and flight heading to maintain to have an efficient and safe flow of air traffic. The Free Flight Initiative would give pilots more flexibility in choosing routes, speeds, and altitudes in real time, resulting in the enroute controller becoming more of a supervisor than an active participant. The pilots, however, will need to be aware of the traffic around them and make adjustments to their flight paths based on this. Thus, it is necessary to examine both the information required by the supervisory interface as well as that required in the cockpit display. What are

the implications of this type of operation for the information and presentation of that information available to pilots and to air traffic controllers?

Case Studies: Support for Free Flight Air Traffic Management. Three studies looked at different aspects of information needed to support free flight air traffic management (Duley, Galster, & Parasuraman, 1998; Knecht, 1998; Morphew & Wickens, 1998). One study (Duley et al., 1998) conducted a survey of 58 air traffic controllers to classify 69 information parameters into the following four categories: should always be present, should be available upon request, should be presented by the system when deemed necessary, and is not needed and should be eliminated. Only 13 information parameters were common across three different levels of free flight. Moreover, there was surprising variability among controllers. This variability must be understood to create safe and efficient interfaces.

Case Study: Pilot Workload and Performance Using Predictive Information Displays. Another study (Morphew & Wickens, 1998) looked at the performance and workload of pilots when predictive information was added to three prototype traffic displays. The workload for pilots decreased monotonically with the addition of predictive information. Also, a greater inferred measure of safety was achieved without a decrease in flight efficiency.

Case Study: Coordination of Airline Ground Traffic. An interesting study (Obradovich et al., 1998) looked at the requirements for tools to coordinate a number of people involved in commercial airline ground traffic. Delays in taxiing to the gate and leaving the gate are extremely costly to the U.S. commercial airlines. Estimates exceed $1 billion per year. The parties involved in coordinating ground movement of aircraft are air traffic control staff, airline ramp control and gate staff, and flight crews. Airline dispatches and traffic managers in FAA traffic management units are also indirectly involved. In studying these problems, researchers used observations, interviews in real time during low-intensity periods of operation, and critical incidents. The major challenges identified were maintaining adequate situation awareness; communicating, coordinating, and accommodating priorities; planning under shifting constraints; and planning under uncertainty.

Military Systems. The design of military systems for use in planning, logistics work, and for training are complex systems. How are HCI principles and methodologies used in the design and development of such systems?

Case Study: Close Combat Tactical Trainer. Mastaglio and Williamson (1995) described a user-centered design and evaluation of the Close Combat Tactical Trainer (CCTT). This system is a distributed network of simulators and workstations with a virtual environment representing real-world terrain. The CCTT was originally targeted for installation at 32 sites throughout the world with 500 simulators and more than 300 workstations constituting the system. In addition to the normal types of interaction (screen and input devices), interactions in this environment

included users responding as if operating actual equipment. The researchers identified four types of interfaces needed:

- Crew stations that are simulator components
- Workstation interfaces
- Visual displays
- Control devices to move the user through the environment

Moreover, three different types of users were identified:

- Soldiers/trainees
- Leaders/trainers
- Contractor operators of the system

For each interface, the tasks performed were based on an assessment of the tasks targeted for training or the tasks that an operator had to perform. Each component in the overall system was designed and evaluated within an operational scenario. Using the results and feedback from these scenarios, the designs were modified. The number of user subjects can quickly become unworkable in a large, complex system, so it is necessary to identify representative users whose feedback is needed to ensure a workable system. Other design issues involved trade-offs in the realism needed in the virtual environment. The cost to produce a completely realistic system was unacceptable; however, it was necessary to identify the relevant portions of the environment where realism was critical to the training.

Case Study: CommandNet Tool. Adkins et al. (2001) worked with the U.S. Navy's Commander Third Fleet to use and evaluate collaborative technology during a 5-week multinational exercise involving 7 nations, more than 22,000 people, 50 ships, and 200 aircraft. The research team provided a collaborative environment both at sea and in an austere environment onshore to establish a forum to exchange relevant information between the military and the various humanitarian organizations. The CommandNet Tool was originally developed as an electronic log book, but the researchers found that it also served as a more general communication tool when radio communications could not be established. Because there were a number of jobs to be performed, including such basics as securing food, water, and shelter for a large number of "refugees," the most frequent users of the technology were those who were closely located with the support personnel for this technology and who could quickly provide assistance. The researchers also noted the need to identify collaborative requirements in humanitarian assistance efforts and to modify the CommandNet Tool to provide more situational awareness.

Case Study: Real-Time Collaboration Tools for Planning Activities. A scenario-based evaluation was conducted on a loosely integrated collaborative system (Damianos et al., 2000). The software was developed under military funding, so the scenario used was that of a humanitarian disaster relief mission. A framework for evaluating collaborative products had been developed as part of this program as well (Drury et al., 1999). Study participants used collaborative software to find an alternative

hospital and an evacuation route for patients who had to be transported from a damaged hospital.

One of the challenges for this type of evaluation was to design training that was sufficient to familiarize participants with the software but fit within a reasonable fraction of the time allocated to the experiment. Therefore, the scenario had to be iteratively revised to encompass only those features that could be trained rapidly. Furthermore, setting up collaborative software is a challenge. The original plan had been to run the experiment in a distributed environment, but firewalls and connectivity issues forced the researchers to "simulate" the distributed nature of the experiment.

HCI STANDARDS AND THE GOVERNMENT'S ROLE

Government agencies often participate in committees and organizations responsible for developing sets of standards. When legislation mandates the adherence to standards, government agencies are often given the task of providing the means for determining compliance.

Section 508 of the Federal Rehabilitation Act (2001) in the United States was discussed earlier in this chapter. As discussed in the section The Government As a Producer of Information, section 508 legislates that Federal agencies' electronic and information technology is accessible to people with disabilities. This section discusses other standards activities, both international and very specific activities for specialized software.

International Standards Organization

A number of standards organizations exist worldwide with the most widely recognized being the ISO (ISO Home Page, 2001). The ISO was founded in 1946 to promote the development of international standards and conformity assessments to facilitate the exchange of goods and services worldwide. ISO currently has members from more than 90 countries, each of which is represented by the national body "most representative of standardization in its country." The American National Standards Institute (ANSI) represents the United States on the ISO committee (American National Standards Institute, 2001). Countries with less fully developed national standards organizations participate as correspondent members.

In 1987, an ISO technical committee published the 9000 series standards on quality management and assurance. A number of these ISO standards are issued in conjunction with the International Electrotechnical Commission (IEC) and are designated accordingly in the numbering scheme. ISO standards also designate the various stages of development. *TS* denotes a technical specification that may develop later into a formal standard. *TR* indicates a technical report that will not be turned into a standard but contains information pertinent to the standard but of a different nature. Draft versions of standards, reports, and technical specifications are indicated by a *D* in the numbering scheme.

Table 40.2 contains a condensed version of the ISO specifications that are of particular interest to the HCI community.

TABLE 40.2. HCI Related ISO Standards

Standard	Topics Covered
ISO/IEC 9126 parts 1-4 ISO/IEC 9126-11 ISO 20282 ISO 9241-2	Software engineering, usability guidelines, quality in use, internal and external metrics, product quality, task requirements
ISO 9241, parts 3–17	Ergonomic requirements for visual display terminals
ISO 13406	Ergonomic requirements for flat panel displays
ISO/ICE 18019	User software documentation process
ISO/ICE 15910	Design guidelines for user documentation
ISO 13407	Humans-centered design process and usability
ISO TR 16982	Methods
ISO/IEC 10741-1	Dialogue interaction
ISO/IEC 11581	Icon symbols and fundtions
ISO 14915	Software ergonomics for multimedia user interfaces
ISO/IEC 14754	Pen-based interfaces
ISO/IEC 18021	User interfaces for mobile tools
ISO/IEC 10075-1	Ergonomic principles for mental workload
ISO DTS 1607	Guidance on accessibility

More detailed descriptions of these standards can be found in Stewart and Travis (chapter 51, this volume), Cakir and Dzida (1997), and at the Serco Usability Services Web Site (2001). The actual standards can be purchased from the ISO Web site.

The various standards come in several varieties. Some standards are guidelines that can be used to guide designers in their considerations of the users' task requirements, work environments, and interaction requirements during design and development of the user interface software. Other ISO standards can be categorized as ergonomic standards. These reflect the acceptable, current state of knowledge in ergonomics and as such should be independent of technical specifications. The ISO standards are developed by consensus of the members of the technical committee and are much less precise than technical standards that are usually agreed upon by participating industries to ensure interoperability and compatibility.

Currently, the European Union (EU) countries have legislation to implement the following three directives.

1. The Display Screen Equipment Directive legislated in 1990 specifies ergonomic requirements for requirements for workstation equipment. Suppliers must also be in compliance with the principals of software ergonomics and requirements for ease of use in the design of the user interface software.
2. The Machine Direction legislated in 1998 requires that machinery must comply with health, safety, and ease-of-use requirements.
3. The Suppliers Direction legislated in 1993 requires that technical specifications used for procurement by public access bodies must reference relevant standards adopted by the European Committee for Standardization (CEN).

Manufacturers desiring to sell their products in a global market must ensure that they are in compliance with the required standards. There are a number of assessment bodies approved by ISO to analyze and certify this compliance. As more and more companies become certified with respect to ISO standards, the marketplace will create demands for other companies to follow suit.

Standards for the Federal Aviation Administration

The Human Factors Division of the FAA provides human factors research for acquisition, certification, regulation, and standardization activities (FAA Human Factors Division, 2001). This is provided through university-sponsored research and the production of technical reports. In 1993, the U.S. Department of Transportation established Order 9550.8 that provided for the policy and coordination of human factors considerations into the FAA program via the Human Factors Coordinating Committee (HFCC). This act was in part based on an assessment of air traffic by the now-defunct Office of Technology Assessment (2001) in the United States. This report noted that long-term improvements will come primarily from human factors solutions. The duties of the HFCC were as follows:

- To conduct human factors research on existing systems and identify problems and solutions necessary for performance improvements
- To acquire the necessary human factors information needed for design and development of the systems
- To apply the necessary human factors information to the acquisition process.

Benefits from this were identified as increased efficiency and effectiveness, improved aviation safety, and a decrease in maintenance costs.

Voting Standards

In the United States, the Federal Election Commission is the agency overseeing the election process. However, states and localities hold the ultimate responsibility for determining the election processes used in their jurisdictions. The Office of Election Administration (under the Federal Election Commission) assists the states and localities in this effort. In 1990, the Office of Election Administration released voluntary standards for computer-based voting systems to aid the local authorities in establishing their voting practices. In 1965, the Voting Rights Act was passed, stating that ballots and other assistance had to be provided in the language of citizens from non-English speaking minorities. No other standards are currently mandatory.

However, a number of efforts are proposed that differ in the amount of standardization and control proposed for voting technology. Fischer (2001) summarized a number of these measures while the complete Congressional Record as well as information about current bills is available at the Library of Congress' Thomas site (Thomas Legislative Information, 2001).

There are currently five types of voting carried out in the United States: hand-counted paper ballots; mechanical lever machines; computer punchcards; marksense forms, which are optically scanned; and direct recording electronic machines. Punchcards are the most common but are on the decline as the machines needed to use these are no longer being manufactured. Marksense forms are the second most frequently used but are increasing in use. Direct recording electronic machines are also on the rise. Internet voting is also being considered.

For each technology, there are three issues to consider: errors made in the process of voting; errors made in the process of counting; and security, privacy, and fraud possibilities. The U.S. 2000 presidential election illustrated errors with punch card technology. Two other issues that are not directly related to the technologies but must be considered are flexibility and cost. Elections have several parts: issues and candidates that are voted on nationally, statewide candidates and issues, and local candidates and issues. If technology is standardized, including ballot design, is it flexible enough to be customized to any region in the country? Cost is also an issue because many smaller communities lack the funds necessary to purchase new voting technologies.

THE GOVERNMENT AS A SOURCE OF FUNDING FOR HCI RESEARCH

Another aspect of government involvement in HCI is the funding of research in this area. In the United States, agencies such as the National Science Foundation (NSF, 2001) fund HCI research that is of general interest, whereas other agencies fund research of a more specialized nature that is relevant to their agency's mission (e.g., Office of Naval Research, Defense Advanced Projects Research Agency, National Aeronautics and Space Administration). Many agencies hold workshops and sponsor studies to outline research needs in various areas. The discussion in this section includes summaries of reports from studies and workshops and some funding initiatives in different agencies. Although the research topics here are derived primarily from studies and funding programs in the United States, there are similar efforts in other areas of the world.

National Research Council Study: Every-Citizen Interface

The study commissioned by the National Research Council (1997) outlined identified three areas of concentration for research to produce every-citizen interfaces. To have interfaces for every citizen, it is necessary to identify new paradigms for human–machine interaction. The next generation Internet with increased capacity will soon be a reality. We will be able to take advantage of these capabilities to enrich the input and output mechanisms used in computer interfaces. In addition to the current types of input, speech, gestures, image recognition, handwriting, and optical character recognition should become commonplace. Control strategies for interaction should be expanded to include more natural forms of communication,

including speech, gesture, and manipulation of physical or virtual objects. Speech systems need to go beyond the simple spoken commands to mixed-initiative dialogues, with the user and the computer engaging in a dialog concerning a task. Similarly, output modalities should be enriched to include tactile and olfactory pathways and to support richer visualization environments including immersive and virtual reality environments.

The second area of concern was focused on component subsystems necessary for every citizen interfaces. The emphasis here is on human and organizational studies to determine current behaviors and the ways that technology can support them. Research on input technologies and output technologies are included in this recommendation. In addition the recommendation includes the necessity of developing modality and medium independence to facilitate the use of an individual system in a variety of situations by a highly diverse population of users.

The third recommendation of the committee is focused on system level research in assembling the various subsystem components. Systems should be developed according to theories of collaboration, communication, and problem solving. Studies on human–centered design methodologies and social science research to determine how well technology is serving the general public are needed. This should include studies and theories of collaboration and work practices in the Networked Information Infrastructure. The report recommends an iterative approach to this research, designing prototype systems based on developed theories, evaluating use, and refining the system based on user feedback.

The National Science Foundation

The Computer and Information Science and Engineering Directorate (2001b) of the NSF has three goals:

- Enable the United States to retain the number-one position in computing, communication, information science, and engineering.
- Promote an understanding of the principles and uses of advanced computing, communications, and information services in service to society.
- Contribute to a universal, transparent, and affordable participation in an information-based society.

Initiatives funded at NSF include Accessibility, Digital Libraries (cosponsored by the Defense Advanced Projects Research Agency, the National Library of Medicine, the Library of Congress, the National Aeronautics and Space Administration, and the National Endowment for the Humanities in cooperation with the Smithsonian Institution, the National Archives and Records Administration, and the Institute of Museum and Library Services), and more general HCI projects within the Information Technology Research program (ITR; National Science Foundation ITR Program, 2001). The ITR program encourages research in people and social groups interacting with computers, scalable information infrastructures and pervasive computing, information management, and systems design and implementation.

In addition, the Digital Government Initiative at NSF (National Science Foundation Digital Government Initiative, 2001) has the goal of broadly addressing the potential improvement of agency, interagency, intergovernmental operations, and government–citizen interactions. Funding initiatives in this area cover such topics as statistics, crisis management, spatial data, community planning, natural resources, and environmental concerns. The technologies being investigated include human–computer interaction, metadata creation, visualization of statistical data, and data mining. Although research is certainly needed in the area of digital government, other barriers must also be removed before practices of digital government will be easily adopted. These include lack of funding for agencies, system integration problems, legacy systems, and the 1-to 3-year planning horizon of most government agencies. In May 2000, the NSF Science Foundation announced a partnership to advance digital government. The DigitalGovernment.Org partnership brings together computer science researchers and federal, state, and local agencies to expand and improve online government services. This new consortium will support researchers in digital government by providing insights into agency needs and acting as a facilitator for technology transfer (National Science Foundation, 2001a).

The European Union Research Initiative

In October of 2000, the European Commission adopted guidelines for guiding research in that community during the 2002–2006 time frame (European Commission, 2000). This initiative focuses on areas in which community attention can provide a "European added value." Funding is to encourage larger projects of longer duration and to strengthen the social dimensions of science. Of particular interest to the HCI community is the need for research to develop an information society.

CHI'99 Research Agenda Workshop

The ACM SIGCHI (Association for Computing Machinery Special Interest Group in Computer Human Interaction) community conducted a deliberative process involving a high-visibility committee, a day-long workshop at CHI'99 conference and a collaborative authoring process. A report was produced based on these activities (Scholtz et al., 1999). The introduction to the report (p. 13) states:

A major research program in Highly Effective Human–Computer Interaction will make the next generation of user interfaces dramatically more useful, usable, and universal. More useful systems will contribute to societal goals of providing quality services, improving education, and fostering a strong economic climate. More usable systems will result in safer, more reliable and more satisfying systems to use, as well as systems that require less maintenance. Universal interfaces will enable increased participation and success for all citizens: novices and experts, young and old, men and women, rich and poor.

Workshop participants identified the following research agendas for each of the areas:

- *Useful*. Research is needed to study, analyze, and reflect on the use of technology, and to incorporate these findings into principles for design and evaluation. The three goals of this research should be to understand the factors that make technology useful, to understand the role of pervasive information technology, and to produce methodologies and metrics to measure and predict the utility of new work practices.

- *Usable*. Good interfaces are difficult to build. Research to support the development of good interfaces should develop engineering methodologies for integrating design, development, and usability evaluation methodologies into the software engineering process and develop user-interface architectures that better support the building of good interfaces.

- *Universal*. Research is urgently needed to reduce the gaps and digital divides through accommodation of diverse abilities and support for mobility and disabling circumstances.

DISCUSSION AND FUTURE DIRECTIONS

The Government As a Provider of Information

As more government services go online, two issues will be of concern: developing interfaces for every citizen and accessibility issues. Web sites and services must be available to a wide range of users with diverse educations, computer experiences, languages, and physical abilities. Moreover, citizens will be accessing these services from a wide range of computing devices. Not only do new handheld technologies need to be accommodated, but so do older, slower computers with limited bandwidth connections. Citizens can easily feel disenfranchised if they are unable to participate in government activities due to accessibility difficulties. Guidelines are emerging for Web sites for accessibility. Government agencies need to be particularly careful to adhere to good design principles and to conduct usability tests with representatives of the target population. In the United States, accessibility is currently legislated for governmental Web sites and guidelines for compliance have been established. It will be important to ensure that these guidelines are kept up to date as technology changes. Moreover, studies should be undertaken to ensure that the current guidelines are sufficient to ensure accessibility.

More research will be needed to facilitate access to large databases and to help users locate relevant information when searching content with unfamiliar terminology. Meta-data, federations of repositories, and automatic tagging of content are topics where HCI expertise is needed. Visualization techniques and other presentation mechanisms are needed to understand complex statistical and tabular data.

Although security and privacy are not normally considered to be HCI related issues, e-government and e-commerce efforts are going to require guarantees concerning information access. Citizens accessing private records over the Internet will want some indication concerning security. Internet users today have concerns with the amount and type of data that is being collected about them as they access Web sites. It is insufficient to merely state security policies or privacy polices. It will be necessary to find methods to continually display this type of information via the user interface (Diller & Masten, chapter 63, this volume).

The Government As a Consumer of Software

The case studies in this section point out the human considerations that need to be taken into account for user interfaces designed for specialized activities. With respect to collaboration activities, the study by Harper and Sellen (1995) pointed out that the type of data being shared was an indicator of whether a human was needed in the loop. A number of other cases studies pointed out the necessity of carefully examining existing work practices to develop requirements for the design of new systems. In the study on commodity trading (Parikh & Lohse, 1995), electronic support actually slowed down the trading activities and adaptability that face to face trading supported. In supporting specialized activities such as this, it is critical to understand what advantages the current processes support and to adequately support these advantages in an electronic medium. The implication is that users of the system must be part of the design team as their specialized knowledge is critical to achieving an effective system design.

Designers and customers need to understand that work practices can be changed by the user interface design. Therefore, it is imperative to have both management and users as representatives on the design teams and to understand and buy into any such changes. To design interfaces for groups performing specialized jobs, it is necessary to thoroughly understand the tasks that the users do, what in the current environment facilitates their jobs, and the obstacles that workers encounter (Scholtz & Salvador, 1998). It is also important to consider the information requirements of all parties when designing collaboration systems. Because many of these jobs are complex, studies must be conducted over reasonable periods of time to ensure that all aspects of the job have been observed and considered. A systematic method for documenting and sharing this information with team members is needed. A number of approaches to task analysis and contextual design are available and should be considered (Holtzblatt, chapter 49; Redish & Wixon, chapter 48, this volume).

The need for better education about usability principles, both on the part of customers and suppliers, is also a key. Customers need to understand that the initial cost for usability engineering can be recovered through increased productivity in the workplace. More studies of the costs and benefits of usability engineering should be performed. Methodologies such as the IUSR project for organizations that purchase software to identify ways to measure their own costs and benefits are needed (Bias et al., chapter 62, this volume).

New technologies and work practices necessitate new informational requirements, hence new displays must be provided, as in the case of the Free Flight Initiative for air traffic control. The variability across participants must also be examined to identify the nature of these discrepancies. Adaptable interfaces are one solution to accommodating differing user expertise, but in critical systems such as air traffic control it is essential to ensure that required information is always readily available.

Evaluation of large-scale system is an issue. How can usability tests be conducted for software used in military operations or in air traffic control? What factors from the complex situations in which these tools are used in must be preserved in the evaluation effort to ensure its validity? Methodologies for evaluating these systems is an area of research that should be pursued in the HCI community.

Another issue in government procurement is to ensure that usability of the software is taken into consideration. The IUSR project and the specification of human-centered design processes and usability requirements in RFP are two ways to accomplish this. These efforts are not unique to government agencies. Parallel efforts in industry and in governmental agencies can be valuable and can leverage off of each other.

HCI Standards and the Government's Role

Standards, such as those issued by the ISO, are currently legislated in some countries, hence many consumer products will adhere to these standards regardless of the country in which it is being sold. It is likely that more countries will adopt legislation concerning these standards once studies can be conducted on the benefits and costs of the current efforts of adherence and compliance.

New efforts in standardization may come from accessibility issues as well as from new technological concerns. Issues concerning voting standards within the United States were discussed in the section in this chapter. As information technologies become more widely available for mobile use, many concerns are being raised about the use of these technologies (cell phones and navigation systems) while driving. Green (chapter 44, this volume) discussed HCI in driving. The U.S. Department of Transportation maintains a Web site to facilitate the interchange of information and opinions on issues such as this (U.S. Department of Transportation, 2001). Assuming that such technologies are found to be distracting, then the question becomes one of regulation. Should the use of these technologies be regulated through design of the devices themselves or through regulation of driver use and behavior? If the devices can be designed to minimize distractions and ill-advised behaviors, then standards can be created to ensure that all such devices comply to these standards. The other approach is to use safety education and awareness and possible regulations to control the use of these devices such as the use of seatbelts while driving today. To study such topics, HCI researchers have some interesting challenges, including these four posed on the Department of Transportation's Web site:

- How can driver distraction be safely and systematically studied during normal driving? Are simulators valid for use in these studies?
- What measures are meaningful for driver distraction? How do these measures correlate to incidents on the roads?
- What technologies, devices, and analysis methods can be used to capture measures of distraction?
- Can models be built to allow prediction of distractions from usage of a specific device?

The issues concerning standardization are complex and cover much territory outside of the current scope of HCI. For the HCI community, a concern with mandatory standards is the issue of compliance checking. If standards are mandated, then there must exist methods for assessing the technologies to ensure that the software has compliance with the standards. Standards must be developed with this aspect in mind. Furthermore, as new technologies become available, new ways to comply with the standards become possible. Standards can become outdated as well. Therefore standardization issues must be considered as an ongoing activity requiring periodic reassessment even after the initial development and implementation.

The Government As a Source of Funding for HCI Research

Software systems and HCI research has progressed to a level where we must go beyond the current scope of HCI and reflect and research the larger issues of how human interactions with technology are affecting work practices, personal lives, and society practices in general. Long-term studies of technology in use such as the studies conducted on HomeNet (Kraut, Scherlis, Mukhopadhyay, Manning, & Kiesler, 1996) found that teenagers were the most active Internet users in a home and that communication in the form of e-mail was the service most heavily used.

Studies of the Blacksburg Electronic Village (Kavanaugh, 1999) found that social networks were able to expand through online communication in the form of Web sites and group discussion forums. Social ties within the community network were strengthened and a increase in resource flow was seen. Members had an increased sense of belonging in the community and felt more closely associated with the organization. However, these effects were seen primarily when there was an existing need to communicate and the resources being exchanged were narrow in scope. The Internet itself does not cause people who are not already prepared to be active in the community to do so; it merely facilitates interaction for those who are already involved in the community.

If technology is going to be useful to society in general, it is essential that more studies of this nature be funded. As stated in Scholtz et al. (1999), we need to identify what makes technology useful.

Other Issues

Two other issues should be noted here that will become increasingly important in the near future: globalization and digital divide issues. Internationalization of software is already a topic within industry (Marcus, chapter 23, this volume). As the diversity of the citizens in our countries grows, we may see that government Web sites will need to accommodate a variety of languages. And as governments collaborate on more issues, software will need to support collaborations in multiple languages across multiple cultures. Moreover, current work practices must change radically to achieve globalization. As O'Hara-Devereaux and Johansen (1994, p. 36) stated, "historically, businesses have

viewed diversity as a chronic problem that had to be minimized and managed. The new challenge of globalization represents an opportunity to take a radically different approach: one that embraces diversity in ways that allow business to grow and profit from the many dramatically different cultural qualities that characterize most of our communities and organizations." Software systems for collaboration in a global environment should facilitate opportunities provided by diversity in the workplace.

As more government transactions become electronic, an issue that should concern all of us is the digital divide. It is not enough that the HCI community develops every-citizen interfaces if every citizen cannot gain access to the technologies needed to participate in digital government activities. Basic findings from studies on digital divide issues include the following:

- Internet access is strongly correlated with income and education (Coley, Cradler, & Engel, 1997).

- Race is also a factor. Whites are significantly more likely than African Americans to have a computer at home or to have computer access at work (Hoffman & Novak, 1998). Education does not explain race differences for home computer ownership. Income does not explain race differences for computer access in the workplace.

- Current students are more likely to have used the Web in the last 6 months than any other income or educational group (Hoffman & Novak, 1998).

- Once access is ensured, especially via computers in the home, race differences disappear. (Hoffman & Novak, 1998).

- Although cost is still a major barrier to acquiring a computer for Hispanics, a second barrier is the lack of information

and understanding of the benefits of technology (Cheskin Research Report, 2000).

- Barriers to community Internet access are not only due to lack of resources but include societal priorities, ambivalence about technology, and lack of political clout (Benton Foundation, 1998).

- Content on the Internet does not serve low-income Americans in four areas: local information, literacy barriers, language barriers, and cultural diversity (Children's Partnership, 2000).

Is a strategic national plan needed to close the digital divide? What types of research and initiatives can be undertaken to eliminate access inequalities? Will new technologies such as high-speed access and mobile wireless access widen or narrow the gap? A combination of funding initiatives to overcome cost barriers plus HCI research to overcome literacy and language barriers along with initiatives to develop relevant content and to provide cultural diversity must be undertaken jointly to close the digital divide. Providing access alone will not be sufficient. The relevant content must also be accessible regardless of language and education.

Clearly HCI community and government agencies would benefit by working more closely in the four roles discussed in this chapter: providing HCI methodologies for developing usable e-government, incorporating HCI into specialized government software, developing HCI standards and compliance testing, and identifying HCI research challenges. Challenging research problems are available in the government domain as well as opportunities to improve the efficiency and effectiveness of government for its citizens.

References

Adkins, M., Kruse, J., Damianos, L., Brooks, J., Younger, B., Rasmussen, E., Rennie, Y., Oshika, B., & Nunamker, J. (2001). Experience using collaborative technology with the United Nations and multinational militaries: Rim of the Pacific 2000 Strong Angel exercise in humanitarian assistance. *Proceedings of the Thirty-Fourth Annual Hawaii International Conference on System Sciences (HICSS-34).* Los Alamitos, CA: IEEE Computer Society.

American National Standards Organization. Home page. Retrieved July 2001 from http://www.ansi.org/

Benton Foundation. (1998). *Losing Ground Bit by Bit: Low-Income Communities in the Information Age.* Retrieved July 2001 from www.benton.org/Library/Low-Income/

Buie, E., & Kreitzberg, C. (1998, June). Tutorial: Using the RFP to get the usability you Need. *Proceedings of Usability Engineering 3: Practical Techniques for Government Systems.* National Institute of Standards and Technology and Usability Professionals' Association.

Cakir, A., & Dzida, W. (1997). International ergonomic HCI standards. In M. Helander, T. K. Landauer, & P. Prabhu (Eds.), *Handbook of human-computer interaction* (2nd ed., pp. 407–430). Amsterdam, The Netherlands: Elsevier Science.

Catarci, T., Matarazzo, G., & Raiss, G. (2000). Driving usability in the public administration: The UWG experience. *Proceedings of the Conference on Universal Usability* (pp. 24–31). New York: ACM Press.

Cheskin Research Report. (2000). *The digital worlds of the U.S. Hispanic.* Retrieved from www.cheskin.com

Children's Partnership. (2000, March). *Online context for low-income and underserved Americans.* Retrieved March 2001 from www.childrenspartnership.org

Clichelli, J. (1998). *Designing government systems for day one job performance.* Retrieved July 2001 from http://www.ctg.albany.edu/research/workshop/background.html

Coley, R., Cradler, J., & Engel, P. (1997). *Computers and classrooms: The status of technology in U.S. schools.* ETS Policy Information Report, ETS Policy Information Center, Princeton, N J. Retrieved July 2001 from ftp://etsis1.ets.org/pub/res/compclss.pdf

The Computer and Information Science and Engineering Directorate of the National Science Foundation. (2001). Home page. Retrieved July 2001 from http://www.cise.nsf.gov

Damianos, L., Drury, J., Fanderclai, T., Hirschman, L., Kurtz, J., & Oshika, B. (2000). Scenario-based evaluation of loosely integrated Collaborative Systems. *Proceedings of the 2000 Conference on Human Factors in Computing Systems (CHI 2000), Extended Abstracts* (pp. 127–128). New York, NY: ACM Press.

Department of Justice. (2001). *Section 508 guidelines*. Retrieved July 2001 from http://www.usdoj.gov/crt/508/web.htm

Duley, J., Galster, S., & Parasuraman, R. (1998). Information manager for determining data presentation preferences in future enroute air traffic management. *Proceedings of the Human Factors and Ergonomics Society 42nd Annual Meeting* (pp. 47–51).

Drury, J., Damianos, L., Fanderclai, T., Kurtz, J., Hirschman, L., & Linton, F. (1999). *Evaluation Working Group: Methodology for evaluation of collaborative systems*, v. 4.0. Retrieved from http://www.nist.gov/nist-icv

E-government Vision for New Zealand, State Service Commission. (2001). Retrieved March 2002 from http://www.e-government.govt.nz/eprogramme/vision.asp

Ellis, J., Rose, A., & Plaisant, C. (1997). Putting visualization to work: Program finder for youth placement. *Proceedings of 1997 Conference on Human Factors in Computing Systems (CHI '97)* (pp. 502–507). New York: ACM Press.

Estonia Government. (2001). Home page. Retrieved July 2001 from http://www.riik.ee/

European Commission. (2000). Research Guidelines. Retrieved July 2001 from http://europa.eu.int/comm/research/nfp.html

Federal Aviation Administration. (2001). Human Factors Division home page. Retrieved July 2001 from http://www.hf.faa.gov

Federal Information Technology Accessibility Initiative. (2001). Home page. Retrieved July 2001 from http://www.section508.gov/aboutfitai.html

Federal Rehabilitation Act, §508. (2001). Retrieved June 25, 2001, from http://www.section508.gov

First Gov. (2001). Home page. Retrieved July 2001 from www.firstgov.gov

Fischer, E. (2001). *The Congressional Research Service Report: RL30773: Voting technologies in the United States: Overview and issues for congress*. Retrieved March 21, 2001, from http://www.cnie.org/nle/rsk-55.html

Free Flight Initiative. (2001). Home page. Retrieved July 2001 from http://ffp1.faa.gov/home/home.asp

Harper, R., & Sellen, A. (1995). Collaborative tools and the practicalities of professional work at the international monetary Fund. *Proceedings of the 1995 Conference on Human Factors in Computing Systems* (pp. 122–129). New York: ACM Press.

The Hart-Teeter Report. (1999). Retrieved July 2001 from http://www.excelgov.org/egovpoll/

Hoffman, D., & Novak, T. (1998, April 17). Bridging the Racial Divide in the Internet. *Science* (pp. 390–391).

International Organization of Standardization. (2001). Home page. Retrieved July 2001 from http://www.iso.ch/iso/en/ISOnlin.frontage

Industry Usability Reporting Project. (2001). Home page. Retrieved July 2001 from http://www.nist.gov/iusr

Johnson, W. (1998). Issues and concerns in the design of cockpit displays of traffic information. *Proceedings of the Human Factors and Ergonomics Society 42nd Annual Meeting* (pp. 40–41). Santa Monica, CA: Human Factors and Ergonomics Society.

Jorna, G., Wouters, M., Gardien, P., Kemp, H., Mama, J., Mavromati, I., McClelland, I., & Vodegel Matzen, L. (1997). The multimedia library: The center of an information rich community. *Proceedings of the 1997 Conference on Human Factors in Computing Systems (CHI '97)* (pp. 510–517). New York: ACM Press.

Kambil, A., & Ginsburg, M. (1996). *Public access web information systems: Lessons from the internet EDGAR project*. Retrieved from http://www.ctg.albany.edu/research/workshop/background.html

Kavanaugh, A. (1999). *The impact of computer networking on community: A social network analysis approach*. Paper presented at the Telecommunications Policy Research Conference. Retrieved March 2002 from http://www.bev.net/project/research/TPRC.UserStudy.Kavanaugh.pdf

Knecht, W. (1998). Requirements for automated collision-avoidance systems for free flight. *Proceedings of the Human Factors and Ergonomics Society 42nd Annual Meeting* (pp. 42–46). Santa Monica, CA: Human Factors and Ergonomics Society.

Kraut, R., Scherlis, W., Mukhopadhyay, R., Manning, J., & Kiesler, S. (1996). The HomeNet field trial of residential internet services. *Communications of the ACM, 39*(12), 55–63.

Lewis, J., Henry, S., & Mack, R. (1990). Integrated office software benchmarks: A case study. In D. Diaper, D. Gilmore, G. Cockton, & B. Schackel (Eds.), *Human-computer interaction, INTERACT 90* (pp. 337–343). Amsterdam: North Holland.

Mastaglio, T., & Williamson, J. (1995). User-centered development of a large-scale complex networked virtual environment. *Proceedings of the 1995 Conference on Human Factors in Computing Systems* (pp. 546–552). New York: ACM Press.

MetaData Research Program. (2001). Retrieved July 2001 from http://www.sims.Berkeley.edu/research/projects/metadata

Moore, G. (1991). *Crossing the chasm*. New York: Harper Business.

Morphew, M., & Wickens, C. (1998). Pilot performance and workload using traffic displays to support free flight. *Proceedings of the Human Factors and Ergonomics Society 42nd Annual Meeting* (pp. 52–56). Santa Monica, CA: Human Factors and Ergonomics Society.

National Committee for Information Technology Standards. (2001). Home page. Retrieved July 2001 from http://www.ncits.org

National Research Council. (1997). *More than screen deep: Toward every-citizen interfaces to the nation's information infrastructure*. Washington, DC: National Academy Press.

National Science Foundation, Digital Government Initiative. (2001). Retrieved July 2001 from http://www.dli2.nsf.gov

National Science Foundation. (2001a). DigitalGovernment.Org. Retrieved July 2001 from http://www.diggov.org

National Science Foundation. (2001b). Home page. Retrieved July 2001 from http://www.nsf.gov

National Science Foundation, Information Technology Research Program. (2001). Retrieved July 2001 from http://www.itr.nsf.gov

O'Hara-Devereaux, M., & Johansen, R. (1994). *GlobalWork: Bridging distance, culture & time*. San Francisco: Jossey-Bass.

Obradovich, J., Smith, P., Denning, R., Chapman, R., Billings, C., McCoy, E., & Woods, D. (1998). Cooperative problem-solving challenges for the movement of aircraft on the ground. *Proceedings of the Human Factors and Ergonomics Society 42nd Annual Meeting* (pp. 57–61). Santa Monica, CA: Human Factors and Ergonomics Society.

Office of the e-Envoy, United Kingdom. (2001). Home page. Retrieved July 2001 from http://www.e-envoy.gov.uk

Office of Technology Assessment in the United States. (2001). Retrieved July 2001 from http://www.wws.Princeton.edu/~ota/

Pankoke-Babatz, U., Mark, G., & Klockner, K. (1997). Design in the POLiTeam Project: Evaluating user needs in real work practice. *Proceedings of Designing Interactive Systems* (pp. 277–287). New York: ACM Press.

Parikh, S., & Lohse, G. (1995). Electronic futures markets versus floor trading: Implications for interface design. *Proceedings of the 1995 Conference on Human Factors in Computing Systems (CHI'95)* (pp. 296–303). New York: ACM Press.

Perera, R. (2000). *Tiny Estonia sets pace in e-government*. International Data Group Report. Retrieved November 20 from http://www.idg.net/crd_government_292948.html

Plaisant, C., Marchionini, G., Bruns, T., Komlodi, A., & Campbell, L. (1997). Bringing treasures to the surface: Iterative design for the

Library of Congress National Digital Library Program. *Proceedings of the 1997 Conference on Human Factors in Computing Systems* (pp. 518-525). New York: ACM Press.

PTI/ICMA. (2001). *2000 Electronic Government Survey. Is Your Local Government Plugged In*? Retrieved from http://www.pti.org/docs/e-gov2000.pdf

Public Technology. (2001). Home page. Retrieved July 2001 from www.pti.org

Scholtz, J., Muller, M., Novick, D., Olsen, D., Shneiderman, B., & Wharton, C. (1999). A research agenda for highly effective human computer interaction: Useful, usable, and universal. *SIGCHI Bulletin*. Retrieved July 2001 from http://www.acm.org/sigchi/bulletin/1999.4/

Scholtz, J., & Salvador, T. (1998). Systematic creativity: A bridge for the gaps in the software development process. In L. Wood (Ed.), *User interface design: Bridging the gap from user requirements to design* (pp. 215-244). Boca Raton, FL: CRC Press.

Serco Usability Services. (2001). User-centered design standards. Retrieved July 2001 from http://www.usability.serco.com/trump/resources/standards.htm

Slaughter, L. (2001). Interfaces for understanding: Improving access to consumer health information. *Extended Abstracts, Conference on Human Factors in Computing Systems* (pp. 89-90). New York: ACM Press.

Thomas Legislative Information on the Internet. (2001). Home page. Retrieved July 2001 from http://thomas.loc.gov/

UK Online Project. (2001). Home page. Retrieved July 2001 from http://ukonline.gov.uk

University of Maryland Human Computer Interaction Laboratory. (2001). *User Interfaces for the U.S. Bureau of Census: Online Survey Interfaces and Data Visualization*. Retrieved July 2001 from http://www.cs.umd.edu/projects/hcil/census/

U.S. Department of Transportation. (2001). *Driver distractions*. Retrieved July 2001 from http://www-nrd.nhtsa.dot.gov/departments/nrd-13/driver-distraction/links.htm

Web Accessibility Forum. (2001). Home page. Retrieved July 2001 from http://www.w3.org/WAI

Winkler, I., & Buie, E. (1995). HCI challenges in government contracting. *CHI '95 Workshop Report*. Retrieved July 2001 from http://www.acm.org/SIGCHI/bulletin/1995.4/buie.html#HDR0

·41·

HUMAN–COMPUTER INTERACTION
IN HEALTH CARE

François Sainfort and Julie A. Jacko
Georgia Institute of Technology

Bridget C. Booske
University of Wisconsin-Madison

INTRODUCTION

U.S. healthcare expenditures are expected to be over $1.5 trillion in 2002. Despite such large spending, many Americans remain uninsured and do not have access to health care services. Furthermore, although our country has the most formidable medical workforce in the world and develops and uses the most modern medical technologies, the World Health Organization (2000) recently rated the quality and performance of the U.S. health care systems worse than most of its counterparts in the Western world. Chassin, Galvin, et al. (1998) documented three types of quality problems: overuse, underuse, and misuse. The results of an extensive review of more than 70 publications covering years 1993 through 2000 provide "abundant evidence that serious and extensive quality problems exist throughout American medicine resulting in harm to many Americans." (Institute of Medicine, 2001, p. 24). In its first report, *To Err Is Human*, the Institute of Medicine (2000) reported serious and widespread errors in health care delivery that resulted in frequent avoidable injuries to patients. The Institute of Medicine (2001) suggests four key underlying reasons for inadequate quality of care in the U.S. health care system: (a) the growing complexity of science and technology, (b) the increase in chronic conditions, (c) a poorly organized delivery system, and (d) constraints on exploiting the revolution in information technology.

Reengineering the delivery of health care services through innovative development, application, and use of information and medical technologies can result in tremendous cost savings, improved access to health care services, as well as improved quality of life for all citizens. Health care professionals in the United States have recognized that both the information revolution and the biological revolution will offer tremendous opportunities—and challenges—for (re)designing the healthcare system of the future. They are aware of the need to utilize in better ways new information and communication technologies and incorporate computing power into care delivery and clinical practice. They are also aware that the widely publicized biological revolution (which includes both advances in genetics and advances in biomedical engineering) will soon bring a large number of screening and diagnostic tests as well as new treatment strategies and disease management tools. It is clear that the combination of biotechnology, computing power, information and communication technologies, distance technology, and sensor technology will make future delivery of health care in the United States unrecognizable from the care we deliver today.

However, unlike other parts of the American economy, the health care system has not yet embraced modern information and communication technology, which is often viewed as too expensive, unusable, and divergent from current practice (Schoen, Davis, Osborn, & Blendon, 2000). In addition, technologies do not typically come "ready-to-use" off the shelf, and it is extremely difficult, if not impossible, for health care provider organizations to have the range of expertise in-house that is needed to design, adapt, and implement technologies to meet an organization's needs. In fact, many health care organizations are often not even fully aware of their own needs, do not know which technologies are available for what, and do not know how modern information and communication technologies can be effectively used to improve and simplify care delivery. Most recent information technology adoptions among health care providers have been driven by federal and state regulations and requirements rather than well-recognized internal needs and growth opportunities.

Health care informatics is a field that can be widely defined as the generation, development, application, and testing of information and communication principles, techniques, theories, and technologies to improve health care delivery. It includes the understanding of data, information, and knowledge used in the delivery of health care and an understanding of how these data are captured, stored, accessed, retrieved, displayed, interpreted, used, and made more efficient. Although health care informatics intersects with the field of medical or bioinformatics, it is different in the sense that it focuses on health care delivery and hence is centered on the patient (or consumer), the clinician (health care professional or "provider"), and, more important, the patient–provider interaction. Human–computer interaction (HCI), from the perspective of both the patient/consumer as well as the provider, is essential to the success of "health care informatics." In this chapter, we first review the characteristics of the health care industry in the United States. We then review information systems used by consumers, patients, and providers and raise HCI issues and challenges associated with both perspectives. Then we propose a framework for evaluating healthcare applications and finally conclude with a discussion of future opportunities and challenges for HCI in health care.

CHARACTERISTICS OF THE HEALTH CARE INDUSTRY

The Health Care Industry

As mentioned in the Introduction, the Institute of Medicine (2001) put forth four key underlying reasons for inadequate quality of care in the U.S. healthcare system today. In addition, a growing trend toward consumerism is becoming a major force in shaping the future organization of the health care industry. These five trends, detailed in this section, are shaping the future of health care in the United States.

Complexity of Science and Technology. The sheer volume of new health care science and technologies—the knowledge, skills, interventions, treatments, drugs, devices—is large today and has advanced much more rapidly that our ability to use and deliver them in a safe, effective, and efficient way. Government as well as private investment in pharmaceutical, medical, and biomedical research and development have increased steadily. The health care delivery system has not kept up with phenomenal advancement in science and technology and with proliferation of knowledge, treatments, drugs, and devices. With current advances in genomics (offering promise in diagnosis and possibly treatment), sensor technologies (offering promise in automated detection, measurement, and monitoring), nanotechnologies (offering promise in diagnosis, treatment, and control), and information and communication technologies (enabling

remote delivery, telemedicine, e-health, and patient empowerment), the complexity of science and technology in healthcare is only going to increase.

Chronic Conditions. As noted by the Institute of Medicine (2001), "because of changing mortality patterns, those aged 65 years and older constitute an increasingly large number and proportion of the U.S. population" (p. 26). As a consequence, there is an increase in both the incidence and prevalence of chronic conditions (defined as conditions lasting more than 3 months and not self-limiting). Hoffman, Rice, and Sung (1996) estimated that patients with chronic conditions make up 80% of all hospital bed days, 83% of prescription drug use, and 55% of emergency room visits. Compared with acute illnesses, effectively treating chronic conditions requires disease management and control over long periods of time, collaborative processes between providers and patient, as well as patient involvement, self-management, and empowerment.

Organization of the Delivery System. The health care delivery system in the United States is a highly complex system that is nonlinear, dynamic, and uncertain. The system is further complicated by a large number of agents, multiple stakeholders, each with multiple, sometimes conflicting, goals, aspirations, and objectives. As a result, the entire system leads to a lack of accountability; it has frequently misaligned reward as well as incentive structures, and it suffers from inefficiencies embedded in multiple layers of processes. The health care "product" or "service" is often ill-defined or difficult to define and evaluate. The processes involved in delivering health care services are complex, ill-specified, and difficult to measure, monitor, and control. Health outcomes are difficult to measure, manage, and analyze. The system experiences growing cost pressures, faces potential insurance premium increases, and is extremely fragmented. Wagner, Austin, and Von Korff (1996) identified five elements needed to improve patient outcomes in a population increasingly afflicted by chronic conditions:

1. Evidence-based, planned care
2. Reorganization of practices to meet the needs of patients who require more time or resources and closer follow-up
3. Systematic attention to patients' need for information and behavioral change
4. Ready access to necessary clinical knowledge and expertise
5. Supportive information systems

Regarding the final point, the Institute of Medicine (2001, p. 30) pointed to the fact that

health care organizations are only beginning to apply information technology to manage and improve patient care. A great deal of medical information is stored on paper. Communication among clinicians and with patients does not generally make use of the Internet or other contemporary information technology. Hospitals and physician groups operate independently of one another, often providing care without the benefit of complete information on the patient's condition or medical history, services provided in other settings, or medications prescribed by other providers.

Information Technology. The revolution in information technology holds great promise in a number of areas for consumers, patients, clinicians, and all organizations involved in the delivery of health care services. A recent report by the National Research Council (2000) of the National Academies identified six major information technology applications domains in health care: consumer health, clinical care, administrative and financial transactions, public health, professional education, and research. Although many applications are currently in use (such as online search for medical information by patients), other, such as remote and virtual surgery and simulation of surgical procedures, are in early stages of development (Institute of Medicine, 2001). The Internet (and intranets) are a driving force of recent changes in the information technology landscape, but not all health care applications are Web-based. Many applications (administrative billing systems, computerized physician order entry systems, etc.) remain on legacy systems, often built around older mainframe systems.

With respect to consumers/patients and providers, the Committee on the Quality of Health Care in America identified five key areas in which information technology could contribute to an improved delivery system (Institute of Medicine, 2001):

1. Access to medical knowledge base
2. Computer-aided decision support systems
3. Collection and sharing of clinical information
4. Reduction in medical errors
5. Enhanced patient and clinician communication

Consumerism. The Internet and other developments in information and communication technologies are contributing to greater consumerism with stronger demands from individuals for information and convenience. People are more demanding; they want timely and easy access to medical information, the latest in technology, and the latest in customer service. Patients are starting to have access to tools that can lead to empowerment and shared decision making regarding their own health care.

There are, however many technical, organizational, behavioral, and social challenges and barriers to greater use of information technology. Technological challenges include the design of optimal, effective, flexible HCIs but also issues of privacy and security of information. Those are discussed in the following section.

The Health Care Regulatory Environment

As Kumar and Chandra (2001) mentioned, the health care industry has unique legislative challenges. Among them, two in particular have implications on the field of health care informatics: the Health Insurance Portability and Accountability Act (HIPAA) and health information and other business data security. The HIPAA was passed and signed into law in 1996. It is designed to improve the portability of health insurance coverage in the group and individual markets, limit health care fraud and abuse, and simplify the administration of health insurance. The act has serious, impending implications for health care providers and information

managers. Of all its mandates, administrative simplification is perhaps the most critical for health care information managers, who will soon be faced with everything from establishing standardized financial and clinical electronic data interchange (EDI) code sets to adopting, assigning, and using unique numerical identifiers for each health care provider, payer, patient, and employer. Both HIPAA and the increasingly growing role of the Internet contribute to creating an even greater and critical concern for data security and privacy.

HIPAA contains five key titles that apply to every provider, payer (including self-insured employers), and health care clearinghouses. Title I pertains to health care access, portability, and renewability, Title II relates to health care fraud and abuse, Title III addresses tax-related provisions and medical savings accounts, Title IV discusses the application and enforcement of group health plan requirements, and finally, Title V relates to revenue offsets.

Currently, the health care industry is focusing its efforts on Title II of the act, also known as Administrative Simplification. Unlike past legislation that has imposed regulations on only Medicare and Medicaid, all health care providers, payers, and clearinghouses must meet HIPAA Administrative Simplification regulations within 26 months of each subsection's finalization. There are four subsections to Title II: transaction standards, coding sets, patient privacy, and security.

Transaction Standards and Coding Sets. The Department of Health and Human Services has adopted national standards for electronic administrative and financial health care transactions. This is one of the most positive attributes of Title II because it will eliminate the conflicting transaction standards, coding sets, and identifiers used by the various players in the industry. There are nine standard transaction definitions defined that must be implemented by October 2002 (2003 for businesses with fewer than 50 employees). The nine standards relate to enrollment, referrals, claims, payments, eligibility for a health plan, payment and remittance advice, premium payments, health claim status, and referral certifications and authorizations. By developing national standards, it is anticipated that EDI of health care data can significantly reduce administrative costs. The Workgroup on Electronic Data Interchange (2000) estimated that EDI could reduce administrative costs by $26 billion per year by streamlining precertification, enrollment status, and reimbursement processes. Code sets have also been defined in which variations are not permitted. Code Sets are based on the following standards:

ICD-9 Volumes 1 and 2: diagnosis coding
ICD-9 Volume 3: inpatient hospital service coding
CPT 4: physician service coding
CDT-2: dental service coding
HCPCS: other health-related coding
DRG: diagnosis-related groups
NDC: national drug coding

HIPAA also establishes unique identifiers for health care providers, health plans and payers, employers, and eventually patients and individuals. Currently the patient/individual identifier is pending because no consensus could be reached.

Privacy and Security. Administrative simplification also addresses patient privacy and security. There are privacy standards for disclosure of patient identifiable information (including demographic data), training of health care workforce, individual's rights to see records, procedures for amending inaccuracies in medical records, maintenance of privacy when patient information is exchanged between business associates, designate a privacy officer, procedures for complaints, sanctions for infractions, duty to mitigate, and document compliance. As a component of disclosure, covered entities (providers, payers, and clearinghouses) must make all reasonable efforts not to use or disclose more than the minimum amount of protected patient information necessary to accomplish the intended purpose of the use of disclosure.

Security standards require establishment of administrative procedures (policies and procedures), physical safeguards (physical access to computers), technical security (individual and network computer access), and electronic signature (optional, but if used, it must be digital). Implementation requirements of the privacy and security standards have undergone a tremendous amount of debate. Although most feel the intent of the standards are good, the cost implications have left many advocating for a longer implementation period. The Department of Health and Human Services estimates the total cost to implement Subtitle II to be approximately $17–$18 billion over the next 10 years. Some industry experts (Kennedy & Blum, 2001) purport implementation and maintenance costs to range from $50–$200 million for a large health care provide or payer.

HEALTH CARE INFORMATICS

As defined earlier, health care informatics comprises the generation, development, application, and testing of information and communication principles, techniques, theories, and technologies to improve the delivery of health care with a focus on the patient/consumer, the provider, and, more important, the patient–provider interaction. Although a number of systems using a number of platforms and technologies have already been developed and are currently being developed, the field itself is still in its infancy. In addition, current systems have been designed to fit within the existing health care delivery system and thus are only marginally or superficially impacting the way health care is being delivered. The true potential of health care informatics is yet to be experienced and will radically transform the way health care is being delivered and managed in the future. The following sections provide background information on current health care applications with an emphasis on two types of users: the consumer/patient and the clinician/provider.

Consumer Health Information

Consumer health information has been defined as "any information that enables individuals to understand their health and make health-related decisions for themselves or their families" (Harris, 1995, p. 23). Patrick and Koss (1995) listed a variety of organizations and entities that produce or disseminate

consumer health information including health-related organizations (involved in provision of, or payment for, health care services and supporting services), libraries, health voluntary organizations (e.g., American Heart Association, American Cancer Association, American Lung Association, etc., and 6000+ other health-interest societies), broadcast and print media, employers, government agencies, community-based organizations (e.g., churches, YMCA, agencies for the elderly), networked computer health information providers, what they called "virtual" communities. Methods of dissemination are diverse and include informal channels, printed text, broadcast electronic media, dial-up services (telephone), nonnetworked computer-based information (e.g., CD-ROM, Kiosk technology), and networked interactive computer-based information. Harris (1995) discussed a number of problems with health information including how to interpret conflicting or differing information, how to judge reliability, how to choose among many alternatives, and how to deal with the vast quantities of information, much of which is superficial or even inaccurate.

Interactive Health Communication

Print and broadcast dissemination of health information leads to a number of problems such as the timing relative to need, single directionality (difficulty of following up, clarifying, and understanding), timeliness and relevance vis-à-vis updates, and not being unique to individuals (Patrick & Koss, 1995). As Harris (1995) pointed out, electronic sources of health information have the potential to be more timely and complete than other media and can become more accessible to all citizens. Consequently, the use of interactive health communication or consumer health informatics has become increasingly popular. Unfortunately, the advantages have not been fully realized. The Science Panel on Interactive Communication and Health defined interactive health communication as "the interaction of an individual—consumer, patient, caregiver, or professional—with or through an electronic device or communication technology to access or transmit health information or to receive guidance and support on health-related issues" (Robinson, Patrick, Eng, & Gustafson, 1998). The panel identified six potential functions of interactive health communication applications: relay information, enable informed decision making, promote healthful behaviors, promote peer information exchange and emotional support, promote self-care, and manage demand for health services.

Ferguson (1997) referred to two types of consumer health informatics. Community consumer health informatics includes resources such as online networks, forums, databases, and Web sites that anyone with a home computer can access. Clinical consumer health informatics includes resources such as programs or systems developed by clinicians, system developers, or health maintenance organizations (HMOs), and provided to selected groups of members or patients. With respect to community consumer health informatics resources, there is a growing concern that the barriers to use of online health resources are becoming similar to those for actual health services. These barriers include cost, geography, literacy, culture, disability, and other factors related to the capacity of people to use services appropriately and effectively. Eng et al. (1998) urged public-and private-sector organizations to collaborate in reducing the gap between the "haves" and "have nots" for health information. They suggested the need for supporting health information technology access in homes and public places, developing applications for the growing diversity of users, funding research on access-related issues, enhancing literacy in health and technology, training health information intermediaries, and ensuring the quality of health information and support.

In a report on consumer health informatics, the U.S. General Accounting Office (1996) described three general categories of consumer health informatics. They include systems that provide health information to user (one-way communication), tailor specific information to the user's unique situation (customized communication), or allow the user to communicate and interact with health care providers or other users (two-way communication). The report also cited a number of issues of concern: access, cost, information quality, security and privacy, computer literacy, copyright, systems development (compatibility, infrastructure, and standardization), and the potential for information overload.

Consumer/Patient Web-Based Applications and E-Health

An estimated 70 million Americans seek health information online (Cain, Mittman, Sarasohn-Kahn, & Wayne, 2000), and there are 10,000 or more health-related Web sites (Benton Foundation, 1999). Consumers and patients desire a range of information and services from getting disease treatment information, obtaining report cards on physicians or hospitals, exchanging information with other patients, to interacting online with their physicians, to managing their own health benefits. The sheer volume of information on the Internet exceeds most expectations yet raises problems: efficient search for information, information retrieval, information visualization, human information processing, understanding, and assimilation. Authors of Internet information need to determine how best to structure the information for use by others. E-health can be described as the transition of healthcare processes and transactions into the Internet-delivered electronic superhighway. There are potential problems specific to patients seeking electronic health-related information (Sonnenberg, 1997): the lack of editorial control of information, conflict of interest for Web site sponsors, and unfiltered information presenting an unbalanced view of medical issues. However, referring primarily to self-help groups, Ferguson (1997) believed that the problem of bad medical information online is not so different from bad medical information at cocktail parties, in the tabloids, advertisements in magazines, and so on. In addition to the content of information, there are technological problems that may affect both access and use of health information on the Internet: slow modems, poor institutional Internet connections, firewalls that interfere with Internet traffic, malfunctioning message routers, and heavy Internet usage in the immediate geographic area (Lindberg & Humphreys, 1998). Most current concerns about interactive health communication center on the fact that these applications have the

potential to both improve health and to cause harm, thus highlighting the need to ensure their accuracy, quality, safety, and effectiveness (Robinson et al., 1998).

Provider Health Care Informatics

Providers and health care provider organizations have long used health information systems to support both administrative and clinical functions of health care delivery and management. However, despite the fact that health care is one of the most information-intensive industries, it has few state-of-the-art information management systems. Health care is fragmented, with hundreds of thousands of payers, hospitals, physicians, laboratories, medical centers, pharmacies, and clinics, each with its own legacy of systems, hardware, software, and platforms. Electronic data interchange and connectivity issues have become critical. Numerous information systems have been developed and implemented, the most noteworthy being integrated electronic medical records and computerized physician order-entry systems. The following describe a sampler of important clinical applications.

Electronic Medical Records. Electronic medical records (EMRs) are slowly being adopted by primary care practices (Ornstein & Bearden, 1994) in the hope that they will reduce costs and be more up to date, accessible, and modifiable than paper-based patient records (Wager, Lee, White, Ward, & Ornstein, 2000). All of this should ultimately improve patient care and reduce medical errors, because physicians will have more accurate patient records and better access to them.

Potential barriers to the adoption and effective use of EMR include cost, a lack of tested systems, problems with data entry, inexperienced vendors, confidentiality concerns, and security concerns (Wager et al., 2000). In addition, EMR may have barriers that are specifically related to the practice of medicine. For example, physician use of EMR while with patients could affect patient perceptions of quality of care or quality of physician–patient interactions. Results of studies examining this issue have not shown this to be the case, however (e.g., Legler & Oates, 1993).

A key factor that will contribute to whether EMR systems begin to become more rapidly adopted in primary care practices is whether physicians themselves perceive the systems to improve quality (of the medical records, patient care, and overall performance). A recent study examined this issue using qualitative methods in five community-based practices that had used EMR for at least 2 years and did not use a duplicate record system (Wager et al., 2000). Results indicated that many physicians and staff members perceived benefits, such as increased access, an ability to search the system, and improved overall quality of medical records. There were, however, several disadvantages mentioned, including the frequency of downtime and the time necessary to develop customized templates.

In a recent study examining the quality of work life of family physicians, Karsh, Beasley, Hagenauer, and Sainfort (2001) collected quantitative data about EMR to compare perceptions of medical records between physicians using EMR and those not using EMR. Specifically, they assessed whether the use of EMR was related to perceptions of improved quality of medical records. The results showed that physicians who used EMR perceived their medical records to be more up to date and accessible. Physicians who used EMR were also more satisfied with the overall quality of their medical records systems. On the other hand, there were no differences in perceptions of whether medical records could be modified to meet individual needs. This suggests that EMR can have positive impacts on medical records.

The results of Karsh et al.'s (2001) study support those of qualitative studies of physicians who use EMR. Wager et al. (2000) found that physicians in primary practice who had used EMR for at least 2 years believed EMR to have many benefits over paper-based systems. The benefits included increased access and availability of patient information to multiple users, the ability to search the system, improved overall quality of patient records, improved quality of documentation, increased efficiency, facilitated cross-training, and improved communication within the practice. Thus, it is clear that EMR has the potential to improve the quality of patient records and therefore, possibly, quality of care. To fully capitalize on such systems, however, they need to be designed to maximize usability, connectivity, and portability while guaranteeing privacy and security.

Computerized Physician Order-Entry Systems. Computerized physician order-entry systems have substantial potential for improving the medication ordering process because they enable physicians to write orders directly online. They ensure complete, unambiguous, and legible orders; they assist physicians at the time of ordering by suggesting appropriate doses and frequencies, by displaying relevant data to assist in prescription decisions, and by checking drugs prescribed for allergies and drug–drug interactions. As such, those systems are believed to potentially help reduce the incidence of medical errors in general, and medication errors in particular. For example, physician order-entry systems and automated medication ordering, dispensing, and administration processes have potential to reduce medication errors (California Health Care Foundation, 2001). In a study of the impact of computerized physician order-entry systems, Bates et al. (1999) found that such systems substantially decreased the rate of non-missed-dose medication errors. For computerized physician order-entry systems to be fully successful, however, physicians need to use them. This leads to issues of human–computer interface design, usability, and integration within the care delivery processes.

Patient Monitoring Systems. Gardner and Shabot (2001) defined patient monitoring as "repeated or continuous observations or measurement of the patient, his or her physiological function, and the function of life support equipment, for the purpose of guiding management decisions" (p. 444). Electronic patient monitors are used to collect, display, store, and interpret physiological data. Increasingly, such data are collected using newly developed sensors from patients in all care settings as well as in patients' own homes. Although such data can be extremely useful for diagnosis, monitoring, alerts, as well as treatment suggestions, the amount, diversity, and complexity of data collected presents challenges to human–computer interface design.

Imaging Systems. Imaging is a central element of the health care process for diagnosis, treatment plan design, image-guided treatment, and treatment response evaluation. Greenes and Brinkley (2001) noted that the proliferation in number and kind of images generated in health care led to the creation of a sub-discipline of medical informatics called "imaging informatics." In their review and summary of the field, they noted that

as processing power and storage have become less expensive, newer, computationally intensive capabilities have been widely adopted. Widespread access to images and reports will be demanded throughout healthcare delivery networks... We will see significant growth in image-guided surgery and advances in image-guided minimally invasive therapy as imaging is integrated in real time with the treatment process. Telesurgery will be feasible (Greenes & Brinkley, 2001, pp. 534–536).

They also highlighted that such ambitious evolution of imaging systems will be in part dependent on continued advances in user interfaces and software functionality.

Information-Retrieval Systems. Hersh, Detmer, and Frisse (2001) defined information retrieval as the "science and practice of indentification and efficient use of recorded media" (p. 539). In their review of the systems they address four elements of the information retrieval process: indexing, query formulation, retrieval, and evaluation and refinement. Indexing is the process by which content is represented and stored and query formulation is the process by which user information needs are translated in terms of an actionable query. Focusing on clinicians' use, Hersh et al. (2001) pointed out that the evolution of the Internet has posed new, so far unmet, challenges regarding the indexing of (essentially loosely and irregularly structured) documents. They call for a unified medical language system. Thinking of novice users such as consumers and patients, another significant challenge resides in query formulation because novice users generally have a difficult time expressing ill-defined information needs because of their lack of domain knowledge.

Other Clinical Decision-Support Systems. Musen, Shahar, and Shortliffe (2001) defined a clinical decision support system as "any computer program designed to help health professionals make clinical decisions" (p. 575). They characterize clinical decision support systems along five dimensions: system function, mode for giving advice, style of communication, underlying decision-making process, and human-computer interaction.

A large number of applications already exist and can be further developed. The opportunities are almost unlimited. Bates et al. (2001) proposed that appropriate increases in the use of information technology in health care, especially the introduction of clinical decision support systems and better linkages in and among systems, could result in substantial reduction in medical errors. However, as noted by Musen et al. (2001, p. 587), "systems can fail... if they require that a practitioner interrupt the normal pattern of patient care." Thus, new technologies (mobile devices, wireless networks, and distance communication technologies) as well as novel human–computer interfaces (based on speech, gestures, and virtual reality) offer huge potential to maximize usability and permit seamless integration of clinical decision-support systems within complex, dynamic work processes.

EVALUATING COMPUTERIZED HEALTH CARE APPLICATIONS

With the proliferation of Web-based health care applications available for consumers and patients, the issue of evaluating applications and guiding users in choosing the best applications has become extremely important. Although evaluation is obviously also important for applications targeted at health care professionals, the issue is not as critical because health care professionals are experts and can exercise their own judgment in the suitability and quality of applications designed to assist them in their work. Consumers and patients, on the other hand, have no or limited basis for exercising such judgment.

Concerned that the growth of the Internet was leading to too much health information with vast chunks of it incomplete, misleading, or inaccurate, Silberg, Lundberg, and Musacchio (1997) proposed four standards for Web sites:

1. *Authorship.* Authors and contributors, their affiliations, and relevant credentials should be provided.
2. *Attribution.* References and sources for all content should be listed clearly, and all relevant copyright information noted.
3. *Disclosure.* Web site "ownership" should be prominently and fully disclosed, as should any sponsorship, advertising, underwriting, commercial funding arrangements or support, or potential conflicts of interest. This includes arrangements in which links to other sites are posted as a result of financial considerations. Similar standards should hold in discussion forums.
4. *Currency.* Dates that content was posted and updated should be indicated.

Murray and Rizzolo (1997) pointed out that "somehow, just the fact that information is traveling quickly through space and being presented on the computer screen lends it an air of authority which may be beyond its due. Sites with official sounding names can dupe the inexperienced or uncritical into unquestioned acceptance of the content." In addition to the standards identified by Silberg et al. (1997), Murray and Rizzolo (1997), and others have cited other criteria for evaluating Web sites. These include the authority of the author/creator, the accuracy of information and comparability with related sources, the workability (user friendliness, connectivity, search access), the purpose of the resource and the nature of the intended users, whether criteria for information inclusion are stated, the scope and comprehensiveness of the materials, and the uniqueness of the resource.

Jadad and Gagliardi (1998) identified a number of instruments used to provide external ratings of Web sites. These ratings are used to produce awards or quality ratings, provide seals of approval, identify sites that are featured as the "best of the Web" or "best" in a given category, or to declare sites as meeting quality standards. They attempted to determine which criteria were used to establish these ratings and to establish the

degree of validation of these rating instruments. However, few organizations listed the criteria behind their ratings, and none provided information on interobserver validity or construct validity. Jadad and Gagliardi (1998) also discussed whether it is desirable to evaluate information on the Internet because of concerns over freedom of expression, excessive regulatory control, and so forth, and whether it is possible to evaluate information on the Internet due to the lack of a gold standard for quality information and the controversy surrounding its definition.

They also pointed out that information on Internet is "different from that found in journals—information is produced and exchanged by groups of people (e.g., health professionals, consumers, vendors, etc.) using multiple formats (e.g., text, video, sound) modified at fast and unpredictable rate, and linked within a highly elaborate and complex network of Internet sites" (p. 611). They concluded that with respect to *external independent evaluation* of sites it is not clear whether evaluation instruments should exist in the first place; whether they measure what they claim to measure; whether they lead to more good than harm; and whether users may ever notice; or if they notice, whether they will ignore evidence in support or against desirability, feasibility or benefits of formal evaluations of health information on the Internet.

Robinson et al. (1998) focused more on internal evaluation of applications, that is self-evaluation by the sponsors or developers of interactive health communication applications. They pointed out a number of barriers to evaluating these applications. These include the fact that the media and infrastructure underlying applications is in a dynamic state; applications themselves change frequently; many applications are used in situations where a variety of influences on health outcomes exist, few of which are subject to easy assessment or experimental controls; developers lack familiarity with evaluation methods and tools; and developers often believe that evaluation will delay development, increase front-end costs, and have limited impact on sales.

Patrick and Koss (1995) suggested that the effectiveness of consumer health information should be measured by how rapidly and completely desired messages are communicated and how completely intended changes in behavior occur. Saying that Silberg et al.'s criteria (explicit authorship and sponsorship, attribution of sources, and dating of materials) are not enough, Wyatt (1997) provided far more specific direction for evaluating Web sites. He believes that evaluation of Web sites should go beyond mere accountability to assess the quality of their content, functions, and likely impact. Evaluating the content should include the following:

- Determining the accuracy of Web material by comparing it to the best evidence (e.g., for effectiveness of treatment, randomized trials; for risk factors, cohort studies; or for diagnostic accuracy, blinded comparisons of test with a standard).
- Determining timeliness by checking the date on Web pages, but recognizing that material may not have been current at that time, so there is a need for independent comparison with the most up-to-date facts.
- Determining if people can read and understand Web material and asking visitors to record satisfaction is unlikely to

reveal problems with comprehension because visitors may not realize they misunderstood something or they may blame themselves. While a minimum reading age for material for the general public is needed, it would be more accurate to ask users questions based on the web content.

Evaluating the functions of Web sites should include determining how easy it is to locate a site, how easy it is to locate material within the site, and whether the site is actually used and by whom. For those investing resources in a Web site, with respect to evaluating the impact, Wyatt (1997) suggested looking at the impact on clinical processes, patient outcomes, and cost-effectiveness compared with other methods of delivering the same information. He recommended using randomized control trials as the most appropriate method for determining impact. Finally, with respect to evaluation methodology, he pointed out the importance of choosing appropriate subjects (not technology enthusiasts) and the need to make reliable and valid measurements. As Wyatt (1997, p. 1880) believes, "Ideally, investigators would have access to a library of previously validated measurement methods, such as those used for quality of life. However, few methods are available for testing the effect of information resources on doctors and patients, so investigators must usually develop their own and conduct studies to explore their validity and reliability."

The Science Panel on Interactive Communication and Health (SciPICH) was convened by the Office of Disease Prevention and Health Promotion of the U.S. Department of Health and Human Services to examine interactive health communication technology and its potential impact on the health of the public. The panel comprised 14 experts from a variety of disciplines related to interactive technologies and health, including medicine, HCI, public health, communication sciences, educational technology, and health promotion. One of the products of the SciPICH products is an evaluation reporting template (Robinson et al., 1998) for developers and evaluators of interactive health communication applications to help them report evaluation results to those who are considering purchasing or using their applications. The template has four main sections: description of the application, formative and process evaluation, outcome evaluation, and background of evaluators. The panel defined the three different types of evaluation as follows:

- *Formative evaluation* is used to assess the nature of the problem and the needs of the target audience with a focus on informing and improving program design before implementation. This is conducted before or during early application development and commonly consists of literature reviews, reviews of existing applications, and interviews or focus groups of "experts" or members of the target audience.

- *Process evaluation* is used to monitor the administrative, organizational, or other operational characteristics of an intervention. It helps developers successfully translate the design into a functional application and is performed during application development. This commonly includes testing the application for functionality and also may be known as alpha and beta testing.

• *Outcome evaluation* is used to examine an intervention's ability to achieve its intended results under ideal conditions (i.e., efficacy) or real-world circumstances (i.e., effectiveness) and also its ability to produce benefits in relation to its costs (i.e., efficiency or cost-effectiveness). This helps developers learn whether the application is successful at achieving its goals and objectives and is performed after the implementation of the application.

Evaluating the effectiveness of Web-based applications whose stated purpose is to relay information and/or enable informed decision making is complicated by the fact that the "success" of these particular types of applications is (a) not necessarily always related to observable behaviors; (b) based on the quality *and* usability of the information within the application; and (c) a function of the application itself as well as the users. In terms of outcome evaluation applications, Robinson et al. (1998) gave examples of the types of questions that such evaluations should address:

• *How much do users like the application?*
• *How helpful or useful do users find the application?*
• *Do users increase their knowledge?*
• *Do users change their beliefs or attitudes (e.g., self-efficacy, perceived importance, intentions to change behavior, and satisfaction)?*
• *Do users change their behaviors (e.g., risk-factor behaviors, interpersonal interactions, compliance, and use of resources)?*
• *Are there changes in morbidity or mortality (e.g., symptoms, missed days of school or work, physiologic indicators)?*
• *Are there effects on cost/resource utilization (e.g., cost-effectiveness analysis)?*
• *Do organizations or systems change (e.g., resource utilization and effects on "culture")?*

However, for Web sites that are designed primarily to provide information or to enable informed decision making, only the first three questions clearly apply. Other potential outcomes related to change in behavior might be applicable depending on the nature of decisions made. Consequently, the evaluation of these types of Web guides needs to focus on the use of the guides (assuming that users who "like" an application will use it more than those who do not), the usefulness of the guides, the usability of the guides, the ability of the guide to increase knowledge, and the contribution of the guide to decision making. These elements can be integrated into a conceptual framework as shown in Fig. 41.1. Part of this framework draws from the Agency for Health Care Policy and Research (1997) report on consumer health informatics and patient decision making, as well as Sainfort and Booske (1996).

The framework suggests that to evaluate the effectiveness of Web-based health applications, at least three perspectives (at the bottom of the figure) can be taken individually or in combination: the consumer, the site sponsor, and outside experts. The framework posits that the characteristics of the system under evaluation primarily influence accessibility and usability of information. Then, accessibility, in conjunction with consumer/patient characteristics, influences actual access to information. In turn, the usability of this information, again in conjunction with consumer characteristics, will influence actual use of information by the consumer. Use of information is a complex construct. The framework emphasizes three main types of use of information: knowledge (inquiry, verify, learn augment, etc.), decision making, and behavior (intentions and actual behavior change). All three uses are generally interconnected, with behavior usually following knowledge or decision making (whether explicitly or implicitly). Also emphasized in the framework is the fact that "use" can be generic (common to most Web-based health applications) or specific (common to the application) and also that use of information accessed can be intended (i.e., the information was sought to accomplish a specific purpose or use) or emergent (i.e., a piece of information that is accessed triggers a new use). Finally, the framework shows, surrounding this entire process, that the context (the situation, time, place, organization, etc.) can influence all key elements: accessibility, access, usability, and use. Consumer evaluations are formed as a result of the experiences they have using the system over time. Expert evaluation usually involves assessment of the system itself and its content, as well as an assessment of its (anticipated) impact on users. Sponsor evaluation involves both consumer and expert evaluations as well as consideration of organizational resources expended in the design, development, implementation, and operation of the system.

It is important to differentiate information access and quantity from use and usefulness (Booske & Sainfort, 1998). Indeed, whereas many studies primarily address attempts to measure the "quantity" of information used, others have acknowledged the need to consider the "quality" of information. The introduction of the Web as a source of information has introduced a number of additional considerations in evaluation of the use and usefulness of information. The most common measure of information use on the Web is the number of "hits" or "visits" to a particular page. Most of these counts do not differentiate between multiple visits by a single user versus multiple users, nor do they consider the amount of time spent on a particular page (i.e., whether the page is merely used as a link to elsewhere or as a source of information itself or even whether the page was accessed inadvertently).

In describing three general approaches to information systems evaluation (i.e., the system's output, behavior, and architecture), Orman (1983) defined the quality and quantity of information as the relevant variables in defining an information system's output. However, he went on to say that both quantity and quality of information were "of considerable theoretical interest but of little practical value since neither can be defined or measured with acceptable precision." Orman defined the quality of information produced by a system in terms of "its contributions to the quality of the decisions it aids" (p. 310) and pointed out that this is "highly influenced by the style and the behavior of the information user and the state of the environment" (p. 311).

This ties in with the concept of the "value" of information, an important part of traditional decision and economic analysis. Within the field of decision science, the primary use of

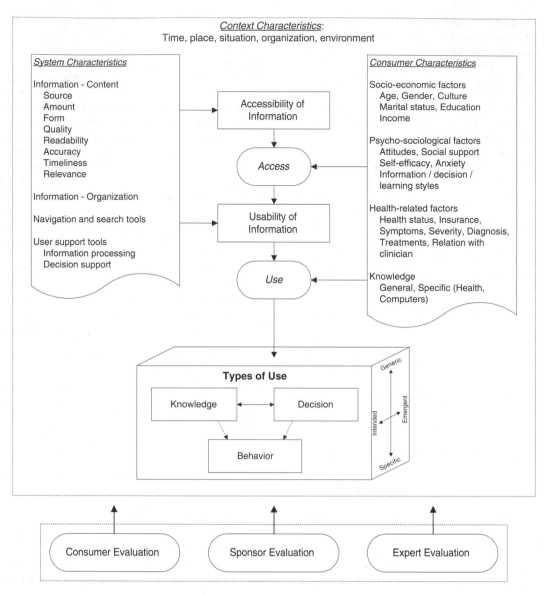

FIGURE 41.1. Conceptual framework.

information is to help make decisions. However, an examination of literature from other disciplines shows that there are other ways to conceptualize the value, and thus the effectiveness of information and its usefulness (Orman, 1983). Although it can be thought that the term *information usefulness* could be an analogue for value of information, within the HCI field, the concept of usefulness includes additional dimensions. The concepts of "perceived usefulness" and "usability" of information *systems* are both receiving increasing attention as more and more systems are developed for novice users. For example, Davis, Bagozzi, and Warshaw (1989) developed scales to measure perceived usefulness and ease of use to evaluate specific software in the work setting. They defined perceived usefulness as "a prospective user's subjective probability that using a specific application system will increase his or her job performance within an organizational context" (p. 983). Ease of use was

defined as "the degree to which the prospective user expects the target system to be free of effort" (p. 985). Using a 7-point Likert scale, potential users of software respond to six items addressing perceived usefulness and six to assess perceived ease of use.

Of particular relevance in considering the usability, and the potential usefulness, of information is the manner of display. Although technological advances continue to increase the number of possible methods for disseminating information, the primary output media for health care information are still print, video, and sound. Print information is often categorized as text, tables, graphs, and figures. There is a quite extensive body of literature comparing the effectiveness of displaying information in tables versus graphs (Jarvenpaa & Dickson, 1988, and Dahlberg, 1991, provide comprehensive summaries of this empirical work). Many of these studies rely on elicited "directed interpretations" through which subjects are given specific questions

about the information in tables and graphs. In contrast, a more recent study by Carswell and Ramzy (1997) elicited spontaneous interpretation of a series of tables, bar graphs, and line graphs to find out "what information subjects choose to take away from a display rather than their ability to extract information when promoted."

Within the context of decision making, Schkade and Kleinmutz (1994) found that different characteristics of information display affected aspects of choice processes. They found that organization of information (such as a matrix versus a list) influenced organization acquisition and form (numeric versus linguistic) influenced information combination and evaluation, whereas sequence had only a limited effect on acquisition. Johnson, Payne, and Bettman (1988) and others have also shown evidence of preference reversals as a result of different information displays.

Many of the issues surrounding the use of the Internet are issues of usability. Usability is a central notion of the HCI field. Simply put, Shackel (1991, p. 42) defined usability as "the capability to be used by humans easily and effectively," whereas his more formal operational and goal-oriented definition says,

for a system to be usable, the following must be achieved (p. 48). *Effectiveness*—the required range of tasks must be accomplished at better than some required level of performance (e.g., in terms of speed and errors); by some required percentage of the specified target range of users; within some required proportion of the range of usage environments
Learnability—within some specified time from commissioning and start of user training; based on some specified amount of user training and support; and within some specified relearning time each time for intermittent users
Flexibility—with flexibility allowing adaptation to some specified percentage variation in tasks and/or environments beyond those first specified
Attitude—and within acceptable levels of human cost in terms of tiredness, discomfort, frustration, and personal effort; so that satisfaction causes continued and enhanced usage of the system.

In an attempt to place system usability in relation to other system concepts, Shackel (1991, p. 50) suggested that

utility (will it do what is needed functionally?), *usability* (will the users actually work it successfully?) and *likeability* (will the users feel it is suitable?) must be balanced in a trade-off against *cost* (what are the capital and running costs and what are the social and organizational consequences?) to arrive at a decision about *acceptability* (on balance the best possible alternative for purchase).

These concepts are all relevant in a decision to purchase or accept an information system but may not be all relevant when the resource or product under consideration is information. As Orman (1983, p. 312) pointed out, the value of an information system is different from the value of its information content, "just as the value of a candy machine is different from the value of the candy it dispenses." Furthermore, usability of information is a necessary but not sufficient criterion for information to be useful.

In the context of informing healthcare consumer decisions, Hibbard, Slovic, and Jewett (1997) asked, How much information is too much? Their answer was that the critical element is the ability to interpret and integrate information items: integration is a difficult cognitive process. They suggest that for a consumer, more information is not necessarily better and that the simple provision is not sufficient when the information is complex. In complex decision situations, it is important to pay particular attention to human information processing capabilities and differences across individuals. For example, specifically addressing information presentation format, Togo and Hood (1992) found a significant interaction effect between gender and format. Personality differences have also been found to contribute to variation in information processing. The way people gather and evaluate information is at least in part based on their psychological type (Slocum & Hellriegel, 1983). Some people are driven to know details before decisions are made, whereas others feel more comfortable assuming what is unknown. Individuals vary on how they receive messages, seek information, organize information, and process information. One approach to categorizing cognitive style, the Myers-Briggs Type Indicator (MBTI), is based on Jung's typology (Myers, 1987).

Other models of learning styles can also be used to differentiate among individuals and their preferred methods or strategies for taking in and processing information. For example, Kolb's learning style model classifies people as having a preference for (a) *concrete experience* or *abstract conceptualization* (how they take information in) and (b) *active experimentation* or *reflective observation* (how they internalize information) (Kolb, 1984). The Hermann Brain Dominance Instrument (HBDI) classifies people in terms of their relative preferences for thinking that are based on the task-specialized functioning of the physical brain (i.e., left brain vs. right brain, cerebral vs. limbic) (Hermann, 1996). The Felder–Silverman Learning Style model classifies learners along five dimensions: sensing or intuitive, visual or verbal, inductive or deductive, active or reflective, and sequential or global (Felder & Silverman, 1988).

In addition to individual differences, other factors that can affect information processing reflect characteristics of the information itself (John & Cole, 1986), for example, information quantity, information source, information format (mode of presentation, organization, order), information complexity, nature of access (e.g., voluntary vs. mandatory), instructions in the use of the information, response formats (e.g., recognition, recall, judgment, choice), as well as the interface itself.

Understanding cognitive processes involved in accessing, processing, interpreting, and using health care information is critical to the successful design and implementation of Web-based health care applications. This particular point extends to applications targeted at providers and other health care professionals. In looking at the impact of computer-based patient record systems on data collection, knowledge organization, and reasoning, Patel, Kushniruk, Yang, and Yale (2000) indicated that exposure to computer-based patient record was associated with changes in physicians' information gathering and reasoning strategies. They concluded that such technology can have profound influence in shaping cognitive behavior. In such systems, the human–computer interface itself can have a strong influence on information gathering, processing, and reasoning strategies. Recently, Patel, Arocha, and Kaufman (2001)

surveyed the literature on aspects of medical cognition and suggest that "cognitive sciences can provide important insights into the nature of the processes involved in human-computer interaction and help improve the design of medical information systems by providing insight into the roles that knowledge, memory, and strategies play in a variety of cognitive activities" (p. 324).

FUTURE OPPORTUNITIES AND CHALLENGES

Information and communication technologies have only begun to impact the health care industry. Regarding the use of the Internet, a number of applications will be at the leading edge. According to Mittman and Cain (1999), those will include consumer health information services, online support groups for patients and caregivers, health care provider information services, provider-patient email, communications infrastructure and transaction services, and electronic medical records. At the same time, they mention that a number of barriers will impede or slow down the development of the Internet in health care: security concerns, weaknesses inherent to Web interfaces (especially browsers, search engine technology, and the inability to interact with legacy systems), mixed or lack of quality of information, physician ambivalence, disarray of current health care information systems, lack of resources for Web development, and lack of unified standards for electronic communications and transactions.

For information technology to fully impact health care delivery with a focus on both patients and providers, these barriers need to be overcome. We believe that the greatest benefits will be reached in the short term by focusing on the following five areas.

Supporting and Enhancing the Patient–Provider Interaction

The Internet will become a critical element of the physician-patient encounter. Organizational Web sites will evolve from publishing generic consumer content to providing personalized, online services to all consumers (individuals, patients, providers, etc.). The Internet needs to provide for technology that truly supports two-way interaction between patients and providers and that has had a direct impact on care delivery.

Effective technologies need to be developed to fully support, enhance, and extend the patient–physician interaction so as to increase the efficiency of care delivery, increase the quality of care received by patients, as well as increase the effectiveness of the work performed by the physician. This latter point is important since a recent study by Linzer et al. (2000) showed that time stress, defined as reports by the physician that they needed more time for patients than they were allotted, was significantly related to burnout, job dissatisfaction, and patient care issues. This study was performed with a national sample ($n = 5,704$) of physicians in primary care specialties and medical and pediatric subspecialties. It demonstrates that the ability to spend sufficient and quality time with each patient is critical to ensure long-term job satisfaction, avoid physician burnout, and increase quality of patient care.

In developing such technologies, as we mentioned earlier, one would need a full understanding of the various cognitive processes involved in information gathering, knowledge acquisition, and organization, reasoning strategies, and decision making. Of particular importance is the recognition of various users with various characteristics potentially using technologies in a variety of situations, contexts, organizations, and environments.

Supporting and Enhancing Collaborative Work Among Providers

The nature of the physician/associate provider (nurse, physician assistant) relationship is evolving from mere interaction to true collaboration with technologies allowing associate providers to increase involvement in clinical decision making and implementation. Technologies to support collaborative work environments in fast-paced, mobile environments are needed.

As mentioned earlier, the health care work environment is complex and involves a variety of health care professionals, all with varying needs for information, knowledge, and support. Designing technologies supporting both individual needs and collaboration among individuals presents a number of challenges. In a recent study, Jacko, Sears, and Sorensen (2001) showed that different health care professionals (physicians, pharmacists, nurses) exhibit different patterns of use of the Internet for clinical purposes and have different perceptions of needed enhancements to support their respective needs. The patterns differed in terms of the range and type of information as well as the depth and specificity of information.

Health care organizations will be faced with technology integration challenges and will make use of Internet as well as Intranet technologies to manage and organize the variety of applications in use by providers and their patients. Of critical importance for health care organizations will be the design of systems to support clinical decision support, knowledge management, organizational learning, administrative transactions, and supply chain.

Developing and Utilizing New Information and Communication Technologies

Wireless, handheld, and nano-technologies will drive higher adoption rates of the Internet in clinical settings, patient monitoring, and disease management. Both the number of interactive wireless device users and the number of devices will increase significantly in the next few years. New applications for such technologies need to be developed, based on fundamental research conducted to investigate usage of such devices.

Mobile computing and wireless technologies will increasingly become an important part of healthcare's information technology. Turisco and Case (2001) reported the following predictions:

• The number of wireless Internet users will reach 83 million by the end of 2005.

- By the end of 2004, there will be 95 million browser-enabled cellular phones and more than 13 million Web-enabled personal digital assistants.
- The wireless local area network (LAN) market is expected to reach $1 billion in 2001, and this figure will double by 2004.

Turisco and Case (2001) reported that mobile computing applications for health care began with reference tools and then moved to transaction-based systems to automate simple clinical and business tasks. They predicted that the next step will provide multiple integrated applications on a single device. In particular, they mentioned the following applications: prescription writing, charge capture and coding, lab order entry and results reporting, clinical documentation, alert messaging and communication, clinical decision support, medication administration, and inpatient care solutions.

Designing and Utilizing Adaptive Human–Computer Interfaces

The development and integration of information and communication technologies designed to support, facilitate, and enhance patient–provider interactions are critical. In particular, research is needed to design optimal human–computer interfaces. A dual focus should be placed on (a) increasing job satisfaction and effectiveness for the medical personnel and (b) increasing quality of care as well as safety for patients. To support these objectives, the human–computer interface must possess at least four qualities:

- It must be multimodal, that is, have the ability to display and accept information in a combination of visual, aural, and haptic modes.
- It must be personalized, that is, tailored to respond in a manner best suited to the current user and his or her needs.
- It must be multisensor, that is, have the ability to detect and transmit changes in the user or situation.
- It must be adaptive, that is, have the ability to change its behavior in real time to accommodate user preferences, user disabilities, and changes in the situation or environment.

Optimal interfaces are critical to virtually all applications connecting people to information technologies and people to people via information and communication technologies. Current research efforts are underway to develop intelligent adaptive multimodal interface systems. In the future, interfaces will automatically adapt themselves to the user (capabilities, disabilities, etc.), task, dialogue, environment, and input–output modes to maximize the effectiveness of HCI.

This is especially critical in health care for both types of users, consumers/patients and providers. Consumers/patients present a number of challenges regarding the interface. In addition to presenting different personal and sociodemographic characteristics, users in health care present varying degrees of health status: healthy consumers, newly diagnosed patients, chronically ill patients, and their caregivers will use the same devices in potentially different ways. Similarly, different providers will have different characteristics and perform a multitude of varying and highly dynamic tasks in different contexts and environments.

Moving to E-Health

The opportunities for improving service and decreasing cost via e-commerce technologies and the supply chain are tremendous. In addition, modern information and communication technologies (including sensors, wireless communication, implant technologies) will enable electronic delivery of health *care*. This is far more comprehensive than the mere electronic delivery of health information to patients and providers and include new developments such as telemedicine and virtual reality. Krapichler et al. (1999) claimed that "virtual environments are likely to be used in the daily clinical routine in [the] medicine of tomorrow" (p. 448). However, a number of barriers will need to be overcome. Organizational barriers to eHealth include infrastructure, organization, culture and strategy, systems integration, and workflow integration. Technological barriers include integration and security, interface design, connectivity, speed, reliability, and usability issues.

The evolution toward eHealth will involve a number of major changes. For example, organizational Web sites will evolve from publishing generic consumer content to providing personalized, online interactive services to profitable patient segments. Delivery organizations will focus on direct-to-patient relationship building, migrating to a health system truly centered on patients and communities. Communities of interest will rapidly expand to become a force in health care navigation. Wireless and handheld technologies will drive higher adoption rates of the Internet in clinical settings. The Internet will become a critical element of the physician–patient encounter and will support and enhance the patient–provider interaction. These changes will lead to a restructuring of the health care industry. Digital health plans will emerge and compete or threaten the traditional health care payer business model. Virtual networks will begin to emerge around specialty services to provide efficient personalized health care.

References

Agency for Healthcare Policy and Research. (1997). Consumer Health Informatics and Patient Decision Making. AHCPR Publication No. 98-N001. Rockville, MD: U.S. Department of Health and Human Services, Public Health Service.

Bates, D. W., Cohen, M., Leape, L. L., Overhage, M., Shabot, M. M., & Sheridan, T. (2001). Reducing the frequency of errors in medicine using information technology. *JAMIA, 8,* 299-308.

Bates, D. W., Teigh, J. M., Lee, J., Seger, D., Kuperman, G. J., Ma'Luf, N., Boyle, D., & Leape, L. (1999). The impact of computerized physician order entry on medication error prevention. *JAMIA, 6*, 313–321.

Benton Foundation. (1999). Networking for better care: Health care in the information age. Retrieved December 5, 2001, from http://www.benton.org/Library/health/

Booske, B. C., & Sainfort, F. (1998). Relationships between quantitative and qualitative measures of information use. *International Journal of Human-Computer Interaction, 10*, 1–21.

Cain, M. M., Mittman, R., Sarasohn-Kahn, J., & Wayne, J. C. (2000). Health e-People: The online consumer experience. Oakland, CA: Institute for The Future, California Health Care Foundation.

California Health Care Foundation. (2001). Addressing medical errors in hospitals: A framework for developing a plan. Oakland, CA: Author.

Carswell, C. M., & Ramzy, C. (1997). Graphing small data sets: should we bother? *Behaviour and Information Technology, 16*, 61–70.

Chassin, M. R., Galvin, R. W., & the National Roundtable on Helath Care Quality. (1998). The urgent need to improve health care quality. *JAMA, 280*, 1000–1005.

Dahlberg, T. (1991). Effectiveness of report format and aggregation: An approach to matching task characteristics and the nature of formats. *Acta Academie Oeconomicae Helsingiensis Series A;76*. Helsinki School of Economics and Business Administration, Finland.

Davis, F. D., Bagozzi, R. P., & Warshaw, P. R. (1989). User acceptance of computer technology: A comparison of two theoretical models. *Management Science, 35*, 982–1003.

Eng, T. R., Maxfield, A., Patrick, K., Deering, M. J., Ratzan, S. C., & Gustafson, D. H. (1998). Access to health information and support: A public highway or a private road? *JAMA, 280*, 1371–1375.

Felder, R. M., & Silverman, L. K. (1988). Learning styles and teaching styles in engineering education. *Engineering Education, 78*, 674–681.

Ferguson, T. (1997). Health online and the empowered medical consumer. *Journal on Quality Improvement, 23*, 251–257.

Gardner, R. M., & Shabot, M. (2001). Patient monitoring systems. In E. H. Shortliffe, L. E. Perreault, G. Wiederhold, & L. Fagan (Eds.), Medical informatics: Computer applications in health care and biomedicine (2nd ed., pp. 443–484). New York: Springer.

Greenes, R. A., & Brinkley, J. F. (2001). Imaging systems. In E. H. Shortliffe, L. E. Perreault, G. Wiederhold, & L. Fagan (Eds.), Medical Informatics: Computer Applications in Health Care and Biomedicine (2nd ed., pp. 485–538). New York: Springer.

Harris, J. (1995, May). Consumer health information demand and delivery: A preliminary assessment. *Partnerships for Networked Health Information for the Public*. Rango Mirage, CA. Summary Conference Report. Washington, DC: Office of Disease Prevention and Health Promotion, U.S. Department of Health and Human Services.

Hermann N. (1996). The whole brain business book. New York, NY: McGraw Hill.

Hersh, W. R., Detmer, W. M., & Frisse, E. H. (2001). Information-retrieval systems. In E. H. Shortliffe, L. E. Perreault, G. Wiederhold, & L. Fagan (Eds.), Medical informatics: Computer applications in health care and biomedicine (2nd ed., pp. 539–572). New York: Springer.

Hibbard, J. H., Slovic, P., & Jewett, J. J. (1997). Informing consumer decisions in health care: Implications from decision-making research. *The Milbank Quarterly, 75*, 395–414.

Hoffman, C., Rice, D. P., & Sung, H. Y. (1996). Persons with chronic conditions. Their prevalence and costs. *JAMA, 276*, 1473–1479.

Institute of Medicine. (2000). To err is human: Building a safer health system. Washington, DC: National Academy Press.

Institute of Medicine. (2001). Crossing the quality chasm: A new health system for the 21st century. Washington, DC: National Academy Press.

Jacko, J. A., Sears, A., & Sorensen, S. J. (2001). A framework for usability: Healthcare professionals and the internet. *Ergonomics, 44*, 989–1007.

Jadad, A. R., & Gagliardi, A. (1998). Rating health information on the Internet: Navigating to knowledge or to Babel? *JAMA, 279*, 611–614.

Jarvenpaa, S. L., & Dickson, G. W. (1988). Graphics and managerial decision making: Research based guidelines. *Communications of the ACM, 31*, 764–774.

John, D. R., & Cole, C. A. (1986). Age differences in information processing: Understanding deficits in young and elder consumers. *Journal of Consumer Research, 13*, 297–315.

Johnson, E. J., Payne, J. W., & Bettman, J. R. (1988) Information displays and preference reversals. *Organizational Behavior and Human Decision Processes, 42*, 1–21.

Karsh, B., Beasley, J. W., Hagenauer, M. E., & Sainfort, F. (2001). Do electronic medical records improve the quality of medical records? In M. J. Smith & G. Salvendy (Eds.), Systems, social and internationalization design aspects of human-computer interaction (pp. 908–912). Mahwah, NJ: Lawrence Erlbaum Associates.

Kennedy, J., & Blum, R. (2001, March 14). *HIPAA: The new network security imperative* (p. 15). Chicago, IL: Lucent Technologies.

Kolb, D. A. (1984). Experiential learning; Experience as the source of learning and development. Englewood Cliffs, NJ: Prentice-Hall.

Krapichler, C., Haubner, M., Losch, A., Schuhmann, D., Seemann, M., & Englmeier, K. H. (1999). Physicians in virtual environments—multimodal human-computer interaction. *Interacting with Computers, 11*, 427–452.

Kumar, S., & Chandra, C. (2001, March). A healthy change. *IIE Solutions, 33*(3), 28–33.

Legler, J. D., & Oates, R. (1993). Patients' reactions to physician use of a computerized medical record system during clinical encounters. *The Journal of Family Practice, 37*, 241–244.

Lindberg, D. A. B., & Humphreys, B. L. (1998). Medicine and health on the Internet: The good, the bad, and the ugly. *JAMA, 280*, 1303–1304.

Linzer, M., Konrad, T. R., Douglas, J., McMurray, J. E., Pathman, D. E., Williams, E. S., Schwartz, M. D., Gerrity, M. S., Scheckler, W. E., Bigby, J. A., & Rhodes, E. (2000). Managed care, time pressure, and physician job satisfaction: Results from the Physician Worklife study. *Journal of General Internal Medicine, 15*, 441–450.

Mittman, R., & Cain, M. (1999). The future of the Internet in health care. Oakland, CA: California HealthCare Foundation.

Murray, P. J., & Rizzolo, M. A. (1997). Reviewing and evaluating Web sites—some suggested guidelines. *Nursing Standard Online, 11*. Retrieved July 30, 1997, from http://www.nursing-standard.co.uk/vol11-45/ol-art.htm

Musen, M. A., Shahar, Y., & Shortliffe, E. H. (2001). Clinical Decision-Support Systems. In E. H. Shortliffe, L. E. Perreault, G. Wiederhold, & L. Fagan (Eds.), Medical Informatics: Computer Applications in Health Care and Biomedicine (2nd ed., pp. 573–609). New York: Springer.

Myers, I. B. (1987). Introduction to Type. Palo Alto, CA: Consulting Psychologists Press.

National Research Council. (2000). Networking Health: Prescriptions for the Internet. Washington, DC: National Academy Press.

Orman, L. (1993). Information independent evaluation of information systems. *Information and Management, 6*, 309–316.

Ornstein, S., & Bearden, A. (1994). Patient perspectives on computer-based medical records. *The Journal of Family Practice, 38*, 606–610.

Patel, V. L., Arocha, J. F., & Kaufman, D. R. (2001). A primer on aspects of cognition for medical informatics. *JAMIA, 8*, 324-343.

Patel, V. L., Kushniruk, A. W., Yang, S., & Yale, J. F. (2000). Impact of a computer-based patient record system on data collection, knowledge organization, and reasoning. *JAMIA, 7*, 569-585.

Patrick, K., & Koss, S. (1995, May 15). *Consumer health information "white paper."* Washington, DC: Consumer Health Information Subgroup, Health Information and Application Working Group, Committee on Applications and Technology, Information Infrastructure Task Force. Working Draft.

Robinson, T. N., Patrick, K., Eng, T. R., & Gustafson, D. (1998). An evidence-based approach to interactive health communication. *JAMA, 280*, 1264-1269.

Sainfort, F., & Booske, B. C. (1996). Role of information in consumer selection of health plans. *Health Care Financing Review, 18*, 31-54.

Schkade, D. A., & Kleinmuntz, D. N. (1994). Information displays and choice processes: Differential effects of organization, form, and sequence. *Organizational Behavior and Human Decision Processes, 57*, 319-337.

Schoen, C., Davis, K., Osborn, R., & Blendon, R. (2000, October). Commonwealth Fund 2000 international health policy survey of physicians' perspectives on quality. New York: Commonwealth Fund.

Shackel, B. (1991). Usability—context, framework, definition, design and evaluation. In B. Shackel & S. Richardson (Eds.), *Human Factors for Informatics Usability* (pp. 41-75). Cambridge, England: Cambridge University Press.

Silberg, W. M., Lundberg, G. D., & Musacchio, R. A. (1997). Assessing, controlling, and assuring the quality of medical information on the Internet: Caveant lector et viewor—let the reader and viewer beware [Editorial]. *JAMA, 277*, 1244-1245.

Slocum, J. W., & Hellriegel, D. (1983, July/August). A look at how managers minds work. *Business Horizons*, 58-68.

Sonnenberg, F. A. (1997). Health information on the Internet: opportunities and pitfalls. *Archives Internal Medicine, 157*, 151-152.

Togo, D. F., & Hood, J. N. (1992). Quantitative information presentation and gender: An interaction effect. *Journal of General Psychology, 119*, 161-167.

Turisco, F., & Case, J. (2001). *Wireless and mobile computing*. Oakland, CA: California HealthCare Foundation.

U.S. General Accounting Office. (1996, July). *Consumer Health Informatics: Emerging Issues*. Report to the chairman, Subcommittee on Human Resources and Intergovernmental Relations, House Committee on Government Reform and Oversight. GAO/AIMD-96-86. Washington, DC: Author.

Wager, K. A., Lee, F. W., White, A. W., Ward, D. M., & Ornstein, S. M. (2000). Impact of an electronic medical record system on community-based primary care practices. *Journal of the American Board of Family Practice, 13*, 338-348.

Wagner, E. H., Austin, B. T., & Von Korff, M. (1996). Organizing care for patients with chronic illness. *Milbank Quarterly, 74*, 511-542.

Workgroup on Electronic Data Interchange. (2000, July/August). *HIPAA: Changing the health care landscape. Oncology Issues* (pp. 21 23). Reston, VA: WEDI Foundation.

World Health Organization. (2000). *The world health report 2000: Health systems: Improving performance*. Retrieved April 11, 2002, from http://www.who.int/whr/2000/index.htm

Wyatt, J. C. (1997). Commentary: Measuring quality and impact of the world wide web. *British Medical Journal, 314*, 1879-1881.

A FRAMEWORK FOR UNDERSTANDING THE DEVELOPMENT OF EDUCATIONAL SOFTWARE

Chris Quintana, Joseph Krajcik, and Elliot Soloway
University of Michigan

Cathleen Norris
University of North Texas

INTRODUCTION

Software usability has been the primary focus for human-computer interaction (HCI) researchers and practitioners. As computers became more mainstream in the 1980s, the HCI community began to realize that ease of use was of utmost importance if larger segments of the population were to fully benefit from using computing technologies. Thus in the last 20 years, the field of HCI has made great strides in taking computers from being arcane devices used only by scientists and engineers to being productive technologies that can be used by many different people.

However, usability is not the only software attribute for HCI to address. Whereas usability issues address how designers can make software use a more efficient and enjoyable experience, there are other aspects of software use that need to be explored by HCI researchers. It is important to see how to make technology more usable, but it is also important to see how the use of technology in different areas can change and affect the user of the technology. One such area is education. A question posed by many researchers in HCI, education, and psychology has been: How can the use of computer technology facilitate learning—not necessarily learning how to use software, but rather using software to learn new, wide-ranging work practices and domains? We consider that question in this chapter by focusing on the use of technology for learning—lifelong learning for people ranging from prekindergarten children to postcollege adults.

Traditional Design Approach: User-Centered Design

To set the context for our discussion on educational technology, we begin by distinguishing between traditional design and designing for learning. Although educational technology certainly must be usable, we need to frame the central design approach for *learner-centered design* (LCD; Soloway, Guzdial, & Hay, 1994) slightly differently than the traditional user-centered design (UCD) approach. By separately considering the LCD approach, we can describe the different ways educational software can be developed to meet the goals of LCD. Let us briefly review the description of the more traditional UCD approach to designing usable software. The concept of "usability" is intuitively understood, but Norman (1986) formulated a specific UCD description that underscores the conceptual design problem: addressing the conceptual distance between a computer user and the computer. Later we use this UCD description to frame the corresponding design problem faced in the design of educational technology.

Audience: Who Are "Users"? In many descriptions of UCD, there is an implicit assumption that the user of a computer tool already possesses some measure of expertise about the work activity they are using the tool to engage in (Mayhew, 1999; Norman, 1986). Specifically, we distinguish "users" as having the following characteristics (Soloway et al., 1994, 1996):

• Users know what activity they want to perform given a tool. Users understand the work domain in which they are working and the work tasks they are completing. Users simply need a computer tool that will help them complete their work tasks easily and efficiently.

• Users engaged in some given work activity share a work culture and therefore can be considered homogenous in a number of important ways. The tasks they perform are often similar from user to user, and a designer can rely on an archetypal user when designing a user-centered tool.

• Users, by the nature of their involvement with their work tasks (e.g., professions, labors of love) have intrinsic motivation for their work so a tool does not have to supply any extra motivating factors.

• Users are not necessarily trying to learn about their work through their tools. Rather, they need tools to help them complete their work. Their work domain familiarity allows them to pursue their tasks without any significant growth in the domain. Work professionals will certainly learn new things about their work, but the tools they use will largely stay the same.

In summary, a "user" is knowledgeable and motivated about their work activity (Fig. 42.1). The goal for the software designer is to design a tool to help users complete their work easily and efficiently.

UCD Problem: The Conceptual Gap Between Learners and Tools. Given the UCD goal of designing a usable system, we need to consider what is meant by "usable." Norman (1986) stated that when users use a tool to complete their work, they will have goals in mind that they need to translate into actions to execute on the tool. Once users have executed an action, they must evaluate the tool's resulting state and interpret that state in terms of their goals. Thus, there exist two important discrepancies between the goals of the user and the physical tool. These discrepancies are represented as "gulfs" between user and tool: a gulf of *execution* and a gulf of *evaluation* (Fig. 42.2).

The gulf of execution is the difference between the goals and intentions of the user and the permissible actions on the tool. The gulf of evaluation reflects the amount of effort the user

FIGURE 42.1. Primary user characteristic.

FIGURE 42.2. Gulfs of execution and evaluation.

FIGURE 42.3. Primary learner characteristic.

must exert to interpret the physical tool state (Norman, 1990). The "size" of the gulfs corresponds to the difficulties users face in understanding and using the tool to complete their tasks. A usable (or user-centered) system must minimize both gulfs, bringing the user and the tool closer together.

Using a Theory of Action to Design Usable Tools. Given the description of the usability gulfs, a "usable" tool is one for which the gulfs have been reduced or removed to make execution and evaluation straightforward. One way of reducing the usability gulfs is to bring the system closer to the user through proper design. Designers can develop usable tools by understanding how people work with tools. Norman (1986) postulated that to develop a design process that realizes usable tools, designers need a "theory of action" explaining "how people generally do things to complete a task." Such a theory can guide a designer in building a system that helps users bridge the usability gulfs.

Norman's theory of action is expressed as a series of execution steps (establish a goal, an intention to achieve the goal, a specific sequence to meet the intention; execute the specific action sequence) and evaluation steps (perceive, interpret, and evaluate the tool state; Norman, 1986). The ramifications of such a "theory of action" for designers is that the theory informs the development of UCD heuristics that work to reduce the gulfs of execution and evaluation. For example, designers should simplify the tool structure, use affordances and metaphors, make permissible actions and the tool state visible, and so on (Norman, 1990, 1986). By analyzing work tasks and understanding how people perform those tasks, designers can design systems that users can operate and understand to successfully engage in their work.

Designing for Learning: Learner-Centered Design

Given the UCD description, we can now frame the design approach for designing educational technology, which we distinguish as a more learner-centered approach (Soloway et al., 1994). Essentially, the goal of LCD is to develop tools that

mediate the development of new understanding by learners of some work activity or domain. Therefore, learner-centered tools need to address the conceptual distance between the audience of learners and the work that they are trying to learn. We can differentiate the LCD perspective by contrasting the dimensions that we used for the UCD description.

Audience: Who Are "Learners"? The audience of learners includes novices in some work practice who are trying to develop a better understanding of that practice (Fig. 42.3). By a "work practice," we mean the responsibilities, tasks, tools, artifacts, terminology, knowledge, and relationships involved in a given work activity. Note that we are not implying that a work practice is activity that only occurs at someone's job. Rather, it is any complex set of activity, artifacts, and so on that is undertaken to meet some goal.

UCD addresses the design of usable tools because of the assumption that users know what they want to do with the tools. By contrast, learners are novices in the target work practice, so even if they are given a usable set of tools, they will not necessarily be able to engage in the underlying activity in which the tools are used. Instead the goal should be to develop tools that learners can use to learn the underlying activity.

Note that although the central goal is to have a learner learn new work, this should not imply that learners are restricted to some specific age group. Learning new work can occur at any point in life. Learners can be students learning new tasks and practices in school, such as beginning business students learning the work of financial analysis. But learners can also be adults in the workforce, such as a new corporate work hire learning her new company's internal consulting practices during a corporate orientation. Whether children or adults, learners need tools that address their lack of expertise and support them in engaging in the new work practice. More specifically, we can characterize learners as having the following characteristics (Soloway & Pryor, 1996):

• Learners do not possess a significant amount of expertise in the work practice. Learners (e.g., the business student or the new consultant) do not share an understanding of the activities,

terminology, etc., of the work practice with their professional counterparts (e.g., financial analysts or the corporate manager). More specifically, learners have an incomplete or naive mental model of the work they are trying to perform.

• Learners are heterogeneous. Traditional design considers a homogenous population consisting of a prototypical user because users share a common work culture. Learners do not necessarily share a common culture nor a common body of expertise in the work practice. Thus, designers need to consider the wide range of diversity in background, development, gender, age, and learning styles in the learner population.

• Learners are not always highly motivated to pursue a new work practice. Experts in the work practice, by the nature of their involvement with their work have an intrinsic motivation that learners may lack. Furthermore, the motivation of learners further suffers from the obstacles that they face in trying to perform the new work (Hogan, Nastasi, & Pressley, 2000). Learners who might have some motivation to learn new work may lose that motivation because of the complexity of the new work.

• A learner's understanding grows as he or she engages in and continues to engage in a new work practice. User-centered tools simply help experts engage in their work easily and do not need to change significantly because experts are not necessarily learning about their work through the tools. Learners, however, will change as they gain new expertise in the work practice. Thus, the tools learners initially use to learn may not be the tools they use when they have developed more expertise in their work.

LCD Problem: The Conceptual Gap Between Learners and Work. Aside from the differences in the target population, the design of learner-centered technology also addresses a different conceptual problem than that for UCD. Certainly, educational software cannot ignore the usability gulfs of execution and evaluation. However, the central problem for educational software involves addressing the conceptual gap—a gulf of expertise—between the learner and the model of expertise embodied by an expert in the work practice (Fig. 42.4).

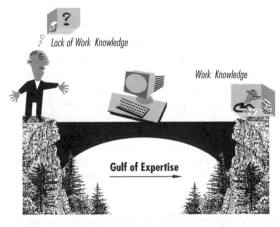

FIGURE 42.4. Gulf of expertise.

Lave and Wegner (1991) stated that learning and knowledge mastery require work practice newcomers to move toward full participation in the practices of the work community. We can describe a work expert as one who can fully participate in the work practice. If we consider the learner to be the "newcomer" in the work practice, then the goal for the learner is to change and grow so that they can fully participate in the work. This movement occurs over the gulf of expertise, that is, over the conceptual gap that lies between the newcomer to the community of practice and the full participant in that community of practice.

For a learner to be able to participate in some work practice, the learner needs to understand what kind of activities are in the practice, facts about the practice, knowledge needed to complete the work activities, and so on. In other words, the learner needs to develop a correct and appropriate conceptual model of the work involved. Thus, we can say that the "size" of the gulf of expertise describes how far a learner is from understanding the work practice. Specifically, the size of the gulf of expertise is proportional to the amount of conceptual change needed in the learner's model of the work practice so that he or she can fully participate with other experts in the work community. Given this description of the conceptual gap addressed by LCD, we can now describe the central design task for educational software: designing software that supports learners in moving across the gulf of expertise.

Overview: Learning Theory and the Design of Educational Software

Given the differences in audience, conceptual gaps, and design goals (summarized in Table 42.1), the final dimension needed to describe educational software design is the underlying learning theory or approach that is used as the basis for the design. Just as the theory of action informs heuristics for the design of usable software, some theoretical approach to learning should be the basis for the development of learner-centered technology. Over the years, there have been different perspectives on how people learn, and those perspectives have led to different design approaches for educational technology. In the remainder of this chapter, we review some different perspectives and the corresponding evolution of educational technology. Although the overall goal of educational software may be to help learners bridge the gulf of expertise, there have been many approaches to helping learners do so over the last 30 years. Specifically, we will review three major perspectives on learning (i.e., behaviorism, information processing, and social constructivism) and the kinds of software that results from each learning perspective.

CROSSING THE GULF OF EXPERTISE: THREE THEORIES OF LEARNING AND THEIR "GAP-CROSSING" STRATEGIES

Given the manner in which we have framed the goals of learner-centered software, the central goal for designers is to develop software that helps learners bridge a gulf of expertise between

TABLE 42.1. Summarizing the Differences Between Users and Learners

Aspects for Design	Designing for Users	Designing for Learners
Primary design goal	Designing highly usable computer tools that allow users to complete their work easily and efficiently	Designing computer tools that support learners in developing an understanding of an unknown work practice (without ignoring tool usability)
Audience characteristics	Users • have a more thorough understanding of the work practice for which they are using tools • are a more homogenous audience because of shared work culture • have more intrinsic motivation in the work practice • will not change significantly as they do their work, meaning that their tools will not change	Learners • have an incomplete or naïve understanding of the given work practice • are a more diverse audience because they do not share a common work culture • have less intrinsic motivation to engage in a work practice because of an incomplete work model and obstacles faced in doing work • will change significantly as they learn new work, necessitating changes in their tools
Conceptual "gaps" to address	Gulfs of execution and evaluation between the user and the tool	Gulf of expertise between the learner and complete work practice expertise (without ignoring the user-centered gulfs between learner and tool)
Underlying approach to bridge the gaps	Using a theory of action that describes how people use tools to complete tasks so designers can minimize the gulfs of execution and evaluation	Using theories of learning so designers can bridge learners across the gulf of expertise • Behaviorism • Information processing • Social constructivism

their current understanding of some work practice to a richer understanding of the practice. There are three prevalent schools of thought that have arisen over the years to address the problem of designing tools for learning: the behaviorist approach, the information processing approach, and the social constructivist approach. Each design approach aims to develop educational software, but the theoretical background for each approach differs, resulting in different kinds of software.

Behaviorist Approach

The behaviorist approach involves looking at a person's external behavior rather than attempting to discern the mental activity or mental models that a person may have. The behaviorist approach is predominately influenced by the work of B. F. Skinner, who advocated an approach in which learning essentially involves "programming" a person's behavior through conditioning that rewards "correct" behavior (or punishes "incorrect" behavior) in a given context. The behaviorist approach can be seen in educational software that adopts this "conditioning behavior" approach. Some early behaviorist software took the form of "teaching machines" that attempted to closely follow Skinnerian thinking. Other software is more loosely tied to classic behaviorism but maintains an emphasis on immediate feedback to students and constant measurement of educational progress to reinforce to students when they are or are not working correctly (Anderson, Reder, & Simon, 1998).

The basic approach for behaviorist software involves having students do activities or answer questions from some target domain. The software then attempts to shape the learner's behavior by rewarding correct answers or punishing incorrect answers. Contemporary examples of software that adopts a basic

behaviorist approach can be seen in many educational games, such as Reader Rabbit, Math Blasters, or Typing Tutor. In each of these cases, learners perform activities from the different domains (e.g., reading, arithmetic, typing). Correct activity is rewarded with game points or by allowing the learners to continue to the next level of the game. Incorrect activity is punished with some sort of negative action on the learner's game character.

The primary focus for developing behaviorist software is on the task analysis. The task analysis involves the detailed breakdown of some complex process into small steps. The learner learns the complex process by training on some predefined criterion for each step of the process. The software monitors the learner's progress as he or she performs the new work activities. By monitoring the learner's progress and following some programmed rubric, the software can inform learners when they have learned some component of the work well enough to proceed with the next work task. Thus behaviorist software transitions a learner across the gulf of expertise through the positive and negative reinforcement. If the learner's behavior continues to be correct, then it is assumed that the learner is learning their new work and is thus continuing forward across the gulf of expertise.

There are many examples of behaviorist software found in many educational games, "flash-card" software, and similar products. However, the behaviorist approach is not a significantly found in current educational software and practices (Anderson et al., 1998). Behaviorist software must be developed for activities that are well defined and that have specifically defined answers. Behaviorist software can be useful for practicing specific skills (i.e., typing, arithmetic), but it can be difficult to develop such software for more exploratory, wide-ranging work that may not have clear answers.

Information Processing Approach

The second major approach to developing educational software is the information processing approach. Whereas the behaviorist approach simply considers the externally visible behavior of a person, the information processing approach takes into consideration models of cognition. A central claim of the information processing approach is that "certain aspects of human cognition involve knowledge that is represented symbolically" (Anderson et al., 1998, p. 232). For example, one symbolic representation that should be familiar to those in HCI is the rule-based approach to human cognition popularized by Card, Moran, and Newell (1983). Here, the mind is seen conceptually as a set of processors and human cognition takes the form of symbolic rules. Thus, the interaction between humans and computers can be seen as a flow of information or symbols between the two sets of processors that use rule sets for cognition or computation.

There are different approaches and theories for formulating rules and rule sets to describe cognitive processes. For example, the GOMS approach (Card et al., 1983) describes procedural knowledge in a set of rules. A more wide-ranging cognitive theory is the ACT* theory (Anderson, 1983), which is a unified theory of the nature, acquisition, and use of human knowledge. The fundamental assumption behind ACT* is that problem-solving knowledge can be externally represented by using sets of production rules. Knowledge begins as declarative information, or different pieces of information that are associatively linked together. Procedural knowledge is formed by making new inferences (or new production rules) from existing declarative knowledge. Aside from the creation of procedural knowledge, ACT* also describes three kinds of learning:

- *Generalization:* Production rules formed in one area are broadened to encompass other areas.
- *Discrimination:* Production rules from a broad set of areas are narrowed to apply in a smaller range of areas.
- *Strengthening:* Certain production rules are strengthened by being applied more often, resulting in material that has been learned "better."

The main outcome of theoretical approaches such as ACT* is that human cognition can be externally represented in some symbolic fashion. The knowledge needed to do some work activity can be formulated and represented. Because cognitive processes can be externally represented, computer simulations of cognitive processes can be created to describe how some work is done and to monitor how people do that work.

In terms of education, the software most commonly associated with the information processing approach includes intelligent tutoring systems (or ITSs). Corbett, Koedinger, and Hadley (2001) described intelligent tutoring systems as problem-solving tools that use expert systems to reason about the target work domain and analyze student activity, reason about the learner's knowledge state, and make decisions about instructional interventions. Examples of ITSs include geometry and algebra tutors (Corbett et al., 2001) and physics tutors (Gertner & VanLehn, 2000; Fig. 42.5). Additionally, there are behavioral approaches to tutoring systems (Emurian & Durham, chapter 34 of this volume) that combine aspects of the behavioral approach and intelligent tutors.

An intelligent tutor is built for a particular work domain (e.g., algebra or geometry) and will contain a specific task model (or cognitive model) describing the problem-solving knowledge needed to do the work in that domain (i.e., knowledge that the learners are trying to acquire). Learners use an ITS to engage in activities from the target work domain. The ITS will monitor learners' work to see how well they are progressing through the work and whether they seem to need help to continue with their tasks. Thus, an ITS also needs a cognitive model of the learner (or the learner's knowledge state) to be able to gauge learners' progress and direct the instruction. Task and learner analysis are necessary for an ITS to develop the specific task and learner models needed by the system. Learners are mediated across the gulf of expertise by practicing on components of the task in the context of the complete task and under the watchful eye of the computer-based tutor using the detailed user model.

From a software engineering/HCI point of view, the main effort for ITS design is the development of the task model and learner's cognitive model. For an ITS to be successful, the software needs to have a detailed description of how to do the underlying work being supported by the software. Thus, one difficulty for ITSs can occur in supporting work practices that are relatively unstructured and complex, such as creative processes involved in design problems. Another difficulty for ITS design is the development of the learner model. The success of a tutor depends on the software being able to identify areas where the learner needs assistance to continue with the material they are learning. However, given the diversity of learners, developing tutors with wide-ranging learner models can be problematic. Given a cognitive model of the task, an ITS can discern the learner's current model of the task from their work as they use the tutor. However, the consideration of learner diversity (e.g., in terms of gender, learning styles, etc.) for the system design could also help improve the kind of intervention offered by ITSs. But developing such specific, yet diverse, learner models can be difficult.

Social Constructivist Approach

The final approach to educational software that we consider is scaffolded software based on a social constructivist perspective of learning. The primary focus for constructivist software is on supporting more open-ended, wide-ranging work practices and the context (i.e., the social arena) in which the learner does the learning. Constructivist software helps learners cross the gulf of expertise by supporting learners to practice and engage in components of the work, many times with collaborators, in some context so that they can begin to develop expertise in the work. Learners try to do their new work in a system consisting of humans and technology, so the constructivist approach calls for scaffolding provided by technology and a human mentor to be the supportive framework that allows the novice learner to engage in and learn the new work.

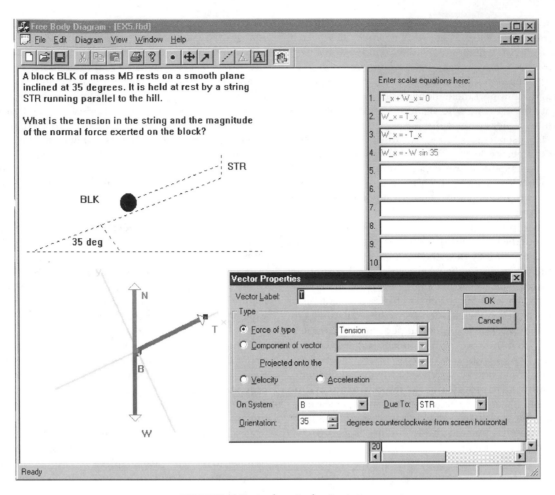

FIGURE 42.5. Andes: A physics tutor.

Social constructivism is a learning perspective that is increasingly gaining favor in education and is the basis for many current educational approaches (Singer, Marx, Krajcik, & Chambers, 2000). Social constructivism proposes that learning and understanding involve active, constructive, generative processes, and enculturation into the work practice being learned (Brown, Collins, & Duguid, 1989; Papert, 1993; Piaget, 1954). Learning is not a simple, passive process of transferring information from expert to novice. Rather, it is an active process, employing a "learning by mindful doing" approach in which learners must cognitively manipulate material to create cognitive links from the new material to their own prior knowledge. Furthermore, learning does not occur in a vacuum; it must occur in some context. As Brown et al. (1989, p. 32) stated, "knowledge is situated, being in part a product of the activity, context, and culture in which it is developed and used." These researchers see knowledge as being contextualized, meaning that learners must build their knowledge within a community of peers and experts. Thus, gaining expertise involves participating in the context of the professional culture to understand the common practices, languages, tools, and values of a professional culture. Singer et al. (2000) summarized four basic tenets of social constructivism:

- *Active construction:* Learners need to actively engage in mindful work to construct their knowledge. The term *mindful* implies that learners should expend some cognitive effort on their work to develop a better understanding of that work. As learners actively construct their knowledge, they develop deep understanding by becoming immersed in the new work practice.

- *Situated cognition:* Learners need to work in a context where they are situated with social and intellectual support so they can see how knowledge is used in the work practice.

- *Community:* Learners need to be exposed to the community (or culture) of the work practice. Learners need to learn the work language and see how professionals discuss and validate their ideas within the culture.

- *Discourse:* Learners need an understanding of the work culture so they can engage in discourse with other members of the work community (whether professionals or other learners) to discuss their work and ideas.

Thus, the social constructivist perspective proposes that learners need to be immersed in and actively participate in representative activity from the new work practice so they can

begin to function as members of the work culture. However, because learners are novices in the work practice, they need a significant amount of structure to mindfully engage in the new work. Software tools can provide this structure in the form of software scaffolding. Software scaffolds can be described as software features that support the mindful performance of a given task or activity by someone who is a novice in that task or activity. Certainly learners cannot use the same tools that work experts use (i.e., user-centered tools) because of the difference in their work expertise levels. Thus, LCD software designers must develop tools modeled on experts' tools, but scaffolded in ways that allow learners to participate in activities similar to those of work experts and allow construction of artifacts used in the work practice (Belamy, 1996).

Metcalf (1999) provided a comprehensive overview of scaffolding from both the educational and software design perspectives, which can be briefly described as follows. From an educational perspective, scaffolding can be described as support provided by a teacher to help a student carry out a previously inaccessible task. For example, teachers provide different kinds of scaffolding for their students, such as coaching, advising, critiquing, or other support to help students perform new complex tasks without doing the task for the student (Collins, Brown, Newman, 1989; Wood, Bruner, & Ross, 1975).

From a software design perspective, scaffolds can be incorporated into software to provide some of the same kinds of support (e.g., coaching, task restructuring, etc.) and help learners engage in new, previously inaccessible work. Scaffolds follow the social constructivist perspective by making aspects of the new work practice accessible and visible (Soloway et al., 1996):

- Scaffolds can make new tasks doable so learners can actively engage in new tasks from the target work practice.

- Scaffolds can situate the learner in a more authentic representation of the work (e.g., software features showing a science lab in the background).

- Scaffolds can make aspects of the work practice and work community (e.g., the terminology used in the work culture) visible and understandable by the students so they can engage in discourse with others.

- Finally, scaffolds should fade from the software when the learner no longer needs the support provided by the scaffold because of increasing expertise. In fact, at that point, the scaffold may now interfere with the task, so scaffolds should fade as directed by a teacher, the learner, or the software itself.

Scaffolds can be defined to provide different dimensions of support, such as process support, management support, or content support. (In fact, the support provided by an ITS can be considered one kind of scaffold among others.) By making different aspects of a target work practice (e.g., science inquiry) accessible to novice learners (e.g., students in school), learner-centered tools can provide the structure and support framework within which learners can see the culture and language of the work practice and engage in different work activities.

The use of software scaffolds to support learners is an ongoing area of research as described by Metcalf (1999). Researchers have discussed scaffolding from an educational standpoint, noting the type of support teachers provide to students, such as coaching, advising, critiquing, etc., to help students perform previously inaccessible tasks (e.g., Collins et al., 1989; Wood et al., 1975). The next step for scaffolding research involved exploring how the scaffolding techniques used by teachers and mentors could be implemented in software. Guzdial (1993) is cited as the first researcher to extensively explore software-based scaffolding methods, identifying areas in which scaffolding could provide learner support via coaching, communicating process, and eliciting articulation. Metcalf (1999) has explored how software-realized scaffolding can "fade" from the software when the learner no longer needs the support to engage in the underlying work tasks. Other researchers have explored scaffolding in the context of science inquiry tools that support specific individual science activities (e.g., Linn, 1996; Tabak et al., 1995).

A key point to remember about scaffolds is that whether or not a software feature is a scaffold depends on whether the feature supports some aspect of the target work practice. A software or user interface feature on its own is not necessarily a scaffold unless the content of that feature supports the learner in a meaningful way. Consider a hypothetical learner-centered word processor that supports the work of writing a funding proposal. Here, scaffolds are defined in the context of writing a complex proposal. One software feature could be the traditional help feature found on most word processors. For example, the feature may provide help for using the word processor itself, such as describing the specific menu commands for changing the page number format (e.g., "Page number format can be changed by going to Format → Paragraph ... → Layout"). In this case, the help feature is not providing any scaffolding to help learners understand aspects of proposal writing. Although page numbers are part of a proposal document, changing the page number format is not really an important task in authoring a proposal. Instead, the help feature is providing help for using the tool. However, that same help feature could also provide help to describe the components of a proposal (e.g., "For a proposal, you will need to write several parts: the research needs, the proposed work, a project team, a project calendar, and a budget."). In this case, the help feature is acting as a scaffold because the help is work-oriented and aimed at relaying information concerning the target work practice.

Constructivist software differs in approach from the intelligent tutoring systems described in the previous section. Constructivist software attempts to convey aspects of the work practice (i.e., in the user interface or in the software structure) in ways that make aspects of the work more accessible and doable by novice learners so that learners can get a foothold into the new work they are learning. For example, Model-It is a system-dynamics modeling tool for science students (Soloway et al., 1996). Students use Model-It to build models of complex systems, such as ecosystems and the atmosphere, to explore scenarios and explain the behavior of such complex systems. One complex task in building system models is defining how different variables in the model relate to each other. For example, in an air-quality model, students may want to relate car exhaust, pollution levels, wind speed, and citizens' health. Professional

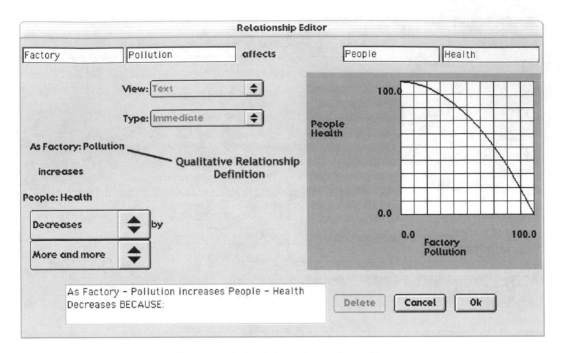

FIGURE 42.6. Model-it relationship editor.

modeling programs can require relationships to be defined by writing differential equations, a task much too difficult for many students. Model-it instead makes relationship definition accessible through a relationship editor (Fig. 42.6) that allows students to define relationships by building qualitative sentences that describe the relationships.

Here, students describe how pollution and health are related by stating that "as pollution increases, people's health decreases." When describing how health decreases, students are presented with qualitative choices (e.g., a little, more and more, the same, a lot) that they understand but that are still useful for defining a working model. Rather than oversimplifying the task, learners can now build relationships using a tool that is simple but that still requires them to reflect on how they believe variables are related. The qualitative, familiar language for describing the relationship acts as a scaffold to support the learner in the new work of model building.

Another example for visualizing work activity comes from Symphony (Fig. 42.7), a scaffolded work environment integrating a range of data collection, visualization, and modeling tools for science inquiry (Quintana, Eng, Carra, wU, & Soloway, 1999). Students new to the complex work of science inquiry will not necessarily understand what activities make up the work. Symphony uses a variety of process maps to visualize the components of the overall science process and individual science activities.

Figure 42.7 shows two kinds of process maps in Symphony. First, the process wheel on the left illustrates the space of activities involved in a science investigation. By visualizing the space of possible activities, learners can begin to see and understand what kinds of activities comprise the work of science inquiry. Second, the right side of Fig. 42.7 shows a directed flow map that describes the procedure for building a graph. Students

step through the flow map, pressing the buttons in the map, which in turn launch the appropriate portion of the graphing tool. Not only does the flow map make graph building accessible to learners, it also explicitly presents the steps of the process so that students can internalize the work involved in graph building.

Other researchers, such as Squires and Preece (1999), are considering the constructivist approach to develop a wider set of design heuristics from the development of constructivist software. The constructivist approach is increasingly making inroads in education, and there is also a shift of emphasis on constructivism to allow learners to engage in more open-ended, exploratory, and personally meaningful construction of knowledge. The primary critique against the constructivist approach, however, is that the social constructivist perspective is just that—a perspective rather than a scientific theory. Many complain that constructivism is too nebulous to form a basis for educational approaches and technology and is not as fully developed and specifically articulated as the cognitive theories used by the information processing approach (Anderson et al., 1998). Still, constructivist approaches have resulted in innovative curricula and technology that have proven to be useful in educational settings.

OPEN ISSUES AND CONCLUDING REMARKS

Although there are different—some might say, competing—approaches to designing educational software, no one approach provides a definitive answer for the development of educational technology. The behaviorist, information processing, and constructivist approaches summarized here have all had successes

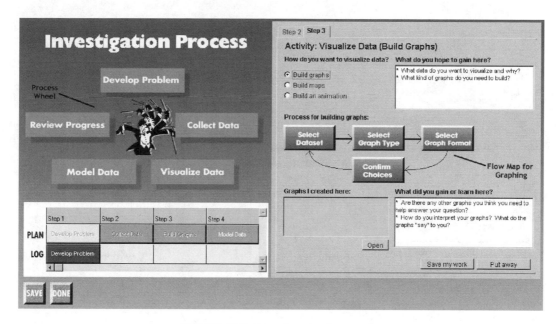

FIGURE 42.7. Symphony main screen.

and failures. Different approaches may better suit different contexts and different kinds of work practices. In the end, what we find is that the interesting part of developing educational software comes from considering the potential impact of having multiple theories of learning at our disposal. Instead of trying to determine the single best paradigm for designing educational technology, the more prudent approach is to consider the central strategies taken by each approach and see which is more applicable to the learning situation at hand. Thus, we see that educational software can incorporate a variety of techniques for transitioning learners across a gulf of expertise, techniques that accomplish the following:

- provide positive and negative reinforcement of correct and incorrect activity as learners engage in new work,
- monitor learner progress through new work activity to provide intelligent intervention to help learners when they encounter difficulty in new work activity,
- visualize different activities and components of new work activity to help learners gain a foothold into the work practice and see the kinds of activities in which they can engage, and
- make complex new work activity doable by novice learners so that they can begin to mindfully engage in and thus better understand a new work practice.

The different approaches we have summarized provide some major techniques that can be employed when developing educational technology. However, the specific design methods that are needed by designers of educational software are not significantly different than those employed in the design of any software. Many of the questions being explored by HCI and software engineering research are found when designing educational technology:

Analysis

As we have seen, task and work analysis is a vital step in the design of educational software, no matter which approach is taken. If software is to support the development of expertise in some domain or work practice, then designers of educational technology much have a strong understanding of that domain or practice. As we have transitioned from the behaviorism to information processing to constructivism, we have seen the need for an increasingly detailed description of the work we want learners to learn and engage in. This parallels the current state in HCI, where the field is transitioning from typical task analysis techniques to cognitive task analysis (Lesgold, 1986) to techniques for analyzing work and context, such as contextual analysis (Beyer & Holtzblatt, 1999), process space analysis (Quintana et al., 1999), and so on. These analysis methods are important to consider for any software design project, but especially for designing educational software.

Design

As designers know, software design in not a solitary endeavor, but an increasingly interdisciplinary effort. When designing educational software, software and HCI experts need to collaborate with experts in the work practice, educational researchers, teachers, and maybe even the learners themselves. Thus, participatory design techniques are especially important to establish common languages and techniques that help a diverse design team work together. Furthermore, HCI researchers need to also be cognizant of issues involving the different areas in which educational software may be used. As we transition from desktop computers to handheld devices and other "ubiquitous computing" environments, a range of design techniques will be

needed to realize the design strategies arising from the different educational approaches.

Evaluation

Software always needs to be evaluated, educational software is no different. However, different evaluation approaches are needed for educational technology. Traditionally, software is evaluated for efficiency and usability. However, educational software also needs to be evaluated in terms of how well learners are learning using the software. Thus, design teams need to evaluate the different cognitive effects resulting from technology use, including the "effects of" technology (i.e., what is the impact of the software as a whole on the learners' understanding of the new work practice?) and the "effects with" technology (i.e., what is the impact of the individual supports in an educational software package?; Salomon, Perkins, & Globerson, 1991). Furthermore, new techniques may be needed for software evaluation given the difficulty of evaluation in "real-world" contexts such as a public school classroom. Sometimes, the use context for the software may make it difficult to carry out traditional controlled studies, so alternative evaluation methods may be needed (Quintana, Fretz, Krajcik, & Soloway, 2000).

The discussion in this chapter has been more technology-centric, focusing on defining the design problem being faced by software designers when developing educational technology and on describing three major approaches for educational software. However, many other social issues must be faced by those developing educational technologies. Educational software does not exist in a vacuum; it is part of an overall educational system consisting of other social aspects. Curriculum design and other supporting materials need to accompany the educational technology, and educators need support for not only learning how the software works, but also how the software can be effectively applied to the current educational setting. Such external support is crucial, for without the approval of educators and teachers, the best educational technology will go unused.

Finally, many critics, such as Stoll and Adcock (2000) and Postman (1993), claim that technology in educational settings is not only unnecessary but can also actually be harmful to learners, especially children. A common criticism of educational technology is that somehow the technology is intended to be a panacea for the current educational ills. Designers of educational technology need to be wary of their claims and intentions for the technology. Educational software is not meant to replace teachers, serve as "one-stop teaching machines," or replace the human–human connections so important to learning. Nonetheless, technology can certainly be a powerful component in an educational system, offering new experiences and new tools for learners.

Will software and technology have a significant impact on education? We still do not fully know the prognosis. There have been some "misses," instances in which educational technology was over hyped, poorly designed, and unused. But this is probably the case in any area where technology is being increasingly used in recent years. There has also been a fair share of "hits," instances in which technology afforded learners the opportunity to investigate, engage in, and understand practices and phenomena that would have been more difficult without the use of technology. In this chapter, we have summarized different approaches to educational technology. But whether one adopts a behaviorist, information processing, constructivist approach—or a combination of the three—the responsibility still lies with the HCI community to effectively design, support, and implement technology that can offer new and valuable educational experiences.

ACKNOWLEDGMENTS

This material is based on work supported by the National Science Foundation under Grant nos. REC 99-80055 and ITR 00-85946.

References

Anderson, J. R. (1983). *The architecture of cognition*. Cambridge, MA: Harvard University Press.

Anderson, J. R., Reder, L. M., & Simon, H. (1998). Radical constructivism and cognitive psychology. In D. Ravitch (Ed.), *Brookings Papers on Education Policy 1998*. Washington, DC: Brookings Institute Press.

Belamy, R. K. E. (1996). Designing educational technology: Computer-mediated change. In B. A. Nardi (Ed.), *Content and consciousness: Activity theory and human-computer interaction* (pp. 123–146). Cambridge, MA: MIT Press.

Beyer, H., & Holtzblatt, K. (1999). Contextual design. *Interactions, 6*, 32–42.

Brown, J. S., Collins, A., & Duguid, P. (1989). Situated cognition and the culture of learning. *Educational Researcher, 18*, 32–42.

Card, S. K., Moran, T. P., & Newell, A. (1983). *The psychology of human-compute interaction*. Mahwah, NJ: Lawrence Erlbaum Associates.

Collins, A., Brown, J. S., & Newman, S. E. (1989). Cognitive apprenticeship: Teaching the crafts of reading, writing, and mathematics. In L. B. Resnick (Ed.), *Knowing, learning, and instruction: Essays in honor of Robert Glaser*. Mahwah, NJ: Lawrence Erlbaum Associates.

Corbett, A. T., Koedinger, K. R., & Hadley, W. H. (2001). Cognitive tutors: From the research classroom to all classrooms. In P. S. Goodman (Ed.), *Technology enhanced learning: Opportunities for change*. Mahwah, NJ: Lawrence Erlbaum Associates.

Gertner, A., & VanLehn, K. (2000). Andes: A coached problem solving environment for physics. In G. Gauthier, C. Frasson, & K. VanLehn (Eds.), *Intelligent Tutoring System: 5th International Conference ITS 2000*. Berlin: Springer.

Guzdial, M. (1993). *Emile: Software-realized scaffolding for science learners programming in mixed media*. Unpublished Ph.D. dissertation, University of Michigan.

Hogan, K., Nastasi, B. K., & Pressley, M. (2000). Discourse patterns and collaborative scientific reasoning in peer and teacher-guided discussions. *Cognition and Instruction, 17*, 379-432.

Lave, J., & Wegner, E. (1991). *Situated learning: Legitimate peripheral participation*. Cambridge, UK: Cambridge University Press.

Lesgold, A. (1986). *Guide to cognitive task analysis*. Pittsburgh, PA: University of Pittsburgh Learning Research and Development Center.

Linn, M. C. (1996). Key to the information highway. *Communications of the ACM, 39*(4), 34-35.

Mayhew, D. J. (1999). *The usability engineering lifecycle*. San Francisco, CA: Morgan Kaufmann.

Metcalf, S. J. (1999). *The design of guided learner-adaptable scaffolding*. Unpublished Ph.D. dissertation, University of Micighan.

Norman, D. A. (1986). Cognitive engineering. In D. A. Norman & S. W. Draper (Eds.), *User-centered system design*. Mahwah, NJ: Lawrence Erlbaum Associates.

Norman, D. A. (1990). *The design of everyday things*. New York: Doubleday-Currency.

Papert, S. (1993). *The children's machine: Rethinking school in the age of the computer*. New York: Basic Books.

Piaget, J. (1954). *The construction of reality in the child*. New York: Basic Books.

Postman, N. (1993). *Technopoly*. Vintage Books.

Quintana, C., Eng, J., Carra, A., Uw, H., & Soloway, E. (1999). Symphony: A case study in extending learner-centered design through process space analysis. *Human Factors in Computing Systems: CHI '99 Conference Proceedings*. New York: ACM Press.

Quintana, C., Fretz, E., Krajcik, J., & Soloway, E. (2000). Assessment strategies for learner-centered software. In B. Fishman &

S. O'Conner-Divilbiss (Eds.), *Proceedings of ICLS 2000*. Mahwah, NJ: Lawrence Erlbaum Associates.

Salomon, G., Perkins, D. N., & Globerson, T. (1991, April). Partners in cognition: Extending human intelligence with intelligent technologies. *Educational Researcher* (pp. 2-9).

Singer, J., Marx, R. W., Krajcik, J., & Clay Chambers, J. (2000). Constructing extended inquiry projects: Curriculum materials for science education reform. *Educational Psychologist, 35*(3), 165-178.

Soloway, E., Guzdial, M., & Hay, K. H. (1994). Learner-centered design: The challenge for HCI in the 21st centrury. *Interactions, 1*(2), 36-48.

Soloway, E., Jackson, S. L., Klein, J., Quintana, C., Reed, J., Spitulnik, J., Stratford, S. J., Studer, S., Eng, J., & Scala, N. (1996). Learning theory in practice: Case studies in learner-centered design. *Human Factors in Computing Systems: CHI '96 Conference Proceedings*. New York: ACM Press.

Soloway, E., & Pryor, A. (1996). The next generation in human-computer interaction. *Communications of the ACM, 39*(4), 16-18.

Squires, D., & Preece, J. (1999). Predicting quality in educational software: Evaluating for learning, usability, and the synergy between them. *Interacting with computers, 11*, 467-483.

Stoll, C., & Adcock, S. (2000). *High-tech heretic: Reflections of a computer contrarian*. New York: Anchor Books.

Tabak, I., Sandoval, W. A., Smith, B. K., Agganis, A., Baumgartner, E., & Reiser, B. J. (1995). Supporting collaborative guided inquiry in a learning environment for biology. *CSCL '95 Proceedings* (pp. 362-366). New York: ACM Press.

Wood, D., Bruner, J. S., & Ross, G. (1975). The role of tutoring in problem-solving. *Journal of child psychology and psychiatry, 17*, 89-100.

·43·

UNDERSTANDING ENTERTAINMENT: STORY AND GAMEPLAY ARE ONE

Jesse Schell
Disney VR Studio

STORY–GAME DUALITY

In 1900, at the dawn of the 20th century, physicists noticed something very strange. They noticed that electromagnetic waves and subatomic particles, which had long been thought to be fairly well-understood phenomena, were interacting in unexpected ways. Years of theorizing, experimenting, and theorizing led to a bizarre conclusion: Waves and particles were the same thing—both manifestations of a singular phenomenon. This "wave–particle duality" challenged the underpinnings of all that was known about matter and energy and made it clear that we didn't understand the universe quite as well as we had thought.

Now it is the dawn of the next century, and storytellers are faced with a similar conundrum. With the advent of computer games, story and gameplay, two age-old enterprises with very different sets of rules, are showing a similar duality. Storytellers are now faced with a medium in which they cannot be certain what path their story will take, just as the physicists found that they could no longer be certain what path their electrons would take. Both groups can now only speak in terms of probabilities.

Historically, stories have been single-threaded experiences that can be enjoyed by an individual, and games have been experiences with many possible outcomes that are enjoyed by a group. The introduction of the single-player computer game challenged these paradigms. Early computer games were simply traditional games, such as tic-tac-toe or chess, with the computer acting as opponent. In the mid-1970s, adventure games with storylines began to appear that let the player become the main character in the story. Thousands of experiments combining story and gameplay began to take place. Some used computers and electronics, others used pencil and paper. Some were brilliant successes, others were dismal failures. The one thing these experiments proved was that experiences could be created that had elements of both story and gameplay. This fact seriously called into question the assumption that stories and games are governed by different sets of rules. In fact, if we look deeply enough, we find common principles that underlie both story- and game-based entertainment. Understanding these fundamental elements provides insight that can help one to create great entertainment of any type.

IS ENTERTAINMENT IMPORTANT?

Before we enter into the dubious endeavor of trying to understand the mechanics of entertainment, we should consider whether such an undertaking is worthwhile. Often, entertainment is considered to be something frivolous, something that we do to pass the time when we are not doing important things. This point of view is part of the reason that the processes of entertainment have remained unexamined for so long. Trying to understand entertainment is worthwhile for several reasons.

1. *Entertainment defines a society's culture.* The shared enjoyment of music, stories, games, and other entertainment experiences help to bind a people together. To better understand the functioning of human society and culture, we would do well to understand the process of entertainment.

2. *Entertainment is a cornerstone of communication.* Much true understanding is achieved through storytelling. Stories are how we communicate events and thought processes. By giving the idea you are trying to convey the context of a story a listener can visualize, it can be much easier to get the point across. The more compelling the story, the more focused the listener will be—and the more likely that true understanding will take place.

3. *Entertainment is a cornerstone of discovery.* Play is how we examine and manipulate things and ideas. Composers play with melody lines, mathematicians play with equations, architects play with designs. Through play, we are able to make both discoveries that lead to understanding and discoveries that lead to invention. By better understanding the process of play, we better understand the processes of both learning and creativity.

4. *We are the most alive when we are entertained.* People who are well and truly entertained are focused, alert, alive, and enjoying themselves. To entertain someone is to fulfill, at least partially, the needs and desires of another person. To understand entertainment is to understand the workings of the human mind.

THE CHALLENGES OF UNDERSTANDING ENTERTAINMENT

Uncovering the processes of entertainment is not without its challenges. Several obstacles stand in our way.

Obstacle 1: Entertainment Is Generally Not Considered a Topic Deserving of Serious Study

The fact that entertainment is, by definition, enjoyable, can make it difficult for entertainment researchers to be taken seriously by their peers, who may feel that research time and effort would be better spent on more "serious" ventures.

Obstacle 2: Being Entertained Is Partially an Unconscious Process

Much of what is happening to us when we are being entertained is not noticed by the conscious mind, because of the very fact that we are so caught up in the entertainment experience. Ironically, the act of trying to focus on exactly what causes you to be entertained by a particular experience effectively "takes you out" of the experience, and it ceases to be entertaining. With practice, one can learn to observe one's own reactions while being entertained, to "step outside one's self" to simultaneously observe one's own reactions, and to determine what about the experience caused these reactions. Still, due to the difficulty of observing one's internal entertainment process, many people take the process for granted or claim that it cannot be effectively analyzed.

Obstacle 3: Honing Entertainment Skill Is Partially an Unconscious Process

Like the process of being entertained, much of the process of improving your skill as an entertainer is also unnoticed by the conscious mind. The act of creating entertainment that is aesthetically pleasing to someone is often very much a right-brain activity. An entertainer or artist often has inspirations that seem to come from out of the blue. Instead of doing a thorough left-brain analysis of how to improve a particular piece of entertainment, an entertainer often just plays around with it until it "feels right," without thinking about (or being able to determine) why it feels that way. Much of the feedback an entertainer gets is in subtle forms of communication from audience members. A sigh at the wrong time, a laugh that is not sincere, a subtle change in posture of an audience member, these are all things that influence an entertainer as to what elements are succeeding or failing. Often, though, the entertainer perceives these subtle communication cues only subliminally and is not consciously aware of them. Entertainers will talk of getting good or bad "vibes" from their audience and use that as important feedback as to how to improve the experience. The fact that some of the most skilled entertainers are often unaware or unable to describe the processes they use to create great entertainment only adds to the challenge of trying to understand the process of entertainment.

Obstacle 4: A Deep Understanding of Human Psychology Is Required to Truly Understand Entertainment

To compound the dual challenge of observing the enjoyment of entertainment and observing its creation, there is the fact that the functional details of the mental processes that allow us to be entertained, such as attention, perception of time, internal visualization, and empathic projection are not well understood. Without a true understanding of the fundamental elements of what makes the mind tick, it is difficult to properly analyze the mechanics of a higher level phenomenon such as entertainment.

These challenges are daunting ones, creating blind spots that justify the fact that the process of entertainment has been able to remain a mystery for so long. However, just being aware of them helps us to see what they hide. Perhaps by trying to understand the process of entertainment, the functioning of more fundamental aspects of the mind may be revealed.

A MODEL FOR ENTERTAINMENT

Every entertainment experience can be thought of as an unfolding sequence of events, experienced by a guest. The term *guest* is used because it applies equally well to both interactive and noninteractive mediums. These sequences of events can be just about any type of experience: movies, books, games, rollercoasters, conversations, songs, jokes, hiking trips—anything that might be considered entertainment.

Guests possess three special abilities that make it possible for them to enjoy these entertainment experiences: focus, empathy and imagination. It is easy to take these abilities for granted, because they are the elements that make up our thought processes.

Focus

The ability to direct attention on what is most interesting at a given moment is called focus. This combination of unconscious desire and conscious will motivates our every action. If an entertainment experience is less interesting than competing experiences, guests will not focus their attention on it, and it is sure to fail.

Empathy

The ability to put one's self in the place of another is called empathy. Sometimes conscious, sometimes unconscious, this uniquely human ability is one of the cornerstones of human communication. When we see someone who is happy, we can feel his or her joy as if it is our own. When we see someone who is sad, we can feel his or her pain. Entertainers use our power of empathy to make us feel we are part of the story world they are creating.

Imagination

The ability to visualize alternate realities is called imagination. Although imagination is usually associated with fantasy or creativity, we use our imagination almost continuously as part of our moment to moment thought and communication processes. Every time we make even the most trivial choice, such as "coffee or tea?," our imagination is at work, visualizing what it would be like to experience the alternate realities that either choice would provide. If you hear a short story, such as "The mailman stole my car," your imagination provides a visualization (although probably a cloudy one) of the event. Skilled entertainers know how to stretch our imaginations, giving us all the right information to make the alternate reality they are creating seem both real and compelling.

Focus, empathy, and imagination are the raw materials that make entertainment possible. How these abilities are manipulated by an entertainment experience is what determines whether the experience is a success or failure. What is needed now is a method for analyzing an entertainment experience to gauge how successfully it makes use of these three abilities.

INTEREST CURVES

The quality of an entertainment experience can be measured by the extent to which its unfolding sequence of events is able to hold a guest's interest. The level of interest over the course of the experience can be plotted out in an interest curve.

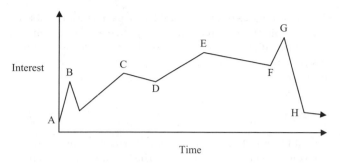

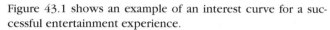

FIGURE 43.1. A good interest curve.

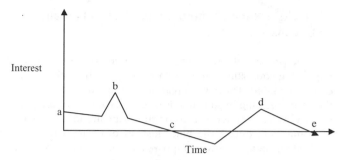

FIGURE 43.2. A bad interest curve.

Figure 43.1 shows an example of an interest curve for a successful entertainment experience.

At Point A, the guests come into the experience with some level of interest; otherwise, they probably wouldn't be there. This initial interest comes from preconceived expectations about how entertaining the experience will be. Depending on the type of experience, these expectations are influenced by the packaging, the advertisements, advice from friends, and so on. While we want this initial interest to be as high as possible to get guests in the door, overinflating it can actually make the overall experience less interesting, as we will discuss later.

Then the experience starts. Quickly we come to Point B, sometimes called "the hook." This is something that really grabs the guests, to get them excited about the experience. In a musical, it is the opening number, or sometimes, the overture. In *Hamlet*, it is the appearance of the ghost. In a video game, it often takes the form of a little movie before the game starts. In this chapter, it was that particle physics story. (Perhaps it worked; after all, you are still reading this.) Having a good hook is important. It gives the guest a hint of what is to come and provides a nice interest spike that will help sustain focus over the less interesting part in which the experience is beginning to unfold, but not much has happened yet.

Once the hook is over, we settle down to business. If the experience is well crafted, the guest's interest will continually rise, temporarily peaking, for example, at Points C and E, occasionally dropping down a bit at Points D and F, only in anticipation of rising again.

Finally, at Point G, there is a climax of some kind, and by Point H, the story is resolved, the guest is satisfied, and the experience is over. Hopefully, guests goes out with some interest remaining, or perhaps with even more than when they arrived. When show business veterans say "leave them wanting more," this is what they mean.

Of course, not every good entertainment experience follows this exact curve. But most successful entertainment experiences will contain some of the elements that Fig. 43.1 displays.

Figure 43.2, on the other hand, shows an interest curve for a less successful entertainment experience. There are lots of possibilities for bad interest curves, but this one is particularly bad, although not as uncommon as one might hope.

As in Fig. 43.1, the guest comes in with some interest at Point a but is immediately disappointed, and because of the lack of a decent hook, the guest's interest begins to wane. Eventually, something somewhat interesting happens, which is good, but

it doesn't last, peaking at Point b, and the guest's interest continues its downhill slide, until it crosses, at Point c, the interest threshold. This is the point where the guests have become so disinterested in the experience that they change the channel, leave the theater, close the book, or shut off the game. This dismal dullness doesn't continue forever, and something interesting does happen later at Point d, but it doesn't last, and instead of coming to a climax, the experience just peters out at Point e—not that it matters, because the guest probably gave up on it some time ago.

Interest curves can be a useful tool when creating an entertainment experience. By charting the level of expected interest over the course of an experience, trouble spots often become clear and can be corrected. Furthermore, when observing guests having the experience, it is useful to compare their level of observed interest to the level of interest that you, as an entertainer, anticipated they would have. Often, plotting different curves for different demographics is a useful exercise. Depending on your experience, it might be great for some groups, but boring for others (e.g., "guy movies" vs. "chick flicks"), or it might be an experience with "something for everyone," meaning well-structured curves for several different demographic groups.

What Comprises Interest?

At this point, you might find your analytical left brain crying out, "I like these charts and graphs, but how can I objectively evaluate how interesting something is to another person? This all seems very touchy-feely!" And it is very touchy-feely. To determine the interest level, you have to experience it with your whole self, using your empathy and imagination, using skills of the right brain as well as the left. Still, your left brain may be happy to know that overall interest can be broken down further, into three factors: inherent interest, poetry of presentation, and psychological proximity.

Inherent Interest. Some events are simply more interesting than others. Generally, risk is more interesting than safety, fancy is more interesting than plain, and the unusual is more interesting than the ordinary. A story about a man wrestling an alligator is probably going to be more interesting than a story about a man eating a cheese sandwich. We simply have internal drives that push us to be more interested in some things than others.

The events don't stand alone, however. They build on one another, creating what is often called the story arc. Part of the inherent interest of events depends on how they relate to one another. For example, in "Goldilocks and the Three Bears," most of the events in the story aren't very interesting: Goldilocks eats porridge, sits in chairs, and takes a nap. But these boring events make possible the more interesting part of the story where the bears discover their home has been disturbed.

Poetry of Presentation

This refers to the aesthetics of the entertainment experience. The more beautiful the artistry used in presenting the experience—whether that artistry be writing, music, dance, acting, comedy, cinematography, graphic design—the more interesting and compelling the guests will find it. Of course, if you can give a beautiful presentation to something that is inherently interesting in the first place, that is all the better.

Psychological Proximity

This is the extent to which the entertainer compels guests to use their powers of empathy and imagination to put themselves into the experience. This is the factor that is crucial to understanding the commonality between story and gameplay and requires some explanation (Fig. 43.3).

Consider the example of winning the lottery (an inherently interesting event). If a stranger wins the lottery, you might be mildly interested in hearing about it. If one of your friends wins the lottery, that is somewhat more interesting. If you win the lottery, you will surely be interested enough to focus your

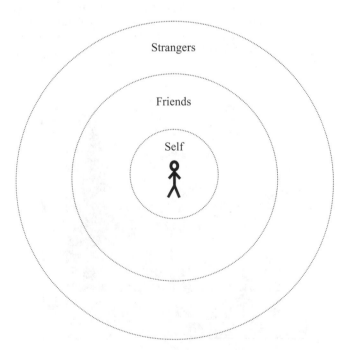

FIGURE 43.3. Psychological proximity.

attention on that fact. Events that happen to us are just more interesting than events that happen to other people.

You would think that this would put storytellers at a disadvantage, because the stories they tell are usually about someone else, often someone you have never heard of, or even someone who doesn't actually exist. However, storytellers know that guests have the power of empathy, the ability to put themselves in the place of another person. An important part of the art of storytelling is to create characters with whom the guests can easily empathize; the more the guests can empathize with the characters, the more interesting the events become that happen to those characters. When you start almost any entertainment experience, the characters in it are strangers. As you get to know them, they become your friends, and you begin to care about what happens to them, and your interest in events involving them grows. At some point, you might even mentally put yourself in their place, bringing you to the height of psychological proximity.

In terms of trying to build psychological proximity, imagination is as important as empathy. Humans exist in two worlds: the outward-facing world of perception and the inward-facing world of imagination. Every entertainment experience creates its own little world in the imagination. This world does not have to be realistic (although it might be), but it does need to be internally consistent. When the world is consistent and compelling, it fills the guest's imagination, and guests mentally enter the world. We often say that guests are "immersed" in the world of the story. This kind of immersion increases psychological proximity, boosting the overall interest of the guest significantly. The suspension of disbelief that keeps guests immersed in the story world is fragile indeed. One small contradiction is all it takes to bring guests back to reality, and "take them out" of the experience.

Episodic forms of entertainment, such as soap operas, sitcoms, and serialized fiction, take advantage of the power of psychological proximity by creating characters and a world that persist from one entertainment experience to the next. Returning guests are already familiar with these persistent characters and settings; each time they experience an episode, their psychological proximity grows, and the fantasy world becomes "more real." This episodic strategy can quickly backfire, however, if the creator fails to carefully maintain the integrity of the characters and the world. If new aspects of the world contradict previously established aspects, or if the regular characters start to do or say things that are "out of character" to serve the story line of some new episode, then not only is the episode compromised, but the integrity of the entire fantasy world, which spans all episodes, past, present, and future. From the guest's point of view, one bad episode can spoil the entire series, because the compromised characters and setting will seem phony from the point of contradiction onward, and it will be difficult for the guest to sustain psychological proximity.

Another way to build psychological proximity in the story world you have created is to provide multiple ways to enter that world. Many people think of toys and games based on popular movies or television shows as nothing but a gimmicky way to make a few extra dollars by riding the coattails of a successful entertainment experience. But these toys and games provide

new ways for children to access an established fantasy world. The toys let them spend more time in that world, and the longer they spend imagining they are in the fantasy world, the greater their psychological proximity for that world and the characters in it becomes. The enduring popularity of *Star Wars* and its sequels is partly due to the fact that excellent toys were marketed along with the movie, making it easier for children to spend their playtime imagining that they are part of the *Star Wars* universe. Of course, if the toys and games meant to complement an entertainment experience are a poor match, or if the primary experience itself is uninteresting to guests, this strategy for boosting psychological proximity is unlikely to succeed.

Interactive entertainment has an even more remarkable advantage, in terms of psychological proximity. The guest can *be* the main character. The events actually happen to the guest and are all the more interesting for that reason. Also, unlike story-based entertainment, in which the story world exists only in the guest's imagination, interactive entertainment creates significant overlap between perception and imagination, allowing the guest to directly manipulate and change the story world. This is why video games can present events with little inherent interest or poetry but still be compelling to guests. What they lack in inherent interest and poetry of presentation, they can often make up for in psychological proximity.

INTEREST FACTOR EXAMPLES

To ensure the relationship between the interest factors is clear, let's compare some different entertainment experiences. Let's say some brave street performers attract attention by juggling running chainsaws (Fig. 43.4). This is an inherently interesting event. It is difficult not to look—at the very least— when it is going on around you. The poetry with which it is presented, however, is usually somewhat limited. There is some psychological proximity, however, because it is easy to imagine what it would be like to catch the wrong end of a chainsaw. When you witness the act in person, the psychological proximity is even greater.

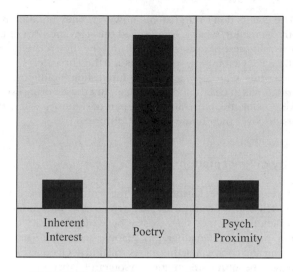

FIGURE 43.5. Violin concerto interest factors.

How about a violin concerto (Fig. 43.5)? The events (two sticks being rubbed together) are not that inherently interesting, and the psychological proximity is usually not notable. In this case, the poetry has to carry the experience. If the music isn't beautifully played, the performance will not be interesting. There are exceptions here. The inherent interest can build when the music is well structured or when the evening's program is well structured. If the music makes you feel as if you are in another place, or if you feel a particular empathy for the musician, there may be significant psychological proximity.

Consider the popular video game, Tetris (Fig. 43.6). The game mainly consists of an endless sequence of falling blocks. This leaves little room for inherent interest or poetry of presentation; however, the psychological proximity can be intense. The guest makes all the decisions, and success or failure is completely contingent on the guest's performance. This is a shortcut that traditional storytelling is unable to take. In terms of an interesting entertainment experience, the large amount of

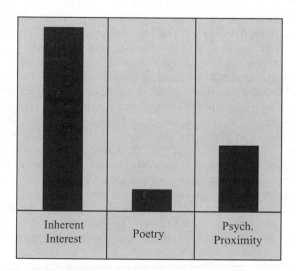

FIGURE 43.4. Chainsaw juggling interest factors.

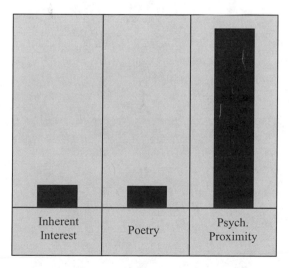

FIGURE 43.6. *Tetris* interest factors.

psychological proximity makes up for what is lacking in poetry or inherent interest.

THE CHALLENGE OF INTERACTIVITY

Although a great deal has been written about the successful design of traditional linear entertainment experiences, relatively little has been said about techniques for interactive entertainment design. To understand better the relationship between story and gameplay, it is worth devoting some attention to these techniques and considering how they relate to guests' focus, empathy, and imagination, and an experience's inherent interest, poetry of presentation, and psychological proximity.

Because interactive entertainment has such an advantage in terms of easily providing high psychological proximity to guests, one might think that most traditional storytellers would prefer to create interactive experiences. This is not generally the case, however, and for a good reason. Despite the advantages of interactive entertainment, it has a serious disadvantage: The storyteller loses control over the sequence of events. Many traditional storytellers cannot even imagine creating a compelling interactive experience, because they are so accustomed to having direct control over each event.

Noninteractive entertainment is challenging to create because although the creator has direct control over the sequence of events, the events are not happening directly to the guest. Interactive entertainment is challenging to create because although the sequence of events is happening directly to the guest, the creator only has indirect control over what those events are going to be.

Many experiments have been conducted in which "gameplay elements" are inserted into a noninteractive experience. This often takes the form of a movie sequence followed by a gameplay sequence, followed by a movie sequence, and so on. These are seldom popular with guests, because continually switching from interactive mode (you are the character) to noninteractive mode (you are watching the character) is jarring to the imagination; it is difficult to stay immersed in either the movie or the game.

Instead of constructing an experience out of inelegant compromises while bemoaning a loss of control, the interactive entertainer must focus on what is controllable, for *the art of interactive entertainment is the art of indirect control*. By influencing the will of the guest, you can influence the sequence of events, and thereby control the interest curve of the experience without wresting control from the guest and thereby jarring the imagination. There are myriad ways to exert influence over a guest, but they can be roughly grouped into four main categories: environment, interface, avatar, and story.

Environment: Influence the Eyes

Anyone who works in an area of the visual arts knows that layout affects where guests will look. This becomes important in an interactive experience, because guests tend to go to what draws their attention. Therefore, if you can control where guests will

FIGURE 43.7. Example of indirect control through graphic design.

look, you can control where they are going to go. Figure 43.7 shows a simple example.

It is difficult, looking at this picture, for your eyes not to be led to the center of the image. A guest looking at this scene in an interactive experience would be likely to focus on the central triangle before considering what might be at the edges of the frame. This is in sharp contrast to Fig. 43.8, where the guest's eyes are compelled to explore the edges of the frame and beyond. If this scene were part of an interactive experience, it would be a good bet that the guest would be trying to find out more about the objects on the edges, rather than the circle in the middle of the scene.

These examples are abstract ones, but there are plenty of real-world examples that illustrate the same thing. Set designers, illustrators, architects, and cinematographers use these principles to guide the eye of their guests and indirectly control

FIGURE 43.8. Another example of indirect control through graphic design.

the focus. One excellent example is the castle at the center of Disneyland. Walt Disney knew there was some risk of guests entering the park and milling about at the entrance, unsure of where to go. The castle is placed such that guests' eyes are immediately drawn to it upon entering the park (similar to Fig. 43.7), and their feet are quick to follow. Soon the guests are at the Disneyland hub, with several visual landmarks beckoning them in different directions (similar to Fig. 43.8). Indirectly, Disney was able to control guests to do just what he wanted them to do: move quickly to the center of Disneyland and then branch out randomly to other parts of the park. Of course, guests are seldom aware of this manipulation. After all, no one told them where to go; all the guests know is that without much thinking, they ended up somewhere interesting and had a fun entertainment experience.

Interface: Influence the Hands

In a good interactive entertainment experience, guests forget that the interface exists. This can only happen if the interface used for the experience is appropriate for what guests try to do with it. However, the interface also exerts indirect control over what the guest will attempt to use it for. For example, if your interactive experience begins with the guest controlling a character driving a car and the interface is a mouse and keyboard, the guest may try to do other things than just drive. They may look for a way that stop the car and get out to walk around. On the other hand, if you use a physical steering wheel and foot pedal for the interface, the guest will focus on the driving task, because the interface has sent the unconscious message that this experience is about driving. A good interface (physical or virtual) sets limits, indirectly controlling the guest to focus on the parts of the experience that are the most entertaining.

Avatar: Influence the Body

Every virtual experience has two components to the interface: the part you control in the real world and the part you control in the virtual world. Often, the virtual component of the experience takes the form of a familiar creature, such as a human being. This virtual puppet is referred to as an *avatar*. Some avatars of note include Mario in *Donkey Kong*, and Lara Croft in *Tomb Raider*. This is a popular form of interface because it takes advantage of the guest's ability to use empathy and imagination to project themselves into the place of the avatar. Many designers have found this notion counterintuitive, believing that greater psychological proximity can be gained by providing a first-person perspective on a scene, with no visible avatar. However, the power of empathy is strong, and when controlling a visible avatar, guests often wince in imagined pain upon seeing their avatar suffer a blow, or sigh in relief upon seeing their avatar escape physical harm. It is almost as if the avatar is a kind of kinesthetic voodoo doll for the guest. Bowlers are another example of this phenomenon, as they try to exert "body English" on a bowling ball as it rolls down the lane toward the pins. These movements are largely subconscious, a result of bowlers

projecting themselves onto the ball. In this sense, the bowling ball serves as a bowler's avatar.

You can exert significant indirect control over what a guest will try to do in an interactive experience by crafting an avatar that is well matched with the sequence of events you have in mind. Because guests project themselves into the avatar, its appearance will indirectly control what guests try to do in the experience.

Story: Influence the Will

Environment, interface, and avatar are all important, but subtle, ways to indirectly control the behavior of a guest in an interactive experience. The most overt form of indirect control that a designer can exert over a guest is that of story. The story gives the guest a goal, a reason for participating in the experience. When a guest is focused on achieving a particular goal, the guest's interest level is high. This is the primary method that game designers use when trying to manage the interest curve for the experience. Most games feature a series of goals, each more challenging than the last, each building on the previous goals. Often, this series of goals takes the form of a linear story, which has the benefit of increasing the guest's interest by featuring a plot, characters, and sequence of events that the guest actually cares about.

There are significant challenges to using story to control a guest's behavior, however. These relate back to the three interest factors described earlier. If guests find a particular goal uninteresting, or too difficult, they may give up on it, putting an end to the entire experience. Similarly, if the goals are not communicated well, players may give up on the experience, which may feel pointless without a clear goal. The other characters in the experience may be computer controlled, and if their behavior and performance is not well crafted (a lack of poetry of presentation), guests may suffer a decrease in psychological proximity and lose interest in the experience.

Some experiences attempt to solve the problem of unrealistic behavior of computer-controlled characters by replacing them with characters controlled by other guests. In these multiplayer experiences, players often feel intense psychological proximity, because the other characters in the experience are real people, often friends of the guest. Although this can be a great boon to the overall interest of the game, it can also be a detriment. In a multiplayer experience, the designer has even less control over the sequence of events, and there is the danger that even though the characters in the experience have real human souls, the sequence of events that occurs may not be exciting enough to sustain their interest. Successful multiplayer experiences can be an even greater challenge to a designer, but the potential payoff can be commensurately great.

CONCLUSION

Entertainment takes thousands of different forms, and new technologies seem to breed new varieties almost faster than we can enjoy them, much less analyze them in detail. Nonetheless, all forms of entertainment share a common basis: They all

take place in the human mind, using the mechanisms of focus, empathy, and imagination. Understanding these mechanisms is important, for the better we understand entertainment, the better we understand ourselves.

Resources

The following selections are excellent ways to further explore the psychology of entertainment.

Alexander, C. (1987). *The timeless way of building*. New York: Oxford University Press.

Arijon, D. (1976). *Grammar of the film language*. Beverly Hills: Silman-James Press.

Arnheim, R. (1974). *Art and visual perception*. Berkeley: University of California Press.

Bang, M. (1991). *Picture this*. Boston: Little, Brown, and Company.

Csikszentmihalyi, M., & Csikszentmihalyi. I. S. (1997). *Optimal experience*. Cambridge, UK: Cambridge University Press.

Dali, S. (1992). *Fifty secrets of magic craftsmanship*. New York: Dover.

Doblin Group. *A model of compelling experiences*. Retrieved from www.doblin.com

Dodsworth, C., Jr. (1998). *Digital illusion* (SIGGRAPH Series). New York: ACM Press.

Flaxon, D. N. *Flaxon alternative interface technologies*. Retrieved from http://www.sonic.net/~dfx/fait/

Henderson, M. (1997). Star Wars: *The magic of myth*. New York: Bantam Books.

Marling, K. A. (1997). *Designing Disney's theme parks: The architecture of reassurance*. Paris: Flammarion.

McCloud, S. (1994). *Understanding comics*. New York: Kitchen Sink Press.

McLuhan, M., & Fiore, Q. (1996). *The medium is the massage*. San Francisco: Hardwired.

McLuhan, M. (1998). *Understanding media*. Cambridge, MA: MIT Press.

Miall, D. S. (1995). Anticipation and feeling in literary response. *Poetics, 23*, 275-298.

Moore, S. (1999). *We love Harry Potter!* New York: St. Martin's Griffin.

Murray, J. H. (1997). *Hamlet on the Holodeck*. New York: The Free Press.

Nelms, H. (1969). *Magic and showmanship*. New York: Dover.

Propp, V. (1998). *Morphology of the folktale*. Austin: University of Texas Press.

·44·

MOTOR VEHICLE DRIVER INTERFACES

Paul Green
University of Michigan Transportation Research Institute

INTRODUCTION

This chapter is written for professionals familiar with human–computer interaction (HCI) but not with the issues and considerations particular to motor vehicles. For non-HCI professionals, reading chapters 1–19 of this text should provide the desired background. For automotive industry interface designers, this chapter should pull together information dispersed in the literature.

The emphasis of this chapter is on how the safety-critical nature of driving changes the implementation of standard HCI practice. Furthermore, this chapter, as one in a reference textbook should, emphasizes design practice over a review of the experimental literature. The design of traditional (noncomputer) motor vehicle driver interfaces (such as switches for the headlights and windshield wipers) is not covered. For information on those interfaces, see Peacock and Karwowski (1993) or the latest edition of the *Society of Automotive Engineers (SAE) Handbook* (SAE, 2000). Readers interested in additional research literature on telematics should see Michon (1993), Parkes and Franzen (1993), Barfield and Dingus (1997), or Noy (1997).

HCI is of interest to motor vehicle designers because of the rapid growth of driver information systems that use computers and communications, a collection of systems referred to as *telematics*. Driver interfaces for telematics applications is the focus of this chapter. The goal of telematics development is to enhance public safety, make transport more efficient (saving time and fuel), make driving more enjoyable, and make drivers more productive while commuting.

What Kinds of Telematics Products Are Likely in the Near Term?

There have been several studies over the years that have used expert opinions to predict the future of automotive electronics, specifically telematics applications (e.g., Ribbens & Cole, 1989; Richardson & Green, 2000; Underwood, 1989, 1992; Underwood, Chen, & Ervin, 1991). The most recent projections appear in Green et al. (2001), summarizing findings from 83 senior executives in the automotive industry. Table 44.1 lists the mean value of the year estimated when each feature would be introduced in 10% of new luxury vehicles in 2004 and 2005. (See that report for information on features in the 2006–2008.)

Because these data were collected in August 2000 and the planning horizon for a new vehicle is 3 years, these estimates should reflect real product intentions, not speculation about the future. For some features (built in personal digital assistants (PDAs), electronic toll tags), there was not complete agreement as to when introduction would occur, because a significant fraction of respondents thought some features would never be introduced. Nonetheless, the table makes the point that the number of systems soon to be introduced is significant. The major categories of interest are driver assistance systems (various collision warning and avoidance systems, night vision), information systems (cell phones, e-mail/Internet, PDAs), and the use of voice technology. In about a decade or less, I believe that wearable computers will be a topic of significant interest for driving (see chapter 32 by Siewiorek and Smailagic.) In aggregate, these systems could potentially reshape the driver's task from one of real-time control of the vehicle (steering the vehicle and using the throttle to adjust speed) to information management.

Chapter Organization

How might one organize information on telematics? In their classic paper on usability, Gould and Lewis (1985) identified three key principles to be followed when designing products for ease of use: early focus on users and tasks, empirical measurement, and iterative design. These principles not only apply to office applications and Web development, but automotive applications as well. In the automotive context, designers need to understand who drive vehicles (users), what in-vehicle tasks they perform, and, most important, the driving task and task context, and the consequence of task failures. These topics are the focus of the first part of this chapter. Secondly, it is also important to be able to measure driver and system performance (empirical measurement). This topic constitutes the second part of the chapter.

Surprisingly, there has been little reporting of the use of iterative design in developing driver interfaces, although it is done. A great deal of automotive design relies on following design guidelines, the third focus of this chapter.

WHAT IS THE DRIVING CONTEXT IN WHICH USERS PERFORM TASKS?

Scott McNealy, the chief executive officer of Sun Microsystems has said, "A car is nothing more than a Java technology-enabled browser with tires" (Kayl, 2000). He is wrong—dead wrong. I have never heard of anyone claiming, "A computer came out of nowhere, hit me, and vanished." Yet such claims are heard for motor vehicles by police officers and insurance adjusters every day. Few people have ever killed as a consequence of operating a computer at their desks, but the loss of life associated with crashes arising from normal motor vehicle operation is huge.

Murray and Lopez (1996a) estimated that 1,391,000 people would die in road traffic crashes in 2000 (see also http://www.unece.org/trans/roadsafe/rsras.htm; http://www.oecd.org/dsti/sti/transpor/road/index.htm). Traffic crashes have been ranked ninth in terms of the percentage of disability-adjusted life-years lost (Murray & Lopez, 1996b), well ahead of war (16th), violence (19th), and alcohol use (20th). If the current trends continue, traffic crashes will become the third largest cause of death and disability after clinical depression and heart disease by the year 2020. Traffic crashes ranked as the ninth biggest killer in the world in 1990.

Additional insights come from crash data for the United States, for which reliable, detailed crash statistics are available. According to the U.S. Department of Transportation (2000), there were 41,611 highway fatalities in the United States in 1999. Of these, 20,818 were passenger-car occupants, 2,472

TABLE 44.1. Estimated Year of Feature Introduction Into Luxury Vehicles

Mean	Never[a]	Feature	Description
2004.3	1	Built-in wireless phone interface	
2004.4	0	GPS navigation	Uses GPS for location (e.g., Hertz NeverLost)
2004.6	1	Automatic collision notification	Calls when airbag deploys (e.g., OnStar)
2004.7	2	Satellite radio	Nationwide satellite broadcast of 100 channels (e.g., XM, Sirius)
2004.8	4	Removable media for entertainment and data	
2004.8	0	E-mail/Internet access	
2004.8	10	Built in PDA (e.g., Palm) docking station	
2004.8	0	Adaptive cruise control	Scans for traffic and adjusts speed to maintain driver set separation
2005.0	1	Rear parking aid	
2005.2	1	MP3 support	Supports MP3 audio file format
2005.2	0	Bluetooth support	Short-range wireless communication allows direct link of cell phones
2005.3	0	Automatic download of traffic and congestion information	
2005.3	2	Blind spot detection and warning	Uses radar and other systems to detect vehicles to the side and rear
2005.5	1	Voice operation of some controls	
2005.6	2	Downloadable software features	
2005.6	2	Downloadable software fixes	
2005.6	3	Forward collision warning	Laser radar systems to detect vehicles ahead
2005.7	4	Forward parking aid	Sonar-based short-range systems to assess distance ahead
2005.7	6	Lane departure warning	Video systems scan for lane markings
2005.7	4	Dual voltage (42/12 V)	Current power is 12-V
2005.8	12	Built-in electronic toll and payment tag	Allows travel through toll booths without stopping
2005.8	0	General purpose text/data speech capability	
2005.9	7	Large general-purpose display	
2005.9	6	Off-board applications via data link	
2005.9	9	Night vision	Infrared-based systems to show objects ahead on a head-up display

[a]Respondents had the option of identifying a particular year or selecting never.
GPS = global positioning systems, PDA = personal digital assistant.

were motorcyclists, 12,001 were truck occupants, 58 were bus occupants, 4,906 were pedestrians, 750 were pedalcyclists, and 606 were in other categories. The truck category includes light trucks (pickups, sport utility vehicles, and vans) that are commonly used for personal transportation in the United States. For additional information, see the latest edition of *Injury Facts*, formerly titled *Accident Facts* (National Safety Council, 2000) and the U.S. Department of Transportation Fatal Analysis Reporting System Web site (http://www-fars.nhtsa.dot.gov/).

What Kinds of Crashes Are Associated With Telematics?

It is widely accepted that some telematics tasks are distracting and distraction can lead to crashes. Crash statistics relating to distraction (Table 44.2) were computed by Wang, Knipling, and Goodman (1996) using the Crashworthiness Data System (CDS). Crashes in CDS are an annual probability sample of approximately 5,000 police-reported crashes involving at least one passenger vehicle that was towed from the scene (out of a population of almost 3.4 million such crashes). Minor crashes (involving property damage only) are not in CDS. CDS crashes are investigated in detail by specially trained teams of professionals who provide much more information than is given in police reports.

Note that distraction-related crashes primarily involved a single vehicle (41%), although rear-end crashes (moving 10%, stopped 22%) were also common. Intersection crashes represented another 18% of the total. Crashes tended to peak in the morning rush hour and, to a much lesser extent, in the evening rush. The overwhelming majority of the crashes occurred in good weather (86% clear, 10% rain, 11% snow/hail/sleet) and many occurred at lower speeds (0–35 miles/hr = 40%; 40–50 miles/hr = 40%; 55–60 = 17%; > 65 miles/hr = 4%). Thus, these data suggest that device test scenarios should emphasize situations in which single-vehicle crashes are likely (often with the vehicle running off the road) as well as those involving rear-end collisions into stopped vehicles.

Other recent evidence appears in Stutts, Reinfurt, Staplin and Rodgman (2001) (see http://www.aaafoundation.org/projects/index.cfm?button=disdrvtest for a summary.) They analyzed

TABLE 44.2. Distraction/Inattention Crashes by Crash Type*

Crash Type Row %, Column %	Sleepy	Distracted	Looked But Did Not See	Unknown	Attentive	Total
Single vehicle	5.8	18.1	0.2	31.8	44.1	100.0
	66.2	41.2	0.7	20.6	45.9	30.0
Rear-end, lead vehicle moving	12.7	21.3	3.4	48.3	14.3	100.0
	27.9	9.6	2.0	6.4	2.9	5.9
Rear-end, lead vehicle stopped		23.9	11.4	52.6	11.8	100.0
	*	21.9	13.8	14.1	4.9	12.1
Intersection/crossing path		7.0	17.9	52.8	22.3	100.0
	*	18.1	63.6	39.8	26.6	34.3
Lane change/merge		5.6	17.2	41.8	35.3	100.0
	*	1.6	6.7	3.4	4.6	3.8
Head-on	1.0	7.0	8.1	46.4	37.5	100.0
	1.7	2.2	3.5	4.3	5.4	4.2
Other		7.3	9.7	53.5	28.9	100.0
	*	5.4	9.7	11.4	9.7	9.7
Total crashes	2.6	13.2	9.7	45.7	28.8	100.0
	100.0	100.0	100.0	100.0	100.0	100.0

*Numbers are percentage.

CDS files from 1995 (the 1st year for which driver attention was examined) until 1999. Table 44.3 shows some of the summary data they reported.

These data, however, should be used with some caution. First, there is a high percentage of unknown and missing data in the set. For example, the driver attention status (1 of 360 variables included) was "unknown" for 36% of the cases. Secondly, 5 years have elapsed from the midpoint of this study until when this chapter is written, and additional time will elapse before it is read. Many driver interfaces have changed considerably since the data were collected. For example, at 25% growth per year (according to data on the Cellular Telecommunications and Internet Association [CTIA] Web site: http://www.wow-com.com/industry/stats/surveys/, the number of cell phones in use has grown by a factor of more than 3 between 1997 and 2001. Given that increase in exposure, the percentage of such crashes is likely to be underreported by a factor of 3. Unlike

TABLE 44.3. Crashes for Which Distraction Was a Known Factor (8.3% of the Crashes)

Distraction	Estimate (%)	Error (%)
Outside person, object, or event	29.4	±4.7
Adjusting radio/cassette/CD	**11.4**	**±7.2**
Other occupant	10.9	±3.3
Moving object in vehicle	4.3	±3.2
Other device/object	2.9	±1.6
Adjusting vehicle/climate controls	**2.8**	**±1.1**
Eating/drinking	1.7	±0.6
Using/dialing cell phone	**1.5**	**±0.9**
Smoking related	0.9	±0.4
Other distractions	25.6	±6.0
Unknown distraction	8.6	±5.3

Note. Distractions associated with in-vehicle controls and displays are shown in boldface.

external distractions, distractions due to vehicle hardware can be reduced or eliminated by appropriate engineering on the part of manufacturers and suppliers.

The Scope of the Cell Phone Problem

Because telematics devices are new, the data on device-related crashes is limited. This situation has hampered progress in understanding the risks of such devices, especially cell phones, the device receiving the most attention. In fact, at the time this chapter was written, only four states recorded whether cell phones were causal factors in crashes, and none were recording information about other types of devices. Cell phone use is not recorded by any of the federal crash data systems. There is, however, relevant crash data from the National Police Agency (NPA) in Japan. Admittedly, there is less evidence to assess the quality of that data than there is for U.S. federal or state data. Furthermore, it is suspected that even when investigated, cell phone use is underreported, because reporting involves drivers admitting to something inappropriate. Table 44.4 shows the annual crash statistics for phones, and for perspective, navigation systems.

Using these data, cell phone–related deaths for 2001 in the United States can be estimated. The average number of cell phone–related deaths in Japan was about 25 per year for the

TABLE 44.4. Cell Phone– and Navigation System–Related Crashes in Japan

Year	Cell Phones		Navigation Systems	
	Injuries	Deaths	Injuries	Deaths
1997	2,095	20	117	1
1998	2,397	28	131	2
1999	2,418	24	205	2

1997–1999 period. In determining this estimate, one must realize that there is some random variation in deaths per year, and the number was increasing over that period. On average, the number of traffic deaths in the United States is about 4.5 times that of Japan. Assuming everything else is equal, this would suggest at least 112 ($= 25 \times 4.5$) cell phone–related deaths per year in the United States for the 1997–1999 period. If the number of calls per driver and the mean call duration are assumed unchanged (both probably are underestimates), then the change in exposure of the population between 1998 and 2001 is simply due to the increase in the number of phones in existence. Using the value of 25% growth per year (from the CTIA data) and 112 deaths in 1998 as the base estimate, the U.S. total for 2001 for cell phone–related crashes is 219 deaths ($= 112 \times 1.25$ cubed across a 3-year period) and should continue to increase at 25% per year. Other sources suggest the number of cell phone–related deaths is much higher (http://www.geocities.com/morganleepena/rebuttal.htm).

By way of comparison, about 120 children and small-statured women were killed in front passenger air bag crashes and about 200 people in Firestone-Explorer rollover crashes over several years. Both of these situations have led to major changes in federal requirements and product engineering changes, as well as significant financial losses to several manufacturers (Green, 2001a). In response to the Firestone situation, the U.S. Senate passed the TREAD Act (http://www.driveusa.net/treadact.htm) 3 weeks after it was introduced, a very short time. The message from Congress is clear: Overwhelming crash statistics (that is, numerous deaths) are not necessary before action is taken on a highway safety matter. This change in philosophy has implications for more than just tires, and congressional action could be just as swift for cell phones. Hence, designing telematics interfaces to ensure safety and usability is a topic worthy of significant effort.

Supporting the crash statistics, the most often-cited study of cell phone use is Redelmeier and Tibshirani (1997; see also Redelmeier & Tibshirani, 2001). They examined data for almost 700 drivers who were cell-phone users and were involved in motor vehicle crashes that resulted in substantial property damage. Each driver's cell-phone records for the day of the crash and the previous week were examined. Redelmeier and Tibshirani reported the risk of a crash was 4.3 times greater when a cell phone was used than when it was not. Interestingly, hands-free units had a greater, although not significant, risk ratio than handheld units (5.9:1 vs. 3.9:1). Other data (Koushki, Ali, & Al-Saleh, 1999; Violanti & Marshall, 1996) suggest similar risk ratios. Other human-factors data show drivers taking longer to respond to brake lights of lead vehicles, departing from the lane more often, and exhibiting other undesired characteristics.

If phone use is a concern, which tasks lead to crashes? Table 44.5 shows task-specific crash statistics for cell phones and, for comparison, navigation systems. This NPA data identifies four primary tasks: receiving a call, dialing, talking, and other. What *other* represents is unknown. Notice that NPA reported the number-one cause of crashes was receiving a call, a surprise to many people. Most people, either at work or at home, immediately answer the phone when it rings, almost regardless of the situation. (An exception is at dinnertime in the United States, when telemarketing calls are common.) For example,

TABLE 44.5. Driver Tasks and Crashes (January–November, 1999)

Cell Phone		Navigation System	
Task	Crashes	Task	Crashes
Receiving call	1077		
Dialing	504	Looking	151
Talking	350	Operating	46
Other	487	Other	8
Total	2418	Total	205

one could be in his or her office talking to a very important person, say the president of the United States; but if the phone rang, many people would ask the president to wait a moment while they answered the incoming call. In fact, many people now have Call Waiting, so they can accept additional incoming calls while on the phone.

The habit of always answering the phone, almost regardless of what else is transpiring, also occurs while driving. Thus, even in hazardous traffic situations, people still answer the phone. Recent work at the University of Michigan Transportation Research Center (UMTRI) (Nowakowski, Friedman, & Green, 2001) shows that drivers answer the phone almost immediately, typically in 1 to 4 s from when it begins to ring. Logically, answering the phone usually should not be more important than driving, but because of habit, people behave otherwise.

Not only is the answering task a problem because of its immediacy, but also sometimes because of its visual demands. At the present time, many drivers use portable phones that might be in a jacket pocket on the seat or in a purse. Thus, in answering the phone, the driver's first task might be to search for it. Although hands-free systems can reduce the duration of the disruption, they do not eliminate the disruption itself or its incorrect prioritization.

In the Japanese data, the second most common cause of crashes was dialing. Clearly, going to a hands-free interface can reduce the visual and manual demands, although not the cognitive demands of dialing. In fact, the early results showed that after November 1999, when Japan limited phone use while driving to hands-free devices, crashes were reduced by 75%.

Finally, a substantial number of calls were associated with simply talking on the phone. Talking on the phone is a cognitive distraction, with drivers focusing on the conversation, not driving, although there are some aspects that are manual (such as holding a handheld phone to one's ear). It is this aspect of cell phone use that is commonly observed by drivers. Current wisdom in the United States is that if drivers are weaving and it is at night, they are drunk. If they are weaving and it is during the day, they are on the phone.

Talking on the phone is different from talking to a passenger, especially an adult in the front seat. Often, that adult behaves as a co-driver, looking both ways at intersections and checking the mirrors during lane changes. Furthermore, passengers may moderate their conversation to fit the driving situation. So, when a driver is moving their head to scan an intersection or expressway entrance or exit, passengers often stop talking. People on the phone have no knowledge of the driving situation and keep talking. Sounds to inform callers of the driving situation could assist in reducing the scope of this problem.

Because the risk to the driver depends on the duration of the call, drivers may shorten their calls if a call timer were in an easy-to-see location. Of course, it is possible that such a display could be distracting to drivers and do more harm than good.

I believe that the best long-term strategy is to implement an in-vehicle workload manager. This system would use (a) data from the navigation system (such as lane width and radius of curvature) to assess the demands due to road geometry, (b) data from the adaptive cruise control system (headway and range rate to vehicles ahead) to assess traffic demands, (c) data from the traction control system to assess road surface friction, and (d) data from the wipers, lights, and clock to assess visibility. This information, along with information on the driver (e.g., age) and the specific visual, auditory, cognitive, and motor demands of each in-vehicle task, could be used to schedule the occurrence of in-vehicle tasks. Thus, when driving on a curving road in heavy traffic in a downpour, incoming cell phone calls would be directed to an answering machine, and 30,000-mile maintenance reminders would be postponed. However, when the driving task demand is low, drivers could have access to a wide range or functions (Green, 2001a; Green, 2001b; Michon, 1993).

Many recognize that excessive visual demand is a significant concern for many telematics tasks, and the driver overload and distraction leading to crashes is a major concern. That has led many who are unfamiliar with human factors to suggest that using voice input and output is "the answer." In addition, there has been a huge growth in cell phone market and an explosion in general purpose phone-based services offered (e.g., voicemail, bank by phone), as well many automotive-specific services (e.g., traffic information). However, there are major public concerns that use of cell phones while driving is not safe. In many countries, only hands-free operation is permitted (Australia, England, Israel, Italy, Switzerland, and Spain), and while this chapter was written, New York was the first state to ban the use of hand-held phones while driving (http://assembly.state.ny.us/leg/?bn=S05400&sh=t). As described earlier, allowing only the use of hands-free cell phones eliminates only some causes of phone-related crashes. Virtually all of these laws exempt 911 and other emergency calls (to police, fire and rescue, etc.) because of their immediate life-threatening importance.

WHAT KIND OF TRIPS DO PEOPLE MAKE AND WHY?

Every 5 years, the U.S. Department of Transportation conducts the Nationwide Personal Transportation Survey (renamed the National Household Travel Survey) to obtain travel data for the U.S. (http://www.fhwa.dot.gov////ohim/nptspage.htm). The 2001 survey was underway at the time of this writing. The most recent data are from 1995 (http://www-cta.ornl.gov/npts/1995/Doc/index.shtml).

In the United States, only a small fraction of all travel is by public transit, and that mode has a significant impact only in large metropolitan areas. The overwhelming majority of travel is via privately owned vehicles (Table 44.6). Thus, aggregate data about trip types and distances reflects motor vehicle travel. Notice that walking is the second most commonly used mode

TABLE 44.6. Transportation Mode Used

Mode	% Person Trips	% Person Miles
Privately owned vehicle	86.1	90.8
Walk	5.4	0.3
Other	4.9	5.5
Transit	1.8	3.1
School bus	1.7	1.3

for trips, but because the trips are short, it accounts for a small fraction of the mileage.

According to the 1995 data (Table 44.7), the most common reason for travel is family and personal business, which includes shopping, running errands, and dropping off and picking up others, accounting for almost half of the trips and one third of the person miles.

Travel to and from work accounts for 1 of every 6 trips and one fifth of the miles traveled. Between 1983 and 1995, there was a 37% increase in commuting distance and a 14% increase in commuting time. The average trip to work is 11.6 miles, takes 13.7 mins, and is driven at 33.6 miles per hour. Trip distances and durations vary considerably. Across all purposes, the average adult spends about 1 hr, and 13 mins per day driving.

The perception that work trips occur during the traditional rush hour is not supported by the data, with only 37% of all trips starting between (6–9 a.m.) and (4–7 p.m.). These data provide background information on the context for which telematics devices should be designed and may suggest scenarios that should be used in evaluating safety and usability.

In contrast to the emerging understanding of the driving task, less is known about the real use of in-vehicle devices while driving, in particular the frequency and duration of use. For example, recent data show that about 3% of all drivers are on the phone at any given moment (Crawford, Manser, Jenkins, Court, & Sepulveda, 2000; http://news.excite.com/news/ap/010723/10/cell-phone-use)

WHO ARE THE USERS?

Almost any adult has the potential to drive, with the determination being made by completing simple driver licensing requirements. Thus, in some ways, the driving population represents the population of candidate users for office computer systems. For most states, the percentage of drivers of either gender is within 1% of being equal (http://www.fhwa.dot.gov/ohim/hs99/dlpage.htm).

TABLE 44.7. Summary of Trip Purposes

Purpose	% Person Trips	% Person Miles
Other family and personal	24.2	19.9
Shopping	20.2	13.5
Work	17.7	22.5
Other social, recreational	16.7	19.5
School, church	8.8	5.7
Visiting	8.2	11.2
Work related	2.6	5.8
Doctor, dental	1.5	1.5
Other	0.2	0.4

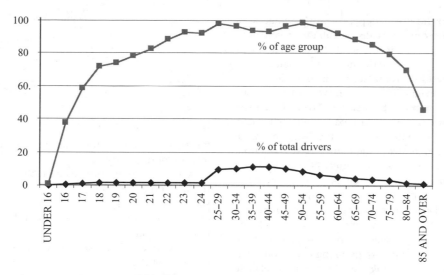

FIGURE 44.1. Distribution of driver age groups. Developed from U.S. Department of Transportation data (http://www.fhwa.dot.gov/ohim/hs99/tables/dl20.pdf). Note that the scale has been expanded at the lower age ranges.

Data with regard to age is shown in Fig. 44.1. Notice that the percentage of the population that is licensed hits 80% at age 21 and remains at that level until the mid-70s. Thus, in designing in-vehicle systems for motor vehicles, few adults can be excluded, and this to some degree differs from the design of office computer systems for which the emphasis is on the working population (generally less than 65 years old). Furthermore, because of a wide age range and other differences, significant differences in individual performance can be expected. For example, in UMTRI telematics studies, older drivers typically required 1.5 to 2 times longer to complete tasks than younger drivers (Green, 2001d).

In contrast to users of ordinary computers, operators of motor vehicles must be licensed. In the United States, the process of becoming a driver begins with obtaining a copy of the driving the state manual and learning the state traffic laws. The candidate then has their vision checked (see http://www.lowvisioncare.com/visionlaws.htm) and takes a test of rules of the road to obtain a learner's permit, often on their 16th birthday in the United States. Consistent with the increasingly common practice of graduated licensing, the learner can drive at restricted times with adult supervision, generally completes a driver's education class (which often includes gory crash movies to convince teenagers not to drink and drive and to wear seatbelts), and, after a few years and an on-the-road test, is licensed to drive. For details, see U.S. Department of Transportation (2000) and the following Web sites:

http://www.highwaysafety.org/safety_facts/qanda/gdl.htm,
http://www.insure.com/auto/teenstates.html,
http://www.nhtsa.dot.gov/people/outreach/stateleg/graddriverlic.htm,
http://www.fhwa.dot.gov/ohim/hs99/dlpage.htm,
http://www.hsrc.unc.edu/pubinfo/grad_overview.htm

The rationale for graduated licensing is to provide new drivers with more experience under less risky conditions. To put this in perspective, the crash rate per million miles is 43 for 16-year-olds, and 16 for 18-year-olds (http://www.hsrc.unc.edu/pubinfo/grad_overview.htm).

In the United States, obtaining a commercial driver's license, needed to drive buses, large trucks, and other vehicles, is a more complex process. Most candidates either obtain on-the-job training, training integrated into their life style (from using machinery on a farm), or attend truck-driving schools (Sloss & Green, 2000). That population tends to be older than the working population as a whole and is predominantly male.

Driver licensing practices vary from country to country. In Japan, for example, the failure rate for the basic licensing exam is much higher than in the United States and there is much greater use of special schools to train drivers.

WHAT KINDS OF VEHICLES DO PEOPLE DRIVE ("THE PLATFORM QUESTION")?

For computers, the concern is the brand of computer a person has, how much memory it has, its processor speed, the capacity of the hard drive, the operating system (and its version) used (Windows, Mac, or Linux) the type of browser (and its version), and so forth. The hardware rarely moves far from its initial location. However, the content of individual computers is in a state of flux, being constantly updated over a life span of just a few years. Fortunately, the physical interface is fairly consistent—a QWERTY keyboard, mouse, and often a 17-inch monitor. The on-screen "desktop" is a more flexible space than the motor vehicle instrument panel.

Unlike personal computers, a motor vehicle is almost completely identified by its make, model, and year and is almost

never updated over a life span averaging 11 years. In most countries (at least where there is left-hand drive), the input devices for the driving task are fairly consistent in type (steering wheel, brake, and throttle), method of operation, and location, as are the primary displays (windshield and mirrors). In contrast, there is no consistency in the controls or displays for telematics tasks. Furthermore, although new motor vehicle models are offered once per year, major changes typically occur every 5 years. Computer software and hardware model upgrades occur almost continually, with a 3-year old computer generally being considered obsolete. Thus, the architecture time frames of the two contexts are quite different.

According to the U.S. Department of Transportation, there were 220 million vehicles registered in the United States in 1999 (http://www.fhwa.dot.gov/ohim/hs99/mvpage.htm). Of these, 60.1% were automobiles, 0.3% were buses, 37.7% were trucks, and 1.9% were motorcycles. In less developed countries, motorcycles make up a much larger fraction of the vehicle fleet. The average miles traveled per vehicle in the United States in 1999 was 11,850 for passenger cars, 10,515 for buses, 11,958 for light trucks, and 26,015 for other trucks (http://www.ita.doc.gov/td/auto/dot.html). For 2001, North American car and truck production as of May 19 totaled 6,395,534 vehicles. Many people do not realize that the United States now produces more trucks (3.3 million) than cars (3.0 million) (http://www.autonews.com/html/main/index.html). The predominance of trucks is due the popularity of minivans, sport utility vehicles (SUVs), and light trucks for personal use, not the production of heavy trucks for commercial use. Table 44.8 summarizes North American production, omitting vehicles manufactured in Japan, vehicles that represent a significant fraction of North American sales. Nonetheless, for the top categories, trucks predominate. Thus, as designers think about packaging telematics applications, trucks are a likely platform.

TABLE 44.8. North America Vehicle Production, Early 2001 (in Excess of 100,000)

Rank	Vehicle	Sales
1	**Chevrolet Silverado C/K**	**224,748**
2	**Ford Explorer**	**210,407**
3	**Ford F series**	**196,165**
4	**Ford F series Super Duty**	**155,760**
5	**Ford Ranger**	**150,049**
6	Honda Accord	137,988
7	**Dodge Grand Cherokee**	**137,064**
8	Ford Focus	133,275
9	**Ford Windstar**	**120,221**
10	Grand Am	111,544
11	Chevrolet Malibu	108,888
12	**Ford Expedition**	**108,173**
13	**Chevrolet S10**	**105,346**
14	Chrysler Neon	104,979
15	**Chevrolet Blazer**	**103,974**
16	**Chevrolet Silverado C/K-Canada**	**102,172**
17	Chevrolet Impala/Lumina	101,304

Note. Source: Automotive News (2001): http://www.autonews.com/html/main/art/prod521.pdf. Trucks are shown in bold.

WHAT MEASURES OF SAFETY AND USABILITY ARE OF INTEREST ("THE EMPIRICAL MEASUREMENT ISSUE")?

As noted earlier, usability is difficult to achieve without empirical measurement. First impressions suggest that the measurement of usability of office computer and Web applications and the measurement of the usability of telematics applications are quite similar. In an office, one measures task completion time, errors, and ratings of ease of use. A typical usability lab has a one-way mirror, cameras, video editing equipment, audio mixers, and at least two rooms, one for the subject and one for an experimenter (see chapter 56).

In a typical laboratory for examining telematics, the same measures may be obtained; however, other measures; as listed in Table 44.9, may also be obtained, especially in driving simulators and on-the-road tests (see Green, 1995a). In many ways, the challenges are similar to those described by Pritchett in the aerospace chapter (chapter 45). The major challenge in assessing the safety and usability of telematics deals with the tradeoffs that drivers naturally make. The impression is that when preoccupied with an in-vehicle task, drivers lose awareness of the driving context, that is, situational awareness. Drivers attempt to compensate by slowing down (to make driving easier), allowing for larger headways, and, if very preoccupied, paying less attention to steering (so lane variance and the number of lane departures increase). However, drivers can respond in strange ways. For example, if asked to use two different in-vehicle systems, one of which is not well designed, they might attempt to maintain equal performance on both, slow down more for the more difficult interface, but compensate by having better steering performance for the poorer interface. Assessment is difficult because the tradeoff functions for all of these measures are unknown. One strategy used to overcome the tradeoff problem is to minimize the opportunity for tradeoffs. For example, this might include using cruise control to fix the speed (and in some cases headway) and provide incentives and feedback

TABLE 44.9. Some Driving-Specific Usability Measures

Lateral
    Number of lane departures
    Mean and standard deviation of lane position
    Standard deviation of steering wheel angle
    Time to line crossing
    Steering entropy
Longitudinal
    Number of collisions
    Time to collision
    Headway (time or distance to lead vehicle)
    Mean and standard deviation of speed
    Speed drop during a task
Visual
    Number of glances
    Mean glance duration
    Maximum glance duration
    Total eyes-off-the-road time

to maintain driving performance, so only task time and errors might tradeoff.

HOW ARE TELEMATICS EVALUATED?

The assessment of mobile applications often occurs in contexts other than simulators and on-the-road, as suggested earlier. Table 44.10 provides a summary of the contexts used and their strengths and weaknesses (see Green, 1995a, b, c for additional details).

Over the last decade there have been enormous improvements in the quality of the tools available for human factors evaluations of in-vehicle evaluations. These include the following:

1. Reductions in the size of video cameras (to that of a postage stamp) and their cost
2. Significant improvements in the quality and reductions in the cost of high-quality, wide field-of-view driving simulators (from millions of dollars to just over $100,000)
3. Other electronic innovations that allow for package of compact instrumentation systems in vehicles
4. Global positioning system systems for precise tracking of vehicle location
5. Digital cell phones for remote downloading of vehicle data.

Development of better, lower cost eye-fixation systems is still needed. Problems remain with their accuracy and ability to hold calibration, dealing with glasses (common for older drivers) and operation in bright sunlight. Until those problems are overcome, studies of driver eye fixations will not be commonplace.

WHAT DESIGN DOCUMENTS EXIST FOR TELEMATICS?

What Types of Documents Are There?

To establish that a product is safe and easy to use, empirical measurement is important. Beyond verification, findings from such measurements can be fed back to the design team to improve the design.

Design is also guided by product and service specifications. In the case of office applications, style guidelines exist for various platforms and applications (see chapter 51, Guidelines, Standards, and Style Guides). For telematics applications, specific platform guidelines do not exist, but there are other types of important written materials (Green, 2001b). Automotive design documents fall into five general classes: principles, information reports, guidelines, recommended practices, and standards. Principles give high level recommendations for design and are similar to found in office HCI applications, such as "design interfaces to minimize learning."

Information report is a term used by the SAE to refer to a compilation of engineering reference data or educational material useful to the technical community. Information reports do not specify how something should be designed, but provide useful background information.

Guidelines give much more specific advice about how to design an interface element. For example, Guideline 9 in chapter 7 of Green, Levison, Paelke, and Serafin (1995, p. 41) states, "Turn displays should show two turns in a row when the turns are in close proximity," where close proximity means 0.1 miles apart or less. The impact of guidelines can depend on the issuing organization. For example, automotive design guidelines written by research organizations have no authority behind them. Guidelines written by the International Organization of Standardization (ISO), although technically voluntary, can become requirements because in some countries, type approval (approval for sale), requires compliance with ISO guidelines. For vehicle models sold worldwide, global manufacturers find building common vehicle systems that comply with ISO standards to less costly than noncomplying, country-specific systems. In Japan, the Japan Automobile Manufacturers Association (JAMA) has a set of guidelines for navigation systems. Although theoretically voluntary, "requests" from the National Police Agency make the JAMA guidelines a requirement for all original equipment manufacturers (OEMs).

Recommended practice, a term used by the SAE, refers to methods, procedures, and technology that are intended as guides to standard engineering practice. The content may be of general nature or may present data that have not yet gained broad engineering acceptance. Common practice is to follow a recommended practice if all possible. It is recognized that a product liability action against a product (especially in the United States) is extraordinarily difficult to defend if the product design deviates from recommended practice.

A *standard* specifies how something must be done and includes broadly accepted engineering practices, or specifications for a material, product, process, procedure, or test method. In the case of the SAE, a standard is technically voluntary because the SAE has no enforcement powers. Even more so than a recommended practice, a product not complying with an SAE standard is almost not defendable in a product liability action and unlikely to be purchased from a supplier by an OEM.

Non-ISO Documents

Table 43.11 provides a summary of the design document activities to date excluding those of the ISO (described later). As indicated in the table, two sets of guidelines are quiet brief (Alliance, European Union [EU]) and are merely statements of general principles (such as interfaces should be simple to operate). Others contain a limited set of specifics (JAMA, SAE J 2364 and J2365, Transport Research Laboratory), and others contain significant details (Battelle, Harmonization of ATT Roadside and Driver Information in Europe, UMTRI). Readers should not take the view that the newer or longer sets of guidelines are necessarily better. The Alliance, EU, JAMA, and SAE guidelines were developed primarily by representatives from industry (with academic involvement for SAE efforts). The Battelle and UMTRI guidelines were developed by teams of contractors for the U.S.

TABLE 44.10. Evaluation Contexts

	Method and Equipment	Advice and Comments
Focus groups	• Groups of 8 to 12 people, demographically similar to customers, sit around a table and discuss a product or service guided by a facilitator • Generally done in multiple cities • Often done by a marketing firm	• Useful in getting ideas for product concepts, but not predictions of the safety or usability of new products because the products are never used • Approach might be used by manufacturers when a usability test might be more appropriate • Generally no quantitative data
Clinics	• Customers in various cities are given the opportunity to experience a new product and the competition, often two or three vehicles, side by side • Customers say which product or feature they prefer	• Only exposes users to a limited number of options • Performance data often are not collected • Approach is commonly used by industry
Part task simulation	• A sample of users operates the device (e.g., computer simulation of a new radio); user task times and errors, as well as comments are recorded • Test facility is not sophisticated	• Not done often
Driving simulator	• Typically, driving simulators are fixed base (no motion) and cost $25,000 to $150,000 each for just hardware and software (facility modifications are an additional expense) • Simulators at manufacturers tend to cost $1–3 million, although some are much more (e.g., Ford is about $10 million). • Usually have 1–3 projectors with 40–120 degree field of view, real vehicle cab, steering system with torque feedback, and realistic sound	• Operation requires considerable experience • Each experiment requires construction of a test road or world and scripting the behavior of vehicles and pedestrians • Facility can require considerable space (e.g., 1,000 square feet) • Generally have a large number of fixed small (lipstick-sized or smaller) cameras • best known U.S. vendors are Systems Technology (about $15,000; http://www.systemstech.com/stidrsm1.htm) and Globalsim Corporation, formerly Hyperion Technologies (>$100,000; http://globalsim.com)
Instrumented vehicle on test track or public roads	• Production vehicle (usually a car and often in the past, a station wagon) is fitted with cameras aimed at driver, forward scene, instrument panel, and lane markings • Sensors for steering wheel angle, brake pressure, speed, headway • Eye-fixation system may be provided • System of interest is installed • In a typical experiment, the driver is asked to follow a test route while a backseat experimenter operates the test equipment	• Typical cost is at least $100,000, although some low-cost systems may use a single camcorder mounted on a vertically mounted curtain rod in the backseat aimed at the instrument panel • Use of DASCAR package has simplified instrumentation, although most organizations use custom hardware, raising cost • Problem in the past has been finding enough space and power for the equipment
Operational field test	• Compact instrumentation is installed in a fleet of vehicles (10–50) • Each vehicle is borrowed by a potential user for 1 week to 1 month • Driving performance is surreptitiously recorded by the vehicle • Unlike an instrumented car, continuous video is not recorded • In addition to data recorded by instrumented vehicle, GPS-determined location is also recorded • Vehicles are periodically polled for data (and data are automatically dumped) by an independent digital cell phone • Test is confined to a single metropolitan area	• Each test requires unique instrumentation • Tests are very expensive and can only be conducted with significant government support • Experiments generally last several years • Planning stage for experiments takes several years • At any given time, there may only one or two operational field tests in progress in the United States

Note. DASCAR = data acquisition system for crash avoidance research; GPS = global positioning system.

Department of Transportation. The TRL and HARDIE guidelines were also contracted efforts (in Europe).

All of these guidelines are voluntary except for the JAMA guidelines, as noted earlier. In the United States, the most important guidelines from the perspective of regulation are SAE Recommended Practices J2364 and J2365, especially J2364. Both SAE J2364 and J2365 are discussed later in this chapter. Although J2634 is still a draft, that document has been approved by the appropriate SAE subcommittees and committees, and eventual approval by the SAE Technical Board is expected. Many

TABLE 44.11. Major Non-ISO Telematics Guidelines and Recommended Practices

Common Document Name	Reference	Size (pages)	Comments
Alliance Guidelines	Alliance of Automobile Manufacturers (2000)	6	Restatement of EU principles; details to be added in 2001 and 2002.
Battelle Guidelines	Campbell, Carney, and Kantowitz (1997)	261	Voluminous document with references to interface design, heavy on trucks; user interface has been said to have a windows flavor, includes physical ergonomics information (e.g., legibility, control sizes) that are not included in the UMTRI guidelines.
EU Guidelines	Commission of the European Communities (1999)	2	Mostly motherhood statements; Some revisions are expected
HARDIE Guidelines	Ross et al., (1996)	480	Early set of European guidelines; less data than UMTRI or Battelle
JAMA Guidelines	Japan Automobile Manufacturers Association (2001)	5	First set of detailed design guidelines for driver interfaces. These guidelines are voluntarily in Japan but followed by all OEMs there and sometimes by aftermarket suppliers; some aspects are particular to Japan
SAE J2364 ("15-Second Rule")	Society of Automotive Engineers (2000)	11	Specifies the maximum allowable task time and test procedures for navigation system tasks performed while driving for systems with visual displays and manual controls
SAE J2365 (SAE Calculation)	Society of Automotive Engineers (2002)	19	Method to compute total task time
TRL Checklist	Quimby (1999)	18	Simple checklist
UMTRI Guidelines	Green, Levison, Paelke, and Serafin (1995).	111	First set of comprehensive design guidelines for the United States; included are principles, general guidelines, and specific design criteria with an emphasis on navigation interfaces.

Note. See http://www.umich.edu/~driving/guidelines.html for electronic copies. UMTRI = University of Michigan Transportation Research Institute; EU = European Union; HARDIE = ; JAMA = Japan Automobile Manufacturers Association; SAE = Society of Automotive Engineers; TRL = Transport Research Laboratory.

manufacturers and suppliers are already complying with both Recommended Practices.

The guidelines developed by the Alliance, EU, JAMA, and SAE are being revised as this chapter is being written, and that process is expected to continue for several years. Readers are therefore advised to see the either the authoring organization or the UMTRI Web site http://www.umich.edu/~driving/guidelines.html for the latest revisions.

ISO Documents

Much of the ISO activity over the last decade has occurred under the auspices of International Standards Organization Technical Committee 22, Subcommittee 13 (ISO TC 22/SC 13-Ergonomics Applicable to Road Vehicles, in particular Working Group 8 (WG8—Transport, Information, and Control Systems or TICS) (Green, 2000a). WG8 has about 50 delegates from the major vehicle producing nations, with the most 15 active members appearing at meetings held two to three times per year, usually in Europe.

Table 44.12 shows the current Working Group 8 work program. Most of the standards in process are quite general, not containing the detail found in the Battelle, HARDIE, or UMTRI guidelines. Such detail is at least a decade away. To promote international harmonization, national standards organizations,

technical societies (e.g., SAE), and government organizations (e.g., U.S. Department of Transportation) often permit ISO standards to supercede their own standards, so ISO standards are important.

SAE J2364 ("The 15-Second Rule")

Of the design documents in the literature, none has generated as much discussion as SAE Recommended Practice J2364 also known as "The 15-Second Total Task Time Rule" or "15-Second Rule." SAE J2364 establishes a design limit for how long drivers can take to complete navigation-system-related tasks involving visual displays and manual controls while driving passenger vehicles (Society of Automotive Engineers, 2000). The practice does not and should not apply to voice interfaces or passenger operation because the task demands are fundamentally different. There is no reason the requirements do not make sense for similar tasks for other systems given its performance basis. The practice applies to OEM and aftermarket products.

In developing this practice, criteria were sought that were related to crash risk, likely to lead to design improvements and easy to measure. Some suggested this document should be design criteria, such as a specification for a maximum number of items on a menu. However, it became clear that such a design practice needed a performance basis and specifying a design

TABLE 44.12. International Organization of Standardization (ISO) Technical Committee 22/SC 13/WG 8 Work Program

Effort	Summary	Status
Dialogue management	Provides high-level ergonomic principles (compatibility with driving, consistency, simplicity, error tolerance, etc.) to be applied in the design of dialogues that take place between the driver of a road vehicle and the TICS while the vehicle is in motion	DIS 15005
Auditory information presentation	Provides requirements for auditory messages including signal levels, appropriateness, coding, along with compliance test procedures	DIS 15006
Measurement of driver visual behavior	Generally describes equipment (cameras, recording procedures, etc.) and procedures (subject descriptions, experiment design parameters, tasks, performance measures, etc.) used to measure driver visual behavior	FDIS 15007
Legibility (visual presentation of information)	Provides requirements for character size, contrast, luminance, and so forth and specifies how they are to be measured	FDIS 15008
Message priority	Prioritization scheme for TICS and other system-initiated and driver-requested messages presented to drivers while driving based on criticality (likelihood of injury if the event occurs and urgency (required response time), both determined on 4-point scales	CD 16951
Suitability of TICS while driving	Generally describes a process for assessing whether a specific TICS, or a combination of TICS with other in-vehicle systems, is suitable for use by drivers while driving. It addresses: (a) user-oriented TICS description and context of use, (b) TICS task description and analysis, (c) assessment, and (d) documentation	DIS 17287
Visual distraction of information and communication systems (formerly navigation accessibility)	Will propose methods and requirements to determine if tasks are too difficult to do while driving	PWI
Intelligent transportation systems safety assurance process for design and development	Provides for a process that incorporates safety impact analysis as an integral part of TICS research and development. It does not establish design or performance requirements per se but does include requirements for safety records and personnel training	PWI

Note. The ISO has a well-defined, 3-year process during which documents go through several stages (Preliminary Work Item, PWI; Committee Draft, CD; Draft International Standard, DIS; Final Draft International Standard, FDIS; International Standard) as they are passed from the Working Group to the Subcommittee to the Technical Committee and finally the secretariat for review and approval. The major hurdles are the Working Group and Subcommittee, where passage requires two thirds of the nations participating. The emphasis of this process is on building a voluntary consensus (http://www.iso.ch/iso/en/stdsdevelopment/whowhenhow/how.html). Because of the limited number of experts available, WG8 is selective in adding items to its work program. TICS = transport information and control systems.

standard would constrain innovation, so a performance-based practice was developed. Many measures were then proposed, with eyes-off-the-road time being most popular. The logic behind this measure is simple. If a driver is not looking at the road while driving, he or she is more likely to crash, with the likelihood of a crash increasing with eyes-off-the-road time. As shown in Green (1999d) (see also Wierwille (1995)), this is reflected in the following equation:

$$\text{\# U.S. deaths in 1989} = \times\text{market.penetration.fraction} \times [-.133 + (.0447 \times (\text{mean glance time})^{1.5} \times (\text{\# of glances}) (\text{frequency}))],$$

where market penetration is the fraction of vehicles with a system (10% → 1), mean glance time is measured in seconds, # of glances is the number of times the device is looked for each use sequence, and frequency is the number of use sequences per week. As an example, a task, say, entering an address, could have a mean glance duration of 2.7 s, require an average of 27.5 glances per entry, and be performed twice per week.

Unfortunately, eyes-off-the-road time is time-consuming and expensive to measure, requires specialized equipment and

skilled personnel, and a fully functional system installed either in a simulator or test vehicle, something only available late in design. Invariably, problems are identified so close to production that few, if any, changes can be made. One of the key lessons from the Gould and Lewis design principles from HCI is that feedback from early in design is critical.

A review of the literature (Green, 1999d) found that task time while driving was highly correlated with eyes-off-the-road time as one would suspect. Looking at in-vehicle systems is time shared while driving, so the more the driver needs to look, the longer the task will take. Furthermore, dynamic (on-the-road) task time and eyes-off-the-road time are correlated with static task time, the time to complete the task when the vehicle is parked (e.g., Green, 1999d; Farber et al., 2000). Static task time is easy to measure and can be done using prototypes available early in design.

The current SAE J2364 test procedure (Society of Automotive Engineers, 2000) requires that 5 to 10 subjects over the age of 45 be tested. Each subject completes five practice trials and three test trials in a parked vehicle, simulator, or laboratory mockup. The test cases to be examined (addresses for destination entry) are to be representative of e.g., is planned

for production. Timing is from when the hand begins to move (typically from the steering wheel to a control) and ends when the goal is achieved. Timing is continuous except for computational interruptions equal to or greater than 1.5 s, a time period during which the device is computing (for example a route). If feedback is provided to the driver, that period is excluded from the 15-s task time limit. The interface complies with J2364 if 5 of the first 5, 6 out of the first 7, or 8 out of the first 10 subjects have a mean task completion time of less than 15 s. Tasks with completion times equal to or greater than 15 s cannot be accessed by drivers in a moving vehicle.

It must be emphasized that the 15-s limit is for a static test. On the road drivers will take 30 to 50% longer overall, and furthermore will be alternating between looking inside the vehicle and looking at the road. The test procedure does not suggest drivers can safely look away from the road continuously for 15 s.

Some have argued that use of static task time fails to identify interfaces requiring long glance durations. However, analysis of real products shows the primary risk is from tasks that take too long to complete. In fact, it is difficult to think of driver tasks for navigation systems task that have short total task times but long glance durations. In real driving, people truncate glances to the interior when the glances become too long but tend to be captured by tasks and complete them, even if the tasks are unacceptably long. In practice, eliminating tasks with long completion tasks (the worst tasks) also eliminates many of the tasks with long glance durations.

Proposals are being reviewed to add a visual occlusion component to J2364. In this method, subjects view the display through liquid crystal display shutter glasses, with the interface being visible for the duration of a glance (say, 1.5 s) and then blocked for a duration somewhat longer (3 to 5 s) than the driver normally looks back at the road (0.5 to 1 s) (Chiang, Brooks, & Weir, 2001). Unlike real driving, drivers may do nothing in the occluded interval. This procedure may be in addition to or an alternative to task time assessment. Current evidence suggests it is unlikely to provide any safety benefits in addition to or replacing the 15-s rule static test and the occlusion procedure unnecessarily complicates the evaluation. However, it may be included in future versions of the standard to obtain a sufficient consensus for passage.

WHAT TOOLS AND ESTIMATION PROCEDURES EXIST TO AID TELEMATICS DESIGN?

SAE J2365 Calculations

Sometimes the results from an SAE J2364 evaluation will be available too late in the design process to have the desired impact. Therefore, SAE J2365 was developed to allow designers and engineers to calculate completion times for in-vehicle tasks involving visual displays and manual controls completed statically, that is, while parked (or in bench-top simulation). The document applies to both OEM and aftermarket equipment.

Although intended for navigation systems, J2365 should provide reasonable estimates for most in-vehicle tasks involving manual controls and visual displays.

The calculation method is based on the goals, operators, methods, and selection rules (GOMS) model described by Card, Moran, and Newell (1983) with task time data from several sources (see chapter 58 of this volume for background information on GOMS). The keystroke data was drawn from UMTRI studies of the Siemens Ali-Scout navigation system (Steinfeld, Manes, Green, & Hunter, 1996; Manes, Green, & Hunter, 1998). Search times were based upon Olson and Nilsen (1987–1988), and the mental time estimates were drawn from Card, Moran, and Newell (1980) and UMTRI Ali-Scout studies. Thus, the times shown in Table 44.13 have been tailored for the automotive context.

The basic approach involves top-down, successive decomposition of a task. The analyst divides the task into logical steps. For each step the analyst identifies the human and device task operators. Sometimes analysts get stuck using this approach because they are not sure how to divide a task into steps. In those cases, using a bottom-up approach may overcome such roadblocks. For each goal, the analyst identifies the method used.

The analyst is advised to use paragraph descriptions of each method to document them and then convert those descriptions into pseudo code. All steps are assumed to occur in series, that is, multiple tasks cannot be completed at the same time. Furthermore, only visible, noncognitively loading shortcuts are assumed to be used by most drivers. Invisible short cuts are likely to be used only by experts.

Next, the pseudo code task description is entered into an Excel spreadsheet. The analyst looks up the associated time for each operator listed in Table 44.10 and sums them to determine total task time. To assist in understanding the process, the practice provides a step-by-step example of entering a street address into a PathMaster/NeverLost navigation system, a popular U.S. product. For background on the calculation method see Green (1999b).

The J2365 approach makes a number of assumptions, many of which are shared with the basic GOMS model. For example, the model assumes error-free performance, which although not true, can be adjusted for (say by increasing the computed value by 25%). Furthermore, activities are assumed to be routine cognitive tasks, with users knowing each step, executed in a serial manner. Again, adjustments in computed time can account for users sometimes forgetting what is next.

Thus, although many of these assumptions are not true, adjustments can be made for them and often the adjustments are small. Furthermore, violations of assumptions tend to affect all interfaces equally, so decisions still hold about which of several interfaces is best. As a practical matter, the estimates are good enough for most engineering decisions. Readers should keep in mind that J2364 only requires the use of 10 subjects at most, so there is some error in those estimates. Those errors are likely to be as large as variability between analysts and between J2365 estimates (Nowakowski & Green, 2000; Nowakowski, Utsui, & Green, 2000).

TABLE 44.13. Operator Times (seconds)

Code	Name	Operator Description	Young Drivers (18–30)	Older Drivers (55–60)
			Time(s)	
Rn	Reach near	From steering wheel to other parts of the wheel, stalks, or pods	0.31	0.53
Rf	Reach far	From steering wheel to center console	0.45	0.77
C1	Cursor once	Press a cursor key once	0.80	1.36
C2	Cursor twice or more	Time/keystroke for the second and each successive cursor keystroke	0.40	0.68
L1	Letter or space once	Press a letter or space key once	1.00	1.70
L2	Letter or space twice times or more	Time/keystroke for the second and each successive cursor keystroke	0.50	0.85
N1	Number once	Press the letter or space key once	0.90	1.53
N2	Number twice or more	Time/keystroke for the second and each successive number key	0.45	0.77
E	Enter	Press the enter key	1.20	2.04
F	Function keys or shift	Press the function keys or shift	1.20	2.04
M	Mental	Time per mental operation	1.50	2.55
S	Search	Search for something on the display	2.30	3.91
Rs	Response time of system-scroll	Time to scroll one line	0.00	0.00
Rm	Response time of system-new menu	Time for new menu to be painted	0.50	0.50

Note. The keystroke times shown include the time to move between keys. System response times to show new menus may be empirically determined.

IVIS Estimates

A more complex estimation procedure, the IVIS DEMAnD model (In-Vehicle Information System Design Evaluation and Model of Attention Demand; described by Hankey, Dingus, Hanowski, Wierwille, & Andrews, 2000a, 2000b). The model, which runs under Windows 98 or Windows NT, allows analysts to calculate a wide range of performance characteristics for proposed user interfaces. (The CD-ROM can be obtained from the U.S. Department of Transportation, Federal Highway Administration, Turner-Fairbank Highway Research Center in McLean, Virginia.)

Consistent with common understanding, the model assumes there are five basic human resources: visual input, auditory input, central processing, manual output, and speech output. Overload of any one of these resources will affect task performance. The more demanding an in-vehicle task, the greater the likelihood of a crash. The model does not include a haptic component because haptic displays are rare in contemporary vehicles. Although many have developed models of human performance that partition human resources more finely, a five-component model is sufficient for most in-vehicle analyses. The data in the model were based on four experiments: Biever (1999), Blanco (1999), Gallagher (2001), and research conducted by Westat, a consulting company. These experiments concentrated on reading visual displays while driving, although there was work on auditory information as well.

To use the model, analysts need to create a description of each task drivers perform. Generally that involves either selecting the task in question from a large library of tasks in the database, modifying an existing description, or creating one from scratch. Tasks are grouped in to seven categories:

conventional, search, search-plan, search-plan-interpret, search-plan-compute, search-compute, and search-plan-interpret-compute. Analysts need to select the driver age category (or specify all ages), the traffic density, the road complexity, the reliance on symbols or labels, the location of the display, and other characteristics.

The output of the model includes about 20 parameters such as the expected number of glances, total task time, ratings of mental demand and frustration, total task demand, and so forth. In addition, the model output specifies if design thresholds are exceeded. The model proposes two sets of thresholds, yellow line and red line. Yellow line thresholds were set of points at which there was a measurable degradation in driving performance ($p < .05$) from baseline driving in the research conducted to support model development. Red line thresholds indicated that a composite group of surrogate safety measures of driving performance was substantially affected. The red line values were determined primarily from the literature and expert opinion. Table 44.14 shows those thresholds.

TABLE 44.14. In-Vehicle Information System Yellow and Red Line Thresholds

Measure	Affected (Yellow)	Substantially Affected (Red)
Single glance time	1.6 s	2.0 s
Number of glances	6 glances	10 glances
Total visual task time (Eyes-off-the-road-time)	7 s	15 s
Total task time (white driving)	12 s	25 s

Note. Data are from Hankey, Dingus, Hanowski, Wierwille, and Andrews (2000a), p. 40.

Independent validation data of the model has yet to appear in the literature but is expected shortly.

CLOSING THOUGHTS

This chapter makes the following key points:

1. A large number of new telematics applications will appear in vehicles over the next few years. They should radically reshape the driver's task, providing the driver with a flood of information.

2. Driving is different from sitting at a desk in an office because of the concern for crash risk.

3. The flood of information from new telematics has the potential to distracting drivers from driving. Distraction is associated with specific types of crashes (single-vehicles running off road, rear-end crashes) that are most common under generally good driving conditions.

4. Relative to other distractions, using figures from the past (when market penetration was low), the number of telematics-related crashes was low. However, current projections are that in excess of 200 people are killed per year in the U.S. in cell-phone-related crashes, and the number is growing. This number of fatalities exceeds that from front air bags and Firestone–Explorer tire rollovers, both of which had major financial impact on industry and led to federal regulatory changes.

5. Workload managers may be a long-term alternative to legislation to reduce distraction that leads to crashes.

6. Trips are made for a wide variety of purposes that need to be considered in assessing telematics applications.

7. Telematics need to be developed for a wide range of driver ages. Well into their retirement years, most people still drive (and will need to be able to use telematics).

8. Although people commonly think of passenger cars as the primary form of personal transportation in the United States, more trucks (especially pickups, SUVs, and vans) are produced in the United States, and they will be the most common platform in the future.

9. In assessing safety and usability, a wide variety of measures of longitudinal and lateral control are used in addition to task completion time, errors, and subjective ratings of ease of use. A major challenge in interface evaluation is dealing with performance tradeoffs between measures.

10. There have been significant advances in driving simulators and instrumented vehicles that have improved their quality and reduced their cost of safety and usability evaluations over the last few years. However, the cost of these systems remains out of the range of most laboratories, especially those in academic settings.

11. Key design documents have been developed by the Alliance, Battelle, EU, HARDIE, JAMA, SAE, TRL, and UMTRI, along with ISO. The SAE documents (especially SAE J2364, "The 15-Second Rule") have the most impact in the United States as JAMA does in Japan. ISO is in the process of developing a large set of guidelines, and other organizations are updating theirs.

12. SAE J2365 and IVIS can assist in predicting task performance time, a simple measure for evaluating telematics safety and usability.

Thus, although the HCI literature provides a framework for test methods and evaluation, there is a great deal that is specific to the motor vehicle context because of the safety-critical nature of the context and the time sharing not found in office activities. To meet the needs of the future, the cost of the tools needs to be reduced, and reliable tools, especially for recording eye fixations, are needed. Significant research is needed to better model driver performance (for the purpose of developing workload managers) and understand how drivers use real telematics applications.

References

Alliance of Automobile Manufacturers. (2000, December 6). *Statement of principles on human-machine interfaces (HMI) for in-vehicle information and communication systems* [draft]. Washington, DC: Author.

Barfield, W., & Dingus, T. A. (1997). *Human factors in intelligent transportation systems.* Mahway, NJ: Lawrence Erlbaum Associates.

Biever, W. J. (1999). *Auditory based supplemental information processing demand effects on driving performance.* Unpublished master's thesis, Virginia Polytechnic Institute and State University, Blacksburg, VA.

Blanco, M. (1999). Effects of in-vehicle information systems (IVIS) tasks on the information processing demands of a commercial vehicle operations (CV)) Driver. Unpublished master's thesis, Virginia Polytechnic Institute and State University, Blacksburg, VA.

Campbell, J. L., Carney, C., & Kantowitz, B. H. (1997). Human factors design guidelines for advanced traveler information systems (ATIS) and commercial vehicle operations (CVO) (Technical Report FHWA-RD-98-057), Washington, DC: U.S. Department of Transportation, Federal Highway Administration.

Card, S. K., Moran, T. P., & Newell, A. (1980). The keystroke-level model for user performance time with interactive systems. *Communications of the ACM, 23,* 396–410.

Card, S. K., Moran, T. P., & Newell, A. (1983). *The psychology of human-computer interaction.* Hillsdale, NJ: Lawrence Erlbaum Associates.

Chiang, D. P., Brooks, A. M., & Weir, D. H. (2001). An experimental study of destination entry with an example automobile navigation system (SAE paper 2001-01-0810). Warrendale, PA: Society of Automotive Engineers.

Commission of the European Communities. (1999). *Statement of principles on human machine interface (HMI) for in-vehicle information and communication systems* (Annex 1 to Commission Recommendation of 21 December 1999 on safe and efficient in-vehicle information and communication systems: A European

statement of principles on human machine interface). Brussels, Belgium: European Union.

Crawford, J. A., Manser, M. P., Jenkins, J. M., Court, C. M., & Sepulveda, E. D. (2000). *Extent and effects of handeld cellular telephone use while driving* (Research Report 167706-1). College Station: Texas Transportation Institute, Texas A&M University.

Farber, E., Blanco, M., Foley, J. P., Curry, R., Greenberg, J., & Serafin, C. (2000). Surrogate measures of visual demand while driving. *Proceedings of the IEA/HFES 2000 Congress* [CD-ROM]. Santa Monica, CA: Human Factors and Ergonomics Society.

Gallagher, J. P. (2001). *An assessment of the attention demand associated with the processing of information for in-vehicle information systems (IVIS)*. Unpublished thesis. Virginia Polytechnic Institute and State University, Department of Industrial and Systems Engineering, Blacksburg, VA.

Gould, J. D., & Lewis, C. (1985). Designing for usability: Key principles and what designers think. *Communications of the ACM, 28*, 300–311.

Green, P. (1995a). Automotive techniques. In J. Weimer (Ed.), *Research techniques in human engineering* (2nd ed., pp. 165–208). New York: Prentice-Hall.

Green, P. (1995b). *Measures and methods used to assess the safety and usability of driver information systems* (Technical Report FHWA-RD-94-088). McLean, VA: U.S. Department of Transportation, Federal Highway Administration.

Green, P. (1995c). *Suggested procedures and acceptance limits for assessing the safety and ease of use of driver information systems* (Technical Report FHWA-RD-94-089). McLean, VA: U.S. Department of Transportation, Federal Highway Administration.

Green, P. (1999a). Estimating compliance with the 15-second rule for driver-interface usability and safety. *Proceedings of the Human Factors and Ergonomics Society 43rd Annual Meeting* [CD-ROM]. Santa Monica, CA: Human Factors and Ergonomics Society.

Green, P. (1999b). *Navigation system data entry: Estimation of task times* (Technical report UMTRI-99-17). Ann Arbor, MI: University of Michigan Transportation Research Institute.

Green, P. (1999c). The 15-second rule for driver information systems. *ITS America Ninth Annual Meeting Conference Proceedings* [CD-ROM]. Washington, DC: Intelligent Transportation Society of America.

Green, P. (1999d). *Visual and task demands of driver information systems* (Technical Report UMTRI-98-16). Ann Arbor, MI: University of Michigan Transportation Research Institute.

Green, P. (2000a, June 8). *The human interface for ITS display and control systems: developing international standards to promote safety and usability*. Paper presented at the International Workshop on ITS Human Interface in Japan, Utsu, Japan.

Green, P. (2000b). Crashes induced by driver information systems and what can be done to reduce them (SAE paper 2000-01-C008). *Convergence 2000 Conference Proceedings* (SAE Publication P-360, pp. 26–36). Warrendale, PA: Society of Automotive Engineers.

Green, P. (2001a, February). *Safeguards for on-board wireless communications*. Paper presented at the Second Annual Plastics in Automotive Safety Conference, Troy, Michigan.

Green, P. (2001b). *Synopsis of driver interface standards and guidelines for telematics as of mid-2001* (Technical Report UMTRI-2001-23). Ann Arbor, MI: University of Michigan Transportation Research Institute.

Green, P. (2001c). *Telematics: Promise, potential, and risks* (Management Briefing Seminar panel session—Traverse City Conference). Ann Arbor, MI: University of Michigan Transportation Research Institute, Office for the Study of Automotive Transportation.

Green, P. (2001d, February). *Variations in task performance between younger and older drivers: UMTRI research on telematics*. Paper presented at the Association for the Advancement of Automotive Medicine Conference on Aging and Driving, Southfield, MI.

Green, P., Flynn, M., Vanderhagen, G., Ziiomek, J., Ullman, E., & Mayer, K. (2001). *Automotive industry of trends in electronics: year 2000 survey of senior executives* (Technical Report UMTRI-2001-15). Ann Arbor, MI: University of Michigan Transportation Research Institute.

Green, P., Levison, W., Paelke, G., & Serafin, C. (1995). *Preliminary human factors guidelines for driver information systems* (Technical Report FHWA-RD-94-087). McLean, VA: U.S. Department of Transportation, Federal Highway Administration.

Hankey, J. M., Dingus, T. A., Hanowski, R. J., Wierwille, W. W., & Andrews, C. (2000a). *In-vehicle information systems behavioral model and design support final report* (Technical Report FHWA-RD-00-135). McClean, VA: U.S. Department of Transportation, Federal Highway Administration. (Available online: http://www.tfhrc.gov/humanfac/00-135.pdf)

Hankey, J. M., Dingus, T. A., Hanowski, R. J., Wierwille, W. W., & Andrews, C. (2000b). *In-vehicle information systems behavioral model and design support: IVIS DEMAnD prototype software user's manual* (Technical Report FHWA-RD-00-136). McClean, VA: U.S. Department of Transportation, Federal Highway Administration. (Available online: http://www.tfhrc.gov/humanfac/00-136.pdf).

Japan Automobile Manufacturers Association. (2001). JAMA guidelines for in-vehicle display systems, version 2.1. Tokyo: Japan Automobile Manufacturers Association.

Kayl, K. (2000). The networked car: Where the rubber meets the road. Retrieved March 2002 from http://java.sun.com/features/2000/10/convergence.html

Koushki, P. A., Ali, S. Y., & Al-Sateh, O. S. (1999). Driving and Using Mobile Phones, *Transportation Research Record No 1694* (paper 99-0064), 27–33.

Manes, D., Green, P., & Hunter, D. (1998). *Prediction of destination entry and retrieval times using keystroke-level models* (Technical Report UMTRI-96-37; also released as EECS-ITS LAB FT97-077). Ann Arbor, MI: University of Michigan Transportation Research Institute.

Michon, J. A. (Ed.). (1993). *Generic intelligent driver support*. London: Taylor and Francis.

Murray, C. J. L., & Lopez, A. D. (1996a). *Global health statistics*. Boston: Harvard School of Public Health.

Murray, C. J. L., & Lopez, A. D. (1996b). *The global burden of disease*. Boston, MA: Harvard School of Public Health.

National Safety Council. (2000). *Injury facts*. Itasca, IL: National Safety Council.

Nowakowski, C., Friedman, D., & Green, P. (2001). *Cell phone ring suppression and HUD caller ID: Effectiveness in reducing momentary driver distraction under varying workload levels* (Technical Report 2001-29). Ann Arbor, MI: University of Michigan Transportation Research Institute.

Nowakowski, C., & Green, P. (2000). *Prediction of menu selection times parked and while driving using the SAE J2365 method* (Technical Report 2000-49). Ann Arbor, MI: University of Michigan Transportation Research Institute.

Nowakowski, C., Utsui, Y., & Green, P. (2000). *Navigation system evaluation: the effects of driver workload and input devices on destination entry time and driving performance and their implications to the SAE recommended practice* (technical report UMTRI-2000-20). Ann Arbor: MI, University of Michigan Transportation Research Institute.

Noy, Y. I. (Ed.). (1997). *Ergonomics and safety of intelligent driver interfaces*. Mahwah, NJ: Lawrence Erlbaum Associates.

Olson, J. R., & Nilsen, E. (1987–1988). Analysis of the cognition involved in spreadsheet software interaction. *Human-Computer Interaction, 3,* 309–349.

Parkes, A. M., & Franzen, S. (1993). *Driving future vehicles.* London: Taylor and Francis.

Peacock, B., & Karwowski, W. (1993). *Automotive ergonomics.* London: Taylor and Francis.

Quimby, A. (1999). A safety checklist for the assessment of in-vehicle information systems: scoring performance (Project Report PA3536-A/99). Crowthorne, UK: Transport Research Laboratory.

Redelmeier, D. A., & Tibshirani, R. J. (1997). Association between cellular-telephone calls and motor vehicle collisions. *New England Journal of Medicine, 336,* 453–458.

Redelmeier, D. A., & Tibshirani, R. J. (2001). Car phones and car crashes: Some popular misconceptions. *Canadian Medical Association Journal, 164,* 1581–1582.

Ribbens, W. B., & Cole, D. E. (1989). *Automotive electronics Delphi.* Ann Arbor, MI: Office for the Study of Automotive Transportation, University of Michigan Transportation Research Institute.

Richardson, B., & Green, P. (2000). *Trends in North American intelligent transportation systems: A year 2000 appraisal* (Technical Report 2000-9). Ann Arbor, MI: University of Michigan Transportation Research Institute.

Ross, T., Midland, K., Fuchs, M., Pauzie, A., Engert, A., Duncan, B., Vaughan, G., Vernet, M., Peters, H., Burnett, G., & May, A. (1996). *HARDIE design guidelines handbook: human factors guidelines for information presentation by ATT systems.* Luxembourg: Commission of the European Communities.

Sloss, D., & Green, P. (2000). National automotive center 21st century truck (21T) dual use safety focus (SAE Paper 2000-01-3426; published in *National Automotive Center Technical Review,* Warren, MI, U.S. Army Tank-Automotive and Armaments Command, National Automotive Center, 63–70). Warrendale, PA: Society of Automotive Engineers.

Society of Automotive Engineers. (2000, January 20). *Navigation and route guidance function accessibility while driving* (SAE draft recommended practice 2364). Warrendale, PA: Author.

Society of Automotive Engineers. (2002). *Calculation of the time to complete in-vehicle navigation and route guidance tasks* (SAE recommended practice J2365), Warrendale, PA: Author.

Society of Automotive Engineers. (2000). *SAE handbook 2000.* Warrendale, PA: Author.

Steinfeld, A., Manes, D., Green, P., & Hunter, D. (1996). *Destination entry and retrieval with the Ali-Scout navigation system* (Technical Report UMTRI-96–30, also released as EECS-ITS LAB FT97-077). Ann Arbor, MI: University of Michigan Transportation Research Institute.

Stutts, J. C., Reinfurt, D. W., Staplin, L., & Rodgman, E. A. (2001). The role of driver distraction in traffic crashes. Washington, DC: AAA Foundation for Traffic Safety. (Available online: http://www.aaafts.org/pdf/distraction.pdf)

U.S. Department of Transportation Bureau of Transportation Statistics. (2000). *National transportation statistics 2000.* Washington, DC: Author.

Underwood, S. E. (1992). *Delphi forecast and analysis of intelligent vehicle-highway systems through 1991: Delphi II.* Ann Arbor Program in Intelligent Vehicle-Highway Systems (IVHS Technical Report-92-17), Ann Arbor, MI: University of Michigan.

Underwood, S. E. (1989). Summary of preliminary results from a Delphi survey on intelligent vehicle-highway systems [Technical report]. Ann Arbor, MI: University of Michigan.

Underwood, S. E., Chen, D., & Ervin, R. D. (1991). Future of intelligent vehicle-highway systems: A Delphi forecast of markets and sociotechnological determinants. *Transportation Research Record No. 1305,* 291–304.

Violanti, J. M., & Marshall, J. R. (1996). Cell phones and traffic accidents: an epidemiological approach, *Accident Analysis and Prevention, 28*(2), 265–270.

Wang, J., Knipling, R. R., & Goodman, M. J. (1996). The role of driver inattention in crashes: New statistics from the 1995 crashworthiness data system. *Association for the Advancement of Automotive Medicine 40th Annual Conference Proceedings* (pp. 377–392). Des Plaines, IL: Association for the Advancement of Automotive Medicine.

Wierwille, W. (1995). Development of an initial model relating driver in-vehicle visual demands to accident rate. *Proceedings of the Third Annual Mid-Atlantic Human Factors Conference* (pp. 1–7). Blacksburg, VA: Virginia Polytechnic Institute and State University.

·45·

HUMAN-COMPUTER INTERACTION IN AEROSPACE

Amy R. Pritchett
Georgia Institute of Technology

INTRODUCTION

Human-computer interaction (HCI) permeates aerospace. Aircraft now use computer-driven displays for cockpit displays, are flown by computers, and communicate via digital datalinks. Air traffic controllers have been using radar screens for 50 years. Flight simulators are sufficiently realistic that they can be used for the complete training of an airline pilot. This chapter describes how the pursuit of effective HCI is an important part of aerospace design and how aerospace has been—and continues to be—an important contributor to the broader HCI community.

This chapter first reviews the history of HCI in aerospace. Then characteristics of aerospace operating environments are discussed to highlight challenges in improving HCI in aerospace. Aerospace is defined broadly as including both atmospheric and space flight and supporting ground systems. This chapter traces the evolution of air transport cockpits (representative of interface and automation designs), flight simulators (representative of virtual reality efforts), and air traffic control (representative of organizational and procedure design). More comprehensive discussions can be found in Wiener and Nagel (1988); Dekker and Hollnagel (1999); and Garland, Wise, and Hopkins (1999) and in journals such as *International Journal of Aviation Psychology*, *Human Factors and Aerospace Safety*, and *Air Traffic Control Quarterly.*

HISTORICAL PERSPECTIVE

Problems in the interaction between human and machine have existed since the first heavier-than-air aircraft. This section traces the evolution of issues with manual control, the display of information, automated and intelligent systems, virtual environments, and communication. This discussion is intended to provide a background on fundamental principles of HCI in aerospace, to illustrate the evolution of designs to current-day standards, and to illustrate contributions of aerospace to the HCI community in general.

Manual Control and Tracking

The difficulties which obstruct the pathway to success in flying machine construction are of three general classes: (1) Those which relate to the construction of the sustaining wings. (2) Those which relate to the generation and application of the power required to drive the machine through the air. (3) Those relating to the balancing and steering of the machine after it is in flight. Of these difficulties [the first] two are already to a certain extent solved.... When this [third] feature has been worked out the age of flying machines will have arrived, for all other difficulties are of minor importance. (Wright, 1901, p. 99)

The most significant obstacle to the first powered, heavier-than-air flight was the stability and controllability of an aircraft by a human pilot. The patent filed by the Wright brothers was for neither engine nor airfoil, but instead for their combined wing-warping mechanism and vertical rudder for banking into a coordinated turn. Improvements soon followed in the form of new control surfaces such as the aileron and better methods

for designing stable aircraft. Despite these improvements, World War I aircraft remained notoriously unstable (Abzug & Larrabee, 1996, pp. 1–5).

Methods of analyzing aircraft stability were developed between World War I and World War II that are still in use; however, these methods did not describe human capabilities. This need to understand the human half of the system spawned research in manual control and human performance in tracking tasks. The earliest studies sought to characterize human behavior in the same terms as used for describing the aircraft, that is, an algebraic representation of output (behavior) when given a specific input (displayed information).

This approach led to several important insights. For example, early research in perceptual motor skills in aviation contributed to models of choice reaction time and response time (e.g., Fitts & Posner, 1967, pp. 93–122). These insights subsequently developed into the Hicks-Lyman and Fitts Laws, which are broadly applicable today (Wickens, 1992, pp. 446–447).

However, as an entity examined separate from the aircraft, human behavior in flying different aircraft is not consistent with a single algebraic representation. This apparent inconsistency was removed by notice that the *combined* behavior of the aircraft and pilot is invariant within a specific operating range, that is, the human adapts his or her control behavior to the aircraft. In the vernacular of system dynamics and control theory, this operating range is centered on the crossover frequency, and hence this effect is called the crossover model. The closed-loop combined behavior comprises an effective time delay largely determined by human information processing delays, a control gain established by the human to create a stable system dynamic, and a first-order lag or integration of error in the signal. This model is best suited to compensatory tracking (i.e., when the human attempts to zero an error signal) but also describes well pursuit and predictive tracking behavior (when the human is tracking the input directly and when the human is tracking the input signal with some predictive indications, respectively; National Research Council [NRC], 1997, pp. 120–125; Wickens, 1992, pp. 468–474).

Both major results from these aviation research programs—Fitts law and the crossover model—continue to be fundamental in the design of aerospace systems, both in establishing vehicles with controllable dynamic behavior and in creating the dynamic presentations (e.g., control cues on cockpit displays such as flight directors) that inform and guide pilots (e.g., Hess, 1995; Stokes & Wickens, 1988).

Fundamentals of Information Display

In World War I era aircraft, the pilot's main references came from outside the cockpit—visual references to the horizon and ground features, the sensation of speed due to wind noise and the hum of vibrating bracing wires, and engine noise. Cockpit instruments were often simple, such as a spring-loaded vane stuck into the airstream next to the cockpit, the deflection of which provided some measure of airspeed. In 1929, J. H. Doolittle demonstrated that, with the correct instrumentation, it is possible to pilot an aircraft based solely on reference to cockpit instrumentation (Doolittle, 1961). This was a major

breakthrough, because it allowed aircraft to be flown in all but the most turbulent weather conditions, when previously low-lying clouds or fog had been absolute obstructions to safe flight.

Subsequent studies created a standard for cockpit instrumentation which is still in use. Several categorizations are possible of the information required to pilot an airplane (e.g., Newman, 1998); using definitions taken from system dynamics (e.g., Blakelock, 1991, p. 229), the basic cockpit instrumentation may be categorized as comprising the following:

- "Flight control" information, indicating the immediate effects of control actions, namely, the pitch and roll angles of the aircraft relative to the horizon, as well as whether the aircraft is slipping or skidding.

- "Guidance" information, indicating the velocity of the aircraft overall (airspeed) and vertically (vertical speed), as well as the heading of the aircraft relative to north.

- "Navigation" information, indicating the position of the aircraft relative to the earth. Basic systems display bearings and ranges to radio beacons; advanced systems either use this data to calculate a more intelligible presentation of position or propagate an estimate of position from knowledge of velocity and attitude.

- "System" information indicating the status of aircraft systems such as engines, hydraulics, fuel, etc.

All of these data sets are highly correlated; for example, a control action rolling the aircraft quickly leads to a change of heading, which then changes the course and position of the aircraft. These categorizations are not absolute; the altimeter, for example, shows vertical position (altitude) but is often treated as guidance information given the sensation of pitch changes and vertical speed provided by the sweep of its needle. These categorizations do serve, however, to identify the most time-critical information, which must be sampled frequently (the control information, because the aircraft can quickly pitch and roll); the information next highest in required sampling rate (the guidance information, which is directly impacted by control actions); and then finally most elements of navigation and system information, which need only be sampled at a rate on the order of minutes.

These temporal and causal relationships led to the development of the "Basic T" cockpit layout, common to during WWII and a standard to this day (e.g., Doolittle, 1961; Green, Self, & Ellifritt, 1995, pp. 1–13; Nijboer, 1998). As shown in Fig. 45.1, the control information is provided by the "artificial horizon" in the upper center of the "T." Guidance information is provided on the left (airspeed indicator), bottom (heading indicator or directional gyro), and right (altimeter). The supplemental vertical speed indicator and indication of slip and skid (turn coordinator) fill in the corners. Navigation and systems information are placed around the Basic T as space is available, usually

FIGURE 45.1. "Basic T" arrangement of cockpit instruments, with engine and navigation instruments surrounding on the right and bottom. Photo courtesy of T. Pritchett, Department of Aviation, Mount Royal College, Calgary, Alberta, Canada.

with similar types of information grouped together. These groupings of information are often emphasized by painted outlines.

This standard cockpit arrangement serves as an early example of display design benefiting from proximity relationships that support good sampling behavior, and from an organization of display information that matches the ecology of the environment. Such considerations remain central to the layout of cockpits and are reinforced by pilot training that teaches a scan pattern centered on control information. In cockpits not using computer screens (often referred to as steam gauge cockpits), this Basic T layout is still standard.

Subsequent research also examined display presentation and formatting. For example, Fitts and Jones (1947) classified types of display interpretation errors. These results guided research into specific display formats, such as the altimeter display formatting guidelines established by Grether in the early 1950s (summarized in Grether, 1995).

Computer Displays

The computerized display of information is not new to aviation; air transport cockpits with cathode ray tubes or active matrix liquid crystal displays have been in operation since 1980, and radar displays were implemented as early as the 1950s. Use of computer displays had several motivations. First, much of the underlying sensing already provided digital output. Second, many displays could only be created by computers, such as radar returns for air traffic controllers and moving map displays for pilots. Third, the number of displays reached critical levels where concerns about the workload and wide visual scan demanded from the pilot were matched with the impossibility of finding more display space. For example, World War II aircraft typically contained 25–35 dials, gauges, and controls (e.g., Nijboer, 1998). Subsequent aircraft with new systems (including jet engines) rapidly added controls and displays. The air transport cockpit touted as having the "most" physical dials and gauges is that in the Concorde, shown in Fig. 45.2; each of the three crew members are faced with a daunting instrument panel, including one gauge per engine measurement for each of the four engines (Sexton, 1988).

In contrast, the most recent Boeing cockpit is shown in Fig. 45.3; most of the dials are replaced by six computer displays: two each for the captain and first officer showing control and guidance information and navigation information; and two central displays providing information about aircraft systems, alerts, and checklists. To present important information in the pilot's primary field of view during maneuvering, takeoff and landing, many cockpits are now also equipped with head-up displays (HUDs) that project information onto a transparent screen through which he or she can also see out the forward windscreen (e.g., Stokes & Wickens, 1988).

These displays originally built on established metaphors. Air Traffic Control (ATC) radar screens mimicked the maps on which controllers had pushed around aircraft markers (colloquially called shrimp boats; Nolan, 1994). Likewise, the computer version of the Basic T is the primary flight display

FIGURE 45.2. Cockpit of Concorde 101 on display at the Imperial War Museum, Duxford, Cambridgeshire.

(PFD), an example of which is shown in Fig. 45.4. Although the PFD displays "tapes" for altitude and airspeed, the same relative positioning of information is used as in the Basic T layout.

Likewise, navigation information is shown in a "moving map" presentation, which uses the same symbology as paper charts, as shown in the electronic horizontal situation indicator (EHSI) in Fig. 45.5.

In selecting the basic elements of the display, most branches of aviation have historically opted for a small set of symbols, colors, and display configurations (some military cockpits are the exception). For example, Boeing air transport aircraft use a common symbology for their computer displays; likewise, their computer displays use only seven colors, each with an assigned semantic meaning: red for danger, yellow for warning, white for general information, and so forth. Although this limited set of design choices restricts the flexibility of the designer in designing the display and then of the operator in configuring the display, it also prevents problems with discrimination between similar features, allows for extensive testing, and enables comprehensive training documents (e.g., Society of Automotive Engineers, 1988).

FIGURE 45.3. B777 flight deck. Photo reproduced with permission of Boeing Management Company, Seattle, WA.

Virtual Environments

The danger of training pilots was quickly recognized, and World War I saw the first recorded use of simple flight simulators. For example, the Antoinette trainer was effectively half a barrel sliced lengthwise in which the student sat; a control column tilted the "aircraft" in response to the student's inputs through pulleys, with instructors at the corners providing simulated turbulence (Rolfe & Staples, 1986).

More sophisticated training devices started appearing in the early 1930s. The most notable was developed by Edwin Link using the knowledge and facilities available to him from the family piano and organ business. Demand for these trainers in teaching instrument flying skills was created by the issuance of air mail contracts to the U.S. Army Air Corps, which immediately required pilots capable of all-weather flying. By the onset of World War II, American, Commonwealth, and German pilots commonly had 50 hours in these devices before active duty; the United States used 10,000 Link trainers during the war years (Moroney & Moroney, 1999). Link trainers, such as the one shown in Fig. 45.6, used simple analogue computers to create both motion and instrument output in response to pilot control inputs. Most had enclosed cockpits excluding external visual references by the pilot. The body of the simulator cab mimicked the shape of an aircraft with moveable control surfaces, allowing the instructor to monitor control activity. Versions for a crew of three also provided scrolling pictures of ground features for a bombardier to aim on, and a rotating star-field for the navigator to practice taking star fixes (Rolfe & Staples, 1986).

Demand for flight simulators capable of providing higher fidelity required fast digital computers capable of solving aircraft equations of motion in real time. This led to the development of digital computers with a range of applications, such as the Whirlwind computer arising out of the Airplane Stability and Control Analyser project started in 1943 and special-purpose digital computers developed by Sylvania and by Link in the 1950s and 1960s (Rolfe & Staples, 1986).

Since that time, flight simulators have become sophisticated, immersive virtual reality environments. Beyond including an exact replica of the cockpit, two components contribute significantly to their realism. The first is the motion system common to the highest fidelity simulators. Six degrees of freedom motion platforms now provide a wide range of motion and fast response rate. Adding to their apparent realism is the knowledge acquired about human proprioceptive sensing. Although motion systems are not well suited to providing a sensation of motion for an extended duration, they can provide earlier indications of acceleration and sudden movements than are detectable from vision alone. To stay within their range of extension, motion systems typically "trick" the pilot by providing an accurate initial indication of acceleration and then slowly returning to a central position at a rate below vestibular thresholds, an effect created through "washout" filters. Additional cues may also be provided by inflating the pilot's g-suit to simulate the onset of g-loads during maneuvering, tightening the pilot's harness to simulate deceleration, and shaking the pilot's seat and instrument panel to simulate turbulence (review provided in Baarspul, 1989, pp. 116–123). Motion platforms are mandated for the highest level of civilian flight trainers; motion systems are less common in military systems because of their cost and concerns about simulator sickness, a phenomena similar to motion sickness (e.g., Moroney & Moroney, 1999; Stoffregen, Hettinger,

FIGURE 45.4. B777 primary flight display. Photo courtesy of Jerome Meriweather.

Haas, Roe, & Smart, 2000). Although the mechanisms underlying simulator sickness are still debated, they may indicate the limit of simulated motion currently achievable with motion platforms, although Stoffregen et al. (2000) also proposed mechanisms to alleviate simulator sickness.

The second component contributing to a flight simulator's fidelity is the vision system creating the out-the-window scene. Early systems simply mounted CRT's at the cockpit windows, often with infinity optics for collimation, each presenting a scene rendered by simple computer-generated imagery or provided by a video camera "flying" over a model of terrain. Modern systems project highly detailed, wide field-of-view scenes on large domes surrounding the simulator cab. Visual systems for air transport aircraft simulators receiving the highest level of certification have, at a minimum, a display resolution from the pilot's eye point of three arc-minutes and create highly detailed visual scenes, including airport layout and lighting as well as incidental features such as cars on roads, buildings, and landmarks (Federal Aviation Regulation 121.409). A wide range of atmospheric conditions are also included, such as the motion of snow in high winds. Vision systems for military simulations can be even more extensive.

Although historically not well documented because of their proprietary nature, these systems have often represented the cutting-edge of computer graphics (e.g., Foley, van Dam, Feiner, & Hughes, 1989, p. 919). This capability has both a corresponding cost (the highest fidelity air transport simulators currently cost between $10 and $15 million (US), with roughly half the cost attributed to the visual system) and a corresponding benefit (simulators receiving the highest level of certification from aviation authorities can be used to legally train and test pilots such that their first flight onboard a real aircraft is on active-duty flying passengers).

As general purpose computing has increased in performance, flight simulation has taken advantage of its increased capability. A range of flight simulators are in use today, including simple software and joysticks intended for personal computers, suitable for practice by pilots at home. The best air

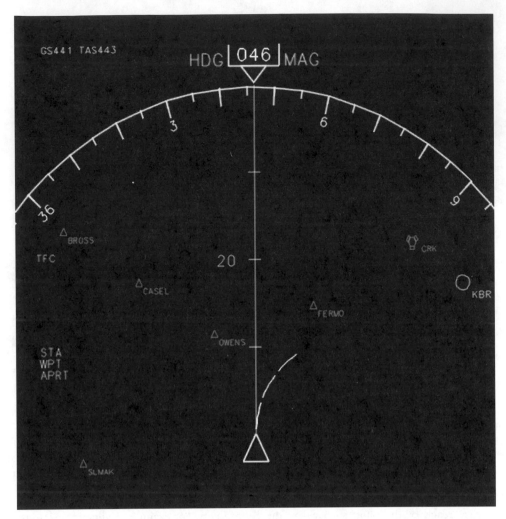

FIGURE 45.5. Electronic horizontal situation indicator.

transport simulators are likely to switch, in the next several years, to workstation-based technology and common programming standards (Nordwall, 2001).

Automated Systems

Paralleling the development of cockpit instruments was the development of automatic control systems or "autopilots" to control aircraft directly, demonstrated as early as 1914 by Sperry (Billings, 1997). Originally, these used analogue connections from sensor to actuator and were used for simple tasks such as maintaining altitude and heading; pilots were still required to perform all the maneuvers to get up to cruise altitude and to descend and land. More advanced systems added flight control modes that create a choice of control behaviors, including tracking heading, track or course, maintaining specific vertical speeds, and capturing altitudes. The most modern aircraft are controlled through flight control computers, which, when not serving as autopilot, measure the pilot's

control inputs and convert them into commands to the control surfaces; these commands are transmitted through the aircraft electronically (in fly-by-wire systems) or optically (in fly-by-light systems), and some aircraft filter the pilot's inputs to make certain that the aircraft stays within its flight envelope. Autopilot modes are typically selected through a control panel placed centrally in the cockpit. For example, the air transport autopilot mode control panel shown in Fig. 45.7 is placed between the two pilots at the top of the instrument panel.

In theory, the pilot of a modern jet transport need only taxi to the runway, turn control over to the autoflight system, and move the flap and gear levers to "up" after takeoff; the system will fly the entire flight, with the pilot required to lower the flap and gear levers before landing and taxi after landing. The flight plan can be entered onboard the aircraft and, in some cases, downloaded from computers in the terminal as part of the preflight briefing. Sheridan noted that these aircraft may be categorized as "telerobots" (1992, p. 240); likewise, designers of automatic flight control systems have stated, "We like

FIGURE 45.6. Link trainer used for World War II pilot training. Photo courtesy of T. Pritchett, Department of Aviation, Mount Royal College, Calgary, Alberta, Canada.

to think of the airplane as just another peripheral" (Tomayko, 1992, p. 19). With such aircraft, the only additional technology required to fly from a ground-based simulator is a datalink with the aircraft; this capability is now being considered for freighters (e.g., Croft, 2001) and is the basis for military uninhabited aerial vehicles (UAVs). A common input device for these systems is the control display unit (CDU) shown in Fig. 45.8, typically placed in easy reach of the pilot, such as by the throttles.

These automated systems can also provide "command information" to the pilot as a guide to follow in flying the aircraft. For example, flight directors indicate (directly on the artificial horizon or PFD) the pitch and bank that will achieve the target flight condition entered into the autoflight system. How to present this command information, and its underlying dynamics, remains an important component of designing both displays and autoflight systems (e.g., Stokes & Wickens, 1988; Verspay et al., 1996).

Intelligent Systems and Decision Support

Intelligent systems in aerospace have extended beyond the automated systems that fully control the aircraft. Some of these systems are intended to help monitor for situations that are so rare, they are rarely sampled or monitored by a human or are so catastrophic that any automatic assistance will contribute to safety. For example, engine temperature sensors have long been used to indicate engine fires, given that human sampling behavior will often not attend to the rarely changing engine temperature gauges (e.g., Wickens & Flach, 1988). Other systems monitor for failures of the human. For example, as early as World War II, aircraft were commonly equipped with alerts warning the pilot if he or she appeared to be landing with the landing gear up. Likewise, American air traffic controllers' screens include a conflict alert that is triggered by aircraft coming too close to each other; this system is commonly called the "snitch patch" because of the punitive consequences of its activation.

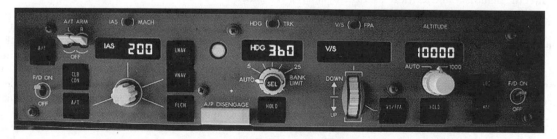

FIGURE 45.7. B777 Mode control panel providing autopilot controls. Photo courtesy of Jerome Meriweather.

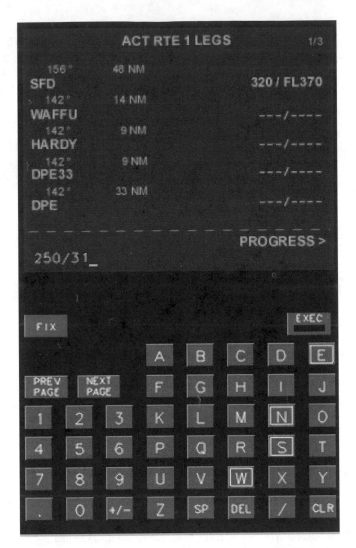

FIGURE 45.8. Schematic of a control display unit (CDU) for reading from and entering data into autoflight systems.

Far more sophisticated systems are now being developed. Some serve as alerting systems, except that they give commands that must be followed by the pilot. For example, the traffic alert and collision avoidance system (TCAS) detects aircraft that may come into conflict and calculates an avoidance maneuver for each; the command given to the pilots, therefore, is essentially an electronic contract that both sets of flight crew must follow to guarantee safe separation (RTCA, 1983). A similar system, the enhanced ground proximity warning system (EGPWS), uses a terrain database to determine whether an aircraft is too close to the ground without being lined up with a runway for landing.

Other systems serve as intelligent agents or decision support systems. In air transportation, for example, aids have been proposed to help pilots decide following engine failures during takeoff whether it is safer to abort the takeoff (and risk overrunning the end of the runway) or continue the takeoff (and risk not getting the aircraft off the ground; e.g., Bove

& Anderson, 1999). Military systems have a more extensive history, such as the Pilot's Associate system for actively managing the information presentation to the pilot and performing planning and battlefield assessment tasks (e.g., Dimitrov & Rippy, 1999; Kupferer, 1999). Decision support tools are also being developed for flight planning and rescheduling in airline operating centers (e.g., Smith, McCoy, & Layton, 1993; see also Smith & Geddes, chapter 33, this volume) and in air traffic control centers (e.g., Green, den Braven, & Williams, 1991; Quinn & Robinson, 2000; Swenson et al., 1997).

Implementation of these systems has not been without problems. Alerting systems, for example, have typically used conservative thresholds; the corresponding prevalence of false alarms has led to both overreliance (where a false alarm is acted on) and overreliance (where correct alert is not trusted) (e.g., Pritchett, 2001; Wiener & Curry, 1980). Studies have also identified situations in which pilots use these alerting systems other unanticipated ways, such as tracking the alerts as aural indicators of the flight envelope during extreme maneuvers (Pritchett, 2001). Likewise, problems with decision support systems have ranged from underreliance to overreliance, as well as "automation bias," in which the operator wishes to use the system appropriately but instead uses its output as a heuristic replacement for effective judgment (e.g., Mosier, Skitka, Dunbar, & McDonnell, 2001; Smith et al., 1993). Greater detail on these issues can be found in Smith and Geddes, chapter 33 in this volume.

Organization, Communication, and Coordination

In 1929, the first air traffic controller, Archie League, stood on the runway of St. Louis's airport and waved flags to communicate to aircraft when it was their turn to land or take off. This was quickly followed by voice communications for air traffic control. Twenty-five years later, civilian air traffic control radar displays were implemented in the United States, providing the information needed for ground control of the movements of aircraft (Nolan, 1994). Recent developments include prototype decision support tools such as the Final Approach Spacing Tool and the Traffic Management Advisor (both being developed and tested at Dallas–Fort Worth) for highlighting the current and near-term traffic situations and for suggesting commands to give to aircraft (e.g., Quinn & Robinson, 2000; Swenson et al., 1997).

As noted, communication has historically been achieved through voice transmissions on a shared frequency monitored by a controller and all aircraft in a control sector. This communication is highly proceduralized, aiding comprehension by placing each word in the context of an ordered, limited-vocabulary message; even so, it still has an estimated error rate of 1% (Cardosi, 1993), with confusion over voice instructions leading to some of the worst accidents in air transport history. This form of communication is also responsible for high controller workload; in high-density control sectors, the voice frequencies are constantly in use with the controller transmitting more than half the time and with aircraft often unable to interrupt to send a message.

Communication is now being expanded to include computerized datalink transmission. Digital transmissions in air

transport started with the Aircraft Communications Addressing and Reporting System (ACARS) system, which allows for transmission of non-flight-critical information between aircraft and their airlines. Messages in the cockpit are printed out by a printer or displayed on a computer display. Although not used to transmit air traffic control communications, this system has demonstrated a long and reliable use of digital communication of operational data for pilots (e.g., text descriptions of weather en route), for cabin crew (e.g., transmitting requirements for cabin servicing), and for passengers (e.g., gate information and the occasional sports score).

Digital communication between controllers and pilots (and between pilots) is also now being implemented in several test conditions, motivated by the current congestion on voice frequencies, the shortage of voice frequencies for air traffic control, and the anticipated benefits of providing a persistent (possibility multimodal) display of important communications (e.g., den Braven, 1992). Controller workload is also anticipated to be reduced in some operations, if provided with good human–computer interfaces (e.g., Talotta et al., 1992). Corresponding cockpit displays have been proposed, including text displays of messages, overlays on graphic navigation displays of commanded routes, and synthesized voice.

CURRENT CHARACTERISTICS OF HCI IN AEROSPACE

The previous sections described elements of HCI in aerospace as they have evolved from the start of heavier-than-air flight to systems considered feasible for near-term implementation. These elements have combined to create task environments for pilots and controllers (and others including airline dispatchers, mechanics, etc.) that are coordinated, automated, complex, information intensive, dynamic, proceduralized, safety-critical, and certified by regulatory agencies. The following sections discuss each of these characteristics and the design challenges they create. Some of these challenges benefit from the knowledge acquired through the history of aerospace and of HCI; others highlight new and emerging issues.

Characteristic: Coordinated

Description. At the time of this writing, new ATC systems are required around the world. In some regions, an air traffic control system is needed where no substantive system has existed before. In other regions, including the United States and Western Europe, dramatic changes will be required to meet the increasing demand for air transportation of people and cargo; initiatives such as NASA's Small Aircraft Transportation System (SATS) additionally seek to replace cars with aircraft for general purpose personal transportation (e.g., Wilson, 2001).

ATC is enabled by three core technologies: Communication, Navigation, and Surveillance (CNS). ATC is also enabled by the procedures, methods and protocols used for the process of air traffic management (ATM). The collective problem of redesigning the ATC system, therefore, involves both technologies and procedural and process changes and is often referred to as CNS/ATM.

Advances in the individual technologies have been demonstrated. For communication, as noted earlier, the potential now exists for digital, high-bandwidth communication, enabling transmission of more information in general, as well as opening up new communication lines between operators who to date have communicated only indirectly or not at all (e.g., allowing systems on different aircraft to communicate directly) and more closely integrating the flight deck and air traffic control station (e.g., Green et al., 1991; Kerns, 1999). Navigation technology has now advanced (primarily through the development of civilian systems using the global positioning system) to the point where aircraft using standard, inexpensive equipment can fly directly to their destination rather than tracking from one ground-based radio beacon to the next and to the point where aircraft separation minimums no longer need to account for navigation error and hence may be reduced. Finally, surveillance technologies, which identify the position of aircraft to the controller, are dramatically changing. Currently performed by radar, surveillance will likely evolve to a combination of navigation (where aircraft know where they are) and communication (where aircraft broadcast their locations); these broadcast methods of surveillance will provide a picture of the traffic situation to all who can receive it.

Redesigning the methods and procedures used for ATM, however, is not as straightforward as upgrading CNS technologies. The current system is both distributed and centralized— distributed in that controllers and pilots are each endowed with authority for a significant set of decisions, centralized in that, at each level of a hierarchy of individual sectors (control centers and nations), a single body is identified as the final authority.

Challenges in Design. Several major trade-offs exist in redesigning ATC, each containing philosophic and legal issues as well as methodological HCI concerns. The first addresses the distribution of cognition within an airspace system. ATM can be conceived as centralized (the "Swiss train" model, in which a central authority exactly specifies all behavior of entities within the system), as distributed (the "TCAS and have at it" model where aircraft fly completely independently, using onboard systems to avoid each other), or as some intermediary between the two. Each conception of the ATC system involves significant difficulty given the scale of the ATC system and the unpredictable effects common to every facet of operation (variations in aircraft performance, changes in weather, unanticipated failures, etc.). A centralized system requires real-time computations of overwhelming scope. Likewise, a distributed conception views the overall dynamic as an emergent behavior that is difficult to analyze a priori for overall performance and for the safety levels expected of air transport. Both create a system behavior of enormous magnitude that humans will need to comprehend and oversee.

Another major trade-off exists in designing the macrolevel concerns (e.g., improving the performance of the ATC system as a whole) and microlevel concerns (e.g., improving the performance and task environment of individuals). In some ways

these concerns are mutually supportive; for example, a safety issue at the microlevel can quickly develop into a safety issue at the macrolevel. In other ways, however, these concerns are contradictory; for example, macrolevel concerns will often impose communication and coordination burdens on a pilot. These concerns may have substantial impact on the future research needs of HCI in aerospace; while HCI often focuses on the microlevel concerns of individuals, their collective behavior must be anticipated and designed for as well.

A final issue centers on the aerospace equivalent of the organizational impact of information technology. Future cockpits will provide pilots with the same big picture information as is currently available only to the air traffic controller, blurring current distinctions between pilot and controller for command authority (e.g., Farley & Hansman, 1999; Kerns, 1999). Therefore, CNS/ATM changes may require—or opportunistically suggest—a shift in the roles, tasks and authority within in the ATC system.

Characteristic: Automated

Description. Aerospace is increasingly automated, with pilots and controllers interacting with decision-support systems, with alerting systems that can detect mechanical failures, with alerting systems that detect human errors, and with completely automated systems that remove the human from direct involvement in system control. An impact of the prevalence of automated systems in aerospace has been to change the human's role, tasks, and required skills. In air transportation, for example, the transition to computerized cockpits has changed the pilot's task significantly; concerns have been raised that pilots mainly sit and monitor computers for most of a flight, whereas pilot control is only required in emergencies or abnormal situations. Corresponding issues include loss of basic flying skill due to lack of reinforcement during normal operations, changes in pilot training to account for the difficulties experienced in understanding and monitoring automated systems, the ability of pilots to monitor and detect faults in systems, the inherent difficulty of stepping into an established situation with little notice, the potential for pilots to not follow or use the automation when it is correct, the potential for pilots to use or follow the automation when it is incorrect, and the impact on pilots situation assessment activities and corresponding situation awareness (e.g., Beringer & Harris, 1999; Dekker & Hollnagel, 1999; Endsley & Kiris, 1995; Mosier, Skitka, Dunbar, & McDonnell, 2001; Rignér & Dekker, 2000; Wiener & Curry, 1980).

Beyond control automation, information automation has also been proposed for helping pilots with the information-intense environment, including the alerting systems and decision aids noted earlier (Billings, 1997; Noyes & Starr, 2000). However, these systems are also prone to problematic interactions between human and computer (e.g., Chen & Pritchett, 2001; Pritchett, 2001; Smith et al., 1993).

Challenges in Design. As in many other domains, these concerns with human-automation interaction have led to increasing recognition that the impact of introducing automation should be considered both for long-term ramifications if it is used as intended and for the potential of well-intended misuse of the automation (e.g., Braune & Graeber, 1992; Tenney, Rogers, & Pew, 1998; Wiener & Curry, 1980). However, specific mechanisms and design guidelines for achieving a desired human-automation interaction in aerospace task environments are not yet widely agreed on. Popular categorizations of behavior use as discriminants who monitors who (machine monitoring human or vice versa?), who closes the control loop, who is responsible for undesired outcomes, and the type of intervention that human or machine can make into each other's activities (e.g., Billings, 1997; Olson & Sarter, 2000; Sheridan, 1992). Methods of allocating functions have included cost-benefit analysis, comparison of human and machine capabilities, dynamic adaptation in operations based on physiologic assessments of pilot stress, and models of the distribution of cognition between teams of pilots and intelligent systems (e.g., Fitts et al., 1951; Howard, 1999; Prinzel, Freeman, Scerbo, Mikulka, & Pope, 2000; Sheridan, 1992). Because automation can often be turned on by the pilot, studies have also examined the automation management strategies provided to pilots through procedures and training (e.g., Billings, 1997; Olson & Sarter, 2000).

Characteristic: Complex

Description. The complexity of the overall air transportation system and of sophisticated aircraft systems is mirrored in the complexity of the interfaces and procedures used by controllers and pilots. Take, for example, the flight management systems (FMSs) common to modern air transport aircraft. These systems can command hundreds of control behaviors; the differences between some behaviors may be subtle in both ensuing system behavior (e.g., does the elevator control speed and the throttle pitch, or vice versa?) and in the feedback to the pilot (e.g., pilots may only have direct indication of flight control mode through flight mode annunciators [FMAs], typically two- or three-letter abbreviations, on the autopilot control panel and PFD). Dealing with this complexity can be difficult, and pilots can no longer be thoroughly trained on each mode and may find their sheer number overwhelming. For example, Sarter and Woods (1992, 1994) highlighted instances in which pilots were confused by transitions between modes and were unable to predict future aircraft behavior.

Recent research has investigated coping mechanisms often used by pilots. In surveys, pilots have reported relying heavily on on-the-job training to understand the FMS, feeling comfortable with it only after months of operation (Vakil & Hansman, 2000). Two coping mechanisms for conceptualizing the FMS have been demonstrated: inferential simplification and frequential simplification (e.g., Javaux & Olivier, 2000). Inferential simplification occurs when the pilot assumes that a mode shares attributes with similar modes, for example, assuming that all FMS modes used to climb the aircraft act the same. Likewise, frequential simplification occurs when some aspect of a mode is rarely observed; for example, some modes are normally (and can only be) triggered in certain flight conditions, with the result that

the flight condition requirement is not reinforced in the pilot's conceptualization. This simplification can lead to surprises when unusual modes are used or rare conditions are experienced.

Dealing with system complexity can also be difficult because of the amount of feedback commonly provided to operators about system functioning (Howard, 1999). Again using the FMS as an example, Vakil and Hansman (2000) found that, although approximately 90% of system modes control aircraft speed and vertical movement, few indications are provided in the cockpit as to the current mode and future behavior in these dimensions beyond the FMA on the PFD (shown in Fig. 45.4) and the autopilot control panel (shown in Fig. 45.7).

Challenges in Design. Methods of providing feedback in the vertical plane are currently being examined, such as vertical profile displays (e.g., Fig. 45.9) providing an equivalent function in the vertical plane as the navigation display (Fig. 45.5) provides in the horizontal. Other methods being examined include tactile and auditory alerts of changes in control behavior (e.g., Nikolic & Sarter, 2001). Similar improvements in feedback are also being examined for alerting and health management and diagnostic systems (e.g., Pritchett & Vándor, 2001).

These studies suggest not only that pilots need sufficient information to understand system functioning for confident, consistent interactions, but also that there may be limits on how complex systems may be. Therefore, another challenge is to eliminate unnecessary complexity within systems and tasks, while retaining essential functionality.

Characteristic: Information Intensive

Description. Historically, pilots and controllers were information starved; for example, to calculate position pilots needed to relate their charts to indications of bearing and distance to radio beacons. In comparison, modern aerospace task environments are information rich; for example, cockpits can now provide a multifunction moving map display that shows position, the programmed flight plan, navigation fixes, location and intensity of turbulent weather, contour maps of terrain, and nearby traffic. Built-in computer screens or carried-in laptops will soon provide "electronic flight bags," computerized presentations of the paper charts, manuals, checklists, and so forth that pilots currently carry in their flight cases (Hansman & Mykityshyn, 1995; In New Cockpits, 2001; Shamo, 2000).

One difficulty with information-intensive environments comes from the data-entry requirements they imply. For example, the advent of automated systems demands data-intensive inputs into these systems, often at unanticipated times. Current systems have attempted to reduce data-entry workload through development of pages and menus that infer and sequence most of the input. For example, rather than requiring entry of all waypoints in a flight plan, CDUs use knowledge of the destination airport to suggest the most likely routes to the pilot. Although this has helped significantly, widespread concern remains that pilots need to "go heads down" too much (i.e., look down at the keyboard rather than at flight instruments and out the window); several airlines have restricted in-flight reprogramming to cruise. In addition, recent research suggests that data-entry

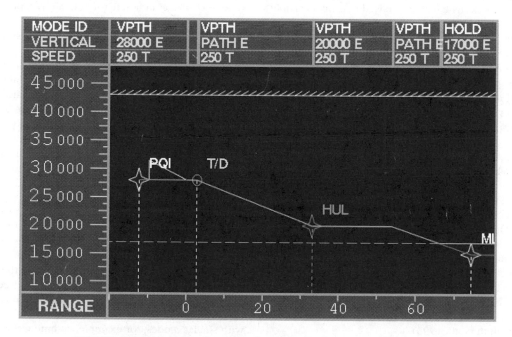

FIGURE 45.9. Example of a vertical profile display. Figure courtesy of S. Vakil and R. J. Hansman, MIT International Center for Air Transportation, Cambridge, Massachusetts.

errors are becoming more prevalent than data interpretation errors (e.g., Sheridan, 1992, p. 244).

Given this information intensive task environment, supporting good situation awareness has received substantial attention (reviewed in Endsley, 1995b). Although the definition of situation awareness and its underlying mechanisms remain the subject of debate, its importance is intuitively and widely agreed on. For example, the development of good situation awareness is considered an important part of pilots' and controllers' duties and is emphasized in training and procedures. Awareness of automation is also increasingly emphasized, with specific terms such as *mode awareness*, *automation awareness*, and *alerting system awareness* (e.g., Pritchett, 2001; Sarter & Woods, 1995).

Several factors may contribute to poor situation awareness. The first is information overload, when the operator faces too much information; for example, a first officer described the cockpit during an emergency as "All the lamps are blinking and there are a lot of warning sounds in the cockpit. It is really a terrible environment. It is not possible to manage all this information. With so many malfunctions you stop analyzing them and concentrate on the flying. That's the only thing to do" (Mårtensson, 1995, p. 315).

The second possible obstruction to good situation awareness is too little information, either because the information is not available, is not selected by the pilot when configuring the display, or is not presented in a sufficiently compelling manner. This is especially a problem at this time with automated cockpit systems (e.g., Johnson & Pritchett, 1995; Pritchett, 2001; Sarter & Woods, 1995).

The third projected obstruction to good situation awareness stems from representational effects. The importance of pilots' internal representation of the environment has been widely demonstrated, with a corresponding emphasis on displays which provide an external representation that efficiently interacts with pilots' internal representations (e.g., Howard, 1999; Zhang, 1997). However, efforts to date have not reached this goal in all cases; studies have also found representational errors can be persistent and difficult to overcome (e.g., Jones & Endsley, 2000).

Challenges in Design. To address problems with data entry, one current challenge is in designing new "command-control" languages and interfaces that shape the underlying functioning of a system around its data-entry requirements, an example of which is shown in Fig. 45.10 (e.g., Faerber, Vogl, & Hartley, 2000; Riley, 2000). Likewise, faster data-entry devices may appear soon in cockpits, such as keyboards and cursor controls for direct manipulation of screen elements.

To address issues with providing the correct information and with supporting situation awareness, research has examined what information pilots and controllers need to know through

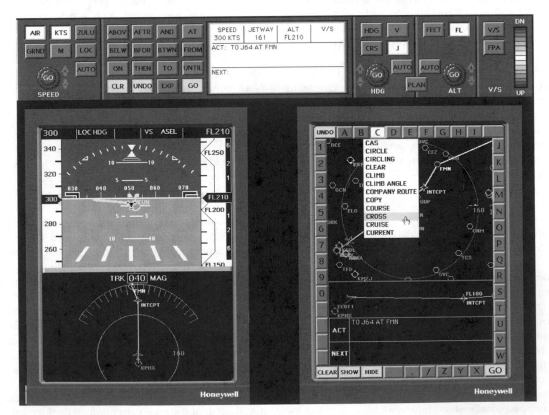

FIGURE 45.10. Command control language interface. Image courtesy of C. Misiak and V. Riley, Honeywell Labs, Minneapolis, Minnesota.

surveys, task analysis, operational observations, and simulator experiments (e.g., Endsley, Farley, Jones, Midkiff, & Hansman, 1998; Fitts et al., 1951, p. 32; Pritchett & Hansman, 1997). Because flight conditions can differ in their information requirements, such studies also examine when information must be shown. A common method of providing pilots and controllers with the correct amount of information relies on their configuring the display; for example, the navigation display on air transport aircraft typically has several different selectable display modes, each suited to a particular task (general navigation, course tracking, flight planning, etc.), with the additional capability to toggle on or off the overlay of additional information such as traffic or weather. Various pop-up mechanisms are also available, often as a precursor to an alert about a developing situation. Other efforts examine relating display content to pilot information needs by intelligent systems that drive the display configuration (e.g., Haas, Nelson, Repperger, Bolia, & Zacharias, 2001; Kupferer, 1999).

Concerns with information overload are generally addressed by careful display design to prevent clutter, by allowing the pilot to select information as needed, and by information management by intelligent agents, including alerting systems and systems that prioritize alerts according to their urgency (e.g., Proctor, 1998). Likewise, concerns with information starvation still exist, especially as studies of automated systems suggest that more information about automated systems' functioning can have significant benefits (e.g., Pritchett & Vandor, 2001; Vakil, Midkiff, & Hansman, 1996). However, realizing these goals can be challenging; some of them can be conflicting (such as showing more information about an automated system while reducing clutter).

The importance of pilots' internal representation of the environment has been widely demonstrated, with a corresponding emphasis on displays that provide an external representation that efficiently interacts with pilots' internal representations (e.g., Howard, 1999; Zhang, 1997). However, efforts to date have not reached this goal in all cases; studies have also found representational errors can be persistent and difficult to overcome (e.g., Jones & Endsley, 2000).

Addressing the design challenge of providing the best display representations may require new display metaphors that integrate information in a manner more directly tied to the operator's internal representation. To date, most computerized information-intense displays have built on historic metaphors. For example, the PFD (shown in Fig. 45.4) builds on the format and presentation of the Basic T (shown in Fig. 45.1); likewise, the new controller screens provided by the display system replacement (DSR) follow formats similar to the radar scopes they replace. These underlying conventions distinguish between horizontal displays (generally a two-dimensional plan-view) and vertical information (generally presented as a text readout of altitude on horizontal displays or, in some recent cockpits, a two-dimensional plan-view in the vertical plane). In cockpits the additional distinction is often made between command, control, guidance, navigation, and systems information.

Recent research and development efforts are examining the use of displays that provide a more-integrated picture of all of these various categories through the use of novel (to aerospace)

display formats such as perspective. The application of perspective displays for both cockpits and air traffic controllers faces several difficulties. First, the resolution required in the horizontal plane is rarely finer than a few hundred feet, whereas the resolution required in the vertical plane is typically within 50 feet. This usually prevents equal scaling in the horizontal and vertical planes, with a commensurate distortion of the display from a true visual representation. In addition, perspective displays allow for the occlusion of far features by near features and ambiguities in perception of location, which is not acceptable in most aerospace displays (e.g., Merwin & Wickens, 1996). Finally, the attributes required of the display vary widely; for example, in developing cockpit perspective displays, different tasks may want an exocentric versus egocentric viewpoint and different fields of view depending on whether fine control around straight and level is desired, whether the aircraft is rapidly maneuvering, or whether the pilot is planning a future flight segment.

Perhaps the most novel uses of integrated displays are being explored in the use of synthetic- and enhanced-vision systems in the cockpit. Some concepts propose using a windowless cockpit containing the visual displays used with flight simulators (described earlier in this chapter), providing outside reference through either a large-format field of view display or a head-mounted display, providing computer generated imagery and sensing technologies such as video camera links and infrared cameras. Other concepts propose a conformal overlay on the view of the outside world, such as that created by projecting onto the inside of cockpit windows or by using a see-through display; enhanced vision may allow a pilot to fly by looking "outside" in all weather conditions without special instrument flight training and may highlight important pieces of information such as hazards that may otherwise not be salient components of the visual scene.

Simple enhanced-vision systems have been in use for many years in the form of HUDs (Green et al., 1995). As shown in Fig. 45.11, these display flight information within the pilot's forward field of view using clear combiners located between the pilot's eyes and the windscreen in a format evocative of Basic T and PFD formats. In the simplest cases, the HUD is simply presenting information in a more convenient location, without providing a direct correlate between its presentation and the outside world. However, newer systems meet the definition of augmented reality in that they present information that is conformal with the outside view (Dix, Finlay, Abowd, & Beale, 1998); for example, if the pilot controls the aircraft such that the HUD's flight path symbol is positioned over the runway threshold, then the aircraft is flying directly to the runway. HUDs have already demonstrated substantial benefits, including providing a less-expensive mechanism for flying all-weather approaches by using the pilot and HUD rather than requiring autoflight systems and can potentially help reduce controlled flight into terrain (CFIT) accidents (e.g., de Muynck & Khatwa, 1999; Proctor, 1997; Wilson, 2001). However, issues have also been raised with localization of the pilot's attention on the forward field of view, visual workload, clutter, obscuration of outside features by HUD symbology, and with difficulties in scanning colocated information presented by a variety of constructs such as text, graphics,

FIGURE 45.11. Heads up display. Photo of BAE Visual Guidance System, courtesy of BAE Systems, Farnborough, United Kingdom.

and out-the-window views (e.g., Martin-Emerson & Wickens, 1997; Newman, 1998; Ververs & Wickens, 1998).

Characteristic: Dynamic

Description. HCI in aerospace has always been driven by the dynamics of flight. Building on the synthetic- and enhanced-vision initiatives described in the previous section, new displays are being proposed that support these dynamics. For example, as shown in Fig. 45.12, "Tunnel-in-the-sky" displays are being developed in which aircraft can be piloted using one integrated display by flying the aircraft down the tunnel. These displays are enabled by both advances in computing power and databases providing a detailed visual scene (e.g., Newman, 1998).

Challenges in Design. For effective control, it is essential that the tunnel give a proper picture of motion. These displays need both to integrate a large amount of information into a small viewing area and to integrate the different types of information into a representation that does not require the pilot to cognitively integrate their various dimensions and time scales (Wickens, 1992, pp. 466–474). Analyses of aircraft motion and human perception have provided a theoretical basis for the design of these displays (e.g., Grunwald, 1996a; Theunissen, 1997).

Prototypes suggest these displays have great potential, with marked decreases in total navigation system error (i.e., pilots are able to track courses more accurately) and favorable ratings by pilots (e.g., Barrows, Enge, Parkinson, & Powell, 1996; Grunwald, 1996b; Theunissen, 1997). Some hypothesize these displays may reduce the difficulty of piloting aircraft sufficiently to reduce the training requirements for general aviation pilots. Likewise, it is hoped that these displays will improve safety by providing pilots with better awareness of terrain and will allow for improved air traffic control procedures involving more intricate maneuvers for better landing rates at airports (e.g., Theunissen, 1997).

Dynamic issues also arise in other tasks such as flight planning and air traffic control. Computer systems can assist in these tasks by extrapolating from the current situation to predict when or if events may occur, up to a time horizon seconds, minutes, or hours away. These predictions implicitly require trade-offs; for example, in detecting loss of aircraft separation, Type I and Type II errors will be significantly reduced by lowering the required time horizon, but alerts too late for an effective avoidance maneuver are of little use. Therefore, trade-offs between time horizon, system performance and the severity of allowable actions to resolve a hazard are constantly required in designing not only intelligent systems, but also in the displays that support predictive and judgment activities.

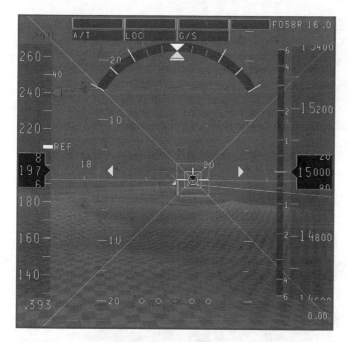

FIGURE 45.12. Tunnel in the sky display used in recent National Aeronautics and Space Administration tests of synthetic vision. Courtesy of E. Theunissen, Technische Universiteit Delft, The Netherlands.

The dynamics of flight also demands that operations can't be suspended for rest or reevaluation and implies that events can occur quickly and then worsen or improve in response to the pilot's actions. Added to the other features of the aerospace environment such as automated and information intense, the dynamics of aerospace operations makes pilot and controller workload a significant concern (e.g., Kantowitz & Casper, 1986).

Characteristic: Standardized and Proceduralized

Description. Aerospace is increasingly a standardized and proceduralized domain. One facet of this standardization focuses on creating standard design features for displays and human interactive systems. Groups such as the Society of Automotive Engineers and RTCA, Inc. (formerly the Radio Technical Commission on Aeronautics) are using teams of academic researchers and industry practitioners to develop standards for cockpit displays and systems (see, for example, RTCA, 1983; Society of Automotive Engineers, 1988). Development of these standards requires a careful balance between reaping the benefits of standardization and stifling innovation.

Substantial effort has also been take to standardize human behavior through personnel selection; the military, airlines, and air traffic control authorities have developed a battery of aptitude tests and minimum qualifications that admit a small, comparatively homogeneous population. This homogeneity is further accented by training that creates a population of experts at their tasks.

The operating environment is also structured by procedures. Taking a broad definition of "procedures," they cover not only the set of actions to be used for standard tasks, but also provide guidelines for when tasks should be executed (e.g., "Do not attempt to program a new route into the FMS below 18,000 feet"), the relative priority of tasks, the range of acceptable actions, and the division of labor. In air traffic control, published procedures dictate the expected routes of flight arriving into and departing from airports, creating a shared set of expectations for both controllers and pilots. Aerospace procedures are rigorously designed and tested and often must be certified by a government authority.

Challenges in Design. Although HCI generally focuses on the design of technologies, HCI in aerospace cannot ignore training and procedures because they are tightly coupled with computer technologies. For example, a flight simulator experiment demonstrated that a change in air traffic control procedures changed the information content pilots wanted in their cockpit traffic displays (Pritchett & Yankosky, 2000). Likewise, a separate simulator experiment demonstrated that the types of diagnostic decision aids best benefiting pilots in detecting, identifying, and resolving onboard system failures depends on the amount of training the pilots have about those failures (Davis & Pritchett, 1999).

Historically, procedures and training followed technologies, that is, if a better technology was invented, training and procedures were then developed to make best use of it. At this time, many aerospace procedures are difficult to change; training programs are also often difficult to change because they are themselves significant investments created through careful research and design. Therefore, implementing new interface and automated technologies often requires their designers to demonstrate that the corresponding procedural and training adjustments are feasible, beneficial, and, in air transport, cost-effective. To this end, researchers of flight deck systems and of training methods have proposed mechanisms for improving communication between designers and trainers, including methods of describing systems and displays that can be understood and used by both communities, up-front specification of how automated systems should be used and how they should fit within cockpit resource management activities, and methods of training pilots on new systems (e.g., Feary, Sherry, Palmer, & Polson, 2000; Olson & Sarter, 2000; Sherry & Polson, 1999; Seamster, 1999).

Characteristic: Safe

Description. Airline transportation in developed countries is one of the safest industries in the world, rivaled only by a few notables such as nuclear power production. Safety has always been the primary driver of improvements to the air traffic control system and continues to be the stated primary concern of manufacturers of air transport aircraft (followed by operating cost and passenger comfort). This concern is motivated by the seriousness of an air transport accident; such an event fits the formal definition of a catastrophic accident, that is, an incident whose costs are orders of magnitude greater than the investment lost. Many militaries are also becoming

more concerned about safety, as standards of acceptable risk become more conservative and as aircraft become fewer and more expensive.

Historical methods for improving safety addressed the human in isolation (e.g., training in flying skills, training in crew resource management, improved personnel selection) and the technologies in isolation (e.g., making the machines more reliable). These efforts have improved to the point that irrecoverable single-point failures rarely occur. Current safety research instead focuses on the chain of events that collectively lead to accidents, where each link in the chain is often recoverable if taken in isolation, with notations of organizational complexity and the role of latent failures in leading to accidents (e.g., Maruino, Reason, Johnston, & Lee, 1995).

While aviation operating environments tend to be structured, they are not always completely predictable; weather changes, mechanical failures, and human error commonly occur and must be resolved during the flight. Dealing with these unpredicted eventualities is the purview of the human operators, sometimes calling for the execution of a rote procedure, and sometimes calling for creativity. Because of this reliance on the human operator, human error has been a major focus in safety-related research in recent years, with some studies implicating human performance as a causal factor in approximately 70% of air transport accidents (e.g., Vakil & Hansman, 2000). Transitioning to computerized cockpits has reduced the air transport accident rate by about half; however, interaction with automation or pilot lack of awareness of information displayed in the cockpit is implicated in many of the remaining accidents, suggesting that computerized cockpits may introduce new forms of error, such as the crash at Nagoya in 1994 in which pilot and aircraft fought each other for control (e.g., FAA Human Factors Team, 1996; Sarter & Woods, 1995; Wiener & Curry, 1980).

Challenges to Design. Improving aerospace safety presents several challenges to HCI. The first is in designing computer-based systems that will always be supportive of good human performance, especially in high-tempo, high-stress, high-time-criticality situations in which other things may be going wrong (e.g., aircraft system failures or the arrival of bad weather). This requires addressing the issues noted throughout this chapter to an extreme: ensuring that a motivated and trained pilot will have the information resources to be aware of what automated systems are doing and why and what they will do next; creating systems that alert pilots and controllers of impending hazards; and creating displays (including synthetic and augmented vision systems) that provide pilots with good situation awareness and support effective decision making at all times. One method of addressing these concerns are formal methods that examine elements of the interface (e.g., Johnson, 1996; Zhang, 1997), the automation's functioning for inconsistencies (e.g., Rushby, 1999), the feedback of the automation's functioning (e.g., Degani, Shafto, & Kirlik, 1999), and the large-scale system design at the level of the entire aircraft or the entire airspace system (e.g., Blom et al., 1998; Sage & Johnson, 1999). Other methods address these concerns by involving the future operator throughout the design (e.g., Scott, 1991) and by identifying

which errors can directly and quickly impact safety (e.g., Sarter & Alexander, 2000).

The second challenge in improving aerospace safety is in measuring HCI in naturalistic environments for unsafe and very rare events. Creating a naturalistic test environment is itself quite difficult; for example, in working with two- and three-person crews, flight simulator and personnel costs are substantial, and crews are quite adept at averting abnormalities before they turn into the desired test cases. Given these difficulties, testing to date has often focused on simple measures of pilot or controller comprehension. For example, pilots may be briefed before flight simulator experiments on what to do when an alert is given, and their reactions recorded for speed and for correct identification of the alert; however, such tests may not captures issues when the pilot does not agree with the alert or chooses a nonstandard resolution to the problem. Therefore, simulator studies have examined methods for examining safety issues, including the use of confederates and methods of briefing pilot-subjects for more naturalistic behavior by encouraging realistic judgments and verbalization of their actions (e.g., Pritchett, Vándor, & Edwards, 1999). Longitudinal studies are also possible in operations in which anonymous reporting systems exist, such as the aviation safety reporting system (ASRS) administered by the National Aeronautics and Space Administration and available to U.S. pilots and controllers, and in which flight data recorders are scanned after every flight for possible safety issues (e.g., O'Leary, 1999).

The third challenge is finding measures of system safety. Aerospace does not rely on one type of measure of a system's effectiveness, but instead typically uses the measures common to HCI, human factors, and systems engineering. None of these measures provide a comprehensive view of system safety. For example, although subjective assessments by pilots in simulator tests are common and carefully considered, they have been found to conflict with objective performance measures in several studies (e.g., Pritchett, 1999; Trujillo, 2001) and cannot be solely relied on. Given the tempo and intensity of aerospace environments, workload and situation awareness measures have been inferred to serve as intermediary metrics of performance and subjective difficulty but are not direct indicators of ultimate performance (e.g., Endsley, 1995a; Hart, Hauser, & Lester, 1984; Kantowitz & Casper, 1988). The need to measure the performance of the entire air transport system has also resulted in a recent emphasis on large-scale analyses through agent-based simulation of human performance within the organization (e.g., Gore & Corker, 2000; Pritchett, Lee, & Goldsman, 2001) and through iterative design and test processes that examine for both macro- and microlevel issues (e.g., Blom et al., 1998).

Characteristic: Regulated and Certified

Description. In addition to the efforts made during design by manufacturers and operators to improve safety, many aspects of aerospace (like other extremely safe domains) are highly regulated, including the certification of technologies, procedures and training programs by regulatory bodies (in the United

States, for example, the Federal Aviation Administration). Certification requires the applicant to demonstrate that a system, training program, or procedure meets minimal safety criteria. This certification process can be lengthy and expensive.

Historically, certification criteria were designed for electromechanical systems, with vague specifications on human-interactive technologies such as the following:

FAR 25.771(a) Each pilot compartment and its equipment must allow the minimum flight crew to perform their duties without unreasonable concentration or fatigue.

FAR 25.777(a) Each cockpit control must be located to provide convenient operation and to prevent confusion and inadvertent operation. (Singer, 1999, p. 91)

These specifications are best suited for older cockpits without HCI issues. Additionally, although certification methods often include extensive simulator and flight tests, they cannot investigate all potential issues, and certification decisions are often criticized as being highly subjective.

Challenges in Design. The need for more-rigorous methods of testing, measuring and describing a system's performance vis-à-vis HCI considerations is being widely recognized. Increased emphasis has been placed on incorporating more direct measures of human performance in certification (e.g., RTCA, 1999); commensurately, there may also be an increasing trend to use the formal methods discussed earlier, standards (e.g., Society of Automotive Engineers, 1988), and basic minimal specifications

for interfaces and automation (e.g., Courteney, 1999; Singer, 1999).

One controversial method of certification includes specifying the design process so that designers are forced to consider human performance issues throughout the design, rather than allowing the interfaces to be designed at the end once system functioning has been finalized. For example, the European Joint Aviation Regulations now require the applicant to submit a plan showing how human factors issues will be addressed during aircraft design. This plan, the regulations emphasize, is not the subject of certification; however, it is intended that the plan require designers to seriously and consistently address human performance from the start of design, with conformance to the plan monitored by the certificating agency (Courteney, 1999).

CONCLUSION AND SUMMARY

This chapter briefly summarized HCI in aerospace. First, its history was tracked, highlighting areas in which aerospace has contributed to HCI through early investigation of topics such as manual control, virtual environments, and information display. Second, the current characteristics of aerospace operating environments were reviewed to highlight current design challenges, as summarized in Table 45.1.

Some of these characteristics and design challenges are common to many domains, such as the difficulty of designing

TABLE 45.1. Summary of Characteristics of Aerospace Operating Environments and Design Challenges

Characteristic	Design Challenges
Coordinated	• Determining the distribution of cognition within the system (e.g., centralized vs. distributed) • Addressing both overall system performance and individual's concerns • Predicting organizational impact of new technologies and information
Automated	• Predicting long-term ramifications of automation • Predicting well-intended misuse of automation and decision support systems • Allocating functions appropriately to automation
Complex	• Developing display representations of complex systems and complex tasks • Eliminating unnecessary complexity in systems and tasks
Information intensive	• Addressing data-entry demands on the pilot • Presenting the appropriate information at appropriate times • Developing new display metaphors for integrated displays and synthetic vision systems
Dynamic	• Capturing essential attributes of flight dynamics in display design • Fostering better predictions of future conditions and planning • Preventing extreme workload
Standardized and proceduralized	• Understanding and capitalizing on relationships between training, procedures, interaction with automation, and information display
Safe	• Supporting human performance in worst-case conditions • Measuring the potential for very rare events in naturalistic environments • Identifying generalizable measures of safety for extremely safe systems
Certified and regulated	• Identifying rigorous measures for certification of human-integrated systems • Identifying enforceable, traceable design processes that promote good, safe designs

displays for information intensive environments, and hence aerospace can learn from the current state of knowledge provided by the HCI community. Other characteristics are comparatively unstudied by the HCI community, such as designing for extremely safe environments, and here aerospace may continue to contribute methods and insights as they are discovered.

ACKNOWLEDGMENTS

Thanks are due to the many people who have volunteered figures, including Ted Pritchett, Sanjay Vakil, John Hansman, Erïk Theunissen, Randy Mumaw, Jerome Meriweather, Chris Misiak, and Victor Riley.

References

Abzug, M. J., & Larrabee, E. E. (1996). *Airplane stability and control: A history of the technologies that made aviation possible*. New York: Cambridge University Press.

Baarspul, M. (1989). *Lecture notes on flight simulation techniques* (LR-596). Delft, The Netherlands: Technische Hogeschool Delft, Faculty of Aerospace Engineering.

Barrows, A. K., Enge, P., Parkinson, B. W., & Powell, J. D. (1996). Flying curved approaches and missed approaches: 3-d display trials onboard a light aircraft. *Proceedings of Institute of Navigation GPS-96*. Alexandria, VA: Institute of Navigation.

Beringer, D. B., & Harris, H. C., Jr. (1999). Automation in general aviation: Two studies of pilot responses to autopilot malfunctions. *The International Journal of Aviation Psychology, 9*, 155–174.

Billings, C. (1997). *Aviation automation: The search for a human-centered approach*. Mahwah, NJ: Lawrence Erlbaum Associates.

Blakelock, J. H. (1991). *Automatic control of aircraft and missiles* (2nd ed.). New York: Wiley.

Blom, H. A. P., Bakker, G. J., Blanker, P. J. G., Daams, J., Everdij, M. H. C., & Klompstra, M. B. (1998). Accident risk assessment for advanced ATM. *Proceedings of the 2nd USA/Europe ATM R&D Seminar*, Orlando, FL.

Bove, T., & Andersen, H. B. (1999, June). *The effect of an advisory system on pilots' go/no-go decision during take-off*. Presented at HESSD-99, Workshop on Human Error, Safety and System Development, Liège.

Braune, R. J., & Graeber, R. C. (1992). Human-centered designs in commercial transport aircraft. *Proceedings of the Human Factors Society 36th Annual Meeting* (1118–1123). Santa Monica, CA: Human Factors & Ergonomics Society.

Cardosi, K. (1993). *An analysis of en route controller-pilot voice communication* (Report No. DOT/FAA/RD-93/11). Washington, DC: U.S. Department of Transportation, Federal Aviation Administration.

Chen, T. L., & Pritchett, A. R. (2001). Cockpit decision aids for emergency flight planning. *Journal of Aircraft, 38*, 935–943.

Courteney, H. (1999). Human factors of automation: The regulator's challenge. In S. Dekker & E. Hollnagel (Eds.), *Coping with computers in the cockpit* (pp. 109–130). Brookfield, VT: Ashgate.

Croft, J. (2001, July 16). Despite human frailties, pilots will remain on scene. *Aviation Week and Space Technology*, 82.

Davis, S. D., & Pritchett, A. R. (1999). Alerting system assertiveness, knowledge, and over-reliance. *Journal of Information Technology Impact, 3*, 119–143.

de Muynck, R. J., & Khatwa, R. (1999). *Flight simulator evaluation of the safety benefits of terrain awareness and warning systems* (NLR-TP-99379). Amsterdam: National Aerospace Laboratory NLR.

Degani, A., Shafto, M., & Kirlik, A. (1999). Modes in human-machine systems: Constructs, representation, and classification. *International Journal of Aviation Psychology, 9*, 125–138.

Dekker, S., & Hollnagel, E. (Eds.). (1999). *Coping with computers in the cockpit*. Brookfield, VT: Ashgate.

den Braven, W. (1992). *Design and evaluation of an advanced air-ground data-link system for air traffic control* (NASA TM 103899). Moffett Field, CA: NASA Ames Research Center.

Dimitrov, G. V., & Rippy, L. O. (1999). Rotorcraft pilot's associate: Technology for the battlefield of tomorrow. *Proceedings of the 55th AHS Vertical Flight Society Annual Forum* (pp. 1358–1363). Alexandria, VA: American Helicopter Society.

Dix, A. J., Finlay, J. E., Abowd, G. D., & Beale, R. (1998). *Human-computer interaction* (2nd ed.). New York: Prentice Hall.

Doolittle, J. H. (1961). Early experiments in instrument flying. *Annual report of the board of regents of the Smithsonian institution* (Smithsonian publication 4478). Washington, DC: Government Printing Office.

Endsley, M. R. (1995a). Measurement of situation awareness in dynamic systems. *Human Factors, 37*, 65–84.

Endsley, M. R. (1995b). Toward a theory of situation awareness. *Human Factors, 37*, 32–64.

Endsley, M. R., Farley, T. C., Jones, W. M., Midkiff, A. H., & Hansman, R. J. (1998). *Situation awareness information requirements for commercial airline pilots* (ICAT-98-1). Cambridge, MA: Massachusetts Institute of Technology, International Center for Air Transportation.

Endsley, M. R., & Kiris, E. O. (1995). The out-of-the-loop performance problem and level of control in automation. *Human Factors, 37*, 381–394.

Faerber, R. A., Vogl, T. L., & Hartley, D. E. (2000). Advanced graphical user interface for next generation flight management systems. *Proceedings of HCI-Aero 2000: International Conference on Human-Computer Interaction in Aeronautics* (pp. 107–112).

Farley, T., & Hansman, R. J., (1999). *An experimental study of the effect of shared information on pilot/controller re-route negotiation* (ICAT-99-1). Cambridge, MA: Massachusetts Institute of Technology, International Center for Air Transportation.

Feary, M., Sherry, L., Palmer, E., & Polson, P. (2000). Evaluation of a formal methodology for developing aircraft vertical flight guidance training material. *Proceedings of HCI-Aero 2000: International Conference on Human-Computer Interaction in Aeronautics* (pp. 123–129).

Federal Aviation Administration Human Factors Team. (1996). *The interfaces between flight crews and modern flight deck systems*. Washington, DC: Author.

Fitts, P. M. (Ed.), Chapanis, A., Frick, F. C., Garner, W. R., Gebhard, J. W., Grether, W. F., Henneman, R. H., Kappauf, W. E., Newman, E. B., & Williams, A. C. Jr. (1951). *Human engineering for an effective air-navigation and traffic control system*. Washington, DC: National Academy of Sciences Archives.

Fitts, P. M., & Jones, R. E. (1947). *Analysis of 270 "pilot error" experiences in reading and interpreting aircraft instruments* (Report

TSEAA-694-12A). Wright-Patterson Air Force Base, OH: Aeromedical Laboratory.

Fitts, P. M., & Posner, M. I. (1967). *Human performance.* Belmont, CA: Brooks/Cole.

Foley, J. D., van Dam, A., Feiner, S. K., & Hughes, J. F. (1989). *Computer graphics: Principles and practice* (2nd ed.). Reading, MA: Addison-Wesley.

Garland, D. J., Wise, J. A., & Hopkin, V. D. (Eds.). (1999). *Handbook of aviation human factors.* Mahwah, NJ: Lawrence Erlbaum Associates.

Gore, B. F., & Corker, K. M. (2000). A systems engineering approach to behavioral predictions of an advanced air traffic management concept. *Proceedings of the 19th IEEE/AIAA Digital Avionics Systems Conference* (pp. 4.B.3.1-8). Piscataway, NJ: Institute of Electrical & Electronics Engineers.

Green, S. M., den Braven, W., & Williams, D. H. (1991). Profile negotiation: A concept for integrating airborne and ground-based automation for managing arrival traffic. *Proceedings of the 1991 RTCA Technical Symposium.* Washington, DC: RTCA Inc.

Green, R. J., Self, H. C., & Ellifritt, T. S. (Eds.). (1995). *50 years of human engineering.* Wright-Patterson Air Force Base, OH: Armstrong Laboratory, Fitts Human Engineering Division.

Grether, W. F. (1995). Human engineering: The first 40 years 1945-1984. In R. J. Green, H. C. Self, & T. S. Ellifritt (Eds.), *50 years of human engineering.* Wright-Patterson Air Force Base, OH: Armstrong Laboratory, Fitts Human Engineering Division.

Grunwald, A. J. (1996a). Improved tunnel display for curved trajectory following: Control considerations. *Journal of Guidance, Control, and Dynamics, 19,* 370-377.

Grunwald, A. J. (1996b). Improved tunnel display for curved trajectory following: Experimental evaluation. *Journal of Guidance, Control, and Dynamics, 19,* 378-384.

Haas, M. W., Nelson, W. T., Repperger, D., Bolia, R., & Zacharias, G. (2001). Applying adaptive control and display characteristics to future air force crew stations. *The International Journal of Aviation Psychology, 11,* 223-235.

Hansman, R. J., & Mykityshyn, M. (1995). *Current issues in the design and information content of instrument approach charts* (DOT/FAA/AAR-95/1). Cambridge, MA: Massachusetts Institute of Technology.

Hart, S. G., Hauser, J. R., & Lester, P. T. (1984). Inflight evaluation of four measures of pilot workload. *Proceedings of the Human Factors Society 28th Annual Meeting* (pp. 945-949). Santa Monica, CA: Human Factors & Ergonomics Society.

Hess, R. A. (1995). Modeling the effects of display quality upon human pilot dynamics and perceived vehicle handling qualities. *IEEE Transactions on Systems, Man and Cybernetics, 25,* 338-344.

Howard, M. (1999). Visualizing automation behavior. In S. Dekker & E. Hollnagel (Eds.), *Coping with computers in the cockpit* (pp. 55-67). Brookfield, VT: Ashgate.

In new cockpits, paper's passé. (2001, July 16). *Aviation Week and Space Technology,* 69.

Javaux, D., & Olivier, E. (2000). Assessing and understanding pilots knowledge of mode transitions on the A340-200/300. *Proceedings of HCI-Aero 2000: International Conference on Human-Computer Interaction in Aeronautics* (pp. 81-86).

Johnson, C. W. (1996). Literate specification: Using design rationale to support formal methods in the development of human-machine interfaces. *Human-Computer Interaction, 11,* 291-320.

Johnson, E. N., & Pritchett, A. R. (1995). Experimental study of vertical flight path mode awareness. *Proceedings of the Sixth IFAC/IFIP/IFORS/IEA. Symposium on Analysis, Design and Evaluation of Man-Machine Systems.* Amsterdam: Elsevier.

Jones, D. G., & Endsley, M. R. (2000). Overcoming representational errors in complex environments. *Human Factors, 42,* 367-378.

Kantowitz, B. H., & Casper, P. A. (1988). Human workload in aviation. In E. L. Wiener & D. C. Nagel (Eds.), *Human factors in aviation* (pp. 157-187). San Diego, CA: Academic Press.

Kerns, K. (1999). Human factors in air traffic control/flight deck integration: Implications of data-link simulation research. In D. J. Garland, J. A. Wise, & V. D. Hopkin (Eds.), *Handbook of aviation human factors* (pp. 519-546). Mahwah, NJ: Lawrence Erlbaum Associates.

Kupferer, R. (1999). Rotorcraft pilot's associate from concept to flight demonstration. *Proceedings of the 55th AHS Vertical Flight Society Annual Forum* (pp. 1303-1311). Alexandria, VA: American Helicopter Society.

Mårtensson, L. (1995). The aircraft crash at Gottrora: Experiences of the cockpit crew. *International Journal of Aviation Psychology, 5,* 305-326.

Martin-Emerson, R., & Wickens, C. D. (1997). Superimposition, symbology, visual attention, and the head-up display. *Human Factors, 39,* 581-601.

Maurino, D. E., Reason, J., Johnston, N., & Lee, R. B. (1995). *Beyond aviation human factors.* Burlington, VT: Ashgate.

Merwin, D. H., & Wickens, C. D. (1996). *Evaluation of perspective and coplanar cockpit displays of traffic information to support hazard awareness in free flight* (ARL-96-5/NASA-96-1). Savoy, IL: University of Illinois at Urbana-Champaign, Aviation Research Laboratory.

Moroney, W. F., & Moroney, B. W. (1999). Flight simulation. In D. J. Garland, J. A. Wise, & V. D. Hopkin (Eds.), *Handbook of aviation human factors* (pp. 355-388). Mahwah, NJ: Lawrence Erlbaum Associates.

Mosier, K. L., Skitka, L. J., Dunbar, M., & McDonnell, L. (2001). Aircrews and automation bias: The advantages of teamwork? *The International Journal of Aviation Psychology, 11,* 1-14.

National Research Council. (1997). *Aviation safety and pilot control: Understanding and preventing unfavorable pilot-vehicle interactions.* Washington, DC: National Academy Press.

Newman, R. L. (1998). Definition of primary flight reference. *Journal of Aircraft, 35,* 497-500.

Nijboer, D. (1998). *Cockpit: An illustrated history of World War II aircraft interiors.* Erin, Ontario: Boston Mills.

Nikolic, M. I., & Sarter, N. B. (2001). Peripheral visual feedback: A powerful means of supporting effective attention allocation in event-driven data-rich environments. *Human Factors, 43,* 30-55.

Nolan, M. S. (1994). *Fundamentals of air traffic control* (2nd ed.). Belmont, CA: Wadsworth.

Nordwall, B. D. (2001, July 16). Virtual reality pending as simulators diversify. *Aviation Week and Space Technology,* 68-69.

Noyes, J. M., & Starr, A. F. (2000). Civil aircraft warning systems: Future directions in information management and presentation. *The International Journal of Aviation Psychology, 10,* 169-188.

Olson, W. A., & Sarter, N. B. (2000). Automation management strategies: Pilot preferences and operational experiences. *The International Journal of Aviation Psychology, 10,* 327-341.

O'Leary, M. (1999). *The British Airways human factors reporting programme.* Presented at HESSD-99, Workshop on Human Error, Safety and System Development, Liège.

Prinzel, L. J., Freeman, F. G., Scerbo, M.W., Mikulka, P. J., & Pope, A. T. (2000). A closed-loop system for examining psychophysiological measures for adaptive task allocation. *The International Journal of Aviation Psychology, 10,* 393-410.

Pritchett, A. R. (1999). Pilot performance at collision avoidance during closely spaced parallel approaches. *Air Traffic Control Quarterly, 7,* 47-75.

Pritchett, A. R. (2001). Reviewing the role of cockpit alerting systems. *Human Factors and Aerospace Safety, 1,* 5–38.

Pritchett, A. R., & Hansman, R. J. (1997). Variations among pilots from different flight operations in party line information requirements for situation awareness. *Air Traffic Control Quarterly, 4,* 29–50.

Pritchett, A. R., Lee, S. M., & Goldsman, D. (2001). Hybrid-system simulation for national airspace system safety analysis. *Journal of Aircraft, 38,* 835–840.

Pritchett, A. R., & Vándor, B. (2001). Designing situation displays to promote conformance to automatic alerts. *Proceedings of the Annual Meeting of the Human Factors and Ergonomics Society.* Santa Monica, CA: HFES.

Pritchett, A. R., Vándor, B., & Edwards, K. E. (1999, June). *Testing and implementing cockpit alerting systems.* Presented at *HESSD-99, Workshop on Human Error, Safety and System Development,* Liège.

Pritchett, A. R., & Yankosky, L. J. (2000). Pilot performance at new ATM operations: Maintaining in-trail separation and arrival sequencing. *Proceedings of the AIAA Guidance, Navigation and Control Conference.* Reston, VA: American Institute of Aeronautics & Astronautics.

Proctor, P. (1997, December 1). Economic, safety gains ignite HUD sales. *Aviation Week and Space Technology,* 54–56.

Proctor, P. (1998, April 6). Integrated cockpit safety system certified. *Aviation Week and Space Technology,* 61.

Quinn, C., & Robinson, J. E., III. (2000). A human factors evaluation of active final approach spacing tool concepts. *Proceedings of the 3rd USA/Europe Air Traffic Management R&D Seminar.*

Rignér, J., & Dekker, S. (2000). Sharing the burden of flight deck automation training. *The International Journal of Aviation Psychology, 10,* 317–326.

Riley, V. (2000). Developing a pilot-centered autoflight interface. *SAE World Aviation Congress.* Warrendale, PA: Society of Automotive Engineers.

Rolfe, J. M, & Staples, K. J. (Eds.). (1986). *Flight simulation.* Cambridge, England: Cambridge University Press.

RTCA. (1983). *Minimum operational performance standards for traffic alert and collision avoidance system (TCAS) airborne equipment* (RTCA/DO-185). Washington, DC: Author.

RTCA. (1999). *Final report of the RTCA Task Force 4: Certification* Washington, DC: RTCA.

Rushby, J. (1999). Using model checking to help discover mode confusions and other automation surprises. Presented at *HESSD-99, Workshop on Human Error, Safety and System Development,* Liège.

Sage, M., & Johnson, C. W. (1999, June). Formally verified, rapid prototyping for air traffic control. Presented at *HESSD-99, Workshop on Human Error, Safety and System Development,* Liège.

Sarter, N. B., & Alexander, H. M. (2000). Error types and related error detection mechanisms in the aviation domain: An analysis of ASRS incident reports. *The International Journal of Aviation Psychology, 10,* 189–206.

Sarter, N. B., & Woods, D. D. (1992). Pilot interaction with cockpit automation: Operational experiences with the flight management system. *The International Journal of Aviation Psychology, 2,* 303–321.

Sarter, N. B., & Woods, D. D. (1994). Pilot interaction with cockpit automation II: An experimental study of pilot's model and awareness of the flight management system. *The International Journal of Aviation Psychology, 4,* 1–28.

Sarter, N. B., & Woods, D. D. (1995). How in the world did we ever get into that mode? Mode error and awareness in supervisory control. *Human Factors, 37,* 5–19.

Scott, W. B. (1991, June 3). 777's flight deck reflects strong operations influence. *Aviation Week and Space Technology,* 52–58.

Seamster, T. (1999). Automation and advanced crew resource management. In S. Dekker & E. Hollnagel (Eds.), *Coping with computers in the cockpit* (pp. 195–213). Brookfield, VT: Ashgate.

Sexton, G. (1988). Crew-cockpit design and integration. In E. L. Wiener & D. C. Nagel (Eds.), *Human factors in aviation* (pp. 495–526). San Diego, CA: Academic Press.

Shamo, M. K. (2000). What is an electronic flight bag and what is it doing in my cockpit? *Proceedings of HCI-Aero 2000: International Conference on Human-Computer Interaction in Aeronautics* (pp. 65–70).

Sheridan, T. B. (1992). *Telerobotics, automation, and human supervisory control.* Cambridge, MA: The MIT Press.

Sheridan, T. B. (1998). Allocating functions rationally between humans and machines. *Ergonomics in Design, 6*(3), 20–25.

Sherry, L., & Polson, P. G. (1999). Shared models of flight management system vertical guidance. *The International Journal of Aviation Psychology, 9,* 139–153.

Singer, G. (1999). Filling the gaps in the human factors certification net. In S. Dekker & E. Hollnagel (Eds.), *Coping with computers in the cockpit* (pp. 87–107). Brookfield, VT: Ashgate.

Smith, P. J., McCoy, C. E., & Layton, C. (1993). The design of cooperative problem-solving systems for flight planning. *Proceedings of the IEEE International Conference on Systems, Man, and Cybernetics* (pp. 701–708). Piscataway, NJ: Institute of Electrical & Electronics Engineers.

Society of Automotive Engineers. (1988). *Design objectives for CRT displays for part 25 (transport) aircraft* (Aerospace Recommended Practice ARP 1874). Warrendale, PA: Author.

Stokes, A. F., & Wickens, C. D. (1988). Aviation displays. In E. L. Wiener & D. C. Nagel (Eds.), *Human factors in aviation* (pp. 387–431). San Diego, CA: Academic Press.

Stoffregen, T. A., Hettinger, L. J., Haas, M. W., Roe, M. M., & Smart, L. J. (2000). Postural instability and motion sickness in a fixed-base flight simulator. *Human Factors, 42,* 458–469.

Swenson, H. N., Hoang, T., Engelland, S., Vincent, D., Sanders, T., Sanford, B., & Heere, K. (1997). Design and operational evaluation of the traffic management advisor at the fort worth air route traffic control center. *1st USA/Europe Air Traffic Management R&D Seminar.*

Talotta, N. J. et al. (1992). *Controller evaluation of initial data link terminal air traffic control service: Mini study 2* (DOT/FAA/CT-92/2). Washington, DC: Federal Aviation Administration.

Tenney, Y. J., Rogers, W. H., & Pew, R. W. (1998). Pilot opinions on cockpit automation issues. *The International Journal of Aviation Psychology, 8,* 103–120.

Theunissen, E. (1997). *Integrated design of man-machine interface for 4-d navigation.* Delft, The Netherlands: Delft University Press.

Tomayko, J. K. (1992). The airplane as computer peripheral. *Invention and Technology,* 19–24.

Trujillo, A. (2001). The effects of history and predictive information on the pilot's ability to predict an alert. *Proceedings of the Annual Meeting of the Human Factors and Ergonomics Society.* Santa Monica, CA: HFES.

Vakil, S. S., & Hansman, R. J. (2000). *Analysis of complexity evolution management and human performance issues in commercial aircraft automation systems* (ICAT-2000-3). Cambridge, MA: Massachusetts Institute of Technology, International Center for Air Transportation.

Vakil, S. S., Midkiff, A. H., & Hansman, R. J. (1996). *Development and evaluation of an electronic vertical situation display* (ASL-96-2). Cambridge, MA: Massachusetts Institute of Technology, Aeronautical Systems Laboratory.

Verspay, J. J. L. H., de Muynck, R. J., Nibbelke, R. J., vd Bosch, J. J., Kolstein, G., & van Gelder, C. A. H. (1996). Simulator evaluation of control and display issues for a future regional aircraft. *Proceedings of the AIAA Atmospheric Flight Mechanics Conference* (pp. 176–187). Reston, VA: American Institute of Aeronautics & Astronautics.

Ververs, P. M., & Wickens, C. D. (1998). Head-up displays: Effects of clutter, display intensity, and display location on pilot performance. *The International Journal of Aviation Psychology, 8*, 377–403.

Wickens, C. D. (1992). *Engineering psychology and human performance* (2nd ed.). New York: HarperCollins.

Wickens, C. D. & Flach, J. M. (1988). Information processing. In E. L. Wiener & D. C. Nagel (Eds.). *Human factors in aviation* (pp. 111–155). San Diego, CA: Academic Press.

Wickens, C. D., Liang, C. C., Prevett, T., & Olmos, O. (1996). Electronic maps for terminal area navigation: Effects of frame of reference and dimensionality. *International Journal of Aviation Psychology, 6*, 241–271.

Wiener, E. L., & Curry, R. E. (1980). Flight-deck automation: promises and problems. *Ergonomics, 23*, 995–1011.

Wiener, E. L., & Nagel, D. C. (Eds.). (1988). *Human factors in aviation*. San Diego, CA: Academic Press.

Wilson, J. R. (2001, July). Creating cars that fly. *Aerospace America*, 52–61.

Wright, W. (1901). Some aeronautical experiments. Lecture delivered to Western Society of Engineers, September 18, 1901. Reprinted in M. W. McFarland (Ed.). (1953). *The Papers of Wilbur and Orville Wright* (Vol. 1, pp. 99–118). New York: McGraw-Hill.

Zhang, J. (1997). Distributed representation as a principle for analysis of cockpit information displays. *The International Journal of Aviation Psychology, 7*, 105–121.

USER-CENTERED DESIGN IN GAMES

Randy J. Pagulayan, Kevin Keeker, Dennis Wixon,
Ramon L. Romero, and Thomas Fuller
Microsoft Game Studios

INTRODUCTION

The intent of this chapter is to review principles and challenges in game design and evaluation and to discuss user-centered techniques that address those challenges. First, we present why games are important, followed by the definitions and differences between games and productivity software. In the next section, we discuss the principles and challenges that are unique to the design of games. That discussion provides a framework for what we believe are the core variables that should be measured to aid in game design and evaluation. The chapter concludes with some examples of how those variables can be operationalized in which we present the methods used by the Microsoft Game Studios User-testing Group.

WHY GAMES ARE IMPORTANT

Computers that appeared commercially in the 1950s created a technological barrier that was not easy to overcome for the greater population. Only scientists, engineers, and highly technical persons were able to use these machines (Preece et al., 1994). As computers became less expensive, more advanced, and more reliable, the technology that was once only available to a small group of people permeated throughout the population and into everyday life. To ease this transition, the need for a well-designed interface between the user and the technology became a necessity. Computers games come from similar origins and potentially may head down a similar path. Early attempts at making commercial video games have failed because of unnecessarily high levels of complexity. The following quote from Nolan Bushnell (cofounder of Atari) states the issue quite succinctly: "No one wants to read an encyclopedia to play a game" (Kent, 2000, p. 28). In retrospect, some of the most successful video games were the ones that were indeed very simple.

Entertainment in general is a field of great importance to everyone (Schell, chapter 43, this volume). Today, the video game industry is one of the fastest growing forms of entertainment to date. According to the Digital Interactive Software Association (IDSA), revenue from computer and console games practically doubled from $3.2 billion in 1994 to $6.02 billion in 2000, when more than 219 million games were sold (IDSA, 2000). To put this into perspective, the number of people in the United States who played video games during 2000 was 5 times that of those who went to America's top five amusement parks combined, and 2 times as many as those who attended all Major League Baseball games (IDSA). Thirty to forty percent of homes in the United States own a console gaming system, with another 10 to 20% renting or sharing consoles (Cassell & Jenkins, 2000). These statistics do not even take into account the international importance of video games.

One of the misconceptions about video games is that they are currently played only by a small segment of the population, that is, younger boys. This misconception may stem from a variety of things, from controversial beliefs about boys and men possessing an innate mathematical superiority (e.g., Williams, 1987), to statements that the electronics industry has traditionally been marketed to boys (e.g., Chaika, 1996), to the fact that one of the most successful companies in the gaming industry (i.e., Nintendo) has indeed targeted the younger male population. Nintendo's positioning has traditionally been targeted toward youth and adolescent age groups, which have clearly been demonstrated by previous marketing campaigns and promotions on such things as cereal boxes (Herz, 1997; Provenzo, 1991). However, the average age of those who play video games is 28 (IDSA, 2000). Also, there is a large movement in the gaming industry that targets the female market as well (Gorriz & Medina, 2000). Purple Moon was founded in 1996, and Girl Games Inc. was founded in 1993, both as an effort to make games dedicated for girls (Cassell & Jenkins, 2000). A recent survey done by Peter D. Hart Research Associates claims that three in five Americans say they play computer or video games, and that 43% of them are female gamers (IDSA).

Currently, many would argue that games play a large role in pushing research and technology forward. This includes such areas hardware innovations (graphics cards, processors, sound cards; e.g., Kroll, 2000), research on the relationships between artificial intelligence and synthetic characters or computer graphics (e.g., Funge, 2000; Laird, 2001), and physics (e.g., Hecker, 2000). These are all areas that previously were confined to engineering and computer science communities.

The gaming industry is becoming as widespread and popular as computers and televisions and is also matching the same levels of complexity and advanced technology. The difference, though, is that user-centered design (UCD) principles have not reached the same level of usage or awareness in game design and development as it has in other electronic applications. To understand the differences and unique challenges brought forth in games, one must appreciate how games fit in relation to other software applications in which UCD principles are present, such as in productivity applications.

DEFINING GAMES

There are many ways to define a class of objects. One popular approach is to define an object by stating the principles that differentiates one class of objects from another. Another is to define a class of objects by enumerating the items in the class. Here, we do both. First, we distinguish games from productivity applications, then we list several types of games.

Games Versus Productivity Applications

The distinction between games and productivity is often clear and straightforward in practice. As with some other forms of art and entertainment, we know a game when we see one. However, articulating a clear and succinct set of principles that capture the differences and similarities between a game and a productivity application is not as straightforward. In other words, the distinction is simple in practice, but more difficult in principle. However, shedding some light on the difference between computer games and productivity may help readers understand both the choice of methods and differences used in

their application. It may also deepen one's understanding of productivity applications. For example, the goal of iterative usability testing on games is to reduce the obstacles to fun, rather than the obstacles to accomplishment (as in productivity applications). Many of the issues of concern in games are virtually identical to those used on other applications in which the obstacles to fun are often similar to obstacles in the way of productive work. These include confusing screen layouts, misleading button labeling, or an inconsistent model of use (to mention a few). Thus, there are both principles and methods that can be successfully applied both to games and productivity applications. At the same time, there are fundamental differences between them, which are sometimes easy to see and articulate and sometimes quite subtle. Let us consider briefly some of these similarities and differences.

Process Versus Results

The purpose of games is different from the purpose of productivity applications. At their root, productivity applications are tools. Similar to tools, the design intentions behind productivity applications are to make tasks easier and quicker, to reduce the likelihood of errors, to increase the quality of the result, and to extend domains of work to larger and larger populations. In this sense, a word processor or spreadsheet differs only from a powered woodworking tool in terms of its complexity and domain of application. Broadly speaking, both the motivation for design and usability work on productivity applications such as a word processor is to make better documents—more quickly, more easily, and with fewer errors—and to extend these capabilities to the widest number of people. Similarly, a power mitre box cuts wood with more precise angles more quickly and reliably than a hand mitre box, which in turn makes work more precise, faster, and more reliable than a hand saw. The focus of design and usability is to produce an improved product or result.

At their root, games are different. Games are intended to be a pleasure to play. Ultimately, games are like movies and literature. They exist to stimulate thinking and feeling. This is not to say that word processors or other tools cannot be a pleasure to use or that people do not think or feel when using them, but that is not their design intention. By and large the outcome goal of games (i.e., winning) serves to enhance the pleasure of participation or the process of playing. Thus, the goal of both design and usability when applied to games is creating a pleasurable process. This fundamental difference leads us to devote more of our effort to collecting user evaluation of games (as opposed to strictly performance) than we would if we were working on productivity applications, where more of our work would measure accomplishing tasks (or productivity).

Defining Goals Versus Importing Goals

Games define their own goals. The goals of productivity applications are defined by the external environment, independent of the application itself.

In some cases, the design goal of productivity applications is simple—complete the checkout process in buying a product. For more general applications, such as word processors or spreadsheets, the goals can be complex and variable. In these situations, one of the most difficult design and usability problems is mapping the interface to the users' goals. When an application presents goals (e.g., Microsoft Works), it presents them in a simple and straightforward way—as a choice of possible work products (e.g., Letter and Labels, Business and Address Cards, Cover Letters; see Fig. 46.1).

Games create an artificial world in which the objectives are set entirely within that world. For example, in chess the objective is to capture the opponent's king. The terms *opponent* and *king* have a different meaning in this situation than they do in other contexts. *Capture* also has a specific meaning here, again different than other contexts. In some sense, this simplifies the problem of the game designer and of the usability practitioner. User's goals are easily identified because they are designed in as the purpose of the game. The resulting concern for game design is to convey the goal to the user in a clear and straightforward way. This is one factor that increases the importance of tutorials and training simulations in games. Often games simply present the user with a goal using an introductory and hopefully entertaining cinematic that establishes the background story and thereby explicitly or implicitly presents the user with a high-level goal. In addition, many games will provide "briefings" to describe a "mission" that outlines immediate goals of this part of the game and may suggest some of the obstacles that a gamer may face.

Few Alternatives Versus Many Alternatives

Whereas choices are limited for productivity applications, competition in the gaming industry is fierce.

Pretend that you have to write a book chapter. For that, you have limited options: handwriting, dictation, a typewriter, or various software applications such as Microsoft Word or Corel WordPerfect. In practice however, users may be even more constrained. Publishers may accept a book chapter only in a given format that excludes any handwritten chapters, regardless of its merit.

Competition within the games industry is intense, more intense than any other software domain. It is conceivable that there are more games produced each year than versions of word processors produced ever. Games also naturally compete with other forms of entertainment. Take, for example, the popular PC fantasy role-playing game *Diablo* (Vivendi, 2000) by Blizzard software. *Diablo* must compete with other role-playing games, with other kinds of games in general, with other entertainment options, and with more practical needs. In a situation in which you want an engaging puzzle that will occupy your mind for a few hours and ultimately give you a sense of accomplishment, you could consult the *New York Times* crossword puzzle, argue politics with your best friend, or watch a mystery film. Or you could stock up with health potions and attack demons in *Diablo*. Games must grab your attention, and each game must be noticeably different than the last.

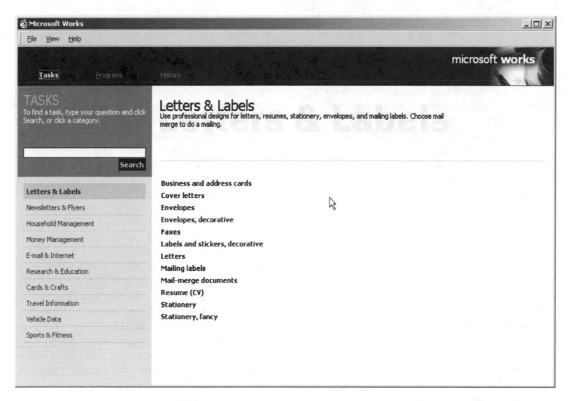

FIGURE 46.1. An example of how a productivity application (Microsoft Works) presents the user with goals.

Being Consistent Versus Generating Variety

Games must provide a variety of experience. In contrast, a user's experience throughout a productivity application strives for consistency.

Variations in users' experience with productivity applications come from differences in their goals. For example, when you need to produce a document with an index, you learn how to do indexing in a word processor. Designers have a goal of making these new functions work in a way that is consistent with the rest of the application while providing the functionality required by the task. In other words, technology can provide significant enhancements for users by presenting new functionality, such as automating routine and tedious tasks. However, an important goal of design is to generate consistency in the user experience.

Games must be different every time because people get bored easily. An important component in this is challenge. Each time the game is played, the user should be invited to learn new rules, or try new strategies to achieve a goal. This learning and exploration is part of the game, which implies that users should have a different experience every time they play.

Imposing Constraints Versus Removing or Structuring Constraints

Games deliberately impose constraints, whereas productivity applications attempt to remove them. Constraints in productivity applications are almost never designed in to intentionally create difficulties. They are most often a reflection of an unsolved (and probably unanticipated) design problem, or a result of the domain in which one is working. A number of successful design approaches attempt to minimize or structure the constraints inherent in task demands. Wizards are a good example. To make a graph, one must select the data to be graphed, choose the type of graph, map the data to axes, and determine display appearance and labeling. This is done effectively by the wizards that structure the task and illustrate the choices to make at each point. The goals of design, usability, and technology are to remove unnecessary constraints because they stand between the user and the result to be obtained, or at least to minimize externally imposed constraints that arise from the interaction between user goals and the environment.

Games, however, often impose artificial constraints because they contribute to the fun of the game. To illustrate this difference, consider the following example. One of our games usability leads, Bill Fulton, has said that the easiest game to use would consist of one button labeled *Push*. When you push it, the display says "YOU WIN." The irony of this example is that this game would have few, if any usability issues, but it clearly would not be much fun. At the same time, making a game difficult does not necessarily make it fun. Consider chess. The constraints consist of rules imposed on moving pieces. However, these constraints combine to yield a complex set of possible combinations of movement that vary through the opening, middle, and end portions of the game. Chess would not be "improved" by introducing more constraints, such as making it harder to distinguish

the pieces or by imposing a new set of bizarre and apparently inconsistent rules of movement, such as "after the 20th move, any bishop left of center, but past the midpoint of the board will now be able to move seven spaces horizontally (diagonal movement is unchanged)." In short, games must impose consistent constraints on user behavior, while making the process of attaining the defined goals interesting and challenging.

Function Versus Mood

Although productivity applications use sound and graphics to convey function, games create environments through the use of sound and graphics.

The use of sound and graphics in productivity applications is relatively small compared to games. Buttons may have drop shadows to produce a three-dimensional effect in an attempt to show that the button affords clicking. Icons may be placed on buttons which afford software functions, such as the disk icon to represent the save function. A "click" sound may be used to provide feedback that a certain function has occurred, such as in selecting a drop-down menu. Even at higher level, overall look of the interface may produce an identity (e.g., buttons that appear to be illuminated from within, or the "lozenge" look of Macintosh OS X).

In contrast, most games attempt to create an "environment" through the use of sound and graphics, which is integral to the game. For example, House of the Dead II (1999) creates a dark, spooky environment through the use of rain effects, compelling depictions of an abandoned city populated by zombies, and a few straggling civilians to be saved (or not). In addition, music and sound effects contribute to the environment by creating a sense of tension.

View of Outcome Versus View of World

Productivity applications rarely have a point of view, or perspective. In contrast, most games need to assume a point of view (first person, third person, Eye of God) or perspective.

Productivity applications may offer alterative views (normal, outline, print, and Web) of the same data. Two of these views (print and Web) are intended to help envision the final product. The others provide ways to see more of the document (normal) or to manipulate the data in specialized ways (outline). It is also common for productivity applications to offer zoom or degrees of magnification, allowing the user to see more or less detail in exchange for a narrower or wider field of view. What is uncommon, is introducing the notion of perspective.

Games often provide perspective, or a point of view. In some games, it attempts to mimic perceived optical flow (i.e., first-person shooters) or center of outflow to specify direction of locomotion. With the increased use of three-dimensional environments, issues often arise in games that are very similar to those experienced in a virtual environment community (e.g., Barfield & Williges, 1998; Smart, 2000). Game designers must appreciate the difference, however, between trying to simulate reality and creating an environment that users perceive as reality (cf. Stoffregen, Bardy, Smart, & Pagulayan, in press). Stanney

(chapter 31, this volume) provided an in-depth discussion of many design and implementation strategies when dealing with three-dimensional environments.

Organization As Buyer Versus Individual As Buyer

Many productivity applications are bought by organizations. Games are almost always bought by individual users. Some productivity applications are sold to home users who are interested in using the same tools both at work and home, but by and large businesses buy productivity applications. In making these purchases, key influential purchasers often spend time evaluating, learning about, and prejustifying their decisions before ever purchasing a product for their business. This changes the marketing of products to focus on direct sales relationships with the influential purchasers rather than consumer advertising.

In contrast, game makers must sell to consumers rather than influential purchasers because the majority of game purchases are made by consumers. This market is large and diverse, which makes it difficult to directly communicate with consumers. Because of this, key features must not only be usable by relatively untrained people, they must be visible, explainable, and appealing to purchasers using minimal media (i.e., box covers and 30-second advertisements).

Form Follows Function Versus Function Follows Form

Users of productivity applications tend to be cautious about adopting innovation while game buyers tend to welcome innovation. Users of productivity applications tend to be reluctant to embrace entirely new and innovative designs. The primary interface of many productivity applications has been relatively stable for many years (take the graphical user interface for example found in current Macintosh and Windows operating systems). Although there are many proffered explanations for this, the most plausible one is that current designs suit user needs relatively well and none of the innovations offered over the years have appeared compelling to users. Innovation does not necessarily equate with greater functionality or ease of use.

Games are compelled by economics to push the technological envelope and to show off exciting features of the system for which they are made. Like other forms of entertainment, games incorporate new and novel features in the hope of attracting a wider audience and not losing their existing audience. Although this is often very exciting for consumers, it can also lead to problems. This issue is discussed further later in the chapter.

Standard Input Devices Versus Novel Input Devices

There is wider variation in game input devices than in productivity applications. Most productivity applications use two input devices, a pointing device and a typing interface. Use of pointing devices or touchscreens can be taught in short and simple tutorials. Typing is a skill that can be learned independently of the applications. In both cases, there is not as much variability in input devices when using productivity applications.

FIGURE 46.2. This is a sample of different input devices used in games, from the traditional keyboard and mouse to different console gamepads.

Games, however, often have unique input devices that are especially designed for a particular games, genres, or platforms. Gamepads, joysticks, steering wheels, aircraft yolks, simulated guns, and microphones all solve particular design problems but often end up creating problems. For example, one could argue that using a joystick should be more satisfying than using a keyboard for controlling an airplane flying through the air. However, it can also be argued that using that same joystick is not very efficient for inputting text on the screen to save your high score. Particular game genres and platforms (see section below) complement certain input devices to make games more or less fun, but also create new issues not seen in productivity application input devices. Figure 46.2 is a sample of different types of input devices.

Summary

We believe the above ten characteristics summarize most of the ways that games differ from productivity applications. Many of the characteristics mentioned are not solely unique to games, however. Many of these characteristics can also be found in areas such as virtual environments, Web applications, or home software applications, such as home publishing software, for example. For purposes of this chapter, however, all these characteristics need to be taken into account and judged appropriately when discussing a game. Later, we discuss how these differences create particular challenges for both the design and evaluation of games.

These principles may appear to be overly general, but this is because as a class, games are very diverse. Now we review some common classification schemes for games and enumerate the various game types.

TYPES OF GAMES

PC Versus Console

Personal computer (PC) versus console (Cassell & Jenkins, 2000) is one of the simplest classifications that can be made. It differentiates those games played on a PC from those played on a console (e.g., Sony PlayStation, Sega Dreamcast, Nintendo GameCube, Microsoft Xbox). There are many characteristics that are associated with each platform. For example, the PC offers different control mechanisms (keyboard, mouse, and other possible game controls). Games usually eschew a keyboard in favor of a controller that provides a series of buttons and joysticks.

Crawford (1982) divides games into even finer categories: arcade (coin-operated), console, PC, mainframe, and handheld. These are all important distinctions. For the purposes of this chapter, however, we often refer to the PC versus console distinction.

Game Genre Classification

The genres of interactive games available are as diverse as those who play them. Some of the more consistent genres used in the games industry include Action, Adventure/Role-Playing, Sports, and Family. Table 46.1 lists the best selling genres in 2001.

TABLE 46.1. Best Selling Game Genres in 2001 for Console Games and PC Games (IDSA 2002)

Console Game Genres	PC Game Genres
Sports	Strategy
Action	Child
Strategy/RPG	Family
Racing	Action
Fighting	RPG
Other Shooters	Sports
Family	Driving
1st Person Shooters	Simulation
Child	Adventure

Note. RPG = role-playing game.
Interactive Digital Software Association (2002). Essential facts about the computer and video game industry. Washington, DC: Interactive Digital Software Association.

PRINCIPLES AND CHALLENGES OF GAME DESIGN

Having differentiated games from other applications, we can look at some of the unique issues in game design and evaluation.

Identifying the Right Kind of Challenges

Games are supposed to be challenging. This requires a clear understanding of the difference between good challenges and frustrating usability problems. Most productivity tools struggle to find a balance between providing a powerful enough tool set for the expert and a gradual enough learning curve for the novice. However, no designer consciously chooses to make a productivity tool more challenging. The ideal tool enables users to experience challenge only in terms of expressing their own creativity. For games, learning the goals, strategies, and tactics to succeed is part of the fun.

Unfortunately, it is not always clear which tasks should be intuitive (i.e., easy to use) and which ones should be challenging. Take a driving game, for example. We are willing to argue that it is not fun to have to discover how to make your car go forward or turn, but that does not mean that learning to drive is not part of the fundamental challenge in the game. Although all cars should use the same basic mechanisms, it may be fun to vary the ways that certain cars respond under certain circumstances. It should be challenging to identify the best car to use on an oval, yet icy, racetrack as opposed to a rally racing track in the middle of the desert.

Learning and mastering these details may involve trial, error, insight, and logic. Furthermore, the challenge level in a game must gradually increase to maintain the interest of the player. If these challenge-level decisions are made correctly, players will have a great time learning to overcome failure. Defining where the basic skills should stop and the challenging skills should start is difficult, so input from users becomes necessary to distinguish good challenges from incomprehensible design.

Addressing Different Skill Levels

Different strategies can be used to regulate the challenge level in the game to support both skilled and unskilled players. Players of all skill levels must do more than simply learn the rules; each one must experience a satisfying degree of challenge. Given that both failure and success can become repetitive quickly, games must address the problem of meeting all players with the correct level of challenge. Although this is complicated in single-player games, it is an even more daunting task when players of different skills are pitted against one another. Tuning a game to the right challenge level is called "game balancing."

There are many ways to balance the difficulty of the game. The most obvious is to let players choose the difficulty themselves. Many games offer the choice of an easy, medium, or hard difficulty level. Although this seems like a simple solution, it is not simple to identify exactly how easy the easiest level should be. Players want to win, but they do not want to be patronized. Too easy is boring; too hard is unfair. Either perception can make a person cease playing.

The environments, characters, and objects in a game provide another possibility for self-regulation. Most games will offer the player some choices regarding their identity, their opponents, and their environment. The better games will provide a variety of choices that allow users to regulate the difficulty of their first experiences. With learning in mind, it is not uncommon for the novice player to choose a champion football team to play against a weak opponent. As long as the player can distinguish the good teams from the bad ones and the teams are balanced appropriately, users will be able to manage their own challenge level.

Another obvious approach to varying skill levels is to require explicit instruction that helps all users become skilled in the game. You might imagine a tutorial in which a professional golfer starts by explaining how to hit the ball and ends by giving instruction on how to shoot out of a sand trap onto a difficult putting green. Instruction, however, need not be presented in a tutorial. It could be as direct as automatically selecting the appropriate golf club to use in a particular situation with no input from the user, similar to the notion of an adaptive interface, where the interface provides the right information at the right time.

Some games take it even further by identifying the skill level of the player and regulating the game difficulty appropriately. In this situation, instruction can be tuned to the skill level of the player by associating it with key behavioral indicators that signifies that the player is having difficulty. If the game does not detect a problem, it does not have to waste the player's time with those instructions.

Productivity tools have implemented similar problem-identification features but often with mixed success because of the open nature of tasks in most productivity applications. Good game tutorials have succeeded by setting clear goals and completely constraining the environment. Doing so focuses the user on the specific skill and simplifies the detection of problem behavior. Other lessons from game tutorial design will be described in later sections of this chapter.

Another in-game approach to regulating the difficulty level requires adjusting the actual challenge level of the opponents

during the game. Evaluating the success of the player and adjusting the opponent difficulty during the game is often called rubber-banding. This tactic is frequently used in racing games. These games monitor the progress of the player and modify the skill of the opponents to ensure that each race is a close one. If the player is not a great driver, the game can detect that and slow down the computer opponents.

Rubber-banding works the same way in multiplayer games. However, restricting players can feel unfair to the restricted player and patronizing to the unskilled player. The popular football game NFL Blitz (Midway Games, 1997) uses this technique in both single and multiplayer modes to keep the challenge level continuously high. Even though this may seem like a good solution, there can be a downside. When players perform skillfully, their performance is moderated by computer-generated bad luck and enhanced opponent attributes. The likelihood of fumbling, throwing an interception, or being sacked increases as a player increases their lead over their opponent. Most people would prefer to play a competitive game (and win) than to constantly trounce a less skilled opponent. But, at the same time, overdeveloped rubber-banding can cheat a skilled player out of the crucial feeling of mastery over the game.

The key for the game designer is to think of ways to maintain challenge, reward, and progress for the unskilled player without severely hampering the skilled player. A final approach focuses on providing tools that maximize the ability of the trailing player to catch up with the leading player. One interesting and explicit example of this is found in Diddy Kong Racing (1997). In this game, the racer can collect bananas along the roadway. Each banana increases the top speed of your car. The player can also collect missiles to fire forward at the leading cars. Each time you hit a car with a missile it not only slows the car's progress, but it jars loose several bananas that one (as the trailer) can pick up. Thus, trailing players have tools that they can use to catch the leaders even if the leader is not making any driving mistakes. The chief distinction between this and rubber-banding is that the game is not modifying skills based on success. Instead, the rules of the game provide the trailing player with known advantages over the leader.

Players Must Be Rewarded Appropriately

Explicit or slow reinforcement schedules may cause users to lose motivation and quit playing a game. Because playing a game is voluntary, games need to quickly grab the user's attention and keep them motivated to come back again and again. One way to accomplish this is to reward players for continued play. Theories of positive reinforcement suggest behaviors that lead to positive consequences tend to be repeated. Thus, it makes sense that positive reinforcement can be closely tied to one's motivation to continue playing a game. However, it is less clear which types of reinforcement schedules are most effective.

Although it's not necessarily the model that should be used for all games, research suggests that continuous reinforcement schedules can establish desired behaviors in the quickest amount of time (Steers & Porter, 1991). Unfortunately, once

continuous reinforcement is removed, desired behaviors extinguish very quickly. Use of partial reinforcement schedules take longer to extinguish desired behaviors but may take too long to capture the interest of gamers. Research suggests that variable ratio schedules are the most effective in sustaining desired behaviors (Jablonsky & DeVries, 1972). This kind of schedule is a staple of casino gambling games, in which a reward is presented after a variable number of desired responses. Overall, there is no clear answer. Creating a game that establishes immediate and continued motivation to continue playing over long periods of time is a complex issue.

Another facet of reinforcement systems that may impact enjoyment of a game is whether the player attributes the fact that they have been playing a game to extrinsic or intrinsic motivations. Intrinsic explanations for behavior postulate that the motivators to perform the behavior come from the personal needs and desires of the person performing the behavior. Whereas extrinsically motivated behaviors are those that people perform to gain a reward from or please other people. In research on children's self-perceptions and motivations, Lepper, Greene, and Nisbett (1973) discovered that children who were given extrinsic rewards for drawing were less likely to continue drawing than those who had only an intrinsic desire to draw. The conclusion that they drew is that children perceived their motivation to draw as coming from extrinsic sources and thus discounted their self-perception that they liked to draw.

The same may be true of reward systems in games (Lepper & Malone, 1987; Malone, 1981). To a certain degree, all reinforcement systems in games are extrinsic because they are created or enabled by game developers, but some reward systems are more obviously extrinsic than others. For instance, imagine the following rewards that could be associated with combat in a fantasy role-playing game (RPG). The player who slays a dragon with the perfect combination of spell casting and sword play may acquire the golden treasure that the dragon was hoarding. In this situation, the personal satisfaction comes from being powerful enough to win and smart enough to choose the correct tactics. The gold is an extrinsic motivator. The satisfaction is intrinsic. By analogy from Lepper et al.'s (1973) research, feelings of being powerful and smart (intrinsic motivators) are more likely to keep people playing than extrinsic rewards.

Collecting and Completing

The chief goal of many games is to acquire all of the available items, rewards, or knowledge contained in the game. It may seem obvious that all gamers like to win. Most racing games allow players to acquire the ability to race new cars and new tracks by winning the basic tracks. These games progress through a series of challenges, unlocking new challenges and opportunities along the way. The ultimate goal is to race with all of the cars and tracks that have been included until new options are exhausted and the game has finally been beaten. In one sense, the goal of the game is to experience everything that is available until all options have been completely identified or uncovered, used, and mastered.

This motivation is even more fully developed in games such as Pokemon Crystal (2000), where the central challenge is to acquire as many of the Pokemon characters as you can and learn all of their skills well enough to outsmart your opponent at selecting the right characters for a head-to-head competition. Not coincidentally, the catch phrase for the Pokemon Crystal game is "Gotta catch 'em all!"

Beyond Interactivity

Game design is not just about adding interactive elements to a narrative. We purport that game designers must use the tools of their medium to channel the thoughts and feelings of their players. Whereas observers are given the freedom of interpretation through their own unique perspective, games offer users more opportunities to change the actual elements and sequences in a story. There have been many great games that are actually linear in structure. In fact, most games are linear in structure, but great games hide this fact from the player. The player feels like they are in control of what happens next when in fact, they are not.

Although most games involve some sort of interactivity, there are as many flavors of these elements as there are game genres. In some cases, authors have suggested that games are inherently more immersive or involving than other media because of the interactive elements (Au, 2001; Lieberman, 1998; Stevenson, 2001). The implication is that button pressing requires or enables a different degree of mental involvement than that required of moviegoers sitting passively in their seats. We propose that button presses and other mechanisms for providing overt interaction are really just physical manifestations of the same types of mental processes that occur during the experience of any work of entertainment or art. Every director or author knows that the audience interacts with their works in a very real way. *The Usual Suspects*, *Casablanca*, and *Psycho* are clear examples of movies that are scripted specifically to involve the user in conscious problem solving by creating expectations, and then violating those expectations.

Thinking of interactivity as a wholly unique phenomenon to games can interfere with effective game design if it causes the designer to think about actions and sequences rather than empathizing with the way that their user will be thinking and feeling as they progress through the game. In most cases, it is not enough to merely keep the player busy at pressing buttons. The designer has to think about the causes and consequences of the player's actions, because that is where the player's attention will be focused. Games give you the opportunity for increased identification, distraction, and physical action. But the most important part of interactivity is not clicking a mouse or moving a joystick. The task of the game designer is to empathize with the viewer and carefully construct an experience that causes them to think clever thoughts and feel profound emotions.

Should There Be a Story?

A story line serves to provide meaning and importance that increases significance, tension, and motivation to succeed in the action sequences. There are those in the games industry that propose that many games neither have nor need a story. It is our contention that all games do have a story and that game designers should approach story line in a similar fashion to the way that authors approach it. The key to understanding narrative in games is to realize that story lines may be embedded in the game, or they may emerge in the course of playing a game.

Most traditional narratives (i.e., books, movies, plays) have a three-act structure. The first act is used to introduce the context and central conflicts. Its purpose is to grab the user's attention and hook them into caring about the characters and their problems. It should make the viewer an active participant and problem solver. The second act is a struggle for success in the primary conflict. At the end of the second act, the conflict changes in some fashion. Success is snatched from the protagonist's hands. Then, the final act often allows the protagonist to regain success. Engagement in the story line is required to get the viewer to suspend their disbelief and become emotionally involved in the trials of the main characters. This immersion is tested each time there is a transition or new element introduced. At each change point, there is a risk to lose viewers, but an opportunity to deepen their involvement in the story.

When most consumers think about story, they think about embedded story lines (Levine, 2001). In games, embedded story lines are often communicated by stopping the game play to watch computer-generated movies. In some games, such as Final Fantasy IX (2000), these embedded stories form a central part of the appeal of the game. In others, they are nearly an afterthought. Embedded narrative need not be presented as computer-generated cut scenes. There may be elements of the story communicated by the action sequences, interactions with characters, or the rewards of the action sequences. This is the easiest form of stories for consumers to recognize and evaluate.

But judgments about this type of narrative may not reflect all of the actual story elements in the game. Ken Levine (2001) pointed out that much of the narrative in a game emerges in the interaction between players throughout the course of the game. This is especially true of a multiplayer game. The story in a multiplayer game often comes not from the game, but from the struggle of multiple opponents. As Levine described it, the story is generated by what happens when people replay an abstract narrative structure and a strict set of rules and possibilities within a novel set of circumstances. In this way, a game of chess or baseball can have the same basic elements every time, but still create opportunities for the players to create epic tales and battles in their minds. The challenge for the game designer is to create a set of rules and opportunities that maximize the ability of users to create unique strategies and anticipate the strategies of others. In this way, the gamers themselves can become the protagonists.

Technological Innovation Drives Design

There is a great deal of pressure on designers to use new technologies that may break old interaction models. As mentioned

earlier, many would argue that games play a large role in pushing research and technology forward in areas that were previously were confined to engineering and computer science communities. On the other side of the coin, in addition to appealing to consumers, games act as showcases for the hardware technology on which they are run. Although this is often exciting for consumers, it can also lead to problems. In the rush to innovate technologically, game makers may not make it a priority to create usable interfaces. In some cases, there may even be strong incentives for game makers to "spruce up" familiar interfaces and break rules of consistency.

Stating that great games are the only thing required to sell the console systems on which they are played would be a dubious claim. The success of a console system depends entirely on whether the games that are played on a particular console are noticeably different from alternative technologies. Because of the rapid pace of innovation, the costs of producing game hardware are quite high. Unfortunately, you cannot sell any games if users do not have access to that video game system. This creates a somewhat dangerous "chicken-and-egg" situation. It would cost the consumer an enormous amount to buy the latest video game hardware to cover production costs. But no games will sell if consumers do not have that hardware. So, Nintendo, Sony, Microsoft, and a handful of other major game system creators sell their hardware at a loss for much of its production run. They make up the loss by collecting royalties from third parties and by publishing games under their own brand.

In his book about the success of the Sony PlayStation, Asakura (2000) noted a series of technological and business decisions that affected the competition between Sony and Nintendo. One example from Asakura's book illustrates how a business decision may influence both customer appeal and usability. Asakura noted that the decision to use relatively low cost CD-ROM technology allowed PlayStation games to be cheaper by reducing the total cost of goods and reducing significant planning and distribution problems. Previous Nintendo games were produced on cartridges with circuit boards. This enabled Nintendo games to minimize loading times to transfer information from the game to the system. It also allowed Nintendo games to save information directly on to the cartridge. Both of these characteristics are positives for consumers because they reduced delays and hassles that were unrelated to playing the game itself. Ultimately, Asakura argued that these consumer benefits were overshadowed by the bottom line, that it cost less for consumers to acquire games, and developers had a higher profit margin on each sale. So this suggests that usability advantages alone are not sufficient for success.

Perceptual-Motor Skill Requirements

The way that functions are mapped onto available input devices can determine the success or failure of a game. A crucial part of the fun in many games comes from performing complex perceptual-motor tasks. Although today's arcade-style games include increasingly more sophisticated strategic elements, a core element of many games is providing the ability to perform extremely dexterous yet satisfying physical behaviors. These behaviors are usually quick and well-timed responses to changes and threats in the environment. If the controls are simple enough to master and the challenges increase at a reasonable difficulty, these mostly physical responses can be extremely satisfying (Csiksentmihalyi, 1990).

Problems can arise when games require unfamiliar input devices. This is a common complication in console game usability research because new console systems usually introduce new input device designs unique to that system (see Fig. 46.2). Furthermore, game designers often experiment with new methods for mapping the features of their games to the unique technical requirements and opportunities of new input devices. Unfortunately, this does not always result in a better gameplay experience.

IMPORTANT FACTORS IN GAME EVALUATION

Typical usability outcome measures, such as task times and errors, can be used to evaluate certain aspects of games, but it is also necessary to measure users' subjective experiences and attitudes about games. This section explores some of the subjective attributes that are common to many games.

Overall Quality (Also Known As "Fun")

Most game genres are subtly different in the experiences that they provoke. It may seem obvious that the point of game design is making a fun game. Some games are so fun that people will travel thousands of miles and spend enormous amounts of money to participate in gaming events. However, we would like to propose a potentially controversial assertion: The fundamental appeal of some games lies in their ability to challenge, to teach, to bring people together, or to simply experience unusual phenomena. Likewise, the definition of fun may be different for every person. When you play simulations games, your ultimate reward may be a combination of learning and mastery. When you play something like the MTV Music Generator (1999), your ultimate reward is the creation of something new. When you go online to play card games with your uncle in Texas, you get to feel connected. Flight Simulator (2000) lets people understand and simulate experiences that they always wished they could have. Although these may be subcomponents of fun in many cases, there may be times when using "fun" as a synonym for overall quality will lead to underestimations of the quality of a game. A fun game is often synonymous with a good game, but researchers are warned to wisely consider which measures best suit the evaluation of each game that they evaluate.

Ease of Use

The ease of use of a game's controls and interface is closely related to fun ratings for that game. Think of this factor as a gatekeeper on the fun of the game. If users must struggle or

cannot adequately translate their intentions into in-game behaviors, they will become frustrated. This frustration can lead users to perceive the game as being unfair or simply inaccessible (or simply not fun). Thus, it becomes clear why usability is important in games. Ease of use should be evaluated with both usability and attitude-measurement methodologies.

Starting a Game. Starting the kind of game that the user wants is an easily definable task with visible criteria for success. This is something we measure in our usability laboratories. Although designers often take game shell (the interface used to start the game) design for granted, a difficult or confusing game shell can limit users' discovery of features and impede their progress toward enjoying the game. The most immediate concern for users can be starting the kind of game that they want to play. Games often provide several modes of play. When the game shell is difficult to navigate, users may become frustrated before they have even begun the game. For example, we have found that many users are unable to use one of the common methods that sports console games use to assign a game controller to a particular team. This has resulted in many users mistakenly starting a computer versus computer game. Depending on the feedback in the in-game interface, users may think that they are playing when, in fact, the computer is playing against itself! In these cases users may even press buttons, develop incorrect theories about how to play the game, and become increasingly confused and frustrated with the game controls. The most effective way to avoid these problems is to identify key user tasks and usability test them.

Tutorials or Instructional Missions. As mentioned earlier, tutorials are sometimes necessary to introduce basic skills to play the game. In this situation, instructional goals are easily translated into the usability labs with comprehension tasks and error rates.

One of the risks of not testing tutorials or instructional missions is inappropriate pacing, which can often result from an ill-conceived learning curve at the start of the game. Many games simply start out at too difficult a challenge level. This is an easy and predictable trap for designers and development teams to fall into because when designers spend months (or even years) developing a game, they risk losing track of the skill level of the new player. A level that is challenging to the development team is likely to be daunting to the beginner.

Unfortunately, the reverse can also be troubling to the new user. Faced with the task of addressing new players, designers may resort to lengthy explanations. Frequently, developers will not budget time to build a ramped learning process into their game. Instead, they may realize late in the development cycle that they need to provide instruction. If this is done too abruptly, the learning process can end up being mostly explanation, and, to be frank, explanation is boring. The last thing that you want to do is to bore your user with a long-winded explanation of what they are supposed to do when they get into your game. It is best to learn in context and at a measured pace or users may just quit the game.

A positive example is the first level of *Banjo Kazooie* (2000). At the start the player is forced to encounter a helpful tutor and listen to a few basic objectives. Then they must complete some basic objectives that teach some of the basic character abilities. Much of the tutorial dialogue may be skipped, but the skills necessary to continue must be demonstrated. In this way, the game teaches new skills but never requires tedious instruction. The player learns primarily by doing. All of this is done in the shadow of a visible path onto the rest of the game so users never loses sight of where they need to go.

In-Game Interfaces. In-game interfaces are used primarily to deliver necessary status feedback and to perform less frequent functions. We measure effectiveness with more traditional lab usability testing techniques and desirability with attitude-measurements such as surveys (see next section for example).

Some PC games make extensive use of in-game interfaces to control the game. For example, simulation and real-time strategy (RTS) games can be controlled by keyboard and mouse presses on interface elements in the game. Usability improvements in these interfaces can broaden the audience for a game by making controls more intuitive and reducing tedious aspects of managing the game play. In-game tutorial feedback can make the difference between confusion and quick progression in learning the basic mechanisms for playing. In this situation, iterative usability evaluations become a key methodology for identifying problems and testing their effectiveness (see next section for example).

Many complex PC and console video games make frequent use of in-game feedback and heads-up displays (HUD) to display unit capabilities and status. For example, most flight combat games provide vital feedback about weapons systems and navigation via in-game displays. Without this feedback, it can be difficult to determine distance and progress toward objectives, unit health, and attack success. This feedback is crucial for player learning and satisfaction with the game. With increasing game complexity and three-dimensional movement capabilities, these displays have become a crucial part of many game genres. Usability testing is required to establish whether users can detect and correctly identify these feedback systems. Attitude measurement is required to assess the utility of these features and gauge whether users have a satisfying amount of status feedback.

Mapping Input Devices to Functions. A learnable mapping of buttons, keys, or other input mechanisms to functions is crucial for enjoying games. We measure effectiveness with usability techniques and desirability with attitude measurements. Without learnable and intuitive controls, the user will make frequent mistakes translating their desires into onscreen actions. We have seen consistently that these kinds of mistakes are enormously frustrating to users, because learning to communicate one's desires through an eight-button input device is not fun. The selection of keys, buttons, and other input mechanisms to activate particular features is often called control mapping. Players tend to feel that learning the control-mapping is the most basic part of learning the game. It is a stepping stone to getting to the fun tasks of avoiding obstacles, developing strategies, and blowing things up.

In contrast with other ease-of-use issues, evaluating the control mapping may involve as much subjective measurement as behavioral observation. Button presses are fast, frequent, and hard to collect automatically in many circumstances. Furthermore, problems with control mappings may not manifest themselves as visible impediments to progress, performance, or task time. Instead, they may directly influence perceptions of enjoyment, control, confidence, or comfort. Differences in experience levels and preferences between participants may create significant variation in attitudes about how to map the controls.

Dissatisfaction with the controller design can also be a central factor that limits enjoyment of all games on a system. For example, the results of one whole set of studies on the games for a particular console game system were heavily influenced by complaints about the system's controller. Grasping the controller firmly was difficult because users' fingers were bunched up and wrists were angled uncomfortably during game play. Ratings of the overall quality of the games were heavily influenced by the controller rather than the quality of the game itself.

Because of these concerns and the importance of optimizing control mappings, we recommend testing them with both usability and attitude assessment methodologies.

Challenge

Challenge is distinct from ease of use and is measured almost exclusively with attitude assessment methodologies. This can be a critical factor to the enjoyment of a game, can be highly individualized, and is rightly considered subjective.

Consumers may have difficulties distinguishing the "appropriate" kinds of challenge that result from calculated level and obstacle design, from the difficulty that is imposed by inscrutable interface elements or poor communication of objectives. In either case, the result is the same. If not designed properly, the player's experience will be poor. Thus, it is up to the user-testing professional to make measurement instruments that evaluate the appropriateness of the challenge level independent of usability concerns.

Pace

We define pace as the rate at which players experience new challenges and novel game details. We measure this with attitude measurement methodologies.

Most designers recognize that appropriate pacing is required to maintain appropriate levels of challenge and tension throughout the game. You might think of this as the sequence of obstacles and rewards that are presented from the start of the game to the end. One group at Microsoft uses a critical juncture analogy to describe pacing. As a metaphor, they suggest that the designer must attend to keeping the user's attention at 10 s, 10 min, 10 hr, and 100 hr, recognizing that the player can always put down the game and play another one. One must

think creatively about giving the user a great experience at these critical junctures. Some games excel at certain points but not others. For example, the massively multiplayer game may have the user's rapt attention at 10 s and 10 min. And the fact that hundreds of thousands pay $10 a month to continue playing indicates that these games are rewarding at the 100 hr mark. But anyone who has played one of these games can tell you that they are extremely difficult and not too fun to play at the 10 hr mark. At this point, you are still getting "killed" repeatedly. This is no fun at all. Arcade games obviously take this approach very seriously. Although they may not scale to 100 hr, good arcade games attract you to drop a quarter and keep you playing for long enough to make you want to spend another quarter to continue.

Over the course of a game, many games take a similar approach to that taken by old-time movie serials. Realizing that users will not complete the game in one sitting, the game becomes a series of punctuated experiences. Each level or section of the game might be expected to build from a casual level of challenge that gradually teaches the user a set of skills that they must master before facing a penultimate challenge. Game magazines often refer to this penultimate challenge or opponent as the "Boss" of a level. This usually manifests itself in a larger or more powerful opponent that must be vanquished before moving on to the next level or set of challenges.

Pacing may also be expressed as a set of interwoven objectives much like the subplots of a movie. Again, *Banjo Kazooie* provides an excellent example of good pacing. Each level in *Banjo Kazooie* contains the tools necessary to complete the major objectives. Finding the tools is an important part of the game. New abilities, objectives, skills, and insights are gradually introduced as the player matures. While progressing toward the ultimate goal (of vanquishing the evil witch and saving the protagonist's sister), the player learns to collect environmental objects that enable them to fly, shoot, become invincible, change shape, gain stamina, add extra lives, and unlock new levels. This interlocking set of objectives keeps the game interesting and rewarding. Even if one is unable to achieve a particular goal, there are always sets of subgoals to work on, some of which may provide cues about how to achieve the major goal.

Summary

Attitude methodologies are better apt to measure factors such as overall fun, graphics and sound, challenge, and pace. The typical iterative usability test is an exploratory exercise designed to uncover problem areas in which the designers intentions don't match the users expectations; as a result, we typically choose to not use usability test to assess "fun" or challenge. When attempting to assess attitudinal issues as "overall fun" and "challenge," we make use of a survey technique that affords testing larger samples. Internally, we have adopted the term Playtest or sometimes Consumer Playtest for this technique. At the same time, we use typical iterative usability methods to determine design elements which contribute to or detract from the experience of fun.

USER RESEARCH IN GAMES

Introduction to Methods at Microsoft Game Studios—Principles in Practice

Thus far, we have introduced a number of variables that we think are important for game development and evaluation. Identifying these issues alone will not result in a better gaming experience, however. To improve the gaming experience, one must be able to accurately measure and improve on many of the aforementioned issues by involving users. In this section, we propose various methodologies and techniques that attempt to do just that. Examples are taken from techniques used by the Microsoft Game Studios User-Testing Group.

Our testing methods can be organized by the type of data being measured. At the most basic level, we categorize our data into two types: *behavioral* and *attitudinal*. Behavioral refers to observable data based on performance or particular actions performed by a participant that one can measure. This is similar to typical measures taken in usability tests (e.g., time it takes to complete a task, number of attempts it takes to successfully complete a task, task completion, etc.). Attitudinal refers to data that represent participant opinions or views, such as subjective ratings from questionnaires or surveys. These are often used to quantify user experiences. Selection of a particular method will depend on what variables are being measured and what questions need to be answered.

Another distinction that is typically made is between *formative* and *summative* evaluations, which we apply to our testing methods as well. Formative refers to testing done on our own products in development. Summative evaluations are benchmark evaluations, either done on our own products or on competitor products. It can be a useful tool for defining metrics or measurable attributes in planning usability tests (Nielsen, 1993) or to evaluate strengths and weaknesses in competitor products for later comparison (Dumas & Redish, 1999).

Facilities

Usability Laboratories. The usability labs at Microsoft are fairly similar to other industry and university usability labs, such as those found at the American Institutes for Research, Lotus Development Corporation, the University of Washington (Dumas & Redish, 1999), Motorola, and Oracle Corporation (R. M. Pagulayan, personal communication, June 26, 2001). Our basic setup includes a participant side and an observer, side divided by a one-way mirror. The observer side contains video-recording equipment, a PC, and a large monitor for observers. The participant side contains two cameras, a PC, and a television for console use.

Playtest Laboratories. The playtest labs at Microsoft were built to allow us to run up to 17 participants at once while minimizing the participants' ability to interact with one another. Our basic setup for each station includes a questionnaire PC, a game PC, a 17″ monitor, and a switchbox to toggle between the two machines as necessary. Two machines are necessary because we often test with early versions of PC games before they are considered stable and do not want to lose data as a result of a crash. Each station also includes a small 13″ stereo television that we use for testing with video game consoles. All participants are given stereo headsets so as to not disturb others around them. Partitions are also set up at each station so they cannot see what is going on at surrounding stations. Data are gathered via online questionnaire data capture.

Usability Techniques. At a certain level, games can possess similar behavioral goals to productivity applications. For example, gamers should be able to navigate through main menu options to change their gamepad settings as easily as a user can navigate a menu structure to change their document settings in their word processor. Gamers should also be able to interact easily with an input device, whether it is a gamepad or a traditional keyboard and mouse. From this perspective, games are the same as other software applications in that they must be easy to use and interact with.

Traditional usability techniques can be used to address a portion of the variables identified as important for game design. Although usability techniques have been around in industry for some time, they are currently evolving to meet many of today's needs (Dumas & Redish, 1999). In addition to measuring performance, we use many standard usability techniques to answer "how" and "why" process-oriented questions. For example, how do users perform an attack, or why are controls so difficult to learn? The following are different user-centered techniques we use to address these issues: (a) structured usability tests, (b) rapid iterative testing and evaluation (RITE; Medlock, Wixon, Terrano, Romero, & Fulton, 2002), and (c) other variations and techniques including open-ended usability tasks, paper prototypes, and empirical guideline documents. In any given usability test, it is common to combine two or more of any of the aforementioned techniques.

Although these methods are useful, they do not allow us to address issues with extended gameplay. We cannot address any issues that may arise after playing the game for a couple of days. This is problematic, because one of the key challenges in game design is longevity. With the competition, the shelf life of a game becomes limited. Usability techniques can only provide information approximately within the first hour of playing the game.

For clarity of presentation, each technique will be discussed separately, followed by a case study. Each case study will only contain information pertinent to a specific technique, thus examples may be taken from a larger usability test.

Structured Usability Test. A structured usability test maintains all the characteristics that Dumas and Redish (1999) proposed as common to all usability tests: (a) goal is to improve usability of the product; (b) participants represent real users; (c) participants do real tasks (however, these tasks may only represent a subset of the capabilities in the game); (d) participant behavior and verbal comments are observed and recorded; and (e) data are analyzed, problems are diagnosed, and changes are

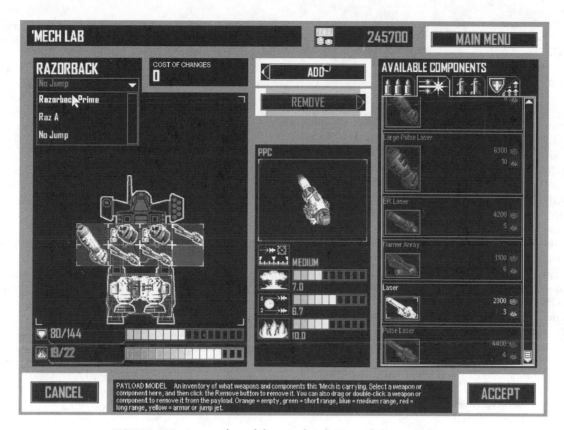

FIGURE 46.3. Screenshot of the 'Mech Lab in MechCommander 2.

recommended. We have found that issues relating to expectancies, efficiency, and performance interaction are well suited for this type of testing. Some common areas of focus for structured usability testing are in game shell screens, or control schemes. The game shell can be defined as the interface where a gamer can determine and or modify particular elements of the game. This may include main menus and options screens (i.e., audio, graphics, controllers, etc.).

An example which uses this method is in the MechCommander 2 (2001) usability test. Portions of the method and content have been omitted.

Case Study: MechCommander 2 Usability Test. MechCommander 2 (MC2) is a PC RTS game in which the gamer takes control of a unit of mechs (i.e., large mechanical robots). One area of focus for this test was on the 'Mech Lab, a game shell screen where mechs can be customized (see Fig. 46.3). Gamers are able to modify weaponry, armor, and other similar features and are limited by constraints such as heat, money, and available slots.

The first step in approaching this test was to define the higher order goals. Overall, the game shell screens had to be easy to navigate, understand, and manipulate, not only for those familiar with mechs and the mech universe, but also for RTS gamers who are not familiar with the mech universe. Our goal was for gamers to be able to modify or customize mechs in the 'Mech Lab.

As mentioned, one of the most important steps in this procedure is defining the participant profile(s). Getting the appropriate users for testing is vital to the success and validation of the data since games are subject to much scrutiny and criticism from its gamers. To reiterate, playing games is a choice. For MC2, we defined two participant profiles that represented all of the variables we wanted to cover. The characteristics of interest included those who were familiar with RTS games (experienced gamers) and those who were not RTS gamers (novice gamers). We also wanted gamers that were familiar with the mech genre, or the mech universe. Overall, we needed a landscape of gamers that had some connection or interest that would make them a potential consumer for this title, whether through RTS experience or mech knowledge.

Tasks and task scenarios were created to simulate situations that a gamer may encounter when playing the game. Most important, tasks were created to address the predefined higher order goals. Participants were instructed to talk aloud, and performance metrics were recorded (i.e., task completion, time). The following are examples from the usability task list.

1. Give the SHOOTIST jumping capabilities. This task allowed us to analyze participant expectations. To succeed in this task, one had to select a "CHANGE WEAPONS" button from a different game shell screen, which brought them into the 'Mech Lab. If the task was to change a weapon on a mech, the terminology would probably have been fine. Thus, this task had uncovered two main issues: (a) could users get to the 'Mech Lab where you modify the mech, and (b) were they able to discover

the process of modifying the mech. It was accurately predicted that gamers would have difficulties with the button terminology. Thus, that button was changed to "MODIFY MECH."

To change the components (i.e., add the jump jets), participants could either select the item, and drag it off the mech, or select the item, and press the "REMOVE" button (see Fig. 46.3). One unexpected issue that arose was that participants unknowingly removed items because the distance required for removing an item was too small. The critical boundary that was implemented was too strict. In addition, participants had difficulties adding items by dragging and dropping because the distance required for adding an item was too large (i.e., the item would not stay on the mech unless it was placed exactly on top of the appropriate location). Appropriate recommendations were made and implemented.

2. Replace the MG Array with the Flamer. One of the constraints presented for modifying a mech was heat limit. Each weapon had a particular heat rating. For example, if the heat limit for a mech is 35 and the current heat rating is 32, only weapons with a rating of 3 or fewer could be added. In this task, the "Flamer" had a heat rating much larger than the "MG Array," thus making impossible to accomplish this task without removing more items. The issues here were the usability of the heat indicator, heat icons, and the discoverability of heat limit concept. None of the participants figured this out. Recommendations included changing the functionality of the Heat Limit Meter and to add better visual cues to weapons that exceed the heat limit. Both of these changes were implemented.

Rapid Iterative Testing and Evaluation Method (RITE). Medlock et al. (2001) have documented another common usability method used by the Microsoft Game User-Testing Group, which they refer to as the RITE method. In this method, fewer participants are used before implementing changes, but more cycles of iteration are performed. With RITE, it is possible to run almost 2 to 3 times the total sample size of a standard usability test. However, only 1 to 3 participants are used per iteration with changes to the prototype immediately implemented before the next iteration (or group of one to three participants). This method has been used by us most commonly for game tutorials, although it is not limited to just tutorials.

The goal of the RITE method is to be able to address as many issues and fixes as possible in a short amount of time in hopes of improving the gamer's experience and satisfaction with the product. However, the utility of this method is entirely dependent on achieving a combination of factors (Medlock et al., 2001). The situation must include (a) a working prototype, (b) the identification of three types of behaviors (critical success behaviors, important but not vital behaviors, and less important behaviors), (c) commitment from the development team to attend tests and immediately review results, (d) time and commitment from development team to implement changes before next round, and (e) the ability to schedule or run new participants as soon as the product has been iterated. Aside from

these unique requirements, planning the usability test is similar to more traditional structured usability tests.

It is helpful to categorize potential usability issues into four categories: (a) clear solution, quick implementation; (b) clear solution, slow implementation; (c) no clear solution; and (d) minor issues. Each category has implications for how to address each issue. In the first category, fixes should be implemented immediately and should be ready for the next iteration of testing. In the second category, fixes should be started, in the hope that it can be tested by later rounds of testing. For the third and fourth category, more data should be collected.

The advantage of using the RITE method is that is allows for immediate evaluation and feedback of recommended fixes that were implemented. Changes are agreed on and made directly to the product. If done correctly, the RITE method affords more fixes in a shorter period of time. In addition, by running multiple iterations over time we are potentially able to watch the number of usability issues decrease. It provides a nice, easily understandable, accessible measure. In general, the more iterations of testing, the better. This method is not without its disadvantages, however. In this situation, we lose the ability to uncover unmet user needs or work practices, we are unable to develop a deeper understanding of gamer behaviors, and we are unable to produce a thorough understanding of user behavior in the context of a given system (Medlock et al., 2002).

The following example demonstrates how the RITE method was used in designing the Age of Empires II: The Age of Kings (1999) tutorial. Again, portions of the method and content have been omitted. See Medlock et al. (2002) for more details.

Case Study: Age of Empires II: The Age of Kings Tutorial. Age of Empires II: The Age of Kings (AoE2) is an RTS game for the PC in which the gamer takes control of a civilization spanning over a thousand years, from the Dark Ages through the late medieval period. In this case study, a working prototype of the tutorial was available, and critical concepts and behaviors were defined. Also, the development team was committed to attending each of the sessions, and they were committed to quickly implementing agreed on changes. Finally, the resources for scheduling were available. The key element in this situation for success was the commitment from the development team to work in conjunction with us.

In the AoE2 tutorial, there were four main sections: (a) marching and fighting (movement, actions, unit selection, the "fog of war"[1]), (b) feeding the army (resources, how to gather, where to find); (c) training the troops (use of minimap, advancing through ages, build and repair buildings, relationship between housing and population, unit creation logic); and (d) research and technology (upgrading through technologies, queuing units, advancing through ages). Each of these sections dealt with particular skills necessary for playing the game. In essence, the tutorial had the full task list built in. In the previous case study, this was not the case.

At a more abstract level, the goals of the tutorial had to be collectively defined (with the development team). In more

---

[1] The *fog of war* refers to the black covering on a minimap or radar that has not been explored yet by the gamer. The fog of war "lifts" once that area has been explored. Use of the fog of war is most commonly seen in RTS games.

TABLE 46.2. Age of Empires Example Concepts and Behaviors Categorized Into Three Concepts Using the RITE Method (Medlock, Wixon, Terrano, Romero, & Fulton, 2002)

Essential Concepts/Behaviors	Important Concepts/Behaviors	Concepts/Behaviors Of Lesser Interest
• Movement	• Queuing up units	• Using hotkeys
• Multiselection of units	• Setting gathering points	• Using minimap modes
• "Fog of war"	• Garrisoning units	• Using trading
• Scrolling main screen via mouse	• Upgrading units through technology	• Understanding sound effects

Note. RITE = rapid iterative testing and evaluation.

concrete terms, specific behaviors and concepts that a gamer should be able to perform after using the tutorial were identified, then categorized into the three levels of importance: (a) essential behaviors that users must be able to perform without exception, (b) behaviors that are important but not vital to product success, and (c) behaviors that were of lesser interest. Table 46.2 lists some examples of concepts and behaviors from each of the three categories. This is an important step because it indirectly sets up a structure for decision rules to be used when deciding what issues should be addressed immediately, and what issues can wait. The general procedure for each participant was similar to other usability tests we perform. If participants did not go to the tutorial on their own, they were instructed to do so by the user-testing engineer.

During the session, errors and failures were recorded. In this situation, an error was defined as anything that caused confusion. A failure was a considered an obstacle that prevented participants from being able to continue. After each session, a discussion ensued among the engineer and the development team to determine what issues (if any) warranted an immediate change at that time.

To do this successfully, certain things had to be considered. For example, how can one gauge how serious an issue is? In typical usability tests, the proportion of participants experiencing the error is a way to estimate its severity. Because changes are made rapidly here, the criteria must change to the intuitively estimated likelihood that users will continue to experience the error. Another thing to consider is clarity of the issue, which was assessed by determining if there is a clear solution. We have often found that if issues do not have an obvious solution then the problem not fully understood. And finally, what errors or failures were essential, important, or of lesser interest. Efforts of the development team should be focused on issues related to the essential category when possible.

At this point we broke down the issues into three groups. The first group included issues with a solution that could be quickly implemented. Every issue in this group was indeed quickly implemented before the next participant was run. The second group consisted of issues with a solution that could not be quickly implemented. The development team began working on these in the hope that they could be implemented for later iterations of the test. Finally, there were issues with no clear solutions. These issues were left untouched because more data were needed to assess the problem at a deeper level (i.e., more participants needed to be run). Any fixes implemented in the

builds were kept as each participant was brought in. Thus, it was possible that many of the participants experienced a different version of the tutorial over the duration of testing.

Overall, seven different iterations were used across 16 participants. Figure 46.4 represents the number of errors and failures recorded over time. The number of errors and failures gradually decreased across participants as new iterations of the build were introduced. By the seventh and final iteration of the build, the errors and failures were reliably reduced to zero.

Although we feel that the *AoE2* tutorial was an enormous success, largely because we used the RITE method, the method does have its disadvantages and should be used with caution. Making changes when issues or solutions are unclear may result in not solving the problem at all and in creating newer usability problems in the interface. We experienced this phenomena a couple of times in the AoE2 study.

Also, making too many changes at once may introduce too many sources of variability and create new problems for users. Deducing specifically the source of the new problem becomes difficult. A related issue is not following up changes with

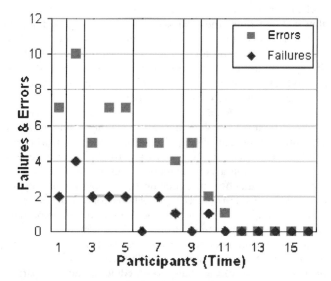

FIGURE 46.4. A record of failures and errors over time for the Age of Empires Tutorial when using the RITE method (Medlock, Wixon, Terrano, Romero, & Fulton, 2002). RITE = rapid iterative testing and evaluation.

enough participants to assess whether or not the solution really addressed the problem. Without this follow-up, there is little evidence supporting that the implementations made were appropriate (which is a problem with traditional usability methods as well). The last thing to consider is that other important usability issues that may surface less frequently are likely to be missed. Using such small samples between iterations allows for the possibility that those less occurring issues may not be detected.

Variations on Usability Methods. Now that we have presented two general types of usability testing, it is worthy to mention some variations on these methods: (a) open-ended tasks, (b) paper prototyping, and (c) empirical guideline documents.

In general, it is often recommended that tasks in usability tests be small, with a specified outcome (e.g., Nielsen, 1993). However, we have found situations where the inclusion of an open-ended task yields important data as well. In many usability studies, we often include an open-ended task where participants are not instructed to perform or achieve anything in particular. In other words, there is no specified outcome to the participant. These tasks can be used to analyze how gamers prioritize certain tasks or goals in a nonlinear environment. These tasks are also useful in situations in which structured tasks may confound the participant experience, or situations where we are interested in elements of discovery. An example of an open-ended task is as follows.

1. *Play the game as if you were at home.* The moderator will tell you when to stop. This example was taken from a usability test on an Xbox sports game. In console sports titles, control schemes are often used repeatedly across different versions of the game. For example, Triple Play Baseball (2001) for PlayStation 2 uses the same control scheme (or very similar) for pitching and batting as many of the Triple Play predecessors (i.e., Triple Play '97 through 2001). Furthermore, similarities often exist across different titles from different developers. To select a pitch in Triple Play Baseball, a given pitch type is mapped to a given button. For example, to select a fast ball, one may have to press the button with the square icon on it. In All-Star Baseball 2002 (2001), the mechanic is the same. To select a fast ball, one may have to press a different button, but the idea is the same. Pitch selection (curveball, fastball, sinker, etc.) occurs via different buttons. The result is that gamers that play baseball console titles begin to expect certain functionalities to work across all baseball console titles.

One of the current sports titles we are working on has introduced a novel control technique that has not been done before. Thus, the reason we introduced the open-ended task was to find out if gamers could discover this new method, assess how they would attack the problem, and assess whether they liked it. At this point, introducing specific tasks would not allow us to make these types of initial discovery conclusions. This task allowed us to see how gamers would attack novel problems that did not match their current expectations. For example, novice gamers may behave differently than more experienced gamers, which has implications for how to introduce novel methods. Again, the attention span of gamers tends to be short. Thus, it

is imperative that we ensure that novel functionality is immediately accessible.

Prototyping, heuristic evaluations, and empirical guideline documents are other techniques we use when more time-consuming testing cannot be done. In practice, these techniques do not differ when used on games. Nielsen (1993) categorized prototyping and heuristic evaluations as "discount usability engineering," and we would agree. We also tend to view empirical guideline documents in a similar manner.

Empirical guideline documents are essentially lists of usability principles for particular content areas based on our collective experience doing user-testing research. Examples of some of these content areas include console game shell design, PC game shell design, PC tutorial design principles, movement, aiming, and camera issues in first/third person shooter games, and online multiplayer interfaces.

Survey Techniques. The use of surveys has been explored in great depth (e.g., Bradburn & Sudman, 1988; Couper, 2000; Labaw, 1981; Payne, 1979; Root & Draper, 1983; Sudman, Bradburn, & Schwarz, 1996) and is considered a valid approach for creating an attitudinal data set as long as you ask questions that users are truly capable of answering (see Root & Draper, 1983). One of the biggest complaints about a survey approach has to do with response rates to mail-based surveys (Armstrong & Lusk, 1988). We avoid this issue by bringing users on site. This also gives us the opportunity to question their attitudes while monitoring some simple behaviors. Even better, it allows us to take more control of the testing situations.

There are a number of advantages to using a technique such as this. We acquire so-called hard numbers that can be used for comparison from time one to time two (the formative approach) and to compare progress against other similar games (the summative approach). The numbers can also be used to give weight to issues. Other advantages include the factors of time and speed. Finally, the data can be presented in a purely descriptive format, meaning that trends of interest can come to light quite easily, even for lay people.

On the other hand, there are a few notable disadvantages. First and foremost, the engineer preparing the test must be intimately aware of what needs to be assessed to include questions in the survey that will answer the development team's questions. In other words, the questionnaire needs to allow for any potential useful user attitudes about the experience. Although there are a number of things we have done to address this (e.g., standardization of general questions to assure coverage of key gameplay perceptions, a question database to refer to) the ability to do this well is mostly a factor of the experience of the engineer and their familiarity with that game as well as the game genre. A high level of game domain knowledge is required. Second, the data is often difficult to interpret (70% believe an aspect of the game is "good," whereas 30% believe it is "bad") without some context. To limit this problem we have built up a comparison database that we can use to answer exactly this kind of question. A third item to consider is that despite the existence of other data sets and reasonable foresight in questionnaire creation, there are still many unpredictable issues that may

arise. Follow-up studies are sometimes necessary to parse out these issues. Finally, the approachable nature of the data is ripe for misinterpretation by the lay person. Careful and consistent presentation formats are called for. It has been rightfully asserted that usability tests do not necessarily require a formal report upon completion (Nielsen, 1993). We believe that that may be less true for using this kind of technique.

Although the use of a large sample can be used for a number of different things there are a few types of playtest studies that we conduct more consistently than others. There are essentially two types of formative studies: the critical facet test and the initial experience test. We also frequently run what we call benchmark (or summative) tests to build our comparison database, and, finally, tests that do not easily fit into any of these categories, including subtle variations on the above, a few cases of large sample observationally based studies, and studies that focus on games from a more conceptual level. In nearly all instances, questioning follows the pattern of a forced choice question (sometimes a set of forced-choice questions) followed by a more open-ended question encouraging the participants to state the reasons behind their response.

As in the usability section, for clarity of presentation, each technique is discussed separately, followed by a case study. Each case study contains only information pertinent to a specific technique, thus examples may be taken from a larger playtest.

Critical Facet Playtest. Games are often a collection of experiences that revolve around particular events. In a baseball game, although there are numerous aspects to the game, such as playing the role of manager, controlling fielders, and controlling runners on the base path, the game revolves around the critical game facet of pitcher versus batter. Similarly, in an air combat game there may be several interesting choices to make with regard to which plane to choose for which mission, perhaps even which weapons to outfit the plane with, but the core gameplay, the critical facets of the game, are (a) being able to control the plane and (b) being able to control a plane well enough to win an air combat battle. In both of these instances and across genres, we have typically conducted tests that focus on critical facets such as these. Although these kinds of game facets can often be assessed in a usability test as well as a playtest, the key is that attitudes about the experience are just as important as the usability (i.e., learnability, performance).

The following example demonstrates how a critical facet test was run in order to determine the optimal point of view behavior (often referred to as "the camera" in the gaming community) for Oddworld: Munch's Oddysee (2001), an Xbox game. Munch's Oddysee is a platform/adventure game that allows you to switch back and forth between the two main characters as they proceed through the increasingly difficult dangers of Oddworld on a quest to save Munch's species from extinction. Its core gameplay includes basic maneuvers including making the characters run, jump, and swim, or making use of their special powers to proceed through the realm. The camera behavior interacts with all components of the game.

Case Study: Munch's Oddysee, Camera. Previous usability testing with the *Munch's Oddysee* had determined that whereas some users indicated dissatisfaction with the behavior of the camera, other participants chose not to mention it at all while engaged in the open-ended usability tasks. The camera's behavior was programmed so as to create maximal cinematic effects (i.e., sometimes zooming out to show the size of an area) and also attempt to enhance gameplay. The camera would often show a specific view with the intent of showing the user what was behind "the next door" or on the other side of the wall while still keeping the main character in view. The most common user complaint was that although they often liked the look, style, and behavior of the camera, they wanted more control over the behavior of the camera as other games in this genre have done. Indeed, some participants would actively say things such as, "That looks really cool right now [after the camera had done something visually interesting], but I want the camera back pointing this way now." The nature of the data from the usability lab was not fully satisfying to the development team. There was feedback related to attitudes of participants that indicated both positive and negative perceptions of the camera. This was weighed against the fact that changing the behavior of the camera represented a major cost to the game in terms of development time and perhaps in the redesign of certain already completed areas of the game.

A critical facet playtest was conducted in order to shed more light on the attitudes related to the camera behavior in Munch's Oddysee. We targeted having at least 25 participants whose previous behavior (i.e., their gaming experience) indicated that they fell within a general definition of the target market for this game.

After having played the game for an hour, participants were asked first for general perceptions of the game followed by more specific questions related to other areas of interest. Questions related to the camera were asked in the latter portion of the questionnaire because previous experience in the usability lab had shown that merely mentioning the camera as part of a task would often cause participants previously silent on the subject to vociferously criticize aspects of the camera's behavior. With the knowledge that we wanted to factor out any priming-related effects, two analyses were conducted.

The first analysis was based on the response to the questions related to the behavior of the camera itself. Nearly half of the participants (46%) indicated that the camera did not give them "enough flexibility" of control. The second analysis was to go back through individual responses and determine if, before we brought up the subject of the camera, any participants (a) mentioned the camera, and (b) if possible, the comments related to the camera were subcategorized as a negative or positive statement. Forty-three percent of the participants were found to have mentioned the camera in a negative fashion before the camera questions. Based on these data and other anecdotal evidence, the development team chose to modify the behavior of the camera in this game to give the players more flexibility of control. The result was more frequent use of a camera behavior that we have termed a third-person follow camera.

The behavior of this camera had the double advantage of being more easily controlled by users and of conforming to a set of behaviors more often expected by users, such as maintaining

focus on the main character without major adjustments to point of view (i.e, to "look over a wall" or "behind a door"). Other camera behaviors (i.e., still camera behaviors that pan with the character) are still a part of the game but have been localized to areas where these alternative camera behaviors can only create an advantage for the user.

Initial Experience Playtest (Formative). As with many things, first impressions are a key component of overall satisfaction with a game. Given that many games are experienced in a certain order (i.e., Mission 1, Mission 2, etc.), there is a lot of value in obtaining attitudinal data related to the first hour(s) of gameplay because the later portions of the game will clearly never be experienced unless the first portions of the game are enjoyed.

As mentioned already, games are entirely a voluntary experience and at any moment a person may become frustrated or bored and put the game down. In a perfect world we would like to be able to identify these moments as they occur to be able to address them. The realities of the manner in which games are created (often times individual portions of a game are completed at different times and in a different order than consumers will actually experience them), and the realities of the time constraints related to testing with users prevent this from happening. If we want to know how users react in the 30th hr of gameplay then we need to conduct a study where the user would experience the first 29 hr of the game. Those 29 hr of gameplay need to be ready and completed before running this 30-hr test. Clearly, this would be difficult, expensive, and time-consuming. As a result, we often focus on the first hour of gameplay because it is easily accessible to us from a research perspective, and the lessons learned can in many cases be applied throughout the game.

The following example explains how a set of formative initial experience tests were run for MechCommander 2, the RTS game described earlier in the chapter. The earlier usability test focused on the choices that users could make before starting a mission, whereas this test focused on in-game components, such as user's ability to take control of their squad and lead them through battles, while experiencing satisfaction and motivation to continue playing.

Case study: MechCommander 2, Initial Missions. A sample of 25 participants, who through previous behavior were identified as representative of the potential target market, were included in the study. The participants were brought on site and asked to begin playing from the first mission of the game. After each mission was completed, the participants were asked to stop playing and fill out a set of questions focusing on their impressions of the "fun," the "excitement," and the "clarity" of objectives, as well as respond to measures designed to assess their "comfort level" with learning the basics of controlling the game. The participants were able to play through as many as three missions before the session ended. For purposes of brevity, this case study will focus on the results related to the first mission.

At the time of this test, the first mission was a quick experience designed to introduce users to some basic control concepts, to allow them to participate in some combat, win that combat, and then move on to the next mission. Also notable was that at the time, the first few missions of the game were intended to act as the tutorial for the game as well as being fun missions. As it happens, this experience was generally not satisfying to these participants, causing a poor initial impression with the game. Approximately one third of the participants felt the "challenge" was "too easy" and that the mission was "not exciting." Furthermore, there were some problems related to clarity of goals, where some participants were observed to move their units into an area where they were not intended to go, disrupting their experience and limiting their ability to proceed quickly. Responses to more open-ended questions indicated that some of the core gameplay components and unique features did not come across to users because they complained about the "standard" nature of the game. Finally several participants complained about the fact that they were "being taught" everything and wanted to "turn the tutorial off."

From the data, the development team, working with feedback from the user-testing engineer, decided to take a number of actions. First, it was decided that a separate set of tutorial missions would be created to give those who wanted to be taught a place to go and learn the basics of controlling the game separate from the actual missions of the game. Second, the scope of the first mission was expanded so that users would have more time in their initial experience with the game. Third, the clarity of objectives was improved via minor interface changes. Fourth, addressing the same issue, the design of the "map" on which the mission took place was revamped such that users were required to take a more linear approach to this mission and be unable to "get lost." Fifth, the amount and challenge of combat was increased. Finally, one of the unique components of the game, the ability to "call in support" from off-map facilities was introduced to the player, demonstrating how they could choose to augment their force on a case by case basis. Despite all these changes, the mission was still targeted to be completed within approximately 10–15 min.

A follow-up playtest was run to verify that the changes we introduced were paying off in terms of generating more user satisfaction with the initial mission of the game. Again, 25 participants (not the same people) representative of the target market were included in the study. Results from this test indicated that the design changes had created a number of payoffs. Far fewer participants felt the mission was "too easy" (13% vs. 33%), only 3% indicated that the mission was "not exciting," measures of "clarity of objectives" improved and, surprisingly, there was no dropoff on ratings of "comfort" with basic controls as a result of the tutorial aspects being improved. Results were not a total success, because some participants were now rating the mission as "too hard," whereas others in response to open-ended questions complained about the "overly linear" nature of the mission (i.e., there was only one way to proceed and few interesting decisions related to how to proceed through the mission). In response to these results, the development team decided to "retune" the first mission to make it a little easier but not to address the comments related to the linearity of the mission. The second mission of the game was far less linear, so it was hoped that, with the major constraints to enjoyment removed, most participants would proceed to the second mission and find the

mission goals to be more engaging. The data from testing with Mission 2 was rated far more "fun" than Mission 1, validating their assumption.

Benchmark Playtests. As indicated earlier, the nature of the data that is supplied using this survey technique is not always clear and easy to interpret. If 10% indicate that learning how to drive a car is "too hard," then is this something we should act on? The question is not easily answered without context. To supply context, benchmark playtests are conducted on every completed Microsoft game, as well as on many competitors. The data are used entirely for comparison purposes. Every benchmark playtest is run using the same protocol. All participants are directed to play as if they are at home, and we stop them to ask a set of questions to gather initial impressions and then let them play for the rest of the session followed by a set of final questions. Both of those question sets are general and concentrate on many of the core gameplay facets (i.e., "fun," "challenge," "pace," etc.) highlighted in previous sections.

The data gathered in these instances are used as comparison for initial experience playtests. Typically, the general questions are followed by a set of more specific questions designed to highlight user impressions about critical facets of a game. The data gathered in response to these questions are often used in comparison to critical facet playtests. Although the choice of comparing data gathered from a benchmark playtest with that from a critical facet playtest (where the experience was likely very different) is certainly controversial, the only goal of the data is to shed light and provide some context, not always to force action based on these comparisons. The decision on how strongly to weight the comparison is entirely a judgment call on the part of the user-testing engineer based on his or her knowledge of how closely the experience in a critical facet playtest models the intended initial experience with the game. As in all good usability testing, the choice of areas to focus on in these critical facet questions is done in close consultation with the development teams.

Qualitative Group Methods. In addition to usability and survey techniques, we employ other UCD methods we categorize as qualitative group methods; deep gameplay and focus groups.

Deep Gameplay. Thus far, the techniques we presented have focused primarily on evaluating the initial experience with a product, or one segmented hour of game play, and are often experienced out of context. Because of the nature of this field, testing portions of game play out of sequence occurs quite often but can be somewhat problematic. Although the initial experience with a game is important, the success or failure of a game often depends on the entire experience. In an attempt to provide game developers with user-centered feedback beyond the first hour of game play, we established the deep gameplay program.

Deep gameplay involves bringing in cohorts of users in to play a product repeatedly throughout the development cycle

and collecting qualitative data from them. The goal of deep gameplay is to provide fairly structured user-centered feedback on gameplay that occurs after the first hour of experience by supplying each team with its own dedicated gameplay team. Deep gameplay teams can be used to (a) evaluate content beyond the first hour in a linear or nonlinear fashion, (b) expose the same group of users to iterations of the same content, (c) test linear based modes of extended play (e.g., a career mode in a snowboarding game or multiple levels of an adventure game), or (d) expose users to a prerelease video game so that they have the experience necessary to evaluate advanced features or features that appear later in the game.

Deep gameplay requires commitment from participants to attend gameplay sessions on a consistent basis (potentially one to three times a month), otherwise the utility of deep gameplay is lost. Participants are exposed to features and game content in various ways. Generally, the order in which they encounter content depends greatly on the development schedule, the type or genre of the game, or the team's specific needs and goals. Commonly, the participants test each level as it becomes available or is significantly improved. The group can also return to previously tested sections that have been iterated to see if it has improved from before. Ideally, participants would be exposed to the game in a linear fashion progressing from the beginning to the end. However, this is only possible to do later in the development cycle because games are usually not developed from start to finish. Generally when this occurs, the team either takes the group back to the beginning of the game or starts a second group from the beginning.

Although there is no single report format for compiling data from these sessions, the report generally includes (a) a list of top issues that were brought up by the participants, along with recommendations; (b) a summarization of the responses to the written questions; (c) and a list of remaining issues from previous sessions.

The purpose of deep gameplay overlaps with other testing methods (viz. usability and playtest), but it has been designed wholly as a supplement, not a replacement. Each of these testing methods has its own unique set of strengths and weaknesses.

There are other methods that exist that attempt to get feedback from potential consumers. These methods are often referred to in the industry as "beta" and "recon". Beta testing usually involves sending a beta or prerelease version of a game to a large number of potential users. Feedback is collected on issues including configuration problems, technical bugs, and gameplay problems. Data collection is often done via newsgroups or electronic bug reports. Beta testing has been successful in getting some useful feedback from a larger number of users, but it does not ensure the amount, focus, or depth of the feedback that our methods provide. Recon's purpose is similar to deep gameplay. The difference is that recon uses professional gamers, which does not truly represent consumer feedback.

Focus Groups. Focus groups provide an additional source of user-centered feedback for members of the project team, including usability practitioners. Focus groups give development

team members a rare opportunity to hear users speak candidly about the game in question. In our group, the interviews are especially useful because the focus group participants often come from a playtest session which gives them the opportunity to use the product before discussing it. Additionally, the focus group interviews offer a qualitative method of data collection that is used to support and supplement data produced using quantitative survey methods.

In this setting, participants can elaborate on their experience with the product and expectations for the finished version of that product. The interviews allow participants to talk about their experiences and share impressions that may not have been elicited by the playtest survey. Allowing users to talk about their experiences can sometimes result in more detailed responses to help validate the quantitative data. This provides the team with a method of delving deeper into gameplay issues and strengths.

By using the qualitative and quantitative methods together, we are sometimes able to use information from focus groups to improve on survey techniques. The focus group data may identify important topics that should be added to the survey, as well as clarify topics that were poorly understood by users. For example, a focus group discussion regarding a sports game revealed that users had varying opinions about what it meant for a game to be a simulation vs. an arcade-style game. Additionally, they had many different ideas as to what game features made up each style of game. From this conversation, we learned that certain playtest survey questions regarding this topic were not reliable and needed revision or exclusion from future surveys.

Besides the general limitations inherent in all focus group studies (Greenbaum, 1988; Krueger, 1994), the focus groups described here include limitations associated with their deviation from validated focus group methods. Generally, focus group studies comprise of a series of 4 to 12 focus groups (Greenbaum, 1988; Krueger, 1994); however, we are only able to run one to three focus groups per topic. This makes it more difficult to confidently identify "real" trends and topics in the target population. Additionally, fewer group interviews make the study more vulnerable to idiosyncrasies of individual groups.

Because all focus group participants have recently completed a survey in playtest, it is also possible that the survey topics confound the issues raised in the general discussion in the focus group. The focus groups are commonly used in conjunction with the survey data, so this is not a serious limitation if used properly.

Because of these limitations, we take many precautions to ensure the quality of the data produced. The data that are produced in focus groups are closely monitored and scrutinized appropriately. Despite these limitations, we have found that focus groups have offered a useful and face-valid method of supplementing user-centered feedback that we provide to game developers.

A Combined Approach. We have provided a number of different techniques that we use in game development and evaluation.

The presentation has been structured in such a way to clearly demonstrate the differences between the techniques we use. The important thing to realize is that no one method exists independent of other methods. All of our techniques are used in conjunction with one another to fully address as many of the game evaluation factors presented earlier in the chapter. The following is a brief case study on an Xbox game called Blood Wake (2001) that demonstrates how some of these techniques can be used together.

Case Study: Blood Wake. Blood Wake is a vehicular combat console game using boats and a combination of powerful weapons. In a game such as this, there are certain game facets that must be done correctly for it to be successful. In Blood Wake, our initial main concern was the aiming mechanism and the controls. Users must also be able to destroy the enemy, and users must be able to easily control their boat. If these two factors were not addressed, the likelihood of a successful game significantly decreases. The next concern was related to the initial missions and gameplay and the general look and feel of the game overall. Users must be able to know what their objectives are, and it has to be fun.

The first problem we addressed was the aiming mechanism. We were presented with two different types of aiming schemes: auto aim and manual stick aim. In the auto aim scheme, the target reticle automatically "snapped" to an enemy that was within a reasonable distance allowing the player to maneuver their boat while shooting the enemy. In the manual stick aim scheme, the user had to manually aim the reticle using the right stick on the game pad with no assistance. In previous usability studies, we have found that many users encounter difficulties with a manual aiming mode. However, this method is used often in popular games, thus the general belief was that it was better or preferred. In addition, many felt the auto aim method would be too easy.

To solve this problem, we decided that a playtest would be most suitable. A usability test would probably allow us to demonstrate errors based on tasks and performance but would not really allow us to make any conclusions about which scheme users preferred. Thus, we ran a within-subjects design playtest to assess which scheme users liked better. Our results showed that 73% of participants liked the auto aim, whereas only 38% liked the manual aim. This was proof enough to convince the developers to focus their efforts on the auto aim method.

Although we solved the aiming problem, we still were not in any position to get feedback on the fun or challenge of the game. The reason for this was that the controls (i.e., handling) of the boat was not really optimized yet. Any conclusions about fun or challenge of the game at this point in development would be a dubious enterprise. The reason for this was because we would not have the ability to separate poor attitudinal ratings due to the gameplay, from the poor attitudinal ratings deriving from the poorly tweaked "feel" of the boat or uncomfortable button mappings. To optimize these controls, we headed to the usability labs.

In the usability test, we presented participants with two control schemes, or button mappings. In one scheme, throttle

and steering were both mapped to the left analog stick (push forward/back for throttle, push left/right for steering). This means that a user would be able to drive and steer with one finger. In another scheme, the throttle was mapped to the right trigger (which exists under the game pad), and steering was on the left analog stick. Previous data suggest that neither scheme was superior, so the goal of this test was to optimize the controls for both schemes.

Participants were presented with a series of tasks that were created to try and simulate common maneuvers performed during the game. The following are examples taken from the usability task list.

1. Once the game starts, you should see a bunch of boxes that form a path in the distance. As fast as you can, try to follow the trail of items and collect as many boxes as you can without turning around if you get off course.
2. Somewhere on this map, there is a straight line of boxes. Find the straight line of boxes then try to slalom (like a skier) through the line of the boxes. Try to do this as fast as you can without collecting any of the boxes.

As you can see from the two examples, the tasks were very performance-based. Participants were instructed to think aloud while we measured task performance and task time. By the end of the study, we were able to provide detailed information to the team about how to optimize the controls. Some of the issues that were addressed included the sensitivity of the steering, controlling boat speed, and difficulties moving the boat in reverse, to just name a few.

At this point, changes were implemented to the controls and the aiming method. The next step would be to validate these changes, but not from a performance-based perspective. We needed to determine if users liked how the controls felt, and we needed to begin getting feedback on missions and game play. Thus, multiple playtests were run focusing on the game play of five of the initial missions, and the controls. Our data on the controls from the playtest validated our earlier recommendations and allowed us to tweak the controls even more.

In addition, by running multiple playtests, we were finally able to get data and feedback on gameplay. After each mission, we elicited feedback on the perceived difficulty, clarity of objectives, attitudes about the environment, and general fun ratings. This allowed us to provide specific details to the development team on where exactly participants were running into problems (such as not knowing where to go next, not understanding how to complete objectives, etc.). This feedback led to dramatic changes design of the missions. For example, after the third mission, participants began to express boredom with the "look" of the environments. They did not know that by Mission 5, more interesting things would be introduced (such as differing the time of day, the weather, and the color of the water). This led to the introduction of more varied environments within the first three missions which helped increase fun ratings.

CONCLUSION

The need for user-centered design methods in video games has indeed arrived. Games drive new technologies and affect an enormous amount of people. In addition, games represent a rich space for research areas involving technology, communication, attention, perceptual-motor skills, social behaviors, virtual environments, to name a few. It is our position that video games will eventually develop an intellectual and critical discipline, like films, which would result in continually evolving theories and methodologies of game design. The result will be an increasing influence on interface design and evaluation. This relationship between theories of game design and traditional HCI evaluation methods has yet to be defined but definitely yields an exciting future.

As mentioned earlier, UCD methods have yet to find their way into the video game industry, at least to the same extent as it has in other fields. However, this is not to say that video games are qualitatively different from other computer applications. There are just as many similarities as differences to other computer fields that have already benefit from current UCD methods. It makes sense to use these methods when applicable but also to adapt methods to the unique requirements that we have identified in video games.

In this chapter, we emphasized the difference between games and productivity applications to illustrate the similarities and differences between these two types of software applications. We also chose to reference many different video games in hopes that these examples would resonate with a number of different readers. Case studies were included to try and demonstrate in practice how we tackle some of the issues and challenges mentioned earlier in the chapter. It is our intention that practitioners in industry, as well as researchers in academia, should be able to take portions of this chapter and adapt them to their particular needs when appropriate, similar to what we have done in creating the actual methods mentioned in this chapter. That said, we acknowledge that most, if not all, of our user-testing methods are not completely novel. They have been structured and refined based on a combination of our applied industry experience, backgrounds in experimental research, and, of course, a passion for video games. This allows us to elicit and use the types of information needed for one simple goal: to make the best video games we can for as many people as possible.

ACKNOWLEDGMENTS

We would like to thank the Microsoft Game Studios User-testing Group. In addition, we would like to express our gratitude to Paolo Malabuyo and Michael Medlock for their insights and input on the creation of this chapter, Kathleen Farrell and Rally Pagulayan for their editing assistance and reviewing of early drafts, and Ed Fries and all of Microsoft Game Studios for their support. The views and opinions contained in this chapter are those of the authors and do not necessarily represent any official views of Microsoft Corporation.

References

Age of Empires II: Age of Kings [Computer software]. (1999). Redmond, WA: Microsoft.

All-Star Baseball 2002 [Computer software]. (2001). Glen Cove, NY: Acclaim Entertainment.

Armstrong, J. S., & Lusk, E. J. (1988). Return postage in mail surveys. *Public Opinion Quarterly, 51*, 233–248.

Asakura, R. (2000). *Revolutionaries at Sony*. New York: McGraw-Hill.

Au, W. J. (2001, April). Playing God. *Salon*. Retrieved from http://www.salon.com/tech/review/2001/04/10/black_and_white/index.html

Banjo Kazooie [Computer software]. (2000). Redmond, WA: Nintendo of America.

Barfield, W., & Williges, R. C. (1998). Virtual environments: Models, methodology, and empirical studies [Special issue]. *Human Factors, 40*(3).

Blood Wake [Computer software]. (2001). Redmond, WA: Microsoft.

Bradburn, N. M., & Sudman, S. (1988). *Polls and surveys: Understanding what they tell us*. San Francisco: Jossey-Bass.

Cassell, J., & Jenkins, H. (2000). *From Barbie to Mortal Kombat: Gender and computer games*. Cambridge, MA: The MIT Press.

Chaika, M. (1996). Computer games marketing bias. *ACM Crossroads*. Retrieved from: www.acm.org/crossroads/xrds3-2/girlgame.html

Couper, M. P. (2000). Web surveys: A review of issues and approaches. *Public Opinion Quarterly, 64*, 464–494.

Crawford, C. (1982). *The art of computer game design*. Berkeley, CA: Osborne/McGraw-Hill.

Csikszentmihalyi, M. (1990). *Flow—The psychology of optimal experience*. New York: Harper & Row.

Diablo [Computer software]. (2000). Paris: Vivendi Universal.

Diddy Kong Racing [Computer software]. (1997). Redmond, WA: Nintendo of America.

Dumas, J. S., & Redish, J. C. (1999). *A practical guide to usability testing* (Rev. ed.). Portland, OR: Intellect Books.

Final Fantasy IX [Computer software]. (2000). Redmond, WA: Electronic Arts.

Flight Simulator [Computer software]. (2000). Redmond, WA: Microsoft.

Funge, J. (2000). Cognitive modeling for games and animation. *Communications of the ACM, 43*(7), 40–48.

Gorriz, C. M., & Medina, C. (2000). Engaging girls with computers through software games. *Communications of the ACM, 43*(1), 42–49.

Greenbaum, T. L. (1988). *The practical handbook and guide to focus group research*. Lexington, MA: D.C. Heath.

Hecker, C. (2000). Physics in computer games. *Communications of the ACM, 43*(7), 35–39.

Herz, J. C. (1997). *Joystick nation: How videogames ate our quarters, won our hearts, and rewired our minds*. New York: Little, Brown.

House of the Dead II [Computer software]. (1999). San Francisco: Sega.

Interactive Digital Software Association. (2000). State of the industry report (2000–2001). Washington, DC: Author.

Jablonsky, S., & DeVries, D. (1972). Operant conditioning principles extrapolated to the theory of management. *Organizational Behavior and Human Performance, 7*, 340–358.

Kent, S. L. (2000). *The first quarter: A 25-year history of video games*. Bothell, WA: BWD Press.

Kroll, K. (2000). Games we play: The new and the old. *Linux Journal, 73es*. Retrieved from www.acm.org/pubs/articles/journals/linux/2000-2000-73es/a26-kroll/a26-kroll.html

Krueger, R. A. (1994). *Focus groups: A practical guide for applied research*. Thousand Oaks, CA: Sage.

Labaw, P. (1981). *Advanced questionnaire design*. Cambridge, MA: Abt Books.

Laird, J. E. (2001). *It knows what you're going to do: Adding anticipation to a Quakebot. Proceedings of the Fifth International Conference on Autonomous Agent* (pp. 385–392). Canada: CSREA Press.

Lepper, M., Greene, D., & Nisbett, R. (1973). Undermining children's intrinsic interest with extrinsic rewards. *Journal of Personality and Social Psychology, 28*, 129–137.

Lepper, M. R., & Malone, T. W. (1987). Intrinsic motivation and instructional effectiveness in computer-based education. In R. E. Snow & M. J. Farr (Eds.), *Aptitude, learning and instruction III: Conative and affective process analyses*. Hillsdale, NJ: Lawrence Erlbaum Associates.

Levine, K. (2001, May). *New opportunities for Storytelling*. Paper presented at the Electronic Entertainment Exposition, Los Angeles, CA.

Lieberman, D. A. (1998, May). *Health education video games for children and adolescents: Theory, design, and research findings*. Paper presented at the annual meeting of the International Communication Association, Jerusalem.

Malone, T. W. (1981). Towards a theory of intrinsic motivation. *Cognitive Science, 4*, 333–369.

MechCommander 2 [Computer software]. (2001). Redmond, WA: Microsoft.

Medlock, M. C., Wixon, D., Terrano, M., Romero, R. L., & Fulton, B. Using the RITE Method to improve products; a definition and a case study. *Humanizing Design, Usability Professional's Association 2002 Annual Conference*. Orlando FL, USA, July 8–12, 2002.

MTV Music Generator [Computer software]. (1999). Warwickshire, United Kingdom: Codemasters Software.

NFL Blitz [Computer software]. (1997). Corsicana, TX: Midway Games.

Nielsen, J. (1993). *Usability engineering*. San Francisco: Morgan Kaufmann.

Oddworld: Munch's Odyssey [Computer software]. (2001). Redmond, WA: Microsoft.

Payne, S. L. (1979). *The art of asking questions*. Princeton, NJ: Princeton University Press.

Pokemon Crystal [Computer software]. (2000). Tokyo: Nintendo Japan.

Preece, J., Rogers, Y., Sharp, H., Benyon, D., Holland, S., & Carey, T. (1994). *Human-computer interaction*. Reading, MA: Addison-Wesley.

Provenzo, E. F., Jr. (1991). *Video kids: Making sense of Nintendo*. Cambridge, MA: Harvard University Press.

Root, R. W., & Draper, S. (1983, December). *Questionnaires as a software evaluation tool*. Paper presented at the ACM Conference on Human Factors in Computing Systems, Boston, MA.

Smart, L. J. (2000). *A comparative analysis of visually induced motion sickness*. Unpublished doctoral dissertation, University of Cincinnati, OH.

Steers, R. M., & Porter, L. W. (1991). *Motivation and work behavior* (5th ed.). New York: McGraw-Hill, Inc.

Stevenson, S. (2001, April). Why are video games for adults

so juvenile? *Slate.* Retrieved from http://slate.msn.com/Culture-Box/entries/01-04-19_104657.asp

Stoffregen, T. A., Bardy, B. G., Smart, L. J., & Pagulayan, R. J. (in press). On the nature and evaluation of fidelity in virtual environments. In L. J. Hettinger & M. W. Haas (Eds.), *Psychological issues in the design and use of virtual environments.* Mahwah, NJ: Lawrence Erlbaum Associates.

Sudman, S., Bradburn, N. M., & Schwarz, N. (1996). *Thinking about answers: The application of cognitive processes to survey methodology.* San Francisco: Jossey-Bass.

Triple Play Baseball [Computer software]. (2001). Redwood City, CA: Electronic Arts Inc.

Williams, J. H. (1987). *Psychology of women: Behavior in a biosocial context* (3rd ed.). New York: W. W. Norton.

THE DEVELOPMENT PROCESS

REQUIREMENTS SPECIFICATION

Michael J. Muller
IBM Research

How do I know what I know?

This question has perplexed and fascinated us for more than 2,000 years, in a variety of cultural, linguistic, and philosophical traditions. It is a question of crucial importance in beginning to work on a product or a service. The authors of the chapters in this section of the handbook (The Development Process) agree that knowledge *of* the users and *from* the users is an essential component of what we need to know before we can design, build, test, and deploy systems.

Although many of us quote with approval the adage, "design is where the action is" (Karat & Dayton, 1995), the authors in this section of the handbook remind us that

- the actions of design take place within a conceptual framework; and
- the terms of that framework are largely given by task analysis and other analytic activities.

Task analysis helps us understand the problems we are to solve, the users and other interested parties for whom we solve those problems, and the constraints and opportunities that inform our solution. Task analysis is concerned with at least three overall activities: learning about the world, discovering possibilities, and communicating with others. In some cases, these three activities occur in *anticipation* of design (see, e.g., the chapters by Mayhew and by Blomberg et al.). In other cases, these three activities take place crucially *during the activities of design* (see, e.g., the chapters by Holtzblatt and by Redish and Wixon, as well as my chapter on participatory design, elsewhere in this handbook). As Redish and Wixon note in their chapter, task analysis can be a set of ongoing activities that inform and reinform other processes throughout the software life cycle.

Task analyses are performed by a variety of workers. Mayhew writes for a usability engineer. Redish and Wixon are concerned more broadly for a diversity of people, from different backgrounds, who may perform this work. Blomberg et al. advocate for specialized training in ethnography. Holtzblatt argues for someone who is trained in her methodology. Readers of these chapters will have their own experiences. I have seen task analyses performed (consciously or unconsciously, formally or informally) by workers with job titles such as designer, usability professional, program manager, user interface developer, user experience researcher, work and usability analyst, and marketer (and my experience is based on fewer than five companies).

Many of these workers bring their own disciplinary backgrounds and their own communities of practice into their task analyses. Indeed, all of us define the quality of our work (in part) in terms of the accepted standards of good work in our discipline, and upon review and approval by other practitioners in our discipline. Differences in disciplinary backgrounds often lead to different strategies of investigation and different definitions of "data" (Chandler, Davidson, & Harootunian, 1994), as well as different analytic frameworks (Klein, 1996)—i.e., what must I do to do a good job in my professional discipline? Differences in communities of co-workers often lead to different implicit audiences for our work (i.e., who will judge whether I have done a good job in my discipline?).

A second major determinant is the problem to be solved. This determinant may be restated in business terms as a matter of scope and risk. Does the problem require an opening to new possibilities or an assurance that a product plan is likely to be successful? A task analysis for the first problem should probably be exploratory and innovative, and even

impressionistic: An in-depth focus on a few people or places may be the best strategy. A task analysis for the second problem should probably be tightly focused and intensely quantitative: Breadth and consistency of data collection are likely to be more important in this case. Blomberg et al. citing Bernard, discuss seven different data collection strategies. Their own work tends to solve the first type of problem (opening a space of possibilities), and their data collection methods tend to provide rich insights into individual and particular cases. Mayhew's approach tends more toward the second question (risk management and mitigation), and her data collection methods tend to favor broader, more formal coverage. Redish and Wixon, and Holtzblatt, offer a number of methods, some of which are appropriate for each of the types of problems that I have listed here.

As Mayhew and Redish and Wixon state clearly, there is no single method (or set of methods) that guarantees successful task analyses in all cases. I hope that readers will find new perspectives, new methods of analysis, and new ways of communicating their results in the chapters in this section.

Who helps me to know what I know?

With whom do we perform our task analysis? All the authors in this section are emphatic that a task analysis is based on information about users and their work (see Mayhew's chapter for illustrative stories about the perils of leaving user information out of the analysis). The role of the user—as data source or as informant or as partner—varies considerably from one chapter to another.[1] Mayhew's contributions have primarily concerned the traditional setting for a usability professional, in which management commissions an expert to describe users' work to inform development work by software professionals. Holtzblatt describes a less traditional version of an expert-based analysis, in which the expert may invite the users to become active members of the analysis team. Blomberg et al. position themselves as experts with strong ties to users. Redish and Wixon provide a more general survey, in which the role of the user depends on the stage in the life cycle, the problem to be solved at that stage, the culture of the organization, and the particular method chosen.

We can ask this question again (*Who helps me?*) in terms of other disciplines in the workplace. Redish and Wixon, and Holtzblatt, discuss methods that can (or even *should*) include other professionals with different backgrounds and different responsibilities. Redish and Wixon explore some of the organizational advantages of exposing desk-bound co-workers to the users' world, through involving those colleagues in task analysis work. By contrast, Blomberg et al. discuss the special contributions that may be made from their specialized perspective, and Mayhew focuses on the modal case in business teams, in which the usability professional acts as a unique, solitary contributor.

How do we know together?

Regardless of the roles of users and co-workers in performing the task analysis, all of the authors emphasize that the outcomes of the task analysis should be effectively communicated to others. The methods of contextual inquiry described by Holtzblatt, and also by Redish and Wixon, are deliberately crafted to support the exchange of views and perspectives, and the gradual deepening of a group's shared understanding. Redish and Wixon also provide pragmatically insightful suggestions about how to make easily understood and inviting displays from the results of other methods (i.e., other than contextual inquiry). All of these methods form a powerful basis for group coherence and collective competence. Blomberg et al. use stories (in an ethnographic context) to make their results clear to other software professionals. In contrast to these three chapters, Mayhew's structured method translates user-related data into the language of the software professionals, at a point in the project life cycle at which Blomberg et al. and Redish and Wixon argue for keeping the user perspective. The tension between these approaches is, of course, due in part to the different kinds of organizations that are being assisted, and the different problems to be solved. I hope that readers will find this contrast informative about the choices that they face.

These two questions, *Who helps me to know?* and *How do we know together?*, become crucially important when we consider the end of the task analysis process. I am using "end" with two distinct meanings. In sense of a *conclusion*, the end of a task analysis is usually a report, presented by a single person who is, by default, the organization's expert on the users' work. It is very rare that a user has this role of analyst and reporter. In almost all cases, the person who writes and presents the task analysis is not the user, but is expected and required to *speak for the users* in the remaining activities of the product life cycle.

In my second sense of *purpose*, the end of a task analysis is to provide information that enables an organization to do what it wants to do more effectively. As the participatory design movement has shown us (e.g., Bjerknes, Ehn, & Kyng, 1987; Schuler & Namioka, 1993), the purposes of the organization (and of its management) may not be the same as the purposes of the end users. (In some cases, the purposes of the end users may not be the same as the purposes of their clients or customers.)

The person who writes and presents the task analysis is thus in a difficult position (Muller, 1997), because s/he is expected and required to speak for people who are not present to speak for themselves (Alcoff, 1991). In a situation with conflicting goals and purposes, whose version of reality should the analyst present? Whose interests should s/he serve? How does the analyst balance among potentially divergent needs of users, managers, and the development team? There are seldom any easy answers to these questions.

Of the chapters in this section of the handbook, Blomberg et al. address these questions most directly. They insist on using the users' stories and the users' language for at least two reasons. First, bringing the users' perspective to the development team will help to inform that team of actual field and usage

---

[1] See also my chapter on methods of participatory analysis, in which users take the role as primary analysts or as full co-analysts of their own work and experiences.

conditions that they need to know about (the other authors agree on this point). Second, using the users' own language provides greater opportunity for users to contribution directly to the analysis and to critique the task analysis as it is being assembled (Redish and Wixon, and Holtzblatt appear to agree on this point). However, a reliance on users' stories in users' own language is a step away from the language of development teams. Mayhew appears to argue for greater effectiveness among those teams by entering the teams' language and value system. Readers will, I hope, make constructive use of these different strategies as they consider their own decisions and loyalties.

CONCLUSIONS

I will close with three questions[2] that remain at the heart of analysis and design of these kinds of systems that have such powerful implications for our work, play, and lives:

- *Who speaks, and who is silent while others speak about them?*
- *Whose interests are served when those who can speak, do speak?*
- *How do I know what I know?*

References

Alcoff, L. (1991). The problem of speaking for others. *Cultural Critique, Winter 1991-1992*, 5-32.

Beyer, H., & Holtzblatt, K. (1998). *Contextual design: Defining customer-centered systems*. San Francisco: Morgan Kaufmann.

Bjerknes, G., Ehn, P., & Kyng, M. (Eds.). (1987). *Computers and democracy: A Scandinavian challenge*. Brookfield, VT: Gower.

Chandler, J., Davidson, A. I., & Harootunian, H. (Eds.). (1994). *Questions of evidence: Proof, practice, and persuasion across the disciplines*. Chicago: University of Chicago Press.

Karat, J., & Dayton, T. (1995). Practical education for improving software usability. *Proceedings of CHI '95* (pp. 162-169).

Klein, J. T. (1996). *Crossing boundaries: Knowledge, disciplinarities, and interdisciplinarities*. Charlottesville, NC: University Press of Virginia.

Krupat, A. (1992). *Ethnocriticism: Ethnography, history, literature*. Berkeley, CA: University of California Press.

Muller, M. J. (1997). Ethnocritical heuristics for reflecting on work with users and other interested parties. In M. Kyng & L. Matthiessen (Eds.), *Computers in context and design* (pp. 349-380). Cambridge, MA: MIT Press.

Rich, A. (1983/1986). North American time. In A. Rich (Ed.), *Your native land, your life* (pp. 33-36). New York: Norton.

Schuler, D., & Namioka, A. (Eds.). (1993). *Participatory design: Principles and practices*. Hillsdale, NJ: Erlbaum.

Wixon, D., & Ramey, J. (Eds.). (1996). *Field methods casebook for software design*. New York: Wiley.

---

[2]These questions are derived from the work of Alcoff (1991), Bjerknes et al. (1987), Krupat (1992), Muller (1997), and Rich (1983/1986), and discussions in the Women's Rights Committee of New York Yearly Meeting of Quakers in 1991-1992.

·47·

REQUIREMENTS SPECIFICATIONS WITHIN THE USABILITY ENGINEERING LIFE CYCLE

Deborah J. Mayhew
Deborah J. Mayhew & Associates

INTRODUCING REQUIREMENTS SPECIFICATIONS

Three key ingredients are necessary to ensure that usability is achieved during product development:

- Application of established design principles and guidelines
- A structured methodology for design
- Managerial and organizational techniques

At this point in the history of the field of human–computer interaction, there are well-established software design principles and guidelines available, based on objective research and reported in the literature. Many of these principles and guidelines are enumerated throughout different chapters in this book. Development organizations need to have staff who are fluent in these design guidelines participate in design efforts, so this general accumulated knowledge will find its way into their products.

However, just having a design guru on board does not guarantee that design principles and guidelines will find their way into products. Design is complex, and there simply is no cookbook approach to design that can rely on general principles and guidelines alone. Development organizations also need *structured methods* for achieving usability in their products.

Similarly, a well-structured and documented design methodology must be introduced and managed—it does not happen by itself. Thus, managerial and organizational techniques must be applied to ensure that the design methodology is followed and includes the application of well-established design principles.

Even when good management practices are being applied, either of the remaining two ingredients alone—design guidelines or design methods—is necessary, but not sufficient. Optimal design cannot be accomplished even by the systematic application of generic guidelines alone, because every product and its intended set of users are unique. Design guidelines must be tailored for and validated against product-unique requirements, and this is what the structured methods accomplish.

Conversely, applying structured methods without also drawing on well-established design principles and guidelines is inefficient at best and may simply fail at worst. Without the benefit of the initial guidance of sound design principles during first passes at design, a particular project with its limited resources may simply never stumble on a design approach that works. For example, formal usability testing is a valuable and objective method for uncovering usability problems. However, without a clear understanding of basic design principles and guidelines, as well as unique requirements data, solving those problems after they have been identified will not be easy or likely.

Previous sections and chapters in this book make reference to a broad variety of design principles and guidelines. This section—Section V: The Development Process—addresses the need for methodology and provides chapters that address a variety of techniques that can be applied during the development process to achieve usability in product design. This chapter sets the stage for this section and, in particular, for the subsection on one stage of the development process: Requirements Specification.

The Usability Engineering Life Cycle (Mayhew, 1999) documents a structured and systematic approach to addressing usability within the product development process. It consists of a set of usability engineering tasks applied in a particular order at specified points in an overall product development life cycle.

Several types of tasks are included in The Usability Engineering Life Cycle:

- Structured usability requirements analysis tasks
- An explicit usability goal-setting task, driven directly from requirements analysis data
- Tasks supporting a structured, tops-down approach to user interface design that is driven directly from usability goals and other requirements data
- Objective usability evaluation tasks for iterating design toward usability goals

Figure 47.1 represents in summary, visual form, The Usability Engineering Life Cycle. The overall life cycle is cast in three phases: Requirements Analysis, Design/Testing/Development, and Installation. Specific usability engineering tasks within each phase are presented in boxes, and arrows show the basic order in which tasks should be carried out. Much of the sequencing of tasks is iterative, and the specific places where iterations would most typically occur are illustrated by arrows returning to earlier points in the life cycle. Brief descriptions of each life cycle task are given below.

Phase 1: Requirements Analysis

User Profile. A description of the specific user characteristics relevant to user interface design (e.g., computer literacy, expected frequency of use, level of job experience) is obtained for the intended user population. This will drive tailored user interface design decisions and also identify major user categories for study in the Contextual Task Analysis task.

Contextual Task Analysis. A study of users' current tasks, workflow patterns, and conceptual frameworks is made, resulting in a description of current tasks and workflow and an understanding and specification of underlying user goals. These will be used to set usability goals and drive Work Reengineering and user interface design.

Usability Goal Setting. Specific, *qualitative* goals reflecting usability requirements extracted from the User Profile and Contextual Task Analysis, and *quantitative* goals defining minimal acceptable user performance and satisfaction criteria based on a subset of high-priority qualitative goals are developed. These usability goals focus later design efforts and form the basis for later iterative usability evaluation.

Platform Capabilities/Constraints. The user interface capabilities and constraints (e.g., windowing, direct manipulation, color, etc.) inherent in the technology platform chosen for the product (e.g., Apple Macintosh, MS Windows, product-unique platforms) are determined and documented. These will define the scope of possibilities for user interface design.

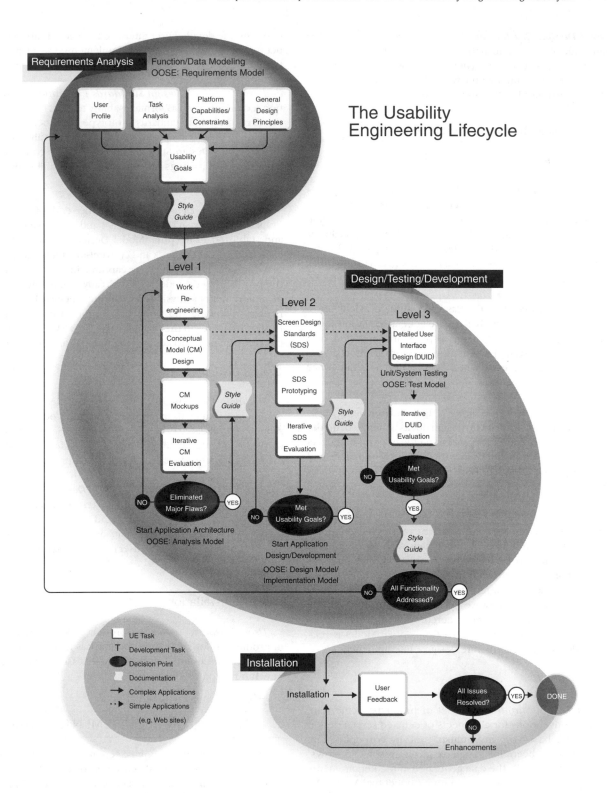

FIGURE 47.1. The Usability Engineering Life Cycle. (Illustration taken from Mayhew, 1999.)

General Design Guidelines. Relevant general user interface design guidelines available in the usability engineering literature are gathered and reviewed. They will be applied during the design process to come, along with all other project-specific information gathered in the previous tasks.

Phase 2: Design/Testing/Development

Level 1 Design

Work Re-engineering. Based on all requirements analysis data and the usability goals extracted from it, user tasks are re-designed at the level of organization and workflow to streamline work and exploit the capabilities of automation. No visual user interface design is involved in this task, just abstract organization of functionality and workflow design. This task is sometimes referred to as Information Architecture.

Conceptual Model Design. Based on all the previous tasks, initial high-level design alternatives are generated. At this level, navigational pathways and major displays are identified, and rules for the consistent presentation of work products, processes and actions are established. Screen design detail is *not* addressed at this design level.

Conceptual Model Mockups. Paper-and-pencil or prototype mockups of high-level design ideas generated in the previous task are prepared, representing ideas about high-level functional organization and Conceptual Model Design. Detailed screen design and complete functional design are *not* in focus here.

Iterative Conceptual Model Evaluation. The mockups are evaluated and modified through iterative evaluation techniques, such as formal usability testing, in which real, representative end users attempt to perform real, representative tasks with minimal training and intervention, imagining that the mockups are a real product user interface. This and the previous two tasks are conducted in iterative cycles until all major usability "bugs" are identified and engineered out of Level 1 (i.e., Conceptual Model) design. Once a Conceptual Model is relatively stable, system architecture design can commence.

Level 2 Design

Screen Design Standards. A set of product-specific standards and conventions for all aspects of detailed screen design is developed, based on any industry and/or corporate standards that have been mandated (e.g., Microsoft Windows, Apple Macintosh, etc.), the data generated in the Requirements Analysis phase, and the product-unique Conceptual Model Design arrived at during Level 1 Design. Screen Design Standards will ensure coherence and consistency—the foundations of usability—across the user interface.

Screen Design Standards Prototyping. The Screen Design Standards (as well as the Conceptual Model Design) are applied to design the detailed user interface to selected subsets of product functionality. This design is implemented as a running prototype.

Iterative Screen Design Standards Evaluation. An evaluation technique, such as formal usability testing, is carried out on the Screen Design Standards prototype, and then redesign/re-evaluate iterations are performed to refine and validate a robust set of Screen Design Standards. Iterations are continued until all major usability bugs are eliminated and usability goals seem within reach.

Style Guide Development. At the end of the design/ evaluate iterations in Design Levels 1 and 2, you have a validated and stabilized Conceptual Model Design, and a validated and stabilized set of standards and conventions for all aspects of detailed Screen Design. These are captured in the document called the product Style Guide, which already documents the results of requirements analysis tasks. During Detailed User Interface Design, following the Conceptual Model Design and Screen Design Standards in the product Style Guide will ensure quality, coherence, and consistency—the foundations of usability.

Level 3 Design

Detailed User Interface Design. Detailed design of the complete product user interface is carried out based on the refined and validated Conceptual Model and Screen Design Standards documented in the product Style Guide. This design then drives detailed product design and development.

Iterative Detailed User Interface Design Evaluation. A technique such as formal usability testing is continued during product development to expand evaluation to previously unassessed subsets of functionality and categories of users, and also to continue to refine the user interface and validate it against usability goals.

Phase 3: Installation

User Feedback. After the product has been installed and in production for some time, feedback is gathered to feed into enhancement design, design of new releases, and/or design of new but related products.

FOCUSING ON REQUIREMENTS ANALYSIS

This section of this book—Section V: The Development Process—provides in-depth coverage of the different phases (and tasks within those phases) of the Usability Engineering Life Cycle. In particular, this subsection, Requirements Specification, provides chapters describing and discussing different tasks and techniques for requirements specification in some depth and from different perspectives. The goal of this chapter is to reinforce the importance of this phase of the life cycle and to provide real-world examples of the benefits of conducting

the kinds of techniques discussed in the later chapters of this subsection.

There are three main things that must be studied and understood to tailor design to support unique requirements:

- The users
- The users' tasks
- The users' work environment

In The Usability Engineering Life Cycle, the first is addressed in a task called The User Profile, and the second two are addressed in a task called Contextual Task Analysis.

The User Profile

There is no single best user interface style or approach for any and all types of users. Specific interface design alternatives that optimize the performance of some types of users may actually degrade the performance of other types of users. For example, an infrequent, casual user needs an *easy-to-learn* and *easy-to-remember* interface, but a high-frequency expert user needs an *efficient*, *powerful*, and *flexible* interface. These are not necessarily the same thing. Similarly, a highly skilled typist might perform better with a *keyboard-oriented* interface, whereas a low-skill typist might do better with a graphical user interface.

Unless designers know the specific characteristics of a population of users (e.g., expected frequency of use, level of typing skill, etc.), they cannot make optimal user interface design decisions for them. The purpose of a User Profile is thus to establish the *general* requirements of a category of users in terms of overall interface style and approach.

The User Profile task fits into the overall Usability Engineering Life Cycle as follows:

- The User Profile task is the first task in The Usability Engineering Life Cycle.
- The User Profile task will feed directly into the Contextual Task Analysis task by identifying categories of users whose tasks and work environment must be studied in that later task.
- The User Profile task will feed directly into the Usability Goal Setting task in that usability goals are in part driven directly by user characteristics (e.g., a low frequency of use indicates a need for ease-of-learning and remembering). Thus, different usability goals will be extracted from the profiles of different categories of users.
- Ultimately, the User Profile task will have a direct impact on all design tasks, which are focused on realizing usability goals, in turn based in part on User Profiles.
- The User Profile task will also drive the selection of usability evaluation issues and test users.
- Output from the User Profile task will be documented in the product Style Guide.

The User Profile task fits into the underlying software development methodology as follows:

- The User Profile task can occur *in parallel with*, *overlapping with*, or *following the development of* the Requirements Model in the Analysis Phase in Object-Oriented Software Engineering (OOSE), or function and data modeling in the requirements phase of a traditional rapid prototyping methodology. It could either define Actors for the Requirements Model or take the definition of user categories from the Requirements Model as its starting point.
- The User Profile task (along with all other Usability Engineering Life Cycle Requirements Analysis tasks) should occur *prior* to the development of the Analysis Model in the Analysis phase of OOSE (or the application architecture design in a traditional rapid prototyping methodology).

Contextual Task Analysis

The purpose of the Contextual Task Analysis task is to obtain a user-centered model of work *as it is currently performed* (i.e., to understand how users currently think about, talk about, and do their work in their actual work environment). Ultimately, the reason for this is that when designing a new product and its user interface, it is important to find an optimal compromise or trade-off between three goals:

- Realizing the power and efficiency that automation makes possible.
- Re-engineering work processes to support more effectively identified business goals.
- Minimizing retraining by having the new product *tap as much as possible into users' existing task knowledge*, and maximizing efficiency and effectiveness by *accommodating human cognitive constraints and capabilities* within the context of their actual tasks.

Traditionally in software development methodologies, the most focus is put on the first goal and some on the second goal, and most of the third is lost, because designers and developers never really understand the users' work and their current work models. A great deal of work re-engineering occurs in application design, and only some of it serves the first two goals. Much of it is unnecessary, unuseful, and results in unnecessary training and usage burdens on the user. The third goal—minimizing the training overhead and maximizing efficiency and effectiveness—cannot be factored in unless designers have a clear picture of users' *current* work models: How they do their work in the realities of their everyday work environment, and how they think about it and talk about it.

Traditional systems analysis usually (but not always) results in the inclusion of all required data and low-level functions, and structures them in a robust *implementation* architecture. Without a truly user-centered approach, however, it often fails to *organize and present* that data and functionality in a manner that *supports and optimizes* the work performance of real users in their real work environment. This missing piece is the whole point of a Contextual Task Analysis.

Thus, the purpose of this task is to supplement more traditional types of systems analyses to define *usability*

requirements and to point toward ways of meeting those requirements. Then, in later tasks, these requirements can be applied directly to making user interface design decisions.

Contextual Task Analysis consists of the following basic steps:

- Gather background information about the work being automated.
- Collect and analyze data from observations of and interviews with users as they do real work in their actual work environment.
- Construct and validate models of how users currently think about and do their work.

A central, key step in the Contextual Task Analysis task is the second step, sometimes referred to as contextual observations/interviews. Here, the idea is that analysts must observe and interview users *in their real life work* context to understand their work and discover their work models. Only then can designers structure and present functionality in an application user interface in a way that taps into users' current work models and optimally supports their tasks.

An *abstract* modeling of users' tasks, which is the focus of more traditional types of systems analysis, does not typically take into consideration key aspects of actual work flow and key aspects of the users and their work environment, and simply does not support this goal. Another way of putting this is that traditional systems analysis models *work in the abstract*, without considering the basic capabilities and constraints of human information processing, the particular characteristics of the intended user population, the unique characteristics of the work environment, and how *users themselves* model, carry out, and talk about their tasks.

Based on an analysis of direct observations, models can be constructed that represent *not* the work in the abstract from a *systems point of view*, but instead represent how users currently think about, talk about, and do their work (i.e., models reflecting the *users' point of view*). These models do not get directly designed into the application or its interface. They feed into only one of the goals referred to earlier—that of tapping into existing user knowledge, habit, and capabilities—and they are juggled with the other two goals of supporting general business goals and exploiting the power of automation. This juggling happens in a later task, called Work Re-engineering in The Usability Engineering Life Cycle, but also often referred to as Information Architecture.

Contextual Task Analysis task fits into the overall Usability Engineering Life Cycle as described:

- The User Profile task will feed directly into the Contextual Task Analysis task by identifying categories of users (i.e., Actors) whose tasks must be studied.
- The Contextual Task Analysis task will feed directly into the Usability Goal Setting task by helping to identify different primary goals for different task types (Use Cases), and by identifying bottlenecks and weaknesses in current work processes that can be reduced through good user interface design.
- The Contextual Task Analysis task will feed directly into the Work Re-engineering task. Current user work models are

re-engineered only as much as necessary to exploit the power of automation and contribute to explicit business goals. Current user knowledge and experience are exploited as much as possible to facilitate ease of learning and use.

- The Contextual Task Analysis task will be documented in the product Style Guide.
- Ultimately, the Contextual Task Analysis task will have a direct impact on all design tasks, and on the selection of usability testing and evaluation issues, as well as on the design of usability testing materials.

Contextual Task Analysis fits into the underlying software development methodology as follows:

- The Contextual Task Analysis task can occur *in parallel with, overlapping with*, or *following the development of* the Requirements Model in the Analysis Phase in OOSE (or function and data modeling in the requirements phase of a traditional rapid prototyping methodology). It could either identify Use Cases for the Requirements Model, or take the definition of Use Cases from the Requirements Model as its starting point for constructing Task Scenarios.
- The Contextual Task Analysis task (along with all other Usability Engineering Life Cycle Requirements Analysis tasks) should occur *prior* to the development of the Analysis Model in the Analysis phase of OOSE (or the application architecture design in a traditional rapid prototyping methodology).

Currently, in the field of Usability Engineering, there is no well-established, universally applied, general, practical, and highly structured technique for performing a Contextual Task Analysis for the purpose of driving the user interface design for a product that has already been identified and scoped. It is more of an art than a science at this point, with each usability practitioner using his or her own informal approach. Very recently, however, some structured techniques have begun to emerge, and experience with them has been reported in the literature (see, e.g., Beyer & Holtzblatt, 1998; Hackos & Redish, 1998). Later chapters in this book represent the latest editions of the thinking of these and other experts on this important phase in The Usability Engineering Life Cycle.

MOTIVATING AND JUSTIFYING REQUIREMENTS SPECIFICATIONS

While the previous section set the stage for Requirements Specifications within the broader perspective of the whole Usability Engineering Life Cycle and underlying development process, this section provides motivation and justification for investing in Requirements Specifications tasks and activities in particular, through the reporting of "war stories" from the experience of myself and other practitioners. The war stories are divided into those relating to user, environment, and task requirements.

User Requirements

Following are two examples from my own experience of the importance of knowing your users.

For two of my clients, I first interviewed project team members (developers) to get a general sense of the user population, to design a User Profile questionnaire that I would later employ to solicit profile information directly from the users themselves. In one case, the project team was convinced that their users would have a generally low level of familiarity with the Microsoft Windows platform they planned for their product. They were thus prepared to depart significantly from the Windows platform user interface standards in their product user interface. The User Profile questionnaire, however, revealed a generally high level of Windows experience. This—and the fact that the Windows user interface standard was a good fit for the application functionality—led me to strongly advise them to adopt the Windows standards as closely as possible. They were still interested in creating their own unique user interface, but early testing of several alternative designs that varied in how faithfully they followed Windows standards clearly showed that users learned much more quickly, the more consistent the design was with Windows standards.

On the other project, the development team felt quite confident that users would generally have high levels of computer literacy. An extensive User Profile questionnaire of more than 800 users, however, revealed that only a very small percentage of potential users had any experience with computer software at all, let alone the Windows user interface standards. In this case, based on this User Profile data, we designed (and validated through testing) a very simplified user interface that departed significantly from the Windows standards.

In both of these cases, two things are clear: That project team members often have serious misconceptions about the important characteristics of their users, and that these misconceptions could have led them to design very inappropriate user interfaces for those users.

Work Environment Requirements

It is important to understand the environment in which users will be utilizing a product to carry out their tasks, because this environment will place constraints on how they work, and how well they work.

Suppose a screwdriver is being designed and all that is known is the size of the screw head. So, something like a traditional screwdriver is designed, with the correct sized blade. But suppose it then turns out that the user needs to apply the screw *from the inside of a narrow pipe* to assemble some piece of equipment. Clearly, a traditional screwdriver will be useless in this work context.

Similarly, suppose a software application is being designed for a set of users, but the designers have never gone to the users' actual work environment. They assume a traditional officelike environment and design software that will work on a traditional workstation. But, suppose it then turns out that in the actual work environment, users are constantly in motion, moving all around the environment to get different parts of an overall job done. If software for a traditional workstation is designed, this will simply not work in this environment. Software that will run on a smaller and more portable device that can be carried around with the user, like the units carried by UPS delivery staff, would be required instead.

In another example, suppose designers have never visited the user's workplace, and they assume users all work in closed offices. A system with voice input and output is designed. But it then turns out that users work in one big open area with desks located right next to one another. The noise from all those talking people and workstations will create an impossible work environment, and most voice recognition systems simply do not work with acceptable accuracy in a noisy environment. The point is, there are many aspects of the actual work environment that will determine how well a tool will work in that environment, and so the environment itself must be studied, and the tool tailored to it.

A real example of the importance of understanding the users' work environment comes from a project I worked on with a large metropolitan police department. Requirements analysis activities revealed that, in the typical police station, the appearance of the interior is dark, run down, and cluttered; the lighting is harsh and artificial; and the air is close and sometimes very hot. The noise level can be high, the work areas are cramped and cluttered, and the overall atmosphere is tense and high pressured at best, chaotic and sometimes riotous at worst. These conditions most likely have a general impact on morale, and certainly will have an impact on cognitive functioning that, in turn, impacts productivity and effectiveness. A user interface must be carefully designed to support the natural and possibly extreme degradations of human performance under these conditions.

In addition, it was observed that in the noisy, stressful, and distracting work environment in a typical police station, users will frequently be interrupted while performing tasks, sometimes by other competing tasks, sometimes by unexpected events, unpredictable prisoners, etc. A user interface in such an environment must constantly maintain enough context information on the screen so that when users' attention is temporarily drawn away from their tasks, they can quickly get reoriented and continue their task without errors, and not have to backup or repeat any work.

Task Requirements

Besides the users themselves and the environments they work in, the tasks they do also have their own inherent requirements.

One compelling example of the need for a thorough understanding of users tasks to achieve usable design comes from one of the earliest books on computer–human interaction (Rubinstein & Hersh, 1984).

For example, recently a water district in Maine installed an on-line billing system so that when a customer called in, the clerk could quickly retrieve a record of usage and billing for that customer. The managers noticed that the clerks were less than happy with the new system. A little investigation revealed that with the manual system, the clerks would write pertinent information in the margins of the records, such as that one person pays on the fifteenth of the month rather than on the first, or that the meter reader should contact the neighbor across the street for access to a house. This informal information was critical to the smooth operation of the office, but because there was no provision for it in an official field of the payment record form, the information was lost in the conversion to the on-line system. Understanding that there is informal as well as official information and recognizing the importance of the former would have reduced the disruption in work style.

Another good example of what can be missed by not understanding users' tasks is found in Coble et al. (1996, p. 231). This report described the use of task analysis techniques to study the functional requirements of physicians for a clinical workstation:

> Before the...session started, a physician explained the purpose and details of each section in his office chart....Later, when he was doing actual work in context, the person performing the...session noticed that the note he was looking at was written with red ink. She probed and the physician said it told him the previous encounter with this patient was a hospital visit. That fact told him he needed to review the hospital discharge summary and hospital laboratory results before entering the patient's exam room. *The physician was surprised that he had not mentioned that need before.* It was so ingrained in how he worked that he did not even process that highly relevant detail consciously anymore. (Italics mine.)

In another example from my own experience, I was performing a requirements analysis in a police department and observing police officers using a system of standardized paper forms to document property. I observed infrequent users of the forms struggling with them (there were many complex forms and many unwritten rules about how to fill them out), and later observed very frequent users using them. The frequent users tended to ignore the forms initially and take free form notes describing the property they had to document in a format very different from that required on the forms. They then transcribed their own notes onto the forms as required. It was clear from this observation (and from follow-up interviews with the frequent users) that the forms were not designed in a way that supported the users' task, and I learned a lot from how frequent users worked around the forms with their own format that helped in designing better online forms. A traditional systems analysis would typically not involve studying users actually using paper forms in their actual work, but would have instead taken the paper forms themselves as a description of the work to be automated. Problems with the forms would have been missed entirely and perpetuated in the online version of the task.

There are always a great many things like this that users simply will not think to report during an offsite interview or "focus group," that only emerge during in-context observations of people doing real work in their workplace. Such things can have major implications for product user interface design.

Here is another example, this time from the context of a customer service application in an insurance company. In the navigational model illustrated in Fig. 47.2, the free text in normal intensity represents the basic user tasks identified from initial in-context observations with users. The bold text in boxes represents the hierarchy of *groups that the design team developed*, with labels they created, again, all based on initial task analysis activities. (This hierarchy represents only a partial set of the functionality required by a Customer Support Rep. For simplicity of the example, not all groups or low level tasks are included.)

In contrast to the *designer*-generated model in Fig. 47.2, in Fig. 47.3 is a *user*-generated model. Again, the free text in normal intensity represents the low-level tasks presented to the user, each one on a separate index card. This time the bold text in boxes represents the *groups that one user formed* from the

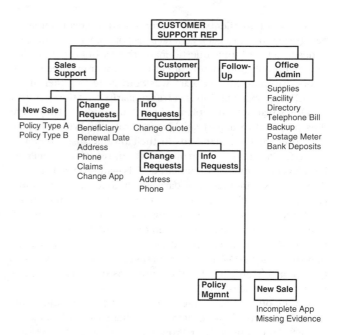

FIGURE 47.2. A designer's navigational model. Admin = administration; App = application; Mgmnt = management.

low-level tasks, with labels they suggested. (Again, this hierarchy represents only a partial set of the functionality required by a Customer Support Rep. For simplicity of the example, not all groups or low-level tasks are included.)

The differences between the designer-generated model and the user-generated model of these tasks are significant.

For example, the design team organized all types of Changes (e.g., Address, Beneficiary) under a single category ("Change Requests") within the category Sales Support, whereas this user distinguished between those changes that have to do with Customer information and those that have to do with Policy details, but located both under the category Customer Support rather than Sales Support. As it turned out, Customer Support staff can make simple changes to Customer information on their own,

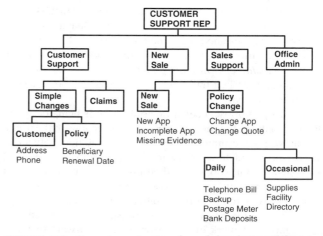

FIGURE 47.3. A user's navigational model. Admin = administration; App = application.

but more complex paperwork and approvals are required for Policy changes. Thus, these are two distinct categories of tasks to users. Even though requests for these types of changes often come from customers via Sales Agents, this user still regarded them as really reflecting Customer Support.

Similarly, the design team lumped all Office Administration tasks in one group, whereas this user distinguished between Daily and Occasional office tasks. Different people get assigned to these types of tasks, so again, this is a meaningful distinction to users.

The designers imagined a category unto itself called Follow-Up, where incomplete tasks of all types (e.g., Incomplete App) are all located, whereas this user considered follow-up activities as belonging with the original task type they are associated with.

Finally, note that the user regarded New Sales (data entering new policy applications) as its own category, wherein the designers had seen this as a subcategory of Sales Support—a very different perception.

One important thing to note in comparing this user's model of tasks with the designers' first pass at a model of tasks is that for any set of low-level tasks, there may be a *large* number of very *different "logical"* task organizations possible, but there will only be a *small* number of rather *similar* ones *that will make sense to users*, given their actual work. A traditional systems analysis will uncover all the low-level tasks, but it typically does not ensure that the *task organization* most consistent with the users' models of their tasks will be presented to the user in the user interface to a product intended to support those tasks. That is the point of a task analysis.

Also note that this user's model is most likely not exactly the same as other users' models, although in most cases, there will be a lot of similarity across individual users' models. It is necessary to *consolidate* all the sampled users' task models and generate one consolidated model (in the same format as the individual users' models, as illustrated in Fig. 47.3) that captures all the across-user commonalties as well as possible.

One last example from my recent experience illustrates the importance of studying users' work in depth to design the user interface to a software application that supports that work.

In this example, the application was a database system intended to support users in a government agency. The mission of the agency was to uncover and prosecute criminal behavior of a specific type. The job of the users of the intended system was to manage publicity for their agency in the media. These users needed to interact with the media to enhance reporting on criminal cases for the purpose of using the media to help discourage criminal behavior and encourage public cooperation in reporting and investigating criminal behavior.

The database of information being provided to these users included information on criminal cases in progress relevant to this agency, prior press releases and news articles on relevant criminal cases, a "phonebook" of contact information for reporters and other contacts in the media, records of past communications with the media, etc. A card sorting exercise revealed that users considered these categories of data types—Cases, Documents, Media Contact Information, Media Communications—to be distinct and meaningful categories that would provide a good foundation for the navigational structure of the application. However, what was *not* revealed by the card sorting exercise, but *did* become clear through other methods of studying users' work in context, was that when users searched for and found a particular item in one of these categories—such as a Case—they then typically wanted to automatically see items in other categories related to the item they had initially looked up.

Based on the card sort data, I had initially designed a navigational structure in which the user first selected a category of data, then searched for items in that category. Assuming from the card sort data that a search for one type of data was independent from searching for another type, I designed the interaction such that searching for something within any category of data was independent of searching for something within any other category of data. That is, if the user searched for and found a Case, and then navigated to the category of Documents, or to the category of Media Contact Information, the initial design assumed that he wanted to start an independent, unrelated new search in that category. However, further discussion with users revealed that in fact when the user searched for and found a Case, it was most likely that she would then want immediate access to Documents *related to that Case*, or Contact Information *related to that Case*. Similarly, if the user initially searched for a particular reporter's contact information, then he most likely would be interested in then seeing all Documents *related to this reporter*, and/or all Cases *related to this reporter*. This required a very different navigational and interaction design. Even though card sorting data had been collected, without the ongoing input of users during early design, it would have been very easy to design the application in such as way as to make the users' most typical type of task very tedious and cumbersome.

Clearly, it is possible—even likely—to design a user interface that does not support users, their tasks, or their work environment if an in-depth requirements analysis is not conducted prior to design and incorporated into design. The chapters that follow describe and discuss a variety of techniques for gathering these all-important requirements specifications.

References

Beyer, H., & Holtzblatt, K. (1998). *Contextual design*. San Francisco, CA: Morgan Kaufmann Publishers, Inc.

Coble, J. M., Karat, J., & Kahn, M. G. (1997). Maintaining a focus on user requirements throughout the development of clinical workstation software. *CHI '97 Proceedings* [On-line]. Available: http://www.acm.org/sigchi/chi97/proceedings/paper/

Hackos, J. T., & Redish, J. C. (1998). *User and task analysis for interface design*. New York: John Wiley & Sons, Inc.

Mayhew, D. J. (1999). *The usability engineering lifecycle*. San Francisco: Morgan Kaufmann Publishers.

Rubinstein, R., & Hersh, H. (1984). *The human factor: Designing computer systems for people* (p. 26). Burlington, MA: Digital Press.

·48·

TASK ANALYSIS

Janice Redish
Redish & Associates

Dennis Wixon
Microsoft

Successful design comes from a marriage of users' goals and (usually) new technologies. Successful design does not necessarily perpetuate users' current ways of working or playing, but it is built on a deep understanding of those ways and of how a new design will change them.

In this chapter, we explore modern interpretations and uses of task analysis.

The first two sections are background:

1. Defining task analysis
2. Considering four principles that underlie our view of task analysis

The next three sections are a practical guide:

3. Planning for a task analysis (issues to consider)
4. Collecting task analysis data
5. Analyzing and presenting the data

In the final section, we present specific ideas for planning a task analysis and for collecting and analyzing task analysis data at different stages of the design and development process.

DEFINING TASK ANALYSIS

Task analysis has different meanings to different authors.

Task Analysis = The Entire Front-End, Predesign Process

Rubenstein and Hersh (1984) have a section in their book with the heading, "Task Analysis." They are referring, not to a specific method, but to doing what it takes to understand "who the people are, what they do now, and what the new system is expected to do for them" (1984, p. 25). They go on to say, "The procedures for task analysis are not yet standardized, but every task analysis should address [these] questions." Their list of 10 questions for a task analysis starts with "Who is the user?" and "What tasks does the user now perform?" Rubenstein and Hersh are using task analysis for the entire predesign process of understanding users and their work.

Task Analysis = One Element of the Front-End Process

Olson and Moran (1996), on the other hand, call the front-end process "defining the problem." The goal of defining the problem in Olson and Moran's process is to understand the users' tasks, but they reserve the name task analysis for one set of methods. Thus, they contrast task analysis with other methods, such as naturalistic observations, interviews, and scenarios or use cases, all of which Rubenstein and Hersh would consider to be part of a task analysis. To Olson and Moran, task analysis is one way to represent the information from observations and interviews. Many of the other techniques used by Rubenstein

and Hersh are also part of Olson and Moran's toolkit of usability activities. They just have names other than task analysis in Olson and Moran's list.

Task Analysis = Many Techniques That Come Into Play at Different Times During the Design and Development Process

Kirwan and Ainsworth (1992) list 41 techniques for task analysis and describe 25 in some detail. Some of the 41, especially *observations* and *charting*, are themselves categories that comprise several techniques. These 41 techniques span the development process from what Kirwan and Ainsworth call the concept stage to what they call the stage of operation and maintenance. Thus, Kirwan and Ainsworth take a broader view of the place of task analysis in the entire design and development process than either Rubenstein and Hersh or Olson and Moran, but they confine task analysis to specific techniques for gathering and presenting information about work. Their focus, also, is on safety critical work, such as nuclear power plants.

Task Analysis = One Element of the Postdesign Process

Other authors (e.g., Card, Moran, & Newell, 1983; Payne & Green, 1989) focus on detailed task analyses for evaluating existing designs. Payne and Green offer TAG (Task Analysis Grammar) as a way to predict the cognitive complexity of an existing design by assessing the number of rules that are embedded in a user's understanding of a design's fundamental operations. Card et al.'s GOMS (goals, operators, methods, and selection) is a method for doing an extremely detailed analysis of how a user would do a task in a specific design.

In contrast to Rubenstein and Hersh's view of task analysis as looking in depth at the user's world before design, these techniques apply to already-designed products and focus only on efficiency of action.

We might note that any technique for evaluating an existing design can be used at the front-end of a new design process as long as there is a previous product to evaluate. In that way, these detailed, efficiency-oriented task analyses could be used before designing a new version of a product.

Task Analysis in This Chapter

Our view (Hackos & Redish, 1998; Whiteside, Bennett, & Holtzblatt, 1988; Wixon, Holtzblatt, & Knox, 1990) fits between Rubenstein and Hersh and Olson and Moran. Task analysis means understanding users' work or play.

Thus, task analysis encompasses all sorts of techniques, including naturalistic observations and interviews, shadowing users or doing "a day in the life of" studies, conducting contextual inquiries, and observing and listening to users doing specific tasks. It includes gathering information that leads to insights about users' life at work or play, to scenarios and use

cases, and sometimes to detailed flowcharts of work processes or specific procedures.

In our view, a major emphasis of task analysis is predesign, and three types of analysis—user, task, and environmental—are necessary input to designing any product. Task analysis is, therefore, an integral part of a triangle that covers users, tasks, and environments. As described in more detail, task analysis goes hand-in-hand with understanding users (user analysis) and understanding the users' physical, technological, cultural, social, and political environments (environmental analysis).

Users are absolutely critical to all three types of analysis. In our view, task analysis requires watching, listening to, and talking with users. Other people, such as managers and supervisors, and other information sources, such as print or online documentation, are useful only secondarily for a task analysis. Relying on them may lead to a false understanding.

Like Kirwan and Ainsworth, we also believe that task analysis does not stop with design. Task analysis continues to be critical at every stage of the design and development process. Task analysis is the major input to use cases and design specifications. Task analysis helps us understand how the emerging product affects users. It is the key to evaluating designs, as scenarios for heuristic evaluations and for usability testing. Task analysis must be the organizing principle for documentation and training.

We recognize that efficiency-oriented, detailed task analyses, such as TAG and GOMS, have a place in evaluating some products, especially those for which efficiency on the order of seconds saved is important (see, for example, Gray, John, & Atwood, 1993). However, that type of task analysis is not the focus of this chapter. The focus here is a broad understanding of the world in which the new product will be used.

CONSIDERING FOUR PRINCIPLES THAT UNDERLIE OUR VIEW OF TASK ANALYSIS

The practical advice for doing task analysis later in this chapter is based on these four principles:

1. Task analysis is an integral part of a broader analysis that also includes understanding users and their environments.
2. Task analysis includes understanding users' goals.
3. Task analysis is relevant at all stages of the design and development process, although the focus, methods, granularity, and presentation of information may differ at different times.
4. The practical reality is that what you do as a task analysis at a given time for a given project depends on many factors.

Task Analysis Is an Integral Part of a Broader Analysis

The first principle is that task analysis by itself is not enough to give you the understanding that you need to design or evaluate a product. The methodology you need here is triangulation—bringing together information about three interwoven elements: users, tasks, and environments.

Figure 48.1 shows how all three of these analyses must come together as you triangulate what you learn to gain the

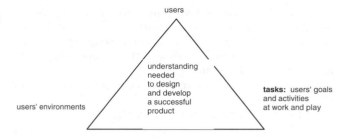

FIGURE 48.1. Triangulating information about users, tasks, and users' environments.

understanding you need. Task analysis is one corner of the triangle of understanding that you need. The other two are:

- Users
  —Who are they?
  —What characteristics are relevant to what you are designing?
  —What do they know about the technology?
  —What do they know about the domain?
  —How motivated are they?
  —What mental models do they have of the activities your product covers?
- Users' environments
  —Physical situation in which the work or play occurs.
  —Technology available to the user (what you might need to find out about could range from what modem speed users have to how often power is interrupted to the cost of upgrading equipment to finding an opportunity for your product because the users' environment favors low technology).
  —Social, cultural, language considerations—what will make the new product acceptable in the users' world? How will the new product change the users' world? How will you help the users make the transition from the old world to the new world?

Task Analysis Includes Understanding Users' Goals

The second principle is that a task is what someone does to achieve a goal.

Considering Norman's Entire Action Cycle as Task Analysis. As Donald Norman explains (1988, p. 46), "to get something done, you have to start with some notion of what is wanted—the goal that is to be achieved." It is true, as he also says (1988, p. 49), that we cannot always articulate our goals clearly. However, in general, we do start with goals, such as,

- making the family happy by getting dinner on the table
- getting a draft of a paper to a co-author in a different city

Norman also describes how we go about trying to meet our goals, that is, how we act. He gives us the seven-stage cycle shown in Fig. 48.2.

The Action Cycle

1. Forming the goal

2. Forming the intention

3. Specifying an action

4. Executing the action

5. Perceiving the state of the world

6. Interpreting the state of the world

7. Evaluating the outcome

FIGURE 48.2. Donald Norman's view of how people get things done (1998, p. 48).

Task analysis is concerned with all the stages of the action cycle.

We can think of forming the intention as deciding what tasks to do to meet a goal. We can think of specifying an action as figuring out what steps and decisions to take to carry out the task. Executing the action would then be actually doing the steps and carrying out the decisions that we have decided are the way to complete the task.

We can think of perceiving the state of the world as seeing the results of the steps and decisions that we took and interpreting the state of the world as relating the results to our understanding of how to carry out the task. Evaluating the outcome would then be judging whether we have successfully met the original goal.

Starting With Task Analysis as Independent of Any Specific Device. An argument within the human–computer interaction (HCI) community has been whether task analysis can be *device independent* (Benyon, 1992; Diaper & Addison, 1992). The problem is one of definition, of which levels you accept as being part of a task analysis.

Benyon calls the goal level (Norman's first stage), the external task, and he would exclude it from the actual task analysis. He would start the task analysis with what he calls the internal task, which begins only after the user has *formed the intention* in Norman's terms. Thus, Preece et al. (1994, p. 411, italics added) define a "task (or internal task) as the activities required, used, or believed to be necessary to achieve a goal *using a particular device*." They thus separate task analysis from work analysis.

However, if we start at Norman's second stage, we are missing a very important part of how users work. If we call Stage 1, work analysis, and the rest, task analysis, we are missing the essential connection between the two. In HCI, if you are designing

a product that you want users to find useful and usable; you must take into account the users' goals, and the entire social context in which those goals are embedded. Part of the design process is to understand what the device must do to help users achieve their goals in ways that give greater value to the user than any available alternative, that is, to understand how to make users want to *form the intention* to use your product.

If we start task analysis only with Norman's second stage, we run the risk of being device-driven in the design rather than being user-centered. Even if the device is predetermined, for example, if we know the solution has to be a software program on a particular platform, working from users' goals is necessary.

Moreover, as Barbara Mirel (1996, 1998) points out, for many users, the issues of usefulness and usability are at a higher level than how to do a specific procedure with a specific device. The problem these users have is that they cannot see the relationship between their problem (their goals) and the solutions (resources, intentions) that are available. They cannot figure out how to form intentions that will lead to tasks that will support their needs. Mirel is talking primarily about managers who have to draw on a variety of information sources to complete tasks "embodying choices, problem-solving, or conditional steps" (1996, p. 16). An example of a user's goal (1996, pp. 16–17) is "to retrieve, relate, and report financial, production, and personnel data in order to persuade [a] manager to allocate effort and resources differently." HCI specialists need to be able to use task analysis to develop solutions for these complex situations, as well as for the more easily decomposable tasks that are traditionally thought of as task analysis.

Therefore, to us, task analysis starts at a device-independent stage and begins by focusing on users' goals.

Task Analysis Is Relevant at All Stages of the Process

Our third principle is that task analysis belongs everywhere in the process of planning, designing, developing, and evaluating a product. Task analysis, like so much else in the user-centered design process, should be done iteratively. The focus, methods, granularity, and presentation may change over time as different questions and different types and levels of information become more or less relevant.

Table 48.1 shows the types of questions that task analysis might help answer at different times in the product life cycle.

Practical Reality Impinges on What We Actually Do

Our fourth principle is that in the fast-paced world of software and web design, in reality, what we can do for a task analysis (or any other aspect of user-centered design) depends on many factors. These factors include:

- time
- budget
- people
- availability of users to observe and talk to
- travel restrictions.

TABLE 48.1. Task Analysis Questions at Different Times

Stage	Examples of Questions That Task Analysis Should Be Used to Answer
Strategic planning	Why would someone or some organization choose to use this product?
	What goals in their world would this product help to meet?
	What benefits are most meaningful and valuable to users?
Predesign	What are the alternatives currently available and technologically possible that would address the "why" listed above?
	How do users achieve relevant goals today?
	How could our product make that easier?
Information architecture	What do users know and what are their environments?
	How do users organize their world?
	What vocabulary do users use today for their goals and tasks?
	How can we incorporate that vocabulary?
Concept design	What metaphors are familiar to users?
Interface design	What do users know about interface conventions?
	How does the task flow of the new product match users' expectations from their current work?
	If we are changing users' task flows, how can we build in help for transitions?
Early prototypes	What tasks should we provide to "heuristic reviewers" and usability testers?
Development	What does the user know that would address the problems we have uncovered?
	What changes should we make to the interface and information to better match users' expectations for work or play?
Postrelease	How well does this release match the user/business needs that we uncovered in the original strategic planning phase?
	Are users now better able to achieve their goals than they were before they had this product?

This chapter is meant to help you decide on the best approaches to consider for whatever situation you are in.

In the next part of this chapter, we consider issues and methods for planning a task analysis, collecting task analysis data, and analyzing and presenting that data.

PLANNING FOR A TASK ANALYSIS (ISSUES TO CONSIDER)

Getting Into the Project Plan

A critical aspect of being able to do task analysis throughout a project is getting this and all other usability activities into the project plan. The extent to which this can happen depends, of course, on other factors, such as whether a project plan exists and in how much detail it is specified. The more strongly a project team uses a formal project plan, the more critical it is to get usability activities, such as task analysis, into the plan. Time, resources, and respect from managers and developers for the information may be dependent on being part of the formal plan.

Another approach that many usability specialists follow is to create a usability project plan that parallels the system design and development project plan. That is fine if the system people understand and respect the parallelism of the plans.

Getting Sign-Off and Meeting Schedules

Whether the usability plan is part of the overall project plan or a parallel track, it is important to get sign-off from the rest of the design and development team with respect to

- activities that the usability team will do
- resources needed for those activities
- information that will be brought back from those activities
- deliverables (formal or informal) that will come from that information
- dates for those deliverables.

As usability specialists, we must acknowledge that, at least in the United States, most software and web development projects are schedule-driven. Task analysis and other usability activities must deliver information in a timely fashion at the right moments in the project schedule.

Getting into the project plan early can help set schedules that allow time and resources to do task analysis and other usability activities. Another way that some usability specialists accomplish this is to take responsibility for elements of the project plan, such as the user interface specification, and use that as the way to integrate usability data into product design and development.

Providing Useful and Usable Data

As usability specialists, we help project teams only if we provide useful and usable data when they need it. Therefore, staying closely aligned with a project team to know the plan and schedule (and when and how that plan and schedule change) is critical.

What else, besides timeliness, makes data useful and usable?

- Data the project team needs—As usability specialists, we should approach any project looking for where the team needs data about users (whether they know it or not) and plan to collect, interpret, and present that data in a way that the team can directly use.
- Data that is credible—Time pressures and limited resources often curtail the extent of any usability activity, including task analysis. In general, we follow the maxim that some data about users and their work is better than no data. Practical

experience tells us that is true, as long as the data are accurate and representative. Limited data can still be good data. Good data is collected in a systematic, careful, rigorous way from appropriate users.

Deciding What the Project Team Needs

The task analysis that you want to do will depend in part on the type of product or Web site that you are working on. Consider at least these four factors as you think about the project for which you are planning task analyses and other usability activities.

Where Is the Product in Its Overall Life Cycle? For example, is the team

- upgrading an existing product without changing the medium (a new software release; a revision of the Web site)?
- changing business processes or medium (going from a "legacy"—DOS-based or green-screen product—to a graphical user interface or to the web)?
- developing something totally new?

How Broad or Specialized Is the User Population for the Product? Is the product for a

- very broad public market?
- niche business market where you can easily define the user population and access to them is through account executives (or similar)?
- special audience? (children? the elderly? persons with disabilities?)

How Widespread Geographically, Culturally, and Linguistically Is the User Population for the Product? Is the product

- Global? How far? How many countries? Cultures? Languages?
- Local? (some products are still used only in a particular country, but even within one country, you are likely to find differences in culture and vocabulary).

How Detailed Must We Specify the Tasks? Is the product

- safety-critical where tasks are very specific and must be done in specified ways? Do users receive training until they prove their competence in completing the tasks accurately and efficiently?
- used by many different types of people for different tasks that they may do in different ways?

Is This a Special Type of Product for Which Traditional Task Analysis May Not Be Useful? Some applications do not fit the traditional approach for task analysis. These are primarily applications that the user does for fun (e.g., games).

Traditional task analysis would not aid much in the design and development of these applications. Games (both technology-based and real-world games) present players with a defined goal

and a set of constraints. The fun is in achieving the goal without violating the constraints. The fun is more in the process than the actual outcome. The outcome exists to motivate the process.

The more interesting the game, the less amenable it seems to be to task analysis. Interesting games have a large number of possible ways to win and present a complex set of constraints. These make the application of task analysis extremely difficult.

Game designers do not typically do task analysis. They are more interested in mood, theme, story, drama, progression, surprise, pacing, and the physical correlates of these experiences. The conceptual space in which they operate and their way of thinking about how to create a fun experience is fundamentally different, perhaps even antithetical to, the analytic approach that is typical of task analysis. (Also see chapter 46 in this handbook, User-Centered Design in Games.)

Deciding on an Appropriate Level of Granularity

Another aspect to consider as you plan a task analysis is the types of analysis to do. Understanding users' goals and their work or play can be done at several different levels. You might be interested in one or more of these types of analysis:

- Analysis of a person's typical day or week ("a day in the life of" or "an evening at home with")—This is probably most needed for the early stages of strategic planning and predesign.
- Job analysis (all the goals and tasks that someone does in a specific role—daily, monthly, or over longer periods)—Again, this is probably most needed for the early stages of strategic planning and predesign.
- Workflow analysis (process analysis, cross-user analysis, how work moves from person to person)—This may be useful in strategic planning and predesign, especially if you have ways of improving workflow among users. It is also needed in concept design and interface design.
- High-level task analysis (the work needed to accomplish a large goal broken down into subgoals and major tasks)—This is needed in information architecture, concept design, and interface design.
- Procedural analysis (the specific steps and decisions the user takes to accomplish a task)—This is needed at the interface design stage and beyond.

Deciding Where to Start

You must first understand how far along in the process the project is.

Unfortunately, by the time usability specialists know about the project, the strategic planning and predesign stages may already be considered closed. The project may be at considerable risk if the strategic planning and predesign questions were never answered or were answered based on speculation or internal discussions without users. However, it may be unproductive for usability specialists to spend time and effort collecting data that speaks to those questions if no one is willing to listen to the answers they bring back. A more productive use of the limited

time and resources for task analysis and other usability activities would be to understand where the project is and how to influence it from that point forward.

Gathering Reusable Data

Time, resources, and costs are likely to limit the number of times you can return to users for task analysis. Also, users do not think in terms of the information needs of a project team or the different stages of design and development of products they might use in the future.

One good approach might be to collect extensive data about users' work in a relatively holistic way, capturing that data on video, audio, or in extensive notes so that you can return to the data—rather than to the users—with different questions in mind at different times.

To do this, an open-ended field study method combined with detailed information gathering is best (Wixon, 1995). Also, having a relatively detailed log of the raw data is necessary (videotape, audiotape, or verbatim transcripts).

Going to Different Users at Different Times

Although the number of times you may go out to users is likely to be limited, we have also said that you are likely to want to do site visits at different times for different questions. Even if you gather extensive data holistically on early site visits and return to the data, you may not have the information that you need. In that case, go out again, and go to different users.

You can use each set of site visits not only to answer the specific immediate issues and questions, but also to enrich the team's general understanding of users, their work, and their environments. Although your immediate focus may be a specific why or what or how question, always drill down so that you are, in fact, seeing the why behind the what or the what and how behind the why.

COLLECTING TASK ANALYSIS DATA

Now that we have considered some issues in planning to do a task analysis, let us discuss ways to work with users to do a task analysis.

A Bit of History

Traditional task analysis had its roots in time and motion studies (Taylor, 1911; see Preece et al., 1994). Observers with stop watches timed workers on assembly lines to find more efficient ways to accomplish the work. The workers were just another part of a system that was part mechanical and part human.

Modern Roots in Ethnography. Modern task analysis, on the contrary, relies more on ethnography and cognitive psychology than on time and motion studies. As Redish writes, "user and task analysis, like ethnography, is about developing as rich an understanding as possible" (Hackos & Redish, 1998, p. 14).

Task analysis today relies primarily on qualitative methods of data collection and analysis. You may report frequency of activities (how many people do the task this way), and at some points in the design and development process, timed studies of tasks may be relevant (how fast users are able to do a task in their current environment). However, the primary methods today are ethnographic—observing, listening, and talking with users in their own environments as they do their own work. These methods were brought into HCI by a group at Digital Equipment Corporation in the late 1980s and early 1990s (Whiteside et al., 1990; Wixon & Comstock, 1994). (Also, see chapter 50 in this handbook, The Ethnographic Approach to Design.)

Of course, a typical user and task analysis is not really ethnography. Redish explains the difference this way:

Ethnography usually means spending a year or two immersed in a different environment. Half a day with each user and a few weeks for the entire visiting time isn't the same. Ethnography, moreover, is usually only about describing the culture. A user and task analysis has a goal that goes beyond description. The point of a user and task analysis is to design a product that may in fact change the culture being observed.... [W]e aren't really doing an ethnographic study . . . when we do user and task analysis. But we are making practical use of ethnographic philosophy and techniques, adapted to the goals of designing products and the constraints of product schedules and budgets (Hackos & Redish, 1998, p. 15).

Modern Roots in Cognitive Psychology. The other source of philosophy and techniques in modern task analysis is cognitive psychology, the study of how people think and learn. Work is not only about action; it is about decisions on when and how to act. The decisions are made by people, and, therefore, people are active parts of any system.

We do user and task analysis because, in fact, it is often the users who are much less predictable and less well understood than the technology. As we now know,

[u]sers come to any new product with preconceived ideas based on their prior experiences. They interpret what they see in an interface and draw their own conclusions about how it works that may be very different from the designers' intentions—and then they act on their conclusions, not on the designers' intentions. Cognitive psychology shows us that we must accept the users as reality because it is they and not the designers (nor their supervisors) who will in the end determine how the product is used (or not used) (Hackos & Redish, 1998, p. 15).

From cognitive psychology, we understand that we must focus on what is happening in users' heads as well as with their hands, and we adapt the technique of think-aloud protocols (having users talk out loud as they work).

Preparing Philosophically to Work With Users

From ethnography and cognitive psychology, we also take our ideas of how to work with users when we are doing a task analysis. Anyone who works with users to gather data for a task analysis should abide by these five principles (inspired by and adapted from the work of Beyer & Holtzblatt, 1995; Holtzblatt & Beyer, 1993; Whiteside et al., 1988; Wixon et al., 1990; Wixon & Jones, 1996; Wixon & Ramey, 1996).

1. Treat the user with respect.

   We are there to understand who users are, what they do and how they do it, and what they value. No matter how differently they do their work (or play) from what we expected, we must respect them, their culture, their actions, and their decisions.

2. Understand that we are visitors (guests) in the users' environment.

   We are there to learn about their reality, not to bring our reality to them.

3. Be in learning mode.

   We are not there to train or teach or demonstrate or sell a new product. (This part is often difficult for engineers, training specialists, marketing specialists, and others when they first start doing site visits. They are used to going to customers for these other reasons. Before going out on a task analysis site visit, they must understand the difference in the reasons for the visit and they must become comfortable with the appropriate frame of mind and behavior.)

4. Let the user be a mentor.

   If we are there to learn, we must watch and listen more than we act or talk. Especially in work contexts, one way to do this is to suggest to the user a relationship of mentor and trainee—user as mentor, site visit observer as trainee. Users are often unsure of how to deal with the strangers in their midst. They, too, have more often been visited by people training them or demonstrating something to them or marketing something to them. Helping them to get into an appropriate relationship is useful, and most users are happy to become the trainer or demonstrator.

5. Be a partner in the task of understanding the user's world.

   Use a site visit to test out your assumptions. Clarify your understanding of what the user is doing and saying as you go along. Capture the users' words. Do not translate as you take notes.

Selecting Users and Environments

Companies have sometimes commissioned large-scale task analyses as part of their long-term strategic planning; that is, outside of the development cycle of a particular product or version (see, e.g., Dray & Mrazek, 1996; Redish & James, 1996). That is terrific when it happens.

However, most task analyses today are small-scale studies, especially when they are part of a product-oriented project. The rapid pace of design and development often does not leave time for site visits to more than a few places. The scale is similar to that used in iterative usability testing—six to eight users per study. As with usability testing, it is better to do a few site visits each time for different purposes at different times in the project than to do many up front and then not have resources to go out for other reasons later. Also, as with iterative usability testing, even though each study is small, over the course of the project, you may see many different sites and users.

Make a Convenience Sample as Representative as Possible. A task analysis study almost always uses a convenience sample. However, care should be taken to make the sample as representative as possible within the constraints of time and budget (see Hackos & Redish, 1998, especially chapters 2 and 7 on defining your users and then selecting appropriate ones for a user and task analysis).

Because it is almost impossible to find one user who truly represents the entire spectrum of users, spending some time with each of a few users is better than spending all the allotted time with just one user. Characteristics on which you might want to base representation include:

- Size of the user's company (which may affect task specialization among users, amount of support available to users, technology available to users, etc.)
- Experience in the domain and with the medium being contemplated for the product
- Gender, age, and background (which may correlate with motivation to learn new ways of working and with interests and values related to technology)

Do Not Limit Task Analysis to Development Partners. Many companies today have "development partners" who agree to work with them throughout the project. Development partners are usually major customers who get to influence the new product in exchange for allowing their people to participate in site visits and usability tests, as well as other activities, such as customer focus groups and beta tests.

Working with development partners is a great idea. However, in many cases, development partners are at the tail end of the distributions for both size and sophistication within the company's market. What works for them may not work well for the company's many other customers. In most cases, restricting site visits to the development partners raises the risk of the product failing to meet the needs of other customers.

If the project you are working on has development partners, but the company also wants the product to be used in many smaller and less sophisticated places, push to include others in usability activities like user and task analysis. In addition to making project managers realize how unrepresentative users in the development partners may be, you can also often make a cost-benefit argument based on plans for implementing the product in different environments. The company is likely to implement the new product for the development partners with a lot of hand-holding, but the wider distribution to the many more but smaller companies must eventually happen without that hand-holding. Getting the product to be usable for the broad majority requires doing task analyses (and other usability activities) with representative users from those smaller companies.

Note that you can, of course, set up a situation in which the development partners or customer partners are in fact representative of the range of customers and users (see Hackos, Elser, & Hammar, 1997; Hackos & Redish, pp. 143–144). However, in many projects, that is just not the case.

Conducting Site Visits

What happens when you actually meet the users? How should you act when working with users?

The two main techniques for collecting task analysis data are observations and interviews. In a typical site visit, you combine both; and, when possible, you do them together—conversing with the user about the work as you observe and listen to the user doing the work.

In certain situations, you cannot converse with the user while the user is working. This may happen when

- the situation is safety-critical (e.g., with air traffic controllers [Means, 1993])
- the user is interacting with a client (e.g., in some situations with retail salespeople or travel agents [Redish & James, 1996])
- the users do not want to be interrupted in their work (e.g., with radiologists [Ramey, Rowberg, & Robinson, 1996])

In those cases, you may be able to talk about the work as soon as the user finishes the task (immediate recall) or at a later time using information from videotape or other artifacts to stimulate recall (cued recall). In situations that depend on the user interacting with a customer, you might have one member of the site visit team act as customer, thus having the user do a real task but giving you the opportunity to talk with the user during the task. (For these and other techniques for task analysis, see Hackos & Redish, 1998, especially chapter 6. For specifics on observing and interviewing during site visits, see chapters 9 and 10 in Hackos & Redish.)

Observing Users. What you take notes on and the level of detail of your notes depend on the stage the project is at and, therefore, the questions and issues you are addressing in this particular study.

In each site visit, you may want to spend all of your time with one user, the person responsible for the tasks you are investigating. You may want to spend time with different users at the same site if several people at the site do the same tasks, especially if they have different levels of domain knowledge, product experience, or technical skill.

You may also want to spend time with different users if you are interested in work analysis or process analysis (how work flows across users who do different parts of a process). You may want to spend time with the people who do each part of the process. If you have already watched them individually and drawn flowcharts of what you see as their process, you may want to bring them together in a conference room to go over your process flow and verify it or discuss it or get answers to questions you have about it.

If you want to get details of steps and decisions to flowchart a procedure, you may want to slow the user down and discuss each step and decision, asking questions about other situations and how they might change the steps or decisions and about the frequency of different situations.

In all cases, as you observe and listen, you must remember the five points we made earlier in the section on preparing philosophically to work with users as well as these four:

1. Be friendly but neutral.
   Be aware of how you give messages with your body, as well as with your words.

2. Be aware of assumptions that you brought with you.
   As you watch and listen, you are verifying those assumptions or changing your understanding of the user's reality.
3. Ask questions to clarify.
   Restate what you think you heard. Your goal must be to bring back the user's understanding of the work. It is very easy to think that you understand when you are in fact filtering what you see and hear through your own view of the work.
4. Note the user's words.
   One goal of a task analysis is to understand the user's vocabulary for objects and actions. Again, it is very easy to translate what you hear into the development team's words without realizing that you are doing it—if you are not attuned to the need to capture the user's words.

Interviewing Users. Most of the talking in a task analysis should be in the context of a conversation with the user about the work during the work. However, there may be more general questions that you want to ask.

In general, interviewing after you have observed is better than before. You will have the context of what you have seen and heard as a frame for the other questions.

The two most important guidelines for successful interviewing are:

1. Ask questions in a neutral manner.
   Do not lead the user. Try to put the questions in a behavioral context rather than as a simple matter of like or dislike. If you say, "We are thinking of adding [feature] to the product. Would you like that?," you are suggesting that the user say, "Yes, I'd like that." If you say, instead, "If the product had [feature], would you use it?" and follow that up with "When and how would you use it?," you are likely to get a more informative answer.
2. Listen far more than you speak.
   You can keep a conversation going with prompts and probes that send the message that you are listening and want the user to talk more. (See Hackos & Redish, 1998, chapter 10, especially pp. 279–291 for more on interviewing and listening skills for user and task analysis.)

Whenever possible, make an interview behavioral rather than attitudinal. For example, if you have brought along prototype screens of the new product to show as part of your site visit, do not just show them and ask what users think. Make that part of the site visit into a mini-usability test even if it is in a conference room and not at the user's desk. Ask the user to walk through the screens doing a realistic scenario.

With interviews, as with observations, clarify what you are hearing, restate so that you know you are getting the user's understanding, try not to filter through your preconceived notions, and do not translate the user's words into your company or product jargon.

Bringing Users to You (Field Studies in the Lab)

The best task analysis, of course, is done in the users' context. However, time constraints, travel restrictions, or security restrictions may make it impossible for you to go to the users.

You can still do task analysis. Have the users bring their tasks to you.

Although you will not get a true environmental analysis (seeing the users work or play in their own settings), you can still get a lot of data about users and tasks. Butler and Tahir (1996), for example, conducted what they called "usability roundtables," in which users (individually or with colleagues from their work) brought their own artifacts and walked and talked the usability specialists through their tasks.

For detailed procedural task analysis (getting down the steps and decisions in completing a specific task), the lab may be the right environment. If you want to capture details, you probably want a videotape record. It is easier to videotape in the lab than to cart equipment to users' sites, set it up, and take it down, although we do recommend videotaping for most task analysis site visits.

If you want quantitative measures, such as time to complete a task under ideal conditions, the lab may again be the best place to do the study. Someone might ask, in that case, whether we are doing a field study or a usability test, but in fact the line between the two techniques is quite blurry. Both can be done in the field or in the lab. Both can be done on old products or new prototypes. Both can be done qualitatively or quantitatively. Both can be done on the user's own work or on scenarios given to the user. Task analysis is a major part of both field studies and usability testing.

ANALYZING AND PRESENTING THE DATA

Once you have collected the data, of course, you need to analyze it and present it. Data is of no value if you do not communicate what you have learned to the people who need the information. To make the data useful, you must bring the data together, think about what you have learned, and draw out implications.

Analyzing the Data

Consider these four principles as you plan how to analyze the data:

1. Involve the design team.
   Making the effort to involve the rest of team pays off handsomely. Involving other team members ensures that they have a stake in the results. It also allows them to work with the raw data, which helps them internalize the "work of the user" more completely, even if they did not get to participate in the site visits. When you involve the team in the analysis, you build a shared understanding of what was seen and heard at the customer or user sites. It also ensures that the questions the team has get answered. Often teams refine and redirect their thinking as they go through an analysis. They may drill down more deeply into the data. They may completely change the questions they have of the data. They may change their thinking about the direction they had planned for the product.
2. Make it traceable.
   Any analysis should include references back to the raw data. There are many advantages. First and foremost, keeping the link ensures the integrity of the analysis. It is important to be able to say to those who were not involved in collecting and analyzing the data that all conclusions are traceable back to statements by users or direct observation of their behavior. Second, as we noted in discussing the first principle, interpretations may change as analysis progresses. During that process, it is important to be able to revisit the data and recall the context of the behavior or the comments. If you set up ways of tracking the data through analysis, the extra time and effort need not be substantial.
3. Make it visible and accessible.
   The analysis may be complex and detailed, but a report laden with text will almost certainly go unread. There is not time in the rush of a project for people to read. Therefore, many teams choose to display their analysis on a wall or in a "war room," where team members can review and add comments to the display (see, e.g., Simpson, 1998). An alternative is a hyperlinked document in which higher level conclusions are linked to more specific analyses. Often, the analysis is graphical so that designers can stand back and see patterns in the data. Some people create multilevel documents with a one-page summary, supported by a three-page overview, which in turn is supported by a 50-page report.
4. Match the form to the questions, the stage, the team's needs.
   The cardinal rule of all documentation is to give users what they need in the form they need it when they need it. That's why most technical communicators have moved from writing extensive tomes that people do not open to helping teams bring communication into the interface. The same principle applies to the internal working of any project team. The best form in which to represent the data depends on many factors, including the questions that were asked, the stage in the project's life cycle (i.e., how the information will be used), the time in which the information is needed, and the team and company culture.

Usability specialists should keep in mind the essential purpose of any analysis. Analysis provides an anchor from which designs can be generated and against which they can be evaluated. The analysis is not an end in itself, and the specialist must always keep in mind the need to keep the design team engaged with the analysis and with the representations of the analysis. In the next section, we describe a few of the possible representations.

Presenting the Data

Here are eight ways to present task analysis data (also, see Hackos & Redish, especially chapter 11. Miles & Huberman [1994] also describe the rich variety of ways to organize qualitative data for interpretation).

Affinity Diagrams. Affinity diagrams are hierarchical pictures of user data. They are produced inductively by grouping similar data elements together into categories and then grouping the categories together. Figure 48.3 is an abstract illustration of an affinity diagram in progress.

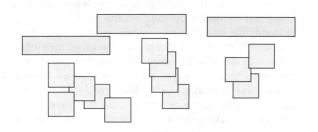

FIGURE 48.3. An abstract illustration of an affinity diagram in progress.

Affinity diagrams derive much of their value from the process that produces them (i.e., a deep engagement with the data combined with recurring reflections on the generalization that best captures a number of data elements). Also, teams often produce "collateral" elements while creating an affinity, such as design ideas and additional questions, that are captured and then used in design or further data gathering (see Beyer & Holtzblatt, 1998, for more on the process).

Figure 48.4 is an example of a small part of the result of doing an affinity diagram. Note how the team has kept the reference to the data with each bullet point (i.e., each note in the affinity diagram).

1.   I must FIND, MAINTAIN, **and** COMMUNICATE **with customers, and sometimes others.**

    1.1. *I need to keep in touch with clients, colleagues, and family.*

        1.1.1.   I must make sure my landline is forwarded--can't miss calls while I'm out.

- P 622 SW 6.2 Seq (Getting ready to leave): Turns off the phone so that the answering machine goes online.

- P 508 SW 5 Seq NOTE: He has a landline, but has forwarded it to his mobile phone so that he always can get calls when he's away from the office and also so that he doesn't forget to forward the phone when he leaves the office.

        1.1.2.   My cell phone is essential during the business day--it must be within easy reach.

- P 349 US 3 Notes He docks his phone on the dash where he can reach it easily.

- P 442 SW 4 Flow NOTE: The phone is everywhere in this flow diagram—as much as he is hitting the pavement, he is also hitting the wire.

- P 552 SW 5 Notes If I leave my phone at home, I've ruined my day. I usually have at least 1 visit, and I need the phone [to help if I get lost, to confirm the visit, and to keep track of other business when I'm traveling].

- P 628 SW 6.1 Seq 6.1 has her bag back there in the trunk too, but her mobile phone is in her hand. (Can't tell if earphone is on; it's not by the time we get out of the car in any case.)

- P 834 SW 8 Seq In the car he puts the mobile phone between his legs, resting on the seat. BREAKDOWN: I normally have the mobile phone put away in my pocket (or somewhere) and an earpiece in my ear so I don't have to mess with it while I drive. (ASSUME: There's usually a switch next to the hanging microphone to let you answer calls.) But this borrowed phone doesn't have an earpiece.

FIGURE 48.4. A partial example of results from an affinity diagram.

FIGURE 48.5. Ellen's appointment book as an example of an artifact.

Artifacts. Some types of tasks are deeply intertwined with their artifacts. For example, the task of making appointments is necessarily interwoven with calendars that show dates and times. As a result, it is often best to begin with existing artifacts (e.g., Ellen's appointment book) and to organize data around that representation.

The team may or may not decide that the representation is in fact the best metaphor to bring into the new product. Even if they do not, the artifacts and representation are often the best way to get an initial understanding for a task analysis. Figure 48.5 shows an example of an artifact that you might want to collect if you were creating a program to help people keep track of time, dates, meetings, tasks, etc.

Flow Diagrams. Flow diagrams answer questions about how information or artifacts flow through a system (process analysis). They illustrate the dependency between system elements or states of the system and what needs to transferred or moved from one part to another. They also show how roles are divided within an organization as data moves from one person or department to another or between the organization and outsiders.

Figure 48.6 is a small part of a flow diagram. The different patterns on the underlying circles represent different people (managers, nonmanagerial professionals, consumers, etc.) who are involved in this flow. Note once again that the references to the raw data are kept with the flow diagram so that the team can return to the data to understand more about the user and the context for each data point.

Personas or Archetypes. Personas or archetypes (user profiles) often describe the user's activities, knowledge, and tasks in some depth. Thus, personas or archetypes may integrate the two elements of users and tasks from the triangle presented

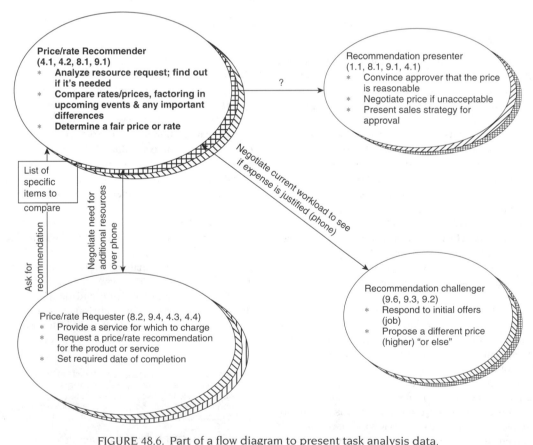

FIGURE 48.6. Part of a flow diagram to present task analysis data.

Picture of Julie at work

Julie Morris

Travel Agent
Globe Travel
Kansas City

- Specialty: Corporate Travel

- Experience: 5 years as travel agent
5 years with PRODUCT

- Technology: PC computer only at work
no computer at home
keyboard savvy
limited mouse experience
computer has Windows
but she spends most of day
in transaction-code based program

- Values: speed of transactions
finding bargains for her clients
knowing lowest fare

About Julie

- 35 years old

- travel agent for 5 years; was in retail sales before

- enjoys the phone contact with clients; likes helping people

- loves to travel, loves parks of being a travel agent

- likes the jobs but it is just a job

- took original required training but has not been to training class since

- no time for new training; after work she has family and other social obligations

- is comfortable with the computer, but thinks she doesn't know a lot about computers: doesn't like it when something goes wrong, for example with printer or fax — doesn't feel she could fix it

Picture of julie and her family skiing (from her desk at work)

Julie's work environment

Picture of Julie's desk and the office

Note the sticky notes all around Julie's computer monitor

- small office; three travel agents

- all help other, but others specialize in leisure and internationals, respectively

- work on salary, not commission

- on fifth floor of office building so get little walk-in traffic

- but keep their desks moderately clean as Julie's partner, Marla. who handles mostly leisure travel has many clients who make appointments to come in

- pleasant atmosphere; travel brochures; nice extra chairs for clients

- printer and fax machine are in small rooms in back to keep noise level down (because they are on the phone so much) — so printing or faxing means getting up from desk.

How Julie works

- clients call; usually know travel dates, destinations, other needs (common to corporate travel — note: leisure travel is different)

- Julie makes notes on paper in spiral notebook she keeps on desk — says it is useful to go back to if issue comes up about not getting arrangements night, she has notes of what person asked for

- knows transaction codes for air travel to most major US cities fast with getting availability

- if booking a car for something special — drop off in another city, getting it someplace other than airport, need something special like ski rack,etc will call car companies

- also book hotel online for major hotels in major cities for clients that repeat common visits; but often calls hotels because needs information about price, room. special request that aren't easy to do online

- asks other agents for help with transaction codes and her colleagues are helpful if they know — but often other person is also on the phone, so Julie can t ask and has to experiment and figure it out herself — or tell client shell do research and call back

- says that a major portion of her time is spent printing tickets and making sure that the right ticket gets in the right envelope and to the right client

FIGURE 48.7. An example of one type of persona (user profile).

in Fig. 48.1 of this chapter. Some people, in fact, also include information about the user's environments (physical, social, cultural, technological) in the description of personas, thus capturing all the triangulated data that we discussed at the beginning of the chapter.

Rich persona descriptions that encompass user, task, and environment information are particularly useful for commercial products, which often begin with market segmentation that classifies and describes potential customers. The task analysis builds on this data by characterizing these users more precisely (Lee & Mikkelson, 2000).

Because personas describe people, they tend to be memorable. Designers can ask "How would X respond to this design that I have just created" (see, also, Cooper, 1999).

Figure 48.7 is one of many ways of showing a persona. The real example would include photographs or sketches where indicated.

Scenarios. A scenario is a short story of a specific situation that is real and relevant to a user. A scenario gives the team the user's goal and specific needs. It often also gives the team the user's names for objects and attributes of those objects. It may give the team information on what the user values. (Is price more important than choice in renting a car, for example.)

Each situation that you observe on a site visit is a scenario. You can also collect scenarios by interviewing users through the critical incident technique (Flanagan, 1954) in which you ask users to recall a specific incident and then to tell you about it.

Figure 48.8 is an example of a scenario from a study of travel agents (Redish & James, 1996). (See, also, chapter 53 in this handbook on Scenario-based Design.)

You can elaborate a scenario with the sequence diagram (flowchart) of the procedure the user went through to accomplish the scenario. If accomplishing the scenario is difficult, and the scenario is important, creating a more efficient procedure could become a requirement for the new product.

Sequence Diagrams. Flow diagrams track work through a system or across people. Sequence diagrams use time to track the actions and decisions that a user takes. Sequence diagrams (procedural analysis) show what users do and when and how they do it. This type and level of information is critical for interface architecture and design because it gives us the functions, objects, and attributes of a system (e.g., menu items and dialogue box design) and the navigation for a Web site.

Client calls the travel agent. Client wants to go to Phoenix for a weekend to see her friend, but she is very concerned about the price. She would like to go sometime this month. She can be flexible about when to go and come back. She wants to know what the best fare is, if it is available this month, and what flights she can get for that fare.

FIGURE 48.8. An example of a scenario.

By laying out the sequence diagram that represents what users do today, you can often see ways to make the product help users be more efficient and effective. Thus, the sequence diagram of the reality of what you find in predesign observations is often elaborate and messy. It may be important for the team to see that reality as they work toward a more useful product.

Figure 48.9 is an example of part of a sequence diagram.

Tables. Tables are an excellent way to show comparisons and so are useful for presenting many types of analysis. Technical communicators, for example, have traditionally used a user/task matrix to understand which tasks are done by which types of users. The user/task matrix becomes a major input to a communication plan—to answer the question of what tasks to include in documentation for people in different roles (e.g., system administrators, end users). Tables can be used to show the relationships between any two (or more) classes of data. (See Table 48.4 later in this chapter for an example of a table with three classes of data.)

User Needs Tables. Kujala, Kauppinen, and Rekola (2001) have recently developed what they call a "user needs table" as a way to present a current task sequence along with the problems and possibilities that the designers should think about for each step in the sequence. They hypothesized that it would be easier to use findings about users' needs in design if the findings were connected to the task sequence that forms the basis of use cases. In their studies, designers found this type of presentation very useful in moving from data to requirements. They also found that writing use cases from this type of presentation keeps the use case in the user's language and keeps the use case focused on the user's point of view. Table 48.2 shows an example of a user needs table.

CONSIDERING TASK ANALYSIS AT DIFFERENT STAGES

In this section, we look in some detail at doing task analysis for different stages of the process. Remember that even if you come in after the first stage, you can still do task analysis. Just as it is relevant at all stages, task analysis is relevant at any stage. The questions, the methods, and the deliverables may vary, but the underlying principles are the same (see Table 48.3).

Strategic Planning and Predesign Stages

Prerequisites for a Task Analysis at This Stage

- Users: You must know or postulate who is likely to want or need to use the product.
- Purpose: You must have at least a broadly conceived understanding of what the promise of this product is.

For example, the product may be aimed at small and medium businesses and provide a systematic way of combining geographic data (maps) with financial information (sales). This is enough information for a usability specialist to begin doing user analysis, task analysis, and site visits with potential customers.

Questions for a Task Analysis at This Stage.
For a task analysis during strategic planning and predesign, you probably want to answer questions like these:

- How do users currently meet the goals and do the tasks that would be fulfilled by this new product?
- How many different ways are there for meeting these goals and doing these tasks in the users' world today?
- Are there any common elements in the way these tasks are done? What are they?
- What are the strengths (from the users' perspective) of the current methods?
- What are the weaknesses (from the users' perspective) of the current methods?
- How connected are these tasks with other user activities?

At the same time, you also want to understand the users themselves (user analysis): who they are, what jobs they have, what matters do them, etc. You also want to understand the contexts in which they would use your product (environmental analysis—physical, technological, social, cultural, political).

Use of the Data From Strategic Planning and Predesign Task Analysis.
The answers to these questions will determine the viability of the initial market definition and product concept. They will also suggest general directions for the product.

Two specific uses of this data might be:

- Assessing the variability in activities related to task performance. This gives the design team some idea of how flexible the design must be. If all users do the tasks in the same way, a "wizard"-like approach may be best. If both the sequence and types of activities are highly variable, a more "toolkit"-oriented approach may be best.
- Deciding the difficult issue of which functions to include in the product.

Presentation of Data From Strategic Planning and Predesign Task Analysis.
At these early stages, the goal is to provide the design and development team with as rich and detailed an understanding of users and their work as possible.

One way to do this is to have other team members go with the usability specialists on site visits. Team members who go on site visits should have at least minimal training in the methods being used so that they observe and listen neutrally, do not become defensive, hear what users say in users' words, and do not take the visit as an opportunity to train or demonstrate. Team members who go with usability specialists on site visits should participate in summarizing, interpreting, and presenting the data—both because their input is valuable and to clarify any differences in interpretation about users and their work.

In addition to actually taking other team members when you are gathering the data, other ways of getting information to designers and developers include:

Triggers, Steps, and Strategies

Identify new potential customers and get their contact information

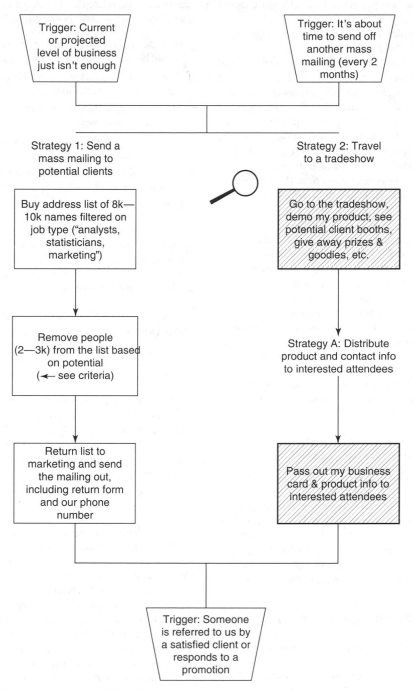

FIGURE 48.9. Part of a sequence diagram.

TABLE 48.2. Example of a User Needs Table

Task Sequence	Problems and Possibilities
Step 1: When trapped in an elevator, passenger makes an emergency alarm.	• Passengers want to get out of the elevator as soon as possible. • All kinds of passengers must be able to make an alarm call (blind, foreigners, etc.). • Sometimes passengers may make false alarms unintentionally. • Passengers may be in panic. • Passengers need instant confirmation that they have created a connection to the service centre operator and that they are going to get help.
Step 2: Unoccupied service centre operator receives the emergency alarm call and asks for information (description of the failure).	• Different versions and types of remote monitoring systems. • Passenger is the only information source. • Service centre operator does not notice the emergency alarm call.
Step 3: Service centre operator completes transmission of information to the system and sends it to the area serviceman.	• Laborious phase for the service centre operator. • Simultaneous calls must be differentiated. • Serviceman cannot see all the information. • Inadequate information from a site system. • Possibility: Instructions as to how to operate the system. • Possibility: Possibility to open phone line from Call Centre to the elevator.
Step 4: Service centre operator calls the serviceman and reads the description of the failure.	• Extra work for the service centre operator.

Note. From "Bridging the Gap Between User Needs and User Requirements," by S. Kujala, M. Kauppinen, and S. Rekola, 2001, *Proceedings of the* PC-HCI 2001 *Conference,* Patras, Greece, December 7–9. Copyright 2001. Reprinted with permission of University of Patras.

- Videos that capture essential issues and provide some understanding of the users environment (e.g., you might have brown bag lunches where you show and discuss the videos).
- Relatively simple, written summaries, such as tables of negative and positive elements of current work practice and relevant, related information.

In all likelihood, the usability engineer will revisit the raw data (videotapes, audiotapes, or transcripts) during subsequent stages of product design.

Reuse of Data Gathered at the Strategic Planning and Predesign Stages.
The data that you gather in this early stage is likely to be very rich. You probably will not use it all for the deliverables the team needs now. Keep the deliverables at this stage to ones that meet the team's need for answers to the immediate questions.

At later stages, you may find it useful to go back to the data you gathered in these visits, especially if you have videotape,

TABLE 48.3. The Different Stages and the Major Thrust of Task Analysis Questions

These Stages	Answer This Type of Question
Strategic planning and predesign	Why?
Information architecture and concept design	What?
Interface design, early prototypes, and development	How?

audiotape, or extensive notes that speak to the more detailed questions of what or how.

An Example of Using Data From an Early Task Analysis: Deciding on Functionality for the Product.
One important use of task analysis data at an early stage may be helping product planners make the difficult decisions of determining the major functionality for a product. Functionality should be based on what will best help users meet their goals or allow users to have goals that meet their needs and values, but that they could not achieve in the world before this product. Functionality that is included only because it is "neat" or "cool" or "possible to include" may hamper the usability and market acceptance of the product.

In most projects, decisions must be made even about functionality that users want and need. Often, time and resources dictate that some functionality must be left out or left to a later release.

Approaches to this problem of choosing which functions should be included in a product vary, but any useful approach must provide a way of

- systematically capturing all users' goals and rating each one in terms of relative importance
- listing all functions under consideration and rating their relative contribution to each user goal
- integrating these two dimensions (users' goals and functions) to make appropriate decisions based on the ratings

A typical approach is to use a matrix with users' goals along one axis and potential product functions along the other axis. Typically, user's goals are scored in terms of their relative

TABLE 48.4. Matrix for Weighing Market Size, Users' Tasks, and Potential Product Functions

Users	Relative Market Size Hi—3 Med—2 Low—1	High-Level Task	Task Relative Importance Hi—3 Med—2 Low—1	Tool Bar Customization	Better Drag and Drop	Progress Meters	Interruption Protection	Total (Sum of Products, How Well the Market Is Served by These Features)
Techno Bob	1	Customize my screen	3	2	1	0	0	9
		Download information	3	0	0	1	2	9
Newbie Ed	1	Customize my screen	1	−1	2	0	0	1
		Download information	3	0	2	3	0	12
Practical Sue	3	Customize my screen	2	2	2	1	0	36
		Download information	2	0	2	1	2	30
Sum				17	35	27	18	

−1—will confuse users
0—users will not use it
1—users mildly positive
2—users strongly positive

importance and potential product functions in terms of their contribution to a specific goal. These results are combined in some way and then summed in either direction with comments (see Table 48.4 for an example).

In Table 48.4, the market size of user groups was rated based on survey data. Customer visits provided data to rate the relative importance of tasks. Each feature was also rated in terms of user reaction. The numbers for market size, importance of task, and anticipated user response are multiplied to form a rating in each cell and then summed across rows and columns. The column totals show how well the feature will serve all the markets. The row totals show how well a given market is served by the features listed.

Matrices like this one provide a way to systematically look at a wide range of features in terms of tasks and market sizes (J. Pruit, person communication, Oct. 2001).

Information Architecture and Concept Design Stages

Prerequisites for a Task Analysis at This Stage. The data analysis from the earlier strategic planning and predesign stages is input to the plans for task analysis at the information architecture and concept design stages. For example, understanding problems that users have with their current tools helps to avoid the trap of designing a system that simply implements the existing work practices with new technology. Although users do not have to learn much to use a new system that mimics the old way of working, there is not much incentive for adopting a new system that has any learning curve and provides no further value.

Many of the critical issues in design revolve around striking the right balance between the way things are done now and the improvements (as seen by the user) that technology brings. Ehn (1988) refers to this as the trade-off between "tradition" and "transcendence."

Questions for a Task Analysis at This Stage. The focus for this stage is what users know and what metaphors work for them. Relevant questions include:

- What are the primary "objects" that users deal with and what are their attributes and organization?
- What are the operations that are performed on these objects?
- What artifacts do users use in performing these tasks?
- How much variability is there in both objects and artifacts among users?

Use of the Data From Information Architecture and Concept Design Task Analysis. The objects, attributes, and operations that users currently work with suggest possible "data models" for the new product. For example, do users now organize information by time in calendars or charts or do they organize information spatially, suggesting that the metaphor should be a map or diagram?

Sometimes the appropriate metaphor is quite literal. For example, a route planning program will probably have street maps and let users get to additional information from the map interface. Sometimes, the appropriate metaphor is not nearly as literal, and we must be cautious about directly carrying over a metaphor where the new technology will cause users to misinterpret the effect of the metaphor (see Carroll & Rosson, 1987).

A related question is whether users have different metaphors for different parts of the task domain in which the software is applied or, for which the software is meant to help users. For

example, when viewing geographical sales data, one may want to compare a map view with a chart or table.

Presentation of Data From Information Architecture.
The information architecture and concept stages broadly define what the system can do. There are a variety of ways to successfully represent this data, including:

- an object hierarchy showing the definition of and the relationship between objects, their types, their attributes, and their features (see Coble, Maffitt, Orland, & Kahn, 1996; Wood, 1996)
- process diagrams showing activities or artifacts and their relationships (see Coble et al., 1996; Graf, 1996)
- diagrams that capture the metaphor that was derived from user data.

Interface Design and Early Prototypes Stages

Prerequisites for a Task Analysis at This Stage. By this point in the product design and development cycle, the team is working on screen layouts, names for menu items, and design elements, such as color and icons. The overall look and feel are largely determined. Now, the usability specialist can contribute most by having specific and detailed data about user work. This detailed information can directly influence designs and prototypes.

Questions for a Task Analysis at This Stage. The focus now shifts from why and what to how. Task analysis questions for these stages include:

- What is the users' workflow?
- What words does the user use for objects, attributes, operations, etc.?
- What computer conventions do users know and feel comfortable using (e.g., right clicking)?
- How long do tasks take users now? What errors do users make? What are users' tolerances for time and errors for specific tasks?

Use of the Data From Interface Design and Early Prototypes Task Analysis. Data from a task analysis at these stages can be used both in design (to generate storyboards and screen sketches) and in evaluation (for heuristic reviews in developing scenarios and metrics and for usability tests).

Presentation of Data From Interface Design and Early Prototypes Task Analysis.
Presentation here is less important because the relationship between data and design is changing. In earlier stages, the primary goal of representation was to influence the creation of the design. At these stages, the primary role of data from users is to evaluate the emerging design.

By this point, data about users, their tasks, and their environments should have been incorporated into the design. The discrepancies between the look and feel of the interface and user's tasks and underlying knowledge should be minimal. The results of task analysis should now provide input into heuristic reviews and inspections of user interfaces (Nielsen & Mack, 1994) and into usability testing (Barnum, 2002; Dumas & Redish, 1999; Rubin, 1994).

CONCLUSIONS

Many usability practitioners do task analysis in one form or another, although often they do not recognize it as such, much as Moliere's character found the fact that he had been speaking prose all his life a revelation.

Some may hesitate to plan for task analysis because they think of it as too complex and time consuming to apply in real-world situations in which time is always too short. In this chapter, we have tried to show that task analysis is a family of flexible and scalable processes that can fit well in almost any development environment.

Some may have thought of task analysis only in terms of highly structured ways of capturing minute details of specific procedures relevant to evaluating already created designs or developing training or documentation for already determined systems. In this chapter, we have tried to show that, although such uses of task analysis continue, the more common use today is in developing a very broad understanding of users' work and is of great use from the earliest strategic planning stages through all the phases of predesign. Task analysis as laid out in this chapter continues to be useful throughout the process, at later stages, being used to develop scenarios for user-oriented evaluations, as well as in inspection methods.

Task analysis is a way to involve the entire team in understanding users. It provides ways to organize the mountain of unstructured data that often comes from field studies or site visits. It is an essential part of the process of creating any product (software, hardware, Web site, document) because products are tools for users to accomplish goals; products are all about doing tasks.

References

Barnum, C. (2002). *Usability testing and research*. New York: Longman.

Benyon, D. (1992). The role of task analysis in systems design. *Interacting with Computers, 4*(1), 102–123.

Beyer, H., & Holtzblatt, K. (1995). Apprenticing with the customer, *Communications of the ACM, 38*(5), 45–52.

Beyer, H., & Holtzblatt, K. (1998). *Contextual design*. San Francisco: Morgan Kaufmann.

Butler, M. B., & Tahir, M. (1996). Bringing the users' work to us: Usability roundtables of Lotus development. In D. Wixon & J. Ramey (Eds.), *Field methods casebook for software design* (pp. 249–267). New York: Wiley.

Card, S. K., Moran, T. P., & Newell, A. (1983). *The psychology of human-computer interaction*. Hillsdale, NJ: Lawrence Erlbaum Associates.

Carroll, J. M., & Rosson, M. B. (1987). The paradox of the active user. In J. M. Carroll (Ed.), *Interfacing thought: Cognitive aspects of*

human-computer interaction (pp. 80-111). Cambridge, MA: MIT Press.

Coble, J., Maffitt, J., Orland, M., & Kahn, M. (1996). Using contextual inquiry to discover physicians true needs. In D. Wixon & J. Ramey (Eds.), *Field methods casebook for software design* (pp. 229-248). New York: Wiley.

Cooper, A. (1999). *The inmates are running the asylum.* New York: Macmillan.

Diaper, D., & Addison, M. (1992). Task analysis and systems analysis for software development. *Interacting with Computers, 4*(1), 124-139.

Dray, S., & Mrazek, D. (1996). A day in the life of a family: An international ethnographic study. In D. Wixon & J. Ramey (Eds.), *Field methods casebook for software design* (pp. 145-156). New York: Wiley.

Dumas, J. S., & Redish, J. C. (1999). *A practical guide to usability testing.* Bristol, UK: Intellect (revised edition of the 1993 book).

Ehn, P. (1988). *Work-oriented design of computer artifacts.* Stockholm: Arbetsliveschentrum.

Flanagan, J. C. (1954). The critical incident technique. *Psychological Bulletin, 51*(4), 327-358.

Graff, R. (1996). Exploring the design of a sales automation workstation using field research. In D. Wixon & J. Ramey (Eds.), *Field methods casebook for software design* (pp. 113-124). New York: Wiley.

Gray, W. D., John, B. E., & Atwood, M. E. (1993). Project Ernestine: Validating a GOMS analysis for predicting and explaining real-world performance. *Human Computer Interaction, 8,* 237-309.

Hackos, J. T., Elser, A., & Hammar, M. (1997). Customer partnering: Data gathering for complex on-line documentation. *IEEE Transactions on Professional Communication, 40*(2), 102-110.

Hackos J. T., & Redish, J. C. (1998). *User and task analysis for interface design.* New York: Wiley.

Holtzblatt, K., & Beyer H. (1993). Making customer-centered designs work for teams. *Communications of the ACM, 35*(5), 93-103.

Kirwan, B., & Ainsworth, L. K. (1992). *A guide to task analysis.* London: Taylor & Francis.

Kujala, S., Kauppinen, M., & Rekola, S. (2001, December 7-9). Bridging the gap between user needs and user requirements. In *Proceedings of PC-HCI 2001 Conference.* Patras, Greece.

Lee, W. O., & Mikkelson, N. (2000). Incorporating user archetypes into scenario-based design. In *Proceedings of the Ninth Annual Conference UPA 2000.* Chicago: Usability Professionals' Association. Available: www.upassoc.org

Means, B. (1993). Cognitive task analysis as a basis for instructional design. In M. Rabinowitz (Ed.), *Cognitive science foundations of instruction* (pp. 97-118). Hillsdale, NJ: Lawrence Erlbaum Associates.

Miles, M. B., & Huberman, A. M. (1994). *Qualitative data analysis: An expanded source book.* New York: Sage.

Mirel, B. (1996). Contextual inquiry and the representation of tasks. *The Journal of Computer Documentation, 20*(1), 14-21.

Mirel, B. (1998). Minimalism for complex tasks. In J. M. Carroll (Ed.), *Minimalism since the Nurnberg funnel.* Cambridge, MA: MIT Press in cooperation with the Society for Technical Communication.

Nielsen, J., & Mack, R. (1994). *Usability inspection methods.* New York: John Wiley.

Norman, D. (1988). *The design of everyday things.* New York: Doubleday. (Originally published as *The Psychology of Everyday Things*; hard cover published by Basic Books.)

Olson, J., & Moran, T. P. (1996). Mapping the method muddle: Guidance for using methods in user interface design. In M. Rudisill, C. Lewis, P. Polson, & T. McKay (Eds.), *Human computer interface design: Success stories, emerging methods and real-world context* (pp. 269-302). New York: Morgan Kaufmann.

Payne, S., & Green, T. R. G. (1989). Task-action grammar: The model for and its development. In D. Diaper (Ed.), *Task analysis for human-computer interaction.* Chichester, UK: Ellis Horwood.

Preece, J., Rogers, Y., Sharp, H., Benyon, D., Holland, S., & Carey, T. (1994). *Human-computer interaction.* Wokinham, England: Addison-Wesley.

Ramey, J., Rowberg, A. H., & Robinson, C. (1996). Adaptation of an ethnographic method for investigation of the task domain in diagnostic radiology. In D. Wixon & J. Ramey (Eds.), *Field methods casebook for software design* (pp. 1-15). New York: Wiley.

Redish, J. C., & James, J. (1996). Going to the users: How to set up, conduct, and use a user and task analysis for (re)designing a major computer system. *Proceedings of the Fifth Annual Conference UPA '96.* Dallas, TX: Usability Professionals' Association. Available: www.upassoc.org

Rubenstein, R., & Hersh, H. (1984). *The human factor: Designing computer systems for people.* Burlington, MA: Digital Press, Digital Equipment Corporation.

Rubin, J. (1994). *Handbook of usability testing.* New York: Wiley.

Simpson, K. T. (1998). The UI war room and design prism: A user interface design approach from multiple perspectives. In L. Wood (Ed.), *User interface design: Bridging the gap from user requirements to design* (pp. 245-274). Boca Raton, FL: CRC Press.

Taylor, F. W. (1911). *Principles of scientific management.* New York: Harper & Row.

Whiteside, J., Bennett, J., & Holtzblatt, K. (1988). Usability engineering: Our experience and evolution. In M. Helander (Ed.), *The handbook of human computer interaction* (pp. 791-817). New York: North-Holland.

Wixon, D. (1995). Qualitative research methods in design and development. *Interactions, 2,* 19-24.

Wixon, D., & Comstock, E. (1995). Evolution of usability at Digital Equipment Corporation. In M. Wicklund (Ed.), *Usability in practice: How companies develop user friendly products* (pp. 147-191). New York: Academic Press.

Wixon, D., Holtzblatt, K., & Knox, S. (1990). Contextual design: An emergent view of system design. *Proceedings of CHI '97* (pp. 329-336). New York: ACM Press.

Wixon, D., & Jones, S. (1996, original workshop held 1991). Usability for fun and profit: A case study of the design of DEC Rally Version 2. In M. Rudisill, C. Lewis, P. Polson, & T. McKay (Eds.), *Human computer interface design: Success stories, emerging methods and real-world context* (pp. 3-36). San Francisco: Morgan Kaufmann.

Wixon, D., & Ramey, J. (1996). Field oriented design techniques— Case studies and organizing dimension. *SIGCHI Bulletin, 28*(3) [On-line]. Available: http://www.acm.org/sigs/sigchi/bulletin/1996. 3/wixon.html

Wood, L. (1996). The ethnographic interview in user-centered work/task analysis. In D. Wixon & J. Ramey (Eds.), *Field methods casebook for software design* (pp. 35-56). New York: Wiley.

·49·

CONTEXTUAL DESIGN

Karen Holtzblatt
Incontext Enterprises

INTRODUCTION

A Brief History of Contextual Design

In the mid-1980s, the industry was looking for ways to make products more usable. Usability as a field was getting underway, testing products in labs and providing a feedback loop at the end of the design process. However, as John Whiteside at Digital Equipment Corporation had predicted, usability testing was only bringing a 15–20% improvement to user experience. It could not address the definition of product function or significantly change the structure of the product because it occurred at the end of the design process. The challenge was how to redress this and really impact the way products fit into people's lives.

Contextual Inquiry, and subsequently Contextual Design, was developed and shaped in response to that challenge. This was an exciting but demanding time, when many in the industry were trying to move engineering-driven organizations to use customer data at the beginning of the product development cycle. Central to this discussion was the question of what kind of data would best drive product concept design and structure.

Engineering groups recognized that the typical marketing data gathered in surveys, focus groups, and user group conferences could target problem areas for customers, but could not give them the detailed data that they needed about how people worked. They implicitly knew that to really define products—to understand what functionality is needed by people and how to structure that functionality within the product to best support work—required a deep understanding of the blow-by-blow details of what people did to get their jobs done. But this information could not be collected with surveys or typical marketing interviews. Marketing techniques provided neither the detail about how the product fit into an overall work practice that involved other tools and manual processes, nor a coherent picture of the user's intent and tasks to guide the structuring of function within the product. Organizationally, this resulted in conflict between marketing, engineering, and usability groups over what to make and how to structure it.

Contextual Inquiry, along with other techniques for going into the field to see how work practice unfolds, was a response to this need. The problem was not that groups could not get along. The problem was that the data needed for design was simply not being collected. To make products more usable, to make products that people really wanted and could use meant understanding what people were really trying to do and designing new technology to support, extend, and transform that practice. But without knowing the practice, missing the target or breaking the practice was probable.

Although initially resisted, by 1990 the idea of user-centered design had spread to the point that people wanted to design with an understanding of work practice. Within companies, evangelists were pushing for site visits, field data collection, and so forth. By the mid-1990s, people were collecting volumes of work practice data and looking for ways to handle it, and the practices of iterative design, rapid prototyping, participatory design with the customer, work modeling, and field research were evolving in parallel. The practices of Contextual Design, the full customer-centered front-end design process incorporating Contextual Inquiry, were designed in support of these changes over the last 13 years by working with teams in companies throughout the industry.

Today, the ideas of user-centered design, user experience, and designing from customer field data have become the norm. Not all companies use all steps of Contextual Design or have good front-end, user-centered design processes. However, the industry has turned the corner from resistance to awareness. Customer-centered design is being pushed in organizations by management and designers alike; people are seeking skills, universities are teaching customer-centered processes, and companies are creating user-centered design methodology groups.

The idea of user-centered design has become central to the consciousness of people who produce products or systems, marking a successful industry change. But, at the level of the individual organization, change is often just beginning. This chapter provides an overview of the steps of Contextual Design (Beyer & Holtzblatt, 1998) and also highlights lessons learned over 13 years spent helping organizations adopt and tailor Contextual Design for organizational use.

Contextual Design Overview

Contextual Design is a full front-end design process that takes a cross-functional team from collecting data about users in the field, through interpretation and consolidation of that data, to the design of product concepts and a tested product structure (see Table 49.1 for the steps). The strength of Contextual Design is that it tells people what to do at each point so that they can move smoothly through the design process from customer data to specific interaction design and code. Contextual Design is often referred to as good scaffolding because it works like a backbone to which other tools and techniques can be easily added. For example, to allow more creative thinking, add additional brainstorming processes before visioning; to incorporate classic market studies, use the vision to structure questionnaires; or to do use cases, add them after the user environment design has stabilized. Also, Contextual Design can be viewed as series of techniques that can be incorporated into a company's standard methodology—it produces artifacts and data that can feed existing requirements specification formats.

Contextual Design therefore can be easily tailored to the needs of companies and adopted one technique at a time. Organizational adoption, which is discussed more at the end of this chapter, depends both on the readiness of people to try something new and the skill with which people alter and incorporate Contextual Design techniques. Piecemeal adoption and alteration of the method to fit the company is a normal part of organizational change. All introduction of user-centered design techniques is about organizational change.

When companies introduce customer-centered design into their organizations, the adoption will change the process itself. Changing the process requires making trade-offs and decisions about what to keep, what to change, and how. However, because each technique serves a specific intent, Contextual Design practitioners need to understand the intention(s) of each step,

TABLE 49.1. The Steps of Contextual Design

Step	Description	Intent
Contextual Inquiry	Collect data using ethnographic techniques by observing and questioning customers while they work.	Get reliable knowledge about what people do, what they care about, and the structure of the work practice.
Interpretation Session and Work Modeling	Capture the key issues of one individual's work practice and model their work using the five work models, each capturing a different dimension of work practice.	Create a multidimensional understanding of the customer shared among cross-functional groups who have to cooperate in the design. Provide a way of capturing complex qualitative data.
Affinity and Work Model Consolidation	Consolidate the individual work models and issues to reveal the structure of the work across a population without losing individual variation.	Create coherent reusable data about a customer population to guide conversations between designers identifying key requirements and stimulating design discussions.
Visioning and Storyboarding	Invent how new technology will address the user work practice by creating a high-level story (the vision) of how work will be changed. Work out the details of this story at the level of the task (storyboard).	Ensure that the technology will fit into the overall work of the user. Guide invention with data at all levels to avoid breaking the work and ensure supporting the work is supported with technology.
User Environment Design	Design a floor plan that shows how the parts of the new system will interrelate. Represent the structure and function clustering of the system independently of consideration of User Interface and implementation.	Ensure that the flow of work within the system the supports the work. Help designers focus on work flow and function within the system instead of considerations of the User Interface and implementation. Create a system representation to support planning.
Paper Prototyping	Test and modify the new system design in partnership with customers using paper mock-ups of the User Interface. Have people do real work tasks in the paper system, not just review it and provide opinions.	Provide a reliable way of talking with users about the proposed system. Verify the design before committing ideas to code. Deepen the requirements of the system. Start testing User Interface ideas and product concepts before shipping.

why the step is included, and what is gained or lost through trade-offs or changes.

For example, central to any customer-centered design process is the collection of rich, reliable data about how people work. People often ask why they cannot work from the "collective" experience of marketers, sales people, business analysts, and others who have years of experience, and then apply the data organization and visioning techniques to that data. But this violates the most basic value of Contextual Design—design from reliable data collected in an agreed-upon way that represents the actual things that people do. As long as the data is coming from someone's opinion—however much experience behind it—its validity and quality are arguable. To create a data-driven organization, the quality and reliability of the customer data must be unquestioned. Challenging this requirement drives teams back into the age-old argument about whose experience is dominant when conflict arises. The Contextual Design methods of field data collection and consolidation ensure that the data is good and the rules of interpreting it clear. Real customer data ensures that team members can create a shared understanding about the data and requirements across business functions. So Contextual Design teams never trade off getting real data from the field to guide the design. As a result, organizations learn to make decisions based on data and to seek data to make decisions. This alone has enormous value within the product development process.

Similarly, each step in Contextual Design was created for a purpose and is multidetermined; there are a number of reasons why and how the exact steps have been assembled. When adopting or redesigning Contextual Design, first understand the fundamental intents of the steps and the parts. When truncating

a step, leaving out a part, or making a change, make sure that the fundamental intent of that step is not violated without purpose. The next section provides an overview of each step (see Table 49.1) identifying its fundamental intent and discussing reasonable trade-offs.

Setting Project Scope

A successful project and successful organizational adoption is as much about defining the project scope as it is about organizational readiness. Organizations and teams are always in the middle of delivering something. Asking people to slow down to do something new or different will always be difficult. Even for people with the best intentions, figuring out how to insert customer-centered processes into an existing organization is daunting. Resistance shows up as questions about the length of time this new process will take, a concern that old knowledge will be ignored and a fear that schedules will slip.

One way to address these concerns is to define the project scope in a way that works for the organization. Because Contextual Design is a set of techniques, it can be tailored in response to issues of time, resources, and place in the life cycle.

The first step of any project is to set the project scope. This involves deciding how wide to open the inquiry and how much change the organization and its people are ready to make. A wide scope requires the team to collect more data, make more models, and create more designs than a small scope. For example, a project to understand how a Web site is performing as a sales site could involve a process to watch eight people

interact with that site, identify the top key issues, and generate some design ideas and quick fixes to improve the site. But a project to define how buying and selling could be reinvented by the web will involve interviewing many people in different contexts: Looking at the sites of multiple companies, observing how people buy in stores and from catalogs, understanding how families make decisions collaboratively, and determining what influences buying decisions.

The wider the scope, the longer the process and the more resources required. The narrower the scope, the shorter the process and the fewer the resources required. A wide scope will produce more change to an existing design. Change to an existing plan is always threatening because it means changing direction and writing more code. The scope of the project will determine the process, but it also must fit corporate goals and peoples' willingness to make change.

Defining the Right Process for the Project

The question for process designers is not whether teams should use Contextual Design, but what teams should do given their goals. A number of factors influence the length of the project: How easy or difficult it is to set up the customer visits, the number of users to be interviewed, the number of work models to be consolidated, the complexity of the design, and the number of people available to do the work. The time to communicate to others and to obtain organizational buy-in will also impact time. Planning the project process means planning all these things, but project scope will affect them all. Consider some of the following differences in process influenced by changing the project scope:

• **What are the top 10 issues I could fix in my existing project?** Talk to 6–8 customers about the current product and its competition, and watch customers using the product. In the interpretation session, capture only affinity notes, not the models. Create the affinity and identify the top issues. This can be done in two weeks. Once the team identifies the top issues, they can use their normal processes to generate new ideas. Or, they can generate new ideas and then mock them up and iterate them in paper prototypes interviews.

• **How is my design, which is already planned?** If the requirements, the vision, the structure, and user interface drawings are already done, no one wants to start from the beginning. Do a Reverse User Environment Design from the user interface drawings and examine the proposed structure. Fix the structure based on User Environment Design principles, make new paper prototypes, and iterate them with 8–10 customers. This gets customer data into the process without asking anyone to start over.

• **What can I do for a next release or stepwise change?** If the company is not open to significant change for the next release, look at the product, its competition, and manual processes. Manual processes may give ideas on what can be improved. Interview three people for each of three significant job titles using the product. Capture affinity notes, sequences, and artifacts. Consolidate only the affinity and sequences. Then

vision new function and storyboard using the sequences to drive the storyboard. Build a Reverse User Environment Design of the product and integrate new function into the existing User Environment Design. Paper prototype and test it with the new system.

• **When do I need to use the full process?** It is only when defining a new product, a major change to a product, opening a new market, or understanding how to use new technology that the full Contextual Design process and all the models are necessary. The full process will push design thinking out much more broadly and bring new significant functionality into the design.

Contextual Design can handle very large scoped projects—the models naturally extend the breadth of ideas to consider and the User Environment Design can capture and record large systems. But even when a company needs a large-scoped project, it is not necessarily the right thing to do. When companies are embarking on their first customer-centered projects, propose projects that are likely to be accepted, to give immediate value, and that will result in actionable changes. This will contribute to the grass-roots movement of introducing customer-centered design.

THE STEPS OF CONTEXTUAL DESIGN

The Contextual Interview—Getting the Right Data

To design a product that meets customers' real needs, designers must understand the customers and their work practice. Yet designers are not usually familiar with or experienced in the work they are supporting. If they operate from their gut feel, they rely on their own experience as a user. But designers generally are more tolerant of technology than average users, so they are not representative of end users.

On the other hand, requirements gathering is not simply a matter of asking people what they need in a system. A product is always part of a larger work practice. It is used in the context of other tools and manual processes. Product design is fundamentally about the redesign of work or life practice, given technological possibility. Work practice cannot be designed well if it is not understood in detail. Any introduction of change into a practice is as likely to make the practice less efficient as more efficient and delightful. So, requirements gathering means understanding how a practice is structured and imagining what technology might do to improve it—those improvements are what is required of the system.

You cannot simply ask people for design requirements, in part because they do not understand what technology is capable of, but more because they are not aware of what they really do. Because the things people do every day become habitual and unconscious, people are usually unable to articulate their work practice. People are conscious of general directions, such as identifying critical problems, and they can say what makes them angry at the system. However, they cannot provide the day-to-day detail about what they are doing to ground designers

in what the practice entails and how it might be augmented with technology.

The challenge of getting design data is being able to get that level of detail about work that is unconscious and tacit. The first step of Contextual Design is Contextual Inquiry, the field data gathering technique that allows designers to go out into the field and talk with people about their work or lives while they are observing them. If designers watch people while they work, the people do not have to articulate their own work practice. If they do blow-by-blow retrospective accounts of things that happened in the recent past, people can stick with the details of a case using artifacts and re-enactment to remind them of what happened. Field data overcomes the difficulties of discovering tacit information.

In Contextual Design, the cross-functional design team conducts one-on-one field interviews with customers in their workplace focusing on the aspects of work that matter for the project scope. The Contextual Interview is based on four principles that guide how to run the interview:

- **Context**—Gather data in the workplace while people are working and focus on what they are doing.
- **Partnership**—Collaborate with customers to understand their work; let them lead the interview by doing their work. Do not come with planned questions.
- **Interpretation**—Determine the meaning of the customer's words and actions together by sharing your interpretations and letting them tune your meaning. When immersed in their real life and real work, people will not let you misconstrue their lives.
- **Focus**—Steer the conversation to meaningful topics by paying attention to what falls within project scope and ignoring things that are outside of it. Let the user know the focus so they can steer, too.

The Contextual Interview starts like a conventional interview, but after a brief overview of the work, it transitions to ongoing observation and discussion with the user about that part of the work that is relevant to the design focus. The interviewer watches the customer for overt actions, verbal clues, and body language. By sharing surprises and understandings with users in the moment, users and designers can enter into a conversation about what is happening, why, and the implications for any supporting system. As much as possible, the interviewer keeps the customer grounded in current work activity, but can also use work artifacts to trigger memories of recent activities.

The fundamental intent of the Contextual Interview is to help designers get design data: Low-level, detailed data about the structure of the practice and the use of technology within that practice.

Choosing Customers Depends on Project Scope.

People are always asking how to shorten the Contextual Design process—time is always critical. Time in a Contextual Design project is directly proportional to the number of interviews conducted. The more interviews, the more time it takes to conduct and interpret them, and the more time to consolidate the data.

So time is something that you can control as you choose how many people to interview.

But the number of people that should be interviewed is directly related to the project scope. The wider the scope, the more people need to be interviewed to cover that scope. A rule of thumb for a small scope, such as top 10 problems, usability improvement, next product release or checking a planned design, is 4-10 customers from 3 to 5 businesses covering 1 or 2 roles. Most large projects will not need more than 15-20 customers distributed over four businesses covering 3-5 roles. Very large, strategic projects might need 20-30 customers.

Choose the number of people to interview by balancing several variables, including the job roles to be supported, industries, and product use. When analyzing the project, first identify the job titles and the people that support the work. The work they do is what counts—if one person is called a system administrator and the other a database administrator, but they do the same work as it relates to the project, it does not matter that their job titles are different. The number of job roles needed to cover the design scope determines how many people must be interviewed. Interview a minimum of three people per job role; four is better.

If the product is to support people across industries, what matters is whether the industries are fundamentally different. Looking at real estate as an example, the work is structured differently within the industry: A group of small, distributed agencies; a large corporate real estate company; and an in-house real estate representative. In each situation, the communication, sharing, and work management are likely to differ, creating three different work practice patterns. So, to cover real estate, collect data in all three contexts. But if there are no changes to work patterns across industries—such as with the document editing component of an office tool suite, where editing is editing whether in banks, insurance companies, or software development organizations—then simply touch multiple industries without worrying about overlap. Try to interview at least two businesses per work pattern.

Geography works the same way. If culture and law affect the work practice, then to be global you need to gather data in multiple geographies. Try to interview two to four companies in a geographic area to ensure reasonable coverage. But if geography does not affect the practice, ensuring coverage is less important and data can simply touch on geographic differences.

Always try to look at customers that use homegrown and competing products, as well as the client's own products. Two to three instances of each of these are usually sufficient.

There is no need to make a grid and fill it all in. Instead, weigh and balance the variables ensuring a selection that has overlap in job title, industry, geography, and type of technology as appropriate to the project scope. The goal is to get enough repetition in the work so that each variable has 3-4 interviews that represent it. One person may represent three of the variables. As long as you have overlap, you will be able to find the common structure and key variations in the work practice. Remember, paper prototype interviews will expand the number of companies and roles represented in the whole project.

So, if time is a driver, cut back the numbers, but know that this will reduce the scope the project can cover.

Interpretation Sessions—Creating a Shared Understanding

Contextual Interviews produce large amounts of customer data, all of which must be shared among the core design team and with the larger, cross-functional team of user interface designers, engineers, documentation people, internal business users, and marketers. Traditional methods of sharing by presentations, in reports, or by e-mail do not allow people to truly process the information or bring their perspectives into a shared understanding. Contextual Design overcomes this by involving the team in interactive sessions to review, analyze, and manipulate the customer data. An interpretation team delivers the best results when its members represent a wide range of job functions and viewpoints. However, if the team is too large, the session will not be managcable. Usually 4–6 people are optimal, with 10 as the absolute maximum.

Although it is good to record interviews for backup, it is not advisable to transcribe the recordings or do videotape analysis. Both transcription and video analysis take too long, and video analysis also limits the perspective to one person. Instead, within 48 hours of each customer interview, gather the team in the design room, where the interviewer can tell the story of the interview from handwritten notes and memory.

Team members ask questions about the interview, drawing out the details of this retrospective account. At the same time, they perform one or more interpretation session roles: Recorder, model builder, moderator, or simple participant. The recorder types notes online, displayed on a monitor or LCD projection panel so that everyone can see them. These notes capture the conversation, including breakdowns and influences, in sequence. They are later transferred to 3M Post-it® Notes and used to build the Affinity Diagram.

Model builders hand sketch up to five work models onto flip charts (Fig. 49.1). Each work model depicts a different perspective of the work:

- The Flow Model depicts people's responsibilities and the communication and coordination required to do a job.
- The Cultural Model reveals influences on a person, whether external to the company (such as dependence on a vendor) or internal company policies.
- The Sequence Model shows each step required to perform a task in order.
- The Physical Model shows the physical layout of the work environment and the constraints it imposes on the design. It also shows the way people physically structure their work environment to support work.
- The Artifact Model shows how artifacts are structured and used, and suggests how their use could be extended.

Each team member brings a different perspective to the data, whereas open discussion enables the team to arrive at a shared understanding. The interactive nature of an interpretation session keeps everyone involved in the task at hand. Team members raise questions in the moment, triggering the interviewer's memory and eliciting more data than would be available from a designer working alone. Design ideas are generated and captured.

The interpretation session is a structured way to ensure that all relevant information from the interview is captured for use in the design process and shared with key team members. It is the context for both understanding and seeing the structure of the data and starting a real design conversation about how to address the user's needs with technology.

The work models structure what the team should look at, and look for, in the data. As such they ensure that the right data are captured. Team members learn that they need to draw these models so they become sensitive to capturing this data when they are in the field. In this way, the work models both provide a way to represent the data and a method to help people get the data they need for design.

Choosing Work Models Depends on Project Scope. The five work models and the affinity support coherent conversations focused on different aspects of the work. Each conversation will drive different design concepts, with some models forcing detailed thinking and others encouraging wide and broad design. The more models captured, the more work both within the interpretation session and during model consolidation. So, the number of models drawn affects resources and time. Depending on the project scope, different models are called for:

- **Identifying key issues:** Creating the affinity diagram will give you top issues, so to identify key issues only capture affinity notes.

- **Designing the next product release:** If you are looking to improve a product, but not reinvent it, you need only capture the affinity notes and the detailed models: The sequence, artifact, and physical layout of the work on the desktop or in the lab.

- **New product, market, or technology direction:** Detailed models only bring stepwise change. For the larger picture of the market and to push more innovative design, you also need to use the broader models: Flow, cultural, and the physical site model. These show you who the people are that could be supported and how they interrelate, what the value proposition is for this population, and how work is distributed across space.

- **Getting reusable corporate data:** Many companies have product lines that address the same work practice. Because the work models represent the fundamental structure of the practice, they tend to be stable over time. The wider data, once collected, can be expanded and reused by the same team and by teams producing related products. Collecting the wide-scoped data at some point and building upon it is efficient in the long run.

What models you capture affect the number of people needed in the interpretation session. To have a smooth interpretation session and capture all the models, you must have an interpretation team of four. A two-person interpretation team is the minimum. At least one outside perspective is needed: One other person to question and prod the interviewer. A three-person team ensures reasonable diversity and group sharing, even when fewer models are used. However, if you only have a two-person

Flow Model

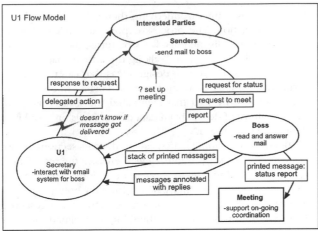

coordination and communication

Sequence Model

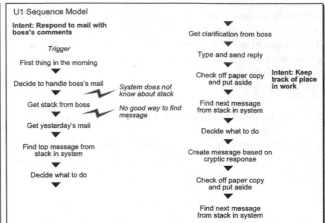

ordered steps to accomplish a task

Artifact Model

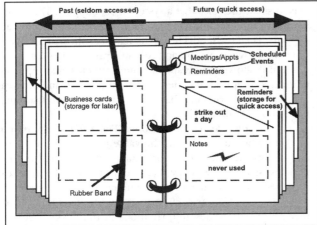

structure and intent of things
used in work

Cultural Model

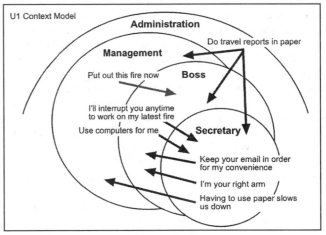

culture and policy

Physical Model

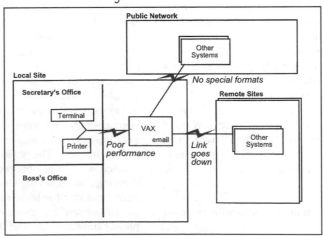

physical environment as it affects work

FIGURE 49.1. The interpretation session results in affinity notes and
five work models.

core team, you can extend your team by rotating in extra people. In this way, you can increase buy-in with limited involvement (2 hours) and benefit from outside perspectives without requiring that these adjunct team members leave their regular jobs for long periods of time.

Consolidation—Creating One Picture of the Customer

Consolidating the work models enables teams to create one representation of a market so that the design addresses a whole population, not just one individual. The most fundamental goal of Contextual Design is to get the team to design from data instead of from the "I." If you walk in the hall and listen to designers talking, you will hear: "I like this feature." "I think the dialogues work best this way." "This user interface component doesn't make sense to me." It is rare to hear, "Our user data says that the work is structured like this, so we need this function." It is natural for people who design to make a system hang together in a way that makes sense to them. But they are not the users, and they are not in any way doing the work of the people they are designing for. Getting designers, marketers, and business analysts out to the field and into interpretation sessions moves them away from design from personal preference, but you do not want them to become attached to their user to the exclusion of the rest.

Product and system design must address a whole market or user population. It must take into consideration the issues of the population as a whole, the structure of work, and the variations natural to that work. The core intent of building consolidated models is to find the issues and the work structure and create a coherent way to see it and talk about it. Because the work models and affinity diagram separate and focus conversation on different aspects of work, consolidation creates six points of view from which to see and discuss the data.

Because consolidated models segment work practice into coherent and meaningful chunks, the team is able to see, capture, and think about very complex and rich information. Complexity often leads to overwhelm, and overwhelm leads to a desire to simplify or reduce data. Data reduction, in the form of identifying the top 10 findings or creating a simple summary, loses the complexity and structure of the work that the interviews revealed. Consolidating the data and systematically walking and discussing each model as a team helps the team think about complex information without getting overwhelmed.

While discussing the key factors of each model, team members will naturally synthesize the findings into a reasonable view of what a product needs to support. Given this data, any good designer will be stimulated to imagine how their product can better support the work. So systematic interaction with consolidated models becomes the basis for design thinking and informal prioritization of customer needs.

The Consolidated Models. Each model focuses the team on a different set of issues to consider.

The affinity diagram (Fig. 49.2) brings issues and insights across all customers together into a wall-sized, hierarchical diagram. Using the Post-it®Notes created in the interpretation session, team members organize all of the data one piece at a

time, finding common underlying themes that cross the customer population. The process exposes and makes concrete common issues, distinctions, work patterns, and needs without losing individual variation. Walking the affinity diagram allows designers to respond with design ideas that extend and transform the work. They write these on Post-it®Notes and stick them right to the data that stimulated the idea. This encourages a culture of design from data over a focus on "cool ideas" generated from the "I."

Building the affinity after a good cross-section of users (usually 6–12) has been interviewed enables the rest of the interviews to be increasingly targeted to address specific issues that may emerge. The affinity is built from the bottom up, which allows the individual notes that come together to suggest groupings instead of trying to force them into predefined categories. Groups are labeled using the voice of the customer—saying what they do and how they think.

The consolidated flow model (Fig. 49.3) identifies the key roles played by individuals and combines them across individuals to show the roles that a system must support. This is your war map, representing the players in the market to target and support. The flow model helps the team see the pattern of relationship between players and reveals opportunities for collaboration; it shows how responsibilities are clustered into roles that can be supported coherently or automated; and it reveals artifacts and information that pass between people that might be supported online.

The consolidated cultural model (Fig. 49.4) is important to marketing because it reveals the value proposition. It collects and shows the influences, constraints, interpersonal friction, policies, standards, and law that people work under. It reveals positive feelings and irritations, including those with the vendor collecting the data. As such, dialogue with the cultural model reveals primary values and irritations in the market being addressed. A product that supports those values or removes those irritations has a good value proposition story. Because cultural differences, policy, and law are also revealed, it shows the design team what they must consider within the design for adoption. Think of it as the "get real" model, because it makes culture and the need to design for the culture real to the team.

Consolidated physical models (Fig. 49.5) reveal the common aspects and variations of physical structure across a customer population. Like the cultural model, the consolidated physical model is a get real model. The physical model reveals the physical constraints of distance, time zone, and overall physical and system access. It reveals the way space is used and how people set up their offices or workspaces to support their work. People's piles are a window into their mind and how they organize the work. The physical model, like the affinity, flow, and cultural models, is a big picture model; it pushes open design thinking and shows the whole social, physical, and cultural environment that the team will have to support. But it is a swing model because it is also a detailed model, revealing how people separate and organize the details of the work. It will guide detailed design and storyboarding, as will the consolidated sequence and artifact models.

The consolidated sequence model (Fig. 49.6) shows the detailed work structure that the system will support or replace.

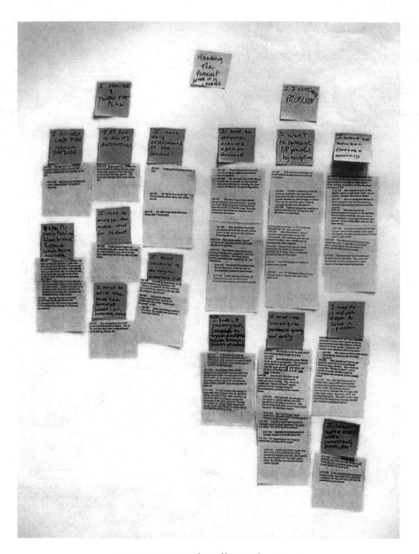

FIGURE 49.2. The affinity diagram.

This is similar to traditional task analysis, showing each step, triggers for the steps, different strategies for achieving each intent, and breakdowns in the ongoing work. Work redesign is ultimately about redesigning the steps in the sequences. Whether the redesign eliminates or changes the steps or eliminates the whole sequence, knowing the steps and intents keeps the team honest. More often than not, technology introduces new problems into the work by failing to consider the fundamental intents that people are trying to achieve. Redesign will better support the work if it accounts for each intent, trigger, and step. This does not mean leaving the work the same; it means that the team has seen the current activity and has completely considered what will happen to it in the new work redesign. The consolidated sequence is critical as a guide for storyboarding.

Consolidated artifact models (Fig. 49.7) reveal the structure of the artifact, how it is used, how it is presented, and the data that it represents. Consolidated artifact models help the team redesign artifacts that have been identified as candidates to be put online. Sometimes teams want to eliminate artifacts they

think serve no useful purpose, so consolidated artifacts also keep the team honest. Any time a team wants to eliminate an artifact, they must ensure that all the intents of that artifact are fulfilled in some other way by the design.

How Can So Little Data Characterize a Whole Market?
Consolidated models create data that is reusable for years. For example, we have been collecting data on system management for 13 years across multiple companies, and the basic flow model roles and high-level activities and intents are the same. The core of any work practice does not change much over time. Technology usually introduces changes at the level of the step in the sequence model. Aspects of the roles may be automated, but the role itself remains, perhaps now played by automation instead of a person. The values that drive the practice, the benefits, persist. Once they are met, they become latent, expected needs instead of unfulfilled needs.

Lasting consolidated data is built anywhere from 15 to 30 field interviews and can characterize markets of millions of

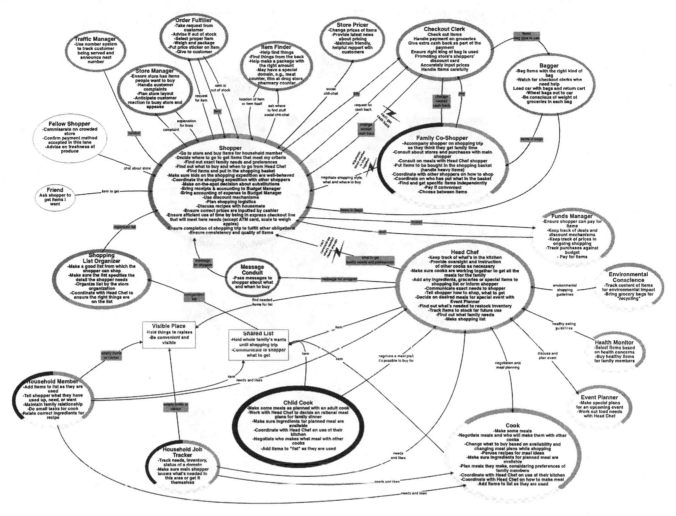

FIGURE 49.3. The consolidated flow model.

people. Even consolidated data from 8 to 10 people can identify a large percentage of the key issues that will eventually be identified in the market. For example, if you consolidate after 10 interviews, you can grow the models and affinity from there. An early affinity can show holes in the data to guide further data collection, but already represents the key areas and distinctions that will grow in detail as more data is added. Similarly, additional data adds depth and detail to all models, but the basic structure, central to the project focus, is identifiable early on. This means that a little bit of data goes a long way. If this data is also stable over time, it becomes a key corporate resource.

Early on in usability testing, people were not comfortable with test results from small numbers of users. Whatever the formal arguments, 15 years later the industry has learned empirically that small numbers are enough. Contextual Design is engaged in market characterization at the level of work practice, so it is compared with marketing studies with random samples and hundreds of data points or to focus groups with 10 people in each session. How can so little data characterize the practice

of so many people? Moreover, how can this data be stable for years?

Contextual data does not claim to show trends. Contextual data cannot be counted because it is based on observing people in the field and each person's activity directs what is seen. You simply do not get the same data from each person. Under these conditions, frequency measures and trends have no meaning. Contextual data shows the structure of the practice. Contextual Design gets its power from designing from an understanding of work practice *structure* without losing variation.

The reason that people get hung up on numbers is that we tend as human beings to focus on our variation, our differences, rather than our basic similarities. A focus on variation leads us to design products with near-infinite customizations. These either destructure a system with ever more customizable and slightly different options or require costly upfront customizations at installation. But, if people are all so different, how could markets exist? Clearly, there is enough similarity of practice between people for the idea of shrinkwrap software and off-the-shelf electronic products to make sense.

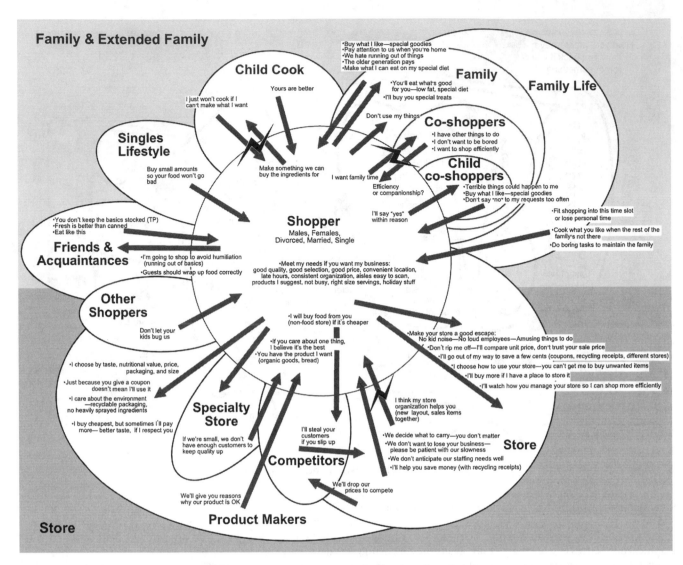

FIGURE 49.4. The consolidated cultural model.

Consolidation helps people see and find the structure in work practice, and this drives successful design. Here is a way to think about it.

We're all different, but we're all alike: Everybody looks different; humans have great variation. People are of different ethnic groups, cultures, and child-rearing practices. Everyone chooses different clothes, hobbies, careers, and lifestyles. So, at one level, we are all unique. But at the same time, we are all alike; we are human beings with one head, two arms, two legs, and so on. From the point of view of clothing manufacturers, we all buy clothes off the racks in department stores; tuning a few preferences in the hemline is enough. Structurally, the variation between people is small and the structure of our bodies is common. So, although we tend to focus on our differences, given the project focus of clothing manufacturing, we are pretty much the same.

There are only so many ways to do work: Any product and system design is really a very narrow focus on the human experience. Within one kind of work, there is only so much variation possible. The roles we play, the intents and goals we have, and the way we do things is common. Variation, once you start looking for structural elements like roles instead of job titles, is small. Experience shows that there are only two to four strategies for any primary task in a work practice. If those are the key strategies for that work practice, the question is not which we support, but how we support them all. Frequency data is not needed to show that these strategies matter; these are the basic elements of the practice. Designs that target a certain domain of work are created for a certain set of people doing a fixed set of work tasks, situated in the larger work culture, using the same set of tools, trying to achieve the same kind of goals. Under these constrained conditions, the work practice will have similar patterns. If this were not the case, there would be custom systems for every single person. Instead, people use the same software systems with some preferences and options. After you have collected data from three to six people doing the

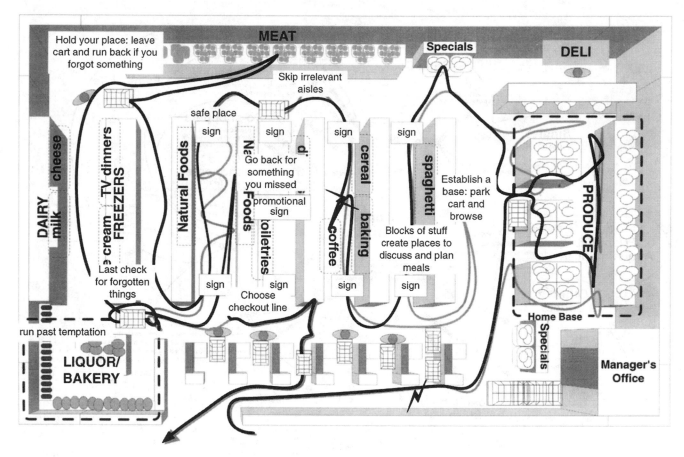

FIGURE 49.5. The consolidated physical model.

same thing, the work pattern and issues start overlapping again and again.

Geographical differences are mostly about law: Even geographical differences in work practice are small. They are mainly about law or lawlike standards, not culture. For example, in Europe, if somebody travels for their job, they are compensated using per diem rates. The actual rates and inclusions vary from country to country, but they are always paid per diem. In contrast, in the United States, they may be compensated per diem or they may need to show their receipts. In any event, because there are only two ways to compensate work travel, there are only two possible interfaces—a receipts-driven and a per diem interface. Cultures are interesting because food, values, personalities, and lifestyles are different. But, from the narrow view of work practice, the variation that affects practice the most tends to be legislated by corporations or government. Other differences between cultures are best understood as variations in emphasis; some cultures emphasize one strategy or cultural value more than another, but most cultures have people in them that represent all the different values and strategies anyway. So, it is possible to sample across cultures and check at the extent of difference before deciding how much global data is needed to characterize the market.

With infinite time and resources, we could satisfy our worry about complete coverage by extensive customer sampling.

Luckily, because a design is targeted at a work or life practice, a small amount of data will let us model the important aspects of the practice and use it to drive design. Start small, grow the data, and reuse it.

The number of interviews affects the timeline of a project. What you consolidate also affects the timeline of a project. You do not need to consolidate models that are not going to help the project, even if you captured them. For example, if you are not redesigning artifacts and the artifacts are not relevant to the design, consolidating the artifacts is a waste of time. You can build the physical model to help the team understand what happened in the interview, but if the arrangement of the details of the model does not impact the problem you are solving, then there is no need to consolidate the physical models. On the other hand, do not postpone consolidation because you have no time now. It will not be possible to recover the context of the interview sufficiently to consolidate data later.

Visioning a New Work Practice

Visioning is about invention. The question is, invention of what? One might argue that the first form of technology was the stick. The stick is an extension of the arm and finger. A good stick is narrow enough to fit in the space that the arm or finger cannot fit.

-Activity 2. -Stop and Buy

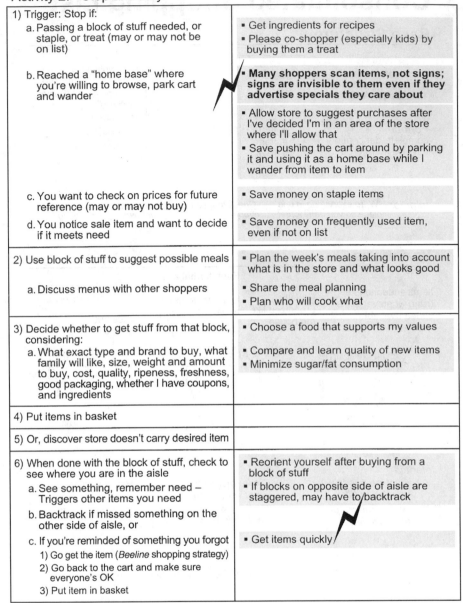

1) Trigger: Stop if:	
a. Passing a block of stuff needed, or staple, or treat (may or may not be on list)	• Get ingredients for recipes • Please co-shopper (especially kids) by buying them a treat
b. Reached a "home base" where you're willing to browse, park cart and wander	• **Many shoppers scan items, not signs; signs are invisible to them even if they advertise specials they care about** • Allow store to suggest purchases after I've decided I'm in an area of the store where I'll allow that • Save pushing the cart around by parking it and using it as a home base while I wander from item to item
c. You want to check on prices for future reference (may or may not buy)	• Save money on staple items
d. You notice sale item and want to decide if it meets need	• Save money on frequently used item, even if not on list
2) Use block of stuff to suggest possible meals	• Plan the week's meals taking into account what is in the store and what looks good
a. Discuss menus with other shoppers	• Share the meal planning • Plan who will cook what
3) Decide whether to get stuff from that block, considering: a. What exact type and brand to buy, what family will like, size, weight and amount to buy, cost, quality, ripeness, freshness, good packaging, whether I have coupons, and ingredients	• Choose a food that supports my values • Compare and learn quality of new items • Minimize sugar/fat consumption
4) Put items in basket	
5) Or, discover store doesn't carry desired item	

6) When done with the block of stuff, check to see where you are in the aisle
a. See something, remember need – Triggers other items you need
b. Backtrack if missed something on the other side of aisle, or
c. If you're reminded of something you forgot
  1) Go get the item (*Beeline* shopping strategy)
  2) Go back to the cart and make sure everyone's OK
  3) Put item in basket
	• Reorient yourself after buying from a block of stuff
	• If blocks on opposite side of aisle are staggered, may have to backtrack
	
	• Get items quickly

FIGURE 49.6. The consolidated sequence model (partial).

A good stick is long enough to reach what the arm cannot. Technology will always need to fit within a larger human practice. So, design is effectively the invention of practice in which technology fits well. Design of technology is first design of the story showing how manual practices, human interactions, and other tools come together with your product to better support the whole practice. Visioning is the Contextual Design technique to help teams tell that story. Visioning is a vehicle to identify needed function in the context of the larger work practice. Visioning ensures that teams postpone lower level decisions about implementation, platform, and user interface until they have a clear picture of how their solution will fit into the whole of the practice. Teams commonly focus too much on low-level details

instead of the full sociotechnical system. This is one cause of breaking the work and failing to create something the market wants. So, the primary intent of visioning is to redesign the work practice, not to design a user interface. Because a visioning session is a group activity, it fosters a shared understanding among team members and helps them use their different points of view to push creativity.

The first step to a visioning session is to "walk the data" model by model, immersing the team in customer data so their inventions will be grounded in the users' work. During the walk, team members compare ideas and begin to get a shared idea of how to respond to the data. Nobody should get to vision unless they have participated in walking the data. Without this rule,

Consolidated Shopping List

Base Structure

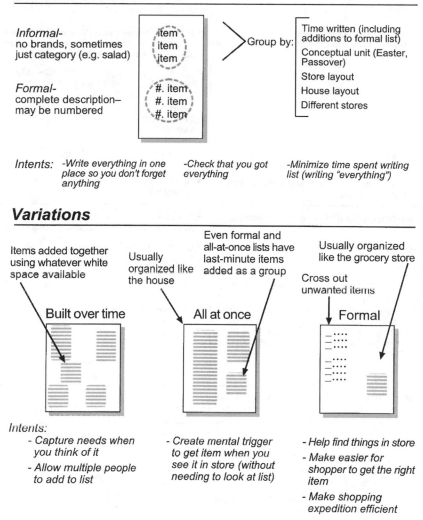

Informal-
no brands, sometimes
just category (e.g. salad)

item
item
item

Formal-
complete description—
may be numbered

\#. item
\#. item
\#. item

Group by:

Time written (including additions to formal list)
Conceptual unit (Easter, Passover)
Store layout
House layout
Different stores

Intents: -Write everything in one place so you don't forget anything   -Check that you got everything   -Minimize time spent writing list (writing "everything")

Variations

Items added together using whatever white space available

Usually organized like the house

Even formal and all-at-once lists have last-minute items added as a group

Usually organized like the grocery store

Cross out unwanted items

Built over time   All at once   Formal

Intents:

- *Capture needs when you think of it*
- *Allow multiple people to add to list*

- *Create mental trigger to get item when you see it in store (without needing to look at list)*

- *Help find things in store*
- *Make easier for shopper to get the right item*
- *Make shopping expedition efficient*

FIGURE 49.7. The consolidated artifact model.

the process is no longer data driven; anyone can walk in and offer their pet design ideas. Simply walking the customer data naturally selects and tailors pre-existing ideas to fit the needs of the population. Because visions are evaluated, in part, on their fit to the data, knowing the data is important.

During the visioning session, the team will pick a starting point and build a story of the new work practice. One person is assigned to be the "pen" who draws the story on a flip chart, fitting ideas called out by the team members into the story as it unfolds. The story describes the new work practice, showing people, roles, systems, and anything else the vision requires. The team does not worry about practicality at this point; all ideas are included. Creating several visions allows the team to consider alternative solutions.

After a set of visions is created, the team evaluates each vision in turn, listing both the positive and negative points of the vision from the point of view of customer value, engineering effort, technical possibility, and corporate value. The negative points are not thrown out, but used to stimulate creative design ideas to overcome objections. When complete, the best parts of each vision and the solutions to objections are brought together into one, synthesized work practice solution (Fig. 49.8).

How Can You Invent From Customer Data?.   Every team raises this question. Customer data tell you what is, not what could be. How can you see the future by looking at the past?

Every invention supports a real need, else why would anyone want it?: Invention is a response to some life or work practice by a designer or technologist who, seeing a need *and* knowing the technology, imagines a new possibility. Edison did not invent the idea of light; he saw candles and gas and invented light bulbs. Dan Bricklin (Beyer, 1994) did not invent spreadsheets;

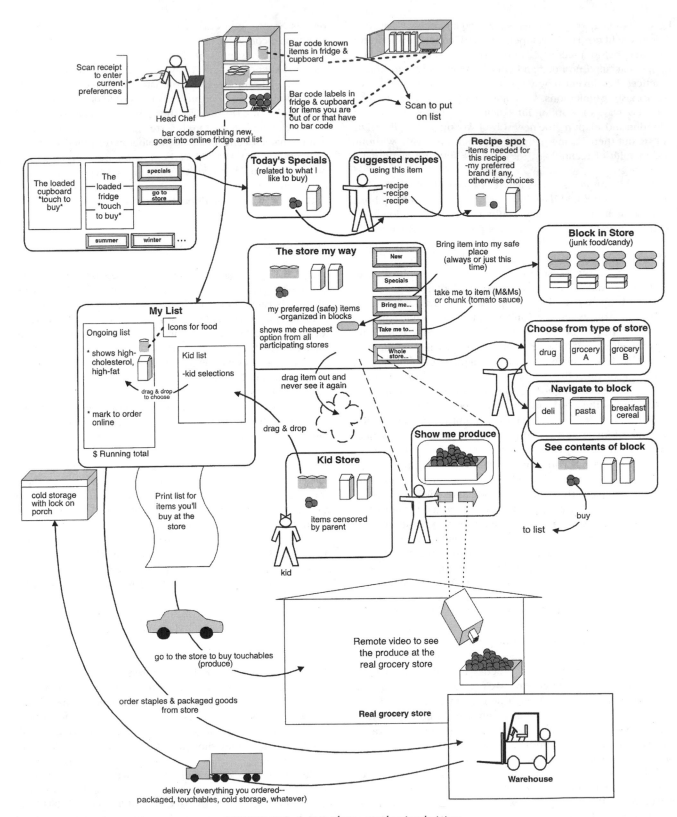

FIGURE 49.8. A complete, synthesized vision.

he saw how paper spreadsheets are used and knew what technology could do. The developers of WordPerfect worked in the basement below a secretarial pool. Nobody invents entirely new things that fulfill no need and contribute in no way to human practice; they invent new ways to fill existing needs and overcome existing limitations. As people incorporate these products and new ways of working into their lives, they reinvent it by adopting and adapting the new way of working. Thus, if designers are out there seeing people living their lives with, without, and in spite of technology, they can see future directions for technology.

A mobile phone businessman tells a great story when asked about how you could have invented the hand-held cell phone by looking at customer data: You would be driving down the road and watching people suddenly pull off the road and park their cars to make a phone call. You would see them pulling into gas stations to make a telephone call. They would be out in their yard in the midst of yard work and run inside to make a telephone call. You would see people looking all over the office and calling over the loud speaker to find a person not at their desk. You would see families call all their child's friends looking for them. Then, knowing phone technology, you would ask yourself what if, when people called, the call went right to them? Gee, how could we, the makers of phones, do that . . . . So, where do you think the idea of cell phones came from, anyway?

A vision is only a good as the team's combined skill: Customer data is the context that stimulates the direction of invention. But there is no invention without understanding the materials of invention: Technology, design, and work practice. The visioning team needs to include people who understand the possibilities and constraints of the technology. If the team is supposed to design web pages and none of them has ever designed a web page, they will not be able to use web technology to design. When the people who always designed mainframes were told to design WYSIWYG interfaces, they replicated the mainframe interface in the WYSIWYG interface. Similarly, if the team has no interaction designers or work practice designers on it, the resulting vision will not be as powerful as if they did. This is why design teams should include people with diverse backgrounds representing all the materials of design and the different functions of the organization. Only then will an innovative design, right for the business to ship, emerge.

Even new technologies can be guided by customer data: Even if you are looking for ways to capitalize on an entirely new technology, there is work practice to observe and gather data about. A designer might say, "We're creating something that has never existed before. There's no customer data to collect." However, there was a time when computers did not have sound and no one knew what it would do. Was there no data to collect to guide this technology? Was this *really* brand new? Sound was once a totally new *technology*, but *sound* is not new. Using sound appropriately in computers is a challenge. What data should you collect? Data on the experience of sound. Walk through the world seeing what people do with sound. Listen, look, and talk with them about random sound, talk, music, sound in the environment, and the role of sound in people's lives. Also pay attention to silence, because you must not violate that. Studying sound and silence, noise and communication,

reveals opportunities and identifies appropriate ways to use sound on a computer.

Don't ask customers what to make; understand what they do: Customers who are not aware of the details of their own practice, who do not know the latest technologies, and who do not know what your business is capable of cannot tell you what to invent. Do not ask them. Instead, understand what they are doing and capture it systematically. Immerse a design team—that *does* understand technology, work practice, and the business—in that data and let them vision. But the vision has to be right for people, so take it out and test it. Let people test drive the future in our paper mock-ups and let their tacit knowledge of their lives shape and direct the vision.

Storyboarding—Working Out the Details

Storyboarding provides the opportunity to work out the detail of the vision. Too often when people design, they break work practice because they jump from their big idea to low-level user interface and implementation design. As soon as designers start focusing on technology, technology and its problems become their central design concern. How technology supports work or life practice is subordinated. This also happens when people design from idealized models of how processes work. A classic story is about an ordering system that broke the work of the corporation. No one knew that the person answering the phone was a good friend of the person prioritizing the orders to be fulfilled. They had created an informal "hot order" channel that was never modeled in the new, planned process.

The steps and strategies of a work practice, being tacit, are easy to overlook. But if you are out in the field watching them, you can see them. This is what is captured in the sequence models. Storyboarding keeps the team honest and the design clean. Guided by the detailed models—the sequence, artifact, and organization of the work place (physical)—the vision is made real in storyboards. Storyboards ensure that the team does not overlook any intents and steps that are critical to the work. Even if the work is changed, think through the details of how it will be changed to ensure that adoption is easy.

Storyboards are like freeze frame movies of the new work practice (Fig. 49.9). Like storyboarding in film, the team draws step-by-step pictures of how people will work in their new world. Storyboards include manual steps, rough user interface components, system activity and automation, and even documentation use. Storyboards are similar in structure to the consolidated sequence model and incorporate ideas from the other detailed models. They show what happens when the roles on the flow interact with the system and each other. Like consolidated sequences, they show how the system will handle multiple strategies.

Because they focus on work practice design, storyboards prevent the design team from prematurely delving into too much detail. They are guided by data, and after each task has been thought through and sketched, the team reviews it to ensure that it remains true to the customer data. This does not mean that no invention happens, but the team must account for the steps and other data elements of the models. They must look at them

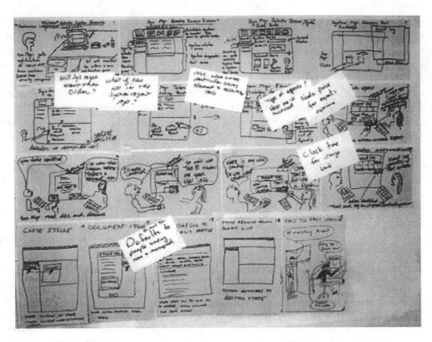

FIGURE 49.9. Storyboards are like freeze frame movies.

and make a conscious decision about how to handle them. They might change all the steps and even eliminate whole sequences; as long as people can still achieve their fundamental intents, the change will work. When teams forget or ignore the user's intent, the design is in trouble.

The vision is a natural "prioritizer." It is not necessary to storyboard every sequence. Instead, identify which sequences are relevant to the vision and storyboard those. The sequence models should represent the core tasks to achieve the work within project scope. Supporting those sequences will be enough for a next product release.

How Is Storyboarding Different From Building Scenarios? The storyboard is fundamentally a future scenario guided by the vision and reined in by the data. Every system implies multiple storyboards of use. Just as a kitchen supports multiple scenarios of cooking (the quick hot dog, the Thanksgiving dinner, the breakfast for the toddler), a system supports multiple scenarios of use. A storyboard is a high-level use case.

Any vision implies multiple scenarios to fulfill the vision. No system should be built around one storyboard or one future scenario. Gathering a set of people together in a room to "invent the future" with future scenarios, returns them to design from the "I." Where do the cases come from that the system must handle? If not from data, they must be generated by the people in the room. If they come from idealized process flows, they are probably unrealistically linear, ignoring the multiple strategies and complexities of real work processes. Good systems are not linear; they do not force or follow a single linear path. Driving scenarios from consolidated data ensures that different strategies for doing a task will be considered and that the design will support the different routes that real people use to achieve their intent and the different roles that people play to get their work done.

Storyboarding consolidated sequences ensures that you maintain the breadth of real work without having to make up cases. Thus, if the team is missing sequence data that supports the vision, they should collect it. But now, knowing what data is needed, data collection can be very focused. Get sequence data from two additional interviews and consolidate just those two sequences. If you can only get one sequence, analyze it as you would for sequence consolidation. Having a guide to the design is always better than making it up. Teams that try to design the future without data get lost, confused, and are prone to overdesign. Data keep teams focused and centered so invention is reasonable and likely to succeed.

User Environment Design

A product really has three layers of design. The user interface is the access mechanism to the function, structure, and flow of the work the user needs to do. The implementation (object model) is the way that the system will make that function, structure, and flow happen. But the core of a product is that middle layer: The explicit work the product is performing. The problem is that the work layer feels invisible. People know what a user interface is and they know what an object model is. But where is the structure of the work of a product? An architect has such a representation for a house design: The floor plan. It shows structure (the rooms each with a purpose and its function), function (what you can do in each room), and flow (the hallways and movement between and within rooms). Having a floor plan gives the architect a way to represent how people will live in the house without worrying about the interior decoration or the wiring system. The User Environment Design is a software floor plan; it represents the structure, function, and flow of a system.

The User Environment Design (Fig. 49.10) separates conversations about the system's underlying structure from conversations about its user interface and implementation. Like the architect's floor plan, it focuses the team on designing the work structure and flow within the system. To support work practice, designers must ensure that the product fits the larger work practice represented by the vision and storyboards, and that the work hangs together within the system itself. The User Environment Design enables the team to view the whole system and the relationships between its parts. It shows how the system will structure the user's work and how it will interface with other systems. It forces the team to focus on the system's function instead of jumping ahead to user interface issues or beneath to implementation issues. This helps ensure a workable and delightful user experience because it keeps the work coherent.

The User Environment Design formalism provides *focus areas* for each activity. Each focus area is a place that defines the functions and objects to be accessed from that place. Links from one focus area to another define the flow within the system that the user can traverse. Because focus areas support each other and naturally cluster related function and related focus areas, the User Environment Design ensures that designers think about the work pattern within the system without being distracted by discussions of icons, placement of function, or words on the user interface.

The User Environment Design was formalized before people were using site maps to guide Web site design. But both web-based and traditional products and systems benefit by having an abstract representation of the whole system that designers can look at, talk about, plan from, and test against scenarios of use. It helps designers and marketers get beyond thinking of the product as a list of functions or features. Because it shows all the parts of the system, it becomes a tool for system planning.

Managers can use the User Environment Design to see how their product interfaces with other products. The User Environment Design helps ensure design from a systems perspective. They can create rollout plans that carve up the system into coherent releases, each of which contributes significant benefit to users. They can assign parts of the system for further design to individual developers who can now see how their part fits into the whole.

Just as consolidated models help you see the structure of the work without getting lost in the variation, the User Environment Design helps teams see the structure of the product without getting lost in its details. Maintaining work coherence within and outside of the system requires this ability to conceptualize structure.

The User Environment Design Works for New Systems: What Do We Do for Existing Systems? Most people can see how building a User Environment Design makes sense for new systems. But most of the time, teams are modifying existing systems, not designing new ones. Existing systems are overwhelming. They are full of legacy features, and they demand that existing users continue to be supported. You cannot just start over.

Implicit in every kind of software is a User Environment Design, just as every house has a floor plan. You just have to find it.

You can redefine and extend an existing system by first building a Reverse User Environment Design to show how that system is structured. Once people see their product structurally, they often realize why people are having a hard time with it. Is your system too leggy? Does it have a star structure requiring the user to go back and forth to a central place to get anywhere? Does your system mirror your database hierarchy? Is the function clustered in your system so that every focus area has a clean purpose, or are functions mixed so the purpose of each place is confusing? These are questions that a Reverse User Environment Design can answer.

With a Reverse User Environment Design, you can plan to watch for these structural problems in the field interview. After the vision, you can determine whether to use the Reverse User Environment Design as the base of the new system, building on it and changing it in response to the storyboards. Used this way, it will constrain the future design and ensure that the design team deals with version-to-version compatibility. Or you can open up the design and create a new User Environment Design, opportunistically taking existing focus areas and functions from the existing system to include in a new design.

Good design is about seeing structure. The User Environment Design helps the design team think about and design for structure in their own and competitive products. Once teams can see structure, they can design for it, whether it is a new product or an existing one.

User Interface Design and Mock-up

At some level, the User Environment Design is a theory. If you were right about how you characterized the market and if you picked the right issues to address and if your vision of the new work practice was a good one and if you translated that vision through the storyboards and into the User Environment Design well, then the system should support and extend people's work. How can you find out?

Users do not understand models. Pelle En (1988) understood this years ago when he realized that holding users hostage in a conference room, squeezing object modeling insight out of their tacit understanding of their work, was painful and inaccurate. Researchers at Aarhus (Kyng, 1988) started using mock-ups of new design artifacts to test their ideas by having people use them in real work situations. They found that, within that context, users could understand the technology and further define the requirements. Users can talk user interface talk, or form factor talk for physical systems, but cannot talk model talk. Contextual Design moves quickly through those phases to get back to the users with our ideas represented as user interfaces.

Mock-up interviews using paper prototypes (Fig. 49.11) help designers understand why design elements work or fail and identify new function. These interviews are based on the principles of Contextual Inquiry given earlier. Test the paper prototype with users in their own context to keep them grounded in their real work practice. Users interact with the prototype by writing in their own content and manipulating and modifying the prototype. The partnership is one of co-design. As the user works with the prototype following a task s/he needs to do or

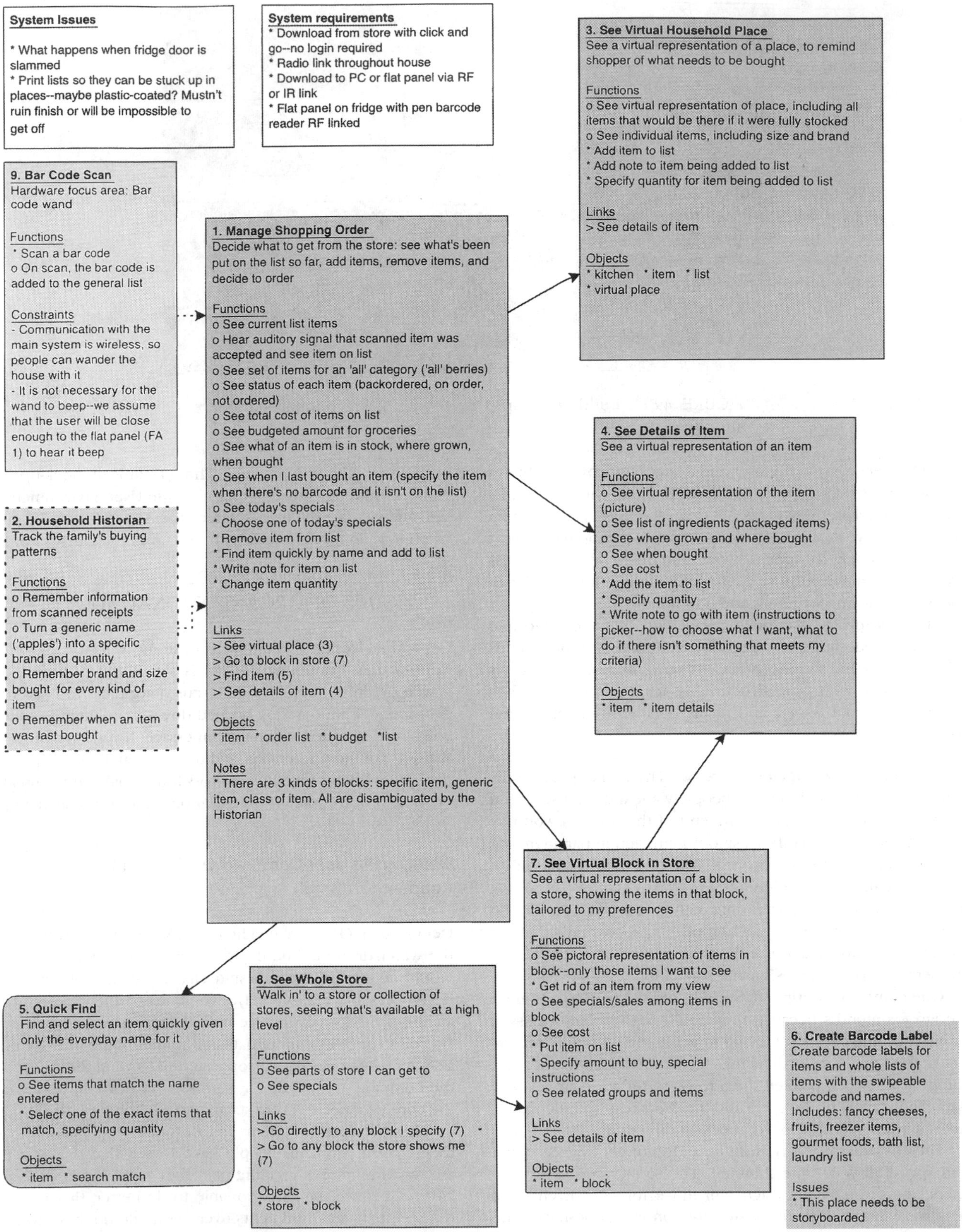

System Issues

* What happens when fridge door is slammed
* Print lists so they can be stuck up in places--maybe plastic-coated? Mustn't ruin finish or will be impossible to get off

System requirements
* Download from store with click and go--no login required
* Radio link throughout house
* Download to PC or flat panel via RF or IR link
* Flat panel on fridge with pen barcode reader RF linked

3. See Virtual Household Place
See a virtual representation of a place, to remind shopper of what needs to be bought

Functions
o See virtual representation of place, including all items that would be there if it were fully stocked
o See individual items, including size and brand
* Add item to list
* Add note to item being added to list
* Specify quantity for item being added to list

Links
> See details of item

Objects
* kitchen * item * list
* virtual place

9. Bar Code Scan
Hardware focus area: Bar code wand

Functions
* Scan a bar code
o On scan, the bar code is added to the general list

Constraints
- Communication with the main system is wireless, so people can wander the house with it
- It is not necessary for the wand to beep--we assume that the user will be close enough to the flat panel (FA 1) to hear it beep

1. Manage Shopping Order
Decide what to get from the store: see what's been put on the list so far, add items, remove items, and decide to order

Functions
o See current list items
o Hear auditory signal that scanned item was accepted and see item on list
o See set of items for an 'all' category ('all' berries)
o See status of each item (backordered, on order, not ordered)
o See total cost of items on list
o See budgeted amount for groceries
o See what of an item is in stock, where grown, when bought
o See when I last bought an item (specify the item when there's no barcode and it isn't on the list)
o See today's specials
* Choose one of today's specials
* Remove item from list
* Find item quickly by name and add to list
* Write note for item on list
* Change item quantity

Links
> See virtual place (3)
> Go to block in store (7)
> Find item (5)
> See details of item (4)

Objects
* item * order list * budget *list

Notes
* There are 3 kinds of blocks: specific item, generic item, class of item. All are disambiguated by the Historian

4. See Details of Item
See a virtual representation of an item

Functions
o See virtual representation of the item (picture)
o See list of ingredients (packaged items)
o See where grown and where bought
o See when bought
o See cost
* Add the item to list
* Specify quantity
* Write note to go with item (instructions to picker--how to choose what I want, what to do if there isn't something that meets my criteria)

Objects
* item * item details

2. Household Historian
Track the family's buying patterns

Functions
o Remember information from scanned receipts
o Turn a generic name ('apples') into a specific brand and quantity
o Remember brand and size bought for every kind of item
o Remember when an item was last bought

7. See Virtual Block in Store
See a virtual representation of a block in a store, showing the items in that block, tailored to my preferences

Functions
o See pictoral representation of items in block--only those items I want to see
* Get rid of an item from my view
o See specials/sales among items in block
o See cost
* Put item on list
* Specify amount to buy, special instructions
o See related groups and items

Links
> See details of item

Objects
* item * block

5. Quick Find
Find and select an item quickly given only the everyday name for it

Functions
o See items that match the name entered
* Select one of the exact items that match, specifying quantity

Objects
* item * search match

8. See Whole Store
'Walk in' to a store or collection of stores, seeing what's available at a high level

Functions
o See parts of store I can get to
o See specials

Links
> Go directly to any block I specify (7)
> Go to any block the store shows me (7)

Objects
* store * block

6. Create Barcode Label
Create barcode labels for items and whole lists of items with the swipeable barcode and names. Includes: fancy cheeses, fruits, freezer items, gourmet foods, bath list, laundry list

Issues
* This place needs to be storyboarded

FIGURE 49.10. User Environment Design for a shopping system. IR = infrared; RF = radiofrequency.

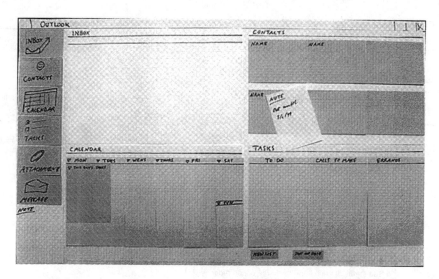

FIGURE 49.11. Build the paper prototype using normal stationary supplies.

did in the recent past, the user and designer uncover problems and adjust the prototype to fix them. Together the user and interviewer interpret what is going on in the usage and come up with alternative designs. Hand-drawn paper prototypes make it clear to the user that icons, layout, and other interface details are not the central purpose of the interviews; it keeps the user focused on testing structure and function.

After the design has been tested with between two and four users, redesign it to reflect the feedback. Multiple rounds of interviews and iterations allow testing in increasing levels of detail, first addressing structural issues, then user interface theme and layout issues, and finally detailed user interaction issues.

Let Customers Speed the Process. The goal of product design is to make something that people want and can use. If you only infuse customer data at the end of the process, you risk creating something that has a smooth interaction paradigm but does not support a valued part of the work. If you only infuse customer data at the beginning of the process, you risk having lots of good function that no one can access well. But, if you spend all your time in the middle of the process trying to get the function, layout, and user interface right before showing it to users, you risk overdesign and stagnation.

One company continued working on its User Environment Design for months because they could not resolve one issue or another, and they kept trying to get it perfect before returning to the field. Finally, their management was ready to cancel the project, which pushed them to go to the users and gather data. They talked to the users and the users resolved the issues, helping them identify the right design directions.

The customer is the bottom line. Do not get bogged down and spend all your time thinking. The sooner you get to the users, the faster your design will move forward. Even for a reasonably big project, all the steps of Contextual Design can be done in 2–3 months. A small project with limited models can be done in 5–6 weeks. After data consolidation, you should be out

in the field within 2–3 weeks. Three rounds of mock-up interviews should be enough to stabilize the User Environment Design and validate the user interface theme and layout. If you are in a hurry, do three rounds with two users each instead of four.

ISSUES OF ORGANIZATIONAL ADOPTION

Contextual Design is about using customer data to produce the right design. Contextual Design is also about making a data-driven organization. If you have customer data, you can make decisions without getting bogged down in arguments and you will have a story for your market message. If you have processes that tell you how to collect and use customer data to produce designs, working together in cross-functional teams quickly is possible. But this means organizational and personal change.

Introducing User-Centered Design Is a Long Road—Start Small

Dennis Allen (1995), when he was at WordPerfect, talked about the water drop method of introducing Contextual Design. He would collect a little data and chat about it at lunch to key developers and managers. He would invite a few people to do an interpretation session. He arranged for a general interest talk. He collected a little more data and told more people. He made a slide show of the data. People started to want the data. Soon, the company was ready for a project. In time, 10 teams from the core product set were starting to use the process.

Change is first about changing attitudes. People have to know they have a problem. People have to see that the way they are working now is creating pain they need not have. People have to believe that it is possible to do better, that confused requirements and arguments over design do not have to come with the job. People have to have the bandwidth to try something new. People have to be willing to change.

Evangelists are the early adopters of new processes. They read about them and try to bring them into their companies. Building awareness through conversation, talks, and grass-roots activities always comes first. Culture and organizations change after their talk changes. Creating the buzz, creating a conversation about process, creating awareness about how people do their work, helps people understand why they should do something as onerous as changing their everyday habits.

This is true even when management is ready for the change. Now that user-centered design is seen as the "right way," managers are trying to change their organizations overnight. Our experience is that executives tend to expect results faster than is reasonable. CEOs often want everybody trained, user-centered processes implemented, and all the products delivered based on customer data immediately. This kind of management initiative can be very disruptive to the day-to-day work of making and selling products.

No matter who it is, grass-roots organizer or CEO, one person acting alone cannot change an organization's processes. Changing a whole organization requires effort at both the decision-making and grass-roots level: Spreading awareness and motivation; setting the new management direction; aligning people's work; changing evaluation criteria and milestone measures; training new skills; dealing with the people who are no good at the new skills; planning projects; tracking success and problems; changing the processes based on feedback; and making the new practice real in the context of a working project.

Change at a huge multinational company, with thousands of people busily working every day, will take years. It will be, like any large cultural movement, won on the ground: Person by person, project by project. So, evangelists: Give yourselves a break. Count every success no matter how small.

The best way to raise awareness is with the water drop technique. Do something small: Collect a little data, interpret it with a buddy, and share it around. Do guerrilla warfare: Find some others to join you, and collect a little data and interpret it without getting permission. Then share it around.

The next step is to identify a test project. Start with a project in an interested part of the company. Do not start with a big project or the most important product in the company. Make sure everyone is trained and set project scope small. Do a subset of the models. Use the data to drive design thinking—then perhaps go to your normal processes. Try the rest of the process on another project. Build success a bit at a time.

In this way, people can try parts of a new process on their own project, see the results, and share them. The Contextual Design process makes projects visible, because it requires a dedicated design room and creates many artifacts hanging on the wall. The environment it creates is so unusual that people frequently will visit the room just to see what is going on. Putting affinity diagrams on an outer wall will lead passersby into the room. A good design room, combined with the water drop method, will help create the buzz you need to get things going.

But remember, failure will kill it. One company had collected data from 100 people. They had taken months and had reams of data—observations cataloged on paper, none of which anyone used. But they had collected too much data, did not communicate it effectively, and did not get any of the results to affect a real shipping design. They lost management's confidence in trying a new process by producing no usable result. Now they will have to wait for new management, people who did not see this wasted work, and try to get them interested again. Pick something small and make sure it is successful. Know your techniques, manage your project scope, take care of each person's needs on the project, and do the painful leg work of organizing the project and getting customers yourself. Then publicize it throughout the company.

By combining upper level managers' pull with grass-roots evangelist push, the evangelist can ferret out pockets of interest inside the organization to initiate action. By setting reasonable expectations, these two parts of the organization will carry along the rest of the people who are engaged, heads-down, in their day-to-day work.

Communicate Out and Give Ideas Away

One team doing a Contextual Design project worried that the 600 coders they worked with would not adopt their designs. So, they identified and targeted friendly coders, asking them for help with prototyping, for their feedback on design ideas, and so forth. Next, in the interest of rolling out the process and the design, these designers gave ownership of the designs to the friendly coders, asking them to write the specifications, prototype the design, and give the prototypes to their managers. Soon, the coders were coming to them because they wanted the designs.

However, the designers did not want to just pass designs to the coders—that would not have achieved the overall awareness or driven customer-centered processes in the company. So, they started working only with coders who agreed to participate in field research, helping to collect and interpret customer work data. Even though this meant the designers gathered far more data than was necessary, they were willing to do this to spread awareness of the value of customer data throughout the company.

Doing something new often looks like a secret project. The team becomes involved with each other, talk a new language, and works a new way. If stakeholders, people on related projects, or the coders who will develop the designs think, "you act like you are *so* special" or if you keep telling them "how great this process is," they will resent it and resist changing. Communicating out and bringing people in is the way to be sure you do not become isolated:

- After you know what you are doing, bring other members of the team out with you on field interviews to take notes.
- Invite others into the interpretation sessions. Add one at a time so there are more people who know how to do the process (3 to 1 is good). It only takes two hours of their time, and then they will know what you are doing "all day in that room" and can help again in the future.
- Build the affinity as a big team event. Invite 10–15 people and involve them in building the data. Get their help in consolidating models, too. Pair each new person with someone on the team.

- Let interested parties and related teams walk the data and vision. Remember affinity wall walking and visioning are ways to hear their design ideas.
- Have bag lunches and open houses in the team room and tell people about your data. Collect their questions and challenges and use them to reset focus.
- Hang your data in the halls to pique interest and demonstrate openness.
- Publish on corporate Web sites and print newsletters to share your work across the organization. Editors and web masters are always looking for new stories.

Communicating out has side benefits as well. A large company had to hang their models up in the same hall that prospective customers for very large sales walked through. The company was able to demonstrate their understanding of the customer: Customers could identify themselves in the models!

Shipping product and changing organizations is all about getting buy-in from the people involved. Communicating, involving, and listening to people are part of the process. It has to be planned and managed by the people introducing the changes.

Ownership Is the Goal; Renovation Is Normal

All companies are already running themselves somehow. They may have very formal processes or they may be small start-ups with informal processes. They may have formal methodology groups in charge of their processes or they may have strong traditions that guide everyday work. Therefore, unless a founder who really wants to do Contextual Design starts a company, all introduction of Contextual Design involves fitting into existing ways of working. Any process, even if adopted by an enlightened CEO with an enlightened set of developers, will have to be changed to fit the company, the people, and their skills. Any user-centered design process, will be renovated and adopted piece by piece.

Part of any adoption process is to make it your own. People will argue that they must change a process to fit their situation or to improve it. Early on in the evolution of Contextual Design, there was a step called "redesigned sequence models," a rewrite of the consolidated sequence steps to reflect the vision. An internal evangelist could not get anyone to use redesigned sequence models. Instead, their people were excited about storyboarding as a technique, which the evangelist realized was just redesigned sequence models but in drawings. So they did storyboarding and were happy. Renaming or modifying a step in Contextual Design makes no difference, as long as the fundamental intents of the step are achieved. Because of the value of visualizing the steps pictorially, storyboarding was incorporated into the Contextual Design process.

When adopting a new process, you cannot make change happen faster than people's awareness, growing skill, and willingness to change. People will resist a new process until they can find a place for themselves within the process and a place for the process within their organization. They have to figure this out themselves.

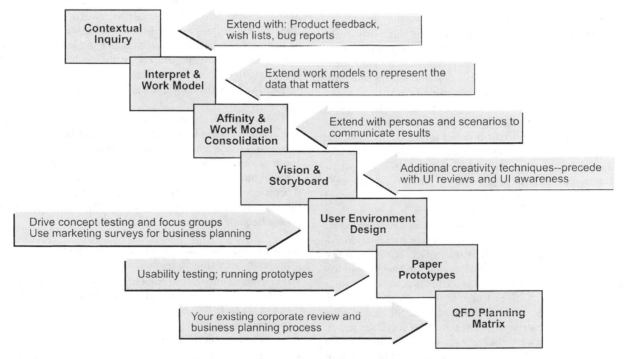

FIGURE 49.12. The steps of Contextual Design act as a process scaffolding upon which complementary activities can be included. QFD = Quality Function Deployment; UI = User Interface.

One company was very excited about collecting customer data, and they started sending hundreds of people out into the field. Soon they were overwhelmed with the amount of data and started looking for a way to handle it. They tried everything—individual reports, group presentations, and postinterviewing data dumps—and still could not handle the data or get people to really think about it. Then their Contextual Design evangelist suggested, "What if we tried this technique called an interpretation session, where you write down all the points and model the data?"

That is simply the second step in Contextual Design; but, until they had tried their own ideas, they could not believe that the interpretation session was efficient. Companies may resist each step of Contextual Design, and then adopt each piece with slight tuning. Companies may lose faith in the process because of an early failure—yet, years later, they are still doing key techniques under different names. There are lots of companies and designers who are trying particular Contextual Design techniques in many diverse projects.

It is not a failure of adoption if people resist or change a particular step—it is part of their adoption process. Training, coaching, and advising will always initiate the adoption process, but it is the people working with it on a daily basis who have moved the industry forward over the last 13 years. These are the people who change and tune the process to make themselves and their organizations comfortable. These people have made customer-centered design and Contextual Design a part of the daily work of companies all around the globe.

Contextual Design is a set of steps (Fig. 49.12) that create scaffolding that organizations can use as a base to build their own approach to understanding the customer and getting that data into their designs. To change, people need to know exactly what to do to be successful. They need to know the techniques, how to work together, and how to fit this work into their daily life. Therefore, change looks like taking on one piece or technique at a time. Most companies start by going to the field to gather customer data and proceed from there. Each step is a success. Each success must be recognized and celebrated.

Looking back, the industry has come a long way. Today, companies all over the world are designing from customer data. The argument is now what customer data to collect and how to collect it, not whether there should be customer data. Using customer data is now a central part of the corporate dialogue on how to make products. I am pleased that Contextual Design has been a part of this revolution.

References

Allen, D. (1995, May). Succeeding as a clandestine change agent. In K. Holtzblatt & H. Beyer (Eds.), *Requirements gathering: The human factor* (Vol. 38, No. 5).

Beyer, H., & Holtzblatt, K. (1998). *Contextual design: Defining customer-centered systems* (1st ed.). San Francisco: Morgan Kaufmann.

Ehn, P. (1988). *Work-oriented design of computer artifacts*. Falkoping, Sweden: Gummessons.

Kyng, M. (1988, September 26–28). Designing for a dollar a day. In *Proceedings of the Conference on Computer-Supported Cooperative Work* (p. 178). Portland, OR.

Beyer, H. (1994, Sept./Oct.). Calling down the lightning. In *IEEE Software* (pp. 106–107).

·50·

AN ETHNOGRAPHIC APPROACH TO DESIGN

Jeanette Blomberg
Blekinge Institute of Technology

Mark Burrell and Greg Guest
Sapient Corporation

INTRODUCTION

In recent years, academic and professional researchers and designers working in the field of Human–Computer Interaction (HCI) have looked to ethnography to provide a perspective on relations between humans and the artifacts they design and use. Within the field of HCI, there are different views among researchers and practitioners on just what constitutes an ethnographic inquiry. For some, ethnography is simply a fashionable term for any form of qualitative research. For others, it is less about method and more about the lens through which human activities are viewed. In this chapter, we will attempt to position the ethnographic approach within historical and contemporary contexts, outline its guiding principles, detail the primary methods and techniques used in ethnographically informed design practice, and provide case examples of ethnography in action.

This chapter provides an introduction to ethnography, primarily as it relates to studies in HCI. We will touch only briefly on some of the more controversial topics current within the field of ethnographic research that have enlivened mainstream academic discourse in recent years. Instead, we will point the reader to books and articles where these topics are discussed in more detail. Our primary aims in this chapter are to provide academics and professionals in the field of HCI with a working understanding of ethnography, an appreciation for its value in designing new technologies and practices, and a discerning eye when it comes to reviewing and evaluating ethnographically informed design studies.

THE RELEVANCE OF ETHNOGRAPHY FOR DESIGN

The turn to ethnography as a resource for design can be traced back to the early 1980s when computer technologies were moving out of the research labs and engineering environments and into mainstream office settings, call centers, manufacturing floors, and educational institutions. There was the realization that the designers and developers of these technologies could no longer rely exclusively on their own experiences as a guide for the user requirements of these new systems. Instead designers and developers needed to find a way to gain an understanding of the everyday realities of people working within these diverse settings (Blomberg, Giacomi, Mosher, & Swenton-Wall, 1991). In many organizations, market research groups were being asked to provide insights into the people and practices that made up these varied settings. However, the techniques most commonly used by market research groups at the time (e.g., attitude surveys, focus groups, telephone interviews, etc.) were not well suited for developing understandings of what people actually do day to day.

Anthropologists and other social scientists had long recognized that what people say and what they do can vary significantly, making reliance on surveys, focus groups, and telephone interviews insufficient for the task at hand. Designers and developers needed a way of getting a firsthand view of the on-the-ground realities—the here and now—of everyday life in these diverse settings. At this time in the early 1980's, social scientists working at the Xerox Palo Alto Research Center were beginning to explore ways of bringing insights from ethnographic research into a productive relationship with the design of new technologies (e.g., Blomberg, 1987, 1988, 1995; Suchman,1983; Suchman et al., 1999a). Not long after, other research labs (e.g., Hewlett-Packard, Apple Computer, and NYNEX) followed suit (e.g., Nardi & Miller, 1990; Sachs, 1995). Today, most industrial research and development labs in the United States have anthropologists and other social scientists on staff (e.g., Intel, AT&T, Kodak, Xerox, and Microsoft to name but a few).

Ethnographically informed design practices also began to take hold in design firms and consulting companies during the early 1990s (e.g., IDEO, Fitch, and the Doblin group). These early explorations culminated in 1993 with the founding of E-Lab, a research and design company that distinguished itself from other design firms at the time by creating an equal partnership between research and design (Wasson, 2000). Ethnographic methods were at the center of E-Lab's research approach, with a commitment to base design recommendations on insights from ethnographic research (Robinson, 1994).

Furthermore, in the mid-1980s the growth in networked applications and devices made possible through the availability of local area networks (LANs) and early Internet implementations created an awareness among designers and developers that they would need to focus beyond the support of single, isolated users interacting with information technologies. What would be needed was a way of exploring the information and communication practices of people interacting with one another, both face-to-face and through mediating technologies. Information technologies were increasingly becoming communication and collaboration technologies that demanded an examination of social interaction across time and space. In response, a group of computer scientists, human factors engineers, and social scientists, somewhat dissatisfied with the dominant perspectives within HCI at the time (e.g., perspectives that emphasized technological possibilities over the uses and users of technology, the interface requirements of standalone applications over networked devices, and human psychology and cognition over social interaction), founded the field of Computer-Supported Cooperative Work (e.g., Greif, 1988; Schmidt & Bannon, 1992). A group of sociologists at Lancaster University and researchers at the Xerox Research Center in Cambridge, England, have played a prominent role in helping to shape the ethnographic research agenda within Computer-Supported Cooperative Work for the last decade (e.g., Bentley et al., 1992; Hughes, Randall, & Shapiro, 1993; Hughes, King, Rodden, & Anderson, 1994; Hughes, Rodden, & Anderson, 1995).

Finally, a more recent trend has redoubled interest in the ethnographic perspective as a valuable tool in the design of new technologies. The explosion of the Internet with its reach into all aspects of people's lives has accelerated the move of information technologies out of the workplace and into homes, recreational environments, and other nonwork-related settings. This has presented a new set of challenges for designers as they design and build applications that leverage powerful, digital technologies for use by people of all ages, engaged in myriad nonwork-related activities. Although the clamor for all that is the Internet may

have temporarily subsided, what has not diminished is the acknowledgment that designing interactive technologies that are useful and engaging will require paying attention to the everyday realities of the people who will use them. Here, the ethnographic perspective is just as relevant.

THE ROOTS OF ETHNOGRAPHY

Ethnography has its historical roots in anthropology, but today is an approach found in most all of the traditional and applied social sciences, and in interdisciplinary fields such as HCI and Human Factors Engineering. In anthropology, ethnography developed as way to explore the everyday realities of people living in small scale, non-western societies and to make understandings of those realities explicit and available to others. The approach relied on the ability of all humans to figure out what is going on through participation in social life. Its techniques bear a close resemblance to the routine ways people make sense of the world in everyday life (e.g., by observing what others do, by participating in social life, and by talking with people). The research techniques and strategies of ethnography developed and evolved over the years to provide ways for the ethnographer to "be present" for the mundane, the exceptional, and the extraordinary events in people's lives.

Over the years within the field of anthropology, both the focus on non-western peoples and the implicit assumptions made about non-western societies (e.g., that they are bounded, closed, and unchanging) have changed. Today, the ethnographic approach is not limited to investigations of small-scale societies, but instead is applied to the study of specific settings within large industrialized societies, such as workplaces, senior centers, and schools; and specific activities such as leisure travel, financial investing, teaching, and energy consumption to name but a few. As a consequence, new techniques and perspectives have been developed and incorporated into anthropology and ethnographic inquiry more generally. However, a few basic principles discussed herein have continued to inform and guide ethnographic practice.

PRINCIPLES OF ETHNOGRAPHY

Natural Settings

Ethnography is anchored in the underlying assumption that to gain an understanding of a world you know little about, you must encounter it firsthand. As such, ethnographic studies always include gathering information in the settings in which the activities of interest normally occur. This does not mean that ethnographic studies never involve techniques that remove people from those everyday settings or that introduce artifacts or activities that would not be present otherwise. The insistence on studying activities in their everyday settings is motivated by the recognition that people have only limited ability to describe what they do and how they do it without immediate access to the social and material aspects of their lives. Furthermore,

some aspects of people's experiences can only be understood through observation of activities as they occur (e.g., people's patterned movements through airports or convenience stores, moment-by-moment shifts in scheduling, etc.).

Holistic

Related to the emphasis on natural settings is the view that activities must be understood within the larger context in which they occur. Historically within anthropology, the notion of holism focused attention on the fact that societies were more than the sum of their parts (however, these parts were specified). The particular aspects of a society (e.g., the court system) could only be understood in relation to the other aspects of the society (e.g., kinship system and religious beliefs). Today, because ethnography is less frequently applied to the study of entire societies, the notion of holism has a somewhat different emphasis. Holism holds that studying an activity in isolation, without reference to the other activities with which it is connected in time and space, provides only a limited and potentially misleading understanding of that activity. So, for example, it would be of dubious value to investigate online search strategies without understanding how these strategies fit into the larger set of activities of which search is but one component (e.g., in the context of online trading, shopping, or report writing).

Descriptive

Ethnographic accounts have always provided a descriptive understanding of people's everyday activities. Ethnographers are concerned first and foremost with understanding events and activities as they occur and not with evaluating the efficacy of people's everyday practices. This is not to say that ethnographic accounts cannot or should not be used to suggest how things could be different or to point out inequities in current ways of doing things. However, there is a strong conviction that to suggest changes or to evaluate a situation, one first needs to understand it as it is. The work practice and technology group at the Xerox Palo Alto Research Center developed a slogan to express this conviction that innovation requires an understanding of the present (Fig. 50.1).

As such, ethnographic accounts strive first and foremost to provide a descriptive and not prescriptive understandings of people's everyday lives. In recent years, there have been many challenges to the idea that a purely descriptive understanding

Innovation
=
Imagination of what could be
based in a knowledge of what is

FIGURE 50.1. Innovation.

is possible. Critics point out that every account is shaped by the perspective of the researcher, the goals of the project, and the dynamics of the relationship between the investigator and those studied, to name but a few factors that shape ethnographic accounts. Although it is hard to argue with this position, in our view this recognition does not diminish the value for design of describing the everyday realities of people engaged in activities that we ourselves might never experience.

Members' Point of View

As already alluded to, ethnographers are interested in gaining an insider's view of a situation. They want to see the world from the perspective of the people studied and are concerned with describing behaviors in terms relevant and meaningful to the study participants. As such, ethnographers are interested in the ways people categorize their world and in the specific language they use to talk about things. This perspective sometimes is at odds with the requirements of quantitative survey research where the relevant categories must be known in advance of the study and where the categories and the language used cannot vary across participant groups. As a consequence, in quantitative social science research, the terms and categories used are likely to be those of the research community and not those of the study participants, which can undermine the validity of the results (see Ethnographic Methods section for further discussion of this topic).

THE POSTMODERN INFLECTION

The scientific paradigm within which ethnography evolved has come under serious questioning over the last decade or so as social studies of science have shown how scientific knowledge production is shaped by the larger social context in which scientific inquiries take place (Latour, 1987; Latour & Woolgar, 1986; Pickering, 1980). As part of this critical discourse, ethnographic accounts have been challenged for their veracity. Likewise, the authority of the ethnographic voice has been questioned (Clifford, 1988; Clifford & Marcus, 1986; Marcus & Fischer, 1986). These challenges have come from a number of fronts, most significantly from study participants who increasingly are able to read ethnographic accounts (Said, 1978) and from feminist theorists who saw in many ethnographic accounts a western, male bias (Harding, 1986; Smith, 1987; Wolf, 1992; Yanagisako & Delaney, 1995). These challenges have made researchers from all fields of inquiry more aware of how their research is shaped by the particular time and place in which it occurs. It is our view that knowledge of the world is always mediated by presuppositions, be they cultural, theoretical, or practical, and as such no ethnographic account is value free. But we also contend that this does not diminish the efficacy of an ethnographic approach as a resource for designing new technologies and practices and that maintaining the illusion of an absolute "truth" is not necessary.

ETHICAL ISSUES

As will be discussed in more detail later, ethnographic research requires developing the trust and participation of the people studied. Without this trust, participants will be reluctant to allow researchers into their homes, boardrooms, and classrooms; and they will not openly share their everyday experiences and concerns. Anthropologists have long realized that such a privileged, trusted position requires reciprocity—if you allow me access to your world, I will protect your interests. This bargain has not always been easy for ethnographers to keep. Over the years, there have been examples of ethnographic research, where wittingly or not, the situation of the people studied has been compromised.[1]

In the context in which ethnographic research is being used to inform the design of new technologies—technologies that will change people's lives—it is critical that the ethnographer reflect on the impact this research could have on study participants (Blomberg, Suchman, & Trigg, 1997). Of course, it is not possible to control all the ways findings from ethnographic research will be used, nor how technologies informed by these studies will be integrated into people's lives. But the ethnographer can work to protect study participants from immediate harm (e.g., that caused by divulging a worker's identity to management) and can inform study participants of possible longer term negative impacts (e.g., job losses brought about by introduction of new technologies). Because ethnographic research has moved into new contexts (e.g., HCI, organizational development), it has been necessary to think creatively about how our ethical guidelines map to these new conditions. However, we cannot lose sight of the importance of protecting the interests of those who have agreed to participate in our studies, be they workers in organizations, traders on wall street, or mothers of special needs children.

ETHNOGRAPHIC METHODS

The ethnographic method is not simply a toolbox of techniques, but a way of looking at a problem, a "theoretically informed practice" (Comaroff & Comaroff, 1992, quoted in Agar, 1996). The methods and techniques outlined have been developed over the years to enable the development of a *descriptive* and *holistic* view of activities as they occur in their *everyday setting* from the *point of view of study participants*. We are not attempting to be exhaustive in our presentation, nor do we want to suggest that there is a fixed set of canonical ethnographic methods and techniques. We encourage researchers

---

[1]To mitigate such negative impacts, the American Anthropological Association has developed a code of ethics that provides guidance for people engaged in ethnographic research. This code outlines the appropriate disclosures and protections that should be given to study participants (see Fluehr-Lobban, 1991, for a discussion of ethical issues in anthropological research).

to develop new techniques as the circumstances require (e.g., studying virtual communities, globally distributed work groups, technologically mediated interactions). What remains constant in the approach taken is a commitment to describe the everyday experiences of people as they occur.

Planning Research

One of the keys to a successful research project is a carefully thought out plan. Research planning can be divided into three general stages—formulating research objectives, devising a sampling strategy, and selecting appropriate methodologies. These activities are interrelated; so, if one of these has not been carefully considered, the entire research is affected.

Research objectives follow from the specific questions to be addressed by the research. It is often useful to develop a research statement, an affirmative sentence that clearly states what one wants to achieve from a given study. This statement acts as a beacon to help keep the research on track through the many twists and turns of a project. If the research is to be used to inform the development of a software application or a Web site that will help doctors manage patients' records, the research statement could be something as simple as, ". . . to understand how doctors manage patient records and the implications this activity has for design." If developing a system that can be customized and used by both doctors and patients is desired, then the statement will change, likely including something about how these groups differ in their practices and in their needs for such a system. Although a research statement might change over the course of a project, this change should also be reflected in changes to the research design (e.g., sampling strategy and selection of research methods and techniques).

Sampling

Once the research objectives have been identified, a sampling strategy should be devised that answers two primary questions: What types of participants best suit the research objectives? And, how many participants should be included in the study to achieve the research objectives? Bernard (1995) identifies seven types of sampling strategies that fall under two main categories: probability and nonprobability. Each sampling type has a specific purpose. Our focus in this chapter is on nonprobability sampling because that is the most commonly used in ethnographic research.[2] The nature of ethnographic work, as well as budget, time, and other recruiting constraints, invariably result in selecting participants based on criteria other than

a strict probability. In most cases, nonprobability sampling is more than adequate to achieve the desired research objectives of ethnographic studies and HCI projects.[3]

Four types of sampling fall under the rubric of nonprobability: quota, purposive, convenience, and snowball (Bernard, 1995). When sampling by quota, the researcher specifies which groups are of interest (e.g., women, teenagers, truck drivers, people who use software X, etc.) and how many individuals will be needed in each group. The number of groups chosen will depend on the research objectives and the amount of time available, but the basic idea is to cover the range of possible variation one would expect across an entire target population. To ensure the desired variability is covered, it is useful to create a "screener,"[4] a questionnaire-like instrument designed to identify characteristics that are appropriate for a given project. Quota sampling is only possible when the desired participants are easy to identify in advance and recruit. There is no statistical basis for determining the number of participants[5] in each group for a quota sample, but a general rule of thumb is between six and ten. If a researcher is not able to specify how many participants will be in each sampled group, a purposive sampling strategy may be called for. This sampling strategy is based on the same principles as quota sampling, but the number of participants for each group is not specified.

Convenience and snowball sampling rely on a sample as you go strategy. This is required in situations where you do not know in advance who will be available to participate or which individuals or groups should participate. Convenience sampling entails selecting people who are available, meet the requirements of the research, and are willing to participate. One might use this strategy, for example, to observe and interview shoppers as they explore a grocery store.

Snowball sampling relies on participants referring others whom they think would be good candidates for the research or on researchers identifying individuals or groups to be included in the study from the ongoing research. Because this method utilizes existing social networks, it is especially valuable when desired participants are initially inaccessible or reluctant to participate (e.g., CEOs, drug users, union leaders) or when the relevant population cannot be known in advance (e.g., Johnson, 1990, for a more detailed discussion of sampling in ethnography).

Gaining Access

Sometimes one of the most difficult tasks for the ethnographer is gaining access to field sites and participants. In more traditional ethnographic studies, the ethnographer's challenge was

---

[2]The intent behind probability sampling, or statistical sampling, is to generalize from the research sample to a larger population with a specified degree of accuracy, measured in terms of probability. All types of probability sampling require a randomly selected and relatively large sample size.
[3]Using nonprobability samples does not mean we cannot make general statements. If participants are chosen carefully, one can obtain reliable data with as few as four or five participants (Nielsen & Landauer, 1993; Romney et al., 1986). Additionally, a recent case study demonstrates that smaller, nonrandomly selected samples can produce the same results as large-scale survey research for as little as 1/100 of the cost (Green, 2001). A nonprobability strategy also does not preclude conducting a statistical analysis or measuring differences between individuals or groups using nonparametric statistics, such as Fisher's Exact Test or nonparametric correlation measures. Their limitation is that they cannot be used to make claims about larger user populations within a specified degree of probability.
[4]Screeners are an essential tool if using an external recruiting agency to locate study participants.
[5]For sampling purposes, participants need not be individuals, but could be families, households, work groups, or other naturally occurring entities.

to gain permission to conduct fieldwork in another country. Once allowed to enter the country, the strategy often was to make friends in the selected village or town and take part in community activities. With the shift in focus away from village or community studies, today it is often necessary to establish more contractual relationships with field sites and study participants. In some cases, recruiting agencies may be used to identify participants, and participants may be offered financial incentives to take part in the study. In addition, gaining access to institutional settings, be they corporations, schools, or government agencies, may require written permission that specifies certain terms and conditions before researchers are given access to the field site. It is important not to underestimate the time (and skill) required to establish these initial relationships and agreements.[6]

Observation

As discussed earlier, ethnographers are interested in understanding human behavior in the context in which it naturally occurs, making observation one of the hallmark methods of the approach. In academic settings, it has been typical for an anthropologist to spend a full year or more in a given field site. Although this continues to be the case for more traditional ethnographic studies, shifts in research focus (e.g., away from studies of entire societies) and in study locations (e.g., less likely in isolated, hard to get to settings) have resulted in more varied research designs that may involve shorter, intermittent fieldwork periods in distributed locations. In corporate and applied settings, the time available for in situ observation may be seriously constrained, sometimes no more than a few days in any one setting. Whereas the amount of time may be relatively brief, observing individuals for even a few days can lead to valuable insights that other methods cannot obtain.

Why Observe? One of the fundamental axioms in the social sciences, and anthropology in particular, is that what people say they do and what they actually do are not always the same. Studies have shown verbal reports to be inconsistent with behavior in a number of areas, including (among many other examples): shopping behavior (Rathje & Murphy, 1991), child rearing (Whiting & Whiting, 1970), recycling (Corral-Verduga, 1997), and health habits (Rich, Lamola, Amory, & Schneider, 2000).

The discrepancies between verbal reports and behavior can be caused by a variety of factors. People may be concerned with their image and so report, consciously or not, behavior that is more socially acceptable. Along these same lines, a participant may respond to a question in a particular way in an attempt to please the researcher. Another source of disparity between behavior and verbal reports is that people are often not aware of their actual behavior because it is so habitual. Such tacit knowledge is often not easily accessible through interview techniques (D'Andrade, 1995).

The limitation of human memory is another reason why interview data can differ from observations. When asking participants about past events, or recurring patterns of behavior, our memory may be selective and skew responses in any number of directions. In many cases, these tendencies occur in predictable patterns (Bernard, 1995).

The complexity of social life is another reason individual accounts of an event may miss certain relevant details. The environments in which humans interact are extremely dynamic and complex—composed of social relationships, artifacts, and physical spaces—which can make it difficult for individuals to fully envision, let alone articulate, after the fact, what is going on.

The Researcher's Observational Role. When it comes to observation, there are varying degrees to which the researcher can become integrated into the scene. At one end of the spectrum, the researcher may become an *observer-participant*. In this role, one attempts to be as unobtrusive as possible, quietly observing events from a discreet, yet strategic, position. At the other end of the spectrum is the *participant-observer*. In this situation, the researcher is actively involved in the events observed.

There are pros and cons associated with each type of role. Although being fully integrated into the action provides the researcher with a first-hand experience of an event and otherwise inaccessible insights, taking good notes in this context is difficult at best. A great deal of energy is spent trying to assimilate rather than on attempting to make sense of the events in the context of the research objectives. One often has to rely on memory, therefore, writing up field notes after the fact. Taking a more observational role affords a wider perspective on events and the time to record and analyze events as they unfold. On the downside, it precludes the opportunity to experience the activity first hand. In many research situations, the ethnography's position moves between these two extremes, sometimes occupying a hybrid position of both partial participant and outside observer.

Structuring Field Observations. Before setting out to observe, decisions need to be made about what, where, and when to observe (Whiting & Whiting, 1970). One might decide to observe individuals as they go about their work and daily routines (person focused), a technique sometimes referred to as "shadowing" (Wasson, 2000). A variation of person-focused observation is the "beeper study," in which participants carry a pager with them during the research period. When the researcher sends a page, the participants record where they are, what they are doing, and the context in which they are doing it. The researcher might also decide to focus on a specific event, such as a meeting or software education class (event focused), or observe the activities that occur over time in a given area, like an office (place focused). One can even shift the subject of observation to an artifact, such as a document, and record its movement from person to person (object focused).

---

[6]Anthropologists have been accused in the past of only studying the disempowered and disenfranchised because these individuals were less likely to feel powerful enough to refuse participation in ethnographic studies. When studying people with more power and ability to say no (Nadar, 1974), it is often necessary to demonstrate how their participation will be of benefit to them, their community or workplace, or the wider society.

Videotaping. Given the complexity of human behavior, it is impossible to notice and record in real time everything of interest to the researcher. This is one of the reasons video cameras are becoming increasingly popular in fieldwork. Video records can be used as a reference to supplement field notes. The ethnographer also has the advantage of being able to watch an event multiple times, so can change levels of analysis or observational focus with subsequent viewings (e.g., interaction between people vs. the movement of one individual in and out of a scene).

Videotaping also allows people not primarily involved in the fieldwork to participate in the analysis and opens up the range of perspectives that can be bought to bear on the analysis (e.g., Blomberg & Trigg, 2000) used video collection tapes in interactions with product developers; also see Brun-Cotton & Wall, 1995; Karasti, 2001; Suchman & Trigg, 1991).

An additional bonus of using video cameras is that they can record events in the absence of the researcher. Not only does this free the researcher to be involved in other activities, but the camera also can be a silent presence[7] in situations where an outsider (even a well-trained participant observer) would be seen as intrusive (e.g., child birth, counselor–student interactions, board room deliberations, etc.). This, however, does not preclude the need to later review the videotapes and incorporate relevant information into the analysis, a time-consuming activity.[8]

In summary, how and what one observes and analyzes depends on the research objectives. In many cases, a combination of techniques is preferable, allowing the ethnographer to view things from a variety of perspectives.[9]

Interviewing

Interviewing is one of the ethnographer's most valuable tools (Gubrium & Holstein, 2002). Interviews can inform research design and observations. They are essential in understanding member's perspective. Interviews are often grouped into three categories: unstructured, semistructured, and structured. In reality, it is more like a continuum, with at one extreme the casual conversation and at the other a formal questionnaire.

In the early stages of fieldwork, interviews are most often open-ended and unstructured. In these early stages, the ethnographer is just beginning to get a perspective on the activities and people studied. An unstructured format gives the researcher the freedom to alter the line of questioning as the interview unfolds. The researcher essentially is learning what questions are important to ask. Unstructured, however, does not mean haphazard or lacking purpose. The researcher will know the topics to be explored when entering the field, and will usually have a loose interview protocol to serve as a guide for the interview.

Although the protocol provides a basic framework for an unstructured interview, the participant plays a major role in the direction the interview takes. As Bernard (1995) puts it, the idea is to "get an informant on to a topic of interest and get out of the way." When the interview moves to a topic of particular interest, the researcher can then probe deeper to elicit more details.[10] Indeed, interviewing is an art, and one of the key skills an ethnographer learns is the art of "interrupting gracefully" (Whyte, 1960).

In an open-ended interview, it is important to avoid what Rhoades (1982) calls the "Joe Friday Syndrome," using an interrogation style of questioning to uncover the "facts." The purpose of keeping the interview open is to allow for a wide range of responses. Using too structured a format at an early stage constrains the range of possible answers and increases the chances of missing critical pieces of information.

As a project progresses and patterns begin to emerge, interviews can become more structured and the line of questioning less broad. The researcher begins to narrow in on topics that are particularly informative and relevant to the research objectives. Questions on the protocol become more focused and specific as answers to previous questions guide the follow-up questioning. A general guideline for determining when enough interviewing has been conducted is the point when responses to questions cease to be novel or surprise the researcher (Blomberg et al., 1991).

Once the range of responses is known and the data begin to show clear patterns and themes, the researcher may want to structure interviews further. A host of structured techniques exist. Some are designed to identify the ways people organize information within a specified domain, such as free listing, card sorts, triad's tests, and paired comparisons (Romney, Batchelder, & Weller, 1986). Other techniques, such as questionnaires and surveys,[11] are used to assess variation between two or more groups or to establish how representative the findings are in a larger population. The main idea behind these techniques is to keep the form and content of the questions consistent for each respondent, thus allowing for differences among the sample population to be ascertained. One advantage of conducting structured interviews after an ethnographic study is completed is that the question structure and language can reflect the way participants talk about and organize experiences, thus increasing the validity of the survey findings.

The Interview As a Communicative Event. The interview has become somewhat ubiquitous in western societies and is viewed as a reliable means of acquiring information of all kinds (e.g., attitudes toward tax increases, the value placed on education, preferences for certain products, basic demographic

---

[7]Videotaping is only ethical with the expressed permission of the participants in the interaction.

[8]A variety of software applications now exists that can help the researcher manage and analyze recorded on video. Caveat, for example, allows the researcher to select and annotate images/events of particular interest. A more sophisticated (although less user-friendly) program is the Observational Coding System that provides for a more quantitative analysis.

[9]For a review of time allocation techniques in ethnographic work, the reader is referred to Gross (1984).

[10]For a brief overview of probing techniques, see Bernard (1995).

[11]A good introductory book on surveys is *How to Conduct Your Own Survey* (Salant & Dillman, 1994). For readers interested in a more advanced treatment of the subject, they are referred to Babbie (1990).

data, etc.). However, as Briggs (1983) points out, what is said in an interview should not be thought of as "...a reflection of what is 'out there,'" but instead must be viewed "...as an interpretation which is jointly produced by the interviewer and respondent." This view compels us to regard the interview as a communicative event in which the structure and context of the interaction condition what the researcher learns. This is no less the case in highly structured interviews (see Jordan & Suchman, 1990, for a critical analysis of the ecological validity of survey research). Briggs recommends that we adopt a wider range of communicative styles in our interactions with study participants, particularly styles that are indigenous to the study population.

Rules of Thumb When Interviewing. Although there are no hard and fast rules for interviewing, a few general guidelines will help facilitate the interview process and increase the chances of obtaining useful information. Some points to remember:

- Interview people in everyday, familiar settings. Not only does this make the participants more comfortable, it also allows them to reference artifacts in the environment that play an integral part in their activities. Moreover, a familiar environment is full of perceptual cues that can help jog the not-so-perfect human memory.

- Establish and maintain good rapport with participants, even if it slows the interview process.

- Do not underestimate the value of casual conversation. Some of the most insightful information comes from informal conversations when social barriers are lowered.

- Assume the respondent is the expert and the researcher the apprentice. This not only shows the participant respect, but also gives them confidence and facilitates conversation. Even if the interviewer happens to be more knowledgeable on the subject, the goal of an ethnographic interview is to get the *participant's* perspective.

- Do not interrupt unnecessarily, complete a participant's sentences, or answer your own questions. Again, the idea is to get the respondent's point of view, not the researcher's.

- When conducting an open-ended interview, avoid asking yes/no questions. Responses to these questions provide less information than questions beginning with what or how.

- Be flexible enough to adapt the line of questioning when necessary. Human behavior is complex and full of surprises.

Connections Between Observation and Interviews. As noted previously, one of the defining qualities of ethnography is its emphasis on holism. To obtain this holistic view, triangulating different data types is useful (Agar, 1996). Observation alone is seldom enough to adequately address research objectives. As such, observation is invariably coupled with interviews. Interviews can extend and deepen one's understand of what has already been observed. Conversely, interviews can be conducted

before observing, giving the researcher a better idea about what is most appropriate to observe. As any experienced ethnographer will attest, much of a researcher's time is consumed with aligning and connecting data from multiple sources.

Interviews can also be conducted in the context of ongoing activities. Instead of setting aside a specific time and place for an interview, the researcher creates an opportunity to ask questions as participants go about their daily activities. It is usually best to get the agreement of the participants ahead of time for this type of interviewing. The strategy can be extremely useful in getting answers to questions that are prompted by observation of ongoing activities.

Self-reporting Techniques

In cases where the event or activity of interest occurs over a long period of time, or requires a significant amount of introspection on the part of the participant, self-reporting techniques can be valuable. This methodology is especially good at revealing patterns in behavior or obtaining data that is otherwise inaccessible (Whyte, 1960, 1984). Although a number of techniques exist, self-reports generally fall under two categories: diaries and visual stories.

Diaries. As implied in the name, diaries consist of a participant's written record, which could include personal thoughts or specific behaviors or accounts of events in which an individual participates. Guest (2000a), for example, asked fishermen to keep daily diaries on their fishing activities and corresponding motivations over a year to establish patterns in behavior that would normally not be observable by a lone researcher. In another recent study, researchers analyzed 180 diaries, written over 80 years ago by young nuns, for the expression of positive and negative content. Interestingly, they found a positive relationship between the proportion of positive statements and life expectancy (Snowdon, 1996).

How diaries are analyzed depends on the research objectives and resource constraints. If researchers and participants have time, it may be useful to have follow-up discussions with participants to clarify points or gain a deeper understanding of the meaning behind the words. The texts can also be coded for themes, key words or phrases, and patterns examined across individuals or between groups.[12]

Visual Stories. Visual stories are essentially pictorial diaries that use a camera in addition to text. They can be particularly valuable when working with nonliterate participants, such as children, or in situations where words alone are inadequate to capture the essence of the subject (Johnson, Ironsmith, Whitcher, Poteat, & Snow, 1997). Much like their textual counterpart, visual diaries can be used in any of a number of ways. In some cases, the researcher may want to structure the visual story to some degree. Wasson (2000), for example, describes giving participants a written guide directing them to take

---

[12]With varying degrees of success, text analysis software has been used to help with large data sets. Some noteworthy programs include: Ethnograph, NUD*IST, E-Z-Text, and NVivo.

photographs of their interaction with a product under study. They were then asked to organize the developed photos into a story that made sense to them, and researchers conducted follow-up interviews over the telephone.

A more open-ended framework can also be informative. Interested in cultural differences between Italian and American fishermen, Johnson and Griffith (1998) instructed participants from both groups to take photographs of whatever they wanted. After developing the film, Johnson coded the pictures based on their content and found significant thematic differences between the groups, which he interpreted as indicating different cultural values.

A more recent derivation of the visual story uses a video camera. The basic technique is the same, but with the added benefit of having the participant's running narrative alongside the visual content. Being able to experience the two sources of information simultaneously provides the researcher with a rich record of an activity. Blomberg, Suchman, and Trigg (1996) set up a stationary video camera in the law office of a study participant and asked him to turn on the camera whenever he had occasion to retrieve documents from his file cabinet. The running narration recorded on videotape provided insights into the everyday use of the file cabinet that helped inform the design of an electronic file cabinet.

Remote Date Collection Techniques and Strategies

Continuing technological developments—in video, audio, wireless, network applications, tracking capabilities, and pervasive computing—have created new (and largely untapped) opportunities to observe and collect rich and dynamic data across geographies in real time, as well as "asynchronously." These technologies increasingly enable ethnographers to remotely collect data in a wide variety of contexts. They also provide a dynamic window into peoples' behaviors and experiences in the digital domain—observing people interact with computer applications, as well as in the context of Internet/network-based social worlds and communities.

Remote Video and Audio Via the Internet. The increasing pervasiveness of the "web cam" is perhaps the simplest illustration of how technology has expanded the observational capabilities of ethnographers. In contrast to standard offline videotape recording, an Internet-enabled digital video camera can stream video in real time. Moreover, simple remote control software is already in use, enabling a viewer (e.g., an ethnographer) to pan, zoom, and focus the camera (within limits of course) from a remote location. This digital video data can be viewed by multiple people in multiple geographies either in real time or by accessing video archives. This capability may prove particularly useful for geographically distributed research and design teams.

Although storage of the digital data provides a challenge, as does bandwidth limitations, continuing developments in broadband, video/audio compression, and streaming and digital storage promise to decrease these challenges.

Observations and Interactions in the Digital Domain. Ethnographers working to support the development or redesign of interactive computer applications obviously need to attend to the ways in which people interact with and through computer technologies. This includes understanding how people traverse the digital domain—their paths, patterns of interaction, and individual experiences in relation to the Internet and other interactive computer applications.

Data derived from increasingly sophisticated tracking technologies—that monitor, gather, collect, and integrate information on peoples' computer-based activities—may prove to be a useful source of information for ethnographers. In this context, some usability professionals recently have begun to assess web site usability by analyzing server log files (Kantner, 2001). This essentially involves looking for patterns in log files—which can be viewed as indicators of online behavior—as clues to problematic interactions with and aspects of web-based interfaces. Although, the quantitative data supplied by log files and similar indicators of online behavior obviously does not enable one to understand the situated nature of the activities, it may highlight interesting patterns that can be subject to further qualitative exploration.

Monitoring and networking technologies that enable one to remotely view what another person sees on their computer screen have existed for several years (Perkins, 2001). New tools are providing an opportunity to interact in real time (e.g., via a "chat" interface or voice), at the same time as online behaviors are captured. These capabilities are being used by businesses providing live online customer service, as well as by usability professionals conducting remote online testing of web applications.[13] Although the current focus of these emerging tools is primarily on assessing the usability of applications, as they increase their flexibility they may become a useful tool for ethnographers working on the design of interactive computer systems.

Artifact Analysis

Ethnographers have long had an interest in the material world of the people they study. The artifacts people make and use can tell us a great deal about how people live their lives.[14] Artifact analysis can be an important part of contemporary ethnographic studies (e.g., Rathje & Murphy, 1991). For example, conducting an artifact analysis of the stuff on people's desks can say a great deal about their work. Or studying the contents of an automobile's glove box can tell a great deal about how the car is used

---

[13] One cannot underestimate the potential for these new tools to create social and political dilemmas concerning personal privacy and the further penetration of corporations into all aspects of people's lives. This will in turn raise important ethical issues for the ethnographer as participant and consumer of this new information.

[14] Archaeologists rely almost exclusively on the artifacts that remain in archaeological sites for their interpretations of the behavior and social organization of past human societies.

(e.g., about the role the car plays in connecting people's work and home lives). Depending on the kinds of research questions asked, it may be useful to include the collection and analysis of specific artifacts.

Record Keeping

In all ethnographic research, it is essential to keep good records. Although the strength of the ethnographic voice derives in part from the fact that the ethnographer is present and witness to events of interest; the ethnographer should not rely exclusively on experiential memory of these events. It is important that field notes be taken either during or soon after observing or interviewing. The specific nature of the notes will depend on the research questions addressed, the research methods used, and the whether audio or video records supplement note taking. Field notes should at least include the date and time when the event or interview took place, the location, and who was present. Beyond that, notes can vary widely, but it is often useful to indicate the difference between descriptions of what is observed, verbatim records of what is said, and personal interpretations or reflections. When working with a team of researchers, field notes need to be understandable to other team members. This is often a good standard for the specificity required of field notes even when working alone. If such a standard is maintained, it will be more likely that the notes will be useful to the researcher months and even years later, in the event re-analysis or a comparative study is undertaken.

Relation Between Qualitative and Quantitative Data

In a previous section, we touched on the complementary nature of observational and interview techniques and the benefit of combining these two approaches. Triangulation of data can serve to connect quantitative and qualitative data as well. It is sometimes the case that prior to the start of a project the only data available is quantitative in nature, usually from one or more surveys. Surveys can tell us something about how representative certain characteristics—typically demographic variables—are of a group of people relative to a larger population. Qualitative data derived from ethnographic research can complement quantitative research in a number of ways. By providing a meaningful context, ethnographic research can inform the content and language of more structured questions, thus making them more meaningful and relevant to the participants. Without any qualitative research to provide context, the validity of a survey can be seriously flawed. We may find out after the fact that we have sent out hundreds of surveys asking the wrong questions or using terminology respondents do not understand.

Qualitative techniques also allow researchers to dig deeper after a survey has been tabulated, and aid in interpreting and explaining trends that the quantitative data might reveal (Guest, 2000b). Numbers show patterns from an aggregate, but invariably they can not explain why such patterns exist. Ethnographic methods can thus illuminate processes and identify the meaning behind the numbers.

COMMUNICATING AND APPLYING THE RESULTS OF ETHNOGRAPHIC RESEARCH AND ANALYSIS

The challenge for ethnographic researchers aiming to inform and inspire the creation of compelling and useful technology solutions is to make the learning derived from ethnographic research and analysis relevant and actionable for development teams. To "connect the dots" between ethnography and technology design, ethnographers must find ways to effectively represent and communicate the insights gained from their research to a broader design team.

This section briefly outlines some of the general ways in which ethnographic research can facilitate the design of useful and compelling technology solutions. In addition, a number of specific ways to represent, communicate, and apply the learning derived from ethnographic research are outlined and illustrated. These modes of representation, communication, and application should not be viewed as prescriptive and/or exhaustive. Instead, they are intended to provide examples that will stimulate thinking about how to apply the results of ethnographic research in the context of a particular design/development processes.

Broadly speaking, ethnographic research and analysis in the context of human computer interface design can:

- Enhance the working models of developers about the people who will interact with technology solutions and the domains and contexts in which they will do so.
- Provide generative tools that support innovation and creativity.
- Provide a critical lens for evaluating and prioritizing design ideas.
- Serve as a guidepost or point of reference for development teams.

These broad purposes are reviewed briefly.

Enriching and Enhancing the Working Models of Developers

To design a technology solution, developers must have at least an implicit working understanding of the people who will interact with the solution. Such working models may include assumptions about a range of essential characteristics of the people who will engage with the solution and the contexts in which they will do so (Newman, 1998). Indeed, some would argue that successful design requires a high degree of empathy with the target population (e.g., Leonard & Rayport, 1997). Implicit and/or explicit models or assumptions about users may be formed through some combination of direct experience (e.g., interacting with and/or observing people in the target population in controlled or noncontrolled settings) and secondary learning (talking with others about the target group, viewing videotapes of target activities, reading, analogy to other directly experienced groups, etc.). However formed, the working models of

developers may be of varying levels of complexity, robustness, coherence, consistency, and viability. The broad, deep, and contextualized understanding that can be derived from ethnographic research can serve to build, enrich and enhance the development team's explicit and implicit working models of end users.

Generative Tools or Frameworks to Support Innovation and Creativity

The design of technology solutions for human beings obviously poses a range of potential creative challenges at varying levels of complexity. What should we build? What kinds of experiences should the technology solution support or enable? What features and functions would be useful, compelling, and satisfying for a particular group of people in a particular domain/context? How can we use existing or emerging technological capabilities to enhance a particular group's experiences in a selected domain or solve a particular human problem? Even if there are clear parameters defining the functionality that will be built (e.g., a set of requirements), a development team must still generate a compelling, easy-to-use, and satisfying way of delivering that functionality that makes sense and is maximally valuable to the target users. The learning derived from ethnographic research and analysis can serve as a generative foundation or tool that can inspire innovative ideas for enhancing the experiences of end users through the use of technology. By providing a deep understanding of a human domain (patterns of relationship, systems of meaning, organizational structure, guiding principles or rules, etc.), ethnography can promote an essential condition for creativity that matters (Robinson & Hackett, 1997)—relevant and actionable innovations that solve problems and create new and realizable opportunities.

Critical Lens for Evaluating and Prioritizing Ideas

Design teams not only face the challenge of generating innovative ideas and concepts, but also the equally important task of evaluating and prioritizing ideas and options that arise from various sources (e.g., business stakeholders, end users, development teams). The learning derived from ethnographic research and analysis can provide a critical experiential lens through which development teams can begin to evaluate and prioritize ideas based on how they may fit into (or not) or enhance the lives of selected groups of people. The need for evaluation and prioritization may occur at various points throughout the development process, ranging from decisions about features and functions, broad directions for design concepts, varying content organizational models, and so forth.

Guideposts for Development Teams

The learning derived from ethnographic analysis, particularly when represented and communicated in compelling visual or narrative forms, can serve as an experiential guidepost or frame of reference for individual designers and design teams throughout the development process. Even though such guidelines may not prescribe or specify what should be done, they can aid developers by focusing attention on essential aspects of an experience, highlighting variations in the experiences of different types of end users, and limiting exploration of experiential "dead-ends." In other words, they can provide a general structure and direction within which a team can focus its creative energies.

MAKING ETHNOGRAPHY MATTER: COMMUNICATING AND APPLYING ETHNOGRAPHIC INSIGHTS TO DESIGN

This section outlines some of the ways in which the insights derived from ethnographic work can be represented and communicated to effectively inspire and guide the design of valuable and compelling human–computer interactive systems. These ways of representing and communicating what is learned are intended as examples of ways in which ethnographers and development teams can make ethnographic work relevant for design.

Experience Models

To help generate potentially valuable ideas, critically evaluate and prioritize design concepts, and guide the actions of development teams, it is important to identify and articulate patterns, principles, and relational frameworks for understanding experience. Simple re-presentations of observational data or lists of disconnected findings are not particularly useful as an aid to design. In other words, it is important to ascertain and visualize patterns and principles that underlie how people create meaning, interact, and organize their experiences in a selected domain. Experience models or frameworks identify, highlight, and visualize relevant patterns of human behavior and experience to guide design.[15] They can address and map dimensions of individual experience (how individuals make decisions, organize information, modify behavior over time, etc.), as well as the dynamics of group behavior (patterns of communication and collaboration, social contexts, and structures). They can be of varying levels of complexity, generality, scope, and specificity.

The collaborative construction of experience models can enable teams to move beyond simplistic and untested assumptions

---

[15] Experience models or frameworks were first used at E-Lab as a tool to connect insights from ethnographic research with design innovations. The frameworks or models, which consist of visual representations depicting key analytic relationships, were used by design teams as they explored and validated design ideas and directions. Frameworks and models have more recently been used to help inform business and brand strategy as well as technology and artifact design.

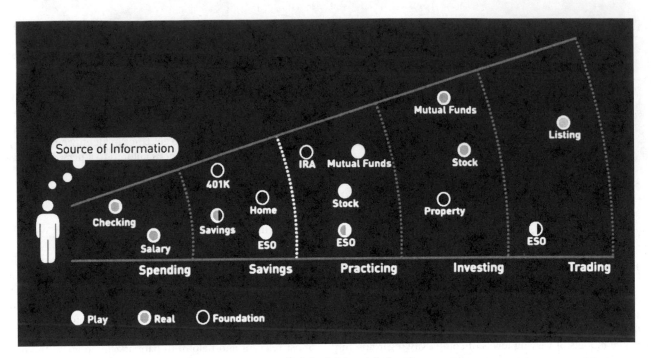

FIGURE 50.2. Experience model of financial development zones. ESO = employee stock options.

regarding domains of experience and end-user characteristics toward the development of a shared, principled understanding of how relevant experiences are structured and work. Once constructed, experience models become dynamic tools or frameworks for focusing a development team's attention on important aspects of experience, generating innovative design concepts, and prioritizing and evaluating concepts/designs, as well as a shared frame of reference throughout the development process.

The model presented in Fig. 50.2, is one of several developed in the context of ethnographic research and analysis for a financial services company serving individual investors. This company aimed to develop web applications that would facilitate customers' active engagement in the investment process with particular financial instruments. The model was intended to articulate and visualize a financial development process, as well as the varied meanings of money. This particular model highlighted the role of practice in developing the confidence and knowledge to become engaged in the investment process, and the iterative/recurrent nature of the process, as people learned to deal with new financial instruments and domains (e.g., securities, bonds, options, etc.). Moreover, it illustrated the distinctions that people make between real, play, and foundational money, and the relationship between these categories, investment behavior, and financial development. To oversimplify a bit, people are more fully engaged and active in the investment process when they view the assets/investments as real (e.g., money that is used to address their current and emerging needs, pay bills, etc.) rather than as play (e.g., stock options that are perceived as intangible and somewhat imaginary) or foundational (e.g., savings for the future that are left untouched). As people

have an opportunity to practice and develop their knowledge, they may move from construing a particular financial instrument or activity as play to real. These notions suggested that web applications in this domain should not be focused on simply providing a wealth of financial information or a plethora of tools. Instead, these patterns helped to foster the generation of numerous ideas of ways to engage people in playful learning in the financial domain, with the aim of facilitating the financial development process.

In addition to being generative tools, experience models can provide frameworks that help teams prioritize and evaluate concepts and goals. For example, a health services company aimed to develop an electronic medical record system (combining client server applications with web based portals). This system would, among other things, increase the efficiency and effectiveness of their medical practice, enable patients to view their health records online, and ultimately empower patients and foster a proactive approach to wellness and health care (both by clinicians and patients). At the outset of the engagement, the health services company had generated a rather long requirements list (several hundred features and functions), and a particular view of the structure and function of the web components of the system. It was clear that the budget for this initiative was not sufficient to build a system that met all of the initial requirements. Perhaps more importantly, it was unclear which components would ultimately add the most value for the various stakeholders (clinicians, patients, the business owners, etc.), and there was no principled way of prioritizing and evaluating potential features, functions, and design concepts. Field research examined the experiences of and relationships between clinicians and patients in context (in clinic settings

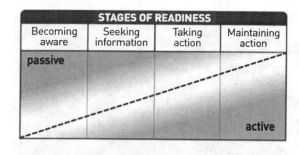

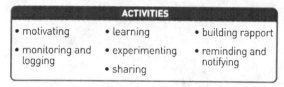

FIGURE 50.3. Experience model of stages of readiness.

and in homes) through a variety of methods (see Case Study 1). Collaborative analysis led to the development of a number of experience models of varying levels of complexity regarding the health management process. For example, one of the simpler models (Fig. 50.3) described how individuals, in the process of adopting an active/proactive stance in relation to health issues, move through varying stages of readiness.

A more comprehensive, integrative model highlighted the ways in which various factors interact in influencing a person to take action in addressing a health issue and mapped the role of various health care-related activities (e.g., monitoring, motivating, learning, sharing, building rapport) in various stages of readiness. The combination of these models enabled the team to identify or map the most important opportunities (Fig. 50.4) for facilitating progression toward a proactive orientation to health, and provided guidance in identifying ways to provide messages and experiences tailored to a person's stage and readiness.

Opportunity Maps

An opportunity map that derives from the creative application of experience models, can serve both generative and evaluative purposes. In the previous example, it provided a principled lens for critically evaluating and prioritizing the initial list of requirements: Which features and functions would have the best chance to really facilitate health-related action and would be most relevant and helpful to patients? In addition, the opportunity map encouraged the team to view the web application from a very different vantage point from the one initially conceptualized as a portal enabling access to an electronic medical record and a library of medical information. Instead, the design

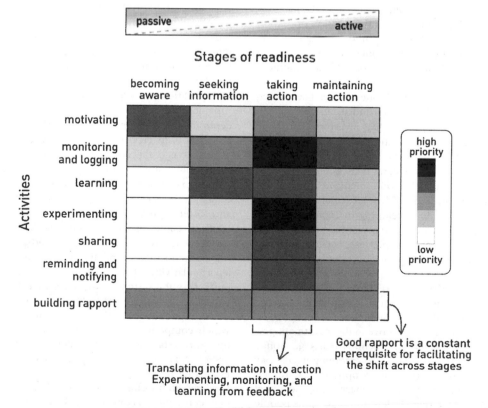

FIGURE 50.4. Opportunity map for stages of readiness.

moved toward a framework for effective and flexible collaboration between clinicians and patients through varying stages of the health management processes.

Profiles

One of the primary challenges in developing interactive systems is to design them so that they meet the needs of varying users, who may play different roles, engage in varied tasks, have different motivations and strategies, and so forth. Profiles identified through ethnographic studies are a simple but valuable tool for representing and communicating some of the experiential variations and similarities within a target population. Profiles may be descriptive of the experiences and characteristics of individuals or composite representations of prototypic users. They highlight a selected range of experiential dimensions and variations of people in the target audience (characteristic strategies, modes of interaction, tools, key relationships, expectations, etc.) to help development teams understand and anticipate how certain types of people may experience and interact with technology solutions. For example, Fig. 50.5a and Fig. 50.5b show composite profiles based on fieldwork conducted for a project concerned with how young people relate to wireless communication devices and applications.

The value of profiles also is enhanced by making them visible and dynamically present for development teams (e.g., profile posters displayed in project rooms, multimedia representations that are reviewed with development teams, role-playing scenarios, and walk-throughs based on profiles, etc.). Rich and dynamic representations of essential characteristics and experiences of individuals can serve as a common frame of reference and reminder to development teams regarding the people for whom they are designing the system. Such profiles may take many forms, including narrative descriptions, matrices/tables, integrated still images, and video snippets.

Although not a substitute for actual feedback from and testing with end users, profiles can function as partial stand-ins or virtual advocates for certain user groups (Beyer & Holtzblatt, 1997). By doing so, they may provide a partial lens and communication vehicle for evaluating design concepts, stimulating and framing questions about how various users might construe or engage with a particular design and enabling teams to frame design decisions.

Scenarios

Scenarios are another way ethnographic research findings can be portrayed. The notion of scenario-based design has become increasingly popular (Carroll, 2000; Rosson & Carroll, chapter 53 in this volume; Nardi, 1992; Sonderegger, Manning, Charron, & Roshan, 2000). Scenarios illustrate how a person's experiences and actions unfold in specific contexts or situations (Fig. 50.6). They may highlight interactions (with computer systems, people, business entities, etc.), decisions processes, activity sequences, influencing factors, and so forth. They may also illustrate the different ways in which varied groups or types of people experience and navigate through similar situations.

Analysis of scenarios can foster the identification of areas of difficulty (pain points) and experiential gaps (or opportunities) that may be addressed or enhanced through technology solutions. When integrated with profiles, they can illustrate how different target audiences navigate through the same situation, which in turn can suggest ways in which solutions can and should be adapted for varying target audiences. Scenarios can serve as a dynamic reminder of the broad range of contexts and situations in which people may engage and interact with technology solutions, fostering thoughtful discussions about how to design solutions that fit into and complement or enhance peoples lives (Fig. 50.6).

Scenarios based in ethnographic research have an advantage (over those created from the designers imaginations alone), in that levels of detail can be added based on the initial research or follow-up visits to the field site can provide missing elements of the scenario. This is important in that it is not always possible to know in advance just what aspects of the activity should be included in the scenario to provide generative or evaluative value for design.

Design Representations: Mock-ups and Prototypes

Representational artifacts, be they paper prototypes, mock-ups, or working prototypes, can play an important mediating role in connecting use requirements and design possibilities. When informed by studies of practice, these design representations respecify practices and activities in ways that are recognizable to practitioners. The prototypes go beyond simple demonstrations of functionality to incorporate materials from the participants' site, embody envisioned new technological possibilities, convey design ideas in relation to existing practices, and reveal requirements for new practices. Prototyping practices as such recover and invent use requirements and technological possibilities that make sense each in relation to the other (Suchman et al., 1999b). In addition, these representational artifacts facilitate the communication of what has been learned about technologies in use to the larger research and technology development communities.

In an ethnographic study of engineering practice at a state Department of Highways, design prototypes critically deepened the researchers' understanding of the requirements of the work of document filing and retrieval (the focus of the study). At each step, from early design discussions with practitioners, to the creation of paper mock-ups of possible interfaces to the online project files, and finally to installing a running system at the work site, the researchers became more aware of the work's exigencies. For example, in recognition of some of the difficulties that engineers experienced with their filing system, various alternative document coding strategies that augmented the existing filing system were designed. Through successive rounds, in which engineers were asked to code documents using mocked-up coding forms (both paper-based and online), the researchers' understanding of the requirements of the work deepened. Eventually, the search and browsing interfaces evolved to be more finely tuned to the requirements of the engineers' work (e.g., Trigg, Blomberg, & Suchman, 1999).

Blake

Background

Blake is a junior at Princeton University. He is majoring in economics, and is heavily involved in his capella singing group. This summer he will travel with the group to perform in Europe. Although American, Blake grew up in Japan, and moved back to the US for college. Blake's father is in international banking, and his older brother works in a management consulting company. Blake is currently interning at an investment bank, and has his sights set on entering the corporate life after he graduates.

Quick mobile stats

Cellular phone user since high school
Nokia 5190
Device and plan paid for by family
Family AT&T plan

Handspring Visor Edge
Present from family
Paid for by family

Walkman
Old Sony model
Paid for by himself

Technology
Blake uses his cell phone to:
- receive calls from his family (mostly his mother)
- make arrangements to meet friends
- locate friends when they are in the city
- occasionally play "snake" when he needs to pass time
Bake uses his Handspring Visor to:
- record all the information (mainly telephone and address) of his friends and work contacts
- keep track of his list of "to-do's"
Blake uses his computer to:
- write papers and do research
- email

Key Insights

• Blake "sells" the idea of a cell phone to his mom

When they lived in Japan, Blake's mother was always concerned about safety. Blake was very much aware of this, and convinced his mother to buy him a cellular phone by "selling" her on the idea that it was a safety device and that would enable her to reach him at any time. In his opinion, framing the phone in this way was a "good move on (his) part."

• Blake was the last in his family to acquire a PDA

Blake's older brother was the first in the family to purchase a Handspring PDA, and soon after, both parents also acquired similar PDAs. Blake was already familiar with the functionality of the PDA before his parents gave him one last year. Because he was the last in the family to get a PDA, he currently has the most advanced model, which is very thin, and has a slick brushed chrome exterior. He likes using his PDA, and feels that carrying it gives him a sense of security.

• Blake needs a new computer

Although Blake has the latest model Handspring Visor, not all of his devices are as up to date. At college, he uses a 90 mHz computer that does not have a USB port. Blake's computer frustrates him because it is slow, and also because he cannot download the programs that enable all the functions on his PDA. Blake would like to get a new computer, because he sees it as a "center" for his collection of devices.

FIGURE 50.5a. Composite profile of Blake in relation to wireless communications.

Bridget

Background

Bridget is a 16-year-old junior at a Catholic school on Long Island. She has a very busy schedule. In addition to the demands of high school, she is intensively involved in Irish dancing, including dance classes, competitions, and considerable commuting time. Bridget has an outgoing disposition, and dedicates hours each day to keeping in touch with three groups of people: friends from her old school, friends from her new school, and friends from dancing. Bridget's father is in finance, her mother works part time, and her 10-year-old brother is in 4th grade.

Quick mobile stats

• Borrows her mother's cellular phone
• Will receive a phone once she starts driving

Technology

Bridget uses her computer for several hours each day to:
  • Instant Message with her friends
  • Email
  • Surf the Internet
  • Look for music
Bridget uses the home phone to:
  • Chat with friends
Bridget uses her mother's cellular phone to:
  • Check in with her mom

Key Insights

• Bridget learns to Instant Message

When Bridget began using AOL Instant Messenger several years ago, she quickly learned that the medium invites and even requires its own form of writing. This is not her English teacher's prose. It is an amalgam of slang, abbreviations and symbols, all delivered rapid-fire to support the conversational feel of Instant Messaging (IM). Bridget is very much aware that not everyone uses IM like she does with her friends. She says, "grown-ups like to type all proper with capitals," and complains about an uncle who types long paragraphs before hitting "send," thus losing the conversational feel of IM.

• Bridget "hangs out" online

Bridget loves to Instant Message. She has a buddy list of over one hundred friends, and often conducts up to six conversations at once. She flips from dialogue box to dialogue box, and thinks of communicating on IM as very similar to having a conversation. While she is online, Bridget also checks and writes email, and surfs the Internet. On most days, Bridget IMs in the evenings, because "that's when all the other kids are online."

• Bridget experiences the pitfalls of Instant Messaging

Bridget has come to understand the merits and drawbacks of communicating on IM. One of her frustrations is the difficulty of conveying the nuances of tone that a telephone conversation supports. There have been several instances when what Bridget "said" had been misunderstood by her friends, which then caused some upsets. Another feature of IM that Bridget has been hurt by is the ability to copy and paste sections of conversations. Unlike a telephone conversation, which is fleeting, IM conversations can be saved and forwarded on to other people. In the complex social landscape of teenhood, this can become the source of many misunderstandings.

FIGURE 50.5b. Composite profile of Bridget in relation to wireless communications.

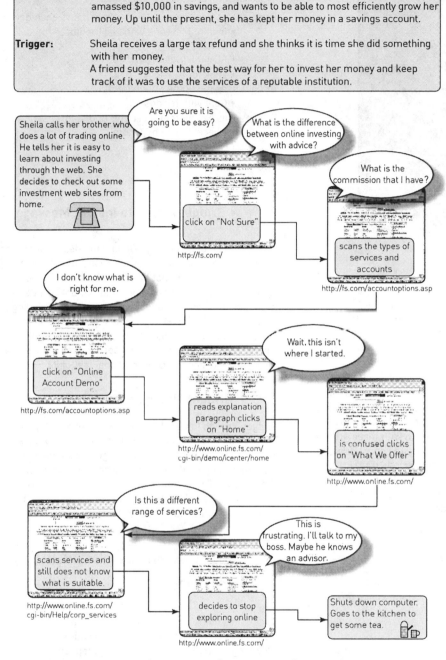

FIGURE 50.6. Simple scenario of a woman exploring online investing resources.

RELATION TO OTHER QUALITATIVE APPROACHES AND PERSPECTIVES

The ethnographic approach has strong connections to and affinities with other approaches that have contributed to the development of the field of HCI, namely distributed cognition, activity theory, ethnomethodology, and participatory design. There is

not space here to go into depth on any of these approaches. Our aim is simply to highlight relations between these approaches and ethnography, and provide a way to distinguish between them.

Distributed cognition (sometimes referred to as social or situated cognition) was first introduced to the HCI community by Lave (1988) and Hutchins (1995). Distributed cognition located cognition in social and material processes. When it was

introduced, it challenged the dominant paradigm within HCI, that cognition primarily involved the psychological and mental processes of individuals. The connection between distributed cognition and ethnography is not only in the insistence that our understanding of human activity be located outside individual mental processes, in human interaction, but also in the conviction that to gain an understanding of human activity, ethnographic, field-based methodologies are required.

Activity theory also shares with ethnography a commitment to field-based research methodologies. In addition, there is the shared view that behavior (activity) should be a primary focus of investigation and theorizing, and a recognition that objects (artifacts) are key components in descriptive and explanatory accounts of human experience (e.g., Engeström, 2000; Nardi, 1996).

Ethnomethodology is often used interchangeably with ethnography in the HCI literature. This is not only because the terms are etymologically similar, but also because many of the social scientists contributing to the field of HCI have adopted an ethnomethodological approach (e.g., Bentley et al., 1992; Button & Harper, 1996; Crabtree, 2000; Hughes et al., 1993, 1994, 1995) with its focus is on locally and interactionally produced accountable phenomena. Ethnomethodology's particular set of commitments (e.g., Heritage, 1984) are not shared, however, by everyone working within the ethnographic paradigm.

Participatory design does not have its roots in qualitative social science research, but instead developed as a political and social movement, and as a design approach committed to directly involving end users in the design of new technologies (see Muller, chapter 54 this volume; also Kensing & Blomberg, 1999; Schuler & Namioka, 1993). Within the HCI context, participatory design has shed much of its political and social action underpinnings, and is viewed primarily as a set of methods and techniques for involving users in design. Its connection to ethnography is in the commitment to involve study participants in the research, and in the value placed on participants' knowledge of their own practices. Also, in recent years, those working within the field of participatory design have incorporated ethnographic techniques (e.g., Crabtree, 1998; Kensing, Simonsen, & Bødker, 1998) as a way of jointly constructing with participants knowledge of local practices.

ETHNOGRAPHY IN ACTION

Case Study 1: Health Care Provider

An onsite health care provider wanted to build a web-based portal that would facilitate communication between employees, employers, and clinicians, and ultimately improve employee health. The research was conducted in employee homes, employer offices, and health clinics, and involved interviews and shadowing of patients as they interacted with health care practitioners in clinics.

One of the main findings, illustrated in the experience model (see Fig. 50.3, p. 34), was that the health experience is composed of several stages and associated activities. Awareness and/or acceptance of a condition or need is the first stage, followed by the desire to search for relevant information. As the search progresses, an individual eventually becomes informed enough to begin to take more direct action toward addressing the situation. Once action is taken, the key is to maintain this behavior for an appropriate length of time, often indefinitely.

The research team found that certain activities were more likely to be associated with each stage, suggesting that understanding where a person stood in relation to the stages was fundamental to developing effective strategies for communicating and motivating healthy behavior. For example, when an individual is in the information-seeking stage, it would be important to provide a means to locate, store, and interpret general information. Once an individual begins taking action, reinforcement of their behavior is crucial and a design solution needs to support such activities as monitoring and logging progress.

During the home interviews, the field team discovered that people use, and respond well to, quantifiable measures of progress such as cholesterol or T-cell counts. Moreover, such indicators serve as positive reinforcement, further motivating healthy behavior. We recommended, therefore, that the proposed system provide functionality that would allow users to monitor and log health-related information that is relevant to them.

Another key finding to come out of the research was that communication between patients and clinicians was extremely important for a patient's health, but it is frequently impeded by a variety of barriers. For a system to be successful, it would have to remove at least some of these barriers to pass through a patient's trust filter (Fig. 50.7).

Combing these two major findings, the research team recommended building a health history tool that allows patients among other things, to monitor and log pertinent information in a web-based platform. The log also would be accessible from the clinician's office, allowing a patient and health care practitioner to review the patient's progress together and make annotations throughout the visit. Such a solution would improve the communication between patient and clinician while simultaneously empowering patients to take charge of their health.

Case Study 2: Department of Highways

The headquarters of a state Department of Highways was the site for a collaborative research and design effort with engineers charged with the design of a bridge, scheduled for completion by the year 2002.[16] The project aimed to design an electronic document management system that was informed by an understanding of the everyday requirements of engineering work at the Department of Highways. The project began with onsite

---

[16]For more on the project with the Department of Highways, see Suchman (1998, 1999).

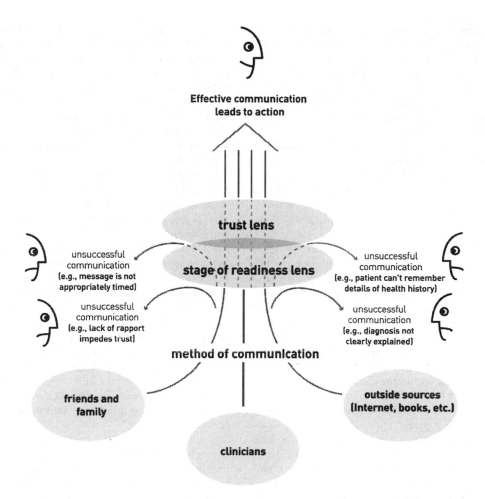

Effective communication leads to action

trust lens

stage of readiness lens

unsuccessful communication (e.g., message is not appropriately timed)

unsuccessful communication (e.g., patient can't remember details of health history)

unsuccessful communication (e.g., lack of rapport impedes trust)

unsuccessful communication (e.g., diagnosis not clearly explained)

method of communication

friends and family

clinicians

outside sources (Internet, books, etc.)

FIGURE 50.7. Communication model.

interviews and observations of engineering practice, with a focus on the document-related work practices.

Based on an initial understanding of the document management requirements of the work, as part of the design process, several alternative paper-based document coding forms were designed. After several iterations, a coding form was settled on that was then incorporated into the electronic document management system, both as a form to be scanned into a document database and as a model for an online coding form. The evolution of the coding form was informed by the prompted use of the form by engineers at the Department of Highways (Fig. 50.8).

One of the key insights that came from the ethnographic study was the need to design continuing connections between the digital and physical document worlds. This included locating familiar ways of organizing documents in the new electronic system, and taking advantage of visual memory in document search and browsing by displaying page images of the documents and not just the text (Fig. 50.9).

The insights gained from the ethnographic study also pointed to challenges that would face engineering teams adopting the new system. First, because members of project teams would no longer be the sole interface to the documents in the project files, team members would need to consider who might view the documents and for what purposes before deciding to add a document to the database. This was not necessary when the project files were paper-based, because the physical location of the documents, in the engineering team's work area, restricted access. Electronic access now meant that users of the system could be located anywhere within the Department of Highways, making explicit access controls necessary. In addition, it would be crucial that an ongoing relation between the paper and digital document renderings be maintained as engineers found it most useful to work with the printouts of large engineering documents. The online renderings were not particularly useful by themselves. The research and design team was able to anticipate these work practice issues, make the highways engineers aware of them, and suggest possible ways they could be addressed.

CONCLUSIONS

Over the last two decades, ethnographic studies have become an important tool for designers and development teams designing new information and communication technologies. Today, in academic, institutional, and corporate settings, there is the realization that understanding the everyday realities of people

FIGURE 50.8. Engineer using mock-up of coding form to code documents.

living and working in a wide range of environments and engaged in myriad activities is essential for creating technologies that provide engaging and productive experiences for their users.[17] Emerging from these last two decades of research and practical experience is the recognition that representational tools (experience models, scenarios, mock-ups, and prototypes, etc.) and design and development practices (collaborative data analysis, joint opportunity mapping, video review sessions, etc.) are

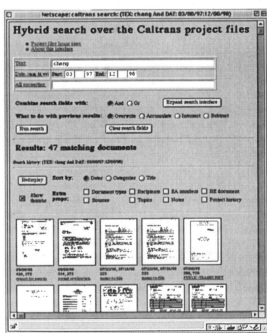

FIGURE 50.9. Components (document scanner, PC, coding forms, etc.) of the designed document management system and document search results page with thumbnails.

---

[17]For a discussion of the relation between ethnography and design, see also Anderson (1994), Grudin and Grintner (1995), Rogers and Belloti (1997), and Shapiro (1994).

necessary for connecting ethnographic studies and technology design. Insights from ethnographic studies do not facilely map directly onto design specifications or straightforwardly generate user requirements, but instead must be actively engaged with design agendas and activities. Those wishing to leverage the potential of ethnographic studies should not only understand what motivates the approach and is at its foundation (e.g., natural settings, holistic, descriptive, members' point of view), but should also recognize the importance of creating the conditions in which design can take advantage of ethnographic insights.

References

Agar, M. (1996). *The professional stranger* (2nd ed.). San Diego, CA: Academic Press.

Anderson, R. J. (1994). Representations and requirements: The value of ethnography in system design. *Human-Computer Interaction, 9*(2), 151–182.

Babbie, E. (1990). *Survey research methods* (2nd ed.). Belmont, CA: Wadsworth Publishing Company.

Bentley, R., Hughes, J. A., Randall, D., Rodden, T., Sawyer, P., Shapiro, D., & Sommerville, I. (1992). Ethnographically-informed system design for air traffic control. *Proceedings of Computer Supported Cooperative Work*. New York: ACM Press.

Bernard, H. R. (1995). *Research methods in anthropology: Qualitative and Quantitative approaches* (2nd ed.). London: Altamira Press.

Beyer, H., & Holtzblatt, K. (1998). *Contextual design: Defining customer-centered systems*. San Francisco, CA: Morgan Kaufmann Publishers.

Blomberg, J. (1987). Social Interaction and office communication: Effects on user's evaluation of new technologies. In R. Kraut (Ed.), *Technology and the transformation of white collar work* (pp. 195–210). Hillsdale, NJ: Lawrence Erlbaum Associates.

Blomberg, J. (1988). The variable impact of computer technologies on the organization of work activities. In I. Greif (Ed.), *Computer-supported cooperative work: A book of readings* (pp. 771–782). San Mateo, CA: Morgan Kaufmann.

Blomberg, J. (1995). Ethnography: Aligning field studies of work and system design. In A. F. Monk & N. Gilbert (Eds.), *Perspectives on HCI: Diverse approaches* (pp. 175–197). London, UK: Academic Press Ltd.

Blomberg, J., Giacomi, J., Mosher, A., & Swenton-Wall, P. (1991). Ethnographic field methods and their relation to design. In D. Schuler & A. Namioka (Eds.), *Participatory design: Perspectives on systems design* (pp. 123–155). Hillsdale, NJ: Lawrence Erlbaum Associates.

Blomberg, J., Suchman, L., & Trigg, R. (1996). Reflections on a work-oriented design project. *Human-Computer Interaction, 11*(3), 237–265.

Blomberg, J., Suchman, L., & Trigg, R. (1997). Back to work: Renewing old agendas for cooperative design. In M. Kyng & L. Mathiassen (Eds.), *Computers and design in context* (pp. 267–287). Cambridge, MA: MIT Press.

Blomberg, J., & Trigg, R. (2000). Co-constructing the relevance of work practice for CSCW design: A case study of translation and mediation. *Occasional Papers from the Work Practice Laboratory, 1*(2), 1–23. Blekinge Institute of Technology.

Briggs, C. (1983). Learning How To Ask.

Brun-Cottan, F., & Wall, P. (1995). Using video to re-present the user. *Communications of the ACM, 38*(5), 61–71.

Button, G., & Harper, R. (1996). The relevance of "work-practice" for design. *Computer-Supported Cooperative Work, 5*, 263–280.

Carroll, J. M. (2000). *Making use: Scenario-based design of human-computer interactions*. Cambridge, MA: MIT Press.

Clifford, J. (1988). *The predicament of culture: Twentieth-century ethnography, literature, and art*. Cambridge, MA: Harvard University Press.

Clifford, J., & Marcus, G. (Eds.). (1986). *Writing culture: The poetics and politics of ethnography*. Berkeley, CA: University of California Press.

Corral-Verduga, V. (1997). Dual 'realities' of conservation behavior: Self reports vs. observations of re-use and recycling behavior. *Journal of Environmental Psychology, 17*, 135–145.

Crabtree, A. (1998, November 12–14). Ethnography in participatory design. *Proceedings of the Participatory Design Conference PDC'98* (pp. 93–105). Seattle, WA.

Crabtree, A. (2000). Ethnomethodologically informed ethnography and information system design. *Journal of the American Society for Information Science, 51*(7), 666–682.

D'Andrade, R. G. (1995). *The development of cognitive anthropology*. Cambridge, MA: Cambridge University Press.

Engeström, Y. (2000). From individual action to collective activity and back: Developmental work research as an interventionist methodology. In P. Luff, J. Hindmarsh, & C. Heath (Eds.), *Workplace studies: Recovering work practice and informing system design* (pp. 150–166). Cambridge, UK: Cambridge University Press.

Fluehr-Lobban, C. (Ed.). (1991). *Ethics and the profession of anthroplogy: Dialogue for a new era*. Philadelphia, PA: University of Pennsylvania Press.

Green, E. C. (2001). Can qualitative research produce reliable quantitative findings? *Field Methods, 13*(1), 1–19.

Greif, I. (Ed.). (1988). *Computer-supported cooperative work: A book of readings*. San Mateo, CA: Morgan Kaufmann.

Gross, D. (1984). Time allocation: A tool for the study of cultural behavior. *Annual Review of Anthropology, 13*, 519–558.

Grudin, J., & Grinter, R. E. (1995). Ethnography and design. *Computer Supported Cooperative Work, 3*(1), 55–59.

Gubrium, J. F., & Holstein, J. A. (Eds.). (2002). *Handbook of interview research: Context and method*. Thousand Oaks, CA: Sage Publication.

Guest, G. (2000a). *Shrimp, poverty, and marine resources on the Ecuadorian coast: A multi-level analysis of fishing effort in the Rio Verde estuary*. Ph.D. dissertation, Department of Anthropology, University of Georgia, Athens, GA.

Guest, G. (2000b). Using Guttman scaling to rank wealth: Integrating quantitative and qualitative data. *Field Methods, 12*(4), 346–357.

Harding, S. (1986). *The science question in feminism*. Ithaca, NY: Cornell University Press.

Heritige, J. (1984). *Garfinkel and ethnomethodology*. Cambridge, MA: Polity Press.

Hughes, J., King, V., Rodden, T., & Anderson, H. (1994). Moving out from the control room: Ethnography in system design. In R. Furuta & C. Neuwirth (Eds.), *Proceedings of the Conference on Computer Supported Cooperative Work (CSCW'94)* (pp. 429–439). New York: ACM Press.

Hughes, J. A., Randall, D., & Shapiro, D. (1993). From ethnographic record to system design: Some experiences from the field. *Computer Supported Cooperative Work, 1*(3), 123-147.

Hughes, J. A., Rodden, T., & Anderson, H. (1995). The role of ethnography in interactive system design. *ACM Interactions, 2*(2), 56-65.

Hutchins, E. (1995). *Cognition in the wild.* Cambridge, MA: MIT Press.

Johnson, J. C. (1990). *Selecting ethnographic informants.* Newbury Park, CA: Sage.

Johnson, J. C., & Griffith, D. C. (1998). Visual data: Collection, analysis, and representation. In V. DeMunck & E. Sobo (Eds.), *Using methods in the field: A practical introduction and casebook* (pp. 211-228). Walnut Creek, CA: Altamira.

Johnson, J. C., Ironsmith, M., Whitcher, A. L., Poteat, G. M., & Snow C. (1997). The development of social networks in preschool children. *Early Education and Development, 8*(4), 389-406.

Jordan, B., & Suchman, L. (1990, March). Interactional troubles in face-to-face survey interviews. *Journal of the American Statistical Association, 85*(409), 232-253.

Kantner, L. (2001). Assessing web site usability from server log files. In R. Branaghan (Ed.), *Design by people, for people: Essays on usability* (pp. 245-262). Chicago, IL: Usability Professionals Association.

Karasti, H. (2001). Bridging work practice and system design—Integrating systemic analysis, appreciative intervention, and practitioner participation. *Computer Supported Cooperative Work—An International Journal, 10*(2), 211-246.

Kensing, F., & Blomberg, J. (1998). Participatory design: Issues and concerns. *Computer Supported Cooperative Work, 7*(3-4), 163-165.

Kensing, F., Simonsen, J., & Bødker, K. (1998). MUST—A Method for Participatory Design. *Human-Computer Interaction, 13*(2), 167-198. Mahwah, NJ: Lawrence Erlbaum.

Latour, B. (1987). *Science in action: How to follow scientists and engineers through society.* Cambridge, MA: Harvard University Press.

Latour, B., & Woolgar, S. (1986). *Laboratory life: The construction of scientific facts.* Princeton, NJ: Princeton University Press.

Lave, J. (1988). *Cognition and practice.* Cambridge, UK: Cambridge University Press.

Leonard, D., & Rayport, J. F. (1997, Nov./Dec.). Sparking innovation through empathic design. *Harvard Business Review*, 102-113.

Marcus, G., & Fischer, M. (1986*). Anthropology as cultural critique: An experimental moment in the human sciences.* Chicago, IL: University of Chicago Press.

Nadar, L. (1974). Up the anthropologist—Perspectives gained from studying up. In D. Hymes (Ed.), *Reinventing anthropology.* New York, Vintage.

Nardi, B. (1992). The use of scenarios in design. *SIGCHI Bulletin, 24*(4), 13-14.

Nardi, B. (1996). *Context and consciousness: Activity theory and human-computer interaction.* Cambridge, MA: MIT Press.

Nardi, B., & Miller, J. (1990). An ethnographic study of distributed problem solving in spreadsheet development. In *Proceedings of CSCW'90* (pp. 197-208). New York: ACM.

Newman, S. E. (1998). Here, there, and nowhere at all: Distribution, negotiation, and virtuality in postmodern engineering and ethnography. *Knowledge and Society, 11*, 235-267.

Nielsen, J., & Landauer, T. K. (1993, April 24-29). A mathematical model of the finding of usability problems. *Proceedings of ACM INTERCHI'93 Conference* (pp. 206-213). Amsterdam, The Netherlands. New York: ACM Press.

Pelto, P. J., & Pelto, G. H. (1978). *Anthropological research: the structure of inquiry.* Cambridge, MA: Cambridge University Press.

Perkins, R. (2001). Remote usability evaluation over the Internet. In R. Branaghan (Ed.), *Design by people, for people: Essays on usability* (pp. 153-162). Chicago, IL: Usability Professionals Association.

Pickering, A. (Ed.). (1980). *Science as practice and culture.* Chicago, IL: University of Chicago Press.

Rathje, W. L., & Cullen Murphy, C. (1991). *Rubbish! The archaeology of garbage.* New York: HarperCollins.

Rhoades, R. E. (1982). *The art of the informal agricultural survey. Training Document 2* (pp. 1-40). Lima, Peru: International Potato Center.

Rich, M., Lamola, S., Amory, C., & Schneider, L. (2000). Asthma in life context: Video intervention/prevention assessment (VIA). *Pediatrics, 105*(3), 469-477.

Robinson, R. E. (1994). The origin of cool things. In *Design that packs a wallop: Understanding the power of strategic design* (pp. 5-10). American Center for Design Conference Proceedings. New York: American Center for Design.

Robinson, R. E., & Hackett, J. P. (1997). Creating the conditions of creativity. *Design Management Journal, 8*(4), 10-16.

Rogers, Y., & Bellotti, V. (1997, May/June). How can ethnography help? *Interactions, 4*, 58-63.

Romney, A. K., Batchelder, W. H., & Weller, S. C. (1986). Culture as consensus: A theory of culture and informant accuracy. *American Anthropologist, 88*, 313-338.

Sachs, P. (1995, September). *Transforming work: Collaboration, learning, and design.* New York: Communications of the ACM.

Said, E. (1978). *Orientalism.* New York, Pantheon.

Salant, P., & Dillman, D. A. (1994). *How to conduct your own survey.* New York: Wiley & Sons, Inc.

Schmidt, K., & Bannon, L. (1992). Taking CSCW seriously: Supporting articulation work. *Computer Supported Cooperative Work, 1*(1-2), 7-40.

Shapiro, D. (1994, October 22-26). The limits of ethnography: Combining social sciences for CSCW. *Proceedings of Computer Supported Cooperative Work (CSCW'94)* (pp. 417-428). Chapel Hill, NC: ACM Press.

Schuler, D., & Namioka, A. (Eds.). (1993). *Participatory design: Principles and practices.* Hillsdale, NJ: Lawrence Erlbaum Associates.

Smith, D. (1987). *The everyday world as problematic: A feminist sociology.* Boston, Northwestern University Press.

Snowdon, D. A. (1996). Aging and Alzheimer's disease: Lessons from the nun study. *The Gerontologist, 37*(2), 150-156.

Sonderegger, P., Manning, H., Charron, C., & Roshan, S. (2000, December). Scenario design. In Forrester Report.

Suchman, L. (1983). Office procedures as practical action: Models of work and system design. *ACM Transactions on Office Information Systems, 1*(4), 320-328.

Suchman, L. (1998, August 9). Organizing alignment: The case of bridge building. *Symposium on Situated Learning, Local Knowledge and Action: Social Approaches to the Study of Knowing in Organizations.* San Diego, CA: Academy of Management Annual Meeting.

Suchman, L. (2000). Embodied practices of engineering work. Special issue of *Mind, Culture and Activity, 7*(1/2), 4-18.

Suchman, L., Blomberg, J., Orr, J., & Trigg, R. (1999a). Reconstructing technologies as social practice. *American Behavioral Scientist, 43*(3), 392-408.

Suchman, L., Blomberg, J., & Trigg, R. (1999b). Working artifacts: Ethnomethods of a prototypes. Presented at the 1998 American Sociological Association in the session Ethnomethodology: Hybrid Studies of the Workplace and Technology, August 22, 1998. San Francisco, CA.

Suchman, L., & Trigg, R. (1991). Understanding practice: Video as a medium for reflection and design. In J. Greenbaum & M. Kyng

(Eds.), *Design at work: Cooperative design of computer systems* (pp. 65–89). Hillsdale, NJ: Lawrence Erlbaum Associates.

Trigg, R., Blomberg, J., & Suchman, L. (1999). Moving document collections online: The evolution of a shared repository. In *Proceedings of the European Conference on Computer-Supported Cooperative Work (ECSCW '99)*. Copenhagen, Denmark.

Wasson, C. (2000). Ethnography in the field of design. *Human Organization, 59*(4), 377–388.

Weller, S. C., & Romney, A. K. (1988). *Systematic data collection.* London: Sage Publications.

Whiting, B., & Whiting, J. (1970). Methods for observing and recording behavior. In R. Naroll & R. Cohen (Eds.), *Handbook of method in cultural anthropology* (pp. 282–315). New York: Columbia University Press.

Whyte, W. F. (1960). Interviewing in field research. In R. Adams & J. Preiss (Eds.), *Human organization research* (pp. 299–314). Homewood, IL: Dorsey.

Whyte, W. F. (1984). *Learning from the field: A guide from experience.* Newbury Park, CA: Sage.

Wolf, M. (1992). *A thrice-told tale: Feminism, postmodernism, and ethnographic responsibility.* Stanford, CA: Stanford University Press.

Yanagisako, S., & Delaney, C. (Eds.). (1995). *Naturalizing power: Essays in feminist cultural analysis.* New York: Routledge.

·VB·

DESIGN AND DEVELOPMENT

Tom Stewart
System Concepts Limited

BACKGROUND

In the past, advertisers made exaggerated claims about the technical capability of their computer systems. Now they also make exaggerated claims about the user friendliness of the user interfaces. Yet when it comes to applying human factors knowledge to the design and development of real products, the reality falls far short of the hype. System designers do not set out to design unfriendly interfaces, but clearly many interfaces are not as well matched to the capabilities and limitations of the users or the requirements of their tasks as the promotional material promises.

There are many reasons for this *usability gap*, including inadequate methods and tools and insuficient understanding and application of the lessons that have been learned from the growing discipline of human–computer interaction (HCI). One of the purposes of this handbook is to help plug this usability gap and ensure that future systems offer improved HCI. In this section, we focus on design and development, which is arguably one of the key areas where the groundwork is laid for ensuring that the resulting systems will underdeliver in terms of usability. One of the most significant problems in design and development lies in the way we develop the kind of large systems that are becoming essential for succesful operation of private and public enterprises. In these large projects, it is difficult to achieve the kind of creative quality that needs to be at the heart of good systems.

Two examples of this creative quality come to mind. The first of these was the Visicalc spreadsheet program, which some would say was responsible for the explosion of personal computers (PCs) in the office environment. The elegance of Visicalc came from a sound understanding of what users would want to use the package for, rather than the application of any particular theory of user interface design. The originator, Dan Bricklin, was a business school student in the early 1980s with real requirements for doing "what if" spreadsheet calculations. He sat down and drew up the specification for the kind of system he would like to use. The two strokes of genius were first to write that specification in terms of what he wanted to achieve and not to worry at that stage about what convolutions the software might have to undergo to achieve this and second to give the programming task to a creative programmer (Bob Frankston) with extensive experience writing computer games.

The result of these two strokes of genius was when the specification said a column of numbers should move to allow a new column to be created, the programmer assumed (correctly) that the best way to do that was for it to happen directly on the screen without the user having to specify all the transformations separately. This style of direct manipulation is now a common feature of many interfaces, and the lasting tribute to Visicalc was the proliferation of "visiclones." Although these were successful as products, it was gurus like Ben Shneiderman who can be given much of the credit for explaining and promoting the merits of direct manipulation as an interface technique (see Shneiderman, 1987). Spreadsheets cannot do everything and are not the best model for all tasks. But Visicalc represented one of the first user interfaces in which users found that once they understood it even to a minor degree, the package was much more useful than they had expected. Up to that time, most computers had promised much but delivered little.

The second major leap was represented by the Apple Macintosh. Of course, the windows, icons, menus, and pointers (WIMP) approach was really pioneered by Xerox at the PARC Research Center, but it was not until the Macintosh that it all came together in an affordable package. Many criticisms can be made of the Mactintosh as a product, and indeed its commercial success is not yet quite what the original reception appeared to promise. Nonetheless, like Visicalc, it has generated a plethora of look-alikes, and no self-respecting user interface today would be complete without some concession to this style of interface.

Although the Macintosh interface as a whole is a joy to play with—and sometimes to use—it really comes into its own in some of the graphics applications.

At a User Interface Workshop we organized under the auspices of the Alvey Programme in the United Kingdom, the designer of MacDraw, Mark Cutter, described the interface design process behind this package and its predecessor, Lisadraw. His description of the process made it clear that considerable ingenuity and creativity had gone into the design. Many of the features were ones that he himself wanted in a drawing aid. He also went to some trouble to have colleagues at Apple try out early versions and incorporate their feedback and comments. This may seem to be a completely different process from the Visicalc example, but I believe there is one common key element knowledge of the user's task requirements.

With Visicalc, it was the user who was the specifier. With Macdraw, the task (i.e., drawing) is one that all children and most adults understand well. The assumptions the designer made about what a user might need or expect at a particular point were therefore based on an implicit model of the user and his or her task that bore some close relation to the real world.

STRUCTURE OF THE SECTION

Neither of the approaches discussed above are easy to scale up to large commercial systems, where design and devlopment project teams may number hundreds of individuals, each with his or her own expertise and viewpoint. It is difficult to find users with the necessary skills and experience to be able to specify buildable systems and equally difficult to find software designers who know business tasks well enough to ensure usable systems. However, there are some promising signs and in this section, each of the five chapters addresses an important component in the design and development phase of the project life cycle.

When asked about their reluctance to apply what is already known about human computer interaction, many designers argue that human factors guidelines are too difficult to use in practice. Back in the 1980s, Mosier and Smith (1986) reported some market research among the users of the Mitre Corporation compilation of 580 user interface guidelines. Many of these guidelines have stood the test of time and have now found their way into national and international standards. As might be anticipated, the guidelines have proved useful but are still too difficult to find and apply in every case. What is required is a skilled interpretation of the general guidelines to make them specific to particular designs. This interpretation requires an understanding of the intended users and their tasks.

In this section, Stewart and Travis (chapter 51) review the history and describe the application of numerous guidelines, standards, and style guides. One of the recurring themes in this chapter is what might be called "Stewart's Law of Usability Standards": the easier it is to formulate the usability standard or guideline, the more difficult it is to apply in practice. By this I mean that a simple guideline such as "allow the user to control the pace and sequence of the interaction" has a great deal of backing as a general guideline. In practice, of course, there are many situations in which the rule fails, in which the answer is "it depends" and the context makes it more appropriate for the system to control some part of the interaction. The designer wishing to follow such a simple guideline needs to interpret the guideline in the context of the system, and this requires insight and thought. The alternative approach, in which the guideline is preceded by statements defining the context—for example, "if the user is x and the task is y and the environment is z then do abc"—become so tedious and confusing that they are quickly ignored. In the ISO [International Organization of Standardization] 9241 series that we describe in the chapter, there has been an attempt to define a middle course that involves giving specific practical examples as an aid to design. In ISO 13407, Human Centered Design Processes for Interactive Systems (which we also describe), the standard is concerned with the process itself. We believe this is one way of ensuring that usability standards remain relevant when technology changes and also offer practical help to designers and developers.

Although there is a growing body of guidance available to designers to help them organize the design process, which provides them with lists of specific points about what does and what does not work well for users, there is still a shortage of effective tools and techniques available to the user interface designer. One of the earliest and most influential HCI tools was Moran's Command Language Grammar (Moran, 1981) which was a representational framework of the conceptual user interface. A key feature of this is the notion of levels of interface (task-related, semantic, syntactic, and interactional) and the representation of users' tasks in terms of goals and subgoals, the operators for acting on the interface, the methods used to achieve subgoals, and selection rules for choosing between two or more alternative methods.

Such tools are generally analytic, however, and offer little help until the designer has already made a creative leap. Beaudouin-Lafon and Mackay (chapter 52) review prototype development and tools, which also address the more creative aspects of the design process. They discuss three classes of tools aimed at idea generation, rapid prototyping, and development. Many of the techniques they review, from brainstorming to rapid prototyping, are not the exclusive preserve of HCI specialists. The authors explain how to integrate the output of these creative methods with software development toolkits, interface builders, application frameworks, models, and environments.

SCENARIO-BASED DESIGN

Much current user interface design seems more akin to art than science. Innovative designers seem prepared to risk much, often based on appealing but substantially untested scenarios of future users. But even with solid user trials evidence behind them, interface designers must be prepared to suffer the disbelief and cynicism of their colleagues. Many user interface innovations were treated sceptically on the basis of their modest performance in laboratory tests before becoming highly successful products. Once they became available to users, their benefits

were so obvious that precise measurement became irrelevant. The pre-Xerox PARC research on the mouse and multiple windows by Englebart, in his pioneering work on augmenting human intelligence in the 1960s, never really pointed to the spectacular success that WIMPs would experience in the 1980s.

Nonetheless, scenario-based design has become a well recognized but often poorly understood part of the system designers repertoire. Rosson and Carroll (chapter 53) review the history and practice of scenario-based design and discuss some of the strengths and limitations of this "solution-first" approach to tackling ill-structured problems.

Unfortunately, even though it is fashionable to talk about involving users in the design process, and no self-respecting design activity claims to do otherwise, taking such involvement seriously is another matter. Although most designers these days accept the wisdom of involving users, the reality is that the process is often far from satisfactory. In many cases, the "users" to which the designers refer are simply senior management representatives of the user organization. Their involvement may involve little more than signing-off specifications and deliverables with much of the emphasis on contractual milestones. In "shrink-wrap" software development, the "users" may be mythical profiles drawn up by marketing specialists and occasionally exposed to features of the system through focus groups or market research surveys.

Muller (chapter 54) describes the more active and direct involvement of users in the process of participatory design (which he calls the third space in HCI). This process involves users playing a full and active part in the design process.

The value of this approach is encouraged by the growth in the number of discretionary users, that is, users who can choose whether to use the system or product. The rise of the PC market illustrates this well. This market really took off when the price of a PC fell to a level where it was within the budgetary discretion of many managers. Thus, they were able to buy PCs from the corner store without having to involve their own computer departments or make extensive financial justifications. Even where the proper approval procedures were followed, the fact that the users were paying directly out of their own budgets gave them a far larger say in the purchasing decision. Now, the price of the PC has fallen to the point where individuals are prepared to pay for serious, professional-quality PCs out of their own pockets. Here, too, we see evidence that improved usability is an important consideration for potential purchasers. This willingness to pay for user interface quality applies just as much to hardware (with users buying high-quality monitors, flat panel displays, etc.) as to software.

In many situations, users are not the purchasers of the information technology products, but here, too, there is an increased willingness to pay for improved HCI to satisfy user demands and expectations, often in the interests of improved acceptability. Involving users need not be more expensive than other design approaches, however. Indeed, as this chapter explains, the key to the process is for designers and users to work together and unite their two worlds during the design process—hence, the third space in Muller's chapter title. Because both sides must compromise to meet in the middle, the resulting process can

be shorter and more efficient than more traditional and adversarial methods.

There is increasing demand to ensure that access to computer technology is made available to all. This greatly increases the challenges facing designers of such universal access interfaces. Stephanidis and Savidis (chapter 55) describe a unified user interface development process allowing the designer to create a system that adapts automatically and that also can be adapted to deal with the diversity of users and environments, which universal access demands.

Adaptive interfaces arose out of early attempts to exploit artificial intelligence (AI) techniques in user interface design. One approach involved building user models, which encapsulated key characteristics of users and act as design tools for interface designers. But AI techniques also proved useful in the creation of intelligent front ends that contain application experts to decouple the user from the vagaries of different services and systems accessed over a network. Both these approaches were combined in a project sponsored by the UK Alvey program in the 1980s. The Adaptive Intelligent Dialogues (AID) project was a joint effort by Standard Telephones and Cables (STC), British Telecom, Data Logic, Essex University, Hull University, and the Strathclyde man–machine interaction (MMI) Unit.

Adaption intuitively seems like a good idea, so a major aim of the project was not so much to test whether this was true as to discover under what circumstances adaption has advantages. Elaborating the advantages is also important as part of the kind of cost–benefit analysis, which will be necessary to justify future investment in adaptive systems. The most obvious disadvantage was the enormous overhead an adaptive system requires to monitor the interaction and change its nature. This overhead is not only in terms of the costs of the extra hardware and software, but also in system complexity (which has an impact on the design process, the ease of modifying and updating the system, and the system response time).

A less obvious consequence of the complexity was that a whole new dimension of error and misunderstanding can occur. It is all too easy to imagine reciprocal misunderstanding that is perpetuated rather than resolved as the iterations progress. The complexity itself gives rise to increased risk of error, but it is the potentially greater consequence of errors that could be more problematic.

Another disadvantage of adaptive systems is the uncertainty for the user about what the system is doing or is about to do. This could make it more difficult for the user to learn the system and may actually inhibit early exploratory learning.

CONCLUSION

Investing in user interface research or consultancy may seem expensive, and many designers find the idea of copying a successful interface attractive. Ample evidence in the marketplace indicates that this approach is popular, and it will work in certain circumstances. A clear example to follow is a powerful teaching device and the number of the significant interface innovations in recent years that have been widely promulgated in this way

with good effect. There are pitfalls for the unwary designer, however. For example, the interface may not actually be good. Just because it has received widespread pubicity or appears attractive to the casual observer, there is no guarantee that it will be the right interface for the user. Indeed, it is easy to misunderstand what makes a good interface, and by copying one the designer may miss the point. Many of the early copies of the Xerox/Macintosh interface focused on the ability to illustrate objects with pretty icons. I believe that what made such interfaces usable had little to do with the graphics and much to do with the ability to alter contexts without the usual computer problems of having to terminate one activity before you could even consider another. Many of the clones tended to miss that point and simply provided menus using icons that delivered few of the real benefits.

In this harsh commercial world, simply copying a competitor will not provide an edge in the marketplace but will relegate a device to a subordinate position. Computer products are becoming increasingly similar in functionality through common components, standards, and so on. The user interface offers the most visible and perhaps most attractive means of differentiating among products. In the worst case, copying an interface solution may result in litigation. There are already some court cases in session over the copyright of icons.

The chapters in this section provide pointers to help interface designers create usable systems, but there are no short cuts to good HCI design. The process must involve understanding of the requirements of the users and their tasks, innovating creative solutions, and testing them. There simply is no substitute for testing proposed solutions with real users.

References

Mosier, J. N., & Smith, S. L. (1986). Application of guidelines for designing user interface software. *Behaviour and Information Technology, 5*, 39–46.

Moran, T. P. (1981). The Command Language Grammar: A representation for the user interface of interactive computer systems. *International Journal of Man-Machine Studies, 15*, 3–50.

Shneiderman, B. (1987). Designing the user interface: *Strategies for effective human-computer interaction*. Reading, MA: Addison-Wesley.

·51·

GUIDELINES, STANDARDS, AND STYLE GUIDES

Tom Stewart and David Travis
System Concepts Ltd.

INTRODUCTION

What Do We Mean by Guidelines, Standards, and Style Guides?

Human–computer interaction (HCI) guidelines, standards, and style guides represent three approaches to improving the usability of systems. They are not mutually exclusive categories. For example, many HCI standards simply provide agreed guidelines rather than specify requirements. Some style guides are implemented in such a way that they have become standards from which designs cannot vary. Generally (and indeed in this chapter), however, we use the terms as follows:

- *Guidelines*—recommendations of good practice that rely on the credibility of their authors for their authority
- *Standards*—formal documents published by standards making bodies that are developed through some form of consensus and formal voting process
- *Style Guides*—sets of recommendations from software providers or agreed within development organizations to increase consistency of design and to promote good practice within a design process of some kind

Why Do Guidelines, Standards, and Style Guides Exist?

Guidelines, standards, and style guides generally exist to improve the consistency of the user interface and to improve the quality of interface components. They help specifiers to procure systems and system components that can be used effectively, efficiently, safely, and comfortably. They also help restrict the unnecessary variety of interface hardware, software, and technology and ensure that the benefits of any variations are fully justified against the costs of incompatibility, loss of efficiency, and increased training time for users. Even standards still under development can have an impact on hardware and software development. The major suppliers play an active part in generating the standards and increasingly are incorporating the guidance on good practice into products before the standards themselves are published.

Although designing usable systems requires far more than simply applying guidelines, standards, and style guides, doing so can make a significant contribution by promoting consistency, good practice, common understanding, and an appropriate prioritization of user interface issues.

Consistency. Anyone who uses computers knows only too well the problems of inconsistency between applications and often even within the same application. Inconsistency, even at the simplest level, can cause problems. Consider three examples:

- Press the <escape> key in one place, and you are safely returned to your previous menu choice. In another place, you are unceremoniously "dropped" to the operating system, the friendly messages disappear, and you lose all your data.

- On the Web, inconsistency is rampant. Even something as straightforward as a hypertext link may be denoted by underlining on one site, by a mouseover on a second site, and by nothing at all on a third site.
- Different and confusing keyboard layouts sit side by side in many offices.

Guidelines, standards, and style guides play an important part in helping address these issues by collating and communicating agreed best practices for user interfaces and for the processes by which they are designed and evaluated. They can provide a consistent reference across design teams or across time to help avoid such unpleasant experiences. Indeed, in other fields, consistency—for example, between components that should interconnect—is the prime motivation for standards. It is certainly a worthwhile target for user interface standards.

Good Practice. In many fields, standards provide definitive statements of good practice. In user interface design, there are many conflicting viewpoints about good practice. Standards, especially those from the International Organization for Standardization (ISO), can provide independent and authoritative guidance. The International standards are developed slowly, by consensus, using extensive consultation and development processes. This has its disadvantages in such a fast-moving field as user interface design, and some have criticized any attempts at standardization as premature. However, there are areas in which a great deal is known that can be made accessible to designers through appropriate standards, and there are approaches to user interface standardization, based on human characteristics, that are relatively independent of specific technologies.

The practical discipline of having to achieve consensus helps moderate some of the wilder claims of user interface enthusiasts and helps ensure that the resulting standards do represent good practice. The slow development process also means that standards can seldom represent the leading edge of design. Nonetheless, properly written, they should not inhibit helpful creativity.

Common Understanding. Standards themselves do not guarantee good design, but they do provide a means for different parties to share a common understanding when specifying interface quality in design, procurement, and use.

- Standards allow users to set appropriate procurement requirements and to evaluate competing suppliers' offerings.
- Standards allow suppliers to check their products during design and manufacture and provide a basis for making claims about the quality of their products.
- Standards allow regulators to assess quality and provide a basis for testing products.

Appropriate Prioritization of User Interface Issues. One of the most significant benefits of standardization is that it places user interface issues squarely on the agenda. Standards are serious business, and whereas many organizations pay

little regard to research findings, few can afford to ignore standards. Indeed, in Europe—and increasingly in other parts of the world—compliance with relevant standards is a mandatory requirement in major contracts.

HISTORY

Deusches Institut für Normung

Thirty years ago, the German national standards organization, Deusches Institut für Normung (DIN) started to publish a series of standards that shook the computer world. These standards, DIN 66-234, were published in a number of parts and collectively addressed the ergonomics problems of visual display terminals and the workplace.

If we ask why there was widespread concern, especially from the computer manufacturers, many of whom happened to be based in the United States, then we receive two answers. First, the standards were based on too little and too recent research. A particular issue that received such criticism was the requirement that the thickness of the keyboard should be restricted to 30 mm. A number of manufacturers reported studies disputing the importance of keyboard thickness, arguing with the proposed dimension and demonstrating that users showed preferences for quite different arrangements.

Of course, it should not be overlooked that the 30-mm keyboard thickness was a difficult target to reach at that time. Most key mechanisms themselves required greater depth, and major manufacturers had substantial investment in tooling keyboards to quite different thicknesses.

The second answer is that the very idea of ergonomics requirements affecting sales directly was completely alien to many of the suppliers. Certainly, the large manufacturers employed ergonomists, human factors engineers, and psychologists in their research and development departments. Certainly, there was growing recognition that the human aspects of computer technology were important. But at that time, price performance was the main objective, and it came as a major culture shock for the computer industry that ergonomics standards could have such a major impact on whether a product would sell.

Note that it was not the DIN standards but their integration into workplace regulations (ZH 618 Safety Regulations for Display Workplaces in the Office Sector, published by the Central Association of Trade Cooperative Associations) that gave the ergonomics requirements such "teeth." Failure to comply with these regulations leaves an employer uninsured against industrial compensation claims.

DIN 66-234 also contained a number of parts that dealt exclusively with software issues. For example, Part 3 dealt with the grouping and formatting of data, Part 5 with the coding of information, and Part 8 with the principles of dialogue design. Although these were more in the form of recommendations, they too were heavily criticized, particularly for their broad scope and their inhibitory effect on interface design.

Indeed, a major criticism of most early standards in this field was that they were based on product design features such as height of characters on the screen. Such standards were specific to current technology, for example, cathode ray tubes (CRT), and did not readily apply to other technologies. They may therefore inhibit innovation and force designers to stick to old solutions.

The International Organization for Standardization

In the late 1970s, the kind of concern about the ergonomics of visual display terminals (also called visual display units) that stimulated the German standards became more widespread, especially in Europe.

The primary concern at that time concerned the possibility that prolonged use (especially of displays with poor image quality) might cause deterioration in users eyesight.[1]

When a new International Standards work item to address this concern was proposed, the information technology committee decided that this was a suitable topic for the recently formed ergonomics committee ISO/TC 159. The work item was allocated to the subcommittee ISO/TC 159/SC4 Signals and Controls, and an inaugural meeting was held at the British Standards Institution (BSI) in Manchester in 1983. The meeting was well attended by delegates from many countries, and a few key decisions were made.

At that time, there was a proliferation of office-based systems, and SC4 decided to focus on office tasks (word processing, spreadsheet, etc.) rather than try to include computer-aided design or process control applications. It also decided that a multipart standard should cover the wide range of ergonomics issues that needed to be addressed to improve the ergonomics of display screen work. A number of working groups were established to carry out the technical work of the subcommittee. Table 51.1 lists the current working groups of ISO/TC159/SC4.

Little did any of those present realize that it would be nearly 7 years before the first parts of ISO 9241 would be published and that it would take to the end of the century to publish all 17 parts. Table 51.2 shows the parts of ISO 9241 and Table 51.3 lists the other standards that are part of the ISO/TC159/SC4 work program.

TABLE 51.1. The Working Groups (WG) of ISO/TC159/SC4

WG1	Fundamentals of controls and signalling methods
WG2	Visual display requirements
WG3	Control, workplace, and environmental requirements
WG4	Task requirements (disbanded)
WG5	Software ergonomics and human–computer dialogue
WG6	Human-centred design processes for interactive systems
WG8	Ergonomics design of control centres

---

[1] Since then several studies have shown that aging causes the main effect on eyesight, and because display screen work can be visually demanding, many people only discover this deterioration when they experience discomfort from intensive display screen use. This can incorrectly lead them to attribute their need for glasses to their use of display screens.

TABLE 51.2. Parts of ISO 9241

Ergonomic Requirements for Office Work With Visual Display Terminals		Responsible Working Group
Part 1	General introduction	WG6
Part 2	Guidance on task requirements	WG4 (finished)
Part 3	Visual display requirements	WG2
Part 4	Keyboard requirements	WG3
Part 5	Workstation layout and postural requirements	WG3
Part 6	Guidance on the work environment	WG3
Part 7	Requirements for displays with reflections	WG2
Part 8	Requirements for displayed colors	WG2
Part 9	Requirements for nonkeyboard input devices	WG3
Part 10	Dialogue principles	WG5
Part 11	Guidance on usability	WG5
Part 12	Presentation of information	WG5
Part 13	User guidance	WG5
Part 14	Menu dialogues	WG5
Part 15	Command dialogues	WG5
Part 16	Direct-manipulation dialogues	WG5
Part 17	Form-filling dialogues	WG5

Note. ISO = International Organization for Standardization.

GUIDELINES

Background

Whereas European interest in the early 1980s seemed to focus on computer hardware ergonomics, in the United States there was a growing interest in user interface software. Computer hardware seemed like a possible target for formal standardization, but the highly contextual nature of best practice made formal standards at best premature and at worst dangerous traps for perpetuating obsolete practices in this rapidly developing technology. Rather than attempt to develop formal standards, several HCI groups and individual researchers started to assemble collections of guidelines, tips, and hints into books and compendiums. We describe three of the most influential in this section.

Smith and Mosier's Guidelines for Designing User Interface Software

In 1984 Sidney Smith and Jane Mosier published *Guidelines for Designing User Interface Software* for the U.S. Air Force. With 944 guidelines, this document remains the largest collection of publicly available user interface guidelines in existence. These guidelines draw extensively from four sources: Brown et al. (1983), Engel and Granda (1975), MIL-STD-1472C (1983), and Pew and Rollins (1975). Their report provides user interface guidelines in six categories: data entry, data display, sequence control, user guidance, data transmission, and data protection. One example of a guideline from this document is shown in Fig. 51.1.

Although focused on character user interfaces (CUIs), many of the guidelines are still relevant to today's user interfaces— especially Web sites.

Shneiderman's User Interface Guidelines

Ben Shneiderman published a landmark text in 1987, *Designing the User Interface*. This book contains many tables of guidelines and includes detailed explanations and background research to

TABLE 51.3. Other Human System Interaction Standards for Which ISO/TC 159/SC4 Is Responsible

ISO Standards		Responsible Working Group
13406-1	Ergonomics requirements for flat panel displays—Introduction	WG2
13406-2	Ergonomics requirements for flat panel displays—Ergonomics requirements	WG2
14915-1	Software ergonomics for multi-media interfaces—Design principles and framework	WG5
14915-2	Multimedia control and navigation	WG5
14915-3	Media selection and combination	WG5
14915-4	Domain-specific multimedia aspects	WG5
TS 16071	Accessibility	WG5
13407	Human centred design processes for interactive systems	WG6
TS 16982	Usability methods supporting HC design	WG6
TR 18529	Human centred lifecycle process descriptions	WG6
11064-1	Principles for the design of control centres	WG8
11064-2	Principles of control suite arrangement	WG8
11064-3	Control room layout	WG8
11064-4	Workstation layout and dimensions	WG8
11064-5	Displays and controls	WG8
11064-6	Environmental requirements for control rooms	WG8
11064-7	Principles for the evaluation of control centres	WG8
11064-8	Ergonomics requirements for specific applications	WG8

Note. ISO = International Organization for Standardization, TC = Technical Committee.

> **1.3 DATA ENTRY: Text**
> *1.3/10 + Upper and Lower Case Equivalent in Search*
>
> Unless otherwise specified by a user, treat upper and lower case letters as equivalent in searching text.
>
> *Example:* "STRING," "String," and "string" should all be recognized/accepted by the computer when searching for that word.
>
> *Comment:* In searching for words, users will generally be indifferent to any distinction between upper and lower case. The computer should not compel a distinction that users do not care about and may find difficult to make. In situations when case actually is important, allow users to specify case as a selectable option in string search.
>
> *Comment:* It may also be useful for the computer to ignore such other features as bolding, underlining, parentheses and quotes when searching text.
>
> See also: 1.0/27 3.0/12

FIGURE 51.1. A guideline from Smith and Mosier (1986).

justify each guideline. He presented "Eight golden rules of dialog design," which he explained as representing underlying principles of design that were applicable to most interactive systems. These principles were as follows:

- *Strive for consistency.* In particular, use consistent sequences of actions, and use the same terminology wherever appropriate. Shneiderman claimed that this was the most frequently violated yet "the easiest one to repair and avoid" (p. 69). Certainly it ought to be easy to ensure consistency, but as we discuss later with respect to style guides, it can prove difficult to ensure that unhelpful inconsistencies do not creep into designs, especially where distributed teams are involved.

- *Enable frequent users to use shortcuts.* Abbreviations, special keys, and macros can all be appreciated by frequent knowledgeable users. One of the traps of WIMPs (windows, icons, mouse, and pop-up menus) interfaces is that they are easy to demonstrate to senior management and superficially appear simple to use. We have seen several examples in which experienced users have become extremely frustrated by mouse-intensive systems—in some cases, to the extent of suffering work-related upper limb disorders.

- *Offer informative feedback.* Visual feedback can be particularly effective. Shneiderman went to discuss the value of direct manipulation in this context.

- *Design dialogs to yield closure.* Organizing actions into groups with a beginning, middle, and end plays to a basic psychological desire for closure and provides a sense of satisfaction when tasks are completed.

- *Offer simple error handling.* Where possible, design systems so that users cannot make serious errors and ensure that error messages really help the users.

- *Permit easy reversal of actions.* It may not always be possible, but there is nothing more reassuring for users than knowing that they can undo an action that has unintended and sometimes dramatic consequences.

- *Support internal locus of control.* Another basic psychological desire is for control; the more systems are able to provide users with control, the more satisfying they will be, especially for experienced users. Some modern office software packages break this rule, and many of us experience extreme frustration when a "clever" piece of software insists on reformatting what we have carefully laid out on a page.

- *Reduce short-term memory load.* Although "seven plus or minus two chunks" may be an oversimplified description of our memory limitations, many interfaces demand extraordinary feats of memory to perform simple tasks.

Nielsen's Usability Heuristics

Rolf Molich and Jakob Nielsen (1990) carried out a factor analysis of 249 usability problems to derive a set of "heuristics" or rules of thumb that would account for all of the problems found. Nielsen (1994a) further revised these heuristics, resulting in the 10 guidelines listed below.

1. *Visibility of system status.* The system should always keep users informed of what is going on, through appropriate feedback within reasonable time.
2. *Match between system and the real world.* The system should speak the users' language, with words, phrases, and concepts familiar to the user, rather than system-oriented terms. Follow real-world conventions, making information appear in a natural and logical order.
3. *User control and freedom.* Users often choose system functions by mistake and will need a clearly marked "emergency exit" to leave the unwanted state without having to go through an extended dialogue. Support undo and redo.
4. *Consistency and standards.* Users should not have to wonder whether different words, situations, or actions mean the same thing. Follow platform conventions.
5. *Error prevention.* Even better than good error messages is a careful design that prevents a problem from occurring in the first place.
6. *Recognition rather than recall.* Make objects, actions, and options visible. The user should not have to remember information from one part of the dialogue to another. Instructions for use of the system should be visible or easily retrievable whenever appropriate.
7. *Flexibility and efficiency of use.* Accelerators—unseen by the novice user—may often speed up the interaction for the expert user such that the system can cater to both inexperienced and experienced users. Allow users to tailor frequent actions.
8. *Aesthetic and minimalist design.* Dialogues should not contain irrelevant or rarely needed information. Every extra unit of information in a dialogue competes with the relevant units of information and diminishes their relative visibility.
9. *Help users recognize, diagnose, and recover from errors.* Error messages should be expressed in plain language (no codes), precisely indicate the problem, and constructively suggest a solution.

10. *Help and documentation.* Even though it is better if the system can be used without documentation, it may be necessary to provide help and documentation. Any such information should be easy to search, focused on the user's task, list concrete steps to be carried out, and not be too large.

Nielsen (1994b) described a method for structuring a guidelines-based user interface review. Nielsen's method uses Molich and Nielsen's user interface heuristics, and indeed Nielsen terms this method an "heuristic evaluation." The method however can be applied using any set of HCI guidelines. (For further information on evaluation techniques, see chapter 57, Inspection-Based Evaluations by Cockton, Lavery, and Alan Woolrych.)

STANDARDS

Sources of HCI Standards

Many standards bodies have been in existence for some time and are organized according to traditional views of technology and trade. Software is used as part of systems that involve a range of technologies. The purpose of this section is to introduce a number of key standards bodies that are working on standards relevant to HCI and describe their main activities briefly.

In most people's minds, one of the most basic and fundamental objectives of standardization is to minimize unnecessary variations. Ideally, for any product category, there is one standard that should be satisfied, and products that meet that standard give their owners or users some reassurance about quality or about what standards makers refer to as "interoperability." Thus, yachting enthusiasts in Europe who buy lifejackets meeting EN 396 might reasonably expect them to keep them afloat if they have the misfortune to fall overboard in the Florida Keys. Similarly, an office manager in the United States who orders A4 paper for a photocopier might reasonably expect paper that meets that standard (ISO 216:1975) to fit, even though it is not the typical size used locally.

This brings us to a rather important point: It is often difficult to achieve a single agreed standard, and a common solution is to have more than one standard. An obvious example concerns paper size where there are the ISO A series (A0, A1 etc.), the ISO B series (B0, B1 etc.) as well as U.S. sizes (legal, letter etc.). Although this solves the standards makers' problems in agreeing on a single standard, it is an endless source of frustration for users of the standard—as anyone who has forgotten to check the paper source in an e-mailed document can testify.

There is another reason why there are more standards than one might imagine, especially when it comes to user interface design issues: Computer technology forms the basis of many different industries, and standards can have an important impact on market success. But it is not just at the international level that there appears to be some duplication. In the United Kingdom, the British Standards Institution mirror committee to SC4 published an early version of the first six parts of ISO 9241 as a British Standard BS 7179: 1990. The primary reason for this was to provide early guidance for employers of users of visual displays, who wanted to use standards to help them select equipment that met the requirements in the Schedule to the Health and Safety (Display Screen Equipment) Regulations 1992. These regulations are the U.K. implementation of a European Community Directive on the minimum safety and health requirements for work with display screen equipment (90/270/EEC). Of course, as a spinoff the British Standards Institution were able to generate revenue by selling these standards several years before the various parts of ISO 9241 became available as British Standards.

A similar process has taken place in the United States with the Human Factors and Ergonomics Society (HFES) developing HFS 100 on Visual Display Terminal Ergonomics as an ANSI-authorized Standards Developing Organisation. More recently, there are two HFES standards development committees working on HFES 100 (a new version of HFS 100) and on HFES 200, which addresses user interface issues. It includes sections on accessibility, voice and telephony applications, color and presentation, and slightly rewritten parts of the software parts of ISO 9241.

As we mentioned earlier, international ergonomics standards in HCI are being developed by the ISO. The work of ISO is important for two reasons. First, the major manufacturers are international, and therefore the best and most effective solutions need to be international. Second, the European Standardisation Organisation (CEN) has opted for a strategy of adopting ISO standards wherever appropriate as part of the creation of the single market. CEN standards replace national standards in the European Union and European Free Trade Area member states.

The ISO comprises national standards bodies from member states. (see www.iso.ch for more information). Its work is conducted by technical and subcommittees that meet every year or so and are attended by formal delegations from participating members of that committee. In practice, the technical work takes place in working groups of experts, nominated by national standards committees but expected to act as independent experts. The standards are developed over a period of several years and in the early stages, the published documents may change dramatically from version to version until consensus is reached (usually within a working group of experts). As the standard becomes more mature (from the Committee Draft Stage onward), formal voting takes place (usually within the parent subcommittee), and the draft documents provide a good indication of what the final standard is likely to look like. Table 51.4 shows the main stages.

Types of HCI Standards

Standards related to HCI fall into two categories: (a) Process-oriented standards specify procedures and processes to be followed and (b) product-oriented standards specify required attributes of the user interface. Some product-oriented standards specify the requirements in terms of performance rather than product attributes. These standards describe the users, tasks, and context of use and assess usability in terms of user performance and satisfaction to be achieved.

TABLE 51.4. The Main Stages of International Standards Development

WI	Work Item—an approved and recognized topic for a working group to be addressing, which should lead to one or more published standards
WD	Working Draft—a partial or complete first draft of the text of the proposed standard
CD	Committee Draft—a document circulated for comment and approval within the committee working on it and the national mirror committees; voting and approval are required for the document to reach the next stage
DIS	Draft International Standard—a draft standard circulated widely for public comment via national standards bodies; voting and approval are required for the draft to reach the final stage
FDIS	Final Draft International Standard—the final draft is circulated for formal voting for adoption as an International Standard
IS	International Standard—The final published standard

Note. Documents may be reissued as further CDs and DISs.

Process-Oriented Standards

ISO 6385: 1981 Ergonomic Principles in the Design of Work Systems. This standard sets out the ergonomic principles that should be applied to the design of work systems. ISO 13407 is also based on these principles and on the description of the aims and objectives of ergonomics which are contained in ISO 6385.

ISO 9241-1: 1993 Ergonomic Requirements for Office Work With Visual Display Terminals (VDTs)—General Introduction. This part introduces the multipart standard ISO 9241 for the ergonomic requirements for the use of visual display terminals for office tasks and explains some of the basic underlying principles. It provides some guidance on how to use the standard and describes how conformance to parts of ISO 9241 should be reported.

ISO 9241-2: 1993 Guidance on Task Requirements. This part deals with the design of tasks and jobs involving work with visual display terminals. It provides guidance on how task requirements may be identified and specified within individual organizations and how task requirements can be incorporated into the system design and implementation process

ISO 9241-11: Guidance on Usability. This part provides the definition of usability used in ISO 13407: "Usability: the extent to which a product can be used by specified users to achieve specified goals with effectiveness, efficiency and satisfaction in a specified context of use." ISO 9241-11 explains how to identify the information that is necessary to consider when specifying or evaluating usability in terms of measures of user performance and satisfaction. Guidance is given on how to describe the context of use of the product (hardware, software, or service) and the required measures of usability in an explicit way. It includes an explanation of how the usability of a product can be specified

and evaluated as part of a quality system, for example, one that conforms to ISO 9001. It also explains how measures of user performance and satisfaction can be used to measure how any component of a work system affects the whole work system in use.

ISO 10075-1: 1994 Ergonomic Principles Related to Mental Workload—General Terms and Definitions. This part of ISO 10075 explains the terminology and provides definitions in the area of mental workload.

ISO/IEC 14598-1: Information Technology—Evaluation of Software Products—General Guide. The concept of quality in use has been used in ISO/IEC 14598-1 to distinguish between quality as an inherent characteristic of a software product and the quality that is achieved when a software product is used under stated conditions, that is, a specified context of use. This definition of quality in use is similar to the definition of usability in ISO/DIS 9241-11. The use of the term *quality in use* therefore implies that it is necessary to take account of human-centered issues in evaluating software products. "Quality in use: the extent to which an entity satisfies stated and implied needs when used under stated conditions."

Product-Oriented Standards. In the product-oriented view, usability is seen as one relatively independent contribution to software quality and is defined in this way in ISO/IEC 9126: 1991 (Information Technology—Software Product Evaluation—Quality characteristics and guidelines for their use): "a set of attributes of software which bear on the effort needed for use and on the individual assessment of such use by a stated or implied set of users."

Usable products can be designed by incorporating product features and attributes known to benefit users in particular contexts of use. ISO 9241 provides requirements and recommendations relating to the attributes of the hardware, software, and environment that contribute to usability, and the ergonomic principles underlying them. Parts 3 to 9 contain hardware design requirements and guidance that can have implications for software. Parts 10 to 17 of ISO 9241 and other standards deal specifically with attributes of the software.

ISO 9241-3: 1993 Visual Display Requirements. This part specifies the ergonomics requirements for display screens that ensure that they can be read comfortably, safely, and efficiently to perform office tasks. Although it deals specifically with displays used in offices, it is appropriate to specify it for most applications that require general purpose displays to be used in an officelike environment.

ISO 9241-4: Keyboard Requirements. This part specifies the ergonomics design characteristics of an alphanumeric keyboard that may be used comfortably, safely, and efficiently to perform office tasks. Keyboard layouts are dealt with separately in various parts of ISO/IEC 9995: 1994 Information Processing—Keyboard Layouts for Text and Office Systems.

ISO 9241-5: Workstation Layout and Postural Require-ments. This part specifies the ergonomics requirements for a visual display terminal workplace that allows the user to adopt a comfortable and efficient posture.

ISO 9241-6: Environmental Requirements. This part specifies the ergonomics requirements for the visual display terminal working environment that provide the user with comfortable, safe, and productive working conditions.

ISO 9241-7: Display Requirements With Reflections. This part specifies methods of measurement of glare and reflections from the surface of display screens, including those with surface treatments. It is aimed at display manufacturers who wish to ensure that antireflection treatments do not detract from image quality.

ISO 9241-8: Requirements for Displayed Colors. This part specifies the requirements for multicolor displays that are largely in addition to the monochrome requirements in Part 3.

ISO 9241-9: Requirements for Nonkeyboard Input Devices. This part specifies the ergonomics requirements for nonkeyboard input devices that may be used in conjunction with a visual display terminal. It covers devices such as a mouse, trackerball, and other pointing devices. It also includes a performance test. It does not address voice input.

ISO 9241-10: 1996 Dialogue Principles. This part deals with general ergonomic principles that apply to the design of dialogues between humans and information systems: suitability for the task, suitability for learning, suitability for individualization, conformity with user expectations, self descriptiveness, controllability, and error tolerance

ISO 9241-12: Presentation of Information. This part contains specific recommendations for presenting and representing information on visual displays. It includes guidance on ways of representing complex information using alphanumeric and graphic/symbolic codes, screen layout, and design, as well as the use of windows.

ISO 9241-13: User Guidance. This part provides recommendations for the design and evaluation of user guidance attributes of software user interfaces including prompts, feedback, status, online help, and error management.

ISO 9241-14: Menu Dialogues. This part provides recommendations for the ergonomic design of menus used in user-computer dialogues. The recommendations cover menu structure, navigation, option selection and execution, and menu presentation (by various techniques including windowing, panels, buttons, fields, etc.). Part 14 is intended to be used by both designers and evaluators of menus, but the focus is primarily toward the designer.

ISO 9241-15: Command Language Dialogues. This part provides recommendations for the ergonomic design of command languages used in user-computer dialogues. The recommendations cover command language structure and syntax, command representations, input and output considerations, and feedback and help. Part 15 is intended to be used by both designers and evaluators of command dialogues, but its focus is primarily toward the designer.

ISO 9241-16: Direct Manipulation Dialogues. This part provides recommendations for the ergonomic design of direct manipulation dialogues and includes the manipulation of objects and the design of metaphors, objects, and attributes. It covers those aspects of graphical user interfaces that are directly manipulated and not covered by other parts of ISO 9241. Part 16 is intended to be used by both designers and evaluators of command dialogues, but its focus is primarily toward the designer.

ISO 9241-17: Form-Filling Dialogues. This part provides recommendations for the ergonomic design of form-filling dialogues. The recommendations cover form structure and output considerations, input considerations, and form navigation. Part 17 is intended to be used by both designers and evaluators of command dialogues, but its focus is primarily toward the designer.

ISO/IEC 10741-1: Dialogue Interaction—Cursor Control for Text Editing. This International Standard specifies how the cursor should move on the screen in response to the use of cursor control keys.

ISO/IEC 11581-1: Icon Symbols and Functions—Part 1: Icons—General. This part contains a framework for the development and design of icons, including general requirements and recommendations applicable to all icons.

ISO/IEC 11581-2: Icon Symbols and Functions—Part 2: Object Icons. This part contains requirements and recommendations for icons that represent functions by association with an object, and that can be moved and opened. It also contains specifications for the function and appearance of 20 icons.

How to Use ISO 9241 Standards

Although it was not made explicit at the time, SC4 had an underlying set of assumptions about HCI design activities and how the standards would support these. These activities included the following:

- Analyzing and defining system requirements
- Designing user–system dialogues and interface navigation
- Designing or selecting displays
- Designing or selecting keyboards and other input devices
- Designing workplaces for display screen users
- Supporting and training users
- Designing jobs and tasks

TABLE 51.5. How Parts of ISO 9241 Were Intended to Be Used in HCI Design Activities

HCI Activity	Relevant Part of ISO 9241
Analyzing and defining system requirements	ISO 9241-11:1998 *Guidance on usability*
Designing user–system dialogues and interface navigation	ISO 9241-10:1996 *Dialogue principles*
	ISO 9241-14:1997 *Menu dialogues*
	ISO 9241-15:1998 *Command dialogues*
	ISO 9241-16:1999 *Direct manipulation dialogues*
	ISO 9241-17:1998 *Form-filling dialogues*
Designing or selecting displays	ISO 9241-3:1992 *Display requirements*
	ISO 9241-7:1998 *Requirements for displays with reflections*
	ISO 9241-8:1997 *Requirements for displayed colors*
	ISO 9241-12:1998 *Presentation of information*
Designing or selecting keyboards and other input devices	ISO 9241-4:1998 *Keyboard requirements.*
	ISO 9241-9: 2000 *Requirements for nonkeyboard input devices*
Designing workplaces for display screen users	ISO 9241-5:1998 *Workstation layout and postural requirements*
	ISO 9241-6:1998 *Guidance on the work environment*
Supporting and training users	ISO 9241-13:1998 *User guidance*
Designing jobs and tasks	ISO 9241-2:1992 *Guidance on task requirements*

Note. ISO = International Organization for Standardization; HCI = human–computer interaction.

Table 51.5 shows how it was anticipated that the standards would be used to support these activities.

Strengths and Limitations of HCI Standards

It is important to be aware of the strengths and limitations of standards; they cannot be understood (nor therefore used effectively) in isolation from the context in which they were developed. It is important to realize the following:

- Standards are developed over an extended period of time
- It is easy to misunderstand the scope and purpose of a particular standard.
- Standard making involves politics as well as science.
- The language of standards can be obscure.

    Nonetheless

- Structure and formality can be a help as well as a hindrance.
- The benefits do not just come from the standards themselves.
- Being international makes it all worthwhile.

Standards Are Developed Over an Extended Period of Time. One of the reasons the standards-making process is slow is that there is an extensive consultation period at each stage of development with time allowed for national member bodies to circulate the documents to mirror committees and then to collate their comments. Another reason is that working group members can spend a great deal of time working on drafts and reaching consensus only to find that the national mirror committees reject their work when it comes to the official vote. It is particularly frustrating for project editors to receive extensive comments (which must be answered) from countries that do not send experts to participate in the work. Of course, the fact that the work is usually voluntary means that it is difficult to get people to agree to work quickly.

There are some benefits that come directly from the slow pace of the process, however. One benefit is that when the technology is moving more quickly than the standards makers can react, it makes it clear that certain types of standards may be premature. For example, ISO 9241-14:1997 Menu Dialogues was originally proposed when character-based, menu-driven systems were a popular style of dialogue design. Its development was delayed considerably for all manner of reasons. But these delays meant that the final standard was relevant to pull-down and pop-up menus that had not even been considered when the standard was first proposed.

Another benefit is that during the development process, those who may be affected have the opportunity to prepare for the standard. Thus, by the time ISO 9241-3:1992 Display requirements was published, many manufacturers were able to claim that they already produced monitors that met the standard. They had not been in that position when the standard was first proposed (although some argued that they would have been improving the design of their displays anyway). Certainly the standards provided a clear target for both demanding consumers and quality manufacturers.

It Is Easy to Misunderstand the Scope and Purpose of a Particular Standard. HCI standards have been criticized for being too generous to manufacturers in some areas and too restrictive in other areas. The "overgenerous" criticism misses the point that most standards are setting minimum requirements, and in ergonomics standards makers must be very cautious about setting such levels. There certainly are areas where being too restrictive is a problem, however. Examples include the following:

- *ISO 9241-3:1992 Ergonomics Requirements for Work With Visual Display Terminals: Display Requirements.* This standard has been successful in setting a minimum standard for display screens which has helped purchasers and manufacturers. However, it is biased toward CRT display technology. An

alternative method of compliance based on a performance test (which is technology independent) has now been published which should help redress the balance.

• *ISO FDIS 9241-9:1999 Ergonomics Requirements for Work With Visual Display Terminals: Nonkeyboard Input Devices.* This standard has suffered because technological developments were faster than either ergonomics research or standards making. Although there has been an urgent need for a standard to help users to be confident in the ergonomic claims made for new designs of mice and other input devices, the lack of reliable data forced the standards makers to slow down or run the risk of prohibiting newer, even better solutions.

Standards Making Involves Politics As Well As Science.

Although ergonomics standards are generally concerned with such mundane topics as keyboard design or menu structures, they nonetheless generate considerable emotion among standards makers. Sometimes this is because the resulting standard could have a major impact on product sales or legal liabilities. Other times the reason for the passion is less clear. Nonetheless, the strong feelings have resulted in painful experiences in the process of standardization. These have included undue influence afforded to major players. Large multinational companies can attempt to exert undue influence by dominating national committees. Although draft standards are usually publicly available from national standards bodies, they are not widely publicized. This means it is relatively easy for large, well-informed companies to provide sufficient experts at the national level to ensure that they can virtually dictate the final vote and comments from a country. Another problem has been the use of "Horse trading" and bargaining to achieve an agreement. End users' requirements can be compromised as part of "horse trading" between conflicting viewpoints. In the interests of reaching agreement, delegates may resort to making political trade-offs largely independent of the technical merits of the issue.

The Language of Standards Can Be Obscure.

In ISO, the formal rules and procedures for operating seem to encourage an elitist atmosphere with standards written for standards enthusiasts. ISO has recognized this and is attempting to make the process more customer focused, but such changes take time. These procedures and rules reinforce elitist tendencies and sometimes resulted in standards that leave much to be desired in terms of brevity, clarity, and usability. There are three contributory factors:

1. *The use of stilted language and boring formats.* The unfriendliness of the language is illustrated by the fact that although the organization is known by the acronym ISO, its full English title is the International Organization for Standardization. The language and style are governed by a set of directives, and these encourage a wordy and impersonal style.

2. *Problems with translation and the use of "Near English."* There are three official languages of the ISO: English, French, and Russian. In practice, much of the work is conducted in English, often by nonnative speakers. The result of this is that the English used in standards is often not quite correct—it is "near English." The words are usually correct, but the combination often makes the exact meaning unclear. These problems are exacerbated when the text is translated.

3. *Confusions between requirements and recommendations.* In ISO standards, there are usually some parts that specify what has to be done to conform to the standard. These are indicated by the use of the word *shall*. In ergonomics standards, however, we often want to make recommendations as well. These are indicated by the use of the word *should*. Such subtleties are often lost on readers of standards, especially those in different countries. For example, in the Nordic countries, they follow recommendations (shoulds) as well as requirements (shalls), so the distinction is diminished. In the United States, they tend to ignore the *shoulds* and only act on the *shalls*.

Structure and Formality Can Be a Help As Well As a Hindrance.

One of the benefits of standards is that they do represent a rather simplified and structured view of the world. There is also a degree (sometimes excessive) of discipline in what a standard can contain and how certain topics can be addressed. Manufacturers (and ergonomists) frequently make wildly different claims about what represents good ergonomics. This is a major weakness for our customers, who may conclude that all claims are equally valid and that there is no sound basis for any of it. Standards force a consensus and therefore have real authority in the minds of our customers. Achieving consensus requires compromises, but then so does life.

The formality of the standards mean that they are suitable for inclusion in formal procurement processes and for demonstrating best practice. In the United Kingdom, parts of ISO 9241 may be used by suppliers to convince their customers that visual display screen equipment and its accessories meet good ergonomic practice. Of course, they can also be "abused" in this way with overeager salespeople misrepresenting the legal status of standards, but this is hardly the fault of the standards makers.

The Benefits Do Not Just Come From the Standards Themselves.

There are several ways in which ergonomics standardization activities can add value to user interface design apart from the standards themselves, which are the end results of the process.

In 1997, the U.S. National Institute of Standards and Technology (NIST) initiated a project (Industry USability Reporting; IUSR) to increase the visibility of software usability. They were helped in this endeavor by prominent suppliers of software and representatives from large consumer organisations. One of the key goals was to develop a common usability reporting format (Common Industry Format; CIF). This is currently being processed as an American National Standards Institute (ANSI) standard through National Committee for Information Technology Standards (NCITS). CIF has been developed to be consistent with ISO 9241 and ISO 13407 and is viewed by the IUSR team as "an implementation of that ISO work." This activity in itself should have a major impact on software usability (for more information, visit http://www.nist.gov/iusr).

In the hardware arena, many people are aware of the TCO 99 sticker that appears on computer monitors and understand that it is an indication of ergonomic and environmental quality. What

TABLE 51.6. Members of ISO/TC159/SC4 Ergonomics of Human–System Interaction

P Members			
Austria	Belgium	Canada	Czech Republic
China	Denmark	Finland	France
Germany	Ireland	Italy	Japan
Korea	Netherlands	Norway	Poland
Slovakia	Spain	Sweden	Thailand
United Kingdom	United States of America		
O Members			
Australia	Hungary	Mexico	Romania
Tanzania			

they may not know is that TCO is the Swedish Confederation of White Collar Trades Unions and that ISO 9241 was used as a major inspiration for its original specification. They publish information in English, and details are available on their Web site (http://www.tco.se/eng/index.htm).

Being International Makes It All Worthwhile. Although there are national and regional differences in populations, the world is becoming a single market with the major suppliers taking a global perspective. Variations in national standards and requirements not only increase costs and complexity, they also tend to compromise individual choice. Making standards international is one way of ensuring that they have impact and can help improve the ergonomics quality of products for everyone. That has to be a worthwhile objective. Table 51.6 shows the member countries of ISO/TC159/SC4.

STYLE GUIDES

Creating a Style Guide

Although few organizations have the history, the infrastructure or the stamina to impose and police rigid user interface standards, style guides can be developed to reduce the unnecessary variation caused by dispersed design teams and extended system development timescales. The main stages in the process involve the following objectives.

Choosing the Right Guidelines. We have already pointed out that there are many good sources of guidelines, including a number of proprietary style guides provided by major vendors. There is no right answer to which guidelines to select, it depends on the specific system under development and the style of interface.

Tailoring the Guidelines Into Specific Design Rules for Your Application. For instance, a guideline stating that displays should be consistently formatted might be translated into design rules that specify where various display features should appear, such as the display title, prompts and other user guidance, error messages, command entries, and so forth. For

maximum effectiveness, guideline tailoring must take place early in the design process before any actual design of user interface software. To tailor guidelines, designers must have a thorough understanding of task requirements and user characteristics. Thus, task analysis is a necessary prerequisite of guidelines tailoring.

The process of developing, reviewing and agreeing style guides can be a positive process for enhancing organizational communication, especially across traditional organizational barriers.

Implementing the Style Guide. Many style guides offer little more than general recommendations on good practice. The problem with these is that they take considerable interpretation and may therefore still result in different parts of a system behaving differently. The mere presence of a style guide does not ensure consistency. Designers have to choose to conform or be disciplined to conform for the benefits to be achieved.

One of the best ways of encouraging them to follow a guide is to support it with a code library and provide lots of good examples for designers to follow. Interactive demonstrations can be particularly valuable.

Policing and Maintaining the Style Guide. In practice, it is better to motivate and encourage designers to follow good examples in style guides than to rely on postdesign monitoring and policing operations. As we pointed out earlier, international standards are slowly being developed that address many aspects of user interface design—both hardware and software. These standards can be used to provide support for in-house measures. In our experience, senior managers are more likely to take style guide and user interface issues seriously if they know that there are public standards that support them.

We have incorporated checklists based on parts of ISO 9241 in some style guides that we have developed. Table 51.7 shows a checklist based on the seven key principles in ISO 9241-10: 1996 Dialogue Principles. Each principle has been rephrased as a question (in bold) with specific questions below. The "correct" answer is usually yes, although there may be occasions where the recommendation is not applicable or not possible; for example, it may not be possible to undo a "commit" action. Table 51.8 shows a similar checklist for user guidance based on ISO 9241-13: 1998 User Guidance. The checklists can be used by designers when reviewing their own work as well as by those involved in signing off on the design.

A Case Study on the Development of a Style Guide

Our client was a European organization with major offices in a number of European cities. Each office was staffed by nationals from throughout Europe, and English was used as the predominant common language, although it was the first language of only a few of the employees. The organization had a unionized workforce, and the work was moving from being paper intensive to an increasing dependence on computer-based systems. These systems were mainly designed by third-party

TABLE 51.7. Task Design Checklist Based on ISO 9241-10:1996
Dialogue Principles

	Yes	No
Is the dialogue suitable for the user's task and skill level?	☐	☐
Does the sequence match the logic of the task?	☐	☐
Are there any unnecessary steps that could be avoided?	☐	☐
Is the terminology familiar to the user?	☐	☐
Does the user have the information they need for the task?	☐	☐
Is extra information available if required (keep dialogue concise)?	☐	☐
Does the dialogue help users perform recurrent tasks?	☐	☐
Does the dialogue make it clear what the user should do next?	☐	☐
Does the dialogue provide feedback for all user actions?	☐	☐
Are users warned about (and asked to confirm) critical actions?	☐	☐
Are all messages constructive and consistent?	☐	☐
Does the dialogue provide feedback on response times?	☐	☐
Can the user control the pace and sequence of the interaction?	☐	☐
Can the user choose how to restart an interrupted dialogue?	☐	☐
Does the dialogue cope with different levels of experience?	☐	☐
Can users control the amount of data displayed at a time?	☐	☐
Is the dialogue consistent?	☐	☐
Are the appearance and behavior of dialogue objects consistent with other parts of the dialogue?	☐	☐
Are similar tasks performed in the same way?	☐	☐
Is the dialogue forgiving?	☐	☐
Does the dialogue provide "undo" (and warn when not available)?	☐	☐
Does the dialogue prevent invalid input?	☐	☐
Are error messages helpful?	☐	☐
Can the dialogue be customized to suit the user?	☐	☐
Does the dialogue offer users different ways of working?	☐	☐
Does the dialogue provide helpful defaults?	☐	☐
Can users choose different levels of explanation?	☐	☐
Can users choose different data representations (e.g., show files as icons or lists)?	☐	☐
Does the dialogue support learning?	☐	☐
Does system feedback help the user learn (e.g., menu items that indicate shortcut key combinations)?	☐	☐
Is context sensitive help provided (where possible)?	☐	☐

suppliers under contract to the different offices, therefore there was a wide variety of different styles and ages of computer systems used by most employees. The increase in screen working was beginning to cause concern for the staff representatives, and the publication of the display screen directive acted as a focus for the organization to start to take ergonomics seriously.

We were invited to support a newly formed ergonomics committee and work with its members to develop an internal infrastructure to promote good ergonomics. Right from the start, we proposed to use international standards to underpin our strategy. The organization was receptive to this approach, and our first task was to brief them on the range of standards that were or soon would be available in this area.

One of the early issues concerned the difference between a user interface *standard* and a user interface *style guide*. Some of the ergonomics committee would have liked to develop user interface standards that would be mandatory across the organization to avoid the kind of inconsistency problem with function keys mentioned earlier. There were others who argued that it was more realistic to aim at a style guide that could

recommend good practice but which could permit deviation if circumstances required it.

In our experience, few organizations have the infrastructure, internal control, or the stamina to enforce strict interface standards. In this organization, it would have been all too easy for individual developers to allow variations to creep into their designs. There was little communication between different development groups, and much of the detailed code writing was carried out offsite by contract programmers. There was also a culture of "not invented here" when anyone tried to introduce new ideas, especially those concerned with the way individuals work. The majority of the workforce were highly qualified individuals with strong views about technical issues. There were already many different systems and subsystems with little obvious similarity or consistency. New systems could easily be developed in isolation with totally contradictory features. Typically, these would only be discovered at the user acceptance testing stage, and by then the time penalties for correcting them would have been difficult to cost justify.

TABLE 51.8. Checklist Based on ISO 9241-13:1998 User Guidance

	Yes	No
General		
Can the user guidance be readily distinguished from other information?	☐	☐
Do system-initiated messages disappear when no longer applicable?	☐	☐
Do user-initiated messages remain until the user dismisses them?	☐	☐
Are messages specific and helpful?	☐	☐
Can the user continue while the guidance is displayed?	☐	☐
Are important messages distinctive?	☐	☐
Can users control the level of guidance they receive?	☐	☐
Wording		
Do messages describe results before actions (e.g., to clear screen, press *del*)?	☐	☐
Are most messages worded positively (i.e., what to do, not what to avoid)?	☐	☐
Are messages worded in a consistent grammatical style?	☐	☐
Is the guidance written in a short, simple sentences?	☐	☐
Are messages written in the active voice?	☐	☐
Is the wording user oriented?	☐	☐
Prompts indicating that the system is waiting for input		
Do prompts indicate the type of input required?	☐	☐
Is online help available to explain prompts (if required)?	☐	☐
Are prompts displayed in consistent positions?	☐	☐
Does the cursor appear automatically at the first prompted field?	☐	☐
Feedback indicating that the system has received input		
Does the system always provide some form of feedback for all user actions?	☐	☐
Is normal feedback unobtrusive and nondistracting?	☐	☐
Are the type and level of feedback suitable for the skills of the users?	☐	☐
Does the system provide clear feedback on system state (e.g., waiting for input)?	☐	☐
Are selected items always highlighted?	☐	☐
If the action requested is not immediate, is there feedback that requests have been accepted (and confirmation when the are complete; e.g., remote printing)?	☐	☐
Does the system show progress indicators when appropriate?	☐	☐
Is system response feedback appropriate (not too fast or too slow)?	☐	☐
Status information indicating what the system is currently doing		
Is appropriate status information available at all times?	☐	☐
Is status information always displayed in a consistent location?	☐	☐
Does the system always indicate when user input is not possible?	☐	☐
Are system modes clearly indicated?	☐	☐
Error prevention and validation		
Are the function keys consistent across the system?	☐	☐
Does the system anticipate problems and warn the user appropriately?	☐	☐
Does the system warn the user when data might be lost by a user action?	☐	☐
Is "undo" provided, where appropriate?	☐	☐
Can users modify or cancel input prior to input?	☐	☐
Can users edit incorrect input (rather than have to reenter the complete field)?	☐	☐
Does the field level validation		
immediately indicate that there is an error?	☐	☐
position the cursor at the beginning of the first incorrect field?	☐	☐
indicate all fields in error (including cross-field errors)?	☐	☐
Error messages		
Can users get more help easily?	☐	☐
Do error messages indicate what is wrong and what should be done?	☐	☐
Can the user tell when an error message has reoccurred?	☐	☐
Do error messages disappear when the error has been corrected?	☐	☐
Can users remove error messages prior to correction if they wish?	☐	☐
Do error messages appear in a consistent location?	☐	☐
Can error messages be moved if they obscure part of the screen?	☐	☐
Do error messages appear as soon as the wrong input has been entered?	☐	☐

(Continued)

TABLE 51.8. (Continued)

	Yes	No
Can users turn off confirmation screens?	☐	☐
Can users adjust the volume of warning tones or messages?	☐	☐
Online help		
Is the online help context sensitive?	☐	☐
Is the system-initiated online help unobtrusive?	☐	☐
Can users turn off system-initiated online help?	☐	☐
Can the users initiate online help by a simple consistent action?	☐	☐
Does the system accept synonyms and close spelling matches when the user searches for the appropriate area of online help?	☐	☐
Is online help clear, understandable, and specific to the users' tasks?	☐	☐
Does online help provide both descriptive and procedural help?	☐	☐
Can the user easily go between the task screens and online help?	☐	☐
Can the users configure online help to suit their preferences?	☐	☐
Can the user easily return to the task?	☐	☐
Are there suitable features to help users find appropriate help?	☐	☐
Are there quick methods for		
going directly to another screen?	☐	☐
browsing?	☐	☐
exploring linkages between topics?	☐	☐
returning to the previous help page?	☐	☐
returning to a home location?	☐	☐
accessing a history of previously consulted topics?	☐	☐
If the online help system is large, are any of the following provided to aid search:		
String search of a list of topics?	☐	☐
Keyword search of online help text?	☐	☐
Hierarchical structure of online help text?	☐	☐
Map of online help topics?	☐	☐
If the online help system has a hierarchical structure, is there		
an overall indication of the structure?	☐	☐
easy access to any level in the hierarchy?	☐	☐
an obvious and consistent method of accessing more detail?	☐	☐
a quick means of accessing the main (parent) topic?	☐	☐
are topics self-contained (i.e., not dependent on reading previous sections)?	☐	☐
If the information is scrollable, does the topic remain clear?	☐	☐
Does the context sensitive help provide information on		
the current dialogue step?	☐	☐
the current task?	☐	☐
the current applications?	☐	☐
the task information presented on the screen?	☐	☐
Does online help explain objects, what they do, and how to use them?	☐	☐
Does online help indicate when it is not available on an object?	☐	☐

It was felt that there would be significant benefits in developing an enterprisewide, high-level user interface style guide for new systems. The key benefits were seen in terms of maintaining best practices, interworking, staff mobility, and efficiency of procurement, design, and maintenance.

The senior management committee responsible for information technology raised the question of the additional costs of such a style guide. We pointed out that the costs associated with the style guide need not be high. All system developments require design decisions. Therefore, there was no reason why good design decisions made using a style guide should be any more expensive than the normal ad hoc design decisions for user system interfaces. Indeed in most cases, there would be savings in time and effort required because some design decisions had already been made.

The guide took the form of a high level statement of principles, a list of approved proprietary platform standards; an agreed, organizationwide subset of these standards; an icon and function library; a glossary of organizationwide terms; and an interactive demonstration to illustrate the various preferred interface styles.

The style guide improved communication in the organization in two ways. First, it was clearly an organizationwide initiative and not the exclusive property of any one site. Everyone therefore felt entitled to contribute to its development. Indeed, it was quickly recognized that not taking part could lead to specific interests being ignored. The process of developing, reviewing, and agreeing the style guide was therefore in itself a positive process in improving communication between sites with different national and cultural viewpoints.

Commenting on the style guide became a relatively safe way of making general comments about the development process. In some cases, small cross-site subgroups would be set up to deal with specific queries, improving working relationships significantly.

The interactive version of the style guide was an extremely useful vehicle for encouraging designers and users to simulate their systems long before hard design decisions were made. Thus, it was possible to bring together mixed groups to review ideas and even to test these ideas with potential users. Whereas previous discussions between these groups had tended to take the form of each side expressing their views and then agreeing to differ, the sessions that involved the demonstration systems and the practical feedback from users focused attention on real differences. These real differences could then be discussed and resolved in a reasonably calm and positive manner.

CONCLUSION

One of the recurring themes in this chapter is what might be called "Stewart's Law of Usability Standards": the easier it is to formulate the usability standard or guideline, the more difficult it is to apply in practice. By this we mean that a simple guideline such as "allow the user to control the pace and sequence of the interaction" has a great deal of backing as a general guideline. In practice, of course, there are many situations in which the rule fails, where the answer is "it depends," and where the context makes it more appropriate for the system to control some part of the interaction. The designer wishing to follow such a simple guideline needs to interpret the guideline in the context of the system, and this requires insight and thought. The alternative approach, where the guideline is preceded by statements defining the context—for example, "if the user is x, the task is y, and the environment is z, then do abc" become so tedious and confusing that they are quickly ignored. In the ISO 9241 series that we have described, there has been an attempt to define a middle course that involves giving specific practical examples as an aid to design. In ISO 13407 Human-Centered Design Processes for Interactive Systems, the standard is concerned with the process itself.

ISO 13407-1999: Human-Centered Design Processes for Interactive Systems provides guidance for project managers to help them follow a human-centered design process. By undertaking the activities and following the principles described in the standard, managers can be confident that the resulting systems will be usable and work well for their users.

The standard describes four principles of human-centered design:

1. Active involvement of users (or those who speak for them)
2. Appropriate allocation of function (making sure human skill is used properly)
3. Iteration of design solutions (allowing time for iteration in project planning)
4. Multidisciplinary design (but beware overly large design teams)

It also involves four key human-centered design activities:

1. Understand and specify the context of use (make it explicit—avoid assuming it is obvious)
2. Specify user and organizational requirements (note there will be a variety of different viewpoints and individual perspectives)
3. Produce design solutions (note plural, multiple designs encourage creativity)
4. Evaluate designs against requirements (involves real user testing, not just convincing demonstrations)

To claim conformance, the standard requires that the procedures used, the information collected, and the use made of results be specified (a checklist is provided as an annex to help). This approach to conformance has been used in a number of parts of ISO 9241 because so many ergonomics recommendations are context specific. Thus, there is often only one *shall* in these standards, which generally prescribes what kind of evidence is required to convince another party that the relevant recommendations in the standard have been identified and followed.

We believe this is one way of ensuring that usability standards remain relevant when technology changes and also offer practical help to designers and developers.

References

Brown, C. M., Brown, D. B., Burkleo, H. V., Mangelsdorf, J. E., Olsen, R. A., & Perkins, R. D. (1983, June 15). *Human Factors Engineering Standards for information processing systems (LMSC-D877141)*. Sunnyvale, CA: Lockheed Missiles and Space Company.

Engel, S. E., & Granda, R. E. (1975, December). *Guidelines for man/display interfaces* (Technical Report TR 00.2720). Poughkeepsie, NY: IBM.

MIL-STD-1472C, Revised. (1983, 1 September). Military standard: Human engineering design criteria for military systems, equipment and facilities. Washington, DC: Department of Defense.

Molich, R., & Nielsen, J. (1990). Improving a human-computer dialogue. *Communications of the ACM, 33*, 338–348.

Nielsen, J. (1994a). Enhancing the explanatory power of usability heuristics. Proceedings of the Association of Computerized Machinery. CHI'94 Conference (pp. 152–158). New York: ACM Press.

Nielsen, J. (1994b). Heuristic evaluation. In J. Nielsen, & R. L., Mack (Eds.), *Usability inspection methods*. New York: John Wiley.

Pew, R. W., & Rollins, A. M. (1975). *Dialog specification procedures* (Report 3129, revised). Cambridge, MA: Bolt Beranek and Newman.

Shneiderman, B. (1987). *Designing the user interface—Strategies for effective human–computer interaction*. Reading, MA: Addison-Wesley

Smith, S. L., & Mosier, J. N. (1984, September). Design guidlines for user-system interface software (Report ESD-TR-84-190). Bedford, MA: The Mitre Corp.

·52·

PROTOTYPING TOOLS AND TECHNIQUES

Michel Beaudouin-Lafon
Université Paris—Sud

Wendy Mackay
Institut National de Recherche en
Informatique et en Automatique (INRIA)

INTRODUCTION

"A good design is better than you think."
—Rex Heftman, cited by Raskin, 2000, p. 143.

Design is about making choices. In many fields that require creativity and engineering skill, such as architecture or automobile design, prototypes both inform the design process and help designers select the best solution. This chapter describes tools and techniques for using prototypes to design interactive systems. The goal is to illustrate how they can help designers generate and share new ideas, get feedback from users or customers, choose among design alternatives, and articulate reasons for their final choices.

We begin with our definition of a prototype and then discuss prototypes as design artifacts, introducing four dimensions for analyzing them. We then discuss the role of prototyping within the design process, in particular, the concept of a design space and how it is expanded and contracted by generating and selecting design ideas. The next three sections describe specific prototyping approaches: rapid prototyping, both offline and online, for early stages of design; iterative prototyping, which uses online development tools; and evolutionary prototyping, which must be based on a sound software architecture.

What Is a Prototype?

We define a prototype as a *concrete representation* of part or all of an interactive system. A prototype is a tangible artifact, not an abstract description that requires interpretation. Designers, as well as managers, developers, customers, and end users, can use these artifacts to envision and reflect on the final system.

Prototypes may be defined differently in other fields. For example, an architectural prototype is a scaled-down model of the final building. This is not possible for interactive system prototypes: The designer may limit the amount of information the prototype can handle, but the actual interface must be presented at full scale. Thus, a prototype interface to a database may handle only a small pseudo-database but must still present a full-size display and interaction techniques. Full-scale, one-of-a-kind models, such as a handmade dress sample, are another type of prototype. These usually require an additional design phase to mass produce the final design. Some interactive system prototypes begin as one-of-a-kind models that are then distributed widely (because the cost of duplicating software is so low); however, most successful software prototypes evolve into the final product and then continue to evolve as new versions of the software are released.

Hardware and software engineers often create prototypes to study the feasibility of a technical process. They conduct systematic, scientific evaluations with respect to predefined benchmarks and, by systematically varying parameters, fine-tune the system. Designers in creative fields, such as typography or graphic design, create prototypes to express ideas and reflect on them. This approach is intuitive, oriented more to discovery and generation of new ideas than to evaluation of existing ideas.

Human–computer interaction (HCI) is a multidisciplinary field that combines elements of science, engineering, and design (Dykstra-Erickson, Mackay, & Arnowitz, 2001; Mackay & Fayard, 1997). Prototyping is primarily a design activity, although we use software engineering to ensure that software prototypes evolve into technically sound working systems and we use scientific methods to study the effectiveness of particular designs.

PROTOTYPES AS DESIGN ARTIFACTS

We can look at prototypes as both concrete artifacts in their own right or as important components of the design process. When viewed as artifacts, successful prototypes have several characteristics: They support *creativity*, helping the developer to capture and generate ideas, facilitate the exploration of a design space, and uncover relevant information about users and their work practices. They encourage *communication*, helping designers, engineers, managers, software developers, customers, and users to discuss options and interact with each other. They also permit *early evaluation* because they can be tested in various ways, including traditional usability studies and informal user feedback, throughout the design process.

We can analyze prototypes and prototyping techniques along four dimensions:

- *Representation* describes the form of the prototype (e.g., sets of paper sketches or computer simulations).
- *Precision* describes the level of detail at which the prototype is to be evaluated (e.g., informal and rough or highly polished).
- *Interactivity* describes the extent to which the user can actually interact with the prototype (e.g., watch only or fully interactive).
- *Evolution* describes the expected life cycle of the prototype (e.g., throw away or iterative).

Representation

Prototypes serve different purposes and thus take different forms. A series of quick sketches on paper can be considered a prototype; so can a detailed computer simulation. Both are useful; both help the designer in different ways. We distinguish between two basic forms of representation: offline and online.

Offline prototypes (also called *paper prototypes*) do not require a computer. They include paper sketches, illustrated storyboards, cardboard mock-ups, and videos. The most salient characteristics of offline prototypes (of interactive systems) is that they are created quickly, usually in the early stages of design, and they are usually thrown away when they have served their purpose.

Online prototypes (also called *software prototypes*) run on a computer. They include computer animations, interactive video presentations, programs written with scripting languages, and applications developed with interface builders. The cost of producing online prototypes is usually higher and may require skilled programmers to implement advanced interaction and/or visualization techniques or to meet tight performance

constraints. Software prototypes are usually more effective in the later stages of design, when the basic design strategy has been decided.

In our experience, programmers often argue in favor of software prototypes even at the earliest stages of design. Because they already are familiar with a programming language, these programmers believe it will be faster and more useful to write code than to "waste time" creating paper prototypes. In 20 years of prototyping, in both research and industrial settings, we have yet to find a situation in which this is true.

First, offline prototypes are inexpensive and quick. This permits a rapid iteration cycle and helps prevent the designer from becoming overly attached to the first possible solution. Offline prototypes make it easier to *explore the design space* (discussed in detail later), examining a variety of design alternatives and choosing the most effective solution. Online prototypes introduce an intermediary between the idea and the implementation, slowing down the design cycle.

Second, offline prototypes are less likely to constrain the designer's thinking. Every programming language or development environment imposes constraints on the interface, limiting creativity and restricting the number of ideas considered. If a particular tool makes it easy to create scroll bars and pull-down menus and difficult to create a zoomable interface, the designer is likely to limit the interface accordingly. Considering a wider range of alternatives, even if the developer ends up using a standard set of interface widgets, usually results in a more creative design.

Finally, and perhaps most important, offline prototypes can be created by a wide range of people, not just programmers. Thus, all types of designers, technical or otherwise, as well as users, managers, and other interested parties, can all contribute on an equal basis. Unlike programming software, modifying a storyboard or cardboard mock-up requires no particular skill. Collaborating on paper prototypes not only increases participation in the design process, but also improves communication among team members and increases the likelihood that the final design solution will be well accepted.

Although we believe strongly in offline prototypes, they are not a panacea. In some situations, they are insufficient to fully evaluate a particular design idea. For example, interfaces requiring rapid feedback to users or complex, dynamic visualizations usually require software prototypes. However, particularly when using video and "Wizard-of-Oz" techniques, which we describe later, offline prototypes can be used to create sophisticated representations of the system.

Prototyping is an iterative process, and all prototypes provide information about some aspects while ignoring others. The designer must consider the purpose of the prototype (Houde & Hill, 1997) at each stage of the design process and choose the representation that is best suited to the current design question.

Precision

Prototypes are explicit representations that help designers, engineers, and users reason about the system being built. By their nature, prototypes require details. A verbal description such as "the user opens the file" or "the system displays the results" provides no information about what the user actually does. Prototypes force designers to *show* the interaction: Just how does the user open the file and what are the specific results that appear on the screen?

Precision refers to the relevance of details with respect to the purpose of the prototype.[1] For example, when sketching a dialogue box, the designer specifies its size, the positions of each field, and the titles of each label. Not all these details are relevant to the goal of the prototype, however. It may be necessary to show where the labels are, but too early to choose the text. The designer can convey this by writing nonsense words or drawing squiggles, which shows the need for labels without specifying their actual content.

Although it may seem contradictory, a detailed representation need not be precise. This is an important characteristic of prototypes: Those parts of the prototype that are not precise are those open for future discussion or for exploration of the design space, yet they need to be incarnated in some form so the prototype can be evaluated and iterated.

The level of precision usually increases as successive prototypes are developed and more and more details are set. The forms of the prototypes reflect their level of precision; sketches tend not to be precise, whereas computer simulations are usually very precise. Graphic designers often prefer using hand sketches for early prototypes because the drawing style can directly reflect what is precise and what is not—the wigglely shape of an object or a squiggle that represents a label are directly perceived as imprecise. This is more difficult to achieve with an online drawing tool or a user interface builder.

The form of the prototype must be adapted to the desired level of precision. Precision defines the tension between what the prototype states (relevant details) and what the prototype leaves open (irrelevant details). What the prototype states is subject to evaluation; what the prototype leaves open is subject to more discussion and design space exploration.

Interactivity

An important characteristic of HCI systems is that they are *interactive*: users both respond to them and act on them. Unfortunately, designing effective interaction is difficult: Many interactive systems (including many Web sites) have a good "look" but a poor "feel." HCI designers can draw from a long tradition in visual design for the former but have relatively little experience with how interactive software systems should be used—personal computers have only been commonplace for about a decade. Another problem is that the quality of interaction is tightly linked to the end users and a deep understanding of their work practices. A word processor designed for professional typographers requires a different interaction design than one designed for secretaries, even though ostensibly they serve similar purposes. Designers must take the context of use into account when designing the details of the interaction.

---

[1] The terms *low-fidelity* and *high-fidelity* prototypes are often used in the literature. We prefer the term *precision* because it refers to the content of the prototype itself, not its relationship to the final, as-yet-undefined system.

A critical role for an interactive system prototype is to illustrate how the user will interact with the system. Although this may seem more natural with online prototypes, in fact it is often easier to explore different interaction strategies with offline prototypes. Note that interactivity and precision are orthogonal dimensions. One can create an imprecise prototype that is highly interactive, such as a series of paper screen images in which one person acts as the user and the other plays the system. Or one may create a precise but noninteractive prototype, such as a detailed animation that shows feedback from a specific action by a user.

Prototypes can support interaction in various ways. For offline prototypes, one person (often with help from others) plays the role of the interactive system, presenting information and responding to the actions of another person playing the role of the user. For online prototypes, parts of the software are implemented, whereas others are "played" by a person (an approach called the "Wizard of Oz" after the character in the 1939 movie of the same name). The key is that the prototype *feels* interactive to the user.

Prototypes can support different levels of interaction. *Fixed prototypes*, such as video clips or precomputed animations, are noninteractive. The user cannot interact, or pretend to interact, with it. Fixed prototypes are often used to illustrate or test scenarios (see chapter 53 by Rosson and Carroll). *Fixed-path prototypes* support limited interaction. The extreme case is a fixed prototype in which each step is triggered by a prespecified user action. For example, the person controlling the prototype might present the user with a screen containing a menu. When the user points to the desired item, she presents the corresponding screen showing a dialogue box. When the user points to the word *OK*, she presents the screen that shows the effect of the command. Even though the position of the click is irrelevant (it is used as a trigger), the person in the role of the user can get a feel for the interaction. Of course, this type of prototype can be much more sophisticated, with multiple options at each step. Fixed-path prototypes are effective with scenarios and can also be used for horizontal and task-based prototypes (discussed in detail in the next section).

Open prototypes support large sets of interactions. Such prototypes work like the real system, with some limitations. They usually only cover part of the system (discussed in the next section) and often have limited error-handling or reduced performance relative to that of the final system.

Prototypes may thus illustrate or test different levels of interactivity. Fixed prototypes simply illustrate what the interaction might look like. Fixed-path prototypes provide designers and users with the experience of what the interaction might be like, but only in prespecified situations. Open prototypes allow designers to test a wide range of examples of how users will interact with the system.

Evolution

Prototypes have different life spans. *Rapid* prototypes are created for a specific purpose and then thrown away. *Iterative* prototypes evolve, either to work out some details (increasing their precision) or to explore various alternatives. *Evolutionary* prototypes are designed to become part of the final system.

Rapid prototypes are especially important in the early stages of design. They must be inexpensive and easy to produce because the goal is to quickly explore a wide variety of possible types of interaction and then throw them away. Note that rapid prototypes may be offline or online. Creating precise software prototypes, even if they must be reimplemented in the final version of the system, is important for detecting and fixing interaction problems. We present specific prototyping techniques, both offline and online, later in the chapter.

Iterative prototypes are developed as a reflection of a design in progress, with the explicit goal of evolving through several design iterations. Designing prototypes that support evolution is sometimes difficult. There is a tension between evolving toward the final solution and exploring an unexpected design direction, which may be adopted or thrown away completely. Each iteration should inform some aspect of the design. Some iterations explore different variations of the same theme. Others may systematically increase precision, working out the finer details of the interaction. We describe tools and techniques for creating iterative prototypes later in the chapter.

Evolutionary prototypes are a special case of iterative prototypes in which the prototype evolves into part or all of the final system (Fig. 52.1). Obviously this only applies to software prototypes. Extreme Programming (Beck, 2000), advocates this approach, tightly coupling design and implementation and building the system through constant evolution of its components. Evolutionary prototypes require more planning and practice than the approaches above because the prototypes are both representations of the final system and the final system itself, making it more difficult to explore alternative designs. We advocate a combined approach, beginning with rapid prototypes and then using iterative or evolutionary prototypes according to the needs of the project. Later in the chapter, we describe how to create evolutionary prototypes by building on software architectures specifically designed to support interactive systems.

PROTOTYPES AND THE DESIGN PROCESS

In the previous section, we looked at prototypes as artifacts (i.e., the results of a design process). Prototypes can also be seen as artifacts *for* design (i.e., as an integral part of the design process). Prototyping helps designers think: Prototypes are the tools they use to solve design problems. In this section, we focus on prototyping as a process and its relationship to the overall design process.

User-Centered Design

The HCI field is both user-centered (Norman & Draper, 1986) and iterative. User-centered design places the user at the center of the design process, from the initial analysis of user requirements (see chapters 48–50 in this volume) to testing and evaluation (see chapters 56–59 in this volume). Prototypes support this goal by allowing users to see and experience the final

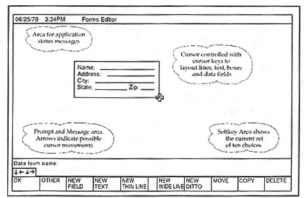
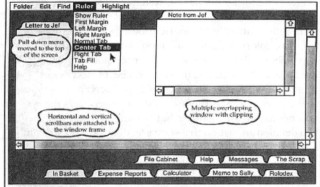

FIGURE 52.1. Evolutionary prototypes of the Apple Lisa: July 1979 (left), October 1980 (right). *Note.* From "Inventing the Lisa User Interface" by R. Perkins, D. S. Keller, and F. Ludolph, 1997, ACM *Interactions*, 4, pp. 43, 47. Copyright 1997 by the Association for Computing Machinery. Reprinted with permission.

system long before it is built. Designers can identify functional requirements, usability problems, and performance issues early and improve the design accordingly.

Iterative design involves multiple design–implement–test loops,[2] enabling the designer to generate different ideas and successively improve on them. Prototypes support this goal by allowing designers to evaluate concrete representations of design ideas and select the best.

Prototypes reveal the strengths as well as the weaknesses of a design. Unlike pure ideas, abstract models, or other representations, they can be *contextualized* to help understand how the real system would be used in a real setting. Because prototypes are concrete and detailed, designers can explore different real-world scenarios, and users can evaluate them with respect to their current needs. Prototypes can be compared directly with existing systems, and designers can learn about the context of use and the work practices of the end users. Prototypes can help designers (re)analyze users' needs during the design process, not abstractly as with traditional requirements analysis, but in the context of the system being built.

Participatory Design

Participatory (also called cooperative) design is a form of user-centered design that actively involves the user in all phases the design process (see Greenbaum & Kyng, 1991, and chapter 54 by Muller in this volume). Users are not simply consulted at the beginning and called in to evaluate the system at the end; they are treated as partners throughout. This early and active involvement of users helps designers avoid unpromising design paths and develop a deeper understanding of the actual design problem. Obtaining user feedback at each phase of the process also changes the nature of the final evaluation, which is used to fine-tune the interface rather than discover major usability problems.

A common misconception about participatory design is that designers are expected to abdicate their responsibilities as designers, leaving the design to the end user. In fact, the goal is for designers and users to work together, each contributing their strengths to clarify the design problem as well as explore design solutions. Designers must understand what users can and cannot contribute. Usually, users are best at understanding the context in which the system will be used and subtle aspects of the problems that must be solved. Innovative ideas can come from both users and designers, but the designer is responsible for considering a wide range of options that might not be known to the user and balancing the trade-offs among them.

Because prototypes are shared, concrete artifacts, they serve as an effective medium for communication within the design team. We have found that collaborating on prototype design is an effective way to involve users in participatory design. Prototypes help users articulate their needs and reflect on the efficacy of design solutions proposed by designers.

Exploring the Design Space

Design is not a natural science. The goal is not to describe and understand existing phenomena but to create something new. Designers do, of course, benefit from scientific research findings, and they may use scientific methods to evaluate interactive systems. But designers also require specific techniques for generating new ideas and balancing complex sets of trade-offs to help them develop and refine design ideas.

Designers from fields such as architecture and graphic design have developed the concept of a *design space*, which constrains design possibilities along some dimensions, while leaving others open for creative exploration. Ideas for the design space come from many sources: existing systems, other designs, other designers, external inspiration, and accidents that

---

[2]Software engineers refer to this as the Spiral model (Boehm, 1988).

prompt new ideas. Designers are responsible for creating a design space specific to a particular design problem. They explore this design space, expanding and contracting it as they add and eliminate ideas. The process is iterative, more cyclic, than reductionist. That is, the designer does not begin with a rough idea and successively add more precise details until the final solution is reached. Instead, she begins with a design problem, which imposes set of constraints, and generates a set of ideas to form the initial design space. She then explores this design space, preferably with the user, and selects a particular design direction to pursue. This closes off part of the design space but opens up new dimensions that can be explored. The designer generates additional ideas along these dimensions, explores the expanded design space, and then makes new design choices. Design principles (e.g., Beaudouin-Lafon & Mackay, 2000) help this process by guiding it both in the exploration and choice phases. The process continues, in a cyclic expansion and contraction of the design space, until a satisfying solution is reached.

All designers work with constraints—not just limited budgets and programming resources, but also design constraints. These are not necessarily bad; one cannot be creative along all dimensions at once. Some constraints are unnecessary, however, derived from poor framing of the original design problem. If we consider a design space as a set of ideas and a set of constraints, the designer has two options. She can modify ideas within the specified constraints or modify the constraints to enable new sets of ideas. Unlike traditional engineering, which treats the design problem as a given, designers are encouraged to challenge, and if necessary, change the initial design problem. If she reaches an impasse, the designer can either generate new ideas or redefine the problem (and thus change the constraints). Some of the most effective design solutions derive from a more careful understanding and reframing of the design brief.

Note that all members of the design team, including users, may contribute ideas to the design space and help select design directions from within it. However, it is essential that these two activities are kept separate. Expanding the design space requires creativity and openness to new ideas. During this phase, everyone should avoid criticizing ideas and concentrate on generating as many as possible. Clever ideas, half-finished ideas, silly ideas, impractical ideas all contribute to the richness of the design space and improve the quality of the final solution. In contrast, contracting the design space requires critical evaluation of ideas. During this phase, everyone should consider the constraints and weigh the trade-offs. Each major design decision must eliminate part of the design space: rejecting ideas is necessary to experiment and refine others and make progress in the design process. Choosing a particular design direction should spark new sets of ideas, and those new ideas are likely to pose new design problems. In summary, exploring a design space is the process of moving back and forth between creativity and choice.

Prototypes aid designers in both aspects of working with a design space: generating concrete representations of new ideas and clarifying specific design directions. The next two subsections describe techniques that have proven most useful in our own prototyping work, both for research and product development.

Expanding the Design Space: Generating Ideas

The most well-known idea generation technique is *brainstorming*, introduced by Osborn (1957). His goal was to create synergy within the members of a group: Ideas suggested by one participant would spark ideas in other participants. Subsequent studies (Collaros & Anderson, 1969; Diehl & Stroebe, 1987) challenged the effectiveness of group brainstorming, finding that aggregates of individuals could produce the same number of ideas as groups. They found certain effects, such as production blocking, free-riding, and evaluation apprehension, were sufficient to outweigh the benefits of synergy in brainstorming groups. Since then, many researchers have explored different strategies for addressing these limitations. For our purposes, the quantity of ideas is not the only important measure: The relationships among members of the group are also important. As de Vreede, Briggs, van Duin, and Enserink (2000) pointed out, one should also consider elaboration of ideas as group members react to each other's ideas.

We have found that brainstorming, including a variety of variants, is an important group-building exercise in participatory design. Designers may, of course, brainstorm ideas by themselves. But brainstorming in a group is more enjoyable and, if it is a recurring part of the design process, plays an important role in helping group members share and develop ideas together.

The simplest form of brainstorming involves a small group of people. The goal is to generate as many ideas as possible on a pre-specified topic; quantity not quality, is important. Brainstorming sessions have two phases: The first generates ideas and the second reflects on those ideas. The initial phase should last no more than an hour. One person should moderate the session, keeping time and ensuring that everyone participates and preventing people from critiquing each other's ideas. Discussion should be limited to clarifying the meaning of a particular idea. A second person records every idea, usually on a flipchart or transparency on an overhead projector. After a short break, participants are asked to reread all the ideas, and each person marks their three favorite ideas.

One variation is designed to ensure that everyone contributes, not just those who are verbally dominant. Participants write their ideas on individual cards notes for a prespecified period of time. The moderator then reads each idea aloud. Authors are encouraged to elaborate (but not justify) their ideas, which are then posted on a whiteboard or flipchart. Group members may continue to generate new ideas, inspired by the others they hear.

We use a variant of brainstorming that involves prototypes called *video brainstorming* (Mackay, 2000): Participants not only write or draw their ideas, they act them out in front of a video camera (Fig. 52.2). The goal is the same as other brainstorming exercises: to create as many new ideas as possible, without critiquing them. The use of video, combined with paper or cardboard mock-ups, encourages participants to actively experience the details of the interaction and to understand each idea from the perspective of the user.

Each video brainstorming idea takes 2 to 5 minutes to generate and capture, allowing participants to simulate a wide variety of ideas quickly. The resulting video clips provide illustrations

FIGURE 52.2. Video brainstorming: One person moves the transparency, projected onto the wall, in response to the actions of the user, who explores how he might interact with an online animated character. Each interaction idea is recorded and videotaped.

of each idea that are easier to understand (and remember) than hand-written notes. (We find that raw notes from brainstorming sessions are not very useful after a few weeks because the participants no longer remember the context in which the ideas were created.)

Video brainstorming requires thinking more deeply about each idea. It is easier to stay abstract when describing an interaction in words or even with a sketch, but acting out the interaction in front of the camera forces the author of the idea (and the other participants) to consider seriously how a user would interact with the idea. It also encourages designers and users to think about new ideas in the context in which they will be used. Video clips from a video brainstorming session, even though rough, are much easier for the design team, including developers, to interpret than ideas from a standard brainstorming session.

We generally run a standard brainstorming session, either oral or with cards, before a video brainstorming session to maximize the number of ideas to be explored. Participants then take their favorite ideas from the previous session and develop them further as video brainstorms. Each person is asked to "direct" at least two ideas, incorporating the hands or voices of other members of the group. We find that, unlike standard brainstorming, video brainstorming encourages even the quietest team members to participate.

Contracting the Design Space: Selecting Alternatives

After expanding the design space by creating new ideas, designers must stop and reflect on the choices available to them. After exploring the design space, designers must evaluate their options and make concrete design decisions—choosing some ideas, specifically rejecting others, and leaving other aspects of the design open to further idea generation activities. Rejecting

good, potentially effective ideas is difficult, but necessary to make progress.

Prototypes often make it easier to evaluate design ideas from the user's perspective. They provide concrete representations that can be compared. Many of the evaluation techniques described elsewhere in this handbook can be applied to prototypes to help focus the design space. The simplest situation is when the designer must choose among several discrete, independent options. Running a simple experiment, using techniques borrowed from psychology (see chapter 56 by Dumas) allows the designer to compare how users respond to each of the alternatives. The designer builds a prototype, with either fully implemented or simulated versions of each option. The next step is to construct tasks or activities that are typical of how the system would be used, and ask people from the user population to try each of the options under controlled conditions. It is important to keep everything the same, except for the options being tested.

Designers should base their evaluations on both quantitative measures, such as speed or error rate, and qualitative measures, such as the user's subjective impressions of each option. Ideally, of course, one design alternative will be clearly faster, prone to fewer errors, and preferred by the majority of users. More often, the results are ambiguous, and the designer must take other factors into account when making the design choice. (Interestingly, running small experiments often highlights other design problems and may help the designer reformulate the design problem or change the design space.)

The more difficult (and common) situation is when the designer faces a complex, interacting set of design alternatives in which each design decision affects a number of others. Designers can use heuristic evaluation techniques, which rely on our understanding of human cognition, memory, and sensory-perception (see chapters 1–6). They can also evaluate their designs with respect to ergonomic criteria (see chapter 51 by Stewart and Trans) or design principles (Beaudouin-Lafon & Mackay, 2000). (See chapters 56–60 for a more thorough discussion of testing and evaluation methods.)

Another strategy is to create one or more scenarios (see chapter 53 by Rosson and Carroll) that illustrate how the combined set of features will be used in a realistic setting. The scenario must identify who is involved, where the activities take place, and what the user does over a specified period of time. Good scenarios involve more than a string of independent tasks; they should incorporate real-world activities, including common or repeated tasks, successful activities, and breakdowns and errors, with both typical and unusual events. The designer then creates a prototype that simulates or implements the aspects of the system necessary to illustrate each set of design alternatives. Such prototypes can be tested by asking users to "walk through" the same scenario several times, once for each design alternative. As with experiments and usability studies, designers can record both quantitative and qualitative data, depending on the level of the prototypes being tested.

The previous section described an idea-generation technique called video brainstorming, which allows designers to generate a variety of ideas about how to interact with the future system. We call the corresponding technique for focusing in on a design

video prototyping. Video prototyping can incorporate any of the rapid-prototyping techniques (offline or online) described later. They are quick to build, force designers to consider the details of how users will react to the design in the context in which it will be used, and provide an inexpensive method of comparing complex sets of design decisions.

To an outsider, video brainstorming and video prototyping techniques look very similar. Both involve small design groups working together, creating rapid prototypes and interacting with them in front of a video camera. Both result in video illustrations that make abstract ideas concrete and help team members communicate with each other. The critical difference is that video brainstorming expands the design space by creating a number of unconnected collections of individual ideas, whereas video prototyping contracts the design space by showing how a specific collection of design choices work together.

Prototyping Strategies

Designers must decide what role prototypes should play with respect to the final system and in which order to create different aspects of the prototype. The next subsections presents four strategies: *horizontal, vertical, task-oriented,* and *scenario-based,* which focus on different design concerns. These strategies can use any of the prototyping techniques covered in the sections that follow.

Horizontal Prototypes. The purpose of a horizontal prototype is to develop an entire layer of the design at the same time. This type of prototyping is most common with large software development teams, where designers with different skill sets address different layers of the software architecture. Horizontal prototypes of the user interface are useful to get an overall picture of the system from the user's perspective and address issues such as consistency (similar functions are accessible through similar user commands), coverage (all required functions are supported), and redundancy (the same function is/is not accessible through different user commands).

User interface horizontal prototypes can begin with rapid prototypes and progress through to working code. Software prototypes can be built with an interface builder (discussed later in the chapter), without creating any of the underlying functionality, making it possible to test how the user will interact with the user interface without worrying about how the rest of the architecture works. Some level of scaffolding or simulation of the rest of the application is often necessary, however, otherwise the prototype cannot be evaluated properly. As a consequence, software horizontal prototypes tend to be evolutionary (i.e., they are progressively transformed into the final system).

Vertical Prototypes. The purpose of a vertical prototype is to ensure that the designer can implement the full, working system from the user interface layer down to the underlying system layer. Vertical prototypes are often built to assess the feasibility of a feature described in a horizontal, task-oriented, or scenario-based prototype. For example, when we developed

the notion of magnetic guidelines in the CPN2000 system to facilitate the alignment of graphical objects (Beaudouin-Lafon & Mackay, 2000), we implemented a vertical prototype to test not only the interaction technique but also the layout algorithm and the performance. We knew that we could only include the particular interaction technique if the we could implement a sufficiently fast response.

Vertical prototypes are generally high precision, software prototypes because their goal is to validate an idea at the system level. They are often thrown away because they are generally created early in the project, before the overall architecture has been decided, and they focus on only one design question. For example, a vertical prototype of a spelling checker for a text editor does not require text editing functions to be implemented and tested. The final version will need to be integrated into the rest of the system, however, which may involve considerable architectural or interface changes.

Task-Oriented Prototypes. Many user interface designers begin with a task analysis (see chapter 48 by Redish and Wixon) to identify the individual tasks that the user must accomplish with the system. Each task requires a corresponding set of functionality from the system. Task-based prototypes are organized as a series of tasks, which allows both designers and users to test each task independently, systematically working through the entire system.

Task-oriented prototypes include only the functions necessary to implement the specified set of tasks. They combine the breadth of horizontal prototypes, to cover the functions required by those tasks, with the depth of vertical prototypes, enabling detailed analysis of how the tasks can be supported. Depending on the goal of the prototype, both offline and online representations can be used for task-oriented prototypes.

Scenario-Based Prototypes. Scenario-based prototypes are similar to task-oriented ones, except that they do not stress individual, independent tasks but rather follow a more realistic scenario of how the system would be used in a real-world setting. Scenarios are stories that describe a sequence of events and how the user reacts (see chapter 53 by Rosson and Carroll). A good scenario includes both common and unusual situations and should explore patterns of activity over time. Bødker, Christiansen, and Thüring (1995) developed checklist, to ensure that no important issues have been left out.

We find it useful to begin with *use scenarios* based on observations of or interviews with real users. Ideally, some of those users should participate in the creation of the specific scenarios, and other users should critique them based on how realistic they are. Use scenarios are then turned into *design scenarios*, in which the same situations are described but with the functionality of the new system. Design scenarios are used, among other things, to create scenario-based video prototypes or software prototypes. Like task-based prototypes, the developer needs to write only the software necessary to illustrate the components of the design scenario. The goal is to create a situation in which the user can experience what the system would be like in a realistic situation, even if it addresses only a subset of the planned functionality.

The following section describes a variety of rapid prototyping techniques that can be used in any of these four prototyping strategies. We begin with offline rapid prototyping techniques, followed by online prototyping techniques.

RAPID PROTOTYPES

The goal of rapid prototyping is to develop prototypes quickly, in a fraction of the time it would take to develop a working system. By shortening the prototype-evaluation cycle, the design team can evaluate more alternatives and iterate the design several times, improving the likelihood of finding a solution that successfully meets the user's needs.

How rapid is rapid depends on the context of the particular project and the stage in the design process. Early prototypes (e.g., sketches) can be created in a few minutes. Later in the design cycle, a prototype produced in less than a week may still be considered "rapid" if the final system is expected to take months or years to build. Precision, interactivity, and evolution all affect the time it takes to create a prototype. Not surprisingly, a precise and interactive prototype takes more time to build than an imprecise or fixed one.

The techniques presented in this section are organized from most rapid to least rapid, according to the representation dimension introduced earlier. Offline techniques are generally more rapid than online ones; however, creating successive iterations of an online prototype may end up being faster than creating new offline prototypes.

Offline Rapid Prototyping Techniques

Offline prototyping techniques range from simple to elaborate. Because they do not involve software, they are usually considered a tool for thinking through the design issues, to be thrown away when they are no longer needed. This section describes simple paper-and-pencil sketches, three-dimensional mock-ups, Wizard-of-Oz simulations, and video prototypes.

Paper and Pencil. The fastest form of prototyping involves paper, transparencies, and post-it notes to represent aspects of an interactive system (for an example, see Muller, 1991). By playing the roles of both the user and the system, designers can get a quick idea of a wide variety of different layout and interaction alternatives in a short period of time.

Designers can create a variety of low-cost "special effects." For example, a tiny triangle drawn at the end of a long strip cut from an overhead transparency makes a handy mouse pointer, which can be moved by a colleague in response to the user's actions. Post-it Notes, with prepared lists, can provide "pop-up menus." An overhead projector pointed at a whiteboard makes it easy to project transparencies (hand-drawn or preprinted, overlaid onto each other as necessary) to create an interactive display on the wall. The user can interact by pointing (Fig. 52.3) or drawing on the whiteboard. One or more people can watch the user and move the transparencies in response to her actions.

FIGURE 52.3. Hand-drawn transparencies can be projected onto a wall, creating an interface a user can respond to.

Everyone in the room gets an immediate impression of how the eventual interface might look and feel.

Note that most paper prototypes begin with quick sketches on paper, then progress to more carefully drawn screen images made with a computer (Fig. 52.4). In the early stages, the goal is to generate a wide range of ideas and expand the design space, not to determine the final solution. Paper-and-pencil prototypes are an excellent starting point for horizontal, task-based, and scenario-based prototyping strategies.

Mock-Ups. Architects use mock-ups or scaled prototypes to provide three-dimensional illustrations of future buildings. Mock-ups are also useful for interactive system designers,

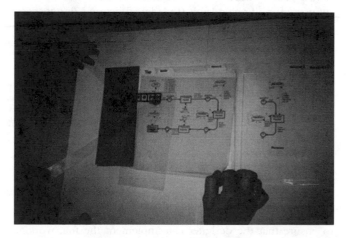

FIGURE 52.4. Several people work together to simulate interacting with this paper prototype. One person moves a transparency with a mouse pointer, while another moves the diagram accordingly.

FIGURE 52.5. Mock-up of a handheld display with carrying handle.

FIGURE 52.6. Scaled mock-up of an air traffic control table, connected to a wall display.

helping them move beyond two-dimensional images drawn on paper or transparencies (see Bødker, Ehn, Knudsen, Kyng, & Madsen, 1988). Generally made of cardboard, foamcore or other found materials, mock-ups are physical prototypes of the new system. Figure 52.5 shows an example of a handheld mock-up showing the interface to a new handheld device. The mock-up provides a deeper understanding of how the interaction will work in real-world situations than possible with sets of screen images.

Mock-ups allow the designer to concentrate on the physical design of the device, such as the position of buttons or the screen. The designer can also create several mock-ups and compare input or output options, such as buttons versus trackballs. Designers and users should run through different scenarios, identifying potential problems with the interface or generating ideas for new functionality. Mock-ups can also help the designer envision how an interactive system will be incorporated into a physical space (Fig. 52.6).

Wizard of Oz. Sometimes it is useful to give users the impression that they are working with a real system, even before it exists. Kelley (1983) dubbed this technique the Wizard of Oz, based on a scene in the 1939 movie of the same name. The heroine, Dorothy, and her companions ask the mysterious Wizard of Oz for help. When they enter the room, they see an enormous green human head, breathing smoke and speaking with a deep, impressive voice. When they return later to see the Wizard, Dorothy's small dog pulls back a curtain, revealing a frail old man pulling levers and making the mechanical Wizard of Oz speak. They realize that the impressive being before them is not a wizard at all, but simply an interactive illusion created by the old man.

The software version of the Wizard of Oz operates on the same principle. A user sits a terminal and interacts with a program. Hidden elsewhere, the software designer (the wizard) watches what the user does and, by responding in different ways, creates the illusion of a working software program. In some cases, the user is unaware that a person, rather than a computer, is operating the system.

The Wizard-of-Oz technique lets users interact with partially functional computer systems. Whenever they encounter something that has not been implemented (or there is a bug), a human developer who is watching the interaction overrides the prototype system and plays the role destined to eventually be played by the computer. A combination of video and software can work well, depending on what needs to be simulated.

The Wizard of Oz was initially used to develop natural language interfaces (e.g., Chapanis, 1982; Good, Whiteside, Wixon, & Jones, 1984). Since then, the technique has been used in a wide variety of situations, particularly those in which rapid responses from users are not critical. Wizard-of-Oz simulations may consist of paper prototypes, fully implemented systems, and everything in between.

Video Prototyping. Video prototypes (Mackay, 1988) use video to illustrate how users will interact with the new system. As explained earlier, they differ from video brainstorming in that the goal is to refine a single design, not generate new ideas. Video prototypes may build on paper-and-pencil prototypes and cardboard mock-ups and can also use existing software and images of real-world settings.

We begin our video prototyping exercises by reviewing relevant data about users and their work practices and then review ideas we video brainstormed. The next step is to create a use scenario, describing the user at work. Once the scenario is described in words, the designer develops a storyboard. Similar to a comic book, the storyboard shows a sequence of rough sketches of each action or event, with accompanying actions and dialogue (or subtitles), with related annotations that explain what is happening in the scene or the type of shot (Fig. 52.7). A paragraph of text in a scenario corresponds to about a page of a storyboard.

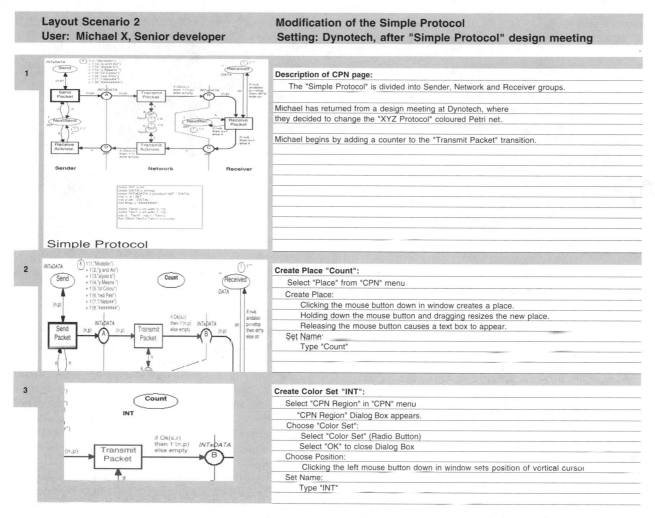

| Layout Scenario 2 | Modification of the Simple Protocol |
| User: Michael X, Senior developer | Setting: Dynotech, after "Simple Protocol" design meeting |

Description of CPN page:

The "Simple Protocol" is divided into Sender, Network and Receiver groups.

Michael has returned from a design meeting at Dynotech, where they decided to change the "XYZ Protocol" coloured Petri net.

Michael begins by adding a counter to the "Transmit Packet" transition.

Create Place "Count":

Select "Place" from "CPN" menu
Create Place:
  Clicking the mouse button down in window creates a place.
  Holding down the mouse button and dragging resizes the new place.
  Releasing the mouse button causes a text box to appear.
Set Name:
  Type "Count"

Create Color Set "INT":

Select "CPN Region" in "CPN" menu
  "CPN Region" Dialog Box appears.
Choose "Color Set":
  Select "Color Set" (Radio Button)
  Select "OK" to close Dialog Box
Choose Position:
  Clicking the left mouse button down in window sets position of vertical cursor
Set Name:
  Type "INT"

FIGURE 52.7. This storyboard is based on observations of real Coloured Petri Net (CPN) users in a small company and illustrates how the CPN developer modifies a particular element of a net, the "Simple Protocol."

Storyboards help designers refine their ideas, generate "what if" scenarios for different approaches to a story, and communicate with the other people who are involved in creating the production. Storyboards may be informal "sketches" of ideas, with only partial information. Others follow a predefined format and are used to direct the production and editing of a video prototype. Designers should jot down notes on storyboards as they think through the details of the interaction.

Storyboards can be used like comic books to communicate with other members of the design team. Designers and users can discuss the proposed system and alternative ideas for interacting with it (Fig. 52.8). Simple videos of each successive frame, with a voiceover to explain what happens, can also be effective. We usually use storyboards to help us shoot video prototypes, which illustrate how a new system will look to a user in a real-world setting. We find that placing the elements of a storyboard on separate cards and arranging them (Mackay & Pagani, 1994) helps the designer experiment with different linear sequences and insert or delete video clips.

The process of creating a video prototype, based on the storyboard, provides an even deeper understanding of the design and how a user will interact with it.

The storyboard guides the shooting of the video. We often use a technique called "editing-in-the-camera" (see Mackay, 2000), which allows us to create the video directly, without editing later. We use title cards, as in a silent movie, to separate the clips and to make it easier to shoot. A narrator explains each event, and several people may be necessary to illustrate the interaction. Team members enjoy playing with special effects, such as "time-lapse photography." For example, we can record a user pressing a button, stop the camera, add a new dialogue box, and then restart the camera to create the illusion of immediate system feedback.

Video is not simply a way to capture events in the real world or to capture design ideas but can also be a tool for sketching and visualizing interactions. We use a second live video camera as a Wizard-of-Oz tool. The wizard should have access to a set of prototyping materials representing screen objects.

FIGURE 52.8. Video prototyping: The Coloured Petri Net (CPN) design team reviews their observations of CPN developers and then discuss several design alternatives. They work out a scenario and storyboard it, then shoot a video prototype that reflects their design.

FIGURE 52.9. Complex Wizard-of-Oz simulation, with projected image from a live video camera and transparencies projected from an overhead projector.

Other team members stand by, ready to help move objects as needed. The live camera is pointed at the wizard's work area, with either a paper prototype or a partially working software simulation. The resulting image is projected onto a screen or monitor in front of the user. One or more people should be situated so that they can observe the actions of the user and manipulate the projected video image accordingly. This is most effective if the wizard is well prepared for a variety of events and can present semiautomated information. The user interacts with the objects on the screen as wizard moves the relevant materials in direct response to each user action. The other camera records the interaction between the user and the simulated software system on the screen or monitor, to create either a video brainstorm (for a quick idea) or a fully storyboarded video prototype.

Figure 52.9 shows a Wizard-of-Oz simulation with a live video camera, video projector, whiteboard, overhead projector, and transparencies. The setup allows two people to experience how they would communicate via a new interactive communication system. One video camera films the woman at left, who can see and talk to the other woman. Her image is projected live onto the left side of the wall. An overhead projector displays hand-drawn transparencies, manipulated by two other people, in response to gestures made by the woman at right. The entire interaction is videotaped by a second video camera.

Combining Wizard-of-Oz and video is a particularly powerful prototyping technique because it gives the person playing the user a real sense of what it might actually feel like to interact with the proposed tool, long before it has been implemented. Seeing a video clip of someone else interacting with a simulated tool is more effective than simply hearing about it, but interacting with it directly is more powerful still. Video prototyping may act as a form of specification for developers, enabling them to build the precise interface, both visually and interactively, created by the design team.

Online Rapid Prototyping Techniques

The goal of online rapid prototyping is to create higher precision prototypes than can be achieved with offline techniques. Such prototypes may prove useful to better communicate ideas to clients, managers, developers, and end users. They are also useful for the design team to fine tune the details of a layout or an interaction. They may exhibit problems in the design that were not apparent in less precise prototypes. Finally, they may be used early on in the design process for low precision prototypes that would be difficult to create offline, such as when dynamic interactions or visualizations are needed.

The techniques presented in this section are sorted by interactivity. We start with noninteractive simulations (i.e., animations), followed by interactive simulations that provide fixed or multiple-paths interactions. We finish with scripting languages that support open interactions.

Noninteractive Simulations. A noninteractive simulation is a computer-generated animation that represents what a person would see of the system if he or she were watching over the user's shoulder. Noninteractive simulations are usually created when offline prototypes, including video, fail to capture a particular aspect of the interaction, and it is important to have a quick prototype to evaluate the idea. It is usually best to start by creating a storyboard to describe the animation, especially if the developer of the prototype is not a member of the design team.

One of the most widely used tools for noninteractive simulations is Macromedia Director. The designer defines graphic objects called *sprites* and defines paths along which to animate them. The succession of events, such as when sprites appear and disappear, is determined with a time line. Sprites are usually created with drawing tools (e.g., Adobe Illustrator or Deneba Canvas), painting tools (e.g., Adobe Photoshop), or even

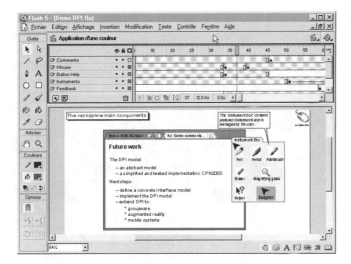

FIGURE 52.10. A noninteractive simulation of a desktop interface created with Macromedia Flash. The time line (top) displays the active sprites and the main window (bottom) shows the animation. Reprinted with permission from O. Beaudoux.

scanned images. Director is a powerful tool; experienced developers can create sophisticated interactive simulations; however, noninteractive simulations are much faster to create. Other similar tools exist on the market, including Abvent Katabounga, Adobe AfterEffects, and Macromedia Flash (Fig. 52.10).

Figure 52.11 shows a set of animation movies created by Dave Curbow to explore the notion of accountability in computer systems (Dourish, 1997). These prototypes explore new ways to inform the user of the progress of a file copy operation. They were created with Macromind Director by combining custom-made sprites with sprites extracted from snapshots of the Macintosh Finder. The simulation features cursor motion, icons being dragged, windows opening and closing, and so on. The result is a realistic prototype that shows how the interface looks and behaves that was created in just a few hours. Note that the simulation also features text annotations to explain each step, which helps document the prototype.

Noninteractive animations can be created with any tool that generates images. For example, many Web designers use Adobe Photoshop to create simulations of their Web sites. Photoshop images are composed of various layers that overlap like transparencies. The visibility and relative position of each layer can be controlled independently. Designers can quickly add or delete visual elements, simply by changing the characteristics of the relevant layer. This permits quick comparisons of alternative designs and helps visualize multiple pages that share a common layout or banner. Skilled Photoshop users find this approach much faster than most Web authoring tools.

We used this technique in the CPN2000 project (Mackay, Ratzer, & Janecek, 2000) to prototype the use of transparency. After several prototyping sessions with transparencies and overhead projectors, we moved to the computer to understand the differences between the physical transparencies and the transparent effect as it would be rendered on a computer screen. We later developed an interactive prototype with OpenGL,

which required an order of magnitude more time to implement than the Photoshop mock-up.

Interactive Simulations. Designers can also use tools such as Adobe Photoshop to create Wizard-of-Oz simulations. For example, the effect of dragging an icon with the mouse can be obtained by placing the icon of a file in one layer and the icon of the cursor in another layer and by moving either or both layers. The visibility of layers, as well as other attributes, can also create more complex effects. Like Wizard-of-Oz and other paper prototyping techniques, the behavior of the interface is generated by the user who is operating the Photoshop interface.

More specialized tools, such as Hypercard and Macromedia Director, can be used to create simulations that the user can directly interact with. Hypercard (Goodman, 1987) is one of the most successful early prototyping tools. It is an authoring environment based on a stack metaphor: A stack contains a set of cards that share a background, including fields and buttons. Each card can also have its own unique contents, including fields and buttons (Fig. 52.12). Stacks, cards, fields, and buttons react to user events (e.g., clicking a button) as well as system events (e.g., when a new card is displayed or about to disappear; Fig. 52.13). Hypercard reacts according to events programmed with a scripting language called Hypertalk. For example, the following script is assigned to a button, which switches to the next card in the stack whenever the button is clicked. If this button is included in the stack background, the user will be able to browse through the entire stack:

```
on click
    goto next card
end click
```

Interfaces can be prototyped quickly with this approach by drawing different states in successive cards and using buttons to switch from one card to the next. Multiple-path interactions can be programmed by using several buttons on each card. More open interactions require more advanced use of the scripting language but are fairly easy to master with a little practice.

Director uses a different metaphor, attaching behaviors to sprites and to frames of the animation. For example, a button can be defined by attaching a behavior to the sprite representing that button. When the sprite is clicked, the animation jumps to a different sequence. This is usually coupled with a behavior attached to the frame containing the button that loops the animation on the same frame. As a result, nothing happens until the user clicks the button, at which point the animation skips to a sequence where, for example, a dialogue box opens. The same technique can be used to make the OK and Cancel buttons of the dialogue box interactive. Typically, the Cancel button would skip to the original frame, whereas the OK button would skip to a third sequence. Director comes with a large library of behaviors to describe such interactions so that prototypes can be created completely interactively. New behaviors can also be defined with a scripting language called Lingo.

Many educational and cultural CD-ROMs are created exclusively with Director. They often feature original visual displays and interaction techniques that would be almost impossible to create with the traditional user interface development tools

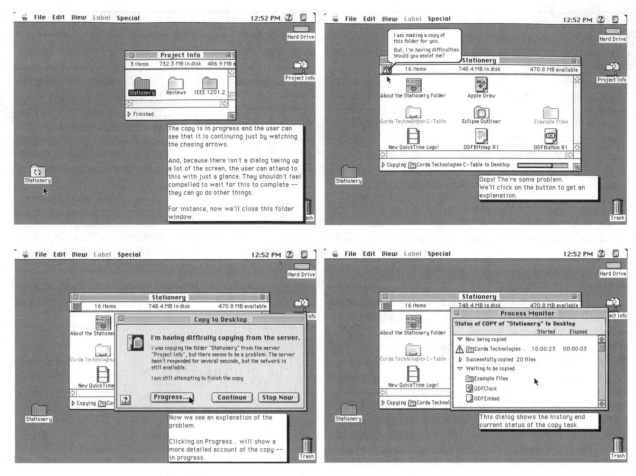

FIGURE 52.11. Frames from an animated simulation created with Macromind Director. Reprinted with permission from D. Curbow.

described in the next section. Designers should consider tools such as Hypercard and Director as user interface builders or user interface development environments. In some situations, they can even be used for evolutionary prototypes.

Scripting Languages. Scripting languages are the most advanced rapid prototyping tools. As with the interactive-simulation tools described above, the distinction between rapid prototyping tools and development tools is not always clear.

Scripting languages make it easy to quickly develop throw-away prototypes (a few hours to a few days), which may or may not be used in the final system for performance or other technical reasons.

A scripting language is a programming language that is both light weight and easy to learn. Most scripting languages are interpreted or semicompiled (i.e., the user does not need to go through a compile-link-run cycle each time the script or program is changed). Scripting languages can be forbidding: They are not

FIGURE 52.12. A Hypercard card (right) is the combination of a background (left) and the card's content (middle). Reprinted with permission from Apple Computer.

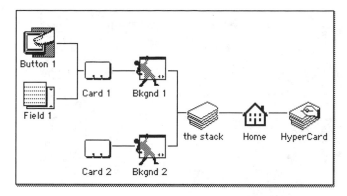

FIGURE 52.13. The hierarchy of objects in Hypercard determines the order (from left to right) in which a handler is looked up for an event. © Apple Computer, Inc. Used with permission.

In summary, domain objects form the basis of the interaction as well as its purpose: Users operate on domain objects by editing their attributes. They also manipulate them as a whole, e.g. to create, move and delete them.

Interaction ⬚instrument⬚s

An interaction ⬚instrument⬚ is a mediator or two-way transducer between the user and domain objects. The user acts on the ▪tool▪, which transforms the user's actions into commands affecting relevant target domain objects. Instruments have reactions enabling users to control their actions on the ▪tool▪, and provide feedback as the command is carried out on target objects (Figure 1).

A scrollbar is a good example of an interaction ⬚instrument⬚. It operates on a whole document by changing the part that is currently visible. When the user clicks on one of the arrows of the scrollbar, the scrollbar sends the document a

| Search string | instrument |
| Replace with | tool |

FIGURE 52.14. Using the Tk text and canvas widgets to prototype a novel search and replace interaction technique. *Note.* From "Instrumental interaction: An interaction model for designing post-WIMP user interfaces" by M. Beaudouin-Lafon, 2000, *Proceedings of the Association for Computing Machinery Human Factors in Computing Systems* (CHI'2000), CHI *Letters* 2(1), 452. Copyright 2000 by the Association of Computing Machinery. Reprinted with permission.

strongly typed, and nonfatal errors are ignored unless explicitly trapped by the programmer. Scripting languages are often used to write small applications for specific purposes and can serve as glue between preexisting applications or software components. Tcl (Ousterhout, 1993) was inspired by the syntax of the Unix shell, it makes it easy to interface existing applications by turning the application programming interface (API) into a set of commands that can be called directly from a Tcl script.

Tcl is particularly suitable for developing user interface prototypes (or small- to medium-sized applications) because of its Tk user interface toolkit. Tk features all the traditional interactive objects (called widgets) of a user interface toolkit: buttons, menus, scroll bars, lists, dialogue boxes, and so on. A widget is typically only one line. For example:

```
button.dialogbox.ok -text OK -command {destroy.dialogbox}
```

This command creates a button, called ".dialogbox.ok" with the label "OK." It deletes its parent window ".dialogbox" when the button is pressed. A traditional programming language and toolkit would take 5 to 20 lines to create the same button.

Tcl also has two advanced, heavily parameterized widgets: the text widget and the canvas widget. The text widget can be used to prototype text-based interfaces. Any character in the text can react to user input through the use of *tags*. For example, it is possible to turn a string of characters into a hypertext link. In Beaudouin-Lafon (2000), the text widget was used to prototype a new method for finding and replacing text. When entering the search string, all occurrences of the string are highlighted in the text (Fig. 52.14). Once a replace string has been entered, clicking an occurrence replaces it (the highlighting changes from yellow to red). Clicking a replaced occurrence returns it to its original value. This example also uses the canvas widget to create a custom scroll bar that displays the positions and status of the occurrences.

The Tk canvas widget is a drawing surface that can contain arbitrary objects: lines, rectangles, ovals, polygons, text strings, and widgets. Tags allow behaviors (i.e., scripts) that are called when the user acts on these objects. For example, an object

that can be dragged will be assigned a tag with three behaviors: button-press, mouse-move, and button-up. Because of the flexibility of the canvas, advanced visualization and interaction techniques can be implemented more quickly and easily than with other tools. For example, Fig. 52.15 shows a prototype exploring new ideas to manage overlapping windows on the screen (Beaudouin-Lafon, 2001). Windows can be stacked and slightly rotated so that it is easier to recognize them, and they can be folded so it is possible to see what is underneath without having to move the window. Even though the prototype is not perfect (for example, folding a window that contains text is not properly supported), it was instrumental in identifying a number of problems with the interaction techniques and finding appropriate solutions through iterative design.

Tcl and Tk can also be used with other programming languages. For example, Pad++ (Bederson & Meyers, 1998) is implemented as an extension to Tcl/Tk: the zoomable interface is implemented in C for performance and accessible from Tk as a new widget. This makes it easy to prototype interfaces that use zooming. It is also a way to develop evolutionary prototypes: a first prototype is implemented completely in Tcl, then parts of it are reimplemented in a compiled language to enhance performance. Ultimately, the complete system may be implemented in another language, although it is more likely that some parts will remain in Tcl.

Software prototypes can also be used in conjunction with hardware prototypes. Figure 52.16 shows an example of a hardware prototype that captures hand-written text from a paper flight strip (using a combination of a graphics tablet and a custom-designed system for detecting the position of the paper strip holder). We used Tk/TCL, in conjunction with C++, to present information on a RADAR screen (tied to an existing air traffic control simulator) and to provide feedback on a touch-sensitive display next to the paper flight strips (Mackay, Fayard,

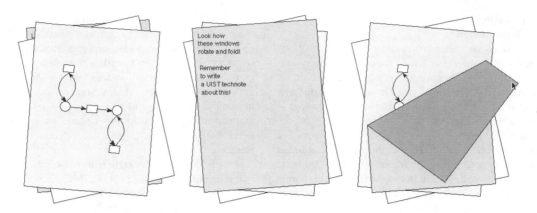

FIGURE 52.15. Using the Tk canvas widget to prototype a novel window manager. *Note*. From "Novel interaction techniques for overlapping windows" by M. Beaudouin-Lafon, 2001, *Proceedings of the Association for Computing Machinery Symposium on User Interface Software and Technology* (UIST 2001). CHI *Letters*, 3(2), 154. Copyright 2001 by the Association for Computing Machinery. Reprinted with permission.

Frobert, & Médini, 1998). The user can write in the ordinary way on the paper flight strip, and the system interprets the gestures according to the location of the writing on the strip. For example, a change in flight level is automatically sent to another controller for confirmation and a physical tap on the strip's ID lights up the corresponding aircraft on the RADAR screen.

ITERATIVE PROTOTYPES

Prototypes may also be developed with traditional software development tools. In particular, high-precision prototypes usually require a level of performance that cannot be achieved with the rapid online prototyping techniques described earlier. Similarly, evolutionary prototypes intended to evolve into the final product require more traditional software development tools. Finally, even shipped products are not "final," because subsequent releases can be viewed as initial designs for prototyping the next release.

Development tools for interactive systems have been in use for more than 20 years and are constantly being refined. Several studies have shown that the part of the development cost of an application spent on the user interface is 50% to 80% of the total cost of the project (Myers & Rosson, 1992). The goal of development tools is to shift this balance by reducing production and maintenance costs. Another goal of development tools is to

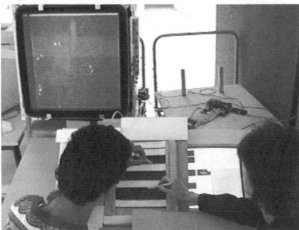

FIGURE 52.16. Caméléon's augmented stripboard (left) is a working hardware prototype that identifies and captures hand writing from paper flight strips. Members of the design team test the system (right), which combines both hardware and software prototypes into a single interactive simulation.

anticipate the evolution of the system over successive releases and support iterative design.

Interactive systems are inherently more powerful than non-interactive ones (see Wegner, 1997, for a theoretical argument). They do not match the traditional, purely algorithmic, type of programming: An interactive system must handle user input and generate output at almost any time, whereas an algorithmic system reads input at the beginning, processes it, and displays results at the end. In addition, interactive systems must process input and output at rates that are compatible with the human perception–action loop (i.e., in time frames of 20 to 200 ms). In practice, interactive systems are both reactive and real-time systems, two active areas in computer science research.

The need to develop interactive systems more efficiently has led to two interrelated streams of work. The first involves creation of software tools, from low-level user interface libraries and toolkits to high-level user interface development environments (UIDE). The second addresses software architectures for interactive systems, or how system functions are mapped onto software modules. The rest of this section presents the most salient contributions of these two streams of work.

Software Tools

Since the advent of graphical user interfaces in the 1980s, a large number of tools have been developed to help with the creation of interactive software, most aimed at visual interfaces. This section presents a collection of tools, from low-level (i.e., requiring a lot of programming) to high-level tools.

The lowest level tools are *graphical libraries* that provide hardware independence for painting pixels on a screen and handling user input, and *window systems* that provide an abstraction (the window) to structure the screen into several "virtual terminals." *User interface toolkits* structure an interface as a tree of interactive objects called widgets, whereas user interface builders provide an interactive application to create and edit those widget trees. *Application frameworks* build on toolkits and UI builders to facilitate creation of typical functions such as cut/copy/paste, undo, help, and interfaces based on editing multiple documents in separate windows. *Model-based tools* semiautomatically derive an interface from a specification of the domain objects and functions to be supported. Finally, *user interface development environments* or UIDEs provide an integrated collection of tools for the development of interactive software.

Before we describe each of these categories in more detail, it is important to understand how they can be used for prototyping. It is not always best to use the highest-level available tool. High-level tools are most valuable in the long term because they make it easier to maintain the system, port it to various platforms, or localize it to different languages. These issues are irrelevant for vertical and throw-away prototypes, so a high-level tool may prove less effective than a lower level one.

The main disadvantage of higher level tools is that they constrain or stereotype the types of interfaces they can implement. User interface toolkits usually contain a limited set of "widgets," and it is expensive to create new ones. If the design must incorporate new interaction techniques, such as bimanual interaction (Kurtenbach, Fitzmaurice, Baudel, & Buxton, 1997) or zoomable interfaces (Bederson & Hollan, 1994), a user interface toolkit will hinder rather than help prototype development. Similarly, application frameworks assume a stereotyped application with a menu bar, several toolbars, a set of windows holding documents, and so on. Such a framework would be inappropriate for developing a game or a multimedia educational CD-ROM that requires a fluid, dynamic, and original user interface.

Finally, developers need to truly master these tools, especially when prototyping in support of a design team. Success depends on the programmer's ability to quickly change the details as well as the overall structure of the prototype. A developer will be more productive when using a familiar tool than if forced to use a more powerful but unknown tool.

Graphical Libraries and Window Systems. Graphical libraries underlie all the other tools presented in this section. Their main purpose is to provide the developer with a hardware-independent, and sometimes cross-platform application programming interface (API) for drawing on the screen. They can be separated into two categories: direct drawing and scene-graph based. Direct drawing libraries provide functions to draw shapes on the screen once their geometry and their graphical attributes are specified. This means that every time something is to be changed on the display, the programmer has to either redraw the whole screen or figure out exactly which parts have changed. Xlib on Unix systems, Quickdraw on MacOS, Win32 GDI on Windows, and OpenGL (Woo, Neider, & Davis, 1997) on all three platforms are all direct drawing libraries. They offer the best compromise between performance and flexibility but are difficult to program.

Scene-graph based libraries explicitly represent the contents of the display by a structure called a scene graph. It can be a simple list (called display list), a tree (as used by many user interface toolkits; see next subsection), or a direct acyclic graph (DAG). Rather than painting on the screen, the developer creates and updates the scene graph, and the library is responsible for updating the screen to reflect the scene graph. Scene graphs are mostly used for three-dimensional graphics (e.g., OpenInventor, Strass, 1993), but in recent years they have also been used for two-dimensional graphics (Beaudouin-Lafon & Lassen, 2000; Bederson et al., 2000). With the advent of hardware-accelerated graphics cards, scene-graph based graphics libraries can offer outstanding performance while easing the task of the developer.

Window systems provide an abstraction to allow multiple client applications to share the same screen. Applications create windows and draw into them. From the application perspective, windows are independent and behave as separate screens. All graphical libraries include or interface with a window system. Window systems also offer a user interface to manipulate windows (move, resize, close, change stacking order, etc.) called the window manager. The window manager may be a separate application (as in X-Windows), it may be built into the window system (as in Windows), or it may be controlled of each application (as in MacOS). Each solution offers a different trade-off between flexibility and programming cost.

Graphical libraries include or are complemented by an input subsystem. The input subsystem is event driven: Each time the user interacts with an input device, an event recording the interaction is added to an input event queue. The input subsystem API lets the programmer query the input queue and remove events from it. This technique is much more flexible than polling the input devices repeatedly or waiting until an input device is activated. To ensure that input events are handled in a timely fashion, the application has to execute an event loop that retrieves the first event in the queue and handles it as fast as possible. Every time an event sits in the queue, there is a delay between the user action and the system reaction. As a consequence, the event loop sits at the heart of almost every interactive system.

Window systems complement the input subsystem by routing events to the appropriate client application based on its focus. The focus may be specified explicitly for a device (e.g., the keyboard) or implicitly through the cursor position (the event goes to the window under the cursor). Scene-graph based libraries usually provide a picking service to identify which objects in the scene graph are under or in the vicinity of the cursor.

Although graphical libraries and window systems are fairly low level, they must often be used when prototyping novel interaction or visualization techniques. Usually, these prototypes are developed when performance is key to the success of a design. For example, a zoomable interface that cannot provide continuous zooming at interactive frame rates is unlikely to be usable. The goal of the prototype is then to measure performance to validate the feasibility of the design.

User Interface Toolkits.   User interface toolkits are probably the most widely used tool nowadays to implement applications. All three major platforms (Unix/Linux, MacOS, and Windows) come with at least one standard user interface toolkit. The main abstraction provided by a toolkit is the *widget*, a software object that has three facets that closely match the Model-View-Controller (MVC) model described later: a presentation, a behavior, and an application interface.

The *presentation* defines the graphical aspect of the widget. Usually, the presentation can be controlled by the application, but also externally. For example, under X-Windows, it is possible to change the appearance of widgets in any application by editing a text file specifying the colors, sizes, and labels of buttons, menu entries, and so on. The overall presentation of an interface is created by assembling widgets into a tree. Widgets such as buttons are the leaves of the tree. Composite widgets constitute the nodes of the tree: A composite widget contains other widgets and controls their arrangement. For example, menu widgets in a menu bar are stacked horizontally, whereas command widgets in a menu are stacked vertically. Widgets in a dialogue box are laid out at fixed positions or relative to each other so that the layout may be recomputed when the window is resized. Such constraint-based layout saves time because the interface does not need to be laid out again when a widget is added or when its size changes as a result of, for example, changing its label.

The *behavior* of a widget defines the interaction methods it supports: a button can be pressed, a scroll bar can be scrolled, a text field can be edited. The behavior also includes the various

FIGURE 52.17. Callback functions.

possible states of a widget. For example, most widgets can be active or inactive, and some can be highlighted. The behavior of a widget is usually hardwired and defines its class (menu, button, list, etc.). It is sometimes parameterized, however (e.g., a list widget may be set to support single or multiple selection).

One limitation of widgets is that their behavior is limited to the widget itself. Interaction techniques that involve multiple widgets, such as drag-and-drop, cannot be supported by the widgets' behavior alone and require a separate support in the user interface toolkit. Some interaction techniques, such as toolglasses or magic lenses (Bier, Stone, Pier, Buxton, & De Rose, 1993), break the widget model both with respect to the presentation and the behavior and cannot be supported by traditional toolkits. In general, prototyping new interaction techniques requires either implementing them within new widget classes, which is not always possible, or not using a toolkit at all. Implementing a new widget class is typically more complicated than implementing the new technique outside the toolkit (e.g., with a graphical library) and is rarely justified for prototyping. Many toolkits provide a "blank" widget (Canvas in Tk, Drawing Area in Motif, JFrame in Java Swing) that can be used by the application to implement its own presentation and behavior. This is usually a good alternative to implementing a new widget class, even for production code.

The *application interface* of a widget defines how it communicates the results of the user interactions to the rest of the application. Three main techniques exist. The first and most common is called a callback function or *callback* for short: When the widget is created, the application registers the name of one or more functions with it. When the widget is activated by the user, it calls the registered functions (Fig. 52.17). The problem with this approach is that the logic of the application is split among many callback functions (Myers, 1991).

The second approach is called *active variables* and consists of associating a widget with a variable of the application program (Fig. 52.18). A controller ensures that when the widget state changes, the variable is updated with a new value and, conversely, when the value of the variable changes, the widget state reflects the new value. This allows the application to change the state of the interface without accessing the widgets directly, therefore decoupling the functional core from the presentation. In addition, the same active variable can be used with multiple widgets, providing an easy way to support multiple views. Finally, it is easier to change the mapping between

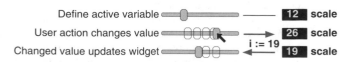

FIGURE 52.18. Active variables.

FIGURE 52.19. Listener objects.

widgets and active variables than it is to change the assignment of callbacks. This is because active variables are more declarative and callbacks more procedural. Active variables work only for widgets that represent data (e.g., a list or a text field) but not for buttons or menus. Therefore, they complement, rather than replace, callbacks. Few user interface toolkits implement active variables. Tcl/Tk (Ousterhout, 1994) is a notable exception.

The third approach for the application interface is based on *listeners*. Rather than registering a callback function with the widget, the application registers a listener object (Fig. 52.19). When the widget is activated, it sends a message to its listener describing the change in state. Typically, the listener of a widget would be its model (using the MVC terminology). The first advantage of this approach is that it matches true more closely the most common architecture models. It is also more true to the object-oriented approach that underlies most user interface toolkits. The second advantage is that it reduces the "spaghetti of callbacks" described above: By attaching a single listener to several widgets, the code is more centralized. A number of recent toolkits are based on the listener model, including Java Swing (Eckstein, Loy, & Wood, 1998).

User interface toolkits have been an active area of research over the past 15 years. InterViews (Linton, Vlissides, & Calder, 1989) has inspired many modern toolkits and user interface builders. A number of toolkits have also been developed for specific applications such as groupware (Roseman & Greenberg, 1996, 1999) or visualization (Schroeder, Martin, & Lorensen, 1997).

Creating an application or a prototype with a user interface toolkit requires a solid knowledge of the toolkit and experience with programming interactive applications. To control the complexity of the interrelations between independent pieces of code (creation of widgets, callbacks, global variables, etc.), it is important to use well-known design patterns. Otherwise the code quickly becomes unmanageable and, in the case of a prototype, unsuitable for design space exploration. Two categories of tools have been designed to ease the task of developers: user interface builders and application frameworks.

User Interface Builders. A user interface builder allows the developer of an interactive system to create the presentation of the user interface (i.e., the tree of widgets) interactively with a graphical editor. The editor features a palette of widgets that the user can use to "draw" the interface in the same way as a graphical editor is used to create diagrams with lines, circles, and rectangles. The presentation attributes of each widget can be edited interactively as well as the overall layout. This saves a lot of time that would otherwise be spent writing and fine-tuning rather dull code that creates widgets and specifies their attributes. It also makes it extremely easy to explore and test design alternatives.

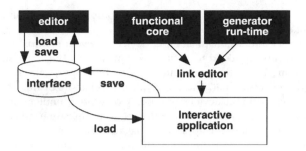

FIGURE 52.20. Iterative user interface builder.

User interface builders focus on the presentation of the interface. They also offer some facilities to describe the behavior of the interface and to test the interaction. Some systems allow the interactive specification of common behaviors such as a menu command opening a dialogue box, a button closing a dialogue box, a scroll bar controlling a list, or text. The user interface builder can then be switched to a "test" mode in which widgets are not passive objects but actually work. This may be enough to test prototypes for simple applications, even though there is no functional core nor application data.

To create an actual application, the part of the interface generated by the user interface builder must be assembled with the missing parts (i.e., the functional core), the application interface code that could not be described from within the builder, and the run-time module of the generator. Most generators save the interface into a file that can be loaded at run-time by the generator's run-time module (Fig. 52.20). With this method, the application need only be regenerated when the functional core changes, not when the user interface changes. This makes it easy to test alternative designs or to iteratively create the interface: Each time a new version of the interface is created, it can be readily tested by rerunning the application.

To make it even easier to modify the interface and test the effects with the real functional core, the interface editor can be built into the target application (Fig. 52.21). Changes to the interface can then be made from within the application and tested without rerunning it. This situation occurs most often with interface builders based on an interpreted language (e.g., Tcl/Tk, Visual Basic).

In either case, a final application can be created by compiling the interface generated by the user interface builder into actual code, linked with the functional core and a minimal run-time

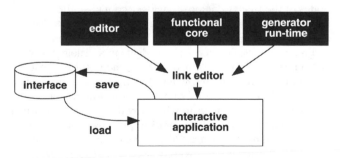

FIGURE 52.21. Interactive user interface builder.

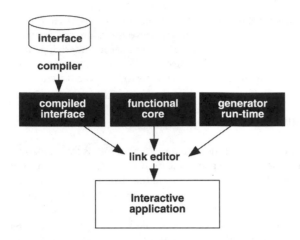

FIGURE 52.22. Generation of the final application.

module. In this situation, the interface is not loaded from a file but directly created by the compiled code (Fig. 52.22). This is both faster and eliminates the need for a separate interface description file.

User interface builders are widely used to develop prototypes as well as final applications. They are easy to use, they make it easy to change the look of the interface, and they hide a lot of the complexity of creating user interfaces with toolkits. Despite their name, however, they do not cover the whole user interface, only the presentation. Therefore, they still require a significant amount of programming, namely, some part of the behavior and all the application interface. Systems such as NeXT's Interface Builder (NeXT, 1991) ease this task by supporting part of the specification of the application objects and their links with the user interface. Still, user interface builders require knowledge of the underlying toolkit and an understanding of their limits, especially when prototyping novel visualization and interaction techniques.

Software Environments

Application Frameworks. Application frameworks address a different problem than user interface builders and are actually complementary. Many applications have a standard form where windows represent documents that can be edited with menu commands and tools from palettes. Each document may be saved into a disk file; standard functions such as copy/paste, undo, and help are supported. Implementing such stereotyped applications with a user interface toolkit or builder requires replicating a significant amount of code to implement the general logic of the application and the basics of the standard functions.

Application frameworks address this issue by providing a shell that the developer fills with the functional core and the actual presentation of the nonstandard parts of the interface. Most frameworks have been inspired by MacApp, a framework developed in the eighties to develop applications for the Macintosh (Apple Computer, 1996). Typical base classes of MacApp include Document, View, Command, and Application. MacApp supports multiple document windows, multiple views of a document, cut/copy/paste, undo, saving documents to files, scripting, and more.

With the advent of object-oriented technology, most application frameworks are implemented as collections of classes. Some classes provide services such as help or drag-and-drop and are used as client classes. Many classes are meant to be derived to add the application functionality through inheritance rather than by changing the actual code of the framework. This makes it easy to support successive versions of the framework and limits the risks of breaking existing code. Some frameworks are more specialized than MacApp. For example, Unidraw (Vlissides & Linton, 1990) is a framework for creating graphical editors in domains such as technical and artistic drawing, music composition, or circuit design. By addressing a smaller set of applications, such a framework can provide more support and significantly reduce implementation time.

Mastering an application framework takes time. It requires knowledge of the underlying toolkit and the design patterns used in the framework, and a good understanding of the design philosophy of the framework. A framework is useful because it provides a number of functions "for free," but at the same time it constrains the design space that can be explored. Frameworks can prove effective for prototyping if their limits are well understood by the design team.

Model-Based Tools. User interface builders and application frameworks approach the development of interactive applications through the presentation side: First the presentation is built, then behavior (i.e., interaction) is added; finally the interface is connected to the functional core. Model-based tools take the other approach, starting with the functional core and domain objects and working their way toward the user interface and the presentation (Szekely, Luo, & Neches, 1992, 1993). The motivation for this approach is that the raison d'être of a user interface is the application data and functions that will be accessed by the user. Therefore, it is important to start with the domain objects and related functions and derive the interface from them. The goal of these tools is to provide a semiautomatic generation of the user interface from the high-level specifications, including specification of the domain objects and functions, specification of user tasks, specification of presentation, and interaction styles.

Despite significant efforts, the model-based approach is still in the realm of research; no commercial tool exists yet. By attempting to define an interface declaratively, model-based tools rely on a knowledge base of user interface design to be used by the generation tools that transform the specifications into an actual interface. In other words, they attempt to do what designers do when they iteratively and painstakingly create an interactive system. This approach can probably work for well-defined problems with well-known solutions (i.e., families of interfaces that address similar problems). For example, it may be the case that interfaces for management information systems (MIS) could be created with model-based tools because these interfaces are fairly similar and well understood.

In their current form, model-based tools may be useful to create early horizontal or task-based prototypes. In particular,

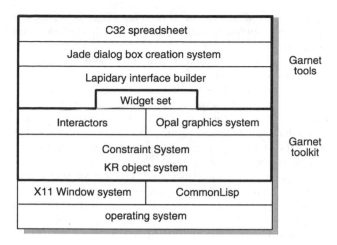

FIGURE 52.23. The Garnet toolkit and tools (Myers et al., 1990).

they can be used to generate a "default" interface that can serve as a starting point for iterative design. Future systems may be more flexible and therefore usable for other types of prototypes.

User Interface Development Environments. Like model-based tools, user interface development environments (UIDE) attempt to support the development of the whole interactive system. The approach is more pragmatic than the model-based approach, however. It consists of assembling a number of tools into an environment where different aspects of an interactive system can be specified and generated separately.

Garnet (Fig. 52.23) and its successor Amulet (Myers et al., 1997) provide a comprehensive set of tools, including a traditional user interface builder, a semiautomatic tool for generating dialogue boxes, a user interface builder based on a demonstration approach, and so on. One particular tool, Silk, is aimed explicitly at prototyping user interfaces.

Silk (Landay & Myers, 2001) is a tool aimed at the early stages of design, when interfaces are sketched rather than prototyped in software. Using Silk, a user can sketch a user interface directly on the screen (Fig. 52.24). Using gesture recognition, Silk interprets the marks as widgets, annotations, and so on. Even in its sketched form, the user interface is functional; for example, buttons can be pressed and tools can be selected in a toolbar. The sketch can also be turned into an actual interface (e.g., using the Motif toolkit). Finally, storyboards can be created to describe and test sequences of interactions. Silk therefore combines some aspects of offline and online prototyping techniques, trying to achieve the best of both worlds. This illustrates a current trend in research where online tools attempt to support not only the development of the final system, but the whole design process.

EVOLUTIONARY PROTOTYPES

Evolutionary prototypes are a special case of iterative prototypes, and are intended to evolve into the final system. Methodologies such as Extreme Programming (Beck, 2000) consist mostly of developing evolutionary prototypes. Because prototypes are rarely robust or complete, it is often impractical and sometimes dangerous to evolve them into the final system. Designers must think carefully about the underlying *software architecture* of the prototype, and developers should use well-documented *design patterns* to implement them.

Software Architectures

The definition of the software architecture is traditionally done after the functional specification is written but before coding starts. The designers design on the structure of the application and how functions will be implemented by software modules. The software architecture is the assignment of functions to modules. Ideally, each function should be implemented by a single module and modules should have minimal dependencies among them. Poor architectures increase development costs (coding, testing, and integration), lower maintainability, and reduce performance. An architecture designed to support prototyping and evolution is crucial to ensure that design alternatives can be tested with maximum flexibility and at a reasonable cost.

Seeheim and Arch. The first generic architecture for interactive systems was devised at a workshop in Seeheim (Germany) in 1985 and is known as the Seeheim model (Pfaff & ten Hagen, 1985). It separates the interactive application into a *user interface* and a *functional core* (then called "application," because the user interface was seen as adding a "coat of paint" on top of an existing application). The user interface is made of three modules: the presentation, the dialogue controller, and the application interface (Fig. 52.25). The *presentation* deals with capturing user's input at a low level (often called lexical level, similar to the lexical, syntactic, and semantic levels of a compiler). The presentation is also responsible for generating output to the user, usually as a visual display. The *dialogue controller* assembles the user input into commands (syntactic level), provides some immediate feedback for the action being carried out, such as an elastic rubber line, and detects errors. Finally, the *application interface* interprets the commands into calls to the functional core (semantic level). It also interprets the results of these calls and turns them into output to be presented to the user.

All architecture models for interactive systems are based on the Seeheim model. They all recognize that there is a part of the system devoted to capturing user actions and presenting output (the presentation) and another part devoted to the functional core (the computational part of the application). In between are one or more modules that transform user actions into functional calls and application data (including results from functional calls) into user output.

A modern version of the Seeheim model is the Arch model (The UIMS Workshop Tool Developers, 1992). The Arch model is made of five components (Fig. 52.26). The *interface toolkit component* is a preexisting library that provides low-level services such as buttons and menus. The *presentation component* provides a level of abstraction over the user interface toolkit. Typically, it implements interaction and visualization techniques that are not already supported by the interface toolkit. It may

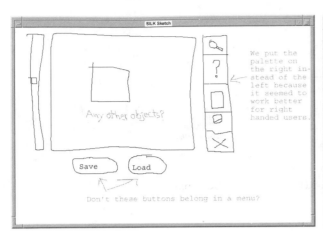
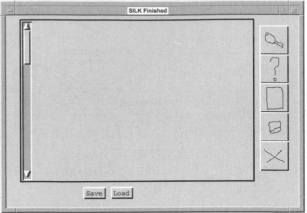
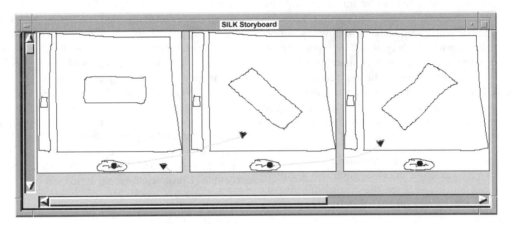

FIGURE 52.24. A sketch created with Silk (top left) and its automatic transformation into a Motif user interface (top right). A storyboard (bottom) used to test sequences of interactions, here a button that rotates an object. Reprinted with permission from J. Landay.

also provide platform independence by supporting different toolkits. The *functional core component* implements the functionality of the system. In some cases, it already exists and cannot be changed. The *domain adapter component* provides additional services to the dialogue component that are not in the functional core. For example, if the functional core is a Unix-like file system and the user interface is a iconic interface similar to the Macintosh Finder, the domain adapter may provide the dialogue controller with a notification service so the presentation can be updated whenever a file is changed. Finally, the *dialogue component* is the keystone of the arch. It handles the translation between the user interface world and the domain world.

The Arch model is also known as the Slinky model because the relative sizes of the components may vary across

applications as well as during the life of the software. For example, the presentation component may be almost empty if the interface toolkit provides all the necessary services and be later expanded to support specific interaction or visualization techniques or multiple platforms. Similarly, early prototypes may have a large domain adapter simulating the functional core of the final system or interfacing to an early version of the functional core; the domain adapter may shrink to almost nothing when the final system is put together.

The separation that Seeheim, Arch, and most other architecture models make between user interface and functional core is a good, pragmatic approach but it may cause problems in some cases. A typical problem is a performance penalty when the interface components (left leg) have to query the domain components (right leg) during an interaction such as drag-and-drop. For example, when dragging the icon of a file over the desktop, icons of folders and applications that can receive the file should highlight. Determining which icons to highlight is a semantic operation that depends on file types and other information and must therefore be carried out by the functional core or domain adapter. If drag-and-drop is implemented in the user interface toolkit, this means that each time the cursor goes over

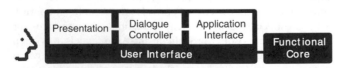

FIGURE 52.25. Seeheim model (Pfaff & ten Hagen, 1985).

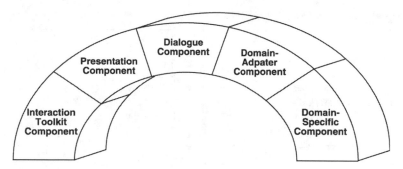

FIGURE 52.26. The Arch Model (The UIMS Workshop Developers Tool, 1992).

a new icon, up to four modules have to be traversed once by the query and once by the reply to find out whether or not to highlight the icon. This is both complicated and slow. A solution to this problem, called *semantic delegation*, involves shifting in the architecture some functions such as matching files for drag-and-drop from the domain leg into the dialogue or presentation component. This may solve the efficiency problem, but at the cost of an added complexity especially when maintaining or evolving the system, because it creates dependencies between modules that should otherwise be independent.

Model-View-Controller and Presentation-Abstraction-Controller Models. Architecture models such as Seeheim and Arch are abstract models and are thus rather imprecise. They deal with *categories* of modules such as presentation or dialogue, when in an actual architecture several modules will deal with presentation and several others with dialogue.

The model-view-controller or MVC model (Krasner & Pope, 1988) is much more concrete. MVC was created for the implementation of the Smalltalk-80 environment (Goldberg & Robson, 1983) and is implemented as a set of Smalltalk classes. The model describes the interface of an application as a collection of triplets of objects. Each triplet contains a model, a view, and a controller. A *model* represents information that needs to be represented and interacted with. It is controlled by applications objects. A *view* displays the information in a model in a certain way. A *controller* interprets user input on the view and transforms it into changes in the model. When a model changes, it notifies its view so the display can be updated.

Views and controllers are tightly coupled and sometimes implemented as a single object. A model is abstract when it has no view and no controller. It is noninteractive if it has a view but no controller. The MVC triplets are usually composed into a tree. For example, an abstract model represents the whole interface, it has several components that are themselves models, such as the menu bar and the document windows, all the way down to individual interface elements such as buttons and scrollbars. MVC supports multiple views fairly easily. The views share a single model; when a controller modifies the model, all the views are notified and update their presentation.

The presentation-abstraction-control or PAC model (Coutaz, 1987) is close to MVC. Like MVC, an architecture based on PAC is made of a set of objects, called PAC agents, organized in a tree. A PAC agent has three facets: The *presentation* takes care of capturing user input and generating output; the *abstraction* holds the application data, like a Model in MVC; the *control* manages the communication between the abstraction and presentation facets of the agent, and with subagents and super-agents in the tree. Like MVC, multiple views are easily supported. Unlike MVC, PAC is an abstract model; there is no reference implementation.

A variant of MVC, called MVP (model-view-presenter), is close to PAC and is used in ObjectArts' Dolphin Smalltalk. Other architecture models have been created for specific purposes such as groupware (Dewan, 1999) or graphical applications (Fekete & Beaudouin-Lafon, 1996).

Design Patterns

Architecture models such as Arch or PAC only address the overall design of interactive software. PAC is more fine-grained than Arch, and MVC is more concrete because it is based on an implementation. Still, a user interface developer has to address many issues to turn an architecture into a working system.

Design patterns have emerged in recent years as a way to capture effective solutions to recurrent software design problems. In their book, Gamma, Helm, Johnson, and Vlissides (1995) presented 23 patterns. Many of these patterns come from interactive software, and most of them can be applied to the design of interactive systems. It is beyond the scope of this chapter to describe these patterns in detail. Nonetheless, it is interesting that most patterns for interactive systems are behavioral patterns, that is, patterns that describe how to implement the control structure of the system.

Indeed, there is a battle for control in interactive software. In traditional, algorithmic software, the algorithm is in control and decides when to read input and write output. In interactive software, the user interface needs to be in control because user input should drive the system's reactions. Unfortunately, more often than not, the functional core also needs to be in control. This is especially common when creating user interfaces for legacy applications. In the Seeheim and Arch models, it is often believed that control is located in the dialogue controller when in fact these architecture models do not explicitly address the issue of control. In MVC, the three basic classes—**Model**, **View**,

and Controller—implement a sophisticated protocol to ensure that user input is taken into account in a timely manner and that changes to a model are properly reflected in the view (or views). Some authors actually describe MVC as a design pattern, not an architecture. In fact, it is both: The inner workings of the three basic classes is a pattern, but the decomposition of the application into a set of MVC triplets is an architectural issue.

It is now widely accepted that interactive software is event-driven: The execution is driven by the user's actions, leading to a control localized in the user interface components. Design patterns such as **Command, Chain of Responsibility, Mediator,** and **Observer** (Gamma et al., 1995) are especially useful to implement the transformation of low-level user event into higher level commands, to find out which object in the architecture responds to the command, and to propagate the changes produced by a command from internal objects of the functional core to user interface objects.

Using design patterns to implement an interactive system not only saves time, it also makes the system more open to changes and easier to maintain. Therefore, software prototypes should be implemented by experienced developers who know their pattern language and who understand the need for flexibility and evolution.

SUMMARY

Prototyping is an essential component of interactive system design. Prototypes may take many forms, from rough sketches to detailed working prototypes. They provide concrete representations of design ideas and give designers, users, developers, and managers an early glimpse into how the new system will look and feel. Prototypes increase creativity, allow early evaluation of design ideas, help designers think through and solve design problems, and support communication within multidisciplinary design teams.

Prototypes, because they are concrete and not abstract, provide a rich medium for exploring a design space. They suggest alternate design paths and reveal important details about particular design decisions. They force designers to be creative and to articulate their design decisions. Prototypes embody design ideas and encourage designers to confront their differences of opinion. The precise aspects of a prototype offer specific design solutions: Designers can then decide to generate and compare alternatives. The imprecise or incomplete aspects of a prototype

highlight the areas that must be refined or require additional ideas.

We begin by defining prototypes and then discuss them as design artifacts. We introduce four dimensions by which they can be analyzed: representation, precision, interactivity, and evolution. We then discuss the role of prototyping within the design process and explain the concepts of creating, exploring, and modifying a design space. We briefly describe techniques for generating new ideas, for expanding the design space, and for choosing among design alternatives to contract the design space.

We describe a variety of rapid prototyping techniques for exploring ideas quickly and inexpensively in the early stages of design, including offline techniques (from paper-and-pencil to video) and online techniques (from fixed to interactive simulations). We then describe iterative prototyping techniques for working out the details of the online interaction, including software development tools and software environments. We conclude with evolutionary prototyping techniques, which are designed to evolve into the final software system, including a discussion of the underlying software architectures, design patterns, and extreme programming.

This chapter has focused mostly on graphical user interfaces that run on traditional workstations. Such applications are dominant today, but this is changing as new devices are being introduced, from cellular phones and personal digital assitants to wall-sized displays. New interaction styles are emerging, such as augmented reality, mixed reality, and ubiquitous computing. Designing new interactive devices and the interactive software that runs on them is becoming ever more challenging: Whether aimed at a wide audience or a small number of specialists, the hardware and software systems must be adapted to their contexts of use. The methods, tools, and techniques presented in this chapter can easily be applied to these new applications.

We view design as an active process of working with a design space, expanding it by generating new ideas and contracting it as design choices are made. Prototypes are flexible tools that help designers envision this design space, reflect on it, and test their design decisions. Prototypes are diverse and can fit within any part of the design process, from the earliest ideas to the final details of the design. Perhaps most important, prototypes provide one of the most effective means for designers to communicate with each other, as well as with users, developers, and managers, throughout the design process.

References

Apple Computer. (1996). Programmer's guide to MacApp. Cupertino, CA: Apple (collective).

Beaudouin-Lafon, M. (2000). Instrumental interaction: An interaction model for designing post-WIMP user interfaces. *Proceedings of the ACM Human Factors in Computing Systems (CHI '2000), CHI Letters, 2*(1), 446–453. New York: ACM Press.

Beaudouin-Lafon, M. (2001). Novel interaction techniques for overlapping windows. *Proceedings of ACM Symposium on User Interface Software and Technology (UIST 2001). CHI Letters, 3*(2), 153–156.

Beaudouin-Lafon, M., & Lassen, M. (2000). The architecture and implementation of a post-WIMP graphical application. *Proceedings of ACM Symposium on User Interface Software and Technology (UIST 2000). CHI Letters, 2*(2), 181–190.

Beaudouin-Lafon, M., & Mackay, W. (2000, May). Reification, polymorphism and reuse: Three principles for designing visual interfaces. *Proceedings of the Conference on Advanced Visual Interfaces (AVI 2000)* (pp. 102–109). Palermo, Italy.

Beck, K. (2000). *Extreme programming explained*. New York: Addison-Wesley.

Bederson, B., & Hollan, J. (1994). Pad++: A zooming graphical interface for exploring alternate interface physics. *Proceedings of ACM Symposium on User Interface Software and Technology (UIST '94)* (pp. 17-26). New York: ACM Press.

Bederson, B., & Meyer, J. (1998). Implementing a zooming interface: Experience Building Pad++. *Software Practice and Experience, 28*(10), 1101-1135.

Bederson, B. B., Meyer, J., & Good, L. (2000). Jazz: An extensible zoomable user interface graphics toolkit in Java. *Proceedings of ACM Symposium on User Interface Software and Technology (UIST 2000). CHI Letters, 2*(2), 171-180.

Bier, E., Stone, M., Pier, K., Buxton, W., & De Rose, T. (1993). Toolglass and magic lenses: The see-through interface. *Proceedings ACM SIGGRAPH '93* (pp. 73-80). New York: ACM Press.

Boehm, B. (1988). A spiral model of software development and enhancement. *IEEE Computer, 21*(5), 61-72.

Bødker, S., Christiansen, E., & Thüring, M. (1995). A conceptual toolbox for designing CSCW applications. *Proceedings of the International Workshop on the Design of Cooperative Systems (COOP '95)* (pp. 266-284).

Bødker, S., Ehn, P., Knudsen, J., Kyng, M., & Madsen, K. (1988). Computer support for cooperative design. In *Proceedings of the CSCW'88 ACM Conference on Computer-Supported Cooperative Work* (pp. 377-393). Portland, OR: ACM Press.

Chapanis, A. (1982). Man/computer research at Johns Hopkins. In R. A. Kasschau, R. Lachman, & K. R. Laughery (Eds.), *Information Technology and Psychology: Prospects for the Future* (pp. 238-249). New York: Praeger.

Collaros, P. A., & Anderson, L. R. (1969). Effect of perceived expertness upon creativity of members of brainstorming groups. *Journal of Applied Psychology, 53*, 159-163.

Coutaz, J. (1987). PAC, an object oriented model for dialog design. In H.-J. Bullinger & B. Shackel (Eds.), *Proceedings of INTERACT '87* (pp. 431-436). Amsterdam: Elsevier Science.

de Vreede, G.-J., Briggs, R. O., van Duin, R., & Enserink, B. (2000). Athletics in electronic brainstorming: Asynchronous brainstorming in very large groups. *Proceedings of HICSS-33*.

Dewan, P. (1999). Architectures for collaborative applications. In M. Beaudouin-Lafon (Ed.), *Computer-supported co-operative work* (Trends in Software Series, pp. 169-193). New York: John Wiley.

Diehl, M., & Strobe, W. (1987). Productivity loss in brainstorming groups: Toward the solution of a riddle. *Journal of Personality and Social Psychology, 53*, 497-509.

Dykstra-Erickson, E., Mackay, W. E., & Arnowitz, J. (2001, March). Trialogue on Design (of). *ACM/Interactions, 8*(2), 109-117.

Dourish, P. (1997). Accounting for system behaviour: Representation, reflection and resourceful action. In M. Kyng & L. Mathiassen (Eds.), *Computers and design in context* (pp. 145-170). Cambridge, MA: MIT Press.

Eckstein, R., Loy, M., & Wood, D. (1998). *Java Swing*. Cambridge, MA: O'Reilly.

Fekete, J.-D., & Beaudouin-Lafon, M. (1996). Using the multi-layer model for building interactive graphical applications. In *Proceedings of ACM Symposium on User Interface Software and Technology (UIST '96)* (pp. 109-118). New York: ACM Press.

Gamma, E., Helm, R., Johnson, R., & Vlissides, J. (1995). *Design patterns, elements of reusable object-oriented software*. Reading, MA: Addison Wesley.

Goldberg, A., & Robson, D. (1983). *Smalltalk—80: The language and its implementation*. Reading, MA: Addison Wesley.

Goodman, D. (1987). *The complete HyperCard handbook*. New York: Bantam Books.

Good, M., Whiteside, J., Wixon, D., & Jones, S. (1984, October). Building a user-derived interface. *Communications of the ACM, 27*(10), 1032-1043.

Greenbaum, J., & Kyng, M. (Eds.). (1991). *Design at work: Cooperative design of computer systems*. Hillsdale, NJ: Lawrence Erlbaum Associates.

Houde, S., & Hill, C. (1997). What do prototypes prototype? In *Handbook of human computer interaction* (2nd ed. rev., pp. 367-381). Amsterdam: North-Holland.

Kelley, J. F. (1983). An empirical methodology for writing user-friendly natural language computer applications. In *Proceedings of CHI '83 Conference on Human Factors in Computing Systems*. New York: ACM Press.

Krasner, E. G., & Pope, S. T. (1988, August/September). A cookbook for using the model-view-controller user interface paradigm in Smalltalk-80. *Journal of Object-Oriented Programming*, 27-49.

Kurtenbach, G., Fitzmaurice, G., Baudel, T., & Buxton, W. (1997). The design of a GUI paradigm based on tablets, two-hands, and transparency. *Proceedings of ACM Human Factors in Computing Systems (CHI '97)* (pp. 35-42). New York: ACM Press.

Landay, J., & Myers, B. A. (2001). Sketching interfaces: Toward more human interface design. *IEEE Computer, 34*, 56-64.

Linton, M. A., Vlissides, J. M., & Calder, P. R. (1989). Composing user interfaces with InterViews. *IEEE Computer, 22*, 8-22.

Mackay, W. E. (1988). Video prototyping: A technique for developing hypermedia systems. *Demonstration, CHI '88, Conference on Human Factors in Computing Systems*. Retrieved from http://www.lri.fr/~mackay/publications.html

Mackay, W. E., & Pagani, D. (1994). Video Mosaic: Laying out time in a physical space. *Proceedings of ACM Multimedia '94* (pp. 165-172). New York: ACM Press.

Mackay, W. E., & Fayard, A.-L. (1997). HCI, natural science and design: A framework for triangulation across disciplines. *Proceedings of ACM DIS '97, Designing Interactive Systems* (pp. 223-234). New York: ACM Press.

Mackay, W., Fayard, A.-L., Frobert, L., & Médini, L. (1998). Reinventing the familiar: Exploring an augmented reality design space for air traffic control. *Proceedings of ACM Conference on Human Factors in Computing Systems (CHI '98)* (pp. 558-565). New York: ACM Press.

Mackay, W. E. (2000). Video techniques for participatory design: Observation, brainstorming & prototyping. *Tutorial Notes, CHI 2000, Human Factors in Computing Systems*. Retrieved from http://www.lri.fr/~mackay/publications.html

Mackay, W., Ratzer, A., & Janecek, P. (2000). Video artifacts for design: Bridging the gap between abstraction and detail. *Proceedings ACM Conference on Designing Interactive Systems (DIS 2000)* (pp. 72-82). New York: ACM Press.

Muller, M. (1991). PICTIVE: An exploration in participatory design. *Proceedings of ACM Conference on Human Factors in Computing Systems (CHI '91)* (pp. 225-231). New York: ACM Press.

Myers, B. A., Giuse, D. A., Dannenberg, R. B., Vander Zander, B., Kosbie, D. S., Pervin, E., Mickish, A., & Marchal, P. (1990). Garnet: Comprehensive support for graphical, highly-interactive user interfaces. *IEEE Computer, 23*(11), 71-85.

Myers, B. A. (1991). Separating application code from toolkits: Eliminating the spaghetti of call-backs. *Proceedings of ACM SIGGRAPH Symposium on User Interface Software and Technology (UIST '91)* (pp. 211-220). New York: ACM Press.

Myers, B. A., & Rosson, M. B. (1992). Survey on user interface programming. *Proceedings of the ACM Conference on Human Factors in Computing Systems (CHI '92)* (pp. 195-202). New York: ACM Press.

Myers, B. A., McDaniel, R. G., Miller, R. C., Ferrency, A. S., Faulring, A., Kyle, B. D., Mickish, A., Klimotivtski, A., & Doane, P. (1997). The Amulet environment. *IEEE Transactions on Software Engineering, 23(6)*, 347-365.

NeXT Corporation. (1991). *NeXT Interface Builder reference manual.* Redwood City, CA.

Norman, D. A., & Draper, S. W. (Eds.). (1986). *User centered system design.* Hillsdale, NJ: Lawrence Erlbaum Associates.

Osborn, A. (1957). *Applied imagination: Principles and procedures of creative thinking* (rev. ed.). New York: Scribner's.

Ousterhout, J. K. (1994). *Tcl and the Tk Toolkit.* Reading, MA: Addison Wesley.

Perkins, R., Keller, D. S., & Ludolph, F. (1997). Inventing the Lisa User Interface. *ACM Interactions, 4(1)*, 40-53.

Pfaff, G. P., & ten Hagen, P. J. W. (Eds.). (1985). *User interface management systems.* Berlin: Springer.

Raskin, J. (2000). *The humane interface.* New York: Addison Wesley.

Roseman, M., & Greenberg, S. (1996). Building real-time groupware with GroupKit, a groupware toolkit. *ACM Transactions on Computer-Human Interaction, 3(1)*, 66-106.

Roseman, M., & Greenberg, S. (1999). Groupware toolkits for synchronous work. In M. Beaudouin-Lafon (Ed.). *Computer-supported co-operative work* (Trends in software series, pp. 135-168). New York: John Wiley.

Schroeder, W., Martin, K., & Lorensen, B. (1997). *The visualization toolkit.* New York: Prentice Hall.

Strass, P. (1993). IRIS Inventor, a 3D graphics toolkit. *Proceedings ACM Conference on Object-Oriented Programming, Systems, Languages and Applications (OOPSLA '93)* (pp. 192-200). New York: ACM Press.

Szekely, P., Luo, P., & Neches, R. (1992). Facilitating the exploration of interface design alternatives: The HUMANOID. *Proceedings of ACM Conference on Human Factors in Computing Systems (CHI '92)* (pp. 507-515). New York: ACM Press.

Szekely, P., Luo, P., & Neches, R. (1993). Beyond interface builders: Model-based interface tools. *Proceedings of ACM/IFIP Conference on Human Factors in Computing Systems (INTERCHI '93)* (pp. 383-390). New York: ACM Press.

The UIMS Workshop Tool Developers. (1992). A metamodel for the run-time architecture of an interactive system. *SIGCHI Bulletin, 24(1)*, 32-37.

Vlissides, J. M., & Linton, M. A. (1990). Unidraw: A framework for building domain-specific graphical editors. *ACM Transactions on Information Systems, 8(3)*, 237-268.

Wegner, P. (1997). Why interaction is more powerful than algorithms. *Communications of the ACM, 40(5)*, 80-91.

Woo, M., Neider, J., & Davis, T. (1997). *OpenGL programming guide.* Reading, MA: Addison-Wesley.

·53·

SCENARIO-BASED DESIGN

Mary Beth Rosson and John M. Carroll
Virginia Tech

THE BASIC IDEA

Scenario-based design is a family of techniques in which the use of a future system is concretely described at an early point in the development process. Narrative descriptions of envisioned usage episodes are then employed in a variety of ways to guide the development of the system that will enable these use experiences.

Like other user-centered approaches, scenario-based design changes the focus of design work from defining system operations (i.e., functional specification) to describing how people will use a system to accomplish work tasks and other activities. However, unlike approaches that consider human behavior and experience through formal analysis and modeling of well-specified tasks, scenario-based design is a relatively lightweight method for envisioning future use possibilities.

A user interaction scenario is a *sketch of use*. It is intended to vividly capture the essence of an interaction design, much as a two-dimensional, paper-and-pencil sketch captures the essence of a physical design.

A SIMPLE EXAMPLE

Scenarios are stories. They consist of a setting, or situation state, one or more actors with personal motivations, knowledge, and capabilities, and various tools and objects that the actors encounter and manipulate. The scenario describes a sequence of actions and events that lead to an outcome. These actions and events are related in a usage context that includes the goals, plans, and reactions of the people taking part in the episode.

Table 53.1 presents three brief scenarios in which a member of a club uses different network tools to interact with club members. In all of these scenarios, the person's goal is to visit a club and interact with her friends. The scenarios contrast three ways that such a goal might be supported by computer network technologies. Each is a potential "solution" to Sharon's needs, but the user experience varies from asynchronous text-based reading and posting, to a real-time graphical simulation of a meeting place.

Designers can quickly construct scenarios like these to make envisioned possibilities more concrete. The example contrasts three contemporary approaches to online interactions, but not as an abstraction, or a list of features or functions. It contrasts three episodes of human–computer interaction (HCI) and personal experience.

Scenarios of envisioned use can be successively detailed to discover and address finer grained design issues. They serve as grist for group brainstorming, to develop further alternatives, or to raise questions about the assumptions behind the scenarios. They can be used to analyze software requirements, as a partial specification of functionality, and to guide the design of user interface layouts and controls. They can be used to identify and plan evaluation tasks that will be performed by usability test participants.

WHY SCENARIO-BASED DESIGN?

One reason that scenarios have become so popular in interactive system design is that they enable rapid communication about usage possibilities and concerns among many different stakeholders. It is easy to write simple scenarios such as those in Table 53.1, and takes only a little more effort to enrich it with a rough sketch or storyboard. When designers are working through ideas, they want to make progress quickly, so that they can obtain feedback and continue to refine their ideas. Scenarios are one way to do this.

The design of an interactive system is an ill-defined problem. Such problems tend to evoke a problem-solving strategy termed *solution-first* (Cross, 2001). In the solution-first strategy, designers generate and analyze a candidate solution as a means of clarifying the problem state, the allowable moves, and the goal. They exploit the concreteness of their own solution proposals to evoke further requirements for analysis.

A solution-first approach to design is energizing, effective, and efficient; it explains the popularity of contemporary system development approaches like rapid prototyping (Wasserman & Shewmake, 1982) and extreme programming (Beck, 1999). But this general strategy also entrains well-known hazards (Cross, 2001): Designers tend to generate solutions too quickly, before they analyze what is already known about the problem and possible moves. Once an approach is envisioned, they may have trouble abandoning it when it is no longer appropriate. Designers may too readily reuse pieces of a solution they have used earlier, one that is familiar and accessible but perhaps inappropriate. They may not analyze their own solutions well, or they may consider too few alternatives when exploring the problem space. In the next three subsections, we consider how scenario-based design may help to minimize these hazards of solution-first problem solving (see Table 53.2).

Scenarios Are Concrete But Rough

Design analysis is always indeterminate, because the act of design changes the world within which people act and experience. Requirements always change (Brooks, 1995). When designs incorporate rapidly evolving technologies, requirements change even more rapidly. The more successful, the more widely adopted, and the more impact a design has, the less possible it will have been to determine its correct design requirements. And in any case, refinements in software technology and new perceived opportunities and requirements propel a new generation of designs every 2 to 3 years.

Design representations that are at once concrete but flexible help to manage ambiguous and dynamic situations. Analysts must be concrete to avoid being swallowed by indeterminacy; they must be flexible to avoid being captured by false steps. Systematic decomposition is a traditional approach to managing ambiguity, but it does not promote flexibility. Instead, designers end up with a set of concrete subsolutions, each of which is fully specified. Unfortunately, by the time the set of subsolutions is specified, the requirements often have changed.

TABLE 53.1. Three Scenarios for a University Student Attending a Club Meeting Online

A. Science Fiction Club in a Web Forum

After 3 years at Virginia Tech, Sharon has learned to take advantage of her free time in between classes. In her hour between her morning classes, she stops by the computer lab to visit the science fiction club. She has been meaning to do this for a few days because she knows she'll miss the next meeting later this week. As she opens a Web browser, she realizes that this computer will not have her bookmarks stored, so she starts at the homepage of the Blacksburg Electronic Village. She sees local news and links to categories of community resources (businesses, town government, civic organizations). She selects "Organizations" and sees an alphabetical list of community groups. She is attracted by a new one, the Orchid Society, so she quickly examines their Web page before going back to select the Science Fiction Club page. When she gets to the club page, she sees that there are two new comments in the discussion on Asimov's *Robots and Empire*, one from Bill and one from Sara. She browses each comment in turn, then submits a reply to Bill's comment, arguing that he has the wrong date associated with discovery of the Zeroth Law.

B. Science Fiction Club in a Community MOO

After 3 years at Virginia Tech, Sharon has learned to take advantage of her free time in between classes. In her hour between her morning classes, she stops by the computer lab to visit the science fiction club. She has been meaning to do this for a few days because she knows she'll miss the next meeting later this week. As she starts up the Blacksburg community MOO, she can see that the last person using this computer must have been interested in orchids, because the welcoming text describes her location as an orchid garden, along with Penny and Alicia, who are discussing some new exotic varieties. The text description mentions an exit to Main Street, so she leaves the garden and starts moving south. Along the street she runs into George, who is working on a banner for the fair. She gives him a quick hello, and continues southward until she sees an eastward exit will take her to Eastenders Pub; this is where the Science Fiction Club meets. She enters the room and is told that Bill and Sara are already there, along with a pitcher of Newcastle Brown. She can tell from their current comments that they have been discussing the time line from Asimov's *Robots and Empire*.

C. Science Fiction Club in a Collaborative Virtual Environment

After 3 years at Virginia Tech, Sharon has learned to take advantage of her free time in between classes. In her hour between her morning classes, she stops by the computer lab to visit the science fiction club. She has been meaning to do this for a few days because she knows she'll miss the next meeting later this week. When she tries to start up the online collaborative environment, she finds that this computer does not have the client, so she waits for a minute or two while it is automatically downloaded and installed. After she logs in, she is taken back to her previous visit location and sees the familiar panoramic view of her livingroom, her to-do lists and sketchpad, and the interactive map of Blacksburg. She positions and zooms in on the map until she can see downtown buildings She enters the Eastenders Pub subspace, where the science fiction club usually meets. She sees a panoramic image of the bar, faces that show Bill and Sara are here, a food and drink menu, and various standard tools. The map updates to show a floorplan of the Pub—the dining room, the darts room, the office, and the main bar. Bill and Sara are using a chat tool and a shared whiteboard to sketch an event time line for Asimov's *Robots and Empire*. Joining Bill and Sara in the chat tool, she types "Based on the Zeroth Law, I'm afraid I must drink some of your beer."

Scenarios of use reconcile concreteness and flexibility. A scenario envisions a concrete design solution, but it can be couched at many levels of detail. Initial scenarios are often extremely rough. They specify a possible design by specifying the tasks users can carry out, but without committing to lower level details describing *how* the tasks will be carried out or *how* the system will present the functionality for the tasks. The examples in Table 53.1 are at an intermediate level, with some detail regarding task flow but little information about individual user-system interactions.

Concrete material is interpreted more easily and more thoroughly than abstract material. For example, people remember a prototypical example far better than they remember the abstract category to which it belongs (Medin & Schaffer, 1978; Rosch, Mervis, Gray, Johnson, & Boyes-Braem, 1976). Incomplete material tends to be elaborated with respect to personal knowledge when it is encountered. This process of elaboration creates more robust and accessible memories, relative to memories for more complete material (Wertheimer, 1939). The combination of concreteness and incompleteness in scenarios engages a powerful variety of constructive cognitive processes.

The fluidity of design situations demands provisional solutions and tentative commitments. Yet if every design decision is suspended, the result will be a design space, not a design. A

TABLE 53.2. Concerns Stemming From the Solution-First Approach to Design and Aspects of Scenario-Based Design That Address Each Concern

Hazards of the Solution-First Approach	How Scenario-Based Design Can Help
Designers want to select a solution approach quickly, which may lead to premature commitment to their first design ideas.	Because they are concrete but rough, scenarios support visible progress but also relax commitment to the ideas expressed in the scenarios.
Designers attempt to quickly simplify the problem space with external constraints, such as the reuse of familiar solutions.	Because they emphasize people and their experiences, scenarios direct attention to the use-appropriateness of design ideas.
Designers are intent on elaborating their current design proposal, resulting in inadequate analysis of other ideas or alternatives.	Because they are evocative and by nature are incomplete, scenarios promote empathy and raise usage questions at many levels.

scenario is a concrete design proposal that a designer can evaluate and refine, but it is also rough, so that it can be easily altered, and many details can be deferred.

Scenarios Maintain an Orientation to People and Their Needs

Designers need constraints; there are just too many things that might be designed. The current state of technology development makes some solutions impossible and others irresistible: On the one hand, designers cannot use technology that does not yet exist. On the other hand, designers are caught up in a technological zeitgeist that biases them toward making use of the latest gadgets and gizmos. They are likely to be biased toward familiar technologies, even when they are aware of limitations in these technologies.

Scenarios are work-oriented design objects. They describe systems in terms of the work that users will try to do when they use those systems, ensuring that design will remain focused on the needs and concerns of users (Carroll & Rosson, 1990). Scenarios address what has been called the "representational bias" in human cognition—people overestimate the relevance of things that are familiar to them (Kahneman & Tversky, 1972; Tversky & Kahneman, 1974). For instance, designers in a Web development company with years of experience in forms-based transactions will see this interaction paradigm as a solution to problems that might be better served by real-time interaction techniques. It is difficult to move beyond the familiar, but generating and sharing a vivid representation of exceptions to the status quo can promote innovative thinking. Scenarios describing unusual but critical circumstances can provide such a perspective.

The reuse of familiar ideas is just one type of constraint that designers may apply in their solution-first problem solving. Other constraints may arise from the organizational structures within which the design work is embedded. Design projects are often chartered with a priori commitments to follow a systematic decomposition process. This makes them easy to manage, but unlikely to succeed with respect to discovering the real requirements of users and clients. Schedules and resources are often assigned in ways that create ongoing conflicts between system designers and usability engineers. Usability engineers need to evaluate scenarios and prototypes at every stage of system development, but if schedules and resources do not provide for this, this work can conflict with software construction and refinement.

Constraints such as these can distract designers with ancillary factors so that they lose sight of what is essential in the design project, namely, the needs and concerns of users. The designer can become "unsituated" with respect to the real design situation, which is not the marketing manager's projections, or the instructional designer's list of steps, or the software engineer's system decomposition. The real design situation is the situation that will be experienced by the user, and designers need to stay focused on that.

Scenarios can be made even more effective as work-oriented design objects when users are directly involved in creating them.

Ackoff (1979) argued that the indeterminacy of design situations makes it imperative that *all* stakeholders participate directly. Scenarios support a simple and natural process of participatory design, where prospective users begin by enacting or relating episodes of current activities, then work iteratively with designers to transform and enrich these scenarios with the opportunities provided by new technologies (Carroll, Chin, Rosson, & Neale, 2000; Chin, Rosson, & Carroll, 1997).

Scenarios Are Evocative, Raising Questions at Many Levels

There is a fundamental tension between thinking and doing: Thinking impedes progress in doing, and doing obstructs thinking. Sometimes this conflict is quite sharp, as when one must stop and think before taking another step. But frequently it is more a matter of trading off priorities. Designers are intelligent people performing complex and open-ended tasks. They want to reflect on their activities, and they routinely do reflect on their activities. However, people take pride not only in what they know and learn, but in what they can do and in what they actually produce.

Donald Schön (1983, 1987) discusses this conflict extensively in his books on reflective practice. For example, he analyzes a coach reacting to an architecture student's design concept for a school building, which included a spiral ramp intended to maintain openness while breaking up lines of sight (the student calls the idea "a Guggenheim"): "when I visited open schools, the one thing they complained about was the warehouse quality—of being able to see for miles. It [the ramp] would visually and acoustically break up the volume" (Schön, 1987, p. 129).

In the episode analyzed by Schön, the coach feels that the student needs to explore and develop her design concept more thoroughly, noting that a ramp has thickness and that this will limit her plans to use the space underneath the ramp; he urges her to draw sections. However, he does not justify this advice; as Schön put it, he does not reveal "the meanings underlying his questions" (Schön, 1987, p. 132). Schön regarded this as a hopeless confrontation in which no progress can be made on the particular design project or on the larger project of understanding how to design. Both the student and the coach are willing to act publicly and to share actions, but they do not reflect enough on their own and one another's values and objectives and on their interpersonal dynamics.

Reflection is not always comfortable; it forces one to consider one's own competence, to open oneself to the possibility of being wrong. Nonetheless, designers create many opportunities for reflection, for example organizing design review meetings, or building prototypes for formative evaluation. Such activities promote identification and integration of different perspectives; they raise concrete and detailed design issues to guide further work. In this way, they help designers to reflect on the work they have already done. But they do not evoke reflection *in the context of doing design*. Design reviews and formative evaluations are ancillary activities that must be coordinated with design itself.

Scenarios help designers to reflect about their ideas in the context of doing design. The narrative is written to evoke an image of people doing things, pursuing goals, using technology in support of these goals. The story enables readers to empathize with the people in the situation, which in turn leads to questions about motivations, intentions, reactions, and satisfaction. For example, in the middle scenario from Table 53.1, is it valuable to Sharon to opportunistically encounter friends on her way to the club meeting? What effect does her recognition of the town's layout have on her success or experience at navigating the online environment?

Scenarios promote reflection and analysis in part because the human mind is adept at overloading meaning in narrative structures, both in generation and interpretation, as illustrated by the remarkable examples of dreams (Freud, 1900), myths (Levi-Strauss, 1967), and folktales (Propp, 1957). It is well-known that when people communicate, they rely on the *given-new contract* (Haviland & Clark, 1974): They assume or allude to relevant background information then present what is novel. This normative structure eases both the generation and interpretation of narratives.

Schön (1983) described design as a "conversation" with a situation comprised of many interdependent elements. The designer makes moves and then "listens" to the design situation to understand their consequences: "In the designer's conversation with the materials of his design, he can never make a move which has only the effects intended for it. His materials are continually talking back to him, causing him to apprehend unanticipated problems and potentials" (p. 101). When a move produces unexpected consequences, and particularly when it produces undesirable consequences, the designer articulates "the theory implicit in the move, criticizes it, restructures it, and tests the new theory by inventing a move consistent with it" (p. 155).

Scenarios often include implicit information about design consequences. Returning to the scenarios in Table 53.1, the archived forum in the first scenario allows Sharon to arrive at a discussion "at any time," find out what has been said, and make a contribution. In the second scenario, it is Sharon's navigation "down Main Street" that sets up her casual encounters with other community residents. At the same time, these features have less positive consequences, for example, the need to browse and read each of the comments in turn, or the requirement that she "walk" to get to the meeting place. These trade-offs are important to the scenarios, but often it is enough to imply them (this is an aspect of the roughness property discussed earlier).

There are times, however, when it is useful to reflect more systematically on these relationships, to make them explicit. In another situation Sharon may find the archived discussion too long or disorganized to browse, or she may be distracted by friends on Main Street and never make it to the club meeting. These alternative scenarios present a failure with respect to the starting goal. To understand, address, and track the variety of both desirable and undesirable consequences of the original annotation design move, the designer might want to make explicit the relevant causal relationships in a scenario. Doing so provides yet another view of the envisioned situation (see Table 53.3).

TABLE 53.3. Design Features of the Scenarios Presented in Table 53.1

A. *Science Fiction Club on a Web forum*
Accessing an online meeting through an Web discussion forum
+ enables convenient browsing of an entire discussion at many points in time or place
− but when a discussion becomes long or complex, it may be difficult to browse or understand

B. *Science Fiction Club in a Community* MOO
Accessing an online meeting room by "walking" through a spatial model of the town
+ allows fortuitous social encounters while moving around the MOO
+ evokes application and development of real-world spatial knowledge about the town
− but navigation to nonimmediate sites may be tedious or awkward
− but the overall spatial model may be poorly evoked by step-by-step navigation

C. *Science Fiction Club in a Collaborative Virtual Environment*
Accessing an online meeting by clicking on a spatial map of the town
+ simplifies navigation through direct pointing in an interactive map
+ suggests a town-oriented spatial context for the meeting within in the community
   but a realistic map of the town may be perceptually complex
− but a data-rich map may require a long time to download or update during use

Note. Features are expanded to consider possible positive and negative consequences for users in the source scenario or closely related alternatives. Each analyzed feature with its consequences is called a *claim*.

Scenarios and analyses such as shown in Table 53.3 can help designers move more deliberately toward specific consequences. For example, it might be decided to provide discussion-summarization support in an online forum, so that the convenience of the forum interaction is obtained, but so that it also scales well to complex discussions. Alternatively, the opportunistic encounters with other residents might be considered desirable enough that efforts would be made to provide "presence" information in the Web forum. As each elaboration is envisioned and proposed, it, too, is explored for further consequences and interdependencies.

Scenarios of use are multifarious design objects; they can describe designs at multiple levels of detail and with respect to multiple perspectives. In this way, they can help designers reflect on several aspects of a problem situation simultaneously. The scenarios in Table 53.1 provide a high-level task view, but can also be elaborated to convey the moment-to-moment thoughts and experiences of the actors to provide a more detailed cognitive view or in terms of moment-to-moment actions to provide a more detailed functional view. Or they might be elaborated in terms of the hardware or software components needed for implementing the envisioned functionality (Rosson & Carroll, 1995; Wirfs-Brock, 1995). Each of these variations in resolution and perspective is a permutation of a single underlying scenario. The permutations are integrated through their roles as complementary views of the same design object.

Using scenarios in this way makes them a more powerful design representation. They allow the designer the flexibility to

develop and analyze key usage scenarios in great detail, for example, to describe core application functionality while merely sketching less critical scenarios. At the same time, designers are able to switch among multiple perspectives, for example, directly integrating usability views with software views. Such a flexible and integrative design object can help designers manage the many interdependent consequences implied by design moves (Rosson & Carroll, 2000).

A FRAMEWORK FOR SCENARIO-BASED DESIGN

The concrete and work-oriented nature of scenarios make them an effective representation for human-centered design activities, particularly when these activities include participation by end users or other stakeholders (Carroll et al., 2000; Chin et al., 1997; Karat, 1995; Karat & Bennett, 1991; Muller, 1992; Muller et al., 1995; Rosson & Carroll, 2002). Scenarios can be quickly developed, shared, and revised; they are easily enriched with sketches, storyboards, or other mock-ups (Erickson, 1995; Kyng, 1995). A scenario of use can be directed at many concerns in system development, including documentation design and object-oriented software design (Carroll, 1995; Carroll, 2000). Given these many virtues, it is no surprise that scenarios are pervasive in software design and development (Rosson, Maass, & Kellogg, 1989; Weidenhaupt, Pohl, Jarke, & Haumer, 1998). But here we expand this generally accepted view of scenarios as a user-centered design representation. We offer a programmatic framework for employing scenarios of use in interactive system design (Carroll, 2000; Rosson & Carroll, 2002).

The framework summarized in this section incorporates scenario-based analysis and design into all phases of system development, from requirements analysis through usability evaluation and iterative development. We exploit the general advantages of scenario-based design described in the previous section but at the same time show how to make the impacts of scenario-based reasoning comprehensive and systematic. The overall process is one of usability engineering, where the scenarios support continual assessment and elaboration of the system's usefulness, ease of use, and user satisfaction. The aim is to develop a rich understanding of current activities and work practices and to use this understanding as a basis for activity transformation.

Figure 53.1 provides an overview of the scenario-based design (SBD) framework. We assume that system development begins with an initial vision or charter, even though this may be quite sketchy or nonbinding. This vision motivates a period of intense analysis during which the current situation is examined for problems and opportunities that might be addressed by available technologies. The analysts' understanding of the current situation is communicated in problem scenarios and claims. Problem scenarios describe prototypical stakeholders engaged in meaningful activities; the claims enumerate features of the current situation that are understood to have important consequences—both positive and negative—for the scenario actors.

The problem scenarios are transformed and elaborated through several phases of iterative design. Design envisioning

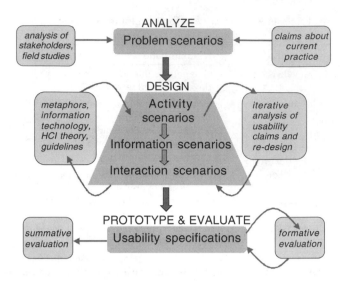

FIGURE 53.1. An overview of the scenario-based design framework. Scenarios serve as a central representation throughout the development cycle, first describing the goals and concerns of current use, and then being successively transformed and refined through an iterative design and evaluation process (from Rosson & Carroll, 2002). HCI = human–computer interaction.

is inspired by metaphors and technology options but at the same time is constrained by the designers' knowledge of interactive system design. Each set of scenarios is complemented by claims that analyze the possible positive and negative consequences of key design features. Claims analysis leads designers to reflect on the usage implications of their design ideas while the ideas are being developed.

SBD is guided by usability evaluation throughout development. Each narrative serves as a test case for analytic evaluation; each claim hypothesizes usability outcomes for one or more test cases. Design scenarios are also evaluated more directly through empirical usability studies. In these the claims analysis structures a mediated evaluation process, wherein the hypothesized usage impacts are operationalized and tested explicitly (Scriven, 1967). The empirical findings are interpreted with respect to the ongoing claims analysis, refining or redirecting the design efforts. We turn now to a brief example illustrating the key elements of the framework.

Requirements Analysis

A challenge for any software development project is identifying the complete and correct set of requirements (Brooks, 1995). Many system requirements are functional, addressed by the actual services and information provided by the final system. Other requirements are nonfunctional, for example, the measured quality of the software implementation or user interactions or pragmatic features of the system development process such as schedule, cost, or delivery platform (Rosson & Carroll, 2000; Sommerville, 1992; Sutcliffe & Minocha, 1998). In SBD,

TABLE 53.4. A Root Concept for Developing Online Activities for a Science Fiction Club

Component	Contributions to the Root Concept
High-level vision	Club members interact anytime, anywhere; develop shared resources
Basic rationale	Network-based interaction overcomes barriers of place and time
	Digital media are convenient to archive, organize, and retrieve over time
Stakeholder group:	
Club officer	Convenient scheduling and posting of shared events and information
Club member	Ongoing access to club activities, persistent recognition of contributions
Prospective member	Self-paced exploration of club vision, history, and membership
Starting assumptions	Open-ended participatory design process
	Members have pervasive access to personal computers and network connections
	Community computing development accomplished via volunteer efforts

we express an initial analysis of requirements as a *root concept* (Table 53.4). The root concept enumerates key aspects of the team's starting vision; it is used to guide further analysis and elaboration of system requirements.

Table 53.4 contains a root concept for the science fiction club example that we use to illustrate the framework. The starting vision and rationale in this case are quite straightforward: There are obvious advantages to meeting with associates in person, but the constraint of same-time, same-place can limit frequency and length of such meetings. Moving some of the club's activities online increases interaction opportunity and flexibility. A side-effect is that digital interactions can be stored and used for other purposes.

The root concept also documents the team's shared beliefs about the project's major stakeholders. A stakeholder is any person or organization who will be affected (either positively or negatively) by the system (Checkland, 1981; Muller, 1992). It is important to take a broad view of stakeholders, particularly early in requirements analysis, so that appropriate individuals and groups will be represented in analysis activities. In the example, we consider several different types of club members because they will have distinct goals with respect to system use—an officer who might find an online system convenient for scheduling and posting information; a "regular" club member who will now have more options for participating, and a prospective member who can learn about the club and its activities in a customized, self-directed fashion.

Although the emphasis of SBD is on analyzing and developing usable functionality, there may be a range of nonfunctional concerns that will constrain development. These are documented as starting assumptions in the root concept. For our example project, we assume that the design of the online club software will involve considerable participation by stakeholders, that club members already have the computing resources needed to access the system, and that use and maintenance of

the final system will take place through members' volunteer efforts.

The root concept lays the groundwork for analyzing the club's current activities. This might involve fieldwork, for example, a visit to a meeting where notes, photographs, or videotaped records are made. It may be more modest, perhaps a survey of current members aimed at eliciting descriptions of activities, or a series of semistructured interviews with club officers, members, and prospective members. A rich source of requirements are activity artifacts—for instance, a club newsletter or calendar, reports, or other shared products created by the group. Such artifacts are an excellent source of implicit information about stakeholders' values and activities (Bødker, 1991; Norman, 1988; Rosson & Carroll, 2001).

Field studies of current practices generate rich and diverse data about needs and opportunities. To direct these data productively toward design efforts, a more abstract representation of themes and relationships is required. An affinity diagram (group members post and organize individual observations; Beyer & Holtzblatt, 1998) is helpful in discovering these themes. Other useful techniques include diagrams of the stakeholder relationships, hierarchical task analysis of central activities, and summaries of how collected artifacts support group activities (see Rosson & Carroll, 2002; chapter 2).

In SBD, a key result of requirements analysis is a set of problem scenarios and claims. A problem scenario is a narrative of current practice that synthesizes actors, themes, relationships, and artifacts discovered in the fieldwork. These scenarios are not design-neutral, however. Even during early analysis activities, we assume that the team has a vision of how technology might enhance current practice. The fieldwork and related analyses will inevitably be colored by this vision (Carroll, Karat, Alpert, van Deusen, & Rosson, 1994). If the team fails to establish a vision, or creates inconsistent or contradictory visions, this, too, will influence requirements analysis, but in a less positive fashion.

An effective technique for generating problem scenarios is to first describe a set of hypothetical stakeholders—individuals who will represent the different sorts of people studied during the fieldwork. It is important to create a rich but realistic view of these individuals, because they form the basis for describing and later transforming current activities and experiences (the technique is similar to the *persona* concept in Cooper, 1999). Our examples focus on the experiences of Sharon, a busy third-year student at a large state university; one of Sharon's interests is science fiction, and she pursues this interest through reading and discussion with friends in a science fiction club.

The scenario in Table 53.5 conveys some aspects of the club's current practice; it enacts a typical activity and simultaneously communicates issues uncovered during fieldwork. Problem scenarios like this may be based directly on an observed episode, or they may be entirely synthetic. The goal is to introduce and contextualize the themes and relationships that will guide subsequent design work. This particular story combines our concept of a typical student club member with issues related to real-world meetings—for example, the need to arrive in a particular place at a particular time, the protocol for being greeted, for entering into a conversation, for proposing new topics, and so on.

TABLE 53.5. A Problem Scenario Describing Sharon's Visit to the Science Fiction Club Meeting

Sharon Joins an Ongoing Science Fiction Club Discussion

Sharon is a busy third-year psychology student at Virginia Tech. Even though she has a biology exam tomorrow morning, she has been looking forward to her science fiction club meeting for several days, so she decides to go and stay up late to study when she gets back. She remembers that they were planning to talk about Asimov's *Robots and Empire*, and she has a new theory about the timeline for first detection of the Zeroth Law.

The meeting is scheduled for 7 p.m. at their usual room in the town library, but she is late getting back from dinner with her roommate, so she misses her regular bus and arrives 15 minutes late. The meeting is already underway; she notes that they have a relatively small group tonight but is happy to see Bill and Sara, who are the real experts on Asimov. She is even more delighted to see that these two are already having a heated discussion about the Zeroth Law. But she is cannot immediately tell what points have been made, so she sits back a while to catch the drift of the conversation. At a break, Bill greets her and asks her what she thinks about Faucian's insight. She replies that she isn't sure about how central he is to the plot, but that she has a new theory about the timeline. They promise to hear her proposal in a few minutes then resume the argument.

The themes and relationships implicit in a scenario can be made more explicit and open for discussion by analyzing them in claims (Table 53.6). Problem claims are analyzed by identifying features of the scenario that have notable consequences for the actors' experience. This is an instance of analytic evaluation and as such is clearly guided by the expectations and biases of the evaluator. A more systematic evaluation can be obtained by asking questions of the scenario that are guided by cognitive or motivational theories of human behavior (Carroll & Rosson, 1992; Polson, Lewis, Rieman, & Wharton, 1992). The first claim captures a key aspect of Sharon's experience in the scenario—meeting in person with other club members creates a rich

TABLE 53.6. Two Claims Analyzed From the Club Meeting Problem Scenario

Face-to-Face Interaction With Club Members at a Meeting
  + ensures that both nonverbal and verbal communication contribute to the interaction
  + leverages many years of experience with communication protocols and conventions
  − but may introduce distracting or irrelevant personal information about partners
  − but inhibits parallel communication activities (i.e., among multiple parties at once)

A *Regular Physical Space Used for Club Meetings*
  + promotes a feeling of familiarity and intimacy among established members
  + simplifies the planning and execution process for arriving at meetings
  − but requires members to travel to the site for interaction
  − but physical locations are valuable resources that must be shared among organizations

Note. The feature of interest appears in the shaded area; hypothesized positive consequences are prefaced with a plus sign and negative consequences with a minus sign.

communication bandwidth; both verbal and nonverbal content can be shared, which helps Sharon know when to jump in with her new topic.

Once an activity feature has been associated with a claim, the analysis is extended through "what if" reasoning that explores other possible positive or negative consequences. For instance, what if Sharon arrives in a T-shirt with a distracting political message? What if the argument goes on for so long that Sharon is never able to raise her topic? Scenarios both evoke and support this sort of questioning; the concrete setting of the scenario invites reasoning about partially overlapping alternatives and outcomes.

An important characteristic of claims analysis is that it considers both positive and negative consequences. During requirements analysis, there is a tendency to focus on the difficulties or concerns of current practice, as apparent in activity breakdowns or contradictions (Bødker, 1991; Kuutti, 1995; Kyng, 1995; Nardi, 1996). In contrast, designers tend to focus on the likely benefits of proposed features. Claims analysis imposes a balanced view of both problems and opportunities. From the perspective of requirements analysis, it ensures that we build on aspects of the current situation that are already working well. From the perspective of design envisionment, it forces us to consider the side effects or other undesired impacts of changes to a situation.

Problem scenarios and associated claims are the central product of requirements analysis in SBD. Note though that they are *not* requirements in the traditional sense of the term—that is, they are not a specification of features required of a system. Instead they serve as requirement criteria; design solutions are expected to address the positive and negative consequences conveyed by the scenarios and claims. For instance, we will "require" that the online club environment reinforce the advantages of in-person meetings as much as possible but at the same time address the disadvantages. This is quite different from specifying that it will have a synchronous communication channel. Individual features of the solution will be identified, elaborated, evaluated, revised, or discarded in an extended iterative process.

Activity Design

Requirements emerge and are refined throughout system development (Brooks, 1995), but at some point a team understands enough about project stakeholders and their needs that they are ready to make specific design proposals. Indeed some projects may be so overdetermined that system functions are specified in advance, and requirements analysis consist simply of analyzing user characteristics and preferences. In SBD, the initial step toward specifying a design solution is made by envisioning how current activities might be enhanced or even completely transformed by available technologies. We deliberately minimize attention to the concrete steps of user interaction at this point, emphasizing the basic goals and motivations of the new activities (see also Constantine & Lockwood, 1999).

SBD is *activity*-oriented—we analyze current practice at the level of meaningful activities and build from this to new activities

TABLE 53.7. Using Metaphors and Available Information Technology to Reason About Activities

Activity Design Features Suggested by Metaphors for an Online Meeting

Reading at the library	Self-paced, individual access to structured information
Hearing a lecture	Large audience; prepared materials; one-way communication
Visiting a museum	Array of artifacts, small groups or individuals examine, discuss
Going to a cocktail party	Friends forming subgroups; social exchange and mingling

Activity Design Features Suggested by Information Technology for an Online Meeting

A hierarchy of Web pages	Mix of text and graphics, category names and links
An e-mail distribution list (listserv)	One-way "push" communication, possibly large audience
A shared whiteboard	Informal sketches
Meeting groupware	Explicit agenda, support for floor control, meeting records

(Kuutti & Arvonen, 1992). A danger in this is that the designers will focus too much on how goals are pursued in the current situation and on understanding and responding to people's current expectations about their tasks and about technology. To encourage consideration of new options and insights, we deliberately expand the "design space" before envisioning the new activities. By design space, we mean the array of possible concepts and technologies that might be relevant to the problem domain, along with some analysis of what these options might bring to the design solution (MacLean, Young, & Moran, 1989; Moran & Carroll, 1996).

Table 53.7 exemplifies two techniques useful in exploring design alternatives. The upper part of the table shows how different conceptual metaphors evoke contrasting views of stakeholder activities. Metaphors are often used deliberately in user interface design, with the hope that users will recruit them in reasoning by analogy about what a system does or how it works (Carroll, Mack, & Kellogg, 1988; Carroll & Thomas, 1982; Madsen, 1994). Here we emphasize the role of metaphors in helping designers "think outside the box" (Erickson, 1990; Verplank, 1988). Addressing real-world activities and concerns is crucial to effective system design, but it is often metaphoric thinking that promotes the insights of truly creative design.

An analysis of available information technology provides a complement to the metaphoric thinking. In a sense, the technology provides another set of metaphors for thinking about the activities to be supported, but in this case the analogy is with classes of software and devices that already exist (e.g., Web information systems, e-mail or database packages). At the same time, a technology-oriented analysis such as this directs the design back to many of the pragmatic concerns of software development, by enumerating possible technology and how it might contribute to the solution. This analysis may also be influenced by the project's starting assumptions (Table 53.3), for instance if the development organization already has developed a shared whiteboard or has considerable experience with Web information systems.

The exploration of metaphors and technology does not generate a new activity design. Rather, it provides a set of lenses for discussion. The team might consider what it would be like if the online science fiction club were designed to be like a cocktail party versus a lecture; they can argue about the relative advantages of using a structured groupware framework versus an open-ended Web site. These divergent discussions form a backdrop for the convergent process of scenario writing.

The generation of activity scenarios is a creative process influenced by many factors. The problem scenario provides a starting point—a realistic context and set of goals for meetings among club members. The associated claims motivate some basic design moves—the general heuristic is to maintain or even enhance the positive consequences for the actors while minimizing or removing the negatives (Carroll & Rosson, 1992). The metaphors and technology exploration provide solution ideas—how our concept of meetings or discussions might be transformed and how existing technology might support these activities. Of course, the designers' knowledge of HCI, and of interactive system design broadly speaking, also provides important guidance—for example, knowing the relative affordances of different computer-mediated communication channels, understanding the motivational challenges of discretionary-use software, and so on.

Two contrasting scenarios for Sharon's interaction with her club interaction are in Table 53.8. Both activities address the goals of the actors in the problem scenario—joining a club discussion and introducing a new topic of personal interest. Both respond to claims analyzed for the problem scenario, for example attempting to maintain the familiarity and intimacy of a real-world meeting in a club, while making participation more flexible and enabling parallel conversations. However, they address these goals and concerns with rather different views of what constitutes an online discussion and what network-based technology might be used to support it.

The first example was influenced by the lecture and library metaphors. Contributions to the online club materials are automatically distributed to group members; this is analogous to sitting in a room and listening to a lecture. The online material is organized into topical categories that can be browsed in a self-paced fashion, just as Sharon might browse stacks of books in a library. These metaphors are easily supported by a combination of e-mail and Web pages.

The second scenario shows an influence of the cocktail party, museum, and library metaphors. It emphasizes social exchange and informal conversation, as well as responses to an assortment of club-specific artifacts in the space. Sharon is able to see which of her fellow members are around and can follow the conversation but also carry out her own exploration in parallel. She jumps in and out of the conversation as her interest in the topic increases or decreases. The club members are engaged in activities that refer to artifacts on display in their room—discussion topics and a bookshelf that displays book titles categorized by theme.

Both of these scenarios have attractive consequences for Sharon and her friends. Both are possible solution approaches, so how do the designers choose? Again, many pragmatic factors contribute to this decision—the kinds of software currently

TABLE 53.8. Two Alternative Activity Designs for the Online Club Meeting

A. *Sharon Visits the Science Fiction Club Online Forum*

Sharon is a third-year psychology student at Virginia Tech, and after 3 years she has learned to take advantage of her free time between classes. In the hour between her morning classes, she stops by the computer lab to visit the science fiction club, because she heard from a friend that they are discussing her favorite book, Asimov's *Robots and Empire*, and she wants to share her new theory about the timeline for the Zeroth Law.

When she logs on to a computer, she first checks her e-mail and sees that as she hoped there are several e-mails from club members proposing and responding to views on this book. Rather than read each e-mail, she follows the convenient link to the club's Web site, which takes her right to the ongoing discussion. As always, the reviews are first, then the discussion topics, where she finds the new discussion thread started by Sara and Bill. She reads the new thread before adding her theory about the Zeroth Law and notes that Bill is also fascinated by this piece of the story. She summarizes her theory, and because she wants the group to focus on this issue, she makes it a first-level topic but links it to Bill's post to acknowledge the relation. When she submits it, she is reminded that an e-mail has been sent to the club listserv with her contribution.

Before leaving, Sharon backs up to the homepage and browses the book categories to look for new books and discussions. Underneath her favorite category of "Artificial Intelligence" (where the Asimov series is placed), she discovers an intriguing new entry, "Brain Evolution." She doesn't recognize any of the authors in this category, so sends herself a reminder to track down a couple of books from the category later that day.

B. *Sharon Goes to the Science Fiction Club's Online Room*

Sharon is a busy third-year psychology student at Virginia Tech. After 3 years, she has learned to take advantage of free time between classes. In the hour between her morning classes, she stops by the computer lab to visit the science fiction club because she heard from her friend that they are discussing her favorite book, Asimov's *Robots and Empire*, and she wants to share her new theory about the timeline for the Zeroth Law.

When she logs onto a computer, she first checks her e-mail and sees that as she hoped there are several e-mails from club members proposing and responding to views on this book. Rather than read each e-mail, she follows the convenient link to the club's online room. She is taken to their regular discussion spot, the bar of a local pub. As she arrives, she sees that Sara, Bill, and Jennifer are already there. She reviews their conversation, and notes that they are discussing Jennifer's new review of Asimov's *Robots and Empire*. Before she joins in, she quickly opens and browses Jennifer's review. She agrees with Jennifer, so she eagerly jumps in to take her side against Bill and Sara. In a few minutes, the chat moves on to plan a group outing that night. She has to study, so she drops out of the conversation to create a new discussion with her theory about the Zeroth Law. She sees that an announcement is sent to all the club members when she has finished creating the object.

Sara keeps an eye on the others' conversation, and when there is a break, she invites them to visit her new topic. They discuss the Zeroth Law for a while, but leave it open for others to visit. On her way out, Bill tells her he has a new "Brain Evolution" grouping he is working on. She hasn't heard of the titles he mentions, so she sends herself a reminder to track down a couple of books from the category later that day.

in use, the team's design expertise, development resources, organizational priorities, and so on. But assuming that both solutions are genuine possibilities, the designers must also evaluate them with respect to their usage implications. One way to do this is with participatory design sessions (Carroll et al., 2000; Chin et al., 1997; Kyng, 1995; Muller, 1992) that focus on how well the alternatives suit stakeholders' needs. In SBD, we also evaluate scenarios through claims analysis, where the positive and negative implications of design features are considered through "what if" discussions.

A sample claim analyzed from each scenario appears in Table 53.9. In this illustrative example, the scenarios were intentionally written to be very similar in many respects; the claims capture one of the basic design contrasts built into the alternative designs. The Web site offers a convenient hierarchical listing of topics, whereas the online room holds a number of different "objects" that people discuss in real time. The analysis helps the designers to see the relative advantages and disadvantages of an organized asynchronous interaction and a more ad hoc synchronous exchange. Such an analysis may or may not be enough to mandate one alternative over the other. But at the least it begins to lay out usage issues that will be the topic of continuing design.

Much progress can be made even at this early level of envisioning activities. The narratives in Table 53.8 are concrete and evocative; designers or their clients can readily understand what is being proposed and begin to consider the relative pros and cons of the design ideas. Yet the scenarios are "just talk"; indeed, if they are shared and discussed over an interactive medium,

they can easily be extended or revised as part of a real-time design review and discussion.

Information Design

As design continues, tentative decisions are made about the design direction. For this continuing example, we will elaborate the second of the two activity scenarios in Table 53.9. We have opted for the "online room" concept because it strongly

TABLE 53.9. Activity Claims That Help to Contrast the Implications of Competing Scenarios

Discussion Archives Organized by Topic and Content Submitters
+ leverages people's familiarity with categorical hierarchies
+ emphasizes the central and permanent recognition of individuals' contributions to the archive
− but browsing extensive stored archives may be tedious or complex
− but people may be disinclined to contribute more transient and informal content to an organized archive

Real-Time Conversation Organized by the People Present in a Space
+ leverages people's familiarity with real world conversational strategies
+ encourages a combination of topic-specific and ad hoc informal exchange
− but requires that conversation participants be present at the same time
− but newcomers may find it hard to interrupt an ongoing conversation

reinforces important strengths of current real-world practice (e.g., familiarity, intimacy, well-learned conversation skills; see Table 53.6), while addressing some of the problems identified in requirements analysis (e.g., the need to travel to a specific place, the possible distraction of irrelevant personal information). We now begin to elaborate the underlying activity with information and interaction details—the user interface.

As for activity design, we first explore the design space with metaphors and technology options. The metaphors applied at to information design may even overlap with those used in activity design but with the emphasis shifted to what the users will see and understand about the system; the concerns of this phase are similar to those in the "gulf of evaluation" discussed by Norman (1986). For instance, the three metaphors of library, museum, and cocktail party suggest these information design ideas:

- *Library.* Documents look like pages in books with title-bearing covers. Objects are arranged in alphabetical or category order on shelves. There are desks and chairs for browsing and note taking.
- *Museum.* The space is broken into relatively small topic-specific rooms. Objects of interest are mounted on the walls. There is space around each object enabling a group of interested parties to form. Descriptive titles and text are attached to each display object.
- *Cocktail party.* There are a number of attractive "seating areas," perhaps including a table and chairs. Visitors are organized into groups and emit a "buzz" of conversation. New arrivals are greeted with waves or smiles.

Technology options explored at this phase might include hyperlinks (icons or other controls are used to navigate or access content objects), MUDs (multiuser domains based on a spatial model), other graphic techniques for rendering a room (a static photograph, a panorama, a three-dimensional model), as well as techniques for representing the people and objects in the space (avatars, buddy lists, texture-mapped forms, and objects).

Ideas such as these are discussed and combined with the design team's experience in information design. Information design possibilities are "tried out" in the activity design scenarios, with attention directed toward claims analyzed in earlier phases. Does the information design further promote intimacy among club members? Will newcomers feel welcome and will they be able to "catch up" and participate? Does the design allow for parallel activity? Design inquiry such as this—and the scenarios that provide a real-world context for the reflective process—is a hallmark of SBD.

Two alternative information designs are presented in Table 53.10. For simplicity just the central actions making up the discussion have been elaborated. Both scenarios offer a view into a virtual room that contains club members and documents. Both assume an information model that is spatial, applying the concepts of rooms, furniture, and so on as a pervasive metaphor. Both also borrow from the other more specific metaphors described above—for example, there is a bookshelf that organizes reviews and discussions, there are notices and other artifacts posted on the walls, the participants form groupings and are in conversation. The technology supporting each scenario is quite different, however; in one case, it consists of a traditional text-based MUD, and in the second, it provides a graphic rendition of the underlying spatial model.

Either of the information designs could be used to represent an online club space, but the two proposals have rather different implications for how club members are likely to experience the space. In the first case, all attention will be directed at a sometimes complex stream of descriptive text. The experience is rather like reading a book or a play, with different people and objects providing the content, but much of the mental experience under the control of the reader. In contrast, the graphic view offers a concrete rendition of the space, and the attention of the participant is instead directed toward a specific activity, in this case a shared discussion of a new review.

These general implications are captured in the claims presented in Table 53.11. As for the earlier design claims, these arguments do not mandate one choice over another; rather, they provoke discussion of each alternative's pros and cons. In this illustrative example, we elect to pursue the graphic view rather than the text-based view, largely because it simplifies the comprehension and participation process. We note as an important negative consequence, however, that the "real-world" view of the pub architecture and contents may dampen the creativity or fantasy of members' contributions. This may become an issue as design continues; for example, we may search for ways to suggest a hybrid approach, inviting both real-world and fantasy content (Cherny, 1995).

Information design comprises all aspects of how the task information is organized and rendered during users' activities. In most design projects, this will include paying special attention to the needs of new or inexperienced users. For instance, suppose that this was Sharon's first visit to the online club. How would she know that the bookshelf held recent reviews and discussions that she could open? In the graphic case, the visual cues provided, along with general experience in using graphic user interfaces, might be enough to cue the behavior described in the scenario. But would she know to use the "toolbox" to create a new discussion object? It is often useful to create "documentation design" versions of other design scenarios, where the hypothetical actor(s) are assumed to have little or no experience using the system. In SBD, the design of supporting documentation (help texts or other guidance) is also inspired by appropriate metaphors (e.g., a coach, a policeman) and technology (e.g., online tutorials, animated demonstrations). More detailed examples of scenario-based documentation design can be found in Rosson and Carroll (2002).

Interaction Design

User interaction design becomes even more detailed when the concrete exchanges between the user(s) and the system are specified. Of course, to some extent this has been foreshadowed by decisions already made—a graphic array of people and objects implies some way of indicating or selecting among them; a text-based chat record implies a mechanism for character or word input. We have also noted several cases of system

TABLE 53.10. Alternative Information Designs for the Online Club Meeting Scenario

A. *Sharon Goes to the Science Fiction Club's Room in the Community MOO*

Sharon's background and goal to share her Zeroth Law theory . . .

When she logs onto a computer, Sharon first checks her e-mail and sees several e-mails marked with a dot indicating that they are new; a quick read of the senders confirms that they are from club members proposing and responding to views on this book. She opens the first one, knowing that it will contain a convenient link to the club's online room. She is taken to their regular discussion spot, and she skims the familiar description of the bar. She grins to see a new seating option someone has added, a snailshell–toadstool combination, and seats herself at this spot; she is told that she is "reclining in a luxurious curl" and ready to join in the activities. She also notes a new exit leading to a "Fractal Immersion Room," and makes a mental note to visit later. Finally, the welcoming text stream concludes with its usual status report, informing her that Sara, Bill, and Jennifer are in the pub, that Jennifer has just added a new review to the bookshelf, and that this review is currently in the possession of Bill.

Text messages from her friends begin appearing, including a quick interleaved hello from Bill, before he comments on a point Jennifer made in her review. Sharon thinks Bill might be mistaken, but before joining in, she asks Bill if she can pick up the review so that she, too, can read Jennifer's comments. She finds the issue Bill is debating and sees that she agrees with Jennifer, so she eagerly jumps in to take her side against Bill and Sara. In a few minutes, the chat moves on to plan a group outing that night. She has to study, so she drops out of the conversation to create a new discussion with her theory about the Zeroth Law. After she issues the commands to instantiate the discussion and then types in a provocative starting premise, the system reports to the group that a new discussion has been added, and that an e-mail announcement has been sent to the club mailing list.

The discussion of Sharon's proposed new topic, her resolution to read the "Brain Evolution" titles . . .

B. *Sharon Goes to the Science Fiction Club's Room in the Collaborative Environment*

Sharon's background and goal to share her Zeroth Law theory . . .

When she logs on to a computer, she first checks her e-mail and sees several e-mails marked with a dot indicating that they are new; a quick read of the senders confirms that they are from club members proposing and responding to views on this book. She opens the first one, knowing that it will contain a convenient link to the club's online room. She is taken to their regular discussion spot, and she sees the familiar panoramic image of Eastenders Pub, with the mirror and bar prints on the wall, the wooden brass-trimmed bar, and the red canvas bar stools. Miniature images of her friends Bill, Sara, and Jennifer are also there, positioned in a close group at one end of the bar. On the club's special bookshelf, she sees all of the reviews and discussions contributed recently, organized as usual by name of author. As usual, the reviews appear as simple text documents, and discussions appear as indented lists. One review on a middle shelf is highlighted in yellow, telling her that this is new since she last visited. She guesses that this review may be what the others are discussing in the chat area, so she opens it to see what all the excitement is about. When she does, she can see that the other three also have the review open, in fact she can see from their named pointers exactly the passage that they are discussing. She finds that she agrees with Jennifer in this case, and eagerly jumps in to take her side in the argument.

In a few minutes, the chat moves on to plan a group outing that night. She has to study, so she drops out of the conversation to create a new discussion with her theory about the Zeroth Law. She uses the room toolbox to create a new discussion object, indicates that this concerns Asimov's *Robots and Empire*, and adds a provocative opening premise about the Zeroth Law. When finished, the discussion object is positioned automatically on the top shelf and given a bright yellow color. She is also provided feedback indicating that an announcement has been sent to the club mailing list.

The discussion of Sharon's proposed new topic, her resolution to read the "Brain Evolution" titles . . .

---

feedback, a critical element of any interaction design. But many details are still open, and the scenarios serve as a usage context for considering interaction options.

As in other aspects of design reasoning, metaphors can be used to inspire interaction design. For instance in thinking about how people navigate to the science fiction club space (i.e., the Eastenders Pub), we contrasted the metaphor of walking down a street and through a door, with that of using a map to point at and access a spatial location directly. In the first case, the metaphor emphasizes the three-dimensionality of the real world and opens up the possibility that other people or objects might be encountered "along the way." The second emphasizes a more two-dimensional view of the world, but leverages people's familiarity with maps, as well as relieving the possible tedium of step-based navigation. In this case, we decided that the convenience and familiarity of map-based access was more desirable than the suggestion of three-dimensionality in the underlying model. But we left as an open issue the possibility of adding other information to the map, which would enable and prompt opportunistic encounters and exchanges.

With respect to our continuing example, we have simplified it even more than in the discussion of information design, addressing a single question of user interaction design: How should club members access and interact with the review and discussion objects (and presumably any other objects that are available for use)? This is a central interaction that will be repeated many times, so it warrants special attention. The abbreviated scenarios in Table 53.12 show the influence of two familiar technologies for interaction with graphic objects—direct manipulation (double click to open) versus a less direct, command-oriented manipulation (select and apply a menu command).

The two scenarios convey some of the trade-offs in deciding whether to provide menu-based interaction with task objects. The support of direct manipulation builds on the pervasive use of such interaction techniques in modern interactive software. But the what-if reasoning applied to this scenario pointed to some usage situations that are not well-supported by this simple technique. What if Sharon did not know the others and did not want to join in on the discussion? What if she wanted to get a quick (private) sense of the issues were

TABLE 53.11. Claims Contrasting Usage Implications of the Alternative Information Designs

Textual Descriptions of People and Objects Present in an Online Space
+ focus participants' attention on a single source of information about the situation and events
+ leverages club members' experience and enjoyment with fantasy-producing textual imagery
− but interleaving many sorts of descriptions and communications may become quite complex
− but it may be impossible to integrate individual text-based fantasies into a coherent mental model

Two-Dimensional Visual Depictions of People and Objects Present in an Online Space
+ leverages club members' familiarity and habits with real-world places and objects
+ enables parallel processing of spatially distributed information
+ allows participants to use spatial cues in organizing activities (e.g., position in review, location of chat area)
− but club members may feel restricted by the constraints of a two-dimensional space
− but objects or people distributed in the space may distract from the activity in focus

before joining the group? A command-based interaction (i.e., through a customizable menu system) offers more possibilities for user control but at the cost of adding more steps to the detailed interaction (see Table 53.13). Developing and analyzing these alternatives helped us as designers to think through the relative benefits of ease of execution versus user control and flexibility.

Note that this concrete and specific interaction design detail has broad implications for what users are able to do with this system. Direct manipulation techniques are simple, familiar, and pervasive but can limit functionality: Different science club documents may implement different meanings of "open," but they will not offer any other options for interaction. In contrast, a menu system is flexible and extensible, even to the extent of admitting new kinds of function for club artifacts not yet invented. We recognized these issues by exploring this scenario and its variants; during claims analysis we considered scenario variants in which other object-specific functions could be useful, ultimately leading us to choose the more complex menu-based interaction technique. This example demonstrates how even small details can have important consequences for the activities being envisioned and supported; in SBD, the scenario context ensures a continuous focus on activities, even during detailed interaction design.

In the science fiction club example, we have considered and incorporated standard user interaction technology—the familiar WIMP paradigm of windows, icons, menus, and pointing. However, SBD can also be used to envision and analyze the implications of more novel user interaction paradigms and devices. For example, we might consider a role for intelligent agents as part of a new user scenario and contrast this to a scenario involving community-generated FAQs (frequently asked questions) repository. Or we could explore the implications of gesture or speech recognition in lieu of (or as a complement to) conventional keyboard and mouse input devices. A key advantage of exploring these ideas within a scenario context is that designers are less likely to be caught up in the new technologies for their own sake; the method leads them to try out their new ideas in usage situations that at the least are believable and that are analyzed explicitly with respect to usability consequences. More detailed examples of SBD activities focused on emerging interaction paradigms are discussed in Rosson and Carroll (2002).

TABLE 53.12. Alternative Interaction Designs for Opening the Review As Part of the Online Club Meeting Scenario

A. Sharon Goes to the Science Fiction Club's Room in the Collaborative Environment
Sharon's background and goal to share her Zeroth Law theory . . .
. . . Sharon arrives in the bar, sees that her friends are talking about a new review . . .
Sharon wants to open the review to see what all the excitement is about. She moves her mouse pointer to the bright yellow icon and clicks twice. A separate window titled "Jennifer's Review" opens to the side, covering the other icons on the bookshelf. Sharon is automatically positioned at the same location as Sara in the review text; she knows from experience that this means that Sara made the last comment in the chat area. This irritates her for a minute because she wanted to see what Bill was talking about, not Sara, but she quickly finds out where Bill is positioned via his colored rectangle in the scroll bar and moves to share his view.
Sharon participates in the argument, then creates her new discussion object . . .
The discussion of Sharon's proposed new topic, her resolution to read the "Brain Evolution" titles . . .

B. Sharon Goes to the Science Fiction Club's Room in the Collaborative Environment
Sharon's background and goal to share her Zeroth Law theory . . .
. . . Sharon arrives in the bar, sees that her friends are talking about a new review . . .
Sharon wants to open the review to see what all the excitement is about. She moves her mouse pointer to the bright yellow icon and clicks the right button to bring up the menu. She recognizes the list of review-specific action choices and selects the second item (join) rather than the first one (browse). A separate window opens to the side, covering the other icons on the bookshelf. Sharon is automatically positioned at the same location as Sara in the review text; she knows from experience that this means that Sara made the last comment in the chat area. This irritates her for a minute because she wanted to see what Bill was talking about, not Sara, but she quickly finds out where Bill is positioned via his colored rectangle in the scroll bar and moves to share his view.
Sharon participates in the argument, then creates her new discussion object . . .
The discussion of Sharon's proposed new topic, her resolution to read the "Brain Evolution" titles . . .

TABLE 53.13. Claims That Capture Some of the Trade-offs Associated With Alternate Interaction Techniques

Double-Clicking to Open the Object Represented by a Visual Icon
+ leverages users' general experience with graphic user interfaces
+ promotes a feeling of direct interaction with the object represented by the icon
− but the semantics of the double click may be hidden and vary as a function of the object that is opened

Selecting and Then Requesting a Menu to Open the Object Represented by a Visual Icon
+ makes explicit the command that is being addressed to the object represented by the icon
+ creates an opportunity to choose among multiple, object-appropriate actions
− but the selection-then-choice operation may seem tedious for frequent actions
− but the menu list of options must be perceived and interpreted and slow down the interaction
− but selecting and addressing a command creates a level of indirection (thus distance) in goal mapping

Usability Evaluation

In SBD, we assume that usability evaluation takes place early and throughout the design and development process. Any representation of a design can be evaluated. Figure 53.1 emphasizes a phase of evaluation that takes place after detailed user interaction scenarios have been developed. This is relatively formal usability testing—users are recruited to carry out representative tasks on early operational prototypes. Such evaluation does require sufficient progress on a design to enable construction of a prototype, although such prototypes may be built using special-purpose languages or tools or may even be low-fidelity prototypes constructed of cardboard and paper (Muller, 1992; Virzi, Sokolov, & Karis, 1996).

Early in design, user feedback may be obtained in rather informal settings, for example a participatory design session (Chin et al., 1997; Muller, 1992). Users can be included in discussion and envisionment of activity scenarios. The design ideas are also subjected constantly to analytic evaluation through claims analysis and other design review activities (e.g., usability inspections or cognitive walkthrough; Nielsen, 1995; Nielsen & Mack, 1994; Polson et al., 1992). All of these activities yield formative evaluation feedback that guides changes and expansion of the design vision.

SBD implements Scriven's (1967) concept of *mediated evaluation*. In mediated evaluation, empirical data are collected (Scriven calls this "pay-off" evaluation), but the materials and methods used in the empirical test are guided by prior analytic evaluation. The analytic evaluation may have many different components, for example, an expert-based inspection or perhaps a cognitive model constructed for a particularly complex or critical interaction sequence (Gray, John, & Atwood, 1992; Kieras, 1997). In SBD, the primary method for analytic evaluation is claims analysis. The claims written during scenario generation and discussion analyze the usability issues most likely to influence the system's success or failure; they are used as a skeleton for constructing and administering empirical usability

tests. One way to view a claims analysis is as a series of usability hypotheses that can be assessed empirically; claims also help to explain *why* a design feature has an observed impact on users' experiences.

Scenarios and their associated claims are combined to create *usability specifications* (Carroll & Rosson, 1985; Good, Spine, Whiteside, & George, 1986). A usability specification is a representative task context that has been analyzed into critical subtasks, with each subtask assigned target usability outcomes. In SBD, the design scenarios provide a realistic task context, and the associated claims provide an analysis of the scenario that is parameterized with expected or desired usability outcomes. When the prototype has sufficient functionality that it can be tested with representative users, the specified tasks are tested and the results compared with the target outcomes.

A sample usability specification developed from the science fiction club scenario appears in Table 53.14. The scenario serves as a usability specification in two ways. Early (or at any point) in design, representative users may be asked to simply explore a rough prototype while pursuing the open-ended goals stated in the task context. Because the actual experience of users would vary considerably in this case, it does not make sense to establish performance outcomes; indeed, these tests are likely to include instructions to think aloud, so as to provide as much feedback as possible about the user experience (Mack, Lewis, & Carroll, 1983). Nonetheless, the system is considered successful in these cases if it satisfies the goals of a user enacting the scenario; collecting general ease of use or satisfaction ratings at the end of each episode can provide a measure of this.

When the prototype is robust enough to measure subtask times, more detailed usability specifications guide empirical testing. In the example, a set of five simple subtasks has been analyzed from the user interaction scenario. These tasks are directly related to claims that have been developed for key design features (only some of these claims have been documented

TABLE 53.14. Usability Specification Developed From the Science Fiction Club Scenario

Task Context: Sharon is a busy university student and a regular member of the science fiction club. During a few free minutes, she sees from her e-mail that new discussions have begun at their online club room. She joins them, planning to share her new idea about her favorite Asimov novel when there is a break in the conversation.

Overall Scenario Outcome: Average rating of at least 4.0 (of 5) on *ease of use* and *satisfaction*

Subtasks	Performance Targets	Satisfaction Targets
Subtask 1: Navigate to the online club room	20 sec, 0 errors	4 on *convenience*
Subtask 2: Identify present club members	5 sec, 1 error	4.5 on *presence*
Subtask 3: Identify and open Jennifer's review	10 sec, 0 errors	4 on *directness*
Subtask 4: Locate and join Bill in review	15 sec, 1 error	0.5 on *confusion*
Subtask 5: Create new discussion object	60 sec, 1 error	4.5 on *feedback*

here). Performance measures are established, based either on the designers' own (expert) experiences with the prototype or on benchmark data collected from comparable systems. Satisfaction measures are constructed to assess one or more of the specific concerns raised in the claims. For example, a negative consequence of menu-based interaction is that it may reduce feelings of directness. The usability specification tracks this issue by requiring that users' perception of this quality be at an acceptably high level (as operationalized by a Likert-type rating scale with a range of 1–5). The satisfaction qualities specified for the other subtasks were similarly derived from advantages or disadvantages hypothesized by claims.

Usability specifications developed in this way have two important roles in evaluation. First, they provide concrete usability objectives that can be serve as a management tool in system development—if a product team accepts these targets, then the team's usability engineers are able to insist that redesign and improvement continue until they are met (Carroll & Rosson, 1985; Good et al., 1986).

Second, the specifications tie the results of empirical evaluation directly to the usability issues raised during design. For instance, our interaction design scenario specified that Sara determines Bill's position from his colored rectangle in the scroll bar (see Table 53.12). A positive consequence is that awareness of others and their activities is enhanced; a negative consequence is that the display becomes more crowded. The time it takes to locate Bill is specified as one measure of the feature's impact, but this performance target is complemented by users' subjective reactions to the feature. For example, users might indicate level of agreement to a statement such as: "The scroll bar with rectangles indicating others' positions was confusing." Problematic results with respect to either of these usability targets raise specific issues for redesign.

As development continues, more complete prototypes or initial working systems are constructed. At this point, new usage scenarios are introduced, with new actors, goals, and activity contexts. These scenarios are also subjected to claims analysis, and the resulting claims used to develop additional usability testing materials. This is an important step in the evaluation, because it ensures that the design has not been optimized for the set of design scenarios that has guided development thus far. Near the end of the development cycle (or at predefined milestones), a carefully orchestrated *summative evaluation* is carried out, with the goal of assessing how well the system performs with respect to its usability specifications.

Scenario-Based Iterative Design

The SBD framework described in this chapter is highly iterative. Although we emphasize the grounding of new activities in a thorough analysis of current practice, we assume that many new possibilities will not be realized until design ideas are made concrete and exposed to actual use. At times, entirely new activities may be envisioned, when breakdowns in the technical or social environment demand a radical transformation of current practice. Even in these cases, an SBD process is valuable—problem

scenarios help designers predict how stakeholders will need to evolve their goals and expectations if they are to learn and adopt the new activities (Orlikowski, 1992).

More commonly, the development process will involve a more gradual coevolution of computing technology and the users' activities, where the technology raises new possibilities for action, and people's creative use of the technology in turn creates new requirements for support (Carroll, Kellogg, & Rosson, 1991). In these cases, user interaction scenarios provide a central activity-centered thread in iterative design, serving as intermediate design products that can be generated, shared, and revised as part of an overarching envisionment-evaluation-refinement cycle.

Scenarios Throughout the System Life Cycle

The SBD framework is aimed at the iterative development of activities that people may pursue with computing support. Our example has focused on the central processes of requirements analysis, design, and usability evaluation, but one of the great strengths of scenario-based methods is that they support of a diverse range of system development goals (Carroll, 1997; see Fig. 53.2). Product planners present day-in-the-life scenarios to managers as design visions (Dubberly & Mitsch, 1992); requirements engineers gather workplace scenarios through direct observation and interviews and analyze scenarios as primary data (Antón, McCracken, & Potts, 1994; Holbrook, 1990; Hsia et al., 1994; Kaindl, 1997; Kuutti, 1995; Potts, 1995).

Even if scenarios are not developed and transformed as described in the SBD framework, they may be used at many points along the way. For instance, task-based user documentation is often structured by scenarios. Minimalist help and training provide many examples of this, such as a "training wheels" system that blocks functions that are not relevant to a paradigmatic novice-use scenario (Carroll & Carrithers, 1983), or a "view matcher" that guides new programmers through a predefined scenario of debugging and modification (Carroll, Singer,

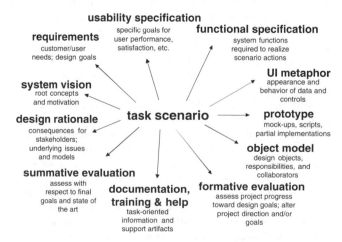

FIGURE 53.2. Scenarios have diverse uses throughout system development life cycle. UI = user interface.

Bellamy, & Alpert, 1990; Carroll & Rosson, 1991; Rosson & Carroll, 1996).

Usage scenarios have also come to play a central role in object-oriented software engineering (Jacobson, 1995; Jacobson, Booch, & Rumbaugh, 1998; Rubin & Goldberg, 1992; Wirfs-Brock, Wilkerson, & Wiener, 1990). A *use case* is a scenario written from a functional point of view, enumerating all of the possible user actions and system reactions that are required to meet a proposed system function (Jacobson, Christersson, Jonsson, & Övergaard, 1992). Use cases can then be analyzed with respect to their requirements for system objects and interrelationships. Wirfs-Brock (1995) described a variant of use case analysis in which she developed a "user-system conversation": Using a two-column format, a scenario is decomposed into a linear sequence of inputs from the user and the corresponding processing and/or output generated by the system. Kaindl (2000) extended this analysis by annotating how scenario steps implement required user goals or system functions.

Scenarios are promising as a mediating representation for analyzing interactions between human-centered and software-centered object-oriented design issues (Rosson & Carroll, 1993, 1995). As we have seen, scenarios can be decomposed with respect to the software objects needed to support the narrated user interactions. These software objects can then be further analyzed with respect to their system responsibilities, identifying the information or services that should be contributed by each computational entity (Wirfs-Brock & Wilkerson, 1989; Beck & Cunningham, 1989; Rosson & Gold, 1989). This analysis (often termed *responsibility-driven design*, Wirfs-Brock et al., 1990) may lead to new ideas about system functionality, for example, initiatives or actions taken by a software object on behalf of the user or another object. Scenarios and claims analysis are useful in describing these new ideas and considering their usability implications *in the context of use* (Rosson, 1999; Rosson & Carroll, 1995; Rosson & Carroll, 2001). For example, a calendar object may be given the responsibility to notify club members of upcoming events; this helps to keep the members informed, but individuals may find the reminders annoying if they are too frequent, or they may come to rely on them too much.

The general accessibility of scenarios makes them an excellent medium for raising and discussing a variety of competing concerns. Software engineers are concerned about issues such as code reuse, programming language or platform, and so on; management is concerned with project resources, scheduling, and so on; a marketing team focuses on issues such as the existing customer base and the product cost. These diverse concerns are *nonfunctional requirements* on system development—concerns about "how" a system will be developed, fielded, and maintained rather than "what" a system will provide (Sommerville, 1992). Usability goals are often specified as nonfunctional requirements, in that they typically focus on the quality of the system rather than its core functions (Mylopoulos, Chung, & Nixon, 1992). The low cost of development, content flexibility, and natural language format of scenarios and claims make them excellent candidates for contrasting and discussing a range of such issues throughout the software development life cycle (Rosson & Carroll, 2000; Sutcliffe & Minocha, 1998).

CURRENT CHALLENGES

When we design interactive systems, we make use. We create possibilities for learning, work, and leisure, for interaction and information. Scenarios—descriptions of meaningful usage episodes—are popular representational tools for making use. They help designers to understand and to create computer systems and applications as artifacts of human activity, as things to learn from, as tools to use in one's work, as media for interacting with other people.

Scenario-based design offers significant and unique leverage on some of the most characteristic and vexing challenges of design work: Scenarios are at once concrete and flexible, helping developers manage the fluidity of design situations. Scenarios emphasize the context of work in the real world; this ensures that design ideas are constantly evaluated in the context of real-world activities, minimizing the risk of introducing features that satisfy other external constraints. The work-oriented character of scenarios also promotes work-oriented communication among stakeholders, helping to make design activities more accessible to many sources of expertise. Finally, scenarios are evocative, raising questions at many levels, not only about the needs of the people in a scenario as written, but also about variants illustrating design tradeoffs.

Scenario-based methods are not a panacea. A project team who complains "We wrote scenarios, but our system still stinks!" must also report how their scenarios were developed, who reviewed them, and what roles they played in system development. If a user interaction scenario is not grounded in what is known about human cognition, social behavior, and work practices, it may well be inspiring and evocative, but it may mislead the team into building the wrong system (Carroll et al., 1998). Scenarios are not a substitute for hard work.

At the same time, *any* work on user interaction scenarios directs a project team to the needs and concerns of the people who will use a system. It is in this sense that scenarios can provide a lightweight approach to human-centered design. Simply writing down and discussing a few key expectations about users' goals and experiences will enhance a shared vision of the problems and opportunities facing system users. Adopting a more systematic framework as described here adds control and organization to the creative process of design and at the same time generates work products (scenarios and claims) that can serve as enduring design rationale during system maintenance and evolution (McKerlie & MacLean, 1994; Moran & Carroll, 1996).

Where are scenarios taking us? The current state of the art in the design of interactive systems is fragmented. Scenarios are used for particular purposes throughout system development, but there is no comprehensive process (Carroll, 1995; Jarke, Bui, & Carroll, 1998; Weidenhaupt et al., 1998). Scenario practices have emerged piecemeal, as local innovations, leading to a

considerable variety of scenario types specialized for particular purposes (Campbell, 1992; Young & Barnard, 1987). A detailed textual narrative of observed workplace practices and interactions, a use case analysis of an object-oriented domain model, a day-in-the-life video envisionment of a future product, and the instructions for test subjects in an evaluation experiment could *all* be scenarios. Recognizing this, and cross-leveraging the many different views of scenarios, is a potential strength of scenario-based design. But much work remains in developing overarching frameworks and methods that exploit this potential advantage.

It is important for us to be ambitious, skeptical, and analytic about scenarios and scenario-based design. Forty years ago, Herman Kahn (1962) expressed puzzlement that scenarios were not more widely used in strategic planning. In the 1990s, scenarios have become so pervasive in interactive system design that younger designers may wonder what would be an alternative to scenario-based design! But there is as yet some strangeness to scenarios. We are not much further than Kahn was in understanding how scenarios work as tools for planning and design or in understanding how to fully exploit their unique strengths as aides to thought.

References

Ackoff, R. L. (1979). Resurrecting the future of operations research. *Journal of the Operations Research Society, 30*, 189–199.

Antón A., McCracken, W. M., & Potts, C. (1994). Goal decomposition and scenario analysis in business process reengineering. *Proceedings of CAiSE '94: Sixth Conference on Advanced Information Systems Engineering* (pp. 94–104). Utrecht, The Netherlands: Springer-Verlag.

Beck, K. (1999) *Extreme Programming explained. Embrace change*. Reading, MA: Addison-Wesley.

Beck, K., & Cunningham, W. (1989). A laboratory for teaching object-oriented thinking. In N. Meyrowitz (Ed.), *Proceedings of Object-Oriented Systems, Languages and Applications: OOPSLA '89* (pp. 1–6). New York: ACM Press.

Beyer, H., & Holtzblatt, K. (1998). *Contextual design: A customer-centered approach to system design*. San Francisco: Morgan Kaufmann.

Bødker, S. (1991). *Through the interface: A human activity approach to user interface design*. Hillsdale, NJ: Lawrence Erlbaum Associates.

Brooks, F. (1995). *The mythical man-month: Essays on software engineering*. Reading, MA: Addison-Wesley. (Original work published 1975)

Campbell, R. L. (1992). Will the real scenario please stand up? *SIGCHI Bulletin, 24*, 6–8.

Carroll, J. M. (Ed.). (1995). *Scenario-based design: Envisioning work and technology in system development*. New York: John Wiley.

Carroll, J. M. (1997). Scenario-based design. In M. Helander & T. K. Landauer (Eds.), *Handbook of human-computer interaction* (2nd ed., pp. 383–406). Amsterdam: North-Holland.

Carroll, J. M. (2000). *Making use: Scenario-based design of human-computer interactions*. Cambridge, MA: MIT Press.

Carroll, J. M., & Carrithers, C. (1983). Blocking errors in a learning environment. Psychonomic Society 24th Annual Meeting (San Diego, CA, November 17). Abstract in Proceedings, p. 356.

Carroll, J. M., Chin, G., Rosson, M. B., & Neale, D. C. (2000). The development of cooperation: Five years of participatory design in the Virtual School. *Proceedings of DIS 2000: Designing Interactive Systems* (pp. 239–251). New York: ACM Press.

Carroll, J. M., Karat, J., Alpert, S. A., van Deusen, M., & Rosson, M. B. (1994). Demonstrating Raison d'Etre: Multimedia design history and rationale. In C. Plaisant (Ed.), *CHI '94 Conference Companion* (pp. 29–30). New York: ACM Press.

Carroll, J. M., Kellogg, W. A., & Rosson, M. B. (1991). The task-artifact cycle. In J. M. Carroll (Ed.), *Designing interaction: Psychology at the human-computer interface* (pp. 74–102). Cambridge, UK: Cambridge University Press.

Carroll, J. M., Mack, R. L., & Kellogg, W. A. (1988). Interface metaphors and user interface design. In M. Helander (Ed.), *Handbook of human-computer interaction* (pp. 67–85). Amsterdam: North Holland.

Carroll, J. M., & Rosson, M. B. (1985). Usability specifications as a tool in iterative development. In H. R. Hartson (Ed.), *Advances in Human-Computer Interaction* (pp. 1–28). Norwood, New Jersey: Ablex.

Carroll, J. M., & Rosson, M. B. (1990). Human-computer interaction scenarios as a design representation. *Proceedings of the 23rd Annual Hawaii International Conference on Systems Sciences* (pp. 555–561). Los Alamitos, CA: IEEE Computer Society Press.

Carroll, J. M., & Rosson, M. B. (1991). Deliberated evolution: Stalking the View Matcher in design space. *Human-Computer Interaction, 6*, 281–318.

Carroll, J. M., & Rosson, M. B. (1992). Getting around the task-artifact cycle: How to make claims and design by scenario. *ACM Transactions on Information Systems, 10*, 181–212.

Carroll, J. M., Rosson, M. B., Chin, G., & Koenemann, J. (1998). Requirements development in scenario-based design. *IEEE Transactions on Software Engineering, 24*, 1156–1170.

Carroll, J. M., Singer, J. A., Bellamy, R. K. E., & Alpert, S. R. (1990). A View Matcher for learning Smalltalk. In J. C. Chew & J. Whiteside (Eds.), *Proceedings of CHI90: Human Factors in Computing Systems* (pp. 431–437). New York: ACM Press.

Carroll, J. M., & Thomas, J. C. (1982). Metaphors and the cognitive representation of computing systems. *IEEE Transactions on Systems, Man, and Cybernetics, 12*, 107–116.

Checkland, P. B. (1981). *Systems thinking, systems practice*. Chichester, UK: John Wiley.

Cherny, L. (1995). Mud community. *The mud register: Conversational modes of action in a text-based virtual reality* (pp. 42–126). Unpublished PhD dissertation, Stanford University, Palo Alto, CA.

Chin, G., Rosson, M. B., & Carroll, J. M. (1997). Participatory analysis: Shared development of requirements from scenarios. In *Proceedings of Human Factors in Computing Systems, CHI '97 Conference* (pp. 162–169). New York: ACM Press.

Constantine, L. L., & Lockwood, L. A. D. (1999). *Software for use: A practical guide to the models and methods of usage-centered design*. Reading, MA: Addison-Wesley.

Cooper, A. (1999). *The inmates are running the asylum: Why high tech products drive us crazy and how to restore the sanity*. Indianapolis, IN: SAMS Press.

Cross, N. (2001). Design cognition: Results from protocol and other empirical studies of design activity. In C. Eastman, M. McCracken, & W. Newstetter (Eds.), *Design knowing and Learning: Cognition in Design Education* (pp. 79–103). Amsterdam: Elsevier.

Dubberly, H., & Mitsch, D. (1992). Knowledge navigator. *CHI '92 Special Video Program: Conference on Human Factors in Computing Systems.* ACM SIGCHI, New York: ACM Press.

Erickson, T. (1990). Working with interface metaphors. In B. Laurel (Ed.), *The art of human-computer interface design* (pp. 65-73). Reading, MA: Addison-Wesley.

Erickson, T. (1995). Notes on design practice: Stories and prototypes as catalysts for communication. In J. M. Carroll (Ed.), *Scenario-based design: Envisioning work and technology in system development* (pp. 37-58). New York: John Wiley.

Freud, S. (1900). *The interpretation of dreams* (standard ed., Vol. IV). London: Hogarth.

Good, M., Spine, T. M., Whiteside, J., & George, P. (1986). User-derived impact analysis as a tool for usability engineering. In M. Mantei & P. Oberton (Eds.), *Proceedings of Human Factors in Computing Systems: CHI '86* (pp. 241-246). New York: ACM Press.

Gray, W. D., John, B. E., & Atwood, M. E. (1992). The precis of Project Ernestine, or an overview of a validation of GOMS. In P. Bauersfeld, J. Bennett, & G. Lynch (Eds.), *Proceedings of Human Factors in Computing Systems: CHI '92* (pp. 307-312). New York: ACM Press.

Haviland, S. E., & Clark, H. H. (1974). What's new? Acquiring new information as a process in comprehension. *Journal of Verbal Learning and Verbal Behavior, 13,* 512-521.

Holbrook, H. (1990). A scenario-based methodology for conducting requirements elicitation. *ACM SIGSOFT Software Engineering Notes, 15,* 95-103.

Hsia, P., Samuel, J., Gao, J., Kung, D., Toyoshima, Y., & Chen, C. (1994, March). Formal approach to scenario analysis. *IEEE Software,* 33-41.

Jacobson, I. (1995). The use-case construct in object-oriented software engineering. In J. M. Carroll (Ed.), *Scenario-based design: Envisioning work and technology in system development* (pp. 309-336). New York: John Wiley.

Jacobson, I., Booch, G., & Rumbaugh, J. (1998). *The unified software development process.* Reading, MA: Addison-Wesley.

Jacobson, I., Christersson, M., Jonsson, P., & Övergaard, G. (1992). *Object-oriented software engineering—A use-case driven approach.* Reading, MA: Addison-Wesley.

Jarke, M., Bui, X. T., & Carroll, J. M. (1998). Scenario management: An interdisciplinary approach. *Requirements Engineering, 3,* 155-173.

Kahn, H. (1962). *Thinking about the unthinkable.* New York: Horizon Press.

Kahneman, D., & Tversky, A. (1972). Subjective probability: A judgement of representativeness. *Cognitive Psychology, 3,* 430-454.

Kaindl, H. (1997). A practical approach to combining requirements definition and object-oriented analysis. *Annals of Software Engineering, 3,* 319-343.

Kaindl, H. (2000). A design process based on a model combining scenarios with goals and functions. *IEEE Transactions on Systems, Man, and Cybernetics, 30,* 537-551.

Karat, J. (1995). Scenario use in the design of a speech recognition system. In J. M. Carroll (Ed.), *Scenario-based design: Envisioning work and technology in system development* (pp. 109-133). New York: John Wiley.

Karat, J., & Bennett, J. B. (1991). Using scenarios in design meetings—A case study example. In J. Karat (Ed.), *Taking design seriously: Practical techniques for human-computer interaction design* (pp. 63-94). Boston: Academic Press.

Kieras, D. (1997). A guide to GOMS model usability evaluation using NGOMSL. In M. G. Helander, T. K. Landauer, & P. V. Pradhu (Eds.), *Handbook of human-computer interaction* (2nd ed., pp. 733-766). Amsterdam: North-Holland.

Kuutti, K. (1995). Work processes: Scenarios as a preliminary vocabulary. In J. M. Carroll (Ed.), *Scenario-based design: Envisioning work and technology in system development* (pp. 19-36). New York: John Wiley.

Kuutti, K., & Arvonen, T. (1992). Identifying potential CSCW applications by means of Activity Theory concepts: A case example. In J. Turner & R. Kraut (Eds.), *Proceedings of Computer-Supported Cooperative Work: CSCW '92* (pp. 233-240). New York: ACM Press.

Kyng, M. (1995). Creating contexts for design. In J. M. Carroll (Ed.), *Scenario-based design: Envisioning work and technology in system development* (pp. 85-107). New York: John Wiley.

Lévi-Strauss, C. (1967). *Structural anthropology.* Garden City, NY: Anchor Books.

Mack, R. L., Lewis, C. H., & Carroll, J. M. (1983). Learning to use office systems: Problems and prospects. *ACM Transactions on Office Information Systems, 1,* 254-271.

MacLean, A., Young, R. M., & Moran, T. P. (1989). Design rationale: The argument behind the artifact. *Proceedings of Human Factors in Computing Systems: CHI '89* (pp. 247-252). New York: ACM Press.

Madsen, K. H. (1994). A guide to metaphorical design. *Communications of the ACM, 37,* 57-62.

McKerlie, D., & MacLean, A. (1994). Reasoning with design rationale: Practical experience with Design Space Analysis. *Design Studies, 15,* 214-226.

Medin, D. L., & Schaffer, M. M. (1978). A context theory of classification learning. *Psychological Review, 85,* 207-238.

Moran, T., & Carroll, J. M. (Eds.). (1996). *Design rationale: Concepts, techniques, and use.* Hillsdale, NJ: Lawrence Erlbaum Associates.

Muller, M. J. (1992). Retrospective on a year of participatory design using the PICTIVE technique. In A. Janda (Ed.), *Proceedings of Human Factors of Computing Systems, CHI '92* (pp. 455-462). New York: ACM Press.

Muller, M. J., Tudor, L. G., Wildman, D. M., White, E. A., Root, R. A., Dayton, T., Carr, R., Diekmann, B., & Dykstra-Erickson, E. (1995). Bifocal tools for scenarios and representations in participatory activities with users. In J. M. Carroll (Ed.), *Scenario-based design: Envisioning work and technology in system development* (pp. 135-163). New York: John Wiley.

Mylopoulos, J., Chung, L., & Nixon, B. (1992). Representing and using nonfunctional requirements: A process-oriented approach. *IEEE Transactions on Software Engineering, 18,* 483-497.

Nardi, B. A. (Ed.). (1996). *Context and consciousness: Activity theory and human-computer interaction.* Cambridge, MA: MIT Press.

Nielsen, J. (1995). Scenarios in discount usability engineering. In J. M. Carroll (Ed.), *Scenario-based design: Envisioning work and technology in system development.* (pp. 59-83). John Wiley.

Nielsen, J., & Mack, R. L. (1994). *Usability inspection methods.* New York: John Wiley.

Norman, D. A. (1986). Cognitive engineering. In D. A. Norman & S. D. Draper (Eds.), *User centered system design* (pp. 31-61). Hillsdale, NJ: Lawrence Erlbaum Associates.

Norman, D. A. (1988). *The psychology of everyday things.* New York: Basic Books.

Orlikowski, W. J. (1992). Learning from notes: Organizational issues in groupware implementation. In J. Turner & R. Kraut (Eds.), *CSCW '92: Proceedings of the Conference on Computer Supported Cooperative Work* (pp. 362-369). New York: ACM Press.

Polson, P. G., Lewis, C., Rieman, J., & Wharton, C. (1992). Cognitive walkthroughs: A method for theory-based evaluation of user interfaces. *International Journal of Man-Machine Studies, 36,* 741-773.

Potts, C. (1995). Using schematic scenarios to understand user needs. *Proceedings of ACM Symposium on Designing Interactive Systems: DIS '95* (pp. 247-256). New York: ACM Press.

Propp, V. (1958). *Morphology of the folktale*. The Hague, The Netherlands: Mouton. (Original work published 1928)

Rosch, E., Mervis, C. B., Gray, W., Johnson, D., & Boyes-Braem, P. (1976). Basic objects in natural categories. *Cognitive Psychology, 7*, 573–605.

Rosson, M. B. (1999). Integrating development of task and object models. *Communications of the ACM, 42*, 49–56.

Rosson, M. B., & Carroll, J. M. (1993). Extending the task-artifact framework. In R. Hartson & D. Hix (Eds.), *Advances in human-computer interaction* (Vol. 4, pp. 31–57). New York: Ablex.

Rosson, M. B., & Carroll, J. M. (1995). Narrowing the gap between specification and implementation in object-oriented development. In J. M. Carroll (Ed.), *Scenario-based design: Envisioning work and technology in system development* (pp. 247–278). New York: John Wiley.

Rosson, M. B., & Carroll, J. M. (1996). The reuse of uses in Smalltalk programming. *ACM Transactions on Computer-Human Interaction, 3*, 219–253.

Rosson, M. B., & Carroll, J. M. (2000). Nonfunctional requirements in scenario-based development. *Proceedings of OZCHI 2000* (pp. 232–239). North Ryde, Australia: CSIRO Mathematical and Information Sciences.

Rosson, M. B., & Carroll, J. M. (2001). Scenarios, objects, and points-of-view in user interface design. In M. van Harmelen (Ed.), *Object modeling and user interface design*. London: Addison Wesley Longman.

Rosson, M. B., & Carroll, J. M. (2002). *Usability engineering: Scenario-based development of human-computer interaction*. San Francisco: Morgan Kaufmann.

Rosson, M. B., & Gold, E. (1989). Problem-solution mapping in object-oriented design. In N. Meyrowitz (Ed.), *Proceedings of OOPSLA '89: Conference on Object-Oriented Programming Systems, Languages, and Applications* (pp. 7–10). New York: ACM Press.

Rosson, M. B., Maass, S., & Kellogg, W. A. (1989). The designer as user: Building requirements for design tools from design practice. *Communications of the ACM, 31*, 1288–1298.

Rubin, K., & Goldberg, A. (1992). Object behavior analysis. *Communications of the ACM, 35*, 48–62.

Schön, D. A. (1967). *Technology and change: The new Heraclitus*. New York: Pergamon Press.

Schön, D. A. (1983). *The reflective practitioner: How professionals think in action*. New York: Basic Books.

Schön, D. A. (1987). *Educating the reflective practitioner*. San Francisco: Jossey-Bass.

Scriven, M. (1967). The methodology of evaluation. In R. Tyler, R. Gagne, & M. Scriven (Eds.), *Perspectives of curriculum evaluation* (pp. 39–83). Chicago: Rand McNally.

Sommerville, I. (1992). *Software engineering* (4th ed.). Reading, MA: Addison-Wesley.

Sutcliffe, A. G., & Minocha, S. (1998, June 8–9). Scenario-based analysis of non-functional requirements. *Workshop paper for Requirements Engineering for Software Quality (REFSQ '98) at CAISE '98*. Pisa, Italy.

Tversky, A., & Kahneman, D. (1974). Judgements under uncertainty: Heuristics and biases. *Science, 185*, 1124–1131.

Verplank, W. L. (1988). Graphic challenges in designing object-oriented user interfaces. In M. Helander (Ed.), *Handbook of human-computer interaction* (pp. 365–376). Amsterdam: North-Holland.

Virzi, R. A., Sokolov, J. L., & Karis, D. (1996). Usability problem identification using both low- and high-fidelity prototypes. In *Proceedings of Human Factors in Computing Systems: CHI '96* (pp. 236–243). New York: ACM Press.

Wasserman, A. I., & Shewmake, D. T. (1982). Rapid prototyping of interactive information Systems. *ACM Software Engineering Notes, 7*, 171–180.

Weidenhaupt, K., Pohl, K., Jarke, M., & Haumer, P. (1998). Scenarios in system development: Current practice. *IEEE Software, 15*, 34–45.

Wertheimer, M. (1939). Laws of organization in perceptual forms. In W. D. Ellis (Ed.), *A Sourcebook of Gestalt Psychology* (pp. 331–363). London: Harcourt, Brace & Company.

Wirfs-Brock, R. (1995). "Designing objects and their interactions: A brief look at responsibility-driven design. In J. M. Carroll (Ed.), *Scenario-based design: Envisioning work and technology in system development* (pp. 337–360). New York: John Wiley.

Wirfs-Brock, R., & Wilkerson, B. (1989). Object-oriented design: A responsibility-driven approach. In N. Meyrowitz (Ed.), *Object-oriented programming: Systems, languages and applications, proceedings of OOPSLA '89* (pp. 71–76). New York: ACM Press.

Wirfs-Brock, R., Wilkerson, B., & Wiener, L. (1990). *Designing object-oriented software*. Englewood Cliffs, NJ: Prentice Hall.

Young, R. M., & Barnard, P. B. (1987). The use of scenarios in human-computer interaction research; Turbocharging the tortoise of cumulative science. *Proceedings of CHI+GI'87: Conference on Human Factors in Computing Systems and Graphics Interface* (pp. 291–296). New York: ACM Press.

·54·

PARTICIPATORY DESIGN: THE THIRD SPACE IN HCI

Michael J. Muller
IBM Research

INTRODUCTION

Participatory design (PD) is a set of theories, practices, and studies related to end users as full participants in activities leading to software and hardware computer products and computer-based activities (Greenbaum & Kyng, 1991; Muller & Kuhn, 1993; Schuler & Namioka, 1993). The field is extraordinarily diverse, drawing on fields such as user-centered design, graphic design, software engineering, architecture, public policy, psychology, anthropology, sociology, labor studies, communication studies, and political science. This diversity has not lent itself to a single theory or paradigm of study or approach to practice (Slater, 1998). Researchers and practitioners are brought together—but are not necessarily brought into unity—by a pervasive concern for the knowledges, voices, and rights of end users, often within the context of software design and development or of other institutional settings (e.g., workers in companies, corporations, universities, hospitals, governments). Many researchers and practitioners in PD (but not all) are motivated in part by a belief in the value of democracy to civic, educational, and commercial settings, a value that can be seen in the strengthening of disempowered groups (including workers), in the improvement of internal processes, and in the combination of diverse knowledges to make better services and products.

PD began in an explicitly political context, as part of the Scandinavian workplace democracy movement (e.g., Ehn & Kyng, 1987; more recently, see Bjerknes & Bratteteig, 1995; Beck, 1996). Early work took the form of experiments conducted by university researchers in alliances with organized labor (for historical overviews, see Ehn, 1993; Levinger, 1998). Subsequent work supplemented the foundational democratic motivation with a need for combining complex knowledge for realistic design problems. Fowles (2000), for example, wrote of transforming the "symmetry of ignorance" (mutual incomprehension between designers and users) into a complementary "symmetry of knowledge" through symmetries of participation and symmetries of learning. Similarly, Holmström (1995) analyzed a "gap in rationalities" among developers and users. Recently, PD has achieved a status as a useful commercial tool in some settings (e.g., McLagan & Nel, 1995), with several major and influential consultancies forming their business identities around participatory methods.[1] This overall corporate and managerial "mainstreaming" of PD has been greeted by some with enthusiasm and by others with dismay.

This chapter primarily addresses methods, techniques, and practices in participatory design, with modest anchoring of those practices in theory. I do not repeat our recent encyclopedic survey of participatory practices (Muller, Hallewell Haslwanter, & Dayton, 1997). Rather, I pursue a trend within those practices that has shown the most growth during the past 5 years, and I motivate my interest in that trend through recent advances in the theory of cultural studies. I focus on participatory practices that fall in the hybrid realm between the two distinct work domains of software professionals and end users. Following a review of work in the area of hybridity in cultural studies, I argue that this in-between domain, or third space, is a good place to look for new insights and understandings, and for syntheses of diverse knowledges into ideas for products and work practices.

I begin with a bibliographic overview of major participatory design resources. I then take a brief look at the concept of hybridity from cultural studies and apply this concept to participatory design and to user-centered design, discussing areas where the world of software professionals overlaps and hybridizes with the world of end users. I argue that participatory design offers a kind of generalized third space within the field of user-centered design and describe a number of specific practices within the field of participatory design that make good use of the qualities of the third space. I conclude with problems and challenges for the future.

Major Bibliographic Sources for Participatory Design

Theory, practice, and experience in participatory design have been published in a series of conference proceedings and several major books. Four important conference series have made major contributions to PD:

- *Computers in Context.* Three conferences have been held, at 10-year intervals, in the Computers in Context series, most recently in 1995. Major papers from the conferences have appeared as two influential books (Bjerknes, Ehn, & Kyng, 1987; Kyng & Matthiessen, 1997).

- *IRIS Conference (Information Systems Research In Scandinavia).* The annual IRIS conference series often includes sessions and individual contributions on participatory topics. Proceedings may be available through the IRIS Association or online.[2]

- *Participatory Design Conference.* This conference has met on even-numbered years since 1990. Proceedings are published by Computer Professionals for Social Responsibility (CPSR).[3] Selected papers from most of the conferences have appeared in edited volumes or special journal issues (e.g., Kensing & Blomberg, 1998; Muller & Kuhn, 1993; Schuler & Namioka, 1993).

- *IFIP Conferences.* A number of conferences and workshops (sponsored by International Federation for Information Processing (IFIP) Technical Committee (TC) 9 have focused on selected topics within participatory design (e.g., Briefs, Ciborra, & Schneider, 1983; Clement, Kolm, & Wagner, 1994; Docherty, Fuchs-Kittowski, Kolm, & Matthiessen, 1987; Gärtner & Wagner, 1995; and van den Besselaar, Clement, & Jaervinen, 1991).[4]

---

[1] In the interest of fairness to other consultancies, I will not provide the names of commercial ventures.
[2] http://iris.informatik.gu.se/
[3] www.cpsr.org
[4] http://www.ifip.or.at/. For TC 9, see http://www.ifip.or.at/bulletin/bulltcs/memtc09.htm

Major papers, panels, and tutorials on participatory design have also appeared in the CHI, CSCW, ECSCW, and DIS conference series, beginning as early as 1988 (Proceedings available through the Association for Computing Machinery[5]). A smaller number of participatory contributions have appeared in *Proceedings* of the Usability Professionals' Association[6] conference series, of the INTERACT conference series, and of the Human Factors and Ergonomics Society conference series. Several papers at the Co-Designing 2000 Conference[7] addressed participatory themes.

In addition to the books cited above, major collections of papers and/or chapters related to participatory design appeared in Carroll's (1995) volume on scenarios in user interaction, Greenbaum and Kyng's (1991) *Design at Work*, and Wixon and Ramey's (1996) collection of papers on field-oriented methods. Individual books that have been influential in the field include Bødker's (1990) application of activity theory to issues of participation, Ehn's (1988) account of work-oriented design, Suchman's (1987) discussion of situated action, and Beyer and Holtzblatt's presentation of contextual inquiry and contextual design (1998; see also chapter 49 by Holtzblatt in this book). Earlier influential works include a series of books on sociotechnical theory and practice by Mumford (e.g., 1983; Mumford & Henshall, 1979/1983), as well as Checkland's (1981) soft systems methodology. Noro and Imada (1991) developed a hybrid ergonomic approach, involving participation and quality programs, which has been influential around the Pacific rim.

Three journals have carried the greatest number of PD papers:

- *Scandinavian Journal of Information Systems*[8]
- *Computer Supported Cooperative Work: The Journal of Collaborative Computing*[9]
- *Human Computer Interaction*[10]

Three special issues of *Communications of the ACM* have addressed participatory topics: Muller and Kuhn (1993) edited a subset of papers from the 1992 Participatory Design Conference; Suchman (1995) edited an issue concerned with issues of representation in software work; and Madsen (1999) edited a set of papers comparing Scandinavian and North American practices. One issue of the *CPSR Newsletter* provided a set participatory practices and experiences from more marginal domains (Muller, 1994).

Computer Professionals for Social Responsibility maintains a set of PD resources (http://www.cpsr.org/program/workplace/PD.html), including a list of PD-related Web sites (http://www.cpsr.org/program/workplace/PD-resources.html).

HYBRIDITY AND THE THIRD SPACE

This chapter is concerned with participatory methods that occur in the hybrid space between software professionals and end users. Why is this hybrid space important?

Bhabha (1994) has made an influential argument that the border or boundary region between two domains—two spaces—is often a region of overlap or hybridity (i.e., a third space that contains an unpredictable and changing combination of attributes of each of the two bordering spaces). His area of concern was colonization, in which some native people find themselves caught between their own traditional culture and the newly imposed culture of the colonizers. Their continual negotiation and creation of their identities, as efforts of survival, creates a new hybrid or third culture (Bhabha, 1994; see also Lyotard, 1984) and even a third language (Bachmann-Medick, 1996).

Within this hybrid third space, the old assumptions of both the colonizers and the colonized are open to question, challenge, reinterpretation, and refutation (Bhabha, 1994). Enhanced knowledge exchange is possible, precisely because of those questions, challenges, reinterpretations, and renegotiations (Bachmann-Medick, 1996). These dialogues across differences and—more important—*within* differences are stronger when engaged in by groups, emphasizing not only a shift from assumptions to reflections, but also from individuals to collectives (Carrillo, 2000).

Bhabha's conception has become highly influential. Bachmann-Medick (1996) applied the concepts to translation theory. Grenfell (1998) interpreted concepts of hybridity in a study of living-at-the-border in multicultural education settings. Evanoff (2000) surveyed a number of theoretical applications of hybridity, from evolutionary biology to constructivist perspectives in sociology to democratic responses to intercultural ethical disagreements. He explored formulations from multiple disciplines, involving "third culture" in intercultural ethics, "third perspective" involving "dynamic in-betweenness" in Asian–Western exchanges, and a psychological "third area" in the development of a "multicultural personality."

A summary of the claims relating to third spaces (or hybridity) appears in Table 54.1.

Hybridity and Human–Computer Interaction

Within human–computer interaction (HCI), there have been many calls for mutual or reciprocal learning within hybrid spaces (e.g., Bødker, Ehn, Kyng, Kammersgaard, & Sundblad, 1987; Bødker, Knudsen, Kyng, Ehn, & Madsen, 1988; Druin, 1999; Druin et al., 2000; Ehn & Sjögren, 1991; Floyd, 1987; Kensing & Madsen, 1991; Mogensen & Trigg, 1992; Muller,

---

[5] www.acm.org
[6] www.upassoc.org
[7] http://vide.coventry.ac.uk/codesigning/
[8] http://www.cs.auc.dk/~sjis/
[9] http://www.wkap.nl/journalhome.htm/
[10] http://hci-journal.com/

TABLE 54.1. Summary of Claims Relating to Third Spaces

- Overlap between two (or more) different regions or fields (in-betweenness)
- Marginal to reference fields
- Novel to reference fields
- Not "owned" by any reference field
- Partaking of selected attributes of reference fields
- Potential site of conflicts between/among reference fields
- Questioning and challenging of assumptions
- Mutual learning
- Synthesis of new ideas

Negotiation and (co-)creation of . . .
- Identities
- Working language
- Working assumptions and dynamics
- Understandings
- Relationships
- Collective actions

Dialogues across and within differences (disciplines)
- Polyvocality
- What is considered to be data?
- What are the rules of evidence?
- How are conclusions drawn?
- Reduced emphasis on authority—increased emphasis on interpretation
- Reduced emphasis on individualism—increased emphasis on collectivism
- Heterogeneity as the norm

Wildman, & White, 1994; Mumford, 1983; Tscheligi et al., 1995). Beeson and Miskelly (2000) appealed to the notion of hybridity ("heterotopia") in describing workers who, like colonized peoples, deal "in a space which is not their own" (p. 2), taking limited and opportunistic actions to preserve "plurality, dissent, and moral space" (p. 1). Maher, Simoff, and Gabriel (2000) described the creation of virtual design spaces for sharing diverse perspectives. In an early formulation, Lanzara (1983) suggested that

a large part of the design process, especially in large-scale projects and organizations involving several actors, is not dedicated to analytical work to achieve a solution but mostly to efforts at reconciling conflicting [conceptual] frames or at translating one frame into another. Much work of the designer is . . . concerned with . . . defining collectively what is the relevant problem, how to see it.

Tscheligi et al. (1995), in a panel on prototyping, considered that the "products" of prototyping include not only artifacts, but also understandings, communications, and relationships—a theme that was echoed in a more recent panel on modeling (Kaindl, Constantine, Karat, & Muller, 2001). Fanderclai (1995, 1996) captured a strong sense of possible new dynamics and new learnings in a hybrid online space. Finally, Thackara (2000) based part of his plenary address at CHI 2000 on the concept of the third space, providing a needed hybridity to HCI studies.

Participatory Design as the Third Space in HCI

In this chapter, I want to extend the HCI analyses surveyed in the preceding paragraphs and make an analogy between Bhabha's concept of two spaces and the problem of HCI methods to bridge between two spaces—the world of the software professionals and the world of the end users (see also Muller, 1997a, 1997b). Each world has its own knowledges and practices; each world has well-defined boundaries. Movement from one world to the other is known to be difficult. We can see this difficulty manifested in our elaborate methods for requirements analysis, design, and evaluation—and in the frequent failures to achieve products and services that meet users' needs or are successful in the marketplace.

Traditional scientific practice in HCI has focused on instruments and interventions that can aid in transferring information between the users' world and the software world. Most of the traditional methods are relatively one-directional (e.g., we analyze the requirements *from* the users; we deliver a system *to* the users; we collect usability data *from* the users). Although there are many specific practices for performing these operations, relatively few of them involve two-way discussions, and fewer still afford opportunities for the software professionals to be surprised, that is, *to learn something that we didn't know we needed to know.*

The PD tradition has, from the outset, emphasized mutuality and reciprocity, often in a hybrid space that enabled new relationships and understandings. Bødker et al. (1988) made specific references to "the mutual validation of diverse perspectives." Floyd (1987) analyzed software practices into two paradigms, which she termed product-oriented (focused on the computer artifact as an end in itself) and process-oriented (focused on the human work process, with the computer artifact as means to a human goal). In her advocacy of balancing these two paradigms, Floyd noted that the process-oriented paradigm required mutual learning among users and developers (see also Segall & Snelling, 1996). Most of PD theories and practices require the combination of multiple perspectives, in part because complex human problems require multiple disciplines (e.g., software expertise and work-domain expertise) for good solutions (e.g., Fowles, 2000; Holmström, 1995) and in part because the workplace democracy tradition requires that all of the interested parties (in the United States, we would say "stakeholders") should have a voice in constructing solutions (e.g., Ehn & Kyng, 1987).

Participatory Design Contains Its Own Third Space

The preceding argument—that PD serves as a kind of third space to HCI—might be interesting but is hardly worth a chapter in a handbook. I now turn to the question of hybridity in methods within the field of PD itself.

In their "tools for the toolbox" approach, Kensing and Munk-Madsen (1993) developed a taxonomy to analyze about 30 participatory methods (see also Kensing, Simonsen, & Bødker, 1996; and, in independent convergences on the same attribute, see Gjersvik & Hepsø, 1998; Luck, 2000; Reid & Reed, 2000). The first dimension of their taxonomy contrasted *abstract* methods (suitable for a software professional's organization) with *concrete* methods (suitable for work with

end users).[11] Muller, White, and Wildman (1993) and Muller et al. (1997) elaborated on this taxonomic dimension by asking *whose work domain served as the basis for the method* (in the United States, we would call this a matter of "turf," as in "on whose turf did the work take place?"). At the *abstract* end of the continuum, the users have to enter the world of the software professionals to participate (e.g., rapid prototyping; Grønbæk, 1989) and quality improvement (Braa, 1996). At the *concrete* end of the continuum, the software professionals have to enter the world of the users to participate (e.g., ethnography; Blomberg, Giacomi, Mosher, & Swenton-Wall, 1993; Crabtree, 1998; Orr & Crowfoot, 1992; Suchman & Trigg, 1991; see also chapter 50 by Blomberg, Burrell, & Guest in this book), ongoing tailoring during usage (Henderson & Kyng, 1991; MacLean, Carter, Lovstrand, & Moran, 1990), and end user "design" by purchasing software for small companies (Krabbel & Wetzel, 1998; Robertson, 1996, 1998).

For the purposes of this chapter, one can now ask: What about the practices that did not occur at the *abstract* or *concrete* end points of the continuum? *What about the practices in between?* These practices turn out to occur in an uncertain, ambiguous, overlapping disciplinary domain that does not "belong" to either the software professionals or the end users (i.e., these practices occur in neither the users' turf nor the software professionals' turf). The practices in between the extremes are hybrid practices and constitute the third space of participatory design. As we explore hybrid methods that occur in this third space, we can look for HCI analogies of the attributes and advantages that were listed for third space studies in Table 54.1.

THIRD SPACE: NEGOTIATION, SHARED CONSTRUCTION, AND COLLECTIVE DISCOVERY IN PD AND HCI

In this, the main section of the chapter, I describe a diversity of participatory design techniques, methods, and practices that provide hybrid experiences or that operate in intermediate, third spaces in HCI. Because my theme is hybridity, I have organized these descriptions in terms strategies and moves that introduce novelty, ambiguity, and renewed awareness of possibilities, occurring at the margins of existing fields or disciplines (see Table 54.1). In several cases, a single report may fall into several categories. For example, Ehn and Sjögren (1991) conducted a workshop (see Workshops) in which a story-telling method (see Stories) provided a space in which people negotiated the naming and defining of workplace activities (see Language). I hope that the strategies and moves of the PD practitioners and researchers will become clear, despite the multiple views onto individual reports.

Sitings

One of the simplest parameters that can be manipulated to influence hybridity is the site of the work. At first, this appears to be a simple issue. As Robins (1999) said, "There are two approaches to participatory design: 1. Bring the designers to the workplace. 2. Bring the workers to the design room." This binary choice reflects the taxonomic distinctions that I reviewed above. However, even within the binary choice, the selection of the site can be important. Fowles (2000), in a discussion of participatory architectural practice, provided insight that applies for HCI as well: "If possible[,] design workshops should be located in the locality of the participating group and in the School of Architecture. Bringing the public into the School helps to demystify the profession, and taking students in the community furthers their understanding of the problem and its context" (p. 65). Pedersen and Buur (2000), in their work on industrial sites, agree (italics in the original):

> When collaborating with users *in our design environment* (e.g., a meeting space at the company), we can invite a number of users from different plants and learn from hearing them exchange work experiences.... Being in a foreign environment (and with other users), users will tend to take a more general view of things.
>
> When collaborating with users *in their work context*, users tend to feel more at ease as they are on their home ground—we are the visitors. Tools and environment are physically present and easy to refer to. This makes for a conversation grounded in concrete and specific work experiences.
>
> The idea was born to create a type of design event with activities in both environments and with two sets of resources to support design collaboration. (p. 95)

In our study of telephone operators' work, we held our sessions at operator service offices and in research offices (Muller et al., 1995a). The work site meetings had the advantages of easy access to equipment on which we could demonstrate or experiment. During those meetings, we had a sense of being strongly tied to practice. The research site meetings were less tied to specific practices and had a tendency to lead to more innovative ideas. Perhaps more subtly, the two different sites enfranchised different marginal participants. At the work site, it was easy to bring in additional work-domain experts (mostly trainers and procedures experts): They became adjunct members of the core analysis team for the duration of those meetings, *and* they became resources for the core team afterward. At the research site, it was easy to bring in more technology experts, as well as the graduate students who later performed data analysis. The research site meetings became an occasion of enfranchisement, contribution, and early commitment for these additional actors. Both core and adjunct members became authors of our report (Muller et al., 1995a).

Brandt and Grunnet (2000) also considered site selection in their Smart Tool and Dynabook projects, which were concerned with working conditions in the office and in the home, respectively. In the Smart Tool case, they conducted dramatic scenarios in the project designers' environment. In the Dynabook case, they asked people at home to create and enact scenarios in their own living areas.

Third Space. In terms of hybridity, the selection of site can be a deliberate strategy to introduce new experiences and

---

[11]Their second dimension was of less interest for the purposes of this chapter.

perspectives to one or more parties in the design process—a de-centering move that can bring people into positions of ambiguity, renegotiation of assumptions, and increased exposure to heterogeneity. Returning to Bhabha's (1994) original argument, site selection initially appears to be a matter of *moving across the boundary* between different work cultures, rather than *living within the boundary*. The use of *common design practices across sites*, however, makes those practices (and the membership of the design group) into a kind of movable third space. The practices and the group membership become stable features that persist across multiple sites. At the same time, the practices, and even the membership, grow and evolve with exposure to new sites and new understandings. In these ways, the practices become an evolutionary embodiment of the knowledge of the learnings of the group (e.g., Floyd, 1987; Muller, 1997a).

Claimed Benefits. What have practitioners gained through site selection, within this deliberately hybrid-oriented work area? Several themes emerge:

- *Improved learning and understanding.* Fowles (2000) described a move from a "symmetry of ignorance" toward a "symmetry of knowledge" as diverse parties educated one another through a "symmetry of learning"—and even a kind of "transformation" through exposure to new ideas. (p. 63) Brandt and Grunnet (2000), Pedersen and Buur (2000), and Muller et al. (1995b) also claimed that the selection of site led to the strengthening of the voices that were comfortable at each site.

- *Greater ownership.* Petersen and Buur (2000) noted that their procedures strengthened user involvement in their project. Fowles (2000) and Muller (1995b; see also Muller et al., 1994) made specific reference to increases in commitment and ownership of the evolving knowledge and design of the group.

Workshops

Workshops may serve as another alternative to the two "standard" sites that most of us think about. In PD, workshops are usually held to help diverse parties ("stakeholders") communicate and commit to shared goals, strategies, and outcomes (e.g., analyses, designs, and evaluations, as well as workplace-change objectives). Workshops are often held at sites that are, in a sense, neutral—they are not part of the software professionals' workplace, and they are not part of the workers' workplace.

More important, workshops usually introduce novel procedures that are not part of conventional working practices. These novel procedures take people outside of their familiar knowledges and activities and must be negotiated and collectively defined by the participants. Workshops are thus a kind of hybrid or third space, in which diverse parties communicate in a mutuality of unfamiliarity and must create shared knowledges and even the procedures for developing those shared knowledges.

The best-known workshop format in PD is the Future Workshop (e.g., Kensing & Madsen, 1991), the overall framework of which proceeds through three stages: *critiquing* the present;

envisioning the future; and *implementing* movement from the present to the future. These three activities involve participants in new perspectives on their work and help to develop new concepts and new initiatives.

Sanders (2000) described a family of "generative tools," activities that are selectively combined into Strategic Design Workshops, under an overall conceptual strategy that combines market research ("what people say"), ethnography ("what people do"), and participatory design ("what people make"). Activities include the construction of collages focused on thinking (e.g., "how do you expect your work to change in the future?"), mapping (e.g., laying out an envisioned work area on paper), feeling ("use pictures and words to show a health-related experience in your past"), and storytelling (see Stories and Making Descriptive Artifacts). Dandavate, Steiner, and William (2000) provided a case study of Sanders's method.

In a different setting, Buur, Binder, and Brandt (2000) developed a workshop in which workers carried a mock-up of a proposed new device (see Making NonFunctional Artifacts) through an industrial plant, recording how it would be used. They then acted out a 5-minute video scenario (see Dramas), which they subsequently presented to other, similar worker teams in a workshop.

Cameron (1998), too, faced a different setting and problem, and chose a workshop solution. This project dealt with safety issues in urban design in Baltimore and—like the METRAC program in Toronto (Nisonen, 1994)—invited community members to contribute their domain expertise as people who lived with safety issues on an everyday basis. Cameron provided a manual, based on a professionally developed set of safety guidelines. Community members became community organizers, bringing the project topic and the proposed guidelines to their own constituencies. Two additional workshops refined the safety audit information from the constituencies, selected priority issues to fix, and adopted an action plan. Cameron observed that

> One of the successful aspects of the Design for Safety workshop is that it provided a forum for a diverse group of people to productively discuss common problems and work through shared solutions and consensus. The workshops also showed that crime and safety were not solely the responsibility of the police, but that public works employees, traffic engineers, and especially residents must work together to envision as well as carry out the plan. . . . Requiring that residents share the workshop information at community association meetings further assisted the transfer of responsibility from the workshop into the neighborhood. (p. 172)

Third Space. The various workshop approaches have several commonalities. Each workshop brings together diverse participants to do common work, to produce common outcomes, and to develop a plan of joint action. They are thus opportunities that require mutual education, negotiation, creation of understanding, and development of shared commitments. Each workshop takes place in an atmosphere and (often) in a site that is not "native" to any of the participants. Thus, all of the participants are at a disadvantage of being outside of their own familiar settings, and they must work together to define their new circumstances and relationships. The combination of diverse voices leads to syntheses of perspectives and knowledges.

Claimed Benefits. Advantages claimed for these experiences in hybridity include

- *Development of new concepts* that have direct, practical value for product design (Dandavate et al., 2000; Kensing & Madsen, 1991; Sanders, 2000) or for community action (Cameron, 1998)
- *Engagement* of the interested parties ("stakeholders") in the process and outcome of the workshop
- *Combinations of different people's ideas* into unified concepts

Stories

Stories and storytelling have played a major role in ethnographic work since before there was a field called "HCI" (for review, see Crabtree, 1998; Suchman & Trigg, 1991; see also chapter 50 by Blomberg et al. in this book). Stories have also had an important history in HCI (see Carroll, 1995; Erickson, 1996; Muller, 1999a; see also Carroll's chapter in this book). I will not attempt to review these areas. Rather, I will focus on those aspects of story collecting and storytelling that involve the construction of third spaces and hybridity.

Stories in participatory work may function in at least three ways. First, they may be used as triggers for conversation, analysis, or feedback (Salvador & Howells, 1998; Salvador & Sato, 1998, 1999). Second, they may be told by end users as part of their contribution to the knowledges required for understanding product or service opportunities and for specifying what products or services should do (Brandt & Grunnet, 2000; Lafreniére, 1996; Muller, 2001; Muller et al., 1995b; Noble & Robinson, 2000; Patton, 2000; Sanders, 2000; Tschudy, Dykstra-Erickson, & Holloway, 1996). Third, they may be used by design teams to present their concept of what a designed service or product will do, how it will be used, and what changes will occur as a result (Druin, 1999; Druin et al., 2000; Ehn & Kyng, 1991; Ehn & Sjögren, 1986, 1991; Gruen, 2001; Muller et al., 1994; Sanders, 2000).

Beeson and Miskelly (1998, 2000) used hypermedia technologies to enable communities to tell their own stories, with the intention that "plurality, dissent, and moral space can be preserved" (Beeson & Miskelly, 2000, p. 1). They were concerned with allowing multiple authors to reuse community materials selectively, telling different stories within a common context. The different accounts were organized according to themes and laid out spatially on the image of a fictitious island for navigation by end users.

Their work entered several areas or aspects of hybridity. First, the authors of the stories (i.e., community members) were using hypermedia technology for the first time and were thus in the role of learners; at the same time, they were the owners of the stories, and thus in the role of experts. Second, the authors wrote from their own perspectives, which were sometimes in strong conflict with one another. Third, the authors could make use of one anothers' materials, effectively moving away from single-author narratives and into a kind of collaborative collage of materials, which conveyed interlinked stories. Fourth, just as

the community members were negotiating and defining their roles as learner-experts, the software professionals/researchers were negotiating and defining their roles as experts-facilitators-students.

A second line of practice and research has emphasized end users telling their stories using a system of paper-and-pencil, cardlike templates. The earliest version was the Collaborative Analysis of Requirements and Design (CARD) technique of Tudor, Muller, Dayton, and Root (1993), later developed into a more general tool in Muller et al. (1995b) and further refined in Muller (2001). Lafreniére (1996) developed a related practice, Collaborative Users' Task Analysis (CUTA), repairing some of the deficits of CARD for his settings. Tschudy, Dykstra-Erickson, and Holloway (1996) developed their own highly visual version, PictureCARD, for a setting in which they had no language in common with the users whose stories they wished to understand.

The card-based practices used pieces of cardboard about the size of playing cards. Each card represented a component of the user's work or life activities, including user interface events (i.e., screen shots); social events (conversations, meetings); and cognitive, motivational, and affective events (e.g., the application of skill, the formation of goals or strategies, surprises and breakdowns, evaluations of work practices). The cards were used by diverse teams in analysis, design, and evaluation of work and technology. Because the cards were novel object to all the participants, they occasioned third-space questionings and negotiations, resulting in new shared understandings and co-constructions. Often, teams used the cards to prepare a kind of storyboard, narrating the flow of work and technology use and annotating or innovating cards to describe that work. The resulting posters formed narratives of the work that were demonstrated to be understandable to end users, corporate officers, and software professionals and which led to insights and decisions of large commercial value (see Sanders, 2000, for a differently constructed example of storyboard posters used to describe work).

Druin (1999; Druin et al., 2000) pursued a third line of storytelling research and practice, with children as design partners in a team that also included computer scientists, graphic designers, and psychologists (for other participatory work with children, see Sanders, 2000; Sanders & Nutter, 1994). Their purpose was to envision new technologies and practices in children's use of computers and related devices. They used both online storyboarding techniques and the construction of prototypes of spaces in which the jointly authored stories could be performed. This work kept everyone learning from everyone else—children learning about technologies and the storyboarding environment, adults learning about children's views and other adults' expertises, and everyone negotiating the meaning of new technological and narrative ideas, as well as their implementations.

So far, this section has addressed primarily the acquisition of stories. But stories are also for telling to others. Sanders (2000) described the construction of storyboards based on users' experiences. Gruen (2000, 2001) described guidelines and practices through which a diverse team could begin with a concept, and then could craft a convincing and engaging story around it. Sanders's and Gruen's procedures led to hybrid experiences, in

the sense that few software professionals or end users think in terms of story-construction or rubrics for effective fictions.

Third Space. Story collecting and storytelling generally require a kind of third space in which to occur. Beeson and Miskelly (1998, 2000) were specifically concerned to create a new space for story writing and story reading and to maintain some of the most important aspects of third spaces in that new space (i.e., preservation and expression of new meanings, relationships, conflicts, multiple perspectives, and "heterotopia"). The three card-based practices use unfamiliar media (the cards) and made those media central to the team's activities, thus requiring conscious attention to shared conceptualizing and defining of those media, as well as the creation of new media when needed. Druin and colleagues created new software environments and new devices to craft and implement stories of futuristic technologies. Finally, Gruen engaged diverse teams in new roles as story writers, guided by expert-derived guidelines, in the writing of professionally structured and paced stories for organizational or commercial use.

Claimed Benefits. The story collecting and story telling practices are diverse, and serve multiple purposes. A brief summary of the claims of their value to projects and products is as follows:

- *Articulation* and preservation of a diverse community's views (Beeson & Miskelly, 1998, 2000)

- *Practical application* to work analysis, task analysis, new technology innovation, and usability evaluation in commercially important products and services (Gruen, 2000, 2001; Lafreniére, 1996; Muller, 2001; Muller et al., 1995b; Sanders, 2000; Tudor et al., 1993; Tschudy et al., 1994)

- *Cocreation of new ideas* and children's articulation and self-advocacy (Druin, 1999; Druin et al., 2000)

Photographs

There are many ways to tell stories. One approach that has informed recent PD work is end-user photography. Patton (2000) noted that both taking pictures and organizing them into albums are familiar activities to most people in affluent countries. These activities allow end users to enter into a kind of native ethnography, documenting their own lives. In keeping with the issues raised in the preceding Stories section, it is important that the informants themselves (the end users) control both the camera and the selection of images (see Bolton, 1989, for a set of discussions of the uses and abuses of documentary photography). They thus become both authors and subjects of photographic accounts of their activities. This dual role leads to one kind of hybridity, in which the photographic activities partake of both the world of common social life, and the world of documenting and reporting on working conditions.

In an exploration of products for mobile knowledge workers, Dandavate et al. (2000) similarly asked their informants to take pictures as part of a documentation of the working lives. In their study, informants were also invited to construct collages of their working lives, selectively reusing the photographs (among other graphic items) in those collages. The collages were, in effect, one type of interpretation by the photographers of their own photographs. Similarly to Patton's work, Dandavate et al. asked their informants to go out of their conventional professional roles as office workers (but well within their roles as members of an affluent culture) in the activity of taking the photographs. Dandavate et al. asked their informants to go even further out of role through the construction of the collages based on their photographs and the interpretation of the collages. The activities were thus marginal, partaking of attributes of informal life and professional life, of familiar and unfamiliar activities. They concluded that the photographic work led to new learnings and understandings that had not been accessible through observational studies, as well as a stronger sense of ownership by their informants in the outcome of the study.

Noble and Robinson (2000) formed an alliance between an undergraduate design class at Massey University and a union of low-status service workers, developing photodocumentaries of service work. The photographs served as a kind of hybrid boundary object (Star & Griesemer, 1989): For the students, the photographs were composed artifacts of design, whereas for the union members, the photographs were common and casually produced snapshots. Discussions between union members and students were rich, conflicted, and productive as they negotiated the status and meaning of these hybrid objects. These discussions—and the exhibits and posters that they produced (i.e., the collective actions of the students and the union members)—could not have been successful without mutual learning and construction of new understandings.

Third Space. End-user photography is an interesting case of hybridity and the production of third spaces. Photography is a good example of an "in-between" medium, one that is part of many people's informal lives (Dandavate et al., 2000; Noble & Robinson, 2000; Patton, 2000) but that is also an intensively studied medium of communication and argumentation (Bolton, 1989; Noble & Robinson, 2000). Photography occurs at the margin of most people's work and yet can easily be incorporated into it.

The resulting photographs in these projects have attributes of their dual worlds; they are partially informal and quotidian and partially formal and documentary. Discussions around the photographs and transformation of the photographs into photo narratives (Patton, 2000) or collages (Dandavate et al., 2000) can lead to mutual learning and new ideas, particularly through the inclusion of the voices of the photographers, the viewers, and especially the people depicted in the photographs (Noble & Robinson, 2000).

Claimed Benefits. The use of end-user photographs appears to be new and experimental, and there are few strongly supported claims of benefits. Informal claims of success and contribution include the following:

- *Richer, contextualized communication medium* between end-users and designers (in some cases, the designers were not, themselves, software professionals)

- *Stronger engagement* of designers with end users' worlds
- *Enhanced sharing* of views and needs among end users, leading to stronger articulation by them as a collective voice

Dramas

Drama provides another way to tell stories—in the form of theatre or of video. One of the important tensions with regard to drama in PD is the question of whether the drama is considered a finished piece or a changeable work in progress.

Many PD drama practitioners make reference to Boal's Theatre of the Oppressed (Boal, 1974/1992). Boal described theatrical techniques whose purpose was explicitly to help a group or a community find its voice(s) and articulate its position(s). The most influential of Boal's ideas was his Forum Theatre, in which a group of nonprofessional actors performs a skit in front of an audience of interested parties. The outcome of the skit is consistent with current events and trends—often to the dissatisfaction of the audience. The audience is then invited to become authors and directors of the drama, changing it until they approve of the outcome.

A second technique of interest involves the staging of a tableau (or a "frozen image," in Brandt & Grunnet, 2000) in which a group of nonprofessional actors positions its members as if they had been stopped in the middle of a play. Each member can tell what s/he is doing, thinking, planning, and hoping.

Forum Theatre was used informally in the UTOPIA project and other early Scandinavian research efforts (Ehn & Kyng, 1991; Ehn & Sjögren, 1991), addressing the question of new technologies in newspaper production. Changes in work patterns and work-group relations were acted out by software professionals in the end users' workplace, using cardboard and plywood prototypes, in anticipation of new technologies. The workers served as the audience and critiqued the envisioned work activities and working arrangements. The drama was carried out iteratively, with changes, until it was more supportive of the skilled work of the people in the affected job titles. The researchers made repeated visits with more detailed prototypes, again using the vehicle of a changeable drama, to continue the design dialogue with the workers. This work was widely credited with protecting skilled work from inappropriate automation, leading to a product that increased productivity while taking full advantage of workers' skills.

Brandt and Grunnet (2000) made a more formal use of Boal's Forum Theatre and "frozen images" in the two projects described earlier (see Sitings). Working with refrigeration technicians in the Smart Tool project, they and the technicians enacted work dramas and tableaux around four fictitious workers, leading to insights about the technicians' work and the technological possibilities for enhanced support of that work. Here is a description of one use of Forum Theatre:

[T]he stage was constructed of cardboard boxes which in a stylized way served as . . . the different locations in the scenario. At first the service mechanics sat as an audience and watched the play. After the first showing of the "performance" the refrigeration technicians were asked to comment and discuss the dramatized scenario critically. . . .

The role of the refrigeration technicians changed from being a passive audience into being directors with an expert knowledge. The users recognized the situations shown in the dramatized scenario. . . . Because of the openness of the scenario there was a lot of "holes" to be filled out. For instance, one . . . technician explained that he preferred to solve the problems himself instead of calling his boss. This information meant that the Smart Tool should be able to help him solve his problems while being in his car. . . . Another [technician] wanted to have personal information that his boss was not allowed . . . [to] access. (p. 14)

Incidents were analyzed through tableaux. The designers positioned themselves in the "frozen image" of the work situation and then led a discussion of (a) the work activities that were captured in the stopped action and (b) the work relations in which each particular tableau was embedded.

Muller et al. (1994) presented a related tutorial demonstration piece called Interface Theatre, with the stated goal of engaging a large number of interested parties (gathered in an auditorium, for example) in a review of requirements and designs. In Interface Theatre, software professionals acted out a user interface "look and feel" using a theatrical stage as the screen, with each actor playing the role of a concrete interface component (e.g., Kim the Cursor, Marty the Menubar, Dana the Dialogue box).

Pedersen and Buur (2000; see also Buur et al., 2000), following previous work of Binder (1999), collaborated with industrial workers to make videos showing proposed new work practices and technologies. After a collaborative analysis of the work (see Games), workers acted out their new ideas and took control of which action sequences were captured on video for subsequent explanation to other workers and management.

Young (1992) made a participatory version of Vertelney's (1989) method of video prototyping. In Vertelney's approach, the designer constructed a stop-action animation of the appearance and dynamics of a user interface, using paper- and pencil-materials (see Low Tech Prototypes) to draw user interface (UI) components. The components were placed under a video camera, and the designer moved the components as they would occur in a software interface. When an event occurred (e.g., a pull-down menu, or a pop-up dialogue box), the designer stopped the camera, placed the new UI component on under the camera, and then continued recording. Young's innovation was to include users as crafters of UI components and as directors of the animated events.

Finally, Salvador and Sato (1998, 1999) used acted-out dramas as triggers for questions in a setting similar to a focus group.

Although all of these practices are loosely tied together through the use of drama, there are important contrasts. One important dimension of difference is the extent to which the drama is improvised in the situation or scripted in advance. Boal's techniques make a crucial use of improvisation by the user audience to change the action and outcome of the drama. This theme is most clearly seen in the work of Brandt and Grunnet (2000), Ehn and Sjögren (1986, 1991), and Muller et al. (1994).

Young's work (1992) takes an intermediate position. Users contribute to the creation of Young's video prototypes and can influence the prototype during its production. Once the prototype has been completed, however, the video itself is relatively

fixed an unchangeable, unless the participants return to the cameras and paper-and-pencil materials to craft a new video. Similarly, Buur and colleagues aided users in constructing relatively unchangeable video descriptions.

At the opposite extreme is the work of Salvador and colleagues (Salvador & Howells, 1998; Salvador & Sato, 1998, 1999). Their work uses live dramas as points of departure for discussions with the audience. Their dramas come *from* the software professionals *to* the users and are left relatively unchanged. The point of the dramas in their work is to trigger discussions, and a critical success component of those discussions is that the actors are members of the discussion and can engage with the end users about their characters' thoughts and actions.

Third Space. Taken as a somewhat diverse participatory genre, the dramatic approaches provide many of the aspects of hybridity reviewed in the cultural studies introduction to this chapter. Drama brings a strong overlap of the world of end users and the world of software developers, showing concrete projections of ideas from one world into the other world and, in most uses, allowing modification of those ideas. Drama is marginal to the work domains of most software professionals and most end users and thus moves all parties into an ambiguous area where they must negotiate meaning and collaboratively construct their understandings. Agreements, conflicts, and new ideas can emerge as their multiple voices and perspectives are articulated through this rich communication medium.

Claimed Benefits. Similarly to end-user photography, most of the theatrical work has the feel of experimentation. It is difficult to find clear statements of advantages or benefits of these practices (see Conclusions). In general, practitioners and researchers made the following claims:

- *Building bridges* between the worlds of software professionals and users
- *Enhancing communication* through the use of embodied (i.e., acted-out) experience and through contextualized narratives
- *Engaging small and large audiences* through direct or actor-mediated participation in shaping the drama (influencing the usage and design of the technology)
- *Increasing designers' empathy* for users and their work
- *Simulating use of not-yet-developed tools* and technologies ("dream tools," Brandt & Grunnet, 2000) to explore new possibilities
- *Fuller understanding* by focus group members, leading to a more informed discussion

Games

From theory to practice, the concept of games has had an important influence in participatory methods and techniques. Ehn's theoretical work emphasized the negotiation of language games in the course of bringing diverse perspectives together in participatory design (Ehn, 1988; for applications of this theory, see Ehn & Kyng, 1991; Ehn & Sjögren, 1986, 1991). In this view, part

of the work of a heterogeneous group is to understand how to communicate with one another—and of course communication isn't really possible on a strict *vocabulary* basis but requires an understanding of the *perspectives* and *disciplinary cultures* behind the words (Bachmann-Medick, 1996; Muller, 1997a, 1997b, 1999b). Thus, the work of heterogeneous teams is, in part, the "mutual validation of diverse perspectives" that Bødker et al. (1988) advocated.

Games have also been an important concept in designing practices, with the convergent strategies of enhanced teamwork and democratic work practices within the team. Muller et al. (1994) explained the concepts as follows:

When properly chosen, games can serve as levelers, in at least two ways. First, games are generally outside of most workers' jobs and tasks. They are therefore less likely to appear to be "owned" by one worker, at the expense of the alienation of the non-owners. Second, . . . [PD] games . . . are likely to be novel to most or all of the participants. Design group members are more likely to learn games at the same rate, without large differences in learning due to rank, authority, or background. . . . This in turn can lead to greater sharing of ideas. . . .

In addition, games . . . can help groups of people to cohere together [and] communicate better. One of the purposes of games is enjoyment—of self and others—and this can both leaven a project and build commitment among project personnel. (pp. 62–63)

Derived from Ehn's (1988) theoretical foundation, Ehn and Sjögren (1986, 1991; see also Bødker, Grønbæk, & Kyng, 1993) adopted a "design-by-playing" approach, introducing several games into PD practice:

- *Carpentopoly*, a board game concerned with business issues in the carpentry industry
- *Specification Game*, a scenario-based game based on a set of "situation cards," each of which described a workplace situation (Players—members of the heterogeneous analysis/design team—took turns drawing a card and leading the discussion of the work situation described on the card)
- *Layout Kit*, a game of floorplans and equipment symbols for a workers' view of how the shop floor should be redesigned (see also Horgan, Joroff, Porter, & Schön, 1998)
- *Organization Kit and Desktop Publishing Game*, a part of the UTOPIA project (Ehn & Kyng, 1991), in which cards illustrating components of work or outcomes of work were placed on posters, with annotations

Petersen and Buur (2000) extended the Layout Kit in new ways. Collaborating with workers at Danfoss, they jointly created a board game for laying out new technologies in an industrial plant:

A map of the plant layout served as the game board. . . . Foam pieces in different colors and shapes worked as game pieces for the team to attach meaning to. . . . Often, in the beginning of the game, the placement of the piece was only accepted when touched by almost everybody. . . . The participants were forced to justify the placement, which fostered a fruitful dialogue about goals, intentions, benefits, and effects. People were asking each other such things as . . . "what if we change this?," "on our plant we do this, because . . . ," "would you benefit from this?." (pp. 96–98)

The games became the foundation of the videos produced in collaboration with the workers (described in Dramas).

Buur et al. (2000) extended the Specification Game, making a game from the outcome of a participatory ethnographic analysis of work at an industrial plant. They first collected video observations from work activities and developed a set of 60 to 70 video excerpts for further discussion. They next constructed a set of cards, one for each video excerpt, with a still-frame image from the video displayed on each card. Game participants then grouped these cards into thematic clusters, organized the clusters, and analyzed the subsets of actions in each cluster (for a related nongame technique, see affinity diagramming in Beyer & Holtzblatt, 1998).

We took the concept of games in a different direction, for use in non-Scandinavian workplaces, by introducing several new games (Muller et al., 1994):

- *CARD*, a card game for laying out and critiquing an existing or proposed work or activity flow (see Stories)
- *PICTIVE*, a paper-and-pencil game for detailed screen design (Muller et al., 1995b)
- *Icon Design Game*, a guessing game for innovating new ideas for icons (this game assumes subsequent refinement by a graphic designer)
- *Interface Theatre*, for design reviews with very large groups of interested parties (see Dramas)

Our games emphasized hands-on, highly conversational approaches to discussing both the user interface concept itself and the work processes that it was intended to support. We attempted to foster an informal and even playful tone, for the reasons sketched in the earlier quotation.

Third Space. Each of these 10 games took all of its players outside of their familiar disciplines and familiar working practices but strategically reduced the anxiety and uncertainty of the situation by using the social scaffolding of games. Each game required its players to work together through mutual learning to understand and define the contents of the game and to interpret those contents to one another in terms of multiple perspectives and disciplines. The conventional authority of the software professionals was thus replaced with a shared interpretation based on contributions from multiple disciplines and perspectives.

Claimed Benefits. Participatory design work with games has been claimed to lead to the following benefits:

- *Enhanced communication* through the combination of diverse perspectives
- *Enhanced teamwork* through shared enjoyment of working in a gamelike setting
- *Improved articulation* of the perspectives, knowledge, and requirements of workers
- *New insights* leading to important new analyses and designs with documented commercial value

Constructions

Preceding sections have considered hybridity in participatory activities, such as sitings, workshops, stories, photography, dramas, and games. This section continues the survey of participatory practices that bring users and software professionals into unfamiliar and ambiguous "third space" settings. In this section, I focus on the collaborative construction of various concrete artifacts:

- *Physical reflections of a cocreated language* of analysis and design
- *Descriptions of work* in unfamiliar media
- *Low-tech prototypes* for analysis and design
- *High-tech prototypes* for design and evaluation

Language. The precedings section noted Ehn's theoretical work on *PD as language games* (Ehn, 1988). Ehn's interest converges with Bhabha's "third space" argument (Bhabha, 1984): Part of the characterization of hybridity was the negotiation and cocreation of working language and meaning. This section takes Ehn's position seriously and considers the role of language creation in participatory practices that lead to hybridity.

Several projects have made physical objects into a kind of vocabulary for work analysis, design, or evaluation. The cards described in the preceding section (Games) are examples (Buur et al., 2000; Ehn & Sjögren, 1986, 1991; Lafreniére, 1996; Muller, 2001; Muller et al., 1995b; Tschudy et al., 1994). In each of these methods, the cards became a kind of "common language" (e.g., Muller et al., 1995b) through which the design team communicated with one another and with their labor and management clients.

In two of the methods, the cards themselves were acknowledged to be incomplete, and part of the work of the team was to develop and refine the cards so as to reflect their growing understanding and their new insights (Lafreniére, 1996; Muller, 2001). Team members (users and others) were encouraged to disregard, if appropriate, the template of information on each card, up to and including the decision to turn the card over and write on the back. In subsequent sessions, the concepts that were written on the blank backs of cards usually became new kinds of cards. The working vocabulary of the team thus grew as the shared understanding of the team grew. This extensibility of the set of cards was observed in nearly all sessions but was particularly important in sessions that were envisioning future technologies or future work practices. The cards thus became a point of hybridity, where assumptions were questioned and challenged, where extensive and polyvocal dialogue was required for the team to assign meaning to the cards, where conflicts were revealed and resolved and where the team had to construct its understanding and its language.

Similarly, the board games of Ehn and Sjögren, and especially of Pedersen and Buur (2000), used deliberately ambiguous playing pieces. The analysis team had to assign meaning to the pieces, and did so in a collaborative way.

Chin, Schuchardt, Myers, and Gracio (2000), working with a community of physical scientists who were not software

professionals, introduced softwarelike flowcharts to their clients (see Kensing & Munk-Madsen, 1993, for a discussion of the relationship between concrete tools and abstract tools). This work shared, with the other work reviewed in this section, aspects of symbol ambiguity and language cocreation:

> To attune scientists to the construction of workflow diagrams, we provided them a simple, informal example of how a meteorologist might diagram his [sic] work in collecting and reporting weather conditions.... Although we used circles and arrows in our example, we did not impose any specific symbology or rules on the scientists' construction of workflow diagrams.... At times, the scientists did struggle in developing some diagrams, but the labor was mostly centered on the elucidation of the research processes rather than the mechanics of diagramming. (p. 32)

Third Space. Common to all of these projects was the cocreation of a physically represented language, both within the team and from the team to its clients and stakeholders. This kind of lay linguistic work requires mutual education and mutual validation for the new language components to have meaning to all of the parties. These negotiations of multiple knowledges are at the heart of the "third space" proposal of Bhabha (1984).

Claimed Benefits. Most of these projects involved a number of activities and a number of aspects of hybridity. It is difficult to determine how much of their successes were specifically due to the language-related components. Benefits that *may* have resulted from the negotiation and cocreation of language include the following:

- *Enhanced understandings* of one anothers' perspectives and needs
- *Critical examinations of assumptions* underlying the ways that each party expressed its perspective
- *Shared ownership of the language* and its physical manifestation (cards, flowcharts, game pieces)
- *Improved communication* within the team and from the team to interested outsiders (clients, stakeholders)

Making Descriptive Artifacts

Another way of moving end users into unfamiliar and hence reflective experiences is to ask them to use "projective" or artistic methods to report on their experiences and needs. In one sense, these methods produce another kind of language of expression and therefore may have been included in the preceding section. Because the outcomes are so distinctively different from the language-oriented work of the preceding section, I thought it best to review this work in its own section.

Sanders employed user-created collage in her participatory practice for a number of years (Sanders, 2000; see also Dandavate et al., 2000; Sanders & Branaghan, 1998; Sanders & Nutter, 1994). The choice of collage is of course strategic: Relatively few people make collages as part of their work activities, and relatively few people interpret their collages to one another as part of their work conversations. Yet the content of the collages is strongly anchored in what people know. The collages

thus become marginal constructions—not part of any defined workplace field or discipline, but informed by familiar knowledges. The novelty of the collage encourages the challenging of assumptions, and the interpretation and presentation of collages encourages mutual learning across the diversity of experiences and knowledges of the participants.

For completeness, I make reference to the work of Noble and Robinson (2000) on collaborative creation of photo documentaries and of Patton (2000) on end user creation of photo collages, reviewed in Photographs. Their work also produced descriptive artifacts that took users and their collaborators into unfamiliar areas.

Third Space. These methods have in common the use of a nonstandard medium for making users' needs known and for developing new insights in a workplace setting. The making of collages may be new for many participants. They are thus in a kind of "third space" between their work culture and the artistic or expressive culture of collages, and they have to reflect on the differences as they construct their approach to making collages of their own experiences.

It is not clear, in Sanders's work, whether the collage is created collaboratively among end users or whether each collage is a solitary production. If the collage creation is done collaboratively, then it might give rise to some of the other attributes of hybridity in Table 54.1 (e.g., challenging assumptions, cocreation of meanings and collective actions, dialogues).

Claimed Benefits. Basing her claims on years of practice with collages and related practices, Sanders (2000) claimed the following benefits:

- *Using visual ways* of sensing, knowing, remembering, and expressing
- *Giving access and expression to emotional side* of experience
- *Acknowledging the subjective perspective* in people's experiences with technologies
- *Revealing unique personal histories* that contribute to the ways that people shape and respond to technologies

Low-tech Prototypes

Beaudouin-Lafon and Mackay provided a chapter in this text on prototyping (chapter 52), including participatory prototyping. Therefore, I include only a brief account in this chapter so as not to duplicate their efforts.

Low-tech prototypes may lead to "third space" experiences because they bring people into new relationships with technologies—relationships that are "new" in at least two important ways. First, the end users are often asked to think about technologies or applications that they have not previously experienced. Second, in *participatory* work with low-tech prototypes, end users are being asked to use the low-tech materials to reshape the technologies—a "design-by-doing" approach (Bødker et al., 1993). In this way, participatory work with low-tech prototypes involves much more user contribution and user

initiative than the more conventional use of "paper prototypes" as surrogates for working systems in usability testing (e.g., Daly-Jones, Bevan, & Thomas, 1999; Rettig, 1994).

The UTOPIA project provided impressive demonstrations of the power of low-tech cardboard and plywood prototypes to help a diverse group to think about new technologies, office layouts, and new working relations that might result from them (Bødker et al., 1987, 1988, 1993; Ehn & Kyng, 1991). Subsequent projects to translate this work to North America led to the PICTIVE method of paper-and-pencil constructions of user interface designs by heterogeneous design teams (Muller et al., 1995b); prototyping of consumer appliances using foam core and hook-and-loop attachments (Sanders & Nutter, 1994); and a more experimental simulation of e-mail, using paper airplanes (Dykstra & Carasik, 1991).

Third Space. Low-tech prototyping has a reputation for bringing new insights through the combination of diverse perspectives. The UTOPIA project is widely credited with mutual education among shop-floor print workers and computer systems researchers. Our experiences with PICTIVE almost always involved mutual education. Understanding and changing the artifact become important arenas for people to explore their understandings of one anothers' positions, to question one anothers' approaches, to discover and resolve conflicts, to engage in combinations of views leading to plans for collective action, and to accommodate heterogeneity of views and interests.

Claimed Benefits. The low-tech participatory prototyping approaches have been extraordinarily influential, with adoption on four continents. Claimed benefits include

- *Enhanced communication and understanding* through grounding discussions in concrete artifacts
- *Enhanced incorporation of new and emergent ideas* through the ability of participants to express their ideas directly via the low-tech materials
- *Enhanced working relations* through a sense of shared ownership of the resulting design
- *Practical application with measured successes* in using low-tech design approaches to real product challenges, achieving consequential business goals

Evolutionary Prototyping and Cooperative Prototyping

This last section on participatory methods is concerned with software prototyping. As noted above, I rely on Beaudouin-Lafon and Mackay's chapter in this book to cover prototyping in greater depth and breadth. I include this brief overview to make my survey of hybridity in participatory practices complete.

Bødker and Grønbæk (1991) and Madsen and Aiken (1993) explored the potential of cooperative prototyping in several projects, using different technology infrastructures. In general, they found that this approach led to enhanced communication with end users, improved incorporation of end-user insights into the prototypes, and stronger collective ownership

and collective action-planning by the team. They also observed time-consuming breakdowns in the design process itself when new ideas required significant programming effort.

In a different prototyping approach, a system is delivered to its end-users as series of iterative prototypes, each of which gradually adds functionality (e.g., Anderson & Crocca, 1993; Bertelsen, 1996; Trigg, 2000). What appears to be critical is that the prototype functions as a *crucial artifact* in the end users' work, for example, a resource of documents for librarians (Anderson & Crocca, 1993), an online event checklist that served as the crucial coordination point for the work of diverse contributions (Bertelson, 1996), or a database supporting funding work in a nonprofit organization (Trigg, 2000). Trigg (2000) provided a series of observations and tactical recommendations about how to engage the users in the evaluations that both they and the software professionals had agreed were needed.

Third Space. This brief survey of cooperative prototyping and "iterative delivery" approaches shows several aspects of hybridity. In the case of cooperative prototyping, the cooperative work may be done in a physical third space that is neither the end users' office nor the software developers' office (Sitings). In the case of the delivery of iterated prototypes, each prototype is presented in the end users' setting but is unusual and only partially functional and thus occasions reflection on its nature, on its role in the end users' work, and thus on the work itself. In both cases, the invitation (or perhaps the necessity) of the end users' actions to help shape the technology becomes an important means of refocusing their attention, as well as the attention of the software developers. The ensuing conversations are concerned with the interlinked feasibility of changes to technology and to work practices, with attributes of hybridity including polyvocal dialogues, challenging one anothers' assumptions, and developing plans for collective actions.

Claimed Benefits. Some of the virtues of the low-tech prototyping approaches have also been claimed for the cooperative prototyping and "iterative delivery" approaches:

- *Enhanced communication and understanding* through grounding discussions in concrete artifacts
- *Enhanced working relations* through a sense of shared ownership of the resulting design

Additional claims for software-based prototypes include

- *Earlier understanding of constraints* posed by the practical limitations of software
- *Improved contextual grounding of the design* in the end users' work practices

CONCLUSION

My theme has been hybridity and the ways in which selected methods in participatory design may bring useful attributes of hybridity or third space approaches into HCI work. I considered eight trends in PD—selection of sites of shared work,

TABLE 54.2. Hybridity in Participatory Practices

Attribute	Sittings	Workshops	Stories	Photography	Dramas	Games	Language	Descriptive	Prototype
Overlap/in-betweenness	?	+	−	+	+	+	+	+	+
Marginality	+	+	−	?	+	+	?	+	?
Novelty	+	+	?	?	+	+	+	+	+
Uncertain/shared "ownership"	?	+	?	−	+	+	+	−	−
Selected attributes	+	?	+	+	−	+	+	−	+
Conflicts	+	+	+	−	+	−	+	−	+
Questioning assumptions	+	?	+	+	+	+	+	?	+
Mutual learning	+	+	+	+	+	+	+	?	+
Synthesis of new ideas	?	+	+	+	+	+	?	+	+
Negotiation/(co-)creation	+	+	+	+	+	+	+	+	+
Identities	−	−	+	+	−	?	?	+	?
Working language	−	?	+	+	−	+	+	+	+
Working assumptions and dynamics	+	?	+	+	+	+	+	?	+
Understandings	+	+	+	+	+	+	+	+	+
Relationships	?	+	+	+	−	+	?	+	?
Collective actions	?	+	?	+	?	?	?	+	+
Dialogues	+	+	+	+	+	+	+	+	+
Polyvocality	+	+	+	+	+	+	+	+	+
What is considered to be data?	−	−	−	+	−	−	+	+	−
What are the rules of evidence?	−	−	−	+	−	−	+		▬
How are conclusions drawn?	−	−	−	?	−	−	+	−	−
↓authority − ↑interpretation	+	?	+	+	+	+	+	?	+
↓individualism − ↑collectivism	?	+	?	+	?	+	?	?	+
Heterogeneity as the norm	+	+	+	+	−	+	+	+	+

Note. ? = not sure.

workshops, stories, end-user photography, dramas, creation of shared languages, descriptive artifacts (low-tech prototypes), and working prototypes—and I explored how each of these categories of practice may contribute to hybridity and what advantages may result. The deliberate and selective use of hybridity has led to powerful methods in PD for increasing communication effectiveness, team coherence, innovation, and quality of outcome. Hybridity is thus at the heart of PD, fostering the critical discussions and reflections necessary to challenge assumptions and to create new knowledges, working practices, and technologies. When we consider HCI as a set of disciplines that lie between the space of work and the space of software development, we see that the hybrid third spaces developed within PD have much to offer HCI in general.

Table 54.2 summarizes the discussion of hybridity in PD, using the criteria derived from cultural studies (Table 54.1) and the experiences described in the eight areas of practice. Table 54.2 shows different patterns of hybridity for different methods, techniques, and practices.

Certain attributes are relatively common across practices (e.g., in-betweenness, questioning assumptions, negotiation, and heterogeneity as the norm). Other attributes are relatively rare (e.g., considerations of what constitutes legitimate data for analysis or design, how those data are analyzed as evidence, and how conclusions are drawn in each of the several fields that are represented in a team). These are difficult questions in the study of disciplinarity (Chandler, Davidson, & Harootunian, 1994; Klein, 1996), so it is perhaps not surprising that there is relatively weak support for their exploration in participatory

practices. For projects in which these are pivotal questions, we may need new methods that leverage hybridity in new ways. I hope that this survey of PD practices for creating third spaces will lead to new practices that strengthen these missing attributes. Conversely, I hope that new work in PD and HCI can help to ground some of the cultural studies discussions in new ways.

This chapter would not be complete without a list of unsolved problems in participatory design:

- *Participation by nonorganized workforce.* The field of PD has long been concerned about how to engage in meaningful participative activities with workers or others who are not organized into a group with collective bargaining power or other collective representation (e.g., Greenbaum, 1993, 1996; van den Besselaar, Greenbaum, & Mambrey, 1996). This has been a particularly difficult problem when we have tried to compare methods from one country (and political culture) to another (e.g., Muller et al., 1991)

- *Evaluation and metrics.* One of the weaknesses of the literature on participatory practices is the dearth of formal evaluations. There is a small set of papers that have examined software engineering projects across companies and have found positive outcomes related to end-user participation (Cotton, Vollrath, Froggatt, Lengnick-Hall, & Jennings, 1988; Saarinen & Saaksjarvi, 1989). I have been unable to discover any formal experiments comparing participatory methods with nonparticipatory methods in a credible workplace context. Indeed, such studies would be difficult to perform, because they would

require that a product be implemented and marketed twice (once with participation, and once without). The problem is made more difficult because measurements and metrics of organizational outcomes, user participation, and user satisfaction are currently vexing research issues (e.g., Garrety & Badham, 1998; Kappelman, 1995; for review, see Gasson, 1995).

• *Universal usability and "universal participation"?* Nearly all of the practices described in this chapter (and in the longer set of methods in Muller et al., 1997) are strongly visual and require hands-on manipulation of materials. These approaches violate the emerging requirements of universal usability for people with visual or motor disabilities (see, e.g., Universal Usability Fellows, 2000, and the *Proceedings of the Conference on Universal Usability*[12]; see also section IIIB in this book). Ironically, participatory design, which was founded on the principle of political inclusion, needs new ideas to be universally inclusive (Luck, 2000).

References

For a more general PD bibliography, see http://www.cpsr.org/conferences/pdc98/bibliography.html.

Anderson, W. L., & Crocca, W. T. (1993). Engineering practice and codevelopment of product prototypes. *Communications of the ACM, 36,* 49–56.

Bachmann-Medick, D. (1996). Cultural misunderstanding in translation: Multicultural coexistence and multicultural conceptions of world literature. *Erfurt Electronic Studies in English, 7.* Retrieved May 2002 from http://webdoc.gwdg.de/edoc/ia/eese/artic96/bachmann/7_96.html

Beck, E. E. (1996). P for political? Some challenges to PD towards 2000. In *Proceedings of the Participatory Design Conference 96* (pp. 117–125). Cambridge, MA: CPSR.

Beeson, I., & Miskelly, C. (1998). Discovery and design in a community story. *Proceedings of the Participatory Design Conference 98* (pp. 1–10). Seattle: CPSR.

Beeson, I., & Miskelly, C. (2000). Dialogue and dissent in stories of community. In *Proceedings of the Participatory Design Conference 2000* (pp. 147–155). New York: CPSR.

Bertelsen, O. W. (1996). The Festival checklist: Design as the transformation of artifacts. *Proceedings of the Participatory Design Conference 96* (pp. 93–102). Cambridge, MA: CPSR.

Beyer, H., & Holtzblatt, K. (1998). *Contextual design: Defining customer-centered systems.* San Francisco: Morgan Kaufmann.

Bhabha, H. K. (1994). *The location of culture.* London: Routledge.

Binder, T. (1999). Setting the scene for improvised video scenarios. In *CHI 99 Extended Abstracts* (pp. 230–231). New York: ACM Press.

Bjerknes, G., & Bratteteig, T. (1995). User participation and democracy: A discussion of Scandinavian research on system development. *Scandinavian Journal of Information Systems, 7,* 73–98.

Bjerknes, G., Ehn, P., & Kyng, M. (Eds.). (1987). *Computers and democracy: A Scandinavian challenge.* Brookfield, VT: Gower.

Blomberg, J., Giacomi, J., Mosher, A., & Swenton-Wall, P. (1993). Ethnographic field methods and their relation to design. In D. Schuler & A. Namioka (Eds.), *Participatory design: Principles and practices* (pp. 123–156). Hillsdale, NJ: Lawrence Erlbaum Associates.

Boal, A. (1992). *Games for actors and non-actors* (A. Jackson, Trans.). London: Routledge. (Original work published 1974)

Bolton, R. (Ed.). (1989). *The contest of meaning: Critical histories of photography.* Cambridge, MA: MIT Press.

Braa, K. (1996). Influencing qualities of information systems—future challenges for participatory design. *Proceedings of the Participatory Design Conference 96* (pp. 163–172). Cambridge, MA: CPSR.

Brandt, E., & Grunnet, C. (2000). Evoking the future: Drama and props in user centered design. In *Proceedings of the Participatory Design Conference 2000* (pp. 11–20). New York: CPSR.

Briefs, U., Ciborra, C., & Schneider, L. (1983). *System design for, with, and by the users.* Amsterdam: North-Holland.

Buur, J., Binder, T., & Brandt, E. (2000). Taking video beyond "hard data" in user centred design. In *Proceedings of the Participatory Design Conference 2000* (pp. 21–29). New York: CPSR.

Bødker, S. (1990). *Through the interface: A human activity approach to user interface design.* Hillsdale, NJ: Lawrence Erlbaum Associates.

Bødker, S., Ehn, P., Kyng, M., Kammersgaard, J., & Sundblad, Y. (1987). A UTOPIAN Experience: On design of powerful computer-based tools for skilled graphic workers. In G. Bjerknes, P. Ehn, & M. Kyng (Eds.), *Computers and democracy: A Scandinavian challenge* (pp. 251–278). Brookfield, VT: Gower.

Bødker, S., & Grønbæk, K. (1991). Design in action: From prototyping by demonstration to cooperative prototyping. In J. Greenbaum & M. Kyng (Eds.), *Design at work: Cooperative design of computer systems* (pp. 197–218). Hillsdale, NJ: Lawrence Erlbaum Associates.

Bødker, S., Grønbæk, K., & Kyng, M. (1993). Cooperative design: Techniques and experiences from the Scandinavian scene. In D. Schuler & A. Namioka (Eds.), *Participatory design: Principles and practices* (pp. 157–176). Hillsdale, NJ: Lawrence Erlbaum Associates.

Bødker, S., Knudsen, J. L., Kyng, M., Ehn, P., & Madsen, K. H. (1988). Computer support for cooperative design. In *CSCW'88: Proceedings of the Conference on Computer Supported Cooperative Work* (pp. 377–394). New York: ACM Press.

Cameron, M. (1998). Design for safety: Working with residents to enhance community livability. *Proceedings of the Participatory Design Conference 98* (pp. 171–172). Seattle: CPSR.

Carrillo, R. (2000). Intersections of official script and learners' script in *Third Space*: A case study on Latino families in an after-school computer program. In *Proceedings of Fourth International Conference of the Learning Sciences.* Mahwah, NJ: Lawrence Erlbaum Associates.

Carroll, J. (Ed.). (1995). *Scenario-based design for human-computer interaction.* New York: John Wiley.

Chandler, J., Davidson, A. I., & Harootunian, H. (Eds.). (1994). *Questions of evidence: Proof, practice, and persuasion across the disciplines.* Chicago: University of Chicago Press.

Checkland, P. (1981). *Systems thinking, systems practice.* New York: John Wiley.

Chin, G., Schuchardt, K., Myers, J., & Gracio, D. (2000). Participatory workflow analysis: Unveiling scientific research processes with physical scientists. In *Proceedings of the Participatory Design Conference 2000* (pp. 30–39). New York: CPSR.

[12]Available through ACM, www.acm.org.

Clement, A., Kolm, P., & Wagner, I. (Eds.). (1994). *NetWORKing: Connecting workers in and between organizations* (IFIP Transactions A-38). Amsterdam: North-Holland.

Cotton, J. L., Vollrath, D. A., Froggatt, K. L., Lengnick-Hall, M. L., & Jennings, K. R. (1988). Employee participation: Diverse forms and different outcomes. *Academy of Management Review, 13*, 8–22.

Crabtree, A. (1998). Ethnography in participatory design. *Proceedings of the Participatory Design Conference 98* (pp. 93–105). Seattle: CPSR.

Daly-Jones, O., Bevan, N., & Thomas, C. (1999). *IN-USE 6.2 handbook of user-centred design*. Teddington, UK: Serco Usability Services. Retrieved May 2000 from http://www.ejeisa.com/nectar/inuse/6.2/index.htm

Dandavate, U., Steiner, D., & William, C. (2000). Working anywhere: Co-Design through participation. In *Proceedings of CoDesigning 2000* (pp. 101–110). London: Springer.

Docherty, P., Fuchs-Kittowski, K., Kolm, P., & Matthiessen, L. (1987). *System design for human development and productivity: Participation and beyond*. Amsterdam: North-Holland.

Druin, A. (1999). Cooperative inquiry: Developing new technologies for children with children. *Proceedings of CHI 99* (pp. 592–599). New York: ACM Press.

Druin, A., Alborzi, H., Boltman, A., Cobb, S., Montemayor, J., Neale, H., Platner, M., Porteous, J., Sherman, L., Simsarian, K., Stanton, D., Sundblad, Y., & Taxen, G. (2000). Participatory design with children: Techniques, challenges, and successes. *Proceedings of PDC 2000* (pp. 226–227). New York: CPSR.

Dykstra, E. A., & Carasik, R. P. (1991). Structure and support in cooperative environments: The Amsterdam Conversation Environment. *International Journal of Man-Machine Studies, 34*, 419–434.

Ehn, P. (1988). *Work-oriented design of computer artifacts*. Falköping, Sweden: Arbetslivcentrum/Almqvist and Wiksell International.

Ehn, P. (1993). Scandinavian design: On participation and skills. In P. S. Adler & T. A. Winograd (Eds.), *Usability: Turning technologies into tools*. New York: Oxford University Press. Retrieved May 2002 from http://www.cpsr.org/confcrences/pdc98/history.html

Ehn, P., & Kyng, M. (1987). The collective resource approach to systems design. In G. Bjerknes, P. Ehn, & M. Kyng (Eds.), *Computers and democracy: A Scandinavian challenge* (pp. 17–58). Brookfield, VT: Gower.

Ehn, P., & Kyng, M. (1991). Cardboard computers: Mocking-it-up or hands-on the future. In J. Greenbaum & M. Kyng (Eds.), *Design at work: Cooperative design of computer systems* (pp. 169–196). Hillsdale, NJ: Lawrence Erlbaum Associates.

Ehn, P., & Sjögren, D. (1986). Typographers and carpenters as designers. In *Proceedings of Skill-Based Automation*.

Ehn, P., & Sjögren, D. (1991). From system descriptions to scripts for action. In J. Greenbaum & M. Kyng (Eds.), *Design at work: Cooperative design of computer systems* (pp. 241–268). Hillsdale, NJ: Lawrence Erlbaum Associates.

Erickson, T. (1996). Design as story-telling. *Interactions*, 30–35.

Evanoff, R. (2000). The concept of "third cultures" in intercultural ethics. *Eubios Journal of Asian and International Bioethics, 10*, 126–129.

Fanderclai, T. (1995). MUDs in education: New environments, new pedagogies. *Computer-Mediated Communication, 2*, 8.

Fanderclai, T. (1996). Like magic, only real. In L. Cherny, & E. R. Weise (Eds.), *Wired women: Gender and new realities in cyberspace*. Seattle: Seal Press.

Floyd, C. (1987). Outline of a paradigm change in software engineering. In G. Bjerknes, P. Ehn, & M. Kyng (Eds.), *Computers and democracy: A Scandinavian challenge* (pp. 191–210). Brookfield, VT: Gower.

Fowles, R. A. (2000). Symmetry in design participation in the built environment: Experiences and insights from education and practice. In *Proceedings of CoDesigning 2000* (pp. 59–70). London: Springer.

Garrety, K., & Badham, R. (1998, June). *The four-dimensional politics of technology, or postmodernising participatory design*. Paper presented at Cultural Politics of Technology workshop, Centre for Technology and Society, Trondheim. Germany. Retrieved May 2002 from http://www.ntnu.no/sts/content/Papers/Four.html

Gasson, S. (1995). User involvement in decision-making in information systems development. In *Proceedings of 18th IRIS*. Gjern, Denmark: IRIS Association. Retrieved May 2002 from http://iris.informatik.gu.se/conference/iris18/iris1826.htm

Gärtner, J., & Wagner, I. (1995). *Political frameworks of systems design from a cross-cultural perspective* (IFIP WG.9.1 Workshop). Aarhus: IFIP.

Gjersvik, R., & Hepsø, V. (1998). Using models of work practice as reflective and communicative devices: Two cases from the Norwegian offshore industry. *Proceedings of the Participatory Design Conference 98* (pp. 107–116). Seattle: CPSR.

Greenbaum, J. (1993). A design of one's own: Towards participatory design in the United States. In D. Schuler & A. Namioka (Eds.), *Participatory design: Principles and practices* (pp. 21–40). Hillsdale, NJ: Lawrence Erlbaum Associates.

Greenbaum, J. (1996). Post modern times: Participation beyond the workplace. *Proceedings of the Participatory Design Conference 96* (pp. 65–72). Cambridge, MA: CPSR.

Greenbaum, J., & Kyng, M. (1991). *Design at work: Cooperative design of computer systems*. Hillsdale, NJ: Lawrence Erlbaum Associates.

Grenfell, M. (1998). *Border-crossing: Cultural hybridity and the rural and small schools practicum*. Presented at the Australian Association for Research in Education conference.

Gruen, D. (2000). Storyboarding for design: An overview of the process. (Lotus Research). Retrieved May 2002 from http://www.research.ibm.com/cambridge, under "Papers".

Gruen, D. (2001). *Stories in design tutorial*. IBM Make It Easy Conference.

Grønbæk, K. (1989). Extending the boundaries of prototyping—Toward cooperative prototyping. In *Proceedings of 12th IRIS* (pp. 219–239). Aarhus: IRIS Association.

Henderson, A., & Kyng, M. (1991). There's no place like home: Continuing design in use. In J. Greenbaum & M. Kyng (Eds.), *Design at work: Cooperative design of computer systems* (pp. 219–240). Hillsdale, NJ: Lawrence Erlbaum Associates.

Holmström, J. (1995). The power of knowledge and the knowledge of power: On the systems designer as a translator of rationalities. In *Proceedings of the 18th IRIS*. Göteborg: IRIS Association. Retrieved May 2002 from http://iris.informatik.gu.se/conference/iris18/iris1829.htm#E21E29

Horgan, T., Joroff, M. L., Porter, W. L., & Schön, D. A. (1998). *Excellence by design—Transforming workplace and work practice*. New York: John Wiley.

Kaindl, H., Constantine, L., Karat, J., & Muller, M. J. (2001). Methods and modeling: Fiction or useful reality? (panel). In *CHI 2001 Extended Abstracts* (pp. 213–214). New York: ACM Press.

Kappelman, L. (1995). Measuring user involvement: A diffusion of innovation perspective. *The DATABASE for Advances in Information Systems, 26*(2&3), 65–86.

Kensing, F., & Blomberg, J. (Eds.). (1998). *Computer Supported Cooperative Work* [special issue on participatory design], 7.

Kensing, F., & Madsen, K. H. (1991). Generating visions: Future workshops and metaphorical design. In J. Greenbaum & M. Kyng (Eds.),

Design at work: Cooperative design of computer systems (pp. 155–168). Hillsdale, NJ: Lawrence Erlbaum Associates.

Kensing, F., & Munk-Madsen, A. (1993). PD: Structure in the toolbox. *Communications of the ACM, 36*, 78–85.

Kensing, F., Simonsen, J., & Bødker, K. (1996). MUST—A method for participatory design. *Proceedings of the Participatory Design Conference 96* (pp. 129–140). Cambridge, MA: CPSR.

Klein, J. T. (1996). *Crossing boundaries: Knowledge, disciplinarities, and interdisciplinarities.* Charlottesville, VA: University Press of Virginia.

Krabbel, A., & Wetzel, I. (1998). The customization process for organizational package information systems: A challenge for participatory design. *Proceedings of the Participatory Design Conference 98* (pp. 45–54). Seattle: CPSR.

Kyng, M., & Matthiessen, L. (Eds.). (1997). *Computers in design and context.* Cambridge, MA: MIT Press.

Lafreniére, D. (1996). CUTA: A simple, practical, and low-cost approach to task analysis. *Interactions, 3*, 35–39.

Lanzara, G. F. (1983). The design process: Frames, metaphors, and games. In U. Briefs, C. Ciborra, & L. Schneider (Eds.), *Systems design for, with, and by the users* (pp. 29–40). Amsterdam: North-Holland.

Levinger, D. (1998). Participatory design history. Retrieved May 2002 from http://www.cpsr.org/conferences/pdc98/history.html

Luck, R. (2000). Does "inclusive design" require an inclusive design process? In *Proceedings of CoDesigning 2000* (pp. 71–79). London: Springer.

Lyotard, J.-F. (1984). *The post-modern condition: A report on knowledge.* Minneapolis: University of Minnesota Press.

MacLean, A., Carter, K., Lovstrand, L., & Moran, T. (1990). User-tailorable systems: Pressing the issues with buttons. *Proceedings of CHI '90* (pp. 175–182). New York: ACM Press.

Madsen, K. H. (1999). *Communications of the ACM* [special issue on usability in Scandinavia and the US], *42.*

Madsen, K. H., & Aiken, P. (1993). Experiences using cooperative interactive storyboard prototyping. *Communications of the ACM, 36*, 57–64.

Maher, M. L., Simoff, S. J., & Gabriel, G. C. (2000). Participatory design and communication in virtual environments. In *Proceedings of the Participatory Design Conference 2000* (pp. 127–134). New York: CPSR.

McLagan, P., & Nel, C. (1995). *The age of participation: New governance for the workplace and the world.* San Francisco: Berrett-Koehler.

Mitchell, W. J. T. (1995). Translator translated: Interview with cultural theorist Homi Bhabha. *Artforum, 33*(7), 80.

Mogensen, P., & Trigg, R. (1992). Artifacts as triggers for participatory analysis. *Proceedings of the Participatory Design Conference '92* (pp. 55–62). Cambridge, MA: CPSR.

Muller, M. J. (Ed.). (1994). *CPSR Newsletter* (3) [participatory design issue], *12.* May 2002 from http://www.cpsr.org/publications/newsletters/issues/1994/Summer1994/

Muller, M. J. (1997a). Ethnocritical heuristics for reflecting on work with users and other interested parties. In M. Kyng & L. Matthiessen (Eds.), *Computers in context and design* (pp. 349–380). Cambridge, MA: MIT Press.

Muller, M. J. (1997b). Translation in HCI: Formal representations for work analysis and collaboration. In *Proceedings of CHI 97* (pp. 544–545). New York: ACM Press.

Muller, M. J. (1999a). *Catalogue of scenario-based methods and methodologies* (Lotus Research Technical Report 99-06). Retrieved May 2002 from http://www.research.ibm.com/cambridge, under "Papers."

Muller, M. J. (1999b). Translation in HCI: Toward a research agenda (Lotus Research Technical Report 99-05). Retrieved May 2002 from http://www.research.ibm.com/cambridge, under "Papers."

Muller, M. J. (2001). Layered participatory analysis: New development in the CARD technique. In *Proceedings of CHI 2001* (pp. 90–97). New York: ACM Press.

Muller, M. J., Blomberg, J. L., Carter, K., Dykstra, E. A., Greenbaum, J., & Halskov Madsen, K. (1991). Participatory design in Britain and North America: Responses to the "Scandinavian challenge" (panel discussion). In *Proceedings of CHI'91* (pp. 389–392). New York: ACM Press.

Muller, M. J., Carr, R., Ashworth, C. A., Diekmann, B., Wharton, C., Eickstaedt, C., & Clonts, J. (1995a). Telephone operators as knowledge workers: Consultants who meet customer needs. In *Proceedings of CHI'95* (pp. 130–137). New York: ACM Press.

Muller, M. J., Hallewell Haslwanter, J. D., & Dayton, T. (1997). Participatory practices in the software lifecycle. In M. Helander, T. Landauer, & P. Prabhu (Ed.), *Handbook of human–computer interaction* (pp. 255–298). Amsterdam: Elsevier.

Muller, M. J., & Kuhn, S. (Eds.). (1993). *Communications of the ACM special issue on participatory design, 36.*

Muller, M. J., Tudor, L. G., Wildman, D. M., White, E. A., Root, R. W., Dayton, T., Carr, R., Diekmann, B., & Dykstra-Erickson, E. A. (1995b). Bifocal tools for scenarios and representations in participatory activities with users. In J. Carroll (Ed.), *Scenario-based design for human–computer interaction* (pp. 135–164). New York: Wiley.

Muller, M. J., White, E. A., & Wildman, D. M. (1993). Taxonomy of PD practices: A brief practitioner's guide. *Communications of the ACM, 36*, 26–28.

Muller, M. J., Wildman, D. M., & White, E. A. (1994). *Participatory design through games and other group exercises.* Tutorial at CHI '94 conference, Boston.

Mumford, E. (1983). *Designing human systems for new technology: The ETHICS method.* Manchester, UK: Manchester Business School.

Mumford, E., & Henshall, D. (1983). *Designing participatively: A participative approach to computer systems design.* Sandbach, UK: Manchester Business School. (Original work published 1979)

Nisonen, E. (1994). Women's safety audit guide: An action plan and a grass roots community development tool. *CPSR Newsletter, 12.* Retrieved May 2002 from http://www.cpsr.org/publications/newsletters/issues/1994/Summer1994/nisonen.html

Noble, A., & Robinson, C. (2000). For the love of the people: Participatory design in a community context. In *Proceedings of CoDesigning 2000* (pp. 81–91). London: Springer.

Noro, K., & Imada, A. S. (Eds.). (1991). *Participatory ergonomics.* London: Taylor and Francis.

Orr, J., & Crowfoot, N. C. (1992). Design by anecdote—the use of ethnography to guide the application of technology to practice. *Proceedings of the Participatory Design Conference '92* (pp. 31–37). Cambridge, MA: CPSR.

Patton, J. W. (2000). Picturing commutes: Informant photography and urban design. *Proceedings of the Participatory Design Conference 2000* (pp. 318–320). New York: CPSR.

Pedersen, J., & Buur, J. (2000). Games and moves: Towards innovative codesign with users. In *Proceedings of CoDesigning 2000* (pp. 93–100). London: Springer.

Reid, F. J. M., & Reed, S. E. (2000). Interaction and entrainment in collaborative design meetings. In *Proceedings of CoDesigning 2000* (pp. 233–241). London: Springer.

Rettig, M. (1994). Prototyping for tiny fingers. *Communications of the ACM, 37*, 21–27.

Robertson, T. (1996). Participatory design and participative practices in small companies. *Proceedings of the Participatory Design Conference 96* (pp. 35–43). Cambridge, MA: CPSR.

Robertson, T. (1998). Shoppers and tailors: Participative practices in small Australian design companies. *Computer Supported Cooperative Work, 7*, 205-221.

Robins, J. (1999). Participatory design (class notes). Champaign, IL: University of Illinois. Retrieved May 2002 from http://www.lis.uiuc.edu/~jrobins/pd/

Saarinen, T., & Saaksjarvi, M. (1989). The missing concepts of user participation: An empirical assessment of user participation and information system success. In *Proceedings of the 12th IRIS (Information System Research in Scandinavia)* (pp. 533-553). Aarhus, Denmark: Aarhus University.

Salvador, T., & Howells, K. (1998). Focus Troupe: Using drama to create common context for new product concept end-user evaluations." In *Proceedings of CHI '98* (pp. 251-252). New York: ACM Press.

Salvador, T., & Sato, S. (1998). Focus troupe: Mini-workshop on using drama to create common context for new product concept end-user evaluations. *Proceedings of the Participatory Design Conference 98* (pp. 197-198). Seattle. CPSR.

Salvador, T., & Sato, S. (1999). Methods tools: Playacting and focus troupes. Theater techniques for creating quick, intense, immersive, and engaging focus group sessions. *Interactions, 6*, 35-41.

Sanders, E. B.-N. (2000). Generative tools for co-designing. In *Proceedings of CoDesigning 2000* (pp. 3-12). London: Springer.

Sanders, E. B.-N., & Branaghan, R. J. (1998). Participatory expression through image collaging: A learning-by-doing experience. *Proceedings of the Participatory Design Conference '98* (p. 199) . Seattle: CPSR.

Sanders, E. B.-N., & Nutter, E. H. (1994). Velcro-modeling and projective expression: Participatory design methods for product development. In *Proceedings of the Participatory Design Conference '94* (p. 125). Chapel Hill, NC: CPSR.

Schuler, D., & Namioka, A. (1993). (Eds.). *Participatory design: Principles and practices*. Hillsdale, NJ: Lawrence Erlbaum Associates.

Segall, P., & Snelling, L. (1996). Achieving worker participation in technological change: The case of the flashing cursor. In *Proceedings of the Participatory Design Conference 96* (pp. 103-109). Cambridge, MA: CPSR.

Slater, J. (1998). Professional misinterpretation: What is participatory design? In *Proceedings of the Participatory Design Conference 98.* Seattle: CPSR.

Star, S. L., & Griesemer, J. R. (1989). Institutional ecology, "translations," and boundary objects: Amateurs and professionals in Berkeley's Museum of Vertebrate Zoology, 1907-1939. *Social Studies of Science, 19*, 387-420.

Suchman, L. (1987). *Plans and situated actions: The problem of human-machine communication.* Trowbridge, England: Cambridge University Press.

Suchman, L. (Ed.). (1995). Representations of work [Special issue]. *Communications of the ACM, 38.*

Suchman, L., & Trigg, R. (1991). Understanding practice: Video as a medium for reflection and design. In J. Greenbaum & M. Kyng (Eds.), *Design at work: Cooperative design of computer systems* (pp. 65-90). Hillsdale, NJ: Lawrence Erlbaum Associates.

Thackara, J. (2000). Edge effects: The design challenge of pervasive interface. Plenary presentation at CHI 2000, The Hague, The Netherlands.

Trigg, R. H. (2000). From sandbox to "fundbox": Weaving participatory design into the fabric of a busy non-profit. In *Proceedings of the Participatory Design Conference 2000* (pp. 174-183). New York: CPSR.

Tscheligi, M., Houde, S., Marcus, A., Mullet, K., Muller, M. J., & Kolli, R. (1995). Creative prototyping tools: What interaction designers really need to produce advanced user interface concepts. In *CHI'95 conference companion* (pp. 170-171). New York: ACM Press.

Tschudy, M. W., Dykstra-Erickson, E. A., & Holloway, M. S. (1996). PictureCARD: A storytelling tool for task analysis. In *Proceedings of the Participatory Design Conference '96* (pp. 183-192). Cambridge, MA: CPSR.

Tudor, L. G., Muller, M. J., Dayton, T., & Root, R. W. (1993). A participatory design technique for high-level task analysis, critique, and redesign: The CARD method. *Proceedings of the HFES '93* (pp. 295-299). Seattle: Human Factors and Ergonomics Society.

Universal Usability Fellows. (2000). Retrieved May 2002 from http://www.universalusability.org/index.html

van den Besselaar, P., Clement, A., & Jaervinen, P. (1991). *Information system, work and organization design* (pp. 199-200). Amsterdam: North-Holland.

van den Besselaar, P., Greenbaum, J., & Mambrey, P. (1996). Unemployment by design: Participatory design and the changing structure of the workforce in the information society. In *Proceedings of the Participatory Design Conference 96* (pp. 199-200). Cambridge, MA: CPSR.

Vertelney, L. (1989). Using video to prototype user interfaces. *SIGCHI Bulletin, 21*, 57-61.

Young, E. (1992). *Participatory video prototyping.* Poster presented at the CHI '92 conference, Monterey, CA, May 1992.

Wixon, D., & Ramey, J. (Eds.). (1996). *Field methods casebook for software design*. New York: John Wiley.

·55·

UNIFIED USER INTERFACE DEVELOPMENT

Constantine Stephanidis and Anthony Savidis
Institute of Computer Science,
Foundation for Research and Technology—Hellas

INTRODUCTION

The ongoing paradigm shift toward a knowledge-intensive information society has brought about radical changes in the way people work and interact with each other and with information. Computer-mediated human activities undergo fundamental changes, and new ones appear continuously, as novel, intelligent, distributed, and highly interactive technological environments emerge, making available concurrent access to heterogeneous information sources and interpersonal communication. The progressive fusion of existing and emerging technologies is transforming the computer from a specialist's device into an information appliance. This dynamic evolution is characterized by several dimensions of diversity that become evident when considering the broad range of user characteristics, the changing nature of human activities, the variety of contexts of use, the increasing availability and diversification of information, knowledge sources and services, the proliferation of diverse technological platforms, and so on.

In this context, the "typical" computer user can no longer be identified: Information artifacts are used by diverse user groups, including people with different cultural, educational, training and employment backgrounds; novice and experienced computer users; the very young and the elderly, and people with disabilities. As a consequence, existing computer-mediated human activities undergo fundamental changes, and a wide variety of new ones appear, such as access to online information, e-communication, digital libraries, e-business, online health services, e-learning, online communities, online public and administrative services, e-democracy, tele-work and tele-presence, and online entertainment. From a specialist's device, the computer is being transformed into an information appliance for the citizen in the information society. Similarly, the context of use is changing. The "traditional" use of computers (i.e., scientific use by the specialist, business use for productivity enhancement) is increasingly being complemented by residential and nomadic use, thus penetrating a wider range of human activities in a broader variety of environments, such as the school, home, marketplace, and other civil and social contexts. Finally, technological proliferation contributes with an increased range of systems or devices to facilitate access to the communitywide pool of information resources. These devices include computers, standard telephones, cellular telephones with built-in displays, television sets, information kiosks, special information appliances, and various other "network-attachable" devices.

The information society has the potential to improve the quality of life of citizens, the efficiency of social and economic organization, and to reinforce cohesion. However, as with all major technological changes, it can also have disadvantages, introducing new barriers, human isolation, and alienation (the so-called "digital divide"), if the diverse requirements of all potential users are not taken seriously into consideration and if an appropriate "connection" to computer applications and services is not guaranteed.

It is in this context that the notion of universal access becomes critically important for ensuring social acceptability of the emerging information society. Universal access implies the accessibility and usability of information society technologies (IST) by anyone, anywhere, anytime (Stephanidis, 2001a). Its aim is to enable equitable access and active participation of potentially all citizens in existing and emerging computer-mediated human activities. Therefore, it is important to develop *universally accessible* and *usable* user interfaces, embodying the capability to interact with the user in all contexts, independently of location, user's primary task, target machine, run-time environment, or the current physical conditions of the external environment. To this end, the needs of the broadest possible end user population must be taken into account in the early design phases of new products and services. The field of human-computer interaction (HCI) plays a critical role toward facilitating universal access, as citizens in the information society experience technology through their contact with the user interface of interactive products, applications, and telematic services.

This chapter presents a principled and systematic approach toward coping with diversity in the target users groups, tasks, and environments of use, named unified user interface development (Stephanidis, 2001c). The theoretical grounds on which unified user interface development methodology is based is provided by the concept of *user interfaces for all*, rooted in the idea of applying *universal design* in the field of HCI (Stephanidis, 1995, 2001a). The underlying principle is to ensure accessibility at design time and to meet the individual requirements of the user population at large.

The remainder of this section discusses the challenges that HCI has to face in the context of universal access, namely accessibility and high quality of interaction, and briefly summarizes past and more recent approaches to these challenges. Subsequently, it introduces and exemplifies the notion of automatic user interface adaptation and defines the concept of unified user interface. The second section presents a methodology for unified user interface design and the third section describes an architecture and an engineering process for the development of unified user interfaces.

HCI for Universal Access

In the context of universal access, accessibility means that for each user task, there is a sequence of input actions and associated feedback, via accessible input—output devices, leading to successful accomplishment. This implies the availability of alternative input—output devices to suit the interaction needs and requirements of different users.

In the past, the term *computer accessibility* was usually associated with access to interactive computer-based systems by people with disabilities. In traditional efforts to improve accessibility, the driving goal has been to devise hardware and software configurations (or alternative access systems) that enable disabled users to access interactive applications originally developed for able users. There have been two possible technical routes to alleviate the lack of accessibility of interactive software products. The first is to treat each application separately and take all the necessary implementation steps to arrive at an alternative accessible version (*product-level adaptation*). The

second alternative is to "intervene" at the level of the particular interactive application environment (e.g., MS Windows, X Windowing System) to provide appropriate software and hardware technology so as to make that environment alternatively accessible (*environment-level adaptation*). The latter option extends the scope of accessibility to cover potentially all applications running under the same interactive environment, rather than a single application.

Product-level adaptation has been practically tackled as redevelopment from scratch. Due to the high costs associated with this strategy, it is considered the less favorable option for providing alternative access, whereas environment-level adaptations, addressing a range of applications, are acknowledged as a more promising strategy (Mynatt & Weber, 1994). In the past, the vast majority of approaches to environment-level adaptations have focused on the issue of accessibility of graphical environments by blind users (Mynatt & Edwards, 1992; Mynatt & Weber, 1994). Through such efforts, it became apparent that any approach to environment-level adaptations should be based on well-documented and operationally reliable software infrastructures, supporting effective and efficient extraction of dialogue primitives during user–computer interaction. Such dynamically extracted dialogue primitives are to be reproduced, at run-time, in alternative input—output forms, directly supporting user access. Recent examples of software infrastructures that satisfy the above requirements are Microsoft Active Accessibility™, and Java Accessibility™ by JavaSoft.

Despite recent progress, the prevailing practices aiming to provide alternative access systems, either at the product or environment level, have been criticized for their essentially reactive nature (Stephanidis & Emiliani, 1999). The critique is grounded on two lines of argumentation. The first is that reactive solutions typically provide limited and low-quality access. This is evident in the context of nonvisual interaction, in which the need has been identified to provide nonvisual user interfaces that are more than automatically generated adaptations of visual dialogues (Savidis & Stephanidis, 1995). The second line tackles the economic feasibility of the reactive paradigm to accessibility. In particular, it is argued that continuing to adopt prevailing practices within the software industry leads to the need to reactively develop solutions for interactive computer-based products that are accessible to users with situational, temporary, or permanent disabilities. Clearly, this is suboptimal for all parties concerned, rendering the reactive approach to accessibility inadequate and inappropriate in the long run. Instead, what is required is a proactive approach to cater for the requirements of the broadest possible end user population (Stephanidis & Emiliani, 1999).

Another important aspect of universal access is interaction quality. In this context, high interaction quality means that all supported paths toward task accomplishment "maximally fit" individual users in the particular context and situation of use. Interaction quality is to be pursued and measured in the interface design and evaluation phases. When designing in a universal access perspective, a large diversity of potential design parameters (i.e., user- and usage-context attributes) may affect interaction quality. Even the same user, enrolled in different contexts and situations, may require different access and quality properties. For example, while driving, any user can be considered as situationally motor and visually impaired; therefore, interaction should require minimum attention, provide simple dialogues and speech input and output, and reduce touch input. On the contrary, in a noisy environment, the same user can be considered as functionally deaf, and therefore visual interaction is to be preferred. Low interaction quality may also reduce accessibility. Even with a "physically" accessible interface, users may be unable to carry out an interaction task if they cannot understand the tasks supported by a system and the action sequences required for performing them or if they cannot interpret interface artifacts and "navigate" in the interface. Finally, what is "good design" for one user may be a "bad design" for another. In this context, it is unrealistic to expect that a single-minded interface design will ensure high-quality interaction for all values of the many design parameters. Instead, alternative design decisions will have to be made, even for the same tasks or subtasks. Consequently, the outcome of such a design process is not a "singular" design, but a design space populated with appropriate dialogue patterns, together with their associated design parameters (e.g., user- and usage-context attribute values).

In the light of these issues, the challenge of accessibility and interaction quality needs to be addressed through more proactive and generic approaches that account for all dimensions and sources of variation. These dimensions range from the characteristics and abilities of users, to the characteristics of technological platforms, to the relevant aspects of the context of use (Stephanidis et al., 1998a).

Automatic Interface Adaptation

A key element elaborated for coping with diversity in the HCI field is the one of adaptation (Stephanidis, 2001b), that is, the capability, on the part of the interaction-specific software, applications, or services, to self-adapt to individual users' characteristics and contexts of use. The distinctive property of a self-adapting interface is that it can realize alternative patterns of interactive behavior by automatically adapting to accommodate specific user- and context-oriented parameters.

An example of practical application employing adaptation techniques to ensure accessibility and high-quality of interaction for all potential users is the AVANTI Web browser (Stephanidis, et al., 1998b; Stephanidis, Paramythis, Styrakis, & Savidis, 2001). This section briefly discusses some representative types of adaptation in the AVANTI browser, with the purpose of illustrating how adaptation takes place in practice.

The AVANTI browser provides an accessible and usable interface to a range of user categories, irrespective of physical abilities or technology expertise, and supports various situations of use. The end user groups targeted in AVANTI, in terms of physical abilities, include (a) "able-bodied" people, assumed to have full use of all their sensory and motor communication "channels"; (b) blind people; and (c) people with different forms of motor impairments in their upper limps causing different degrees of difficulty in employing traditional computer input devices. In particular, in the case of people with motor

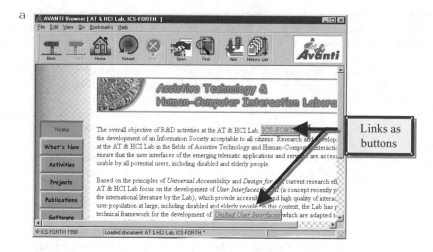

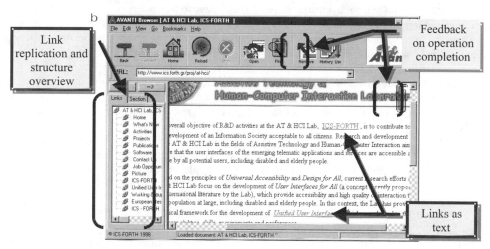

FIGURE 55.1. Adapting to the user and the context of use. (a) Conventional, simplified instance of the interface. (b) Adapted instance for an experienced user.

impairments, two coarse levels of impairment are considered: light motor impairments (i.e., users have limited use of their upper limps but can operate traditional input devices or equivalents with adequate support) and severe motor impairments (i.e., users cannot operate traditional input devices at all). Furthermore, because the AVANTI system is intended to be used both by professionals (e.g., travel agents) and the general public (e.g., citizens, tourists), the users' experience in the use of, and interaction with, technology is another major parameter taken into account in the design of the user interface. The system supports users with any level of computer expertise and with or without previous experience in the use of Web-based software. In terms of usage context, the system is intended to be used both by individuals in their personal settings (e.g., home, office) and by the population at large through public information terminals (e.g., information kiosks at a railway station or airport). In the case of private use, the front end of AVANTI should be appropriate for general Web browsing, allowing users to make use of the accessibility facilities beyond the context of a particular information system. Additionally, users are to be continuously supported as their communication and interaction requirements change over time due to personal or environmental reasons (e.g., stress, tiredness, system configuration). This entails the capability, on the part of the system to detect dynamic changes in the characteristics of the user and the context of use (either of temporary or of permanent nature) and cater for these changes by modifying itself appropriately.

Figures 55.1–3 show some instances of the AVANTI browser illustrating some of the available categories of adaptation. Figure 55.1 contains two instances of the interface that demonstrate adaptation based on the characteristics of the user and the usage context. Specifically, Fig. 55.1a presents a simplified instance intended for use by a user unfamiliar with Web browsing. Note the "minimalist" user interface with which the user is presented, as well as the fact that links are presented as buttons, arguably increasing their affordance (at least in terms of functionality) for users familiar with windowing applications in general. The second instance, Fig. 55.1b, the interface has been

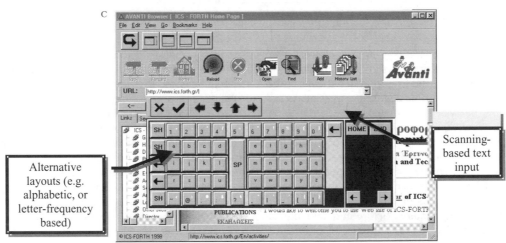

FIGURE 55.2. Instances for users with motor impairments. (a) Scanning for switch-based interaction. (b) Window manipulation toolbar. (c) On-screen, "virtual" keyboard.

adapted for an experienced user. Note the additional functionality that is available to the user (e.g., a pane where the user can access an overview of the document itself, or of the links contained therein, an edit field for entering the uniform resource locators (URLs) of local or remote HTML documents). Figure 55.2 contains some examples demonstrating disability-oriented adaptations in the browser's interface. The instance in Fig. 55.2a presents the interface when a special interaction technique for users with motor impairments is activated, namely hierarchical interface scanning (either manually or automatically activated). Scanning is a mechanism allowing the user to "isolate" each interactive object in the interface and to interact with it thorough binary switches (Stephanidis & Savidis, 2001). Note the scanning highlighter over an image-link in the HTML document and the additional toolbar that was automatically added in the user interface. The latter is a "window manipulation" toolbar, containing three sets of controls enabling the user to perform typical actions on the browser's window (e.g., resizing and moving). Figure 55.2b illustrates the three sets of controls in the toolbar,

as well as the "rotation" sequence between the sets (the three sets occupy the same space on the toolbar, to utilize screen real estate more effectively and to speed up interaction; the user can switch between them by selecting the first of the controls). Figure 55.2c presents an instance of the same interface with an on-screen, "virtual" keyboard activated for text input. Interaction with the on-screen keyboard is also scanning-based. Finally, the single interface instance in Fig. 55.3 illustrates a case of adaptive prompting (Stephanidis et al., 1998b). Specifically, this particular instance is displayed in those cases in which there exists high probability that the user is unable to initiate the "open location" task (as would be the case if there were adequate evidence that the user is attempting to load an external document with unsupported means, e.g., using "drag and drop"). In this case, adaptive prompting is achieved through the activation of a "tip" dialogue (i.e., a dialogue notifying the user about the existence of the "open location" functionality and offering some preliminary indications of the steps involved in completing the task).

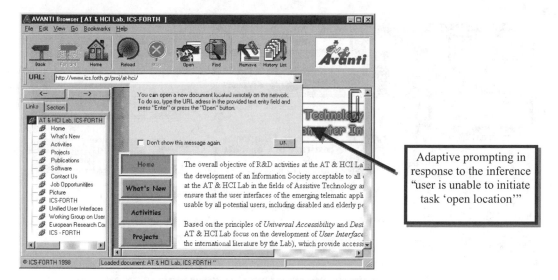

FIGURE 55.3. Awareness prompting.

The primary target for adaptation in the AVANTI Web browser is to ensure that each of the system's potential users is presented with an instance of the user interface that has been tailored to offer the highest possible degree of accessibility (limited, of course, by the system's knowledge about the user). Secondly, adaptation is employed to tailor the system further to the inferred needs or preferences of the user, so as to achieve the desired levels of interaction quality.

In general, the overall adapted interface behavior can be viewed as the result of two complementary classes of system initiated actions: (a) adaptations driven from initial user and context information, usually acquired before initiating interaction, and (b) adaptations decided on the basis of information inferred or extracted by monitoring interaction.

The former behavior is referred to as adaptability, reflecting the interface's capability to automatically tailor itself initially to each individual end user. The latter behavior is referred to as adaptivity and characterizes the interface's capability to cope with the dynamically changing and evolving situation of use. It should be noted that adaptability is crucial to guarantee accessibility, because it is essential to initially (i.e., before initiation of interaction) provide a fully accessible interface instance to each end user. Furthermore, adaptivity can be applied only on accessible running interface instances (i.e., ones with which the user is capable of performing interaction), because interaction monitoring is required for the identification of changing or emerging decision parameters that may drive dynamic interface enhancements. Referring to the examples of adaptation illustrated above, Figs. 55.1 and 55.2 represent instances of adaptability, whereas Fig. 55.3 represents an instance of adaptivity. The complementary roles of adaptability and adaptivity are depicted in the diagram of Fig. 55.4, whereas the key differences between these two types of adaptation behavior are illustrated in Table 55.1.

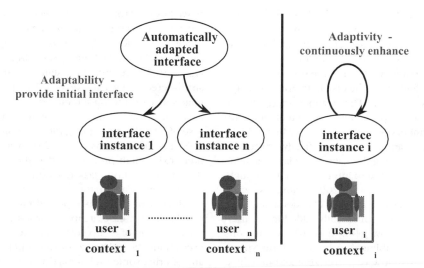

FIGURE 55.4. Complementary roles of adaptability and adaptivity.

TABLE 55.1. Key differences between adaptability and adaptivity in the context of Unified User Interfaces

Adaptability	Adaptivity
1. User and usage-context attributes are considered known before interaction	1. User and usage-context attributes are dynamically inferred or detected
2. "Assembles" an appropriate initial interface instance for a particular end user and usage context	2. Enhances the initial interface already "assembled" for a particular end user and usage context
3. Works before interaction is initiated	3. Works after interaction is initiated
4. Provides a user-accessible interface	4. Requires a user-accessible interface

Clearly, the development of a self-adapting interactive application such as the AVANTI browser cannot be supported by traditional user interface development methods because they lack design and implementation means for adaptation (Stephanidis & Savidis, 2001). Methods capable of supporting adaptation should capture the adaptation behavior in the user interface design, and encapsulate it accordingly in the implementation (Stephanidis, 2001c). Because adaptation implies providing dialogue according to user- and context-related factors, suitable methods and tools should provide means for capturing user and context characteristics and their interrelationships with alternative interactive behaviors, as well as appropriate mechanisms for deciding adaptations on the basis of those parameters and relationships. Additionally, suitable methods and tools should provide appropriate techniques for managing alternative interactive behaviors and applying adaptation decisions at run-time. This is not the case in currently available interface development tools, which are mainly targeted to the provision of advanced support for implementing physical aspects of the interface via different techniques (e.g., visual construction, task based, demonstration based; Savidis & Stephanidis, 2001b).

Previous work on adaptation has addressed various key technical issues. Early approaches to adaptability include (Koller, 1993; Lai & Malone, 1988; MacLean, Carter, Lovstrand, & Moran, 1990; Robertson, Henderson, & Cards, 1991; Sherman & Shortliffe, 1993). All these systems allow the user to modify certain aspects of their interactive behavior while working with them.

Although concrete software architectures for adaptive user interfaces have not been defined in the past, various proposals as to what should be incorporated (in a computable form) into an adaptive interactive system have been formulated. In (Dieterich, Malinowski, Kühme, & Schneider-Hufschmidt, 1993), these categories of computable artifacts are summarized as being the types of models that are required within a structural model of adaptive interactive software. Other relevant aspects of adaptivity that have been investigated in previous efforts include user modeling (Kobsa & Wahlster, 1989); the need for explicit design, as well as run-time availability of design alternatives (Browne, Norman, & Adhami, 1990); abstraction (Cockton, 1987); dialogue modeling (Benyon, 1984; Cote Muñoz, 1993); and interface updates for adaptivity (Cockton, 1993; Short, 1997). It can be concluded from the above that the incorporation of

adaptation capabilities into interactive software is far from trivial, and there is a genuine requirement for the definition of an overall framework that can accommodate adaptation-oriented requirements.

The unified user interface development framework (Stephanidis, 2001c), of which the AVANTI browser constitutes an application (Stephanidis et al., 2001; Stephanidis & Savidis, 2001), provides a design and engineering methodology supporting the development of adaptable and adaptive user interfaces.

The Concept of Unified User Interfaces

From a user perspective, a unified user interface can be considered as an interface tailored to personal attributes and to the particular context of use, whereas from the designer perspective, it can be seen as an interface design populated with alternative designs, each alternative addressing specific user and usage-context parameter values. Finally, from an engineering perspective, a unified user interface is a repository of implemented dialogue artifacts from which to select, by means of a decision-making mechanism, the most appropriate at run-time. The distinctive property of a unified user interface is that it can realize alternative patterns of interactive behavior, at the physical, syntactic or semantic level of interaction, by automatically adapting to accommodate specific user- and context-oriented parameters. Typically, such alternative interactive behaviors encompass interaction elements available in different interaction environments, for example, Windows™, OSF/Motif™, and HAWK toolkit for nonvisual interaction (Savidis et al., 1997), suitable for different target user groups (i.e., sighted and blind users, respectively).

In developing unified user interfaces, the primary challenge is to envision, model, and deploy the required type and range of adapted interactive behaviors. In the emerging interactive paradigm (characterized by nomadicity, ubiquitousness, seamless interactivity, etc), context-oriented assessments are likely to determine both the type and range of adapted behavior that best suits a given human task or activity. For example, a unified user interface may adapt depending on the type of interaction technology available at the end user's site. Thus, for instance, in the case of low screen resolution and presence of audio output hardware, the interface may "decide" to provide fewer visual effects while emphasizing audio feedback. The context of use is also important to the extent that it provides an account of primary and secondary tasks. For example, when driving, visual attention is primarily devoted to driving tasks, and any type of secondary interaction should not disturb the user's focus of attention. From the above, it follows that the proposed concept of automatically adapted behavior denotes context-sensitive adaptation. This in turn may be distinguished into (a) *user-adapted behavior*, in which the interface is capable of automatically selecting an interaction approach more appropriate to the particular end user (i.e., user awareness), and (b) *usage-context adapted behavior*, in which the interface is capable of automatically selecting an interaction approach more appropriate to the particular situation of use (i.e., usage context awareness).

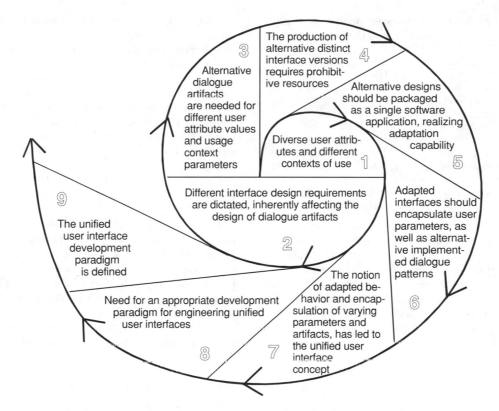

FIGURE 55.5. From diverse user requirements and usage contexts, to the unified user interface engineering paradigm.

A unified user interface comprises a single (unified) interface specification, targeted to potentially all user categories, and is defined by the following key engineering properties: (a) it encompasses user- and context-specific information; (b) it contains alternative dialogue artifacts in an implemented form, each appropriate for different user- and context-specific parameters; and (c) it applies adaptation decisions, activates dialogue artifacts, and is capable of interaction monitoring.

Figure 55.5 depicts the trajectory that led to the unified user interface concept and the respective design and development paradigms. The starting point of this trajectory is that for designing interactive software applications and telematic services accommodating the requirements of the widest possible target population users in different contexts of use, it is necessary to take into consideration the diverse attributes that characterize the users and the envisaged contexts of use (Step 1). These varying user and usage-context attribute values give rise to different design requirements, which, in turn, affect the design of dialogue artifacts (Step 2). As a result, alternative dialogue artifacts have to be constructed, at various points of the interface design process, as dictated by the differing user- and usage-context attribute values (Step 3).

When trying to map the outcomes of such design processes into an implemented interactive application, a key issue is how the various alternative dialogue artifacts are "packaged." Because the production of alternative interface versions requires prohibitive resources for development, maintenance, upgrading, and distribution (all distinct versions should potentially be made available for concurrent "execution"), it turns out to be unrealistic (Step 4). Consequently, the "packaging" of the alternative dialogue patterns into a single software application is the most promising approach. In this context, packaging may not necessarily imply the construction of a monolithic software system incorporating all alternative dialogue artifacts; rather, it can be instantiated as a logical collection within a single resource. For example, a repository can be made directly accessible by a single software application that encompasses adaptation capabilities, thus being able to select the most appropriate dialogue patterns for a particular end user and target usage context (Step 5).

To facilitate such a capability, interactive applications should encompass information about individual users, as well as alternative dialogue patterns in an implemented form (Step 6). This notion of encapsulation has led to the definition of the unified user interface concept, which proposes the realization of user- and usage-context-adapted behavior by encapsulating all the varying design parameters and alternative dialogue artifacts (Step 7). The need for an appropriate development strategy for unified user interfaces (Step 8) leads to the introduction of the unified user interface development paradigm, targeted to the development of unified user interfaces (Step 9).

The unified user interface development paradigm requires interdisciplinary processes driving the production of automatically adapted software applications and services. It is general

enough so as not to exclude particular design and implementation practices; at the same time, it offers sufficient details to drive the engineering process of unified user interface software. As with any new development paradigm, it naturally requires some initial investment to be effectively adopted, assimilated, and applied. However, if the constructed software products are intended to be used by user populations with diverse requirements and operated in different usage contexts, it is argued that the gains will outweigh the overhead of additional resources that need to be invested (Stephanidis, 2001c; Stephanidis & Emiliani, 1999).

UNIFIED USER INTERFACE DESIGN

Unified user interfaces encapsulate automatically adapted behaviors and provide end users with appropriately individualized interaction facilities. Hence, the process of designing unified user interfaces does not lead to a single design outcome (i.e., a particular design instance for a particular end user). Rather, it collects and appropriately represents alternative design solutions as well as the conditions under which each of these should be instantiated (i.e., a design rationale). Two major challenges that can be identified in this respect concern the process for the production of the alternative designs and the organization of all potential design instances into a single design structure.

Clearly, producing and enumerating distinct designs through the execution of multiple design processes is not a practical solution. Ideally, a single design process is desirable, leading to a design outcome that may directly be mapped to a single (i.e., unified) software system implementation.

On these grounds, the *Unified User Interface design method* (Savidis et al., 2001) has been defined so as to address two objectives:

1. enabling the "fusion" of all potentially distinct design alternatives into a single unified form, without requiring multiple design phases;
2. producing a design structure that can be easily translated by user interface developers into an implementation form.

The unified user interface design method is characterized by two properties that distinguish both the conduct of the method and its respective outcomes. First, the method adopts an analytical design perspective in the sense that it requires insight into how users perform tasks in existing task models as well as what design alternatives and underpinning rationale should be embedded in the envisioned and reengineered task models. In this context, the method links with other analytical perspectives into HCI design, such as design rationale (Moran & Carroll, 1996) and ethnography (Simonsen & Kensing, 1997), to obtain the real-world insight that is required, while extending the traditional design inquiry by focusing explicitly on the concept of *polymorphism* as an aid to designing and implementing user- and use-adapted behaviors.

In the unified user interface design method, a polymorphic design artifact constitutes a collection of alternative solutions for a single design problem, each alternative addressing different parameters. In this context, the main problem can be identified as optimally designing artifacts for end users and contexts of use, while the parameters of the problem are the various attributes characterizing the users and the contexts of use. The method supports a disciplined hierarchical approach to populating and articulating rationalized design spaces. This entails a middle-out design perspective, whereby enumerated design alternatives are fused into design abstractions and subsequently transformed into polymorphic artifacts to facilitate automatic realization of alternative interactive behavior. In terms of conduct, the method is related to hierarchical task analysis, with the distinction that alternative decomposition schemes can be employed at any point of the hierarchical task analysis process. Each decomposition seeks to address different values of the driving design parameters. This approach leads to the notion of design polymorphism, which is characterized by the pluralism of plausible design options consolidated in the resulting task hierarchy.

Overall, the method introduces the notion of *polymorphic task decomposition*, through which any task (or subtask) may be decomposed in an arbitrary number of alternative sub-hierarchies (Savidis et al., 2001). The design process realizes an exhaustive hierarchical decomposition of various task categories, starting from the abstract level, by incrementally specializing in a polymorphic fashion (because different design alternatives are likely to be associated with differing user- and usage-context attribute values), toward the physical level of interaction. The outcomes of the method include the following: (a) a design space that is populated by collecting and enumerating design alternatives, (b) a polymorphic task hierarchy that comprises alternative concrete artifacts, and (c) for each artifact produced, a recorded design rationale underlying its introduction.

Polymorphic Task Hierarchies

Design alternatives are necessitated by the different contexts of use and provide a global view of task execution. This is to say that design alternatives offer rich insight into how a particular task may be accomplished by different users in different contexts of use. Because users differ with regard to ability, skill, requirements, and preferences, tentative designs should aim to accommodate the broadest possible range of capabilities across different contexts of use. Thus, instead of restricting the design activity to producing a single outcome, designers should strive to compile design spaces containing plausible alternatives.

A polymorphic task hierarchy combines three fundamental properties: (a) *hierarchical organization*, (b) *polymorphism*, and (c) *task operators*. The hierarchical decomposition adopts the original properties of hierarchical task analysis (Johnson, Johnson, Waddington, & Shouls, 1988) for incremental decomposition of user tasks to lower level actions. The polymorphism property provides the design differentiation capability at any level of the task hierarchy, according to particular user- and usage-context attribute values. Finally, task operators, which

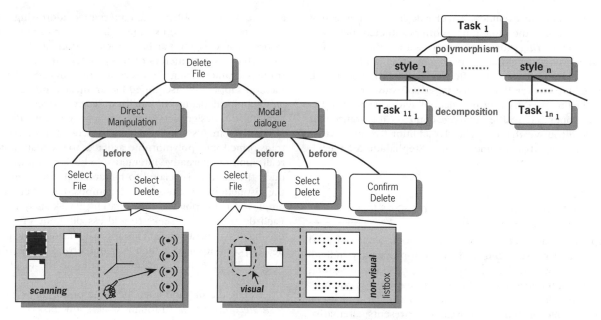

FIGURE 55.6. Example of polymorphic task hierarchy.

are based on the powerful CSP (communicating sequential processes) language for describing the behavior of reactive systems (Hoare, 1978), enable the expression of dialogue control flow formulae for task accomplishment. The concept of polymorphic task hierarchies is illustrated in Fig. 55.6.

Each alternative task decomposition is called a decomposition style, or simply a style, and is given an arbitrary name; the alternative task subhierarchies are attached to their respective styles. The example in Fig. 55.6 shows how two alternative dialogue styles for a "Delete File" task can be designed; one exhibiting direct manipulation properties with object-function syntax (i.e., the file object is selected before operation to be applied) with no confirmation, and another realizing modal dialogue with a function-object syntax (i.e., the delete function is selected, followed by the identification of the target file) and confirmation.

Additionally, the example demonstrates the case of physical specialization. Because "selection" is an abstract task, it is possible to design alternative ways for physically instantiating the selection dialogue: via scanning techniques for users with motor impairments, via three-dimensional (3D) hand pointing on 3D-auditory cues for blind people, via enclosing areas (e.g., irregular "rubber banding") for sighted users, and via Braille output and keyboard input for deaf-blind users. The unified user interface design method does not require the designer to follow the polymorphic task decomposition all the way down the user-task hierarchy until primitive actions are met. A nonpolymorphic task can be specialized at any level, following any design method the interface designer chooses. For example, in Fig. 55.6 (lower portion), graphic illustrations are used to describe each of the alternative physical instantiations of the abstract "selection" task.

It should be noted that the interface designer is not constrained to use a particular model, such as CSP operators, to describe user actions for device-level interaction (e.g., drawing, drag and drop, concurrent input). Instead, an alternative may be preferred, such as an event-based representation, for example, event representation language (ERL) (Hill, 1986) or user action notation (UAN) (Hartson, Siochi, & Hix, 1990).

In the unified user interface design method, it is not sufficient to represent alternative task hierarchies by means of the traditional task model, employing the *xor* operator among alternatives and thus requiring the polymorphic task decomposition approach. This is primarily because the *xor* operator in the traditional task model means that the user is allowed to perform any, but only one, of the *N* subtasks. This means that the physical interaction context (i.e., interface components) for performing any of the subtasks is made available to the user, although the user is required to accomplish only one of those subtasks. However, if alternative subhierarchies are related via polymorphism, it is implied that a particular end user will be provided with the design (out of the *N* alternative ones) that maximally matches the specific user's characteristics. Clearly, it is meaningless to provide all designed artifacts (which are likely to address diverse user characteristics) concurrently to end users, and force users to work with only one of those. Hence, the *xor* operator is not the appropriate way to organize alternative dialogue patterns. As discussed in more detail later, design polymorphism entails a decision-making capability for context-sensitive selection among alternative artifacts so as to assemble a suitable interface instance, whereas task operators support temporal relationships and access restrictions applied to the interactive facilities of a particular interface instance.

In the unified user interface design method, there are three categories of design artifacts, all of which are subject to polymorphism on the basis of varying user- and usage-context parameter values. These three categories are as follows:

- *User tasks*, relating to what the user has to do; user tasks are the center of the polymorphic task decomposition process;
- *System tasks*, representing what the system has to do or how it responds to particular user actions (e.g., feedback); in the polymorphic task decomposition process, they are treated in the same manner as user tasks;
- *Physical design*, which concerns the various interface components on which user actions are to be performed; physical structure may also be subject to polymorphism.

System tasks and user tasks may be freely combined within task "formulas," defining how sequences of user-initiated actions and system-driven actions interrelate. The physical design, providing the interaction context, is always associated with a particular user task. It provides the physical dialogue pattern associated to a task-structure definition. Hence, it simply plays the role of annotating the task hierarchy with physical design information.

In some cases, given a particular user task, there is a need for differentiated physical interaction contexts, depending on user and usage-context parameter values. Hence, even though the task decomposition is not affected (i.e., the same user actions are to be performed), the physical design may have to be altered.

There are also cases in which the alternative physical designs are dictated due to alternative task structures (i.e., polymorphic tasks). In such situations, each alternative physical design is directly attached to its respective *style* (i.e., subhierarchy).

In summary, the rules for identifying polymorphic artifacts are as follows:

- If alternative designs are assigned to the same task, then attach a polymorphic physical design artifact to this task; the various alternative designs depict the styles of this polymorphic artifact.
- If alternative designs are needed due to alternative task structures (i.e., task-level polymorphism), then each alternative physical design should be assigned to its respective style.

User tasks, and in certain cases system tasks, need not always be related to physical interaction but may represent abstraction on either user or system actions. For instance, if the user has to perform a selection task, then clearly the physical means of performing the selection are not explicitly defined, unless the dialogue steps to perform selection are further decomposed. This notion of continuous refinement and hierarchical analysis— starting from higher level abstract artifacts and incrementally specializing toward the physical level of interaction—is fundamental in the context of hierarchical behavior analysis, both for tasks that humans have to perform (Johnson et al., 1988) and for functional system design (Saldarini, 1989).

The polymorphic task decomposition process follows the methodology of abstract task definition and incremental specialization, in which tasks may be hierarchically analyzed through various alternative schemes. In such a recursive process, involving tasks ranging from the abstract task level, to specific physical actions, decomposition is applied either in a traditional

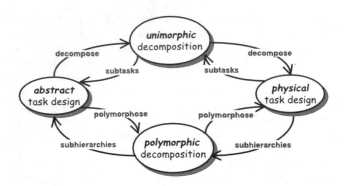

FIGURE 55.7. Polymorphic task decomposition: process overview.

unimorphic fashion or by means of alternative styles. The overall process is illustrated in Fig. 55.7; the decomposition starts from abstract or physical task design, depending on whether top-level user tasks can be defined as being abstract. Next follows the description of the various transitions (i.e., design specialization steps) from each of the four states illustrated in the process state diagram of Fig. 55.7.

An abstract task can be decomposed either in a polymorphic fashion, if user- and usage-context attribute values require alternative dialogue patterns, or in a traditional manner, following a unimorphic decomposition scheme. In the latter case, the transition is realized via a decomposition action, leading to the task hierarchy decomposition state. In the case of a polymorphic decomposition, the transition is realized via a "polymorphose" action, leading to the design alternative subhierarchies state.

Reaching this state means that the required alternative dialogue styles have been identified, each initiating a distinct subhierarchy decomposition process. Hence, each subhierarchy initiates its own instance of polymorphic task decomposition process. While initiating each distinct process, the designer may either start from the abstract task design state or from the physical task design state. The former is pursued if the top-level task of the particular subhierarchy is an abstract one. In contrast, the latter option is relevant in case the top-level task explicitly engages physical interaction issues.

From this state, the subtasks identified need to be further decomposed. For each subtask at the abstract level, there is a subtask transition to the abstract task design state. Otherwise, if the subtask explicitly engages physical interaction means, a sub task transition is taken to the physical task design state.

Physical tasks may be further decomposed either in a unimorphic fashion or in a polymorphic fashion. These two alternative design possibilities are indicated by the "decompose" and "polymorphose" transitions respectively.

The polymorphic task model provides the design structure for organizing the various alternative dialogue patterns of automatically adapted interfaces into a unified form. Such a hierarchical structure realizes the fusion of all potential distinct designs, which may be explicitly enumerated given a particular unified user interface. Apart from the polymorphic organization model, the following primary issues need to be addressed as well: (a) when polymorphism should be applied, (b) which

are the user- and usage-context attributes that need to be considered, (c) what are the run-time relationships among alternative styles, and (d) how the adaptation rationale, connecting the designed styles with particular user- and usage-context attribute values, is documented.

In the context of the unified user interface design method, and as part of the polymorphic task decomposition process, designers should always assert that every decomposition step (i.e., those realized either via the polymorphose or through the decompose transitions of Fig. 55.7) satisfies all constraints imposed by the combination of target user- and usage-context attribute values. These two classes of parameters will be referred to collectively as decision parameters/attributes. An "accessibility gap" is usually encountered when there is a particular decomposition (for user or system tasks, as well as for physical design) that does not address some combination(s) of the decision attribute values. Such a design gap can be remedied by constructing the necessary alternative subhierarchy(ies) addressing the excluded decision attribute values.

In the unified user interface design method, the representation of end user characteristics may be developed in terms of attribute–value pairs, using any suitable formalism. There is no predefined or fixed set of attribute categories. Some examples of attribute classes are general computer-use expertise, domain-specific knowledge, role in an organizational context, motor abilities, sensory abilities, and mental abilities.

The value domains for each attribute class are chosen as part of the design process (e.g., by interface designers or human factors experts), whereas the value sets need not be finite. The broader the set of values, the higher the differentiation capability among various individual end users. For instance, commercial systems realizing a single design for an "average" user have no differentiation capability at all. The unified user interface design method does not pose any restrictions on the attribute categories considered relevant or on the target value domains of such attributes. Instead, it seeks to provide only the framework in which the role of user- and usage-context attributes constitutes an explicit part of the design process. It is the responsibility of interface designers to choose appropriate attributes and corresponding value ranges, as well as to define appropriate design alternatives. For simplicity, designers may choose to elicit only those attributes from which differentiated design decisions are likely to emerge. The construction of context attributes may follow the same representation approach as users' characteristics.

Recording Design Rationale for Alternative Styles

When a particular task is subject to polymorphism, alternative subhierarchies are designed, each associated to different user and usage-context parameter values. A running interface implementing such alternative artifacts should encompass decision-making capability so that, before initiating interaction with a particular end user, the most appropriate of those artifacts are activated for all polymorphic tasks. Hence, polymorphism can be seen as a technique potentially increasing the number of alternative interface instances represented by a typical hierarchical task model. If polymorphism is not applied, a task model merely represents a single interface design instance, in which further run-time adaptation is restricted; in other words, there is a fundamental link between adaptation capability and polymorphic design artifacts.

Consider the case where the design process reveals the necessity of having multiple alternative subdialogues available concurrently to the user for performing a particular task. This scenario is related to the notion of multimodality, which can be more specifically called *task-level* multimodality, in analogy to the notion of multimodal input, which emphasizes pluralism at the input-device level. This issue is further clarified with an example using the physical design artifact of Fig. 55.8, which depicts two alternative dialogue patterns for file management: one providing direct manipulation facilities and the other employing command-based dialogue. Both artifacts can be represented as part of the task-based design, in two ways:

1. through *polymorphism*, in which each of the two dialogue artifacts is defined as a distinct style: The two resulting styles are defined as being compatible, which implies that they may coexist at run-time (i.e., the end user may freely use the command line or the interactive file manager interchangeably).

Task: *Delete File*		
Style :	Direct Manipulation	Modal Dialogue
Targets :	Speed, naturalness, flexibility	Safety, guided steps
Parameters :	User (expert, frequent, average)	User (casual, naive)
Properties :	Object first, function next	Function first, object next
Relationships :	Exclusive	Exclusive

FIGURE 55.8. Recording design rationale for alternative styles.

2. via *decomposition*, in which the two artifacts are defined to be concurrently available to the user, within the same interface instance, via the or operator. In this case, the interface design is "hard-coded," representing a single interface instance, without need for further decision making.

The advantages of the polymorphic approach are as follows. (a) It is possible to make only one of the two artifacts available to the user, depending on user parameters. (b) Even if both artifacts are initially provided to end users, when a particular preference is dynamically detected, the alternative artifact can be dynamically disabled. (c) If more alternative artifacts are designed for the same task, the polymorphic design is directly extensible, whereas the decomposition-based design would need to be turned into a polymorphic one (except in the unlikely case in which it is still desirable to provide concurrently to the user all defined subdialogues).

The need for alternative styles emerges during the design process, when it is identified that some particular user or usage-context attribute values are not addressed by the various dialogue artifacts that have already been designed. Starting from this observation, one could argue that all alternative styles, for a particular polymorphic artifact, are mutually exclusive to each other (in this context, *exclusive* means that, at run-time, only one of those styles may be "active"). There exist, however, cases in which it is meaningful to make artifacts belonging to alternative styles concurrently available in a single adapted interface instance. For example, we discussed earlier how two alternative artifacts for file management tasks, a direct-manipulation and a command-based artifact, can both be present at run-time. In the unified user interface design method, four design relationships between alternative styles are distinguished, defining whether alternative styles may be concurrently present at run-time. We will now show how these four fundamental relationships reflect pragmatic, real-world design scenarios.

Exclusion. The exclusion relationship is applied when the various alternative styles are deemed to be usable only within the space of their target user and usage-context attribute values. For instance, assume that two alternative artifacts for a particular subtask are being designed, aiming to address the "user expertise" attribute. One is targeted to users considered "novice," and the other is targeted to "expert" users. Next, these two are defined to be mutually exclusive to each other, because it is probably meaningless to concurrently activate both dialogue patterns. For example, at run-time a novice user might be offered a functionally "simple" alternative of a task, whereas an expert user would be provided with additional functionality and greater "freedom" in selecting different ways to accomplish the same task.

Compatibility. Compatibility is useful among alternative styles for which the concurrent presence during interaction allows the user to perform certain actions in alternative ways, without introducing usability problems. The most important application of compatibility is for *task-multimodality*, as it discussed earlier (see Fig. 55.9, where the design artifact provides two alternative styles for interactive file management).

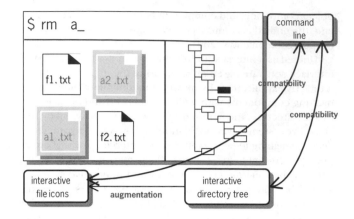

FIGURE 55.9. Relationships among alternative styles.

Substitution. Substitution has a strong connection with adaptivity techniques. It is applied in cases during interaction in which some dialogue patterns need to be substituted by others. For instance, the ordering and the arrangement of certain operations may change on the basis of monitoring data collected during interaction, through which information such as frequency of use and repeated usage patterns can be extracted. Hence, particular physical design styles would need to be "cancelled," whereas appropriate alternatives would need to be "activated." This sequence of actions (i.e., "cancellation" followed by "activation") is the realization of substitution. Thus, in the general case, substitution involves two groups of styles: some styles are "cancelled," or substituted by other styles, which are then "activated."

Augmentation. Augmentation aims to enhance the interaction with a particular style that is found to be valid but insufficient to facilitate the user's task. To illustrate this point, let us assume that during interaction, the user interface detects that the user is unable to perform a certain task. This would trigger an adaptation (in the form of adaptive action) aiming to provide task-sensitive guidance to the user. Such an action should not aim to invalidate the active style (by means of style substitution), but rather to augment the user's capability to accomplish the task more effectively, by providing informative feedback. Such feedback can be realized through a separate but compatible style. It follows, therefore, that the augmentation relationship can be assigned to two styles when one can be used to enhance the interaction while the other is active. Thus, for instance, the adaptive prompting dialogue pattern, which provides task-oriented help, may be related via an augmentation relationship with all alternative styles (of a specific task), provided that it is compatible with them.

UNIFIED INTERFACE ENGINEERING

The unified user interface development process provides a new technical framework to enable the construction of self-adapting software products, which are open, expandable, and compliant to recent popular development paradigms, such as distributed

object computing, and component-based development. To this end, the unified user interface architecture has been elaborated (Savidis & Stephanidis, 2001a).

Unified user interface architecture promotes insight into user interface software based on structuring the implementation of interactive applications by means of independent intercommunicating components with well-defined roles and behaviors. In this architecture, the notion of *encapsulation* plays a key role: To realize system-driven adaptations, all the various parameters, decision-making logic, and alternative interface artifacts are explicitly represented in a computable form, constituting integral parts of the run-time environment of an interactive system. It should be noted that the architectural properties of the unified user interface software architecture comply with the definition of architecture provided by the object management group (Mowbray & Zahavi, 1995; Jacobson, Griss, & Johnson, 1997). In accordance with these definitions, an architecture should supply components, description of functional role(s) per component, communication protocols among components, or application programming interfaces (APIs), as well as implementation and interoperability issues.

A unified user interface performs run-time adaptation to meet context-specific requirements as designated by a particular instance of the design space or the global task execution context. In this manner, it can practically attain the target of providing different interface instances for different end users and usage contexts. Figure 55.10 depicts the basic orthogonal components of the unified user interface architecture on the vertical dimension. These are (a) the dialogue patterns component, (b) the decision-making component, (c) the user information server, and (d) the context information server.

Unified user interface architecture enables the deployment of existing interactive software, requiring only some additional implemented modules, mainly serving coordination, control, and communication purposes. In this context, the dialogue control module of a typical arch-based interactive system may be expanded into a dialogue patterns component without affecting its original functional role. This capability facilitates the vertical

growth of existing interactive applications, following the unified user interface architectural paradigm, to accomplish automatically adapted interaction. The rest of this section provides an account of the characteristic properties and implementation issues of each of the main components of the unified user interface architecture. Communication protocols among the components are described in detail in (Savidis & Stephanidis, 2001a).

Context Information Server

This component encompasses information regarding the usage environment and interaction-relevant machine parameters. During the interface design process, the identification of important parameters relevant to the context(s) of use needs to be carried out. The context information server is not intended to support device independence but to provide device awareness, thus enabling the decision-making component to select those interaction patterns, which, apart from fitting the particular end user attributes, are also appropriate for the type of equipment available on the end user machine.

The usage-context attribute values are communicated through simple message passing to the decision-making component before the initiation of interaction. Additionally, during interaction, some dynamically changing usage-context parameters may also be fed to the decision-making component for decisions regarding *adaptive* behavior. For instance, assuming that the initial decision for selecting feedback leads to the use of audio effects, the dynamic detection of an increase in environmental noise may result in a run-time decision to switch to visual feedback (the underlying assumption being that such a decision does not conflict with other constraints).

This component is expected to be at the level of system software, accessing external peripheral equipment and providing information on the available input–output devices (i.e., some kind of a simple "registry" for installed peripheral equipment). This component is also responsible for usage-context monitoring. The idea of monitoring the environment within which the user interacts has been practically applied in the context of smart-home technology, mainly to identify emergency situations, by processing the input from various types of special-purpose sensors. Because of the lower-level functional requirements of these components, a typical third-generation programming language is considered suitable for their implementation.

User Information Server

This module maintains the individual profiles of end users, which, together with usage-context information (provided by the context information server), constitute the primary input source for the adaptation decision-making process (which is carried out by the decision-making component). An initial end user identification mechanism (e.g., a password at start-up) is required by the user information server for retrieving the corresponding user profile or making any deductions based on such knowledge. In the case of desktop systems with dedicated users, all interactive applications may have a hard-wired "user ID"; otherwise explicit user identification is needed. Alternative

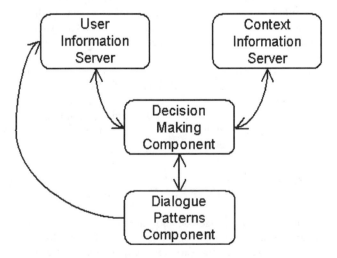

FIGURE 55.10. The unified user interface architecture.

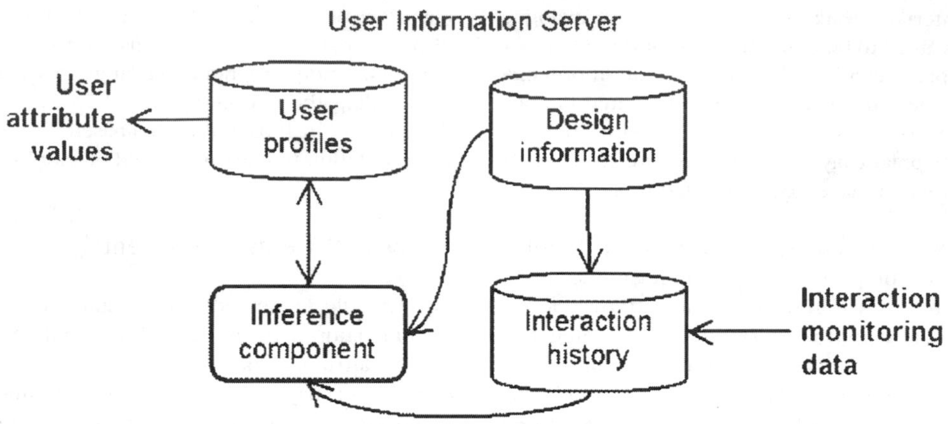

FIGURE 55.11. User information server.

means of identification, such as smart cards, can be employed for systems deployed in public information points.

From a knowledge representation point of view, static or preexisting user knowledge may be encoded in any appropriate form, depending on the type of information the user information server should feed to the decision-making process. Moreover, additional knowledge-based components may be employed for processing retrieved user profiles, drawing assumptions about the user, or even updating the original user profiles. Systems such as BGP-MS (Kobsa, 1990), PROTUM (Vergara, 1994), or USE-IT (Akoumianakis & Stephanidis, 2001) may be employed for such intelligent processing purposes.

Apart from such initial (i.e., before initiation of interaction) manipulation of user profiles, the user information server may also collect and process run-time interaction events to draw (additional) inferences about the end user. Such inferences may result in the identification of dynamic user preferences, loss of orientation in performing certain tasks, fatigue, inability to complete a task, and so on. In the communication with the rest of the architectural components, user-oriented information does not pose any special requirements because it can be conveyed through message passing of sequences of attribute values.

A typical knowledge representation approach may be employed for representing user models and profiles, as well as for drawing assumptions about particular user attributes at run-time (i.e., during interaction). An appropriate, local (to the user information server component) user representation method may be defined and adopted, both for storage of user information and for manipulation via inference engines. However, this representation, when communicated to the "external world" (i.e., the rest of the architectural components), should always be converted to the general form of attribute value pairs. In some cases, the user information server may merely play the role of a user profile repository. In such situations, a minimalistic implementation approach can be taken in which a database of user profiles is maintained and a small implementation shell is added to access the database by processing communication requests. If more than one repository is needed, the user information server should be at least capable of receiving and processing (i.e., reasoning about) interaction monitoring information. The

best candidate for this purpose is the employment of a logic programming language (e.g., Prolog), with support for interprocess communication.

From the analysis of various aspects of existing adaptive systems (Dieterich et al., 1993) with respect to the employment and utilization of user modeling methods, an appropriate architectural design pattern for the user information server has been derived. This architectural pattern is illustrated in Fig. 55.11 and consists of the following four logical components:

1. The *interaction history*, which monitors and stores interaction events; the modules performing such monitoring reside in the interactive software.
2. The *user profiles*, needed to drive an adaptation-oriented decision making process; the individual user profiles may be locally stored within this component.
3. The *design information*, providing design-related knowledge, which is necessary for associating interaction monitoring data with design context, as well as for supporting dynamic user attribute detection.
4. The *inference component*, which encompasses all the necessary knowledge to derive user attribute values during interaction (e.g., preferences, disorientation, expertise).

Decision-Making Component

This module encompasses the logic for deciding the necessary adaptation actions on the basis of the user and context attribute values received from the user information server and the context information server, respectively. Such attribute values are supplied to the decision-making component before the initiation of interaction (i.e., initial values resulting in initial interface adaptation, referred to as *adaptability*) as well as during interaction (i.e., changes in particular values or detection of new values resulting in dynamic interface adaptations, referred to as *adaptivity*). The decision-making component is only responsible for deciding the necessary adaptation actions, which are then directly communicated to, and subsequently performed by, the dialogue patterns component.

As part of the interface design process, alternative dialogue design artifacts may need to be constructed to fit different user and usage-context parameters. In the implementation process, these distinct dialogue artifacts become implemented interactive components. The run-time adaptation process for a particular situation of use is practically a context-sensitive selection of those components, which have been designed to address that particular situation.

To perform such a context-sensitive selection, the decision-making component encompasses information regarding all the dialogue patterns present within the dialogue patterns component and their specific design role. There are two categories of adaptation actions which are decided and communicated to the dialogue patterns component: (a) activation of specific dialogue components and (b) cancellation of previously activated dialogue components.

These two categories of adaptation actions suffice to express the interface component manipulation requirements for realizing either adaptability or adaptivity. Substitution is modeled by a message containing a *cancellation* action (i.e., the dialogue component to be substituted), followed by the necessary number of *activation* actions (i.e., which dialogue components to activate in place of a cancelled component). The need to send in the same message information regarding the component to be cancelled and the component to be used instead emerges when the implemented interface requires knowledge of all (or some) of the newly created components during interaction. For instance, if the new components include a container and the various contained components, and, if upon the creation of the container, information on the number and type of the particular contained components is required, it is necessary to ensure that all the relevant information (i.e., all engaged components) is received as a single message.

This component encompasses the representation of design logic in a form that, for each subtask, appropriately associates user and context attribute values with the various implemented alternative styles (i.e., dialogue patterns), if any. Typical logic programming languages and knowledge representation formalism are most appropriate for implementing this kind of functionality. For applications with simple decision-making logic, common programming languages may also suffice for hard-coding the decision process. In cases of more complex design relationships and design-based reasoning, a knowledge representation framework should be employed.

Dialogue Patterns Component

This module implements all the alternative dialogue patterns identified during the design process on the basis of the user and context attribute values (see Fig. 55.12). The dialogue patterns component may employ predeveloped interactive software in combination with additional interactive components. The latter case normally occurs when the directly deployed interactive software does not provide all the patterns required for addressing the target user and context attribute values.

The dialogue patterns component should be capable of applying pattern activation and cancellation decisions originated from the decision-making component. Additionally, interaction monitoring components may be attached to implemented dialogue patterns, providing monitoring information to the user information server for further processing (e.g., key strokes, notifications for use of interaction objects, task-level monitoring). The particular level of detail and frequency of monitoring are to be requested at run-time by the user information server.

This component should encompass all dialogue patterns in an implementation form. In this sense, typical development methods for building interactive software may be freely employed. The unified user interface architecture poses no restrictions on the type of interface tool to be used. Additionally, some special purpose functionality needs to be introduced "on top" of the interactive software implementing the designed dialogue patterns, realizing the "packaging" (of interactive software) as the dialogue patterns component of the unified user interface architecture.

Dialogue Patterns Component

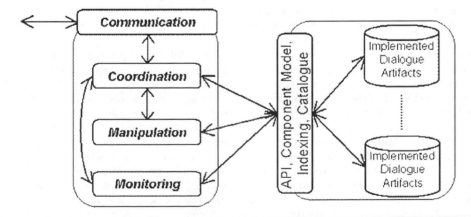

FIGURE 55.12. Dialogue patterns component.

Such extra required functionality is split into three categories: (a) communication with the rest of the architectural components (i.e., receiving and sending messages, following the intercomponent communication semantics); (b) monitoring and encompassing interaction monitoring code, which is to be attached to various interface components; and (c) coordination, capable of applying activation and cancellation decisions on interface components (as received from decision-making component), and enabling and disabling interaction monitoring components (as received from user information server).

This is the only type of functionality required to make existing interactive software available in the context of the unified user interface architecture. It should be noted that the employed interactive software may internally encompass architectural links in the context of a particular software framework, such as an Arch-based interactive structure (UIMS, 1992). In practice, there may be some lower level issues that need to be addressed:

• Attaching interaction monitoring software as independent software modules may present implementation barriers, requiring monitoring code to be "injected" within the original software implementation. This implementation barrier was recently overcome, because all widely available user interface development tools are now built on top of software toolkits (e.g., Java AWT/Swing, Microsoft Foundation Classes, or Microsoft ActiveX™) that explicitly provide monitoring application programming interfaces (e.g., Microsoft Active Accessibility™ and Java Accessibility™ by JavaSoft).

• When expanding interactive software to encapsulate adapted behaviors, some designed dialogue patterns may not be available in an implemented form. This is naturally expected when new user or usage-context parameter values (not originally foreseen) are taken into consideration, thus resulting in the construction of new design artifacts. In such cases, a dedicated implementation process should be initiated to implement the required extra dialogue patterns.

• It may be the case that some user interface components are built with non-programming-oriented tools, thus "lacking" an application programming interface through which run-time activation and cancellation could be performed. This issue would be resolved if widely available tools in this category (e.g., Visual Basic, Java Beans interactive construction tools, ActiveX component builders) generated code through which programmatic access to interface components would be possible.

Adaptability and Adaptivity Cycles

In unified user interfaces, the completion of an adaptation cycle, being either adaptability or adaptivity, is realized in a specific number of distributed processing stages performed by the various components of the architecture. During those stages, the components communicate with each other, requesting or delivering specific pieces of information. The overall communication requirements among the components are illustrated in Fig. 55.13. The final outcome of an adaptation cycle, the activation and cancellation of decisions, is emphasized with a large dashed arrow.

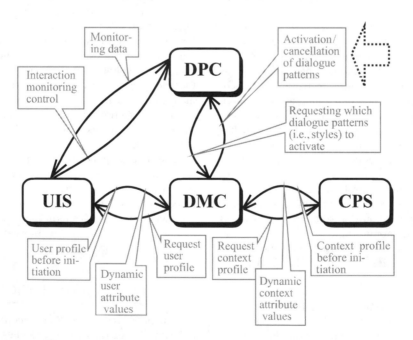

FIGURE 55.13. Communication requirements among the components of the unified user interface architecture to perform interface adaptation cycles. UIS = user information server; DPC = dialogue patterns component; DMC = decision-making component; CPS = context parameters server.

The communication protocol among the architecture components presented here mainly emphasizes the rules that govern the exchange of information among various communicating parties, as opposed to a strict message syntax description. Hence, the primary focus is on the semantics of communication regarding the type of information communicated, the content it conveys, and the usefulness of the communicated information at the recipient component side.

In the unified user interface architecture, there are four bidirectional communication channels, each engaging a pair of communicating components. For instance, one such channel concerns the communication between the user information server and the decision-making component. Each channel defines two protocol classes (for example, user information server → decision-making component, i.e., type of messages sent from the first to the latter, and decision-making component → user information server, i.e., type of messages sent in the opposite direction). The four communication channels therefore define four pairs of protocol categories.

Communication Between the User Information Server and the Decision Making Component. In this communication channel, there are two communication rounds: (a) before initiation of interaction, the decision-making component requests the user profile from the user information server, which replies directly with the corresponding profile (as a sequence of attribute values); and (b) after initiation of interaction, each time the user information server detects some dynamic user attribute values (on the basis of interaction monitoring), it communicates those values immediately to the decision-making component.

Communication Between the User Information Server and the Dialogue Patterns Component. The communication among these two components aims to enable the user information server to collect interaction monitoring information, as well as to control the type of monitoring to be performed. The user information server may request monitoring at three levels: (a) *task* (i.e., when initiated or completed); (b) *method* for interaction objects (i.e., which logical object action has been accomplished, such as "pressing" a "button" object); and (c) *input event* (i.e., a specific device event, such as moving the mouse or pressing a key).

In response to monitoring control messages, the dialogue patterns component takes care of (a) activating and disabling the appropriate interaction monitoring software modules and (b) continuously exporting monitoring data, according to the monitoring levels requested, back to the user information server (initially, no monitoring modules are activated by the dialogue patterns component).

Communication Between the Decision-Making Component and the Dialogue Patterns Component. Dialogue patterns are organized within the dialogue patterns component in the following manner. Suppose that alternative dialogue patterns are designed for a particular user subtask, addressing distinct values of the decision parameters (i.e., user and context parameters). Then, as part of the software implementation, each such alternative dialogue pattern is associated with its respective subtask, while it is given an arbitrary name, unique among the dialogue patterns of its corresponding subtask. Each implemented dialogue pattern for a particular subtask is referred to as *style*. Styles can thus be identified by reference to their designated name.

At start-up, before initiation of interaction, the dialogue patterns component requests (for each user task) the names of the styles (i.e., implemented dialogue patterns) to be activated, thus realizing the adaptability behavior. The decision-making component triggers the adaptability cycle and responds accordingly.

After initiation of interaction, the decision-making component may "take the initiative" to communicate dynamic style activation and cancellation messages to the dialogue patterns component. Such a communication always occurs at the end of each adaptivity cycle.

Communication Between the Decision-Making Component and the Context Parameters Server. The communication between these two components is simple: The decision-making component requests the context parameter values (i.e., usage context profile), and the context parameters server responds accordingly. During interaction, dynamic updates on certain context property values are to be communicated to the decision-making component for further processing (i.e., possibly new inferences will be made).

As previously mentioned, the context parameters server aims to support two levels of functionality: (a) It encompasses information regarding the available input–output facilities (i.e., a type of system "registry") on the end user machine, and (b) it monitors particular environment parameters, such as environment noise or user presence in front of the terminal (e.g., via infrared sensors). The first category of information is necessary for selecting, at the decision-making component level, those interaction techniques that, while conforming to user attribute values, can be fully supported via the peripheral equipment at the end user terminal. The second category is necessary for many purposes. For instance, it may be used to provide interaction not conflicting with the particular environment state (such as avoiding audio feedback, if high environment noise is detected). Another scenario concerns supporting the inference process of making dynamic assumptions about the user. For example, if notification is received that the user moves away from the terminal, then that particular interaction session is terminated; otherwise, if the device remains idle for a long period of time while the user is still in front of the terminal, this could be interpreted as potential "confusion," "loss of orientation," or "inability to perform the task."

CONCLUSIONS

This chapter has discussed the concept of unified user interface, reflecting on the notion of user and usage-context self-adapting interfaces as a promising technical vehicle toward achieving universal access. We have elaborated on the unified user interface design method and engineering paradigm, putting forward a profile of a new technology for developing universally accessible interactions.

There are a number of issues that emerge from this chapter. Because accessibility and interaction quality have become global requirements in the information society, adaptation needs to be "designed into" the system rather than decided on and implemented a posteriori. This entails design techniques that can capture alternative design options and design representations that can be incrementally extended (i.e., design pluralism) to encapsulate evolving or new artifacts. Another important issue concerning design that has been learned through practice and experience is that universal access means breaking away from the traditional perspective of "typical" users interacting with a desktop machine in a business environment, moving toward interactive systems accessible at anytime, anywhere, and by anyone. This requires that future systems embody the capability for context-sensitive processing so as to present their users with a suitable computational embodiment or metaphor depending on user-, situation-, and context-specific attributes.

The unified user interface development methodology is the first systematic effort in this direction that has been thoroughly developed and applied in practice. The results of such an effort and the experience gained justify the argument that developing universally accessible and usable user interfaces is more of a challenge than a utopia (Stephanidis & Emiliani, 1999). Nonetheless, unified user interfaces should not be considered the only possible approach for addressing universal access in the information society. The unified user interfaces development framework augments and enhances traditional approaches to developing user interfaces in the following dimensions: (a) It is oriented toward iteratively capturing and satisfying multiple user- and context-oriented requirements; (b) it employs adaptation-based techniques to facilitate and enhance personalized interaction; furthermore, unified user interfaces development is characterized by the need to identify appropriate evaluation methods and process to assess the effectiveness of the self-adapting behavior of the interface.

The AVANTI browser has provided a concrete example of the unified user interface development process. The AVANTI unified interface is capable of adapting itself to suit the requirements of users who are able bodied and those who have visual or motor impairments in a variety of contexts of use. Adaptability and adaptivity are used extensively to tailor and enhance the interface respectively, to effectively and efficiently meet the target of interface individualization for end users. The experience acquired has demonstrated the applicability of the methodologies, techniques, and tools comprising the unified user interfaces development paradigm in the construction of a large-scale, "real-world" self-adapting interactive application.

The unified user interface development process is a resource-demanding design and engineering process. It requires the systematic classification and organization of both design as well as engineering artifacts, according to the unified design and engineering practices, thus exhibiting a relatively high entry barrier for development. Multidisciplinary expertise is required, especially within the implementation phase. The experience gained so far has clearly indicated the considerable development overhead of crafting unified user interfaces, which is proportional to the targeted degree of diversity in users and usage contexts. Additionally, it has been observed that an initial "start-up" investment is required to set up a unified user interface architecture, starting from the preliminary stages before the actual embedding of components in a running system (e.g., implemented dialogues, decision-making logic, or user information models). Finally, an additional need has emerged for an automated integration process to expand the system's adaptation behavior and address additional user and context parameter values, and subsequently to implement decision-making logic and dialogue patterns. Ongoing and future work seeks to further advance the unified user interface development approach to take account of recent developments in enterprise computing (e.g., distributed and component-based computing) and novel user interface software technologies (e.g., augmented and virtual realities, wearable equipment).

Despite the recent rise of interest in the topic of universal access and the successful and indisputable progress in research and development, many challenges lie ahead. Effective use of adaptation in user interfaces implies a deep and thorough knowledge of the user requirements, and this implies identifying diversity in the user population, mapping diversity-related requirements to concrete designs, and evaluating the related adaptation behavior. The same holds for the study of context, which, as we pointed out in previous sections, is critical for the quality of a system's adaptable and adaptive behavior (Stephanidis et al., 1999). The latter issue introduces another challenge to be addressed, namely, the compelling need to assess the implications of universal access on digital content, functionality, and interaction. This entails, among other things, a renewed account of the properties of adaptation and how they can be embodied in interactive products and services. Clearly, the traditional view that adaptable and adaptive behavior is a characteristic property of the interactive software may no longer suffice given the trends toward ubiquitous access, mobile computing, and Internet appliances.

Finally, theoretical work must be supported with large-scale case studies, which can provide the instruments for experimentation, thus ultimately improving the empirical basis of the field. Such case studies should not only aim to demonstrate technical feasibility of research and development propositions, but also to assess the economic efficiency and efficacy of competing technological options in the longer term.

ACKNOWLEDGMENTS

Part of the research and technological development (RTD) work reviewed in this chapter was carried out in the context of the following European Commission funded projects: The ACCESS TP1001 (Development platform for unified ACCESS to enabling environments) project was partially funded by the TIDE Programme of the European Commission and lasted 36 months (January 1, 1994, to December 31, 1996). The partners of the ACCESS consortium are CNR-IROE (Italy), Prime contractor; ICS-FORTH (Greece); University of Hertfordshire (United Kingdom); University of Athens (Greece); NAWH (Finland); VTT (Finland); Hereward College (United Kingdom); RNIB (United Kingdom); Seleco (Italy); MA Systems & Control

(United Kingdom); and PIKOMED (Finland). The AVANTI AC042 (Adaptable and Adaptive Interaction in Multimedia Telecommunications Applications) project was partially funded by the Advanced Communications Technologies & Services (ACTS) Programme of the European Commission, and lasted 36 months (September 1, 1995, to August 31, 1998). The partners of the AVANTI consortium are ALCATEL Italia, Siette division (Italy), Prime Contractor; IROE-CNR (Italy); ICS-FORTH (Greece); GMD (Germany), VTT (Finland); University of Siena (Italy); MA Systems and Control (UK); ECG (Italy); MATHEMA (Italy); University of Linz (Austria); EUROGICIEL (France); TELECOM (Italy); TECO (Italy); and ADR Study (Italy).

References

Akoumianakis, D., & Stephanidis, C. (2001). USE-IT: a tool for lexical design assistance. In C. Stephanidis (Ed.), *User Interfaces for All—Concepts, Methods and Tools* (pp. 469–487). Mahwah, NJ: Lawrence Earlbaum Associates.

Benyon, D. (1984). MONITOR: A self-adaptive user-interface. In *Proceedings of IFIP Conference on Human–Computer Interaction: INTERACT '84* (Vol. 1, pp. 335–341). Amsterdam: North-Holland, Elsevier Science.

Browne, D., Norman, M., & Adhami, E. (1990). Methods for building adaptive systems. In D. Browne, M. Totterdell, & M. Norman (Eds.), *Adaptive user interfaces* (pp. 85–130). London: Academic Press.

Cockton, G. (1987). Some critical remarks on abstractions for adaptable dialogue managers. In *Proceedings of the 3rd Conference of the British Computer Society, People & Computers III, HCI Specialist Group*, University of Exeter (pp. 325–343). Cambridge, UK: Cambridge University Press.

Cockton, G. (1993). Spaces and distances—software architecture and abstraction and their relation to adaptation. In M. Schneider-Hufschmidt, T. Kühme, & U. Malinowski (Eds.), *Adaptive user interfaces—principles and practice* (pp. 79–108). Amsterdam: North-Holland, Elsevier Science.

Cote Muñoz, J. (1993). AIDA—an adaptive system for interactive drafting and CAD applications. In M. Schneider-Hufschmidt, T. Kühme, & U. Malinowski (Eds.), *Adaptive user interfaces—principles and practice* (pp. 225–240). Amsterdam: North-Holland, Elsevier Science.

Dieterich, H., Malinowski, U., Kühme, T., & Schneider-Hufschmidt, M. (1993). State of the art in adaptive user iinterfaces. In M. Schneider-Hufschmidt, T. Kühme, & U. Malinowski (Eds.), *Adaptive user interfaces—principles and practice* (pp. 13–48). Amsterdam: North-Holland, Elsevier Science.

Hartson, H. R., Siochi, A. C., & Hix, D. (1990). The UAN: A user-oriented representation for direct manipulation interface design. *ACM Transactions on Information Systems, 8*, 181–203.

Hill, R. (1986). Supporting concurrency, communication and synchronisation in human–computer interaction—the sassafras UIMS. *ACM Transactions on Graphics, 5*, 289–320.

Hoare, C. A. R. (1978). Communicating sequential processes. *Communications of the ACM, 21*, 666–677.

Jacobson, I., Griss, M., & Johnson, P. (1997). Making the reuse business work. *IEEE Computer, 10*, 36–42.

Johnson, P., Johnson, H., Waddington, P., & Shouls, A. (1988). Task-related knowledge structures: analysis, modeling, and applications. In D. M. Jones & R. Winder (Eds.), *People and computers: From research to implementation—Proceedings of HCI '88* (pp. 35–62). Cambridge, England: Cambridge University Press.

Kobsa, A. (1990). Modeling the user's conceptual knowledge in BGP-MS, a user modeling shell system. *Computational Intelligence, 6*, 193–208.

Kobsa, A., & Wahlster, W. (Eds.). (1989). *User models in dialog systems.* New York: Springer Symbolic Computation.

Koller, F. (1993). A demonstrator based investigation of adaptability. In M. Schneider-Hufschmidt, T. Kuhme, & U. Mallinowski, (Eds.), *Adaptive user interfaces* (pp. 183–196). Amsterdam: Elsevier.

Lai, K., & Malone, T. (1988). Object Lens: A spreadsheet for co-operative work. In *Proceedings of the Conference on Computer Supported Co-operative Work* (pp. 115–124). New York: ACM Press.

MacLean, A., Carter, K., Lovstrand, L., & Moran, T. (1990). User-tailorable systems: Pressing the issues with buttons. In *Proceedings of the ACM Conference on Human Factors in Computing Systems* (pp. 175–182). New York: ACM Press.

Moran, T. P., & Carroll, J. M. (1996). *Design rationale: Concepts, techniques, and use.* Hillsdale, NJ: Lawrence Erlbaum Associates.

Mynatt, E. D., & Edwards, W. K (1992). *The Mercator environment: A nonvisual interface to the X window system.* (Technical Report GIT-GVU-92-05, February) Graphics Visualization & Usability Center. Atlanta, GA: Georgia Institute of Technology.

Mynatt, E. D., & Weber, G. (1994). Nonvisual presentation of graphical user interfaces: Contrasting two approaches. In the *Proceedings of the ACM Conference on Human Factors in Computing Systems (CHI '94)* (pp. 166–172). New York: ACM Press.

Mowbray, T. J., & Zahavi, R. (1995). *The essential CORBA: Systems integration using distributed objects.* New York: John Wiley.

Robertson, G., Henderson, D., & Card, S. (1991). Buttons as first class objects on an Xdesktop. In *Proceedings of the Symposium on User Interface Software and Technology* (pp. 35–44). New York: ACM Press.

Saldarini, R. (1989). Analysis and design of business information systems. *Structured Systems Analysis* (pp. 22–23). New York: MacMillan.

Savidis, A., Akoumianakis, D., & Stephanidis, C. (2001). The unified user interface design method. In C. Stephanidis (Ed.), *User interfaces for all—concepts, methods and tools* (pp. 417–440). Mahwah, NJ: Lawrence Erlbaum Associates.

Savidis, A., & Stephanidis, C. (1995). Developing dual user interfaces for integrating blind and sighted users: The HOMER UIMS. In the *Proceedings of the ACM Conference on Human Factors in Computing Systems (CHI '95)* (pp. 106–113). New York: ACM Press.

Savidis, A., & Stephanidis, C. (2001a). The unified user interface software architecture. In C. Stephanidis (Ed.), *User interfaces for all—concepts, methods and tools* (pp. 389–415). Mahwah, NJ: Lawrence Erlbaum Associates.

Savidis, A., & Stephanidis, C. (2001b). The I-GET UIMS for Unified User Interface Implementation. In C. Stephanidis (Ed.), *User interfaces for all—concepts, methods and tools* (pp. 489–523). Mahwah, NJ: Lawrence Erlbaum Associates.

Savidis, A., Stephanidis, C., & Akoumianakis, D. (1997a). Unifying toolkit programming layers: A multi-purpose toolkit integration module. In M. D. Harrison & J. C. Torres (Eds.), *Conference Proceedings of the 4th Eurographics Workshop on Design, Specification and Verification of Interactive Systems (DSV-IS '97)* (pp. 177–192). Vienna: Springer Verlag.

Savidis, A., Stergiou, A., & Stephanidis, C. (1997b, May 28-30). Generic containers for metaphor fusion in non-visual interaction: The HAWK Interface toolkit. *Proceedings of the Interfaces '97 Conference* (pp. 194-196). Montpellier, France: EC2 & Development.

Sherman, H. E., & Shortliffe, H. E. (1993). A user-adaptable interface to predict users' needs. In M. Schneider-Hufschmidt, T. Kuhme, & U. Mallinowski (Eds.), *Adaptive user interfaces* (pp. 285-315). Amsterdam: Elsevier.

Short, K. (1997). Component based development and object modeling, version 1.0 [Computer software]. Dallas, TX: Texas Instruments.

Simonsen, J., & Kensing, F. (1997). Using ethnography in contextual design. *Communications of the ACM, 40,* 82-88.

Stephanidis, C. (1995). Towards user interfaces for all: Some critical issues [Panel session]. In Y. Anzai, K. Ogawa, & H. Mori (Eds.), *Symbiosis of human and artifact—future computing and design for human-computer interaction. Proceedings of the 6th International Conference on Human-Computer Interaction* (pp. 137-142). Amsterdam: Elsevier.

Stephanidis, C. (2001a). User interfaces for all: New perspectives into HCI. In C. Stephanidis (Ed.), *User interfaces for all—concepts, methods and tools* (pp. 3-17). Mahwah, NJ: Lawrence Erlbaum Associates.

Stephanidis, C. (2001b). Adaptive techniques for universal access. *User Modelling and User Adapted Interaction International Journal, 11,* 159-179. Retrieved from http://www.wkap.nl/issuetoc.htm/0924-1868+11+1/2+2001

Stephanidis, C. (2001c). The concept of unified user interfaces. In C. Stephanidis (Ed.), *User interfaces for all—concepts, methods and tools* (pp. 371-388). Mahwah, NJ: Lawrence Erlbaum Associates.

Stephanidis, C., & Emiliani, P. L. (1999). Connecting to the information society: A European perspective. *Technology and Disability Journal, 10,* 21-44. Available from http://www.ics.forth.gr/proj/at-hci/files/TDJ_paper.PDF

Stephanidis, C., Paramythis, A., Sfyrakis, M., & Savidis, A. (2001). A case study in unified user interface development: The AVANTI web browser. In C. Stephanidis (Ed.), *User interfaces for all—concepts, methods and tools* (pp. 525-568). Mahwah, NJ: Lawrence Erlbaum Associates.

Stephanidis, C., Paramythis, A., Sfyrakis, M., Stergiou, A., Maou, N., Leventis, A., Paparoulis, G., & Karagiannidis, C. (1998b). Adaptable and adaptive user interfaces for disabled users in the AVANTI project. In S. Trigila, A. Mullery, M. Campolargo, H. Vanderstraeten, & M. Mampaey (Eds.), *Lecture notes in computer science, Vol. 1430: Proceedings of the 5th International Conference on Intelligence in Services and Networks (IS&N '98), Technology for Ubiquitous Telecommunication Services* (pp. 153-166). Haidelberg, Germany: Springer-Verlag.

Stephanidis, C., Salvendy, G., Akoumianakis, D., Bevan, N., Brewer, J., Emiliani, P. L., Galetsas, A., Haataja, S., Iakovidis, I., Jacko, J., Jenkins, P., Karshmer, A., Korn, P., Marcus, A., Murphy, H., Stary, C., Vanderheiden, G., Weber, G., & Ziegler, J. (1998a). Toward an information society for all: An international R&D agenda. *International Journal of Human-Computer Interaction, 10,* 107-134. Retrieved from http://www.ics.forth.gr/proj/at-hci/files/white_paper_1998.pdf

Stephanidis, C., & Savidis, A. (2001). Universal access in the information society: methods, tools and interaction technologies. *Universal Access in the Information Society, 1.*

UIMS Developers Workshop. (1992). A meta-model for the run-time architecture of an interactive system. *SIGCHI Bulletin, 24,* 32-37.

Vergara, H. (1994). *PROTUM—A Prolog based tool for user modeling* (Bericht Nr. 55/94; WIS-Memo 10). University of Kostanz, Germany.

Part

·VC·

TESTING AND EVALUATION

Andrew Sears
UMBC

Developing computing technologies is an iterative process, and evaluation is a critical component of any effective iterative process. In this context, an effective evaluation will not only identify problems, but also will provide guidance for the next iteration of design activities. An effective evaluation will be based on knowledge of the intended users, their goals, and the environment in which they will interact with the system.

Different approaches are useful throughout the development process, depending on the purpose of the evaluation and the resources available. The first three chapters in this section discuss what have become the three standard approaches for evaluating interfaces: user-, inspection-, and model-based evaluations. The fourth chapter focuses on the increasing importance of evaluating the affective, or emotional, aspects of the interactions between people and computers. Because many decisions during the development process are driven by the benefits expected as compared with the costs incurred, the reader is also referred to the cost justification chapter by Bias in the next section of this handbook.

In *User-Based Evaluations*, Dumas explores a variety of techniques that depend on the involvement of individuals representative of those who will ultimately use the system. Three approaches are discussed: questionnaires, observation, and usability testing. The chapter begins with a brief overview of the issues involved in developing questionnaires followed by review of several questionnaires that have been used in a variety of development efforts, including the software usability scale (SUS), computer user satisfaction inventory (CUSI), questionnaire for user interaction satisfaction (QUIS), and software usability measurement inventory (SUMI) questionnaires. Next, Dumas discusses the use of observation as an evaluation technique, and several important limitations are highlighted. The focus of this chapter is clearly on empirical usability testing. A brief history of usability testing is provided, as well as a definition to guide the reader through the remainder of the chapter. This is followed by a detailed discussion of six characteristics of a valid usability test:

focusing on usability, including appropriate participants, having something specific to evaluate, having participants think aloud, recording and analyzing data, and communicating the results. Dumas provides an interesting argument regarding the need to have participants think aloud during evaluation sessions while providing detailed guidance about selecting participants, selecting tasks, providing instructions, executing a usability test, collecting and analyzing data, and communicating the results of the evaluation. Additional issues, including variations on standard usability testing and concerns about the validity of usability tests are also discussed.

In *Inspection-Based Evaluations*, Cockton, Lavery, and Woolrych shift the focus to techniques that rely on the judgments of experts rather than input from potential users. They begin by discussing a generic process that effectively describes all inspection-based techniques. Basic issues are discussed, including candidate problem discovery, confirmation and elimination of candidate problems, and recommendation generation, many of which apply to all evaluation techniques. After highlighting the different types of inspection-based techniques that exist, heuristic evaluations, guidelines-based evaluations, cognitive walkthroughs, and heuristic walkthroughs are presented in more detail. In next section, Cockton et al. revisit Gray and Salmon's critique of studies involving evaluation techniques with a focus on how these concerns can be addressed in future research. The generic process presented early in the chapter, and the issues raised through the review of Gray and Salmon's critique, provide a framework for reviewing the existing literature. Through this review, the reader learns of the limitations of these studies while gaining insight into the efficacy of various evaluation techniques. The chapter concludes with a review of the major issues, directions for future research, and recommendations for practitioners.

Kieras focuses on *Model-Based Evaluations* as a supplement to conventional usability evaluation techniques. He begins by discussing the benefits model-based evaluations can

provide and how model-based techniques can fit into the standard development process. Brief descriptions are provided for the three most common approaches to model-based evaluation: task network models, cognitive architecture models (discussed in detail by Byrne in chapter 5 of this handbook), and GOMS models. Next, he discusses key issues affecting the development and use of model based evaluation techniques. This may prove to be the most useful part of the chapter because it provides practical guidance for both practitioners and researchers. Kieras describes GOMS models, including a brief review of why GOMS models work and the limitations of these models. The chapter concludes by providing a set of "if–then" scenarios to guide those interested in applying the most appropriate model-based evaluation techniques to their situation.

This section concludes with *Beyond Task Completion: Evaluation of Affective Components of Use* by Karat. For years the focus has been on usability. In this chapter, Karat argues that computing technologies are now being used for a wider variety of activities and that for many of these activities, productivity is not necessarily the primary concern. He argues that an effective evaluation may no longer focus on traditional measures such as task completion times and error rates. Instead, it may be more important to assess the users' feelings about their interactions and the technologies. Much of this chapter focuses on the importance of shifting attention from productivity to perceptions and the corresponding challenges. As a result, it is unclear how well existing techniques will perform if the focus is on designing experiences that people enjoy or value. The chapter explores these issues and concludes by discussing various tools and techniques that may be used to more effectively assess affective aspects of interactions between people and computing technologies.

The first three chapters in this section will provide practitioners and researchers with a solid foundation with respect to the most common approaches for evaluating computing technologies: user-, inspection-, and model-based evaluations. The fourth chapter looks beyond the state-of-the-art in evaluation techniques, suggesting that affective aspects of use will be increasingly important as the use of computing technologies continues to grow. Practitioners will find practical advice to guide their selection and use of evaluation techniques and the interpretation of the resulting data. These chapters provide guidance to researchers by highlighting methodological issues that must be addressed when conducting research in this area and identifying numerous research opportunities.

USER-BASED EVALUATIONS

Joseph S. Dumas
Oracle Corporation

INTRODUCTION

Over the past 20 years, there has been a revolution in the way products, especially high-tech products, are developed. It is no longer accepted practice to wait until the end of development to evaluate a product. That revolution applies to evaluating usability. As the other chapters in this handbook show, evaluation and design now are integrated. Prototyping software and the acceptance of paper prototyping make it possible to evaluate designs as early concepts, then throughout the detailed design phases. User participation is no longer postponed until just before the product is in its final form. Early user involvement has blurred the distinction between design and evaluation. Brief usability tests are often part of participatory *design* sessions, and users are sometimes asked to participate in early user interface design walkthroughs. Although the focus of this chapter is on user-based evaluation methods, I concede that the boundary between design methods and evaluation methods grows less distinct with time.

In this chapter, I focus on user-based evaluations, which are evaluations in which users directly participate. But the boundary between user-based and other methods is also becoming less distinct. Occasionally, usability inspection methods and user-based methods merge, such as in the pluralistic walkthrough (Bias, 1994). In this chapter, I maintain the somewhat artificial distinction between user-based and other evaluation methods to treat user-based evaluations thoroughly. I describe three user-based methods: user-administered questionnaires, observing users, and empirical usability testing. In the final section of the chapter, I describe when to use each method.

USER-ADMINISTERED QUESTIONNAIRES

A questionnaire can be used as a stand-alone measure of usability, or it can be used along with other measures. For example, a questionnaire can be used at the end of a usability test to measure the subjective reactions of the participant to the product tested, or it can be used as a stand-alone usability measure of the product. Over the past 20 years, there have been questionnaires that

- Measure attitudes toward individual products
- Break attitudes down into several smaller components, such as ease of learning
- Measure just one aspect of usability (Spenkelink, Beuijen, & Brok, 1993)
- Measure attitudes that are restricted to a particular technology, such as computer software
- Measure more general attitudes toward technology or computers (Igbaria & Parasuraman, 1991)
- Are filled out after using a product only once (Doll & Torkzadeh, 1988)
- Assume repeated use of a product
- Require a psychometrician for interpretation of results (Kirakowski, 1996)

- Come with published validation studies
- Provide comparison norms to which one can compare results

Throughout this history, there have been two objectives for questionnaires developed to measure usability: (a) create a short questionnaire to measure users' subjective evaluation of a product, usually as part of another evaluation method, and (b) create a questionnaire to provide an absolute measure of the subjective usability of a product. This second objective parallels the effort in usability testing to find an absolute measure of usability, that is, a numerical measure of the usability of a product that is independent of its relationship to any other product.

Creating a valid and reliable questionnaire to evaluate usability takes considerable effort and specialized skills, skills in which most usability professionals don't receive training. The steps involved in creating an effective questionnaire include the following:

- Create a number of questions or ratings that appear to tap attitudes or opinions that you want to measure. For example, the questions might focus on a product's overall ease of use. At the beginning of the process, the more questions you can create, the better.
- Use item analysis techniques to eliminate the poor questions and keep the effective ones. For example, if you asked a sample of users to use a product and then answer the questions, you could compute the correlation between each question and the total score of all of the questions. You would eliminate questions with low correlations. You would also eliminate questions with small variances because nearly all of the respondents are selecting the same rating value or answer. You would also look for high correlations between two questions because this indicates that the questions may be measuring the same thing. You could then eliminate one of the two.
- Assess the reliability of the questionnaire. For example, you could measure test–retest reliability by administering the questionnaire twice to the same respondents, but far enough apart in time that respondents would be unlikely to remember their answers from the first time. You could also measure split half reliability by randomly assigning each question to one of two sets of questions, then administering both sets and computing the correlation between them (Gage & Berliner, 1991).
- Assess the validity of the questionnaire. Validity is the most difficult aspect to measure but is an essential characteristic of a questionnaire (Chignell, 1990). A questionnaire is valid when it measures what it is suppose to measure, so a questionnaire created to measure the usability of a product should do just that. Demonstrating that it is valid takes some ingenuity. For example, if the questionnaire is applied to two products that are known to differ on usability, the test scores should reflect that difference. Or test scores from users should correlate with usability judgments of experts about a product. If the correlations are low, either the test is not valid or the users and the experts are not using the same process. Finally, if the test is valid, it should correlate highly with questionnaires with known validity, such as the usability questionnaires discussed in the following subsections.

Off-the-Shelf Questionnaires

Because an effective questionnaire takes time and special skills to develop, usability specialists have been interested in using off-the-shelf questionnaires that they can borrow or purchase. The advantages of using a professionally developed questionnaire are substantial. These questionnaires usually have been developed by measurement specialists who assess the validity and reliability of the instrument as well as the contribution of each question.

Historically, there have been two types of questionnaires developed: (a) short questionnaires that can be used to obtain a quick measure of users' subjective reactions, usually to a product that they have just used for the first time, and (b) longer questionnaires that can be used alone as an evaluation method and that may be broken out into more specific subscales.

Short Questionnaires. There have been a number of published short questionnaires. A three-item questionnaire was developed by Lewis (1991). The three questions measure the users' judgment of how easily and quickly tasks were completed. The Software Usability Scale (SUS) has 10 questions (Brooke, 1996). It can be used as a stand-alone evaluation or as part of a user test. It can be applied to any product, not just software. It was created by a group of professionals then working at Digital Equipment Corporation. The 10 SUS questions have a Likert scale format—a statement followed by a five-level agreement scale. For example,

I think that I would like to use this system frequently.

Strongly	————————→	Strongly
Disagree		Agree
1	2	3	4	5

Brooke (1996) described the scale and the scoring system, which yields a single, 100-point scale.

A somewhat longer questionnaire is the Computer User Satisfaction Inventory (CUSI). It was developed to measure attitudes toward software applications (Kirakowski & Corbett, 1988). It has 22 questions that break into two subscales: affect (the degree to which respondents like the software) and competence (the degree to which respondents feel they can complete tasks with the product).

Stand-Alone Questionnaires. These questionnaires were developed to measure usability as a stand-alone method. They have many questions and attempt to break users' attitudes into a number of subscales. The Questionnaire for User Interaction Satisfaction (QUIS) was developed at the Human–Computer Interaction Lab (HCIL) at the University of Maryland at College Park (Chin, Diehl, & Norman, 1988). QUIS was designed to assess users' subjective satisfaction with several aspects of the human–computer interface. It has been used by many evaluators over the past 10 years, in part because of its inclusion in Shneiderman's (1997) editions. It consists of a set of general questions, which provide an overall assessment of a product, and a set of detailed questions about interface components. Version 7.0 of the questionnaire contains a set of demographic questions, a measure of overall system satisfaction, and

hierarchically organized measures of 11 specific interface factors: screen factors, terminology and system feedback, learning factors, system capabilities, technical manuals, online tutorials, multimedia, voice recognition, virtual environments, Internet access, and software installation.

Because QUIS's factors are not always relevant to every product, practitioners often select a subset of the questions to use or use only the general questions. There is a long form of QUIS (71 questions) and a short form (26 questions). Each question uses a 9-point rating scale, with the end points labeled with adjectives. For example,

Characters on the screen are:

Hard to read ————————→ *Easy to read*
1 2 3 4 5 6 7 8 9

There is a Web site for QUIS (www.lap.umd.edu/QUIS/index. html). Licenses for use are available for a few hundred dollars. The site also contains references to evaluations that have used QUIS.

The Software Usability Measurement Inventory (SUMI) was developed to evaluate software only (Kirakowski, 1996). It is a well-constructed instrument that breaks the answers into six subscales: global, efficiency, affect, helpfulness, control, and learnability.

The Global subscale is similar to QUIS's general questions. The SUMI questionnaire consists of 50 statements to which users reply that they either *agree, are undecided,* or *disagree.* For example:

- This software responds too slowly to inputs.
- The instructions and prompts are helpful.
- The way that system information is presented is clear and understandable.
- I would not like to use this software every day.

Despite it length, SUMI can be completed in about 5 minutes. It does assume that the respondents have had several sessions working with the software. SUMI has been applied not only to new software under development, but also to compare software products and to establish a usability baseline. SUMI has been used in development environments to set quantitative goals, track achievement of goals during product development, and highlight good and bad aspects of a product.

SUMI's strengths come from its thorough development. Its validity and reliability have been established. In addition, its developers have created norms for the subscales so that you can compare your software against similar products. For example, you could show that the product you are evaluating scored higher than similar products on all of the subscales. The norms come from several thousand respondents.

The questionnaire comes with a manual for scoring the questions and using the norms. The developers recommend that the test be scored by a trained psychometrician. For a fee, the developer will do the scoring and the comparison with norms. The license comes with 50 questionnaires in the language of your choice, a manual, and software for scoring the results and

creating reports. The Web site for SUMI is http://www.ucc.ie/hfrg/questionnaires/sumi/index.html.

Questionnaires can play an important role in a toolkit of usability evaluation methods. It is difficult to create a good one, but there are several that have been well constructed and extensively used. The short ones can be used as part of other evaluation methods, and for most usability specialists, using them is preferable to creating their own. The longer ones can be used to establish a usability baseline and to track progress over time. Even the longest questionnaires can be completed in 10 minutes or less. Whether any of these questionnaires can provide an absolute measure of usability remains to be demonstrated.

OBSERVING USERS

Although observing users is a component of many evaluation methods, such as watching users through a one-way mirror during a usability test, this section focuses on observation as a stand-alone evaluation method. Some products can only be evaluated in their use environment, where the most an evaluator can do is watch the participants. Indeed, one could evaluate any product by observing its use and recording what happens. For example, if you were evaluating new software for stock trading, you could implement it and then watch trading activity as it occurs.

Unfortunately, observation has several limitations when used alone (Baber & Stanton, 1996), including the following:

- *It is difficult to infer causality while observing any behavior.* Because the observer is not manipulating the events that occur, it is not always clear what caused a behavior.

- *The observer is unable to control when events occur.* Hence, important events may never occur while the observer is watching. A corollary to this limitation is that it may take a long time to observe what you are looking for.

- *Participants change their behavior when they know they are being observed.* This problem is not unique to observation; in fact it is a problem with any user-based evaluation method.

- *Observers often see what they want to see, which is a direct challenge to the validity of observation.*

Baber and Stanton provide guidelines for using observation as an evaluation method.

A method related to both observation and user testing is private camera conversation (DeVries, Hartevelt, & Oosterholt, 1996). Its advocates claim that participants enjoy this method and that it yields a great deal of useful data. The method requires only a private room and a video camera with a microphone. It can be implemented in a closed booth at a professional meeting, for example. The participant is given a product and asked to go into the room and, when ready, turn on the camera and talk. The instructions on what to talk about are quite general, such as asking them to talk about what they like and dislike about the product. The sessions are self-paced but quite short (5–10 minutes). As with usability testing, the richness of the verbal protocol is enhanced when two or more people who know each other participate together.

The product of the sessions is a videotape that must be watched and analyzed. Because the participants are allowed to be creative and do not have to follow a session protocol, it is difficult to evaluate the usability of a product with this method.

A related method has been described by Bauersfeld and Halgren (1996). The rationale behind this passive video observation is the assumption that a video camera can be less intrusive than a usability specialist but more vigilant. In this method, a video camera is set up in the user's work environment. There is a second camera or a scan converter that shows what is on the user's screen or desk surface. The two images are mixed and recorded. Participants are told to ignore the cameras as much as possible and to work as they normally would. Participants are shown how to turn on the equipment and told to do so whenever they work. This method can be used during any stage of product development and not just for evaluation.

Although passive video capture is done without a usability specialist present, we still don't know whether participants act differently because they know they are being taped. In addition, the data must be extracted from the videotapes, which takes as much time to watch as if the participant were being observed directly. Still, this method can be used in situations in which an observer can't be present when users are working.

EMPIRICAL USABILITY TESTING

Usability testing began in the early 1980s at a time when computer software was beginning to reach a wider audience than just computing professionals. The explosion of end user computing was made possible by new hardware and software in the form of both the mini- and microcomputer and expansion of communications technology, which moved computing from the isolated computer room to the desktop. The advent of the cathode ray tube (CRT) and communications technology made it possible to interact directly with the computer in real time.

The 1982 conference, Human Factors in Computer Systems, held at Gaithersburg, Maryland, brought together for the first time professionals interested in studying and understanding human–computer interaction. Subsequent meetings of this group became known as the Computer-Human Interaction (CHI) Conference. At that first meeting, there was a session on evaluating text editors that described early usability tests (Ledgard, 1982). The reports of these studies were written in the style of experimental psychology reports, including sections titled "Experimental Design" and "Data Analysis," in which the computation of inferential statistics was described.

But the reliance on psychological research experiments as a model for usability testing was challenged early. Young and Barnard (1987) proposed the concept of scenarios instead of experiments, and 2 years later, CHI Conference writers were discussing issues such as "The Role of Laboratory Experiments in HCI: Help, Hindrance or Ho-Hum?" (Wolf, 1989).

The first books on HCI began appearing at this time. Perhaps the most influential book on usability, Shneiderman's (1987) first edition of *Designing the User Interface*, did not have a

section or index item for usability testing but did have one on quantitative evaluations. In that section, Shneiderman wrote the following:

> Scientific and engineering progress is often stimulated by improved techniques for precise measurement. Rapid progress in interactive systems design will occur as soon as researchers and practitioners evolve suitable human performance measures and techniques.... Academic and industrial researchers are discovering that the power of traditional scientific methods can be fruitfully employed in studying interactive systems. (p. 411)

In the 1992 edition, there again was no entry in the index for usability testing, but there is one for usability laboratories. Shneiderman described usability tests but called them "pilot tests." These tests "can be run to compare design alternatives, to contrast the new system with current manual procedures, or to evaluate competitive products" (p. 479).

In the 1997 edition, there is a chapter section on usability testing and laboratories. Shneiderman wrote:

> Usability-laboratory advocates split from their academic roots as these practitioners developed innovative approaches that were influenced by advertising and market research. While academics were developing controlled experiments to test hypotheses and support theories, practitioners developed usability-testing methods to refine user interfaces rapidly. (p. 128)

This brief history shows that usability testing has been an established evaluation method for only about 10 years. The research studies by Virzi (1990, 1992) on the relatively small number of participants needed in a usability test gave legitimacy to the notion that a usability test could identify usability problems quickly. Both of the book length descriptions of usability testing (Dumas & Redish, 1993; Rubin, 1994) explicitly presented usability testing as a method separate from psychological research. Yet, as discussed later in this chapter, comparisons between usability testing and research continue. The remaining sections on usability testing cover usability testing basics, important variations on the essentials, challenges to the validity of user testing, and additional issues.

Valid usability tests have the following six characteristics.

- The focus is on usability.
- The participants are end users or potential end users.
- There is some artifact to evaluate, such as a product design, a system, or a prototype of either.
- The participants think aloud as they perform tasks.
- The data are recorded and analyzed.
- The results of the test are communicated to appropriate audiences.

The Focus Is on Usability

It may seem like an obvious point that a usability test should be about usability, but sometimes people try to use a test for other, inappropriate purposes or call other methods usability tests. Perhaps the most common mismatch is between usability and marketing and promotional issues, such as adding a question to a posttest questionnaire asking participants if they would buy the product they just used. If the purpose of the question is to provide an opportunity for the participant to talk about his or her reactions to the test session, the question is appropriate. But if the question is added to see if customers would buy the product, the question is not appropriate. A six-participant usability test is not an appropriate method for estimating sales or market share. Obviously, a company would not base its sales projections on the results of such a question, but people who read the test report may draw inappropriate conclusions, for example, when the product has several severe usability problems, but five of the six participants say that they would buy it. The participants' answers could provide an excuse for ignoring the usability problems. It is best not to include such questions or related ones about whether customers would use the manual.

The other common misconception about the purpose of a test is to view it as a research experiment. The fact is, a usability test *looks* like research. It often is done in a "lab," and watching participants think out loud fits a stereotype some people have about what a research study looks like. But a usability test is not a research study (Dumas, 1999).

A Usability Test Is Not a Focus Group. Usability testing sometimes is mistaken for a focus group, perhaps the most used and abused empirical method of all time. People new to user-based evaluation jump to the conclusion that talking with users during a test is like talking with participants in a focus group. But a usability test is not a group technique, although two participants are sometimes paired, and a focus group is not a usability test unless it contains the six essential components of a test. The two components of a usability test that are most often missing from a focus group are (a) a primary emphasis is not on usability and (b) the participants do not perform tasks during the session.

The most common objective for a usability test is the diagnosis of usability problems. When testers use the term *usability test* with no qualifier, most often they are referring to a diagnostic test. When the test has another purpose, it has a qualifier such as *comparison* or *baseline*.

When Informal Really Means Invalid. One of the difficulties in discussing usability testing is finding a way to describe a test that is somewhat different from a complete diagnostic usability test. A word that is often used to qualify a test is *informal*, but it is difficult to know what informal really means. Thomas (1996) described a method, called "quick and dirty" and "informal," in which the participants are not intended users of the product and in which time and other measures of efficiency are not recorded. Such a test may be informal in some sense of that word, but it is certainly invalid and should not be called a usability test. It is missing one of the essentials: potential users. It is not an informal usability test because it is not a usability test at all. Still we need words to describe diagnostic tests that differ from each other in important ways. In addition, tests that are performed quickly and with minimal resources are best called "quick and clean" rather than "informal" or "quick and dirty" (Wichansky, 2000).

The Participants Are End Users or Potential End Users

A valid usability test must test people who are part of the target market for the product. Testing with other populations may be useful, that is, it may find usability problems. But the results cannot be generalized to the relevant population—the people for whom it is intended.

The key to finding people who are potential candidates for the test is a user profile (Branaghan, 1997). In developing a profile of users, testers want to capture two types of characteristics: those that the users share and those that might make a difference among users. For example, in a test of an upgrade to a design for a cellular phone, participants could be people who now own a cell phone or who would consider buying one. Of the people who own a phone, you may want to include people who owned the previous version of the manufacturer's phone and people who own other manufacturers' phones. These characteristics build a user profile. It is from that profile that you create a recruiting screener to select the participants.

A common issue at this stage of planning is that there are more relevant groups to test than there are resources to test them. This situation forces the test team to decide on which group or groups to focus. This decision should be based on the product management's priorities not on how easy it might be to recruit participants. There is almost always a way to find the few people needed for a valid usability test.

A Small Sample Size Is Still the Norm. The fact that usability testing uncovers usability problems quickly remains one of its most compelling properties. Testers know from experience that in a diagnostic test, the sessions begin to get repetitive after running about five participants in a group. The early research studies by Virzi (1990, 1992; see Fig. 56.1) showing

that 80% of the problems are uncovered with about five participants and 90% with about 10 continue to be confirmed (Law & Vanderheiden, 2000).

What theses studies mean for practitioners is that, given a sample of tasks and a sample of participants, just about all of the problems testers will find appear with the first 5 to 10 participants. This research does not mean that all of the *possible* problems with a product appear with 5 or 10 participants, but most of the problems that are going to show up with one sample of tasks and one group of participants will occur early.

There are some studies that do not support the finding that small samples quickly converge on the same problems. Lewis (1994) found that for a very large product, a suite of office productivity tools, 5 to 10 participants was not enough to find nearly all of the problems. The studies by Molich et al. (1998, 2001) also do not favor convergence on a common set of problems.

As I discuss later, the issue of how well usability testing uncovers the most severe usability problems is clouded by the unreliability of severity judgments.

Recruiting Participants and Getting Them to Show Up. To run the test you plan, you will need to find candidates and qualify them for inclusion in the test. Usually, there are inclusion and exclusion criteria. For example, from a user profile for a test of an instruction sheet that accompanies a ground fault circuit interrupter (GFCI)—the kind of plug installed in a bathroom or near a swimming pool—a test team might want to include people who consider themselves "do-it-yourselfers" and who would be willing to attempt the installation of the GFCI but exclude people who actually had installed one before or who were licensed electricians. The way the testers would qualify candidates is to create a screening questionnaire containing the specific questions to use to qualify each candidate. Then they

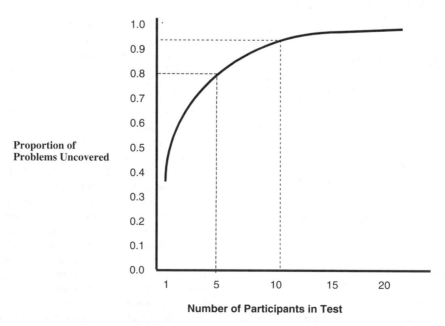

FIGURE 56.1. An idealized curve showing the number of participants needed to find various proportions of usability problems.

have to find candidates to recruit. It often takes some social skills and a lot of persistence to recruit for a usability test. It takes a full day to recruit about six participants. For the test of the GFCI instruction sheet, they may have to go to a hardware store and approach people who are buying electrical equipment to find the relevant "do-it-yourselfers."

Many organizations use recruiting firms to find test participants. Firms charge about $100 (in year 2000 U.S. dollars) for each recruited participant. But the testers create the screening questions and test them to see if the people who qualify fit the user profile.

To get participants to show up, the testers or the recruiting firm need to do the following:

- Be enthusiastic with them on the phone.
- Offer them some incentive. Nothing works better than money, about $50 to $75 an hour (in year 2000 dollars) for participants without any unusual qualifications. Some testing organizations use gift certificates or free products as incentives. For participants with unusual qualifications, such as anesthesiologists or computer network managers, the recruiter may need to emphasize what the candidates are contributing to their profession by participating in the test.
- As soon as participants are qualified, send or fax or e-mail citing the particulars discussed on the phone and a map with instructions for getting to the site.
- Contact participants one or two days before the test as a reminder.
- Give participants a phone number to call if they can't make the session, need to reschedule, or will be late.

If testers follow all of these steps, they will still have a no-show rate of about 10%. Some organizations over recruit for a test, qualifying some extra candidates to be backups in case another participant is a no-show. A useful strategy can be to recruit two participants for a session and, if both show up, run a codiscovery session with both participants. See below for a description of codiscovery.

There Is a Product or System to Evaluate

Usability testing can be performed with most any technology. The range includes the following:

- Products with user interfaces that are all software (e.g., a database management system), all hardware (a high-quality pen), and those that are both (a cell phone, a clock radio, a hospital patient monitor, a circuit board tester, etc.)
- Products intended for different types of users (such as consumers, medical personnel, engineers, network managers, high school students, computer programmers, etc.)
- Products that are used together by groups of users, such as cooperative work software (Scholtz & Bouchette, 1995)
- Products in various stages of development (such as user-interface concept drawings; early, low-tech prototypes; more

fully functioning, high-fidelity prototypes; products in beta testing, and completed products)
- Components that are imbedded in or accompany a product (such as print manuals, instruction sheets that are packaged with a product, tutorials, quick-start programs, online help, etc.)

Testing Methods Work Even With Prototypes. One of the major advances in human–computer interaction over the last 15 years is the use of prototypes to evaluate user interface designs. The confidence that evaluators have in the validity of prototypes has made it possible to move evaluation sooner and sooner in the development process. Evaluating prototypes has been facilitated by two developments: (a) using paper prototypes and (b) using software specifically developed for prototyping. Ten or more years ago, usability specialists wanted to create prototyping tools that make it possible to save the code from the prototype to use it in the final product. Developers soon realized that a prototyped version of a design is seldom so close to the final design that it is worth saving the code. Consequently, the focus of new prototyping tools has been on speed of creating an interactive design. In addition, the speed with which these software tools and paper prototypes can be created makes it possible to evaluate user interface concepts before the development team gets so enamored with a concept that they won't discard it.

There have been several studies that have looked at the validity of user testing using prototypes. These studies compare paper or relatively rough drawings with more interactive and polished renderings. The studies all show that there are few differences between high- and low-fidelity prototypes in terms of the number or types of problems identified in a usability test or in the ratings of usability that participants give to the designs (Cantani & Biers, 1998; Landay & Myers, 1995; Virzi, Sokolov, & Karis, 1996; Wiklund, Dumas, & Thurrot, 1992).

The Participants Think Aloud As They Perform Tasks

This is the execution phase of the test. It is where the test participant and the test administrator interact and it is where the data are collected. Before the test session starts, the administrator gives a set of pretest instructions. The instructions tell the participant how the test will proceed and that the test probes the usability of the product, not their skills or experience.

One of the important parts of the pretest activities is the instructions on thinking aloud. The administrator tells the participants to say out loud what they are experiencing as they work. Interest in thinking aloud was revived with the recent publication of two articles that were written independently at almost the same time (Boren & Ramey, 2000; Dumas, 2001). Both of these articles question assumptions about the similarity between the think-aloud method as used in usability testing and the think-aloud method used in cognitive psychology research. Until these articles were published, most discussions of the think-aloud method used in usability testing automatically noted its superficial resemblance to the method described by Ericsson and Simon (1993) to study human problem solving.

In both methods, the participants are taught to think aloud by providing instructions on how to do it, showing an example of the think aloud by giving a brief demonstration of it, and by having the participant practice thinking aloud. But there ends the similarity to research.

In cognitive psychology research, thinking aloud is used to study what is in participants' short-term memory. Called Level 1 thinking aloud, the research method focuses on having the participant say out loud what is in the participants short-term memory, which can only occur when the participants' describe what they are thinking as they perform cognitive tasks such as multiplying two numbers. Participants are discouraged from reporting any interpretations of what is happening, any emotions that accompany the task, and their expectations or violations of them. The research method is thought, by its proponents, to provide entree only into the short-term memory of the participants in the research.

In usability testing, the focus is on interactions with the object being tested and with reporting not only thoughts, but also expectations, feelings, and whatever the participants want to report. Reports of experiences other than thoughts are important because they often are indicators of usability problems.

This discrepancy between the two sets of think-aloud instructions led Boren and Ramey (2000) to look at how thinking aloud is used in user testing as well as practices related to thinking aloud, such as how and when to encourage participants to continue to do it. They reported the results of observing test administrators implementing these practices and how little consistency there is among them. Boren and Ramey explored other aspects of the verbal communication between administrators and participants, including how to keep the participants talking without interfering with the think-aloud process.

Bowers and Snyder (1990) conducted a research study to compare the advantages and disadvantages of having test participants think out loud as they work, called *concurrent* thinking aloud, with thinking out loud after the session, called *retrospective* thinking aloud. In the retrospective condition, the participants performed tasks in silence then watched a videotape of the session while they thought aloud. This is an interesting study because of its implications for usability testing. The group of participants who performed concurrent thinking aloud were not given typical think-aloud instructions for a usability test. Instead, their instructions were typical of a think aloud research study. Participants were told to "describe aloud what they are doing and thinking." They were not told to report any other internal experiences. In addition, they were never interrupted during a task. There was no probing. Any encouragement they needed to keep talking was only done between tasks. The retrospective participants were told that they would be watching the videotape of the session after the tasks and would be asked to think aloud then.

There were several interesting results. First, there were no differences between the concurrent and retrospective groups in task performance or in task difficulty ratings. The thinking aloud during the session did not cause the concurrent group to take more time to complete tasks, to complete fewer tasks, or to rate tasks as more difficult in comparison with the performance of the retrospective participants. These findings are consistent with the results from other think-aloud research studies, although in some studies thinking aloud does take longer.

The differences between the groups were in the *types* of statements the participants made when they thought out loud. The concurrent group verbalized about 4 times as many statements as the retrospective group, but the statements were almost all descriptions of what the participants were doing or reading from the screen. The participants who did concurrent thinking aloud were doing exactly as they were instructed; they were attending to the tasks and verbalizing a kind of "play-by-play" of what they were doing. The participants in the retrospective condition made only about one fourth as many statements while watching the tape, but many more of the statements were explanations of what they had been doing or comments on the user interface design. "The retrospective subjects...can give their full attention to the verbalizations and in doing so give richer information" (Bowers & Snyder, 1990, p. 1274).

This study shows us what would happen if we tried to get participants in usability tests to report only Level 1 verbalizations and did no probing of what there were doing and thinking. Their verbalizations would be much less informative. The study does show that retrospective thinking aloud yields more diagnostic verbalizations, but it takes 80% longer to have the participants do the tasks silently then to think out loud as they watch the tape.

There have been two fairly recent studies that compared a condition in which the tester could not hear the think aloud of participants with a condition in which they could (Lesaigle & Biers, 2000; Virzi, Sorce, & Herbert, 1993). In the Virzi et al. study, usability professionals who recorded usability problems from a videotape of test participants thinking aloud, were compared with usability professionals who could see only the performance data of the test participants. Those who had only the performance data uncovered 46% of the problems with the product, whereas those seeing the think aloud condition uncovered 69%. In the Lesaigle and Biers study, usability professionals who could see a video of only the screens the participants could see were compared with comparable professionals who could see the screens and hear the participants think aloud. The results showed that there were fewer problems uncovered in the screen only condition compared with the screen plus think-aloud condition. Both of these studies suggest that in many cases participants' think-aloud protocols provide evidence of usability problems that do not otherwise show up in the data.

Conflict in Roles. Dumas (2001) explored how difficult it can be for test administrators to keep from encouraging or discouraging participants' positive or negative statements. He saw a conflict in two roles that administrators play: (a) the friendly facilitator of the test and (b) the neutral observer of the interaction between the participant and the product.

The friendly facilitator role and the neutral observer role come into conflict when participants make strong statements expressing an emotion such as, "I hate this program!" Almost anything the test administrator says at that point can influence whether the participants will report more or fewer of these negative feelings. Consider the following statements:

"Tell me more about that": relatively neutral in content but could be interpreted as encouraging more negative statements

"That's great feedback": again relatively neutral to someone who has training in test administration but, I believe, sounds evasive to participants

"Those are the kinds of statements that really help us to understand how to improve the product": reinforcing the negative

"I really appreciate your effort to help us today": says nothing about the content of what the participant said and is part of playing the friendly role with participants. Will the participant hear it that way?

Silence: Neutral in content, but how will it be interpreted? In human interaction, one person's silence after another's strong statement is almost always interpreted as disagreement or disapproval. Without any other instructions, the participant is left to interpret the test administrator's silence—you don't care, you don't want negative comments, strong feelings are inappropriate in this kind of test, and so on.

Not all of the biasing responses to emotional statements are verbal. If the tester is in the room with the participant and takes notes when participants make an emotional statement, he or she may be reinforcing them to make more. Any of these responses could push test participants to utter more or fewer strong feelings. Dumas suggested that one way to avoid this conflict in roles is to tell participants what the two roles are in the pretest instructions.

The Special Case of Speech-Based Products. For the most part, the basic techniques of user testing apply to speech applications. There are a few areas, however, where testers may need to modify their methods (Dobroth, 1999):

• It is not possible for test participants to think aloud while they are using a speech recognition application because talking interferes with using the application. If participants were to speak aloud, the verbalized thoughts may be mistaken for input by the speech recognizer. Moreover, if participants think aloud while using a speech application, they may not be able to hear spoken prompts. One way to get around this problem is to have participants comment on the task immediately after finishing it. This works well for tasks that are short and uncomplicated. If tasks are longer, however, participants will begin to forget exactly what happened in the early parts of the task. In this case, the test administrator can make a recording of the participants' interaction with the system as they complete the task. At the end of the task, participants listen to the recording and stop it to comment on parts of the interaction that they found either clear or confusing. Both of these solutions provide useful information but add a substantial amount of time to test sessions.

• Evaluating speech-based interfaces often is complicated by the presence of a recognizer in the product. The recognizer interprets what the test participant says. Often the recognizer can't be changed; it is the software that surrounds it that is being tested. Using a poor recognizer often clouds the evaluation of the rest of the software. In a "Wizard-of-Oz" test (see chapter 52 for more on this technique), the test administrator creates the impression in participants that they are interacting with a

voice response system. In reality, the flow and logic of each interaction is controlled by the test administrator, who interprets participants' responses to prompts, and responds with the next prompt in the interaction. Using this method also allows the administrator to be sure that error paths are tested. In a speech interface, much of the design skill is in dealing with the types of errors that recognizers often make.

• In the past, it was much more difficult to create a prototype of a speech-based product, but several options are now available. Another option is to use the speech capabilities of office tools such as Microsoft's PowerPoint.

Selecting Tasks. One of the essential requirements of every usability test is that the test participants attempt tasks that users of the product will want to do. When a product of even modest complexity is tested, however, there are more tasks than there is time available to test them. Hence the need to select a sample of tasks. Although not often recognized as a liability of testing, the sample of tasks is a limitation to the scope of a test. Components of a design that are not touched by the tasks the participants perform are not evaluated. This limitation in thoroughness is often why testing is combined with usability inspection methods, which have thoroughness as one of their strengths.

In a diagnostic test, testers select tasks for several reasons:

• They include important tasks, that is, tasks that are performed frequently or are basic to the job users will want to accomplish, and tasks, such as log in or installation, that are critical, if infrequent, because they affect other tasks. With almost any product there is a set of basic tasks. *Basic* means tasks that tap into the core functionality of the product. For example, a nurse using a patient monitor will frequently look to see the vital sign values of the patient and will want to silence any alarms once she or he determines the cause. In addition, the nurse will want to adjust the alarm limits, even though the limit adjustment may be done infrequently. Consequently, viewing vital signs, silencing alarms, and adjusting alarm limits are basic tasks.

• They include tasks that probe areas where usability problems are likely. For example, if testers think that users will have difficulty knowing when to save their work, they may add saving work to several other tasks. Selecting these kinds of tasks makes it more likely that usability problems will be uncovered by the test, an important goal of a diagnostic test. But including these kinds of tasks makes it likely that a diagnostic test will uncover additional usability problems. In effect, these tasks pose a more difficult challenge to a product than if just commonly done or critical tasks are included. These tasks can make a product look less usable than if they were not included. As we will see below, this is one of the reasons why a diagnostic test does not provide an accurate measure of a product's usability.

• They include tasks that probe the components of a design. For example, tasks that force the user to navigate to the lowest level of the menus or tasks that have toolbar shortcuts. The goal is to include tasks that increase thoroughness at uncovering problems. When testing other components of a product, such as a print manual, testers may include tasks that focus on what

is in the manual, such as a task that just asks the participant to locate a number of items (Branaghan, 1998).

Some additional reasons for selecting tasks are:

- They may be easy to do because they have been redesigned in response to the results of a previous test.
- They may be new to the product line, such as sending an order for a drug to the hospital pharmacy.
- They may cause interference from old habits, such as a task that has been changed from a previous release of the product.

With so many reasons for selecting tasks, paring the task list to the time available is an important part of test planning. Typically testers and developers get together in the early stages of test planning to create a task list. In addition to including tasks in the list, the testers need to make some preliminary estimate of how long each task will take. The time estimate is important for deciding how many tasks to include, and it may also be useful for setting time limits for each task. Even in a diagnostic test, time limits are useful because testers want participants to get through most of the tasks. Setting time limits is always a bit of a guess. Until you conduct a pilot test, it is difficult to make accurate estimates of time limits, but some estimate is necessary for planning purposes.

The Tasks Are Presented in Task Scenarios. Almost without exception, testers present the tasks that the participants do in the form of a task scenario. For example:

You've just bought a new combination telephone and answering machine. It is in the box on the table. Take the product out of the box and set it up so that you can make and receive calls.

A good scenario is short, in the user's words not the product's, unambiguous, and gives participants enough information to do the task. It never tells the participant *how* to do the task. From the beginning, usability testers recognized the artificiality of the testing environment. The task scenario is an attempt to bring a flavor of the way the product will be used into the test. In most cases, the scenario is the only mechanism for introducing the operational environment into the test situation. Rubin (1994, p. 125) describes task scenarios as adding context and the participant's rationale and motivation to perform tasks. "The context of the scenarios will also help them to evaluate elements in your product's design that simply do not *jibe with reality*" and "The closer that the scenarios represent *reality*, the more *reliable* the test results" (emphasis added). Dumas and Redish (1999, p. 174) said, "The whole point of usability testing is to predict what will happen when people use the product on their own.... The participants should feel as if the scenario matches what they would have to do and what they would know when they are doing that task *in their actual jobs*" (emphasis added).

During test planning, testers work on the wording of each scenario. The scenario needs to be carefully worded so as not to mislead the participant to try to perform a different task. Testers also try to avoid using terms in the scenario that give the participants clues about how to perform the task, such as using the name of a menu option in the scenario.

In addition to the wording of the task scenarios, their order may also be important. It is common for scenarios to have dependencies. For example, in testing a cellular phone there may be a task to enter a phone number into memory and a later task to change it. A problem with dependencies happens when the participant can't complete the first task. Testers have developed strategies to handle this situation, such as putting a phone number in another memory location that the test administrator can direct the participants to when they could not complete the earlier task.

Testers continue to believe in the importance of scenarios and always use them. There is no research, however, showing that describing tasks as scenarios rather than simple task statements makes any difference to the performance or subjective judgments of participants. But taking note of the product's use environment may be important, as described in the next section.

The Participants Are Observed, and Data Are Recorded and Analyzed

Capturing Data As They Occur. Recording data during the session remains a challenge. All agree that testers need to plan how they will record what occurs. There are too many events happening too quickly to be able to record them in free-form notes. The goal is to record key events while they happen rather than having to take valuable time to watch videotapes later. There are three ways that testers deal with the complexity of recording data:

- Create data collection forms for events that can be anticipated (Kantner, 2001a).
- Create or purchase data logging software (Philips & Dumas, 1990).
- Automatically capture participant actions in log files (Kantner, 2001b) or with specialized software (Lister, 2001).

Figure 56.2 shows a sample data collection form for the task of saving a file to a diskette in Microsoft Windows. Notice that it is set up to capture both paths to success and paths to failure. The form also allows for capturing a Reject, which is a task that a participant considers complete but the data collector knows is not. Rejects are important to note because, although they are failures, they often have task times that are faster than even successful tasks.

The use of data logging software continues at many of the larger testing facilities. Most have created their own logging software. Logging test activities in real time, however, continues to be a messy process. Testers almost always have to edit the data log after each session to remove errors and misunderstandings, such as when a task was really over.

It is difficult to use forms or software when the administrator is sitting in the test room beside the participant and is conducting the test alone. Without a data recorder, it is difficult, but still possible, to sit in the test room with the test participant and to record data at the same time.

Task 1. Copy a Word file to a diskette

Pass   (Time_____)

Explorer:
☐ Dragged file from one Explorer pane to another with ____left ____ right button
☐ File: Send to: Floppy A
☐ Copied and Pasted in Explorer with _____ Toolbar _____ Edit menu ____ with Keyboard

My Documents:
☐ Dragged to Desktop then back with ____left _____ right button _____ CTRL
☐ File: Send to: Floppy A
☐ Copied and Pasted with __ Toolbar __ Edit menu __ with Keyboard

Word
☐ Opened Word and chose File: Save as

_____
_____
_____

Fail or Reject (Time_____)

Word

☐ Chose Save

_____
_____
_____

Help _____Windows _____ Word _____Topic:_____

FIGURE 56.2. Sample data collection form.

Collecting data is a special challenge with Web-based products. There are so many links and controls on a typical Web page that it is difficult to record what is happening short of watching the session again on videotape. This difficulty has renewed interest in automatic data collection. But the tools to do this capture usually record data that are at too low a level to uncover usability problems. Most usability problems don't need to be diagnosed at the mouse click or key press level.

Getting Developers and Managers to Watch Test Sessions. One of the important assets of testing is that it sells itself. Watching even a few minutes of live testing can be very persuasive. There are two reasons why testers need to get key project staff and decision makers to come to watch a test session:

• When people see their first live test session they are almost always fascinated by what they see. They gain understanding of the value of the method. Watching a videotape of a session does not provide the same experience. Expend whatever effort it takes to get these people to attend test sessions.

• When developers see live sessions, it is much easier to communicate the results to them. When they have seen some of the usability problems themselves, they are much less likely

to resist agreeing on what the most important problems are. Some of them will even become advocates for testing.

Even though testing is known and accepted by a much wider circle of people than it was 10 years ago, the experience of watching a user work at a task while thinking aloud still converts more people to accept usability practices than any other development tool.

Participant and Administrator Sit Together. Most usability tests are run with a single test participant. Studies show that when two participants work together, sometimes called the codiscovery method (Kennedy, 1989), they make more utterances. The nature of the utterances also is different, with codiscovery participants making more *evaluative*, as opposed to *descriptive* statements and making more statements that developers view as useful (Hackman & Biers, 1992). But using codiscovery does require recruiting twice as many participants. A related method is to have one participant teach another how to do a task (Vora, 1994).

The Usability Lab Is Now Ubiquitous. Usability labs continue to be built, and there is a brisk business in selling lab

equipment. The demand for labs is driven by the advantages of having recording equipment and the ability to allow stakeholders to view the test sessions. In essence, the method sells itself in the sense that developers and managers find compelling the experience of watching a live test session. A testing facility, especially one with a one-way mirror, adds a sense of scientific credibility to testing, which, as we will discuss below, may be a false sense.

The basic makeup of a suite of usability test equipment has not changed much with time. It consists of video and audio recording equipment and video mixing equipment. For testing products that run on general-purpose computer equipment, a common setup is a scan converter showing what is on the test participant's screen and a video camera focused on the face or head and shoulders of the participant.

There are some recent innovations in lab equipment that are enhancing measurement. Miniaturization continues to shrink the size of almost all lab equipment, hence the arrival of portable lab setups that fit in airplane overhead compartments. Relatively inexpensive eye-tracking equipment has made it possible to know where participants are looking as they work.

The quality of video images recorded during sessions has always been poor. Second-generation copies, which are often used in highlight tapes, make the quality of highlight tapes even poorer. Scan converters selling for under $2,000 produce surprisingly poor images, making it difficult to see screen details.

The move to digital video and inexpensive writeable CDs promises to improve recordings and to make it substantially easier to find and edit video segments.

Mimicking the Operational Environment. Testers often make changes to the setup of the test room. Rubin (1994, p. 95) describes the requirements for the testing environment as follows: "Make the testing environment as realistic as possible. As much as possible, try to maintain a testing environment that mimics the actual working environment in which the product will be used." But is putting a couch in the test room to make it look more like a room in a home simulating the use environment? It may not be, but going to the participant's home for testing is a complex process (Mitropoulos-Rundus & Muzak, 1997).

The literature on product evaluation, when viewed from the perspective of 50 years, shows that in complex operational environments, researchers and practitioners have used software simulations or hardware–software simulators to mimic that operational environment. For example, aircraft and automobile simulators are used to study interactions with cockpits and dashboards as well as for operator training. More recently, hospital operating-room simulators have been developed to study equipment interaction issues in anesthesiology (Gaba, 1994).

A variable, usually called *fidelity*, is used to describe the degree to which simulations or simulators mimic the operational environment. In those interactions between users and aircraft, automobiles, and operating rooms, the environment is so important that simulations are needed to mimic it. There may be other environments that influence the usability of the products we test, and we need to think more about the fidelity of our testing environments (Wichansky, 2000). I discuss this issue further in the section Challenges to the Validity of Testing.

The Impact of the Testing Equipment. An issue that has been debated throughout the history of usability testing is the impact of one-way mirrors and recording equipment on the test participants. This debate comes to a head in discussions about whether the test administrator should sit with participants as they work or stay behind the one-way mirror and talk over an intercom. Some testing groups always sit with the participant, believing that it reduces the participants' anxiety about being in the test and makes it easier to manage the session (Rubin, 1994). Other testing groups normally do not sit with the participants, believing that it makes it easier to remain objective and frees the administrator to record the actions of the participants (Dumas & Redish, 1999). There is one study that partially addressed this issue. Barker and Biers (1994) conducted an experiment in which they varied whether there was a one-way mirror and cameras in the test room. They found that the presence of the equipment did not affect the participants' performance or ratings of usability of the product.

Remote Testing. Remote usability testing refers to situations in which the test administrator and the test participant are not at the same location (Hartson, Castillo, Kelso, Kamler, & Neale, 1996). This can happen for a number of reasons, such as testing products used by only a few users who are spread throughout the country or the world. Products such as Net-Meeting software make it possible for the tester to see what is on the participant's screen and, with a phone connection, hear the participants think aloud.

There are other technologies that can provide testers with even more information, but they often require both parties to have special video cards and software. There are also technologies for having ratings or preference questions pop up while participants are working remotely (Abelow, 1992). But Lesaigle and Biers' (2000) study showed that uncovering problems only through participants' questionnaire data had the least overlap with conditions in which testers uncovered problems by watching the participants work or seeing the screens on which participants. They concluded that "the present authors are skeptical about using feedback provided by the user through online questionnaires as the sole source of information" (p. 587). Still, no one would disagree that some remote testing is better than no testing at all.

The primary advantages of remote testing include the following:

- Participants are tested in an environment in which they are comfortable and familiar.
- Participants are tested using their own equipment environment.
- Test costs are reduced because participants are easier to recruit, do not have to travel, and often do not have to be compensated. In addition, there are no test facility costs.

But there can be disadvantages to remote testing:

- With live testing, the viewer or conferencing software can slow down the product being tested.
- Company firewalls can prevent live testing. Most viewers and meeting software often cannot be used if there is a firewall.

Perkins (2001) described a range of remote usability testing options that includes user-reported critical incidents, embedded survey questions, live remote testing with a viewer, and live remote testing with conferencing software.

Measures and Data Analysis

In this section, I discuss test measures, discrepancies between measures, and data analysis.

Test Measures. There are several ways to categorize the measures taken in a usability test. One is to break them into two groups: (a) performance measures, such as task time and task completion, and (b) subjective measures, such as ratings of usability or participants' comments. Another common breakdown uses three categories: (a) efficiency measures (primarily task time), (b) effectiveness measures (such as task success), and (c) satisfaction measures (such as rating scales and preferences).

Most performance measures involve time or simple counts of events. The most common time measure is time to complete each task. Other time measures include the time to reach intermediate goals such as the time for events such as to finding an item in Help. The counts of events in addition to task completion include the number of various types of errors, especially repeated errors, and the number of assists. An assist happens when the test administrator decides that the participant is not making progress toward task completion, but the administrator can continue to learn more about the product by keeping the participant working on the task. An assist is important because it indicates that there is a usability problem that will keep participants from completing a task. The way assists are given to participants by test administrators, part of the art of running a test, is not consistent from one usability testing organization to another (Boren & Ramey, 2000).

There are some complex measures that are not often used in diagnostic tests but are sometimes used in comparison tests. These measures include the time the participant works toward the task goal divided by the total task time (sometimes called *task efficiency*) and the task time for a participant divided by the average time for some referent person or group, such as an expert or an average user. It seems only natural that an important measure of the usability of a product should be the test participants' opinions and judgments about the ease or difficulty of using it. The end of the test session is a good time to ask for those opinions. The participant has spent an hour or two using the product and probably has as much experience with it as he or she is likely to have. Consequently, a posttest interview or a brief questionnaire is a common subjective measure (see, however, the discussion about the discrepancies between measures in the next subsection).

Eye tracking is a relatively new measure in user testing. Eye-tracking equipment has come down in price in recent years. You can purchase a system for about $40,000, plus or minus $10,000, depending on accessories and data reduction software. The new systems are head mounted and allow the test participant a good deal of movement without losing track of where the eye is looking.

Interest in where participants are looking has increased with the proliferation of Web software. There are so many links and controls on a Web page that it can be difficult to know exactly where participants are looking. An eye tracker helps solve that problem. Not all test participants can be calibrated on an eye tracker; up to 20 percent of typical user populations cannot be calibrated because of eye abnormalities.

The data from the tracker is broken into fixations—300-millisecond periods during which the point of regard doesn't move more than 1° of visual angle. Each fixation has a start time, duration, point of gaze coordinates, and average pupil diameter. Eye movement analysis involves looking at fixations within an area of interest (AOI), which is a tester-defined area on a screen. These regions usually define some object or control on a page. There are AOIs on each page and testers use eye-tracking software to compute statistics such as the average amount of time and the number and duration of fixations in each AOI. Then there are statistics that measure eye movements from one AOI to another and plots of scan paths.

Eye tracking systems produce a great deal of data; a 60-Hz tracker produces 3,600 records a minute. Consequently, data reduction becomes a major task. Consequently, eye tracking isn't something that is used without a specific need. Goldberg (2000) identified evaluation criteria that can benefit most from eye tracking data, with visual clarity getting the most benefit. But there are other evaluation areas that are likely to benefit as tracking technology becomes cheaper and easier to manage.

Discrepancies Between Measures. Some investigators find only a weak correlation between efficiency measures and effectiveness measures. Frokjaer, Hertzum, and Hornbaek (2000) described a study in which they found such a weak correlation. They then went back and looked at several years of usability test reports in the proceedings of the annual CHI conference. They noted that it is common for testers to report only one category of performance measure and caution not to expect different types of measures to be related.

There is a vocal minority of people writing about usability testing measures who argue against the use of quantitative performance measures in favor of a qualitative analysis of test data. Hughes (1999) argued that qualitative measures can be just as reliable and valid as qualitative measures.

A common finding in the literature is that performance measures and subjective measures are often weakly correlated. Lesaigle and Biers (2000) compared how well testers uncovered usability problems under a number of conditions:

- They only can see the screen the participant sees.
- They can see the screen and hear the participants think aloud.
- They can see screens and hear the think aloud and see the participants face.
- They see only the responses to questionnaire items.

The results show that uncovering problems only through participants' questionnaire data had the least overlap with the other three conditions. The authors concluded that "questionnaire data taps a somewhat different problem set," and "the questionnaire data was less likely to reveal the most severe problems

(p. 587). Bailey (1993) and Ground and Ensing (1999) both reported cases in which participants perform better with products that they don't prefer and vice versa. Bailey recommended using only performance measures and not using subjective measures when there is a choice.

One of the difficulties with test questions is that they are influenced by factors outside of the experience that participants have during the test session. There are at least three sources of distortions or errors in survey or interview data: (a) the characteristics of the participants, (b) the characteristics of the interviewer or the way the interviewer interacts with the participant, and (c) the characteristics of the task situation itself. Task-based distortions include such factors as the format of questions and answers, how participants interpret the questions, and how sensitive or threatening the questions are (Bradburn, 1983). In general, the characteristics of the task situation produce larger distortions than the characteristics of the interviewer or the participant. Orne (1969) called these task characteristics the "demand characteristics of the situation." (See Dumas, 1998b, 1998c, for a discussion of these issues in a usability testing context.) In addition to the demand characteristics, subjective measures can be distorted by events in the test, such as one key event, especially one that occurs late in the session.

Creating closed-ended questions or rating scales that probe what the tester is interested in is one of the most difficult challenges in usability test methodology. Test administrators seldom have any training in question development or interpretation. Unfortunately, measuring subjective states is not a knowledge area where testers' intuition is enough. It is difficult to create valid questions, that is, questions that measure what we want to measure. Testers without training in question development can use open-ended questions and consider questions as an opportunity to stimulate participants to talk about their opinions and preferences.

Testers often talk about the common finding that the way participants perform using a product is at odds with the way the testers themselves would rate the usability of the product. There are several explanations for why participants might say they liked a product that, in the testers eyes, was difficult to use. Most explanations point to a number of factors that all push user ratings toward the positive end of the scale. Some of the factors have to do with the demand characteristics of the testing situation, for example, participants' need to be viewed as positive rather than negative people or their desire to please the test administrator. Other factors include the tendency of participants to blame themselves rather than the product and the influence of one positive experience during the test, especially when it occurs late in the session.

Test participants continue to blame themselves for problems that usability specialists would blame on the user interface. This tendency seems to be a deep-seated cultural phenomenon that doesn't go away just because a test administrator tells the participant during the pretest instructions that the session is not a test of the participants' knowledge or ability. These positive ratings and comments from participants often put testers in a situation in which they feel they have to explain away participants' positive judgments with the product. Testers always feel

that the performance measures are true indicators of usability, whereas subjective statements are unreliable. For example, a very long task time or a failure to complete a task is a true measure of usability, whereas a positive rating of six out of seven on usability is inflated by demand characteristics.

Data Analysis. Triangulation of measures is critical. It is rare that a usability problem affects only one measure. For example, a poorly constructed icon toolbar will generate errors (especially picking the wrong icon on the toolbar), slow task times (during which participants hesitate over each icon and frequently click through them looking for the one they want) and statements of frustration (participants express their feelings about not being able to learn how the icons are organized or be able to guess what an icon will do from the tool tip).

Much of the data analysis involves building a case for a usability problem by combining several measures, a process that is called *triangulation* (Dumas & Redish, 1999). The case building is driven by the problem list created during the test sessions. It is surprising how much of this analysis is dependant on the think-aloud protocol. We depend on what participants say to help us understand what the problem is.

Identifying usability problems is key. Most usability problems do not emerge from the analysis of the data after the test. The problems are observed during the sessions and are recorded on problem sheets or data logs. Later, the problem sheet or log drives the data analysis. The problem sheet is usually created by the test administrator during the test sessions or immediately afterward. The sheet is organized by participant and by task. What gets recorded on the sheet are observations, such as "didn't see the option," and interpretations, such as "doesn't understand the graphic." When the same problem appears again, it is noted.

Experienced usability testers see the basic causes of problems. From individual instances of problems, the experienced tester sees patterns that point to more general problems. For example, a tester might see instances of participants spending time looking around the screen and aimlessly looking through menu options and conclude that "the participants were overwhelmed with the amount of information on the screen." From a number of instances of participants not understanding terms, the tester might conclude "the interface has too much technical and computer jargon." From a number of instances of participants doing a task twice to make sure it was completed, the tester might conclude that "there is not enough feedback about what the system is doing with the participant's actions." Seeing the underlying causes of individual problem tokens is one of the important skills that a usability tester develops. It is not entirely clear that such skills can be taught quickly. Testers often have years of experience studying and practicing problem identification skills. But do experienced testers see the same problems and causes? As I discuss later, there is some doubt about the consistency of problem labeling.

While watching a test session, a product developer will see the same events or tokens as the test administrator. But developers tend to see all problems as local. Instead of seeing that there needs to be a general review of the language in the interface, the developer sees problems with individual words. This

conflict often doesn't appear until the testers and developers sit down to discuss what they saw and what to do about it. Usability professionals believe that this conflict over what "really" happened during the test remains a major barrier to improving a product's usability. Handling this conflict takes some diplomacy. Developers don't like to be told that they have tunnel vision and can't see the underlying causes of individual tokens, and usability professions don't like hearing that the local fix will solve the problem. This conflict continues to limit the impact of testing on product improvement.

There have been several research studies that have looked at how many usability problems are uncovered by different populations. These studies consistently show that usability specialists find more problems than product developers or computer scientists. But all of the studies have used inspection evaluation methods not user-based evaluation methods.

One of the issues still being debated about usability problems is whether to place them into a company's software bug tracking system (Wilson & Coyne, 2001). Putting them into the system can be effective if the bugs are more likely to be fixed. But fitting the bugs into a bug severity rating scale often is difficult, and there is always a risk that the fix will solve only the local impact of the problem not its basic structural cause. Some bug tracking systems require that a bug be assigned only one cause, which would not adequately describe many usability problems.

One way to call attention to important problems is to put them into a measurement tool such as a problem severity scale. These scales determine which problems are the most severe and, presumably, more likely to be candidates to be fixed. There have been several recent research studies that have looked at the validity the reliability of these scales.

A disappointing aspect of this research is the lack of consistency in severity judgments. This lack appears in all forms of usability evaluation, inspection and user based, and is one of the most important challenges to usability methodology. Several practitioners have proposed severity rating schemes: Nielsen (1992), Dumas and Redish (1999), Rubin (1994), and Wilson and Coyne (2001). The schemes have three properties in common:

1. They all use a rating scale that is derived from software bug reporting scales. The most severe category usually involves loss of data or task failure and the least severe category involves problems that are so unimportant that they don't need an immediate fix. All of the authors assume that the measurement level of their scale is at least ordinal, that is, the problems gets worse as the scale value increases. The middle levels between the extremes are usually difficult to interpret and are stated in words that are hard to apply to specific cases. For example, Dumas and Redish proposed two middle levels: (a) problems that create significant delay and frustration and (b) problems that have a minor effect on usability. Nielson's middle levels are (a) major usability problem (important to fix and so should be given high priority) and (b) minor usability problem (fixing is given low priority). Practitioners are not given any guidance on how problems fit into the scale levels, especially the middle ones.

2. All of the authors admit, at least indirectly, that their scales alone are not enough to assess severity. The authors propose one or more additional factors for the tester to consider in judging severity. For example, Nielsen (1992) described four factors in addition of the severity rating itself: frequency, impact, persistence, and something called "market impact." Rubin (1994) proposed multiplying the rating by the number of users who have the problem. Dumas and Redish (1999) added a second dimension: the scope of the problem from local to global, with no levels in between. With the exception of Rubin's multiplication rule, none of these other factors are described in enough detail to indicate how their combination with the severity scale would work, which is, perhaps, an indicator of the weakness of the severity scales themselves.

3. None of the scales indicate how to treat individual differences. For example, what does one do if only two of eight participants cannot complete a task because of a usability problem. Is that problem in the most severe category or does it move down a level? If a problem is global rather than local, does that change its severity? The authors of these scales provide little guidance.

There have been a number of research studies investigating the consistency of severity ratings. These studies all show that the degree of consistency is not encouraging. Most studies have looked at the inconsistencies among experts using severity scales with inspection methods such as heuristic evaluation. But Jacobsen and John (1998) showed that it also applies to usability testing. They asked four experienced usability testers to watch tapes of the same usability test and then identify problems, including the top-10 problems in terms of severity. Of the 93 problems identified with the product, only 20% were detected by all evaluators, whereas 46% were only found by a single evaluator. None of the top-10 severe problems appeared on all four evaluators' lists.

Lesaigle and Biers (2000) reported a disappointing correlation coefficient (0.16) among professional testers' ratings of the severity of the same usability problems in a usability test. They used Nielsen's severity rating scale. Cantani and Biers (1998) found that heuristic evaluation and user testing did not uncover the same problems, and that severity ratings of usability professionals did not agree with each other.

The results of these studies cast doubt on one of the most often-mentioned assets of usability testing: its touted ability to uncover the most severe usability problems.

Communicating Test Results

In the early days of user testing, there almost always was a formal test report. Testers needed reports to communicate what they did, what they found, and what testing was all about. Now it is more common for the results of a test to be communicated more informally, such as at a meeting held soon after the last test session. Communication at these meetings is facilitated when the product team has attended at least some of the test sessions.

One of the important reasons for the change in reporting style for diagnostic usability tests is the confidence organizations have in the user testing process. It now is less often necessary to write a report to justify conducting the test. Organizations with

active usability programs have come to accept user testing as a valid and useful evaluation tool. They don't feel that they need to know the details of the test method and the data analysis procedures. They want to know the bottom line: What problems surfaced, and what should they do about them? In these organizations, a written report may still have value but as a means of documenting the test.

The Value of Highlight Tapes. A highlight tape is a short, visual illustration of the 4 or 5 most important results of a test. In the early days of testing, almost every test had a highlight tape, especially a tape aimed at important decision makers who could not attend the sessions. These tapes had two purposes: to show what happed during the test in an interesting way and to illustrate what a usability test is and what it can reveal.

As usability testing has become an accepted evaluation tool, the second purpose for highlight tapes has become less necessary. One of the disappointing aspects of highlight tapes is that watching them does not have the same impact as seeing the sessions live. Unless the action moves quickly, even highlight tapes can be boring. This characteristic makes careful editing of the highlights a must. But if the editing system is not digital, it takes about 1 hour to create 1 minute of finished tape. A 15 minute tape can take 2 days to create, even by an experienced editor. Most of that time is taken finding appropriate segments to illustrate key findings. The emergence of digital video will make highlight tapes less time-consuming.

Some testers use the capabilities of tools such as PowerPoint to put selections from a videotape next to a bullet in a slide presentation rather than having a separate highlight tape. Others have begun to store and replay video in a different way. There are video cards for personal computers that will take a feed from a camera and store images in mpeg format on a compact disk (CD). Each CD stores about an hour of taping. A tester can then show an audience the highlights by showing segments of the CDs in sequence, thus eliminating the need for editing. Because the cost of blank CDs is only about a dollar, they are cheaper to buy than videotapes and take up less storage space.

VARIATIONS ON THE ESSENTIALS

In this section, I discuss aspects of usability testing that go beyond the basics of a simple diagnostic test. The section includes measuring and comparing usability, baseline usability test, and allowing free exploration.

Measuring and Comparing Usability

A diagnostic usability test is not intended to measure usability as much as to uncover as many usability problems as it can. It doesn't directly answer the question, "How usable is this product?" It would be wonderful to be able to answer that question with a precise, absolute statement such as, "It's very usable" or better, "Its 85% usable." But there is no absolute measure of usability, and without a comparative yardstick it is difficult to pinpoint a product's usability.

It would be ideal if we could say that a product is usable if participants complete 80% of their tasks and if they give it an average ease-of-use rating of 5.5 out of 7, with 7 being *very usable*. But all tasks and tests are not equal. One of the limiting factors in measuring usability is the makeup of the diagnostic test itself. It typically tests a very small sample of participants; it encourages those participants to take the time to think aloud and to made useful verbal diversions as they work; it allows the test administrator the freedom to probe interesting issues and to take such actions as skipping tasks that won't be informative; and it deals with a product that might be in prototype form and, consequently, will occasionally malfunction. Those qualities make a diagnostic test good at exploring problems, but limited at measuring usability.

Historically, human factors professionals have made a distinction between formative and summative measurement. A formative test is done early in development to contribute to a product's design; a summative test is performed late in development to evaluate the design. A diagnostic test is clearly a formative test. But what specifically is a summative usability test?

At the present time, without a comparison product, we are left with the judgment of a usability specialist about how usable a product is based on their interpretation of a summative usability test. Experienced usability professionals believe that they can make a relatively accurate and reliable assessment of a product's usability given data from a test designed to measure usability, that is, a test with a stable product and a larger sample than is typical and one in which participants are discouraged from making verbal diversions, and the administrator makes minimal interruptions to the flow of tasks. This expert judgment is the basis of the common industry format (CIF) I describe below. Perhaps someday, we will be able to make more precise measurements based directly on the measures that are not filtered through the judgment of usability professional. In the meantime, those judgments are the best estimate we have.

Comparing the Usability of Products

An important variation on the purpose of a usability test is one that focuses primarily on comparing usability. Here the intention is to measure how usable a product is relative to some other product or to an earlier version of itself. There are two types of comparison tests: (a) an internal usability test focused on finding as much as possible about a product's usability relative to a comparison product (a comparative test or a diagnostic, comparative test) and (b) a test intended to produce results that will be used to measure comparative usability or to promote the winner over the others (a competitive usability test).

In both types of comparison tests, there are two important considerations: (a) The test design must provide a valid comparison between the products and (b) the selection of test participants, the tasks, and the way the test administrator interacts with participants must not favor any of the products.

Designing Comparison Tests. As soon as the purpose of the test moves from diagnosis to comparison measurement, the test design moves toward becoming more like a research design. To

demonstrate that one product is better on some measure, you need a design that will validly measure the comparison. The design issues usually focus on two questions:

- Will each participant use all of the products, some of the products, or only one product?
- How many participants are enough to detect a statistically significant difference?

In the research methods literature, a design in which participants use all of the products is called a "within-subjects" design, whereas in a "between-subjects" design each participant uses only one product. If testers use a between-subjects design, they avoid having any contamination from product to product, but they need to make sure that the groups who use each product are equivalent in important ways. For example, in a typical between-subject design, members of one group are recruited because they have experience with Product A, whereas a second group is recruited because they have experience with Product B. Each group then uses the product they know. But the two groups need to have equivalent levels of experience with the product they use. They also need to have equivalent skills and knowledge with related variables, such as job titles and time worked, general computer literacy, and so on.

Because it is difficult to match groups on all of the relevant variables, between-subjects designs need to have enough participants in each group to wash out any minor differences. An important concern to beware of in the between-subjects design is the situation in which one of the participants in a group is especially good or bad at performing tasks. Gray and Salzman (1998) called this the "wildcard effect." If the group sizes are small, one superstar or dud could dramatically affect the comparison. With larger numbers of participants in a group, the wildcard has a smaller impact on the overall results. This phenomenon is one of the reasons that competitive tests have larger sample sizes than diagnostic tests. The exact number of participants depends on the design and the variability in the data. Sample sizes in competitive tests are closer to 20 in a group than the 5 to 8 that is common in diagnostic tests.

If testers use a within-subjects design in which each participant uses all of the products, they eliminate the effect of groups not being equivalent but then have to worry about other problems, the most important of which are order and sequence effects and the length of the test session. Because within-subjects statistical comparisons are not influenced by inequalities between groups, they are statistically more powerful than between-subjects designs, which means testers need fewer participants to detect a difference. To eliminate effects due to order and the interaction of the product with each other, you need to counterbalance the order and sequence of the products. (See Fisher & Yates, 1963, and Dumas, 1998a, for rules for counterbalancing.) They also have to be concerned about the test session becoming so long that participants get tired.

There are some designs that are hybrids because they use within-subjects comparisons but don't include all of the combinations. For example, if tester's are comparing their product to two of their competitors, they might not care about how the two competitors compare with each other. In that case, each participant would use the testers' product and one of the others, but no one would use both of the competitors' products. This design allows the statistical power of a within-subjects design for some comparisons—those involving your product. In addition, the test sessions are shorter than with the complete within-subjects design.

Eliminating Bias in Comparisons. For a comparison test to be valid, it must be fair to all of the products. There are at least three potential sources of bias: the selection of participants, the selection and wording of tasks, and the interactions between the test administrator and the participants during the sessions.

The selection of participants can be biased in both a between- and a within-subjects design. In a between-subjects design, the bias can come directly from selecting participants who have more knowledge or experience with one product. The bias can be indirect if the participants selected to use one product are more skilled at some auxiliary tasks, such as the operating system, or are more computer literate. In a competitive test using a between-subjects design, it is almost always necessary to provide evidence showing that the groups are equivalent, such as by having them attain similar average scores in a qualification test or by assigning them to the products by some random process. In a within-subjects design, the bias can come from having the participants have more knowledge or skill with one product. Again, a qualification test could provide evidence that they know each product equally well.

Establishing the fairness of the tasks is usually one of the most difficult activities in a comparison test, even more so in a competitive test. One product can be made to look better than any other product by carefully selecting tasks. Every user interface has strengths and weaknesses. The tasks need to be selected because they are typical for the sample of users and the tasks they normally do. Unlike a diagnostic test, the tasks in a competitive test should not be selected because they are likely to uncover a usability problem or because they probe some aspect of one of the products.

Even more difficult to establish than lack of bias in task selection is apparent bias. If people who work for the company that makes one of the products select the tasks, it is difficult to counter the charge of bias even if there is no bias. This problem is why most organizations will hire an outside company or consultant to select the tasks and run the test. But often the consultant doesn't know enough about the product area to be able to select tasks that are typical for end users. One solution is to hire an industry expert to select or approve the selection of tasks. Another is to conduct a survey of end users, asking them to list the tasks they do.

The wording of the task scenarios can also be a source of bias, for example, because they describe tasks in the terminology used by one of the products. The scenarios need to be scrubbed of biasing terminology.

Finally, the test administrator who interacts with each test participant must do so without biasing the participants. The interaction in a competitive test must be as minimal as possible. The test administrator should not provide any guidance in performing tasks and should be careful not to give participants rewarding feedback after task success. If participants are to be told

when they complete a task, it should be done after every complete task for all products. Because of the variability of task times it causes, participant should not be thinking aloud and should be discouraged from making verbal tangents during the tasks.

Baseline Usability Tests

One of the ways to measure progress in user interface design is by comparing the results of a test to a usability baseline. Without a baseline, it can be difficult to interpret quantitative measures from a test and put them in context. For example, if it takes a sample of participants 7 minutes to complete a task with an average of two errors, how does a tester interpret that result? One way is to compare it to a usability goal for the task. Another is to compare it to the results of the same task in an earlier version of the product.

But establishing a baseline of data takes care. Average measures from a diagnostic usability test with a few participants can be highly variable for two reasons. First, because of the small number of participants, average scores can be distorted by a wildcard. Because of this variability, it is best to use a sample size closer to those from a comparison test than those from a diagnostic test. Second, the thinking-aloud procedure typically used in diagnostic tests adds to the variability in performing the task. It is best not to have participants think aloud in a baseline test, which makes the data cleaner but also lessens its value as a diagnostic tool.

Allowing Free Exploration

An important issue in user testing is what the participant does first. For example, if all users will have some training before they use the product, the tester might want to provide this training. There is often a preamble to the first task scenario that puts the test and the tasks into some context. Most often, the preamble leads to the first task scenario. Using this procedure immediately throws the participant into product use. Some testers argue that this procedure is unrealistic, that in the "real world" people don't work that way but spend a few minutes exploring the product before they start doing tasks. Others argue that going directly to tasks without training or much of a preamble puts stress on the product to stand on its own, stress that is beneficial in making the product more usable.

Should testers consider allowing the test participants 5 to 10 minutes of exploration before they begin the task scenarios? Those in favor of free exploration argue that without it, the product is getting a difficult evaluation and that the testing situation is not simulating the real use environment, especially for Web-based products. Users must know something about the product to buy it, or their company might give them some orientation to it. Those against free exploration argue that it introduces added variability into the test; some participants will find information that helps them do the tasks, but others won't find the same information. Furthermore, nobody really knows what users do when no one is watching. A usability test session is a constructed event that does not attempt to simulate every component of the real use environment. Finally, the test is intended to be a difficult evaluation for the product to pass. This debate continues, but most testers do not allow free exploration.

CHALLENGES TO THE VALIDITY OF USABILITY TESTING

For most of its short history, user testing has been remarkably free from criticism. Part of the reason for this freedom is the high face validity of user testing, which means that it appears to measure usability. User testing easily wins converts. When visitors watch a test for the first time, they think they are seeing a "real" user spontaneously providing their inner experiences through their think-aloud protocol. Visitors often conclude that they are seeing what really happens when no one is there to watch customers. When a usability problem appears in the performance of a test participant, it is easy to believe that every user will have that problem.

But some impressions of user testing can be wrong. A test session is hardly a spontaneous activity. On the contrary, a user test is a very constructed event. Each task and each word in each scenario has been carefully chosen for a specific purpose. And unfortunately, we don't know what really happens when no one is watching.

In the past 5 years, researchers and practitioners have begun to ask tough questions about the validity of user testing as part of a wider examination of all usability evaluation methods. This skepticism is healthy for the usability profession. Here I discuss four challenges to validity:

1. How do we evaluate usability testing?
2. Why can't we map usability measures to user interface components?
3. Are we ignoring the operational environment?
4. Why don't usability specialists see the same usability problems?

How Do We Evaluate Usability Testing?

One of the consequences of making a distinction between usability testing and research is that it becomes unclear how to evaluate the quality and validity of a usability test, especially a diagnostic test. As I have noted, usability professionals who write about testing agree that a usability test is not a research study. Consequently, it is not clear whether the principles of research design should be applied to a diagnostic usability test. Principles, such as isolating an independent variable and having enough test participants to compute a statistical test, do not apply to diagnostic usability testing. The six essential characteristics of user testing described above set the minimum conditions for a valid usability test but do not provide any further guidance. For example, are all samples of tasks equal in terms of ensuring the valid of a test? Are some better than others? Would some samples be so bad as to invalidate the test and its results? Would any reasonable sample of tasks uncover the global usability

problems? Is a test that misses uncovering a severe usability problem just imperfect, or is it invalid?

Dumas (1999) explored other ways to judge the validity of a user test. For example, Skinner (1956) invented a design in which causality between independent and dependent variables was established with only one animal. By turning the independent variable on and off several times with the same animal, he was able to establish a causal relationship between, for example, a reinforcement schedule and the frequency and variability of bar pressing or pecking. In some ways, Skinner's method is similar to having the same usability problem show up many times both between and within participants in a usability test. In this analogy, usability problems that repeat would establish a causal relationship between the presentation of the same tasks with the same product and the response of the participants. This relationship is exactly why a tester becomes confident that problems that repeat are caused by a flawed design. But should we end there? Should we only fix repeating problems? And what if, as often happens, some participants don't have the problem? It is not clear where to draw the repetition line.

Hassenzahl (1999) agued that a usability tester is like a clinician trying to diagnose a psychological illness. An effective tester is one who is good at tying symptoms, that is, usability problems, with a cause—a poor design. In this analogy, a goal for the profession is to create a diagnostic taxonomy to make problem interpretations more consistent. Gray and Salzman (1998) and Lund (1998) have made similar points. Until that happens, however, we are left looking for good clinicians (testers), but we have little guidance about that makes a valid test.

Why Can't We Map Usability Measures to User Interface Components?

An important—Gray and Salzman (1998) said the most important—challenge to the validity of usability testing is the difficulty of relating usability test measures to components of the user interface. Practitioners typically use their intuition and experience to make such connections. For example, a long task time along with several errors in performing a task may be attributed to a poorly organized menu structure. Would other testers make the same connection? Do these two measures always point to the same problem? Do these measures only point to this one problem? Is the problem restricted to one menu or several? Are some parts of the menu structure effective?

As we have seen, common practice in test reporting is to group problems into more general categories. For example, difficulties with several words in an interface might be grouped under a "terminology" or a "jargon" category. Unfortunately, there is no standardized set of these categories. Each test team can roll their own categories. This makes the connection from design component to measures even more difficult to make. Landauer (1995) urged usability professions and researchers to link measures such as the variability in task times to specific cognitive strategies people use to perform tasks. Virzi et al. (1993) compared the results of a performance analysis of objective measures with the results of a typical think aloud protocol analysis. They identified many fewer problems using performance analysis.

That study and others suggest that many problems identified in a usability test come from the think-aloud protocol alone. Could some of these problems be false alarms, which is to say they are not usability problems at all? Expert review, an inspection evaluation method, has been criticized for proliferating false alarms. Bailey, Allan, and Raiello (1992) claimed that most of the problems identified by experts are false alarms. But they used the problems that they identified from user testing as the comparison. If Bailey et al. are correct, most of the problems identified by user testing also are false alarms. Their study suggests that the only practice that makes any difference is to fix the one or two most serious problems found by user testing.

Without a consistent connection between measures and user interface components, the identification of problems in a user test looks suspiciously like an ad hoc fishing expedition.

Are We Ignoring the Operational Environment?

Meister (1999) took the human factors profession to task for largely ignoring the environment within which products and systems are used (see Fig. 56.3). He asserted that in human factors, the influence of the environment on the human-technology interaction is critical to the validity of any evaluation. He proposed that human factors researchers have chosen erroneously to study the interaction of people and technology largely in a laboratory environment. He noted that "any environment in which phenomena are recreated, other than the one for which it was intended, is artificial and unnatural" (p. 66). Although Meister did not address usability testing directly, he presumably would have the same criticism of the use of testing laboratories to evaluate product usability.

Those who believe that it is important to work with users in their operational environment as the usability specialists gather requirements also believe that at least early prototype testing should be conducted in the work environment (Beyer & Holtzblatt, 1997). The assumption these advocates make is that testing results will be different if the test is done in the work

FIGURE 56.3. The scope of human factors. From "Usability testing methods: When does a usability test become a research experiment?" by J. Dumas, 2000, *Common Ground*, 10. Reprinted with permission.

environment rather than in a usability lab. These differences will lead to designs that are less effective if the richness of the work environment is ignored. The proponents of testing in the work environment offer examples to support their belief, but to date there are no research studies that speak to this issue.

One could imagine a continuum on which to place the influence of the operational environment on product use. For some products, such as office productivity tools, it seems unlikely that the operational environment would influence the usability of a product. For some other products, such as factory floor operational software, the physical and social environments definitely influence product use; and then there is a wide range of products that fall in between. For example, would an evaluation of a design for a clock radio be complete if test participants didn't have to read the time from across a dark room? Or shut the alarm off with one hand while lying down in a dark room?

Meister admitted that it is often difficult to create or simulate the operational environment. The classic case is an accident in a power plant that happens once in 10 years. One of the reasons the operational environment is not considered more often in usability evaluation is that it is inconvenient and sometimes difficult to simulate. When we list convenience as a quality of a usability lab, we need to keep in mind that for some products, the lab environment may be insufficient for uncovering all of the usability problems in products.

Why Don't Usability Specialists See the Same Usability Problems?

Earlier I discussed the fact that usability specialists who viewed sessions from the same test had little agreement about which problems they saw and which ones were the most serious (Jabobsen & John, 1998). There are two additional studies that also speak to this point (Molich et al., 1998, 2001). These studies both had the same structure. A number of usability labs were asked to test the same product. They were given broad instructions about the user population and told that they were to do a "normal" usability test. In the first study, four labs were included; in the second, there were seven. The results of these studies were not encouraging. There were many differences in how the labs went about their testing. It is clear form these studies that there is little commonality in testing methods. But even with that caveat, one would expect these labs staffed by usability professionals to find the same usability problems. In the first study, there were 141 problems identified by the four labs. Only one problem was identified by all of the labs. Ninety-one percent of the problems were identified by only one lab. In the second study, there were 310 problems identified by the seven teams. Again, only one problem was identified by all seven teams, and 75% of the problems were identified by only one team.

Our assumption that usability testing is good method for finding the important problems quickly has to be questioned by the results of these studies. It is not clear why there is so little overlap in problems. Are slight variations in method the cause? Are the problems really the same but just described differently? We look to further research to sort out the possibilities.

ADDITIONAL ISSUES

In this final section on usability testing, I discuss five final issues:

1. How do we evaluate ease of use?
2. How does user testing compare with other evaluation methods?
3. Is it time to standardize methods?
4. Are there ethical issues in user-based evaluation?
5. Is testing Web-based products different?

How Do We Evaluate Ease of Use?

Usability testing is especially good at assessing initial ease of learning issues. In many cases, a usability test probes the first hour or two of use of a product. Testers see this characteristic as an asset because getting started with a new product is often a key issue. If users can't get by initial usability barriers, they may never use the product again, or they may use only a small part of it.

Longer term usability issues are more difficult to evaluate. Product developers often would like to know what usability will be like after users learn how to use a product. Will users become frustrated by the very affordances that help them learn the product in the first place? How productive will power users be after 6 months of use?

Although there is no magic potion that will tell developers what usability will be like for a new product after 6 months, there are some techniques that address some long-term concerns:

- Repeating the same tasks one or more times during the session—this method gets at whether usability problems persist when users see them again.
- Repeating the test—a few weeks in between tests provides some estimate of long-term use.
- Providing training to participants who will have it when the product is released—establishing a proficiency criterion that participants have to reach before they are tested is a way to control for variations in experience.

Although these techniques sometimes are useful, assessing the ease of use for a new product is difficult with any evaluation method.

How Does Usability Testing Compare With Other Evaluation Methods?

In the early 1990s, there were several research studies that looked at the ability of user testing to uncover usability problems and compared testing with other evaluation methods, especially expert reviews and cognitive walkthroughs (Desurvire, 1994; Jeffries, Miller, Wharton, & Uyeda, 1991; Karat, Campbell, & Fiegel, 1992; Nielsen & Phillips, 1993). The evaluation methods

together are now called UEMs—usability evaluation methods. In these studies, testing generally came out quite well in comparison with the other methods. Its strengths were in finding severe usability problems quickly and finding unique problems, that is, problems not uncovered by other UEMs.

Jeffries et al. (1991) found that usability testing didn't uncover as many problems as an expert review and that no one expert found more than 40% of the problems. Furthermore, when the authors segmented the problems by severity, usability testing found the smallest number of the least severe problems and the expert reviewers found the most. Karat et al. (1992) compared usability testing to two kinds of walkthroughs and found that testing found more problems and more severe problems. In addition, usability testing uncovered more unique problems than walkthroughs. Desurvire (1994) compared usability testing to both expert reviews and walkthroughs and found that usability testing uncovered the most problems, the most severe problems, and the most unique problems.

Dumas and Redish (1993), reviewing these studies from a usability testing perspective, summarized the strengths of usability testing as uncovering more severe problems than the other methods. Since that time, this clear-cut depiction of usability testing has been challenged. All of these studies and more were reviewed by Gray and Salzman (1998) and Andre, Williges, and Hartson (1999) in a meta-analysis of the comparison research. In Gray and Salzman's view, all of the studies are flawed, being deficient in one or more of five types of validity. Their analysis makes it difficult to be sure what conclusions to draw from the comparison studies. Andre et al. proposed three criteria to evaluate UEMs: thoroughness (finding the most problems), validity (finding the true problems), and reliability (repeatedly finding the same problems). They found that they only could compare UEM studies on thoroughness, with inspection methods being higher on it than usability testing. Andre et al. could not find sufficient data to compare UEMs on validity or reliability. Fu, Salvendy, and Turley (1998) proposed that usability testing and expert reviews find different kinds of problems.

As we have described above, it now appears that the ability of experts or researchers to consistently agree on whether problems are severe or not makes it difficult to tout usability testing's purported strength at uncovering severe problems quickly. Even the conclusion that usability testing finds unique problems is suspect because those problems might be false alarms. Andre et al. proposed that usability testing be held up as the yardstick against which to compare other UEMs. But the assumption that usability testing uncovers the true problems has not been established.

Gray and Salzman's analysis was criticized by usability practitioners (Olson & Moran, 1998). The practitioners were not ready to abandon their confidence in the conclusions of the comparison studies and continue to apply them to evaluate the products they develop. To date, no one has shown that any of Gray and Salzman's or Andre et al.'s criticisms of the lack of validity of the UEM studies is incorrect. At present, the available research leaves us in doubt about the advantages and disadvantages of usability testing relative to other UEMs.

Is It Time to Standardize Methods?

Several standards setting organizations have included user-based evaluation as one of the methods they recommend or require for assessing the usability of products. These efforts usually take a long time to gestate and their recommendations are sometimes not up to date, but the trends are often indicative of a method's acceptance in professional circles.

The International Organization of Standardization (ISO) Standard ISO 9241, "Ergonomic requirements for office work with visual display terminals (VDTs)," describes the ergonomic requirements for the use of visual display terminals for office tasks. Part 11 provides the definition of usability, explains how to identify the information which is necessary to take into account when evaluating usability, and describes required measures of usability. Part 11 also includes an explanation of how the usability of a product can be evaluated as part of a quality system. It explains how measures of user performance and satisfaction, when gathered in methods such as usability testing, can be used to measure product usability.

ISO/DIS 13407, "Human-centered design processes for interactive systems," provides guidance on human-centered design including user-based evaluation throughout the life cycle of interactive systems. It also provides guidance on sources of information and standards relevant to the human-centered approach. It describes human-centered design as a multidisciplinary activity, which incorporates human factors and ergonomics methods such as user testing. These methods can enhance the effectiveness and efficiency of working conditions and counteract possible adverse effects of use on human health, safety, and performance.

One of the most interesting efforts to promote usability methods has been conducted by the U.S. Food and Drug Administration (FDA), specifically the Office of Health and Industrial Programs, which approves new medical devices. In a report titled "Do It by Design" (http://www.fda.gov/cdrh/humfac/doit. html), the FDA described what it considers best practices in human factors methods that can be used to design and evaluate devices. Usability testing plays a prominent part in that description. The FDA stops short of requiring specific methods but does require that device manufacturers prove that they have an established human factors program. The FDA effort is an example of the U.S. Governments' relatively recent but enthusiastic interest in usability (http://www.usability.gov).

The most relevant standards-setting effort to those who conduct user-based evaluations is the National Institute of Standards and Technology's (NIST) Industry Usability Reporting Project (IURP). This project has been underway since 1997. It consists of more than 50 representatives of industry, government, and consulting who are interested in developing standardized methods and reporting formats for quantifying usability (http://zing.ncsl.nist.gov/iusr/).

One of the purposes of the NIST IUSR project is to provide mechanisms for dialogue between large customers, who would like to have usability test data factored into the procurement decision for buying software, and vendors, who may have usability data available.

NIST worked with a committee of usability experts from around the world to develop a format for usability test reports, called the common industry format (CIF). The goal of the CIF is to facilitate communication about product usability between large companies who want to buy software and providers who want to sell it. The CIF provides a way to evaluate the usability of the products buyers are considering on a common basis. It specifies what should go into a report that conforms to the CIF, including what is to be included about the test method, the analysis of data, and the conclusions that can be drawn from the analysis. The CIF is intended to be written by usability specialists and read by usability specialists. One of its assumptions is that given the appropriate data specified by CIF, a usability specialist can measure the usability of a product their company is considering buying.

The CIF is not intended to apply to all usability tests. It applies to a summative test done late in the development process to measure the usability of a software product, not to diagnostic usability tests conducted earlier in development.

The American National Standards Institute (ANSI) has created CIF as one of its standards (ANCI/NCITS 354-2001). The CIF document is available from http://techstreet.com. It is difficult to know how this standard will be used, but it could mean that in the near future vendors who are selling products to large companies could be required to submit a test report in CIF format.

Are There Ethical Issues in User Testing?

Every organization that does user testing needs a set of policies and procedures for the treatment of test participants. Most organizations with policies use the federal government or American Psychological Association policies for the treatment of participants in research. At the heart of the policies are the concepts of informed consent and minimal risk. Minimal risk means that "the probability and magnitude of harm or discomfort anticipated in the test are not greater, in and of themselves, than those ordinarily encountered in daily life or during the performance of routine physical or psychological examination or tests."[1] Most usability tests do not put participants at more than minimal risk. If the test director feels that there may be more than minimal risk, he or she should follow the procedures described in the Notice of Proposed Rule Making in the Federal Register, 1988, Vol. 53, No. 218, pp. 45661–45682.

Even if the test does not expose participants to more than minimal risk, testers should have participants read and sign an informed consent form, which should describe the purpose of the test; what will happen during the test, including the recording of the session, what will be done with the recording, and who will be watching the session; and the participants' right to ask questions and withdraw from the test at any time. Participants need to have the chance to give their consent voluntarily. For most tests, that means giving them time to read the form and asking them to sign it as an indication of their acceptance of what is in the form. For an excellent discussion of how to create and use consent forms see Waters, Carswell, Stephens, and Selwitz (2001).

The special situation in which voluntariness may be in question can happen when testers sample participants from their own organizations. The participants have a right to know who will be watching the session and what will be done with the videotape. If the participants' bosses or another senior members of the organization will be watching the sessions, it is difficult to determine when the consent is voluntary. Withdrawing from the session may be negatively perceived. The test director needs to be especially careful in this case to protect participants' rights to give voluntary consent. The same issue arises when the results of a test with internal participants are shown in a highlight tape. It that case, the participants need to know before the test that the tape of the session might be viewed by people beyond the development team. Test directors should resist making a highlight tape of any test done with internal participants. If that can't be avoided, the person who makes the highlight tape needs to be careful about showing segments of tape that place the participant in a negative light, even if only in the eyes of the participant.

The names of test participants also need to be kept in confidence for all tests. Only the test director should be able to match data with the name of a participant. The participants' names should not be written on data forms or on videotapes. Use numbers or some other code to match the participant with their data. It is the test director's responsibility to refuse to match names with data, especially when the participants are internal employees.

This discussion should make it clear that it may be difficult to interpret subjective measures of usability when the participants are internal employees. Their incentive to give positive ratings to the product may be increased when they believe that people from their company may be able to match their rating with their name.

Is Testing Web-Based Products Different?

There is nothing fundamentally different about testing Web products, but the logistics of such tests can be a challenge (Grouse, Jean-Pierre, Miller, & Goff, 1999). Often the users of Web-based products are geographically dispersed and may be more heterogeneous in their characteristics than users of other technologies. The most important challenge in testing these products is the speed with which they are developed (Wichansky, 2000). Unlike products with traditional cyclic development processes, Web products often do not have released versions. They are changed on a weekly, if not daily, basis. For testing, this means gaining some control over the product being tested. It needs to be stable while it is tested, not a moving target.

With control comes the pressure to produce results quickly. Conducting a test in 8 to 12 weeks is no longer possible in fast-paced development environments. Testing in 1 or 2 weeks is more often the norm now. Testing with such speed is only possible in environments where the validity of testing is not questioned and the test team is experienced.

---

[1] Notice of Proposed Rule Making in the Federal Register, 1988, Vol. 53, No. 218, p. 45663.

The Future of Usability Testing

Usability testing is clearly the most complex usability evaluation method, and we are only beginning to understand the implications of that complexity. It appears that usability testing has entered into a new phase in which its strengths and weaknesses are being seriously debated, although it remains very popular and new usability labs continue to open. Before 1995 the validity of testing was seldom challenged. The recent research has opened up a healthy debate about our assumptions about this method. We can never go back to our earlier innocence about this method, which looks so simple in execution but whose subtleties we are only beginning to understand.

WHICH USER-BASED METHOD TO USE?

Deciding which of the user-based evaluation methods to use should be done in the context of the strengths and weaknesses of all of the usability inspection methods discussed in Chapter 57. Among the user-based methods, direct or video observation is useful in special situations. It allows usability specialists to observe populations of users who cannot otherwise be seen or who can only be observed through the medium of videotape. Questionnaires are a useful way to evaluate a broad sample of users, to measure the usability of a product that has been used by the same people over a long period of time, and to sample repeatedly the same user population. The best questionnaires also have the potential to allow usability comparisons across products and, perhaps, to provide an absolute measure of usability. Usability testing can be used throughout the product development cycle to diagnose usability problems. Its findings have the most credibility with developers of all of the evaluation methods. As currently practiced, tests can be conducted quickly and allow retesting to check whether solutions to usability problems are effective. Using testing to compare products or to provide an absolute measure of usability requires more time and resources and testers who have knowledge of research design and statistics.

References

Abelow, D. (1992). Could usability testing become a built-in product feature? *Common Ground, 2*, 1-2.

Andre, T., Williges, R., & Hartson, H. (1999). The effectiveness of usability evaluation methods: Determining the appropriate criteria. *Proceedings of the Human Factors and Ergonomics Society, 43rd Annual Meeting* (pp. 1090-1094). Santa Monica, CA: Human Factors and Ergonomics Society.

Baber, C., & Stanton, N. (1996). Observation as a technique for usability evaluation. In P. Jordan, B. Thomas, B. Weerdmeester, & I. McClelland (Eds.), *Usability evaluation in industry* (pp. 85-94). London: Taylor & Francis.

Bailey, R. W. (1993). Performance vs. preference. *Proceedings of the Human Factors and Ergonomics Society, 37th Annual Meeting* (pp. 282-286). Santa Monica, CA: Human Factors and Ergonomics Society.

Bailey, R. W., Allan, R. W., & Raiello, P. (1992). Usability testing vs. heuristic evaluation: A head-to-head comparison. *Proceedings of the Human Factors and Ergonomics Society, 36th Annual Meeting* (pp. 409-413). Santa Monica, CA: Human Factors and Ergonomics Society.

Barker, R. T., & Biers, D. W. (1994). Software usability testing: Do user self-consciousness and the laboratory environment make any difference? *Proceedings of the Human Factors Society, 38th Annual Meeting* (pp. 1131-1134). Santa Monica, CA: Human Factors and Ergonomics Society.

Bauersfeld, K., & Halgren, S. (1996). "You've got three days!" Case studies in field techniques for the time-challenged. In D. Wixon & J. Ramey (Eds.), *Field methods casebook for software design* (pp. 177-196). New York: John Wiley.

Beyer, H., & Holtzblatt, K. (1997). *Contextual design: Designing customer-centered systems.* San Francisco: Morgan Kaufmann.

Bias, R. (1994). The pluralistic usability walkthrough: Coordinated empathies. In J. Nielsen & R. Mack (Eds.), *Usability inspection methods* (pp. 63-76). New York: John Wiley.

Boren, M., & Ramey, J. (2000, September). Thinking aloud: Reconciling theory and practice. *IEEE Transactions on Professional Communication*, 1-23.

Bowers, V., & Snyder, H. (1990). Concurrent versus retrospective verbal protocols for comparing window usability. *Proceedings of the Human Factors Society, 34th Annual Meeting* (pp. 1270-1274). Santa Monica, CA: Human Factors and Ergonomics Society.

Bradburn, N. (1983). Response effects. In R. Rossi, M. Wright, & J. Anderson (Eds.), *The handbook of survey research* (pp. 289-328). New York: Academic Press.

Branaghan, R. (1997). Ten tips for selecting usability test participants. *Common Ground, 7*, 3-6.

Branaghan, R. (1998). Tasks for testing documentation usability. *Common Ground, 8*, 10-11.

Brooke, J. (1996). SUS: A quick and dirty usability scale. In P. Jordan, B. Thomas, B. Weerdmeester, & I. McClelland (Eds.), *Usability evaluation in industry* (pp. 189-194). London: Taylor & Francis.

Cantani, M. B., & Biers, D. W. (1998). Usability evaluation and prototype fidelity: Users and usability professionals. *Proceedings of the Human Factors Society, 42nd Annual Meeting* (pp. 1331-1335). Santa Monica, CA: Human Factors and Ergonomics Society.

Chignell, M. (1990). A taxonomy of user interface terminology. *SIGCHI Bulletin, 21*, 27-34.

Chin, J. P., Diehl, V. A., & Norman, K. L. (1988). Development of an instrument measuring user satisfaction of the human-computer interface. *Proceedings of Human Factors in Computing Systems '88*, 213-218.

Desurvire, H. W. (1994). Faster, cheaper! Are usability inspection methods as effective as empirical testing? In J. Nielsen & R. Mack (Eds.), *Usability inspection methods* (pp. 173-202). New York: John Wiley.

DeVries, C., Hartevelt, M., & Oosterholt, R. (1996). Private camera conversation: A new method for eliciting user responses. In P. Jordan, B. Thomas, B. Weerdmeester, & I. McClelland (Eds.), *Usability evaluation in industry* (pp. 147-156). London: Taylor & Francis.

Dobroth, K. (1999, May). Practical guidance for conducting usability tests of speech applications. Paper presented at the annual meeting of the American Voice I/O Society (AVIOS). San Diego, CA.

Doll, W., & Torkzadeh, G. (1988). The measurement of end-user computing satisfaction. *MIS Quarterly, 12*, 259-374.

Dumas, J. (1998a). Usability testing methods: Using test participants as their own controls. *Common Ground, 8,* 3-5.

Dumas, J. (1998b). Usability testing methods: Subjective measures Part I—Creating effective questions and answers. *Common Ground, 8,* 5-10.

Dumas, J. (1998c). Usability testing methods: Subjective measures Part II—Measuring attitudes and opinions. *Common Ground, 8,* 4-8.

Dumas, J. (1999). Usability testing methods: When does a usability test become a research experiment? *Common Ground, 9,* 1-5.

Dumas, J. (2000). Usability testing methods: The fidelity of the testing environment. *Common Ground, 10,* 3-5.

Dumas, J. (2001). Usability testing methods: Think-aloud protocols. In R. Branghan (Ed.), *Design by people for people: Essays on usability.* Chicago: Usability Professionals' Association.

Dumas, J., & Redish, G. (1993). *A practical guide to usability testing.* NJ: Ablex.

Dumas, J., & Redish, G. (1999). *A practical guide to usability testing* (Rev. ed.). London: Intellect Books.

Ericsson, K. A., & Simon, H. A. (1993). *Protocol Analysis: Verbal Reports as Data.* Cambridge, MA: MIT Press.

Fisher, R. A., & Yates, F. (1963). *Statistical tables for biological, agricultural and medical research.* Edinburgh, Scotland: Oliver & Boyd.

Frokjaer, E., Hertzum, M., & Hornbaek, K. (2000). Measuring usability: Are effectiveness, efficiency, and satisfaction really correlated? *Proceedings of Human Factors in Computing Systems 2000,* 45-52.

Fu, L., Salvendy, G., & Turley, L. (1998). Who finds what in usability evaluation. *Proceedings of the Human Factors and Ergonomics Society, 42nd Annual Meeting* (pp. 1341-1345). Santa Monica, CA: Human Factors and Ergonomics Society.

Gaba, D. M. (1994). Human performance in dynamic medical domains. In M. S. Bogner (Ed.), *Human error in medicine* (pp. 197-224). Hillsdale, NJ: Lawrence Erlbaum Associates.

Gage, N., & Berliner, D. (1991). *Educational psychology* (5th ed.). New York: Houghton Mifflin.

Goldberg, J. H. (2000). Eye movement-based interface evaluation: What can and cannot be assessed? *Proceedings of the IEA 2000/HFES 2000 Congress (44th Annual Meeting of the Human Factors and Ergonomics Society)* (pp. 625 628). Santa Monica, CA: Human Factors and Ergonomics Society.

Gray, W., & Salzman, M. (1998). Damaged merchandise? A review of experiments that compare usability methods [Special Issue]. *Human-Computer Interaction, 13,* 203-261.

Grouse, E., Jean-Pierre, S., Miller, D., & Goff, R. (1999). Applying usability methods to a large intranet site. *Proceedings of the Human Factors and Ergonomics Society, 43rd Annual Meeting* (pp. 782-786). Santa Monica, CA: Human Factors and Ergonomics Society.

Ground, C., & Ensing, A. (1999). Apple pie a-la-mode: Combining subjective and performance data in human-computer interaction tasks. *Proceedings of the Human Factors and Ergonomics Society, 43rd Annual Meeting* (pp. 1085-1089). Santa Monica, CA: Human Factors and Ergonomics Society.

Hackman, G. S., & Biers, D. W. (1992). Team usability testing: Are two heads better than one? *Proceedings of the Human Factors Society, 36th Annual Meeting* (pp. 1205-1209). Santa Monica, CA: Human Factors and Ergonomics Society.

Hartson, H. R., Castillo, J. C., Kelso, J., Kamler, J., & Neale, W. C. (1996). Remote evaluation: The network as an extension of the usability laboratory. *Proceedings of Human Factors in Computing Systems '96,* 228-235.

Hassenzahl, M. (1999). Usability engineers as clinicians. *Common Ground, 9,* 12-13.

Hughes, M. (1999). Rigor in usability testing. *Technical Communication, 46,* 488-494.

Igbaria, M., & Parasuraman, S. (1991). Attitudes towards microcomputers: Development and construct validation of a measure. *International Journal of Man-Machine Studies, 34,* 553-573.

Jacobsen, N., & John, B. (1998). The evaluator effect in usability studies: Problem detection and severity judgments. *Proceedings of the Human Factors and Ergonomics Society, 42nd Annual Meeting* (pp. 1336-1340). Santa Monica, CA: Human Factors and Ergonomics Society.

Jeffries, R., Miller, J., Wharton, C., & Uyeda, K. (1991). User interface evaluation in the real world: A comparison of four techniques. *Proceedings of Human Factors in Computing Systems '91,* 119-124.

Kantner, L. (2001a). Following a fast-moving target: Recording user behavior in Web usability testing. In R. Branaghan (Ed.), *Design by people for people: Essays on usability* (pp. 235-244). Chicago: Usability Professional's Association.

Kantner, L. (2001b). Assessing Web site usability from server log files. In R. Branaghan (Ed.), *Design by people for people: Essays on usability* (pp. 245-261). Chicago: Usability Professional's Association.

Karat, C. M., Campbell, R., & Fiegel, T. (1992). Comparison of empirical testing and walk-through methods in user-interface evaluation. *Proceedings of Human Factors in Computing Systems '92,* 397-404.

Kennedy, S. (1989). Using video in the BNR usability lab. *SIGCHI Bulletin, 21,* 92-95.

Kirakowski, J. (1996). The software usability measurement inventory (SUMI): Background and usage. In P. Jordan, B. Thomas, B. Weerdmeester, & I. McClelland (Eds.), *Usability evaluation in industry* (pp. 169-177). London: Taylor & Francis.

Kirakowski, J., & Corbett, M. (1988). Measuring user satisfaction. In D. Jones & R. Winder (Eds.), *People and computers* (Vol. IV, pp. 189-217). Cambridge, England: Cambridge University Press.

Landauer, T. K. (1995). *The trouble with computers.* Cambridge, MA: MIT Press.

Landay, J. A., & Myers, B. (1995). Interactive sketching for the early stages of user interface design. *Proceedings of Human Factors in Computing Systems '95,* 43-50.

Law, C. M., & Vanderheiden, G. C. (2000). Reducing sample sizes when user testing with people who have, and who are simulating disabilities—experiences with blindness and public information kiosks. *Proceedings of the IDEA 2000/HFES 2000 Congress, 4,* 157-160. Santa Monica, CA: Human Factors and Ergonomics Society.

Ledgard, H. (1982). Evaluating text editors. *Proceedings of Human Factors in Computer System,* 135-156.

Lesaigle, E. M., & Biers, D. W. (2000). Effect of type of information or real-time usability evaluation: Implications for remote usability testing. *Proceedings of the IEA 2000/HFES 2000 Congress, 6,* 585-588. Santa Monica, CA: Human Factors and Ergonomics Society.

Lewis, J. (1991). Psychometric evaluation of an after-scenario questionnaire for computer usability studies: The ASQ. *SICCHI Bulletin, 23,* 78-81.

Lewis, J. (1994). Sample size for usability studies: Additional considerations. *Human Factors, 36,* 368-378.

Lewis, J. R. (1995). IBM computer usability satisfaction questionnaires: Psychometric evaluation and instructions for use. *International Journal of Human-Computer Interaction, 7,* 57-78.

Lister, M. (2001). Usability testing software for the Internet. *Proceedings of Human Factors in Computing Systems 2001, 3,* 17-18.

Lund, A. M. (1998). The need for a standardized set of usability metrics. *Proceedings of the Human Factors and Ergonomics Society, 42nd Annual Meeting* (pp. 688-691). Santa Monica, CA: Human Factors and Ergonomics Society.

Meister, D. (1999). *The history of human factors and ergonomics*. Mahwah, NJ: Lawrence Erlbaum Associates.

Mitropoulos-Rundus, D., & Muzak, J. (1997). How to design and conduct a consumer in-home usability test. *Common Ground, 7*, 10–12.

Molich, R., Bevan, N., Curson, I., Butler, S., Kindlund, E., Miller, D., & Kirakowski, J. (1998). Comparative evaluation of usability tests. *Proceedings of the Usability Professionals' Association* (pp. 1–12). Dallas, TX: Usability Professionals' Association.

Molich, R., Kindlund, E., Seeley, J., Norman, K., Kaasgaard, K., Karyukina, B., Schmidt, L., Ede, M., van Oel, W., & Kahmann, R. (2002). Comparative usability evaluation. In press.

Nielsen, J. (1992). Finding usability problems through heuristic evaluation. *Proceedings of Human Factors in Computing Systems '92* (pp. 373–380).

Nielsen, J., & Phillips, V. L. (1993). Estimating the relative usability of two interfaces: Heuristic, formal, and empirical methods compared. *Proceedings of the Association of Computerized Machinery INTERCHI '93 Conference on Human Factors in Computing Systems* (pp. 214–221). New York: ACM Press.

Olson, G., & Moran, R. (1998). Damaged merchandise? A review of experiments that compare usability methods [Special Issue]. *Human-Computer Interaction, 13*, 203–261.

Orne, M. (1969). Demand characteristics and the concept of quasi-controls. In R. Rosenthal & R. Rosnow (Eds.), *Artifact in behavioral research* (pp. 143–179). New York: Academic Press.

Perkins, R. (2001). Remote usability evaluation over the Internet. In R. Branaghan (Ed.), *Design by people for people: Essays on usability* (pp. 153–162). Chicago: Usability Professional's Association.

Philips, B., & Dumas, J. (1990). Usability testing: Functional requirements for data logging software. *Proceedings of the Human Factors Society, 34th Annual Meeting* (pp. 295–299). Santa Monica, CA: Human Factors and Ergonomics Society.

Rubin, J. (1994). *Handbook of usability testing*. New York: John Wiley.

Scholtz, J., & Bouchette, D. (1995). Usability testing and group-based software: Lessons from the field. *Common Ground, 5*, 1–11.

Shneiderman, B. (1987). *Designing the user interface: Strategies for effective human computer interaction*. Reading, MA: Addison-Wesley.

Shneiderman, B. (1992). *Designing the user interface: Strategies for effective human computer interaction* (2nd ed.). Reading, MA: Addison-Wesley.

Shneiderman, B. (1997). *Designing the user interface: Strategies for effective human computer interaction* (3rd ed.). Reading, MA: Addison-Wesley.

Skinner, B. F. (1956). A case history in scientific method. *American Psychologist, 11*, 221–233.

Spenkelink, G., Beuijen, K., & Brok, J. (1993). An instrument for measurement of the visual quality of displays. *Behaviour and Information Technology, 12*, 249–260.

Thomas, B. (1996). Quick and dirty usability tests. In P. Jordan, B. Thomas, B. Weerdmeester, & I. McClelland (Eds.), *Usability evaluation in industry* (pp. 107–114). London: Taylor & Francis.

Virzi, R. A. (1990). Streamlining the design process: Running fewer subjects. *Proceedings of the Human Factors Society, 34th Annual Meeting* (pp. 291–294). Santa Monica, CA: Human Factors and Ergonomics Society.

Virzi, R. A. (1992). Refining the test phase of usability evaluation: How many subjects is enough? *Human Factors, 34*, 457–468.

Virzi, R. A., Sokolov, J. L., & Karis, D. (1996). Usability problem identification using both low and high fidelity prototypes. *Proceedings of Human Factors in Computing Systems '96*, 236–243.

Virzi, R. A., Sorce, J. F., & Herbert, L. B. (1993). A comparison of three usability evaluation methods: Heuristic, think-aloud, and performance testing. *Proceedings of the Human Factors and Ergonomics Society, 37th Annual Meeting*, 309–313.

Vora, P. (1994). Using teaching methods for usability evaluations. *Common Ground, 4*, 5–9.

Waters, S., Carswell, M., Stephens, R., & Selwitz, A. (2001). Research ethics meets usability testing. *Ergonomics in Design, 9*, 14–20.

Wichansky, A. (2000). Usability testing in 2000 and beyond. *Ergonomics, 43*, 998–1006.

Wiklund, M., Dumas, J., & Thurrott, C. (1992). Does the fidelity of software prototypes affect the perception of usability? *Proceedings of the Human Factors Society, 36th Annual Meeting* (pp. 1207–1212). Santa Monica, CA: Human Factors and Ergonomics Society.

Wilson, C. E., & Coyne, K. P. (2001). Tracking usability issues: To bug or not to bug? *Interactions, 8*, 15–19.

Wolf, C. G. (1989). The role of laboratory experiments in HCI: Help, hindrance or ho-hum? *Proceedings of Human Factors in Computing Systems '89*, 265–268.

Young, R., & Barnard, P. (1987). The use of scenarios in HCI research: Turbo charging the tortoise of cumulative science. *Proceedings of Human Factors in Computing Systems '87*, 291–296.

·57·

INSPECTION-BASED EVALUATIONS

Gilbert Cockton
University of Sunderland

Darryn Lavery
Microsoft Corporation

Alan Woolrych
University of Sunderland

USABILITY, INSPECTION, AND COMPLEMENTARY EVALUATION METHODS

A usability inspection method (UIM) is applied to an interactive system to assess its usability. An analyst thus inspects the system (or a specification or sketch) and reports on its usability. To be effective, a UIM must take account of the nature of usability, a concept that is more subtle than naïve ideas of "user friendliness" would suggest.

Defining Usability: Implications for Evaluation

Usability is a key quality of interactive systems, typically encountered negatively as system features and behaviors that obstruct usage, learning, or satisfaction. Positively, a usable system is one that can be used effectively, efficiently, and enjoyably. By effective, we understand that a user can achieve goals that the system was intended to support. By efficient, we understand that goals can be achieved with acceptable levels of resource, be this mental energy, physical effort, or time. By enjoyable, we understand that usage delivers levels of enjoyment appropriate to the context of use.

No elaboration of "effectively, efficiently, and enjoyably" involves absolute properties; indeed, it is a matter for philosophical debate whether any object can have intrinsic absolute properties. Rather, attributes of objects may only result from their relation to some measuring or observing device. If so, we cannot talk of attributes or properties independently of ways of determining them. Usability is not just a question of measurement, however, any more than, for example, the robustness of physical objects. Thus, when buying a chair, the issue is not whether it is strong, but whether it is strong *enough*, and in turn, this requires further elaboration: strong enough for what? A chair may be strong enough for most humans to sit on but would break if a person climbed onto it with a heavy load—few chairs make good small stepladders. An antique chair may not be strong enough to sit on, but it needs to stay in one piece when being moved during cleaning. Even a chair with a spike in the seat is not absolutely unusable, especially where the intended purpose is to stop it from being sat on. Such a chair will be unusable (and downright dangerous) for almost all usage contexts typically associated with chairs, however.

Usability is thus always relative to contexts of use. A system is only intended for some contexts of use. As applications become more generic (e.g., word processors, spreadsheets), they must be usable enough across many potential contexts of use. Conversely, as usability problems become more severe, they can be fatal in almost any context; a digital equivalent of spiky chairs is a Web-page stuffed full of frenetic animations, hidden controls, and cryptic self-absorbed text (in small type and poorly contrasting colors).

A UIM must be able to cope with the full range of usability problems, from ones that would be near universal to those that are specific to tightly defined contexts of use. The justifications for predicting problems can thus be based on universals of human perception and cognition (or human flesh in the case

of spiky chairs!) at one extreme and at the other extreme, on specific sociocultural factors in an expected context of use (e.g, antique chairs in a museum). The nature of usability thus sets requirements for UIMs. Those that cannot take account of intended contexts of use can only predict problems that arise from psychological universals.

The Phases of Usability Inspection

Analysts' requirements for supporting resources from a UIM arise from the high-level structure of usability inspection. Analysts must begin by studying the UIM. Each time a UIM is applied, analysts must study the target system in whatever form it is provided. Once a system has been studied, a UIM can be applied. This is essentially a search problem—analysts look first for possible problems. Analyst behavior could be modelled by the *generate and test* strategy from early artificial intelligence research, in which possibilities are first generated (e.g., moves in a chess game) and then tested to identify possibilities with specific attributes (e.g., the "best" next move). Finally, analysts must move from reporting on problems to generating recommendations on solutions. We can thus model usability inspection as having four distinct phases: (a) analyst preparation, (b) candidate problem discovery, (c) confirmation or elimination of candidate problems, and (d) recommendation generation.

Analyst Preparation. Analysts must first understand a UIM. They must then understand a system and its expected context of use. Gaps in any understandings will translate into a poor set of predictions (and ultimately recommendations).

UIMs are still developing and have yet to mature. Analysts thus are always researchers who have much to contribute to UIM development. The oldest UIM, guidelines conformance, is more than 30 years old, but no set of guidelines has survived (unchanged). Moreover, guidelines conformance is more of an approach than a method; analysts must decide how to assess conformance. The most structured UIM, Cognitive Walkthrough, has a much shorter history and underwent many changes in its first 4 years, but it has been largely undeveloped since then (work at Microsoft (Spencer, 2000) is a notable exception).

Such short and turbulent histories give analysts little solid to draw on when learning UIMs. Methods tend to be reported in short conference papers, with the occasional journal article or a chapter in a student text or practitioner's book. Unpublished tutorial notes often present the best accounts of methods. In reality, analysts must consult a small set of sources and then work out how to apply a UIM, taking into account any available evaluation. For the authors' studies, we have developed and used compact tutorial material with self-assessment for three UIMs (Lavery, Cockton, & Atkinson, 1996a, 1996b, 1996c, 1996d; Lavery & Cockton, 1997a).

Novice analysts should practice UIMs on familiar systems to quickly establish the scope and accuracy of a UIM relative to their experience. Once analysts are confident that they understand a UIM, they can then apply it, but they must be

well briefed about systems and their intended contexts of use. Without knowledge of the latter, analysts must guess how and when potential problems would really be severe. They can thus only trap features and behaviors that are so incompatible with general human capabilities that usage contexts become irrelevant. Not surprisingly, relatively few developers are so unaware of their own limitations as humans that extreme features get introduced; for example, we have yet to encounter pale lemon text on a white background. Typical usability problems tend to be more subtle and closely related to contextually specific factors such as user knowledge (e.g., of similar systems), semiotic expectations (i.e., expected meanings of system elements), domain factors (e.g., number formats for accounting), human activity structures (e.g., you don't pay before you buy), and resources (e.g., there simply isn't time available for a task like that). Such knowledge of expected contexts of use let analysts make sound judgements about the likely severity or frequency of a problem.

Finally, analysts must know about and understand the systems they are analyzing. Some UIM developers think otherwise, preferring to keep analysts in holy ignorance, supposedly inducing user empathy. Such empathy can only be truly grounded in knowledge of usage contexts. Assumptions about users that go beyond the specified context of use are likely to be highly unreliable (and often insulting to real users' intelligence). The result of analyst ignorance about system features is not better empathy with the user, but incorrect claims that features are missing or poorly supported. Although there may be evidence of usability problems in such analyst errors, they are still errors. Properly informed analysts are more likely to note that a feature is difficult to find (rather than absent) or that a design rationale has overlooked some critical criteria. Such problem reports are far more useful than bogus ones.

Candidate Problem Discovery. Two forms of prediction failure must be avoided with UIMs: predictions must be valid (no false positives) and comprehensive (few false negatives). False negatives are by definition due to a failure to identify possible candidate problems in initial analysis. UIMs can reduce false negatives by adequate provision of *discovery resources*. Such resources begin with knowledge of the system, but this is generally beyond a UIM unless it requires system specifications in a specific form. Applying UIMs to undocumented designs is always a risky enterprise. More powerful resources come from representations of usage contexts, however, especially tasks and scenarios. Representations that cover domain knowledge are also valuable.

Analysts who understand a UIM, a system and its application domain, its design rationales and its expected context of use are well placed to find possible problems. Knowledge of the system is vital.

Confirmation and Elimination of Candidate Problems. Having generated possible problems by the application of system, interaction, domain, and contextual knowledge, analysts must further consider these candidate problems to eliminate or confirm them as (im)probable. This should improve prediction *validity*.

UIMs can reduce false positives by adequate provision of *analysis resources*, which overlap with discovery resources. Knowledge of the system is far less important, but knowledge of expected contexts of use, human capabilities, and key properties of human–computer interaction (HCI) are critical. Contextual knowledge may eliminate a possible problem as improbable. Knowledge of display-based interaction may eliminate possible problems that overlook users' abilities to discover information and to explore interactive behaviours. Knowledge of human capabilities such as visual attention may either confirm possible problems (e.g., key information in the wrong place) or eliminate them (e.g., misleading information in the wrong place!).

Recommendation Generation. Current UIMs provide little, if any, support for the generation of recommendations for fixing designs to avoid predicted problems. However, it is vital that UIMs are properly applied in preceding phases. Recommendations that are based on UIM misapplications are likely to be ill founded. It is thus important that UIMs are properly applied and not just that problems can be successfully predicted by analysts who use them.

Relation to Other Methods

UIMs are the oldest HCI method, beginning with the development of design guidelines. Checking for conformance with these guidelines was the first form of inspection method. This became less favored in the 1980s as new styles of interaction multiplied the number of design guidelines to the point where guideline collections became so large that they became unusable for most practitioners. User testing became the dominant approach in HCI, with model-based approaches developing as theory-based complements. In the late 1980s, the cost of user testing and doubts about the cost–benefit ratios for model-based methods resulted in an interest in discount usability methods. Inspection methods had a renaissance in the late 1980s and early 1990s, being seen as cost-effective alternatives to user testing that required less expertise than model-based methods. Since the mid-1990s, the emphasis in HCI has been on contextual analysis and evaluating systems in context. As a result, inspection methods have changed little. Studies of the scope and accuracy of UIMs have continued and breakthroughs in UIMs remain possible, however.

Model-Based Evaluations. Model-based approaches draw on the formality of computer science or theoretical models from cognitive psychology (see chapter 58 of this handbook). Models of systems, tasks and users are used individually or in combination to support analyses of system quality. Model-based methods, by definition, require a model of some or all of a human–computer system. Inspection methods by contrast require no such models, although task models are used with some UIMs.

User-Based Evaluations. Model- and inspection-based methods are both analytical, that is, they arrive at conclusions

through the use of calculation, deduction, or judgement. User-based evaluations by contrast are empirical and arrive at conclusions through observation, analysis, and inference. Model- and inspection-based methods thus rely critically on the quality of their underlying theories of HCI, which are explicit and formal for model-based methods and implicit and informal for inspection-based methods. User-based evaluations do not rely on predictive theories but simply find out what happens during some usage. However, user-based methods are also highly dependent on prior analyses of expected contexts of use. User tests must be planned and results interpreted. Neither can be performed reliably without good contextual knowledge.

User testing is no less of a search problem than are other evaluation methods. Systems induce potential usability problems that are only exposed within certain usage scenarios. Model- and inspection-based methods must simulate such scenarios to avoid false negatives, which requires powerful discovery resources. User-based methods also require discovery resources, in the form of appropriate test users, test activities, and problem extraction procedures (to "find" usability problems in test data). User-based methods remain more reliable because, once selected and briefed, test participants are substantial discovery resources in themselves. Hence, user testing will expose problems through user behaviors that are difficult to anticipate. User testing can only avoid false negatives through careful planning and data analysis, however. There is no guarantee that user-based approaches will always outperform other evaluation methods. Still, real human participants significantly reduce the chances of both oversights and misjudgement—seeing, after all, is believing. Once problems have emerged, there is no need to predict them.

The Need for Inspection Methods. Heuristic evaluation was the most used usability method in a recent major survey (Rosenbaum, Rohn, & Humburg, 2000). Usability professionals take risks with UIMs for valid reasons. Given this, it is vital that a UIMs' strengths and weaknesses are thoroughly understood.

UIMs are used because model-based approaches remain limited or immature, are expensive to apply, and their use is largely restricted to research teams (e.g., Gray, John, & Atwood, 1992). UIMs also are cheap to apply, have been applied to many commercial designs by practitioners, and are seen as a low-cost and low-skill method. In addition, development resources may rule out user testing, leaving UIMs as the best available practical approach—some usability is better then no usability. UIMs also can be used before a testable prototype has been implemented and can be iterated without exhausting or biasing a group of test participants. Finally, UIMs also serve as a *discovery resource* for user testing, which can be designed to focus on potential problems predicted by UIMs.

A SURVEY OF INSPECTION METHODS

With UIMs, analysts examine (representations of) a system, following the method to draw conclusions about probable usability problems. "Following the method" takes one of two distinct forms. The predominant form is rule-based conformance assessment, in which a usable system adheres to a set of "rules." The other form is the *walkthrough*, in which a usable system is one in which no likely user difficulties emerge during step-by-step analysis of a hypothetical interaction.

Rule-based methods operate at different levels of abstraction. The most abstract methods look for breaches of principles that express properties of systems or interaction. Often known as "-ities," such principles include browsability, predictability, and flexibility. The least abstract methods look for breaches of "style rules" that prescribe design decisions at a level of details such as the structure and content of window title bars. In between principles and style guides are heuristics (less abstract than principles) followed by guidelines (more abstract than style guides). Levels of abstraction are not necessarily homogeneous. They are generally homogeneous at the extremes of abstraction, but heuristics and guidelines can be expressed with varying specificity. Heuristics as rules of thumb tend toward the general but can also be quite specific ("Provide clearly marked exits"). Guidelines range from detailed design advice as specific instructions to general wisdom ("Know the user").

The homogeneity and abstraction level of UIMs have immediate consequences for analysts. More abstract rules require extensive translation either to *narrow down* a rule to an instance or the opposite. The latter, *generalizing* from an instance to a rule (breach) tends to be made more readily and with more confidence, as indicated in a pilot study on problem matching (Lavery, Cockton, & Atkinson, 1997). Inspection using this generalizing tactic must proceed feature by feature, scanning all rules to see if a feature breaks any rule. In contrast, the narrowing tactic proceeds rule by rule, scanning all features to see if a feature breaks a rule. Decision making will thus become quicker as rules become less abstract, because less translation between rules and system features is required. However, more detail results in much larger rule sets that introduce further search problems. Not only must analysts search the system for features to analyze, but they must also search large rule sets for rules that features could breach.

There are further concerns beyond the narrowing/ generalizing dilemma with rule-based methods. Although rules can be applied to interaction, the overwhelming tendency is for rules to be applied to system features and not to the integrated use of several features within a hypothetical interaction. Rule-based methods thus tend to be *system-centric*, incorrectly assuming that usability can be an inherent property of the system alone. This introduces the risk of both false positives (by overestimating the impact of a rogue feature on users of normal intelligence) and false negatives (by failing to examine adverse interactions between individually conformant features). Rule-based methods can thus best identify problems that arise at the extremes of cognitive and perceptual overload that would require exceptional (and highly improbable) contexts of use to ameliorate them (for a spiked chair, such a context would be a knight in armour).

In contrast, procedural inspection methods are better placed to consider interaction beyond brief encounters with individual system features. In principle, they can bring any aspect of

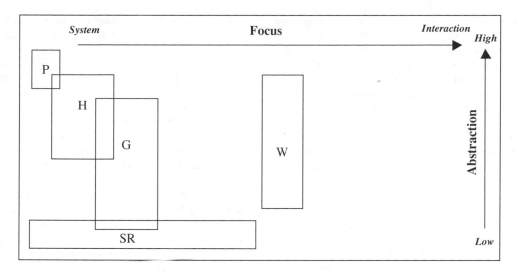

FIGURE 57.1. Relationship of types of inspection-based method by level of abstraction and focus.

context into consideration within the hypothetical interaction that forms the focus of the walk-through. Walk-through methods can also support heterogeneity in levels of abstraction, ranging from a user's motives for being in a role (which requires them to carry out a task) to the human information processing of keystroke level interaction. As with rule-based methods, however, an increase in the set of concerns that must be addressed at each step of the walkthrough increases the chances of error (assuming the method is practical enough for any use at all to be attempted).

Figure 57.1 relates the five main types of inspection-based method to each other within a space formed by level of abstraction (vertical axis, top is high) and focus (horizontal axis; middle is neutral, left is system-centered, and right is interaction-centered). Heterogeneity increases with the area of the rectangle for each method type. It is primarily determined by abstraction range, but it can also be extended by range of focus. Method types are represented by letters: P(rinciples), H(euristics), G(uidelines), S(tyle) R(ules) and W(alkthrough). The four rule-based types overlap, with heuristics spanning abstraction levels within principles and guidelines, and guidelines overlapping with style rules. Heuristics, guidelines and style rules can all be specialized for particular application domains, and thus they span from almost wholly system-centered toward a balance of system- and interaction-centered.

Having related the five types of inspection-based methods, we cover four specific methods in detail. For each method, the emphasis is initially on description, with brief analyses based on the framework presented above. The next section presents assessments of the methods. Space limitations have prevented us covering several worthy methods in detail, notably cognitive dimensions (Green & Petre, 1996), ergonomic criteria (Bastien & Scapin, 1995), platform-specific guides (Martin, Chapman, & Leben, 1991; Apple Computer, 1992; Microsoft Corp., 1999) and software engineering principles (Gram & Cockton, 1996). The methods covered are the best known and most commonly used and thus the most evaluated.

Heuristic Evaluation

Heuristic-based methods involve the evaluation of an interface against rules of thumb. These heuristics or "golden rules" are general rules derived from more extensive collections of interface guidelines (for example, 944 guidelines in the case of Smith & Mosier, 1986).

Using a heuristic evaluation (for example, Nielsen, 1992; Nielsen & Molich, 1990), the evaluator uses a small set of "rules" to find usability problems by inspecting the user interface for violations, either using task scenarios or free exploration of the system. For example, systems that require the user to type in the name of a file would breach a "prevent errors" heuristic, while a system that allowed the user to choose the name would not. UNIX command line programs are notoriously bad for either producing no feedback ("provide feedback" rule) or providing cryptic error messages ("good error messages" and "speak the user's language" rules).

The most recent version of heuristic evaluation (HE) (Nielsen, 1994) was derived from a database of 249 usability problems from evaluations of 11 interactive systems. Nielsen collected seven sets of guidelines, including his own original heuristics (Nielsen, 1993b). He then, for each problem in the database, assigned a rating of 1 to 5 to each individual guideline according to how well it explained the usability problem. Average ratings for each guideline were calculated.

A principal components analysis was undertaken to determine if a few factors accounted for most variability of the usability problems. Nielsen found that seven factors could only account for 30% of the variability, and 53 factors were need to account for 90%, which is too many for an HE. The names of the seven factors, which were chosen by Nielsen included "visibility of system status," "match between system and real world," and "error prevention."

Nielsen then attempted to pick the principles that provide the best explanation of the usability problems. The principle which found the most usability problems with a rating of 3 or

TABLE 57.1. Revised Heuristics From Nielson (1994)

1. Visibility of system status
2. Match between system and the real world
3. User control and freedom
4. Consistency and standards
5. Error prevention
6. Recognition rather than recall
7. Flexibility and efficiency of use
8. Aesthetic and minimalist design
9. Help users recognize, diagnose, and recover from errors
10. Help and documentation

more ("explains a major part of the problem, but there are some aspects of the problem that are not explained") was chosen and this process was repeated with the remaining usability problems. This supported the claim that the seven factors could form a basis for a new set of 10 heuristics (Table 57.1). Nielsen noted, however, that the new set is good at explaining existing problems but does not know how effective they will be at *finding* usability problems. The current version of HE thus comes with known limitations in its discovery resources, and it is no surprise that others have attempted to fill the gap.

Muller et al. (1995) claimed that the heuristics do not consider the context in which the proposed system will be used. Three new heuristics were added to assess how well the interactive system meets the needs of its users and their work environment (Table 57.2).

The value of these extra heuristics was demonstrated for an application in which 247 usability problems were found, resulting in 89 recommendations, of which 72% were implemented. Each problem and recommendation was scored for their connection to each of the 13 heuristics. The three new heuristics were solely responsible for identifying 15% of the problems and 10% of the recommendations. By rating problems and recommendations using a severity scale (from 1–5), they found that *problems* revealed with the new heuristics were rated as slightly less important than those identified using the old heuristics, but there was *no difference in importance for the recommendations*. This reinforces the need to consider all aspects of inspection method usage (i.e., analyst preparation and recommendation generation as well as problem discovery and confirmation/elimination).

Overall, HE is limited in its support for analyst preparation and recommendation generation. Its main possible strengths lie in discovery resources, analysis resources, or both. An assessment of HE's actual strengths is given later in the chapter.

Guideline-Based Methods

Guidelines provide advice at a range of levels of abstraction and contextual sensitivity to developers. They take a wide range of

TABLE 57.2. Additional Heuristics From Muller et al. (1995)

1. Respect the user and his or her skills
2. Promote a pleasurable experience with the system
3. Support quality work.

TABLE 57.3. The Six Sections of Smith and Mosier's (1986) Guidelines

Section	Functional Area	Guidelines
1	Data entry	199
2	Data display	298
3	Sequence control	184
4	User guidance	110
5	Data transmission	83
6	Data protection	70

forms. Early collections quickly became monumental, so current international standards now separate guidelines for specific dialogue styles.

The best known collection of design guidelines was prepared for the U.S. Air Force by Mitre Corporation (Smith & Mosier, 1986). More than 15 years old at the time of this writing, its contents largely address text interfaces and high-level issues of dialogue and functionality. Nevertheless, its most basic advice remains highly relevant. For example, many forms on Web sites would benefit immensely from following Smith and Mosier's guidelines on data input. Smith and Mosier organized the guidelines into six areas, as shown in Table 57.3. Clearly, such a collection cannot be used "rule by rule" in a usability inspection, even with computer-support for search and retrieval.

An example guideline that remains valid to this day is for data entry (p. 56):

1.4/20 Data Format Cueing in Labels
Include in a field label additional cueing of data format when that seems helpful.
"Example"
|DATE (MM/DD/YY): __/__/__|

More recent guidelines are provided by ISO (International Organization of Standardization) 9241, a multipart standard on ergonomics requirements for office work with visual display terminals. Guidelines are provided for a range of dialogue styles: menu (Part 14), command (Part 15), direct manipulation (Part 16), and form-filling dialogues (Part 17). There is not space to cover any of these standards in detail, which are available from the standards body of individual countries. See also chapter 51 of this handbook (Stewart and Travis).

Guidelines vary significantly in their support for analyst preparation and provision of discovery and analysis resources. Overall, the first two stages of UIM uses are well supported, with detailed presentations of guidelines that match well to specific system features. This also provides good support for recommendation generation. Guidelines tend to be weak on analysis resources, however, and tend to provide little support for distinguishing between contexts where they do (not) apply.

Cognitive Walk-Through

Cognitive walk-through (Lewis & Wharton, 1997) is a *procedural* inspection method. It assesses the learnability of "walk-up-and-use" systems. One or more evaluators choose a set of tasks for analysis. Analysts step through each step of the task

answering a set of questions backing up their answers with empirical evidence, experience, or scientific evidence. A negative answer to any of the questions indicates potential usability problems with the system. Using the Cognitive Walkthrough (CW), analysts assess if the user can (or cannot) form appropriate goals and then choose appropriate actions to satisfy these goals from previous knowledge and system cues.

CW is underpinned by CE+, a theory of learning (Polson & Lewis, 1990). For example, the theory predicts that a user new to a system will choose an action with a good match to the user's current goal.

The role of theory in methods was not covered in the initial framework presented earlier, but it is clearly the basis for further key differences between UIMs. However, there seems to be a relation between interaction-centeredness and theory-based methods in that the most system-centered methods are generally atheoretical.

There have been two main forms of CW presented in the literature: those that explicitly consider users' goals (e.g., Lewis, Polson, Wharton, & Rieman, 1990; Polson, Lewis, Rieman, & Wharton, 1992) and those later versions that do not (e.g., Wharton, Rieman, Lewis, & Polson, 1994).

Early Versions of Cognitive Walkthrough. In early versions (for example, Lewis et al., 1990; Polson et al., 1992), users' goals were explicitly considered. Each version works in roughly the same way, the major differences being the number of questions asked and the wording of these questions. The walkthrough starts with the evaluator(s) choosing a set of tasks for analysis. The actions needed to achieve these tasks are described. The evaluators step through each action answering a set of questions: would the user form the right goal, choose the correct action, and form new goals where appropriate?

In Polson et al. (1992) three pages of questions are asked for each action. The first addresses user goals. The evaluator is asked to describe the correct goal structure for that action. This goal structure is compared with the goal hierarchy carried over from the last action. For the first action, the user's assumed initial goals are those that they bring to the system. A problem is noted if there is any mismatch between the two hierarchies.

The second page addresses problems associated with actions. These questions assess if the user would choose the correct action and have any problems executing it. When answering these questions, the evaluator assumes the user has generated the correct goal structure needed for the particular action under analysis.

The final page of questions assumes that the user has chosen the correct action. They assess the effect of system feedback on the users' goal hierarchy. The evaluator assesses if users will perceive progress toward any of their goals, and if they will form any new goals which may or may not be appropriate.

Because these three pages present a massive set of decisions for analysts, tool support (in the form of a Hypercard stack; Rieman et al., 1991) was developed and used in several studies (Cuomo & Bowen, 1992; Desurvire, Kondziela, & Atwood,

1992). An alternative to tool support is simplification, as in Cognitive "Jogthrough" (Rowley & Rhoades, 1992), which makes CW less formal and collapses interaction steps, but the cost-benefit of such simplifications has not been assessed.

The choice of task evaluated is critical to the success of the method, yet the method offers analysts no support here (Wharton, Bradford, Jeffries, & Franzke, 1992). Without support, analysts may fail to consider different task methods. Also, usability problems may be missed if subtasks are not analyzed in the context of different superordinate goals. A task definition methodology is needed (Jeffries, Miller, Wharton, & Uyeda, 1991).

Overall, the first versions of CW provided extensive learning materials for analysts, but these proved to be unwieldy as discovery sources, limiting the benefits that arise with interaction-centered methods. The success case in CW is a rare example of an analysis resource for problem elimination, letting analysts argue that some apparent design flaws would not automatically result in usability problems. Recommendation generation is also partially supported by the CE+ theory.

Later Versions of Cognitive Walkthrough. In response to criticisms, a simpler version of CW was developed (Wharton et al., 1994), which "de-emphasises the explicit consideration of the user's goal structure" (p. 129) thus reducing the number of questions needed and making the method easier to use.

The inputs to this version are a task scenario, a set of actions needed to solve that task, and assumptions about the target users of the interface. Analysts step through each task action trying to generate a success and failure story by considering four questions (Table 57.4).

During inspection, analysts are recommended to record any user assumptions and knowledge requirements, but the method does not suggest how this information could be used. Lewis and Polson taught this version in a tutorial at the ACM CHI '92[1] conference where "The response received was dramatically different from the CHI'91 experience [where an older version was taught] and quite favorable" (Wharton et al., 1994, p. 129).

Franzke (1995) extended the later version of CW by gathering empirical evidence both to support and extend CE+, the underlying theory of learning. In her study, she investigated three factors that make people choose actions given a goal: quality of match, type of action (for example, radio button, selecting a menu, and dragging and dropping) and the number of actions available for consideration. By examining usage data with

TABLE 57.4. Cognitive Walkthrough Questions (Wharton et al., 1994)

1. Will the user try to achieve the right effect?
2. Will the user notice that the correct action is available?
3. Will the user associate the correct action with the effect that the user is trying to achieve?
4. If the correct action is performed, will the user see that progress is being made toward solution of the task?

---

[1]ACM CHI is the world's largest annual HCI conference organized by the Special Interest Group on Computer-Human-Interaction (SIGCHI) of the Association for Computing Machinery (ACM).

two graph tools, she found that if an action has a good match with the user's goal, then it will always be picked quickly regardless of how many competing actions are available. In the absence of a good match, then all actions must be considered. The amount of time needed to find the correct action appears to be proportional to the number of conflicting actions which are available. Finally, Franzke found that the users under study tended to select visible commands first before later trying direct manipulation on interaction objects. These insights have been used to provide guidance for determining whether the user will notice the correct action is available. We know of no studies to determine if this advice improves the quality of an evaluator's walkthrough.

Overall, later versions of CW reduced extensive learning materials for analysts, making them less unwieldy as discovery resources but perhaps losing some benefits that arise with interaction-centered methods, although many should be preserved by task descriptions. Neither analysis resources nor recommendation generation is improved by the later versions (both may actually be reduced, but the real impact here needs to be determined).

A Hybrid Method: Heuristic Walkthrough

Heuristic Walkthrough (Sears, 1997) is a cross between HE and CW. The input to the method is a prioritised list of user tasks. These tasks should include frequent or critical tasks and may include tasks designed purely to ensure coverage of the system.

There are two phases to the evaluation: a task based phase and a free-form phase. In the first phase, the evaluators explore the tasks using a set of thought-provoking questions derived from CW. They are free to explore the tasks in any order, spending as long as they need, but they should be guided by the priorities. In the second phase, evaluators are free to explore the system. They use the set of thought of provoking-questions and HE.

Sears has compared Heuristic Walkthrough with CW and HE. Heuristic Walkthrough found more actual problems than CW and found fewer false positives than HE. Overall, Sear's merge of the two methods results in general improvements to analyst preparation and discovery resources, but there is no sign of improvements in analysis resources, other than the resulting reduction in false positives, which is most likely due to the initial CW inhibiting problem discovery.

SCOPING AND ASSESSING INSPECTION METHODS

In the previous section, a framework was developed and applied to inspection-based methods, highlighting potential limitations of different methods on the basis of their level and homogeneity of abstraction, their position on a system- and interaction-centered spectrum, and additionally, their underpinning by explicit theories. In this section, we review UIM assessments to see which potential limitations emerge in practice.

Gray and Salzman's Critique

In the previous section, we noted the complexity of the inspection task. This inevitably extends to evaluating methods. Ignoring complexity and its implications results in flaws that seriously diminishes a study's credibility (Gray & Salzman, 1998). Gray and Salzman reviewed the validity of five major studies in the literature (including Jeffries et al., 1991; Desurvire et al., 1992; Nielsen, 1992). They cited Cook and Campbell (1979, cited by Gray and Salzman), who suggested four main types of validity: statistical conclusion validity, internal validity, construct validity, and external validity. They add a fifth type of validity: conclusion validity. Gray and Salzman found validity problems with each study reviewed. We now describe the forms of (in)validity and examples.

Statistical Conclusion Validity. Was the change to the dependent variable caused by the manipulation of the independent variable? Statistical tests are used to determine if the change did not occur by chance, for example, by what Gray and Salzman (1998) called wildcard participants, who do significantly better or worse than the average participant despite the conditions. Recruiting more participants reduces the wildcard effect.

For example, Jeffries et al. (1991) compared CW, HE (actually expert review), guidelines and user testing and found HE superior. The study used few evaluators and thus it is not clear if the effects of the study (for example the superiority of HE) occurred by chance.

Internal Validity. Was the effect caused by the variable under manipulation or by a third unknown confounding variable? There is no simple mechanical test for internal validity, but study design must consider the instrumentation, selection of participants and the study setting, and the use of controls where appropriate.

One study where HE (expert review) was shown to be superior potentially suffers from internal validity as the HE group were given a 2-week period at their own pace to complete the evaluation, whereas evaluators in the other conditions had far less time. Also, the worst performing methods were used by software engineers with less usability expertise than the HE group.

Construct Validity. Does the study manipulate the conditions it claims to manipulate (causal construct validity) and does it measure the changes it claims to measure (effect construct validity)?

One example of causal construct invalidity would occur when participants in a study used an evaluation method differently from practitioners. Gray and Salzman (1998) commented that there are different ways of using methods, and so it is difficult to generalize. However, practitioners need to be informed how different ways of employing an evaluation method affect the number of problems found and which evaluation methods are most effective for which types of software.

One study (Karat, Campbell, & Fiegel, 1992) attempted to address causal construct validity, addressing "mono-operation" bias by comparing inspections by individuals with those by

pairs. Similarly, they addressed "mono-method" bias by evaluating two different software systems.

Unfortunately they did not specify the differences in the software being evaluated, reducing the usefulness of this study for practitioners. Also, the evaluation had two phases: self-guided exploration and prescribed scenarios. Gray and Salzman (1998) claimed that this study's "confounding" construct validity was threatened because they were able in the scenario phase to draw on their experience gained in the self-guided exploration phase.

For "effect construct validity," different evaluation methods focus on either *intrinsic attributes* of the interface or *payoff measures* of usability (for example, performance time, errors). However, "None of the studies we reviewed report systematic ways of relating pay-off problems to intrinsic features, all apparently rely upon some form of expert judgement" (Gray & Salzman, 1998, p. 17). Although they "believe that effect construct validity is the single most important issue facing HCI researchers and practitioners," Gray and Salzman unfortunately gave this type of validity light coverage.

External Validity. Do the results of the study generalize to a subpopulation or across a subpopulation? An external validity problem would occur if the results are generalized beyond the settings in the experiment.

Conclusion Validity. Gray and Salzman (1998) introduced a fifth type of validity, conclusion validity. A study is *conclusion invalid* if either the claims were not investigated in the study or that the data presented in the study contradicted their claim. One study (Nielsen, 1992) investigated the effects of evaluator expertise on the number of problems identified using HE and claims that "usability specialists were much better than those without usability expertise by finding usability problems with heuristic evaluation" (Nielsen, 1992, p. 380). Gray and Salzman claimed this study has conclusion invalidity because it is not clear what the effect of HE was on the evaluator's ability to find usability problems or how many problems evaluators would find through expert review. They suggest the data supported the more modest claim that "experts named more problems than non-experts."

Gray and Salzman (1998) found that many studies presented conclusions that were not supported by the data. They did not argue against presenting advice based on experience rather than experimental evidence, but that the source of such advice should be made explicit.

Improving the Assessment of UIMs

Gray and Salzman's critique means that we cannot trust existing simple *comparisons* of UIMs, and thus claims that HE in some way finds more problems than CW cannot be accepted on the basis of these studies. However, the studies are still relevant to the extent that they (strongly) suggest ways in which specific discovery and analysis resources influence outcomes. Still, it would be better if we had assessments that we could trust, so we first consider what is required here. Two of Gray and Salzman's (1998) challenges require little to meet them:

(statistical) *conclusion validity* is a fundamental scientific concern. Researchers should not make unsupported claims (and referees and editors should spot them). Due caution and candor are all that is required to avoid writing invalid conclusions. Due vigilance is required to prevent their publication. For *statistical conclusion validity*, seeking competent statistical advice *before* carrying out studies is enough.

In contrast, *external validity* is a challenge for most psychological experiments. This follows from the controls that ensure internal and construct validity. However, if these are not properly addressed, then there can be no external validity, because this can result in unrealistic settings that cannot generalize. Internal and construct validity are logically before external validity. Safe generalizations cannot be made from unsound experiments, but sound experiments may allow few generalizations. This tension from experimental psychology is extremely challenging.

Causal construct validity can be addressed by ensuring that UIMs under assessment are actually applied by analysts and result in predictions. At the very least, UIM use should be compared with unstructured expert judgement, but this is rare (Bastien & Scapin, 1995, is an exception). Where this is not done, it may still be possible to establish whether the assessed method actually resulted in the prediction. For example, HE cannot be claimed to have supported a prediction if cited heuristics for it have no logical connection with the problem (Cockton & Woolrych, 2001).

Effect construct validity is a question of good experimental design, ensuring that instruments and measures in a study really do report the true value of theoretical dependent variables. In too many studies, the main dependent variable has been the percentage of accurate predictions. However, the result, for example, that a method predicts 70% of known problems is a spurious measure. First, it is not possible to know with confidence 100% of the usability problems associated with a system. Second, such figures are misleading unless they are balanced by the percentage of false positives (i.e., predictions that cannot be matched to known problems). An *effectiveness* metric (Sears, 1997) penalizes a method for excessively high false alarms:

$$\text{Effectiveness} = \text{Thoroughness} \times \text{Validity},$$

where

$$\text{Thoroughness} = \frac{\text{number of real usability problems found by UIM}}{\text{number of real problems that exist}}$$

and

$$\text{Validity} = \frac{\text{number of real usability problems found by UIM}}{\text{number of problems predicted by UIM}}.$$

However, the *thoroughness* metric is only reliable if causal construct validity is achieved (i.e., the number of real usability problems must be shown to have been found by the UIM and not by expert judgement). Finally, internal validity requires control of five key confounding variables, which we now describe.

Analyst Training and Preparation. This confound has already been addressed. Analysts who do not understand a method, the system under evaluation, or the expected context of its use cannot be expected to generate sensible predictions. Analysts who do not understand a UIM can readily both predict many false positives and also fail to predict problems. Both would misrepresent a UIM's quality. Studies must carefully control and assess analysts' understanding. It should not be possible to wholly attribute inappropriate method use and unpredicted problems to analysts' *misunderstanding.*

Reliability of Prediction Merging. When assessing UIMs, it is essential to use a range of analysts. Their predictions must be merged to produce a single problem set. This procedure must be reliable and carefully controlled (Lavery et al., 1997). Errors in forming a single set of predictions for each analyst (group) and then all analysts will corrupt the problem count, as well as distorting the counts of (in)appropriate UIM applications. This problem has been addressed by Connell and Hammond (1999), who demonstrated how it can greatly change the outcome of a UIM assessment. Cockton and Woolrych (2001) applied a structured problem report format (Lavery & Cockton, 1997b) to an extensive assessment of HE. The format has been shown to increase control over prediction merging.

Systematic matching rules require structured usability problems. The individual components of a usability problem report must be reported separately; however, earlier problem report formats (Jeffries, 1994; John & Packer, 1995; Mack & Montaniz, 1994) either restricted the types of problems that can be recorded, missed some crucial aspect of a usability problem, or were unstructured and hence offer no guidance (Lavery et al., 1997). One study (van Schaik, Edwards, & Petrie, 1995) did develop a report that separated causes and difficulties (but provided no prompting on the nature of difficulties). Overall, problem matching in existing studies has been too "liberal" and sometimes based on such crude tactics as matching any problems that concern the same design feature (Lavery, Cockton, & Atkinson, 1997).

Ability of User Testing to Expose Actual Problems. The construction of a known problem set is best achieved through user testing. This brings challenges. Test design will affect the number and types of usability problems discovered. By missing usability problems, both UIMs and user testing may misreport product quality, depending upon the metrics and methods employed by the study. When user test results are used to validate UIMs, failures in user test design will result in miscoding (as false positives) problems that a UIM can in fact predict.

Reliability of Known Problem Extraction. Covering a wide range of usability problems in user testing does not ensure that such problems will make it to the final documented problem set. Errors in forming the set of actual problems from user testing risk miscoding predicted problems as false positives. The SUPEX method (Cockton & Lavery, 1999) addresses this extraction problem. Informed interpretation of problem (type) counts requires explicit levels of problem generalisation. Similarly, confidence in coding predictions as (un)successful also requires

knowledge of the abstraction level for the actual problem set. SUPEX also addresses these issues of abstraction and generalization.

Reliability of Matching Predicted to Actual Problems. To assess a UIM, predictions must be matched against actual problems. Errors in matching will risk miscoding predicted problems as false alarms or vice versa. Existing approaches to problem matching have either been ad hoc or specialized. Ad hoc approaches rely on subjective judgment and hence are unlikely to be replicable, and therefore a true measure of the method's quality may not be reported. Existing specialised approaches (Lewis et al., 1990; Mack & Montaniz, 1994) either are intended for one UIM only (and so do not generalize) or ignore some crucial aspects of usability problems (or both). An investigation of the effects of independent problem matching found evidence of matcher bias (Lavery et al., 1997) and recommended structured problem reports as a way of reducing this bias. Without such reports, problem counts are likely to be so unreliable that few calculations in a UIM assessment can be trusted. This view is given extensive support by Connell and Hammond (1999) in their assessment of HE, where they show that original actual problem counts from one assessment (Nielsen, 1994) are lower than they could be. Replicable and robust assessments of usability inspection methods thus require a set of matching rules applicable to a wide range of methods.

Controlling Confounding Variables: An Example. A recent study (Cockton & Woolrych, 2001) made systematic attempts to control the five confounding variables just described. Control of all the variables can be further improved, but the study shows what is possible for improving study quality. Three forms of training addressed analysts' understanding of HE. Controlling this variable aimed to improve causal construct and internal validity. By constraining analysts to use a common report format (Lavery & Cockton, 1997b), the study eased merging of analyst predictions. By controlling this variable, they aimed to improve effect construct validity (errors in problem set construction result in errors in effect measurement). The ability of user testing to expose actual problems was addressed by systematically deriving task sets for user testing from analyst predictions, again improving effect construct validity. Reliability of actual problem extraction was addressed by use of general principles from the SUPEX method (Cockton & Lavery, 1999) such as transcription, segmentation, and difficulty isolation and generalisation. Finally, reliability of matching predicted to actual problems was addressed by directly associating instances of actual problems with predicted problems. The report format eased such associations. Using the report format for both predictions and actuals, as recommended in Lavery and Cockton (1997b), would have further improved effect construct validity.

Improving External Validity: An Example. As a result of improving internal validity, external validity is improved by supporting conclusions that are focused and accurate. Further support, however, is required from extensive coding of accurate predictions, unpredicted problems, and false alarms. A recent study (Cockton & Woolrych, 2001) coded predicted and actual

problems in several ways to identify how and why HE succeeds and fails as a UIM. Frequency was coded as *high* (more than 20% of users), *medium*, or *low* (one user). Problem impact was classified as *severe* (task failure, more than 2 minutes wasted or major impact on task quality), *nuisance* (under 2 minutes wasted or minor quality impact), and *minor* (immediate recovery or no task impact).

Problem *discoverability* was also coded as an analytical variable (rather than by actual frequency of prediction). The easiest problems to discover are *perceivable*, as they can be seen "at a glance." Next come *actionable* problems, which can be discovered after a few simple actions at most (e.g., a mouse click). Hardest to find are *constructable problems*, for which only several interaction steps involving multiple application objects may reveal a problem. This coding allowed investigation of the relationship between predicted and actual discoverability. Significantly, 80% of missed problems were constructable, but only 7% of successful ones (*p* < .008, Fisher's exact test). Heuristic applications for nonbogus predictions were further coded as being (in)appropriate, using explicit criteria from the HE training manual provided for analysts (Lavery, Cockton, & Atkinson, 1996c), which provides conformance questions that state, "What the system should do, or users should be able to do, to satisfy the heuristic." Such questions are answered with conformance evidence (i e , the "design features or lack of design features that indicate partial satisfaction or breaches of the heuristic"). For many heuristic applications, these criteria were clearly ignored.

All coding schemes used key concepts from usability practice to provide practitioners with *scoping* as well as scoring of methods, that is, the problems that are (not) predicted must be analyzed for regularities in their frequency, impact, discovery requirements, and other relevant effect constructs. In addition, predictions that are due to the assessed method must be separated from those that are not. Measures of effectiveness (Sears, 1997) are thus not enough. To achieve causal construct validity, successful predictions must be shown to be due to the assessed UIM. To support practitioners in their professional practice, missed problems and false alarms must be analyzed across a range of coding dimensions to establish why they arise and what can be done to avoid them.

Software developers need to know the potential benefits of using a particular method (e.g., the number of usability problems predicted), resources needed (e.g., time to evaluate the system, training time for evaluators, evaluators with particular skills) and risks (e.g., missing certain types of problems or predicting "false" usability problems). The factors that may affect the quality of the predictions (e.g., maturation of the prototype, skill level of the evaluator, or choice of different prototyping mediums) also need to be understood and communicated through guidelines.

A Survey of Inspection Method Assessment

The most common effect construct is a count of problems predicted for a given interactive system (for example, Jeffries et al., 1991; Cuomo & Bowen, 1992; Dutt, Johnson, & Johnson, 1994).

For example, one study (Cuomo & Bowen, 1992) found that evaluators using guidelines (i.e., Smith & Mosier, 1986) predicted more usability problems than evaluators using CW and HE. Such assessments indicate something about the *discovery resources* associated with a method's use but cannot distinguish between predictions that are due to method use and those that are due to analysts' existing expertise. Worse still, this measure does not distinguish between successful and unsuccessful predictions (false alarms).

Other studies have compared predicted problems against known usability problems, for example, those revealed during user testing. This effect construct can only distinguish between successful predictions and false alarms if the procedural confounds described in earlier are properly controlled. Although one study (Cuomo & Bowen, 1994) found that guidelines (Smith & Mosier, 1986) predicted the most usability problems, the method found the fewest problems that were confirmed by user testing. Most practitioners prefer to avoid such invalid methods.

The strengths and weaknesses of the particular methods can only be assessed by using more sophisticated measures (effect constructs). Studies have thus classified usability problems by severity (e.g., Jeffries et al., 1991) and by type (e.g., Cuomo & Bowen, 1992, 1994; John & Mashyna, 1997). Types may be based on a range of constructs, for example, by relation to stages in Norman's theory of action (Norman, 1986). Using this coding scheme, a study (Cuomo & Bowen, 1992, 1994) found that an early version of CW tended to focus on problems related to the formation of actions, and was poor at finding problems concerned with evaluating the display. However, effect constructs need to extend beyond measures of problem discovery as dependent variables. Some studies (e.g., John & Marks, 1997; Johnson, 1997; Sawyer, Flanders, & Wixon, 1996) have thus examined the impact (or persuasiveness) and ability of methods to improve the usability of the systems as the number of usability problems (or committed to be) fixed divided by the number of problems reported. However, although the impact of an inspection may be high, there is no guarantee that by addressing its recommendations that the system would be improved—indeed it could be made worse (for example, Nielsen, 1993a). A study (John & Marks, 1997) thus measured *design-change (in)effectiveness* by whether an implemented recommendation increases, leaves unaffected, or reduces usability problems. Their study found that only 11% of predicted problems led to an effective change.

Another study (Johnson, 1997) used too simple a measure of quality: the number of design requirements suggested when using the inspection method. Problem identification precedes requirement proposal, however, and thus reporting such measures may hide how these requirements were proposed. The effectiveness of implemented requirements was also not investigated.

Control over more sophisticated effect measures improves the information available about a method's scope and effectiveness; however, other measures are required to determine the factors that may affect the quality of the predictions. Thus, one study (Nielsen, 1992) found that usability experts predicted more problems than nonexperts when using HE, which was confirmed by another study (Desurvire et al., 1992). Analyst

expertise is thus not surprisingly a major determinant of the discovery resources that are applied during usability inspections. Other studies have identified additional discovery resources, such as the prototype medium (paper and running prototype), which influenced the problems predicted using HE (Nielsen, 1990).

There are thus a range of effect measures of increasing sophistication that can be applied to UIM assessment, as well as a range of causal construct measures that can separate the impact of the method from the influence of other factors such as analyst expertise or the assessed medium (e.g., paper or running prototype). We now survey these measures in practice, beginning with assessments of HE. We then survey five comparative studies.

Heuristic Evaluation. Jakob Nielsen has undertaken several studies to assess the effectiveness of HE. Earlier studies were carried out in coordination with Rolf Molich. Although the first study (Molich & Nielsen, 1990) studied the effectiveness of expert evaluation (and not HE), it did provide the justification for HE and thus will be described. It aimed "to investigate whether industrial data processing professionals would be able to recognize serious interface problems in simple but realistic dialogues" (p. 336). They posed a competition in a Danish computer magazine to find as many usability problems with a telephone number enquiry system for home use, which used an ASCII display. The competition asked readers to find as many problems as possible with the system. They were also free to suggest improvements to the dialogue that would avoid usability problems. The article stressed that the reader's primary aim was to "articulate the usability problems you have identified, instead of merely indicating them implicitly through subtle changes in an alternate design." A sample screen shot and specification were supplied. Seventy-seven readers submitted entries; 51 were from industry (mostly designers and programmers), 10 were teachers and students from university or high schools; and the remaining 16 did not specify their job.

Molich and Nielsen (1990) developed a model solution by checking against a short checklist of usability considerations for dialogue design; these later became heuristics (Nielsen & Molich, 1990). The model solution contained a total of 30 known problems, 2 of which were added when reviewing the submissions (which indicates the quality of the initial "known" problem set).

There was surprise and disappointment at the results. On average, analysts found 11.2 of the problems (37%). The best submission found 18 of the 30 problems (60%). On average, analysts did slightly better at finding serious problems (3.5 out of 8 or 44%). Still, the failure to find problems caused concern: "Dialogue design appears deceptively simple, yet it is full of pitfalls" (Molich & Nielsen, 1990, p. 342).

Following this initial study, four studies of HE were reported (Nielsen & Molich, 1990). The second was the study above, although its reported average yield later rose to 38%.

In the first study, 37 computer science students on a user interface design course applied HE to Teledata (the Danish

videotext system). The students were given a lecture about the heuristics and were provided with a set of ten screen problems. They found 51% of the known problems. In the third and fourth studies, 34 computer science students from the same population as the first study evaluated two running telephone-based systems. The first, a savings application, had 48 known usability problems. The second, a transport application, had 34. Before the evaluation, students were given a lecture about the heuristics. Students found an average of 26% and 20% of problems for each application. Because the same evaluators evaluated both systems, Nielsen and Molich tried to compare the ability of individual evaluators. They found that "while some people are better than others at doing heuristic evaluation of user interfaces, this tendency is not very strong" (Nielsen & Molich, 1990, p. 251).

In summary, evaluators found 51%, 38%, 26%, and 20% of known usability problems for the four systems. These results were judged to be poor, but Nielsen and Molich did remark that being able to find some problems is better than nothing. By combing analysts at random, they found that a higher proportion of usability problems would be found. For example, with the Teledata system (Study 1) two evaluators would find on average 81% of the known problems. With three evaluators, 90% of problems would be found. However, with the transport application (Study 4), five evaluators would only find 55% of problems on average. Nielsen and Molich recommended that heuristic evaluation be carried out with groups of three to five. However this recommendation does not seem to be based on a thorough analysis of the data, nor does it consider acceptable levels of thoroughness—is 55% really good enough? Why did these differences in discoverability exist? Such questions must be answered before practitioners can use UIMs with confidence. The next study (Nielsen, 1990) assessed the effect of prototype medium (paper and running prototype) on analyst's predictions on a mock-up of the Danish Teledata videotext system. The paper prototype consisted of 10 screen shots and a specification. Both versions had the same dialogue structure and screen design. Both versions had 50 known problems[2] of which 15 were major and 35 were minor. Thirty-seven evaluators from a course on user interface design evaluated the paper prototype in the first study described in (Nielsen & Molich, 1990). Another 34 evaluators on the course later assessed the running prototype.

For a particular problem, difficulty was defined as the difference between the probability of it being found minus the average probability for all problems. Thus, a positive rating for a usability problem meant it was easier to find than average, whereas a negative rating meant more difficult. Most problems range from easy to difficult to find in both conditions, but some of the problems were easier to identify in the paper than in the running prototype and vice versa. No analysis of differences in discovered problem types by inspection medium was presented. Instead, for each condition Nielsen identified the seven most common problems (two major problems for the paper system and five major problems for running prototype). He examined each major problem, giving reasons

---

[2]Although Nielsen and Molich (1990) claimed it has 52 problems, implementing it as a running prototype removed two problems.

why it was easier or more difficult to find the problem in a particular medium. Nielsen concluded that the study "indicates that the computer mockup tended to focus the evaluators more on the major problems while the paper mockup tended to focus evaluators more on minor problems" (p. 318). He recommended using both for an evaluation. This follows from Cockton and Woolych's (2001) analysis of discoverability: Constructable problems, which tend to be severe, can only be readily found with computer prototypes.

A later study (Nielsen, 1992) investigated the effects of evaluator expertise. It found that the best analysts were usability experts with knowledge of the application domain. The system under study was a telephone-based interface with voice response known as the "Banking System." The interface took the form of a printed dialogue from the system that had been earlier recorded. This dialogue contained a user error.

Three different types of evaluators evaluated the interface. The first group (novices) consisted of 31 computer science students "who had completed their first programming course" and had "no formal knowledge of user interface design principles" (Nielsen, 1992, p. 375). The second group, usability specialists were "people with graduate degrees and/or several years of job experience in the usability area." The final group were double experts who were usability specialists and had experience of voice response systems.

On average, novices found 22% of problems, experts (usability experts) found 41% of problems, and double experts (both domain and usability experts) found 60% of known problems. The difference between experience level was statistically significant. By grouping together analysts, a higher number of problems would be found, for example, between three and five regular usability specialists would find between 74% and 87% of the known problems (compare this with the 55% and 90% in earlier studies). It would seem that the 3 to 5 analysts rule may only work for usability experts (and even then, is 87% enough?). Nielsen's explanation was as follows: "These detailed results indicate that the double specialists found more problems, not because they were necessarily better usability specialists in general, but because they had specific experience with usability issues for the kind of user interface that was being evaluated" (Nielsen, 1992, p. 377).

Given this, it is unfortunate that the study did not assess the effect of possessing domain expertise but without usability expertise. Still, Nielsen had clearly identified domain knowledge as a critical *discovery resource*. Previous evaluations of HE identified similar resources for successful inspection. Thus paper and running prototypes proved to be different discovery resources that influenced the problems that were found. For HE, these appear to be the only effective resources beyond those that analysts bring to an inspection.

Overall, HE as presented on paper (and presumably in Nielsen's HE lectures) does little to prepare analysts for inspection. The most effective discovery resources are brought by the analysts themselves. As for analysis resources, none of Nielsen's studies provide data on the confirmation or elimination of possible problems. This focus on discovery at the expense of confirmation or elimination eases Nielsen's recommendation of

multiple analysts as a way of increasing the discovery resources that are applied during inspection. However, it is also clear that the use of multiple analysts will *dilute* the impact of *analysis resources* used to eliminate improbable problems. The result will be both more true *and* false positives (i.e., for every extra true problem found by multiple analysts, improbable ones excluded by most analysts will be preserved through the oversight of a single analyst). Assessments of UIMs must thus use Sears' measure of *effectiveness* (Sears, 1997) and not just the measure of *thoroughness* used in most of Nielsen's studies.

Nielsen's assessments of HE are thus misleading in that they fail to consider both thoroughness *and* validity, but also because they fail to establish that HE actually changes the predictions that untrained analysts would otherwise have made and thus fails to establish that HE actually improves on analysts' existing expertise. Also, these studies have not assessed the quality of recommendations made as a result of applying HE. If HE is poorly applied to genuine usability problems, then inappropriate recommendations may well result.

These limitations in existing studies of HE were addressed in a recent thorough assessment of HE (Cockton & Woolrych, 2001) that paid particular attention to the research methods and procedures applied. Special attention was paid to five confounding procedural variables as described earlier. Given the conditions under which HE could be said to be effective and efficient, only *superficial thoroughness* was achieved (74%), with poor *validity* (32%). These results come from a large study involving 99 analysts working in groups and two phases of user testing to confirm analyst predictions. Matches between predicted and actual usability problems do suggest that HE is thorough, but analysis of the actual problems and the successful predictions indicate only a limited role for HE. First, the missed problems tended to be severe, but too complex in origin for HE to predict. Second, 61% of the heuristics applied during successful prediction were inappropriate. Also, HE cannot be said to be efficient because 65% of predictions turned out to be false. Analysis of unpredicted problems revealed essential *discovery resources* that HE lacked, notably a focus on task execution and complex domain goals that are the origin of subtle interaction problems. Analysis of false predictions revealed analysis resources that HE lacked, notably a respect for users' intelligence and an understanding of display based interaction that eliminates logically possible problems as empirically improbable. In short, HE leads analysts to underestimate user capabilities as enhanced within the distribution cognition of HCI.

Cognitive Walkthrough. A number of studies (e.g., Lewis et al., 1990; Jeffries et al., 1991; Cuomo & Bowen, 1992; Wharton et al., 1992) have assessed CW, but given the different versions of the method, one must be cautious when comparing findings.

The first assessment was undertaken on, and is described with, the original CW (Lewis et al., 1990). The four developers of the method each performed a walkthrough on two tasks on four telephone interfaces and recorded the problems found. Twenty unique problems were found, eighteen of which were found by two or more of the four evaluators. The problems found were

compared against existing empirical usage data for the four systems converted into action trees (i.e., "graphic representations of the paths taken by subjects or groups of subjects through the space of actions afforded by the interface" (Lewis et al., 1990, p. 239). They then "looked for points where the user had deviated from a direct path to a solution and where that derivation seemed to reflect one of the problems found in the walkthroughs." Of the 18 nonunique predictions, they found evidence for 15. In a second validation of the predicted problems, two evaluators created action trees for the walkthrough results. The empirical and predicted action trees were then compared. They found that 13 of the 44 observed paths leading to errors (29.5%) were predicted. By weighting against the number of users committing the error, 51 out of 105 (48.5%) of observed error paths were predicted. The proportion of predicted paths ending in error that were not observed was not reported in this paper, although a later account reported that "The false alarm rate . . . was high, almost 75%" (Polson et al., 1992, p. 763). Clearly, CW provides insufficient analysis resources for eliminating improbable problems (as evidenced in CW's lack of support for forming reasoned conclusions about success or failure cases).

A later version of the method (Polson et al., 1992) also reported that students of user interface design found 50% of errors for a voicemail application. In a further example, although analysts found only 30% of errors with a text editor, they did find 70% of problems that had been rated as serious. Finally, with a "single, fairly simple task" (Polson et al., 1992, p. 763), all problems were found with no false alarms for a document routing application. The study also reported that "The method has also been used in a number of design studies . . . commercial software, student projects, research tools and both telephone and graphical interfaces. Without exception participants have indicated that the method was useful in identifying potential problems in the designs" (p. 763).

Other studies (for example, Desurvire et al., 1992) have directly compared CW against other usability inspection methods and user testing and found a potential to predict usability problems, although generally fewer than HE. Other independent studies however have criticized CW's scope, for example, found that the CW tended to miss general ("flaw that affects several parts of the interface", p. 123) and recurring (i.e., those that always interfere with interaction) problems (Jeffries et al., 1991). Another study further refined CW's scope (Cuomo & Bowen, 1992) by mapping problems predicted by CW onto Norman's model of interaction (Norman, 1986). This model splits interaction into two phases of execution and evaluation. The execution phase has four stages: establishing a goal, forming an intention for the goal, specifying an action sequence, and executing the actions. The evaluation has three stages: perceiving the system state, interpreting the state, and evaluating it with respect to the original goals and intentions. This study used the automated CW (Rieman et al., 1991), which mostly predicted problems concerned with action formation. CW was extremely poor at detecting evaluation related problems.

Changes to CW were first driven by the challenges of creating the goal structures required for analysis but later by the perceived cost of using the method. This led Wharton (1992) to develop and extend the original CW questions. Meanwhile, Polson et al. (1992) independently developed their own extended method.

Wharton's first revisions (Wharton, 1992) were used in two studies (Jeffries et al., 1991; Wharton et al., 1992), in which evaluators found the method too tedious. This is perhaps not surprising because analysts had to fill in six pages of questions for each action. Both studies failed to report key details—Jeffries et al. even failed to describe which CW version was being used.

In response to this feedback, the latest version (Wharton et al., 1994) was developed. Although CW's creators believe that most criticisms have been addressed in the latest revisions, the sole assessment of its effectiveness (John & Mashyna, 1997; John & Packer, 1995) suggests that problems remain.

In the first part of this case study (John & Packer, 1995), a single analyst evaluated a paper specification of a multimedia authoring system with CW. The analyst kept a diary of his activities and kept notes of any experiences learning and using the method. After the researchers had removed duplicates, the analyst had predicted a total of 46 problems. In the second part of the study, the predicted problems were compared against usability problems revealed during user testing of an implemented version of the authoring system. Four undergraduate students were given a set of tasks to achieve with the system and asked to think aloud their thoughts. The evaluation was videotaped.

Each researcher independently used a computer tool (MacSHAPA, University of Illinois)[3] to mark segments of videotapes where the test users had experienced difficulties, adhering to fixed criteria for marking usability problems (e.g., the user expressing surprise). The two problem sets were merged to form a total of 60 usability problems.

To match problems, both predicted and actual problem sets had to be filtered. Because only part of the system was built, the implemented version differed from the analyzed specifications, so problems that could not be predicted from the specification had to removed., including two run-time bugs that could not be predicted by CW, leaving 37 potentially predictable actual problems.

The 46 predicted problems were then filtered by removing 16 that concerned unimplemented (and thus untested) system features and 12 that were invalidated by changes during implementation by developers who found the specification incomplete or ambiguous. This left 18 matchable predictions.

The researchers compared the two filtered problem sets. The analyst using CW had only precisely predicted 2 (5%) of the problems found during user testing and a further "two were specified vaguely by a single CW PDR [problem description report] (5%)." Of the total problems, 90% were thus not predicted. John and Mashyna (1997) compared these disappointing results against previous assessments of CW. One study (Desurvire et al., 1992) reported a similar thoroughness for CW

---

[3] Available from www.csse.swin.edu.au/macshapa/index.html

of 8%. Another (Cuomo & Bowen, 1994) reported a *validity* of 58% (but no thoroughness measure). In the John studies, only 11% (2 out of 18 problems) were valid, but John and Mashyna (1997) noted that Cuomo and Bowen (1994) were not comparing usability problems but problem types, so it is unclear how many actual problems were verified.

To explain the poor thoroughness result, actual problems were associated, where possible, with the CW question that could have predicted them. Nearly half (16/37) could have been predicted by Question 1 ("Will the user try to achieve the right effect?"). Seven could be associated with no CW question, however, and thus were beyond CW's scope. Despite the potential thoroughness of CW, due to problems within the scope of Question 1, John Rieman (one of CW's original developers who reviewed predictions) commented that one analyst "often failed to recognize steps where the user might not have the right goal" (John & Packer, 1995, p. 435). Analyst effectiveness may thus be improved by expert review of walkthroughs, although such coaching may not be readily available.

The researchers also examined the unverified predictions. This analysis was inconclusive, largely because the study could not control key confounding variables, especially the ability of user testing to expose predicted problems.

A more recent study exercised more controls within its comparison, although fewer than (Cockton & Woolrych, 2001). The study is probably the first analyst-centered study of CW (Sears & Hess, 1999). Instead of focusing on some supposed inherent quality of a UIM's rules or procedures, this study examined the impact of short and (very) detailed task descriptions on analyst performance. The study only reports thoroughness measures; however, it demonstrates the importance of discovery resources for easing the analyst's task, and furthermore, it measures the impact of these resources by coding problems by the CW question relevant to the prediction. Short task descriptions resulted in relatively more predictions related to finding actions (see Table 57.4). Detailed descriptions resulted in relatively more predictions related to feedback (see Table 57.4). Sears and Hess based their explanations on analysis of the analyst's task. Detailed (to the level of naming and locating controls) task descriptions led to oversights in finding action specification problems, but left analysts with the energy to look for feedback problems. Short descriptions forced analysts to adopt a user perspective, which may have better prepared them for eliminating improbable feedback related predictions. The study demonstrates the importance of discovery resources, but further studies are required to establish the validity of predictions that are made or eliminated by detailed or short descriptions respectively.

CW's poor performance in assessments suggests that not only is it low on discovery resources, but that it suppresses analysts' existing expertise in a way that HE does not. This suggests that discovery resources are not solely positive in their effects but bring with them risks of de-skilling. CW's analysis resources are poor, as its low validity scores suggest, but these are not as poor as HE's in some studies. Recommendation generation for CW has not been assessed.

Guidelines. There are too many assessments of guidelines to review in the space available here (for a recent survey, see Reed et al., 1999) and also relevant comparative studies in the next section, Comparative Studies). One study, however, is interesting for the light it throws on discovery resources. The study (deSouza & Bevan, 1990) evaluated a draft standard for ISO 9241 Part 14 (Menu design). Three designers were given a week to study the guidelines before spending a day using the guidelines to redesign a menu interface, which they had to justify in terms of the guidelines. Designers made errors or had difficulties with 91% of the guidelines. Errors and difficulties were analyzed, enabling recommendations for improvements to be made to the draft standard. Designers appeared to find it difficult to integrate detailed design guidelines with their existing experience. Despite difficulties with guideline interpretation, the resulting interfaces only violated an average of 11% of assessable guidelines. This study suggests that analysts can still perform well despite their misunderstanding of a method. Analysts' existing skills can compensate for limitations in effective discovery resources within a method. This does not appear to hold for CW, however, so the interaction between a method's discovery resources and analyst skill is clearly complex. Research is required to better understand how and why discovery resources can supplement, repress, or have no impact on analysts' ability to find problems

Comparative Studies

Gray and Salzman (1998) were particularly critical of comparative UIM assessments. This is not surprising. It is difficult enough to design and implement assessments of single methods. Comparative assessments of multiple methods merely compound the opportunities for compromising all five forms of validity within one study, and in multiple guises. Existing conclusions about the relative effectiveness of different methods can rarely be trusted, and even where studies were sound, they are often now too old to apply to current versions of UIMs. At best, comparative studies can be used to recruit additional evidence to support indications about effective discovery and analysis resources. This section is written on this basis. For clarity, the mention of a study is not an endorsement of any of its conclusions. The value of these studies is now limited to specific insights that need to be further tested in more rigorous settings with current methods.

Expert Review (HE) versus CW versus Guidelines versus User Testing. This study (Jeffries et al., 1991) compared four usability evaluation methods on a user interface with some challenge (it was not "walk up and use"). A CW (materials used were later published as Wharton, 1992) was performed by a group of three software engineers using tasks chosen by the experimenters. A set of 62 Hewlett Packard guidelines were used by a group of three software engineers. Expert review was performed by four usability specialists; the researchers claim HE was used, but no heuristics were given to the evaluators (according to Karat et al., 1992; Virzi, Sorce, & Herbert, 1993). Finally, a human factors engineer conducted a usability test using six participants. This was a performance test (according to Virzi et al., 1993). Analysts were prompted to record any usability problems even if they did not use their technique but were asked to record how they found a particular problem.

The study claimed that expert review was the best performing method, followed by user testing and last joint guidelines and CW. The study lacks internal and statistical conclusion validity, however. It also lacks effect construct validity, because the simple problem counts for inspection methods were not translated into thoroughness or validity measures. Still, CW's poor performance in this study is consistent with individual assessments of CW, but the study contributes nothing to our understanding of HE as a UIM, because HE was not applied. The main reason for drawing attention to this study is thus to warn practitioners about drawing conclusions from studies that lack validity in almost all guises.

HE versus CW. Another study (Desurvire et al., 1992) compared the effectiveness of CW and HE for evaluating a telephone-based interface. Three different groups of three analysts used each method: human factors experts with more than 3 years HCI experience, software engineers who were the original designers of the system being analyzed, and nonexperts who had some experience with computer systems.

The researchers gave analysts a set of tasks in the form of flow charts. HE analysts first worked independently and then as a group discussed the problems found, either adding or deleting problems. This hints at group merging as a valuable discovery and analysis resource. CW analysts worked in groups and used the Automated Cognitive Walkthrough (Rieman et al., 1991). A problem was only reported if two or more of the group agreed on it. Note that this approach (overly?) amplifies the elimination potential of any analysis resources, in contrast to Nielsen's (1992) multiple analyst approach that dilutes critical resources for suppressing false alarms.

A usability evaluation using 18 users performing the same tasks was conducted. Severity ratings (minor annoyance or confusion, problem-caused error, or problem-caused task failure) and user attitude ratings (still content with system, frustrated with system, or wants to throw system out of window) were recorded for each problem.

The predicted problems were then filtered into problems that did occur in tests, those which did not occur, and statements that were classified as improvements to the interface. Note that a lack of control over user testing means that the second group cannot be assumed to be false alarms, as noted earlier, and thus missed problems could not be reliably analyzed. No validity measures are thus reported.

HE carried out by experts was the best for predicting usability problems observed in tests (thoroughness of 44%), followed by experts using CW (28%). Furthermore, for both methods, experts were better at predicting observed problems than were the software engineers, who in turn were better than the nonexperts. This suggests that typical software developers may bring better discovery resources to inspection than typical end users.

The study also counted recommendations that arose from UIM use. HE analysts suggested the most improvements (24). CW analysts were the next best, but only suggested five improvements. Experts from both groups were best at finding observed severe problems (those causing task failure) but still found only 29% of serious observed problems.

This study reports a narrower gap than most studies on thoroughness between HE and CW, which at its most extreme is an order of magnitude (74% vs. 8%). This may be due to the resources that expert analysts bring to inspection, but without a proper analysis of method application, it is not clear that either HE or CW were actually responsible for any predictions.

Individual HE versus Paired HE versus User Testing. This study (Karat et al., 1992) compared the effectiveness of user testing, individuals using HE, and pairs using HE when evaluating two office systems with graphic user interfaces. Twelve heuristics were used, mostly from the set in Nielsen (1992).

Each evaluation had two sections: self-guided exploration and prescribed scenarios. Analysts, who were either user interface specialists, developers, or end users, only evaluated one of two systems. For each system, six participants were used in a nonstandard user test. The participants used the system and reported the problems they experienced to a human factors professional who was responsible for recording them. For each system, six analysts applied HE individually and 12 analysts in pairs evaluated only one of the two systems.

Problem sets for a particular condition (for example, pairs using HE) were merged and duplicates removed to form problem types. User testing found more than twice the number of problem types found by HE pairs, and more than 3 times the number found by independent HE. Unfortunately, details of individual analyst (pair) performance (e.g., average number of problem types identified) were not reported. Variations in analyst performance with a UIM can provide vital information to practitioners, so it is unfortunate that this information was not reported. The amount of time needed to identify each problem type was calculated by dividing the number of problem types found by a method divided by the time taken for the evaluation. For each unique usability problem type, user testing took nearly half the time for the first system and less than half the time for the second, compared with using HE either in pairs or individually.

The researchers classified the usability problems found using a model of potential usability problem areas: The first system had 47 areas and the second 43. Again user testing was more efficient in locating problem areas than HE, but HE identified a couple of areas that user testing did not find. Once again, however, it is not clear how the simple problem counts should be interpreted. No thoroughness or validity measures are presented for HE, nor was the user testing sufficiently controlled to allow confident coding of false alarms; however, the use of pairs allows issues associated with analysis resources and problem elimination to be explored. Future studies could investigate whether pairs of analysts produce more valid predictions than individuals.

HE versus CW versus Guidelines. Cuomo and Bowen compared guidelines (i.e., Smith & Mosier, 1986), HE, and CW for an interactive Gantt chart prototype. The first part of the study is documented in two separate papers (Cuomo & Bowen, 1992, 1994). As with the previous study, analysis time was included as a dependent variable. Guideline analysis took longest, whereas for this study HE took the least. In the first part of the study, independent analysts identified usability problems. One

analyst used guidelines, with which they were already familiar, so only a checklist was needed. Two analysts independently evaluated the user interface using HE. Two analysts (apparently as a team) used the automated CW (Rieman et al., 1991); one was an independent analyst and the other was a member of the design team for the prototype. All analysts were human factors professionals with experience of usability evaluation.

The usability problems identified were filtered for redundancies, and related instances of problems were grouped into problem types. For example, instances of disabling nonactive button or menu items were grouped as one problem type. Finally, problem types found by the evaluators using HE were grouped together. Guidelines found more problem types (47) than HE (32), which in turn found more problem types than CW (24). Again, although this performance is consistent with results from other multimethod studies, simple problem counts are no substitute for more informative measures of thoroughness and validity. The range of dependent variables does provide some useful insights into each method's effective scope, however.

Predictions were associated with one or more of seven stages of Norman's (1986) theory of interaction and also with one of Smith and Mosier's (1986) six functional areas of an interface. Norman described interaction with a system as having two distinct phases, execution and evaluation, which are in turn split into a number of distinct stages. The execution phases starts with establishing a goal, forming an intention for the goal, specifying an action sequence, and finally executing the actions. Evaluation starts with perceiving the system state, interpreting the state, and then evaluating it with respect to the original goals and intentions. Problem types were mapped to each stage; sometimes a problem type was mapped to more than one stage. Three problem types, which mapped to no stage, were discarded.

Guidelines and HE identified problem types for all stages of interaction, but CW was best at finding action formation related problems, and very poor at finding evaluation related problems—only one such problem was found. Also, problem types identified by CW were rarely found with other methods.

Smith and Mosier identified six functional areas in a user interface (see Table 57.3). The last two were irrelevant for the prototype under study. CW found no problems with the data display and user guidance areas, but HE and guidelines did (overlapping in their results), although most problem types concerned data-entry and sequence control areas. These scoping analyses are useful to practitioners, although there is a risk that the proportions of problem types found are distorted by false alarms. Method scope as indicated by this study should thus be treated with some caution.

In the second part of the study (documented in Cuomo & Bowen, 1994), predicted problem types were associated with results from user testing by looking "for evidence in the usability data that the predicted problem type did indeed cause the users a problem" (Cuomo & Bowen, 1994, p. 100). The authors remarked that "the specific instance of the problem type may or may not have been predicted" (Cuomo & Bowen, 1994, p. 100).

Of those problem types causing noticeable difficulty, HE found the most with 16 problem types (a validity of 46% of its predicted problem types), CW found 14 (a validity of 58%) and guidelines found 11 (22% validity).[4] CW was claimed to be best "at predicting problems that cause users noticeable difficulty (as observed during a usability study)" (Cuomo & Bowen, 1994, p. 86).

All problems identified by user testing were also coded using Norman's model; 71% of problem relating to action formation were predicted, but only between 20 and 33% of problems relating to execution, perception, interpretation, and evaluation. It thus appears that most UIMs are weak at identifying problems relating to the evaluation phase of Norman's theory of interaction.

Overall, the analysis of (un)successful predictions in this study does provide useful indications of gaps in the scope of specific methods, and the benefits of combining methods to achieve better coverage. The study does not, however, inform the effective combination of methods.

Summary: The Need to Improve Method Assessment and Scoping

Existing method assessments are inconclusive. Too many studies report simple problem counts and not enough report reliable thoroughness and validity measures. Still fewer have used a wide enough range of dependent measures to reliably determine the scope of specific UIMs. The urgent requirement in UIM assessment is thus to improve the internal validity of studies (and related construct validity issues), as well as to improve the external validity by applying a relevant range of dependent variables based on informative and useful coding dimensions.

The real test of method effectiveness is some combination of uptake and trust. As already reported, HE was the most used usability method in a recent major survey (Rosenbaum et al., 2000). It was not well trusted, however, such practitioner surveys provide vital triangulation for the results of formal studies. They also present a more positive picture than earlier studies, such as Nielsen's (1995) survey of industrialists about the take-up and perceived usefulness of particular inspection methods or Bellotti's (1990) investigation of the industrial adoption of HCI methods, which pinpointed software developers' concerns about using a particular method and methods in general. However, so much remains to accomplish via formal studies to improve methods that there is still limited value in large evaluations in the field.

CONCLUSION

Inspection methods will always be an important part of the usability tool set. They are an essential complement to user testing, eliminating usability problems before user testing and thus making better use of fewer test participants as well as reducing analysis and reporting time. They would be an equally

---

[4]These figures will not tally with the original number of problem types found because some types are assigned to more than one step in Norman's theory of interaction.

good complement to model-based methods, should these ever improve to the point where they could replace much user testing. Inspection methods could also be used during model development, resulting in better models before applying more expensive analysis methods.

UIMs are thus unlikely to disappear, and their flaws must be addressed, which is not possible until those flaws are better understood. Currently, most assessments of UIMs contain too many flaws to let UIM flaws be reliably revealed. It is not enough to know what percentage of problems were (in)correctly predicted by some analysts when using some method. The reliability of such percentages depends completely on the quality of the method assessment.

Acceptable assessment quality is difficult to achieve. The known problem set must be rigorously derived from user test data, and the limitations of this data in terms of potential missed problems must be clearly stated. Analysts used in method assessments must be adequately trained before using a UIM. Missed problems must be analyzed to reveal the types of problems that fall outside its scope. Successful predictions must be shown to be due to the application of the UIM and not to an analyst's existing knowledge: The UIM must be shown to make a difference. Also, unsuccessful predictions need to be analyzed to reveal risks. It must be further demonstrated that such problems have not failed to find their way into the known problem set due to data analysis errors or flaws in user test design. Finally, it should be clearly unreasonable to match such problems to any problem in the known set. Gray and Salzman (1998) formalized some of these procedural errors as leading to specific validity problems for method assessment.

Few UIMs provide good resources for candidate problem discovery. Only CW could systematically cover the search space, but to do this analysts must have a thorough understanding of the full range of tasks that a system is intended to support. CW does not support such understandings, and thus will only ever be as good as the discovery resources that contextual methods (e.g., Beyer & Holtzblatt, 1998) bring to usability inspection. Critical gaps in task coverage could well explain the poor performance of CW relative to HE. Until UIMs are improved, analysts need to make full use of methods for contextual analysis and system specification to maximize their effective search space during usability inspections. Such *discovery resources* are critical to reducing missed problems in UIM applications. Discovery resources are wide ranging. As well as tasks and scenarios, domain and product knowledge are vital to successful prediction.

For methods such as HE with their high rates of false positives, *analysis resources* are as critical as discovery resources. CW's *success cases*, although vague in formation, structure, and content, do at least let analysts discount the impact of some logically possible problems, thus improving CW's validity. Other methods are less discriminating. Until UIMs are improved, analysts need to make full use of the research literature in HCI, as well as methods for contextual analysis. Knowledge of tasks, scenarios, and the application domain can be used to eliminate bogus predictions, as can more general knowledge of human behavior, especially when interacting with display-based computer systems. On a far more mundane level, analysts with good product knowledge will not claim that implemented features don't exist. Similarly, analysts with good design knowledge will not argue that a problem exists because they mistakenly think that a design should be different.

We are currently at the stage where improvements can be made on analyst preparation and the provision of discovery and analysis resources for UIMs. It will be interesting to see how these develop, and indeed whether they can develop without essentially turning inspection methods into model-based methods. This follows from the nature of desirable discovery resources, which are essentially models of the system, of user interaction, or of its intended context of use. Similarly, as analysis resources become more sophisticated, they are likely to become similar to the analysis procedures applied in model-based methods.

We cannot yet expect major improvements to recommendation generation. Clearly, missed problems must always result in gaps in recommendations. Also, false alarms are a serious nuisance, because recommendations associated with them could result in unwise design changes. The most dangerous flaws in a UIM, however, are expert predictions that are then rationalized to fit the method. What tends to happen here is that a prediction is modified to bring it within the method's scope. This distorts problem reports and in turn results in inappropriate change recommendations. Thus by improving application of methods, we will inevitably improve recommendation generation by reducing missed problems, false positives, and prediction distortion. Only progress here can let UIMs begin to deliver reliable problem sets, and only then can the step from problem confirmation to recommendation generation be systematically studied. Usability practitioners thus currently face dilemmas in UIM choice. A chosen UIM should provide adequate analyst preparation, but there are currently no reliable method comparisons. Clearly, the resources (time, training, developer liaison) needed to apply a method will greatly influence its uptake among usability practitioners. Accurate and reliable estimates are not available for UIMs, however, nor are indications of the relative demands that different methods place on practitioners. Also, although it is clear that all UIMs fail to find many problems, the nature of missed problems for each method has yet to be thoroughly established. This prevents practitioners from choosing a well-informed combination of methods for a particular project, especially when UIMs are known to miss some serious forms of usability problem.

So, *caveat emptor*—let the buyer beware—is the fairest advice that can be given on selecting and applying UIMs. It is not enough to know how to apply a method in outline. Practitioners need much better assessments of UIMs. Until these are in place, practitioners should thoroughly examine the assessments that do exist, bearing in mind any methodological flaws. Even the best current UIM assessments are provisional, and all practitioners must thus think of themselves as researchers who can take little on trust when using UIMs. Every UIM application is an experiment and should be treated as such, so that practitioners can learn from their experiences. There are, as yet, no giants' shoulders to stand on, but every usability practitioner should be capable of strapping together their own scaffolding for effective UIM assessment and application. Such scaffolding must rest on foundations of thorough analyst preparation and be built up through careful exploitation of a wide range of discovery resources. It is made rigid and safe through painstaking application of analysis resources. Without them, the flimsy

tower of wild speculation will come crashing down around the analyst in a heap of invalid predictions. Only with such strong scaffolding in place can analysts reliably recommend repairs for an inspected system. UIMs are currently small components in building and using such scaffolding. Expert analysts and professional development environments provide far more effective resources from their own project knowledge and common sense.

Methods are currently the weakest link in usability inspection. They cannot be expected to bear the full weight of usability engineering and must be combined with other approaches to ensure that usability targets are met. Inspection-based methods are not wholly ineffective, however. If invalid predictions are carefully culled, then the predictions generated by careful and expert method application can provide highly valuable and effective insights into potential usability problems. The authors speak here from their own experience, having regularly produced predictions from inspection of commercial products that were confirmed in subsequent user testing. Nonetheless, much success is due to our insights from extensive assessments of existing methods. Use methods carelessly, and they will let you down. Assess them rigorously, and they will improve with every use. Even the bluntest chisel can be ground into the sharpest tool in the box.

Note: The views and opinions contained in this chapter are those of the authors. In the case of the second author, they do not necessarily represent any official views of Microsoft Corporation.

References

Apple Computer, I. (1992). *Macintosh human interface guidelines.* Reading, MA: Addison-Wesley.

Bastien, J. M. C., & Scapin, D. L. (1995). Evaluating a user interface with ergonomic criteria. *International Journal of Human-Computer Interaction, 7,* 105–121.

Bellotti, V. M. E. (1990). A framework for assessing applicability of HCI techniques. In D. Diaper, D. Gilmore, G. Cockton, & B. Shackel (Eds.), *Proceedings of IFIP INTERACT '90: Human-Computer Interaction* (pp. 213–218). Amsterdam: North-Holland.

Beyer, H., & Holtzblatt, K. (1998). *Contextual design: Defining customer-centered systems.* San Francisco: Morgan Kaufmann.

Cockton, G., & Lavery, D. (1999). A framework for usability problem extraction. In M. A. Sasse & C. Johnson (Eds.), *IFIP INTERACT '99: Human-Computer Interaction* (pp. 344–352). Amsterdam: IOS Press.

Cockton, G., & Woolrych, A. (2001). Understanding inspection methods: Lessons from an assessment of heuristic evaluation. In A. Blandford & J. Vanderdonckt (Eds.), *People and computers XV* (pp. 171–192). Berlin: Springer-Verlag.

Connell, I. W., & Hammond, N. V. (1999). Comparing usability evaluation principles with heuristics: Problem instances vs. problem types. In M. A. Sasse & C. Johnson (Eds.), *IFIP INTERACT '99: Human-Computer Interaction* (pp. 621–629). Amsterdam: IOS Press.

Cook, T. D., & Campbell, D. T. (1979). *Quasi-experimentation: Design and analysis issues for field settings.* Chicago, IL: Rand McNally.

Cuomo, D. L., & Bowen, C. D. (1992). Stages of user activity model as a basis for user-system interface evaluations. *Proceedings of the Human Factors Society 36th Annual Meeting* (pp. 1254–1258). Santa Monica, CA: Human Factors Society.

Cuomo, D. L., & Bowen, C. D. (1994). Understanding usability issues addressed by three user-system interface evaluation techniques. *Interacting with Computers, 6,* 86–108.

deSouza, F. D., & Bevan, N. (1990). The use of guidelines in menu interface design: Evaluation of a draft standard. In D. Diaper, D. Gilmore, G. Cockton, & B. Shackel (Eds.), *Proceedings of IFIP INTERACT '90: Human-Computer Interaction* (pp. 435–440). Amsterdam: North-Holland.

Desurvire, H. W., Kondziela, J. M., & Atwood, M. E. (1992). What is gained and lost when using evaluation methods other than empirical testing. In A. Monk, D. Diaper, & M. D. Harrison (Eds.) *Proceedings of the HCI '92 Conference on People and Computers VII* (pp. 89–102). Cambridge, England: Cambridge University Press.

Dutt, A., Johnson, H., & Johnson, P. (1994). Evaluating evaluation methods. In G. Cockton, S. W. Draper, & G. R. S. Weir (Eds.), *People and computers IX. Proceedings of BCS HCI 94* (pp. 109–121). Cambridge, England: Cambridge University Press.

Franzke, M. (1995). Turning research into practice: Characteristics of display-based interaction. In I. Katz, R. Mack, & L. Marks (Eds.), *Proceedings of ACM CHI '95 Conference on Human Factors in Computing Systems* (pp. 421–428). New York: ACM Press.

Gram, C., & Cockton, G. (1996). *Design principles for interactive software.* London: Chapman & Hall.

Gray, W. D., John, B. E., & Atwood, M. E. (1992). The precis of Project Ernestine, or, an overview of a validation of GOMS. In P. Bauersfeld, J. Bennett, & G. Lynch (Eds.), *Proceedings of ACM CHI '92 Conference on Human Factors in Computing Systems* (pp. 307–312). New York: ACM Press.

Gray, W. D., & Salzman, M. (1998). Damaged merchandise? A review of experiments that compare usability evaluation methods. *Human-Computer Interaction, 13,* 203–261.

Green, T. R. G., & Petre, M. (1996). Usability analysis of visual programming environments: A "cognitive dimensions" framework. *Journal of Visual Languages and Computing, 7,* 131–174.

Jeffries, R. (1994). Usability problem reports: Helping evaluators communicate effectively with developers. In J. Nielsen & R. L. Mack (Eds.), *Usability inspection methods* (pp. 273–294). New York: John Wiley.

Jeffries, R., Miller, J. R., Wharton, C., & Uyeda, K. M. (1991). User interface evaluation in the real world: A comparison of four techniques. In S. P. Robertson, G. M. Olson, & J. S. Olson (Eds.), *Proceedings of ACM CHI '91 Conference on Human Factors in Computing Systems* (pp. 119–124). New York: ACM Press.

John, B. E., & Marks, S. J. (1997). Tracking the effectiveness of usability evaluation methods. *Behaviour and Information Technology, 16,* 188–202.

John, B. E., & Mashyna, M. M. (1997). Evaluating a multimedia authoring tool. *Journal of the American Society of Information Systems, 48,* 1004–1022.

John, B. E., & Packer, H. (1995). Learning and using the cognitive walk-through method: A case study approach. In I. Katz, R. Mack, & L. Marks (Eds.), *Proceedings of ACM CHI '95 Conference on Human Factors in Computing Systems* (pp. 429–436). New York: ACM Press.

Johnson, H. (1997). Generating user requirements from discount usability evaluations. In D. Harris (Ed.), *Engineering psychology and cognitive ergonomics* (Vol. 2, pp. 339-357). Aldershot, UK: Ashgate Publishing.

Karat, C., Campbell, R., & Fiegel, T. (1992). Comparison of empirical testing and walkthrough methods in user interface evaluation. In P. Bauersfeld, J. Bennett, & G. Lynch (Eds.), *Proceedings of ACM CHI '92 Conference on Human Factors in Computing Systems* (pp. 397-404). New York: ACM Press.

Lavery, D., & Cockton, G. (1997a). Cognitive Walkthrough: Usability evaluation materials (Technical Report TR-1997-20). Department of Computing Science, University of Glasgow, Scotland.

Lavery, D., & Cockton, G. (1997b). Representing predicted and actual usability problems. In H. Johnson, P. Johnson, & E. O'Neill (Eds.), *Proceedings of International Workshop on Representations in Interactive Software Development* (pp. 97-108). London: Queen Mary and Westfield College, University of London.

Lavery, D., Cockton, G., & Atkinson, M. (1996a). Analytic usability evaluation materials. Retrieved May 2002 from http://www.cet. sunderland.ac.uk/~csOgco/asp.htm

Lavery, D., Cockton, G., & Atkinson, M. (1996b). Cognitive dimensions: Usability evaluation materials (Technical Report TR-1996-17). Department of Computing Science, University of Glasgow, Scotland.

Lavery, D., Cockton, G., & Atkinson, M. (1996c). Heuristic evaluation for software visualisation: Usability evaluation materials (Technical Report TR-1996-16). Department of Computing Science, University of Glasgow, Scotland.

Lavery, D., Cockton, G., & Atkinson, M. (1996d). Heuristic evaluation: Usability evaluation materials (Technical Report TR-1996-15). Department of Computing Science, University of Glasgow, Scotland.

Lavery, D., Cockton, G., & Atkinson, M. (1997). Comparison of evaluation methods using structured usability problem reports. *Behaviour and Information Technology, 16*, 246-266.

Lewis, C., Polson, P., Wharton, C., & Rieman, J. (1990). Testing a walkthrough methodology for theory-based design of walk-up-and-use interfaces. In J. Carrasco & J. Whiteside (Eds.), *Proceedings of ACM CHI '90 Conference on Human Factors in Computing Systems* (pp. 235-242). New York: ACM Press.

Lewis, C., & Wharton, C. (1997). Cognitive walkthroughs. In M. Helander, T. K. Landauer, & P. Prabhu (Eds.), *Handbook of human-computer interaction* (2nd ed., pp. 717-732). Amsterdam: Elsevier Science.

Mack, R., & Montaniz, F. (1994). Observing, predicting and analyzing usability problems. In J. Nielsen & R. L. Mack (Eds.), *Usability inspection methods* (pp. 295-339). New York: John Wiley.

Martin, J., Chapman, K. K., & Leben, J. (1991). *Systems application architecture: Common user access*. Englewood Cliffs, NJ: Prentice-Hall.

Microsoft Corp. (1999). *Microsoft Windows user experience*. Redmond, WA: Microsoft Press.

Molich, R., & Nielsen, J. (1990). Improving a human-computer dialogue. *Communications of the ACM, 33*, 338-342.

Muller, M. J., McClard, A., Bell, B., Dooley, S., Meiskey, L., Meskill, J. A., Sparks, R., & Tellam, D. (1995). Validating an extension to participatory heuristic evaluation: Quality of work and quality of life. In I. Katz, R. Mack, & L. Marks (Eds.), *Proceedings of ACM CHI '95 Conference on Human Factors in Computing Systems (Conference Companion)* (pp. 115-116). New York: ACM Press.

Nielsen, J. (1990). Paper versus computer implementations as mockup scenarios for heuristic evaluation. In D. Diaper, D. Gilmore, G. Cockton, & B. Shackel (Eds.), *Proceedings of IFIP INTERACT '90: Human-Computer Interaction* (pp. 315-320). Amsterdam: North-Holland.

Nielsen, J. (1992). Finding usability problems through heuristic evaluation. In P. Bauersfeld, J. Bennett, & G. Lynch (Eds.), *Proceedings of ACM CHI '92 Conference on Human Factors in Computing Systems* (pp. 373-380). New York: ACM Press.

Nielsen, J. (1993a). Iterative user-interface design. *IEEE Computer, 26*, 32-41.

Nielsen, J. (1993b). *Usability engineering*. New York: Academic Press.

Nielsen, J. (1994). Enhancing the explanatory power of usability heuristics. In B. Adelson, S. Dumais, & J. Olson (Eds.), *Proceedings of ACM CHI '94 Conference on Human Factors in Computing Systems* (pp. 152-158). New York: ACM Press.

Nielsen, J. (1995). Getting usability used. In K. Nordby, P. Helmersen, D. Gilmore, & S. Arnesen (Eds.), *Proceedings of INTERACT '95: IFIP TC13 Fifth International Conference on Human-Computer Interaction* (pp. 3-12). London: Chapman & Hall.

Nielsen, J., & Molich, R. (1990). Heuristic evaluation of user interfaces. In J. Carrasco & J. Whiteside (Eds.), *Proceedings of ACM CHI '90 Conference on Human Factors in Computing Systems* (pp. 249-256). New York: ACM Press.

Norman, D. A. (1986). Cognitive engineering. In D. A. Norman & S. W. Draper (Eds.), *User centered system design: New perspectives on human-computer interaction* (pp. 31-61). Mahwah, NJ: Lawrence Erlbaum Associates.

Polson, P. G., & Lewis, C. H. (1990). Theory-based design for easily learned interfaces. *Human-Computer Interaction, 5*, 191-220.

Polson, P. G., Lewis, C., Rieman, J., & Wharton, C. (1992). Cognitive Walkthroughs: A method for theory-based evaluation of user interfaces. *International Journal of Man-Machine Studies, 36*, 741-773.

Reed, P., Holdaway, K., Isensee, S., Buie, E., Fox, J., Williams, J., & Lund, A. (1999). User interface guidelines and standards: Progress, issues, and prospects. *Interacting With Computers, 12*, 119-142.

Rieman, J., Davies, S., Hair, D. C., Esemplare, M., Polson, P., & Lewis, C. (1991). An automated Cognitive Walkthrough. In S. P. Robertson, G. M. Olson, & J. S. Olson (Eds.), *Proceedings of ACM CHI '91 Conference on Human Factors in Computing Systems* (pp. 427-428). New York: ACM Press.

Rosenbaum, S., Rohn, J. A., & Humburg, J. (2000). A toolkit for strategic usability: Results from workshops, panels, and surveys. In R. Little & L. Nigay (Eds.), *Proceedings of ACM CHI 2000 Conference on Human Factors in Computing Systems* (pp. 337-344). New York: ACM Press.

Rowley, D. E., & Rhoades, D. G. (1992). The Cognitive Jogthrough: A fast-paced user interface evaluation procedure. In P. Bauersfeld, J. Bennett, & G. Lynch (Eds.), *Proceedings of ACM CHI '92 Conference on Human Factors in Computing Systems* (pp. 389-395). New York: ACM Press.

Sawyer, P., Flanders, A., & Wixon, D. (1996). Making a difference—the impact of inspections. In M. J. Tauber (Ed.), *Proceedings of ACM CHI '96 Conference on Human Factors in Computing Systems— Conference Companion* (pp. 376-282). New York: ACM Press.

Sears, A. (1997). Heuristic Walkthroughs: Finding the problems without the noise. *International Journal of Human-Computer Interaction, 9*, 213-234.

Sears, A., & Hess, D. (1999). Cognitive Walkthroughs: Understanding the effect of task description detail on evaluator performance. *International Journal of Human-Computer Interaction, 11*, 185-200.

Smith, S. L., & Mosier, J. N. (1986). Guidelines for designing user interface software (Technical Report ESD-TR-86-278). Bedford, MA: MITRE Corporation.

Spencer, R. (2000). The streamlined Cognitive Walkthrough method, working around social constraints encountered in a software development company. In R. Little & L. Nigay (Eds.), *Proceedings of ACM*

CHI 2000 Conference on Human Factors in Computing Systems (pp. 353–359). New York: ACM Press.

van Schaik, P., Edwards, J., & Petrie, H. (1995). SATURN Project. Deliverable 3c: An initial evaluation of an existing ATM. Technical Report, Psychology Division, University of Hertfordshire, England.

Virzi, R. A., Sorce, J. F., & Herbert, L. B. (1993). A comparison of three usability evaluation methods: Heuristic, think-aloud, and performance testing. *Proceedings of the Human Factors and Ergonomics Society 37th Annual Meeting* (pp. 309–313). Human Factors and Ergonomics Society.

Wharton, C. (1992). Cognitive Walkthroughs: Instructions, forms and examples (Technical Report CU-ICS-#92-17). Institute of Cognitive Science, University of Colorado, Boulder.

Wharton, C., Bradford, J., Jeffries, R., & Franzke, M. (1992). Applying Cognitive Walkthroughs to more complex user interfaces: Experiences, issues, and recommendations. In P. Bauersfeld, J. Bennett, & G. Lynch (Eds.), *Proceedings of ACM CHI '92 Conference on Human Factors in Computing Systems* (pp. 381–388). New York: ACM Press.

Wharton, C., Rieman, J., Lewis, C., & Polson, P. (1994). The Cognitive Walkthrough: A practitioner's guide. In J. Nielsen & R. L. Mack (Eds.), *Usability inspection methods* (pp. 105–140). New York: John Wiley & Sons.

·58·

MODEL-BASED EVALUATION

David Kieras
University of Michigan

INTRODUCTION

What Is Model-Based Evaluation?

Model-based evaluation is using a *model* of how a human would use a proposed system to obtain predicted usability measures by calculation or simulation. These predictions can replace or supplement empirical measurements obtained by user testing. In addition, the content of the model itself conveys useful information about the relationship between the user's task and the system design.

Organization of This Chapter

This chapter first argues that model-based evaluation is a valuable supplement to conventional usability evaluation and then surveys the current approaches for performing model-based evaluation. Because of the considerable technical detail involved in applying model-based evaluation techniques, this chapter cannot include "how-to" guides on the specific modeling methods, but they are all well-documented elsewhere. Instead, this chapter presents several high-level issues in constructing and using models for interface evaluation and comments on the current approaches in the context of those issues. This assists the reader in deciding whether to apply a model-based technique, which one to use, what problems to avoid, and what benefits to expect. More detail is presented about one form of model-based evaluation, GOMS (goals, operators, methods, and selection rules) models, which is a well-developed, relatively simple and "ready-to-use" methodology applicable to many interface design problems. A set of concluding recommendations summarizes the practical advice.

Why Use Model-Based Evaluation?

Model-based evaluation can be best viewed as an alternative way to implement an iterative process for developing a usable system. This section summarizes the standard usability process and contrasts it with a process using model-based evaluation.

Standard Usability Design Process. In simplified and idealized form, the standard process for developing a usable system centers on empirical user testing of prototypes and seeks to compare user performance to a specification or identify problems that impair learning or performance. After performing a task analysis and choosing a set of benchmark tasks, an interface design is specified based on intuition and guidelines both for the platform/application style and usability. A prototype of some sort is implemented, and then a sample of representative users attempts to complete the benchmark tasks with the prototype. Usability problems are noted, such as excessive task completion time or errors, being unable to complete a task, or confusion over what to do next. If the problems are serious enough, the prototype is revised and a new user test conducted. At some point, the process is terminated and the product completed,

either because no more serious problems have been detected or there is not enough time or money for further development. See Dumas (chapter 56, this volume) for a complete presentation.

The standard process is a straightforward, well-documented methodology with a proven record of success (Landauer, 1995). The guidelines for user interface design, together with knowledge possessed by those experienced in interface design and user testing, adds up to a substantial accumulation of wisdom on developing usable systems. There is no doubt that if this process were applied more widely and thoroughly, the result would be a tremendous improvement in software quality. User testing has always been considered the "gold standard" for usability assessment. However, it has some serious limitations, some practical and others theoretical.

Practical Limitations of User Testing. A major practical problem is that user testing can be too slow and expensive to be compatible with current software development schedules, so a focus of human–computer interaction (HCI) research for many years has been to uncover ways to tighten the iterative design loop. For example, better prototyping tools allow prototypes to be developed and modified more rapidly. Clever use of paper mock-ups or other early user input techniques allows important issues to be addressed before making the substantial investment in programming a prototype. So-called inspection evaluation methods seek to replace user testing with other forms of evaluation, such as expert surveys of the design or techniques such as Cognitive Walkthroughs (see Cockton, Lavery, & Woolrych, chapter 57, this volume).

If user testing is really the best method for usability assessment, then it is necessary to come to terms with the unavoidable time and cost demands of collecting behavioral data and analyzing it, even in the rather informal manner that normally suffices for user testing. For example, if the system design is substantially altered on an iteration, it would be necessary to retest the design with a new set of test users. Although it is hoped that the testing process finds fewer important problems with each iteration, the process does not get any faster with each iteration—the same adequate number of test users must perform the same adequate number of representative tasks, and their performance must be assessed.

The cost of user testing is especially pronounced in expert-use domains, where the user is, for example, a physician, a petroleum geologist, or an engineer. Such users are few, and their time is valuable. This may make relying on user testing too costly to adequately refine an interface. A related problem is evaluating software that is intended to serve experienced users especially well. Assessing the quality of the interface requires an almost complete prototype that can be used in a realistic way for an extended period of time so that the test users can become experienced. This drives up the cost of each iteration because the new version of the highly functional prototype must be developed, and the lengthy training process has to be repeated. Other design goals can also make user testing problematic: Consider developing a pair of products for which skill is supposed to transfer from one to the other. Assessing such transfer requires prototyping both products fully enough to train users on the first, and then training them on the second, to see if the

savings in training time are adequate. Any design change in either of the products might affect the transfer and thus requires a repeat test of the two systems. This double dose of development and testing effort is probably impractical except in critical domains, where the additional problem of testing with expert users will probably appear.

Theoretical Limitations of User Testing. From the perspective of scientific psychology the user testing approach takes very little advantage of what is known about human psychology and thus lacks grounding in psychological theory. Although scientific psychology has been underway since the the late 1800s, the only concepts that user testing relies on are a few basic concepts of how to collect behavioral data. Surely more is known about human psychology than this! The fact is that user testing methodology would work even if there were no systematic scientific knowledge of human psychology at all—as long as the designer's intuition leads in a reasonable direction on each iteration, it suffices merely to revise and retest until no more problems are found. Although this is undoubtedly an advantage, it does suggest that user testing may be a relatively inefficient way to develop a good interface.

This lack of grounding in psychological principles is related to the most profound limitation of user testing: It lacks a systematic and explicit representation of the knowledge developed during the design experience; such a representation could allow design knowledge to be accumulated, documented, and systematically reused. After a successful user testing process, there is no representation of how the design "works" psychologically to ensure usability—there is only the final design itself, as described in specifications or in the implementation code. These descriptions normally have no theoretical relationship to the user's task or the psychological characteristics of the user. Any change to the design, or to the user's tasks, might produce a new and different usability situation, but there is no way to tell what aspects of the design are still relevant or valid. The information on why the design is good, or how it works for users, resides only in the intuitions of the designers. Designers often have outstanding intuitions, and we know from the history of creations such as the medieval cathedrals that intuitive design is capable of producing magnificent results, but it is also routinely guilty of costly overengineering or disastrous failures.

The Model-Based Approach. The goal of model-based evaluation is to get some usability results before implementing a prototype or testing with human subjects. The approach uses a model of the human-computer interaction situation to represent the interface design and produce predicted measurements of the usability of the interface. Such models are also termed *engineering models* or *analytic models* for usability. The model is based on a detailed description of the proposed design and a detailed task analysis; it explains how the users will accomplish the tasks by interacting with the proposed interface and uses psychological theory and parametric data to generate the predicted usability metrics. Once the model is built, the usability predictions can be quickly and easily obtained by calculation or by running a simulation. Moreover, the implications of variations

on the design can be quickly explored by making the corresponding changes in the model. Because most variations are relatively small, a circuit around the revise–evaluate iterative design loop is typically quite fast once the initial model-building investment is made. Thus, unlike user testing, iterations generally get faster and easier as the design is refined.

In addition, the model itself summarizes the design and can be inspected for insight into how the design supports (or fails to support) the user in performing the tasks. Depending on the type of model, components of it may be reusable not only in different versions of the system under development, but in other systems as well. Such a reusable model component captures a stable feature of human performance, task structures, or interaction techniques; characterizing them contributes to our scientific understanding of HCI.

The basic scheme for using model-based evaluation in the overall design process is that iterative design is done first using the model and then by user testing. In this way, many design decisions can be worked out before investing in prototype construction or user testing. The final user testing process is required for two reasons: First, the available modeling methods only cover certain aspects of usability; at this time, they are limited to predicting the sequence of actions, the time required to execute the task, and certain aspects of the time required to learn how to use the system. Thus, user testing is required to cover the remaining aspects. Second, because the modeling process is necessarily imperfect, empirical testing is required to ensure that some critical issue has not been overlooked. If the user testing reveals major problems, along the lines of a fundamental error in the basic concept of the interface, it will be necessary to go back and reconsider the entire design; again, model-based iterations can help address some of the issues quickly. Thus, the purpose of the model-based evaluation is to perform some of the design iterations in a lower cost, higher speed mode before the relatively slow and expensive user testing.

What "Interface Engineering" Should Be. Model-based evaluation is not the dominant approach to user interface development; most practitioners and academics seem to favor some combination of user testing and inspection methods. Some have tagged this majority approach as a form of "engineering." However, even a cursory comparison to established engineering disciplines makes it clear that conventional approaches to user interface design and evaluation has little resemblance to an engineering discipline. In fact, model-based evaluation is a deliberate attempt to develop and apply true engineering methods for user interface design. The following, somewhat extended analogy will help clarify the distinction, as well as explain the need for further research in modeling techniques.

If civil engineering were done with iterative empirical testing, bridges would be built by erecting a bridge according to an intuitively appealing design and then driving heavy trucks over it to see if it cracks or collapses. If it does, it would be rebuilt in a new version (e.g., with thicker columns) and the trial repeated; the iterative process continues with additional guesses until a satisfactory result is obtained. Over time, experienced bridge builders would develop an intuitive feel for good designs and how strong the structural members need to be, and so would

often guess correctly. Time and cost pressures, however, would probably lead to cutting the process short by favoring conservative designs that are likely to work, even though they might be unnecessarily clumsy and costly.

Although early bridge building undoubtedly proceeded in this fashion, modern civil engineers do not build bridges by iterative testing of trial structures. Rather, under the stimulus of design failures (Petrosky, 1985), they developed a body of scientific theory on the behaviors of structures and forces and a body of principles and parametric data on the strengths and limitations of bridge-building materials. From this theory and data, they can quickly construct models in the form of equations or computer simulations that allow them to evaluate the quality of a proposed design without having to physically construct a bridge. Thus, an investment in theory development and measurement enables engineers to replace an empirical iterative process with a theoretical iterative process that is much faster and cheaper per iteration. The bridge is not built until the design has been tested and evaluated based on the models, and the new bridge almost always performs correctly. Of course, the modeling process is fallible, so the completed bridge is tested before it is opened to the public, and occasionally the model for a new design is found to be seriously inaccurate, and a spectacular and deadly design failure is the result. The claim is not that using engineering models is perfect or infallible, only that it saves time and money and thus allows designs to be more highly refined. In short, more design iterations result in better designs, and better designs are possible if some of the iterations can be done very cheaply using models.

Moreover, the theory and the model summarizes the design and explains why the design works well or works poorly. The theoretical analysis identifies the weak and strong points of the design, giving guidance to the designer in where intuition can be applied to improve the design; a new analysis can then test whether the design has actually been improved. Engineering analysis does not result in simply static repetition of proven ideas. Rather, it enables more creativity because it is now possible to cheaply and quickly determine whether a new concept will work. Thus, novel and creative concepts for bridge structures have steadily appeared once the engineering models were developed.

Correspondingly, model-based evaluation of user interfaces is simply the rigorous and science-based techniques for how to evaluate user interfaces without empirical user testing; it likewise relies on a body of theory and parametric data to generate predictions of the performance of an engineered artifact, and explain why the artifact behaves as it does. Although true interface engineering is nowhere as advanced as bridge engineering, useful techniques have been available for some time and should be more widely used. As model-based evaluation becomes more developed, it will become possible to rely on true engineering methods to handle most of the routine problems in user interface design, with considerable savings in cost and time and with reliably higher quality. As has happened in other branches of engineering, the availability of powerful analysis tools means that the designer's energy and creativity can be unleashed to explore fundamentally new applications and design concepts.

THREE CURRENT APPROACHES

Research in HCI and allied fields has resulted in many models of human-computer interaction at many levels of analysis. This chapter restricts attention to approaches that have developed to the point that they have some claim, either practical or scientific, to being suitable for actual application in design problems. This section identifies three current approaches to modeling human performance that are the most relevant to model-based evaluation for system and interface design. These are task network models, cognitive architecture models, and GOMS models.

Task Network Models

In task network models, task performance is modeled in terms of a network of processes. Each process starts when its prerequisite processes have been completed and has an assumed distribution of completion times. This basic model can be augmented with arbitrary computations to determine the completion time and what its symbolic or numeric inputs and outputs should be. Note that the processes are usually termed "tasks," but they need not be human-performed at all, but can be machine processes instead. In addition, other information, such as workload or resource parameters can be attached to each process. Performance predictions are obtained by running a Monte-Carlo simulation of the model activity, in which the triggering input events are generated either by random variables or by task scenarios. A variety of statistical results, including aggregations of workload or resource usage values can be readily produced. The classic SAINT (Systems Analysis of Integrated Networks of Tasks, Chubb, 1981) and the commercial MicroSaint tool (Laughery, 1989) are prime examples. These systems originated in applied human factors and systems engineering and are heavily used in system design, especially for military systems.

Cognitive Architecture Models

Cognitive architecture systems are surveyed by Byrne (chapter 5, this volume). These systems consist of a set of hypothetical interacting perceptual, cognitive, and motor components assumed to be present in the human, the properties of which are based on empirical and theoretical results from scientific research in psychology and allied fields. The functioning of the components and their interactions are typically simulated with a computer program, which effectively produces a simulated human performing in a simulated task environment that supplies inputs (stimuli) to the simulated human and reacts to the outputs (responses) produced by the simulated human. Tasks are modeled primarily by programming the cognitive component according to a task analysis, and then performance predictions are obtained by running the simulation using selected scenarios to generate the input events in the task. Because these systems are serious attempts to represent a theory of human psychological functions, they tend to be rather complex and are primarily used in basic research projects; there has been very limited experience in using them in actual design settings.

GOMS Models

GOMS models are the original approach to model-based evaluation in the computer user interface field; both the model-based evaluation approach and GOMS models were presented as methods for user interface design in the seminal Card, Moran, and Newell (1983) presentation of the psychology of human-computer interaction. They based the GOMS concept on the theory of human problem-solving and skill acquisition. In brief, GOMS models describe the knowledge of procedures that a user must have in order to operate a system. The acronym and the approach can be summarized as follows: The user can accomplish certain *goals* (G) with the system; *operators* (O) are the basic actions that can be performed on the system such as striking a key or finding an icon on the screen; *methods* (M) are sequences of *operators* that, when executed, accomplish a *goal; selection rules* (S) describe which *method* should be used in which situation to accomplish a *goal*, if there is more than one available. Constructing a GOMS model involves writing out the methods for accomplishing the task goals of interest and then calculating predicted usability metrics from the method representation.

There are different forms of GOMS models, systematized by John and Kieras (1996a, 1996b), which represent the methods at different levels of detail, and whose calculations can range in complexity from simple hand calculations to full-fledged simulations. John and Kieras pointed out that the different forms can be viewed as being based on different simplified cognitive architectures, which make the models easy to apply to typical interface design problems and insulate the model builder from many difficult theoretical issues. More so than any other model-based approach, GOMS models have a long and well-established track record of success in user interface design, although they are not used as widely as their simplicity and record would justify. Although still under development by researchers, GOMS models are emphasized in this chapter because in some forms they are a "ready-to-use" modeling methodology. A later section will describe their rationale more completely, but the reader is referred to John and Kieras (1996a, 1996b) for a thorough discussion.

ISSUES IN MODEL-BASED EVALUATION

This section presents several key issues in model-based evaluation that the potential user of these techniques should consider and take into account. In the context of each issue, the three basic approaches to model-based evaluation are commented on as appropriate. Advice is given to both the user of model-based evaluation and the developer of model-based techniques.

Psychological Constraints Are Essential

The concept of model-based evaluation in system design has a long history and many proposed methods (for early surveys see Elkind, Card, Hochberg, & Huey, 1989; MacMillan et al., 1989; Pew, Baron, Feehrer, & Miller, 1977). However, the necessary scientific basis for genuinely powerful models has been slow to develop. The key requirement for model-based evaluation is that building a model to evaluate a design must be a routine, production, or engineering activity and not a piece of basic scientific research on how human psychological factors are involved in a particular computer usage situation. This means that the relevant psychological science must not only be developed first, but also then systematized and encapsulated in the modeling methodology itself. That is, a modeling methodology must provide *constraints* on the content and form of the model, and these constraints must provide the psychological validity of the model as a predictor of human performance. In other words, if the model builder can do essentially anything in the modeling system, then the only way the resulting model can be psychologically valid is if the model builder does all of the work to construct a valid psychological theory of human cognition and behavior in the task and then ensure that the constructed model accurately reflects this theory.

Of course, it takes tremendous time, effort, and training to construct original psychological theory, far more than should be necessary for most interface design situations. Although the decisions in truly novel or critical design situations might require some fundamental psychological research, most interface design situations are rather routine: The problem is to match a computer system to the user's tasks using conventional interface design concepts and techniques. It should not be necessary to be an expert researcher in human cognition and performance to carry this out.

Thus, the key role of a modeling system is to provide constraints based on the psychological science, so that a model constructed within the system has a useful degree of predictive validity. In essence, simply by using the modeling system according to its rules, the designer must be able to construct a scientifically plausible and usefully accurate model "automatically."

A simple series of examples will help make the point: Computer user interfaces involve typing arbitrary strings of text on the keyboard and pointing with a mouse. The time required to type on the keyboard and to point with a mouse are fairly well documented. If task execution times are of interest, an acceptable modeling system should include these human performance parameters so that the interface designer does not have to collect or guess them.

Furthermore, because both hands are involved in typing strings of text, users cannot type at the same time as they move the mouse cursor; these operations must be performed sequentially, taking more time than if they could be done simultaneously. A modeling system should make it impossible to construct a model of an interface that overzealously optimizes execution speed by assuming that the user could type strings and point simultaneously; the sequential constraint should be enforced automatically. A high-quality modeling system would not only enforce this constraint, but also automatically include the time costs of switching between typing and pointing, such as the time to move the hand between the mouse and the keyboard. There are many such constraints on human performance, some of them quite obvious, as in these examples, and some very subtle. A good modeling system will represent these constraints in such a way that they are automatically taken into account in how the model can be constructed and used. Because of the

subtleties involved, computational tools are especially valuable for constructing and using models because they can help enforce the psychological constraints and make it easier for the model-builder to work within them.

A Brief History of Constraints in Modern Psychological Theory

Theoretical constraints are not easy to represent or incorporate; a coherent and rigorous theoretical foundation is required to serve as the substrate for the network of constraints, and suitable foundations were not constructed until fairly recently. Through most of second half of the 20th century, psychological theory was mired in a rather crude form of information-processing theory, in which human activity was divided into information-processing stages, such as perception, memory, decision making, and action, usually depicted as a flowchart with a box for each stage, various connections between the boxes, and perhaps with some fairly simple equations that described the time required for each stage or the accuracy of its processing. However, there was little constraint on the possible data contained in each box or on the operations performed there; a box could be of arbitrary complexity, and no actual explicit mechanism had to be provided for any of them. Such models were little more than a "visual aid" for theories posed in the dominant forms of informal verbal statements or rather abstract mathematical equations. Later, many researchers began to construct computer simulations of these "box models," which provided more flexibility than traditional mathematical models and also contributed more explicitness and rigor than traditional verbal models. Still, the operations performed in each box were generally unstructured and arbitrary.

An early effort at model-based evaluation in this theoretical mode appears in the famous Human Operator Simulator (HOS) system (see Glenn, Zaklad, & Wherry, 1982; Harris, Iavecchia, & Bittner, 1988; Harris, Iavecchia, & Dick, 1989; Lane, Strieb, Glenn, & Wherry, 1981; Pew et al., 1977; Strieb & Wherry, 1979; Wherry, 1976). HOS contained of a set of *micromodels* for low-level perceptual, cognitive, and motor activities, invoked by task-specific programs written in a special-purpose procedural programming language called HOPROC (Human Operator Procedures language). The micromodels included such things as Hick's and Fitts' Law, formulas for visual recognition time, a model of short-term memory retention, and formulas for calculating the time required for various motor actions such as pushing buttons and walking. The effort was ambitious and the results impressive, but in a real sense, HOS was ahead of its time. The problem was that psychological theory was not well enough developed at the time to provide a sound foundation for such a tool; the developers were basically trying to invent a cognitive architecture good enough for practical application before the concept had been developed in the scientific community. Interestingly, the spirit of the HOPROC language lives on in the independently developed notations for some forms of GOMS models. In addition, the scientific base for the micromodels was in fact very sparse at the time, and many of them are currently out of date empirically and theoretically. HOS appears to have been subsumed into some commercial modeling systems; for example, a task network version of HOS is available from Micro Analysis and Design, Inc. (http://www.maad.com/), and its micromodels are used in their Integrated Performance Modeling Environment (IPME), as well as CHI System's COGNET/IGEN produced by CHI Systems (http://www.chiinc.com/).

The task network models also originated in this box-model mode of psychology theory and show it in their lack of psychological constraints; their very generality means they contribute little built-in psychological validity. Even if the HOS micromodels are used, the flexibility of the modeling system means that model builders themselves must identify the psychological processes and constraints involved in the task being modeled and program them into the model explicitly.

Led by Anderson (1983) and Newell (1990), researchers in human cognition and performance began to construct models using a *cognitive architecture* (see Byrne, chapter 5, this volume). Cognitive architecture parallels the concept of computer architecture: a cognitive architecture specifies a set of fixed mechanisms, the "hardware," that comprise the human mind. To construct a model for a specific task, the researcher "programs" the architecture by specifying a psychological strategy for doing the task, the "software." (Parameter value settings and other auxiliary information might be involved as well.) The architecture provides the coherent theoretical framework within which processes and constraints can be proposed and given an explicit and rigorous definition. Several proposed cognitive architectures exist in the form of computer simulation packages in which programming the architecture is done in the form of production systems, collections of modular if–then rules, that have proved to be an especially good theoretical model of human procedural knowledge. Developing these architectures, and demonstrating their utility, is a continuing research activity (see Byrne, chapter 5, this volume); not surprisingly, they all have a long way to go before they accurately incorporate even the subset of human abilities and limitations that appear in an HCI design context.

The psychological validity of a model constructed with a cognitive architecture depends on the validity of both the architecture and the task-specific programming, so it can be difficult to assign credit or blame for success or failure in modeling an individual task. However, the fixed architecture and its associated parameters are supposed to be based on fundamental psychological mechanisms that are required to be invariant across all tasks, while the task-specific programming is free to vary with a particular modeled task. To the extent that the architecture is correct, one should be able to model any task simply by programming the architecture using only task-analytic information and supplying a few task-specific parameters. The value of such architectures lies in this clear division between universal and task-specific features of human cognition; the model builder should be free to focus solely on the specific task and system under design and let the architecture handle the psychology.

Achieving this goal in psychological research is a daunting challenge. What about the practical sphere? In fact, the role of architectural constraints in some of the extant commercial modeling systems is problematic. The task network models basically have such an abstract representation that there is no straightforward way for architectural assumptions to constrain

the modeling system. Once one has opted for representing human activity as a set of arbitrary interconnected task processes, there is no easy way to somehow impose more constrained structure and mechanism on the system. Attempting to do so simply creates more complexity in the modeling problem—the modeler must figure out how to *underuse* the *overgeneral* capabilities of the system in just the right way.

Another commercial system, COGNET/IGEN (see Zachary, Santarelli, Ryder, & Stokes, 2000, for a recent and relatively complete description), is in the form of a cognitive architecture, but a very complex one that incorporates a multitude of ideas about human cognition and performance, so many that it appears to be rather difficult to understand how it works. However, the essence of a cognitive architecture is the insistence on a small number of fundamental mechanisms that provide a comprehensive and coherent system. For example, several of the scientifically successful cognitive architectures require that all cognitive processing must be expressed in the form of production rules that can include only certain things in their conditions and actions. These rules control all of the other components in the architecture, which in turn have strictly defined and highly limited capabilities. These highly constrained systems have been successful in a wide range of modeling problems, so it is difficult to see why a very complex architecture is a better starting point. Again, to be useful in both scientific and practical prediction, the possible models must be constrained—too many possibilities are not helpful, but harmful.

From the point of view of cognitive architectures and the constraints supplied by the architecture, the modeling approaches described in this chapter, as currently implemented, span the range from little or no architectural content or constraints (the task network systems) to considerable architectural complexity and constraints (the cognitive-architecture systems). GOMS models occupy an intermediate position: They assume a simplified, but definitely constraining, cognitive architecture that allows them to be applied easily by interface designers and still produce usefully accurate results. But at the same time, they are less flexible than the modeling systems at the other extremes.

Modeling Cognitive Versus Perceptual-Motor Aspects of a Design

As Byrne (chapter 5, this volume) points out, cognitive architectures have lately begun to incorporate not just proposed cognitive mechanisms, but also proposals for perceptual and motor mechanisms that act as additional sources of constraint on performance. Calling these a "cognitive" architecture is something of a misnomer, because perceptual and motor mechanisms are normally distinguished from cognitive ones. However, including perceptual and motor constraints is actually a critical requirement for modeling user interfaces; this follows from the traditional characterization of human-computer interaction in terms of the interactive cycle (Norman, 1986). The user sees something on the screen if they are looking in the right place and can sense and recognize it, involving the perceptual system and associated motor processes such as eye movements. The user

decides what to do, an exclusively cognitive activity, and then carries out the decision by performing motor actions that are determined by the physical interaction devices that are present; this may also involve the perceptual system, such as visual guidance for mouse pointing.

Occasionally, the cognitive processes of deciding what to do next can dominate the perceptual and motor activities. For example, one mouse click might bring up a screen containing a single number, such as a stock price, and the user might think about it for many minutes before simply clicking on a "buy" or "sell" button. But much of the time, users engage in a stream of routine activities that require only relatively simple cognitive processing, and so the perceptual and motor actions take up most of the time and determine most of the task structure. Two implications follow from this thumbnail analysis.

Modeling Purely Cognitive Tasks Is Generally Impractical. Trying to model purely cognitive tasks such as human problem-solving, reasoning, or decision-making processes is extremely difficult because they are so open-ended and unconstrained (see also Landauer, 1995). For example, there are a myriad possible ways in which people could decide to buy or sell a stock, and the nature of the task does not set any substantial or observable constraints on how people might make such decisions—stock decisions are based on everything from gut feelings, to transient financial situations, to detailed long-term analysis of market trends and individual corporate strategies. Trying to identify the strategy that a user population will follow in such tasks is not a routine interface design problem, but a scientific research problem, or at least a very difficult task analysis problem. Fortunately, a routine task analysis may produce enough information to allow the designer to *finesse* the problem, that is, sidestep it or avoid having to confront it. For example, if one could determine what information the stock trader needs to make the decisions and then make that information available in an effective and usable manner, the result would be a highly useful and usable system without having to understand exactly how users make their decisions.

Modeling Perceptual-Motor Activities Is Critical. A good modeling approach at a minimum must explicitly represent the perceptual and motor operations involved in a task. For most systems, the perceptual and motor activities involved in interacting with a computer take relatively well-defined amounts of time are heavily determined by the system design and frequently dominate the user's activity; leaving them out of the picture means that the resulting model is likely to be seriously inaccurate. For example, if two interface designs differ in how many visual searches or mouse points they logically require to complete a task, the one requiring fewer is almost certainly going to be faster to execute and will probably have a simpler task structure as well, meaning it will probably be easier to learn and less error prone. This means that any modeling approach that represents the basic timing and the structure of perceptual and motor activity entailed by an interface is likely to provide a good approximation to the basic usability characteristics of the interface.

This conclusion is both good news and bad news. The good news is that because perceptual-motor activities are relatively

easy to model, it can be easy to get fairly reliable and robust model-based evaluation information in many cases. One reason why GOMS models work so well is that they allow the modeler to easily represent perceptual-motor activity fairly completely, but with a minimum of complications to represent the cognitive activity. The bad news is that different modeling approaches that include perceptual-motor operations are likely to produce similar results in many situations, making it difficult to tell which approach is the most accurate.

This does not mean continuing the effort to develop modeling systems is futile; rather, the point is that trying to verify or compare models in complex tasks is quite difficult because of practical difficulties in both applied and basic research. Despite the considerable effort and expense to collect it, data on actual real-world task performance are often lacking in detail, coverage, and adequate sample sizes, Even in the laboratory, collecting highly precise, detailed, and complete data about task performance is difficult, and researchers are typically trapped into using tasks that are artificial, performed by nonexpert subjects, or trivial relative to actual tasks. There is no easy or affordable resolution to this dilemma, so the practitioner who seeks to use models must be cautious about claims made by rival modeling approaches and look first at how they handle perceptual-motor activities. The theorist seeking to improve modeling approaches must be constantly iterating and integrating over both laboratory and actual applications of modeling methods.

The Science Base Must Be Visible

Even though the modeling methodology encapsulates the constraints provided by psychological theory, it is critical that the psychological assumptions be accessible, justified, and intelligible. An architecture is the best way to do this, because the psychological assumptions are either hard-wired into the modeling system architecture or are explicitly stated in the task-specific programming supplied by the modeler. The basis for the task-specific programming is the task analysis obtained during the overall design process, and the basis for the architecture is a documented synthesis of the scientific literature.

The importance of the documented synthesis of the scientific literature cannot be overstated. The science of human cognition and performance that is relevant to system design is not at all "finished"; important new results are constantly appearing, and many long-documented phenomena are incompletely understood. Thus, any modeling system will have to be updated repeatedly as these theoretical and empirical issues are thrashed out, and it will have to be kept clear which results it incorporates and which it does not.

The current commercial modeling tools have seriously lagged the scientific literature; although some conservatism would be desirable to damp out some of the volatility in scientific work, the problem is not just conservatism, but rather obsolescence, as in the case of the micromodels inherited from HOS. Perhaps these systems would still be adequate for practical work, but unfortunately, it is difficult to get a scientific perspective on their adequacy because they have been neither described nor tested in forums and under ground rules similar to those used for mainstream scientific work in human cognition and performance. Thus, they have not been subject to the full presentation, strict review, criticism, and evolution that is characteristic of the cognitive architecture and GOMS model work. The practitioner should therefore greet the claims of commercial modeling system with healthy skepticism, and developers of modeling systems should participate more completely in the open scientific process.

The Role of Detail

In the initial presentation above, the reader may have noticed the emphasis on the role of detailed description, both of the user's task and the proposed interface design. Modeling has sometimes been criticized because it appears to be unduly labor intensive, especially with GOMS. Building a model, and using it to obtain predictions may indeed involve substantial detail work. However, working out the details about the user's task and the interface design is, or should be, a necessary part of any interface design approach. If the user's task has not been described in detail, chances are that the task analysis is inadequate, and a successful interface will be more difficult to achieve; extra design iterations may be required to discover and correct deficiencies in the original understanding of the user's needs. If the interface design has not been worked out in detail by the interface designer, the prospects of success are especially poor. The final form of an interface reflects a mass of detailed design decisions; these should have been explicitly made by an interface designer whose focus is on the user, rather than the programmers who happen to write the interface code. So the designer has to develop this detail as part of any successful design effort. In short, using model-based evaluation does not require any more detail than should be available anyway; it just requires that this detail be developed more explicitly and perhaps earlier than is often the case.

Cognitive Architectures Are Committed to Detail. The cognitive architecture systems are primarily research systems dedicated to synthesizing and testing basic psychological theory. Because they have a heavy commitment to characterizing the human cognitive architecture in detail, they naturally work at a extremely detailed level. The current cognitive architecture systems differ widely in the extent to which they incorporate the most potent source of practical constraints, namely, perceptual-motor constraints, but at the same time, they are committed to enabling the representation of a comprehensive range of complex cognitive processes, ranging from multitask performance to problem solving and learning. Thus, these systems are generally very flexible in what cognitive processes they can represent within their otherwise very constrained architectures.

However, the detail has a downside. The cognitive architectures are typically difficult to program, even for simple tasks, and have the further drawback that as a consequence of their detail, currently unresolved psychological issues can become exposed to the modeler for resolution. For example, the nature of visual short-term memory is rather poorly understood at

this time, and no current architecture has an empirically sound representation of it. Using one of the current architectures to model a task in which visual short-term memory appears to be prominent might require many detailed assumptions about how it works and is used in the task, and these assumptions typically cannot be tested within the modeling project itself. One reason why is the difficulty discussed above of getting high-precision data for complex tasks. But the more serious reason is that in a design context, data to test the model is normally not available because there is not yet a system to collect the data with! Less detailed modeling approaches such as GOMS may not be any more accurate, but they at least have the virtue of not sidetracking the modeler into time-consuming detailed guesswork or speculation about fundamental issues.

Task Networks Can Be Used Before Detailed Design. Although model-based evaluation works best for detailed designs, the task network modeling techniques were developed to assist in design stages before detailed design, especially for complex military systems. For example, task network modeling was used to determine how many human operators would be required to properly man a new combat helicopter. Too many operators drastically increases the cost and size of the aircraft; too few means the helicopter could not be operated successfully or safely. Thus, questions at these stages of design are what capacity (in terms of the number of people or machines) is needed to handle the workload and what kinds of work needs to be performed by the each person or machine.

In outline, these early design stages involve first selecting a *mission profile*, essentially a high-level scenario that describes what the system and its operators must accomplish in a typical mission, then developing a basic *functional analysis* that determines the functions (large-scale operations) that must be performed to accomplish the mission and what their interactions and dependencies are. Then the candidate high-level design consists of a tentative *function allocation* to determine which human operator or machine will perform each function (see Beevis et al., 1992). The task network model can then be set up to include the tasks and their dependencies and simulations run to determine execution times and compute workload metrics based on the workload characteristics of each task.

Clearly, entertaining detailed designs for each operator's controls or workstation is pointless until such a high-level analysis determines how many operators there will be and what tasks they are responsible for. Note that the cognitive architecture and GOMS models are inherently limited to predicting performance in detailed designs, because their basic logic is to use the exact sequence of activities required in a task to determine the sequence of primitive operations. Thus, for high-level design modeling, the task-network models appear to be the best, or only, choice.

However, there are limitations that must be clearly understood. The ability of the task network models to represent a design at these earliest stages is a direct consequence of the fact that these modeling methods do not have any detailed mechanisms or constraints for representing human cognition and performance. Recall that the tasks in the network can consist of any arbitrary process whose execution characteristics can follow any desired distribution. Thus, the tasks and their parameters can be freely chosen without any regard to how a human will be actually do them in the final version of the system. Hence this early-design capability is a result of a lack of theoretical content in the modeling system itself.

While the choice of tasks in a network model is based on a task analysis, the time distribution parameters are more problematic: How does one estimate the time required for a human to perform a process specified only in the most general terms? One way is to rely on empirical measurements of similar tasks performed in similar systems, but this requires that the new system must be similar to a previous system not only at the task-function level, but at least roughly at the level of design details.

Given the difficulty of arriving at task parameter estimates rigorously, a commonly applied technique is ask a subject matter expert to supply *subjective estimates* of task time means and standard deviations and workload parameters. When used in this way, a task-network model is essentially a mathematically straightforward way to start with estimates of individual subtask performance, with no restrictions on the origin or quality of these estimates, and then combine them to arrive at performance estimates for the entire task and system.

Clearly, basing major design decisions on an aggregation of mere subjective estimates is hardly ideal, but as long as a detailed design or preexisting system is not available, there is really no alternative to guide early design. In the absence of such analyses, system developers would have to choose an early design based on "gut feel" about the entire design, which is surely more dangerous.

Note that if there is a detailed design available, the task-network modeler could decompose the task structure down to a fine enough level to make use of basic human performance parameters, similar to those used in the cognitive-architecture and GOMS models. For example, some commercial tools supply the HOS micromodels. It is difficult to see the advantage in using task network models for detailed design, however. The networks and their supplementary executable code do not seem to be a superior way to represent task procedures compared to the computer- program-like format of some GOMS models or the highly flexible and modular structure of production systems.

Another option would be to construct GOMS or cognitive architecture models to produce time estimates for the individual tasks and use these in the network model instead of subjective estimates. This might be useful if only part of the design has been detailed, but otherwise staying with a single modeling approach would surely be simpler. If one believes that interface usability is mostly a matter of getting the details right, along the lines originally argued by Whiteside, Jones, Levy, and Wixon (1985) and verified by many experiences in user testing, modeling approaches that naturally and conveniently work at a detailed design level will be especially valuable.

The Value of Generativity

It is useful if a modeling method is *generative*, meaning that a single model can generate predicted human behavior for a whole class of scenarios consisting of just the input events

or specifications for a task situation. Many familiar modeling methods, including the *Keystroke-Level* type of GOMS model, are nongenerative in that they start with a specific scenario in which the model builder has specified, usually manually, what the user's actions are supposed to be for the specified inputs. A nongenerative model predicts metrics defined only over this particular input–output sequence. To see what the results would be for a different scenario, a whole new model must be constructed (although parts might be duplicated). Because nongenerative modeling methods are typically labor intensive, involving a manual assignment of user actions to each input–output event, they tend to sharply limit how many scenarios are considered, which can be risky in complex or critical design problems.

An example of a sophisticated nongenerative modeling method is the CPM-GOMS models developed by Gray, John, and Atwood (1993) to model telephone operator tasks. These models decomposed each task scenario into a set of operations performed by perceptual, cognitive, and motor processors like those proposed in the Card, Moran, and Newell (1983) *Model Human Processor*. The sequential dependencies and time durations of these operations were represented with a PERT[1] chart, which then specified the total task time, and whose critical path revealed the processing bottlenecks in task performance. Such models are nongenerative in that a different scenario with a different pattern of events requires a different PERT chart to represent the different set of process dependencies. Because there is a chart for each scenario, predicting the time for a new scenario, or different interface design, requires creating a new chart to fit the new sequence of events. However, a new chart can often be assembled from templates or portions of previous charts, saving considerable effort (see John & Kieras, 1996a, 1996b for more detail).

In contrast, if a model is generative, a single model can produce predicted usability results for any relevant scenario, just like a computer program for calculating the mean of a set of numbers can be applied to any specific set of values. A typical *Hierarchical Task Analysis* (HTA, see Annett, Duncan, Stammers, & Gray, 1971; Kirwan & Ainsworth, 1992) results in a generative representation in that the HTA chart can be followed to perform the task in any subsumed situation. The forms of GOMS models that explicitly represent methods (see John & Kieras, 1996a, 1996b) are also generative. The typical cognitive architecture model is generative in that it is programmed to perform the cognitive processes necessary to decide how to respond appropriately to any possible input that might occur in the task. In essence, the model programming expresses the general procedural knowledge required to perform the task, and the architecture, when executing this procedural knowledge, supplies all of the details; the result is that the model responds with a different specific time sequence of actions to different specific situations.

For example, Kieras, Wood, and Meyer (1997) used a cognitive architecture to construct a production-rule model of some of the telephone operator tasks studied by Gray et al. (1993).

Because the model consisted of a general "program" for doing the tasks, it would behave differently depending on the details of the input events; for example, greeting a customer differently depending on information on the display and punching function keys and entering data depending on what the customer requires. Thus, the specific behavior and its time course of the model depends on the specific inputs, in a way expressed by a single set of general procedures.

A generative model is typically more difficult to construct initially, but because it is not bound to a specific scenario, it can be directly applied to a large selection of scenarios to provide a comprehensive analysis of complex tasks. The technique is especially powerful if the model runs as a computer simulation; the different scenarios are just the input data for the program, and it produces the predicted behavior for each one. Furthermore, because generative models represent the procedural knowledge explicitly, they readily satisfy the desirable property of models described above: The content of a generative model can be inspected to see how a design "works" and what procedures the user must know and execute.

GOMS MODELS: A READY-TO-USE APPROACH

As summarized above, GOMS is an approach to describing the knowledge of procedures that a user must have in order to operate a system. The different types of GOMS models differ in the specifics of how the methods and sequences of operators are represented. The aforementioned CPM-GOMS model represents a specific sequence of activity in terms of the cognitive, perceptual, and motor operators performed in the context of a simple model of human information processing. At the other extreme of detail, the *Keystroke-Level Model* (Card et al., 1980) is likewise based on a specific sequence of activities, but these are limited to the overt keystroke-level operators (i.e., easily observable actions at the level of keystrokes, mouse moves, finding something on the screen, turning a page, and so forth). The task execution time can be predicted by simply looking up a standardized time estimate for each operator and then summing the times. The Keystroke-Level Model has a long string of successes to its credit (see John & Kieras, 1996a). Without a doubt, if the design question involves which alternative design is faster in fairly simple situations, there is no excuse for measuring or guessing when a few simple calculations will produce a usefully accurate answer.

It is easy to generalize the Keystroke-Level Model somewhat to apply to more than one specific sequence of operators. For example, if the scenario calls for typing in some variable number of strings of text, the model can be parameterized by the number of strings and their length. However, if the situation calls for complex branching or iteration, and clearly involves some kind of hierarchy of task procedures, such sequence-based models become quite awkward, and a more generative form of model is required.

---

[1]The acronym stands for Program Evaluation and Review Technique, which is attributed to the U.S. Navy. According to some sources, this technique, developed during the 1950s, was later combined with the *critical path method* to produce the project planning tool commonly known as PERT charts.

The generative forms of GOMS models are those in which the procedural knowledge is represented in a form resembling an ordinary computer programming language and are written in a fairly general sort of way. This form of GOMS model can be applied to many conventional desktop computing interface design situations. It was originally presented in Card et al. (1983, chapter 5), and further developed by Kieras, Polson, and Bovair (Bovair, Kieras, & Polson, 1990; Kieras & Polson, 1985; Polson, 1987), who provided a translation between GOMS models and the production-rule representations popular in several cognitive architectures and demonstrated how these models could be used to predict learning and execution times. Kieras (1988, 1997a) proposed a structured-natural-language notation, NGOMSL ("Natural" GOMS Language), which preserved the empirical content of the production-rule representation, but resembled a conventional procedural programming language. This notation was later formalized into a fully executable form, GOMSL (GOMS Language), for use in computer simulation tools that implement a simplified cognitive architecture that incorporates a simple hierarchical-sequential flow of control (Kieras, 1999; Kieras, Wood, Abotel, & Hornof, 1995). See Baumeister, John, and Byrne (2000) for a survey of computer tools for GOMS modeling.

Continuing the analogy with conventional computer programming languages, in generative GOMS models, the operators are like the primitive operations in a programming language; Methods are like functions or subroutines that are called to accomplish a particular goal, with individual steps or statements containing the operators, which are executed one at a time, as in a conventional programming language. Methods can assert a subgoal, which amounts to a call of a submethod, in a conventional hierarchical flow of control. When procedural knowledge is represented explicitly in this way, and in a format that enforces a uniform "grain size" of the operators and steps in a method, then there are characteristics of the representation that relate to usability metrics in straightforward ways.

For example, the collection of methods represents "how to use the system" to accomplish goals. If a system requires a large number of lengthy methods, then it will be difficult to learn; there is literally more required knowledge than for a system with a smaller number or simpler methods. If the methods for similar goals are similar, or in fact the same method can be used to accomplish different but similar goals, then the system is "consistent" in a certain, easily characterized sense: In a procedurally consistent system, fewer methods, or unique steps in methods, must be learned to cover a set of goals compared to an inconsistent system, and so it is easier to learn. One can literally count the amount of overlap between the methods to measure procedural consistency.

Finally, by starting with a goal and the information about the specific situation, one can follow the sequence of operators specified by the methods and submethods to accomplish the goal. This generates the sequence of operators required to accomplish the goal under that specific situation; if the methods were written to be adequately general, they should suffice to generate the correct sequence of operators for any relevant task situation. The times for the operators in the trace can be summed, as in the Keystroke-Level Model, to obtain a predicted execution time. Details of the timing can be examined or "profiled" to see where the processing bottlenecks are.

Why GOMS Models Work

The reasons why GOMS models have useful predictive and heuristic power in interface design can be summarized under three principles: The *rationality principle* (cf. Card et al., 1983) asserts that humans attempt to be efficient given the constraints on their knowledge, ability, and the task situation. Generally, when people attempt to accomplish a goal with a computer system, they do not engage in behavior that they know is irrelevant or superfluous—they are focused on getting the job done. Although they might perform suboptimally due to poor training (see Bhavnani & John, 1996) they generally try to work as efficiently as they know how, given the system they are working with. How they accomplish a goal depends on the design of the system and its interface; for example, in a word-processing system, there are only a certain number of sensible ways to delete a word, and the user has some basis for choosing between these that minimizes effort along some dimension. Between these two sets of constraints—the user's desire to get the job done easily and efficiently and the computer system's design—there is considerable constraint on the possible user actions. This means that we can predict user behavior and performance at a useful level of accuracy *just from the design of the system and an analysis of the user's task goals and situation*. A GOMS model is one way of combining this information to produce predicted performance.

Procedural primacy is the claim that regardless of what else is involved in using a system, at some level the user must infer, learn, and execute procedures to accomplish goals using the system. That is, computers are not used purely passively; the user has to do something with them, and this activity takes the form of a procedure that the user must acquire and execute. Note that even display-only systems still require some procedural knowledge for visual search. For example, making use of the flight status displays at an airport requires choosing and following some procedure for finding one's flight and extracting the desired information; different airlines use different display organizations, some of which are probably more usable than others. Because the user must always acquire and follow procedures, the complexity of the procedures entailed by an interface design is therefore related to the difficulty of using the interface. Although other aspects of usability are important, the procedural aspect is always present. Therefore, analyzing the procedural requirements of an interface design with a technique such as GOMS will provide critical information on the usability of the design.

Explicit representation refers to the fact that any attempt to assess something benefits from being explicit and clear and relying on some form of written formalized expression; thus, all task analysis techniques (Beevis et al., 1992; Kirwan & Ainsworth, 1992) involve some way to express aspects of a user's task. Likewise, capturing the procedural implications of an interface design will benefit from representing the procedures explicitly in a form that allows them to be inspected and manipulated. Hence, GOMS models involve writing out user

procedures in a complete, accurate, and detailed form. By doing so in a specified format, it becomes possible to define metrics over the representation (e.g., counting the number of statements) that can be calibrated against empirical measurements to provide predictions of usability. Moreover, by making user procedures explicit, the designer can then apply the same kinds of intuition and heuristics used in the design of software: clumsy, convoluted, inconsistent, and "ugly" user procedures can often be spotted and corrected just like poorly written computer code. Thus, by writing out user procedures in a notation like GOMS, the designer can often detect and correct usability problems without even performing the calculations. This approach can be applied immediately after a task analysis to help choose the functionality behind the interface, as well as to help in the initial design decisions (Kieras, 1997b).

Limitations of GOMS Models

GOMS models address only the procedural aspects of an computer interface design. This means that they do not address a variety of nonprocedural aspects of usability, such as the readability of displayed text, the discriminability of color codes, or memorability of command strings. Fortunately, these properties of usability are directly addressed by standard methods in human factors.

Within the procedural aspect, user activity can be divided into the open-ended "creative" parts of the task (such as composing the content of a document or thinking of the concept for the design of an electronic circuit) on the one hand, and the routine parts of the task on the other (which consist of simply manipulating the computer to accept the information that the user has created, and then to supply new information that the user needs). For example, the creator of a document has to input specific strings of words into the computer, rearrange them, format them, spell check them, and then print them out. The creator of an electronic device design has to specify the circuit and its components to a CAD system and then obtain measures of its performance. If the user is reasonably skilled, these activities take the form of executing routine procedures involving little or no creativity.

The bulk of time spent working with a computer is in this routine activity, and the goal of computer system design should be to minimize the difficulty and time cost of this routine activity so as to free up time and energy for the creative activity. GOMS models are easy to construct for the routine parts of a task, because, as described above, the user's procedures are constrained by the task requirements and the design of the system, and these models can then be used to improve the ability of the system to support the user. However, the creative parts of task activity are purely cognitive tasks, and as discussed above, attempting to formulate a GOMS model for them is highly speculative at best and would generally be impractical. Applying GOMS thus takes some task analysis skill to identify and separate the creative and routine procedural portions of the user's overall task situation.

Finally, although a GOMS model is often a useful way to express the results of a task analysis, similar to the popular Hierarchical Task Analysis technique (Annett et al., 1971; Kirwan & Ainsworth, 1992), building a GOMS model does not "do" a task analysis. The designer must first engage in task analysis work to understand the user's task before a GOMS model for the task can be constructed. In particular, identifying the top-level goals of the user and selecting relevant task scenarios are all logically prior to constructing a GOMS model.

CONCLUDING RECOMMENDATIONS

• If you need to predict the performance of a system prior to detailed design, when overall system structure and functions are being considered, use a task network model.

• If you are developing a detailed design and want immediate intuitive feedback on how well it supports the user's tasks, write out and inspect a high-level or informal GOMS model for the user procedures while you are making the design decisions.

• If your design criterion is the execution speed for a discrete selected task, use a Keystroke-Level model.

• If your design criteria include the learnability, consistency, or execution speed of a whole set of task procedures, use a generative GOMS model such as CMN-GOMS or NGOMSL. If numerous or complex task scenarios are involved, use a GOMS model simulation system.

• If the design issues hinge on understanding detailed or subtle interactions of human cognitive, perceptual, and motor processing and their effect on execution speed and only a few scenarios need to be analyzed, use a CPM-GOMS model.

• If the resources for a research-level activity are available, and a detailed analysis is needed of the cognitive, perceptual, and motor interactions for a complex task or many task scenarios, use a model built with the simplest cognitive architecture that incorporates the relevant scientific phenomena.

References

Anderson, J. R. (1983). *The architecture of cognition*. Cambridge, MA: Harvard University Press.

Annett, J., Duncan, K. D., Stammers, R. B., & Gray, M. J. (1971). *Task analysis*. London: Her Majesty's Stationery Office.

Baumeister, L. K., John, B. E., Byrne, M. D. (2000). A comparison of tools for building GOMS models. In *Proceedings of CHI 2000* (pp. 503–509). New York: ACM Press.

Bhavnani, S. K., & John, B. E. (1996). Exploring the unrealized potential of computer-aided drafting. In *Proceedings of the CHI'96 Conference on Human Factors in Computing Systems*. New York: ACM Press.

Beevis, D., Bost, R., Doering, B., Nordo, E., Oberman, F., Papin, J-P., Schuffel, I. H., & Streets, D. (1992). *Analysis techniques for man-machine system design* (Report AC/243(P8)TR/7). Brussels, Belgium: Defense Research Group, NATO Headquarters.

Bovair, S., Kieras, D. E., & Polson, P. G. (1990). The acquisition and performance of text editing skill: A cognitive complexity analysis. *Human-Computer Interaction, 5*, 1-48.

Card, S. K., Moran, T. P., & Newell, A. (1980). The keystroke-level model for user performance time with interactive systems. *Communications of the ACM, 23*, 396-410.

Card, S., Moran, T., & Newell, A. (1983). *The psychology of human-computer interaction*. Hillsdale, NJ: Lawrence Erlbaum Associates.

Chubb, G. P. (1981). SAINT, a digital simulation language for the study of manned systems. In J. Moraal & K. F. Kraas (Eds.), *Manned system design* (pp. 153-179). New York: Plenum.

Elkind, J. I., Card, S. K., Hochberg, J., & Huey, B. M. (Eds.). (1989). *Human performance models for computer-aided engineering* (Committee on Human Factors, National Research Council). Washington, DC: National Academy Press.

Glenn, F. A., Zaklad, A. L., & Wherry, R. J. (1982). Human operator simulation in the cognitive domain. In *Proceedings of the Human Factors Society* (pp. 964-969). Santa Monica, CA: Human Factors and Ergonomics Society.

Gray, W. D., John, B. E., & Atwood, M. E. (1993). Project Ernestine: A validation of GOMS for prediction and explanation of real-world task performance. *Human-Computer Interaction, 8*, 237-209.

Harris, R. M., Iavecchia, H. P., & Bittner, A. C. (1988). Everything you always wanted to know about HOS micromodels but were afraid to ask. In *Proceedings of the Human Factors Society* (pp. 1051-1055).

Harris, R., Iavecchia, H. P., & Dick, A. O. (1989). The Human Operator Simulator (HOS-IV). In G. R. McMillan, D. Beevis, E. Salas, M. H. Strub, R. Sutton, & L. Van Breda (Eds.), *Applications of human performance models to system design* (pp. 275-280). New York: Plenum.

John, B. E., & Kieras, D. E. (1996a). Using GOMS for user interface design and evaluation: Which technique? *ACM Transactions on Computer-Human Interaction, 3*, 287-319.

John, B. E., & Kieras, D. E. (1996b). The GOMS family of user interface analysis techniques: Comparison and contrast. *ACM Transactions on Computer-Human Interaction, 3*, 320-351.

Kieras, D. E. (1988). Towards a practical GOMS model methodology for user interface design. In M. Helander (Ed.), *Handbook of human-computer interaction* (pp. 135-158). Amsterdam: North-Holland/ Elsevier.

Kieras, D. E. (1997a). A guide to GOMS model usability evaluation using NGOMSL. In M. Helander, T. Landauer, & P. Prabhu (Eds.), *Handbook of human-computer interaction* (2nd ed., pp. 733-766). Amsterdam: North-Holland.

Kieras, D. E. (1997b). Task analysis and the design of functionality. In A. Tucker (Ed.), *The computer science and engineering handbook* (pp. 1401-1423). Boca Raton, FL: CRC.

Kieras, D. E. (1999). *A guide to GOMS model usability evaluation using GOMSL and GLEAN3*. Retrieved from ftp://www.eecs.umich.edu people/kieras.

Kieras, D. E., & Polson, P. G. (1985). An approach to the formal analysis of user complexity. *International Journal of Man-Machine Studies, 22*, 365-394.

Kieras, D. E., Wood, S. D., Abotel, K., & Hornof, A. (1995). GLEAN: A computer-based tool for rapid GOMS model usability evaluation of user interface designs. In *Proceeding of UIST* (pp. 91-100). New York: ACM Press.

Kieras, D. E., Wood, S. D., & Meyer, D. E. (1997). Predictive engineering models based on the EPIC architecture for a multimodal high-performance human-computer interaction task. *ACM Transactions on Computer-Human Interaction, 4*, 230-275.

Kirwan, B., & Ainsworth, L. K. (1992). *A guide to task analysis*. London: Taylor and Francis.

Landauer, T. (1995). *The trouble with computers: Usefulness, usability, and productivity*. Cambridge, MA: MIT Press.

Lane, N. E., Strieb, M. I., Glenn, F. A., & Wherry, R. J. (1981). The human operator simulator: An overview. In J. Moraal & K. F. Kraas (Eds.), *Manned system design* (pp. 121-152). New York: Plenum.

Laughery, K. R. (1989). Micro SAINT—A tool for modeling human performance in systems. In G. R. McMillan, D. Beevis, E. Salas, M. H. Strub, R. Sutton, & L. Van Breda (Eds.), *Applications of human performance models to system design* (pp. 219-230). New York: Plenum Press. (See also Web site of Micro Analysis and Design: http://www.maad.com)

McMillan, G. R., Beevis, D., Salas, E., Strub, M. H., Sutton, R., & Van Breda, L. (1989). *Applications of human performance models to system design*. New York: Plenum.

Newell, A. (1990). *Unified theories of cognition*. Cambridge, MA: Harvard University Press.

Norman, D. A. (1986). Cognitive engineering. In D. A. Norman, & S. W. Draper (Eds.), *User centered system design*. Hillsdale, NJ: Lawrence Erlbaum Associates.

Petrosky, H. (1985). *To engineer is human: The role of failure in successful design*. New York: St. Martin's Press.

Pew, R. W., Baron, S., Feehrer, C. E., & Miller, D. C. (1977). Critical review and analysis of performance models applicable to man–machine systems operation (Technical Report No. 3446). Cambridge, MA: Bolt, Beranek and Newman.

Polson, P. G. (1987). A quantitative model of human-computer interaction. In J. M. Carroll (Ed.), *Interfacing thought: Cognitive aspects of human-computer interaction*. Cambridge, MA: Bradford, MIT Press.

Strieb, M. I., & Wherry, R. J. (1979). An introduction to the human operator simulator (Technical Report 1400.02-D). Willow Grove, PA: Analytics Inc.

Wherry, R. J. (1976). The human operator simulator—HOS. In T. B. Sheridan, & G. Johannsen (Eds.), *Monitoring behavior and supervisory control*. New York: Plenum.

Whiteside, J., Jones, S., Levy, P. S., & Wixon, D. (1985). User performance with command, menu, and iconic interfaces. In *Proceedings of CHI '85*. New York: ACM Press.

Zachary, W., Santarelli, T., Ryder, J., & Stokes, J. (2000). Developing a multi-tasking cognitive agent using the COGNET/IGEN integrative architecture (Technical Report No. 001004.9915), Lower Gwynedd, PA: CHI Systems. (See also Web site for CHI Systems: http://www.chiinc.com/)

BEYOND TASK COMPLETION: EVALUATION OF AFFECTIVE COMPONENTS OF USE

John Karat
IBM T. J. Watson Research Center

INTRODUCTION

Evaluation is a fundamental part of human–computer interaction (HCI) and interface design. We attempt to understand the needs of some audience and then formulate system designs to meet those needs. Along the way, those responsible for the design carry out a wide range of evaluations of their design ideas. These may range from informal discussions in the hallway at work to extensive user based studies (Karat, 1997). Design doesn't just happen at once; it develops as we test ideas and use feedback to inform future design decisions. Although some design decisions are made based on guidance that comes from standards or guidelines that might apply to a range of user tasks and contexts, the main challenge rests in trying to make design decisions that will have a positive impact for a user in a task context. Understanding the impact of any design decision on someone trying to use the system to do something is beyond existing theory or guidelines. The question of how the designer knows any particular design choice is an acceptable choice for a given purpose requires some sort of evaluation.

The focus of almost all evaluation in HCI has been on how well someone can complete a specified task using the technology being evaluated. We can measure time on task, observe error rates, and note task completion rates to provide information about whether the technology is satisfactory for its intended purpose as long as there is a clear intended purpose. Although sometimes it might not be easy to say exactly what the purpose of a system is (e.g., what is the purpose of a word-processing or spreadsheet program), we can generally identify users who have tasks in specified contexts that the system is intended to support. This might involve identifying critical scenarios—those situations which we will make facilitating as much as possible the primary goals in design. Measuring the effectiveness, efficiency, and satisfaction of users trying to carry out tasks in particular contexts is how we define measuring usability (ISO 9241 from the International Organization of Standardization). Although there is an affective component of this measure—user satisfaction—it is generally regarded as a less powerful measure; it is "more subjective" and, in the productivity-oriented end, just not as critical.

Nonetheless, it is increasingly clear that the uses of computing technology are reaching far beyond office productivity tools. Various authors have pointed out this trend (e.g., Furnas, 2001; Grudin, 1990) and suggested that the HCI field needs to adapt techniques to a larger scope of human activities and interests. This includes moving beyond the desktop to considering less productivity-centered aspects of human life such as providing rewarding social interactions and addressing concerns of society as a whole. As this trend continues, I expect we will shift our focus in HCI from people interacting with computers, through people interacting with information, toward people interacting with people through technology. Looking beyond the role of technology to complete a task to technology's role in making a better world takes some expansion of the focus of the HCI field. We need to be able to ask whether people value this technology, not just whether they find this technology useful. Measuring such value goes beyond economic notions. It should consider all aspects of a system that a user might feel makes owning and using it important. Such benefits can be identified, measured, and given a role in HCI design (Landauer, 1995; Putnam, 2000; Resnick, 2001).

As we move forward, the specific goals a user has in mind when approaching the technology will become more varied and less easily identified. Topics such as appearance and aesthetics will play an increasingly important role in HCI design (Karvonen, 2000; Laurel, 1993). HCI professionals will need to be able to respond with something other than a blank stare when asked to assist in the design of artifacts that provide emotionally satisfying user experiences. Style (e.g., form factor, color, and materials used) is becoming increasingly important to the design of standard personal computers. Devices for entertainment that "connect to the net" are becoming widespread. Newer uses of technology call for a better understanding of how people will view the technology as acceptable. It calls for approaches that are not all oriented toward the efficiency of task completion. Questions that we should consider include the following:

- How can we design a user experience that is engaging?
- When will people buy something because of its image?
- How can we measure the social value of technology?

Although it would be helpful if the techniques developed in defining usability engineering could be directly applied to all technology, there are reasons to be aware of the limitations. At the core, usability engineering is oriented to the design of functional tools, and HCI design is rapidly moving into domains other than functional tools. I want to focus on how we need to extend the framework if we want to be a part of the design and development of technology for more varied use. Before we conclude that moving from designing applications that people use to designing things that people value is a simple step, we need to be aware that design approaches other than usability engineering have been involved in designing artifacts for a long time and that "function" is not universally held as the main driver of design.

The HCI field has evolved along with the technology since the 1980s. With the exception of a few visionaries in the field, almost all of the work carried out under the broad banner of HCI has been oriented toward developing and refining technology that serves as a tool for people. The efforts were (and still are) generally oriented at helping people carry out tasks. With this in mind, HCI specialists and technologists shared a common goal: to make the new technology systems as efficient as possible. For the technologists, smaller, faster, and more powerful technology have been the goals. For the HCI specialist, the components of usability, effectiveness, efficiency, and user satisfaction have provided the focus. Within the bounds of work task-oriented systems, there is little different now from 20 years ago in this regard. The fundamental principles of usability engineering put forward in the 1980s (Gould, 1988; Whiteside, Holtzblat, & Bennett, 1988) and the various extensions in the family of user-centered design techniques still provide the best guidance for understanding how to approach the task of making better systems for people. New technology for accomplishing

the same tasks requires creativity in understanding how to match people to the new form factors, but to a large extent the basic work tasks haven't changed. A good task analysis of "creating an office memo" carried out 20 years ago would still provide useful information for usability work on today's (or tomorrow's) office systems.

Two interesting trends have created new opportunities for behaviorally oriented work. One is the movement of the technology out of the workplace and into the home and everyday lives of people. Although we certainly do spend some resource on labor-saving devices for the home such as dishwashers and such, few of us consider such economic issues when making purchases of most of the things we surround ourselves with outside of the work environment. Did you really consider the effectiveness and efficiency of items such as lighting fixtures, furniture, home entertainment equipment when deciding on these purchases? The second trend is to consider a wider role for technology systems in the workplace than we have traditionally assigned to our tools. Can technology enhance collaboration? Can it change what it means to work together on common goals? Can technology improve my quality of life?

These trends bring a new challenge to HCI research and practice. Try as we might, it just isn't possible to always find a measurable objective task to orient our engineering methods toward making easier or more efficient. It will take more than just asking if someone is "satisfied" with an experience to trying to understand more fully why they might "value" it—where value can suggest a number of considerations (e.g., ethical issues in Nardi & O'Day, 1999). What makes someone like playing a game? Why do people spend hours in chat rooms? If we are to consider answering these questions as a part of the HCI field—and this certainly seems like a topic that HCI should consider—we must become better equipped to address the affective side of value. Although we might still have a long way to go in making user-centered design common practice, the basic tools for task analysis and performance measurement are already well covered, as can be seen in many chapters in this book and the general focus in the usability field. It is possible that many in the field feel that there are aspects of personal choice and value that cannot be easily subjected to the usability engineering approaches of task-based systems. I do not argue that furthering the practice of use-based evaluation is the only means for advancing such systems, only that it is an area that could benefit their design and development through "normal science" (Kuhn, 1996). One alternative would seem to be to declare such issues as outside of the scope of the field. This is certainly an option for anyone who feels that the challenges of designing usable functional systems sufficiently challenging. Adopting evaluation techniques to cover affective areas certainly offers significant challenge and demand attention, but the HCI field seems to be the right home among these new challenges, and many within it will be interested in this direction.

LEANING FORWARD AND LEANING BACK

Certainly usability is important. The ability of the user of a system to accomplish a desired goal with acceptable effort is a primary means for determinig how good a system is. Although this is certainly true for tools that we *need to use*, it seems less true for systems that we might *want to use*. The distinction between what someone might need to do and what they might choose to do as a matter of personal choice might not always be clear. Ask yourself if you need a wristwatch or a new car. Now ask if you make decisions about the purchase of these things based on usability alone.

Let me inject a story from my personal experience. Basically, I am a dedicated usability engineer—I believe that good design can result from the hard work of understanding users combined with creative problem solving to try and address the needs of people. I would like to believe that usability engineering is at least sufficient for good design, although I will allow that it might not be necessary (it is difficult to account for random acts of good design or the possibility of insight from other sources). As long as the sort of systems we are talking about are strictly functional—designed to assist in specific tasks and valued by users for their suitability to the task—I think this is an appropriate approach. I waver when I think of the future uses of computing technology, however. I need look no further than my wrist to find an example of technology that I think of as having "good design" but which would not be the product of usability engineering design. I am speaking here of an analog wristwatch, one with two hands for hour and minute designation and no markings on the dial except a gold circle where the 12 would normally appear. I would argue that this watch is adequately functional but certainly less than optimal for a number of time-related tasks (e.g., I can't do the sort of timing that second hands or digital readouts provide for event duration). In fact, the watch costs considerably more than watches with "more functionality." Still, I am a satisfied user. I like my watch.

Is this behavior irrational? Only in a system that does not allow for preferences based on something other than function. Is it idiosyncratic? Not completely because to some extent it is possible for me to find others who share my preference (and also find those who do not). Could I use the principles of usability engineering or user-centered design to produce a design like the one on my wrist? I do not think so.

There is a metaphor that describes two environments fairly well, although not perfectly. It is to think about things that one does with a computer "leaning forward" versus things that one does while "leaning back." In most situations, leaning forward is associated with a task focus, whereas leaning back is associated with recreational or discretionary behavior. When I am at work to accomplish a specific task, I think of leaning forward to complete it. When at home, I lean back and watch television. This is not a perfect metaphor; we can certainly see people playing video games hunched over a system, focused on the goal of the game. But goal attainment in a lean-forward task involves doing something to complete a task. Goal attainment in a lean-back task is more indirect—we don't watch a program or play a game just to complete it; we do so because we are engaged or entertained.

I want to move forward by looking at two primary features of usability engineering: the use of observations of users and use situations as a data gathering activity to inform design and the use of measurable behavioral objectives in design iteration. This addresses the major areas of concern in extending usability engineering to environments in which there is no specific task.

Although we might not think of "leaning back" activities such as creating entertainment "content" or aesthetic design as being influenced by usability engineering, they are generally activities that do attend to their audiences in the creative process. Test screenings are carried out for theater, movies, and television and "user observations" are collected to inform the final product. The presentation of a design concept for an artifact (ranging from a product to a building) is done to judge the adequacy of the current design and contribute to the final version. The main difference is not whether user testing takes place, although this term might not be used in other domains. The difference is more in how the data-gathering activities are set up and what kind of observations are sought. "Experience this artifact and tell me if you like the experience" replaces "Try to carry out this task using this system" as the main guide for what the session will involve. Clear usability objectives—a fundamental part of usability engineering—are rarely present in non-task-oriented artifacts or design.

It might be useful to address some terminology that will be. This chapter addresses *experiences involving technology that people value*. It will be fairly broad in talking about "experiences"—broader than talking about "use." All situations in which *computing technology delivers content which people experience* offer the potential for evaluation to inform the design of the technology. So if someone asks whether this includes the design of a television, the answer is that it does as long as changing the technology can change the experience in some way. It should be clear that the experience is a function of content, technology, and context.

In some cases, we assume that both content and context are givens—that we are focused on evaluating the technology. Some of what is described here addresses specifically trying to understand better the role of content in understanding the value of the experience of it. This can be seen as related to efforts that might be directed at understanding context in determining system usability. Context is used here to include all aspects of the situation in which the experience takes place. "On the subway," "with friends," and "alone on a tropical island" are all elements of context. A criticism of some work in HCI evaluation is that is oversimplifies context and thus can lead to incorrect evaluations. This is certainly a valid concern, and one that is even more important when considering affective components of use. Discussions about content are intended to be of a similar flavor—if you do not examine it, conclusions about technology delivering it can be misguided.

The standard terminology of HCI is to talk about users of systems, user goals, context of use, and usability. For the most part, goals in using systems intended to provide pleasurable experiences rather than to solve problems are not explicitly held by the users of the technology. We can certainly say people are using this because they have the goal of reaching a pleasurable state, but this is awkward and has not proven useful as a guiding approach in design. Partly this is because of the difficulty in objectively defining the goal state; without this, there isn't much we can say about the path to the goal. The real problem is that for usability engineering to work, it needs to be guided by models of goals and how people work to achieve them. Rather than pretending that valuable user experiences are something that we understand like we understand more explicit goal-directed

behavior, user goals will not be defined. The science of understanding enjoyment is not capable of defining a goal-directed approach. What is clear is that people can value experiences that involve technology beyond the contribution of those experiences to task-oriented activities. The focus here will be on the value people place on the experience with a system. Although we might view usability as just a name for the measure of the quality of interaction with a system, it is difficult for people to break from the task-oriented focus unless different terminology is employed. So although addressing how techniques from usability engineering does apply to entertainment, and although it is possible to address enjoyment within the usability framework, different terms will be employed here. The act of experiencing content—whether it is looking at an image, reading some text, or watching a movie—is defined as "valuing" the content. Unless a particular mode is identified, "systems which people value" addresses a wide range of possible reasons (sensory or cognitive) for valuing the experience.

MEASUREMENT TO INFORM DESIGN

An assumption will be carried forward in to the discussion that follows, that is, that we will look for ways in which the design of systems that are intended to be seen as valuable to people can be informed by data gathered from potential users of the technology. This is the central tenant of user-centered design (UCD; Karat, 1997). In this view, the main concern becomes not whether to use potential users of the technology during the design process to help in iterative design, but how to use them and how to use measurement to inform design. The question to be addressed here concerning the involvement of user based data in the design cycle is what kind of guidance is possible for the developers and designers of the next generation of technology that will address issues beyond productivity aids.

There are a number of possibilities. We can state that this is a problem that UCD was not developed to handle. In its original formulation (Norman & Draper, 1986) and in work that followed (Whiteside et al., 1988), there is a clear focus on providing measurable objectives for measuring design progress and success. Although the examples provided are all drawn from the class of applications known as "productivity applications," it is not clear whether the procedures outlined can be extended to systems that people value independently of their productivity value. The method has some basic guidance—determine what the system is for, set measures of usability for those tasks, and then iteratively design and evaluate to those criteria. This is illustrated in Fig. 59.1. Not fundamental to the method, but a major practical factor, is that most of the systems that we have designed in the past 20 years are oriented to improving on the productivity of past systems as the primary measurable criteria. Although we might not have done as well as we might in following this guidance (Landauer, 1995), we still are aware of the basic directions. We examine *tasks* people carry out and aim to improve them by decreasing effort to carry them out. At issue is whether we can find an analog for tasks in which there is no specific goal to be achieved and for which eliminating user effort might not be a desirable goal.

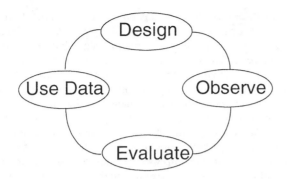

FIGURE 59.1. Design iteration cycle.

One position to take is that it is not possible to directly extend the method, that systems which people value are not subject to the same iterative process that systems with usability goals are. It is easy to find examples of artifacts that are not iteratively designed with user input in the traditional UCD sense—works of art and culture, for example. Picasso and Beethoven did not sit down with representative user samples before creating their next works, although they may have been influenced by collegues and others, They, and other artists and craftspeople, have certainly listened to their audience, customers, or patrons but not in the joint design sense that we associate with UCD or participatory design. Could design in music, art, or architecture proceed in such a fashion? Perhaps, but the fact is that it hasn't. In fact, there is a strong culture of individual creativity in many, if not most, of the traditional design fields. Is HCI design different because we discovered something in computer systems technology that other fields have not known? The answer is most likely no. We might have formulated it well to suit our design environment—the kind of systems we build, the kind of culture associated with engineering design, and the kind of pace at which we design work systems. UCD is not something new. Instead, HCI design is moving from a domain that is primarily function driven in which the guidance of our methods clearly applies toward a domain in which the applicability is less obvious but largely applicable. Furthermore, it is less obvious precisely because we have less clear guidance on what constitutes success.

It is important to distinguish between not having clear objectives and having no possible clear objectives. Not having clear objectives is often associated with design failure—we can't design successfully because we are not sure what we want to design. This suggests a lack of effort in trying to identify objectives. For systems that people value, it is not that we are unclear about the objective, it is that we are unsure how to go about reaching it. We can learn a lot by looking; a closer examination of people and what they value in life can lead to a better sense of how to address those needs. This might be thought of as simply a "task analysis will prevail" view. If we look closely enough at what people think of as contributing to their quality of life, we can figure out how to measure it—perhaps creatively—in ways that will facilitate informing design in much the same way as we track productivity in design now. There is a bit of experience and a bit of blind faith in this statement.

In domains in which productivity was not the primary objective, basic UCD processes have proven useful. There have been failures and uncertainty about how to proceed in such areas. The advice of contextual design (Holtzblatt & Beyer, 1998) is even more important in these domains, and the contributions of standards, guidelines, and heuristics may be even more limited.

There are alternatives to this optimistic view about the adaptability of user-based techniques. We could view the creation of artifacts that people value as an artistic endeavor, one that either cannot or should not be viewed as subject to engineering design (Wroblewski, 1991). It is not claimed that all design should follow UCD principles, at least not at this stage in the maturity of these principles. There are fields with rich design traditions that are far closer to craft than to engineering (e.g., see the discussion in Schon, 1983). One point to keep in mind in discussing different approaches, though, is that design as craft and design as data-based usability engineering are not necessarily in opposition. Whether design happens in the head of one creative individual or in a team, it represents a synthesis of data observed in the world. Just because the data was explicitly sought (as in UCD) or not (as perhaps in artistic design), the designer still uses observations to drive design. There is still a great deal to be learned about what makes for good design, and good designers that can inform design practice.

The studio model of design practiced in fields such as architecture or music is a legitimate competing model to UCD for developing systems that people value. It is completely possible that training in "sensitivity to the form" is at least a partner to "training in the ability to observe" (Schon, 1983). The engineering-oriented designer might ask, "Which technique is more effective?" That is, which technique produces the better result with a higher probability? Experiments to address this issue might be conceived but would seem impractical to carry out. The logistics would be simply overwhelming.

EXPERIENCES PEOPLE VALUE

Without trying to be complete in discussing what people might value outside of tools for helping in carrying out tasks, I identify several general classes of systems for which the nature of HCI activities can be discussed that might be useful in the design of. systems people value.

First, there are systems for which technology can deliver content that people find valuable. At the most general level, we can view a Web browser as this sort of system. Although the Web can certainly be viewed as a system that helps people complete "tasks" such as obtaining driving directions or finding out what the weather will be like this week in Paris, it also can be viewed as providing content that people simply enjoy browsing. For these types of systems, the two major features that one can explore with HCI methods are the simplicity of the interactions and the quality of the content. Simplicity of interaction is a fairly common theme in HCI and will not be covered in this chapter. Quality of content—from the perspective

of providing valuable experiences for the reader or viewer—is a different issue. Quality of content is not generally explored within task-based HCI (outside of education applications) but is something that is easily accommodated within behavioral observation methods employed in HCI. Here it seems that the value comes primarily from *what* the system gives the user access to, and these systems will be called *content* driven.

Second, there are systems that support activities people find valuable and will use without strong task-based requirements. The need to communicate is an example of this type of activity, and e-mail and chat systems will be addressed. Certainly we can look at e-mail and chat systems through a task-oriented perspective, and we have done this in the HCI field. In doing so, however, we are missing the point that these systems have a great deal of use outside of the task perspective. E-mail has enabled new forms of social communication and provided use that cannot be accounted for in terms of goal achievement. Chat systems have become wildly popular, and although we are certainly finding business uses for them, their popularity rises from their support for social activity. A system that is developed to support social communication might be a lot more powerful than one designed to support a particular work communication task. Systems that can fulfill other basic human needs could be researched in similar ways. Here it seems that the value comes primarily from *who* the system gives the user access to, and these systems will be called *communication* driven.

Third, there are systems that derive value from interaction and the results the interaction can produce. Games are the most common technology artifact, but the piano or other musical instrument may be more of the extended prototype. The goal might not necessarily be ease of use; there is something in acquiring a skill that many people value. Mastery of something complex can be a goal people will work for, whether it is being able to execute a triple axle, play a Beethoven concerto, or blast the enemy in a game to smithereens. Again, although HCI techniques might not generally be seen as employed in the design of such systems, it is easy to see how they might be. Here it seems that the value comes primarily from the *interactive experience* the system gives the user access to, and these systems will be called *experience* driven.

All three of these domains offer challenges appropriate to the skills of an HCI practitioner or researcher. Whether it is in the area of understanding the basic phenomena of why these systems are valued when their purpose is less obvious than our transaction-oriented approaches might assume or in iteratively trying to design and develop systems which "meet needs." It is not important that the classification presented here be complete or nonoverlapping. For example, interactive dolls or animals can be seen as artifacts that are valued as surrogate communicative playmates or as interactive toys. Advice will not be presented in separate sections for each of the three cateories because the HCI field has not yet sufficiently differentiated HCI methods which would call for applicability to one category (say content driven) over another (say social). In all cases, the advice is similar: Consider what the value might be, observe people in the appropriate class of interaction, and use the observations to inform future design decisions.

What to Measure?

How can we measure the value of a user experience? The problem is related to how we measure any sort of behavioral item. There are aspects of human behavior that lend themselves to measurement because they can be directly observed—things such as task completion or the amount of time spent on a task. Affective measures—such as user satisfaction or value of an experience—are not among these items. Measurement relies on indirect measures—techniques such as interviews or surveys— or on psychometric measures that are either known to be or can be shown to be associated with the quality we seek to measure. Just because these reactions are not easily measured does not mean that they are not real or that they are somehow random. We know that emotional reactions in humans are very real. We also know that they can be very reliable, that there are stimuli (like a laughing or a crying baby) that will produce consistent responses in the population. Like many of the more easily measured attributes of behavior, we also know that there are individual differences in reactions or evaluations of value. Again, this variability should not be confused with randomness or unpredictability.

Although some work has been done in measuring or monitoring arousal states through various physiological measures such as heart rate or skin resistance, these will not be discussed in detail here. The focus is on measures that in one form or another rely on self-reports from the holder of the experience. These can be questionnaires, in-context observations, or interview sessions. The relative merits of the various techniques will not be discussed here because there is evidence that all can be useful in the domain of measuring the value of a user experience. It is suggested that practitioners select a particular tool based on the demands of the situation and that, if possible, data are collected using two methods to achieve cross-validation.

It is interesting to look at some products outside of the technology sphere and ask about their methods for gathering data to inform design. One could argue that there are many consumer products that aim for success by providing a valuable user experience—not simply to succeed by providing a needed product or function. In product categories from automobiles to snack foods, one can find the use of self-reports and interviews with a range of clever observation techniques. For example, consumer product manufacturers have monitored shopper behavior in stores with cameras to see how long people stop in front of a display as a way of determining whether products grab the attention of shoppers. In general, however, it is still difficult to determine whether a product would be considered as giving the user a valuable experience without observing the actual use of the product. "Grabs attention" or "interests" might be reasonable goals for a marketing campaign, but it is unclear what role these would play in an overall effort to provide people with valuable experiences. Although companies such as Proctor and Gamble use homelike settings to evaluate products, indirect measures are still relied on to a great extent. The benefits might be substantial, but the logistics and costs of observation in context are just too great to make it the technique of choice. Observing a few people in real-life settings coupled with other

methods of data collection might be the cost-effective compromise in many situations.

Another place to look for experiences in evaluation of an experience is to the entertainment industry. The question "What do people enjoy?" is certainly key here, and the industry looks not only at the content (e.g., the movies) but also at the setting (the theaters themselves) as components of this experience. A common technique here is the use of screenings before a movie is released, where people are invited to see a movie and asked to provide feedback (generally in filling out a questionnaire) following the viewing. Such methods can actually provide information about the viewers experience, in contrast to monitoring systems for television viewership that simply monitor what people are watching and draw conclusions from this data about what they might have enjoyed.

Other Challenges in Adopting Affective Evaluation to Technology Design

There are some other issues to keep in mind in evaluating user experience that might seem counterintuitive to the practitioner who has worked long under the goal of "making it easy." There are many activities in which people engage where acquiring and demonstrating skill can be important parts of the experience. Games that are trivial pass from being interesting to boring. Consider the game of tic-tac-toe (also known as naughts and crosses). For children, or generally for people unfamiliar with the game, the rules are simple. Interest in the game—and the value to someone of playing it—can vanish once one has learned the basic strategies (this is a game that will always result in a draw between two moderately skilled players). Other domains to consider are those of more complex games (such as chess) or of musical instruments, domains that share the property that it takes years (even decades) to master the skill and yet people do persist with such "hard-to-learn" devices which might deliver little functional value. Understanding the balance between ease of use and the value of possessing a skill poses an interesting challenge to HCI in the future.

As long as the technology we are considering has some form—from a desktop system to a computer embedded in jewelry (Mountford, 1994; Picard, 1997), another factor to consider is the value of the aesthetics of the form. Visual appearance is the most obvious form to consider, but people have also considered other sensual dimensions such as sound, touch, and even smell. That form is important is not new; functionally oriented tools are designed with appearance as a part of usability, even if this is mostly considered in the second-class satisfaction measurements. But one important point here is that aesthetic elements of a system (beyond those subject to industrial engineering guidelines or brand appearances) are rarely addressed in the same design tradition as usability issues. One comparison that has been made between the HCI design tradition and the visual design traditions is in the area of accountability: How do we judge if it is good? William Gaver (2000) referred to HCI design as driven by "empirical accountability"—evaluation by data about the fit of the system to the task. He went on to describe visual design as driven by "peer accountability"—evaluation by reference to

what other experts think of the design. In some way, of course, all product design is driven by marketplace accountability: What people will eventually purchase? But the fact is that we first design systems and then see how they do in the marketplace—we integrate in our processes for designing things we think will relate to market success. For HCI and visual design, these are often different criteria. As we merge technology and aesthetics, we are likely to experience some gaps in communication between different design traditions.

A final point to mention here is the consideration of impact on the overall quality of life. This is an issue often considered beyond the scope of a project when we are considering productivity tools because we have economic incentive to make work tasks easier and faster. Even when there are possible quality-of-life issues in workplace technology (such as impacts on the skill level or number of people employed), they are rarely attended to outside of countries where worker participation in design of technology that impacts them is required by law. The issue of whether the greater good is served will play a larger role when we address developing technology that is not required by work, but that influences what we do with our free time. The focus is not on the philosophical issues here, for the responsibility of those who develop technology to the people of the world is an important topic addressed elsewhere and is in the domain of codes of ethics of professional societies like the Human Factors and Ergonomics Society or the Association of Computing Machinery. Approaching the development of discretionary technology with the assumption of "if people will buy it, it must be good" is simplistic. The HCI field should debate and consider seriously quality-of-life impacts related to new technology (Kubey & Csikszentmihalyi, 1990).

AN EXAMPLE—HCI AND ENTERTAINMENT

Content systems will now be discussed as an example of the content, communication, and experience categories defined above. Content-driven systems are those valued for the content that they can provide access to, more than for their own features. The paradox of the video cassette recorder (VCR) helps illustrate a point about the distinction between usability and value. People in the HCI field have joked for years about the difficulty of using VCRs. Still, people buy them in large quantities. This is clearly not for the joy of playing with the buttons or trying to figure out how to use the features. People own VCRs because they give them access to content they enjoy. Although there are certainly educational videos that might provide task-oriented training, VCR purchase is likely driven far more by entertainment goals than by practical task-oriented goals.

There are three primary aspects that need to be considered: the quality of the content itself, how the content might be accessed, and the experience of browsing, viewing, or reading the content in a possibly new context. These are the primary aspects that need to be considered from a design and evaluation perspective to understand why people would value a content system. Content quality issues are discussed by focusing on the how various content might be seen as valuable to someone when viewed in traditional passive modes such as reading and

viewing or watching (Yamaguchi, Hosomi, & Miyashita, 1997). Because a great deal of current and proposed technology deals with simply "delivering content," the contribution of the content to a valuable user experience and how HCI techniques can be employed in content must be addressed.

Access to content is discussed by focusing on new technologies for accessing traditional media. Issues here are more in line with traditional HCI approaches, in which "see the picture" might be seen as an explicit user goal, and we might simply focus on making the viewing task easy. The point here is that simple access is not the only story in looking to support valuable user experiences. Going to the movies is a part of enjoying a movie. The value does not always come from the content alone. This will lead in to the third issue—the discussion of content integrated with viewing context. The focus is on viewing systems for pleasurable experiences in as much a complete system context as possible. When people are looking for pleasurable experiences, they are often looking beyond a single content or access element and are also looking for social interaction. Although this certainly complicates the design problem, it highlights the full range of how such problems need to be addressed and the perils of simplification of the context particularly in discretionary use situations.

Content Quality

What makes a movie, play, book, work of art, or other item something valuable to someone to experience? Is it within the domain of the HCI designer to consider the quality of the content? To some extent, particularly the wide world of preexisting content in the form of movies, text, and images, the user experience designer simply takes the content for granted as a starting point. For some HCI work, work in which the practitioner is responsible for a system that delivers preexisting content, this is the correct thing to do. If people value access to content, and technology can provide them easier or more convenient access, then focusing on access and leaving the content to stand on its own is a reasonable approach. There is plenty to do in designing easy-to-use technology for new ways of accessing existing content.

But we often overlook the impact of the technology on the perceived quality of the content. In the world of such new technology as electronic books and multimedia computers, it is perfectly reasonable to ask about what makes good content for different viewing technology. There seem to be two basic questions. First, we can ask about evaluating content—answering the question about whether something is perceived as worthy of attention or not. Clearly, there are tools used in HCI that can be useful for this purpose. These include field and laboratory observation methods and techniques such as questionnaires and focus groups. In selecting exactly how one will observe and what one wants to look for, the first advice that is necessary to give is to consider how you want to operationalize "quality" or "value." It is easy to become discouraged if you are tied to objective measures, but can be viewed as an exciting challenge if you view the task as attempting to add some clear definition to a soft attribute like value. Although "goal attainment" is not obvious if we try to observe someone seeking to be entertained, we

can generally operationalize our measurements of it for specific purposes. We can measure the time someone spends looking at something as a measure of interest. We can count the number of laughs or tears shed watching a movie. We can ask directly if an experience was enjoyable. No specific recommendation is offered here for whether to use contextual observation or questionnaire as the method for data colleciton. Both of these general methods have value for exploring content quality. Generally contextual methods have a greater role in early evaluation and survey methods in later stages. Techniques for collecting these kinds of data are discussed in more detail in other chapters of this book.

Second, we can ask whether there is design advice that can be given to content designers by HCI research. Here we look for practical theories related to human emotional states. Rather than detailing the range of research activities on this topic, we can summarize by saying that they are a long way from being in a form that would provide practical guidance for content designers. Laurel's work (1993) is one exception to the case in this area. It is useful to now examine the general topic of entertainment and mention the sort of areas and guidance that one might expect.

Defining entertainment value is by no means an easy task. Although the term is often used in everyday language, it is actually difficult to define (see Dyer, 1992, for a discussion about the issue). Entertainment has been described as being "any activity without direct physical aim, anything that people attend to simply because it interests them" (Langer, 1953). For the purposes of the discussion here, and in looking at how the methods of the HCI field can be used to address artifacts designed to be "entertaining," we can say that people are entertained when they are voluntarily undergoing an experience that interests them and gives them some amount of satisfaction.

Using this description, we can see that entertainment covers a large number of activities, many of which can be seen in applications in common use such as games or even chat programs. For example, using technology to enhance storytelling and listening is one are that has received some attention (Pausch, Snoddy, Taylor, Watson, & Haseltine, 1996). And although making learning fun has been a goal of educators for a long time, there is still relatively little research contributing a solid understanding of how to do this (see Malone, 1980, and Jacques, Preece, & Carey, 1995, for good examples of work in this area). Talking and gossiping have a forum in electronic chat rooms; reading news on the Web is becoming increasingly popular; the previously solitary video game experience has found new meaning in the networked game era; shopping has reached major proportions on the Web, newly augmented by the thrills of online auction; and the growth of gambling and pornography industries on the Web seem to be stoppable only by legislation.

A major element of the entertainment topic and its relation to HCI would seem to be whether the artifact is interactive or primarily passive. The distinction is not pure; content such as a DVD movie can certainly be interacted with as part of the experience, and games can be appealing because of the graphics. But it is helpful to make the distinction between "watching" and "interactive" appeal in examining the role of HCI in entertainment applications.

There has been some effort to advance engagement—holding the attentional resources of the viewer—as an important concept for design of multimedia systems, particularly in education (e.g., Chapman, Selvarajah, & Webster, 1999; Jacques et al., 1995; Malone & Lepper, 1987; Webster & Ho, 1997). Much of this work is aimed at tapping into the effort that people are willing to give to activities such as games. Certainly if we can make required activities—such as learning in school—more enjoyable, we might find that people learn better. These efforts have met with mixed success, and I will not attempt to address them here. For the most part, their goal is to provide more attractive interfaces for task-oriented systems, and while they might succeed to some extent, they also might fail because they are diverting by trying to mix "business and pleasure." If there are task-related benefits that occur in entertaining activities, such as incidental learning that might take place while viewing a movie or play, the task behaviors tend to become the major focus. For the study and advancement of positive experiences with technology, the role of engagement in education must be a major focus.

This is a brief overview of the value of passive experience of content such as reading or viewing content. It serves as an invitation to HCI professionals to consider this topic as open to contributions and to suggest that content is not something that is outside of the design of the total user experience.

Access to Content

At some level, it could be easy to think of the role of HCI in content access as simply to make the task of getting the content as usable (meaning effective, efficient, and satisfactory) as possible. Accessing the content is the assumed goal, and making it easy is the usability engineer's task. Certainly, there may be times when this is an appropriate approach to take—after all, we are primarily interested in the music not the player, right? Not always. Certainly we don't want the access device to get in the way of our enjoyment, but sometimes it can be a part of it. When all mechanisms for access are essentially equally easy and functional, and sometimes even when they aren't, aesthetic attributes of the content player can become distinguishing features that people value. The appropriate metric for looking at access for nontask systems is whether the interaction seems engaging to the user rather than whether it is easy.

An example for the audio product world illustrates this point. Stereo equipment is not highly differentiated by ease-of-use considerations. Although there is a fair amount of product differentiation in the labeling and placement of controls, the fact that all systems have very similar controls for playing the tapes, disks, and the like. People are also attracted to a particular device by other factors such as price, technical performance, and design. There is a line of products, for example, that do differentiate themselves in design in general and user experience in particular—the products of Bang and Olafeson. The issue isn't that everyone demands this particular brand; price factors keep this from being the case at a minimum, and the design style might not be for everyone. The point is that this is a product that offers a different user experience, and it is one that some people value enough to pay a substantial premium on the base value of the technology. Such designs do not emerge from a focus on efficiency or traditional usability. They emerge from a culture aimed at producing things that people will value. Does Bang and Olafeson use user-centered design? Probably not by that name; actually, they operate largely through the designs of a single individual. As an organization, they are aware of the value they are trying to create and that they take steps to advance it that can be seen as "evaluation methods"—sessions in which user reactions to design ideas are examined in detail much greater than a check on a satisfaction scale.

There are lots of ways to structure such evaluations—in focus groups, in home settings, in product laboratories—but all require two leaps for the usability engineer. First, we must get over a fundamental mistrust of subjective data. Second, we must invest in more than a surface picture of satisfaction. It is difficult to iterate on design—to know how to improve on a previous version—if all we have is a number that represents a user's satisfaction with a particular instance of design. This requires a "craft" view of the design process to supplement the engineering view (Wroblewski, 1991).

Aesthetics should not play the only role in designing systems for access to content, but affective components do have a role here beyond the normal usability perspective. Work is work, and play is an activity in which people are generally freer to make decisions on feelings rather than efficiency. This doesn't mean that people will give priority to other design features, but it does mean that they are more likely to do so than in situations in which a specific goal is the reason for using a system.

User Experience of Content in a Context

If everyone could see or hear all of the music, plays, sporting events, and cultural exhibits that they wanted in the comfort of their home, would anyone go to the movie, to a museum, or to a stadium to experience these things? Is smelling the popcorn or hearing the crowd roar part of the experience that makes an event enjoyable?

Any technology use takes place in some context. By far the most pervasive assumption that we have had in HCI in designing our tools is that they are used by someone in a nice quiet office. As we moved from individual work tools to organizational work environments, the focus on computer-supported cooperative work has at least expanded our boundaries of consideration to include other people in a work environment as part of the context of use. For systems to fill leisure time, one might consider the various contexts in which people like to spend their leisure time. In some cases, solitude is what people value and seek. For example, there are ways in which someone might view having other people around to be a nuisance (e.g., when reading a book or enjoying a painting in a museum). There are times when reading together with others or being part of a museum tour might be a preferred experience, however. Although it might not be possible to determine a single best possible viewing context for any particular content, it is possible to consider whether or how a system might be used in solitary or social situations

and to include evaluations of user experiences with a system in these different contexts (Karat, Pinhanez, Karat, Arora, & Vergo, 2001; Pinhanez et al., 2001). How might we do this? Observation of current and proposed practice *in context* would serve as the best advice here. Although there is a good deal of emphasis on techniques such as heuristic and laboratory evaluation in usability engineering, these techniques are simply not adequate for discovering the value of social context, particularly for casual use systems (Putnam, 2000). In these evaluations, one should attempt to observe the interactions between the users and their environment and how the technology might enhance or diminish such interactions. Noting such interactions, even if it is just in a simple list of pluses and minuses, can help in gaining a perspective on how the technology will fit in a real context.

Again, not everyone values social contact to the same extent, but in general people are social creatures who do value being around other people at times. Considering the social context of use for systems that are intended to be a part of filling our free time should be a part of the design. It is just as easy to imagine a hall of failed usable products, products that were perfectly suitable for delivering desired content in a usable package that were not widely accepted because they did not provide some equivalent quality of the "real experience." It would seem embarrassing if usability engineering were associated with the design process that ignored such elements.

Communication- and Experience-Driven Systems

What do you need a cell phone for, really? Are there tasks that we could identify that would predict that this technology would be as pervasive worldwide as it is becoming? What is the value of communication limited by small-screen displays and numeric keypad message entry? What makes a game fun? Why are many successful games minor variations of a basic hunt-and-destroy theme? We need more evaluation-based information on to be able to answer these questions. What do we know about how to approach these questions? For the most part, we are left with the basic advice posed above in Fig. 59.1. We need to consider the domain and the technology, make guesses about how to apply the technology, and then look to be informed by people attempting to use it. For designing new communication technology that people would value or new games, we have even less formal guidance than we do for content-driven systems. We have schools and design traditions for content areas such as film, theatre, literature, new media, and so on. We do not have game design schools or chat-room design schools. For these reasons, the advice offered here is more basic and subject to development by the practitioner based on context. If you think you have technology to support communication, ask questions about why people might value it.

HCI designers are encouraged to think of communication as more than just a task to be facilitated in a functional system. The functional view cannot easily account for the proliferation of communication devices available (from cell phones to pagers to various e-mail and chat programs), or for the value or preferences that people have for different modalities of communication. The development of communication-driven technologies needs to consider value issues.

As a contrast, an analysis of a different class of popular applications—games—suggests that success for these applications can depend to a high extent on content quality, can depend to a fairly low extent on access demands from an ease of use perspective, and can rely in varying degrees on context issues (e.g., Solitaire is extremely popular solitary game, whereas other interactive games might form the social clubs of the future).

A peek in to a video arcade will provide some interesting information for anyone interested in game design. Relative to the population as a whole, one will find that the population in the arcade is younger and composed of a higher proportion of male players. Does this mean that games are only popular—or possibly only interesting—to young males? It certainly suggests that games of a certain type, those in which someone is blasting something on a mission to find or save something, seem attractive to a particular audience. This is evidence that we have some knowledge of how to succeed in providing a valuable use experience for at least some segment of the population but would stop short of concluding that this is the only audience interested in game experiences. Brenda Laurel (1993), among others, looked at gender in game preference and concluded that there are fundamental differences—possibly genetic—in preferences exhibited by males and females. Males enjoy highly competitive "hunting"-related games, whereas females prefer social play as a free-time activity. This certainly reflects the type of games that seem popular with males. Although the advice seems appropriate in designing games for young females, the success based on this advice has not been as apparent. It is easier to design a computer-based experience that is consistent with a desire to compete and hunt than it is to design one that provides a satisfying social experience.

TOOLS FOR THE HCI EVALUATOR

I would like to offer a summary of tools and techniques that might be used in the evaluation of the value of a user experience with a system. These are all drawn from the wide range of techniques that have been described within this. The goal is to give some guidance on what tools to think of drawing from in planning a project. As always in our field, new twists or applications might call for creatively modifying the general tools to suit the particular situation. This is not weakness in the definition of the tool sets or methodologies but the reality of the kind of innovative environment we find ourselves in.

The basic areas for evaluation are content, access and interaction, and context of experience. The assumption is that the value of an experience is a function of these—not necessarily a linear combination, but at least that the value is likely to increase with increases in the values of the components. So that, all other things held equal, we can expect the value of an experience to be better if the content is improved, the interaction and access made more engaging, or the context more appealing. There is some danger in this assumption—as there are in all attempts to decompose overall system quality measures like usability into elements that we can work on in engineering teams. But the risks

TABLE 59.1. System Category by Value Aspect Matrix

	Content Driven	Communication Driven	Experience Driven
Content quality	Primary importance—seek ways to evaluate content	Content generally assumed to be developed by users; give ways to develop valuable content	Content will interact with interaction; difficult to evaluate in isolation
Access and interaction	Ease of use is a secondary issue	Explore pervasive versus context-specific use	Challenge level can be a major factor
Context of experience	Explore individual and group use	Availability a major issue; explore synchronous and asynchronous use	Explore individual and group use

seem no greater here than in other activities done in human–computer interaction, and the more holistic approaches seem to offer no real solutions.

Table 59.1 provides an overview of the experience component (content, access and interaction, and context of experience), the primary evaluation domain target for each component, and the high-level evaluation technique appropriate to that target. For content evaluation, subjective measures such as enjoyment and engagement are primary targets. Techniques for evaluating these components are a mix of objective measures (such as how long people will spend watching a video when they are free to do something else) and self-report measures to interviews or questionnaires. In general, there is a trade-off between questionnaire methods and interview methods, with questionnaires generally offering the possibility of greater response numbers (they can more easily be analyzed and administered outside of face-to-face situations) and interviews offering greater depth of information about the possible reasons for positive or negative evaluations. As with all of the guidelines here, it is not possible to recommend a single best approach for all situations. Suggesting a mix of approaches in any project is generally considered "best practice," not just a suggestion to cover lack of specific guidance.

How might we go about evaluating aesthetic appeal of a system and its contribution to how a system is valued? The answer here is to give advice that is not very satisfactory to usability engineers or to HCI researchers but is familiar to individuals involved in many design practices. Although the efficiency aspects of usability are suitable for empirical evaluation, the aesthetic aspects are not part of our known science base. One method that everyone agrees on as a good metric of appeal is simply market success or large sample polls, neither of which are particularly useful for informing iterative design cycles (although such market information can be useful in narrowing initial design ideas and directions). Commonly used in design fields such as graphic design, architecture, or even music composition is expert or peer evaluation (Schon, 1983). Although perhaps "unscientific," extended critical reviews by people with experience in the particular domain (often termed "crits" for short) is the accountability metric most often employed in evaluating design.

How might this be done in a usability engineering environment? No radical cultural change is advocated in which the user-centered methods of one set of practitioners be abandoned in place of the wisdom of experts. It is, however, useful to emphasize that searches for the objective measures that engineering design seeks are likely to fail. Using the voices of experienced designers, even when they might not be as experienced with evaluating interactive systems, can provide input and guidance where no other source will be directly useful. Forming a design panel and reviewing designs periodically with them is a reasonable strategy for development of systems with important user experience characteristics. If this can be done in an atmosphere of learning and cooperation, rather than in one of who is in control, it is more likely to succeed.

This is not a provable position; the reader cannot be provided a study of N design projects in which advice offered by "experts" was found to be of significant value. The study probably would never be carried out for cost–benefit reasons. It would be too expensive and its conclusions too narrow to be practical. Rather, it is an offering of experience of my own and others in the design process community within HCI (see the proceedings of the Association for Computing Machinery Designing Interactive Systems [DIS] conference) in attempting to understand different design traditions and their applicability to the design of computational devices. One conclusion, possibly controversial, is that aesthetic-based design has more to offer discretionary use systems than task-oriented systems.

One final area to mention involves methods to evaluate the value of a system based on its possible contribution to quality-of-life issues. This means both how a system might be viewed as enhancing the well-being of an individual (e.g., through stress reduction or being viewed as making life for the family better) or be viewed as benefiting society as a whole (e.g., through facilitation of worldwide communication). Such factors are often viewed as intangibles, as not subject to careful evaluation either because we don't know how to approach the evaluation or because we don't think we can measure the benefit economically. The advice here is simply to attempt formulating best guesses. Because many people in HCI come from behavioral science backgrounds, we have little experience in making suggestions based on evidence we know is incomplete. In the engineering context of building systems, people make judgments based on incomplete knowledge all the time; there is no deceit in this as long as one is open about the limitations of the evaluations. Taken this way, suggestions that a system might have quality-of-life value becomes a topic of discussion in design and development to be considered, rather than something nobody dares talk about. Again, the behaviorally oriented usability

engineer is the best person to take up the challenge and advance evaluations in this area.

As computing continues to make its way into all aspects of human life, it becomes important to take up the challenge of how it impacts quality of life more seriously. Producers of film and television programming have had many years to refine techniques for engaging the human brain's attention-fixing mechanisms, but there is reason to believe that providing passive experiences alone do not provide the emotional rewards found in more actively engaging activities such as sports, games, or hobbies (Csiklszentmihalyi, 1992).

CONCLUSIONS

The issues in designing systems to provide users with valuable experiences are different in a number of ways than designing systems that are usable for some task set. One can ask if the two are necessarily orthogonal goals. Is it a question of either/or—either usable or entertaining? As discussed above, usability does play an important role, at least in the access to content that people find valuable. Can the role go beyond that to include making our tools more entertaining (making our experiences with them more valued?) or our games more productive? This is yet another design challenge to consider, one that may become more realistic once we get a better sense of what contributes to valuable experiences in nontask systems. For now, we have to focus on creating the balance between different qualities, functionality and aesthetics, and clearly understanding aesthetics is not as well known (in the technology field at least) as

improving usability. Questions such as "Can word processing be entertaining and productive?" or "Do we care if games are easy to use?" must wait to be answered until we have a better understanding of the synergy between different kinds of design goals and design traditions. For some systems, we will focus on productivity-oriented techniques, and for others we will focus on fun. There is great potential in, but not a quick path to, merging the two.

Perhaps it is asking too much in asking that the science of HCI reach out beyond functional tools more broadly to the design of systems that people value. The skills involved in the design of aesthetically pleasing or entertaining systems might best be left to people who focus on these areas. The only problem with this is that as our technology becomes more pervasive, the distinction between tool and valued artifact becomes terribly blurred. After all, isn't my watch basically a tool for telling time? The more correct question is whether the design of technology artifacts needs to become increasingly interdisciplinary (meaning that additional skills need to be brought to it) or do the people responsible for usability of a system need to broaden their perspectives to better accommodate human enjoyment? Until we have a better sense of what it means to design for "leaning back," this question will be difficult to answer. For now, I simply encourage HCI designers to take the challenge seriously. It is fairly clear that observation of potential use and interactions with users will continue to be a main activity. It is not as clear whether a second principle—setting measurable objectives—will be as important to successful design. Insight, intuition, or design sense may be found to be of greater importance than we thought 10 years ago.

References

Chapman, P., Selvarajah, S., & Webster, J. (1999). Engagement in multimedia training systems. *Proceedings of the 32nd Hawaii International Conference on System Sciences*.

Csikszentmihalyi, M. (1992). *Flow: The psychology of happiness*. New York: Harper & Row.

Dyer, R. (1992). *Only entertainment*. London: Routledge.

Furnas, G. W. (2001). Designing for the MoRAS. In J. M. Carroll (Ed.), *Human-computer interaction in the new millennium*. Reading, MA: Addison Wesley.

Gaver, W. (2000). Looking and leaping. In D. Boyarski & W. Kellogg (Eds.), *Proceedings of DIS 2000—Designing Interactive Systems*. New York: ACM Press.

Gould, J. (1988). How to design usable systems. In M. Helander (Ed.), *Handbook of human-computer interaction*. Amsterdam: North-Holland.

Grudin, J. (1990). The computer reaches out: The historical continuity of interface design evolution and practice in user interface engineering. *Proceedings of ACM CHI '90 Conference on Human Factors in Computing Systems* (pp. 261-268). New York: ACM Press.

Holtzblatt, K., & Beyer, H. (1998). *Contextual design*. San Francisco: Morgan Kauffmann.

Jacques, R., Preece, J., & Carey, T. (1995). Engagement as a design concept for multimedia. *Canadian Journal of Educational Communication, 24*, 49-59.

Karat, C.-M., Pinhanez, C., Karat, J., Arora, R., & Vergo, J. (2001). Less clicking, more watching: Results of the iterative design and evaluation of entertaining web experiences. In Michitaka Hirose (Ed.), *Human-Computer Interaction—INTERACT '01* (pp. 455-463). Amsterdam: IOS Press.

Karat, J. (1997). Evolving the scope of user centered design. *Communications of the ACM, 40*, 33-38.

Karat, J. (1997). Software valuation methodologies. In M. Helander (Ed.), *Handbook of human-computer interaction* (pp. 689-704). North-Holland: Amsterdam.

Karvonen, K. (2000). The beauty of simplicity. In *Conference on Universal Usability: CUU 2000* (pp. 85-90). New York: ACM Press.

Kubey, R., & Csikszentmihalyi, M. (1990). Television and the quality of life. Hillsdale, NJ: Lawrence Erlbaum Associates.

Kuhn, T. S. (1996). *The structure of scientific revolutions*. Chicago: University of Chicago Press.

Laurel, B. (1993). *Computers as theatre*. Reading, MA: Addison-Wesley.

Landauer, T. K. (1995). *The trouble with computers: Usefulness, usability, and productivity*. Cambridge, MA: MIT Press.

Langer, S. K. (1953). *Feeling and form*. New York: Schribner & Sons.

Malone, T. W. (1980). *What makes things fun to learn? A study of intrinsically motivating computer games*. Unpublished Ph.D. Thesis, Stanford University, Department of Psycvhology, Stanford CA.

Malone, T., & Lepper, M. (1987). Making learning fun: A taxonomy of intrinsic motivatons for learning. In R. E. Snow & M. J. Farr (Eds.),

Aptitude, learning and instruction: Cognitive and affective process analyses. Hillsdale, NJ: Lawrence Erlbaum Associates.

Mountford, J. (1994). Constructing new interface frameworks. In B. Adelson, S. Dumais, & Olson, J. (Eds.), *Proceedings of ACM CHI'94 Conference on Human Factors in Computing Systems* (pp. 239-240). New York: ACM Press.

Nardi, B., & O'Day, V. (1999). Information ecologies: Using technology with heart. Cambridge, MA: MIT Press.

Norman, D., & Draper, R. (1986). User-centered system design: New perspectives in human-computer interaction. Hillsdale, NJ: Lawrence Erlbaum Associates.

Pausch, R., Snoddy, J., Taylor, R., Watson, S., & Haseltine, E. (1996). Disney's Aladdin: First sterps toward storytelling in virtual reality. *Proceedings of SIGGRAPH 96* (pp. 193-203). New York: ACM Press.

Picard, R. (1997). *Affective computing.* Cambridge, MA: MIT Press.

Pinhanez, C., Karat, C.-M., Vergo, J., Karat, J., Arora, R., Riecken, D., & Cofino, T. (2001). Can Web entertainment be passive? *Proceedings on Multimedia and the Web.* New York: IEEE Press.

Putnam, R. (2000). *Bowling alone: America's declining social capital.* New York: Simon & Schuster.

Resnick, P. (2001). Beyond bowling together: SocioTechnical Capital. In J. M. Carroll (Ed.), *Human-computer interaction in the new millennium.* Reading, MA: Addison Wesley.

Schon, D. (1983). *The reflective practitioner.* New York: Basic Books.

Tsukahara, W., & Ward, N. (2001). Responding to subtle, fleting changes in the user's internal state. In *Proceedings of ACM CHI '2001 Conference on Human Factors in Computing Systems.* (pp. 77-84). New York: ACM Press.

Webster, J., & Ho, H. (1997). Audience engagement in multimedia presentations. *The DATA BASE for Advances in Information Systems, 28,* 63-77.

Whiteside, J., Holtzblatt, K., & Bennett, J. (1988). Usability engineering: Our experience and evolution. In M. Helander (Ed.), *Handbook of human-computer interaction* (pp. 791-817). Amsterdam: North-Holland.

Wroblewski, D. (1991). Interface design as craft. In J. Karat (Ed.), Taking software design seriously: Practical techniques for HCI design (pp. 1-19). New York: Academic Press.

Yamaguchi, T., Hosomi, I., & Miyashita, T. (1997). WebStage: An active media enhanced world wide web browser. In *Proceedings of ACM CHI '97 Conference on Human Factors in Computing Systems* (pp. 391-398). New York: ACM Press.

·VI·

MANAGING HUMAN–COMPUTER INTERACTION AND EMERGING ISSUES

Hans-Jörg Bullinger and Jürgen Ziegler
Fraunhofer IAO

With the rapid dispersal of information and communication technologies (ICT) into all areas of working and living, the challenges for managing both the development and the application of ICT are changing. The scope, as well as the depth, of the impact of ICT in a modern, e-business-oriented enterprise are significantly larger than used to be the case with specialized, stand-alone applications or typical office tools. Human–computer interaction (HCI) and usability must be considered key factors for achieving the goals that are underlying the transition to electronic business, online government, and similar far-reaching concepts. In a recent study on Internet use (Forrester 1999), for instance, usability was named only second to high-quality content, preceding factors like download time or topicality. Awareness for these issues is rising in management, but it is often unclear how usability and user-centered design should be addressed in a concrete organizational setting.

THE CHANGING CONTEXT FOR HUMAN–COMPUTER INTERACTION

To understand the full scale and potential of user-centered approaches, HCI issues need to be positioned in a larger framework as shown, for instance, in Fig. VI.1. The changing business context, like innovation speed, global networking, customer orientation, or virtualization, requires new, highly flexible forms of systems and applications. Only widely accepted, standards-based developments will have the impact to satisfy these requirements. In conjunction with the organizational changes taking place in such an environment, work is becoming much more knowledge-intensive and cooperative. Although in the past, systems used to be mainly tailored to operative tasks, in the future creative, team-oriented, projectlike activities will become a predominant form of work. An often-discussed phenomenon is the so-called information technology (IT) productivity paradox: whereas IT investments went considerably up over the past decades, the productivity of information workers increased only slightly, as studies in the United States and other countries show (for an overview, see Landauer, 1996). A considerable portion of this effect may be attributed to utility and usability issues. With new types of tasks requiring much more flexible and creative work processes, there is the danger that IT will not be able to deliver sufficient benefits unless functionality and interaction are developed in a much more user-centered manner.

Another important determinant is the broadening range of people expected to use IT. This development pertains to all social strata and requires the support of a wide sociodemographic

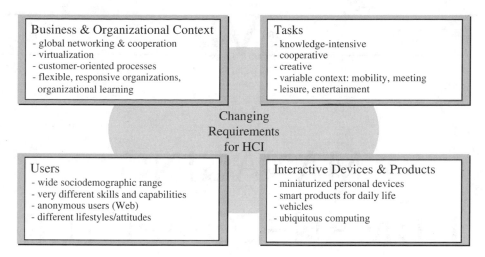

FIGURE VI.1. Challenges for human–computer interaction (HCI).

range, and users with very different skills and capabilities. More and more services are offered exclusively in the World Wide Web. Potential user groups encompass, for instance, all citizens of a city in e-government applications, all members of an organization using an intranet or the residents of a building equipped with "Smart Housing" functions. Consequently, the user groups addressed are characterized by an extreme heterogeneity of skills, knowledge, age, gender, disabilities, literacy, culture, income, etc. (e.g., Shneiderman, 2000; Stephanidis, 2001). Therefore, many new questions arise concerning the acceptance and accessibility, which go beyond the traditional issues of software ergonomics. If, for example, certain user groups like older persons and people with disabilities are excluded from the use of new technologies by inappropriate user interfaces, it is not merely a matter of neglecting important market segments. In addition, there will be the danger of splitting the society into users and nonusers with various social consequences related to education, job market, and other areas.

Finally, we see that IT is increasingly integrated into all sorts of devices and products, and used in professional and personal lives. Being able to master this growing complexity in our environment will be an important aspect for technology acceptance and quality of life in the future.

HUMAN–COMPUTER INTERACTION AND MANAGEMENT

Managing ICT and HCI aspects, in particular, needs to address the challenges arising from these developments in a variety of ways. The most important aspect is to recognize the importance of HCI-related issues and to understand them as an integral part of the productive potential of ICT infrastructures, applications, and services. The objectives and approaches for integrating HCI have traditionally been different for organizations developing software products, in-house application development, and user organizations. The importance of HCI has typically been

recognized earlier and to a broader extent in the software products business. The driving factor here is the immediate impact of a product's usability on its market success. In-house developers and user organizations typically had a more limited view of the importance of the user interface and usability related issues. The main concern here was often to satisfy minimum ergonomic requirements for the design of workplaces. This view has been particularly predominant in countries with strong workplace regulations. The notion of user interfaces as part of the workplace is about to change dramatically, however. With the increase of electronic networking with customers and suppliers, external users suddenly play an important role for the enterprise and its e-business systems. Many companies are discovering that their customers are at the same time users. Managing HCI-related developments in these different contexts requires a clear understanding of what the different characteristics of these user groups are, and with what approaches and methods their requirements can be captured.

A major concern in managing HCI activities has been to look at the cost of including user-centered design in the overall development process. Although cost factors are relatively easy to determine, benefits typically are much harder to define and measure (Karat, 1997). In the few cases where concrete cost-benefit analyses were done, benefits were mainly calculated on the basis of increased productivity (e.g., as determined by the decrease in learning and execution times needed for performing tasks, or in more indirect measures such as reduction of help desk calls). A merely productivity-oriented justification of HCI, however, fails to recognize important trends in management and organizational strategies. A general development resulting from the growing globalization and increased competition is the focus on value as a major determinant of a company's future success. Value-oriented management places higher emphasis on strategies, knowledge, and agility of the enterprise. In this view, it will be important to understand that the ability to develop and deploy highly usable ICT constitutes an important aspect of this overall value scheme. Transferring this notion into practice implies not only the implementation of

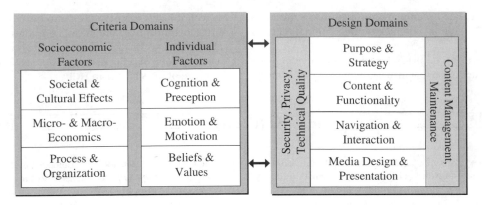

FIGURE VI.2. Reference framework for evaluating web applications.

user-centered development processes with the typical activities of usability engineering and testing, but also the management of usability-related knowledge and organizational learning processes ensuring the take-up and further development of such knowledge.

Another important challenge lies in the recognition and understanding of the wide range of factors influencing usability and acceptability of new technologies. The effectiveness, acceptance, and ultimately, the success of new ICT is to a considerable degree dependent on factors that lie outside the technology itself. Socioeconomic aspects of information technologies are, in many instances, more relevant than functional capabilities or even cost of the technology alone. The scope of issues addressed creates the need for evaluation approaches that take this diversity into account. In Fig. VI.2, for instance, a reference model (WebSCORE; Bullinger, Ziegler, & Bauer, 2000) for evaluating web-based applications is shown that provides a structured set of evaluation factors and criteria. This model tries to address the full range of aspects that are critical for the acceptance and success of interactive ITs.

One of the most challenging aspects of managing ICT is the fact that the application of information systems is inextricably intertwined with the organizational structure and processes of an enterprise. It has long been recognized in research and industrial practice that ICT should not merely be used for automating existing processes, but rather as enablers for new organizational forms. Re-engineering of processes, virtualization of organizational structures, and the focus on core competencies are factors that should inform and direct the development and deployment of new technical infrastructures and applications. HCI, as well as computer-supported cooperative work (CSCW), have an important role to play in this context. If insufficient usability creates barriers for the effective use of a system, the intended organizational effects often cannot be achieved. Introducing ICT in conjunction with new organizational patterns also requires a thorough understanding of how people cooperate through ICT and what obstacles can be expected (cf., e.g., Duane & Finnegan, 2000). Failing to address these issues when managing development and introduction processes can seriously endanger the realization of new organizational strategies and business processes.

SECTION OVERVIEW

The chapters in this section of the handbook address a variety of issues related to managing ICT development and use with respect to HCI. In the chapter on Technology Transfer, Schofield compares the perspectives of researchers versus practitioners in the area of technology transfer. He discusses internal transfer issues, giving a thorough explanation of the processes and structures with special attention to funding, development, implementation, and integration in existing systems. Also, important factors concerning external technology transfer are presented.

The chapter by Bias, Mayhew, and Upmanyu discusses different aspects of cost-justifying usability activities. Methods for quantifying costs and benefits are demonstrated. The main question is which usability engineering method is the most beneficial for the various approaches of product development. Examples for building a cost-justification tool are given. Taking a broader perspective, Friedman and Kahn discuss the relevance of human values and ethics in the context of system design. The chapter begins with a review of the arguments for embedding values in the design of technologies. The authors provide criteria for evaluating how human values and ethical issues can be distinguished from usability aspects. Special ethical responsibilities for professionals and HCI professionals in particular are discussed.

The notions of security, privacy and trust—that play an increasingly important role in a digitized world—are provided in the chapter authored by Diller, Lin, and Tashjian. Arguments concerning the relevance of trust, different trust models, and an overview on the history of trust in commercial relationships are stated. The principle of Relative Trust is introduced and elaborated.

Finally, Karwowski provides an entirely new perspective on *achieving compatibility in design and evaluation* in the final chapter of this section. This chapter discusses the computer–human system compatibility in the context of design and evaluation of HCIs. The concept of the computer–human system compatibility is introduced, and the methodology for quantitative assessment of such compatibility is proposed. Two numerical examples are used to illustrate the proposed methodology.

A modeling framework for defining and quantifying the measures of ergonomic system complexity and compatibility is discussed.

Overall, this section couples scientific advancement with practical implementation so that management issues and emerging technologies can be woven into the fabric of the discipline of HCI in appropriate and meaningful ways. By addressing critical issues like technology transfer, ethics, cost, security, privacy, trust, and compatibility in design, this section serves to expand the range of possibilities in HCI.

References

Bullinger, H.-J., Ziegler, J., & Bauer, W. (2002, in press). Intuitive human-computer interaction—Towards a user-friendly information society. *International Journal of Human-Computer Interaction*.

Duane, A., & Finnegan, P. (2000). Managing intranet technology in an organizational context: Toward a "stages of growth" model for balancing empowerment and control. In *Proceedings of the 21st International Conference on Information Systems* (pp. 242–258).

Forrester Research, Inc. (1999, January). Media field study.

Karat, C.-M. (1997). Cost-justifying usability engineering in the software life cycle. In M. Helander, Th. Landauer, & P. Prabhu (Eds.), *Handbook of human-computer interaction* (2nd completely revised ed.). Amsterdam: Elsevier.

Landauer, Th. (1996). *The trouble with computers*. Cambridge, MA: MIT Press.

Shneiderman, B. (2000). Universal usability. *Communications of the ACM, 43*(5), 85–91.

Stephanidis, C. (Ed.). (2001). *User interfaces for all. Concepts, methods and tools*. Mahwah, NJ: Lawrence Erlbaum.

·60·

TECHNOLOGY TRANSFER

Kevin M. Schofield
Microsoft Corporation

INTRODUCTION

This chapter is intended to provide guidance on how to approach technology transfer of human–computer interaction (HCI)-related research. Admittedly, there are many different perspectives one could take in looking at this often difficult problem: Researcher vs. practitioner, industry vs. government vs. academia. There is also the extra-added dimension of whether the transfer is within the boundaries of a corporation or across corporate lines.

In this chapter, we will discuss technology transfer both internally and externally to a company, then at the end discuss commonalities across the two. Primarily, we will look at it from the perspective of the researcher, because in practice the burden is more on the researcher to justify the transfer and to make it work. Practitioners, however, will also gain value as well from reading this, because it will help them to understand what researchers are going through.

Although any technology transfer has its challenges, HCI tends to be a particularly difficult one. This is because of many factors, but the two main ones are that it is often more about abstract ideas than specific implementations, and because we, in the HCI community, are still fighting the (wrong) impression that HCI is an afterthought and not the "real meat" of research and development efforts.

INTERNAL TECHNOLOGY TRANSFER

I am constantly surprised by the number of people who believe that technology transfer is some sort of Rube Goldberg machine,[1] where technology is inserted in one end of the contraption, strange things happen in the middle that usually involve people in uncomfortable and contorted positions, and it magically pops out on the other end. Countless books and articles have been written (Lesko, Nicolai, & Steve, 1998; Mock, Kenkeremath, & Janis, 1993) in an attempt to document the perfect mechanical process for technology transfer. And yet, despite the fact that nearly everyone has had painful experiences trying to define a mechanical process for technology transfer, they still try to do it and complain bitterly when it fails (Butler, 1990; Hiltzik, 1999; Isaacs & Tang, 1996; Singh, 1993).

Technology Transfer as a Social Process

Technology transfer is not a mechanical or logistical process; it is fundamentally a social process. It succeeds when people build a relationship between the provider and the recipient that fosters trust and communication. Manning (1974) recognizes that successful technology transfer centers on viewpoints and perspectives, and fundamentally on communications. Foley (1996) speaks to this point most directly—that technology transfer is a "full-contact sport" that centers on the people.

Successful product organizations understand that risk is their mortal enemy. They work hard to proactively manage the risk in their development process, or to remove the risk factors altogether. One of the most prevalent and difficult-to-manage risk factors is an external dependency, and let us be honest: An external dependency from a research organization looks about as risky as it gets. As long as your counterparts in product organizations think of a research organization that way, technology transfer is difficult at best and often outright impossible.

To succeed with technology transfer, we need to mitigate the risk—or at least the perceived risk—in the minds of the people we wish to receive our technology.

Up to this point, none of this is particularly controversial, but this is where the paths diverge. Many people will tell you that you succeed in mitigating the risk by creating well-defined, step-by-step processes through which you and your industry partner will enact the technology transfer. I argue that this approach fails more often than it succeeds, for two main reasons:

1. The only experience that people in industry have with external dependencies is the occasional dependence on an external contractor or supplier to deliver a finished component ready for integration. They inevitably use this as the model for defining their technology transfer process from research, and it is fundamentally incompatible. Research technologies are not finished components, and any product organization that expects a research group to deliver a finished component fundamentally misunderstands the role, expertise, and hiring practices of a research organization. Research groups almost never understand the development and test practices of a commercial product organization; even if they did, those practices vary so widely between organizations that past experience does not guarantee that they could successfully deliver a finished component. Moreover, it is not the goal of a research organization to develop technologies into finished components; its goal is to discover and prove solutions to previously unsolved problems. To do so requires different skills and expertise, and quite frankly different development and testing methodologies. Both research and product organizations need to understand this fundamental difference and embrace it as a way to complement each others' strengths, rather than ignore it and delude themselves about a theoretical capability that practically speaking is not there.

2. People who do not understand each other cannot communicate, do not trust each other, and cannot be expected to work cooperatively to a shared goal, even within the best-defined process. The trust and communication must come first. If the two people on opposite sides of a table trust each other, then the two of them can accomplish anything; if they do not, they will never accomplish anything of value.

Building Trust in Steps

So, then comes the catch: How does one build trust? By working side by side, of course! This means that one must start with

---

[1] For those in the UK, that would be "Heath Robinson machine."

the kind of activities that are initially low in risk, but high in communication, and build on one's successes to build more trust and overcome successively higher levels of risk.

Step 1 is to establish trust that one is an expert in the domain. Technical people, whether in research or in industry, are almost universally avid readers and understand the importance of staying up to date in their field. But, we all suffer from a lack of time to weed through the volume of less-than-useful information to find the truly valuable nuggets of wisdom. If someone in a product organization expresses an interest in one's field, an offer to forward them a set of papers, articles, and books that represent the seminal reading is a great first response. Granted, doing a literature search is not the most glamorous work, but it fundamentally demonstrates a working knowledge of the domain and an ability to provide guidance. Equally important, it shows a healthy respect for their intellect and a flattering assumption that they will be able to read and digest the material. One of two things will happen: Either they will actually read the materials sent to them, in which case they have not only made an initial investment in seeing a technology transfer happen, but also provided a great topic of further conversation; or, they will not read the material and most likely conclude that it is simply easier to rely on the researcher as their expert rather than to become an expert themselves. Either way, it is a foot in the door. They will ask endless questions as they try to decide for themselves what is within the realm of possibility and, more importantly, practicality. It is essential to ask them just as many questions to understand as completely as possible their constraints and make clear recommendations on what they can expect to build.

Step 2 is to move from simply giving domain guidance on the state of the art to providing feedback on product design plans. This involves offering to review specifications and to provide timely feedback. Timeliness is absolutely critical—schedules are the rules of the game, and an ability to stay within their stated schedule reflects an understanding of the rules, an appreciation for their importance, and a commitment to the success of their project. This is also a critical test of a researcher's ability to think practically; in their distrust, they might expect suggestions of wildly impractical things that would push out their schedule or require resources out of proportion to the relative importance of the technology to the overall product. It is the researcher's job to show once again an understanding of their constraints and the value brought to their team effort. Success will be apparent when a subtle but important shift happens: Instead of the researcher asking to review their design documents, they will ask the researcher to review them.

Step 3 is a significant one indeed: When they ask the researcher to help write the specification for the product. Do not expect this to happen until there has been a clear success at Step 2 that established credibility; but, when it happens, the product organization is making a loud and unambiguous statement that they now think of the researcher as part of the product team. This is an enormous step for a product group to take in their relationship, and it is a heavy responsibility to take on.

At first, they will probably only delegate small parts, and often they might ask the researcher to co-write design documents; but, regardless of the size of the assignment, the key to success is the same: Whatever is designed must be easily buildable and testable. If there is any significant disagreement on whether it can be built or tested, the product organization will not take the risk. Development organizations (at least the successful ones) are inherently conservative and will overstate the costs to build new technologies.[2] This is not only another test of whether the researcher understands their constraints, but also equally whether he or she understands their development process. I encourage "aiming low" initially and looking for indications from the team that they would like to work together to design something more aggressive. If there is success in co-designing components, they will loosen the reins and delegate more responsibility (with more autonomy).

Step 4 is where one (finally) gets involved in implementation. It has taken enormous patience on the part of the researcher to get here, and there are still land mines everywhere. No two development organizations are alike; they all have different practices for creating, documenting, integrating, accepting, testing, and deploying new work. It is impossible to understand all of their processes, and it is very unlikely that they will all be written down; yet, every one of them is an opportunity to break form and cause a rift. I would encourage asking the group manager how to bring a new employee into the group, and what training and mentoring that person would go through; furthermore, see if there are opportunities to take advantage of such a process to help to get up to speed. If one has made it this far, the product team wants to see a success as much as the researcher, because one's success is the other's success; they will be very reasonable about doing things to help to understand them, especially if it is clear to them that is the goal: To understand them. All development groups fall in love with their own processes, and one earns their cooperation by showing equal respect for those processes, no matter how silly they might seem to an outsider.

The key to success in a development process is to underpromise and over-deliver. Set expectations reasonably low; they should think that you can carry your own weight, but not that you are God's gift to engineering. Promise metrics for work and for the technologies created that are achievable but not overly aggressive, then exceed those metrics. By doing that, it is possible to fully gain their trust and move on to discuss with them more aggressive technology transfers.

It is important to note that in this "pyramid" of sorts that we are building with increasing levels of risk and corresponding trust, it is possible to peak at any level. For instance, if the researcher does not have the development skills to co-develop components with a real product team, then do not try to do it! By all measures, every step in this process can be considered technology transfer. Product organizations need knowledge, understanding, and ideas about technology just as much as they need finished technology components; they need to understand what cannot be built just as much as what can. And, most important, they need researchers to tell them honestly what they

---

[2]Ironically, it has been my experience that development organizations tend to estimate wrong not in the new technologies, but in the incremental improvements to legacy components and particularly in the "integration" work in making multiple components work together.

are, and are not, capable of doing for them. Even without ever delivering a finished component to a product organization, one can still have a litany of technology transfer successes for which the product organization will sing praises. It is more important to proceed in measured steps built on past successes and build the trust and the lines of communication that will guarantee future successes.

Thinking for the Future

It is also important to be thinking to the future, to be thinking about what comes next. There is always a desire to simply throw a technology over the wall and then to move on to the next research project, but this is unrealistic. It never really works that way and, even if it did, "throwing it over the wall" would end the relationship and any opportunities for future technology transfer; the ongoing relationship after the transfer is an opportunity to carry on a dialogue about the next great technology.

From the product organization perspective, there is rarely a "clean" way to integrate a component. The overwhelming majority of development work is revisions to existing products; very rarely are new products started. Revision work means that new components need to be integrated into an existing legacy framework; this almost always requires development work on both sides of the integration to ensure the optimal match.

Applying the Approach to Human–Computer Interaction Technology Transfer

As if this was not difficult enough, applying this approach to HCI-related technology transfer introduces its own challenges. One can read *The Psychology of Human-Computer Interaction* (Card, Moran, & Newell, 1983) and learn that, at a very fundamental level, there is a set of scientific principles that hold very broadly. However, we in the HCI community have also learned that interactive systems must be designed within the context of a particular task and human, and that this very fact makes it tricky at best and misleading or impossible at worst to try to generalize specific designs to other contexts. Even with the best of intentions and the most thorough usability testing, there are no clear guidelines about how much of an HCI-related research technology can actually be transferred, and particularly for integration into an existing product. So "throwing over the wall" is especially difficult, because it calls for potential redesign, as well as further development and integration, test, localization, support, and operations (it is increasingly a service world, after all).

In the traditional view of technology transfer, this would be fatal. In the "relationship" view, however, this is an opportunity to build a working model that allows you to overcome the challenges and work side by side with a product organization to guide the transfer of your work.

Beyond simply moving up the pyramid, there are other important things one can do to deepen the relationship. Scheduling regular "maintenance" conversations can help to maintain communication channels and to keep abreast of activity in the product organization; it is also an opportunity to continue to update them on progress on new work. The relationship can

also be used to improve the researcher's own work by learning about critical real-world issues. Good product organizations have a wealth of information about their customers; by using their access to real customers (and aggregate information about them), the relevance of research activities can be improved. It is an opportunity to ask key questions, learn about critical product strategic direction, conduct user studies, and find out what difficult HCI and technical issues are about to become critical issues for real customers. This is the golden opportunity to get out of the ivory tower.

EXTERNAL TECHNOLOGY TRANSFER

External technology transfers, regardless of whether they originate in a government agency, academia, or an industrial lab, are almost by definition crisper types of transfers, because things must cross the boundaries of legal entities. This means that one will be participating in a transaction involving the sale or licensing of technology, or contracting to provide some service, or both. So the first order of business is negotiating the "deal."

Know What You Are Selling

It is critical to know and understand what is being sold: Outright ownership of intellectual property, or a license to it? Is it a complete solution that has been developed or just pieces? Those pieces might include any or all of the following:

- User interface design
- System specifications
- A working prototype
- An actual implementation, tested to some level of quality
- The source code for a software implementation
- Copyrights
- Patents
- Working time as a commitment to support ongoing productization
- A running service that you host.

Know What You Are Not Selling

It is equally important to know what is not being sold. If this is work that one would like to continue, he or she will need to make sure that to preserve rights and ownership to continue that work; otherwise, one could very well put oneself out of business by selling complete ownership to a pride and joy, or by signing an exclusive license that precludes licenses with any other company.

Know the People Involved

Deals inevitably involve lawyers as well as what are known as "business development" people: Those whose job is to negotiate deals that further the business interests of their employer.

One can assume that the company negotiating will have both business development people and lawyers; it makes good sense for the researcher to have them, too. This is a situation in which we need to put our egos and high-minded notions about doing business "on a handshake" aside. Although the overwhelming majority of companies are not in the business of stealing from people like us, and will not try to do so, to get the most value in return for what is being offered, one needs someone on his or her side who understands what is customary in intellectual property deals and how to negotiate for it. Even a brilliant and an excellent debater who does not know what to reasonably ask for is at a serious disadvantage. The bottom line: Find someone with good business development skills and experience to negotiate the deal.

Likewise, once the terms of a deal have been negotiated, one needs a lawyer to write it up and make it legally binding. Do not even think about self-representation in the drafting of an intellectual property agreement; The laws are changing too quickly (which is not the fault of the lawyers) for a researcher to understand which ones apply to the situation and should be factored into the drafting of an agreement. Mack et al. (1993) lay out the basics of existing laws and how they relate to technology transfer.

The key to success is to understand the defined role of each of the three people on a team: Oneself, the business development person, and the lawyer. The researcher's role is to be the technical expert and to place a value on the work as well as what the people on the other side of the table are offering in return; one's role as "client" is to decide what is needed to be successful and beyond the needs what additionally is desired but negotiable. The role of the business development person is to take the articulated needs and desires, and try to structure the terms of an agreement that will work for both parties. He or she understands business risks and will help to understand them and make informed decisions about how much risk can be tolerated. The role of the lawyer is to take the terms and write them down in words that both parties understand and that can be interpreted under the law to protect the client's interests. The lawyer also understands and can articulate the legal risks; laws are often subtle and ambiguous things that can be interpreted in many ways (in fact, nations have an entire branch of government that does nothing but interpret the laws). Any contractual obligation runs the risk of being interpreted in a way other than how it was intended, and a lawyer can help to understand how likely that is based on the language of the law and similar previous cases where the law has been interpreted by the courts. Just as there is always business risk, there is always legal risk, and in the end it will fall on the researcher to make the decision as to whether the risk is acceptable.

Where do deals go wrong? In my personal experience, they often go wrong when these three roles become confused, and when the business development person and the lawyer start to make the key decisions. The researcher must live with the result of the deal, not them. On one end of the spectrum, there is no such thing as a risk-free deal; on the other end, even a high-risk deal could be worth doing if the reward is high enough. Those decisions are the client's, not theirs, and the researcher as client should insist on making them.

There are many good sources for both business development people and lawyers. Venture capitalists will often have a "short list" of ones they trust to do their business transactions. The Chamber of Commerce for a local area can also provide recommendations, and often will track complaints registered against specific ones. Many regions also have associations of entrepreneurs, inventors, and small business owners, with great resources to draw from.

Of course, if one already works in a research lab, the institution most likely has a technology licensing office that will negotiate and draft deals on the researcher's behalf (and are likely required to do so if the researcher's employment agreement assigns ownership of inventions to the employer). In that case, one will still need to stay involved to make sure that the researcher's needs are met in whatever deal is negotiated.

Crafting a Deal

The most difficult process of negotiating a deal is crafting an arrangement that meets the needs of both sides. I have seen many negotiations take much longer to conclude, and in many cases fail to conclude successfully, because either or both sides did not bother to try to understand the other side's business needs. Business partnerships are always about finding a way to help both parties be more successful. The best way to do that is to understand what one's prospective partner's business is about, and likewise to share enough information about the business, openly and honestly, so that together a combination can be found that works for both sides. Find out everything available about their current business situation:

- their revenues and profits
- their competition
- their most important customers
- what customers are saying about their product
- where they say they want to take their business in the future
- the problems and challenges they are facing.

Steinberg (1998) describes his experiences in negotiating deals and his own well-known and well-respected philosophy for how to structure deals makes good business sense for all parties.

Companies will pay for the value that can be delivered to them. They will pay a certain amount of money to make an even larger reduction in their costs (because in the end they save money). They will of course pay to help them make even more money. Finally, they will pay if one solves a problem for them. As a deal is structured, try to cast it in terms of what it does for them; those are terms that they can understand and, more importantly, quantify in a valuation.

Take it further: Help them in any way possible to place a value on what is brought to the table. For example: Do a user study on their existing product, and another showing how the technology being offered will improve their product (if it happens to address a key customer complaint, all the better and that should definitely be brought to their attention). During the negotiations, show them a smart, effective professional

who can work with them. This does not mean that you need to negotiate hard to the very last item; contrary to popular belief, the tough negotiators are not always the most respected, and, in fact, they are often the ones that create their own reputation for being difficult to work with. It is much more important that to demonstrate an understanding of them and an ability to speak their language, as well as a willingness to help them make the case to the decision makers in their organization's senior management.

Confidentiality and Nondisclosure Agreements

One large challenge in negotiating deals is the issue of confidentiality, which usually rears its head first in the often-dreaded nondisclosure agreement (NDA). NDAs are signed prior to revealing confidential information to ensure that the information will not be disclosed to competitors. That part is good business and a natural, noncontroversial part of good faith bargaining: We all need to be able to keep secrets. The difficult part of nondisclosures is the issue of residuals: By looking at confidential information, I learn things, and then I carry that learning around in my head for the rest of my career. What am I allowed to do with that information in my head, and in fact who owns it? From the discloser's point of view, one wants to make sure that someone cannot use the confidential information to compete with. From the other side, it is impossible to know exactly what will be disclosed, or what business opportunities are going to come one's way tomorrow, so it is deeply problematic to sign away the ability to enter certain businesses simply for the privilege of looking at confidential information. Both sides sound very reasonable, and they are, which is why NDAs are no trivial matter and often become the stopping point in negotiations.

Whenever possible, try to complete as much of the negotiations as possible without entering into an NDA, because it simplifies matters and prevents the trust issues from rising to the surface too early. The downside is that this makes the early negotiations a precarious dance, in which researcher needs to show the other party enough to convince them that the technology is real and solves their problem, without giving away key secrets. There are things that still can be done: Tell them "what" it does, instead of "how" it does it, and show them the system working. The goal is to make them crave it enough that they will want to sign an NDA under the researcher's terms to complete the technical due diligence required for them to close the deal.

It is absolutely critical to think this all through before getting to the negotiation table: these are not decisions to make—ever—under time and social pressure. It is also critical to realize that the whole issue of confidentiality and NDAs is one more business risk; admit it and decide for oneself whether (and when) the potential reward outweighs the risk. This is one of the clear cases in which a lawyer will be extremely conservative and protective, and describe in great detail everything that could be lost by entering into an NDA (or by showing technology without an NDA). But in the end, the decision is the researcher's.

Going to the Negotiating Table

Negotiating a deal is probably the most hyped and feared part of this whole process; perhaps we have all had too many nightmares about slick car salesmen tricking us into paying too much for too little in return. The reason we fear car salesmen is that the salesman has all of the information about what the car is really worth and shares none of it with us; we are forced to blindly trust him, and many of us do not.

Steinberg (1998) once again shares his wisdom on a sound and ethical approach to negotiation in his "twelve essential rules for negotiating." Not nearly as ambitious as Steinberg, I have only three basic rules for negotiating:

1. I obtain as much information about each side's position as possible, before arriving at the table.
2. I have a list of what I really need in order to succeed and a separate list of what I want to have in addition. I hold firm to my needs, and I am willing to compromise on my wants.
3. I always negotiate a deal where both sides win.

Understanding a potential business partner's position is absolutely critical to negotiating success, for a number of reasons. First, it tells what they are looking for. Ask the same list of questions we discussed with internal technology transfer. Who are their customers? What are those customers saying about their products? Who are their competitors? What is the company looking for that will help them to be more successful? What are their strengths and weaknesses? Second, and very much related, it tells how they will value what they are offered. Successful business deals involve an exchange of value that both sides view as fair and equitable, but value is of course relative to the company and its context. Understand that to find an equal trade. Reading the company's annual report is an excellent source of information about a company (if it is a publicly traded company). Reading news articles and competitive reviews also provides invaluable information about the business pressures the company is under, as well as the assets that they bring to the table.

Having the list of the things that one really needs and the things that one wants in addition is a valuable step in preparing for negotiation. I have seen many people come to the table unwilling to compromise on anything; They ask for too much in the beginning and believe strongly that compromising on anything is a sign of weakness. Negotiations like those always take longer than they should and are very frustrating. In some cases, they spend more money in lawyers' fees for fruitless negotiations than the value of the small items that they refuse to compromise on. I recommend starting with a basic negotiation on the core needs of both sides; not only does that keep you focused on the heart of the deal, but it tends to simplify things just by taking all the peripheral items off the table. Once the heart of the deal is done, and both sides feel comfortable that they can be successful because they are getting what they need, adding additional pieces is much easier with a lower stress level and a structure in place.

Remember to be honest with yourself as you detail all this information, because lying to yourself is the surest path to failure. Assume that they will have an accurate valuation of what

is brought to the table (regardless of whether they are willing to tell you what it is), so insisting that something is worth more than its real value is foolish. Understand one's own strengths and weaknesses; be honest about what is needed to be successful, and do not promise things that cannot be delivered.

All of this brings us to the last rule: Always negotiate a win-win. Always negotiate a deal where both sides feel that they are receiving what they needed and can be successful. Beyond the obvious ethical reasons for doing this, there is also the very practical consideration that the two parties will need to continue to work together. Many people go to the negotiating table believing that signing a deal is the end of the process, when in fact it is just the beginning. Once the paperwork is signed, there is a relationship between the parties that essentially lasts forever. This relationship often makes itself known in unanticipated ways; for example, if one is licensing a patent to a company, the company has a vested interest in ensuring that the patent maintenance fees continue to be paid to the Patent and Trademark Office to keep the patent valid, and will want to have regular information to confirm that the payments are being made. Almost every deal one can imagine, no matter how cut-and-dry, has some aspect that will require communication between the parties on an ongoing basis after the deal is done. If one licenses a technology to a company, the company may be sending royalty checks, and the licensor may want some way to audit their sales to ensure that they are accurately paying. They, in turn, may want technical support, including important bug fixes and updates. They may additionally have negotiated the option to license future upgrades, and as a paying customer they will likely want to provide their input on features and enhancements to the technology that would be of most help to them.

Embracing the Relationship

By admitting from the beginning that there is an ongoing relationship, one can embrace this notion and turn it to an advantage. In fact, I encourage its use to build a future revenue stream. A researcher can build design and consulting services into the deal, which is particularly helpful for HCI-related technology transfer, because as we discussed earlier they often need reworking to fit into a larger context. One can use the ongoing relationship as a "foot in the door" to be able to offer future sales and deals as new technologies are developed. In fact, viewed as a "strategic partner" one might even want to offer them the right of first refusal on future offerings, as a way of demonstrating a commitment to them. Finally, as in contemplating future growth of the researcher's business, one may need additional sources of funding, and a partner who has a vested interest in the researcher's success can be a great source for funding. Even if none of this is true, assume that one day there will be a need for a good reference or recommendation from them; that alone is reason enough to want to have a great ongoing relationship.

COMMONALITIES FOR INTERNAL AND EXTERNAL TECHNOLOGY TRANSFER

Intellectual Property

One of the issues that is common to both internal and external technology transfer is intellectual property (IP). I am not a lawyer, and so I obviously cannot give legal advice on how to protect IP or how to treat others' IP. What I can do is point out some places where the IP issues get thorny and make some business recommendations about how to deal with them.

Anything received from a third party may carry restrictions on how it may be used, and more importantly whether it can be redistributed in its original or modified form, or combined with some other components. This includes libraries of software routines, data, copyrighted works and designs, and patents. These restrictions can come from explicit license agreements that accompany the third-party components, or they could come from any of a number of different laws, including patent, copyright, trademark, trade secret, and export. Any time a third-party component is used in one's work, that creates a business risk of constraining the ability to transfer the work to another party, either because one does not possess the right to do so, or the rights that are possessed are not sufficient to the needs of the party that wants to license it. This question fits into the larger scheme of what it traditionally called the "build or buy" decision: Whether it makes more business sense to build something oneself, or to buy or license it from a third party.[3]

I strongly recommend that whenever possible, the IP issues should be dealt with at the time a third-party component is acquired, rather than waiting until an opportunity to transfer it. This accomplishes two things. First, it allows one to negotiate and make business decisions about acquiring the component before there is a commitment and dependence on the component built into your technology—once the dependency is there, the "switching cost" is much higher for moving to an alternate and one could be forced to pay a much larger licensing fee than before. Second, it simplifies the technology transfer process. Any company worth its salt will perform a "due diligence" on the technology before they close a licensing deal. Part of that will be an analysis of who really owns the technology, or whether the licensor has acquired the right to further license it. In essence, one will need to prove the right to license your work to them. Do not be surprised if they also ask the researcher to "warrant" the work—to guarantee that one has the right to license all of it to them, and that right will be defended in court if necessary. The bottom line: Clear it up front and save a lot of trouble later.

It All Comes Down to the Relationship

The most significant common aspect of internal and external transfer comes back to the notion that we began with: Technology transfer is a social process that succeeds or fails based on the

---

[3]It is important to point out, though, that building from scratch does not necessarily mean an automatic escape from third parties; for example, one can still violate someone else's patent even with code written from scratch.

relationships that have been built. Technology transfer partners need to trust each other, and that trust is built with communication and follow-through. Researchers should understand clearly and honestly what value they bring to the table at the various stages of the relationship, and they should make it their business to know how their partners see and value what they bring. After all, business is fundamentally the exchange of value between partners who need each others' competencies, and the truly successful companies are the ones that build relationships that last across a continuous series of business transactions.

When beginning a technology transfer effort, assume that what is started that day is the beginning of a working relationship that will last forever. Build the relationship from the ground up with the expectation of ever-increasing levels of cooperation and trust that will allow the partnership to take on ever more challenging technology transfers, in whatever manner is most appropriate to the needs of the business.

There are far too many stories of companies that have struggled with the transfer of technologies that could change the world and failed. As Buderi (2000) describes, the next chapter of this story is being written now. By rewriting the rules to focus on the social side of the process, we can ensure that our best work will see the light of day and this story will have a happy ending.

References

Buderi, R. (2000). *Engines of tomorrow: How the world's best companies are using their research labs to win the future.* New York: Simon & Schuster.

Butler, K. (1990). Collaboration for technology transfer—Or "how do so many promising ideas get lost?" In *Proceedings of the CHI '90 Conference on Human Factors in Computing Systems* (pp. 349-351). New York: ACM Press.

Card, S. K., Moran, T. P., & Newell, A. (1983). *The psychology of human-computer interaction.* Hillsdale, NJ: Lawrence Erlbaum Associates.

Foley, J. (1996). Technology transfer from university to industry. *Communications of the ACM, 39*(9), 30-31.

Hiltzik, M. (1999). *Dealers of lightning.* New York: Harper Collins Publishers.

Isaacs, E. A., & Tang, J. C. (1996). Technology transfer: So much research, so few good products. In *Proceedings of the CHI 96 Conference on Human Factors in Computing Systems* (Vol. 2, pp. 155-156). New York: ACM Press.

Lesko, J., Nicolai, P., & Steve, M. (1998). *Technology exchange in the information age.* Columbus, OH: Battelle Press.

Manning, G. K. (Ed.). (1974). *Technology transfer: Successes and failures.* San Francisco: San Francisco Press.

Mock, J. E., Kenkeremath, D. C., & Janis, F. T. (1993). *Moving R&D to the marketplace: A guidebook for technology transfer managers.* Falls Church, VA: Technology Prospects, Inc.

Singh, G. (1993). From research prototypes to usable, useful systems: lessons learned in the trenches. In *Proceedings of the ACM Symposium on User Interface Software and Technology* (pp. 139-143). New York: ACM Press.

Steinberg, L. (1998). *Winning with integrity.* New York: Villard Books.

·61·

HUMAN VALUES, ETHICS, AND DESIGN

Batya Friedman and Peter H. Kahn, Jr.
University of Washington

INTRODUCTION

Human values and ethical considerations no longer stand apart from the human–computer interaction (HCI) community—perhaps in some separate field called "computer ethics"—but are fundamentally part of our practice.

This shift reflects, at least in part, the increasing impact and visibility that computer technologies have had on human lives. Computer viruses have destroyed data on millions of machines. Large linked medical databases can, and often do, infringe on individuals' privacy. Accountability becomes a problem when large computer systems malfunction and result in human deaths. The outcome of the 2000 presidential elections in the United States may have hinged on the poor design of the computerized Florida election ballot (the "butterfly ballot"). On and on, the media portray such problems. Presidential reports (PITAC, 1999) ask that such problems be addressed. Corresponding research agendas on human values have been integrated into recent government-funded programs, such as the National Science Foundation's funding initiative on Information Technology Research. Thus, the writing decades ago by cyberneticists like Norbert Wiener (1953/1985) and computer scientists like Joseph Weizenbaum (1972) now seem prescient. They argued that humans fundamentally control technology, and that we must make wise and humane choices about its design and use.

But, if human values, especially those with ethical import—such as rights to privacy and property, physical welfare, informed consent, trust, and accountability, to name but a few—are important, they are no less controversial. What values count? Who decides? Are values relative? Or are some values universal, but expressed differently across culture and context? Does technology even have values? If not, how do values become implicated in design? Also, it is clear that values can conflict. For example, shared online calendars can enhance trust within an organization, but at the expense of individual privacy. Thus, on what basis do some values override others in the design of, say, hardware, algorithms, databases, and interfaces? Finally, how can HCI designers working within a corporate structure and with a mandate to generate revenue bring values and ethics into their designs?

In this chapter, we review how the field has addressed such questions. The second section reviews positions on how values become implicated in technological designs. The third section distinguishes usability from human values with ethical import. The fourth section reviews foundational issues that make ethics so controversial. We also show how scholars have bounded the moral controversies in ways that make it possible for practitioners to shape their work from an ethical stance. The fifth section reviews the major HCI approaches to the study of human values, ethics, and design: Computer Ethics, Social Informatics, Computer-Supported Cooperative Work, Participatory Design, and Value Sensitive Design. We discuss the strengths (and limitations) of each approach by drawing on exemplar projects. The sixth section calls attention to key values relevant for design: Human welfare, ownership and property, privacy, freedom from bias, universal usability, trust, autonomy, informed consent, accountability, identity, calmness, and environmental sustainability. Each of these values could merit its own chapter; and, indeed, in this *Handbook*, some do. Yet by reviewing them here, however briefly, we highlight their ethical status and thereby suggest that they have a distinctive claim on resources in the design process. The seventh section reviews the special ethical responsibilities that HCI professionals incur by virtue of their professional standing. Finally, in the eighth section, we highlight a few suggestions for moving the field forward.

HOW VALUES BECOME IMPLICATED IN TECHNOLOGICAL DESIGN

Technological innovations implicate human values. Consider, for example, that in the early 1900s, missionaries introduced a technological innovation—steel ax heads—to the Yir Yoront of Australia, a native people. The missionaries did so without regard for traditional restrictions on ownership, and indiscriminately distributed the ax heads to men, women, old people, and young adults alike. In so doing, they altered relationships of dependence among family members and reshaped conceptions of property within the culture (Sharp, 1952/1980). Another example: About four decades ago, snowmobiles were introduced into the Inuit communities of the Arctic and have now largely replaced travel by dog sleds. This technological innovation thereby altered not only patterns of transportation, but symbols of social status, and moved the Inuit toward a dependence on a money economy (Houston, 1995; Pelto, 1973). Or a computer example. Electronic mail rarely displays the sender's status. Is the sender perhaps a curious layperson, system analyst, full professor, journalist, assistant professor, entry-level programmer, senior scientist, or high school student? Who knows until the e-mail is read, and maybe not even then. This design feature (and associated conventions) has thereby played a significant role in allowing electronic communication to cross traditional hierarchical boundaries and to contribute to the restructuring of organizations (Sproull & Kiesler, 1991).

But how exactly do values become implicated in technological designs? Three types of positions have been offered in the literature: Embodied, exogenous, and interactional.

The Embodied Position

The embodied position holds that designers inscribe their own intentions and values into the technology; and once developed and deployed, the resulting technology determines specific kinds of human behavior. Indeed, in the history of science literature (Smith & Marx, 1994), this position is sometimes referred to as technological determinism. To illustrate the idea, consider Latour's (1992) description of the "Berliner lock and key": A device designed to deal with people who forget to lock the door behind them. The device initially works as a normal lock. You put your key into the lock, turn the key, and the door unlocks. But then to remove the key, it is necessary to push it through the key hole to the other side. Then, by moving to the other side of the door, you rotate the key one more turn,

thereby relocking the door. Thus, when used, this technology "determines" that people will lock their doors.

There is a hard and soft version of the embodied position. In the hard version, it is argued that the very meaning and intentions that designers and builders bring to their task literally become a part of the technology (cf. Appadurai, 1988; Cole, 1991). Many people have trouble with this version because it imputes mental states to things that do not seem to have the capacity to have them. As Smith and Marx (1994) write: "Critics of 'hard' determinism question the plausibility of imputing agency to 'technology'.... How can we reasonably think of this abstract, disembodied, quasi-metaphysical entity [technology], or of one of its artifactual stand-ins (e.g., the computer), as the initiator of actions capable of controlling human destiny?" (p. xii). Yet, it is possible that as the field of Artificial Intelligence moves forward—for example, as computational learning systems increasingly mimic human agency—the strong version of the embodied position will gain some currency.

In the more common soft version, it is recognized that objects themselves do not literally embody an intention or value. It is also recognized that designers themselves are shaped by organizational, political, and economic forces; and that because of such forces, a particular technology may never take hold in a society. But, it is argued, if the technology takes hold, then it becomes very difficult for an individual or society to override the values driven by the technology. Think about how the design and deployment of passenger-side front air bags have in a sense held children hostage in the back seat of millions of cars. Or consider that a virus program readily destroys data or grabs CPU cycles (or whatever it is programmed to do). Granted, you could perhaps use the program to teach about self-replicating code, but such activity would in effect be secondary to its primary function. In other words, to execute the computer virus code likely leads to the destruction of data. Similarly, if your goal in online interactions is to keep data secure, you can design technical security mechanisms, such as mandatory encryption, certificates, and anonymizers. Such designs ensure behavior. Thus, according to Akrich (1992), a large part of a designer's work is that of "inscribing" or embodying his or her vision of "the world in the technical content of the new object" (p. 208).

The Exogenous Position

The exogenous position holds that societal forces—that involve, for example, economics, politics, race, class, gender, and religion—significantly shape how a deployed technology will be used. Consider, for example, the Internet. In many ways, its basic file-sharing functionality remains unchanged from its initial inception. Files are still broken into packets, and packets are routed and then reassembled. Why is it, then, that this underlying technology was primarily used for nearly a quarter of a century by the scientific and educational communities before becoming appropriated by the private sectors for commerce? Once appropriated, why did the development of e-commerce occur so rapidly? According to the exogenous theorists, answers to such questions cannot be found in the supposedly embodied values of the technology itself. Rather:

To understand the origin of a particular kind of technological power, we must first learn about the actors. Who were they? What were their circumstances? ... Why was this innovation made by these people and not others? Why was it possible at this time and this place rather than another time and place? ... Instead of treating "technology" per se as the locus of historical agency... [these advocates] locate it in a far more various and complex, social, economic, political and cultural matrix (Smith & Marx, 1994, p. xiii).

In turn, the exogenous position has been used and developed in various ways. Bulliet (1994), for example, examines three technologies that emerged in the Islamic world in the fifth century: Block printing, the harnessing of draft animals, and wheeled transport. Each technology failed to have an immediate transforming social and economic impact, not because of its lack of economic advantage, but because of "social filters" that involved race, class, and lifestyle. Perdue (1994) explores the comparative history of agrarian societies (medieval Western European, eighteenth-century Russian, and fourteenth-century Chinese). He argues that studies of technology need contextual accounts that integrate environmental, technological, social, and cultural elements. Scranton (1994) draws on postmodern theory to argue that local contingencies, diversities, disjunctions, multiple oppositions, and contrasting norms should be central to exogenous explanations. Hughes (1994) suggests that younger developing technologies tend to be more open to sociocultural influences, whereas older, more mature technologies tend to be more deterministic. Through many such studies, a recurring view is that technological systems are not value neutral, but invariably favor the interests of people with economic and political power (Noble, 1991; Smith, 1994; Winner, 1986).

The Interactional Position

The interactional position holds that whereas the features or properties that people design into technologies more readily support certain values and hinder others, the technology's actual use depends on the goals of the people interacting with it. A screwdriver, after all, is well-suited for turning screws, and yet amenable as a poker, pry bar, nail set, cutting device, and tool to dig up weeds. Moreover, through human interaction, technology itself changes over time. On occasion, such changes (as emphasized in the exogenous position) can mean the societal rejection of a technology, or that its acceptance is delayed. But more often it entails an iterative process whereby technologies get invented and then redesigned based on user interactions, which then are reintroduced to users, further interactions occur, and further redesigns implemented. Typical software updates (e.g., of word processors, browsers, and operating systems) epitomize this iterative process.

Two sides of the interactional position have been emphasized in the literature. One side emphasizes the properties designed into the technology (Friedman, 1997a). For example, let us for the moment agree that disabled people in the work place should be able to access technology, just as they should be able to access a public building (Perry, Macken, Scott, & McKinley, 1997). As system designers, we can make the choice

to try to construct a technological infrastructure that disabled people can access. If we do not make this choice, then we single-handedly undermine the human value of universal access.

The other side of the interactional position emphasizes how users use the technology in the context of social-organizational structures. Orlikowski (2000), for example, studied the use of Lotus Development Corporation's *Notes* software by two groups within a large multinational consulting firm: Technologists and consultants. The technologists used *Notes* extensively. They used e-mail, maintained electronic discussions with *Notes* databases, and created their own database designs. Moreover, "supported by the cooperative norms of technical support, the technologists used many of the properties of *Notes* to promote their collective technical work, and to cooperate with each other" (p. 415). In contrast, the consultants used *Notes* minimally, sometimes even begrudgingly. Orlikowski found that what she calls such "technology-in-practice" was enacted for at least three different reasons. First, some consultants had doubts about the value of *Notes* for their performance. Second, in contrast to the technologists, the consultants were under a time-based billing structure. "Because many consultants did not see using *Notes* as an activity that could be billed to clients, they were unwilling to spend time learning or using it" (p. 416). Third, consultants feared that the collaborative properties of *Notes* would threaten their status within the company. Thus, Orlikowski proposes "a view of technology structures, not as embodied in given technological artifacts, but as enacted by the recurrent social practices of a community of users" (p. 421).

Regardless of emphasis, from the interactional position it should be clear that design and social context matter, dialectically. Moreover, users are not always powerless when faced with unwelcomed value-oriented features of a technology.

DISTINGUISHING USABILITY FROM HUMAN VALUES WITH ETHICAL IMPORT

The language and conceptualizations within HCI currently provide solid means by which to pursue issues of usability (Adler & Winograd, 1992; Nielsen, 1993; Norman, 1988). Usability refers to characteristics of a system that make it work in a functional sense, including that it is easy to use, easy to learn, consistent, and recovers easily from errors.

Some HCI designers, however, have a tendency to conflate usability with human values with ethical import. (In the next section, we discuss what exactly is meant by the term moral. Here, it can be understood to involve issues of fairness, justice, human welfare, and virtue.) This conflation arises insofar as usability is itself a human value, although not always a moral one. But when it is, both can be addressed by the same design. For example, systems that can be modified by users to meet the needs of specific individuals or organizations can both (a) enhance usability and (b) help users to realize their goals and intentions: The moral value of autonomy. Other times, however, usability can conflict with human values with ethical import. Nielsen (1993), for example, asks us to imagine a computer system that checks for fraudulent applications of people who are applying for unemployment benefits. Specifically, the system asks applicants numerous personal questions, and then checks for inconsistencies in their responses. Nielsen's point is that even if the system receives high usability scores, some people may not find the system socially acceptable, based on the moral value of privacy.

In terms of a general framework, four pairwise relationships exist between usability and human values with ethical import.

1. *A design is good for usability and independently good for human values with ethical import.* This relationship is exemplified previously, where adaptable systems can also promote user autonomy. Another example involves browser designs that offer users more efficient cookie management than currently offered in Netscape or Internet Explorer and also promote values of informed consent and privacy (Friedman, Howe, & Felten, 2001).

2. *A design is good for usability, but at the expense of human values with ethical import.* This relationship is exemplified in the previous example offered by Nielsen, where a highly usable system undermines the value of privacy. Tang (1997) provides another example by means of a case study of a team designing a workstation. At one point, the team was trying to decide how to power a microphone, and finally decided to power it directly from the workstation. Consequently they eliminated a separate hardware on/off switch on the microphone. From a usability perspective, Tang points to benefits of this design. Users, for example, no longer had to remember to turn the microphone off when it was not in use (to conserve battery power), or be inconvenienced by replacing a dead battery. However, despite these usability advantages, Tang found that some users thought the design—by not allowing them direct control over the microphone—undermined their privacy and security.

3. *A design is good for human values with ethical import, but at the expense of usability.* This relationship is exemplified by choosing a browser preference with the setting "accept or decline each cookie individually." The values of privacy and informed consent are well supported; but, for most people, the nuisance factor is too high. Another example (the flip side of an earlier one) arises when a system is purposefully made adaptable to promote autonomy, but in the process the system becomes unwieldy and difficult to use.

4. *A design good for usability is necessary to support human values with ethical import.* This relationship is exemplified by security systems that are not so cumbersome to use that either critical features of the security system are disabled or individuals invent workarounds that compromise the security. For example, if a security system requires 3 passwords and 10 steps to login, a user's workaround might involve taping a yellow sticky to the side of the computer screen with the passwords. Other instances of this relationship arise when morality requires that all individuals in a specified group be able to successfully use the system. For example, to have a fair national election using a computerized voting system, all citizens of voting age must be able to use the system.

HCI professionals are often responsible for usability. Accordingly, it is important that we be aware of the possible relationships between usability and human values with ethical import.

At times, the two support one another. Other times we need to give ground on usability to promote human values with ethical import, or, conversely, give ground on human values with ethical import to promote usability. Such optimizations require judicious decisions, carefully weighing and coordinating the advantages of each.

MORAL, PHILOSOPHICAL, AND PSYCHOLOGICAL FOUNDATIONS

HCI professionals—like people in many fields—may sometimes wonder if morality is too controversial to be integrated in a principled way into their work. After all, who is to say for another person what is right or wrong, or good or bad? How do we make sense of the seemingly different moralities among people? Is not morality relative to person or at least culture?

Such questions form the backdrop to a wide range of issues that moral philosophers and psychologists have pursued. In this section, we review relevant literature to help bound moral controversies such that HCI researchers and practitioners can, with legitimate grounding, move forward proactively in shaping their work from a moral stance.

Moral Theory. One common starting point in moral discussions is to raise the question, "What do you mean by morality?" Or, "How do you define it?" In response, moral philosophers have offered what can be viewed as three overarching approaches toward developing moral theory: Consequentialist, deontological, and virtue-based. Briefly stated, consequentialist theories maintain that a moral agent must always act to produce the best available outcomes overall (see Scheffler, 1982, for an analysis). Utilitarianism is the most common form of consequentialism, wherein a moral agent should act to bring about the greatest amount of utility (e.g., happiness) for the greatest number of people. In contrast, deontological theories maintain that there are some actions that a moral agent is forbidden to do, or, in turn, must do, regardless of consequences (Dworkin, 1978; Gewirth, 1978; Kant, 1785/1964; Rawls, 1971). For example, a deontologist might maintain that it is immoral to torture an innocent child even if such an act would bring about great good (e.g., to prevent a bombing). Along similar lines, it can be argued that vendors who collect and sell information about individuals' web-browsing activity without individuals' knowledge and consent violate individuals' right to privacy. Indeed, most any argument for a basic right makes a deontological move.

Both consequentialist and deontological theories are centrally concerned with answering the fundamental question, "What ought I to do?" In turn, answers are often viewed to be morally obligatory, meaning that the action is required of every moral agent "regardless of whether he wants to accept them or their results, and regardless also of the requirements of any other institutions such as law or etiquette" (Gewirth, 1978, p. 1). Moreover, such obligatory prescriptions are often framed in the negative (as in "Thou shall not steal"), such that it is possible to fulfill the prescription (e.g., just don't steal). In turn, virtue-based theories are centrally concerned with answering the fundamental question, "What sort of person ought I to be?," in which the focus is on long-term character traits and personality (see Louden, 1984). This tradition dates back to Aristotle's delineation in *Nichomachean Ethics* of the ethical virtues (e.g., courage, temperance, friendship, wisdom, and justice), and developed, for instance, by Foot (1978), MacIntyre (1984), and Campbell and Christopher (1996). In virtue ethics, the prescription is often viewed to be morally discretionary, meaning that whereas an action is not morally required, it is conceived of as morally worthy and admirable based on considerations of welfare and virtue (Kahn, 1992). Moreover, such discretionary prescriptions are often framed in the positive (as in "be charitable"), such that it is not possible to completely satisfy the declaration. After all, one cannot continuously practice charity without soon becoming destitute oneself (Fishkin, 1982).

With these core distinctions in mind—between a theory of the right (both consequentialism and deontology) that is largely viewed as obligatory, and a theory of the good (virtue theory) that is largely viewed to be discretionary—it is possible to characterize a wide range of HCI designs. For example, consider Mattel Corporation's "Barbie doll" Web site that was up and running during the latest census collection in the United States. On the Web site, Barbie posed as a census taker and asked children to provide information about their family. Presumably, Mattel sought to use the resulting information to market their products more effectively. Is this action moral? Most people would presumably say no. But why? From a deontological position, such a deceptive web design violates a moral obligation (e.g., the obligation not to intentionally deceive others, let alone to deceive children for a corporation's material gain), and should not have been designed and deployed.

Other human values articulated in HCI fit within a virtue orientation. For example, SeniorNet is an organization that brings seniors together via computer networks. Mynatt, Adler, Ito, Linde, and O'Day (1999) found that "SeniorNetters repeatedly commented on the warmth and friendliness of the community as something that differentiated SeniorNet from other net communities, and as a reason for their participation and comfort with the community" (p. 232). Are values such as warmth and friendliness moral values? For a consequentialist and even more so for a deontologist the answer is probably no; a person, after all, is presumably not under a moral obligation to engage in "warm-hearted" actions. But for a virtue ethicist the answer is fundamentally yes insofar as such values are conceived to be central to moral personhood. Or consider Web sites (such as Alfie.com) that offer children a menu of computer games, like "pinball" or "miniature golf," that children can play. Most of these games do not promote such virtues as friendship, caring, or compassion. Must they? Presumably, even virtue theorists would say that not every game has to. Rather, promoting such virtues is usually conceived as discretionary—perhaps morally praiseworthy if done, and perhaps contributing to an account of human flourishing—but not morally obligatory.

Moral Epistemology. This discussion raises the question, "Who is to say what's moral?" or "How do we know?" Such questions move us into the field of moral epistemology: The study of the limits and validity of moral knowledge. Often at

stake is whether it is even possible for a moral statement to be objectively true or false, and for a moral value to be objectively right or wrong, or good or bad. A wide variety of positions have been taken. For instance, some believe that moral knowledge corresponds to or approaches a correspondence with a moral reality that exists independent of human means of knowing (Boyd, 1988; Sturgeon, 1988). Others believe moral knowledge can be objectively grounded by constructing rational principles that strive for coherence and consistency while building on the common ground and specific circumstances of a society (Dworkin, 1978). Others believe that the only thing that can be said of moral knowledge is that it can be true subjectively for an individual or culture depending on that individual's or group of individuals' desires, preferences, and goals (Mackie, 1977; Rorty, 1982). Finally, others believe that any moral knowledge is unattainable, even in a weak sense (the full skeptic's position; see Nagel, 1986, for a characterization).

The skeptic's position—that moral knowledge is unattainable—reflects the morally relativistic position described previously: That no one can say what is really right or wrong and so ethics and values become, at best, a personal choice in terms of HCI design. But consider: No one can prove that at this moment you are not really just a brain in a vat being stimulated by electrical impulses to think that you are reading this chapter. Still, you have compelling reasons to think that proposition false. Similarly, with morality. Although the skeptic's position cannot be proved conclusively false, moral philosophers have provided compelling reasons not to believe it (Nagel, 1986; Williams, 1985).

Moral Variability and Universality. Anthropologists often document moral differences between cultures. For example, from some anthropological accounts, we learn that devout Hindus believe that it is immoral for a widow to eat fish, or for a menstruating woman to sleep in the same bed with her husband (Shweder, Mahapatra, & Miller, 1987). Other accounts document that members of the Yanomamo tribe of Brazil at times practice infanticide and that the women are "occasionally beaten, shot with barbed arrows, chopped with machetes or axes, and burned with firebrands" (Hatch, 1983, p. 91). Per our discussion, there are three ideas to understand about such examples.

1. *Variability in human practice does not prove or disprove the moral skeptic's position.* Imagine going to a culture where the people there did not believe in logical transitivity. (If A = B and B = C, then A = C.) That finding would be interesting psychologically and culturally, but presumably has no bearing on whether logical transitivity is true or false. It is or it is not. If it is true, then people who think otherwise are simply wrong. So, too, with the moral life. People can believe that a certain act (such as shooting women with barbed arrows) is moral; but documenting such a belief does not make it so.

2. *When moral differences occur between peoples, it is not necessarily the case that the practices are believed legitimate by the victims.* For example, in Hatch's (1983) report on the Yanomamo, he also reports that the women did not appear to enjoy such physically abusive treatment, and were seen running

in apparent fear from such assaults. Psychological data of a similar kind can be found in a study by Wainryb (1995) on the Druze population in Israel. The Druze largely live in segregated villages, are of Islamic religious orientation, and organized socially around patriarchal relationships. The father, as well as brothers, uncles, and other male relatives, and eventually a woman's husband, exercise considerable authority over women and girls in the family, and restrict their activities to a large degree. However, when these women were interviewed, a majority of them (78%) unequivocally stated that the husband's or father's demands and restrictions were unfair. Thus, Yanomamo and Druze women— like many women in Western societies—are often unwilling victims within what they themselves perceive to be an uncaring or unjust society. In such situations, it is less the case that societies differ on moral ground, and more that some societies (Western societies included) are involved explicitly in immoral practices.

3. *Moral variability may be much less pervasive than many people suppose.* Reconsider, for example, the Shweder et al. (1987) report that devout Hindus believe it is immoral for a widow to eat fish, or for a menstruating woman to sleep in the same bed as her husband (but two of many dozens of their examples). At first glance, for Western eyes at least, such moral beliefs do seem different. However, when Shweder et al.'s data were reanalyzed by Turiel, Killen, and Helwig (1987), they showed bases for not only difference, but also moral commonality. In their reanalysis, for example, Turiel et al. found that devout Hindus believed that harmful consequences would follow from a widow who ate fish (the act would offend her husband's spirit and cause the widow to suffer greatly), and from a menstruating woman who sleeps in the same bed with her husband (the menstrual blood is believed poisonous and can hurt the husband). Although such beliefs, themselves, differ from those in Western culture, the underlying concern for the welfare of others is congruent with it. More generally, Turiel et al.'s claim is that when researchers differentiate informational and metaphysical assumptions about the world from moral judgments based on those assumptions, then the moral life often takes on a greater universal cast (cf. Turiel, 1998, in press).

When analyzing moral variability, conceptualizations of morality that entail abstract characterizations of justice and welfare tend to highlight moral universals, whereas definitions that entail specific behaviors or rigid moral rules tend to highlight moral cross-cultural variation. Typically, theorists who strive to uncover moral universals believe they are wrestling with the essence of morality, with its deepest and most meaningful attributes. In contrast, theorists who strive for characterizing moral variation argue that, by the time you have a common moral feature that cuts across cultures, you have so disembodied the idea into an abstract form that it loses virtually all meaning and utility. For instance, in the example of devout Hindus who believe that by eating fish a widow hurts her dead husband's spirit, is the interesting moral phenomenon that Hindus, like ourselves, are concerned with not causing others harm? Or is the interesting moral phenomenon that Hindus believe in spirits that can be harmed by earthly activity?

In our view, both questions have merit, and a middle ground provides a more sensible and powerful approach for the HCI

community: One that allows for an analysis of universal moral values, as well as allowing for these values to play out differently in a particular culture at a particular point in time (Friedman, 1997b; Kahn, 1991, 1999; Kahn & Lourenço, 1999).

Designing for Diversity. In this chapter, we will continue to draw on the preceding analysis to help us review the HCI literature, and to offer morally principled design methods that respect culture and context. As a case in point, imagine you are designing a computerized voting booth. At what height would you place the electronic ballot? A reasonable answer might go something like: "It depends—tell me, how tall are the people who vote? Moreover, at a later time will the voting booth be used by other people of different height? After all, a voting booth designed only for players of the National Basketball Association will disenfranchise most voters in Japan." In other words, the universal value is to enfranchise all voters, but the specific mechanism by which to do so may need to be adaptable to specific contexts and cultures. The general principle then is that designs need to be robust enough to substantiate the value under consideration and yet adaptable enough so that different cultures (or subcultures) can use the designs in their own way.

Many problems occur when this principle is ignored or when unanticipated users of a system emerge. For example, as described by Friedman and Nissenbaum (1996), since the early 1970s, the computerized National Resident Medical Match Program has placed most medical students in their first jobs. In the system's original design, it was assumed that only one individual in a family would be looking for a residency. At the time, such an assumption was perhaps not out of line, because there were few women residents. But as women have increasingly made their way into the medical profession, marriages between residents have become more frequent, and bias against placing couples emerged. Another example involved a dog icon that was used to indicate the printing orientation of a printer, landscape or vertical. When the printer was shipped to Islamic countries, the vendor discovered that people in such countries often considered dogs as offensive animals. Or consider the case of competitive educational software that was exported to Micro Asia, with dismal results because the value of competition promoted by the software clashed too strongly with the culture's emphasis on cooperation. Or consider that data protection laws and policies differ across national boundaries (Agre & Rotenberg, 1998; Rotenberg, 2000). For example, Bennett (1998) describes how most of the European countries have applied the same data-protection policies to both the public and private sector. However, the United States, Canada, Australia, and Japan have preferred "to regulate only the public sector's practices and to leave the private sector governed by a few sectoral laws and voluntary codes of practice" (p. 100). Thus, lenient designs from the private sector of these latter countries need to be adaptable to transfer readily to European countries.

Granted, building value adaptability into systems requires additional time and financial resources. Also, from a user's perspective, additional options add further complexity and challenges. Thus value adaptability is not always the best way to go. But it often is, especially when we can anticipate that similar values will play out in different ways for different users. In such cases, not only are a larger number of human lives enhanced from an ethical standpoint, but also from an economic standpoint such systems increase market share and generate profits.

APPROACHES TO HUMAN VALUES AND ETHICS IN DESIGN

The computing field has addressed issues of human values and ethics by means of a handful of approaches. To some extent, these approaches overlap with one another. For example, sociotechnical analyses, which are central to Social Informatics, often form the front end of efforts in Participatory Design, and are incorporated into the empirical investigations of Value-Sensitive Design. Both Computer-Supported Cooperative Work and Participatory Design share particular interests in collaboration in the work place. But, that said, the approaches differ significantly when considering how each integrates their respective positions on moral epistemology, methods, and contexts studied.

In this section, our goal is not to review each approach comprehensively (which is beyond the scope of this chapter). Rather, we seek to show how each approach contributes to our understanding of how to integrate human values and ethics in design.

Computer Ethics

When applied moral philosophers brought their talents and energies to bear on understanding the impact of computer technologies on social life, the field of Computer Ethics was born (Bynum, 1985; Johnson, 1985; Moor, 1985).

In their resulting examinations, computer ethicists have embraced two complimentary goals. One goal has been to utilize existing moral theory to bring clarity to issues at hand, and—at appropriate times—to proscribe norms of behavior. For example, in the computer science literature, the term trust has often been used synonymously with security (Schneider, 1999). Yet, drawing on ethical theory, Nissenbaum (1999) has shown that these two terms need to be distinguished. For example, if, as HCI professionals, our goal is to create a place where people feel safe in their online interactions, we can move in two design directions. We can move forward with technical solutions that involve security features like locks, keys, passwords, and encryption. Or we can understand how trusting relationships are created and fostered, and design them into our online systems. Each direction leads to a very different online environment. The strength of this type of philosophical contribution is that it helps translate moral abstractions into crisp working conceptualizations that HCI professionals can use.

A second goal of computer ethicists builds on the innovations of the technology itself. For the technologies have not only generated new entities (computer programs, the Internet, Web pages), but have also enlarged the scale of activities (data mining), increased the power and pervasiveness of its effects (ease of communication), and often become invisible to human purview (using computers for surveillance). Thus, computer ethicists have been interested in understanding how such

innovations extend the boundaries of traditional ethical concepts. Moor (1985), for example, examines the ways in which the invisibility of computers affect human lives. *Invisible abuse* is the intentional use of the invisible operations of a computer to engage in unethical conduct, such as the invasion of the property and privacy of others. *Invisible programming values* are those values that are embedded in a computer program, such as bias. *Invisible complex calculations* refer to calculations that are too complex for human inspection and understanding. In all three situations, Moor argues that computer technologies raise special ethical issues, different from other forms of technology, and that the task of computer ethics is to fill in what he refers to as the resulting "conceptual vacuum" and "policy vacuum" (p. 266).

One theoretical debate in this literature is whether or not the technological innovations fundamentally challenge ethical theory (Johnson & Miller, 1997; Spinello & Tavani, 2001). In other words, is it the case that the innovations are so qualitatively different and far-reaching from past ones that traditional ethical theories have to be, if not abandoned, at least significantly revised? Or is it the case that the technology simply offers a new domain within which traditional ethical theory works? Toward understanding this debate, a helpful analogy can be made to literature, wherein there exist the categories (genres) of fiction and nonfiction. These categories seem clear enough until we encounter a new form of writing that blurs the boundaries, such as historical novels. Does that new form undermine the traditional categories? Philosophers like Searle (1983) argue no: That the categories of fiction and nonfiction fundamentally remain, and what the new literary form shows is that writing can embody aspects of both categories. Similarly, it can be argued that whereas the technology can blur traditional ethical boundaries, and demand further refinements and clarity of moral theory, that ethics itself has not changed, nor its fundamental precepts.

Regardless of how one views this debate, it is clear that the field of Computer Ethics advances our understanding of key values that lie at the intersection of computer technology and human lives. However, there are some limitations to the field's contributions to HCI. For one, Computer Ethics often remains too divorced from technical implementations. How exactly, for example, can HCI professionals build interfaces that enhance trust within a community of users? Or how exactly do we address the value problems that arise through invisible computing? In addition, Computer Ethics has focused too often on a single value at a time. Yet, as HCI professionals, we commonly wrestle with design trade-offs between competing moral values. Collaborations between computer ethicists and HCI professionals would be a fruitful way to address these limitations.

Social Informatics

In the second section, we reviewed embodied, exogenous, and interactional positions on how values become implicated in technology. Recall that the embodied position holds that designers inscribe their own intentions and values into the technology; and once developed and deployed, the resulting technology is said to determine specific kinds of human behavior. As also

noted, many researchers find such a position highly untenable, and instead emphasize the social context in which information systems are used. Recall, for example, Orlikowski's (2000) study, reviewed earlier, where she investigated the effects when Lotus Development Corporation's *Notes* was introduced into a large corporation. She found that, at least in part, the incentive systems in the corporation did more to influence how, and whether, *Notes* was used than the mere capability of the software itself.

Over the last 25 years, this emphasis on the social context of information technologies has been the subject of systematic research (Attewell, 1987; Borgman, 2000; Iacono & Kling, 1987; Kiesler, 1997; King, 1983; Kling, 1980; Orlikowski, 1993; Poltrock & Grudin, 1994). The research has appeared under many labels, including social analysis of computing, social impacts of computing, information systems, sociotechnical systems, and behavioral information systems. In more recent years, this overarching enterprise has begun to coalesce within a new field called Social Informatics (Kling, Rosenbaum, & Hert, 1998; Kling & Star, 1998; Sawyer & Rosenbaum, 2000). As defined by participants at the 1997 National Science Foundation-sponsored workshop on this topic, Social Informatics is the interdisciplinary study of the design, uses, and consequences of information technologies that takes into account their interaction with institutional and cultural contexts.

To illustrate the value of Social Informatics, Kling (1999) contrasts the design and functioning of two electronic journals: *Electronic Transactions of Artificial Intelligence* (ETAI) and *Electronic Journal of Cognitive and Brains Sciences* (EJCBS). Both journals envision attracting high-quality papers. Both journals also have implemented technology that works effectively. However, Kling argues that differences in their sociotechnical designs lead ETAI to prosper and EJCBS to wane. For example, articles submitted to ETAI are reviewed in a two-phase process. In phase 1, the article is open to public online discussion for a period of 3 months. Based on the resulting discussions, authors have an opportunity to revise their papers. In phase 2, the article undergoes a quick confidential peer review. In contrast, articles submitted to EJCBS are evaluated by their general online readership. Articles that receive a minimum score are then transferred to an archive of accepted papers. EJCBS has been designed as much as possible as an autonomous system that would run on its own after it was programmed. It removes editorial attention from the publishing process, and instead relies on a readers' plebiscite. But such a design, according to Kling, misconstrues the social context of successful academic journal publishing, one that requires a lively group of authors and readers, and attention from senior scientists in the field.

Work in Social Informatics has been successful in providing sociotechnical analyses of deployed technologies. Yet, in terms of its contributions to HCI, Johnson (2001), for example, writes: "One aspect that still confounds me is how to reconcile the basic premise of social informatics—that it is critical to gain knowledge of the social practices and values of the intended users—with the basic work of system developers. How, if at all, can programmers practice and apply social informatics?" (p. 18). Granted, Kling (1999) recognizes the importance, for example, of "workplace ethnography, focus groups, user participation in design teams, and participatory design strategies." But then he

says "to discuss it [these methods] in detail here would lead us away from our focus on the structural elements of a sociotechnical analysis." Thus, it appears that Social Informatics can move in at least two directions. One direction leads to developing the sociotechnical analyses, and viewing this work as complimenting work in design (and other areas, such as Computer Ethics). Another direction leads to an expansion of Social Informatics such that it fundamentally embraces design (and other areas) into its theoretical framework.

Computer-Supported Cooperative Work

Although social informatics has emphasized the sociotechnical analyses of deployed technologies, the field of Computer-Supported Cooperative Work (CSCW) has, for some time, focused on the design of new technologies to help people collaborate effectively in the workplace (Galegher, Kraut, & Egido, 1990; Grief, 1988; Grudin, 1988). Groupware is the name often used for software that seeks to facilitate CSCW goals. The history of CSCW goes back to early work in various research laboratories, like Xerox PARC, EuroPARC, Bell Lab, and IBM. There, computer professionals, working within relatively small groups themselves, and sometimes remotely, sought to improve their collaborations.

Typically, then, the values considered in CSCW designs have been closely tied to group activities and workplace issues. Cooperation has been, of course, an overarching value. But, in addition, the field has paid attention to such values as privacy, autonomy, ownership, commitment, security, and trust. Isaacs, Tang, and Morris (1996), for example, designed a system to support informal interactions in the workplace, using the piazza (the plaza) as the metaphor. The application provided means for workers to opt out of piazza interactions, thus protecting workers' privacy and autonomy. Olson and Teasley (1996) report on the planning, implementation, and use of groupware tools over the course of a year in a real group with remote members. One of their key findings was that "social responsibility and commitment appeared diminished or missing when people did not meet face-to-face" (p. 425). Dewan and Shen (1998) discuss an access-control model that accounts for joint ownership of shared objects, different ownership rights for different types of users, and the delegation of access rights (security). Hudson and Smith (1996) sought after methods that allow workers to share video information about themselves to their colleagues while providing protections for privacy. One solution was to shadow images on the video screen so that colleagues can tell that you are in your office, but not what you are doing. Fuchs (1999) developed a notification service for awareness information that also addresses potential conflicts between awareness and privacy. Van House, Butler, and Schiff (1998) examined how in a workplace with environmental planning data sets that trust is created and assessed, and "how changes in technology interact with those practices of trust" (p. 335).

CSCW has traditionally focused on the workplace, as its name implies. Yet if the recent Conference Proceedings of CSCW are any indication, then the field is quickly expanding to include non-workplace settings. For example, in the CSCW 1996

Proceedings, about 6% of the papers involved a non-workplace setting. In 1998, it was 13%. In 2000, it was 31%. The CSCW 2000 conference co-organizers note: "the fact that these [work] place technologies are available not only in working settings but also in homes means that the focus of our attention [in the CSCW community] has broadened to encompass a much wider range of activities that we could have imagined when the CSCW conference series began in 1986" (Dourish & Kiesler, 2000, p. v).

Correspondingly, the range of values that the CSCW field has begun to analyze has also started to expand. For example, Mynatt et al. (1999) delineate the values of safety, civility, warmth, and friendship that are fostered by SeniorNet, an organization that brings seniors together through computer networking technologies. O'Neill and Gomez (1998) describe a project that links middle and high school science students with working scientists as mentors for the students' science projects. The researchers "illustrate the unique dynamics of these relationships, consider their technical and social demands, and discuss the potential for CSCW systems to help sustain long-term help relationships by better accommodating their needs" (p. 325). McCarthy and Anagnost (1998) explore the social ramification of a group preference agent for music in a shared environment, a fitness center. By technical means, they thus seek to democratize the music selection process. Palen, Salzman, and Youngs (2000) tracked 19 new cell mobile phone users for 6 weeks. In their discussion, they call attention to issues of unfairness that arose in the context of use.

It remains unclear how much more the field of CSCW can expand to include non-workplace settings before its very name (that has "work" in its title) becomes an historical artifact rather than a description of its current activity. But that is a quibble with nomenclature. The direction of the field seems intellectually vibrant and exciting. Thus, as CSCW continues to expand into a broader range of human activity, it will increasingly encounter and thereby take hold of a broader range of human values with ethical import.

A substantive question thus emerges: How should we understand the epistemic standing of moral values within a CSCW framework? One way of currently reading the field is that moral values are simply relative to culture (and to the "culture" of any particular work group). Harrison and Dourish (1996), for example, argue that "privacy is relative, not a set of psychological primitives" (p. 71). As such, these values should be respected only if the workers themselves think the values are important. Yet, as the field continues to broaden, the tensions embedded in this perspective will continue to become cause of concern. Greenbaum (1996), for example, argues that CSCW needs to consider the political dimensions of labor and not assume that work is cooperative. Suchman (1994), too, emphasizes the power hierarchies within social organizational groups. In such situations—which presumably arise within every organization, at least to some extent—then values often will conflict. Perhaps management (which holds the power) seeks efficiency over privacy, whereas workers seek the converse. What does a designer do? Thus, once CSCW analyses move beyond largely homogeneous groups and into organizational structures, then potentially a principled position on the moral standing of human values will be required.

Participatory Design

In Norway, in the early 1970s, there was a general consensus that computer systems should not deskill workers, but enhance skill, protect crafts, and foster meaningful work. At that time, strong labor unions also helped enact into law a national *codetermination agreement*. This agreement entitled workers along with management to determine which technologies are introduced into the workplace (Kuhn & Winograd, 1996). Thus, emerging from this social structure was a new approach to system design—Participatory Design—that fundamentally sought to integrate workers' knowledge and a sense of work practice into the system design process (Bjerknes & Bratteteig, 1995; Bodker, 1990; Ehn, 1989; Greenbaum & Kyng, 1991; Kyng & Mathiassen, 1997).

In light of value considerations, then, Participatory Design has embedded within it a commitment to democratization of the workplace and human welfare. It also has what can be viewed as virtue-based moral commitments insofar as it seeks to account for meaningful activity in everyday lives.

At least five important methods have emerged from, or have been elaborated by, the field of Participatory Design. (1) *Identifying Stakeholders*—Toward achieving designs that work, it is often necessary to identify the people they directly and indirectly effect (cf. Korpela et al., 1998). (2) *Workplace Ethnography*—Ethnographies document the practices and beliefs of a group from the group's perspective. Methods include analyzing documents and artifacts in the group's environment, participant observation, field observations, surveys, and formal and informal interviews (cf. Kensing, Simonsen, & Bodker, 1998; Mackay & Fayard, 1999). (3) *Future Workshops*—A future workshop is a method to uncover common problems in the workplace and to solve them. This method is divided into three phases. As described by Kensing and Masden (1991), "the Critique phase is designed to draw out specific issues about current work practice; the Fantasy phase allows participants the freedom to imagine 'what if' the workplace could be different; and the Implementation phase focuses on what resources would be needed to make realistic changes" (p. 157). (4) *User Participation in Design Teams*—Four ways users participate in design teams can be characterized (Kuhn & Winograd, 1996): Directness of interaction with designers; length of involvement in the design process; scope of the participation; and the degree of control over the design decisions. User participation has been central to Participatory Design projects conducted in Europe and North America. In addition, Korpela et al. (1998) write that their 7-year experience in Nigeria suggests that user participation is also a must in developing countries. (5) *Mock-ups and Prototypes*—Both mock-ups and prototypes create physical representations of technological designs. A mock-up looks roughly like the artifact it represents, but completely lacks the artifact's functionality. Mock-ups usually occur very early in the design process. As noted by Ehn and Kyng (1991), mock-ups encourage "hands-on" experience, and are understandable to the end-user, cheap to build, and fun to work with. In turn, prototypes incrementally embed functionality into the artifact through successive iterations. Both methods help end-users envision the potential for the proposed technology and the resulting changes in work practice (Greenbaum & Kyng, 1991).

Some HCI practice in the United States has followed closely in the Scandinavian style of Participatory Design. But what has been embraced even more commonly is *Pragmatic Participatory Design*, a term we use to refer to the above rich constellation of methods and design techniques, but largely stripped of their moral commitments.

Two reasons may help account for why the HCI community in the United States has been resistant to embrace the moral commitments of Participatory Design, while embracing many of its methods. One reason is that although the United States is politically committed to the value of participatory democracy, its capitalistic business culture is not. Thus, the moral values that lie at the core of Participatory Design (participation, democracy, and moral personhood) run counter to values in the U.S. workplace. A second reason can be viewed in light of the cultural homogeneity in the Scandinavian countries. Compared with the United States, for example, these countries historically have been more homogeneous in terms of race, ethnicity, and religion, and thus have encountered fewer opportunities for corresponding prejudices and hostilities. Thus, when applied in more diverse contexts, Participatory Design has too little to say when divisive constituencies argue on the basis of narrowly conceived self interests and hostile prejudices. After all, at least in theory, Participatory Design values each participant's voice, even those that appear uncaring and unjust. This problem has been of concern within the field (Gross, Parker, & Elliott, 1997; Muller, 1997).

Value-Sensitive Design

Given the limitations of the other approaches in integrating ethics and sociotechnical analyses with actual design, another approach has recently emerged called Value-Sensitive Design. This approach seeks to design technology that accounts for human values in a principled and comprehensive manner throughout the design process (Friedman, 1997a). Value-Sensitive Design is primarily concerned with values that center on human well-being, human dignity, justice, welfare, and human rights. This approach is principled in that it maintains that such values have moral epistemic standing independent of whether a particular person or group upholds such values. At the same time, Value-Sensitive Design maintains that how such values play out in a particular culture at a particular point in time can vary, sometimes considerably.

Value-Sensitive Design articulates an interactional position for how values become implicated in technological designs. From this position, Friedman and Nissenbaum (1996) have analyzed bias in computer systems. Cranor and Resnick (2000) have analyzed anonymity in ecommerce. Agre and Mailloux (1997) have analyzed privacy and computerized vehicle-highway systems. Thomas (1997) has analyzed universal access within a communications company. Ackerman and Cranor (1999) have analyzed interface components to safeguard users' privacy on the Internet. Abowd and Jacobs (2001) have called attention to how the design of advanced sensing technologies within and

outside the home can impinge on the individual's right (as established by the Fourth Amendment) to be protected against illegal search and seizure by the government. Shneiderman and Rose (1997) have proposed social impact statements for information systems.

Methodologically, at the core of Value-Sensitive Design lies an iterative process that integrates conceptual, empirical, and technical investigations (Friedman & Kahn, in press). *Conceptual investigations* involve philosophically informed analyses of the central constructs and issues under investigation. Questions include: How are values supported or diminished by particular technological designs? Who is affected? How should we engage in trade-offs among competing values in the design, implementation, and use of information systems? *Empirical investigations* involve social-scientific research on the understandings, contexts, and experiences of the people affected by the technological designs. *Technical investigations* involve analyzing current technical mechanisms and designs to assess how well they support particular values, and, conversely, identifying values, and then identifying and/or developing technical mechanisms and designs that can support those values. As mentioned, these investigations are iterative and integrative. For example, results from the empirical investigations may reveal values initially overlooked in the conceptual investigations, or help to prioritize competing values in the design trade-offs between technical mechanisms and values considerations.

To illustrate this methodology, consider a current project by Friedman, Felten, and their colleagues as they have sought to understand how to design web-based interactions to respect informed consent (Friedman, Howe, & Felten, 2001; Friedman, Millett, & Felten, 2000; Millett, Friedman, & Felten, 2001). They began their project with a conceptual investigation of informed consent itself. What is it? How can it be garnered in diverse online interactions in general, and web-based interactions in particular? To validate and refine their resulting conceptual analysis, and initiate their technical investigation, they conducted a retrospective analysis of existing technology. Namely, they examined how the cookie and web-browser technology embedded in Netscape Navigator and Internet Explorer changed—with respect to informed consent—over a 5-year period, beginning in 1995. (These results are summarized in Informed Consent.) Then, the design work began. Their design improvements are being implemented in the Mozilla browser (the open-source code for Netscape Navigator) and undergoing empirical investigations (usability studies and formative evaluation), which will then reshape the initial technical and conceptual work.

The National Science Foundation recently sponsored two workshops to help shape a research agenda for Value-Sensitive Design. The recommendations from the workshops' final reports (Friedman, 1999; Friedman & Borning, 2001) have focused on the need for (1) *theoretical and conceptual analyses* that study not only particular values in the online context, but also complexities that arise when trade-offs among competing values are required in a design; (2) *translations* of well-analyzed values into technical implementations; (3) *proof-of-concept projects* in which multidisciplinary teams apply Value-Sensitive Design to a particular technology, domain, or design

problem; (4) *contextual analyses* that investigate the impact of different stakeholders who influence the design and use of a technology, and who may have different goals and priorities that, in turn, lead to different value trade-offs; (5) *integrating the methodology* into organizational structures and work practices; and (6) *criteria and metrics*—both qualitative and quantitative—that can guide the design process and assess the success of particular designs.

HUMAN VALUES WITH ETHICAL IMPORT

We review and discuss 12 specific human values with ethical import. Some of these values have garnered individual chapters in this handbook. But, by including these values here, we highlight their ethical status and thereby suggest that they have a distinctive claim on resources in the design process.

Two caveats. First, not all the values we review are fundamentally distinct from one another. Nonetheless, each value has its own language and conceptualizations within their respective fields, and thus warrants separate treatment here. Second, this list is not comprehensive. Perhaps no list could be, at least within the confines of a chapter. Peacefulness, compassion, love, warmth, creativity, humor, originality, vision, friendship, cooperation, collaboration, purposefulness, devotion, diplomacy, kindness, musicality, harmony—the list of other possible values could get very long very fast. Our particular list comprises many of the traditional values that hinge on the deontological and consequentialist moral orientations reviewed previously: Human welfare, ownership and property, privacy, freedom from bias, universal usability, trust, autonomy, informed consent, and accountability. In addition, we have chosen several nontraditional values within the HCI community: Identity, calmness, and environmental sustainability. Our goal here is not only to point to important areas of future inquiry, but also to illustrate how an overarching framework for human values and ethics in design can move one quickly and substantively into new territory.

Human Welfare

Perhaps no value is more directly salient to individuals at large as their own welfare, and the welfare of other human beings. Moreover, societal interest in the moral dimensions of computing often arises when such harms occur. The faulty computer technology and interface design of Therac-25 led to the physical suffering and death of cancer patients (Leveson & Turner, 1993). Faulty computer technologies are implicated in nuclear accidents like Chernobyl. Indeed, the Risks literature (Neumann, 1995) is full of almost daily examples.

Some people in the HCI community, like Leveson (1991) refer broadly to harms that impact people and objects as harms to physical welfare. But there is merit in demarcating three categories of welfare claims (Friedman, 1997b; Kahn, 1992, 1999; Turiel, 1983). *Physical welfare* appeals to the well-being of individuals' biological selves, which is harmed by injury, sickness, and death. Many of the previous examples—that involve

Therac-25, "smart" missiles gone astray, nuclear accidents, and so forth—involve harms of this kind. In many such instances, physical harms occur because of faulty hardware or software. But it is important to recognize that technological designs working correctly can themselves lead to corresponding harms or benefits. For example, by means of the connectivity of the web, stalkers can more easily find their "prey," and pedophiles can more easily find unwitting children to contact and potentially abuse. On the positive side, the mediating quality of the web can enhance anonymity (e.g., in a Chat Room) and buffer individuals from physical harm. *Material Welfare* appeals to physical objects that humans value and human economic interests. It goes hardly without saying that a major driving force behind computer technologies in general has been to enhance the material welfare of humans. In turn, a good deal of effort is spent protecting against material harms that computers can engender, such as damaged data from computer viruses and stolen financial information. *Psychological Welfare* refers to the higher order emotional states of human beings, including comfort, peace, and mental health. The connectivity offered by the web again provides a clear example of psychological benefits and harms. A benefit accrues when the web enhances friendships; and a harm occurs when the web allows for new forms of betrayal (e.g., the "friend" you thought you found in a chat room turns out to be a Bot).

Ownership and Property

From the anthropological literature, it would appear that all cultures embrace the idea that people can own property, although the form of such conceptions can vary considerably (Herskovits, 1952). In perhaps its most stringent form, ownership can be understood as a general right to property, which, in turn, entails a group of specific rights, including the right to possess an object, use it, manage it, derive income from it, and bequeath it (Becker, 1977).

This basic concept of a property right seems simple enough when applied to tangible objects. If you own a table, for example, you can keep it in your house, eat on it, let others eat on it, rent it, and give it away. But current computational technologies blur the boundaries between the tangible and intangible. Herein lie difficult questions. Can users, for example, legitimately copy software for their personal use? After all, to do so involves no loss of a physical object from the original seller. Or can a programmer modify part of the code from an operating system and then sell the modified system? Provisional legislative answers in the United States have hinged on nuanced distinctions between copyright, patent, licensing, and trade secrets. For example, unlike a patent, a copyright does not protect ideas, but only the expression of an idea once it is fixed in a tangible medium. The courts have largely granted developers copyrights (but not patents) to their software. But even here further controversies arise. Let us say that a developer does not copy any code of a competitor's software, but copies its "look and feel"? Is that an infringement? The courts have said yes if other technical means readily exist for implementing a different "look and feel" for the interface (Lipinski & Britz, 2000).

While case law is solidifying for issues that focus on computer software, it is hardly keeping pace with other computational developments that impact ownership and property. Consider, for example, a workplace discussion group. Who gets to decide whether the comments from the online discussion forum are made accessible to the public at large? In other words, do the workers have equal ownership of the compiled contributions? Or is ownership divided by level of participation? Or is the owner the organizer of the discussion group? Or the president of the company? Can one member delete the contribution from another member (if, for example, he or she finds the contribution offensive)? Or consider Goldberg's (2000) Telegarden, an installation that combines robotics and a real-time web camcorder, such that users interact from a remote distance with a real physical environment (a garden), and plant seeds and water them. Who owns each actual live plant? The user? The owner of the installation? Goldberg himself? Who decides how to respond to an infestation of pests, and whether pesticides will be used? If the garden becomes overcrowded, do the people who physically maintain the site have the right to uproot all the plants that were planted and cared for remotely by others? Or consider whether a cookie on a user's machine belongs to the user or to the Web site that set the cookie. If the cookie belongs to the user, and the browser does not allow the average user to delete the cookie, then by virtue of a technical mechanism HCI designers have in effect deprived an average user of the capability to exercise one of his or her specific property rights.

Our point is that through conventions and technical mechanisms, HCI designers shape answers to questions of ownership and property that lie at the forefront of social discourse and legal codification.

Privacy

Privacy refers to a claim, an entitlement, or a right of an individual to determine what information about himself or herself can be communicated to others (Schoeman, 1984). Historically, a good deal of our privacy protections arose because it was simply too much effort to collect and sort through relevant information about other individuals. But as computer technologies increasingly garner large amounts of information about specific individuals, and increasingly link information databases, our historical protections for privacy are being eroded.

In the HCI community, three general approaches have emerged for privacy protections. One approach informs "people when and what information about them is being captured and to whom the information is being made available" (Bellotti, 1998, p. 70). For example, a video monitor next to a surveillance camera in a convenience story can inform customers of the information being recorded. A second approach allows "people to stipulate what information they project and who can get hold of it" (Bellotti, 1998, p. 70). Tang's (1997) example of an on/off switch for a video conferencing workstation (noted earlier) maps onto this approach. A third approach applies privacy-enhancing technologies (PETs) that prevent sensitive data from being tagged to a specific individual in the first place. PETs have become "extraordinarily successful" in taxing

"the capacities of even the most powerful surveillance institutions" (Phillips, 1998, p. 245), and thus have provoked sharp conflicts in attempts to disseminate them. Indeed, Agre (1998) argues that PETs "disrupt the conventional pessimistic association between technology and social control. No longer are privacy advocates in the position of resisting technology as such" (p. 4).

As we move toward an era of ubiquitous computing, where embedded computation in our everyday physical objects are linked, the amount of information known about us will increase by many orders of magnitude. Thus, here, as in other domains of applications, we in the HCI community face difficult questions. How do we inform users about the risks to privacy they incur through the use of various information systems? How do we inform users about the technical protections for privacy available to them? How do we help users to understand the risks from aggregated vs. individual data? How much protection is afforded by anonymity? What should the default on systems be?—toward greater privacy protection? Or toward greater access to information? Many of these questions form part of current inquiries in HCI and, particularly CSCW, that seek to balance individual privacy with group awareness (Boyle, Edwards, & Greenberg, 2000; Cohen, Cash, & Muller, 2000; Godefroid, Herbsleb, Jagadeesan, & Li, 2000; Jancke, Venolia, Grudin, Cadiz, & Gupta, 2001; Milewski & Smith, 2000; Shoemaker & Inkpen, 2001; Svensson, Hook, Laaksolahti, & Waern, 2001).

Freedom From Bias

Bias refers to systematic unfairness perpetrated on individuals or groups. Three forms of bias in computer systems have been identified (Friedman & Nissenbaum, 1996): pre-existing social bias, technical bias, and emergent social bias.

Pre-existing social bias has its roots in social institutions, practices, and attitudes. It occurs when computer systems embody biases that exist independently of, and usually prior to, the creation of the software. For instance, the work by Nass and his colleagues (Nass & Gong, 2000; Reeves & Nass, 1996) has shown that people respond to a computer's "gender" along stereotypical lines. Male voice interfaces are rated more competent and persuasive, and more knowledgeable about technical subjects. Female voice interfaces are viewed as more knowledgeable about topics such as love and relationships. As Nass points out, as a designer it is all too easy to inadvertently build on these pre-existing social biases when building interfaces. Thus, toward minimizing pre-existing bias designers must not only scrutinize the design specifications, but must couple this scrutiny with a good understanding of relevant biases out in the world. In addition, it can prove useful to identify potential user populations that might otherwise be overlooked and include representative individuals in the field test groups. Rapid prototyping, formative evaluation, and field testing with such well-conceived populations of users can be an effective means to detect unintentional biases throughout the design process.

Technical bias occurs in the resolution of technical design problems. For example, imagine a database for matching organ donors with potential transplant recipients. If certain individuals retrieved and displayed on initial screens are favored systematically for a match over individuals displayed on later screens, technical bias occurs. Technical bias also originates from attempts to make human constructs such as discourse, judgments, or intuitions amenable to computers. For example, consider a legal expert system that advises defendants on whether or not to plea bargain by assuming that law can be spelled out in an unambiguous manner and is not subject to human interpretations in context. Toward minimizing technical bias, designers often need to look beyond the features internal to a system and envision the design, algorithms, and interfaces in use.

Emergent social bias emerges in the context of the computer system's use, often when societal knowledge or cultural values change, or the system is used with a different population. Toward minimizing emergent bias, designers need to plan for not only a system's intended contexts of use, but also its potentially emergent contexts. Yet, given limited resources, such a proposal cannot be pursued in an unbounded manner. Thus, three practical suggestions are as follows: First, designers should reasonably anticipate probable contexts of use and design for these. Second, where it is not possible to design for extended contexts of use, designers should communicate to users the contextual constraints. As with other media, we may need to develop conventions for communicating the perspectives and audience(s) assumed in the design. Third, system designers and administrators can take responsible action if bias emerges with changes in context. The National Resident Medical Match Program offers a good example. Although the original design of the Admissions Algorithm did not deal well with the changing social conditions (when significant numbers of dual-career couples participated in the match), those responsible for maintaining the system responded conscientiously to this societal change and modified the system's algorithm to place couples more fairly (Roth, 1990).

Universal Usability

Universal usability refers to making all people successful users of information technology. Or, if this requirement of "all people" is too stringent, it can be reframed as "all people who so desire," or in some other way. Shneiderman (1999), for example, says that universal usability means "having more than 90 percent of all households as successful users of information and communication services at least once a week." In many respects, universal usability comprises a special case of freedom from bias, one that focuses on usability as a means to systematically prevent unfair access to information systems.

Three challenges are often addressed by proponents of universal usability: (1) *technological variety*—supporting a broad range of hardware, software, and network access; (2) *user diversity*—accommodating users with differences in, say, skills, knowledge, age, gender, disabilities, literacy, culture, and income; and (3) *gaps in user knowledge*—bridging the gap between what users know and what they need to know (Shneiderman, 1999, 2000b). Toward addressing these challenges, current work includes Aberg and Shahmehri (2001);

Beard and Korn (2001); Cooper and Rejmer (2001); Jacko, Dixon, Rosa, Scott, and Pappas (1999); and Oviatt (1999). Stephanidis (2001) provides an anthology of current concepts, methods, and tools.

Universal usability, of course, is not necessarily always a moral good insofar as it depends on what is being used or accessed. Virtually no one would suggest, for example, that there is a moral imperative to provide universal access to re-runs of "I Love Lucy." Nonetheless, universal usability often is a moral good. Hert and her colleagues (Hert, Marchionini, Liddy, & Shneiderman, 2000), for example, shows that if all U.S. citizens have the right to access federal statistics, and if those statistics are only available in an online format, then to be able to exercise their right, the online federal statistics system must be usable by all U.S. citizens. Moreover, often universal access with ethical import provides increased value to a company. For example, based on his case study within a large communications corporation (NYNEX), Thomas (1997) shows how making communications systems more accessible leads to three direct corporate benefits. One, increasing access makes a communications device more pervasive in social life, and thereby more valuable for everyone. Second, increasing access increases market share. Third, increasing access "forces technologists, developers, marketers, and executives to think 'out of the box'" (p. 271).

Trust

People sometimes use the term trust broadly to include expectations of natural phenomena or machine performance. It is in this sense that people trust that "the sun will rise tomorrow" or that "brakes will stop a car." Indeed, the Computer Science and Telecommunications Board in their thoughtful publication *Trust in Cyberspace* (Schneider, 1999) adopted the terms "trust" and "trustworthy" to describe systems that perform as expected along the dimensions of correctness, security, reliability, safety, and survivability. However, equating the term "trust" with expectations for machine performance (or physical phenomena) misconstrues fundamental characteristics of this value. Specifically, trust is said to exist between people who can experience good will, extend good will toward others, feel vulnerable, and experience betrayal (Baier, 1986; Friedman, Kahn, & Howe, 2000; Kahn & Turiel, 1988). Moreover, on the societal level trust enhances our social capital (Putnam, Leonardi, & Nanetti, 1993).

In their analysis of trust online, Friedman et al. (2000) suggest that it is important for designers to distinguish two overarching contexts: e-commerce and interpersonal relationships. In e-commerce, for example, certain characteristics of the technology—such as those that concern security, anonymity, accountability, and performance history—can make it difficult for consumers to assess their possible financial harms and the potential good will of the company. One solution that has emerged has been a form of insurance (usually through the credit card companies) that limits a person's financial liability. Trust in e-commerce has received a good deal of attention in the last several years in the HCI and larger community (Dieberger, Hook, Svensson, & Lonnqvist, 2001; Egger, 2000; Fogg & Tseng, 1999; Greenspan, Goldberg, Weimer, & Basso, 2000; Jones, Wilkens, Morris, & Masera, 2000; Olson & Olson, 2000; Shneiderman, 2000a).

In online interpersonal interactions, violations of trust make us most vulnerable psychologically: for example, hurt feelings or embarrassment. Rocco (1998) suggests that interpersonal trust online succeeds best when preceded by face-to-face interaction. In response, Zheng, Box, Olson, and Olson (2001) suggest that when users do not meet physically but engage in online chat to get to know one another that they can establish the same kinds of interpersonal trust that are established in face-to-face interactions (cf. Bos, Gergle, Olson, & Olson, 2000). Interestingly, whereas the characteristic of anonymity works against establishing financial trust in e-commerce, it is double-edged in interpersonal relationships. On the negative side, online anonymity can limit the depth of interpersonal interactions in so far as we engage in a singular means of expression (written). On the positive side, online anonymity opens up new interpersonal opportunities. For example, a gay teenager in an intolerant family and community might rely on the anonymous characteristics of the web to find and interact with like-minded peers. Thus, toward enhancing interpersonal trust online, we need to build tools that allow users to have easy and refined control about what personal information is made known to others.

Autonomy

People decide, plan, and act in ways that they believe will help them to achieve their goals. In this sense people value their autonomy. It might appear relatively easy for HCI designers to support user autonomy. The idea would be that whenever possible provide users with the greatest possible control over the technology. However, the task is harder than that. After all, most users of a word processor have little interest, say, in controlling how the editor executes a search and replace operation or embeds formatting commands. In other words, autonomy is protected when users are given control over the right things at the right time. The hard work arises in deciding what those features are and when those conditions occur.

Toward this end, Friedman and Nissenbaum (1997) identify four aspects of systems that can promote or undermine user autonomy. The first involves system capability. Recall Isaacs, Tang, and Morris's (1996) design of a system to support informal interactions in the workplace, using the piazza as the metaphor. Their application provided a means for workers to opt out of piazza interactions, and thus supported worker autonomy. The second involves system complexity. For example, as previously noted, software programs can proliferate features at the expense of usability and user autonomy suffers. The third involves misrepresentation of the system. For example, hyperbolic advertising claims can lead users to develop inaccurate expectations of the system and thereby frustrate users' goals. The fourth involves system fluidity. Over time, user's goals often change. Thus, systems need to provide ready mechanisms for users to review and adapt their systems.

Many systems depend on categorization, and it is here that Suchman (1994) worries that organizations can run roughshod

of user autonomy. For, according to Suchman, whomever determines the categories—and how those categories can be used—imputes their own personal values into the system and has power over the user. In response, Winograd (1994) points to the frequent need for socially coordinated activity through which groups of people seek to share information and technology. In such activity, Winograd argues, we need some degree of standardization wherein designers impose categories. That is their job. The key, according to Winograd, is to cultivate regularized activity without becoming oppressive. In turn, Malone (1994) responds to Suchman by arguing that not only are categories often useful, but to some extent they are necessary given the structure of human cognition. At the same time, Malone suggests that no category system is complete and that designs need to be adaptable, often by their users.

Informed Consent

Informed consent provides a critical protection for privacy, and supports other human values such as trust and autonomy. Yet, currently, there is a mismatch between industry practice and the public's interest. According to a recent report from the Federal Trade Commission (2000), for example, 59% of Web sites that collect personal identifying information neither inform Internet users that they are collecting such information nor seek the user's consent. Yet, according to a Harris Poll (2000), 88% of users want Web sites to garner their consent in such situations. The Federal Trade Commission (2000, p. iv) hopes that industry will continue to make progress on this problem, in conjunction with its proposed legislation. In turn, the HCI community has been recognizing the need to understand better what constitutes informed consent, and to realize it in online interactions.

Friedman, Millett, & Felten (2000) offer an analysis of what constitutes informed consent, and show how both words—"informed" and "consent"—have import (Faden & Beauchamp, 1986; *The Belmont Report*, 1978). The idea of "informed" encompasses disclosure and comprehension. Disclosure refers to providing accurate information about the benefits and harms that might reasonably be expected from the action under consideration. Comprehension refers to the individual's accurate interpretation of what is being disclosed. In turn, the idea of "consent" encompasses voluntariness, competence, and agreement. Voluntariness refers to ensuring that the action is not controlled or coerced, and that an individual could reasonably resist participation should he or she wish to. Competence refers to possessing the mental, emotional, and physical capabilities needed to give informed consent. Agreement refers to a reasonably clear opportunity to accept or decline to participate. Based on this account, Friedman, Millett, and Felten (2000) offer general design principles for informed consent online. Namely: (1) Decide whether the capability is exempt from informed consent. (2) Be particularly careful when invoking the sanction of implicit consent. (3) Defaults settings should err on the side of preserving informed consent. (4) Put the user in control of the "nuisance factor." (5) Avoid technical jargon. (6) Provide the user with choices in terms of potential effects rather than in terms of technical mechanisms. (7) Field test to help ensure

adequate comprehension and opportunities for agreement. (8) Design proactively.

In conjunction with this analysis, Millett et al. (2001) have examined how cookie technology and web browser designs have responded to concerns about informed consent. Specifically, they document relevant design changes in Netscape Navigator and Internet Explorer over a 5-year period, starting in 1995. Their retrospective examination led them to conclude that while cookie technology has improved over time regarding informed consent, some startling problems remain. They specify six problems and offer design remedies. For example, in both Netscape Navigator and Internet Explorer, the information disclosed about a cookie still does not adequately specify what the information will be used for or how the user might benefit or be harmed by its use. One remedy is to redesign the browser's cookie dialog box to include three additional fields, one for stating the purpose for setting the cookie, one for a brief statement of benefits, and one for a brief statement of risks.

Accountability

Medical expert systems. Automated pilots. Loan approval software. Computer-guided missiles. Increasingly, computers participate in decisions that affect human lives. In cases of computer failure, there is a common response to blame the computer—"it's the computer's fault." Indeed, Friedman and Millett (1995) found that 83% of undergraduate computer science majors she interviewed attributed aspects of agency—either decision making and/or intentions—to computers. In addition, 21% of the students consistently held computers morally responsible for error. Yet, if we accept that humans, but not computational systems, are capable of being moral agents, then such blame is fundamentally misplaced (Searle, 1980).

How can HCI designers minimize this tendency of users to attribute blame to computational systems? Suggestions have been offered based on minimizing two types of distortions (Friedman & Kahn, 1992). In the first type of distortion, the computational system diminishes or undermines the human user's sense of his or her own moral agency. In such systems, human users are placed into largely mechanical roles, either mentally or physically, and frequently have little understanding of the larger purpose or meaning of their individual actions. To the extent that humans experience a diminished sense of agency, human dignity is eroded and individuals may consider themselves to be largely unaccountable for the consequences of their computer use. Conversely, in the second type of distortion the computational system masquerades as an agent by projecting intentions, desires, and volition. Strikingly, even when computer interfaces only minimally mimic human agency, people appear predisposed—at least in certain regards—to treat computers as social agents. For example, as Nass and his colleagues have shown, people respond to multiple voices from a single computer as though they were separate entities, respond to a computer's "gender" along stereotypical lines, are less likely to criticize a computer directly (i.e., if the computer itself asks for an evaluation) than to criticize the computer to a third party (a different computer or human), and respond to computer flattery

(Nass & Gong, 2000; Reeves & Nass, 1996). Thus, to the extent that humans inappropriately attribute agency to computational systems, humans may well consider the computational systems, at least in part, to be morally responsible for the effects of computer-mediated or computer-controlled actions.

Identity

The idea of personal identity embraces two seemingly contradictory ideas. On the one side is the obvious fact that each one of us has many roles. A single person can be, for example, a father, lover, poker player, gourmet cook, computer geek, and animal lover. Indeed, William James says that a person "has many social selves as there are individuals who recognize him" (quoted in Rosenberg, 1997, p. 23). On the other side, virtually all of us feel like we live reasonably coherent lives, and that the person we are today is pretty much whom we were yesterday and last week, if not last year. Thus, identity appears to be multiple and unified, and both aspects are essential to human flourishing. Too far toward multiplicity and we end up schizophrenic; too far toward being unified and there are too few mechanisms for psychological growth, and too little basis for healthy social functioning (e.g., it makes little sense to present the same "persona" to one's boss as one does to one's child).

In terms of HCI, it is important that the field as a whole substantively support both manifestations of identity, and that in our designs we not swing too far one direction or the other. To date, the networked personal computer has easily supported multiplicity. Thus, the same person can easily communicate with many unrelated groups (chats, list serves, etc.), easily establish a different identity within many of these groups, or even multiple identities within a single group (Turkle, 1996). This trend is being checked in interesting ways. For example, individuals in some online communities link their comments to their personal homepage ("to find out more about me, click here"). Some chats (e.g., SeniorNet) require single identities. Indeed, with the advent of increasingly linked online databases coupled with ubiquitous computing, we believe that the pendulum with soon shift powerfully and pervasively toward the unification of identity. If so, following Bers, Gonzalo-Heydrich, and DeMaso (2001) and Schiano and White (1998), the challenge for HCI designers will be to find ways such that individuals have flexibility to establish and reveal not only an integrated self, but a multiplicity of identities.

Calmness

Weiser and Brown (1997) suggest that in the last 50 years, there have been two great trends in the human relationship with computing: The first was with the mainframe computer and the second is currently with the personal computer. They suggest that the next great trend is toward ubiquitous computing, characterized by deeply embedded computation in the world. In turn, they argue that the "most potentially interesting, challenging, and profound change implied by the ubiquitous computing era is a focus on *calm*. If computers are everywhere, they had better

stay out of the way, and that means designing them so that the people being shared by the computers remain serene and in control" (p. 79).

The central design mechanism Weiser and Brown put forward is that which conveys information in our peripherial awareness, but wherein the information is moved to the center of our awareness at appropriate times. One architectural example they provide is of a glass window between offices and hallways. Such a window, they suggest, "extends our periphery by creating a two-way channel for clues about the environment. Whether it is motion of other people down the hall (it's time for lunch; the big meeting is starting) or noticing the same person peeking in for the third time while you are on the phone (they really want to see me; I forgot an appointment), the window connects the person inside to the nearby world" (pp. 81–82). A computational example involves an Internet multicast, a continuous video from another location that provides not so much videoconferencing, but "more like a window of awareness" of another location (p. 82). Another example involves a "dangling string," "an eight-foot piece of plastic spaghetti that hangs from a small electric motor mounted in a ceiling. The motor is electronically connected to a nearby Ethernet cable, so that each bit of information that goes past causes a tiny twitch of the motor. A very busy network causes a madly whirling string with a characteristic noise; a quiet network causes only a small twitch every few seconds" (p. 83).

The idea that ubiquitous computing will need to take up the challenge of how to preserve calmness in human lives seems on the mark. There are, however, some limitations in the solution Weiser and Brown put forward in terms of designing for the periphery. Perhaps the most notable is that very quickly the periphery itself can become overloaded with information, especially if work place conventions expect workers to be continuously aware of such information. Imagine, for example, working in an office with a window out onto the hallway, with four Internet multicasts playing in the background, a dangling string in the corner, five other peripheral information streams on each of several pieces of software you are using, and then add in peripheral information streams on one's regular phone, cell phone, personal digital assistant, and any and all other pieces of embedded computation in the office of today (or the future). Then try to get some work done. It will not be easy.

An alternative design approach (Kahn & Friedman, in preparation; cf. Farrell, 2001) toward preserving calmness has as its starting point the psychological literature that shows that direct experiences with nature have beneficial effects on people's physical, cognitive, and emotional well-being. For example, Ulrich (1984) found that postoperative recovery improved when patients were assigned to a room with a view of a natural setting (a small stand of deciduous trees) versus a view of a brown brick wall. More generally, studies have shown that even minimal connection with nature—such as looking at a natural landscape—can reduce immediate and long-term stress, reduce sickness of prisoners, and calm patients before and during surgery (see, for reviews, Beck & Katcher, 1996; Kahn, 1999; Ulrich, 1993). Building on this literature, the question becomes in what ways computer technologies can augment the human relationship with the natural world with beneficial effects. One

implementation might involve video plasma display "windows" in inside offices that stream in real-time views of a psychologically restorative local nature scene. Another implementation might involve robotic "pets" as possible companions for the elderly. The point is that while such augmented natural interactions may not be as psychologically beneficial as real nature, it may be more so than no nature, in which case it becomes a plausible area for design applications.

Environmental Sustainability

We pollute air and water, deplete soil, deforest, create toxic wastes, and through human activity are extinguishing over 27,000 species each year (a conservative estimate). Such problems have generated a tremendous amount of attention among the populace, and have slowly been coming under the purview of the computing community (cf. IEEE and ACM: Software Engineering Code of Ethics and Professional Practice, 1998). On the production end, there is concern about the resources used in producing computer technologies, as well as the resulting toxic wastes (e.g., of making computer chips). On the consumption end, there is concern about the resources computer technologies use. The electrical demands, for example, are particularly high, especially when energy sources are scarce.

Another question arises of how we can design systems that foster a healthier and more life-affirming connection with the natural world (cf. Kahn & Kellert, 2002). The more straightforward approach has been simply to harness computer technologies in the service of environmental science, in computer models, for example, of global warming or earth tectonics. Environmental educators have also been using networked computers such that students can share environmental data or narratives with other students (or scientists) in diverse geographical locations. Or consider an approach being taken by Borning and his colleagues (Noth, Borning, & Waddell, 2000) where he is using a computer simulation ("UrbanSim") of an urban environment to help residents, politicians, and planners visualize the effects of their proposed land-use plans. Another approach might involve designing calming technologies, as previously discussed, that integrate restorative aspects of nature into human lives.

PROFESSIONAL ETHICS

Computer professionals not uncommonly experience ethical conflicts in the workplace. Consider, for example, the situation where, by means of his or her specialized knowledge, a computer professional knows of a harm that can result from the implementation of a computer technology. Should he or she inform others of such impeding harm even if it will jeopardize his or her job security? Or consider the situation where a client asks a software developer to develop an expert system that recommends against loans in certain neighborhoods. Although such a policy (called "redlining") may serve the economic interests of the loan-granting organization, it unfairly discriminates against individuals on the basis of where they live. Should a computer

professional deny such a request? Or consider the situation offered by Quesenbery (Molich, 2001):

You have set up early usability tests of a paper prototype with nurses at a medical facility. The test was difficult to schedule because nurses' time is guarded carefully and marketing carefully guards the relationship with customers. The nursing managers insist on 'taking the test' themselves first, and then on being present in the room during the test with the other participants to 'be sure they do it right.' You believe that the manger's presence will be intimidating to the nurses, altering the results of the test. Do you continue with the usability tests? (p. 218)

Quesenbery offers two possible answers. One answer is yes. She argues that if you turn down this opportunity, another one might not arise; and although you would have to carefully evaluate the test results for bias caused by the managers, valuable data about the prototype would still be generated. The other answer is no. She argues that it is unethical to put nurses into such a stressful situation where they are studied in front of their obviously critical managers.

In response to ethical issues that arise within the computing professions, numerous organizations have developed ethical codes of conduct. Codes have been developed, for example, by the ACM, IEEE Computer Society, DPMA, and ICCP. Berleur and Brunnstein (1997) provide a comparison of 30 different codes of ethics. At the end of 1998, the IEEE and the ACM together adopted a revised Software Engineering Code of Ethics and Professional Practice (http://www-cs.etsu.edu/seeri/secode.htm). At the highest level of abstraction (see Gotterbarn, 1999, for a review), the code states eight principles wherein software engineers shall (1) act consistently with the public interest; (2) act in a manner that is in the best interests of their client and employer consistent with the public interest; (3) ensure that their products and related modifications meet the highest professional standards possible; (4) maintain integrity and independence in their professional judgment; (5) subscribe to and promote an ethical approach to the management of software development and maintenance; (6) advance the integrity and reputation of the profession consistent with the public interest; (7) be fair to and supportive of their colleagues; and (8) participate in life-long learning regarding the practice of their profession and shall promote an ethical approach to the practice of their profession. In turn, each principle is elaborated upon with specific guidelines. For example, under the first principle (to act consistently with the public interest), the code states that software engineers should "disclose to appropriate persons or authorities an actual or potential danger to the user, the public, or the environment, that they reasonably believe to be associated with the software or related documents."

Before moving forward with this topic, it is important to take hold of a question often discussed in the literature, of whether the computer profession is even a profession (Weckert & Adeney, 1997). If it is not, then concerns about professional ethics—special moral requirements above and beyond what are applied to ordinary people—disappear.

What, then, is a profession? According to Bayles (1981), a profession is comprised of members who (1) have extensive training, (2) have an intellectual component such that the professional's primary role is to advise the client about things the

client does not know, and (3) provide an important service to society. Bennion (1969) also points out that professionals are typically self-employed, and establish a client–agent relationship based on trust. In addition, as Johnson and Miller (1997) write: "Being a professional means more than just having a job. The difference is commitment to doing the right thing because you are a member of a group that has taken on responsibility for a domain of social activity—a social function. The group is accountable to society for this domain, and for this reason, professionals must behave in ways that are worthy of public trust" (p. 22).

Based on such criteria, in some respects the computer field constitutes a profession. Computer personnel have intellectual knowledge and extensive training, and they provide an important service to society. Computer personnel are also especially well positioned to understand how new technologies may impact human lives; accordingly, many people argue that they incur the responsibility to communicate such impacts to the general public. Too, the media, industry, and academy widely use the term "computer professionals." That said, the computer field certainly does not constitute a canonical profession, like medicine or law. After all, many computer personnel work for businesses rather than as a consultant, and create artifacts (e.g., software, hardware, algorithms, interface designs) rather than offer advice. Too, computer personnel vary greatly in the amount and type of their training. Some computer personnel are self-taught, without even a high-school diploma. Other computer personnel complete 6-month to 2-year programs, or a bachelor's, master's, or doctoral program.

If we accept—as most people do—that in at least some important regards the computer field constitutes a profession, then how has the profession understood and conveyed its correlative ethical responsibilities? One approach has been through codes of ethics, noted previously. Such codes serve multiple purposes (Gotterbarn, 1999; Anderson, Johnson, Gotterbarn, & Perrolle, 1993). They serve to educate computer employees and managers. They help garner the trust of the general public. They provide computer employees with a formal document to turn to (and appeal to) when they face conflicts that pit ethical decisions against the economic benefit of their company. They can function as a means of deterrence and discipline. And they can enhance a profession's public standing.

Another approach for advancing the profession's ethical responsibilities involves using hypothetical scenarios to flesh out the meaning of the codes (Parker, 1979; Parker, Swope, & Baker, 1990). For example, Anderson et al. (1993) provide an analysis of nine scenarios that illustrate how the 1992 ACM Code of Ethics bear out in practice. They offer, for example, the hypothetical situation of a consultant, named Diane, who is designing the database management system for the personnel office of a medium-sized company. The database will store sensitive information, including performance evaluations, medical records, and salaries. In an effort to cut costs, the CEO of the company rejects Diane's suggestions and opts for a less secure system. Diane remains convinced that a more secure system is needed. What should she do? Anderson et al. (1993) then point to specific passages in the code that pertain to the importance of maintaining privacy and confidentiality. Based on the code, Diane's first obligation is to try to educate the company officials on

the ethics of the situation. "If that fails, then Diane needs to consider her contractual obligations . . . on honoring assigned responsibilities" (p. 100).

Another possible approach toward advancing the profession's ethical responsibilities involves that of licensing its members. The model draws from the American Medical Association or the American Bar Association. Both associations can revoke their members' licenses given serious violations of their respective professional organization's code of ethics. But there are good reasons why most professional computing organizations have resisted this move. Computing is an enormously diverse field; and it is difficult to formally demarcate all its areas, let alone the relevancy of one area to others. What exactly must an independent Web site designer know, for example, beyond whatever the designer and employer believe necessary? Does the field really want to exclude people all along its fringes? There has been concern that through licensing the profession might become a closed shop and thereby enhance the status and incomes of those admitted at the expense of those excluded. There has also been concern that licensing could stifle creativity, originality, and excellence (see the discussion by Neumann, 1991). Licensing also runs counter to the history of computing that has emphasized ease of entry, entrepreneurship, and a culture not often found in professional licensing organizations.

Setting aside licensing, and recognizing the value of codes of ethics and analyses of hypothetical dilemmas, what else can the computing profession do to advance its ethical responsibilities? Johnson and Miller (1997) offer two suggestions. One is that corporations could have ombudspersons to whom computing professionals could report their concerns, anonymously if desired. Another is that professional societies (like the ACM) could maintain "hotlines that professionals could call for advice on how to get their concerns addressed" (p. 23). Both suggestions, if implemented, could prevent many problems from escalating into "whistle-blowing" affairs and help integrate ethical considerations into the workplace.

The computing profession is also recognizing the importance of analyzing not only hypothetical scenarios, but real life events. Neumann (1995) is a pioneer in this regard, publishing ongoing chronicles of the risks that arise through computing. For the more interesting cases, Neumann's approach could be extended. What we have in mind here builds on a leading mountaineering journal that each year publishes brief accounts of the major mountaineering accidents of the previous year. To read these accounts, year in and year out, is to build up a rich repertoire of how mistakes happen in the field, and how they can be avoided. Similarly, it would be possible to provide comprehensive accounts of the major ethical problems that have arisen each year in the computing profession. Moreover, these accounts could be classified based on type of ethical value (e.g., privacy, trust, human welfare, security, or universal usability), user population (e.g., workers, elderly, children, or people in developing nations), and applications (e.g., online communities, information visualization, augmented reality, wearable computing, or groupware). With such a searchable database, computing professionals could gain ready access to real-life ethical case studies directly relevant to their own endeavors.

Most of the ethical issues of HCI professionals are subsumed under those encountered by computer professionals in general.

But HCI professionals incur unique responsibilities whenever they involve human subjects. At universities, such studies come under the purview of a Human Subjects Institutional Review Board (an IRB) that seeks to protect the rights and welfare of human subjects. But, in industry, such oversight is almost always lacking. Consider the following questions:

• You conduct usability studies with people within your company. The vice-President later asks you which people did well. Do you tell (Molich, 2001)?

• You collect video footage of people engaged in your usability study. Can you show the footage at a conference (Mackay, 1995)? Have you obtained written consent from the subjects? What if the subject initially gives consent but then comes off looking foolish in the video. Have you given the subject a chance to review the video footage and to opt out?

• You obtain informed consent, and then collect keystroke data on employees over a period of months. How do you keep your subjects aware that your data collection is on-going?

• You collect data from subjects living in an "aware home" (that has ubiquitous computation embedded throughout the living environment). How do you convey to subjects the extent of personal information (e.g., time spent in the bathroom) that can be mined from the resulting database?

• You collect data from employees in your company, and promise to keep identifying information confidential. But if there is only one disabled person in a given department, how can you maintain confidentiality?

• You are involved in a safety critical situation. You recognize that often usability studies in your company involve only a handful of subjects, which allows for fast prototyping and more quickly bringing products to market. But, in this situation, is a more rigorous psychological study of usability needed, even given that it will require additional resources and time (Molich, 2001)?

• You collect information from an online chat focused on sexual abuse, and then publish the information widely, quoting key emotional passages that now become visible on a societal level. Even though you report the passages without identifying names, others can now go to this chat, search on the text, and easily identify the people of interest. Have you adequately protected the privacy of these online "subjects"? How does using subjects garnered from the Internet differ from those garnered from more traditional subject pools (Frankel & Siang, 1999)?

These questions and many other related issues can be addressed by drawing on the rich literature on protecting the rights and welfare of human subjects (*The Belmont Report*, 1978) and its application to the internet (Frankel & Siang, 1999). Indeed, often HCI professionals may be the only source for giving voice to these ethical considerations.

CONCLUSIONS

During the early periods of computerization, around the 1950's, cyberneticist Norbert Wiener (1953/1985) argued that technology could help make us better human beings and create a more just society. But for it to do so, he argued, we have to take control of the technology. We have to reject the "worshiping [of] the new gadgets which are our own creation as if they were our masters" (p. 678). Similarly, a few decades later, computer scientist Joseph Weizenbaum (1972) wrote:

What is wrong, I think, is that we have permitted technological metaphors...and technique itself to so thoroughly pervade our thought processes that we have finally abdicated to technology the very duty to formulate questions....Where a simple man might ask: "Do we need these things?," technology asks "what electronic wizardry will make them safe? Where a simple man will ask "is it good?," technology asks "will it work?" (pp. 611–612).

As HCI professionals, we have profound opportunities to shape the designs and implementations of computer technologies from an ethical stance.

In this chapter, we have reviewed varying approaches, projects, and ideas that offer us important ways of bringing human values and ethics into our design practice. Several ideas are worth emphasizing. First, as is well known, it is much easier to design systems right initially than to attempt to retrofit poor systems after they have become entrenched within organizations and other social systems. Thus, it is imperative that we take a proactive stance on human values, ethics, and design. Second, many of the difficult problems in this area require multidisciplinary collaborations. Third, we need to hold out human values with ethical import as a design criterion—along with the traditional criteria of reliability, efficiency, and correctness—by which systems may be judged poor and designers negligent. As with the traditional criteria, we need not require perfection, but commitment.

This *Handbook* has an entire section devoted to issues of diversity. Individual chapters include, for example, designing for gender differences, children, the elderly, internationalization, and motor, perceptual, and cognitive impairments. Yet, in our review of the literature, current HCI approaches to human values and ethics do not always fare well when used in diverse contexts. For example, Participatory Design can be particularly effective when a community shares many deeply held sensibilities, such as a commitment to participatory democracy in the workplace and to the idea of meaningful work itself. But this approach is more difficult to apply when divisive constituencies argue on the basis of narrowly conceived self-interests or hostile prejudices. What happens, for example, when a manager values accountability over a worker's privacy? Or efficiency over a worker's autonomy? If each value has equal weight, on what basis does a designer move forward with a particular design?

As the field of HCI seeks to bring moral commitments into diverse contexts, it needs a principled moral means to adjudicate competing value claims. It is for this reason that we provided early on a moral philosophical and psychological framework for approaching this problem. We suggested, for example, that when moral conflicts occur between diverse groups, that such variability by itself does not prove or disprove the moral relativist's position. People can believe that a certain act (such as racial discrimination) is moral; but documenting such a belief does not make it so. We also suggested that when moral differences appear in another culture, that a question to ask

is whether there are victims within that culture. If there are, then it is probably less the case that societies differ on moral ground, and more that some societies (Western societies included) may be involved explicitly in immoral practices. We suggested that close attention must be paid to the level of moral analysis. Namely, definitions of morality that entail abstract characterizations of justice and welfare tend to highlight moral universals, whereas definitions that entail specific behaviors or rigid moral rules tend to highlight moral cross-cultural variation. In our view, both levels of analysis have merit, and a middle ground provides an epistemically sensible and powerful approach in HCI research: One that allows for an analysis of universal moral constructs (such as justice, rights, welfare, and virtue), as well as allowing for the ways in which these constructs play out in a particular culture at a particular point in time. By embracing this moral theoretical framework, we are not saying that all moral problems can be solved. But such problems will be discussed (and argued about) in ways that respect diversity and prevent oppression.

In the future, certain trends—technologically and societally—will pose particular challenges in terms of human values, ethics, and design. From our view, three trends stand out. First, computational technologies will increasingly allow for the erosion of personal privacy. Even today, for example, surveillance cameras capture our images in banks and airports, and in many stores, malls, and even streets. In cars, GPS navigation systems not only receive positioning data, but can broadcast one's position. Businesses can (and sometimes do) monitor workers' electronic communications. Indeed, think of perhaps the last bastion of the private space, the home; and recognize that "aware homes" of the future will have the potential to record virtually every movement an individual makes within his or her home, and to link that data to large networked databases. Thus, protections of individual privacy will need to become an even more central concern to the HCI community. The second trend is that computational technologies will increasingly provide means for government to erode civil liberties. Moreover, the public may unwittingly accept such consequences based on the assumption that these technologies will substantially enhance our nation's physical security. But such an assumption is not always warranted. Imagine, for example, if the government required each of us to obtain biometric national identification cards, and to use these cards for all of our financial transactions, air travel, entry into government buildings, and so forth. Secure? Hardly. For the systems that store this data can be cracked. Indeed, once biometric data is stolen, it is not so easy to get a new face, fingerprint, or DNA. Third, in our increasingly linked communicative technological infrastructure, the pervasiveness and speed of information can undermine people's psychological well-being and quality of life. How can we respond to this problem of information overload? One solution is that we need, at times, to turn away from our technological devices and thus to check their encroachment into human lives. Another solution is that, in our designs, we need to find ways to increase the ratio of quality over quantity of information. We can also profit by designing "calming" technologies (that go beyond placing information in our peripheral awareness, which itself can become overloaded).

In the last decade, the HCI community has made tremendous progress in integrating human values and ethics into the practice of design. As the field continues to move forward, the challenge remains how to design technology wisely, ethically, so as to create essential conditions by which humans live and flourish.

ACKNOWLEDGMENT

We thank Alan Borning and Gaetano Borriello for sustained discussions on human values and computational technologies. Valerie Wonder assisted with library research. The ideas presented in this paper are, in part, based on work supported by the National Science Foundation under Award Nos. IIS-0000567, IIS-9911185, and SES-0096131.

References

Aberg, J., & Shahmehri, N. (2001). An empirical study of human web assistants: Implications for user support in web information systems. In *Proceedings of the Conference on Human Factors in Computing Systems (CHI 2000)* (pp. 404–411). New York: Association for Computing Machinery Press.

Abowd, G. D., & Jacobs, A. (2001, September/October). The impact of awareness technologies on privacy litigation. *SIGCHI Bulletin*, 9.

Ackerman, M. S., & Cranor, L. (1999). Privacy critics: UI components to safeguard users' privacy. In *Extended Abstracts Conference on Human Factors in Computing Systems (CHI 99)* (pp. 258–259). New York: Association for Computing Machinery Press.

Adler, P. S., & Winograd, T. (Eds.). (1992). *Usability: turning technologies into tools*. Oxford: Oxford University Press.

Agre, P. E. (1998). Introduction. In P. E. Agre & M. Rotenberg (Eds.), *Technology and privacy: The new landscape* (pp. 1–28). Cambridge, MA: The MIT Press.

Agre, P. E., & Mailloux, C. A. (1997). Social choice about privacy: Intelligent vehicle-highway systems in the United States. In B. Friedman (Ed.), *Human values and the design of computer systems* (pp. 289–310). New York: Cambridge University Press.

Agre, P. E., & Rotenberg, M. (1998). *Technology and privacy: The new landscape*. Cambridge, MA: The MIT Press.

Anderson, R. E., Johnson, D. G., Gotterbarn, D., & Perrolle, J. (1993). Using the ACM code of ethics for decision making. *Communications of the ACM, 36*(2), 98–107.

Appadurai, A. (Ed.). (1988). *The social life of things*. New York: Cambridge University Press.

Akrich, M. (1992). The de-scription of technical objects. In W. E. Bijker & J. Law (Eds.), *Shaping technology/building society: Studies in sociotechnical change* (pp. 205–224). Cambridge, MA: The MIT Press.

Attewell, P. (1987). The deskilling controversy. *Work and Occupation, 14*(3), 323–346.

Baier, A. (1986). Trust and antitrust. *Ethics, 96*, 231–260.

Bayles, M. D. (1981). *Professional ethics*. Belmont, CA: Wadsworth.

Beard, M., & Korn, P. (2001). What I need is what I get: Downloadable user interfaces vis Jini and Java. In *Extended Abstracts of the Conference on Human Factors in Computing Systems (CHI 2001)* (pp. 15–16). New York: Association for Computing Machinery Press.

Beck, A., & Katcher, A. (1996). *Between pets and people*. West Lafayette, IN: Purdue University Press.

Becker, L. C. (1977). *Property rights: Philosophical foundations*. London: Routledge & Kegan Paul.

The Belmont Report. (1978). Ethical principles and guidelines for the protection of human subjects of research. Washington, DC: The National Commission for the Protection of Human Subjects of Biomedical and Behavioral Research.

Bellotti, V. (1998). Design for privacy in multimedia computing and communications environments. In P. E. Agre & M. Rotenberg (Eds.), *Technology and privacy: The new landscape* (pp. 63–98). Cambridge, MA: The MIT Press.

Bennett, C. J. (1998). Convergence revisited: Toward a global policy for the protection of personal data? In P. E. Agre & M. Rotenberg (Eds.), *Technology and privacy: The new landscape* (pp. 99–123). Cambridge, MA: The MIT Press.

Bennion, F. A. R. (1969). *Professional ethics*. London: Charles Knight & Co.

Berleur, J., & Brunnstein, K. (Eds.). (1997). *Ethics of computing: Codes, spaces for discussion, and law*. Dordrecht, The Netherlands: Kluwer Academic Publishers.

Bers, M. U., Gonzalez-Heydrich, J., & DeMaso, D. R. (2001). Identity construction environments: Supporting a virtual therapeutic community of pediatric patients undergoing dialysis. In *Proceedings of the Conference of Human Factors in Computing Systems (CHI 2001)* (pp. 380–387). New York: Association for Computing Machinery.

Bjerknes, G., & Bratteteig, T. (1995). User participation and democracy: A discussion of Scandinavian research on system development. *Scandinavian Journal of Information Systems, 7*(1), 73–97.

Bodker, S. (1990). *Through the interface—A human activity approach to user interface design*. Hillsdale, NJ: Lawrence Erlbaum Associates.

Borgman, C. L. (2000). *From Gutenberg to the global information infrastructure: Access to information in the networked world*. Cambridge, MA: The MIT Press.

Bos, N., Gergle, D., Olson, J. S., & Olson, G. M. (2001). Being there versus seeing there: Trust via video. In *Extended Abstracts of the Conference of Human Factors in Computing Systems (CHI 2001)* (pp. 291–292). New York: Association for Computing Machinery.

Boyd, R. A. (1988). How to be a moral realist. In G. Sayre-McCord (Ed.), *Essays on moral realism* (pp. 181–228). Ithaca, NY: Cornell University Press.

Boyle, M., Edwards, C., & Greenberg, S. (2000). The effects of filtered video on awareness and privacy. In *Proceedings of the Conference on Computer Supported Cooperative Work (CSCW 2000)* (pp. 1–10). New York: Association for Computing Machinery.

Bulliet, R. (1994). Determinism and pre-industrial technology. In M. R. Smith & L. Marx (Eds.), *Does technology drive history? The dilemma of technological determinism* (pp. 201–216). Cambridge, MA: The MIT Press.

Bynum, T. W. (Ed.). (1985). [Entire issue]. *Metaphilosophy, 16*(4).

Campbell, R. L., & Christopher, J. C. (1996). Moral development theory: A critique of its Kantian presuppositions. *Developmental Review, 16*, 1–47.

Cohen, A. L., Cash, D., & Muller, M. J. (2000). Designing to support adversarial collaboration. In *Proceedings of the Conference on Computer Supported Cooperative Work (CSCW 2000)* (pp. 31–39). New York: Association for Computing Machinery Press.

Cole, M. (1991, April). Discussant to B. Friedman & P. H. Kahn, Jr. (Presenters) "Who is responsible for what? And can what be responsible? The psychological boundaries of moral responsibility." Paper session at the biennial meeting of the Society for Research in Child Development, Seattle, WA.

Cooper, M., & Rejmer, P. (2001). Case study: Localization of an accessibility evaluation. In *Extended Abstracts of the Conference on Human Factors in Computing Systems (CHI 2001)* (pp. 141–142). New York: Association for Computing Machinery Press.

Cranor, L., & Resnick, P. (2000). Protocols for automated negotiations with buyer anonymity and seller reputations. *Netnomics 2*(1), 1–23.

Dewan, P., & Shen, H. (1998). Flexible meta access-control for collaborative applications. In *Proceedings of the Conference on Computer Supported Cooperative Work (CSCW 1998)* (pp. 247–256). New York: Association for Computing Machinery Press.

Dieberger, A., Hook, K., Svensson, M., & Lonnqvist, P. (2001). Social navigation research agenda. In *Extended Abstracts of the Conference on Human Factors in Computing Systems (CHI 2001)* (pp. 107–108). New York: Association of Computing Machinery Press.

Dourish, P., & Kiesler, S. (2000). From the papers co-chairs. In *Proceedings of the Conference on Computer Supported Cooperative Work (CSCW 2000)* (p. v). New York: Association for Computing Machinery Press.

Dworkin, R. (1978). *Taking rights seriously*. Cambridge, MA: Harvard University Press.

Egger, F. N. (2000). "Trust me, I'm an online vendor": Towards a model of trust for e-commerce system design. In *Extended Abstracts of the Conference of Human Factors in Computing Systems (CHI 2000)* (pp. 101–102). New York: Association for Computing Machinery.

Ehn, P. (1989). *Work-oriented design of computer artifacts*. Hillsdale, NJ: Lawrence Erlbaum Associates.

Ehn, P., & Kyng, M. (1991). Cardboard computers: Mocking-it-up or hands-on the future. In J. Greenbaum & M. Kyng (Eds.), *Design at work: Cooperative design of computer systems* (pp. 169–195). Hillsdale, NJ: Lawrence Erlbaum Associates.

Faden, R., & Beauchamp, T. (1986). *A history and theory of informed consent*. New York: Oxford University Press.

Farrell, S. (2001). Social and informational proxies in a fishtank. In *Extended Abstracts of the Conference of Human Factors in Computing Systems (CHI 2001)* (pp. 365–366). New York: Association for Computing Machinery.

Federal Trade Commission. (2000, May). *Privacy online: Fair information practices in the electronic marketplace* (A Report to Congress). Washington, DC: Author.

Fishkin, J. S. (1982). *The limits of obligation*. New Haven, CT: Yale University Press.

Fogg, B. J., & Tseng, H. (1999). The elements of computer credibility. In *Proceedings of the Conference of Human Factors in Computing Systems (CHI 1999)* (pp. 80–87). New York: Association for Computing Machinery.

Foot, P. (1978). *Virtues and vices*. Berkeley, CA: University of California Press.

Frankel, M. S., & Siang, S. (1999, November). *Ethical and legal aspects of human subjects research on the internet*. Washington, DC: American Association for the Advancement of Science. Workshop report [On-line]. Available: http://www.aaas.org./spp/dspp/sfrl/projects/inters/main.htm

Friedman, B. (Ed.). (1997a). *Human values and the design of computer technology*. New York: Cambridge University Press.

Friedman, B. (1997b). Social judgments and technological innovation: Adolescents' understanding of property, privacy, and electronic information. *Computers in Human Behavior, 13*(3), 327–351.

Friedman, B. (1999). *Value-sensitive design: A research agenda for information technology* (Contract No.: SBR-9729633). Report to the National Science Foundation [On-line]. Available: http://ischool.washington.edu/vsd

Friedman, B., & Borning, A. (2001). Value-sensitive design: Cultivating research and community (Contract No.: IIS-0000567). Unpublished report to the National Science Foundation.

Friedman, B., Howe, D. C., & Felten, E. (2002). Informed consent in the Mozilla browser: Implementing value-sensitive design. *Proceedings of the Thirty-Fifth Annual Hawai'i International Conference on Systems Science.* Abstract, p. 247; CD-ROM of full-paper, OSPE101. Los Alamitos, CA: IEEE Computer Society.

Friedman, B., & Kahn, P. H., Jr. (1992). Human agency and responsible computing: Implications for computer system design. *Journal of Systems Software, 17,* 7-14.

Friedman, B., & Kahn, P. H., Jr. (in press). A value-sensitive design approach to augmented reality. In W. Mackay (Ed.), *Design of augmented reality environments.* Cambridge, MA: The MIT Press.

Friedman, B., Kahn, P. H., Jr., & Howe, D. C. (2000). Trust online. *Communications of the ACM, 43*(12), 34-40.

Friedman, B., & Millett, L. (1995). "It's the computer's fault"—Reasoning about computers as moral agents. In *Conference Companion of the Conference on Human Factors in Computing Systems (CHI 95)* (pp. 226-227). New York: Association for Computing Machinery Press.

Friedman, B., Millett, L., & Felten, E. (2000). *Informed consent online: A conceptual model and design principles.* UW-CSE Technical Report 00-12-2. University of Washington, Department of Computer Science & Engineering. Seattle, WA.

Friedman, B., & Nissenbaum, H. (1996). Bias in computer systems. *ACM Transactions on Information Systems, 14*(3), 330-347.

Friedman, B., & Nissenbaum, H. (1997). Software agents and user autonomy. *Proceedings of the First International Conference on Autonomous Agents* (pp. 466-469). New York: Association for Computing Machinery Press.

Fuchs, L. (1999). AREA: A cross-application notification service for groupware. In S. Bodker, M. Kyng, & K. Schmidt (Eds.), *Proceedings of the Sixth European Conference on Computer Supported Cooperative Work* (pp. 61-80). Dordrecht, The Netherlands: Kluwer Academic Publishers.

Galegher, J., Kraut, R. E., & Egido, C. (Eds.). (1990). *Intellectual teamwork: Social and technological foundations of cooperative work.* Hillsdale, NJ: Lawrence Erlbaum Associates.

Gewirth, A. (1978). *Reason and morality.* Chicago, IL: University of Chicago Press.

Godefroid, P., Herbsleb, J. D., Jagadeesan, L., & Li, D. (2000). Ensuring privacy in presence awareness systems: Automated verification approach. In *Proceedings of the Conference on Computer Supported Cooperative Work (CSCW 2000)* (pp. 59-68). New York: Association for Computing Machinery Press.

Goldberg, K. (Ed.). (2000). *The robot in the garden: Telerobotics and telepistemology on the Internet.* Cambridge, MA: The MIT Press.

Gotterbarn, D. (1999, November/December). How the new software engineering code of ethics affects you. *IEEE Software, 16*(6), 58-64.

Greenbaum, J. (1996). Back to labor: Returning to labor process discussions in the study of work. In *Proceedings of the Conference on Computer Supported Cooperative Work (CSCW 1996)* (pp. 315-324). New York: Association for Computing Machinery Press.

Greenbaum, J., & Kyng, M. (Eds.). (1991). *Design at work: Cooperative design of computer systems.* Hillsdale, NJ: Lawrence Erlbaum Associates.

Greenspan, S., Goldberg, D., Weimer, D., & Basso, A. (2000). Interpersonal trust and common ground in electronically mediated communication. In *Proceedings of the Conference on Computer Supported Cooperative Work (CSCW 2000)* (pp. 251-260). New York: Association for Computing Machinery Press.

Grief, I. (Ed.). (1988). *Computer-supported cooperative work: A book of readings.* San Mateo, CA: Morgan Kaufmann.

Gross, M. D., Parker, L., & Elliot, A. M. (1997). MUD: Exploring tradeoffs in urban design. In *Proceedings of the 7th International Conference on Computer Aided Architectural Design Futures* (pp. 373-387). Dordrecht, The Netherlands: Kluwer Academic Publishers.

Grudin, J. (1988). Why CSCW applications fail: Problems in the design and evaluation of organizational interfaces. In *Proceedings of the Conference on Computer Supported Cooperative Work (CSCW '88)* (pp. 85-93). New York: Association for Computing Machinery Press.

Harris Poll/*Business Week.* (2000). *A growing threat.* http://www.businessweek.come/2000/00_12/b3673010.htm

Harrison, S., & Dourish, P. (1996). Re-place-ing space: The roles of place and space in collaborative systems. In *Proceedings of the Conference on Computer Supported Cooperative Work (CSCW 96)* (pp. 67-76). New York: Association for Computing Machinery Press.

Hatch, E. (1983). *Culture and morality.* New York: Columbia University Press.

Herskovits, M. J. (1952). *Economic anthropology: A study of comparative economics.* New York: Alfred A. Knopf.

Hert, C., Marchionini, G., Liddy, L., & Shneiderman, B. (2000). Extending understanding of federal statistics in tables: Integrating technology and user behavior in support of university usability [On-line]. Available: http://istweb.syr.edu/~tables

Houston, J. (1995). *Confessions of an igloo dweller.* New York: Houghton Mifflin.

Hudson, S. E., & Smith, I. (1996). Techniques for addressing fundamental privacy and disruption tradeoffs in awareness support systems. In *Proceedings of the ACM 1996 Conference on Computer Supported Cooperative Work (CSCW 96)* (pp. 248-257). New York: Association for Computing Machinery Press.

Hughes, T. P. (1994). Technological momentum. In M. R. Smith & L. Marx (Eds.), *Does technology drive history? The dilemma of technological determinism* (pp. 101-113). Cambridge, MA: The MIT Press.

Iacono, S., & Kling, R. (1987). Changing office technologies and the transformation of clerical jobs. In R. Kraut (Ed.), *Technology and the transformation of white collar work.* Hillsdale, NJ: Lawrence Erlbaum.

Isaacs, E. A., Tang, J. C., & Morris, T. (1996). Piazza: A desktop environment supporting impromptu and planned interactions. In *Proceedings of the Conference on Computer Supported Cooperative Work (CSCW 96)* (pp. 315-324). New York: Association for Computing Machinery Press.

Jacko, J. A., Dixon, M. A., Rosa, R. H., Jr., Scott, I. U., & Pappas, C. J. (1999). Visual profiles: A critical component of universal access. In *Proceedings of the Conference on Human Factors in Computing Systems (CHI 99)* (pp. 330-337). New York: Association for Computing Machinery Press.

Jancke, G., Venolia, G. D., Grudin, J., Cadiz, J. J., & Gupta, A. (2001). Linking public spaces: Technical and social issues. In *Proceedings of the Conference of Human Factors in Computing Systems (CHI 2001)* (pp. 530-537). New York: Association for Computing Machinery.

Johnson, D. G. (1985). *Computer ethics* (1st ed.). Englewood Cliffs, NJ: Prentice Hall.

Johnson, D. G. (2001). *Computer ethics* (3rd ed.). Upper Saddle River, NJ: Prentice Hall.

Johnson, D. G., & Miller, K. (1997). Ethical issues for computer scientists and engineers. In A. B. Tucker, Jr. (Ed.), *The computer science and engineering handbook* (pp. 16-26). Boca Raton, FL: CRC Press.

Johnson, E. H. (2000). Getting beyond the simple assumptions of organization impact [social informatics]. *Bulletin of the American Society for Information Science, 26*(3), 18-19.

Jones, S., Wilkens, M., Morris, P., & Masera, M. (2000). Trust requirements in e-business. *Communications of the ACM, 43*(12), 80-87.

Kahn, P. H., Jr. (1991). Bounding the controversies: Foundational issues in the study of moral development. *Human Development, 34*, 325-340.

Kahn, P. H., Jr. (1992). Children's obligatory and discretionary moral judgments. *Child Development, 63*, 416-430.

Kahn, P. H., Jr. (1999). *The human relationship with nature: Development and culture.* Cambridge, MA: MIT Press.

Kahn, P. H., Jr., & Friedman, B. (in preparation). Augmented reality of the natural world and its psychological effects: A value-sensitive design approach.

Kahn, P. H., Jr., & Kellert, S. R. (Eds.). (2002). *Children and nature: Psychological, sociocultural, and evolutionary investigations.* Cambridge, MA: MIT Press.

Kahn, P. H., Jr., & Lourenço, O. (1999). Reinstating modernity in social science research—or—The status of Bullwinkle in a post-postmodern era. *Human Development, 42*, 92-108.

Kahn, P. H., Jr., & Turiel, E. (1988). Children's conceptions of trust in the context of social expectations. *Merrill-Palmer Quarterly, 34*, 403-419.

Kant, I. (1785/1964). *Groundwork of the metaphysic of morals* (H. J. Paton, Trans.). New York: Harper Torchbooks. (Original work published 1785.)

Kensing, F., & Madsen, K. H. (1991). Generating visions: Future workshops and metaphorical design. In J. Greenbaum & M. Kyng (Eds.), *Design at work: Cooperative design of computer systems* (pp. 155-168). Hillsdale, NJ: Lawrence Erlbaum Associates.

Kensing, F., Simonsen, J., & Bodker, K. (1998). MUST: A method for participatory design. *Human-Computer Interaction, 13*(2), 167-198.

Kiesler, S. (Ed.). (1997). *The culture of the Internet.* Mahwah, NJ: Lawrence Erlbaum Associates.

King, J. L. (1983). Centralized versus decentralized computing: Organizational considerations and management options. *Computing Surveys, 15*(4), 320-349.

Kling, R. (1980). Social analyses of computing: Theoretical perspectives in recent empirical research. *Computing Surveys, 12*(1), 61-110.

Kling, R. (1999). What is social informatics and why does it matter? *D-Lib Magazine, 5*(1) [On-line]. Available: http://www.dlib.org/dlib/january99/kling/oikling.html

Kling, R., Rosenbaum, H., & Hert, C. (1998). Social informatics in information science: An introduction. *Journal of the American Society for Information Science, 49*(12), 1047-1052.

Kling, R., & Star, S. L. (1998). Human centered systems in the perspective of organizational and social informatics. *Computers and Society, 28*(1), 22-29.

Korpela, M., Soriyan, H. A., Olufokunbi, K. C., Onayade, A. A., Davies-Adetugbo, A., & Adesanmi, D. (1998). Community participation in health informatics in Africa: An experiment in tripartite partnership in Ile-Ife, Nigeria. *CSCW Journal, 7*(3-4), 339-358.

Kuhn, S., & Winograd, T. (1996). Profile 14: Participatory design. In T. Winograd (Ed.), *Bringing design to software.* Reading, MA: Addison-Wesley.

Kyng, M., & Mathiassen, L. (Eds.). (1997). *Computers and design in context.* Cambridge, MA: The MIT Press.

Latour, B. (1992). Where are the missing masses? The sociology of a few mundane artifacts. In W. E. Bijker & J. Law (Eds.), *Shaping technology/building society: Studies in sociotechnical change* (pp. 225-258). Cambridge, MA: The MIT Press.

Leveson, N. G. (1991). Software safety in embedded computer systems. *Communications of the ACM, 34*(2), 34-46.

Leveson, N. G., & Turner, C. S. (1993). An investigation of the Therac-25 accidents. *IEEE Computer, 26*(7), 18-41.

Lipinski, T. A., & Britz, J. J. (2000). Rethinking the ownership of information in the 21st century: Ethical implications. *Ethics and Information Technology, 2*(1), 49-71.

Louden, R. B. (1984). On some vices of virtue ethics. *American Philosophical Quarterly, 21*, 227-235.

MacIntyre, A. (1984). *After virtue.* Notre Dame, IN: University of Notre Dame Press.

Mackay, W. E. (1995). Ethics, lies and videotape. In *Proceedings of the Conference on Human Factors in Computing Systems (CHI 95)* (pp. 138-145). New York: Association for Computing Machinery Press.

Mackay, W. E., & Fayard, A. L. (1999). Designing interactive paper: Lessons from three augmented reality projects. In *Proceedings of IWAR '98* (pp. 81-90). Natick, MA: A. K. Peters.

Mackie, J. L. (1977). *Ethics: Inventing right and wrong.* New York: Penguin Books.

Malone, T. W. (1994). Commentary on Suchman article and Winograd response. *CSCW Journal, 3*, 37-38.

McCarthy, J. F., & Anagnost, T. D. (1998). MusicFX: An arbiter of group preferences for computer supported collaborative workouts. In *Proceedings of the Conference on Computer Supported Cooperative Work (CSCW 98)* (pp. 363-372). New York: Association for Computing Machinery Press.

Milewski, A. E., & Smith, T. M. (2000). Providing presence cues to telephone users. In *Proceedings of the Conference on Computer Supported Cooperative Work (CSCW 2000)* (pp. 89-96). New York: Association for Computing Machinery Press.

Millett, L., Friedman, B., & Felten, E. (2001). Cookies and Web browser design: Toward realizing informed consent online. In *Proceedings of the Conference on Human Factors in Computer Systems (CHI 2001)* (pp. 46-52). New York: Association for Computing Machinery Press.

Molich, R. (2001). Ethics in HCI. In *Extended Abstracts of the Conference on Human Factors in Computing Systems (CHI 2001)* (pp. 231-232). New York: Association for Computing Machinery Press.

Moor, J. H. (1985). What is computer ethics? *Metaphilosophy, 16*(4), 266-275.

Muller, M. J. (1997). Ethnocritical heuristics for reflecting on work with users and other interested parties. In M. Kyng & L. Mathiassen (Eds.), *Computers and design in context* (pp. 349-380). Cambridge, MA: The MIT Press.

Mynatt, E. D., Adler, A., Ito, M., Linde, C., & O'Day, V. L. (1999). The network communities of SeniorNet. In *Proceedings of the Sixth European Conference on Computer Supported Cooperative Work* (pp. 219-238). Dordrecht, The Netherlands: Kluwer Academic Publishers.

Nagel, T. (1986). *The view from nowhere.* Oxford, England: Oxford University Press.

Nass, C., & Gong, L. (2000). Speech interfaces from an evolutionary perspective. *Communications of the ACM, 43*(9), 36-43.

Neumann, P. G. (1991). Inside risks: Certifying professionals. *Communications of the ACM, 34*(2), 130.

Neumann, P. G. (1995). *Computer related risks.* New York: Association for Computing Machinery Press.

Nielsen, J. (1993). *Usability engineering.* Boston, MA: AP Professional.

Nissenbaum, H. (1999). Can trust be secured online? A theoretical perspective. *Etica e Politca* [On-line], 2. Available: http://www.univ.trieste.it/~dipfilo/etica_e_politica/199_2/homepage.html

Noble, D. D. (1991). *The classroom arsenal: Military research, information technology and public education.* London, England: Falmer Press.

Norman, D. A. (1988). *The psychology of everyday things.* New York: Basic Books.

Noth, M., Borning, A., & Waddell, P. (2000). An extensible, modular architecture for simulating urban development, transportation, and environmental impacts (UW CSE TR 2000-12-01) [On-line]. Available: http://www.urbansim.org

Olson, J. S., & Olson, G. M. (2000). i2i trust in e-commerce. *Communications of the ACM, 43*(12), 41–44.

Olson, J. S., & Teasley, S. (1996). Groupware in the wild: Lessons learned from a year of virtual collaboration. In *Proceedings of the Conference on Computer Supported Cooperative Work (CSCW 96)* (pp. 419–427). New York: Association for Computing Machinery Press.

O'Neill, D. K., & Gomez, L. M. (1998). Sustaining mentoring relationships on-line. In *Proceedings of the Conference on Computer Supported Cooperative Work (CSCW 98)* (pp. 325–334). New York: Association for Computing Machinery Press.

Orlikowski, W. J. (1993). Learning from *Notes*: Organizational issues in groupware implementation. *The Information Society, 9*(3), 237–250.

Orlikowski, W. J. (2000). Using technology and constituting structures: A practice lens for studying technology in organizations. *Organization Science, 11*(4), 404–428.

Oviatt, S. (1999). Mutual disambiguation of recognition errors in a multimodal architecture. In *Proceedings of the Conference on Human Factors in Computing Systems (CHI 99)* (pp. 576–583). New York: Association for Computing Machinery Press.

Palen, L., Salzman, M., & Youngs, E. (2000). Going wireless: Behavior & practice of new mobile phone users. In *Proceedings of the Conference on Computer Supported Cooperative Work (CSCW 2000)* (pp. 201–210). New York: Association for Computing Machinery Press.

Parker, D. (1979). *Ethical conflicts in computer science and technology.* Arlington, VA: AFIPS Press.

Parker, D., Swope, S., & Baker, B. (1990). *Ethical conflicts in information and computer science, technology, and business.* Wellesley, MA: QED Information Sciences.

Pelto, P. J. (1973). *The snowmobile revolution: Technology and social change in the arctic.* Menlo Park, CA: Cummings Publishing Company.

Perdue, P. C. (1994). Technological determinism in agrarian societies. In M. R. Smith & L. Marx (Eds.), *Does technology drive history? The dilemma of technological determinism* (pp. 169–200). Cambridge, MA: The MIT Press.

Perry, J., Macken, E., Scott, N., & McKinley, J. L. (1997). Disability, inability and cyberspace. In B. Friedman (Ed.), *Human values and the design of computer technology* (pp. 65–89). New York: Cambridge University Press.

Phillips, D. J. (1998). Cryptography, secrets, and structuring of trust. In P. E. Agre & M. Rotenberg (Eds.), *Technology and privacy: The new landscape* (pp. 243–276). Cambridge, MA: The MIT Press.

Poltrock, S. E., & Grudin, J. (1994). Organizational obstacles to interface design and development. *ACM Transactions on Computer-Human Interaction, 1,* 52–80.

President's Information Technology Advisory Committee (PITAC). (1999, February 24). *Information technology research: Investing in our future* (Advance Copy). National Coordination Office for Computing, Information and Communications. Washington, DC [On-line]. Available: http://www.itrd.gov/ac/report/

Putnam, R., Leonardi, R., & Nanetti, R. (1993). *Making democracy work: Civic traditions in modern Italy.* Princeton, NJ: Princeton University Press.

Rawls, J. (1971). *A theory of justice.* Cambridge, MA: Harvard University Press.

Reeves, B., & Nass, C. (1996). *The media equation: How people treat computers, television, and new media like real people and places.* New York: Cambridge University Press.

Rocco, E. (1998). Trust breaks down in electronic contexts but can be repaired by some initial face-to-face contact. In *Proceedings of the Conference of Human Factors in Computing Systems (CHI 98)* (pp. 496–502). New York: Association for Computing Machinery.

Rorty, R. (1982). *Consequences of pragmatism.* Minneapolis, MN: University of Minnesota Press.

Rosenberg, S. (1977). Multiplicity of selves. In R. D. Ashmore & L. Jussim (Eds.), *Self and identity* (pp. 23–45). New York: Oxford University Press.

Rotenberg, M. (2000). *The privacy law sourcebook 2000: United States law, international law, and recent developments.* Washington, DC: Electronic Privacy Information Center.

Roth, A. E. (1990). New physicians: A natural experiment in market organization. *Science, 250,* 1524–1528.

Sawyer, S., & Rosenbaum, H. (2000). Social informatics in the information sciences: Current activities and emerging direction. *Informing Science, 3*(2), 89–95.

Scheffler, S. (1982). *The rejection of consequentialism.* Oxford, England: Oxford University Press.

Schiano, D. J., & White, S. (1998). The first noble truth of cyberspace: People are people (even when they MOO). In *Proceedings of the Conference of Human Factors in Computing Systems (CHI 98)* (pp. 352–359). New York: Association for Computing Machinery.

Schneider, F. B. (Ed.). (1999). *Trust in cyberspace.* Washington, DC: National Academy Press.

Schoeman, F. D. (Ed.). (1984). *Philosophical dimensions of privacy: An anthology.* Cambridge, England: Cambridge University Press.

Scranton, P. (1994). Determinism and indeterminacy in the history of technology. In M. R. Smith & L. Marx (Eds.), *Does technology drive history? The dilemma of technological determinism* (pp. 143–168). Cambridge, MA: The MIT Press.

Searle, J. R. (1980). Minds, brains, and programs. *Behavioral and Brain Sciences, 3,* 417–458.

Searle, J. R. (1983, October 27). The word turned upside down. *New York Review of Books,* 74–79.

Sharp, L. (1952/1980). Steel axes for stone-age Australians. In J. P. Spradley & D. W. McCurdy (Eds.), *Conformity and conflict* (pp. 345–359). Boston: Little, Brown, & Company. (Reprinted from *Human Organization,* 1952, *11,* 17–22.)

Shneiderman, B. (1999). Universal usability: Pushing human-computer interaction research to empower every citizen. ISR Technical Report 99-72. University of Maryland, Institute for Systems Research. College Park, MD.

Shneiderman, B. (2000a). Designing trust into online experience. *Communications of the ACM, 43*(12), 57–59.

Shneiderman, B. (2000b). Universal usability. *Communications of the ACM, 43*(5), 84–91.

Shneiderman, B., & Rose, A. (1997). Social impact statements: Engaging public participation in information technology design. In B. Friedman (Ed.), *Human values and the design of computer systems* (pp. 117–133). New York: Cambridge University Press.

Shoemaker, G. B. D., & Inkpen, K. M. (2001). Single display privacyware: Augmenting public displays with private information. In *Proceedings of the Conference of Human Factors in Computing Systems*

(CHI 2001) (pp. 522-529). New York: Association for Computing Machinery.

Shweder, R. A., Mahapatra, M., & Miller, J. B. (1987). Culture and moral development. In J. Kagan & S. Lamb (Eds.), *The emergence of morality in young children* (pp. 1-82). Chicago: University of Chicago Press.

Smith, M. R. (1994). Technological determinism in American culture. In M. R. Smith & L. Marx (Eds.), *Does technology drive history? The dilemma of technological determinism* (pp. 1-35). Cambridge, MA: The MIT Press.

Smith, M. R., & Marx, L. (Eds.). (1994). *Does technology drive history? The dilemma of technological determinism*. Cambridge, MA: The MIT Press.

Software engineering code of ethics and professional practice. (1998). [On-line]. Available http://www.acm.org/serving/se/code.htm

Spinello, R. A., & Tavani, H. T. (2001). The Internet, ethical values, and conceptual frameworks: An introduction to cyberethics. *Computers and Society, 31*(2), 5-7.

Sproull, L., & Kiesler, S. (1991). *Connections: New ways of working in the networked organization*. Cambridge, MA: The MIT Press.

Stephanidis, C. (Ed.) (2001). *User interfaces for all: Concepts, methods, and tools*. Mahwah, NJ: Lawrence Erlbaum Associates.

Sturgeon, N. L. (1988). Moral explanations. In G. Sayre-McCord (Ed.), *Essays on moral realism* (pp. 229-255). Ithaca, NY: Cornell University Press.

Suchman, L. (1994). Do categories have politics? The language/action perspective reconsidered. *CSCW Journal, 2*(3), 177-190.

Svensson, M., Hook, K., Laaksolahti, J., & Waern, A. (2001). Social navigation of food recipes. In *Proceedings of the Conference of Human Factors in Computing Systems (CHI 2001)* (pp. 341-348). New York: Association for Computing Machinery.

Tang, J. C. (1997). Eliminating a hardware switch: Weighing economics and values in a design decision. In B. Friedman (Ed.), *Human values and the design of computer technology* (pp. 259-269). New York: Cambridge University Press.

Thomas, J. C. (1997). Steps toward universal access within a communications company. In B. Friedman (Ed.), *Human values and the design of computer systems* (pp. 271-287). New York: Cambridge University Press.

Turiel, E. (1983). *The development of social knowledge*. Cambridge, England: Cambridge University Press.

Turiel, E. (1998). Moral development. In N. Eisenberg (Ed.), *Social, emotional, and personality development* (pp. 863-932). Volume 3 of W. Damon (Ed.), *Handbook of child psychology* (5th ed). New York, NY: Wiley.

Turiel, E. (in press). *The Culture of Morality: Social Development and Social Opposition*. Cambridge, England: Cambridge University Press.

Turiel, E., Killen, M., & Helwig, C. C. (1987). Morality: Its structure, functions and vagaries. In J. Kagan & S. Lamb (Eds.), *The emergence of morality in young children* (pp. 155-244). Chicago: University of Chicago Press.

Turkle, S. (1996). *Life on the screen: Identify in the age of the Internet*. New York: Simon and Schuster.

Ulrich, R. S. (1984). View through a window may influence recovery from surgery. *Science, 224*, 420-421.

Ulrich, R. S. (1993). Biophilia, biophobia, and natural landscapes. In S. R. Kellert & E. O. Wilson (Eds.), *The biophilia hypothesis* (pp. 73-137). Washington, DC: Island Press.

Van House, N. A., Butler, M. H., & Schiff, L. R. (1998). Cooperative knowledge work and practices of trust: Sharing environmental planning data sets. In *Proceedings of the Conference on Computer Supported Cooperative Work (CSCW 98)* (pp. 335-343). New York: Association for Computing Machinery Press.

Wainryb, C. (1995). Reasoning about social conflicts in different cultures: Druze and Jewish children in Israel. *Child Development, 66*, 390-401.

Weckert, J., & Adeney, D. (1997). *Computer and information ethics*. Westport, CT: Greenwood Press.

Weiser, M., & Brown, J. S. (1997). The coming age of calm technology. In P. Denning & B. Metcalfe (Eds.), *Beyond calculation: The next 50 years of computing* (pp. 75-85). New York: Springer-Verlag.

Weizenbaum, J. (1972, May 12). On the impact of the computer on society: How does one insult a machine? *Science, 178*, 609-614.

Wiener, N. (1953/1985). The machine as threat and promise. In P. Masani (Ed.), *Norbert Wiener: Collected works and commentaries* (Vol. IV, pp. 673-678). Cambridge, MA: The MIT Press. (Reprinted from *St. Louis Post Dispatch*, December 13, 1953.)

Williams, B. (1985). *Ethics and the limits of philosophy*. Cambridge, MA: Harvard University Press.

Winner, L. (1986). *The whale and the reactor: A search for limits in an age of high technology*. Chicago: University of Chicago Press.

Winograd, T. (1994). Categories, disciplines, and social coordination. *CSCW Journal, 2*(3), 191-197.

Zheng, J., Bos, N., Olson, J., & Olson, G. M. (2001). Trust with*out* touch: Jump-start trust with social chat. In *Extended Abstracts of the Conference of Human Factors in Computing Systems (CHI 2001)* (pp. 293-294). New York: Association for Computing Machinery.

· 62 ·

COST JUSTIFICATION

Randolph G. Bias
Austin Usability

Deborah J. Mayhew
Deborah J. Mayhew and Associates

Dilip Upmanyu
Tivoli Systems

A friend decided she would try one of those new grocery e-tail companies—order her family's groceries online, and then just have to be home during a prescribed 2-hour period for the delivery. It sounded like a great idea, a new trend, the way grocery shopping would be done forevermore. She found the Web site, a competitor for the grocery chain she usually patronized, and completed the registration form, entering all sorts of demographic data to enable the company to provide her customized service. She was asked to choose an ID and a password. This sounded like a good idea, ensuring that no one would be able to order food in her name. She clicked the "Submit" button, eager to begin her shopping experience. She was dismayed to receive an error message; it seems the ID she selected was already in the system, and the message asked her to pick a new one. Alas, the entire form had been blanked out, and my friend had to start from scratch. Not being particularly computer-savvy, my friend figured she was at fault. She went through the process again, this time using her last name as the ID. With the same result. She gritted her teeth, went through the same process a third time, put in her whole name as her ID, and, perhaps surprisingly, but certainly frustratingly, she got the same error message. Despite all encouragement from her spouse, and assurances that this sort of thing happens, she has refused to try again this or any online shopping service. How long will it be before she overcomes this well-earned aversion? How much revenue will the grocery e-tailer lose from this one family because of this one usability problem? More interestingly, is the news in today's newspaper that this e-tailer is going out of business, directly related to this usability problem, or something else altogether?

JUSTIFICATION, OF COST JUSTIFICATION

"The usability engineering of web sites is *always* carried out under sub-optimal conditions. Development constraints... demand that we pursue a quicker, less thorough course of usability engineering than we would like" (Bias & Keough, 2000). In times of tight budgets, it is absolutely necessary to be able to demonstrate, with real quantitative data, that any money invested in usability engineering yield a positive return. Otherwise, the usability engineer is dependent on the faith and inspiration of the stakeholders, when trying to get the funding to carry out a user-centered design approach. Plus, said usability engineer will be competing for those funds with others who *do* have quantitative figures—lines of code, sales projections, market research, defect rates, customer support calls, person-months.

Consider, too, that much of the software development world sees usability still as "nice-to-have." To compete for those constrained development resources, the usability engineer must be able to cost-justify any expenditures on usability. This chapter is intended to give to the usability engineer, the usability engineering manager, or anyone advocating for the implementation of a user-centered design approach, the tools to demonstrate that a particular program of usability engineering will yield a particular return on investment (ROI).

THE RETURN ON INVESTMENT IS WHAT IT IS

The ROI for any effort is what it is—the trick is to project or calculate it. Theoretically, after each new marketing campaign or addition in sales force or added product feature, we could calculate effects on revenues and costs. (This is not necessarily easy, because there are usually multiple variables with no experimental control, and thus it is hard to attribute a certain percentage of any changes to one variable.) Given the state of web design, and our experiences, usability engineers are confident that just about all usability engineering efforts are well cost-justified. Clare-Marie Karat (1994) reported a range of ROI ratios of from 2:1 to 100:1 for traditional software user interface designs. With the web development tools enabling more and more folks to play the role of designer, and with the broad access to the web making for an ever-widening circle of users, there is every reason to suspect that usability engineering for Web sites yields an even *higher* positive return.

So, assuming that any change in a company's product or process will yield *some* change in that company's revenues or expenses or both, the trick is to estimate what that ROI will be.

THE EMPIRICAL APPROACH TO USER-CENTERED DESIGN

Although it is hard to attribute exact fluctuations in costs and benefits to certain methods employed in a user-centered design approach, it is possible to collect empirical data to drive projections and calculations of ROI. In this chapter, we will endeavor to encourage and equip the usability engineer, or the software development manager, to collect the data and make the calculations to justify expenditures on usability engineering. We start by considering the wide range of possible beneficiaries of a user-centered design approach.

TO WHOM DO BENEFITS ACCRUE?

It is important to realize, and to communicate to all the product/Web site stakeholders, all of the different benefits that can result from a program of usability engineering. Done right (and early in the development cycle), usability engineering can yield efficiency gains for the *site/product development team*, because problems are discovered early when they are relatively inexpensive to repair. The *customer support team* will require a small staff, because more users can carry out their tasks without help. Similarly, the *documentation team* may need to generate fewer user aids, and will do a better job of creating their help screens and other hard-copy and online documentation. A more intuitive product means that fewer resources need to be devoted to *training* materials, or perhaps the training products can focus more on advanced usage because the fundamentals are more intuitive. Of course, a more usable product will mean that *the user* will enjoy increased throughput, both because the interaction design allows quicker completion of tasks through fewer displays and fewer keystrokes/mouse clicks, and because

less time is spent making and recovering from errors. Indeed, *the customer* may be able to hire a less sophisticated or less experienced employee to carry out tasks previously handled only by higher paid employees—all of which will lead to increased customer satisfaction, increased trade press joy, and increased sales.

QUANTIFYING THE COSTS (EASY!)

It is a simple matter to quantify the *costs*. For in-house usability teams, it is a simple matter to calculate the costs of the usability engineers—their salaries and additional costs, such as benefits, a desk, a computer, and so forth. ("Commonly, the fully loaded cost of an employee is at least twice his or her salary" [Nielsen, 2000, p. 276].) For a particular project, or even a particular data collection effort, a fractional cost of a usability professional can be calculated. For usability engineering contractors or consultants, the costs will be easy to calculate, simply by reading the invoice. Add to these costs the costs of engaging the software developers in the usability evaluations or in the implementation of the corrections to the interface that arose during testing. That is, perhaps the designer/developers will be asked to generate a user interface prototype for testing. Or perhaps they will be observing end-user testing in the lab. Certainly the development team will need to invest some time in helping build a user profile and test scenarios.

Recognize that there will be *some* costs, even without usability engineering. That is, when calculating the comparison between the costs of pursuing a usability engineering approach and not, it is important to include the designer/developer time to fix software or design bugs that may be discovered late in the process anyway, say, during system test.

QUANTIFYING THE BENEFITS (HARDER)

Quantifying the benefits is harder, but still possible. One reasonable approach is to make projections based on usability testing. Perhaps the crispest arena is reduced call support burden. Imagine a usability test that revealed that, of the test participants who carried out a task, 30% had to call the (simulated) help desk. Then (we are still imagining), in light of the usability test results, the product is redesigned, and subsequent testing reveals that only 5% of the test participants need help to complete the task. It is a relatively simple matter to project the reduction in the number of calls to the help desk, using the projected number of licenses (for a product) or visits (for a Web site). Then multiply that number of *avoided* help line calls times the average cost of a call to the company to find the amount of money saved thanks to that one usability test. (We will go through a more thorough example in a later section.)

Another way to project the benefits is by using historical data. If you have experience with a similar sort of interface, at a similar stage of development, you might reasonably expect the same ROI as you earned the last time. Or, if there are data in the research literature, from a project that is similar to yours, you could use those ROI data as an estimate. Of course, the closer

your current situation matches that of the other development effort from which you are extrapolating—in complexity of the user interface, sophistication and experience level of the development team, processes followed by the development team, resources available to the development team (e.g., machines for testing), domain expertise in house—the better the estimate will be. As with sales projections, it is always best to have some customer data to drive the estimates.

It is important to return to the "actuals" after the project, to assess how good your ROI estimates were. This is a difficult thing to budget time for, as you will be off on the next project. But the continuous improvement of your financial modeling, and your empirically based validation of your usability effort, in the eyes of your stakeholders and funders, depends on your ability to demonstrate that your projections were good (and conservative!), and that the hour and dollar spent on usability engineering was well worth it.

THE VALUE OF MONEY—VARIOUS FINANCIAL METHODS

Raw Dollars

One of the frustrations with usability engineering of human computer interfaces, circa 2002, is there is no acknowledged metric with which all interfaces can be compared. Perhaps there will come a time when we will have an agreed upon scale—that particular user interface is a "6," but making those improvements makes it a "9." The joy of a cost-benefit analysis approach is the reliance on a universally understood metric: the dollar. (Or the franc, or the yen.) It enables managers, engineers, writers, marketers, and even psychologists to communicate on even ground: If we spend X dollars on usability engineering, we expect to save Y dollars thanks to gained development efficiencies and reduced support call burden, and to earn Z additional dollars in revenue, thanks to increased sales. As stated earlier, the benefits of usability engineering tend to be so robust, that attending only to raw dollars will usually be sufficient to justify dedicating resources to a systematic, user-centered design approach.

Three different but related factors, however, may motivate individuals today, and the whole discipline in the near future to employ more sophisticated financial models:

1. tight development budgets and the need for usability engineers to compete with other product-related disciplines,
2. relatedly, the need to be smart about *which* usability engineering methods to employ (more in the next section), and
3. the (assumed) ever-improving accuracy of our projections of the benefits of usability engineering.

These three factors may drive the usability engineer intent on being as accurate as possible in projecting the relative value of usability to employ an accounting technique, such as one of the following three. (See Karat [1994] for more detail, including formulas.)

Payback Period

The payback period is the amount of time it takes to recover through incremental cash flows (or subsequent cost reductions) a project's net investment cash outflow. The calculation locates the point at which the cash outflow equals the cash inflow. In simple terms, the shorter the payback period, the better the investment. An organization may have a standard payback period that any investment must beat, in order to be funded. Alternately, two different courses of usability engineering (say, a "basic" course vs. a "complete" course) can be compared against each other, to drive a decision. Payback period is not a measure of profitability, but rather a measure of speed of capital recovery. (*Note*: It is *not* necessarily the case that the more expensive, "full" course will yield a longer payback period. For a passionate and well-reasoned argument for the joys of a "time to break even" approach, see Conklin, 1996.)

Because payback period ignores the time value of money, many organizations use discounted cash flow techniques to make investment decisions. Two of the most popular are internal rate of return (IRR) and net present value (NPV).

Internal Rate of Return

The IRR on a project is the company's rate of return on invested capital. For example, if a particular project costs $1,000 and earns $100 in Years 1 and 2 plus returning the initial investment at the end, then the IRR is 10%.

IRR

	Time 0	Year 1	Year 2
Cash Flow	($1,000)	$100	$1,100
	10% IRR		

The IRR in this case is compared against a company's cost of capital to make an investment decision. Cost of capital in simple terms is the company's cost of obtaining funding whether through the debt or equity markets. Most companies use both sources. For example, if a company's capital structure is 50% debt with an average cost of 10% and 50% through equity with an average return of 20%, the cost of capital is 15% (.5*.1) + (.5*.2). In summary, the decision rule for IRR is as follows: If the IRR is greater than the cost of capital, then the project should be accepted; but, if the opposite is true, then the project should be rejected. NPV is an alternative to IRR. With IRR, we calculate a percentage, whereas with NPV we calculate a dollar amount.

Net Present Value

NPV is today's value of the cash inflows less the cash outflows. In this technique, the future cash inflows are discounted using the company's cost of capital. From a finance view, any project with a positive NPV should be accepted. In the above example, with a cost of capital of 10%, the NPV is zero and we would be indifferent regarding the project.

NPV

	Time 0	Year 1	Year 2
Cash Flow	($1,000)	$100	$1,100
	10% Cost of Capital		
	($ 0.00) NPV		

The challenge to any usability engineer is to demonstrate that the benefits in the form of cash inflows (today's dollars) outweigh the cost of the project as measured against the organization's cost of capital. In many companies, 15–20% is the hurdle rate used to determine whether or not a project is acceptable.

NOT "IF," BUT "WHICH"

As suggested in the last section, one can apply a cost-justification approach *not only* to justify (or not) usability engineering of a product interface, but (*more likely, these days*) to steer the product development team toward a *particular set* of usability engineering methods. The field of usability engineering has come to a point of sophistication where we need not ask the question, "Shall we employ usability engineering?" Rather, we should ask, "Which methods will yield us the best return-on-investment for our usability engineering dollar and hour?" It would be folly for a web or other software development team to ask the question, "Shall we test the software before 'going live,' or before shipping?" Obviously, the sensible question is, "How much software testing shall we employ, and what type of testing?"

Likewise, given the importance of usability to the success of Web sites and just about all software, the sensible question is not "Shall we employ usability engineering in the development of our product/site?" Rather, "How much usability engineering should we invest in, which methods should we employ, and when?" Clearly, $20,000 spent on a prototype test, early in the development cycle, may yield a greater ROI than a $20,000 end-user test after the product is developed, given the relative costs of repairing the problems discovered, even given the increased time-adjusted cost of the earlier test.

WHY A COST-JUSTIFICATION APPROACH IS MORE IMPORTANT FOR WEB INTERFACES

Another anecdote will serve to illustrate the increased importance of usability on Web sites as compared with traditional office software.

A contract development team was building a Web site for a client organization. The Web site was to include up-to-date drug information and was intended to be used by physicians as a substitute for the standard desk references they currently use to look up such drug information as side effects, interactions, appropriate uses, and data from clinical trials. The business model for the site was an advertising model. Physicians would visit the site regularly because more current information would be available, and that information would be more readily findable on the site, relative to in published desk references. Pharmaceutical

companies would buy advertising for their drug products on the site because the visitors to the site (physicians) represent their target market. Regular and increasing traffic caused by repeat visitors, and new visitors joining based on word-of-mouth among physicians, would drive up the value of advertising, generating a profit for the client.

The development team generated a prototype design that the client would use to pursue venture capital to support the full-blown development and initial launch and maintenance of the site. The client paid for this prototype development.

Once the prototype was ready, a usability engineering staff was brought on board the project to design and conduct a usability test. Eight physicians were paid to fly in to the development center and participate in the test as test users. Several basic search tasks were designed for the physicians to perform. They were pointed to the prototype's Home Page, and left on their own to try to successfully find the drug information that was requested in the first task.

Within 45 seconds of starting their first search task, seven of the eight physicians in effect gave up, and announced, unsolicited, that the site was unusable and if it were a real site, they would abandon it at that point and never return.

Clearly, if the site had launched as it was prior to this test, not only would an optimal ROI not have been realized, but in fact, the site would have failed all together and a complete loss of the clients' investment would have resulted. If seven-eighths of all visitors never returned, enough traffic would not have been generated to have motivated advertisers to buy advertising. The entire investment would have been lost.

Instead, the test users were asked to continue with the entire test protocol, and the data generated revealed insights into the problem that was a show-stopper on the first task, as well as other problems revealed in other test tasks. The site was redesigned to eliminate the identified problems. Clearly, the usability test, which had an associated cost, was worth the investment in this case.

This anecdote illustrates something that distinguishes Web sites from commercial software products. In a commercial software product, the buyers discover the usability problems only after they have paid for the product. Often they cannot return it once they have opened the shrink wrap and installed the product. Even if it has a money-back guarantee, they are not likely to return it, and anyway, perhaps there are not many alternative products on the market with noticeably greater usability.

On a Web site, on the other hand, it costs the visitor nothing to make an initial visit. On a Web site based on an advertising model, such as the one described, the site investor makes nothing at all unless there is sufficient ongoing traffic to attract and keep advertisers. On an e-commerce site, the investor makes nothing at all unless the visitors actually find and successfully purchase products.

A Web site is not a product and the user does not have to buy it to use it. The Web site is just a channel, like a TV show, magazine, or catalog, and if users do not find and repeatedly and successfully use the channel, the investor gets no ROI for having developed the channel. Thus, usability can absolutely make or break the ROI for a Web site even more so than for traditional software products.

It is also true that competition is even more dependent on relative usability on Web sites than is the case with traditional software products or sales channels. Someone wishing to buy a book may be inclined to buy from a particular bookstore that is easy to get to, even if it is not the best bookstore around. On the other hand, if a customer cannot easily find the desired book through the barnesandnoble.com Web site, they need not even get out of their chair to shop at a competitor's site instead. It is not enough to simply have a Web site that supports direct sales; your site must be more usable than the competition's site or you will lose business based on the relative usability of the selling channel alone. For example, 60% of a sample of consumers shopping for travel online stated that if they cannot find what they are looking for quickly and easily on one travel site, they will simply leave and try a competitor's site ("Travel Data Overview," Forrester Research, Inc., August, 2000).

In addition, if you are a catalog order company, such as LL Bean or Lands' End, and your product is good but your Web site is bad, customers will not use your Web site and will resort to traditional sales channels (fax, phone) instead. This will result in a poor ROI for the Web site that was intended to be justified by relatively low costs, compared with that of those more traditional channels.

Forrester reports a buy-to-look ratio of only about 2% on e-commerce Web sites ("The Best of Retail Site Design," Forrester Research, Inc., October, 2000), and although part of this may be caused by other factors, usability is surely a large factor in this poor performance of Web sites as a sales channel.

Another Forrester Report ("Get ROI From Design," Forrester Research, Inc., June, 2001) cites other compelling statistics from a variety of studies. In one study, 65% of shopping attempts at a set of prominent e-commerce sites ended in failure. In another, 42% of users were unable to complete the job application process on a set of corporate sites. Another study showed that journalists could find answers to only 60% of their questions on corporate public relation sites. Still another study showed that adding certain product information to e-commerce sites reduced product-related inquiries by more than 20%. Finally, Forrester points out that when amazon.com let customer service slip in 2000, they fell behind their competitors in three categories for the first time. In this same report, Forrester notes that the companies they interviewed spent between $100,000 and $1 million on site redesigns, but few had any sense of which specific design changes paid off, if any. Clearly, usability engineering programs that are specifically aimed at such goals as increased success rate of shopping attempts, job application completion, and information requests could had a profound impact on the ROI of these redesign projects.

As illustrated by all these examples, usability engineering can have a profound impact on ROI in the case of Web sites, and even more dramatic cost-justification cases can be made for usability engineering programs during web development, compared with during traditional software development projects. A profit is made whenever a commercial software product is sold, regardless of how productive users are once they start to use that software product, but there simply is no ROI until and unless visitors to Web sites complete their tasks efficiently and successfully.

TABLE 62.1. Usability Cost-Justification Software

Manufacturer	Year 1	Year 2	Year 3	Total
Cost				
Usability lab	($33,000)			
Usability lab projects	($179,974)			
Total cost	($212,974)	$0	$0	($212,974)
Cost savings				
Customer support	$168,000	$168,000	$168,000	
Software development	$15,050			
Additional sales				
Total cost savings	$183,050	$168,000	$168,000	$519,050
Net	($29,926)	$168,000	$168,000	$306,074

A SPREADSHEET MODEL FOR CUSTOM-BUILDING YOUR OWN COST-JUSTIFICATION TOOL

The following is an example of a cost-justification tool, built on a simple spreadsheet tool with the addition of some fairly simple macros. It is offered as a model for how the reader might generate a cost-justification tool to project his or her own.

The following section of the spreadsheet, in Table 62.1, represents the summary cost-benefit analysis. According to this summary, we expect to spend $212,974 in Year 1. These expenses include building a $33,000 usability lab, plus about $180,000 spent on usability engineering support on a total of three products. The summary reveals that by the end of Year 2, we expect a net gain, and by the end of Year 3, we expect a total ROI of $519,048, for a net gain of $306,074. And this before any additional sales from improved usability are factored in.

The details that fuel these summary figures are offered in Table 62.1.

Our example illustrates the cost benefit from employing usability in a software development environment. First let us examine the costs. Tables 62.2 and 62.3 project the cost of a "basic" program of usability engineering and a "complete" program

TABLE 62.2. Cost of "Basic" Usability Engineering Support

Activity	Cost Item	Units (Hours)	Unit Cost	Total Cost
Generate wish list	Developer	3	70	$210
	Customer support	3	56	168
User survey	Developer	16	70	1,120
	Postage			50
Specification review	Usability engineer	8	70	560
Scope review	Usability engineer	4	70	280
Contents review	Usability engineer	8	70	560
Create usability plan	Usability engineer	8	70	560
Feature review	Usability engineer	8	70	560
Function-task analysis	Usability engineer	40	70	2,800
	Developer	40	70	2,800
	Travel			5,000
Detailed feature review	Usability engineer	8	70	560
QA review	Usability engineer	4	70	280
	QA engineers	32	70	2,240
Usability walkthrough	Usability engineer	32	70	2,240
	Developer	104	70	7,280
	Customers	48	30	1,440
	Travel			6,000
	Miscellaneous			100
Beta survey	Usability engineer	128	70	8,960
	Developer	104	70	7,280
	Postage			50
Product ship meeting	Usability engineer	4	70	280
Total				$51,378

Note. QA = quality assurance.

TABLE 62.3. Cost of "Complete" Usability Engineering Support

Activity	Cost Item	Units (Hours)	Unit Cost	Total Cost
Generate wish list	Developer	3	70	$210
	Customer support	3	56	168
User survey	Developer	16	70	1,120
	Postage			50
Generate UI prototype	Developer	160	70	11,200
Test UI prototype	Usability engineer	96	70	6,720
	Test subjects	80	30	2,400
	Developer	72	70	5,040
	Miscellaneous			200
Specification review	Usability engineer	8	70	560
Scope review	Usability engineer	4	70	280
Contents review	Usability engineer	8	70	560
Create usability plan	Usability engineer	8	70	560
Feature review	Usability engineer	8	70	560
Function-task analysis	Usability engineer	40	70	2,800
	Developer	40	70	2,800
	Travel			5,000
Detailed feature review	Usability engineer	8	70	560
QA review	Usability engineer	4	70	280
	QA engineers	32	70	2,240
Usability walkthrough	Usability engineer	32	70	2,240
	Developer	104	70	7,280
	Customers	48	30	1,440
	Travel			6,000
	Miscellaneous			100
Beta survey	Usability engineer	128	70	8,960
	Developer	104	70	7,280
	Postage			50
Product ship meeting	Usability engineer	4	70	280
Include usability survey	Usability engineer	4	70	280
Total				$77,218

Notes. UI = user interface; QA = quality assurance.

of usability engineering, respectively. The "complete" program has all the "basic" program has, plus a prototype usability test and the inclusion of a usability survey, along with the product itself. In the interest of making the usability cost-benefit analysis as conservative as possible, an attempt is made to be as thorough as possible, accounting for all the meetings, the developer hours, and anything else that would *not* have been a cost had a program of usability engineering not been pursued.

The hourly cost for each development team member (Developer, Usability Engineer, Quality Assurance Engineer, Customer Support Professional) is intended to be a "fully loaded" salary, including benefits, computer support, etc.

Table 62.4 represents the costs of building a usability lab.

There are two main elements of costs in our example; the lab build-out and the conduct of a program of usability engineering. This example assumes that existing space is used for the lab, thus the incremental cost consists of lab design, build-out, and hardware. The cost of the conduct of a program of usability engineering represents the cost to run two basic and one expanded or "complete" usability project. That is, we are going to assume that in Year 1 we are going to use the lab, and the usability engineers, to provide "basic" support to two product development efforts (say, Web sites) and "complete" support to one product development effort. Thus, adding the lab cost

($33,000) to the two basic support efforts (at $51,378 each, for a total of $102,756) and one complete support effort ($77,218) yields a total Year 1 cost for usability engineering of $212,974, which we saw in Table 62.1. (Note, when you are building your own cost-justification spreadsheet, you could add macros to dynamically change the costs as you increase the number of basic and complete programs you undertake.)

Now, let us address the potential benefits. In this example, we consider only the cost savings based on market costs for software development and customer support personnel. The customer support benefit is derived from reduced call support

TABLE 62.4. Cost of Building a Usability Lab

Lab Feature	Units	Unit Cost	Total Cost
Lab design and equipment selection	160	35	$5,600
Carpenters and electricians	25	80	2,000
One-way mirror	1	1,900	1,900
Video cameras/VCR combination	3	3,500	10,500
Computers	4	2,000	8,000
Overhead projector	1	5,000	5,000
Total			$33,000

TABLE 62.5. Financial Metrics

	Year				Cost of Captial 20%
	0	1	2	3	
Cash flow	($212,974)	$183,048	$168,000	$168,000	
Discounted cash flow	($212,974)	$152,540	$116,667	$97,222	
Net present value	$153,455				
Internal rate of return	64%				
Payback period	2 Years				

burden and the software development savings are a result of reduced rework or engineering hours. This example illustrates that usability would result in a positive NPV of $153,455 or a payback period of 2 years, just considering the customer support and software development savings.

Note that our spreadsheet example has a macro in it that allows the user (of the spreadsheet) to change the "cost of capital" (basically, how much it costs to spend the money vs. keeping it and earning interest on it) in increments of 1%, which automatically yields different calculations of the time value of the expenditures and savings (see Table 62.5).

Now, let us look at that savings in more detail. We determine that because of problems discovered early (when they are cheaper to fix) rather than late, the software development team will realize a savings of $15,050 in the first year (as seen in Table 62.1). That is, based on historical data, we expect that it will take software developers 215 fewer hours (at $70 per hour) to fix the problems.

As for the detail for the reduced call support burden, our product, an e-tail site, expects 20,000 visitors a year. (Note the conservative nature of the estimates. If our estimates are low, and our projections yield a positive ROI, then actual numbers that exceed the estimates will yield even higher actual ROI.) By working with the customer support team and reviewing their (very detailed!) call logs, we calculate that 70% of the calls are for nondefect problems. That is, the code works "as spec'd," and users just cannot figure out how to carry out their task. The majority of customer support calls a software manufacturer will receive tend to be these how-to/nondefect issues. Table 62.6 reveals a further breakdown of these usage problems *that usability engineering testing is likely to help resolve*.

The customer support team also knows (and knows well!) the cost of a call. In our example, that cost is $120. Given the number of visits, and the cost of a call, and the percentage of "usage" problems, we can calculate that these calls cost the company $1,680,000 a year. If usability testing was able to reveal problems and be an agent of change to drive fixes just some of them, enough to reduce the "usage" calls by 10%, the company would realize a savings of $168,000 a year.

As in Table 62.5, Table 62.6 has a macro that allows the usability engineer, based on ever-improving data, to change the "usability benefit" (i.e., the percentage of usage calls avoided) by 1% increments.

This completes the first example. As we saw in Table 62.1, we expect to spend $212,974 in the first year, and at the end of 3 years to have realized savings of $519,048, for a net savings of $306,074. In Table 62.5, we see that this investment has a payback period of 2 years, with an internal rate of return of 64%. If we learn, during the course of this project, that the cost of capital goes down, or that the usability benefit is actually higher than 10%, our projected benefit of the usability engineering work will go up.

SAVINGS TO THE CUSTOMER

The illustration above was exceedingly conservative because it did not take into account all the benefits that might possibly derive (not even increased sales). In the following example, we illustrate how to calculate potential benefits to the customer.

Let us say we have a customer company that is buying from us a software product, to be used by 500 employees. If this customer receives a more usable product, then the users of that product will require less training, spend less time making and recovering from errors, and will be quicker at traversing the correct path through the user interface (see Table 62.7).

TABLE 62.6. Savings From Reduced Call Support Burden

Product	Total Cases	Documentation Change	Environment Problem	Documentation Questions	Installation Questions	Product Usage	Usability Subtotal
E-tailer	20,000	1%	9%	1%	10%	49%	70%
Usability Benefit		10%					

Product	Cost Per Call	Total Cost	Usability Support Questions	Cost of Usability Related Cases	Usability Engineering Savings	Revised Total Cost
E-tailer	$120	$2,400,000	70%	$1,680,000	$168,000	$2,232,000

TABLE 62.7. Usability Cost Justification

Customer	Year			
Cost Savings	1	2	3	Total
Productivity improvement	$33,542	$33,542	$33,542	
User error reduction	$16,771	$16,771	$16,771	
Reduced training	$17,500			
Total cost savings	$67,813	$50,313	$50,313	$168,439

Thanks to the program of usability engineering, productivity increases because users were able to navigate the product menus/Web site more efficiently. Conservatively (again), based on historical findings, we expect to improve performance using this software tool by 30 seconds. A 30-second benefit for 500 users would result in $33,542 worth of savings in the course of a year. Error reduction is the savings associated with crashes or redoing any given task. A 15-second benefit, thanks to one less error to correct, would results in savings of $16,771 for the same 500 users. Naturally a more user-friendly product is more intuitive and would result in less required training. At a training cost of $55/hour, a 1-hour reduction would result in a $17,500 one time savings. So, one customer with 500 "scats" will enjoy a savings of more than $50,000 a year, thanks to our usability engineering effort.

INCREASED CONVERSION RATES

Here is another example. Critical to an e-tailer's success is its conversion rate and ability to generate repeat business. Many of us have had unpleasant web experiences, whether it be difficulty in navigating the Web site (to buy groceries online!) or conflicts with the user's browser environment. In most of these situations, the potential customer does not call/e-mail. Instead, the potential customer reverts to a traditional route or alternatively moves to a competitor's Web site. In the example of a physicians' reference Web site the failure to employ usability

would result in a 13% conversion rate (1 of 8 users returning), or 87% of potential users deciding not to use the site.

What if we project that via usability testing, we would yield a 50% increase in the conversion rate, raising it from 13% to 19%? The macro in the spreadsheet allows you to adjust the projection based on a different projected benefit. Furthermore, the spreadsheet in Table 62.8 reveals that we expect to gain $40 per active user, in advertising dollars. A small (say, $10,000) usability test would yield a $25,000 increase in advertising dollars. In our previous example, the usability testing and subsequent site redesign increased the conversion rate 700%! Clearly, the ROI projections based on that testing would be much, much higher, and the usability work well cost-justified.

SOME FINAL POINTS

What of the Not-For-Sale, Internal-Use Software or Web Site?

In this chapter, our examples have focused on vended software and e-commerce sites. Similar cost-benefit analyses are possible for intranet sites of informational sites. "For intranet sites, efficiency, memorability, and error reduction become the most important usability attributes" (Nielsen, 2000, p. 274). Thus, employees' time spent locating documents on the company intranet site can be calculated, and the resultant improvements from a well-designed, usable site demonstrated.

Are We Worried About Achieving a Particular Return on Investment, or About Avoiding a Disaster (and a Law Suit)?

We have worried about numbers—dollars in particular. But there are intangible costs and benefits to be considered that may or may not translate into dollars. Could another cost of poor usability engineering be a bad review in the trade press? Or it could be a lawsuit, if someone thinks an actual or implied contract was not fulfilled because people could not use a (perhaps defect-free) user interface effectively? Or it could mean a picket line of angry users, picked up by the 6:00 o'clock news? When evaluating whether or not to pursue a course of usability

TABLE 62.8. Conversion Rate Benefits

Users	Conversion Rate	Active Users	Advertising Dollars Per User	Dollar Value of Subscribers	Lost Advertising Dollars
10,000	13%	1,250	$40	$50,000	$350,000
Usability Benefit		50%			

Users	Conversion Rate	Active Users	Advertising Dollars Per User	Dollar Value of Subscribers	Lost Advertising Dollars
10,000	19%	1,875	$40	$75,000	$325,000

engineering, it is important to consider not only what you will get that you like, but also what you will not get that you would not like.

The "Developers Are Users" Inhibitor—How Shall We Learn *Not* to Depend on Our Own Intuitions?

Why has usability not been more warmly and universally embraced as a vital aspect of software user interface development? Why is it that "... usability is one of the first things cut in a budget" (Rhodes, 2000, p. 4)? Perhaps it is because all developers are also users. Every software developer is also a software user, and so reflexively feels justified to depend on his or her own intuitions as to what is usable. Although the typical usability engineer would not think of telling a software developer how to structure his or her code, the typical developer naturally feels qualified to make usability pronouncements. The oh-so-vital distinction, between being a "user" and a "representative user," tends to get lost. The experienced usability professional has had much practice trying to convince a product developer that the target audience for this product/site is *likely not* someone with a master's degree in computer science or electrical engineering, and *certainly not* someone with 18 months of experience working on *this project*. The best way to drive this point home is to encourage software developers to observe usability testing (or participate directly in the usability evaluation; see, e.g., Bias, 1994). Previously unable to comprehend that someone could *not* understand how to navigate a particular path through the software user interface, the product developer becomes convinced when he or she sees two or three test participants (the first one was just "stupid") repeat a problem. In lieu of participating in or observing a usability test, being able to view a highlight videotape from a usability test that shows several smart-looking people repeating a problem can also be very convincing.

Finding the Stakeholder in Which the Cost and Benefit Lines Cross—What to Do When the Costs Outweigh the Benefits for a Particular Suborganization

Imagine the following scenario. Armed with your usability cost-justification projections, you eagerly and confidently stride into a room of decision makers at your software company. Having done your homework and generated an empirical, cost-benefit analysis, you propose a course of usability engineering that, while costing the development team an extra $20,000, will save a whopping $100,000 in the development of training materials. "Way to go," right? Well, if your audience consists of the company's director of product development and the director of training, they may well be underwhelmed by your proposal. Now, there is likely someone higher up the organizational structure, who worries about *both* the development budget and the training budget who will be more impressed and more likely to be convinced by your cost-benefit analysis.

CONCLUSIONS

In case study after case study, usability is proving its worth. In a wide variety of domains, using a wide variety of usability engineering methods, those who spend money on professional usability engineering, and take the time to calculate the ROI for those expenditures, are finding that they are glad they spent the money.

- "An investment in user and task analysis early in the design and development process will reap considerable benefits in cost savings later in the process" (Hackos & Redish, 1998, p. 122).
- "... An executive development manager observed that reducing call volume by even 25% would cover the hiring of three additional support technicians" (Montaniz & Kissel, 1995).
- "Sun Microsystems has shown how spending $20,000 could yield a savings of $152 million" (Rhodes, 2000, p. 4).
- "Performing usability cost-benefit analyses within one's own company may be a first step toward introducing usability engineering techniques into a company" (Donahue, Weinschenk, & Nowicki, 1999).

Above, we have offered encouragement and a plan for gathering cost and benefit projections, and actual calculations, and using a metric everyone can understand—dollars—to cost-justify a usability engineering effort. We have offered some examples of spreadsheets to model this sort of cost-justification approach. Readers will have to derive their own values, for cost of a developer hour, cost of a "basic" program of usability engineering, likely reduction in customer support calls due to usability engineering, and so forth. Furthermore, at no point in the above was the whole package put together. That is, if we pursue a certain program of usability engineering, what would be the grand total of all the benefits, including (but likely not limited to):

- reduced costs due to development efficiency
- reduced costs due to fewer calls to the customer support line
- reduced costs due to less need for documentation
- reduced costs due to fewer training materials needed
- increased sales/visits
- increased return clients/visitors
- increased user throughput
- decreased customer training costs
- decreased customer costs due to ability to hire less sophisticated user audience
- reduced chance of lawsuits
- reduced chance of public relations disaster
- increased chance of trade press praise
- increased customer satisfaction.

Our choice not to build one "grand total" cost-benefit analysis (showing, certainly, an ROI of hundreds of thousands of dollars to one) is intended to emphasize the importance of knowing the "value proposition" for your audience; the director of training may be unmoved by your argument to reduce training needs.

(Indeed, this can be parried by arguing for those training dollars being spent on training the more sophisticated functionality of the product and *not* the day-to-day tasks.)

Given the state of the software development world today, though, usability engineering almost certainly will yield a robust ROI, in almost any user interface/Web site development effort. By coming up with reasonable projections in advance, the usability advocate will gain the resources necessary to carry out that first program of usability engineering. By following up and calculating the "actuals," after the product ships or "goes live," the usability advocate will likely demonstrate how conserva-tive the original projections were, and thus earn the confidence of the stakeholders for quite a long time. Thereafter, the cost-justification effort will become a way to continue to validate the usability expense, and to fine-tune the program of usability engi-neering, demonstrating which usability methods were the most cost-effective for which types of products, at various points in the development cycle.

(Portions of this chapter will also appear in a chapter by Deborah J. Mayhew and Randolph G. Bias in Ratner, J., Ed., *Human Factors and Web Development* (2nd ed.), Lawrence Erlbaum, to be published in 2003.)

References

Bias, R. G. (1994). The pluralistic usability walkthrough: Coordinated empathies. In J. Nielsen & R. L. Mack (Eds.), *Usability inspection methods*. New York: John Wiley & Sons, Inc.

Bias, R. G., & Keough, K. (2000, June). Usability triage for web sites. *Human factors and the web*.

Conklin, P. F. (1996). Bringing usability effectively into product de-velopment. In M. Rudisill, C. Lewis, P. G. Polson, & McKay, T. D. (Eds.), *Human-computer interface design: Success stories, emerging methods, and real-world context*. San Francisco: Morgan Kaufmann Publishers, Inc.

Donahue, G. M., Weinschenk, S., & Nowicki, J. (1999). Usability is good business [On-line]. Available: http://www.compuware.com/intelligence/articles/usability.htm

Forrester Research, Inc. (2000, October). *The best of retail site design*. R. K. Souza.

Forrester Research, Inc. (2000, August). *Travel data overview*. H. H. Harteveldt.

Forrester Research, Inc. (2001, June). *Get ROI from design*. R. K. Souza.

Hackos, J. T., & Redish, J. C. (1998). *User and task analysis for inter-face design*. New York: Wiley Computer Publishing.

Karat, C. (1994). A business case approach to usability cost-justification. In R. G. Bias & D. J. Mayhew (Eds.), *Cost-justifying usability*. Cambridge: Academic Press.

Montaniz, F., & Kissel, G. V. (1995). Reversing the charges. *Interactions, 2*, 29–33.

Nielsen, J. (2000). *Designing web usability: The practice of simplicity*. Indianapolis, IN: New Riders Publishing.

Rhodes, J. S. (2000). Usability can save your company [On-line]. Available: http://webword.com/moving/savecompany.html

· 63 ·

THE EVOLVING ROLE OF SECURITY, PRIVACY, AND TRUST IN A DIGITIZED WORLD

Steve Diller, Lynn Lin, and Vania Tashjian
Cheskin

Recently, spam was unleashed that offered a novel possibility—"new software [that] lets you find out almost anything about anyone." By "harnessing the power of the Internet," one could find long-lost friends and family, collect detailed financial information on entertainers, learn "the truth" about one's spouse's fidelity, or determine the sexual proclivities of politicians. All for the remarkably low price of $29.95.

Although digging up dirt on others may be the true oldest profession, the claims made by the spam suggest that the methods available have greatly increased in their power and ease of use. Additionally, the spam implies that these methods are somehow intrinsic to the web. When it comes to the central defining concept of the Internet, a dystopic "ubiquitous spying" competes with the utopian "ubiquitous computing" for dominance in the public mind, and for good reason. In a sense, they are the two sides of the human–computer interaction coin. If the network is everywhere, then not only are we as individuals everywhere, but, simultaneously, everyone else is everywhere too, including in "our" files, if they choose to be.

If the concept behind the spam is correct, then the implications are enormous. Concepts like "security," "privacy," and "trust" may be severely challenged. Because these concepts operate at the heart of our economy, governance, and interactions technologies that threaten their operation are, by definition, revolutionary in their potential impact. Although a kind of "salvation of the world" remains a central feature of the technoelite's sense of self, few of the pioneers of human–computer interaction ever imagined that digital networks would force a radical re-envisioning of the central political concept of our age, "individuality," and all that flows from that concept. However, that may be a fundamental result of their innovations.

THE IDEA OF "INDIVIDUALITY"

If we want to make sense of the challenges digital technology make to many of our central organizing concepts, perhaps the best place to start is to delve deeply into the nature of them, starting with the idea of "individuality." The *American Heritage Dictionary* defines "individuality" as "a single, distinct entity." In modern, Western parlance, however, the word has taken on additional meaning. For example, in the United States, the idea serves as a central organizing principle in ideologies seeking to determine the ideal functioning of society. More Cartesian theorists, particularly classical Liberals and Libertarians, see "the person," and his freedom from governmental interference, as a necessary precondition to a well-run society. Thinkers on the liberal Left also stress the primacy of "the person," although much of their efforts focus on freedom from corporate interference and domination.

It is important to note that the idea of "individuality" has always been a universal human concept. The main bone of contention has concerned what the unit of measurement should be for determining the "distinct entity" of greatest concern. That entity can just as easily be "the family," "the clan," "the corporation," or "the nation" as "the person."

The victory in the West of free-market liberal democracy, for the most part, can be seen as the triumph of the Liberal/ Libertarian strains of thinking about individualism, over more collectively oriented ideologies such as Communism and Fascism. Since the end of the Cold War, the more person-oriented ideologies have increasingly operated center stage in the emerging world system. Still, more traditional cultures, whether in the developing world or inside the boundaries of the more developed nations, continue to hold to more collectivist notions. This has major ramifications for their response to the web, relative to the individualists.

Scientific thinking has also bolstered the ideas behind person-centered individuality. Evolutionary theory, for instance, has tended to focus on individual members of species, and their particular survival strategies. Psychology, too, has played a role here, developing person-centered therapeutic strategies, and by developing a conceptual framework centered on ideas such as "the ego," as the unit of individualism in modern culture. The idea that humans instinctively protect their individual existence against threats to it remains a powerful ideological support for the concept of the person as the indivisible, and most significant, unit of autonomous action in society.

THE INSTITUTIONAL SUPPORTS OF INDIVIDUALISM

To support the institution of "the person," many governments around the world have developed legal and other socially supported protections. These protections are designed to maintain the primacy of the person within society and come in three distinct classes.

First and foremost is the protection of persons' bodies from interference by other persons or institutions. These rights take two forms. The more developed of the two protects persons from physical attack from other persons. The other, somewhat less explicitly protected right, at least in the U.S. Constitution, seeks to maintain the physical autonomy of individual persons. Laws against slavery and in support of abortion freedoms fall under this category. Second are civil rights, such as freedom of speech, religion, and voting rights. These rights are designed to make it possible for persons to act as political beings with a minimum of interference by larger entities.

Third, and most important for this discussion, are property rights. Protection of property represents an extension of civil rights and bodily integrity rights, in that the ownership and free use of property makes more effective personal control of one's immediate environment possible. The general idea is that private property makes the other two categories of freedoms possible.

The Case of Business Rights

In many societies, at least some of the institutional protections for individuals have been extended to corporations. For example, the U.S. Supreme Court has ruled that corporations are to be treated, for many legal purposes, as "persons." Although citizenship, bodily, and voting rights were not extended, property and speech rights were. As a result, the range of concerns individual

persons face vis-à-vis the web, for the most part, also apply to corporations.

WHERE SECURITY, PRIVACY, AND TRUST COME IN

For the institutional protections of individualism to meet their goals, certain operational concepts are required.

Security

First and foremost among these concepts is "security." Institutionally, "security" refers to the range of functional supports of civil, bodily, and property rights, including the police, the rule of law, the right to bear arms, etc. From an emotional standpoint, "security" also refers to the sensation that one's rights are protected, and, therefore, one's individuality can remain intact. Individual persons, in this society at least, seem to possess an intuitive understanding of when the supports of their individuality are under threat.

This understanding generally kicks in at the bodily level first. For instance, a person walking into a "dangerous" neighborhood generally pays great attention to the sensory cues that tell them when other individuals may not respect their bodily rights. Without that fundamental respect, individual personhood is impossible.

Many people see property rights as the next most fundamental right to secure. Security of property, particularly in more developed societies, rarely will be forgotten, perhaps because the dynamism of modern capitalism results in a constantly shifting economic environment that requires constant attention to navigate successfully. In developing societies, in contrast, property rights are generally less secure, but the general thrust of public policy, worldwide, is to ensure greater property protections.

In Western societies, not only does a huge governmental apparatus exist to support property rights, but large portions of "civil society" also function to enhance these rights. The daily functioning of corporations, worldwide, provides considerable influence to ensure that property rights serve as the bedrock concept underlying transactions.

TACTICS TO SECURE PROPERTY RIGHTS

The tactics utilized by many governments and societies to manage and secure property rights, typically flow from two essential concepts: labeling and privacy. As we will see, digital technology threatens the smooth functioning of these techniques. Each is discussed herein.

Security of Labels of Individual Identity

One tactic that most governments use to protect property rights is to establish a set of labels that are legally attached to property to determine which individuals own it. These labels, such as icons or unique marks or names and numerical equivalents, such as social security numbers used in the United States, provide the primary basis for determining which person or corporation owns specific property. If the right to control the use of one's labels of identity is not maintained, then other individuals might be able to access one's property. Labels are at the heart of the security of property.

"Branding" in the marketing realm is a significant use of this concept. Corporations need to retain control of their brands, just as individuals need to control their names to maintain positive reputations. Reputation, in a fundamental sense, is as valuable a property right as ownership of land or stock, and most governmental systems seeks to protect this right by upholding the security of labels.

The Special Tactic of "Privacy"

In addition to governmental coercion and the support of civil society, the three classes of rights previously described also depend on "privacy." Privacy, which the *American Heritage Dictionary* defines as "secluded from the sight, presence, or intrusion of others," and "of or confined to one person," operates as a vital tactic in the support of security.

With privacy in place, all three classes of rights can more effectively function. For instance, in the bodily rights area, privacy makes it possible for individuals to make whatever use they want of their own bodies without interference of others. In the area of property rights, shielding others from knowledge of our assets and how to access them serves to secure those rights. In the case of civil rights, the privacy of the ballot makes a free voting choice, unaffected by the preferences of those more powerful than ourselves, much more likely.

As in the case of security in general, most persons intuitively grasp the power of privacy to support the exercise of their rights. Privacy may be the most powerful technique of all, because it is effective even when governmental or civil supports weaken or fail. Consequently, it is a technique that individuals greatly appreciate. When they feel its absence, they understand that the exercise of their rights is at great risk.

Most contemporary rights theorists seem to believe that privacy is an essential, indispensable technique in the securing of individuality. Privacy as a support to security can be a double-edged sword, however; it can act to prevent all actions from scrutiny, even when those actions can have as their goal the subversion of other's rights. Consequently, a certain vagueness as to the limits of privacy suits the needs of those seeking to secure individual rights. Privacy must function in some form for individuals to exist. Consequently, the tension between privacy as guarantor and privacy as subverter of rights will also continue to exist as long as the individual as a concept exists. Finally, privacy will be a concern wherever individuals act, be it in the brick and mortar world or cyberspace.

THE ROLE OF TRUST

Although privacy and security secure the autonomy of persons and corporations, they do not provide the resources needed for such persons to actually function in the world. To obtain these

resources, persons and corporations require some sort of mechanism that helps them "let down the walls" of their solitude, selectively, to obtain what they need from others. The decision to engage in interactions with others, of course, is a necessity for all persons and corporations. The question then is, how do they get what they need and retain their independent existence?

Two primary mechanisms provide this capability. The first is law, both criminal and civil. Persons and corporations assume, to varying degrees, that other persons and institutions will refrain from physically assaulting them or stealing their property. The standard may be fairly low, as in contemporary Russia, or high, as in contemporary Japan. But, in most societies, law does function to ease the risk of interactions between persons and companies.

The other primary mechanism, trust, is psychological in nature. "Trust," defined by the *American Heritage Dictionary* as a "firm reliance on the integrity, ability, or character of a person or thing," makes it possible, more than any other mechanism, for persons to let down their security systems and open themselves to the world. For digital networks to function in the context of an individualized world, trust must function there as well.

Trust and Risk

When a person or corporation considers lowering the walls to act in the world, they have historically faced six types of risk, four of which often involve direct economic transactions:

- financial (risk of losing money or paying too much);
- functional (risk of receiving the wrong or a malfunctioning product);
- social (risk of embarrassment and/or loss of reputation); and
- physical (risk that we might be physically harmed).

In addition, engaging with the world puts us at risk in two other ways:

- emotional (risk that one might be emotionally hurt by an interaction); and
- identity (risk that others may impersonate us for financial or other types of gain).

To effectively act, a person or corporation will engage in the process of developing trust, be it with other persons or corporations. At the heart of this process is the goal of risk reduction.

Trust as Process

First and foremost, it is important to keep in mind that "trust" is a dynamic process. Trust deepens or retreats based on experience. The trusting process begins when an individual perceives indications—"forms"—that suggest a person or firm may be worthy of trust. These indications can include behaviors, such as manners, professionalism, and sensitivity.

Both persons and firms understand that these forms are designed to represent trustworthiness. These formal claims to trustworthiness become strengthened over time and are eventually transformed into "character traits," such as dependability, reliability, and honesty.

As it becomes clear that "character" underlies the forms, one will be willing to participate in more informal transactions. When only forms are known, one is likely to prefer formal, written contracts with others. However, as one begins to rely on a sense that a "trustworthy character" underlies the other's behavior, one will require progressively less new information.

In a study conducted by Studio Archetype/Sapient (U.S.-based design firm) and Cheskin (U.S.-based research and consulting firm) in 1999, most respondents felt that commercial relationships require far less knowledge of trustworthiness than loving relationships. In large part, this is because trust in more intimate relationships involves more valued personal assets than money, such as self-respect, desirability, and worthiness as a lover or spouse.

Because less valuable assets are at stake in a commercial relationship, consumers generally do not necessarily expect to ever know if a firm possesses the "character" that might make it worthy of deeper levels of trust. Still, experience over time in a commercial relationship is vitally important in making transactions smoother, simpler, and more likely to become habitual.

Figure 63.1 suggests an archetypal process of trust development over time. In this model, individuals begin the process by not trusting. They then enter an "extrinsic" level of trust (i.e., conditional trust based on the use of formal characteristics of trustworthiness by the other). If the transaction goes smoothly, after a time, the person shifts to an "intrinsic" level of trust, depending on experience as the basis for engaging comfortably in transactions. Getting to this "intrinsic" level is important to all parties in transactions, because, once there, transactions can occur more easily and at lower cost. Still, even in this advanced state, the maintenance of high levels of trust is contingent on the continued trustworthy behavior of both parties. In that sense, it always remains conditional.

SECURITY AND PRIVACY IN THE ONLINE WORLD

For individuals and corporations, developing a sense of "intrinsic" trust vis-à-vis the web has been only partially achieved. One reason is that two different "others" are involved in the trust-building process. On the one hand, persons must interact with a network, and the network must be seen as supportive of security and privacy. On the other hand, those with whom we interact on the web (Web sites, bulletin board systems, etc.), must also support security and privacy.

Because digital information can "flow" more easily from device to device than information stored on paper, the functioning of privacy can be impaired, and the protection of labels can be more difficult as well. This "flow" has also led to new ways of thinking about the nature of disparate transactions, with major implications for the ways governments and persons think about privacy and security.

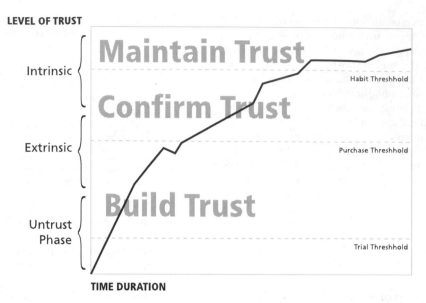

FIGURE 63.1. Trust development over time. © 1999 Copyrights of Studio Archetype/Sapient and Cheskin. All Rights Reserved.

PRIVACY ISSUES ONLINE: THE "THREAT OF THE NET"

As noted earlier, in the case of privacy, protection involves the prevention of information flow that others can use at the expense of the individual. Historically, the full range of behaviors an individual engaged in could not be collected, evaluated, and acted on. Consequently, such transactions were not generally viewed as being in need of privacy protection.

The ability to more easily access and track transactional behavior that digital information makes possible has resulted in a change in perspective. At the same time, the vulnerability of databases of such information to hackers makes protection of privacy a key concern of persons and corporations. As a result, discussions have been taking place in civil liberties, technology, and marketing circles about offering technology-based solutions to protecting personal data.

"Secure" Servers

Currently, billions of dollars are being spent by corporations and persons to protect private information stored in computers and servers. Although the software that has been developed provides some protection, most experts in the field recognize that a determined hacker can generally find his way into most systems.

It appears that the primary reason for the difficulty to protect digital data lies in the double-edged nature of such data. Its great strength is its ease of use and manipulation. Its greatest weakness is its accessibility via digital techniques. The struggle between those who build digital security systems and those who seek to break them (on occasion the same people) makes this an arms race that neither side can win.

Consequently, persons and corporations need to make hard decisions about how (or whether) certain types of data should be accessible by anyone through networked computers.

Re-establishing Control of "Private" Information

In addition to concerns about the security of servers, individuals have noticed an unexpected new threat to their privacy—the collection of vast amounts of information about discrete transactions that, in the past, would never be aggregated. With aggregation, made possible by digital technology, others can infer our likely future preferences and behavior in ways not previously imagined.

The result of this new capability has been an attempt to develop new tactics to secure privacy, tactics that take varying approaches to the entire question of aggregation of data. The core idea of a few start-up companies' business plans is to develop software that would allow consumers more control of their personal information, and because personal information is in great demand, consumers should charge a price for releasing it. In time, individuals could be managing and trading their personal information the way they now manage and trade stocks, retirement funds, and other financial assets.

However, no consensus yet exists about the need or even appropriateness of protecting transactional histories as an extension of privacy rights. The contrast between how badly marketers want personal information and how concerned consumers are of their privacy is striking. The vast majority of Web sites collect and aggregate personally identifiable information for marketing purposes. In the Georgetown Internet Privacy Policy Survey of 1999, 364 dot-com Web sites, with an estimated reach of 98.8%, were evaluated for the way in which they handled transactional information. Among these Web sites,

92.9% were collecting personally identifiable information, but only 9.5% sites had an adequate privacy policy that addressed the important issues of notice, choice, access, and security and had contact information for the site. Meanwhile, in another survey done by *Business Week* in March 2000, 82% of Americans are not at all comfortable with online activities being merged with personally identifiable information, such as income, driver's license, credit data, and marital status.

As a result, some startup dot-coms have been trying to use technology to create a market-based solution, by making it possible for individuals to receive payment for their behavioral data. The hope is that, in time, people may realize the economic value of their personal information and trade their personal data for cash. If so, it may be seen as a modernized version of credit bureau activities, which individuals usually see as advantageous, because they can lower the cost of credit to deserving individuals.

Struggle Over Ownership of Health Data

In the United States, among the various types of personal data, consumers are especially concerned with their health information. A perceived lack of privacy has led people to withdraw from full participation in their own health care because they are afraid their most sensitive health records will fall into the wrong hands, and lead to discrimination, loss of benefits, stigma and unwanted exposure.

The public's fear about improper release of personal health information, particularly when accessed via the Internet, is widespread, as has been shown by many recent studies. In a 2000 survey funded by the California HealthCare Foundation, 55% of Internet users had no problem with information about what they had purchased online being shared; 48% were willing to have information about what ads they had clicked on shared; but only 3% were willing to have their personal health information shared.

Similarly, 60% of Americans studied by Princeton Survey Research in 2000 indicated they were unwilling to have their medical records shared with hospitals that were offering relevant preventive medicine programs; 61% were unwilling to have a new employer granted access to their medical records; and 70% were unwilling to have their records shared with pharmaceutical manufacturers, even if those manufacturers were providing information about new drugs that might help the patients.

In Table 63.1, from a study conducted by Cheskin in 2000, we find a middling expectation that aggregated information is available to commercial interests.

Whether the issue is health records or transactions, the protection of aggregated data remains a key, unresolved aspect of online trust. The challenge for all those who interact on the web—in other words, everyone—is to determine the cost-benefit ratio for that interaction, in a situation in which such a determination simply is not possible. The resulting inability to calculate the degree of risk will likely stunt the use of the web as long as privacy cannot be assured, or as long as individuals think as individuals.

TABLE 63.1. Attitudes Toward Online Personal Information Use

	Total Sample
Personal information I give on a Web site may be sold for marketing purposes	6.3
The information may be sold to another company or Web site	6.5
A great deal of information about me is gathered without my permission	6.2
Abuses of personal information are rampant	5.7

Note. Mean scores are based on an 11-point scale, where 0 means "strongly disagree" and 10 means "strongly agree."

Protection of Individual Labels

In the case of security, the protection at issue is the integrity of the labels of individuals as well as corporations. Historically, labels have always been at some risk. For instance, forgery of signatures has existed for as long as labels have been used to secure property rights.

However, because digitized information can be more easily collected, from many different sources, than analog information, a person's credit card numbers, names, addresses, and other labels can be acquired more easily than in the past. Generally speaking, aggregation of multiple labels makes it easier for a thief to impersonate an individual or corporation, allowing them to then digitally access individuals' assets, which are also, increasingly available in digital form (see Tables 63.2 and 63.3).

In the study conducted by Cheskin in 2000 across the Western Hemisphere, fear of "identity theft" was the most serious concern of persons in every country. This fear was far greater than concern with the misuse of behavioral/transactional data by corporations and was greater than concerns that governments would make use of online data to observe the actions of private citizens.

The perceived threat from hackers is substantially greater than the perceived threat from corporations and governmental institutions. This threat primarily involves a risk to one's personal information, although financial risk is close behind.

Forrester Research, in a 1999 study, estimated that $2.8 billion in online transactions did not occur because of such concerns, and Jupiter Communications estimated that, by 2002, upwards of $18 billion would not be spent online for the same reason. As of the writing of this chapter, no one has come up

TABLE 63.2. Predatory Individuals

The Biggest Threat Is From:	Total Sample
Hackers who will steal my personal information	8.1
Hackers who will steal my money	7.6
Dishonest companies *or people* who might cheat or con me	7.5
Politicians who will get personal information about me	4.8

Note. Mean scores are based on an 11-point scale, where 0 means "strongly disagree" and 10 means "strongly agree."

TABLE 63.3. Predatory Institutions

The Biggest Threat Is From:	Total Sample
Dishonest *companies* or people who might cheat or con me	7.5
Marketers and advertisers who will sell my name	6.4
"Unfulfillment"—cannot return an item you bought	5.9
Insurance companies who will get personal information about me	5.4

Note. Mean scores are based on an 11-point scale, where 0 means "strongly disagree" and 10 means "strongly agree."

with an effective method for completely protecting identity labels from potential misuse.

ONLINE TRUST DEVELOPMENT BETWEEN TWO PARTIES

Although concerns about privacy and security remain significant vis-à-vis the network as a whole, the four types of risk of greatest historical concern to transactions—financial, social, functional, and physical—can and have been addressed by e-commerce sites. Given the rapid development of e-commerce, apprehensions about privacy and identity protections are not necessarily deal breakers.

In a sense, the process of trust development for Web sites follows a similar pattern to development of trust in general. An archetypal example is depicted herein. In the model (Fig. 63.2), the individual interacts with a site over time, gradually coming to trust the site to deliver on its promises.

Although this model describes the overall process, it begs the question—what formal qualities (or, in the case of Web

sites, design attributes, etc.) allow individuals to feel comfortable making that first transaction with an online entity?

Building Blocks of Web Site Trustworthiness

In the Studio Archetype/Cheskin study in 1999, six primary components of Web sites were identified that play a major role in communicating trustworthiness when well-designed:

- Seals of approval (symbols of "guarantors" of trustworthiness)
- Brand
- Technological sophistication
- Navigation
- Presentation
- Fulfillment

The six primary components breakdown into a total of 28 different ways in which trustworthiness may be established. Each is briefly addressed in Table 63.4.

The six different major components that communicate trustworthiness interact with each other in complex ways. It is important to note that these six components, when handled correctly, will largely defuse the four traditional risks of commerce. They do not, however, necessarily deal with the issue of identity risk or privacy concerns for the network as a whole.

INTERACTIONS BETWEEN THE ELEMENTS

While all six elements are important in the development of online trustworthiness, some are more significant than others, and the interaction between them needs to be understood

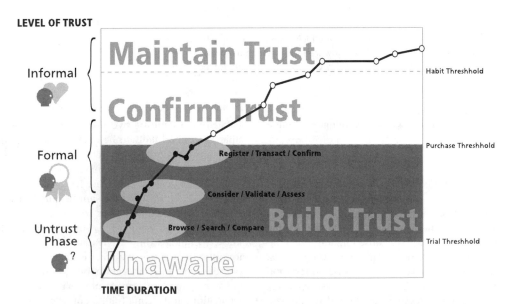

FIGURE 63.2. A model to understand e-commerce trust. © 1999 Copyright of Studio Archetype and Cheskin Research. All Rights Reserved.

TABLE 63.4. Establishing Trustworthiness

Seals of Approval	Information About Other Companies That Specialize in Ensuring the Safety of Web Sites
Network Level 1	Icons symbolizing security of the computer network as a whole, such as TRUSTe or VeriSign
Network Level 2	Text accompanying the icons
Technology Level 1	Icons symbolizing commerce-enabling functions, such as MS Commerce Server, ICAT, IBM e.business mark, and Browser compatibility marks
Technology Level 2	Text accompanying the icons
Merchant Level 1	Icons symbolizing merchant service security like MasterCard, VISA, Amex
Merchant Level 2	Text accompanying the icons
Brand	Importance of the Company's Reputation in Choosing to Do Business With Them
Overall Brand Equity	Consumer awareness of what this company does for consumers outside of the web
Web Brand Equity	How well the company's Web site fits with the consumers' sense of what the company is about generally
Benefit Clarity	On one's first visit to the site, how easy it is to discern what the site is promising to deliver
Portal/Aggregator Affiliations	Mention of an affiliation to portals and aggregators, such as Yahoo!, eXcite, iVillage, Lycos, etc.
Co-op Third Party Brands	Promotion of "third-party" quality brands
Relationship Marketing	Sending updates and other notices to consumers
Community Building	Facilitating interactions between individual shoppers
Depth of Product Offering on the Site	How many varieties of product types the site contains
Breadth of Product Offering on the Site	How many types of products the site contains
Navigation	Ease of Finding What the Visitor Seeks
Navigation Clarity	Terminologies for navigation and content are apparent for the user to differentiate
Navigation Access	The navigation system placement is consistent, persistent, and easy to find
Navigation Reinforcement	There are prompts, guides, tutorials, and instructions to aid and inform the user to perform transaction and or search task on the site
Fulfillment	The Process One Works Through From the Time a Purchase Process Is Initiated Until the Product Is Received
Protection of Personal Information	The information one provides is guaranteed to be used for no purpose other than what one gave it for, without their approval
Tracking	The site provides feedback or a confirmation number once the order is placed
Recourse	The transaction process allows for recourse if one has a problem at any time during the process
Return Policy	How clearly the return policy is explained
Simplicity of Process	How simple it is to buy something
Presentation	Ways in Which the Look of the Site, In and of Itself, Communicates Meaningful Information to You
Clarity of Purpose	The visuals/layout effectively convey the idea and the purpose of the site. Consumers would know they can purchase products when they get to the site
Craftsmanship	The degree to which, when one first views the homepage, one believes that the Web site developers were skilled in their efforts
Resembles Other Trusted Sites	How much the site resembles others consumers have come to trust
Technology	The Ways in Which the Site Technically Functions
Functionality	Overall, how well the site seems to work
Speed	How quickly each page, text, and images appears

to ease concerns of individuals when engaging in online transactions.

The Centrality of Effective Navigation and Branding

Effective navigation and a well-known brand, when viewed as isolated elements, both communicate trustworthiness. Fulfillment, viewed in isolation, has relatively little impact.

Although, when they are viewed as interacting elements, the picture changes. Strong navigation can best be understood as the foundation of communicating trustworthiness. Generally speaking, effective navigation needs to be joined to either a well-known brand or effective fulfillment if consumers are going to perceive the site as trustworthy. As long as effective navigation is one of two components in place, a site is significantly more likely to be considered trustworthy than a site with only one component in place, or a well-known brand with strong fulfillment but weak navigation.

However, even when a site has a well-known brand, is easily navigable, and offers a simple transaction process, it still may not be considered more trustworthy than sites without all three components in place. For instance, a site with a well-known brand, strong navigation, and strong fulfillment was found to be less trustworthy than a site of another well-known brand with poor navigation but strong fulfillment. In short, even if a company can combine a well-known brand, strong navigation, and strong fulfillment, it cannot ensure that its site will be perceived as trustworthy if its brand is not considered trustworthy.

Lesser-Known Brands Must Build Sites With Strong Navigation and Fulfillment

Because newer brands, by definition, are lesser-known, the only way they can compete with better-known brands is to make sure that both navigation and fulfillment work well for visitors. For these brands, navigation and fulfillment are equally important in building trust. As navigation or fulfillment improves, so does trust.

CULTURAL ASPECTS OF ONLINE TRUST DEVELOPMENT

Collectivism Vs. Individualism

Jarvenpaa, Tractinsky, Saarinen, and Vitale, in their study, Consumer's Trust in an Internet Store: A Cross-Cultural Validation, have also constructed a model pertaining to the online trust issue. The model includes seven hypotheses as listed here, addressing factors such as perceived size of the store, perceived store reputation, trust in store, consumer attitude and perceived risk, as well as the interrelations between these factors.

- H1: Higher consumer trust toward an Internet store will reduce the perceived risks associated with buying from that store.
- H2: The store's perceived reputation is positively associated with a consumer's trust in an Internet store.
- H3: The store's perceived size is positively associated with a consumer's trust in an Internet store.
- H4: Higher consumer trust toward an Internet store will generate more favorable attitudes toward shopping at that store.
- H5: The lower the consumer's perceived risk associated with buying from an Internet store, the more favorable the consumer's attitudes toward shopping at that store.
- H6: Favorable attitudes toward an Internet store will increase the consumer's willingness to purchase from that Internet store.
- H7: Reduced perceived risks associated with buying from an Internet store will increase a consumer's willingness to purchase from that Internet store.

Figure 63.3 (the Internet Consumer Trust Model: Jarvenpaa et al., 1999) summarizes the hypothesized relations between the variables. Plus (+) denotes a positive relation, and minus (−) denotes a negative relation between variables.

More interestingly, beyond these conventional factors, the study also examines the relationship between trust and consumers' cultural background.

Not surprisingly, cultural attitudes about individualism vs. collectivism play a key role in how individuals think about online trust, and sociologists have studied how these cultural norms play out in the online world. Hofstede (1980) found this dimension to have the strongest variation across cultures. Those cultures high on the individualism scale are characterized as self-reliant, competitive, trusting of others, and focused on utilitarian views of exchange and competence (Bhawuk & Brislin, 1992). Because of the utilitarian view, others are trusted if the circumstances suggest that it is in the other's own interest to behave well.

Individualism also promotes a trusting stance; one gets better outcomes assuming that others are reliable. Hence, individualists are more likely to trust others until they are given some reason not to trust. By contrast, those high on collectivism are more likely to base their trust on relationships with first-hand knowledge. Because of the emphasis on social relatedness and interdependence, collectivists are sensitive to the ingroup-outgroup boundary (Triandis, 1989). Members of collectivist cultures are less likely to trust someone who is not part of their ingroup (Yamagishi & Yamagishi, 1994). Three more hypotheses are proposed addressing the above arguments.

- H8: Consumers from individualistic cultures exhibit higher trust in specific Internet stores.
- H9: Consumers from individualistic countries exhibit a lower perception of risk in specific Internet stores.
- H10: In an individualistic culture, size and reputation will have a stronger effect on trust than in a collectivist culture.

Although these last three hypotheses regarding cultural factors were not validated in the study, the argument may have particular value in understanding the emergence of a true worldwide online commercial environment, and authors of the research project suggested further studies to refine the model.

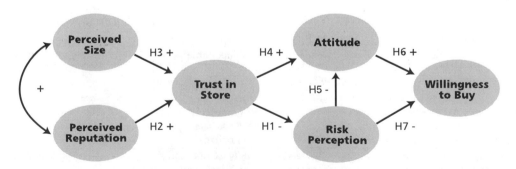

FIGURE 63.3. Internet consumer trust model.

TABLE 63.5. Online Monitoring and Tracking Perceptions

	U.S.	Latin America	Brazil	Total Sample
The government is able to track where we go on the Internet	7.2	5.3	6.0	6.7
When you are online, your activities are monitored	5.9	5.1	5.5	5.7

Note. Mean scores are based on an 11-point scale, where 0 means "strongly disagree" and 10 means "strongly agree."

Other Cultural Factors

Other culture-related factors could also affect consumers' perception and attitudes regarding trust on the Internet, including government activity, media, business structure, etc. These factors may or may not approve hypotheses of Jarvenpaa's cross-cultural study. For example, in the study done by Cheskin in July 2000, U.S. consumers, compared with those in Latin America and Brazil, seemed more alert to the possibilities of how their online activities might be tracked or monitored. Table 63.5 addresses the question.

The differences found between people online in the United States, Spanish-speaking Latin America and Brazil probably represent basic cultural differences simply extended to the Internet. Among those in the United States, the belief that the government can track online activities is significantly higher than among those in Latin America or Brazil.

The Effect of Experience and Use of Earlier Technological Metaphors

People in the United States have far more exposure to the Internet environment, from its early days to today's rapid growth, than other societies. Media reports in the United States have often highlighted examples of online monitoring and abuses of personal information. This may affect perceptions among users in the United States and may lead to a sense of being "jaded" about the probability of online surveillance.

People in the United States do not believe that e-mail and instant messages are as private as a phone call. But among Spanish-speaking Latin Americans and, to a slightly lesser extent, those in Brazil, e-mail and instant messages are believed to approach the privacy level of a phone call (see Table 63.6).

TABLE 63.6. Email and Instant Message Privacy Perceptions

	U.S.	Latin America	Brazil	Total Sample
E-mail is just as private as a phone call	4.8	6.8	5.8	5.3
Instant message is just a private as a phone call	4.8	6.9	5.8	5.3

Note. Mean scores are based on an 11-point scale, where 0 means "strongly disagree" and 10 means "strongly agree."

Telephone companies are largely providing Internet access in Latin America. The recent joint venture between Microsoft and Telmex, to expand Internet access in Mexico, may help explain why Latin Americans associate the Internet more closely with the telephone than do people in the United States. Another explanation may be that Latin Americans are less exposed to the media hype surrounding Internet and e-mail privacy than are people in the United States.

Perception Vs. Behavior

If American culture is seen as one that favors individualism, whereas Latin American societies seem to promote collectivism, the findings from Cheskin's Trust in the Wired Americas study seem to disapprove Jarvenpaa's study.

However, when it comes to comfort when purchasing online, the U.S. audience is significantly more comfortable purchasing online than Latin American and Brazilian audiences as indicated in Table 63.7.

This contrast between U.S. consumers' lack of trust on the Internet in their perceptions vs. their actual purchasing behavior can possibly be explained by one factor that has been missing in Jarvenpaa's model: Consumer expectations. If people in a culture rating high on individualism are more trusting of others, they may also tend to have higher expectations in their interactions with business, government, and other individuals, and therefore demand a higher level of security, privacy, and trust. Thus, Jarvenpaa's culture-related hypotheses can be modified as such:

- H11: Consumers from individualistic cultures have a higher level of expectations in terms of their security and privacy being protected in an online transaction.
- H12: Consumers from individualistic cultures have a lower level of tolerance of risks involved in an online transaction.
- H13: Consumers from individualistic cultures are given more control of their personal information, and therefore feel more comfortable about making purchases online.

These hypotheses, however, are yet to be tested. Besides, there may be more culture-related factors that affect people's trust level, as well as their attitudes toward online security and privacy, such as history, offline lifestyle, economic system, etc.

TABLE 63.7. Online Purchase Comfort

I Am Comfortable:	U.S.	Latin America	Brazil	Total Sample
Getting information	8.7	9.1	8.4	8.8
Researching products	8.8	8.7	8.4	8.7
Purchasing online	5.9	4.6	4.4	5.5

Note. Mean scores are based on an 11-point scale, where 0 means "strongly disagree" and 10 means "strongly agree."

TABLE 63.8. Personal Information Protection

I Believe That:	Total Sample
Protection of my personal information is impossible	6.1
Web sites are free to do what they like with my personal information	4.0[a]
Personal information is protected by privacy laws	4.7
There are international laws governing how personal information is handled	4.1
The legal system protects my personal information online	3.4

Note. Mean scores are based on an 11-point scale, where 0 means "strongly disagree" and 10 means "strongly agree."
[a]The belief that Web sites are free to do what they want with personal info is higher (4.4) among people in the United States.

OTHER WAYS WEB SITES AND OTHER PLAYERS TRY TO AFFECT WEB SITE TRUSTWORTHINESS

Within the fields of Web site development and e-commerce, a wide variety of efforts have been made to reassure consumers about all six types of perceived risk in the online world. The consumer responses to those efforts, from the Cheskin 2000 study, are discussed here.

LAWS AND GOVERNMENT

There is little faith by most people that the legal system or the government offers protection to individuals online. However, as Table 63.8 indicates, the primary reason for such low expectations of government is a belief that protection simply is not possible.

Private Institutions

Although there is little evidence that people trust the government or the legal system to protect them online, many respondents feel that private institutions, namely, credit card companies, lower financial risk if not other types of risk (see Table 63.9).

Less than a quarter of online users strongly believed they would be liable for 100% of the amount if someone used their credit card online without their permission. There were no significant differences in this belief based on age, gender, expertise, or region.

Recognition of "Seals of Approval"

Of the five seals tested, the Visa symbol was most recognized overall, followed by MasterCard, TRUSTe, VeriSign, and BBBOnline.

Visa and MasterCard were universally recognized throughout the Americas. TRUSTe and VeriSign, well-known "security symbols," were more highly recognized in the United States than in Latin America, and VeriSign was more highly recognized than TRUSTe in Latin America. BBBOnline was also more highly recognized in the United States than in Latin America; however, it was the least-recognized symbol in all the Americas.

The numbers in Table 63.10 represent the percentage of individuals who recognize seals of approval. As Table 63.10 indicates, overall recognition of seals of approval increased rapidly from the 1999 Studio Archetype/Cheskin study to the 2000 Cheskin study.

Privacy Statements Irrelevant

Very few respondents claimed to have read the privacy statements associated with each symbol. Visa's statement recognition scored highest overall, with TRUSTe following. Credit card privacy statements are read more in Latin American countries than security symbols' statements. This may be due to the higher familiarity of these logos and institutions in Latin America.

Table 63.11 indicates the relative insignificance of the statements attached to specific seals of approval.

Effects on Perceived Trustworthiness

As of this writing, it seems likely that seals of approval are playing a significant role in enhancing trust between persons and Web sites. Trust in the network, however, is another matter. Significant differences are seen between the United States and Latin America in perceptions of symbol trustworthiness:

- *Security Symbols Matter in the US:* Security symbols increase trust more than credit card symbols in the United States. Of these, TRUSTe rated highest, whereas MasterCard rated lowest.
- *Credit Card Symbols Matter in Latin America:* Visa is significantly the most trusted symbol in Spanish-speaking Latin America and Brazil. Security symbols have far less significance than credit card symbols. Of the three security symbols tested, VeriSign did rate significantly higher than TRUSTe. This may be due to the word itself, which in Spanish can appear related to the word "verdad" ("verdade" in Portuguese), or the concept of truth.

Table 63.12 indicates the percentage of people who feel the presence of a particular symbol increases the trustworthiness of a site.

TABLE 63.9. Credit Card Perceptions

	Internet Use			Have Purchased Online		
	Light	Medium	Heavy	Yes	No	Total Sample
If someone uses my credit card online without my permission, I will be liable for 100% of the total spent	4.3	3.3	3.4	3.3	3.9	3.4

Note. Means are based on an 11-point scale, where 0 means "strongly disagree" and 10 means "strongly agree."

TABLE 63.10. Seals of Approval Recognition

Have Seen	U.S. 1999[a] (%)	U.S. 2000 (%)	Latin America (%)	Brazil (%)	Total Sample (%)
Visa	70	89	90	85	89
MasterCard[b]	79	63	77	67	66
TRUSTe	10	69	26	20	56
VeriSign	36	59	39	40	53
BBBOnline	18	37	16	12	31

[a]Data from the 1999 Trust Study conducted by Cheskin and Studio Archetype/Sapient.
[b]The Visa and MasterCard symbols from year to year were not identical.

TABLE 63.11. Statement Familiarity

Have Read Statement of:	U.S. (%)	Latin America (%)	Brazil (%)	Total Sample (%)
Visa	27	29	18	27
TRUSTe	25	5	2	19
VeriSign	21	11	10	18
MasterCard	15	22	13	16
BBBOnline	10	3	0	8

TABLE 63.12. Percent Reporting Increased Trust

Symbol	U.S. 1999[a] (%)	U.S. 2000 (%)	Latin America (%)	Brazil (%)	Total Sample (%)
VISA	11	38	80	78	50
TRUSTe	9	55	19	19	45
MasterCard	13	27	72	65	40
VeriSign	25	38	25	32	35
BBBOnline	16	40	04	05	30

[a]Data derived from the 1999 e-Commerce Trust Study conducted by Cheskin (1999) and Studio Archetype/Sapient (1999).

The Greater the Experience, the Less the Need for Seals of Approval

As illustrated in Figs. 63.4 and 63.5, one's experience online tends to be related to the amount of trust a seal of approval offers. Light users, those with less expertise, and those who have never purchased online agree that credit card symbols, especially VISA, would increase their trust. Additionally, younger users, age 25 or less, report that credit card symbols inspire more trust than security symbols. On the other hand, heavy users of the Internet agree that security symbols would increase their trust, more so than for light or medium users.

DATABASE SECURITY

Although there is continuing concern about security of the network, there are significant differences by region. This may point to perceptual differences in what a database is, and how information is stored, as well as cultural differences vis-à-vis individualism (see Table 63.13).

This finding suggests that a Web site can boost trust among users by clearly addressing how it handles database security. Given the generally low level of trust in the security of databases, Web sites that offer the security of a "banker's vault" can go a long way toward promoting trust.

THE FUTURE OF ONLINE TRUST

Given the uncertainties of privacy and security online, and the unlikelihood that either will be secured anytime soon, particularly with respect to the network as a whole, should one conclude that online trust will remain largely stunted? To some extent, yes. However, when engaging in individual transactions, lack of trust in the network seems to be less significant with each passing year. As has been noted, the relative increase in trust between players on the web has been made possible through the use of effective design, the involvement of seals of approval, simple experience, and a belief that credit cards limit one's liability if a transaction should go wrong.

Although transactional trust looks likely to continue to improve, the same cannot be said for security of the network. As long as this insecurity exists, persons and corporations will continue to experience a general unease with the entire structure. The nature of digital information makes it likely that all players on the web will continue to suffer from identity theft and aggregation of transactional and other personal information, and, because no one can know where such information is stored, controlling its use will likely always be difficult.

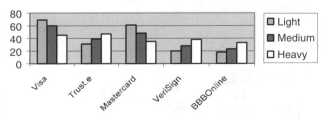

FIGURE 63.4. % increased trust by level of use.

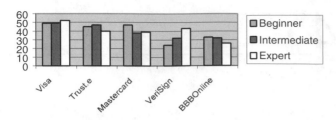

FIGURE 63.5. % increased trust by level of expertise.

TABLE 63.13. Database Security Perceptions

	U.S.	Latin America	Brazil	Total Sample
"Personal information I give on a Web site resides on a server or in a database that is secure from unauthorized people, such as hackers."	3.8	5.3	3.3	4.1

Note. Mean scores are based on an 11-point scale, where 0 means "strongly disagree" and 10 means "strongly agree."

THE END OF PRIVACY?

Do digital networks mean the end of private property? More intriguing than even this question, do they mean the end of the individual as the essential unit of society? At this point, it seems unlikely that either will come about. However, individuals will likely continue to experience an increased insecurity, which the protectors of network security cannot completely undo. Over time, most persons and corporations will simply factor that insecurity into their calculations and actions. A general erosion of individual autonomy may well occur, raising the cost of transactions, and changing the psychological make-up of those involved in the network in ways we cannot predict at this early stage. A "conditional" trust of the network may be the end result, though, in which individuals simply assume, for convenience's sake, that most of the time they are reasonably protected. For the foreseeable future, that may be the way individualism lives on in the digital age.

References

Bhawuk, D. P. S., & Brislin, R. (1992). The measurement of intercultural sensitivity using the concepts of individualism and collectivism. *International Journal of Intercultural Relations, 16*, 413–436.

Business Week Survey. (2000, March 20). Georgetown Internet Privacy Policy Survey, May, 1999 [On-line]. Available: http://www.cdt.org/privacy/survey/findings/

Cheskin. (2000). Trust in the wired Americas [On-line]. Available: www.cheskin.com

Forrester Research, Inc. (1999). Study on the web, as cited in Digitrends, "Keep a close eye on the world of online privacy," Kevin Jeys, Spring 2001, p. 28.

Hofstede, G. (1980). *Culture's consequences: International differences in work related values.* Beverly Hills, CA: Sage Publications.

Jarvenpaa, S. L., Tractinsky, N., Saarinen, L., & Vitale, M. (1999). Consumer trust in an Internet store: A cross-cultural validation [On-line]. Available: http:///www.ascusc.org/jcmc/vol5/issue2/jarvenpaa.html

Studio Archetype/Sapient and Cheskin. (1999). *eCommerce Trust Study.* Available: http://www.cheskin.com

Triandis, H. C. (1989). The self and social behavior in differing cultural contexts. *Psychological Review, 96*, 506–520.

Yamagishi, T., & Yamagishi, M. (1994). Trust and commitment in the United States and Japan. *Motivation and Emotion, 18*, 129–165.

ACHIEVING COMPATIBILITY IN HUMAN–COMPUTER INTERFACE DESIGN AND EVALUATION

Waldemar Karwowski
University of Louisville

THE CONCEPT OF COMPATIBILITY

The human factors and ergonomics of computing systems advocates the systematic use of knowledge concerning relevant human characteristics to achieve compatibility in the design of interactive systems among people, computers, and outside environments. This is done to achieve other specific goals, such as (system) effectiveness, safety, ease of performance, and to contribute to overall human well-being. The term *compatibility* is used in a narrow context, typically in relation to the problem of information processing (Karwowski, 2000). For example, it is often used in relation to design of displays and controls, such as classical spatial (location) compatibility or the intention-response-stimulus compatibility related to movement of controls (for example, see Wickens & Carswell, 1987). Cognitive compatibility is often expressed and quantified through some other measures, such as mental workload.

The *American Heritage Dictionary of English Language* (Morris, 1978) defines *compatible* as follows: (a) capable of living or performing in harmonious, agreeable, or congenial combination with another or others; and (b) capable of orderly, efficient integration and operation with other elements in a system. Probing further, the word *congenial* is defined as "suited to one's needs, agreeable," whereas *harmony* (from Greek *harmos*: joint) is defined as (a) agreement in feeling, approach, action disposition, or the like, sympathy, accord; and (b) the pleasing interaction or appropriate combination of elements in a whole.

Today the design criteria for the computer-human system compatibility are not well-defined, and the degree of desired or required compatibility cannot be easily specified given the lack of the universal matrix for compatibility measurements (Karwowski, 2000). Computer-human compatibility is usually considered as a static and linear concept that can be intuitively understood. The use of compatibility in a context of *ergonomics systems* has been advocated by Karwowski and his coworkers (Karwowski, 1991; Karwowski, Marek, & Noworol, 1988). Recently, Karwowski (1997) introduced the term *human-compatible systems* to focus on the need for comprehensive treatment of compatibility in the human factors discipline.

To survive, complex systems must be adaptable, that is, capable of functioning in an uncertain environment (Conrad, 1983). The same applies to the computer-human systems (CHS). Adaptation means alterations or adjustments by which a CHS improves its condition in relationship to its environment. Adaptation implies changes that allow for a system to become suitable to a new or special use or situation. This notion proposes that optimal compatibility may be achieved by adaptation (primarily design or learning) of both CHS subsystems.

Recently, Karwowski (2000) proposed the development of a new science that would aim to discover laws of the artifact-human compatibility, propose theories of the artifact-human compatibility, and develop quantitative matrix for measurement of such compatibility. The name of this new science, *symvatology*, was coined by using two Greek words: *symvatotis* (compatibility) and *logos* (logic, or reasoning about). Specifically, the new word is the result of joining the *symvato-* (from *symvatosis*) and *logy* (a combining form denoting "the science of"). Symvatology was proposed as the science of system-human compatibility. To optimize system performance, the system-human compatibility should be considered at all levels, including the physical, perceptual, cognitive, emotional, social, organizational, environmental, and so on. This requires a means to measure the input and output that characterize the set of CHS interactions (Karwowski, 1991).

CHS complexity

The American Heritage Dictionary of English Language (1978) defines *complex* as consisting of interconnected or interwoven parts. According to Zuse (1999), complexity can be defined as "a function of the relationships among the components of an object." This definition proposes a link between complexity and multiple relationships or interactions. Certain features can be attributed to complex systems such as a large number of elements, high dimensionality, an extended space of possibilities (Kokol, Brest, & Umer, 1996), understandability, and metric of difficulty with respect to understanding multiple relationships or interactions among two or more components of an object or entity (Coskun & Grabowski, 2001).

Different disciplines have defined software complexity in different ways. Software engineers, for instance, have defined complexity based on the degree of presence or absence of certain software characteristics—numbers of errors in the code, development cost and time, or size of the code (Kokol et al., 1996). Mathematical approaches considered source code length, volume, difficulty of programming, and effort to program (Halstead, 1977), dimensionality, number of components, and number of relationships among components (Kokol et al., 1996; Zuse, 1993) as complexity measures for software products.

From the psychological point of view, software complexity was defined in terms of its understandability by human developers, and users, as well as in terms of the workload, stress, and fatigue that software complexity can induce (Alford, 1994). Social scientists have described complexity in terms of degrees of interaction and coupling. Interactions in sociotechnical systems can be complex or linear, depending on whether events are unpredicted or unexpected. Coupling is the degree to which "reciprocal interdependence exists across many units and levels" and is a measure of the degree of slack or redundancy in the system (Perrow, 1984).

Marr (1977) pointed out that intelligent systems complexity should be addressed at the following complexity levels: *task complexity* to describe the intricacies of the system's functions; *representation complexity* to capture the intricacies of an intelligent system's knowledge and reasoning models; and *implementation complexity* to describe the richness and robustness of the implemented intelligent system. Coskun and Grabowski (2001) distinguished four types of computer software complexity in safety-critical, large-scale systems: (a) *architectural/structural complexity*, which refers primarily to the structure of the software code, its functions, and utilities; (b) *data processing/reasoning/functionality complexity*, which refers to the measurement of complexity associated with

intelligence-related functions such as data processing, reasoning, and functionality; (c) *user interface complexity*, which refers to the complexity associated with the users' view of the information displays and understandability of user screen, which is the interface between the user and the system; and (d) *decision support/explanation complexity*, which refers to the complexity associated with the decision support and explanations provided by system.

CHS compatibility is a dynamic phenomenon affected by the system's structure, its inherent complexity, and its *ergonomic entropy*, or the irreducible level of incompatibility between the system's elements (Karwowski, 1995). The structure of system interactions determines the complexity and related compatibility relationships in a given system. Therefore, compatibility should always be considered in relation to the system's complexity.

A FRAMEWORK OF THE GENERIC ERGONOMIC SYSTEM

Based on our earlier work (see Karwowski, 1991; Karwowski et al., 1988), any generic artifact–human system (S) can be represented as a construct that contains the following:

- a human subsystem (H),
- an artifact subsystem (A),
- an environmental subsystem (N), and
- a set of interactions (I) occurring between different elements of these subsystems over time (T).

The set of interactions is viewed here as a set of all possible interactions between people, artifacts, and various environments that are present in a given state of the system [H, A, N, T]. These interactions (see Fig. 64.1) reflect the existence (or nonexistence) of the relationships between the subset of all relevant human characteristics, such as anatomic, physiologic, biomechanical, or psychological; the subset of characteristics of the artifact subsystem (for example the computer–computer task system); and the elements of the environmental subsystem, representing the subset of environmental conditions (physical environment, social support, organizational structure, etc.). These interactions also impact the system complexity.

THE COMPLEXITY–COMPATIBILITY PARADIGM

To express the innate relationship between systems' complexity and compatibility, Karwowski et al. (1988, 1994) originally proposed the *complexity-incompatibility principle*, which is stated as follows:

As the (artifact-human-environment) system complexity increases, the incompatibility between the system elements, as expressed through their ergonomic interactions at all system levels, also increases, leading to greater ergonomic (nonreducible) entropy of the system and decreasing the potential for effective ergonomic intervention.

This principle reflects the natural phenomena that others in the field have described in terms of difficulties encountered in humans interacting with consumer products and technology in general. For example, according to Norman (1988), the paradox of technology is that added functionality to an artifact typically comes with the trade-off of increased complexity. These added complexities often lead to increased difficulty and frustration when interacting with various artifacts. One of the reasons for this is that technology that has more features also provide less feedback. Moreover, Norman (1988) argued that the added complexity cannot be avoided when functions are added and can only be minimized with good design that follows natural mapping between the system elements (i.e., control–display compatibility).

The complexity-compatibility paradigm representation for the CHS is shown in Fig. 64.2.

The system space, denoted here as an ordered set (complexity, compatibility) are defined by the four pairs as follows: (high, high), (high, low), (low, high), (low, low). Under the best scenario (i.e., under the most optimal state of system design), the CHS exhibits high compatibility and low complexity levels. It should be noted that the transition from high to low level of system complexity does not necessarily lead to improved (higher) level of system compatibility. Also, it is often the case in most CHSs, that an improved (higher) system's compatibility

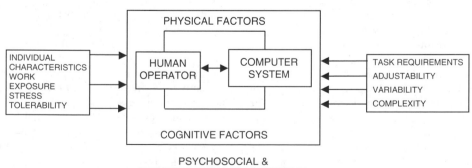

FIGURE 64.1. Generic framework for a human–computer interaction system.

COMPLEXITY

COMPATIBILITY	LOW, HIGH	HIGH, HIGH
	LOW, LOW	HIGH, LOW

FIGURE 64.2. The compatibility–complexity paradigm representation.

can only be achieved at the expense of increasing system's complexity.

In context of a CHS, the previously stated general principle describing the relationship between the systems' complexity and compatibility (Karwowski et al. 1988, 1994), can now be rephrased as follows:

As the computer-human system complexity increases, the system compatibility expressed through the set of relevant interactions at all system levels decreases, leading to diminished potential for effective system design.

The above principle reflects the natural phenomenon that others in the ergonomics field have described in terms of difficulties encountered by humans when interacting with the computers as consumer products or with computer-based technology in general. The above principle is illustrated below based on two examples: (a) the design of a computer display and (b) design of an office chair, using the framework of the ergonomic system and entropy.

ENTROPY OF CHSs

Entropy, related to the probability of an order–disorder, is a useful concept for analysis of the CHS because changes in entropy (disorder) can be quantified. Natural processes spontaneously move from the more ordered to the less ordered (more probable) state. Low entropy (or well-ordered systems) can be found in an open or closed system, however. The entropy of the universe, which is an isolated system, is continually increasing, that is, the disorder of the universe is increasing.

Of great importance to the CHS is that the dynamic equilibrium is more ordered and, therefore, less probable than the normal state. This is because the system in the state of dynamic equilibrium has lower entropy. Depending on how widely open or isolated the system is, the maintenance of the lower entropy level over time may be more or less difficult. Isolated systems are closed to the transfer of matter or heat. Such systems move toward states of higher entropy (lower order) and stop changing when they reach an equilibrium state, characterized by the maximum entropy. This does not preclude local islands of decreasing entropy, but the overall entropy of the closed system (like the universe) always increases (Silver, 1998).

A CHS can be open, isolated or closed, depending on its purpose, structure, and dynamic state of interactions with the outside environment; this is not always intentional and the intentions can change. As such, these systems show changing levels of entropy (or order) over time. It should be noted that a change (increase) in incompatibility, which I denote here as (change in) an ergonomic disorder, follows nature by tending to increase the disorder or increase the system entropy as much as possible. Incompatibility is considered an "attractor," which means a state or phenomenon to which the system will gravitate whatever the initial conditions or a steady-state point at which the system starts (if poorly designed to begin with) and where it remains.

For example, while working with a video display terminals, the human operator engages in a controlled interaction with the surfaces of support (chair seat, backrest, floor, work surface) with the goal of maintaining dynamic equilibrium (Dainoff & Baillett, 1991). The structure of this interaction is determined by four classes of constraints: (a) system goals, (b) workstation characteristics, (c) operator characteristics, and (d) chair characteristics. The nature of the task may require different postural orientations depending on the viewing distance to the copy or screen, physical dimensions of the human operator, and the nature and range of computer workstation adjustments. Within these constraints, it is assumed that an operator will try to minimize muscular efforts to maintain an equilibrium.

In view of the complexity–compatibility paradigm (Karwowski, 2000), the human operators are trying to minimize locally the incompatibilities at different levels of the system (i.e., at those interactions that are possible for them to use and modify). This process of human adaptation to the task can be considered the last-resort solution when the system incompatibility cannot be further reduced. In this case, human operators attempt to change the structure of their physical interactions with the CHS (i.e., the workstation and chair components or computer subsystem). From the operator point of view (human subsystem), the easiest interaction to modify is the postural interaction, aimed at control of the body.

Entropy of the Generic Ergonomic System

As discussed by Karwowski (1995), the entropy of the artifact-human system (S) (system entropy: $E(S)$ can be modeled using the entropies (E) due to the human-related interactions (*H-subsystem*), artifact-related interactions (*A-subsystem*), environment-related interactions (*N-subsystem*), and time (T). In view of this, the system entropy can be defined as a function of entropy of the set of ergonomic interactions $E(I)$ and time (T) as follows: $E(S) = f [E(I), T]$, where $E(S)$ is an expression of the total system incompatibility with the human.

Because the set of ergonomic interactions (I subsystem) consists of the possible relations between the subsystem elements of $\{H\}$, $\{A\}$, and $\{N\}$, the $E(I)$ can be expressed as follows:

$$E(I) = \{E(H), E(A), E(N)\},$$

where $E(H)$ = contributing entropy due to human subsystem, $E(A)$ = contributing entropy due to artifact subsystem, and $E(N)$ = contributing entropy due to environmental subsystem.

It should be noted that $E(H)$, or the contributing entropy due to human subsystem, is interpreted here as the entropy

due to natural human limitations (in relation and as relevant to a given system), which adversely affect the compatibility of an artifact system with the human characteristics defined by a subset (*H*).

The entropy of an artifact-human system (*S*) depends on the variety and complexity of all relevant interactions for the subsystems *H*, *A* and *N*. Any manipulation of elements of the artifact-human system may result in subsequent changes in the number or structure (or both) of other interactions, and therefore changes in system entropy.

COMPATIBILITY REQUIREMENTS AND SYSTEM REGULATION

In design of human-compatible systems, one must consider the structure of system interactions (*I*) at the given system level, which induces complexity of that system. It is the complexity of the {*A*} and {*N*} subsystem interactions that defines the ergonomic compatibility requirements (*CR*) of the system. In general, the greater the system compatibility requirements, the greater the need for ergonomic intervention efforts, called here the *system regulation*. Therefore, it follows that an entropy of the artifact-environment {*A*, *N*} system, or the system regulator, *E*(*R*), is also defined by the system's structure complexity. The optimal (ergonomic) system has minimal ergonomic compatibility requirements (i.e., minimal system incompatibility). In other words, the optimally designed artifact-human system satisfies most (if not all) of the compatibility requirements.

The artifact and environmental subsystems can be regulated through the design efforts (at least to some degree) and are jointly called the system regulator (*R*). Consequently, the sum of their respective entropies defines the entropy of the system regulator:

$$E(R) = E(A) + E(N),$$

or equivalently

$$E(R) = E\{(A) + (N)\}, \text{ denoted as } E(R) = E(A, N)$$

It follows from this discussion that an entropy of the regulator also defines the system's compatibility requirements. This is because at any given level of human entropy *E*(*H*), the *E*(*R*) is directly related to the artifact-human system entropy *E*(*S*), which is the outcome of such regulation. This can be illustrated as the process system regulation, with *E*(*H*) as the system input, *E*(*R*) as the system regulator, and the artifact-human system entropy *E*(*S*) as the output, with the regulation structure as follows:

$$\begin{array}{c|c} & E(R) \\ \hline E(H) & E(S) \end{array}.$$

According to Ashby's (1964) law of requisite variety, for the system shown above, the outcome, (i.e., the entropy) can be defined as follows:

$$E(S) \geq E(H) + E(H(R)) - E(R),$$

where *E*(*H*(*R*)) is the entropy of the regulator (*R*) when the state of human subsystem (*H*) is known. When *E*(*H*(*R*)) = 0 (i.e., when the regulator subsystem *R* = {*A*, *N*} is a determinate function of the human subsystem (*H*) functioning under *R*, the above equation transforms as follows:

$$E(S) \geq E(H) - E(R),$$

showing that system entropy can be defined as a difference between the human entropy and entropy of the regulator. This equation also indicates that system entropy, if minimal at a given design stage (and given the specific value of the human entropy), can only be further reduced by increasing entropy of the regulator. This means that human limitations can be overcome by improved system design. In the context of ergonomics, this is typically done by redesign efforts leading to an increase in the system complexity. On the other hand, this also implies a need for increasing the system's compatibility requirements.

Following Ashby's (1964) law of requisite variety, Karwowski (1995) proposed the corresponding law, called the **law of requisite complexity**, which states that only design complexity can reduce system complexity. This means that only added complexity of the regulator (R = redesign), expressed by the system compatibility requirements, can be used to reduce the overall system complexity and consequently reduce the ergonomics system entropy (i.e., reduce the overall artifact-human system incompatibility).

It should be noted here that the minimum value of the artifact-human system's entropy *E*(*S*)$_{\text{MIN}}$ is equal to the human entropy *E*(*H*) and occurs when the value of *E*(*R*) = 0, indicating a system state in which no further system regulation is possible. This is called the ergonomic entropy of the system:

$$E(S)_{\text{MIN}} \geq E(H), \quad \text{when } E(R) = 0.$$

As discussed by Karwowski et al. (1994), the ergonomic entropy *E*(*S*)$_{\text{MIN}}$ is the "nonreducible" level of system entropy with respect to ergonomic intervention (regulation) efforts. It should be noted, however, that this entropy can be modified by the nonergonomic means, that is, through the human, rather than system, adaptation efforts. This can be done by reducing the value of human entropy *E*(*H*), through improvements in human performance and reliability by training, education, motivation, and so on.

THE HUMAN–COMPUTER INTERACTION (HCI) SYSTEM

Based on our earlier work (see Karwowski et al., 1988, 1995), the CHS (*S*) can be represented as a construct that contains the human subsystem (*H*), a computer subsystem (*A*), an environmental subsystem (*E*), and a set of interactions (*I*) occurring between different elements of these subsystems over time (*T*). The set (*I*) is as a set of all possible interactions between the human, computer, and various environmental conditions (*N*) that are present at a given state of the system.

These interactions reflect the existence (or nonexistence) of the relationships between the subset of all relevant human characteristics, such as anatomic, physiologic, biomechanical, or psychological factors; the subset of characteristics of the computer subsystem, and the elements of N, representing the subset of environmental conditions (physical environment, social support, organizational structure, etc.). These interactions also impact the system complexity. As discussed by Karwowski et al. (1985), the lack of compatibility or ergonomic incompatibility (*EI*), defined as system degradation (disintegration), is reflected in the system's measurable inefficiency and associated human losses.

The HCI system (*HCI-S*) can be formally defined as follows:

$$[\textit{HCI-S}] = \{(H), (R = \{K, M, N\}), (I), (T)\},$$

where H = a set of human operator characteristics (perceptual, physical, and cognitive, etc.), K = a set of task requirements (physical, sensory, perceptual, cognitive, psychosocial), M = a set of computer characteristics (hardware and software, including computer interfaces), N = an environment, I = a set of interactions between H and R, and T = time. The set of interactions embodies all possible interactions between H, K, and M of the regulator in a given environment subsystem, regardless of their nature or strength of association. Such interactions can be elemental (i.e., one-to-one association) or complex (such as an interaction between the human operator, particular software that is used to achieve the desired computing task, and the available computer interface).

CHS Entropy Determination: Example 1—Visual Display Design

The following example relates to the perceptual/cognitive interactions of the human operator with the visual display terminal (VDT) subsystem. Let the ergonomic entropy of the VDT-human operator system be defined as follows:

$$E(S) = \{\text{user}(H), R = (\text{task}(K),$$
$$M = \{\text{workstation}(W), \text{visual display}(V)\}, (T)\},$$

where K and M subsets of the system regulator (R) are the ergonomic design variables. Let us consider the perceptual/cognitive demands at the VDT work due to various human operator interactions with the visual display terminal. If the values of K and W are constants, then the characteristics of V define R. The design parameters of $\{R\}$ directly affect the interactions between the system and the user. These interactions (between H and K), define the current compatibility requirements $\{CR\}$ of the system.

According to the formula presented here $[E(S) = E(H) - E(R)]$, the entropy of the user-VDT system $E(S)$ equals the human entropy (($E(H)$), defined by the perceptual and cognitive interactions between the human operator and the visual display terminal (V), which are due to specific information processing requirements (K), minus an entropy of the regulator $E(R)$, which is determined by the specific screen and display design (see Fig. 64.3).

It should be noted that the higher the $E(R)$ at the given value of $E(H)$, the lower the $E(S)$. Therefore, the value of system

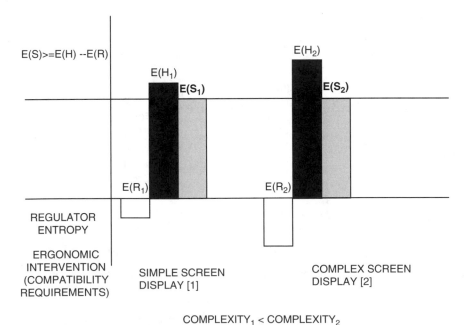

FIGURE 64.3. An illustration of hypothetical entropy changes for the screen display–human operator system at two levels of computer screen complexity.

entropy $E(S)$ is determined by system's compatibility requirements (CR), and the consequent (resultant) level of complexity of the system regulator (R) with an increased entropy $E(R)$, due to some new design concept or solution. To satisfy the compatibility requirements of the information processing task (K) (i.e., to make the extraction and processing of information more compatible with the user), one has to design more sophisticated display by increasing the number of different screen features, adopting variety of windows, application of different color schemes, and so on. These ergonomic intervention efforts (system regulation) lead to an increase in the visual display complexity. Such increased complexity, and the increase in corresponding entropy, must in turn be dealt with by the human operator.

CHS Entropy Determination: Example 2—Chair Design

Another example of system regulation in CHS is the physical interactions of the chair–human operator subsystem. Let the ergonomic entropy $E(S)$ of the CHS system be defined as follows:

$$E(S) = \{\text{user}(H), R = \{\text{task}(K),$$
$$M = \{\text{workstation}(W), (\text{computer}), \text{chair}(C)\}\}, (T)\}$$

where K and M subsets of the system regulator (R) are the ergonomic design variables.

Let us consider the postural strain at the computer work due to physical interactions (Grieco, 1986). If K and W are assumed to be constants (i.e., their values are set before the human operator starts working with the system), then the chair characteristics (C) can be considered as the system regulator (R), which design parameters directly affect the interactions between the system and the user. The complexity of interactions (between H and C) defines the current compatibility requirements {CR} of the system. The above allows to simplify the considered entropy of the examined ergonomic system as follows: $E(S) = E \{H, R = \text{chair}\}$

According to equation; $E(S) = E(H) - E(R)$, the user-chair system entropy $E(S)$ equals the human entropy of the system (defined by the physical interactions between the human operator and the chair) minus the entropy due to specific chair design or regulator $E(R)$. The value of $E(R)$ is determined by the chair compatibility requirements, and the consequent level of complexity of the regulator (new system design concept). To satisfy the compatibility requirements of the chair (i.e., to make the chair more compatible with the user, or simply to make the chair fit the user better and ensure more comfort, the designers typically increase the number of adjustable chair features (degrees of freedom), thereby increasing chair's complexity [and entropy $E(R)$]. In other words, to reduce the system entropy $E(S)$, one needs to increase the entropy of the system regulator $E(R)$. This increased system complexity may significantly reduce effectiveness of the ergonomic intervention, because the user may be unable to deal with the added system complexity (i.e., may not be able to properly adjust variety of chair features (adjustments) to ensure his or hers comfort (see Fig. 64.4). Remember that the ergonomic compatibility requirements are subject to changes over time.

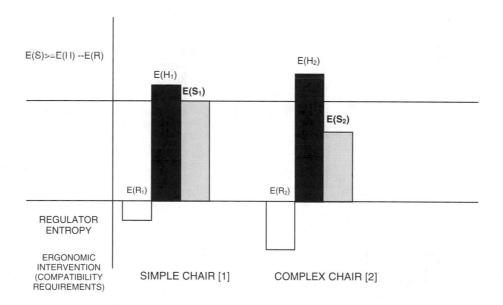

FIGURE 64.4. An illustration of hypothetical entropy changes for the office chair–human operator system at two levels of chair complexity (degrees of freedom for chair adjustments).

A FRAMEWORK FOR QUANTIFICATION OF COMPLEXITY AND COMPATIBILITY

To maximize their usability, viable products and systems should be designed to maximize their compatibility with the human operator while minimizing their complexity. Jamaldin and Karwowski (1997) proposed a framework for defining and quantifying system complexity and the related measure of system compatibility. The proposed modeling framework includes compatibility matrix as a function of the system components' significance, the strength of system interactions and local compatibilities between the system elements (i.e., human–computer and task, environment subsystems, and relevant interactions), and influence between the system components.

Modeling of CHS Complexity

The complexity of the CHS can be defined by the structure and extent of interactions among the subsystem elements. Such a complexity, which is introduced into the system by the human (H), computer (M), and environment (EN) components, can be defined by the structure of interactions among all of the system elements. Karwowski et al. (1994) pointed out that according to the laws of system adaptation, any manipulation of elements of the human-computer-environment system may result in subsequent changes in the number or structures of the interactions. It is the complexity of the M and EN resulting interactions that defines the extent of ergonomic intervention efforts required to achieve (and regulate) the desired system compatibility.

The complexity of any system $\Psi(S)$ is manifested in the complexity of the interactions observed between subsystem components Ψ_i: $\Psi(S) = f(\Psi_i)$. On the other hand, the complexity of an ergonomic system is influenced by:

1. the number of components subcomponents of the ergonomic system (N_c),
2. the number of interactions within the system (N_I), and
3. the strength of these interactions (k_I), where $\Psi(S) = f(N_c, N_I, k_I)$.

The complexity of system interactions is a function of the inherent complexity of the interacting components (Ψ_C), and therefore $\Psi(S) = f(\Psi_I) = f(\Psi_C)$, which implies that complexity of interactions in the human-computer system is largely influenced by the original design process, because the complexity of the interacting components is embedded into the original design of the system. It should be noted here that according to Suh (1990), the design objectives for any system must be defined in terms of specific requirements, called *functional requirements*.

A Structure of Ergonomic System Interactions

As described earlier, any ergonomic system (ES) can be defined in terms of three main elements and their interactions at a given time as follows:

$$ES = (H, M, EN, I, T)$$

All system elements can be defined in terms of their components (Karwowski, 1991) as follows: $H[H = \{b_1, b_2 \ldots b_I\}]$ is the human subsystem, $M[M = \{m_1, m_2 \ldots m_j\}]$ is the computer subsystem, $EN[EN = \{e_1, e_2 \ldots e_k\}]$ is the environmental subsystem, and $I[I = \{i_{pq}\}]$ is the subset of interactions between all subelements (or components) within the system, and where p and q take all the indices from $\{b_1, b_2, .., b_l, m_1, m_2, .., m_j, e_1, e_2, e_k\}$; and where $i_{pq} = \{\emptyset\}$ if no interaction exists between the system components.

The interactive nature of the structure of the ergonomic system at a given time (t_0) can be represented by three characteristic matrices each of size $[p \times q]$; the adjacency matrix (A_{t0}), the reachability matrix (R_{t0}), and connectivity matrix (C_{t0}) which are defined as follows:

$$A_{t0} = \{a_{pq}\}, \quad a_{pq} = 1 \quad \text{if there is an immediate (simple) interaction (direct arc) from } p \text{ to } q$$
$$= 0 \quad \text{otherwise}$$
$$R_{t0} = \{r_{pq}\}, \quad r_{pq} = 1 \quad \text{if component } q \text{ is reachable from } q$$
$$= x \quad \text{otherwise}$$
$$C_{t0} = \{c_{pq}\}, \quad c_{pq} = nd \quad \text{if component } p \text{ reaches component } q$$
$$= x \quad \text{otherwise}$$

where $a_{pq} \cdot r_{pq}$, and c_{pq} are adjacency, reachability, and connectivity entries, an x is some undefined distance, and nd is the total number of nonidentical interactions along all possible, but nonidentical reachability paths between p and q, originating at, but not returning, to p. When calculating (nd), the interactions that are common to more than one reachability path between p and q must only be counted once.

The structures of the adjacency matrix (A_{t0}), reachability (R_{t0}), and connectivity (C_{t0}) are shown on Fig. 64.5. From these defined matrices, three new column matrices of size $[N_C \times I]$ are denoted as elemental vectors. These include the elemental adjacency vector ($\mathbf{A_T}$), elemental reachability vector ($\mathbf{R_T}$), and elemental connectivity vector ($\mathbf{C_T}$), that are calculated by summing the values across each individual row as follows:

$$\mathbf{A_T} = \{u_p\}, u_p = \sum_{q=b_1}^{e_k} a_{pq}; \quad \mathbf{R_T} = \{v_p\}, v_p = \sum_{q=b_1}^{e_k} r_{pq};$$
$$\mathbf{C_T} = \{w_p\}, w_p = \sum_{q=b_1}^{e_k} c_{pq}$$

where $p \oplus \{b_1, b_2, .., b_l, m_1, m_2 \ldots, m_j, e_1, e_2, .., e_k\}$.

The total adjacency (T_A), total reachability (T_R), and total connectivity (T_c) of the system are derived by summing all elemental values within each individual matrix. Thus, the total adjacency of the system (T_A) represents the total number of direct interactions within the system. The total reachability of the system (T_R) represents the total number of possible interactions (direct and indirect). The total connectivity of the system (T_A)

$$
\begin{array}{c}
\text{System Adjacency}\\
\text{(Structure of simple}\\
\text{interactions):}
\end{array}
\quad
A_{to} =
\begin{array}{c}
\\
h_1\\ h_2\\ h_3\\ h_4\\ \\ m_1\\ m_2\\ m_3
\end{array}
\begin{array}{ccccccc}
h_1 & h_2 & h_3 & h_4 & m_1 & m_2 & m_3\\
0 & 1 & 0 & 0 & 0 & 0 & 0\\
1 & 0 & 0 & 0 & 0 & 0 & 0\\
0 & 0 & 0 & 1 & 0 & 0 & 0\\
0 & 0 & 0 & 0 & 0 & 0 & 0\\
1 & 1 & 1 & 0 & 0 & 0 & 0\\
1 & 1 & 0 & 1 & 0 & 0 & 0\\
1 & 1 & 0 & 0 & 0 & 0 & 0
\end{array}
\quad : \quad A_T =
\begin{array}{c}
1\\1\\1\\0\\3\\3\\2
\end{array}
$$

$$TA = 11$$

$$
\begin{array}{c}
\text{System Reachability}\\
\text{(Structure of all}\\
\text{interactions):}
\end{array}
\quad
R_{to} =
\begin{array}{c}
\\
h_1\\ h_2\\ h_3\\ h_4\\ \\ m_1\\ m_2\\ m_3
\end{array}
\begin{array}{ccccccc}
h_1 & h_2 & h_3 & h_4 & m_1 & m_2 & m_3\\
1 & 1 & 0 & 0 & 0 & 0 & 0\\
1 & 1 & 0 & 0 & 0 & 0 & 0\\
0 & 0 & 1 & 1 & 0 & 0 & 0\\
0 & 0 & 0 & 1 & 0 & 0 & 0\\
1 & 1 & 1 & 1 & 1 & 0 & 0\\
1 & 1 & 0 & 1 & 0 & 1 & 0\\
1 & 1 & 0 & 0 & 0 & 0 & 1
\end{array}
\quad : \quad R_T =
\begin{array}{c}
2\\2\\2\\1\\5\\4\\3
\end{array}
$$

$$TR - 19$$

$$
\begin{array}{c}
\text{System Connectivity}\\
\text{(Directness of all}\\
\text{interactions):}
\end{array}
\quad
C_{to} =
\begin{array}{c}
\\
h_1\\ h_2\\ h_3\\ h_4\\ \\ m_1\\ m_2\\ m_3
\end{array}
\begin{array}{ccccccc}
h_1 & h_2 & h_3 & h_4 & m_1 & m_2 & m_3\\
0 & 1 & x & x & x & x & x\\
1 & 0 & x & x & x & x & x\\
x & x & 0 & 1 & x & x & x\\
x & x & x & 0 & x & x & x\\
1 & 1 & 1 & 2 & 0 & x & x\\
1 & 1 & x & 1 & x & 0 & x\\
1 & 1 & x & x & x & x & 0
\end{array}
\quad : \quad R_T =
\begin{array}{c}
1\\1\\1\\0\\5\\3\\2
\end{array}
$$

$$TC = 13$$

FIGURE 64.5. The structure of system adjacency, reachability, and connectivity matrices in the quantification of complexity and compatibility framework.

represents the total number of interactions along nonidentical reachability paths.

The structures of T_A, T_R, and T_c are illustrated using an example of the complex ergonomics (human–chair system) system shown in Fig. 64.6 below.

$$
T_A = \sum_{p=b_1}^{e_k} u_p \qquad T_R = \sum_{p=b_1}^{e_k} v_p
$$
$$
T_C = \sum_{p=b_1}^{e_k} w_p
$$

Quantification of Compatibility

A compatible system is one that requires a minimal level of human interaction(s) to achieve the required level of fit (i.e., the level of desirable user comfort and safety; Karwowski, 1991). Accordingly, an ergonomic system consisting of a human and a nonadjustable chair is less compatible with the human than a similar system where the chair is adjustable. It should be noted here that the second (adjustable) system could be more complex depending on the complexity of the adjustable chair and extent of interactions between its components. System compatibility, which is the target of any design or ergonomic intervention effort, hinges on the notion of complexity because, in most cases, system complexity must be increased to increase compatibility (Karwowski et al., 1994).

The notions of compatibility and complexity are similar in form in terms of their dependence on connectivity (c_{pq}) and the strength of interaction (k_{pq}) in their basic structure. Unlike complexity, however, the notion of compatibility is tied to two

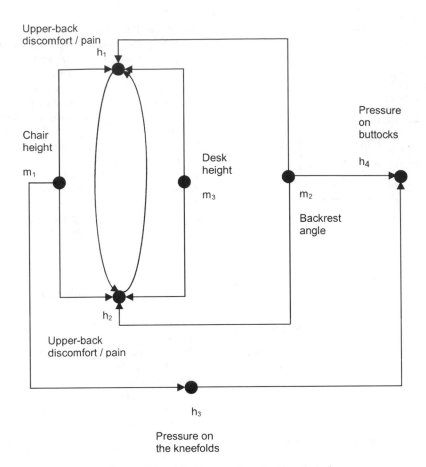

FIGURE 64.6. Illustration of the interactions for the chair–human operator system in the quantification of complexity and compatibility framework.

additional factors, the significance of components (a) and the inherent compatibilities (q) between them.

Collectively, all factors that impact compatibility are collectively expressed as the rate of influence of compatibility $\theta = f[\Psi] = f(I)$. Altering the influence of a given component (I_p) on other components within the system may alter system compatibility by making the system more (or less) compatible based on the nature of the interaction(s) involved.

The compatibility of the ergonomic system is a function of the inherent incompatibility between system components (I_p), the strength of their interactions (k_{pq}), and their significance (α_p). Thus, $I_p = f(\alpha_p, k_{pq}, i_{pq}, c_{pq})$ and because $c_{pq} = f(k_{pq})$, therefore $I_p = f(\alpha_p, c_{pq}, i_{pq})$, which allows definition of the system compatibility as follows: $\theta = f(I_p) = f(\alpha_p, c_{pq}, i_{pq})$.

Complexity Versus Compatibility

The value of compatibility between two system components (qpq) is described on a scale of 0 (*fully incompatible*) to 1 (*fully compatible*). Conceptually, compatibility (i_{pq}) and incompatibility (h_{pq}) between two components are complementary sets, thus $[i_{pq} + b_{pq} = 1]$. This indicates that when the system is fully compatible with the human, all individual compatibilities must be at their maximum, therefore $i_{pq} = 1$,

or $b_{pq} = 0$. In real systems, however, the value of zero incompatibility is not achievable because, according to Karwowski's (1991) framework, an ergonomic system has a nonreducible (ergonomic) entropy below which no further reduction is possible. The notion that entropy and incompatibility are similar in form and pattern leads to conclude that incompatibility also reaches a minimum nonreducible value.

The interdependence between system complexity (Ψ) and compatibility (θ) can be seen in the fact that compatibility is dependent, among other things, on system connectivity (c_{pq}) and strength of interactions (k_{pq}), factors that are inherent to the structure of complexity. This is essential to the design of a system of higher compatibility. Clearly, the best design is one with the highest system compatibility and lowest complexity, and because both Ψ and θ are measured on a scale of 0 to 1, their boundaries form a square denoted by the square of compatibility (see Fig. 64.7). Therefore, each ergonomic system (ES) at a given state of compatibility and complexity can he represented by its compatibility and complexity coordinates on the square: $ES = (\Psi, \theta)$.

It can be seen from Fig. 64.4 that system states located within quadrant (I) and closer to corner (1,0) are ideally the best case design scenario, where compatibility is high and complexity is low. Quadrant (II) provides the second best case, most

COMPLEXITY [Ψ]

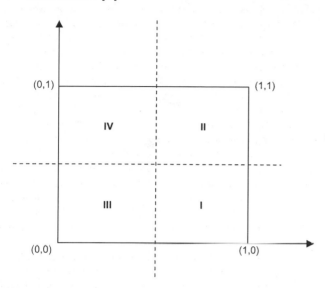

COMPATIBILITY [θ]

FIGURE 64.7. The complexity–compatibility quantification framework.

commonly centered on ergonomic intervention where complexity and compatibility are closer to their higher boundaries (1,1), that is, high system complexity is often increased to increase system compatibility with the human subsystem.

FUZZINESS IN COMPUTER-HUMAN SYSTEMS

Ergonomics aims to ensure compatibility in the computer system-human functioning with respect to complex and uncertain (cognitive) interrelationships between system users, computers, and the environment. Such analysis must account for natural fuzziness and nonlinear dynamics (chaos) (Jee Feingenbaum, 1980; Gleick, 1987) of human cognitive processes (Guastello, 1995; Karwowski, 2000; Karwowski, Grobelny, Yang, & Lee, 1999; Smithson, 1982). Fuzziness describes an event ambiguity and measures the degree to which an event occurs, not *whether* it occurs (Kosko, 1992; Zadeh, 1973).

Fuzzy systems provide a useful framework for modeling variety of complex tasks, situations, systems, CHSs and their environments, and their interactions with people. Because fuzziness occurs at all levels of human interactions with outside environments, ranging from physical to cognitive tasks, it can be used as the natural model of human sensory, information processing, communication, or physiological functioning (Karwowski, 1991, 1992). The cognitive models use formal structures to represent the computer-human environment as perceived by the human operator. However, such perception is biased by the natural fuzziness due to the human (human fuzziness) and by the related uncertainty of the system under observation (CHS fuzziness) (Ukita, Karwowski, & Salvendy, 1994). For example, the measures of mental workload, such as those of perceptual-cognitive effort or response loading, are fuzzy phenomena, by the very nature of the underlying

processes (Karwowski, 1992). Because fuzziness occurs at many levels of cognitive processes, people understand and apply vague and uncertain concepts, perceive natural categories as fuzzy objects with no clear boundaries separating members from nonmembers, and are able to comprehend vague concepts of the natural language and manipulate them according to rules of fuzzy logic (Karwowski et al., 1999).

Karwowski, Kosiba, Benabdallah, and Salvendy (1990) and Karwowski and Salvendy (1992) discussed the problems of fuzziness due to high complexity of an artifact-human systems and the nature of user's perception and information processing. Because the interaction between people and computers reflects the cognitive fuzziness of the data, as well as user's uncertainty exhibited in perception of computing environment, HCI focuses also on developing an effective fuzzy-based communication tools.

Experimental Verification of the Fuzzy GOMS Model

According to the model of computer users' information processing (GOMS model), the user's cognitive structure consists of the following four components (Card, Moran, & Newell 1983): (a) a set of Goals, (b) a set of Operators, (c) a set of Methods for achieving the goals, and (d) a set of Selection Rules for choosing among competing methods for goals. In 1983, Card et al. also devised a text-editing experiment to show the validity of GOMS model and concluded that GOMS knowledge representations were valid for such tasks. Karwowski et al. (1990) proposed enhancements to the GOMS model to account for uncertainty within selection rules and to generalize the GOMS model, where goals, operators and methods components can be either precise or fuzzy and where the selection rules can be expressed in either probabilistic or fuzzy manner.

The example presented below refers to the generalized GOMS structure, with the sets of goals and operators precisely defined, whereas the (predicted) methods used by subjects, as well as specific selection rules applied to accomplish the editing task, are based on fuzzy concepts. The fuzzy concepts used included application of linguistic descriptors, fuzzy connectives and fuzzy logic, and possibilistic measures of uncertainty. Such model was defined to as the Fuzzy GOMS model. Karwowski et al. (1989) reported an experiment designed to validate the fuzzy-based GOMS model for text editing task. The experiment was a variation of the manuscript editing experiment by Card et al. (1983), with the subject verbalizing five cursor placement rules.

For example, the editing rules used by the subject included the following fuzzy statements: (a) If the word is more than half of a screen from the cursor and on the same screen or if the word is more than half of a screen from the cursor and across the printed page, then use Method 1. (b) If the word is more than 70 lines and the pattern is not distinct, then use Method 2. (c) If the word is less than half of a screen and on the left half of the page, use Method 3. (d) If the word is less than half of a screen and on the right half of the page, use Method 4. Because the subject did not have the perfect cognitive ability to divide a screen directly in half, he elicited the knowledge as fuzzy knowledge. The membership functions of the linguistic

descriptors were derived based on perceived numbers of lines or characters. Once all the rules, methods, and corresponding membership functions have been elicited, the theory of possibility (Zadeh, 1978) was used to model the subject's rule selection process. For this purpose, each of the potential rules was assigned a possibility measure equal to the membership value(s) associated with it during the elicitation phase of experiment.

The possibility measure $p(A)$ was defined after Zadeh (1978) as follows:

$$p(A) = \text{Poss}\{X \text{ is } A\} \geq \sup \min \{A(u), p_x(u)\},$$

where $p_x(u)$ is the possibility distribution induced by the proposition $(X \text{ is } Z)$, and A is a fuzzy set in the universe U.

The subtask "move down 27 lines to a position in column 20," illustrates the process of predicting the rule selection based on the linguistic ambiguity of the subject's actions. For example, the rules (R) that applied in the above task were as follows:

Rule 1: Membership value of more than a half of the screen = 0.4 (the possibility that the rule applies is 0.4).
Rule 2: Membership value of more than 70 lines = 0 (the possibility that the rule applies is 0).
Rule 3: Membership value of less than half of the screen = 0.3, and membership value of left hand side of the line = 0.4 (the possibility that the rule applies is 0.3 and 0.4).
Rule 4: membership value of right half of line = 0.9, and membership value of less than half of the screen = 0.3 (the possibility that the rule applies is 0.3 and 0.9).
Rule 5: membership value of more than 70 lines = 0 (the possibility that the rule applies is 0).

The possibility measure of the possibility distribution of X that the subject would select a given rule from the universe of available rules R was defined after Zadeh (1978). In case of the example cited above, the most applicable rule was derived based on possibility measure of $\{X \text{ is Rule \#}\}$ as follows:

$$\text{Poss}\{X \text{ is Rule \#}\} = \max[\{(\text{rule } 1, 0.4)\}, \{(\text{rule } 2, 0)\},$$
$$\min\{(\text{rule } 3, 0.3), (\text{rule } 3, 0.4)\}, \min\{(\text{rule } 4, 0.3),$$
$$(\text{rule } 4, 0.9)\}, \{(\text{rule } 5, 0)\}]; \max[\{(\text{rule } 1, 0.4)\},$$
$$\{(\text{rule } 2, 0)\}, \{(\text{rule } 3, 0.3)\}, \{(\text{rule } 4, 0.3)\}, \{(\text{rule } 5, 0)\}];$$

$$\text{Poss}\{X \text{ is Rule\#}\} = \{(Rule\#1, 0.4)\}.$$

Because the possibility of selecting Rule 1 as the most applicable one out of the five rules (R) was 0.4, it was predicted based on the possibilistic measure of uncertainty that the subject would use Rule 1 (i.e., the "control d" method). All fuzzy model predictions in the experiment were checked against the actual selection rule decisions made by the subjects. In the pilot study (Karwowski et al., 1989), a model was run using the fuzzy GOMS approach to the cursor placement task. The fuzzy GOMS model predicted 13 out of 17 decisions, or 76%, correctly. The nonfuzzy GOMS model predicted only 8, or 47% of the cursor placement decisions correctly. It was reported that the use of fuzzy concepts seemed natural within the knowledge elicitation process. It was much easier to ask for membership values in the linguistic terms than to ascertain exact cut-offs for selection rules. This observation supports the results of the study by Kochen (1975) who found a higher degree of consistency in subjects responses when allowed to give imprecise (verbal) descriptors.

In the follow-up study reported by Karwowski et al. (1990), five subjects were asked to perform a word placement task while explaining why they were choosing their particular methods and verbalizing the associated selection rules. If the selection rules appeared to have fuzzy components, these components were quantified by asking the subjects to verbalize a membership value for the applicability of the rule. Overall, across all subjects and trials, the nonfuzzy GOMS model successfully predicted 58.7% of the responses, whereas the fuzzy GOMS model predicted significantly more correct responses (i.e., 82.3% of all subjects decisions).

CONCLUSIONS

One of the main shortcomings of the human factors and ergonomics of computing systems is lack of methodological tools for quantification and measurement of the degree of fit between the humans and computer-task-environment subsystems. This chapter develops a framework for quantifying the measures of computer-human system complexity and compatibility. The introduced concepts of complexity and compatibility should allow for quantitative evaluation of the impact of redesign changes on the complex computer-human systems.

References

Alford, M. (1994, April 18-22). Attacking requirements complexity using a separation of concern. In *Proceedings of the first international conference on requirements engineering* (pp. 2-5). Colorado Springs, CO.

Ashby, W. R. (1964). *An introduction to cybernetics*. London: Methuen & Co.

Cambel, A. B. (1993). *Applied Chaos Theory*. Boston: Academic Press.

Card, S. K., Moran, T. P., & Newell, A. (1983). *The psychology of human-computer interaction*. London: Lawrence Erlbaum Associates.

Chidamber, S. R., & Kemerer, C. F. (1994). A metrics suite for object-oriented design. *IEEE Transaction on Software Engineering, 20*(6), 476-493.

Conrad, M. (1983). *Adaptability*. New York: Plenum.

Coskun, E., & Grabowski, M. (2001). An interdisciplinary model of complexity in embedded intelligent real-time systems. *Information and Software Technology, 43*, 527-537.

Dainoff, M. J., & Balliett, J. (1991). Seated posture and workstation configuration. In M. Kumashiro & E. D. Megaw (Eds.), *Towards human work: Solutions to problems in occupational health and safety* (pp. 156-163). London: Taylor & Francis.

Feigenbaum, M. J. (1980). Universal behavior in nonlinear systems. *Los Alamos Science, 1*, 4.

Gleick, J. (1987). *Chaos: Making a new science*. New York: Penguin Books.

Grieco, A. (1986). Sitting posture: an old problem and a new one. *Ergonomics, 39*, 345.

Guastello, S. J. (1995). *Chaos, catastrophe, and human affairs*. Mahwah, NJ: Lawrence Erlbaum Associates.

Halstead, M. H. (1977). *Elements of software science*. New York: Elsevier.

Jamaldin, B., & Karwowski, W. (1997). Quantification of human–system compatibility (HUSYC): an application to analysis of the bhopal accident. In P. Seppala, T. Luopajarvi, C-H. Nygard, & M. Mattila (Eds.), *From experience to innovation: Proceedings of the 13th Triennial Congress of the International Ergonomics Association* (Vol. 3, pp. 46–48). Tampere, Finland: Tampere University of Technology Press.

Karwowski, W. (1991). Complexity, fuzziness and ergonomic incompatibility issues in the control of dynamic work environments. *Ergonomics, 34*, 671–686.

Karwowski, W. (1992). The human world of fuzziness, human entropy, and the need for general fuzzy systems theory. *Journal of Japan Society for Fuzzy Theory and Systems, 4*, 591–609.

Karwowski, W. (1995). A general modeling framework for the human-computer interaction based on the principles of ergonomic compatibility requirements and human entropy. In A. Grieco, G. Molteni, E. Occhipinti, & B. Piccoli (Eds.), *Work with display units* (Vol. 94, pp. 473–478), Amsterdam: North-Holland.

Karwowski, W. (1997). Ancient wisdom and future technology: The old tradition and the new science of human factors/ergonomics. In *Proceedings of the Human Factors and Ergonomics Society 4th Annual Meeting* (pp. 875–877). Santa Monica, CA: Human Factors and Ergonomics Society.

Karwowski, W. (2000). Symvatology: The science of an artifact-human compatiblty. *Theoretical Issues in Ergonomics Science, 1*, 76–91.

Karwowski, W., & Salvendy, G. (1992). Fuzzy-set-theoretic applications in modeling of man-machine interactions. In R. R. Yager & L. A. Zadeh (Eds.), *An introduction to fuzzy logic applications in intelligent systems* (pp. 201–220). Boston: Kluwer Academic.

Karwowski, W., Grobelny, J., Yang, Y., & Lee, W. G. (1999). Applications of fuzzy systems in human factors. In H. Zimmermman (Ed.), *Handbook of fuzzy sets and possibility theory* (pp. 589–620). Boston: Kluwer Academic.

Karwowski, W., Kosiba, E., Benabdallah, S., & Salvendy, G (1990). A framework for development of fuzzy GOMS model for human-computer interaction. *International Journal of Human-Computer Interaction, 2*, 287–305.

Karwowski, W., Marek, T., & Noworol, C. (1988). Theoretical basis of the science of ergonomics. *Proceedings of the 10th Congress of the International Ergonomics Association* (pp. 756–758). London: Taylor & Francis.

Karwowski, W., Marek, T., & Noworol, C. (1994). The complexity-incompatibility principle and the science of ergonomics. In F. Aghazadeh (Ed.), *Advances in industrial ergonomics and safety* (Vol. VI, pp. 37–40). London: Taylor & Francis.

Kochen, M. (1975). *Applications of Fuzzy Sets in Psychology*, L. A Zadeh, K. S. Fu, K. Tanaka, & M. Shimuro (Eds.), Fuzzy Sets and their Applications to Cognitive and Decision Processes, Academic Press, New York, 395–408.

Kokol, P., Brest, J., & Umer, V. (1996). Software complexity—an alternative view. *ACM SIGPLAN notices, 31(2)*, 35–41.

Kosko, B. (1992). *Neural networks and fuzzy systems: a dynamical systems approach to machine intelligence*. Englewood Cliffs' NJ: Prentice Hall.

Marr, D. (1977). *Vision: A computational investigation into the human representation and processing of visual information*. San Francisco: Freeman.

McCabe, T. J. (2000). *Software matrics*. Retrieved from http://www.mccabe.com/products/software_matrics.htm

Morris, W. (Ed.). (1978). *The American Heritage Dictionary Of English Language*. Boston: Houghton Mifflin.

Norman, D. (1988). *The design of everyday things*. New York: Doubleday.

Perrow, C. (1984). *Normal accidents: Living with high-risk technologies*. New York: Basic Books.

Silver, B. L. (1998). *The ascent of science*. Oxford, England: Oxford University Press.

Smithson, M. (1982). Applications of fuzzy set concepts to behavioral sciences. *Mathematical Social Sciences, 2*, 257–274.

Such, N. P. (1990). The Principles of Design. Oxford University Press, New York.

Ukita, A., Karwowski, W., & Salvendy, G. (1994). Aggregation of evidence in a fuzzy knowledge-based method for automated tuning of microwave electric circuits. *Journal of Intelligent and Fuzzy Systems, 2*, 299–313.

Wickens, C. D., & Carswell, C. M. (1987). *Information Processing, Decision-Making, and Cognition*, G. Salvendy (Ed.). Handbook of Human Factors, John Wiley & Sons, New York, 72–107.

Zadeh, L. A. (1973). Outline of a new approach to the analysis of complex systems and decision processes. *IEEE Transactions on Systems, Man and Cybernetics, SMC-3*, 28–44.

Zadeh, L. A. (1978). Fuzzy sets as a basis for a theory of possibility. *Fuzzy Sets and Systems, 1*, 3–28.

Zuse, H. (1999). Software metrics and objects-oriented systems. Lecture Notes in Computer Science, No 1743, 329, Springer Verlag.

·Conclusion·

PERSPECTIVES ON HUMAN–COMPUTER INTERACTION

Gavriel Salvendy

School of Industrial Engineering, Purdue University and Department
of Industrial Engineering, Tsinghua University, Beijing China

OVERVIEW

Initially computers evolved to crunch numbers; no interaction took place between the computer and the human. In the early 1970s, more and more computers were designed to aid in human task performance. In these cases, engineers and designers determined whether a given task was being performed effectively, which human skills were required to perform the task, and in what ways humans and computers could interact to improve task performance. They considered *what* activity was being performed, *how* it was performed, and whether the sequence of task activities was effective. In establishing *how* the task was performed, the design of computer interfaces began to play an important role, thus determining the job content of those who interact with computers on the job. And this has an impact on job satisfaction and evaluation, quality and quantity of work, and industrial competitiveness (Salvendy, 1997, 2001).

JOB SATISFACTION: SIMPLIFIED VERSUS ENRICHED JOBS

Job satisfaction is a function of the individual's likes and dislikes, difficulties in performing a given task, and the values placed on specific attributes and actions by the individual. There is some indication that older or less educated individuals are more satisfied performing a task in a simplified way, whereas younger and more educated people are more satisfied in performing the task in an enriched way. In simplified jobs, the task is decomposed to its smallest denomination, whereas in enriched tasks, it typically includes the combined performance of a number of simplified tasks.

In one study (Salvendy, 1977) of 359 industrial workers, 45% of the labor force who were both older (over 45 years) and less educated were more satisfied doing a simplified job, 45% who were younger (less than 45 years) and more educated were more satisfied in performing an enriched task, and 10% of workers did not like to perform the job in any form.

Whether the job is performed in a simplified or an enriched way, it has major impact on personnel selection, personnel training, group effectiveness, and job evaluation. The performance of an enriched job is usually requires a greater variety and a higher level of abilities than does a simplified task. Because of this, a greater number of people have the ability to perform simplified tasks than have the ability to perform enriched ones.

Of course, the increased diversity and depth of ability required of enriched versus simplified jobs means that it takes longer to train people on enriched jobs, which results in an initial increased loss in production. It is easy to create a situation in which the same function can be performed both in simplified and enriched mode, as illustrated by Cook and Salvendy (1999) for the student selection and admission process at a major university. Simplified versus enriched job design also has implications for group work. When the job of an individual in group work is enriched, then that individual understands better the cognitive thought processes of the other team members, which results in more effective team performance of the group

than when the member's job design is simplified, because for simplified jobs, members do not understand well the mental model of the other group members (Eberts, Majchrzak, Payne, & Salvendy, 1990).

Job evaluation determines the relative amount of money one job merits in relation to another. The factor method of job evaluation (Hannon, Newman, Milkovich, & Brakefield, 2001) is among the most widely used methods of job evaluation. Using this method, the higher and more diversified the abilities required to perform a job, the higher the job evaluation score and the higher the pay. Thus, job enrichment commands higher pay than performing the same activities in a simplified job setting.

DIMENSIONS OF HUMAN–COMPUTER INTERACTION (HCI)

The quality and quantity of work output in an HCI setting is a function of many variables. Probably the four most important are as follows: First, does the function or activity need to be performed? Frequently activities are associated with task performance that do not add overall value. These functions need to be eliminated from the task performance and from the interface design. Second, are all functions and activities that add value to the overall task performance incorporated in the interface design? Third, in complex HCI, it is frequently helpful to provide decision support systems to the user (Sage, 2001). For example, in the supervision of flexible manufacturing systems (FMSs) (Sheridan, 1997), this is widely practiced. Each piece of manufacturing equipment in an FMS has a microprocessor, and each machine is connected via both an information system and a conveyor to the other machines in the system. The supervisor monitors the machines and decides which part should be processed where and by which machine so that FMS efficiency is maximized. In this case, the decision support to the supervisor may include animation, which shows each component of the system and how many parts are queued for each machine. Another example for decision support in an FMS may be in providing real-time information, on the remaining life of machine components. This information is useful to determine which parts should be routed to which machine to minimize use of a machine on which the components need to be changed. Fourth, when the human can be taken out of the loop in interaction with computers, major increases in quality and output can be achieved. An example of this is the use of electronic tuning machines. A major international corporation approached me to help redesign an HCI to reduce performance time and increase the quality of the electronic tuning machines.

In the original system it took more than 30 minutes to tune each machine, and the quality was judged to be very good only in about 40% of the tuning. In performing the classical task analysis and applying HCI rules and guidelines, no major reduction in performance or increase in the percentage of good-quality tuning could be achieved. When we decided to automate the process by emulating the thought processes of the top experts in machine tuning using fuzzy neural networks, we were able to devise a method that resulted in high-quality tuning, achieved

without humans, in 1 to 5 minutes, depending on the complexity of the tuning (Ukita, Karwowski, & Salvendy, 1994). This illustrates the case that, in HCI design and evaluation, we should consider each of the following options:

1. Improve HCI as an independent discipline
2. Improve HCI with the inclusion of decision support
3. Improve operation by removing humans from systems and automating operations processes
4. Consider system HCI

SYSTEM HCI

An example that illustrates this approach well is a project I spearheaded for the U.S. Postal Service to design, implement, and evaluate video-encoding systems to sort letters. The way the system works is that addresses are automatically encoded into a computer system (without human intervention), and then transmitted to a computer screen for sorting by a human. The HCI issue was what form the address should be displayed on the computer screen, how to display it, and how the operator should decode it.

The systems approach in this HCI scenario included the following:

1. Based on available guidelines and standards, 11 possible interface design scenarios were identified.
2. Based on 1-day, controlled laboratory experiments, 3 of the 11 scenarios were identified as providing the best results.
3. Three groups of postal operators performed the three design scenarios, sorting letters for 6 weeks; two scenarios were identified as being superior to the third.
4. Each of the two scenarios were performed at two geographic locations by a large group of individuals for 1 year. Based on the information derived from this study, the best scenario emerged.

During the 1-year study, the following were accomplished:

1. Appropriate equipment (e.g., tables and chairs) were finalized, and optimal group (e.g., number of operators) and facility size were determined.
2. A short videotape was developed to train operators on proper use of equipment, describing correct posture to minimize fatigue and discomfort.
3. A job evaluation method was devised to determine the level of payment for operators; organizational structure was determined for group leaders.
4. Physical layout for the computer workstations within the facility was established, including special design features for operator rest periods.
5. Methods for selecting operators were developed and validated, as were personnel training and job design.

It was determined that operators' satisfaction, work output, and quality were influenced by all of these factors and that optimizing merely HCI design results in a suboptimal performance. The overall system and each of its components must be designed and evaluated as an integrated system to achieve optimal work performance and satisfaction.

USABILITY

How successful a product will be in the marketplace and how many units of it will be sold depends on the price, the need for the product, its reliability and usability, and the effectiveness of its marketing. Because usability is directly within the domain of the HCI professional, I concentrate on some aspects of it here. For a product to be usable, both inherent and apparent usability is critical; both need to be carefully designed, tested, and evaluated (Fu & Salvendy, 2002).

An easy and effective way to compare the usability of products is to use an index system as advocated by Lin, Choong, and Salvendy (1997). In this way, the relative usability of a product can be compared with the usability of other products. This leads to interest in establishing a national independent usability laboratory for evaluating the usability of all products before they are released by a company to the marketplace. This could be analogous to the operation of the Underwriters Laboratory in the United States.

In this proposed laboratory, the usability evaluation method should be imbedded in a software package; with computer control systems, this software would automatically evaluate the usability of HCI products. This evaluation tool should not only provide an index of usability, but also indicate what needs to be done and how to increase the usability index of a specific HCI product. This unbiased information would empower the customer to make informed decisions in choosing and purchasing software interface products and services.

With the emergence of e-Business, the role and function of usability has significantly increased to include customer relations management in both commercial Web sites and e-Business. Customer relations management is a well-established discipline in traditional retailing and is gaining interest in e-business settings (Yoo & Donthu, 2001). Development of guidelines, standards, and usability evaluation methods for customer relations management in e-business may be one of the most economically valuable HCI contributions. Increased customer loyalty makes repeated shopping at a site, more likely which results in increased sales without spending funds on marketing and advertisements.

THE FUTURE IS NOW

Major opportunities exist to be at the forefront of HCI in the following areas:

1. *Push rather than pull.* The last 20 years have seen information technology dictating the nature of HCI. In the next 20 years, HCI professionals and researchers will determine which technologies need to be developed to push HCI to new professional and scientific heights.

2. *International center for usability testing.* To provide consistent and reliable knowledge to consumers so that they

can make informed decisions in business globalization, an international center for usability testing should be established. For this purpose, software packages with control systems will have to be developed to provide intelligent systems for evaluating and improving usability for each information technology product or service.

3. *HCI will broaden.* HCI will encourage the design of customer relations management jobs and the invention of new information technology products or services, such as store navigation aids that would tell customers where to find which products on their shopping list. These approaches will change from the narrow HCI emphasis of ease or joy of use to ease and joy of accomplishing objectives. To this end, many HCI functions can be automated, and thus objectives can be more effectively achieved. This was evident in recent HCI studies and was achieved in one project aimed at improving ease of use of electronic tuning machines (Ukita et al., 1994). In this case a major improvement occurred not by making the product easier to use but by automating the electronic tuning machine using fuzzy neuro net and eliminating the need for human interaction.

References

Cook, J., & Salvendy, G. (1999). Job enrichment and mental workload in computer-based with implications for adaptive job design. *International Journal of Industrial Ergonomics, 24,* 13–23.

Eberts, R., Majchrzak, A., Payne, P., & Salvendy, G. (1990). Integrating social and cognitive factors in the design of human–computer interactive communication. *International Journal of Human Computer Interaction, 2,* 1–27.

Fu, L., & Salvendy, G. (2002). The contribution of apparent and inherent usability to a user's satisfaction in searching and browsing task on the web, *ergonomics,* in press.

Hannon, J., Newman, J., Milkovich, G., & Brakefield, J. (2001). Job evaluation in organizations. In G. Salvendy (Ed.), *Handbook of industrial engineering: Technology and operations management* (3rd ed.). New York: John Wiley.

Lin, H., Choong, Y.-Y., & Salvendy, G. (1997). A proposed index of usability: A method for comparing the relative usability of different software systems. *Behaviour and Information Technology, 16,* 267–278.

Sage, A. (2001). Decision support systems. In G. Salvendy (Ed.), *Handbook of industrial engineering: Technology and operations management* (3rd ed.). New York: John Wiley.

Salvendy, G. (1977). An industrial engineering dilemma: Simplified vs. enlarged jobs (Keynote address). *Proceedings, 4th International Conference on Production Research.*

Salvendy, G. (Ed.). (1997). *Handbook of human factors and ergonomics* (2nd ed.). New York: John Wiley.

Salvendy, G. (Ed.). (2001). *Handbook of industrial engineering: Technology and operations management* (3rd ed.). New York: John Wiley.

Sheridan, T. B. (1997). Supervisory control. In G. Salvendy (Ed.), *Handbook of human factors and ergonomics* (2nd ed.). New York: John Wiley.

Ukita, A., Karwowski, W., & Salvendy, G. (1994). Aggregation of evidence in a fuzzy knowledge-based method for automated tuning of microwave electric circuits. *Journal of Intelligent and Fuzzy Systems, 2,* 299–313.

Yoo, B., & Donthu, N. (2001). Developing a scale to measure the perceived quality of our Internet shopping site (SITEQUAL). *Quarterly Journal of Electronic Commerce, 2,* 31–45.

AUTHOR/REFERENCE INDEX

Reiser, B. J., 834
Reiswig, G., 520
Reithinger, N., 302
Rejmer, P., 1197
Rekimoto, J., 167, 655
Rekola, S., 940
Remde, J. R., 789
Remondeau, C., 303
Rempel, D., 165, 166, 167
Rennels, G., 674
Rennie, Y., 805
Repenning, A., 440
Repperger, D., 880
Resnick, M., 439, 440
Resnick, P., 357, 594, 1164, 1197
Resnikoff, H. L., 582
Rettig, M., 541, 1067
Reuter-Lorenz, P., 95
Revelle, G., 439
Rex, D. B., 582
Reynard, G., 593
Reynolds, P. L., 696
Rheingold, H., 17, 619
Rhoades, D. G., 1137
Rhoades, J. A., 593
Rhoades, R. E., 985
Rhodes, B., 330
Rhodes, E., 821
Rhodes, J. S., 1212
Ribbens, W. B., 860
Rice, D. R., 921
Rice, R. E., 593, 619
Rich, A., 911
Rich, E., 330
Rich, M., 985
Richards, J., 95
Richards, J. T., 593, 617
Richardson, B., 860
Richardson, C., 165
Rickard, T. C., 695
Ricker, K. L., 33
Rideout, V., 412
Riding, R., 541
Riecken, D., 771, 1164
Riedel, O., 634
Riedl, J., 328, 330, 581, 593, 594
Riedl, M. O., 117
Rieman, J., 16, 117, 1049, 1137, 1138
Riesbeck, C. K., 675
Rigas, D. I., 237, 239
Rignér, J., 881
Riley, V., 675, 881
Ringle, M. D., 185
Rippy, L. P., 879
Risch, J. S., 582
Risden, K., 439
Rist, T., 261, 328
Ritter, F. E., 117
Riva, G., 370
Rivera, K., 95
Riviere, C. N., 427
Rizzolo, M. A., 821
Roads, C., 239
Robbins, D., 165
Robbins, M. A., 426
Robert-Ribes, J., 303
Roberts, B., 17
Roberts, D., 412, 540, 771
Roberts, L. A., 789
Roberts, S. D., 368

Roberts, T. L., 15, 619
Robertson, G., 164, 165, 167, 304, 522, 581, 582, 1088
Robertson, T., 1068
Robertson, T. S., 369
Robin, H., 582
Robins, J., 1068
Robinson, C., 940, 1067
Robinson, E., 751
Robinson, J. E. III, 881
Robinson, P., 501, 729
Robinson, R. E., 985
Robinson, T. N., 822
Robson, D., 1030
Rocco, E., 594, 1200
Rockley, A., 539
Rockwell, T., 676
Rodden, T., 80, 355, 356, 357, 501, 984, 985
Rodgman, E. A., 860
Rodney, R. C., 695
Rodvold, M., 676
Roe, D. B., 185
Roe, M. M., 881
Roediger, H. L., 50, 696
Roessler, A., 368
Rogers, D., 438
Rogers, K. J. S., 395
Rogers, M., 771
Rogers, W. A., 426, 427
Rogers, W. H., 881
Rogers, Y., 17, 262, 440, 525, 619, 675, 906, 940, 985
Rogozan, A., 303
Rohn, J. A., 1137
Rolfe, J. M., 881
Rolland, J. P., 634
Rollins, A. M., 17, 1005
Rolls, E. T., 696
Romero, R. L., 905
Rommelse, K., 329
Romney, A. K., 985, 986
Root, R. A., 1049
Root, R. W., 593, 771, 789, 906, 1067, 1068
Rorty, R., 1200
Rosa, R. H. Jr., 520, 1198
Rosch, E., 1050
Rose, A., 806, 1200
Rose, K., 439
Roseman, I. J., 96
Roseman, M., 1031
Rosen, B., 426
Rosenbaum, H., 1199, 1200
Rosenbaum, S., 1137
Rosenberg, R., 521
Rosenberg, S., 1200
Rosenblitt, D. A., 594
Rosenbloom, P. S., 50, 675
Rosenfeld, A., 96
Rosenfeld, L., 284, 730
Rosengren, K. E., 96
Rosenstock, I. M., 370
Rosenthal, D. J., 520
Rosenthal, M., 502
Rosenzweig, A. S., 94
Roshan, S., 985
Roskos, E., 219
Ross, G., 834
Ross, L., 370
Ross, T., 860

Rosson, M. B., 616, 940, 1031, 1048, 1050
Roston, G. P., 219
Rotenberg, M., 1196, 1200
Roth, A. E., 1200
Roth, D. L., 426
Roth, E. M., 146, 675
Roth, S. F., 581, 582
Rouch, H., 411
Roukos, S., 185
Rouse, W. B., 674, 675, 676
Rousseau, G. K., 426
Roussel, N., 357
Rowan, M., 730
Rowberg, A. H., 940
Rowley, D. E., 1137
Rowley, P., 164
Roy, D., 521
Roy, D. M., 502
Roy, E. A., 34
Rössel, M., 330
Rubenstein, R., 940
Rubert, M., 426
Rubichi, S., 34
Rubin, A., 412
Rubin, A. M., 96
Rubin, J., 440, 541, 728, 940, 1117
Rubin, K., 1050
Rubin, K. S., 675
Rubin, P., 303
Rubine, D., 167
Rubinson, H., 675
Rubinstein, G., 789
Rubinstein, R., 921
Rudell, A. P., 34
Rudisill, M., 49
Rudmann, S., 674, 676
Rudnicky, A., 303, 713
Ruef, A. M., 618
Rugg, M. D., 50
Rumbaugh, J., 1049
Rummer, R., 329
Rushby, J., 881
Rusk, N., 440
Russell, D. M., 582
Russell, J. A., 93, 94
Russell, L., 619
Russell, M. J., 304
Russell, S., 675
Russo, M., 412
Ruthsatz, J. M., 480
Rutkowski, C., 34
Rutledge, J., 167
Ryall, K., 730
Ryan, S. D., 696
Ryder, J., 1151
Ryder, J. W., 521
Ryokai, K., 411

S

Saaksjarvi, M., 1068
Saarinen, L., 1225
Saarinen, T., 1068
Saberi, K., 238
Sacerdoti, E. D., 676
Sacher, H., 730
Sachs, L., 674, 676
Sachs, P., 985
Sack, W., 619

SUBJECT INDEX